Capute and Accardo's Neurodevelopmental Disabilities in Infancy and Childhood

Capute and Accardo's Neurodevelopmental Disabilities in Infancy and Childhood

Fourth Edition

Edited by

Fatima Y. Ismail
Pediatric Neurology and Developmental Medicine, Department of Pediatrics, College of Medicine and Health Sciences, United Arab Emirates University, United Arab Emirates; Department of Neurology, Johns Hopkins University School of Medicine, MD, United States

Pasquale J. Accardo
Department of Pediatrics, Virginia Commonwealth University School of Medicine, Richmond, VA, United States

Bruce K. Shapiro
Department of Neurology and Developmental Medicine, Kennedy Krieger Institute, Baltimore, MD, United States; Department of Pediatrics, Johns Hopkins University School of Medicine, Baltimore, MD, United States

ELSEVIER

ACADEMIC PRESS
An imprint of Elsevier

Academic Press is an imprint of Elsevier
125 London Wall, London EC2Y 5AS, United Kingdom
525 B Street, Suite 1650, San Diego, CA 92101, United States
50 Hampshire Street, 5th Floor, Cambridge, MA 02139, United States

ISBN: 978-0-12-824060-1

For Information on all Academic Press publications
visit our website at https://www.elsevier.com/books-and-journals

Publisher: Stacy Masucci
Acquisitions Editor: Megan McManus
Editorial Project Manager: Sam Young
Production Project Manager: Gomathi Sugumar
Cover Designer: Miles Hitchen

Typeset by MPS Limited, Chennai, India

Dedication

The editors would like to dedicate this work to their families, mentors, colleagues, and trainees.

—Fatima Y. Ismail
—Pasquale J. Accardo
—Bruce K. Shapiro

Contents

List of contributors

Christina Briscoe Abath Department of Neurology, Boston Children's Hospital, Harvard Medical School, Boston, MA, United States

Jennifer Accardo Departments of Pediatrics and Neurology, Virginia Commonwealth University, Richmond, VA, United States

Pasquale J. Accardo Department of Pediatrics, Virginia Commonwealth University School of Medicine, Richmond, VA, United States

Marilee C. Allen Department of Neurology and Developmental Medicine, Kennedy Krieger Institute, Baltimore, MD, United States; Department of Pediatrics, Johns Hopkins University School of Medicine, Baltimore, MD, United States

Miya R. Asato Department of Neurology and Developmental Medicine, Kennedy Krieger Institute, Baltimore, MD, United States; Departments of Neurology and Pediatrics, Johns Hopkins University School of Medicine, Baltimore, MD, United States

Kadee D. Atkinson Center for Diversity in Public Health Leadership Training, Kennedy Krieger Institute, Baltimore, MD, United States

Elizabeth Barkoudah Department of Neurology, Boston Children's Hospital, Harvard Medical School, Boston, MA, United States

Jeremy P. Bauer Shriners Children's Hospital, Oregon Health and Science University, Portland, OR, United States

Harolyn M.E. Belcher Office for Health, Equity, Inclusion, and Diversity, Kennedy Krieger Institute, Department of Pediatrics, Johns Hopkins University School of Medicine, Department of Mental Health, Johns Hopkins Bloomberg School of Public Health, Baltimore, MD, United States

Peter A. Blasco Department of Pediatrics, Division of Neurodevelopmental Pediatrics, Institute on Development and Disability, Oregon Health and Science University, Portland, OR, United States

Vera Joanna Burton Department of Neurology and Developmental Medicine, Kennedy Krieger Institute, Baltimore, MD, United States; Department of Neurology, Johns Hopkins University School of Medicine, Baltimore, MD, United States

Philippines Cabahug Kennedy Krieger Institute, International Center for Spinal Cord Injury, Baltimore, MD, United States; Department of Physical Medicine and Rehabilitation, Kennedy Krieger Institute and Johns Hopkins School of Medicine, Baltimore, MD, United States

Meeta Cardon Division of Child Neurology, Department of Neurology, University of New Mexico, Albuquerque, NM, United States

Thomas D. Challman Geisinger Autism & Developmental Medicine Institute, Danville, PA, United States; Geisinger Health System and Geisinger Commonwealth School of Medicine, Lewisburg, PA, United States

Eric M. Chin Department of Neurodevelopmental Medicine, Kennedy Krieger Institute, Baltimore, MD, United States; Departments of Neurology and Pediatrics, Johns Hopkins University School of Medicine, Baltimore, MD, United States

Celanie K. Christensen Department of Pediatrics, Indiana University School of Medicine, Indianapolis, IN, United States

Lindsay Cirincione Department of Behavioral Psychology, Kennedy Krieger Institute, Baltimore, MD, United States; Department of Physical Medicine & Rehabilitation, Johns Hopkins University School of Medicine, Baltimore, MD, United States

Julie S. Cohen Department of Neurology and Developmental Medicine, Kennedy Krieger Institute, Baltimore, MD, United States; Department of Neurology, Johns Hopkins University School of Medicine, Baltimore, MD, United States

Kimberly C. Davis Department of Pediatrics, Sections of Psychology and Neurology, Texas Children's Hospital, Houston, TX, United States

Janet Dean Kennedy Krieger Institute, International Center for Spinal Cord Injury, Baltimore, MD, United States; Department of Physical Medicine and Rehabilitation, Kennedy Krieger Institute and Johns Hopkins School of Medicine, Baltimore, MD, United States

Shannon L. Dean Department of Neurology and Developmental Medicine, Kennedy Krieger Institute, Department of Neurology, Johns Hopkins University School of Medicine, Baltimore, MD, United States

Jessica Durkee-Shock Medical Virology Section, National Institute of Allergy and Infectious Diseases, National Institutes of Health, Bethesda, MD, United States

Darius Ebrahimi-Fakhari Department of Neurology, Movement Disorders Program, Boston Children's Hospital, Harvard Medical School, Boston, MA, United States

Karen Evankovich Department of Pediatrics, Sections of Psychology and Neurology, Texas Children's Hospital, Houston, TX, United States

Patricia Ann Evans Departments of Neurology and Neurotherapeutics, Pediatrics, and Psychiatry, University of Texas Southwestern School of Medicine, Children's Medical Center, Dallas, TX, United States

John Michael Falligant Department of Behavioral Psychology Kennedy Krieger Institute, Johns Hopkins University School of Medicine, Baltimore MD, United States

Kristin Fauntleroy-Love Department of Pediatrics, Indiana University School of Medicine, Indianapolis, IN, United States

Kristen Fisher Division of Pediatric Neurology and Developmental Neuroscience, Texas Children's Hospital, Baylor College of Medicine, Houston, TX, United States

Scott Frampton Department of Occupational Therapy, Kennedy Krieger Institute, Baltimore, MD, United States

Michelle A. Frank-Crawford Kennedy Krieger Institute, and Johns Hopkins University School of Medicine, Baltimore, MD, United States

Alexis L. Franks Department of Pediatrics, Division of Child Neurology and Child Development, University of Pittsburgh School of Medicine, Pittsburgh, PA, United States

Sheryl Lynn Frierson Luke Waites Center for Dyslexia and Learning Disorders, Scottish Rite for Children, Dallas, TX, United States

Gwendolyn J. Gerner Department of Psychology, Kennedy Krieger Institute, Baltimore, MD, United States; Department of Psychiatry and Behavioral Sciences, Johns Hopkins University School of Medicine, Baltimore, MD, United States

Ryan E. Gill Department of Neurology and Developmental Medicine, Kennedy Krieger Institute, Baltimore, MD, United States; Department of Neurology, Johns Hopkins University School of Medicine, Baltimore, MD, United States

Tanjala T. Gipson The University of Tennessee Health Science Center, Boling Center for Developmental Disabilities, Le Bonheur Children's Hospital, Memphis, TN, United States

Andres Jimenez Gomez Neurodevelopmental Disabilities Program, Joe Dimaggio Children's Hospital, Hollywood, FL, United States; Stiles-Nicholson Brain Institute, Florida Atlantic University, Jupiter, FL, United States

Eliza Gordon-Lipkin Metabolism, Infection and Immunity Section, National Human Genome Research Institute, National Institutes of Health, Bethesda, MD, United States

Andrea Lynne Gropman Neurometabolic Translational Research, Center for Experimental Neurotherapeutics, St. Jude's Pediatric Research Hospital, Memphis, TN, United States

Darrah Haffner Division of Pediatric Neurology, Nationwide Children's Hospital, The Ohio State University, Columbus, OH, United States

Louis P. Hagopian Kennedy Krieger Institute, and Johns Hopkins University School of Medicine, Baltimore, MD, United States; Department of Psychiatry and Behavioral Sciences, Johns Hopkins University School of Medicine, Baltimore MD, United States

Samantha L. Hardesty Kennedy Krieger Institute, and Johns Hopkins University School of Medicine, Baltimore, MD, United States

Jennifer Harmon Department of Neurology, Wake Forest University School of Medicine, Winston Salem, NC, United States

Jacqueline Harris Department of Neurology and Developmental Medicine, Kennedy Krieger Institute, Departments of Neurology, Pediatrics and Genetics, Johns Hopkins Medical Institution, Baltimore, MD, United States

Mariam Hull Division of Pediatric Neurology and Developmental Neuroscience, Texas Children's Hospital, Baylor College of Medicine, Houston, TX, United States

Ayra Hussain Departments of Biological Sciences and Modern Languages, Linguistics & Intercultural Communication, University of Maryland, Baltimore County, Baltimore, MD, United States

Samantha R. Hutchison Office for Health, Equity, Inclusion, and Diversity, Kennedy Krieger Institute, Baltimore, MD, United States

William Ide Kennedy Krieger Institute, Baltimore, MD, United States; Department of Neurology and Developmental Medicine, Kennedy Krieger Institute, Department of Physical Medicine & Rehabilitation, Johns Hopkins University School of Medicine, Baltimore, MD, United States

Fatima Y. Ismail Pediatric Neurology and Developmental Medicine, Department of Pediatrics, College of Medicine and Health Sciences, United Arab Emirates University, United Arab Emirates; Department of Neurology, Johns Hopkins University School of Medicine, MD, United States

Wendy A. Jennejahn Outpatient Speech-Language Clinic, Speech and Language Department, Kennedy Krieger Institute, Baltimore, MD, United States; Speech and Language Department, Kennedy Krieger Institute, Baltimore, MD, United States

Deanna Johnson Kennedy Krieger Institute, Department of Physical Therapy, Baltimore, MD, United States

Tara Johnson Arkansas Children's Biomedical Research Institute, Arkansas Children's Hospital, Department of Pediatrics, University of Arkansas for Medical Sciences, Little Rock, AR, United States

Mikael Guzman Karlsson Division of Pediatric Neurology and Developmental Neuroscience, Baylor College of Medicine, Houston, TX, United States

Akshat Katyayan Division of Pediatric Neurology and Developmental Neuroscience, Texas Children's Hospital, Baylor College of Medicine, Houston, TX, United States

Kate Kielty Department of Pediatrics, Division of Child Neurology, UPMC Children's Hospital of Pittsburgh, University of Pittsburgh School of Medicine, Pittsburgh, PA, United States

Aram Kim Division of Neurodevelopmental Medicine, Thompson Autism and Neurodevelopmental Center - Children's Hospital of Orange County, Orange, CA, United States

Brooke Kimbrell Department of Neurology and Developmental Medicine, The Kennedy Krieger Institute, Baltimore, MD, United States; Department of Pediatrics, Johns Hopkins University School of Medicine, Baltimore, MD, United States

Jenna Klotz Department of Neurology, Division of Child Neurology and Neuromuscular Medicine, Stanford University School of Medicine, Palo Alto, CA, United States

Sarah Ann Korth Kennedy Krieger Institute, International Center for Spinal Cord Injury, Baltimore, MD, United States; Department of Physical Medicine and Rehabilitation, Kennedy Krieger Institute and Johns Hopkins School of Medicine, Baltimore, MD, United States

Eboni Lance Department of Neurology and Developmental Medicine, Kennedy Krieger Institute, Baltimore, MD, United States; Department of Neurology, Johns Hopkins University School of Medicine, Baltimore, MD, United States

Janice Laux Kennedy Krieger Institute, Department of Physical Therapy, Baltimore, MD, United States

Steven M. Lazar Section of Pediatric Neurology and Developmental Neuroscience, Meyer Center for Developmental Pediatrics & Autism, Baylor College of Medicine, Texas Children's Hospital, Houston, TX, United States

Ryan Lee Department of Pediatrics, University of Hawaii - Manoa, Hawaii, HI, United States; Shriners Hospitals for Children - Honolulu, Hawaii, HI, United States

Mary L. Leppert Department of Neurology and Developmental Medicine, Kennedy Krieger Institute and Department of Pediatrics, Johns Hopkins University School of Medicine, Baltimore, MD, United States

Wilfreda Lindsey Department of Neurology and Developmental Medicine, Kennedy Krieger Institute, Johns Hopkins University School of Medicine, Baltimore, MD, United States

Paul H. Lipkin Division of Neurology and Developmental Medicine, Kennedy Krieger Institute, Department of Pediatrics, Johns Hopkins School of Medicine, Baltimore, MD, United States

Milos R. Ljubisavljevic Department of Physiology, College of Medicine and Health Sciences, United Arab Emirates University, United Arab Emirates

Thomas Lock Department of Pediatrics, Indiana University School of Medicine, Indianapolis, IN, United States

Michael Alonzo Lopez Division of Pediatric Neurology, Department of Pediatrics, Heersink School of Medicine, University of Alabama at Birmingham, Birmingham, AL, United States

Carmen Lopez-Arvizu Center for Developmental Behavioral Health, Kennedy Krieger Institute, Department of Psychiatry and Behavioral Science, Johns Hopkins School of Medicine, Baltimore, MD, United States

Natasha N. Ludwig Department of Neuropsychology, Kennedy Krieger Institute, Baltimore, MD, United States; Department of Psychiatry and Behavioral Sciences, Johns Hopkins School of Medicine, Baltimore, MD, United States

Kirby Matthews Speech Language Department, Kennedy Krieger Institute, Baltimore, MD, United States

Meghan McAnally National Institute of Neurological Disorders and Stroke, NIH, Bethesda, MD, United States

Alexandria E. Melendez-Zaidi Department of Neurology, Texas Tech Health Sciences Center, El Paso, El Paso, TX, United States

Deepa Unnikrishnan Menon Department of Neurology & Developmental Medicine, Johns Hopkins University School of Medicine, Kennedy Krieger Institute, Baltimore, MD, United States

Amy Jo Merchant Newmeyer Department of Pediatrics, Baylor College of Medicine-Temple Campus, McLane Children's Hospital, Temple TX, United States

Alisa Mo Department of Neurology, Boston Children's Hospital, Harvard Medical School, Boston, MA, United States

Scott M. Myers Geisinger Autism & Developmental Medicine Institute, Danville, PA, United States; Geisinger Health System and Geisinger Commonwealth School of Medicine, Lewisburg, PA, United States

Kavita Nadendla Department of Neurology and Developmental Medicine, Kennedy Krieger Institute, Department of Physical Medicine & Rehabilitation, Johns Hopkins University School of Medicine, Baltimore, MD, United States

Archana Nelliot Department of Neurology and Developmental Medicine, Kennedy Krieger Institute, Department of Pediatrics, Johns Hopkins University School of Medicine, Baltimore, MD, United States

Michael Oanea Department of Neurology, Boston Children's Hospital, Harvard Medical School, Boston, MA, United States

Meghan O'Neill Department of Developmental Behavioral Pediatrics, Ann & Robert H. Lurie Children's Hospital of Chicago, Chicago, IL, United States

Valerie Paasch Department of Behavioral Psychology, Kennedy Krieger Institute, Baltimore, MD, United States; Department of Physical Medicine & Rehabilitation, Johns Hopkins University School of Medicine, Baltimore, MD, United States

Mered Parnes Division of Pediatric Neurology and Developmental Neuroscience, Texas Children's Hospital, Baylor College of Medicine, Houston, TX, United States

Rajapillai L.I. Pillai Department of Neurology, Boston Children's Hospital, Harvard Medical School, Boston, MA, United States

Meghna Rajaprakash Department of Neurology and Developmental Medicine, Kennedy Krieger Institute, Department of Neurology Johns Hopkins University School of Medicine, Baltimore, MD, United States

Michael Regner Department of Radiology, Oregon Health & Science University, Portland, OR, United States

Sarah Risen Department of Pediatrics, Section of Neurology and Developmental Neuroscience, Baylor College of Medicine, Texas Children's Hospital, Houston, TX, United States

Lance Rodan Department of Neurology, Boston Children's Hospital, Boston, MA, United States; Division of Genetics and Genomics, Boston Children's Hospital, Harvard Medical School, Boston, MA, United States

Cristina Lavinia Sadowsky Kennedy Krieger Institute, International Center for Spinal Cord Injury, Baltimore, MD, United States; Department of Physical Medicine and Rehabilitation, Kennedy Krieger Institute and Johns Hopkins School of Medicine, Baltimore, MD, United States

Alexandra Santana Department of Neurology, Movement Disorders Program, Boston Children's Hospital, Harvard Medical School, Boston, MA, United States

Benjamin Joffe Schindel Department of Neurology and Developmental Medicine, Kennedy Krieger Instittute, Johns Hopkins University School of Medicine, Baltimore, MD, United States

Natalie Schmitt Speech Language Department, Kennedy Krieger Institute, Baltimore, MD, United States

Kevin Senn Department of Pediatrics, Institute on Development and Disability, Oregon Health and Sciences University, Portland, OR, United States

J. Seth Van Zant Department of Pediatrics/Division of Pediatric Neurology, Baylor College of Medicine, Houston, TX, United States; Department of Pediatric Neurology/Neurodevelopment, Christus Children's Hospital, San Antonio, TX, United States

Bruce K. Shapiro Department of Neurology and Developmental Medicine, Kennedy Krieger Institute, Baltimore, MD, United States; Department of Pediatrics, Johns Hopkins University School of Medicine, Baltimore, MD, United States

Keith J. Slifer Department of Behavioral Psychology, Kennedy Krieger Institute, Baltimore, MD, United States; Department of Physical Medicine & Rehabilitation, Johns Hopkins University School of Medicine, Baltimore, MD, United States

Clay Smith Department of Neurology and Developmental Medicine, Kennedy Krieger Institute, Baltimore, MD, United States; Department of Pediatrics, Johns Hopkins University School of Medicine, Baltimore, MD, United States

Siddharth Srivastava Department of Neurology, Boston Children's Hospital, Harvard Medical School, Boston, MA, United States; Rosamund Stone Zander Translational Neuroscience Center, Department of Neurology, Boston Children's Hospital, Boston, MA, United States

Carl E. Stafstrom Departments of Neurology and Pediatrics, Division of Pediatric Neurology, Johns Hopkins University School of Medicine, Baltimore, MD, United States

Melissa Svoboda The Children's Hospital of San Antonio, Baylor College of Medicine, San Antonio, TX, United States

Danielle Schwartzenburg Takacs Division of Pediatric Neurology and Developmental Neuroscience, Texas Children's Hospital, Baylor College of Medicine, Houston, TX, United States

Anisa Tatini Department of Neurology and Developmental Medicine, Kennedy Krieger Institute, Department of Physical Medicine & Rehabilitation, Johns Hopkins University School of Medicine, Baltimore, MD, United States

Elaine Tierney Department of Psychiatry and Behavioral Sciences, Johns Hopkins School of Medicine, Kennedy Krieger Institute, Baltimore, MD, United States

Melissa Trovato Kennedy Krieger Institute, Baltimore, MD, United States; Department of Neurology and Developmental Medicine, Kennedy Krieger Institute, Department of Physical Medicine & Rehabilitation, Johns Hopkins University School of Medicine, Baltimore, MD, United States

Margaret A. Tunney Department of Behavioral Psychology, Kennedy Krieger Institute, Baltimore, MD, United States; Department of Physical Medicine & Rehabilitation, Johns Hopkins University School of Medicine, Baltimore, MD, United States

Sherry Vinson Department of Pediatrics, Baylor College of Medicine, Houston, TX, United States

Robert G. Voigt Michael R. Boh Centers for Child Development, Developmental Pediatrics, Ochsner Children's Hospital, New Orleans, LA, United States; Faculty of Medicine The University of Queensland Medical School, Brisbane, Australia

Toni M. Whitaker Department of Pediatrics/Division of Developmental Pediatrics, University of Tennessee Health Science Center, Le Bonheur Children's Hospital, Memphis, TN, United States

Kerry M. White Department of Pediatrics, Indiana University School of Medicine, Indianapolis, IN, United States

Matthew Whitehead Department of Radiology, Children's Hospital of Philadelphia, PA, United States; Perelman School of Medicine, Philadelphia, PA, United States

Jonny Wilson Department of Pediatrics, Division of Pediatric Neurology, Oregon Health & Science University, Portland, OR, United States

Kirkland A. Wilson Rare Disease Institute, Children's National Hospital, Washington, DC, United States; National Human Genome Research Institute, National Institutes of Health, Bethesda, MD, United States

Zainab Zaman Department of Neurology, Movement Disorders Program, Boston Children's Hospital, Harvard Medical School, Boston, MA, United States

Arnold J. Capute, MD, MPH (1923–2003)

Preface

In the late 1970s, Drs. Capute and Accardo traveled from Baltimore to New York City to meet with an international medical publisher to propose a reference volume on developmental disabilities. Since the outline called for three volumes, they were uncertain whether the work would be a manual, a textbook, an encyclopedia, or a handbook. The usage of these terms varied between America and Europe. Since most of the contributors were to be recruited from Johns Hopkins faculty, the publisher was interested but doubtful whether the editors could actually carry out the proposal. For a variety of reasons, the project idled and then died a quiet death.

For a number of years (1978–82), the Kennedy group had published a short-lived series—*Monographs in Developmental Pediatrics*—with University Park Press in Baltimore. Later, the annual Spectrum in Developmental Disabilities course was used to generate the Spectrum Monographs series with Baltimore's York Press (volumes 1–7 from 1994 to 2005). The last two volumes in this series (8 and 9, 2010) were published by Paul H. Brookes in Baltimore. The series title was then changed to the Capute Monographs.

In the 1980s, discussions with the American Board of Pediatrics with regard to the creation of a new pediatric subspecialty suggested that a textbook that would delineate the scope of the field would be a significant plus to granting recognition. The one-volume *Arnold J. Capute and Pasquale J. Accardo (editors) Developmental Disabilities in Infancy and Childhood* (Baltimore, MD: Paul H. Brookes, 1991) was the result. This edition was dedicated to Louise and Patricia. In 1996 a two-volume second edition was published (dedicated to Louise, Patricia, and children), and after Dr. Capute's death,[1] Dr. Accardo put out a third edition in 2008 (dedicated to Arnold and Patricia, with grandchildren not included).

There are five major changes to the present edition.

First, the editors have reverted to a more portable single-volume edition.

A second change is the focus on neuropathologic, biochemical, and genetic mechanisms. The clinical presentation for the spectrum of developmental disorders has changed very little, but the knowledge base for the underlying science has virtually exploded since the first edition. Dr. Capute (and Dr. Accardo) would find it difficult to absorb these dramatic changes. The field is both recognizable and radically new.

A third change is the development of the new training program in Neurodevelopmental Disabilities and Child Neurology under the combined auspices of the American Board of Pediatrics and the American Board of Psychiatry and Neurology. While these were in existence when the third edition was published, they were still too young to adequately impact the chapter presentations. Many of the contributors to the present volume are graduates of the new subspecialty training or faculty active in teaching it.

A fourth change relates to the evolution and expansion of spectrum disorders—what a half century ago was characterized as early infantile autism.

A fifth change is the addition of Drs. Fatima Y. Ismail and Bruce K. Shapiro as coeditors to carry out the bulk of the preparation of this fourth edition. They have succeeded in shortening production time from the 4 years that it took to finish each of the previous three editions to an almost unbelievable 2 years.

Although they have aged much, many of the current leaders in this field have been direct recipients of Dr. Capute's teaching. To those of us who knew him, he remains il maestro di color che sanno.

Fatima Y. Ismail
Pasquale J. Accardo
Bruce K. Shapiro

1. P.J. Accardo: The Father of Developmental Pediatrics: Arnold J. Capute MD, MPH (1923-2003). *Journal of Child Neurology* (December 2004, number 12), 19: 978–981.

Section I

Neurodevelopmental disorders

Chapter 1

The evolution of neurodevelopmental disabilities: an updated view of the spectrum

Bruce K. Shapiro[1,2]

[1]*Department of Neurology and Developmental Medicine, Kennedy Krieger Instittute, Baltimore, MD, United States,* [2]*Department of Pediatrics, Johns Hopkins University School of Medicine, Baltimore, MD, United States*

Introduction

The modern era of neurodevelopmental disabilities (NDDs) began in the early 20th century. Initial efforts focused on identification. Binet and Simon developed an instrument to detect school-aged children who could not be accommodated in traditional schools. This instrument was refined and evolved into the Stanford Binet Intelligence Test. Children with significant motor impairment were the objects of interest in the nascent discipline of neurology, with a classic neuropathological text by Sigmund Freud, the term "cerebral palsy" being coined by Sir William Osler, and attempts at nonsurgical approaches to treatments commenced.

Of particular note is the work of Arnold Gesell [1]. Gesell, like many of the early investigators of child development, observed children as they matured. Gesell's observations resulted in the following findings:

1. **Development was ordered**. Based on his observations, Gesell defined the sequences of development and noted that the components of these sequences were not achieved in a random fashion. For example, he noted that children rolled, sat, crawled, and then walked. He called these components "milestones."

   Gesell's empirical observations were the foundation of a quantitative approach to child development. If a child met their milestones on time, they would be doing 4-month activities at 4 months, 9-month activities at 9 months, 1-year activities at 1 year, etc. Consequently, these achievements could be plotted with age on the abscissa and function on the ordinate. Development would be the rate of change in function over time, as described by the slope of the line. Fig. 1.1 demonstrates this relationship. Given that it is assumed that most cases start at (0,0) this relationship can be reduced to Functional Age divided by Chronological Age times 100. Developmental quotient (DQ) = developmental age (DA)/chronologic age (CA) × 100.
2. **The developmental process precedes birth**. In contrast to other theories of child development, studies of infants who were born prematurely identified an orderly progression of development [2]. In monitoring the development of infants who were born prematurely, most people attempt to "correct" for prematurity. This is an incomplete approach because (1) it assumes that prematurity acts in a uniform fashion, (2) it does not account for the severity of illness encountered by prematurely born infants, and (3) it attempts to capture a dynamic longitudinal process with a cross-sectional measure. Evaluating a child's function at two points in time and then determining a developmental rate may prove to be a more accurate approach.
3. **Children who were delayed followed the same order but did so at a slower pace**. This relationship holds for children who are delayed as well. For example, a 3-year-old child who is functioning at an 18-month-old level would have a DQ of 50. Thus, the function/age relationship can be used for identification, diagnosis, prognosis, and monitoring the progress of children who are delayed.

Capute and Accardo's Neurodevelopmental Disabilities in Infancy and Childhood. DOI: https://doi.org/10.1016/B978-0-12-824060-1.00058-4

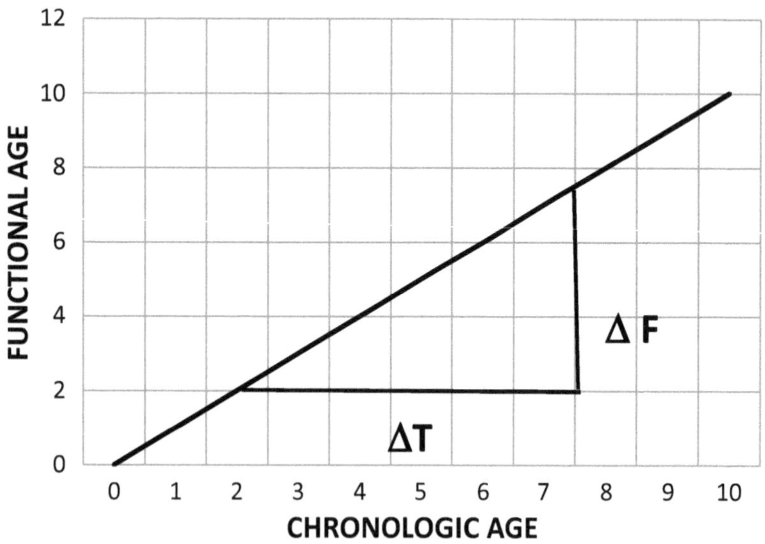

FIGURE 1.1 Developmental quotient (DQ) = developmental (functional age)/chronological age x 100.

3. **Development was the functional outcome of complex neurological processes.** If a child's development was proceeding normally, it could be inferred that the brain was functioning normally or that compensatory mechanisms were effective. Thus, the developmental assessment was a sensitive extension of the neurological examination.

Subsequent efforts focused on children with these disorders. The treatment of motor disorders received a major focus due to the polio diagnosis of President Franklin D. Roosevelt and efforts in rehabilitation of impairments suffered by soldiers during the Second World War. The work of the National Society for Crippled Children, later to become Easter Seals, and the funding of the March of Dimes brought the issues relating to children with motor impairments to the public's attention. In 1947, six members of the Cerebral Palsy Advisory Medical Council of the National Society for Crippled Children and Adults founded the American Academy of Cerebral Palsy, the precursor to the American Academy of Cerebral Palsy and Developmental Medicine, to provide interdisciplinary physician education about cerebral palsy. Shortly thereafter, parent groups that wanted to maintain their children's ability to live fulfilling lives in the community established the National Association of Parents and Friends of Retarded Children, which is now known as the Arc. Research into intellectual disability, legislation that enabled training teachers to teach children with disabilities, the growth of advocacy groups, and the recognition of the limitations of institutionalization resulted in major initiatives to address the needs of children with disabilities. In 1961, President John Kennedy appointed a special President's Panel on Mental Retardation, which resulted in 112 recommendations. In 1963, landmark legislation (public law (PL) 88–164) was passed that established a nation-wide system of university-affiliated training programs that sought to train professionals to meet the current and future needs of people with intellectual disabilities and related disabilities [3]. These university-affiliated facilities have evolved into the Maternal and Child Health Bureau Leadership Education in Neurodevelopmental and Related Disabilities Programs. The same legislation established a nationwide system of university-based research facilities, under the purview of the Eunice Kennedy Shriver National Institute of Child Health and Human Development, to develop the knowledge that would be transmitted to the trainees. Over the next 20 years, Congress passed 116 acts or amendments that provided support for people with intellectual disabilities and their families in the areas of education, employment, health, housing, income maintenance, nutrition, rights, social services, transportation, and vocational rehabilitation. Specific Learning Disability and Attention Deficit Hyperactivity Disorder (previously called Minimal Brain Dysfunction) were codified in recognition of school-aged children who evidenced academic underachievement but did not meet criteria for intellectual disability.

In 1970, the Developmental Disability Service and Facilities Construction Act (PL 91–517) defined developmental disability as "a disability attributable to mental retardation, cerebral palsy, actually epilepsy, or other neurologically handicapping condition found to be related to mental retardation requiring treatment similar to that for mentally retarded individuals." Developmental disability had to originate before age 18 and was expected to continue indefinitely and constitute a substantial handicap [4].

The Amendments to the Developmental Disability Services and Facilities Construction Act (PL 95–602) expanded the scope of the 1970 Act and, in addition to statewide planning councils, established state protection and advocacy programs to protect the rights of people with developmental disabilities and funding to assist in the operation and administration of the University Affiliated Facilities. It also included a revised definition of developmental disability. A severe, chronic disability of a person which is attributable to a mental or physical impairment or combination of mental and physical impairments; is manifested before the person attains age 22; is likely to continue indefinitely; results in substantial functional limitations in three or more of the following areas of major life activity: self-care, receptive and expressive language, learning, mobility, self-direction, capacity for independent living, and economic self-sufficiency; and reflects the person's need for confirmation of a sequence of special interdisciplinary, or generic care, treatment, or other services that are of lifelong or extended duration and are individually planned [5]. Because this definition allowed for other conditions such as chronic renal disease, cystic fibrosis, and juvenile diabetes to meet the criteria, the term NDDs was used to refer to the brain-based disorders of the initial definition.

The rights of individuals with disabilities became a major focus that built on the foundational work of the civil rights movement of the 1960s. Institutional care was no longer acceptable, and the movement to community-based services accelerated with two important legal cases: Halderman vs Pennhurst State School and Hospital and Olmstead vs L.C. The Pennhurst case was the first time that a federal court ruled that an institution must be closed based on a constitutional right to community services. The Olmstead decision (1999) found that unjustified segregation of persons with disabilities constituted discrimination in violation of the Americans with Disabilities Act and held that public entities must provide community-based services to persons with disabilities.

The Education for All Handicapped Children Act (PL 94−142) guaranteed all children with disabilities the right to a free and appropriate education in the least restrictive environment. This act required that all states provide education for children aged 3−21 with disabilities. It was also a major factor in moving children out of institutional settings and into the mainstream of society. This law was amended in 1986 (PL 99−457) and expanded to include early intervention services for children from birth to three years. Ultimately, the two laws were amalgamated into the Individuals with Disabilities Education Act (IDEA) [6].

In 1990, the Americans with Disabilities Act (ADA) was enacted [7]. This was a borad piece of legislation that addresses discrimination in employment, transportation, public accommodations, access to state and local government services, and telecommunications. Perhaps the best known aspect of ADA is the required architectural modifications, such as curb cuts, wheelchair ramps, accessible pedestrian signals, and tactile warning surfaces, that enable people with disabilities to more fully participate in society. The ADA has improved access to public services, increased understanding of the abilities of people with disabilities, established a right to equal access to public services, and demonstrated the contributions that people with disabilities can make to the economy.

The concept of NDDs continues to expand in the current era. The quest to secure the rights of individuals with disabilities has expanded to self-advocacy, where no decisions are made about me without me, and neurodiversity, which questions the construct of disability. The expansion of the fields of neuroimaging and genetics has increased our ability to determine the etiology of NDDs. Determining the etiology of NDDs gives insight into the biology of these disorders and holds promise for prevention, amelioration, or cure. At the same time, this new knowledge has fragmented the parent groups and holds potential for less effective advocacy.

Assumptions that underpin the construct

NDD is a group of disorders that are uniquely pediatric in onset and affect multiple areas of life function. The rationale for this is multidimensional and is based on a number of assumptions:

1. *Brain dysfunction in childhood is different from that in adults.* Most brain dysfunction in childhood results from different etiologies than in adults. The attached Table 1.1 lists etiologies for childhood brain dysfunction and provides examples. These disorders are not selective in their expression and affect the entire brain. Consequently, in contrast to the traditional stroke model of brain dysfunction in adults, childhood brain dysfunction usually does not localize using traditional approaches. The clinical implication is that multiple functions are impaired and that multiple neurodevelopmental diagnoses and comorbidities are the rule.
2. *Children respond to the same brain injury as adults in different ways.* Maturation, plasticity, learning, and compensatory mechanisms have been put forth as modifiers of the effects of the injury. For example, in some cases of severe brain injury, the language outcomes are good, but in others, the youngest children have the worst outcomes [8,9]. In some areas, new learning occurs, while in others, compensatory mechanisms are employed.

TABLE 1.1 Etiologies of brain dysfunction in childhood.

Genetic	Down syndrome Fragile X Williams syndrome Prader Willi syndrome Rett syndrome
Hypoxic-ischemic Injury	Perinatal asphyxia Abruptio placentae Cyanotic congenital heart disease Epiglottitis
Iatrogenic	Medical providers Non-medical providers Lack of access to health care Medication errors
Immunologic	Ataxia telangiectasia Acute flaccid myelitis
Infectious	Congenital infections Meningitis Encephalitis
Metabolic	Aminoacidopathies Organic acidurias Lysosomal disorders Hypothyroidism
Nutritional	Vitamin A deficiency Iodine deficiency
Oncologic (neurocutaneous disorders)	Neurofibromatosis Tuberous sclerosis
Structural malformations	Migrational defects
Trauma	Traumatic brain injury (accidental and non-accidental)
Toxins	Lead, alcohol, tobacco, radiation

In the developing brain, there is an immediate injury, but a secondary injury may result from the response to the initial insult. There are tertiary effects from the initial injury that result from disruption of normal developmental processes. These are often seen as deficits in cognition, academics, behavior, and attention.

3. *The treatment of NDDs differs from the treatment of disabilities in adults.* In adults, the therapeutic goal is to return the person to their previous level of function. For some children with acquired injuries, the goal is the same as for adults. In children with congenitally acquired dysfunction, determining success is a nebulous process. This is because we do not have the ability to forecast the ultimate functional outcome, and therapies that seek to develop prerequisite skills are often unproven.

Diagnosis by function

NDDs are defined by function. There are advantages to grouping NDDs by *function.* It focuses on similarities, facilitates planning, and may create efficiencies in service provision. However, defining disorders by function has important limitations in the clinical realm:

1. *Functional definitions do not address etiology.* Each of these disorders may have multiple etiologies. For example, cerebral palsy may result from hypoxic-ischemic injury in a neonate, a genetic disorder, or brain trauma. This has led to some calling the neurodevelopmental disorders "wastebasket diagnoses" [10].

   In addition, some have suggested that, when an etiology has been established, the neurodevelopmental diagnosis should no longer be applied. Most recently, there has been a debate as to whether to include genetic disorders leading to the cerebral palsy phenotype under the neurodevelopmental diagnosis umbrella [11]. This presents a false choice between etiology and function. Both provide important information about the patient. Similarly, we do not exclude children with Trisomy 21 from intellectual disability.

2. *Diagnoses by function do not address mechanisms.* NDDs may have multiple mechanisms that result in functional limitations. Understanding how etiology results in brain dysfunction is necessary if we are to progress beyond symptomatic therapy. Our current level of understanding does not allow for a direct link between disordered function and the underlying biological mechanisms. For example, how does a duplication or a deletion in a chromosome result in intellectual disability? Often, we are confronted with multiple potential causes and mechanisms of the functional deficit: e.g., the extremely low birthweight infant (immaturity) who had low Apgar scores, required respiratory support (hypoxic ischemia), and had multiple episodes of presumed infection (inflammatory damage), as well as individual differences in brain plasticity, compensation, and recovery mechanisms.
3. *The definition of developmental disabilities does not state the degree of impairment required to be considered disabled.* This has created substantial diagnostic confusion. How much motor impairment is needed to move a diagnosis from clumsiness to a specific developmental disorder of motor function (developmental coordination disorder) to cerebral palsy? How much social impairment is required to change a diagnosis from language disorder or ADHD to autism spectrum disorder?
4. *The diagnoses of NDDs are not useful predictors of outcome.* It is not possible to predict the outcome for an individual child with a neurodevelopmental disability given the wide expression of these disorders. For example, diagnosing a child with cerebral palsy does not provide information about the child's function other than to note that there is significant motor impairment. If additional information is provided about the nature of the motor impairment, diagnosing spastic hemiplegia may direct attention to the affected arm, but prognosis is not possible. Adding the child's resistant epilepsy and associated dysfunctions in the cognitive, academic, and attentional areas gives a more specific prognosis. However, until familial and environmental factors are considered, the prognosis is tentative at best.

Spectra of neurodevelopmental disabilities

NDDs are defined by function and are called by their most prominent manifestations. Originally, NDDs were considered to be categorical disorders with sharp diagnostic boundaries. These disorders were established without considering overlapping symptoms and signs. Each was considered unique. ADHD and autism could not coexist. Moderate intellectual disability excluded the diagnosis of ADHD.

More recently, there has been an increased movement to consider these disorders as spectra with indistinct diagnostic margins. This approach may have addressed some of the limitations of categorical diagnosis, but it has substantively altered the diagnostic process. How many fidgets are "very much"? Where does inefficiency stop and noticeable differences start? How do you define impairment? If you go to the food court at a shopping mall to observe gait, you will see a wide range of variation that is "typical."

Studies of families have shown high heritability for ADHD and for ASD [12,13]. However, studies also show much higher rates of ASD in siblings of children with ADHD and higher rates of ADHD in siblings of children with ASD. This raises the question of whether the observed variation is the result of one entity having multiple disorders (pleiotropy) or whether there is one disorder that has multiple expressions.

The NDDs are called by their most prominent manifestations. The functions—motor, cognitive (language), and behavioral—that form the basis for NDDs are the result of coordinated, complex neurological interactions. The etiologies of NDDs result in diffuse neurologic dysfunction. Often, this dysfunction results in additional impairments, and multiple developmental diagnoses are the rule. However, this dysfunction may not rise to the level of a coexisting diagnosis in all cases but still impacts the child's function in cognitive (language), motor, or behavioral domains (reduced penetrance).

The iceberg model of NDDs (Fig. 1.2) captures the relationship between the spectrum of the identified disorders and the underlying continuum of CNS dysfunction. The neurodevelopmental diagnoses represent the tip of the iceberg and focus on diagnosis. The continuum of brain dysfunction is below the waterline and must be addressed for successful management.

If one views NDDs as the functional result of a continuum of neurological dysfunction, the case can be made that every child has every diagnosis. The difference is only in the degree to which the dysfunction is expressed. This has led some authors to recast the older construct of minimal brain dysfunction as developmental brain disorder [14], ESSENCE (Early Symptomatic Syndromes Eliciting Neurodevelopmental Clinical Examinations) [15], static

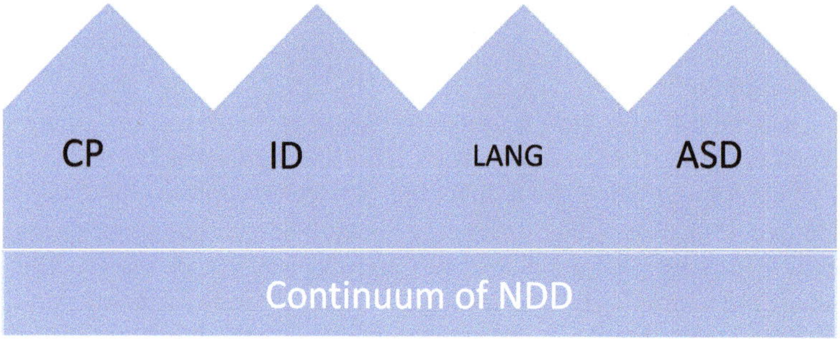

FIGURE 1.2 The Iceberg of NDD: The diagnoses represent the tip of the iceberg but there is an underlying continuum of neural dysfunction that may not rise to the level of an additional diagnosis but may impact the child's function in significant ways. *CP*, Cerebral palsy; *ID*, intellectual disability; *LANG*, language disorder; *ASD*, autism spectrum disorder.

encephalopathy, or global developmental delay. Recent findings from genetic analysis have supported this view and challenged the uniqueness of the NDDs.

When confronting the continuum, the symptoms of one entity may be overemphasized or underemphasized, leading to an over or underdiagnosis. Diagnostic overshadowing may be seen in cases where intellectual disability and autism spectrum disorder are present, and the intellectual disability may be overlooked or not addressed in favor of the ASD diagnosis. Attempts to distinguish language disorders from specific reading disabilities in young school-aged children have not been successful. Specific learning disabilities may be neglected after the diagnosis of ADHD, and vice versa. Specific developmental disorders of written expression are often overlooked when there are other diagnoses associated with academic underachievement.

In addition to having difficulty defining NDDs by severity, the spectral approach to diagnosis is further complicated by the age of the child. Age is a major complication in considering a diagnosis. The expression of developmental dysfunction may evolve as the child matures. For example, early motor delays may resolve by the time the child is two years old, but language delays become evident. Ultimately, the child attains communicative competence, but difficulty with academic achievement, organization, attention, or behavior may later become prominent expressions of the developmental dysfunction. Some dysfunctions may not appear to be atypical at a young age, but they become evident when the child reaches the age when they must integrate and apply the skills they have attained.

Some diagnostic criteria are being applied to children for whom they were not developed. For example, ADHD diagnostic criteria were developed for preadolescent school-aged children and have required modification/reinterpretation for adolescents and preschoolers. Some disorders are not diagnosable in very young children, such as ADHD in toddlers. Some are no longer impaired in adults, such as specific reading disabilities in adults who have attained literacy.

Spectral diagnosis that considers only the behavior and not the mechanism leads to misdiagnosis. "Aloofness" may be the result of the child being socially unaware or withdrawing due to stranger anxiety or a "slow to warm up" temperament. Flapping may be a stereotypy or synkinesis (overflow) associated with a specific developmental disorder of motor function. The terminology we use to describe the behaviors may also affect the diagnosis. One person's "stick to itivemess" is another's "persistence" and someone else's "perseveration." Spontaneity and impulsivity are in the eye of the beholder.

The continuum of neurological dysfunction extends from function to genetics (Fig. 1.3). Current research foci attempt to define how these points link to each other. Diagnosis, test performance, research domain criteria, endophenotypes, physiology, anatomy, and biochemistry are intermediate points on the continuum. The mediating effects on the function of environment, learning, epigenetics, and maturation remain largely unknown.

The impact of neurodevelopmental disabilities

The functional limitations that define NDDs restrict the individual's ability to participate in society as a whole. The way society approaches people with NDDs has changed over the past 50 years. Parents are no longer counseled to place their children in large residential institutions, and to "forget about them." Current thinking suggests that segregating people with developmental disabilities inhibits the individual's ability to cope with the demands of daily living. Children with disabilities interact with typically developing peers in mainstream classes and receive necessary support services in the classroom. Vocational training is no longer focused on teaching prerequisite skills. Supported employment techniques teach a specific job and work to enable the integration of a person with a disability into the workplace. Adaptive technology facilitates participation by circumventing functional limitations. The neurodiversity movement rejects the medical approach to disability which holds that NDDs are pathological conditions that need to be "treated." They espouse the viewpoint that brain differences are a normal part of the human condition and posit that societal

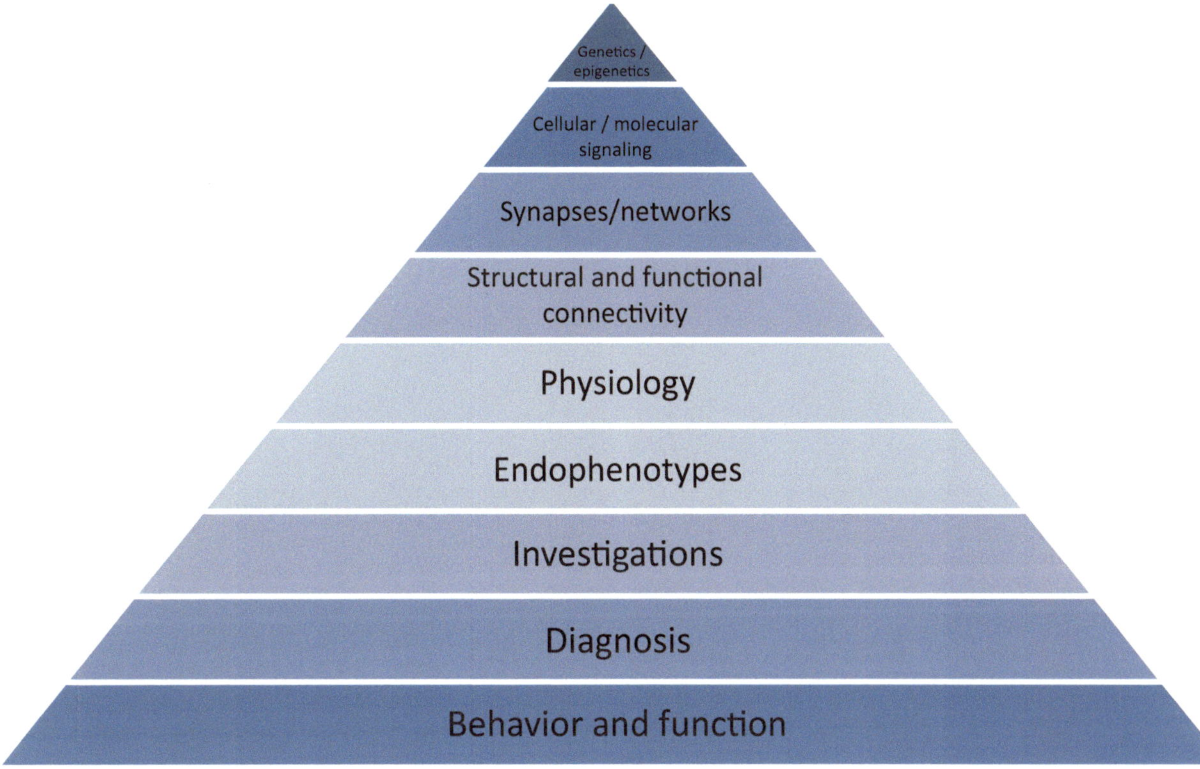

FIGURE 1.3 The continuum of neurological dysfunction extends from function to genetics.

barriers are the main contributing factor that disables people. The neurodiversity movement celebrates the unique qualities that people have and the contributions they make to society.

In composite, NDDs are common. Approximately 17.8% of children have a developmental disability [16]. Depending on the source, developmental disabilities may affect as many as 7%–19% of school children. NDDs may include: ADHD, Autism, Blindness, Cerebral Palsy, Deafness, Epilepsy, Intellectual Disability, Language Disorder, and Specific Developmental Disabilities (Reading, Math, Written Expression, and Motor Function). Table 1.2 lists the relative prevalence of these disorders.

Clinical implications

The clinical implications of etiologies that result in childhood brain dysfunction are important in the evaluation, diagnosis, and management of children with NDDs. These etiologies are not localized and usually affect multiple brain areas or single areas that are responsible for multiple functions. Because NDDs are called by the functions that are involved, multiple diagnoses are the rule.

The clinical corollaries of this are that dysfunction in one area predicts dysfunction in other areas. A complete evaluation of a child with a motor delay should also include an evaluation of their language abilities. Children with developmentally inappropriate levels of hyperactivity, impulsivity, or inattention should have their fine motor and cognitive abilities assessed.

A related corollary is: "If the treatment program is not working, reevaluation is needed because it is likely that something is being missed." Misdiagnosis is not a major cause of the failure of a management program. Incomplete diagnosis is much more common.

If the multiple diagnoses of a child with a neurodevelopmental disorder are thought to be independent of each other, then they are appropriately considered comorbid conditions. On the other hand, if they are considered to be part of a single syndrome, then calling them coexisting conditions is more appropriate. While this may be thought to be a matter of semantics, there are major implications for management. If the conditions are comorbid, parents may become overwhelmed by a long list of diagnoses and exhausted by trying to treat each one. If the conditions are coexisting, then they can be more easily understood as manifestations of an overarching neurological entity, for example, developmental brain disorder, ESSENCE, or encephalopathy, and a management program that is comprehensive yet selective can result.

TABLE 1.2 Prevalence of developmental disabilities based on the National Health Interview Survey.

ADHD	5%–10%
Autism	0.05%–1.7%
Cerebral Plasy	0.1%–0.3%
Deafness	0.1%
Moderate to Severe Hearing Loss	0.064%
Developmental Coordination Disorder	5%–6%
Epilepsy	1%
Intellectual Disability	1.11%–3%
Receptive Expressive Langauge Disorder	6%
Specific Learning Disabilities	4%–10%

Data From : McGuire DO, Tian LH, Yeargin-Allsopp M, Dowling NF, Christensen DL. Prevalence of cerebral palsy, intellectual disability, hearing loss, and blindness, National Health Interview Survey, 2009–2016. Disabil Health J 2019 Jul;12(3):443–451.

Neurodevelopmental disabilities continues to evolve

Initial formulations of NDD focused on a restricted set of functionally-based disorders. Diagnosis by function was born out of practical necessity and a limited understanding of the underlying mechanisms. NDDs remain a useful construct, but they are not complete. Cure remains an elusive goal for many NDDs, although there is substantial progress in prevention.

Traditionally, most treatments for NDDs have fallen outside the purview of medicine, focusing on addressing the expressions of brain dysfunction to limit functional limitations and increase the child's ability to participate more fully in society. However, advances in genetics and neurobiology have opened windows to understanding the mechanisms of these disorders. With this increased understanding, there is potential for the development of agents that prevent, ameliorate, and may lead to a cure for NDDs.

Neurodevelopmental disorders may arise from synaptopathies, dendritopathies, disorders of neurotransmission and intracellular signaling, neurodegeneration, defective genes, and impaired epigenetic machinery, as well as acquired brain injuries in the form of infections, inflammation, ischemia, and metabolic disturbances occurring early in life. Advances in neuroscience, imaging, molecular biology, and genetics have advanced the understanding of the biological mechanisms that may be associated with the dysfunctions that constitute NDDs. There is a debate about whether one system should replace the other. Are we ready as a field to move beyond the functional diagnosis of certain phenotypes where the underlying mechanism is important for improving treatments and outcomes? In reality, the systems are complementary, with one focusing on the biologic mechanisms while the other focuses on the functional outcomes of the disordered biology.

As the care of children with chronic medical conditions progressed, the interactions between functional and physiological disorders expanded and complicated the management of NDDs. The prevalence of NDDs in chronic diseases is unknown. The common occurrence of NDDs has only recently been appreciated in children with congenital heart disease, chronic renal disease, epilepsy, or sickle cell anemia, among others. The prevalence of NDDs in children with acquired disorders may be underestimated because families of children with acquired disorders from cancer or trauma may not consider them to have neurodevelopmental disorders but instead choose to identify with their etiologic diagnosis. Chronic conditions may have intermittent periods of physiologic decompensation that may impact cognitive, motor, or behavioral development and thereby complicate the management of neurodevelopmental disorders.

The recognition of neurodevelopmental dysfunction in chronic disease raises the question of where the boundaries of NDD should lie. While every person with an NDD would meet the criteria for a developmental disability, not every person with a developmental disability would have their functional limitations due to CNS dysfunction. Should every systemic/chronic disease of childhood that impacts brain function be classified as an NDD?

The question of whether NDD should be expansive or restrictive is made more complex in the cases of children with chronic neurologic disorders such as spina bifida, epilepsy, or muscular dystrophy. While some of them can have coexisting cognitive, language, or behavioral disabilities, most do not. However, the adult outcomes for many of these individuals are suboptimal.

Some of the initial criteria that underpin the construct of developmental disabilities are subject to question. For example, there are exceptions to the requirement that three or more areas of life function be affected before a diagnosis of developmental disability can be made. The amendments to the 2000 DD Act state, "An individual from birth to age 9, inclusive, who has a substantial developmental delay or specific congenital or acquired condition, may be considered to have a developmental disability without meeting 3 or more of the criteria described ... if the individual, without services and supports, has a high probability of meeting those criteria later in life" [17].

The requirement for onset in childhood to diagnose NDD may be challenged. New knowledge has enabled us to detect and diagnose disorders in childhood that do not have their full expression until adulthood. Adrenomyeloneuropathy and Huntington's disease are two examples where the abnormality is present at birth but the disorder is not evident until adulthood. Adult-onset ADHD, or ASD raises an additional question. Are these disorders temporal extensions of childhood disorders, or are they distinctive? Do they share the same biological origins?

Finally, the success in the establishment of community-based services for children with NDDs has brought to the fore issues related to aging. Service systems for those who "age out" of the services guaranteed by IDEA are infrequent, poorly coordinated, and variable. Issues relating to aging, what services are needed, and how to provide them are among the challenges confronting the field of NDD.

One may ask, "Are the current models sufficient to propel the field?" Cure and significant amelioration are distant objectives for the majority of these disorders. However, the current constructs recognize the contributions of individuals to a greater society, regardless of ability. As a result, continuing efforts that seek to diminish functional limitations and increase the ability of the individual to participate in a greater society lead to new ways to view childhood-onset disability and to help families.

References

[1] Gesell A, Amatruda CS. Developmental diagnosis. 2nd ed. New York, NY: Harper and Row; 1941.

[2] de Vries JIP, Fong BF. Normal fetal motility: an overview. Ultrasound Obstet Gynecol 2006;27(6):701–11. Available from: https://doi.org/10.1002/uog.2740.

[3] Mental Retardation and Community Mental Health Centers Construction Act of 1963, PL 88–164, 42 USC, p. 292–2921.

[4] The Developmental Disabilities Services and Facilities Construction Act of 1970, PL 91–517, 42 USC p. 2661–2678.

[5] Rehabilitation, Comprehensive Services, and Developmental Disabilities Amendments of 1978, PL 95–602, 29 USC, p. 701 et seq.

[6] Individuals with Disabilities Education Improvement Act (IDEA) of 2004, PL 108–446, 20 USC, p. 1400 et seq.

[7] Americans with Disabilities (ADA) of 1990, PL 101–336, 42 USC, p. 12101 et seq.

[8] Verger K, Junqué C, Jurado MA, Tresserras P, Bartumeus F, Nogués P, et al. Age effects on long-term neuropsychological outcome in paediatric traumatic brain injury. Brain Inj [Internet] 2000;14(6):495–503. Available from: https://doi.org/10.1080/026990500120411.

[9] Sariaslan A, Sharp DJ, D'Onofrio BM, Larsson H, Fazel S. Long-term outcomes associated with traumatic brain injury in childhood and adolescence: a nationwide Swedish cohort study of a wide range of medical and social outcomes. PLoS Med [Internet] 2016;13(8):e1002103. Available from: https://doi.org/10.1371/journal.pmed.1002103.

[10] Sanger TD. Is cerebral palsy a wastebasket diagnosis? J Child Neurol [Internet] 2008;23(7):726–8. Available from: https://doi.org/10.1177/0883073808314963.

[11] MacLennan AH, Lewis S, Moreno-De-Luca A, Fahey M, Leventer RJ, McIntyre S, et al. Genetic or other causation should not change the clinical diagnosis of cerebral palsy. J Child Neurol [Internet] 2019;34(8):472–6. Available from: https://doi.org/10.1177/0883073819840449.

[12] Tick B, Bolton P, Happé F, Rutter M, Rijsdijk F. Heritability of autism spectrum disorders: a meta-analysis of twin studies. J Child Psychol Psychiatry [Internet] 2016;57(5):585–95. Available from: https://doi.org/10.1111/jcpp.12499.

[13] Larsson H, Chang Z, D'Onofrio BM, Lichtenstein P. The heritability of clinically diagnosed attention deficit hyperactivity disorder across the lifespan. Psychol Med [Internet] 2014;44(10):2223–9. Available from: https://doi.org/10.1017/S0033291713002493.

[14] Moreno-De-Luca A, Myers SM, Challman TD, Moreno-De-Luca D, Evans DW, Ledbetter DH. Developmental brain dysfunction: revival and expansion of old concepts based on new genetic evidence. Lancet Neurol [Internet] 2013;12(4):406–14. Available from: https://doi.org/10.1016/S1474-4422(13)70011-5.

[15] Gillberg C. The ESSENCE in child psychiatry: early symptomatic syndromes eliciting neurodevelopmental clinical examinations. Res Dev Disabil [Internet] 2010;31(6):1543–51. Available from: https://doi.org/10.1016/j.ridd.2010.06.002.

[16] Zablotsky B, Black LI, Maenner MJ, Schieve LA, Danielson ML, Bitsko RH, et al. Prevalence and trends of developmental disabilities among children in the United States: 2009–2017. Pediatrics [Internet] 2019;144(4):e20190811. Available from: https://doi.org/10.1542/peds.2019-0811.

[17] Developmental Disabilities Assistance and Bill of Rights Act of 2000, PL 106–402, 42 USC, p. 1501 et seq.

Chapter 2

Patterns of typical and atypical neurodevelopment

Steven M. Lazar[1], Melissa Svoboda[2], Sarah Risen[3] and Scott M. Myers[4]

[1]*Section of Pediatric Neurology and Developmental Neuroscience, Meyer Center for Developmental Pediatrics & Autism, Baylor College of Medicine, Texas Children's Hospital, Houston, TX, United States,* [2]*The Children's Hospital of San Antonio, Baylor College of Medicine, San Antonio, TX, United States,* [3]*Department of Pediatrics, Section of Neurology and Developmental Neuroscience, Baylor College of Medicine, Texas Children's Hospital, Houston, TX, United States,* [4]*Geisinger Autism & Developmental Medicine Institute, Geisinger Commonwealth School of Medicine, Lewisburg, PA, United States*

Introduction

Child development is the basic science of pediatrics.

Julius B. Richmond [1].

All disciplines of pediatric medicine share the common goal of guiding patients to their best developmental outcome. The term *development* can be defined as the process of growth and change in an expected and orderly fashion to a typical endpoint of mature independent function. Often, this process does not go as expected, which leads to significant concern and consternation for patients and their families. Human development has gone through many iterations of inquiry, and multiple theories have arisen trying to answer the question: how does an infant (or even a fetus) go from needing complete care and physiologic support to an independent adult [2]? Most formulations on the process of human development have been defined as either approaching through discreet stages or a continuous pattern with reliable markers of achievement and level of development known as developmental milestones.

Throughout these stages of development, consistent and discreet behaviors, skills, or abilities that are met in a typical, expected, and sequential order within a preordained timeframe are described as developmental milestones. Most milestones are defined by behaviors that can be observed by others. For a developmental milestone to be considered mastered, the behavior or skill must be consistent, reproducible, and generalizable [3,4].

Across cultures and geography, a consistent order or sequence that nearly all individuals experience emerges from a cohesive outline that transcends individual experience [5–8]. While varied, typical human development is predicable. For example, a typically developing infant's first independent steps occur within a relatively small window, and the skills leading up to mastery occur in a consistent sequence. When one's developmental progress does not adhere to this pattern, their development is determined to be abnormal or atypical. Given the negative connotations commonly ascribed to the terminology of normal and abnormal, the authors of this chapter have determined to use the terminology of typical and atypical in the discussion of developmental patterns. Furthermore, the use of the term *typical* and *atypical* allows one to focus on the current level of abilities and behaviors (what the child can do) as opposed to their disabilities (what the child cannot do).

An historical perspective on the study of child development

Historical references to the study of child development prior to the 20th century were quite limited and relegated to theories of education and learning. Some of the earliest descriptions of child development were reported within physicians and scientists' personal observations of their own children. In 1907, Alfred Binet, through a lens of injustice seen in placing children in mental institutions, created objective measures of intellect independent of observer bias [9,10].

Capute and Accardo's Neurodevelopmental Disabilities in Infancy and Childhood. DOI: https://doi.org/10.1016/B978-0-12-824060-1.00026-2

This signaled a change in theory that academic underachievement was due to voluntary character qualities such as poor working habits, moral fiber, and willingness to accept more innate factors of capacity [11,12].

Many theories of human development involve a series of conflicts. Amongst developmental theories, they can be categorized among those with an emphasis on biology and maturation, cognition, behavior, environment, social and cultural, or adaptation. Of the historical perspectives on development, Sigmund Freud's (1856–1939) theory of psychosexual development is one of the most well-known, infamous, yet influential. He theorized that development was a series of conflicts between innate biological drives and societal restrictions [13]. While most of the particulars of his theory are no longer accepted, his concept of development as a series of stages has influenced many later theories. As a counter to this theory, Erik Erikson proposed the psychosocial stage theory, in which development is modeled as a series of existential crises to be overcome [14,15].

Jean Piaget (1896–1980) proposed the cognitive theory of development, which states that interactions with the environment are guided by adaptation and reaction to our environment [16–19]. His theory suggested that early life was driven by reactions to sensory and motor responses, while later stages of life were driven by an increasingly complex and abstract manner of responding to our environment. While Piaget focused on an individual's response and understanding of their environment, others have emphasized the influence of one's environment on developmental change. Foremost, B. F. Skinner's (1904–90) proposal of an extreme behavioral model of development considered children and their behavior to be contingent on their response to environmental conditions [20,21]. While the current understanding of development has moved away from this radical behavioral model, the principles have had a lasting impact on clinical practice, particularly with applied behavioral analysis therapy [22].

Other theories, such as Lev Vygotsky's (1896–1934) sociocultural theory [23] and Urie Bronfenbrenner's (1917–2005) ecological systems theory [24], emphasized the importance of social environment and culture on learning and development. Vygotsky's theory supplies a framework in which the gap between a child's current skills and those yet to be learned allows the support of others to learn new skills. This theory has held a prominent position in educational practice.

While these early theories focused on internal strife and adaptation through a psychologic lens or environmental influences, other theories focused on the biological basis of development. Notable theories include the attachment theory proposed by John Bowlby (1907–90) [25], in which infant/child caregiver bonds and interactions drive the course of development, while ethologic theories emphasize evolutionary pressures and perspectives comparative to other primate species. Within the biologic models of development, Arnold Gesell (1880–1961) proposed a maturational model of development that saw genetics and neurology as a blueprint for developmental change that can be mapped by defined milestones [26–30].

Neurobiological perspective

While most neurodevelopmental assessments and conclusions are based upon observation of behaviors, the origin of all behavior and development belongs to the brain [31]. While the behaviors of a child outwardly change and develop, the underlying central and peripheral nervous system matures (through neuronal growth and pruning) and adapts (through plasticity) in a predictable, sequential, and reliable pattern throughout the human lifespan [32] (see Chapter Biological basis of neurodevelopmental disabilities for a more complete discussion). Neuroscience research over the last few decades has provided empirical evidence linking functional and neuroanatomic maturation and changes to key areas of behavioral development [33].

Neuroconstructivism is a more recent theoretical framework that associates cognitive development with the development of neural structures within and between brain regions. It proposes that by understanding the constraints that govern the development of neural structures and the interplay between neural structures and functions through interaction within brain regions and with the environment, a developmental trajectory can be explained [34]. The constraints can be at the cellular level, such as genes, epigenetic modifications, neural activity, and oscillations, or at the network level, such as experience-dependent plasticity and remodeling and cross-modal and cross-regional plasticity, which are all subjected to the constraining effect of critical and sensitive periods of brain development. The environment can exert a constraining effect. This framework postulates that atypical brain development arises from abnormal constraints [35].

While developmental theories allow for the evaluation and discussion of the process of change, that is, the dynamics of brain development, developmental milestones allow for the operationalization of theory to create and analyze data related to patients. Given the variation in the expected age of skill achievement, it is important to understand that each developmental milestone is typically achieved within a range of ages and not at a specific timeframe. Many developmental milestone charts and tables have listed the expected months as either the 50th or 75th percentile of achievement, and this can lead to a wide discrepancy in published lists of developmental milestones [36].

Developmental monitoring, screening, and assessment are an integral part of the clinical practice of general pediatrics as well as the management of children with concerns for neurodevelopmental disability. While there are many developmental milestones and skills that have been described, the most clinically useful milestones are ones that emerge within a reasonably small and predicable age range, are easily defined and observed, do not vary significantly from person to person, and are functionally relevant and memorable [2,3,37].

While milestone charts and tables can be quite useful, there are significant limitations as well as risk for misunderstanding of the dynamic process underlying the numbers. Principally, the main limitation of milestone charts, as they are typically presented, is the significant underlying variation in precision and accuracy of the data collected to collate the charts [3,4,36,38].

Developmental milestones and typical development

Developmental milestones are commonly categorized amongst one or several streams of development being: motor (including both gross and fine motor skills), cognitive (with subdivisions of receptive and expressive language skills and problem solving), and neurobehavioral (adaptive and emotional behaviors). The domains of developmental milestones can be variably combined or subdivided into further descriptive domains via a process of lumping or splitting. Convergence (or lumping) of various streams of development leads to areas such as adaptive/self-care (fine motor skills, communication skills), social (communication and motor skills), and play (communication, motor, and neurobehavioral/emotional skills). Similarly, individual domains such as language can be subdivided into receptive, expressive, and pragmatic languages. Furthermore, the relative influence of a particular stream of development is dependent upon the age of the individual being evaluated (i.e., fine motor skills and visual-perceptual skills are intimately linked in infancy and toddlerhood [39,40] but can diverge with higher-level communication skills at older ages).

Typical developmental trajectories

To understand the nature of atypical development, the typical progression of developmental milestones must be understood. The following section will provide a narrative of the neuro-maturational sequence of development, touching on key milestones that are reliable indicators of developmental skills by history or observation. The ages listed are approximations to the 50th percentile of acquisition compiled through a variety of sources [2–4,36–38,41–48]. Note should be made that the exact timing of skill acquisition is less significant than the overall trajectory given typical temporal variation. A more complete listing of milestones can be seen in Table 2.1.

Gross motor

Gross motor developmental milestones relate to the coordinated use of large muscle groups that pertain to positioning, posture, station, and mobility. They progress in a sequential order with corresponding appearance and resolution of primitive protective postural responses. Gross motor milestones can be categorized into various streams of positioning, including proneness arising in early infancy, sitting arising in mid-infancy, ambulation arising in late infancy, and stair use arising in early toddlerhood.

In the prone position, an infant's gross motor skills progress in a proximal to distal fashion and will begin to lift its chin off the surface at 1 month, list its chest at 2 months, support weight on forearms at 3 months, then begin to roll from prone to supine at 4 months. From a neutral-prone position, an infant must progress through each of these stages in order to complete the more complex skill (i.e., in order to roll, an infant must prop onto forearms/wrists, then asymmetrically push to roll into a supine position). Infants then begin to roll supine to prone at 5 months, requiring lower extremity rotational force. By 7 months, an infant will be able to support weight on hands as well as begin to creep, or "army crawl." At 8 months, these skills are integrated, which gives rise to reciprocal crawling raised on both hands and knees.

When placed in the seated position, at 4 months of age, an infant will hold its head steady, while at 5 months, an infant will support its head without a lag when raised from a supine to seated position. At 5 months, an anterior protective response emerges, where an infant's arms will extend when flexed or tilted forward. This protective response then integrates into volitional anterior propping for independent sitting at 6 months. By 7 months, the lateral protective response has developed which allows for better truncal support and independent sitting without propping. At 8 months,

TABLE 2.1 Developmental milestones acquired at approximately 50th–75th percentile.

	Gross motor	Fine motor	Adaptive/ self-help	Visual problem-solving	Receptive language	Expressive language	Social/emotional
1 month	**Chin up in prone position**[a] **Turns head in supine position**	**Hands fisted near face**	**Sucks well**	**Gazes at high contrast objects**[a] **Follows face**	**Alerts to voice/sound**[a]	Throaty noises	Discriminates mother's voice Cries out of distress Soothes when picked up[a]
Red flags	*Muscle tone too low to feed*				*No response to loud sounds*		
2 months	**Chest up in prone position**[a] **Head bobs when held in sitting position**	Will unfist hands at times **Retains rattle if placed in hand** Holds hands together	Opens mouth at sight of breast or bottle	**Follows large, highly contrasting objects horizontally and vertically**[a] Recognizes mother	**Alerts to voice/sound**	**Social smile (6 weeks)**[a] Vowel-like noises	Reciprocal smiling Responds to adult voice and smile
Red flags	*Cannot raise head in prone*				*Does not alert to voice*		*Does not fixate on face*
3 months	**Props on forearms in prone position**[a] Rolls to side	Hands unfisted 50% of the time Inspects fingers Bats at objects	Brings hands to mouth	**Responds to visual threat**[a] **Reaches for face** **Follows objects in circle (in supine position)**[a] **Regards toys**	**Regards speaker**	Coos[a] Vocalizes when talked to	Expression of disgust (sour taste, loud sound) Visually follows person who is moving across a room
4 months	**Sits with trunk support** **No head lag when pulled to sit** **Props on wrists in prone**[a] Rolls front to back	Hands held predominately open[a] Clutches at clothes **Reaches persistently** **Plays with rattle**	Briefly holds onto breast or bottle	Mouths objects Manipulates fingers at midline[a] Stares longer at novel faces than familiar **Shakes rattle** **Reaches for ring/rattle**	Orients head in direction of a voice[a] Stops crying to soothing voice	Laughs out loud[a] Vocalizes when alone	Smiles spontaneously at pleasurable sight/sound Stops crying at parent voice To and fro alternating vocalizations
Red flags	*Does not bring hands to midline*				*No cooing or gurgling sounds*		*Does not smile*
5 months	**Sits with pelvic support** Rolls back to front **Puts arms out front when falling** **Sits with arms supporting trunk**	**Transfers objects: hand-mouth-hand**[a] **Reaches/grasps dangling ring**[a]	Gums/ mouths pureed food	Turns head to look for dropped spoon **Regards pellet or small cracker**[a]	Begins to respond to name **Orients to bell laterally**[a]	Says "Ah-goo"[a] Razzes[a] Expresses anger with sounds other than crying	Recognizes caregiver visually Forms attachment relationship to caregiver
6 months	**Sits momentarily propped on hands** Pivots in prone In prone position, bears weight on one hand	**Transfers hand-hand** **Rakes pellet**[a] **Palmar grasps cube**[a] **Lifts cup**[a] **Reaches with one hand**	Feeds self crackers Places hands on bottle	Touches reflection and vocalizes **Removes cloth on face** **Bangs and shakes toys**	Stops momentarily to "no" Gestures for "up"	Reduplicative babble with consonants[a] Listens, then vocalizes when adult stops **Smiles/vocalizes to mirror**	Stranger anxiety (familiar versus unfamiliar people)

Red flags		*Does not transfer objects*			*Does not turn to voice, does not smile, laugh, or express self*		
7 months	**Bounces when held** **Sits without support steadily** **Lateral protection** **Puts arms out to sides for balance**	**Radial-palmar grasp** **Pulls peg out of board**[a]	Refuses excess food	**Explores different aspects of toy**[a] Observes cube in each hand Finds partially hidden object	Looks toward familiar object when named Attends to music **Orients indirectly to bell**[a]	Increasing variety of syllables	Looks from object to parent and back when wanting help (eg, with a wind-up toy)
8 months	**Gets into sitting position** **Commando crawls** **Pulls to sitting/ kneeling position**	Bangs spoon after demonstration **Scissor grasp of cube** **Takes cube out of cup**	Holds own bottle Finger feeds small cereals or crackers **Secures small pellet**[a]	**Pulls string to obtain ring**[a] **Inspects bell**[a]	Responds to "Come here" Looks for family members, "Where's mama?"... etc	Says "Dada/Mama" (nonspecific)[a] Echolalia begins Shakes head for "no"	Lets parents know when happy versus upset Engages in gaze monitoring: adult looks away and child follows adult glance with own eyes
9 months	**"Stands" on feet and hands** Begins creeping **Pulls to stand** Bear walks (all four limbs straight)	**Immature scissor grasp**[a]	Bites, chews cookie	**Rings bell**[a] **Seeks object after it falls silently to the floor**[a]	Enjoys gesture games[a] Orients to name well **Orients directly to bell**[a]	Waves "bye-bye"[a] Nonreduplicative babble Imitates sounds	Uses sounds to get attention Separation anxiety **Follows a point, "Oh look at..."** Recognizes familiar people visually
Red flags		*Does not sit or roll*			*Does not babble, does not respond in back-and-forth fashion*		
10 months	Creeps well **Cruises around furniture using two hands** Stands with one hand held **Walks with two hands held**	Clumsy release of cube **Inferior pincer grasp of pellet** Isolates index finger and pokes	Drinks from cup held for child	**Uncovers toy under cloth**[a] **Pokes at pellet in bottle or probes small holes in pegboard**[a] **Tries to put cube in cup, but may not be able to let go**[a]	Enjoys peek-a-boo Understands "no"[a]	Says "Dada/Mama" (specific)	Experiences fear Looks preferentially when name is called
11 months	Pivots in sitting position Cruises furniture using one hand **Stands for a few seconds** **Walks with one hand held**	**Throws objects** Stirs with spoon **Fine pincer grasp of pellet**[a]	Cooperates with dressing	**Finds toy under cup**[a] Looks at pictures in book	Bounces to music	Says first word other than a name[a] Vocalizes to songs	Gives objects to adult for action after demonstration (lets adult know he or she needs help)
12 months	Stands well with arms high, legs splayed Posterior protection Independent steps	**Makes crayon mark**[a] **Releases one cube in cup**[a] **Holds crayon** Attempts tower of two cubes	Finger feeds part of meal Takes off hat	Rattles spoon in cup Lifts box lid to find toy	**Follows one-step command with gesture**[a] Recognizes names of two objects and looks when named	Points to get desired object (proto-imperative pointing) Uses several gestures with vocalizing (e.g., waving, reaching) Uses two words other than names[a]	Shows objects to parent to share interest Points to get desired object (proto-imperative pointing)

(Continued)

TABLE 2.1 (Continued)

	Gross motor	Fine motor	Adaptive/ self-help	Visual problem-solving	Receptive language	Expressive language	Social/emotional
Red flags	*Does not stand or bear weight with support*				*Does not respond to name or understand "no." Does not follow a point. Indifference to caregiver.*		
14 months	Walks with arms high and out (high guard) **Stands without pulling up** Falls by collapse Walks well	Attempts to release pellet in bottle Imitates back and forth scribble **Adds third cube to a two-cube tower** **Puts round peg in and out of hole**[a]	Drinks from cup with some spilling Removes socks/shoes Chews well Puts spoon in mouth (turns over)	Dangles ring by string **Reaches around clear barrier to obtain object**[a] Unwraps toy in cloth **Dumps pellet out of bottle after demonstration**[a]	Looks appropriately when asked, "Where's the ball?" **Follows one-step command without gesture**[a]	Uses three words other than names[a] Immature jargoning[a] Names one object Points at object to express interest (proto-declarative pointing)	Shows desire to please caregiver Solitary play Functional play Points at object to express interest (proto-declarative pointing) Purposeful exploration of toys through trial and error
16 months	Stoops to pick up toy Creeps up stairs Runs stiff-legged Walks carrying toy Climbs on furniture Stands on one foot with slight support Walks backwards **Walks up stairs with one hand held**	**Builds three-to four-cube tower** **Places 10 cubes in cup** Releases pellet into bottle **Puts several round pegs in board with urging** **Scribbles spontaneously**[a]	Uses spoon with some spilling Attempts to brush own hair Fusses to be changed Picks up and drinks from cup Fetches and carries objects (same room)	Turns pages in book Places circle in single-shape puzzle **Dumps pellet out without demonstration**[a] Finds toy observed to be hidden under layers of covers **Places circle in form board**[a]	**Points to one body part**[a] Points to one object of three when named Gets object from another room upon demand Understands simple commands, "Bring to mommy" **Points to one picture when named**[a]	Uses 4–6 words[a] Mature jargoning with real words[a]	Shows empathy (someone else cries, child looks sad) Hugs adult in reciprocation Recognizes without a demonstration that a toy requires activation; hands it to adult if can't operate Kisses by touching lips to skin Periodically visually relocates caregiver Self-conscious; embarrassed when aware of people observing
Red flags	*Does not have pincer grasp*				*Does not use "mama/dada/papa"*		*No protoimperative pointing*
18 months	**Creeps down stairs** Runs well Seats self in small chair Throws ball while standing	**Makes four-cube tower** **Crudely imitates vertical stroke** **Completes round peg board without urging**[a] **Places 10 cubes in a cup**[a]	Removes garment Gets onto adult chair unaided Moves about house without adult	Matches pairs of objects **Replaces circle in form board after it has been turned around (usually with trial and error)**[a]	Points to two of three objects when named **Points to 2–3 body parts**[a] Points to self Understands "mine" Points to familiar people when named	Uses 7–10 words[a] Uses giant words (all gone, stop that) Imitates environmental sounds (e.g., animals) Names one picture on demand	Passes M-CHAT Engages in pretend play with other people (e.g., tea party, birthday party) Begins to show shame (when does wrong) and possessiveness
Red flags	*Does not walk independently*				*Has <6 Words*		*No protodeclarative pointing*
20 months	Squats in play Carries large object **Walks downstairs with one hand held**	Makes five- to six-cube tower Completes a square peg board	Places only edibles in mouth Feeds self with spoon entire meal **Uses a tool to reach toy out of reach**[a]	Deduces location of hidden object **Places square in form board**[a]	**Points to 2–3 pictures when named**[a] Begins to understand her/him/me	Holophrases ("Mommy?" and points to keys, meaning: "These are Mommy's keys.") Two-word combinations[a] Answers requests with "no"	Begins to have thoughts about feelings Engages in tea party with stuffed animals Kisses with pucker

22 months	**Walks up stairs holding rail, putting both feet on each step** **Kicks ball with demonstration** Walks with one foot on walking board	Closes box with lid **Imitates vertical line and circular scribble**	Uses spoon well Drinks from cup well Unzips zippers Puts shoes on partway	**Makes tower of 3 cubes**[a]	**Points to 4–5 pictures when named** Points to five to six body parts Points to four pieces of clothing when named	Uses 25 to 50 words[a] Asks for more Adds one to two words/ week	Watches other children intensely Begins to show defiant behavior
24 months	**Walks down stairs holding rail, both feet on each step** **Kicks ball without demonstration** **Throws overhand**	**Makes a single-line "train" of cubes**[a] **Imitates circle** **Imitates horizontal line** **Attempts to fold paper (either crumping or fold without crease)** [a]	Opens door using knob Sucks through a straw Takes off clothes without buttons Pulls off pants	Sorts objects Matches objects to pictures Shows use of familiar objects **Completes form board**[a]	**Follows two-step command**[a] Understands me/you **Points to 5–10 pictures**[a]	Two-word sentence (noun + verb)* Telegraphic speech Uses 50 + words[a] 50% intelligibility[a] Refers to self by name **Names three pictures**	Parallel play Begins to mask emotions for social etiquette
Red flags		*Unable to walk well*			*Unable to follow simple commands*	*Lack of words, two word phrases*	*Poor eye contact, lack of imitation of actions/words*
28 Months	Jumps from bottom step with one foot leading Walks on toes after demonstration Walks backward 10 steps	Strings large beads awkwardly Unscrews jar lid Turns paper pages (often several at once)	Holds self and verbalizes toilet needs Pulls pants up with assistance	**Matches shapes** **Matches colors**	**Understands "just one"**[a]	**Repeats two digits**[a] Begins to use pronouns (I, me, you) Names 10–15 pictures	Reduction in separation anxiety
30 Months	**Walks up stairs with rail, alternating feet** **Jumps in place** Stands with both feet on balance beam Walks with one foot on balance beam	**Makes eight-cube tower** **Makes a "train" of cubes and includes a stack**[a] **Folds paper with definite crease**[a]	Washes hands Puts things away Brushes teeth with assistance	**Replaces circle in form board after it has been turned around (little or no trial and error)**[a] Points to small details in pictures	**Follows two prepositions: "put block in...on box"**[a] Understands actions words: "playing ... washing ... blowing"	Echolalia and jargoning ceases Names objects by use Refers to self with correct pronoun[a] Recites parts of well-known story/ fills in words	Imitates adult activities (e.g., sweeping, talking on phone)
33 months	**Walks swinging arms opposite of legs (synchronous gait)**	**Makes 9- to 10-cube tower**[a] Puts six square pegs in pegboard **Imitates cross**[a]	Toilet trained Puts on coat unassisted	Points to self in photos Points to body parts based on function "What do you hear with?"	Understands three prepositions Understands dirty, wet Points to objects by use: "ride in...put on feet...write with"	Gives first and last name Counts to 3 Begins to use past tense Enjoys being read to (short books)	Begins to take turns Tries to help with household tasks
3 years	Balances on one foot for 3 seconds Goes up stairs, alternating feet, no rail Pedals tricycle Walks heel to toe Catches ball with stiff arms	**Copies circle**[a] Cuts with scissors: side-to-side (awkwardly) Strings small beads well **Imitates bridge of cubes**[a]	Independent eating Pours liquid from one container to another Puts on shoes without laces Unbuttons	**Draws a two- to three-part person**[a] Understands long/short, big/ small, more/less Knows own gender Knows own age Matches letters/ numerals	Points to parts of pictures (nose of cow, door of car) Names body parts with function Understands negatives Groups objects (foods, toys) **Follows two-step prepositional command**[a]	Uses 250 + words Three-word sentences[a] Uses pronouns correctly 75% intelligibility[a] Uses plurals Names body parts by use Asks to be read to	Starts to share with/without prompt Fears imaginary things Imaginative play Uses words to describe what someone else is thinking "Mom thought I was asleep"

(*Continued*)

TABLE 2.1 (Continued)

	Gross motor	Fine motor	Adaptive/self-help	Visual problem-solving	Receptive language	Expressive language	Social/emotional
Red flags	*Frequent falling, difficulty with stairs*					*Does not use 3-word sentences*	*Lack of pretend play*
4 years	Balances on one foot, 4–8 seconds Hops on one foot two to three times Standing broad jump: 1–2 feet Gallops Throws ball overhand 10 feet Catches bounced ball (4½ yrs)	**Copies square**[a] Ties single knot Cuts 5-inch circle Uses tongs to transfer Writes part of first name **Imitates gate with cubes**[a]	Goes to toilet alone Wipes after bowel movement Washes face/ hands Brushes teeth alone Buttons Uses fork well	**Draws a four- to six-part person** Can give amounts (usually less than 5) correctly Understands simple analogies: Points to five to six colors Points to letters/numerals when named Rote counts to 4 "Reads" several common signs/ store names	Follows three-step commands Points to things that are the same versus different Names things when actions are described (e.g., swims in water, you cut with it, it's something you read, it tells time. . .) Understands adjectives: bushy, long, thin, pointed	Uses 300–1000 words Tells stories[a] **100% intelligibility**[a] Uses "feeling" words Uses words that tell about time	Deception: interested in "tricking" others and concerned about being tricked by others Has a preferred friend Labels happiness sadness, fear, and anger in self Group play
Red flags	*Cannot jump in place*				*Unintelligible speech, does not answer questions*		*Ignores other children*
5 years	**Walks down stairs with rail, alternating feet** Balances on 1 foot >8 seconds Hops on one foot 15 times **Skips** **Running broad jump 2–3 feet** Walks backward heel-toe Jumps backward	**Copies triangle**[a] Puts paper clip on paper Can use clothes-pins to transfer small objects Cuts with scissors **Writes first name** **Builds stairs of blocks from model**[a]	Spreads with knife Independent dressing Bathes independently	**Draws an 8–10-part person** Gives amounts (<10) Identifies coins Names letters/ numerals out of order **Rote counts to 10** Names 10 colors Uses letter names as sounds to invent spelling Knows sounds of consonants and short vowels by end of kindergarten Reads 25 words	Knows right and left on self Points to different one in a series Understands "er" endings (e.g., batter, skater) Understands adjectives: busy, long, thin, pointed Enjoys rhyming words and alliterations Produces words that rhyme Points correctly to "side," "middle," "corner"	Repeats six- to eight-word sentence Defines simple words Uses 2000 words Knows telephone number Responds to "why" questions Retells story with clear beginning, middle, end	Has group of friends Apologizes for mistakes Responds verbally to good fortune of others
Red flags	*Resistance to dressing, sleeping, toileting*			*Cannot recognize shapes, letters, colors*		*Unable to rhyme*	*Inappropriately fearful, shy, or angry.*
6 years	Tandem walks	**Builds stairs of blocks from memory**[a] **Copies diamond**[a] **Copies drawing of flag (divided rectangle)**[a] **Writes first and last name**[a] Creates and writes short sentences Forms letters with downgoing and counterclockwise strokes	Ties shoes Combs hair Looks both ways at street Remembers to bring belongings	**Draws a 12–14-part person** Number concepts to 20 Simple addition/subtraction Understands seasons Sounds out regularly spelled words Reads 250 words by end of first grade	Asks what unfamiliar words mean Can tell which words do not belong in a group	Repeats 8–10-word sentences Describes events in order Knows days of the week 10,000 word vocabulary	Has best friend of same sex Plays board games Distinguishes fantasy from reality Wants to be like friends and please them Enjoys school
Red flags	*Does not skip or hop on one foot*			*Does not write name*			*Does not name friends or recognize emotions in others*

Bolded items represent readily reproducible items within an office visit.
[a]*Milestones used on validated developmental testing measures including CAT and CLAMS and Gesell Developmental Observation Scale – revised.*

an infant will come to a seated position independently as the ability to roll and bear weight on hands and knees integrates with postural protective responses. At 10 months, the posterior protective response develops and allows the child to pivot. Concurrently, infants will develop functional mobility through creeping at 7 months (forward propulsion in prone position using the upper extremities with the lower extremities dragged behind) and crawling at 8–9 months (forward propulsion on hands and knees utilizing both upper and lower extremities in a left-right alternating fashion). These mobility skills start to diminish as the infant matures in their upright bipedal ambulatory skills.

Once the skills in the prone and seated positions have started to mature, the early stages of bipedal ambulation emerge. At 8 months of age, an infant can grasp and pull from a seated or kneeling position to stand with support. At 9 months, the infant will start to move his legs in a lateral fashion around a fixed support. The infant will start to move away from a fixed support with both hands (10 months) or one hand (11 months) held. Between 11 and 12 months, the child will begin to take independent steps, and by 13 months, he will abandon crawling. After 15 months, the toddler no longer falls and will begin to stiffly run with his knees straight. At 21 months, running matures with bending of the knees in a more fluid movement with brief periods of both feet off the ground. At 24 months, the child can jump from a standing position with both feet of the ground straight up, and by 36 months, they will be able to jump forward. These skills mature further as the child can gallop with a single foot forward at 4 years and skip an alternating lead foot at 5 years of age.

Fine motor and visual perceptual skills

In early infancy, extending to childhood, fine motor and visual perceptual skills are inextricably linked. Practical assessment of these skills can be completed with a simple set of toys and readily obtainable objects. The child's interaction with each object allows for a detailed assessment of the maturing visual-perceptual and fine motor systems. The most common tools include a bright-colored ring with a string, a pellet or small piece of rolled paper, drawing utensils and paper, a pegboard, a formboard with a circle or square, and a set of 10 blocks.

Using a brightly colored ring suspended by a string, an examiner can elicit a visual focus at midline on the ring at 1 month, a visual track in the horizonal and vertical planes at 2 months, and a circular motion at 3 months. As the palmar grasp reflex begins to wane, the infant will hold hands primarily unfisted, which allows the infant at 5 months to reach out and grasp at the ring and will begin to transfer from one hand to the other. At 7 months, the infant will begin to inspect the ring, while at 8 months, the infant will pull the ring to them by the string when it is out of reach.

Using a small pellet, the examiner can monitor the proximal to distal maturation of fine motor hand movements. At 5 months, the infant will begin to regard the pellet, and by 7 months, they will attempt to grasp the pellet, typically with a raking motion on their wrist, and succeed by 8 months. This grasping motion then transitions radially with the thumb and first two fingers grasping with an inferior pincer grasp at 10 months. By 11–12 months, this pincer grasp matures with isolation of the thumb and first finger as well as dropping with purpose.

While an infant's grasp matures, at around 12 months, they can fully grasp a crayon or make utensils. At 16 months, the child can mirror another who scribbles, and by 18 months, they will start to scribble spontaneously. Once they reach 24 months, they begin to copy single strokes and then copy in the appropriate orientation at 30 months. Additionally, at 30 months, the child will create a circle with repetitive circular motions and create a single circle by 36 months. Progressively more complex figure copying can help track visual-motor integration reliably up to 12 years of age (Fig. 2.1) [28,47].

Alongside drawing, the integration of visual-motor skills can be reliably tacked using a series of 1-inch blocks [47]. At 6 months, the infant can grasp a block with a radial-palmar grasp with one hand, and by 8 months, he can hold a block in each hand. Using a cup or receptacle, an infant will combine them without an intentional release at 10 months and, by 12 months, will spontaneously release the block into the cup. At 18 months, the child will spontaneously drop up to 10 blocks into a cup. At 18–21 months, the child will begin to stack blocks up to 3 high. Progressively more complex block constructs can be imitated up to a 6-year-old level (Fig. 2.1).

Adaptive skills

While the above can accurately assess fine motor skills, they are intrinsically related to the development of functional adaptive skills. At 4 months, an infant will hold hands mostly unfisted. By 8 months, the infant will begin to hold its own bottle as well as finger-feed using early grasps. At 14 months, the infant will be able to grasp and bring a spoon to their mouth to feed, while at 21 months, they will be able to use a fork. These skills also cross into dressing, where a child will assist with dressing at 12 months, unzip at 21 months, unbutton at 3 years, button at 4 years, and tie shoes at 5 years.

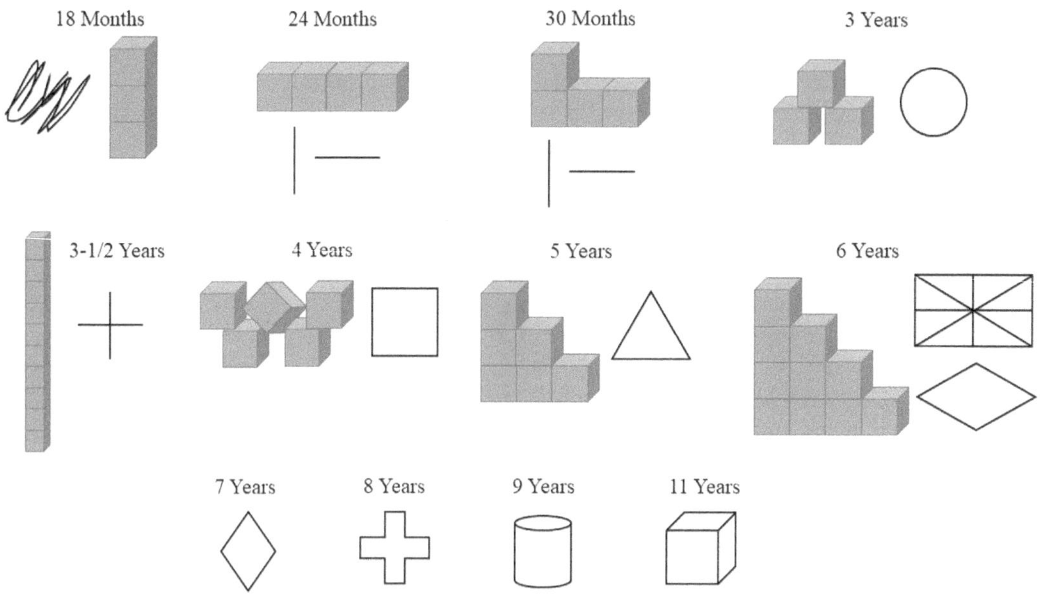

FIGURE 2.1 Typical block constructs and figure copying by typical age of attainment.

Speech and language

Speech and language development is commonly divided into multiple subdomains. Speech is the physical production of sounds that convey meaning. In this chapter, we will discuss the typical progression of spoken language. Language itself is a system that ascribes meaning to symbols in a manner in which information can be shared amongst multiple organisms. Language is traditionally categorized as the understanding of language (receptive) or the production of language (expressive).

Receptive language develops in infants initially as responsiveness to sounds. At 1 month, an infant will be alert to a sound. At 4 months, the infant can localize their gaze and head to a familiar voice. Infants mature in their localization of nonsocial sounds, such as that of a bell, with lateral orientation at 5 months, indirect orientation (lateral axis then vertical axis) at 7 months, and a mature direct localization at 9 months. By 10 months of age, the infant starts to understand the word "no." Note should be made that this is the word itself and not the typical association of higher pitch and volume accompanied. At a year of age, the child will begin to follow gestured commands ("give me" with the hand outstretched), while at 16 months they no longer require the associated visual cue. At 18 months, the infant starts to ascribe words to up to two personal body parts and a pictural representation of an object. By 21 months, the child should be able to point to two pictures by command. At 24 months, the child can extend command following two unlinked steps ("take your shoes off and wash your hands"). After 30 months, the child begins to understand "one" from a greater number of objects. At this stage, understanding of prepositions begins.

Parallel to receptive language, expressive language follows an ordered sequence of prelinguistic skills that progress to sequentially more complex utterances. An infant will progress through stages of cooing at 3 months, laughing at 4 months, and razzing (pursed lip blowing) and vowel combinations with some guttural consonants, such as "ah-goo," at 5 months. When an infant reaches 6 months, they begin to babble by making reduplicative consonant-vowel pairings ("babababa" or "dadadada"). These utterances shorted into nonspecific doublets such as "mama" and "dada" without specific meaning at 8 months. These doublets take meaning for specific people at 10 months. By 11 months, the infant begins to build a lexicon, starting with one word, which expands to 2 words at 12 months, 3 words at 14 months, 4–6 words at 16 months, 7–10 words at 18 months, and 20 words at 21 months.

Through this period of lexicon building, the child will initiate string nonsense sounds with prosody and pitch changes of mature sentences (immature jargoning) at 14 months. This then matures with the inclusion of whole words into utterances at 16 months. At 18 months, echolalia becomes prominent with the rise in rate of word acquisition. At 21 months, the child starts to string two words together, first into phrases combining concepts ("Daddy up"), which then start to include verbs at 24 months ("I want milk"). Incorporation of appropriate pronouns begins at 30 months, while echolalia ceases. Between the ages of 24 and 36 months, an "explosion" of language is typically noted, which corresponds to maturing higher-order cortical language networks [49].

At 3 years of age, the child begins to string three words into sentences with a word count of up to 250 words. At this time, they also begin to ask questions, answer their gender and age, and provide their full name when prompted. Between 3 and 4 years, the child will begin to answer questions related to adaptive skills, such as "What do you do when you are hungry? Cold? Tired?." At 4 years old, the child will string multiple sentences together to relay a personal experience or story.

Intelligibility of speech progresses from 50% at 2 years, 75% at 3 years, and 100% at 4 years of age.

Social communication and play

Social development begins at 1 month, when the infant will visually focus on a face. At 2 months, the infant will smile in response to another's face reliably. At 6 months, the infant begins to develop stranger anxiety, which leads to separation anxiety at 9 months as the caregiver-child bond grows. At 8 months, following another's gaze becomes an apparent skill, while following a point begins at 9 months. At 9 months, the infant will begin to engage in imitative games and wave "bye-bye." Expression of wants through protoimperative pointing (isolation of finger to indicate a need, want, or request) emerges at 12 months, while expression of interest through protodeclarative pointing (isolation of finger to indicate interest or request shared attention) emerges at 14 months.

At 15 months, the child will begin to show signs of empathy and hug in reciprocity. Play begins to mature with pretend play and imitation of household chores at 18 months. At 2 years of age, the child will play next to peers without persistent interaction (parallel play). At 3 years, interactive group play becomes prominent amongst peers.

Streams of neurodevelopment

Neurodevelopment can be described in the form of multiple interrelated developmental streams, or areas of development that follow an expected trajectory for one portion of developmental functioning. This active process suggests ongoing and constant progress as the child's neural networks grow and change. Dr. Arnold Capute (1923–2003) is recognized as the "father of neurodevelopmental pediatrics" and has posited three core streams of development: motor, cognitive, and neurobehavioral [50–52]. These 3 larger categories can be broken down into more specific streams of development, which can be individually assessed (Fig. 2.2).

The motor stream includes gross motor, fine motor, as well as oromotor development. The cognitive stream is made up of language, problem-solving, visual-motor (aka visual-perceptual), and adaptive functioning domains. The

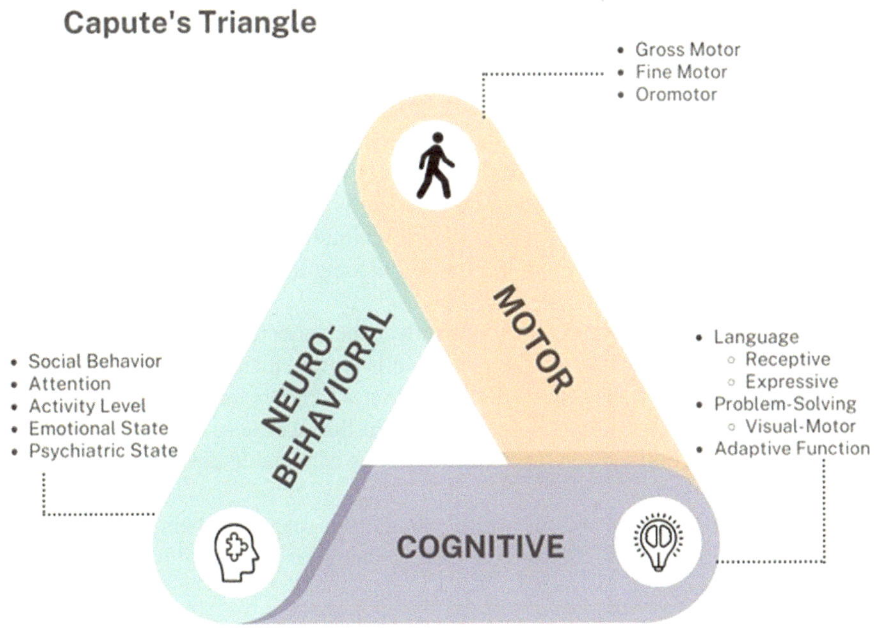

FIGURE 2.2 Capute's triangle of neurodevelopment. The triangle is a conceptual framework that illustrates the dynamic interplay among three fundamental domains of child development: cognition, motor skills, and communication, which influence each other's growth and progress as children develop. The domains do not evolve in isolation but rather as a cohesive system.

neurobehavioral stream includes social behavior, attention/activity level, and emotional/psychiatric states [53–55]. These streams are not rigid pipes and are the result of complex interactions. Examples of the overlap occurring are obvious visual-motor (vision, motor, as well as a cognitive component) or less obvious, such as social (which encompasses language, cognitive, motor, and neurobehavioral components). In addition, developmental milestone charts typically display these smaller streams of development as one skill precedes the next in a natural flow of brain development.

The developmental quotient

Developmental delays for each stream of development are diagnosed when the child's current milestone level is significantly behind what is expected for their chronological age (see below). Typically, this is defined as current milestones that are 1.5–2 standard deviations or more below the mean for that child's age [40,46,53]. For children over 6 years old, more formal neuropsychological testing is typically used to find how far a child is from the norm. However, in younger children, the developmental quotient (DQ) can be used to provide an objective assessment of the extent of developmental delay. The DQ is calculated by dividing the developmental age equivalent (DA) for a certain stream of development by the child's chronological age (CA) multiplied by 100 ($DQ = DA \div CA \times 100$). This percentage can be thought of in terms of a standard deviation (15) around a mean (100), such as how an intelligence quotient (IQ) is commonly represented. For example, a DQ of 70 is 2 standard deviations below the mean and therefore qualifies as a significant delay in that stream of development. Additionally, a $DQ < 50$ generally means there is very little chance of the child catching up, and this is likely to be a true deficit throughout life. There may be different, or dissociated, DQs for different streams of development, and this one number is not meant to encompass an entire child's developmental profile.

$$DQ = \frac{\text{Developmental Age (DA)}}{\text{Chronologic Age (CA)}} \times 100$$

The developmental age can be found from a referenced milestone chart (such as Table 2.1 included in this text) or calculated from certain developmental tests made for young children, such as the Capute Scales [40,46]. The chronological age can also be corrected for prematurity if desired and within the age range to do so. In addition, a corrected and uncorrected DQ may be helpful to understand a premature child [56].

The DQ quantifies the degree of a child's delay. For example, for a 20-month-old boy who just started independent walking, we would calculate his gross motor DQ as follows: his developmental level would be 12 months over his chronological age (20 months), giving you a DQ of 60 ($12 \div 20 \times 100 = 60\%$). To interpret this, a DQ of 60 is less than 2 standard deviations below the mean ($DQ < 70$) and therefore definitely qualifies as a gross motor delay.

Limitations of the global developmental quotient

While the developmental quotient (DQ) is easy to use, there are limitations. Caution should be used in the predictive value of the DQ the younger a child is. For example, if a child is 4 months old and gross motor skills are at a 2-month-old level, $DQ = 50$. However, this child has only been in the world for 4 months and is only 2 months behind where he should be. Over time, he may gain many skills or maybe even stay just 2 months behind his chronological age for some time, so his DQ will improve over time. The DQ is less predictive of the future outcome the younger the child is. This predictive value improves as the child reaches 6 years old.

In addition, a global DQ that is taken at a single point in time does not reflect the trajectory of a child's development. It also does not account for splinter skills or other nuances of the child and is simply a single percentage of skill attainment in one stream of development at a single point in time. A DQ of 50 on a single evaluation does not fully reflect the nuance needed for neurodevelopmental diagnosis and management, given that there are multiple trajectories of development that can lead to this point in time. See Fig. 2.4 for further descriptions of variable developmental trajectories within the neurodevelopmental assessment.

Despite these pitfalls, the DQ is an excellent tool that assesses the child's skills in a snapshot of time. If the DQ is less than 70, the pediatrician can be assured the child is behind and should refer to therapies and possibly begin a work-up for etiology.

Clinical application of developmental milestones

Assessment using developmental milestones is the cornerstone of a clinical neurodevelopmental evaluation. Current clinical guidelines recommend developmental surveillance—the process of recognizing children who may be at risk for developmental delays—at every well-child check until 5 years of age. In addition, more detailed developmental screening—the use of standardized tools to identify and refine that recognized risk—is advised via standardized developmental screening tests anytime a parent has concerns and at the 9-, 18-, and 30-month visits [57].

Surveillance is the close observation of development over time and is typically done by the primary care physician. Essentially, this translates to asking about development at these visits, taking a good developmental history, and addressing any parental concerns regarding development. The developmental history is divided into distinct streams of development, and discussion of each one as compared to normed developmental milestones is key. If a parent or clinician reveals any developmental concerns, a formal screening test is recommended and/or a direct referral to services should be made. Screening is a more formal process whereby a more formal test is administered to evaluate development. These tests more formally and systematically identify areas of concern.

A formal neurodevelopmental assessment includes a detailed developmental history for each stream of development. It is important to understand the trajectory of development rather than just obtaining the child's current milestone attainment. As children get older, the recall bias of the caregiver becomes a barrier, and it can then be helpful to ask about a more memorable past milestone (e.g., first walking/taking independent steps) and then current functioning to understand the gross motor trajectory. These trajectories are important for assessing the overall pattern of development (see pattern graphs in the following section). For example, regression, or a loss of previously mastered skills, can be seen in communication and social domains in a child with autism spectrum disorder. In addition, the pattern of developmental delays is important to assist in diagnosis. For example, receptive language that is worse than expressive language as well as social deficits may be indicative of an autism spectrum disorder. Lastly, formal neuropsychological assessments are used to assess direct observations of the child more objectively.

Patterns of atypical neurodevelopment

As the achievement of developmental milestones is typically orderly and sequential, atypical patterns of development occur when a child does not achieve a milestone as expected compared to age-matched peers. Through a thorough developmental history, including the timing and rate of developmental milestone achievement in and across each domain, the clinician can recognize atypical patterns of development. Neurodevelopmental dysfunction is reflected by four patterns of abnormal developmental milestone achievement: delay, dissociation, deviance, and regression (Table 2.2) [52]. These patterns are not mutually exclusive. In fact, certain combinations of abnormal brain development patterns may suggest specific neurodevelopmental disorders.

Developmental delay is a significant lag in milestone achievement, though still within the typical sequence, in one or more streams of development. A significant delay is considered a DQ $<70\%-75\%$, corresponding to performance at $<1.5-2$ standard deviations below the mean. Dissociation is a significant asynchrony in one stream of development when compared to other developmental areas. A difference in developmental rates of 15% or more between two streams is consistent with dissociation. Deviation is the atypical, nonsequential attainment of milestones within a developmental stream, where more advanced age-level skills are achieved before lower-level

TABLE 2.2 Definitions of atypical patterns of neurodevelopment.

Delay	A significant temporal lag in the sequential attainment of milestones in one or more developmental streams (i.e., isolated gross motor delay in setting of muscular dystrophy)
Dissociation	A significant difference in the sequential temporal attainment of milestones between two or more major developmental streams (i.e., expressive language delay with intact receptive language)
Deviance	A violation of the developmental sequence; nonsequential unevenness in attainment of milestones within one or more developmental streams (i.e., performing higher level skills before mastering foundational skills or skipping a milestone)
Regression	A persistent loss of previously attained milestones within one or more developmental streams (i.e., leukoencephalopathy with progressive neurologic deterioration)

foundational skills. While developmental delay and dissociation reflect abnormal rates of achievement in typical, sequential development (skills that are expected but in a younger child), deviance always reflects underlying atypical neurodevelopment. As development is typically progressive, developmental regression, or the loss of previously attained developmental milestones, is always worrisome. Milestones are successfully achieved when a child can perform that skill independently, spontaneously, and appropriately. A careful developmental history is needed to diagnose regression, considering potential parental misperceptions of achieved milestones (e.g., a parent reporting their child had three words when the words were only said in repetition). Atypical patterns of development are clinical manifestations of dysfunction of the developing brain and require further etiological exploration as well as longitudinal neurodevelopmental monitoring.

Classic examples of atypical neurodevelopment

Developmental delay occurring in one stream of development results in specific diagnoses. For instance, cerebral palsy is a significant delay in motor skills, while receptive-expressive language disorder is a significant delay in the acquisition of language milestones. Attention-deficit/hyperactivity disorder can be considered a delay in neurobehavioral development.

Global developmental delay (GDD) reflects a significant lag in two or more streams of development in children younger than 5 years, with or without dissociation across streams, and typically projects a higher risk of later diagnosed cognitive impairments and intellectual developmental disability (IDD). The impact of one neurodevelopmental disability on performance may affect several other streams of development (e.g., spasticity in cerebral palsy can affect gross-, fine-, and oro-motor skills) [58]. Reflecting widespread deficits of neuromaturation in young children, GDD can also be considered a significant delay in both language and nonverbal/visual-motor problem-solving skills with associated deficits in adaptive behavior without clear specific motor deficits.

After identifying a delay in one developmental stream, it is important to consider the child's developmental achievement and trajectory across all streams to identify global patterns of atypical development and further define developmental diagnoses. Developmental delay is commonly associated with dissociation, and these relationships create a neurodevelopmental framework to guide diagnosis and management. Receptive-Expressive Language Disorder is a delay in language acquisition that is also dissociated from nonverbal or problem-solving skills; however, dissociation may also occur within a stream of development, such as expressive language being more profoundly delayed relative to receptive skills in expressive language disorder. Sensory impairments (vision or hearing) typically result in the dissociation of development. A child with visual impairment often has associated delays in gross and fine motor skills alongside visual-motor problem-solving skills, but often with typical language development. Whereas a child with hearing impairment typically presents with isolated language delay dissociated from development in other areas. A specific learning disability with impairment in reading, math, and/or written expression is classically defined as a developmental delay in a specific skill as well as dissociation between performance of that skill relative to the child's cognitive or academic ability. Our understanding and definition of specific learning disabilities have evolved from low academic performance—IQ discrepancy (or dissociation) —to recognition of a discrepancy (or dissociation) between the actual and expected performance of a skill with a more global understanding of the child's abilities and contextual factors [59,60]. Consideration of the functional impact of a developmental delay and dissociation can guide neurodevelopmental diagnosis throughout the spectrum of severity.

Children with autism spectrum disorder (ASD) often present with a combination of developmental delays in language and social skills, dissociated from other developmental streams, with deviations in language and/or visuomotor problem solving milestone achievements. An uneven, developmentally delayed, dissociated, deviated, and communicatively disordered developmental profile, along with neurobehavioral manifestations of restricted interests and repetitive behaviors, is a typical neurodevelopmental profile observed in children with autism spectrum disorder. An example of language deviance is a child who can label numerous objects, colors, letters, and numbers but does not use more foundational functionally communicative skills such as gesture language, verbal requests, or the use of "mama or dada" specifically. This specific pattern of deviance results from strong rote memory skills, despite significantly impaired social and pragmatic language skills, which increases suspicion for a diagnosis of ASD [61]. Developmental deviance may be readily apparent through a comprehensive neurodevelopmental history but may be missed in assessment if the deviance is beyond the ceiling of a psychometric evaluation (or the testing is discontinued prior to reaching the true ceiling of the deviant skills). Recent measures have been evaluated to quantify developmental deviance and significant response "scatter" on standardized neurodevelopmental assessments [62]. Developmental deviance does not specifically imply autism spectrum disorder and can be seen in other neurodevelopmental disabilities. For instance, a child with spastic diplegia may cruise along furniture before being able to sit independently. Thus, the presence of deviance in any stream is likely a harbinger of atypical neurodevelopment and should prompt a thorough evaluation.

Developmental regression is another abnormal pattern that should prompt thorough history, testing, and evaluation. Regression requires the loss of a skill previously mastered (that the child performed independently, accurately, spontaneously, and consistently for a period of 3–6 months). Developmental regression may become apparent abruptly or progress slowly over time and may impact only one or all streams of development. Furthermore, regression can occur in a child with prior normal development or in a child with prior atypical development. Etiologies for regression include genetic and metabolic disorders, including neurodegenerative and epileptic encephalopathies, autoimmune disorders, toxins, or the presence of a tumor or other structural abnormality impacting function. It is critical to understand the child's developmental profile and trajectory of milestone achievement prior to the onset of regression, the rate of developmental regression, and the developmental streams and skills involved in the regression. Furthermore, the timeframe of a regression (i.e., chronic or subacute, often seen in metabolic and neurodegenerative disorders, versus acute, often seen in traumatic brain injury and infectious meningoencephalitides) can help guide diagnostic evaluation. Considering the differential diagnosis list includes some etiologies with specific or emerging treatment and/or an expected clinical course, a diagnostic evaluation is always recommended in a child with a true history of regression.

"Classic autistic regression" [63–67] is sometimes evoked by clinicians as a reason to not pursue further diagnostic evaluation. However, as knowledge of genetic, metabolic, and neuroimmune (infectious or autoimmune) disorders continues to evolve and considering the risk of missing a treatable etiology, even "classic autistic regression" warrants an evaluation. A strong history and neurodevelopmental exam can help identify certain patterns of regression associated with epileptic, toxic, genetic, metabolic, and autoimmune etiologies of regression (see Part III for discussion of specific clinical disorders). While language regression is more common in children with a diagnosis of autism compared to children with either specific language impairment or global developmental delay, estimates of autistic regression are quite variable, depending on the study definition and methodology [66,67]. Noting significant heterogeneity in current literature, a 2021 metaanalysis estimates 30% of children with autism experience language and/or social regression with an average age of onset at ~20 months (15–30 months range) [63]. Interestingly, few cases of regression are reported in prospective studies of infants at high risk for developing autism [68].

Atypical neurodevelopment and developmental diagnoses

Mapping atypical patterns, or profiles, of neurodevelopment (delay, dissociation, and deviation) onto the developmental streams can highlight specific neurodevelopmental diagnoses. A child with global developmental delay typically presents with delays, though to varying degrees, in all streams of development. Mixed receptive-expressive language disorders present with broad language delays and a variable associated impact on social, adaptive, or self-help skills. In contrast, a child with an expressive language disorder presents with significant delay and dissociation in only expressive language. As "spectrum" diagnoses, ASD and cerebral palsy can be associated with variable patterns of developmental delay, dissociation, and deviation. Classically, children with ASD have a clear dissociation in development with language and social skills significantly more impaired than visuomotor problem-solving skills. Developmental deviance within one or more of these areas is also typical in children with ASD. As a disorder of motor function, CP presents with more impaired gross and fine motor skills compared to the other streams of development, though motor deficits can adversely affect visuomotor, oromotor, and adaptive/self-help skills.

Alongside the neurodevelopmental profile that is present at the time of evaluation, each child comes with a unique temporal pattern of development. Tracking developmental trajectories (Fig. 2.3) can assist in etiologic considerations, guide diagnostics, and help to counsel on outcomes and prognostication. Historical milestone achievement by the patient is necessary for accurate diagnosis and management, as strict reliance on psychometric measures of DQs can be misleading. Fig. 2.3 details four sample developmental trajectories that each could present with a DQ of 50. Congenital static encephalopathy represents the child who has persistently developed at a rate of 50% of what was expected since birth and is expected to continue on this trajectory (Fig. 2.3A). Early life and perinatal insults can show a brief, typical developmental trajectory, which then leads to a change in course to a DQ of 50 (Fig. 2.3B). An acquired brain injury later in childhood may show typical development until the inciting event and lead to a pattern of regression. Multiple predictive lines are included to represent the guarded prognosis and outcomes of acquired brain injury recovery (Fig. 2.3C). Neurodegeneration is modeled as an initially delayed developmental trajectory that later in life shows a further regression and progressive loss of skills (Fig. 2.3D). This trajectory highlights the guarded nature of neurodevelopmental prognostication, especially with an incomplete etiologic evaluation.

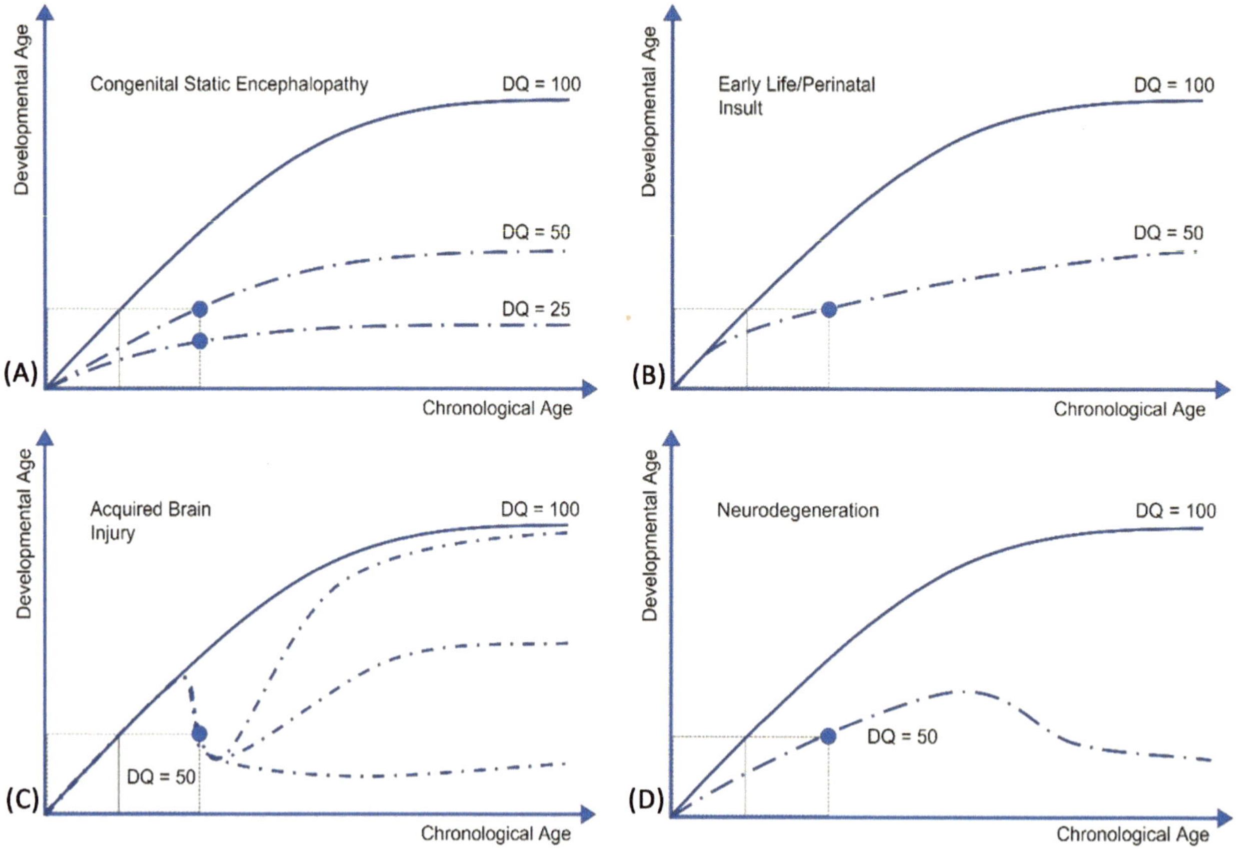

FIGURE 2.3 Variable atypical neurodevelopmental trajectories of children with a DQ of 50 inclduing congential static encephalopathy (A), early life or perinatal insult (B), acquired brian injury in childhood (C), and neurodegeneration (D).

The spectrum and continuum of neurodevelopmental disabilities in context

Importantly, given that child development reflects underlying brain development, skills across neurodevelopmental streams are often interconnected, and difficulties are rarely present in complete isolation. While certain patterns of developmental impairments are diagnostic of specific neurodevelopmental disabilities, associated deficits that may not reach categorical diagnostic criteria may still be functionally impairing and are equally important to recognize. Frequently, the presenting complaint or primary neurodevelopmental diagnosis is only the "tip of the iceberg," indicating a broader underlying continuum of neuronal dysfunction.

Despite this continuum of neurodisability, clinically, children often present with only one "chief complaint." Along the spectrum of neurodevelopmental disabilities, gross motor delays are the most common presentation in infants and early toddlers, while concerns with language development are the most common clinical presentation thereafter. Considering the spectrum and continuum of neurodevelopmental disabilities, the evaluation of all children with a developmental concern should include a history and assessment of skills within and across all streams of development, even if the chief complaint is only within one area. Commonly, deficits associated with the primary neurodevelopmental disorder may be equally or more functionally impairing. While more severe and profound neurodevelopmental disorders present earlier clinically, milder deficits can also be significantly impairing. In fact, less obviously severe impairments, such as attention-deficit/hyperactivity disorder (ADHD) or learning disabilities, are still variants within the continuum of neurodevelopmental disabilities resulting from the same underlying neurological dysfunction. It can be helpful to parents to connect the various disabilities to the continuum of neurodevelopmental disabilities, as one broad etiology manifests across the spectrum of development.

Neurodevelopmental disabilities are the result of abnormal brain development and neural dysfunction, manifesting clinically as developmental delay, dissociation, and deviance, and neurodevelopmental disabilities persist and/or evolve over time. Understanding these developmental principles and atypical developmental patterns is critical for accurate neurodevelopmental assessment and diagnosis, which then serve as the foundation for ongoing biomedical evaluations and rehabilitation planning.

Conceptualizing patterns of atypical neurodevelopment

Human neurodevelopment is a complex, dynamic process involving orderly and predictable maturation of the central nervous system (CNS), with the concomitant emergence of behavior that is similarly orderly and predictable: quantifiable developmental milestones and qualitative features within each stream of development [28,36,69]. Independent assessment of skills within each stream of development identifies an individual's performance relative to age-specific expectations and a profile of relative strengths and weaknesses, which facilitates the recognition of patterns of atypical neurodevelopment. Careful gathering of historical information and serial evaluations over time allow for consideration of developmental trajectory within and across domains.

Neurodevelopmental disabilities are chronic disorders involving cognition, communication, motor skills, or behavior that arise from CNS dysfunction during the developmental period and result in substantial limitations in multiple aspects of adaptive functioning [70,71]. Traditionally, a spectrum of milder, higher prevalence disorders (e.g., language disorders, learning disorders, developmental coordination disorder, ADHD) and more severe, but less common, disorders (e.g., intellectual disability, cerebral palsy, autism spectrum disorder, epilepsy) have been included within the concept of neurodevelopmental disabilities. Other low-frequency, high-severity conditions such as spinal cord disorders (e.g., myelomeningocele and other congenital malformations of the neural tube) and severe sensory impairments (i.e., blindness and deafness) are sometimes included within the neurodevelopmental disabilities concept. These are clinically defined conditions; there are currently no biomarkers that can be used to make the diagnoses or to differentiate among them. Although subject to maturational changes, neurodevelopmental disabilities are typically nonprogressive and tend to follow a relatively steady trajectory rather than a pattern of remitting and relapsing. Tic disorders are an exception to the latter since they are characterized by the waxing and waning of core symptoms.

The co-occurrence of neurodevelopmental disabilities is the rule rather than the exception [71–73]. Historically, the extensive cooccurrence and neural origin of childhood-onset neurodevelopmental disabilities have been acknowledged and emphasized by terminology such as minimal brain dysfunction [74], deficits in attention, motor control, and perception (DAMP) [75], early symptomatic syndromes eliciting neurodevelopmental clinical examinations (ESSENCE) [76], and the continuum aspect of Capute and colleagues' spectrum and continuum of developmental disabilities [51,52]. See Chapter The evolution of neurodevelopmental disabilities: an updated view of the spectrum of this text for a more thorough discussion of the theoretical constructs behind the spectrum and continuum of developmental disabilities.

It was recognized in the mid-19th century that epilepsy, intellectual impairment, and various psychiatric disorders often coexisted in individuals and within families, and this heritable diathesis became known as the neuropathic taint [77]. However, it wasn't until a decade ago that the neurodevelopmental continuum [78,79] and developmental brain dysfunction (DBD) [53,72] models conceptualized certain psychiatric disorders (including schizophrenia and bipolar disorder) as existing on the same pathogenetic continuum as the traditional neurodevelopmental disabilities, largely based on compelling evidence of shared genetic architecture. Rather than being etiologically discrete entities, these psychiatric disorders and neurodevelopmental disabilities represent a range of outcomes of altered neural development.

Developmental brain dysfunction

The developmental brain dysfunction (DBD) conceptual framework is illustrated in Fig. 2.4 [53,72]. In this conceptual model, "developmental brain dysfunction" serves as an umbrella term for the constellation of pathophysiologic processes underlying neurodevelopmental disabilities, most of which are yet to be elucidated, and represents the common denominator of categorical neurodevelopmental diagnoses and their component signs and symptoms.

Neurodevelopmental disabilities result from multiple genomic and genetic determinants interacting in the context of prenatal and postnatal environmental factors, selective vulnerability over time, and stochastic developmental variation to give rise to clinically diverse phenotypes. In contrast to focal neurologic deficits that occur in the mature adult nervous system due to insults such as cerebrovascular accidents, dysfunction of the developing brain tends to be diffuse, resulting in varying degrees of impairment in cognitive, motor, and neurobehavioral functioning and sometimes observable CNS morphological anomalies (e.g., microcephaly, macrocephaly, structural malformations) or neurophysiological abnormalities (e.g., seizures, EEG abnormalities) [53,72]. Clinical diagnoses are determined by the profile of signs and symptoms and specific impairments using operationalized diagnostic criteria (e.g., threshold number and duration of

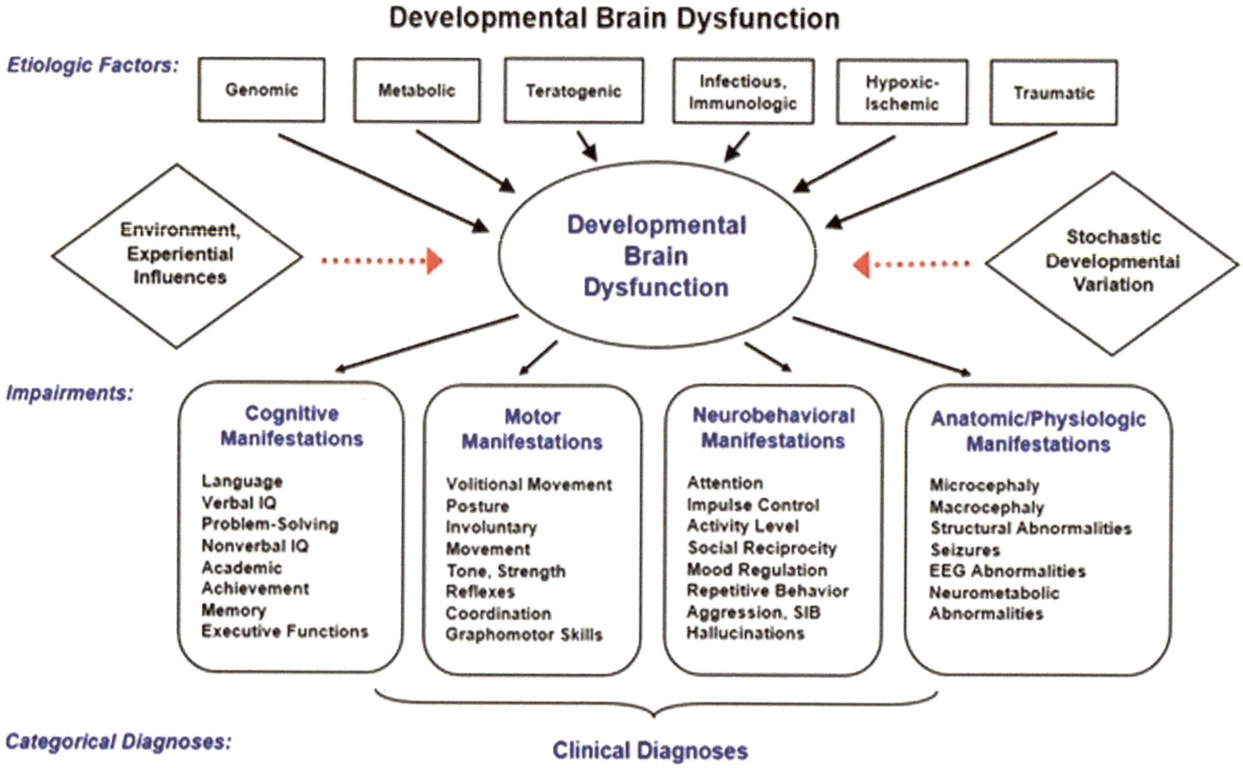

FIGURE 2.4 Model of developmental brain dysfunction. *Adapted from Myers SM. In: Batshaw ML, et al., editors. Children with disabilities, 7th ed. (2013) & 8th ed. (2019). Baltimore, MD: Brookes Publishing.*

symptoms and behaviors, presence of significant functional impairment, exclusion criteria). There is not a one-to-one correspondence between etiologic diagnosis and categorical disorder diagnosis; any neurodevelopmental disorder can result from multiple different causes (etiologic heterogeneity), and any genetic or other etiologic insult can result in multiple different neurodevelopmental disorders (variable phenotypic expression).

The observed frequent cooccurrence of different disorders within individuals and families (including those with identical pathogenic variants), overlap of signs and symptoms across disorders, within-disorder etiologic heterogeneity, and sharing of etiologies and risk factors across disorders are all consistent with the DBD conceptual model [72,80]. For example, the DBD concept fits very well with advances in understanding of the genetic architecture of neurodevelopmental disabilities, which includes rare de novo and inherited deleterious variants (e.g., copy number variants, single-gene sequence variants) of large effect size [81–85] and polygenic risk conferred by many common variants (single-nucleotide polymorphisms) of individually small effect size but collective importance [82,86,87]. It has become clear that genetic risk is not disorder-specific; pathogenic variants in the same genes are identified across diagnostic categories, including ASD, ID, epilepsy, ADHD, schizophrenia, and bipolar disorder [82,84,88], and polygenic risk is similarly shared across disorders [82,89,90]. Based on these findings and the DBD model, it has been proposed that neurodevelopmental disabilities should be considered a group for gene discovery research, assessing the pathogenicity of variants, determining penetrance, and exploring the origins of variation in phenotypic expression, as opposed to evaluating genotype-phenotype relationships and curating variants for individual diagnoses [72,80,91].

Diagnosing atypical neurodevelopment

The diagnostic evaluation culminates in a diagnostic formulation, which should include not only the categorical diagnoses but also delineation of the specific impairments (which are typically the primary targets of intervention), assessment of severity and prognosis, etiologic diagnoses (or potential etiologic diagnoses and plans for investigation), and a treatment plan.

Diagnostic classification

Unlike most areas of medicine in which conditions are classified systematically as *diseases* based on etiology and pathophysiology, in developmental medicine and psychiatry, conditions are classified phenomenologically as *disorders*, or syndromes defined by clusters of attributes or symptoms [92,93]. This is the primary way that neurodevelopmental disability diagnoses are classified using the *International Classification of Diseases* (ICD-11) [94] and *Diagnostic and Statistical Manual of Mental Disorders* (DSM-5) [95] diagnostic classification systems. Although phenomenological categorical diagnoses are effective heuristics that enhance interrater reliability, they often do not align well with genomic, neuroimaging, and other neurobiological findings, and their validity has been challenged [92,96,97]. However, diagnostic classification is important not only for treatment planning, prognostication, patient/caregiver education, and other aspects of clinical care for individuals, but also for effective communication and transfer of information among clinicians and researchers, etiologic and outcome research, and allocation of resources at the societal level [53,93]. Current diagnostic practices help clinicians make sense of complex observations and fulfill certain practical decision-making needs, and they are firmly entrenched in administrative systems such as those involved in insurance reimbursement and drug approval. Neither elimination of the use of categorical diagnoses nor rigid adherence to their criteria is likely to be a helpful strategy for clinicians in the short term. While awaiting empirically supported advances in diagnostic classification that overcome the limitations of existing systems, clinicians can strive to identify and treat prominent symptoms regardless of whether they are core features of the categorical diagnoses, incorporate existing valid dimensional assessments, and treat diagnostic criteria not as fundamental truths but as a somewhat flexible framework [93,98].

Levels of diagnosis

Disorders of development and behavior can be described on three different levels, each of which has value. This is depicted in Fig. 2.5 as an inverted triangle, with level 1 (specific impairments) at the top, level 2 (categorical diagnoses) in the middle, and level 3 (etiologic diagnoses) at the bottom. Level 1 is broad, signifying that an individual may have many specific impairments or symptoms. Analysis of the profile of impairments leads to the identification of a smaller number of categorical disorder diagnoses, depicted as the narrower level 2. Underlying the categorical diagnoses and specific impairments are one or more etiologies (level 3), which can sometimes be diagnosed. The relative importance of each level of diagnosis for guiding pharmacologic and nonpharmacologic intervention, genetic counseling, identification of associated medical risks, prognostication, and resource allocation is depicted in the lower panel of Fig. 2.5.

Specific impairments

The first level of diagnosis, which is primarily descriptive rather than integrative, involves the identification of specific deficits in function by caregiver interviews, standardized behavior rating scales, direct observation and elicitation, and developmental/psychological testing [53]. In addition to being a prerequisite for formulating categorical diagnoses, delineation of the specific impairments is essential for treatment planning since educational, behavioral, and psychopharmacologic interventions most often address these deficits rather than the categorical or etiologic diagnoses [53,99].

It is the specific deficits in function that are addressed, for example, by educational and habilitative interventions, with the goal of building developmentally appropriate skills and maximizing functional independence. Problem behaviors associated with neurodevelopmental disabilities may cause harm to the individual or others, result in property damage, interfere with learning and adaptive functioning, and contribute substantially to the burden on families. In most cases, problem-focused behavioral interventions informed by functional behavior assessment are an important and effective component of treatment [100–102]. These interventions address specific observable behaviors, not the diagnostic classification.

Even psychopharmacology is transdiagnostic; medication choices are often determined more by the target symptoms than by categorical or etiologic diagnoses [101,103–105]. The traditional psychiatric model of psychopharmacologic decision-making involves first making the diagnosis of a specific categorical disorder, such as ADHD or bipolar disorder, and then choosing a medication based on that diagnosis. However, because symptoms exist on a continuum and occur across different disorders, and modifications of diagnostic criteria are necessary to apply many psychiatric diagnoses to individuals with more severe disabilities, the value of categorical distinctions in selecting pharmacologic treatment is often unclear. Therefore, a target-symptom approach is often the most appropriate treatment strategy [99,101,103,104].

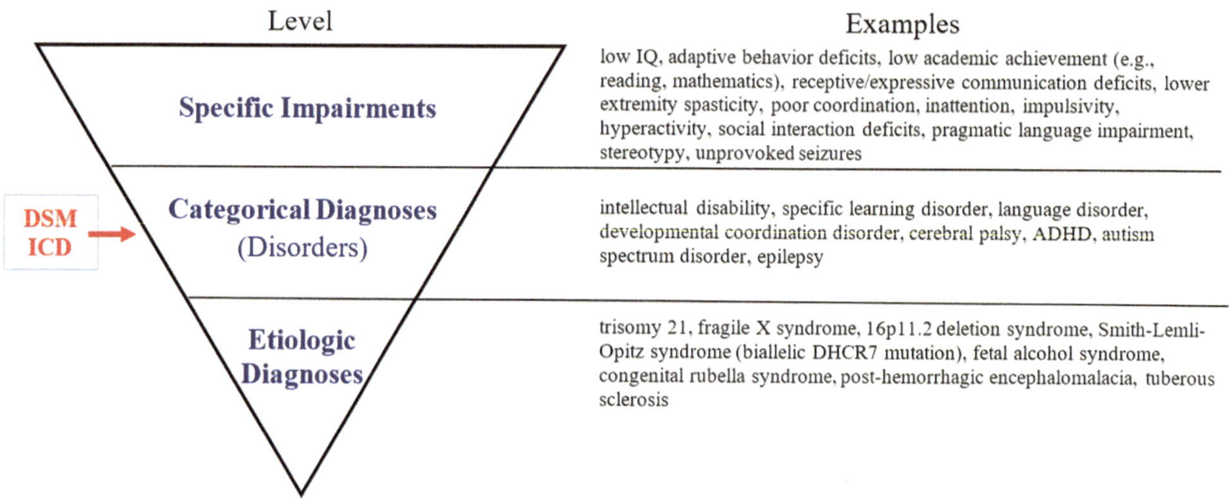

Significance of Each Level of Diagnosis

Level of Diagnosis	Importance					
	Treatment		Genetic Counseling	Identification of Associated Medical Problems	Prognostication	Resource Allocation
	Behavioral, Educational	Psychopharmacologic				
Specific Impairments	++	++	-	-/+	+/++	+
Categorical Diagnoses (Disorders)	+	+/++	+	+	++	++
Etiologic Diagnoses	-	-/+*	++	++	+	-/+

Key: - minor or no importance, + important, ++ very important
*potential/emerging importance (may become very important with future advances)

FIGURE 2.5 Levels of neurodevelopmental diagnoses and their significance. *Adapted from Myers SM. In: Batshaw ML, et al., editors. Children with disabilities, 7th ed. (2013) & 8th ed. (2019). Baltimore, MD: Brookes Publishing.*

Categorical diagnoses

The second level of diagnosis involves assigning categorical disorder diagnoses based on patterns of specific deficits and other signs and symptoms according to established diagnostic criteria (e.g., ICD-10, DSM-5). Although there are significant weaknesses, as described previously, categorical diagnoses are reasonably reliable and very useful for communication. Identification of the categorical diagnoses is important for providing a prognosis, attempting to standardize and guide treatment, and communicating information to patients, families, and treatment providers, including educators, habilitative therapists, psychologists, and healthcare providers (Fig. 2.6). Epidemiologic information is also most often collected at this level of diagnosis, and this information can be used to guide resource allocation.

Etiologic diagnoses

The third level of diagnosis, the etiologic diagnosis, involves identifying the specific underlying cause(s) of the developmental disorder, which may be genetic, metabolic, teratogenic, infectious/immunologic, hypoxic-ischemic, traumatic, a combination of more than one of these, or unknown/idiopathic. The etiologic level of diagnosis can currently be determined in only a minority of individuals with brain-based developmental disabilities, but much more commonly in those with disorders of the spinal cord or sensory deficits (blindness or deafness). In general, the more severe the neurodevelopmental disability, the more likely a specific etiologic diagnosis can be identified. For example, with the development of chromosomal microarray analysis and next-generation sequencing, the molecular diagnostic yield of genetic testing has increased to 55%–70% for severe IDD [81].

Recommendations regarding etiologic investigations in clinical practice depend on the specific diagnoses, physical examination findings, and family history. Lower-severity disabilities, such as language disorders, specific learning disorders, and ADHD, usually do not require etiologic investigation, such as laboratory or imaging studies, unless the history and/or physical examination suggest the possibility of a specific etiology that requires further evaluation (e.g., vision or hearing impairment, obstructive sleep apnea, thyroid dysfunction, lead toxicity, or physical stigmata of a genetic syndrome). In contrast, genetic testing is routinely warranted for individuals with IDD, or global developmental delay, ASD, epilepsy, and probably cerebral palsy [72,106].

The etiologic diagnosis is important, especially in the case of genetic disorders, for which accurate counseling regarding recurrence risk can be provided. In some cases, associated medical problems or undesired outcomes can be identified, treated, and perhaps prevented with early recognition of the genetic abnormality. For example, a child who is found to have a recurrent 22q11.2 deletion as the cause of his or her IDD or ASD is at increased risk for cardiac, palate, and kidney anomalies as well as immunologic and psychiatric disorders, and there is a 6%–10% chance that one of the parents has the deletion [107]. An affected parent has a 50% chance of passing the deletion on to each of their children. In the case of certain genetic epilepsies, knowing the etiologic molecular diagnosis allows for important refinement of antiepileptic therapy decisions [108]. The value of determining an etiologic diagnosis is not limited to genetic disorders that directly affect neurological function, and genetic tests are not the only valuable studies. For example, magnetic resonance imaging of the brain may reveal a cerebral infarction as the cause of a young child's hemiplegia, prompting testing for a constitutional coagulopathy for which additional treatment, stroke recurrence risk counseling, and family genetic counseling would be available. An infant with congenital cytomegalovirus (CMV) infection requires serial assessments for hearing loss, which may be delayed in onset and/or progressive, and, in the case of certain inborn errors of metabolism, neuroprotective or even curative treatment can be provided if the diagnosis is made early. All these important aspects of medical management are more closely linked to etiologic diagnosis than to categorical diagnoses or specific impairments (Fig. 2.6) [53]. Although current treatments are based mainly on target symptoms, advances in the identification of genetic etiologies, protein functions, and temporospatial expression patterns and their convergence on core molecular pathways hold promise for the development of genetics-based therapeutics for neurodevelopmental disabilities [109–111].

References

[1] Richmond JB. Child development: a basic science for pediatrics. Pediatrics 1967;39:649–58.

[2] Pellegrino L. Child development. In: Batshaw ML, Roizen NJ, Pellegrino L, editors. Children with disabilities. 8th ed. Baltimore, MD: Brooks Publishing; 2019, p. 177–98.

[3] Sheldrick RC, Perrin EC. Evidence-based milestones for surveillance of cognitive, language, and motor development, 13. Academic Pediatrics; 2013, p. 577–86. Available from: https://doi.org/10.1016/J.ACAP.2013.07.001.

[4] Sheldrick RC, Schlichting LE, Berger B, Clyne A, Ni P, Perrin EC, et al. Establishing new norms for developmental milestones. Pediatrics 2019;144. Available from: https://doi.org/10.1542/peds.2019-0374.

[5] Shoaibi A, Neelon B, Østbye T, Benjamin-Neelon SE. Longitudinal associations of gross motor development, motor milestone achievement and weight-for-length z score in a racially diverse cohort of US infants. BMJ Open 2019;9:1–8. Available from: https://doi.org/10.1136/bmjopen-2018-024440.

[6] Villar J, Fernandes M, Purwar M, Staines-Urias E, di Nicola P, Cheikh Ismail L, et al. Neurodevelopmental milestones and associated behaviours are similar among healthy children across diverse geographical locations. Nat Commun 2019;10:1–10. Available from: https://doi.org/10.1038/s41467-018-07983-4.

[7] Ertem IO, Krishnamurthy V, Mulaudzi MC, Sguassero Y, Balta H, Gulumser O, et al. Similarities and differences in child development from birth to age 3 years by sex and across four countries: a cross-sectional, observational study. Lancet Glob Health 2018;6:e279–91. Available from: https://doi.org/10.1016/S2214-109X(18)30003-2.

[8] McCoy DC, Waldman M, Fink G. Measuring early childhood development at a global scale: evidence from the caregiver-reported early development instruments. Early Child Res Q 2018;45:58–68. Available from: https://doi.org/10.1016/j.ecresq.2018.05.002.

[9] Binet A, Simon T. Le développement de l'intelligence chez les enfants. L'année Psychologique 1907;14:1–94.

[10] Binet A, Simon T. The development of intelligence in children (The Binet-Simon Scale). Baltimore, MD, US: Williams & Wilkins Co; 1916. Available from: https://doi.org/10.1037/11069-000.

[11] Sears RR. Your ancients revisited: a history of child development. Rev Child Dev Res 1975.

[12] Anderson JE. Child development: an historical perspective. Child Dev 1956;27:181. Available from: https://doi.org/10.2307/1126088.

[13] Freud S. Three essays on the theory of sexuality (1905). The standard edition of the complete psychological works of Sigmund Freud, volume VII (1901–1905): a case of hysteria, three essays on sexuality and other works, 1953, p. 123–246.

[14] Erikson EH. Identity: youth and crisis. WW Norton & Company; 1968.
[15] Maree JG. The psychosocial development theory of Erik Erikson: critical overview. Early Child Dev Care 2021;191:1107–21. Available from: https://doi.org/10.1080/03004430.2020.1845163.
[16] Piaget J, Cook MT. The origins of intelligence in children. 1952.
[17] Piaget J. The child's construction of reality. Routledge & Kegan Paul; 1955.
[18] Piaget J. Play, dreams and imitation in childhood. New York, NY, US: W W Norton & Co; 1952.
[19] Piaget J. La naissance de l'intelligence chez l'enfant. Paris: Delachaux et Niestlé Neuchatel; 1935.
[20] Ferster CB, Skinner BF. Schedules of reinforcement. Appleton-Century-Crofts; 1957. Available from: https://doi.org/10.1037/10627-000.
[21] Gewirtz JL, Peláez-Nogueras MBF. Skinner's legacy to human infant behavior and development. Am Psychol 1992;47:1411–22. Available from: https://doi.org/10.1037/0003-066X.47.11.1411.
[22] Proctor RW, Weeks DJ. The goal of BF Skinner and behavior analysis. Springer Science & Business Media; 2012.
[23] Vygotsky LS. Mind in society: the development of higher psychological processes. Harvard University Press; 1980.
[24] Bronfenbrenner U. The ecology of human development: experiments by nature and design. Harvard University Press; 1979.
[25] Bowlby J. The bowlby-ainsworth attachment theory. Behav Brain Sci 1979;2:637–8.
[26] Gesell A. Maturation and infant behavior pattern. Psychological Rev 1929;36:307–19. Available from: https://doi.org/10.1037/H0075379.
[27] Gesell A. 2nd rev. ed. Maturation and the patterning of behavior. A handbook of child psychology, Vol 1. Russell & Russell: Atheneum Publishers; 1933, p. 209–35. Available from: https://doi.org/10.1037/11552-004.
[28] Gesell A. The first five years of life. Harper & Brothers Publishers; 1940.
[29] Gesell A, Ilg FL, Ames LB, Bullis GE. The child from five to ten. Oxford, England: Harper; 1946.
[30] Gesell A, Ilg FL, Ames LB. Youth; the years from ten to sixteen. Oxford, England: Harper & Brothers; 1956.
[31] Accardo PJ, Shapiro BK, Capute AJ, editors. Behavior belongs in the brain: neurobehavioral syndromes. York Press; 1997.
[32] Volpe JJ, Inder TE, Darras BT, de Vries LS, du Plessis AJ, Neil JJ, et al., editors. Volpe's neurology of the newborn. 6th ed. Elsevier; 2018. Available from: https://doi.org/10.1016/C2010-0-68825-0.
[33] Nketia J, Amso D, Brito NH. Towards a more inclusive and equitable developmental cognitive neuroscience. Develop Cognit Neurosci 2021;52:101014. Available from: https://doi.org/10.1016/J.DCN.2021.101014.
[34] Westermann G, Mareschal D, Johnson MH, Sirois S, Spratling MW, Thomas MSC. Neuroconstructivism. Dev Sci 2007;10:75–83. Available from: https://doi.org/10.1111/J.1467-7687.2007.00567.X.
[35] Sirois S, Spratling M, Thomas MSC, Westermann G, Mareschal D, Johnson MH. Précis of neuroconstructivism: how the brain constructs cognition. Behav Brain Sci 2008;31. Available from: https://doi.org/10.1017/S0140525X0800407X.
[36] Dosman C, Andrews D, Goulden K. Evidence-based milestone ages as a framework for developmental surveillance. Paediatr Child Health 2012;17:561–8.
[37] Scharf RJ, Scharf GJ, Stroustrup A. Developmental milestones. Pediatr Rev 2016;37:25–38. Available from: https://doi.org/10.1542/PIR.2014-0103.
[38] De M, Drachler L, Marshall T, Carlos J, Leite C. A continuous-scale measure of child development for population-based epidemiological surveys: a preliminary study using item response theory for the Denver test. Paediatr Perinat Epidemiol 2007;21:138–53.
[39] Carlson AG, Rowe E, Curby TW. Disentangling fine motor skills' relations to academic achievement: the relative contributions of visual-spatial integration and visual-motor coordination. J Genet Psychol 2013;174:514–33. Available from: https://doi.org/10.1080/00221325.2012.717122.
[40] O'Connor Leppert ML, Shank TP, Shapiro BK, Capute AJ. The Capute scales: CAT/CLAMS – a pediatric assessment tool for the early detection of mental retardation and communicative disorders. Ment Retard Dev Disabil Res Rev 1998;4:14–19. Available from: https://doi.org/10.1002/(SICI)1098-2779.
[41] Wilks T, Gerber RJ, Erdie-Lalena C. Developmental milestones: cognitive development. Pediatr Rev 2010;31:364–7. Available from: https://doi.org/10.1542/PIR.31-9-364.
[42] Gerber RJ, Wilks T, Erdie-Lalena C. Developmental milestones: motor development. Pediatr Rev 2010;31:267–77. Available from: https://doi.org/10.1542/PIR.31-7-267.
[43] Gerber RJ, Wilks T, Erdie-Lalena C. Developmental milestones 3: social-emotional development. Pediatr Rev 2011;32:533–6. Available from: https://doi.org/10.1542/PIR.32-12-533.
[44] Capute AJ, Accardo PJ. The infant neurodevelopmental assessment: a clinical interpretive manual for CAT-CLAMS in the first two years of life, Part 1. Curr Probl Pediatr 1996;26:279–306. Available from: https://doi.org/10.1016/s0045-9380(96)80008-9.
[45] Capute AJ, Accardo PJ. The infant neurodevelopmental assessement: a clinical interpretive manual for CAT-CLAMS in the first Two years of life, part 2. Curr Probl Pediatr 1996;26:265–306. Available from: https://doi.org/10.1016/S0045-9380(96)80008-9.
[46] Wachtel RC, Shapiro BK, Palmer FB, Allen MC, Capute AJ. CAT/CLAMS: a tool for the pediatric evaluation of infants and young children with developmental delay. Clin Pediatr 1994;33:410–15. Available from: https://doi.org/10.1177/000992289403300706.
[47] Guddemi M, Sambrook A, Wells S, Randel B, Fite K, Selva G, et al. Arnold Gesell's developmental assessment revalidation substantiates child-oriented curriculum. SAGE Open 2014;4. Available from: https://doi.org/10.1177/2158244014528918.
[48] Capute A, Palmer FB, Shapiro BK, Wachel RC, Ross A, Accardo PJ. Primitive reflex profile: a quantitation of primitive reflexes in infancy. Dev Med Child Neurol 1984;26:375–83. Available from: https://doi.org/10.1111/J.1469-8749.1984.TB04456.X.
[49] Kuhl PK. Brain mechanisms in early language acquisition. Neuron 2010;67:713–27. Available from: https://doi.org/10.1016/J.NEURON.2010.08.038.

[50] Accardo PJ. The father of developmental pediatrics: Arnold J. Capute, MD, MPH (1923–2003). J Child Neurol 2004;19:978–81. Available from: https://doi.org/10.1177/08830738040190121201.

[51] Capute AJ, Accardo PJ. A neurodevelopmental perspective on the continuum of developmental disabilities. In: Capute AJ, Accardo PJ, editors. Developmental disabilities in infancy and childhood. 1st ed. Baltimore, MD: Brookes Publishing; 1990, p. 7–42.

[52] Capute AJ, Palmer FB. A pediatric overview of the spectrum of developmental disabilities. J Dev Behav Pediatr 1980;1:66–9.

[53] Myers SM. Diagnosing developmental disabilities. In: Batshaw ML, Roizen NJ, Pellegrino L, editors. Children with disabilities. 8th ed. Baltimore, MD: Brooks Publishing; 2019, p. 199–224.

[54] Voigt RG. Making developmental-behavioral diagnoses. In: Voigt RG, Macias MM, Myers SM, Tapia CD, editors. AAP developmental and behavioral pediatrics. 2nd Edition American Academy of Pediatrics; 2018, p. 187–222.

[55] Voigt RG. Developmental-behavioral pediatric diagnosis. In: 23rd ed. Kline MW, editor. Rudolph's pediatrics, vol. 1. New York, NY: McGraw-Hill Education; 2018.

[56] Miller G, Dubowitz LMS, Palmer P. Follow-up of pre-term infants: is correction of the developmental quotient for prematurity helpful. Early Hum Dev 1984;9:137–44. Available from: https://doi.org/10.1016/0378-3782(84)90094-X.

[57] Lipkin PH, Macias MM, Norwood KW, Brei TJ, Davidson LF, Davis BE, et al. Promoting optimal development: identifying infants and young children with developmental disorders through developmental surveillance and screening. Pediatrics 2020;145. Available from: https://doi.org/10.1542/peds.2019-3449.

[58] Riou EM, Ghosh S, Francoeur E, Shevell MI. Global developmental delay and its relationship to cognitive skills. Dev Med Child Neurol 2009;51:600–6. Available from: https://doi.org/10.1111/j.1469-8749.2008.03197.x.

[59] Miciak J, Fletcher JM. The critical role of instructional response for identifying dyslexia and other learning disabilities. J Learn Disabil 2020;53:343–53. Available from: https://doi.org/10.1177/0022219420906801.

[60] Fletcher JM, Francis DJ, Morris RD, Lyon GR. Evidence-based assessment of learning disabilities in children and adolescents. J Clin Child Adolesc Psychol 2005;34:506–22. Available from: https://doi.org/10.1207/s15374424jccp3403_7.

[61] Myers SM, Challman TC. Autism spectrum disorder. In: Voigt RG, Macias MM, Myers SM, Tapia CD, editors. Developmental-behavioral pediatrics. Isaca, IL: American Academy of Pediatrics; 2018, p. 407–75.

[62] Hare-Harris AE, Mitchel MW, Myers SM, Mitchel AD, King BR, Ruocco BG, et al. Within-task variability on standardized language tests predicts autism spectrum disorder: a pilot study of the response dispersion index. J Neurodev Disord 2019;11:21. Available from: https://doi.org/10.1186/S11689-019-9283-Z.

[63] Tan C, Frewer V, Cox G, Williams K, Ure A. Prevalence and age of onset of regression in children with autism spectrum disorder: a systematic review and meta-analytical update. Autism Res 2021;14:582–98. Available from: https://doi.org/10.1002/aur.2463.

[64] Barger BD, Campbell JM, McDonough JD. Prevalence and onset of regression within autism spectrum disorders: a meta-analytic review. J Autism Dev Disord 2013;43:817–28. Available from: https://doi.org/10.1007/s10803-012-1621-x.

[65] Hansen RL, Ozonoff S, Krakowiak P, Angkustsiri K, Jones C, Deprey LJ, et al. Regression in autism: prevalence and associated factors in the CHARGE study. Ambul Pediatr 2008;8:25–31. Available from: https://doi.org/10.1016/j.ambp.2007.08.006.

[66] Baird G, Charman T, Pickles A, Chandler S, Loucas T, Meldrum D, et al. Regression, developmental trajectory and associated problems in disorders in the autism spectrum: the SNAP study. J Autism Dev Disord 2008;38:1827–36. Available from: https://doi.org/10.1007/s10803-008-0571-9.

[67] Pickles A, Simonoff E, Conti-Ramsden G, Falcaro M, Simkin Z, Charman T, et al. Loss of language in early development of autism and specific language impairment. J Child Psychol Psychiatry 2009;50:843–52. Available from: https://doi.org/10.1111/j.1469-7610.2008.02032.x.

[68] Jones EJH, Gliga T, Bedford R, Charman T, Johnson MH. Developmental pathways to autism: a review of prospective studies of infants at risk. Neurosci Biobehav Rev 2014;39:1–33. Available from: https://doi.org/10.1016/J.NEUBIOREV.2013.12.001.

[69] Silbereis JC, Pochareddy S, Zhu Y, Li M, Sestan N. The cellular and molecular landscapes of the developing human central nervous system. Neuron 2016;89:248. Available from: https://doi.org/10.1016/j.neuron.2015.12.008.

[70] Mullin AP, Gokhale A, Moreno-De-Luca A, Sanyal S, Waddington JL, Faundez V. Neurodevelopmental disorders: mechanisms and boundary definitions from genomes, interactomes and proteomes. Transl Psychiatry 2013;3. Available from: https://doi.org/10.1038/tp.2013.108.

[71] Ismail FY, Shapiro BK. What are neurodevelopmental disorders. Curr OpNeurol 2019;32:611–16. Available from: https://doi.org/10.1097/WCO.0000000000000710.

[72] Moreno-De-Luca A, Myers SM, Challman TD, Moreno-De-Luca D, Evans DW, Ledbetter DH. Developmental brain dysfunction: revival and expansion of old concepts based on new genetic evidence. Lancet Neurol 2013;12:406–14. Available from: https://doi.org/10.1016/S1474-4422(13)70011-5.

[73] Willcutt EG. Behavior and molecular genetic approaches to comorbidity. Curr Dev Disord Rep 2019;6:31–6. Available from: https://doi.org/10.1007/s40474-019-00162-0.

[74] Clements SD, Peters JE. Minimal brain dysfunctions in the school-age child: diagnosis and treatment. Arch Gen Psychiatry 1962;6:185–97. Available from: https://doi.org/10.1001/archpsyc.1962.01710210001001.

[75] Gillberg C. Deficits in attention, motor control, and perception: a brief review. Arch Dis Child 2003;88:904–10. Available from: https://doi.org/10.1136/adc.88.10.904.

[76] Gillberg C. The ESSENCE in child psychiatry: early symptomatic syndromes eliciting neurodevelopmental clinical examinations. Res Dev Disabil 2010;31:1543–51. Available from: https://doi.org/10.1016/j.ridd.2010.06.002.

[77] Johnson MR, Shorvon SD. Heredity in epilepsy: neurodevelopment, comorbidity, and the neurological trait. Epilepsy Behav 2011;22:421–7. Available from: https://doi.org/10.1016/j.yebeh.2011.07.031.

[78] Owen MJ, O'Donovan MC, Thapar A, Craddock N. Neurodevelopmental hypothesis of schizophrenia. Br J Psychiatry 2011;198:173–5. Available from: https://doi.org/10.1192/bjp.bp.110.084384.

[79] Owen MJ, O'Donovan MC. Schizophrenia and the neurodevelopmental continuum:evidence from genomics. World Psychiatry 2017;16:227–35. Available from: https://doi.org/10.1002/wps.20440.

[80] Myers SM, Challman TD, Bernier R, Bourgeron T, Chung WK, Constantino JN, et al. Insufficient evidence for "Autism-Specific" genes. Am J Hum Genet 2020;106:587–95. Available from: https://doi.org/10.1016/j.ajhg.2020.04.004.

[81] Vissers LELM, Gilissen C, Veltman JA. Genetic studies in intellectual disability and related disorders. Nat Rev Genet 2016;17:9–18. Available from: https://doi.org/10.1038/nrg3999.

[82] Sullivan PF, Geschwind DH. Defining the genetic, genomic, cellular, and diagnostic architectures of psychiatric disorders. Cell 2019;177:162–83. Available from: https://doi.org/10.1016/j.cell.2019.01.015.

[83] Satterstrom FK, Kosmicki JA, Wang J, Breen MS, de Rubeis S, An JY, et al. Large-scale exome sequencing study implicates both developmental and functional changes in the neurobiology of autism. Cell 2020;180:568–84. Available from: https://doi.org/10.1016/j.cell.2019.12.036 e23.

[84] Rees E, Kirov G. Copy number variation and neuropsychiatric illness. Curr OpGenet Dev 2021;68:57–63. Available from: https://doi.org/10.1016/j.gde.2021.02.014.

[85] Fernández IS, Loddenkemper T, Gaínza-Lein M, Sheidley BR, Poduri A. Diagnostic yield of genetic tests in epilepsy: a meta-analysis and cost-effectiveness study. Neurology 2019;92:E418–28. Available from: https://doi.org/10.1212/WNL.0000000000006850.

[86] Grove J, Ripke S, Als TD, Mattheisen M, Walters RK, Won H, et al. Identification of common genetic risk variants for autism spectrum disorder. Nat Genet 2019;51:431–44. Available from: https://doi.org/10.1038/s41588-019-0344-8.

[87] Pardiñas AF, Holmans P, Pocklington AJ, Escott-Price V, Ripke S, Carrera N, et al. Common schizophrenia alleles are enriched in mutation-intolerant genes and in regions under strong background selection. Nat Genet 2018;50:381–9. Available from: https://doi.org/10.1038/s41588-018-0059-2.

[88] Sanders SJ, Sahin M, Hostyk J, Thurm A, Jacquemont S, Avillach P, et al. A framework for the investigation of rare genetic disorders in neuropsychiatry. Nat Med 2019;25:1477–87. Available from: https://doi.org/10.1038/s41591-019-0581-5.

[89] Lee PH, Feng YCA, Smoller JW. Pleiotropy and cross-disorder genetics among psychiatric disorders. Biol Psychiatry 2021;89:20–31. Available from: https://doi.org/10.1016/j.biopsych.2020.09.026.

[90] Lee PH, Anttila V, Won H, Feng YCA, Rosenthal J, Zhu Z, et al. Genomic relationships, novel loci, and pleiotropic mechanisms across eight psychiatric disorders. Cell 2019;179:1469–82. Available from: https://doi.org/10.1016/j.cell.2019.11.020 e11.

[91] Gonzalez-Mantilla AJ, Moreno-De-Luca A, Ledbetter DH, Martin CL. A cross-disorder method to identify novel candidate genes for developmental brain disorders. JAMA Psychiatry 2016;73:275–83. Available from: https://doi.org/10.1001/jamapsychiatry.2015.2692.

[92] Hyman SE. The diagnosis of mental disorders: The problem of reification. Annu Rev Clin Psychol 2010;6:155–79. Available from: https://doi.org/10.1146/annurev.clinpsy.3.022806.091532.

[93] Clark LA, Cuthbert B, Lewis-Fernández R, Narrow WE, Reed GM. Three approaches to understanding and classifying mental disorder: ICD-11, DSM-5, and the National Institute of Mental Health's Research Domain Criteria (RDoC). Psychol Sci Publ Int 2017;18:72–145. Available from: https://doi.org/10.1177/1529100617727266.

[94] World Health Organization. International statistical classification of diseases and related health problems (11th Ed.) 2019.

[95] American Psychiatric Association. Diagnostic and statistical manual of mental disorders: DSM-5. Diagnostic and statistical manual of mental disorders: DSM-5™, 5th Ed. https://doi.org/10.1176/appi.books.9780890425596; 2013.

[96] Lilienfeld SO, Treadway MT. Clashing diagnostic approaches: DSM-ICD versus RDoC. Annu Rev Clin Psychol 2016;12:435–63. Available from: https://doi.org/10.1146/annurev-clinpsy-021815-093122.

[97] Insel TR, Cuthbert BN. Brain disorders? Precisely: precision medicine comes to psychiatry. Science 1979;348:499–500. Available from: https://doi.org/10.1126/science.aab2358 2015.

[98] Thapar A, Cooper M, Rutter M. Neurodevelopmental disorders. Lancet Psychiatry 2017;4:339–46. Available from: https://doi.org/10.1016/S2215-0366(16)30376-5.

[99] Myers SM, Johnson CP, Lipkin PH, Cartwright JD, Desch LW, Duby JC, et al. Management of children with autism spectrum disorders. Pediatrics 2007;120:1162–82. Available from: https://doi.org/10.1542/peds.2007-2362.

[100] Beavers GA, Iwata BA, Lerman DC. Thirty years of research on the functional analysis of problem behavior. J Appl Behav Anal 2013;46:1–21. Available from: https://doi.org/10.1002/jaba.30.

[101] McGuire K, Fung LK, Hagopian L, Vasa RA, Mahajan R, Bernal P, et al. Irritability and problem behavior in autism spectrum disorder: a practice pathway for pediatric primary care. Pediatrics 2016;137:S136–48. Available from: https://doi.org/10.1542/peds.2015-2851L.

[102] Kurtz PF, Leoni M, Hagopian LP. Behavioral approaches to assessment and early intervention for severe problem behavior in intellectual and developmental disabilities. Pediatr Clin North Am 2020;67:499–511. Available from: https://doi.org/10.1016/j.pcl.2020.02.005.

[103] Bostic JQ, Rho Y. Target-symptom psychopharmacology: between the forest and the trees. Child Adolesc Psychiatr Clin North Am 2006;15:289–302. Available from: https://doi.org/10.1016/j.chc.2005.08.003.

[104] Accordino RE, Kidd C, Politte LC, Henry CA, McDougle CJ. Psychopharmacological interventions in autism spectrum disorder. Expert Opin Pharmacother 2016;17:937–52. Available from: https://doi.org/10.1517/14656566.2016.1154536.

[105] Waszczuk MA, Zimmerman M, Ruggero C, Li K, MacNamara A, Weinberg A, et al. What do clinicians treat: diagnoses or symptoms? The incremental validity of a symptom-based, dimensional characterization of emotional disorders in predicting medication prescription patterns. Compr Psychiatry 2017;79:80–8. Available from: https://doi.org/10.1016/j.comppsych.2017.04.004.

[106] Savatt JM, Myers SM. Genetic testing in neurodevelopmental disorders. Front Pediatr 2021;9:1–24. Available from: https://doi.org/10.3389/fped.2021.526779.

[107] McDonald-McGinn DM, Sullivan KE, Marino B, Philip N, Swillen A, Vorstman JAS, et al. 22q11.2 deletion syndrome. Nat Rev Dis Prim 2015;1:1–19. Available from: https://doi.org/10.1038/nrdp.2015.71.

[108] Poduri A. When should genetic testing be performed in epilepsy patients? Epilepsy Curr 2017;17:16–22. Available from: https://doi.org/10.5698/1535-7511-17.1.16.

[109] Parenti I, Rabaneda LG, Schoen H, Novarino G. Neurodevelopmental disorders: from genetics to functional pathways. Trends Neurosci 2020;43:608–21. Available from: https://doi.org/10.1016/j.tins.2020.05.004.

[110] Arnett AB, Wang T, Eichler EE, Bernier RA. Reflections on the genetics-first approach to advancements in molecular genetic and neurobiological research on neurodevelopmental disorders. J Neurodev Disord 2021;13:1–10. Available from: https://doi.org/10.1186/s11689-021-09371-4.

[111] van Karnebeek CDM, Bowden K, Berry-Kravis E. Treatment of neurogenetic developmental conditions: from 2016 into the future. Pediatr Neurol 2016;65:1–13. Available from: https://doi.org/10.1016/j.pediatrneurol.2016.07.010.

Chapter 3

Biological basis of neurodevelopmental disabilities

Kate Kielty[1], Sarah Risen[2] and Alexandria E. Melendez-Zaidi[3]

[1]*Department of Pediatrics, Division of Child Neurology, UPMC Children's Hospital of Pittsburgh, University of Pittsburgh School of Medicine, Pittsburgh, PA, United States,* [2]*Department of Pediatrics, Section of Neurology and Developmental Neuroscience, Baylor College of Medicine, Texas Children's Hospital, Houston, TX, United States,* [3]*Department of Neurology, Texas Tech Health Sciences Center, El Paso, El Paso, TX, United States*

Introduction

Neurodevelopmental disabilities (NDD) are derived from internal and external mechanisms that disrupt formation and maturation of the developing brain [1]. Appreciation and understanding must be given to the development of the brain and spinal cord in order to contextualize the etiologies and implications of specific disorders. Outcomes of disruption in development are influenced by injury-specific factors including timing of when the initial insult occurs, the mechanism (or mechanisms) of injury, the severity of the injury, and the downstream biochemical, molecular, and genetic responses of the system. The variety of combinations in which these factors can unfold and interact contribute to the vast spectrum of NDD globally as well as within specific etiologies. In this section, we will review development of the central nervous system (CNS) as well as congenital and acquired factors which can interrupt the typical development process.

Development of the central nervous system

CNS development begins rapidly in the first days of embryogenesis and continues throughout life with the earlier periods (birth through childhood) showing larger developmental gains relative to adolescence and adulthood, where the process eventually plateau. Conducted by intrinsic genetic and epigenetic regulation with cues from external stimuli, development is orchestrated through a series of phases which overlap in harmony to produce the complex networks of structure and function which comprise the CNS. Knowledge of these processes has been derived from animal models, fetal autopsy, and, in more recent decades, advances in genetics and neuroimaging, providing a framework for understanding the basis of congenital malformations and neurological conditions. Full comprehension of CNS development entails a review of processes occurring at the molecular, biochemical, and genetic levels in detail that are beyond the scope of this chapter. With a particular focus on the cerebral cortex, we will describe distinct phases of development at the cellular and structural level including neurulation, neural proliferation and migration, connectivity, and myelination [2]. For a more comprehensive review of nervous system development (see Volpe) [3].

Neurulation

In the first 2–4 weeks of gestation, the initial circular embryo undergoes gastrulation to differentiate its three germ layers: the ectoderm, mesoderm, and endoderm. Two poles emerge at the cranial and caudal positions of a newly cylindrical form, beginning the process of neurulation in which the primordial CNS is created, as seen in Fig. 3.1. The neural plate develops from the ectoderm between these poles, with a depression along the plate forming the neural groove. Lying

Capute and Accardo's Neurodevelopmental Disabilities in Infancy and Childhood. DOI: https://doi.org/10.1016/B978-0-12-824060-1.00003-1

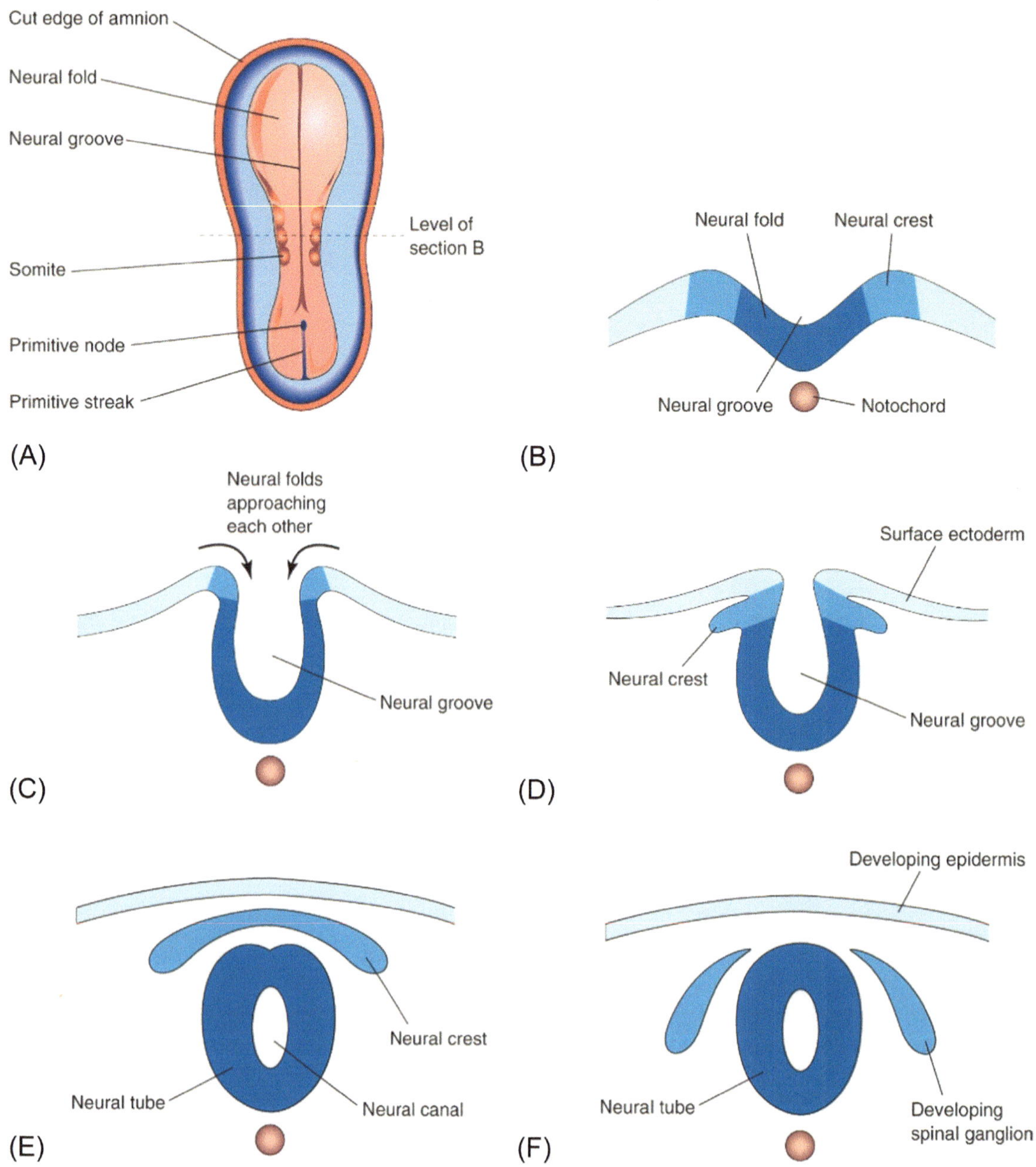

FIGURE 3.1 The process of neurulation. *From Moore KL, et al. Third week of human development. The developing human: clinically oriented embryology. Edinburgh: Elsevier; 2020.*

bilaterally adjacent to the groove is the neural fold which elevates and extends to fuse together. This process of folding creates the beginnings of a tubal structure as the coalescence extends both caudally and rostrally, elongating the tube. As this enveloping occurs, neuropores at the cranial and caudal ends are formed and close at approximately 24- and 26-day gestation, respectively [3]. The neural tube, lined by neuroepithelium, is eventually separated from the initial layer of ectoderm as the primordial CNS. The remaining ectoderm further demarcates into the neural crest that later develops structures including components of the peripheral nervous system; cranial nerves V, VII, IX, and X; autonomic ganglia; and meninges. The surface ectoderm contributes to portions of the cutaneous system including the epidermis, hair, nails, cutaneous glands, the anterior part of the pituitary gland, enamel of the teeth, and the lens of the eye [4].

The caudal eminence of the neural tube develops into the primordial spinal cord as the cranial eminence becomes the brain, with the closed fluid-filled tubal cavity the beginning of the ventricular system and the central canal of the spinal cord [4]. Three vesicles are initially formed: the prosencephalon, mesencephalon, and rhombencephalon, which then further differentiate into five main vesicles. At this point, cell lines are committed to forming the structures of the

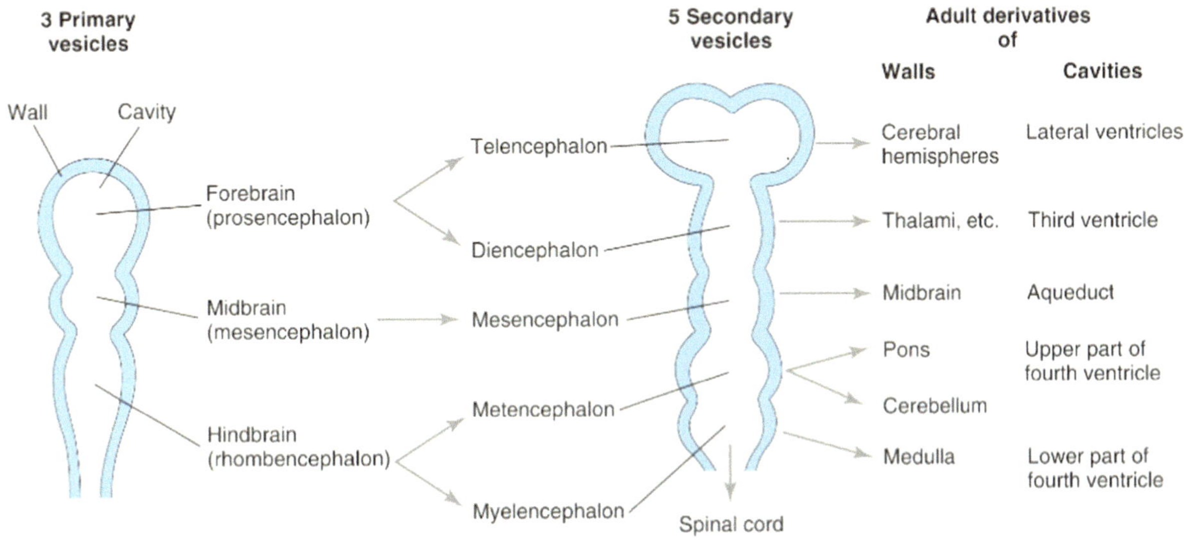

FIGURE 3.2 Ventral Induction. *From Moore KL, et al. Nervous system. The developing human: clinically oriented embryology. Edinburgh: Elsevier; 2020.*

adult brain as shown in Fig. 3.2. Ventral induction occurs around the fifth week of gestation, during which the prosencephalon is cleaved into two separate cerebral hemispheres, and subsequent midline development begins to form structures including the corpus callosum, optic and olfactory vesicles, and pituitary gland [3].

Neural proliferation

With formation of the main structural subsets of the CNS, neural proliferation begins, lasting from approximately 6–27 weeks' gestation, though some structures continue to proliferate into the postnatal period, as is the case of the dentate gyrus of the hippocampus [5,6]. The period of neural proliferation involves rapid cellular divisions of progenitor cell lines and neurogenesis, initially starting in the spinal cord and brainstem before involving the ventricular zone to create the neurons of the cerebral cortex at an exponential rate [2]. At the time of birth, a newborn infant is predicted to have approximately 20 billion neurons within the neocortex [7]. Paralleling the formation of neurons, glial cells are also undergoing proliferation which extends postnatally into the second year of life and may continue at a slower rate thereafter. Glial cells ultimately account for over half of the cells in the mature brain and play essential roles in facilitating the next steps in development [3].

Neural migration

As neurons are formed through neurogenesis and proliferation, glia create a web to guide neural migration [8,9]. Following neurogenesis, neurons climb radial glia outwards, perpendicular to the ventricular system to form the six layers of the cerebral cortex. As this migration occurs, new layers begin to form, surpassing the layers previously created behind them, until the outermost of six distinctive cortical layers are created, at which point migration is arrested with the production of reelin. In addition to radial migration, a separate process of tangential migration parallel to the surface of the brain and perpendicular to the radial glia takes place to further thread the neurons to create an intricate structural network in the cerebral cortex [3]. Within the later stages of this process, the morphology of the cerebral cortex changes from a flat surface to the topographical landscape of sulci and gyri as demonstrated in Fig. 3.3.

While this process is most impressive in the cerebral cortex, the cerebellum is simultaneously formed beginning in the first trimester by a separate cascade of proliferation and migration at the midbrain–hindbrain junction, with continued maturation of the structure into the second year of life [10].

Neural differentiation, organization, and connectivity

With distinctive neurons in their new structural homes at approximately 6 months of gestation, there begins the establishment of neural networks, which provide the functionality of the CNS. Phenotypic differentiation of neurons from both a biochemical and morphological perspective creates neuronal subtypes throughout the CNS. Synaptogenesis initiates at

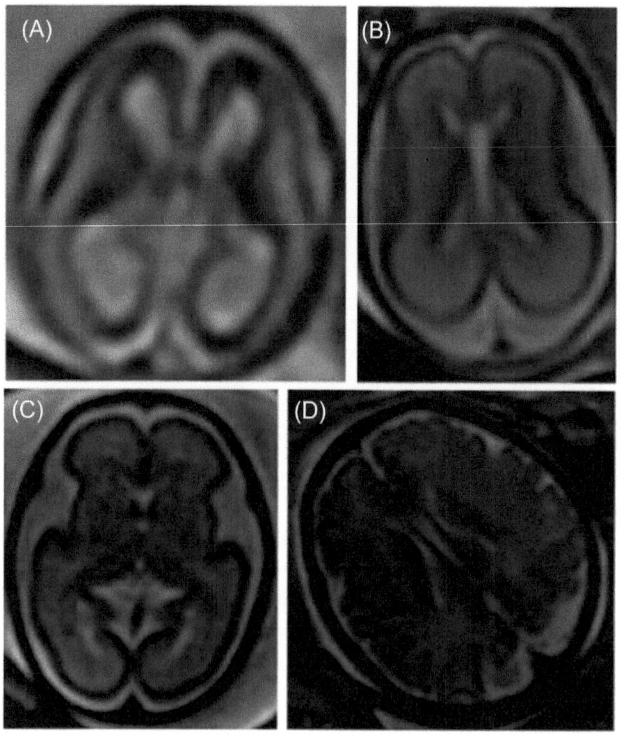

FIGURE 3.3 Axial single-shot fast spin echo (ssFSE) sequences of four different normal fetuses demonstrating the expected changes in fetal brain parenchymal growth and formation during the second and third trimester of pregnancy. (A) A 16-week-old fetus, (B) 20-week-old fetus, (C) 26-week-old fetus, and (D) 35-week-old fetus. *From Guimaraes CVA, Dahmoush HM. Fetal brain anatomy. Neuroimaging Clin N Am 2022;32(3):663–81. doi:10.1016/j.nic.2022.04.009. PMID: 35843668.*

27 weeks' gestation creating electrochemical connections between neurons. Synaptic density continues to increase until 2–3 years of life with synaptic pruning contributing to the formation of functional circuits [11]. The expression of behavior in the developing brain reflects complex interplay of brain structural maturation, neurotransmitter expression and regulation, formation of neural circuits and functional specification. The changes at all levels of CNS hierarchy are tightly regulated by the genetic blueprint and epigenetic modifications derived from internal events and external experiences. With advances in functional magnetic resonance imaging, resting-state neural hubs and network connectivity are apparent as early as the second trimester in utero [12,13]. Neural organization, synapse formation, and pruning are further covered in Chapter: Neuroplastcity in the neurodevelopmental disorders.

Myelination

The final step of CNS development involves the maturation of newly created networks through a process of myelination. Myelin, wrapped around neuronal axons by oligodendrocytes, increases the speed of conduction and enhances the efficiency of neural circuits. Myelination begins in the midbrain and cerebellum following fiber growth and organization [14]. At 40 weeks' gestation, approximately the time of a full-term birth, myelination has begun at the level of the brainstem, cerebellum, and the internal capsule, though it is not until the end of the first year of life that the white matter is formed throughout the cerebrum. Further myelination of fibers continues in specific structures asynchronously and at different rates, with maturation continuing into adulthood for selective regions [15].

Timing of disruption

The wide timespan of CNS development allows a unique susceptibility to a multitude of disruptions. Insults or genetic changes can influence neural development at any point from embryogenesis to later stages in myelination, with outcomes ranging from negligible to hindering survival. Examples of disruption to development during embryogenesis and fetal CNS development are highlighted in Table 3.1.

The concepts of *critical* and *sensitive periods* are described to highlight how external influences can alter development. *Critical periods* are defined as "the time during which the action of a specific external or internal condition or stimulus is required for the normal progress of development [16]." A prime example of a critical period is the stage in

TABLE 3.1 Examples of associated outcomes in early central nervous system (CNS) development disruption.

Age	Developmental phase	Process of development	Examples of outcomes of disruption
2–5 weeks' gestation	Neurulation, dorsal induction	Closure of cranial neuropore	Anencephaly Encephalocele
		Closure of caudal neuropore	Spina bifida Caudal regression syndrome
	Neurulation, ventral induction	Ontogenesis of the cerebral midline	Holoprosencephaly Fetal alcohol syndrome Smith–Lemli–Opitz syndrome
		Formation of midline structures	Agenesis of the corpus callosum Septo-optic dysplasia Agenesis of the septum pellucidum
6–22 weeks' gestation	Neural proliferation	Proliferation in the cerebral cortex	Microcephaly Megalencephaly Focal cortical dysplasia
		Proliferation in cerebellar-brainstem development	Pontocerebellar hypoplasia Dandy–Walker malformation Joubert syndrome
12–20 weeks' gestation	Neural migration	Germinative zone formation	Schizencephaly
		Formation of cortical laminae	Lissencephaly Pachygyria Polymicrogyria
		Late neuronal migration	Neuronal heterotopias Focal cortical dysplasias (types I, III)
6 months of gestation into adolescence	Neural organizational and connectivity	Synaptogenesis Synaptic pruning Network formation	Rett syndrome Autism spectrum disorder ADHD Phelan–McDermid syndrome Fragile X syndrome
6 months into adulthood	Myelination	Myelin production	Pelizaeus–Merzbacher disease and other hypomyelinating leukodystrophies

ADHD, Attention-deficit hyperactive disorder.

which amblyopia occurs due to abnormal visual input in early life [17]. Similarly, *sensitive periods* reflect "the time period in development during which the nervous system is highly susceptible to the effects of harmful internal or external conditions [16]." Critical periods are, by definition, sensitive periods, though not all sensitive periods can be considered critical [11]. As these concepts are at risk of being oversimplified, it is important to appreciate that the temporal and spatial heterogeneity of neurodevelopment blurs between critical and sensitive epochs and is best understood as a series of overlapping, interrelated events.

Within early periods of neurological development, structural malformations can occur by perturbation in the initial formation of a particular structure (a primary malformation), or through destruction of a previously developed structure (a secondary malformation). Take, for example, the case of white matter diseases: the ability to create myelin is impacted in hypomyelination disorders such Pelizaeus–Merzbacher disease, a primary process, whereas the destruction of previously formed fibers, as is the case in focal periventricular necrosis from hypoperfusion leading to periventricular leukomalacia (PVL), is considered a secondary process [18]. In addition to a singular brain insult, progressive conditions in neurodevelopmental disabilities are characterized by continuous decline from ongoing processes. In these disorders, continuing insults or insufficiencies impair further aspects of development and lead to regression. Metachromatic leukodystrophy is one of many examples of progressive white matter disorders, where myelin is damaged by accumulation of sulfatides secondary to deficiency in the lysosomal enzyme arylsulfatase A [19].

Perturbations that occur early may subsequently impact the development of other organ systems or threaten the ability for survival. In the case of holoprosencephaly, where the prosencephalon fails to separate into two distinct hemispheres, midline facial features can also fail to form—ranging from hypotelorism or midline cleft palate—considered mild cases—to anophthalmic cyclopia in severe cases which are associated with lower survival [20]. Certain conditions may also extend beyond the nervous system due to the underlying genetic mechanisms involved. This is the case with neurocutaneous disorders, in which individuals have both neurological manifestations as well as distinctive cutaneous and ophthalmologic sequalae, highlighting the underlying ectodermal derivation of the nervous system [21].

In medicine, conditions are generally considered "congenital" when they are present from the time of birth. This terminology may lead to some confusion in neurodevelopment where there is significant overlap in pathology during the weeks before and after birth. The perinatal period is a particularly vulnerable time for the developing fetus, exemplified by the case of prematurity, which an most important risk factor for cerebral palsy [22]. The premature neonate, born with immature organs, must continue to develop despite systemic threats to survival. Inability to effectively regulate cerebral perfusion predisposes the premature brain to injury secondary to hypoxia or systemic hypotension [18]. Cerebrovascular fluctuations may also injure the delicate germinal matrix, causing intraventricular hemorrhage [23]. Beyond vascular susceptibilities, the premature neonate is at risk of exposure to inflammation, infection, hyperbilirubinemia, anemia, hypoglycemia, and electrolyte imbalances, all of which can also impact neurodevelopmental outcomes [5]. Furthermore, underlying genetic or metabolic conditions may predispose a fetus to early delivery or birth complications (e.g., in the case of conditions with hypotonia) and further contribute to neural disruption beyond the systemic and environmental factors. Advances in neonatal medical care have significantly increased survival but introduced further risks, such as sensorineural hearing loss secondary to aminoglycosides exposure in treatment of neonatal sepsis [24], or retinopathy of prematurity which has an association with high oxygen-saturation levels in premature infants receiving respiratory support [25,26].

While congenital conditions and perinatal insults account for a broad range of neurodevelopmental disabilities, acquired injuries (such as trauma and infection) in childhood also lead to long-term disability from direct harm and disruption to ongoing neurodevelopment. Injuries may be focal in nature, such as in stroke, and/or may have diffuse impact, such as anoxic brain injury due to submersion.

Mechanisms of disruption

Disruption to CNS development may come from a variety of endogenous and exogenous sources. Though categories of insults exist, the developing brain may be affected by multiple types of insults, either as a consequence of an initial mechanism or as unrelated events. Each insult's impact on the developing brain is also varied depending on severity of the insult in addition to the timing within brain development, and the system's ability to respond. Table 3.2 provides a broad categorization of etiologies that result in disruption to the CNS.

Genetic, epigenetic, and metabolic conditions

Specific genetic conditions form a large cohort of the neurodevelopmental disabilities population with most malformations of cortical development thought to be caused by underlying genetic mutations [6]. While some of these are well recognized—such as Fragile X syndrome (FXS) and tuberous sclerosis complex (TSC)—advancements in availability and efficiency of genetic testing have led to the discovery of rarer genetic disorders, with over 1500 genes identified as related to neurodevelopmental disabilities [27]. Identification of underlying genetic mechanisms opens the door for interventions and treatments, such as the case of dietary manipulation in phenylketonuria, or with strategies from more recent advancements in research, such as enzyme replacement or gene therapies.

Infection/inflammation

Congenital and perinatal infections occur due to the immature immune system's inability to combat viruses or bacteria transmitted from the maternal host. Transmission may be transplacental, ascending, or from exposure during passage of the birth canal. The leading cause of congenital infections worldwide is cytomegalovirus (CMV), which is most detrimental when acquired during the first trimester and accounts for 20% of sensorineural hearing loss in the developed world [28,29]. In addition to their direct teratogenic impacts on development, maternal infections and the inflammatory response increase the risk of prematurity [30], which further places the neonate at increased susceptibility as previously

TABLE 3.2 Categorization and examples of mechanisms in central nervous system (CNS) disruption and corresponding chapters.

Mechanism etiology	Examples
Genetic/epigenetic including metabolic and neurodegenerative disorders	• Disorders in methylation ◦ Prader–Willi syndrome • Trinucleotide repeat expansion disorders ◦ Fragile X syndrome • Mitochondrial disorders ◦ MELAS • Single gene disorders ◦ CHARGE syndrome • Chromosomal abnormalities ◦ Down syndrome ◦ Turner syndrome ◦ Microdeletion syndromes
Infection	• Prenatal infection ◦ Cytomegalovirus ◦ Toxoplasmosis ◦ Zika virus ◦ Syphilis • Neonatal meningitis ◦ Group B Streptococcus ◦ *Escherichia coli* ◦ *Listeria monocytogenes* ◦ *Staphylococcus aureus* • Meningitis and encephalitis in childhood ◦ Herpes simplex virus ◦ Enterovirus ◦ *Streptococcus pneumoniae*
Exogenous substances and environmental factors	• Medications and substances ◦ Thalidomide ◦ Valproic acid ◦ Tobacco ◦ Alcohol ◦ Marijuana • Environmental exposures ◦ Organophosphate pesticides ◦ Lead ◦ Methylmercury ◦ Polychlorinated biphenyls (PCBs) ◦ Radiation
Vascular	Hypoxia/ischemia hemorrhage
Trauma	Nonaccidental trauma Accidental trauma
Inflammatory/autoimmune	
Oncologic	
Systemic illness	
Psychosocial adversity	Prenatal maternal stress Malnutrition Neglect

discussed. Bacterial infections leading to neonatal meningitis are particularly harmful to brain development with high morbidity and must be diagnosed and treated early for best outcomes [31].

Following the perinatal period, infants and children remain susceptible to bacterial and viral meningitis and encephalitis. Children who are immunocompromised or unimmunized against the most common bacterial and viral culprits are particularly susceptible to severe infection. Sequalae of these neurological infections lead to long-term disability, such as hearing loss, epilepsy, cognitive impairment, motor deficits, and lower adaptive functioning [32].

Exogenous substances and environmental toxins

In the late 1950s and early 1960s, babies were being born with alarming rates of congenital malformations including limb abnormalities such as phocomelia. These birth defects were later connected to prenatal use of a medication marketed toward pregnant women, thalidomide. This devastating man-made health crisis impacted at least 10,000 neonates, 40% of whom died in the first year [33]. In the following half-century, pharmaceutical regulation increased and knowledge on the impact of maternal medication usage on the developing fetus now allows for better informed guidance to expecting mothers. While awareness has increased, maternal medications and substance use continue to impact neurodevelopment, resulting in conditions such as fetal alcohol syndrome disorders [34].

Neurotoxic effects may also occur from a variety of environmental exposures both prenatally and in childhood. Young children are especially at risk for injury from these environmental toxins due to oral exploration, increased gastrointestinal absorption rates, and a more permeable blood–brain barrier [5]. For over half a century, research of lead exposure and intoxication has demonstrated association with intellectual disability and neuropsychiatric problems. Public health efforts continue to limit lead exposures in developing and developed countries, which the World Health Organization estimates has cost over 20 million disability-adjusted life years worldwide [35].

Vascular insults

As previously discussed, the perinatal period is uniquely susceptible to neurological injury, especially to hemorrhage and ischemic stroke. Fragility and lack of autoregulation of cerebral vasculature, particularly in premature and very low birth weight neonates, can lead to hemorrhage, most commonly from the germinal matrix into the ventricular system. In severe intraventricular hemorrhage, marked by brain parenchyma involvement, neurodevelopment is most likely to be impaired [3]. Focal infarcts may occur from either arterial or venous thromboses. A more diffuse or global ischemic injury may occur from hypoxic events, leading to hypoxic-ischemic encephalopathy (HIE) in the neonate. Neurodevelopmental outcomes range by the severity of the injury and the vascular territories which are affected, and can include future motor, language, cognitive, and behavioral impairments. Depending on the region involved, these injuries may cause neonatal seizures and/or future epilepsy, both of which may further negatively impact development [36].

Cerebrovascular injury can also occur throughout childhood, though stroke is significantly less common than in the adult population. Risk factors for stroke in the pediatric population include cardiac disease, prothrombotic state, and arteriopathies (including Moyamoya disease), with an increased risk in certain genetic conditions such as sickle cell disease (SCD) [37].

Trauma

The leading cause of mortality in children aged 1–19 years old in the United States is due to unintentional injury. This category includes mild to severe traumatic brain injury related to sports and recreational activities as well as motor vehicle crashes, drownings, and suffocation [38]. Abusive head trauma is more common in infants but may occur across all age groups. Severity of injury depends upon the mechanism of the injury and subsequent deficits evolve over time, with behavioral and cognitive problems the most common sequelae in mild-moderate injury [39]. Ongoing management strives to maximize functional outcomes through optimal neurorehabilitation and education strategies.

Autoimmune and inflammatory conditions

The field of neuroimmunology has grown over the past decade with increased research on inflammatory and autoimmune conditions. Autoimmune encephalitis is associated with rapid onset of neuropsychiatric symptoms and encephalopathy in adults. However, consistent with the constructs of neurodevelopmental stage at time of injury, autoimmune encephalitis in young children typically presents with loss of developmental skills as a focal deficit (such as impaired use of a limb or difficulty talking mimicking a stroke) and/or behavior problems (such as impaired sleep or excessive irritability) [40,41]. Autoimmune encephalitis has been associated with antibodies to N-methyl-D-aspartate (NMDA), myelin oligodendrocyte glycoprotein (MOG), glutamic acid decarboxylase 65 (GAD65), and several other cytosolic and membrane-bound motifs. Many cases remain seronegative, however, as there are likely further neural autoantibodies to be discovered. Postinfectious and inflammatory syndromes such as febrile infection–related epilepsy syndrome (FIRES) are also included in this category, with need for further understanding of neurodevelopmental impacts of these conditions following the acute phase of injury [42].

Oncologic

The field of oncology has made some of the greatest leaps in the last century, with nearly 90% of children with a diagnosis of any cancer surviving at least 5 years from date of diagnosis. This increase in survival has forced the field to assess additional outcome measures such as quality of life and developmental outcomes following treatment. Children who undergo treatment for cancer are well documented to have higher rates of neurodevelopmental deficits across domains including motor function, cognitive functioning, behavior, and regulation of mood [43]. This is the case for *all* cancers, not only those in which the primary tumor is located within the CNS. Systemic chemotherapy is directly neurotoxic causing a range of pathology, from peripheral neuropathies to leukoencephalopathies to weakening the blood–brain barrier, therefore allowing enhanced inflammation and production of damaging reactive oxygen species (ROS). Intrathecal chemotherapy causes similar but more severe effects. Radiation therapy additionally causes significant necrosis to the parenchyma and vasculature. Beyond direct physiological affects to neurodevelopmental skills, children with cancer spend much less time in school during their treatment, which causes them to be behind in the class material and may also contribute to compromised social and behavioral skills [44].

In children with primary tumors located in the CNS, developmental outcomes are further influenced by the tumor itself or because of necessary treatment. A prime example of the latter is the case of posterior fossa syndrome following gross surgical resection of medulloblastoma. Similarly, development may be compromised by unavoidable removal of surrounding "normal" tissue. Complications of surgery—including stroke, intracerebral hemorrhage, infection—may occur and further contribute to impaired development in ways already discussed above. Though advances now allow for more targeted treatment, radiation of the complete CNS axis may be required (in metastatic medulloblastoma, for example). The impact of radiation treatment on cognition is well documented and its consequences so severe that it is contraindicated in children less than 3 years of age, given the profound neurodevelopmental complications that may result [45].

Finally, several brain tumors are seen in the context of genetic diseases causing a predisposition for cancer including neurofibromatosis (associated with low-grade gliomas particularly in the optic pathway), tuberous sclerosis complex (associated with subependymal giant cell astrocytoma), or Sturge–Weber syndrome. Evidence suggests that the genetic conditions themselves may have impaired development, potentiated by intracranial tumors and related treatment, should they occur.

Systemic and chronic illness

In adults, chronic diseases such as hypertension, diabetes mellitus, and obesity have been associated with more rapid cognitive decline. A similar phenomenon occurs in children with chronic illness who experience higher rates of developmental impairments [46].

Asthma, bronchopulmonary dysplasia

The mechanisms by which asthma impacts neurodevelopment are not well defined, though the relationship is strong. It is hypothesized that the driver of asthma and associated NDDs (most common of which is attention-deficit hyperactive disorder [ADHD], in this population) is inflammation and, potentially, shared genetic and epigenetic regulation and/or mitochondrial dysfunction [47–50]. In addition to this, children with severe asthma are more likely to come from low socioeconomic backgrounds and have higher exposure to environmental pollutants, both of which are, independently, risk factors for impaired development (see below). Children with bronchopulmonary dysplasia (BPD) have been well described to have higher rates of neurodevelopmental deficits even when prematurity and birth weight are accounted for (which is a clear risk factor for BPD) including deficits in language acquisition, cognition including general intelligence, and potentially visuospatial or visuomotor abilities. It is hypothesized that this may be due to recurrent periods of hypoxia, hypercapnia, and respiratory acidosis in addition to metabolic and epigenetic factors [51–53].

Congenital heart disease

Children and, now even adults, with congenital heart disease (CHD) have become more common in the setting of advances in cardiopulmonary bypass, cardiac catheterization, intensive care, and open-heart surgery. The prevalence is high at 9 per 1000 [54]. Children have many reasons for risk of neurodevelopmental disabilities which include altered cerebral blood flow with impaired oxygen delivery (prenatally and postnatally), decreased brain development causing a phenotype akin to prematurity, chronic hypoxia, and the risks inherent to surgical interventions which include coagulopathies and associated ischemic events (e.g., stroke), prolonged hospitalization, frequent exposure to anesthesia, episodes of cardiac arrest requiring cardiopulmonary resuscitation and global hypoxia, and increased PVL (though the mechanism of PVL formation in this setting is less clear) [55–57]. The cardiac intensive care unit environment and stressors to the infant brain can also contribute to neurodevelopmental outcomes [58]. Children with CHD may also have comorbid conditions that themselves cause NDD (such as prematurity or genetic syndromes) [59,60].

Obesity, malnutrition

Obesity and malnutrition are not mutually exclusive. The increased rates of both obesity and malnutrition in socioeconomically disadvantaged populations may contribute to already compromised neurodevelopmental outcomes seen in children. Malnutrition causes definitive stunting of linear growth and, in some cases, neurodevelopmental delays though this relationship is not as straightforward as may be presumed. However, malnutrition in the form of micronutrient shortages is well known to impact development. Specifically, deficiencies in iodine, B12, and iron may cause early problems in CNS development given their requirement for processes such as neuronal migration, synapse formation, DNA methylation, and myelination to mention a few. Though NDD are common comorbidities in children with obesity, there is not a cause-effect established between obesity in infancy/childhood and NDD. A growing body of evidence, however, supports a role for maternal obesity in the development of NDDs with mechanisms centered in a heightened inflammatory state and/or aberrant hormonal signaling to influence fetal brain growth [61–63].

Malnutrition has also been associated with worse physical and mental health as well as the full spectrum of neurodevelopmental disabilities [64,65]. Balanced levels of macro- and micronutrients are required for brain development (neurogenesis, neuronal migration, synaptogenesis, myelination) and ongoing brain function. Malnutrition can also impair synaptic plasticity, which is the ability of synapses to strengthen or weaken over time in response to increases or decreases in their activity [66–68]. This can impact learning and memory processes. This has been shown in animal models, where protein malnutrition led to impaired long-term potentiation (LTP), a mechanism related to synaptic plasticity. Malnutrition can alter the synthesis and function of neurotransmitters, such as serotonin and dopamine, which are critical for mood regulation and cognitive function. For example, iron deficiency can disrupt the production and function of certain neurotransmitters, altering brain development and functioning resulting in a variety of associated neurodevelopmental disorders [69,70]. Severe malnutrition can lead to structural changes in the brain. Studies have shown that malnourished children often have smaller brain volumes with a significant reduction in gray and white matter [71]. Chronic malnutrition is an independent risk factor for impairments in cognition and behavioral issues [72–74]. While malnutrition may be initially considered as a concern in only low-income countries, developed countries are also affected, such as the United States where 16 million children are considered to be living with significant food insecurity [75].

Psychosocial factors

Over the past decades, there has been increased study and understanding of the impact of the psychosocial environment on the developing brain. Maternal stressors and demographics during pregnancy impact outcomes in neurodevelopment and may reflect systemic racial and socioeconomic disadvantage. As an example, when compared to other racial groups, non-Hispanic black women in the United States have 48% higher rates of spontaneous preterm birth [76]. The impact of race in healthcare has been historically underrepresented in the literature and is an area in ongoing need of research.

Furthermore, adverse childhood events, including neglect and abuse, have been shown to directly impact structural and functional development of the brain [77]. These negative early life influences have an association with poor long-term outcomes, particularly in relation to cognition and emotional regulation separate from any physical manifestations of injury [78]. Socioeconomic status has been directly linked to cortical and subcortical brain development, white matter microstructure, and neural function, which was directly associated with impaired neurocognitive processing and the spectrum of developmental functions [79–81]. The physical environmental and home stressors even impact adolescent neurodevelopmental, with reduced prefrontal cortical thickness associated with learning deficits [82]. Importantly, evidence also suggests these changes may be reversed or minimized through early environmental enrichment and positive interactions [83]. While psychosocial factors play a role in the etiology of neurodevelopmental disabilities, providers and caregivers who work in this population must also be aware that children with disabilities are at an increased risk of child abuse themselves, impacting the potential for optimal growth and development [84].

Response to injury

The initial insult may not always account for the entire extent of injury to the developing brain, as there are downstream effects at the biochemical and cellular levels. These cascades are moderated by both genetic and inflammatory influences and have been best defined in the literature in the context of traumatic brain injury and hypoxic-ischemic injury.

Using traumatic brain injury as an example, the primary insult leads to direct cell death while a secondary phase of injury ensues: Disruption of the blood–brain barrier and mechanosensitive release of glutamate leads to sodium and calcium influx which in turn triggers signaling molecules and apoptotic proteins, leading to mitochondrial damage and ROS accumulation. Further apoptotic cell death results and an environment of neuroinflammation is created with release of pro-inflammatory cytokines and persistent immune cell activation. Repair and recovery of the CNS are impacted by these processes and contribute to neurodevelopmental morbidity [85,86].

Clinical outcomes following disruption to neurodevelopment

Considering the timing, mechanism, severity, and systemic response to disruptions to neurodevelopment, clinical outcomes are significantly varied across individuals. This variation contributes to the spectrum and continuum of neurodevelopmental disabilities.

Though some neurodevelopmental disabilities can be predicted prenatally, presentations in childhood are more common and may reflect influences of events that are temporally far removed. For example, a perinatal hemorrhage leading to porencephaly might not be detected until a mother observes a left-hand preference in her 5-month-old infant, a time when motor development is expected to be symmetrical. Intellectual disability as a result of lead exposure in early life might not be apparent until a comprehensive neuropsychological evaluation is performed in a school-age child. Children with mild HIE are more prone to ADHD and autism, both of which typically require an older age of observation and diagnosis given the inherent "higher-level" deficits in the disorders.

While a comprehensive history and physical, neuroimaging, plus laboratory and genetic testing may provide an etiology to the underlying neurodevelopmental diagnosis, the complexity and multitude of mechanisms at play is not often fully understood in each individual. Some etiologies may have a specific supplement or treatment to improve outcomes. Independent of etiology, early intervention with therapies and support is essential to improving patient function. From advances in intensive neonatal and pediatric care, to comprehensive metabolic evaluations in treatable intellectual disability and epilepsy, to identification of food insecurity, to neuropsychological testing revealing dyslexia or subtle learning disabilities; new opportunities continue to arise in conditions previously thought unpreventable and/or untreatable.

Conclusion

At the core of all neurodevelopmental disorders is disruption to the fundamental biological processes underlying brain maturation and development. The timing, type, and severity of the disruption, along with the molecular and genetic and epigenetic response of the system, impact brain recovery and neurodevelopmental outcomes. As knowledge of these processes continues to grow through research, further understanding of medical and therapeutic interventions arises, with opportunity for improved care for individuals with neurodevelopmental disabilities.

References

[1] Meredith RM. Sensitive and critical periods during neurotypical and aberrant neurodevelopment: a framework for neurodevelopmental disorders. Neurosci Biobehav Rev 2015;50:180–8. Available from: https://doi.org/10.1016/j.neubiorev.2014.12.001.

[2] Accardo PJ, Capute AJ. Capute & 'Accardo's neurodevelopmental disabilities in infancy and childhood. Baltimore, MD: Paul H. Brookes Pub; 2008.

[3] Volpe JJ. 'Volpe's neurology of the newborn. Philadelphia, PA: Elsevier; 2018.

[4] Moore KL, N. PTV, Torchia MG. The developing human: clinically oriented embryology. 11th ed. Elsevier; 2020.

[5] Andescavage NN. Fetal Development. In: Batshaw ML, Roizen NJ, Pellegrino L, editors. Children with Disabilities, 8th Edition. Baltimore, MD: Paul H Brookes. Publishing Co.; 2019.

[6] Juric-Sekhar G, Hevner RF. Malformations of cerebral cortex development: molecules and mechanisms. Annu Rev Pathol 2019;14 (1):293–318. Available from: https://doi.org/10.1146/annurev-pathmechdis-012418-012927.
[7] Larsen CC, Bonde Larsen K, Bogdanovic N, Laursen H, Græm N, Badsberg Samuelsen G, et al. Total number of cells in the human newborn telencephalic wall. Neuroscience 2006;139(3):999–1003. Available from: https://doi.org/10.1016/j.neuroscience.2006.01.005.
[8] Allen NJ, Lyons DA. Glia as architects of central nervous system formation and function. Science 2018;362(6411):181–5. Available from: https://doi.org/10.1126/science.aat0473.
[9] Sild M, Ruthazer ES. Radial glia: progenitor, pathway, and partner. Neuroscientist 2011;17(3):288–302. Available from: https://doi.org/10.1177/1073858410385870.
[10] Haldipur P, Millen KJ. What cerebellar malformations tell us about cerebellar development. Neurosci Lett 2019;688:14–25. Available from: https://doi.org/10.1016/j.neulet.2018.05.032.
[11] Ismail FY, Fatemi A, Johnston MV. Cerebral plasticity: windows of opportunity in the developing brain. Eur J Paediatric Neurol 2017;21 (1):23–48. Available from: https://doi.org/10.1016/j.ejpn.2016.07.007.
[12] Thomason ME, Grove LE, Lozon Jr TA, Vila AM, Ye Y, Nye MJ, et al. Age-related increases in long-range connectivity in fetal functional neural connectivity networks in utero. Dev Cogn Neurosci 2015;11:96–104. Available from: https://doi.org/10.1016/j.dcn.2014.09.001.
[13] van den Heuvel MI, Turk E, Manning JH, Hect J, Hernandez-Andrade E, Hassan SS, et al. Hubs in the human fetal brain network. Dev Cogn Neurosci 2018;30:108–15. Available from: https://doi.org/10.1016/j.dcn.2018.02.001.
[14] Dubois J, Dehaene-Lambertz G, Kulikova S, Poupon C, Hüppi PS, Hertz-Pannier L. The early development of Brain White matter: a review of imaging studies in fetuses, newborns and infants. Neuroscience 2014;276:48–71. Available from: https://doi.org/10.1016/j.neuroscience.2013.12.044.
[15] Ropper AH, Samuels MA, Klein J, Prasad S. Adams and 'Victor's principles of neurology. 11th ed. McGraw-Hill Education; 2019.
[16] Erzurumlu RS, Killackey HP. Critical and sensitive periods in neurobiology. Curr Top Developmental Biol 1982;207–40. Available from: https://doi.org/10.1016/s0070-2153(08)60522-0.
[17] Hensch TK, Quinlan EM. Critical periods in amblyopia. Vis Neurosci 2018;35. Available from: https://doi.org/10.1017/s0952523817000219.
[18] Novak CM, Ozen M, Burd I. Perinatal brain injury: mechanisms, prevention, and outcomes. Clin Perinatol 2018;45(2):357–75. Available from: https://doi.org/10.1016/j.clp.2018.01.015.
[19] van der Knaap MS, Bugiani M. Leukodystrophies: a proposed classification system based on pathological changes and pathogenetic mechanisms. Acta Neuropathol 2017;134(3):351–82. Available from: https://doi.org/10.1007/s00401-017-1739-1.
[20] Dubourg C, Bendavid C, Pasquier L, Henry C, Odent S, David V. Holoprosencephaly. Orphanet J Rare Dis 2007;2(1). Available from: https://doi.org/10.1186/1750-1172-2-8.
[21] Rosser T. Neurocutaneous disorders. CONTINUUM 2018;24(1):96–129. Available from: https://doi.org/10.1212/con.0000000000000562.
[22] Graham HK, Rosenbaum P, Paneth N, Dan B, Lin J-P, Damiano DL, et al. Cerebral palsy. Nat Rev Dis Prim 2016;2(1). Available from: https://doi.org/10.1038/nrdp.2015.83.
[23] Valdez Sandoval P, Hernández Rosales P, Quiñones Hernández DG, Chavana Naranjo EA, García Navarro V. Intraventricular hemorrhage and posthemorrhagic hydrocephalus in preterm infants: diagnosis, classification, and treatment options. 'Childs Nerv Syst 2019;35(6):917–27. Available from: https://doi.org/10.1007/s00381-019-04127-x.
[24] Khairy MA, Abuelhamed WA, Ahmed RS, El Fouly HE, Elhawary IM. Hearing loss among high-risk newborns admitted to a tertiary neonatal intensive care unit. J Matern Fetal Neonatal Med 2017;31(13):1756–61. Available from: https://doi.org/10.1080/14767058.2017.1326902.
[25] BOOST II United Kingdom Collaborative Group, BOOST II Australia Collaborative Group, BOOST II New Zealand Collaborative Group, et al. Oxygen saturation and outcomes in preterm infants. N Engl J Med 2013;368(22):2094–104. Available from: https://doi.org/10.1056/NEJMoa1302298.
[26] Bashinsky AL. Retinopathy of prematurity. North Carol Med J 2017;78(2):124–8. Available from: https://doi.org/10.18043/ncm.78.2.124.
[27] Leblond CS, Le T-L, Malesys S, Cliquet F, Tabet A-C, Delorme R, et al. Operative list of genes associated with autism and neurodevelopmental disorders based on database review. Mol Cell Neurosci 2021;113:103623. Available from: https://doi.org/10.1016/j.mcn.2021.103623.
[28] Leruez-Ville M, Foulon I, Pass R, Ville Y. Cytomegalovirus infection during pregnancy: state of the science. Am J Obstet Gynecol 2020;223 (3):330–49. Available from: https://doi.org/10.1016/j.ajog.2020.02.018.
[29] Manicklal S, Emery VC, Lazzarotto T, Boppana SB, Gupta RK. The "silent" global burden of congenital cytomegalovirus. Clin Microbiol Rev 2013;26(1):86–102. Available from: https://doi.org/10.1128/cmr.00062-12.
[30] Goldenberg RL, Culhane JF, Iams JD, Romero R. Epidemiology and causes of preterm birth. Lancet 2008;371(9606):75–84. Available from: https://doi.org/10.1016/s0140-6736(08)60074-4.
[31] Ku LC, Boggess KA, Cohen-Wolkowiez M. Bacterial meningitis in infants. Clin Perinatol 2015;42(1):29–45. Available from: https://doi.org/10.1016/j.clp.2014.10.004.
[32] Christie D, Viner RM, Knox K, Coen PG, Wang H, El Bashir H, et al. Long-term outcomes of pneumococcal meningitis in childhood and adolescence. Eur J Pediatrics 2011;170(8):997–1006. Available from: https://doi.org/10.1007/s00431-010-1390-5.
[33] Vargesson N, Stephens T. Thalidomide: history, withdrawal, Renaissance, and safety concerns. Expert Opin Drug Saf 2021;20(12):1455–7. Available from: https://doi.org/10.1080/14740338.2021.1991307.
[34] Wozniak JR, Riley EP, Charness ME. Clinical presentation, diagnosis, and management of fetal alcohol spectrum disorder. Lancet Neurol 2019;18(8):760–70. Available from: https://doi.org/10.1016/s1474-4422(19)30150-4.

[35] Naranjo VI, Hendricks M, Jones KS. Lead toxicity in children: an unremitting public health problem. Pediatr Neurol 2020;113:51–5. Available from: https://doi.org/10.1016/j.pediatrneurol.2020.08.005.

[36] Dunbar M, Kirton A. Perinatal stroke. Semin Pediatr Neurol 2019;32:100767. Available from: https://doi.org/10.1016/j.spen.2019.08.003.

[37] Malone LA, Felling RJ. Pediatric stroke: unique implications of the immature brain on injury and recovery. Pediatr Neurol 2020;102:3–9. Available from: https://doi.org/10.1016/j.pediatrneurol.2019.06.016.

[38] West BA, Rudd RA, Sauber-Schatz EK, Ballesteros MF. Unintentional injury deaths in children and youth, 2010–2019. J Saf Res 2021;78:322–30. Available from: https://doi.org/10.1016/j.jsr.2021.07.01.

[39] Chevignard M, Câmara-Costa H, Dellatolas G. Pediatric traumatic brain injury and abusive head trauma. Handb Clin Neurol 2020;451–84. Available from: https://doi.org/10.1016/b978-0-444-64150-2.00032-0.

[40] Goldberg EM, Titulaer M, de Blank PM, Sievert A, Ryan N. Anti-N-methyl-D-aspartate receptor-mediated encephalitis in infants and toddlers: case report and review of the literature. Pediatr Neurol 2014;50(2):181–4. Available from: https://doi.org/10.1016/j.pediatrneurol.2013.10.002.

[41] Hardy D. Autoimmune encephalitis in children. Pediatr Neurol 2022;132:56–66. Available from: https://doi.org/10.1016/j.pediatrneurol.2022.05.004.

[42] Cellucci T, Van Mater H, Graus F, Muscal E, Gallentine W, Klein-Gitelman MS, et al. Clinical approach to the diagnosis of autoimmune encephalitis in the pediatric patient. Neurol Neuroimmunol Neuroinflamm 2020;7(2). Available from: https://doi.org/10.1212/nxi.0000000000000663.

[43] National Academies of Sciences, Engineering, and Medicine; Health and Medicine Division; Board on Health Care Services; Committee on Childhood Cancers and Disability; Aiuppa L., Cartaxo T., Spicer C.M., et al., editors. Childhood Cancer and Functional Impacts Across the Care Continuum. Washington (DC): National Academies Press (US); 2020 Dec 9. 4, Childhood Cancers and Function. Available from: https://www.ncbi.nlm.nih.gov/books/NBK569402/.

[44] Martinez-Santos AE, Fernandez-De-La-Iglesia JDC, Sheaf G, Coyne I. A systematic review of the educational experiences and needs of children with cancer returning to school. J Adv Nurs 2021;77(7):2971–94. Available from: https://doi.org/10.1111/jan.14784.

[45] Millard NE, De Braganca KC. Medulloblastoma. J Child Neurol 2016;31(12):1341–53. Available from: https://doi.org/10.1177/0883073815600866.

[46] Goldberg S. Chronic illness and early development: parent-child relationships. Pediatr Ann 1990;19(1):35–41. Available from: https://doi.org/10.3928/0090-4481-19900101-07.

[47] Zhu Z, Zhu X, Liu CL, Shi H, Shen S, Yang Y, et al. Shared genetics of asthma and mental health disorders: a large-scale genome-wide cross-trait analysis. Eur Respir J 2019;54(6):1901507. Available from: https://doi.org/10.1183/13993003.01507-2019.

[48] Arrondo G, Solmi M, Dragioti E, Eudave L, Ruiz-Goikoetxea M, Ciaurriz-Larraz AM, et al. Associations between mental and physical conditions in children and adolescents: an umbrella review. Neurosci Biobehav Rev 2022;137:104662. Available from: https://doi.org/10.1016/j.neubiorev.2022.104662.

[49] Holmberg K, Lundholm C, Anckarsäter H, Larsson H, Almqvist C. Impact of asthma medication and familial factors on the association between childhood asthma and attention-deficit/hyperactivity disorder: a combined twin- and register-based study: epidemiology of allergic disease. Clin Exp Allergy 2015;45(5):964–73. Available from: https://doi.org/10.1111/cea.12529.

[50] Leffa DT, Caye A, Santos I, Matijasevich A, Menezes A, Wehrmeister FC, et al. Attention-deficit/hyperactivity disorder has a state-dependent association with asthma: the role of systemic inflammation in a population-based birth cohort followed from childhood to adulthood. Brain Behav Immun 2021;97:239–49. Available from: https://doi.org/10.1016/j.bbi.2021.08.004.

[51] Cheong JLY, Doyle LW. An update on pulmonary and neurodevelopmental outcomes of bronchopulmonary dysplasia. Semin Perinatol 2018;42(7):478–84. Available from: https://doi.org/10.1053/j.semperi.2018.09.013.

[52] Gou X, Yang L, Pan L, Xiao D. Association between bronchopulmonary dysplasia and cerebral palsy in children: a meta-analysis. BMJ Open 2018;8(9):e020735. Available from: https://doi.org/10.1136/bmjopen-2017-020735.

[53] Kramer BW, Niklas V, Abman S. Bronchopulmonary dysplasia and impaired neurodevelopment-what may be the missing link? Am J Perinatol 2022;39(S 01):S14–17. Available from: https://doi.org/10.1055/s-0042-1756677.

[54] Liu Y, Chen S, Zühlke L, Black GC, Choy MK, Li N, et al. Global birth prevalence of congenital heart defects 1970–2017: updated systematic review and meta-analysis of 260 studies. Int J Epidemiol 2019;48(2):455–63. Available from: https://doi.org/10.1093/ije/dyz009.

[55] Howell HB, Zaccario M, Kazmi SH, Desai P, Sklamberg FE, Mally P. Neurodevelopmental outcomes of children with congenital heart disease: a review. Curr Probl Pediatr Adolesc Health Care 2019;49(10):100685. Available from: https://doi.org/10.1016/j.cppeds.2019.100685.

[56] Ortinau CM, Smyser CD, Arthur L, Gordon EE, Heydarian HC, Wolovits J, et al. Optimizing neurodevelopmental outcomes in neonates with congenital heart disease. Pediatrics 2022;150(Suppl 2). Available from: https://doi.org/10.1542/peds.2022-056415L e2022056415L.

[57] Peyvandi S, Latal B, Miller SP, McQuillen PS. The neonatal brain in critical congenital heart disease: insights and future directions. NeuroImage 2019;185:776–82. Available from: https://doi.org/10.1016/j.neuroimage.2018.05.045.

[58] Smith LM, Harrison TM. Neurodevelopment in the congenital heart disease population as framed by the life course health development framework. J Cardiovasc Nurs 2023;. Available from: https://doi.org/10.1097/JCN.0000000000000977 Advance online publication.

[59] Katz JA, Levy PT, Butler SC, Sadhwani A, Lakshminrusimha S, Morton SU, et al. Preterm congenital heart disease and neurodevelopment: the importance of looking beyond the initial hospitalization. J Perinatol 2023;43(7):958–62. Available from: https://doi.org/10.1038/s41372-023-01687-4.

[60] Patt E, Singhania A, Roberts AE, Morton SU. The genetics of neurodevelopment in congenital heart disease. Can J Cardiol 2023;39(2):97–114. Available from: https://doi.org/10.1016/j.cjca.2022.09.026.

[61] Tong L, Kalish BT. The impact of maternal obesity on childhood neurodevelopment. J Perinatol 2021;41(5):928–39. Available from: https://doi.org/10.1038/s41372-020-00871-0.

[62] van der Burg JW, Sen S, Chomitz VR, Seidell JC, Leviton A, Dammann O. The role of systemic inflammation linking maternal BMI to neurodevelopment in children. Pediatric Res 2016;79(1-1):3–12. Available from: https://doi.org/10.1038/pr.2015.179.

[63] Sanchez CE, Barry C, Sabhlok A, Russell K, Majors A, Kollins SH, et al. Maternal pre-pregnancy obesity and child neurodevelopmental outcomes: a meta-analysis. Obes Rev 2018;19(4):464–84. Available from: https://doi.org/10.1111/obr.12643.

[64] Cohen Kadosh K, Muhardi L, Parikh P, Basso M, Jan Mohamed HJ, Prawitasari T, et al. Nutritional support of neurodevelopment and cognitive function in infants and young children-an update and novel insights. Nutrients 2021;13(1):199. Available from: https://doi.org/10.3390/nu13010199.

[65] Prado EL, Dewey KG. Nutrition and brain development in early life. Nutr Rev 2014;72(4):267–84. Available from: https://doi.org/10.1111/nure.12102.

[66] Ciancarelli I, Morone G, Iosa M, Cerasa A, Calabrò RS, Iolascon G, et al. Influence of oxidative stress and inflammation on nutritional status and neural plasticity: new perspectives on post-stroke neurorehabilitative outcome. Nutrients 2022;15(1):108. Available from: https://doi.org/10.3390/nu15010108.

[67] Flores O, Pérez H, Valladares L, Morgan C, Gatica A, Burgos H, et al. Hidden prenatal malnutrition in the rat: role of β_1-adrenoceptors on synaptic plasticity in the frontal cortex. J Neurochem 2011;119(2):314–23. Available from: https://doi.org/10.1111/j.1471-4159.2011.07429.x.

[68] Barra R, Morgan C, Sáez-Briones P, Reyes-Parada M, Burgos H, Morales B, et al. Facts and hypotheses about the programming of neuroplastic deficits by prenatal malnutrition. Nutr Rev 2019;77(2):65–80. Available from: https://doi.org/10.1093/nutrit/nuy047.

[69] Lukowski AF, Koss M, Burden MJ, Jonides J, Nelson CA, Kaciroti N, et al. Iron deficiency in infancy and neurocognitive functioning at 19 years: evidence of long-term deficits in executive function and recognition memory. Nutr Neurosci 2010;13(2):54–70. Available from: https://doi.org/10.1179/147683010X12611460763689.

[70] Shah HE, Bhawnani N, Ethirajulu A, Alkasabera A, Onyali CB, Anim-Koranteng C, et al. Iron deficiency-induced changes in the hippocampus, corpus striatum, and monoamines levels that lead to anxiety, depression, sleep disorders, and psychotic disorders. Cureus 2021;13(9):e18138. Available from: https://doi.org/10.7759/cureus.18138.

[71] Ayaz A, Nisar I, Muhammad A, Ahmed K, Chand P, Jehan F. Structural changes in the brain on magnetic resonance imaging in malnourished children: a scoping review of the literature. Pediatric Neurol 2023;149:151–8. Available from: https://doi.org/10.1016/j.pediatrneurol.2023.08.020.

[72] Kirolos A, Goyheneix M, Kalmus Eliasz M, Chisala M, Lissauer S, Gladstone M, et al. Neurodevelopmental, cognitive, behavioural and mental health impairments following childhood malnutrition: a systematic review. BMJ Glob Health 2022;7(7):e009330. Available from: https://doi.org/10.1136/bmjgh-2022-009330.

[73] Liu J, Raine A, Venables PH, Dalais C, Mednick SA. Malnutrition at age 3 years and lower cognitive ability at age 11 years: independence from psychosocial adversity. Arch Pediatr Adolesc Med 2003;157(6):593–600. Available from: https://doi.org/10.1001/archpedi.157.6.593.

[74] Strauß H, Venables P, Zentner M. Associations between early childhood poverty and cognitive functioning throughout childhood and adolescence: a 14-year prospective longitudinal analysis of the Mauritius Child Health Project. PLoS ONE 2023;18(2):e0278618. Available from: https://doi.org/10.1371/journal.pone.0278618.

[75] Cusick SE, Georgieff MK. The role of nutrition in brain development: the golden opportunity of the first 1000 days. J Pediatr 2016;175:16–21. Available from: https://doi.org/10.1016/j.jpeds.2016.05.013.

[76] Purisch SE, Gyamfi-Bannerman C. Epidemiology of preterm birth. Semin Perinatol 2017;41(7):387–91. Available from: https://doi.org/10.1053/j.semperi.2017.07.009.

[77] Herzog JI, Schmahl C. Adverse childhood experiences and the consequences on neurobiological, psychosocial, and somatic conditions across the lifespan. Front Psychiatry 2018;9:420. Available from: https://doi.org/10.3389/fpsyt.2018.00420.

[78] Bick J, Nelson C. Early adverse experiences and the developing brain. Neuropsychopharmacology 2016;41(1):177–96. Available from: https://doi.org/10.1038/npp.2015.252.

[79] Thomas MSC, Coecke S. Associations between socioeconomic status, cognition, and brain structure: evaluating potential causal pathways through mechanistic models of development. Cognit Sci 2023;47(1):e13217. Available from: https://doi.org/10.1111/cogs.13217.

[80] McDermott CL, Seidlitz J, Nadig A, Liu S, Clasen LS, Blumenthal JD, et al. Longitudinally mapping childhood socioeconomic status associations with cortical and subcortical morphology. J Neurosci 2019;39(8):1365–73. Available from: https://doi.org/10.1523/JNEUROSCI.1808-18.2018.

[81] Rosen ML, Sheridan MA, Sambrook KA, Meltzoff AN, McLaughlin KA. Socioeconomic disparities in academic achievement: a multi-modal investigation of neural mechanisms in children and adolescents. NeuroImage 2018;173:298–310. Available from: https://doi.org/10.1016/j.neuroimage.2018.02.043.

[82] Uy JP, Goldenberg D, Tashjian SM, Do KT, Galván A. Physical home environment is associated with prefrontal cortical thickness in adolescents. Dev Sci 2019;22(6):e12834. Available from: https://doi.org/10.1111/desc.12834.

[83] Miguel PM, Pereira LO, Silveira PP, Meaney MJ. Early environmental influences on the development of 'children's brain structure and function. Dev Med Child Neurol 2019;61(10):1127–33. Available from: https://doi.org/10.1111/dmcn.14182.

[84] Legano LA, Desch LW, Messner SA, Idzerda S, Flaherty EG, Haney SB, et al. Maltreatment of children with disabilities. Pediatrics 2021;147 (5). Available from: https://doi.org/10.1542/peds.2021-050920.

[85] Thapa K, Khan H, Singh TG, Kaur A. Traumatic brain injury: mechanistic insight on pathophysiology and potential therapeutic targets. J Mol Neurosci 2021;71(9):1725–42. Available from: https://doi.org/10.1007/s12031-021-01841-7.

[86] Chakkarapani AA, et al. Therapies for neonatal encephalopathy: targeting the latent, secondary and tertiary phases of evolving brain injury. Semin Fetal Neonatal Med 2021;26(5):101256. Available from: https://doi.org/10.1016/j.siny.2021.101256.

Chapter 4

Epigenetics in neurodevelopmental disabilities

Mikael Guzman Karlsson[1] and Jacqueline Harris[2]

[1]Division of Pediatric Neurology and Developmental Neuroscience, Baylor College of Medicine, Houston, TX, United States, [2]Department of Neurology and Developmental Medicine, Kennedy Krieger Institute, Departments of Neurology, Pediatrics and Genetics, Johns Hopkins Medical Institution, Baltimore, MD, United States

In recent years, epigenetics has emerged as an important area of biology, shedding light on the intricate interplay between genetics and the environment and differential expression of genes in shaping an individual's traits and characteristics. This chapter delves into the role of epigenetics in neurodevelopmental disabilities by providing an in-depth overview of the field, discussing key epigenetic mechanisms, and exploring their influence on neurodevelopment and behavior. In addition, the chapter examines how epigenetics serves as a mechanism by which environmental stimuli contribute to neurodevelopmental disabilities, outlines various epigenetic disorders including key examples of imprinting disorders and Mendelian disorders of the epigenetic machinery, and discusses the potential of epigenetic-based therapeutics for treating neurodevelopmental disabilities.

Introduction

Epigenetics is a rapidly evolving field of study that seeks to understand the molecular mechanisms underlying the regulation of gene expression. While the DNA sequence (the genotype) provides the basis for an organism's traits and characteristics (the phenotype), it is becoming increasingly evident that there is more to the story than the genetic code.

The term "epigenetics" was coined by the British developmental biologist Conrad Waddington in the 1940s to describe the intricate interactions between genes and their environment that ultimately give rise to an organism's phenotype [1]. Epigenetic changes refer to changes in gene expression that are not caused by alterations in the DNA sequence itself but rather by chemical modifications to DNA or histone proteins or by the activity of noncoding RNA (ncRNA) molecules. These modifications can be stable and passed down through generations, making epigenetics a critical component of the complex relationship between genetics and environment in determining an individual's traits. In essence, epigenetic mechanisms provide a molecular framework for understanding how environmental factors can impact gene expression and how this can affect an organism's development, health, and behavior. As our knowledge of epigenetics expands, so does our appreciation of the intricate and dynamic relationship between our genes and the world around us.

Our genome is comparable to the blueprint of a highly complex building, containing detailed plans for every structural component, from the foundation to the roof. However, the actual construction of each building section requires unique materials and construction methods. Acting as the blueprint reader, epigenetic mechanisms determine which genes to utilize and how to employ them. This is similar to selecting appropriate building materials for the furnace (analogous to the liver) or the electrical system (representing the nervous system). The true power of epigenetic mechanisms lies in their dynamic ability to interpret the genetic blueprint and regulate gene expression profiles accordingly, a critical process in human development. This ability enables the creation of over 200 diverse cell types, each with a unique function, from a single genetic blueprint. Similar to how the choice of materials and construction methods can define the specific operation of a room, epigenetic modifications determine the precise transcriptional profile and subsequent function of each cell type by stably expressing specific genes while silencing others. As a result, a diverse, functional building with various rooms (cell types) is constructed from the same original genetic blueprint.

Capute and Accardo's Neurodevelopmental Disabilities in Infancy and Childhood. DOI: https://doi.org/10.1016/B978-0-12-824060-1.00022-5

Epigenetic mechanisms

Eukaryotic cells store DNA in a compact and structured form called chromatin. This highly organized DNA–protein structure is vital for efficiently storing genetic information and regulating gene expression. The process of transcription, which converts genetic information into RNA, depends on chromatin's three-dimensional restructuring. Chromatin is dynamic and can shift between a condensed, transcriptionally repressed state known as heterochromatin and a more relaxed, transcriptionally active state known as euchromatin. The following sections discuss key epigenetic mechanisms, including DNA methylation, histone modifications and variants, ATP-dependent chromatin remodeling, and regulatory ncRNAs. Together these mechanisms modulate chromatin states, ultimately dictating whether a specific gene is expressed or silenced.

DNA methylation

One of the most extensively studied epigenetic mechanisms involves adding a methyl group to the cytosine base of DNA, a process called DNA methylation. This process is primarily observed at CpG dinucleotides and is facilitated by enzymes known as DNA methyltransferases (DNMTs), including DNMT1, DNMT3A, and DNMT3B. DNA methylation typically represses gene expression by blocking transcription factor binding or recruiting methyl-binding domain (MBD) proteins, which interact with other repressive complexes to inhibit transcription. On the other hand, the ten-eleven translocation (TET) family proteins can promote gene expression by removing methyl groups or DNA demethylation, making DNA more accessible to the transcription machinery.

Histone modifications and variants

Histones are proteins that package and order DNA into structural units called nucleosomes, which together form chromatin. Histone modifications, such as acetylation, methylation, phosphorylation, and ubiquitination, occur on the N-terminal tails of histone proteins and play a crucial role in regulating gene expression. For example, histone acetyltransferases (HATs) add acetyl groups to histone tails, promoting an open chromatin structure and increased gene expression. In contrast, histone deacetylases (HDACs) remove acetyl groups, resulting in a more compact chromatin structure and reduced gene expression. Histone methylation, mediated by histone methyltransferases (HMTs), can either activate or repress gene expression, depending on the modified histone residue. Additionally, incorporating histone variants into nucleosomes can influence chromatin structure and gene expression.

ATP-dependent chromatin remodeling

ATP-dependent chromatin remodeling is another crucial epigenetic mechanism that modulates gene expression by altering chromatin structure. Chromatin remodelers are large multisubunit complexes that use the energy derived from ATP hydrolysis to reposition, evict, or exchange histones within nucleosomes. This process can create or eliminate access to DNA-binding sites for transcription factors and other regulatory proteins, thereby influencing gene expression. Chromatin remodelers can be classified into four leading families: SWI/SNF, ISWI, CHD, and INO80. Each family has distinct target sequences and effects on chromatin structure, with some promoting gene expression and others repressing it, depending on the context.

Regulatory noncoding RNAs

Regulatory ncRNAs represent a diverse class of RNA molecules that do not code for proteins but can influence gene expression through various epigenetic mechanisms. MicroRNAs (miRNAs), small interfering RNAs (siRNAs), and long noncoding RNAs (lncRNAs) are examples of regulatory ncRNAs that play crucial roles in gene regulation. MiRNAs and siRNAs primarily bind to complementary sequences in target messenger RNAs (mRNAs), leading to mRNA degradation or translational repression. lncRNAs can regulate gene expression through various mechanisms, such as acting as scaffolds for the recruitment of chromatin-modifying complexes, serving as decoys for transcription factors, or forming RNA–DNA triplex structures that modulate the binding of regulatory proteins to DNA. The functional outcomes of ncRNA-mediated regulation can be either activation or repression of gene expression, depending on the specific ncRNA and target genes involved.

Epigenetic modifications can be stable or dynamic and can be influenced by genetic variation or by environmental stimuli. For example, DNA methylation is about 20% heritable genome-wide but individual sites can be between 0% and 99% heritable [2]. All of the heritability in humans seems to be due to inherited genetic variations as opposed to direct transgenerational transmission of DNA methylation [2]. DNA methylation at certain sites is so stable across time that distinctive DNA methylation "signatures" can be used to diagnose genetic syndromes [3] while at other sites, DNA methylation can change across developmental time points. Histone modifications can even change moment to moment to regulate transcription.

Epigenetics in neurodevelopment

The development of the mammalian brain is a complex process consisting of highly orchestrated development, differentiation, migration, and pruning of billions of morphologically and functionally distinct neuron and glia populations to create functional circuits. Intrinsic transcriptional and epigenetic networks closely regulate progenitor proliferation and differentiation, cellular migration, morphological maturation, synaptogenesis, and programmed cell death. For example, DNA methylation at non-CpG sites becomes progressively more enriched in brain tissues throughout development and has been shown to be essential for neurogenesis and maturation of neural progenitor cells [4]. The process of 3D chromatin structure alteration and remodeling is also a process that dynamically changes to facilitate neural development and differentiation with certain regions of chromatin become increasingly more compressed as nervous system development progresses causing certain genes to become permanently silenced, thus dictating neural progenitor cells' fate [5,6]. What is more, heterozygous knockouts of the chromatin remodeler proteins lead to specific alterations in neuronal differentiation and cortex organization. Specifically, in mice, heterozygous loss of Chd2 leads to a distinct deficiency of GABAergic neurons [7] whereas heterozygous loss of Ctcf leads to excessive neuronal apoptosis and disorganization of the forebrain and telencephalon [8]. All types of histone modifications also appear to be important for neural development and differentiation. For example, in mouse brain, deletion of Dpy30 or Ezh2—both common components of histone methyltransferases—leads to neurogenic and gliogenic deficits including suppression of neurogenesis and differentiation of NPCs into neurons and glia [9–11]. Several studies have shown that delicately orchestrated increases and decreases in H3K9 acetylation are necessary for early and accurate neural differentiation [12,13].

In addition to affecting the development of the nervous system, epigenetics offers a unique perspective on how environmental stimuli can contribute to the development of NDDs. Environmental factors such as prenatal exposure to toxins, maternal stress, and early life experiences can have lasting effects on brain development and function. These environmental stimuli can interact with an individual's genetic makeup, leading to alterations in the epigenome, which in turn can modulate gene expression and contribute to the development of NDDs [14].

One critical insight from social science research is the role of early life experiences in shaping the epigenome. Adverse childhood experiences (ACEs), such as maltreatment, neglect, or exposure to domestic violence, have been linked to lasting changes in the epigenetic regulation of genes involved in stress response, immune function, and neuronal plasticity [15]. These epigenetic alterations can potentially mediate the relationship between ACEs and the increased risk of developing NDDs, such as ADHD, autism spectrum disorder (ASD), and learning disabilities [16,17].

Another important area of social science research focuses on the effects of socioeconomic factors on the epigenome and the development of NDDs. Socioeconomic disparities have been associated with differences in DNA methylation patterns, which can in turn influence cognitive function and susceptibility to NDDs [18]. For instance, lower socioeconomic status has been linked to alterations in the methylation of genes involved in neurodevelopment, stress response, and immune function in those born preterm, potentially contributing to disparities in cognitive abilities and the prevalence of NDDs among different populations [19].

A compelling example of how environmental factors can induce lasting epigenetic changes is the Dutch Hunger Winter, a famine that occurred in the Netherlands during the winter of 1944–45. Children who were exposed to the famine during prenatal development showed persistent epigenetic modifications, such as changes in DNA methylation, that persisted even six decades after the famine [20]. These epigenetic changes were associated with long-term health consequences, including increased risk for obesity, type 2 diabetes, and other metabolic disorders [21]. The Dutch Hunger Winter serves as a historical example of how environmental stressors during critical periods of development can lead to lasting epigenetic changes with potential implications for neurodevelopmental outcomes.

In summary, social science research provides valuable insights into how environmental stimuli, such as early life experiences and socioeconomic factors, can contribute to the development of NDDs through epigenetic mechanisms. While these disorders will be covered in further detail in later chapters, understanding these complex interactions between environment, epigenetics, and neurodevelopment can inform prevention and intervention strategies to reduce the impact of NDDs on individuals and society.

Epigenetic disorders

While it is clear that epigenetic regulation of gene expression plays a role in neurodevelopmental and neurophysiological processes, specific neurodevelopmental disorders also result from disruption of normal epigenetic processes. These epigenetic disorders are categorized as either imprinting disorders or Mendelian disorders of the epigenetic machinery (MDEMs), which result when a genetic variant disrupts a specific component of the epigenetic machinery.

Imprinting disorders introduction

Imprinting is the process by which only one allele of an autosomal gene is expressed in a tissue depending on the allele's parent of origin. This process of silencing or activating alleles based on parent of origin is done through specific patterns of DNA methylation. Imprinting disorders occur when the expression of imprinted genes is disrupted by sequence change including deletion or duplication or by epigenetic regulation. Imprinting occurs in less than 1% of human genes but disruption of imprinting in these few regions leads to significant symptoms. There are now at least 14 imprinting disorders known in humans and several of them cause significant neurodevelopmental disabilities (Table 4.1). Here we review several of the imprinting disorders with neurodevelopmental features as prominent parts of the phenotype.

Angelman syndrome

Clinical presentation

Angelman syndrome (AS) occurs in 1:12,000 to 1:24,000 individuals and is characterized by intellectual disability with speech and language most affected, hypotonia, ataxia, seizures with a characteristic electroencephalography (EEG) pattern, gastrointestinal issues, microcephaly, sleep dysfunction, and a specific behavioral pattern [22,23]. This specific behavior pattern includes a happy demeanor, paroxysmal laughter, anxiety, excitability with hand flapping stereotypies, hyperactivity with disruptive and impulsive and aggressive behavior, and an obsessive interest in water [24,25]. There are also progressive neurologic features including distal lower extremity hypertonia, upper extremity tremor, and worsening of the ataxia whereas seizures can get better with age [22,23]. The vast majority of individuals with AS fall in the severe to profound range of intellectual disability and most are completely or functionally nonverbal, although research indicates that individuals have a wide array of nonverbal communicative methods that go beyond their abilities with speech production [26].

TABLE 4.1 Neurodevelopmental disabilities in imprinting disorders.

Imprinting disorder	Intellectual disability	Other neurodevelopmental features
Angelman syndrome	Present	Seizures, hypotonia, language impairment, ataxia, sleep impairment, anxiety, stereotypies, hyperactivity, impulsivity, aggression, lower extremity hypertonia, upper extremity tremor
Beckwith–Wiedemann syndrome		Expressive speech delay (likely due to big tongue), ADHD
Birk–Barel syndrome	Present	Hypotonia, hyperactivity, aggression, ASD, seizures
Central precocious puberty		
Central precocious puberty 2		
Kagami–Ogata syndrome	Present	
Mulchandani–Bhoj–Conlin syndrome		
Prader–Willi syndrome	Present	Hypotonia, hyperphagia, obsessive-compulsive disorder, poor executive function, aggression, impulsivity, anxiety, skin picking
Pseudohypoparathyroidism type 1A		
Pseudohypoparathyroidism type 1B		
Schaaf–Yang syndrome	Present	ASD, areas of hypertonia, language problems
Silver–Russell syndrome		Autistic traits, attention problems
Temple syndrome	Present	Hypotonia
Transient neonatal diabetes mellitus		

ADHD, Attention-deficit hyperactive disorder; *ASD*, autism spectrum disorder.

Molecular genetics/diagnosis

AS is caused by loss of the maternal copy of the *UBE3A* (ubiquitin-protein ligase 3A) gene. Only the maternal copy is expressed in the brain. This loss of the maternal copy can happen through one of four mechanisms. These mechanisms, in order of frequency with which they are the cause of AS, are maternal deletion of chromosome 15q11–13, pathogenic sequence variant in maternally inherited copy of *UBE3A*, paternal uniparental disomy (UPD) of chromosome 15q11–13, and imprinting defect at the chromosome 15q11–13 region. Broadly speaking, individuals with the deletion have the most severe and classic phenotype [25].

When AS is suspected, diagnosis can be made by DNA methylation analysis in the majority of cases. However, this analysis will not distinguish between the specific genetic mechanisms. First-line testing is generally chromosomal microarray (CMA), which will detect deletions and the majority of cases of UPD along with DNA methylation analysis or methylation-specific multiplex ligation-dependent probe amplification (MS-MLPA) to confirm parent of origin. If these steps do not yield the genetic mechanism, next step is *UBE3A* sequencing. If CMA and *UBE3A* sequencing are both normal but there is a clinical diagnosis of AS with an abnormal DNA methylation analysis, then an imprinting defect is presumed to be the cause unless research-based testing is pursued to confirm [22]. Establishing a molecular diagnosis is important for prognosis, reproductive recurrence counseling, and also for eligibility for specific clinical trials to treat neurodevelopmental manifestations of AS. Parental testing is important, especially for those with a deletion to counsel about reproductive recurrence [27].

Management

Currently, management of AS is entirely symptomatic. Seizures should be treated, when present, and there is some literature on preferences for certain antiepileptic treatments over others [28]. Appropriate management of specific neurodevelopmental and behavioral manifestations should be done. Additionally, there are current clinical trials looking at potential disease-modifying treatments for AS.

Prader–Willi syndrome

Clinical presentation

Prader–Willi syndrome (PWS) affects around 1 in every 10,000–30,000 births. It is characterized by intellectual disability, significant hypotonia, early feeding problems with a switch to hyperphagia and obesity, short stature with small hands and feet and genitals, and a distinct behavioral profile [29,30]. Individuals with PWS generally fall in the mild intellectual disability range but they have poor executive function and behaviors that make their adaptive functioning worse than would be expected for their IQ [31]. Behaviorally, individuals suffer from anxiety throughout their lifespan with a significant obsessive-compulsive component which contributes to the food obsession and food-seeking behaviors [32].

Molecular genetics/diagnosis

PWS is caused by loss of paternal chromosome 15q11–13 region, which is the opposite molecular defect of Angelman syndrome (AS). An individual can have PWS through any of three different molecular genetic mechanisms which are in order of how commonly they occur—deletion of the paternal copy of chromosome 15q11–13, UPD, and an imprinting defect [29]. Testing for PWS should start with DNA methylation analysis followed by a SNP array if positive for PWS. Further testing may be pursued as necessary [27].

Management

Individuals with PWS should be started on growth hormone as soon as possible after the diagnosis is made, if there are no absolute contraindications. Growth hormone can improve lean muscle mass and hypotonia and reduce obesity in individuals with PWS beyond just linear growth [33]. Beyond growth hormone, management of PWS is symptomatic. Appropriate management of specific neurodevelopmental and behavioral manifestations should be done. Behavioral therapy and nutrition should work together to prevent overeating and obesity. Several current clinical trials are underway to more specifically treat the behavioral and endocrine manifestations of PWS.

Maternal 15q duplication syndrome

Maternally derived duplication of chromosome 15q11.2–13.3 leads to hypotonia from infancy with a weak cry, developmental delay leading to usually severe intellectual disability with speech and language more affected, autism spectrum disorder, seizures, and schizophrenic features. Diagnosis can be made with SNP array and methylation testing to confirm maternal derivation (or reverse order). Individuals with an isodicentric chromosome 15 usually have a more severe developmental phenotype and worse seizure disorder. As with AS, maternal testing is important for appropriate genetic counseling to families [27]. Management is symptom based including developmental and behavioral services and treatment of epilepsy.

Kagami–Ogata syndrome

Clinical presentation

Kagami–Ogata syndrome (KOS) is an extremely rare syndrome with estimates of less than 1 in 1 million births that is characterized by distinctive features including full cheeks, a long and prominent philtrum, and a small thorax with coat hanger ribs [34]. Individuals also have polyhydramnios prenatally, respiratory problems, feeding problems, and frequent abdominal wall defects. Many individuals die before the age of 4 and there have only been a few reported cases in adulthood [34]. Nearly all individuals reported with KOS who survived have developmental delay leading to intellectual disability and many of these individuals have severe ID [35]. More than one-third also have reported hypotonia with delayed age at walking [34,35]. A small percentage have seizures as well [34]. Very little is known about the specific cognitive or behavioral profile in KOS.

Molecular genetics/diagnosis

Like many of the other imprinting disorders, KOS has multiple molecular mechanisms that can lead to it. KOS can be caused by a microdeletion on the maternal copy of the long arm of chromosome 14, by paternal UPD of chromosome 14, or by an epimutation at the imprinting center—specifically the *MEG3* differentially methylated region. In suspected cases of KOS, methylation analysis should be done first and, if abnormal, should be followed by analysis for UPD and chromosomal SNP array [36]. More comprehensive methylation analysis beyond chromosome 14, specifically Episign, can also be diagnostic.

Management

Like most other disorders in this chapter, current treatment is symptomatic and supportive only. Respiratory interventions such as mechanical ventilation are often needed and difficult respiratory status may contribute to developmental outcomes. Feeding interventions including tube feedings are also often needed. Individuals are at higher risk of hepatoblastoma and so should be screened regularly [36]. Individuals with KOS should receive appropriate developmental services and therapies, although this can be challenging because they are frequently medically unstable in early life.

Temple syndrome

Temple syndrome is a syndrome caused by loss of the paternally imprinted region of chromosome 14. It has been seen in cases of paternal deletion of 14q region, in cases of maternal UPD of chromosome 14, and in imprinting defects—specifically loss of methylation at the 14q32 imprinting region. It is clinically characterized by low birth weight leading to short stature, hypotonia, small hands and feet, and precocious puberty. While these features are nearly universal, many other features are present in nearly or greater than half of individuals with Temple syndrome including obesity and development of diabetes mellitus type 2 and intellectual impairment [37]. Some describe the phenotype as similar to PWS [38]. Just under half of individuals with Temple syndrome are in the borderline cognitive impairment to mild intellectual disability range [37] with the remainder classified as average cognition. Little is known about associated behavioral issues. Chromosomal microarray can detect deletion of the paternal copy of the 14q region and cases of extensive maternal UPD of chromosome 14. Methylation analysis will detect all types of Temple syndrome and can also be used to confirm that the cases suspected based on a deletion or UPD seen on microarray are actually Temple syndrome.

Mendelian disorders of the epigenetic machinery

Alterations in the normal biallelic state of genes that encode proteins that play a direct role in epigenetic functions are an increasingly recognized cause of genetic intellectual disability. While individually rare, together these MDEMs account for a large proportion of genetic intellectual disability. Further underscoring the importance of these epigenetic machinery proteins in neurodevelopment and neurologic functioning, the vast majority of known MDEMs has significant neurodevelopmental disability as a core part of the phenotype (Table 4.2) [39–41].

TABLE 4.2 Mendelian disorders of the epigenetic machinery (MDEM) genes associated with neurodevelopmental disabilities and reported phenotypes (originally in Ref. [41]).

Gene	Primary mechanism	Function	Dual function?	ID	ADHD	Anxiety	Hypotonia	Other behavior problems
ALG13	Reader	HMR		✓			✓	Limited reports of SIB
ASH1L	Writer	HMT/HMR and HAR	✓	✓	✓	✓	✓	ASD, SIB
ASXL1	Reader	HMR		✓			✓	
ASXL2	Reader	HMR		✓	✓		✓	ASD
ASXL3	Reader	HMR		✓			✓	
ATRX	Remodeler	RE/HMR	✓	✓	✓		✓	
BPTF	Reader	HMR and HAR		✓		✓	✓	
BRPF1	Reader	HMR and HAR		✓	✓		✓	
BRWD3	Reader	HMR and HAR		✓			✓	Autistic features
CHD1	Remodeler	RE/HMR	✓	✓			✓	Autistic features
CHD2	Remodeler	RE/HMR	✓	✓	✓			
CHD3	Remodeler	RE/HMR	✓	✓			✓	Stereotypies
CHD4	Remodeler	RE/HMR	✓	✓				
CHD5	Remodeler	RE/HMR	✓	✓	✓		✓	Aggression, disruption
CHD7	Remodeler	RE/HMR	✓	✓			✓	ASD, aggression, SIB, OCD
CHD8	Remodeler	RE/HMR	✓	✓	✓	✓	✓	ASD, aggression, SIB
CREBBP	Writer	HAT/HAR	✓	✓	✓	✓	✓	Repetitive behavior, aggression, SIB, OCD
DNMT3A	Writer	DNMT/HMR	✓	✓	✓	✓	✓	ASD
DPF2	Reader	HAR		✓			✓	Behavioral prob NOS
EED	Reader	HMR		✓			✓	
EHMT1	Writer	HMT/HMR	✓	✓		✓	✓	ASD, aggression
EP300	Writer	HAT/HAR	✓	✓	✓	✓	✓	Repetitive behavior, aggression, SIB, OCD
EZH2	Writer	HMT		✓		✓	✓	ASD
HDAC4	Eraser	HDAC		✓			✓	
HDAC6	Eraser	HDAC		✓				
HDAC8	Eraser	HDAC		✓	✓			ASD, aggression, SIB
KAT5	Writer	HAT		✓	✓			Stereotypies
KAT6A	Writer	HAT/HAR	✓	✓			✓	
KAT6B	Writer	HAT/HAR	✓	✓			✓	
KAT8	Writer	HAT		✓	✓		✓	ASD
KDM1A	Eraser	HDM		✓			✓	
KDM3B	Eraser	HDM		✓	✓		✓	ASD
KDM4B	Eraser	HDM/HMR	✓	✓	✓		✓	OCD

(Continued)

TABLE 4.2 (Continued)

Gene	Primary mechanism	Function	Dual function?	ID	ADHD	Anxiety	Hypotonia	Other behavior problems
KDM5B	Eraser	HDM/HMR	✓	✓	✓		✓	
KDM5C	Eraser	HDM/HMR	✓	✓	✓	✓	✓	Aggression, repetitive behaviors, stereotypies
KDM6A	Eraser	HDM		✓	✓	✓	✓	
KDM6B	Eraser	HDM		✓	✓		✓	ASD
KMT2A	Writer	HMT/HMR & HAR and DNUMR	✓	✓	✓	✓	✓	Repetitive behaviors, OCD, aggression
KMT2B	Writer	HMT/HMR and DNUMR	✓	✓	✓		✓	ASD, stereotypies
KMT2C	Writer	HMT/HMR	✓	✓	✓		✓	ASD, aggression
KMT2D	Writer	HMT/HMR	✓	✓	✓	✓	✓	OCD, stereotypies
KMT2E	Writer	HMT/HMR	✓	✓	✓	✓	✓	ASD, aggression, SIB, stereotypies
KMT5B	Writer	HMT		✓	✓		✓	ASD
MBD5	Reader	HMR and DNMR		✓	✓	✓	✓	ASD, aggression, SIB, stereotypies
MECP2	Reader	DNMR		✓		✓	✓	ASD, aggression, SIB, repetitive behavior, stereotypies
MORC2	Reader	HMR		✓			✓	
MSL3	Reader	HMR		✓			✓	ASD
NSD1	Writer	HMT/HMR	✓	✓	✓	✓	✓	ASD, aggression, repetitive behavior, OCD
PHF21A	Reader	HMR		✓	✓	✓	✓	ASD, SIB
PHF6	Reader	HMR		✓			✓	
PHF8	Eraser	HDM/HMR	✓	✓				ASD
PHIP	Reader	HMR and HAR		✓	✓	✓	✓	ASD, aggression, stereotypies
PRDM13	Writer	HMT		✓				
PRDM8	Writer	HMT		✓				
RAI1	Reader	HMR		✓	✓	✓	✓	Aggression, OCD
RERE	Reader	HMR		✓			✓	ASD, behavior problems NOS
SETD1A	Writer	HMT		✓	✓	✓	✓	Aggression, OCD
SETD1B	Writer	HMT		✓		✓		ASD
SETD2	Writer	HMT		✓	✓	✓	✓	ASD, aggression, OCD

(Continued)

TABLE 4.2 (Continued)

Gene	Primary mechanism	Function	Dual function?	ID	ADHD	Anxiety	Hypotonia	Other behavior problems
SETD5	Writer	HMT		✓		✓	✓	ASD, stereotypies, OCD
SMARCA2	Remodeler	RE/HAR	✓	✓				
SMARCA4	Remodeler	RE/HAR	✓	✓	✓		✓	Repetitive behaviors, stereotypies
SRCAP	Remodeler	RE		✓	✓	✓	✓	ASD, aggression, mood disorders
TCF20	Reader	HMR		✓	✓	✓	✓	ASD, mood disorders
TET3	Eraser	DNME		✓	✓	✓	✓	ASD, stereotypies
UBR7	Reader	HMR		✓			✓	
WHSC1 (NSD2)	Writer	HMT/HMR	✓	✓	✓	✓	✓	ASD, stereotypies
ZMYND11	Reader	HMR and HAR		✓	✓		✓	ASD, aggression

ASD: Autism; *NOS*: not otherwise specified; *OCD*: obsessive compulsive disorder; *SIB*: self-injurious behavior.

MDEMs can be divided into categories based on the function the protein encoded by the gene with the pathogenic variant(s). The categories are (1) writers, proteins that place chemical groups on DNA or histones; (2) erasers, proteins that remove chemical groups from DNA or histones; (3) readers, proteins which recognize these chemical groups on DNA or histones and recruit appropriate enzymes or complexes; and (4) remodelers, which alter the physical conformation of chromatin. All of these functions lead to activation or repression of gene expression [39]. The majority of MDEMs are inherited in an autosomal-dominant fashion and are caused by heterozygous loss of function variants in components of the machinery that perform this writing, erasing, reading, and remodeling. Writers, erasers, and readers can be further subdivided based on the type of chemical group they write, erase, or read and whether they write, erase, or read marks on DNA or histones. Many genes that cause MDEMs encode proteins that have dual functions (e.g., writer and reader) [42]. The most widely studied chemical groups that are written, erased, and read are methyl groups and acetyl groups. Methylation, or the addition of methyl groups, can occur on both DNA—catalyzed by DNMTs and typically associated with silencing of gene expression—and on histones—catalyzed by HMTs (PRMTs or KMTs) which can be activating or repressing [43]. Acetylation occurs only on histones and is carried out by HATs and is typically associated with transcriptional activation [43]. Erasers include TET enzymes which demethylate DNA, histone demethylases which demethylate histones, and HDACs, which remove the acetyl group from histones [44]. Readers have specialized domains that recognize epigenetic marks. Some readers recognize methyl groups, some acetyl groups, and some both. Some readers bind to and recruit complexes directly to DNA and some to histones [41]. All known MDEMs that cause neurodevelopmental disorders are listed in Table 4.2. Following is a detailed description of several MDEMs that are phenotypically or molecularly representative of the larger group.

Tatton–Brown–Rahman syndrome

Clinical presentation

Tatton–Brown–Rahman syndrome (TBRS) is a disorder characterized by overgrowth, hypotonia, mild to moderate intellectual disability, and a specific behavioral phenotype [45]. Nearly all described individuals have cognitive impairment with IQs ranging from 39 to 76. Limited case studies suggest verbal reasoning skills are a strength compared to spatial processing skills [46,47]. Behavioral concerns are also extremely prominent in TBRS. Frequently described behaviors include aggression, stereotypies, obsessive-compulsive behaviors, anxiety, impulsivity, hyperactivity, and sensory processing issues [47–49]. Neuropsychiatric disorders are extremely common. More than half of individuals with TBRS have ASD and/or psychotic disorders such as schizophrenia [47,50]. A few individuals reported in the literature seem to have regressed after childhood but it is unclear whether this is due to onset of schizophrenia as opposed to true regression or neurodegeneration of any kind.

Molecular genetics/diagnosis

TBRS is caused by heterozygous loss of function variants in *DNMT3A* which is a writer of DNA methylation [51]. The vast majority of cases are de novo. Diagnosis is made through sequencing, most commonly either gene panel testing or whole exome sequencing. TBRS also has a known and specific DNA methylation signature that can be identified using DNA methylation array such as Episign [3] and is diagnostic of TBRS. This can be used as a primary diagnostic tool or to clarify variants of uncertain pathogenicity in *DNMT3A*.

Management

Currently management of TBRS is entirely symptom based and should include appropriate therapies, including physical therapy, occupational therapy, and speech/language therapy, appropriate educational interventions, and multimodal therapies for the behavioral and psychiatric issues including behavioral therapy and medications.

CHARGE syndrome

Clinical presentation

CHARGE syndrome is characterized by growth retardation in addition to multiple organ anomalies that includes the heart, choanae, genital/urinary systems, ear, and ocular coloboma, and less commonly vertebral/limb, renal, and tracheal abnormalities [52–54]. Cognitive functioning is also impaired in the majority of individuals CHARGE syndrome with more than half with ID although there is a large range with some individuals displaying average intellect [55,56]. Hearing loss is very common in this syndrome which contributes to the speech/language delay that is commonly seen [56]. Clinical studies involving patients with CHARGE syndrome report primary weaknesses in visuospatial construction, sequential processing, and selective attention juxtaposed with relative strengths in semantic skills, logical reasoning, and planning [55]. Notably, brain malformations, microcephaly, and low vision were prognostic predictors of intellectual functioning [57].

The neurobehavioral profile associated with CHARGE syndrome includes increased incidence of psychiatric illnesses including obsessive-compulsive disorder (43%–49%) and anxiety disorder (37%–53%) [58,59] with potential high comorbidity of the two psychopathology [60]. Depression (8%–24%) [59,61,62] and ADHD (26%–34%) [60,61], self-injurious behaviors (40%–54%) [59,61–63], and aggressive behaviors (38%–53%) [59,61,62,64] are also commonly found in those with this syndrome. Those with dual sensory impairment (deaf–blind) generally present with more challenging behaviors [61]. Autistic behaviors are commonly reported (26%–43%) [58,59,61], although some researches suggest these features may be due to sensory deficits, such as hearing loss and visual impairment, and are qualitatively different from the social impairment seen in idiopathic ASD [65].

Molecular genetics/diagnosis

CHARGE syndrome is caused by heterozygous mutations in CHD7 [52]. CHD7 is a chromatin remodeler which binds to elements on DNA and changes the conformation of chromatin which changes the availability of certain genes for expression. CHD7 variants that cause CHARGE syndrome are thought to be loss of function variants. Genetic variants are identified through sequencing of CHD7. Very rarely, deletions of chromosome 8q that impact CHD7 can be the cause and these are diagnosed typically through microarray or sometimes also through sequencing of CHD7. In addition, there is a specific DNA methylation episignature associated with CHARGE syndrome that can be used for diagnosis through clinical Episign testing [3].

Management

Management of CHARGE syndrome is symptomatic. Given the large number of different organs that can be affected by CHARGE, a multidisciplinary team is optimal. Vision and hearing should be checked and monitored and any impairments should be addressed. Other organ system anomalies should be treated by the relevant medical specialist [66]. Multiple modalities of therapies should be engaged from an early age. Educational needs for children with CHARGE are complex and varied and guidelines for appropriate IEPs have been published [67].

Rubinstein–Taybi syndrome

Clinical presentation

Rubinstein–Taybi syndrome (RTS) is a syndrome resulting in facial dysmorphisms, postnatal growth retardation, broad thumbs and great toes, and intellectual disability. A host of other anomalies including hypotonia, cardiac, kidney, GI and feeding issues, immune deficiencies, dental, and skin issues are also common in this syndrome. RTS occurs in about 1 in every 125,000 births [68].

Cognitively, the largest group of individuals with RTS are reported to fall in the moderate intellectual disability range with individuals scattering into the mild and severe range as well. Little work has been done to characterize specific domains of cognition that are more and less impaired in RTS, the few studies that do exist suggest verbal skills are relatively stronger than nonverbal and spatial skills among those with RTS but these studies have very small sample sizes and often inconsistent measures [69,70]. Neurobehaviorally, individuals with RTS have been described to show relative strengths in social communication and affinity for interacting with others [71], regardless of severity of intellectual impairment [72]. However, individuals with RTS often meet criteria for autism because of the high prevalence of restricted and repetitive behaviors [73]. Other behavioral problems that are present include attention problems, hyperactivity, and motor stereotypies [70,71]. Externalizing problems are less consistently documented in RTS, whereas elevated rates of anxiety, obsessive-compulsive disorder, and depression are common and seem to increase with age [73,74].

Molecular genetics/diagnosis

RTS is caused by heterozygous variants in either *CREBBP* or *EP300* genes—both essential components of histone acetylation complexes. Pathogenic variants in *CREBBP* cause RTS type 1 (RTS 1) while those in *EP300* cause RTS type 2 (RTS 2). Among those with clinically diagnosed RTS, about 60% have molecularly confirmed RTS 1 and about 10% RTS 2. The remaining one-third of patients with a clinical diagnosis have no identifiable variants in either gene [68]. However, since there is currently no consensus diagnostic criteria for RTS, it is uncertain whether these individuals are misdiagnosed as RTS and actually have a different syndrome, or whether there is a yet unidentified genetic cause of RTS.

Diagnosis of RTS is typically made by sequencing and deletion/duplication analysis of CREBBP and EP300. Deletion/duplication analysis either specifically of those two genes or via chromosomal SNP microarray is important as many individuals have been identified with RTS caused by deletions on chromosome 16 including the *CREBBP* region. In addition, RTS has a known DNA methylation episignature that can be used to diagnose RTS through Episign.

Individuals with RTS 2 seem to present with milder cognitive impairment than patients with RTS 1, with more clustering in the mild ID range. However, individuals with RTS 2 seem to have more impairment from certain features of ADHD and anxiety than those with RTS 1 [74,75].

Management

Management of RTS includes a multidisciplinary medical team of subspecialists for all the different medical complications. Individuals with RTS should also receive appropriate therapies from an early age and ongoing including physical therapy, occupational therapy, and speech/language therapy. Tailored educational interventions are essential and, when necessary, interventions for the behavioral and psychiatric issues including behavioral therapy and medications.

Kabuki syndrome

Clinical presentation

Kabuki syndrome (KS) is a genetic disorder occurring in about 1 in every 32,000 births characterized by dysmorphic facial features, postnatal growth deficiency, hypotonia, and multiple congenital skeletal and other organ system anomalies. These include congenital heart defects, structural kidney issues, skeletal anomalies, immune deficiencies, GI and feeding issues, cleft palate, and persistence of the fetal finger pads [76,77].

KS is associated with a very specific neurodevelopmental profile. Most individuals have mild intellectual disability with some scattered into more impairment and some in the low average IQ range. However, almost all individuals with KS have disproportionate and significant weaknesses in visuospatial construction, perception, and memory with many having relative strengths in verbal and language skills [78], although speech and language impairments still exist when compared to typically developing individuals [79]. Executive functioning and working memory are generally less affected than visuospatial skills [78,80]. The neurobehavioral profile of KS is characterized by early

ADHD features and very prevalent anxiety that increases with age [41]. Individuals are very prosocial with what is typically described as a happy demeanor and more preserved social adaptive skills [81,82] but with some challenges in pragmatic language [79,82].

Molecular genetics/diagnosis

KS results from heterozygous variants in *KMT2D* (80%) which result in KS type 1 (KS 1) or *KMD6A* (5%–10%) which result in KS type 2 (KS 2) [83]. *KMT2D* encodes a HMT that places mono-, di-, and tri-methylation at the fourth lysine of histone 3 (H3K4) which promotes open chromatin. *KDM6A* removes methylation at H3K27; when present, this methylation promotes closed chromatin. As such, loss of KMT2D or KDM6A both lead to a chromatin configuration that is too closed. As such, KS is a model example of MDEMs as variants in two different genes with two different functions that lead to similar transcriptional and phenotypic outcomes [41].

Diagnosis of KS is made by clinical consensus criteria [77] or by sequence analysis or deletion/duplication analysis of *KMT2D* and *KDM6A*, although sequence variants are by far the most common. As with many of the other epigenetic disorders outlined in this chapter, KS also has a distinctive DNA methylation episignature [84] that can be used for diagnosis through clinical Episign.

Management

Currently, management of KS is symptomatic only although there are clinical trials investigating disease-modifying therapies underway. Individuals with KS should receive care from the relevant medical subspecialists, therapies including physical, occupational, and speech therapy and behavioral interventions that may include therapy and medications [85].

The future of epigenetic therapies for neurodevelopmental disabilities

Given the important role of epigenetic marks and machinery in neurodevelopment, cognition, and behavior, there may also be a role for epigenetic modification as treatment for neurodevelopmental disorders. Several studies have demonstrated promise in using epigenetic modification to treat epigenetic disorders and similar therapies may have use for NDDs more broadly. Several antisense oligonucleotides which serve to reactivate the silenced paternal allele of *UBE3A* are currently in clinical trials to treat AS and this kind of therapy could be used in other imprinting disorders as well [86–89]. In mouse models of RTS, HDAC inhibitors and phosphodiesterase inhibitors reverse the behavioral and neurophysiologic phenotype of these animals [90,91]. In mouse models of KS, several epigenetic therapies—HDAC inhibitors, ketogenic diet (thought to be working through a mechanism of HDAC inhibition by beta-hydroxybutyrate), and KDM1A inhibition (KDM1A is the primary demethylase at H3K4 where KMT2D acts as a methyltransferase)—all reverse the behavioral and neurohistopathologic phenotype of these animals [92–94]. Clinical trials are now also underway in KS to test some of these therapies in humans. Beyond these therapies for specific rare genetic syndromes, several epigenetic therapies are in development to potentially treat broader categories of neurodevelopmental disorders. KDM1A inhibitors are being trialed in adult psychiatric conditions with plans to begin clinical trials in pediatrics in autism, aggressive behaviors, and more. Several pharmaceutical companies and academic centers are also looking to improve existing HDAC inhibitors to trial them in autism, intellectual disability, and anxiety among other things.

In conclusion, epigenetics is a rapidly developing field that is demonstrating itself to have immense importance to brain development and neurodevelopmental disorders with potential for further understanding underlying mechanisms of NDDs as well as potential treatments. For clinicians working with individuals with NDDs, an epigenetic disorder should be considered in individuals with NDDs—particularly intellectual disability with or without ADHD and anxiety—and hypotonia or history of early life hypotonia and/or growth abnormalities (microcephaly, significant macrocephaly, short stature, or overgrowth). However, given the increasing percentage of genetic ID accounted by MDEMs, as well as the broadening clinical spectrum of many MDEMs with more mild cases being described with NDDs but without ID, it is reasonable to consider an epigenetic disorder in any individuals with NDDs who meet existing American College of Medical Genetics or American Academy of Pediatrics criteria for broader genetic testing. The testing toolbox to diagnose epigenetic disorders is actively expanding. It has significant overlap with the general toolbox of genetic testing in NDDs but also has some tests that are newer and unique (Table 4.3). For example, in an individual with an NDD in whom an epigenetic disorder is suspected but other diagnoses are suspected as well, chromosomal microarray and whole exome sequencing are still first-line tests. However, more and more DNA methylation profiles for specific disorders are being published, and tests like whole genome DNA methylation arrays like Episign have a number of advantages like giving clarity on pathogenicity of variants of uncertain significance and diagnosing methylation abnormalities like in types of AS that would not be picked up by sequencing. In the future, this type of testing could become first line in certain clinical scenarios.

TABLE 4.3 Genetic testing for epigenetic disorders.

Type of test	When to use
Chromosomal microarray	First-line testing for individuals with ID, ASD, NDD + other anomalies
Whole exome sequencing	First-line testing for individuals with ID, ASD, NDD + other anomalies
Single gene testing	When a specific disorder is strongly suspected (clinical features, family history, etc.) or if a diagnostic DNA methylation signature is identified and specific variant info is desired
DNA methylation testing	Can be a specific region (e.g., Angelman/Prader–Willi region methylation) or whole genome. Use when a specific diagnosis (e.g., PWS) is strongly suspected but sequencing has been negative or is not yet done. When all or most disorders on the differential diagnosis have a known signature especially if some of them like Temple syndrome might be missed by exome sequencing alone. To clarify variants of uncertain significance in epigenetic genes or regions that are known to cause epigenetic syndromes
Whole genome sequencing	Not currently a first-line test, still difficult to interpret. Might use in individuals in whom all other testing has been negative

Generally do not recommend doing gene panel testing in any scenario when epigenetic disorder is suspected, prefer to do exome sequencing.
ASD, Autism spectrum disorder; *NDD*, neurodevelopmental disabilities; *PWS*, Prader–Willi syndrome.

References

[1] Waddington CH: The epigenotype. Int J Epidemiol 1942;41(1):10–13.

[2] Villicaña S, Bell JT. Genetic impacts on DNA methylation: research findings and future perspectives. Genome Biol 2021;22(1):127.

[3] Aref-Eshghi E, Kerkhof J, Pedro VP, Groupe DI FranceBarat-Houari M, Ruiz-Pallares N, Andrau J, Lacombe D, et al. Evaluation of DNA methylation episignatures for diagnosis and phenotype correlations in 42 Mendelian neurodevelopmental disorders. Am J Hum Genet 2020; 106(3):356–70.

[4] Lister R, Mukamel EA, Nery JR, Urich M, Puddifoot CA, Johnson ND, et al. Global epigenomic reconfiguration during mammalian brain development. Science 2013;341(6146):1237905.

[5] Kishi Y, Fujii Y, Hirabayashi Y, Gotoh Y. HMGA regulates the global chromatin state and neurogenic potential in neocortical precursor cells. Nat Neurosci 2012;15(8):1127–33.

[6] Meshorer E, Yellajoshula D, George E, Scambler P, Brown D, Misteli T. Hyperdynamic plasticity of chromatin proteins in pluripotent embryonic stem cells. Dev Cell 2012;22(1):233–4.

[7] Kim YJ, Khoshkhoo S, Frankowski JC, Zhu B, Abbasi S, Lee S, et al. Chd2 is necessary for neural circuit development and long-term memory. Neuron 2018;100(5):1180–93.

[8] Watson LA, Wang X, Elbert A, Kernohan KD, Galjart N, Berube NG. Dual effect of CTCF loss on neuroprogenitor differentiation and survival. J Neurosci 2014;34(8):2860–70.

[9] Shah K, King GD, Jiang H. A chromatin modulator sustains self-renewal and enables differentiation of postnatal neural stem and progenitor cells. J Mol Cell Biol 2020;12(1):4–16.

[10] Pereira JD, Sansom SN, Smith J, Dobenecker M, Tarakhovsky A, Livesey FJ, et al. Ezh2, the histone methyltransferase of PRC2, regulates the balance between self-renewal and differentiation in the cerebral cortex. Proc Natl Acad Sci 2010;107(36):15957–62.

[11] Hwang WW, Salinas RD, Siu JJ, Kelley KW, Delgado RN, Paredes MF, et al. Distinct and separable roles for EZH2 in neurogenic astroglia. eLife 2014;3:e02439.

[12] Qiao Y, Wang R, Yang X, Tang K, Jing N. Dual roles of histone H3 lysine 9 acetylation in human embryonic stem cell pluripotency and neural differentiation. J Biol Chem 2015;290(4):2508–20.

[13] Nothof SA, Magdinier F, Van-Gils J. Chromatin structure and dynamics: focus on neuronal differentiation and pathological implication. Genes 2022;13(4):639. Available from: https://doi.org/10.3390/genes13040639.

[14] Stolzenberg DS, Champagne FA. Hormonal and non-hormonal bases of maternal behavior: the role of experience and epigenetic mechanisms. Hormones Behav 2016;77:204–10.

[15] Lutz P, Chay M, Pacis A, Chen GG, Aouabed Z, Maffioletti E, et al. Non-CG methylation and multiple histone profiles associate child abuse with immune and small GTPase dysregulation. Nat Commun 2021;12(1):1132.

[16] Levy S, Muench J. The epigenetic impact of adverse childhood experiences through the lens of personalized medicine. Epigenomics 2022;14(8):425–9.

[17] Chaste P, Leboyer M. Autism risk factors: genes, environment, and gene-environment interactions. Dialogues Clin Neurosci 2012; 14(3):281–92.

[18] McDade TW, Ryan CP, Jones MJ, Hoke MK, Borja J, Miller GE, et al. Genome-wide analysis of DNA methylation in relation to socioeconomic status during development and early adulthood. Am J Phys Anthropol 2019;169(1):3–11.

[19] Oldenburg KS, O'Shea TM, Fry RC. Genetic and epigenetic factors and early life inflammation as predictors of neurodevelopmental outcomes. Semfetal Neonatal Med 2020;25(3):101115.

[20] Tobi EW, Goeman JJ, Monajemi R, Gu H, Putter H, Zhang Y, et al. DNA methylation signatures link prenatal famine exposure to growth and metabolism. Nat Commun 2014;5(1):5592.

[21] Lumey LH, Stein AD, Susser E. Prenatal famine and adult health. Annu Rev Public Health 2011;32(1):237–62.

[22] Dagli AI, Mathews J, Williams CA. Angelman syndrome GeneReviews is a registered trademark of the University of Washington, Seattle. All rights reserved In: Adam MP, Mirzaa GM, Pagon RA, et al., editors. GeneReviews(®). Seattle (WA): University of Washington; 1993.

[23] Buiting K, Williams C, Horsthemke B. Angelman syndrome - insights into a rare neurogenetic disorder. Nat Rev Neurol 2016;12(10):584–93.

[24] Grebe SC, Limon DL, McNeel MM, Guzick A, Peters SU, Tan W, et al. Anxiety in Angelman syndrome. Am J Intellect Dev Disabil 2022;127(1):1–10.

[25] Williams CA, Beaudet AL, Clayton-Smith J, Knoll JH, Kyllerman M, Laan LA, et al. Angelman syndrome 2005: updated consensus for diagnostic criteria. Am J Med Genet A 2006;140(5):413–18.

[26] Pearson E, Wilde L, Heald M, Royston R, Oliver C. Communication in Angelman syndrome: a scoping review. Dev Med Child Neurol 2019;61(11):1266–74.

[27] Butler MG, Duis J. Chromosome 15 imprinting disorders: genetic laboratory methodology and approaches. Front Pediatr 2020;8:154.

[28] Samanta D. Epilepsy in Angelman syndrome: a scoping review. Brain Dev 2021;43(1):32–44.

[29] Driscoll DJ, Miller JL, Cassidy SB. Prader-Willi syndrome GeneReviews is a registered trademark of the University of Washington, Seattle. All rights reserved In: Adam MP, Mirzaa GM, Pagon RA, et al., editors. GeneReviews(®). Seattle (WA): University of Washington; 1993.

[30] Butler MG, Miller JL, Forster JL. Prader-Willi syndrome - clinical genetics, diagnosis and treatment approaches: an update. Curr Pediatr Rev 2019;15(4):207–44.

[31] Whittington J, Holland A. Cognition in people with Prader-Willi syndrome: insights into genetic influences on cognitive and social development. Neurosci Biobehav Rev 2017;72:153–67.

[32] Shelkowitz E, Gantz MG, Ridenour TA, Scheimann AO, Strong T, Bohonowych J, et al. Neuropsychiatric features of Prader-Willi syndrome. Am J Med Genet A 2022;188(5):1457–63.

[33] Deal CL, Tony M, Höybye C, Allen DB, Tauber M, Christiansen JS. Growth hormone in Prader-Willi syndrome clinical care guidelines workshop participants: Growth Hormone Research Society workshop summary: consensus guidelines for recombinant human growth hormone therapy in Prader-Willi syndrome. J Clin Endocrinol Metab 2011;98(6):1072.

[34] Sakaria RP, Mostafavi R, Miller S, Ward JC, Pivnick EK, Talati AJ. Kagami-Ogata syndrome: case series and review of literature. AJP Rep 2021;11(2):e65–75.

[35] Kagami M, Kurosawa K, Miyazaki O, Ishino F, Matsuoka K, Ogata T. Comprehensive clinical studies in 34 patients with molecularly defined UPD(14)pat and related conditions (Kagami-Ogata syndrome). Eur J Hum Genet 2015;23(11):1488–98.

[36] Ogata T, Kagami M. Kagami-Ogata syndrome: a clinically recognizable upd(14)pat and related disorder affecting the chromosome 14q32.2 imprinted region. J Hum Genet 2016;61(2):87–94.

[37] Ioannides Y, Lokulo-Sodipe K, Mackay DJG, Davies JH, Temple IK. Temple syndrome: improving the recognition of an underdiagnosed chromosome 14 imprinting disorder: an analysis of 51 published cases. J Med Genet 2014;51(8):495–501.

[38] Kagami M, Nagasaki K, Kosaki R, Horikawa R, Naiki Y, Saitoh S, et al. Temple syndrome: comprehensive molecular and clinical findings in 32 Japanese patients. Genet Med 2017;19(12):1356–66.

[39] Fahrner JA, Bjornsson HT. Mendelian disorders of the epigenetic machinery: postnatal malleability and therapeutic prospects. Hum Mol Genet 2019;28(R2):R254–64.

[40] Larizza L, Finelli P. Developmental disorders with intellectual disability driven by chromatin dysregulation: clinical overlaps and molecular mechanisms. Clin Genet 2019;95(2):231–40.

[41] Ng R, Kalinousky A, Harris J. Epigenetics of cognition and behavior: insights from Mendelian disorders of epigenetic machinery. J Neurodev Disord 2023;15(1):16.

[42] Boukas L, Havrilla JM, Hickey PF, Quinlan AR, Bjornsson HT, Hansen KD. Coexpression patterns define epigenetic regulators associated with neurological dysfunction. Genome Res 2019;29(4):532–42.

[43] Biswas S, Rao CM. Epigenetic tools (the writers, the readers and the erasers) and their implications in cancer therapy. Eur J Pharmacol 2018;837:8–24.

[44] Yang AY, Kim H, Li W, Kong AT. Natural compound-derived epigenetic regulators targeting epigenetic readers, writers and erasers. Curr Top Med Chem 2016;16(7):697–713.

[45] Ostrowski PJ, Tatton-Brown K. Tatton-Brown-Rahman syndrome GeneReviews is a registered trademark of the University of Washington, Seattle. All rights reserved In: Adam MP, Mirzaa GM, Pagon RA, et al., editors. GeneReviews(®). Seattle (WA): University of Washington; 1993.

[46] Lane C, Tatton-Brown K, Freeth M. Tatton-Brown-Rahman syndrome: cognitive and behavioural phenotypes. Dev Med child Neurol 2020; 62(8):993–8.

[47] Tatton-Brown K, Zachariou A, Loveday C, Renwick A, Mahamdallie S, Aksglaede L, et al. The Tatton-Brown-Rahman syndrome: a clinical study of 55 individuals with de novo constitutive DNMT3A variants. Wellcome Open Res 2018;3:46.

[48] Balci TB, Strong A, Kalish JM, Zackai E, Maris JM, Reilly A, et al. Tatton-Brown-Rahman syndrome: six individuals with novel features. Am J Med Genet Part A 2020;182(4):673–80.

[49] Smith H, Lane C, Al-Jawahiri R, Freeth M. Sensory processing in Sotos syndrome and Tatton-Brown-Rahman syndrome. J Abnorm Psychol 2023;132(6):768–78.

[50] Tenorio J, Alarcón P, Arias P, Dapía I, García-Miñaur S, Palomares Bralo M, et al. Further delineation of neuropsychiatric findings in Tatton-Brown-Rahman syndrome due to disease-causing variants in DNMT3A: seven new patients. Eur J Hum Genet 2020;28(4):469–79.
[51] Tatton-Brown K, Seal S, Ruark E, Harmer J, Ramsay E, Del Vecchio Duarte S, et al. Mutations in the DNA methyltransferase gene DNMT3A cause an overgrowth syndrome with intellectual disability. Nat Genet 2014;46(4):385–8.
[52] Janssen N, Bergman JEH, Swertz MA, Tranebjaerg L, Lodahl M, Schoots J, et al. Mutation update on the CHD7 gene involved in CHARGE syndrome. Hum Mutat 2012;33(8):1149–60.
[53] Vissers LELM, Ravenswaaij-Arts CMAV, Admiraal RJC, Hurst JA, Vries LBAD, Janssen IM, et al. Mutations in a new member of the chromodomain gene family cause CHARGE syndrome. Nat Genet 2004;36(9):955–7.
[54] Pagon RA, Graham JM, Zonana J, Yong S. Coloboma, congenital heart disease, and choanal atresia with multiple anomalies: CHARGE association. J Pediatr 1981;99(2):223–7.
[55] Lasserre E, Vaivre-Douret L, Abadie V. Psychomotor and cognitive impairments of children with CHARGE syndrome: common and variable features. Child Neuropsychol 2013;19(5):449–65.
[56] Vesseur A, Langereis M, Free R, Snik A, van Ravenswaaij-Arts C, Mylanus E. Influence of hearing loss and cognitive abilities on language development in CHARGE Syndrome. Am J Med Genet Part A 2016;170(8):2022–30.
[57] Raqbi F, Le Bihan C, Morisseau-Durand M, Dureau P, Lyonnet S, Abadie V. Early prognostic factors for intellectual outcome in CHARGE syndrome. Dev Med Child Neurol 2003;45(7):483–8.
[58] Wachtel LE, Hartshorne TS, Dailor AN. Psychiatric diagnoses and psychotropic medications in CHARGE syndrome: a pediatric survey. J Dev Phys Disabil 2007;19(5):471–83.
[59] Blake KD, Salem-Hartshorne N, Daoud MA, Gradstein J. Adolescent and adult issues in charge syndrome. Clin Pediatr 2005;44(2):151–9.
[60] Hartshorne N, Hudson A, MacCuspie J, Kennert B, Nacarato T, Hartshorne T, et al. Quality of life in adolescents and adults with CHARGE syndrome. Am J Med Genet. Part A 2016;170A(8):2012–21.
[61] Hartshorne TS, Grialou TL, Parker KR. Autistic-like behavior in CHARGE syndrome. Am J Med Genet Part A 2005;133A(3):257–61.
[62] Souriau J, Gimenes M, Blouin C, Benbrik I, Benbrik E, Churakowskyi A, et al. CHARGE syndrome: developmental and behavioral data. Am J Med Genet Part A 2005;133A(3):278–81.
[63] van Dijk JPM, de Kort A. Reducing challenging behaviors and fostering efficient learning of children with CHARGE syndrome. Am J Med Genet Part A 2005;133A(3):273–7.
[64] Thelin JW, Fussner JC. Factors related to the development of communication in CHARGE syndrome. Am J Med Genet Part A 2005; 133A(3):282–90.
[65] Graham Jr JM, Rosner B, Dykens E, Visootsak J. Behavioral features of CHARGE syndrome (Hall-Hittner syndrome) comparison with Down syndrome, Prader-Willi syndrome, and Williams syndrome. Am J Med Genet Part A 2005;133A(3):240–7.
[66] Trider C, Arra-Robar A, van Ravenswaaij-Arts C, Blake K. Developing a CHARGE syndrome checklist: health supervision across the lifespan (from head to toe). Am J Med Genet A 2017;173(3):684–91.
[67] Slavin LJ, Hartshorne TS. The development of an educational checklist for individuals with CHARGE syndrome. Int J Dev Disabil 2019;67(4):256–62.
[68] Stevens CA. Rubinstein-Taybi syndrome GeneReviews is a registered trademark of the University of Washington, Seattle. All rights reserved In: Adam MP, Mirzaa GM, Pagon RA, et al., editors. GeneReviews(®). Seattle (WA): University of Washington; 1993.
[69] Adrien J, Taupiac E, Thiébaut E, Paulais M, Van-Gils J, Kaye K, et al. A comparative study of cognitive and socio-emotional development in children with Rubinstein-Taybi syndrome and children with autism spectrum disorder associated with a severe intellectual disability, and in young typically developing children with matched developmental ages. Res Dev Disabil 2021;116:104029.
[70] Waite J, Beck SR, Heald M, Powis L, Oliver C. Dissociation of cross-sectional trajectories for verbal and visuo-spatial working memory development in Rubinstein-Taybi syndrome. J Autism Dev Disord 2016;46(6):2064–71.
[71] Galéra C, Taupiac E, Fraisse S, Naudion S, Toussaint E, Rooryck-Thambo C, et al. Socio-behavioral characteristics of children with Rubinstein-Taybi syndrome. J Autism Dev Disord 2009;39(9):1252–60.
[72] Taupiac E, Lacombe D, Thiébaut E, Van-Gils J, Michel G, Fergelot P, et al. Psychomotor, cognitive, and socio-emotional developmental profiles of children with Rubinstein-Taybi syndrome and a severe intellectual disability. J Intellect Dev Disabil 2021;46(1):80–9.
[73] Awan N, Pearson E, Shelley L, Greenhill C, Tarver J, Waite J. The behavioral phenotype of Rubinstein–Taybi syndrome: a scoping review of the literature. Am J Med Genet Part A 2022;188(9):2536–54.
[74] Qu'd D, Schmitt LM, Leston A, Harris JR, Slavotinek A, Riddle I, et al. Behavioral and neuropsychiatric challenges across the lifespan in individuals with Rubinstein-Taybi syndrome. Front Genet 2023;14:1116919.
[75] López M, García-Oguiza A, Armstrong J, García-Cobaleda I, García-Miñaur S, Santos-Simarro F, et al. Rubinstein-Taybi 2 associated to novel EP300 mutations: deepening the clinical and genetic spectrum. BMC Med Genet 2018;19(1):36.
[76] Adam MP, Hudgins L. Kabuki syndrome: a review. Clin Genet 2005;67(3):209–19.
[77] Adam MP, Banka S, Bjornsson HT, Bodamer O, Chudley AE, Harris J, et al. Kabuki syndrome: international consensus diagnostic criteria. J Med Genet 2019;56(2):89–95.
[78] Harris J, Mahone EM, Bjornsson HT. Molecularly confirmed Kabuki (Niikawa-Kuroki) syndrome patients demonstrate a specific cognitive profile with extensive visuospatial abnormalities. J Intellect Disabil Res 2019;63(6):489–97.
[79] Morgan AT, Mei C, Da Costa A, Fifer J, Lederer D, Benoit V, et al. Speech and language in a genotyped cohort of individuals with Kabuki syndrome. Am J Med Genet A 2015;167(7):1483–92.
[80] Dongen LCMV, Wingbermühle PAM, Veld WMVD, Stumpel CTRM, Kleefstra T, Egger JIM. Exploring the cognitive phenotype of Kabuki (Niikawa-Kuroki) syndrome. J Intellect Disabil Res 2019;63(6):498–506.

[81] Mervis CB, Becerra AM, Rowe ML, Hersh JH, Morris CA. Intellectual abilities and adaptive behavior of children and adolescents with Kabuki syndrome: a preliminary study. Am J Med Genet A 2005;132A(3):248–55.

[82] White SM, Thompson EM, Kidd A, Savarirayan R, Turner A, Amor D, et al. Growth, behavior, and clinical findings in 27 patients with Kabuki (Niikawa-Kuroki) syndrome. Am J Med Genet 2004;127A(2):118–27.

[83] Boniel S, Szymańska K, Śmigiel R, Szczałuba K. Kabuki syndrome—clinical review with molecular aspects. Genes 2021;12(4):468.

[84] Sobreira N, Brucato M, Zhang L, Ladd-Acosta C, Ongaco C, Romm J, et al. Patients with a Kabuki syndrome phenotype demonstrate DNA methylation abnormalities. Eur J Hum Genet 2017;25(12):1335–44.

[85] Barry KK, Tsaparlis M, Hoffman D, Hartman D, Adam MP, Hung C, et al. From genotype to phenotype—a review of Kabuki syndrome. Genes 2022;13(10):1761.

[86] Milazzo C, Mientjes EJ, Wallaard I, Rasmussen SV, Erichsen KD, Kakunuri T, et al. Antisense oligonucleotide treatment rescues UBE3A expression and multiple phenotypes of an Angelman syndrome mouse model. JCI Insight 2021;6(15):e145991. Available from: https://doi.org/10.1172/jci.insight.145991.

[87] Meng L, Ward AJ, Chun S, Bennett CF, Beaudet AL, Rigo F. Towards a therapy for Angelman syndrome by targeting a long non-coding RNA. Nature 2015;518(7539):409–12.

[88] Amanat M, Nemeth CL, Fine AS, Leung DG, Fatemi A. Antisense oligonucleotide therapy for the nervous system: from bench to bedside with emphasis on pediatric neurology. Pharmaceutics 2022;14(11):2389. Available from: https://doi.org/10.3390/pharmaceutics14112389.

[89] Dindot SV, Christian S, Murphy WJ, Berent A, Panagoulias J, Schlafer A, et al. FIRE consortium: an ASO therapy for Angelman syndrome that targets an evolutionarily conserved region at the start of the UBE3A-AS transcript. Sci Transl Med 2023;15(688):eabf4077.

[90] Bourtchouladze R, Lidge R, Catapano R, Stanley J, Gossweiler S, Romashko D, et al. Model of Rubinstein-Taybi syndrome: defective long-term memory is ameliorated by inhibitors of phosphodiesterase 4. Proc Natl Acad Sci 2003;100(18):10518–22.

[91] Alarcón JM, Malleret G, Touzani K, Vronskaya S, Ishii S, Kandel ER, et al. Chromatin acetylation, memory, and LTP are impaired in CBP +/− mice: a model for the cognitive deficit in Rubinstein-Taybi syndrome and its amelioration. Neuron 2004;42(6):947–59.

[92] Benjamin JS, Pilarowski GO, Carosso GA, Zhang L, Huso DL, Goff LA, et al. A ketogenic diet rescues hippocampal memory defects in a mouse model of Kabuki syndrome. Proc Natl Acad Sci 2017;114(1):125–30.

[93] Bjornsson HT, Benjamin JS, Zhang L, Weissman J, Gerber EE, Chen YC, et al. Histone deacetylase inhibition rescues structural and functional brain deficits in a mouse model of Kabuki syndrome. Sci Transl Med 2014;6(256):256ra135.

[94] Zhang L, Pilarowski G, Pich EM, Nakatani A, Dunlop J, Baba R, et al. Inhibition of KDM1A activity restores adult neurogenesis and improves hippocampal memory in a mouse model of Kabuki syndrome. Mol Ther Methods Clin Dev 2021;20:779–91.

Chapter 5

Neuroplasticity in neurodevelopmental disorders

Fatima Y. Ismail[1,2] **and Milos R. Ljubisavljevic**[3]
[1]*Pediatric Neurology and Developmental Medicine, Department of Pediatrics, College of Medicine and Health Sciences, United Arab Emirates University, United Arab Emirates,* [2]*Department of Neurology, Johns Hopkins University School of Medicine, Baltimore, MD, United States,* [3]*Department of Physiology, College of Medicine and Health Sciences, United Arab Emirates University, United Arab Emirates*

Introduction

Neuroplasticity refers to the central nervous system's inherent biological capacity to change, both structurally and functionally, in a preprogrammed manner, in response to experiences and perturbations. This malleability is primed, mediated, and regulated by a genetic blueprint and epigenetic influences through molecular and cellular processes ensuring the appropriate degree of change occurs at the adequate time, in response to specific inputs and for the correct duration. The primary purpose of neuroplasticity in the developing brain is to support structural formation of the nervous system, and functional maturation and adaptation—whether by losing previous functions (e.g., suppression of primitive reflexes), or by acquiring new skills (e.g., language and higher cognitive abilities).

The term "plasticity" is often used interchangeably with brain development, maturation, and learning. However, it is essential to distinguish between these terms. In neuroscience, plasticity refers to the mechanistic framework through which brain development and maturation are achieved. While development and maturation are interlinked, they are distinct processes. Development involves forming the foundational structure and organization of the nervous system, beginning in utero with neural tube formation and continuing through childhood to achieve complete structural construction of the brain. Maturation refers to the growth, fine tuning, and functional specialization of the nervous system, ultimately leading to proper adult-like neurological function. Maturation extends into early adulthood, particularly for higher order cognitive skills. Learning, on the other hand, is an experience-dependent process of acquiring skills or behaviors, occurring at any age.

Neuroplasticity can be conceptualized through various frameworks that are interconnected, although not mutually exclusive. Table 5.1 summarizes these different frameworks, which elucidate the functions of neuroplasticity, the biological processes underlying its mechanisms, the hierarchical organizational levels at which it operates to achieve specific functions, and the intrinsic and extrinsic factors that modulate and regulate neuroplasticity and, consequently, an organism's behavior. The choice of framework depends on the scientific question, the observed biological or behavioral phenomenon, and the required level of analysis.

In this chapter, we focus on the functional framework and the clinical contexts of neuroplasticity, in both typical and atypical brain development, with a specific focus on neurodevelopmental disorders and early brain injuries.

Mechanisms of plasticity in the developing brain

Brain development is a dynamic and iterative process facilitated by various plasticity mechanisms that operate across different temporal and spatial scales. Table 5.1 ("structural frameworks" and "neurophysiological frameworks") and Table 5.2 list examples of mechanisms underlying brain plasticity at different levels of CNS organization. Discussing the expansive list of mechanisms of neuroplasticity is beyond the scope of this chapter. Here, we focus on the core mechanisms underpinning structural and functional plasticity in normal brain development.

Capute and Accardo's Neurodevelopmental Disabilities in Infancy and Childhood. DOI: https://doi.org/10.1016/B978-0-12-824060-1.00010-9

TABLE 5.1 Conceptual frameworks of neuroplasticity.

Category	Description	
Structural framework	• Examines changes in brain structure ◦ Neurogenesis and proliferation, differentiation and migration ◦ Dendritic spine remodeling ◦ Axonal sprouting and regeneration ◦ Synaptic pruning ◦ Corticospinal tract lateralization	
Temporal framework	Represents periods during development when the brain is particularly responsive to inputs, including age-related plasticity, critical and sensitive periods.	
Neurophysiological framework	Explores the underlying biological and molecular processes and mechanisms driving neuroplasticity at the level of the synapse.	
	Hebbian mechanisms	• Forms of synaptic plasticity that depend on the timing of presynaptic and postsynaptic activities: ◦ Spike-timing-dependent plasticity: The relative timing of spikes in the presynaptic and postsynaptic neurons determines the direction of plasticity. Specific mechanisms include long term potentiation (LTP) and long term depression (LTD). ◦ Cell assemblies and neural ensembles: Groups of neurons that become activated together in response to specific stimuli.
	Non-Hebbian mechanisms	• Forms of synaptic plasticity that do not strictly depend on the timing of presynaptic and postsynaptic activities: ◦ Spike-timing independent plasticity: Repetitive activity may strengthen or weaken the synaptic strength. ◦ Homeostatic plasticity: Explores plasticity's role in maintaining neural circuit balance and stability. ◦ Synaptic scaling: Global adjustment of synaptic strengths across a network to maintain a stable level of activity. ◦ Heterosynaptic plasticity: Involves changes at synapses that were not directly involved in a specific pattern of activity. ◦ Metaplasticity (plasticity of plasticity): Examines how the history of synaptic activity can influence a synapse's response to subsequent activity.
	Neuromodulation mechanisms	Effect of neuromodulatory systems (neurotransmitters, hormones) on synapsis and networks
Cognitive and behavioral frameworks	Explores the interactions between behavior and neuroplasticity through various overlapping mechanisms	
	Experience-dependent framework	Focuses on how sensory and environmental experiences drive plasticity
	Skill acquisition framework	Focuses on plastic changes induced by skill acquisition and practice
	Educational neuroscience framework	Integrates neuroscience findings with education to optimize learning experiences
	Emotion–cognition interaction framework	Explores the interplay between emotional experiences, cognition, and neuroplasticity
	Social interaction framework	Investigates social interactions influence on brain plasticity and socio-emotional development
	Neurofeedback framework	Uses real-time neuroimaging to data enhance self-regulation and inducing plastic changes
Functional framework	• Focuses on the specific functions that neuroplasticity serves in the brain ◦ Developmental maturational plasticity ◦ Reactive plasticity to injury ◦ Excessive/destabilizing plasticity ◦ Plasticity that increases susceptibility to further injuries [2] ◦ Adaptive and maladaptive plasticity [2] ◦ Cross-modal plasticity	
Intervention framework	• Focuses on plasticity induced by interventions ◦ CNS active medications ◦ Training programs (physical therapy, intense exercises) ◦ Brain stimulation techniques (invasive and noninvasive)	

TABLE 5.2 Examples of mechanisms underlying neuroplasticity.

CNS organization level	Mechanisms
Intracellular	Synaptic protein expression Molecular signaling pathways Protein synthesis and degradation
Synaptic	Dendritic spine morphology and synaptic function Synapse formation and elimination Synaptic scaling Receptor dynamics (e.g., NMDA and AMPA receptors)
Cellular	Neuronal excitability Synaptic facilitation Synaptic pruning Synaptic potentiation Synaptic depression Astrocytes and gliotransmitters
Circuit	Neural circuit formation Functional connectivity within circuits Synaptic transmission within circuits
Network	Large-scale network dynamics Cortical maps and representations Information processing across brain regions

Structural mechanisms of plasticity, such as neurogenesis, neuronal differentiation, and migration are the earliest and the most predominant mechanisms of brain plasticity during early fetal development. Around 27 weeks of gestation, synaptogenesis progressively increases in a heterochronous cortex-specific maturational pattern. In this process, the primary sensorimotor cortices, along with the frontal and occipital poles, undergo maturation earlier than parietal and frontal regions. The visual and auditory areas reach their synaptogenesis peaks earlier (around 3 months of age) than the prefrontal areas (around 18 months of age). Activity-dependent synaptic pruning in the auditory cortex precedes that observed in the prefrontal cortex [1].

Activity-dependent mechanisms of plasticity are primarily observed at the synaptic level. Synaptic plasticity is induced by the repetitive firing of neighboring neurons. The nature of the presynaptic neuron dictates the direction of plasticity: when adjacent neurons fire simultaneously, they strengthen or reinforce the synaptic connection in the case of an excitatory presynaptic neuron (Hebbian rule). Conversely, if the presynaptic neuron is inhibitory, the connection weakens (anti-Hebbian rule). These fundamental principles underpins models of synaptic plasticity, such as long term potentiation (LTP) and long-term depression (LTD), which are critical to associative learning and memory. The strength of synaptic plasticity is influenced by the frequency and order of presynaptic and postsynaptic firing. This spike-timing-dependent plasticity (STDP) supports the refinement of sensory topographical maps, mediates associative learning, and facilitates network remodeling [3].

Activity-dependent synaptic plasticity can modify neuronal structure and function by changing the number and morphology of dendritic spines, altering neurotransmitter release, and modifying postsynaptic receptor trafficking, among other mechanisms. At the network level, synaptic plasticity influences cortical map organization [4], region-specific progressive myelination [5], and functional coupling between cortical network hubs [6].

It is important to recognize that the mechanisms of neuroplasticity operate in synchrony. A change in one mechanism (e.g., synaptic activity) may have upstream and downstream effects on other levels of CNS organization, with time scales ranging from seconds to minutes (e.g., induction and consolidation of synaptic plasticity) to days (e.g., structural changes in synapses and circuits) (Fig. 5.1).

These structural and functional hetero-modal changes at various levels of CNS organization result in a predetermined and sequential pattern of maturation of brain functions. Primary sensory areas (e.g., vision and hearing) develop adult-like patterns well before secondary association areas (e.g., language and cognition). Clinically, this progression in reflected in the emergence and consolidation of developmental skills, as captured by the assessment of developmental milestones (discussed in chapter: Typical and Atypical Brain Development).

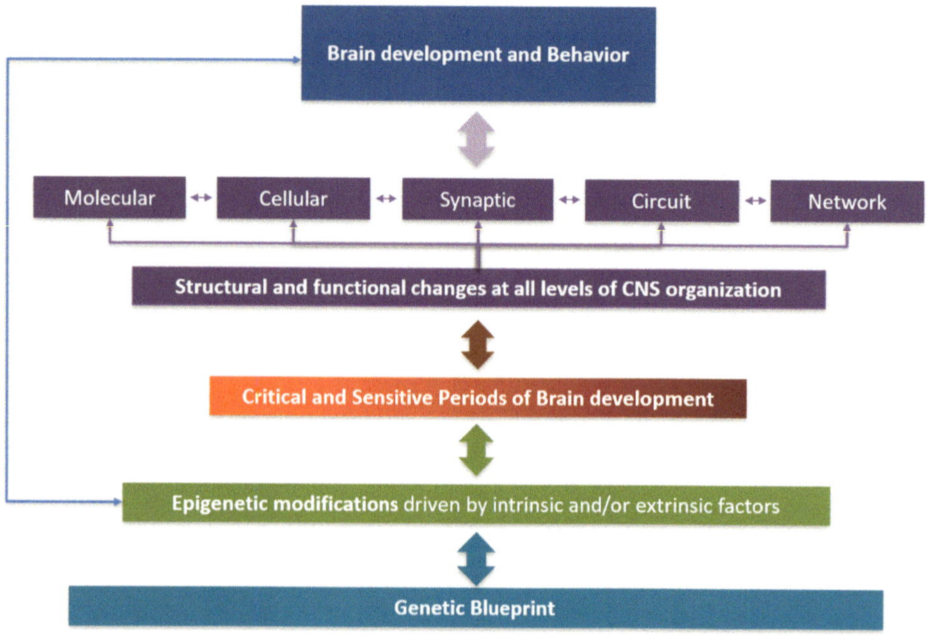

FIGURE 5.1 Plasticity mechanisms at different levels of central nervous system (CNS) organization. Abnormal plasticity mechanisms at one level may trigger changes in other levels of CNS organization, ultimately affecting brain maturation, function, and behavior.

Plasticity mechanisms are subject to alterations—both enhancement and suppression—by a variety of intrinsic factors, including neurotransmitters and hormones, as well as extrinsic factors, such as personal experiences or injuries. These modifying agents can significantly dictate the magnitude and trajectory of the plasticity responses, thereby influencing the eventual outcomes. A more detailed exploration of these mechanisms will be undertaken later, focusing on their relevance in specific clinical conditions.

The concepts of critical and sensitive periods

Plasticity in early brain development is heightened during genetically programmed, time-limited periods, known as *critical and sensitive periods*, during which the brain is most amenable to change. Critical and sensitive periods of plasticity differ in their spatial and temporal properties and can be circuit- or modality-specific.

Critical periods facilitate the construction and consolidation of experience-dependent structural and functional brain connections. In certain brain networks, only a few genetically predetermined patterns of connectivity exist, awaiting consolidation by specific stimuli. Consequently, the presence or absence of experience or input results in permanent changes to the network (critical period for network development). If the stimulus is restored or reintroduced later, likelihood of altering the network remains low. Molecular mechanisms regulate the onset and offset of the critical period, acting as "initiators" or "brakes" to stabilize the connectivity pattern and minimize further drastic changes to the network.

Through experience-dependent plasticity, synaptic connections are fine-tuned and refined based on their activity patterns (Hebbian plasticity). To maintain the balance and stability of synapses and neuronal networks, homeostatic plasticity mechanisms regulate and counterbalance synaptic strength, ensuring normal activity patterns. This balance is crucial during development for the proper specialization of network functions. For example, disruption to visual input early in life may lead to permanent changes in the visual cortex that hinder normal binocular vision [7,8]. Therefore the presence of visual input is critical for the normal development of the visual cortex. In children with amblyopia, if correction is not performed before the age of 7–8 years, the likelihood of establishing normal vision decreases [9].

Other brain networks may experience broader and more "flexible" periods of heightened plasticity (*sensitive periods*), during which stimuli or experiences strongly influence network maturation and function. However, the absence of these stimuli or experiences may not be detrimental to the network. Such networks can develop multiple connectivity patterns depending on the stimulus [10]. There age-related difference in plasticity potential during sensitive periods likely explains the young age advantage in acquiring a second language [11] or mastering a musical instrument [12]. Table 5.3 summarizes the conceptual differences between critical and sensitive periods of plasticity.

TABLE 5.3 Differences between critical and sensitive periods.

	Critical period	Sensitive period
Definition	A *specific and restricted time window* during development when the brain is highly sensitive to environmental input, allowing for optimal learning and development occur.	A *broader and longer developmental period* during which the brain is more receptive to certain types of experiences or stimuli, though the effects may be subtler than during critical periods.
Timing	Often coincide with periods of rapid synaptic growth and pruning, particularly during early childhood.	Can extend throughout development, with some sensitive periods occurring during childhood, adolescence, and even adulthood depending on the specific network.
Outcomes	Perturbation during critical periods might result in permanent deficits.	Perturbation during a sensitive period may not result in permanent deficits.
Mechanisms	Often involve strong experience-dependent plasticity and synaptic remodeling.	Mechanisms include a balance between experience-dependent and experience-expectant plasticity. The brain is still receptive to input but may require more intensive or specific experiences to induce plastic changes.
Examples	Visual system development, where lack of visual input during the critical period can result in permanent vision deficits.	Language acquisition, where exposure to language during sensitive periods enhances fluency but can still occur later in life with varying degrees of effort.
Clinical implications	Mitigate aberrant plasticity by interventions to (re)-open, extend, or close a critical/sensitive period of brain development.	Early (re)habilitation interventions.
References	[13–15]	[16–18]

The concepts of sensitive periods is gaining attention as these may represent windows of opportunity in the developing brain for therapeutic interventions. Tracking epochs of enhanced plasticity (sensitive periods) and identifying the molecular mechanisms controlling their onset and offset may have significant clinical implications. These implications include the potential to modulate these periods, by reopening them, extending their duration, or precociously closing them. Such modulation could serve as a tool to limit aberrant plasticity and potentially restore normal network connectivity and function.

Plasticity in the developing brain: clinical implications

The developing brain exhibits a complex array of plasticity patterns, each contributing to distinct developmental functions. Fig. 5.2 delineates a functional schema that organizes the spectrum of plasticity forms evident in brain maturation. It is crucial to recognize that these patterns and their underlying mechanisms do not operate in isolation but often they converge and interact. Moreover, while the developing brain is thought to have a remarkable propensity or capacity for neuroplasticity, it is imperative to recognize that not all plasticity outcomes are beneficial. Gaining insights into the mechanisms that govern neuroplasticity can enhance our understanding of atypical brain development, as seen in neurodevelopmental disorders and the aftermath of early brain injuries.

Impaired plasticity machinery at the core of neurogenetic disorders

Many neurological and neurodevelopmental diseases can be considered as disorders of plasticity, supported by evidence of aberrant plasticity machinery at various levels of CNS organization. Impairment or disruption of neuroplasticity mechanisms can alter the developmental trajectories, resulting in delays, plateaus, or regression in brain functions (refer to chapter: Typical and Atypical Patterns of Development).

A single disease may cause or be associated with multiple impaired plasticity mechanisms. Similarly, specific impaired mechanisms of plasticity can be observed in various diseases. Table 5.4 lists examples of impaired plasticity mechanisms in selected genetic disorders associated with neurodevelopmental disabilities.

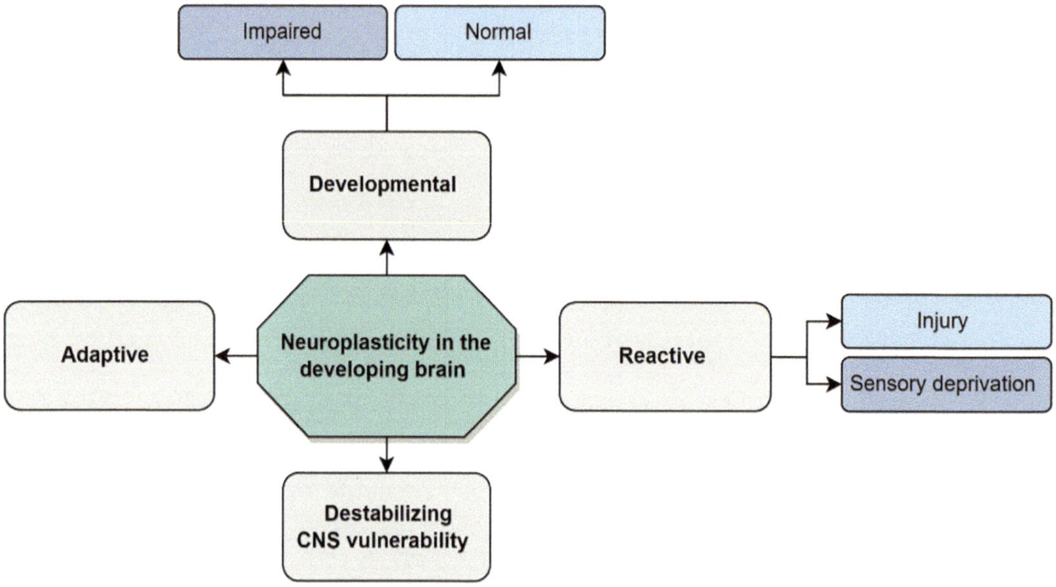

FIGURE 5.2 Clinical and functional patterns of plasticity in the developing brain. This framework conceptualizes plasticity based on its various functions and clinical manifestations in the developing brain, both in health and disease. These functions and their driving mechanisms are not mutually exclusive. For example, the young brain matures and adapts to the environment, but if the developing brain is injured, reactive plasticity occurs in conjunction with baseline developmental plasticity. The developmental trajectory may be altered based on the net effect of developmental plasticity, reactive plasticity, and intervention-related plasticity.

TABLE 5.4 Selected examples of distorted plasticity mechanisms in genetic disorders associated with neurodevelopmental disabilities.

Mechanism(s) of plasticity impairment	Clinical disorder	Genetic defect	Upstream/downstream effects on plasticity	Treatments based on targeting impaired plasticity mechanisms	References
Molecular level: Defect in signaling and transcription pathways					
Impaired synaptic proteins expression and activity-dependent protein translation	Fragile X syndrome (FXS)	CGG trinucleotide expansion in FMR1 gene	Alters dendritic spine morphology and synaptic function Dysregulation of mGluR signaling Prolonged LTD Excitatory ("mGluR" excess)/inhibitory ("GABAergic" deficit) Low cAMP levels	GABAa agonist mGluR5 antagonist Phosphodiesterase-4D (PDE4D) inhibitor	[19,20]
Dysregulated mammalian target of rapamycin (mTOR) signaling	Tuberous sclerosis complex (TSC) (increased mTOR signaling)	Mutations in TSC1 or TSC2 genes	Abnormal cellular growth, dysplasia, epileptogenesis	mTOR inhibitors (everolimus rapamycin)	[21,22]
Neuronal level: Defects in neurogenesis and dendritogenesis					
Abnormal neural migration	Disorders of cortical development	Mutations in DCX, LIS1 genes (e.g., lissencephaly)	LIS1 protein degradation Impairment in microtubule-cytoskeleton Abnormal neural migration	Calpain inhibitors to prevent LIS1 protein degradation	[23,24]

(Continued)

TABLE 5.4 (Continued)

Mechanism(s) of plasticity impairment	Clinical disorder	Genetic defect	Upstream/downstream effects on plasticity	Treatments based on targeting impaired plasticity mechanisms	References
Impaired dendritic spine morphology and density	Rett syndrome	Mutations in MECP2 gene	Reduced spine density, reduced neuronal cell packing density, immature dendritic spines	Strategies to target MECP2 mutation or pathways downstream of MECP2	[25–27]
Synaptic level					
Impaired homeostatic synaptic scaling	Angelman syndrome	Deficiency in ubiquitin protein ligase E3 (UBE3A) gene (maternally inherited)	Abnormal dendritic spine growth and maturation Abnormal internalization of glutamate-mediated AMPA receptors in response to experience-dependent stimulation Impaired LTP and LTD and abnormal GABAergic function	Strategies enhancing GABAergic transmission via small molecules Targeted gene therapy to reinstate UBE3A Antisense oligonucleotide (ASO) (parental unsilencing)	[28,29]
Circuit and network level					
Abnormal functional connectivity	Nonsyndromic autism spectrum disorder	Various genetic and environmental factors	Multiple mechanisms at different levels of CNS organization leading to altered information processing and integration on network levels	Targeting excitatory/inhibitory balance, local and long-range network connectivity and synchronization (medications, brain stimulation)	[30,31]

These mechanisms are not mutually exclusive; a single mechanism can be impaired in many diseases (e.g., mTOR signaling dysfunction in FXS, TSC, and NF), while a single disease may involve multiple impaired mechanisms (e.g., Rett syndrome).

For example, in Rett syndrome, mechanisms of homeostatic synaptic scaling and synaptic overshoot are impaired. Synaptic scaling is a form of homeostatic plasticity that adjusts the strength of all of a neuron's synapses up or down to stabilize neuronal output. Synaptic overshoot refers to a temporary increase in synaptic strength that exceeds the level needed to achieve a desired activity level. In Rett syndrome, there is an increase in glutamate signaling accompanied by an inappropriate increase, rather than a decrease, in glutamate receptors. This leads to an excitatory/inhibitory imbalance, exposing the developing brain to cytotoxic levels of glutamate, which interfere with normal neuronal transmission and the activity-dependent neuronal plasticity required for learning and memory [32–34].

In Fragile X syndrome (FXS), epigenetic silencing of the fragile X mental retardation 1 (FMR1) gene through DNA methylation of the CGG repeats is associated with abnormal differentiation of neural progenitors cells, impaired neuronal migration, and defective maturation and stabilization of activity-dependent dendritic spine plasticity. These changes also defects in spine density, shape, and turnover [35,36]. Impaired homeostatic plasticity, due to blocking retinoic acid-mediated regulation of synaptic strength, has been identified in neurons derived from induced pluripotent stem cells from FXS patients [37]. Altered neuronal excitability due to a defect in presynaptic control of calcium influx, and abnormal expression of GABAergic receptor subunits have been described in FXS [38,39].

Abnormal hyperactivation of the mammalian target of rapamycin (mTOR) signaling pathway in TSC is linked to dysregulated of cell proliferation and growth, impaired neurogenesis, and abnormal synaptic excitability [40]. While FXS and TSC share some features of impaired plasticity such as defective dendritic spine pruning resulting in abnormal age-dependent increases in dendritic spine density, mTOR, unlike FMR1 protein, it downregulates activity-dependent protein synthesis, leading to increased LTP and impaired mGluR-mediated LTD [41], and excitatory/inhibitory synaptic imbalance which cause hyperexcitability, epileptogenesis, and defective activity-dependent modulation of cortical circuits and cortico-cortical connectivity [42,43].

The clinical relevance of identifying abnormal plasticity mechanisms lies in the potential to target core mechanisms for therapeutic interventions using genetic and molecular therapies. Examples of disease-specific, plasticity-informed therapeutic targets are listed in Table 5.2. The ability to restore normal plasticity in some of these disorders or to curb aberrant plasticity during early brain development represents a promising new frontier in the field of neurodevelopmental disabilities.

Excessive or destabilizing plasticity in epilepsy

When plasticity mechanisms escape homeostatic regulation, they can become destabilizing, impairing normal brain functions. Homeostatic dysregulation of synaptic plasticity can result from genetic or acquired factors, leading to destabilizing plasticity as observed in epilepsy.

Epileptogenesis refers to the cascade of molecular and cellular events that transform a normal neuronal circuit into a spontaneous generator of recurrent seizures, leading to the emergence of an epileptic condition or refractoriness of existing epilepsy [44]. Clinically, the hallmark of an epileptogenic disorder is the presence of a latent period between the insult and the emergence of the first unprovoked seizures. During this latent period, processes such as neuroinflammation, neurodegeneration, gliosis, abnormal axonal sprouting, altered gene expression, and epigenetic regulation take place [45]. These aberrant mechanisms of plasticity, whether occuring individually or collectively, lead to an excitatory/inhibitory imbalance in neuronal excitability. This imbalance promotes depolarization through the formation of new excitatory glutamatergic synapses and weakening of inhibitory tonic input due to the loss of GABAergic synapses, culminating in a decreased seizure threshold [46].

At both local and large-scale network levels, the functional reorganization of networks during seizures can lead to the consolidation of aberrant networks. This process may contribute to seizure progression and refractoriness, as well as neuropsychiatric comorbidities.

In acquired causes of epilepsy, there is interindividual variability in the latency period between CNS injury and the onset of unprovoked seizures. Factors contributing to this variability include genetic and epigenetic predisposition, the nature of the CNS insult (traumatic, vascular, metabolic), and age at which CNS insult occurs. Epileptogenesis can be thought of as a threat to *critical periods* of circuit development. If not controlled, a naive circuit may be permanently transformed into an epileptic one. Growing evidence suggests that early life seizures, during critical periods of hippocampal development, can lead to long-term structural and functional changes, potentially affecting hippocampal plasticity into adulthood [47]. This evidence underscores the importance of aggressive seizure control during early brain development to prevent long-term dysfunction. It is postulated that interventions during epileptogenesis may decrease the risk for developing epilepsy [48].

Age-dependent selective vulnerability to injury

The immature brain undergoes dynamic alterations in the expression and composition of excitatory (glutamate) and inhibitory (GABA) neurotransmitters and receptor subunits. These changes occur in distinct regions and during specific developmental windows, triggering a cascade of genetic, epigenetic, molecular, and cellular events that can render the developing brain more susceptible to perturbations and injuries.

In the embryonic brain, GluN2B glutamate receptors are predominantly expressed. However, during the first few weeks of life, a developmental switch to the GluN2A subunit occurs. This switch helps to curb premature synaptic maturation, a crucial factor for experience-dependent synaptic plasticity [49].

Similarly, GABA receptors undergo a developmental switch in function during early postnatal life. In contrast to adult neurons, where GABA typically causes hyperpolarization, immature neurons depolarize in response to GABAA activation. This activation leads to depolarization of the postsynaptic potential and initiation of action potentials. Additionally, GABAA activation coactivates glutamatergic receptors by removing the NMDA blockade, generating "primitive currents" known as giant depolarizing potentials (GDP). These GDPs play a pivotal role in neuronal assembly, synaptogenesis, and the formation of functional units. GDPs are characterized by recurrent bursts with large polysynaptic currents, laying the foundation for the hyperexcitability of the immature brain and its low thresholds for epileptogenesis and seizures. Glutamate transporters play a role in counterbalancing GABA-mediated hyperexcitability until GABAA switches to its inhibitory adult pattern [50].

The dynamic nature of CNS maturation can influence the pattern of injury. Hypoxic-ischemic injury (HII) can result in two distinct patterns of injury depending on gestational age at the time of insult [51,52]. Periventricular leukomalacia (PVL), often described in premature infants, reflects the selective vulnerability of the abundant late oligodendrocyte progenitor cells, located in the periventricular region, to HII. In contrast, term infants subjected to HII exhibit a different injury profile, primarily affecting the perirolandic area, posterior putamen, and ventrolateral thalamus. These regions are connected by glutamate-rich pathways, making them more susceptible to glutamate-mediated excitotoxicity, often observed in early phases of hypoxic-ischemic injury [53,54]. Studies have shown that cholinergic neurons are more sensitive to hypoxia-ischemia compared to dopaminergic and GABAergic-specific neurons [55]. The selective heightened vulnerability of cholinergic neurons is thought to be the result of their biological immaturity relative to dopaminergic and GABAergic neurons at the time of injury, highlighting selective heightened vulnerability as a function of maturation.

While perinatal HII is often considered a one-time event, it triggers a cascade of primary, secondary, and tertiary injuries that impact long-term brain development. Within minutes of ischemia and hypoxia, a primary energy failure and depletion of ATP occur, causing dysfunction of the Na/K ATP pump. This leads to an increase in intracellular Na concentration resulting in cytotoxic edema. Overactivation of AMPA and NMDA receptors, along with the failure to reuptake glutamate from the synapse, contributes to glutamate excitotoxicity, ultimately causing necrosis and apoptosis. A secondary energy failure occurs between 6 and 72 hours, propagated by mitochondrial dysfunction, inflammation, and glial activation. This leads to decreased protein synthesis and loss of neurotrophic factors, further worsening necrosis and apoptosis. In the days to weeks following the injury, tertiary injury in the form of anterograde degeneration, hypomyelination, and abnormal neurovascular remodeling occurs, setting the stage for long-term dysfunction. The institution of therapeutic hypothermia within 6 hours of HII is considered the therapeutic window for mitigating the injury cascade and improving long-term outcomes.

Following focal brain lesions the pattern of cortical damage and reorganization depends on several factors, including the timing of the injury in relation to postconception age, the stage of developmental maturity of brain systems, the nature, location, and severity of the injury, and the involvement of neuroplasticity mechanisms in reorganization and remodeling, referred to as *reactive plasticity*.

The pattern of upper limb weakness in patients with perinatal stroke and hemiplegic cerebral palsy is largely affected by the timing of corticospinal tract (CST) development at the time of injury. CSTs (the major motor output from the brain to the contralateral motor neurons in the spinal cord) show an early maturation profile during the third trimester of pregnancy and first 2 years of life, making it particularly vulnerable to perinatal injury. During the early stages of development, the motor cortex sends axonal projections to spinal motor neurons bilaterally. Through activity-dependent competitive elimination and axonal pruning at the spinal cord level, the ipsilateral projections from each hemisphere gradually retract, while the contralateral projections are strengthened. This process leads to the establishment of crossed CST from each motor cortex, ultimately resulting in the lateralization of dominant motor control.

When early unilateral focal brain injury occurs before the competitive elimination of ipsilateral CST fibers is complete, it interferes with the normal process of lateralization. The ipsilateral CST fibers from the unaffected hemisphere then engage in strong competition with those from the injured side. This competition can ultimately lead to the elimination of CST fibers from the damaged hemisphere. As a result, the healthy hemisphere assumes bilateral control of movements. This bilateral control can diminish the functional use of the affected hand and lead to the development of mirror movements, which are typically seen as clinically maladaptive. However, if a unilateral brain lesion occurs after the competitive elimination of ipsilateral fibers has taken place, no such reorganization occurs, and the potential for the recovery of useful hand function is limited.

Reorganization and remodeling following hemispherectomy

While rare, functional hemispherectomy is an effective procedure for intractable childhood epilepsy and serves as an interesting model to study early brain plasticity. In functional hemispherectomy, the lesioned hemisphere is completely disconnected, with varying degrees of cortical excision depending on the case. Many patients attain relatively favorable neurological recovery.

The extent of remodeling and restoration of motor and cognitive function varies. Language development following early life hemispherectomy depends on the side of surgery and the etiology of epilepsy. In cases of left hemispherectomy, all aspects of language development are impaired except for sentence comprehension. Following right hemispherectomy, language development typically falls within a borderline range. Regardless of the surgical side, children with cortical dysplasia, as an etiology for their epilepsy, tend have less favorable outcomes, whereas those with a vascular etiology in the left hemisphere tend to have the most favorable outcomes [56].

In the motor domain, interesting case reports describe remarkable motor recovery. For example, a 7-year-old patient who underwent a complete right-sided anatomical hemispherectomy showed normal intellect. A follow-up after 11 years revealed that she had independent mobility with residual spasticity in the upper and lower limbs. Intriguingly, a functional MRI study conducted a few years after the surgery revealed that movements of the left foot generated activity within the primary motor cortex of the left precentral gyrus, confirming complete relocation of left-sided motor control to the remaining (unlesioned) left hemisphere [57]. The neural mechanisms responsible for this enhanced functional recovery remain unclear, but both human studies and animal models have reported structural and functional remodeling in the lesioned and contralesional hemispheres at the cortical and subcortical levels [58].

Cochlear implants and sensitive periods of speech and language development

Injury to the brain in the form of congenital sensory deprivation (e.g., congenital deafness) challenges the developing brain to reorganize its "default" structural and functional plasticity patterns [59]. In infants born with bilateral congenital sensorineural deafness, a cross-modal plasticity occurs in which the auditory cortex get "repurposed" to receive and process other sensory inputs (due to lack of auditory input). Understanding critical and sensitive periods of auditory cortex development and their relationship to speech and language development has underscored the importance of timely restoration of auditory input to minimize impairment in speech and language development.

Cochlear implants are the most effective neural prosthesis in modulating cortical plasticity, bypassing the damaged hair cells in the cochlea and directly stimulating the neurons in the auditory nerve. The best clinical outcomes are observed when implantation occurs within the first 12–18 months of life. However, there are limitations and constraints to language development in children with cochlear implants [60]. For prelingual deaf children, early implantation supports speech and language development; whereas late implantation may improve hearing but may not help with the discrimination and recognition of complex sounds. This discrepancy in outcomes emphasizes the importance of delivering therapy within an early therapeutic window.

This is further substantiated by animal models, which have shown hyperexcitability in the auditory cortex due to a decrease in inhibitory synaptic strength. Additionally, there is a loss of tonotopy in the auditory cortex which can be reversed with auditory restoration within 6 months [61]. The significant impact of cochlear implants on speech and language further substantiates the importance of population-wide newborn hearing screening programs [62] (see Chapter Developmental surveillance and screening).

Toward a plasticity-informed approach in neurodevelopmental disorders

The dynamic nature of brain development mandates a careful understanding of the interrelated, time-dependent aspects of aberrant plasticity in neurodevelopmental disorders. In this section, we discuss cerebral palsy as a neurodevelopmental disorder that may benefit from plasticity-informed approaches to diagnosis and treatment.

While the brain insult in cerebral palsy (CP) is often described as a static one-time "hit" or lesion during early brain development, the injury profile and pathological consequences are dynamic and progressive. This progression is driven by cascading changes at all levels of neural organization from synapses to networks, which alter normal brain development. The phenotype of CP varies among patients, but the cardinal feature is a disorder of movement (an essential criterion) that includes spasticity, dyskinesia, hypotonia, and weakness. Associated features may include cognitive and speech impairment, epilepsy, chronic pain, musculoskeletal deformities, sleep, and behavioral disturbances, reflecting the widespread nature of damage and impact on normal brain development (see Chapter Cerebral palsy).

CP is a clinical diagnosis confirmed through behavioral assessments of motor function (the current gold standard) in infants and young children with a high-risk medical history or abnormal neuroimaging profile. The average age of CP diagnosis is 12–24 months in most clinical settings. Barriers to early and accurate diagnosis may include (1) incomplete manifestation of motor impairment at the time of clinical evaluation, (2) subtle motor deficits that can evade detection at early stages of the disease, (3) clinical assessment tools that suffer from interrater variability, potentially leading to delayed or inaccurate diagnosis, and (4) changes in motor function with age as a result of brain plasticity and maturation.

Below we explore two plasticity-relevant questions related to diagnosing and treating CP.

Can a patient at high risk for cerebral palsy be identified early enough before the closing of the critical window for motor system plasticity?

A "watchful waiting" approach, where clinicians serially evaluate the infant or young child for clear manifestation of motor impairment before confirming a diagnosis of CP is common in practice. However, the complex interplay between brain structure and function cannot be fully captured by serial clinical examinations alone. It can be argued that the findings of the clinical examination are late manifestations of already consolidated structural and functional remodeling. Waiting for the full manifestation of motor dysfunction may put patients at a disadvantage, as the critical window for neuromuscular development may be closing or already closed, making meaningful restoration or compensation for the impaired function difficult to achieve.

The rationale for the earliest accurate diagnosis of CP is supported by a growing body of experimental and translational evidence showing the young brain has a heightened capacity for neuroplasticity, making it crucial to begin expropriate interventions as soon as possible [63].

Can our understanding of developmental and intervention-induced neuroplasticity be leveraged to improve outcomes in patients with cerebral palsy?

Motor impairment in CP is a target for many interventions. Perhaps the most studied is constraint-induced movement therapy (CIMT) for children with hemiplegic CP. CIMT is based on compensatory reliance on the paretic limb, where the stronger hand is immobilized with a soft cast to force the use of the paretic limb and promote new learn new motor learning strategies through intensive, repetitive task practice. The addition of bimanual training further facilitates motor learning and improves paretic hand function. The effectiveness of CIMT is variable [64]. Recent evidence suggests a correlation between the size, location, and severity of damage to the motor system (e.g., CSTs and transcallosal fibers) and the degree of improvement in hand function following CIMT. The plasticity mechanisms by which CIMT improves paretic hand function are not fully understood. It has been suggested that CIMT induces structural and functional experience-dependent plasticity and reorganization in sensorimotor tracts [65,66]. CIMT is also thought to modify inter-hemispheric inhibition; by constraining the unaffected hand, the abnormal inhibition from the unaffected hemisphere to the damaged hemisphere decreases. On a cellular level, in animal model of stroke, CIMT has been shown to promote changes in dendrites and dendritic spines in the ipsilateral and contralateral sensorimotor cortex and increase the expression of synaptic GluR2 in the ipsilateral sensorimotor cortex [67].

The pattern of CST lateralization after brain injury may influence the outcomes of CIMT. Studies of small sample size have reported that patients with ipsilateral CST projections may not benefit from CIMT [68,69]. As a result, mapping of CST lateralization was thought to be the primary biomarker of response to CIMT. However, in a recent prospective randomized control trial of 83 children with unilateral CP, the pattern of CST organization (alone) did not correlate with outcomes of CIMT or bimanual therapy. Children with ipsilateral, bilateral, and contralateral projections, all improved after therapy. The study did not examine regions such as somatosensory system, transcallosal fibers, thalamocortical projections that are involved in motor control.

The findings of this study highlight the importance of creating profiles of individual structural and functional patterns of sensorimotor system (re)organization beyond the corticospinal system. Early efforts for such multimodal assessments with CP, of brain plasticity show promising results in older children. In a pilot study, a multimodal assessment of 15 children aged 9.4 ± 2.5 years, the use of EEG, structural and functional magnetic resonance imaging (sMRI, fMRI), diffusion

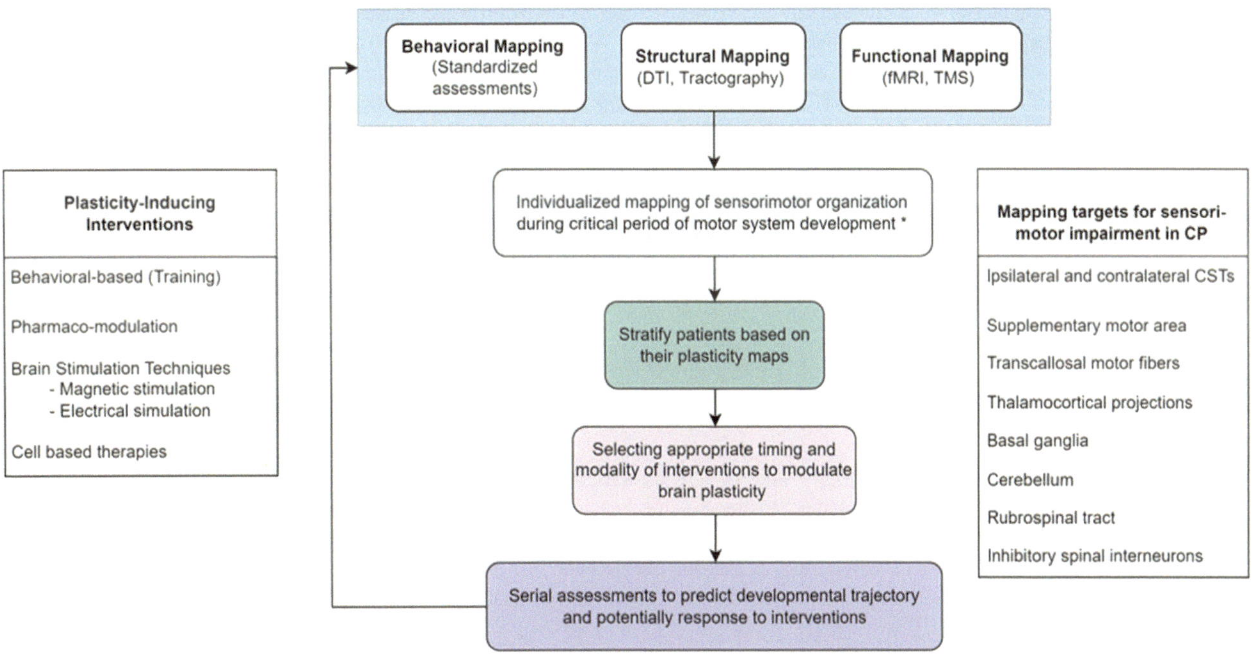

FIGURE 5.3 **The quest for an individualized, multimodal, multisystem mapping of neuroplasticity in children with perinatal brain injury with or at risk for cerebral palsy.** Mapping patient-specific cortical-subcortical (re)organization plasticity patterns following early brain injury may be key to explaining interindividual variability in phenotype, identifing sensitive periods for plasticity-inducing interventions. *While corticospinal tracts (CST) are the most studied targets for plasticity-based interventions in cerebral palsy (CP), other targets should be considered due to their involvement in sensorimotor impairment in CP. DTI, Diffusion tensor imaging; fMRI, functional MRI; TMS, transcranial magnetic stimulation.

tensor imaging (DTI), transcranial magnetic stimulation (TMS), electromyography (EMG), and hand function assessments identified multiple patient-specific patterns of brain changes that correlated with individual motor abilities. Another pilot study reported the feasibility of multimodal assessment (DTI, TMS, and fMRI) in two children aged 6 and 9 years with unilateral CP, revealing individual cortical reorganization patterns in response to motor rehabilitation [66].

The knowledge derived from multimodal and multisystem mapping of brain plasticity can be used to individualize interventions that induce plasticity and promote adaptive function. Fig. 5.3 illustrates a potential plasticity-based framework for evaluating a patient at high risk for developing CP.

Concluding remarks

Viewing neurodevelopmental disorders in children as disorders of brain plasticity highlights the dynamic changes in the developing brain over time, whether following injury or perturbation, and in response to interventions. Understanding how brain plasticity mechanisms affect the clinical manifestations of brain dysfunction is a crucial step in designing interventions. This perspective fosters a paradigm shift from passive symptom management strategy to active engagement in therapeutic approaches that attenuate the effects of the primary pathology, thereby supporting more favorable trajectories of brain development and maturation.

References

[1] Tau GZ, Peterson BS. Normal development of brain circuits. Neuropsychopharmacology 2010;35(1):147–68. Available from: https://doi.org/10.1038/npp.2009.115, PMID: 19794405; PMCID: PMC3055433.

[2] Nava E, Röder B. Adaptation and maladaptation insights from brain plasticity. Prog Brain Res 2011;191:177–94. Available from: https://doi.org/10.1016/B978-0-444-53752-2.00005-9, PMID: 21741552.

[3] Feldman DE. The spike-timing dependence of plasticity. Neuron 2012;75(4):556–71. Available from: https://doi.org/10.1016/j.neuron.2012.08.001, PMID: 22920249; PMCID: PMC3431193.

[4] Batschelett M, Gibbs S, Holder CM, Holcombe B, Wheless JW, Narayana S. Plasticity in the developing brain: neurophysiological basis for lesion-induced motor reorganization. Brain Commun 2021;4(1):fcab300. Available from: https://doi.org/10.1093/braincomms/fcab300.

[5] Buyanova IS, Arsalidou M. Cerebral White Matter Myelination and Relations to Age, Gender, and Cognition: A Selective Review. Front Hum Neurosci 2021;15:662031. Available from: https://doi.org/10.3389/fnhum.2021.662031, PMID: 34295229; PMCID: PMC8290169.

[6] Oldham S, Ball G, Fornito A. Early and late development of hub connectivity in the human brain. Curr Opin Psychol 2022;44:321–9. Available from: https://doi.org/10.1016/j.copsyc.2021.10.010.

[7] Voss P, Thomas ME, Cisneros-Franco JM, de Villers-Sidani E. Dynamic brains and the changing rules of neuroplasticity: implications for learning and recovery. Front Psychol 2017;8:1657. Available from: https://doi.org/10.3389/fpsyg.2017.01657.

[8] Takesian AE, Hensch TK. Balancing plasticity/stability across brain development. Prog Brain Res 2013;207:3–34. Available from: https://doi.org/10.1016/B978-0-444-63327-9.00001-1.

[9] Thompson B, Concetta Morrone M, Bex P, Lozama A, Sabel BA. Harnessing brain plasticity to improve binocular vision in amblyopia: an evidence-based update. Eur J Ophthalmol 2023;0(0). Available from: https://doi.org/10.1177/11206721231187426.

[10] Dehorter N, Del Pino I. Shifting developmental trajectories during critical periods of brain formation. Front Cell Neurosci 2020;14:283. Available from: https://doi.org/10.3389/fncel.2020.00283.

[11] Birdsong D. Plasticity, variability and age in second language acquisition and bilingualism. Front Psychol 2018;9:81. Available from: https://doi.org/10.3389/fpsyg.2018.00081.

[12] White EJ, Hutka SA, Williams LJ, Moreno S. Learning, neural plasticity and sensitive periods: implications for language acquisition, music training and transfer across the lifespan. Front Syst Neurosci 2013;7:90. Available from: https://doi.org/10.3389/fnsys.2013.00090.

[13] Hubel DH, Wiesel TN. The period of susceptibility to the physiological effects of unilateral eye closure in kittens. J Physiol 1970;206(2):419–36.

[14] Hensch TK. Critical period plasticity in local cortical circuits. Nat Rev Neurosci 2004;5(11):877–88.

[15] Merzenich MM, Jenkins WM. Reorganization of cortical representations of the hand following alterations of skin inputs induced by nerve injury, skin island transfers, and experience. J Hand Ther 1993;6(2):89–104.

[16] Knudsen EI. Sensitive periods in the development of the brain and behavior. J Cognit Neurosci 2004;16(8):1412–25.

[17] Johnson MH. Interactive specialization: A domain-general framework for human functional brain development? Developmental Cognit Neurosci 2011;1(1):7–21.

[18] Werker JF, Hensch TK. Critical periods in speech perception: new directions. Annu Rev Psychol 2015;66:173–96. Available from: https://doi.org/10.1146/annurev-psych-010814-015104 Epub 2014 Sep 17. PMID: 25251488.

[19] Huber KM, Gallagher SM, Warren ST, Bear MF. Altered synaptic plasticity in a mouse model of fragile X mental retardation. Proc Natl Acad Sci USA 2002;99(11):7746–50. Available from: https://doi.org/10.1073/pnas.122205699.

[20] Protic DD, Aishworiya R, Salcedo-Arellano MJ, et al. Fragile X syndrome: from molecular aspect to clinical treatment. Int J Mol Sci 2022;23(4):1935. Available from: https://doi.org/10.3390/ijms23041935.

[21] Feliciano DM, Lin TV, Hartman NW, Bartley CM, Kubera C, Hsieh L, et al. A circuitry and biochemical basis for tuberous sclerosis symptoms: from epilepsy to neurocognitive deficits. Int J Dev Neurosci 2013;31(7):667–78. Available from: https://doi.org/10.1016/j.ijdevneu.2013.02.008.

[22] Luo C, Ye WR, Shi W, et al. Perfect match: mTOR inhibitors and tuberous sclerosis complex. Orphanet J Rare Dis 2022;17(1):106. Available from: https://doi.org/10.1186/s13023-022-02266-0.

[23] Kerjan G, Gleeson JG. Genetic mechanisms underlying abnormal neuronal migration in classical lissencephaly. Trends Genet 2007;23 (12):623–30. Available from: https://doi.org/10.1016/j.tig.2007.09.003.

[24] Yamada M, Hirotsune S, Wynshaw-Boris A. A novel strategy for therapeutic intervention for the genetic disease: preventing proteolytic cleavage using small chemical compound. Int J Biochem Cell Biol 2010;42(9):1401–7. Available from: https://doi.org/10.1016/j.biocel.2010.05.017.

[25] Asaka Y, Jugloff DG, Zhang L, Eubanks JH, Fitzsimonds RM. Hippocampal synaptic plasticity is impaired in the Mecp2-null mouse model of Rett syndrome. Neurobiol Dis 2006;21(1):217–27. Available from: https://doi.org/10.1016/j.nbd.2005.07.005.

[26] Tropea D, Giacometti E, Wilson NR, Beard C, McCurry C, Fu DD, et al. Partial reversal of Rett Syndrome-like symptoms in MeCP2 mutant mice. Proc Natl Acad Sci USA 2009;106(6):2029–34. Available from: https://doi.org/10.1073/pnas.0812394106.

[27] Vashi N, Justice MJ. Treating Rett syndrome: from mouse models to human therapies. Mamm Genome 2019;30(5-6):90–110. Available from: https://doi.org/10.1007/s00335-019-09793-5.

[28] Yang X. Characterizing spine issues: if offers novel therapeutics to Angelman syndrome. Dev Neurobiol 2020;80(5-6):200–9. Available from: https://doi.org/10.1002/dneu.22757.

[29] Copping NA, McTighe SM, Fink KD, Silverman JL. Emerging gene and small molecule therapies for the neurodevelopmental disorder Angelman syndrome. Neurotherapeutics 2021;18(3):1535–47. Available from: https://doi.org/10.1007/s13311-021-01082-x.

[30] Krol A, Wimmer RD, Halassa MM, Feng G. Thalamic reticular dysfunction as a circuit endophenotype in neurodevelopmental disorders. Neuron 2018;98(2):282–95. Available from: https://doi.org/10.1016/j.neuron.2018.03.021.

[31] Jiang CC, Lin LS, Long S, et al. Signalling pathways in autism spectrum disorder: mechanisms and therapeutic implications. Signal Transduct Target Ther 2022;7(1):229. Available from: https://doi.org/10.1038/s41392-022-01081-0.

[32] Johnston M, Blue ME, Naidu S. Recent advances in understanding synaptic abnormalities in Rett syndrome. F1000Res 2015;4. Available from: https://doi.org/10.12688/f1000research.6987.1.

[33] Blue ME, Naidu S, Johnston MV. Altered development of glutamate and GABA receptors in the basal ganglia of girls with Rett syndrome. ExpNeurol 1999;156(2):345–52. Available from: https://doi.org/10.1006/exnr.1999.703.

[34] Krishnan K, Wang BS, Lu J, Wang L, Maffei A, Cang J, et al. MeCP2 regulates the timing of critical period plasticity that shapes functional connectivity in primary visual cortex. Proc Natl Acad Sci USA 2015;112(34):E4782–91. Available from: https://doi.org/10.1073/pnas.1506499112.

[35] Castren ML, Castren E. BDNF in fragile X syndrome. Neuropharmacology 2014;76(Pt C)):729–36. Available from: https://doi.org/10.1016/j.neuropharm.2013.05.018.

[36] He CX, Portera-Cailliau C. The trouble with spines in fragile X syndrome: density, maturity and plasticity. Neuroscience 2013;251:120–8. Available from: https://doi.org/10.1016/j.neuroscience.2012.03.049.

[37] Zhang Z, Marro SG, Zhang Y, Arendt KL, Patzke C, Zhou B, et al. The fragile X mutation impairs homeostatic plasticity in human neurons by blocking synaptic retinoic acid signaling. Sci Transl Med 2018;10(452):eaar4338. Available from: https://doi.org/10.1126/scitranslmed.aar4338.

[38] Paluszkiewicz SM, Martin BS, Huntsman MM. Fragile X syndrome: the GABAergic system and circuit dysfunction. Dev Neurosci 2011;33 (5):349–64. Available from: https://doi.org/10.1159/000329420.

[39] Telias M. Molecular mechanisms of synaptic dysregulation in fragile X syndrome and autism spectrum disorders. Front Mol Neurosci 2019;12:51. Available from: https://doi.org/10.3389/fnmol.2019.00051.

[40] Bateup HS, Johnson CA, Denefrio CL, Saulnier JL, Sabatini BL. Hyperexcitability in mouse models of tuberous sclerosis. Neuron 2013;78 (3):510–22. Available from: https://doi.org/10.1016/j.neuron.2013.03.017.

[41] Auerbach BD, Osterweil EK, Bear MF. Mutations causing syndromic autism define an axis of synaptic pathophysiology. Nature 2011;480 AU:24(7375):63–8. Available from: https://doi.org/10.1038/nature10658.

[42] Curatolo P. Mechanistic target of rapamycin (mTOR) in tuberous sclerosis complex-associated epilepsy. Pediatr Neurol 2015;52(3):281–9. Available from: https://doi.org/10.1016/j.pediatrneurol.2014.10.028.

[43] D'Argenzio L, Koch G, Bombardieri R, Mori F, Moavero R, Centonze D, et al. Abnormal parieto-motor connectivity in tuberous sclerosis complex. Epilepsy Res 2009;87(1):102–5. Available from: https://doi.org/10.1016/j.eplepsyres.2009.07.009.

[44] Pitkänen A, Engel J. Past and present definitions of epileptogenesis and its biomarkers. Neurotherapeutics 2014;11(2):231–41. Available from: https://doi.org/10.1007/s13311-014-0257-2.

[45] Rakhade SN, Jensen FE. Epileptogenesis in the immature brain: emerging mechanisms. Nat Rev Neurol 2009;5(7):380–91. Available from: https://doi.org/10.1038/nrneurol.2009.80, PMID: 19578345; PMCID: PMC2822660.

[46] Wasterlain CG, Gloss DS, Niquet J, Wasterlain AS. Epileptogenesis in the developing brain. Handb Clin Neurol 2013;111:427–39. Available from: https://doi.org/10.1016/B978-0-444-52891-9.00046-4.

[47] Sayin U, Hutchinson E, Meyerand ME, Sutula T. Age-dependent long-term structural and functional effects of early-life seizures: evidence for a hippocampal critical period influencing plasticity in adulthood. Neuroscience 2015;288:120–34. Available from: https://doi.org/10.1016/j.neuroscience.2014.12.017.

[48] Younus I, Reddy DS. Epigenetic interventions for epileptogenesis: a new frontier for curing epilepsy. Pharmacol Ther 2017;177:108–22. Available from: https://doi.org/10.1016/j.pharmthera.2017.03.002.

[49] Paoletti P, Bellone C, Zhou Q. NMDA receptor subunit diversity: impact on receptor properties, synaptic plasticity and disease. Nat Rev Neurosci 2013;14:383–400.
[50] Khalilov I, Minlebaev M, Mukhtarov M, Khazipov R. Dynamic changes from depolarizing to hyperpolarizing GABAergic actions during giant depolarizing potentials in the neonatal rat hippocampus. J Neurosci 2015;35(37):12635–42. Available from: https://doi.org/10.1523/JNEUROSCI.1922-15.2015.
[51] Johnson MH. Interactive specialization: a domain-general framework for human functional brain development? Dev Cogn Neurosci 2011; 1(1):7–21.
[52] Johnston MV, Hoon AH. Possible mechanisms in infants for selective basal ganglia damage from asphyxia, kernicterus, or Mitochondrial encephalopathies. J Child Neurol 2000;15(9):588–91. Available from: https://doi.org/10.1177/088307380001500904.
[53] Baxter P. Markers of perinatal hypoxia-ischaemia and neurological injury: assessing the impact of insult duration. Dev Med Child Neurol 2020;62(5):563–8. Available from: https://doi.org/10.1111/dmcn.14421.
[54] Johnston MV. Excitotoxicity in neonatal hypoxia. Ment Retard Dev Disabil 2001;7:229–34.
[55] Johnston MV. Neurotransmitter alterations in a model of perinatal hypoxic-ischemic brain injury. Ann Neurol 1983;13:511–18.
[56] Nahum AS, Liégeois FJ. Language after childhood hemispherectomy: a systematic review. Neurology 2020;95(23):1043–56. Available from: https://doi.org/10.1212/WNL.0000000000011073.
[57] Foster MT, Das K, May P. Motor cortex relocation after complete anatomical hemispherectomy for intractable epilepsy secondary to 'Rasmussen's encephalitis. Br J Neurosurg 2019;33(2):234–6. Available from: https://doi.org/10.1080/02688697.2018.1549316.
[58] Sebastianelli L, Versace V, Taylor A, Brigo F, Nothdurfter W, Saltuari L, et al. Functional reorganization after hemispherectomy in humans and animal models: what can we learn about the brain's resilience to extensive unilateral lesions? Brain Res Bull 2017;131:156–67. Available from: https://doi.org/10.1016/j.brainresbull.2017.04.005.
[59] Merabet LB, Pascual-Leone A. Neural reorganization following sensory loss: the opportunity of change. Nat Rev Neurosci 2010;11(1):44–52. Available from: https://doi.org/10.1038/nrn2758.
[60] Tomblin JB, Barker BA, Hubbs S. Developmental constraints on language development in children with cochlear implants. Int J Audiol 2007;46(9):512–23. Available from: https://doi.org/10.1080/14992020701383043.
[61] Glennon E, Svirsky MA, Froemke RC. Auditory cortical plasticity in cochlear implant users. Curr Opin Neurobiol 2020;60:108–14. Available from: https://doi.org/10.1016/j.conb.2019.11.003.
[62] Kral A, Dorman MF, Wilson BS. Neuronal development of hearing and language: cochlear implants and critical periods. Annu Rev Neurosci 2019;42:47–65. Available from: https://doi.org/10.1146/annurev-neuro-080317-061513.
[63] Novak I, Morgan C, Adde L, et al. Early, accurate diagnosis and early intervention in cerebral palsy: advances in diagnosis and treatment. JAMA Pediatr 2017;171(9):897–907. Available from: https://doi.org/10.1001/jamapediatrics.2017.1689.
[64] Novak I, Mcintyre S, Morgan C, Campbell L, Dark L, Morton N, et al. A systematic review of interventions for children with cerebral palsy: state of the evidence. Dev Med Child Neurol 2013;55:885–910. Available from: https://doi.org/10.1111/dmcn.12246.
[65] Inguaggiato E, Sgandurra G, Perazza S, Guzzetta A, Cioni G. Brain reorganization following intervention in children with congenital hemiplegia: a systematic review. Neural Plasticity 2013;2013. Available from: https://doi.org/10.1155/2013/356275.
[66] Bleyenheuft Y, Dricot L, Gilis N, Kuo H-C, Grandin C, Bleyenheuft C, et al. Capturing neuroplastic changes after bimanual intensive rehabilitation in children with unilateral spastic cerebral palsy: a combined DTI, TMS and fMRI pilot study. Res Dev Disabil 2015;43:136–49. Available from: https://doi.org/10.1016/j.ridd.2015.06.014.
[67] Hu J, Li C, Hua Y, Liu P, Gao B, Wang Y, et al. Constraint-induced movement therapy improves functional recovery after ischemic stroke and its impacts on synaptic plasticity in sensorimotor cortex and hippocampus. Brain Res Bull 2020;160:8–23. Available from: https://doi.org/10.1016/j.brainresbull.2020.04.006.
[68] Juenger H, Kuhnke N, Braun C, Ummenhofer F, Wilke M, Walther M, et al. Two types of exercise-induced neuroplasticity in congenital hemiparesis: a transcranial magnetic stimulation, functional MRI, and magnetoencephalography study. Dev Med Child Neurol 2013;55(10):941–51. Available from: https://doi.org/10.1111/dmcn.12209.
[69] Kuhnke N, Juenger H, Walther M, Berweck S, Mall V, Staudt M. Do patients with congenital hemiparesis and ipsilateral corticospinal projections respond differently to constraint-induced movement therapy? Dev Med Child Neurol 2008;50:898–903. Available from: https://doi.org/10.1111/j.1469-8749.2008.03119.x.

Section II

The developmental evaluation and diagnosis

Chapter 6

Developmental surveillance and screening

Toni M. Whitaker[1] and Paul H. Lipkin[2]

[1]*Department of Pediatrics/Division of Developmental Pediatrics, University of Tennessee Health Science Center, Le Bonheur Children's Hospital, Memphis, TN, United States,* [2]*Division of Neurology and Developmental Medicine, Kennedy Krieger Institute, Department of Pediatrics, Johns Hopkins School of Medicine, Baltimore, MD, United States*

Introduction

Neurodevelopmental disabilities (NDD) represent common chronic child health conditions, affecting up to 17% or one in six children in the United States. The NDD of highest prevalence, attention-deficit/hyperactivity disorder, represents approximately 9% of children, with autism spectrum disorder (ASD) 1%–2%, and intellectual disability estimated at 1% of children [1–3]. Cerebral palsy (CP) affects 3 per 1000 children [4]. As lifelong health conditions with significant effects on everyday function at home, school, work, and in the community, they have a large impact on resources in health care, educational services, and community services, with health care expenses in adults with intellectual and developmental disabilities (IDD) totaling nearly $5 billion in 2011 [5]. Amongst children with IDD, health care expenses also were significantly higher in children with IDD compared to their peers, with inpatient and emergency department use nearly twice that of other children, with per capita expense 100% higher [6,7]. For those with ASD, the annual direct medical, direct nonmedical, and productivity costs associated with this NDD are substantial and are estimated to represent approximately 1%-3.5% of the US Gross Domestic Product in 2025 [8,9]. Children with CP have health care expenditures 10 times higher than other children, with more than $1 million in costs in a lifetime and more than $12 billion in lifetime costs for all those affected by CP in the United States [10].

Given the prevalence of these medical conditions and the high impact they have on health care and the general economy, identification and management of these disorders in the general population of children by pediatric health care clinicians is indicated through systems of surveillance and screening in health care settings. As with other health conditions, surveillance and screening are key components of health care and are aimed at early identification and intervention in children for the identified problem. Common examples of health *surveillance* include serial measurement of weight and vital signs, medical history review, and new symptom review, provided at each health supervision visit during childhood to every child. It is typically performed or reviewed by the physician or other child health clinician. In contrast, health *screening* of such conditions is usually diagnosis-specific, episodic, and is performed at key ages. It is typically implemented by clinic support staff on every child with interpretation by the medical professional. Screening usually involves performance of a standardized test, is rapid, and generally of low cost. Pediatric examples include newborn screening for inborn errors and anemia screening at 12 months of age.

For neurodevelopmental and related disorders, *developmental surveillance* and standardized *developmental screening* have been developed, rooted in the same principles used in general health surveillance and screening [11]. The practice is most often embedded in pediatric primary care. However, one can also think of developmental surveillance and screening as beginning in the newborn period and continuing into primary care as well as the community. In the newborn period, infants are uniformly screened for conditions that, if untreated, will result in NDD. There are several dozen inborn errors of metabolism screened in newborns throughout the US, including amino acid disorders, organic acid disorders, disorders of fatty acid oxidation, disorders of carbohydrate metabolism, and endocrine disorders [12–14], each if untreated may lead to delays in motor, intellectual, behavior, or social development. In addition, newborns are routinely screened for congenital hearing loss [15], which also if untreated leads to delays in language development and which may be accompanied by other developmental disabilities. Screening is now also routine for congenital heart disease, which is now known to frequently be associated with NDD, particularly in those with complex cardiac conditions [16].

Capute and Accardo's Neurodevelopmental Disabilities in Infancy and Childhood. DOI: https://doi.org/10.1016/B978-0-12-824060-1.00042-0

After newborn screening, developmental surveillance and screening occur in pediatric primary care as well as in early childhood programs. Rooted in the primary health care delivery system for all children, developmental surveillance and screening are a starting point for specialized care for children with known or at high risk for developmental disability. The guidance for this practice has been developed by expert panels convened by the American Academy of Pediatrics (AAP) aiming for the early identification and intervention of children with developmental disorders and includes general developmental screening, including delays in motor development and autism screening [11], as well as specific guidance for ASDs [17], vision screening [18,19], motor delays [20], and behavioral and emotional conditions [21].

The child who is identified with significant delays or with a specific disabling condition becomes a candidate for special interventions in learning, motor skills, language development, or behavior. Such interventions are provided as a legal right in the United States within the Individuals with Disabilities Education Act (IDEA) [22,23]. Developmental screening only identifies "risk" for specific conditions and therefore require further evaluation of the child who has concerns identified on screening in order to identify specific disabilities (e.g., intellectual, autism spectrum) and/or specific medical conditions (e.g., CP, neuromuscular disorders, genetic disorders). The benefits and efficacy of early identification and developmental intervention for NDD are accepted in special populations of NDD, such as ASD [24–26], high-risk neonates [27], and those with motor delays/deficits [28]. Despite the complexity of the research, systematic reviews of outcome, including long-term outcomes into teens and adulthood, have continued to support the overall benefit of early childhood education in improving a variety of developmental, educational, and behavioral outcomes as well as promoting resilience for children who may be at additional risk of poor outcomes [29–33]. In medical conditions with established interventions and treatments, delays in identification have been reported despite demonstrated benefits of earlier identification (e.g., Duchenne Muscular Dystrophy (DMD), Fragile X syndrome, CP) [27,28,34,35]. It is therefore at these diagnostic and therapeutic points in health and developmental care that the medical and developmental specialist typically provide short- and long-term specialty care.

The screening process: from primary care to specialty care and treatment

The key principles for the early identification and treatment of the child with a developmental disability begin in pediatric primary care with *developmental surveillance* and *developmental screening* of every child. Care for the child who is identified with either method as having a verified concern then shifts out of primary care with referral to developmental specialty care for comprehensive *developmental evaluation*, provided through medical specialists and/or other early childhood professionals. Specialty medical care also ensues alongside special therapeutic intervention, with treatment for identified disabilities and associated medical conditions through developmental interventions and medical care. Chronic condition management is initiated for those with identified NDD in the primary pediatric home, the specialty pediatric home, and community-based treatment services in schools, developmental specialty centers, and therapy centers. (See Fig. 6.1 for the developmental screening process.)

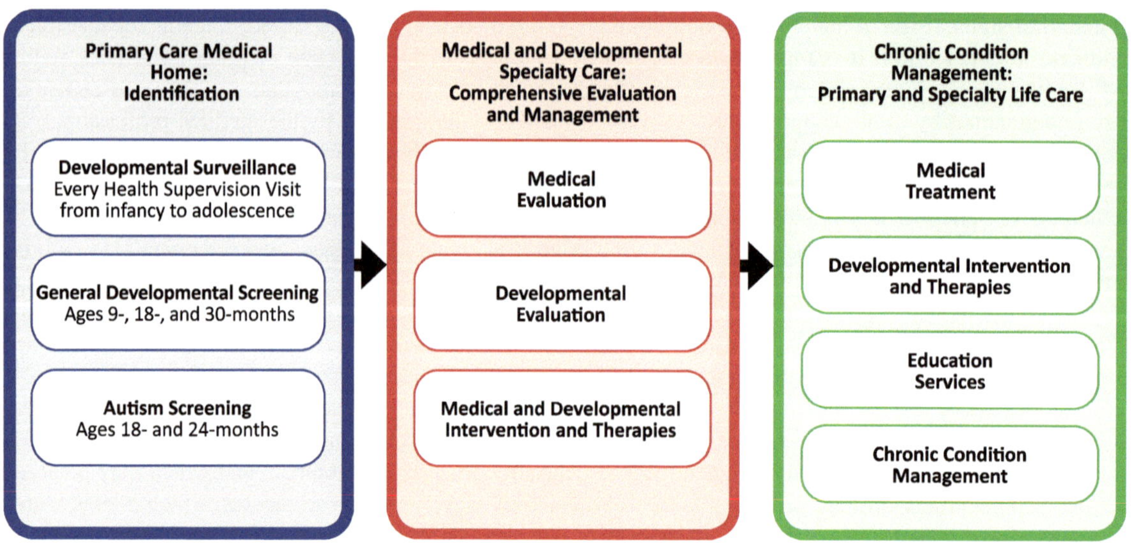

FIGURE 6.1 The developmental screening process from primary care to specialty care and intervention to lifelong management.

Identification in the medical home

Developmental surveillance

Developmental surveillance is defined by the AAP [11] as a flexible, longitudinal, continuous, and cumulative process aimed at identifying children who may have developmental problems. It is performed at every health supervision visit (previously referred to as well-child visit or preventive care visit) in pediatric primary care beginning in infancy and continuing into adolescence and young adulthood. It includes obtaining historical information on the child and family as well as observing the child in the office visit. There are six key components obtained by history:

1. Eliciting and attending to the parents' concerns about their child's development.
2. Documenting and maintaining a developmental history.
3. Making accurate observations of the child.
4. Identifying risk and protective factors.
5. Maintaining an accurate record of documenting the process and findings.
6. Sharing and obtaining opinions and findings with other professionals, such as child care providers, home visitors, preschool teachers, and developmental therapists, especially when concerns arise.

While developmental surveillance has been defined for use in primary care in the early identification of the child with developmental delays and disabilities, key components overlap with components of a thorough medical specialty evaluation. With its performance at every health supervision visit, surveillance provides a longitudinal evaluation of a child's developmental progress. In the repeated review of a child's developmental milestones over multiple pediatric primary care visits, such as described in the Bright Futures Guidelines [36] and the Centers for Disease Control and Prevention's (CDC's) *Learn the Signs. Act Early.* program (https://www.cdc.gov/ncbddd/actearly/index.html), pediatric surveillance may identify delays or other developmental problems such as regression, dissociation, or deviation. The medical and family history review may identify health and social risk factors for developmental disability as well as protective factors for development. Further discussion regarding risk factors and relevance of social determinants of health are later in this chapter under "Other Considerations in Screening, Identification, and Management." The serial observations of the child's development afforded by the health supervision visits, along with related documentation, provide the primary care pediatric clinician a unique view and chronology of the child's development and patterns across the streams of gross motor, fine motor, speech and language, social engagement and response. Developmental surveillance coupled with direct physical and neurological examination provides the primary care clinician insights into a child's development to share with a parent and other professionals for consideration of each child's pattern of development and any concerns that may be indicative of a developmental disability. In continuing from infancy through adolescence, surveillance offers opportunities for recognition of a developmental disability throughout childhood and into adolescence. Particular attention to surveillance beyond the infant and preschool years is recommended at the 4- to 5-year-old visits, due to the child's entry into elementary education at this time and the opportunity for identification of previously unrecognized problems in learning, language, attention or social behaviors.

When concerns arise during longitudinal developmental surveillance, the primary care clinician should perform standardized developmental screening or refer a child for comprehensive developmental evaluation. Referral for intervention and therapy services may be considered at this time as well, depending on resources in the local community and the duration of time expected between referral and full neurodevelopmental evaluation. The surveillance process also includes regular discussions with the parents of their child's developmental progress and can support the positive ongoing relationship between the primary care clinician and the parent or caregiver. Behavioral and emotional problems are also common in children, including those with developmental disabilities. Discussions with parents during developmental surveillance may elicit these concerns as well, creating opportunities to identify and intervene with these problems [21].

Increasingly, developmental surveillance and screening are recognized as complementary and mutually beneficial strategies for improving early identification of developmental concerns and receipt of early developmental intervention services [37–40].

Developmental screening

In contrast to developmental surveillance, developmental screening involves the administration of a standardized instrument designed for such screening. Any screening test used should meet professional measurement standards including reliability, validity, sensitivity, and specificity [11]. It is typically brief for incorporation into the typical health supervision visit and most commonly uses a parent-report questionnaire. Its aim is the identification of children at risk of a developmental disorder and is not diagnostic. Like other forms of pediatric screening, developmental screening is to be performed universally in all children at specific recommended ages in the pediatric periodicity table, based on known patterns of development and practical implementation opportunities [41]. The AAP currently recommends that each child have formal general developmental screening at 9, 18, and 30 months of age during the first 3 years of life, using a general developmental screening test that measures all developmental domains. Autism-specific screening is also recommended at the 18- and 24-month visits [11,17]. Screening is also recommended when surveillance concerns arise regarding a child's pattern of development from the parent or caregiver, the primary pediatric health care professional, or other professionals such as child care providers, home visitors, preschool teachers, and developmental therapists. (See Table 6.1 for developmental screening visits.)

The ages for screening present opportunities for identification of specific developmental disabilities during health supervision visits when few other screenings or preventive care actions are required. These screening visits also allow the progress of developmental skills to be highlighted. (See Table 6.2 for examples of developmental emphasis at typical health supervision visits.) At the 9-month visit, delays in motor development, such as seen in CP, may be evident, as well as problems in hearing that were not identified in newborn screening and problems in vision. Milder motor disabilities, including mild CP, hemiplegia, and mild abnormalities in gait or coordination, as well as early manifestations of muscular dystrophies, may be identified on the 18-month screening visit. With both general and autism screening at this visit, screening offers opportunities for identifying delays seen in autism, language disabilities, and intellectual disabilities. The 24-month visit is focused on autism-specific screening to target that disorder characterized by problems in social language and related behaviors. This autism screening visit is rooted in the predictive validity of the autism-specific screening tests. The 30-month visit offers one more opportunity for screening before the publicly-mandated early intervention system of developmental care transitions over to the preschools in the public education system. Like the 18-month visit, the 30-month visit is focused on screening for language, social, or intellectual disabilities, as well as milder problems in speech and language, attention, learning, and related behaviors. It also may identify children with mild motor delays or signs of the motor problems including regression seen in neuromuscular disorders.

In screening all children three times before 3 years of age, those with disabilities, whether mild or severe, will have an opportunity for earliest identification and treatment of a specific disability and related medical conditions, including the publicly mandated early intervention treatments. When a child is identified with a disability or at high risk, they are legally entitled to therapeutic and developmental interventions through local early intervention agencies, provided in the United States through the Individuals with Disability Education Act (IDEA) [22,23]. Medical care tied to this early identification includes treatments of genetic or neurological disorders and also allows for parental counseling, including

TABLE 6.1 Developmental screening visits.[a,b]

Age of visit	Screening tool category	Developmental areas screened	Identifiable neurodevelopmental and related disorders
Newborn	Blood audiology	General hearing	Inborn errors of metabolism, deafness and hearing loss
9 months	General	Gross motor, fine motor, receptive language, vision, hearing	Cerebral palsy and other neuromotor disorders, vision impairment, deafness and hearing loss
18 months	General and autism	Gross motor, fine motor, verbal language (expressive and receptive), social language and related behaviors	Cerebral palsy (mild to moderate), mild motor or neuromuscular disorders, autism spectrum disorders, language disorders, intellectual disability
24 months	Autism	Social language and related behaviors	Autism spectrum disorder
30 months	General	Verbal language (expressive and receptive), social language, behavior	Language disorders, autism spectrum disorder, intellectual and other learning disabilities, attention disorders, disruptive behavior disorders, mild motor delays or loss

[a]*Lipkin PH, Macias MM; Council on Children with Disabilities, Section on Developmental and Behavioral Pediatrics. Promoting optimal development: identifying infants and young children with developmental disorders through developmental surveillance and screening. Pediatrics. 2020 Jan;145(1): e20193449. https://doi.org/10.1542/peds.2019–3449. Epub 2019 Dec 16. PMID: 31843861.*
[b]*Committee on Practice and Ambulatory Medicine and B.F.P.S. Workgroup. "2021 Recommendations for preventive pediatric health care." Pediatrics. 2021;147(3): e2020049776.*

TABLE 6.2 Examples of developmental emphasis at typical health supervision visits.

Age of visit	Developmental areas screened	Identifiable developmental disabilities
9 months	Gross motor Fine motor Receptive language Hearing Vision	Cerebral palsy and other neuromotor disorders vision impairment deafness and hearing loss
18 months	Gross motor and fine motor Cognition Language (expressive and receptive) Social communication	Cerebral palsy (mild to moderate), mild neuromotor or neuromuscular disorders, intellectual disability, language disorders, autism spectrum disorder
24 months	Social communication Language (expressive and receptive)	Intellectual disability, language disorders, autism spectrum disorder
30 months	Gross and fine motor Social communication Language (expressive and receptive) Cognition Attention Behavior	Mild and moderate motor delays language disorders autism spectrum disorder intellectual and other learning disabilities attention disorders disruptive behavior disorders

genetic counseling. As previously described, evidence exists demonstrating the benefits of both early therapeutic interventions and medical interventions for the early diagnosis and initiation of such services, with improved short- and long-term outcomes. Treatments may include medical treatments, such as medications to treat the motor manifestations and tone abnormalities in CP or new gene therapies targeting muscular dystrophy [24–34].

In order to assure children receive appropriate diagnosis and treatment, the primary care clinician should refer those children who show concerns on screening or surveillance for comprehensive developmental evaluation, as well as medical evaluation and testing aimed at identifying specific medical conditions. The primary care clinician should also conduct a historical review of the child's prior newborn metabolic and hearing screenings, growth, and an update of environmental, medical, family, and social history for additional risk factors. The clinician should refer the child with language delays for audiological evaluation. Vision screening or referral should be considered for the child with abnormal visual regard or attention problems. For the child with delayed motor development, a full neurologic examination should also be performed by the primary clinician. When increased tone, microcephaly, or macrocephaly are noted, neuroradiologic examination should be considered to identify congenital or acquired brain anomalies. If tone is normal or decreased, the primary clinician should order plasma creatine kinase levels for the early identification and treatment of the muscular dystrophies or related disorders. Thyroid-stimulating hormone is also recommended to expedite treatment of an unidentified thyroid disorder [20]. The child with suspected global developmental delay, intellectual disability, or ASD should have chromosomal microarray and fragile X testing completed by the primary or a specialty physician. For the primary care clinician who seeks subspecialist assistance, a comanagement plan and care-coordination should be developed with the specialist and the family. (See Table 6.3 for common medical testing within the primary care setting.)

Screening tests

Developmental screening tests or tools are designed and validated for screening in primary care for all children and not for developmental evaluation by a medical or developmental specialist. Therefore, specialty clinicians do not typically use screening tests directly but will receive referrals from primary care clinicians and other professionals who identify concerns on screening. Specialty providers can often provide helpful guidance to a primary care clinician on appropriate choice and use of screening tests in addition to advice on planning for specific evaluation and treatment, thereby maximizing the value of the specialty consultation. Thus, specialty providers should be familiar with commonly used screening tests as well as tips and potential pitfalls of their use. Regardless of the stage during which collaboration occurs,

TABLE 6.3 Primary care medical testing of the child with developmental delay.[a]

Area of delay	Common examination findings	Medical test	Suspected neurodevelopmental or related disorders
Gross motor	Increased tone	Neuroimaging	Cerebral palsy and related neuromotor disorders
Gross motor	Normal or low tone	Plasma creatine kinase Thyroid stimulating hormone	Duchenne muscular dystrophy Hypothyroidism
Speech-language		Audiologic evaluation	Deafness and hearing loss
Global, including severe language and/or motor processing delays	Short or tall stature; micro- or macro-cephaly; Dysmorphology	Chromosomal microarray Whole exome sequencing Fragile X assay	Intellectual or other learning disabilities
Autism	Language and social delays	Chromosomal microarray Whole exome sequencing Fragile X assay	Autism spectrum disorder

[a] *Lipkin PH, Macias MM; Council on Children with Disabilities, Section on Developmental and Behavioral Pediatrics. Promoting optimal development: identifying infants and young children with developmental disorders through developmental surveillance and screening. Pediatrics. 2020 Jan;145(1): e20193449. https://doi.org/10.1542/peds.2019–3449. Epub 2019 Dec 16. PMID: 31843861.*

team-based care, which is promoted by the AAP, provides opportunities to enhance communication and cooperation among various entities involved in the care of the child and support to the family [42].

Primary care clinicians and other professionals may choose from many existing screening tests. The options are dynamic since the tests may change editions, new tests are created, validation studies are added, as well as other factors. It is therefore important for the clinician to access current information about screening tests. The AAP's *Screening Technical Assistance & Resource Center* (STAR Center; https://screeningtime.org/star-center/#/screening-tools) [43] has maintained an online resource of commonly used screening tests since 2017 that may be used to review basic features of current screening tests. This resource provides specific information on the tests, including type of screening and topics covered (developmental, autism, social-emotional, etc.), number of items, usual time needed to administer, price, literacy level, and validation studies (with sensitivity and specificity). The AAP does not endorse individual tests so the clinician will need to judge which tests meet their unique practice needs. While each clinician may experience challenges incorporating screening into office work-flow, this AAP resource and others exist to support the clinician in implementation [39]. The AAP offers additional online resources including training and family support materials to facilitate screening [44].

It is important to consider whether screens are general, covering more than one domain of development or type of behavior, or are more specific to a condition. If condition-specific screens are chosen, more than one screening test may be needed to ensure completeness of overall screening. Screens may ask parents for their own concerns about the child's development, while others may be rooted in developmental skills and milestones and give continuous scores for skills that are compared to normed "cut" points, indicating "pass" or "fail" for risk stratification. (See Table 6.4 for a list of common developmental screening tests now in use.)

As an example of differences among general developmental screening tests, the *Ages and Stages Questionnaire, Third Edition* (ASQ-3) [51] is designed to gather information on specific developmental skills from the parent while the *Parents' Evaluation of Developmental Status* (PEDS) [52] asks the parent about concerns in multiple developmental domains. Both are validated general developmental screening tests but provide different information to the primary care clinician, and thus, to the specialist. The *Modified Checklist for Autism in Toddlers, Revised, with Follow-Up* (M-CHAT-R/F) [48] is a validated screening test for ASD, updated from a prior version, with high sensitivity, but requires use of a standardized follow-up interview to reduce high rates of false-positive results. While this popular screening test is commonly used, the follow-up interview is often not performed, thus results may be suboptimal if not used as validated [53,54].

Even when screening tests themselves are appropriately validated, care should be taken to consider whether the screening test is used according to the guidelines on which validation studies were based (e.g., for the appropriate age

TABLE 6.4 Common developmental screening tests for use in primary care (Note: all are parent report).

General name	Ages	Administration time (minutes)	Publication information	References
Ages & Stages Questionnaires, Third Edition (ASQ-3)	2–66 months	10–15	Paul H. Brookes Publishing	[45]
Parents Evaluation of Developmental Status (PEDS)	0–8 years	2–10	Ellsworth & Vandermeer Press	[46]
Parents Evaluation of Developmental Status: Developmental Milestones (PEDS:DM)	0–8 years	4–6	Ellsworth & Vandermeer Press	[46]
Survey of Well-Being of Young Children (SWYC)	1–65 months	Less than 5	http://www.theswyc.org	[47]
Autism				
Modified Checklist for Autism in Toddlers, Revised, with Follow-up (M-CHAT-R/F)	16–48 months	5–10	http://www.mchatscreen.com	[48]
Social Communication Questionnaire (SCQ)	4 years and over	5–10	Western Psychological Services	[49]
Language				
Communication and Symbolic Behavior Scales Developmental Profile (CSBS DP): Infant-Toddler Checklist	6 months–6 years	5–10	Paul H. Brookes Publishing Co.	[50]

child; for translated tests, whether validated studies have been done in multiple languages). Developmental screening questions or tests embedded in electronic health records systems should be reviewed for similar relevant factors since inclusion does not necessarily guarantee that the screening test meets all desired criteria. Some tests provide handouts for parent education and encouragement of promotion of developmental skills.

There remains a significant role for clinical decision-making, particularly in the developmental surveillance process where additional screenings may be employed when a concern is identified or referrals may be made without specific screening test [55–57]. Specific screening tests are designed to increase accuracy of clinical decision-making and avoid delays in identification of delays and disabilities [58].

Gaps remain in surveillance and screening. Some pediatric primary care clinicians may be unaware of appropriate tests or may be unfamiliar with appropriate use despite care guidelines recommending consistent use, thus continued efforts to encourage screening is important [59]. Further, even when screening is completed and results show concerns, not all children receive appropriate referrals for further evaluation and intervention [53,54,60]. When screening is "passed," or does not show concerns, the role of surveillance remains relevant as this longitudinal process is intended to support monitoring over time [54].

Disparities in health equity exist in all stages of developmental surveillance, screening, referral, evaluation, and intervention. Black, Indigenous and children of color are less likely to be diagnosed with developmental or behavioral disorders and are diagnosed at later ages [1,61,62]. Racial, ethnic, and language disparities have been described in early childhood developmental/behavioral evaluations [63,64]. While gains have been made in identifying developmental delays and disabilities in children of non-white race and from households with lower median income, both with historical concerns of inadequate identification, attention should be paid to ensuring appropriate and equitable early identification and intervention [65,66].

Multiple state and national educational efforts, including those that encourage and support quality improvement projects, have sought to increase rates of developmental surveillance and screening in the medical home and by other early childhood professionals. The CDC's *Learn the Signs. Act Early.* program promotes parent-engaged monitoring and targets various professionals and caregivers with its health education campaign [56,67]. The *Act Early* program updated specific milestones used in developmental checklists in early 2022 in conjunction with experts from the AAP and other representatives. The new milestones are intended to reflect those that most children (>75%) typically can do by select ages rather than 50% in order to improve identification of those with significant delays. Milestone checklists help support parents and complement the overall process of surveillance by developmental professionals, and are not intended to replace or alter other steps in the process, including standardized screening or diagnostic evaluation [56]. Additional partnerships with other national programs have further extended outreach to child care, home visiting, and educational systems, where there is increasing likelihood a child may receive developmental screening. The AAP provides resources and education to health care providers as well as platforms serving parents and families directly to promote developmental surveillance and screening [11,68].

From developmental surveillance and screening to comprehensive evaluation and referral

Developmental evaluation

A developmental evaluation must be distinguished from developmental surveillance and screening. As defined by the AAP [11], developmental evaluation is a complex process to identify a specific developmental disorder or multiple disorders that affect the child. The evaluation is typically performed by a specialized medical or early childhood professional and provides the child and parent with information regarding best treatments and interventions for the disorder along with prognostic information. It can also assist in identification of related medical conditions or co-occurring risk for developmental or behavioral disorders which may be present at that time or arise later in life. Medical specialists include neurodevelopmental pediatricians or neurologists, developmental-behavioral pediatricians, general pediatric neurologists, pediatric physiatrists, or related advanced practice nurses. Other professionals with related expertise include early childhood educators, child psychologists, speech-language pathologists, physical therapists and occupational therapists working in medical centers, early intervention programs, or community practices. Ideally, the developmental evaluation is multidisciplinary, with the team working with the family in a coordinated manner around care and treatment. The description of comprehensive evaluations is beyond the scope of this chapter, but is described in other chapters in Section II. The developmental evaluation and diagnosis.

Referral

Pediatric clinicians should consider the degree to which additional evaluation is warranted when concerns on surveillance or screening are found. For relatively mild delays in a single area, referral for single intervention services (such as physical therapy for a mild motor delay or speech therapy for a mild phonologic disorder) with monitoring within the medical home may be appropriate as a first step. However, diagnostic clarity may be immediately sought for more complex diagnoses such as ASD. Consideration of available resources in the local area is often a significant determining factor. Many experts recommend a simultaneous approach to referral for further evaluation and initiation of intervention services based on functional abilities, rather than diagnosis alone. An accurate and sustainable diagnosis may take some time to assign and intervention and evaluation may need to occur concurrently to ensure prompt receipt of therapeutic services by the child [11,23,69,70].

Both at the local and national levels, innovative strategies to address need for evaluations and services may be needed when faced with shortages in appropriate resources. Primary care clinicians may need additional guidance from specialists who are familiar with local availability of resources for evaluation and intervention to determine most efficient referral patterns. For example, specialty providers with extensive waiting lists for new patient evaluations may offer interim guidance to primary care clinicians about a child's treatment needs and offer suggestions for other professionals and resources for initiating treatment for the child. For patients joining a waiting list for evaluations, parents can be provided with basic resources such as early intervention or public school services. Additionally, the referring clinician may help families secure some intervention services before additional formal evaluation is completed. There may be a relative risk in waiting for a diagnostic label if interventions could proceed sooner based on functional limitations rather than a specific diagnosis.

Pediatric workforce shortages include professionals with training in NDD, child neurology, developmental-behavioral pediatrics, and others who typically provide further evaluation for developmental delay and disability. Therefore, creative solutions at a community level may need to be employed [71–76]. Even when specialists exist, practice models differ and assessing availability of certain diagnostic services may be difficult [77]. Innovative strategies to address workforce shortages at multiple steps may be employed to reduce delays in identification and intervention [69,78]. For example, delays in interventions associated with bottlenecks in the formal diagnostic process may be reduced by changing the initial emphasis to therapeutic interventions based on function [69,78]. Comprehensive and coordinated care for children with special needs requires ongoing input and participation for primary care and specialty providers, as well as advocacy for systemic improvements [79].

Referral processes, whether for further evaluation or for treatment services may sometimes be complex, with reliance on data that often must be curated within the medical home. Without frequent verification of accuracy of information about community providers, errors may add challenges to this process for families. Families may have had to follow-up with a primary care clinician multiple times before receiving further evaluation. While a coherent team approach is often touted as optimal, this is rare in many communities and alternate team-building may be needed to ensure that families receive relevant information. It is important for medical specialists and intervention providers to be in reciprocal contact with primary care clinicians in the medical home to facilitate bidirectional flow of information about the child's diagnosis, recommended services, and eligibility status.

Once a primary care clinician sends a referral request for further evaluation or therapeutic intervention services, families may experience a variety of outcomes, ranging from minimal to optimal contact, evaluation, and/or interventions. In some cases, although a primary care clinician or other provider may recommend services or other therapeutic interventions, the burden may fall to the family to initiate contact and follow through with securing the services. For many families, the service and evaluation systems are complex and difficult to navigate. Specialty providers may therefore be connected with parents who are in various stages of a process to determine appropriate diagnoses and interventions. Further complicating the known challenge of inadequate numbers of professionals in many areas is the complex system of financing that is often burdensome for families to navigate. Many relevant services may be provided through public systems without charge to families, though accessing such services is sometimes challenging. Medical insurance coverage for developmental concerns varies among plans, and in-network providers may be sparse in some cases. Because of the complexities and state and local variations, it is critical that both primary care and specialty providers understand the systems that serve children and families in order to facilitate optimal care.

Both specialty and primary care clinicians should be familiar with the eligibility guidelines for programs to which children are referred. Children with developmental delays and disabilities and some with risk for such are often eligible by federal law through Part C of IDEA for early intervention services through a state early intervention program [22,23]. However, since specific eligibility rules vary by state and may change over time, it is imperative that clinicians have current and correct information to give appropriate anticipatory guidance to families. Clinicians should counsel parents on specific steps to seek early intervention services as well as alternate plans if there are problems in accessing them. When needed, consideration should be given to provision of specific services by alternate means, such as a medical order for therapy for a child who does not meet eligibility criteria for a public early intervention program.

Quality improvement initiatives have found persistent difficulty with referral processes, particularly since these often involve complex systems out of the control of the clinician [80–82]. However, such systematic appraisals of gaps also lead to improvements in autism and developmental screening and referral [37,64,82]. Specialty clinician support to the primary care clinician, including feedback on referral outcomes and recommendations for specific services and continued monitoring, not only helps the child to receive appropriate direct services, but helps the primary care clinician to improve surveillance and screening processes overall.

Other considerations in screening, identification, and management

Special populations

Universal screening is recommended for all children but select children will benefit from additional surveillance and screening due to additional associated risks for developmental disability based on presence of other identified medical conditions or environmental or social circumstances. Many of the additional risk conditions are described elsewhere in this text (Section III, Clinical disorders), including many risk factors that should be elicited in the developmental history (Chapter: The Developmental History). Evidence on risk factors is vast and select conditions that warrant additional surveillance and screening are described below, though this brief review is not intended to be exhaustive [83]. (See Table 6.5 for examples for special populations with increased developmental risk.)

A model for developmental surveillance, screening and evaluation for children with health conditions placing them at higher risk for developmental delays and disabilities has been created for those children with complex congenital heart disease [16]. In this model, children are risk stratified into those without significant additional risk and those with such risk. For those with risks similar to the general population, the usual system of surveillance and screening used in the general population is recommended. Any medical condition or socioeconomic situation that affords additional risk for developmental delay

TABLE 6.5 Examples of developmental surveillance, screening, and evaluation for special populations with increased developmental risk.

Setting and developmental practice	Condition with increased risk for developmental delay or disability

Primary care: additional developmental surveillance and screening recommended	• Complex congenital heart disease
	• High-risk neonatal conditions:
	  ◦ Preterm birth
	  ◦ Other neonatal medical conditions (infection, congenital anomalies, neurologic insults, and others)
	  ◦ Neonatal risk with increased socioeconomic or environmental risk
	  ◦ Intrauterine risks (medication, illicit drugs, infections, environmental toxins, maternal health and the uterine environment, and others)
	• Treatments for critical illnesses (e.g., need for extracorporeal membrane oxygenation [ECMO])
	• Traumatic brain injury
	• Post-natal or childhood meningitis
	• Treatment for certain childhood cancers (e.g., Leukemia)
	• Chronic health conditions, particularly those with known genetic origins
	• Environmental toxins (e.g., lead toxicity)
	• Family history of developmental disability (especially a sibling)
	• Adverse childhood experiences, social and environmental considerations (including poverty and systemic racism)

Specialty care: developmental evaluation due to additional risk status	• Complex congenital heart disease • High-risk neonatal conditions (as above) • Sickle cell disease • those already identified with specific developmental delays, disability, and/or behavioral conditions

or disability should be considered for additional surveillance and screening beyond that typically recommended universally for all children. Primary care clinicians may recognize this additional risk in association with select conditions or may benefit from the specialist's highlighting of the need for additional surveillance and screening where applicable. Communication between the primary and specialty providers about these goals benefits the family and enhances collaboration among clinicians. Those children at higher risk than the general population should have vigilant surveillance by their primary care and specialty care clinicians, with condition-specific ages and frequencies for comprehensive developmental evaluation based upon specific known risks and can bypass the usual screening guidelines. For example, children with a known risk for motor disability such those with a high-risk neonatal condition should have formal developmental evaluation in infancy and early childhood as indicated. Children at high risk for disorders affecting learning and attention, such as sickle cell disease, should have developmental evaluation in the late preschool or early elementary education period.

Neonatal risk

Children born preterm and those with other neonatal medical conditions and complications may have increased risk for developmental disability. Risk registries or systematic monitoring is available for some of these conditions and varies by location, though more common risk factors are more likely to be monitored. Some state early developmental intervention programs employ programs of surveillance and treatment for affected children. Graduates of neonatal intensive care units also are often monitored in dedicated clinics (some with attention to research on outcomes) and/or are referred for developmental surveillance to a variety of specialists. It is well known that preterm birth; as well as commonly associated medical conditions such as systemic or central nervous system infections [84–86], congenital anomalies (particularly those likely to affect major and/or multiple organ systems) [16,84,87], and direct neurologic insults (such as ventricular hemorrhage or stroke) [88] pose risks for significant developmental delay and/or disability. More significant degrees of prematurity are associated with higher degree of risk [89,90]. Though etiologic mechanisms are not established, there is a demonstrated higher risk for ASD in children who were born preterm [91]. Even when the medical aspects of preterm birth are considered, there is likely to be an additional overlay of socioeconomic or environmental risk [92].

In-utero risks may include exposure to medication, illicit drugs, infections, and environmental toxins among others [93]. Maternal health and the uterine environment may affect pregnancy outcomes, with some risk factors more completely identified than others and some newer associations not yet fully understood [94–98]. Certain treatments for critical illness, while life-saving, may be associated with increased risk for developmental concerns (e.g., need for extracorporeal membrane oxygenation [ECMO] is a risk factor for adverse neurodevelopmental outcomes) [99,100], while others are felt to be protective (e.g., the association of antenatal steroids with reduced risk of adverse neurodevelopmental outcomes in preterm infants) [101,102].

Other chronic health conditions and risk factors

Traumatic brain injury, post-natal or childhood meningitis, and treatment for certain childhood cancers may directly affect the central nervous system. Children with other chronic health conditions, particularly those with known genetic origins such as complex congenital heart disease also have known risk for significant developmental disabilities and related disorders [16]. Environmental toxins and increased lead toxicity may also affect the central nervous system and adversely affect developmental outcomes [93]. In addition, those already identified with specific developmental delays, disability, or behavioral conditions benefit from closer surveillance and screening of other developmental domains to ensure that they are not experiencing more global deficits [20]. Known sensory impairment (vision, hearing) may also contribute to deficits in progress in related domains [18].

Family history of developmental disability, especially having a sibling with identified developmental delay or disability, confers potential risk regardless of whether the etiology has been identified [103]. There is increasing evidence of potential deleterious effects of social and environmental considerations, though there is also recognition that strengths and protective factors may reduce the burden and afford more optimal outcomes [104,105]. Adverse childhood experiences, poverty, and systemic racism have been increasingly shown to be associated with poorer developmental outcomes [106–108].

The chronic medical, social, and environmental factors may be cumulative and convey additional risk. The interplay among genetic, medical, environmental and social effects is difficult to disentangle [32,33,109]. Whether a child has a single or multiple risk factors, the primary and specialty clinicians should be prepared to discuss appropriate monitoring and anticipatory guidance and intervention for these children when applicable. Caution is encouraged to ensure that developmental and behavioral aspects are fully considered by the child's treating team, particularly for those children with multiple medical conditions or disabilities.

Payment

Specific discussion on billing and coding for developmental screening is outside the scope of this chapter, though are important components to the primary care clinician in a medical home since appropriate payment for completion of these critical tests and the professional judgement involved in interpretation and management facilitates the next step of early identification. Developmental surveillance is part of the medical visit and not separately paid, though can be counted when billing by time. Developmental screening is a billable service with payment for use of a standardized screening test with related medical record documentation. Billing codes are available for components of a more comprehensive developmental evaluation. Comprehensive guides are available from trusted sources such as the AAP for additional guidance [43,110].

Summary

As laid out through the course of this chapter, a comprehensive system of care aimed at the early identification and the early medical and developmental intervention of the child with neurodevelopmental disability is rooted and begins in the primary care medical home with the primary care clinician performing developmental surveillance and screening on every child during the periodic and usual health supervision visits. (See Fig. 6.1 for the developmental screening process.) This developmental screening process flows from the primary care clinician to the neurodevelopmental specialty care physician and the developmental specialists (medical and related early childhood professionals) for comprehensive medical and developmental evaluation and diagnosis of the child with significant delays in development. These professionals are also charged with initiating referral and treatment of the child for indicated medical and developmental interventions and therapies and beginning a system of lifelong management of the developmental disabilities and related disorders affecting the child. Needs of the children include ongoing primary and specialty medical management, special educational services and supports, developmental interventions and therapies, and chronic condition management commonly in place for those with other chronic health problems. Systems aimed at community integration of the individual as well as individual and family support are indicated and require ongoing assistance and linkage from the medical and other professionals treating the children and adults with developmental disabilities throughout their lifespan.

References

[1] Zablotsky B, Black LI, Maenner MJ, Schieve LA, Danielson ML, Bitsko RH, et al. Prevalence and trends of developmental disabilities among children in the United States: 2009–2017. Pediatrics. 2019;144(4).

[2] Maenner MJ, Warren Z, Williams AR, Amoakohene E, Bakian AV, Bilder DA, et al. Prevalence and Characteristics of Autism Spectrum Disorder Among Children Aged 8 Years — Autism and Developmental Disabilities Monitoring Network, 11 Sites, United States, 2020.. MMWR Surveill Summ 2023;72(SS-2):1–14. Available from: http://dx.doi.org/10.15585/mmwr.ss7202a1.

[3] Patrick ME, Shaw KA, Dietz PM, Baio J, Yeargin-Allsopp M, Bilder DA, et al. Prevalence of intellectual disability among eight-year-old children from selected communities in the United States, 2014. Disabil Health J 2021;14(2):101023.

[4] Durkin MS, Benedict RE, Christensen D, Dubois LA, Fitzgerald RT, Kirby RS, et al. Prevalence of cerebral palsy among 8-year-old children in 2010 and preliminary evidence of trends in its relationship to low birthweight. Paediatr Perinat Epidemiol 2016;30(5):496–510.

[5] Fujiura GT, Li H, Magana S. Health services use and costs for americans with intellectual and developmental disabilities: a national analysis. Intellect Dev Disabil 2018;56(2):101–18.

[6] Kim J, Stevens P, Carbone PS, Jones KB. Health care use and spending of pediatric patients with an intellectual or developmental disability. Med Care 2020;58(5):468–73.

[7] Lindgren S, Lauer E, Momany E, Cope T, Royer J, Cogan L, et al. Disability, hospital care, and cost: utilization of emergency and inpatient care by a cohort of children with intellectual and developmental disabilities. J Pediatr 2021;229:259–66.

[8] Leigh JP, Du J. Brief Report: forecasting the economic burden of autism in 2015 and 2025 in the United States. J Autism Dev Disord 2015;45 (12):4135–9.

[9] Zuvekas SH, Grosse SD, Lavelle TA, Maenner MJ, Dietz P, Ji X. Healthcare costs of pediatric autism spectrum disorder in the United States, 2003–2015. J Autism Dev Disord 2021;51(8):2950–8.

[10] Kancherla V, Amendah DD, Grosse SD, Yeargin-Allsopp M, Van Naarden Braun K. Medical expenditures attributable to cerebral palsy and intellectual disability among medicaid-enrolled children. Res Dev Disabil 2012;33(3):832–40.

[11] Lipkin PH, Macias MM. Promoting optimal development: identifying infants and young children with developmental disorders through developmental surveillance and screening. Pediatrics 2020;145(1).

[12] Children HRaSAACoHDiNa. Recommended uniform screening panel. 2018. Available from: https://www.hrsa.gov/advisory-committees/heritable-disorders/rusp/index.html.

[13] Centers for Disease C, Prevention. CDC grand rounds: newborn screening and improved outcomes. Mmwr. 2012;61(21):390–3.

[14] Almannai M, Marom R, Sutton VR. Newborn screening: a review of history, recent advancements, and future perspectives in the era of next generation sequencing. Curr Opin Pediatr 2016;28(6):694–9.

[15] Hearing TJCoI, Position statement: principles and guidelines for early hearing detection and intervention programs. J Early Hear Detect Interv. 2019;4(2):1–44.

[16] Marino BS, Lipkin PH, Newburger JW, Peacock G, Gerdes M, Gaynor JW, et al. Neurodevelopmental outcomes in children with congenital heart disease: evaluation and management: a scientific statement from the American Heart Association. Circulation. 2012;126(9):1143–72.

[17] Hyman SL, Levy SE, Myers SM. Council on children with disabilities SOD, behavioral P. Identification, evaluation, and management of children with autism spectrum disorder. Pediatrics. 2020;145(1).

[18] Donahue SP, Baker CN, Committee on P, Ambulatory Medicine AAP, Section on Ophthalmology AAP, American Association of Certified O, et al. Procedures for the evaluation of the visual system by pediatricians. Pediatrics. 2016;137(1).

[19] Loh AR, Chiang MF. Pediatric vision screening. Pediatr Rev 2018;39(5):225–34.

[20] Noritz GH, Murphy NA. Neuromotor screening expert P. Motor delays: early identification and evaluation. Pediatrics 2013;131(6):e2016–27.

[21] Weitzman C, Wegner L, Section on D, Behavioral P, Committee on Psychosocial Aspects of C, Family H, et al. Promoting optimal development: screening for behavioral and emotional problems. Pediatrics 2015;135(2):384–95.

[22] The individuals with disabilities education act. Sect. 1400–1415 (2004).

[23] Lipkin PH, Okamoto J, Council on Children with D, Council on School H. The individuals with disabilities education act (IDEA) for children with special educational needs. Pediatrics 2015;136(6):e1650–62.

[24] Landa RJ. Efficacy of early interventions for infants and young children with, and at risk for, autism spectrum disorders. Int Rev Psychiatry 2018;30(1):25–39.

[25] Sacrey LA, Bennett JA, Zwaigenbaum L. Early infant development and intervention for autism spectrum disorder. J Child Neurol 2015;30 (14):1921–9.

[26] Sallows GO, Graupner TD. Intensive behavioral treatment for children with autism: four-year outcome and predictors. Am J Ment Retard 2005;110(6):417–38.

[27] Khurana S, Kane AE, Brown SE, Tarver T, Dusing SC. Effect of neonatal therapy on the motor, cognitive, and behavioral development of infants born preterm: a systematic review. Dev Med Child Neurol 2020;62(6):684–92.

[28] Blauw-Hospers CH, Hadders-Algra M. A systematic review of the effects of early intervention on motor development. Dev Med Child Neurol 2005;47(6):421–32.

[29] McCoy DC, Yoshikawa H, Ziol-Guest KM, Duncan GJ, Schindler HS, Magnuson K, et al. Impacts of early childhood education on medium- and long-term educational outcomes. Educ Res 2017;46(8):474–87.

[30] Magnuson KA, Kelchen R, Duncan GJ, Schindler HS, Shager H, Yoshikawa H. Do the effects of early childhood education programs differ by gender? A meta-analysis. Early Child Res Q 2016;36:521–36.

[31] Schindler HS, Kholoptseva J, Oh SS, Yoshikawa H, Duncan GJ, Magnuson KA, et al. Maximizing the potential of early childhood education to prevent externalizing behavior problems: a meta-analysis. J Sch Psychol 2015;53(3):243–63.
[32] Boyce WT, Levitt P, Martinez FD, McEwen BS, Shonkoff JP. Genes, environments, and time: the biology of adversity and resilience. Pediatrics. 2021;147(2).
[33] Shonkoff JP, Slopen N, Williams DR. Early childhood adversity, toxic stress, and the impacts of racism on the foundations of health. Annu Rev Public Health 2021;42:115–34.
[34] Thomas S, Conway KM, Fapo O, Street N, Mathews KD, Mann JR, et al.Muscular Dystrophy Surveillance, Tracking, and Research Network (MD STARnet) Time to diagnosis of Duchenne muscular dystrophy remains unchanged: findings from the muscular dystrophy surveillance, tracking, and research network, 2000–2015. Muscle Nerve 2022. Available from: https://doi.org/10.1002/mus.27532. Epub ahead of print. PMID: 35312090.
[35] Centers for Disease C, Prevention. Delayed diagnosis of fragile X syndrome—United States, 1990–1999. Mmwr. 2002;51(33):740–2.
[36] Lipkin P, Macias M. Developmental milestones for developmental surveillance at preventive care visits. In: Hagan JFSJ, Duncan PM, editors. Bright futures: guidelines for health supervision of infants, children, and adolescents. 4th ed. Elk Grove Village, IL, USA: American Academy of Pediatrics; 2017.
[37] Barger B, Rice C, Wolf R, Roach A. Better together: developmental screening and monitoring best identify children who need early intervention. Disabil Health J 2018;11(3):420–6.
[38] Barger B, Rice C, Roach A. Socioemotional developmental surveillance in young children: monitoring and screening best identify young children that require mental health treatment. Child Adolesc Ment Health 2018;23(3):206–13.
[39] Morelli DL, Pati S, Butler A, Blum NJ, Gerdes M, Pinto-Martin J, et al. Challenges to implementation of developmental screening in urban primary care: a mixed methods study. BMC Pediatr 2014;14:16.
[40] Guevara JP, Gerdes M, Localio R, Huang YV, Pinto-Martin J, Minkovitz CS, et al. Effectiveness of developmental screening in an urban setting. Pediatrics. 2013;131(1):30–7.
[41] Committee on Practice and Ambulatory Medicine, Workgroup BFPS. Recommendations for preventive pediatric health care. Pediatrics. 2021;147(3) e2020049776.
[42] Katkin JP, Kressly SJ, Edwards AR, Perrin JM, Kraft CA, Richerson JE, et al. Guiding principles for team-based pediatric care. Pediatrics. 2017;140(2).
[43] Pediatrics A.A. American Academy of Pediatrics Screening Technical Assistance & Resource Center, Screening Tools 2017 Available from: https://screeningtime.org/star-center/#/screening-tools.
[44] Pediatrics A.A. Developmental Surveillance and Screening Resources for Pediatricians 2021 Available from: https://www.aap.org/en/patient-care/developmental-surveillance-and-screening-patient-care/developmental-surveillance-resources-for-pediatricians/.
[45] Squires J, Potter L, Bricker D, editors. The ASQ user's guide. 3rd ed. Baltimore, MD: Paul H Brookes Publishing; 2009.
[46] Glascoe FP, Marks KP, Poon JK, Macias MM. Identifying and addressing developmental-behavioral problems: a practical guide for medical and non-medical professionals, trainees, researchers and advocates. Nolensville, TN: PEDS test; 2013.
[47] Sheldrick RC, Perrin EC. Evidence-based milestones for surveillance of cognitive, language, and motor development. Acad Pediatr 2013;13 (6):577–86.
[48] Robins DL, Casagrande K, Barton M, Chen CM, Dumont-Mathieu T, Fein D. Validation of the modified checklist for Autism in toddlers, revised with follow-up (M-CHAT-R/F). Pediatrics 2014;133(1):37–45.
[49] Rutter M, Bailey A, Lord C. The social communication questionnaire (SCQ) manual. Los Angeles: Western Psychological Services; 2003.
[50] Wetherby AM, Prizant BM. Communication and symbolic behavior scales: developmental profile. Baltimore, MD: Paul H Brookes Publishing; 2002.
[51] Squires J, Bricker D. Ages & stages questionnaires®, third edition (ASQ®-3): a parent-completed child monitoring system. Baltimore: Paul H. Brookes Publishing Co., Inc.; 2009.
[52] Glascoe FP. PhD. Collaborating with parents. 2nd ed. Nolensville, Tennessee: PEDSTest.com, LLC. 2013. Available from: https://pedstest.com/research/supporting/.
[53] Guthrie W, Wallis K, Bennett A, Brooks E, Dudley J, Gerdes M, et al. Accuracy of autism screening in a large pediatric network. Pediatrics. 2019;144(4).
[54] Carbone PS, Campbell K, Wilkes J, Stoddard GJ, Huynh K, Young PC, et al. Primary care autism screening and later autism diagnosis. Pediatrics. 2020;146(2).
[55] Sheldrick RC, Garfinkel D. Is a positive developmental-behavioral screening score sufficient to justify referral? A review of evidence and theory. Acad Pediatr. 2017;17(5):64–470.
[56] Zubler JM, Wiggins L, Macias MM, Whitaker TM, Shaw JS, Squires JK, et al. Evidence-informed milestones for developmental surveillance tools. Pediatrics. 2022;149(3) e2021052138.
[57] Sheldrick RC, Schlichting LE, Berger B, Clyne A, Ni P, Perrin EC, et al. Establishing new norms for developmental milestones. Pediatrics. 2019;144(6).
[58] Sheldrick RC, Marakovitz S, Garfinkel D, Carter AS, Perrin EC. Comparative accuracy of developmental screening questionnaires. JAMA Pediatr 2020;174(4):366–74.
[59] Fenikile TS, Ellerbeck K, Filippi MK, Daley CM. Barriers to autism screening in family medicine practice: a qualitative study. Prim Health Care Res Dev 2015;16(4):356–66.
[60] Monteiro SA, Dempsey J, Berry LN, Voigt RG, Goin-Kochel RP. Screening and referral practices for autism spectrum disorder in primary pediatric care. Pediatrics 2019;144(4).

[61] Maenner MJ, Shaw KA, Baio J, EdS, Washington A, Patrick M, et al. Prevalence of autism spectrum disorder among children aged 8 years – autism and developmental disabilities monitoring network, 11 sites, United States, 2016. MMWR Surveill Summ 2020;69(4):1–12.
[62] Zablotsky B, Black LI. Prevalence of children aged 3–17 years with developmental disabilities, by urbanicity: United States, 2015–2018. Natl Health Stat Rep 2020;139:1–7.
[63] Zuckerman KE, Mattox KM, Sinche BK, Blaschke GS, Bethell C. Racial, ethnic, and language disparities in early childhood developmental/behavioral evaluations: a narrative review. Clin Pediatr (Phila) 2014;53(7):619–31.
[64] Zuckerman KE, Chavez AE, Wilson L, Unger K, Reuland C, Ramsey K, et al. Improving autism and developmental screening and referral in US primary care practices serving Latinos. Autism. 2021;25(1):288–99.
[65] Shaw KA, McArthur D, Hughes MM, Bakian AV, Lee LC, Pettygrove S, et al. Progress and Disparities in early identification of autism spectrum disorder: autism and developmental disabilities monitoring network, 2002–2016. J Am Acad Child Adolesc Psychiatry 2021;.
[66] Wallis KE, Davis Rivera LB, Guthrie W, Bennett AE, Mandell DS, Miller JS. Provider responses to positive developmental screening: disparities in referral practices? J Dev Behav Pediatr 2021;42(1):23–31.
[67] Centers for Disease Control and Prevention. NCoBDaDD. Learn the signs. Act early. 2021 Available from: https://www.cdc.gov/ncbddd/actearly/index.html.
[68] Pediatrics A.A. Developmental surveillance and screening resources for pediatricians: American Academy of Pediatrics; 2021 [updated 10/06/202110/28/2021]. Available from: https://www.aap.org/en/patient-care/developmental-surveillance-and-screening-patient-care/developmental-surveillance-resources-for-pediatricians/.
[69] Gordon-Lipkin E, Foster J, Peacock G. Whittling down the wait time: exploring models to minimize the delay from initial concern to diagnosis and treatment of autism spectrum disorder. Pediatr Clin North Am 2016;63(5):851–9.
[70] Adams RC, Tapia C. Council on children with D. Early intervention, IDEA Part C services, and the medical home: collaboration for best practice and best outcomes. Pediatrics. 2013;132(4):e1073–88.
[71] Association CsH. Pediatric Workforce Shortages Persist2018 [updated January 19, 2018]. Available from: https://www.childrenshospitals.org/Issues-and-Advocacy/Graduate-Medical-Education/Fact-Sheets/2018/Pediatric-Workforce-Shortages-Persist.
[72] Bridgemohan C, Bauer NS, Nielsen BA, DeBattista A, Ruch-Ross HS, Paul LB, et al. A Workforce survey on developmental-behavioral pediatrics. Pediatrics. 2018;141(3).
[73] Basco WT, Rimsza ME. Committee on pediatric W, American Academy of P. Pediatrician workforce policy statement. Pediatrics. 2013;132(2):390–7.
[74] Roizen NJ, Ruch-Ross HS, Bauer NS, Nielsen BA, DeBattista A, Paul LB, et al. Developmental-behavioral pediatrics 13 years after the first board certification: evolving subspecialty. J Dev Behav Pediatr 2021;42(2):83–90.
[75] Roberts G, Efron D, Price A, Hiscock H, Wake M. The time and practice challenges of developmental-behavioral pediatrics: an Australian national study. J Dev Behav Pediatr 2011;32(5):368–74.
[76] American Academy of Pediatrics Committee on Pediatric W. Pediatrician workforce statement. Pediatrics. 2005;116(1):263–9.
[77] Jimenez ME, Martinez Alcaraz E, Williams J, Strom BL. Access to developmental pediatrics evaluations for at-risk children. J Dev Behav Pediatr 2017;38(3):228–32.
[78] Dworkin PH, Kelly DP, Schonfeld DJ. Strategic principles to inform workforce development in developmental and behavioral pediatrics. J Dev Behav Pediatr 2020;41(4):316–18.
[79] Council on Children with D, Medical Home Implementation Project Advisory C. Patient- and family-centered care coordination: a framework for integrating care for children and youth across multiple systems. Pediatrics. 2014;133(5):e1451–60.
[80] Bright MA, Zubler J, Boothby C, Whitaker TM. Improving developmental screening, discussion, and referral in pediatric practice. Clin Pediatr (Phila) 2019;58(9):941–8.
[81] Flower KB, Massie S, Janies K, Bassewitz JB, Coker TR, Gillespie RJ, et al. Increasing early childhood screening in primary care through a quality improvement collaborative. Pediatrics. 2020;146(3).
[82] Campbell K, Carbone PS, Liu D, Stipelman CH. Improving autism screening and referrals with electronic support and evaluations in primary care. Pediatrics. 2021;147(3).
[83] Bear LM. Early identification of infants at risk for developmental disabilities. Pediatr Clin North Am 2004;51(3):685–701.
[84] Pappas A, Shankaran S, Hansen NI, Bell EF, Stoll BJ, Laptook AR, et al. Outcome of extremely preterm infants (<1000 g) with congenital heart defects from the National Institute of Child Health and Human Development Neonatal Research Network. Pediatr Cardiol 2012;33(8):1415–26.
[85] Gordon-Lipkin E, Hoon A, Pardo CA. Prenatal cytomegalovirus, rubella, and Zika virus infections associated with developmental disabilities: past, present, and future. Dev Med Child Neurol 2021;63(2):135–43.
[86] Mukhopadhyay S, Puopolo KM, Hansen NI, Lorch SA, DeMauro SB, Greenberg RG, et al. Neurodevelopmental outcomes following neonatal late-onset sepsis and blood culture-negative conditions. Arch Dis Child Fetal Neonatal Ed 2021;106(5):467–73.
[87] Brosig CL, Bear L, Allen S, Hoffmann RG, Pan A, Frommelt M, et al. Preschool Neurodevelopmental outcomes in children with congenital heart disease. J Pediatr 2017;183:80–6 e1.
[88] Mukerji A, Shah V, Shah PS. Periventricular/Intraventricular hemorrhage and neurodevelopmental outcomes: a meta-analysis. Pediatrics 2015;136(6):1132–43.
[89] Luu TM, Rehman Mian MO, Nuyt AM. Long-term impact of preterm birth: neurodevelopmental and physical health outcomes. Clin Perinatol 2017;44(2):305–14.

[90] Brumbaugh JE, Hansen NI, Bell EF, Sridhar A, Carlo WA, Hintz SR, et al. Outcomes of extremely preterm infants with birth weight less than 400 g. JAMA Pediatr 2019;173(5):434–45.

[91] Crump C, Sundquist J, Sundquist K. Preterm or early term birth and risk of autism. Pediatrics 2021;148(3).

[92] Hoffman L, Bann C, Higgins R, Vohr B. Eunice Kennedy Shriver National Institute of Child H, Human Development Neonatal Research N. Developmental outcomes of extremely preterm infants born to adolescent mothers. Pediatrics 2015;135(6):1082–92.

[93] Council On Environmental H. Prevention of childhood lead toxicity. Pediatrics 2016;138(1).

[94] Matias SL, Pearl M, Lyall K, Croen LA, Kral TVE, Fallin D, et al. Maternal prepregnancy weight and gestational weight gain in association with autism and developmental disorders in offspring. Obes (Silver Spring) 2021;29(9):1554–64.

[95] Cordero C, Schieve LA, Croen LA, Engel SM, Maria Siega-Riz A, Herring AH, et al. Neonatal jaundice in association with autism spectrum disorder and developmental disorder. J Perinatol 2020;40(2):219–25.

[96] Cordero C, Windham GC, Schieve LA, Fallin MD, Croen LA, Siega-Riz AM, et al. Maternal diabetes and hypertensive disorders in association with autism spectrum disorder. Autism Res 2019;12(6):967–75.

[97] Croen LA, Qian Y, Ashwood P, Zerbo O, Schendel D, Pinto-Martin J, et al. Infection and fever in pregnancy and autism spectrum disorders: findings from the study to explore early development. Autism Res 2019;12(10):1551–61.

[98] McGuinn LA, Windham GC, Messer LC, Di Q, Schwartz J, Croen LA, et al. Air pollution, neighborhood deprivation, and autism spectrum disorder in the Study to Explore Early Development. Env Epidemiol 2019;3(5).

[99] Sadhwani A, Cheng H, Stopp C, Rollins CK, Jolley MA, Dunbar-Masterson C, et al. Early neurodevelopmental outcomes in children supported with ECMO for cardiac indications. Pediatr Cardiol 2019;40(5):1072–83.

[100] Kim F, Bernbaum J, Connelly J, Gerdes M, Hedrick HL, Hoffman C, et al. Survival and developmental outcomes of neonates treated with extracorporeal membrane oxygenation: a 10-year single-center experience. J Pediatr 2021;229:134–40 e3.

[101] Chawla S, Natarajan G, Shankaran S, Pappas A, Stoll BJ, Carlo WA, et al. Association of neurodevelopmental outcomes and neonatal morbidities of extremely premature infants with differential exposure to antenatal steroids. JAMA Pediatr 2016;170(12):1164–72.

[102] Sotiriadis A, Tsiami A, Papatheodorou S, Baschat AA, Sarafidis K, Makrydimas G. Neurodevelopmental outcome after a single course of antenatal steroids in children born preterm: a systematic review and meta-analysis. Obstet Gynecol 2015;125(6):1385–96.

[103] Hansen SN, Schendel DE, Francis RW, Windham GC, Bresnahan M, Levine SZ, et al. Recurrence risk of autism in siblings and cousins: a multinational, population-based study. J Am Acad Child Adolesc Psychiatry 2019;58(9):866–75.

[104] Duffee JH, Mendelsohn AL, Kuo AA, Legano LA, Earls MF, Council On Community P, et al. Early childhood home visiting. Pediatrics. 2017;140(3).

[105] Glascoe FP, Leew S. Parenting behaviors, perceptions, and psychosocial risk: impacts on young children's development. Pediatrics 2010;125(2):313–19.

[106] Garner AS, Shonkoff JP, Committee on Psychosocial Aspects of C, Family H, Committee on Early Childhood A, Dependent C, et al. Early childhood adversity, toxic stress, and the role of the pediatrician: translating developmental science into lifelong health. Pediatrics 2012;129(1):e224–31.

[107] Council On Community P. Poverty and child health in the United States. Pediatrics 2016;137(4).

[108] Council On Early C, Council On School H. The pediatrician's role in optimizing school readiness. Pediatrics. 2016;138(3).

[109] Shonkoff JP, Boyce WT, Levitt P, Martinez FD, McEwen B. Leveraging the biology of adversity and resilience to transform pediatric practice. Pediatrics 2021;147(2).

[110] Pediatrics A.A. Standardized Screening/Testing Coding Fact Sheet for Primary Care Pediatricians: Developmental/Emotional/Behavioral (January 2020) 2020 [updated January 1, 2020]. Available from: https://downloads.aap.org/AAP/PDF/coding_factsheet_developmentalscreening testingandEmotionalBehvioraassessment.pdf.

Chapter 7

Developmental history

Pasquale J. Accardo
Department of Pediatrics, Virginia Commonwealth University School of Medicine, Richmond, VA, United States

Introduction

In medicine, traditionally, 90% of the diagnosis was to be achieved by the history, with the physical examination contributing perhaps 10% to the process. Laboratory tests and imaging studies should only verify the conclusions already reached by history and physical. In fact, however, many areas of medicine increasingly rely on laboratory and imaging studies to the exclusion of a detailed history. Child development remains one of the few medical specialties where history remains paramount for diagnosis.

The developmental history can be obtained either orally or in writing with open-ended or objective questionnaires. Such questionnaires present a useful way of collecting a large amount of information, some of which can be quantified in a standardized manner. But developmental and behavioral signs and symptoms are complex enough to limit the utility of many questionnaires. For example, the Vanderbilt ADHD Diagnostic Rating Scale uses a Likert scoring system that contains a significant subjective component. When different observers rate each behavior, they will tend to produce the same overall pattern of strengths and weaknesses, but some observers will be "easy markers" while others will grade the same behavior more severely. The resulting pattern may be more revealing than the composite score. When available, an open-ended description of a child's strengths and weaknesses in the classroom or other structured setting can be more helpful than such a questionnaire. The results from developmental questionnaires should always be reviewed in a face-to-face parent interview. When exploring some developmental milestones and behaviors, it can be helpful to ask several questions and use descriptors not usually found in standardized questionnaires, symptom checklists, and behavior rating scales. This can act as a validity check on the responses by seeking the same or similar information via different pathways.

A question like "Does your child make eye contact?" can be answered "yes" or "no." Alternatively, the response can be graded on a Likert scale from "always," "often," "sometimes," "rarely," to "never." However, when asking such a question, one should note any parental hesitation, uncertainty, or confusion as indicated by the tone of voice, pause, or facial expression. If there is more than one informant (parent, stepparent, grandparent, aunt, child-care worker, or therapist), discrepancies in the responses will need to be further explored. Discrepancies may relate to the accuracy of the individual observer or to differences in the observational environment.

The value of a developmental history will be limited by the reliability of the informant and the skill of the examiner. Parents are typically the most reliable reporters of their children's current developmental function and behavior; mothers are usually more reliable than fathers. Developmental information may be recorded in baby books or bolstered by linking milestones to significant family events, such as holidays and birthdays. Reliability is not related to parent education, socio-economic status, or the number of previous children.

While the history is being elicited, several mild developmental differences or behavior problems might be reported. It would be appropriate to mention to the parent if these are of little significance or even recommend simple approaches to address them. In this way, they will be removed from the final formulation. An example would be the fact that the child never crawled. Parents can obsess over such insignificant variations in development and behavior and become anxious if they are not specifically addressed.

Typically, the younger child can remain in the room while the history is being obtained. This allows prolonged observation of free play behavior as well as parent-child interaction. Parents should be allowed the option of having the child out of the room to discuss what they consider sensitive issues.

Capute and Accardo's Neurodevelopmental Disabilities in Infancy and Childhood. DOI: https://doi.org/10.1016/B978-0-12-824060-1.00020-1

Parents may jump ahead of the interviewer with an assumed diagnosis or etiology in mind; they will then answer questions in a manner to lead the interviewer in the "right" direction. The answer to the question of whether a child exhibits a specific behavior will instead be an answer to the question of whether the parent thinks that behavior is of any importance to the diagnosis. The parent may feel that it is none of the interviewer's business to know the answer to certain questions since they do not see the information as relating to the problem at hand. Behaviors that might be considered to reflect poor parenting will frequently be denied. It is not uncommon for a parent to fail to report a milestone or behavior that the child is noted to exhibit during the evaluation. Since parents generally tend to at least slightly overestimate their child's abilities, any obvious underreporting should be carefully investigated.

The use of an interpreter can have an impact on the reliability of the results. There are two components relevant to the use of an interpreter: the difference in language and the impact of the culture that the language reflects. Mere awareness of cultural differences cannot automatically surmount certain cultural barriers. Cultural differences in parental expectations are more noticeable with personal and social milestones. Parents and other historians usually do not lie, but they do tend to answer the question they think the interviewer asked (or should have asked) rather than the question that actually was asked.

The developmental history has six goals: (1) obtain the information needed to formulate a developmental diagnosis; (2) allow the parent to feel that the interviewer has learned enough about the whole child to make such a diagnosis; (3) establish a foundation for prognosis and the monitoring of progress; (4) identify coexisting conditions; (5) provide clues to etiology; and (6) obtain the information necessary for the development of a comprehensive management program. Just because a diagnosis can be made quickly does not mean that it should be made quickly. A detailed history that is not rushed can help to establish rapport with the family. The interviewer's interest in the details of the child's history will be interpreted as an interest in the child and the family.

The developmental history typically starts out with the identification of the parent's main concerns, their onset, duration, and intensity; in medicine, this is referred to as the chief complaint. The reported onset will identify when the parents first became concerned; the interviewer will then need to search for earlier signs of problems in this area. The sudden onset of a developmental problem with no earlier detected difficulties in the area should raise concern for a subacute or acute disorder (neurodevelopmental disorders are by definition chronic disorders). Previous evaluations, videos, and diagnostic evaluations can provide documentation of earlier functions. They are of particular importance when considering the possibility of regression. Even when such records are unavailable, the fact that the child previously may have been found ineligible for specific services is revealing. Parental agreement with or concerns about such prior assessments need to be explored. Generally, the earlier the first parental concerns, the more severe the developmental problem. A chronological review of the child's medical and developmental history serves a dual purpose: (1) depending on their timing and severity, acute and chronic medical conditions can have an impact on a child's development, and (2) the time spent on such a history gives the final diagnostic summation a greater force. A comprehensive developmental evaluation not only focuses on development but also includes a detailed review of the child's medical, family, social, educational, treatment, and behavioral histories. The history can also briefly review the child's daily routine, including feeding, play, school placement or other out-of-home experiences, and finally sleep.

Birth history and early infancy

The pregnancy, labor, and delivery histories do not contribute directly to the developmental diagnosis but may provide information that contributes to an etiologic diagnosis. Most children with developmental disorders do not have abnormal birth histories, and most children with abnormal birth histories do not have developmental disorders. Among the areas to review are:

- Parental ages—older parental ages can be associated with point mutations (father) or chromosomal anomalies (mother)
- Pregnancy history—fetal wastage is an important marker for genetic as well as obstetric risk
- Multiple gestations
- Fetal activity—fetal hyperactivity is associated with later attention deficit hyperactivity disorder, and fetal hypoactivity is associated with intellectual disability and several specific genetic syndromes
- Exposure to drugs, alcohol, and tobacco during the pregnancy
- Physical or emotional trauma during the pregnancy
- Duration of the pregnancy—prematurity is associated with a wide variety of negative developmental outcomes. Prematurity may be the result of an abnormal central nervous system and not the cause. Typically, the more severe

the prematurity, the more severe the neurodevelopmental disability. Late preterm delivery has been associated with an increase in developmental morbidity.
- Atypical presentation such as breech
- Low birth weight not due to prematurity (intrauterine growth retardation) may indicate genetic disorders or other negative influences on fetal growth
- Type of delivery—emergency Caesarian section often reflects the presence of adverse influences on the developing fetus
- Length of neonatal intensive care unit NICU stay including duration of respiratory support and the occurrence of seizures or intracranial bleeds
- Assistive reproductive technology is associated with at least a mild increase for a variety of neurodevelopmental diagnoses

The infant's temperament can be a marker for future neurobehavioral difficulties. Typical "3-month colic" lasts 3 months. Prolonged colic or severe irritability throughout the first year of life may reflect a damaged—and therefore irritable—central nervous system. An infant who is unusually quiet during the first year of life may have significantly delayed cognition. The infant who is described as "aloof" and almost seems to dislike being cuddled may develop autism spectrum disorder (although most infants with this presentation will not be autistic).

When investigating the impact of various perinatal events, it is important to differentiate the mother's reported subjective discomfort from objective factors having a potentially negative impact on the baby.

Medical history

A medical history of hospitalizations, surgery, head injuries, and unusual behavioral responses to medications in early childhood can be revealing. Children who have had pressure equalization (PE) tubes placed will have had follow-up formal audiological assessments other than their newborn hearing screening. Many children who have tonsillectomy and adenoidectomy for obstructive sleep apnea will then stop snoring and exhibit more restful sleep; their daytime behavior can often be noted to improve along with their attention span. It seems obvious that children who have had neurosurgical procedures or significant head trauma would be at higher risk for later neurodevelopmental problems, but the linkage is not always appreciated. Exposure to prolonged anesthesia prior to age 3 is a problematic etiology for later neurodevelopmental problems. Paradoxical responses to anesthetics and sedatives are frequent in children with attention deficit hyperactivity disorder. For example, a baby who is increasingly hyperactive on diphenhydramine will often later exhibit an attentional disorder that has a positive behavioral response to stimulant medication.

A review of systems can help identify visual and hearing problems that might cause or intensify developmental delays. Growth disturbances of short stature, obesity, or organ malformations can be components of genetic syndromes. Heart and lung disease can produce chronic hypoxia, which contributes to motor weakness, hypotonia, and speech and motor delays. Allergies can produce chronic fatigue that may seasonally exacerbate developmental and behavioral problems. Dietary history may indicate a nutritional deficiency. Drooling, choking, or vomiting may be associated with dysphagia. Diurnal and nocturnal enuresis are not uncommon in developmental disorders. Constipation (sometimes with bleeding) may be related to restrictive diets (as in autism spectrum disorders), insufficient toilet time (as in attentional deficits), or sphincter dyssynergy (as in cerebral palsy). Seizures and tics can be secondary to more generalized brain problems that also involve neurodevelopmental disorders.

Feeding and sleep

Feeding problems can be severe enough to impact growth, and this in turn can affect the energy available for the achievement of developmental milestones. Some degree of malnutrition can impact muscle strength and delay developmental milestones that are dependent on motor abilities. Feeding problems related to oral motor coordination difficulties can follow prolonged intubation in the NICU. Similarly, these oral motor delays can contribute to delayed speech milestones. Very selective menus can be part of a sensory processing disorder that may respond to feeding therapy or occupational therapy; such restricted diets tend to be most severe in children with autism spectrum disorder.

Sleep disturbances (insomnia) can have a global impact on a child's behavior. They rarely cause developmental and behavioral problems, but they often exacerbate their severity. The two most common sleep disturbances are late-onset and sleep interruption. Sleep-onset disorders typically respond to sleep hygiene measures. Melatonin, Clonidine, and Gabapentin may be needed to supplement this approach. Sleep interruption disorders can be addressed with a scheduled waking pattern. Chronic snoring raises the suspicion of obstructive sleep apnea and should be investigated further.

Children with attentional disorders are often restless sleepers who toss and turn throughout the night; they can also have a history of falling out of bed. This nocturnal sign of hyperactivity is often neglected. Children with neurodevelopmental disorders have a higher prevalence of nocturnal enuresis. While this may be a minor or side issue, when the child is no longer wetting every night, the use of medication to address the enuresis should be considered for its positive emotional impact.

Streams of development

The developmental history focuses on the streams of development to delineate a pattern of milestone achievement. Delays in each stream can be compared and may exhibit a dissociation pattern that supports a specific developmental diagnosis. Such a "patternological diagnosis" identifies the predominant disorder on the spectrum of developmental disorders. A careful review of the signs that do not automatically belong to the spectrum diagnosis can identify a second developmental diagnosis or items on the continuum of developmental disorders that, although not diagnostic of another disorder, nevertheless need to be addressed for the effectiveness of the treatment program.

The spectrum of developmental disorders focuses on what distinguishes each developmental diagnosis from the others. The continuum of developmental disorders emphasizes the signs and symptoms that each disorder shares with the others. Delay occurs within a stream; dissociation occurs between streams. Deviance is an older term that referred to atypical behaviors that did not occupy a place in a particular stream of development. While these atypical behaviors did not seem to occur in the sequence of normal development, it is now recognized that they do appear, although early on and briefly. For example, echoing is observed in typically developing children for several weeks late in the second year of life.

Motor

Gross motor achievement is the primary developmental focus for children under 9 months. While motor delay is often very concerning to the parent, such milestones have the widest range of attainment. For example, while most children walk around 12–13 months, the age for independent walking can range from 7 to 24 months. Delays within this range are significant only when accompanied by other delays or neuromotor findings such as tone abnormalities, exaggerated deep tendon reflexes, and persistent primitive reflexes.

Sitting and walking should receive more emphasis than rolling over and crawling. Sitting stages range from the early development of head control through to propping reactions and are the best indicators for readiness to walk. Walking requires the ability to take a step while holding on (cruising), the absence of the plantar grasp, and the presence of an anterior forward parachute. The child who has these prerequisites is ready to walk and will do so eventually.

In the absence of other neuromotor signs, toe walking (equinus or toe-heel gait) for longer than 3 months after the onset of independent ambulation (or with a later onset) can be associated with language delay and autism spectrum disorder. If the toe walking contributes to frequent falls, a referral for management may be in order. Running typically has its onset around 3 months after the onset of walking; the child who is reported to have started running at the same time as walking may have an attentional disorder. Such children will not wait for the full development of the anterior parachute reaction needed to protect the head when an ambulating child falls forward. Motor milestones such as swimming and riding a bicycle are useful when they are achieved at the expected age level, but their failure often depends on extrinsic factors such as exposure, expectations, and teaching. When inquiring about motor clumsiness, it is important to have the parent distinguish clumsiness related to motor inability from clumsiness related to the child's rushing and not paying attention.

Language

Language is the best predictor of developmental dysfunction. Delayed attainment of language milestones is associated with intellectual disability, language disorders, autism spectrum disorders, and hearing loss. It is also common in cerebral palsy because of the associated cognitive dysfunction and may be the precursor of language-based learning disorders.

For the infant, early vocalizations such as cooing and babbling are important. The child who is noted to have been very quiet in infancy may present later with an intellectual disability or autism spectrum disorder. Perhaps the biggest red flag in language development is language regression, the loss of words previously used to communicate. This would typically occur between 18 and 30 months of age and is often a presenting sign for autism spectrum disorder. Other causes need to be ruled out. When the language regression is accompanied by a loss of motor skills, the underlying problem may be a neurodegenerative disease. Significant emotional trauma, that occurs with child abuse, can also precipitate language loss and the onset of mutism.

Language milestone attainment is usually obtained from the parent history because infants and toddlers are unlikely to speak to strangers. The Clinical Linguistic and Auditory Milestones Scale, which is part of the Capute Scales, measures expressive and receptive language development for children under 3 years of age. One point that can be especially helpful while obtaining this language history is the fact that some language milestones are neurodevelopmentally linked to each other. Thus a 50-word vocabulary is always associated with the use of two-word phrases: a child with less than 50 words should not be reported to be using two-word phrases, and a child with a vocabulary above 50 words should be using two-word phrases. A disconnect between these two milestones should lead the examiner to question the accuracy of the informant. The most common reporting error in early language involves determining the actual size of the child's vocabulary (their dictionary). Many people have difficulty with numerical estimates. When there is doubt, the parent can be asked to record every single different word the child uses over a several-week period and report back with that vocabulary list. Alternatively, they can be provided with a checklist of the 200 most commonly used early words.

An isolated expressive language delay is the single most common language disorder in children. When there is a discrepancy between expressive and receptive language milestone achievement, it is almost always expressive that is more delayed. With severe language delays, both expressive and receptive language will be equally delayed. When expressive language is reported to be ahead of receptive language, this is probably an artifact secondary to some degree of echolalia being misconstrued as expressive language. Echolalia is a component of several language delay patterns and is not pathognomonic for autism spectrum disorder. Programmed speech and "single word" phrases (such as "thank you," "all done," "bye-bye," and "ice cream") may inflate vocabulary estimates. Conversely, parents may underestimate vocabulary size by not counting mispronounced words or word approximations (such as "ba" for bottle). If used consistently by the child, word approximations and incorrect pronunciations should be counted when estimating vocabulary size.

For children 2–6 years of age, a useful rule of thumb is that the child should be using sentences with a word length one greater than their age in years. For example, 2-year-olds use 2–3-word sentences, and 3-year-old children use 3–4-word sentences. This rule falls apart when children start to string together clauses with "and" and other conjunctions. With older children, the history of their early language development is an especially important marker for later problems. The history should address whether a child can tell a coherent story with a beginning, middle, and end about an incident and whether there is enough detail for an uninvolved person to understand what is being related. Humor is a marker for the development of higher language abilities. "Can the child tell a joke?" is easily ascertained. Five-year-old children recognize the need for novelty in jokes, but the punch line usually does not relate to the set-up. Whether a child understands a joke provides additional information. Impulsive children may not wait for the punch line. Some children laugh because others are laughing but do not actually understand the joke.

Apart from the description of their current language usage, it is always helpful to obtain a language sample by engaging the child in conversation. Ascertaining the child's interests from the parent beforehand can provide a friendly opening topic. Discussing school performance with the child should usually be reserved for the end of the conversation. As a rule, at least 80% of a child's utterances should be readily intelligible by age 4 years; a lower degree of intelligibility should raise concern for a phonological disorder. Even when resolved by school age, articulation difficulties (phonological disorders) can be early markers for later reading (phonological processing) disorders. Note should be made of the sequence leading from prolonged intubation in the NICU through oral motor/feeding difficulties to pronunciation difficulties and later reading decoding disorders.

While conversing with a child, a number of different areas can be explored. The neurodevelopmental pediatrician should be able to estimate the age-appropriateness of a child's vocabulary and syntax. Verbally bright children should be readily distinguishable from children with language disorders and cognitive delays. There are also separate and distinct language markers for other neurodevelopmental conditions, such as autism spectrum disorder and attentional disorders.

Nonverbal problem-solving

Nonverbal problem-solving builds on earlier visual, fine motor, and cognitive skills. The best way to assess the nonverbal problem-solving skills of infants via history is to ask about the way they most commonly interact (play) with objects. From birth to 3 months, the child visually tracks objects of interest. From 3 to 6 months, the infant swats at objects, tries to grab them, and transfers them between hands. From 6 to 9 months, the child brings the grasped object to the mouth; from 9 to 12 months, the baby prefers to bang objects either together or against surfaces to make noise.

By 1 year of age, the child acquires a voluntary casting release (throwing). A typical 1-year-old can "clear the table" of a highchair after eating or discard all the toys in a play pen and then wail for a mother to replace them—and then repeat the performance. At 15 months, the child starts to stack one block on top of another, and at 18 months, one puzzle piece can be placed. At 16 months, the child imitates a scribble, does it spontaneously at 18 months, and imitates a stroke (a line) by age 2 years. The infantile developmental sequence of look, grab, mouth, bang, and release is one that most parents can readily describe. Interactions with toys in older children depend on exposure, and it is better to switch over to paper and pencil tasks that parents can report and that can be readily confirmed in the office. Most parents can describe whether their child can draw a line, a circle, a face, or a person.

Self-help

While self-help milestones such as feeding, toileting, and dressing skills depend on the previous mastery of certain motor and cognitive skills, their appearance will exhibit significant variability depending on the demands that the family exerts on the child. With different expectations, there can be wide variation in the achievement of toileting and self-feeding. Some skills, such as swimming and bicycle riding, are never achieved. Many self-help skills are acquired through social imitative learning; that is, the child is not specifically taught by adults but masters the task by spontaneously imitating adult behavior. Thus most children are toilet trained around 3 years of age with minimal to no effort by the parent. If desired, typically developing children could be toilet trained around 15 months after they have achieved the three prerequisites for toilet training—walking, pulling pants down, and the ability to consistently follow one-step commands. It is the absence of the last that is the cause of most delays in toilet training.

Socialization

Babies develop increasing social interactions with their mothers throughout the first several years of life. Both separation and stranger anxiety come in by 8 months of age. These reactions might be muted in families where the child is constantly exposed to many adults in the home. Up until age 2, most children are content to play alone. Between 2 and 3 years of age, children will increasingly observe and even imitate the play of other children, with no interest in joining or sharing with them. They will at best engage in parallel play, sitting next to the other child and playing—sometimes with the same toys—but not interacting with that child; their interest is more with the toys than with the child. From 3 years on, play becomes associated with more complex interactions among children. The increasingly lower ages at which children are being exposed to electronic games, devices, and media can be expected to further delay the development of social interaction and play.

With older children, it is important to ask, "With what aged children do they play best?" Children tend to prefer to play with children who are at similar developmental levels. Asking about the type of activities they undertake, outdoor large-muscle activities versus more quiet activities that require sustained attention, may be useful. How well they play with peers may be related to cognition, language, or motor abilities. Finally, determining whether the child is invited to social activities such as birthday parties and sleepovers indicates social acceptance by peers.

Attention

The challenging behaviors associated with attentional disorders are often explained away as part of the primary diagnosis. Doing so can divert the physician from focusing on an important component of the treatment regimen.

"Attention" can be a confusing term. Some parents will describe a child as "attention-seeking" (continually demanding attention from others); this is not what is meant by an attention deficit. Attention spans can vary across settings and situations. A child who cannot sit still in class for a lesson may be able to watch an action movie for 2 hours or play a video game even longer. Interest level can override a short attention span, but not for repeated and prolonged periods, and especially not for nonfavored tasks such as schoolwork. This is why it is always helpful to obtain teacher observations: in the classroom situation, the child's behavior will automatically be comparable to peers at a similar developmental level. Any child or adult with a strong level of interest in a specific topic or activity (especially video games and electronic media) will be able to lengthen an otherwise short attention span. While a short attention span is the hallmark of attention deficit disorder, some children (especially girls described as daydreamers) will exhibit overfocused behaviors.

Attentional deficits can also impact speech performance. While the following behaviors can occur with language disorders, they are often quite prominent in children with attentional disorders. Such children talk so fast that it can impact their speech intelligibility. They often exhibit a "motor mouth" ("chatterbox") behavior that parents will describe as "The child's mouth runs ahead of the child's brain" or "The child's brain is faster than the child's mouth." The presence of word finding problems, word substitutions or circumlocutions, and word opposites (saying the reverse of what was intended, e.g., yes/no, up/down) can also be part of the presentation. Children with attentional disorders may need to have instructions repeated and often respond with "Huh" or "What did you say?" After a while they, may develop the habit of just tuning out such requests and "acting deaf." They are often restricted to responding to at most single-step commands and cannot handle multiple-step commands at the expected age level. Children who have had repeat audiological assessments to rule out "hearing loss" often have "listening" problems instead; many parents can easily discriminate between the two on questioning.

Hyperactivity is another relative term that needs descriptive support. Apart from excessive gross motor activity, hyperactivity also includes fidgetiness, body parts in constant motion (such as bouncing in one's seat and foot tapping), and extraneous noises (such as humming and tongue clicking). Such children will often be out of or under their seats in class, and with their excess motor activity, they will have trouble keeping their hands to themselves so that they do not respect personal space. This may irritate or antagonize other children and contribute to socialization difficulties.

The impulsivity component can be especially important with younger preschool children, where this behavior can represent a threat to their safety. Such children will elope, run out into traffic, and sometimes jump from dangerous heights. When they run away, some children will constantly look back to make certain that the parent is in their line of sight or following them; others will simply take off and never look back. This latter behavior represents a more serious safety risk and suggests the use of a locator device.

The ability to understand cause-and-effect relationships may be unimpaired, but it is almost never brought to bear in the heat of the moment. Impulsivity also contributes to other problematic behaviors, such as lying and stealing. The child is aware of the rules and the consequences but simply never applies them in time. The child ignores recognized and accepted consequences—but not deliberately so. Even while being observed to misbehave, a child will deny culpability with the shortest answer to get the parent out of their face. They answer the perceived question, "Did you do that deliberately and intend to get punished for it?" with a truthful "No." When children with attentional disorders are asked why they misbehave, a typical response is "I don't know." Such a lack of even attempting an explanation or excuse is almost pathognomonic for an attentional deficit.

Children with attentional disorders often have problems with sleep onset and with marked overactivity while asleep. Hyperactivity persisting into bedtime can delay sleep onset. Not necessarily counterintuitively, the continued influence of stimulant medication can help with sleep onset by settling the child down. Thrashing around during sleep—to the point of falling out of bed—indicates sleep that is less restful and may leave the child tired and less attentive the next day. Sleep disorders do not cause attentional problems, but they certainly may exacerbate them.

Since the behaviors that comprise attention deficit hyperactivity disorder are well known from various diagnostic checklists, it can be helpful for the neurodevelopmental interview to focus on the impact of such behaviors on specific areas. In the area of school performance, the hallmark sign of an attentional disorder is inconsistency. One day the child knows the material; the next day it seems forgotten, only to "return" later. A child may study for a test, completely master the material before bedtime, and the next day at school either ace the quiz or fail it. If the child fails the test, later that evening the child can nevertheless be shown to know the material. Such inconsistency is often misinterpreted by teachers and parents as stupidity, laziness, or even deliberate, willful misbehavior. When a child has a learning disability in a certain area, the child consistently does poorly in, for example, reading, while doing well in mathematics and other subjects less dependent on reading. The child with an attentional disorder will do well in reading and mathematics one day and poorly in both the next day. The same inconsistency can be noted in daily classroom deportment. Parents can often predict what kind of day the child will have in school when the child first gets up in the morning.

Attentional disorders tend to impact mathematics more than reading because the relative arbitrariness of pure numbers doesn't provide additional clues to correct calculations (until the introduction of word problems). For obvious reasons, attention affects reading decoding less than reading comprehension. Finally, ADHD impacts handwriting because of the tendency of the child to rush; handwriting that improves when the child is encouraged to slow down is not usually due to dysgraphia and will often dramatically improve when the child is placed on stimulant medication. Children with attentional deficits who have more difficulty with reading decoding than with mathematics probably have an associated specific learning disability impacting reading (possibly dyslexia).

Learning

School is the major developmental task for school-aged children. The differential diagnosis for academic underachievement is long and detailed. The history should address the following:

- Onset of school problems
- Are the problems academic, behavioral, or both?
- Are academic problems in a single subject area or generalized?
- Are the problems confined to school?
- Have there been any previous evaluations of the problems?
- Is the child receiving any interventions (tutoring, 504 Plan, Individualized educational plan IEP)?
- How effective are the interventions?
- Are there issues with homework?

While attention and learning problems both tend to improve over time, worsening or even regression can be observed at key turning points in the child's school career. Whereas the first four grades (kindergarten through third) focus on the mastery of phonics and elementary calculations, the fourth grade emphasizes the rapid application of these rote skills. Children with attentional or learning problems who have mastered such skills with some difficulty and can apply them only laboriously may start to fail—to be overloaded—when expected to use them to manage longer assignments. This is what has been referred to as high-output failure. The transition to middle school presents the child with the added burden of multiple teachers and diverse expectations—all while moving from one class to another. Learning problems can also be characterized by the grade level at which the child started to experience trouble. There is often a difference between the grade when trouble is first experienced and the grade when the child actually begins to fail. Some learning disabilities affect all academic achievement areas, but many processing disorders impact specific subjects more severely. Thus a child with learning disabilities can present with greater deficits in reading (dyslexia), mathematics (dyscalculia), or written language (dysgraphia) than in other subjects. Some processing problems impact the way a child acquires information (input) or the way the child demonstrates their learning (output). With regard to the latter, some children will perform better with short-answer tests (true/false or fill-in-the blank), while others will do better with essay tests or on oral exams. Apart from formal psychoeducational testing, parent and teacher descriptions of a child's strengths and weaknesses can often help identify the specific learning pattern. The parent who helps the child with their homework can often provide a detailed description of the types of errors and confusions that characterize the child's learning style.

A significant challenge in distinguishing academic failure or underachievement due to attentional or learning problems is the frequent coexistence of these two disorders in the same child. In neurodevelopmental disabilities, diagnoses are more likely to be incomplete than incorrect. Some children diagnosed with a learning disability are given specific educational interventions and show improvement, but then plateau and no longer seem to be responding to treatment. They may then be found to have an associated attentional disorder. Similarly, children who have a significant attentional disorder and show dramatic improvement on stimulant medication but later plateau in their academics may be found to have associated learning disabilities. The severity of the one disability at first overshadows the presence of the associated disorder (diagnostic overshadowing).

Another important fact to be gleaned from daycare, preschool, and school reports is whether behavioral problems reported at school are also being observed in the home setting and, if so, to what degree of severity. Attentional behaviors typically present more of a problem in a classroom than in a home setting; this relates to the more structured demands that are a routine part of the educational system. Detention, suspension, and expulsions can be markers for serious behavior disorders.

The quality of school multidisciplinary evaluations used to address children's learning and behavior problems can vary significantly across state, county, and school district lines. It is most helpful to have reviewed such records prior to obtaining the developmental history. Familiarity with the quality of reports from different schools can also be helpful.

Emotional history

Children with neurodevelopmental disorders may exhibit the full continuum of behavioral dysfunction, so the presence of emotional problems needs to be explored. Symptoms of anxiety and depression may not meet criteria for a separate diagnosis but often accompany neurodevelopmental disorders, especially when developmental treatment modalities are insufficiently effective. When children begin to realize they are different, they become sad about not meeting expectations. With time, they can develop a poor self-image and become increasingly anxious about situations in which they can expect to do poorly. A "failure set" (the expectation of failure and the worthlessness of trying) is easier to avoid when a child is first noted to be struggling in school than to wait for it to develop fully.

Tantrums and violent outbursts are a frequent cause of parental concern in children, both with and without neurodevelopmental disorders. Such behaviors often first become prominent around age two (the "terrible twos") when the child starts to use the word "no." The older psychiatric literature highlighted the onset of "no" as the beginning of individuation, the first emergence of the independent ego, and the child's separation from the mother. Tantrums should be described in terms of frequency, duration, and components such as screaming, falling out, biting, hitting, punching, kicking, pinching, scratching, head banging, throwing things, and property destruction. Identifying precipitating factors is the single most important step in their treatment. Rather than being overly concerned with punishment (the response, such as spanking), the focus should be on averting the episodes by identifying and then avoiding or removing the triggers (the stimulus). Children can be described as explosive, aggressive, destructive, and oppositional. Their moods of sadness and anxiety may suggest other psychiatric diagnoses.

Does the child like going to school? Did the child ever like going to school? Does the child talk down about themselves either academically ("I'm dumb or stupid") or socially ("I don't have any friends"; "Nobody likes me")? When interacting with peers, is this child a leader, a follower, gullible, or a loner? Does the child participate in any group activities such as clubs, scouts, or team sports? Children with attentional deficits will have trouble following the rules in games, while children with autism spectrum disorder may be so overly rule-bound and nitpicky as to irritate their teammates. Children with attentional deficits perform best as pitchers or catchers when playing baseball; having constant interaction with the ball helps keep them focused on the game. Does the child get into trouble with peers by being too bossy, by not respecting personal space, or by being too touch-feely? Is the child socially inappropriate in public, for example, by blurting out embarrassing statements? Does the child exhibit common sense, empathy, and remorse? If the child does show remorse, what is their response when asked why they misbehaved? As noted above, "I don't know" is the classic attentional deficit response. Recurrent somatic complaints such as headaches, stomach aches, and leg pains can indicate emotional stress rather than physical disease.

Certain behavior patterns are so closely related to the neurodevelopmental disorder that they have been deemed behavioral phenotypes. Tics and obsessive-compulsive disorder are seen in Tourette's syndrome. Overeating and skin picking are prominent in Prader-Willi syndrome. Self-hugging and pulling out fingernails are part of Smith-Magenis syndrome.

When confronting concerns about certain behaviors, history should focus on trying to distinguish whether a child is willfully not doing a task or whether the child is unable to perform the required action. Does the child seem angry, or are they frustrated? Is the oppositionality generalized, or is it related to specific demands? Has the child been making self-deprecating remarks? If the child has been thought to do the task before, can the parent describe the conditions under which the behavior was accomplished? Are behaviors such as tics, humming, or vocalizing under the child's control?

Atypical behaviors

Atypical behaviors are characteristic of autism spectrum disorder. Some relate to the social communication aspect of the disorder, while others reflect more on repetitive behaviors and fixations. In the past, it was thought that such "deviant" behaviors did not occur in typically developing children—they were not to be found in the sequential listing of normal milestones, so they could not be delayed or dissociated. Closer observation has identified many of these behaviors—at least briefly—in typically developing children. They are only more intense and prolonged in children on the autism spectrum.

In the area of language, echoing occurs in typically developing children between 18 and 24 months, where it is characterized by the briefness of its duration. More prolonged echoing passing into the third year of life is seen in some language disorders but is most prominent and prolonged in autism spectrum disorder. The echolalia can be immediate or delayed. When children repeat entire movie scripts, it may be done in a monotone manner or with different voices and accompanying sound effects. (A related behavior is echopraxia, in which the child mimics another's bodily posture and movements like a mime.) The echoing contributes, in turn, to the tendency of these children to reverse pronouns and call themselves "you." Alternatively, they will refer to themselves by name or in the third person ("he/she"). Children with autism spectrum disorder often talk to themselves or to no one in particular; they converse more "at" you than "with" you. Eye contact tends to be poor. Rather than the hyperkinetic saccades observed in children with attentional disorders, children with autism spectrum disorder almost seem to avoid eye contact deliberately; they often "peer" out of the corner of their eyes.

Part of the problem with eye contact may relate to the fact that they have difficulty "reading faces"—interpreting and understanding the emotional expressions of others. This impacts social reciprocity, and usually their default interpretation of a neutral facial expression is one of anger and hostility. Thus it can be helpful to greet a child with autism spectrum disorder with an unmistakable, if not exaggerated, smile. Some children with attentional deficits have a similar negative default for neutral facial expressions, and this can contribute to an almost paranoid affect.

Inappropriate laughter and a high threshold for pain may reflect the child's not knowing how to express sensations of pleasure or pain. Children with language disorders may have difficulty relating to peers, but they often attempt to do so. Children with autism spectrum disorder, on the other hand, may be hyperverbal and still show little to no interest in interacting with their peers. They will engage at best in parallel play and be more interested in the toys of other children than in the children themselves. Taking another child's toys and engaging in large motor activities such as running around on the playground should not be credited as appropriate social interaction. Their lack of social interest can also be reflected in their climbing over people as if they were objects of furniture. These children prefer to be alone and often find the very presence of others annoying. When talking a lot, children with attentional deficits may talk tangentially (with a constant change of subject), while children with autism spectrum disorder may overly focus on a topic and constantly repeat the same "lecture" on a subject of their interest, like "little professors." Parents can thus become experts in the details of video games that they have never played.

At the severe end of the spectrum, children with autism will treat all people like "blocks of wood." Children with less severe presentations may exhibit social interaction with family members, other adults, and other teenagers. In the mildest presentations, one may see play with age-peers. It can be difficult to assess such social interaction in a toddler who is home with a mother and does not have any structured peer group exposure, such as play groups, daycare, or preschool.

Repetitive behaviors, obsessions, and a restrictive range of interests are seen in autism spectrum disorder and severe intellectual disability. Some of the more prominent repetitive behaviors include rocking, going in circles, spinning wheels (and anything else that can spin), watching fans (or a cycling washer or dryer), flapping, looking at one's hands (especially against a light), perusing details of objects, flipping light switches on and off, opening and closing doors and drawers (or sometimes just obsessively closing all doors), turning faucets on and off, and repeatedly flushing toilets, as well as rituals relating to a daily schedule that must be completed in minute detail. The flapping can be replaced by other repetitive hand and finger movements. Parents will often report that the child does not seem to know how to play with toys. They either prefer household objects or do "odd things" with toys. Sometimes pretend play that may appear to be imaginative is actually the rote repetition of a scene from a television show or a movie.

Children with autism spectrum disorder, intellectual disability, or anxiety prefer routine and can become upset with the slightest change in schedule or other detail in their daily lives. Those with autism spectrum disorder or intellectual disability often have a fascination with sorting and lining things up that extends beyond the normal age for such behavior. The principle for the sorting is not always readily discernible, but the children will become very upset if the order is changed. The preference for lining up toys such as cars is reflected in a fascination with trains, especially Thomas the Tank Engine (and friends). In the absence of social communication disorder, intellectual disability, or anxiety disorder, such behaviors might qualify for a diagnosis of obsessive-compulsive disorder.

Sensory issues

Sensory issues frequently accompany neurodevelopmental disorders. While such sensory symptoms are reported most commonly in children with autism spectrum disorder, they can be associated with other neurodevelopmental disorders such as attention deficit hyperactivity disorder, speech language disorder, and learning disabilities.

One of the most commonly involved senses is taste. Food selectivity with a very restrictive diet is often present in children with autism spectrum disorder, with their preferences often being related to texture, odor, and even the position of the food on the plate. Sometimes the selectivity extends further to the use of specific dishes and eating utensils. Many young children with autism spectrum disorder will restrict their diet to under 10 food items. This selectivity can be refined to the point that they will eat French fries or chicken nuggets from only one fast food chain and not from the others. While many parents report texture as a significant factor in food selectivity, more often the preferred foods occupy a place on a color spectrum going from yellow to brown. Surprisingly, despite the severe dietary restrictions, many of these children maintain growth at or above their expected percentiles, so malnutrition is rarely a concern. Some children need to smell all their food and may be so sensitive to odors that they go around smelling everything. Therefore food selectivity may reflect several senses in addition to taste. The restrictive menu probably contributes to the constipation reported in these children. In addition to being a sensory issue, the food restriction may also be categorized as a repetitive behavior.

Some children with neurodevelopmental disorders can be overly sensitive to loud noises, high-pitched sounds, babies' crying, and machine noises such as the vacuum cleaner, fireworks, sirens, hair and hand dryers, and lawnmowers. The same children who are very sensitive to "loud" noises will often be oblivious to their own high volume ("outdoor voice"). Some children with autism spectrum disorder will have a particular affinity for or severe dislike of a specific type of music (e.g., Indian, rap, country, etc.).

Children who intensely dislike being touched (tactile defensiveness) may not respect the personal space of others and hug or touch them inappropriately. Some tactile sensitivities will relate to food texture (such as hard, soft, mushy, and crunchy). In addition to food, texture can also affect which clothes are preferred. Soft (such as silk or linen) is often preferred to rough (such as jeans, wool, and anything "itchy" or 'scratchy'); tags and seams (in socks) can be unusually annoying.

With regard to the vestibular sense, some children with autism spectrum disorder like the sensation of circular movement: they may spin themselves until they become dizzy, run in repetitive circles, spin wheels on toy cars and bikes, and sit and stare at fans. Such sensory issues may act as triggers for negative behaviors.

Family history

The family medical history should focus less on the diseases of the elderly and more on conditions found in relatives that are children or that were childhood diagnoses. The history should cover three generations—siblings, parents, aunts and uncles, cousins, and grandparents. Information should be sought about early motor delays, language disorders (including those that resolved), intellectual disability, autism spectrum disorder, academic underachievement, parental level of education, and childhood attentional deficits. The history may need to use older, non-politically correct terminology to avoid missing important information. Psychiatric conditions, including depression, bipolar disorder, obsessive-compulsive disorder, and anxiety disorder, should be noted because of their coexistence with neurodevelopmental disorders and the possibility of diagnostic overshadowing.

The history should be charted against a timeline that correlates particular family stresses with the onset, worsening, or improvement of behavioral symptoms. Adverse childhood experiences such as physical, emotional, and sexual abuse, domestic violence, substance abuse, parental imprisonment, homelessness, and the suicide and death of close family members can have a significant short- and long-term impact on a child's developmental course. Other stresses such as divorce, relocation, unemployment, bankruptcy, and other legal and financial problems typically impact all children's behavior and development, but more so on those with neurodevelopmental disorders. Occasionally, a child's limitations may protect them from the impact of such family stressors. However, even infants can be extremely sensitive "emotional barometers" to such unspoken stresses. They may have no comprehension of the details of the family's problems, but they sense the significant adult's emotional turmoil and can exhibit a confused and sometimes exaggerated response.

Telemedicine

The advent of and increasing reliance on telemedicine pose significant questions for history-taking. An in-person interview allows the examiner to rely on four of the five senses—sight, sound, touch, and odor. A telephone interview restricts the examiner to a single sense—sound—while videoconferencing allows two—sight and sound. Rather than restricting the child and family to the waiting area and examining room, videoconferencing can allow a brief glance at the home setting, a fragment of a house call. All the information obtained from eye contact is distorted by such electronic media. A child with autism who would not make eye contact in person may be fascinated by staring at the image on the electronic screen. Telemedicine presents a situation analogous to, or inverse of, distance learning. Some students thrive with distance learning, while others have difficulties with it. The pluses and minuses of telemedicine do not make it better or worse; they are just different. Its effective application will require refinements to the interviewing technique.

Conclusion

The developmental history is the first step in the process of establishing a diagnosis and comprehensive management plan for the child who presents with concerns about developmental dysfunction. The differential diagnosis is large, so the history must be comprehensive. Previous medical issues, developmental attainments, behavioral, social, and academic function, and family and social information are all necessary components of a comprehensive developmental history.

The accuracy of a parent's description of their child's current functional level can be verified by direct observation and testing. The more important task of interpreting their description of the child's developmental course over time is more challenging. Internal consistency and increased detail are helpful characteristics of an accurate report. Family photographs, videos, and even baby books can assist in identifying significant achievements and transitions.

A developmental diagnosis attempts to place the child somewhere along a neurodevelopmental spectrum that goes from cerebral palsy to attentional deficits. With a reliable informant, it should be possible to diagnose the child with one or more of the six major disability categories. The history obtained by the physician can then be correlated with the histories obtained by the other members of the interdisciplinary team.

Attempts to assess the accuracy of the developmental history typically compare its results to quantitative measures of development or intelligence. The problem with using a developmental age or intelligence quotient as the outcome variable is that neither a developmental age nor an intelligence quotient is a developmental diagnosis; they are merely pieces of data that are used in formulating a diagnosis. Giving such data preferential status simply because it is quantitative does not assist in the diagnostic process.

The developmental history provides the information needed to formulate a developmental diagnosis, establishes a foundation for prognosis and monitoring of progress, identifies coexisting conditions, yields clues to etiology, and provides information about the child, family, and community necessary for the development of a comprehensive management program.

Chapter 8

The physical evaluation

Eboni Lance[1,2] and Bruce K. Shapiro[1,3]
[1]Department of Neurology and Developmental Medicine, Kennedy Krieger Institute, Baltimore, MD, United States, [2]Department of Neurology, Johns Hopkins University School of Medicine, Baltimore, MD, United States, [3]Department of Pediatrics, Johns Hopkins University School of Medicine, Baltimore, MD, United States

Past, present, and future

The physical examination remains a hallmark of medical practice. Physical diagnosis began with the discovery of percussion in 1760 by Auenbrugger, followed by the invention of the stethoscope in 1816 by Laennec. These developments led to the widespread implementation of physical diagnosis by the French School in the 1800s, led by Pierre Louis [1]. William Osler incorporated the physical examination to standard medical education at the Johns Hopkins University medical clinic in 1893 [1]. In 2004 the United States Medical Licensing Examination added Step 2, Clinical Skills, an evaluation of a physician's physical examination skills before they are permitted to practice medicine without supervision. For neurologists in particular, the physical examination is used as a tool to localize brain injury.

There are at least five justifications for performing a physical examination as part of the evaluation of neurodevelopmental dysfunction. Perhaps the most fundamental purpose of the physical examination is to ensure physiologic stability. Physiologic instability may arise from many causes. Undetected major organ dysfunction, silent aspiration, thyroid disease, malnutrition, epilepsy, drug toxicity, or a previous insult or injury can all impede a reliable diagnosis. Similarly, prognosis is not possible without physiologic stability because the child's presentation may change significantly once stability is achieved.

Equally important, the physical evaluation clarifies and confirms the hypotheses generated by the history. While the history establishes a differential diagnosis, the physical evaluation revises the initial considerations. For example, poor eye contact is commonly associated with autism spectrum disorder. However, poor eye contact may be seen in blindness or severe visual impairment, cerebral visual impairment, hyperactivity, social anxiety, avoidance, or profound intellectual disability. Motor delays in infancy may be related to cerebral palsy, spina bifida, muscular dystrophy, spinal muscular atrophy, nutritional deficiency, or a benign variant.

The physical evaluation contributes significantly to the etiologic evaluation. Establishing an etiology may provide valuable information to families, influence the management program, or offer potential interventions that may ameliorate or cure the disorder. While practice guidelines endorse genetic evaluations for global developmental delay, intellectual disability, and autism, the diagnostic yields are higher if there are associated dysmorphisms. Some recommendations, such as for brain magnetic resonance imaging (MRI), are based on the presence of physical examination findings, including abnormalities in head size, asymmetries, seizures, or epilepsy. Some etiologies, such as toxin exposure or nutritional disorders, may not be detected by genetic evaluations, neuroimaging, or newborn screening but may be noted on physical evaluation.

History and physical evaluation serve as the basis for a comprehensive, coordinated management program. Documenting the skills the child possesses and estimating the skills required for adult functioning help set the objectives for the management program. This provides a "road map" to track progress and guide families toward achievable outcomes. It also prioritizes therapeutic interventions, thereby, preventing families from becoming overburdened.

The physical evaluation serves as a baseline to quantify changes across time. This is particularly true when measuring growth (including head circumference), mental status, or motor function. The physical assessment may reveal acceleration of growth or development, steady progress, plateau, or degeneration/regression.

Capute and Accardo's Neurodevelopmental Disabilities in Infancy and Childhood. DOI: https://doi.org/10.1016/B978-0-12-824060-1.00027-4

Neurodevelopmental disorders are the functional expressions of static brain injury. While the expression of the disorders may change over time, the primary injury, lesion or defect remains stable. Progressive disorders, such as neurodegenerative diseases or unstable hydrocephalus, may be associated with functional impairments, and the appropriate diagnosis may be delayed until the progressive nature of the child's dysfunction is recognized.

Neurodevelopmental disorders are not limited to a single organ system and are all-encompassing. Consequently, there is no "standard" physical assessment. The physical assessment is guided by the presenting complaint, the child's age, the developmental level, and the history.

The physical assessment of children with neurodevelopmental dysfunction addresses the following areas: growth, age-appropriate general physical assessment with a focus on dysmorphisms (Table 8.1), neurologic evaluation, including mental status assessment and other evaluations relevant to the presenting problem. It also addresses considerations for the treatment and follow-up plans. (Table 8.2).

However, in the current technological era, is the physical examination still a necessary part of the evaluation of children with neurodevelopmental dysfunction? Telehealth can obtain a history and bypass the constraints on access imposed by geography and time. Brain MRI and ultrasound allow us to assess anatomical differences and measure cerebral metabolism directly. Next-generation genetic studies not only identify new syndromes but also expand the spectrum of classic syndromes. Despite these innovations, physical evaluation remains an efficient, cost-effective, and detailed means of quantifying structures and functions.

Certain physical examination findings still require in-person assessment for differential etiological or diagnostic considerations. While many vital signs and general physical exam assessments can be evaluated using smartphones or other electronic devices, the reliability of these measurements is not guaranteed and may be user-dependent. Virtual assessment of skin birthmarks is possible, but careful examination for a sacral dimple may be missed on telehealth. Quality of breathing (shallow vs deep, labored, accessory muscle use, etc.), cardiac murmurs, and pulse (regular vs irregular, bounding, etc.) are best assessed in person. A thorough dysmorphology examination, including measurements and comparison to relatives, can lead to an efficient differential diagnosis and targeted genetic testing.

In terms of the neurological examination, a thorough mental status examination can be done via telehealth, though the participant may be able to use visual cues to answer orientation questions. Many of the commonly tested cranial nerves can be assessed via telehealth including eye, face, and tongue movements. However, assessing the smell

TABLE 8.1 Areas of focus in infant, preschool, and school-age neurodevelopmental physical evaluation.

Infant	General physical examination	Birthmarks Growth parameters Cardiovascular status (pulse, heart rate, respirations)
	Mental status	Level of alertness Interaction with caregiver and provider
	Motor examination	Spontaneous movement of extremities Asymmetry in movement or tone
	Reflexes	Presence or absence of primitive and postural reflexes
Preschool	General physical examination	Nutrition and growth
	Mental status	Communication Ability to follow commands Interaction with caregiver and provider
	Motor examination	Coordinated movements and balance Demonstration of motor milestones Excessive activity level
School age	Mental status	Mood Understanding of complex language
	Motor examination	Strength testing Muscle tone and bulk Subtle neurological signs Coordinated movements Excessive activity level

TABLE 8.2 Physical examination areas of focus in monitoring visits.

ADHD	General physical examination	Nutrition and growth, cardiovascular examination, activity level
	Mental status	Affect, attention, following commands
	Motor examination	Subtle neurological signs, "motor on" behaviors (fidgeting, constant movement), coordinated movement difficulties, repetitive motor movements
Autism spectrum disorder	General physical examination	Dysmorphology examination, nutrition and growth, abdominal examination, activity level, cutaneous examination
	Mental Status	Interaction with examiner, attention, use of language (including higher level skills if applicable), following commands
	Motor examination	Subtle neurological signs, coordinated movement difficulties, repetitive motor movements
Developmental delay/ intellectual disability	General physical examination	Dysmorphology examination, nutrition and growth, activity level, cutaneous examination
	Mental status	Attention, use of language, following commands
	Motor examination	Muscle tone and bulk

(olfactory nerve), facial sensation (trigeminal nerve), gagging, and swallowing (glossopharyngeal, vagus nerves) is more accurate if done in person. Children with language disorders require hearing screening, which can be partially assessed dvia vestibulocochlear nerve testing. For the motor examination, the ability to move limbs against gravity and pronator drift can certainly be assessed via telehealth; however, more subtle deficits in grading of muscle strength require examiner assessment.

Different subtypes of cerebral palsy can have subtle differences in tone type (i.e., "clasp knife spasticity") and location (i.e., unilateral tone differences in hemiplegia, upper extremity tone differences in spastic diplegia). Decreased axial tone and subtle neurological signs are also common in neurodevelopmental disorders like ADHD and autism spectrum disorder and require in-person assessment; the absence of these findings may lead to other etiologic and diagnostic considerations for the patient's symptoms [2–4]. Reflexes and sensory assessment require in-person evaluation, but careful gait evaluation can be done with the right space and equipment via telehealth. Additional resources regarding telehealth examinations are included in Table 8.3.

Growth parameters and vital signs

Each physical evaluation should begin with an assessment of the child's growth parameters and vital signs. Height and weight should be measured correctly and compared against appropriate percentiles depending on the child's age, sex, and, if needed, diagnoses. Certain genetic diagnoses, such as Down syndrome [5] and Marfan's syndrome [6], have specialized growth charts based on population data. Growth rates in preterm infants can predict neurodevelopmental outcomes later in life [7]. Children on stimulant medication, a first-line treatment for ADHD, have their weight closely tracked to avoid growth suppression from appetite suppression, a known side effect of the medication [8]. Head circumference in particular is an essential vital sign in the neurodevelopmental disability population, determining if a child is microcephalic, normocephalic, or macrocephalic and further refining a differential diagnosis, including overgrowth syndromes or genetic disorders [9–11]. An abnormal head circumference should trigger additional physical examination maneuvers: examination of the head shape and suture status; evaluation for signs of increased intracranial pressure, such as a bulging fontanel in an infant; and inquiring about the head sizes of immediate family members as a consideration for familial megaloencephaly. While most electronic medical records contain head circumference percentiles [12] for children up to 2 years of age, online calculators can provide parameters for older children, adolescents, and adults as well [13]. Appropriate measurement of the head circumference should be done around the largest part of the head, from just above the eyebrows around the occiput, with a second measurement to confirm reliability [14].

TABLE 8.3 Telehealth resources for neurological examination.

Child Neurology Society pediatric neurology examination via telehealth	https://www.childneurologysociety.org/wp-content/uploads/2021/08/Pediatric-Neuro-Exam-via-Telemedicine-Oct2020-v2.pdf
American Academy of Neurology telehealth and remote care website	https://www.aan.com/practice/telemedicine/
Articles	**Dysmorphology** Del Campo, M., Beach, D., Wells, A., & Jones, K. L. (2021). Use of telemedicine for the physical examination of children with fetal alcohol spectrum disorders. *Alcoholism, Clinical and Experimental Research, 45*(2), 409–417. https://doi.org/10.1111/acer.14533. Rizvi, I., Fallah, K. N., Cepeda, A., Jr., Turner, K., Nguyen, P. D., & Greives, M. R. (2021). Where are we headed? Diagnosing abnormal head shapes through virtual care visits. *The Journal of Craniofacial Surgery*, 10.1097/SCS.0000000000008118. Advance online publication. https://doi.org/10.1097/SCS.0000000000008118. **Neurology Examination** Harper, K. A., Butler, E. C., Hacker, M. L., Naik, A., Eoff, B. R., Phibbs, F. T., Isaacs, D. A., Gallion, S. J., Thomas, E. P., Scott, J. L., Ploucher, S., Meystedt, J. C., McLeod, M. E., & Charles, P. D. (2021). A comparative evaluation of telehealth and direct assessment when screening for spasticity in residents of two long-term care facilities. *Clinical Rehabilitation, 35*(4), 589–594. https://doi.org/10.1177/0269215520963845. Meinhold, W., Yamakawa, Y., Honda, H., Mori, T., Izumi, S. I., & Ueda, J. (2021). A smart tendon hammer system for remote neurological examination. *Frontiers in Robotics and AI, 8*, 618656. https://doi.org/10.3389/frobt.2021.618656. Al Hussona, M., Maher, M., Chan, D., Micieli, J. A., Jain, J. D., Khosravani, H., Izenberg, A., Kassardjian, C. D., & Mitchell, S. B. (2020). The virtual neurologic exam: instructional videos and guidance for the COVID-19 Era. *The Canadian Journal of Neurological Sciences. Le journal canadien des sciences neurologiques, 47*(5), 598–603. https://doi.org/10.1017/cjn.2020.96. Cohen AB, Nahed BV. The digital neurologic examination. *Digital Biomarkers* 2021;*5*(1):114–126. Published 2021 Apr 26. https://doi.org/10.1159/000515577. Tarolli, C. G., Biernot, J. M., Creigh, P. D., Moukheiber, E., Salas, R., Dorsey, E. R., & Cohen, A. B. (2021). Practicing in a pandemic: a clinician's guide to remote neurologic care. Neurology. *Clinical Practice, 11*(2), e179–e188. https://doi.org/10.1212/CPJ.0000000000000882. Basil, G., Luther, E., Burks, J. D., Govindarajan, V., Urakov, T., Komotar, R. J., Wang, M. Y., & Levi, A. D. (2021). The focused neurosurgical examination during telehealth visits: guidelines during the COVID-19 pandemic and beyond. *Cureus, 13*(2), e13503. https://doi.org/10.7759/cureus.13503. **Developmental and behavioral pediatrics** Perez, D. L., Biffi, A., Camprodon, J. A., Caplan, D. N., Chemali, Z., Kritzer, M. D., Moo, L. R., Newhouse, A. L., Ramirez-Gomez, L., Razafsha, M., Tanev, K. S., & Schmahmann, J. D. (2020). Telemedicine in behavioral neurology-neuropsychiatry: opportunities and challenges catalyzed by COVID-19. *Cognitive and Behavioral Neurology: Official Journal of the Society for Behavioral and Cognitive Neurology, 33*(3), 226–229. https://doi.org/10.1097/WNN.0000000000000239. Soares, Neelkamal S. MD*; Langkamp, Diane L. MD, MPH† (2012). Telehealth in developmental-behavioral pediatrics, *Journal of Developmental & Behavioral Pediatrics, 33*(8), 656–665. https://doi.org/10.1097/DBP.0b013e3182690741.

Physical assessment

A standard physical examination including cardiovascular, respiratory, and abdominal assessments should be completed in all patients. Although these systems are not the central focus of a neurodevelopmental physical examination, findings could be associated with certain genetic conditions. Certain symptoms may also signal a need for urgent medical care or follow-up with a primary care provider. Physiological status could impact the patient's performance throughout the rest of the full evaluation; careful documentation of any changes in performance tied to medical status may be useful for school accommodation considerations and recommendations for other medical providers.

TABLE 8.4 Common dysmorphology findings.

Finding	Description/definition	Examples of associated conditions
Epicanthal folds	"A skinfold originating in the upper lid extends over the medial end of the upper lid, the medial canthus, and the caruncle, and ends in the skin of the lower lid"	Down syndrome, Fetal alcohol syndrome, Turner syndrome, Phenylketonuria, Williams syndrome, Noonan syndrome, Rubinstein-Taybi syndrome, Blepharophimosis syndrome
Low-set ears	"When the helix of the ear meets the cranium at a level below that of a horizontal plane through both inner canthi"	Down syndrome, Turner syndrome, Beckwith-Wiedemann syndrome, Rubinstein-Taybi syndrome, Smith-Lemli-Opitz syndrome, Trisomy 13, Trisomy 18
Hypertelorism	"Increased separation between the bony orbits, usually greater than two standard deviations above the mean"	Craniosynostosis syndromes (Apert, Carpenter, Crouzon syndromes), median facial clefting, Bohring-Opitz syndrome, Greig cephalopolysyndactyly, Noonan syndrome, Joubert syndrome, Trisomy 18, 22q11 syndrome, Neurofibromatosis type 1, Aarskog syndrome, Cat-Eye syndrome, CHARGE association, Loeys-Dietz syndrome, Wolf-Hirschhorn syndrome, Morquio syndrome, and Hurler syndrome
2–3 Toe syndactyly	"Toes two and three fused together"	22q13.3 deletion syndrome, Burn-McKeown syndrome, Cornelia de Lange syndrome 1, Feingold syndrome 1, Hunter-MacDonald syndrome, Joubert syndrome, Silver-russel syndrome 2, Smith-Lemli-Opitz syndrome
5th Finger clinodactyly	"Congenital curvature of a digit distal to the metacarpal phalangeal joint in the coronal plane"	Down syndrome, Rubinstein-Taybi syndrome, Cenani-Lenz syndactyly, Klinefelter syndrome, Turner syndrome, Fanconi anemia

Notes: Information on epicanthal folds from Jones, Kenneth Lyons, Marilyn C Jones, and Miguel del Campo. *Smith's Recognizable Patterns of Human Malformation*. Eighth edition. Philadelphia, PA: Elsevier, Inc., 2022 and Medline Plus (https://medlinepulus.gov).
Information on low-set ears from Lewis ML. A comprehensive newborn exam: part I. General, head and neck, cardiopulmonary. Am Fam Physician. 2014 Sep 1;90(5):289–96. PMID: 25251088 and Medline Plus (https://medlinepulus.gov).
Information on hypertelorism from Jones, Kenneth Lyons, Marilyn C Jones, and Miguel del Campo. *Smith's Recognizable Patterns of Human Malformation*. Eighth edition. Philadelphia, PA: Elsevier, Inc., 2022 and Sirkek B, Sood G. Hypertelorism. [Updated 2021 Jul 30]. In: StatPearls [Internet]. Treasure Island (FL): StatPearls Publishing; 2021 Jan-. Available from: https://www.ncbi.nlm.nih.gov/books/NBK560705/.
Information on 2–3 toe syndactyly from NCBI MedGen UID: 1645640
Information on 5th finger clinodactyly from Solomon BD, Muenke M. When to suspect a genetic syndrome. Am Fam Physician. 2012 Nov 1;86(9):826–33 and Banda DN, Yaish AM. Clinodactyly. [Updated 2021 Aug 11]. In: StatPearls [Internet]. Treasure Island (FL): StatPearls Publishing; 2021 Jan. Available from: https://www.ncbi.nlm.nih.gov/books/NBK556068/.
https://elementsofmorphology.nih.gov/ website is another resource for the above information.

The physical assessment of each patient should include a careful dysmorphology examination. Facial dysmorphisms include upslanting or downslanting palpebral fissures, low-set ears, a broad nasal bridge, and a smooth philtrum. Certain syndromes are associated with specific facial features that can aid in diagnosis [15]. Technological advances, including computer algorithms, can also support and improve diagnostic accuracy [16,17]. Common dysmorphology findings are presented in Table 8.4; less common findings could support targeted genetic testing. During the physical assessment, discussion with the family about dysmorphisms can lead to clues about inherited traits or even reveal parentage information that may not have been previously discussed.

Mental status

The mental status examination is an integral component of the neurological examination. All aspects of the examination have clinical relevance for neurodevelopmental assessment: appearance and behavior, speech and language, mood, thoughts and perceptions, and cognitive function [18]. Direct communication and interaction with patients of all ages can yield information regarding both their developmental and neurological status. While caregivers can provide a clear history of developmental milestones over the lifespan, the mental status examination allows the provider to objectively evaluate the children's pragmatic, receptive, and expressive language skills. Examples for every age group include an infant's ability to vocalize or laugh, a preschool child's understanding of their name and age, a school-age child's ability to tell or understand a joke, and a teenager's affect and ability to follow directions.

The history provided by the family prior to the physical examination can help guide the type of focused questioning or mental status evaluation performed with the patient. For example, nonverbal patients may be able to engage in interactive games and/or repeat skills or movements once demonstrated by the examiner. Patients with limited language skills answer yes-or-no questions, point to colors, and sign answers to questions. Patients who initially appear shy or have a slow-to-warm-up temperament can sit on their caregiver's lap during the initial mental status examination, and hopefully, be transitioned to the examination table once they've become more familiar with the examiner. Patients with attention difficulties struggle with attention tasks like digit span and spelling words backwards; patients with impulsive behaviors may quickly follow two-step commands without listening to key words, exemplified by an out-of-order response to the command "Point to the ceiling **after** you point to the floor."

The Mini-Mental Status Examination [19] remains the gold standard for cognitive dysfunction screening in adults as part of the mental status examination. The pediatric version, validated in children ages 3–14 years [20], has some evidence for use in neurodevelopmental disorders such as cerebral palsy [21] and learning difficulties [22]. Other brief or very brief cognitive screening tools [23] have been studied in both children and adults with various conditions showing promising results. The Montreal Cognitive Assessment [24] (MoCA) is the most widely used in clinical settings, with an administration time of 10 minutes, both in-person and remote testing options, and simple online training tools. The MoCA has not been widely studied in the pediatric population [25,26] at this time, though there is literature regarding specific neurodevelopmental disabilities such as intellectual disability [27] and ADHD [28].

Cranial nerves

A lot of useful information can be gleaned from a thorough cranial nerve assessment as part of a neurodevelopmental evaluation. The optic, oculomotor, trochlear, and abducens nerves cover vision and eye movements, which can be assessed at multiple points in the lifespan. Infants begin to track moving objects at 3 months of age; even in the absence of tracking, a blink in response to a visual threat should be able to be elicitable. Absence of horizontal or vertical ocular motility could indicate congenital ophthalmologic disorders (e.g., congenital ocular motor apraxia) or hydrocephalus (e.g., Parinaud's syndrome). A careful examination of a developmentally mature patient can identify visual field deficits that can be associated with certain cerebral palsy subtypes. A detailed fundoscopic examination could identify potential etiologic markers such as cataracts, macular cherry red spots, chorioretinitis, colobama, or signs of elevated intracranial pressure such as papilledema. Patients with feeding and speech articulation difficulties should have a careful evaluation of the trigeminal, facial, glossopharyngeal, vagal, and hypoglossal nerves to determine if there are contributing factors to their symptoms. Abnormal vestibulocochlear examinations should trigger referral to audiology for a thorough, developmentally appropriate hearing evaluation.

Localized cranial injuries should also be considered during a neurodevelopmental evaluation. Moebius syndrome involves the congenital absence of the abducens and facial nerves. Different cranial nerves travel together through structures such as the cavernous sinus, internal auditory meatus, and jugular foramen, to name a few, making them susceptible to specific patterns of injury. Certain constellations of symptoms, such as miosis, ptosis, and anhidrosis in Horner's syndrome, can suggest acute disruption of pathways due to infection, inflammation, or abnormal lesions.

Motor assessment

The first part of a motor assessment in a neurodevelopmental evaluation should simply be observation. While collecting a history, physicians should observe whether the patient can sit quietly throughout the evaluation without constant attention or redirection. They should look for signs of excess movement, such as 'motor on' behaviors associated with ADHD, like fidgeting, eye-blinking, or facial grimacing associated with tics, as well as uncontrolled involuntary movements indicative of choreoathetosis. Similarly, decreased movement on one side of the body or in one limb or decreased muscle bulk due to atrophy could add to diagnostic and therapeutic considerations.

Each limb should be assessed for strength using a standard 5-point scale, range of motion at each joint, and quality/quantity of tone, being careful to note head positioning at the time of manipulation in case of asymmetrical tonic neck reflex activation. This standard motor assessment of tone and strength in the upper and lower extremities is almost always indicated and performed, but just as important for neurodevelopmental evaluation is the documentation of functional use. Questions include:

- Does the child have functional use of their hands? Are they able to pick up and transfer a toy?
- Can they reliably push buttons or operate a switch, allowing for consideration of a communication device?

- Is their increased tone causing difficulties with care, creating a need for orthopedic or neurosurgical consultation?
- Is their decreased tone or poor coordination leading to delayed motor milestones, impaired athletic performance, or poor posture?
- Are the deficits symmetric and bilateral, lateralized, or limited to just the upper or lower extremities?
- Does the child have a limited or excessive range of motion, suggesting various treatment options or alternative diagnostic evaluations?

A careful cerebellar examination involves the evaluation of coordinated movements through sequenced motor tasks requiring understanding of complex instructions or pattern repetition. Often, a cerebellar evaluation is limited by the child's receptive language and motor skills. In addition, children with developmental coordination disorder, often associated with other neurodevelopmental disorders, must undergo these cerebellar maneuvers to ensure their difficulties aren't related to a structural issue requiring additional diagnostic evaluations, such as a cerebellar stroke or vestibular disorder.

Cerebellar dysfunction is associated with multiple neurodevelopmental disorders. Autism spectrum disorders are associated with cerebellar abnormalities across multiple animal and human studies, and ADHD and specific learning disabilities in reading have similar associations [29,30]. Differentiating cerebellar ataxia from sensory ataxia using the Romberg test is an important clinical distinction. Recognizing the differences between cerebellar ataxic gait, sensory ataxic gait, and antalgic gait is important for differential diagnosis purposes. Gait should also be assessed in general, particularly symmetry in arm swing and stride, stress gait abilities with accompanying posture, and caution/carefulness of steps.

Sensory assessment

A thorough sensory assessment is commonly overlooked in a neurodevelopmental evaluation. The task requires evaluating multiple sensory modalities, including light touch, pain, temperature, position and vibration sense, and discriminative sensation [18]. The exam can be limited by the availability of necessary tools, such as cotton swabs, a sharp-edged utensil, ice or cold water, and a tuning fork—important equipment for any neurodevelopmental provider to have on hand. If these items are not readily available readily in an examination room, they should be included in the provider's a personal equipment arsenal.

While a careful sensory evaluation is required for patients with spinal cord injuries, it can also provide diagnostic clues for patients with more common neurodevelopmental issues. A child with toilet training difficulties should receive a careful sensory and reflex genitourinary evaluation if a parent reports that the child can't seem to tell if their bladder is full. Parents of children with developmental delays and autism spectrum disorders may report that their child doesn't seem to feel pain, recalling their propensity for touching hot objects or showing minimal reactions to having their extremities stepped on or crushed. A sensory evaluation can provide parental relief or support for these observations, which may be unexplored by other providers. Children with balance and coordination difficulties who struggle with the Romberg test may warrant neuroimaging and additional workups.

Reflexes

Reflexes can support for a variety of diagnoses over the lifespan in different ways, from the persistence of primitive reflexes in cerebral palsy to the change from hyperreflexia to hyporeflexia over time in spinal cord injury. The infant examination primarily consists of evaluating primitive and postural reflexs, with special consideration for head position. Upper and lower extremities should be assessed at all ages, as subtle asymmetries may warrant consideration alongside muscle tone and strength changes. Children with various neurodevelopmental disorders frequently require distraction to elicit patellar reflexes, which can be done using various maneuvers, such as asking them to press their palms together or clap on the count of three while tapping on the tendon [31]. If the reflex is still unable to be elicited despite these techniques, the ankle reflex can be assessed to ensure the reflex arc is intact.

Summary

A physical and neurological evaluation is an integral part of assessing a child with any developmental concern. Positive and negative findings can help localize the dysfunction, generate a relevant differential diagnosis, and guide further assessment. While technology and telehealth may make some components of the traditional physical evaluation challenging to execute remotely, a great deal of valuable information can still be gleaned.

References

[1] Walker H.K. The origins of the history and physical examination. In: Walker H.K., Hall W.D., Hurst J.W., (Eds.). *Clinical Methods: The History, Physical, and Laboratory Examinations*. 3rd edition. Boston: Butterworths; 1990. Chapter 1. Available from: https://www.ncbi.nlm.nih.gov/books/NBK458/.

[2] Sweeney KL, Ryan M, Schneider H, Ferenc L, Denckla MB, Mahone EM. Developmental trajectory of motor deficits in preschool children with ADHD. Dev Neuropsychol 2018;43(5):419. Available from: https://doi.org/10.1080/87565641.2018.1466888.

[3] Jansiewicz EM, Goldberg MC, Newschaffer CJ, Denckla MB, Landa R, Mostofsky SH. Motor signs distinguish children with high functioning autism and Asperger's syndrome from controls. J Autism Dev Disord 2006;36(5):613–21. Available from: https://doi.org/10.1007/s10803-006-0109-y.

[4] Vitiello B, Ricciuti AJ, Stoff DM, Behar D, Denckla MB. Reliability of subtle (soft) neurological signs in children. J Am Acad Child Adolesc Psychiatry 1989;28(5):749–53. Available from: https://doi.org/10.1097/00004583-198909000-00017.

[5] Zemel BS, Pipan M, Stallings VA, et al. Growth charts for children with down syndrome in the United States. Pediatrics 2015;136(5):e1204–11. Available from: https://doi.org/10.1542/peds.2015-1652.

[6] Erkula G, Jones KB, Sponseller PD, Dietz HC, Pyeritz RE. Growth and maturation in Marfan syndrome. Am J Med Genet 2002;109(2):100–15. Available from: https://doi.org/10.1002/ajmg.10312.

[7] Yitayew M, Chahin N, Rustom S, Thacker LR, Hendricks-Muñoz KD. Fenton vs. intergrowth-21st: postnatal growth assessment and prediction of neurodevelopment in preterm infants. Nutrients 2021;13(8). Available from: https://doi.org/10.3390/nu13082841.

[8] Baweja R, Hale DE, Waxmonsky JG. Impact of CNS stimulants for attention-deficit/hyperactivity disorder on growth: epidemiology and approaches to management in children and adolescents. CNS Drugs 2021;35(8):839–59. Available from: https://doi.org/10.1007/s40263-021-00841-w.

[9] Jones, S.G., Samanta, D. Macrocephaly. [Updated 2023 Jul 24]. In: StatPearls [Internet]. Treasure Island (FL): StatPearls Publishing; 2024 Jan-. Available from: https://www.ncbi.nlm.nih.gov/books/NBK560786/

[10] Perenc L, Guzik A, Podgórska-Bednarz J, Drużbicki M. Abnormal head size in children and adolescents with congenital nervous system disorders or neurological syndromes with one or more neurodysfunction visible since infancy. J Clin Med 2020;9(11). Available from: https://doi.org/10.3390/jcm9113739.

[11] Becerra-Solano LE, Mateos-Sánchez L, López-Muñoz E. Microcephaly, an etiopathogenic vision. Pediatr Neonatol 2021;62(4):354–60. Available from: https://doi.org/10.1016/j.pedneo.2021.05.008.

[12] Kuczmarski R.J., Ogden C.L., Guo S.S., Grummer-Strawn L.M., Flegal K.M., Mei Z., et al. 2000 CDC growth charts for the united states: methods and development. Vital Health Stat 11. 2002;(246):1-190.

[13] Segal M., Rapin I. Head circumference calculator. Chestnut Hill MA: Simulconsult, Inc.

[14] Tanski S., Garfunkel L.C. Assessing Growth and Nutrition in Tanski S., Garfunkel L.C., Duncan P.M., Weitzman M. (Eds.). *Performing Preventive Services: A Bright Futures Handbook*. Elk Grove Village, IL. American Academy of Pediatrics 2010. Chapter 2.

[15] Jones K, Jones M, del Campo M. Smith's recognizable patterns of human malformation. 8th ed. Amsterdam, Netherlands: Elsevier Health Sciences; 2021.

[16] Marwaha A, Chitayat D, Meyn MS, Mendoza-Londono R, Chad L. The point-of-care use of a facial phenotyping tool in the genetics clinic: enhancing diagnosis and education with machine learning. Am J Med Genet A 2021;185(4):1151–8. Available from: https://doi.org/10.1002/ajmg.a.62092.

[17] Javitt MJ, Vanner EA, Grajewski AL, Chang TC. Evaluation of a computer-based facial dysmorphology analysis algorithm (Face2Gene) using standardized textbook photos. Eye (London). 2021;. Available from: https://doi.org/10.1038/s41433-021-01563-5.

[18] Bickley LS, Szilagyi PG, Hoffman RM, Soriano RP. Bates' guide to physical examination and history taking. 13th ed. Philadelphia PA: Wolters Kluwer; 2021.

[19] Creavin ST, Wisniewski S, Noel-Storr AH, Trevelyan CM, Hampton T, Rayment D, et al. Mini-Mental State Examination (MMSE) for the detection of dementia in clinically unevaluated people aged 65 and over in community and primary care populations. Cochrane Database Syst Rev 2016;2016(1):CD011145. Available from: https://doi.org/10.1002/14651858.CD011145.pub2.

[20] Ouvrier RA, Goldsmith RF, Ouvrier S, Williams IC. The value of the mini-mental state examination in childhood: a preliminary study. J Child Neurol 1993;8(2):145–8. Available from: https://doi.org/10.1177/088307389300800206.

[21] Moura R, Andrade PMO, Fontes PLB, Ferreira FO, Salvador LS, Carvalho MRS, et al. Mini-mental state exam for children (MMC) in children with hemiplegic cerebral palsy. Dement Neuropsychol 2017;11(3):287–96. Available from: https://doi.org/10.1590/1980-57642016dn11-030011.

[22] Salvador LS, Moura R, Ferreira FO, Andrade PMO, Carvalho MRS, Haase VG. The Mini-Mental Examination for Children (MMC): evidence of validity for children with learning difficulties. Dement Neuropsychol 2019;13(4):427–35. Available from: https://doi.org/10.1590/1980-57642018dn13-040010.

[23] Patnode CD, Perdue LA, Rossom RC, Rushkin MC, Redmond N, Thomas RG, et al. Screening for cognitive impairment in older adults: an evidence update for the U.S. Preventive Services Task Force. JAMA 2020;323(8):764–85. Available from: https://doi.org/10.1001/jama.2019.22258.

[24] Nasreddine ZS, Phillips NA, Bédirian V, Charbonneau S, Whitehead V, Collin I, et al. The Montreal Cognitive Assessment, MoCA: a brief screening tool for mild cognitive impairment. J Am Geriatr Soc 2005;53(4):695–9. Available from: https://doi.org/10.1111/j.1532-5415.2005.53221.x.

[25] Pike NA, Roy B, Gupta R, Singh S, Woo MA, Halnon NJ, et al. Brain abnormalities in cognition, anxiety, and depression regulatory regions in adolescents with single ventricle heart disease. J Neurosci Res 2018;96(6):1104–18. Available from: https://doi.org/10.1002/jnr.24215 06.

[26] Pike NA, Poulsen MK, Woo MA. Validity of the Montreal Cognitive Assessment screener in adolescents and young adults with and without congenital heart disease. Nurs Res 2017;66(3):222–30. Available from: https://doi.org/10.1097/NNR.0000000000000192.

[27] Edge D, Ewing L. The predictive validity of the MoCA-LD for assessing mental capacity in adults with intellectual disabilities. J Appl Res Intellect Disabil 2019;32(5):1280–7. Available from: https://doi.org/10.1111/jar.12621.

[28] Helene Bergly T, Julius Sømhovd M. The relation between ADHD medication and mild cognitive impairment, as assessed by the montreal cognitive assessment (MoCA), in patients entering substance use disorder inpatient treatment. J Dual Diagn 2018;14(4):228–36. Available from: https://doi.org/10.1080/15504263.2018.1496305.

[29] Stoodley CJ. The cerebellum and neurodevelopmental disorders. Cerebellum 2016;15(1):34–7. Available from: https://doi.org/10.1007/s12311-015-0715-3.

[30] Sathyanesan A, Zhou J, Scafidi J, Heck DH, Sillitoe RV, Gallo V. Emerging connections between cerebellar development, behaviour and complex brain disorders. Nat Rev Neurosci 2019;20(5):298–313. Available from: https://doi.org/10.1038/s41583-019-0152-2.

[31] Delwaide PJ, Toulouse P. Facilitation of monosynaptic reflexes by voluntary contraction of muscle in remote parts of the body. Mechanisms involved in the Jendrassik Manoeuvre. Brain 1981;104(Pt 4):701–9. Available from: https://doi.org/10.1093/brain/104.4.701.

Chapter 9

Developmental testing for the physician

Chapter 9.1

The neuromotor evaluation

Eric M. Chin[1,2], Ayra Hussain[3] and Peter A. Blasco[4]

[1]*Department of Neurodevelopmental Medicine, Kennedy Krieger Institute, Baltimore, MD, United States,* [2]*Departments of Neurology and Pediatrics, Johns Hopkins University School of Medicine, Baltimore, MD, United States,* [3]*Departments of Biological Sciences and Modern Languages, Linguistics & Intercultural Communication, University of Maryland, Baltimore County, Baltimore, MD, United States,* [4]*Department of Pediatrics, Division of Neurodevelopmental Pediatrics, Institute on Development and Disability, Oregon Health and Science University, Portland, OR, United States*

Concerns about the development of infants and young children fall into three broad areas—language/cognitive problems, neuromotor and orthopedic concerns, and behavioral issues. Developmental screens are useful for identifying neuromotor abnormalities in low-risk populations; this section, however, focuses on the definitive, multifaceted assessment that should follow referral for motor concerns, covering elements of the clinical evaluation, the classification of motor dysfunction, and a description of commonly used motor performance tools.

Neuromotor evaluation of the infant

In order to make a meaningful statement about an infant's motor competence, the clinician should organize data gathered from the history, physical examination, and neurodevelopmental examination according to the following schema:

- motor developmental milestones
- the classic neurological examination
- markers of cerebral neuromotor maturation (primitive reflexes and postural reactions)

Motor milestones are extracted from the developmental history as well as from observations during the neurodevelopmental examination. Use of a reference table of sequential gross and fine motor milestones allows individual milestone acquisition profiles to be viewed in terms of motor age equivalents (Tables 9.1.1 and 9.1.2). Milestone assessment is best summarized as a single (or narrow) motor age. This approach allows one to think of the child in terms of his or her level of motor function. The motor age can be converted to a motor quotient ($MQ = 100 \times \frac{\text{Motor age}}{\text{Chronological or adjusted age}}$), which provides a useful expression of deviation from the norm [1–5]. It is also convenient to adjust for prematurity at this point. An MQ less than 70 is considered abnormal, and less than 50 is particularly concerning [5]. Values above 100 are not of particular significance except for the pleasure of identifying above-average performers. Whether truly gifted athletes can be recognized early by this method is thought-provoking but unknown. The motor age and the MQ are each good summary descriptors of the child and have more meaning than trying to analyze each individual milestone. Normed milestone-based assessments such as the Developmental Assessment of Young Children (DAYC-2) [6] additionally allow calculation of a standard score for age, complementing use of the MQ. There is also value in comparing the individual child's gross motor and fine motor performance. Dissociation of these developmental ages by more than a couple of months raises a red flag for neurological dysfunction in the lower domain, which should be addressed via neurological examination. A gross motor dissociation pattern, for example, would be typical for spastic diplegia or paraplegia. Isolated gross motor delays are

Capute and Accardo's Neurodevelopmental Disabilities in Infancy and Childhood. DOI: https://doi.org/10.1016/B978-0-12-824060-1.00033-X

TABLE 9.1.1 Gross motor milestones [1,2].

Prone:	
Head up	1 month
Chest up	2 months
Up on elbows	3 months
Up on hands	4 months
Roll:	
Front to back	3–4 months
Back to front	4–5 months
Sit with support ("tripod" sitting)	5 months
Sit without support	6–7 months
Creep (in prone, "belly crawl")	6–7 months
Come to sit (unassisted)	8 months
Crawl (quadruped—up on all fours)	8 months
Pull to stand	8–9 months
Cruise	9–10 months
Walk with two hands held	10 months
Walk with one hand held	11 months
Walk alone	12 months
Run (stiff-legged)	15 months
Walk upstairs (with rail)	21 months
Jump in place	24 months
Pedal tricycle	30 months
Walk downstairs, alternating feet	3 years

Source: Adapted from Capute AJ, Accardo PJ. The infant neurodevelopmental assessment: a clinical interpretive manual for CAT-CLAMS in the first two years of life, part 2. Curr Probl Pediatr 1996;26(8):279–306; Knobloch H, Stevens FM, Malone AF. Manual of developmental diagnosis: the administration and interpretation of the revised Gesell and Amatruda developmental and neurologic examination. Harper Collins Publishers; 1980. p. 286.

TABLE 9.1.2 Fine motor milestones [2–4].

Retains ring (rattle)	1 month
Hands unfisted	3 months
Reaches	3–4 months
Hands to midline	3–4 months
Transfer	5 months
Takes 1-inch cube	5–6 months
Takes pellet (grasp)	6–7 months
Immature pincer	7–8 months
Mature pincer	10 months
Release	12 months

Source: Adapted from Knobloch H, Stevens FM, Malone AF. Manual of developmental diagnosis: the administration and interpretation of the revised Gesell and Amatruda developmental and neurologic examination. Harper Collins Publishers; 1980. p. 286; Capute AJ, Accardo PJ. The infant neurodevelopmental assessment: a clinical interpretive manual for CAT-CLAMS in the first two years of life, part 1. Curr Probl Pediatr 1996;26(7):238–57; Johnson CP, Blasco PA. Infant growth and development. Pediatr Rev 1997;18(7):224–42.

sometimes associated with mild hypotonia and/or later cognitive deficits [7]. In contrast, children with hemiplegic cerebral palsy frequently function better in gross motor than fine motor domains due to impaired functioning in bimanual tasks [8].

Thus if a 15-month-old boy is not walking alone but pulls up to stand (age equivalent 8–9 months), cruises around furniture (9–10 months), and uses a mature pincer grasp (10 months), the motor age indicates that he is behaving like an average 10-month-old. His MQ for chronological age would be 67—concerning for significant delays in both domains. If he had been born at 28 weeks gestation, his adjusted age would be 12 months, and the MQ is a more reassuring 83. Standard scores from DAYC-2 lookup tables further reassure us that gross motor performance at a 10-month age equivalent is consistent with a standard score of 90 at age 12 months—that is, less than one standard deviation below average.

Motor milestones do not take into account the *quality* of a child's movement. The child with hemiplegia may sit, stand, and walk "on time" despite obvious differences between the right and left side. The motor portion of the neurological examination takes qualitative features into better account. The neurological assessments of tone, strength, deep tendon reflexes, and coordination are challenging in the infant because of their subjective nature, often compounded by limited cooperation. The elicitation of reflexes requires patience and often repeated trials (The tendon you are swinging at is usually a moving target.) Muscle tone (resistance to passive movements), strength (active power), and musculoskeletal joint laxity are difficult to disentangle in the contrary subject. The best clues may come from observation, not handling. Spontaneous or prompted motor activities (e.g., weight bearing in sitting or standing) require adequate strength. Weakness may be best appreciated from observing stationary postures and transition movements. The Gowers sign (arising from floor-sitting to standing using the hands to "walk up" one's legs) is a classic example, indicative of pelvic girdle and quadriceps muscular weakness frequently seen in myopathies. Spontaneous postures such as frog-legs and scissoring (Fig. 9.1.1) provide visual clues to hypotonia/weakness and spastic hypertonus, respectively. After 2–3 years of age, as cooperation improves, the neurological examination becomes easier and often more meaningful. Station and gait also need to be assessed. Station refers to the posture assumed in sitting or standing and should be viewed from anterior, lateral, and posterior perspectives looking for body alignment. Gait, of course, can only be examined in progress. Initially, the typical toddler walks on a wide base, slightly crouched, with the arms abducted and elevated (often termed "high guard"). Forward progression is more staccato than smooth. Movements gradually become more fluid, the base narrows, and arm swing evolves, leading to a mostly adult pattern of walking by 3 years of age.

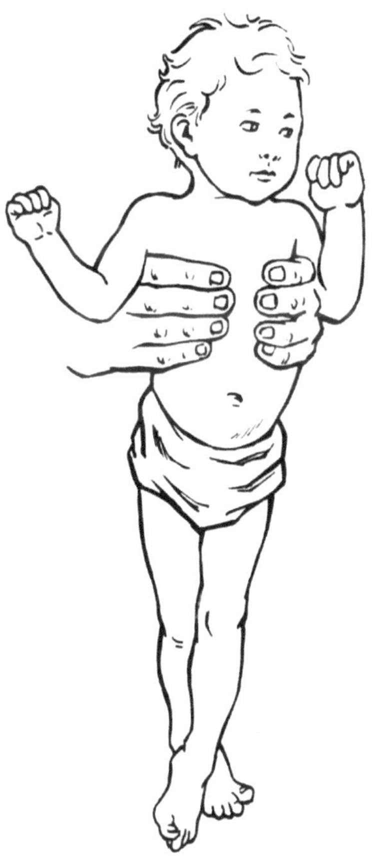

FIGURE 9.1.1 Scissoring. In vertical suspension, the ankles cross one another. Note the fisting and tonic labyrinthine ("surrender") posture.

The motor neuromaturational markers are the primitive reflexes, which develop during gestation and generally disappear between the third and sixth months after birth, and the postural reactions, which are not present at birth but develop sequentially between 3 and 10 months of age (Fig. 9.1.2). The most clinically useful primitive reflexes are the Moro, tonic labyrinthine, asymmetric tonic neck, and positive support (Figs. 9.1.3–9.1.5). As with all reflexes, each requires a specific sensory stimulus to generate the stereotyped motor response that follows. Normal babies and infants demonstrate these postures inconsistently and transiently whereas those with neurological injuries show stronger and more sustained primitive reflex posturing [9,10].

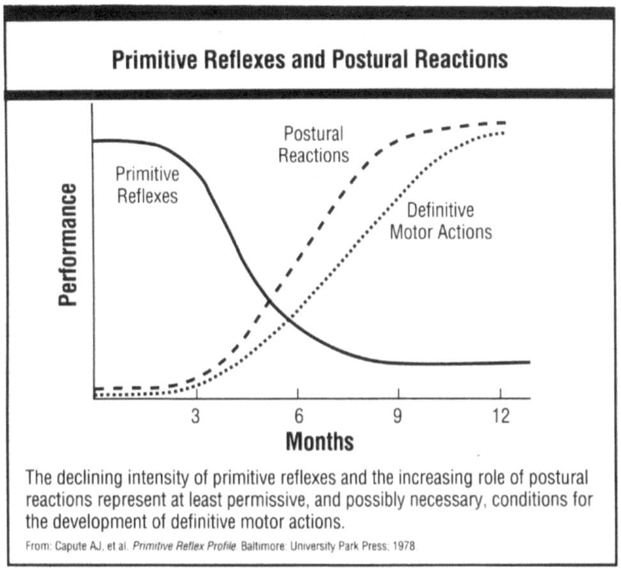

FIGURE 9.1.2 Primitive reflexes and postural reactions. The declining intensity of primitive reflexes and the increasing strength of postural reactions represent at least permissive, and possibly necessary, conditions for the development of definitive motor actions [10]. *From Blasco PA. Primitive reflexes [Internet]. Clin Pediatr 1994;38897. Available from: http://doi.org/10.1177/000992289403300703. Reprinted with permission.*

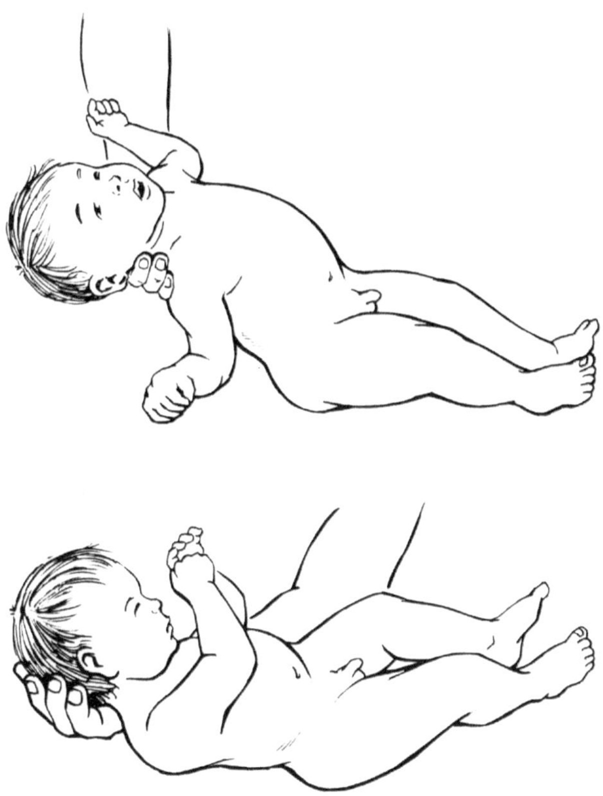

FIGURE 9.1.3 Tonic labyrinthine reflex. In the supine position, the baby's head is mildly extended (about 45 degrees) by gentle support under the shoulders. This yields relative extension of the legs and retraction of the shoulders, producing the upper extremity "surrender posture." With active or passive neck flexion (ideally to 45 degrees above the plane of the body), the arms come forward to the midline and the legs flex.

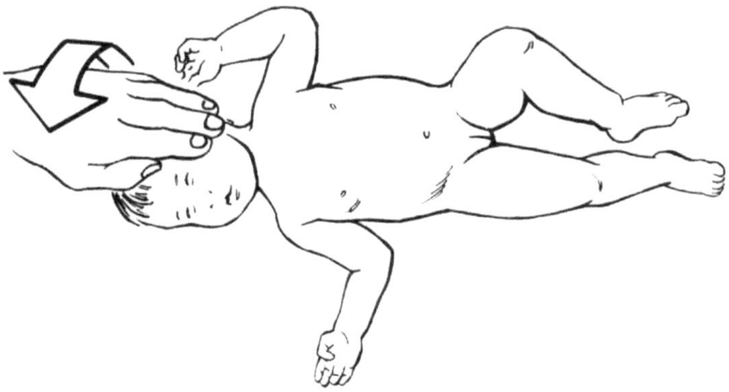

FIGURE 9.1.4 Asymmetric tonic neck reflex. The sensory limb of the Asymmetric Tonic Neck Reflex (ATNR) probably involves cervical proprioceptors. With either active or passive turning to the side, the baby reflexively extends the fingers, arm, and leg on the face side and flexes the extremities on the occiput side. There is also some mild paraspinous muscle tightening on the occiput side, producing subtle trunk curvature.

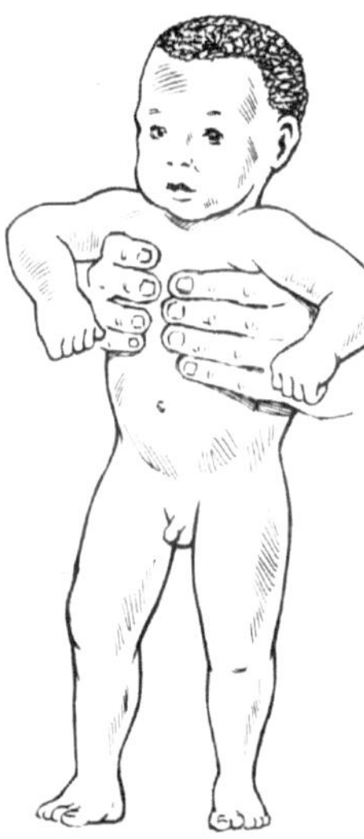

FIGURE 9.1.5 Positive support reflex. With support around the trunk, the child is suspended and lowered to pat the feet gently on a flat surface. This stimulus produces reflex extension at the hips, knees, and ankles so the baby stands up straight. Often children go up on their toes initially but should come down flat within 20–30 s before sagging back down toward a sitting position.

FIGURE 9.1.6 Parachute reaction. The examiner has suspended the child horizontally by the waist and lowered him face down toward a flat surface. The arms extend in front, slightly abducted at the shoulders, and the fingers spread as if to break a fall.

Primitive reflexes are somewhat tricky to gauge, even in expert hands. The appearance of postural reactions after 2–3 months of age is easier to recognize clinically and can provide insight into the motor potential of young infants. Postural reactions are sought in each of the three major categories: righting, protection, and equilibrium (Figs. 9.1.6 and 9.1.7). These movements are much less stereotyped than the primitive reflexes, and they require a complex interplay of cerebral, cerebellar, and brainstem adjustments to a barrage of sensory inputs (proprioceptive, visual, vestibular). They are markedly slowed in their appearance in the baby with nervous system damage [9] and may even help in the early differentiation of subtypes of cerebral palsy [11].

FIGURE 9.1.7 Postural mechanisms. The infant is comfortably seated, supported about the waist (if necessary) and distracted by a toy or parent. The examiner gently tilts the child to one side, noting deviation of the head back toward the midline (head-righting), lateral extension of one arm toward the side of the fall (protection), and upward counter movements of the arm and leg (equilibrium reactions).

Individual therapists, teachers, physicians, and others trained in child development often disagree on what should be considered significant in terms of motor delay and neurological examination findings. This is where clinical experience and clinical judgment come into play. Some evaluation elements are fairly subjective, like muscle tone and coordination. The calculation of a motor quotient is objective and useful in determining the significance of motor delay. This ratio is most meaningful when applied in children under 3 or 4 years of age and, by itself, is not an absolute indicator of normality or abnormality. When added with other aspects of the neurological examination, the combined elements become the basis for determining motor competence and identifying the motor-impaired infant.

Classification of motor impairment

Once neuromotor abnormality has been established, further assessment as to its exact nature and etiology is essential. The differential diagnosis of motor disability is divided into four general categories: (1) static encephalopathies; (2) progressive neurological and neuromuscular diseases; (3) spinal cord and peripheral nerve disorders; and (4) structural (musculoskeletal) abnormalities. Based on clinical examination and history, one can usually get a good sense of which category the patient falls into.

Static encephalopathies, by definition, indicate nonprogressive brain dysfunction. An insult that occurs during early fetal brain development may produce a central nervous system anomaly, which could be the result of a genetic variant or a very early biochemical or mechanical insult that permanently alters anatomical development of the brain. Alternatively, a brain developing in a normal fashion can be damaged before, during, or after birth by a wide variety of insults—infection (meningitis, encephalitis), trauma, ischemia, toxins (such as lead or via inborn errors of metabolism), and others.

Brain injury resulting in neuron or axon destruction is permanent with little to no true neuronal repair or regrowth. However, intact areas continue to develop during childhood so that functional improvements may occur for a time. The gross architecture of sensorimotor networks is typically established by age 2 [12], although improvements in motor performance are known to continue through adolescence [13]. When motor impairment is due to an anomaly or to an insult that involves the immature (i.e., developing) brain, we refer to the disorder as cerebral palsy.

Progressive diseases of the brain, spinal cord, nerves, or muscles produce motor impairment that worsens with time. Although the number of disorders in this category is large, each individual disease is extremely rare. Combined, the prevalence of the neurogenetic disorders (many of which exhibit neuromotor progression over time) appear to be on the order of 1 in 1000 individuals [14]. Degenerative diseases can present in the neonate (e.g., Zellweger syndrome, nemaline myopathy), early infancy (spinal muscular atrophy), school age (Duchenne muscular dystrophy, X-linked adrenoleukodystrophy), adolescence (spinal muscular atrophy type IV, facioscapulohumeral dystrophy), or adulthood (Huntington disease, adrenomyeloneuropathy). Knowing the specific diagnosis is important because it helps anticipate the rate of progression, determine other prognostic implications, provide meaningful genetic counseling, and, increasingly, identify diagnosis-specific treatments.

Children with progressive conditions may initially experience a period of normal or near-normal development, obscuring the fact that a slow disease process is going on. In Duchenne muscular dystrophy, for example, the diagnosis is usually not made until age 4–6 when increasing clumsiness and difficulty keeping up with peers brings the child to attention, yet astronomic elevations of creatine kinase (CK) enzyme indicating muscle destruction are present from birth. The importance of diagnosing such diseases and their genetic underpinnings early cannot be overstated. Early diagnosis is crucial to providing interdisciplinary treatment, access to new treatments, and proper genetic counseling.

Nervous system tumors would fall in this category, at least before treatment. As long as the tumor expands in the brain or spinal cord, the patient experiences progressive loss of neurologic function. If the tumor is successfully treated, the individual may be cured with no residual deficits or may be left with a static motor impairment. Once again, early diagnosis and treatment are critical to maximizing survival with a minimum of disability.

Spinal cord and peripheral nerve disorders are almost all static conditions. In this category, traumatic injuries and spina bifida (meningomyelocele) are most frequently encountered. The loss of motor and sensory function varies depending on where along the spinal cord the injury lies. There is almost always another anomaly, the Chiari II malformation, present at the brainstem in children with meningomyelocele. This anomaly leads to hydrocephalus in most (not all) cases. Once again, early involvement of an interdisciplinary diagnostic and treatment team is emphasized.

Structural (musculoskeletal) abnormalities refer to situations in which some anatomical structure (e.g., a limb) is missing or deformed or in which support tissues for the nerves and muscles are inadequate (e.g., connective tissue defects and biochemically abnormal bone). Some structural defects are progressive in the sense of ongoing tissue damage, and many are extremely complex to manage. As examples, osteogenesis imperfecta and the many varieties of childhood arthritis fit in this category. Repeated fractures and degenerative joint changes can lead to progressive loss of motor function. On the milder end of the spectrum, there exist a wide variety of fairly common orthopedic deformities, which may or may not impact motor development or motor function (e.g., club feet, developmental hip dysplasia). They present at birth or in early infancy, they cause aberrations in the quality of movement, and their early recognition hinges on physical examination. Early treatment can make a big difference in outcome and in the extent of intervention required to achieve correction.

Timing of the diagnosis

The ability to make an early diagnosis depends, in large measure, on the type of disability. The diagnosis of spina bifida or club foot is readily made at birth. In the case of cerebral palsy, diagnosis is delayed until a problem with motor development is recognized and investigated. The child with severe impairments is likely to be identified earlier (usually well before 12 months of age) than the child with milder delays because the former will fail to achieve developmental milestones sooner and will demonstrate more neurological examination signs. The key to early identification hinges on (1) appropriate monitoring of infants with risk factors; (2) skill with the pediatric neuromotor evaluation, as elucidated above; and (3) use of adjunctive assessments (such as neuroimaging) when appropriate [15].

Formal motor assessments

Formal assessments of motor development vary in form, in purpose, and in target population. Effective usage requires judicious test selection and appropriate interpretation of results. The forms of assessments range from caregiver report of milestones to direct examination of motor function, of overall neurologic status, or of movement patterns and qualities [16]. Examinations themselves range from evaluations a physician can complete at bedside to detailed assessments requiring specifically trained therapists and equipment. Full descriptions of instruments are beyond the scope of this chapter, but general principles will be reviewed.

As with cognitive assessments, psychometric test properties are key considerations when using motor assessments and interpreting their results (see Bannigan and Watson for a review of concepts [17]). In brief, the reliability of a tool describes measurement error and is generally evaluated in terms of consistency across repeated measurements—performed either by the same evaluator (intrarater reliability) or by different evaluators (interrater reliability). Validity describes whether a tool measures what it is intended to (along one or more dimensions of interest while minimizing the influence of irrelevant dimensions). Finally, utility reflects the practicality of a tool in real-world clinical or research settings (e.g., considering burdens in terms of time, personnel, and costs). For developmental motor evaluations, additional considerations in test selection include the age/functional level of the patient as well as the type(s) of conditions/movement disorders involved.

Motor assessments for infants

Structured assessments of infant motor functioning may be considered for three main purposes: discrimination (differentiating groups of typically developing infants from those with motor impairments), prediction (of an individual's developmental motor trajectory), and evaluating an individual's changes over time (e.g., in response to therapy) [18]. While structured assessments can additionally help inform diagnosis, consideration of medical history and of possible etiologies is generally also needed to, for example, differentiate cerebral palsy from progressive neurologic disorders with a similar presentation [15]. Motor assessment in infants can be particularly challenging due to the rapid development of motor phenotypes [19] including the gradual replacement of sporadic involuntary movements with more clearly purposeful movements [20].

Many established infant motor assessments are highly reliable; bedside neuromotor evaluations are almost always adequate to detect motor abnormalities, but more specialized or more detailed instruments are typically needed to assess response to intervention or to predict motor outcomes. Normed milestone-based assessments (such as the DAYC-2 referenced above) typically rely upon caregiver report. In contrast, structured neurological examinations (e.g., the Hammersmith Infant Neurodevelopmental Examination, or HINE [21]) and therapist-administered motor batteries (e.g., the Test of Infant Motor Performance, or TIMP [22]) rely upon in-person assessment of tone, primitive reflexes, and/or postural reactions. Other assessments rely on distinct conceptual frameworks (e.g., the observational Prechtl General Movements Assessment [23], which predicts motor outcome based on the presence and typicality/atypicality of "writhing" and "fidgety" movements thought to presage the emergence of precise voluntary motor development).

Assessments from any of these categories can achieve "good" or "very good" interobserver agreement [16]. Further, all of these types of instruments are able to discriminate between typical and atypical motor development [16]. In contrast, to date, only therapist- or psychologist-administered assessments have been designed and validated to capture longitudinal changes [16]. There is some evidence, for example, of improvements in TIMP scores following therapy referrals in infants born preterm [24]. Therapist-administered assessments, however, often require specialized training and may require up to an hour to administer [25]. While individual assessments have shown promise for prediction of motor outcomes (e.g., the General Movements Assessment demonstrating nearly 100% sensitivity in specific populations for a later diagnosis of cerebral palsy [25]), specificity requires interpretation within clinical context. As such, current guidelines [15] recommend combining standardized motor assessments with a medical history, standardized neurological examination, and neuroimaging for making an early, accurate diagnosis of cerebral palsy.

Motor assessments for toddlers/school-aged children

Goals of evaluating motor abilities, discriminating between typical and atypical motor patterns, and predicting future levels of motor functioning pertain also to child-aged assessments. However, the broader and better-defined motor phenotypes seen in children (as compared to infants) permit more fine-grained assessment:

- Increasingly refined motor abilities enable the performance of increasingly sophisticated and diverse tasks; as such, functional measures for children are often oriented toward more specific motor-related domains (e.g., on functional mobility, on self-care skills, on classroom-relevant fine motor dexterity, and/or on specific patient- or family-oriented goals).
- Motor pathology also evolves with age—often with disorder-specific trajectories. Functional motor assessments (those assessing activities and participation in society as well as those assessing body systems and functions) and disorder-specific assessments each merit consideration in children with motor concerns.

Motor assessments can examine functioning broadly (i.e., defining function with respect to developmental norms) or with respect to more specific goals:

- Developmental motor assessments: Comprehensive developmental assessment batteries (whether milestone-based interview scales or observational assessments) frequently include gross motor and fine motor domains as subtests (e.g., Gross Motor and Fine Motor domains of the Bayley Scale of Infant and Toddler Development [26] or the Developmental Assessment of Young Children [6]). Dedicated developmental motor batteries are also available (e.g., the Peabody Developmental Motor Scales [27], the Movement Assessment Battery for Children [28], or the Bruininks-Oseretsky Test of Motor Proficiency [29]). Gross motor assessment using well-established comprehensive developmental batteries or using well-established motor-specific batteries generally achieves good-to-excellent test-retest reliability and good-to-excellent content and construct validity; however, their responsiveness to clinically important change over time is uncertain [30]. These multidomain "snapshots" of developmental achievement are helpful for diagnostic purposes.
- Activities-based assessments: Other assessments focus upon specific functional domains (e.g., mobility or self-care) instead of using a developmental model. These include normed scales (e.g., the Pediatric Evaluation of Disability Inventory [PEDI] [31], the Functional Independence Measure for Children [WeeFIM] [32], the Pediatric Outcomes Data Collection Inventory [PODCI] [33], and the Activities Scale for Kids [ASK] [34]). Alternatively, patients and/or families—in collaboration with therapists and medical providers—may select specific goals and accompanying benchmarks using tools such as the Canadian Occupational Performance Measure (COPM) [35] or Goal Attainment Scale (GAS) [36]. These assessments may be particularly helpful for evaluating responses to interventions (e.g., is a course of therapy impacting the targeted function as anticipated?).

More focused motor assessments are available for specific motor disorders (e.g., spasticity, dystonia, or choreoathetosis) and in selected clinical populations (e.g., in children with cerebral palsy or, more specifically, in children with holoprosencephaly). Several assessments are examination-based instruments designed to quantify the severity of a single motor feature across joints/body regions (e.g., the Modified Ashworth Scale [37] for muscle tone in general or the Tardieu Scale [38] for spasticity; the Barry−Albright Dystonia Scale [39] for dystonia; and the Scale for the Assessment and Rating of Ataxia [40] or the International Cooperative Ataxia Rating Scale [41] for ataxia). Most such assessments appear to have good relative reliability (consistency across measurements) [42,43] in children with cerebral palsy, but concerns have been raised regarding their content and construct validity [44]—particularly as consensus definitions of pediatric motor disorders have changed over time [45]. Use of discriminative tools such as the Hypertonia Assessment Tool (HAT) [46] may also be helpful for determining which motor disorder(s) are present in a particular child. It should also be noted that some measurement scales have primarily been developed and validated within specific clinical populations (e.g., for children with cerebral palsy); validity will likely be weaker when used outside of the specified clinical population. That said, diagnosis-specific assessments can be quite powerful:

- The focus on developmental stages corresponding to age equivalents <12 months in the Carter Neurocognitive Assessment, for example, is well suited to capture the slow, incremental developmental gains often seen in children with holoprosencephaly [47];
- The Gross Motor Functional Classification System, Expanded and Revised (GMFCS E&R) [48] and the Gross Motor Function Measure (GMFM) [49] can offer prognostic information for children with cerebral palsy. Beyond age 2, the five-tiered GMFCS E&R classification scheme is stable for most children over time [50]—facilitating communication as well as long-term planning. Developmental trajectories for the more detailed therapist-administered GMFM have been defined longitudinally [51]—permitting more individualized monitoring and treatment planning. Analogous to the GMFCS, systems are also available to classify function in other domains (e.g., the Manual Abilities Classification System [MACS] [8] for manual dexterity).

Summary and future directions

The diagnostic motor evaluation takes into account the child's medical history, developmental trajectory, and neuromotor examination. The presence of pathological motor signs (e.g., spasticity) and the absence of expected motor progress (e.g., emergence of postural responses) are both powerful predictors of motor disorders. Normed formal assessments can quantify abnormalities seen on clinical examination, but diagnosis depends upon clinical judgment—integrating these complementary streams of data.

The clinical motor evaluation continues to evolve as understanding of motor development (typical and disordered) continues to advance. The increasing use of normed formal assessments reflects the power of quantitative measures to aid in differentiating typical from disordered motor patterns, in monitoring change over time, and in prognosticating motor potential. However, as the natural history of motor disorders becomes better understood, as treatment paradigms change, and as new instruments are developed, we anticipate that recommendations regarding the use of specific formal assessments will continue to evolve and to be disorder specific. Further, as understanding of basic mechanisms underlying motor development advance, new qualitative phenomenology may also continue to be added to the providers' toolkit (Prechtl's General Movements as an example).

References

[1] Capute AJ, Accardo PJ. The infant neurodevelopmental assessment: a clinical interpretive manual for CAT-CLAMS in the first two years of life, part 2. Curr Probl Pediatr 1996;26(8):279−306.

[2] Knobloch H, Stevens FM, Malone AF. Manual of developmental diagnosis: the administration and interpretation of the revised Gesell and Amatruda developmental and neurologic examination. HarperCollins Publishers; 1980. p. 286.

[3] Capute AJ, Accardo PJ. The infant neurodevelopmental assessment: a clinical interpretive manual for CAT-CLAMS in the first two years of life, part 1. Curr Probl Pediatr 1996;26(7):238−57.

[4] Johnson CP, Blasco PA. Infant growth and development. Pediatr Rev 1997;18(7):224−42.

[5] Capute AJ, Shapiro BK. The motor quotient. A method for the early detection of motor delay. Am J Dis Child 1985;139(9):940−2.

[6] Dumont R., Willis J.O. Developmental assessment of young children [Internet]. Encyclopedia of special education. 2008. http://doi.org/10.1002/9780470373699.speced0636

[7] Lundberg A. Dissociated motor development – developmental patterns, clinical characteristics, causal factors and outcome, with special reference to late walking children [Internet]. 10, Neuropediatrics 1979. pp. 161−182. http://doi.org/10.1055/s-0028-1085323

[8] Eliasson AC, Krumlinde-Sundholm L, Rösblad B, Beckung E, Arner M, Öhrvall AM, et al. The Manual Ability Classification System (MACS) for children with cerebral palsy: scale development and evidence of validity and reliability. Dev Med Child Neurol 2006;48(7):549–54.
[9] Blasco PA. Primitive reflexes [Internet]. Clin Pediatr 1994;388–97. Available from: https://doi.org/10.1177/000992289403300703.
[10] Capute AJ, Accardo PJ, Vining EP, Rubenstein JE, Walcher JR, Harryman S, et al. Primitive reflex profile. A pilot study. Phys Ther 1978;58 (9):1061–5.
[11] Futagi Y, Tagawa T, Otani K. Primitive reflex profiles in infants: differences based on categories of neurological abnormality. Brain Dev 1992;14(5):294–8.
[12] Gao W, Gilmore JH, Giovanello KS, Smith JK, Shen D, Zhu H, et al. Temporal and spatial evolution of brain network topology during the first two years of life. PLoS ONE 2011;6(9):e25278.
[13] Gasser T, Rousson V, Caflisch J, Jenni OG. Development of motor speed and associated movements from 5 to 18 years. Dev Med Child Neurol 2010;52(3):256–63.
[14] Bargiela D, Yu-Wai-Man P, Keogh M, Horvath R, Chinnery PF. Prevalence of neurogenetic disorders in the North of England. Neurology 2015;85(14):1195–201.
[15] Novak I, Morgan C, Adde L, Blackman J, Boyd RN, Brunstrom-Hernandez J, et al. Early, accurate diagnosis and early intervention in cerebral palsy: advances in diagnosis and treatment. JAMA Pediatr. 2017;171(9):897–907.
[16] Heineman KR, Hadders-Algra M. Evaluation of neuromotor function in infancy-a systematic review of available methods. J Dev Behav Pediatr 2008;29(4):315–23.
[17] Bannigan K, Watson R. Reliability and validity in a nutshell [Internet]. J Clin Nurs 2009;3237–43. Available from: https://doi.org/10.1111/j.1365-2702.2009.02939.x.
[18] Spittle AJ, Doyle LW, Boyd RN. A systematic review of the clinimetric properties of neuromotor assessments for preterm infants during the first year of life. Dev Med Child Neurol 2008;50(4):254–66.
[19] AACAP. Assessment of young children [Internet]. [cited Apr 5, 2022]. https://www.aacap.org/AACAP/Member_Resources/AACAP_Committees/Infant_and_Preschool_Committee/Assessment_of_Young_Children.aspx
[20] Caesar R, Boyd RN, Colditz P, Cioni G, Ware RS, Salthouse K, et al. Early prediction of typical outcome and mild developmental delay for prioritisation of service delivery for very preterm and very low birthweight infants: a study protocol. BMJ Open 2016;6(7). Available from: https://www.ncbi.nlm.nih.gov/pmc/articles/PMC4947748/.
[21] Haataja L, Mercuri E, Regev R, Cowan F, Rutherford M, Dubowitz V, et al. Optimality score for the neurologic examination of the infant at 12 and 18 months of age. J Pediatr 1999;135(2 Pt 1):153–61.
[22] Campbell SK. Test-retest reliability of the test of infant motor performance [Internet]. Pediatric Phys Therapy 1999;60–6. Available from: https://doi.org/10.1097/00001577-199901120-00002.
[23] Einspieler C, Prechtl HRF, Bos A, Ferrari F, Cioni G. Prechtl's method on the qualitative assessment of general movements in preterm, term and young infants. Mac Keith Press; 2008. p. 104.
[24] Rose RU, Westcott SL. Responsiveness of the Test of Infant Motor Performance (TIMP) in infants born preterm. Pediatr Phys Ther 2005;17 (3):219–24.
[25] Noble Y, Boyd R. Neonatal assessments for the preterm infant up to 4 months corrected age: a systematic review. Dev Med Child Neurol 2012;54(2):129–39.
[26] Bayley N. Bayley scales of infant and toddler development. 3rd ed. [Internet]. PsycTESTS Dataset; 2012. http://doi.org/10.1037/t14978-000.
[27] Rhonda Folio M, Fewell RR. Peabody developmental motor scales (Pdms-2). 2000.
[28] Movement Assessment Battery for Children, Second Edition [Internet]. Encyclopedia of autism spectrum disorders; 2021. p. 3000. http://doi.org/10.1007/978-3-319-91280-6_301041.
[29] Bruininks RH, Bruininks BD. Bruininks-Oseretsky test of motor proficiency. 2nd ed. [Internet]. PsycTESTS Dataset; 2012. http://doi.org/10.1037/t14991-000.
[30] Griffiths A, Toovey R, Morgan PE, Spittle AJ. Psychometric properties of gross motor assessment tools for children: a systematic review [Internet]. BMJ Open 2018;e021734. Available from: https://doi.org/10.1136/bmjopen-2018-021734.
[31] Pediatric evaluation of disability inventory [Internet]. Shirley Ryan AbilityLab. [cited Apr 20, 2022]. https://www.sralab.org/rehabilitation-measures/pediatric-evaluation-disability-inventory.
[32] Wong V, Wong S, Chan K, Wong W. Functional Independence Measure (WeeFIM) for Chinese children: Hong Kong cohort. Pediatrics 2002;109(2):E36.
[33] Gates PE, Campbell SR. Effects of age, sex, and comorbidities on the pediatric outcomes data collection instrument (PODCI) in the general population. J Pediatr Orthop 2015;35(2):203–9.
[34] Tests & measures: a resource for pediatric physical therapy practitioners/activities scale for kids-performance version [Internet]. [cited Apr 20, 2022]. https://blogs.elon.edu/ptkids/2015/03/16/activities-scale-for-kids-performance-version/
[35] Law M, Baptiste S, McColl M, Opzoomer A, Polatajko H, Pollock N. The Canadian occupational performance measure: an outcome measure for occupational therapy. Canadian Journal of Occupational Therapy 1990;57(2):82–7.
[36] Graham F, Kennedy-Behr A, Ziviani J. Occupational performance coaching: a manual for practitioners and researchers. Routledge; 2020. p. 222.
[37] Modified Ashworth Scale (MAS) for multiple sclerosis (MS) [Internet]. APTA. [cited Apr 20, 2022]. https://www.apta.org/patient-care/evidence-based-practice-resources/test-measures/modified-ashworth-scale-mas-for-multiple-sclerosis-ms.

[38] Tardieu Scale and Modified Tardieu Scale (MTS) [Internet]. [cited Apr 20, 2022]. https://strokengine.ca/en/assessments/tardieu-scale-and-modified-tardieu-scale-mts/

[39] Bijlsma JWJ. Eular compendium on rheumatic diseases. Wiley-Blackwell; 2009. p. 824.

[40] Scale for the Assessment and Rating of Ataxia (SARA) [Internet]. Physiopedia. [cited Apr 20, 2022]. https://www.physio-pedia.com/Scale_for_the_Assessment_and_Rating_of_Ataxia_(SARA)

[41] International Cooperative Ataxia Rating Scale [Internet]. Shirley Ryan AbilityLab. [cited Apr 20, 2022]. https://www.sralab.org/rehabilitation-measures/international-cooperative-ataxia-rating-scale.

[42] Flamand VH, Massé-Alarie H, Schneider C. Psychometric evidence of spasticity measurement tools in cerebral palsy children and adolescents: a systematic review. J Rehabil Med 2013;45(1):14–23.

[43] Stewart K, Harvey A, Johnston LM. A systematic review of scales to measure dystonia and choreoathetosis in children with dyskinetic cerebral palsy. Dev Med Child Neurol 2017;59(8):786–95.

[44] Scholtes VAB, Becher JG, Beelen A, Lankhorst GJ. Clinical assessment of spasticity in children with cerebral palsy: a critical review of available instruments. Dev Med Child Neurol 2006;48(1):64–73.

[45] Sanger TD, Delgado MR, Gaebler-Spira D, Hallett M, Mink JW. Task force on childhood motor disorders. Classification and definition of disorders causing hypertonia in childhood. Pediatrics 2003;111(1):e89–97.

[46] Jethwa A, Mink J, Macarthur C, Knights S, Fehlings T, Fehlings D. Development of the Hypertonia Assessment Tool (HAT): a discriminative tool for hypertonia in children. Dev Med Child Neurol 2010;52(5):e83–7.

[47] Leevers HJ, Roesler CP, Flax J, Benasich AA. The Carter Neurocognitive Assessment for children with severely compromised expressive language and motor skills. J Child Psychol Psychiatry 2005;46(3):287–303.

[48] McDowell B. The gross motor function classification system - expanded and revised [Internet]. Dev Med Child Neurol 2008;725. Available from: https://doi.org/10.1111/j.1469-8749.2008.03104.x.

[49] Kenyon LK. Gross motor function measure (GMFM-66 and GMFM-88) users' manual [Internet]. Phys Occup Ther Pediatr 2014;341–2. Available from: https://doi.org/10.3109/01942638.2014.931744.

[50] Alriksson-Schmidt A, Nordmark E, Czuba T, Westbom L. Stability of the gross motor function classification system in children and adolescents with cerebral palsy: a retrospective cohort registry study. Dev Med Child Neurol 2017;59(6):641–6.

[51] Hanna SE, Rosenbaum PL, Bartlett DJ, Palisano RJ, Walter SD, Avery L, et al. Stability and decline in gross motor function among children and youth with cerebral palsy aged 2 to 21 years. Dev Med Child Neurol 2009;51(4):295–302.

Chapter 9.2

Language and communication assessment

Mary L. Leppert[1] and Wendy A. Jennejahn[2,3]

[1]*Department of Neurology and Developmental Medicine, Kennedy Krieger Institute and Department of Pediatrics, Johns Hopkins University School of Medicine, Baltimore, MD, United States,* [2]*Outpatient Speech-Language Clinic, Speech and Language Department, Kennedy Krieger Institute, Baltimore, MD, United States,* [3]*Speech and Language Department, Kennedy Krieger Institute, Baltimore, MD, United States*

The acquisition of speech and language use and understanding is one of the most critical and impressive processes in human development. It requires the neurologic integrity required to hear, understand, and attend to verbal communication, the desire to establish rapport with others, and the natural curiosity that drives learning. Typical language development is remarkably fluid, building in complexity of use and understanding with tremendous synchrony (Table 9.2.1). The complexity of speech and language disorders make atypical language difficult to characterize, which has led to a host of overlapping diagnostic terms. Recently, the CATALISE-2 Consortium [3] has consensus-guided terms employed in the description of disordered language development. Broadly, language disorders are problems that are "likely to endure to middle childhood and beyond, with significant impact on everyday social interaction or educational progress" [3]. The consortium further subcategorized language disorders as language disorders of "differentiating conditions" and "developmental language disorders." Language disorders that occur as part of the overall constellation of findings in a specific underlying condition such as fragile X, other genetic syndromes, or cerebral palsy are those of differentiating conditions. Developmental language disorder is the term used to describe language disorders not associated with underlying conditions. The assessment of language presently requires the use of older nomenclature, so the following text will refer broadly to speech, language, receptive, expressive, and pragmatic skills.

Speech is the production of sound used to communicate verbally. It requires good articulation, phonation, voice, and fluency. Language is the art of communicating. Language is broadly categorized as expressive, receptive, and pragmatic communication. Expressive language is the ability to communicate with words, and is assessed in quantitative, for example, vocabulary size, mean length of utterance, or in qualitative measures of language form, content, or use (Table 9.2.2) [4]. Receptive language is the capacity to understand verbal communication accurately and efficiently. Pragmatic language is everything that goes into communication that is not at the word level. Pragmatic skills allow the interpretation of tone of voice, cadence, prosody, facial expression, body language, and inference. Disorders of speech and language may occur independently, or may cooccur. Disorders of expressive, receptive, and pragmatic language often cooccur in patterns; expressive language may occur in isolation, receptive language disorders are generally accompanied by expressive language disorders, and pragmatic deficits may accompany receptive and expressive language disorder or may occur in isolation. Given the complexities of speech and language development, it is no wonder that disorders of communication are one of the most common presenting complaints of early childhood. In early childhood, approximately 16% of children have language delays, and almost half persist into childhood [5]. The prevalence of children with language delay is estimated at 7.5% in school-aged children [6]. The assessment of speech and language requires a firm understanding of normal milestones acquisition and an appreciation of common variations of development.

The clinical presentation of speech and language impairments differs with age. In early childhood, a paucity of word use relative to age expectancy is the most common presentation (Table 9.2.3). In later childhood, persisting speech sound disorders are a frequent reason for children to present for evaluation. Rarely in early childhood or preschool years do children present with concerns about understanding of language. Once children enter school, language assessments are most often prompted by academic underachievement, and less frequently by behavioral concerns. Older children with language disorders may present with social or behavioral issues. Language regression is an uncommon presentation, but warrants additional consideration in assessment.

The assessment of speech and language concerns begins with determining if either process is normal, delayed, atypical, or simply a developmentally acceptable variation of normal. Once normal and developmental variations of normal are ruled out, the evaluation turns to determining if speech and language concerns are primary or secondary in nature (Fig. 9.2.1). Speech and language development may be secondary to differentiating conditions such as hearing impairment, cognitive impairments, autism spectrum disorder (ASD), genetic or motor disorders or differences in the anatomy, respiratory physiology, or neurologic capacities. Each of these diagnoses must be considered and excluded before conferring a communication

TABLE 9.2.1 Typical acquisition of speech and language.

Age	Speech sounds Rossetti, 2005 [1] Crowe and McLeod, 2020 [2]	Expressive	Receptive	Pragmatic/ communicative intent
3 months		Coos	Attends to speaker's mouth or eyes	Social smile
6 months	puh, buh, ma, da,	Ahgoos Babbles Takes turns vocalizing	Orients directly to familiar voice Recognizes own name	Maintains eye contact Imitates facial expressions
9 months	Four different syllables	Mama/Dada (nonspecific) Gestures	Orients directly to bellStops when name is called	Shouts or vocalizes to gain attention
12 months	Babbles longer strings of sounds mamama	Mama and Dada (specific) Says one to two words spontaneously Imitates consonant - vowel combinations	Understands "no" Follows command with gesture	Extends arm to show objects Points to objects Vocalizes when another person calls
18 months		Mature jargon 7–10 words Asks for "more"	Points to picture Identifies two or more facial body parts Understands 50 words	Uses words to protest Requests assistance
24 months	/p, b, d, m, n, h, w/	Uses 50 words Uses two-word sentences frequently	Follows two-step command	Gestures to request toileting needs
30 months		Begins to use pronouns Points to pictures of familiar objects	Responds to simple questions Understands concept of "1"	
36 months	/t, k, g, f/ ng, y	250 words Uses plurals Three-word sentence States gender and first and last name	Follows a three-step unrelated command Understands prepositions Names colors	
48 months	/v,sz/l/ sh, ch, j			
60 months	Voiced th, zh, r			
72 months	Voiceless th			

disorder diagnosis, as intervention strategies and prognoses vary with each different etiology. Sufficient hearing is required to reproduce accurate speech sounds. The identification of hearing impairment is particularly time sensitive, as language outcomes may be mitigated by early identification and intervention [7], which has driven the national Early Hearing Detection and Intervention (EHDI) program. Less common primary causes of speech impairment are seen in disorders stemming from physiological, anatomical, or neuromuscular differences, such as cerebral palsy, cleft palate, or vocal cord injury, as oral, pharyngeal, and respiratory integrity are necessary to coordinate the production of speech appropriately.

Language may also be delayed in children with cognitive impairments as the ability to build a repository of vocabulary, access it effectively, summarize it appropriately, and to understand verbally presented information with accuracy and efficiency are often commensurate with intellectual ability. The CATALISE-2 Consortium states that low nonverbal skills do not preclude the diagnosis of a language disorder [3]. Communication also presumes an innate ability to establish rapport with others, a characteristic that is often compromised in children with ASD.

The primary responsibility of an examiner then, is to identify atypical language development, determine if it is a primary developmental language disorder or a consequence of a differentiating condition. Newborn hearing screening is now a universal practice in the United States [8,9] but approximately 25% of hearing impairment are postnatally

TABLE 9.2.2 Components of language.

Form—the structure of language			
Phonology	Rules for combining sounds	Speech	
Morphology	Rules for combining meaningful units	Word endings such as plurals, -ing	
Syntax	Grammar, sequencing units	Word order	
Content			
Semantics	Meaning	Word meanings (vocabulary)	Synonyms/antonyms Figurative language
Use			
Pragmatics	Social aspects of language	Interpretation and use of language in a social context	Discourse-level language (conversation, storytelling)

Source: Adapted from Bloom L, Lahey M. Language development and language disorders. New York: Wiley; 1978.

TABLE 9.2.3 Red flags for speech and language impairment.

Age (years)	Speech	Expressive	Receptive	Pragmatic
<2	No intelligible utterance	• Not babbling • No words	• No response to speech • Does not understand "no"	No attempt to communicate
2–3		• No words • Single word use	• Cannot follow commands	Does not respond to communication

acquired [10]. Thus the first assessment measure for every child with speech or language concerns is an audiology assessment. Once hearing is ruled out as a contributor to speech or language disorders, an assessment of general development and social skills are necessary to rule out intellectual disability or ASDs as underlying causes of atypical or delayed development.

The office assessment of speech

The informal assessment of speech sound production in the office setting is generally made as simple as an estimate of the portion of speech which is clearly understood by the evaluator. Typically, 25% of a child's utterances are understood by the evaluator by 18 months, 50% by 24 months, 75% by 36 months, and 100% by 4 years. These speech intelligibility benchmarks are a good way to determine the need for further assessment by a speech-language pathologist (SLP). Early identification and treatment can mitigate other problems including frustration/maladaptive behaviors, expressive language delay, and future difficulties with phonological awareness needed for reading.

Common developmental variations of speech and language in early childhood

Common variants of speech production in early childhood include reduplication, assimilation, sound omissions, substitutions, phoneme rearrangements, and developmental dysfluency. Vowels are generally acquired early; vowel errors are rare in normal development. Reduplication is when a syllable, or part of a syllable, is repeated, for example, "wawa" for water and "baba" for blanket. Assimilation is when one sound in a word is influenced by a nearby sound and becomes similar or identical. This especially occurs with velar sounds saying "kaek" for cat or "gaw-gee" for doggie and with labial sounds "fwim" for swim or "bay-ble" for table. Assimilation can persist until age 2 and half in typical development. Sound omissions,

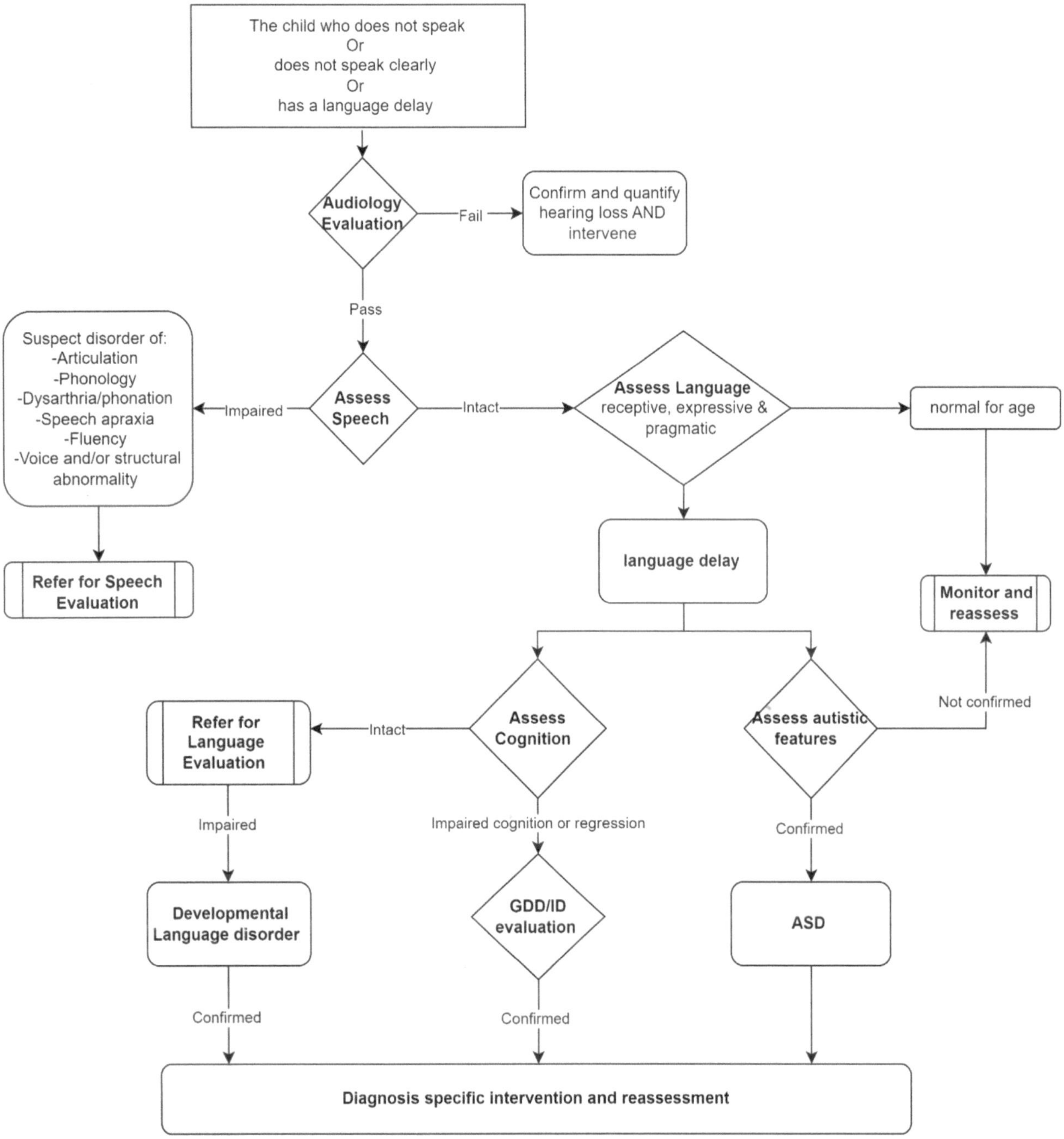

FIGURE 9.2.1 Assessment algorithm for a child who does not speak or has a language delay.

substitutions, and rearrangements may be thought of collectively as word approximations. Omissions are seen when the first or last phonemes of a word are dropped, for example, "ottle" for bottle, or "Schoo bus" for "school bus." Substitutions are common as children acquire specific phonemes in sequential manner that may extend to 7 or 8 years for sounds like "r," "th." Substitutions such as "wed woom" stand in place of "red room," and "fought" replaces "thought." Phonemic rearrangements include "pisghetti," for spaghetti or "hopspitle" for hospital. In each of these scenarios, a child who demonstrates consistent and appropriate use of the word approximation, the word should be counted toward the child's vocabulary estimate.

Developmental dysfluency is a common observation in children between 18 and about 48 months of life. Often confused with stuttering, developmental dysfluency is not a disorder of motor output of sound, but a strategy used to stall communication as children with rapidly growing vocabulary repositories search for the correct word. Fluent speech requires the ability to summon the precise word for the context of the communication, which is only achieved with continual language production. Examples of developmental dysfluency include repetitions "mommy, mommy, mommy...a squirrel is on the porch," filler sounds "mommy there is a um um um squirrel on the porch," or prolonged pauses "mommy a.......................squirrel is on the porch." Developmental dysfluency lessens as children gain the facility of vocabulary and should be seen only infrequently after 4 years of age, at which time it is associated only with more rare word use.

These early childhood speech production variants should improve or become suppressed as the child's motor and linguistic systems mature. SLPs are knowledgeable about stuttering risks that are likely to persist, normative data for the acquisition of sounds and consonant-vowel syllable shapes, and when common variants should no longer be present in the child's speech production.

Disorders of speech sound production

"Pediatric speech sound disorders" is a broad term that encompasses all of the difficulties or combination of difficulties a child could experience with speech production. Developmental sound variations that have no known cause and persist beyond age norms may be related to the motor aspects of speech for individual sounds, called "articulation," or the linguistic aspect, called "phonology" where there are patterns of errors. Functionally, these speech sound changes are part of the child's developing motor and linguistic systems that may become familiar and predictable to listeners, allowing the listener to understand most or all of the child's message depending on familiarity with the child or the context (e.g., topic and listener's background knowledge) of the child's message. Atypical sound changes warranting a referral for further evaluation include the persistent or unusually prevalent deletion of initial consonants "oi" for toy, replacement of a consonant sound by a glottal stop (e.g., a sound made by rapidly closing the space between the vocal folds interrupting the airflow, then releasing while breathing out as in the pause between "ut-oh." It is often used to represent the "t" sound as in "bo-el" for "bottle"), backing sounds produced in the front of the mouth such as "key" for tea, "go" for dough, "goo" for zoo, or "feger" for feather, and deleting the easier to produce element of a cluster "sop" for stop, "sow" for snow, "lu" for glue.

Speech sound disorders that are acquired or persist beyond the typical developmental timeframes fall into the general areas of motor/neurologic aspects, structural abnormalities (e.g., craniofacial anomalies and cleft palate), syndrome related (i.e., genetic syndromes such as veliocardiofacial syndrome or CHARGE syndrome), and sensory and perceptually based due to hearing loss. The pediatric neuromotor speech disorders of apraxia and dysarthria are commonly referred to as "developmental motor speech disorders" [11] because they do not manifest the same way as adults where the disorders are acquired. Though the speech output can be categorized as a neuromotor speech disorder, the etiology is unknown. A hallmark of these disorders is significantly reduced speech intelligibility. Features of childhood apraxia [12] (inconsistent voicing errors, trial and error, vowel or consonant distortions) and dysarthria (imprecise, unclear sound production) often overlap, making differential diagnosis between the two challenging. Though the term "developmental" is commonly used, pediatric motor speech disorders that are not acquired are idiopathic and often persist beyond the developmental period.

Childhood apraxia of speech (CAS) presents with no abnormalities in reflexes or muscle tone. The core impairment in planning and/or programming spatiotemporal parameters of movement sequences results in errors in speech sound production and prosody [13].

Dysarthria is characterized by "abnormalities in the strength, speed, range, steadiness, tone, or accuracy of movements required for breathing, phonatory, resonatory, articulatory, or prosodic aspects of speech production" [14]. This definition captures the adult presentation which is superimposed on the pediatric neurodevelopmental population. Adult dysarthria is acquired and therefore a change from previous speech function. In children, speech development is not complete, changing with maturation, resulting in overlapping characteristics between dysarthria and developing speech [15]. Despite diagnostic challenges, dysarthria is fundamentally a neurologic motor condition that results in weakness of the muscles for speech production. Thus children who have a disorder with associated muscle weakness, such as cerebral palsy and Down syndrome, necessitate the diagnosis of dysarthria when the speech characteristics are present.

A different classification system proposes four motor speech disorders in children: speech motor delay (SMD), childhood dysarthria (CD), CAS, and the combination of CD + CAS [16]. SMD is a childhood motor speech disorder characterized by imprecise and unstable speech, prosody, and voice that does not meet criteria for either childhood dysarthria or childhood apraxia of speech [17]. The protocol for diagnosing SMD is not clinically feasible at the outset, but offers a classification for children with delayed development of speech motor control that has been lacking [18], thus potentially filling a void for a previously undescribed group of speech-impaired children when key components of SMD become operationally defined.

There is no one comprehensive speech assessment making it challenging for physicians and SLPs to sort out speech production deficits in children. The reasons for this are multifactorial. Listening for and describing the speech production symptoms are paramount for making differential diagnosis, but consensus for established diagnostic criteria is lacking, especially for childhood apraxia where the theories are highly debated [19]. Further, many clinical features overlap between the speech sound disorders themselves, normal speech development, and other speech-language deficits such as poor language generation, word retrieval deficits, and dysfluency/stuttering (Table 9.2.4). As such, it is important to refer children at risk for severe speech disorders, or with significantly reduced intelligibility to a SLP for a comprehensive speech *and* language assessment that includes receptive and expressive language testing. The speech production assessment should include a hearing screening, case history of motor development and feeding skills, oral-mechanism examination of structures and function including feeding, and baseline data of the child's speech production skills (e.g., inventory of sounds, syllable shapes, omissions, substitutions, distortions, rate, and prosody). To look at clinical symptoms of speech, SLPs phonetically transcribe the child's speech either spontaneously, through imitation, or standardized word-level picture naming tasks, for example, the *Goldman-Fristoe Test of Articulation-third edition* [20] or the *Arizona-fourth edition* [21]. SLPs also listen to and analyze connected language, determine percent of speech intelligibility, and perform dynamic assessment to determine if cueing techniques are beneficial. Consideration of a differential diagnosis or a dual diagnosis for speech production requires that SLPs assess children longitudinally. Speech sound production may also be influenced by the acquisition or exposure of two or more languages simultaneously in early childhood. In simultaneous language exposures, the influences of each language are taken into consideration. Further, accents and dialects are noted during evaluation, but not considered as errors.

It is inappropriate to attempt to define the type of speech production disorder in nonverbal or minimally verbal children. Currently, there is a trend to diagnose children who are slow to produce babbling or only have a few sounds as having childhood apraxia of speech. In older children, we sometimes see the inappropriate diagnosis of apraxia when the underlying problem is language based, such as children experiencing word-retrieval and/or language formulation deficits. Physicians should refer to SLPs for comprehensive assessment and help educate parents that making the diagnosis is often a process. This process is made easier when children have language output to analyze, or minimally, when they can imitate. SLPs look to physicians for information about overall motor functioning, observations of tone, and ideally, a benchmark of speech intelligibility. SLPs may also refer back to the physician for confirmation of tethered oral tissue or other structural concerns (e.g., cleft palate).

The office assessment of language

Expressive language: Infants and young toddlers rarely demonstrate their language skills in the medical office, so evaluators are reliant on parent history of language for the first 18 months of life. In the first year, there are three phases of language development: guttural, prelinguistic, and linguistic. The guttural phase lasts only the first 2 months, and is marked by crying, burping, and sneezing—sounds made without social intent. Social intent characterizes the nonword level vocalizations of the prelinguistic phase. This phase begins in the third month with cooing, and is followed by razzing, ah-gooing babbling, and the nonspecific use of "mama" and "dada." The linguistic phase of language development begins at about 11 months, with the utterance of the first recognizable word, and the specific use of "mama" and "dada." Demonstration of word acquisition

TABLE 9.2.4 Components of spoken speech and language.

Speech	Expressive-output	Receptive-input	Pragmatic-input and output
Articulation	Vocabulary/ Lexicon	Understanding word meaning	Tone of voice
Phonology	Length of utterance	Interpreting sentence structures	Facial expression
Voice	Morphology	Temporally efficient understanding	Body language
Fluency	Syntax	Inference understanding	Figurative language
			Interpreting verbal and nonverbal components of conversation or discourse
			Adapting communication to the situation or context

begins slowly, with children using 7–10 words by 18 months, the rate of acquisition grows exponentially as children consistently use 20 words by 21 months, 50 words by 24 months, and 250 by 3 years. Children's vocabulary is initially associative, they use new vocabulary to label an image in their visual field, but with repetition, the words become part of the reference lexicon and able to be retrieved without visual prompts. Nouns are the first parts of speech learned, followed by adjectives and then verbs, so as vocabulary grows, so too does the length of utterance. Once the vocabulary size exceeds about 20 words, children go from single word utterances to two-word phrases, "truck" becomes "big truck," "daddy's truck," or "fire truck." Verb use elevates the utterances to two-word sentences, which come in at about 24 months, with a vocabulary size of 50 or more words. Pronoun use begins at 30 months, and plurals and prepositions by 36 months. In the office evaluation of expressive language in young children, one listens to the number of words used, and the length of utterance the child typically uses, and confirms vocabulary estimates and mean length of utterances with parents. When there is a discrepancy between the vocabulary size, the parts of speech employed, or the typical length of utterance, further assessment is recommended. Common patterns of discrepancy include a larger vocabulary size than is anticipated in a child using single word utterances or two-word phrases, or the report of sentence use in a child with a small vocabulary. Inflated vocabulary size is often due to crediting a child for words that are repetitions of others' utterances, or echolalia. Echolalia is a typical developmental phenomenon in which a child repeats the last word, or several words uttered by another. Echolalia typically begins at 18 months of life, peaks at 24 months, and should be gone by 30 months. Vocabulary size is also overestimated when parents "count" words that their child utters only once. Vocabulary that are genuinely part of the child's lexicon must be used specifically, consistently, and repeatedly. The use of two-word phrases is overestimated if one-word concepts are counted as two-word phrases, for example, "thank you," "more please," "good bye." Two-word phrases should be novel combinations of words that are not used rotely. Length of utterance is easily observed in children over 3 years of age, but the estimate of vocabulary size becomes more challenging, as words are often too numerous for parents to count. Expressive language delays may be suspected in the older child when they appear to have difficulty with word retrieval. Developmental dysfluency should be absent in children after 3–4 years of age, so children who stall to search for words are likely having difficulty accessing their reference vocabulary fluently.

Language regression, generally marked by the cessation of word use, requires additional consideration. The first consideration is to determine if the language regression is significant. Significant regression is the cessation of word of phrase use, after a child has demonstrated consistent and specific use of words and phrases. The use of words once or twice, which are not heard again, does not constitute regression. Language regression occurs in approximately 32% of children with ASD, and generally occurs at a mean age of 21 months [22]. Regression in children with ASD may also be accompanied by social skill regression. If other streams of development, motor, fine motor, adaptive, or cognitive skills share the regressive pattern; the examiner is advised to consider neurodegenerative disorders.

Expressive language in children who are simultaneously learning two or more languages is often delayed for a brief period as children are focused on listening and comprehension. This delayed period is generally between the ages of 18 and 36 months. When assessing language in bilingual children, it is important to inventory the number and use of unique words in each language. Frequently, the sum of the two lexicons are appropriate for the age of the child, but the number of words in each language may be insufficient. Bilingual children also go through a brief period in which they intermix the two languages, but they quickly learn to employ the appropriate language for the listener or the venue. Receptive language skills should be at age expectation in each of the languages the child is learning. Delays in receptive language necessitate further assessment.

Receptive language: The assessment of receptive language begins with parent report of the understanding of the word "No" in children, typically at 10 months of age. By a year, children can follow one-step commands with a gesture; parents point to the cup or extend an open hand to request a cup from their child. By 16 months, most children can follow familiar commands without a gesture, so long as the request includes words in their reference vocabulary. At 18 months, children begin to point to pictures, and to body parts of the face, by age 24 months they follow two-step, unrelated commands and at 36 months, they follow two-step prepositional commands, for example, "put the block under the cup, and put the pencil on top of the cup." Receptive language may be overestimated when children are credited for two-step commands are closely related or part of a routine, for example, "Take the dirty diaper and put it in the trash" or "put on your pajamas and brush your teeth." As children age, the success with which they follow multiple-step commands must also consider their ability to attend to the commands, and to remain on task.

Pragmatic language: In early childhood, pragmatics are simply evaluated by the child's ability to show affection, and to reciprocate visual attention, facial expression, and conversation. Children begin to point to make requests in the second year of life, and follow a pointing gesture by looking at the object of attention, and back at the person initiating the request, or joint attention. As children age, pragmatics are evaluated in the office by the child's ability to understand age-appropriate jokes or idioms. The linguistic demands of inferential thought become more taxing as children age, and sarcasm, idioms, and irony become an imperative for successful social dynamics. The social difficulties of children with pragmatic deficits may be very distressing to the child or the family, as the children with pragmatic disorders often have the desire to socialize, but may not have the finesse to do so in an age-appropriate manner.

Suspected delay in receptive, expressive, or pragmatic language prompts the evaluation of hearing, and nonverbal skills to rule out hearing or cognitive impairment as an underlying etiology of language delay. Confirmed expressive, receptive and expressive, and or pragmatic deficits beg the assessment of associated developmental or behavioral diagnoses, as the frequency of cooccurring diagnoses is high. Behavioral disruption is particularly common in young children with expressive language delay and is often a consequence of frustration from the inability to effectively convey wants or needs. Developmental diagnoses, including attention-deficit hyperactivity disorder (ADHD) and language-based learning disability, are very commonly seen in children with developmental language disorders, as are mental health concerns such as anxiety and mood disorders.

Language assessment

Assessment of language in a subspecialty practice such as developmental and behavioral pediatrics or neurodevelopmental disabilities also includes standardized measures to quantify delays in each component of language. Examiner awareness of the component tested by any particular measure is necessary to characterize the type of language disorder. Tests may measure reference or productive lexical breadth, semantics (word meaning), grammar use (including morphology and tense), the quality of discourse (longer utterances), or the comprehension of single words or discourse. Optimally, the tests employed should have good psychometric properties, be age appropriate, and give quantifiable results.

Expressive language measures of the size of vocabulary production in early childhood often rely on parent report of vocabulary size using inventories such as the language development survey (LDS) [23] or the MacArthur–Bates Communicative Developmental Inventories (MB-CDIs) [24]. Tests such as the Capute Scales [25] accept parent report of vocabulary size without inventorying words, but the evaluator must match the parental estimate of vocabulary with the other measures of expressive language, for example, the presence of immature versus mature jargon, the length of utterance, and word type use, to optimize reliability. The Expressive Vocabulary Test-third edition (EVT-3) [26] is a clinician-administered assessment of vocabulary available for all ages after 2 and half years, and is often paired with receptive vocabulary tests to measure disparities between reference vocabulary and word use. Fluharty-2 [27] subtests on describing actions and sequencing events may provide insight into expressive measures of semantics and discourse, but more comprehensive assessment of expressive language skills beyond lexical capacity is optimally performed by a SLP.

Receptive language measures differ from expressive assessments, in that they are less reliant on parent-provided data. The Receptive-Expressive Emergent Language Test, fourth edition (REEL-4) [28] is one measure that relies on parent interview for assessment of receptive as well as expressive language. Evaluating receptive language requires the awareness that some tests measure reference vocabulary, while others measure comprehension of discourse or connected language. The Peabody Picture Vocabulary Test-5 [29], for example, is a measure of reference vocabulary. It does not require a child to summon a word from their lexicon, only to accurately identify pictorial representations of words within their lexicon. Assessment of retention and comprehension of language discourse may be measured in the Capute Scales [25] in young children, or in the Fluharty- 2 [27] subtests on responding to directives and answering questions.

The characterization and quantification of language delays in the developmental evaluation allows the evaluator to confidently discuss the findings and implications with families, direct families to appropriate services, and define the probability of cooccurring or consequential secondary disorders of development or behavior. It behooves the evaluator to refer children with identified speech or language disorders to SLPs who provide expert refinement of the characterization of language deficits and recommendations for the most appropriate therapeutic interventions.

The speech-language assessment by the SLP broadly compares a child's spoken and written communication skills to that of typically developing peers using a variety of means. A complete assessment includes a review of case history, and the clinical judgment of evaluator observations are employed in conjunction with parent report and responses to parent/teacher questionnaires, and results of criterion-referenced, standardized, and/or dynamic assessments. The ultimate goal of the evaluation is to determine a delay versus a disorder, consider prognosis, and to provide treatment recommendations and goals; however, the SLP is also often tasked with qualifying a child for public early intervention or school-based services where criteria and cutoffs vary widely and arbitrarily from state to state. Public schools often rely on results of standardized assessments using standard scores that range from 1 to 2.5 standard deviations below the mean as their criteria for eligibility for school-based speech-language treatment services [30].

From the perspective of the SLP, the importance of a school-aged child receiving necessary school-based speech-language treatment services should not be underestimated. Services provided in the school setting address goals that are both meaningful to the child's education by incorporating their curriculum and to the child's overall success in achieving mastery generalizable to everyday routines. This is in contrast to the outpatient setting where SLPs strive to make the context of the therapy room functional when it is not a naturalistic setting. Thus the SLP bears the burden of making a case for eligibility of school-based services within the constraints of the standardized tests available and school eligibility criteria. Service delivery recommendations, including frequency of treatment, are best made by the SLP in conjunction with results of comprehensive evaluation that includes more than results of standardized tests.

Limitations of standardized assessments

Most standardized language assessments look at receptive and expressive language skills, or comprehension and use of vocabulary, semantics often including meaning relationships among words, morphology (i.e., the smaller units of meaning), and syntax. More recently, tests include pragmatic language (i.e., language in a situational context) checklists, supralinguistics (i.e., complex language where meaning is not obvious from the lexical or grammatical information), and preliteracy or literacy. Table 9.2.5 reviews a list of commonly used measures for the speech-language assessment. Standardized speech-language tests are far from perfect at identifying speech-language disorders (i.e., sensitivity) or translating to treatment goals. They are often heavily vocabulary based which relies on the child's exposure and experiences [56]. Standardized assessments also often fall short with considerations of English as second language learners, multilingualism, race/ethnicity, dialectical considerations, geographical region, and influence of the presence of subgroups (e.g., children with disabilities and gifted) in the normative sample. Further, revisions of tests, or new editions, do not always equate to new normative data. Currently, the fifth editions of Test of Language Development—Primary (TOLD-P5) [35] and—Intermediate (TOLD-I5) [41] have new normative data. The Test of Integrated Language and Literacy (TILLS) [45] does not include children with disabilities in their normative sample and the test avoids words with "th" to minimize African American Vernacular English (AAVE) dialectical considerations.

TABLE 9.2.5 Speech and language assessment tools.

Test, authors, date published	Age group	Language components
Language Development Survey Rescorla, 1989 [23]		Expressive vocabulary
MacArthur–Bates Communicative Developmental Inventories (MB-CDIs) Fenson et al., 2007 [24]	8–37 months	Vocabulary comprehension Vocabulary expression Gesture use Early grammar use
The Capute Scales Capute and Accardo, 2005[25]	0–36 months	Receptive/expressive language cognition
The Mullen Scales of Early Learning (MSEL) Mullen, 1995 [1]	0–68 months	Receptive/expressive language motor development—gross and fine Perceptual abilities Cognitive abilities
Receptive—Expressive Emergent Language Test—4th edition (REEL-4) Brown, Bzoch, and League, 2020 [28]	0–36 months	Criterion-referenced measure of receptive and expressive language including some preverbal skills Vocabulary inventory
Rossetti Infant Toddler Language Scale (RI-TLS) Rossetti, 2005 [31]	0–36 months— can be used for older children whose skills fall in this developmental age range	Criterion-referenced measure of preverbal (interaction/attachment, gesture use, play) and verbal (receptive, expressive, pragmatic) aspects of communication
Peabody Picture Vocabulary Test-5th edition (PPVT-5) Dunn, 2018 [29]	2.6–90 years	Receptive vocabulary—identification in a field of 4
Expressive Vocabulary Test-3rd edition (EVT-3) Williams, 2018 [26]	2.6–90 years	Expressive vocabulary—naming

(Continued)

TABLE 9.2.5 (Continued)

Test, authors, date published	Age group	Language components
Fluharty Preschool Speech and Language Screening Test-2nd edition (FLUHARTY-2) Fluharty, 2001 [27]	3–6.11 years	Articulation expressive receptive—connected language skills
Preschool Language Assessment Instrument-2 (PLAI-2) Blank, Rose, and Berlin, 2003 [32]	3.0–5.11 years	Receptive/expressive language Pragmatic language
Preschool Language Scales-5th edition (PLS-5) Zimmerman, 2001 [33]	0–7.11 years	Preverbal Receptive/expressive language Early literacy
Clinical Evaluation of Language Fundamentals—Preschool 3rd edition (CELF-P3) Wiig, Secord, and Semel, 2020 [34]	3.0–6.11 years	Receptive/expressive language Language content Language structure Academic language readiness Early literacy Story organization Pragmatics activity checklist
Test of Language Development—Primary: 5th edition (TOLD-5) Newcomer and Hammill, 2019 [35]	4.0–8.11 years	Receptive/expressive language Phonology-supplemental
Test of Auditory Comprehension of Language-4th edition (TACL-4) Carrow-Woolfolk, 2014 [36]	3.0–12.11 years	Comprehension of spoken: • Vocabulary • Grammatical morphemes • Elaborated phrases and sentences
Test of Expressive Language (TEXL) Carrow-Woolfolk and Allen, 2014 [37]	3.0–12.11 years	Spoken language abilities in: • Vocabulary • Grammatical morphemes • Elaborated phrases and sentences
Structured Photographic Expressive Language Test—3rd edition (SPELT-3) Dawson, Stout, and Eyer, 2003 [38]	4.0–9.11 years	Use of specific morphological and syntactical structures
Rice Wexler Test of Early Grammatical Impairment (Rice Wexler TEGI) Rice and Wexler, 2001 [39]	3.0–8.0 years	Phonology Grammar—third person singular, past tense, etc. Grammaticality judgment
Clinical Evaluation of Language Fundamentals-5th edition (CELF-5) Wiig, Secord, and Semel, 2013 [40]	5.0–21.0 years	Receptive/expressive language Language content Language structure Pragmatics Reading comprehension Structured writing
Test of Language Development: Intermediate-5th edition (TOLD:I-5) Hammill and Newcomer, 2020 [41]	8.0–17.11 years	Receptive/expressive language Listening Organizing Speaking
Oral and Written Language Scales-2nd edition (OWLS-II) Carrow-Woolfolk, 2011 [42]	3.0–21.11 years for spoken language5.0–21.11 years for written language	Receptive/expressive language Pragmatics Supralinguistics Reading comprehension Written expression
Comprehensive Assessment of Spoken Language, 2nd edition (CASL-2) Carrow-Woolfolk, 2017 [43]	3.0–21.11 years	Receptive/expressive language Supralinguistics

(Continued)

TABLE 9.2.5 (Continued)

Test, authors, date published	Age group	Language components
Oral Passage Understanding Scale (OPUS) Carrow-Woolfolk and Klein, 2016 [44]	5.0–21.0 years	Listening Comprehension—oral responses to questions about passages read aloud by the examiner
Test of Integrated Language and Literacy Skills (TILLS) Nelson, Plante, Helm-Estabrooks, and Hotz, 2016 [45]	6.0–18.0 years	Comprehensive test of language and literacy Notable that the standardization sample had no students with language disorder, other disabilities, or parent concern about language impairment
Clinical Evaluation of Language Fundamentals-5th edition Metalinguistics (CELF-5 Metalinguistics) Wiig and Secord, 2014 [46]	9.0–21.11 years	Higher level language skills embedded in upper grade curricula Making inferences Conversational skills Multiple meanings Figurative language Metalinguistics profile
Test of Narrative Language-2nd edition (TNL-2) Gillam and Pearson, 2017 [47]	4.0–15.11 years	Assessment of important aspects of narrative development without the need to transcribe the examinee's story
School-Age Language Assessment Measures (SLAM) Crowley and Baigorri, 2015 [48] https://www.leadersproject.org/disability-evaluation/school-age-language-assessment-measures-slam/	Preschool to high school	Dynamic assessment—not norm referenced Uses inferential stories requiring higher-order thinking Can be used with or without picture support, story retell or written expression Available in multiple languages
Narrative Language Measure (NLM) Petersen and Spencer, 2012 [49] https://www.languagedynamicsgroup.com/cubed/cubed_download/	Grade based, pre-K to third grade	Norm referenced by grade Listening Evaluates story grammar elements, language complexity, and episodic structure Informally assesses vocabulary
The Narrative Assessment Protocol-2 (NAP-2) Bowles, Justice, Khan, Piasta, Skibbe, and Foster, 2020 [50] http://www.narrativeassessment.com/	3.0–6.0 years	Provides a comprehensive summary of a child's narrative skills with information about how often the child uses each of 20 different narrative skills
Clinical Assessment of Pragmatics (CAPs) Lavi, 2019 [51]	7.0–18.0 years	Uses video scenes of real people in real-life social situations to assess ability to understand and use pragmatic language, including nonverbal cues and overall dynamics of social context
Social Emotional Evaluation (SEE) Wiig, 2008 [53]	6.0–12.11 years	Evaluates the social skills and higher-level language that students need to interact successfully in everyday situations
Social Language Development Test—Adolescent: Normative Update (SLDT-A:NU) Bowers, Huisingh, and LoGiudice, 2017 [54]	12.0–17.11 years	Measures students' abilities to make inferences, and interpret and respond to social interaction
Social Language Development Test—Elementary: Normative Update (SLDT-A:NU) Bowers, Huisingh, and LoGiudice, 2016 [55]	6.0–11.11 years	Measures language-based skills of social interpretation and interaction with friends to differentiate typically developing children from those diagnosed with autism spectrum disorder
Multidimensional Social Competence Scale (MSCS) Trevisan, Tafreshi, Slaney, Yager, and Iarocci, 2018 [56]	17–25 years	Criterion referenced across seven domains that represent social competence

Service eligibility and goal development should not be determined from one standardized assessment alone [57,58]. Incorporating "dynamic assessment," or nonstandardized assessment is an important step to taking a deeper dive into the child's areas of deficits (e.g., the efficiency of vocabulary acquisition, morphology, and discourse) [59]. It is also useful for multilanguage learners. A comprehensive speech-language assessment should also include the area of pragmatics or social communication (Table 9.2.5). Standardized tests struggle to capture the nuances of social communication because language is complex and social interactions are highly nuanced. A number of pragmatic assessments are reviewed in Table 9.2.5. While they provide some useful data, they should not be used in isolation to diagnose a social communication disorder [2]. Informal assessment looks at functional use of language in context. Language sampling, conversational, and/or narrative discourse analysis is unique to a comprehensive speech-language evaluation and provides rich information that translates to meaningful goal development. In sum, to support families in obtaining appropriate speech-language services, the first step is to seek a comprehensive evaluation that includes more than results of a standardized test. From there, the SLP seeks to make recommendations that result in a functional and meaningful treatment plan with the goal of achieving better outcomes for the child and the family.

The acquisition of language in early childhood is dynamic and fluid. The rapidity with which children identify, learn, and conjure vocabulary, compose, understand discourse, and make inferences is a marvel. The complexity of language development, however, provides many opportunities for atypical development, which are often difficult to identify or categorize. The office assessment of language is a starting point for preliminary or broad diagnoses, but specific characterization of the speech or language disorder(s) require assessment of the form, content, and use of language and its many components, which are often best accomplished with specific test measures and dynamic assessment by experts of speech and language pathology. Longitudinal assessment requires periodic re-evaluation to measure the efficacy of treatment, the progress toward language goals, and the emergence of associated disorders.

References

[1] Mullen EM. Mullen scales of early learning. Bloomington, MN: Pearson Inc.; 1995.

[2] Crowe K, McLeod S. 'Children's English consonant acquisition in the United States: a review. Am J Speech Lang Pathol 2020;1–15. Available from: https://doi.org/10.1044/2020_AJSLP-19-00168.

[3] Bishop DVM, Snowling MJ, Thompson PA, Greehalgh TCATALISE-2 Consortium. Phase 2 of CATALISE: a multinational and multidisciplinary Delphi consensus study of problems with language development: terminology. J Child Psychol Psychiatry 2017;58(10):1068–80.

[4] Bloom L, Lahey M. *Language development and language disorders*. New York: Wiley; 1978.

[5] Feldman HM. How young children learn language and speech. Pediatr Rev 2019;40(8):398–411.

[6] Paul R. Language disorders. In: Gallagher A, Bulteau C, Cohen D, editors. Handbook of child neurology. Neurocognitive development: disorders and disabilities, 174. Elsevier Press; 2020. p. 21–35.

[7] Ching TYC, Dillon H, Button L, Seeto M, Van Buynder P, Marnane V, et al. Age at intervention for permanent hearing loss and 5-year language outcomes. Pediatrics 2017;140(3)e20164274. Available from: https://doi.org/10.1542/peds.2016-4274.

[8] American Academy of Pediatrics Joint Committee on Infant Hearing Detection and Intervention Programs. Year 2007 position statement: principles and guidelines for early hearing detection and intervention programs. Pediatrics 2007;120:898. Available from: https://doi.org/10.1542/peds.2007-2333.

[9] U.S. Preventive Services Task Force. Universal screening for hearing loss in newborns; US preventive task force recommendation statement. Pediatrics 2008;122:143–8.

[10] Weichbold V, Nekahm-Heis D, Welzl-Mueller K. Universal newborn hearing screening and postnatal hearing loss. Pediatrics 2006;117(4):e631–6.

[11] Crary MA. Developmental motor speech disorders. San Diego, CA: Singular Publishing Group, Inc; 1993.

[12] Strand EA, McCauley RJ. Dynamic evaluation of motor speech skill (DEMMS). Baltimore MD: Brookes Publishing; 2019.

[13] American Speech-Language-Hearing Association. Childhood apraxia of speech (Technical report). 2007b. Retrieved from http://www.asha.org/public/speech/disorders/ChildhoodApraxia/doi:10.1094/PDIS-91-4-0467B

[14] Duffy JR. Motor speech disorders: substrates, differential diagnosis, and management. St. Louis, MO: Elsevier; 2013.

[15] Allison KM, Hustad KC. Data-driven classification of dysarthria profiles in children with cerebral palsy. J Speech Lang Hear Res 2018;61(12):2837–53.

[16] Shriberg LD, Kwiatkowski J, Mabie HL. Estimates of the prevalence of motor speech disorders in children with idiopathic speech delay. Clin Linguist Phon 2019.

[17] Shriberg LD, Campbell TF, Mabie HL, McGlothlin JH. Initial studies of the phenotype and persistence of speech motor delay (SMD). Clin Linguist Phon 2019.

[18] Shriberg LD, Wren YE. A frequent acoustic sign of speech motor delay (SMD). Clin Linguist Phon 2019.

[19] Allison KM, Cordella C, Iuzzini-Seigel J, Green JR. Differential diagnosis of apraxia of speech in children and adults: a scoping review. J Speech Lang Hear Res 2020;63(9):2952–94.

[20] Goldman R, Fristoe M. Goldman Fristoe test of articulation-3 (GFTA-3). Bloomington, MN: NCS Pearson Inc; 2015.
[21] Fudala JB, Stegall S. Arizona articulation and phonology scale. 4th ed. Torrance CA: Western Psychological Services; 2017.
[22] Barger BD, Campbell JM, McDonough JD. Prevalence and onset of regression within autism spectrum disorders: a meta-analytic review. J Autism Dev Disord 2013;43:817–28.
[23] Rescorla L. The language development survey: a screening tool for delayed language in toddlers. J Speech Hear Disord 1989;54:587–99.
[24] Fenson L, Bates E, Dale PS, Marchman VA, Reznick JS, Thal DJ. MacArthur–Bates communicative development inventories. Baltimore, MD: Brookes Publishing; 2007.
[25] Accardo PJ, Capute AJ. The Capute scales. Baltimore MD: Brookes Publishing Company; 2005.
[26] Williams KT. Expressive vocabulary test. 3rd ed. Bloomington, MN: NCS Pearson; 2018.
[27] Fluharty NB. Fluharty preschool speech and language screening test. 2nd ed. Austin, TX: Pre-Ed Inc; 2001.
[28] Brown VL, Bzoch KR, League R. Receptive-Expressive Emergent Language 4th edition (REEL-4). 2020.
[29] Dunn DM. The peabody picture vocabulary test–Fifth Edition (PPVT-5). Toronto, Ontario, Canada: Pearson Canada Assessment, Inc.; 2018.
[30] Spaulding T, Szulga M, Figueroa C. Using norm-referenced tests to determine severity of language impairment in children: disconnect between US policy makers and test developers. Lang Speech Hear Serv Sch 2012;43:176–90. Available from: https://doi.org/10.1044/0161-1461(2011/10-0103).
[31] Rossetti L. The Rossetti infant-toddler language scale. East Moline, IL: Lingui Systems Inc.; 2005.
[32] Blank M, Rose SA, et al. Preschool language assessment instrument. 2nd ed. Austin, TX: PRO-ED Inc.; 2003.
[33] Zimmerman IL. Preschool language scales. 5th ed. Bloomington, MN: NCS Pearson; 2001.
[34] Wiig EH, Secord WA, Semel E. Clinical evaluation of language fundamentals – preschool. 3rd ed. Bloomington, MN: NCS Pearson; 2020.
[35] Newcomer PL, Hammill DD. Test of language development – primary. 5th ed. Austin, TX: Pro-Ed Inc.; 2019.
[36] Carrow-Woolfolk E. Test for auditory comprehension of language. 4th ed. Austin, TX: Pro-Ed Inc.; 2014.
[37] Carrow-Woolfolk E, Allen E. Test of expressive language. Austin, TX: Pro-Ed Inc.; 2014.
[38] Dawson JI, Stout CE, Eyer JA. Structured photographic expressive language test. 3rd ed. Austin, TX: Pro-Ed Inc.; 2003.
[39] Rice ML, Wexler K. Rice Wexler test of early grammatical impairment. USA: The Psychological Corporation; 2001.
[40] Wiig EH, Semel E, Secord WA. Clinical evaluation of language fundamentals. 5th ed. Bloomington, MN: NCS Pearson; 2013.
[41] Hammill DD, Newcomer PL. Test of language development–intermediate. 5th ed. Austin, TX: Pro-Ed Inc.; 2020.
[42] Carrow-Woolfolk E. Oral and written language scales. 2nd ed. Torrance, CA: Western Psychological Services; 2011.
[43] Carrow-Woolfolk E. Comprehensive assessment of spoken language. 2nd ed. Torrance, CA: Western Psychological Services; 2017.
[44] Carrow-Woolfolk E, Klein AM. Oral passage understanding scale. Torrance, CA: Western Psychological Services; 2016.
[45] Nelson N, Plante E, Helm-Estabrooks N, Hotz G. Test of integrated language an literacy skills. Baltimore, MD: Brookes Publishing; 2016.
[46] Wiig EH, Secord WA. Clinical evaluation of language fundamentals-metalinguistics. 5th ed. Bloomington, MN: NCS Pearson; 2014.
[47] Gillam RB, Pearson NA. Test of narrative language. 2nd ed. Austin TX: Pro-Ed Inc; 2017.
[48] Crowley, C. School-age language assessment measures. Leaders Project, Teachers College, Columbia University. 2015. http://www.leaderspro-ject.org
[49] Petersen D, Spencer T. The narrative language measures: tools for language screening, progress monitoring, and intervention planning. Perspect Lang Learn Educ 2012;19:119. Available from: https://doi.org/10.1044/lle19.4.119.
[50] Bowles RP, Justice LM, Khan KS, Piasta SB, Skibbe LE, Foster TD. Development of the narrative assessment protocol-2: a tool for examining young 'children's narrative skill. Lang Speech Hear Serv Sch 2020;51(2):390–404. Available from: https://doi.org/10.1044/2019_LSHSS-19-00038.
[51] Lavi A. Clinical assessment of pragmatics. Torrance, CA: Western Psychological Services; 2019.
[52] Wiig E. Social emotional evaluation. Austin, TX: Pro-Ed Inc; 2008.
[53] Bowers L, Huisingh R, LoGiudice CM. Social language development test – adolescent: normative update. 2017.
[54] Bowers L, Huisingh R, LoGiudice CM. Social language development test – elementary: normative update. 2016.
[55] Trevisan DA, Tafreshi D, Slaney KL, Yager J, Iarocci G. A psychometric evaluation of the Multidimensional Social Competence Scale (MSCS) for young adults. PLoS ONE 2018;13(11):e0206800. Available from: https://doi.org/10.1371/journal.pone.0206800.
[56] Campbell T, Dollaghan C, Needleman H, Janosky J. Reducing bias in language assessment: processing-dependent measures. J Speech Lang Hear Res 1997;40(3):519–25. Available from: https://doi.org/10.1044/jslhr.4003.519.
[57] Denman D, Speyer R, Munro N, Pearce WM, Chen YW, Cordier R. Psychometric properties of language assessments for children aged 4–12 years: a systematic review. Front Psychol 2017;8:1515.
[58] Spaulding TJ, Plante E, Farinella KA. Eligibility criteria for language impairment: is the low end of normal always appropriate? Lang Speech Hear Serv Sch 2006;37(1):61–72. Available from: https://doi.org/10.1044/0161-1461(2006/007).
[59] Crowley C. Dynamic assessment: how does it work in the real world of preschool evaluations? Leader live. 2014. https://leader.pubs.asha.org/do/10.1044/dynamic-assessment-how-does-it-work-in-the-real-world-of-preschool-evaluations/full/#comment-8588.

Chapter 9.3

Cognitive assessment

Amy Jo Merchant Newmeyer

Department of Pediatrics, Baylor College of Medicine-Temple Campus, McLane Chidren's Hospital, Temple, TX, United States

History of cognitive assessment

In the early 1800s, Jean Esquirol first proposed the concept of varying levels of intellectual functioning, and that language skills function as a criterion for distinguishing these levels. French psychologist Albert Binet subsequently published the Binet–Simon Scale in 1904, which was the first intelligence scale for children. It was designed to measure cognitive ability related to performance in academic skills, with the intention of screening children in the French public schools. Albert Binet claimed that this was a way to differentiate which children would not perform well in a typical classroom setting due to intellectual delay, versus those who were performing poorly for other reasons [1].

In 1910 the Binet–Simon scale was translated into English and administered to school children in the United States. The 1937 revision of the instrument officially became known as the Stanford-Binet [1]. Other intelligence tests were also developed around this time, including the Weschler scales. David Weschler developed the Wechsler Bellevue scale in 1939, which was designed for use with adults [1]. At that time, he felt the Stanford-Binet was too heavily based on verbal skills and sought to assess nonverbal intelligence skills. The Stanford-Binet has been revised over the years, and now includes both verbal and nonverbal intelligence scores. The Stanford-Binet and the Weschler scales were the predominant tests of cognition for decades. In the 1980s, Alan Kaufman raised criticism of previous intelligence tests, including their limited ability to assess children who were nonverbal or non-English speaking. In 1983 he published the Kaufman Assessment Battery for Children (KABC) [2].

Although definitions of intellectual disability have broadened over time to include adaptive behavior and the need for supports, there is still a focus on obtaining objective data through the administration of a standardized IQ test in addition to measures of behavior and adaptive functioning [3]. Standardized IQ assessments can help with the design of educational interventions and with transition planning for older children and adolescents [1].

The nature and purpose of cognitive assessment

Standards for educational and psychological testing have been released by the American Educational Research Association (AERA), American Psychological Association (APA), and National Council on Measurement in Education (NCME), which define assessment more broadly as a procedure for integrating test results with information from multiple sources that is used to make inferences or decisions about the characteristics of a person [4]. Other sources of information used in the assessment process could include additional standardized assessments, structured interviews, standardized questionnaires, and patient observations. Assessment is a continual practice that occurs throughout the provider–patient relationship [5], rather than relying on a one-time test to make diagnostic inferences. The fact that assessment can continue without testing is one way to distinguish between these two terms [6].

The purposes of screening and assessment also differ and can impact the choice of tests or questionnaires that are used. A screening tool is designed to determine the risk of a particular condition and is not meant to replace a more in-depth assessment. To ensure that all cases of a particular condition are identified, a screening test should have a high false positive and a low false negative rate.

Ideally, the assessment process encompasses information from multiple sources, including the patient, parents/caregivers, teachers, medical providers, and other professionals that interact with the patient. It is not ideal to base the diagnosis of a patient on the results of a single test. The choice of sources for assessment is dependent upon the purpose or reason for the assessment.

Most commonly, assessment is undertaken to initially identify or diagnose a disorder or other medical condition. The process of diagnosis typically involves using a series of assessment tools or other strategies to identify areas to target for intervention. Criteria for diagnoses of many psychological and developmental conditions are contained in the *Diagnostic and Statistical Manual of Mental Disorders* (DSM-5) [7]. Standardized tests and other instruments such as standardized questionnaires are often developed around the criteria outlined in this manual.

Once an initial diagnosis is established, assessments are often performed periodically thereafter to monitor an individual's development and learning. For instance, the Bayley Scales of Infant Development may be utilized to routinely follow children at risk for developmental delay based on prematurity. Once a developmental diagnosis is made, ongoing assessment with the same instrument may be used to monitor progress over time. Developmental and cognitive assessments are often used to identify children who might benefit from services, guide decision-making about a child's educational services, and report and communicate with families and other professionals.

Dynamic assessment

The goal of standardized assessment is to classify and group students for differential instruction. In contrast, dynamic assessment is characterized by a pretest-intervene-posttest model that focuses on learning processes over time. If the overall goal is to understand how a child learns, then using dynamic assessment and engaging in a real-life learning process would best meet this goal [8].

The basis for dynamic assessment is in the concept of "zone of proximal development" theorized by Vygotsky [9]. He theorizes that the development of higher intellectual functioning requires social interactions within zones of development at gradually more complex levels [1]. Identifying a child's developmental abilities and limitations in relationship to their learning environment is a core concept of dynamic assessment. The assessor actively engages the child in learning and promotes positive changes in their cognitive development. The concept of dynamic assessment includes the motivational and instructional factors that contribute to a child's development, in addition to the results of a traditional standardized cognitive assessment [10].

Psychometrics of cognitive assessment

When determining the ranges of assessments to administer, the examiner must choose an instrument with psychometric properties that are acceptable to measure the specific construct in question. One of the most important factors to consider is *standardization*. Standardization refers to the adherence to uniform administration procedures, time limits, and directions. Scoring should also be uniform across settings [11].

Normatization refers to the selection of a large, representative population of individuals when developing the scoring norms when the instrument is developed. Adequate normatization permits accurate comparison of the individual's test results with those of the representative group. Accurate statements can then be made about a child's level of performance compared to peers, or level of performance on one measure when compared to another measure of intelligence [11].

The *reliability* of a test is related to the consistency and predictability of the test. The scores on the test should be similar on two different administrations of the test to the same individual [12]. High reliability should include both test-retest reliability and interrater reliability when comparing one examiner to another. There should also be reliability between different forms of the same test if applicable.

Because there is likely some degree of measurement error in a test, scores are presented with a confidence interval based on the standard error of measurement. Patient fatigue, anxiety, or hunger can impact measurement error, along with the general environmental conditions when the test is administered. Using uniform testing conditions can decrease the variance in scores. Confidence intervals are usually reported at 90%–95%, meaning that an individual's true score on the test would be found in the same range 90%–95% of the time on repeat administrations [11].

Validity assesses whether a test measures what it is purported to measure. There are different types of validity:

1. *Content validity* refers to whether the test measures the construct it is purported to measure. In developing a test, the experts must determine that the items measure all the main areas of the content area in the correct proportion. One of the limitations of content validity is that it relies on the subjective judgment of the experts [12].
2. *Criterion validity, or predictive validity*, compares the performance of the test in question to performance on another outside measure. For instance, comparing a new intelligence test to an established measure such as the Stanford-Binet. This relationship is usually expressed as a correlation coefficient. This can be further divided into *concurrent validity* and *predictive validity*. *Concurrent validity* refers to measurements taken at around the same time as the test. For instance, taking a test of intelligence and a test of academic achievement. *Predictive validity* refers to outside measurements after the test in question is scored. For instance, taking a test of intelligence and then a test of academic achievement a year later. Predictive validity is preferred for selection of certain groups, such as

identifying which individuals are most likely to benefit from specific treatment approaches. The most difficult problem for criterion validity is finding the most acceptable and definable outside criterion [12].

3. *Construct validity* refers to whether a test measures the desired theoretical construct or trait. To determine construct validity, the test developer must consider how this construct relates to other identified variables or traits, and then confirm a correlation with the other traits. Once these relationships are verified by published studies, the degree of confidence that can be placed in the test increases. Even if the validity of a test is determined during the construction of the testing instrument, the test may not be valid in a specific situation with an individual patient [12].

Scoring: The raw score on a standardized test must be converted to a standard score which compares the individual test-takers test performance to that of their peers. The distribution of scores is typically viewed on a bell-shaped curve, where 50% of the scores fall below average or the mean of 100, and another 50% of the scores fall above the mean of 100. The average standard score is 100, with a standard deviation of 15. Thus approximately 68% of the population will fall within one standard deviation from the mean (85–115). Standard scores may also be presented as a scaled score, with a mean of 10 and standard deviation of 3. Raw scores can also be transformed into percentile rank and t-scores. T-scores have a mean of 50 and standard deviation of 10 and are most often used for behavioral checklist scores [11].

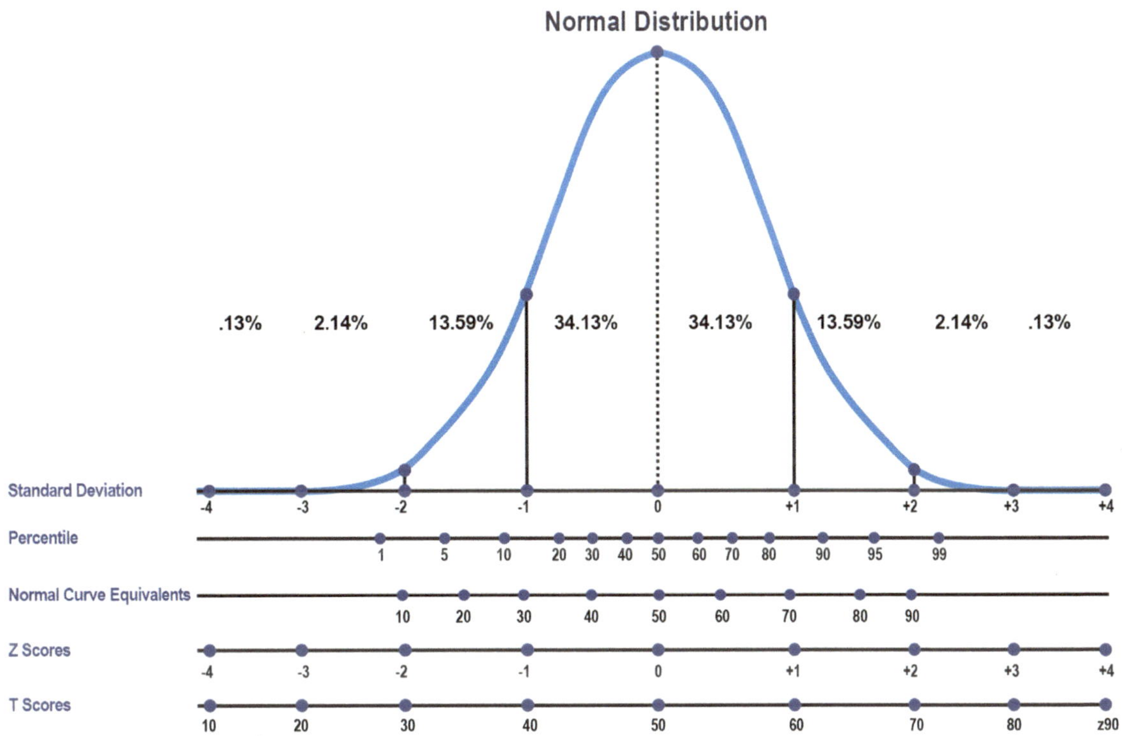

Choosing the right test/instrument

The first step to choosing a cognitive test is to determine what specific cognitive construct needs measured. The provider should determine whether the evaluation of the child requires a full cognitive assessment, a short measure of IQ, or the measurement of a specific cognitive area, such as memory or attention. Once that is determined, the next consideration is to choose a measure with high reliability and validity to assess that particular construct. Other factors that may influence patient performance on a cognitive test include communication skills, presence of vision or hearing impairments, presence of motor impairment, or cultural factors that may influence the outcome of the test. The provider should also ensure that a standardized cognitive test is only one part of a comprehensive assessment, which may include questionnaires, parent interview, patient interview, and observations of the patient.

Common cognitive assessment tools

Stanford-Binet

The Stanford-Binet was first introduced into the United States in 1910 and remains one of the most popular intelligence tests. It has undergone several revisions over time, with the latest version (Stanford Binet 5) released in 2003. It is one of the few tests that can be administered from age 2 through adulthood. It assesses five factors (fluid reasoning, knowledge, quantitative reasoning, visual-spatial processing, and working memory), and includes both verbal and nonverbal domains. It is more sensitive in measuring cognitive function in younger children and high-risk individuals than previous versions [1].

Wechsler scales

The original Wechsler-Bellevue scale was released in 1939 as a measure of adult intelligence [13]. In 1949 the Weschler Intelligence Scale for Children was published, and this extension placed it in direct competition with the Stanford-Binet. The Wechsler Preschool and Primary Scale of Intelligence was subsequently published in 1967 [14], extending the age range for children to ages 2.5–7.5 years. The various Wechsler scales have been revised several times since their inception. They are currently the most widely used standardized IQ tests. They are preferred due to the presence of tests specific to certain age groups, and the reliability and stability of IQ scores over time [1].

Kaufman battery

KABC was originally developed with a focus on gaining information about a child's learning potential, in order to apply that information to contextualized learning settings. The KABC is based on neuropsychological theory [15]. The most recent edition, the KABC-II NU was developed to provide an assessment that is sensitive to ethnic minorities and children with disabilities. It encompasses five scales, including simultaneous processing, sequential processing, planning, learning, and knowledge. It has a nonverbal option for children who have limited expressive language ability. The test has good reliability and was standardized on 3025 3- to 18-year-old individuals across 39 states. It was conormed with the Kaufman Test of Educational Achievement-II [16].

Bayley Scales of Infant Development

The Bayley Scales of Infant Development was originally published in 1969 and is rooted in theories of infant development rather than intelligence. As a result, it is only moderately predictive of later IQ scores [17]. The Bayley is sensitive to identifying both developmental delay and developmental disability. The Bayley Scales of Infant Development-IV, which assesses children ages 1–42 months, was published in 2019 and provides scores in five domains: cognition, language, social-emotional, motor, and adaptive behavior [18].

Leiter International Performance Scale-3

The Leiter International Performance Scale was originally developed in 1929 to assess individuals with hearing impairment and those who are non-English speaking. The Leiter International Performance Scale-3 was published in 2013. It was normed on 1600 + children that represent the US population on ethnicity/race, gender, and age. The Leiter consists of 20 subtests and is normed on individuals ages 3 to 75 + years. One unique feature of the Leiter International Performance Scale is the ability to administer the test without any verbal instructions, and without requiring any verbal responses. The Leiter has only nonverbal subtests and is often used to test children with ASD or those with significant language deficits [19].

Differential Abilities Scales

The Differential Abilities Scales (DAS) test was originally developed in Great Britain as the British Abilities Scales. The US version of the British test was published in 2007 as DAS-II.

DAS-II can be administered to children 2 and half to 17 years of age. There is a separate test battery for children ages 2 years 6 months to 6 years 11 months, and a school age battery for children ages 7 years 0 months to 17 years 11 months. The DAS provides an overall General Conceptual Abilities score which is similar to an IQ score. It also provides a diagnostic profile of a child's intellectual ability. In addition, there is a nonverbal composite for non-English–speaking children or children with communication disorders. Fewer subtests are required for scoring, which can be an advantage for children who have difficulty with attention or disruptive behaviors.

The DAS-II was normed on 3480 children ages 2 years 6 months to 17 years 11 months stratified by age, gender, race/ethnicity, parent education, and geographic region based on 2002 US Census data [20].

Kaufman Brief Intelligence Test-2

The Kaufman Brief Intelligence Test (KBIT) was originally published in 1990 and the second edition was published in 2004. It is normed for individuals ages 4–90 years. It provides a verbal and nonverbal standard score, along with a composite score and percentile ranks by age. One of the advantages of the KBIT-2 is the short administration time of 20 minutes [21].

NEPSY-II

The NEPSY was initially published in 2007 [22] as a measure of neuropsychological functioning of young children. The NEPSY-II was released in 2007 for children ages 3–16. It covers six domains, including executive function and attention, language, memory and learning, sensorimotor, visual-spatial processing, and social perception. Results are expressed in percentile ranks and scaled scores. The patterns of the scores can be compared to those of children with various conditions, including ASD and ADHD. Qualitative behavioral observations of the child can also be completed to the normative sample. The full battery can be given as a comprehensive neuropsychological test, or selected subtests can be administered in areas of interest (memory, attention, etc.).

Raven's progressive matrices clinical edition (Raven's 2)

The Raven's 2 was released in 2018. It is a measure of reasoning ability, observational skills, and nonverbal intelligence. It can be administered to those with communication disorders, or with language limitations. The age range for the Raven's 2 is 4–85 years [23].

There are other less common cognitive tests that may also be administered, based upon the specific needs of the patient. Table 9.3.1 contains additional information about cognitive tests that may be useful tools for the neurodevelopmental specialist.

Memory and attention

In addition to tests of general cognitive functioning, tests of individual cognitive functions may also be helpful in differential diagnosis and in developing recommendations for treatment. For example, children with ADHD may have average IQ on a standard cognitive test, but struggle in school due to deficits in memory, attention, and processing speed. Executive functioning is a set of processes and skills that help a person plan and complete tasks. These functions include attentional control, verbal and nonverbal working memory, and problem-solving. Tests of memory and attention are outlined in Table 9.3.1.

Special considerations

Multicultural assessment

It is important to recognize that some standardized cognitive tests may be suboptimal for use in diverse populations, particularly if they have been normed to a more specific population. The examiner should ascertain whether the standardized test and testing procedures are appropriate, fair, and useful for individuals in a certain culture or group. Factors to consider include test bias, psychometrics of the test, examiner bias, and acculturation [5]. When working with children with developmental disabilities, it is also essential to consider what accommodations may be needed to minimize the impact of the individual's attributes that are not relevant to the construct measured. For instance, a child with communication impairment may not be able to complete a test that requires verbal answers [5]. Examples of test modification include the following:

1. Modifying the presentation form
2. Modifying response format

TABLE 9.3.1 Cognitive assessment tools.

Instrument	Variables assessed	Age range	Administration time	Training requirements	Strengths	Digital	References
Bayley Scales of Infant and Toddler Development-4	Cognitive, language, motor, social emotional, and adaptive	16 days to 42 months	30–70 min	Level B	Nonverbal option available Normed to diverse populations (language difficulties or cultural differences)	Q global for administration of adaptive and social-emotional subtests	[18]
CTONI-2	Nonverbal cognitive: Analogical thinking; categorical formulation; sequential reasoning	6–89 years	40–60 min	Level B	Oral instructions available for common non-English languages	No digital versions	[24]
Differential Ability Scales-II	Cognitive two batteries: Early years and school age	30 months to 18 years	45–60 min	Level C	Spanish version available	Scoring software available	[20]
Kaufman Assessment Battery for Children-II	Simultaneous, sequential, planning, learning, and knowledge	3–18 years	25–75 min	Level C	Subtests for children with expressive language and motor difficulties; conormed with Kaufman Test of Educational Achievement (KTEA)	Q-global scoring	[16]
Leiter International Performance Scale-3	Cognitive/attention: Nonverbal figure ground form completion classification/analogies order	3 to 75 + years	20–45 min	Level C	Can be used for non-English speaking individuals	Scoring software available	[19]
Mullen Scales of Early Learning (MSEL)	Gross motor, visual reception, fine motor, expressive language, and receptive language	Birth to 68 months	15–60 min	Level B	Pinpoints strengths and weaknesses	No digital scoring or administration	[25]
NEPSY-II	Executive functioning/attention; language; memory/learning; sensorimotor functioning; visuospatial processing; social perception	3–16 years	45 min to 3 h	Level C	Vary subtests administered based on needed areas of evaluation	Select subtests on Q interactive	[22]
Raven's Progressive Matrices (Raven's 2)	Nonverbal intelligence	4–90 years	40–60 min	Level B	Can be administered to nonverbal children and adults	Digital administration and scoring available	[23]
Stanford-Binet Intelligence Scales-Fifth Edition	Verbal and nonverbal IQ Fluid Knowledge Quantitative Visual-spatial Working memory	2–85 years	45–60 min	Level C	Single test for all age ranges	No digital versions	[26]

Wechsler Preschool and Primary Scale of Intelligence (WPPSI)-IV	Cognitive (verbal comprehension, visual spatial), fluid reasoning (working memory), Processing speed for 30–47 months	30–91 months	30–45 min	Level C	Large normative sample; comprehensive test of intelligence	Digital scoring assistant	[27]
WISC-V	Cognitive primary scales: Verbal comprehension index (VCI), visual spatial index (VSI), working memory index (WMI), fluid reasoning index (FRI), processing speed index (PSI)	6–17 years	60 min	Level C	Large normative sample; comprehensive test of intelligence; strong reliability and validity	Digital version; Q-interactive; scoring assistant	[28]
WAIS-IV	Cognitive primary scales: Verbal comprehension Perceptual reasoning Working memory Processing speed	16–91 years	60–90 min	Level C	Updated normative data; use for assessing older adults	Digital version; Q-interactive; scoring assistant	[29]
Executive functioning							
Behavior Rating Inventory of Executive Function (BRIEF-2)	Behavior regulation; emotion regulation; cognitive regulation	5–18 years	10 min	Level B	Parent, teacher, and self-report forms for screening Spanish version available	Online administration and scoring PARiConnect	[30]
Children's Memory Scale	Attention; working memory; verbal and visual memory; short-and long-delay memory; recall and recognition; learning characteristics	5–16 years	30 min	Level C	Can be used to assess for learning disabilities	Digital scoring	[31]
Conners' Kiddie Continuous Performance Test- 2nd edition	Inattentiveness; impulsivity; sustained attention; vigilance	4–7 years	8 min	Level B	Computer game-based format	Online or; digital administration and scoring	[32]
Conners' Continuous Performance Test-3rd edition	Inattentiveness; impulsivity; sustained attention; vigilance	8 + years	15 min	Level B	Computer game-based format	Online or digital administration and scoring	[33]
Rey Complex Figure Test and Recognition Trial (RCFT)	Visuospatial ability; visuospatial memory	6–89 years	45 min	Level C	Identifies possible causes of memory deficits	No digital form available	[34]
Wide Range Assessment of Memory and Learning (WRAML-3)	Short- and long-term memory; attention/concentration	5–90 years	70–90 min	Level C	Includes assessment of ability to learn new material	Q-global for scoring	[35]

3. Modifying timing
4. Modifying test setting
5. Using only portions of a test
6. Using substitute or alternate tests [4]

The test administrator should review the testing manual to determine what modifications are valid and reliable for the group in question. Special norms may be appropriate to compare the individual's performance with that of peers with the same condition [5].

Computer-based assessment

Computer-based tests are now available in most areas of assessment, including tests of cognitive functioning. These tools provide better visuals, user interaction, and real-time scoring and which can result in time and cost savings to the provider and patient [5]. These tests also offer ability to adapt testing based on the individual test-taker response. Score reports and written narratives can be easily generated without separately inputting test results. However, these narratives should be carefully reviewed by the individual administering the test, as cutting and pasting a computer-generated report may lead to errors in interpretation of testing results [36]. Computer-based assessments can be instrumental in the provision of assessments over telehealth, which has expanded rapidly over the last few years.

Test user qualifications

Professionals who administer standardized tests should ensure they have the necessary knowledge, skills, training, and credentials to administer the test (international test commission 2013). The purchase of tests is usually restricted to those who meet certain minimum qualifications. Most test publishers use a three-tiered system that was first developed by the APA in 1950. It was later dropped by the APA in 1974, but a similar system continues to be used by publishers in determining who can purchase the tests [5].

A level: Test users are not required to have advanced training in test administration and interpretation. They may have training or practical experience in use of tests, a bachelor's degree in human services, education, psychology or related disciplines; and/or training or certification relevant to the particular assessment.

B level: Test users typically have a graduate degree in education, counseling, psychology or related disciplines, have licensure or certification documenting training in test administration and interpretation, or have completed coursework in test administration. Being a member of a professional organization such as the American Speech-Language-Hearing Association (ASHA) may also make one eligible to purchase B-level products.

C level: Test users must have B-level qualifications and a doctoral degree in a discipline that provides training in test administration and interpretation, such as psychology. Test users may also be working under the direct supervision of a qualified professional in psychology or a related field.

Limitations of cognitive testing

Basing assessment on traditional cognitive tests fails to address the responsiveness of an individual to instruction and practice of certain skills. This can lead to an underestimate of what an individual could achieve in the future. Traditional cognitive tests test a child at one point in time, and do not usually contain information on how a child responds in a learning environment. This may underestimate a child's true performance in an environment with educational supports [10].

Some critics of cognitive assessment state that tests may be biased against certain diverse groups of individuals, particularly if they are not normed in a broad population. In addition, cognitive testing can be too heavily relied on for determining services, without taking into account how an individual's adaptive skills and behavior impact their daily functioning [5].

References

[1] Borkowski JG, Carothers SS, Howard K, Schatz J, Farris JR. Intellectual assessment and intellectual disability. Handbook of intellectual and developmental disabilities. Boston, MA: Springer; 2007. p. 261–77.

[2] Kaufman AS. K-ABC: Kaufman assessment battery for children: administration and scoring manual. American Guidance Service; 1983.

[3] Schalock RL, Borthwick-Duffy SA, Bradley VJ, Buntinx WH, Coulter DL, Craig EM, et al. Intellectual disability: definition, classification, and systems of supports. Washington, DC: American Association on Intellectual and Developmental Disabilities; 2010.

[4] American Psychological Association. American Educational Research, Association, & National Council on Measurement in Education. Standards for educational and psychological testing. *Washington, D. C: American Psychological Association*; 2014.
[5] Sheperis, C., Drummond, R., and Jones, K. (2020). Assessment procedures for counselors and helping professionals (9th ed.). Pearson.
[6] Graham JR, Naglieri JA, Weiner IB. Handbook of psychology: assessment psychology, 10. John Wiley & Sons, Inc.; 2013. p. 3–25.
[7] American Psychiatric Association. Diagnostic and statistical manual of mental disorders: DSM-5. United States; 2013.
[8] Lidz CS, Gindis B. Dynamic assessment of the evolving cognitive. Vygotsky's educational theory in cultural context 2003;99.
[9] Rutland AF, Campbell RN. The relevance of 'Vygotsky's theory of the 'zone of proximal development'to the assessment of children with intellectual disabilities. J Intellect Disabil Res 1996;40(2):151–8.
[10] Jitendra AK, Kameenui EJ. Dynamic assessment as a compensatory assessment approach: a description and analysis. Remedial Spec Educ 1993;14(5):6–18.
[11] Kral M. Interpreting psychoeducational testing reports, individualized family service plans (IFSP), and individualized education program plans (IEP). In: Voigt RG, Macias MM, Myers SM, editors. AAP developmental and behavioral pediatrics. 2010. p. 477–87.
[12] Groth-Marnat, G and Wright J. (2016). Handbook of psychological assessment. John Wiley & Sons.
[13] Wechsler, D. (1939). The Measurement of Adult Intelligence. Williams & Wilkins.
[14] Wechsler, D. (1967). Wechsler Preschool and Primary Scale of Intelligence (WPPSI). The Psychological Corporation.
[15] Mays KL, Kamphaus RW, Reynolds CR. Applications of the Kaufman Assessment Battery for Children, in neuropsychological assessment. Handbook of clinical child neuropsychology. Boston, MA: Springer; 2009. p. 281–96.
[16] Kaufaman AS, Kaufman NL. Kaufman Test of Educational Achievement 3rd Edition. Pearson. 2014.
[17] Black MM, Black MM, Matula K. Essentials of Bayley scales of infant development II assessment, 5. John Wiley & Sons Incorporated; 2000.
[18] Bayley N, Aylward GP. Bayley scales of infant and toddler development-fourth edition. San Antonio, TX: Pearson; 2019.
[19] Roid GH, Miller LJ, Pomplun M, Koch C. Leiter international performance scale-third edition. Wood Dale, IL: Stoelting Company; 2013.
[20] Elliott CD. Differential abilities scales-second edition. San Antonio, TX: Harcourt Assessment; 2007.
[21] Kaufman AS. Kaufman brief intelligence test: KBIT. Circle Pines, MN: AGS, American Guidance Service; 1990.
[22] Korkman M, Kirk U, Kemp S. NEPSY-II. 2nd ed. San Antonio, TX: Harcourt Assessment; 2007.
[23] Raven J. Raven's progressive matrices 2, clinical edition. San Antonio, TX: Pearson; 2018.
[24] Hammill DD, Pearson NA, Wiederholt JL. Comprehensive test of nonverbal intelligence-second edition. Austin, TX: Pro-ed; 2009.
[25] Mullen E. Mullen scales of early learning. San Antonio, TX: Pearson; 1995.
[26] Roid GH. Stanford-Binet intelligence scales. 5th ed. Rolling Meadows, IL: Riverside Publishing; 2003.
[27] Weschler D. Weschler preschool and primary intelligence scale for children-fourth edition. San Antonio, TX: Pearson; 2013.
[28] Weschler D. Weschler intelligence scale for children-fifth edition. San Antonio, TX: Pearson; 2014.
[29] Weschler D. Weschler adult intelligence scale-fourth edition. San Antonio, TX: Pearson; 2008.
[30] Gioia GA, Isquith PK, Guy SC, Kenworthy L. Behavior rating inventory of executive function-second edition. Lutz, FL: PAR; 2015.
[31] Cohen MJ. Children's memory scale. San Antonio, TX: The Psychological Corporation; 1997.
[32] Conners K. Conners" kiddie continuous performance test-second edition. New York, NY: Multi-Health Systems, Inc; 2015.
[33] Conners K. Conners' continuous performance test-third edition. New York, NY: Multi-Health Systems, Inc; 2014.
[34] Meyers JE, Meyers KR. Rey complex figure test and recognition trial. Lutz, FL: Psychological Assessment Resource Inc; 1995.
[35] Adams W, Sheslow D. Wide range assessment of memory and learning-third edition. San Antonio, TX: Pearson; 2021.
[36] Butcher JN, Perry J, Hahn J. Computers in clinical assessment: historical developments, present status, and future challenges. J Clin Psychol 2004;60(3):331–45.

Chapter 9.4

Psychoeducational assessment

Amy Jo Merchant Newmeyer

Department of Pediatrics, Baylor College of Medicine-Temple Campus, McLane Chidren's Hospital, Temple, TX, United States

Neurodevelopmental pediatricians often play a key role in the initial identification of a specific learning disability. They should be aware of the process of psychoeducational assessment process, and the validity and appropriateness of commonly used standardized tests of educational skills. Working in collaboration with other educational and medical professionals, they can help guide the evaluation process and provide input on any needed educational services and/or classroom accommodations.

Psychoeducational assessments should be comprehensive and include more than one set of measures, which are selected based on the student's unique needs. The selection of the assessment protocol is dependent on the referral question and what additional information is needed to guide the diagnostic process. Many genetic conditions, such as Turner syndrome or neurofibromatosis type 1, are associated with specific learning profiles. Knowledge of these profiles can inform psychoeducational assessment strategies, as many children with neurodevelopmental diagnoses have comorbid learning difficulties. In addition, the presence of other comorbid conditions such as anxiety, ADHD, depression, and hearing or visual impairment can also affect classroom functioning [1].

Psychoeducational assessments are an essential component of the initial identification of a specific learning disability or intellectual disability. These assessments often include measures of cognitive and behavioral characteristics and observations from the classroom. Federal law surrounding educational services for a learning disability requires that the child undergo a comprehensive initial evaluation prior to the development of an individualized education plan (IEP). This evaluation is necessary to document the presence of a disability, and to determine if the disability is affecting the child's learning in the classroom setting. This process is governed by federal special education laws in all public schools. In addition to initial diagnosis, the psychoeducational assessment results of a psychoeducational assessment can be utilized for intervention planning and measurement of student progress over time [1].

The environment where testing is conducted can affect student performance on standardized assessments. Special considerations during the standardized assessment of children with neurodevelopmental disorders include the following:

1. Evaluate the environment for distractors such as sounds and harsh light.
2. Ensure that the student utilizes any necessary adaptive equipment, such as an augmentative communication device.
3. Take time to establish rapport at the beginning of the testing session.
4. Incorporate positive reinforcement in the testing setting, taking into account parent and child input of what types of reinforcers are most effective.
5. Monitor for fatigue, attention, and frustration, and take testing breaks if needed [1].

Special education law

Elementary and Secondary Education Act of 1965

Prior to the 1960s, few children with disabilities were educated in the public school district. In fact, some states had specific laws barring children with disabilities from the general education classroom. In 1965 the Elementary and Secondary Education Act was passed, which provided federal aid for the education of disadvantaged students. In 1966 an amendment was passed with the addition of grant funding for programs that served children with disabilities. Subsequent laws have expanded available services [2].

The Education of the Handicapped Act of 1970 (public law 91-230)

Public law (PL) 91-230 first introduced the concept of an appropriate education for all children with disabilities. In 1970, prior to the passage of PL 91-230, only one in five children with a disability was educated in the public school district [3].

Education for All Handicapped Children Act (public law 94-142)

In 1975 Congress passed the Education for All Handicapped Children Act (PL 94-142). PL 94-142 authorized funding for special education services for children ages 6–17 years. Specific provisions of part B of the bill include the following:

1. States must make a reasonable effort to identify students with a disability that would be eligible for special education services.
2. Each child should be placed in the least restrictive environment possible to meet their educational needs.
3. The child's progress must be periodically evaluated, with changes made to the educational program when needed in conjunction with the parents and school specialists.
4. Each child's learning needs must be evaluated in consultation with parents and educational professionals to develop an IEP that meets those needs.
5. All children, regardless of disability, have access to a free and appropriate education.
6. A process for appeal and due process must be developed that allows parents to challenge the school decisions (Boyer)

In 1986 the Education for All Handicapped Children Act was reauthorized as PL 99-457. This re-authorization added additional provision for early intervention services, and mandated states to provide services to children with disabilities and their families from the time of birth. These mandates are included in Part C of the law [2].

Individuals with Disabilities Education Act (public law 110-476)

In 1990 PL 110-476 was authorized, which changed the name of the law from the Education for All Handicapped Children Act to the Individuals with Disabilities Education Act (IDEA). The law contained the new disability categories of autism and traumatic brain injury. Another new mandate of this act was the addition of an Individual Transition Plan to assist students in transitioning to postsecondary life [3].

PL 105-17 was passed in 1997 as the Individuals with Disabilities Education Act Amendments of 1997. PL 105-17 included an emphasis on access to the general curriculum for students with disabilities. This law also provided parents with a process for mediation to resolve disputes with the schools and local educational agencies. The most recent reauthorization of IDEA was in 2004 [2].

No Child Left Behind Act

The No Child Left Behind Act was finalized in 2002 and replaced the Elementary and Secondary Education Act of 1965. In the prior legislation, funding was directed to schools with the largest number of children whose family income was near the federal poverty level. The No Child Left Behind Act changed this to distribute funding to schools based on students' academic performance. It required all students to be assessed yearly between grades 3 and 8 in both mathematics and reading. Many students with disabilities were expected to participate in the standardized assessment process, as only 1% of the student body could be assessed with a different standard [2].

Every Student Succeeds Act (public law 114-95)

PL 114-95, the Every Student Succeeds Act, was passed in December 2015. This law replaced the No Child Left Behind Act and set new standards for annual standardized testing, while placing more responsibility on the individual states to oversee this process. The Every Student Succeeds Act also set new standards for students with disabilities. Only 1% of students with disabilities can now be exempt from standardized testing. Each state is also allowed to use their own definition of disability. The Every Student Succeeds Act also recognizes that students with disabilities are often bullied in school, and requires the states to design and implement plans for reducing bullying in the public schools [4].

Identification of learning disabilities

Identifying learning disabilities in the public school system is based upon the criteria outlined in the IDEA. The criteria for an educational classification of a specific learning disability implies that other diagnoses and factors that could account for poor academic performance must be ruled out prior to this classification being used.

IDEA's definition of specific learning disability:

1. Specific learning disability means a disorder in one or more of the basic psychological processes involved in understanding or in using language, spoken or written, that may manifest itself in the imperfect ability to listen, think,

speak, read, write, spell, or to do mathematical calculations, including conditions such as perceptual disabilities, brain injury, minimal brain dysfunction, dyslexia, and developmental aphasia.

2. Disorders not included. Specific learning disability does not include learning problems that are primarily the result of visual, hearing, or motor disabilities, of intellectual disability, of emotional disturbance, or of environmental, cultural, or economic disadvantage. IDEA Section 300.8(c) (10) [5].

Other definitions of learning disability exist in the general literature, which may differ from the definition in the IDEA. The DSM-5 definition of specific learning disorder is based upon the compilation of the individual's history (developmental, medical, family, educational), school reports, and psychoeducational assessment.

1. Difficulties learning and using academic skills, as indicated by the presence of at least one of the following symptoms that have persisted for at least 6 months, despite the provision of interventions that target those difficulties:
   i. Inaccurate or slow and effortful word reading (e.g., reads single words aloud incorrectly or slowly and hesitantly, frequently guesses words, has difficulty sounding out words).
   ii. Difficulty understanding the meaning of what is read (e.g., may read text accurately but not understand the sequence, relationships, inferences, or deeper meanings of what is read).
   iii. Difficulties with spelling (e.g., may add, omit, or substitute vowels or consonants).
   iv. Difficulties with written expression (e.g., makes multiple grammatical or punctuation errors within sentences; employs poor paragraph organization; written expression of ideas lacks clarity).
   v. Difficulties mastering number sense, number facts, or calculation (e.g., has poor understanding of numbers, their magnitude, and relationships; counts on fingers to add single-digit numbers instead of recalling the math fact as peers do; gets lost in the midst of arithmetic computation and may switch procedures).
   vi. Difficulties with mathematical reasoning (e.g., has severe difficulty applying mathematical concepts, facts, or procedures to solve quantitative problems).
2. The affected academic skills are substantially and quantifiably below those expected for the individual's chronological age, and cause significant interference with academic or occupational performance, or with activities of daily living, as confirmed by individually administered standardized achievement measures and comprehensive clinical assessment. For individuals age 17 years and older, a documented history of impairing learning difficulties may be substituted for the standardized assessment.
3. The learning difficulties begin during school-age years but may not become fully manifest until the demands for those affected academic skills exceed the individual's limited capacities (e.g., as in timed tests, reading or writing lengthy complex reports for a tight deadline, excessively heavy academic loads).
4. The learning difficulties are not better accounted for by intellectual disabilities, uncorrected visual or auditory acuity, other mental or neurological disorders, psychosocial adversity, lack of proficiency in the language of academic instruction, or inadequate educational instruction [6].

IQ-achievement discrepancy model

The IQ-achievement discrepancy model is often utilized to determine the presence of a specific learning disability, and to determine eligibility for special education services. In the discrepancy model, the standard score of the overall IQ on a cognitive assessment is compared with the standard score on a standardized educational assessment. In this model, a significant discrepancy is defined as a greater than one standard deviation between the cognitive and educational assessment scores. However, this model has been criticized for a variety of reasons. First, this model may overidentify a learning disability in those whose IQ is above the mean and underidentify learning disability for those with an IQ under the mean. This approach often results in the "wait to fail" model, where children may not receive services until a severe discrepancy exists. Due to these concerns, this model has fallen out of favor and is no longer required to determine eligibility for special education services. The IQ-achievement discrepancy may not be present until children are older, leading to delayed identification and intervention [7].

Response to intervention

The response to intervention (RtI) model allows for more immediate instructional intervention for children who are struggling academically and provides children with educational supports that are matched to their learning needs. RtI is intended to reduce the number of referrals for special education services by providing intervention within the general education setting. It was first described in the Individuals with Disabilities Education Improvement Act of 2004 [8,9].

The main components of the RtI model include the following:

1. *Universal screening*: Universal screening allows for classification of students into three groups, which include those who are making adequate progress, those who are at some risk of academic failure and may need assistance, and those who are at high risk of academic failure without additional support.
2. *Continuous progress monitoring*: Progress is monitored frequently to identify response to the level of intervention provided, and to assist in determining when there is a need to increase the level of support.
3. *Continuum of evidence-based interventions*: Validated interventions may initially occur at a group level, and then progress to an individual level of support if there is inadequate response to the group intervention.
4. *Data-based decision making*: Decisions about the type and intensity of intervention needed are made based on student performance and assessment of the impact of an earlier intervention.
5. *Implementation fidelity*: There should be consistent documentation of the type and level of implementation, and the student response to intervention [9].

Assessment tools (criterion-referenced assessment and norm-referenced assessment)

The purpose of a criterion-referenced assessment is to determine whether individual students have mastered a specific set of skills or concepts drawn from a specific curriculum. Students are compared to an overall minimum standard of performance, rather than being compared to the knowledge of other students. Items are designed to have similar difficulty, rather than varying from easy to hard. Scores from a criterion-referenced assessment are usually expressed as a percentage. Advantages of criterion-referenced assessments is that students will know what information they need to master to pass a certain course or curriculum, and that teachers will know what content needs to be covered in a certain class or course. The scores on a criterion-based assessment can be used for screening purposes or to identify target areas for future educational interventions [10]. Curriculum-based assessments can be used to monitor a student's performance in comparison to peers receiving the same instruction. However, criterion-referenced assessments are limited in assisting with decisions about eligibility and placement [10].

Norm-referenced assessments are the most used method of evaluating students. The results are used to compare a student's performance to that of other children in a broad area of knowledge. Test administration procedures and scoring processes are standardized, regardless of ability or other medical conditions. Any changes to the administration of the test will limit the validity of the scores [1]. The items in these types of assessments will vary in difficulty to discriminate between high-scoring and low-scoring children. Scores are usually expressed as a percentile, grade equivalent, or standard score [10].

Each type of assessment has strengths and weaknesses, and should be chosen based on the needs of the individual patient. Using a varied battery of tests that combine both criterion-based and norm-referenced, approaches should be considered by the assessing clinician.

Testing accommodations

Under IDEA, Section 504 of the Rehabilitation Act of 1973 (PL 93-112), and the Americans with Disabilities Act of 1990 (PL 101-336), testing accommodations are frequently provided to students with learning disabilities on standardized testing and in-class examinations. Some examples of testing accommodations include extended time, testing in a less distracting environment, use of a calculator during a math assessment, or having a test read aloud [2].

Psychoeducational assessment tools

Early Learning Accomplishment Profile-third edition

The Early Learning Accomplishment Profile (ELAP)-3rd edition was published in 2002 and is a criterion-referenced assessment that provides a systematic method to observe skill development of children ages birth to 36 months. It includes six domains: gross motor, fine motor, cognition, language, self-help, and social emotional. Effective for communicating and collaborating with parents about their child's progress; use for children with special needs. Age ranges are given for items, but are approximations based on previous research. Age equivalents are not provided as it is not norm referenced [11].

Learning Accomplishment Profile-D

The Learning Accomplishment Profile (LAP)-D third edition is a comprehensive, norm-referenced measure designed to assess the development of young children ages 30–72 months. The results of the LAP-D can be utilized to plan specific, individualized activities for use in the classroom or home setting. English and Spanish versions of the test are available. The LAP-D was normed on 2099 children, 77 of whom had disabilities. A stratified 45 sampling procedure was used based on language, geographic region, age, race, gender, and type of setting in four geographic regions in the United States. The scoring on the LAP-D provides standard scores, percentile ranks, Z-scores, and age equivalent scores. Specific developmental skill data can also be obtained [12,13].

Brigance Inventory of Early Development-III

The Brigance Inventory of Early Development-III was published in 2013 and is a criterion-referenced assessment to assess the skill mastery of children ages birth to 7 years. Domains assessed include physical development, language development, academic/cognitive: literacy, academic/cognitive: mathematics and science, daily living, and social and emotional development. It was normed on a study sample of over 2400 children. Results assist with development of IEP goals, and help to target intervention, and monitor individual progress in skill development [14].

Developmental Assessment of Young Children-2

The Developmental Assessment of Young Children (DAYC)-2nd edition was published in 2012. It is based upon observation of the child in their natural settings during their interactions with familiar people and aligns with the five mandated areas of assessment for the IDEA law. Information is obtained through observation, caregiver report, and direct assessment. It was normed on a representative sample of 1832 children. It is normed for children ages birth to 5 years 11 months. Standard scores, percentile ranks, age equivalents, and an overall general development score is provided [15].

Battelle Developmental Inventory-third edition

The Battelle Developmental Inventory was developed by Jean Newborg. It is designed to assess children birth to age 7 years 11 months in five domains: adaptive, social-emotional, motor, cognitive, and communication. It is a norm-referenced test which provides standard scores, scaled scores, percentile rank, and age equivalent. It was normed with a sample of 2500 children in 20 different age groups. Special group studies were also performed for children with speech and language delay, autism, prematurity, cognitive delay, and motor delay. English and Spanish versions are available [16].

Woodcock Johnson IV Tests of Achievement

The Woodcock Johnson Tests of Achievement is a widely used, comprehensive battery of tests of achievement for ages 2–90 years. There are two complementary batteries: the Tests of Cognitive Abilities and the Tests of Oral Language. It was normed on 7416 individuals, employing 2010 Census data to stratify the sample. The standard battery includes letter-word identification, applied problems, spelling, passage comprehension, calculation, writing samples, word attack, oral reading, sentence reading fluency, math facts fluency, and sentence writing fluency. An extended battery adds additional tests of reading, spelling, science, social studies, and humanities [17].

Wechsler Individualized Achievement Test-fourth edition

The Wechsler Individual Achievement Test fourth edition was published in 2020. It is a comprehensive test of academic achievement for individuals ages 4 years 0 months to 50 years 11 months. It contains dyslexia index scores to support reliable screening for dyslexia. Results can be presented as Fall, Winter, and Spring grade-based standard scores, age-based standard scores, percentile ranks, age and grade equivalents, and growth scale values [19].

Wide Range Achievement Test-fifth edition

The Wide Range Achievement Test-5th edition (WRAT-5) was published in 2017 as a test of academic achievement for individuals range ages 5–85. Subtests include word reading, spelling, math computation, and sentence comprehension. Scores are reported as standard scores by age and grade level (Fall, Winter, Spring), percentile ranks, and grade equivalents (Wilkinson and Robertson) (Table 9.4.1) [20].

TABLE 9.4.1 Psychoeducational assessment tools.

Instrument	Variables assessed	Criterion or norm referenced	Age range	Administration time	Training requirements	Strengths	Digital options	References
Battelle Developmental Inventory-3rd edition	Adaptive, personal-social, communication, motor, cognitive	Norm referenced	Birth to 7 years 11 months	60–90 min (screener 10–30 min)	Level A[a]	English and Spanish	Online and offline electronic scoring available	[17]
Beery-Buktenica Developmental Test of Visual Motor Integration Sixth Edition	Visual perception and motor coordination	Norm referenced	2 years to 99 years 11 months	5–10 min	Level B	Short-format for ages 2–7 years	Manual scoring only	[21]
Bracken School Readiness Assessment Third Edition	Colors, letters, numbers/counting, size/comparison	Norm referenced	3 years to 6 years 11 months	10–15 min	Level A	Spanish version available	Q-global; scoring software available	[22]
Brigance Inventory of Early Development III	Physical development, language development, literacy, mathematics and science, daily living, social and emotional development	Norm- and criterion-referenced versions available	Birth to 7 years	20–60 min	Level A[a]	Alignment with state educational standards	Online scoring available	[15]
Comprehensive Test of Phonological Processing (C-TOPP) second edition	Phonological awareness, phonological memory, naming	Norm referenced	4 years to 24 years 11 months	40 min	Level B	Early identification of phonological awareness skills	Print version only	[23]
Developmental Assessment of Young Children, second edition	Cognition, communication, social-emotional, physical, development, adaptive behavior	Norm referenced	Birth to 5 years 11 months	10–20 minutes per domain	Level B	Allows for caregiver interview and observation; can test one or more domains	Print only	[16]

(Continued)

TABLE 9.4.1 (Continued)

Instrument	Variables assessed	Criterion or norm referenced	Age range	Administration time	Training requirements	Strengths	Digital options	References
Developmental Indicators for the Assessment of Learning (DIAL-4)	Cognitive/basic concepts; language; motor; self-help; social-emotional development	Norm referenced	3 years to 5 years 11 months	30–45 min	Level B	English and Spanish material available; teacher and parent questionnaire available	Q global scoring available	[24]
Early Learning Accomplishment Profile 3	Gross motor, fine motor, cognition, language, self-help, and social emotional	Criterion referenced	Birth to 36 months	45–90 min	Level A[a]	Spanish version available; use of natural observation for scoring; able to score three times per year; aligned with State Standards and Head Start Outcomes	Online software version; online ongoing assessment	[12]
Gray Oral Reading Test-fifth edition	Oral reading fluency; comprehension		6 years to 23 years 11 months	20–30 min	Level B	Two equivalent forms available	Print version only	[25]
Kaufman Test of Educational Achievement Third Edition	Reading, math, written language, oral language	Norm referenced	4 years to 25 years 11 months	50–85 min	Level B	Dyslexia index score available	Q interactive administration; Q global scoring	[26]
Learning Accomplishment Profile LAP-3	Gross motor, fine motor, prewriting, cognition, language, self-help, social-emotional	Criterion referenced	36–72 months	45–90 min	Level A[a]	See ELAP-3	Online ongoing assessment	[27]
Learning Accomplishment Profile-D Third edition	Gross motor, fine motor, cognitive, and language	Norm referenced	30–72 months	45–90 min	Level A[a]	Spanish norms available; designed for children with typical and atypical development		[14]
Learning Accomplishment Profile Birth to Kindergarten	Gross motor, fine motor/prewriting, cognitive, language, social-emotional, executive function	Criterion referenced	Birth to 72 months	45–90 min	Level A[a]	Combines E-LAP and LAP-3 into one assessment	Online assessment using computer, phone or tablet	[13]

Test of Early Written Language Third Edition (TEWL-3)	Basic writing, contextual writing, normative writing	Norm referenced	4 years to 11 years 11 months	30–50 min	Level B	Two forms available	Print version only	[28]
Test of Mathematical Abilities Third edition (TOMA-3)	Mathematical symbols and concepts; computation; mathematics in everyday life; word problems; attitude toward math	Norm referenced	8 years to 18 years 11 months	60–90 min	Level A	Covers broad range of math concepts	Print version only	[29]
Test of Written Language Fourth Edition (TOWL-4)	Overall writing, contrived writing, spontaneous writing	Norm referenced	9 years to 17 years 11 months	60–90 min	Level B	Determine strengths and weaknesses in various aspects of writing	Print version only	[30]
Test of Word Reading Efficiency Section Edition (TOWRE-2)	Pronunciation of printed words and phonemically regular nonwords	Norm referenced	6 years to 24 years 11 months	5–10 min	Level B	Used to monitor three to four times a year	Print version only	[31]
Individual Achievement Test Fourth edition	Reading, written expression, mathematics	Both	4–50 years	Varies by grade level	Level B	Dyslexia index score	Online administration and scoring in Q-interactive; digital scoring and report generation	[19]
Wide Range Achievement Test Fifth Edition (WRAT-5)	Word reading; sentence comprehension; spelling; math computation	Norm referenced	5 to 85 + years	15–45 min	Level B	Can administer in small groups	Q global scoring: Q interactive scoring and administration	[20]
Woodcock-Johnson Tests of Achievement IV	Eleven subtest standard batteries; nine additional subtests for extended battery	Norm referenced	2–90 years	5–10 min per subtest	Level B	Parallel forms for retesting	Web-based scoring and reports	[18]

[a]*No level given for qualifications by publisher—classified as level A.*

Summary

The choice of which test to utilize for a patient evaluation is based on a variety of factors, including:

1. The referring provider's request
2. The family's chief concern
3. Need for documentation for services at the school or in the community
4. The child's unique developmental needs—for instance, hearing impairment or visual impairment

In some cases, a focused test may be needed—for instance, an executive functioning assessment for a child with ADHD. In other cases, a broad measure assessing a wide range of abilities would be most effective. Children with neurodevelopmental disabilities may also face challenges that limit their ability to complete an assessment, such as executive function difficulty with ADHD, or receptive language impairment. In such cases, the scores obtained should be interpreted with the understanding that the underlying neurodevelopmental diagnosis may impact the overall score and may not reflect their true ability.

This chapter reviews some of the more common assessments utilized in neurodevelopmental practice. There may be additional assessments that are not listed in this chapter that could be utilized in psychoeducational assessment of children. It is also important to consider the role of the psychoeducational assessments in a broader context. The most accurate diagnoses and recommendations will be made by combining the information in the patient history, observations, test scores, and results of cognitive, educational, and adaptive assessments.

References

[1] Magyar CI, Pandolfi V, Peterson CR. Psychoeducational assessment. Handbook of intellectual and developmental disabilities. 2007. p. 333–51.

[2] Hurwitz KA. A review of special education law. Pediatr Neurol 2008;39(3):147–54.

[3] US Department of Education. A history of the Individuals with Disabilities Education Act. 2020. https://sites.ed.gov/IDEA-History.

[4] Penuel W, Meyer E, Valladares MR. Making the most of the Every Student Succeeds Act (ESSA): helping states focus on school equity, quality and climate. Natl Educ Policy Center 2016;.

[5] US Department of Education. (May 25, 2018). Sec. 300.8 (c) (10). https://sites.ed.gov/idea/regs/b/a/300.8/c/10.

[6] American Psychiatric Association. Diagnostic and statistical manual of mental disorders: DSM-5. 2013. United States.

[7] Restori AF, Katz GS, Lee HB. A critique of the IQ/achievement discrepancy model for identifying specific learning disabilities. Eur J Psychol 2009;5(4):128–45.

[8] Fox L, Carta J, Strain P, Dunlap G, Hemmeter ML. Response to intervention and the pyramid model. Tampa, Florida: University of South Florida, Technical Assistance Center on Social Emotional Intervention for Young Children; 2009.

[9] Individuals With Disabilities Education Act, 20 U.S.C. § 1400. 2004.

[10] Knight P. A briefing on key concepts: formative and summative, criterion and norm-referenced assessment. Learning and Teaching Support Network; 2001.

[11] Fessler MA, Plourde PA. Psychoeducational assessment. Capute & Accardo's eurodevelopmental disabilities in infancy and childhood: neurodevelopmental diagnosis and treatment. 1. Brookes Pub 2007;591608.

[12] Glover ME, Preminger JL, Sanford AR. Early learning accomplishment profile-third edition. Lewisville, NC: Kaplan Early Learning Company; 2001.

[13] Crowell NA, Ramey CT, Ramey SL. Learning accomplishment profile- birth to kindergarten. Lewisville, NC: Kaplan Early Learning Company; 2004.

[14] Hardin BJ, Pesner-Feingberg ES, Weeks SW. The learning accomplishment profile-diagnostic-third edition. Lewisville, NC: Kaplan Early Learning Company; 2005.

[15] French BF. Brigance inventory of early development-third edition. North Billerica, MA: Curriculum Associates; 2013.

[16] Voress JK, Maddox T. Developmental assessment of young children-second edition. Austin, TX: Pro-Ed; 2012.

[17] Newborg J. Battelle developmental inventory-third edition. Itasca, IL: Riverside Insights; 2020.

[18] Schrank FA, Mather N, McGrew KS. Woodcock Johnson IV tests of achievement. IL: Riverside: Rolling Meadows; 2014.

[19] Weschler D. Weschler individual achievement test-fourth edition. San Antonio, TX: Pearson; 2020.

[20] Wilkinson GS, Robertson GJ. Wide range achievement test-fifth edition. San Antonio, TX: Pearson; 2017.

[21] Beery KE, Buktenica MA, Beery NA. Beery-Buktenica developmental test of visual motor integration-sixth edition. Bloomington, MN: Pearson; 2010.

[22] Bracken BA. Bracken school readiness assessment-third edition. San Antonio, TX: Pearson; 2007.

[23] Wagner R, Torgesen J, Rashotte C, Pearson NA. Comprehensive test of phonological processing-second edition. Austin, TX: Pro-Ed; 2013.

[24] Mardell C, Goldenberg DS. Developmental indicators for the assessment of learning-fourth edition. San Antonio, TX: Pearson; 2011.

[25] Widerholt JL, Bryant BR. Gray oral reading tests-fifth edition. Austin, TX: Pro-Ed; 2012.

[26] Kaufman AS, Kaufman NL. Kaufman test of educational achievement-third edition. San Antonio, TX: Pearson; 2014.

[27] Hardin BJ, Pesner-Feingberg ES. The learning accomplishment profile-third edition. Lewisville, NC: Kaplan Early Learning Company; 2004.
[28] Hresko W, Herron S, Peak P, Hicks D. Test of Early Written Language, Third Edition. Austin, Texas:ProED;2012.
[29] Brown, V. L., Cronin, M. E., & Bryant, D. P.. Test of mathematical abilities (3rd ed.). Austin, TX: PRO-ED;2012.
[30] Hammill DD, Larsen SC. Test of written language-fourth edition. Austin, TX: Pro-Ed; 2009.
[31] Torgesen, J, Wagner, R, and Rashotte, C. Test of Word Reading Efficiency-Second Edition (TOWRE-2). Austin, TX: Pro-Ed;2012.

Chapter 9.5

Functional and behavioral assessment

John Michael Falligant[1] and Louis P. Hagopian[2]

[1]Department of Behavioral Psychology Kennedy Krieger Institute, Johns Hopkins University School of Medicine, Baltimore MD, United States,

[2]Department of Psychiatry and Behavioral Sciences, Johns Hopkins University School of Medicine, Baltimore MD, United States

Individuals with neurodevelopmental disabilities including ASD and other intellectual and developmental disabilities (IDD) are at increased risk for engaging in problem behavior (e.g., aggression, property destruction, self-injury [1]). Problem behavior is a heterogeneous phenomenon with respect to its onset, presentation, and causal variables. Problem behavior can emerge at any time in the developmental period and into adulthood in some cases. Individuals may present with a wide range of behaviors that differ along the dimensions of topography, frequency, and severity. Individuals may present with a single form of problem behavior (e.g., aggression) or multiple forms (e.g., aggression, elopement, pica, motor stereotypies). These behaviors may occur episodically or throughout the day up to hundreds or even thousands of times per day. When problem behavior occurs regularly and with high intensity, it is likely to produce injuries to self and others and greatly impair functioning. The consequences of these behaviors may range from relatively minor to highly significant and even life-threatening [2].

Although estimates vary, approximately 50% of individuals with neurodevelopmental disabilities display some form of problem behavior, with a subset (5%–10%) of these individuals exhibiting very severe topographies of problem behavior [3]. There are a number of variables associated with problem behavior among this population, some of which are thought to increase risk for its development. For example, the presence of severe problem behavior is more common among individuals with IDD who also have ASD. More pronounced deficits in intellectual functioning and communication skills (e.g., ability to convey wants and needs to others), the presence of genetic conditions, sensory impairments, and the presence of repetitive and restrictive behaviors/interests are associated with increased prevalence problem behavior among these individuals as well [4–6].

Problem behavior in this population is thought to be the product of interactions between deficits related to the primary disability (e.g., limited communication repertoires and adaptive skills, affective dysregulation) and learning experiences that reinforce problem behavior. Deficits in functioning and adaptive behavior may increase the frequency and intensity of frustrative experiences that set the stage for the occurrence of problem behavior [7]. As the occurrence of problem behavior is generally concerning, socially unacceptable, and often disruptive or dangerous, caregivers understandably react to problem behavior in an attempt to interrupt it. Common reactions include redirecting the individual to another activity, consoling them, or removing them from the event that thought to occasion the behavior. Although well intended, and sometimes effective, these reactions also have the potential to inadvertently reinforce the behavior via operant conditioning processes—thus increase the probability of problem behavior in the future. For example, if a child engages in aggressive behavior when presented with instructional demands, the teacher may provide the child a break from work to calm the child and prevent injury or disruption in the environment. Likewise, if a child engages in self-injury when the caregiver's attention is diverted elsewhere, the caregiver may respond with statements of concern (e.g., "stop hitting yourself, you're going to get hurt!") accompanied by physical attention (e.g., rubbing child's back, holding child's hand). These patterns of interaction can increase the future probability of problem behavior under similar conditions and contribute to its maintenance across time and settings. Additionally, caregivers may also work to actively prevent situations that might cause distress or occasion problem behavior. Although potentially beneficial, making such accommodations can broaden and result in the establishment routines that become unsustainable and a source of chronic stress for caregivers [7,8].

Diagnostic and behavioral assessment

As problem behavior has multiple dimensions, its assessment and management must consider a variety of factors including caregiver–child interactions that occasion and reinforce the behavior, other skills deficits, developmental variables, the environmental context, psychiatric comorbidities, and familial variables—as well as the broader sociocultural context in which the child lives. Assessment also includes the use of standardized rating scales to document characterize the type and severity of the problem for purpose of measuring response to treatment over time. Diagnostic assessment is designed to accurately classify and diagnose the psychiatric disorder(s) that best characterizes the problem behavior (e.g., stereotypic movement disorder with self-injurious behavior), or other conditions thought to contribute to it (e.g., emotion dysregulation and impulsivity). In contrast to diagnostic assessment, behavioral assessment is designed to identify the variables that occasion and reinforce problem behavior (i.e., the operant reinforcing function of problem behavior) which also requires analysis of caregiver–child interactions and skills deficits. This involves the use of a variety of functional behavioral assessment methods [9]. Prior to discussing specific functional behavioral analysis (FBA) strategies, a number of other broader considerations relevant to assessment and management of problem behavior are discussed.

Surveillance and risk assessment

Pediatricians play a critical role with respect to ongoing monitoring and surveillance for the emergence and development of problem behavior (see Ref. [10]). Analysis of the family's capacity to effectively manage problem behavior—which includes the availability of social supports, healthcare resources, and caregivers' physical ability to respond to challenging behavior—is necessary to determine the level of additional supports needed through educational and social service organizations. Additionally, as described above, the occurrence of problem behavior can be a significant source of both acute and chronic stress for caregivers [11]. Thus assessment of caregiver stress and psychiatric issues is necessary, and may also require referral for mental health resources for caregivers.

A first step of surveillance for high-risk populations entails routinely inquiring about problem behavior at regular visits so that emerging problem behavior or maladaptive caregiver–child interactions that could lead to the development or maintenance of problem behavior are identified early [12]. Directly observing the child's problem behavior (and caregivers' responses to problem behavior) in the examination room can be informative, although some children (and/or caregivers) may be inhibited and fail to behave as they typically do in the natural environment. Therefore caregiver report of problem behavior, and how they respond to it may be a primary source of information for the pediatric provider [7,13].

There is significant variability in problem behavior regarding its form, and the risks it may pose. Risk can be determined by the presence of injuries and by the inherent physical risks of the behavior itself. For example, problem behavior such as head-directed self-injury, head banging, self-biting, pica, and elopement impose direct and immediate risks to the individual. These behaviors can produce concussions, lacerations, tissue loss resulting in disfigurement, permanent sensory impairment (e.g., loss of vision), or even death. Additionally, other forms of problem behavior, such as aggression and property destruction, present considerable risk to others (e.g., parents, siblings, and teachers), and can result in out-of-home residential placement, hospitalization, and/or use of chemical and physical restraint. Safety planning may be indicated as a primary step to mitigate imminent risk to the patient, including psychoeducation regarding emergency services and interactions with first responders [10].

Assessment of cooccurring medical and psychiatric conditions

Pediatric management of problem behavior in ASD has been described elsewhere and will only be summarized here (readers are encouraged to access these resources regularly as guidance continues to evolve; see [10,12]. Pediatricians should evaluate whether cooccurring medical conditions may be contributing to problem behavior, especially in cases in which the onset of problem behavior is sudden or there are known medical problems. Considering the increased prevalence of communication difficulties among this population, a thorough review of symptoms and physical examination may be necessary to adequately assess for the presence of contributing medical problems that the child may be unable to report. This assessment may entail review of supplements and medications, gastrointestinal issues, dental pain, sleep problems, and neurological changes [14]. In addition to medical conditions, cooccurring psychiatric conditions appear to be highly prevalent among individuals with ASD and IDD and may contribute to irritability and behavioral dysregulation [10,12,15]. Irritability is an associated clinical feature of many psychiatric disorders and may be the primary symptom or characteristic of a contributing psychiatric disorder and behavioral dysfunction in this population (particularly if they have limited communication repertoires). Therefore when a patient presents with a new or worsening problem behavior with irritable features, a differential psychiatric diagnosis may be necessary to determine whether psychiatric conditions are contributing to the occurrence of challenging behavior. These include but are not limited to problems related to impulse control, agitation, and mood instability [10].

Assessment of psychosocial and environmental stressors

Ongoing surveillance also requires an assessment of potential psychosocial stressors and their relation to the emergence of problem behavior. Children with neurodevelopmental disabilities are at increased risk for physical and sexual abuse, victimization, and bullying. Additional environmental stressors may lead to problem behavior when there is a poor match between the needs of the child and their environment, such as when there is insufficient structure or failure to meet the emotional needs of the child-based developmental age and other contextual factors [10]. Examples of this would include home environments with limited environmental stimulation and/or structure, or school environments in which there is a poor match between the developmental and educational level of the child and their classmates (e.g., patient is verbal but is placed in classroom with exclusively nonverbal classmates [7]).

Norm referenced behavior rating scales

Norm referenced behavior rating scales, such as the Aberrant Behavior Checklist (ABC [16]) and Child Behavior Checklist (CBCL [17]) are used to characterize the type and severity of problem behavior; and as outcome measures quantify the effectiveness of treatment. The ABC is a 58-item rating Likert-style rating scale that evaluates the subdomains of irritability, hyperactivity/noncompliance, lethargy/social withdrawal, stereotypic behavior, and inappropriate speech. There is strong psychometric support for the validity and reliability of the ABC in a variety of clinical populations, including individuals with neurodevelopmental disabilities. For example, the irritability subscale of the ABC has been used to measure changes in irritability in a number of clinical trials for risperidone and aripiprazole (the only two U.S. Food and Drug Administration (FDA) approved medications for children with ASD [15]). The CBCL also has extensive empirical support for assessing clinically elevated behavioral and emotional symptoms among individuals with IDD and ASD (e.g., [18]). The CBCL is a Likert-style scale consisting of eight DSM-oriented scales, eight syndrome scales, two composite scale scores for (one for internalizing symptoms and one for externalizing symptoms), and a total score.

The Repetitive Behavior Scale-Revised (RBS-R [19,20]) is another useful instrument for assessing the variety of restrictive and repetitive behaviors (RBBs), which are characteristic of ASD and may cause considerable impairment in adaptive functioning [21]. The 43 items are evaluated on a Likert-style scale to measure RBBs across six subscales, including stereotypic behavior, self-injurious behavior, compulsive behavior, ritualistic behavior, insistence on sameness, and restricted behavior.

In both clinical and research contexts, these instruments are also used to quantify the effectiveness of behavioral and pharmacological interventions targeting problem behavior or other clinical targets (e.g., irritability). Ideally, response to treatment would be determined by gathering information from multiple sources including: caregiver report, direct observation, and longitudinal behavioral measuring by teachers or other providers relevant to contexts in which the problem behavior is the most problematic and most likely to occur [10]. Clear and measurable treatment goals should be established (e.g., reduction in frequency of aggressive outbursts by a specific number or reduction in duration of outbursts by a specific time). For example, an 80% reduction in problem behavior from baseline to treatment is a common benchmark used in the domain of applied behavior analysis for assessing treatment effectiveness [22]. However, such measures of efficacy are only meant for the purpose of capturing a change in a single dimension of behavior (i.e., frequency) and do not always reflect broader changes in quality of life and other clinical targets.

Assessment of the operant reinforcing function of problem behavior

Although standard rating scales such as the ABC [16] are excellent for characterizing the type and severity of problem behavior, they are not designed to identify the events that occasion and reinforce the behavior (these are described further below). As noted, problem behavior in this population is thought to be the product of interactions between individual deficits in adaptive behavior secondary to neurodevelopmental disabilities (e.g., limited communication skills, poor frustration tolerance, impulsivity, chronic irritability) and environmental consequences that reinforce problem behavior. Events in the environment may initially occasion an instance of problem behavior that is subsequently reinforced and strengthened based on the type of consequence it produces. It is very useful to identify the different types of consequences that serve to strengthen and maintain problem behavior because this yields information about the function of the problem behavior. The most commonly reported operant reinforcing functions for problem behavior include access to attention, access to tangible rewards, escape from unpleasant tasks/demands/situations, and sensory reinforcement

[23]. Behavioral interventions are more effective when they are matched to the underlying function of problem behavior.

That is, it allows clinicians to understand why problem behavior is occurring, and then educate and guide caregivers to respond in ways that are more effective. Pediatricians are in an excellent position to identify and educate caregivers on how to respond appropriately. However, referral to specialized services may be needed when first line interventions fail, when problem behavior is more severe, or when maladaptive interactions are well established and require more intensive intervention.

Interviews and rating scales to assess operant reinforcing function

Extant research indicates that assessment methods that rely on the reports or accounts of others (e.g., parents, teachers) typically have limited validity for identifying reinforcing function of problem behavior—relative to methods involving direct observation of behavior [9,24,25]. However, this indirect assessment may yield valuable information to inform preliminary hypotheses about the potential function of problem behavior and guide initial intervention strategies. Apart from primary caregivers (e.g., parents), it is helpful to interview other individuals that interact with the child including teachers, other caregivers (e.g., grandparents), and/or other professionals (e.g., SLPs) when possible [13]. Information from these sources may corroborate information provided from the primary caregiver, identify contextual factors associated with problem behavior, and help determine the degree to which challenging behavior is trans-situational and occurs across different settings and individuals—or if it is highly context specific.

The quality and usefulness of information derived from interviews can vary depending on a number of factors, including the recall and reliability of the informant, whether the interview is open-ended or close-ended, and the experience of the interviewer [13]. Open-ended interview aids provide a general structure for the interview while permitting more flexibility than close-ended, structured interview methods (described below). These open-ended aids help determine what questions to ask, and how to phrase the questions, in a manner that encourages informants to provide accurate and complete information to help the clinician identify environmental variables contributing to the patient's problem behavior. Table 9.5.1 briefly summarizes the main components of a functional assessment interview (FAI), the objectives, and some example questions for illustrative purposes. More detailed open-ended FAIs have been described in the literature [26,27]. For example, the FAI [28] breaks the interview into nine different sections addressing the form, frequency, timing, intensity, environmental variables, and specific antecedent/consequent events surrounding

TABLE 9.5.1 Components, objectives, and example questions for a brief functional assessment interview.

Component	Objectives and example questions
Structural dimensions of problem behavior	Describe the behaviors of concern. For each behavior, determine its approximate frequency and patterning over time.
	What are the main forms of problem behavior your child engages in? What do the behaviors look like?
	In general, how often does each behavior occur per day? Do the behaviors occur in bouts?
	When did you first notice these behaviors emerge?
Antecedents to problem behavior	Identify events in the environment that occasion or increase the probability of problem behavior occurring. Do this for each behavior.
	Can you tell when each behavior is about to occur? If so, how?
	What seems to trigger each behavior of concern?
	In what situations does each behavior almost always happen; and almost never happen?
Consequences that follow problem behavior	Identify the consequences produced by problem behavior that may inadvertently strengthen or maintain its occurrence. Do this for each behavior.
	What do you do when each behavior occurs? How do others respond?
	What seems to settle things down or stop each behavior when it is occurring?
	Have you found anything to be effective in decreasing the behavior or preventing it from occurring? What have you tried?

problem behavior. It also addresses patient-specific factors (e.g., communication level, skill repertoires) that may contribute to problem behavior or impact the effectiveness of potential treatments.

Although more flexible than the close-ended interview approach, a drawback to open-ended interview tactics is that the lack of continuity and consistency of specific interview questions may prevent clinicians from obtaining all of the information necessary to develop hypotheses about the potential function(s) of problem behavior or begin formalizing a treatment approach [13]. Accordingly, a number of structured interview tools have been developed to provide a degree of standardization to the interview process and ensure that information specific to a number of important dimensions of problem behavior is obtained. Structured behavioral assessment instruments are useful for gathering information with respect to the frequency, severity, and topography of problem behavior in a variety of contexts (e.g., home, school, community). These tools allow clinicians to quantify the degree to which these behaviors are present, identify situations in which these behaviors are more or less likely to occur, and provide an index of impairment across an array of behavioral and psychological domains. Two commonly used and empirically supported interview tools are the Question about Behavioral Function (QABF [29]) and Functional Analysis Screening Tool (FAST [30]). The QABF is a 15-item Likert-style rating scale designed to aid in identifying the function of problem behavior. Similarly, the FAST is a 16-item measure that includes 16 yes/no items, along with questions pertinent to demographic and topographical variables, which can be used to develop hypothesis surrounding the operant reinforcing function.

Functional assessment of problem behavior with direct observation

A common type of structured behavioral observation involves ABC recording, in which the observer records the antecedent and consequent events that preceded and followed each instance of problem behavior, respectively. These events can be written down in narrative fashion or can be recorded using an ABC checklist [31]. For example, during a behavioral observation, an observer may note that every time aggression occurred, the antecedent involved the presentation of a task or demand by a caregiver (e.g., "stand up," "come here," "put on your jacket"), and the consequence involved temporarily removing the directive and allowing the patient to have a break from the task (e.g., "okay you don't have to, take a minute to calm down"). These types of antecedent-behavior-consequence sequences may be indicative of a potential function (e.g., escape from demands) and can subsequently be used to form preliminary treatment recommendations (e.g., psychoeducation and behavioral parent training and time-out). However, such methods are correlational at best, and rely on the observer to accurately discriminate which events occasioned and reinforced the behavior.

Functional analysis of problem behavior

A formal functional analysis (FA) is a controlled assessment in which environmental conditions are systematically manipulated to determine their relation to problem behavior [32,33]. FA is most valid and scientifically rigorous method of assessment [23,34]. FA involves examining how problem behavior changes as function of specific antecedent conditions, and consequences for problem behavior arranged by a trained therapist. Thus this methodology involves placing the child in different analog contexts designed to simulate various situations. For example, if a child's problem behavior is hypothesized to be maintained by attention from a caregiver, the "test condition" involves arranging a situation in which the caregiver provides a form of attention (e.g., telling child to stop, physically touching the child) contingent on each instance of the problem behavior. In the control condition, which serves as a basis for comparison, the caregiver interacts with the child continuously and does not provide any attention contingent upon the occurrence of problem behavior. The difference in the amount of problem behavior occurring in the test and control conditions determines whether or not the specific function is indicated. FA represents the gold standard of behavioral assessment in the clinical practice of applied behavior analysis, as it is highly controlled, and yields objective data based on the frequency of problem behavior. As this assessment method requires significant time, staff, and technological resources, it is usually only available in highly specialized clinical and school settings. Although federal law mandates a functional assessment to be conducted for students who have a formal behavior intervention plan, it does not specify the methodology. Many public school settings rely primarily on interviews, rating scales, and naturalistic observations.

As described above, problem behavior is classified based on its function, which include two broad classes: (1) socially maintained (occasioned and reinforced through the interactions of others [e.g., attention, escape, access to preferred items/activities]) and (2) automatically maintained (occurs independent of social contingencies). Automatically maintained problem behavior occurs independent of interactions with others and is thought to be maintained by the sensory products it produces via some unknown biological process. Certain topographies of problem behavior may be more likely to belong to functional class of behavior of another. For example, aggression is most often socially maintained, whereas pica or rumination is most often automatically maintained. Self-injurious behavior is socially maintained in approximately three-fourths of cases and automatically maintained in one-fourth of cases [7].

Service coordination and referral to specialists

Pediatricians are in an excellent position to detect the emergence of problem behavior, and to identify caregiver–child interactions of concern though interviews, observations, and rating scales. As noted, behavioral assessment encompasses an array of techniques that range from minimally intensive assessment procedures, including rating scales, interviews, and behavioral observations, to more resource-intensive assessments including formal functional analysis of problem behavior. The former techniques (e.g., rating scales, interviews, behavioral observations) are common first-line assessment approaches employed in school and community settings, and by pediatricians or other healthcare professionals. For patients that require more resource-intensive methods including functional analysis, referral for more intensive services may be necessary. Intensive behavioral services are available in some school and community settings, however, when those services are insufficient to meet the needs of the patient; referral to specialized behavioral providers would be indicated. Practitioners with the appropriate level of training in behavior analysis to perform such analyses are typically Board-Certified Behavior Analysts (BCBAs) or licensed psychologists with specialized training in behavior analysis. The BCBA is a graduate-level certification for independent practitioners who provide behavior-analytic services (those with doctoral level training possess the BCBA-D credential to signify this). Psychologists are licensed professionals with highly specialized training in the assessment and treatment of serious mental, emotional, and behavioral disorders. Some psychologists are qualified to work with individuals with IDD and have training in applied behavior analysis (and often possess the BCBA credential). When comprehensive assessment findings indicate other factors are contributing to the presentation of problem behavior, this would necessitate referrals for specialized psychiatric services (to a psychiatrist with specialized training in working with individuals with NDD), educational services, or services of other health disciplines (clinical psychology, speech and language, occupational therapy, etc.; see [10]).

References

[1] Strand RC, Vister OM, Eldevik S, Eikeseth S. Nature, prevalence, and characteristics of challenging behaviors in functional assessment. Functional assessment for challenging behaviors and mental health disorders. 2021. p. 153–81.

[2] Kalb LG, Vasa RA, Ballard ED, Woods S, Goldstein M, Wilcox HC. Epidemiology of injury-related emergency department visits in the US among youth with autism spectrum disorder. J Autism Dev Disord 2016;46(8):2756–63.

[3] Emerson E, Kiernan C, Alborz A, Reeves D, Mason H, Swarbrick R, et al. The prevalence of challenging behaviors: a total population study. Res Dev Disabil 2001;22(1):77–93.

[4] Petty JL, Bacarese-Hamilton M, Davies LE, Oliver C. Correlates of self-injurious, aggressive and destructive behaviour in children under five who are at risk of developmental delay. Res Dev Disabil 2014;35(1):36–45.

[5] Soke GN, Rosenberg SA, Hamman RF, Fingerlin TE, Rosenberg CR, Carpenter L, et al. Prenatal, perinatal, and neonatal factors associated with self-injurious behaviors in children with autism spectrum disorder. Res Autism Spectr Discord 2019;61:1–9.

[6] Soke GN, Rosenberg SA, Hamman RF, Fingerlin T, Rosenberg CR, Carpenter L, et al. Factors associated with self-injurious behaviors in children with autism spectrum disorder: findings from two large national samples. J Autism Dev Disord 2017;47(2):285–96.

[7] Kurtz PF, Leoni M, Hagopian LP. Behavioral approaches to assessment and early intervention for severe problem behavior in intellectual and developmental disabilities. Pediatric Clinics 2020;67(3):499–511.

[8] Feldman I, Koller J, Lebowitz ER, Shulman C, Itzchak EB, Zachor DA. Family accommodation in autism spectrum disorder. J Autism Dev Disord 2019;49(9):3602–10.

[9] Pence ST, Roscoe EM, Bourret JC, Ahearn WH. Relative contributions of three descriptive methods: implications for behavioral assessment. J Appl Behav Anal 2009;42(2):425–46.

[10] McGuire K, Fung LK, Hagopian L, Vasa RA, Mahajan R, Bernal P, et al. Irritability and problem behavior in autism spectrum disorder: a practice pathway for pediatric primary care. Pediatrics 2016;137(Supplement 2):S136–48.

[11] Kurtz PF, Strohmeier CW, Becraft JL, Chin MD. Collateral effects of behavioral treatment for problem behavior on caregiver stress. J Autism Dev Disord 2021;51(8):2852–65. Available from: https://doi.org/10.1007/s10803-020-04694-y.

[12] Hyman SL, Levy SE, Myers SM. Identification, evaluation, and management of children with autism spectrum disorder. Pediatrics 2020;145(1):e20193447.

[13] Kranak MP, Gregory MK, Rooker GW. Interviews and report writing in the context of functional assessment. In: Matson JL, editor. Functional assessment for challenging behaviors and mental health disorders. Springer; 2021. p. 261–72.

[14] Myers SM, Johnson CP. Management of children with autism spectrum disorders. Pediatrics 2007;120(5):1162–82.

[15] Lamy M, Pedapati EV, Dominick KL, Wink LK, Erickson CA. Recent advances in the pharmacological management of behavioral disturbances associated with autism spectrum disorder in children and adolescents. Pediatric Drugs 2020;22:473–83.

[16] Aman MG, Singh NN, Stewart AW, Field CJ. The aberrant behavior checklist: a behavior rating scale for the assessment of treatment effects. Am J Ment Defic 1985;89(5):485–91.

[17] Achenbach TM. Manual for the CBCL/4–18 and 1991 profile. Burlington, VT: University of Vermont, Department of Psychiatry; 1991.

[18] Arias AA, Rea MM, Adler EJ, Haendel AD, Van Hecke AV. Utilizing the Child Behavior Checklist (CBCL) as an autism spectrum disorder preliminary screener and outcome measure for the PEERS® intervention for autistic adolescents. J Autism Dev Disord 2021;. Available from: https://doi.org/10.1007/s10803-021-05103-8.

[19] Bodfish JW, Symons FW, Lewis MH. The repetitive behavior scale. Western Carolina Center research reports. 1999.

[20] Bodfish JW, Symons FJ, Parker DE, Lewis MH. Varieties of repetitive behavior in autism: comparisons to mental retardation. J Autism Dev Disord 2000;30:237–43.

[21] Lam KS, Aman MG. The repetitive behavior scale-revised: independent validation in individuals with autism spectrum disorders. J Autism Dev Disord 2007;37(5):855–66.

[22] Falligant JM, Hagopian LP. Further extensions of precision medicine to behavior analysis: a demonstration using functional communication training. J Appl Behav Anal 2020;53(4):1961–81.

[23] Beavers GA, Iwata BA, Lerman DC. Thirty years of research on the functional analysis of problem behavior. J Appl Behav Anal 2013;46 (1):1–21.

[24] Floyd RG, Phaneuf RL, Wilczynski SM. Measurement properties of indirect assessment methods for functional behavioral assessment: a review of research. School Psych Rev 2005;34(1):58–73.

[25] Wiggins HC, Roscoe EM. Evaluation of an indirect assessment for identifying tasks for functional analysis. J Appl Behav Anal 2020;53 (2):997–1012.

[26] Miltenberger RG, Valbuena D, Sanchez S. Functional assessment of challenging behavior. Curr Develop Disord Rep 2019;6(4):202–8.

[27] Dufrene BA, Kazmerski JS, Labrot Z. The current status of indirect functional assessment instruments. Psychol Sch 2017;54(4):331–50.

[28] O'Neill RE, Horner RH, Albin RW, Sprague JR, Storey K, Newton JS. Functional assessment and program development for problem behavior: a practical handbook. New York, NY: Cole Publishing; 1997.

[29] Matson JL, Vollmer TR. The questions about behavioral function (QABF) user's guide. *Baton Rouge, LA: Scientific Publishers*; 1995.

[30] Iwata BA, DeLeon IG, Roscoe EM. Reliability and validity of the functional analysis screening tool. J Appl Behav Anal 2013;46(1):271–84.

[31] Ellingson SA, Miltenberger RG, Long ES. A survey of the use of functional assessment procedures in agencies serving individuals with developmental disabilities. Behav Interv 1999;4(4):187–98.

[32] Iwata BA, Dozier CL. Clinical application of functional analysis methodology. Behav Anal Pract 2008;1(1):3–9.

[33] Iwata BA, Dorsey MF, Slifer KJ, Bauman KE, Richman GS. Toward a functional analysis of self-injury. J Appl Behav Anal 1994;27:197–209.

[34] Didden R, Korzilius H, van Oorsouw W, Sturmey P. Behavioral treatment of challenging behaviors in individuals with mild mental retardation: meta-analysis of single-subject research. Am J Mental Retard 2006;111(4):290–8.

Chapter 10

The etiologic evaluation

Chapter 10.1

Neuroimaging in neurodevelopmental disabilities

Jenny Wilson[1] and Michael Regner[2]

[1]*Department of Pediatrics, Division of Pediatric Neurology, Oregon Health & Science University, Portland, OR, United States,* [2]*Department of Radiology, Oregon Health & Science University, Portland, OR, United States*

Introduction

Neurodevelopmental disabilities are clinically diagnosed, although improvements in technology have expanded the role of neuroimaging in our understanding of these disorders. Cerebral palsy (CP) and significant intellectual disability (ID) associated with other neurologic features (e.g., epilepsy and microcephaly) are likely to have structural abnormalities on imaging. Imaging is one of the strongest predictors of a clinical diagnosis of CP and may provide evidence for an acquired, genetic, or metabolic cause. For some neurodevelopmental disabilities, imaging may not play a central role in clinical management, but holds research value and the potential for development of clinically useful imaging markers. Advanced neuroimaging modalities in combination with computational modeling has elucidated mechanisms of disease, with the ultimate goal of supporting diagnoses or identifying treatments.

Description of imaging modalities

Routine brain imaging includes three different major modalities: head ultrasonography, computed tomography (CT), and magnetic resonance imaging (MRI), which vary by contrast mechanism (i.e., what bright and dark represent), protocols, advantages, disadvantages, and limitations. The clinician may consider consultation with a pediatric radiologist or neuroradiologist to optimize the imaging protocol and interpretation.

Ultrasound of the head

Mechanism: Head ultrasound (US) uses high-frequency sounds (2–10 MHz transmit bandwidth) and near-simultaneous acoustic measurement of the echo (receiver bandwidth) to produce a two-dimensional cross-sectional image of a field-of-view. Because bone causes dense acoustic shadowing, head US requires an acoustic window—typically the fontanels—to successfully transmit acoustic signals intracranially and assess brain parenchyma.

Advantages: Head US is noninvasive (no ionizing radiation), does not require sedation, is performed at the bedside, and is inexpensive. Head US is sensitive to intraventricular hemorrhage, periventricular leukomalacia, and hydrocephalus. Accordingly, US is often the first study of choice for the fetal and neonatal brain.

Disadvantages and limitations: The major limitation of head US is the requirement for an open fontanel; it is thus routinely used only in neonates and infants. Head US also has a limited field-of-view (i.e., poor or no brainstem visualization), relatively poor spatial resolution, artifacts that block US assessment (acoustic shadowing), and lack of tissue characterization (i.e., no specific modalities to detect ischemia or hemorrhage). Cortical malformations, polymicrogyria, and disorders of myelination (leukodystrophies and inborn errors of metabolism) can be difficult to be detected by US.

Capute and Accardo's Neurodevelopmental Disabilities in Infancy and Childhood. DOI: https://doi.org/10.1016/B978-0-12-824060-1.00045-6

Furthermore, the examinations are highly operator dependent; the sonographer must be fluent in the use of the machine, recognize and navigate intracranial anatomy, and optimally image tissue-specific echotextures.

Computed tomography of the head

Mechanism: Head CT uses a polychromatic beam of high-frequency photons (X-rays) in a rotating frame, measuring photon attenuation to produce a three-dimensional image of the head. Brightness on a CT image represents radiodensity at a given point in space and is measured in Hounsfield units (HU, Fig. 10.1.1). Brightness is primarily determined by the physical tissue density and the atomic number. For instance, iodine contrast and other heavy metals are bright on CT out of proportion to physical density because of orbital electron configuration. Most materials and tissues have attenuation characteristics in proportion to their physical density: water is 0 HU, air is −1000 HU (dark), and cortical bone is +1000 HU (bright).

Interpretation of CT examinations of the pediatric brain requires experience at a high-volume pediatric hospital and across developmental ages. For instance, unmyelinated white matter appears hypodense (dark) on CT and does not reach the mature adult radiodensity across the brain until 2 years of age. Myelination occurs in a predictable spatiotemporal pattern: central before peripheral (inside to outside), caudal before rostral (bottom to top), dorsal before ventral (back to front), and sensory before motor [1]. At term, only the corticospinal tracts appear myelinated. Because of the complex nature of spatio-temporal myelination, use of a normative atlas or reference text can be particularly helpful for imaging interpretation.

Advantages: CT provides a rapid, widely accessible, and relatively high-resolution image of intracranial anatomy. CT is often the first-choice study in the setting of trauma given the high sensitivity for hemorrhage and fracture. CT also detects masses, calcifying pathologies (Fig. 10.1.1), and skeletal abnormalities including craniosynostosis. Sedation is not required in most cases. While structural abnormalities from remote injury can be detected, MRI is generally the preferred modality.

Disadvantages and limitations: A major disadvantage of head CT is exposure to ionizing radiation, which is carcinogenic. Compared to adults, children are both more sensitive to radiation-induced carcinogenesis and have more years to develop cancer. For children under 5 years, it is estimated that 570−1350 head CTs would result in one case of radiation-induced cancer, defined as solid-organ malignancy or leukemia [2]. Therefore MRI is the preferred initial imaging study in the nonacute assessment of children with neurodevelopmental disability.

If a head CT is needed, it should be performed with the lowest possible dose that provides diagnostic quality images for that specific scanner. As an alternative to a head CT, many institutions have pediatric "quick brain" and "quick stroke" MRI protocols that take just minutes and are designed to be motion-robust so that sedation is not required.

Magnetic resonance imaging of the brain

Mechanism: Brain MRI uses a strong magnetic field to create an environment in which proton spins are magnetized. Applied radiofrequency (RF) pulses "tip" the protons out of magnetic alignment, and special RF receiving coils "listen"

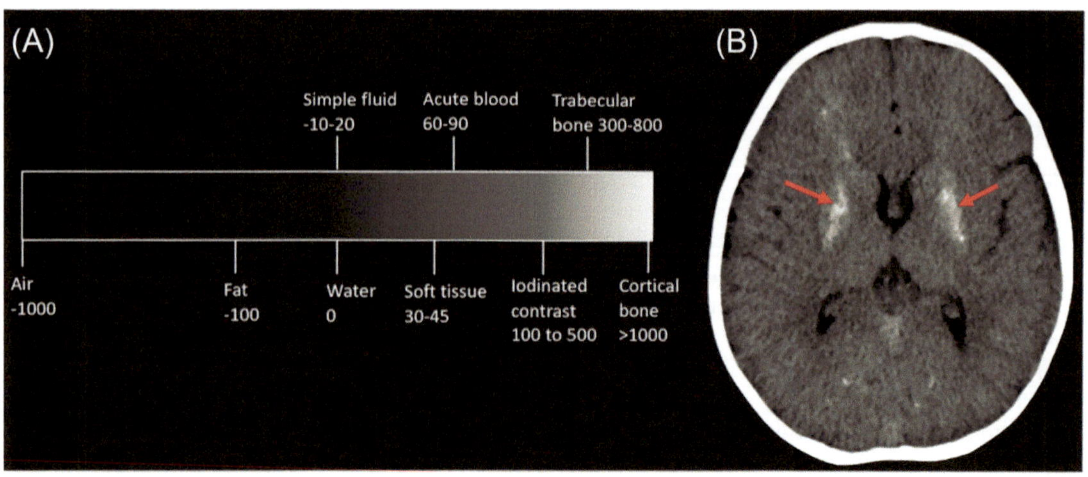

FIGURE 10.1.1 Computed tomography. (A) Hounsfield scale of radiodensity for computed tomography. (B) Head computed tomography identifies basal ganglia calcifications (red arrows) leading to a diagnosis of Aicardi–Goutières syndrome due to IFIH1 mutation.

to the proton spins as they realign with the magnetic field. Different tissue contrast weighting depends upon the local molecular quantum environment and RF pulse sequence used. Major categories of tissue contrast weighting include T1, T2, FLAIR, diffusion-weighted (DWI), apparent diffusion coefficient (ADC), susceptibility-weighted (SWI), time of flight angiography, perfusion imaging, diffusion tensor tractography, magnetic resonance spectroscopy (MRS), and blood-oxygen level-dependent functional MRI (BOLD fMRI).

Interpretation of MRI requires extensive experience with normal and abnormal pediatric brain appearance across development. The mature adult-type appearance of white matter is not reached until ages 9–12 months and 18–24 months for T1- and T2-weighted images, respectively. An MRI myelination atlas (e.g., https://www.myelinationmriatlas.com/) is useful in interpreting exams on children under 2 years of age (Fig. 10.1.2).

Advantages: MRI has several major advantages that make it the preferred neuroimaging modality for children. First, it does not use ionizing radiation and is not carcinogenic. Second, MRI offers excellent spatial resolution to 1 mm or less on most sequences. Third, multiple different tissue-intensity maps evaluate anatomy and tissue characteristics of the brain.

MRI can characterize tissue because intrinsic microscopic magnetic properties of tissue differ. For instance, some substances that are intrinsically T1 hyperintense (bright) include blood, fat, melanin, protein, gadolinium, copper, and manganese. In general, T1-weighted images show lipid and proteinaceous tissues brightly while FLAIR- and T2-weighted images show fluid and edematous tissues brightly. FLAIR imaging is usually T2 weighted and illustrates edema with cerebrospinal fluid appearing dark. Diffusion characteristics can discriminate the cytotoxic edema that occurs in stroke (ADC dark) from vasogenic edema (ADC bright). Susceptibility-weighted images are sensitive for detecting blood products which are seen as areas of hypointensity (dark) where the iron distorts the local magnetic field.

Disadvantages and limitations: MRI is expensive and requires a sophisticated team effort to make it practical and useful. Because individual MRI sequences can require minutes to acquire, patients must be able to remain still within the scanner. In young children, sedation may be necessary to obtain diagnostic-quality MRI images. In older children, anxiolysis can be helpful in appropriate patients as the scanner itself is narrowly enclosed, noisy, and uncomfortable. MRI has limitations in osseous evaluation compared to CT, and metallic artifacts can sometimes be mitigated better on CT than MRI. Lastly, the MRI scanner is a powerful magnet that produces fields strong enough to convert everyday objects into lethal projectiles, produce burns, disable pacemakers and electronic devices, and induce electrical current.

Advanced and investigational imaging of the brain

Advanced magnetic resonance imaging

Diffusion tensor imaging

Diffusion tensor imaging (DTI) measures the mean diffusivity along different directional orientations, thought to reflect the preferential movement of water along axons. DTI uses a measurement termed fractional anisotropy, which takes on a value between 0 (diffusion occurs unrestricted in all directions) and 1 (diffusion of water is completed restricted along one axis). Tractography is the probabilistic prediction of the orientation of white matter tracts and is based upon fractional anisotrophy (Fig. 10.1.3).

Magnetic resonance perfusion

Cerebral blood flow can be inferred using a variety of MRI pulse sequences, most commonly either arterial spin labeling (ASL) or dynamic susceptibility contrast (DSC). MR perfusion is primarily used in stroke and tumor imaging.

Magnetic resonance spectroscopy

MRI is a spatially encoded version of nuclear magnetic resonance (NMR), which is routinely used in the analysis of organic compounds. MRS quantifies different compounds in the brain in preselected regions (voxels placed by the radiologist) with utility in diagnosing tumors, leukodystrophies, mitochondrial disorders, creatine disorders, and in prognosis after hypoxic-ischemic encephalopathy (HIE).

Functional magnetic resonance imaging

fMRI detects alterations in blood flow using BOLD imaging in regions of the brain while the patient is exposed to a stimulus or requested to complete a task called a "paradigm" (Fig. 10.1.3). fMRI requires a cooperative patient with limited movement. It is routinely employed for presurgical mapping of cortical functions including language and motor control. fMRI applications in patients with neurodevelopmental disorders is primarily in the research realm.

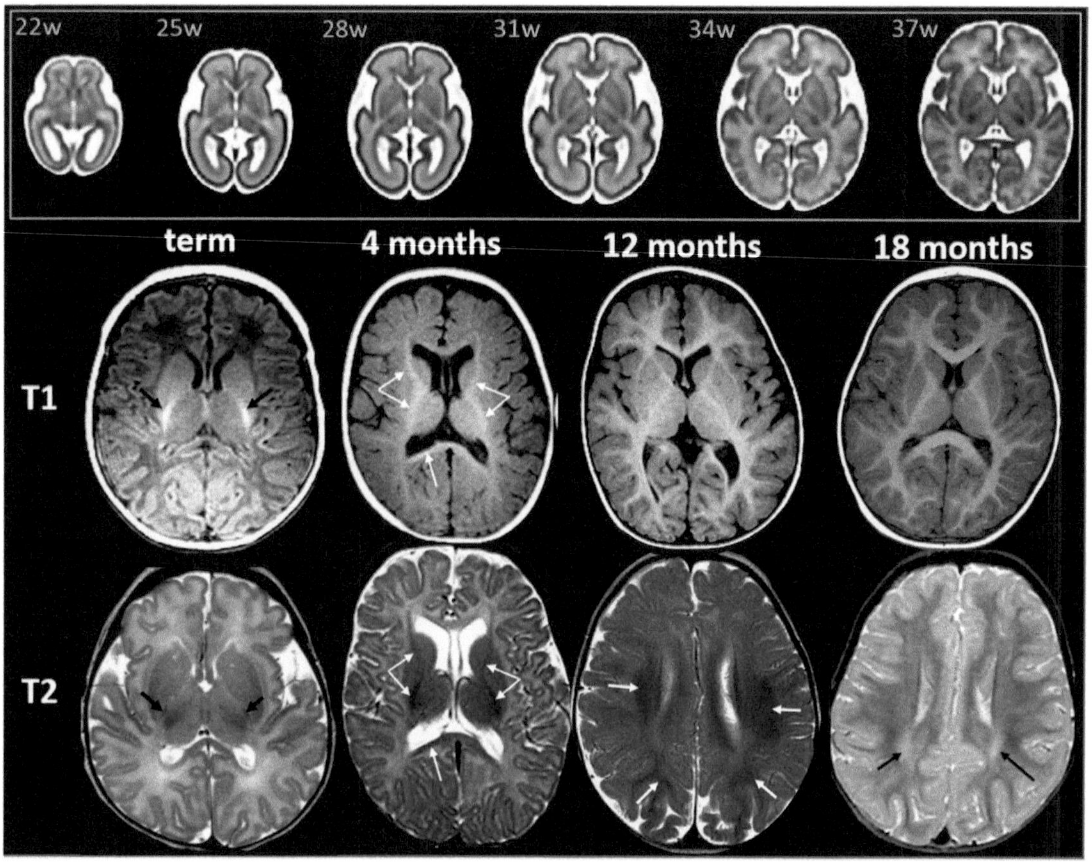

FIGURE 10.1.2 Magnetic resonance imaging (MRI) of the developing brain. Normal MRI findings of fetal and postnatal development. Top panel: spatiotemporal atlas of axial T2 images from a composite of 81 normal fetuses demonstrating increasing sulcation with age. Bottom two rows: progression of myelination from term to 18 months. On T1 imaging, unmyelinated fibers are hypointense to gray matter and myelinated fibers appear hyperintense to gray matter. On T2 imaging, the reverse is true. At any point in time, myelination appears more advanced on T1 compared to T2 imaging. Term age: myelinated fibers are apparent in the posterior limb of the internal capsule (black arrows). 4 months: myelination now apparent in the anterior limb and corpus callosum (white arrows). 12 months: the centrum semiovale and posterior white matter are myelinated on T2 (white arrows) and myelination appears complete on T1. 18 months: mostly complete myelination pattern on T1 and T2. On T2, residual hyperintensity adjacent to the ventricles, or "terminal zones of myelination" (black arrows), are a normal finding from 16 months to 10 years of age and should not be confused with gliotic changes from periventricular leukomalacia. *W*, Weeks. *Preterm imaging from Gholipour A, et al. A normative spatiotemporal MRI atlas of the fetal brain for automatic segmentation and analysis of early brain growth. Sci Rep 2017;7(1):476; term age and older images from Branson, HM. Normal myelination: a practical pictorial review. Neuroimaging Clin N Am 2013;23(2):183–95; reprinted with permission.*

High-field magnetic resonance imaging

While standard MRI uses magnet strengths of 1.5 or 3 Tesla (T), high-field MRI refers to use of magnet field strengths of 7 T, allowing for greater spatial resolution and tissue contrast, although with increased risk of side effects such as dizziness. The 7 T MRI was FDA-approved for use in humans 66 pounds or greater in 2017 and has clinical use in some patients with epilepsy [3].

Functional near-infrared spectroscopy

Functional near-infrared spectroscopy (fNIRS) evaluates regions of cerebral activation by quantifying changes in oxygenated hemoglobin and deoxygenated hemoglobin through light reflected from specific regions in the brain (Fig. 10.1.3). Although fNIRS has limited spatial resolution (2–3 cm), it is portable and does not require the patient be still, which makes it an appealing modality for use in children and individuals with developmental disabilities.

Positron emission tomography

Positron emission tomography (PET) localizes and quantifies a radiotracer that is taken up by a tissue. Because of limited resolution and need to correct for natural changes in tissue attenuation, the region of uptake is often

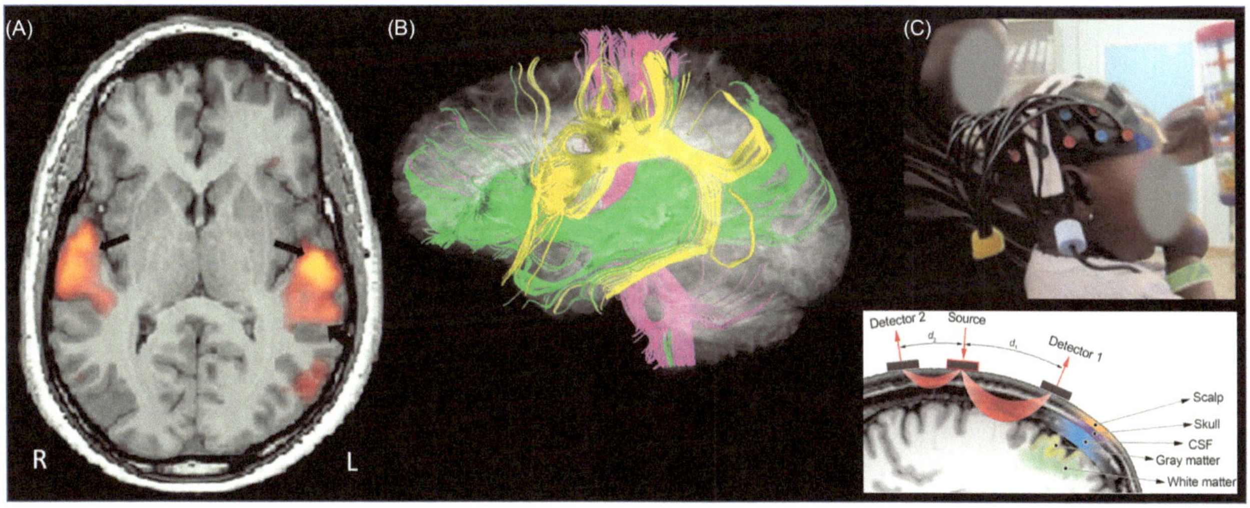

FIGURE 10.1.3 Advanced imaging techniques. (A) Functional MRI of a 10-year-old showing activation of the primary auditory areas (thin arrows) and Wernicke's area (thick arrow) while listening to a story. (B) Tractography using diffusion tensor imaging (DTI). Pink: corticospinal tract, yellow: superior longitudinal fasciculus (including arcuate fasciculus connecting Wernicke to Broca's area), green: inferior longitudinal fasciculus. (C) Functional near-infrared spectroscopy (fNIRS) on a child, and schematic demonstrating the path of light from the source to detectors through the head. *CST*, Corticospinal tract; *fMRI*, functional magnetic resonance imaging; *MCP*, middle cerebellar peduncle; *PCF*, pontine cerebellar fibers; *SCP*, superior cerebellar peduncle; *SLF*, superior longitudinal fasciculus. *A:From Altman NR, Bernal B. Pediatric applications of functional magnetic resonance imaging. Pediatr Radiol. 2015;45(Suppl 3):S382−96, reprinted with permission; B: Courtesy Dr. David Pettersson; C: Top panel, from Lloyd-Fox S, Begus K, Halliday D, et al. Cortical specialisation to social stimuli from the first days to the second year of life: a rural Gambian cohort. Dev Cogn Neurosci 2017;25:92−104, reprinted with permission; bottom panel, from Pinti P, Tachtsidis I, Hamilton A, et al. The present and future use of functional near-infrared spectroscopy (fNIRS) for cognitive neuroscience. Ann N Y Acad Sci 2020;1464(1):5−29, reprinted with permission.*

colocalized with CT. PET has a variety of applications depending on the tracer and body region visualized. 18F-Fluorodeoxyglucose (FDG) is the radiotracer used to evaluate glucose metabolism and is commonly used in tumor evaluation. However, newer radiotracers targeting neurotransmitters such as serotonin, gamma-aminobutyric acid, dopamine, and synaptic vesicles are being used as research tools in neurodevelopmental disabilities including autism and attention-deficit hyperactivity disorder (ADHD) [4].

Clinical guidelines

Cerebral palsy

A noncontrast MRI brain is typically the first imaging study performed to evaluate the etiology of CP [5,6]. The MRI sequences typically performed include high-resolution 3D images in T1, T2, FLAIR, diffusion, and susceptibility-weighted pulse sequences. Children who have previously had imaging demonstrating an abnormality consistent with a diagnosis of CP may not need to have repeat imaging. A CT scan is not preferred as it exposes the child to ionizing radiation and has lower tissue sensitivity. Imaging also serves an important role in the early diagnosis of an infant at risk for CP. An abnormal term-age MRI combined with an abnormal standardized exam such as the General Movement Assessment, the Hammersmith Infant Neurological Examination (HINE), or the Developmental Assessment of Young Children predicts CP in infants [7] (Table 10.1.1).

Global developmental delay and intellectual disability

Neuroimaging may be useful for select children with global developmental delay (GDD) or ID. The overall reported positivity rate by MRI for this indication is estimated to be 38%, but varies widely (0%−98%) depending on the study and population [8]. Higher positivity rates were associated with moderate to profound ID or abnormal neurologic exam [9,10]. However, even if the MRI is abnormal, it is not likely to provide a precise etiology: one review found that an abnormal MRI led to an etiologic diagnosis in just 8% of patients [8]. Clinical guidelines generally acknowledge the utility of neuroimaging in children with GDD/ID who have neurologic abnormalities on exam [11−13]. The risks of sedation must be balanced against the pretest probability for a structural abnormality (Table 10.1.2).

TABLE 10.1.1 Guidelines on imaging in cerebral palsy.

	Year	Status	Timing	Recommendation
NICE [5]	2017	Active	When deciding on imaging timing, take into account: • subtle abnormalities may not be apparent until 2 years of age • presence of "red flags" for a progressive disorder • need for sedation in a younger child • the views of the child and their parents and/or caregivers	"Offer MRI to investigate etiology in a child or young person with suspected or known cerebral palsy if this is not clear from: • antenatal, perinatal, and postnatal history • their developmental progress • findings on clinical examination • results of cranial ultrasound examinations"
AAN [6]	2004	Guideline being updated. Last reaffirmed on July 21, 2007	Not discussed	"Neuroimaging is recommended in the evaluation of a child with CP if the etiology has not been established, for example, by perinatal imaging"

AAN, American Academy of Neurology; *CP*, cerebral palsy; *MRI*, magnetic resonance imaging; *NICE*, National Institute for Health and Care Excellence.

TABLE 10.1.2 Guidelines on imaging in global developmental delay/intellectual disability.

	Year	Status	Tier of testing	Recommendation
AAP [12]	2014	Active	After genetic and metabolic testing	"MRI if (presence of) microcephaly, macrocephaly, or abnormal findings on neurologic examination (focal motor findings, pyramidal signs, extrapyramidal signs, intractable epilepsy, or focal seizures)"
AAN [13]	2003	Retired	First tier if no features of a specific diagnosis	"Neuroimaging is recommended as a part of the diagnostic evaluation. As the presence of physical findings (e.g., microcephaly, focal motor findings) increases the yield of making a specific neuroimaging diagnosis, physicians can more readily consider obtaining a scan in this population"
ACMG [11]	1997	Active	Following a thorough patient assessment, tier left up to the practitioner	Neuroimaging should be considered in patients without a known diagnosis especially in the presence of neurologic symptoms, cranial contour abnormalities, microcephaly, or macrocephaly

AAN, American Academy of Neurology; *AAP*, American Academy of Pediatrics; *ACMG*, American College of Medical Genetics; *MRI*, magnetic resonance imaging.

Autism spectrum disorder

The reported yield of neuroimaging in autism spectrum disorder (ASD) varies. Some report relatively low yield, particularly in the absence of other neurologic signs/symptoms (6.5%–10%) [14,15]. Others have found a higher yield (40%–55%), although mainly of nonactionable findings such as volume loss, white matter changes, dilated Virchow–Robin spaces, or mega cisterna magna [16–19]. Guidelines generally state that imaging is not routinely indicated for children with ASD unless other features such as seizures or focal neurologic deficits are present [20–22] (Table 10.1.3).

TABLE 10.1.3 Guidelines on imaging in autism spectrum disorder.

	Year	Status	Tier of testing	Recommendation
AAP [21]	2020	Active		"MRI may be indicated in the evaluation of atypical regression, microcephaly, macrocephaly, seizures, intracranial manifestations of genetic disorders, abnormal neurologic examination, or other clinical indications."
AACAP [22]	2014	Active		"Genetic or neurologic consultation, neuroimaging, EEG, and additional laboratory tests should be obtained when relevant, based on examination or history."
AAN [20]	2000; affirmed in 2014	Active		"There is no clinical evidence to support the role of routine clinical neuroimaging in the diagnostic evaluation of autism, even in the presence of megalencephaly."

AACAP, American Academy of Child and Adolescent Psychiatry; *AAN*, American Academy of Neurology; *EEG*, electroencephalogram; *MRI*, magnetic resonance imaging.

Developmental regression

While a child with typical autistic regression does not require neuroimaging, developmental regression and/or progressive neurologic symptoms outside of that setting warrants a thoughtful and thorough investigation which may include neuroimaging. The approach should be tailored to the child [23,24].

Hearing disorders

In unilateral, asymmetric, or mixed hearing loss, comprehensive temporal bone imaging should be considered. Imaging is low yield for bilateral symmetric hearing loss. Children with profound hearing loss who are cochlear implant candidates benefit from cochlear imaging for prosthesis planning and confirming implant electrode location. It is important to be aware that many cochlear implants preclude an MRI scan. Temporal bone imaging and brain MRI are recommended for evaluating unilateral auditory neuropathy spectrum disorder (ANSD) and should be considered for bilateral ANSD. Temporal bone imaging in neonates is not recommended unless imaging is indicated for other reasons. MRI is the preferred modality for most peripheral auditory abnormalities [25].

Imaging findings by etiology

Among developmental disabilities, CP has the highest rate of abnormal neuroimaging (90%) [26]. The Surveillance of Cerebral Palsy in Europe (SCPE) classifies neuroimaging abnormalities in CP by timing of insult [27], which serves as a classification framework when considering structural causes of neurodevelopmental disabilities. As the categorization is based on timing rather than etiology, a number of disorders are simply included under "miscellaneous," and some etiologies span more than one category (e.g., HIE may affect white matter and gray matter, and a child with congenital cytomegalovirus [CMV] may have multiple malformations such as polymicrogyria, heterotopic gray matter, and calcifications) (Table 10.1.4).

Malformations

Malformations of cortical development

The European Network Neuro-MIG initiative on Brain Malformations published consensus definitions and classifications of malformations of cortical development (MCD) [28]. MCD describe abnormally developed gray matter or gray matter heterotopias, which may be associated with abnormal brain size. MCD result from disruption of cell proliferation and apoptosis, cell migration or postmigrational development [29]. MCD subtypes include:

- Microcephaly
- Megalencephaly/hemimegalencephaly
- Focal cortical dysplasia: disordered cortical lamination with or without abnormal cell types

TABLE 10.1.4 The Surveillance of Cerebral Palsy in Europe (SCPE) MRI classification system for children with cerebral palsy.

Timing	Category	Example
First and early second trimester	1. Maldevelopments[a]	
	a. Disorders of cortical malformation	Polymicrogyria
	b. Other maldevelopments	Holoprosencephaly, Dandy–Walker malformation, cerebellar hypoplasia
Early third trimester	2. Predominant white matter injury	
	a. Periventricular leukomalacia	
	b. Sequelae of IVH or periventricular venous infarction	
	c. Combination of PVL and IVH sequelae	
Late third trimester/term	3. Predominant gray matter injury	
	a. Basal ganglia/thalamus lesions	Hypoxic-ischemic encephalopathy
	b. Cortico-subcortical lesions only	Watershed ischemia
	c. Arterial infarctions	Middle cerebral artery territory stroke
	4. Miscellaneous	Cerebral atrophy, parenchymal hemorrhage, infection, trauma
	5. Normal	

Classification system for neuroimaging findings in cerebral palsy [27]. *IVH*, Intraventricular hemorrhage; *MRI*, magnetic resonance imaging; *PVL*, periventricular leukomalacia.
[a]*Used interchangeably with "malformations of cortical development."*

- Gray matter heterotopia: clusters of neurons in abnormal locations
- Lissencephaly: thick cortex with reduced gyration which may be agyria or pachygyria
- Cobblestone malformation: irregular, pebbled, internal and external cortical surface
- Polymicrogyria: excessive number of small gyri
- Schizencephaly: cleft lined by abnormal gray matter extending across the hemisphere
- Dysgyria: nonspecific dysmorphic cortex

Other malformations

This category includes holoprosencephaly, agenesis of the corpus callosum, Dandy–Walker malformation, and cerebellar hypoplasia (Fig. 10.1.4).

Predominant white matter injury

Periventricular leukomalacia

White matter injury, also referred to as periventricular leukomalacia (PVL), is the most common cerebral injury in very low birth weight (VLBW) infants [30]. Cystic PVL, one of the most important contributors to spastic CP, has become less common with improvements in neonatal care. Noncystic forms, which may be diffuse or focal (punctate white matter lesions, PWML), now predominate, and are associated with adverse developmental outcomes. Diffuse white matter injury is associated white matter volume loss and thinning of the corpus callosum. An anterior/posterior pattern of PWML is also seen in term infants with congenital heart disease, possibly due to delayed cerebral development predisposing to injury patterns more typical of a premature infant [31].

The findings of PVL evolve over time on ultrasound, beginning with increased echogenicity of the white matter, which may become cystic in 2–5 weeks. Subsequently, the cysts may coalesce with the ventricle so that later scans may just show ventricular enlargement. Taking this evolution into account, there are standardized recommendations for head ultrasound screening for premature infants. For example, the American Academy of Pediatrics (AAP) recommends head ultrasound

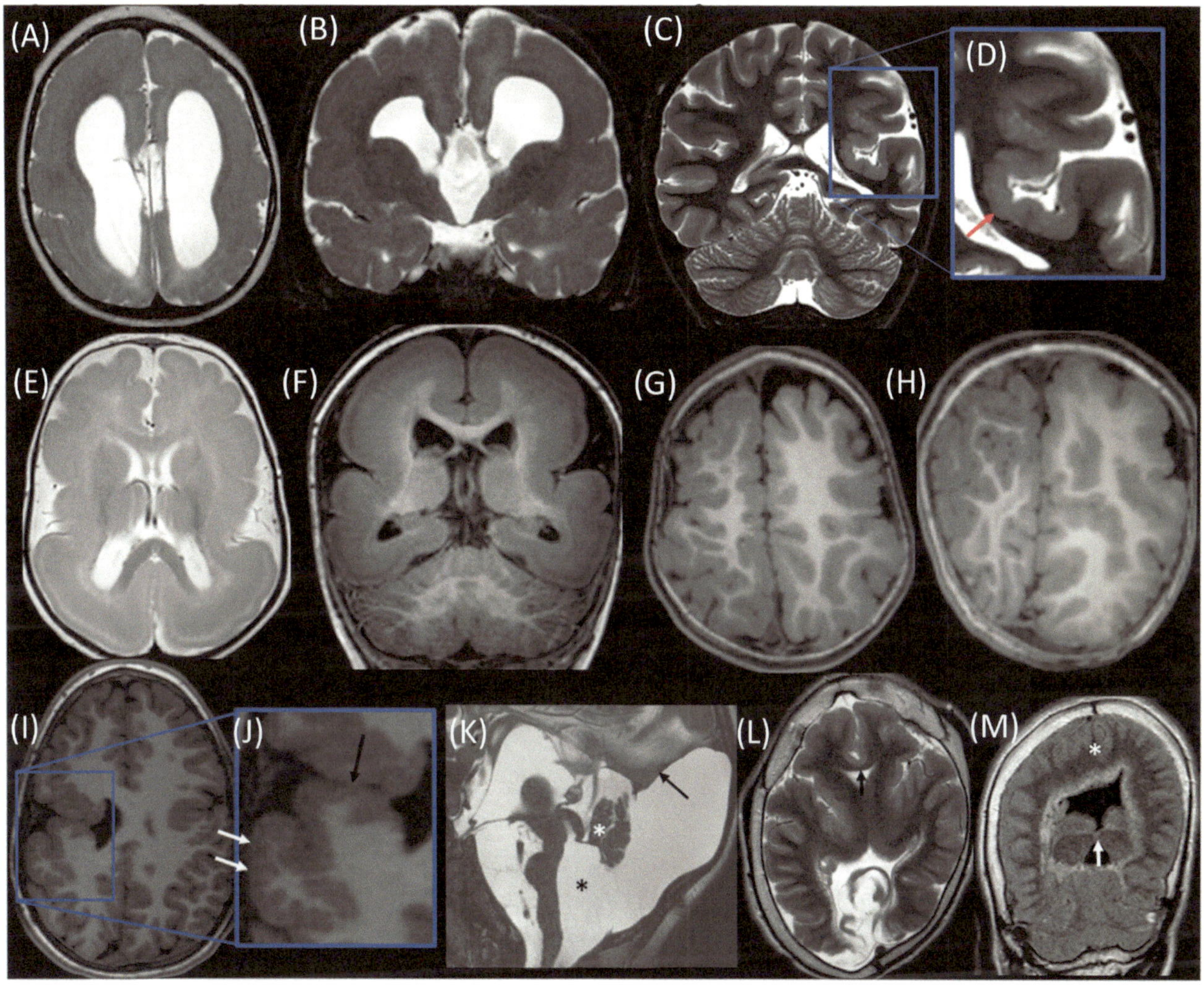

FIGURE 10.1.4 Imaging of malformations of cortical development. (A and B) T2 axial and coronal MRI at 1 year demonstrating pachygyria (spastic/dystonic quadriplegic CP, GMFCS V, ID, epilepsy, *TUBA1A* gene mutation). (C and D) Left perisylvian polymicrogyria at 11 years (T1 coronal, unilateral dystonic CP, GMFCS I). (E and F) Axial T2 and coronal T1 MRI at 6 months showing lissencephaly (hypotonic CP GMFCS V, ID, epilepsy, Miller–Dieker lissencephaly syndrome). (G and H) Axial and coronal T1 MRI at 1 year showing left hemimegalencephaly (epilepsy, hemiplegic CP). (I and J) T1 axial MRI at 11 years showing closed lip schizencephaly (black arrow), with polymicrogyria lining cleft and extending along cortex (white arrows) (left spastic hemiplegic CP, GMFCS I, learning disability). (K) Dandy–Walker malformation with hypoplastic, upturned vermis (white asterisk), enlarged fourth ventricle (black asterisk), enlarged posterior fossa and elevated tentorium cerebelli (black arrow). (L and M) T2 axial and T1 coronal MRI at 20 years showing holoprosencephaly with fusion of a portion of the frontal lobe across the midline (black arrow), absent interhemispheric fissure (asterisk), rudimentary ventricular shape and fusion of the deep gray nuclei (white arrow) (spastic/dystonic quadriplegic CP GMFCS V, intellectual disability). *CP*, Cerebral palsy; *GMFCS*, gross motor function classification system; *ID*, intellectual disability; *MRI*, magnetic resonance imaging.

screening in infants born at <30 weeks or >30 weeks if risk factors are present, with a scan by 7–10 days, then at 4–6 weeks of age, and near term [32]. Of note, while cystic PVL is easily detected on head ultrasound, diffuse PVL may only be visible on MRI [33]. An MRI at term equivalent age may be considered in infants at high risk, to inform prognosis [32]. Scoring systems grading the severity of PVL on ultrasound and MRI predict neurodevelopmental outcomes [34].

Sequelae of intraventricular hemorrhage or periventricular venous infarction

Lower gestational age and birth weight put premature infants at increasing risk for intraventricular hemorrhage (IVH), which is classified as grade I: blood limited to the germinal matrix, grade II: IVH without ventricular dilation, grade III: IVH encompassing >50% of ventricular area and/or ventricular dilation present, and parenchymal hemorrhage (previously grade IV). Parenchymal hemorrhage in this setting is due to venous infarction in the white matter adjacent to the ventricle from blood obstructing flow in the medullary veins. Blood in the ventricular system, particularly higher grades of IVH, predispose to hydrocephalus and need for neurosurgical intervention. Grade III IVH, parenchymal hemorrhage, and posthemorrhagic hydrocephalus increase the risk of neurodevelopmental disabilities [35].

Cerebellar injury

Cerebellar injury, while not included in the SCPE classification, is common in preterm infants and an important risk factor for adverse neurodevelopmental outcomes including CP, ID, and autism. Specifically, VLBW infants are at greater risk of cerebellar hemorrhage or ischemia which may be seen using mastoid windows on head US and is readily visualized on MRI [36] (Fig. 10.1.5).

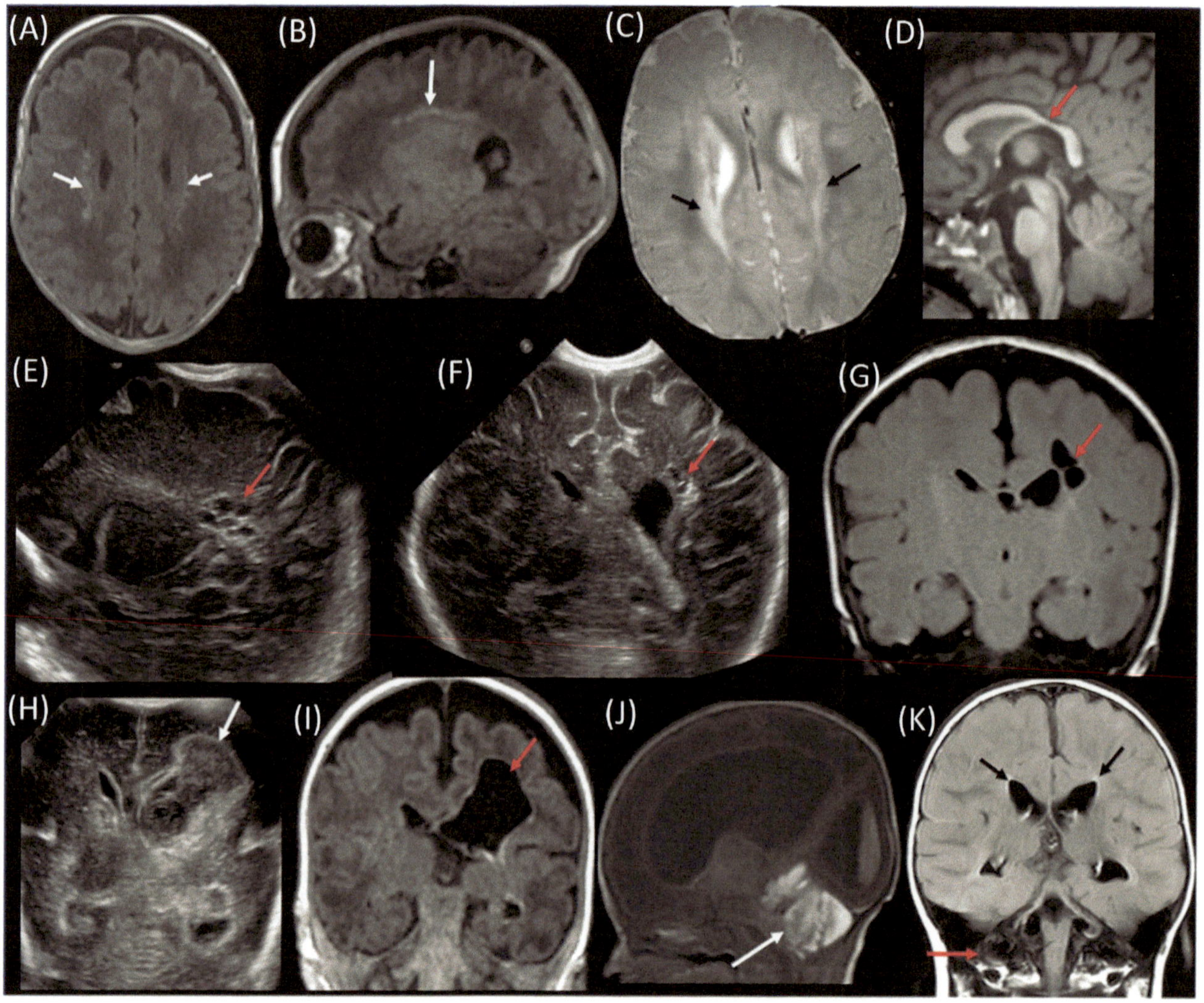

FIGURE 10.1.5 Imaging of white matter injury and cerebellar injury. (A–D) White matter injury of prematurity. (A and B) Axial and sagittal T1 MRI showing punctate white matter lesions in an ex-31-week child at term age (white arrows). (C and D) MRI of the same child at 1 year showing more gliosis on axial T2 (black arrows) and posterior thinning of the corpus callosum on sagittal T1 (red arrow) (spastic diplegic CP, GMFCS II). (E and F) Sagittal and coronal ultrasound demonstrating left cystic PVL in an 8-week-old ex-35-week infant (red arrows) (right hemiplegic CP, GMFCS I). (G) Left cystic PVL on coronal flair MRI in an 8-week-old ex-29-week infant (red arrow) (right hemiplegic CP, GMFCS I). (H and I) Left-sided periventricular venous infarction (IVH grade IV) in an ex-24-week infant at 2 weeks on ultrasound (H) and resulting porencephalic cyst on coronal T1 MRI at 36 weeks CGA (red arrow) (spastic hemiplegic CP, GMFCS II, autism, intellectual disability). (J and K) Cerebellar hemorrhage, IVH, and hydrocephalus in a 28-week premature infant on sagittal T1 MRI (J, white arrow), with follow-up coronal flair MRI at 2 years (K) showing encephalomalacia of the majority of the cerebellum (red arrow) and mild PVL as evidenced by ventricular enlargement and periventricular hyperintensity (black arrows) (spastic and ataxic diplegic cerebral palsy, GMFCS III, dysarthria, normal cognition). *CGA*, Corrected gestational age; *CP*, cerebral palsy; *GMFCS*, gross motor function classification system; *ID*, intellectual disability; *IVH*, intraventricular hemorrhage; *MRI*, magnetic resonance imaging; *PVL*, periventricular leukomalacia.

Predominant gray matter injury

Hypoxic-ischemic encephalopathy

Basal ganglia and thalamic injuries are common sequelae of abrupt HIE and predict dyskinetic CP. Other HIE patterns include watershed or a combination of basal ganglia/thalamic and watershed [37]. Characteristic acute findings of HIE are restricted diffusion on DWI/ADC and T1 hyperintensity in the bilateral putamen and thalami with absence of the posterior limb of the internal capsule bright spot (the bright spot represents myelinating fibers present after 36 weeks gestation) [38]. Basal ganglia/thalamic involvement, abnormalities of the posterior limb of the internal capsule, and presence of a lactate peak on MR spectroscopy predict adverse neurodevelopmental outcomes [39]. The imaging findings evolve with time, with the highest sensitivity of MRI findings at 3–7 days. Imaging is typically performed after rewarming as hypothermia delays the DWI changes [39]. Therapeutic hypothermia decreases the severity of MRI abnormalities without impacting the prognostic value of imaging [40] (Fig. 10.1.6).

Arterial infarctions

Neonatal arterial ischemic stroke is the most common cause of hemiplegic CP and an important cause of other developmental disabilities. On MRI, the ischemic region demonstrates increased signal on DWI within minutes of onset of ischemia, lasting about 3 weeks, which represents cytotoxic edema. There may also be reduced diffusion in the denervated white matter connected to the involved cortex (e.g., cortical spinal tract in the posterior limb of the internal capsule or cerebral peduncle, corpus callosum) or in other functionally connected structures. This dysfunction of brain regions remote from but connected to an area of injury is termed "diaschisis" [41]. Hyperintensity on T2 and flair imaging begins within hours of the stroke and remains for about a month. If the stroke goes undetected in the neonatal period, it may be discovered as an area of encephalomalacia in a vascular territory when imaging is performed for new onset seizures or hemiplegic CP ("presumed perinatal arterial ischemic stroke") [42].

Imaging findings are strongly predictive of outcomes: essentially every neonate with a complete MCA territory stroke will develop hemiplegic CP [43]. Also predictive of CP are basal ganglia involvement and corticospinal tract DWI changes (e.g., cerebral peduncle). Complete MCA territory stroke or involvement of the basal ganglia predicts cognitive and behavior deficits [43]. After the acute phase, decreased volumes in structures impacted by diaschisis including the cerebellum and thalamus predict outcomes [44].

Miscellaneous

Venous

Cerebral venous sinus thrombosis

Cerebral venous sinus thrombosis (CVST) may result in venous congestion and infarction, often with hemorrhage. About 60% of children with CVST develop sequelae which may include cognitive deficits and/or motor disability [45]. Venous structures which may be involved include the venous sinuses, smaller cortical veins, superficial or deep medullary veins, or the deep sinuses [46]. When the smaller veins and sinuses become involved, infarction may ensue. Thrombosis of the medullary veins is a relatively common and probably underrecognized injury pattern in the neonate detected on T2 and blood-sensitive sequences as a fan-shaped pattern of linear blooming within the white matter, often with associated areas of ischemia on DWI, and sometimes associated with parenchymal or subpial hemorrhage (Fig. 10.1.7) [46].

Genetic

Imaging correlates of genetic abnormalities are diverse and may include maldevelopments, white matter or gray matter atrophy, delayed myelination, and thinning of the corpus callosum.

Infections

Prenatal, perinatal, and postnatal infections are an important cause of developmental disabilities with imaging findings that may overlap with other categories (e.g., ischemia, white matter injury, or maldevelopments in the setting of prenatal infection). Additional findings such as atrophy, calcifications, and hydrocephalus may be seen.

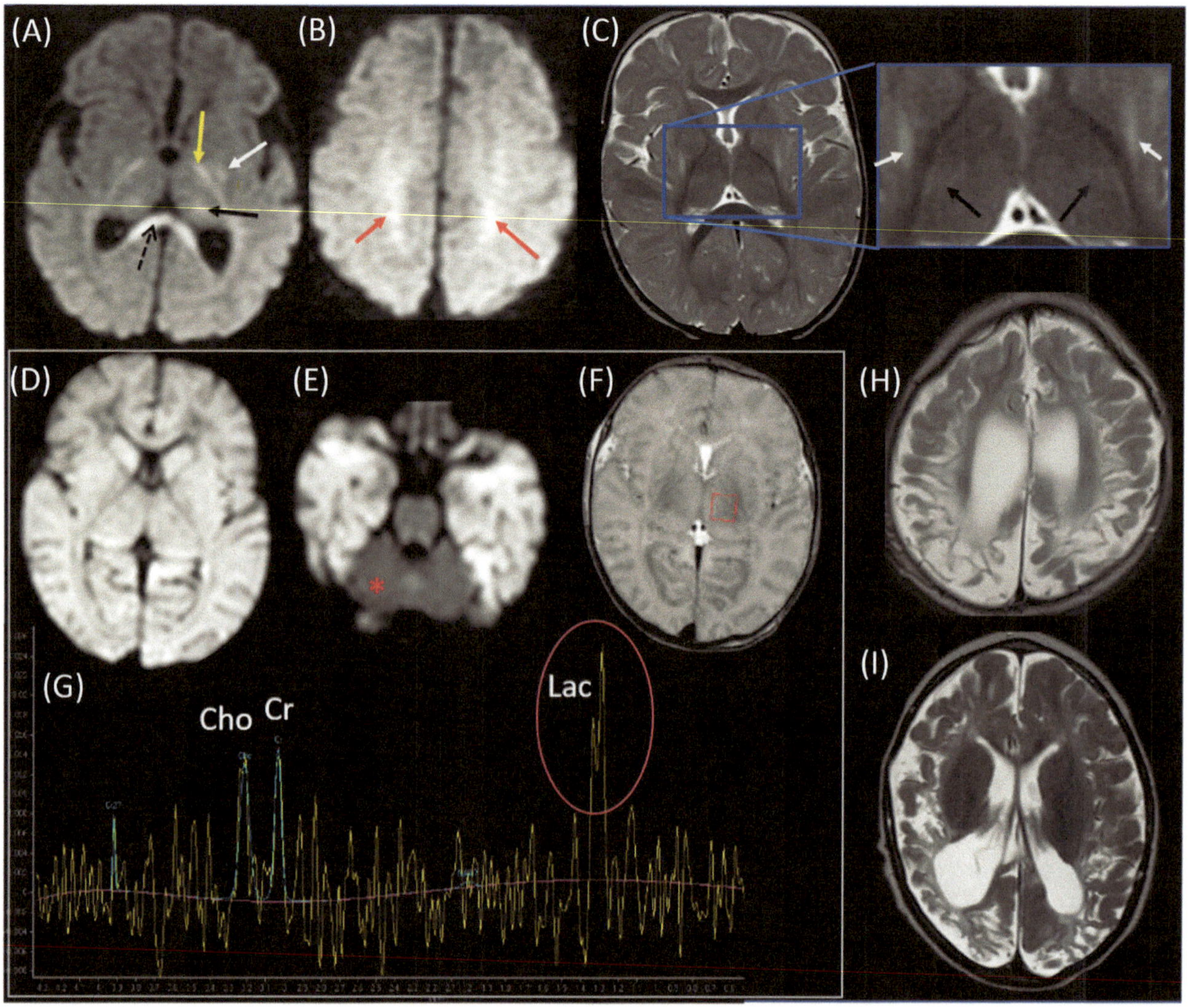

FIGURE 10.1.6 Imaging of neonatal hypoxic-ischemic injury. (A–C) Basal ganglia/thalamic injury pattern of hypoxic-ischemic encephalopathy. (A and B) Axial DWI MRI of a 13-day old with HIE showing restricted diffusion due to direct injury of the bilateral putamen (white arrow), thalami (black arrow), and deep white matter (red arrows). There is also restricted diffusion in associated white matter tracts including the internal capsule of the corticospinal tract (yellow arrow) and corpus callosum (black dashed arrow) due to diaschisis* (mixed CP, GMFCS IV). (C) Axial T2 MRI showing putamen (white arrows) and thalamic (black arrows) gliosis, sequelae of HIE in an 8-month old (dystonic CP, GMFCS III). (D–G) Severe global pattern HIE. (D and E) Global restricted diffusion on DWI imaging of a 4-day-old infant with severe HIE could be misinterpreted as normal if not compared to the normal signal intensity of the cerebellum (red asterisk). (G) ^{1}H-MRS with voxel placed on the thalamus (F) showing a doublet at 1.33 ppm (red circle) corresponding to the presence of lactate (hypotonia, epilepsy, deceased at 9 months). (H and I) Axial T2 MRI showing encephalomalacia and ex vacuo ventricular enlargement in a 3-year old, sequelae of watershed pattern HIE (spastic quadriplegic CP GMFCS IV, intellectual disability, cortical visual impairment). *Cho*, Choline; *Cr*, creatine; *DWI*, diffusion-weighted imaging; *GMFCS*, gross motor function classification system; *HIE*, hypoxic-ischemic encephalopathy; *Lac*, lactate; *MRS*, magnetic resonance spectroscopy; *MRI*, magnetic resonance imaging. **Diaschisis*—a Greek word meaning "shocked through," describes the dysfunction of brain regions remote from but connected to an area of injury. The three forms are diaschisis corticospinalis (impacting cortical spinal tract fibers), diaschisis commissuralis (effect on corpus callosum fibers), and diaschisis associative (involvement of neighboring cortical areas).

Neurodevelopmental disabilities imaging research

Autism

Numerous studies have documented volumetric brain differences in individuals with ASD as compared to controls. Children with autism tend to have larger brain volumes and greater surface area by 2 years of age, however, those differences are not present in older children, suggesting a period of overgrowth followed by decreased growth compared

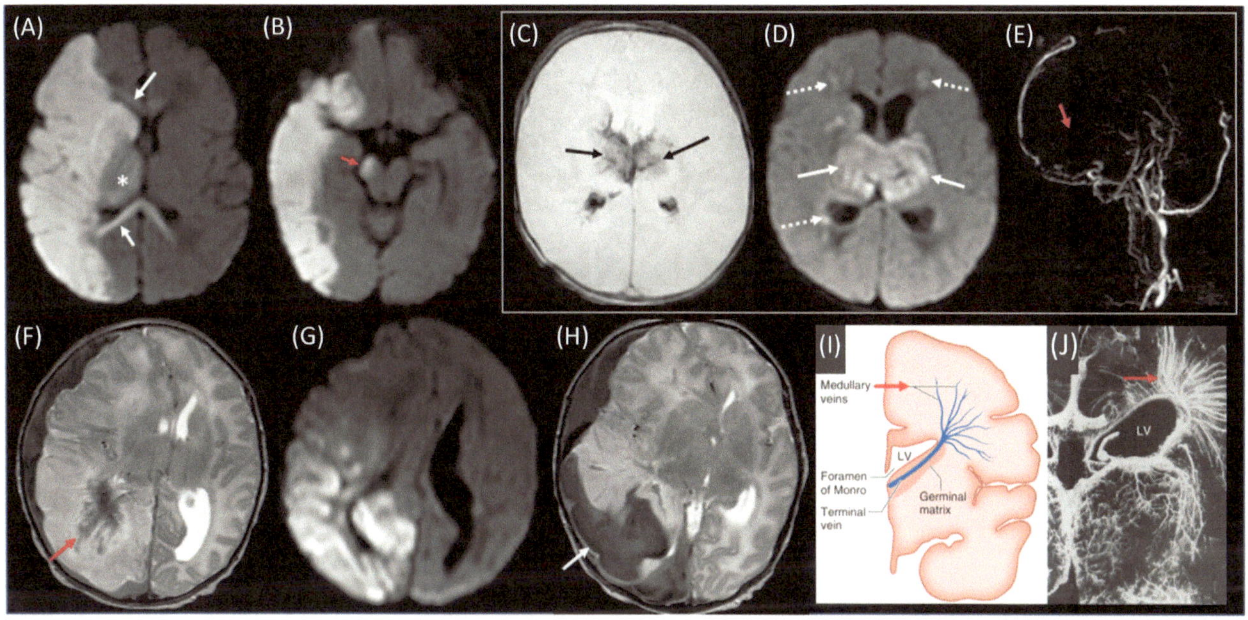

FIGURE 10.1.7 Imaging of stroke. (A and B) Axial DWI MRI of a neonate showing complete right MCA acute arterial ischemic stroke with diaschisis of corpus callosum (white arrows), thalamus (asterisk), and cortical spinal tract in the cerebral peduncle (red arrow), predictive of cerebral palsy (hemiplegic CP GMFCS 1). (C–E) Deep venous thrombosis. (C) Venous congestion apparent as hypointensity on axial SWI MRI within the thalami (black arrows), with (D) acute venous ischemia on axial DWI imaging of the deep gray (solid white arrows) and white matter (dashed white arrows), with (E) absent flow in the deep veins on MRV (red arrow) (dystonic CP GMFCS I, ID). (F–H) Medullary vein thrombosis in a neonate. (F) Axial T2 MRI showing characteristic fan-shaped linear hypointensities of clot within the medullary veins (red arrow), associated with and acute venous ischemia on DWI (G), and subpial hemorrhage on axial T2 (H, white arrow) (left hemianopsia, epilepsy). (I) Schematic of the medullary veins (red arrow) alongside postmortem venogram (J) illustrating the drainage pattern and appearance of the medullary veins. *CP*: Cerebral palsy; *DWI*, diffusion-weighted imaging; *GMFCS*, gross motor function classification system; *MCA*, middle cerebral artery; *MR*, magnetic resonance; *MRV*, magnetic resonance venogram; *SWI*, susceptibility-weighted imaging. *Schematic from Inder TE, Perlman JM, Volpe JJ. Chapter 28 – Preterm Intraventricular Hemorrhage/Posthemorrhagic Hydrocephalus. In: Volpe JJ, ed. Volpe's Neurology of the Newborn (Seventh Edition). Content Repository Only! 2025:777–846.e24, reprinted with permission from Elsevier. Postmortem venogram from Takashima S, Mito T, Ando Y. Pathogenesis of periventricular white matter hemorrhages in preterm infants. Brain Dev 1986;8:25–30, reprinted with permission.*

to typically developing children. The Enhancing Neuroimaging Genetics Through Meta-Analysis (ENIGMA) ASD working group has elucidated more nuanced differences in brain structure by analyzing 1571 patients with ASD and 1651 healthy controls, showing that individuals with autism had greater overall cortical thickness, greater frontal cortex thickness but decreased temporal cortex thickness, and decreased volumes of some deep gray matter structures [47].

fMRI studies of children with autism have shown decreased task-related activation across the "social brain network" including the amygdala, insula, inferior frontal cortex, language circuitry, and reward circuitry [48]. Resting-state fMRI data (obtained when no task is being performed) show a rightward shift of asymmetry involving various functional networks of brain regions [49]. Individuals with ASD generally show reduction in fNIRS activation in a number of regions including the prefrontal cortex, inferior frontal gyrus, and superior temporal cortex [4], and decreased connectivity compared to controls [50]. PET studies have demonstrated abnormalities in serotonin, gamma-aminobutyric acid, and glutamate neurotransmitter function in individuals with ASD. PET tracers targeting synaptic membranes are a promising new avenue to explore synaptic function within ASD [4].

The cerebellum is recognized to play an important role in the development of ASD. Children with cerebellar injury due to prematurity have up to a 40% risk of autism [36], imaging studies demonstrate structural differences in the cerebellum in children with ASD such as decreased cerebellar complexity [51], and functional studies are revealing the cerebellar circuit disruptions that result in ASD [52].

Attention-deficit hyperactivity disorder

Structural imaging studies consistently demonstrate decreased total brain and gray matter volume in children with ADHD compared to controls [53], as well as decreased volumes of subcortical structure including the basal ganglia, amygdala, nucleus accumbens, and hippocampus. These differences are present in childhood but not in adulthood, suggesting a maturational delay [54,55]. Structural differences do not appear to be impacted by stimulant use [55]. Connectivity studies using DTI have demonstrated complex differences in white matter tracts of individuals with ADHD, particularly in the region of the corpus callosum [54].

PET and single-photon emission computed tomography studies have focused on changes in dopamine, generally finding higher dopamine transporter density in the striatum in individuals with ADHD as compared to controls; however, this may be attributable to stimulant use [54]. fMRI studies have evaluated differences in brain activation during tasks that are typically impaired in individuals with ADHD. One example is a "go/no-go" task in which children are told to press a button, then occasionally told to not press the button. Children with ADHD have more difficulty initiating and inhibiting a motor response. Studies have shown decreased activity in the frontal lobe, the supplementary motor area, and the basal ganglia, as well as increased activity of the default mode network (which typically shows reduced activity during tasks, and increased activity in a rest state) during these tasks in individuals with ADHD [54]. fMRI studies have also shown differences in brain activation during tasks assessing memory, attention, reward processing, and time perception [54]. Some studies have shown a decrease in these differences with stimulant medication [56,57]. Studies of individuals with ADHD using fNIRS have demonstrated abnormalities in the right hemisphere, particularly the right frontal regions, during a number of tasks, supporting the theory that the ADHD results from dysfunction of the right hemisphere's role in attentional control and visuo-spatial processing [58].

Tourette syndrome

In structural imaging studies, individuals with Tourette syndrome show cortical thinning in the motor cortex, premotor cortex, prefrontal and parietal regions as well as altered basal ganglia volume compared to controls [59,60]. Tractography studies show an increase in connectivity between the motor cortex and insula in individuals with Tourette syndrome and premonitory urge [59–61].

Multimodal research including functional imaging studies implicate abnormalities in the cortico-striato-thalamo-cortical network in individuals with Tourette syndrome, resulting in loss of the network's inhibitory function [62]. These studies have identified a temporal series of neural activations resulting in tic generation, with onset in the supplementary motor area, activating the primary motor area, which activates the cortico-basal ganglia-thalamo-cortical leading to the tic (Fig. 10.1.8) [63]. This knowledge has informed targets for repetitive transcranial magnetic stimulation as an investigational therapeutic modality in Tourette syndrome [64].

Cerebral palsy

Imaging studies have begun to elucidate how motor tracts develop, reorganize after early injury, and respond to therapeutic interventions. At birth, corticospinal tract neurons innervate contralateral and ipsilateral lower motor neurons in equal proportion. Over the first 2 years of life with active function of the limbs, about 85% of ipsilateral innervations withdraw and the contralateral innervations dominate [65]. After a unilateral corticospinal tract injury, fMRI studies have demonstrated that

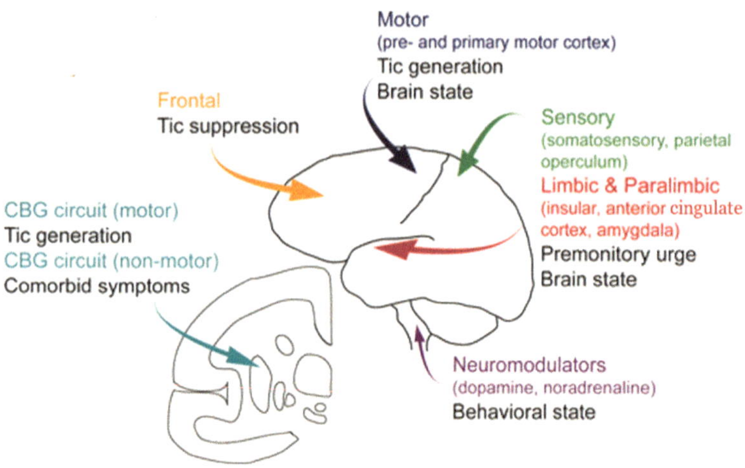

FIGURE 10.1.8 Neural areas involved in tic expression. Neural regions involved in tics. *From Yael D, Vinner E, Bar-Gad I. Pathophysiology of tic disorders. Mov Disord 2015;30(9):1171–8, reprinted with permission.*

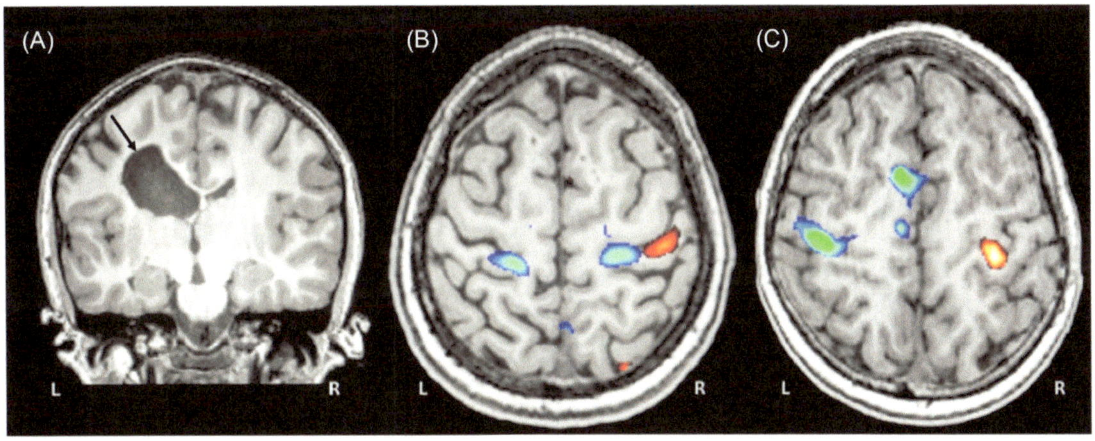

FIGURE 10.1.9 fMRI patterns after periventricular venous infarction. (A) Coronal T1 MRI showing right-sided porencephalic cyst due to periventricular venous infarction (black arrow). (B) Task fMRI activation using ball squeeze task of the paretic hand (blue) and nonparetic hand (red) showing bilateral reorganization for the paretic hand. (C) Typically developing control. *From Craig B, Hilderley A, Kirton A, Carlson H. Imaging developmental and interventional plasticity following perinatal stroke. Can J Neurol Sci 2021;48(2):157–71, reprinted with permission.*

activation patterns can be ipsilesional (recruiting surviving brain tissue by the lesion), contralesional (recruitment of the nonlesioned hemisphere), or bilateral [66,67]. Ipsilesional activation is associated with better motor outcomes, but may not be possible with large lesions [68]. Furthermore, fMRI studies have suggested that efficacy of constraint-induced movement therapy (CIMT) in children with hemiplegic CP results from enhancing ipsilesional activation [69] (Fig. 10.1.9).

Novel approaches in neuroimaging in neurodevelopmental disabilities

Connectivity

While early research focused on differences in the size and activation of brain regions, the recognition that the brain is organized into networks reframed neurodevelopmental disabilities as "dysconnectivity syndromes" [70]. With the ability to perform neuroimaging on large numbers of children and analyze and share that data, endeavors such as the Developing Human Connectome Project (DHCP), an Open Science project funded by the European Research Council, are mapping the connectome in developing infants and children [71]. This project uses fMRI data from large numbers of infants to map the resting state network. In adults, there are a number of well-characterized resting-state networks including sensorimotor, auditory, visual, dorsal attention, default-mode, and others [72]. The DHCP established the presence of sensorimotor, visual, and auditory resting state networks in infants at term age and demonstrated profound abnormalities in the resting-state networks in premature infants [71].

Deep learning/machine learning

Imaging techniques show increasing promise as clinical tools in evaluating developmental disabilities on an individual patient level. Machine learning is a form of artificial intelligence that enables computers to learn from existing data without explicit instruction. Deep learning is a form of machine learning that uses a specific architecture which parallels the structure of the nervous system with input, output, and hidden voxels analogous to neurons (Fig. 10.1.10) [73].

Deep learning has been applied to imaging and clinical data to predict a variety of neurodevelopmental outcomes on an individual level with good success. One study used DTI with whole brain connectome analysis to differentiate isolated speech delay from GDD with greater than 83% accuracy in 3-year-old children [74]. Another study used fNIRS functional connectivity data with deep learning for a >90% sensitivity and specificity for an ASD diagnosis [75]. Predicting outcomes in the preterm population is another area of active research. One study used deep learning from DTI data obtained from preterm infants at term age to generate a model that predicted cognitive abilities in an independent sample of infants at 2 years with >80% accuracy [76]. In another study, a model that included anatomical MRI, DTI, fMRI, and clinical data applied to premature infants had >85% accuracy in predicting cognitive, language, and motor deficits at 2 years of age (Fig. 10.1.11) [70]. Taken together, these studies demonstrate the potential for imaging data to be used as a powerful clinical tool, and highlight the importance of the connectome in developmental outcomes.

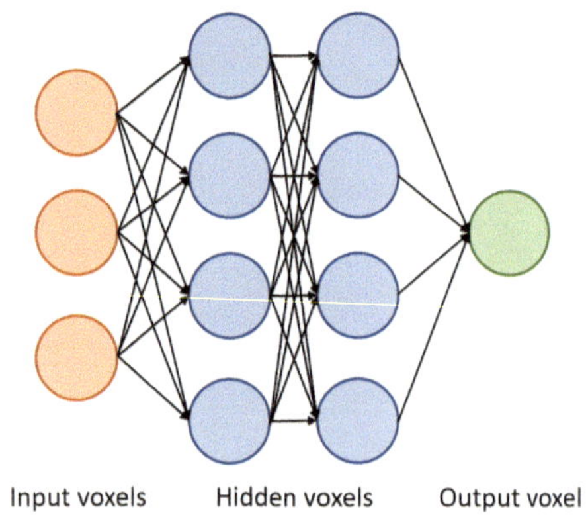

FIGURE 10.1.10 Deep learning network schematic. Schematic of a deep learning network. Information enters input voxels, is processed through the hidden voxels, and then an output is produced. There are parallels to the structure of the nervous system. *Adapted from Zaharchuk G, Gong E, Wintermark M, Rubin D, Langlotz CP. Deep learning in neuroradiology. Am J Neuroradiol 2018;39(10):1776–84.*

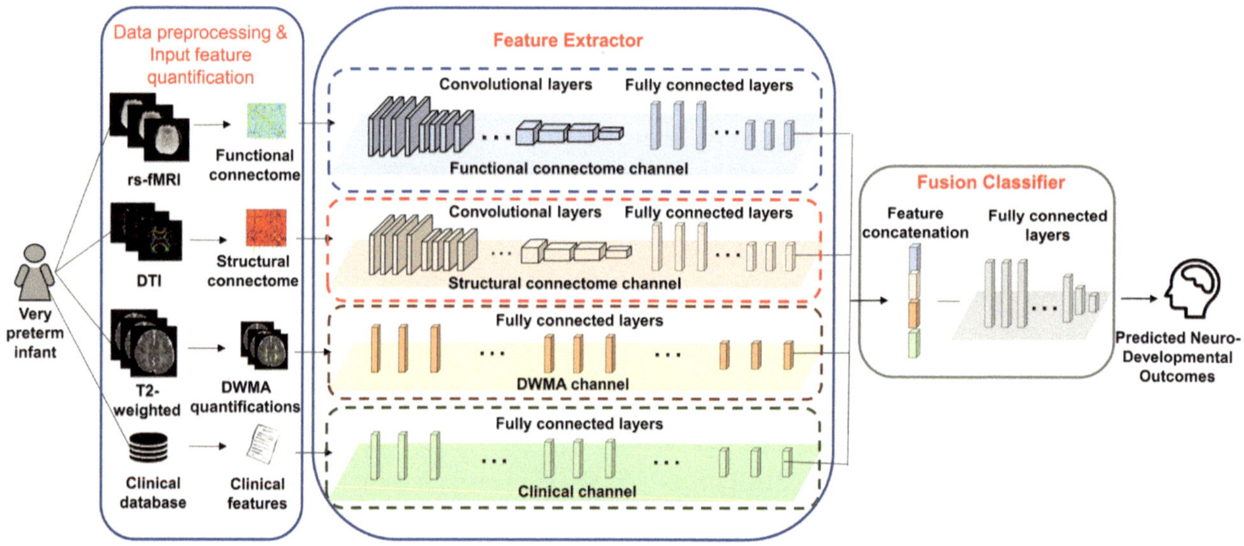

FIGURE 10.1.11 Deep multimodal learning model. Model incorporating data from fMRI, structural MRI, DTI, and clinical factors from very preterm infants is used to predict neurodevelopmental outcomes at 2 years of age. Data are preprocessed and inputted into the model. The feature extractor pulls out high-level features from the four channels which are then integrated by the fusion classifier using a fully connected layer with one output neuron. *DTI*, Diffusion tensor imaging; *DWMA*, diffuse white matter abnormality; *fMRI*, functional magnetic resonance imaging; *MRI*, magnetic resonance imaging; *rs-fMRI*, resting-state functional MRI. *From He L, Li H, Chen M, Wang J, Altaye M, Dillman JR, et al. Deep multimodal learning from MRI and clinical data for early prediction of neurodevelopmental deficits in very preterm infants. Front Neurosci 2021;15:753033, reprinted with permission.*

References

[1] Barkovich AJ. Pediatric neuroimaging. Lippincott Williams & Wilkins; 2005.

[2] Miglioretti DL, Johnson E, Williams A, et al. The use of computed tomography in pediatrics and the associated radiation exposure and estimated cancer risk. JAMA Pediatr 2013;167(8):700–7. Available from: https://doi.org/10.1001/jamapediatrics.2013.311.

[3] Opheim G, van der Kolk A, Markenroth Bloch K, et al. 7T epilepsy task force consensus recommendations on the use of 7T MRI in clinical practice. Neurology 2021;96(7):327–41. Available from: https://doi.org/10.1212/wnl.0000000000011413.

[4] McPartland JC, Lerner MD, Bhat A, et al. Looking back at the next 40 years of ASD neuroscience research. J Autism Dev Disord 2021;. Available from: https://doi.org/10.1007/s10803-021-05095-5.

[5] National Guideline A. National Institute for Health and Care Excellence: clinical guidelines. Cerebral palsy in under 25s: assessment and management. UK: National Institute for Health and Care Excellence; 2017.

[6] Ashwal S, Russman BS, Blasco PA, et al. Practice parameter: diagnostic assessment of the child with cerebral palsy: report of the Quality Standards Subcommittee of the American Academy of Neurology and the Practice Committee of the Child Neurology Society. Neurology 2004;62(6):851–63. Available from: https://doi.org/10.1212/01.wnl.0000117981.35364.1b.

[7] Novak I, Morgan C, Adde L, et al. Early, accurate diagnosis and early intervention in cerebral palsy: advances in diagnosis and treatment. JAMA Pediatr 2017;171(9):897–907. Available from: https://doi.org/10.1001/jamapediatrics.2017.1689.

[8] Murias K, Moir A, Myers KA, Liu I, Wei XC. Systematic review of MRI findings in children with developmental delay or cognitive impairment. Brain Dev 2017;39(8):644–55. Available from: https://doi.org/10.1016/j.braindev.2017.04.006.

[9] van Karnebeek CD, Jansweijer MC, Leenders AG, Offringa M, Hennekam RC. Diagnostic investigations in individuals with mental retardation: a systematic literature review of their usefulness. Eur J Human Genet 2005;13(1):6–25. Available from: https://doi.org/10.1038/sj.ejhg.5201279.

[10] Shevell MI, Majnemer A, Rosenbaum P, Abrahamowicz M. Etiologic yield of subspecialists' evaluation of young children with global developmental delay. J Pediatr 2000;136(5):593–8. Available from: https://doi.org/10.1067/mpd.2000.104817.

[11] Curry CJ, Stevenson RE, Aughton D, et al. Evaluation of mental retardation: recommendations of a consensus conference: American College of Medical Genetics. Am J Med Genet 1997;72(4):468–77 doi: 10.1002/(sici)1096–8628(19971112)72:4 <468::aid-ajmg18 >3.0.co;2-p.

[12] Moeschler JB, Shevell M. Comprehensive evaluation of the child with intellectual disability or global developmental delays. Pediatrics 2014;134(3):e903–18. Available from: https://doi.org/10.1542/peds.2014-1839.

[13] Shevell M, Ashwal S, Donley D, et al. Practice parameter: evaluation of the child with global developmental delay: report of the Quality Standards Subcommittee of the American Academy of Neurology and The Practice Committee of the Child Neurology Society. Neurology 2003;60(3):367–80. Available from: https://doi.org/10.1212/01.wnl.0000031431.81555.16.

[14] Cooper AS, Friedlaender E, Levy SE, et al. The implications of brain MRI in autism spectrum disorder. J Child Neurol 2016;31(14):1611–16. Available from: https://doi.org/10.1177/0883073816665548.

[15] Vasa RA, Ranta M, Huisman TA, Pinto PS, Tillman RM, Mostofsky SH. Normal rates of neuroradiological findings in children with high functioning autism. J Autism Dev Disord 2012;42(8):1662–70. Available from: https://doi.org/10.1007/s10803-011-1407-6.

[16] Boddaert N, Zilbovicius M, Philipe A, et al. MRI findings in 77 children with non-syndromic autistic disorder. PLoS ONE 2009;4(2):e4415. Available from: https://doi.org/10.1371/journal.pone.0004415.

[17] Erbetta A, Bulgheroni S, Contarino V, et al. Neuroimaging findings in 41 low-functioning children with autism spectrum disorder: a single-center experience. J Child Neurol 2014;29(12):1626–31. Available from: https://doi.org/10.1177/0883073813511856.

[18] Rochat MJ, Distefano G, Maffei M, et al. Brain magnetic resonance findings in 117 children with autism spectrum disorder under 5 years old. Brain Sci 2020;10(10). Available from: https://doi.org/10.3390/brainsci10100741.

[19] Zeegers M, Van Der Grond J, Durston S, et al. Radiological findings in autistic and developmentally delayed children. Brain Dev 2006;28 (8):495–9. Available from: https://doi.org/10.1016/j.braindev.2006.02.006.

[20] Filipek PA, Accardo PJ, Ashwal S, et al. Practice parameter: screening and diagnosis of autism: report of the Quality Standards Subcommittee of the American Academy of Neurology and the Child Neurology Society. Neurology 2000;55(4):468–79. Available from: https://doi.org/10.1212/wnl.55.4.468.

[21] Hyman SL, Levy SE, Myers SM. Identification, evaluation, and management of children with autism spectrum disorder. Pediatrics 2020;145(1). Available from: https://doi.org/10.1542/peds.2019-3447.

[22] Volkmar F, Siegel M, Woodbury-Smith M, King B, McCracken J, State M. Practice parameter for the assessment and treatment of children and adolescents with autism spectrum disorder. J Am Acad Child Adolesc Psychiatry 2014;53(2):237–57. Available from: https://doi.org/10.1016/j.jaac.2013.10.013.

[23] Smith T, Ram D. Fifteen-minute consultation: a practical approach to developmental regression in children. Arch Dis Child Educ Pract Ed 2019;104(4):173–7. Available from: https://doi.org/10.1136/archdischild-2017-313792.

[24] Verity C, Baker E, Maunder P, Pal S, Winstone AM. Differential diagnosis of progressive intellectual and neurological deterioration in children. Dev Med Child Neurol 2021;63(3):287–94. Available from: https://doi.org/10.1111/dmcn.14691.

[25] Liming BJ, Carter J, Cheng A, et al. International Pediatric Otolaryngology Group (IPOG) consensus recommendations: hearing loss in the pediatric patient. Int J Pediatr Otorhinolaryngol 2016;90:251–8. Available from: https://doi.org/10.1016/j.ijporl.2016.09.016.

[26] Springer A, Dyck Holzinger S, Andersen J, et al. Profile of children with cerebral palsy spectrum disorder and a normal MRI study. Neurology 2019;93(1):e88–96. Available from: https://doi.org/10.1212/wnl.0000000000007726.

[27] Himmelmann K, Horber V, De La Cruz J, et al. MRI classification system (MRICS) for children with cerebral palsy: development, reliability, and recommendations. Dev Med Child Neurol 2017;59(1):57–64. Available from: https://doi.org/10.1111/dmcn.13166.

[28] Severino M, Geraldo AF, Utz N, et al. Definitions and classification of malformations of cortical development: practical guidelines. Brain 2020;143(10):2874–94. Available from: https://doi.org/10.1093/brain/awaa174.

[29] Raybaud C, Widjaja E. Development and dysgenesis of the cerebral cortex: malformations of cortical development. Neuroimaging Clin N Am 2011;21(3):483–543. Available from: https://doi.org/10.1016/j.nic.2011.05.014.

[30] Volpe JJ. The encephalopathy of prematurity–brain injury and impaired brain development inextricably intertwined. Semin Pediatr Neurol 2009;16(4):167–78. Available from: https://doi.org/10.1016/j.spen.2009.09.005.

[31] Guo T, Chau V, Peyvandi S, et al. White matter injury in term neonates with congenital heart diseases: topology & comparison with preterm newborns. Neuroimage 2019;185:742–9. Available from: https://doi.org/10.1016/j.neuroimage.2018.06.004.

[32] Hand IL, Shellhaas RA, Milla SS. Routine neuroimaging of the preterm brain. Pediatrics 2020;146(5). Available from: https://doi.org/10.1542/peds.2020-029082.

[33] Benders MJ, Kersbergen KJ, de Vries LS. Neuroimaging of white matter injury, intraventricular and cerebellar hemorrhage. Clin Perinatol 2014;41(1):69–82. Available from: https://doi.org/10.1016/j.clp.2013.09.005.

[34] Martinez-Biarge M, Groenendaal F, Kersbergen KJ, et al. Neurodevelopmental outcomes in preterm infants with white matter injury using a new MRI classification. Neonatology 2019;116(3):227–35. Available from: https://doi.org/10.1159/000499346.

[35] Inder E, Perlman J, Volpe JJ. Preterm intraventricular hemorrhage/posthemorrhagic hydrocephalus. In: Volpe JJ, editor. Volpe's neurology of the newborn.. 6th ed. Elsevier; 2018. Chapter 24.

[36] Limperopoulos C, Bassan H, Gauvreau K, et al. Does cerebellar injury in premature infants contribute to the high prevalence of long-term cognitive, learning, and behavioral disability in survivors? Pediatrics 2007;120(3):584–93. Available from: https://doi.org/10.1542/peds.2007-1041.

[37] Barkovich AJ, Hajnal BL, Vigneron D, et al. Prediction of neuromotor outcome in perinatal asphyxia: evaluation of MR scoring systems. Am J Neuroradiol 1998;19(1):143–9.

[38] Rutherford M, Biarge MM, Allsop J, Counsell S, Cowan F. MRI of perinatal brain injury. Pediatr Radiol 2010;40(6):819–33. Available from: https://doi.org/10.1007/s00247-010-1620-z.

[39] Ouwehand S, Smidt LCA, Dudink J, et al. Predictors of outcomes in hypoxic-ischemic encephalopathy following hypothermia: a meta-analysis. Neonatology 2020;117(4):411–27. Available from: https://doi.org/10.1159/000505519.

[40] Rutherford M, Ramenghi LA, Edwards AD, et al. Assessment of brain tissue injury after moderate hypothermia in neonates with hypoxic-ischaemic encephalopathy: a nested substudy of a randomised controlled trial. Lancet Neurol 2010;9(1):39–45. Available from: https://doi.org/10.1016/s1474-4422(09)70295-9.

[41] Srivastava R, Rajapakse T, Carlson HL, Keess J, Wei XC, Kirton A. Diffusion imaging of cerebral diaschisis in neonatal arterial ischemic stroke. Pediatr Neurol 2019;100:49–54. Available from: https://doi.org/10.1016/j.pediatrneurol.2019.04.012.

[42] Lee S, Mirsky DM, Beslow LA, et al. Pathways for neuroimaging of neonatal stroke. Pediatr Neurol 2017;69:37–48. Available from: https://doi.org/10.1016/j.pediatrneurol.2016.12.008.

[43] Wagenaar N, Martinez-Biarge M, van der Aa NE, et al. Neurodevelopment after perinatal arterial ischemic stroke. Pediatrics 2018;142(3). Available from: https://doi.org/10.1542/peds.2017-4164.

[44] Craig BT, Olsen C, Mah S, Carlson HL, Wei XC, Kirton A. Crossed cerebellar atrophy in perinatal stroke. Stroke 2018;. Available from: https://doi.org/10.1161/strokeaha.118.022423.

[45] Sébire G, Tabarki B, Saunders DE, et al. Cerebral venous sinus thrombosis in children: risk factors, presentation, diagnosis and outcome. Brain 2005;128(Pt 3):477–89. Available from: https://doi.org/10.1093/brain/awh412.

[46] Khalatbari H, Wright JN, Ishak GE, Perez FA, Amlie-Lefond CM, Shaw DWW. Deep medullary vein engorgement and superficial medullary vein engorgement: two patterns of perinatal venous stroke. Pediatr Radiol 2021;51(5):675–85. Available from: https://doi.org/10.1007/s00247-020-04846-3.

[47] van Rooij D, Anagnostou E, Arango C, et al. Cortical and subcortical brain morphometry differences between patients with autism spectrum disorder and healthy individuals across the lifespan: results from the ENIGMA ASD working group. Am J Psychiatry 2018;175(4):359–69. Available from: https://doi.org/10.1176/appi.ajp.2017.17010100.

[48] Ecker C, Bookheimer SY, Murphy DG. Neuroimaging in autism spectrum disorder: brain structure and function across the lifespan. Lancet Neurol 2015;14(11):1121–34. Available from: https://doi.org/10.1016/s1474-4422(15)00050-2.

[49] Cardinale RC, Shih P, Fishman I, Ford LM, Müller RA. Pervasive rightward asymmetry shifts of functional networks in autism spectrum disorder. JAMA Psychiatry 2013;70(9):975–82. Available from: https://doi.org/10.1001/jamapsychiatry.2013.382.

[50] Zhu H, Fan Y, Guo H, Huang D, He S. Reduced interhemispheric functional connectivity of children with autism spectrum disorder: evidence from functional near infrared spectroscopy studies. Biomed Optics Exp 2014;5(4):1262–74. Available from: https://doi.org/10.1364/boe.5.001262.

[51] Zhao G, Walsh K, Long J, Gui W, Denisova K. Reduced structural complexity of the right cerebellar cortex in male children with autism spectrum disorder. PLoS ONE 2018;13(7):e0196964. Available from: https://doi.org/10.1371/journal.pone.0196964.

[52] Kelly E, Meng F, Fujita H, et al. Regulation of autism-relevant behaviors by cerebellar-prefrontal cortical circuits. Nat Neurosci 2020;23(9):1102–10. Available from: https://doi.org/10.1038/s41593-020-0665-z.

[53] Greven CU, Bralten J, Mennes M, et al. Developmentally stable whole-brain volume reductions and developmentally sensitive caudate and putamen volume alterations in those with attention-deficit/hyperactivity disorder and their unaffected siblings. JAMA Psychiatry 2015;72(5):490–9. Available from: https://doi.org/10.1001/jamapsychiatry.2014.3162.

[54] Albajara Sáenz A, Villemonteix T, Massat I. Structural and functional neuroimaging in attention-deficit/hyperactivity disorder. Dev Med Child Neurol 2019;61(4):399–405. Available from: https://doi.org/10.1111/dmcn.14050.

[55] Hoogman M, Bralten J, Hibar DP, et al. Subcortical brain volume differences in participants with attention deficit hyperactivity disorder in children and adults: a cross-sectional mega-analysis. Lancet Psychiatry 2017;4(4):310–19. Available from: https://doi.org/10.1016/s2215-0366(17)30049-4.

[56] Hart H, Radua J, Nakao T, Mataix-Cols D, Rubia K. Meta-analysis of functional magnetic resonance imaging studies of inhibition and attention in attention-deficit/hyperactivity disorder: exploring task-specific, stimulant medication, and age effects. JAMA Psychiatry 2013;70(2):185–98. Available from: https://doi.org/10.1001/jamapsychiatry.2013.277.

[57] Rubia K, Alegria AA, Cubillo AI, Smith AB, Brammer MJ, Radua J. Effects of stimulants on brain function in attention-deficit/hyperactivity disorder: a systematic review and meta-analysis. Biol Psychiatry 2014;76(8):616–28. Available from: https://doi.org/10.1016/j.biopsych.2013.10.016.

[58] Doi H, Shinohara K. fNIRS studies on hemispheric asymmetry in atypical neural function in developmental disorders. Front Human Neurosci 2017;11:137. Available from: https://doi.org/10.3389/fnhum.2017.00137.

[59] Worbe Y, Gerardin E, Hartmann A, et al. Distinct structural changes underpin clinical phenotypes in patients with Gilles de la Tourette syndrome. Brain 2010;133(Pt 12):3649–60. Available from: https://doi.org/10.1093/brain/awq293.

[60] Sigurdsson HP, Pépés SE, Jackson GM, Draper A, Morgan PS, Jackson SR. Alterations in the microstructure of white matter in children and adolescents with Tourette syndrome measured using tract-based spatial statistics and probabilistic tractography. Cortex 2018;104:75–89. Available from: https://doi.org/10.1016/j.cortex.2018.04.004.

[61] Bruce AB, Yuan W, Gilbert DL, et al. Altered frontal-mediated inhibition and white matter connectivity in pediatric chronic tic disorders. Exp Brain Res 2021;239(3):955–65. Available from: https://doi.org/10.1007/s00221-020-06017-0.

[62] Mink JW. Neurobiology of basal ganglia circuits in Tourette syndrome: faulty inhibition of unwanted motor patterns? Adv Neurol 2001;85:113–22.

[63] Yael D, Vinner E, Bar-Gad I. Pathophysiology of tic disorders. Mov Disord 2015;30(9):1171–8. Available from: https://doi.org/10.1002/mds.26304.

[64] Kleimaker M, Kleimaker A, Weissbach A, et al. Non-invasive brain stimulation for the treatment of Gilles de la Tourette syndrome. Front Neurol 2020;11:592258. Available from: https://doi.org/10.3389/fneur.2020.592258.

[65] Craig BT, Hilderley A, Kirton A, Carlson HL. Imaging developmental and interventional plasticity following perinatal stroke. Can J Neurol Sci 2021;48(2):157–71. Available from: https://doi.org/10.1017/cjn.2020.166.

[66] Weinstein M, Green D, Rudisch J, et al. Understanding the relationship between brain and upper limb function in children with unilateral motor impairments: a multimodal approach. Eur J Paediatr Neurol 2018;22(1):143–54. Available from: https://doi.org/10.1016/j.ejpn.2017.09.012.

[67] Kuo HC, Ferre CL, Carmel JB, et al. Using diffusion tensor imaging to identify corticospinal tract projection patterns in children with unilateral spastic cerebral palsy. Dev Med Child Neurol 2017;59(1):65–71. Available from: https://doi.org/10.1111/dmcn.13192.

[68] Chu D, Huttenlocher PR, Levin DN, Towle VL. Reorganization of the hand somatosensory cortex following perinatal unilateral brain injury. Neuropediatrics 2000;31(2):63–9. Available from: https://doi.org/10.1055/s-2000-7475.

[69] Sutcliffe TL, Logan WJ, Fehlings DL. Pediatric constraint-induced movement therapy is associated with increased contralateral cortical activity on functional magnetic resonance imaging. J Child Neurol 2009;24(10):1230–5. Available from: https://doi.org/10.1177/0883073809341268.

[70] He L, Li H, Chen M, et al. Deep multimodal learning from MRI and Clinical Data For Early Prediction Of Neurodevelopmental Deficits In Very Preterm Infants. Front Neurosci 2021;15:753033. Available from: https://doi.org/10.3389/fnins.2021.753033.

[71] Eyre M, Fitzgibbon SP, Ciarrusta J, et al. The developing human connectome project: typical and disrupted perinatal functional connectivity. Brain 2021;144(7):2199–213. Available from: https://doi.org/10.1093/brain/awab118.

[72] Damoiseaux JS, Rombouts SA, Barkhof F, et al. Consistent resting-state networks across healthy subjects. Proc Natl Acad Sci USA 2006;103 (37):13848–53. Available from: https://doi.org/10.1073/pnas.0601417103.

[73] Zaharchuk G, Gong E, Wintermark M, Rubin D, Langlotz CP. Deep learning in neuroradiology. Am J Neuroradiol 2018;39(10):1776–84. Available from: https://doi.org/10.3174/ajnr.A5543.

[74] Jeong JW, Sundaram S, Behen ME, Chugani HT. Differentiation of speech delay and global developmental delay in children using DTI tractography-based connectome. Am J Neuroradiol 2016;37(6):1170–7. Available from: https://doi.org/10.3174/ajnr.A4662.

[75] Xu L, Liu Y, Yu J, et al. Characterizing autism spectrum disorder by deep learning spontaneous brain activity from functional near-infrared spectroscopy. J Neurosci Methods 2020;331:108538. Available from: https://doi.org/10.1016/j.jneumeth.2019.108538.

[76] Girault JB, Munsell BC, Puechmaille D, et al. White matter connectomes at birth accurately predict cognitive abilities at age 2. Neuroimage 2019;192:145–55. Available from: https://doi.org/10.1016/j.neuroimage.2019.02.060.

Chapter 10.2

Neurophysiology

Alexandria E. Melendez-Zaidi[1], Akshat Katyayan[2] and Danielle Schwartzenburg Takacs[2]

[1]Department of Neurology, Texas Tech Health Sciences Center, El Paso, El Paso, TX, United States, [2]Division of Pediatric Neurology and Developmental Neuroscience, Texas Children's Hospital, Baylor College of Medicine, Houston, TX, United States

Introduction

The 20th century saw the discovery of the two most common imaging modalities in clinical neurosciences: CT and MRI. The importance of these tools in the evaluation of patients cannot be emphasized enough. However, decades before these were available, clinicians had discovered that the electrical features of the brain allowed for a novel means of evaluation, eventually known as the electroencephalogram (EEG). The history of this tool is robust, but its beginnings are best traced to the contributions of Hans Berger, who published the first report of EEG use in humans in 1929 and demonstrated the first electroclinical seizure in 1930 [1–3]. Collaboration with Frederic and Erna Gibbs brought about improvements in instrumentation and further developments, and the Gibbs' along with William Lennox, were able to demonstrate consistent findings in 12 children with absence epilepsy in 1935, in addition to studying EEG patterns of generalized and focal seizures (Fig. 10.2.1) [4,5].

The applicability of this tool is broad and is used in myriad conditions beyond monitoring of seizure activity. In this section, we will review basic concepts of EEG and how the study is evaluated; we will present common clinical scenarios in which EEG is or is not indicated; typical EEG indications and/or findings in common neurodevelopmental disorders; and finally, review ongoing research in which the tool may be helpful for use in understanding pathophysiology as well as developing electrographic "biomarkers."

The biology underlying electroencephalogram

At first glance, the EEG appears to be a collection of unrelated waves whose amplitudes (the y-axis) are tracked over time (x-axis). These waves, however, are a representation of voltage potentials which themselves represent the communication between groups of neurons within and across different brain regions. The voltages measured by EEG are those that are within the extracellular space, often referred to as local field potentials (LFPs). LFPs are generated as ions move between the extracellular space and cell cytoplasm by way of ion channels and, to a lesser degree, ion pumps on the cell membranes of neurons and some categories of glial cells [6]. Though action potentials are how cells engage in communication, the speed of these events and the integrity of the signal make it difficult to capture. Rather, it is the postsynaptic potentials—the consequences of action potential firing—that generate LFPs that can be amplified and read. Because the time scale of postsynaptic potentials is longer—at least an order of magnitude longer than that of action potentials—and the amplitude can be tuned up or down depending on the relative contribution from excitatory and inhibitory inputs, postsynaptic potentials are a more informative read out of what is happening over time. Neurons in similar alignment and/or involved in the same electrical networks may augment this signal further allowing for it to be read at the scalp, as scalp potentials (Fig. 10.2.2) [7,8]. LFPs differ in time and region in which they are recorded. The number of scalp electrodes helps to determine the degree of temporospatial resolution making EEG a potentially powerful tool in localization of abnormalities.

The electroencephalogram setup

EEG electrodes are placed in a standard fashion according to the 10–20 (international 10–20) system (Fig. 10.2.3). This system was formalized by Herbert H. Jasper in 1949 following a recommendation for improved standardization made at the first international EEG congress in 1947 [9,10]. It was subsequently widely adopted, allowing for more direct comparison of reports between institutions and regions [11]. The 10–20 system uses 21 individual electrodes placed on the skull at relative distances apart. Though improvements in hardware have allowed for the use of additional electrodes and, consequently, improved spatial resolution, these configurations are not typically used outside of the epilepsy community and so the 10–20 system remains the norm. In understanding the EEG, it is important to remember

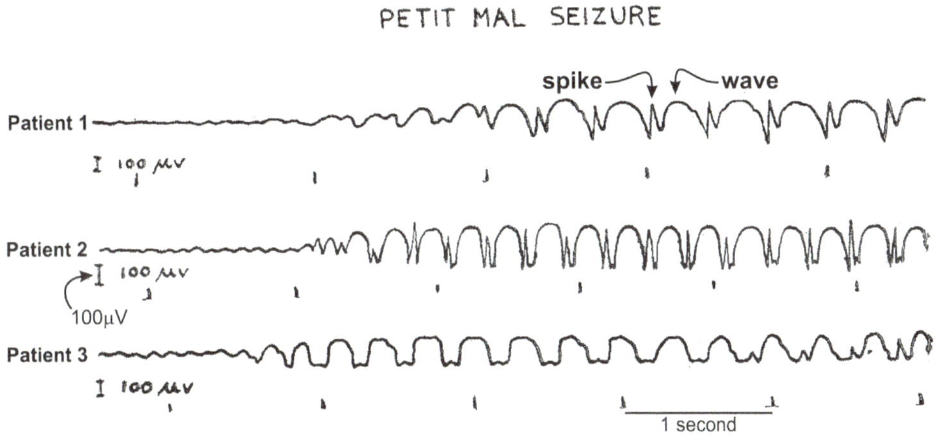

FIGURE 10.2.1 Original electroencephalogram (EEG) recordings and clinical characteristics described in the Gibbs cohort. Representative EEG tracings of patients with "petit mal" and "petit mal variant" epilepsy showing slight variability in speed of spike-and-wave discharges. The paper demonstrates that children with slow spike-and-wave discharges are more likely to have certain diagnoses than age-matched controls with normal EEGs. These diagnoses include cerebral palsy intellectual disabilites, epilepsy, behavioral disorders, and impaired speech. *From Salvati KA, Beenhakker MP. Out of thin air: hyperventilation-triggered seizures. Brain Res 2019;1703:41–52.*

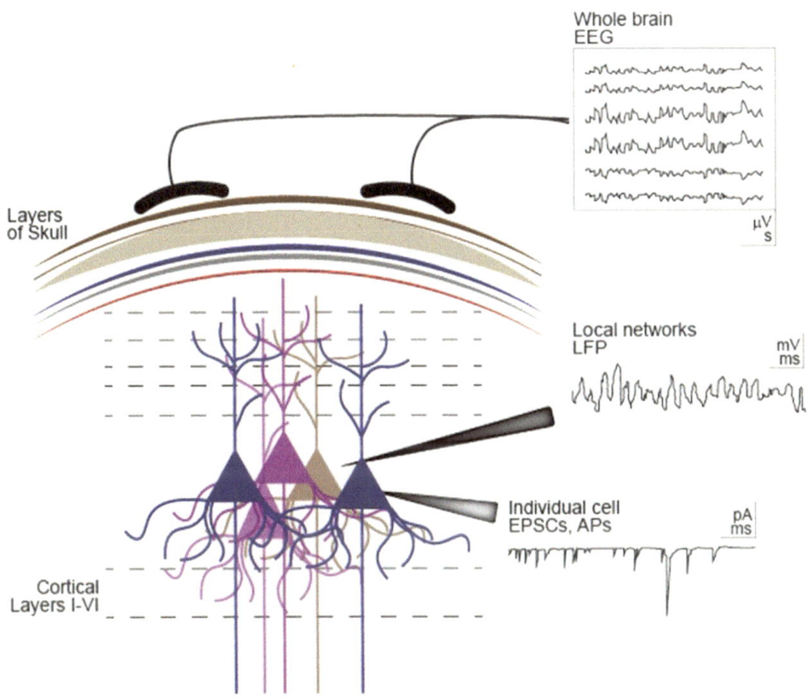

FIGURE 10.2.2 Levels of neurophysiological monitoring. Excitatory or inhibitory postsynaptic currents (EPSCs, shown in the figure, IPSCs), excitatory or inhibitory postsynaptic potentials (EPSPs, IPSPs) and/or action potentials (APs) can be measured in individual neurons. Local field potentials (LFPs) can be measured in groups of close, similarly oriented neurons. Electroencephalogram (EEG) allows for measurements of thousands of neurons providing a more comprehensive view of whole brain activity.

that absolute voltages are not being calculated. The voltage measured at each electrode is being compared to another electrode (the reference electrode) which indicates the direction in which a wave of depolarization occurs [8]. This allows the EEG reader to determine whether there are focal regions or diffuse regions of the brain in initiating epileptic activity. Most EEG studies are short, 20–40 minutes in duration and can be scheduled rapidly. A notable exception to this is neonatal studies which the American Clinical Neurophysiology Society (ACNS) mandates include at least 60 minutes of recording time in which awake, quiet sleep, and active sleep states are captured [11]. The utility of the interictal EEG is not to capture an event, but to evaluate for evidence of epilepsy, features suggestive of an increased risk of epilepsy, and structural and/or functional brain abnormalities.

Longer EEG studies, "continuous EEG" (cEEG), are indicated in many situations, and its use has expanded rapidly in the last decade with wider availability of cEEG monitoring and data demonstrating adverse neurodevelopmental outcomes because of poor control of seizures, especially in critical care settings [12]. Studies may be ordered for the following indications: to characterize an event suspected to be epileptic or to prove an event is nonepileptic; to quantify

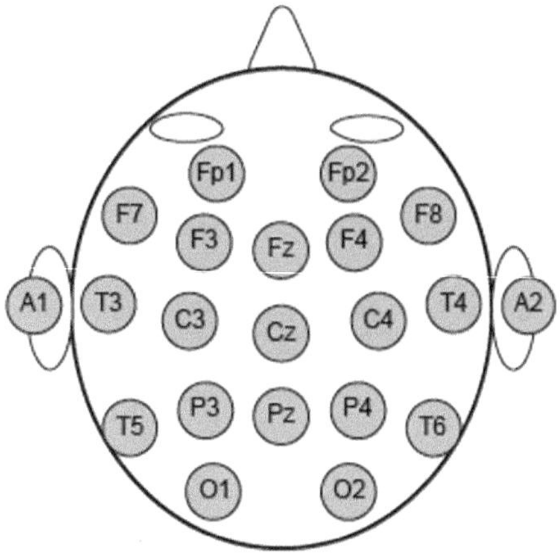

FIGURE 10.2.3 Electroencephalogram (EEG) electrode placement. EEG electrodes are placed in the pictured format, the so-called international 10–20 method. This refers to the distance between electrodes which is either 10% or 20% of the distance of the skull as measured from the front-to-back or left-to-right axes. *Fp*, Prefrontal; *F*, frontal; *T*, temporal; *P*, parietal; *O*, occipital; *C*, central; *z*, zero (a reference electrode placed in the midline sagittal plane of the skull).

seizure burden and medication efficacy in a patient with known epilepsy; to evaluate potential for epilepsy surgery; to monitor critically ill patients for changes in mental status, seizure burden, significant background changes, or new regions of focality which may point to acute changes such as bleeding or ischemia [13].

Report content

The ACNS and many others have published precise guidelines on the method of reporting EEG findings [11,14]. The content of the report can be broken down into the following sections: history, technical description, EEG description, impression, and clinical correlation. While clinicians typically "free text" in these sections, it is important to note that there is ongoing effort to further standardize EEG reporting such that the EEG reader would choose descriptors from predefined lists when describing EEG waveforms, clinical events of interest, seizure type, etc. [15]. In the last decade, this idea has been championed by many experts in the field including the International Federation of Clinical Neurophysiology (IFCN) and International League Against Epilepsy (ILAE). While this method—standardized computer-based organized reporting of EEG (SCORE)—is not yet the norm, it has growing support and may soon usurp the current system described below [15,16].

The EEG description reports features of the EEG that contribute to the resting state of the brain and includes the background and presence or absence of abnormal background waves, organization, and amplitude of the waves, reactivity to external stimuli or lack thereof, internal state changes or lack thereof (e.g., wake vs sleep), and any asymmetry in the reported waveforms. Activation procedures are techniques which may provoke specific seizure types or epileptiform discharges. These procedures are hyperventilation, photic stimulation, and sleep induction. Epileptiform activity may be pointed out in a separate section in which the presence of spikes, sharps, or spike-and-wave discharges is pointed out. The clinical significance of such abnormalities increases the risk for seizures and in the context of clinical activity concerning for seizures can support a suspected diagnosis of epilepsy. Seizures captured during the study will be described clinically and electrographically. Examples of the typical jargon used when reporting EEG abnormalities are depicted in Table 10.2.1.

The impression is the synthesis of the above information reports whether the study was abnormal or not and, if so, the features which are abnormal. The clinical correlation attempts to answer the referral question using the information provided in the impression and addresses whether the question is even answerable with the given data. It is important to remember that a normal EEG does not rule out epilepsy.

Other techniques in neurophysiology: In addition to the EEG, there are several techniques within the field of neurophysiology that may assist in diagnosing and determining the etiology of neurodevelopmental disorders. These include electromyography (EMG), nerve conduction studies (NCS), and evoked potentials (e.g., visual, somatosensory). However, these are typically used if there is a suspicion for concurrent muscle disease (such as myopathy), disease of the peripheral nerves (such as neuropathy), or concern for defects in vision or visual processing [17]. These are less commonly employed in the broad workup and evaluation of Neurodevelopmental disorders (NDDs) compared to use of EEG.

Electroencephalogram indications

One of the most common indications for obtaining an EEG is toward answering the question of "seizure" or "not seizure". The clinical scenarios in which the tool can be used are numerous and include classification of a seizure (general vs focal), localization of a seizure focus via traditional scalp and/or invasive depth electrodes, and evaluation of

TABLE 10.2.1 Epilepsy syndromes with absence seizures.

Syndrome	Seizure semiologies*
Childhood absence epilepsy	Typical absence (all patients) **Generalized tonic clonic seizure (GTCS) (in some patients, at older ages)
Juvenile absence epilepsy	Typical absence (all patients) GTCS Myoclonic jerks (one-third total patients)
Juvenile myoclonic epilepsy	Myoclonic jerks (early morning predominant) GTCS Typical absence (one-third total patients)
Epilepsy with myoclonic absences	Myoclonic absence seizures GTCS
Lennox–Gastaut syndrome	Tonic (most common), focal, myoclonic, atypical absence GTCS, atonic seizure
Myoclonic-astatic epilepsy	Atonic or myoclonic jerks, atonic, GTCS Typical absence, myoclonus, tonic
Eyelid myoclonia with absence	Eyelid myoclonia Typical absence (two-thirds total patients) GTCS, myoclonus

**In descending order of most to least common.*
***Generalized Tonic Clonic Seizures.*

medication or treatment response, to name a few. Because these questions are often addressed by epileptologists, we will limit this section to discussion of chief complaints that may necessitate EEG testing to determine whether epileptiform activity and/or seizures are present or not, with the caveat that a seizure remains a clinical diagnosis and therefore, the presence or absence of abnormalities on EEG can neither confirm nor exclude the diagnosis.

The utility of a routine EEG is highly dependent on the scenario under which it is being ordered, or the "pretest probability." This pretest probability is, however, undefined for most clinical scenarios and heavily dependent on individual heuristics. In general, EEG studies can have relatively high sensitivity and specificity when the event(s) of concern are captured or with longer recording periods [18–20]. For example, serial routine EEGs have a 92% chance or capturing interictal, epileptic abnormalities by the fourth recording period and ambulatory studies have a 61%–100% chance of answering clinical questions depending on the nature of the inquiry (e.g., epileptic vs nonepileptic event or presence or absence of epileptic abnormalities; [18,21]). This sensitivity even translates to the critical care setting where 24–48 hours of cEEG can capture 92% of seizure activity in high-risk populations [22].

In the case of a developmentally typical child with a first-time, unprovoked seizure, 66% will have a recurrent episode [21]. In this specific scenario, routine EEG has not been demonstrated to be predictive of seizure recurrence given the intermediate sensitivity and specificity (57.8% and 69.6%, respectively) that the finding of "epileptic discharges" has in a pediatric recording [23]. Returning to earlier commentary on the heuristics of pretest probability, the impact of an EEG in a child with abnormal brain structure or function (HIE, cortical malformation, genetic diagnosis) is almost certainly higher, though exact numbers are not available and hugely variable [20]. We ask readers to keep individual patient features in mind in the following discussion on clinical scenarios in which concern for seizures is high.

Staring spells have a broad differential that ranges from behavioral to epileptic. The parent typically provides a history of recurrent episodes during which the child is seen staring and generally "nonresponsive" to voice until some time has passed. In history taking, several details can assist in determining the etiology of the episode of concern and are quite helpful in diagnostic certainty [24]. It is critical to determine whether a parent has tried touching the child during the episodes or doing some other stimulatory maneuver to get their attention, outside of just calling the name. Additionally, caregivers will often provide vague details regarding the context of when the episodes occur (e.g., "he was in the middle of playing"). Here, one must determine whether the patient was in a period of rest or stillness when they began staring or if there was a more abrupt behavioral arrest which occurred in mid-speech or mid-movement. One should always ask about other facial or appendicular moves during the episodes; stereotypic lip smacking, tongue thrusting, or rapid blinking are suggestive of automatisms which may be seen in focal motor seizures with impaired awareness [25,26]. Sudden behavioral arrest with staring alone is commonly described in absence seizures, though these seizures frequently also have subtle automatisms as described above [27,28]. Occurrence of the episodes in multiple settings (e.g., home, school, sports practice) and short duration are also more common to epileptic causes of staring versus behavioral events.

Behavioral (nonepileptic) staring episodes can be further classified into daydreaming, childhood preoccupation, migraine-like events, parasomnias, and psychogenic events [25,29]. A decreased blink rate may cause the illusion of staring episodes which can be provoked by concentrating on a computer or other electronic device screen [30]. While adults may also have decreased blink secondary to pathology (e.g., Parkinson's disease), this is not seen in pediatrics, aside from the rare situation in which there is damage to the trigeminal sensory nerve endings [31]. The distinctions between etiologies of behavioral staring episodes are at times difficult to appreciate, and thus EEG is a useful tool to characterize such events. In the setting of a normal EEG and features of (1) distractibility, (2) normal neurological development, and (3) appreciation of episodes by a nonparent (e.g., teacher), an episode is unlikely epileptic [29].

Epileptic staring episodes fall into one of two categories: focal seizures with impaired awareness or generalized absence seizures. As noted above, there are numerous features that distinguish the two, however, given the significant overlap in clinical features the most objective identification can be made with EEG. Once the seizure type is determined, it is appropriate to consider which epilepsy syndrome the patient may fall into. Absence seizures can be seen in a variety of syndromes alongside other seizure types that typically follow the onset of absence seizures. Table 10.2.2 outlines these syndromes and seizure types.

While clinical suspicion may be high for one etiology over the other, a routine EEG is indicated in the work up of this complaint. Children with typical absence seizures have normal background architecture with spontaneous or hyperventilation and sleep potentiated 3 Hz generalized but frontal predominant spike-and-wave discharges with or without clinical features of behavioral arrest [27,28,32]. The 3 Hz frequency is the characteristic electrophysiologic signature of typical absence seizures, as is the ability to activate discharges during the hyperventilation portion of the study. Focal seizures with impaired awareness do not share these EEG signatures, though may reveal focal slowing or discharges

TABLE 10.2.2 Movement disorders and their clinical characteristics.

Hyperkinetic	
Tics	Abnormal movements or sounds, complex or simple. Abrupt, sudden, isolated movement, ± repetitive. Coordinated patterns of movements
Functional tic disorders	Like the above but psychological in origin
Stereotypies	Repetitive, purposeful movement, self-soothing/sensory feedback
Dystonia	Repetitive, patterned twisting involving opposing muscle groups (action and overflow) generalized or focal
Myoclonus	Contraction of opposition muscle groups blocking motion
Tremor	Increased muscle tone during movement and rest
Ballismus	Repetitive, variable, large amplitude movements
Chorea	Irregular, purposeless, nonrhythmic, abrupt, nonsustained flowing movements
Athetosis	Slow, continuous, writhing movements
Hypokinetic	
Akinesia	Absence of movement
Bradykinesia	Slowness of movement
Dysarthria	Weakness in muscles of speech causing abnormal features of speech
Dystonia	As described above
Freezing	Contraction of opposition muscle groups blocking motion
Rigidity	Increased muscle tone during movement and rest

[27,28,33]. Neuroimaging, typically with brain MRI, may be indicated in patients with a strong suspicion for focal seizures, especially with EEG findings of focal epileptiform discharges or a postictal focal deficit [33].

Even when an episode may sound obvious as being nonepileptic, certain families may be more apt to accept a behavioral diagnosis if this study is reassuring. In this situation, it is recommended to review with parents that small, nonsignificant abnormalities may be present which would still not support a diagnosis of seizure. Specifically, EEGs in children with ADHD may have occasional benign focal spikes in the Rolandic/centro-temporal regions [34]. Children with typical migraines may show nonspecific focal regions of cortical irritability or slowing [34]. And children with underlying developmental or intellectual delay may demonstrate diffuse slowing, lack of typical waking/sleep background architecture, and/or incidental interictal epileptiform discharges. This is particularly notable in the autism population in which greater than 50% of patients have abnormal features on their EEG (such as slowing, asymmetry, and spikes) whereas only 15%–20% have epilepsy.

Paroxysmal episodes or movements can be difficult to diagnose even with the assistance of prolonged video EEG. We use the words "episodes" and "movements" to speak of two different phenomena. Paroxysmal movements fall under the category of movement disorders which are broadly defined as conditions in which involuntary movements are abnormal secondary to initiation, implementation, velocity, frequency, or posture [35]. These phenomena include hyperkinetic—chorea, tremor, akathisia, myoclonus, and tics—and hypokinetic—dystonia, bradykinesia—disorders. Movement disorders are nonepileptic and are mentioned here because the clinician should be familiar with these conditions and aware that they are common in children with certain neurodevelopmental disorders (Table 10.2.3).

Contrast this to paroxysmal episodes which are generally more complex. These may have a motor component (such as repetitive movement of a body part, prolonged stiffening, or generalized shaking) or may be more subtle. Within this latter category falls episodes of dysarthric speech or speech difficulty; episodes of paresis without loss of consciousness; aura-like episodes in which a recurrent feeling, image, sound, or paresthesia is noted over some time; or a series of complex movements and/or vocalizations that are difficult to classify in traditional categories. For simplicity's sake, we discuss common clinical scenarios of the paroxysmal episodes above, arranged by age of onset (Table 10.2.4).

In neonates and infants, paroxysmal episodes tend to fall into one of two extreme conditions, the first being self-limited seizures (more "benign") and the other, progressive neurological conditions and/or epilepsies (severe). Because seizures are common in neonates (1–5 per 1000 live births) and can have devastating consequences on development if untreated, any recurrent abnormal movement or episode should raise concern for an epileptic etiology and prompt urgent evaluation [36]. Clinical features which make movements more concerning for seizure include a history of brain injury (HIE, infarction, hemorrhage), known brain malformation, history of meningitis, suspected current infection, or a known metabolic or genetic disorder with neurological involvement. If movements persist in periods of sleep, cannot be interrupted, are stereotyped, or if there has been concurrent decline in development, epileptic etiologies should be considered more likely. Neonates and infants may have movements that are typical for seizures in older populations—including regular, repetitive limb jerking, body stiffening, myoclonus, behavioral arrest, and automatisms; however, infants are also more likely to have subclinical seizures or subtle clinical seizures, the identification of which may escape even an experienced clinician. Newborns do not have generalized tonic clonic seizures or absence seizures, so an episode resembling either should prompt consideration of other etiologies [37]. Therefore based on the present

TABLE 10.2.3 Differential diagnosis of paroxysmal episodes by age.

Infant/Neonate	Young child	Adolescent
Benign neonatal sleep myoclonus Physiological startle response Sandifer syndrome Shuddering/shivering attacks Benign paroxysmal torticollis Infant gratification disorder Epileptic spasms	Syncope (cardiogenic and noncardiogenic) Seizure Panick attack Narcolepsy/cataplexy Migraine, basilar migraine Transient ischemic attack (TIA) Hypoglycemia Daydreaming Infant gratification disorder Gratification disorder Breath holding spells	Syncope (cardiogenic and noncardiogenic) Seizure Panick attack Narcolepsy/cataplexy Migraine, basilar migraine TIA Hypoglycemia Daydreaming Psychogenic nonepileptiform seizures (PNES)

TABLE 10.2.4 Pediatric disease in which polysomnography is indicated or recommended in work up.

Conditions predisposing to obstructive sleep apnea (OSA)	
Obesity	Diet, genetic, Prader–Willi syndrome
Craniofacial abnormalities	Retrognathia, hyperglossia, midface hypoplasia
Neuromuscular disease	Duchenne muscular dystrophy, spinal muscular atrophy, myotonic dystrophy
Medication usage	Alcohol, benzodiazepines, general anesthetics
Upper airway inflammation	Recent upper respiratory infection, smoke exposure
Adenotonsillar hypertrophy	Natural phenomena of childhood
Neuromuscular junction disease	Myasthenia gravis
Spinal cord disease, diaphragm paralysis/injury	Trauma, cervical or brachial plexus neuropathy

guidelines, whenever possible, the evaluation of an infant with suspected seizures should include EEG in order to assist in determining if the events are epileptic in origin. If an episode is captured and found to correlate with ictal activity on the EEG, the event is verified a seizure. However, this is not necessary for diagnosis, nor do abnormalities need to be present (i.e., spikes, sharp waves). Importantly, some features which are "abnormal" in adults may be within the realm of normal in certain infant populations; for this reason, an EEG reader with specialization in pediatric populations, or sufficient experience, should be utilized when able.

Continuous video EEG is the gold standard for diagnosis (evidence level 1). If cEEG is not available, an experienced clinician may provide a probable (level 2b) or possible (level 3) diagnosis of seizure depending on the typicality of the clinical movements. While the goal of cEEG is to capture the movements, characterization of background waveforms may also be helpful in determining risk for seizures [11,13].

Whether abnormal movements in an infant are first being evaluated inpatient or outpatient, urgent initiation of recording is recommended; transfer to a higher level of care or presentation to the emergency room for hospital admission are both acceptable next steps. Beyond seizures, a differential in this age group typically includes normal baby movements, Sandifer's syndrome (body arching or stiffening due to gastroesophageal reflux after feeds), exaggerated startle, provoked tremor (e.g., hypoglycemia or excess caffeine consumption in a breastfeeding mother), benign neonatal sleep myoclonus (BNSM), benign myoclonus of infancy, or shuddering attacks. Caregivers often provide videos and, if this is not available, they should be encouraged to capture the episode. Even when the clinical history is consistent with BNSM, most clinicians will obtain at least routine EEG to ensure no background abnormality is present in the interictal recording.

In preadolescent, older children (ages 1–10) paroxysmal movements/episodes expand from those described above to also including breath-holding spells, syncope, and benign paroxysmal vertigo. Children 6 years or older tend to have seizure types that more closely resemble those seen in adults. In this population, special consideration should be given to those less than 12 months of age; while generalized tonic clonic seizures and absence seizures may occur once past the neonatal stage, these are rare semiologies to observe before the age of 1 year [37]. A history inclusive of either should prompt additional questions to better understand the clinical context of these features. Additionally, commentary must be made regarding febrile seizures: it is common for fever to provoke onset of seizure(s) in the acute period of illness in children ages 6 months to 6 years. This occurrence will not necessarily confer a diagnosis of epilepsy at that time or in the future. Most febrile seizures are brief, generalized events, though more prolonged (> 15 minutes) events or seizures with focal predominance (unilateral motor symptoms) may warrant further diagnostic evaluation and closer monitoring for additional, unprovoked seizures thus, meeting criteria for an epilepsy diagnosis.

Though a formal ACNS recommendation, the utility of EEG in establishing a definitive diagnosis during the work up of a paroxysmal episode is somewhat limited. A normal routine EEG can be found in approximately half of those with an epileptic seizure [20], while rare epileptic waveforms (such as spike-and-wave discharges, spikes) may appear on the EEG of a child without epilepsy [38]. Therefore the most important task of the clinician is procuring a detailed history and diagnosing the episode as epileptic or "other" based on their clinical suspicion, prior to ordering an EEG.

Features of an episode which make a seizure highly likely include: incontinence of bowel or bladder, involuntary self-injury (biting of the tongue or head injury from a fall), or a period of altered consciousness after the seizure known as the "postictal period." This latter feature is quite common in epilepsy in general (children and adults) with close to 75% of patients with epilepsy experiencing a postictal period [39]. A survey-based study at Alberta Children's Hospital found that nearly 90% of children experienced a postictal period and, in 78% of these cases, the episode was readily distinguished by the child's inablity to was unable to move [40].

In this clinical setting, the best use of EEG is to guide the clinician in determining the likelihood of seizure recurrence in a child who has had a clinical event that is deemed consistent with a seizure episode. While antiseizure medication is not recommended after a single episode, if a follow-up EEG demonstrates abnormalities that are deemed to portend a 60% or greater likelihood of seizure recurrence over the next 10 years, this would commonly result in the diagnosis of epilepsy and thus early initiation of antiseizure medication per ILAE definition of epilepsy. Though not an official recommendation in management, there are web-based epilepsy prediction tools which are able to assist the physician in calculating risk of epilepsy based on clinical presentation and neurophysiologic features on EEG [41]. In children with acquired, metabolic, or genetic neurodevelopmental diseases with potential for CNS involvement, a one-time seizure-like event should prompt EEG, and even if the result is normal, some clinicians may still choose to initiate antiseizure medications if there is known high risk.

In other clinical scenarios, alternative testing may be recommended prior to EEG. An episode that was preceded by tunnel vision, nausea, lightheadedness, hearing decrease, or palpitations should raise concern for syncope and an EKG should be obtained, potentially alongside an evaluation by cardiology. Headaches surrounding the event should raise the consideration of atypical migraine. If the episode was preceded by a period of agitation, anger, and/or screaming, it is reasonable to consider breath-holding spells as a cause.

In adolescent children and young adults (early 20s), there is an additional diagnosis that should be kept in mind, psychogenic nonepileptiform seizures (PNES) previously known as pseudoseizures. PNES events are paroxysmal episodes that may look remarkably similar to epileptic seizures, but do not stem from concurrent epileptiform activity in the brain [42]. These episodes are thought to derive from a complex relationship between some combination of psychological disposition, internal and external stressors, trauma, and several psychiatric disorders such as anxiety, panic attacks, or obsessive compulsive disorder (OCD). It is critical to distinguish between epileptic seizures and nonepileptic events, doing so can prevent unnecessary diagnostic testing and healthcare costs but, more importantly, may save the patient from potentially dangerous medications or clinical interventions. PNES does not respond to preventative or abortive antiseizure drugs except that excess of the latter will eventually make the patient unconscious and can decrease respiratory drive, potentially leading to intubation and need for critical care management.

The ILAE has come out with numerous papers regarding best methods to diagnosis PNES. Gold standard diagnosis is capture of the episodes of concern on video EEG to demonstrate absence of correlating epileptic activity before, during, or after the event. The diagnosis of PNES in the absence of video EEG is also possible provided certain clinical criteria are met, though the diagnosis is often not accepted by the patient or caregivers without objective EEG data [43]. History that is suggestive of PNES includes: a fluctuating course, long duration, forced eye closure, crying during the event, pelvic thrusting, or recall of the event, to name a few [44]. When possible, prolonged ambulatory or inpatient EEG should be obtained if PNES is suspected, and the diagnostic study is available. Most critically, the diagnosis of epilepsy does not exclude an episode as PNES. It is estimated that up to 7%–15% of patients diagnosed with PNES also have a known epilepsy diagnosis and between 1% and 10% of children with epilepsy also have nonepileptic episodes [43,45,46]. Thus management of patients with events of both types can be challenging, and both medical interventions and psychological interventions are needed. It is generally held that children with ID are less likely to produce symptoms consistent with PNES compared to neurotypical children. However, as has been previously discussed, many behavioral events in this patient population may be mistaken for seizure. Furthermore, those with intellectual deficits may have inadequate coping skills in response to psychosocial or physical stressors and may manifest turmoil outwardly with a somatoform response as seen in PNES.

Regression of speech may manifest as new-onset problems in expressive language, language comprehension, difficulties with pronunciation, enunciation, or the inability to form words, to name a few. If a clinician suspects or confirms a central etiology for the language abnormalities, the more familiar term "aphasia" may be used, which is formally defined as the inability to comprehend or formulate language because of some insult to or abnormality of the brain. In children that are typically developing, or who only have mild speech delay, the development of progressive aphasia may be further specified as acquired aphasia [47,48]. In some cases, the chief complaint of "language regression" may be preceded by or accompanied by a complaint of hearing problems. Many children with epileptic acquired aphasias will first manifest as auditory agnosia which may appear as though a child is ignoring common sounds with specific associations (e.g., ignoring the ringing of the doorbell or knocking on the door) or ignoring aspects of conversation [49].

In baseline neurotypical children, regression of previously acquired language skills should immediately raise concern for an epileptic or metabolic/degenerative etiology, and rapid evaluation is necessary as earlier treatment with steroids and/or antiseizure medications as well as speech therapy provides the best chance for speech and developmental recovery. However, it is important to expand the differential to etiologies that are more common—such as progressive hearing loss, Rett syndrome, and ASD—or rare—including several metabolic crises or neurodegenerative diseases in which language regression with or without accompanying hearing loss may be an initial symptom, as well as acute behavioral or psychiatric disorder (including selective mutism) [50,51]. There are several clinical features which may suggest one etiology over the other and help guide additional work up in addition to EEG.

In line with recent ILAE guidelines, we refer to epileptic acquired aphasias as developmental and/or epileptic encephalopathy with spike-wave activation in sleep (D/EE-SWAS) rather than their eponymous syndrome names (e.g., Landau–Kleffner syndrome or pseudo-Lennox syndrome) [52]. D/EE-SWAS emphasizes the electrophysiologic and clinical findings that characterize the disorder: persistent spike-wave activation in sleep with notable developmental regression. If D/EE-SWAS is suspected, a routine EEG is an appropriate first study, though a multihour or overnight study may be needed as a follow-up exam to best quantify the spike-wave burden. This is because it is often challenging to get children to sleep during the procedure and, while there is no formal data, this is expected to be even more difficult in patients with developmental concerns. Fortunately, an EEG that captures sleep, even briefly, may provide significant information in diagnosis as the transition from drowsiness to Stage I sleep is where spike-wave activation tends to be the most prominent [52]. Despite this, most epileptologists will want to assess all stages of sleep to confirm the diagnosis above requiring a follow-up overnight EEG. It is generally much faster to arrange for a routine EEG as a means of screening for this diagnosis so this should not be delayed out of preference for a long study which can be expedited if concern is present.

Regarding diagnostic criteria, there is no universally agreed up duration of the EEG study, or percentage of sleep which must exhibit spike-wave discharges. Mandatory criteria are few and include (1) the presence of slow (less than 2 Hz) spike-wave discharges during at least 50% of nonrapid eye movement (REM) sleep; (2) the discharges are activated in sleep [52]. Other commonly used markers for diagnosis used include >85% spike burden in non-REM sleep, or >50% increase in discharges when comparing wakefulness to non-REM sleep states [53,54]. The awake EEG findings are variable and may be of normal background or may demonstrate focal or multifocal discharges and/or focal or diffuse slowing. Normal sleep features are likely to be absent or significantly obscured by the predominance of spike wave discharges.

A normal routine or prolonged awake and asleep EEG that captures several hours of sleep essentially excludes an established or evolving epileptic encephalopathy as the cause of the child's regression. In such a case, genetic, metabolic, or neurodegenerative diseases should be considered. The list of conditions that may first present with language regression, however, is vast but will be addressed in chapters: Developmental and Epileptic encephalopathies; neurodegenerative, neurometabolic, appendices.

Regression of motor skills is a clinical complaint that can be evaluated similarly to the above approach regarding language regression. It is important to note that language regression is more likely to present as an isolated finding in older children whereas younger children are more likely to show concomitant regression in other domains. Isolated motor regression should raise suspicion for other causes including infectious (chronic HIV, progressive toxoplasma), structural (muscle disease such as myopathies or disorders of bone formation causing osteoporosis), metabolic (hypothyroidism, vitamin deficiency), or genetic and neurodegenerative disease [55]. Because a majority of neurodegenerative or metabolic disorders have increased risk for seizures, an EEG as one of numerous first-line tests is advised.

Sleep disorders are one of the most prevalent complaints in the general pediatrics clinic with an estimated 20%–30% of children meeting criteria for these conditions at least once in childhood. Children with neurodevelopmental disabilities such as intellectual delay or autism, or related psychiatric diagnoses (anxiety, depression, and ADHD) are more likely to have sleeping difficulties than the general population, sometimes with rates as high as 83% as in [56–58].

Evaluation in either population is difficult but is often more challenging in children with NDD secondary to sensory sensitivity to electrodes and enhanced anxiety to new environments [59,60]. Given this, it is important to be prudent in ordering testing particularly because the gold standard test—polysomnography (PSG)—is difficult for most to tolerate.

Fortunately, a thorough history and parental data collection is sufficient for diagnosis in most cases and PSG is only indicated in select clinical scenarios (see Table 10.2.5). In any child with a known epilepsy, expanded EEG on a PSG is recommended. PSG is not recommended for several common sleep disorders including restless leg syndrome, insomnia, circadian rhythm disorders, or simple parasomnias.

Sleep-disordered breathing may refer to obstructive sleep apnea (OSA), central sleep apnea, or sleep-related hypoventilation. Among these, OSA is the most common condition. Clinical complaints suggestive of OSA are variable but

TABLE 10.2.5 Conditions predisposing children to obstructive sleep apnea (OSA).

Disease process	Notes
Periodic limb movement disorder (PLMD)	Parental report is typically too unreliable on its own
Narcolepsy, suspected	Requires 6-h polysomnography (PSG) followed by multiple sleep latency test (MSLT)
Sleep-disordered breathing	–
Hypersomnia	MSLT is used to differentiate from narcolepsy
Parasomnias, atypical or injurious	Electroencephalogram (EEG) is used to distinguish or differentiate from seizure, consider expanded montage
Restless leg syndrome (RLS), suspected but unclear	Diagnosis is clinical but poor positive predictive value by parent report
Congenital central alveolar hypoventilation	–
Sleep-related hypoventilation	If present in the setting of neuromuscular disease or chest wall abnormalities
Obstructive sleep apnea (OSA), suspected	PSG should be used to titrate non-invasive positive pressure ventilation
Sleep-related epilepsy, suspected	EEG component may be used to differentiate from parasomnias, recommend expanded montage

generally include snoring with witnessed apnea, excessive daytime sleepiness, daytime hypoxemia, mood irritability, multiple nighttime awakenings, or new onset hypertension. Suspicion for OSA should be high in children with clinical report of snoring if there are predisposing conditions including hypotonia or neuromuscular disease, obesity, and midbrain malformations, several which are named in Table 10.2.6. Children who are between the ages of 2 and 8 years when there is expected, physiologic adenotonsillar hypertrophy causing OSA in some, should be considered at higher risk. In these cases, in addition to PSG, evaluation by Otolaryngologist (ENT) or pulmonology is recommended [61,62]. See chapter: Sleep disorder.

The AAP has provided standard guidelines for health supervision of children with trisomy 21, which includes monitoring for signs of sleep apnea. If snoring, daytime sleepiness, or other sleep-related concerns arise, this population is recommended to see a sleep specialist and undergo a sleep study or PSG; most children below the age of 4 will have had this accomplished [61]. Children with Prader–Willi syndrome should also be observed for sleep-disordered breathing, particularly given the incidence of obesity in this population; they are also prone to a narcolepsy-like condition and hypersomnolence for which PSG may assist in diagnosing at least the former [63,64]. Finally, children with Rett syndrome are documented as having significant sleeping abnormalities including sleep-disordered breathing and disorders initiating sleep and maintaining sleep [56,65].

Parasomnias are sleep disorders occurring during transitions in sleep stages and may involve abnormal behaviors, dreams, movements, or thoughts. They are classified based on their occurrence in REM sleep or non-REM sleep. Common examples of non-REM disorders are sleepwalking, sleep talking, sleep terrors, or confusional arousals. While REM disorders are uncommon in children, the most well known are nightmare disorder and REM sleep behavior disorder (RBD)—the latter of which is seen most often in men over 50 years of age with parkinsonian disease. To meet criteria for nightmare disorder, one must experience recurrent, vivid dreams that result in daytime impairments, and is often linked to posttraumatic stress disorder (PTSD; Lloyd et al.). In general, most diagnoses of parasomnias do not require PSG and it is the recommendation of the American Academy for Sleep Medicine (AASM) that PSG only be ordered when there is concern for comorbid OSA and/or narcolepsy OR there is insufficient clinical history to make a diagnosis [66,67]. Most of the time, history alone is sufficient after parents are asked to keep a sleep diary and record episodes of concern on video. One exception is the diagnosis of periodic limb movement disorder (PLMD). PLMD requires PSG particularly as parental reporting has poor sensitivity.

RBD is best confirmed via PSG given its atypicality in childhood. Suspicion for RBD should arise in children sustaining frequent injuries during sleep, or injuring bedpartners (such as parents or siblings). The disorder is known to be more prevalent in patients with ASD and there is also a correlation in those with ADHD [68]. There is also a recognized association with other neurodevelopmental disorders including general developmental delay and

TABLE 10.2.6 Elements of the EEG report and their significance.

Category	Electroencephalogram (EEG) finding	Subcategory	Correlation	Example
Ictal	Electrographic or electro-clinical seizure	Focal or generalized	Confirms the diagnosis of epilepsy	Absence seizures with typical 3 Hz generalized spike waves confirm the diagnosis of absence epilepsy in a child with frequent staring spells.
Interictal	Spike waves or sharp waves (*epileptiform*)	Focal or generalized	Supports the diagnosis of epilepsy	Right temporal spike waves in a patient with episodes of staring, behavioral arrest, and oral automatisms supports the diagnosis of right temporal epilepsy.
	Slowing (*nonepileptiform*)	Typically, focal	Suggests underlying cortical or subcortical dysfunction	Left parieto-occipital slowing in a child with frequent, worsening headaches and new onset seizures should prompt neuroimaging to rule out a space-occupying lesion.
Background	Slowing	Typically, generalized	Suggests diffuse cerebral dysfunction or encephalopathy	Children with developmental delays will frequently have background slowing compared to neurotypical patients of the same age.
	Asymmetry	Typically, regional/ hemispheric	Suggests lateralized dysfunction	Suppression of sleep features and fast activity on the left in a child with early left-hand dominance and concern for seizures should prompt neuroimaging to rule out perinatal insult (e.g., arterial ischemic stroke).

Smith–Magenis syndrome. It is well known that girls with Rett syndrome have abnormal nocturnal episodes including laughing or screaming, however, these do not meet the criteria for RBD, nor should they prompt PSG unless other concerns are present.

Narcolepsy is rare in childhood though it is known to occur in up to 0.05% of the population, with peak incidence between age 10 and 19 years (though often with delay of childhood diagnosis) [57]. Clinical suspicion for narcolepsy based on detailed history and screening questionnaire tools such as the Epworth Sleepiness Scale or Pediatric Daytime Sleepiness Scale warrants specific diagnostic testing, best accomplished through combination of PSG and the multiple-sleep latency test (MSLT) [66]. Clinical features concerning for this disorder include greater than 3 months of excessive daytime sleepiness and frequent "sleep attacks" in the presence of adequate sleep duration, as well as cataplexy, which are periods of weakness, clumsiness, or falls often provoked by episodes of heightened emotion. Cataplexy alone is often seen in children with Coffin–Lowry syndrome, Norrie disease, and Niemann–Pick type C disease [69]. Secondary narcolepsy is associated with several central nervous system disorders, especially with hypothalamic involvement in stroke, tumors, neuroinflammatory conditions, or brain injury [70]. PSG will reveal rapid entry into REM sleep (decreased REM latency), and MSLT confirms narcolepsy diagnosis if patient enters REM sleep within 8 minutes or less in at least two of five nap opportunities, after having had a full night of normal sleep preceding this test. A normal REM latency (in adults) is 90 minutes.

Sleep-related epilepsy includes several epilepsy syndromes. In the clinic, one should be wary of epileptic events in the child with repetitive or stereotyped movements [66,71]. Like the cases above, PSG is not indicated in the work up, though a standard EEG to look for interictal findings is advised. The semiology of seizures may prompt the clinician to suspect certain nocturnal epilepsy syndromes. If a description is lacking, parents should be asked to document or videotape episodes and report back. Episodes of hemifacial clonic movements with or without evolution to a generalized tonic clonic seizure should raise concern for benign or self-limited epilepsy of childhood with centrotemporal spikes (SeLECTS, formally known as BECTS, or Rolandic seizures). Episodes of myoclonus at the end stages of sleep or upon waking are features common to juvenile myoclonic epilepsy (JME). Language or motor regression is discussed above and associated with the spectrum of conditions falling within D/EE-SWAS. Paroxysmal arousals with complex motor movements or vocalizations or episodes of aura followed by amnesia are concerning for frontal lobe seizures (such as autosomal dominant nocturnal frontal lobe epilepsy, ADNFLE) or temporal lobe seizures. If an overnight study is indicated, generally a video EEG has more clinical utility rather than PSG unless there are concurrent episodes of hypoxia or snoring [66,71,72].

Children with epilepsy have been shown to have significant sleep-related disorders, estimated at between 25% and 80% in prevalence [66,73]. It is well known that OSA and resultant poor sleep efficiency can significantly worsen seizure control through a reduced seizure threshold. Therefore any child with unexpected increase in seizures and concurrent sleep difficulty should be evaluated for sleep disorders through clinical history and potentially, PSG versus cEEG depending on the etiology of sleeping concerns.

Electroencephalogram techniques

The importance of electrophysiological tools—the premier of which is the EEG—has only grown since their advent in the mid-20th century. In parallel to their roles in the clinic, these tools are widely used and investigated in the research space. As is not uncommon, translating these efforts into meaningful clinical applicability has not been easy. In this section, we will briefly discuss the most common techniques and the neurodevelopmental conditions in which they have been most studied and end with a discussion of their potential use as biomarkers in several different contexts.

Quantitative electroencephalogram—defining the spectrum

Clinically, the routine EEG is used to obtain information on the brain resting state and its response to stimulation in the form of "activating procedures" or those which may provoke hypersynchrony, and thereby seizures. The interpretation of the EEG is qualitative and, generally, limited to what the EEG reader can appreciate on visual inspection whereby certain abnormalities are identified with an approximation of their frequencies. Though the availability of commercial software is more commonplace, its use in the clinical realm is limited. For example, Persyst is a software package that may aid reading EEG studies, however, other software packages (such as open source packages available from the Antrobus EEG laboratory or the Swartz Center for Computational Neuroscience) are almost exclusively used in the research space. EEG data, however, can be more quantitatively analyzed with the assistance of commercial and custom-written software when grossly broken down into one of the following domains:

Frequency domain: The most common way to quantify an EEG is by reviewing the contributions of individual oscillatory frequencies, also referred to as spectral analysis. This is done by breaking the EEG up into so-called frequency bands (typically, delta, theta, alpha, beta, and gamma). The contribution of these frequency bands can then be calculated and reported relative to one another, or absolutely during a predetermined period (see Fig. 10.2.4) [74]. Spectral analysis can be performed on a resting (spontaneous) EEG or during specific tasks. Additionally, this analysis can help measure relationships or connectivity between different brain regions by comparing their individual frequency spectra which is a feature named coherence.

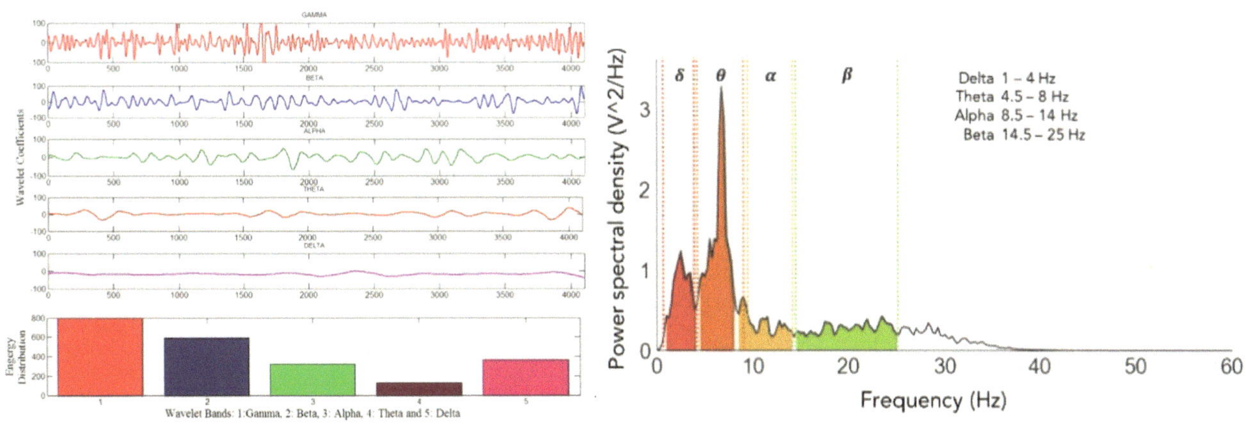

FIGURE 10.2.4 Quantitative electroencephalogram (EEG) methodology. The left panel demonstrates the most common frequency domains of interest, recorded from the EEG. These rhythms (from bottom to top) are delta, theta, alpha, beta, and gamma activity with frequency ranges of 0.4–4, 4–8, 8–13, 13–30, and 30–60, respectively. The right panel demonstrates an example of the EEG power spectrum which plots the frequencies (x-axis) as a function of their power. *From Harpale VK, Bairagi V. Brain seizure detection and classification using EEG signals. Chapter 2; Cox R, Fell J. Analyzing human sleep EEG: a methodological primer with code implementation. Sleep Med Rev 2020;54:101353.*

Time domain: A pitfall of frequency domain analysis is the loss of temporal information, as the signal is generated from a discrete time period. Time domain analysis focuses on EEG changes with respect to time. In this way, the subject can be presented with specific information, or be asked to perform specific tasks, and a small period of the EEG can be reviewed. The signals elicited during this period are called event-related potential(s) (ERPs) [75,76]. Notably, ERPs can be generated in response to many stimuli including motor and sensory events or actions as well as cognitive processes. Because this response is a function of brain networks, an ERP may serve as a surrogate measure of those networks which would be expected to be impacted in several neurodevelopmental disorders. Other more complex measurements, such as network complexity, can be obtained over a time domain though the specifics of this analysis are beyond the scope of this text [77].

Time-frequency domain: In general, this is a more complex means of measuring EEG signal whereby oscillations are considered within the distinct frequencies and the changes in phases of the oscillations and/or patterns of oscillations are assessed across time [78].

Though the United States Food and Drug Administration (FDA) have approved several devices or software for the post hoc analysis of EEG, it is notable that it has not approved quantitative EEG methods in the diagnosis or evaluation of any condition. Additionally, no major professional medical organizations (including the American Academy of Neurology, the International League Against Epilepsy, and the ACNS) have supported its use in official guidelines and has, in fact, published at least one formal recommendation against its use in traumatic brain injury diagnosis [79,80].

Quantifying electroencephalogram differences in children with NDDs

Children with NDDs are more likely to have epilepsy and vice versa; consequently, these children are more likely to undergo EEG evaluation [81]. Given this, it is reasonable that many clinicians and researchers have gravitated toward examining these data for features outside of seizures. Indeed, the scientific literature is full of examples of EEG parameters (such as those described above) that have been measured and found altered in several conditions including ASD, fragile X syndrome (FXS), attention-deficit disorder ADHD, Angelman's syndrome (AS) [82], *STXBP1* encephalopathy syndrome [83], Rett syndrome [84], several psychiatric disorders, and general learning disorders. As such, we will limit commentary to three of the more well studied which are ASD, FXS, and ADHD.

ASD: Research and clinical groups have described several changes in the EEGs of children with autism which indicate inappropriately low, long-range connectivity [75,85,86]. However, because many of these results were obtained in small populations, were inconsistent across research groups, or were obtained using variable methods, the premier organization for studying EEG in autism, the "Autism Biomarkers Consortium for Clinical Trials" [87], rigorously repeated these measures with the goal of describing reproducible changes in these patients [85,86]. This work found that resting-state EEG, visually evoked potentials, and a face-sensitive, ERP, "N170," were reasonably stable over several weeks. However, only features of N170 (latency) and resting-state (slope) had any ability to discriminate between the ASD and control groups [88].

FXS: Both ERPs and resting-state EEG have been explored in FXS with the findings summarized in several reviews [89,90]. A reduction in resting-state theta power has been reported in several studies, thought to represent aberrant glutamatergic signaling in the disease. An auditory ERP, "P300" associated with novelty detection, has been found to be consistently altered though this measure is sensitive to attentional state at the time of study.

ADD/ADHD: For years, numerous groups reported similar abnormalities in the EEGs of patients with ADD/ADHD—including greater power in the theta band at rest, decreased absolute beta with normal gamma band frequencies and higher theta:beta ratios, all suggestive of cortical hypoarousal in tasks necessitating these circuits [91–93]. In 2015 an FDA-certifying report and companion journal article were released demonstrating that an EEG "biomarker" for ADHD, the ratio of theta to beta waves, in conjunction with a confirmatory clinical diagnosis could improve diagnostic accuracy from 61% to 88% [74,94]. Notably, the AAN published a practice advisory soon after this announcement in which they stated they found insufficient evidence to use it in clinical practice [95,96].

Electroencephalogram-based biomarkers

Biomarkers are a measurable unit of some biological process, pathological process, or a response to intervention [97,98]. The FDA approves biomarkers for several reasons, their purpose best described by the category in which they are classified which are defined as one of the following categories: diagnostic, safety, monitoring, response, prognostication, predication, and susceptibility (Fig. 10.2.5; [98]). Qualified biomarkers are those that have been approved by the FDA after additional testing of their validity, reliability, and reproducibility [95] (Fig. 10.2.6).

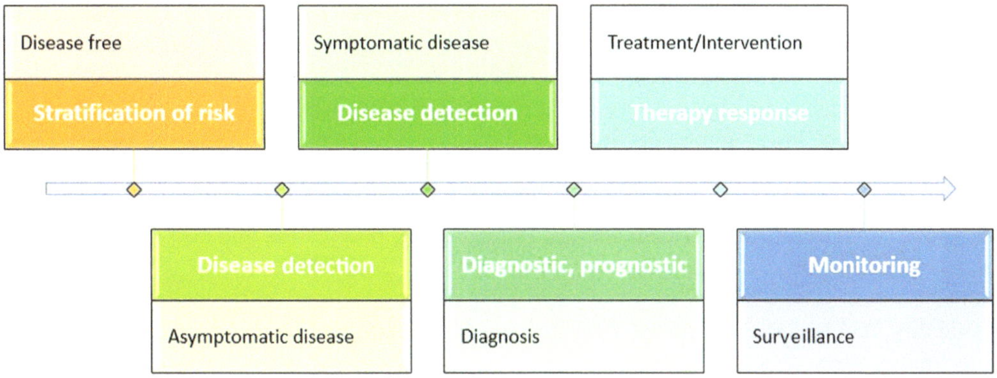

FIGURE 10.2.5 Biomarker categories of use and developmental pipeline. The FDA has defined several categories of biomarker which are: diagnostic, monitoring, prognostic, predictive, pharmacodynamic/response, safety, and susceptibility/risk.

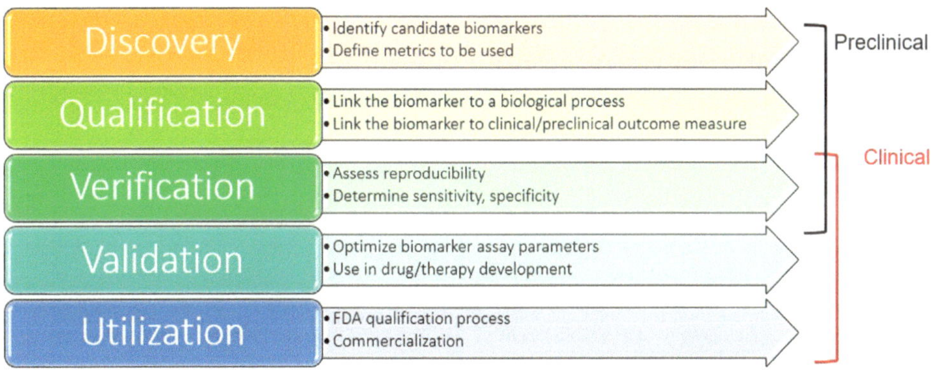

FIGURE 10.2.6 Biomarker development. General stages of biomarker development can be divided into preclinical and clinical stages with the end goal of FDA approval through the Biomarker Qualification Program.

The importance of biomarkers cannot be understated. Within the context of NDDs, discovery and utilization of biomarkers is most relevant in the development of appropriate treatments. Pediatric neurological disorders can especially benefit from validated biomarkers because it is hypothesized that many or most disorders have an age-limit after which time interventions are less likely to be successful, therefore, speed of therapy development and approval is more critical. Biomarkers have the potential to expediate this process if they can predict drug efficacy, replacing traditional, lengthy clinical trial outcomes measurements. Furthermore, robust biomarkers may decrease clinical trial failure: outcome measures for neurological conditions are often features such as cognitive function, emotional regulation, or psychiatric symptoms, all of which are difficult to measure in animal models in which preclinical studies are performed [90,99]. Supportive of this are several reviews which have demonstrated decreased trial failure in oncology, with use of biomarkers as outcome measures, and the now infamous, failed clinical trials in FXS despite very promising preclinical data [100–102].

For these reasons, development of an EEG-based biomarker has gained substantial attention, particularly in the last two decades [74,103–105]. While tools are being developed to improve interrater reliability in EEG reading in specific conditions, there remains substantial deficiencies in this when these tools are not available or utilized. For this reason, quantitative EEG becomes more appealing. As of today, the FDA does not recognize any EEG-related, qualified biomarkers though one is under review in the case of ASD; the N170 being an evoked potentials in response to human faces (NCT05294705).

Despite the above, EEG is frequently used as a clinical outcome measure (COA) in several contexts. One example may be found in infantile spasms. The FDA has authorized use of the EEG finding of "hypsarrhythmia" (a characteristic EEG pattern in severe infantile spasms) as a clinical outcome measure in several trials assessing the effectiveness of the medications vigabatrin and prednisolone [106–108]. While earlier studies suffered from the interrater reliability problem noted above, more recent literature suggests that this problem has been addressed through incorporation of the validated BASED (burden of amplitude and epileptiform discharges) score for measuring the presence or absence of "epileptic encephalopathy" as opposed to hypsarrhythmia [109−111]. Other examples of EEG-based outcome measures which have been proposed for use include:

In the field of congenital heart disease and preterm infants, traditional EEG is being discussed as one of many tools used in prediction algorithms to determine neurodevelopmental outcomes while still hospitalized [112,113].

In children with dyslexia, lower amplitude signals in the alpha frequency band in the parietal, temporal, and occipital regions as compared to controls may correlate with interventions in teaching/learning [114].

In Alzheimer's disease (AD), several changes in the EEG have been demonstrated to evolve with progression of disease allowing its use, alongside fMRI and CSF studies, in monitoring progression and as a marker of treatment response in clinical trials [115].

EEG has several benefits to other measures of neurological function (e.g., MRI) including superior spatiotemporal resolution, portability, and that it is relatively inexpensive. As research evolves, the application of EEG in the clinical realm is sure to follow, albeit, at a slower pace.

In summary, neurophysiological studies, the most employed being EEG, are important tools in caring for children with NDDs. Though informative, their utility is limited to the clinical context in which they are being utilized and are restricted by real-world shortcomings such as availability and quality of study. While there exists an abundance of literature detailing their usage in the research space, it is important to recognize that at present, the utility of these observations in patient management is unclear and is best limited to assisting with the diagnosis and management of seizure disorders.

Despite this, the future is promising. Hope that systematically obtained measures can be used as biomarkers remains strong and the results of corresponding clinical trials are eagerly awaited by clinicians and caregivers alike.

References

[1] Brigo F, Trinka E, Lattanzi S, Bragazzi NL, Nardone R, Martini M. A brief history of typical absence seizures - petit mal revisited. Epilepsy Behav 2018;80:346−53. Available from: https://doi.org/10.1016/j.yebeh.2018.01.007.

[2] Gibbs FA. Petit mal variant revisited. Epilepsia 1971;12(1):89−96. Available from: https://doi.org/10.1111/j.1528-1157.1971.tb03918.x.

[3] Stone JL, Hughes JR. Early history of electroencephalography and establishment of the American Clinical Neurophysiology Society. J Clin Neurophysiol 2013;30(1):28−44. Available from: https://doi.org/10.1097/WNP.0b013e31827edb2d.

[4] Gibbs FA, Gibbs EL. The Gibbs' Boston years: early developments in epilepsy research and electroencephalography at Harvard. Interview by James L. Stone and John R. Hughes. Clin Electroencephalogr 1990;21(4):175−82.

[5] Appendix 6. A brief history of EEG. In: St Louis EK, Frey LC, editors. Electroencephalography: an introductory text and atlas of normal and abnormal findings in adults, children, and infants. Allen Press, Inc; 2016. p. 95.

[6] Ebersole E. Current practice of clinical electroencephalography. 4th ed. Woltser Kluwer Health; 2014. p. 1−27.

[7] Chen H, Koubeissi M. Electroencephalography in Epilepsy evaluation. Continuum 2019;25:431−53.

[8] Marcuse LV, Fields MC, Yoo J. Origin and technical aspects of the EEG. Rowan's primer of EEG. Elsevier Health Services; 2016. p. 1−37.

[9] Jasper HH. The 10-20 electrode system of the International Federation. Electroencephalogr Clin Neurophysiol 1958;10:367−80.

[10] Klem GH, Lüders HO, Jasper HH, Elger C. The ten-twenty electrode system of the International Federation. The International Federation of Clinical Neurophysiology. Electroencephalogr Clin Neurophysiol Suppl 1999;52:3−6.

[11] American Clinical Neurophysiology Society. Guideline twelve: guidelines for long-term monitoring for epilepsy. J Clin Neurophysiol 2008;25(3):170−80. Available from: https://doi.org/10.1097/WNP.0b013e318175d472.

[12] Payne ET, Zhao XY, Frndova H, McBain K, Sharma R, Hutchison JS, et al. Seizure burden is independently associated with short term outcome in critically ill children. Brain 2014;137(Pt 5):1429−38. Available from: https://doi.org/10.1093/brain/awu042.

[13] Herman ST, Abend NS, Bleck TP, Chapman KE, Drislane FW, Emerson RG, et al.Critical Care Continuous EEG Task Force of the American Clinical Neurophysiology Society Consensus statement on continuous EEG in critically ill adults and children, part I: indications. J Clin Neurophysiol 2015;32(2):87−95. Available from: https://doi.org/10.1097/WNP.0000000000000166.

[14] Tatum WO, Olga S, Ochoa JG, Munger Clary H, Cheek J, Drislane F, et al. American Clinical Neurophysiology Society Guideline 7: guidelines for EEG reporting. J Clin Neurophysiol 2016;33(4):328−32. Available from: https://doi.org/10.1097/WNP.0000000000000319.

[15] Beniczky S, Aurlien H, Brøgger JC, Fuglsang-Frederiksen A, Martins-da-Silva A, Trinka E, et al. Standardized computer-based organized reporting of EEG: SCORE. Epilepsia 2013;54(6):1112−24. Available from: https://doi.org/10.1111/epi.12135.

[16] Beniczky S, Rubboli G, Aurlien H, Hirsch LJ, Trinka E, Schomer DLSCORE Consortium. The new ILAE seizure classification: 63 seizure types? Epilepsia 2017;58(7):1298−300. Available from: https://doi.org/10.1111/epi.13799.

[17] Kang PB, McMillan HJ, Kuntz NL, Lehky TJ, Alter KE, Fitzpatrick KF, et al.Professional Practice Committee of the American Association of Neuromuscular & Electrodiagnostic Medicine Utility and practice of electrodiagnostic testing in the pediatric population: an AANEM consensus statement. Muscle Nerve 2020;61(2):143−55. Available from: https://doi.org/10.1002/mus.26752.

[18] Wirrell E, Kozlik S, Tellez J, Wiebe S, Hamiwka L. Ambulatory electroencephalography (EEG) in children: diagnostic yield and tolerability. J Child Neurol 2008;23(6):655−62. Available from: https://doi.org/10.1177/0883073808314158.

[19] Salinsky M, Kanter R, Dasheiff RM. Effectiveness of multiple EEGs in supporting the diagnosis of epilepsy: an operational curve. Epilepsia 1987;28(4):331–4. Available from: https://doi.org/10.1111/j.1528-1157.1987.tb03652.x.

[20] Smith SJM. EEG in the diagnosis, classification, and management of patients with epilepsy. J Neurol Neurosurg Psychiatry 2005;76:ii2–7.

[21] Bouma HK, Labos C, Gore GC, Wolfson C, Keezer MR. The diagnostic accuracy of routine electroencephalography after a first unprovoked seizure. Eur J Neurol 2016;23(3):455–63. Available from: https://doi.org/10.1111/ene.12739.

[22] Claassen J, Mayer SA, Kowalski RG, Emerson RG, Hirsch LJ. Detection of electrographic seizures with continuous EEG monitoring in critically ill patients. Neurology 2004;62(10):1743–8.

[23] Arthur TM, deGrauw TJ, Johnson CS, Perkins SM, Kalnin A, Austin JK, et al. Seizure recurrence risk following a first seizure in neurologically normal children. Epilepsia 2008;49(11):1950–4. Available from: https://doi.org/10.1111/j.1528-1167.2008.01775.x.

[24] Stroink H, van Donselaar CA, Geerts AT, Peters AC, Brouwer OF, Arts WF. The accuracy of the diagnosis of paroxysmal events in children. Neurology 2003;60(6):979–82. Available from: https://doi.org/10.1212/01.wnl.0000049914.25434.72.

[25] Benbadis S. The differential diagnosis of epilepsy: a critical review. Epilepsy Behav 2009;15(1):15–21. Available from: https://doi.org/10.1016/j.yebeh.2009.02.024.

[26] Montenegro MA, Sproule D, Mandel A, Cappell J, Chiriboga CA, Jacob S, et al. The frequency of non-epileptic spells in children: results of video-EEG monitoring in a tertiary care center. Seizure 2008;17(7):583–7. Available from: https://doi.org/10.1016/j.seizure.2008.02.010.

[27] Sadleir LG, Scheffer IE, Smith S, Carstensen B, Farrell K, Connolly MB. EEG features of absence seizures in idiopathic generalized epilepsy: impact of syndrome, age, and state. Epilepsia 2009;50(6):1572–8. Available from: https://doi.org/10.1111/j.1528-1167.2008.02001.x.

[28] Sadleir LG, Scheffer IE, Smith S, Connolly MB, Farrell K. Automatisms in absence seizures in children with idiopathic generalized epilepsy. Arch Neurol 2009;66(6):729–34. Available from: https://doi.org/10.1001/archneurol.2009.108.

[29] Rosenow F, Wyllie E, Kotagal P, Mascha E, Wolgamuth BR, Hamer H. Staring spells in children: descriptive features distinguishing epileptic and nonepileptic events. J Pediatr 1998;133(5):660–3. Available from: https://doi.org/10.1016/s0022-3476(98)70108-8.

[30] Chu CA, Rosenfield M, Portello JK. Blink patterns: reading from a computer screen versus hard copy. Optom Vis Sci 2014;91(3):297–302. Available from: https://doi.org/10.1097/OPX.0000000000000157.

[31] Shaheen BS, Bakir M, Jain S. Corneal nerves in health and disease. Surv Ophthalmol 2014;59(3):263–85. Available from: https://doi.org/10.1016/j.survophthal.2013.09.002.

[32] Hedström A, Olsson I. Epidemiology of absence epilepsy: EEG findings and their predictive value. Pediatr Neurol 1991;7(2):100–4. Available from: https://doi.org/10.1016/0887-8994(91)90004-5.

[33] Hirtz D, Ashwal S, Berg A, Bettis D, Camfield C, Camfield P, et al. Practice parameter: evaluating a first nonfebrile seizure in children: report of the quality standards subcommittee of the American Academy of Neurology, The Child Neurology Society, and The American Epilepsy Society. Neurology 2000;55(5):616–23. Available from: https://doi.org/10.1212/wnl.55.5.616.

[34] Riviello JJ. Pediatric EEG abnormalities. Blum & Rutkove's the clinical neurophysiology primer. Humana Press; 2007. p. 179–204.

[35] Pearson TS, Pons R. Movement disorders in children. Continuum 2019;25(4):1099–120. Available from: https://doi.org/10.1212/CON.0000000000000756.

[36] Pisani F, Facini C, Bianchi E, Giussani G, Piccolo B, Beghi E. Incidence of neonatal seizures, perinatal risk factors for epilepsy and mortality after neonatal seizures in the province of Parma, Italy. Epilepsia 2018;59(9):1764–73. Available from: https://doi.org/10.1111/epi.14537.

[37] Korff C, Nordli Jr. DR. Do generalized tonic-clonic seizures in infancy exist? Neurology 2005;65(11):1750–3. Available from: https://doi.org/10.1212/01.wnl.0000187125.87414.f3.

[38] Cavazzuti GB, Cappella L, Nalin A. Longitudinal study of epileptiform EEG patterns in normal children. Epilepsia 1980;21(1):43–55. Available from: https://doi.org/10.1111/j.1528-1157.1980.tb04043.x.

[39] Josephson CB, Engbers JD, Sajobi TT, Jette N, Agha-Khani Y, Federico P, et al. An investigation into the psychosocial effects of the postictal state. Neurology 2016;86(8):723–30. Available from: https://doi.org/10.1212/WNL.0000000000002398.

[40] MacEachern SJ, D'Alfonso S, McDonald RJ, Thornton N, Forkert ND, Buchhalter JR. Most children with epilepsy experience postictal phenomena, often preventing a return to normal activities of childhood. Pediatr Neurol 2017;72:42–50. Available from: https://doi.org/10.1016/j.pediatrneurol.2017.03.002.

[41] van Diessen E, Lamberink HJ, Otte WM, Doornebal N, Brouwer OF, Jansen FE, et al. A prediction model to determine childhood epilepsy after 1 or more paroxysmal events. Pediatrics 2018;142(6)e20180931. Available from: https://doi.org/10.1542/peds.2018-0931.

[42] LaFrance Jr WC, Devinsky O. The treatment of nonepileptic seizures: historical perspectives and future directions. Epilepsia 2004;45(Suppl 2):15–21. Available from: https://doi.org/10.1111/j.0013-9580.2004.452002.x.

[43] LaFrance Jr WC, Baker GA, Duncan R, Goldstein LH, Reuber M. Minimum requirements for the diagnosis of psychogenic nonepileptic seizures: a staged approach: a report from the International League Against Epilepsy Nonepileptic Seizures Task Force. Epilepsia 2013;54(11):2005–18. Available from: https://doi.org/10.1111/epi.12356.

[44] LaFrance Jr WC, Reuber M, Goldstein LH. Management of psychogenic nonepileptic seizures. Epilepsia 2013;54(Suppl 1):53–67. Available from: https://doi.org/10.1111/epi.12106.

[45] Kutluay E, Selwa L, Minecan D, Edwards J, Beydoun A. Nonepileptic paroxysmal events in a pediatric population. Epilepsy Behav 2010;17(2):272–5. Available from: https://doi.org/10.1016/j.yebeh.2009.12.020.

[46] Asadi-Pooya AA, Emami M. Demographic and clinical manifestations of psychogenic non-epileptic seizures: the impact of co-existing epilepsy in patients or their family members. Epilepsy Behav 2013;27(1):1–3. Available from: https://doi.org/10.1016/j.yebeh.2012.12.010.

[47] Bhardwaj P, Sharma VK, Sharma R, Gautam P. Acquired epileptic aphasia: Landau-Kleffner syndrome. J Pediatr Neurosci 2009;4(1):52–3. Available from: https://doi.org/10.4103/1817-1745.49114.

[48] Landau WM, Kleffner FR. Syndrome of acquired aphasia with convulsive disorder in children. Neurology 1957;7:523–30.

[49] Caraballo RH, Cejas N, Chamorro N, Kaltenmeier MC, Fortini S, Soprano AM. Landau-Kleffner syndrome: a study of 29 patients. Seizure 2014;23(2):98–104. Available from: https://doi.org/10.1016/j.seizure.2013.09.016.

[50] Pearl PL, Carrazana EJ, Holmes GL. The Landau-Kleffner syndrome. Epilepsy Curr 2001;1(2):39–45. Available from: https://doi.org/10.1046/j.1535-7597.2001.00012.x.

[51] Shinnar S, Rapin I, Arnold S, Tuchman RF, Shulman L, Ballaban-Gil K, et al. Language regression in childhood. Pediatr Neurol 2001;24(3):183–9. Available from: https://doi.org/10.1016/s0887-8994(00)00266-6.

[52] International League Against Epilepsy (ILAE). ILAE classification and definition of epilepsy syndromes with onset in childhood: position paper by the ILAE task force on nosology and definitions. 2021. Retrieved from ilae.org/guidelines.

[53] Beenhakker MP, Huguenard JR. Neurons that fire together also conspire together: is normal sleep circuitry hijacked to generate epilepsy? Neuron 2009;62(5):612–32. Available from: https://doi.org/10.1016/j.neuron.2009.05.015.

[54] Singhal NS, Sullivan JE. Continuous spike-wave during slow wave sleep and related conditions. ISRN Neurol 2014;2014619079. Available from: https://doi.org/10.1155/2014/619079.

[55] Smith T, Ram D. Fifteen-minute consultation: a practical approach to developmental regression in children. Arch Dis Child Educ Pract Ed 2019;104(4):173–7. Available from: https://doi.org/10.1136/archdischild-2017-313792.

[56] Veatch OJ, Malow BA, Lee HS, Knight A, Barrish JO, Neul JL, et al. Evaluating sleep disturbances in children with rare genetic neurodevelopmental syndromes. Pediatr Neurol 2021;123:30–7. Available from: https://doi.org/10.1016/j.pediatrneurol.2021.07.009.

[57] Maski K, Owens J. Pediatric sleep disorders. Continuum 2018;24(1, Child Neurology):210–27. Available from: https://doi.org/10.1212/CON.0000000000000566.

[58] Reynolds AM, Soke GN, Sabourin KR, Hepburn S, Katz T, Wiggins LD, et al. Sleep problems in 2- to 5-year-olds with autism spectrum disorder and other developmental delays. Pediatrics 2019;143(3)e20180492. Available from: https://doi.org/10.1542/peds.2018-0492.

[59] Paasch V, Hoosier TM, Accardo J, Ewen JB, Slifer KJ. Technical tips: performing EEGs and polysomnograms on children with neurodevelopmental disabilities. Neurodiagn J 2012;52(4):333–48.

[60] Stowe RC, Afolabi-Brown O. Pediatric polysomnography-a review of indications, technical aspects, and interpretation. Paediatr Respir Rev 2020;34:9–17. Available from: https://doi.org/10.1016/j.prrv.2019.09.009.

[61] Bull MJCommittee on Genetics. Health supervision for children with Down syndrome. Pediatrics 2011;128(2):393–406. Available from: https://doi.org/10.1542/peds.2011-1605.

[62] Greenfeld M, Tauman R, DeRowe A, Sivan Y. Obstructive sleep apnea syndrome due to adenotonsillar hypertrophy in infants. Int J Pediatr Otorhinolaryngol 2003;67(10):1055–60. Available from: https://doi.org/10.1016/s0165-5876(03)00182-4.

[63] Ghergan A, Coupaye M, Leu-Semenescu S, Attali V, Oppert JM, Arnulf I, et al. Prevalence and phenotype of sleep disorders in 60 adults with Prader-Willi syndrome. Sleep 2017;40(12). Available from: https://doi.org/10.1093/sleep/zsx162.

[64] McCandless SECommittee on Genetics. Clinical report—health supervision for children with Prader-Willi syndrome. Pediatrics 2011;127(1):195–204. Available from: https://doi.org/10.1542/peds.2010-2820.

[65] Amaddeo A, De Sanctis L, Arroyo JO, Khirani S, Bahi-Buisson N, Fauroux B. Polysomnographic findings in Rett syndrome. Eur J Paediatr Neurol 2019;23(1):214–21. Available from: https://doi.org/10.1016/j.ejpn.2018.09.003.

[66] Aurora RN, Lamm CI, Zak RS, Kristo DA, Bista SR, Rowley JA, et al. Practice parameters for the non-respiratory indications for polysomnography and multiple sleep latency testing for children. Sleep 2012;35(11):1467–73. Available from: https://doi.org/10.5665/sleep.2190.

[67] Aurora RN, Zak RS, Karippot A, Lamm CI, Morgenthaler TI, Auerbach SH, et al. Practice parameters for the respiratory indications for polysomnography in children. Sleep 2011;34(3):379–88.

[68] Lloyd R, Tippmann-Peikert M, Slocumb N, Kotagal S. Characteristics of REM sleep behavior disorder in childhood. J Clin Sleep Med 2012;8(2):127–31. Available from: https://doi.org/10.5664/jcsm.1760.

[69] Khan Z, Trotti LM. Central disorders of hypersomnolence: focus on the narcolepsies and idiopathic hypersomnia. Chest 2015;148(1):262–73. Available from: https://doi.org/10.1378/chest.14-1304.

[70] Yassin W, Sugihara G, Oishi N, Kubota M, Ubukata S, Murai T, et al. Hypothalamic-amygdalar-brainstem volume reduction in a patient with narcolepsy secondary to diffuse axonal injury. J Clin Sleep Med 2015;11(5):581–2. Available from: https://doi.org/10.5664/jcsm.4710.

[71] Tinuper P, Bisulli F, Cross JH, Hesdorffer D, Kahane P, Nobili L, et al. Definition and diagnostic criteria of sleep-related hypermotor epilepsy. Neurology 2016;86(19):1834–42. Available from: https://doi.org/10.1212/WNL.0000000000002666.

[72] Fisher RS, Cross JH, D'Souza C, French JA, Haut SR, Higurashi N, et al. Instruction manual for the ILAE 2017 operational classification of seizure types. Epilepsia 2017;58(4):531–42. Available from: https://doi.org/10.1111/epi.13671.

[73] Al-Biltagi MA. Childhood epilepsy and sleep. World J Clin Pediatr. 2014;3(3):45–53. Available from: https://doi.org/10.5409/wjcp.v3.i3.45.

[74] Loo SK, Lenartowicz A, Makeig S. Research review: use of EEG biomarkers in child psychiatry research - current state and future directions. J Child Psychol Psychiatry 2016;57(1):4–17. Available from: https://doi.org/10.1111/jcpp.12435.

[75] De Stefano LA, Schmitt LM, White SP, Mosconi MW, Sweeney JA, Ethridge LE. Developmental effects on auditory neural oscillatory synchronization abnormalities in autism spectrum disorder. Front Integr Neurosci 2019;13:34. Available from: https://doi.org/10.3389/fnint.2019.00034.

[76] Jeste SS, Nelson 3rd CA. Event related potentials in the understanding of autism spectrum disorders: an analytical review. J Autism Dev Disord 2009;39(3):495–510. Available from: https://doi.org/10.1007/s10803-008-0652-9.

[77] Gasser T, Molinari L. The analysis of the EEG. Stat Methods Med Res 1996;5(1):67–99. Available from: https://doi.org/10.1177/096228029600500105.

[78] Coben R, Mohammad-Rezazadeh I, Cannon RL. Using quantitative and analytic EEG methods in the understanding of connectivity in autism spectrum disorders: a theory of mixed over- and under-connectivity. Front Hum Neurosci 2014;8:45. Available from: https://doi.org/10.3389/fnhum.2014.00045.

[79] Tenney JR, Gloss D, Arya R, et al. Practice guideline: use of quantitative EEG for the diagnosis of mild traumatic brain injury: report of the Guideline Committee of the American Clinical Neurophysiology Society. J Clin Neurophysiol 2021;38(4):287–92. Available from: https://doi.org/10.1097/WNP.0000000000000853.

[80] Nuwer M. Assessment of digital EEG, quantitative EEG, and EEG brain mapping: report of the American Academy of Neurology and the American Clinical Neurophysiology Society. Neurology 1997;49(1):277–92. Available from: https://doi.org/10.1212/wnl.49.1.277.

[81] Brooks-Kayal A. Molecular mechanisms of cognitive and behavioral comorbidities of epilepsy in children. Epilepsia 2011;52(Suppl 1):13–20. Available from: https://doi.org/10.1111/j.1528-1167.2010.02906.x.

[82] Martinez LA, Born HA, Harris S, et al. Quantitative EEG analysis in Angelman syndrome: candidate method for assessing therapeutics. Clin EEG Neurosci 2020;. Available from: https://doi.org/10.1177/1550059420973095.

[83] Houtman SJ, Lammertse HCA, van Berkel AA, et al. STXBP1 syndrome is characterized by inhibition-dominated dynamics of resting-state EEG. Front Physiol. 2021;12775172. Available from: https://doi.org/10.3389/fphys.2021.775172.

[84] Smirnov K, Stroganova T, Molholm S, Sysoeva O. Reviewing evidence for the relationship of EEG abnormalities and RTT phenotype paralleled by insights from animal studies. Int J Mol Sci 2021;22(10):5308. Available from: https://doi.org/10.3390/ijms22105308.

[85] Webb SJ, Naples AJ, Levin AR, et al. The Autism Biomarkers Consortium for clinical trials: initial evaluation of a battery of candidate EEG biomarkers. Am J Psychiatry 2022;. Available from: https://doi.org/10.1176/appi.ajp.21050485.

[86] O'Reilly C, Lewis JD, Elsabbagh M. Is functional brain connectivity atypical in autism? A systematic review of EEG and MEG studies. PLoS ONE 2017;12(5):e0175870. Available from: https://doi.org/10.1371/journal.pone.0175870.

[87] The Autism Biomarkers Consortium for Clinical Trials. The Autism Biomarkers Consortium for clinical trials confirmation study (ABC-CT CS). NCT05294705. Updated June 21, 2022. Accessed November 1, 2022. https://www.clinicaltrials.gov/ct2/show/NCT05294705

[88] Wang J, Barstein J, Ethridge LE, Mosconi MW, Takarae Y, Sweeney JA. Resting state EEG abnormalities in autism spectrum disorders. J Neurodev Disord 2013;5(1):24. Available from: https://doi.org/10.1186/1866-1955-5-24.

[89] Devitt NM, Gallagher L, Reilly RB. Autism spectrum disorder (ASD) and fragile X syndrome (FXS): two overlapping disorders reviewed through electroencephalography-what can be interpreted from the available information? Brain Sci 2015;5(2):92–117. Available from: https://doi.org/10.3390/brainsci5020092.

[90] Kenny A, Wright D, Stanfield AC. EEG as a translational biomarker and outcome measure in fragile X syndrome. Transl Psychiatry 2022;12(1):34. Available from: https://doi.org/10.1038/s41398-022-01796-2.

[91] Tye C, Rijsdijk F, McLoughlin G. Genetic overlap between ADHD symptoms and EEG theta power. Brain Cogn 2014;87:168–72. Available from: https://doi.org/10.1016/j.bandc.2014.03.010.

[92] Barry RJ, Clarke AR, Hajos M, McCarthy R, Selikowitz M, Dupuy FE. Resting-state EEG gamma activity in children with attention-deficit/hyperactivity disorder. Clin Neurophysiol 2010;121(11):1871–7. Available from: https://doi.org/10.1016/j.clinph.2010.04.022.

[93] Shi T, Li X, Song J, Zhao N, Sun C, Xia W, et al. EEG characteristics and visual cognitive function of children with attention deficit hyperactivity disorder (ADHD). Brain Dev 2012;34(10):806–11. Available from: https://doi.org/10.1016/j.braindev.2012.02.013.

[94] Snyder SM, Rugino TA, Hornig M, Stein MA. Integration of an EEG biomarker with a 'clinician's ADHD evaluation. Brain Behav 2015;5(4) e00330. Available from: https://doi.org/10.1002/brb3.330.

[95] Ewen JB. The eternal promise of EEG-based biomarkers: getting closer? Neurology 2016;87(22):2288–9. Available from: https://doi.org/10.1212/WNL.0000000000003275.

[96] Gloss D, Varma JK, Pringsheim T, Nuwer MR. Practice advisory: the utility of EEG theta/beta power ratio in ADHD diagnosis: report of the Guideline Development, Dissemination, and Implementation Subcommittee of the American Academy of Neurology. Neurology 2016;87(22):2375–9. Available from: https://doi.org/10.1212/WNL.0000000000003265.

[97] Ewen JB, Sweeney JA, Potter WZ. Conceptual, regulatory and strategic imperatives in the early days of EEG-based biomarker validation for neurodevelopmental disabilities. Front Integr Neurosci 2019;13:45. Available from: https://doi.org/10.3389/fnint.2019.00045.

[98] Amur S, LaVange L, Zineh I, Buckman-Garner S, Woodcock J. Biomarker qualification: toward a multiple stakeholder framework for biomarker development, regulatory acceptance, and utilization. Clin Pharmacol Ther 2015;98(1):34–46. Available from: https://doi.org/10.1002/cpt.136.

[99] Budimirovic DB, Berry-Kravis E, Erickson CA, Hall SS, Hessl D, Reiss AL, et al. Updated report on tools to measure outcomes of clinical trials in fragile X syndrome. J Neurodev Disord 2017;9:14. Available from: https://doi.org/10.1186/s11689-017-9193-x.

[100] Parker, JL, Kuzulugil, SS, Pereverzev, K, Mac, S, Lopes, G, Shah, Z, Weerasinghe, A, Rubinger, D, Falconi, A, Bener, A, Caglayan, B, Tangri, R, & Mitsakakis, N. (2021). Does biomarker use in oncology improve clinical trial failure risk? A large-scale analysis. Cancer Med. 2021;10(6):1955–1963. Available from: https://doi.org/10.1002/cam4.3732.

[101] Antoniou, P, Ziogas, DE, Mitsis, M, & Roukos, DH. Precision oncology in patients with nonmetastatic disease: emerging reality or illusion. Future Oncol. 2019;15(16):1805–1810. Available from: https://doi.org/10.2217/fon-2019-0132.

[102] Erickson, CA, Davenport, MH, Schaefer, TL, Wink, LK, Pedapati, EV, Sweeney, JA, Fitzpatrick, SE, Brown, WT, Budimirovic, D, Hagerman, RJ, Hessl, D, Kaufmann, WE, & Berry-Kravis, E. Fragile X targeted pharmacotherapy: lessons learned and future directions. J Neurodev Disord. 2017;9:7. Available from: https://doi.org/10.1186/s11689-017-9186-9.

[103] Heunis TM, Aldrich C, de Vries PJ. Recent advances in resting-state electroencephalography biomarkers for autism spectrum disorder-a review of methodological and clinical challenges. Pediatr Neurol 2016;61:28–37. Available from: https://doi.org/10.1016/j.pediatrneurol.2016.03.010.

[104] Kanda PAM, Anghinah R, Smidth MT, Silva JM. The clinical use of quantitative EEG in cognitive disorders. Dement Neuropsychol 2009;3(3):195–203. Available from: https://doi.org/10.1590/S1980-57642009DN30300004.

[105] Babiloni C, Barry RJ, Başar E, Blinowska KJ, Cichocki A, Drinkenburg WHIM, et al. International Federation of Clinical Neurophysiology (IFCN) - EEG research workgroup: recommendations on frequency and topographic analysis of resting state EEG rhythms. Part 1: applications in clinical research studies. Clin Neurophysiol 2020;131(1):285–307. Available from: https://doi.org/10.1016/j.clinph.2019.06.234.

[106] U.S. Federal Drug Agency 2021, Center for drug evaluation and research. Clinical outcome assessment compendium, accessed 1 September 2022 <https://www.fda.gov/drugs/development-resources/clinical-outcome-assessment-compendium>.

[107] Lux AL, Osborne JP. A proposal for case definitions and outcome measures in studies of infantile spasms and West syndrome: consensus statement of the West Delphi group. Epilepsia 2004;45(11):1416–28. Available from: https://doi.org/10.1111/j.0013-9580.2004.02404.x.

[108] Messer R, Knupp KG. Infantile spasms: opportunities to improve care. Semin Neurol 2020;40(2):236–45. Available from: https://doi.org/10.1055/s-0040-1705121.

[109] Hussain SA, Kwong G, Millichap JJ, Mytinger JR, Ryan N, Matsumoto JH, et al. Hypsarrhythmia assessment exhibits poor interrater reliability: a threat to clinical trial validity. Epilepsia 2015;56(1):77–81. Available from: https://doi.org/10.1111/epi.12861.

[110] Mytinger JR, Hussain SA, Islam MP, Millichap JJ, Patel AD, Ryan NR, Twanow JD, Heyer GL. Improving the inter-rater agreement of hypsarrhythmia using a simplified EEG grading scale for children with infantile spasms. Epilepsy Res. 2015;116:93–98. Available from: https://doi.org/10.1016/j.eplepsyres.2015.07.008. Epub 2015 Jul 28. PMID: 26280806.

[111] Mytinger JR, Vidaurre J, Moore-Clingenpeel M, Stanek JR, Albert DVF. A reliable interictal EEG grading scale for children with infantile spasms - The 2021 BASED score. Epilepsy Res. 2021;173:106631. Available from: https://doi.org/10.1016/j.eplepsyres.2021.106631. Epub 2021 Apr 2. PMID: 33839516.

[112] Marino BS. New concepts in predicting, evaluating, and managing neurodevelopmental outcomes in children with congenital heart disease. Curr Opin Pediatr 2013;25(5):574–84. Available from: https://doi.org/10.1097/MOP.0b013e328365342e.

[113] Pavlidis E, Lloyd RO, Mathieson S, Boylan GB. A review of important electroencephalogram features for the assessment of brain maturation in premature infants. Acta Paediatr 2017;106(9):1394–408. Available from: https://doi.org/10.1111/apa.13956.

[114] Babiloni C, Stella G, Buffo P, Vecchio F, Onorati P, Muratori C, et al. Cortical sources of resting state EEG rhythms are abnormal in dyslexic children. Clin Neurophysiol 2012;123(12):2384–91. Available from: https://doi.org/10.1016/j.clinph.2012.05.002.

[115] Colom-Cadena M, Spires-Jones T, Zetterberg H, Blennow K, Caggiano A, DeKosky ST, et al.Synaptic Health Endpoints Working Group The clinical promise of biomarkers of synapse damage or loss in 'Alzheimer's disease. Alzheimers Res Ther 2020;12(1):21. Available from: https://doi.org/10.1186/s13195-020-00588-4.

Section III

Clinical disorders

Chapter 11

Intrauterine alcohol and drug exposure: preventable causes of intellectual and behavioral disorders

Carmen Lopez-Arvizu[1], Samantha R. Hutchison[2], Kadee D. Atkinson[3] and Harolyn M.E. Belcher[4]

[1]*Center for Developmental Behavioral Health, Kennedy Krieger Institute, Department of Psychiatry and Behavioral Science, Johns Hopkins School of Medicine, Baltimore, MD, United States,* [2]*Office for Health, Equity, Inclusion, and Diversity, Kennedy Krieger Institute, Baltimore, MD, United States,* [3]*Center for Diversity in Public Health Leadership Training, Kennedy Krieger Institute, Baltimore, MD, United States,* [4]*Office for Health, Equity, Inclusion, and Diversity, Kennedy Krieger Institute, Department of Pediatrics, Johns Hopkins University School of Medicine, Department of Mental Health, Johns Hopkins Bloomberg School of Public Health, Baltimore, MD, United States*

Toxins are common causes of neurodevelopmental disabilities (NDDs). Most of the toxins that affect prenatal development are associated with structural malformations. Warfarin, thalidomide, and methotrexate were identified early. Antiepileptic agents such as phenytoin, carbamazepine, and valproate have distinct syndromes associated with intellectual disability. Retinoic acid has been identified more recently.

Postnatal toxins may not be associated with structural abnormality but may have cognitive, academic, or behavioral consequences. Linking an adverse outcome to a specific agent is difficult in the absence of structural abnormality. Exposures may be toxic when occurring with a genetic variation, such as the case with aminoglycosides and hearing loss. The toxicity of an agent only may be expressed in conjunction with immaturity of a metabolic pathway or with neurologic immaturity. Quantifying the amount of exposure necessary to cause a functional impairment is extremely difficult. Exposures are rarely unique, and the effect may result from multiple, interacting exposures.

This chapter will discuss the effects of intrauterine alcohol and drug exposure on neurodevelopment and address their unique neurobiology, age-dependent clinical expressions, and management.

Intrauterine alcohol exposure

Fetal alcohol spectrum disorders (FASD) are a combination of a wide range of nonspecific morphologic, cognitive, and behavioral sequelae that result from prenatal exposure to ethanol. FASD includes fetal alcohol syndrome (FAS), alcohol-related birth defects (ARBD), and alcohol-related neurodevelopmental disorder (ARND).

When alcohol use in pregnancy is confirmed, three physical criteria are required for the diagnosis of FAS: growth retardation, a distinct facial phenotype (including short palpebral fissures, thin upper vermillion border, flat philtrum, low set ears), and evidence of central nervous system (CNS) abnormalities (decreased cranial size, structural brain abnormalities, and/or cognitive disabilities) [1–3].

Epidemiology

Gestational ethanol consumption is the leading cause of preventable intellectual disability in the United States [4]. It is estimated that 3.1%–9.9% of children have some form of FASD [5,6] and a higher prevalence in Native Americans and African-Americans, inner-city populations, and those with low socioeconomic status [7].

Alcohol use among pregnant women is common throughout the world. A 2020 Center for Disease Control and Prevention survey of alcohol use among pregnant women reported 64.7% used alcohol in the previous 12 months, and

Capute and Accardo's Neurodevelopmental Disabilities in Infancy and Childhood. DOI: https://doi.org/10.1016/B978-0-12-824060-1.00021-3

use within the past month was found in 19.6% of women in their first month of pregnancy, decreasing to 4.7% in the second and third trimester. Binge drinking (four or more drinks on one occasion) was reported by 10.5% of first trimester respondents and 1.4% of second or third-trimester respondents [8].

Neurobiology

Structural changes that have been detected by MRI studies include reduction in overall brain volume, corpus callosum hypo/dysgenesis, and marked volume reductions in the frontal and parietal lobes, cerebellum, caudate nucleus, and hippocampus in children and adults with history of prenatal exposure to alcohol [9]. Diffusion tensor imaging studies identified microstructural white matter abnormalities in the corpus callosum, cerebellum, and corticospinal tracts [10]. While some structural abnormalities were observed in individuals with prenatal ethanol exposure, the absence of structural abnormalities does not preclude damage at the cellular level.

Ethanol toxicity to the fetus has been associated with abnormal cellular function, survival, migration, and myelination. Ethanol-induced effects of neurotransmission identified abnormalities in N-methyl-D-aspartate (NMDA), serotonergic, cholinergic, glycine receptors, and L-type calcium channels [11]. Specific to gamma-aminobutyric acid (GABA) neurotransmission, ethanol exposure reduces NMDA receptors' number and function during brain development. In fetuses of mothers who binge drink, the period of alcohol withdrawal that follows is thought to induce neuronal excitotoxicity. Another neurotransmitter system that plays a significant role in neurodevelopment is the serotonergic system. Ethanol exposure impairs interaction between serotonergic neurons and nearby astrocytes, disrupting target neuronal networks' brain organization and function.

Beyond the direct insults to fetal physiology that ethanol produces, indirect mechanisms of teratogenicity result from damage to the mother's physiology. Heavy alcohol consumption is often accompanied by nutritional deficiencies that include folate, choline, and vitamin A. Ethanol impairs placental function, acting as a vasoconstrictor of placental and umbilical cord vessels [12,13]. Amniotic fluid can function as a reservoir for ethanol; prolonging the exposure to ethanol for the fetus [14].

Clinical presentation and age-dependent neurobehavioral outcomes

Initially, FAS came to attention because of the physical stigmata that were thought to be the only outcome of intrauterine alcohol exposure. As a larger spectrum of disorders were appreciated, it was recognized that the heterogeneous phenotype of FASD resulted from multiple factors that affect fetal development besides ethanol exposure. These factors include genetic susceptibility, maternal age, polydrug consumption, nutritional status, and the dose, timing, and pattern of ethanol intake during gestation [11]. Studies also have shown heterogeneity in the clinical presentation depending on exposure during development. The diagnosis of the other disorders included in FASD may be more difficult because they lack the physical findings and are dependent on history. Lack of awareness of FASD, poor access to specialty care, and inconsistent approaches to diagnosis as well as parental stigma all delay the identification and early diagnosis of FASD.

FAS is associated with early *growth retardation* that cannot be explained by other causes such as malnutrition, malabsorption, and genetic syndromes. Diagnostic criterion requires weight, length, and body mass index below the 10th percentile [15].

Classic *facial dysmorphism* of FAS, based on racial norms, are smooth philtrum, thin upper lip, and short palpebral fissures (<3rd percentile) [16,17]. This compilation of facial features has high sensitivity and specificity and serve as the basis for automated screening using standard photographs of children [16].

CNS abnormalities attributable to fetal alcohol effects fall within three categories (1) structural, for example, microcephaly below the 10th percentile, (2) neurological problems not attributable to other causes, for example, hypoxia; and (3) functional: IQ two standard deviations below the mean or significant developmental delay in children less than 2 years old; or performance less than one standard deviation below the mean in three of the following areas: speech and language, fine motor, visuospatial perception, spatial-constructive skills, learning ability, executive function, arithmetic, attention, social skills, or behavior [15].

Beyond the CNS disturbances that qualify for diagnosis of FAS, this patient population struggles with *mental health conditions* that severely impact their quality of life. Ninety-four percent of patients report having some mental health problem [18]. The inability to control impulses leads to an array of psychosocial behavior problems that can have devastating consequences. Lack of social judgment and difficulty comprehending the mental states and emotions of others [19–21] may have a daily impact on interaction with others. Inappropriate sexual behavior was reported in 45%–50%

of FAS patients, the most common being sexual advances (18%) [18]. The increase in inappropriate sexual activity noted in the population of individuals with FAS often results in criminal consequences. One child with FAS stated, "*(It's) hard to keep out of trouble, and I am not that smart. I have to think. Like when I want to push a little kid. I have to stop and think.*" [22]. Up to 35% of adult patients with FAS have been incarcerated for a crime. The first crime that usually presents in patients with FAS is shoplifting or theft [18].

Mental health problems reported in individuals with FAS may include depression and suicide. People with FAS commonly state that they feel different from their peers [22]. Studies also demonstrate that children with FASD have increased exposure to maltreatment and exhibit fewer solutions to social dilemmas compared to peers without prenatal ethanol exposure. They are at increased risk of alcohol and substance abuse and are less likely to reach independence as adults. These psychosocial challenges are compounded by various medical problems, including feeding difficulties, spasticity, ocular, and hearing problems. Overall, FAS is a life-long diagnosis with life-long physical and mental health challenges. The majority of individuals (83%) with FAS will require supervised living environments as adults [18].

More recently diagnostic and statistical manual (DSM) of mental disorders fifth edition, the designation of neurobehavioral disorder associated with prenatal alcohol exposure (ND-PAE) has been included in the DSM-5 as a condition of further study. They propose a diagnosis that includes impaired neurocognitive functioning, self-regulation, and adaptive functioning in the context of confirmation of (more than minimal) prenatal alcohol exposure. It is also a diagnosis of exclusion in the sense that the symptoms cannot be due to a general medical condition, another teratogen, genetic disorder, or environmental neglect (DSM-5). The most significant repercussion of the addition of ND-PAE is that it can include all exposed individuals with or without physical effects of prenatal alcohol exposure (facial dysmorphism, growth impairment).

The behavioral phenotype for FASD has been described as a combination of cognitive and behavioral deficits. Cognitive difficulties include impairment in coordination disorder (impaired gross/fine motor skills), academic achievement, memory, attention, executive function, adaptive functioning, social deficits, language disorders, and intellectual disability [23].

The cognitive difficulties reflect what longitudinal MRI studies have shown, changes in cortical (volume, thickness) and white matter in childhood and adolescence [24].

Behavioral difficulties are related to self-regulation. These present frequently as externalizing behaviors which are usually the reason for pursuing medical care in a way the subtler cognitive symptoms are not and might be missed in their entirety. Subtler symptoms are often missed in children and youth from historically marginalized populations [25].

The self-regulation deficits are expressed as behavioral difficulties, mood lability, attention deficit, and poor impulse control. This definition overlaps with attention deficit hyperactivity disorder (ADHD) diagnosis, making it difficult to separate from each other as affected domains overlap. It is also frequent to see overlap in other exposures besides alcohol in the prenatal period, often other toxic teratogens in combination with suboptimal prenatal conditions which complicates the diagnostic clarity.

Clinical presentation in FASD has involved description of hyperactivity and impulsivity but also more significant externalizing behavior descriptions such as disruptive, oppositionality, defiant behavior, and even delinquency. Comorbidity of ADHD, oppositional defiant disorder, and conduct disorder are more prevalent than in the general population [26].

Long-term outcome of individuals with FASD in transition-age youth and adults report having high service needs due to their difficulties in daily living, including difficulty with independent living, legal problems, employment, and substance abuse. Individuals with lower intellectual functioning and lack of primary support had poorer outcomes [27].

Social and adaptive functioning impairment have significant effects on decision-making ability and navigating the world with significant consequences, especially in the legal system as these individuals might be vulnerable to suggestibility [28,29].

Environmental events including prematurity, continuing exposure to parental alcohol abuse, and early neglect all increase risk of ADHD, so it is difficult to determine the causality of the behavioral phenotype, creating lack of consensus in treatment approaches [30,31].

Laboratory test

The diagnosis of FAS is clinical. Diagnosis continues to be challenging due to lack of identified biomarkers.

Differential diagnosis

FASD is a diagnosis of exclusion, particularly, in the absence of confirmed alcohol exposure [32] and, consequently, is often a challenge to establish. Cases of FASD may go undetected from birth well into adulthood, especially when no phenotypical features are present [16,32]. Up to 80% are not diagnosed or misdiagnosed (7%) [33].

The differential diagnosis of FAS includes genetic syndromes that share any of the three facial features required for FAS diagnosis. A smooth philtrum can suggest Cornelia de Lange syndrome, Opitz syndrome, and toluene embryopathy. Other neurogenic disorders like Miller-Dieker (Lissencephaly) syndrome or fetal valproate syndrome share lip features. Genetic disorders to consider include Williams syndrome, Velocardiofacial syndrome, duplication of 10q or 15q, Trisomy 18 syndrome, or maternal phenylketonuria fetal effects. Noonan syndrome, Dubowitz syndrome, and fetal hydantoin syndrome should also be considered [34,35].

Resources

https://www.cdc.gov/ncbddd/fasd/index.html
https://www.cdc.gov/ncbddd/fasd/facts.html
https://www.cdc.gov/ncbddd/fasd/research.html
https://store.samhsa.gov/product/TIP-58-Addressing-Fetal-Alcohol-Spectrum-Disorders-FASD-/SMA13-4803
https://store.samhsa.gov/sites/default/files/d7/priv/sma13-4803.pdf

Intrauterine tobacco exposure

Tobacco contains the alkaloid nicotine that is responsible for tobacco's addictive properties. In addition, smoke from tobacco contains over 7000 different chemicals, 70 of which are known carcinogens [36]. Biomarkers of exposure to tobacco and nicotine include nicotine and nicotine metabolites, volatile organic compounds, benzene, toluene, cyanide, polycyclic aromatic hydrocarbons such as fluorine metabolites, and aromatic amines such as aniline and propylene glycol [37]. Volatile organic compounds are of special concern as they may adversely affect smokers and nonsmokers [37].

Epidemiology

Tobacco use is associated with four of the leading five causes of death in the United States, namely: heart disease (#1), cancer (#2), chronic lower respiratory diseases (#4), and cerebrovascular diseases (#5) [38]. Collectively tobacco use is associated with conditions that make up 55% of total deaths in the United States [38]. In 2019, 34 million US citizens (14%) were current cigarette smokers, 15.3% adult US men, and 12.7% adult US women [39]. Based on data from the Centers for Disease Control and Prevention's National Center for Health Statistics Vital Statistics Cooperative Program, the rate of tobacco smoking during pregnancy was 8.1% [40]. The highest rates of cigarette use among pregnant women were noted in women ages 20–24 years (34.0%), who were White (83.6%) and had a high school graduate equivalency degree (GED) (40.4%).

Neurobiology

In the fetus, intrauterine tobacco exposure is associated with alterations of neuronal migration, synaptic function, and activation of dopaminergic neurons of the mesocorticolimbic pathway [41]. Smoking during the first trimester may also alter the expression of nicotinic acetylcholine receptors and muscarinic acetylcholine receptors, thereby interfering with the complex interactions between growth factors and neurotransmitters regulating cell growth and differentiation [42]. Intrauterine nicotine exposure modified apical dendritic morphology and nicotinic acetylcholine receptor currents in the medial prefrontal cortex [43]. Long lasting changes were noted in nicotinic acetylcholine receptors were associated with *chrna5* genotype [43]. These alterations of neurogenesis in the prefrontal cortex may be related to the development of attention deficits.

Clinical presentation and age-dependent neurobehavioral outcomes

Prenatal tobacco exposure is a leading preventable cause of infant morbidity and mortality [44,45]. Multiple studies have demonstrated the association between intrauterine tobacco exposure and lower mean birth weights and head circumference measures [46–48]. Prenatal tobacco exposure is associated with alterations in α-fetoprotein [49] and lower maternal serum folate and vitamin B12 [50] and related to fetal growth retardation. Lower birth weights were associated with increased cigarette use [50]. A 2020 study of 83 infants examined the neurobehavioral effects of prenatally exposed tobacco and e-cigarettes [51]. Using the Neonatal Behavioral Assessment Scale, the researchers found less optimal performance on motor maturity and regulation for 1-month-old infants with prenatal tobacco exposure than nonexposed infants [51]. There was an association between the years the mother smoked before pregnancy and motor maturity, reflexes, and regulation [51]. The neurobehavioral profile of the infants with prenatal tobacco exposure showed more irritability and decreased consolability.

The clinical findings in older children who are exposed to tobacco prenatally are also variable. In a study of 357 6-year-old children born to teenage mothers, offspring with prenatal tobacco exposure demonstrated increased activity, impulsivity, aggression, and externalizing behaviors [52]. After adjustment for maternal psychological status, home environment, and environmental tobacco exposure, increased activity was still significantly associated with prenatal tobacco exposure [52]. In a study of 593 10-year-olds, verbal learning and memory deficits associated with prenatal tobacco exposure persisted after controlling for maternal demographic characteristics, current tobacco use, another drug use, and other prenatal drug exposure [53].

By contrast, the National Longitudinal Survey of Youth Study of 704 mothers and 1752 offspring, ages 4–10 years, found no effect of prenatal tobacco exposure risk in regards to conduct problems and oppositional defiant disorder when comparing offspring with prenatal tobacco exposure to their siblings who differed in their exposure to prenatal nicotine [54]. There was a small association between prenatal tobacco exposure and ADHD. Genetic studies suggest that significant, unidentified environmental factors, rather than the teratogenic effects of tobacco, are linked to prenatal tobacco exposure and offspring externalizing problems [54]. The National Collaborative Perinatal Project found that after adjusting for measured and unmeasured confounders, the hypothesized effects of maternal smoking during pregnancy on these outcomes are either nonexistent or indistinguishable from a broader range of familial factors associated with maternal smoking [55].

In youth, the long-term consequences of prenatal tobacco exposure are also evident. A longitudinal study of 131 urban high school students who had teachers complete the Behavior Rating Inventory of Executive Functioning-Teacher Form (BRIEF-TF) found that prenatal tobacco exposure was associated with less optimal behavior regulation controlling for intrauterine substance exposures, violence exposure, sex, birth weight, IQ, mothers' country of birth, age at BRIEF teacher assessment, and lead exposure [56]. MRI studies in adolescence demonstrated reduced volumes in the orbitofrontal, middle frontal, and parahippocampal areas in youth with prenatal tobacco exposure compared to nonexposed peers [57]. A study of 608 individuals, followed since birth and assessed at 22 years of age, found that prenatal tobacco exposure was associated with total behavior problem scores, internalizing and externalizing behavior problem scores, as well as aggression, anxiety/depression, attention, and somatic problems. First-trimester tobacco exposure was associated with more arrests in adult offspring; there was no association between an arrest record and second or third-trimester prenatal tobacco exposure [58].

Treatment and prevention

The goal of "treatment" for pregnant women is smoking cessation rather than smoking reduction. Studies suggest that ceasing tobacco use during pregnancy may reduce the risk of adverse neurobehavioral outcomes. In a study of over 18,000 infants, women who quit smoking had infants who scored more optimally on the easy temperament scale on the Carey Infant Temperament Scale [59]. Studies have documented that even smoking fewer than 5 tobacco cigarettes per day is associated with decreases in birth weight [46].

Resources

https://www.samhsa.gov/national-center-excellence-tobacco-free-recovery
https://www.samhsa.gov/newsroom/press-announcements/201809251100
https://ncsacw.samhsa.gov/files/special_issues_pregnancy_factsheets.pdf
https://www.cdc.gov/tobacco/basic_information/health_effects/pregnancy/index.htm
https://www.cdc.gov/tobacco/campaign/tips/diseases/pregnancy.html
https://www.cdc.gov/tobacco/basic_information/youth/index.htm

Substance use

Substance use and substance use disorders result from a complex interplay between the individual, the environment, and the drug. Substance use among pregnant women in the United States has increased over the last decade. Polysubstance use during pregnancy is common and has adverse maternal and fetal health outcomes [60]. In 2019 among pregnant women aged 15–44, 18.4% reported using illicit drugs, tobacco products, or alcohol in the past month; almost 1/3 reported illicit drug use, marijuana use, and opioid use.

Substance use in pregnancy is associated with harmful risks and effects for the mother and infant that vary by drug of choice, the timing of exposure, dose, and psychosocial conditions. Intrauterine drug exposure (IUDE) may be accompanied by neonatal withdrawal symptoms, as with intrauterine opiate exposure [61]. IUDE may also be associated with numerous adverse neurobehavioral and neurodevelopmental outcomes [60,62].

Intrauterine cannabis (marijuana) exposure

Epidemiology

Marijuana is the most commonly used federally illicit drug [63]. In the past year, marijuana was used by an estimated 49.6 million (17.9%) people ages 12 and older [63]. Young adults, 18–25 years, had the highest percentage of marijuana use (34.5%) [63]. In the past month, 8.3% of pregnant women reported using marijuana [64]. A Centers for Disease Control and Prevention study of eight states (Alaska, Illinois, Maine, New Mexico, New York, North Dakota, Pennsylvania, and West Virginia) participating in the 2017 Pregnancy Risk Assessment Monitoring System found that non-Hispanic White or other race/ethnicity compared to non-Hispanic Black, unmarried, with less than high school education status, and cigarette users were more likely to use marijuana [65].

Neurobiology

Tetrahydrocannabinol binds to natural endogenous endocannabinoid receptors that mediate the direct responses of cannabinoids and endocannabinoid lipid ligands and associated enzymes [66]. Stimulation of the cannabinoid receptor subtype 1 located in multiple brain regions, including the amygdala, basal ganglia, brainstem, cerebellum, cerebral cortex, hippocampus, and hypothalamus, mediates the psychoactive effects of cannabis [66]. Cannabinoid use downregulates cannabinoid receptor subtype 1 in the neocortex and limbic cortices which regulates cognition and emotional processing [66,67]. Cannabis use also acts on the cannabinoid subtype 1 receptors in the ventral striatum, releasing dopamine that controls reward and goal-directed behaviors [68].

Of the 538 natural compounds found in *Cannabis sativa*, more than 100 are phytocannabinoids. Cannabidiol, the most common phytocannabanoid, interacts with multiple receptors, including adenosine, glycine, opioid, serotonin, nonendocannabinoid G protein-coupled, nicotinic acetylcholine, and proliferator-activated receptors. Cannabidiol's medicinal uses include anticonvulsant, anxiolytic, and antinausea for cancer patients.

Tetrahydrocannabinol is the major psychoactive compound in *C. sativa*. Tetrahydrocannabinol acts as a partial agonist of the two primary endocannabinoid receptors, namely cannabinoid type-1 (CB1) receptor and cannabinoid type-2 (CB-2). Tetrahydrocannabinol CB1 receptor modifies psychoactive effects such as anxiety, paranoia, perceptual alternations, and cognitive deficits [69]. Transient CB1 receptor effects result from alterations in the GABA/glutamatergic transmission and dopamine release [70].

Clinical presentation and age-dependent neurobehavioral outcomes

In a study of over 4000 women who delivered live full-term babies, intrauterine cannabis exposure was not associated with a significant reduction in birth weight or head circumference z-scores [46]. Infants enrolled in longitudinal studies of the effects of intrauterine cannabis exposure may also have intrauterine exposure to alcohol, tobacco, and other drugs. The use of both tobacco and cannabis during pregnancy resulted in significantly lower birth weight and smaller head circumference compared to tobacco use alone [46]. Although cannabis exposure alone was not associated with decreased infant weight and head circumference, cannabis may compound the effects of tobacco exposure given cannabis' higher carbon monoxide production compared to tobacco smoke and postulated adverse impact on fetal oxygenation. A meta-analysis of 24 randomized controlled trials, case-control, and cohort studies examining the effects of cannabis use during pregnancy found evidence of an average 109 g reduced birth weight and 77% increased odds of low birth weight infants [71]. No significant differences in neonatal length, head circumference, preterm birth incidence, and small gestational age births were associated with intrauterine cannabis exposure [71].

Neurobehavioral findings during infancy found decreased visual habituation and automatic stability and increased tremors, irritability, startle response, and reflexes [71,72]. At 9 months, infants with intrauterine cannabis scored within the typical range; however, exposed infants had lower scores on the Bayley Scales of Infant Development Mental Developmental Index compared to their nonexposed peers [73]. Inconsistent results from studies of cognition were noted across studies on infants with intrauterine cannabis exposure through age 24 months.

In older children, cognitive studies of intrauterine cannabis exposure found that 3-year-olds had better motor performance on the McCarthy Scales of Children's Abilities but at 4 years, decreased verbal and memory subtests scores were noted [74]. At 6 years, intrauterine cannabis exposure was associated with poorer performance on the Gordon Vigilance test [72].

Comparing children with light versus heavy exposure to cannabis showed an advantage for the light exposure group on the Stanford–Binet composite score, verbal reasoning, quantitative reasoning, and short-term memory. The heavy cannabis exposure group scored in the low average range. However, using stepwise multiple regression modeling to predict Stanford–Binet subscales and controlling for significant sociodemographic and maternal variables revealed that only 1% of the variation was associated with intrauterine cannabis exposure.

Findings of cognitive outcomes for children 9–12 years of age with intrauterine cannabis exposure are variable. In a longitudinal study of children with and without cocaine exposure, 9-year-old children with intrauterine cannabis exposure (with and without intrauterine cocaine, alcohol, and other drugs) had poorer performance on the Wechsler Intelligence scale for children, fourth edition (WISC-IV) processing speed summary score [75]. Notably, this study found that the quality of the home environment and birth mother's vocabulary were associated with many other cognitive measures [75].

Timing of exposure continued to be relevant later for elementary school-aged children. First-trimester cannabis exposure was associated with higher depression and anxiety scores compared to nonexposed peers, although mean scores were not in the clinically significant range [76]. In a longitudinal cohort of 10-year-olds with intrauterine cannabis exposure, second-trimester exposure was associated with a higher percentage of school underachievement and lower (although within the average range) reading comprehension scores compared to nonexposed peers [76].

In adolescence, a longitudinal study of 16-year-old youth with intrauterine cannabis exposure found no differences on composite measures representing general or overall intelligence on unadjusted and adjusted multivariate models using stepwise discriminant function analysis [77].

Variable results continue into older childhood. Spelling [77,78] and reaction time [79] are two examples that demonstrate that different tests that purport to measure the same construct yield different results. Maternal and sociodemographic variables continue to influence the cognitive testing of children who have been exposed to cannabis.

Torres et al., in their extensive systematic review of the extant literature on intrauterine cannabis exposure, caution that the associations between cognitive functioning and cannabis exposure should consider the complex interplay between marijuana exposure, maternal psychological and family status, environment, and sociodemographic factors. Prudent interpretation of study findings and resultant policies should take care not to lead to "punitive policies and enhancing unwarranted stigma....that do more harm than the drug itself" for pregnant women and their children [80].

Resources

https://www.nccih.nih.gov/health/cannabis-marijuana-and-cannabinoids-what-you-need-to-know
https://www.cdc.gov/marijuana/index.htm
https://www.cdc.gov/marijuana/fact-sheets.htm
https://www.samhsa.gov/marijuana
https://store.samhsa.gov/product/preventing-marijuana-use-among-youth/PEP21-06-01-001
https://www.samhsa.gov/marijuana/marijuana-pregnancy

Intrauterine cocaine exposure

Epidemiology

Cocaine among persons aged 12 and older increased from 2.0% in 2002 to 2.5% in 2019 [81]. Among pregnant women aged 15–44, 0.2% reported cocaine use in the past month. While overall substance use in pregnant women has increased in recent years, cocaine and crack cocaine use has decreased.

Neurobiology

Cocaine base and hydrochloride have similar physiological and psychological effects. The psychoactive effects of cocaine include prolonged adrenergic stimulation by presynaptic uptake of the sympathomimetic neurotransmitters—serotonin, dopamine, and norepinephrine. The psychoactive impact of cocaine stems from the extended effect of dopamine on the brain's limbic system and cerebral cortex [82–84]. Women may be disproportionately susceptible to developing stimulant misuse due to hormonal factors and potential impact from socialization and reinforcement of gender constructs. While male and female drug use tends to be similar in prevalence, females may be more sensitive to the reinforcing effects of stimulants [85].

Intrauterine cocaine exposure can negatively affect mothers and infants throughout the perinatal period. Maternal use of cocaine can result in cardiovascular problems, neurological complications, and obstetric complications such as hypertension, myocardial infarction, arrhythmia, sudden death, convulsions, postpartum intracerebral hemorrhage, and fetal death [86]. Cocaine can significantly affect structural development during early gestation in the embryonic stage and subtly influence the developmental process. The teratogenic effects of IUDE involve abnormal growth and maturation, alterations in neurotransmitters and receptors, executive functioning, and brain organization [60,82,87,88]. Neurologic abnormalities in cocaine-exposed infants include smaller head circumference at birth, poor visual attention, delayed intrauterine growth and motor development, tone and movement abnormalities, and smaller occipitofrontal circumference [88,89].

Diagnostic evaluation

Toxicology and self-report measures are typically used to identify maternal cocaine and opiate use [86,88,90]. Urine toxicology screens are typically reliable; however, false negatives are common for cocaine use [91]. The collection of additional biological specimens such as hair samples and meconium analysis can also help determine IUDE [91,92].

The NICU Network Neurobehavioral Scale (NNNS) was developed for infants with exposures that place them at biological and medical risk for low birth weight, prematurity, illness, or injury at birth, as well as socio-environmental risks such as maternal stress, IUDE, and poor prenatal care [93]. It can be used to identify and assess the neurobehavioral outcomes of intrauterine cocaine exposure. The NNNS is a valuable screening tool for neurodevelopment and neurobehavioral assessment but is not recommended to evaluate treatment needs [93].

Clinical presentation and age-dependent neurobehavioral outcomes

Prenatal cocaine exposure can directly affect a neonate's neurobehavioral development through the effects of cocaine and its metabolites, and indirectly through cerebral hypoxia, cerebral vascular compromise, fetal malnutrition, or prematurity [90]. Early signs of behavioral and autonomic disturbance resulting from prenatal cocaine exposure using the Brazelton Newborn Behavioral Assessment Scale include increased irritability and lability, reduced behavioral and autonomic regulation, and lack of orientation and alertness [87]. Additional neurobehavioral symptoms include hyperreflexia, excessive tremors, poor feeding, and sleep disturbance [86,88,90,94]. Prenatal cocaine exposure has also been associated with abnormalities in temperament and social reactivity within infants. Cocaine-exposed infants were found to have low affectivity and flatter temperament styles, including decreased reactivity to nonsocial contexts and increased reactivity to more social contexts. Infants with cocaine exposure also tended to need more soothing time in response to distress than nonexposed infants [95].

In addition to the neurobiological consequences of prenatal cocaine exposure, several biopsychosocial consequences result from maternal-infant interaction and environmental exposures connected to maternal drug use [96–98]. Women who engage in substance abuse display atypical parenting and behavioral patterns that impact the biopsychosocial development of the child [96,99,100]. Several studies have explored the impact of prenatal cocaine exposure on maternal-infant attachment and interaction. Infants with IUDE are more likely to experience disruptions in maternal care than nonexposed infants [96]. The presence of high-risk behavior, environmental instability, reduced inhibitory control, psychological stress, and emotion dysregulation can lead to maladaptive mother-infant interaction [97,98].

In childhood, due to structural and functional alterations resulting from perinatal cocaine use, cognitive deficits such as inattention, impulsivity, self-regulation, and other executive functioning problems can appear through internalizing and externalizing behaviors [89,101–104]. Delayed mental/developmental outcomes have been noted [89,105] as have language delays [106]. Internalizing behavior patterns have also been identified in children with prenatal cocaine exposure.

On the one hand, the quality of parenting, maternal harshness, and environmental risks mediate the association between prenatal cocaine exposure and internalizing behaviors in kindergarten children [102]. The culmination of early life adversity and sociopolitical determinants disproportionately place children of women with substance use disorder at risk for childhood maltreatment. Maternal childhood trauma, quality of parenting, environmental risk and exposures, and cooccurring psychological disorders are common among women with substance use disorders and pose additional risks for child maltreatment for their children [99].

Prenatal cocaine exposure may indirectly be associated with increased neurobehavioral challenges in children through exposure to violence due to the large proportion of children in high-risk neighborhoods with substance-using parents [101]. Exposure to violence among young children is also common through maternal violence exposure (i.e., intimate partner violence, community violence). Increased maternal harshness and decreased sensitivity/warmth are higher in women using cocaine with early life trauma and adversity [99,101]. Lower sociability, withdrawn behaviors, and depressive symptoms have been reported among children with prenatal cocaine exposure and violence exposure [104]. Children with adoption history or involvement in the child welfare system as a result of prenatal cocaine use and environmental risks were found to have higher rates of aggression, delinquent behavior, socioemotional challenges, cognitive deficits, and attention challenges than children with kinship at birth or noncocaine exposed children [106].

In youth, given the impact of cocaine exposure on stress response and the likelihood of these youth growing up in unstable environments, the interaction between neurobehavioral disinhibition and allostatic load resulting from early adversity may increase vulnerability to high-risk behavioral trajectories [107]. Previous research has demonstrated a relationship between prenatal cocaine exposure and substance use in adolescence and emerging adulthood [108–110]. A study of 365 adolescents found that prenatal cocaine exposure and environmental risk factors are independently associated with substance use.

Intrauterine opiate exposure

Epidemiology

In recent years, the use of prescription and illicit opioids has increased to epidemic levels [111]. Among persons ages 12 and older in 2019, 3.7% misused opioids in the past year. The majority are users of prescription pain relievers, 92.6% of opioid misusers report pain reliever misuse only, 3.4% of opioid misusers report heroin use only, and 4.0% of opioid misusers report both heroin and pain reliever misuse. Opioid abuse for pregnant women has increased exponentially over the years. Between 1998 and 2011, the rate of pregnant women with an opioid disorder increased by 127% [112]. Overall, illicit drug use in the past month for pregnant women in 2019 was 5.8%. Among pregnant women between the ages of 15 and 44, 0.9% actively used opiates in the last 30 days. Pregnant women who use opiates are at risk for several issues accompanying dependence, including deleterious health outcomes for the infant and mother [113].

Neurobiology

Fetal effects from opiate use may consist of intrauterine growth restriction, low birth weight, preterm birth, and a higher risk for mortality and morbidity for both the infant and the mother [113,114]. Additional maternal effects may include cardiac arrest, placental abruption, toxemia, third-trimester bleeding, and transfusion requirements [81,112,115]. Pregnant women with substance use disorder are more likely to have a history of early life trauma and/or comorbid psychiatric conditions and involvement in polysubstance use. The combined effects of maternal opiate use and other substance can lead to several adverse outcomes for the infant [115].

Clinical presentation: age-dependent neurobehavioral outcomes

Prenatal opiate exposure can have a long-lasting impact on development from infancy through adulthood. Maternal opiate use can have a negative effect on caregiver sensitivity and responsiveness [116]. Infants with prenatal opiate exposure are likely to be born with withdrawal symptoms that lead to neurobehavioral dysfunction. While much of the research on infancy focuses on neonatal withdrawal, the biopsychosocial aspects of maternal drug use can lead to behavioral and psychiatric consequences for the developing child [98]. It is important to note that polysubstance abuse is a common occurrence with opiate use. Much of the literature has examined the impact of pregnant women using heroin which may have differential environmental risk and neonatal outcomes than prescription opioids [117].

Congenital teratogenic effects of opiates and polysubstance use include oral cleft, clubfoot, and atrial septal defects. While congenital effects are rare, more research is needed to determine the extent to which opioids are associated with structural defects [115,118].

Neonatal abstinence syndrome (NAS) is a behavioral syndrome associated with withdrawal experienced by infants exposed to prenatal opiate use [116]. NAS is characterized by autonomic nervous system disruption, gastrointestinal upset, and CNS hyperirritability [119,120]. Infant withdrawal symptoms can include hyperthermia, sleep disturbance, tremors, irritability, excessive crying, yawning, sweating, nasal congestion and sneezing, and softened stool [115,119,121]. Additional neurologic effects include difficulty with gait, reflex, coordination, balance, and left and right orientation [122].

The Finnegan Neonatal Abstinence Score is the most commonly used assessment tool to determine the severity of NAS in opiate-exposed infants. The scoring system includes 31 items used for assessment throughout early infancy. Three or more scores higher than 8 or a single score higher than 12 necessitates pharmacologic treatment. The Lipsitz scoring system is also commonly used for assessment. The Lipsitz includes 11 items and requires a score of 4 or higher for pharmacologic therapy. Though less frequently used, it is reliable in identifying pharmacologic treatment needs [123]. Meconium urine screens are used to determine newborn substance exposure [119].

Infants of mothers who engage in drug use are at higher risk for child maltreatment and adversity [99].

Children with prenatal opiate exposure are more likely to develop disruptive behavior, attention deficit disorders, and other psychiatric disorders. Opiate exposed children have experienced reduced cognitive ability and increased behavior problems and are known to have smaller brains and thinner cortex [119]. Opiate-exposed children are more likely to have challenges with executive functioning, dysregulated behavioral response, psychomotor dysfunction, sensory processing, language, and cognitive abilities. School-aged children exposed to opiates prenatally experienced increased anxiety and aggression compared to nonexposed children [119,124]. Similar to cocaine exposure, children with prenatal opiate exposure may also experience internalizing and externalizing behavior patterns. Frequent changes in caregivers, environmental exposures of risk or violence, and quality of caregiving also contribute to the neurobehavioral effects of opiate use in children [86,124,125].

While there is minimal research examining the neurobehavioral outcomes of prenatal opiate exposure alone for youth and young adults, much of the research demonstrates that prenatal exposure to drug use increases the vulnerability for substance use in adolescence and early adulthood [126]. Similar to children with cocaine exposure, opiate exposure youth may continue to have difficulty with self-regulation, disinhibition, and executive functioning [122]. These socio-emotional, behavioral, and cognitive deficits can increase the risk of engaging in substance use or other high-risk behaviors [107]. Drug-exposed youth may be more vulnerable to engaging in high-risk behavior, including truancy, early-onset sexual activity, and fighting. Exposures such as poverty, community violence, familial stress, parental separation, and other environmental risks can increase the likelihood of psychiatric morbidity, substance use, and different adverse health outcomes [107,127].

Management of intrauterine alcohol and drug exposure

Early intervention for children who are at risk for or have already been exposed prenatally begins with the mother. Continuing even small exposures have been shown to be associated with adverse outcomes [46]. Mothers who engage use of prenatal toxins may be reluctant to engage in treatment due to lack of access to information and fear of legal action that would risk the loss of children to the social services. Stigma, negative experiences within the healthcare system, lack of childcare, lack of housing, and other social and structural barriers reduce access to proper treatment [128]. A comprehensive model of care, including incentives, review of comorbidities, and trauma-informed and gender-specific treatment modalities, can improve health outcomes for mothers and diminish continuing exposures [15,82,128].

Cessation of in utero exposure to toxins seems to have the best outcomes. Studies suggest that ceasing tobacco use during pregnancy may reduce the risk of adverse neurobehavioral outcomes. There are several treatment methods for mothers with opioid use disorder during the prenatal and postnatal periods [129].

Establishing a stable social climate for the family is a major goal of an overall management program. Having a supportive and nurturing family is key to the outcome of a child with FAS [4]. This might entail enlisting the aid of other family members, church, and social service agencies to address the issues listed above. Referrals of parents for mental health assessments and therapy may be needed. Screening of siblings for developmental and behavioral dysfunction with referral for further evaluation and therapy for children who screen positive.

Addressing the medical needs of the infant and child who has been exposed to prenatal toxins is a necessary part of a comprehensive program. NAS may be seen in neonatal period with symptoms extending into the first year of life. Structural and physiologic impairments may require consultations from cardiology, ophthalmology, otolaryngology, audiology, neurology, and mental health professionals.

Behavioral and education interventions are a major component of a comprehensive program. Toxin-exposed infants should be referred for Individuals with Disabilities Education Act (IDEA) Part C services. Case management services provided through Part C will support parents in establishing a stable social climate by making the parents aware of existing programs to enable the family to reach its goals, making appropriate referrals, assisting with the application processes, and by providing useful information. Speech and language, occupational, and physical therapy services may be required as the child ages. Part C programs also provide ongoing monitoring of the child's developmental progress.

Continued monitoring of academic performance is needed. As the educational demands increase through school, the child with prenatal toxic exposures may have increasing difficulty. ADHD, specific learning disabilities, intellectual disability, social (pragmatic) language disorders, autism spectrum, and anxiety disorders mood and other regulatory disorders may manifest. The school-based Individualized Education Plan (IEP) team may need to collaborate with the medical team (primary care provider, mental health team) and the social services providers to develop an effective program.

Symptom management is a primary focus of pharmacologic treatment, often targeting hyperactivity, impulsivity, mood lability, and attention difficulties. Stimulants, antidepressants, and anxiolytic medication can help treat a patient with hyperactivity, depression, or anxiety, respectively, while neuroleptic medication can treat mood lability or aggressive behavior [130]. As in most neurodevelopmental disorders, the indications of psychopharmacological interventions should not be the only choice for treatment. Pharmacologic treatment should be used as part of a comprehensive management program.

Summary

Many children with FASD go unidentified or are misdiagnosed. Children with FASD who do not display phenotypic features associated with FASD are most at risk for delayed or misdiagnosis. It is imperative for pediatric healthcare providers to obtain a complete family, perinatal, and psychosocial history, including questions about alcohol, prescription drug use, and illicit drug use in both biological parents and family, regardless of the family's socioeconomic status. If the child is adopted or in foster care, efforts should be made to obtain birth and pregnancy records.

Mothers with a history of alcohol and drug use benefit from comprehensive integrated psychosocial and substance abuse treatment services, including childcare resources when possible. Children with FASD and IUDE benefit from early identification and referrals to early intervention services for evaluation and treatment tailored to the child's individual developmental and behavior needs, or close monitoring if the child is growing and meeting developmental milestones. Monitoring of academic skills should continue through elementary school to identify later onset learning disabilities or behavior disorders.

The most important intervention is prevention! Prevention strategies may occur at the policy, community, clinic, and family level. Access to enriched structured early education services and childcare benefit all children. Safe neighborhoods, with affordable housing, livable wages, and strong education systems are protective factors against alcohol and drug use. Prevention of child neglect and maltreatment by providing trauma-informed evidence-based culturally responsive social supports, interventions, and mental health services supports recovery, health, and wellness for families with identified needs and may reduce the likelihood of intergenerational alcohol and drug dependence.

References

[1] Ray O, Ksir CD. Society and human behavior. 7th ed. Boston, MA: McGraw-Hill; 1996.

[2] Chudley AE, Conry J, Cook JL, Loock C, Rosales T, LeBlanc N. Fetal alcohol spectrum disorder: Canadian guidelines for diagnosis. Can Med Assoc J 2005;172(5 suppl):S1.

[3] Centers for Disease Control and Prevention. Facts about fetal alcohol spectrum disorders; 2010 [Available from: http://www.cdc.gov/ncbddd/fasd/facts.html].

[4] Centers for Disease Control and Prevention. Alcohol use and binge drinking among women of childbearing age – United States, 2006–2010. MMWR Morbidity Mortal Wkly Rep 2012;61(28):534–8.

[5] Popova S, Lange S, Probst C, Gmel G, Rehm J. Estimation of national, regional, and global prevalence of alcohol use during pregnancy and fetal alcohol syndrome: a systematic review and meta-analysis. Lancet Glob Health 2017;5(3):e290–9.

[6] Sampson PD, Streissguth AP, Bookstein FL, Little RE, Clarren SK, Dehaene P, et al. Incidence of fetal alcohol syndrome and prevalence of alcohol-related neurodevelopmental disorder. Teratology. 1997;56(5):317–26.

[7] May PA, Gossage JP. Estimating the prevalence of fetal alcohol syndrome. A summary. Alcohol Res Health 2001;25(3):159–67.

[8] England LJ, Bennett C, Denny CH, Honein MA, Gilboa SM, Kim SY, et al. Alcohol use and co-use of other substances among pregnant females aged 12-44 years – United States, 2015–2018. MMWR Morb Mortal Wkly Rep 2020;69(31):1009–14.

[9] Norman AL, Crocker N, Mattson SN, Riley EP. Neuroimaging and fetal alcohol spectrum disorders. Dev Disabil Res Rev 2009;15(3):209–17.

[10] O'Leary-Moore SK, Parnell SE, Lipinski RJ, Sulik KK. Magnetic resonance-based imaging in animal models of fetal alcohol spectrum disorder. Neuropsychol Rev 2011;21(2):167–85.

[11] Goodlett CR, Horn KH. Mechanisms of alcohol-induced damage to the developing nervous system. Alcohol Res Health 2001;25(3):175–84.

[12] Acevedo CG, Carrasco G, Burotto M, Rojas S, Bravo I. Ethanol inhibits L-arginine uptake and enhances NO formation in human placenta. Life Sci 2001;68(26):2893–903.

[13] Kay HH, Grindle KM, Magness RR. Ethanol exposure induces oxidative stress and impairs nitric oxide availability in the human placental villi: a possible mechanism of toxicity. Am J Obstet Gynecol 2000;182(3):682–8.

[14] Underwood MA, Gilbert WM, Sherman MP. Amniotic fluid: not just fetal urine anymore. J Perinatol 2005;25(5):341–8.

[15] Landgraf MN, Nothacker M, Kopp IB, Heinen F. The diagnosis of fetal alcohol syndrome. Dtsch Arzteblatt Int 2013;110(42):703–10.

[16] Astley S. Profile of the first 1,400 patients receiving diagnostic evaluations for fetal alcohol spectrum disorder at the Washington State Fetal Alcohol Syndrome Diagnostic and Prevention Network. Can J Clin Pharmacol 2010;17(1):e132–64.

[17] Astley SJ. Canadian palpebral fissure length growth charts reflect a good fit for two school and FASD clinic-based U.S. populations. J Popul Ther Clin Pharmacol 2011;18(2):e231.

[18] Streissguth AP, Finnegan LP. Effects of prenatal alcohol and drugs. In: Kinney J, editor. Clinical manual of substance abuse. 2nd ed. St. Louis: Mosby-Year Book Inc; 1996. p. 254–71.

[19] Stevens SA, Major D, Rovet J, Koren G, Fantus E, Nulman I, et al. Social problem solving in children with fetal alcohol spectrum disorders. J Popul Ther Clin Pharmacol 2012;19(1):e99–e110.

[20] Greenbaum RL, Stevens SA, Nash K, Koren G, Rovet J. Social cognitive and emotion processing abilities of children with fetal alcohol spectrum disorders: a comparison with attention deficit hyperactivity disorder. Alcohol Clin Exp Res 2009;33(10):1656–70. Available from: https://doi.org/10.1111/j.1530-0277.2009.01003.x.

[21] Kodituwakku PW. Defining the behavioral phenotype in children with fetal alcohol spectrum disorders: a review. Neurosci Biobehav Rev 2007;31(2):192–201.

[22] Stade B, Beyene J, Buller K, Ross S, Patterson K, Stevens B, et al. Feeling different: the experience of living with fetal alcohol spectrum disorder. J Popul Ther Clin Pharmacol 2011;18(3):e475–85.

[23] Cook JL, Green CR, Lilley CM, Anderson SM, Baldwin ME, Chudley AE, et al. Fetal alcohol spectrum disorder: a guideline for diagnosis across the lifespan. CMAJ 2016;188(3):191–7. Available from: https://doi.org/10.1503/cmaj.141593.

[24] Moore EM, Glass L, Infante MA, Coles CD, Kable JA, Jones KL, et al. Cross-sectional analysis of spatial working memory development in children with histories of heavy prenatal alcohol exposure. Alcohol Clin Exp Res 2021;45(1):215–23. Available from: https://doi.org/10.1111/acer.14506.

[25] Ergun G, Schultz MS, Rettig EK. Fetal alcohol spectrum disorder-issues of misdiagnosis and missed diagnosis in black youth: a case report. Innov Clin Neurosci 2021;18(4-6):20–3.

[26] Lange S, Rovet J, Rehm J, Popova S. Neurodevelopmental profile of fetal alcohol spectrum disorder: a systematic review. BMC Psychol 2017;5(1):22. Available from: https://doi.org/10.1186/s40359-017-0191-2.

[27] McLachlan K, Flannigan K, Temple V, Unsworth K, Cook JL. Difficulties in daily living experienced by adolescents, transition-aged youth, and adults with fetal alcohol spectrum disorder. Alcohol Clin Exp Res 2020;44(8):1609–24. Available from: https://doi.org/10.1111/acer.14385.

[28] Greenspan S, Novick Brown N, Edwards W. FASD and the concept of intellectual disability equivalence. In: Nelson M, Trussler M, editors. Fetal alcohol spectrum disorders in adults: ethical and legal perspectives. Springer; 2016. p. 241–66.

[29] Novick Brown N, Greenspan S. Diminished culpability in fetal alcohol spectrum disorders (FASD). Behav Sci Law 2022;40(1):1–13. Available from: https://doi.org/10.1002/bsl.2535.

[30] Wetherill L, Foroud T, Goodlett C. Meta-analyses of externalizing disorders: genetics or prenatal alcohol exposure? Alcohol Clin Exp Res 2018;42(1):162–72. Available from: https://doi.org/10.1111/acer.13535.

[31] Coles CD, Kalberg W, Kable JA, Tabachnick B, May PA, Chambers CD. Characterizing alcohol-related neurodevelopmental disorder: prenatal alcohol exposure and the spectrum of outcomes. Alcohol Clin Exp Res 2020;44(6):1245–60. Available from: https://doi.org/10.1111/acer.14325.

[32] Hoyme HE, May PA, Kalberg WO, Kodituwakku P, Gossage JP, Trujillo PM, et al. A practical clinical approach to diagnosis of fetal alcohol spectrum disorders: clarification of the 1996 institute of medicine criteria. Pediatrics. 2005;115(1):39–47.

[33] May PA, Tabachnick B, Hasken JM, Marais A-S, de Vries MM, Barnard R, et al. Who is most affected by prenatal alcohol exposure: boys or girls? Drug Alcohol Dependence 2017;177:258–67.

[34] Bertrand J, Floyd R, Weber M, O'Connor M, Riley E, Johnson K, et al. National Task Force on FAS/FAE. Fetal alcohol syndrome: guidelines for referral and diagnosis Atlanta, GA. Cent Dis Control Prev 2004;4.

[35] Jones KL. Fetal alcohol syndrome. Smith's recognizable patterns of human malformation. 5th ed. Philadelphia: W.B. Saunders; 1997.

[36] Drope J, Schluger N, Cahn Z, Drope J, Hamill S, Islami F, et al. The tobacco atlas. 6th ed. Atlanta: American Cancer Society, Inc and Vital Strategies; 2018Available from. Available from: https://tobaccoatlas.org/2018/12/14/what-is-tobacco/.

[37] Habibagahi A, Alderman N, Kubwabo C. A review of the analysis of biomarkers of exposure to tobacco and vaping products. Anal Methods 2020;12(35):4276–302.

[38] Heron M. Deaths: leading causes for 2019. Natl Vital Stat Rep 2021;70(9):1–114.

[39] Cornelius ME, Wang TW, Jamal A, Loretan CG, Neff LJ. Tobacco product use among adults – United States, 2019. MMWR Morb Mortal Wkly Rep 2020;69(46):1736–42.

[40] Azagba S, Manzione L, Shan L, King J. Trends in smoking during pregnancy by socioeconomic characteristics in the United States, 2010–2017. BMC Pregnancy Childbirth 2020;20(1):52.

[41] Muhammad A, Mychasiuk R, Nakahashi A, Hossain SR, Gibb R, Kolb B. Prenatal nicotine exposure alters neuroanatomical organization of the developing brain. Synapse. 2012;66(11):950–4.

[42] Falk L, Nordberg A, Seiger A, Kjaeldgaard A, Hellstrom-Lindahl E. Smoking during early pregnancy affects the expression pattern of both nicotinic and muscarinic acetylcholine receptors in human first trimester brainstem and cerebellum. Neuroscience. 2005;132(2):389–97.

[43] Bailey CD, Tian MK, Kang L, O'Reilly R, Lambe EK. Chrna5 genotype determines the long-lasting effects of developmental in vivo nicotine exposure on prefrontal attention circuitry. Neuropharmacology. 2014;77:145–55.

[44] Crume T. Tobacco use during pregnancy. Clin Obstet Gynecol 2019;62(1):128–41.

[45] Dietz PM, England LJ, Shapiro-Mendoza CK, Tong VT, Farr SL, Callaghan WM. Infant morbidity and mortality attributable to prenatal smoking in the U.S. Am J Prevent Med 2010;39(1):45–52.

[46] Sturrock S, Williams E, Ambulkar H, Dassios T, Greenough A. Maternal smoking and cannabis use during pregnancy and infant outcomes. J Perinat Med 2020;48(2):168–72.

[47] Hammoud AO, Bujold E, Sorokin Y, Schild C, Krapp M, Baumann P. Smoking in pregnancy revisited: findings from a large population-based study. Am J Obstet Gynecol 2005;192(6):1856–62.

[48] U.S. Department of Health and Human Services. The health consequences of smoking: a report of the surgeon general, Chapter 5. Reproductive effects. Atlanta, GA; 2004.

[49] Colquhoun DR, Goldman LR, Cole RN, Gucek M, Mansharamani M, Witter FR, et al. Global screening of human cord blood proteomes for biomarkers of toxic exposure and effect. Env Health Perspect 2009;117(5):832–8.

[50] Abel EL. Smoking during pregnancy: a review of effects on growth and development of offspring. Hum Biol 1980;52(4):593–625.

[51] Froggatt S, Reissland N, Covey J. The effects of prenatal cigarette and e-cigarette exposure on infant neurobehaviour: a comparison to a control group. EClinicalMedicine. 2020;28:100602.

[52] Cornelius MD, Goldschmidt L, DeGenna N, Day NL. Smoking during teenage pregnancies: effects on behavioral problems in offspring. Nicotine Tob Res 2007;9(7):739–50.

[53] Cornelius MD, Ryan CM, Day NL, Goldschmidt L, Willford JA. Prenatal tobacco effects on neuropsychological outcomes among preadolescents. J Dev Behav Pediatr 2001;22(4):217–25.

[54] D'Onofrio BM, Van Hulle CA, Waldman ID, Rodgers JL, Harden KP, Rathouz PJ, et al. Smoking during pregnancy and offspring externalizing problems: an exploration of genetic and environmental confounds. Dev Psychopathol 2008;20(1):139–64.

[55] Gilman SE, Gardener H, Buka SL. Maternal smoking during pregnancy and children's cognitive and physical development: a causal risk factor? Am J Epidemiol 2008;168(5):522–31.

[56] Rose-Jacobs R, Richardson MA, Buchanan-Howland K, Chen CA, Cabral H, Heeren TC, et al. Intrauterine exposure to tobacco and executive functioning in high school. Drug Alcohol Depend 2017;176:169–75.

[57] Toro R, Leonard G, Lerner JV, Lerner RM, Perron M, Pike GB, et al. Prenatal exposure to maternal cigarette smoking and the adolescent cerebral cortex. Neuropsychopharmacology. 2008;33(5):1019–27.

[58] Cornelius MD, Goldschmidt L, Day NL. Prenatal cigarette smoking: long-term effects on young adult behavior problems and smoking behavior. Neurotoxicol Teratol 2012;34(6):554–9.

[59] Pickett KE, Wood C, Adamson J, D'Souza L, Wakschlag LS. Meaningful differences in maternal smoking behaviour during pregnancy: implications for infant behavioural vulnerability. J Epidemiol Community Health 2008;62(4):318–24.

[60] Forray A, Foster D. Substance use in the perinatal period. Curr Psychiatry Rep 2015;17(11):91.

[61] World Health Organization. KD35 neonatal withdrawal syndrome from maternal use of drugs of addiction. Geneva, Switzerland; 2019.

[62] Butz AM, Pulsifer MB, Belcher HME, Leppert M, Donithan M, Zeger S. Infant head growth and cognitive status at 36 months in children with in-utero drug exposure. J Child Adolesc Substance Abuse 2005;14(4):15–39.

[63] Substance Abuse and Mental Health Services Administration. Key substance use and mental health indicators in the United States: results from the 2020 national survey on drug use and health. Rockville, MD; 2021. Report no. HHS, Publication no. PEP21-07-01-003, NSDUH Series H-56.

[64] Center for Behavioral Health Statistics and Quality National Survey on Drug Use and Health. Results from the 2020 national survey on drug use and health: detailed tables; 2021, Substance Abuse and Mental. Rockville MD: Health Services Administration.

[65] Ko JY, Coy KC, Haight SC, Haegerich TM, Williams L, Cox S, et al. Characteristics of marijuana use during pregnancy – eight states, pregnancy risk assessment monitoring system, 2017. MMWR Morb Mortal Wkly Rep 2020;69(32):1058–63.

[66] Ferland JN, Hurd YL. Deconstructing the neurobiology of cannabis use disorder. Nat Neurosci 2020;23(5):600–10.

[67] Ceccarini J, Kuepper R, Kemels D, van Os J, Henquet C, Van Laere K. [18F]MK-9470 PET measurement of cannabinoid CB1 receptor availability in chronic cannabis users. Addict Biol 2015;20(2):357–67.

[68] Wijayendran SB, O'Neill A, Bhattacharyya S. The effects of cannabis use on salience attribution: a systematic review. Acta Neuropsychiatr 2018;30(1):43–57.

[69] Bonini SA, Premoli M, Tambaro S, Kumar A, Maccarinelli G, Memo M, et al. Cannabis sativa: a comprehensive ethnopharmacological review of a medicinal plant with a long history. J Ethnopharmacol 2018;227:300–15.

[70] Pertwee RG. The diverse CB1 and CB2 receptor pharmacology of three plant cannabinoids: delta9-tetrahydrocannabinol, cannabidiol and delta9-tetrahydrocannabivarin. Br J pharmacology 2008;153(2):199–215.

[71] Gunn JK, Rosales CB, Center KE, Nunez A, Gibson SJ, Christ C, et al. Prenatal exposure to cannabis and maternal and child health outcomes: a systematic review and meta-analysis. BMJ Open 2016;6(4):e009986.

[72] Fried PA, Watkinson B, Dillon RF, Dulberg CS. Neonatal neurological status in a low-risk population after prenatal exposure to cigarettes, marijuana, and alcohol. J Dev Behav Pediatr 1987;8(6):318–26.

[73] Richardson GA, Day NL, Goldschmidt L. Prenatal alcohol, marijuana, and tobacco use: infant mental and motor development. Neurotoxicol Teratol. 1995;17(4):479–87.

[74] Fried PA, Watkinson B. 36- and 48-month neurobehavioral follow-up of children prenatally exposed to marijuana, cigarettes, and alcohol. J Dev Behav Pediatr 1990;11(2):49–58.

[75] Singer LT, Nelson S, Short E, Min MO, Lewis B, Russ S, et al. Prenatal cocaine exposure: drug and environmental effects at 9 years. J Pediatr 2008;153(1):105–11.

[76] Goldschmidt L, Richardson GA, Cornelius MD, Day NL. Prenatal marijuana and alcohol exposure and academic achievement at age 10. Neurotoxicol Teratol 2004;26(4):521–32.

[77] Fried PA, Watkinson B, Gray R. Differential effects on cognitive functioning in 13- to 16-year-olds prenatally exposed to cigarettes and marihuana. Neurotoxicol Teratol 2003;25(4):427–36.

[78] Rose-Jacobs R, Augustyna M, Beeghly M, Martin B, Cabral HJ, Heeren TC, et al. Intrauterine substance exposures and Wechsler Individual Achievement Test-II scores at 11 years of age. Vulnerable children and youth studies: an international interdisciplinary journal for research, policy care. 2012. p. 1–12.

[79] Willford JA, Chandler LS, Goldschmidt L, Day NL. Effects of prenatal tobacco, alcohol and marijuana exposure on processing speed, visual-motor coordination, and interhemispheric transfer. Neurotoxicol Teratol 2010;32(6):580–8.

[80] Torres CA, Medina-Kirchner C, O'Malley KY, Hart CL. Totality of the evidence suggests prenatal cannabis exposure does not lead to cognitive impairments: a systematic and critical review. Front Psychol 2020;11:816.

[81] Substance Abuse and Mental Health Services Administration. Results from the 2019 National Survey on Drug Use and Health: Detailed Table; 2019 Available from: https://www.samhsa.gov/data/release/2019-national-survey-drug-use-and-health-nsduh-releases

[82] Smid MC, Metz TD, Gordon AJ. Stimulant use in pregnancy: an under-recognized epidemic among pregnant women. Clin Obstet Gynecol 2019;62(1):168–84.

[83] Ciccarone D. Stimulant abuse: pharmacology, cocaine, methamphetamine, treatment, attempts at pharmacotherapy. Prim Care 2011;38(1):41–58.

[84] Marwah J. Frontiers in cocaine research. Can J Physiol Pharmacol 1989;67(9):1153.

[85] Roth ME, Cosgrove KP, Carroll ME. Sex differences in the vulnerability to drug abuse: a review of preclinical studies. Neurosci Biobehav Rev 2004;28(6):533–46.

[86] Hetea A, Cosconel C, Stanescu AAM, Simionescu AA. Alcohol and psychoactive drugs in pregnancy. Maedica 2019;14(4):397–401.

[87] Behnke M, Smith VC, Committee On Substance A, Committee On F, Newborn, Behnke M, et al. Prenatal substance abuse: short- and long-term effects on the exposed fetus. Pediatrics. 2013;131(3):e1009–24.

[88] Chiriboga CA, Bateman DA, Brust JCM, Allen Hauser W. Neurologic findings in neonates with intrauterine cocaine exposure. Pediatric Neurol 1993;9(2):115–19.

[89] Min MO, Minnes S, Lang A, Yoon S, Singer LT. Effects of prenatal cocaine exposure on early sexual behavior: gender difference in externalizing behavior as a mediator. Drug Alcohol Depend 2015;153:59–65.

[90] Neuspiel DR, Hamel SC. Cocaine and infant behavior. J Dev Behav Pediatr 1991;12(1):55–64.

[91] Pesce A, Rosenthal M, West R, West C, Crews B, Mikel C, et al. An evaluation of the diagnostic accuracy of liquid chromatography-tandem mass spectrometry versus immunoassay drug testing in pain patients. Pain Physician 2010;13(3):273–81.

[92] González-Colmenero E, Concheiro-Guisán A, Lorenzo-Martínez M, Concheiro M, Lendoiro E, de-Castro-Ríos A, et al. Drug testing in biological samples vs. maternal surveys for the detection of substance use during whole pregnancy. J Addict Dis 2021;39(2):175–82.

[93] Sucharew H, Khoury JC, Xu Y, Succop P, Yolton K. NICU network neurobehavioral scale profiles predict developmental outcomes in a low-risk sample. Paediatr Perinat Epidemiol 2012;26(4):344–52.

[94] LaGasse LL, Messinger D, Lester BM, Seifer R, Tronick EZ, Bauer CR, et al. Prenatal drug exposure and maternal and infant feeding behaviour. Arch Dis Child - Fetal Neonatal Ed 2003;88(5):F391.

[95] Locke RL, Lagasse LL, Seifer R, Lester BM, Shankaran S, Bada HS, et al. Effects of prenatal substance exposure on infant temperament vary by context. Dev Psychopathol 2016;28(2):309–26.

[96] Beeghly M, Frank DA, Rose-Jacobs R, Cabral H, Tronick E. Level of prenatal cocaine exposure and infant–caregiver attachment behavior. Neurotoxicol Teratol 2003;25(1):23–38.

[97] Porreca A, De Palo F, Simonelli A, Capra N. Attachment representations and early interactions in drug addicted mothers: a case study of four women with distinct adult attachment interview classifications. Front Psychol 2016;7(346).

[98] Cataldo I, Azhari A, Coppola A, Bornstein MH, Esposito G. The influences of drug abuse on mother-infant interaction through the lens of the biopsychosocial model of health and illness: a review. Front Public Health 2019;7(45).

[99] Hatzis D, Dawe S, Harnett P, Loxton N. An investigation of the impact of childhood trauma on quality of caregiving in high risk mothers: does maternal substance misuse confer additional risk? Child Psychiatry Hum Dev 2019;50(5):835–45.

[100] Richardson GA, Goldschmidt L, Willford J. The effects of prenatal cocaine use on infant development. Neurotoxicol Teratol 2008;30(2):96–106.

[101] Veira Y, Finger B, Schuetze P, Colder CR, Godleski S, Eiden RD. Child behavior problems: role of cocaine use, parenting, and child exposure to violence. Psychol Violence 2014;4(3):266–80.

[102] Eiden RD, Godleski S, Colder CR, Schuetze P. Prenatal cocaine exposure: the role of cumulative environmental risk and maternal harshness in the development of child internalizing behavior problems in kindergarten. Neurotoxicol Teratol 2014;44:1–10.

[103] Min MO, Albert JM, Lorincz-Comi N, Minnes S, Lester B, Momotaz H, et al. Prenatal substance exposure and developmental trajectories of internalizing symptoms: toddlerhood to preadolescence. Drug Alcohol Depend 2021;218:108411.
[104] Richardson JB, Brown J, Brakle MV. Pathways to early violent death: the voices of serious violent youth offenders. Am J Public Health 2013;103(7):e5–e16.
[105] Min MO, Minnes S, Kim J-Y, Yoon M, Singer LT. Association of prenatal cocaine exposure, childhood maltreatment, and responses to stress in adolescence. Drug Alcohol Depend 2017;177:93–100.
[106] Singer LT, Minnes S, Min MO, Lewis BA, Short EJ. Prenatal cocaine exposure and child outcomes: a conference report based on a prospective study from Cleveland. Hum Psychopharmacol 2015;30(4):285–9.
[107] Fisher PA, Lester BM, DeGarmo DS, Lagasse LL, Lin H, Shankaran S, et al. The combined effects of prenatal drug exposure and early adversity on neurobehavioral disinhibition in childhood and adolescence. Dev Psychopathol 2011;23(3):777–88.
[108] Rose-Jacobs R, Soenksen S, Appugliese DP, Cabral HJ, Richardson MA, Beeghly M, et al. Early adolescent executive functioning, intrauterine exposures and own drug use. Neurotoxicol Teratol 2011;33(3):379–92.
[109] Bennett DS, Lewis M. Does prenatal cocaine exposure predict adolescent substance use? Neurotoxicol Teratol 2020;81:106906.
[110] Delaney-Black V, Chiodo LM, Hannigan JH, Greenwald MK, Janisse J, Patterson G, et al. Prenatal and postnatal cocaine exposure predict teen cocaine use. Neurotoxicol Teratol 2011;33(1):110–19.
[111] Rudd RA, Seth P, David F, Scholl L. Increases in drug and opioid-involved overdose deaths—United States, 2010–2015. MMWR Morbidity Mortal Wkly Rep 2016;65(50–51):1445–52.
[112] Maeda A, Bateman BT, Clancy CR, Creanga AA, Leffert LR. Opioid abuse and dependence during pregnancy: temporal trends and obstetrical outcomes. Anesthesiology. 2014;121(6):1158–65.
[113] Shatil B, Landau R. Opioid use and misuse in pregnancy. Clin Perinatol 2020;47(4):769–77.
[114] Rizk AH, Simonsen SE, Roberts L, Taylor-Swanson L, Lemoine JB, Smid M. Maternity care for pregnant women with opioid use disorder: a review. J Midwifery Womens Health 2019;64(5):532–44.
[115] Tobon AL, Habecker E, Forray A. Opioid use in pregnancy. Curr Psychiatry Rep 2019;21(12):118.
[116] Daigle KM, Heller NA, Sulinski EJ, Shim J, Lindblad W, Brown MS, et al. Maternal responsivity and oxytocin in opioid-dependent mothers. Dev Psychobiol 2020;62(1):21–35.
[117] Minnes S, Lang A, Singer L. Prenatal tobacco, marijuana, stimulant, and opiate exposure: outcomes and practice implications. Addict Sci Clin Pract 2011;6(1):57–70.
[118] McDaniel CE, Test M, Deodhar P, Jennings R, Bove J, Carlin K. Effects of polysubstance exposure on neonatal outcomes for infants with intrauterine opioid exposure. J Perinatol 2020;40(10):1489–96.
[119] Stover MW, Davis JM. Opioids in pregnancy and neonatal abstinence syndrome. Semin Perinatol 2015;39(7):561–5.
[120] Winklbaur B, Kopf N, Ebner N, Jung E, Thau K, Fischer G. Treating pregnant women dependent on opioids is not the same as treating pregnancy and opioid dependence: a knowledge synthesis for better treatment for women and neonates. Addiction. 2008;103(9):1429–40.
[121] Lester BM, Tronick EZ, LaGasse L, Seifer R, Bauer CR, Shankaran S, et al. The maternal lifestyle study: effects of substance exposure during pregnancy on neurodevelopmental outcome in 1-month-old infants. Pediatrics. 2002;110(6):1182–92.
[122] Bauer CR, Langer J, Lambert-Brown B, Shankaran S, Bada HS, Lester B, et al. Association of prenatal opiate exposure with youth outcomes assessed from infancy through adolescence. J Perinatol 2020;40(7):1056–65.
[123] Lipsitz PJ. A proposed narcotic withdrawal score for use with newborn infants: a pragmatic evaluation of its efficacy. Clin Pediatr 1975;14(6):592–4.
[124] Nygaard E, Slinning K, Moe V, Walhovd KB. Behavior and attention problems in eight-year-old children with prenatal opiate and polysubstance exposure: a longitudinal study. PLoS One 2016;11(6):e0158054.
[125] Baldacchino A, Arbuckle K, Petrie DJ, McCowan C. Neurobehavioral consequences of chronic intrauterine opioid exposure in infants and preschool children: a systematic review and meta-analysis. BMC Psychiatry 2014;14(1):104.
[126] Lambert BL, Bann CM, Bauer CR, Shankaran S, Bada HS, Lester BM, et al. Risk-taking behavior among adolescents with prenatal drug exposure and extrauterine environmental adversity. J Dev Behav Pediatr 2013;34(9):669–79.
[127] Liebschutz JM, Crooks D, Rose-Jacobs R, Cabral HJ, Heeren TC, Gerteis J, et al. Prenatal substance exposure: what predicts behavioral resilience by early adolescence? Psychol Addict Behav 2015;29(2):329–37.
[128] Frazer Z, McConnell K, Jansson LM. Treatment for substance use disorders in pregnant women: motivators and barriers. Drug Alcohol Depend 2019;205:107652.
[129] Klaman SL, Isaacs K, Leopold A, Perpich J, Hayashi S, Vender J, et al. Treating women who are pregnant and parenting for opioid use disorder and the concurrent care of their infants and children: literature review to support national guidance. J Addict Med 2017;11(3):178–90.
[130] Centers for Disease Control and Prevention. Fetal alcohol spectrum disorders (FASDs); 2021. Available from: https://www.cdc.gov/ncbddd/fasd/facts.html.

Chapter 12

Early brain injury

Vera Joanna Burton[1,2], Gwendolyn J. Gerner[3,4] and Marilee C. Allen[1,5]
[1]*Department of Neurology and Developmental Medicine, Kennedy Krieger Institute, Baltimore, MD, United States,* [2]*Department of Neurology, Johns Hopkins University School of Medicine, Baltimore, MD, United States,* [3]*Department of Psychology, Kennedy Krieger Institute, Baltimore, MD, United States,* [4]*Department of Psychiatry and Behavioral Sciences, Johns Hopkins University School of Medicine, Baltimore, MD, United States,* [5]*Department of Pediatrics, Johns Hopkins University School of Medicine, Baltimore, MD, United States*

Introduction

Early brain injury encompasses a large group of disorders that result in damage to the developing brain prenatally, perinatally, and early in postnatal life. Effects of prematurity, limited intrauterine resources due to maternoplacental dysfunction, or direct injury to the developing brain may be recognized in newborns or may not manifest until later, when a child presents with neurologic abnormalities, a delay in developmental milestone attainment, or school difficulties. Although no individual biomarker can currently predict with certainty the impact of a brain injury on a child's functional outcome, a combination of neuroimaging, genetic, and neurobehavioral biomarkers can stratify risk for neurodevelopmental impairments. Risk stratification can be used to identify those most likely to benefit from comprehensive neurodevelopmental follow-up and early intervention services. Early diagnosis and the implementation of targeted interventions can take advantage of key periods of neuroplasticity. This approach focuses on specific interventions to improve function and on providing support for parents during what is often a period of elevated family stress.

Brief overview of epidemiology

The epidemiology of early brain injury is complex. Despite advances in obstetric care, the incidence of prematurity (defined as birth prior to 37 weeks gestation) has remained relatively consistent (10% of United States live births), but varies widely across ethnic groups and geographic regions. Worldwide, premature birth is the most frequent cause of neonatal death and the second most frequent cause of death in children aged less than 5 years old [1,2]. In the US, preterm births have been estimated to account for 64%–75% of infant deaths, 42%–47% of children with cerebral palsy (CP), 27% of children with intellectual disability (ID), 37% of children with visual impairment, and 23% of children with hearing impairment [3].

The risk for early brain injury and neurodevelopmental disability increases with decreasing gestational age (GA) and birthweight. Low birthweight (LBW, defined as having birthweights below 2500 g or 5 lb 8 oz), a standard public health measure, is associated with increased perinatal, neonatal, and neurodevelopmental morbidity. It occurs in 15%–20% of global livebirths and 7%–9% of North American livebirths [4]. A third are full-term infants born small for their gestational age (SGA); the remainder are premature. SGA infants who are small due to constitutional factors (e.g., parental race or ethnicity) do well. However, intrauterine growth restriction (IUGR) can adversely influence fetal brain development and is a common cause of stillbirths (30%) and intrapartum asphyxia [5]. Asphyxia leading to neonatal encephalopathy and hypoxic ischemic encephalopathy (HIE) occurs in approximately 1–3 per 1000 livebirths and is a leading cause of neonatal death and of brain injury at term [6]. It is also associated with an increased risk of neurodevelopmental disability. Perinatal stroke, which can present in neonates, is discussed in Chapter: Acquired brain injury.

Signs and symptoms

The newborn infant is not a passive but an active participant in their development. The rapidly maturing central nervous system results in development that generally proceeds in a sequential and organized manner but is vulnerable to and

Capute and Accardo's Neurodevelopmental Disabilities in Infancy and Childhood. DOI: https://doi.org/10.1016/B978-0-12-824060-1.00004-3

can be modified by environmental and perinatal insults, especially during sensitive or critical periods that vary with brain region and stage of development.

Fetal monitoring

Routine obstetric ultrasounds are used to evaluate fetal anatomy, growth, and wellbeing. Anomalies found on these scans can lead to transfer to a maternal fetal medicine specialist and additional testing and imaging, which may include fetal MRI. Prenatal counseling may include consultation with a neonatologist, geneticist, or pediatric neurologist and can provide families with information regarding risks for medical and developmental concerns. Detection of lagging fetal growth based on clinical findings or ultrasound warrants close monitoring and may include Doppler studies of flow in uterine and fetal vessels.

Neonates with prenatal diagnoses, intrauterine growth restriction (IUGR), prematurity, and suspected brain injury warrant neurodevelopmental assessment. In the early neonatal period, neurodevelopmental abnormalities rarely localize but instead manifest with nonspecific findings (e.g., encephalopathy, seizure, dysregulation of responses to sensory input, or abnormal muscle tone, movement patterns, primitive reflexes). Signs of early brain injury can include irritability, feeding difficulties, and disruptions in parent-infant interactions, which all increase parental stress. Later, signs may more closely mirror classic brain injury patterns. Given the developing brain's inherent neuroplasticity, signs of injury may not manifest until cognitive demands increase, sometimes not until school age.

Neuroimaging

One of the best predictors of neurodevelopmental outcomes is neuroimaging evidence of structural brain injury. The combination of sequential head ultrasounds and MRI at term age equivalent provides the best prediction for motor outcomes in children born prematurely [7]. However, brain imaging cannot be used to diagnose impairments with certainty. For example, in one study of children with a birth weight of less than 1000 g, 30% with intraparenchymal hemorrhage had no impairments at 18–24 months [8]. Another study found that 40% of preterm children who had normal neonatal head ultrasounds showed neurodevelopmental impairment at 2 years [9]. Children born prematurely who had no evidence of structural differences on standard neuroimaging have evidence of altered resting state connectivity at term age equivalent [10]. Further, adolescents and even adults born prematurely continue to demonstrate evidence of more recruitment of additional brain regions during cognitive tasks than those born at term [11–15].

Neurodevelopment and neurobehavior

A number of neonatal neurobehavioral assessments have been designed to evaluate the integrity of an infant's nervous system (Table 12.1). Most include assessing muscle tone, primitive reflexes, and postural responses for abnormal patterns or asymmetries, as well as assessing an infant's state regulation and responses to sensory stimulation (e.g., the Amiel-Tison Neurological Assessment at Term, the Dubowitz Neurological Examination of the Full-Term Infant, the Neonatal Network Neurobehavioral Scale, and Hammersmith Neonatal Neurologic Exam) [16–20]. By closely observing awake, quiet infants, Prechtl's General Movement Assessment evaluates their type and quality of spontaneous movements, which reflect the development of higher cortical control over central pattern generator networks in the brain stem and spinal cord [21]. Abnormal movement patterns such as the persistence of cramped-synchronized movements between 32 and 46 weeks postmenstrual age and the absence of fidgety movements between 47 and 60 weeks postmenstrual age are associated with early brain injury and later diagnosis of CP [22]. Abnormal cramped synchronized movements have been found to have a sensitivity of 93% (CI 86–96) and specificity of 59% (CI 45–71), and the fidgety period has a sensitivity of 97% (CI 93–99) and specificity of 89% (CI 83–93) for a diagnosis of CP [23]. Combining findings from neonatal neuroimaging studies, neonatal assessments, and serial assessments during early infancy improves the prediction of neurodevelopmental outcomes.

Early diagnosis

The use of neonatal bedside ultrasounds and unsedated brain MRI enables earlier identification of brain injury and greater precision in determining its severity. More expanded use of early neuroimaging also means an increased number of findings that do not have clear neurological or functional consequences. Additionally, the ability of the neonatal brain to reorganize in response to environmental input makes certain predictions of neurodevelopmental outcomes

TABLE 12.1 Prediction of neurodevelopmental outcome with neonatal assessments.

Exam	Population	Range of ages of follow-up studies	Outcome
Allen neurodevelopmental exam	Premature	1–5 years	Cerebral palsy; Motor dysfunction; Developmental disability
Amiel-Tison neurological assessment	Premature Term IVH HIE High risk	11 months–2 years	Neurologic;General developmental domains;Developmental disability
Dubowitz neurological assessment of the preterm and full-term infant	Premature	1–11 years	Cerebral palsy; Minor neurological dysfunction; General developmental domains
Brazelton neonatal behavioral assessment scale (NBAS)	Premature; PE	18 months	General developmental domains
Prechtl's assessment of general movements (GMA)	Premature Term HIE Brain injury PE	1–5 years	Cerebral palsy Neurologic; General developmental domains; Motor; Early cognition Feeding impairment
Hammersmith neonatal neurologic examination (HNNE)	Premature Term High risk HIE	1–2 years	Neurodevelopmental disability; General developmental domains; Feeding impairment
Neonatal intensive care unit network neurobehavioural scale (NNNS)	PE; Premature; HIE; PHH;	1–2 years	Motor General developmental domains; Feeding impairment; Atypical sensory profile
Neurobehavioural assessment of the preterm infant (NAPI)	Premature	12–30 months	General developmental domains
Test of infant motor performance (TIMP)	High risk Premature Low birth weight	2–5 years	Motor; Cerebral palsy; General developmental domains

IVH, Intraventricular hemorrhage; *HIE*, hypoxic-ischemic encephalopathy; *PHH*, posthemorrhagic hydrocephalus; *PE*, perinatal exposure.

challenging. Prediction is greatly enhanced by using a combination of tools to support clinical judgment. For example, guidelines for the early diagnosis of Cerebral Palsy (CP) rely on a combination of history, neuroimaging, and standardized neurobehavioral biomarkers that should be used to predict risk and allow for diagnoses of CP within the first year (sometimes as early as 3 months) [24]. Finally, cognitive and neurobehavioral deficits may not be initially apparent in early infancy but manifest as children "grow" into their deficits as expectations and demands increase in each developmental domain. This should be discussed with parents, and the child's development should be closely monitored during the preschool and school years.

Neurobiology

Stages of brain development and neuromaturation

The nervous system begins with the formation of a primitive streak during gastrulation 15–17 days after fertilization. Its anatomical and functional development follows a precise sequence with neuronal proliferation, migration, differentiation, and orientation; synaptic formation, pruning, and reorganization; and myelination [25]. Embryonic sensory organs are linked to neural architecture as they are formed. By 10 weeks GA (8 weeks after fertilization), the brain with cerebral hemispheres has been formed and linked to the body by nerves extending into the spinal cord and periphery; brain waves and a variety of spontaneously generated movements can be detected. Neuromaturation is the functional development of the central nervous system, a dynamic process mediated by the formation and refinement of neuronal circuits based on continuous interactions between genes, epigenetic markings, and first the intrauterine, then the extrauterine environment [26]. Although driven by genetic and epigenetic processes, fetal and infant movement, sensory input, and responses play a critical role in neurogenesis by directing the programmed cell death of inactive neurons, forming, maturing, and pruning synapses, and refining neural networks. The integral relationship linking structure and function facilitates the coordination and functional integration required for optimal development (see Chapter: Typical and atypical neurodevelopment).

Neuroplasticity

Neuroplasticity is the inherent capacity of the central nervous system to undergo neuromaturation while making structural and adaptive functional changes in response to environmental circumstances or injury. This dynamic process is achieved by modulating genetic, molecular, and cellular mechanisms that influence synaptic connections and neural circuitry formation [27]. Given a marked redundancy of neurons and synapses in the fetus and infant, over time inactive neurons are targeted for programmed cell death as the overabundance is pruned according to use. In this way, neuronal networks are shaped by environmental input to facilitate function in the individual's specific environment. Disruption of this process can result in functional deficits. Adaptive neuroplasticity enables a child to perform a specific function despite brain injury, but frequently with less efficient task performance and increased response times.

This inherently dynamic biological capacity for change in brain structure and function persists into early adulthood. Periods of development when the brain is most amenable to change with ongoing neurogenesis, synaptogenesis, and subsequent activity-dependent pruning are thought to be orchestrated by the genetically regulated sequential expression of GABA, NMDA, and AMPA receptors that are sequentially expressed. Different brain regions have critical or sensitive periods in development when they are most amenable to change, and changes may be adaptive or maladaptive. Metabolically active periods create windows of vulnerability in different parts of the brain at different time points [28]. Following a brain injury, neuroplasticity provides an opportunity to improve neurodevelopmental function. The richer and more nurturing the environment and its input, the greater, faster, and more flexible the neuronal circuitry becomes (see Chapter Neuroplasticity in neurodevelopmental disabilities).

Functional development

Assessment of a child's functional abilities involves careful evaluation of developmental milestones, including quality of movement and language, throughout infancy and into childhood [29]. Developmental trajectories are often altered due to reductions or abnormal patterns of synaptic connections and neuronal networks associated with changes in physiological stability and subsequent injury to a critical brain region during sensitive periods of brain development [30]. Extreme immaturity of all organ systems, with subsequent physiological instability after birth, leads to the death or injury of many cells required for establishing neural connections. Infants who experience acute and severe asphyxia at term tend to have injuries in specific regions that are especially vulnerable when that region is developing [31]. Different patterns of brain injury influence the type and severity of later neurodevelopmental disabilities.

Preterm delivery removes the fetus from the nurturing intrauterine environment designed to precisely support fetal movements, nutrition, and sensory input. Following preterm transition to extrauterine life and for term infants requiring intensive medical support, often in a Neonatal Intensive Care Unit (NICU), where the primary goal is physiological stabilization to sustain life [32]. A healthy newborn is exposed to a unique multisensory environment that supports the developing nervous system. For a critically ill neonate, the NICU is a less than ideal environment, which is simultaneously hyperstimulating and hypostimulating [33]. A NICU generally provides increased stimulatory input from ventilators and alarms, continuous lighting, invasive and painful procedures, and less freedom of movement. With reduced parental exposure and complex NICU care, the neonate receives quite different tactile and less rhythmic stimuli, painful procedures, altered positioning in an isolette, and less language exposure. Providing the sensory and motor experience necessary for promoting optimal neuromaturation is quite challenging while maintaining a critical infant's physiological stability and providing life-saving procedures as needed.

Etiology

Prematurity

Prematurity is a complex multifactorial condition resulting from a complex interplay of maternal, fetal, and environmental factors leading to the delivery of a newborn whose immature organ systems are not fully functional or prepared for sustaining extrauterine life. Prematurity is associated with numerous risk factors, ranging from the molecular to the neighborhood environmental level, that can individually or cumulatively lead to a pathway to prematurity: preterm labor, premature rupture of membranes, or indicated preterm birth for the safety of the mother (preeclampsia, infection) or the fetus (poor growth, fetal distress) [3]. Ongoing research is providing a more nuanced understanding of the biological impact of social determinants of health, genetic predisposition, developmental origins of health and disease, and weathering/allostatic load to better understand the cumulative effects of these factors on premature birth [3,34,35]. An increasing number of candidate genes are being identified as associated with prematurity, including Factor V (F5) gene haplotype (blood clotting abnormality) as well as genes involved with cytokine production and immune system function. Gene-environmental interactions are important to explore as contributing factors. Gene polymorphisms (e.g., TNF-α-2 with bacterial vaginosis; CYP1A1 and GSTT1 gene variants with maternal smoking) and epigenetic modifications by specific environmental exposures during critical periods are associated with conditions leading to preterm birth.

Preterm populations were initially categorized by birth weight (BW), but gestational age (GA) is a better proxy for degree of immaturity, which is now subdivided into moderate/late preterm birth (32 weeks to <37 weeks gestation), very preterm birth (28 weeks to <32 weeks gestation), and extremely preterm birth (<28 weeks gestation) [3,36]. The greater the degree of immaturity at birth, the greater the vulnerability and the higher the risk for complications involving any or all of the major organ systems. The risk of rehospitalization, chronic medical problems, and neurodevelopmental disability increases with decreasing gestational age and more neonatal complications. Some specific complications of prematurity, including chronic lung disease (CLD), necrotizing enterocolitis (NEC), retinopathy of prematurity (ROP), infection, and structural brain injury, are independent risk factors for later neurodevelopmental disability [37].

Structural brain injury

Brain injuries seen on neuroimaging in premature infants include periventricular leukomalacia (PVL, white matter injury that can lead to cysts), intraventricular hemorrhage (IVH), with more severe IVH associated with hydrocephalus (due to obstruction of the flow of cerebrospinal fluid by blood products with subsequent inflammation), and periventricular hemorrhagic infarction that results from bleeding around ischemic brain tissue [38]. Acute injury leads to a microglial response with necrosis, the release of cytokines, subsequent inflammation, injury, and often the death of surrounding healthy cells, including neurons and preoligodentrocytes. This can lead to widespread patterns of white and gray matter injuries that alter neurodevelopmental trajectories. The associated impairments in motor, cognitive, behavioral, and socioemotional functioning have collectively been referred to as encephalopathy of prematurity [39,40]. Some preterm infants show periventricular high echogenicity (brightness) on serial ultrasounds; infection or sepsis and a higher risk of neurodevelopmental impairment are associated with persistence beyond 14 days [41]. The risk of neurodevelopmental disability varies with the type, location, and severity of the brain injury.

Intrauterine and fetal growth restriction

Although the terms small for gestational age (SGA), intrauterine growth restriction (IUGR), and fetal growth restriction (FGR) have been used interchangeably, there are important distinctions. All are defined in terms of weight below the 10th percentile for their GA; SGA relies on *postnatal* birthweight, while IUGR and FGR are *prenatal* diagnoses based on a variety of fetal growth parameters on ultrasound (e.g., biparietal diameter, abdominal circumference, femur length), which are then used to estimate fetal weight. Definitions for both IUGR and FGR include reduced fetal growth velocity, which results in a fetus failing to attain its full growth potential. FGR is often reserved for infants who fail to attain their full potential because of compromised placental function. SGA infants who are small based on constitutional factors (e.g., ethnicity, maternal height, gender) are generally not included in many studies of growth restriction. Ideally, comparative fetal growth curves are customized based on constitutional factors.

Studies of SGA subpopulations provide clues to the effects of etiology (generally multifactorial), pathophysiology, intrauterine insults, and adaptations of the developing nervous system. Higher rates of perinatal and neonatal complications (fetal intolerance of labor and delivery, encephalopathy, hypothermia, hypoglycemia, pulmonary hypertension, feeding difficulties) can injure the developing brain beyond the effects of IUGR, and for preterm SGA infants, the effects of complications of prematurity often outweigh those of IUGR [42–44]. Infants have been classified with symmetric, asymmetric, or mixed growth restrictions based on the presence or absence of preserved head growth. Fetal factors, including chromosomal abnormalities, genetic syndromes, congenital anomalies, or early intrauterine infections (especially rubella, CMV, toxoplasma, and syphilis), are associated with symmetric IUGR that can often be recognized early. Each entails a specified pathophysiology with an expected outcome. GA at onset predicts growth patterns and risks associated with most maternal causes of IUGR, including chronic disease (hypertension, cardiovascular disease, renal failure, vasculopathies, autoimmune disorders), malnutrition, exposure to teratogens (antineoplastic drugs, folic acid antagonists), and drug use (smoking, alcohol). Uteroplacental causes (multiple gestations, chronic partial abruption, single umbilical artery, velamentous cord, chronic inflammation, multiple placental infarctions or hemangiomas, syncytial knots, placenta previa) are often associated with later-onset asymmetric IUGR. Maternal vascular disease with reduced uteroplacental perfusion is associated with 25%–30% of FGR cases. Onset of FGR before 32 weeks gestation is associated with placental implantation abnormalities, increased uterine artery resistance, hypoxia that leads to adaptations in the fetal cardiovascular system, symmetric growth restriction, increased risk of preeclampsia, and higher infant mortality and morbidity rates. Pre-eclampsia, a maternal systemic syndrome that originates with inadequate trophoblast invasion of the uterine spiral arteries, generally presents after 20 weeks of gestation with maternal systemic endothelial dysfunction, hypertension, and proteinuria. It is a common cause of FGR, prematurity (delivery is the definitive treatment), intrapartum asphyxia, perinatal mortality (30% of stillbirths), and maternal morbidity (renal or liver failure, thrombocytopenia, hemolysis, seizures) and mortality.

With late-onset growth restriction due to chronic uteroplacental hypoperfusion, the fetus has a limited supply of nutrients, chronic hypoxia, a limited ability to make cardiovascular adaptations, and often poor tolerance for the stresses of uterine contractions, labor, and delivery. A fetal adaptation shunts blood flow toward the major organs, resulting in asymmetric FGR with preferential growth of the brain. Hormonal responses to increased stress may accelerate fetal brain and lung maturation. Despite these survival strategies, FGR is associated with increased rates of stillbirth, perinatal mortality, postnatal complications, reduced total brain and gray matter volume, and adverse neurodevelopmental outcomes. A prospective study of late-onset IUGR observed lower IQ scores and academic achievement and more difficulties with language, learning, and executive function than controls by age 9 years [45]. Another study examining FGR with brain sparing highlighted elevated rates of parent-reported social and attention problems by age 12 years compared to peers despite otherwise comparable cognitive abilities [46]. In these studies, risk factors for adverse neurodevelopmental outcomes included gestational age at delivery, birthweight, signs of worsening uteroplacental function (e.g., abnormal Doppler blood flow in uterine and fetal vessels), neonatal complications, and poor growth. Managing mothers with FGR fetuses involves closely following measures of fetal growth, wellbeing, and uteroplacental function. Determining optimal timing for delivery entails weighing the worsening fetal status in an inhospitable intrauterine environment against the many complications of prematurity. Postnatal management can also be challenging: good postnatal catch-up growth, especially of head circumference, is associated with higher cognitive scores in childhood but also increased obesity and metabolic syndrome [45,47–50].

Hypoxic ischemic encephalopathy

Asphyxia leading to hypoxic ischemic injury is an important cause of neonatal encephalopathy and severe neurodevelopmental disability [51–53]. HIE varies in severity from mild to severe, reflecting the duration and severity of deprivation of blood flow and oxygen to the brain. Risk factors for HIE include IUGR, GA longer than 39 weeks, and acute interruption of placental blood flow (due to maternal hypotension, uterine rupture, placental abruption, or cord prolapse). Chronic severe deprivation is associated with severe injury to the cerebral cortex, deep gray structures (basal ganglia, thalamus), and brainstem, while acute severe deprivation is associated with injury to the deep gray structures and the brainstem with relative sparing of the cortex. Moderate, prolonged, or intermittent HIE is associated with injury to the cerebral cortex and deep gray structures. Injury to the cerebral white matter is seen with mild to moderate HIE while more prolonged HIE is associated with infection and hypoglycemia [54]. Before therapeutic hypothermia was available, 37%–80% of infants with severe HIE died, and 40%–100% of survivors had severe disability; 5% died, and 6%–21% with moderate HIE had disability [53,55,56].

Therapeutic hypothermia has been the standard treatment for moderate-to-severe neonatal HIE following clinical trials that showed improvement in outcomes (death, major disability, or combined death or major disability at 18 months), with a number needed to treat of 7 [6]. Lowering the infant's body and brain temperature soon after a brain injury reduces the secondary injury caused by cascades of metabolic processes that result in cell death [27].

Hypothermia should be initiated within 6 hours of birth with close monitoring to keep infant body temperature within a narrow range for 3 days and then during slow rewarming. Neonatal seizures should be promptly treated. With therapeutic hypothermia, mortality due to moderate to severe HIE is 24%–50%, and among survivors, 7% have CP, 2%–32% have ID, 4% have hearing impairment, and 7% have visual impairment [51–53]. At school age, 55% have persistent neurologic abnormalities, and 35% have moderate-to-severe disabilities [57,58]. The severity of HIE is the major risk factor for neurodevelopmental impairments.

Even without major disability, school-age children who had HIE report they have more difficulties than peers with nonverbal reasoning, verbal skills, reading, academics, fine motor function, visual spatial skills, attention, executive function, behavioral regulation, anxiety, depression, and social relationships [56,59,60]. A study of 2–3-year-olds found that those who had mild HIE had significantly lower mean cognitive scores (97.6 vs 103.6) than controls (with an adjusted mean difference of −5.2); there was no significant difference between mean scores for untreated children with mild HIE and treated children with moderate HIE [61]. A systematic review and metaanalysis of 5 trials of therapeutic hypothermia that inadvertently randomized some infants with mild HIE found that 20% had adverse neurodevelopmental outcomes [62].

Natural history

The etiology of perinatal and neonatal brain injuries is multifactorial. It includes postnatal complications seen in preterm and term NICU infants, which can lead to chronic medical conditions, a higher risk of re-hospitalization, infection, respiratory illness, feeding difficulties, poor growth, and an increased risk of neurodevelopmental disabilities. Medical complications make it difficult to assess neurodevelopment and contribute to functional impairment. Early brain injury is associated with neurodevelopmental disabilities that range from severe cerebral palsy and intellectual disability to subthreshold impairments in multiple domains that adversely influence function into later childhood and adolescence.

Correlation with development and functional status

The environment continues to play a key role in brain development and neuromaturation in these high-risk populations. Neuromaturation continues despite brain injury, although it can alter the rate of development and the role of neuroplasticity in brain recovery. Although difficult to provide in a NICU, nurturing environmental interventions can support neuromaturation and adaptive neuroplasticity, with the goal of shaping brain structural development to improve function.

Associated neurodevelopmental disabilities

The risk of developing epilepsy is higher in preterm than full-term infants, highest in those born before 28 weeks GA, and epilepsy increases the risk of cognitive impairment [63]. Maternal smoking, SGA, Apgar score of less than 4 at 1 minute, need for a ventilator, intracranial hemorrhage, and presence of a congenital abnormality increase a preterm infant's risk for epilepsy, while female sex seems to be protective. Structural brain injury also increases the risk of seizures in term and preterm infants, including the risk of infantile spasms [63].

Early brain injury increases risk of mild to severe motor and cognitive impairment [64]. Children who were born extremely preterm or survived moderate to severe HIE have high prevalence rates for both cerebral palsy and intellectual disability [3,8,59,65,66]. In the general population, the prevalence of CP is 1–3 per 1000 and of ID is 1%–3%; rates for each increase with decreasing gestational age and birth weight and with increasing severity of HIE. Up to 20% of infants born before 28 weeks and 10%–15% with very low birthweight (VLBW, <1500 g) develop CP. Children who survived severe HIE have a high incidence of severe CP and ID; 20%–30% develop CP and 10%–15% develop ID [51–53], although the neuroprotection provided by therapeutic hypothermia has lowered this incidence with current estimates at less than 10% [58].

Most studies of neonates born extremely preterm or with IUGR or HIE report mean cognitive scores that are lower than controls but within the normal range. Preterm or term children with brain injury but *not* CP are more likely to experience more subtle motor delay and coordination difficulties. As many as 6% of children in the general population have developmental coordination disorder (DCD), while children born early are 8 times more likely to have DCD (up to 75% of children born prematurely) [67]. School-age children born preterm or with HIE have an increased prevalence of low-severity disorders, including learning disability, language delay, and difficulties with visual spatial skills, attention, executive function, behavioral regulation, anxiety, depression, and social relationships, compared to their peers [57,60,68–71].

In the 85%–90% of children who were born preterm and do *not* have intellectual disability, studies show mean Full Scale IQ (FSIQ) scores are 8–12 points below those of peers born at term [3,68]. Beyond ID, encephalopathy of prematurity has an impact on the speed and efficiency of processing and responding to information, attention, executive function, memory, and higher-order language, as well as more difficulties with adaptive function, academics, and behavioral problems [57,66,69,71]. A study of 2½-year old children born extremely preterm found significantly higher cognitive, expressive, and receptive language and fine motor skills scores in girls compared to boys [72]. Neonatal diagnosis of ROP is associated with worse cognitive outcomes in preterm males, while chronic lung disease predicts worse cognitive scores in preterm females. Preterm outcome studies show a strong association between higher socioeconomic status and parental education with language development and higher cognitive scores.

SGA outcome studies vary widely in inclusion and exclusion criteria, etiologies, age at outcome assessment, outcome measures, and adjustment for confounding variables. Reviews report higher rates of abnormal neuroimaging, infant cognitive, language, and motor deficits, and later chronic health problems (e.g., obesity, cardiovascular and renal disease, metabolic syndrome) in SGA children or adolescents compared to peers born appropriate for gestational age (AGA) [5,43,48,49]. A metaanalysis of 52,822 children found that SGA and IUGR children had lower mean cognitive scores and higher rates of cognitive scores greater than 1 standard deviation below the mean than AGA controls [46]. A study of SGA young adults found lower mean full scale IQ scores, as well as six Wechsler Adult Intelligence Scale subtest scores, than in AGA controls [73]. Despite these statistically significant lower scores, for all these studies, the *mean* cognitive scores in the SGA groups were within the normal range. A prospective study of late-onset IUGR observed lower IQ scores and academic achievement and more difficulties with language, learning, and executive function than controls by age 9 years [45]. Another study examining FGR with brain sparing highlighted elevated rates of parent-reported social and attention problems by age 12 years compared to peers despite otherwise comparable cognitive abilities [46]. Children born with extremely low birth weight have a higher frequency of learning disabilities in reading, math, and written expression than those born full-term [74]. Many research studies have demonstrated "catch-up" in expressive and receptive lexicons by adolescence among those born preterm [15,75]; however, children born extremely and very preterm are at increased risk of persistent higher-order language problems. Specifically, complex language deficits tend to persist in word fluency, synonym production, and pragmatic language into adolescence. Speech disorders are more frequent [76–79].

Attention deficit hyperactivity disorder (ADHD) is among the most frequent outcomes following premature birth (2–6 times increased); however, the prevalence follows a different pattern than DSM defined ADHD without the male predominance, fewer comorbid conduct disorders, and less influence of socioeconomics and family risk factors [80]. Slow processing speed and impaired working memory drive a larger burden of severity of executive function and academic achievement for children born early [81]. There is conflicting information; however, most studies have not found an association between ADHD and birthweight, gestational age, or neonatal complications [66,80]. Children and adolescents born preterm also tend to perform worse than term controls on measures of inhibition, shifting, planning, and parent ratings of global executive function, metacognition, initiation, and working memory [82–85].

Children born preterm are at increased risk of autism spectrum disorder (ASD), but they are also the group that is most likely to lose the ASD diagnosis. A positive screen on the M-CHAT occurs in 21%–41% and is more common in those with lower BW, lower GA, male gender, chorioamnionitis, acute intrapartum hemorrhage, and abnormal MRI findings [86–88]. High-risk children with other neurodevelopmental disabilities are more likely to have a positive screen for autism; however, this decreases with age [87]. Children screened at 18 months were more likely to screen positive than children screened at age 3 [89]. Using more comprehensive or additional screening tools decreased the likelihood of screening positive for an ASD diagnosis [90]. With longer-term follow-up between the ages of 7 and 11, the rates of autism were between 2.2% and 8% [80,91]. Taken together, this area of research demonstrates that while children born early are at a higher risk for autism, they also have a higher rate of false positive screening, so care should be given when making this diagnosis.

A "preterm behavioral phenotype" that includes difficulty with inattention, emotional regulation (often manifesting with anxiety), and socialization, as well as an increased risk for internalizing rather than externalizing problems, has been proposed for children born prematurely. Parents of children with this profile report difficulties in verbal communication, reciprocal social interactions, and repetitive or persistant behaviors that are often verbal. Many have subclinical symptomatology in multiple social-emotional areas. The proposed "preterm behavioral phenotype" is associated with decreasing gestational age and cognitive impairment but not with neurological or sensory impairments [80,92]. The cumulative effects of slightly lower cognitive potential, increased executive dysfunction, and psychiatric features have significant overall social, adaptive function, and academic consequences.

Early diagnosis initiatives

Evidence-based assessments of infants born prematurely or with IUGR or perinatal brain injury can provide a timely diagnosis, which is critical for targeting appropriate interventions and providing for family and child support [24]. Up to 86% of parents with a child with CP suspected it before the diagnosis was made; many parents experienced depression and lasting anger with a prolonged diagnostic process [93,94]. Parents rarely report seeing the diagnosis as too early and generally report the benefits of having an early diagnosis. Interestingly, they also report remembering prognostic discussions as underestimating the functional outcome of their children [95]. Early diagnosis helps to foster family acceptance, confidence in their infant's medical team, access to early intervention, and more efficient use of resources. In high-risk follow-up settings, initiatives to follow early diagnosis guidelines for CP using a combination of history, neuroimaging, and standardized neurological and neuromotor assessments have decreased the age of diagnosis to less than 1 year of age [96].

Similar initiatives have been suggested for other neurodevelopmental disorders. Despite the push for early diagnosis, it is important to note that many available developmental screening tools have not been validated for use in infants with a high-risk history, such as perinatal brain injury. Therefore, longitudinal evaluation in a setting well-versed in outcomes following premature birth and perinatal brain injury is critical when providing diagnosis for this population and guidance for parents.

Very early therapies

In addition to appropriate medical care in the NICU and during acute hospitalization, there is a new emphasis on developmentally appropriate care for infants while hospitalized. There are a number of standardized neuromotor and neurobehavioral measures, such as the Neurobehavioral Assessment of the Premature Infant (NBAS) and the Assessment of Preterm Infants' Behavior (APIB), available that could be used to guide these early interventions [97,98]. Even when a definitive diagnosis is not yet appropriate, indicating *high risk* facilitates early interventions for associated neurodevelopmental disabilities during peak periods of neuroplasticity.

In the United States, the Individuals with Disabilities Education Act provides rehabilitation services for children from birth to 3 years old. This provision is based on the knowledge that the brain experiences rapid growth during the first 3 years and that supporting development early may promote long-term cost savings. However, in most of the United States, early intervention occurs after the age of 2 years. With the advent of validated early diagnosis algorithms for children at risk and more comprehensive newborn screening panels, children can be identified for therapy interventions during peak periods of neuroplasticity. A 2015 Cochrane review of early intervention for premature infants found evidence that early developmental interventions improve cognitive outcomes up to preschool age. There is also evidence of improved motor outcomes during infancy, but with smaller effects; however, it is clear that the targets and types of intervention are important [99]. There is growing literature identifying effective interventions that promote more functional motor patterns and early problem-solving skills both prior to NICU discharge and during the first few years [100].

Parent support following perinatal brain injury

Across diverse ethnic and cultural groups, parents of infants in the NICU report feelings of guilt and shame, high levels of stress, and mood and anxiety symptoms during and after NICU hospitalizations. A study found that more than one-third of the mothers screened positively for posttraumatic stress and depression symptoms and that these symptoms were associated with negative changes in infant development as well as reports of lower maternal quality of life [101]. Screening parental mental health should be part of routine NICU care, as the use of specific coping strategies and social supports can improve their mental health [102].

Similar themes emerge around diagnosis and health care delivery for infants and toddlers after a NICU stay. A study comparing the thoughts of parents and providers in an early diagnosis of cerebral palsy initiative found that 100% of the parents but only 50% of the providers felt that an early diagnosis or high-risk classification was beneficial. While parents and providers had matching priorities for discussions regarding neuroimaging timing/risk/

benefit, cognition, primary care, motor, and feeding/nutrition, parents wanted more education and options regarding participation, parental well-being, infant pain, and vision [103]. The relationship between developmental progress and parental mental health seen in the NICU continues, with parents of children with moderate or severe conditions having poorer outcomes on measures of depression, marital satisfaction, quality of life, and family functioning. Mothers and fathers generally had similar outcomes, but mothers demonstrated a greater burden of guilt and higher levels of anxiety [94]. Parents report that direct and ongoing communication leads to greater family satisfaction. A study of a preventative care program reported that while developmental outcomes did not differ, parents of children in the intervention group reported improved behavioral outcomes, including less externalizing and dysregulation behaviors and increased competence, compared with reports of parents of children in the control group. According to the report, primary caregivers themselves had less anxiety and depression [104]. Overall, families desire experiences around diagnosis to be informative and timely, with early follow-up support and assistance navigating the health care system [105].

Both in the NICU and after discharge, parents with an infant in or discharged from a NICU are often overwhelmed by anxiety and feelings of helplessness regarding their infant's future. They want to know as early as possible about their child's risk for neurodevelopmental disabilities and what they can do about it. Neurodevelopmental follow-up as well as targeted early intervention programs benefit not only infants at risk for neurodevelopmental disability by taking advantage of windows of neuroplasticity, but they also benefit parents by providing them with support and education about what they can do to support their child's development and behavior. Clinical, translational, and basic research programs endeavor to support the care provided by follow-up and early intervention programs that improve outcomes and quality of life for both parents and infants.

High-risk follow-up programs

High-risk infant follow-up is an important aspect of outpatient care following discharge from the NICU, as it provides unique opportunities to monitor for impairments as they first emerge and quickly triage children to early intervention services, thus maximizing the efficacy of the interventions during peak periods of neuroplasticity [106]. Additionally, it provides an opportunity to educate caregivers about the development and risks associated with their child's history, as well as provide ongoing support for the family following NICU discharge [106].

Many children who meet certain criteria are automatically referred to a high-risk infant follow-up program upon discharge from the NICU, and specific criteria for referral may include the degree of prematurity, diagnosis of HIE, and other perinatal and neonatal complications that place a child at high risk for developmental disability. In some instances, states have preestablished and state run high-risk infant follow-up programs, and all states offer birth-to-three early intervention services as a component of Part C of the Individuals with Disabilities Education Act (IDEA) [107]. Such early intervention services may include physical therapy, occupational therapy, speech and language pathology, vision and hearing services, and education services. High-risk infant follow-up clinics tend to focus on identifying medical complications and/or developmental delays and disabilities. Some clinics address all of these areas, while others may only address medical or developmental needs. In addition to addressing medical and developmental issues, the value of a multidisciplinary approach is that the team can focus on identifying specific interventions tailored to each child's needs, as well as discussing and demonstrating some recommended approaches parents can use when working with their child.

The variation in high-risk infant follow-up clinic design may be dictated by healthcare policies in each state and the available resources [106]. There is a unique opportunity in high-risk infant follow-up care for a multidisciplinary or interdisciplinary care approach. In programs that employ a multidisciplinary or interdisciplinary approach, services such as neonatology, developmental medicine, neurology, and rehab expertise in conjunction with related services such as physical therapy, occupational therapy, speech and language pathology, psychology/neuropsychology, social work, nursing, and nutrition may be integrated and available at the time of each scheduled visit. This provides the opportunity for a more comprehensive evaluation and follow-up care and reduces the number of additional appointments a child and their caregivers have to attend, as multiple services are provided at the time of each follow-up care visit. Presently, there are no standards established as to which measures should be utilized to monitor development. Most often, neurodevelopmental testing tools are selected based on the expertise of the providers and the specific clinical questions to be answered. The information gathered during each evaluation, including specific developmental test scores, is an important aspect to high-risk follow-up care for counseling parents, guiding referrals for intervention, and documenting the child's response to intervention.

References

[1] Harrison MS, Goldenberg RL. Global burden of prematurity. Semin Fetal Neonatal Med 2015;21:74–9. Available from: https://doi.org/10.1016/j.siny.2015.12.007.

[2] Frey HA, Klebanoff MA. The epidemiology, etiology, and costs of preterm birth. Semin Fetal Neonatal Med 2016;21:68–73. Available from: https://doi.org/10.1016/j.siny.2015.12.011.

[3] Preterm birth: causes, consequences, and prevention. In: Behrman RE, Butler AS, editors. Institute of Medicine (US) Committee on understanding premature birth and assuring healthy outcomes. Washington (DC): National Academies Press (US); 2007.

[4] United Nations Children's Fund and World Health Organization. Low birthweight: country, regional and global estimates. UNICEF, New York, 2004.

[5] Nardozza LM, Caetano AC, Zamarian AC, Mazzola JB, Silva CP, Marcal VM, et al. Fetal growth restriction: current knowledge. Arch Gynecol Obstet 2017;295:1061–77. Available from: https://doi.org/10.1007/s00404-017-4341-9.

[6] Jacobs SE, Morley CJ, Inder TE, Stewart MJ, Smith KR, McNamara PJ, et al. Whole-body hypothermia for term and near-term newborns with hypoxic-ischemic encephalopathy: a randomized controlled trial. Arch Pediatr Adolesc Med 2011;165:692–700. Available from: https://doi.org/10.1001/archpediatrics.2011.43.

[7] de Vries LS, van Haastert IC, Benders MJ, Groenendaal F. Myth: cerebral palsy cannot be predicted by neonatal brain imaging. Semin Fetal Neonatal Med 2011;16:279–87. Available from: https://doi.org/10.1016/j.siny.2011.04.004.

[8] Broitman E, Ambalavanan N, Higgins RD, Vohr BR, Das A, Bhaskar B, et al. Clinical data predict neurodevelopmental outcome better than head ultrasound in extremely low birth weight infants. J Pediatr 2007;151 500–5, 505.e1–2.

[9] Laptook AR, O'Shea TM, Shankaran S, Bhaskar B, Network NN. Adverse neurodevelopmental outcomes among extremely low birth weight infants with a normal head ultrasound: prevalence and antecedents. Pediatrics 2005;115:673–80.

[10] Rogers CE, Lean RE, Wheelock MD, Smyser CD. Aberrant structural and functional connectivity and neurodevelopmental impairment in preterm children. J Neurodev Disord 2018;10. Available from: https://doi.org/10.1186/s11689-018-9253-x.

[11] Constable RT, Vohr BR, Scheinost D, Benjamin JR, Fulbright RK, Lacadie C, et al. A left cerebellar pathway mediates language in prematurely-born young adults. Neuroimage 2013;64:371–8. Available from: https://doi.org/10.1016/j.neuroimage.2012.09.008.

[12] Salvan P, Froudist Walsh S, Allin MP, Walshe M, Murray RM, Bhattacharyya S, et al. Road work on memory lane—functional and structural alterations to the learning and memory circuit in adults born very preterm. Neuroimage 2014;152–61. Available from: https://doi.org/10.1016/j.neuroimage.2013.12.031.

[13] Smyser CD, Snyder AZ, Shimony JS, Blazey TM, Inder TE, Neil JJ. Effects of white matter injury on resting state fMRI measures in prematurely born infants. PLOS One 2013;8:e68098. Available from: https://doi.org/10.1371/journal.pone.0068098.

[14] Mullen KM, Vohr BR, Katz KH, Schneider KC, Lacadie C, Hampson M, et al. Preterm birth results in alterations in neural connectivity at age 16 years. Neuroimage 2011;54:2563–70. Available from: https://doi.org/10.1016/j.neuroimage.2010.11.019.

[15] Myers EH, Hampson M, Vohr B, Lacadie C, Frost SJ, Pugh KR, et al. Functional connectivity to a right hemisphere language center in prematurely born adolescents. Neuroimage 2010;51:1445–52. Available from: https://doi.org/10.1016/j.neuroimage.2010.03.049.

[16] Lester BM, Tronick EZ. History and description of the neonatal intensive care unit network neurobehavioral scale. Pediatrics 2004;113:634–40.

[17] Dubowitz L, Ricciw D, Mercuri E. The Dubowitz neurological examination of the full-term newborn. Ment Retard Dev Disabil Res Rev 2005;11:52–60. Available from: https://doi.org/10.1002/mrdd.20048.

[18] Dubowitz L.M.S., Dubowitz V., Mercuri E. The neurological assessment of the preterm and full-term newborn infant. Issue 148 of Clinics in Developmental Medicine, Cambridge University Press; 1999.

[19] Gosselin J, Gahagan S, Amiel-Tison C. The Amiel-Tison neurological assessment at term: conceptual and methodological continuity in the course of follow-up. Ment Retard Dev Disabil Res Rev 2005;11:34–51. Available from: https://doi.org/10.1002/mrdd.20049.

[20] Lester BM, Tronick EZ, Brazelton TB. The neonatal intensive care unit network neurobehavioral scale procedures. Pediatrics 2004;113:641–67.

[21] Einspieler C, Prechtl HF. Prechtl's assessment of general movements: a diagnostic tool for the functional assessment of the young nervous system. Ment Retard Dev Disabil Res Rev 2005;11:61–7. Available from: https://doi.org/10.1002/mrdd.20051.

[22] Ferrari F, Cioni G, Einspieler C, Roversi MF, Bos AF, Paolicelli PB, et al. Cramped synchronized general movements in preterm infants as an early marker for cerebral palsy. Arch Pediatr Adolesc Med 2002;156:460–7.

[23] Kwong AKL, Fitzgerald TL, Doyle LW, Cheong JLY, Spittle AJ. Predictive validity of spontaneous early infant movement for later cerebral palsy: a systematic review. Dev Med Child Neurol 2018;60:480–9. Available from: https://doi.org/10.1111/dmcn.13697.

[24] Novak I, Morgan C, Adde L, Blackman J, Boyd RN, Brunstrom-Hernandez J, et al. Early, accurate diagnosis and early intervention in cerebral palsy: advances in diagnosis and treatment. JAMA Pediatr 2017;171:897–907. Available from: https://doi.org/10.1001/jamapediatrics.2017.1689.

[25] Luhmann HJ, Sinning A, Yang JW, Reyes-Puerta V, Stuttgen MC, Kirischuk S, et al. Spontaneous neuronal activity in developing neocortical networks: from single cells to large-scale interactions. Front Neural Circuits 2016;10:40. Available from: https://doi.org/10.3389/fncir.2016.00040.

[26] Allen MC. Assessment of gestational age and neuromaturation. Ment Retard Dev Disabil Res Rev 2005;11:21–33. Available from: https://doi.org/10.1002/mrdd.20059.

[27] Johnston MV, Fatemi A, Wilson MA, Northington F. Treatment advances in neonatal neuroprotection and neurointensive care. Lancet Neurol 2011;10:372–82. Available from: https://doi.org/10.1016/S1474-4422(11)70016-3.

[28] Sannia A, Natalizia AR, Parodi A, Malova M, Fumagalli M, Rossi A, et al. Different gestational ages and changing vulnerability of the premature brain. J Matern-Fetal Neonatal Med 2015;2268–72. Available from: https://doi.org/10.3109/14767058.2013.796166.

[29] Illingworth RS. The development of the infant and young child–normal and abnormal. Baltimore: Williams & Wilkins; 1972.

[30] Ismail FY, Fatemi A, Johnston MV. Cerebral plasticity: windows of opportunity in the developing brain. Eur J Paediatr Neurol 2017;21:23–48. Available from: https://doi.org/10.1016/j.ejpn.2016.07.007.

[31] Johnston MV, Hoon Jr. AH. Possible mechanisms in infants for selective basal ganglia damage from asphyxia, kernicterus, or mitochondrial encephalopathies. J Child Neurol 2000;15:588–91. Available from: https://doi.org/10.1177/088307380001500904.

[32] El-Metwally DE, Medina AE. The potential effects of NICU environment and multisensory stimulation in prematurity. Pediatr Res 2020;88:161–2. Available from: https://doi.org/10.1038/s41390-019-0738-4.

[33] Vitale FM, Chirico G, Lentini C. Sensory stimulation in the NICU environment: devices, systems, and procedures to protect and stimulate premature babies. Children 2021;8:334. Available from: https://doi.org/10.3390/children8050334.

[34] Geronimus AT. The weathering hypothesis and the health of African–American women and infants: evidence and speculations. Ethn Dis 1992;2:207–21.

[35] Manuck TA. Racial and ethnic differences in preterm birth: a complex, multifactorial problem. Semin Perinatol 2017;41:511–18.

[36] Blencowe H, Cousens S, Oestergaard MZ, Chou D, Moller AB, Narwal R, et al. National, regional, and worldwide estimates of preterm birth rates in the year 2010 with time trends since 1990 for selected countries: a systematic analysis and implications. Lancet 2012;379:2162–72. Available from: https://doi.org/10.1016/S0140-6736(12)60820-4.

[37] Spittle AJ, Thompson DK, Olsen JE, Kwong A, Treyvaud K. Predictors of long-term neurodevelopmental outcomes of children born extremely preterm. Semin Perinatol 2021;45:151482.

[38] Volpe JJ. Brain injury in premature infants: a complex amalgam of destructive and developmental disturbances. Lancet Neurol 2009;8:110–24. Available from: https://doi.org/10.1016/S1474-4422(08)70294-1.

[39] Kinney HC, Volpe JJ. Modeling the encephalopathy of prematurity in animals: the important role of translational research. Neurol Res Int 2012;2012. Available from: https://doi.org/10.1155/2012/295389 295389–17.

[40] Volpe JJ. The encephalopathy of prematurity–brain injury and impaired brain development inextricably intertwined. Semin Pediatr Neurol 2009;16:167–78. Available from: https://doi.org/10.1016/j.spen.2009.09.005.

[41] Bae MH, Jang HJ, Lee NR, Han YM, Byun SY, Park KH. The clinical characteristics and neurodevelopmental outcome of preterm infants with persistent periventricular echogenicity. Pediatr Neonatol 2018;59:606–10. Available from: https://doi.org/10.1016/j.pedneo.2018.02.006.

[42] Sacchi C, Marino C, Nosarti C, Vieno A, Visentin S, Simonelli A. Association of intrauterine growth restriction and small for gestational age status with childhood cognitive outcomes: a systematic review and meta-analysis. JAMA Pediatr 2020;174:772–81. Available from: https://doi.org/10.1001/jamapediatrics.2020.1097.

[43] Jensen EA, Foglia EE, Dysart KC, Simmons RA, Aghai ZH, Cook A, et al. Adverse effects of small for gestational age differ by gestational week among very preterm infants. Arch Dis Child Fetal Neonat Ed 2019;104:F192–8. Available from: https://doi.org/10.1136/archdischild-2017-314171.

[44] Sharma D, Shastri S, Sharma P. Intrauterine growth restriction: antenatal and postnatal aspects. Clin Med Insights Pediatr 2016;10:67–83. Available from: https://doi.org/10.4137/CMPed.S40070.

[45] Geva R, Eshel R, Leitner Y, Valevski AF, Harel S. Neuropsychological outcome of children with intrauterine growth restriction: a 9-year prospective study. Pediatics 2006;118:91–100.

[46] Beukers F, Aarnoudse-Moens CSH, van Weissenbruch MM, Ganzevoort W, van Goudoever JB, van Wassenaer-Leemhuis AG. Fetal growth restriction with brain sparing: neurocognitive and behavioral outcomes at 12 years of age. J Pediatr 2017;188:103–109.e2.

[47] Pels A, Mensing van Charante N, Vollgraff Heidweiller-Schreurs C, Limpens J, Wolf H, de Boer M, et al. The prognostic accuracy of short term variation of fetal heart rate in early-onset fetal growth restriction: a systematic review. Eur J Obstetr Gynecol Reprod Biol 2019;234:179–84. Available from: https://doi.org/10.1016/j.ejogrb.2019.01.005.

[48] Levine TA, Grunau RE, McAuliffe FM, Pinnamaneni R, Foran A, Alderdice FA. Early childhood neurodevelopment after intrauterine growth restriction: a systematic review. Pediatrics 2015;135:126–41. Available from: https://doi.org/10.1542/peds.2014-1143.

[49] Hwang IT. Long-term care, from neonatal period to adulthood, of children born small for gestational age. Clin Pediatr Endocrinol 2019;28:97–103. Available from: https://doi.org/10.1297/cpe.28.97.

[50] de Bie HM, Oostrom KJ, Delemarre-van de Waal HA. Brain development, intelligence and cognitive outcome in children born small for gestational age. Horm Res Paediatr 2010;73:6–14. Available from: https://doi.org/10.1159/000271911.

[51] Shankaran S, Laptook AR, Ehrenkranz RA, Tyson JE, McDonald SA, Donovan EF, et al. Whole-body hypothermia for neonates with hypoxic-ischemic encephalopathy. N Engl J Med 2005;353:1574–84. Available from: https://doi.org/10.1056/NEJMcps050929.

[52] Azzopardi DV, Strohm B, Edwards AD, Dyet L, Halliday HL, Juszczak E, et al. Moderate hypothermia to treat perinatal asphyxial encephalopathy. N Engl J Med 2009;361:1349–58. Available from: https://doi.org/10.1056/NEJMoa0900854.

[53] Gluckman PD, Wyatt JS, Azzopardi D, Ballard R, Edwards AD, Ferriero DM, et al. Selective head cooling with mild systemic hypothermia after neonatal encephalopathy: multicentre randomised trial. Lancet 2025;365:663–70. Available from: https://doi.org/10.1016/S0140-6736(05)17946-X.

[54] Volpe JJ. Neonatal encephalopathy: an inadequate term for hypoxic-ischemic encephalopathy. Ann Neurol 2012;72:156–66. Available from: https://doi.org/10.1002/ana.23647.

[55] Robertson CM, Finer NN, Grace MG. School performance of survivors of neonatal encephalopathy associated with birth asphyxia at term. J Pediatr 1989;114:753–60. Available from: https://doi.org/10.1016/s0022-3476(89)80132-5.

[56] Natarajan G, Pappas A, Shankaran S. Outcomes in childhood following therapeutic hypothermia for neonatal hypoxic-ischemic encephalopathy (HIE. Semin Perinatol 2016;40(8):549–55. Available from: https://doi.org/10.1053/j.semperi.2016.09.007.

[57] Pappas A, Shankaran S, McDonald SA, Vohr BR, Hintz SR, Ehrenkranz RA, et al. Cognitive outcomes after neonatal encephalopathy. Pediatrics 2015;135:e624–34. Available from: https://doi.org/10.1542/peds.2014-1566.

[58] Azzopardi D, Strohm B, Marlow N, Brocklehurst P, Deierl A, Eddama O, et al. Effects of hypothermia for perinatal asphyxia on childhood outcomes. N Engl J Med 2014;371(2):140–9. Available from: https://doi.org/10.1056/NEJMoa1315788.

[59] Marlow N, Rose AS, Rands CE, Draper ES. Neuropsychological and educational problems at school age associated with neonatal encephalopathy. Arch Dis Child Fetal Neonatal Ed 2005;90:F380–7. Available from: https://doi.org/10.1136/adc.2004.067520.

[60] van Handel M, Swaab H, de Vries LS, Jongmans MJ. Behavioral outcome in children with a history of neonatal encephalopathy following perinatal asphyxia. J Pediatr Psychol 2010;35:286–95. Available from: https://doi.org/10.1093/jpepsy/jsp049.

[61] Finder M, Boylan GB, Twomey D, Ahearne C, Murray DM, Hallberg B. Two-year neurodevelopmental outcomes after mild hypoxic ischemic encephalopathy in the era of therapeutic hypothermia. JAMA Pediatr 2020;174:48–55. Available from: https://doi.org/10.1001/jamapediatrics.2019.4011.

[62] Kariholu U, Montaldo P, Markati T, Lally PJ, Pryce R, Teiserskas J, et al. Therapeutic hypothermia for mild neonatal encephalopathy: a systematic review and meta-analysis. Arch Dis Child Fetal Neonatal Ed 2020;105:225–8. Available from: https://doi.org/10.1136/archdischild-2018-315711.

[63] Rantala H, Ingalsuo H. Occurrence and outcome of epilepsy in children younger than 2 years. J Pediatr 1996;135:761. Available from: https://doi.org/10.1016/s0022-3476(99)70098-3.

[64] Hirvonen M, Ojala R, Korhonen P, Haataja P, Eriksson K, Rantanen K, et al. Intellectual disability in children aged less than seven years born moderately and late preterm compared with very preterm and term-born children – a nationwide birth cohort study. J Intellect Disabil Res 2017;61:1034–54. Available from: https://doi.org/10.1111/jir.12394.

[65] Jacobs SE, Berg M, Hunt R, Tarnow-Mordi WO, Inder TE, Davis PG, et al. Cooling for newborns with hypoxic ischaemic encephalopathy. Cochrane Database Syst Rev 2013;2013:CD003311. Available from: https://doi.org/10.1002/14651858.CD003311.pub.

[66] Bhutta AT, Cleves MA, Casey PH, Cradock MM, Anand KJ. Cognitive and behavioral outcomes of school-aged children who were born preterm: a meta-analysis. JAMA 2002;288:728–37. Available from: https://doi.org/10.1001/jama.288.6.728.

[67] Edwards J, Berube M, Erlandson K, Haug S, Johnstone H, Meagher M, et al. Developmental coordination disorder in school-aged children born very preterm and/or at very low birth weight: a systematic review. J Dev Behav Pediatr 2011;32:678–87. Available from: https://doi.org/10.1097/DBP.0b013e31822a396a.

[68] Aarnoudse-Moens CSH, Weisglas-Kuperus N, van Goudoever JB, Oosterlaan J. Meta-analysis of neurobehavioral outcomes in very preterm and/or very low birth weight children. Pediatrics 2009;124:717–28. Available from: https://doi.org/10.1542/peds.2008-2816.

[69] Sansavini A, Annalisa G, Caselli MC. Preterm birth: neuropsychological profiles and atypical developmental pathways. Dev Disabil Res Rev 2011;17:102–13.

[70] Hutchinson EA, De Luca CR, Doyle LW, Roberts G, Anderson PJ. School-age outcomes of extremely preterm or extremely low birth weight children. Pediatrics 2013;131:e1053–61. Available from: https://doi.org/10.1542/peds.2012-2311.

[71] Kerr-Wilson CO, Mackay DF, Smith GC, Pell JP. Meta-analysis of the association between preterm delivery and intelligence. J Public Health (Oxf) 2012;34(2):209–16. Available from: https://doi.org/10.1093/pubmed/fdr024.

[72] Wood NS, Costeloe K, Gibson AT, Hennessy EM, Marlow N, Wilkinson AR. The EPICure study: associations and antecedents of neurological and developmental disability at 30 months of age following extremely preterm birth. Archives of Disease in Childhood-Fetal and Neonatal Edition. 2005;90(2):134–40.

[73] Lohaugen GC, Ostgard HF, Andreassen S, Jacobsen GW, Vik T, Brubakk AM, et al. Small for gestational age and intrauterine growth restriction decreases cognitive function in young adults. J Pediatr 2013;163:447–53. Available from: https://doi.org/10.1016/j.jpeds.2013.01.060.

[74] Grunau RE, Whitfield MF, Davis C. Pattern of learning disabilities in children with extremely low birth weight and broadly average intelligence. Arch Pediatr Adolesc Med 2002;156:615–20. Available from: https://doi.org/10.1001/archpedi.156.6.615.

[75] Wolke D, Samara M, Bracewell MA, Marlow N. Specific language difficulties and school achievement in children born at 25 weeks of gestation or less. J Pediatr 2008;256–62. Available from: https://doi.org/10.1016/j.jpeds.2007.06.043.

[76] Luu TM, Vohr BR, Schneider KC, Katz KH, Tucker R, Allan WC, et al. Trajectories of receptive language development from 3 to 12 years of age for very preterm children. Pediatrics 2009;124:333–41. Available from: https://doi.org/10.1542/peds.2008-2587.

[77] Taylor HG, Minich NM, Klein N, Hack M. Longitudinal outcomes of very low birth weight: neuropsychological findings. J Int Neuropsychol Soc 2004;10:149–63. Available from: https://doi.org/10.1017/S1355617704102038.

[78] Schirmer CR, Portuguez MW, Nunes ML. Clinical assessment of language development in children at age 3 years that were born preterm. Arq Neuropsiquiatr 2006;64:926–31. Available from: https://doi.org/10.1590/S0004-282X2006000600007.

[79] van Noort-van der Spek IL, Franken MC, Weisglas-Kuperus N. Language functions in preterm-born children: a systematic review and meta-analysis. Pediatrics 2012;129(4):745–54. Available from: https://doi.org/10.1542/peds.2011-1728.

[80] Johnson S, Marlow N. Preterm birth and childhood psychiatric disorders. Pediatr Res 2011;69:11R–8R. Available from: https://doi.org/10.1203/PDR.0b013e318212faa0.

[81] Mulder H, Pitchford NJ, Marlow N. Processing speed and working memory underlie academic attainment in very preterm children. Arch Dis Child Fetal Neonatal Ed 2010;95:F267–72. Available from: https://doi.org/10.1136/adc.2009.167965.

[82] Allin M, Walshe M, Fern A, Nosarti C, Cuddy M, Rifkin L, et al. Cognitive maturation in preterm and term born adolescents. J Neurol Neurosurg Psychiatry 2008;79:381–6. Available from: https://doi.org/10.1136/jnnp.2006.110858.

[83] Marlow N, Hennessy EM, Bracewell MA, Wolke D, EPICure Study Group. Motor and executive function at 6 years of age after extremely preterm birth. Pediatrics 2007;120(4):793–804.

[84] Anderson PJ, Doyle LW, Victorian Infant Collaborative Study Group. Executive functioning in school-aged children who were born very preterm or with extremely low birth weight in the 1990s. Pediatrics 2004;114(1):50–7. Available from: https://doi.org/10.1542/peds.114.1.50.

[85] Bayless S, Stevenson J. Executive functions in school-age children born very prematurely. Early Hum Dev 2006;83:247–54. Available from: https://doi.org/10.1016/j.earlhumdev.2006.05.021.

[86] Limperopoulos C, Bassan H, Sullivan NR, Soul JS, Robertson Jr RL, Moore M, et al. Positive screening for autism in ex-preterm infants: prevalence and risk factors. Pediatrics 2008;121(4):758–65.

[87] Kuban KC, O'Shea TM, Allred EN, Tager-Flusberg H, Goldstein DJ, Leviton A. Positive screening on the modified checklist for autism in toddlers (M-CHAT) in extremely low gestational age newborns. J Pediatr 2009;154(4):535–40. Available from: https://doi.org/10.1016/j.jpeds.2008.10.011.

[88] Moore GS, Kneitel AW, Walker CK, Gilbert WM, Xing G. Autism risk in small- and large-for-gestational-age infants. Am J Obstet Gynecol 2012;206(4):314.e1–9. Available from: https://doi.org/10.1016/j.ajog.2012.01.044.

[89] Harel-Gadassi A, Friedlander E, Yaari M, Bar-Oz B, Eventov-Friedman S, Mankuta D, et al. Risk for ASD in preterm infants: a three-year follow-up study. Autism Res Treat 2018;2018:8316212–19. Available from: https://doi.org/10.1155/2018/8316212.

[90] Stephens BE, Bann CM, Acarregui MJ, Pappas A, Adams-Chapman I, McGowan EC, et al. Screening for autism spectrum disorders in extremely preterm infants. J Dev Behav Pediatr 2012;33:535–41. Available from: https://doi.org/10.1097/DBP.0b013e31825fd0af.

[91] Hirschberger RG, Kuban KCK, O'Shea TM, Joseph RM, Heeren T, Douglass LM, et al. Co-occurrence and severity of neurodevelopmental burden (cognitive impairment, cerebral palsy, autism spectrum disorder, and epilepsy) at age ten years in children born extremely preterm. Pediatr Neurol 2018;79:45–52. Available from: https://doi.org/10.1016/j.pediatrneurol.2017.11.002.

[92] Korzeniewski SJ, Joseph RM, Kim SH, Allred EN, O'Shea TM, Leviton A, et al. Social Responsiveness scale assessment of the preterm behavioral phenotype in 10-year-olds born extremely preterm. J Dev Behav Pediatr 2017;38:697–705.

[93] Baird G, McConachie H, Scrutton D. Parents' perceptions of disclosure of the diagnosis of cerebral palsy. Arch Dis Child 2000;83:475–80. Available from: https://doi.org/10.1136/adc.83.6.475.

[94] Bemister TB, Brooks BL, Dyck RH, Kirton A. Parent and family impact of raising a child with perinatal stroke. BMC Pediatr 2014;14:182–243. Available from: https://doi.org/10.1186/1471-2431-14-182.

[95] Guttmann K, Flibotte J, DeMauro SB. Parental perspectives on diagnosis and prognosis of neonatal intensive care unit graduates with cerebral palsy. J Pediatr 2018;203:156–62. Available from: https://doi.org/10.1016/j.jpeds.2018.07.089.

[96] Maitre NL, Burton VJ, Duncan AF, Iyer S, Ostrander B, Winter S, et al. Network implementation of guideline for early detection decreases age at cerebral palsy diagnosis. Pediatrics 2020;145. Available from: https://doi.org/10.1542/peds.2019-2126.

[97] Als H, Butler S, Kosta S, McAnulty G. The assessment of preterm infants' behavior (APIB): furthering the understanding and measurement of neurodevelopmental competence in preterm and full-term infants. Ment Retard Dev Disabil Res Rev 2005;11:94–102. Available from: https://doi.org/10.1002/mrdd.20053.

[98] Carmichael K, Burns Y, Gray P, O'Callaghan M. Neuromotor behavioural assessment of preterm infants at risk for impaired development. Aust J Physiother 1997;43:101–7. Available from: https://doi.org/10.1016/S0004-9514(14)60404-9.

[99] Spittle A, Orton J, Anderson P, Boyd R, Doyle L. Early developmental intervention programmes provided post hospital discharge to prevent motor and cognitive impairment in preterm infants. Cochrane Database Syst Rev 2015;2015. Available from: https://doi.org/10.1002/14651858.CD005495.pub4.

[100] Dusing SC, Tripathi T, Marcinowski EC, Thacker LR, Brown LF, Hendricks-Munoz KD. Supporting play exploration and early developmental intervention versus usual care to enhance development outcomes during the transition from the neonatal intensive care unit to home: a pilot randomized controlled trial. BMC Pediatr 2018;18:46. Available from: https://doi.org/10.1186/s12887-018-1011-4.

[101] Gateau K, Song A, Vanderbilt DL, Gong C, Friedlich P, Kipke M, et al. Maternal post-traumatic stress and depression symptoms and outcomes after NICU discharge in a low-income sample: a cross-sectional study. BMC Pregnancy Childb 2021;21:48. Available from: https://doi.org/10.1186/s12884-020-03536-0.

[102] Roque ATF, Lasiuk GC, Radunz V, Hegadoren K. Scoping review of the mental health of parents of infants in the NICU. J Obstet Gynecol Neonatal Nurs 2017;46:576–87. Available from: https://doi.org/10.1016/j.jogn.2017.02.005.

[103] Byrne R, Duncan A, Pickar T, Burkhardt S, Boyd RN, Neel ML, et al. Comparing parent and provider priorities in discussions of early detection and intervention for infants with and at risk of cerebral palsy. Child Care Health Dev 2019;45:799–807. Available from: https://doi.org/10.1111/cch.12707.

[104] Spittle AJ, Anderson PJ, Lee KJ, Ferretti C, Eeles A, Orton J, et al. Preventive care at home for very preterm infants improves infant and caregiver outcomes at 2 years. Pediatrics 2010;126:e171–8. Available from: https://doi.org/10.1542/peds.2009-3137.

[105] Williams SA, Alzaher W, Mackey A, Hogan A, Battin M, Sorhage A, et al. "It should have been given sooner, and we should not have to fight for it": a mixed-methods study of the experience of diagnosis and early management of cerebral palsy. J Clin Med 2021;10. Available from: https://doi.org/10.3390/jcm10071398.

[106] Allen MC, Dorner RA, Burton VJ. Follow-up of high-risk infants. In: 8th Edition Gomella TL, Eyal FG, Bany-Mohammed F, editors. Gomella's neonatology: management, procedure, on-call problems, disease, and drugs, 2020. Chicago: McGraw-Hill Publishing; 2020.

[107] Part 303 (part C)—early intervention program for infants and toddlers with disabilities. Individuals with Disabilities Education Act. Published May 2, 2017 [accessed 05.07.2023]. https://sites.ed.gov/idea/regs/c.

Chapter 13

Pediatric acquired brain injury

Archana Nelliot[1], Karen Evankovich[2], Kristen Fisher[3], Kimberly C. Davis[2], Eliza Gordon-Lipkin[4] and Sarah Risen[3]

[1]Department of Neurology and Developmental Medicine, Kennedy Krieger Institute, Department of Pediatrics, Johns Hopkins University School of Medicine, Baltimore, MD, United States, [2]Department of Pediatrics, Sections of Psychology and Neurology, Texas Children's Hospital, Houston, TX, United States , [3]Department of Pediatrics, Section of Neurology and Developmental Neuroscience, Baylor College of Medicine, Texas Children's Hospital, Houston, TX, United States, [4]Metabolism, Infection and Immunity Section, National Human Genome Research Institute, National Institutes of Health, Bethesda, MD, United States

Introduction

Acquired brain injury (ABI) results from a broad spectrum of traumatic and nontraumatic etiologies and is associated with an array of secondary neurodevelopmental disabilities. ABI is a nonprogressive insult to the brain resulting in acute and chronic disruption of neurological function. In children, traumatic brain injury (TBI) is the most common cause of ABI. Nontraumatic etiologies of ABI include anoxic/hypoxic-ischemic brain injury, infectious encephalitis/meningitis, neuroimmune and postinfectious disorders, and cerebrovascular disorders. While there are other causes of pediatric ABI, several specific etiologies of brain injury are included in other chapters and will not be the focus of this chapter [prematurity/neonatal hypoxic-ischemic encephalopathy (see Chapter Early brain injury), neurological manifestations of systemic diseases including neurooncological processes (see Chapter Systemic diseases with increased risk for neurodevelopmental disorders), and toxic-metabolic exposures. Despite wide-ranging etiologies and underlying pathophysiology, the long-term neurodevelopmental outcomes and the necessary medical, therapeutic, and educational interventions converge on a common final pathway.

The variability in pediatric ABI sequelae and recovery results from (1) etiology, (2) affected brain regions, (3) injury severity, (4) developmental stage in which the insult occurs, (5) underlying genetic and metabolic factors affecting the secondary responses to ABI, and (6) social environmental features such as family functioning, parental practices, and stressors and resources in the home [1–6]. Compared to adults, brain injury in children occurs in the context of underlying neurodevelopment, a meticulously orchestrated series of physiological processes occurring throughout childhood (Fig. 13.1) [7]. Focal or diffuse brain insults disrupt this ongoing neuromaturation, resulting in long-term neurodevelopmental impairments.

As functional skills become more complex, functional impairments may only become apparent over time. In other words, a brain insult that occurs at only one point in time may impact neural connectivity and neural maturation, potentially impairing typical brain developmental processes throughout the remainder of life. Finally, there are familial, environmental, epigenetic, and social variables that influence both injury and outcomes in children. Thus each brain injury is unique, and children with ABI require long-term neurodevelopmental monitoring.

Depending on the etiology, location and extent of brain injury, and age at injury, children with ABI are at risk for long-term impairments in fine and gross motor functioning, the sensory abilities (vision and hearing), cognitive and language abilities, behavioral and emotion regulation, social and adaptive skills, endocrine function, and sleep regulation. These ABI-related impairments, either in isolation or combination, not only adversely affect a child's functioning at home, in school, and in the community but also impact the whole family. Therefore a comprehensive multidisciplinary team with longitudinal developmental follow-up is essential for identifying problems and providing interventions to optimize the child's ability to participate in all environments while also providing support for the family.

This chapter will first review several etiologies of ABI in children, then describe the neurodevelopmental outcomes common to all causes of ABI, and finally will briefly summarize outpatient care strategies unique to children and their families following brain injury.

Capute and Accardo's Neurodevelopmental Disabilities in Infancy and Childhood. DOI: https://doi.org/10.1016/B978-0-12-824060-1.00052-3

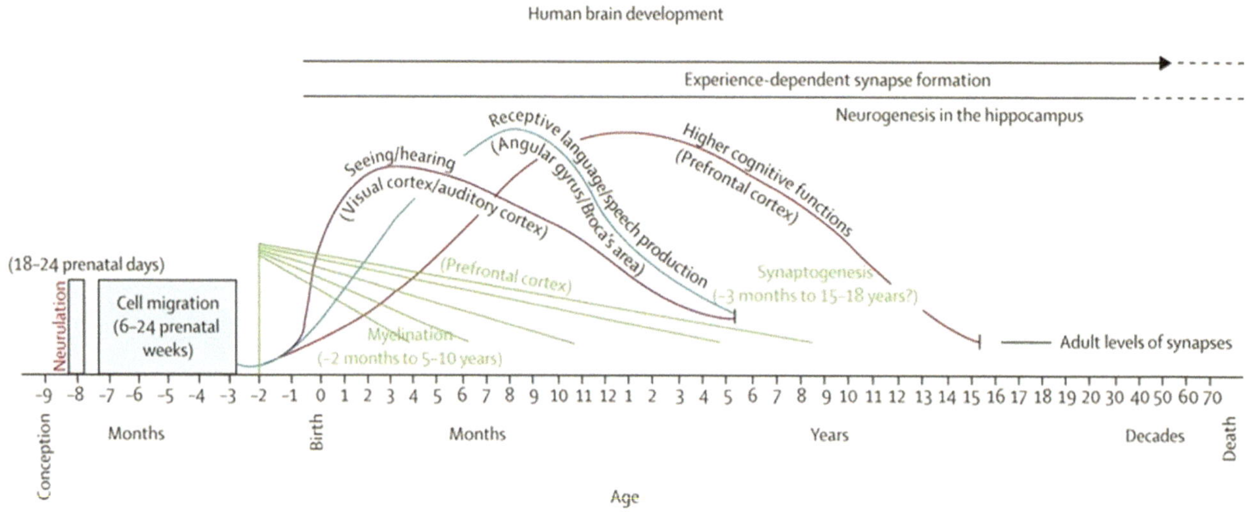

FIGURE 13.1 Early brain development. Depiction of (1) the neurodevelopmental physiology that occurs prenatally and following birth including myelination and synaptogenesis and (2) the overlap of critical periods for developmental skills with the underlying pathophysiology. A brain injury at any age impacts both already acquired functional skills as well as the development of future skills secondary to disruption of these developmental pathophysiological processes. Source: *From Grantham McGregor et al, [7] adapted from Thompson R.A., Nelson C.A. Developmental science and the media. Early brain development. Am Psychol. 2001;56(1):5-15. https://doi.org/10.1037/0003-066x.56.1.5.*

Etiologies of acquired brain injury

Traumatic brain injury, including abusive head trauma

Epidemiology

Approximately 2.8 million Americans sustained a TBI in 2013, with the highest incidence of TBI occurring in the infant/toddler, adolescent, and elderly populations [8]. In children, there are more than a half-million TBI-related emergency department (ED) visits consistent with mild injury (concussion). Worldwide, the incidence of pediatric TBI varies greatly by country and ranges between 47 and 280 children per 100,000; the incidence in the United States is 70–75 children per 100,000 [9].

The mechanism of injury for pediatric TBIs varies by age. Globally, pediatric TBIs result from motor vehicle accidents, especially in adolescents. In children younger than 1 year of age, abusive head trauma (AHT) is the leading cause of injury with the highest incidence in the first 1–2 months of life [8]. The rate of AHT in infants/toddlers is consistent worldwide. In the United States, the incidence of AHT is ~ 8 per 100,000 children of all ages, though the incidence is as high as 40 per 100,000 in infants [10–12]. Falls are the leading cause of TBI in children 5–14 years of age. Sports injuries (especially football, lacrosse, and hockey), recreational activities (e.g., skiing, surfing), and assault are also important causes of head injury [13,14]. Independent of etiology, a substantial number of children that survive moderate and severe TBIs are left with permanent disability, associated with high health care resource utilization as well as significant economic burden [15,16].

Neurobiology

TBI results from a direct or indirect biomechanical force exerted on the brain. Primary brain injury occurs at impact and can include direct tissue damage in penetrating injuries, skull fractures, cerebral contusions (coup and contrecoup), vascular injury resulting in hemorrhage in or around the brain, and diffuse axonal injury (shearing of the nerves that send information between different areas of the brain). A neurometabolic cascade is also triggered at the moment of impact, including widespread release of glutamate and other neurotransmitters, ionic changes, and metabolic alterations. TBI also results in impaired cerebral blood flow and glucose metabolism [17]. These changes continue to evolve over days and contribute to additional (secondary) brain injury.

Secondary brain injury refers to a complex, multifaceted cascade of events in the hours to days after the primary insult, resulting in further brain injury [18–23]. On a macro level, secondary injuries include brain swelling/edema, increased intracranial pressure, ischemia, seizures, and vascular dysregulation. At a molecular level ongoing excitotoxicity, mitochondrial dysfunction/damage, blood–brain barrier disruption, neuroinflammation, oxidative stress, reduced glucose metabolism, alteration of synapses, and ongoing axonal degeneration ultimately culminate in apoptotic cell death. Additional systemic alterations, such as elevated body temperature, hyper- or hypoglycemia, or hypotension, have been associated with poorer outcomes, likely compounding the effects of secondary brain injury [24,25]. Secondary brain injuries are more common in children with moderate to severe TBIs and especially in children with AHT compared to accidental TBI [26–28].

Presentation and diagnosis

Most children with a TBI present after a mechanical injury to the head, which may be followed by loss of consciousness and/or posttraumatic amnesia (PTA). For all severities of TBI, symptoms present immediately; however, the number, severity, and type of symptoms may evolve over several hours to days. Epidural hematoma most often presents with loss of consciousness but one-third will have normal exams and be asymptomatic or only report subtle symptoms initially [29]. In severe TBI, children are more likely than adults to have delayed increased intracranial hypertension [30].

Clinical history detailed neurological exam and imaging [head computed tomography (CT), X-ray and/or brain magnetic resonance imaging (MRI)] are cornerstones of TBI evaluation. Decision making on the need for head CT in an ED is often guided by Pediatric Emergency Care Applied Research Network guidelines [31] to identify intracranial bleeds and determine if there is need for emergent neurosurgical intervention. However, acute head CT imaging has limited reliability in determining the severity or extent of the TBI and associated morbidity (Table 13.1).

There are two common ways to categorize the severity of TBI: the Glasgow Coma Scale (GCS) [32,33] and the duration of PTA (Table 13.1). The GCS assesses motor response (scored from 1 to 6), verbal response (1–5), and eye opening (1–4) within the first 24 hours after brain injury. Following acute resuscitation, a GCS score ≤ 8 indicates a severe TBI; scores 9–12 indicate a moderate TBI; and scores 13–15 indicate a mild TBI. While lower GCS is associated with increased mortality and morbidity, GCS alone has poor prognostic ability, and children with TBI and children with TBI and GCS 3–5 can still have meaningful recovery with functional independence [34–36].

In school-aged children, the duration of PTA can be measured using the Children's Orientation and Amnesia Test [37]. Duration of PTA as well as the time to follow commands (TFC) have been shown to be the most consistent and reliable, though weak, predictors of long-term outcome [38,39]. Accurate determination of TFC requires a child to be unimpaired by sedating medications though may be intubated; however, assessment of PTA is more challenging to assess when a child is intubated as testing requires verbal responses to assess orientation and memory (Table 13.2).

TABLE 13.1 Categorization of traumatic brain injury severity.

Severity	Glasgow Coma Scale	Time to follow commands	Duration of posttraumatic amnesia
Mild	13–15	<30 min	< 24 h
Moderate	9–12	1–6 h	1–7 days
Severe	3–8	> 6 h	> 7 days

TABLE 13.2 Reliability of measures of injury severity in traumatic brain injury.

Measure of injury severity	Reliability
Loss of consciousness	Weakest measure due to inconsistent documentation.
Neurologic examination	Minimal assistance while sedated or in coma.When examination possible, helpful to detect focal deficits.
Head CT	Good to detect skull fractures, bleeding, and focal lesions. Inadequate to assess diffuse axonal injury.
GCS	Evidence primarily supports use to predict mortality.
PTA and TFC	Best evidence to predict morbidity.

CT, Computed tomography; *GCS*, Glasgow Coma Scale; *PTA*, posttraumatic amnesia; *TFC*, time to follow commands.

Clinical course

Mild traumatic brain injury Mild traumatic brain injury (mTBI), also known as a concussion, is the most common form of TBI and represents roughly 75% of all TBIs in the United States [40]. As described in Table 13.1, mTBIs are classified as head injuries involving loss of consciousness of 30 minutes or less (and the majority of children have no loss of consciousness), with a GCS of 13–15 and PTA lasting less than 24 hours. Complicated mTBIs have a similar clinical presentation but with evidence of intracranial injury including subarachnoid hemorrhages, intracranial contusions, or extra axial hematomas, which may require intervention.

The course of recovery after mTBI is variable and associated with alterations in brain function for a period of time after the immediate injury. Importantly, the expected time frames for normal clinical recovery in children (4 weeks) is different than the typical course of recovery in adults (within 2 weeks) [41–43]. While the most common concussion symptoms vary by the age of the child at time of injury, concussive symptoms typically fall into four domains: physical (e.g., headache, dizziness), cognitive (e.g., inattention, feeling slowed down, executive function difficulties), sleep (e.g., poor sleep, difficulty falling asleep), and mood (e.g., irritability, sadness, nervousness). Preschool children may present with atypical postconcussive symptoms including stomachaches, appetite changes, enuresis, clinginess/increased dependence, disengagement, and behavioral dysregulation [44]. While concussive symptoms severity may wax and wane over the course of recovery, ~75% of children with mTBI recover 6–8 weeks after injury, though high school athletes recover more quickly, within 2–4 weeks on average [45]. A small percentage of children (2%–3%) report symptoms >1 year after their initial injury.

Postconcussion syndrome describes the 10%–25% of patients in whom concussive symptoms persist for over 6 weeks [46]. Factors most strongly associated with a prolonged recovery include family factors (Hispanic ethnicity, lower socioeconomic status, parental distress, and anxiety) and a premorbid neurological or psychiatric diagnoses including learning disabilities, behavioral problems, premorbid anxiety, and lower resilience [40,47–49]. Other associated risk factors include female sex, injury occurring in adolescence, prior concussions, early presentation of symptoms, and carriers of an apolipoprotein E4 allele [43,48]. The presence of prolonged postconcussive symptoms is associated with significant decreases in health care-related quality of life in all domains of social and emotional functioning and particularly impacting school functioning [50].

Given the significant and wide-ranging impact of persistent concussive symptoms, medical care requires an interdisciplinary team focused on longitudinal monitoring of symptoms and a comprehensive management plan, including school accommodations, monitored exercise, and emotional support, and treatment for ongoing physical symptoms such as sleep disruption is necessary. Providing the child and parents with early education on the expected concussive symptoms, the typical course of recovery, and lifestyle factors to implement immediately to support concussion recovery can significantly reduce the likelihood of developing postconcussion syndrome [51]. There is an overall paucity of evidence for pharmacologic intervention in the management of acute symptoms and postconcussive syndrome. Hypertonic saline administration in the acute period after injury has shown some benefit in treating and preventing headaches [52], as has short-term use of non steroidal antiinflammatory (NSAIDs) and acetaminophen. Although used frequently to treat nausea, there is limited evidence that early use of antiemetics results in early symptom reduction. Early return to physical activity and subsymptomatic exercise protocols have also been shown to support concussion recovery, both in reduced number and shorter duration of symptoms [53].

A very brief period of complete physical and cognitive rest, generally of 24–48 hours, followed by return to both cognitive and physical activity is the mainstay of management after a concussion. School accommodations are critical in the care plan for children with concussion. Returning to school is important for numerous reasons, including maintaining a routine, keeping up with school work, and social interactions; however, a child with a concussion should have concussion-related accommodations immediately upon return to school. A "return to learn" plan is recommended by the Centers for Disease Control (CDC) [54] and American Academy of Pediatrics [55,56], both offering guidelines and materials covering common accommodations [54]. There has been consistent evidence, including randomized controlled trials, that graded, subsymptomatic levels of exercise actually speed up recovery [57]. Medical management is often symptom-based (e.g., to reduce the burden of symptoms such as posttraumatic headaches, mood disorders, sleep issues, and neurocognitive deficits) and should be monitored by a physician to avoid secondary complications that impede recovery, such as medication overuse headaches. In children with mTBI, a tailored, individualized approach is generally recommended with close initial follow-up.

Moderate and severe traumatic brain injury The criteria for diagnosis of moderate and severe TBI are listed in Table 13.1 and include longer periods of inability to follow commands (1–6 hours for moderate TBI and more than 6 hours for severe TBI) and longer periods of PTA (up to 7 days in moderate TBI and more than 7 days in severe TBI).

The course of recovery can be variable. Severe TBI with more extensive primary injury, more frequent and significant secondary injuries, a longer intensive care unit (ICU) stay, and a longer hospital stay [58,59]. These children often require the aggressive use of neuroprotective measures to reduce and prevent the impact of secondary brain injury; therefore children with severe TBI are typically transferred to a level 1 Pediatric Trauma Center. In 2019 the Brain Trauma Foundation published the third edition of the Guidelines on the Management of Severe Pediatric TBI [60,61], providing comprehensive recommendations on the critical care, neurosurgical interventions, clinical parameters, and neuromonitoring in children with severe TBI.

Following severe TBI, recovery is most rapid during the first year and slows during the following years [62,63]. However, children with TBI can continue to gain new functional skills over their lifetime. Recovery is further influenced by family factors, including socioeconomic status, resources, coping skills, and support systems [35,58,64]. Additionally, variability in the rate of recovery and the differing effects on neurodevelopmental domains in children who sustain TBI at different ages suggest an age-dependent neuro-pathophysiological influence on neurodevelopmental outcome [65]. For many years it was felt that the younger the age of injury, the better the outcome due to the plasticity of the brain. However, the influence of age on TBI recovery is much more nuanced. In some, though not all, studies, an earlier age (infants and toddlers) of injury has been associated with increased risk for lasting impairments across all areas compared to older children who sustain a brain injury [66–68]. Timing of injury relative to a child's stage of neurodevelopment will impact different developmental domains that are more vulnerable during that period of development. For example, a child may return to the previous developmental level following injury but may have trouble progressing past that stage [65,69]. This difficulty with progressing into new developmental levels or attaining new developmental milestones has been described as a neurocognitive stall [65]. This may be noted in all age groups and is typically manifest as a slowing or lack of further progression in the intellectual, motor, and social development of the child 1 year after injury. Beyond this period of cognitive stall, children will continue to gain skills, though likely at a slower developmental rate.

Abusive head trauma AHT has a high mortality rate and is also a major cause of morbidity in young children [70,71]. In fact, given the unique differences in the infant and young child's nervous system and the major neurodevelopmental changes that occur throughout the first years of life, infants and very young children appear to be particularly vulnerable to effects of brain injury, especially cognitive development [67,72]. Furthermore, secondary brain injuries (including seizures, increased intracranial pressure, diffuse hypoxic-ischemic injury, and stroke) are more common in AHT compared to accidental TBI [26–28]. Infants and children with AHT are also more likely to present with clinical seizures; additionally, this population has a very high risk for electrographic-only seizures (nonclinical or subclinical seizures, 50%–60%), and of those, 30% develop nonconvulsive status epilepticus [73,74]. Given the potentiation of secondary brain injury from seizure activity, continuous EEG monitoring, if available, is recommended for severe TBI, especially due to AHT. Seizures may be triggered by the neurometabolic cascade, blood products, hypoxemia, and neuroinflammation. The presence of increased mortality, morbidity, and secondary injuries in AHT emphasizes the importance of strict adherence to the severe TBI guidelines and the use of aggressive neuroprotective strategies (treatment strategies designed to reduce secondary injury) in this population [28]. Additionally, children with AHT make similar functional gains during inpatient rehabilitation and following discharge as do children with nonabusive (accidental) head trauma, underscoring the need to ensure ongoing rehabilitation services [27].

Hypoxic/anoxic ischemic encephalopathy

Epidemiology

Anoxic or hypoxic (absence of or decreased oxygen delivery) and ischemic (reduced blood flow) injury are a common cause of ABI in children. The most common etiologies of this kind of injury vary by age. The epidemiology of *perinatal* hypoxic ischemic encephalopathy (HIE) is described in Chapter: Early brain injury. In young children, cardiac arrest is a frequent etiology of hypoxic/anoxic brain injury (H/AnBI); however, there are few published reports on the overall incidence and epidemiology of nonfatal cardiac arrests in the general pediatric population [75]. While cardiac arrest secondary to primary cardiac dysfunction is common in high-risk populations such as children with congenital heart disease and cardiac arrhythmias, the most common cause of cardiac arrest in children is respiratory arrest. Cardiac arrest may occur in a hospital setting (e.g., in an emergency room or ICU), or it may occur in a community setting. Other causes of H/AnBI include nonfatal drownings, which occur in 10/100,000 children (16/100,000 in children under age 4) [76], and asphyxiation, which may be unintentional (e.g., sudden infant death syndrome and accidental suffocation) or intentional (e.g., hanging and strangulation are the most common mechanism of suicide in adolescents) [77].

Neurobiology

Regardless of etiology, a decrease in cerebral oxygenation and perfusion triggers a global cascade of metabolic changes, and injury evolves over time after the initial insult [78]. There is a switch to anaerobic metabolism with corresponding decreases in adenosine triphosphate (ATP) production, transcellular transportation, and accumulation of toxic metabolites. Excitotoxicity and oxidative stress result in ongoing neuronal damage, neuroinflammatory reactions, and eventually neuronal cell death via necrosis and apoptosis. Similar to TBI, there is a secondary phase of injury characterized by cerebral edema, increased intracranial pressure, seizures, and further neuronal cell death. The tertiary stage which may extend months to years after the original injury is characterized by neuronal repair, remodeling, and gliosis.

Depending on the duration of the insult and the stage of brain development (i.e., brain myelination), different patterns of injury may be seen in children with H/AnBI [79]. Given the significant impact on energy reserves, the brain tissues most vulnerable to injury are the most metabolically active such as the deep gray structures, the perirolandic cortex, and areas that are actively myelinating. Areas of the brain with a high density of postsynaptic glutamate receptors are especially vulnerable to excitotoxic damage. Watershed injury to the subcortical white matter may occur in some cases, and animal studies suggest that this may be due to the shunting of blood supply from these regions to the deep gray structures, including the brainstem, thalami, basal ganglia, hippocampi, and cerebellum. As children approach 2 years of age and their myelination patterns mature to resemble those of adults, the pattern of H/AnBI also begins to mirror that of adults. In mild to moderate injury, the damage primarily affects watershed areas. In more severe cases, the injury extends to deep gray matter structures such as the thalami and basal ganglia, as well as the diffuse cerebral cortex—particularly the sensorimotor and visual cortices. The cerebellum and hippocampi are also commonly involved in these more severe injuries.

Presentation and diagnosis

Diagnosis is based on a combination of history (and particularly of one of the mechanisms described above), a detailed neurologic assessment (which may not show any focal signs but rather an encephalopathy), and imaging (MRI, CT, and cranial ultrasound in neonates and infants). MRI is the most sensitive neuroimaging modality and demonstrates the evolution of hypoxic-ischemic injury [79]. Within the first few hours, areas of diffusion restriction on imaging are seen on diffusion weighted imaging sequences. The abnormalities tend to peak at around 3–5 days and then seemingly improve ("pseudonormalize") over 7–12 days following injury. T1 and T2 weighted images can often be normal in the first day or two and gradually develop persistent signal intensity differences at 7–10 days. In the chronic phase of injury, there is atrophy of the affected areas with a persistent T2 hyperintensity. It is important to note that in several cases of H/AnBI, acute MRI may be normal but subacute and chronic changes are revealed on subsequent imaging. Additionally, MRI findings do not consistently relate to neurological function [80], thus following the child's clinical examination and monitoring development over time is ultimately the best predictor of outcomes.

Clinical course

The course of recovery for children is complex and multifaceted, and a comprehensive systematic approach is more likely to mitigate ongoing injury. More recently, with the advent of therapeutic hypothermia for the treatment after neonatal HIE and cardiac arrest, there has been significant improvement in overall mortality associated with hypoxic-ischemic injury in children [81,82]. During the acute period following injury, in addition to therapeutic hypothermia, it is crucial to minimize secondary brain injury Continuous EEG monitoring should be considered in children with hypoxic ischemic injury given the increased risk for clinical and subclinical seizures in the acute period following the insult. Furthermore, during the subacute period, children with H/AnBI are more likely to experience paroxysmal sympathetic hyperactivity compared to those with TBI. Rehabilitation should begin as early as safely possible, with mobilization starting once the child is stable in the ICU. A comprehensive rehabilitation team includes physical therapy, occupational therapy, speech therapy, and educational support; additional specialty services such as nutrition may also be required. For children who develop disorders of consciousness (see below), the postacute management is somewhat more complicated, prolonged with less favorable outcomes.

Infectious meningitis and encephalitis

Epidemiology

Infectious meningitis (inflammation of meninges) and encephalitis (inflammation of the brain parenchyma) remain important and often preventable sources of ABI in children; these etiologies exist on a spectrum and the terms are often use interchangeably. The incidence varies across the world, with vaccination rates and resource availability significantly impacting etiology, mortality, and morbidity.

The overall rates of bacterial meningitis have increased worldwide, largely due to increases in incidence in lower- and middle-income countries [83,84]. In contrast, in the United States, the rates of bacterial meningitis declined by 55% from 1986 to 1995 and by another 31% since 1998; this trend holds true for other high-income countries. In children under 2 months, the incidence of bacterial meningitis is 81 in 100,000 cases, while it is 0.4 in 100,000 in adolescents [85]. Globally, the mortality associated with bacterial meningitis has decreased by 21%, but in lower resource settings, it can still be as high as 30% of cases. In the United States, the mortality rate has remained fairly consistent since the 1980s at ~15%. Although vaccines have influenced the aforementioned trends, *Haemophilus influenzae* type B (HiB) and *Streptococcus pneumoniae* are still primary causative agents for bacterial meningitis in areas with low vaccination rates. Group B streptococcus, *Listeria monocytogenes* and *Escherichia coli* are the most common pathogens in newborns. *S. pneumoniae* and *Neisseria meningitidis* are the most common pathogens in other age groups.

Viral encephalitis most commonly occurs in children under the one year of age; however, the incidence is difficult to determine due to underreporting, with estimates ranging from 4 to 10 cases per 100,000 per year [86,87]. In areas with higher vaccination rates, viral meningitis occurs more often than bacterial meningitis. There are over a hundred viruses that are known to cause meningoencephalitis. Vaccines have greatly curtailed the rates of mumps, measles, rubella, and polio, and the most common causative agents are currently nonpolio enteroviruses and herpesviruses.

Neurobiology

Whether bacterial or viral, the infectious agent typically spreads to cause a secondary central nervous system (CNS) infection. The most common route of spread is hematogenous, where inflammation leads to the breakdown of the blood–brain barrier. However, infectious agents can also access the CNS via direct infection of the choroid plexus epithelium and retrograde axonal transport [88,89]. Once in the CNS, the presence of pathogens triggers a significant immune response. Neuronal damage occurs due to the direct invasion of brain tissue by the infectious agent, the release of exo- and endotoxins, and the neuroinflammatory response. This can result in secondary brain injury through seizures, edema, and increased intracranial pressure. Abscesses, subdural effusions, and communicating hydrocephalus may also develop over time. Additionally, pneumococcal and HiB infections are associated with direct damage to the cochlea and hair cells, resulting in hearing impairment.

Presentation and diagnosis

The classic symptoms of meningitis (fever, headache, neck stiffness, and photophobia) are often seen in older children and adolescents. Younger children, especially neonates, may present with either hyperthermia or hypothermia, lethargy and poor feeding, irritability or seizures. The diagnosis relies on patient history, analysis of cell counts in the cerebral spinal fluid (CSF), and either culture or polymerase chain reaction (PCR) testing. Despite extensive testing, the causative pathogen may remain elusive, especially in cases of viral meningoencephalitis. Imaging (CT and MRI) can be helpful for evaluating complications such as edema, subdural collections, abscesses, and hydrocephalus, as well as for assessing the extent of secondary brain injury. Diagnostic algorithms have been outlined by the International Encephalitis Consortium, incorporating clinical, laboratory, and imaging features suggestive of specific infections in children [90].

Clinical course

Outcomes after meningitis are dependent on several factors: the host (i.e., neonate vs child, immunocompromised vs not), the pathogen (e.g., virus vs bacteria), availablity of a pathogen-sepcific treatment, as well as the timing of presentation to medical care and access to appropriate treatment. Early initiation of pathogen-specific antimicrobials is critical and the mainstay of treatment. Concurrent and early steroid treatment is also indicated for pneumococcal and Haemophilus influenzae type b (HiB) infections to decrease the risk of hearing loss [91].

Long-term neurodevelopmental sequelae are common and occur in up to half of cases. These include epilepsy, motor and cognitive impairments, and behavioral dysfunction [92,93]. Bacterial meningitis remains an important cause of acquired sensorineural hearing impairment in childhood [94].

Postinfectious disorders

Epidemiology

Postinfectious brain disorders occur shortly after a known or presumed infection and include Sydenham's chorea, acute cerebellar ataxia (ACA), and acute disseminated encephalomyelitis (ADEM). Postherpes simplex virus (HSV) autoimmune encephalitis has also recently been described and will be covered below with autoimmune encephalitis.

Postinfectious brain disorders are fairly rare. ADEM has an estimated incidence of 0.4–0.6 per 100,000 children per year and ACA has a similar reported incidence [95,96]. Sydenham's chorea, a neurologic disorder associated with rheumatic fever, while once fairly common, has declined precipitously over the last century with the introduction of antibiotics for streptococcal infections. All of these disorders typically present in early childhood, between 2 and 8 years of age, and are often monophasic. However, cognitive sequelae, particularly in ADEM, may persist long term [97].

Pediatric acute-onset neuropsychiatric syndrome or pediatric autoimmune neuropsychiatric disorder associated with streptococcal infection is a somewhat controversial diagnosis involving the explosive onset of obsessive-compulsive disorder symptoms and/or tics, after a Group A streptococcal infection. Several studies have failed to consistently demonstrate a definitive set of diagnostic criteria, pathophysiologic process, and response to proposed treatments, and as such, we will not be discussing this syndrome in this chapter [98]. [See chapter Developmental movement disorders: Tics and stereotypies].

Neurobiology

These disorders are thought to be caused by a secondary immune response to the original infectious pathogen. In some patients, the primary infection may be asymptomatic or oligosymptomatic and while comprehensive investigation for acute underlying infection should always be performed, in many cases, a causative pathogen is not identified. Secondary immune responses are thought to result in the formation of autoantibodies against various brain structures (e.g., basal ganglia in Sydenham's chorea and myelin in ADEM) or may occur secondary to molecular mimicry by the causative organism. Just as the original infectious pathogen can be difficult to identify, autoantibodies associated with these syndromes may also be difficult to detect consistently.

Sydenham's chorea Described at one time as "St Vitus Dance," children commonly present with new acute onset unilateral or bilateral choreiform movements weeks to months after group A streptococcal pharyngitis [99]. Chorea is considered part of the clinical spectrum of acute rheumatic fever as a "Major Criteria" in the Jones Diagnostic Criteria, occuring in up to 40% of patients with acute rheumatic fever [100]. Although anti-ASO and anti-DNAse antibodies may be detected in these patients, indicating a recent streptococcal infection, they are not believed to cause brain disease. Instead, the underlying mechanism is thought to involve both B-cell and T-cell responses triggered by molecular mimicry of cells within the basal ganglia [101]. Neuroimaging studies have revealed T2-hyperintensities in the basal ganglia on brain MRI in some patients, as well as volume differences in the basal ganglia compared to age-matched controls [102]. Antibodies against the dopamine-2 receptor have also been detected in up to one-third of patients [103].

The natural history of this disorder typically involves the resolution of symptoms in approximately 8 months, although symptoms may persist up to 2 years in some cases [104]. The standard treatment for Sydenham's chorea includes penicillin prophylaxis until the age of 21 as part of the standard care for rheumatic fever. For motor symptoms, valproic acid or other antiepileptic medications have shown some efficacy as symptomatic treatments. Although the use of immunomodulatory medications has been reported in case studies, they are not considered standard care. In addition to chorea, behavioral changes including anxiety, attention problems, and mood changes have been reported in 20%–30% of patients [105]. Medications can also be used to manage psychiatric and behavioral symptoms as needed [105].

Acute cerebellar ataxia Similar to Syndenham's chorea, ACA presents as acute onset of abnormal movements in a young child. In cases of ACA, a child may wake up with gait changes and falls, and on examination, a pattern of ataxia consistent with cerebellar dysfunction is observed. There is often a history of a viral illness occurring days or weeks before the onset of neurological symptoms. Specifically in ACA, a child may wake up with gait changes and falls, and on examination, a pattern of ataxia consistent with cerebellar dysfunction is observed. A history of a viral illness either days or weeks prior to their neurologic symptoms may be reported [106]. However, because these acute neurologic changes may overlap with other neurologic disorders, including infectious cerebellitis, neoplasms, stroke, and inborn errors of metabolism, ACA is considered a diagnosis of exclusion. A comprehensive workup, including neuroimaging, metabolic labs, and lumbar puncture, should be performed to rule our alternative causes. In ACA, this workup is often normal, though isolated pleocytosis may be seen in the CSF [107,108].

ACA is thought to be a postinfectious autoimmune phenomenon, supported by clinical history and its association with various infectious pathogens such as enterovirus, varicella, Epstein-Barr virus, and others. This association is primarily based on epidemiologic cooccurrence or the detection of viral markers in the CSF [109]. However, no single pathogen has been definitively implicated. Similarly, no specific autoimmune biomarker has been identified. Although several small studies have isolated a variety of autoantibodies, including those targeting neuronal proteins, none have been consistently associated with ACA. [109].

The prognosis for ACA is generally favorable, with over 90% of children achieving full recovery within months or even weeks. While rehabilitation services may be beneficial for managing ataxia symptoms during the recovery process, no other interventions are typically required. Although there have been case reports on the use of intravenous immunoglobulin (IVIG) or steroids for ACA, there are no large clinical trials supporting the use of immunomodulatory medications for this condition.

Acute disseminated encephalomyelopathy ADEM is a neuroinflammatory phenomenon marked by multifocal neurologic deficits and neuroimaging abnormalities that primarily affect the white matter of the CNS. The clinical presentation is variable, with an acute or subacute onset of a broad spectrum of neurological deficits, depending on the location of CNS lesions. Diagnostic criteria are described in Table 13.3. It is commonly considered to be a postinfectious phenomenon, as an infectious prodrome is reported in 75% of cases [110,111]. ADEM has been known to occur following a variety of viral infections, including coronavirus, EBV, varicella, and Zika virus. The hypothesis of an autoinflammatory pathogenesis is supported by pathological studies showing microglial activation and demyelination in the white matter, similar to other CNS autoimmune diseases including multiple sclerosis [112].

While blood tests (e.g., elevated ESR and WBC) and CSF biomarkers (e.g., pleocytosis, elevated protein, and the presence of oligoclonal bands) may support the diagnosis, they are not required for it [113]. Anti-myelin oligodendrocyte glycoprotein (MOG) antibodies are present in approximately half of cases, but in monophasic ADEM, these titers diminish and disappear over time [114–116].

Immunotherapy is the mainstay of treatment for ADEM. While there is no standardized protocol, intravenous corticosteroids and/or IVIG are typically the first line of treatment. In severe or refractory cases of ADEM, plasma exchange or rituximab may be used, especially when an identifiable autoantibody is known or suspcted. Otherwise, treatment for ADEM is supportive and symptomatic, depending on the specific manifestations of the patient, which may include seizures, neuropsychiatric symptoms, motor/sensory impairment, or vision impairment. Rehabilitation services should be introduced as soon as the patient is stabilized.

ADEM is a monophasic event in the vast majority of patients (95%). However, in the event of a second occurrence, especially in the absence of encephalopathy and in postpubertal patients more than 3 months after initial onset, recurrent neuroinflammatory diseases should be considered. Other demyelinating diseases that affect the CNS white matter include multiple sclerosis, neuromyelitis optica spectrum disorder, anti-MOG associated demyelinating disease, and patients with recurrence of ADEM.

TABLE 13.3 Diagnostic criteria for ADEM [117].

Monophasic ADEM	Multiphasic ADEM
A first polyfocal clinical CNS event with presumed inflammatory cause characterized by: – Encephalopathy not explained by fever – Brain MRI during the acute phase (within 3 months) with diffuse, poorly demarcated, large >1–2 cm lesions involving predominantly the cerebral white matter; T1 hypointense white matter lesions are rare; deep gray matter lesions (e.g., thalamus or basal ganglia) can be present. – No new symptoms, signs or MRI findings after 3 months of the initial ADEM.	New event, meeting criteria for monophasic ADEM, that occurs ≥3 months after the initial event and can be associated with new or reemergence of prior clinical and MRI findings.

ADEM, Acute disseminated encephalomyelitis; *CNS*, central nervous system; *MRI*, magnetic resonance imaging.
Source: Adapted from Krupp LB, Tardieu M, Amato MP, et al. International Pediatric Multiple Sclerosis Study Group criteria for pediatric multiple sclerosis and immune-mediated central nervous system demyelinating disorders: revisions to the 2007 definitions. Mult Scler Houndmills Basingstoke Engl 2013;19 (10):1261–67.

Most patients with ADEM respond well to immunotherapy and full neurologic recovery occurs in over 70% [110,118]. In 20%–30% of children with ADEM, cognitive and behavior deficits may persist long term [119,120]. Deficits in attention and executive function have been commonly reported [118,121]. A longitudinal study of 83 children with ADEM also reported reduced age-expected brain growth [122].

Autoimmune encephalitis

Autoimmune encephalitides (AE) are a group of inflammatory brain disorders manifesting as acute or subacute onset of encephalopathy presumed to be of an autoimmune etiology. Presenting symptoms in children may include mood or behavior changes, fluctuations in consciousness, seizures, sleep disturbance, or movement disorders [123,124]. Some forms of AE may also present with systemic symptoms like diarrhea or dysautonomia. While AE is more common in adulthood, cases are increasingly being identified in children, and occur more frequently than any viral encephalitides [125,126].

As an autoimmune phenomenon, AE is presumed to be caused by a misdirected immune response that attacks an individual's CNS, causing neurologic dysfunction. AEs may arise as a paraneoplastic phenomenon in the presence of a known or yet to be identified neoplasm (e.g., anti-NMDA, N-methyl-D-aspartate, receptor encephalitis in the setting of ovarian teratoma, opsoclonus myoclonus syndrome in the setting of neuroblastoma), as a postinfectious phenomenon (e.g., anti-NMDA receptor encephalitis after HSV encephalitis), or by an unknown/idiopathic mechanism [123,127,128]. While in some cases of AE, autoantibodies are identified in the blood or CSF, other cases may be "antibody-negative." The most common and widely recognized form of AE is anti-NMDA receptor encephalitis where antibodies are produced against the NMDA receptor within the CNS. This form of AE accounts for 40% of pediatrics cases [126]. Over 15 other autoantibodies have also been implicated in AE, with the most common in children being anti-MOG, and anti-GAD65. Additionally, new antibodies are continousely being being identified [124].

Diagnosis of AE is primarily based on clinical history and examination, and exclusion of alternative diagnoses. In recognition that diagnostic testing for autoantibodies in AE has limitations (i.e., time for specialized laboratory processing, some antibodies have yet to be discovered), clinical criteria to diagnose AE have recently been established and adapted for pediatric patients (Table 13.4) along with a diagnostic algorithm [124,129]. Diagnosis is further subcategorized by certainty of diagnosis based on paraclinical evidence of CNS inflammation and serologies.

TABLE 13.4 Diagnostic criteria for autoimmune encephalitis in children.

	Possible AE	Probable antibody-negative AE	Definite antibody-positive AE
1. Symptoms: Onset of neurologic and/or psychiatric symptoms over ≤3 months in previously healthy child		Yes	
2. Clinical evidence: • Altered mental status or abnormal EEG • Focal neurologic deficits • Cognitive difficulties • Acute developmental regression • Movement disorder (except tics) • Psychiatric symptoms • Seizures not explained by known seizure disorder or other condition		≥2 features present	
3. Paraclinical evidence • CSF inflammatory changes (leukocytosis or oligoclonal bands) • MRI features of encephalitis • Brain biopsy with inflammatory infiltrates excluding other disorders	Not available	≥1 feature present	
4. Serology: Presence of serum or CSF autoantibody associated with AE	Not available	No	Yes
5. Reasonable exclusion of alternative causes		Yes	

AE, Autoimmune encephalitis; *EEG*, electroencephalography; *MRI*, magnetic resonance imaging; *CSF*, cerebral spinal fluid.
Source: Adapted from Cellucci T, Van Mater H, Graus F, et al. Clinical approach to the diagnosis of autoimmune encephalitis in the pediatric patient. Neurol Neuroimmunol Neuroinflamm 2020;7(2):e663.

Treatment of patients with AE includes immunotherapy and workup/treatment for underlying neoplasm or infection. While there is not yet a standardized protocol for all types of AE, international consensus recommendations have recently been established for the treatment of anti-NMDA receptor encephalitis in pediatric patients [130,131]. Escalation of treatment from first-line immunotherapies (IV corticosteroids, IVIG, or plasmapheresis) to second-line immunotherapies (cyclophosphamide or rituximab) and duration of treatment are dictated by disease severity [130]. Symptomatic management and rehabilitative services for psychiatric manifestations, seizures, feeding, or autonomic issues should be initiated simultaneously in all patients, similar to other forms of ABI [132].

Many patients improve significantly with treatment and some may return to baseline at 2 years [123,133–135]. However, the long-term effects of AE are increasingly being recognized. While motor deficits typically resolve, subtle cognitive changes have been identified by neuropsychological testing, including deficits in processing speed, memory, executive function, and adaptive behavior [136,137]. Clinical factors which may portend poorer outcomes include younger age of onset, ICU requirement, and longer time from diagnosis to treatment initiation [136,138,139]. While longitudnal studies to better understand the long-term outcomes of AE are underway, close neuropsychological monitoring is currently recommended.

Pediatric stroke

Epidemiology

Pediatric stroke is a relatively rare but significant cause of ABI. Perinatal strokes (in infants less than 28 days old) are more common and occur in roughly 1 in 2300 live births, whereas childhood strokes (in children 29 days to 18 years of age) occur less commonly in around 2–13 per 100,000 [140].

Neurobiology

Strokes can be further categorized into three subtypes by etiology: arterial ischemic stroke (AIS; loss of blood flow in arteries supplying the brain), venous stroke (secondary to a clot in the brain's venous system), or hemorrhagic stroke (bleeding into the brain substance, ventricles, or subarachnoid space of the brain). AIS is the most common type of pediatric stroke. Children under the age of 1 year are at higher overall risk of AIS. AIS accounts for at least 80% of neonatal strokes [141].

Unlike adult stroke, hypertension and coronary artery disease are not the primary contributors to increased risk in pediatric stroke. Pediatric AIS risk factors include cardiac (congenital heart) disease, acquired and inherited thrombophilias, inborn errors of metabolism, arteriopathies (such as Moyamoya disease), sickle cell disease, rheumatologic disease (such as SLE and systemic vasculitides), and malignancy. The overall hypercoagulable state of pregnancy in addition to prenatal exposures (e.g., infection, placental insufficiency, preeclampsia, prolonged rupture of membranes) contributes to the risk of perinatal stroke [142]. It should be noted that pediatric stroke is often multifactorial, and it is a combination of risk factors and a precipitating event that leads to the presentation of stroke. Despite this, 10%–25% of childhood strokes have no identifiable etiology [143].

Presentation and diagnosis

The presenting signs and symptoms of stroke vary in age and arterial territory affected. In younger children and neonates, there may be few or subtle focal neurologic signs of stroke, while hemiparesis, hemisensory loss, speech and language difficulties, or ataxia are common presenting symptoms of stroke in older children. In contrast to the adult stroke population, roughly a third of strokes in children present with seizures and a change in mental status. Importantly, 94% of neonates with strokes present with seizures. However, in some infants, sequelae of a perinatal stroke may not be evident until many months later when a child develops asymmetric motor skills [144]. In children with arteriopathies, there is often a history of transient ischemic attacks (TIAs) prior to the first stroke [145]. Sometimes the onset of symptoms is slow or stuttering. In the case of venous thromboses, signs of increased intracranial pressure, including headaches, nausea, and vomiting may be the presenting symptoms. Hemorrhagic strokes tend to present acute severe headaches in addition to seizures, changes in mental status, and focal neurologic signs.

Clinical course

Mortality is higher in hemorrhagic strokes and cerebral sinovenous thrombosis in comparison to AIS [146]. Since pediatric stroke is not a common occurrence, there are few randomized controlled trials on treatment and management. Thus, most interventions are extrapolated from adult treatment trials. A guideline for the management of pediatric stroke was published by the American Heart Association [147]. Head CT is often performed emergently to rule out findings that require urgent neurosurgical intervention. MRI brain with perfusion sequencing is the preferred imaging modality to identify the extent of the stroke. CT angiography with vessel imaging of the head and neck can optimally evaluate large vessel occlusions, dissection, or other vascular anomalies. As in adults, thrombolysis and mechanical thrombectomy are the mainstay of acute stroke management and are safe to use in children [147]. In children with sickle cell disease, the American Society of Hematology recommends that management focus on hydration, supplemental oxygen, and urgent exchange transfusion [148].

Stroke prevention depends on a detailed evaluation for risk factors as detailed in the guideline published by the American Heart Association [147]. The steps include an evaluation for congenital heart disease, thrombophilias, and blood work. Stroke recurrence, while relatively rare, is still significant in neonates and children with congenital heart disease, arteriopathies, and systemic diseases such as sickle cell disease. Surgical correction of the congenital heart anomaly (e.g., patent foramen ovale (PFO)), antithrombotic agents, and anticoagulation are recommended for use in these high-risk groups. In children with arteriopathies like Moyamoya, surgical revascularization is recommended as soon as feasible to prevent ongoing TIAs and strokes. For children with sickle cell disease, management of their underlying disease using hydroxyurea and transfusions is recommended.

Recovery from stroke also varies by age, etiology, and extent of the stroke. Outcomes described in the literature estimate that between 10 and 30% of children with pediatric stroke develop moderate to severe disability [146,149]. In children with perinatal stroke, cerebral palsy occurs in 30%–40%, and particularly hemiparetic cerebral palsy (CP). Hemiparesis after childhood AIS occurs at higher rates, 40%–90% of the time. Cognitive deficits including lower IQs and executive functioning difficulties are commonly seen in children with strokes, with larger volume strokes involving the cortical and subcortical areas associated with worse cognitive outcomes [146]. Poststroke epilepsy is common in all etiologies of pediatric stroke, occurring in 7%–20% of children [146]. However, children are believed to fare better than adults with similar strokes due to the young brain's ability to harness developmental plasticity and more efficient regenerative mechanisms to reorganize neural networks [149,150]. As in adults, rehabilitation therapy after a stroke affords these children the most chance for recovery.

Postintensive care unit syndrome

With advances in critical care medicine, a rise in long-term morbidity has paralleled reduced mortality. Following a significant critical illness (PICU admission >48 hours, one or more organ dysfunction), 82% of children experience a significant functional decline in at least one developmental domain [151]. Postintensive care syndrome-pediatrics (PICS-p) (Fig. 13.2) is a more recent clinical construct that systematically organizes the range of physical, cognitive, psychological, and social symptoms that emerge in a child and their family members following a critical illness [152,153]. Of children admitted to a neurocritical care unit, nearly 90% reported new deficits at the first follow-up clinic visit, with cognitive and behavioral impairments being the most common [154]. Furthermore, the number of neurodevelopmental morbidities increases over several years following PICU discharge [155].

PICS-p not only recognizes the breadth of impact critical illness has on a child's function but importantly also emphasizes the traumatic nature of the PICU experience within the child and the family. Nearly half of parents of children with critical illness report significant stress symptoms even 12 months following hospital discharge [156]. Importantly, parental stress does not seem to be associated specifically with their child's functional outcome [157].

A child with a critical illness affecting the central nervous system must be considered as having an ABI. Whether resulting from traumatic or nontraumatic causes, such as infections or cerebrovascular events, all etiologies of pediatric acquired brain injury (pABI) can result in significant challenges for the child and their family. Beyond direct brain injury, critical care measures to save and comfort the child during the critical illness (i.e., intubation and ventilation, use of sedation) are also important factors influencing outcomes. Insults and injury to the brain at any point risk disruption to critical, specialized, and complicated neurodevelopmental pathophysiological processes; thus it is not surprising that children have a high incidence of functional deficits following neurocritical care.

Neurodevelopmental disabilities providers are particularly well trained to care for sequelae following critical illness in children and PICS-p. Physical sequelae range from poor coordination to spastic quadriplegia. Other physical symptoms include sleep dysregulation, headaches, dizziness, and tone or spasms. Similarly, cognitive impairments in PICS-p include a wide spectrum, though typically with a more variable profile than developmental cognitive deficits. Cognitive impairments range from inattention, slow processing, and executive dysfunction to intellectual disability. Emotional health in PICS-p includes mood and behavior disorders as well as the traumatic stress of the PICU experience on the child and family. Finally, PICS-p incorporates social health into the construct including returning to school and

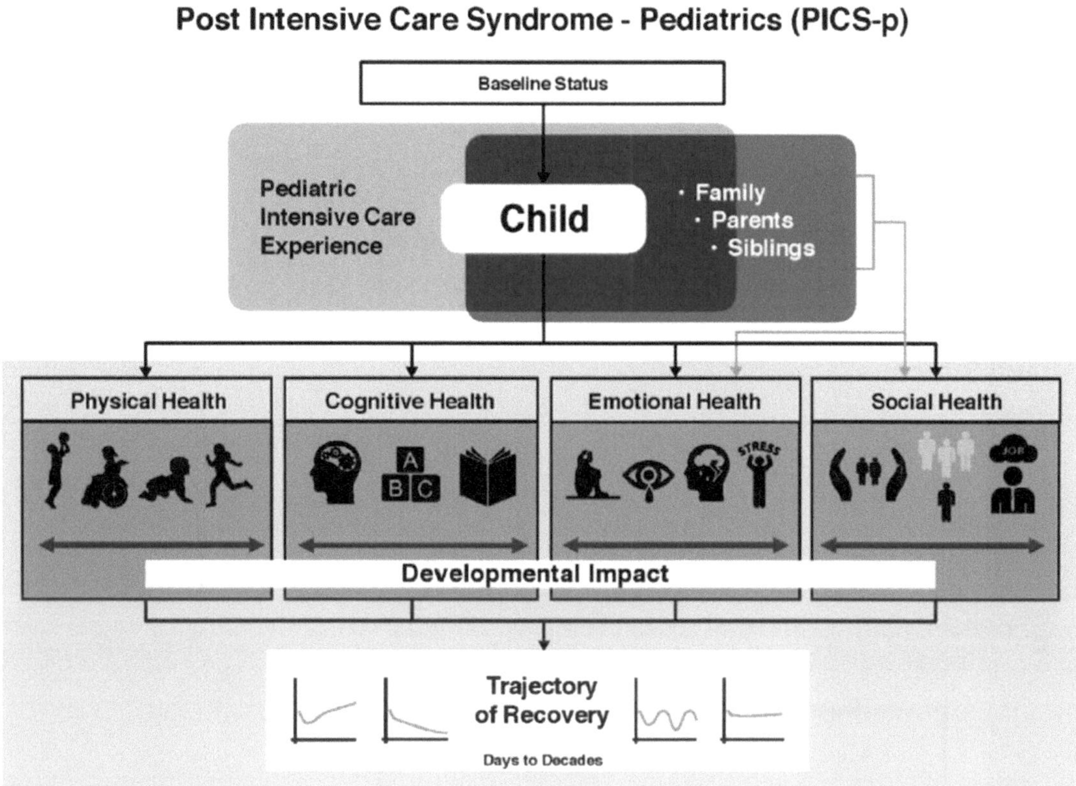

FIGURE 13.2 The postintensive care syndrome-pediatrics (PICS-p) is a clinical construct that recognizes the physical, cognitive, emotional, and social sequelae following critical illness in a child; additionally, PICS-p recognizes the impact on the child as well as parents, siblings, and the whole family [152,153].

community activities. Through the comprehensive, holistic training in neurodevelopmental disabilities, an neurodevelopmental disabilities (NDD) provider can easily apply their knowledge and skills to the evaluation and interdisciplinary management of each of these domains in children following critical illness.

Outcomes and rehabilitative strategies

After brain injury and medical stabilization, some patients exhibit an extended state of altered consciousness collectively termed disorders of consciousness (DoC). Children with ABI who have relatively intact consciousness or who emerge rapidly from DoC are at risk for a constellation of cognitive, sensorimotor, and behavioral sequelae. Recovery from ABI occurs over many years, though pABI recovery may appear paradoxical in that children gain functional skills and have improved overall outcomes while the number of neurodevelopmental morbidities and diagnoses also increases. This paradox results from (1) natural underlying neurodevelopmental pathophysiological processes, (2) more complex and nuanced skills needed as a child ages, and (3) the potential for ongoing neuroplasticity and recovery inherent following an ABI. Therefore a child with an ABI requires an extended time period of neurodevelopmental monitoring, potentially lasting until the child transitions to adult medical care.

The needs of children and their families following pABI have been explored (Fig. 13.3) [158], with common themes emerging, including identification and management of comorbidities and sequelae, specific support returning to school, the emotional impact, and need for whole family support. McKinlay et al. [159] provide 14 comprehensive practice recommendations regarding services for children and families following pABI.

The severity of disability can vary within and among different streams of development, often related to regions of brain involvement, though localization of deficits based on imaging can be less precise in young children. In the overall ABI population, global reduction gray matter volume and thinning for certain cortical regions correlate with motor, cognitive, visual, and communication deficits compared to children without brain injury [160]. Typically, children with the most profound developmental outcomes have severe global impairments that are clinically present immediately and persist following the injury and are consistent with imaging or other markers of severe brain injury. Profound-severe

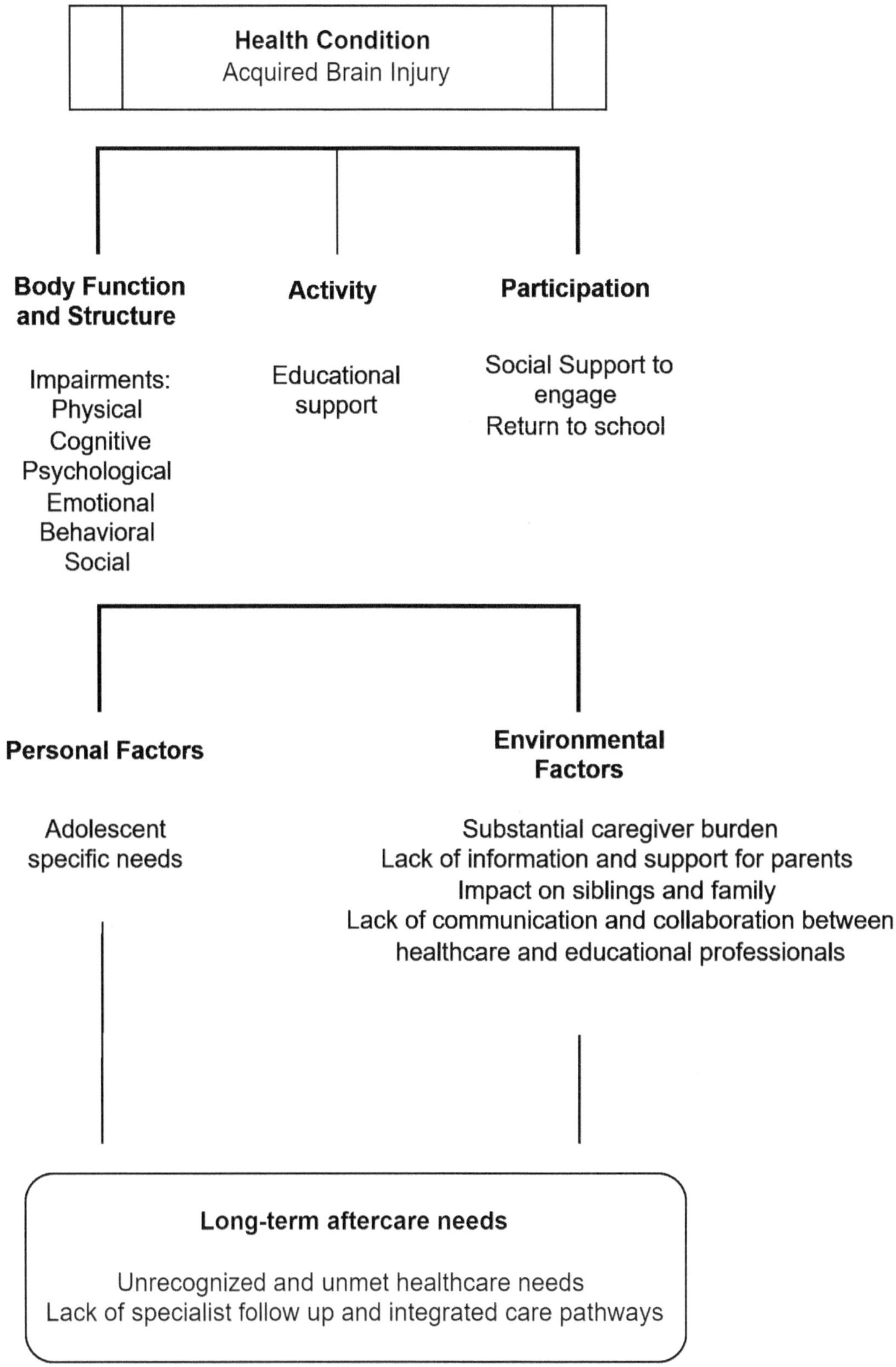

FIGURE 13.3 Needs assessment of child and parents after pediatric acquired brain injury mapped onto the International Classification of Function framework, adapted from [158].

deficits are more common in nontraumatic etiologies [161]. Other than the most severe injury outcomes, which are fortunately the least common, children with brain injury are expected to have improvement in overall function and to gain developmental skills over time. The most rapid and extensive gain in skills occurs over the first year following injury; however, the extent of recovery following injury is dependent on many variables and a child's outcome is difficult to predict in the acute or subacute setting. While profound deficits in overall developmental streams are not expected to emerge over time, functionally impairing deficits in high-level skills certainly can develop at a later stage. As developmental skills become increasingly more complex, new impairments may be diagnosed over time following an ABI.

Based on the severity of the sequelae, children and adolescents will be admitted to an intensive inpatient rehabilitation program, enrolled in intensive outpatient rehabilitative therapies, or transitioned to school with educational support. For children with ABI, optimal rehabilitation facilities have: (1) expertise in pediatric rehabilitation, (2) developmentally appropriate treatments, (3) family-centered care, and (4) a focus on education and school reintegration. The goals of rehabilitation are to: (1) mitigate complications resulting from immobilization and disuse; (2) promote recovery of cognitive, behavioral, and sensorimotor function; (3) teach compensatory strategies for impaired or lost function; and (4) reduce the impact of chronic disability on future development. Interdisciplinary intervention including physical therapy (PT), occupational therapy (OT), speech therapy, neuropsychology, behavior psychology, special education, and social work may be necessary, depending on the nature and severity of the injury and ensuing impairments.

Disorders of consciousness

DoC represents the state of wakefulness and awareness of a patient following brain injury. Stages of DoC include coma (complete lack of both arousal and awareness), vegetative state (now referred to as an unresponsive wakefulness syndrome VS/UWS, arousal without awareness), and minimally conscious state (MCS, arousal with inconsistent awareness) (Table 13.5). MCS is subcategorized into MCS− and MCS + . MCS− is typically the first stage when purposeful responses are more inconsistent and when patients do not yet have inconsistent language. When in the MCS + state of

TABLE 13.5 Characteristics of levels of consciousness.

Function	Coma	VS/UWS	MCS−	MCS +
Sleep-wake cycle (eye opening)	Absent	Present	Present	Present
Awareness	Absent	Absent	Present	Present
Response to noxious stimuli	+/−generalized/ reflexive	+/−generalized/ reflexive	Present	Present
Motor function	No purposeful movement	No purposeful movement	Inconsistent purposeful movement	Purposeful movement
Visual/ Auditory responses	Absent	Startle	Inconsistent localization	localization
Language	Absent	Absent	Absent	Response to commands, yes/no questions or verbalizing words
Affect	Absent	May laugh, smile, cry, grimace but not in direct relation to stimuli	+/−Affective responses congruent with stimuli	Affective responses congruent with stimuli
Emergence (to the next level of consciousness)	Return of sleep-wake cycle	Clear repeatable response to stimuli	Clear repeatable verbalization of words or response to commands	Functional object use for two distinct objects/reliable, consistent, and functional interactive communication

Summary of clinical features that describe different states of consciousness.
VS/UWS, vegetative state/Unresponsive wakefulness state; *MCS*, minimally conscious state, subcategorized as MCS− and MCS + based on specific clinical characteristics.

wakefulness and awareness, patients have more consistent purposeful movements and language though they do not yet demonstrate functional motor or language skills to be categorized as being in a conscious state. DoCs can occur following the sudden onset of an ABI, from which the individual may or may not regain consciousness, or from a terminal decline of consciousness toward the end of life [162].

The assessment of DoC is challenging, since consciousness tends to wax and wane over the short term (second to hours) and long term (days). Additionally, behaviors in patients with DoC are often subtle, inconsistent, and variable, thus it is challenging to definitively know a child's level of consciousness, especially early in recovery. Underlying central and peripheral impairments such as sensory deficits, neuromuscular abnormalities, and aphasia may also mask conscious awareness. In pediatrics, a number of standardized behavioral assessment tools have been described in the literature: Rappaport Coma/Near Coma Scale [163], Level of Cognitive Functioning Assessment Scale [164], Western Neuro Sensory Stimulation Profile [165], and Post-Acute Level of Consciousness Scale [166]. The Coma Recovery Scale-Revised is considered the gold standard assessment tool for adults given its evidence for adequate reliability and validity and was revised and validated (preliminarily) in typically developing young children [167,168]. Although EEG, PET, and fMRI are sometimes used clinically in the assessment of levels of consciousness in DoCs, there is insufficient evidence to support their routine clinical use at this time [162].

A clinical practice guideline published in 2018 as a collaboration between the AAN, ACRM, and NIDILRR [162] recommends an interdisciplinary team with expertise in brain injury rehabilitation to address the complex care needs of a patient with DoC to facilitate arousal, purposeful behavior, communication, and self-care using standard rehabilitative practices described below.

Motor impairments

Following ABI, impairments may be noted in gross motor functions such as strength, balance, speed, and coordination [169] as well as decreased fine motor speed and coordination [170]. Studies have shown that functional outcomes of early motor impairments due to ABI are typically good. Seventy percent of nearly 500 children with ABI admitted to inpatient rehabilitation regained mobility (7.6 months after injury on average) with self-care and mobility continuing to improve over 5 years following injury [169]. Certain etiologies of ABI, such as acute anoxic brain injury, are more likely to result in motor impairment due to underlying physiology with frequent involvement of the metabolically active perirolandic cortex. Fine motor function may also be impaired, impacting a child in activities of daily living (ADLs) and school. In fact, children make great gains in gross motor skills but have more persistent fine motor impairments following brain injury [170,171].

Sensory impairments

Vision may be affected by ABI and necessitate an ophthalmological or neuro-ophthalmological evaluation. Visual impairments may arise from optic nerve injury, cranial nerve damage limiting ocular movements, and cortical damage resulting in impaired visual processing. Vision impairments may include nystagmus, decreased visual acuity, difficulty with tracking, and visual field cuts. Double vision is common following ABI because of cranial nerve injury. Severe injury to the optic nerve is less common but is irreversible and can cause blindness [172,173]. Occipital lobe injury can cause cortical visual impairment (CVI). CVI includes a range of deficits, from mild vision loss to impaired acuity with functional vision to complete vision perception loss (cortical blindness). Assessment of CVI extends beyond visual acuity, including visual fields, contrast sensitivity, motion detection, visual search, and visual processing measures. CVI can improve over time. Vision therapy is an option, especially for infants and toddlers following ABI. Interestingly, improvement in CVI has also been associated with enrollment in early childhood intervention (ECI) including nonvision-related therapies [174].

Sensorineural hearing loss or conductive hearing loss can result from ABI (especially bacterial meningitis); however, there is an overall paucity of studies on hearing and auditory processing following TBI and many other etiologies of brain injury in children [175]. Sensorineural loss is more pronounced at the higher hearing frequencies [176]. Given these concerns, audiological evaluation is recommended following ABI.

Feeding disorders

Dysphagia is a common complication of severe ABI (10%–20%) [177,178]; however swallowing and feeding skills may improve over time, especially in TBI where dysphagia can resolve within 3 months [179]. Untreated dysphagia risks aspiration and complications such as aspiration pneumonia. Furthermore, poor feeding can impair adequate oral nutrition and hydration, which adversely impacts brain injury recovery. Associated impairments include atypical tongue movements, poor jaw stability, inefficient chewing, impaired lip closure, and reduced attention and impulsivity during feeding. For a child with ABI requiring intubation and critical care, a formal swallow evaluation is essential. In children with less critical injury, clinical signs of aspiration can include coughing after a swallow, wet voice after swallowing, and delayed swallow initiation.

Nutrition has been shown to be an important factor in recovery after ABI. Thus a feeding tube, either nasogastric tube (short term) or a percutaneous gastrostomy tube (long term) is often necessary to provide adequate nutrition and hydration to children with dysphagia. As metabolic needs of a child vary by age and level of function, nutritional monitoring is beneficial to assure the appropriate caloric intake, and nutritional content is provided to ensure proper growth [See chapter: Dysphagia and feeding disorders].

Communication skills and language impairments

A wide array of communication and language deficits present following ABI in children including dysarthria, receptive and/or expressive language disorder including aphasias, and social communication impairment. Overall language development may be adversely affected by both timing and severity of TBI. There is evidence that children who sustain moderate TBIs between the ages of 4 and 6 years have significantly worse verbal abilities and language skills than their counterparts with mild TBI [66]. In young children with TBI, basic receptive and expressive deficits improve over time, though many remain more impaired than same aged peers. Older children and adolescents with brain injury to regions of the brain related to language may struggle with receptive, expressive, or mixed aphasia. A child with aphasia may retain some basic social courtesy responses or answer basic questions in a simplistic manner, thus without an astute provider and language assessment, aphasia may be missed. Furthermore, complex and pragmatic language impairments in children with ABI contribute to deficits in both social and adaptive function [180,181], associated with a higher rate of autism, social isolation, and externalizing behavior problems [182–184]. Thus speech and language assessment is an important aspect of ongoing neurodevelopmental monitoring following ABI.

Cognitive sequelae

Changes in cognitive functioning are a common consequence of ABI regardless of etiology, age of injury, or injury severity. While they may vary in severity, the most common and persistent cognitive sequelae associated with ABI are in the areas of sustained attention, processing speed, memory, and executive functioning (i.e., initiation, planning and organization, impulse control, emotional regulation) [65,185]. Long-term cognitive outcome is associated with injury severity and factors such as age of ABI, preinjury adaptive functioning, and behavior [72,186].

The cognitive sequelae and recovery of function following ABI may not be readily detectable by medical providers in a limited outpatient clinic visit. For this reason, initial impairments and recovery of function can be identified and monitored with data obtained from serial neuropsychological evaluations, which include direct testing of the child's cognitive and academic functioning as well as the collecting collateral information from caregivers and teachers via standardized questionnaires. The neuropsychological evaluation may be used to estimate premorbid functioning, to assess cognitive changes resulting from the ABI, to monitor recovery in the context of ongoing development, and to predict future areas of concern. As the child advances in age, even subtle cognitive sequelae may become more impairing as academic and social demands increase.

Neuropsychological evaluations during the acute period of recovery are used to guide therapeutic interventions and to make recommendations for return to school planning. Results are shared with the child's medical team to guide therapies to include speech therapy, cognitive rehabilitation, executive functioning coaching, psychotherapy, and psychopharmacological intervention. In addition, the neuropsychological evaluation is shared with the child's school district to help determine eligibility for educational support and provide classroom-based learning and hospital-to-school transition recommendations. The child's education plan may include a more restrictive educational setting as the child returns to school, which can become less restrictive with ongoing recovery. In addition, the child may be able to access a wide array of instructional modifications and accommodations, as well as related services such as speech, OT, PT, and behavioral interventions.

Behavioral sequelae

Behavior changes as a function of an ABI are common and frequently experienced in a biphasic timeline: during acute injury and evolution of long-term sequelae.

Initially, in the acute phase, children might display immediate cognitive impairments, emotional lability, and behavioral challenges resulting from the direct impact of the injury on the brain [187]. These behavioral changes can be profound and include agitation, confusion, and emotional dysregulation related to the brain's response to the trauma and the swelling and disruption of neural networks. The severity and persistence of these acute symptoms can vary based on the type and extent of the injury, with some children experiencing transient behavioral changes that gradually resolve as the brain heals [188].

Following this acute phase, a recovery and rehabilitation period typically ensues. During this phase, many children show improvements in certain behavioral domains, particularly in cognitive functioning and emotional regulation. Some children, however, exhibit persistent behavioral challenges, which can manifest as difficulties in attention, executive functioning, impulsivity, and emotional dysregulation [35]. These behavioral challenges, in turn, negatively affect the child's adaptive functioning, academic development and academic performance, and social relationships.

Long-term trajectories of behavioral changes after pediatric brain injury can be variable. Some children may show remarkable resilience and significant improvements over time, while others might continue to experience persistent behavioral difficulties that affect their long-term educational, social, and vocational outcomes [189]. Factors such as the type and severity of the injury, the age of the child at the time of injury, the location of the brain damage, and the effectiveness of rehabilitation and support services all contribute to shaping the trajectory of postinjury behavioral changes. Frontal lobe injuries often manifest as persistent deficits in sustained attention, behavioral inhibition, and planning/organization, while injuries affecting the limbic system may result in emotional dysregulation and mood disturbances [188]. Social impairments, including difficulties in interpreting social cues and maintaining peer relationships, are common sequelae of ABIs in children [187]. These behavioral changes can persist long term, impacting the child's psychosocial adjustment and quality of life [188]. It is noteworthy, however, that some children may continue to improve even years after the initial injury.

Across the spectrum of ABI, the most commonly reported long-term neurodevelopmental sequelae is that of a secondary attention deficit hyperactivity disorder (SADHD). SADHD differs from that of a primary/developmental ADHD as the symptoms emerge following brain injury. The spectrum and severity of symptoms can vary and do not always correlate with the degree of injury. The treatment of SADHD is congruent to that of primary/developmental ADHD, but studies have shown that treatment of SADHD with stimulant medications provides less response than that seen in primary ADHD [190].

Aspects of outpatient care strategies specific to acquired brain injury

There has been an increase in clinical practice guidelines and consensus-based recommendation for close follow-up care and neurodevelopmental monitoring in high-risk infants, including congenital heart disease [191], very preterm birth [192], and neonatal encephalopathy [193], as they are susceptible to long-term neurodevelopmental sequelae. While children with ABI experience similar neurodevelopmental consequences, clinical guidelines do not currently exist. Unfortunately, the wide variety of etiologies and severity of injury in pABI contributes to disparate acute-care pathways and fractured, inconsistent follow-up. Without guidelines specific to the long-term management of children with ABI, ongoing care strategies are extrapolated from other etiologies of brain injury, needs assessment and consensus papers from groups with expertise caring for children with pABI, as well as from general management strategies of children with developmental disorders and similar impairments/diagnoses which are discussed throughout this book.

Up to 31% of children hospitalized with brain injury have unmet or unrecognized needs for health care services 1 year after injury [194–196]. Cognitive, language, physiatry, and educational services are the most commonly reported unmet needs by caregivers. Moreover, estimates suggest that close to 40% of caregivers report that their child did not see a physician at all during the first year after injury [197]. Children with specified clinical care pathways, longer hospitalizations, or inpatient rehabilitation admissions are most likely to be connected to ancillary services that extend into the chronic phase of recovery [198,199]. Social and environmental factors influencing access to follow-up care in pABI include insurance type, preexisting psychosocial condition, and nonwhite racial background [197].

The initial steps of outpatient care for children with ABI should begin prior to hospital discharge. Families with a clear understanding of the child's brain injury, level of cognitive and behavioral functioning, and the related therapeutic and educational needs are more likely to follow through with recommendations and follow-up visits [200]. Ideally, initial steps in long-term management should begin with the establishment of an outpatient care team familiar with pediatric ABI. This may include a range of specialties including physicians, therapists, neuropsychologists, and/or social workers who can collectively collaborate with the child's educational team and family. Each child should be evaluated by the outpatient team and an individualized program consisting of targeted goals to promote recovery should be developed. In the early stages of recovery, children and families typically require more frequent medical follow-up, intensive therapies, and/or a more restrictive educational plan. As goals are met, the frequency and intensity of services will decrease. At this time, the child's care will transition to a focus on chronic, long-term disability monitoring and transition planning, which may consist of short bursts of therapy, or phase-in care to address new challenges as they arise.

In addition to outpatient care for the child, caregivers and families often need support following brain injury. The initial phase postdischarge often requires a significant adjustment to a "new normal" by the child and the family. Parents may benefit from counseling to assist them in mourning the changes in their child and adjusting to the "normal." They may also need explicit instruction in developing and implementing a behavioral management program at home. The particular needs of the child and the family may also vary over time based on other psychosocial stressors, including barriers to obtaining the necessary psychosocial and educational support.

Conclusion

Acquired brain injuries in children are a challenging group of disorders with potential long-term motor, cognitive, and behavioral sequelae that impact the adaptive functioning of the child and may also impact the family. The unique vulnerability of the developing brain throughout childhood magnifies the impact of these injuries, often leading to intricate and long-term consequences that extend beyond the initial recovery phase. The complex interplay between neurological insult, developmental stage, and individualized factors, such as genetic predispositions and socio-environmental influences, contributes to the broad spectrum and variability of neurodevelopmental outcomes. While advances in acute management have improved survival rates in recent decades, the persistent challenges in motor skills, cognition, behavior, and socio-emotional well-being underscore the importance of ongoing interdisciplinary monitoring and the application of comprehensive, coordinated, and tailored interventions to meet the evolving needs of the child as they age, as well as those of their family. Concerted efforts in research, advocacy, and interdisciplinary care can mitigate the enduring burdens associated with pediatric acquired brain injuries, ultimately enhancing the quality of life for affected children and their families.

References

[1] Forsyth R, Kirkham F. Predicting outcome after childhood brain injury. CMAJ 2012;184(11):1257–64. Available from: https://doi.org/10.1503/cmaj.111045.

[2] Wilson M, Montgomery H. Impact of genetic factors on outcome from brain injury. Br J Anaesth 2007;99(1):43–8. Available from: https://doi.org/10.1093/bja/aem142.

[3] Zeiler FA, McFadyen C, Newcombe VFJ, et al. Genetic influences on patient-oriented outcomes in traumatic brain injury: a living systematic review of non-apolipoprotein E single-nucleotide polymorphisms. J Neurotrauma 2021;38(8):1107–23. Available from: https://doi.org/10.1089/neu.2017.5583.

[4] Carmichael J, Hicks AJ, Spitz G, Gould KR, Ponsford J. Moderators of gene-outcome associations following traumatic brain injury. Neurosci Biobehav Rev 2021;130:107–24. Available from: https://doi.org/10.1016/j.neubiorev.2021.08.015.

[5] Griffiths H, Goyal MS, Pineda JA. Brain metabolism and severe pediatric traumatic brain injury. Childs Nerv Syst 2017;33(10):1719–26. Available from: https://doi.org/10.1007/s00381-017-3514-y.

[6] Wade SL, Zhang N, Yeates KO, Stancin T, Taylor HG. Social environmental moderators of long-term functional outcomes of early childhood brain injury. JAMA Pediatr 2016;170(4):343–9. Available from: https://doi.org/10.1001/jamapediatrics.2015.4485.

[7] Grantham-McGregor S, Cheung YB, Cueto S, et al. Developmental potential in the first 5 years for children in developing countries. Lancet 2007;369(9555):60–70. Available from: https://doi.org/10.1016/S0140-6736(07)60032-4.

[8] Taylor CA, Bell JM, Breiding MJ, Xu L. Traumatic brain injury-related emergency department visits, hospitalizations, and deaths – United States, 2007 and 2013. MMWR Surveill Summ 2017;66(9):1–16. Available from: https://doi.org/10.15585/mmwr.ss6609a1.

[9] Dewan MC, Mummareddy N, Wellons JC, Bonfield CM. Epidemiology of global pediatric traumatic brain injury: qualitative review. World Neurosurg 2016;91:497–509. Available from: https://doi.org/10.1016/j.wneu.2016.03.045.

[10] Parks S, Sugerman D, Xu L, Coronado V. Characteristics of non-fatal abusive head trauma among children in the USA, 2003–2008: application of the CDC operational case definition to national hospital inpatient data. Inj Prev J Int Soc Child Adolesc Inj Prev 2012;18(6):392–8. Available from: https://doi.org/10.1136/injuryprev-2011-040234.

[11] Niederkrotenthaler T, Xu L, Parks SE, Sugerman DE. Descriptive factors of abusive head trauma in young children – United States, 2000–2009. Child Abuse Negl 2013;37(7):446–55. Available from: https://doi.org/10.1016/j.chiabu.2013.02.002.

[12] Nuño M, Shelley CD, Ugiliweneza B, Schmidt AJ, Magaña JN. Differences in incidence and case fatality of abusive head trauma. Child Abuse Negl 2020;104:104488. Available from: https://doi.org/10.1016/j.chiabu.2020.104488.

[13] QuickStats CDC. Injury and traumatic brain injury-related death rates by age – United States, 2006. MMWR 2010;59:303.

[14] Faul M, Coronado V. Epidemiology of traumatic brain injury. Handb Clin Neurol 2015;127:3–13. Available from: https://doi.org/10.1016/B978-0-444-52892-6.00001-5.

[15] Schneier AJ, Shields BJ, Hostetler SG, Xiang H, Smith GA. Incidence of pediatric traumatic brain injury and associated hospital resource utilization in the United States. Pediatrics. 2006;118(2):483–92. Available from: https://doi.org/10.1542/peds.2005-2588.

[16] Graves JM, Rivara FP, Vavilala MS. Health care costs 1 year after pediatric traumatic brain injury. Am J Public Health 2015;105(10):e35–41. Available from: https://doi.org/10.2105/AJPH.2015.302744.

[17] Giza CC, Hovda DA. The new neurometabolic cascade of concussion. Neurosurgery. 2014;75(Suppl 4):S24–33. Available from: https://doi.org/10.1227/NEU.0000000000000505 (0 4).

[18] Thapa K, Khan H, Singh TG, Kaur A. Traumatic brain injury: mechanistic insight on pathophysiology and potential therapeutic targets. J Mol Neurosci MN 2021;71(9):1725–42. Available from: https://doi.org/10.1007/s12031-021-01841-7.

[19] Hill CS, Coleman MP, Menon DK. Traumatic axonal injury: mechanisms and translational opportunities. Trends Neurosci 2016;39(5):311–24. Available from: https://doi.org/10.1016/j.tins.2016.03.002.

[20] Bruggeman GF, Haitsma IK, Dirven CMF, Volovici V. Traumatic axonal injury (TAI): definitions, pathophysiology and imaging-a narrative review. Acta Neurochir 2021;163(1):31–44. Available from: https://doi.org/10.1007/s00701-020-04594-1.

[21] Jamjoom AAB, Rhodes J, Andrews PJD, Grant SGN. The synapse in traumatic brain injury. Brain J Neurol 2021;144(1):18–31. Available from: https://doi.org/10.1093/brain/awaa321.

[22] Jassam YN, Izzy S, Whalen M, McGavern DB, El Khoury J. Neuroimmunology of traumatic brain injury: time for a paradigm shift. Neuron. 2017;95(6):1246–65. Available from: https://doi.org/10.1016/j.neuron.2017.07.010.

[23] Lifshitz J, Sullivan PG, Hovda DA, Wieloch T, McIntosh TK. Mitochondrial damage and dysfunction in traumatic brain injury. Mitochondrion. 2004;4(5–6):705–13. Available from: https://doi.org/10.1016/j.mito.2004.07.021.

[24] Natale JE, Joseph JG, Helfaer MA, Shaffner DH. Early hyperthermia after traumatic brain injury in children: risk factors, influence on length of stay, and effect on short-term neurologic status. Crit Care Med 2000;28(7):2608–15. Available from: https://doi.org/10.1097/00003246-200007000-00071.

[25] Vavilala MS, Bowen A, Lam AM, et al. Blood pressure and outcome after severe pediatric traumatic brain injury. J Trauma 2003;55 (6):1039–44. Available from: https://doi.org/10.1097/01.TA.0000101759.23607.57.

[26] Hymel KP, Makoroff KL, Laskey AL, Conaway MR, Blackman JA. Mechanisms, clinical presentations, injuries, and outcomes from inflicted versus noninflicted head trauma during infancy: results of a prospective, multicentered, comparative study. Pediatrics. 2007;119(5):922–9. Available from: https://doi.org/10.1542/peds.2006-3111.

[27] Risen SR, Suskauer SJ, Dematt EJ, Slomine BS, Salorio CF. Functional outcomes in children with abusive head trauma receiving inpatient rehabilitation compared with children with nonabusive head trauma. J Pediatr 2014;164(3):613–19. Available from: https://doi.org/10.1016/j.jpeds.2013.10.075.

[28] Miller Ferguson N, Sarnaik A, Miles D, et al. Abusive head trauma and mortality —- an analysis from an international comparative effectiveness study of children with severe traumatic brain injury. Crit Care Med 2017;45(8):1398–407. Available from: https://doi.org/10.1097/CCM.0000000000002378.

[29] Cremonini C, Lewis M, Wong MD, Benjamin ER, Inaba K, Demetriades D. Traumatic epidural hematomas in the pediatric population: clinical characteristics and diagnostic pitfalls. J Pediatr Surg 2020;55(9):1773–8. Available from: https://doi.org/10.1016/j.jpedsurg.2020.05.011.

[30] Bennett Colomer C, Solari Vergara F, Tapia Perez F, et al. Delayed intracranial hypertension and cerebral edema in severe pediatric head injury: risk factor analysis. Pediatr Neurosurg 2012;48(4):205–9. Available from: https://doi.org/10.1159/000343385.

[31] Kuppermann N, Holmes JF, Dayan PS, et al. Identification of children at very low risk of clinically-important brain injuries after head trauma: a prospective cohort study. Lancet Lond Engl 2009;374(9696):1160–70. Available from: https://doi.org/10.1016/S0140-6736(09)61558-0.

[32] Teasdale G, Jennett B. Assessment of coma and impaired consciousness. A practical scale. Lancet 1974;2(7872):81–4. Available from: https://doi.org/10.1016/s0140-6736(74)91639-0.

[33] Teasdale G, Maas A, Lecky F, Manley G, Stocchetti N, Murray G. The Glasgow Coma Scale at 40 years: standing the test of time. Lancet Neurol 2014;13(8):844–54. Available from: https://doi.org/10.1016/S1474-4422(14)70120-6.

[34] Lieh-Lai MW, Theodorou AA, Sarnaik AP, Meert KL, Moylan PM, Canady AI. Limitations of the Glasgow Coma Scale in predicting outcome in children with traumatic brain injury. J Pediatr 1992;120(2 Pt 1):195–9. Available from: https://doi.org/10.1016/s0022-3476(05)80426-3.

[35] Catroppa C, Godfrey C, Rosenfeld JV, Hearps SSJC, Anderson VA. Functional recovery ten years after pediatric traumatic brain injury: outcomes and predictors. J Neurotrauma 2012;29(16):2539–47. Available from: https://doi.org/10.1089/neu.2012.2403.

[36] Emami P, Czorlich P, Fritzsche FS, et al. Impact of Glasgow Coma Scale score and pupil parameters on mortality rate and outcome in pediatric and adult severe traumatic brain injury: a retrospective, multicenter cohort study. J Neurosurg 2017;126(3):760–7. Available from: https://doi.org/10.3171/2016.1.JNS152385.

[37] Ewing-Cobbs L, Levin HS, Fletcher JM, Miner ME, Eisenberg HM. The Children's Orientation and Amnesia Test: relationship to severity of acute head injury and to recovery of memory. Neurosurgery. 1990;27(5):683–91 discussion 691.

[38] Davis KC, Slomine BS, Salorio CF, Suskauer SJ. Time to follow commands and duration of posttraumatic amnesia predict GOS-E peds scores 1 to 2 years after TBI in children requiring inpatient rehabilitation. J Head Trauma Rehabil 2016;31(2):E39–47. Available from: https://doi.org/10.1097/HTR.0000000000000159.

[39] Wilson JT, Teasdale GM, Hadley DM, Wiedmann KD, Lang D. Post-traumatic amnesia: still a valuable yardstick. J Neurol Neurosurg Psychiatry 1994;57(2):198–201. Available from: https://doi.org/10.1136/jnnp.57.2.198.

[40] Lumba-Brown A, Yeates KO, Sarmiento K, et al. Diagnosis and management of mild traumatic brain injury in children: a systematic review. JAMA Pediatr 2018;172(11):e182847. Available from: https://doi.org/10.1001/jamapediatrics.2018.2847.

[41] McCrory P, Meeuwisse W, Dvořák J, et al. Consensus statement on concussion in sport-the 5(th) international conference on concussion in sport held in Berlin, October 2016. Br J Sports Med 2017;51(11):838–47. Available from: https://doi.org/10.1136/bjsports-2017-097699.

[42] Davis GA, Anderson V, Babl FE, et al. What is the difference in concussion management in children as compared with adults? A systematic review. Br J Sports Med 2017;51(12):949–57. Available from: https://doi.org/10.1136/bjsports-2016-097415.

[43] Risen SR, Reesman J, Yenokyan G, Slomine BS, Suskauer SJ. The course of concussion recovery in children 6–12 years of age: experience from an interdisciplinary rehabilitation clinic. PM R 2017;9(9):874–83. Available from: https://doi.org/10.1016/j.pmrj.2016.12.005.

[44] Suskauer SJ, Rane S, Reesman J, Slomine BS. Caregiver-report of symptoms following traumatic brain injury in a small clinical sample of preschool-aged children. J Pediatr Rehabil Med 2018;11(1):7–14. Available from: https://doi.org/10.3233/PRM-160424.

[45] Williams RM, Puetz TW, Giza CC, Broglio SP. Concussion recovery time among high school and collegiate athletes: a systematic review and meta-analysis. Sports Med 2015;45(6):893–903. Available from: https://doi.org/10.1007/s40279-015-0325-8.

[46] Makdissi M, Schneider KJ, Feddermann-Demont N, et al. Approach to investigation and treatment of persistent symptoms following sport-related concussion: a systematic review. Br J Sports Med 2017;51(12):958–68. Available from: https://doi.org/10.1136/bjsports-2016-097470.

[47] Iverson GL, Karr JE, Maxwell B, Zafonte R, Berkner PD, Cook NE. Examining criteria for defining persistent post-concussion symptoms in children and adolescents. Front Neurol 2021;12:614648. Available from: https://doi.org/10.3389/fneur.2021.614648.

[48] Olsson KA, Lloyd OT, Lebrocque RM, McKinlay L, Anderson VA, Kenardy JA. Predictors of child post-concussion symptoms at 6 and 18 months following mild traumatic brain injury. Brain Inj 2013;27(2):145–57. Available from: https://doi.org/10.3109/02699052.2012.729286.

[49] Laliberté Durish C, Yeates KO, Brooks BL. Psychological resilience as a predictor of persistent post-concussive symptoms in children with single and multiple concussion. J Int Neuropsychol Soc JINS 2018;24(8):759–68. Available from: https://doi.org/10.1017/S1355617718000437.

[50] Russell K, Selci E, Chu S, Fineblit S, Ritchie L, Ellis MJ. Longitudinal assessment of health-related quality of life following adolescent sports-related concussion. J Neurotrauma 2017;34(13):2147–53. Available from: https://doi.org/10.1089/neu.2016.4704.

[51] Ponsford J, Nguyen S, Downing M, et al. Factors associated with persistent post-concussion symptoms following mild traumatic brain injury in adults. J Rehabil Med 2019;51(1):32–9. Available from: https://doi.org/10.2340/16501977-2492.

[52] Hanalioglu D, Hanalioglu S, Arango JI, Adelson PD. Current evidence for pharmacological management of pediatric concussion: a systematic review. Childs Nerv Syst 2023;39(7):1831–49. Available from: https://doi.org/10.1007/s00381-023-05960-x.

[53] Leddy JJ, Master CL, Mannix R, et al. Early targeted heart rate aerobic exercise versus placebo stretching for sport-related concussion in adolescents: a randomised controlled trial. Lancet Child Adolesc Health 2021;5(11):792–9. Available from: https://doi.org/10.1016/S2352-4642(21)00267-4.

[54] CDC HEADS UP Campaign I HEADS UP I CDC Injury Center, https://www.cdc.gov/headsup/about/index.html; [accessed 19.12.23].

[55] Dichiaro M, Baker D, Tlustos SJ. Return to learn after traumatic brain injury. Pediatr Clin North Am 2023;70(3):445–60. Available from: https://doi.org/10.1016/j.pcl.2023.01.004.

[56] Halstead ME, McAvoy K, Devore CD, et al. Returning to learning following a concussion. Pediatrics. 2013;132(5):948–57. Available from: https://doi.org/10.1542/peds.2013-2867.

[57] Leddy JJ, Baker JG, Willer B. Active rehabilitation of concussion and post-concussion syndrome. Phys Med Rehabil Clin N Am 2016;27(2):437–54. Available from: https://doi.org/10.1016/j.pmr.2015.12.003.

[58] Anderson V, Catroppa C. Advances in postacute rehabilitation after childhood-acquired brain injury: a focus on cognitive, behavioral, and social domains. Am J Phys Med Rehabil 2006;85(9):767–78. Available from: https://doi.org/10.1097/01.phm.0000233176.08480.22.

[59] Campbell CGN, Kuehn SM, Richards PMP, Ventureyra E, Hutchison JS. Medical and cognitive outcome in children with traumatic brain injury. Can J Neurol Sci 2004;31(2):213–19. Available from: https://doi.org/10.1017/s0317167100053853.

[60] Kochanek PM, Tasker RC, Carney N, et al. Guidelines for the management of pediatric severe traumatic brain injury, third edition: update of the brain trauma foundation guidelines. Pediatr Crit Care Med 2019;20(3S Suppl 1):S1–82. Available from: https://doi.org/10.1097/PCC.0000000000001735.

[61] Kochanek PM, Tasker RC, Bell MJ, et al. Management of pediatric severe traumatic brain injury: 2019 consensus and guidelines-based algorithm for first and second tier therapies. Pediatr Crit Care Med 2019;20(3):269–79. Available from: https://doi.org/10.1097/PCC.0000000000001737.

[62] Taylor HG, Yeates KO, Wade SL, Drotar D, Klein SK, Stancin T. Influences on first-year recovery from traumatic brain injury in children. Neuropsychology 1999;13(1):76–89. Available from: https://doi.org/10.1037//0894-4105.13.1.76.

[63] Gorman S, Barnes MA, Swank PR, Ewing-Cobbs L. Recovery of working memory following pediatric traumatic brain injury: a longitudinal analysis. Dev Neuropsychol 2017;42(3):127–45. Available from: https://doi.org/10.1080/87565641.2017.1315581.

[64] Keenan HT, Bratton SL. Epidemiology and outcomes of pediatric traumatic brain injury. Dev Neurosci 2006;28(4-5):256–63. Available from: https://doi.org/10.1159/000094152.

[65] Babikian T, Merkley T, Savage RC, Giza CC, Levin H. Chronic aspects of pediatric traumatic brain injury: review of the literature. J Neurotrauma 2015;32(23):1849–60. Available from: https://doi.org/10.1089/neu.2015.3971.

[66] Crowe LM, Anderson V, Barton S, Babl FE, Catroppa C. Verbal ability and language outcome following traumatic brain injury in early childhood. J Head Trauma Rehabil 2014;29(3):217–23. Available from: https://doi.org/10.1097/HTR.0b013e31829babfd.

[67] Anderson V, Ylvisaker M. Executive function and the frontal lobes: themes for child development, brain insult and rehabilitation. Dev Neurorehabil 2009;12(5):253–4. Available from: https://doi.org/10.3109/17518420903086899.

[68] Andruszkow H, Deniz E, Urner J, et al. Physical and psychological long-term outcome after traumatic brain injury in children and adult patients. Health Qual Life Outcomes 2014;12:26. Available from: https://doi.org/10.1186/1477-7525-12-26.

[69] Giza CC, Kolb B, Harris NG, Asarnow RF, Prins ML. Hitting a moving target: basic mechanisms of recovery from acquired developmental brain injury. Dev Neurorehabil 2009;12(5):255–68. Available from: https://doi.org/10.3109/17518420903087558.

[70] Duhaime AC, Christian CW, Rorke LB, Zimmerman RA. Nonaccidental head injury in infants – the "shaken-baby syndrome. N Engl J Med 1998;338(25):1822–9. Available from: https://doi.org/10.1056/NEJM199806183382507.

[71] Chevignard MP, Lind K. Long-term outcome of abusive head trauma. Pediatr Radiol 2014;44(Suppl 4):S548–58. Available from: https://doi.org/10.1007/s00247-014-3169-8.

[72] Anderson V, Catroppa C, Morse S, Haritou F, Rosenfeld J. Functional plasticity or vulnerability after early brain injury? Pediatrics. 2005;116 (6):1374–82. Available from: https://doi.org/10.1542/peds.2004-1728.

[73] Oh A, Olson LD, Chern JJ, Kim H. Clinical characteristics and nonconvulsive seizures in young children with abusive head trauma. J Child Neurol 2019;34(12):713–19. Available from: https://doi.org/10.1177/0883073819853973.

[74] Dingman AL, Stence NV, O'Neill BR, Sillau SH, Chapman KE. Seizure severity is correlated with severity of hypoxic-ischemic injury in abusive head trauma. Pediatr Neurol 2018;82:29–35. Available from: https://doi.org/10.1016/j.pediatrneurol.2017.12.003.

[75] Topjian AA, de Caen A, Wainwright MS, et al. Pediatric post-cardiac arrest care: a scientific statement from the American Heart Association. Circulation. 2019;140(6):e194–233. Available from: https://doi.org/10.1161/CIR.0000000000000697.

[76] Felton H, Myers J, Liu G, Davis DW. Unintentional, non-fatal drowning of children: US trends and racial/ethnic disparities. BMJ Open 2015;5 (12):e008444. Available from: https://doi.org/10.1136/bmjopen-2015-008444.

[77] Cyr CCanadian Paediatric Society, Injury Prevention Committee. Preventing choking and suffocation in children. Paediatr Child Health 2012;17(2):91–4. Available from: https://doi.org/10.1093/pch/17.2.91.

[78] Douglas-Escobar M, Weiss MD. Hypoxic-ischemic encephalopathy: a review for the clinician. JAMA Pediatr 2015;169(4):397–403. Available from: https://doi.org/10.1001/jamapediatrics.2014.3269.

[79] Huang BY, Castillo M. Hypoxic-ischemic brain injury: imaging findings from birth to adulthood. Radiogr Rev Publ Radiol Soc N Am Inc 2008;28(2):417–39. Available from: https://doi.org/10.1148/rg.282075066 quiz617.

[80] Christophe C, Fonteyne C, Ziereisen F, et al. Value of MR imaging of the brain in children with hypoxic coma. AJNR Am J Neuroradiol 2002;23(4):716–23.

[81] Doherty DR, Parshuram CS, Gaboury I, et al. Hypothermia therapy after pediatric cardiac arrest. Circulation. 2009;119(11):1492–500. Available from: https://doi.org/10.1161/CIRCULATIONAHA.108.791384.

[82] Finder M, Boylan GB, Twomey D, Ahearne C, Murray DM, Hallberg B. Two-year neurodevelopmental outcomes after mild hypoxic ischemic encephalopathy in the era of therapeutic hypothermia. JAMA Pediatr 2020;174(1):48–55. Available from: https://doi.org/10.1001/jamapediatrics.2019.4011.

[83] Kwambana-Adams B. Global burden of meningitis and implications for strategy. Lancet Neurol 2023;22(8):646–8. Available from: https://doi.org/10.1016/S1474-4422(23)00244-2.

[84] Nakamura T, Cohen AL, Schwartz S, et al. The global landscape of pediatric bacterial meningitis data reported to the World Health Organization-Coordinated Invasive Bacterial Vaccine-Preventable Disease Surveillance Network, 2014–2019. J Infect Dis 2021;224(12Suppl 2):S161–73. Available from: https://doi.org/10.1093/infdis/jiab217.

[85] GBD 2016 Neurology Collaborators. . Global, regional, and national burden of neurological disorders, 1990–2016: a systematic analysis for the Global Burden of Disease Study 2016. Lancet Neurol 2019;18(5):459–80. Available from: https://doi.org/10.1016/S1474-4422(18)30499-X.

[86] George BP, Schneider EB, Venkatesan A. Encephalitis hospitalization rates and inpatient mortality in the United States, 2000–2010. PLoS One 2014;9(9):e104169. Available from: https://doi.org/10.1371/journal.pone.0104169.

[87] Messacar K, Fischer M, Dominguez SR, Tyler KL, Abzug MJ. Encephalitis in US children. Infect Dis Clin North Am 2018;32(1):145–62. Available from: https://doi.org/10.1016/j.idc.2017.10.007.

[88] Kohil A, Jemmieh S, Smatti MK, Yassine HM. Viral meningitis: an overview. Arch Virol 2021;166(2):335–45. Available from: https://doi.org/10.1007/s00705-020-04891-1.

[89] Hoffman O, Weber RJ. Pathophysiology and treatment of bacterial meningitis. Ther Adv Neurol Disord 2009;2(6):1–7. Available from: https://doi.org/10.1177/1756285609337975.

[90] Venkatesan A, Tunkel AR, Bloch KC, et al. Case definitions, diagnostic algorithms, and priorities in encephalitis: consensus statement of the international encephalitis consortium. Clin Infect Dis 2013;57(8):1114–28. Available from: https://doi.org/10.1093/cid/cit458.

[91] Wang Y, Liu X, Wang Y, Liu Q, Kong C, Xu G. Meta-analysis of adjunctive dexamethasone to improve clinical outcome of bacterial meningitis in children. Childs Nerv Syst 2018;34(2):217–23. Available from: https://doi.org/10.1007/s00381-017-3667-8.

[92] Chandran A, Herbert H, Misurski D, Santosham M. Long-term sequelae of childhood bacterial meningitis: an underappreciated problem. Pediatr Infect Dis J 2011;30(1):3–6. Available from: https://doi.org/10.1097/INF.0b013e3181ef25f7.

[93] Snoek L, Gonçalves BP, Horváth-Puhó E, et al. Short-term and long-term risk of mortality and neurodevelopmental impairments after bacterial meningitis during infancy in children in Denmark and the Netherlands: a nationwide matched cohort study. Lancet Child Adolesc Health 2022;6(9):633–42. Available from: https://doi.org/10.1016/S2352-4642(22)00155-9.

[94] Rodenburg-Vlot MBA, Ruytjens L, Oostenbrink R, Goedegebure A, van der Schroeff MP. Systematic review: incidence and course of hearing loss caused by bacterial meningitis: in search of an optimal timed audiological follow-up. Otol Neurotol 2016;37(1):1–8. Available from: https://doi.org/10.1097/MAO.0000000000000922.

[95] Leake JAD, Albani S, Kao AS, et al. Acute disseminated encephalomyelitis in childhood: epidemiologic, clinical and laboratory features. Pediatr Infect Dis J 2004;23(8):756–64. Available from: https://doi.org/10.1097/01.inf.0000133048.75452.dd.

[96] Desai J, Mitchell WG. Acute cerebellar ataxia, acute cerebellitis, and opsoclonus-myoclonus syndrome. J Child Neurol 2012;27(11):1482–8. Available from: https://doi.org/10.1177/0883073812450318.

[97] Gordon-Lipkin E, Banwell B. Acute disseminated encephalomyelitis. In: Fox RJ, Rae-Grant AD, Bethoux F, editors. Multiple sclerosis and related disorders. Springer Publishing Company; 2018. p. 363–71. Available from: 10.1891/9780826125941.0040.

[98] Singer HS. Autoantibody-associated movement disorders in children: proven and proposed. Semin Pediatr Neurol 2017;24(3):168–79. Available from: https://doi.org/10.1016/j.spen.2017.08.003.

[99] Taranta A, Stollerman GH. The relationship of Sydenham's chorea to infection with group A streptococci. Am J Med 1956;20(2):170–5. Available from: https://doi.org/10.1016/0002-9343(56)90186-3.

[100] Walker KG, Wilmshurst JM. An update on the treatment of Sydenham's chorea: the evidence for established and evolving interventions. Ther Adv Neurol Disord 2010;3(5):301–9. Available from: https://doi.org/10.1177/1756285610382063.

[101] Kirvan CA, Swedo SE, Heuser JS, Cunningham MW. Mimicry and autoantibody-mediated neuronal cell signaling in Sydenham chorea. Nat Med 2003;9(7):914–20. Available from: https://doi.org/10.1038/nm892.

[102] Giedd JN, Rapoport JL, Kruesi MJ, et al. Sydenham's chorea: magnetic resonance imaging of the basal ganglia. Neurology. 1995;45 (12):2199–202. Available from: https://doi.org/10.1212/wnl.45.12.2199.

[103] Dale RC, Merheb V, Pillai S, et al. Antibodies to surface dopamine-2 receptor in autoimmune movement and psychiatric disorders. Brain J Neurol 2012;135(Pt 11):3453–68. Available from: https://doi.org/10.1093/brain/aws256.

[104] Cardoso F. Autoimmune choreas. J Neurol Neurosurg Psychiatry 2017;88(5):412–17. Available from: https://doi.org/10.1136/jnnp-2016-314475.

[105] Punukollu M, Mushet N, Linney M, Hennessy C, Morton M. Neuropsychiatric manifestations of Sydenham's chorea: a systematic review. Dev Med Child Neurol 2016;58(1):16–28. Available from: https://doi.org/10.1111/dmcn.12786.

[106] Connolly AM, Dodson WE, Prensky AL, Rust RS. Course and outcome of acute cerebellar ataxia. Ann Neurol 1994;35(6):673–9. Available from: https://doi.org/10.1002/ana.410350607.

[107] Poretti A, Benson JE, Huisman TAGM, Boltshauser E. Acute ataxia in children: approach to clinical presentation and role of additional investigations. Neuropediatrics. 2013;44(3):127–41. Available from: https://doi.org/10.1055/s-0032-1329909.

[108] Thakkar K, Maricich SM, Alper G. Acute Ataxia in childhood: 11-year experience at a major pediatric neurology referral center. J Child Neurol 2016;31(9):1156–60. Available from: https://doi.org/10.1177/0883073816643407.

[109] Yeshokumar A, Gordon-Lipkin E, Banwell B. Post-viral syndromes. Clinical neurovirology.. 2nd ed. CRC Press; 2020.

[110] Tenembaum S, Chamoles N, Fejerman N. Acute disseminated encephalomyelitis: a long-term follow-up study of 84 pediatric patients. Neurology. 2002;59(8):1224–31. Available from: https://doi.org/10.1212/wnl.59.8.1224.

[111] Murthy SNK, Faden HS, Cohen ME, Bakshi R. Acute disseminated encephalomyelitis in children. Pediatrics. 2002;110(2 Pt 1):e21. Available from: https://doi.org/10.1542/peds.110.2.e21.

[112] Young NP, Weinshenker BG, Parisi JE, et al. Perivenous demyelination: association with clinically defined acute disseminated encephalomyelitis and comparison with pathologically confirmed multiple sclerosis. Brain J Neurol 2010;133(Pt 2):333–48. Available from: https://doi.org/10.1093/brain/awp321.

[113] Koelman DLH, Chahin S, Mar SS, et al. Acute disseminated encephalomyelitis in 228 patients: a retrospective, multicenter US study. Neurology. 2016;86(22):2085–93. Available from: https://doi.org/10.1212/WNL.0000000000002723.

[114] Hacohen Y, Banwell B. Treatment approaches for MOG-Ab-associated demyelination in children. Curr Treat Options Neurol 2019;21(1):2. Available from: https://doi.org/10.1007/s11940-019-0541-x.

[115] Hennes EM, Baumann M, Schanda K, et al. Prognostic relevance of MOG antibodies in children with an acquired demyelinating syndrome. Neurology. 2017;89(9):900–8. Available from: https://doi.org/10.1212/WNL.0000000000004312.

[116] Lopez-Chiriboga AS, Sechi E, Buciuc M, et al. Long-term outcomes in patients with myelin oligodendrocyte glycoprotein immunoglobulin G-associated disorder. JAMA Neurol 2020;77(12):1575–7. Available from: https://doi.org/10.1001/jamaneurol.2020.3115.

[117] Krupp LB, Tardieu M, Amato MP, et al. International Pediatric Multiple Sclerosis Study Group criteria for pediatric multiple sclerosis and immune-mediated central nervous system demyelinating disorders: revisions to the 2007 definitions. Mult Scler 2013;19(10):1261–7. Available from: https://doi.org/10.1177/1352458513484547.

[118] Burton KLO, Williams TA, Catchpoole SE, Brunsdon RK. Long-term neuropsychological outcomes of childhood onset acute disseminated encephalomyelitis (ADEM): a meta-analysis. Neuropsychol Rev 2017;27(2):124–33. Available from: https://doi.org/10.1007/s11065-017-9343-7.

[119] Beatty C, Bowler RA, Farooq O, et al. Long-Term neurocognitive, psychosocial, and magnetic resonance imaging outcomes in pediatric-onset acute disseminated encephalomyelitis. Pediatr Neurol 2016;57:64–73. Available from: https://doi.org/10.1016/j.pediatrneurol.2016.01.003.

[120] Suppiej A, Cainelli E, Casara G, Cappellari A, Nosadini M, Sartori S. Long-term neurocognitive outcome and quality of life in pediatric acute disseminated encephalomyelitis. Pediatr Neurol 2014;50(4):363–7. Available from: https://doi.org/10.1016/j.pediatrneurol.2013.12.006.

[121] Shilo S, Michaeli O, Shahar E, Ravid S. Long-term motor, cognitive and behavioral outcome of acute disseminated encephalomyelitis. Eur J Paediatr Neurol 2016;20(3):361–7. Available from: https://doi.org/10.1016/j.ejpn.2016.01.008.

[122] Aubert-Broche B, Weier K, Longoni G, et al. Monophasic demyelination reduces brain growth in children. Neurology. 2017;88(18):1744–50. Available from: https://doi.org/10.1212/WNL.0000000000003884.

[123] Florance NR, Davis RL, Lam C, et al. Anti-N-methyl-D-aspartate receptor (NMDAR) encephalitis in children and adolescents. Ann Neurol 2009;66(1):11–18. Available from: https://doi.org/10.1002/ana.21756.

[124] Cellucci T, Van Mater H, Graus F, et al. Clinical approach to the diagnosis of autoimmune encephalitis in the pediatric patient. Neurol Neuroimmunol Neuroinflamm 2020;7(2):e663. Available from: https://doi.org/10.1212/NXI.0000000000000663.

[125] Gable MS, Gavali S, Radner A, et al. Anti-NMDA receptor encephalitis: report of ten cases and comparison with viral encephalitis. Eur J Clin Microbiol Infect Dis 2009;28(12):1421–9. Available from: https://doi.org/10.1007/s10096-009-0799-0.

[126] Gable MS, Sheriff H, Dalmau J, Tilley DH, Glaser CA. The frequency of autoimmune N-methyl-D-aspartate receptor encephalitis surpasses that of individual viral etiologies in young individuals enrolled in the California Encephalitis Project. Clin Infect Dis 2012;54(7):899–904. Available from: https://doi.org/10.1093/cid/cir1038.

[127] Sun Q, Wang Y, Xie Y, Wu P, Li S, Zhao W. Long-term neurological outcomes of children with neuroblastoma with opsoclonus-myoclonus syndrome. Transl Pediatr 2022;11(3):368–74. Available from: https://doi.org/10.21037/tp-21-519.

[128] Kumar PA, Chowdhury D. Anti-NMDAR encephalitis preceded by infection: Systematic literature review with comparison of non-HSV and HSV related anti-NMDAR encephalitis. J Neurol Sci 2021;429. Available from: https://doi.org/10.1016/j.jns.2021.118838.

[129] Graus F, Titulaer MJ, Balu R, et al. A clinical approach to diagnosis of autoimmune encephalitis. Lancet Neurol 2016;15(4):391–404. Available from: https://doi.org/10.1016/S1474-4422(15)00401-9.

[130] Nosadini M, Thomas T, Eyre M, et al. International consensus recommendations for the treatment of pediatric NMDAR antibody encephalitis. Neurol Neuroimmunol Neuroinflamm 2021;8(5):e1052. Available from: https://doi.org/10.1212/NXI.0000000000001052.

[131] Nosadini M, Eyre M, Molteni E, et al. Use and safety of immunotherapeutic management of N-methyl-D-aspartate receptor antibody encephalitis: a meta-analysis. JAMA Neurol 2021;78(11):1333–44. Available from: https://doi.org/10.1001/jamaneurol.2021.3188.

[132] Hutchinson ML, Yeshokumar AK, Armangue T. Antibody-mediated encephalitis in children: focus on diagnostic clues and acute symptom management. Semin Pediatr Neurol 2021;37:100873. Available from: https://doi.org/10.1016/j.spen.2021.100873.

[133] Hageboutros K, Hattiangadi Thomas N, Hutchinson M, Banwell B, Baum KT. Neuropsychological functioning in children and adolescents with anti-NMDA receptor encephalitis (anti-NMDARE). J Neurol 2023;270(1):402–12. Available from: https://doi.org/10.1007/s00415-022-11372-9.

[134] Wilkinson-Smith A, Blackwell LS, Howarth RA. [Formula: see text]Neuropsychological outcomes in children and adolescents following anti-NMDA receptor encephalitis. Child Neuropsychol 2022;28(2):212–23. Available from: https://doi.org/10.1080/09297049.2021.1965110.

[135] Loerinc LB, Blackwell L, Howarth R, Gombolay G. Evaluation of the anti-N-methyl-D-aspartate receptor encephalitis one-year functional status score in predicting functional outcomes in pediatric patients with anti-N-methyl-D-aspartate receptor encephalitis. Pediatr Neurol 2021;124:21–3. Available from: https://doi.org/10.1016/j.pediatrneurol.2021.08.004.

[136] McKeon GL, Robinson GA, Ryan AE, et al. Cognitive outcomes following anti-N-methyl-D-aspartate receptor encephalitis: a systematic review. J Clin Exp Neuropsychol 2018;40(3):234–52. Available from: https://doi.org/10.1080/13803395.2017.1329408.

[137] Yeshokumar AK, Gordon-Lipkin E, Arenivas A, et al. Neurobehavioral outcomes in autoimmune encephalitis. J Neuroimmunol 2017;312:8–14. Available from: https://doi.org/10.1016/j.jneuroim.2017.08.010.

[138] Gordon-Lipkin E, Yeshokumar AK, Saylor D, Arenivas A, Probasco JC. Comparative outcomes in children and adults with anti-N-methyl-D-aspartate (anti-NMDA) receptor encephalitis. J Child Neurol. 2017;32(11):930–5. Available from: https://doi.org/10.1177/0883073817720340.

[139] Matricardi S, Patrini M, Freri E, et al. Cognitive and neuropsychological evolution in children with anti-NMDAR encephalitis. J Neurol 2016;263(4):765–71. Available from: https://doi.org/10.1007/s00415-016-8056-9.

[140] Felling RJ, Sun LR, Maxwell EC, Goldenberg N, Bernard T. Pediatric arterial ischemic stroke: epidemiology, risk factors, and management. Blood Cell Mol Dis 2017;67:23–33. Available from: https://doi.org/10.1016/j.bcmd.2017.03.003.

[141] Oleske DM, Cheng X, Jeong A, Arndt TJ. Pediatric acute ischemic stroke by age-group: a systematic review and meta-analysis of published studies and hospitalization records. Neuroepidemiology. 2021;55(5):331–41. Available from: https://doi.org/10.1159/000518281.

[142] Lehman LL, Rivkin MJ. Perinatal arterial ischemic stroke: presentation, risk factors, evaluation, and outcome. Pediatr Neurol 2014;51 (6):760–8. Available from: https://doi.org/10.1016/j.pediatrneurol.2014.07.031.

[143] Mittal SO, ThatiGanganna S, Kuhns B, Strbian D, Sundararajan S. Acute ischemic stroke in pediatric patients. Stroke. 2015;46(2):e32–4. Available from: https://doi.org/10.1161/STROKEAHA.114.007681.

[144] Grunt S, Mazenauer L, Buerki SE, et al. Incidence and outcomes of symptomatic neonatal arterial ischemic stroke. Pediatrics 2015;135(5): e1220–8. Available from: https://doi.org/10.1542/peds.2014-1520.

[145] Amlie-Lefond C, Bernard TJ, Sébire G, et al. Predictors of cerebral arteriopathy in children with arterial ischemic stroke: results of the International Pediatric Stroke Study. Circulation. 2009;119(10):1417–23. Available from: https://doi.org/10.1161/CIRCULATIONAHA.108.806307.

[146] Malone LA, Levy TJ, Peterson RK, Felling RJ, Beslow LA. Neurological and functional outcomes after pediatric stroke. Semin Pediatr Neurol 2022;44:100991. Available from: https://doi.org/10.1016/j.spen.2022.100991.

[147] Ferriero DM, Fullerton HJ, Bernard TJ, et al. Management of stroke in neonates and children: a scientific statement from the American Heart Association/American Stroke Association. Stroke. 2019;50(3):e51–96. Available from: https://doi.org/10.1161/STR.0000000000000183.

[148] DeBaun MR, Jordan LC, King AA, et al. American Society of Hematology 2020 guidelines for sickle cell disease: prevention, diagnosis, and treatment of cerebrovascular disease in children and adults. Blood Adv 2020;4(8):1554–88. Available from: https://doi.org/10.1182/bloodadvances.2019001142.

[149] Malone LA, Felling RJ. Pediatric stroke: unique implications of the immature brain on injury and recovery. Pediatr Neurol 2020;102:3–9. Available from: https://doi.org/10.1016/j.pediatrneurol.2019.06.016.

[150] Kornfeld S, Delgado Rodríguez JA, Everts R, et al. Cortical reorganisation of cerebral networks after childhood stroke: impact on outcome. BMC Neurol 2015;15:90. Available from: https://doi.org/10.1186/s12883-015-0309-1.

[151] Choong K, Fraser D, Al-Harbi S, et al. Functional recovery in critically ill children, the "WeeCover" multicenter study. Pediatr Crit Care Med 2018;19(2):145–54. Available from: https://doi.org/10.1097/PCC.0000000000001421.

[152] Manning JC, Pinto NP, Rennick JE, Colville G, Curley MAQ. Conceptualizing post intensive care syndrome in children – the PICS-p framework. Pediatr Crit Care Med 2018;19(4):298–300. Available from: https://doi.org/10.1097/PCC.0000000000001476.

[153] Hartman ME, Williams CN, Hall TA, Bosworth CC, Piantino JA. Post-intensive-care syndrome for the pediatric neurologist. Pediatr Neurol 2020;108:47–53. Available from: https://doi.org/10.1016/j.pediatrneurol.2020.02.003.

[154] Williams CN, Kirby A, Piantino J. If you build it, they will come: initial experience with a multi-disciplinary pediatric neurocritical care follow-up clinic. Child 2017;4(9). Available from: https://doi.org/10.3390/children4090083.

[155] Pinto NP, Rhinesmith EW, Kim TY, Ladner PH, Pollack MM. Long-term function after pediatric critical illness: results from the survivor outcomes study. Pediatr Crit Care Med 2017;18(3):e122–30. Available from: https://doi.org/10.1097/PCC.0000000000001070.

[156] Colville G, Pierce C. Patterns of post-traumatic stress symptoms in families after paediatric intensive care. Intensive Care Med 2012;38 (9):1523–31. Available from: https://doi.org/10.1007/s00134-012-2612-2.

[157] Yagiela LM, Carlton EF, Meert KL, Odetola FO, Cousino MK. Parent medical traumatic stress and associated family outcomes after pediatric critical illness: a systematic review. Pediatr Crit Care Med 2019;20(8):759–68. Available from: https://doi.org/10.1097/PCC.0000000000001985.

[158] Keetley R, Radford K, Manning JC. A scoping review of the needs of children and young people with acquired brain injuries and their families. Brain Inj 2019;33(9):1117–28. Available from: https://doi.org/10.1080/02699052.2019.1637542.

[159] McKinlay A, Linden M, DePompei R, Aaro Jonsson C, Anderson V, Braga L, et al. Service provision for children and young people with acquired brain injury: Practice recommendations. Brain Inj 2016;30(13–14):1656–64. Available from: http://doi.org/10.1080/02699052.2016.1201592. Epub 2016 Oct 14. PMID: 27740853.

[160] Pagnozzi AM, Pannek K, Fripp J, Fiori S, Boyd RN, Rose S. Predicting motor and cognitive outcomes from MRIs of brain structure in children with acquired brain injury: a pilot study. Pediatr Neonatol 2023;64(3):297–305. Available from: https://doi.org/10.1016/j.pedneo.2022.07.014.

[161] Lambregts SAM, Van Markus-Doornbosch F, Catsman-Berrevoets CE, et al. Neurological outcome in children and youth with acquired brain injury 2-year post-injury. Dev Neurorehabil 2018;21(7):465–74. Available from: https://doi.org/10.1080/17518423.2018.1460770.

[162] Giacino JT, Katz DI, Schiff ND, et al. Practice guideline update recommendations summary: disorders of consciousness: report of the guideline development, dissemination, and implementation subcommittee of the American Academy of Neurology; the American Congress of Rehabilitation Medicine; and the National Institute on Disability, Independent Living, and Rehabilitation Research. Neurology. 2018;91 (10):450–60. Available from: https://doi.org/10.1212/WNL.0000000000005926.

[163] Molteni E, Colombo K, Pastore V, et al. Joint neuropsychological assessment through coma/near coma and level of cognitive functioning assessment scales reduces negative findings in pediatric disorders of consciousness. Brain Sci 2020;10(3):162. Available from: https://doi.org/10.3390/brainsci10030162.

[164] Villa F, Colombo K, Pastore V, et al. LOCFAS-assessed evolution of cognitive and behavioral functioning in a sample of pediatric patients with severe acquired brain injury in the postacute phase. J Child Neurol 2015;30(9):1125–34. Available from: https://doi.org/10.1177/0883073814553798.

[165] Ansell BJ, Keenan JE. The Western Neuro Sensory Stimulation Profile: a tool for assessing slow-to-recover head-injured patients. Arch Phys Med Rehabil 1989;70(2):104–8.

[166] Eilander HJ, van de Wiel M, Wijers M, et al. The reliability and validity of the PALOC-s: a post-acute level of consciousness scale for assessment of young patients with prolonged disturbed consciousness after brain injury. Neuropsychol Rehabil 2009;19(1):1–27. Available from: https://doi.org/10.1080/09602010701694822.

[167] American Congress of Rehabilitation Medicine, Brain Injury-Interdisciplinary Special Interest Group, Disorders of Consciousness Task ForceSeel RT, Sherer M, et al. Assessment scales for disorders of consciousness: evidence-based recommendations for clinical practice and research. Arch Phys Med Rehabil 2010;91(12):1795–813. Available from: https://doi.org/10.1016/j.apmr.2010.07.218.

[168] Slomine BS, Suskauer SJ, Nicholson R, Giacino JT. Preliminary validation of the coma recovery scale for pediatrics in typically developing young children. Brain Inj 2019;33(13-14):1640–5. Available from: https://doi.org/10.1080/02699052.2019.1658221.

[169] Beretta E, Molteni E, Galbiati S, Stefanoni G, Strazzer S. Five-year motor functional outcome in children with acquired brain injury. Yet to the end of the story? Dev Neurorehabil 2018;21(7):449–56. Available from: https://doi.org/10.1080/17518423.2017.1360408.

[170] Kuhtz-Buschbeck JP, Hoppe B, Gölge M, Dreesmann M, Damm-Stünitz U, Ritz A. Sensorimotor recovery in children after traumatic brain injury: analyses of gait, gross motor, and fine motor skills. Dev Med Child Neurol 2003;45(12):821–8. Available from: https://doi.org/10.1017/s001216220300152x.

[171] Mimouni-Bloch A, Shaklai S, Levin M, et al. Developmental and acquired brain injury have opposite effects on finger coordination in children. Front Hum Neurosci 2023;17:1083304. Available from: https://doi.org/10.3389/fnhum.2023.1083304.

[172] Bodack MI. Pediatric acquired brain injury. Optometry 2010;81(10):516–27. Available from: https://doi.org/10.1016/j.optm.2010.06.004.

[173] Cockerham GC, Goodrich GL, Weichel ED, et al. Eye and visual function in traumatic brain injury. J Rehabil Res Dev 2009;46(6):811–18. Available from: https://doi.org/10.1682/jrrd.2008.08.0109.

[174] Jimenez-Gomez A, Fisher KS, Zhang KX, Liu C, Sun Q, Shah VS. Longitudinal neurological analysis of moderate and severe pediatric cerebral visual impairment. Front Hum Neurosci 2022;16:772353. Available from: https://doi.org/10.3389/fnhum.2022.772353.

[175] Riccardi JS. A scoping review of auditory dysfunction after childhood traumatic brain injury. Am J Audiol 2023;32(2):469–76. Available from: https://doi.org/10.1044/2022_AJA-22-00198.

[176] Munjal SK, Panda NK, Pathak A. Audiological deficits after closed head injury. J Trauma 2010;68(1):13–18. Available from: https://doi.org/10.1097/TA.0b013e3181c9f274. PMID: 20065751.

[177] Alighieri C, De Maere K, Poncelet G, et al. Occurrence of speech-language disorders in the acute phase following pediatric acquired brain injury: results from the Ghent University Hospital. Brain Inj 2021;35(8):907–21. Available from: https://doi.org/10.1080/02699052.2021.1927185.

[178] Huang CT, Lin WC, Ho CH, et al. Incidence of severe dysphagia after brain surgery in pediatric traumatic brain injury: a nationwide population-based retrospective study. J Head Trauma Rehabil 2014;29(6):E31–6. Available from: https://doi.org/10.1097/HTR.0000000000000032.

[179] Morgan AT. Dysphagia in childhood traumatic brain injury: a reflection on the evidence and its implications for practice. Dev Neurorehabil 2010;13(3):192–203. Available from: https://doi.org/10.3109/17518420903289535.

[180] Hendrix CL, King TZ, Wise J, Haarbauer-Krupa J. Early brain injury and adaptive functioning in middle childhood: the mediating role of pragmatic language. J Int Neuropsychol Soc 2020;26(9):835–50. Available from: https://doi.org/10.1017/S1355617720000399.

[181] Deighton S, Ju N, Graham SA, Yeates KO. Pragmatic language comprehension after pediatric traumatic brain injury: a scoping review. J Head Trauma Rehabil 2020;35(2):E113–26. Available from: https://doi.org/10.1097/HTR.0000000000000515.

[182] Greene RK, Rich-Wimmer N, Williams CN, Hall TA. Social functioning and autistic behaviors in youth following acquired brain injury. Child 2022;9(11):1648. Available from: https://doi.org/10.3390/children9111648.

[183] On ZX, Ryan NP, Konjarski M, Catroppa C, Stargatt R. Social cognition in paediatric traumatic brain injury: a systematic review and meta-analysis. Neuropsychol Rev 2022;32(1):127–48. Available from: https://doi.org/10.1007/s11065-021-09488-2.

[184] Ryan NP, Catroppa C, Beare R, et al. Predictors of longitudinal outcome and recovery of pragmatic language and its relation to externalizing behaviour after pediatric traumatic brain injury. Brain Lang 2015;142:86–95. Available from: https://doi.org/10.1016/j.bandl.2015.01.007.

[185] Dikmen SS, Corrigan JD, Levin HS, Machamer J, Stiers W, Weisskopf MG. Cognitive outcome following traumatic brain injury. J Head Trauma Rehabil 2009;24(6):430–8. Available from: https://doi.org/10.1097/HTR.0b013e3181c133e9.

[186] Anderson VA, Morse SA, Catroppa C, Haritou F, Rosenfeld JV. Thirty month outcome from early childhood head injury: a prospective analysis of neurobehavioural recovery. Brain J Neurol 2004;127(Pt 12):2608–20. Available from: https://doi.org/10.1093/brain/awh320.

[187] Ganesalingam K, Sanson A, Anderson V, Yeates KO. Self-regulation as a mediator of the effects of childhood traumatic brain injury on social and behavioral functioning. J Int Neuropsychol Soc 2007;13(2):298–311. Available from: https://doi.org/10.1017/S1355617707070324.

[188] Anderson V, Godfrey C, Rosenfeld JV, Catroppa C. 10 years outcome from childhood traumatic brain injury. Int J Dev Neurosci 2012;30(3):217–24. Available from: https://doi.org/10.1016/j.ijdevneu.2011.09.008.

[189] Yeates KO, Swift E, Taylor HG, et al. Short- and long-term social outcomes following pediatric traumatic brain injury. J Int Neuropsychol Soc 2004;10(3):412–26. Available from: https://doi.org/10.1017/S1355617704103093.

[190] Jin C, Schachar R. Methylphenidate treatment of attention-deficit/hyperactivity disorder secondary to traumatic brain injury: a critical appraisal of treatment studies. CNS Spectr 2004;9(3):217–26. Available from: https://doi.org/10.1017/s1092852900009019.

[191] Ware J, Butcher JL, Latal B, et al. Neurodevelopmental evaluation strategies for children with congenital heart disease aged birth through 5 years: recommendations from the cardiac neurodevelopmental outcome collaborative. Cardiol Young 2020;30(11):1609–22. Available from: https://doi.org/10.1017/S1047951120003534.

[192] Litt JS, Campbell DE. High-risk infant follow-up after NICU discharge: current care models and future considerations. Clin Perinatol 2023;50(1):225–38. Available from: https://doi.org/10.1016/j.clp.2022.11.004.

[193] Merchant N, Shekhar L, Huertas-Ceballos A, Johnson S, Arasu A. Neurodevelopmental care following neonatal discharge. Paediatr Child Health 2022;32(9):324–31. Available from: https://doi.org/10.1016/j.paed.2022.07.002.

[194] Greenspan AI, MacKenzie EJ. Use and need for post-acute services following paediatric head injury. Brain Inj 2000;14(5):417–29. Available from: https://doi.org/10.1080/026990500120529.

[195] Fuentes MM, Wang J, Haarbauer-Krupa J, et al. Unmet rehabilitation needs after hospitalization for traumatic brain injury. Pediatrics. 2018;141(5). Available from: https://doi.org/10.1542/peds.2017-2859.

[196] Diener ML, Kirby AV, Sumsion F, Canary HE, Green MM. Community reintegration needs following paediatric brain injury: perspectives of caregivers and service providers. Disabil Rehabil 2022;44(19):5592–602. Available from: https://doi.org/10.1080/09638288.2021.1946176.

[197] Slomine BS, McCarthy ML, Ding R, et al. Health care utilization and needs after pediatric traumatic brain injury. Pediatrics. 2006;117(4):e663–74. Available from: https://doi.org/10.1542/peds.2005-1892.

[198] Glang A, Todis B, Thomas CW, Hood D, Bedell G, Cockrell J. Return to school following childhood TBI: who gets services? NeuroRehabilitation 2008;23(6):477–86.

[199] Bennett TD, Niedzwecki CM, Korgenski EK, Bratton SL. Initiation of physical, occupational, and speech therapy in children with traumatic brain injury. Arch Phys Med Rehabil 2013;94(7):1268–76. Available from: https://doi.org/10.1016/j.apmr.2013.02.021.

[200] Keenan RP, Lovanio K, Lapidus G, Chenard D, Smith S. Improved concussion discharge instructions in a pediatric emergency department. Adv Emerg Nurs J 2020;42(1):63–70. Available from: https://doi.org/10.1097/TME.0000000000000280.

Chapter 14

Developmental and epileptic encephalopathies

Ryan E. Gill[1,2], Natasha N. Ludwig[3,4] and Carl E. Stafstrom[5]

[1]*Department of Neurology and Developmental Medicine, Kennedy Krieger Institute, Baltimore, MD, United States,* [2]*Department of Neurology, Johns Hopkins University School of Medicine, Baltimore, MD, United States,* [3]*Department of Neuropsychology, Kennedy Krieger Institute, Baltimore, MD, United States,* [4]*Department of Psychiatry and Behavioral Sciences, Johns Hopkins School of Medicine, Baltimore, MD, United States,* [5]*Departments of Neurology and Pediatrics, Division of Pediatric Neurology, Johns Hopkins University School of Medicine, Baltimore, MD, United States*

Introduction

Children are highly prone to seizures and epilepsy, and those with developmental impairments have an even greater susceptibility. Approximately 1.2% of the US population has epilepsy, including about 3 million adults and 470,000 children [1]. It is estimated that 69 million people worldwide live with epilepsy, with the majority residing in developing countries [2]. Epilepsy is more common in individuals with neurodevelopmental disorders than in the general population. About 22% of those with intellectual disability (ID) also have epilepsy, with the prevalence increasing with more severe ID [3]. A wide range of epilepsy prevalence has been reported in autism spectrum disorder (ASD), ranging from 2.4% to 26% [4]. In children with cerebral palsy (CP), epilepsy is present in up to 25%–57%, and the epilepsy is drug-resistant in 19%–28% [5].

Definitions

Definitions of seizure, epilepsy, and related terms have evolved over the past decade, so it is critical to begin our discussion with the most updated nomenclature (Table 14.1). A seizure is a neurological event consisting of the "transient occurrence of signs and/or symptoms due to abnormal, excessive, or synchronous neuronal activity in the brain" [6]. A seizure is either provoked or unprovoked. A provoked seizure has an acute cause, such as head trauma, stroke, infection, or metabolic derangement. An unprovoked seizure has no acute precipitant, though there could be a history of a remote insult such as a stroke or trauma. A provoked seizure, therefore, can be thought of as occurring outside of the context of epilepsy, whereas a seizure with an inciting factor in someone with epilepsy could be described as "triggered," for example, by sleep deprivation, alcohol, medications that lower the seizure threshold, etc.

Epilepsy is the result of recurrent, unprovoked seizuresand is defined as (1) the occurrence of at least two unprovoked seizures occurring over 24 hours apart, (2) the occurrence of one unprovoked seizure and a presumed recurrence risk of over 60% over the next 10 years (i.e., the presence of a structural brain abnormality), or (3) the diagnosis of an epilepsy syndrome [7].

The International League Against Epilepsy's (ILAE) most recent classification of the epilepsies, published in 2017, introduced three levels of classification, taking into account etiology and the robust scientific advances in the field. The first level of classification is *seizure type* based on onset site: focal, generalized, or unknown.

The second level of classification is *epilepsy type*: focal, generalized, combined focal and generalized, or unknown. Epilepsy is classified on clinical grounds and is supported by EEG findings. Generalized epilepsies may include a range of seizure types, including absence, myoclonic, atonic, tonic, and tonic-clonic seizures. Focal epilepsies include unifocal, multifocal, and hemispheric. When seizures or EEG findings demonstrate both generalized and focal features, combined generalized and focal epilepsy is the appropriate classification. An unknown epilepsy type is used when a patient has a diagnosis of epilepsy but there is insufficient information to define the epilepsy onset as focal or generalized.

Capute and Accardo's Neurodevelopmental Disabilities in Infancy and Childhood. DOI: https://doi.org/10.1016/B978-0-12-824060-1.00009-2

TABLE 14.1 Definitions.

Seizure	Transient occurrence of signs or symptoms due to abnormal excessive or synchronous neuronal activity in the brain. Signs may be motor, sensory, cognitive, autonomic, or a combination of these.
Epilepsy	The condition of recurrent, unprovoked seizures.
Epilepsy syndrome	Cluster of features (e.g., seizure type, age of onset, EEG features, imaging findings, prognosis) that tend to cooccur.
Developmental encephalopathy (DE)	Syndrome with aberrant neurodevelopment causing cognitive impairment.
Epileptic encephalopathy (EE)	Syndrome in which epilepsy itself contributes to severe cognitive and behavioral impairments, above and beyond what might be expected from the underlying pathology alone.
Developmental epileptic encephalopathy (DEE)	Syndrome combining EE and DE in which the developmental impairment is the result of both the underlying etiology and the epileptiform activity, independently.

Epilepsy syndrome is the third level of diagnostic classification and refers to a cluster of features that tend to cooccur (e.g., seizure type, EEG features, etiology, age of onset, genetics, imaging findings, prognosis, and natural history). Diurnal occurrence, seizure triggers, and comorbidities may also contribute to the identification of an epilepsy syndrome. Many epilepsy syndromes are not pathognomonic of etiology; for example, a wide array of etiologies can underlie epilepsy syndromes such as West syndrome/infantile epileptic spasms syndrome, Lennox–Gastaut syndrome, and Dravet syndrome [8].

Beyond these three levels of classification, epilepsy is further classified according to the etiologic categories of structural, genetic, infectious, metabolic, immune, or unknown [8]. Based on advances in understanding the genetic and structural brain underpinnings of the various epilepsy syndromes, the former term "idiopathic" is no longer used. Idiopathic previously referred epilepsy with no known etiology other than suspected hereditary susceptibility. Now, the term Genetic Generalized Epilepsies (GGE) is preferred; GGE includes Childhood Absence Epilepsy (CAE), Juvenile Absence Epilepsy (JAE), Juvenile Myoclonic Epilepsy (JME), and Generalized Tonic-Clonic Seizures Alone [8]. The use of the term "genetic" is more precise, as it would include inherited, de novo, and complex etiologies. The 2017 ILAE classification of the epilepsies position paper also introduced the term "developmental and epileptic encephalopathy," which brings us to the crux of this review.

Epilepsy and encephalopathy

Encephalopathy is an altered mental state associated with a presumed brain-based etiology. Developmental encephalopathy (DE) occurs when developmental impairments are associated with an atypical course of brain development, such as developmental delay (DD), Intellectual disabilities (ID), cerebral palsy (CP), and autism spectrum disorder (ASD). In contrast, epileptic encephalopathy (EE) occurs when observed cognitive and behavioral impairments are associated with the impact of epilepsy on brain functioning and that these impairments are "above and beyond what might be expected from the underlying pathology of epilepsy alone" [9]. In epileptic encephalopathy (EE), the epileptic activity itself is believed to be the nidus of cognitive and behavioral deterioration, and within the context of development, EE can present as a developmental plateau or regression and/or the emergence of behavioral and psychiatric features. Importantly, while individuals with DE are at higher risk of cooccurring epilepsy, it does not mean that they harbor an EE if the epileptiform activity is not having a specific deleterious effect on cognition and behavior.

There are several possible mechanisms by which epileptic activity contributes to encephalopathy in EE. Early-life seizures often begin during phases of critical brain development and may lead to synaptic reorganization, reduced neurogenesis, and alterations in the brain's excitatory and inhibitory balance [10,11]. Not only seizures but also interictal (in-between seizures) epileptiform activity (i.e., spikes, sharp waves) can contribute to encephalopathy. Interictal spikes are abnormal neuronal discharges resulting from synchronous depolarizations that occur when a patient is not having a seizure. There is evidence that interictal spikes lead to transient effects on behavior and cognition in children and adults and can lead to chronic effects in the developing brains of animal models [12]. It is important to note that while seizures typically present as discrete events, interictal discharges are more likely to be omnipresent. In EE, the amelioration of the epileptiform activity may lead to improvements in the cognitive and behavioral presentation [8], although such improvement is not required for diagnosis, and seizures in EE are often drug-resistant.

A direct causal relationship between epileptiform activity and cognitive function is implied in EE, but in some patients, cognitive impairment is present prior to seizure onset, indicating that a preexisting neurological framework may be involved [13]. In others, cognitive impairment or deterioration may ensue even if seizures and epileptiform activity are completely ameliorated. These observations suggest that the underlying disorder may independently contribute to the neurodevelopmental phenotype [14]. Developmental and epileptic encephalopathy (DEE) describes developmental impairment that is the result of *both* the underlying etiology and epileptiform activity, functioning independently of each other [8] (Fig. 14.1 and Table 14.2). Several different clinical presentations illustrate this concept. A child with premorbid delays in development could experience plateau or regression with the onset of epileptiform activity. A child with typical development could experience a slowed rate of development prior to the onset of significant epileptiform activity on the EEG, such as in Dravet syndrome. Alternatively, a severe early-life epilepsy could abate relatively early in life, but the child may continue to have profound developmental impairments, as seen in DEE due to variants in genes such as *STXBP1* or *KCNQ2*.

Genetic underpinnings and investigations in developmental and epileptic encephalopathy

Genetic variants are the primary disruptors that lead to DEE. Genetic etiologies of DEE include single-gene disorders, syndromes with significant genetic heterogeneity (i.e., West Syndrome), or known clinical syndromes in which genes are being newly identified (i.e., *GRIN2A* in epilepsy with continuous spike-waves during slow-wave sleep [CSWS]). DEE can also be the result of acquired causes (HIE, stroke) or associated with cortical malformation, which could have a genetic or acquired etiology [8].

In a patient with DEE, next-generation sequencing (NGS) techniques (whole exome sequencing, whole genome sequencing, and targeted gene panels) are the most appropriate avenues for genetic etiologic investigation [15]. NGS allows for the identification of single nucleotide variants (SNV)—changes in a single nucleotide that can lead to a change in protein structure or function. In contrast, chromosomal microarrays (CMA) detect copy number variants (CNV), which are larger deletions or duplications of genetic material that affect one or more genes. SNVs can cause a premature stop codon that leads to a truncated protein. Missense variants occur when a single nucleotide is replaced with another codon that can increase or decrease the function of a protein. While NGS yields a diagnosis in 17% of patients with epilepsy and neurodevelopmental disorders, the rate for DEE is ~50% [16]. Up to 50% of DEE occurs as the result of de novo pathogenic variants in single genes and can be identified by NGS, and a general recurrence risk of 1% is typical. However, parental somatic mosaicism may occur in up to 10% of parents of children with DEE, significantly increasing recurrence risk.

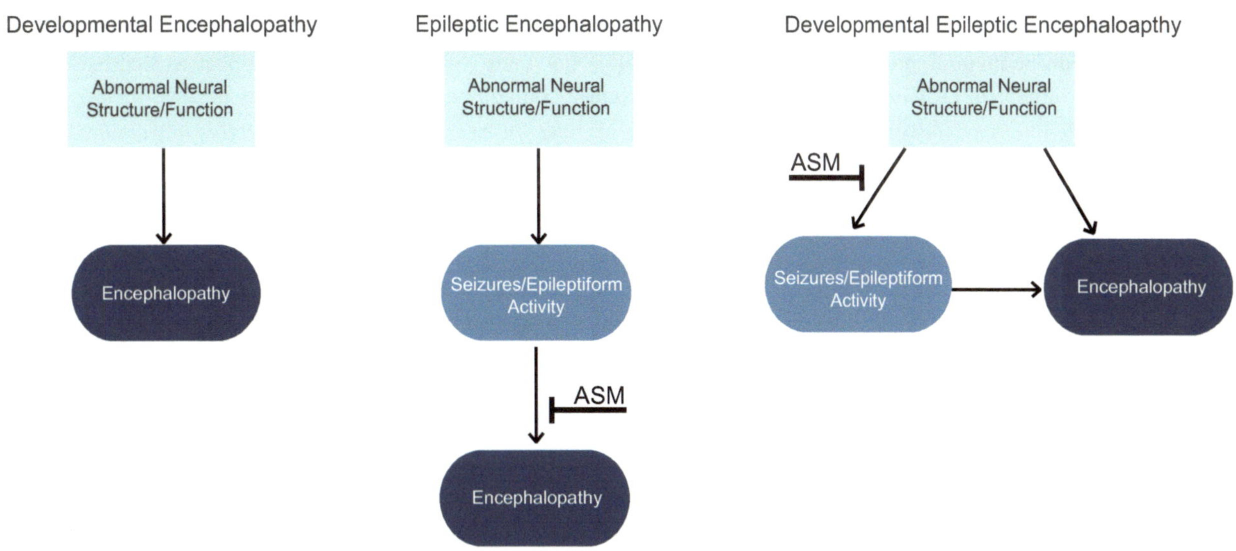

FIGURE 14.1 Schematization of mechanisms of developmental encephalopathy, epileptic encephalopathy, and developmental epileptic encephalopathy. ASM, anti-seizure medication.

TABLE 14.2 Epileptic encephalopathy versus developmental epileptic encephalopathy.

	Epileptic encephalopathy	Developmental epileptic encephalopathy
Pre-onset developmental functioning	Age-expected function	Delays in development may precede onset of epileptiform activity
Cause of developmental impairments	Seizures or interictal activity	Underlying disorder and seizures or interictal activity
Treatment	Aggressive antiseizure medication	May benefit from precision medicine approach
Response to treatment	Possible return to prior or near-prior cognition/function	Developmental impairments likely persist even with adequate control of seizures
Example	Epileptic encephalopathy with spike-wave activation in sleep	Dravet syndrome

Genetic identification of a cause for DEE can have a significant positive impact on clinical care, particularly with early access to testing, and can direct treatment. Clinical management may be directly impacted by identifying a genetic diagnosis, leading to the use or avoidance of certain antiseizure medications (ASM); for example, sodium channel blockers are contraindicated in *SCN1A*-related epilepsies. For families, the conclusion of the diagnostic odyssey can also have significant personal utility, often increasing knowledge, connection, and communication while decreasing feelings of blame and guilt [17].

Etiology and treatment implications

Recognizing the presence of an epileptic component to an encephalopathy is crucial, as it suggests there may be a treatable component that could lead to improvement in cognition, behavior, or daily function [18]. It may not be clear whether epileptic activity or developmental underpinning is the more impactful contributor in a presentation. This is a complex endeavor, as the occurrence and severity of encephalopathy associated with epilepsy vary. For example, encephalopathy is essential to the diagnosis of Lennox–Gastaut syndrome, usually present in West syndrome, and is present in about half of patients with epilepsy with myoclonic-atonic seizures (EMAtS, formerly called Doose syndrome) [14].

Epilepsy management may be more conservative in a patient with a developmental encephalopathy with comorbid epilepsy, compared with that in a patient with a DEE [19], and may be most aggressive in a purely EE. When a genetic mutation is responsible for both the neurobehavioral impact and epilepsy (DEE), treatment of epilepsy is expected to improve the neurobehavioral presentation, at least partially, but there would still be an impact associated with the genetic etiology alone. In DEE with a more impactful epileptic component, the potential risks of gene therapy altering the underlying pathology contributing to the encephalopathy may outweigh the potential benefit, and management of epilepsy may be the priority.

Developmental and epileptic encephalopathies—selected examples

The identification of specific syndromes can be beneficial in treatment when precision medicine approaches are available. Additionally, some diagnoses are indications for specific treatments that are not widely approved. The ensuing sections will discuss some specific examples of DEE. While not intended to be an exhaustive list, understanding these methods of classification can aid in diagnosis and inform treatment.

Neonatal/infantile onset developmental and epileptic encephalopathies (as defined by clinical syndrome)

DEEs with onset in neonates or infants include four major clinical presentations: EIDEE, EIMFS, IESS, and DS.

Early infantile developmental and epileptic encephalopathy

Early infantile developmental and epileptic encephalopathy (EIDEE) includes infants previously classified as having Ohtahara Syndrome (Early Infantile Epileptic Encephalopathy, or EIEE) and Early Myoclonic Epilepsy (EME). Onset is within the early infantile period (0–3 months). One or more of the following seizure types is required for diagnosis: tonic seizures, myoclonic seizures, epileptic spasms, or sequential seizures. Sequential seizures involve a sequence of several features during a seizure (e.g., focal tonic → focal clonic → spasms). Tonic seizures are frequent and can occur in multiple daily clusters. Features that can help to differentiate tonic seizures from spasms include duration >3 seconds and younger age of occurrence [20]. Myoclonic seizures can be quite variable, ranging from occasional to near-continuous, and can occur during wakefulness, sleep, or both. The neurological examination may show severe abnormalities of tone and posture, and patients may have concurrent movement disorders. Differentiating seizures from paroxysmal movements in the neonatal period can be challenging. Some characteristics more indicative of a benign movement disorder in infancy include: brief (3–4 seconds) paroxysmal events occurring multiple times daily, sometimes in clusters separated by a few minutes; onset in the mid-first year of life (~4–7 months), no alteration of consciousness during an event, specific trigger such as emotion or postural change, normal development, normal neurological exam, normal EEG during and in between events, self-limited nature, and typical resolution in toddlerhood [21].

The cardinal EEG feature of EIDEE is burst-suppression with medium-to-high voltage bursts lasting 1–5 seconds, alternating with intervals of diffuse attenuation for 5–10 seconds. Ictal EEG varies according to seizure type. Brain imaging often shows structural brain abnormalities. Metabolic studies should be considered if a structural cause is not identified on neuroimaging [22]. Genetic pathogenic variants have been identified in over half of patients with EIDEE [23]; *STXBP1*, *KCNQ2*, and *SCN2A* are among the more commonly identified genes. There are unique features of some specific EIDEE that can help predict the genotype (Table 14.3).

Pathogenic variants of several genes can lead to distinct phenotypes. For example, pathogenic variants in *KCNQ2* are associated with two distinct electroclinical phenotypes in the neonatal period. First, benign familial neonatal epilepsy (BFNE) is an autosomal dominant condition of recurrent seizures in an otherwise healthy newborn that begin in the first few days of life and typically remit by 4 months of age. BFNE can be caused by loss-of-function (LOF) variants in *KCNQ2* or *KCNQ3*. While truncating variants in *KCNQ2* typically cause BFNE, missense variants in *KCNQ2* may lead to an encephalopathy that typically presents with tonic seizures in the first week of life and progresses to a refractory DEE.

Epilepsy of infancy with migrating focal seizures

Epilepsy of infancy with migrating focal seizures (EIMFS) was previously called Malignant Migrating Partial Seizures of Infancy. Seizures start sporadically between 2 and 7 months of age with a mixture of spasms, tonic seizures, myoclonic seizures, and focal seizures, arising from independent foci in both hemispheres. The seizures evolve, and by the second phase of the syndrome, which can occur as early as 1 month of age, seizures become more frequent, may occur in multiple daily clusters, or may be nearly continuous for days to weeks. Focal seizures may become bilateral and migrate from limb to limb. Hypomotor seizures are also common, as are autonomic symptoms including drooling, cyanosis, and flushing, as well as apnea and desaturations that may be misdiagnosed as gastroesophageal reflux. On examination, hypotonia is common, and acquired microcephaly may occur [40].

The typical EEG background of EIMFS is diffuse or focal slowing with multifocal spikes; rarely, a burst-suppression pattern occurs [41]. MRI typically shows atrophy, white matter abnormalities, and underdevelopment of the corpus callosum. Many genetic causes of EIMFS have been reported, the most common of which is a de novo gain-of-function (GOF) mutation in *KCNT1*. Variants in *SCN2A*, *SCN1A*, and several others have also been identified as causes of EIMFS. *KCNT1* encodes a sodium-activated potassium channel that regulates neuronal hyperexcitability through the enhancement of slow hyperpolarization. Counterintuitively, *KCNT1* GOF variants lead to a severe phenotype, likely via decreased activity of inhibitory interneurons. The prognosis is poor, seizures are typically refractory to ASMs, and life expectancy is reduced. There are reports of responses to levetiracetam or potassium bromide [42]. Some patients with *KCNT1* variants respond to quinidine, an inhibitor of the *KCNT1* channel, reflecting a potential example of targeted therapy. However, treatment failures have also been reported, and high serum levels of quinidine may result in cardiac toxicity [42]. Even when significant seizure control is obtained in EIMFS, there is usually little to no improvement in the severe developmental deficits [43], further illustrating the challenge of differentiating the contribution of epileptic activity versus that of the underlying neurobiological dysfunction in DEE.

TABLE 14.3 Etiologic specific presentations of early infantile developmental and epileptic encephalopathy.

EIDEE	Typical genetic variant	Gene function and implications	Seizure onset and features	Clinical pearls	Unique or precise treatment approaches	Phenotypic spectrum
KCNQ2-DEE	LOF in *KCNQ2* (Kv7.2 subunit)	Responsible for M-current (regulates neuronal excitability by stabilizing resting potential and suppressing repetitive action potential firing) Malfunction → ↑ network excitability	1st week of life; multiple daily seizures; mostly focal tonic seizures; myoclonic, focal; autonomic changes may occur	Burst-suppression pattern on EEG in >60%	SCBs [24,25]; Ezogabine shown to improve seizure frequency, controlled trials needed. XEN496 is ezogabine like compound currently in clinical trial in KCNQ2-related DEE [26].	Self-limited (familial) neonatal epilepsy (SeLNE)—*KCNQ2* LOF (*KCNQ3* and *SCN2A* less frequent causes)
SCN2A-DEE	GOF in *SCN2A*; (Nav1.2 subunit)	Generation and propagation of action potentials (particularly in early development)	0–3 months; (those after 3 months may have LOF and IS); Focal tonic or clonic; apnea, desaturations	Hypotonia, microcephaly, dysautonomia, feeding intolerance, enhanced startle, paroxysmal movements (ataxia, chorea, dystonia) [27]	SCBs, ketogenic diet, high-dose steroids [28];	EIMFS; self-limited familial infantile epilepsy, episodic ataxia, autism intellectual disability, and schizophrenia [29]
SCN8A-DEE	GOF in *SCN8A*; (Nav1.6 α subunit)	Regulates initiation and propagation of action potential; Impairment of channel inactivation → neuronal hyperexcitability	Within first 18 months (average 4 months); heterogeneous; GTCs with autonomic features; Status epilepticus; SUDEP in 1.6% [30]; Febrile seizures uncommon	Hypotonia, movement disorders (ataxia); GI disorders; increased risk for premature death including SUDEP	High-dose SCBs (CBZ, OXC, PHT), benzodiazepines; worsening of seizures or development with LEV in some [31]	Range from Self-limited familial infantile epilepsy to EIDEE; LOF → ID, movement disorders, ASD with or without seizures
STXBP1-DEE [32]	LOF in *STXBP1*	Synaptic vesicle docking → neurotransmitter release	0–3 months; focal or ES at onset.; Asymmetric tonic, or sequential seizures (tonic, autonomic, clonic, spasms) spasms, axial myoclonic	severe cognitive impairment; hypotonia; later may have ataxic gait, movement disorders	No definite ASM superiority [33]; some report dramatic LEV effect	Some presentations with epilepsy, but no clear genotype-phenotype correlation [34]

CDKL5-DEE (CDD)	LOF in *CDKL5* (on short arm of X-chromosome)	Protein deficiency of a serine threonine kinase protein with roles in neuronal proliferation, migration, formation growth; synaptic development and function	0–3 months; tonic seizures; "hyperkinetic-tonic-spasms" sequence; Stage 1- early epilepsy w brief tonic seizures; Stage 2- epileptic encephalopathy, epileptic spasms without hypsarrhythmia; Stage 3- late multifocal and myoclonic epilepsy	4:1 female predominance; hypotonia; CVI in >75%; GI issues (constipation > reflux > air swallowing), hand stereotypies; deceleration of head growth; sleep disruption, dysautonomia; phenotypic similarities with Rett Syndrome (common metabolic pathway w/ *MECP2*)	"honeymoon" periods with new ASM of max 6 mos;[35] ASM effectiveness limited, wanes with time; spasms more resistant to first-line treatments than in other etiologies; KDT and palliative surgeries mixed results; neurosteroid ganaxolone approved; fenfluramine, soticlestat being investigated [36]	A less severe phenotype without epilepsy, head growth deceleration, or CVI associated with mostly missense variants or skewed XCI (possible presence of some CDKL5 activity in early brain development) [37]
KCNT1-DEE [38]	GOF in *KCNT1* Na-gated K channel (slack)	Slow hyperpolarization; Possible greater dampening of interneuron excitability than pyramidal neuron excitability [39]	neonatal or early infancy; multiple seizure types	Motor impairment, respiratory problems, ND delay or regression after seizure onset	Quinidine, phenobarbital, clobazam, ketogenic diet	EIMFS; ADNFLE

ADHD, Attention deficit hyperactivity disorder; *ADNFLE*, autosomal dominant nocturnal frontal lobe epilepsy; *ASD*, autism spectrum disorder; *ASO*, anti-sense oligonucleotide; *ASM*, antiseizure medication; *CBZ*, carbamazepine; *CDKL5*, cyclin-dependent kinase-like 5; *CLB*, clobazam; *CVI*, cerebral visual impairment; *DD*, developmental delay; *EIDEE*, early infantile developmental and epileptic encephalopathy; *EIMFS*, epilepsy of infancy with migrating focal seizures; *ES*, epileptic spasms; ESX, ethosuximide; *GI*, gastrointestinal; *GLUT1 DS*, GLUT1 deficiency syndrome; *GOF*, gain-of-function; *GTC*, generalized tonic-clonic seizure; *ID*, intellectual disability; *IPS*, intermittent photic stimulation; *KDT*, ketogenic diet therapy; *LEV*, levetiracetam; *LGS*, Lennox–Gastaut syndrome; *LOF*, loss-of-function; *LTG*, lamotrigine; *MRE*, medically-refractory epilepsy; *OXC*, oxcarbazepine; *PHT*, phenytoin; *SCB*, sodium channel blocker; *STXBP1*, syntaxin1a binding protein; *SUDEP*, sudden unexpected death in epilepsy; *SZ*, seizure; *TC*, tonic-clonic; *VPA*, valproic acid; *WS*, West syndrome; *XCI*, X-chromosome inactivation.

Like *KCNQ2* variants, there are distinct presentations that can occur as the result of variants in *KCNT1*. The range of phenotypic expression of pathogenic variants in *KCNT1* includes EIMFS, non-EIMFS *KCNT1* DEE (see Table 14.3), later-onset DEE, and autosomal dominant hypermotor epilepsy.

Infantile epileptic spasms syndrome/West syndrome

West syndrome (WS) comprises the triad of (1) epileptic (or infantile) spasms, (2) hypsarrhythmia on the EEG, and (3) developmental impairment (often regression). Infantile Epileptic Spasms Syndrome (IESS) is a term proposed to encompass West Syndrome (WS) and infants with spasms not meeting full criteria for West Syndrome [20]. IESS is a phenotypically homogenous syndrome (in terms of seizure semiology with flexion, extension, or mixed spasms) with a range of etiologies that may include structural, metabolic, infectious (CMV), or genetic underpinnings. Tuberous Sclerosis Complex (TSC), for example, represents a clinical entity that can progress to IESS with both genetic and structural causes. Adrenocorticotropic hormone (ACTH) and glucocorticoids (GC) are the mainstays of treatment for IESS, with efficacy in controlling seizures as well as improving the marked EEG background abnormalities. In TSC specifically, however, the first-line treatment for infantile spasms is vigabatrin. ACTH may influence epileptic activity by several proposed mechanisms, including reduced secretion of corticotropin-releasing hormone (CRH), influencing GABA homeostasis, and numerous other factors that affect excitation/inhibition balance in the developing brain [10]. Given the relationship between the synthesis and release of glucocorticoid and spasm control, hypotheses have implicated neuroinflammation in the pathogenesis of IESS [44].

Dravet syndrome

Dravet syndrome (DS) (previously severe myoclonic epilepsy of infancy) is a DEE typically resulting from pathogenic mutations in *SCN1A*, which encodes the alpha subunit of the sodium channel (Nav1.1). Truncating, missense, or copy number variants in *SCN1A* lead to loss of channel function, primarily affecting inhibitory interneurons [45]. The core phenotype of DS is epilepsy onset during the first year of life in a previously typically developing infant. This timing erroneously led to the attribution of vaccine administration to seizures and encephalopathy in some infants who had Dravet syndrome, which was triggered after a vaccine, but the seizures were actually predestined by the underlying genetic variant. In DS, children present with hemiclonic or generalized tonic-clonic seizures, often with fever, around 6 months of age, and subsequently develop other seizure types [46]. Initial EEG and brain imaging are typically normal, which may delay diagnosis.

Avoidance of seizure triggers, especially hyperthermia, is an important treatment approach in DS. Valproic acid and clobazam are the most commonly utilized initial ASMs in DS [47]. Sodium channel blockers such as carbamazepine, oxcarbazepine, lamotrigine, and phenytoin should be avoided, as they are likely ineffective and may even worsen seizures [48]. Longer use of sodium channel blockers during the first 5 years of life may have a negative impact on cognitive outcomes in DS [49]. Newer treatment approaches for DS include stiripentol, cannabidiol, and fenfluramine [50]. Epilepsy in DS almost always becomes drug resistant, and this disorder carries a high risk for sudden, unexplained death in epilepsy (SUDEP) [51].

Development in DS is initially normal, but regression or plateau becomes evident between 2 and 3 years of age, a time at which the EEG shows bursts of epileptiform abnormalities and seizures are prolonged but infrequent. Later in childhood, when ictal activity is more prominent, treatment with ASM may improve cognition [52], suggesting features of EE. While DS has historically been classified as an EE, there is increasing evidence that the underlying sodium channel mutation plays an important role in the development of encephalopathy in DS. LOF pathogenic variants in *SCN1A* account for the vast majority of cases of DS, and patients with mosaic LOF variants demonstrate milder phenotypes on average as compared to those with heterozygous truncating variants [53]. In contrast to LOF variants in *SCN1A* resulting in DS, GOF variants yield a different phenotypic spectrum. *SCN1A* developmental and epileptic encephalopathy with early onset, profound impairment, and movement disorder is caused by a GOF variant in *SCN1A* and may actually benefit from sodium channel blockade [54].

Pathogenic variants in *SCN1A* are also seen in milder epilepsy phenotypes, including genetic (formerly generalized) epilepsy with febrile seizures plus (GEFS +) and febrile seizures. GEFS + includes a spectrum of both generalized and focal epilepsy phenotypes. While the most commonly identified cause of GEFS + is a pathogenic variant in *SCN1A* [55], other linked genes include *GABRA1*, *GABRB3*, *CABRG2*, *SCN1B*, and *STX1B* [56].

Childhood onset developmental and epileptic encephalopathies (as defined by clinical syndrome)

While many DEEs involve onset in infancy, others are present during childhood. Such DEEs are categorized by clinical presentation and include: epilepsy with myoclonic-atonic seizures (EMAtS, formerly called myoclonic-astatic epilepsy (MAE) or Doose syndrome), Lennox–Gastaut syndrome (LGS), the epilepsy-aphasia spectrum, febrile infection-related epilepsy syndrome (FIRES), and hemiconvulsion-hemiplegia-epilepsy syndrome (HHE).

Epilepsy with myoclonic-atonic seizures

Epilepsy with myoclonic-atonic seizures (EMAtS) is a generalized epilepsy with the onset of drop attacks (atonic seizures) during the preschool years in children with normal development. Febrile seizures precede the drop seizures in about two-thirds of patients; other seizure types, including absence and generalized tonic-clonic seizures, can occur. A developmental plateau or regression usually occurs in the setting of frequent seizures and may improve with the achievement of seizure control. EMAtS is one of the epilepsy syndromes that can be seen in families with GEFS + ; a family history of seizures is reported in up to half of patients [57]. Children with febrile seizures related to *SCN1A* variants experience an earlier onset of febrile seizures [58]. The initial EEG may be normal but later evolve into generalized epileptiform activity. There is no clear treatment consensus; initial treatment is often levetiracetam, followed by valproic acid. Other medications used may include ethosuximide, benzodiazepines, lamotrigine, topiramate, or zonisamide. Diet therapy (ketogenic diet or modified Atkins diet) may be the most effective treatment, with a greater reduction in seizure burden than with the first 3 ASMs [57]. About half of patients achieve seizure freedom; failure to do so is associated with persistent developmental deficits [57].

Lennox–Gastaut syndrome

Lennox–Gastaut syndrome (LGS) is a DEE that includes a wide range of etiologies. About 20% of cases evolve from IESS. The cardinal features of LGS are multiple seizure types, intellectual impairment, and a generalized slow-spike-wave discharge (<3 Hz) EEG pattern. LGS accounts for 1%–2% of all children with epilepsy and 6%–7% of children with refractory epilepsy. The peak onset of LGS is between 3 and 5 years, with a range of 18 months to 8 years.

Tonic seizures are the most common seizure type in LGS and usually occur during sleep. Tonic seizures in LGS consist of sustained muscle contraction lasting a few seconds to a minute (average 10 seconds). While this tonic contraction may include most muscles of the body, it can present with just flexion of the head and trunk or as abduction and elevation of the arms [59]. Atypical absence seizures (staring spells of variable duration, usually tens of seconds, with diminished responsiveness) and atonic seizures are also common. It is important to differentiate LGS from EMAtS, as treatment and prognosis differ. Both phenotypes may include seizures associated with falls. In EMAtS, atonic seizures include sudden loss of muscle tone resulting in a drop attack. In LGS, tonic seizures involve a sudden increase in muscle tone and can result in a fall if standing.

Since seizure freedom is unlikely, the goal of treatment is to balance seizure control with medication side effects to uphold or improve quality of life. Nonpharmacologic treatment options are also available (e.g., neuromodulation, dietary therapy, surgical resection). For patients with an identified seizure focus, surgical resection should be considered and may lead to a dramatic improvement in seizure frequency and quality of life. Corpus callosotomy (CC) may be more effective than vagal nerve stimulation (VNS) in reducing tonic seizures, but VNS may lead to improved quality of life with lower risk. Resective surgery in combination with CC or deep brain stimulation (DBS) may also lead to meaningful reductions in seizure frequency [60].

Epilepsy-aphasia spectrum

DEE with Spike-Wave Activation in Sleep (DEE-SWAS) and Epileptic Encephalopathy with Spike-Wave Activation in Sleep (EE-SWAS) are part of a spectrum of conditions with varying degrees of cognitive, language, behavioral, and motor regression in the setting of marked activation of interictal spike-and-wave discharges in sleep, known as the epilepsy-aphasia spectrum. Electrical status epilepticus of sleep (ESES) is defined as sleep-activated spike-wave discharges occupying 85% or more of NREM sleep associated with cognitive problems during wakefulness. The strict ESES criteria are not validated biologically and do not fully encompass the reality that disturbances in cognitive and behavioral function can be seen in patients with lower percentages of sleep affected by these EEG changes. Therefore, the 2022 ILAE classification notes that DEE-SWAS/EE-SWAS replaces ESES and its associated EEG findings (CSWS, or continuous spike-and-wave in slow-wave sleep). DEE-SWAS occurs in those with preexisting developmental impairments with evident worsening of function in the setting of significant activation of spikes and waves in sleep. EE-SWAS denotes patients with previously typical development with an activation of spike-and-wave complexes in slow-wave sleep. The Landau-Kleffner Syndrome (LKS) designation is maintained to describe a specific subtype of EE-SWAS characterized by language regression with acquired auditory agnosia [61].

Together, DEE/EE-SWAS are rare epilepsy presentations, with peak seizure onset around 4–5 years (range 2–12 years), affecting boys and girls equally. The sleep activation of spikes and waves typically occurs with the developmental/behavioral regression or plateau (which can include any domain), about 1–2 years after seizure onset, although some patients may not have any clinical seizures. Neuroimaging is recommended given associations with early-life thalamic injury and cortical malformations (e.g., bilateral perisylvian polymicrogyria). Most often, infrequent pharmacoresponsive focal motor or focal-to-bilateral tonic-clonic seizures are seen, usually early in presentation, around 2–5 years of age. A family history of seizures is present in up to 50% of children. Pathogenic variants in *GRIN2A*, which encodes the NMDA glutamate receptor alpha-2 subunit, are a major monogenic cause, but a number of others have been associated.

There are four general approaches to treatment: high-dose oral benzodiazepines before sleep, steroids, other ASMs (valproic acid, ethosuximide, levetiracetam), or epilepsy surgery. None is ideal, and the management approach varies widely across centers. Benzodiazepines and steroids demonstrate better efficacy than ASMs. Generally, measured outcomes include improvement in neurocognitive function, seizures, or EEG (i.e., reduction in SWI) [62].

Clinical seizures usually remit, and the EEG improves around puberty. While cognitive and behavioral changes often improve, some impairments may remain, and function is limited in about half of patients. Etiology and duration of disease are predictors of cognitive outcome. Poorer outcomes are seen in children with an earlier onset and longer duration of disease (>2 years) [61]. It is thought that the neuropsychological impairments persist because disrupted neural networks are unable to support higher-function organization as abnormal epileptiform activity impedes the consolidation of synaptic connections necessary for learning [63].

LKS, also known as acquired epileptic aphasia, occurs in approximately 1 in 350,000 children and has peak onset at 3–7 years (range 2–14 years) with a 2:1 male:female predominance. The hallmark feature is language regression in a child with previously typical language development, often with behavioral changes and mild seizures. There are no randomized clinical trials evaluating pharmacological treatment of LKS [64]. In severe cases, high-dose steroids are often used as first-line treatment and improvement is typically seen in about half of these patients. Other treatments include high-dose benzodiazepines including clobazam and diazepam. Levetiracteam, lamotrigine, valproic acid, and sulthiame may be effective. There are reports of improvement with the ketogenic diet or immunoglobins in children who are not responsive to steroids or conventional ASMs. Although seizures typically remit, the behavioral and language outcomes are variable and language function may not achieve premorbid level of functioning [65].

Self-Limited Epilepsy with Centrotemporal Spikes (SeLECTS), formerly called benign Rolandic epilepsy or benign epilepsy with centrotemporal spikes (BECTS), is the most common epilepsy syndrome. SeLECTS is classified as a self-limited focal epilepsy of childhood; it can be considered the mildest syndrome on the epilepsy-aphasia spectrum. Centrotemporal spikes are generated from the rolandic region, the area of cortex surrounding the central sulcus, consisting of the pre and postcentral gyri, which leads to the distinct clinical semiology of the seizures. SeLECTS typically occurs in children with a normal neurological examination and cognitive function, although language impairments and neuropsychiatric concerns may be seen even prior to seizure onset [66] and, when present, predict higher total seizure frequency and epilepsy duration [67]. A natural history study of children with SeLECTS indicated a higher risk of neuropsychiatric symptoms (e.g., ADHD, mood disorders, and behavioral disorders) [67]. Intellectual profiles are generally within normal limits, but epilepsy-related variables, including earlier age of onset, greater frequency of seizures, use of ASMs, longer treatment duration, and the presence of neurodevelopmental disorders, are related to lower intellectual performance [68].

Febrile infection-related epilepsy syndrome

Febrile infection-related epilepsy syndrome (FIRES), a subcategory of new-onset refractory status epilepticus (NORSE), is a rare, severe EE that occurs in previously healthy children. FIRES is marked by new-onset refractory status epilepticus (SE) in children and adolescents, preceded by a febrile infection 24 hours to 2 weeks prior to the onset of SE, at which time fever is not necessarily present. In contrast to febrile SE, SE of FIRES begins days after the onset of fever. Morbidity and mortality are high during the acute phase, which can last up to 12 weeks and is marked by a high seizure burden of up to hundreds of seizures per day. During the ensuing chronic phase, drug-resistant multifocal epilepsy and cognitive impairments are common sequelae. FIRES is felt to likely represent an immune-inflammatory-mediated process [69].

Hemiconvulsion-hemiplegia-epilepsy syndrome

Hemiconvulsion-hemiplegia-epilepsy (HHE) is a rare syndrome in infants and children under 4 years old that occurs after focal motor SE in the setting of febrile illness with unilateral hemispheric cerebral edema. Hemiplegia will be present immediately. Aphasia may be present acutely if the dominant hemisphere is affected. Atrophy of the affected hemisphere ensues, and many will have permanent hemiplegia and drug-resistant epilepsy, as well as variable degrees of intellectual disability [61]. The pathogenesis is not well understood; hyperthermia, a proinflammatory state, and cytotoxic edema have been hypothesized [70].

Other childhood epilepsy syndromes

Some childhood epilepsy syndromes are self-limited, meaning that they tend to resolve with age. Several of these epilepsies were considered "benign," but that qualifier is no longer favored. In recent years, neuropsychological deficits, often in language function, have been identified in some affected children, suggesting the occurrence of a broader spectrum of developmental impairments that cooccur or are directly impacted by epilepsy [71]. While the DEE presentations discussed previously demonstrate well-recognized encephalopathy, childhood epilepsy syndromes and their increasingly recognized neurocognitive associations are presented in Table 14.4.

Precision medicine approaches in developmental and epileptic encephalopathies

The vast heterogeneity of the etiology of DEEs, with over 400 monogenic causes identified [86], makes precision medicine challenging. Precision therapy for a DEE may start with a rationally chosen ASM based on the genetic variant and the ASM's mechanism of action (e.g., use of sodium channel blockers for seizures associated with GOF sodium channelopathies). While some ASMs may be more effective than others in certain DEEs, some ASMs may actually exacerbate seizures (e.g., sodium channel blockers may worsen seizures in LOF sodium channelopathies (*SCN1A*)). In some cases, developmental gains may occur simply by removing an agent that exacerbates seizures and epileptiform activity. Precision medicine holds great promise in the treatment of DEE. This concept will be explored through the example of Tuberous Sclerosis Complex (TSC), a multisystem genetic disorder occurring in about 1 in 6000 children. TSC is caused by pathogenic variants in one of two genes, *TSC1* or *TSC2*, which encode for the proteins hamartin and tuberin, respectively. Absence or deficiency of either protein leads to unchecked activation of the mechanistic target of the rapamycin (mTOR) pathway, with subsequent excess cell growth and tumor formation in several body systems, including the central nervous system. Epilepsy affects up to 85% of individuals with TSC, and in most cases, seizures become refractory to ASM treatment. Seizures begin early in TSC, typically within the first year of life. Infantile spasms affect approximately one-third of children with TSC, and a greater proportion of patients with TSC-related IS are refractory to treatment as compared with IS from other causes.

TSC may be a disease model for epilepsy prevention. In TSC, earlier seizure onset is associated with a negative impact on neurodevelopment [87,88]. Early control of seizures with vigabatrin is associated with improved developmental outcomes [89]. To improve long-term developmental outcomes in TSC, early identification and treatment of epilepsy in TSC are tantamount. To facilitate early seizure recognition, TSC management guidelines recommend surveillance EEGs every 6 weeks up to age 12 months, followed by every 3 months up to age 24 months [90]. The EPISTOP trial compared conventional to preventative treatment [91]. In the conventional group, vigabatrin was started after the onset of a seizure, as compared to the preventative group, in which patients without a prior seizure were started on vigabatrin as soon as epileptiform activity was detected on the EEG. Preventative treatment was associated with a delayed time to the onset of seizures, a reduced risk of seizures, and a decreased severity of epilepsy. In the United States, the PREVeNT trial included infants with TSC but without seizures who underwent monthly EEGs. Once epileptiform activity was detected, the participants were randomized to receive preventive treatment with vigabatrin or placebo [92]. Results indicate that vigabatrin treatment did not decrease seizures or impove cognition [93].

Early treatment or preventative treatment with ASM are promising approaches, but the potential for disease-modifying therapy is also being explored. The use of mTOR inhibitors (sirolimus and everolimus) has emerged as the first ASM to target the neurobiology of the underlying disease in TSC. mTOR inhibitors have already proven successful in the treatment of several manifestations of TSC, including angiofibromas, angiomyolipomas (AML), lymphangioleiomyomatosis (LAM), cardiac rhabdomyomas, and subependymal giant cell astrocytomas (SEGA). The potential for prevention and early intervention via mTOR inhibition holds promise and is being investigated in a trial using sirolimus in infants with TSC to prevent or delay seizures.

TABLE 14.4 Selected childhood epilepsy syndromes.

	Previous name	Known genetic etiologies	Peak age onset	Seizures	Clinical pearls	Neurocognitive comorbidities	Treatment	Prognosis
Genetic generalized epilepsies								
Childhood Absence Epilepsy (CAE)	Petit mal epilepsy; pyknolepsy	Most genes associated are ion channels	6–7 years	staring spells (seconds) many times daily, sometimes with eyeblinking or motor automatisms; immediate return to baseline mental status	If onset <4 years → consider GLUT1 DS	Significant cognitive, behavioral, psychiatric comorbidities	ETX for initial monotherapy if only absences seizures; if persistent absences seizures or GTC use VPA >LTG [72]	Remission (in many, but not all) usually at 10–14 years; 15% develop JME
Epilepsy with Eyelid Myoclonia (EEM) [73]	Jeavons syndrome	In a minority: SYNGAP1, NEXMIF, RORB, CHD2	6–8 years	Triad: (1) Eyelid myoclonia (jerking of the eyelids and upward eye and head deviation immediately following eye closure)– which may be associated with brief absence seizure (2) eye closure induced seizures (3) photosensitivity; Other seizures may occur: absence, GTCs, myoclonic	2:1 male: female; ID and psychiatry comorbidities described	Cognitive outcomes have historically been regarded as normal, there is increasing recognition of dysfunction [74], including an association between earlier seizure onset and higher rates of neurocognitive comorbidities and medically-refractory seizures [75]	Limited evidence; broad spectrum ASMs: VPA, LEV, LTG, ETX' benzodiazepines; (CBZ, PHT, OXC may worsen seizures)	GTCs medication-responsive, but EM frequently treatment resistant → lifelong epilepsy
Epilepsy with myoclonic absence (EMA)	Tassinari syndrome	reported associations with variants in the following single genes: SYNGAP1, GLUD1, SCN1A, SLC2A1, SETD1B [76] and FOXP1 [77].	7 years	Myoclonic absence = repetitive (3 per second) prominent myoclonic jerks of upper limbs > lower limbs or face; absence duration 10–60 seconds (longer than CAE) Triggers include hyperventilation, photic stimulation, and awakening.	Males 70%; sometimes with family history of febrile seizures or epilepsy	50% have developmental impairment before seizure onset; learning disabilities, academic underperformance in 2/3 [78] Most eventually show ID (70%)	Some respond to VPA or VPA + LTG	Seizures persist in >50% [79] Some progress to LGS. Favorable outcomes with only myoclonic absence or simple absence seizures, whereas the concurrence of other seizures types (tonic-clonic seizures, drop seizure, absence status) → less favorable outcome

Self-limited focal epilepsies								
Self-limited epilepsy with centrotemporal spikes (SeLECTS)	Childhood epilepsy with centrotemporal spikes, (benign) Rolandic epilepsy, (benign) epilepsy with centrotemporal spikes	Likely complex in heritance; no pathogenic variant identified in most; variants in GRIN2A may show evolution to EE-SWAS [61]	7 years	Unilateral facial sensorimotor or oropharyngeal symptoms, speech arrest, and/or hypersalivation; Brief, infrequent, in sleep or at or just before waking; 50% may progress to hemi or bilateral convulsions	Slight male predominance	↑ risk ADHD, mood disorders, behavioral disorders [67]	Variable prevalence of ASM treatment	Seizures usually remit by puberty
Self-limited epilepsy with autonomic seizures (SeLEAS)	Panayiotopoulus syndrome	No causative genes; rare cases of *SCN1A* variants [61]	3–6 years	Prolonged > 30 minutes, occur in sleep, associated with autonomic features: color change (pallor > flushing or cyanosis), pupillary changes (mydriasis > miosis), incontinence (urine > fecal); temperature changes; hypersalivation Consciousness initially intact, but becomes progressively impaired; Deviation of head or eyes after	Can imitate migraines, sleep disorders, cyclic vomiting syndrome, gastroenteritis, GERD [81]	Neuropsychological assessment typically reveals normal IQ. Some studies have reported deficits in visual and visual perceptual functions, less often minor attention and memory deficits [82]	Not always recommended	Seizures typically remit within 1–2 years of onset; nearly 50% only have 1 seizure

(Continued)

TABLE 14.4 (Continued)

	Previous name	Known genetic etiologies	Peak age onset	Seizures	Clinical pearls	Neurocognitive comorbidities	Treatment	Prognosis
				vomiting in >50% [80]				
Childhood occipital visual epilepsy (COVE)	Late-onset (benign) occipital epilepsy; Idiopathic childhood occipital epilepsy-Gastaut type	Likely complex/ polygenic inheritance; no causative genes identified [61]	8–9 years	Brief, abrupt onset, focal sensory visual seizures in wakefulness. Elementary visual hallucinations in peripheral fields, multiple, move horizontally; complex may develop, but less commonly [83]. ± ipsilateral head or eye deviation. ~50% have postictal migraine-like headache, ± nausea, vomiting [61].	Obtain MRI to rule-out structural cause of epilepsy	Usually normal development; mild cognitive impairment reported	ASM recommended; dramatic response to CBZ in >90%	Long-term prognosis variable; some may have seizures into adulthood, including other seizure types [84]
Photosensitive Occipital Lobe Epilepsy (POLE) [85]	Idiopathic photosensitive occipital lobe epilepsy	*GRIN2A* considered as a candidate gene	4–17 years	Diagnosis requires: Focal visual seizures triggered by photic stimulus and occipital spike-waves on EEG, facilitated by eye closure and IPS	EEG photosensitivity may remain even with clinical seizure remission; over 25% have morbid migraines	Variable	ASM recommended; monotherapy typically sufficient	Remission rate about 80%; but half of those will relapse

Abbreviations: *ADHD*, Attention deficit hyperactivity disorder; *ADNFLE*, autosomal dominant nocturnal frontal lobe epilepsy; *ASD*, autism spectrum disorder; *ASO*, anti-sense oligonucleotide; *ASM*, antiseizure medication; *CBZ*, carbamazepine; *CDKL5*, cyclin-dependent kinase-like 5; *CLB*, clobazam; *CVI*, cerebral visual impairment; *DD*, developmental delay; *EIDEE*, early infantile developmental and epileptic encephalopathy; *EIMFS*, epilepsy of infancy with migrating focal seizures; *ES*, epileptic spasms; ESX, ethosuximide; *GI*, gastrointestinal; *GLUT1 DS*, GLUT1 deficiency syndrome; *GOF*, gain-of-function; *GTC*, generalized tonic-clonic seizure; *ID*, intellectual disability; *IPS*, intermittent photic stimulation; *KDT*, ketogenic diet therapy; *LEV*, levetiracetam; *LGS*, Lennox–Gastaut syndrome; *LOF*, loss of function; *LTG*, lamotrigine; *MRE*, medically-refractory epilepsy; *OXC*, oxcarbazepine; *PHT*, phenytoin; *SCB*, sodium channel blocker; *STXBP1*, syntaxin1a binding protein; *SUDEP*, sudden unexpected death in epilepsy; *SZ*, seizure; *TC*, tonic-clonic; *VPA*, valproic acid; *WS*, West syndrome; *XCI*, X-chromosome inactivation.

TSC-Associated Neuropsychiatric Disorders (TAND) is an umbrella term used to encompass the multitude of neuropsychiatric manifestations that occur in approximately 90% of TSC patients [94]. Anxiety, depression, attention deficits, self-injury, aggression, autism spectrum disorder, intellectual disability, learning disabilities, and disordered sleep are among the disturbances that are not uncommon and can have a profound impact on the lives of patients and their families [95]. Treatment with everolimus has not been associated with improvement in cognitive function in children and adolescents with TSC, but other studies are ongoing [96]. The potential for disease-modifying therapy for epilepsy and TAND holds great promise for patients with TSC.

Quality-of-life and family functioning in developmental and epileptic encephalopathie

Children with epilepsy have reduced quality-of-life (QoL) [97]. Not only does epilepsy impact the child's well-being, but it also has a significant impact on family functioning. Notably, Rozensztrauch and Koltuniuk [97] found that for children with epilepsy, the negative impact of child health on caregiver and family functioning, as reported by the caregiver, is greater than for children with cancer undergoing active treatment [98] and adolescents with chronic pain conditions [99]. Little is known about QoL outcomes in individuals with DEEs specifically. However, recent work examining a novel caregiver-reported QoL measure designed for tracking QoL in individuals with neurodevelopmental disabilities in response to intervention (i.e., the Quality-of-Life Inventory-Disability, [100]) suggests that children with DEE (i.e., *SCN2A*, *SLC6A1*, *SCN1A*, and *KCNQ2*) have lower QoL compared to those with Rett syndrome, cerebral palsy, autism spectrum disorder, and Down syndrome [101]. A greater number of days disrupted by seizures was strongly associated with lower QoL in this group. Taken together, the impact of DEEs on the child and family is an area of clinical concern and an important consideration for management.

Conclusion

Developmental impairment in DEE results from two factors: the underlying neurobiology and the superimposed epileptiform activity, whereas in EE, the ictal and interictal activity alone is believed to be the driving force behind developmental impairments. While they are largely viewed as separate entities, deciphering causation is challenging. What separates DE from comorbid epilepsy and EE from DEE? As technology advances and additional specific genetic etiologies are uncovered, understanding of the role of a neurobiological predestination and the degree to which it is the result of or exacerbated by epileptic activity will continue to evolve. DEE, viewed as a spectrum, may be the most appropriate term to capture this idea.

Regardless of semantics, the essence remains the same—early recognition of DEE is crucial to informing counseling, treatment approaches, and prognosis [15]. Treatment should follow precision medicine-based approaches that are unique to an individual's needs. Attention should be given to the delicate balance between optimizing seizure control and avoiding medication side effects that impair quality of life.

References

[1] Zack MM. National and state estimates of the numbers of adults and children with active epilepsy—United States, 2015. Morb Mortal Wkly Rep 2017;66. Available from: https://doi.org/10.15585/mmwr.mm6631a1.

[2] Ngugi AK, Bottomley C, Kleinschmidt I, Sander JW, Newton CR. Estimation of the burden of active and life-time epilepsy: a meta-analytic approach. Epilepsia 2010;51(5):883–90. Available from: https://doi.org/10.1111/j.1528-1167.2009.02481.x.

[3] Robertson J, Hatton C, Emerson E, Baines S. Prevalence of epilepsy among people with intellectual disabilities: a systematic review. Seizure 2015;29:46–62. Available from: https://doi.org/10.1016/j.seizure.2015.03.016.

[4] El Achkar CM, Spence SJ. Clinical characteristics of children and young adults with co-occurring autism spectrum disorder and epilepsy. Epilepsy Behav 2015;47:183–90. Available from: https://doi.org/10.1016/j.yebeh.2014.12.022.

[5] Hanci F, Türay S, Dilek M, Kabakuş N. Epilepsy and drug-resistant epilepsy in children with cerebral palsy: a retrospective observational study. Epilepsy Behav 2020;112:107357. Available from: https://doi.org/10.1016/j.yebeh.2020.107357.

[6] Fisher RS, Boas W, van E, Blume W, Elger C, Genton P, et al. Epileptic seizures and epilepsy: definitions proposed by the International League Against Epilepsy (ILAE) and the International Bureau for Epilepsy (IBE). Epilepsia 2005;46(4):470–2. Available from: https://doi.org/10.1111/j.0013-9580.2005.66104.x.

[7] Fisher RS, Acevedo C, Arzimanoglou A, Bogacz A, Cross JH, Elger CE, et al. ILAE official report: a practical clinical definition of epilepsy. Epilepsia 2014;55(4):475–82. Available from: https://doi.org/10.1111/epi.12550.

[8] Scheffer IE, Berkovic S, Capovilla G, Connolly MB, French J, Guilhoto L, et al. ILAE classification of the epilepsies: position paper of the ILAE Commission for Classification and Terminology. Epilepsia 2017;58(4):512–21. Available from: https://doi.org/10.1111/epi.13709.

[9] Berg AT, Berkovic SF, Brodie MJ, Buchhalter J, Cross JH, Van Emde Boas W, et al. Revised terminology and concepts for organization of seizures and epilepsies: report of the ILAE Commission on Classification and Terminology, 2005–2009. Epilepsia 2010;51(4):676–85. Available from: https://doi.org/10.1111/j.1528-1167.2010.02522.x.

[10] Shao L-R, Habela CW, Stafstrom CE. Pediatric epilepsy mechanisms: expanding the paradigm of excitation/inhibition imbalance. Children 2019;6(2). Available from: https://doi.org/10.3390/children6020023 Article 2.

[11] Barry JM, Holmes GL. Why are children with epileptic encephalopathies encephalopathic? J Child Neurol 2016;31(13):1495–504. Available from: https://doi.org/10.1177/0883073816662140.

[12] Shao L-R, Stafstrom CE. Pediatric epileptic encephalopathies: pathophysiology and animal models. Sem Pediatr Neurol 2016;23(2):98–107. Available from: https://doi.org/10.1016/j.spen.2016.05.004.

[13] Lin JJ, Mula M, Hermann BP. Uncovering the lifespan neurobehavioral comorbidities of epilepsy. Lancet 2012;380(9848). Available from: https://doi.org/10.1016/S0140-6736(12)61455-X.

[14] Specchio N, Curatolo P. Developmental and epileptic encephalopathies: what we do and do not know. Brain 2021;144(1):32–43. Available from: https://doi.org/10.1093/brain/awaa371.

[15] Raga S, Specchio N, Rheims S, Wilmshurst JM. Developmental and epileptic encephalopathies: recognition and approaches to care. Epileptic Disord 2021;23(1):40–52. Available from: https://doi.org/10.1684/epd.2021.1244.

[16] Scheffer IE. How gene discovery has transformed management of people with epilepsy. Eur J Paediatr Neurol 2020;24:2. Available from: https://doi.org/10.1016/j.ejpn.2020.01.010.

[17] Jeffrey JS, Leathem J, King C, Mefford HC, Ross K, Sadleir LG. Developmental and epileptic encephalopathy: personal utility of a genetic diagnosis for families. Epilepsia Open 2021;6(1):149–59. Available from: https://doi.org/10.1002/epi4.12458.

[18] Lin K, Stafstrom CE. Cognition, behavior, and psychosocial effects of seizures in the developing brain. In: Jones NC, Kanner AM, editors. Psychiatric and behavioral aspects of epilepsy: current perspectives and mechanisms. Springer International Publishing; 2022. p. 3–15. Available from: https://doi.org/10.1007/7854_2020_189.

[19] Scheffer IE, Liao J. Deciphering the concepts behind "epileptic encephalopathy" and "developmental and epileptic encephalopathy. Eur J Paediatr Neurol 2020;24:11–14. Available from: https://doi.org/10.1016/j.ejpn.2019.12.023.

[20] Zuberi SM, Wirrell E, Yozawitz E, Wilmshurst JM, Specchio N, Riney K, et al. ILAE classification and definition of epilepsy syndromes with onset in neonates and infants: position statement by the ILAE task force on nosology and definitions. Epilepsia 2022;63(6):1349–97. Available from: https://doi.org/10.1111/epi.17239.

[21] Fernández-Alvarez E. Transient benign paroxysmal movement disorders in infancy. Eur J Paediatr Neurol 2018;22(2):230–7. Available from: https://doi.org/10.1016/j.ejpn.2018.01.003.

[22] Pearl PL. Amenable treatable severe pediatric epilepsies. Sem Pediatr Neurol 2016;23(2):158–66. Available from: https://doi.org/10.1016/j.spen.2016.06.004.

[23] Symonds JD, Elliott KS, Shetty J, Armstrong M, Brunklaus A, Cutcutache I, et al. Early childhood epilepsies: epidemiology, classification, aetiology, and socio-economic determinants. Brain 2021;144(9):2879–91. Available from: https://doi.org/10.1093/brain/awab162.

[24] Casas-Alba D, Aguilar A, Alonso I, et al. Relationship between epileptic activity and developmental outcome in KCNQ2-related epilepsy. Pediatr Neurol 2023;144:11–15. Available from: https://doi.org/10.1016/j.pediatrneurol.2023.03.004.

[25] Kuersten M, Tacke M, Gerstl L, Hoelz H, Stülpnagel Cv, Borggraefe I. Antiepileptic therapy approaches in KCNQ2 related epilepsy: a systematic review. Eur J Med Genet 2020;63(1):103628. Available from: https://doi.org/10.1016/j.ejmg.2019.02.001.

[26] Knight D, Mahida S, Kelly M, Poduri A, Olson HE. Ezogabine impacts seizures and development in patients with KCNQ2 developmental and epileptic encephalopathy. Epilepsia 2023;. Available from: https://doi.org/10.1111/epi.17627.

[27] Morrison-Levy N, Borlot F, Jain P, Whitney R. Early-onset developmental and epileptic encephalopathies of infancy: an overview of the genetic basis and clinical features. Pediatr Neurol 2021;116:85–94. Available from: https://doi.org/10.1016/j.pediatrneurol.2020.12.001.

[28] Kim HJ, Yang D, Kim SH, et al. The phenotype and treatment of SCN2A-related developmental and epileptic encephalopathy. Epileptic Disord 2020;22(5):563–70. Available from: https://doi.org/10.1684/epd.2020.1199.

[29] Richardson R, Baralle D, Bennett C, et al. Further delineation of phenotypic spectrum of SCN2A-related disorder. Am J Med Genet A 2022;188(3):867–77. Available from: https://doi.org/10.1002/ajmg.a.62595.

[30] Sahly AN, Shevell M, Sadleir LG, Myers KA. SUDEP risk and autonomic dysfunction in genetic epilepsies. Auton Neurosci 2022;237:102907. Available from: https://doi.org/10.1016/j.autneu.2021.102907.

[31] Schreiber JM, Tochen L, Brown M, et al. A multi-disciplinary clinic for SCN8A-related epilepsy. Epilepsy Res 2020;159:106261. Available from: https://doi.org/10.1016/j.eplepsyres.2019.106261.

[32] Stamberger H, Crosiers D, Balagura G, et al. Natural history study of STXBP1-developmental and epileptic encephalopathy into adulthood. Neurology. 2022;99(3):e221–33. Available from: https://doi.org/10.1212/WNL.0000000000200715.

[33] Balagura G, Xian J, Riva A, et al. Epilepsy course and developmental trajectories in *STXBP1*-DEE. Neurol Genet 2022;8(3):e676. Available from: https://doi.org/10.1212/NXG.0000000000000676.

[34] Abramov D, Guiberson NGL, Burré J. STXBP1 encephalopathies: clinical spectrum, disease mechanisms, and therapeutic strategies. J Neurochem 2020;. Available from: https://doi.org/10.1111/jnc.15120 Published online July 8.

[35] Jakimiec M, Paprocka J, Śmigiel R. CDKL5 deficiency disorder—a complex epileptic encephalopathy. Brain Sci 2020;10(2):107. Available from: https://doi.org/10.3390/brainsci10020107.

[36] Hong W, Haviland I, Pestana-Knight E, et al. CDKL5 deficiency disorder-related epilepsy: a review of current and emerging treatment. CNS Drugs 2022;36(6):591–604. Available from: https://doi.org/10.1007/s40263-022-00921-5.

[37] Aznar-Laín G, Fernández-Mayoralas DM, Caicoya AG, Rocamora R, Pérez-Jurado LA. CDKL5 deficiency disorder without epilepsy. Pediatr Neurol 2023;144:84–9. Available from: https://doi.org/10.1016/j.pediatrneurol.2023.04.015.

[38] Bonardi CM, Heyne HO, Fiannacca M, et al. KCNT1-related epilepsies and epileptic encephalopathies: phenotypic and mutational spectrum. Brain 2021;144(12):3635–50. Available from: https://doi.org/10.1093/brain/awab219.

[39] Gertler TS, Cherian S, DeKeyser JM, Kearney JA, George AL. $K_{Na}1.1$ gain-of-function preferentially dampens excitability of murine parvalbumin-positive interneurons. Neurobiol Dis 2022;168:105713. Available from: https://doi.org/10.1016/j.nbd.2022.105713.

[40] Marsh E, Melamed SE, Barron T, Clancy RR. Migrating partial seizures in infancy: expanding the phenotype of a rare seizure syndrome. Epilepsia 2005;46(4):568–72. Available from: https://doi.org/10.1111/j.0013-9580.2005.34104.x.

[41] Yoshitomi S, Takahashi Y, Imai K, Koshimizu E, Miyatake S, Nakashima M, et al. Different types of suppression-burst patterns in patients with epilepsy of infancy with migrating focal seizures (EIMFS). Seizure 2019;65:118–23. Available from: https://doi.org/10.1016/j.seizure.2019.01.009.

[42] El Kosseifi C, Cornet M-C, Cilio MR. Neonatal developmental and epileptic encephalopathies. Sem Pediatr Neurol 2019;32:100770. Available from: https://doi.org/10.1016/j.spen.2019.08.006.

[43] Dilena R, DiFrancesco JC, Soldovieri MV, Giacobbe A, Ambrosino P, Mosca I, et al. Early treatment with quinidine in 2 patients with epilepsy of infancy with migrating focal seizures (EIMFS) due to gain-of-function KCNT1 mutations: functional studies, clinical responses, and critical issues for personalized therapy. Neurotherapeutics 2018;15(4):1112–26. Available from: https://doi.org/10.1007/s13311-018-0657-9.

[44] Orsini A, Foiadelli T, Costagliola G, Michev A, Consolini R, Vinci F, et al. The role of inflammatory mediators in epilepsy: focus on developmental and epileptic encephalopathies and therapeutic implications. Epilepsy Res 2021;172:106588. Available from: https://doi.org/10.1016/j.eplepsyres.2021.106588.

[45] De Jonghe P. Molecular genetics of Dravet syndrome. Dev Med Child Neurol 2011;53(s2):7–10. Available from: https://doi.org/10.1111/j.1469-8749.2011.03965.x.

[46] Dravet C. The core Dravet syndrome phenotype. Epilepsia 2011;52(s2):3–9. Available from: https://doi.org/10.1111/j.1528-1167.2011.02994.x.

[47] Wirrell EC, Laux L, Donner E, Jette N, Knupp K, Meskis MA, et al. Optimizing the diagnosis and management of dravet syndrome: recommendations from a North American consensus panel. Pediatr Neurol 2017;68:18–34. Available from: https://doi.org/10.1016/j.pediatrneurol.2017.01.025 e3.

[48] Chiron C, Dulac O. The pharmacologic treatment of Dravet syndrome. Epilepsia 2011;52(s2):72–5. Available from: https://doi.org/10.1111/j.1528-1167.2011.03007.x.

[49] de Lange IM, Gunning B, Sonsma ACM, van Gemert L, van Kempen M, Verbeek NE, et al. Influence of contraindicated medication use on cognitive outcome in Dravet syndrome and age at first afebrile seizure as a clinical predictor in SCN1A-related seizure phenotypes. Epilepsia 2018;59(6):1154–65. Available from: https://doi.org/10.1111/epi.14191.

[50] Lagae L, Sullivan J, Knupp K, Laux L, Polster T, Nikanorova M, et al. Fenfluramine hydrochloride for the treatment of seizures in Dravet syndrome: a randomised, double-blind, placebo-controlled trial. Lancet 2019;394(10216):2243–54. Available from: https://doi.org/10.1016/S0140-6736(19)32500-0.

[51] Shmuely S, Sisodiya SM, Gunning WB, Sander JW, Thijs RD. Mortality in Dravet syndrome: a review. Epilepsy Behav 2016;64:69–74. Available from: https://doi.org/10.1016/j.yebeh.2016.09.007.

[52] Catarino CB, Liu JY, Liagkouras I, Gibbons VS, Labrum RW, Ellis R, et al. Dravet syndrome as epileptic encephalopathy: evidence from long-term course and neuropathology. Brain 2011;134(Pt 10):2982–3010. Available from: https://doi.org/10.1093/brain/awr129.

[53] de Lange IM, Koudijs MJ, van 't Slot R, Gunning B, Sonsma ACM, van Gemert LJJM, et al. Mosaicism of de novo pathogenic SCN1A variants in epilepsy is a frequent phenomenon that correlates with variable phenotypes. Epilepsia 2018;59(3):690–703. Available from: https://doi.org/10.1111/epi.14021.

[54] Berecki G, Bryson A, Terhag J, Maljevic S, Gazina EV, Hill SL, et al. SCN1A gain of function in early infantile encephalopathy. Ann Neurol 2019;85(4):514–25. Available from: https://doi.org/10.1002/ana.25438.

[55] Myers KA, Scheffer IE, Berkovic SFon behalf of the ILAE Genetics Commission. Genetic literacy series: genetic epilepsy with febrile seizures plus. Epileptic Disord 2018;20(4):232–8. Available from: https://doi.org/10.1684/epd.2018.0985.

[56] Bayat A, Bayat M, Rubboli G, Møller RS. Epilepsy syndromes in the first year of life and usefulness of genetic testing for precision therapy. Genes 2021;12(7):1051. Available from: https://doi.org/10.3390/genes12071051.

[57] Nickels K, Kossoff EH, Eschbach K, Joshi C. Epilepsy with myoclonic-atonic seizures (Doose syndrome): clarification of diagnosis and treatment options through a large retrospective multicenter cohort. Epilepsia 2021;62(1):120–7. Available from: https://doi.org/10.1111/epi.16752.

[58] Sijben AEJ, Sithinamsuwan P, Radhakrishnan A, Badawy RAB, Dibbens L, Mazarib A, et al. Does a SCN1A gene mutation confer earlier age of onset of febrile seizures in GEFS + ? Epilepsia 2009;50(4):953–6. Available from: https://doi.org/10.1111/j.1528-1167.2009.02023.x.

[59] Arzimanoglou A, Resnick T. All children who experience epileptic falls do not necessarily have Lennox-Gastaut syndrome... But many do. Epileptic Disord 2011;13(S1):3–13. Available from: https://doi.org/10.1684/epd.2011.0422.

[60] Thirunavu V, Du R, Wu JY, Berg AT, Lam SK. The role of surgery in the management of Lennox–Gastaut syndrome: a systematic review and meta-analysis of the clinical evidence. Epilepsia 2021;62(4):888–907. Available from: https://doi.org/10.1111/epi.16851.

[61] Specchio N, Wirrell EC, Scheffer IE, Nabbout R, Riney K, Samia P, et al. International League Against Epilepsy classification and definition of epilepsy syndromes with onset in childhood: position paper by the ILAE task force on nosology and definitions. Epilepsia 2022;63(6):1398–442. Available from: https://doi.org/10.1111/epi.17241.

[62] Baumer FM, McNamara NA, Fine AL, Pestana-Knight E, Shellhaas RA, He Z, et al. Treatment practices and outcomes in continuous spike and wave during slow wave sleep: a multicenter collaboration. J Pediatr 2021;232:220–8. Available from: https://doi.org/10.1016/j.jpeds.2021.01.032 e3.

[63] Filippini M, Arzimanoglou A, Gobbi G. Neuropsychological approaches to epileptic encephalopathies. Epilepsia 2013;54(s8):38–44. Available from: https://doi.org/10.1111/epi.12422.

[64] Pangalila R, van der Meulen I. What is the effect of pharmacological treatment for continuous spike-wave during slow wave sleep syndrome and Landau-Kleffner syndrome? A cochrane review summary with commentary. Dev Med Child Neurol 2022;64(4):411–12. Available from: https://doi.org/10.1111/dmcn.15137.

[65] Clark M, Holmes H, Ngoh A, Siyani V, Wilson G. Overview of Landau–Kleffner syndrome: early treatment, tailored education and therapy improve outcome. Paediatr Child Health 2021;31(5):207–19. Available from: https://doi.org/10.1016/j.paed.2021.02.005.

[66] Overvliet GM, Aldenkamp AP, Klinkenberg S, Vles JSH, Hendriksen J. Impaired language performance as a precursor or consequence of Rolandic epilepsy? J Neurol Sci 2011;304(1):71–4. Available from: https://doi.org/10.1016/j.jns.2011.02.009.

[67] Ross EE, Stoyell SM, Kramer MA, Berg AT, Chu CJ. The natural history of seizures and neuropsychiatric symptoms in childhood epilepsy with centrotemporal spikes (CECTS). Epilepsy Behav 2020;103:106437. Available from: https://doi.org/10.1016/j.yebeh.2019.07.038.

[68] Zanaboni MP, Pasca L, Bova SM, Chiappedi MA, Filippini M, Giordano L, et al. WISC-IV intellectual profiles in Italian children with self-limited epilepsy with centrotemporal spikes. Epileptic Disord 2023;25(2):160–72. Available from: https://doi.org/10.1002/epd2.20003.

[69] Specchio N, Pietrafusa N. New-onset refractory status epilepticus and febrile infection-related epilepsy syndrome. Dev Med Child Neurol 2020;62(8):897–905. Available from: https://doi.org/10.1111/dmcn.14553.

[70] Auvin S, Bellavoine V, Merdariu D, Delanoë C, Elmaleh-Bergés M, Gressens P, et al. Hemiconvulsion–hemiplegia–epilepsy syndrome: current understandings. Eur J Paediatr Neurol 2012;16(5):413–21. Available from: https://doi.org/10.1016/j.ejpn.2012.01.007.

[71] Jackson DC, Dabbs K, Walker NM, Jones JE, Hsu DA, Stafstrom CE, et al. The neuropsychological and academic substrate of new/recent-onset epilepsies. J Pediatr 2013;162(5):1047–53. Available from: https://doi.org/10.1016/j.jpeds.2012.10.046 e1.

[72] Kessler SK, McGinnis E. A practical guide to treatment of childhood absence epilepsy. Pediatr Drugs 2019;21(1):15–24. Available from: https://doi.org/10.1007/s40272-019-00325-x.

[73] Smith KM, Wirrell EC, Andrade DM, et al. A comprehensive narrative review of epilepsy with eyelid myoclonia. Epilepsy Res 2023;193:107147. Available from: https://doi.org/10.1016/j.eplepsyres.2023.107147.

[74] Arvio M, Sauna-aho O, Nyrke T, Bjelogrlic-Laakso N. Intellectual disability in patients with epilepsy with eyelid myoclonias. SAGE Open Med Case Rep 2018;6. Available from: https://doi.org/10.1177/2050313X18777951 2050313X18777951.

[75] Cerulli Irelli E, Cocchi E, Ramantani G, et al. Spectrum of epilepsy with eyelid myoclonia: delineation of disease subtypes from a large multicenter study. Epilepsia 2023;64(1):196–207. Available from: https://doi.org/10.1111/epi.17450.

[76] Hiraide T, Hattori A, Ieda D, et al. De novo variants in SETD1B cause intellectual disability, autism spectrum disorder, and epilepsy with myoclonic absences. Epilepsia Open 2019;4(3):476–81. Available from: https://doi.org/10.1002/epi4.12339.

[77] Frydson I, Avula S, Ahmad SF. A case of a seven-year-old boy with epilepsy with myoclonic absence: importance of seizure semiology, genetic etiology, and electroencephalogram correlation for timely intervention. Child Neurol Open 2022;9. Available from: https://doi.org/10.1177/2329048X221131738 2329048X221131738.

[78] Jara Jde la, Vásquez-Hernández C, Ramírez-Rojo E, Moya-Vilches J. Uncommon epileptic syndromes in children: a review. Seizure – Eur J Epilepsy 2021;90:17–27. Available from: https://doi.org/10.1016/j.seizure.2021.05.005.

[79] Bureau M, Tassinari CA. Epilepsy with Myoclonic absences. Brain Dev 2005;27(3):178–84. Available from: https://doi.org/10.1016/j.braindev.2004.01.008.

[80] Ferrie C, Caraballo R, Covanis A, et al. Panayiotopoulos syndrome: a consensus view. Dev Med Child Neurol 2006;48(3):236–40. Available from: https://doi.org/10.1017/S0012162206000508.

[81] Graziosi A, Pellegrino N, Di Stefano V, Raucci U, Luchetti A, Parisi P. Misdiagnosis and pitfalls in Panayiotopoulos syndrome. Epilepsy Behav 2019;98:124–8. Available from: https://doi.org/10.1016/j.yebeh.2019.07.016.

[82] Specchio N, Trivisano M, Di Ciommo V, et al. Panayiotopoulos syndrome: a clinical, EEG, and neuropsychological study of 93 consecutive patients. Epilepsia 2010;51(10):2098–107. Available from: https://doi.org/10.1111/j.1528-1167.2010.02639.x.

[83] Caraballo R, Koutroumanidis M, Panayiotopoulos CP, Fejerman N. Idiopathic childhood occipital epilepsy of gastaut: a review and differentiation from migraine and other epilepsies. J Child Neurol 2009;24(12):1536–42. Available from: https://doi.org/10.1177/0883073809332395.

[84] Wyllie E. Wyllie's treatment of epilepsy. 7th ed. Wolters Kluwer Health; 2020. p. 216–7. Available from: https://wolterskluwer.vitalsource.com/books/9781496397713.

[85] Cerrahoğlu Şirin T, Yılmaz T, Elmalı AD, Bebek N, Demirbilek V, Baykan B. Photosensitive occipital lobe epilepsy: delineation of an underrecognized reflex epilepsy syndrome according to the new ILAE criteria and long-term follow-up. Epileptic Disord 2023;25(2):187–99. Available from: https://doi.org/10.1002/epd2.20011.

[86] Palmer EE, Howell K, Scheffer IE. Natural history studies and clinical trial readiness for genetic developmental and epileptic encephalopathies. Neurotherapeutics 2021;18(3):1432–44. Available from: https://doi.org/10.1007/s13311-021-01133-3.

[87] Capal JK, Bernardino-Cuesta B, Horn PS, Murray D, Byars AW, Bing NM, et al. Influence of seizures on early development in tuberous sclerosis complex. Epilepsy Behavior: E&B 2017;70(Pt A):245–52. Available from: https://doi.org/10.1016/j.yebeh.2017.02.007.

[88] Chu-Shore CJ, Major P, Camposano S, Muzykewicz D, Thiele EA. The natural history of epilepsy in tuberous sclerosis complex. Epilepsia 2010;51(7):1236–41. Available from: https://doi.org/10.1111/j.1528-1167.2009.02474.x.

[89] Bombardieri R, Pinci M, Moavero R, Cerminara C, Curatolo P. Early control of seizures improves long-term outcome in children with tuberous sclerosis complex. Eur J Paediatr Neurol 2010;14(2):146–9. Available from: https://doi.org/10.1016/j.ejpn.2009.03.003.

[90] Northrup H, Aronow ME, Bebin EM, Bissler J, Darling TN, de Vries PJ, et al. Updated international tuberous sclerosis complex diagnostic criteria and surveillance and management recommendations. Pediatr Neurol 2021;123:50–66. Available from: https://doi.org/10.1016/j.pediatrneurol.2021.07.011.

[91] Kotulska K, Kwiatkowski DJ, Curatolo P, Weschke B, Riney K, Jansen F, et al. Prevention of epilepsy in infants with tuberous sclerosis complex in the EPISTOP trial. Ann Neurol 2021;89(2):304–14. Available from: https://doi.org/10.1002/ana.25956.

[92] van der Poest Clement E, Jansen FE, Braun KPJ, Peters JM. Update on Drug management of refractory epilepsy in tuberous sclerosis complex. Pediatr Drugs 2020;22(1):73–84. Available from: https://doi.org/10.1007/s40272-019-00376-0.

[93] Bebin EM, Peters JM, Porter BE, McPherson TO, O'Kelley S, Sahin M, et al. Early Treatment with vigabatrin does not decrease focal seizures or improve cognition in tuberous sclerosis complex: The PREVeNT Trial. Ann Neurol 2023;28. Available from: https://doi.org/10.1002/ana.26778.

[94] de Vries PJ, Wilde L, de Vries MC, Moavero R, Pearson DA, Curatolo P. A clinical update on tuberous sclerosis complex-associated neuropsychiatric disorders (TAND). Am J Med Genet Part C: SemMed Genet 2018;178(3):309–20. Available from: https://doi.org/10.1002/ajmg.c.31637.

[95] Curatolo P, Moavero R, de Vries PJ. Neurological and neuropsychiatric aspects of tuberous sclerosis complex. Lancet Neurol 2015;14(7):733–45. Available from: https://doi.org/10.1016/S1474-4422(15)00069-1.

[96] Luo C, Ye W-R, Shi W, Yin P, Chen C, He Y-B, et al. Perfect match: mTOR inhibitors and tuberous sclerosis complex. Orphanet J Rare Dis 2022;17(1):106. Available from: https://doi.org/10.1186/s13023-022-02266-0.

[97] Rozensztrauch A, Kołtuniuk A. The quality of life of children with epilepsy and the impact of the disease on the family functioning. Int J Environ Res Public Health 2022;17:2277. Available from: https://doi.org/10.3390/ijerph19042277.

[98] Scarpelli AC, Paiva SM, Pordeus IA, Varni JW, Viegas CM, Allison PJ. The pediatric quality of life inventory (PedsQL) family impact module: reliability and validity of the Brazilian version. Health Qual Life Outcomes 2008;20:35. Available from: https://doi.org/10.1186/1477-7525-6-35.

[99] Parmar A, Yeh EA, Korczak DJ, Weiss SK, Lu Z, Zweerink A, et al. Family functioning among adolescents with narcolepsy. Paediatr Child Health 2019;24(8):490–4. Available from: https://doi.org/10.1093/pch/pxy192.

[100] Downs J, Jacoby P, Leonard H, Epstein A, Murphy N, Davis E, et al. Psychometric properties of the Quality of Life Inventory-Disability (QI-Disability) measure. Qual Life Res 2019 Mar;28(3):783–94. Available from: https://doi.org/10.1007/s11136-018-2057-3.

[101] Cohen SR, Helbig I, Kaufman MC, Schust Myers L, Conway L, Helbig KL. Caregiver assessment of quality of life in individuals with genetic developmental and epileptic encephalopathies. Dev Med Child Neurol 2022;64(8):957–64. Available from: https://doi.org/10.1111/dmcn.15187.

Chapter 15

Cerebral palsy

Elizabeth Barkoudah[1], Jenny Wilson[2], Eric M. Chin[3,4] and Peter A. Blasco[5]

[1]*Department of Neurology, Boston Children's Hospital, Harvard University, Boston, MA, United States,* [2]*Department of Pediatrics, Division of Pediatric Neurology, Oregon Health & Science University, Portland, OR, United States,* [3]*Department of Neurodevelopmental Medicine, Kennedy Krieger Institute, Baltimore, MD, United States,* [4]*Departments of Neurology and Pediatrics, Johns Hopkins University School of Medicine, Baltimore, MD, United States,* [5]*Department of Pediatrics, Division of Neurodevelopmental Pediatrics, Institute on Development and Disability, Oregon Health and Science University, Portland, OR, United States*

Introduction

Cerebral palsy (CP) is a complex and heterogeneous disorder in the sense that its antecedents are numerous and its manifestations, both primary (i.e., motor dysfunction) and associated (i.e., cognitive deficits, seizures, etc.), are protean. The characteristics of the CP patient population encompass the purview of almost all of neurodevelopmental medicine plus substantial portions of general pediatrics, genetics, orthopedics, rehabilitation medicine, and behavioral pediatrics. Thorough care of patients with CP requires insights from all of these disciplines. Addressing the needs, medical/surgical and psychosocial, of children with CP and their families in a comprehensive and cohesive way extends beyond the scope of any single discipline and will only be accomplished in the context of a structured and thorough understanding of the disorder and its complexities. The intent of this chapter is to provide that.

Background and history

William John Little (1810–94), whose name used to be applied to this disorder (Little's disease), was an English orthopedist who first described CP and associated it with difficult birth [1]. His 1861 publication was the culmination of more than 20 years of clinical experience and thought given to the relationship between orthopedic deformities and early neurological insult. He presented 63 case abstracts, of which at least 34 were individuals with spastic diplegia, and associated these syndromes with mechanical birth injury, asphyxia neonatorum and prematurity. He also made note of the difficulty determining whether neurological damage occurred before or during parturition. His fundamental point was that a third "termination" (i.e., outcome) from difficult birth processes existed between the extremes of death and normal recovery, that is, survival with "other diseases" (by which he meant disability). This represented a remarkably novel viewpoint at a time when teething was considered the primary cause of most paralytic syndromes. The prevailing notion was that teething produced convulsions and associated cerebral damage. Little presented his thesis on orthopedic deformity resulting from abnormal birth processes to the Obstetrical Society of London in 1861 and was met with disbelief, if not derision.

Sigmund Freud (1856–1939) in his neuropathologic treatises on the classification of CP recognized spastic diplegia and the close association between the diplegic phenotype and premature birth [2]. His decade of work on these childhood motor syndromes failed to elucidate a clear correlation between neuropathology and clinical classification features. Freud sided with Little in rejecting the commonly accepted belief that CP was caused by teething seizures, and he was more skeptical than Little about the role of abnormal birth processes, pointing out that these events might be a symptom of underlying fetal abnormality. Despite their cautions, both Freud and Little are commonly cited as creating an undue emphasis on birth insults as the cause of CP [3].

In 1888 Sir William Osler popularized use of the term "cerebral palsy" with his review of the clinical details of 151 cases of the "cerebral palsies of children" most of which were spastic hemiplegia [4]. Osler's meticulous review of the pathology literature existing at that time emphasized the varied nature of the lesions encountered and the unsatisfactory state of knowledge.

Capute and Accardo's Neurodevelopmental Disabilities in Infancy and Childhood. DOI: https://doi.org/10.1016/B978-0-12-824060-1.00041-9

Crothers and Paine [5], in their classic book *The Natural History of Cerebral Palsy*, reviewed and contrasted the literature on pathology available prior to 1955, going back to the time of Little and Osler. The major point of contention through the latter half of this period was not so much one of prenatal versus perinatal timing for the prime cause of CP but rather whether the insult was primarily vascular/hemorrhagic due to trauma during delivery or was a primary infectious process involving the motor cortex (polioencephalitis), analogous to poliomyelitis involving the spinal cord.

Definition

CP is defined as a disorder of movement and posture due to a nonprogressive defect or lesion of the immature brain [6,7]. The lesion may be the result of an insult to a normally developing brain or to anomalous brain development from an early stage. In 2004 Bax's original definition was "reconsidered" by an international work group of the American Academy for Cerebral Palsy and Developmental Medicine (AACPDM) and was expanded [8]. The current definition is contained within the "Report on the Definition and Classification of Cerebral Palsy, April 2006":

> *"A group of permanent disorders of the development of movement and posture, causing limitation of activity, that are attributed to non-progressive disturbances that occurred in the developing fetal or infant brain. The motor disorders of cerebral palsy are often accompanied by disturbances of sensation, perception, cognition, communication, and behavior, by epilepsy, and by secondary musculoskeletal problems." [9]*

The terminology used in the definition was deliberate:

"*group*"—CP is not a homogenous condition.

"*permanent*"—While children with CP make developmental progress over time, they are on a distinct motor developmental curve that will not normalize.

"*disorders of movement and posture*"—CP does not describe a delay in motor development, but rather aberrant motor development. Furthermore, CP does not refer to cognitive disability.

"*causing limitation of activity*"—CP is a disability. Occasionally a child may have spasticity that does not cause any activity limitation, which is not CP.

"*nonprogressive*"—This term distinguishes CP from progressive neurologic conditions such as leukodystrophies. The term nonprogressive refers to the cerebral process, not the clinical manifestations. Generally, children with CP do not experience progressive decline; however, there can be clinical worsening with time. Specifically, children may develop orthopedic complications such as hip dislocations, contractures and scoliosis; the onset of dystonia may occur later in childhood [10]; adolescents with limited ambulation may lose gross motor function as they gain height and weight and become less involved with therapies [11]; and adults may experience further mobility loss as well as pain and fatigue [12]. Teasing out the clinical decline that accompanies CP from true neurodegeneration is one of the challenges faced by the practitioner.

"*disturbances*"—There is a misconception that CP is solely due to anoxic brain injury, when in fact hypoxic-ischemic encephalopathy represents a minority among causes of CP. Other destructive processes such as periventricular leukomalacia (PVL) and intraventricular hemorrhage (IVH) due to prematurity are common, but so are early changes such as malformations of cortical development. A "disturbance" may not be apparent on neuroimaging, often in the setting of a genetic cause.

"*developing fetal or infant brain*"—CP is a manifestation of the abnormality that occurs in the developing brain. The precise age cut-off of is not clearly defined, however some suggest "2 or 3 years" of age.

"*accompanied by*"—While CP refers to the motor problem, the other disturbances listed are included in the definition to emphasize their frequency and importance (see *Associated Problems*).

Epidemiology

CP is a clinical diagnosis and is therefore subject to interobserver variability [13]. Consequently, epidemiologic studies of CP will be affected by inconsistency in diagnosis. Some reports include only "disabling" cases of CP, thereby excluding the mildest end of the spectrum. These may be the most numerous cases and also will have the most diagnostic variability. On the opposite end of the spectrum, the most severely involved infants may be undercounted because of death before being assigned a diagnosis. Since there is no agreed upon age of onset criterion, some do not acknowledge postnatal CP. Others report "non-CP motor impairment" as a distinct group. Whether this refers to developmental coordination disorder (DCD), mild CP, or something else is unclear. In sum, CP epidemiology is imprecise. Population surveillance data on CP likely represent underestimates of true prevalence, which would take into account the mildest cases, missed diagnoses, and an extended age of onset.

CP is a relatively common cause of motor disability in children. Within high-income countries, systematic meta-analysis estimate prevalence of CP as between 1.4–1.6 per 1000 live births [14–16]. As with other neurodevelopmental disabilities, the risk of CP varies based on sex, race, and socioeconomic status. There is a 30%–40% greater prevalence in males [17], and the disparity appears to extend across subtypes and across levels of functional impairment [18]. In the United States, children with black mothers have approximately a 30% greater risk of CP than those with white mothers—a difference that appears to be attributable to a greater incidence of premature birth and low birth weight [19]. Prevalence of CP appears to be 70% higher in low- and middle-income US communities than in high-income communities.

Within low- and middle-income countries, systematic meta-analysis of available data estimate prevalence of CP as between 3.0 and 3.9 per 1000 live births though national registries as of 2022 have only recently begun collecting data and harmonizing definitions [20–23]. Perinatal and postnatal etiologies (e.g., meningitis, kernicterus, trauma, and hypoxic/ischemic injury) appear to represent a greater proportion of cases as do cases of greater severity (quadriplegia) [24,25]. Jahan et al. [24] provide data from four low- and middle-income countries though prevalence could not be calculated and case ascertainment is suspect. Mean age at diagnosis ranged from 2½ to 6½ years. Postnatal etiology was 13.5%. Mild cases must be underreported as moderate-severe involvement (GMFCS levels III–V) accounted for 55%–84%. In a study from Uganda only 2% were associated with preterm birth and 25% were postnatal [26]. Implications for prevention strategies seem evident in these countries. Gaps in prenatal care, in perinatal care, and in rehabilitative resources have been identified as opportunities to improve outcomes [24].

Rates of CP have only recently begun to decrease in developed countries, although direct interpretation of trends is complicated by changing patterns of neonatal care and survivorship. Rates in Western Australia fluctuated between 1.5 and 3.1 per 1000 live births in the 1960s until 1994 [27], but two registries have reported recent declines (net −0.13/1000 European live births from 1980 to 2003 [28] and −0.43/1000 Western Australian live births from 1995 to 2009 [29], respectively). While overall prevalence over time has fluctuated, prevalence of specific etiologies and injury patterns has markedly shifted [29]:

- In the 2000s, children with extremely low birth weight (<1000 g) [28] and those born extremely preterm (<28 weeks gestation) [30] have comprised increasing proportions of individuals with CP (6% and 11% in those studies, respectively), which appears to reflect increased survival in these groups [30].
- Patterns of preterm brain injury may be changing. For example, the incidence of cystic periventricular leukomalacia (PVL) in very/extremely low birth weight infants (<1500 g) declined sixfold at one medical center from 1992 to 2002—a change that was significantly predicted by a decrease in mechanical ventilator days [31]. Of note, no corresponding decrease in CP incidence was seen in that cohort over the study period.
- Hypoxic-ischemic encephalopathy (HIE) may be decreasing as an apparent cause of CP in developed countries (e.g., a nonsignificant decrease from 10% of cases for 2003–06 to 6% for 2007–10 in Sweden) [30]. Neonatal hypothermia was introduced within this time frame, which may reduce the risk of CP by 38% in appropriate settings and cohorts [32]. However changes in the case definition of HIE, other advances in maternal/neonatal care, and broader demographic shifts may also have contributed to weakening the links between HIE and CP [30,33,34].
- Kernicterus is now rare as a cause of CP in developed countries, although cases (particularly in very preterm infants) do occasionally occur [35,36].

Etiology

Study of the etiology and pathology of this disorder are closely intertwined. The neuroanatomy of the condition explains the nature of the lesion which produces motor dysfunction. Etiology addresses the question of causation from the aspect of the mechanism underlying the neuroanatomical change. Indeed, the two provide information about each other. For example, with lissencephaly, maldevelopment of the nervous system has been present since the first trimester of prenatal life. The child may have experienced a very difficult birth, but the anatomy indicates that little or none of the motor or cognitive dysfunction is a consequence of birth insult. On the other hand, for a child with hemiplegia who had a normal birth history and a postnatal cerebrovascular accident resulting in porencephaly, the neuroanatomy and the etiologic history are consistent with an insult of postnatal origin altering a previously normal nervous system.

Etiologies are categorized by timing: prenatal, perinatal and postnatal. The many factors associated with increased risk for CP generally predispose to specific etiologies. For example, preterm birth, a risk factor seen in up to 35% [37],

TABLE 15.1 Risk factors for cerebral palsy in children born at term.

Risk factor	Range of point estimates
Placental abnormalities	2.4–4.9
Birth defects	2.2–27.4
Low birth weight	3.4–8.2
Meconium aspiration	25–27
Delivery mode:	
Emergency C-section	1.8–8.1
Instrumented delivery	1.9–2.0
Breech	3.3–3.4
Birth asphyxia	4.6–140
Neonatal seizures	3.7–63
Respiratory distress syndrome	2.1–18.1
Neonatal hypoglycemia	2.7–11.6
Neonatal infections	1.8–29

Source: Adapted from McIntyre S, Taitz D, Keogh J, Goldsmith S, Badawi N, Blair E. A systematic review of risk factors for cerebral palsy in children born at term in developed countries. Dev Med Child Neurol 2013;55:499–507.

can predispose to PVL or IVH, but the underlying precipitant of earlier birth and the etiology for the CP could equally be a central nervous system (CNS) malformation. In some cases, etiology is associated with a number of risk factors. A meta-analysis found 38 significant risk factors for all children with CP and 10 risk factors associated with CP in children born at term (Table 15.1) [38]. In about 40%–50% of individuals who develop CP there were no concerns about them in the neonatal period: they were born at term without apparent risk factors and appeared "healthy," requiring only routine neonatal care [39]. Early recognition of CP in infants without risk factors depends on detection of motor delay and referral for neuromotor evaluation.

With advances in genetic testing, it has become evident that genetic factors play an important role in the etiology of CP. There are more than 100 genes associated with CP [40]. Whole exome sequencing indicates a genetic cause in 9%–26% of unselected individuals with CP [40–43], in 14%–58% with cryptogenic CP (no risk factors) [40,41,44,45], and in 8%–15% with CP risk factors [40,41]. The genetic abnormality may result in CP by leading to structural brain change. As examples, TUBA1A causes malformations of cortical development; COL4A1 predisposes to stroke. In other cases, the genetic change may produce an abnormality at the microscopic or molecular level sufficient to explain the motor disability. An example is CTNNB1, which causes spasticity or dystonia without structural brain changes on MRI. Genetic and metabolic etiologies may respond to specific treatments, impact prognosis, require monitoring for other medical problems, and inform family planning [46,47].

The discovery of genetic causes of CP has raised the question of which individuals warrant genetic testing (see chapters: Movement Disorders and Genetic Mimics of Cerebral Palsy; Neurometabolic disorders; Neurodegenerative disorders). Additionally, the current definition of CP does not address the diagnostic challenges that have presented themselves in the genomic era. It is not always clear if a genetic condition results in a "nonprogressive" disturbance in the brain. In these situations, if the clinical picture is not one of neurodegeneration, then the CP diagnosis is appropriate.

The following are important etiological examples of CP:

Prenatal

Infection—Intrauterine infections are important risk factors for and causes of CP. Epidemiologic data indicate that maternal genitourinary tract and respiratory infections as well as intra-amniotic infections (chorioamnionitis) increase the risk of CP [48,49]. Some infectious agents such as cytomegalovirus or Zika virus may directly cause malformations of cortical development, ischemia, hemorrhage, and calcifications resulting in CP. One study identified CMV DNA in the newborn screening cards of 9% of children with CP, perhaps suggesting a greater role for this infection than previously recognized [50].

Malformations of cortical development (MCD)—Approximately 10% of children with CP have MCDs, a severely impacted group. These children are more likely to have GMFCS level IV–V involvement (45%), moderate to profound intellectual disability (52%), and epilepsy (55%) than children with CP who do not have MCDs [51]. However, children with schizencephaly often have spastic hemiplegic CP, GMFCS levels I–II [52]. MCDs such as lissencephaly and cobblestone malformation are often genetic in etiology (yield of genetic testing 75%–80%). Symmetric bilateral malformations suggest a genetic etiology [53]. Other MCDs such as polymicrogyria and microcephaly are frequently infectious in origin (yield of genetic testing is 20%) [54].

Perinatal

Periventricular leukomalacia (PVL)—PVL is a white matter injury of premature infants due to susceptibility of pre-oligodendrocytes to ischemia and inflammation. It occurs in up to 50% of VLBW infants and is the most common cause of CP in prematurely-born individuals [55]. Children with CP due to PVL often have spastic diplegia because the corticospinal tracts to the legs are more impacted; however, other distributions and dystonia can occur [56]. PVL can also occur in term infants, a group comprising about 7% of children with CP [57]. The mechanism may be similar to that of PVL in the preterm infant, occurring in utero in the late second or early third trimester, although in some cases the injury may have occurred perinatally [58].

Periventricular hemorrhagic infarction—Preterm birth increases the risk for intraventricular hemorrhage. In 10%–15% of such cases, blood in the ventricle causes venous infarction of the periventricular white matter, which results in a porencephalic cyst. This complication is most likely to occur in extremely premature infants and results in CP when the motor fibers are involved [59].

Perinatal stroke—Stroke occurs in about 1/1100 live births and is the most common cause of hemiplegic CP. Of individuals with CP born at term, about 22% have stroke as the etiology [60]. Motor disability occurs in 30%–60% of individuals with neonatal stroke, although complete MCA territory stroke is 100% predictive of CP [61].

Hypoxic-ischemic encephalopathy (HIE)—Neonatal encephalopathy is a broad term to describe the neonate with depressed neurologic function. HIE is one cause of neonatal encephalopathy. Overall, about 13% of children with HIE go on to develop CP [62]. Therapeutic hypothermia decreases that risk (pooled risk ratio 0.63; 95% CI = 0.50, 0.78) [63].

Postnatal

Kernicterus—Kernicterus (chronic bilirubin encephalopathy) is a preventable neurologic injury due to neonatal hyperbilirubinemia, in the classic form characterized by the tetrad of dyskinetic CP, impaired upgaze due to oculomotor apraxia, sensorineural hearing loss, and dysplasia and staining of the baby teeth [64]. Kernicterus has become exceedingly rare in developed countries but is still a common cause of dyskinetic CP in the developing world [60,65].

Infections—Central nervous system infection (meningitis/encephalitis) is an important cause of CP. Approximately 10% of infants with neonatal meningitis develop CP. Sepsis and other systemic infections also increase risk of CP in term and preterm infants, possibly through a mechanism of white matter injury [66].

Traumatic brain injury (TBI)—TBI is a leading cause of pediatric mortality and disability. In young children, abusive head trauma is an important condition that results in neurologic disability in 65% and CP in 45% of individuals [67,68].

Classification

There are several classification systems to describe CP, reflecting the heterogeneity of causes, distributions, types of motor involvement, and severity. Classification aids in determining cause, coordinating care, monitoring comorbidities, selecting treatments, and predicting long-term outcomes. One approach is based on the type of neuromotor involvement: spastic or extrapyramidal.

Spasticity is characterized by a velocity-dependent increase in muscle tone with exaggerated deep tendon reflexes resulting from hyperexcitability of the stretch reflex [69]. Presynaptic inhibition is reduced due to the inability to synthesize or transport gamma-aminobutyric acid (GABA) to the anterior horn cell. Upper motor neuron (UMN) lesions cause spasticity primarily due to the loss of this inhibitory control. UMN lesions are heterogeneous in terms of type, location, and extent, and the subsequent spinal network adaptations vary considerably among individuals. Thus spasticity is not caused by a single mechanism but by an intricate chain of alterations in different interdependent networks [70]. The modified Ashworth Scale is the most widely used clinical tool to "objectively" measure muscle tone and to evaluate the efficacy of interventions for hypertonicity [71]. It relies on the examiner's "feel" for the resistance of a muscle during passive range of motion. The Tardieu Scale and the Modified Tardieu Scale are similar clinical measures of spastic muscle tone accomplished by assessing

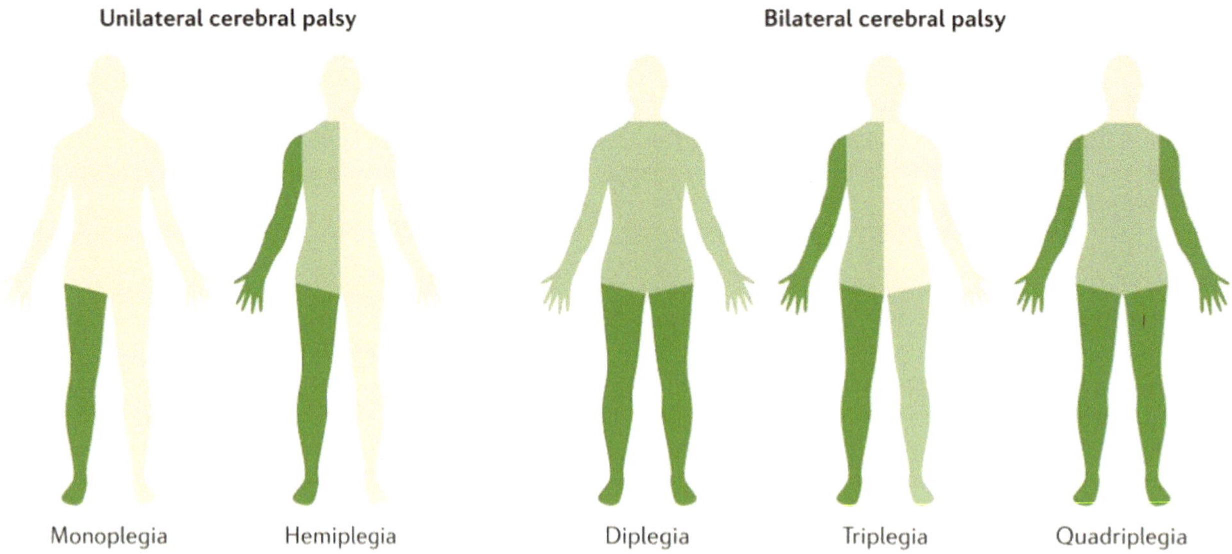

FIGURE 15.1 Topographical description in spastic cerebral palsy. Spastic cerebral palsy accounts for 70%–80% of cases and is due to an injury of the pyramidal tracts affecting voluntary movement [74]. The neurophysiology of spasticity is due to a chronic reduction in presynaptic inhibition and hyperexcitability of motor neurons through alterations in motor neuron membrane and synaptic input. Monoplegia and hemiplegia affect one side of the body; in monoplegia one limb is affected whereas in hemiplegia both arm and leg on one side are affected. Hemiplegia can be asymmetric affecting the arm or leg greater than the other extremity. Diplegia, triplegia and quadriplegia affect both sides of the body. In diplegia, the predominant picture is involvement of the lower extremities. However, the arms can be affected though not to the same degree. In triplegia, both lower extremities and one arm are affected. A common picture is diplegia from periventricular leukomalacia and a hemiplegia from an intraventricular hemorrhage. This results in one lower extremity being more severely affected due to a dual mechanism of injury. In quadriplegia, all extremities are involved.

the muscle's response to stretch applied at given velocities. The Tardieu Scale has gained favor because it may be more reliable than the modified Ashworth Scale [72]. Computerized gait analysis has become a standard component of the evaluation of gait deviations in children with CP and aids in determining the best therapeutic approach by providing quantification of gait characteristics [73].

Spastic (pyramidal) CP is described topographically based on which limbs are affected (Fig. 15.1). When used with a measure of severity, the Gross Motor Function Classification System (GMFCS), the combination provides an excellent description of where and to what extent a child is affected by CP:

- **Monoplegia:** only one limb is affected; the least common type, sometimes mistakenly applied in infancy to one arm before leg involvement also becomes apparent (i.e., a hemiplegia).
- **Diplegia:** all four limbs are involved but the legs to a markedly greater degree than the arms.
- **Hemiplegia:** the arm and leg on one side of the body are involved. The arm typically is more affected than the leg, although this can vary.
- **Paraplegia:** the lower half of the body is affected; the arms are normal. Although almost always indicative of a spinal cord insult, paraplegic CP does rarely occur.
- **Triplegia:** all four limbs are affected, but one is dramatically spared relative to the other three. Most commonly both legs and an arm are affected, suggesting spastic hemiplegia superimposed on spastic diplegia. Much less often one leg is spared (cf: *My Left Foot* by Christy Brown).
- **Quadriplegia**: all four limbs are involved. It is the most disabling form of spastic CP.

In extrapyramidal CP, the brain damage spares the pyramidal tracts but results in other tone abnormalities, movement disorders, coordination deficits, and balance problems. All four extremities are involved with the face and the arms usually more affected than the legs. Tone abnormalities (hypotonia, lead pipe rigidity, cogwheel rigidity) and extrapyramidal movement patterns (described below) result from an imbalance between the direct and indirect striatal output systems and impaired cerebellar motor control [75]. Abnormal movements may increase with stress, excitement, or purposeful activity and typically disappear during sleep. Children with extrapyramidal CP not uncommonly have normal intelligence, but their abilities can be underestimated due to the severity of their motor and communication deficits [76]. Scales designed to classify movement disorders are based on body functions and structures. The Burke-Fahn-Marsden Dystonia Rating Scale is a 120-point rating scale used to describe the severity of dystonia in nine body regions. The scale takes into account severity and frequency of the dystonic movements. The Barry-Albright Dystonia Scale was developed to address the concerns of evaluating secondary dystonias more commonly seen in children [77]. It is the most commonly reported scale for CP and easiest scale to use clinically. The Dyskinesia Impairment Scale is the only tool that considers both dystonia and choreoathetosis in CP.

- **Ataxic:** features involve lurching, bobbing, imprecise and dysmetric movements. Ataxic CP is rare and should suggest a genetic etiology.
- **Dystonic:** involuntary, sustained muscle contraction resulting in twisting and repetitive movements or postures.
- **Choreoathetoid:** a combination of chorea (rapid, irregular, unpredictable jerks) and athetosis (slow twisting, writhing movements).
- **Hypotonic:** markedly low muscle tone resulting in significant functional impairment. There is controversy as to whether it should be included as CP diagnostically despite fulfilling all the criteria. Hypotonia is a common feature in preterm infants and also in a number of syndromes or diagnoses that have historically been excluded from the CP group, such as Down syndrome, Prader-Willi syndrome, and global developmental delay. In the Surveillance of CP in Europe (SCPE) decision tree, children with generalized hypotonia are excluded under the umbrella of CP. In Australia, the most recent CP register data show 1.9% with hypotonic CP [78]. Levy et al. did a descriptive review of cases of ataxic and hypotonic CP in the Canadian Cerebral Palsy Registry [79]. They exhibited more in the way of prenatal features (intrauterine growth restriction (IUGR), malformations), and their MRIs were more likely to be either normal or demonstrate a CNS malformation.

There are issues and controversies in applying the diagnoses of ataxic and hypotonic CP, the least frequent subtypes with prevalence estimates <5%. The authors feel that individuals with central hypotonia fulfilling diagnostic criteria for CP should in fact be diagnosed as such [80]. A mindset to be cautious, acknowledge uncertainty, and consider alternative diagnoses is in order in both instances.

The classification of mixed CP is used when the child displays a combination of spastic and extrapyramidal features. Mixed CP often gets left out of epidemiological studies when "predominant motor type" is reported, but co-occurrence of spasticity and dystonia is quite common, as revealed by careful motor phenotyping [81].

When classified according to the predominant neuromotor feature, spastic CP accounts for 80%–95% of cases, while extrapyramidal makes up 5%–20% (4%–17% dyskinetic, 5% ataxic, and 2% hypotonic) [81]. Mixed CP increases as motor function worsens, regardless of which neuromotor subtype predominates. The frequency of hypotonic CP is not clear because of inconsistent classification practices, in particular automatic exclusion of the diagnosis. Many infants who are initially hypotonic progress to other CP subtypes [76]. Hypotonia can be seen in minor neuromotor disorders, not classified as CP. See Chapter Developmental Coordination Disorder and Minor Neuromotor Disabilities for further discussion regarding the spectrum of minor neuromotor disorders.

Historically, CP has also been classified based on degree of involvement—mild, moderate, and severe—without specified criteria for each group. The GMFCS was developed to categorize severity based on abilities and limitations in motor functioning [82,83] (Fig. 15.2). It consists of five levels that are applied to five consecutive age bands and describes functional changes over time. The functions in each level focuses on usual, rather than best, motor performance in a variety of settings: home, school, community. The intent was to improve communication related to treatment decisions and research into treatment outcomes, provide a more precise developmental picture of the child with CP, and anticipate motor progress and needs overtime [82] (See Fig. 15.3).

Similar tools have been designed for evaluating upper extremity function—the Manual Ability Classification System (MACS)—and for speech/communication—the Communication Function Classification System (CFCS). The MACS describes hand use in daily activities [84] and is divided into five levels based on the child's typical performance, not maximal capacity. Knowing a child's MACS level can help parents, teachers, and others to understand situations in which a child is independent or in need of support or adaptations. The CFCS classifies the everyday communication of an individual with CP into one of five levels according to effectiveness of communication [85].

Diagnosis

General approach

The diagnosis of NDDs in general is made at two levels: first identification of the clinical problems (CP, intellectual disability, autism, etc.) and, secondly, determination of the underlying etiology. The diagnosis of CP is made on clinical grounds by establishing that a motor disability is present, the motor problem is localized to the brain, and the course is

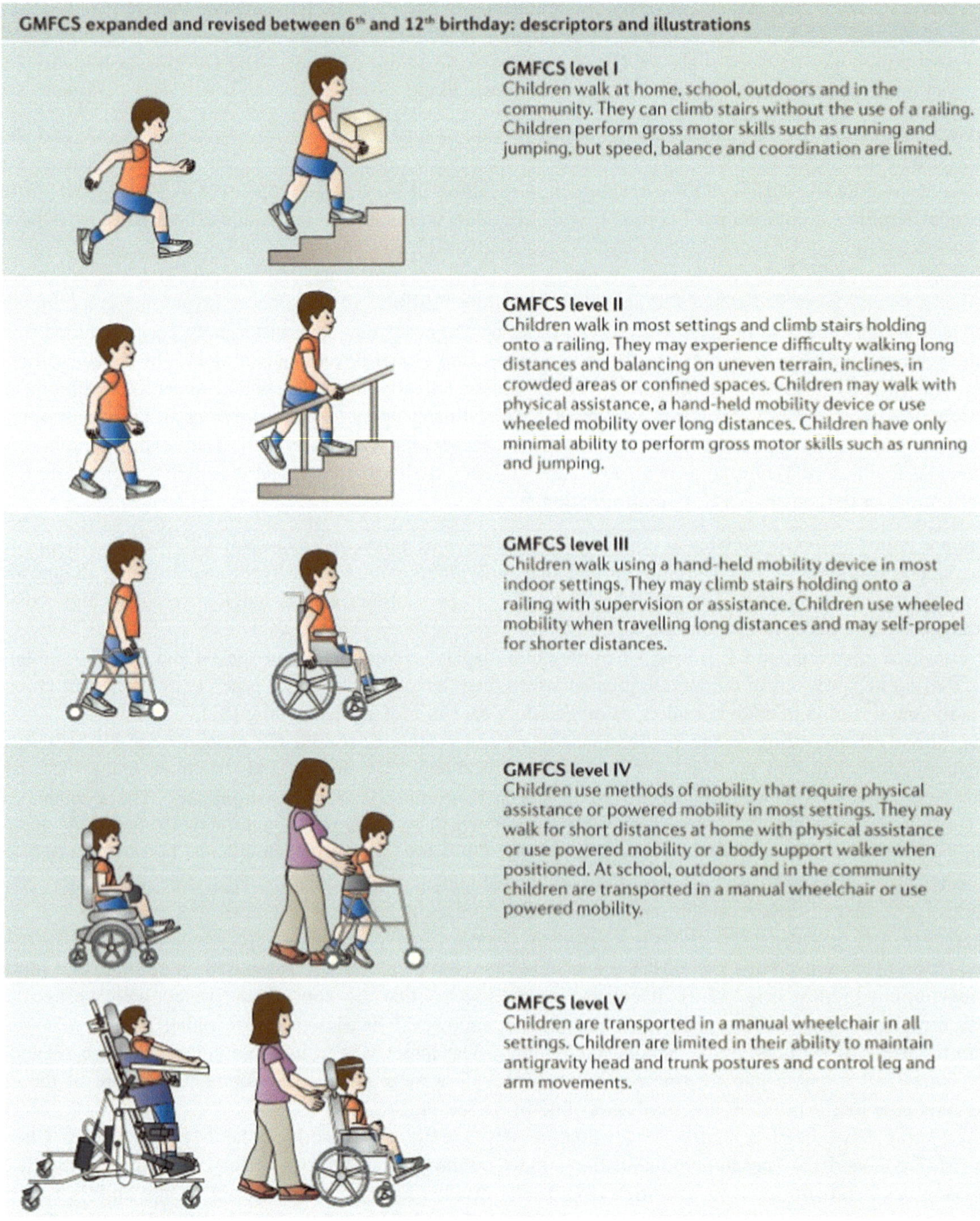

FIGURE 15.2 Gross motor function classification system (GMFCS) expanded and revised for children with cerebral palsy, 6–12 years of age. The GMFCS uses a five-level rating system that corresponds to ability and impairment of daily motor functioning based on self-initiated movement with emphasis on sitting, walking and wheeled mobility. Shown above are for ages 6–12 with an additional age band for youth 12–18 years of age reflective of environmental and personal factors on mobility methods at those ages. The GMFCS-E&R descriptors were devised by Palisano et al. [82,83]. *Courtesy: Reid B, Harvery A, Graham HK. The royal children's hospital, Melbourne, Victoria, Australia.*

not progressive. A child will typically present with a concern about motor development: delay, tone abnormality, weakness, abnormal movements, abnormal gait, asymmetry. Historical features, such as a history of prematurity or head injury, increase the level of suspicion. The developmental history helps establish whether motor development is significantly delayed and/or disordered and whether the developmental trajectory is consistent with a nonprogressive,

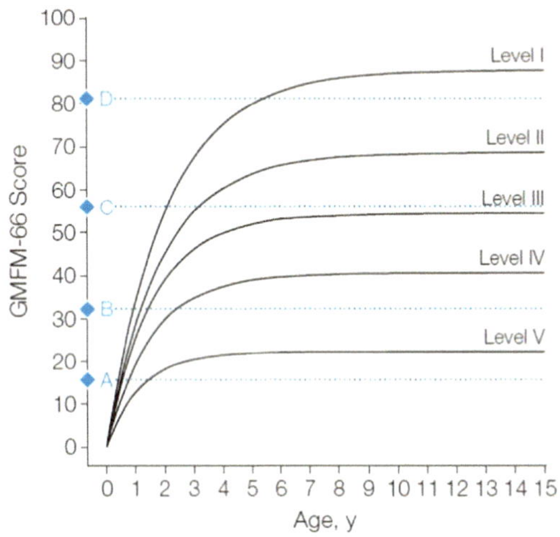

FIGURE 15.3 Predicted average development by the gross motor function classification system levels. The diamonds on the vertical axis identify 4 Gross Motor Function Measure-66 (GMFM-66) items that predict when a child is expected to have a 50% chance of completing that item successfully. Diamond A—whether a child can lift and maintain vertical head position with trunk supported while sitting. Diamond B—whether when in a seated position a child can maintain sitting unsupported by his/her arms for 3 s. Diamond C—measures a child's ability to walk forward 10 steps unsupported. Diamond D—ability to walk down 4 steps alternating feet with arms free. *Courtesy: Rosenbaum PL, Walter SD, Hanna SE, Palisano RJ, Russell DJ, Raina P, et al. Prognosis for gross motor function in cerebral palsy: creation of motor development curves. JAMA 2002;288(11):1357–1363. Available from: https://jamanetwork.com/.*

neurologic disorder. In addition, it alerts one to significant comorbid neurodevelopmental conditions. The physical and neurological examinations establish the presence of neuromotor abnormalities attributable to a cerebral disorder. The practitioner should be reasonably satisfied that the child is not manifesting a neurodegenerative process. Once the diagnosis is made or strongly suspected, it must be compassionately and accurately relayed to the family. An attempt at classification is made, although this may not be entirely clear at younger ages. The evaluation for the underlying etiology is undertaken, typically starting with MRI of the brain [86]. Although an MRI with abnormalities consistent with CP supports the diagnosis and increases the level of certainty, the diagnosis remains a clinical one (i.e., *the practitioner makes the diagnosis, not the MRI*). Then the child is evaluated for associated problems. Referrals for rehabilitation services are placed, and resources are provided to the family.

Timing of diagnosis

Early diagnosis of CP is desirable in order to initiate CP-specific therapies and to provide a definitive answer for families. A diagnosis of CP or "high-risk for CP" can be made most of the time under 1 year and even under 6 months of age [21]. Standardized assessments are sometimes incorporated into the evaluation of the high-risk infant and may aid in diagnosis:

- Prechtl's General Movement Assessment (GMA): an observational exam which assesses the movement patterns in young infants.
- Hammersmith Infant Neurological Examination (HINE): a standardized neuromotor examination.
- Developmental Assessment of Young Children (DAYC): an evaluation of motor, cognitive, and adaptive milestones in infants and preschool-aged children.

Some examples of how children with CP present and are diagnosed follow:

Hemiplegic cerebral palsy

Hemiplegia most often becomes evident around 4–6 months of age, when parents may note that the infant has a hand preference when reaching or maintains fisting on one side. There may be decreased spontaneous movement and/or crossing over the midline to grab an object. On examination postural responses may demonstrate asymmetry, as with the parachute reaction. There may be noticeably increased tone and hyperreflexia. Often leg involvement is so subtle, especially prior to standing and ambulation, that a monoplegia is suspected. If the leg is affected more than the arm, asymmetric crawling may be an early tip-off, or the asymmetry may not come to attention until after a year of age when the child is noted to be toe-walking on one side. An ipsilateral visual field defect may be demonstrable. The clinical findings plus confirmatory evidence on neuroimaging establish the diagnosis of CP.

Diplegic cerebral palsy

The child with spastic diplegia may begin to develop increased tone in the legs after the first few months of life and may show more stereotyped, less variable leg movements. In later infancy scissoring is often present, and parents may note difficulty with diaper changes. The child may "log roll" rather exhibit derotational/segmental rolling. Instead of reciprocal crawling, the child will usually drag his or her legs in extension. The mildest cases may go unappreciated until the child is up walking. On examination spasticity, extensor posturing, hyperreflexia and decreased variability of movements are noted. Upper extremity involvement is typically not apparent at this age. A history of prematurity and lack of progressive symptoms raises the suspicion for CP, as opposed to other disorders in the differential such as leukodystrophies. If no pathology is demonstrated on MRI scan, the possibility of hereditary spastic paraplegia (HSP) needs to be considered, and additional diagnostic work-up is warranted. See chapter: Movement disorders and genetic mimics of cerebral palsy.

Quadriplegic cerebral palsy

The child with quadriplegic involvement may have early feeding difficulties, gastroesophageal reflux, and irritability. Low trunk tone in the face of extremity hypertonicity and global developmental delays may be apparent early, and parents may comment on exaggerated "startle" responses. Visual deficits and/or strabismus, microcephaly, dystonic mouth opening, and poor head control may be present on examination. Primitive reflexes will persist beyond the expected time and may be obligatory, and postural reactions will be absent or delayed in appearance. Tone will gradually increase over the first year of life. Signs of dystonia or a movement disorder may start to appear after 18 months of age.

Differential diagnosis

There are a number of disorders that share clinical characteristics of CP but do not satisfy all aspects of the definition. The following are examples of disorders in the differential diagnosis of the child with significant neuromotor abnormality:

Hereditary spastic paraplegia

HSP is due to length-dependent degeneration of the corticospinal tracts, predominantly affecting the spinal cord. The "pure" form is characterized by lower extremity spasticity and weakness. In "complex" forms, there may be arm and bulbar involvement and other neurologic features [87] such as seizures, intellectual disability, and peripheral neuropathy. More than 80 genetic types have been identified [88]. Most forms of HSP have a variably progressive course. Distinguishing early-onset HSP from CP is challenging. HSP typically presents at a later age than CP with a decline after a period of normal motor development. Gait analysis may be helpful [89,90]. Neuroimaging may show decreased size of the spinal cord at the cervical and thoracic levels in adults [91]. Genetic testing establishes the diagnosis in 30%–50% of cases [87]. While treatment differs only somewhat from CP, prognosis is substantially different [See chapter: Movement disorders and genetic mimics of cerebral palsy].

Neurodegenerative disorders

The developmental trajectory of neurodegenerative disorders is typically a period of normal or near-normal development followed by a slowing of developmental progress, then plateau for a period of time, and finally functional deterioration. They can mimic CP, usually declaring at the stage of functional decline. Common mimics include leukodystrophies, Rett syndrome, CNS tumors, mitochondrial disorders, ataxia-telangiectasia, neurodegeneration with brain iron accumulation disorders [See chapter: Neurodegenerative disorders].

Spinal cord disorders

Spinal cord disorders are excluded from the definition of CP. A spinal cord disorder would be suspected in spastic paraplegia or spastic quadriplegia with complete absence of bulbar signs or other brain-related symptoms (e.g., microcephaly, epilepsy, intellectual disability). Tethering of the spinal cord may present with ankle tightness and can mimic CP, or it can be an associated problem with CP. Spinal cord tether typically does not cause proximal lower extremity spasticity [See chapter: Spinal cord disorders].

Developmental coordination disorder

A number of descriptions have been used to describe "the clumsy child" [92], including dyspraxia, minimal brain (cerebral) dysfunction and DCD [93]. The preferred term is now DCD, a common disorder affecting 5%–6% of children and characterized by significant motor skill deficits and difficulty with coordination [94]. The current consensus definition is:

1. Motor performance that is substantially below expected levels given the child's chronological age and appropriate opportunities for skill acquisition.

2. The disturbance significantly interferes with activities of daily living or academic achievement.
3. The motor impairment is not solely explainable by intellectual disability.
4. The disturbance cannot be explained by any specific congenital or acquired neurological disorder or a severe psychosocial problem.

While considered a distinct category, DCD is essentially the mildest end of the CP spectrum [44]. As with CP, children with DCD frequently exhibit other neurodevelopmental problems: language impairments, learning disabilities, and ADHD. The motor deficits are central in origin, supported by the presence of associated neurological deficits and by research demonstrating microstructural white matter alterations on neuroimaging [95]. The boundaries of the definition of DCD are likely to be refined over time. Etiology has not been adequately addressed: genetic associations, preterm birth, and environmental factors are given only cursory mention [94]. Excluding neurological causes from the definition is debatable: children with syndromes or acquired brain injuries can have neuromotor residua—more obvious in the form of CP or milder in the form of DCD [See chapter: Developmental coordination disorder and minor neuromotor disabilities].

Neuromuscular disorders

Peripheral neuromuscular disorders may mimic CP in some circumstances. Monoplegic CP may be misdiagnosed in unrecognized peripheral nerve or brachial plexus injury. However, brachial plexus injury should present at birth with hyporeflexia and weakness and should never have findings of increased tone, in contrast to CP. Neuromuscular disorders such as myopathies or spinal muscular atrophy may mimic hypotonic CP in infancy. The key distinctions are areflexia/hyporeflexia and true weakness. As a rule cognitive and language development are normal, although in certain disorders (e.g., Duchenne muscular dystrophy) neurocognitive deficits are frequently exhibited [96] (See chapter: Neuromuscular disorders).

Structural disorders

Occasionally structural musculoskeletal disorders, such as connective tissue disorders and ligamentous laxity, may initially be mistaken for CP. A child with joint hypermobility may be perceived to be on the spectrum of hypotonic CP. Joint hypermobility is a common finding in childhood and may be benign. It may also be associated with genetic etiologies such as the Ehlers-Danlos syndromes, Marfan syndrome and skeletal dysplasias [97]. The infant with decreased mobility and irritability due to unrecognized fractures in osteogenesis imperfecta can be misdiagnosed as having a central cause for the symptoms.

Etiologic work-up

The etiologic underpinnings of a case of CP should be viewed as a causal chain, the entirety of which warrants full explanation. The chain should be revisited periodically, as medical knowledge evolves, if any link remains unspecified. For example, for a child with spastic diplegic CP and a history of prematurity, MRI findings consistent with PVL would offer a satisfactory explanation for the diagnosis. If that child were born at term, the same white matter changes might prompt a genetic or metabolic evaluation. For a child with hemiplegic CP brain, MRI revealing a porencephalic cyst would explain the clinical features. However, genetic testing may reveal the cause of the porencephaly to be a mutation in COL4A1. (Fig. 15.4) As a caution, while understanding of causal pathways to CP is advancing, parents and practitioners should not expect to find a single cause in every case. Teasing out risk factors, associations, and causes can be confusing and challenging [98].

Imaging

Once the clinical diagnosis of CP or highly suspected CP is established, brain MRI scan is indicated [86,99]. It will be abnormal in about 90% of children with CP [100]. In a child with neonatal imaging that explains the diagnosis, such as ultrasound showing periventricular echodensities, repeat imaging may not be necessary. Timing of imaging is pertinent: an early scan can be helpful in establishing etiology, however the risks of sedation are slightly higher in a younger child, and subtle findings on imaging may not become clear until the brain becomes more fully myelinated after 2 years of age [99]. Computed tomography of the head is less desirable because of lower sensitivity and exposure to ionizing radiation [86]. In children with spastic diplegia without a history of prematurity or with quadriplegia but absence of bulbar and cerebral cortical involvement, an MRI of the spine should be considered. See chapter: Neuroimaging in neurodevelopmental disabilities.

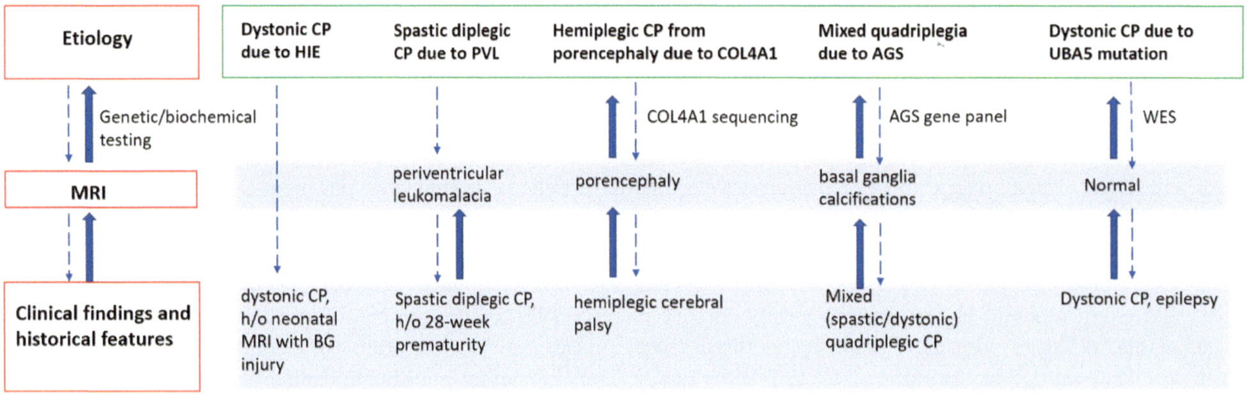

FIGURE 15.4 Cerebral palsy causal pathway. Example cases of cerebral palsy, clinical presentations, and etiologies. The bottom row describes the clinical presentation, middle row the MRI findings, and top row the etiology. For example, a child with dystonic CP with evidence of basal ganglia injury on neonatal MRI has an established causal pathway with no further work-up needed (far left). A child with dystonic cerebral palsy and epilepsy with no risk factors who has a normal MRI, may then undergo whole exome sequencing uncovering a UBA5 mutation which explains the clinical picture (far right). *AGS*, Aicardi-Goutières syndrome; *BG*, basal ganglia; *COL4A1*, collagen type IV, alpha-1 subunit; *CP*, cerebral palsy; *HIE*, hypoxic-ischemic encephalopathy; *h/o*, history of; *MRI*, magnetic resonance imaging; *PVL*, periventricular leukomalacia; *UBA5*, ubiquitin like modifier activating enzyme 5; *WES*, whole exome sequencing.

Genetic/Metabolic testing

Genetic and metabolic testing have become increasingly incorporated into the etiologic evaluation of the child with CP. Genetic causes underlie 14%–33% of CP cases based on studies of broad testing [42,101]. Genetic and metabolic disorders may respond to specific treatments [46,47], impact prognosis and monitoring for other medical problems, and inform family planning. One recommended approach is to pursue genetic testing in individuals with suspected CP who have the following:

- Lack of perinatal risk factors.
- Mismatch between history and clinical presentation.
- Consanguinity or family history of CP.
- Possible degenerative course.
- Congenital anomalies or dysmorphic features.
- Normal MRI or MRI not matching history or clinical presentation [41].

Testing to be pursued depends on the clinical features, availability, and cost. It may include biochemical testing for conditions such as peroxisomal disorders and mitochondrial disorders, various gene panels, or whole exome/genome sequencing. Involvement of a geneticist can be helpful (Fig. 15.5). The practitioner ordering the genetic test must have the expertise to counsel families on the risk, benefits, and possible outcomes of genetic testing, and be able to provide or refer for genetic counseling once results are received.

Associated problems

While motor impairment is the core deficit in CP, most individuals with CP have additional neurological, developmental, medical, and/or behavioral manifestations. [102] These comorbidities can be considered in terms of their causes [103]:

1. Other neurologic features of the same brain injury that caused CP (termed "co-causal" disorders).
2. Secondary disorders representing sequelae of CP (termed "complications").
3. Disorders occurring independently of the causes and complications of CP (termed "co-occurring" disorders).

We focus the discussion here on cocausal disorders and complications. Causal pathways can be complex for some coexisting disorders (e.g., some mental health diagnoses likely have co-causal roots as well as contributions secondary to CP-related disability) [102] (See Table 15.2 associated problems).

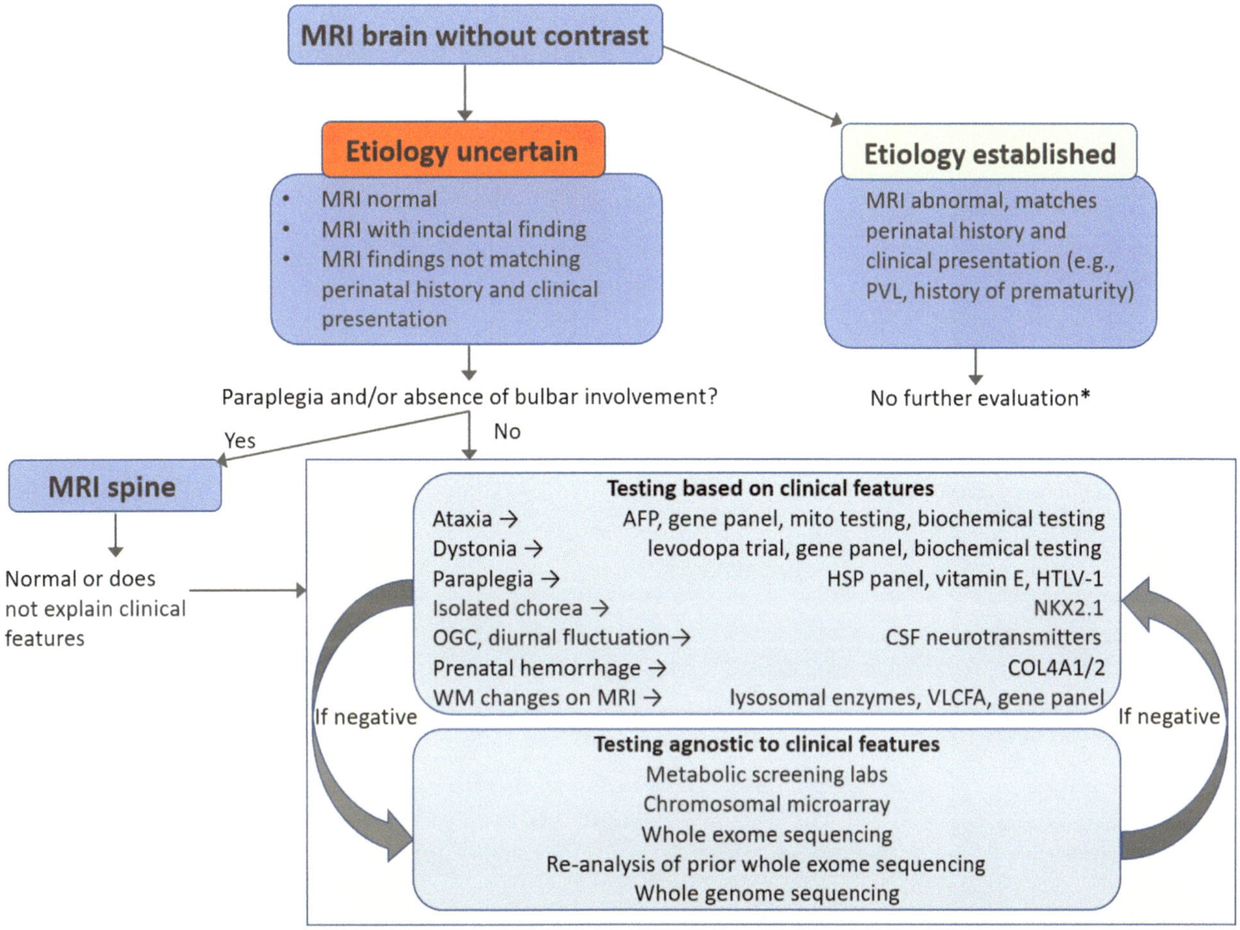

FIGURE 15.5 Suggested evaluation for the etiology of cerebral palsy. The evaluation for the cause of cerebral palsy starts with an MRI of the brain. If the MRI result is normal or the findings do not fit the perinatal history and clinical picture, genetic testing is indicated. This testing may be directed by the clinical features, or through broad testing (e.g., whole exome sequencing). Metabolic/biochemical testing may include serum amino acids, urine organic acids, serum lactate, serum ammonia, serum uric acid, serum biotinidase, serum lactate, urine purines/pyrimidines, urine creatinine, GAA, and serum acylcarnitines. *If the imaging findings explain the clinical findings and fit the perinatal history, genetic testing is probably not indicated. However, one study demonstrated that a small number of patients with risk factors for cerebral palsy still had a genetic cause. The presence of risk factors for cerebral palsy may not decrease the likelihood of an underlying genetic cause [40]. *AFP*, Alpha fetoprotein; *COL4A1/2*, collagen type IV alpha-1 and 2 Chain; *CSF*, cerebrospinal fluid; *mito*, mitochondrial; *HTLV-1*, human T-cell lymphotropic virus type 1; *NKX2.1*, NK2 homeobox 1; *OGC*, oculogyric crisis; *PVL*, periventricular leukomalacia; *VLCFA*, very long chain fatty acids; *WM*, white matter.

Co-causal neurological impairments

In a Norwegian CP registry 61% of individuals carried a coexisting neurological diagnosis [102]. CP subtypes (and their underlying brain injury patterns) are associated with patterns of cognitive and other neurological deficits, although there is substantial interindividual variability.

Seizures and epilepsy

35% of individuals with CP have been diagnosed with epilepsy. The long-term trajectory of epilepsy in individuals with CP varies substantially. While seizures can be severe and refractory in some, a majority of children with nonsyndromic CP and epilepsy appear to achieve seizure freedom by adolescence [104]. Epilepsy is more common in individuals with brain malformations and individuals with severe gray matter injury; its occurrence is linked to more severe cognitive and physical disability [105].

TABLE 15.2 Associated problems.

Orthopedic deformities – Muscle/tendon contracture – Bone deformities/malalignments – Joint dislocation/degeneration – Scoliosis – Osteoporosis and fracture
Cognitive deficits – Intellectual disability – Learning disability
Sensory deficits – Sensation loss – Visual impairment – Oculomotor disturbance – Hearing loss: sensorineural and/or conductive (recurrent otitis)
Oral-motor performance impairments – Speech deficits (dysarthria) – Feeding and swallowing dysfunction (dysphagia) – Drooling – Aspiration
Poor growth/Undernutrition
Pain Gastroesophageal reflux Bowel and bladder dysfunction
Seizures Cervical myelopathy/Radiculopathy Lumbar myelopathy/Radiculopathy Autonomic dysfunction Sleep disturbances
Behavioral/Emotional disturbances – ADHD, autism, depression, anxiety

Cognitive, language, and learning disabilities

About 49% of individuals with CP have intellectual disability; 28% have a full scale IQ less than 50 [106]. Prevalence and severity of intellectual disability correlate with the degree of gross motor impairment although correlation is imperfect even at the ends of the spectrum (13% of individuals with GMFCS I-II have intellectual disability, and 21% of individuals with GMFCS V have typical intelligence [107]). The degree of gross motor impact (rather than physiologic subtype) appears to be the better predictor of cognitive functioning [108].

Inter-domain discrepancies are frequently seen in the cognitive profiles of individuals with CP, for example, verbal versus performance IQ. However, there is no CP profile. On standardized cognitive batteries, studies of individuals with CP frequently demonstrate lower performance than verbal IQ scores, particularly in individuals with bilateral spastic CP and those born preterm [108]. Between preschool-age and school-age testing, verbal IQ may further increase [109], and performance IQ may decline [110], thereby widening the discrepancy. Of note, performance IQ subscores may be influenced by fine motor, attention, and visual deficits [111]. Among subdomains visual-spatial skills may be the most frequently (~90%) impaired in individuals with bilateral CP [112].

Receptive and expressive language skills may also be discrepant. Receptive language skills are closely associated with general cognitive ability (and, neurobiologically, with preservation of white matter [113]). While expressive communication is most closely related to the type and severity of motor disorder [114], verbal expression may be impacted by severe dysarthria. The distinction needs to be made between children with poor expressive language as opposed to those with dysarthria. While both are strongly associated with poor gross motor function, the former predominantly have moderate-to-severe intellectual disability, while the latter predominantly have typical cognition or mild intellectual disability [115]. Non-speaking children with age-appropriate comprehension skills often have dyskinetic rather than spastic CP [113].

Poor academic achievement due to specific learning disabilities is not uncommon. In an Australian study, approximately 30% of school-aged children with CP scored within the typical range for reading, and approximately 30% scored within the typical range for mathematics skills [116]. Using a discrepancy formula, Frampton et al. estimated that 19% of children with hemiplegia have a specific reading disability and 25% have a specific mathematics disability [117]. Global intellectual deficits as well as more specific working memory/executive function deficits have been associated with both reading and mathematics skill deficiencies in children with CP, and reading and mathematics achievement share substantial variance (40% in Jenks et al.'s cohort) [118]. Most children with CP appear to have mild-to-moderate deficits in executive function (reviewed in Craig et al. [119] and Fluss and Lidzba [111]). When prospectively employing standardized clinical assessments, one population-based study estimated the rate of comorbidity with ADHD to be 30%. Prevalence was higher among children with intellectual disability or autism [78]. Impairments in pragmatic communication skills have been reported in children with CP [120], and estimates of autism spectrum disorder comorbidity range from 2% to 30% [119]. Påhlman et al. found 30% of school-aged children with CP in western Sweden had autism, and prevalence of autism was higher in children with intellectual disability, ADHD, or a history of preterm birth [121].

Visual impairment

About 11% of individuals with CP are "functionally blind," and one in four have some degree of visual impairment [106]. Among children with CP and visual impairment, ocular refraction error, altered eye movements, ocular misalignment (strabismus), and reduced visual acuity occur across all subtypes of CP and frequently co-occur, although children with quadriplegic CP have the most frequent and severe visual and oculomotor impairments [122]. Visual field limitations (whether a dense field cut or a less-defined deficit) appear to be common in children with hemiplegia and diplegia; they may be difficult to assess in children with quadriplegia [122]. While structural ocular etiologies can occur, visual impairments in individuals with CP are frequently neurogenic (termed cerebral visual impairment or cortical visual impairment; CVI). CVI should be distinguished from ocular visual impairment as CVI presents specific challenges (e.g., particularly impacting perception of faces, symbols, or moving objects) that can be addressed using specific accommodations (e.g., judicious use of color, white space, and lines for guidance) [123].

Hearing loss

About 4% of children with CP have a severe sensorineural hearing impairment/deafness [106]. Rates may be higher in nonambulatory children (18% in Shevell et al. [124]). Infants with severe neonatal bilirubin encephalopathy requiring exchange transfusion in particular may go on to have severe hearing loss [125]. Conductive hearing loss appears to be uncommon in children with CP [126].

Dysautonomia

Analogous to the clinical entity "paroxysmal sympathetic hyperactivity" defined in individuals with acquired brain injury [127], similar episodic irritability with associated extreme changes in vital signs may be seen in some children with CP—particularly those with severe neurologic impairment [128]. Clinical manifestations of dysautonomia are myriad and not uncommon, for example, cutaneous vascular changes—cyanotic and/or cold, mottled distal extremities, especially the feet—due to disturbed sympathovagal balance [129]. Prevalence rates have not been established. Objective measures (such as assessments of heart rate variability) have suggested more widespread differences in autonomic functioning in individuals with CP [130], but studies to date lack consistent methodology, and clinical significance is uncertain.

Medical complications of cerebral palsy

In the Norwegian CP cohort 77% carried a coexisting diagnosis thought to be a complication of CP [102]. Systems most commonly affected were musculoskeletal (50% of individuals) and digestive (39% of individuals). As expected, rates of comorbidities were highest for CP subtypes with the greatest degree of motor impairment ;95% for those with spastic quadriplegic CP and 86% for those with dyskinetic CP as compared to 65% for those with spastic hemiplegia.

Musculoskeletal complications

Abnormalities of muscle tone, particularly spasticity, and asymmetries of motor involvement produce abnormal forces on the muscles, tendons, joints, and bones. Tendon contractures, joint dislocations, bony deformities of the limbs, and scoliosis are prone to develop. Although many interventions can reduce spastic hypertonus, the imbalance between agonist and antagonist muscle groups remains, so the propensity to contracture and deformity persists.

Osteopenia is common among children with CP who are nonambulatory and/or have complex medical problems. This results in bone fragility, and pathologic fractures can occur [131]. They are generally seen in the context of risk factors—poor nutrition, seizure medications, decreased sun exposure, and decreased weight-bearing. Prevention strategies include: (1) optimization of nutrition with attention to calcium intake, (2) vitamin D supplementation, (3) weight-bearing [132]. Among children who have experienced a pathologic fracture, some advocate treatment with bisphosphonates [133]. Calcium supplementation and vitamin D, standing frames and gait trainers for weight-bearing, and bisphosphonates to improve bone mineralization and decrease fracture rate have some but limited evidence to support their use [134].

Gastrointestinal/Feeding and swallowing issues

Feeding and nutrition: Inadequate nutrition is common in severely affected children. In a study from Norway 21% of children with CP were completely dependent on assistance for feeding and another 10% required some assistance [135]. Of those at GMFCS levels IV and V 45% had a gastrostomy tube, which can lead to improved growth [135] as well as longer life expectancy [136]. Severe motor impairment is also associated with a higher prevalence of overweight/obesity (16%) [135]. Worsening of cardiometabolic capacity and direct adipose impairment of muscle function appear to contribute to a cycle of obesity and functional immobility. Swallowing dysfunction is associated with chronic congestion and pulmonary aspiration (41%) [91] and chronic drooling ($\sim$10%) [137].

Gastrointestinal: Children with CP have ubiquitous gastrointestinal symptoms (92% overall, including 74% with chronic constipation, 32% with reflux and vomiting, and 32% with abdominal pain) [138,139]. Decreased mobility is thought to impact gastrointestinal motility to some degree. Gastrointestinal issues often have a major impact on the comfort and nutritional status of individuals with CP [See chapter: Dysphagia and feeding disorders].

Respiratory disorders

Respiratory diseases represent the leading cause of death for individuals with CP. Recurrent aspiration, impaired airway clearance, and thoracic spine deformity appear to contribute to respiratory compromise [140].

Sleep disorders

An estimated 23% of children with CP have a sleep disorder [106]. Reported sleep disorders include difficulties initiating and maintaining sleep, sleep-wake transition problems, sleep-disordered breathing, excessive somnolence, arousal difficulty, and hyperhidrosis [141]. Many medical (epilepsy, severe visual impairment, CP subtype) and environmental (living in a single-parent household, bed-sharing) and behavioral factors are associated with these chronic sleep problems [141]. Sleep disorders in individuals with CP can be multifactorial, challenging to address, often contribute to behavioral challenges, and may limit productive daytime activities [See chapter Sleep disorders].

Pain

Seventy percent of adults with CP report pain [142], the culmination of a steady rise in prevalence with age [143]. Acute and chronic musculoskeletal pain are well-described in individuals with CP, but neuropathic pain, visceral pain, headaches, and primary pain conditions also contribute [144]. The lower extremities are the most common sites; in particular, hip dislocation can cause substantial pain [143].

Genitourinary health

Over half of children and adults with CP have lower urinary tract symptoms (most commonly incontinence). Other frequent disorders include urinary urgency, frequency, hesitancy, and urinary tract infections [145]. Data are limited regarding reproductive health in individuals with CP. Among those who become pregnant, more than half lose a degree of mobility, highlighting the importance of having OT and PT services available during and after pregnancy. Pregnancy outcomes show the same number of miscarriages and stillbirths as in the general population but higher rates of Caesarian section and preterm birth [146]. For women with a broadly defined spectrum of disabilities, neonatal complications occur at mildly higher rates, including low birth weight, small for gestational age (SGA), neonatal morbidities, and need for neonatal intensive care unit admission [147]. Data specific to women with CP are not available, nor is there information on the ultimate neurodevelopmental outcome of the newborns.

Psychiatric and behavioral disorders

Among individuals with CP, 26% of children are reported to have "abnormal behavior" [106]; 20% have a mental health diagnosis, and 28%–35% have mental health symptoms [148]. Among adults with CP, mood and anxiety disorders are reported in 30% of women and 20% of men [148,149].

Treatment

Although CP is a lifelong disability, there are many interventions that can reduce the impact on the individual and improve their quality of life. Aims of treatment need careful discussion with the individual and the family, beginning with an evaluation of the individual's current level of functioning in order to ascertain achievable short-term goals. Treatment planning should take into account potential drawbacks of intervention; for example, muscle tone reduction (see below). Appropriate methods of measuring outcomes are necessary to determine if treatment aims are being met and whether treatments need to be altered. Common goals of CP treatment include improvement in motor function, pain reduction, ease of caregiving, improved access to adaptive technology, and prevention of secondary problems such as orthopedic deformities. Benefits include assistance with social interactions, improvement in education and learning, diminished maladaptive behaviors, and promoting independence. This chapter will focus on the primary treatment categories for motor dysfunction in CP including pharmacotherapy, habilitative therapy strategies, and surgery. See chapters: Speech and language therapy, Intellectual disabiltiy, Specific learning disability, Cognitive and behavioral therapy, Circumvention therapy, Classroom for treatment of non-motor disorders. Integration of these treatments provides the most benefit; it is not a sequential approach. CP treatment is tailored to each individual to target the specific symptoms and requires a multidisciplinary team in order to develop coordinated, comprehensive treatment plans [See chapter Medical home].

Pharmacotherapy

A detailed evaluation to determine whether and to what degree muscle tone is interfering with motor function is an important early step. In some instances a child's increased tone is actually aiding function, so reducing it would be counterproductive [150]. Enteral medications are commonly used as first line agents for generalized spasticity and dyskinesias given their ease of usage and known side effect profiles. These medications are initiated on a trial basis due to limited evidence-based guidelines [151] and variability of response among individual patients. While baclofen is generally favored, other antispastic medications such as tizandine, dantrolene, and benzodiazepines are acceptable alternatives [151–153]. Second line medications such as clonidine or gabapentin may provide dual benefit for tone management and other neurologic associations including sleep disruption, dysautonomia, pain, and neuroirritability [154,155]. There is emerging evidence that cannabinoids are beneficial for spasticity management in other neurological disorders (e.g., multiple sclerosis) and may be applicable to patients with CP [156]. However a recent randomized, controlled trial showed no benefit and some significant adverse events [157].

The treatment approach for extrapyramidal features should be chosen according to the movement phenomenology. Dystonia is said to have a greater impact on several functional domains, activity, participation, and quality of life when compared to choreoathetosis [158,159]. Most treatment modalities addressing dystonia in CP have not been studied in a controlled fashion, and treatment is still largely determined by expert opinion. Resources exist to review available evidence [159,160]. Medications used to treat dystonia include oral baclofen, benzodiazepines, trihexyphenidyl, clonidine, gabapentin, dantrolene, and tetrabenazine, but practice varies widely [159,161–163]. Several antiepileptic medications (carbamazapine, phenytoin, levetiracetam, and topiramate) can be beneficial for management of dyskinesias while allowing co-management of seizures and dysautonomia [164]. Many of these medications can be used in the acute setting of status dystonicus [165,166] with more recent interest in high-dose clonidine and intranasal options such as dexmedetomidine [167,168]. One should always search for triggers of status dystonicus such as infection, GI issues, or sources of pain (e.g., fracture).

Chemodenervation agents such as botulinum toxin (btx), phenol, and ethyl alcohol are ideal for management of focal or segmental spasticity or dystonia. [151]. Targeted injections combined with rehabilitative therapies can allow for improved motor functioning and delay (but not avoidance) of orthopedic surgery [169,170]. These injections require repeat administration, typically every 4–6 months, with timing, dose, and location dependent on needs [169]. Risks should be taken into account: individuals with a history of dysphagia or aspiration pneumonia may be at greater risk for systemic adverse events [171], and there are data demonstrating long-term structural muscle changes after injections [172,173].

Stem cell transplantation is a regenerative therapy that has the potential to replace damaged, nonfunctional cells in the brain, as well as to provide support to the remaining neurons and oligodendrocytes (by trophic and/or anti-inflammatory mechanisms) [174,175]. There are many kinds of stem cells, each with unique characteristics. In a systematic review, stem cell therapy for CP compared with traditional management modalities showed a significant positive effect on gross motor function, although the magnitude of improvement was small [176]. Stem cell therapy is an experimental technique that is not ready to be adopted as the standard of care for patients with CP as high quality randomized controlled trials (RCTs) are still lacking [See Chapter Innovative therapies].

Habilitative therapies

Physical therapy (PT) is an established treatment modality for CP and is an integral and vital part of management as it promotes range of motion, positioning, stamina, and coordination, all of which can directly impact motor function. Although widely used and recommended, the effectiveness of PT is limited, in part due to lack of guidelines for optimal timing, intensity, and duration of treatment. [177]. Intensive function-based programs can benefit CP regardless of type [178]. PT following btx helps improve range of motion and motor functioning [179] [See chapter: Physical therapy].

Occupational therapy (OT) is equally important and focuses on fine motor skill development to improve capacity for self-care (feeding, toileting, dressing, personal hygiene) and for fine motor abilities needed for school success. Constraint-induced movement therapy (CIMT), whereby use of the unaffected limb is inhibited, improves the functioning of a hemiplegic or monoplegic arm [178]. Similarly, bimanual upper limb therapy involving carefully planned, repeated practice of two-handed games and activities improves the use and coordination of both arms in daily tasks [180] [See chapter: Occupational therapy].

PT and OT should be implemented as soon as possible and carried over by parents or caregivers at home and in the school setting when appropriate [181]. Early intervention capitalizes on the brain's capacity to grow and change during critical developmental periods. Session frequency and duration are determined based on the child's needs and the goals established during evaluation. Services can be rendered in the community or in school as part of an individual education plan.

In contrast to medically-based PT and OT, school-based PT and OT are designed to support the student's ability to access their educational environment. School-based therapy is not intended to meet all of the therapeutic needs of a student; rather it is provided for educational purposes only. Goals for intervention in the school setting address the child's functional needs for accessing all areas of the school curriculum. PT interventions are designed to enable the student to travel throughout the school environment, participate in classroom activities, maintain and change positions in the classroom, as well as manage stairs, restrooms, and the cafeteria.

Therapy programs can be built around a number of techniques or frameworks. A few have good evidence for benefit in individuals with CP, a few have substantial evidence for lack of benefit, and most have little evidence for or against their efficacy [182]. Therapies with established efficacy in children with CP include constraint-induced or bimanual therapies supporting hand use in children with unilateral CP, goal-directed therapy, treadmill training and strength training, as well as modalities such as equine-assisted therapies [183]. Therapies with evidence against use for motor outcomes include passive neurodevelopmental therapy and sensory integration therapy.

Orthoses play an important role in the physical management of children with CP. They are designed with one of two primary aims: either to affect the body structure or to assist in function. Lower extremity orthotics are designed to correct or prevent deformity, provide a stable base of support, facilitate training in skills, and improve gait efficiency [184]. There are a variety of lower extremity orthotics including shoe inserts, ankle-foot orthotics, and knee orthotics. Scoliosis occurs at higher rates in those with more severe motor involvement (GMFCS levels IV and V). Rigid plastic thoracolumbosacral orthoses (TLSOs) may reduce spinal curvature and improve sitting ability while the orthosis is worn [185]; however, TLSOs are unlikely to alter the rate of progression of spine deformity [186].

Serial casting is the process of applying and then removing a cast on an arm or leg every seven days for a period of several weeks. With each cast change, the joint is held in position to provide constant passive stretch and thereby increase range of motion, reduce hypertonicity, and improve gait. Serial casting on its own can have significant beneficial effects [187], and the benefits are enhanced when combined with btx [188].

Adaptive equipment essentially represents a circumvention strategy to improve mobility, comfort, and/or ability to engage in activities. Wheelchairs and strollers facilitate long-distance transportation (whether self-propelled, caregiver-propelled, or motorized). For individuals with limited head/neck/trunk control, inclusion of hip and trunk support may be particularly important when fitting adaptive seating [189]. For such individuals, mechanical supports can also be implemented in stationary systems to facilitate standing, sitting, and/or upper extremity use.

Surgical

The mainstay of orthopedic intervention is nonoperative with the goals of maintaining and gaining function and preserving joint motion and mobility. Surgical treatment is undertaken in concert with other treatment modalities: bracing, casting, assistive devices, tone management, and developmental therapies. Orthotic devices are often prescribed to prevent deformity (i.e., through control of spasticity), to improve function, and occasionally to impose partial postoperative immobilization. The intent of surgery is to correct or prevent deformity and to diminish muscle imbalance. Timing of surgical intervention varies on the pathology being treated. For hip and spine pathology the decision to intervene is based more on progression of deformity and less on actual age. In the child who can walk, surgery is generally timed at an age where there is less chance of recurrence. Computerized gait analysis has greatly enhanced the ability to pinpoint muscles to be altered. Surgical interventions are consolidated when possible, commonly referred to as Single Event Multilevel Surgery (SEMLS) [190], thereby keeping surgery and anesthesia episodes to a minimum.

Risk for hip displacement increases with CP severity; for GMFCS level V it is 90%. Indicators of progressive hip subluxation are easy to recognize [191,192], and frank hip dislocation is avoidable with containment surgery. Nonoperative measures have not been found to prevent subluxation. [170] The incidence of scoliosis is directly related to GMFCS level. In ambulatory children scoliosis may develop but rarely comes to surgery. Children at GMFCS level V have a 75% chance of developing a scoliosis curve >40 degrees by age 20 [193]. Curves tend to progress most rapidly during the adolescent growth spurt. Details regarding the specific surgical procedures about the hip, knee, foot, and spine along with their indications and outcomes are provided in Subchapter: surgical interventions.

Upper extremity deformities in CP are complex, and there is no consensus on how to best manage them. Upper extremity surgery is infrequent compared to the lower extremities. There are two main categories. The first is children with severe elbow flexion, forearm pronation, wrist flexion, and a thumb-in-palm deformity. The main goal of treatment of these deformities is to improve the resting position of the hand for hygiene, dressing, and pain. The second category is spastic hemiplegia or mild spastic quadriplegia, where the surgical goal is to improve the function of the involved hand. Surgery may involve releasing tight tendons or their muscle bellies, transferring tendons, or fusion of the wrist bones [194,195]. If sensation is relatively preserved, surgical treatment may produce substantial functional gains. In the face of severe sensory loss, the lack of sensory feedback largely nullifies functional benefits from surgical procedures [196].

Neurosurgical techniques to improve motor function had little application to children with CP until the surge of interest in selective dorsal rhizotomy (SDR) aimed at eliminating spasticity in the lower limbs. Warwick Peacock in South Africa, who pioneered the application of SDR to children in the 1980s, pointed out that the ideal candidate for rhizotomy appeared to be the child with pure spastic diplegia. Rhizotomy effectively eliminates or reduces lower extremity spasticity in the short term [197], and there has been very little recurrence in the long term with notable functional benefits. [198,199] Cohorts of individuals with spastic diplegia who underwent SDR in childhood have now been followed for over 25 years. More recently, as an alternative to intrathecal baclofen, SDR has been applied to nonambulatory quadriplegic children including those with dystonia [200,201]. In children with severe mixed CP (GMFCS levels III and IV) , using a combined dorsal and ventral rootlet rhizotomy technique, Abdel Ghany et al. reported marked tone improvement in all and functional gains in up to 80%. On the other hand, among children with spastic quadriplegia undergoing traditional SDR, Kim et al. encountered a "poor outcome" in 25% [202].

Neurosurgical treatment of movement disorders in children, specifically dystonia, utilizing deep brain stimulation (DBS) evolved around the turn of the century. The universal experience has been that acquired dystonias with no structural change (previously termed "primary dystonias") respond much better than secondary dystonias [203,204]. The globus pallidus internus has become the standard target for treatment of secondary dystonia, and bilateral stimulation seems to be necessary [204]. At present DBS is reserved for children with major functional deficits who have not responded to other treatments, including oral medications and intrathecal baclofen [205]. See chapter: Surgical interventions for details of neurosurgical treatments in the CP population.

Prognosis

Motor function over time

When parents receive a new diagnosis of CP, their first two questions almost always are:

1. Why does my child have CP? (i.e., etiology), and
2. Will he (or she) walk?

Prognosis in terms of ambulation potential is relatively straightforward [206,207]. Molnar highlights the importance of CP subtype: almost all children with diplegia and hemiplegia will walk. Looking at early milestones, her guideline states that (1) every child who maintains independent sitting by age 2 will walk and (2) no child who fails to sit by age

4 will ever walk. The in-between group, those who achieve sitting between 2 and 4 years old, may or may not walk depending on several variables that influence the prediction:

- Primitive reflexes, especially if obligatory.
- Seizures, especially if difficult to control.
- Anticonvulsant drugs, especially multiple.
- Intellectual disability, especially if severe to profound.
- Blindness.
- Intercurrent and chronic medical problems, for example, respiratory failure, obesity.

A fundamental premise of CP is that the neurological insult is static and therefore motor function, although disordered, should improve with time. For the most part this is true, excepting those children who are the most profoundly impaired. Longitudinal studies of the GMFCS, a measure of CP severity, demonstrate that the great majority of children assigned a level at early age (most reliably after age 4) stay in the same level (i.e., make motor developmental progress) as they get older [208,209]. Among the few who do change a level, as often as not they make better progress rather than worse [209]. Children in GMFCS level V have little to no independent motor functional capacity, and that does not change with time. In other words, they make minimal, if any, motor progress, although some may become more mobile using assistive devices.

In older children and adults there can be too much focus on ambulatory function with failure to recognize the value of limited standing activity, like standing up to reach for a high object, bearing weight and taking a step to facilitate transfer, etc. Preserving even limited functional weight-bearing can be precious in terms of promoting independence [210].

For children who are severely involved and appear to lose skills over time (especially in later childhood) and who do not have a clear underlying etiology, it can be extremely difficult to sort out whether one is dealing with a static or a progressive neurological disorder. In this scenario the possibility that a child with a neurodegenerative disorder has been misdiagnosed with CP must be considered. However, motor deterioration (or pseudodegeneration) can occur in CP when other factors come into play. Physical growth—both linear and weight gain—can work against motor skills in those who have marginal capabilities. Intercurrent illnesses, uncontrolled seizures, need for multiple and/or high-dose antiseizure medications, and progressive orthopedic deformities (contractures, bony malalignment, degenerative arthritis) [211] can cause motor setbacks. These functional losses may or may not be regained depending on the success of treatments.

For children who are marginal ambulators in early childhood (GMFCS levels III and IV) loss of ambulatory or weight-bearing skills is not uncommon as they grow into adolescence and adulthood due to fatigue, joint pain, and joint degeneration [212,213]. Osteoarthritis develops prematurely in adults with CP [214] due to muscle imbalance and abnormal joint forces which promote thinning of articular cartilage. The process is gradual and starts in childhood [215].

Adults have increased risk of cervical and lumbar myelopathy. Cervical stenosis, cervical and lumbar disc degeneration or herniation, and lumbosacral spondylolysis causing myelopathy and/or radiculopathy can produce chronic pain and loss of function [211]. Scoliosis, hyperlordosis and crouch gait are contributing factors. Scoliosis pain can be a new experience for the adult. It is rarely an issue in children [210]. Loss of bladder or bowel control may be a warning of spinal cord impingement. Deep tendon reflexes can be a tip-off on examination if they dramatically increase (myelopathy) or decrease (radiculopathy). Significant anterior pelvic tilt may be a promoting factor in the lower spine, and long-standing dystonic neck twisting may be a key factor in the cervical spine.

When adults experience functional deterioration, it will often be secondary to something other than their CP—unrelated co-occurring disorders like Alzheimer disease, Parkinson disease, stroke or indirectly associated disorders like depression, excessive weight gain, and arthritis. Too often symptoms and loss of function in these individuals are simply attributed to the CP without seeking other, possibly treatable, explanations.

Rehabilitation programs may benefit adults with specific goals. Botulinum toxin continues to have a role in adults for relieving painful spasms, improving upright posture, and treating cervical dystonia. Intrathecal baclofen with high (cervical) placement and phenol nerve or motor point blocks also may have a place. Surgical interventions to relieve pain including hip arthroplasty [216] and lumbar or cervical spine fusion [217] can be successfully undertaken.

Transition

Most children with CP survive to be adults with CP. They and their families must negotiate the same transitions throughout childhood and into adulthood as children with other NDDs (see Chapter : Transition and aging). Guidelines for the process of transition have been developed—the mechanics of when to start, what information to gather, and how to organize things. The issue is finding adult healthcare providers who are comfortable and familiar with CP [218]. New or changing neurologic symptoms are of particular concern [12]. Services and expertise are usually plentiful for children, leading to high hopes for adult services which more often than not fail to be met [219]. Additionally, patients and parents bemoan the loss of long and trusted relationships with their pediatric providers. Roadblocks to accessing adequate adult healthcare include lack of clinician education, inadequate examination rooms and equipment, and time allotted to care. More time is necessary, especially when there are communication barriers as with individuals who are non-verbal and use a communication device [12].

Many people with NDDs have decreased participation in society. Independent living and employment rates for adults with CP are low. In a study of ambulatory adults, Shrader et al. [220] uncovered two distinct subgroups: self-reporters and those who required proxy report. Among self-reporting adults high school graduation and bachelor's degree achievement was at or above national levels, yet their employment rate was lower (67%). The proxy report group had substantially lower educational achievement and even less employment (36%). Looking at all adults in the Swedish CP registry, 21.2% were employed and 13.4% had a partner or spouse [221]. Reduced ability to communicate was the greatest risk factor associated with unemployment.

Through supported employment programs in the United States, people with CP and other developmental disabilities have for decades been paid to work but for subminimum wage. Legislation to eliminate this section of the Fair Labor and Standards Act is currently under way in an attempt to foster better access to fully paid, competitive, and integrated employment [222]. Efforts to increase opportunities for employment and to change societal attitudes about disabilities in general are every bit as important as the traditional hands-on therapies and medical/surgical interventions employed in childhood.

Life expectancy

Life expectancy of children and adults with CP is shortened compared to the general population and is highly influenced by risk factors: severity of motor involvement, degree of intellectual disability, nutrition and oral-motor dysfunction, seizures and antiseizure medication, residential environment. Outcome varies by type of CP; for example, individuals with hemiplegia and diplegia do better than those with quadriplegia. Estimating life expectancy at any point in time is a challenge. Using data available from many longevity studies, Day et al. have provided a rigorous methodology for calculating life expectancy in individual CP cases [223]. Reliable information on each of the risk factors mentioned above is essential to applying the method.

However, people with CP are living longer than they used to. Rimmer showed that regular exercise improves functional status for adults and thus higher survival rates [224]. Higher rates of mortality due to cancers, other organ system diseases like diabetes and heart disease, and accidents suggest that inadequate medical surveillance for common adult conditions that are unrelated to CP is a factor in early mortality.

Prevention

Improvements in CP morbidity would appear to be predicated on two conditions. First of all, decreasing the incidence of preterm births should be accomplishable by changes primarily of a social and public health policy nature aimed at prenatal and prepregnancy care and education [225,226]. Secondly, identifying and improving the perinatal and neonatal events that are damaging to the brain via neuroprotective strategies should result in better outcomes [227]. Antenatal magnesium sulfate and postnatal therapeutic hypothermia decrease CP [227,228]. Other interventions are under investigation and may have promise [229,230]. See Chapter: Early brain injury.

Resources

Parents

Families raising a child with a disability face more than the usual challenges and stresses. To help navigate these challenges they rank "more information" as a top priority [231], and developmental disabilities professionals are ideally positioned to direct parents to high quality informational and support resources. Particularly valuable resources for networking interactions among parents and professionals are listed in Table 15.3. Equipment clearinghouses, respite care, adaptive sporting opportunities, adaptive clothing and toys, feeding supplies and diet strategies, and other resources are typically discovered via these networks. Political and legislative activities may also be included. Connecting parents to someone who knows available community resources of high quality and how to obtain them, is an essential first step to fruitful networking activities. Parent support groups can provide valuable service to physicians as well in terms of identifying the best resources.

The best information resource often turns out to be other parents of special needs children—"someone who has been there." Parents commonly connect with one another via social media/Facebook groups. Every state has at least one Parent Training and Information Center (PTIC) and Family to Family (F2F) program. Their purpose is to support the family as collaborators, decision makers, problem solvers and active partners with healthcare professionals and resource

TABLE 15.3 Family resources.

United Cerebral Palsy (UCP) Association
https://www.ucp.org
Cerebral Palsy Foundation (CPF)
https://www.yourcpf.org
Child Neurology Foundation (CNF)
https://www.childneurologyfoundation.org
American Academy for Cerebral Palsy and Developmental Medicine (AACPDM)
https://www.aacpdm.org
The Cerebral Palsy Research Network
https://www.cprn.org
National Organization of Rare Disorders (NORD)
https://www.rarediseases.org
Parent Training and Information Centers (PTIC)
https://www.understood.org/articles/en/parent-training-centers-a-free-resource
https://www.parentcenterhub.org/find-your-center
Family to Family (F2F) Health Information Centers [formerly Family Voices]
https://www.familyvoices.org/lfpp/f2fs
The Arc
https://thearc.org
Parent Advocacy Center for Educational Rights (PACER)
https://www.pacer.org/about
Exceptional Parent
https://www.epmagazine.com
Public school system Special Education or Part H coordinator
Municipal or county Health and Human Services department

agency personnel. They facilitate networking opportunities and provide guidance for advocacy activities at state and national levels. Some diagnosis-specific organizations, like the United Cerebral Palsy Association (UCP), have affiliate offices in most large cities. In addition to being familiar with local resources UCP is known for putting on excellent conferences, and some affiliates are involved with research endeavors. A listing of support groups specific to uncommon disease diagnoses can be found in the *Exceptional Parent Resource Guide* and through NORD (the National Organization of Rare Disorders). *Exceptional Parent* magazine provides practical, up-to-date informational materials for families of children and adults with special healthcare needs through their monthly print publication and website.

Professionals

Physicians can provide guidance to families to start the support-finding process by referring a parent to just one or two of the agencies listed in Table 15.3. We should keep telephone numbers, brochures, and contact information handy in the clinic for quick access to give families. In addition to directly assisting families, NDD physicians can help families with whom they are not directly involved through the training of medical students, residents, fellows, practicing pediatricians, and other healthcare disciplines simply by being role models for listening to parents and partnering with them in medical decision-making. Another avenue is to serve as a consultant or board member for community support organizations and assist with technical support and/or training.

Just as with parents, professionals need resources—networking and information are fundamental to what we do: new ideas, new treatments, new applications of therapy techniques, adaptive equipment resources, support group resources for parents [232]. We need to have our own resources for information and networking—our "professional home":

- The American Academy for Cerebral Palsy and Developmental Medicine (AACPDM) and their journal *Developmental Medicine and Child Neurology* (DMCN).
- The American Academy of Pediatrics and its Council on Children with Disabilities (COCWD).
- The Child Neurology Society, which has a CP special interest group.
- The Cerebral Palsy Research Network and their CP Toolkit.

The Child Neurology Society and AACPDM annual scientific meetings both take place in the fall. The annual Spectrum of Developmental Disabilities conference sponsored and organized by the Kennedy Fellows Association takes place in the spring [https://hopkinscme.cloud-cme.com].

References

[1] William John AP. Little and cerebral palsy in the nineteenth century. J Hist Med Allied Sci 1989;44(1):56–71. Available from: https://academic.oup.com/jhmas/article/44/1/56/759499.

[2] Accardo P. Freud on diplegia. Commentary and translation. Am J Dis Child 1982;136(5):452–6.

[3] Paneth N. Birth and the origins of cerebral palsyAvailable from: https://www.nejm.org/doi/full/10.1056/NEJM198607103150209.

[4] Osler SW. The cerebral palsies of children : a clinical study from the Infirmary for Nervous. Diseases, Phila 1987;92.

[5] Crothers B. The natural history of cerebral palsy. Cambridge Mass: Harvard Univ. Press; 1959.

[6] Bax M. Terminology and classification of cerebral palsy. Dev Med Child Neurol 1964;295–307.

[7] Mac Keith RC, Mackenzie ICK, Polani PE. The little club: memorandum on terminology and classification of "cerebral palsy.". Cereb Palsy Bull 1959;27–35.

[8] Baxter P, Morris C, Rosenbaum P, Paneth N. The definition and classification of cerebral palsy. Dev Med Child Neurol 2007;49:1–44.

[9] Rosenbaum P, Paneth N, Leviton A, Goldstein M, Bax M, Damiano D, et al. A report: the definition and classification of cerebral palsy April 2006. Dev Med Child Neurol, 49. 2007. p. 8–14.

[10] Cerovac N, Petrović I, Klein C, Kostić VS. Delayed-onset dystonia due to perinatal asphyxia: a prospective study. Mov Disord 2007;22(16):2426–9. Available from: https://onlinelibrary.wiley.com/doi/full/10.1002/mds.21747.

[11] Hanna SE, Rosenbaum PL, Bartlett DJ, Palisano RJ, Walter SD, Avery L, et al. Stability and decline in gross motor function among children and youth with cerebral palsy aged 2 to 21 years. Dev Med Child Neurol 2009;51(4):295–302. Available from: https://pubmed.ncbi.nlm.nih.gov/19391185/.

[12] Smith SS, Gannotti PT, Hurvitz EA, Jensen FE, Krach LE, Kruer MC, et al. Adults with cerebral palsy require ongoing neurologic care: a systematic review. Ann Neurol 2021;89:860–71.

[13] Aravamuthan BR, Fehlings D, Shetty S, Fahey M, Gilbert L, Tilton A, et al. Variability in cerebral palsy diagnosis. Pediatrics 2021;147(2):76. Available from: https://doi.org/10.1542/peds.2020-010066.

[14] Paneth N, Hong T, Korzeniewski S. The descriptive epidemiology of cerebral palsy. Clin Perinatol 2006;33(2):251–67.

[15] Maenner MJ, Blumberg SJ, Kogan MD, Christensen D, Yeargin-Allsopp M, Schieve LA, et al. Prevalence of cerebral palsy and intellectual disability among children identified in two U.S. National Surveys, 2011–2013. Ann Epidemiol 2011;26(3):222–6.

[16] Oskoui M, Coutinho F, Dykeman J, Jett N, Pringsheim T. An update on the prevalence of cerebral palsy: a systematic review and meta-analysis. Dev Med Child Neurol 2013;55(6):509–19.

[17] Chounti A, Hägglund G, Wagner P, Westbom L. Sex differences in cerebral palsy incidence and functional ability: a total population study. Acta Paediatr 2013;102(7):712–17. Available from: https://onlinelibrary.wiley.com/doi/full/10.1111/apa.12240.

[18] Romeo D.M., Sini F., Brogna C., Albamonte E., Ricci D., Mercuri E. Sex differences in cerebral palsy on neuromotor outcome: a critical review. 2016.

[19] Wu YW, Xing G, Fuentes-Afflick E, Danielson B, Smith LH, Gilbert WM. Racial, ethnic, and socioeconomic disparities in the prevalence of cerebral palsye. Pediatrics 2011;127:674. Available from: http://www.pediatrics.org/cgi/doi/10.1542/peds.2010-1656.

[20] Khandaker G, Muhit M, Karim T, Smithers-Sheedy H, Novak I, Jones C, et al. Epidemiology of cerebral palsy in Bangladesh: a population-based surveillance study. Dev Med Child Neurol 2019;61(5):601–9.

[21] Novak I, Morgan C, Adde L, Blackman J, Boyd RN, Brunstrom-Hernandez J, et al. Early, accurate diagnosis and early intervention in cerebral palsy: advances in diagnosis and treatment. JAMA Pediatr 2017;171(9):897–907. Available from: https://jamanetwork.com/journals/jamapediatrics/fullarticle/2636588.

[22] Blair E., Cans C., Sellier E. Epidemiology of the cerebral palsies. Cereb palsy A multidiscip approach, 4th ed. 2018;19–28. Available from: https://link.springer.com/chapter/10.1007/978-3-319-67858-0_3.

[23] McIntyre S, Goldsmith S, Webb A, Ehlinger V, Hollung SJ, McConnell K, et al. Global prevalence ofcerebral palsy: a systematic analysis. Dev Med Child Neurol 2022;62(12):1494–506.

[24] Jahan I, Muhit M, Hardianto D, Laryea F, Chhetri AB, Smithers-Sheedy H, et al. Epidemiology ofcerebral palsy in low- and middle-income countries: preliminary findings from an international multi-centrecerebral palsy register. Dev Med Child Neurol 2021;63(11):1327–36.
[25] Gladstone M. A review of the incidence and prevalence, types and aetiology of childhoodcerebral palsy in resource-poor settings. Ann Trop Paediatr 2010;30(3):181–96.
[26] Kakooza-Mwesige A, Andrews C, Peterson S, Mangen FW, Eliasson AC, Forssberg H. Prevalence ofcerebral palsy in Uganda: a population-based study. Lancet Glob Heal 2017;5(12):e1275–82.
[27] Galea C, Mcintyre S, Smithers-Sheedy H, Reid SM, Gibson C, Delacy M, et al. Cerebral palsy trends in Australia (1995–2009): a population-based observational study. Dev Med Child Neurol 2019;61(2):186–93. Available from: https://onlinelibrary.wiley.com/doi/full/10.1111/dmcn.14011.
[28] Sellier E, Platt MJ, Andersen GL, Krägeloh-Mann I, De La Cruz J, Cans C, et al. Decreasing prevalence in cerebral palsy: a multi-site European population-based study, 1980 to 2003. Dev Med Child Neurol 2016;58(1):85–92. Available from: https://onlinelibrary.wiley.com/doi/full/10.1111/dmcn.12865.
[29] McIntyre S. The continually changing epidemiology of cerebral palsy. Acta Paediatr 2018;3(107):374–5.
[30] Himmelmann K, Uvebrant P. The panorama of cerebral palsy in Sweden part XII shows that patterns changed in the birth years2007–2010. Acta Paediatr 2018;107(3):462–8. Available from: https://onlinelibrary.wiley.com/doi/full/10.1111/apa.14147.
[31] Hamrick SEG, Miller SP, Leonard C, Glidden DV, Goldstein R, Ramaswamy V, et al. Trends in severe brain injury and neurodevelopmental outcome in premature newborn infants: the role of cystic periventricular leukomalacia. J Pediatr 2004;145(5):593–9. Available from: http://www.jpeds.com/article/S002234760400455X/fulltext.
[32] Tagin MA, Bch MB, Woolcott CG, Vincer MJ, Whyte RK, Stinson DA. Hypothermia for neonatal hypoxic ischemic encephalopathy an updated systematic review and meta-analysis. Arch Pediatr Adolesc Med 2012;166(6):558–66. Available from: http://www.archpediatrics.com.
[33] MacLennan A. A template for defining a causal relation between acute intrapartum events and cerebral palsy: international consensus statement. BMJBr Med J 1999;319(7216):1054. Available from: http://pmc/articles/PMC1116849/.
[34] Neonatal Encephalopathy and Neurologic Outcome, 2nd ed. Pediatrics. 2014 May 1 [cited 2022 Apr 14];133(5):e1482–8. Available from: /pediatrics/article/133/5/e1482/32738/Neonatal-Encephalopathy-and-Neurologic-Outcome.
[35] Kitai Y, Hirai S, Okuyama N, Hirotsune M, Mizutani S, Ogura K, et al. Diagnosis of bilirubin encephalopathy in preterm infants with dyskinetic cerebral palsy. Neonatology 2020;117(1):73–9. Available from: https://www.karger.com/Article/FullText/502777.
[36] Maisels MJ. Neonatal hyperbilirubinemia and kernicterus—not gone but sometimes forgotten. Early Hum Dev 2009;85(11):727–32.
[37] Thorngren-Jerneck K, Herbst A. Perinatal factors associated with cerebral palsy in children born in Sweden. Obstet Gynecol 2006;108(6).
[38] McIntyre S, Taitz D, Keogh J, Goldsmith S, Badawi N, Blair E. A systematic review of risk factors for cerebral palsy in children born at term in developed countries. Developmental Med Child Neurol 2013;55:499–507.
[39] McIntyre S, Morgan C, Walker K, Novak I. Cerebral palsy–don't delay. Dev Disabil Res Rev 2011;17(2):114–29.
[40] May H, Fasheun J, Bain J, Baugh E, Bier L, Revah-Politi A, et al. Genetic testing in individuals with cerebral palsy. Dev Med Child Neurol 2021;63(12):1448–55.
[41] Chopra M, Gable DL, Love-Nichols J, Tsao A, Rockowitz S, Sliz P, et al. Mendelian etiologies identified with whole exome sequencing in cerebral palsy. Ann Clin Transl Neurol 2022;9(2):193–205.
[42] Jin SC, Lewis SA, Bakhtiari S, Zeng X, Sierant MC, Shetty S, et al. Mutations disrupting neuritogenesis genes confer risk for cerebral palsy. Nat Genet 2020;52(10):1046–56.
[43] van Eyk CL, Webber DL, Minoche AE, Pérez-Jurado LA, Corbett MA, Gardner AE, et al. Yield of clinically reportable genetic variants in unselected cerebral palsy by whole genome sequencing. Genomic Med 2021;6(1).
[44] Takezawa Y, Kikuchi A, Haginoya K, Niihori T, Numata-Uematsu Y, Inui T, et al. Genomic analysis identifies masqueraders of full-term cerebral palsy. Ann Clin Transl Neurol 2018;5(5):538–51.
[45] Yechieli M, Gulsuner S, Ben-Pazi H, Fattal A, Aran A, Kuzminsky A, et al. Diagnostic yield of chromosomal microarray and trio whole exome sequencing in cryptogenic cerebral palsy. J Med Genet 2022;59(8).
[46] Leach EL, Shevell M, Bowden K, Stockler-Ipsiroglu S, van Karnebeek CDM. Treatable inborn errors of metabolism presenting as cerebral palsy mimics: systematic literature review Vol Orphanet J Rare Dis 2014;197.
[47] Méneret A., Roze E. GLUT1 deficiency in a patient diagnosed as cerebral palsy: is NGS a valuable tool to be considered in all cases of cp to detect underlying genetic disorders? Vol. 6, Movement disorders clinical practice. Wiley-Blackwell; 2019. p. 277–9.
[48] Bear JJ, Wu YW. Maternal infections during pregnancy and cerebral palsy in the child. Pediatr Neurol 2016;57:74–9.
[49] Miller JE, Pedersen LH, Streja E, Bech BH, Yeargin-Allsopp M, Van Naarden Braun K, et al. Maternal infections during pregnancy and cerebral palsy: a population-based cohort study. Paediatr Perinat Epidemiol 2013;27(6).
[50] Smithers-Sheedy H, Raynes-Greenow C, Badawi N, Fernandez MA, Kesson A, McIntyre S, et al. Congenital cytomegalovirus among children with cerebral palsy. J Pediatr 2017;181.
[51] Goldsmith S, Mcintyre S, Andersen GL, Gibson C, Himmelmann K, Blair E, et al. Congenital anomalies in children with pre- or perinatally acquired cerebral palsy: an international data linkage study. Dev Med Child Neurol 2021;63(4):413–20.
[52] Kulak, et al. Adv Med Sci 2011; PMID 21444273.
[53] Himmelmann K, Horber V, De La Cruz J, Horridge K, Mejaski-Bosnjak V, Hollody K, et al. MRI classification system (MRICS) for children with cerebral palsy: development, reliability, and recommendations. Dev Med Child Neurol 2017;59(1).
[54] Oegema R, Barakat TS, Wilke M, Stouffs K, Amrom D, Aronica E, et al. International consensus recommendations on the diagnostic work-up for malformations of cortical development. Nat Rev Neurol 2020;16(11):618–35.

[55] Volpe JJ. Brain injury in premature infants: a complex amalgam of destructive and developmental disturbances Vol Lancet Neurol 2009;.
[56] Papadimitriou I, Dalivigka Z, Outsika C, Scarmeas N, Pons R. Dystonia assessment in children with cerebral palsy and periventricular leukomalacia. Eur J Paediatr Neurol 2021;32:8–15.
[57] Marefi, et al. Neurology 2022; PMID: 36041870.
[58] Lasry, et al. Neurology 2010; PMID: 20421583.
[59] Inder T, Perlman J, Volpe J. Preterm intraventricular hemorrhage/posthemorrhagic hydrocephalus. Volpe's neurology of the newborn. 6th ed. Elsevier; 2018. p. 637–98.
[60] Wu YW, Croen LA, Shah SJ, Newman TB, Najjar DV. Cerebral palsy in a term population: risk factors and neuroimaging findings. Pediatrics. 2006;118(2).
[61] Wagenaar N, Martinez-Biarge M, Van Der Aa NE, Van Haastert IC, Groenendaal F. Benders MJNL, et al. Neurodevelopment after perinatal arterial ischemic stroke. Pediatrics. 2018;142:3.
[62] Martinello K., Hart A.R., Yap S., Mitra S., Robertson N.J. Management and investigation of neonatal encephalopathy: 2017 update. Vol. 102, archives of disease in childhood: fetal and neonatal edition. BMJ Publishing Group; 2017. p. F346–58.
[63] Mathew JL, Kaur N, Dsouza JM. Therapeutic hypothermia in neonatal hypoxic encephalopathy: a systematic review and meta-analysis. J Glob Health 2022;12.
[64] Shapiro SM. Chronic bilirubin encephalopathy: diagnosis and outcome. Semin Fetal Neonatal Med 2010;15(3).
[65] Alkén J, Håkansson S, Ekéus C, Gustafson P, Norman M. Rates of extreme neonatal hyperbilirubinemia and kernicterus in children and adherence to national guidelines for screening, diagnosis, and treatment in Sweden. JAMA Netw open 2019;2(3).
[66] Smilga AS, Garfinkle J, Ng P, Andersen J, Buckley D, Fehlings D, et al. Neonatal infection in children with cerebral palsy: a registry-based Cohort study. Pediatr Neurol 2018;80.
[67] Hung KL. Pediatric abusive head trauma. Biomed J, Vol. 43. Elsevier B.V.; 2020. p. 240–50.
[68] Lind K, Toure H, Brugel D, Meyer P, Laurent-Vannier A, Chevignard M. Extended follow-up of neurological, cognitive, behavioral and academic outcomes after severe abusive head trauma. Child Abus Negl 2016;51.
[69] Lance JW. What is spasticity? Lancet 1990;335(8689):606. Available from: http://www.thelancet.com/article/014067369090389M/fulltext.
[70] Mukherjee A, Chakravarty A, Misra UK, Post SG. Spasticity mechanisms-for the clinician. Front Neurol 2010;17(1):149. Available from: http://www.frontiersin.org.
[71] Meseguer-Henarejos AB, Sańchez-Meca J, López-Pina JA, Carles-Hernańdez R. Inter-and intra-rater reliability of the modified ashworth scale: a systematic review and meta-analysis. Eur J Phys Rehabil Med 2018;54(4):576–90.
[72] Morris S. Ashworth and tardieu scales: their clinical relevance for measuring spasticity in adult and paediatric neurological populations. Phys Ther Rev 2002;7(1):53–62. Available from: https://www.tandfonline.com/action/journalInformation?journalCode = yptr20.
[73] Armand S, Decoulon G, Bonnefoy-Mazure A. Gait analysis in children with cerebral palsy. EFORT Open Rev 2016;1(12):448. Available from: https://pmc/articles/PMC5489760/.
[74] Krigger K. Cerebral palsy: an overview. Am Fam Phys 2006;73(1):91–100.
[75] Filloux FM. Neuropathophysiology of movement disorders in cerebral palsy. J Child Neurol 1996;11(Suppl 1):S5 –12.
[76] Jan MMS. Cerebral palsy: comprehensive review and update. Ann Saudi Med 2006;26(2):123. Available from: https://pmc/articles/PMC6074141/.
[77] Stewart K, Harvey A, Johnston LM. A systematic review of scales to measure dystonia and choreoathetosis in children with dyskinetic cerebral palsy. Dev Med Child Neurol 2017;59(8):786–95. Available from: https://onlinelibrary.wiley.com/doi/full/10.1111/dmcn.13452.
[78] Cans C. Surveillance of cerebral palsy in Europe: a collaboration of cerebral palsy surveys and registers. Surveillance of Cerebral Palsy in Europe (SCPE). Dev Med Child Neurol 2000;42(12):816–24 Dev Med Child Neurol. 2000;42(12).
[79] Report from the Australian Cerebral Palsy Register (ACPR): Birth prevalence of CP is declining. J Paediatr Child Health. 2017;53.
[80] Levy JP, Oskoui M, Ng P, Andersen J, Buckley D, Fehlings D, et al. Ataxic-hypotonic cerebral palsy in a cerebral palsy registry: insights into a distinct subtype. Neurol Clin Pract 2020;10:2.
[81] Rice J, Skuza P, Felicity B, Russo R, Fehlings D. Identification and measurement of dystonia in cerebral palsy BAD Barry-Albright Dystonia scale CFCS Communication Functional Classification System HAT Hypertonia Assessment Tool MACS Manual Ability Classification System PMT Predominant motor type SCPE Sur. Dev Med Child Neurol 2017;59(12):1249–55. Available from: https://research.hollandbloorvie.
[82] Palisano R, Rosenbaum P, Walter S, Russell D, Wood EGB. Development and reliability of a system to classify gross motor function in children with cerebral palsy. Dev Med Child Neurol 1997;39(4):214–23.
[83] Palisano RJ, Rosenbaum P, Bartlett D, Livingston MH. Content validity of the expanded and revised gross motor function classification system. Dev Med Child Neurol 2008;50(10):744–50.
[84] Eliasson A-C, Rösblad BP, Arner M, Rosenbaum FRCPP. The manual ability classification system (MACS) for children with cerebral palsy: scale development and evidence of validity and reliability. Dev Med Child Neurol 2006;48(7):549–54.
[85] Hidecker MJC, Paneth N, Rosenbaum PL, Kent RD, Lillie J, Eulenberg JB, et al. Developing and validating the communication function classification system for individuals with cerebral palsy. Dev Med Child Neurol 2011;53(8):704. Available from: https://pmc/articles/PMC3130799/.
[86] Ashwal S., Russman B.S., Blasco P.A., Miller G., Sandler A., Shevell M., et al. Practice Parameter: diagnostic assessment of the child with cerebral palsy report of the quality standards subcommittee of the American Academy of Neurology and the Practice Committee of the Child Neurology Society. 2004.
[87] Shribman S, Reid E, Crosby AH, Houlden H, Warner TT. Hereditary spastic paraplegia: from diagnosis to emerging therapeutic approaches Vol Lancet Neurol 2019;.

[88] Hedera P. Hereditary Spastic Paraplegia Overview. 2000 Aug 15 [updated 2021 Feb 11]. In: Adam. MP, Everman. DB, Mirzaa. GM, Pagon R, Wallace S, Bean L, et al., editors. GeneReviews®. Seattle (WA): University of Washington, Seattle; 1993–2022; 2000.
[89] Ounpuu S. The importance of understanding gait features in hereditary spastic paraplegia: accomplishments and next steps. Dev Med Child Neurol 2016;58.
[90] Wolf SI, Braatz F, Metaxiotis D, Armbrust P, Dreher T, Döderlein L, et al. Gait analysis may help to distinguish hereditary spastic paraplegia from cerebral palsy. Gait Posture 2011;33(4):556–61.
[91] Hedera P, Eldevik OP, Maly P, Rainier S, Fink JK. Spinal cord magnetic resonance imaging in autosomal dominant hereditary spastic paraplegia. Neuroradiology. 2005;47(10).
[92] Taft L, Barowsky E. Clumsy child. Pediatr Rev 1989;10:247–53.
[93] Hadders-Algra M. Developmental coordination disorder: is clumsy motor behavior caused by a lesion of the brain at early age? Neural Plast 2003;10(1–2):39. Available from: http://pmc/articles/PMC2565415/?report = abstract.
[94] Blank R, Smits-Engelsman B, Polatajko H, Wilson P. European academy for childhood disability (EACD): recommendations on the definition, diagnosis and intervention of developmental coordination disorder (long version)*. Dev Med Child Neurol 2012;54(1):54–93. Available from: https://onlinelibrary.wiley.com/doi/full/10.1111/j.1469-8749.2011.04171.x.
[95] Zwicker JG, Brown-Lum M, Izadi-Najafabadi S, Oberlander TF, Rauscher A, Zwicker JG. Differences in white matter microstructure among children with developmental coordination disorder. JAMA Netw Open 2020;3(3).
[96] Banihani R, Smile S, Yoon G, Dupuis A, Mosleh M, Snider A, et al. Cognitive and neurobehavioral profile in boys with duchenne muscular dystrophy. J Child Neurol 2015;30(11).
[97] Tofts LJ, Elliott EJ, Munns C, Pacey V, Sillence DO. The differential diagnosis of children with joint hypermobility: a review of the literature Vol PediatricRheumatol. 2009;.
[98] Blair E, Stanley F. Causal pathways to cerebral palsy. Curr Pediatr 2002;12(3):179–85.
[99] Hickie IB, Scott EM, Cross SP, Iorfino F, Davenport TA, Guastella AJ, et al. Right care, first time: a highly personalised and measurement-based care model to manage youth mental health. Med J Aust 2019;211(S9). Available from: https://onlinelibrary.wiley.com/doi/abs/10.5694/mja2.50383.
[100] Springer A, Dyck Holzinger S, Andersen J, Buckley D, Fehlings D, Kirton A, et al. Profile of children with cerebral palsy spectrum disorder and a normal MRI study. Neurology 2019;93(1):e88–96. Available from: https://n.neurology.org/content/93/1/e88.
[101] Moreno-De-Luca A, Millan F, Pesacreta DR, Elloumi HZ, Oetjens MT, Teigen C, et al. Molecular diagnostic yield of exome sequencing in patients with cerebral palsy. JAMA 2021;325(5):467–75. Available from: https://europepmc.org/articles/PMC7856544.
[102] Hollung SJ, Bakken IJ, Torstein V, Stian L, Wiik RK, et al. Comorbidities in cerebral palsy: a patient registry study. Dev Med Child Neurol 2020;62(1):97–103. Available from: https://www.ssb.no.
[103] Bax M, Gillberg C. Comorbidities in developmental disorders. John Wiley & Sons; 2010. p. 176.
[104] Cooper M, Mackay M, Dagia C, Fahey M, Howell K, Reddihough D, et al. Epilepsy syndromes in cerebral palsy: varied, evolving and mostly self-limited. Brain. 2022;24.
[105] Carlsson M, Hagberg G, Bm BA, Olsson I. Clinical and aetiological aspects of epilepsy in children with cerebral palsy Vol Developmental Med & Child Neurol 2003;.
[106] Novak I., Hines M., Goldsmith S., Barclay R. Clinical prognostic messages from a systematic review on cerebral palsy. Pediatrics. 2012 Nov 1 [cited 2022 Apr 23];130(5):e1285–312. Available from: /pediatrics/article/130/5/e1285/32499/Clinical-Prognostic-Messages-From-a-Systematic.
[107] Gabis LV, Tsubary NM, Leon O, Ashkenasi A, Shefer S. Assessment of abilities and comorbidities in children with cerebral palsy. J Child Neurol 2015;30(12):1640–5. Available from: https://journals.sagepub.com/doi/10.1177/0883073815576792?url_ver = Z39.88-2003&rfr_id = ori%3Arid%3Acrossref.org&rfr_dat = cr_pub + +0pubmed.
[108] Sigurdardottir S, Eiriksdottir A, Gunnarsdottir E, Meintema M, Arnadottir U, Vik T. Cognitive profile in young Icelandic children with cerebral palsy. Devel Med Child Neurol 2008;50(5):357–62.
[109] Ito J, Araki A, Tanaka H, Tasaki T, Cho K. Intellectual status of children with cerebral palsy after elementary education. Pediatr Rehabil 1997;1(4):199–206.
[110] Gonzalez-Monge S, Baya B, Ritz Lp A, Fatima, Msc A-C, Rabilloud M, et al. A 7-year longitudinal follow-up of intellectual development in children with congenital hemiplegia. Dev Med Child Neurol 2009;51(12):959–67.
[111] Fluss J, Lidzba K. Cognitive and academic profiles in children with cerebral palsy: a narrative review. Ann Phys Rehabil Med 2020;63(5):447–56. Available from: https://doi.org/10.1016/j.rehab.2020.01.005.
[112] Pueyo R, Junqué C, Vendrell P, Narberhaus A, Segarra D. Neuropsychologic impairment in bilateral cerebral palsy. Pediatr Neurol 2009;40(1):19–26. Available from: http://www.pedneur.com/article/S0887899408003913/fulltext.
[113] Geytenbeek JJ, Oostrom KJ, Harlaar L, Becher JG, Knol DL, Barkhof F, et al. Language comprehension in nonspeaking children with severe cerebral palsy: neuroanatomical substrate? Eur J Paediatr Neurol 2015;19(5):510–20. Available from: http://www.ejpn-journal.com/article/S1090379815001087/fulltext.
[114] Vos RC, Dallmeijer AJ, Verhoef M, Van Schie EM P, et al. Developmental trajectories of receptive and expressive communication in children and young adults with cerebral palsy. Dev Med Child Neurol 2014;56(10):951–9.
[115] Sigurdardottir S, Torstein V. Speech, expressive language, and verbal cognition of preschool children with cerebral palsy in Iceland. Dev Med Child Neurol 2011;53(1):74–80.
[116] Gillies MB, Bowen JR, Patterson JA, Roberts CL, Siranda T. Educational outcomes for children with cerebral palsy: a linked data cohort study. Dev Med Child Neurol 2017;60(4):397–401.

[117] Frampton I, Yude C, Goodman RM. The prevalence and correlates of specific learning difficulties in a representative sample of children with hemiplegia. Br J Educ Psychol 1998;68(1):39–51. Available from: https://onlinelibrary.wiley.com/doi/full/10.1111/j.2044-8279.1998.tb01273.x.

[118] Jenks KM, van Lieshout ECDM, de Moor JMH. Cognitive correlates of mathematical achievement in children with cerebral palsy and typically developing children. Br J Educ Psychol 2012;82(1):120–35. Available from: https://onlinelibrary.wiley.com/doi/full/10.1111/j.2044-8279.2011.02034.x.

[119] Craig F, Savino R, Trabacca A. A systematic review of comorbidity between cerebral palsy, autism spectrum disorders and attention deficit hyperactivity disorder. Eur J Paediatr Neurol 2019;23(1):31–42. Available from: http://www.ejpn-journal.com/article/S1090379818301533/fulltext.

[120] Holck P, Nettelbladt U, Sandberg AD. Children with cerebral palsy, spina bifida and pragmatic language impairment: differences and similarities in pragmatic ability. Res Dev Disabil 2009;30(5):942–51.

[121] Påhlman M, Gillberg C, Himmelmann K. Autism and attention-deficit/hyperactivity disorder in children with cerebral palsy: high prevalence rates in a population-based study. Dev Med Child Neurol 2021;63(3).

[122] Fazzi E, Signorini SG, Piana RL, Chiara B, Misefari W, Galli J, et al. Neuro-ophthalmological disorders in cerebral palsy: ophthalmological, oculomotor, and visual aspects CVI cerebral visual impairment GVPQ general visuoperceptual quotient. Dev Med Child Neurol 2012;54 (8):730–6.

[123] Philip S, Dutton G. Identifying and characterising cerebral visual impairment in children: a review. Clin Exp Optom 2014;97(3):196–208.

[124] Shevell MI, Dagenais L, Hall N. Comorbidities in cerebral palsy and their relationship to neurologic subtype and GMFCS level. Neurology 2009;72(24):2090–6. Available from: https://n.neurology.org/content/72/24/2090.

[125] Dufresne D, Dagenais L, Shevell MI. Epidemiology of severe hearing impairment in a population-based cerebral palsy cohort. Pediatr Neurol 2014;51(5):641–4. Available from: http://www.pedneur.com/article/S0887899414004032/fulltext.

[126] Reid SM, Modak MB, Berkowitz RG, Reddihough DS. A population-based study and systematic review of hearing loss in children with cerebral palsy. Dev Med Child Neurol 2011;53:11.

[127] Baguley I.J., Perkes I.E., Fernandez-Ortega J.F., Rabinstein A.A., Dolce G., Hendricks H.T. Paroxysmal sympathetic hyperactivity after acquired brain injury: consensus on conceptual definition, nomenclature, and diagnostic criteria. https://home.liebertpub.com/neu. 2014 Aug 27 [cited 2022 Apr 23];31(17):1515–20. Available from: https://www.liebertpub.com/doi/abs/10.1089/neu.2013.3301.

[128] Hills T, Bachrach S. Managing Irritability and nonoperative pain in the noncommunicative child with cerebral palsy. Cereb Palsy 2020;395–412.

[129] Freeman R, Waldorf HA, Dover JS. Autonomic neurodermatology (part II): disorders of sweating and flushing Vol SemNeurol 1992;.

[130] Asior JSG, Zamunér AR, Silva LEV, Williams CA, Baranowski R, Sacha J, et al. Clinical medicine heart rate variability in children and adolescents with cerebral palsy-a systematic literature review. J Clin Med 2020;1141:1141. Available from: http://www.mdpi.com/journal/jcm.

[131] Henderson RC, Lark RK, Gurka MJ, Worley G, Fung EB, Conaway M, et al. Bone density and metabolism in children and adolescents with moderate to severe cerebral palsy. Pediatrics 2002;110(1):e5. Available from: http://pediatrics/article/110/1/e5/64097/Bone-Density-and-Metabolism-in-Children-and.

[132] Fehlings D., Switzer L., Stevenson R., Gaebler-Spira D., Dalziel B., Ozel S. AACPDM care pathways: osteoporosis. 2016 [cited 2022 Apr 24]; Available from: http://www.aacpdm.org/publications/care-pathways/osteoporosis.

[133] Bachrach SJ, Kecskemethy HH, Harcke HT, Lark RK, Miller F, Henderson RC. Pamidronate treatment and posttreatment bone density in children with spastic quadriplegic cerebral palsy. J Clin Densitom 2006;9(2):167–74.

[134] Ozel S, Switzer L, Macintosh A, Fehlings D. Informing evidence-based clinical practice guidelines for children with cerebral palsy at risk of osteoporosis: an update Vol Devel Med Child Neurol 2016;.

[135] Dahlseng MO, Finbråten AK, Júlíusson PB, Skranes J, Andersen G, Vik T. Feeding problems, growth and nutritional status in children with cerebral palsy. Acta Paediatr 2012;101(1):92–8. Available from: https://onlinelibrary.wiley.com/doi/full/10.1111/j.1651-2227.2011.02412.x.

[136] Strauss D, Fasa P, Rosenbloom L, Shavelle R. Life expectancy in cerebral palsy: an update. Dev Med Child Neurol 2008;50:487.

[137] Blasco P, Allaire J. Drooling in the developmentally disabled: management practices and recommendations. Consortium on drooling. Dev Med Child Neurol 1992;34(10):849–62.

[138] Del Giudice E, Staiano A, Capano G, Romano A, Florimonte L, Miele E, et al. Gastrointestinal manifestations in children with cerebral palsy. Brain Dev 1999;21(5):307–11. Available from: http://www.brainanddevelopment.com/article/S038776049900025X/fulltext.

[139] Peterson MD, Gordon PM, Hurvitz EA. Chronic disease risk among adults with cerebral palsy: the role of premature sarcopoenia, obesity and sedentary behaviour. Obes Rev 2013;14(2):171–82. Available from: https://onlinelibrary.wiley.com/doi/full/10.1111/j.1467-789X.2012.01052.x.

[140] Boel L, Pernet K, Toussaint M, Ides K, Leemans G, Jurn H, et al. Respiratory morbidity in children with cerebral palsy: an overview. Dev Med Child Neurol 2018;61(6):646–53.

[141] Newman CJ, O' M, Msc R. Sleep disorders in children with cerebral palsy. Dev Med Child Neurol 2006;48(7):564–8.

[142] van der Slot WMA, Benner JL, Brunton L, Engel JM, Gallien P, Hilberink SR, et al. Pain in adults with cerebral palsy: a systematic review and meta-analysis of individual participant data. Ann Phys Rehabil Med 2021;64(3):101359.

[143] Mckinnon CT, Meehan EM, Harvey AR, Giuliana, Antolovich C, Morgan PE. Prevalence and characteristics of pain in children and young adults with cerebral palsy: a systematic review. Dev Med Child Neurol 2019;61(3):305–14.

[144] Vinkel MN, Rackauskaite G, Finnerup NB. Classification of pain in children with cerebral palsy. Dev Med Child Neurol 2022;64(4):447–52. Available from: https://onlinelibrary.wiley.com/doi/full/10.1111/dmcn.15102.

[145] Samijn B, Van Laecke E, Renson C, Hoebeke P, Plasschaert F, Vande Walle J, et al. Lower urinary tract symptoms and urodynamic findings in children and adults with cerebral palsy: a systematic review. Neurourol Urodyn 2017;36(3):541–9. Available from: https://onlinelibrary.wiley.com/doi/full/10.1002/nau.22982.

[146] Hayward K, Chen AY, Forbes E, Byrne R, Greenberg MB, Fowler EG. Reproductive healthcare experiences of women with cerebral palsy. Disabil Health J 2017;10(3):413–18.

[147] Brown HK, Chen S, Guttmann A, Havercamp SM, Parish SL, Ray JG, et al. Neonatal outcomes of mothers with a disability. Pediatrics 2022;150(3).

[148] Downs J, Blackmore AM, Epstein A, Skoss R, Langdon K, Jacoby P, et al. The prevalence of mental health disorders and symptoms in children and adolescents with cerebral palsy: a systematic review and meta-analysis. Developmental medicine and child neurology, Vol. 60. Blackwell Publishing Ltd.; 2018. p. 30–8.

[149] Whitney DG, Warschausky SA, Ng S, Hurvitz EA, Kamdar NS, Peterson MD. Prevalence of mental health disorders among adults with cerebral palsy. Ann Intern Med 2019;171(5):328–33.

[150] Tilton AH. Management of spasticity in children with cerebral palsy. Semin Pediatr Neurol 2004;11(1):58–65.

[151] Chung C, Chen C, Wong AM. Pharmacotherapy of spasticity in children with cerebral palsy. J Formos Med Assoc 2011;110(4):215–22. Available from: https://doi.org/10.1016/S0929-6646(11)60033-8.

[152] Verrotti A, Greco R, Spalice A, Chiarelli F, Iannetti P. Pharmacotherapy of spasticity in children with cerebral palsy. Pediatr Neurol 2006;34(1):1–6. Available from: http://www.ncbi.nlm.nih.gov/pubmed/16376270.

[153] Delgado MR, Hirtz D, Aisen M, Ashwal S, Fehlings DL, Mclaughlin J, et al. Practice parameter : pharmacologic treatment of spasticity in children and adolescents with cerebral palsy (an evidence-based review) neurology and the practice committee of the child neurology society. Neurology. 2010;74:336–43.

[154] Lapeyre E, Kuks JBM, Meijler WJ. Spasticity: revisiting the role and the individual value of several pharmacological treatments. NeuroRehabilitation 2010;27(2):193–200. Available from: http://www.ncbi.nlm.nih.gov/pubmed/20871149.

[155] Collins A, Mannion R, Broderick A, Hussey S, Devins M, Bourke B. Gabapentin for the treatment of pain manifestations in children with severe neurological impairment: a single-centre retrospective review. BMJ Paediatr Open 2019;3. Available from: http://bmjpaedsopen.bmj.com/.

[156] Malfitano AM, Proto MC, Bifulco M. Cannabinoids in the management of spasticity associated with multiple sclerosis Vol Neuropsychiatric Dis Treat 2008;.

[157] Fairhurst C, Kumar R, Checketts D, Bola T, Turner S. Efficacy and safety of nabiximols cannabinoid medicine for paediatric spasticity in cerebral palsy or traumatic brain injury: a randomized controlled trial. Dev Med Child Neurol 2020;62(9):1031–9.

[158] Monbaliu E, Himmelmann K, Lin JP, Ortibus E, Bonouvrié L, Feys H, et al. Clinical presentation and management of dyskinetic cerebral palsy. Lancet Neurol 2017;16(9):741–9. Available from: http://www.thelancet.com/article/S1474442217302521/fulltext.

[159] Fehlings D, Brown L, Harvey A, Himmelmann K, Lin JP, Macintosh A, et al. Pharmacological and neurosurgical interventions for managing dystonia in cerebral palsy: a systematic review. Dev Med Child Neurol 2018;60(4):356–66. Available from: https://pubmed.ncbi.nlm.nih.gov/29405267/.

[160] Brown L, Harvey A, Himmelmann K, Lin J, MacIntosh A, Mink J, et al. Recommendations for the management of dystonia in individuals with cerebral palsy. AACPDM Care Pathw 2016;. Available from: http://www.aacpdm.org/publications/care-pathways/dystonia.

[161] Harvey AR, Baker LB, Reddihough DS, Scheinberg A, Williams K. Trihexyphenidyl for dystonia in cerebral palsy. Cochrane Database Syst Rev 2018;2018(5). Available from: http://pmc/articles/PMC6494536/.

[162] Sayer C, Lumsden DE, Kaminska M, Lin JP. Clonidine use in the outpatient management of severe secondary dystonia. Eur J Paediatr Neurol 2017;21(4):621–6. Available from: http://www.ejpn-journal.com/article/S1090379817301630/fulltext.

[163] Klopstock T, Gahlot Saini A, Ramirez-Zamora A, Battini R, Scalise R, Sgandurra G, et al. A retrospective longitudinal study in a cohort of children with dyskinetic cerebral palsy treated with tetrabenazine. Front Neurol 2021;12:612429. Available from: https://eprovide.mapi-trust.org.

[164] Siniscalchi A, Gallelli L, De Sarro G. Use of antiepileptic drugs for hyperkinetic movement disorders Vol Curr Neuropharmacol 2010;.

[165] Iodice A, Pisani F. Status dystonicus: management and prevention in children at high risk. Acta Biomed 2019;90:207–12. Available from: http://www.actabiomedica.it.

[166] Allen NM, Lin JP, Lynch T, King MD. Status dystonicus: a practice guide. Dev Med Child Neurol 2014;56(2):105–12.

[167] De Zen L, Della Paolera S, Del Rizzo I, Taucar V, Skabar A, Barbi E. Home intranasal dexmedetomidine for refractory dystonia in pediatric palliative care. J Pain Symptom Manage 2020;59(6):e3–5. Available from: http://www.jpsmjournal.com/article/S0885392420301251/fulltext.

[168] Nakou V, Williamson K, Arichi T, Lumsden D, Tomlin S, Kaminska M, et al. Safety and efficacy of high-dose enteral, intravenous, and transdermal clonidine for the acute management of severe intractable childhood dystonia and status dystonicus: an illustrative case-series. Eur J Paediatr Neurol 2017;21(6):823–32.

[169] Jr PRF, Calhes R, Moura FD, Galli M, Oliveira CS. Effect of physiotherapeutic intervention on the gait after the application of botulinum toxin in children with cerebral palsy : systematic review. Eur J Phys Rehabil Med 2018;54(5):757–65.

[170] Miller SD, Juricic M, Hesketh K, Lynore M, Magnuson S, Sherylin G, et al. Prevention of hip displacement in children with cerebral palsy: a systematic review. Dev Med Child Neurol 2017;58(Supplement 5):62–3. Available from: http://www.otseeker.com.

[171] Paget SP, Swinney CM, Karen, Burton LO, Bau K, Stephen, et al. Systemic adverse events after botulinum neurotoxin A injections in children with cerebral palsy. Dev Med Child Neurolol 2018;60(11):1172–7.

[172] Van Campenhout A, Verhaegen A, Pans S, Molenaers G. Botulinum toxin type A injections in the psoas muscle of children with cerebral palsy: muscle atrophy after motor end plate-targeted injections. Res Dev Disabil 2014;34(3):1052–8.

[173] De Beukelaer N, Weide G, Huyghe E, Vandekerckhove I, Hanssen B, Peeters N, et al. Reduced cross-sectional muscle growth six months after botulinum toxin type-a injection in children with spastic cerebral palsy. Toxins (Basel) 2022;2022:139.

[174] Jiao Y, Li X-Y, Liu J. A new approach to cerebral palsy treatment: discussion of the effective components of umbilical cord blood and its mechanisms of action. Child's Nerv Syst 2019;37:1729–40. Available from: https://us.sagepub.com/en-us/nam/open-access-at-sage.

[175] Sun JM, Case LE, McLaughlin C, Burgess A, Skergan N, Crane S, et al. Motor function and safety after allogeneic cord blood and cord tissue-derived mesenchymal stromal cells in cerebral palsy: An open-label, randomized trial. Dev Med Child Neurol 2022;64(12):1477–86.

[176] Eggenberger S, Boucard C, Schoeberlein A, Guzman R, Limacher A, Surbek D, et al. Stem cell treatment and cerebral palsy: systemic review and meta-analysis conflict-of-interest statement: PRISMA 2009 Checklist statement. World J Stem Cell 2019;11(10):891–903. Available from: https://doi.org/10.4252/wjsc.v11.i10.891.

[177] Sewell MD, Eastwood DM, Wimalasundera N. Managing common symptoms of cerebral palsy in children. BMJ 2014;349. Available from: https://www.bmj.com/content/349/bmj.g5474.

[178] Shankar Ganesh MG, Das SP, Shankar Ganesh G. Evidence-based approach to physical therapy in cerebral palsy. Indian J Orthop 2019;53 (1):20–34. Available from: http://www.ijoonline.com.

[179] Kinnear BZ. Physical therapies as an adjunct to Botulinum toxin-A injection of the upper or lower limb in adults following neurological impairment. Syst Rev 2012;26:1–29. Available from: http://www.systematicreviewsjournal.com/content/1/1/29.

[180] Hung YC, Casertano L, Hillman A, Gordon AM. The effect of intensive bimanual training on coordination of the hands in children with congenital hemiplegia. Res Dev Disabil 2011;32(6):2724–31.

[181] Morgan C, Fetters L, Adde L, Badawi N, Bancale A, Boyd RN, et al. Early intervention for children aged 0 to 2 years with or at high risk of cerebral palsy: international clinical practice guideline based on systematic reviews. JAMA Pediatr 2021;175(8):846–58. Available from: https://jamanetwork.com/journals/jamapediatrics/fullarticle/2780012.

[182] Novak I, Morgan C, Fahey M, Finch-Edmondson M, Galea C, Hines A, et al. State of the evidence traffic lights 2019: systematic review of interventions for preventing and treating children with cerebral palsy Vol Curr Neurol Neurosci Rep 2020;.

[183] Häusler M, Heussen N. Protocol for a systematic review and meta-analysis on the effect of hippotherapy and related equine-assisted therapies on motor capabilities in children with cerebral palsy. Syst Rev 2020;9(1).

[184] Morris C. Orthotic management of children with cerebral palsy. J Prosthet Orthot 2002;14(4):150–8.

[185] Terjesen T, Lange JE, Steen H. Treatment of scoliosis with spinal bracing in quadriplegic cerebral palsy Vol Developmental Med & Child Neurol 2000;.

[186] Miller A, Temple T, Miller F. Impact of orthoses on the rate of scoliosis progression in children with cerebral palsy. J Pediatr Orthop 1996;16(3):332–5.

[187] Milne N, Miao M, Beattie E. The effects of serial casting on lower limb function for children with cerebral palsy: a systematic review with meta-analysis. BMC Pediatr 2020;20(1):1–23. Available from: https://bmcpediatr.biomedcentral.com/articles/10.1186/s12887-020-02122-9.

[188] Dai AI, Demiryürek AT. Serial casting as an adjunct to botulinum toxin type a treatment in children with cerebral palsy and spastic paraparesis with scissoring of the lower extremities. J Child Neurol 2017;32(7):671–5. Available from: https://journals.sagepub.com/doi/10.1177/0883073817701526?url_ver = Z39.88-2003&rfr_id = ori%3Arid%3Acrossref.org&rfr_dat = cr_pub++0pubmed.

[189] Angsupaisal M, Maathuis CGB, Hadders-Algra M. Adaptive seating systems in children with severe cerebral palsy across International classification of functioning, disability and health for children and youth version domains: a systematic review Vol Developmental Med Child Neurol 2015;.

[190] Mcginley JL, Dobson F, Ganeshalingam R, Shore BJ, Rutz E, Kerr Graham H. Single-event multilevel surgery for children with cerebral palsy: a systematic review CHQ Child Health Questionnaire FAQ Gillette Functional Assessment Questionnaire FMS Functional Mobility Scale GDI Gait Deviation Index GGI Gillette Gait Index GPS Gait P. Dev Med Child Neurol 2012;54(2):117–28.

[191] Shore BJ, Mulpuri K, Shrader MW. Radiographic hip screening in cerebral palsy: developing POSNA wide consensus JPOSNA®. J POSNA 2021;3:1–13. Available from: https://www.jposna.org/ojs/index.php/jposna/article/view/231.

[192] O'Donnell M, Mayson T, Miller S, Cairns R, Graham K, Love S, et al. AACPDM care pathways: hip surveill 2017;. Available from: http://www.aacpdm.org/publications/care-pathways/hip-surveillance.

[193] Hägglund G, Pettersson K, Czuba T, Persson-Bunke M, Rodby-Bousquet E. Incidence of scoliosis in cerebral palsy A population-based study of 962 young individuals. Acta Orthop 2018;89(4):443–7.

[194] Koman L, Gelberman R, Toby E, Poehling GG. Cerebral palsy. Manag Up Extremity Clin Orthop Relat Res 1990;253:62–74.

[195] Louwers A, Warnink-Kavelaars J, Daams J, Beelen A. Effects of upper extremity surgery on activities and participation of children with cerebral palsy: a systematic review. Dev Med Child Neurol 2020;62(1):21–7.

[196] Goldner JL. Upper extremity tendon transfers in cerebral palsy. Orthop Clin North Am 1974;5(2):389–414. Available from: https://linkinghub.elsevier.com/retrieve/pii/S0030589820310816.

[197] McLaughlin J, Bjornson PTKM, Steinbok P, Wright V, Drake MBBCh J, Rosenbaum P, et al. Selective dorsal rhizotomy: meta-analysis of three randomized controlled trials. Dev Med Child Neurol 2002;44:17–25.

[198] Munger ME, Aldahondo N, Krach LE, Novacheck TF, Schwartz MH. Long-term outcomes after selective dorsal rhizotomy: a retrospective matched cohort study. Dev Med Child Neurol 2017;59(11):1196–203.

[199] Veerbeek B.E., Lamberts R.P., Fieggen A.G., Verkoeijen P.P.J.L., Langerak N.G. Daily activities, participation, satisfaction, and functional mobility of adults with cerebral palsy more than 25 years after selective dorsal rhizotomy: a long-term follow-up during adulthood. Available from: https://doi.org/101080/0963828820191695001. 2019;43(15):2191–9, https://www.tandfonline.com/doi/abs/10.1080/09638288.2019.1695001

[200] Abdel Ghany W, Nada M, Mahran M, Aboud A, Mahran M, Nasef M, et al. Combined anterior and posterior lumbar rhizotomy for treatment of mixed dystonia and spasticity in children with cerebral palsy. Neurosurgery 2016;79:336–44. Available from: http://www.neurosurgery-online.com.

[201] Scott Gillespie C, Matthew George A, Hall B, Toh S, Ismail Islim A, Hennigan D, et al. The effect of GMFCS level, age, sex, and dystonia on multi-dimensional outcomes after selective dorsal rhizotomy: prospective observational study. Child's Nerv Syst 2021;1729–40. Available from: https://doi.org/10.1007/s00381-021-05076-0.

[202] Kim H, Steinbok P, Wickenheiser D. Predictors of poor outcome after selective dorsal rhizotomy in treatment of spastic cerebral palsy. Childs Nerv Syst 2006;22(1):60–6.

[203] Elkaim LM, Alotaibi NM, Sigal A, Alotaibi HM, Lipsman N, Kalia SK, et al. North American Pediatric DBS Collaboration Deep brain stimulation for pediatric dystonia: a meta-analysis with individual participant data. Devel Med Child Neurol 2018;61:49–56.

[204] Sanger TD. Deep brain stimulation for cerebral palsy: where are we now? Devel Med Child Neurol 2020;62:28–33.

[205] Tsering D, Tochen L, Lavenstein B, Reddy SK, Granader Y, Keating RF, et al. Considerations in deep brain stimulation (DBS) for pediatric secondary dystonia. Child's Nerv Syst 2017;33(4):631–7. Available from: https://link.springer.com/article/10.1007/s00381-017-3361-x.

[206] Molnar GE. Cerebral palsy: prognosis and how to judge it. Pediatr Ann 1979;8(10).

[207] Wu YW, Day SM, Strauss DJ, Shavelle RM. Prognosis for ambulation in cerebral palsy: a population-based study. Pediatrics 2004;114(5).

[208] Palisano RJ, Cameron D, Rosenbaum PL, Walter SD, Russell D. Stability of the gross motor function classification system. Dev Med Child Neurol 2006;48(6):424–8. Available from: https://onlinelibrary.wiley.com/doi/full/10.1111/j.1469-8749.2006.tb01290.x.

[209] Palisano RJ, Avery L, Gorter JW, Galuppi B, McCoy SW. Stability of the gross motor function classification system, manual ability classification system, and communication function classification system. Dev Med Child Neurol 2018;60((10):1026–32. Available from: https://onlinelibrary.wiley.com/doi/full/10.1111/dmcn.13903.

[210] Murphy KP. The adult with cerebral palsy. Orthop Clin North Am 2010;41(4):595–605.

[211] Murphy KP. Cerebral palsy lifetime care-four musculoskeletal conditions. Dev Med Child Neurol 2009;51(Suppl.):30–7.

[212] Morgan P, McGinley J. Gait function and decline in adults with cerebral palsy: a systematic review. Disabil Rehabil 2014;36:1–9.

[213] Bartlett DJ, Hanna SE, Avery L, Stevenson RD, Galuppi B. Correlates of decline in gross motor capacity in adolescents with cerebral palsy in gross motor function classification system levels III to V: an exploratory study. Dev Med Child Neurol 2010;52(7):155–60.

[214] Boldingh EJ, Jacobs-van der Bruggen MA, Bos CF, Lankhorst GJ, Bouter LM. Determinants of hip pain in adult patients with severe cerebral palsy. J Pediatr Orthop B 2005;14(2):120–5.

[215] Carter DR, Bse BT. The pathogenesis of osteoarthritis in cerebral palsy. Med Child Neurol 2009;51:79–83.

[216] Houdek MT, Watts CD, Wyles CC, Trousdale RT, Milbrandt TA, Taunton MJ. Total hip arthroplasty in patients with cerebral palsy: a cohort study matched to patients with osteoarthritis. J Bone Jt Surg - Am Vol 2017;99(6):488–93. Available from: https://journals.lww.com/jbjsjournal/Fulltext/2017/03150/Total_Hip_Arthroplasty_in_Patients_with_Cerebral.6.aspx.

[217] Kim KN, Ahn PG, Ryu MJ, Shin DA, Yi S, Yoon DH, et al. Long-term surgical outcomes of cervical myelopathy with athetoid cerebral palsy. Eur Spine J 2013;23(7):1464–71. Available from: https://link.springer.com/article/10.1007/s00586-013-3119-4.

[218] McLaughlin SE, Machan J, Fournier P, Chang T, Even K, Sadof M. Transition of adolescents with chronic health conditions to adult primary care: factors associated with physician acceptance. J Pediatr Rehabil Med 2014;7(1):63–70.

[219] Manikandan M, Kerr C, Lavelle G, Walsh M, Aisling W, Ryan JM. Health service use among adults with cerebral palsy: a mixed-methods systematic review. Dev Med Child Neurol 2021;64(4):429–46. Available from: https://zenodo.org/record/4730672#.YWAbdtrMJPY.

[220] Rodby-Bousquet E., Graham H.K., Thorpe D.E., Shrader M.W., Church C., Lennon N., et al. Well-being of ambulatory adults with cerebral palsy: education, employment, and physical function of a cohort who received specialized pediatric care. Front Neurol. 2021;12:732906. Available from: http://www.frontiersin.org.

[221] Pizzighello S., Medea E., Langhammer B., Pettersson K., Rodby-Bousquet E. Living conditions and social outcomes in adults with cerebral palsy. Front Neurol. 2021;12:749389. Available from: http://www.frontiersin.org.

[222] September 2020 Subminimum Wages Impacts on the Civil Rights Of People With Disabilities. [cited 2022 Apr 24]; Available from: http://www.usccr.gov.

[223] Day SM, Robert RJ, Scott KJ. Extrapolating published survival curves to obtain evidence-based estimates of life expectancy in cerebral palsy. Dev Med Child Neurol 2015;57(12):1105–18.

[224] Rimmer JH. Physical fitness levels of persons with cerebral palsy Vol Developmental Med Child Neurol 2001;.

[225] Maas VYF, Poels M, Lamain-de Ruiter M, Kwee A, Bekker MN, Franx A, et al. Associations between periconceptional lifestyle behaviours and adverse pregnancy outcomes. BMC Pregnancy Childbirth 2021;21(1).

[226] Burris HH, Baccarelli AA, Wright RO, Wright RJ. Epigenetics: linking social and environmental exposures to preterm birth. Pediatric Research, Vol. 79. Nature Publishing Group; 2016. p. 136–40.

[227] Shepherd E, Salam RA, Middleton P, Han S, Makrides M, Mcintyre S, et al. Neonatal interventions for preventing cerebral palsy: an overview of cochrane systematic reviews Vol Cochrane Database Syst Rev 2018;.

[228] Doyle L, Crowther C, Middleton P, Marret S, Rouse D. Magnesium sulphate for women at risk of preterm birth for neuroprotection of the fetus. Cochrane Database Syst Rev 2009;21(1).

[229] Juul SE, Ferriero DM. Pharmacologic neuroprotective strategies in neonatal brain injury Vol ClPerinatology 2014;119–31.

[230] Zhou KQ, Davidson JO, Bennet L, Gunn AJ. Combination treatments with therapeutic hypothermia for hypoxic-ischemic neuroprotection Vol Dev Med Child Neurol 2020;1131–7.

[231] Liptak G, Revell G. Community physician's role in case management of children with chronic illnesses. Pediatrics. 1989;84(3):465–71.

[232] Johnson CP, Blasco PA. Community resources for children with special healthcare needs. Pediatr Ann 1997;26(11):679–86. Available from: https://ohsu.pure.elsevier.com/en/publications/community-resources-for-children-with-special-healthcare-needs-2.

Chapter 16

Developmental coordination disorder and minor neuromotor dysfunction

Elizabeth Barkoudah[1], Christina Briscoe Abath[1], Meghan McAnally[2] and Rajapillai L.I. Pillai[1]
[1]*Department of Neurology, Boston Children's Hospital, Harvard Medical School, Boston, MA, United States,* [2]*National Institute of Neurological Disorders and Stroke, NIH, Bethesda, MD, United States*

Introduction

The term neuromotor disabilities encompasses a broad array of predominantly motor deficits of neurologic origin. Prototypical disorders include cerebral palsy and spina bifida. This chapter discusses three childhood neuromotor disorders: developmental coordination disorder, developmental dyspraxia, and benign congenital hypotonia. These diagnoses are often considered minor neuromotor disabilities, diagnosed in infants and children with persistent motor delays who nonetheless achieve important milestones such as walking by 2 years of age [1].

Though labeled minor neuromotor disabilities, these disorders carry the risk of accompanying coordination difficulties, motor planning problems, fine motor incoordination, and sensorimotor integration challenges, therefore the implications for children and their families can be quite significant [2]. Early recognition of these subtle deficits allows for a discussion of expectations and prognosis, optimization of care, greater gains with therapies, and appropriate school and community supports to maximize functioning and prevent secondary consequences, including poor self-esteem and negative peer relationships.

We outline the history, definition, epidemiology, clinical manifestations, neurobiology, and treatment for developmental coordination disorder, developmental dyspraxia, and benign congenital hypotonia. Though they are common disorders, much of the pathophysiology, etiology, and treatment of these disorders remain unknown. We conclude this chapter with opportunities for future study.

Developmental coordination disorder

Case 1. Part A.

A 6-year-old right-handed boy with no significant past medical history presents to the neurology clinic due to significant issues with clumsiness and bumping into objects. His prenatal and birth histories are unremarkable. There were no significant intrauterine medication or substance exposures. His family states that he met all his developmental milestones but has always been clumsy compared with other children his age. They do not feel that these issues have gotten worse over time, but they have been more of a problem since he started school. His teacher has noted that learning to tie his shoes is an issue for him, and he has trouble keeping up with other children during recess. He additionally has challenges with reading and is behind grade level. His vital signs and general and neurologic examinations were unremarkable. A diagnosis of developmental coordination disorder (DCD) is made.

History

The first description of DCD was by Collier in the early 1900s, when he described "congenital maladroitness" [3]. This term continued to be used through the 1930s [4]. By the 1960s, the term "minimal brain dysfunction" began to be used for this subset of children with normal intellect but challenges with coordination. Through the 1970s, a variety of terms were used, including sensory integration disorder, clumsy child syndrome, and developmental dyspraxia [5–7].

Capute and Accardo's Neurodevelopmental Disabilities in Infancy and Childhood. DOI: https://doi.org/10.1016/B978-0-12-824060-1.00055-9

The Diagnostic and Statistical Manual from 1980 was the first publication to formally outline a set of criteria to describe developmental incoordination under the term DCD (APA 1980). In 1994, a conference held in London, Ontario, determined the term DCD should be used to improve consistency in the literature; this was known as the "London Consensus." Other statements have reaffirmed this term for both clinical and research use [8]. This term has continued to be used through the DSM-5 without significant alterations. Some authors have argued for the use of the term developmental dyspraxia due to the recognition of cognitive implications associated with the early onset of coordination concerns, though these samples may have been biased as they were from children referred for testing in a developmental medicine clinic [9]. This controversy will be further described in the section on developmental dyspraxia. The ICD-10 term "specific developmental disorder of motor function" is not used routinely by researchers or clinicians.

Definition and epidemiology

DCD is formally defined by the *Diagnostic and Statistical Manual of Mental Disorders* (5th ed.; DSM-5 [10]) as a nonprogressive impaired coordination or clumsiness with onset in the early developmental period that results in functional limitations not better explained by other neurodevelopmental disorders (see Table 16.1). It is typically diagnosed after the age of 5 years old, as an earlier diagnosis may not represent a stable or persistent finding.

DCD is common, estimated to affect approximately 6% of primary school-aged children, though estimates have ranged between 1% and 13% [11]. Variants in estimates largely appear to stem from how strict criteria are applied and methodological differences in diagnosis. Boys are at least two times more likely than girls to meet criteria for DCD in most studies [12,13]. There have not been racial or socioeconomic disparities reported in the literature, though this has not been specifically examined. Documented risk factors include prematurity, in-utero toxic exposures, and the presence of other neurodevelopmental disabilities, including ADHD and autism spectrum disorder [12].

Current DSM-5 criteria allow for the diagnosis of DCD in adults and adolescents; however, appropriate tools for diagnosis, evaluation, and intervention for these age groups are an emerging area of study.

Signs and symptoms

As outlined in the definition, DCD is marked by challenges with coordination and dexterity without other neurologic, motor, or cognitive dysfunction that better explains those difficulties. Symptoms typically include challenges with higher-level gross and fine motor tasks. Fine motor tasks can include school-based challenges like handwriting and play-based tasks such as using scissors or completing art projects. Common gross motor challenges can include bike riding, playing sports, and clumsiness or frequent falls.

TABLE 16.1 Developmental coordination disorder is often characterized as a skill acquisition deficit disorder. The DSM-5 is the 2013 update to the Diagnostic and Statistical Manual of the Mental Disorders published by the American Psychiatric Association. The DSM-5 is a tool and reference guide for clinicians to diagnose, classify, and identify mental health conditions.

DSM-5 diagnostic criteria for developmental coordination disorder
1. The acquisition and execution of coordinated motor skills are substantially below what is expected given the individual's chronological age and opportunity for skill learning and use. Difficulties manifest as clumsiness (e.g., dropping or bumping into objects) as well as slowness and inaccuracy of performance of motor skills (e.g., catching an object, using scissors or cutlery, handwriting, riding a bike, or participating in sports).
2. The motor skills deficit in Criterion A significantly and persistently interferes with activities of daily living appropriate to chronological age (e.g., self-care and self-maintenance) and impacts academic and school productivity, prevocational and vocational activities, leisure, and play.
3. The onset of symptoms is in the early developmental period.
4. The motor skills deficits are not better explained by intellectual disability (intellectual developmental disorder) or visual impairment and are not attributable to a neurological condition affecting movement (e.g., cerebral palsy, muscular dystrophy, or degenerative disorder).

As there are no deficits of power, tone, or cerebellar findings in these patients to physically explain their inability to effectively learn and perform motor tasks, these deficits are generally related to higher-level visual-motor planning and integration rather than corticospinal tract, motor nerve, or muscle pathology. Symptoms of DCD tend to become more apparent in tasks that require precision, advance planning, or multitasking [11]. While executive functioning challenges are common, authors disagree as to whether these are cooccuring or a core deficit of DCD.

Clinical significance and prognosis

DCD is associated with a significantly increased risk of social isolation, low self-worth, and psychiatric comorbidities such as anxiety and depression; these effects are presumed to be due to DCD's impact on activities of daily living, play, school, and social interactions [1,14,15]. Parents of children with DCD tend to worry about their long-term functioning and societal integration. Children with DCD are at risk for bullying [16].

In addition to psychological consequences, DCD is thought to be a risk factor for poor fitness and resultant cardiovascular illness/obesity [17]. One study of 118 children from the U.S. and 43 children from Israel showed statistically significant differences in obesity between children with (45%) and without (23%) DCD [18]. Another study of 578 children found a statistically significant difference in obesity for boys but not girls with DCD [19].

Outcomes of DCD in adolescence and adulthood have not been well studied, but most studies indicate that a significant proportion of adults (estimates have ranged between 50% and 70%) with a history of childhood DCD continue to have challenges [20,21]. The persistence of deficits would seem to suggest that DCD is not merely a developmental delay but a neurodevelopmental disability. Beyond pure motor tasks, nonmotor implications of DCD such as language and communication challenges, poor academic performance, and depression or low self-esteem may persist throughout the lifetime; as such, these should additionally be targets for monitoring and intervention [11].

Etiology and neurobiology

The etiology of DCD remains unclear, with both genetic and environmental factors being potentially implicated by different authors without conclusive evidence. There have not been specific genes identified, though one study noted that nonspecific copy-number variations were more common among children with DCD [22].

The mirror neuron system has been theorized to be involved in DCD through impairment in imitation and internal representation, but evidence for this has been mixed [23]. There is also some evidence of neuroconnectivity differences based on functional imaging studies. However, these have been hampered by sample size, a lack of standardization of diagnoses and populations, and variable methodology and results. For example, some functional neuroimaging studies have demonstrated decreased activation of the dominant parietal lobe with visual-motor tasks in children with DCD compared to controls, as well as decreased activation of tracts connecting the cerebellum and parietal lobe during trail-tracing tasks [24,25]; another showed increased activation of the postcentral gyrus [26]. White matter differences on diffusion tensor imaging have also been suggestive of organizational differences connecting sensory and motor cortices and subcortical regions [27]. The nature and implications of these candidate connectivity differences will require further study.

Diagnostic process

DCD is often diagnosed by developmental pediatricians, neurologists, and neurodevelopmental disabilities specialists based on a consistent history and examination in keeping with DSM-5 criteria (see Table 16.1). However, a lack of recognition of the importance of this diagnosis, challenges with access to specialists, and relatively unfamiliarity with DCD may hinder this approach, as many children with DCD may not be referred to specialist care absent other neurodevelopmental concerns.

There are no specific or definitive tests for DCD at present, and most recent guidelines suggest a team-based approach to diagnosis [11] (See Fig. 16.1). One possible tool is the Developmental Coordination Disorder Questionnaire (DCDQ), a freely available parent and child-report questionnaire that screens for motor impairment compared with other age-matched children between 5 and 15 years old [28]. Scores are summed, with higher scores indicating better coordination (range: 15–75). Cutoff values for concern for diagnosis range between the 5th and 10th percentiles [29]. The survey has been used and validated in varied populations, including in large groups of otherwise healthy children [30,31].

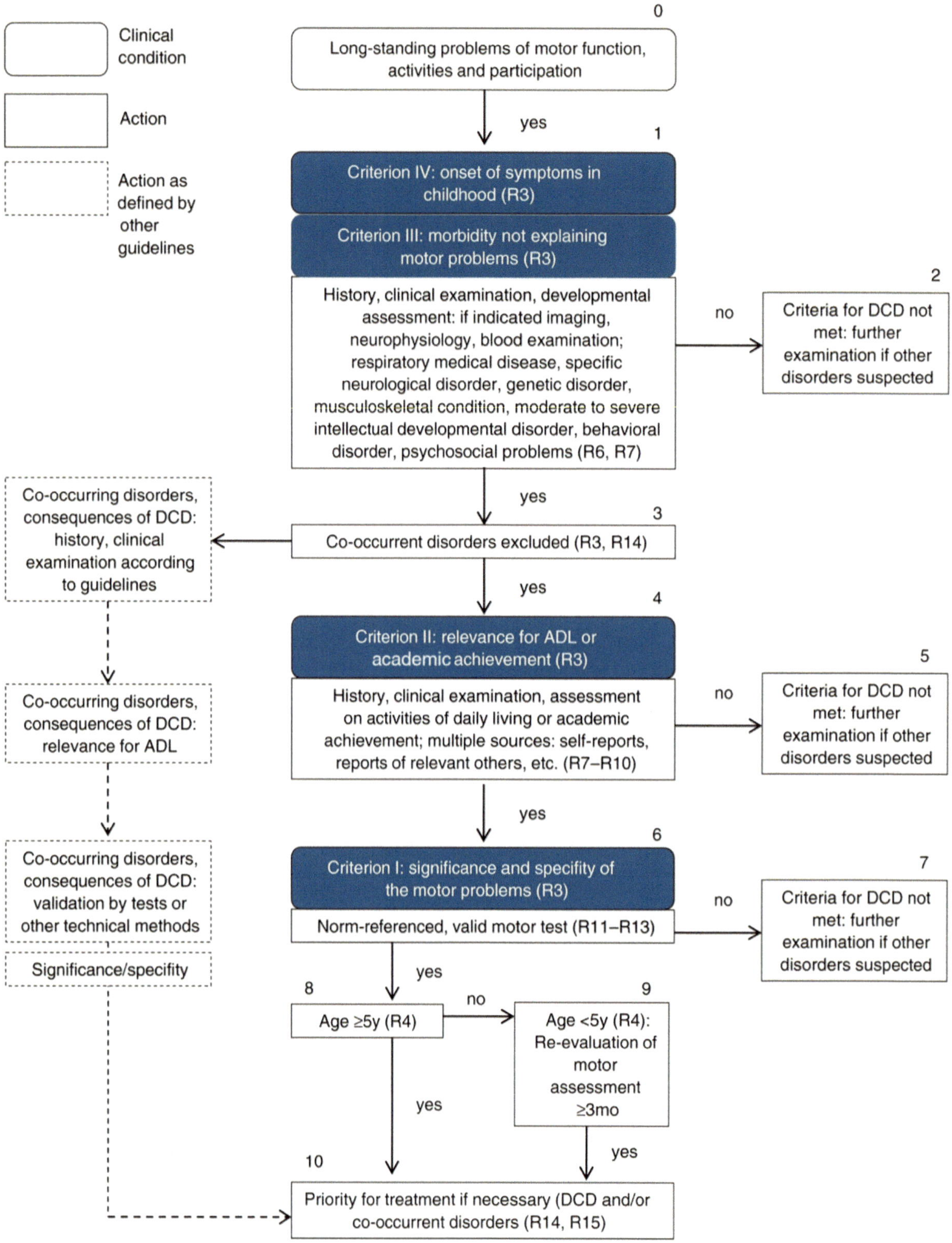

FIGURE 16.1 Flow chart assessment, treatment indication, and planning for DCD. *From Blank R, Barnett A, John C, Dido G, Kirby A, Polatajko H, et al. International clinical practice recommendations on the definition, diagnosis, assessment, intervention, and psychosocial aspects of developmental coordination disorder. Dev Med Child Neurol 2019;61.3: 242–285.*

While the most recent European Academy of Childhood Disability (EACD) guidelines do not recommend the DCDQ as a population-based screening tool due to concerns about sensitivity, the authors of this chapter argue that it is a reasonable first tier assessment tool for primary care physicians or other providers concerned about DCD to use in conjunction with clinical judgment (in a fashion similar to the MCHAT – Modified Checklist for Autism in Toddlers) given the very significant delays in access to developmental or neurodevelopmental-trained specialists in most settings. A DCD diagnosis does, however, require the exclusion of other explanatory pathologies through medical history and examination; similar to the MCHAT. A questionnaire cannot replace the clinical examination and history in diagnosis but may be a useful adjunctive tool.

Other questionnaires like the Movement Assessment Battery for Children, 2nd edition (MABC-2-C), the Motor Observation Questionnaire for Teachers, and the DCD Daily Questionnaire (DCDDailyQ) have been developed but currently lack the level of evidence to recommend their use. They can be useful, however, as they prioritize the evaluation of functional activities of daily living.

Differential diagnosis and cooccuring disorders

The presence of tone, power, bulk, reflex, cerebellar, or other abnormalities on a neurologic examination should cause providers to consider other diagnoses. Similarly, developmental regression is not consistent with DCD, which by definition is nonprogressive. New incoordination should raise the possibility of other diagnoses like neurodegenerative conditions, including some white matter disorders with specific treatments, beyond the scope of this chapter [See chapter Neurodegenerative disorders]. When global developmental delays, severe intellectual disability (ID), or significant gross or fine motor delays are present, a diagnosis of DCD may not be indicated, as coordination difficulties are better explained by other disorders. Some clinicians will diagnose DCD in individuals with ID when coordination difficulties are out of proportion to IQ deficits and cause significant adaptive challenges; however, this has not been well studied or systematized to date [11].

Attention deficit disorder (ADHD) is an important differential consideration for DCD, though these disorders can cooccur. Without comorbid DCD, ADHD can lead to challenges with handwriting, sports, or other complex tasks due to inattention, impulsivity, or hyperactivity. A careful history, including coordination during preferred activities, can be helpful to clarify concomitant DCD in a child with ADHD. The persistence of deficits with effective treatment of ADHD may additionally be suggestive.

Cooccurring disorders

Children with DCD are more likely to have other neurodevelopmental and psychiatric disorders, including ADHD, autism spectrum disorder (ASD), social anxiety, and behavioral challenges. In specialist clinics in neurology or developmental medicine, the cooccurrence of many of these disorders merits specific consideration of DCD in children with other neurodevelopmental disorders. Conversely, a diagnosis of DCD should prompt consideration of other disorders.

Some reports have estimated that up to half of children with DCD have ADHD, though these populations were from developmental clinics and therefore may not be representative [32,33]. The presence of both diagnoses predicts more significant lifetime disability than either alone, including risk for substance use, reading disorders, and lower educational achievement [34].

Challenges in spatial awareness and motor abilities are well recognized in children with ASD, contributing to challenges in adaptive functioning; a diagnosis of cooccuring DCD is allowed by the DSM-5 [35]. A study of over 16,000 children with ASD without intellectual disability or significant neuromotor delays showed that, though only 32% of children with ASD were receiving physical or occupational therapy and only 15% also had an established DCD diagnosis, 89% had an indication of or suspicion for DCD on parent questionnaires (DCDQ) [36]. A population-based study of children with DCD found that 10 out of 122 with severe DCD had ASD [12]. Therefore, despite ongoing significant controversy over whether shared etiology and neuropathology lead to combined ASD and DCD phenotypes or whether motor impairment itself is a feature of ASD (therefore not separate diagnostically), recognition and treatment of motor impairments in children with ASD remain as important for long-term social inclusion, participation, and physical health as they are for children with DCD alone.

The cooccurrence of specific language impairments has been shown, though precise studies in this domain are lacking. Most studies have assessed populations of children with language disorders for previously undiagnosed motor challenges (DCD); they have shown that between 50% and 70% of children with specific language impairments also met criteria for DCD [37,38].

Finally, as previously alluded to, children with DCD are at greater risk of lower self-concept, depression, social anxiety, and low self-esteem [11].

Therapies and management

Treatment for DCD should concentrate on improvements in activities of daily living (including school and leisure), participation and social integration, coping and quality of life, and personal satisfaction [1,11]. The focus of therapies should be on the functional and emotional aspects of motor impairment rather than resolving the impairment, as this is more effective [39]. Therapies should be child-centered, goal-oriented, aim for functionality, not normality, and actively involve children and caregivers. Given cardiovascular risk, physical fitness should be incorporated into therapeutic action plans for DCD.

Therapies can be activity-oriented (designed to improve the performance of an activity) or participation-oriented (improve participation in everyday life activities). Activity-oriented approaches (such as neuromotor task training (NTT) and cognitive orientation to daily occupational performance approach (CO-OP)) have been found to be beneficial for gross and fine motor difficulties [40,41]. Activity-oriented interventions can improve skill performance and should be individualized to the child's strengths and weaknesses [42]. Videogame-based therapies have been shown to improve balance and some functional tasks, but their generalizability to everyday life has not been well studied [43].

Relatively short interventions can show improvements, with as few as 10 sessions in many trials. Comparative studies to assess the relative effectiveness of interventions have not been done, therefore there are not clear recommendations on the intensity or duration of therapies. However, the expert recommendation following any specific DCD-oriented intervention is to provide opportunities for the child to continue to practice the skills learned. There is no evidence for the effectiveness of supplements, such as vitamin E, fatty acids, or other medications for DCD. Cooccuring disorders, such as ADHD, should be treated according to their separate guidelines.

Case 1. Part B.

The patient is sent for cognitive and academic testing, which reveals a specific learning disorder in reading. Conners and Vanderbilt Rating Scales were performed with his family and school; he did not meet criteria for ADHD. There are no signs or symptoms concerning autism spectrum disorder. An individualized education plan with specific accommodations and supports for his learning disability and DCD is put into place, including occupational and physical therapy for his DCD. His therapists craft an individualized plan to target his activity participation and fine motor challenges. His family additionally enrolls him in a martial arts activity; they intermittently check in with his teachers to ensure there is no concern for bullying.

Developmental dyspraxia

Ambiguity and controversy surrounding dyspraxia

Developmental dyspraxia, in contrast to DCD, continues to be poorly defined in the literature, in large part due to imprecision in both generalizability and specificity of the term. From a generalizability perspective, the term dyspraxia has been applied to various conditions, both with and a without known neurological injury. With regards to specificity, a variety of terms have been used in the literature to describe dyspraxia, including developmental coordination disorder, disorder of attention and motor perception (DAMP), clumsiness or clumsy child syndrome, minimal brain dysfunction, developmental apraxia, perceptuomotor dysfunction, motor learning difficulty, sensory integration disorder, and developmental awkwardness [7,44].

The history of developmental dyspraxia runs parallel to that of DCD, as the two remained undifferentiated along with other definitions until the International Consensus Meeting on Children and Clumsiness in October of 1994 established DCD as the formal label for this condition [45]. Despite these efforts, considerable confusion remained, with child health professionals and educators using DCD and dyspraxia inconsistently and ascribing cognitive problems even outside the motor domain to dyspraxia [46]. Even within a single organization, terminology is not consistent: the United Kingdom Dyspraxia Foundation states that dyspraxia is another name for DCD (https://dyspraxiafoundation.org.uk/about-dyspraxia/dyspraxia-glance/), while the United States website (https://dyspraxiausa.org/) states that DCD is a type of dyspraxia.

Given the ambiguity in the use of these terms both among medical professions and in publicly available resources, discussion of dyspraxia with patients and colleagues can be fraught with confusion. While DCD and developmental dyspraxia are understood by many to be synonymous, there may be utility in the term dyspraxia that merits further discussion.

Definitions of dyspraxia

In discussing the term dyspraxia, it is necessary to have a clear definition of normal praxis. Praxis has been defined as the ability to produce a gesture and includes ideation or conceptualization, motor planning, and execution of that movement [44,47]. Developmentally, praxis has been described to start with object manipulation (e.g., grabbing one's tooth), to body-part-as-tool (e.g., using a finger to brush teeth), to holding errors (e.g., representing the toothbrush as having an overly large handle), and finally to accurate pantomime [48,49]. Functional networks involving the dominant frontal and parietal lobes and basal ganglia are involved in the generation of normal praxis, with areas of the parietal cortex such as the inferior parietal sulcus and inferior parietal lobule particularly implicated in the ability to pantomime [50].

The term dyspraxia has been used to signify a disorder of praxis development. In contrast to apraxia (acquired loss of previously normally developed praxis), localization of dyspraxia is generally considered a diffuse process that may involve both motor and cognitive domains [51]. One study found that children with dyspraxia have more difficulty in speech planning and visuospatial reasoning compared to children with dyslexia but no dyspraxia; this group hypothesized that dyspraxia may reflect a difficulty generating an abstract representation as a basis for planning [9]. Another study found patients with congenital left hemisphere lesions to have more dyspraxic symptoms [52]. However, few studies have been done to investigate this, likely in part due to fundamental difficulties in defining the term.

Some have argued that dyspraxia itself is not a neurological disorder but a specific neurological sign indicative of a deficit in performing skilled movements [49]. This position is strengthened by the fact that dyspraxia has been described in several other, more clearly defined disorders. For example, one large study found clinically defined dyspraxia to be present in 6.9% of children with autism spectrum disorder (ASD) and only 0.8% in children without ASD [53]. In this context, dyspraxia can be used simply to describe the signs of dysfunctional development of praxis in the same way one uses dysdiadochokinesia or dysmetria in assessing cerebellar function—that is, as a neurologic sign indicating a disorder, but not the disorder itself.

Dyspraxia in the clinical context

The evaluation for dyspraxia requires testing under multiple conditions and the exclusion of other etiologies, as well as collaboration with other members of a child's care team. First, other sensory and motor conditions that may account for difficulty in skilled, learned movements should be ruled out. Then, skilled learned movement should be tested with real tool use, pantomiming in response to verbal command, and pantomiming as imitation. Testing should include transitive (i.e., involving use of a tool/object) and intransitive gestures (i.e., not involving an object such as a thumbs-up or beckoning), as well as representational (e.g., pretending to brush one's teeth) and symbolic (i.e., waving goodbye) gestures. Error types, if present, can help refine the exam [49]. Some tools exist for testing dyspraxia, such as the Test of Ideational Praxis, though currently there is no standardized set of tests for dyspraxia used in neurologic evaluation [54]. Corroboration or collaboration with occupational, physical, and speech therapists can be essential to defining the scope and impact of suspected dyspraxia.

Treatments for dyspraxia

The treatment for developmental dyspraxia is largely the same as for DCD, involving a team-based approach with therapists and clear goals. Novel apraxia-specific treatments could also be considered, though not well studied in dyspraxia as a distinct entity, such as augmentative virtual reality technologies, mirror boxes, and noninvasive transcranial magnetic stimulation [55–57]. However, it must be emphasized that these studies are preliminary, even for apraxia, and well-powered randomized controlled trials have not yet been performed.

Benign congenital hypotonia

History

Benign congenital hypotonia (BCH), a term introduced in 1960, was initially defined as isolated hypotonia of unclear etiology during the newborn or infant period, with or without associated delays in motor milestones; it implies a good prognosis with recovery often before the age of 2 years [58]. A variety of terms have been used in the literature to describe this transient period of infantile hypotonia, including idiopathic hypotonia, benign essential hypotonia, and transient tone abnormality [59–61]. More recently, the term has been used to describe children with mild central hypotonia, with or without mild motor delay, that improves with time. BCH remains a diagnosis of exclusion, confirmed only retrospectively.

In this section, a framework for the initial clinical approach to evaluation for clinicians considering a diagnosis of BCH is laid out based on the current understanding of epidemiology and underlying etiologies. Previously proposed diagnostic criteria for BCH are presented, and appropriate monitoring and therapies for suspected BCH are discussed.

Definition and epidemiology

Hypotonia is a common pediatric complaint that can reflect pathology anywhere along the neuroaxis. Therefore understanding presenting signs and symptoms and common causes is critical in clinical practice to guide diagnostic workup, monitoring, and management. Hypotonia is defined as decreased resistance to passive manipulation or range of motion of the body. It is a qualitative and subjective finding based on a series of physical exam maneuvers and observations. Tone itself permits a person to maintain desired body postures (i.e., upright sitting or standing). Depending on the underlying etiology, patients can have varying physical manifestations of hypotonia, including axial, appendicular, or global distributions.

Hypotonia often presents in the neonatal or infantile period, before the age of 12 months, as a feature of over 500 different genetic disorders [62]. The incidence of hypotonia itself is difficult to determine, given that it is a feature of many conditions or disorders [62]. One study estimated that genetic and metabolic disorders account for more than 50% of all disorders leading to hypotonia in early life [61].

A major distinction must be made as to whether an infant has signs of a central or peripheral process, which is accomplished by assessing whether the child is weak and determining the presence or absence of deep tendon reflexes. At the most basic level, hypotonia and weakness coupled with depressed or absent deep tendon reflexes are suggestive of a peripheral process. A child with normal strength and brisk reflexes, especially when accompanied by other domains of developmental delay, suggests a central process [63]. One pitfall of this approach is that early in life, incomplete myelination of the developing nervous system limits the expression of upper motor neuron signs, therefore localization of infantile hypotonia must be approached with caution [64,65]. Spasticity replaces neonatal hypotonia in these children towards the end of the first year of life as myelination progresses [66]. This consideration of broad localization and the multitude of pathologies that cause infantile hypotonia is important when considering a diagnosis of exclusion, such as BCH.

BCH as an entity is poorly studied and described in the literature, with ill-defined clinical presentation, familial occurrence, and long-term developmental outcomes. There are currently no DSM-5 criteria for the diagnosis of BCH in neonates and infants. Shuper et al. [67] proposed the first and only diagnostic criteria for BCH in 1987, which lay forward several basic relevant features present in suspected BCH. Of note, the diagnostic criteria require improvement over a period of months and therefore cannot be made at the initial presentation. Due to the paucity of data, the true incidence of BCH is unclear.

Proposed diagnostic criteria for benign congenital hypotonia [68].

1. Generalized hypotonia with the presence of deep tendon reflexes.
2. Lack of evidence for neonatal asphyxia.
3. Fine motor and mental development are within the normal range for the infant's age.
4. Absence of coexisting systemic diseases.
5. Normal screening labs (levels of muscle enzymes, that is, creatine phosphokinase, thyroid studies).
6. Progressive improvement in muscle function was observed during several months of follow-up.

Clinical evaluation for suspected benign congenital hypotonia

Children often present with hypotonia during their first year of life, typically as the "floppy infant" in the newborn period or as the older infant with delays in achieving gross motor milestones [68,69]. Clinicians should follow a systematic approach to determine if any signs of an underlying pathologic etiology are present, including the typical components of the clinical assessment (vital signs, anthropomorphic measurements, a general and neurologic exam, and a comprehensive history, including birth, developmental, and family histories). Treatable causes of hypotonia, such as toxins, infection, spinal muscular atrophy, or endocrine dysfunction, must be specifically considered and excluded based on clinical presentation (see Table 16.2. As BCH can only be diagnosed with improvement over time and in the absence of pathologic causes of hypotonia, serial evaluations and examinations for patients with suspected BCH are required for diagnosis.

Currently, there is no standardized motor assessment utilized to evaluate for hypotonia, which also varies based on an infant's state and level of alertness. Although there are no validated hypotonia scoring systems and the finding is primarily subjective, the utilization of a standardized neurologic assessment of resting, phasic, and postural tone assessment provides consistency in clinical practice (described and depicted in Figs. 16.2–16.4) [70]. Using either the Hammersmith Neonatal Neurological Examination (HNNE) or the Hammersmith Infant Neurological Examination (HINE) could be considered based on patient age. These tools have been widely utilized in the literature, and are freely available (http://hammersmith-neuro-exam.com) [70,71]. Measurement of the popliteal angle is a useful method for documenting tone in the lower extremities, which can be compared to reported norms for age [72,73].

Once hypotonia is identified, the remainder of the general and neurologic exams can provide helpful diagnostic clues if abnormalities are present, suggesting alternative etiologies (see Table 16.2) . Salient features that must be noted indicative of a potential central nervous system (CNS) etiology include macro- or microcephaly, dysmorphic facial features, an abnormal level of consciousness, hyperreflexia, and early signs of spasticity, including a tightly fisted hand with the thumb persistently adducted and enclosed by the other fingers with minimal spontaneous opening (cortical thumb) and thigh adduction with leg crossing on vertical suspension (scissoring) [74]. Features concerning a peripheral

TABLE 16.2 Differentiation of benign congenital hypotonia, peripheral, and central hypotonia in infants.

	Benign congenital hypotonia	Central hypotonia	Peripheral hypotonia
General exam	Normal	Can be normal. Abnormal head circumference, growth parameters, dysmorphic features, organomegaly, and neurocutaneous stigmata are suggestive.	Can be normal. Arthrogryposis, club feet, or other sequelae of decreased intrauterine movement can be present. Respiratory compromise is more common. Some disorders can be associated with cardiomyopathies or other systemic involvement.
Mental status	Normal	Can be normal, but if other domains of development are affected, it is suggestive of central etiology.	Normal
Cranial nerves	Normal	Can be normal. When present, abnormal eye movement or light responses are strongly suggestive of a central etiology.	Usually normal. Myopathic facies, high-arched palate, and swallowing difficulty can be present.
Motor strength and bulk	Normal	Usually normal. Scissoring and cortical thumb can be present.	Can be normal, but when weakness or paucity of movement is present, it is suggestive.
Reflexes	Normal or mildly decreased.	Can be elevated, normal, or depressed.	Can be normal or depressed. Absent reflexes are suggestive.
Other historical features	Normal intrauterine movements, normal pregnancy and normal delivery. No other domains of development were affected.	Can include abnormal intrauterine movements, perinatal distress, seizures, movement disorders, family history of consanguinity, and history of miscarriages.	Some parental features can be suggestive of specific diagnoses (i.e., frontal balding and maternal myotonia for congenital myotonic dystrophy).

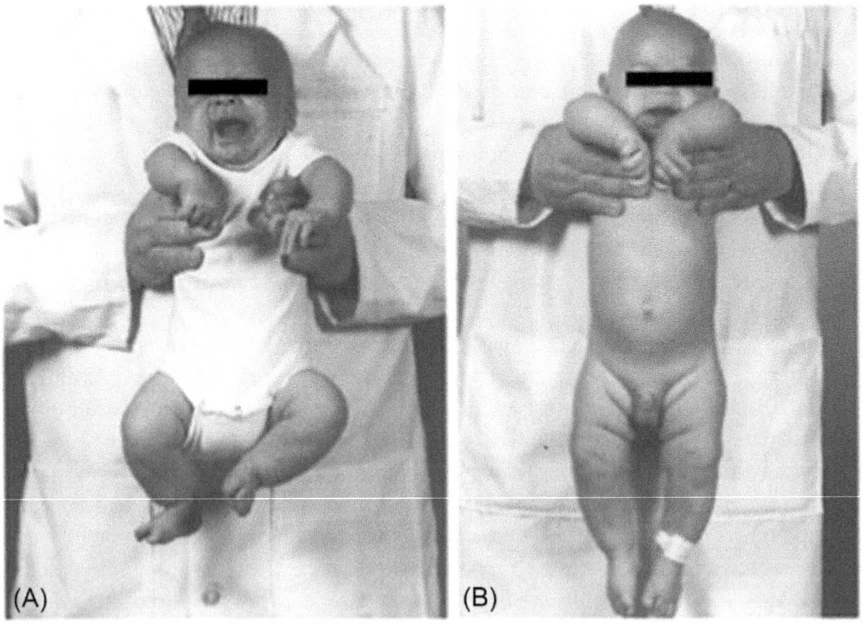

FIGURE 16.2 Vertical suspension. (A) The arms help provide resistance against the examiner's hands, and the legs flex against gravity. (B) The arms and shoulders fail to provide support, and the legs do not resist gravity. *Reprinted with permission from Fenichel G.M. The neurological consultation. Neonatal neurology. 4th ed. Philadelphia, PA: Elsevier; 2007. p. 12.*

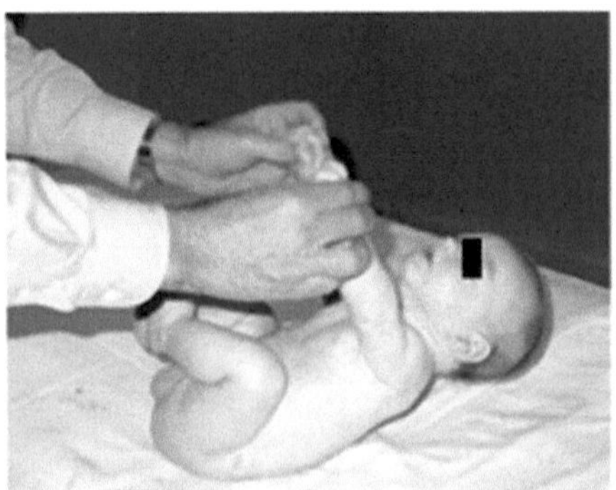

FIGURE 16.3 Traction response. When a newborn is pulled by the arms from a lying to a sitting position, the head lags at first. The infant then flexes, lifting the head to the midline of the rest of the body before it falls forward. *Reprinted with permission from Fenichel G.M. The neurological consultation. Neonatal neurology. 4th ed. Philadelphia, PA: Elsevier; 2007. p. 11.*

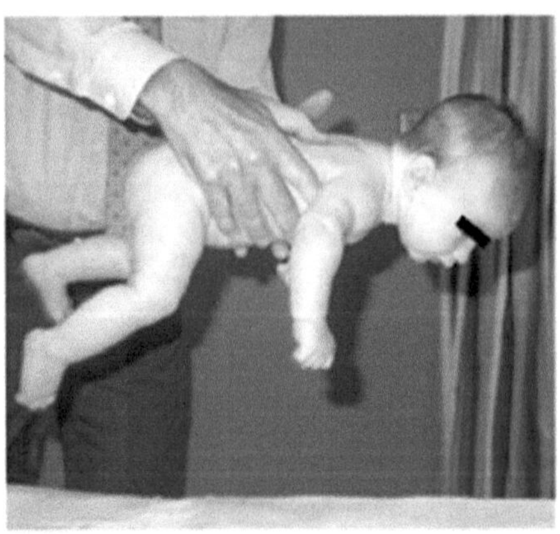

FIGURE 16.4 Horizontal suspension. The head rises intermittently while the head and extremities lift antigravity. *Reprinted with permission from Fenichel GM. Neonatal Neurology, 4th ed. Philadelphia: Elsevier; 2007.*

nervous system (PNS) etiology include weakness, muscle atrophy, fasciculations (especially tongue), diminished or absent deep tendon reflexes, and normal mental status. There can be overlap in features, with some conditions presenting with mixed central and peripheral signs due to the underlying pathology. Any systemic features, including cardiac murmurs, skin lesions, and organomegaly can help narrow the differential and guide the next steps in the diagnostic workup. The second step in the evaluation of an infant or child with hypotonia is obtaining a comprehensive pregnancy and delivery, medical, and three-generation family history [62]. Any abnormalities in the history or exam should alert the physician to a potential alternative diagnosis and prompt further diagnostic workup. A careful pregnancy history can ascertain any prenatal complications, including abnormalities on prenatal ultrasound, abnormalities in the quality of fetal movements, polyhydramnios (which can suggest reduced fetal swallowing), or maternal exposures. Delivery history can elucidate any concerns for birth asphyxia or a sentinel event, as hypoxic-ischemic encephalopathy (HIE) is a common cause of hypotonia. Medical history, including early neonatal history and evolution of mental status, feeding, and tone, is critical to determining the age of onset and trajectory (improving or worsening) of hypotonia. A comprehensive review of the systems should be completed. A three-generation family history should be obtained, specifically inquiring about a history of motor or sensory issues, skeletal muscle disease, cardiac history, birth defects, intellectual disability, or consanguinity [75].

As signs and symptoms often become more apparent and easier to detect in older children, serial exams should be performed to observe the evolution of exam findings, specifically to monitor for improvement, detect abnormalities, and assess for other developmental delays that should prompt further workup. The presence of any other abnormalities in the history or exam should prompt alternate diagnoses to be considered.

Clinical significance and prognosis

When appropriately made, the diagnosis of benign congenital hypotonia can be reassuring to families and prevent unnecessary testing. However, there may be danger in implying that a diagnosis has been made, as benign congenital hypotonia remains a clinical entity only by exclusion [76]. Owing to the advancement of diagnostics (including genetic testing and imaging) and therapeutics, the importance of making a specific diagnosis in a child with hypotonia cannot be emphasized enough. This allows the clinician to initiate timely directed therapies (for instance, in SMA, or inborn errors of metabolism), reduce the risk of adverse events (i.e., malignant hyperthermia with some neuromuscular disorders), and begin multidisciplinary surveillance for multiorgan involvement in some diseases. For this reason, some authors advocate for benign congenital hypotonia to be considered a provisional diagnosis only [77].

While some infants with BCH may be delayed in achieving motor milestones (i.e., sitting, crawling, and walking) in the setting of low tone, many achieve typical motor milestones and, by definition, must have normal motor milestones by 24–30 months. As with DCD, developmental stagnation or regression is not consistent with BCH. When global developmental delays or significant gross or fine motor delays are present, a diagnosis of BCH is not indicated, as hypotonia is better explained by other disorders.

From a prognostic standpoint, by definition, a trajectory of hypotonia improvement and then resolution must be present to generate a diagnosis of BCH. The challenge in providing anticipatory guidance to families regarding finer points, however, remains that BCH is poorly described in the literature, with many investigators not following these children past two years of age and limited evidence reported for long-term outcomes [59]. Some studies have shown that a portion of these children may have additional motor deficits on long-term follow-up. One prospective observational study described the clinical features and short-term outcomes of 43 children diagnosed with presumed benign congenital hypotonia in infancy, noting that approximately half were found to have minor neurological abnormalities such as hypotonia, clumsy gait, and speech delay when followed until the end of their third year of life [67]. Parush et al. evaluated a small sample of six- to eight-year-old children who had been diagnosed with BCH as infants and found that these children had inferior gross motor performance on several measures when they reached school age relative to matched controls [78]. Further longitudinal studies are needed in patients with presumed benign congenital hypotonia to better describe the natural history, including developmental outcomes, and understand the risk for associated sequelae.

Etiology and neurobiology

The etiology of BCH remains unclear, with both genetic and environmental factors being potentially implicated without conclusive evidence. There have not been specific genes identified. Some have proposed a delayed developmental process, such as delayed myelination, that improves with maturation, without studies to indicate the validity of this proposed mechanism. The high familial incidence of BCH in a retrospective chart review demonstrating a family history of delayed motor achievements in 30% of the cases suggests a potential genetic link [78]. Further investigation into the genetic underpinnings and neurobiology of this disorder is warranted.

Diagnostic process

A presumed diagnosis of benign congenital hypotonia is made by a clinical synthesis of physical examination and the history (medical, developmental, and family), as previously outlined. Some argue that benign congenital hypotonia should no longer be considered a specific diagnostic entity, as clinicians must first eliminate other causes of hypotonia [62]. If there are any features potentially concerning for an alternative process, a staged approach to diagnostic workup should be completed.

The evaluation of a hypotonic infant should be comprehensive and, on the basis of current evidence, require a multidisciplinary approach by a team of developmental specialists, including pediatricians, medical geneticists, child neurologists, and physical or occupational therapists [79]. With the ever-evolving scientific landscape leading to more genetic diagnoses and emerging directed treatments and therapies, it remains more important than ever to utilize appropriate tools for diagnosis, evaluation, and intervention. One retrospective chart review analyzed the usefulness of different procedures in the diagnostic approach to hypotonic newborn infants. While 50% of all diagnosed cases were solved by history, clinical, and neurologic exam alone, neuroimaging (i.e., MRI/MR spectroscopy) techniques contributed to the diagnosis in 13% of cases, with the remainder identified through further genetic and biochemical testing or specific nerve and muscle investigations [80]. Proposed diagnostic algorithms for the underlying cause of hypotonia in neonatal and pediatric populations exist and are beyond the scope of this chapter [62].

As the diagnosis of benign congenital hypotonia can ultimately only be confirmed retrospectively, some argue that the gold standard test for this diagnosis is the "test of time" by serially assessing the developmental outcomes [81]. Therefore close monitoring and clinical follow-up are essential in assessing these hypotonic infants for continued progress in their motor skills and normal development in other areas.

Differential diagnosis and cooccuring disorders

Hypotonia itself is a frequent presenting symptom of other disorders, and therefore discerning the etiology is critical to assigning a relevant genetic diagnosis, appropriate prognosis, and initiating appropriate and timely therapeutic management.

Therapies and management

Given the lack of clarity of the neurobiological and genetic basis of BCH, the mainstay of management remains supportive versus continued observation. Physical therapy, occupational therapy, and other therapies tailored to the patient's specific needs can provide significant improvement in functionality in the short-term, but the long-term benefits are poorly characterized for BCH. Nonetheless, as BCH diagnosis cannot be made at onset, we recommend referral of all infants with hypotonia for therapeutic evaluation and treatment.

In one study, the benefit of frequent physical therapy sessions and home treatment implemented by caregivers was demonstrated in a prospective randomized trial aimed at evaluating the effect of frequency of physical therapy on the outcome of infants referred for delayed motor development due to joint hypermobility and benign hypotonia. Infants were randomly placed into monthly and weekly treatment groups. All infants underwent a gross motor developmental assessment using the Alberta Infant Motor Scale, evaluated at 9, 12, 15, and 18 months, or at the onset of independent walking. Assessment of walking at 15 months revealed a clear advantage for the infants who were treated weekly, with 45.5% of the infants from the weekly group independently walking by the age of 15 months compared with 9% of the infants from the monthly group. However, this benefit was no longer appreciated at 18 months, when 82% of children from the weekly session group walked independently compared with 100% from the monthly session group, demonstrating that physical therapy combined with a home treatment protocol implemented by the primary caregivers is sufficient to achieve motor catch-up [82].

Summary

Minor neuromotor dysfunction, despite its name, can have major effects on a child's development and functioning. Though developmental dyspraxia and DCD have often been used interchangeably, it is better to use DCD as a disorder of coordination and dyspraxia as a neurologic sign that may be present in DCD. Given its history, researchers seeking to find more information on DCD may need to keep in mind its alternate titles, such as developmental dyspraxia, in their reviews of the literature. Benign neonatal hypotonia is a diagnosis of exclusion that can only be made definitively after resolution and requires careful evaluation. Revised diagnostic criteria in the era of neurogenetics are needed. For all "minor" neuromotor disabilities, further high-quality evidence on etiology, neurobiology, treatment, and prognosis is needed.

References

[1] Institute of Medicine (U.S.), Committee on Understanding Premature Birth and Assuring Healthy Outcomes, Behrman R.E., Butler A.S. Preterm birth: causes, consequences, and prevention. National Academies Press; 2007. 772

[2] Wilson BN, Neil K, Kamps PH, Babcock S. Awareness and knowledge of developmental co-ordination disorder among physicians, teachers and parents. Child Care Health Dev 2013;39(2):296–300.

[3] Ford FR. In: Thomas CC, editor. Diseases of the nervous system in infancy, childhood and adolescence. 5th ed. C.C. Thomas; 1966. p. 1–1416.

[4] Orton ST. Reading, writing and speech problems in children: a presentation of certain types of disorders in the development of the language faculty. 1st ed. New York: W.W. Norton & Company, Inc.; 1937. p. 1–215.

[5] Ayres AJ. Types of sensory integrative dysfunction among disabled learners. Am J Occup Ther 1972;26(1):13–18.

[6] Gubbay SS. The management of developmental apraxia. Dev Med Child Neurol 1978;20(5):643–6.

[7] Gibbs J, Appleton J, Appleton R. Dyspraxia or developmental coordination disorder? Unravelling the enigma. Arch Dis Child 2007;534–9.

[8] Sugden D. Development coordination disorder as a specific learning difficulty. Leeds Consens Statement ESRC 2006;2004–5.

[9] Waber DP, Boiselle EC, Yakut AD, Peek CP, Strand KE, Bernstein JH. Developmental dyspraxia in children with learning disorders: four-year experience in a referred sample. J Child Neurol 2021;36(3):210–21.

[10] American Psychiatric Association. Developmental coordination disorder. In: Diagnostic and statistical manual of mental disorders, 5th ed. Arlington, VA: American Psychiatric Publishing; 2013. p. 74–5. Available from: https://doi.org/10.1176/appi.books.9780890425596.

[11] Blank R, Barnett A, John C, Dido G, Kirby A, Polatajko H, et al. International clinical practice recommendations on the definition, diagnosis, assessment, intervention, and psychosocial aspects of developmental coordination disorder. Dev Med Child Neurol [Internet] 2019;61 (3):242–85.

[12] Lingam R, Hunt L, Golding J, Jongmans M, Emond A. Prevalence of developmental coordination disorder using the DSM-IV at 7 years of age: a UK population-based study. Pediatrics 2009;123(4):693–700.

[13] Kadesjö BGC. Attention deficits and clumsiness in Swedish 7-year-old children. Dev Med Child Neurol 1998;40(12):796–804.

[14] Piek J, Dworcan M, Barrett N, Coleman R. Determinants of self-worth in children with and without developmental coordination disorder. Int J Disabil Dev Educ 2000;47(3):259–72.

[15] Francis M., Piek J. The effects of perceived social support and self-worth on depressive symptomatology in children with and without developmental coordination disorder. In: 38th APS annual conference. Perth Western Australia: The Australian Psychological Society Ltd.; 2003: p. 70–4.

[16] Campbell WN, Missiuna C, Vaillancourt T. Peer victimization and depression in children with and without motor coordination difficulties. Psychol Sch 2012;49(4):328–41.

[17] Rivilis I, Hay J, Cairney J, Klentrou P, Liu J, Faught BE. Physical activity and fitness in children with developmental coordination disorder: a systematic review. Res Dev Disabil 2011;894–910.

[18] Cermak S, Katz N, Weintraub N, Steinhart S, Raz-Silbiger S, Munoz M, et al. Participation in physical activity, fitness, and risk for obesity in children with developmental coordination disorder a cross-cultural study. Occup Ther Int 2015;22(4):163–7.

[19] Cairney J, Hay J, Faught B, Hawes R. Developmental coordination disorder and overweightand obesity in children aged 9–14 y. Int J Obes 2005;29(4):369–72.

[20] Visser J, Geuze RH, Kalverboer AF. The relationship between physical growth, the level of activity and the development of motor skills in adolescence: differences between children with DCD and controls. Hum Mov Sci 1988;17(4–5):573–608.

[21] Losse A, Henderson SE, Elliman D, Hall D, Knight E, Jongmans M. Clumsiness in children—do they grow out of it? A 10-year follow-up study. Dev Med Child Neurol 1991;33(1):55–68.

[22] Mosca SJ, Langevin LM, Dewey D, Innes AM, Lionel AC, Marshall CC, et al. Copy-number variations are enriched for neurodevelopmental genes in children with developmental coordination disorder. J Med Genet 2016;53(12):812–19.

[23] Reynolds JE, Billington J, Kerrigan S, Williams J, Elliott C, Winsor AM, et al. Mirror neuron system activation in children with developmental coordination disorder: a replication functional MRI study. Res Dev Disabil 2019;84:16–27.

[24] Gomez A, Sirigu A. Developmental coordination disorder: core sensori-motor deficits, neurobiology and etiology. Neuropsychologia 2015;79:272–87.

[25] Kashiwagi M, Tamai H. Brain mapping of developmental coordination disorder. Functional brain mapping and the endeavor to understand the working brain. InTech; 2013.

[26] Licari M, Billington J, Reid S, Wann J, Elliott C, Winsor A, et al. Cortical functioning in children with developmental coordination disorder: a motor overflow study. Exp Brain Res 2015;233(6):1703–10.

[27] Caeyenberghs K, Taymans T, Wilson PH, Vanderstraeten G, Hosseini H, van Waelvelde H. Neural signature of developmental coordination disorder in the structural connectome independent of comorbid autism. Dev Sci 2016;19(4):599–612.

[28] Wilson B, Kaplan B, Crawford S, Campbell A, Dewey D. Reliability and validity of a parent questionnaire on childhood motor skills. Am J Occup Ther 2000;54(5):484–93.
[29] Wilson B, Crawford S, Green D, Roberts G, Aylott A, Kaplan B. Psychometric properties of the revised developmental coordination disorder questionnaire. Phys Occup Ther Pediatr 2009;29(2):182–202.
[30] Missiuna C, Gaines R, Soucie H, McLean J. Parental questions about developmental coordination disorder: a synopsis of current evidence. Paediatr Child Health 2006;.
[31] Nakai A, Miyachi T, Okada R, Tani I, Nakajima S, Onishi M, et al. Evaluation of the Japanese version of the developmental coordination disorder questionnaire as a screening tool for clumsiness of Japanese children. Res Dev Disabil 2011;32(5):1615–22.
[32] Green D, Baird G, Sugden D. A pilot study of psychopathology in developmental coordination disorder. Child Care Health Dev 2006;32 (6):741–50.
[33] Miyahara M, Möbs I, Doll-Tepper G. Severity of hyperactivity and the comorbidity of hyperactivity with clumsiness in three sample sources: school, support group and hospital. Child Care Heal Dev 2001;27(5):413–24.
[34] Rasmussen P, Gillberg C. Natural outcome of ADHD with developmental coordination disorder at age 22 years: a controlled, longitudinal, community-based study. J Am Acad Child Adolesc Psychiatry 2000;39(11):1424–31.
[35] Baumer N, Spence S. Evaluation and management of the child with autism spectrum disorder. Child Neurol 2018;24(1):248–75.
[36] Bhat A. Is motor impairment in autism spectrum disorder distinct from developmental coordination disorder? A report from the SPARK study. Phys Ther 2020;100(4):633–44.
[37] Rintala P, Pienimäki K, Ahonen T, Cantell M, Kooistra L. The effects of a psychomotor training programme on motor skill development in children with developmental language disorders. Hum Mov Sci 1998;17(4–5):721–37.
[38] Visscher C, Houwen S, Scherder E, Moolenaar B, Hartman E. Motor profile of children with developmental speech and language disorders. Pediatrics 2007;120(1):158–63.
[39] Smits-Engelsman B, Blank R, van der Kaay A-C, van der Meijs R, van den Brand E, Polatajko H, et al. Efficacy of interventions to improve motor performance in children with developmental coordination disorder: a combined systematic review and meta-analysis. Dev Med Child Neurol 2012;55(3):229–37.
[40] Offor N, Ossom Williamson P, Caçola P. Effectiveness of interventions for children with developmental coordination disorder in physical therapy contexts: a systematic literature review and meta-analysis. J Mot Learn Dev 2016;4(2):169–96.
[41] Preston N, Magallón S, Hill LJB, Andrews E, Ahern SM, Mon-Williams M. A systematic review of high quality randomized controlled trials investigating motor skill programmes for children with developmental coordination disorder. Clinical Rehabilitation, Vol. 31. SAGE Publications Ltd.; 2017. p. 857–70.
[42] Smits-Engelsman BCM, Jelsma LD, Ferguson GD, Geuze RH. Motor learning: an analysis of 100 trials of a ski slalom game in children with and without developmental coordination disorder. PLoS One 2015;10(10).
[43] Au MK, Chan WM, Lee L, Chen TMK, Chau RMW, Pang MYC. Core stability exercise is as effective as task-oriented motor training in improving motor proficiency in children with developmental coordination disorder: a randomized controlled pilot study. Clin Rehabil 2014;28 (10):992–1003.
[44] Dewey D. What is developmental dyspraxia. Brain Cogn 1995;29(3):254–74.
[45] Polatajko H, Fox M, Missiona C. An international consensus on children with developmental coordination disorder. Can J Occup Ther 1995;62 (1):3–6.
[46] Peters J, Barnett A, Henderson S. Clumsiness, dyspraxia and developmental co-ordination disorder: how do health and educational professionals in the UK define the terms? Child Care Health Dev 2008;27(5):399–412.
[47] Njiokiktjien C, Verschoor C, Vranken M, Vroklage L. Development of ideomotor praxis representation. Dev Med Child Neurol 2000;42 (4):253–7.
[48] Kaplan E. Gestural representation of implement usage: an organismic-developmental study. 1968.
[49] Steinman KJ, Mostofsky SH, Denckla MB. Toward a narrower, more pragmatic view of developmental dyspraxia. J Child Neurol 2010;71–81.
[50] Niessen E, Fink G, Weiss P. Apraxia, pantomime and the parietal cortex. NeuroImage Clin 2014;5:42–52.
[51] Heilman K. Apraxia. Behav Neurol 2010;16(4):86–108.
[52] Nass R. Ontogenesis of hemispheric specialization: apraxia associated with congenital left hemisphere lesions. Percept Mot Ski 1983;57 (3):775–82.
[53] Cassidy S, Hannant P, Tavassoli T, Allison C, Smith P, Baron-Cohen S. Dyspraxia and autistic traits in adults with and without autism spectrum conditions. Mol Autism 2016;7(1).
[54] Lane S, Ivey C, May-Benson T. Test of ideational praxis (TIP): preliminary findings and interrater and test-retest reliability with preschoolers. Am J Occup Ther 2014;68(5):555–61.
[55] Park W, Kim J, Kim MY. Efficacy of virtual reality therapy in ideomotor apraxia rehabilitation: a case report. Med (Baltim) 2021;100(28): e26657.
[56] Romano D, Tosi G, Gobbetto V, Pizzagalli P, Avesani R, Moro V, et al. Back in control of intentional action: improvement of ideomotor apraxia by mirror box treatment. Neuropsychologia 2021;160.
[57] Pastore-Wapp M, Nyffeler T, Nef T, Bohlhalter S, Vanbellingen T. Non-invasive brain stimulation in limb praxis and apraxia: a scoping review in healthy subjects and patients with stroke. Cortex, Vol. 138. Masson SpA; 2021. p. 152–64.
[58] Walton JN. The ‘floppy’ infant. Dev Med Child Neurol 1960;2(1):10–18.

[59] Strubhar AJ, Meranda K, Morgan A. Outcomes of infants with idiopathic hypotonia. Pediatr Phys Ther [Internet] 2007;19(3):227–35.
[60] Chaudhari S, Bhalerao M, Chitale A, Patil B, Pandit A, Hoge M. Transient tone abnormalities in "high risk" infants and cognitive outcome at five years. Indian Pediatr 2010;47(11):931–5.
[61] Prasad AN, Prasad C. Genetic evaluation of the floppy infant. Semin Fetal Neonatal Med [Internet] 2011;16(2):99–108.
[62] Lisi EC, Cohn RD. Genetic evaluation of the pediatric patient with hypotonia: perspective from a hypotonia specialty clinic and review of the literature. Dev Med Child Neurol [Internet] 2011;53(7):586–99.
[63] Marcdante KJ, Kliegman RM. Weakness and hypotonia. Nelson Essentials of Pediatrics. 2019. p. 692–700.
[64] Singer HS, Mink JW, Gilbert DL, Jankovic J. Cerebral palsy. Mov Disord Child 2016;453–75.
[65] Iyer KM. Cerebral palsy. General principles of orthopedics and trauma. Springer International Publishing; 2019. p. 309–21.
[66] Jan MMS. The hypotonic infant: clinical approach. J Pediatr Neurol 2007;5:181–7.
[67] Shuper A, Weitz R, Varsano I, Mimouni M. Benign congenital hypotonia. A clinical study in 43 children. Eur J Pediatr 1987;146(4):360–4.
[68] Bodensteiner JB. The evaluation of the hypotonic infant. Semin Pediatr Neurol 2008;15(1):10–20.
[69] Segal I, Peylan T, Sucre J, Levi L, Bassan H. Relationship between central hypotonia and motor development in infants attending a high-risk neonatal neurology clinic. Pediatr Phys Ther [Internet] 2016;28(3):332–6.
[70] Haataja L, Mercuri E, Regev R, Cowan F, Rutherford M, Dubowitz V, et al. Optimality score for the neurologic examination of the infant at 12 and 18 months of age. J Pediatr [Internet] 1999;135(2):153–61 Aug 1 [cited 13.04.22].
[71] Dubawitz L, Mercuri E, Dubowitz V. An optimality score for the neurologic examination of the term newborn. J Pediatr [Internet] 1998;133 (3):406–16 [cited 13.04.22].
[72] Reade E, Hom L, Hallum A, Lopopolo R. Changes in popliteal angle measurement in infants up to one year of age. Dev Med Child Neurol [Internet] 1984;26(6):774–80 [cited 13.04.22].
[73] Katz K, Rosenthal A, Yosipovitch Z. Normal ranges of popliteal angle in children. J Pediatr Orthop 1992;12(2):229–31.
[74] Piña-Garza JE. The hypotonic infant. Fenichel's Clin Pediatr Neurol 2019;8:150–71.
[75] Miralles R, Panjwani D. Neonatal hypotonia. Emerging topics and controversies in neonatology. Springer International Publishing; 2020. p. 71–101.
[76] Gordon N. Benign congenital hypotonia: a syndrome or a disease. Dev Med Child Neurol 1966;8(3):330–5.
[77] Thompson CE. Benign congenital hypotonia is not a diagnosis. Dev Med Child Neurol [Internet] 2002;44(4):283 [cited 13.04.22].
[78] Parush S, Yehezkehel I, Tenenbaum A, Tekuzener E, Bar-Efrat I, Jessel A, et al. Developmental correlates of school-age children with a history of benign congenital hypotonia. Dev Med Child Neurol 1998;40(7):448–52.
[79] Harris SR. Congenital hypotonia: clinical and developmental assessment. Dev Med Child Neurol [Internet] 2008;50(12):889–92 [cited 13.04.22].
[80] Paro-Panjan D, Neubauer D. Congenital hypotonia: is there an algorithm? J Child Neurol [Internet] 2004;19(6):439–42 [cited 13.04.22].
[81] Crawford TO. Clinical evaluation of the floppy infant. Pediatr Ann [Internet] 1992;21(6):348–54 [cited 13.04.22].
[82] Mintz-Itkin R, Lerman-Sagie T, Zuk L, Itkin-Webman T, Davidovitch M. Does physical therapy improve outcome in infants with joint hypermobility and benign hypotonia? J Child Neurol [Internet] 2009;24(6):714–19 [cited 13.04.22].

Chapter 17

Movement disorders and genetic mimics of cerebral palsy

Zainab Zaman, Alexandra Santana and Darius Ebrahimi-Fakhari
Department of Neurology, Movement Disorders Program, Boston Children's Hospital, Harvard Medical School, Boston, MA, United States

Introduction to childhood-onset movement disorders

The realm of movement disorders in children encompasses a diverse array of neurological syndromes that manifest with impaired voluntary movements, postural dysfunction, abnormal involuntary movements, or the execution of seemingly normal movements at inappropriate or unintended times. It is crucial to note that these disturbances in movement are not rooted in muscle weakness or abnormal tone, though they may sometimes coexist with such conditions. Movement disorders can be broadly categorized into two main groups: those characterized by an excess of involuntary movements, known as *hyperkinetic movement disorders*, and those distinguished by a reduction in both automatic and voluntary movements, referred to as *hypokinetic movement disorders*. Hyperkinetic movement disorders include dystonia, chorea, myoclonus, and tremor. Meanwhile, the quintessential hypokinetic movement disorder is parkinsonism, often described as an akinetic-rigid syndrome in children. It is important to note that hypokinetic disorders are relatively infrequent in children, with most pediatric movement disorders exhibiting a combination of different hyperkinetic disorders. In addition, spasticity frequently cooccurs with movement disorders in children. The presence of nonmovement disorder symptoms, both neurological and nonneurological, often holds valuable diagnostic significance in childhood-onset movement disorders.

Approaching pediatric movement disorders

In the context of an outpatient consultation, a systematic approach should be employed to evaluate movement disorders in children. This approach hinges on five fundamental questions:

1. Is the presenting symptom indicative of a movement disorder?
2. What are the leading phenomenologies and time courses?
3. Could this be a functional neurological disorder?
4. Could the movement disorder be attributed to an acquired cause?
5. Does the disorder align with a recognizable clinical syndrome, such as a treatable monogenic movement disorder or an inborn error of metabolism?

The first question, although seemingly trivial, is important as there are several other conditions that may be mistaken for movement disorders. This includes paroxysmal conditions such as seizures or behavioral mannerism, as well as more static presentations such as musculoskeletal disorders. When the clinical presentation aligns with the characteristics of a movement disorder, the foremost step is to define the predominant phenomenology and establish the temporal evolution. These two pieces of information lay the foundation for constructing a comprehensive differential diagnosis. Given the prevalence of functional movement disorders, it is also important to consider this differential, especially in an outpatient setting. While reaching a definitive diagnosis of a functional neurological disorder in a single encounter is often a challenge, acknowledging the potential of this differential diagnosis constitutes an important initial step. The next question should focus on whether the movement disorder could potentially be attributed to an acquired cause. Major categories include medication-related movement disorders and movement disorders resulting from infectious,

Capute and Accardo's Neurodevelopmental Disabilities in Infancy and Childhood. DOI: https://doi.org/10.1016/B978-0-12-824060-1.00034-1

inflammatory, paraneoplastic, traumatic, or vascular central nervous system disorders. In the absence of significant suspicion for an acquired cause, the focus then shifts to evaluating the possibility of a genetic movement disorder, and identifying symptoms indicative of a readily treatable disorder should be prioritized [1–5].

Characterizing the phenomenology and time course

When confronted with a movement disorder, the initial step is to characterize the phenomenology and the temporal evolution of the disorder. This is a pivotal step, as it lays the foundation for constructing a differential diagnosis. Key aspects to consider include:

- Is the pattern of movements within the range of normal or aberrant?
- Are there excessive movements or a paucity of movements?
- Is the movement paroxysmal (sudden onset and cessation), continual (repetitive), or continuous (uninterrupted)?
- Has the movement disorder evolved or changed over time?
- Are there external factors or emotional states that influence the movement disorder?
- Can the movements be voluntarily suppressed?
- Does the movement disorder diminish during sleep?
- Are there any examination findings suggestive of focal neurological deficits or underlying systemic diseases?
- Is there a family history of a similar or related condition?

Terminology and classification

In clinical practice, accurate diagnosis of a movement disorder relies on a qualitative appreciation of the movement type and its context. Multiple classification systems have been developed to provide a structured taxonomy for the diverse array of movement disorders. Utilizing the correct terminology to describe the phenomenology of a movement disorder is essential, as it not only aids in diagnosis and treatment but also facilitates effective communication between specialized and nonspecialized clinicians.

Ataxia

The term "ataxia" originates from the Greek "without order," and it denotes an inability to execute a typical or expected voluntary movement trajectory without attributing the issue to muscle weakness or involuntary muscle activity around the affected joints. Ataxia may stem from a disruption in the spatial pattern of muscle activity, the timing of such activity, or a combination of both. Notable associated deficits encompass dysmetria (inaccurate movement, often involving undershooting or overshooting a target), dyssynergia (breakdown of coordinated multijoint movements), and dysdiadochokinesis (impaired rhythmicity in rapid alternating movements).

Athetosis

Athetosis, taken from the Greek term meaning "without fixed position," is characterized by slow, writhing, and continuous involuntary movements. There is ongoing debate regarding whether athetosis should be treated as a distinct entity or regarded as part of the broader spectrum of chorea. Many movement disorder specialists opt to use the term "athetosis" primarily to describe a form of cerebral palsy.

Ballismus

Ballismus, or "ballism," refers to involuntary, high-amplitude, flinging movements, typically originating proximally. These movements can be brief or sustained and may occur in conjunction with chorea. Hemiballism, affecting one side of the body, is a common presentation. Over time, hemiballism sometimes becomes less severe and evolves into chorea. Some experts argue that because the same lesions can produce both ballismus and chorea, ballismus is likely an integral part of the chorea spectrum rather than a separate entity.

Chorea

Chorea, derived from the Greek term for "dance-like," characterizes an involuntary, continual, irregular hyperkinetic disorder where movements or movement fragments manifest at unpredictable rates and directions. Movements may range from flowing, akin to athetosis, to rapid. Chorea can affect all body parts, with certain distributions being more indicative of specific diseases or disorders.

Dystonia

Dystonia is characterized by sustained muscle contractions, frequently resulting in twisting and repetitive movements or abnormal postures [6]. Dystonia is often provoked or worsened by voluntary movements and is associated with overflow muscle activation. It may be further described by its anatomic distribution, whether involving a single body region (focal), two or more contiguous (segmental) or noncontiguous (multifocal) body regions, or two or more body regions along with the trunk (generalized).

Myoclonus

Myoclonus denotes quick, shock-like movements in one or more muscles. Myoclonus may be characterized by its distribution (focal, multifocal, or generalized) and whether it occurs at rest or is precipitated by movement (action myoclonus) or other external stimuli (stimulus-sensitive or reflex myoclonus). It can also be classified electrophysiologically based on its origin within the central nervous system (e.g., cortical, subcortical/reticular, spinal cord) and rarely in the peripheral nervous system (e.g., plexus, peripheral nerve).

Parkinsonism

Parkinsonism is a neurological syndrome defined by the presence of two or more cardinal features of Parkinson's disease, encompassing resting tremor, bradykinesia, rigidity, and postural instability. In children, parkinsonism is usually defined by bradykinesia (slowness of movement), usually in combination with rigidity.

Spasticity

Spasticity is a motor abnormality characterized by increased muscular tone, manifested as a velocity-dependent resistance to passive movement in an affected limb, often referred to as a "spastic catch." Typically, spasticity is accompanied by both positive signs, such as pyramidal signs including hyperreflexia, the Babinski sign, and ankle clonus, as well as negative signs, which encompass weakness and impaired motor control in the affected muscle groups. The presence and severity of spasticity depend on the location of the lesion within the central nervous system and may manifest in various patterns, including both legs (referred to as diparesis or paraparesis), all four limbs (quadriparesis or tetraparesis), or one side of the body (hemiparesis).

Stereotypies

Stereotypies encompass involuntary, patterned, coordinated, repetitive, nonreflexive movements that recur in the same manner with each repetition, often exhibiting a rhythmic quality.

Tics

Tics are sudden, involuntary, rapid, abrupt, repetitive, nonrhythmic movements or vocalizations. They are categorized into motor and phonic tics, each further subdivided into simple and complex types. Tics are typically preceded by an uncomfortable sensation or urge that is relieved by performing the movement.

Tremor

Tremor refers to oscillating, rhythmic movements around a fixed point, axis, or plane, occurring when antagonist muscles contract alternately. This oscillation often involves movement around a joint, resulting in visible motion. Some authors also include rhythmic palatal myoclonus as a form of tremor.

Cerebral palsy as a movement disorder in the era of genomics

In recent years, the landscape of neurodevelopmental disorders has undergone a transformation, thanks to remarkable advancements in genetic testing technologies. Cerebral palsy (CP), a heterogeneous group of motor disorders traditionally associated with nongenetic factors like prenatal brain injury or perinatal asphyxia, is now under the scrutiny of emerging evidence that underscores the role of genetic factors in a substantial proportion of cases. This paradigm shift in our understanding of cerebral palsy has been made possible by the advent of molecular diagnostic tools, including next-generation sequencing, which have revolutionized the field by enabling the detection of genetic variants linked to cerebral palsy.

Cerebral palsy is a neurodevelopmental disorder that exerts its impact on motor development and functioning, leading to various abnormalities in movement, coordination, muscle tone, reflexes, posture, and balance. A generally accepted definition of "cerebral palsy" is that of a permanent disorder of movement or posture caused by a nonprogressive lesion in the developing fetal or infant brain. It is vital to note that CP is a clinical diagnosis, and its etiology can vary widely. Acquired risk factors for CP encompass conditions like prematurity, hypoxic-ischemic encephalopathy, maternal and fetal infections, and perinatal stroke. The term "cryptogenic CP" is employed to describe cases where no compelling risk factors are evident. An increasing proportion of conditions formerly classified under the cerebral palsy umbrella now have a recognized genetic basis. This encompasses conditions that arise from single nucleotide variants, structural variants, and copy number variants. Depending on the definition of "lesion," cerebral palsy may encompass some genetic disorders, but this is not the common usage, and we advocate for the use of the term "genetic mimics of cerebral palsy" instead.

The identification of genetic causes behind disorders resembling cerebral palsy holds significant implications [7–10]. It facilitates the potential for disease-modifying treatments in specific instances, allows for a more accurate prognosis, provides a basis for genetic counseling, offers relief for families on a long diagnostic journey, and forms the cornerstone for research that aims at developing precision therapies.

Based on recent metaanalyses, which now include well over 2500 individuals with cerebral palsy from over 15 cohorts, the diagnostic yield of clinical exome sequencing is estimated at around 30% [11,12]. The yield is higher among studies that excluded cerebral palsy risk factors (history of prematurity, low birth weight, perinatal complications, neuroimaging findings of periventricular leukomalacia, hypoxic-ischemic encephalopathy, infarcts, intracranial hemorrhage, and brain malformations) for patient selection compared with studies that included all patients with cerebral palsy, regardless of risk factors or neuroimaging findings. Importantly, in a recent metaanalysis by Gonzalez-Mantilla et al., 36.8% of individuals with pathogenic and likely pathogenic variants benefited from genomic-informed changes in their medical management, including surveillance of gene-related conditions and genomics-informed interventions.

Red flag symptoms for genetic mimics of cerebral palsy

Discerning cerebral palsy from its genetic mimics presents a significant challenge, primarily due to several overlapping characteristics:

- Clinical and radiological similarities: Clinical and radiological features in both cerebral palsy and genetic mimics may be similar, making it difficult to distinguish between them on clinical grounds alone.
- Slowness of progressive features: Notably, the presence of progressive features is often not discernible during the initial presentation, further complicating the diagnostic process.
- Fluctuation of motor symptoms: The fluctuation of symptoms, which is commonly observed in acquired cerebral palsy, can also be encountered in cases where genetic factors are at play, adding an additional layer of complexity.
- Expanding spectrum of genetic disorders: The number of genetic disorders that mimic cerebral palsy is continually expanding, encompassing a broad phenotypic spectrum that, at times, remains incompletely understood.
- Genetic complexity: The landscape of genetically unresolved disorders remains vast, reinforcing the complexity of the genetic mimic of cerebral palsy.

These challenges emphasize the need for comprehensive assessments and advanced genetic diagnostic tools to differentiate between cerebral palsy and its genetic mimics. There are several clinical and radiographic "red flag" symptoms that can help guide clinicians with triaging towards genetic testing (Table 17.1). We argue that next-generation sequencing-based molecular testing, that is, exome or genome sequencing, should be considered in all children with a cerebral palsy-like presentation.

Hereditary spastic paraplegia in the differential diagnosis of cerebral palsy

Hereditary spastic paraplegia (HSP) stands as a genetically diverse group of over 80 monogenic disorders, all characterized by the progressive onset of spasticity resulting from dysfunction in the corticospinal tract [13–15]. The global prevalence is estimated at approximately 5–10 individuals per 100,000 worldwide, although this is likely an underestimation [16,17]. The clinical presentation of HSP may manifest in its "pure" form, characterized by isolated lower extremity spasticity, weakness, and neurogenic bladder, or in its "complex" form, where additional symptoms such as

TABLE 17.1 Red flag symptoms that point to a genetic mimic of cerebral palsy.

I: History:
- Parental consanguinity or family history of similarly affected individuals.
- Full-term birth.
- No documented pre or perinatal risk factors for brain injury.
- Onset of abnormal motor symptoms after a period of normal development.
- Progression or fluctuation of symptoms/severity.
- Episodic worsening with illness, fever, fasting, and/or dietary changes.

II: Examination:
- Dysmorphic features.
- Abnormal eye movements.
- Isolated hypotonia or prominent ataxia.
- Evidence of peripheral neuromuscular disease.
- Multiorgan systemic involvement not explained by the sequelae of prematurity.

III: Neuroimaging:
- Normal brain MRI or changes that are not consistent with an acquired brain injury.
- Progressive MRI changes.

developmental delay/intellectual disability, peripheral neuropathy, seizures, and cerebellar dysfunction are evident [14,18].

As of now, a remarkable >85 distinct forms of HSP have been described, and >70 specific genes have been identified. The inheritance patterns of HSP encompass all modes of Mendelian inheritance. These affected genes implicate various molecular pathways, including organelle shaping, biogenesis, and membrane and cargo trafficking, in their pathogenesis [19]. One of the notable challenges in clinical practice is the frequent initial misdiagnosis of HSP as cerebral palsy. This underscores the importance of detailed clinical evaluation and genetic testing to accurately differentiate between these two conditions. In studies evaluating the diagnostic yield of exome sequencing in individuals with cerebral palsy, HSP accounted for about 25% of genetic mimics of cerebral palsy [11,12] and *SPAST* and *ATL1*, both associated dominant forms of HSP, were the second and third most common genes identified in a recent meta-analysis [12].

When confronted with patients presenting with spasticity, conducting a comprehensive neurologic evaluation becomes paramount. This should include a thorough neurologic history, with an emphasis on eliciting information about potential risk factors for perinatal injury and associated neurological symptoms, such as developmental delay, ataxia, dystonia, neuropathy, urinary symptoms, seizures, and a family history of neurologic disorders. A detailed neurologic examination is essential, focusing on assessing the distribution of spasticity, the presence of pyramidal or extrapyramidal signs, cerebellar manifestations, weakness, and signs of neuropathy. Furthermore, in the evaluation of spasticity, it is advisable to obtain brain magnetic resonance imaging (MRI) of the brain and consider spinal imaging. It is important to consider that progressive symptoms, an initial period of normal development followed by regression, the absence of a history suggestive of perinatal injury, or a normal brain MRI (or an MRI lacking typical findings associated with conditions like prematurity, hypoxic ischemic injury during birth, or neonatal stroke) should raise concerns about an alternative etiology, such as HSP.

When the initial evaluation raises suspicion of HSP, genetic testing should be pursued without delay. This can involve a multiple-gene panel or more comprehensive genetic testing methods such as exome or genome sequencing. The significance of obtaining an accurate diagnosis cannot be overstated, as it facilitates referral to subspecialists, offers genetic counseling to family members, and enables more precise prognostication.

Listed below are several forms of childhood-onset HSP that should be considered in the differential diagnosis of cerebral palsy:

Autosomal-dominant forms of hereditary spastic paraplegia

*Spastic paraplegia 4—*SPAST

Spastic paraplegia 4 (SPG4), also known as *SPAST*-HSP, is the most common form of autosomal-dominant HSP and accounts for about 50% of all HSP cases [20]. It typically presents as a pure form of HSP, although complex forms have been reported. Age of onset is variable, and a bimodal distribution has been described, with a first major peak before the first decade and a second peak between the third and fifth decades [21–24]. SPG4 typically presents with insidiously progressive bilateral lower limb spasticity. Leg muscle spasms are frequent and may even be present before the onset of spasticity. Neurogenic bladder dysfunction may develop. Disease severity usually worsens with the duration of the disease, although some individuals remain mildly affected throughout life. Less common complex forms, which often present during childhood and are associated with de novo variants in *SPAST*, may present with global developmental delay, dystonia, and/or epilepsy [25,26]. Brain and spinal cord MRIs are typically normal, although spinal cord atrophy can be seen. Mild vermis atrophy, thin corpus callosum, subtle white matter changes, and/or cerebellar atrophy have been reported in a small subset of cases.

Spastic paraplegia 3A—ATL1

Spastic paraplegia 3A (SPG3A), also known as *ATL1*-HSP, is typically an autosomal-dominant disorder [27]. It is classically a pure form of HSP; however, complex forms have been reported with axonal motor neuropathy and/or distal amyotrophy with lower motor involvement, and genotype-phenotype correlations are emerging [28]. *ALT1* variants are among the most common causes of autosomal-dominant HSP in all age groups. The average age of onset is about 4 years of age. Motor manifestations consist of progressive, symmetric lower extremity spasticity and weakness.

Spastic paraplegia 31—REEP1

Spastic paraplegia 31 (SPG31), caused by variants in the *REEP1* gene, is an autosomal-dominant disorder [29]. It is typically a pure form of HSP, although complex forms have been reported. Peak onset is either in the second or fourth decades of life. In its pure form, it presents with slowly progressive lower limb spasticity, gait difficulties, and hyperreflexia. In its complex form, it may include additional symptoms such as peripheral neuropathy, cerebellar ataxia, postural tremor, and urinary symptoms [30].

Spastic paraplegia 72—REEP2

Spastic paraplegia 72 (SPG72), caused by variants in *REEP2*, can be an autosomal-dominant or autosomal-recessive disorder [31,32]. SPG72 is a pure form of HSP. Onset is in infancy or early childhood. It presents with slowly progressive lower extremity spasticity, hyperreflexia, and difficulty walking. Affected individuals might also have *pes cavus* or sphincter disturbances [31].

Spastic paraplegia 10—KIF5A

Spastic paraplegia 10 (SPG10), caused by a variant in *KIF5A*, is an autosomal-dominant disorder [33,34]. It can be present in both pure and complex forms. Onset ranges from childhood to adulthood. Pure forms present with lower limb spasticity and hyperreflexia. Affected individuals can also have upper limb involvement, urinary urgency, and distal sensory impairment. Complex forms can present with additional symptoms, including axonal sensorimotor peripheral neuropathy, amyotrophy of the hands, parkinsonism, and cognitive decline [35].

Autosomal-recessive forms of hereditary spastic paraplegia

AP-4 associated HSP (SPG47—AP4B1, *SPG50*—AP4M1, *SPG51*—AP4E1, *and SPG52*—AP4S1*)*

AP-4-associated HSP, also known as AP-4 deficiency syndrome, is a group of four autosomal-recessive disorders [36]. Genes associated with AP-4-HSP include *AP4B1* (SPG47), *AP4E1* (SPG51), *AP4M1* (SPG50), and *AP4S1* (SPG52) [36]. Onset is usually in infancy with hypotonia and global developmental delay [37,38]. Hypotonia evolves into progressive lower extremity spasticity that eventually progresses to involve the upper extremities in the second or third decade of life in a subset of patients. The majority of children with AP-4-HSP become full-time wheelchair users over time [38]. Associated complications include contractures, foot deformities, bowel and bladder dysregulation, and a pseudobulbar affect. Around 60% have epilepsy, often first manifesting with febrile status epilepticus in the first years of life. Secondary microcephaly is common and often in a mild-to-moderate range. All individuals with AP-4-HSP have global developmental delays, with motor and expressive language development being most significantly impaired. Characteristic brain MRI findings include thinning of the corpus callosum with prominent posterior thinning, delayed myelination, nonspecific loss of the periventricular white matter, and exvacuo ventriculomegaly with prominent enlargement of the posterior horns of the lateral ventricles [39].

Spastic paraplegia 26—B4GALNT1

Spastic paraplegia 26 (SPG26), caused by variants in *B4GALNT1*, is an autosomal-recessive complex form of HSP [40–42]. Onset is usually in the first and second decades of life. It presents with lower limb spasticity and muscle weakness. Some patients manifest upper limp involvement, intellectual disability, ataxia, dysarthria, dystonia, and peripheral sensory neuropathy. Brain MRI is usually normal.

Spastic paraplegia 69—RAB3GAP2

Spastic paraplegia 69 (SPG69), caused by variants in *RAB3GAP2*, is an autosomal-recessive disorder complex form of HSP with onset in infancy and features overlapping with Warburg Micro syndrome [43]. It presents with progressive lower extremity spasticity, global developmental delay, hyperreflexia, and clonus, and might have associated dysarthria, cataracts, and hearing impairment. Congenital cataracts and microphthalmia are the important diagnostic clues. Brain MR typically shows agenesis of the corpus callosum.

Spastic paraplegia 56—CYP2U1

Spastic paraplegia 56 (SPG56), caused by variants in *CYP2U1*, is an autosomal-recessive disorder complex form of HSP with onset in infancy [44,45]. It presents with progressive lower limb spasticity and commonly has upper limb involvement. Affected individuals have developmental delay, axonal neuropathy, and pseudoxanthoma elasticum. Brain MR can show white matter lesions, globus pallidus calcifications, and a thin corpus callosum.

Spastic paraplegia 54—DDHD2

Spastic paraplegia 54 (SPG54), caused by variants in *DDHD2*, is an autosomal-recessive complex form of HSP with early onset [46,47]. It presents with spasticity of the lower limbs and developmental delay. Dysarthria, dysphagia, and strabismus can also be seen. Brain MRI can show a thin corpus callosum and periventricular white matter lesions. Brain MRIs can show an abnormal lipid peak [48].

Spastic paraplegia 18—ERLIN2

Spastic paraplegia 18 (SPG18), caused by variants in *ERLIN2*, is an autosomal-recessive disorder [49–51]. It is typically a complex form of HSP, although pure forms have been reported. Onset is in early childhood. It presents with progressive lower-extremity spasticity. Affected individuals might also have developmental delays, joint contractures, and epilepsy. Brain MRIs of the reported cases have shown no significant abnormalities.

Spastic paraplegia 49—TECPR2

Spastic paraplegia 49 (SPG49), caused by variants in *TECPR2*, has now been reclassified as a subtype of hereditary sensory and autonomic neuropathy (HSAN9) [52,53]. It is an autosomal-recessive disorder and was previously known as a complex form of HSP. Onset is in childhood. It presents with developmental delay, behavioral abnormalities, hypotonia, hyporeflexia, dysarthria, neuropathy, ataxic gait, and autonomic dysfunction. Affected individuals may also have epilepsy, as well as hearing and visual impairments. Brain imaging may show a thin or dysplastic corpus callosum, mild ventriculomegaly, delayed myelination, and/or mild cerebral or cerebellar atrophy.

Spastic paraplegia 84—PI4KA

Spastic paraplegia 84 (SPG84), caused by variants in *PI4KA*, is an autosomal-recessive disorder [54,55]. Onset is in childhood or adolescence. It presents with slowly progressive lower limb weakness, spasticity, and walking difficulties. Affected individuals can also have nystagmus, urinary urgency, joint contractures, and learning difficulties.

Other genetic conditions mimicking cerebral palsy

There are several treatable monogenic conditions that can resemble cerebral palsy. Highlighted here are conditions that fall into the categories of monogenic movement disorders, inflammatory diseases, and inborn errors of metabolism. Others are covered in several recent review papers [2,3].

Aicardi-Goutières syndrome

Aicardi-Goutières syndrome is a group of rare genetic autoinflammatory disorders, classified as interferonopathies, that can manifest with movement disorders similar to those seen in CP [56–60]. Affected children present at variable times after the first few months of life, frequently after a period of apparently normal development. Individuals with Aicardi-Goutières syndrome often present with a combination of dystonia and spasticity, and less commonly, chorea. Intellectual disability, seizures, and microcephaly are additional neurological features associated with Aicardi-Goutières syndrome [61]. Calcifications in the basal ganglia and periventricular white matter, as well as leukodystrophic changes, can be seen. Systemic features include recurrent fevers, chill-blain-like rashes, and joint inflammation [62]. The condition arises from variants in an evolving list of genes related to immune response, including biallelic pathogenic variants in *ADAR*, *RNASEH2A*, *RNASEH2B*, *RNASEH2C*, *SAMHD1*, or *TREX1* [56–59].

Creatine metabolism disorders

Creatine metabolism disorders encompass a group of genetic conditions that can present with a variety of movement disorders [63]. These disorders result from impairments in the synthesis, transport, or utilization of creatine—an essential molecule for brain and muscle energy metabolism. Pathogenic variants in GAMT, *AGAT*, or *SLC6A8* disrupt creatine homeostasis, causing neurological symptoms resembling cerebral palsy. Motor symptoms include ataxia, dystonia, or chorea. Early diagnosis through genetic testing and metabolic evaluation is crucial to initiating tailored management strategies, including dietary interventions and creatine supplementation.

Hyperornithinemia-hyperammonemia-homocitrullinuria syndrome

Hyperornithinemia-hyperammonemia-homocitrullinuria syndrome is a disorder of the urea cycle and ornithine degradation pathway [64]. Symptoms encompass chronic neurocognitive deficits, developmental delays, ataxia, spasticity, seizures, and acute encephalopathy due to hyperammonemic crises. The condition necessitates a multifaceted approach to management, including dietary measures, ammonia-scavenging medications, and symptomatic management of neurological manifestations such as seizures.

Arginase deficiency

Arginase deficiency is a rare inherited urea cycle disorder characterized by persistent elevation of plasma arginine [65]. It can present with a progressive spastic paraplegia, often misdiagnosed as cerebral palsy [66]. Clinical manifestations are diverse, encompassing motor deficits, developmental delays, and epilepsy [67,68]. Diagnosis relies on a testing of plasma arginine levels, genetic testing, or reduced arginase enzyme activity in red blood cells. Treatment aims to reduce nitrogenous waste production through dietary restrictions including a low-protein diet, and supplementation with citrulline and/or arginine.

Biotinidase deficiency

Biotinidase deficiency is a rare autosomal-recessive form of multiple carboxylase deficiency that is included in newborn screening in many countries [69]. It may also be diagnosed by biotinidase enzyme activity levels or by genetic testing detecting pathogenic variants in the *BTD* gene. In infancy, patients can present with seizures, developmental delay, and ataxia, as well as optic atrophy and dermatologic manifestations. Patients with delayed onset may be misdiagnosed as having cerebral palsy due to manifestations of spastic paraparesis and lower limb weakness. Early treatment with biotin may improve cutaneous and neurological manifestations.

Glut1 deficiency syndrome

Bi-allelic variants in the glucose transporter gene, *SLC2A1*, lead to a disorder of brain energy metabolism known as glucose transporter type 1, or Glut-1, deficiency [70]. A broad range of phenotypes is associated with Glut-1 deficiency [71]. The majority of patients develop the classic phenotype of early-onset seizures, developmental delay, microcephaly, and a variety of movement disorders (ataxia, spasticity, and dystonia). The nonclassic phenotype may be mistaken for cerebral palsy. The diagnosis is confirmed by a low CSF glucose concentration in the absence of low blood sugar and analysis of the *SLC2A1* gene. Treatment consists mainly of dietary modifications with ketogenic or modified Atkins diets, which can rapidly mitigate symptoms.

Aromatic L-amino acid decarboxylase deficiency

Aromatic L-amino acid decarboxylase (AADC) deficiency is an autosomal-recessive condition caused by mutations in the DDC gene, disrupting the synthesis of critical neurotransmitters dopamine and serotonin [72]. Symptoms vary, but typically involve severe developmental delay, movement disorders, oculogyric crises, and autonomic dysfunction [73–76]. Treatment primarily involves providing precursor molecules like levodopa and 5-hydroxytryptophan to bypass the enzymatic blockage. Intraparenchymal AAV-mediated gene therapy has shown positive results [77–79].

Segawa disease (dopa-responsive dystonia)

Dopa-responsive dystonia, specifically Segawa disease, is a form of dopa-responsive dystonia caused by autosomal-dominant GTPCH deficiency [80]. While often mistaken for cerebral palsy, it features characteristic postural dystonia of the extremities, especially the legs, leading to walking and balance impairments, with significant diurnal fluctuation at least initially. Prompt diagnosis and therapy with levodopa can yield significant improvement in motor symptoms. The mean age of onset typically falls between 5 and 10 years, with a female preponderance.

Sepiapterin reductase deficiency

Sepiapterin reductase deficiency is an autosomal-recessive disorder caused by variants in the *SPR* gene [81]. Symptoms may include nonspecific dystonia or hypotonia and may be mistaken for CP. The classic triad consists of oculogyric crises, paroxysmal stiffening, and hypotonia, with diurnal fluctuation and improvement in symptoms with sleep. Diagnosis is confirmed by analysis of CSF neurotransmitters and genetic testing. Treatment with L-dopa and 5-hydroxytryptophan results in improved neurologic outcomes.

Glutaric aciduria type 1

Glutaric aciduria type 1 is an organic aciduria caused by bi-allelic variants in the glutaryl-CoA dehydrogenase gene (*GCDH*) [82]. Affected children classically present with acute encephalopathic crises in the setting of an intercurrent illness. These metabolic crises lead to irreversible damage to the basal ganglia, with the subsequent sudden onset of movement disorder. Patients may also have delayed milestones, macrocephaly, hypotonia, choreoathetosis, or dystonia. Diagnosis can be made by urine organic acid analysis and confirmed by genetic testing. Treatment with a low-lysine diet, carnitine supplementation, and aggressive emergency treatment during periods of catabolism can improve outcomes.

Adenylate cyclase 5-related dyskinesia

Adenylate cyclase 5 (ADCY5)-related dyskinesia is an autosomal dominant disease with increasing prevalence around the world [83,84]. Children with ADCY5-related dyskinesia are often misdiagnosed as having hypotonic or dyskinetic cerebral palsy, given the combination of developmental delays, hypotonia and dystonia, chorea, or choreoathetosis [83]. Exacerbations of dyskinesia during sleep should prompt evaluation for ADCY5, with episodic or paroxysmal exacerbations lasting for a few minutes up to a few days. Brain MRI is normal. Treatment is supportive. Caffeine has been found to ameliorate dyskinesia in some patients [85]. Deep-brain stimulation is a treatment option with favorable outcomes [86].

Conclusion

There have been remarkable advances in the understanding of the clinical and molecular spectrum of childhood-onset movement disorders. Much of this is fueled by the advent and increasing availability of next-generation sequencing. In clinical practice, it is important to distinguish cerebral palsy from HSP and to identify treatable monogenic movement disorders and inborn errors of metabolism that mimic both. Recognition of symptoms beyond spasticity is crucial, and "red flag" features in history and examination can guide further diagnostic steps, including biochemical testing, brain imaging, and molecular testing. Attention needs to be paid to atypical presentations of treatable disorders including those discussed here. A correct clinical and molecular diagnosis and a better understanding of the clinical spectrum and natural history are of paramount importance as gene-specific disease-modifying therapies emerge.

References

[1] Ortigoza-Escobar JD. Catching the culprit: how chorea may signal an inborn error of metabolism. Tremor Other Hyperkinet Mov (N Y) 2023;13:36.
[2] Ortigoza-Escobar JD. A proposed diagnostic algorithm for inborn errors of metabolism presenting with movements disorders. Front Neurol 2020;11:582160.
[3] Ebrahimi-Fakhari D, Van Karnebeek C, Munchau A. Movement disorders in treatable inborn errors of metabolism. Mov Disord 2019;34 (5):598–613.
[4] Movement disorders and inherited metabolic disorders. In: Ebrahimi-Fakhari D, Pearl PL, editors. Movement disorders and inherited metabolic disorders: recognition, understanding, improving outcomes. Cambridge: Cambridge University Press; 2020. p. i–ii.
[5] Ebrahimi-Fakhari D, Saffari A, Pearl PL. Childhood-onset hereditary spastic paraplegia and its treatable mimics. Mol Genet Metab 2022;137 (4):436–44.
[6] Albanese A, Bhatia K, Bressman SB, Delong MR, Fahn S, Fung VS, et al. Phenomenology and classification of dystonia: a consensus update. Mov Disord 2013;28(7):863–73.
[7] Pavelekova P, Necpal J, Jech R, Havrankova P, Svantnerova J, Jurkova V, et al. Predictors of whole exome sequencing in dystonic cerebral palsy and cerebral palsy-like disorders. Parkinsonism Relat Disord 2023;111:105352.
[8] Moreno-De-Luca A, Millan F, Pesacreta DR, Elloumi HZ, Oetjens MT, Teigen C, et al. Molecular diagnostic yield of exome sequencing in patients with cerebral palsy. JAMA 2021;325(5):467–75.
[9] Yechieli M, Gulsuner S, Ben-Pazi H, Fattal A, Aran A, Kuzminsky A, et al. Diagnostic yield of chromosomal microarray and trio whole exome sequencing in cryptogenic cerebral palsy. J Med Genet 2022;59(8):759–67.
[10] Chopra M, Gable DL, Love-Nichols J, Tsao A, Rockowitz S, Sliz P, et al. Mendelian etiologies identified with whole exome sequencing in cerebral palsy. Ann Clin Transl Neurol 2022;9(2):193–205.
[11] Gonzalez-Mantilla PJ, Hu Y, Myers SM, Finucane BM, Ledbetter DH, Martin CL, et al. Diagnostic yield of exome sequencing in cerebral palsy and implications for genetic testing guidelines: a systematic review and meta-analysis. JAMA Pediatr 2023;177(5):472–8.
[12] Srivastava S, Lewis SA, Cohen JS, Zhang B, Aravamuthan BR, Chopra M, et al. Molecular diagnostic yield of exome sequencing and chromosomal microarray in cerebral palsy: a systematic review and meta-analysis. JAMA Neurol 2022;79(12):1287–95.
[13] Kumar KR, Blair NF, Sue CM. An update on the hereditary spastic paraplegias: new genes and new disease models. Mov Disord Clin Pract 2015;2(3):213–23.
[14] Shribman S, Reid E, Crosby AH, Houlden H, Warner TT. Hereditary spastic paraplegia: from diagnosis to emerging therapeutic approaches. Lancet Neurol 2019;18(12):1136–46.
[15] Blackstone C. Hereditary spastic paraplegia. Handb Clin Neurol 2018;148:633–52.
[16] Coutinho P, Ruano L, Loureiro JL, Cruz VT, Barros J, Tuna A, et al. Hereditary ataxia and spastic paraplegia in Portugal: a population-based prevalence study. JAMA Neurol 2013;70(6):746–55.
[17] Erichsen AK, Koht J, Stray-Pedersen A, Abdelnoor M, Tallaksen CM. Prevalence of hereditary ataxia and spastic paraplegia in southeast Norway: a population-based study. Brain 2009;132(Pt 6):1577–88.
[18] Elsayed LEO, Eltazi IZ, Ahmed AE, Stevanin G. Insights into clinical, genetic, and pathological aspects of hereditary spastic paraplegias: a comprehensive overview. Front Mol Biosci 2021;8:690899.
[19] Blackstone C. Converging cellular themes for the hereditary spastic paraplegias. Curr Opin Neurobiol 2018;51:139–46.
[20] Parodi L., Rydning S.L., Tallaksen C., Durr A. Spastic paraplegia 4. In: Adam MP, Feldman J, Mirzaa GM, Pagon RA, Wallace SE, Bean LJH, et al., editors. GeneReviews®. Seattle (WA): University of Washington, Seattle; 1993-2024.
[21] Varghaei P, Estiar MA, Ashtiani S, Veyron S, Mufti K, Leveille E, et al. Genetic, structural and clinical analysis of spastic paraplegia 4. Parkinsonism Relat Disord 2022;98:62–9.
[22] Rattay TW, Volker M, Rautenberg M, Kessler C, Wurster I, Winter N, et al. The prodromal phase of hereditary spastic paraplegia type 4: the preSPG4 cohort study. Brain 2023;146(3):1093–102.
[23] Rossi S, Rubegni A, Riso V, Barghigiani M, Bassi MT, Battini R, et al. Clinical-genetic features influencing disability in spastic paraplegia type 4: a cross-sectional study by the Italian DAISY network. Neurol Genet 2022;8(2):e664.
[24] Parodi L, Fenu S, Barbier M, Banneau G, Duyckaerts C, Tezenas du Montcel S, et al. Spastic paraplegia due to SPAST mutations is modified by the underlying mutation and sex. Brain 2018;141(12):3331–42.
[25] Mo A, Saffari A, Kellner M, Dobler-Neumann M, Jordan C, Srivastava S, et al. Early-onset and severe complex hereditary spastic paraplegia caused by de novo variants in SPAST. Mov Disord 2022;37(12):2440–6.
[26] Schieving JH, de Bot ST, van de Pol LA, Wolf NI, Brilstra EH, Frints SG, et al. De novo SPAST mutations may cause a complex SPG4 phenotype. Brain 2019;142(7):e31.
[27] Hedera P. Spastic paraplegia 3A. In: Adam MP, Feldman J, Mirzaa GM, Pagon RA, Wallace SE, Bean LJH, et al., editors. GeneReviews®. Seattle (WA): University of Washington, Seattle; 1993-2024.

[28] Alecu JE, Saffari A, Jordan C, Srivastava S, Blackstone C, Ebrahimi-Fakhari D. De novo variants cause complex symptoms in HSP-ATL1 (SPG3A) and uncover genotype-phenotype correlations. Hum Mol Genet 2023;32(1):93–103.

[29] Zuchner S, Wang G, Tran-Viet KN, Nance MA, Gaskell PC, Vance JM, et al. Mutations in the novel mitochondrial protein REEP1 cause hereditary spastic paraplegia type 31. Am J Hum Genet 2006;79(2):365–9.

[30] Beetz C, Schule R, Deconinck T, Tran-Viet KN, Zhu H, Kremer BP, et al. REEP1 mutation spectrum and genotype/phenotype correlation in hereditary spastic paraplegia type 31. Brain 2008;131(Pt 4):1078–86.

[31] Esteves T, Durr A, Mundwiller E, Loureiro JL, Boutry M, Gonzalez MA, et al. Loss of association of REEP2 with membranes leads to hereditary spastic paraplegia. Am J Hum Genet 2014;94(2):268–77.

[32] Roda RH, Schindler AB, Blackstone C. De novo REEP2 missense mutation in pure hereditary spastic paraplegia. Ann Clin Transl Neurol 2017;4(5):347–50.

[33] Liu YT, Laura M, Hersheson J, Horga A, Jaunmuktane Z, Brandner S, et al. Extended phenotypic spectrum of KIF5A mutations: From spastic paraplegia to axonal neuropathy. Neurology 2014;83(7):612–19.

[34] Reid E, Kloos M, Ashley-Koch A, Hughes L, Bevan S, Svenson IK, et al. A kinesin heavy chain (KIF5A) mutation in hereditary spastic paraplegia (SPG10). Am J Hum Genet 2002;71(5):1189–94.

[35] Carosi L, Lo Giudice T, Di Lullo M, Lombardi F, Babalini C, Gaudiello F, et al. Hereditary spastic paraplegia: a novel mutation and expansion of the phenotype variability in SPG10. J Neurol Neurosurg Psychiatry 2015;86(6):702–4.

[36] Ebrahimi-Fakhari D., Behne R., Davies A.K., Hirst J. AP-4-associated hereditary spastic paraplegia. In: Adam MP, Feldman J, Mirzaa GM, Pagon RA, Wallace SE, Bean LJH, et al., editors. GeneReviews®. Seattle (WA): University of Washington, Seattle; 1993-2024.

[37] Ebrahimi-Fakhari D, Cheng C, Dies K, Diplock A, Pier DB, Ryan CS, et al. Clinical and genetic characterization of AP4B1-associated SPG47. Am J Med Genet A 2018;176(2):311–18.

[38] Ebrahimi-Fakhari D, Teinert J, Behne R, Wimmer M, D'Amore A, Eberhardt K, et al. Defining the clinical, molecular and imaging spectrum of adaptor protein complex 4-associated hereditary spastic paraplegia. Brain 2020;143(10):2929–44.

[39] Ebrahimi-Fakhari D, Alecu JE, Ziegler M, Geisel G, Jordan C, D'Amore A, et al. Systematic analysis of brain MRI findings in adaptor protein complex 4-associated hereditary spastic paraplegia. Neurology 2021;97(19):e1942–54.

[40] Boukhris A, Schule R, Loureiro JL, Lourenco CM, Mundwiller E, Gonzalez MA, et al. Alteration of ganglioside biosynthesis responsible for complex hereditary spastic paraplegia. Am J Hum Genet 2013;93(1):118–23.

[41] Harlalka GV, Lehman A, Chioza B, Baple EL, Maroofian R, Cross H, et al. Mutations in B4GALNT1 (GM2 synthase) underlie a new disorder of ganglioside biosynthesis. Brain 2013;136(Pt 12):3618–24.

[42] Alecu JE, Ohmi Y, Bhuiyan RH, Inamori KI, Nitta T, Saffari A, et al. Functional validation of novel variants in B4GALNT1 associated with early-onset complex hereditary spastic paraplegia with impaired ganglioside synthesis. Am J Med Genet A 2022;188(9):2590–8.

[43] Handley M., Sheridan E. RAB18 deficiency. In: Adam MP, Feldman J, Mirzaa GM, Pagon RA, Wallace SE, Bean LJH, et al., editors. GeneReviews®. Seattle (WA): University of Washington, Seattle; 1993-2024.

[44] Tesson C, Nawara M, Salih MA, Rossignol R, Zaki MS, Al Balwi M, et al. Alteration of fatty-acid-metabolizing enzymes affects mitochondrial form and function in hereditary spastic paraplegia. Am J Hum Genet 2012;91(6):1051–64.

[45] Masciullo M, Tessa A, Perazza S, Santorelli FM, Perna A, Silvestri G. Hereditary spastic paraplegia: novel mutations and expansion of the phenotype variability in SPG56. Eur J Paediatr Neurol 2016;20(3):444–8.

[46] Schuurs-Hoeijmakers JH, Geraghty MT, Kamsteeg EJ, Ben-Salem S, de Bot ST, Nijhof B, et al. Mutations in DDHD2, encoding an intracellular phospholipase A(1), cause a recessive form of complex hereditary spastic paraplegia. Am J Hum Genet 2012;91(6):1073–81.

[47] Gonzalez M, Nampoothiri S, Kornblum C, Oteyza AC, Walter J, Konidari I, et al. Mutations in phospholipase DDHD2 cause autosomal recessive hereditary spastic paraplegia (SPG54). Eur J Hum Genet 2013;21(11):1214–18.

[48] Thabet F, Tlili-Graiess K, Tabarki B. Distinct neuroimaging features of DDHD2 gene-related spastic paraplegia, a mimicker of cerebral palsy. Arch Dis Child 2020;105(5):482.

[49] Alazami AM, Adly N, Al Dhalaan H, Alkuraya FS. A nullimorphic ERLIN2 mutation defines a complicated hereditary spastic paraplegia locus (SPG18). Neurogenetics 2011;12(4):333–6.

[50] Rydning SL, Dudesek A, Rimmele F, Funke C, Kruger S, Biskup S, et al. A novel heterozygous variant in ERLIN2 causes autosomal dominant pure hereditary spastic paraplegia. Eur J Neurol 2018;25(7):943–e71.

[51] Park JM, Lee B, Kim JH, Park SY, Yu J, Kim UK, et al. An autosomal dominant ERLIN2 mutation leads to a pure HSP phenotype distinct from the autosomal recessive ERLIN2 mutations (SPG18). Sci Rep 2020;10(1):3295.

[52] Heimer G, Neuser S, Ben-Zeev B, Ebrahimi-Fakhari D. TECPR2-related hereditary sensory and autonomic neuropathy with intellectual disability. GeneReviews®. Seattle (WA): University of Washington, Seattle; 1993-2024.

[53] Neuser S, Brechmann B, Heimer G, Brosse I, Schubert S, O'Grady L, et al. Clinical, neuroimaging, and molecular spectrum of TECPR2-associated hereditary sensory and autonomic neuropathy with intellectual disability. Hum Mutat 2021;42(6):762–76.

[54] Baple E.L., Salter C., Uhlig H., Wolf N.I., Crosby A.H. PI4KA-related disorder. In: Adam MP, Feldman J, Mirzaa GM, Pagon RA, Wallace SE, Bean LJH, et al., editors. GeneReviews®. Seattle (WA): University of Washington, Seattle; 1993-2024.

[55] Verdura E, Rodriguez-Palmero A, Velez-Santamaria V, Planas-Serra L, de la Calle I, Raspall-Chaure M, et al. Biallelic PI4KA variants cause a novel neurodevelopmental syndrome with hypomyelinating leukodystrophy. Brain 2021;144(9):2659–69.

[56] Crow YJ, Hayward BE, Parmar R, Robins P, Leitch A, Ali M, et al. Mutations in the gene encoding the 3′-5′ DNA exonuclease TREX1 cause Aicardi-Goutieres syndrome at the AGS1 locus. Nat Genet 2006;38(8):917–20.

[57] Crow YJ, Leitch A, Hayward BE, Garner A, Parmar R, Griffith E, et al. Mutations in genes encoding ribonuclease H2 subunits cause Aicardi-Goutieres syndrome and mimic congenital viral brain infection. Nat Genet 2006;38(8):910–16.

[58] Rice GI, Bond J, Asipu A, Brunette RL, Manfield IW, Carr IM, et al. Mutations involved in Aicardi-Goutieres syndrome implicate SAMHD1 as regulator of the innate immune response. Nat Genet 2009;41(7):829–32.

[59] Rice GI, Kasher PR, Forte GM, Mannion NM, Greenwood SM, Szynkiewicz M, et al. Mutations in ADAR1 cause Aicardi-Goutieres syndrome associated with a type I interferon signature. Nat Genet 2012;44(11):1243–8.

[60] Beysen D, De Cordt C, Dielman C, Ogunjimi B, Dandelooy J, Reyniers E, et al. Genetic testing contributes to diagnosis in cerebral palsy: Aicardi-Goutieres syndrome as an example. Front Neurol 2021;12:617813.

[61] Wang W, Wang W, He TY, Zou LP, Li WD, Yu ZX, et al. Analysis of clinical characteristics of children with Aicardi-Goutieres syndrome in China. World J Pediatr 2022;18(7):490–7.

[62] Orcesi S, La Piana R, Fazzi E. Aicardi-Goutieres syndrome. Br Med Bull 2009;89:183–201.

[63] Mercimek-Andrews S., Salomons G.S. Creatine deficiency disorders. In: Adam MP, Feldman J, Mirzaa GM, Pagon RA, Wallace SE, Bean LJH, et al., editors. GeneReviews®. Seattle (WA): University of Washington, Seattle; 1993-2024.

[64] Camacho J., Rioseco-Camacho N. Hyperornithinemia-hyperammonemia-homocitrullinuria syndrome. In: Adam MP, Feldman J, Mirzaa GM, Pagon RA, Wallace SE, Bean LJH, et al., editors. GeneReviews®. Seattle (WA): University of Washington, Seattle; 1993-2024.

[65] Scheuerle AE, McVie R, Beaudet AL, Shapira SK. Arginase deficiency presenting as cerebral palsy. Pediatrics 1993;91(5):995–6.

[66] Jichlinski A, Clarke L, Whitehead MT, Gropman A. "Cerebral palsy" in a patient with arginase deficiency. Semin Pediatr Neurol 2018;26:110–14.

[67] Bin Sawad A, Jackimiec J, Bechter M, Trucillo A, Lindsley K, Bhagat A, et al. Epidemiology, methods of diagnosis, and clinical management of patients with arginase 1 deficiency (ARG1-D): a systematic review. Mol Genet Metab 2022;137(1–2):153–63.

[68] Carvalho DR, Brum JM, Speck-Martins CE, Ventura FD, Navarro MM, Coelho KE, et al. Clinical features and neurologic progression of hyperargininemia. Pediatr Neurol 2012;46(6):369–74.

[69] Wolf B. Biotinidase deficiency. In: Adam MP, Feldman J, Mirzaa GM, Pagon RA, Wallace SE, Bean LJH, et al., editors. GeneReviews®. Seattle (WA): University of Washington, Seattle; 1993-2024.

[70] Wang D., Pascual J.M., De Vivo D. Glucose transporter type 1 deficiency syndrome. In: Adam MP, Feldman J, Mirzaa GM, Pagon RA, Wallace SE, Bean LJH, et al., editors. GeneReviews®. Seattle (WA): University of Washington, Seattle; 1993-2024.

[71] Klepper J, Akman C, Armeno M, Auvin S, Cervenka M, Cross HJ, et al. Glut1 deficiency Syndrome (Glut1DS): state of the art in 2020 and recommendations of the international Glut1DS study group. Epilepsia Open 2020;5(3):354–65.

[72] Blau N., Pearson T.S., Kurian M.A., Elsea S.H. Aromatic L-amino acid decarboxylase deficiency. In: Adam MP, Feldman J, Mirzaa GM, Pagon RA, Wallace SE, Bean LJH, et al., editors. GeneReviews®. Seattle (WA): University of Washington, Seattle; 1993-2024.

[73] Hwu WL, Hsu RH, Li MH, Lee HM, Chen HA, Lee NC, et al. Aromatic l-amino acid decarboxylase deficiency in Taiwan. JIMD Rep 2023;64(5):387–92.

[74] Himmelreich N, Bertoldi M, Alfadhel M, Alghamdi MA, Anikster Y, Bao X, et al. Prevalence of DDC genotypes in patients with aromatic L-amino acid decarboxylase (AADC) deficiency and in silico prediction of structural protein changes. Mol Genet Metab 2023;139(3):107624.

[75] Abukhaled M, Al Muqbil M, Alghamdi MA, Hundallah K, Suleiman J, Ben-Omran T, et al. Aromatic L-amino acid decarboxylase deficiency in countries in the Middle East: a case series and literature review. Eur J Pediatr 2023;182(6):2535–45.

[76] Pearson TS, Gilbert L, Opladen T, Garcia-Cazorla A, Mastrangelo M, Leuzzi V, et al. AADC deficiency from infancy to adulthood: symptoms and developmental outcome in an international cohort of 63 patients. J Inherit Metab Dis 2020;43(5):1121–30.

[77] Pearson TS, Gupta N, San Sebastian W, Imamura-Ching J, Viehoever A, Grijalvo-Perez A, et al. Gene therapy for aromatic L-amino acid decarboxylase deficiency by MR-guided direct delivery of AAV2-AADC to midbrain dopaminergic neurons. Nat Commun 2021;12(1):4251.

[78] Francois-Heude MC, Poulen G, Flamand Roze E, Nguyen Morel MA, Gras D, Roch-Toreilles I, et al. Intraputaminal gene delivery in two patients with aromatic L-amino acid decarboxylase deficiency. Mov Disord Clin Pract 2023;10(5):811–18.

[79] Tai CH, Lee NC, Chien YH, Byrne BJ, Muramatsu SI, Tseng SH, et al. Long-term efficacy and safety of eladocagene exuparvovec in patients with AADC deficiency. Mol Ther 2022;30(2):509–18.

[80] Furukawa Y. GTP cyclohydrolase 1-deficient dopa-responsive dystonia. In: Adam MP, Feldman J, Mirzaa GM, Pagon RA, Wallace SE, Bean LJH, et al., editors. GeneReviews®. Seattle (WA): University of Washington, Seattle; 1993-2024.

[81] Friedman J. Sepiapterin reductase deficiency. In: Adam MP, Feldman J, Mirzaa GM, Pagon RA, Wallace SE, Bean LJH, et al., editors. GeneReviews®. Seattle (WA): University of Washington, Seattle; 1993-2024.

[82] Larson A., Goodman S. Glutaric acidemia type 1. In: Adam MP, Feldman J, Mirzaa GM, Pagon RA, Wallace SE, Bean LJH, et al., editors. GeneReviews®. Seattle (WA): University of Washington, Seattle; 1993-2024.

[83] Menon PJ, Nilles C, Silveira-Moriyama L, Yuan R, de Gusmao CM, Munchau A, et al. Scoping review on ADCY5-related movement disorders. Mov Disord Clin Pract 2023;10(7):1048–59.

[84] Ferrini A, Steel D, Barwick K, Kurian MA. An update on the phenotype, genotype and neurobiology of ADCY5-related disease. Mov Disord 2021;36(5):1104–14.

[85] Meneret A, Mohammad SS, Cif L, Doummar D, DeGusmao C, Anheim M, et al. Efficacy of caffeine in ADCY5-related dyskinesia: a retrospective study. Mov Disord 2022;37(6):1294–8.

[86] Cif L, Demailly D, Gehin C, Chan Seng E, Dornadic M, Huby S, et al. Deep brain stimulation effect in genetic dyskinetic cerebral palsy: the case of ADCY5- related disease. Mol Genet Metab 2023;138(1):106970.

Chapter 18

The neurodevelopmental movement disorders: tics and stereotypies

Shannon L. Dean
Department of Neurology and Developmental Medicine, Kennedy Krieger Institute, Department of Neurology, Johns Hopkins University School of Medicine, Baltimore, MD, United States

Introduction

Excessively repetitive movements are one of the earliest signs of abnormal development [1–3]. In early infancy, a comparatively restricted repertoire of repetitive arm and leg movements is seen in typically developing infants, which may metaphorically be thought of as part of the normal "boot-up" phase of motor development, as motor circuits refine their connections [4–6]. In normal development, these movements become less complex and more stereotyped as early as 2 months [7], and infants show the ability to learn to use their bodies in task-specific, goal-directed motor learning by 3 months [8]. Movements that continue to be excessively stereotyped are an early sign of limited motor repertoire and future dysfunction [1,2,9].

Tics and stereotypies are common pediatric movement disorders that appear early in development and consist of repeated stereotyped abnormal movements. These developmental movement disorders can be seen in the context of an overarching developmental disorder, such as autism [10,11]. They can also be seen in otherwise normally developing children as an isolated primary movement disorder [12].

In general, developmental movement disorders have a benign course. They do not always require symptomatic treatment unless the movements are clearly negatively affecting the patient, and they sometimes improve spontaneously with age. Many patients with developmental movement disorders come to clinical attention because the presence of comorbidities has a greater impact on their movements than the movements themselves [13]. These comorbidities range from other developmental disorders such as autism and attention-deficit/hyperactivity disorder (ADHD) to psychiatric comorbidities such as anxiety and obsessive-compulsive disorder (OCD) [13,14]. Thus even when developmental movement disorders are not themselves a cause for alarm or even treatment, they should serve as a prompt for further developmental screening.

Stereotypies

Motor stereotypies are repetitive, usually rhythmic, movements that appear purposeful but are actually purposeless. They begin before age three in the vast majority of patients [15]. When refering to stereotypies as a developmental movement disorder, we are generally referring to more complex and seemingly bizarre movements. Simple unilateral stereotypies, such as thumb sucking in younger children and hair twirling, foot tapping, or pen tapping in older children and adults, are commonly seen but rarely remarked on in the general population [16]. Body rocking is also generally considered a simple motor stereotypy, and is seen commonly either in very young children or in older individuals with other neurodevelopmental disorders. When these common movements are included, the rate of stereotypies in the general population is up to 20%.

Motor stereotypies can be distinguished from tics by their younger age of onset, more rhythmic appearance, their lack of waxing and waning, and their constant semiology over days to weeks (Table 18.1). They can be distinguished from motor overflow again by their characteristic rhythmic appearance, and by their triggers. While motor overflow is an unintentional movement triggered by volitional movements elsewhere in the body, motor stereotypies are instead triggered most commonly by level of arousal and are often, in fact, reduced during volitional movements. Of note,

Capute and Accardo's Neurodevelopmental Disabilities in Infancy and Childhood. DOI: https://doi.org/10.1016/B978-0-12-824060-1.00044-4

TABLE 18.1 A comparison of stereotypies versus tics.

	Stereotypies	Tics
Age of onset	Before 3	3–9 years
Movement appearance	Rhythmic, sometimes with fixed posturing, may go on for minutes.	Usually short and abrupt, but occasionally head shaking is seen; generally, individual movements do not occur for more than a few seconds.
Body parts involved	Often hands or feet, sometimes fixed unusual body posture or facial expression	Rostral to caudal gradient, frequently involving facial muscles
Temporal pattern	Foreseeable fixed frequency: movements look the same over years.	Wax and wane, movement appearance often changes or is replaced by new movements over weeks to months.
Triggers	Excitement, boredom, engrossment, fatigue, and stress	Stress, excitement, illness, and fatigue
Subjective experience/answer to "Why do you do that?"	May not notice the movements or may report they are enjoyable.	More likely to notice movements and feel they "have" to do them.
Voluntary suppression	Stop spontaneously with distraction; can stop when asked but may be upset.	Can stop briefly with effort but find it difficult and uncomfortable to do so, often rebounding of movements after "holding it in."
Treatment	Behavioral therapy only, neuroleptics may be helpful in children with autism and self-injurious stereotypies resistant to behavioral therapy.	Behavioral therapy or medications
Patient distress due to symptoms	Usually low	Variable

while these movement abnormalities are considered distinct problems, they often occur in the same patients [14,17]. This may reflect a failure of motor inhibition as a common underlying mechanism in all of these findings [12,18].

Complex motor stereotypies

Complex motor stereotypies include shaking, posturing, and waving movements in the joints of the upper extremities or, less commonly, the lower extremities. In addition to rhythmic extremity movements, some children will simultaneously vocalize, grimace, or rock their torsos. Movements generally occur for seconds to minutes [16], multiple times a day at a stable frequency in a given individual. They remain constant in semiology for prolonged periods of time, although over the course of months or years they may change in intensity, and some fragments of complex movements may fade away. They are usually not painful, although exceptions such as self-scratching or hitting movements do occur. They are most commonly triggered by feelings of happiness and excitement, but movements may also be triggered when the child is engrossed in an activity, bored, anxious, or tired. Some patients may have their movements when daydreaming, but mental status is always normal during these movements. They should stop with distraction or when the child needs the affected body parts to complete a task; however, children may resume their movements once they are no longer actively using their limbs or attending to the parent or examiner. Some children may be upset by a request to stop their movements, but they should not find it difficult to comply.

Etiology

These movements are thought to be mediated by habitual, rather than goal-directed or voluntary, motor pathways [12]. An EEG study looking at the Bereitschaftspotential (readiness potential) supports this hypothesis [19]. The Bereitschaftspotential can be measured 1.2 seconds immediately prior to planned volitional movements over the supplementary motor area and the motor cortex. It is thought to reflect motor planning activity. Because of its small amplitude, it cannot be appreciated when viewing an EEG with the naked eye and instead only emerges when EEG tracings are averaged and correlated with movement onset. When research participants with complex motor stereotypies are asked to deliberately perform their movements, an Bereitschaftspotential can be measured. It can likewise be measured when healthy control participants are asked to deliberately produce stereotypy-like movements. However, this signal is not measured before spontaneously occurring stereotypies in patients [19]. Notably, this study did not include patients with autism or intellectual disabilities and stereotypies.

The exact cause of these movements is not known, although given the family history of 25%–30% of patients, there is likely a genetic component [12]. Some have argued that stereotypies may be a remnant of prior early motor patterning that is not successfully suppressed in the brains of some individuals based on the observation that similar rhythmic movements are normal in well-developed infants [5,20]. Others have argued that while the movements may not be done purposefully, they may still serve a purpose within the brain to help regulate arousal levels by providing somatosensory input [21]. This is based on the observation that stereotypies are more frequent during periods of very high (agitation, excitement) or low (fatigue, boredom) arousal, and that stereotypies occur more frequently in children with sensory deficits or in low stimulation environments and may improve in high stimulation environments [22]. It also fits with the subjective reports of some patients with autism who experience their movements as soothing or report that they help them pay attention [23].

While the brains of individuals with stereotypies alone look essentially normal, we have some preliminary evidence that hints at the underlying pathophysiology of stereotypies. Small volumetric changes have been reported in the basal ganglia [24], frontal lobes [25], and cerebellum [26] of children with isolated "primary" stereotypies: those that have grossly normal development aside from their movements. Lower GABA levels have been reported in the striatum and cingulate cortex of these children [27]. There is one case report of new onset stereotypies in an adult after a stroke in the putamen [28], and stereotypies are also seen in frontotemporal dementia [29]. In autism, stereotypy severity correlates with smaller volumes in cerebellar lobules VI–VII and higher frontal lobe volumes [26]. The involvement of a circuit involving the cerebellum, striatum, and frontal cortex seems highly likely based on these findings, but much remains to be learned.

Primary complex motor stereotypies

Primary complex motor stereotypies are those that occur in otherwise typically developing children. In this context, children with mild developmental differences such as ADHD or transient mild language or motor delay but ultimately normal IQ are considered to be part of the "typical" group. These children may have a slightly lower IQ and worse motor performance and executive function compared to age-matched controls without these movements, but both their motor and intellectual performance should still fall in the normal range [17]. Moreover, overall motor performance does not correlate with the severity of the stereotypies. In contrast, these movements are considered secondary in children with autism, intellectual disability, Rett syndrome, or sensory abnormalities such as blindness [16]. These movements occur in up to 3%–4% of "typically" developing children [16], although of note, far fewer than 3%–4% of families will bring these movements to medical attention as a cause for clinical concern. Children old enough to answer questions generally report that they often do not notice doing their movements until their attention is brought to them, or less commonly, that the movements are enjoyable [15].

Evaluation

A general neurologic exam is expected to be essentially normal in children with primary motor stereotypies. These movements are not associated with injuries to the central nervous system, so a workup is generally not needed for the movements themselves. However, it is worth evaluating the common comorbidities of stereotypies. Current literature suggests that most patients with primary complex motor stereotypies eventually have at least one comorbid condition such as ADHD, anxiety, OCD, or Tourette syndrome [14]. An increased level of concern for these disorders is warranted. However, due to the young age at which these movements typically manifest, these comorbidities are often not evident at the initial presentation for movement evaluation.

Treatment and outcome

Previously, there was an expectation that primary stereotypies would resolve on their own, generally by age 8 [16]. More recent studies suggest that while stereotypies improve in 77% of patients, they rarely fully resolve [14]. This improvement may take the form of: a reduced frequency of movements in public settings, or a reduction in the complex rhythmic hand movements, with the maintenance of motor fragments (e.g., maintained teeth clenching or grimacing with excitement but no hand movements). For many children, these movements are not particularly bothersome and do not interfere with activities, so treatment is not always necessary. In cases where treatment is desired, either because the child finds the movements embarrassing, or, in rare cases, because they are painful, habit reversal therapy is an effective treatment [30]. Online videos that teach parents how to deliver treatment are available for children who do not have access to in-person therapy and can reduce stereotypies [31].

Secondary stereotypies in autism

It is estimated that about 90% of children with autism exhibit complex motor stereotypies. Literature suggests that children with autism tend to have more frequent and more complex movements, with a greater tendency to include objects or vocalization [32]. Increased frequency and severity of stereotypies are associated with lower IQ in children with autism [33], as well as greater severity of autistic symptoms [34]. However, there is significant clinical overlap in the appearance of the movements themselves between typically developing children and children with autism. Genetic evidence suggests that some genetic changes in 'typically developing' children with primary stereotypies overlap with risk factor genes for autism [35]. This is not to say that children with primary stereotypies meet criteria for autism, but rather to illustrate that autism manifests on a spectrum and that traits commonly associated with autism extend into the "typical" population.

Like children with primary stereotypies, many children with autism report that they do not notice their movements until their attention is drawn to them, though some will report that they will deliberately continue the movements once noticed as they find them soothing or otherwise enjoyable [23].

Evaluation

Given the frequency of stereotypies in autism, they generally do not merit further evaluation beyond appropriate genetic screening for the cause of autism itself. However, there are certain clinical scenarios that do require additional workup.

For example, stereotypies occur frequently in catatonia as well as autism and are part of the diagnostic criteria for both. There is a great deal of overlap in the diagnostic criteria for catatonia and autism [36] (Table 18.2). Clinicians should keep in mind that, in addition to this baseline overlap, individuals with preexisting autism can also develop catatonia as a secondary diagnosis, often in adolescence or early adulthood [37–39]. Therefore, an increase in stereotypies combined with other concerning changes such as decreased speech, changes in general motor activity, increased echolalia/echopraxia, increased exaggerated mannerisms, or increased agitation should prompt referral to neuropsychiatry for further diagnosis and treatment. Patients with catatonia in the context of autism may be slower to respond to standard treatment for catatonia but can improve or resolve their symptoms with high-dose benzodiazepines or electroconvulsive therapy [40,41]. Other specific genetic and neurodegenerative disorders that may have both stereotypies and autistic features are discussed below.

Treatment and outcome

In children with autism, stereotypies are generally grouped together with other repetitive behaviors, the frequency of which is correlated with the severity of autistic symptoms [42]. Increased rates of stereotypies are also associated with lower IQ in children with autism [33,42,43]. When severe, some experts argue that the movements themselves may interfere with education or therapies or increase autism stigma [44–46], thereby warranting a lower threshold for intervention. However, many older patients with autism, capable of expressing their experiences, report instead that the movements help them concentrate or deal with anxiety [23]. Indeed, some researchers have proposed that stereotypies may help regulate arousal levels in some patients [47]. Therefore. It may be better, to consider stereotypies as a result of decreased attention in the classroom, raher than the cause. Further, many older individuals with higher functioning autism express lingering distress about attempts to force them to stop their stereotypies during childhood [23]. For these reasons, we generally recommend against aggressive treatment of stereotypies with behavioral therapy in children with autism unless they are resulting in self-injury, teasing, or other demonstrable harm. On the other hand, more severe self-harming stereotypies are more common in individuals with autism than in children with primary stereotypies [42,48]. Behavioral therapy is effective for stereotypies causing them demonstrable harm [46]. Neuroleptics such as risperidone may be considered for self-harming stereotypies that are refractory to behavioral therapy [49,50]. Before beginning treatment, however, physicians should assess whether the self-harming movements are automatic and involuntary, or a volitional response to frustration or attempt to elicit a specific response from their caregivers [51].

TABLE 18.2 A comparison of autism and catatonia, two disorders that include stereotypies in their diagnostic criteria

Criteria for Diagnosis of Autism (overlap with catatonia in bold)	Criteria for Diagnosis of Catatonia (overlap with autism in bold)
1. Persistent deficits in social communication and social interaction across multiple contexts, as manifested by all of the following:	1. Stupor (i.e. - no psychomotor activity; not actively relating to environment)
A. Deficits in social-emotional reciprocity, ranging, for example, from abnormal social approach and failure of normal back-and-forth conversation; to reduced sharing of interests, emotions, or affect; to **failure to initiate or respond to social interactions.**	2. Catalepsy (i.e. - passive induction of a posture held against gravity; or muscular rigidity and fixity of posture regardless of external stimuli; catalepsy is a motor symptom of schizophrenia similar to waxy flexibility)
B. Deficits in nonverbal communicative behaviors used for social interaction, ranging, for example, from poorly integrated verbal and nonverbal communication; to abnormalities in eye contact and body language or deficits in understanding and use of gestures; to a total lack of facial expressions and nonverbal communication.	3. Waxy flexibility (i.e. - slight, even resistance to positioning by examiner; the limb can be placed in an awkward posture and remain fixed in position for long time despite asking the individual to relax)
C. Deficits in developing, maintaining, and understanding relationships, ranging, for example, from difficulties adjusting behavior to suit various social contexts; to difficulties in sharing imaginative play or in making friends; to absence of interest in peers.	4. Mutism (i.e. - no, or very little, verbal response [exclude if known aphasia])
2. Restricted, repetitive patterns of behavior, interests, or activities, as manifested by at least two of the following:	5. Negativism (i.e. - opposition or no response to instructions or external stimuli; see also *Gegenhalten*)
A. **Stereotyped or repetitive motor movements**, use of objects, or speech (e.g., simple motor stereotypies, lining up toys or flipping objects, echolalia, idiosyncratic phrases).	6. Posturing (i.e. - spontaneous and active maintenance of a posture against gravity)
B. **Insistence on sameness, inflexible adherence to routines, or ritualized patterns of verbal or nonverbal behavior**	7. Mannerism (i.e. - odd, circumstantial caricature of normal actions)
C. **Highly restricted, fixated interests that are abnormal in intensity or focus** (e.g., strong attachment to or preoccupation with unusual objects, excessively circumscribed or perseverative interests).	8. Stereotypy (i.e. - repetitive, abnormally frequent, non-goal-directed movements)
D. **Hyper- or hyporeactivity to sensory input or unusual interest in sensory aspects of the environment** (e.g. apparent indifference to pain/temperature, adverse response to specific sounds or textures, excessive smelling or touching of objects, visual fascination with lights or movement).	9. Agitation (not influenced by external stimuli)
	10. Grimacing
	11. Echolalia (i.e. - mimicking another's speech)
	12. Echopraxia (i.e. - mimicking another's movements)

Complex motor stereotypies in other contexts

Stereotypies occur not only in children with normal development or autism but in many other clinical contexts as well. They are found in 73% of children with blindness [52]. They are also more prevalent in children in early institutional care compared to those in foster care [22]. In both cases, stereotypies are associated with restricted environments, whereas increased appropriate sensory stimulation tends to decrease their occurance [22,52].

Stereotypies are common in children with a variety of neurodevelopmental disabilities, occurring in about 60% of nonautistic children with intellectual disabilities [10]. IQ negatively correlates with the frequency of stereotypies [33]. Similar to children with autism, while these movements are often benign, self-injurious movements such as self-hitting or head banging are more common in children with intellectual disabilities than in those with primary stereotypies [42,48]. Of course, children with these conditions may also exhibit self-injurious behaviors as a more goal-directed response to frustration or attempt to elicit a specific response from their caregivers [51]. Careful observation of the appearance of the movements themselves in terms of rhythmicity and stereotyped nature, as well as whether they are context-dependent or context-independent, can help clarify the nature of the movement and the appropriate treatment strategy [51].

Stereotypies involving the hands occur in 70% to 80% of children with fragile X syndrome, the most common cause of inherited intellectual disability [53]. These movements often start extremely early in these children, sometimes before

6 months of age [53]. Although this has been investigated as a potential screening feature for early identification of Fragile X syndrome, direct observation by clinicians indicates that parents often do not appreciate these movements, which limits their clinical utility.

Rett syndrome is a progressive X-linked dominant developmental disorder occurring almost exclusively in females due to mutations in the MECP2 gene. Patients with Rett syndrome generally have normal development until 6–18 months. Then, they begin to lose previously acquired language and fine motor skills, and they develop characteristic hand stereotypies. Classically, stereotypies in Rett syndrome include a characteristic hand-wringing movement, observed in 63% of patients [54]. However, patients may exhibit other hand movements such as clapping, mouthing, flapping, squeezing, tapping, and "piano playing" finger movements [55]. Hand stereotypies may appear during a developmental plateau phase before the regression truly begins [56]. Other features of the disorder include autism-like features, irregular breathing, epilepsy, and, in some cases, acquired microcephaly. In the later stages of the disease, children with Rett syndrome may develop Parkinsonism, which may mask their characteristic hand movements. Rett syndrome should be considered in girls with language regression, hand stereotypies, and loss of purposeful hand use emergence of between 6 and 18 months. Diagnosis is confirmed by genetic testing.

Neuronal ceroid lipofuscinosis is a group of autosomal recessive neurodegenerative lysosomal storage disorders characterized by progressive intellectual deterioration, vision loss, and epilepsy. Some of these disorders can also include stereotypies early in clinical presentation. Type 1 (CLN1), due to a mutation in PPT1 (palmitoyl-protein thioesterase), presents in infancy with stereotypies as well as global developmental deterioration, loss of interest in toys, retinal blindness, and myoclonus. Other movement disorders, such as dystonia, ataxia, and Parkinsonism, frequently develop.

Stereotypies may occur in other developmental neurogenetic disorders, but not necessarily as one of the most prominent early features. Patients with Cornelia de Lange syndrome, caused by mutations in various genes vital to the function of the cohesion complex during cellular division, develop stereotypies and self-injurious behaviors, intellectual disability, autism spectrum disorder, as well as hirsutism, gastrointestinal dysfunction, and abnormalities in a variety of organ systems, including the eyes, heart, genitourinary tract, and bones [57]. Children with Angelman syndrome typically exhibit happy personalities, elfin features, intellectual disabilities, and occasionally stereotypies [58]. Children with Smith-Magenis syndrome have stereotypies including a characteristic self-hugging movement [58]. In infancy, they present with hypotonia, failure to thrive, and lethargy, and later on they develop coarsening facial features, abdominal obesity, and intellectual impairment [59]. Self-injurious repetitive movements are associated with Lesch-Nyhan syndrome, although they do not typically manifest as classical rhythmic complex motor stereotypies [60].

Head nodding stereotypies

Head nodding stereotypies generally refer to rhythmic movements of the head, including yes–yes, no–no, and sometimes more complex movements, such as figure-of-eight movements. Some patients may have movements of their hands or feet at the same time, or upward gaze deviation. These stereotypies are generally observed less frequently in clinical practice compared to classic complex motor stereotypies (personal observation), although exact prevalence data for this subtype is lacking in the literature.

The need for workup in patients with head nodding stereotypies depends on the clinical context. Head nodding stereotypies do occur in otherwise typically developing children with normal neurologic exams. Like the complex motor stereotypies discussed above, their prognosis is benign in otherwise well children, and the limited data available suggests that there may be a higher rate of complete spontaneous resolution compared to classic complex motor stereotypies involving the extremities [15]. Head nodding and body rocking are also relatively common in children with intellectual disabilities and/or autism [33]. Head nodding and body rocking may be slightly more common in children with intellectual disabilities but not autism compared to children with autism, but of note, these children commonly have multiple stereotypies that may include both head nodding and finger or arm movements [33]. The workup in these cases is for the underlying cause of autism or intellectual disability.

However, certain stereotyped rhythmic head movements do prompt additional workup. Figure-of-eight stereotypies, in which the head moves in a complex figure-of-eight pattern, have been reported frequently in children with rhombencephalosynapsis in particular [61] as well as other abnormalities of the posterior fossa [62]. When this stereotypy is seen, a brain MRI should be considered if the patient has a developmental delay or signs of cerebellar dysfunction, such as abnormal saccades. Likewise, a 2–3 Hz, somewhat suppressible yes–yes head nodding tremor is characteristic of bobble-head doll syndrome, a movement disorder seen in children with lesions that cause compression of the third ventricle [63]. Although this movement is often classified as a tremor—the ability of patients to suppress these movements, suppression in sleep, reduction of the movement during volitional activities, and increase with strong emotions—could easily lead to them being mistaken for a head nodding stereotypy. Additional features that may guide the clinician to this rare diagnosis include a slowly progressive course, tremors in the trunk, and papilledema. Prompt imaging should be obtained if this movement disorder is suspected. Finally, spasmus nutans presents with a triad of torticollis, intermittent head movements, and a low-amplitude, high-frequency (> 5 Hz) nystagmus. While this disorder can be an idiopathic benign finding, it can also be associated with serious other diseases, such as retinal abnormalities or gliomas. Brain imaging and referral to neuro-ophthalmology are recommended.

Tics

Stereotypies are not the only common developmental movement disorder. Tics are also common in children and share common comorbidities with stereotypies [12]. Tics are a distinct movement disorder separate from stereotypies [64]. However, in clinical practice, they are sometimes confused by providers unfamiliar with both disorders. Even for physicians experienced with both tics and stereotypies, distinguishing between the two can be challenging, especially in cases of atypical onset or when a child cannot accurately describe their movements. The reader is encourged to refer to Table 18.1 to help distinguish between these movements.

Motor tics are brief, usually nonrhythmic, repetitive movements that are difficult to control. They generally follow a rostral to caudal distribution, with tics involving the face more common than those involving the hands or feet. Simple motor tics are brief and meaningless movements involving a single muscle group. Common examples include prominent blinking, eye rolling, grimacing, head jerking, or shoulder shrugging. Complex tics are clusters of longer movements that sometimes appear more purposeful. This might take the form of a stereotyped chain of motor and/or vocal tics, such as a blink followed by a grimace and a shrug. Other complex tics might be repetitively touching objects, mimicking the gestures of others (echopraxia), or making foul or inappropriate gestures (copropraxia). Of note, OCD is a common comorbidity of tic disorders, and it may be difficult to distinguish a complex tic from a simple compulsion. Although in both cases, the patient may report that they "have" to do the movements, the reason why is often different. Patients with complex tics will report that they are physically uncomfortable until they do the movements, whereas those with compulsions will generally report that the movement in some way relieves anxiety or stops unwanted obsessive thoughts.

Vocal tics are brief, repeated sounds rather than movements. Simple vocal tics are meaningless nonword sounds such as sniffing, coughing, throat clearing, grunts, gasps, and squeals. Complex vocal tics generally involve words or phrases. This may include repeating one's own words (palilalia), repeating other people's words (echolalia), repeating words after reading them (lexilalia), or foul/inappropriate words (coprolalia). It is important to distinguish true coprolalia from purposeful, goal-directed use of foul language. Coprolalia usually involves repetition of a stereotyped word or phrase that is often uttered in an atypical volume and cadence and is often out of the appropriate social context during which the word might normally be uttered. Very rarely, some patients with Tourette syndrome can have coprolalia that manifest in a more socially appropriate context, such as racial or ethnic slurs, in the presence of individuals who might be the most hurt or offended by the words. When this occurs, it is generally distressing and embarrassing for the patients.

Tics generally begin by age 3–9, although if they are mild, they may not be recognized initially. Eye blinking, for example, may lead to a trip to the ophthalmologist rather than a neurologist or psychiatrist, and repetitive sniffing may be mistaken for a symptom of seasonal allergies. Many, but not all, children with tics experience a "premonitory urge" that is relieved by the performance of the tic, or when the tic is finally performed "just right" [65]. The premonitory urge can be thought of as the "itch" that the tic "scratches." Although not all patients are able to put this experience into words, some describe discomfort in the affected body part or a buildup of energy that needs an outlet [65]. Tics can be suppressed for brief periods of time, but doing so is generally uncomfortable. A rebound period of increased tics is often observed after voluntary suppression. Tics, by their nature, wax and wane over time, sometimes quite dramatically. They are often worse with stress, excitement, fatigue, or illness. Unlike most neurologic disorders, tics are often suggestible, meaning they may worsen when they are under discussion. They are often better when fully engaged in a task [66]. Tics also frequently change over time. A prominent eye blink, for example, might resolve and be replaced by a repetitive grimace.

Provisional tic disorder

Provisional tic disorder is characterized by tics lasting less than 1 year. Multiple studies have reported tics for less than 1 year in about 20% of the population [67–69], but estimates of prevalence vary widely from 5%–47% depending on the methodology used [70]. Of note, the highest estimate, 47%, was based on a direct observation and a longitudinal, year-long study of 1st graders [71]. This is significantly lower than the rate based on parent or teacher reports [72], and it is thought that overall lifetime tic prevalence is likely underestimated and may be extrapolated to be up to 70% of the population [70]. The lifetime incidence of tics is significantly higher than the point incidence, and most children with tics are likely to have these repetitive sounds and movements resolved, often without ever having come to clinical attention.

One of the main questions clinicians may get from parents of a child with a recent onset of tics is the likelihood of developing a chronic tic disorder. First, it is important to determine whether the tics are truly recent by asking questions about prior repetitive sounds and movements that may not have been recognized at the time. Although it is not possible to know with certainty which children will develop chronic tics, there are a few risk factors that make it more likely. Tics are often more severe and bothersome in children with chronic tic disorders compared to those who will only have transient tics [73]. Boys are more likely to have chronic tic disorders than girls, and evidence suggests greater persistence of tics in boys than girls [70]. For example, in one study that observed children monthly, the ratio of boys to girls was 1.6:1 in children with tics spanning only 1–2 consecutive months, but 7.6:1 in children observed to have tics for 3 consecutive months or more [71].

Tourette syndrome/complex motor tics

According to DSM5 criteria, patients with Tourette syndrome (TS) must have had at least two different motor tics and one vocal tic over the course of more than a year. Unlike previous editions of the DSM, the current criteria do not specify frequency or severity. The CDC specifies that tics should occur every day or most days [74], although this requirement is not used in common clinical practice and most movement disorder specialists would argue that criteria can be met even when there are no tics appreciated for weeks at a time [70]. Diagnosis becomes more difficult when there are years of separate bouts of tics. In this case, it becomes unclear whether the diagnosis of Tourette syndrome or two bouts of transient tics is more appropriate [70]. The prevalence of Tourette syndrome is about 0.6% in children [67] but closer to 0.01% by adulthood [75].

If patients have exclusively motor tics or exclusively vocal tics, they are diagnosed with a chronic motor tic disorder or a chronic vocal tic disorder, respectively. Another 0.6% of children qualify for these disorders collectively [67]. The meaning of this distinction is limited. While it is true that children who have both vocal and motor tics have in general an increased complexity and frequency of tics [76], it is this increased complexity and frequency that is clinically and mechanistically meaningful, rather than just the presence of tics that fall into two different categories. Many authors argue that it is better to think of all of the above disorders as part of the broader class of "tic spectrum disorders" [76,77].

Chronic tic disorders generally begin between the ages of 3 and 9 [78]. In one study of grade school children, the highest overall prevalence of tics was in 1st graders, expected to be between the ages of 6 and 7 [71]. In individuals with chronic tic disorders, tics tend to peak between the ages of 9 and 11 [79]. They improve for about 85% of patients somewhere in the teens or early 20s [79,80]. The chance of improvement is impacted by tic severity and degree of comorbidity, with children with more comorbidities and more severe tics being more likely to be in the minority that do not experience significant improvement [80]. It is worth noting that improvement is entirely independent of any symptomatic treatments used to suppress tics, and thus the decision to treat tics should be based on their impact on the child right now rather than on concerns for the future.

Etiology and evaluation

For most children presenting with tics at a typical age and with an otherwise unremarkable neurologic exam, no further workup is needed apart from appropriate screening for comorbidities. 85% of patients with Tourette syndrome will have at least one comorbidity, the average number being 2 [81]. The most common disorders are ADHD (60%–80%), anxiety (30%–80%), and OCD (11%–80%), followed by anger management issues (25%–75%), depression (13%–76%), sleep problems (12%–62%), migraines (25%), and learning disabilities (23%) [82,83]. Tics are also found in 22% of patients with autism [11], and physicians caring for patients with autism should keep in mind that stereotypies are not the only potentially problematic repetitive developmental movement disorder that occurs in this population [84].

Brain imaging is expected to be essentially normal in a typical patient with an otherwise unremarkable neurologic exam and is not routinely indicated. Some research studies have suggested reduced caudate volume or cortical thinning, but these differences are not large enough to be appreciated in routine studies, and the interpretation of these reported findings should be cautious due to the small sample size and the potential for motion artifacts [78]. Increased dopaminergic innervation in the striatum in patients with Tourette syndrome on positron emission tomography (PET) has also been reported [85], and has increased glutamine in the striatum on magnetic resonance spectroscopy (MRS) [18].

Tic disorders are thought to be multifactorial but have a strong genetic component. The risk of tic disorders is 15 times higher in siblings of patients with Tourette syndrome [78,86,87]. The heritability of tic disorders is estimated at 0.77, where a value of 1 would be consistent with 100% heritability [86]. This is significantly higher than the heritability of related psychiatric disorders such as OCD [87]. In general, tics are thought to be a polygenic disorder, with most patients having an aggregation of multiple common risk factor genetic variants. The greater the burden of risk factor variants an individual has, the more prolonged and severe the tics are expected to be [78]. Risk factor genes for tic disorders also likely increase for common comorbidities, including OCD, ADHD, and autism [88,89]. However, even in identical twins, the concordance rate is only 53%, indicating that other factors likely affect how or whether tics are expressed [90]. Prenatal smoking and low birth weight are the most commonly reported very early risk factors for Tourette syndrome, and there have been scattered reports of other factors such as perinatal complications and maternal stress, but on the whole, the evidence is limited by a few studies relying on mostly retrospective data [91]. Evidence of microglia activation in the postmortem tissue of patients with Tourette syndrome does suggest that inflammation could be a factor in tic disorders [92].

While there may be an inflammatory component to Tourette syndrome and many other neurologic and psychiatric disorders, this does not mean that Tourette syndrome is a primarily autoimmune or postinfectious disorder. Tics with severe periods of exacerbations in prepubertal children have been proposed to be part of the criteria of a post-Group A strep phenomenon called Pediatric Autoimmune Neuropsychiatric Disorders Associated with Streptococcal infections, or PANDAS. This diagnostic was proposed based on the observation that children with Sydenham's chorea, a known post-streptococcal disorder, often appear suddenly and often have neuropsychiatric symptoms beyond chorea, including sudden onset of tics, anxiety, obsessions, compulsions, and executive dysfunction [93]. Proponents of PANDAS argue that these other symptoms can occur as a post-streptococcal disease without chorea [94].

This is a controversial diagnosis, and the preponderance of evidence argues against this connection [95]. First, the diagnosis fails to account for the fact that waxing and waning is an expected part of tic disorders, and that many patients diagnosed with PANDAS have a strong family history of tic disorders and common comorbidities [95]. Various antibodies, including anti-tubulin, anti-DR1, anti-D2R, and anti-D1R, and lysoganglioside-GM1, have been reported to be elevated in both patients with Sydenham's chorea, and in patients diagnosed with PANDAS. However, when patients diagnosed with PANDAS who do not have choreiform movements have serial measurements of their antibodies taken, exacerbations in tics do not correlate with elevations in any of these autoantibodies [96–98]. Furthermore, more recently recruited control groups of healthy asymptomatic children in Baltimore have levels of these same autoantibodies similar to PANDAS patients, which argues against their specificity in this diagnosis [96]. Randomized control trials of antibiotics [99] and tonsillectomy [100], both expected to reduce streptococcal infections, have failed to demonstrate any amelioration of tic symptoms in PANDAS patients compared to placebo. The only small, partially blinded control trial of IVIG, plasmapheresis, and placebo had a small sample size and showed significant improvement in tics only in the plasmapheresis group [101]. Of note, no sham version of plasmapheresis was given. This is not to say that there is no inflammatory component to tics, as discussed above. Inflammation may play a role in the genesis of many neuropsychiatric disorders, from schizophrenia [102] to autism [103]. However, while there may be an inflammatory component to all of these disorders, it does not follow that autoimmune activity is the primary driver of these disorders or that immunomodulation is an effective treatment strategy [95].

While the average patient with tics does not require any routine studies, in certain cases, further workup is appropriate. Certainly, a child who presents with acute to subacute chorea in addition to tics should receive appropriate workup for Sydenham's chorea and receive penicillin prophylaxis to prevent the consequences of rheumatic fever [104]. Likewise, any child presenting with an examination or history concerning new-onset encephalopathy, weakness, ataxia, or seizures should receive the appropriate workup, as tics have rarely been reported secondary to head trauma, malignancy, vascular insults, or infection [105]. When tics occur secondary to a focal neurological insult, the most common area involved is the basal ganglia, although tics after peripheral nerve injury have also rarely been reported [105].

Particular care should be taken with the onset of tics beyond the typical expected age range. Although Tourette syndrome can be diagnosed in any patient presenting with tics before age 18, in clinical practice, true Tourette syndrome with tics starting after age 10 is rare [106]. In addition to rare focal injuries and insults, physicians should be on the lookout for functional neurologic symptom disorders presenting with tic-like movements. Although in some older reports functional tic-like movements make up only 10% of secondary tics [105], the COVID-19 pandemic brought obut a mini-pandemic of functional tic-like movements and brought attention to this particular manifestation of functional neurologic symptom disorder [106–109]. Functional tics and true tics share overlapping features, such as suggestibility, distractibility, and comorbidity with anxiety disorders [108]. However, there are key differences that help with the diagnosis. Functional tics are more likely to be diagnosed in young females, whereas Tourette syndrome is more common in males [107,109]. Other suggestive features include onset in the teen or early adult years, lack of the typical rostral-caudal gradient of movements, particularly bizarre movements or vocal tics, coprophenomenon at onset, lack of suppressibility, absence of a waxing or waning course, and presence of other functional symptom disorders. Table 18.3 reviews these differences.

In addition to excluding atypical causes of tics and tic-like movements, tics themselves can rarely cause serious injuries that require additional workups. Myelopathy [110,111], arterial dissection [112], and retinal detachment [113] have all been reported as secondary to violent, painful tics. New-onset focal neurologic signs or symptoms in a child with a chronic tic disorder should be investigated with appropriate imaging.

Treatment

Not all tics require treatment. In children with mild simple tics that do not cause pain or interfere with school or social activities, the only treatment indicated is education and reassurance for the patient and their families [114]. Tics may be misinterpreted as willful behavior in the classroom, and educating the teacher or school personnel about the neurologic basis of these disorders may be necessary. However, in other cases, tics can cause pain or even significant injury. In some literature, about 10%–20% of children with Tourette syndrome will report tics that are painful at some point [115]. Tics may also cause embarrassment, bullying, or social ostracism [116]. One study estimates that coprophenomenon (coprolalia and copropraxia) occurs in approximately 20% of individuals with Tourette syndrome at some point in their lives [117]. Tics can also be distracting and thus interfere with learning or work [118]. Of note, the goal of treatment should be to reduce tics until they are not bothersome, rather than the complete cessation of tics. This approach decreases the risk of deleterious side effects with excessively high medication doses.

When tics are bothersome, physicians should consider not just the tics themselves but also the comorbidities present, as many treatments for tics may also help to treat these other disorders. When two or more symptoms are present and a medication with efficacy for all is not available, the disorder with the most bothersome symptoms should be focused on first. Of note, there has been a historic reluctance to use stimulants when needed in children who have both ADHD and tics. This is based on the fear that stimulants will worsen or even cause tics, given that they act by blocking dopamine re-uptake. However, a double-blind controlled trial of clonidine and methylphenidate in children with both Tourette syndrome and ADHD found that methylphenidate did not increase the severity of tics [119]. Therefore while some argue that α agonists may be the first choice for children with both ADHD and tics due to the dual action provided [114], stimulants do not need to be avoided in children who would benefit from them. Of note, given that α agonists are rarely sufficient as monotherapy

TABLE 18.3 A comparison of true versus functional tics.

	True tics	Functional tics	Areas of overlap
Age of onset	3–9 years	Teens and young adults	
Sex	Male predominant	Female predominant	
Movement qualities	Rostral to caudal gradient, consistency of muscles used in a given tic.	More likely to involve arms/legs, more likely to be "bizarre" (e.g., foreign accents).	Coprophenomenon can occur in both but are unlikely to appear at symptom onset in true tics.
Time course	Waxing and waning	Continuous worsening or abrupt onset	Functional tics may also wax and wane
Urge/suppressibility	Usually present	Usually absent	More patients with functional tics are reporting these findings, likely secondary to greater awareness of typical tic qualities due to social media.
Other features of the history	Less bothered by the movements than the parents.	Social media use, prior history of functional neurologic symptom	Anxiety, depression, OCD, and ADHD
Family history	Tourette syndrome or other tic disorders		Anxiety, depression, OCD, and ADHD
Response to treatment	Response often good	Response usually poor	

ADHD, Attention-deficit/hyperactivity disorder. *OCD*, obsessive compulsive disorder.

for ADHD, many providers would recommend starting stimulants prior to α agonists in children with both ADHD and tics, provided that the tics are mild and ADHD is causing the majority of dysfunction.

For patients with bothersome tics, the ideal first treatment is Comprehensive Behavioral Intervention for Tics, a form of habit reversal therapy. This treatment is effective and avoids the potential side effects inherent to pharmacologic treatment [120]. For this reason, it is generally the first choice for treatment when available [114]. However, like all treatments, it does have its drawbacks. First, habit reversal therapy starts with recognition of the pre-tic urge to allow patients to instead perform a competing physical response that is incompatible with the tic while they wait for the urge to reach its maximum and then ultimately fade away. It is therefore more difficult for patients who do not have a significant prodromal urge [121]. Second, the technique requires patient engagement and practice. It works best in engaged patients with tics bothersome enough that the hope of reducing them provides incentive to master the technique. Finally, while the number of therapists who can use this therapy technique is growing, wait lists are still often long, which may limit its use in patients in urgent need of tic relief. Online therapy is helpful, but families are more likely to stick with it if they have at least initial contact with a therapist [122].

When therapy is not available or is not effective, medications should be considered. In general, the best evidence is for neuroleptics, but given their side effects, they are not generally the first choice in clinical practice. There is reasonably good evidence for the α2 adrenergic agonist clonidine [114,119,123] in tic disorders. Clonidine has the advantage of a comparably favorable side effect profile and utility in treating common disorders such as ADHD and anxiety. It can also help with insomnia. The starting dose is 0.05 mg nightly, and it may be increased slowly up to 0.4 mg, divided into two to three doses if tolerated. A double-blind placebo-controlled trial of the clonidine patch also showed improvement in tic frequency and severity [124]. Dosing was weight-based, with children less than 40 kg receiving a 1 mg patch, those greater than 60 kg receiving a 2 mg patch, and those weighing 40–60 kg receiving a 1.5 mg patch. The literature on the alpha-agonist guanfacine is more mixed [114,123,125,126] but it has the advantage of causing less sedation at higher doses and is thus commonly the first choice in clinical practice. The starting dose is 0.25–0.5 mg, and it is increased slowly up to 4 mg, divided into two doses as needed and tolerated. For both alpha-agonists, decreased blood pressure is a possible but not common side effect, and long-acting guanfacine can additionally prolong the QTc interval. Both medications are weaned slowly to prevent the chance of rebound hypertension.

Topiramate is another possible medication for the treatment of tics [127]. It may be an ideal choice for children with both tics and migraine headaches. Dosing generally starts at 0.5 mg/kg/day or total daily dose of 12.5 mg/day at night, and then is increased to twice a day dosing, weekly as needed, to a maximum dose of 4 mg/kg/day or 200 mg total daily dose. At higher doses, patients should be warned about possible side effects such as cognitive effects, weight loss, and kidney stones.

When these medications are ineffective or cause very severe tics, the next line of treatment is neuroleptics. Atypical neuroleptics, such as aripiprazole and risperidone, are commonly used in clinical practice today. Aripiprazole in particular is the first choice of many providers given its favorable side effect profile in terms of both extrapyramidal and metabolic side effects [128]. Dosing starts at 2 mg and is increased weekly as needed to a maximum dose of 10 mg in children under 50 kg or 20 mg in children over 50 kg. Possible side effects include weight gain, sedation, and extrapyramidal side effects, including dystonic reactions and potentially permanent tardive dyskinesia. Medication is tapered slowly to avoid withdrawal dyskinesia. Risperidone is likewise effective [129] but it carries an increased risk of weight gain and the added risk of prolonged QTc compared to aripiprazole. It is a reasonable choice if aripiprazole is not tolerated or effective. It also has the added benefit of being helpful for episodes of rage that occur in some individuals with Tourette syndrome [130]. Dosing starts at 0.5 mg per day and is increased weekly as needed up to a maximum of 3 mg per day.

First-generation typical neuroleptics are still used when other treatments fail or in some countries where they may be the only treatment available [123]. Haloperidol [131] is the oldest treatment in this class and, like aripiprazole, is FDA-approved for this indication [132]. The starting dose is 0.5 mg, and it is increased weekly as needed by 0.5 mg to a maximum of 3 mg/day in 2–3 divided doses. While the risk of weight gain is low, haloperidol also carries a higher risk of extrapyramidal side effects and prolonged QTc. Pimozide may be slightly effective, but with a lower risk of extrapyramidal side effects [131]. It is also FDA-approved for this indication. The usual starting dose is 0.05 mg/kg nightly, increasing weekly as needed to a maximum dose of 0.2 mg/kg/day or 10 mg/day. Prolonged QTc is a potential concern, as with other neuroleptics apart from aripiprazole.

The majority of patients will respond to these treatments. However, a small percentage will have disabling refractory tics despite the treatments listed above. Botulinum toxin may be effective for simple motor tics [133,134]. Risks include allergic reactions and overweakness, which may last up to 3 months. Deep brain stimulation (DBS) can likewise be considered for superrefractory Tourette syndrome. Open-label studies strongly suggest it is effective [135]. However, the only double-blinded placebo-controlled trial failed [136], possibly due to an inadequate dose during the double-blinded portion becuase of concerns that an adequate dose would unblind participants and study personnel. This difficulty in blinding is a legitimate practical consideration but is worrisome in a procedure where the placebo effect is known to be significant, despite rare but life-threatening side effects [137]. Risks include intracranial bleeding and infections. Nonetheless, when all other agents have failed in severe, disabling tics, the benefits may outweigh the risks. There is also increasing patient interest in the use of cannabinoids for tic treatment, but no high-quality evidence is yet available [114]. Fig. 18.1 provides a possible treatment algorithm.

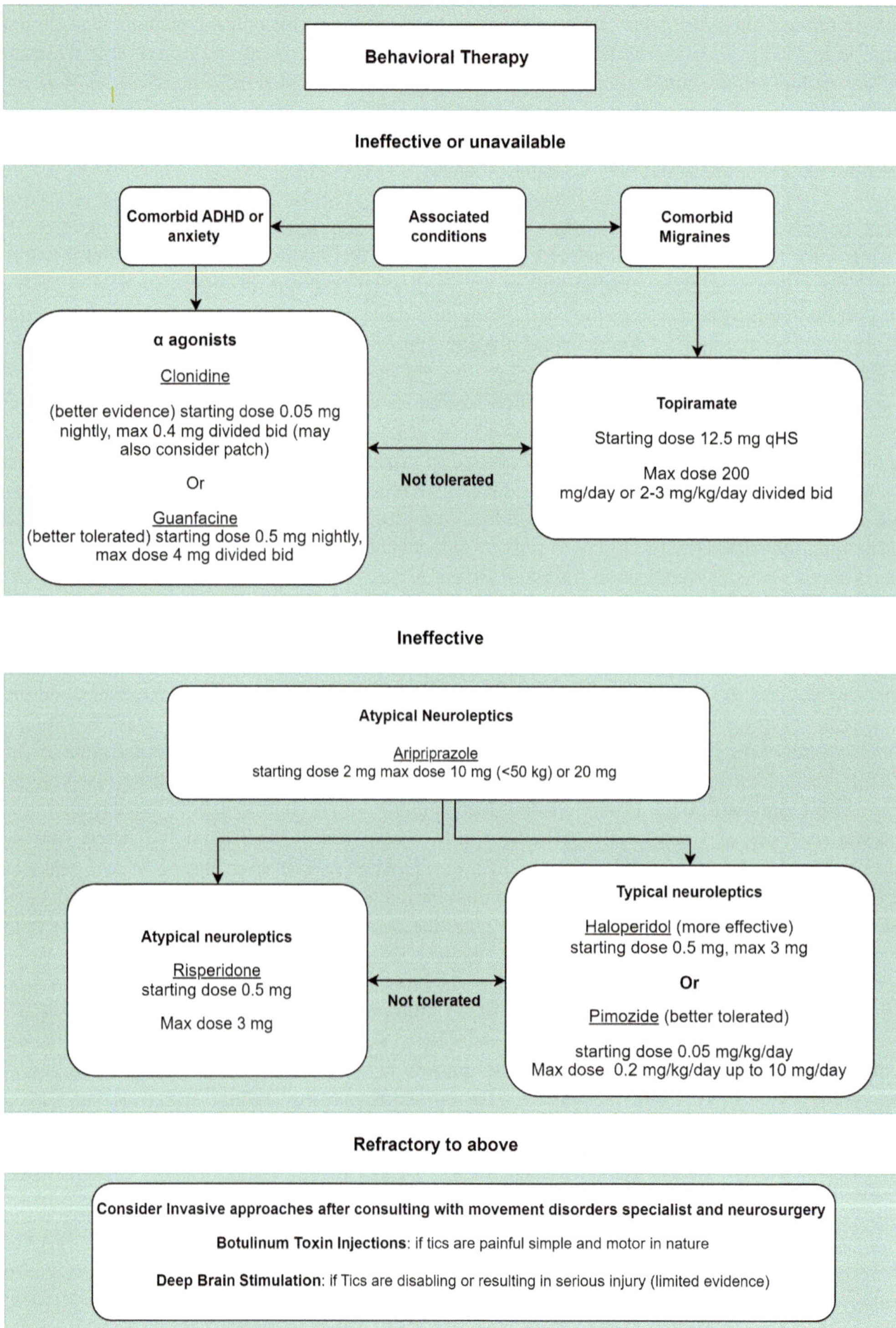

FIGURE 18.1 Treatment algorithm for bothersome tics.

Conclusion

Tics and stereotypies are relatively common neurodevelopmental movement disorders that can occur either as an isolated finding in children with otherwise typical development or in the context of other neurodevelopmental movement disorders. Both conditions are characterized by repetitive, stereotyped movements and share a common set of comorbidities, but they differ in age of onset and appearance. When these movements appear at a typical age of onset in an otherwise normally developing child or in a child with another known developmental disorder in which such movements are expected, no further workup is generally required. Behavioral therapy is effective for problematic stereotypies, although frequently these movements are not particularly bothersome to patients. Both behavioral therapy and a wide range of pharmaceutical therapies are available for tics. Regardless of whether treatment is offered, most children will experience a spontaneous decrease in these movements with age.

References

[1] LeBarton ES, Landa RJ. Infant motor skill predicts later expressive language and autism spectrum disorder diagnosis. Infant Behav Dev 2019;54:37–47.

[2] Wilson RB, Enticott PG, Rinehart NJ. Motor development and delay: advances in assessment of motor skills in autism spectrum disorders. Curr Opin Neurol 2018;31(2):134–9.

[3] Wilson RB, McCracken JT, Rinehart NJ, Jeste SS. What's missing in autism spectrum disorder motor assessments? J Neurodev Disord 2018;10 (1):33.

[4] Dewolf AH, Sylos Labini F, Ivanenko Y, Lacquaniti F. Development of locomotor-related movements in early infancy. Front Cell Neurosci 2020;14:623759.

[5] Thelen E. Kicking, rocking, and waving: contextual analysis of rhythmical stereotypies in normal human infants. Anim Behav 1981;29 (1):3–11.

[6] Thelen E, Bradshaw G, Ward JA. Spontaneous kicking in month-old infants: manifestation of a human central locomotor program. Behav Neural Biol 1981;32(1):45–53.

[7] Sargent B, Scholz J, Reimann H, Kubo M, Fetters L. Development of infant leg coordination: exploiting passive torques. Infant Behav Dev 2015;40:108–21.

[8] Sargent B, Schweighofer N, Kubo M, Fetters L. Infant exploratory learning: influence on leg joint coordination. PLoS One 2014;9(3):e91500.

[9] Hadders-Algra M. General movements: a window for early identification of children at high risk for developmental disorders. J Pediatr 2004;145(2 Suppl):S12–18.

[10] Saade S, Martin V, Lanovaz M. Prevalence of stereotypy in individuals with developmental disabilities: a systematic review. Review Journal of Autism and Developmental Disorders 2016;3. Available from: https://doi.org/10.1007/s40489-016-0069-x.

[11] Canitano R, Vivanti G. Tics and Tourette syndrome in autism spectrum disorders. Autism 2007;11(1):19–28.

[12] Singer HS. Motor control, habits, complex motor stereotypies, and Tourette syndrome. Ann N Y Acad Sci 2013;1304:22–31.

[13] Huisman-van Dijk HM, Matthijssen SJMA, Stockmann RTS, Fritz AV, Cath DC. Effects of comorbidity on Tourette's tic severity and quality of life. Acta Neurol Scand 2019;140(6):390–8.

[14] Oakley C, Mahone EM, Morris-Berry C, Kline T, Singer HS. Primary complex motor stereotypies in older children and adolescents: clinical features and longitudinal follow-up. Pediatr Neurol 2015;52(4):398–403 e1.

[15] Harris KM, Mahone EM, Singer HS. Nonautistic motor stereotypies: clinical features and longitudinal follow-up. Pediatr Neurol 2008;38 (4):267–72.

[16] Singer HS. Motor stereotypies. Semin Pediatr Neurol 2009;16(2):77–81.

[17] Mahone EM, Ryan M, Ferenc L, Morris-Berry C, Singer HS. Neuropsychological function in children with primary complex motor stereotypies. Dev Med Child Neurol 2014;56(10):1001–8.

[18] Mahone EM, Puts NA, Edden RAE, Ryan M, Singer HS. GABA and glutamate in children with Tourette syndrome: A ^{1}H MR spectroscopy study at 7T. Psychiatry Res Neuroimaging 2018;273:46–53.

[19] Houdayer E, Walthall J, Belluscio BA, Vorbach S, Singer HS, Hallett M. Absent movement-related cortical potentials in children with primary motor stereotypies. Mov Disord 2014;29(9):1134–40.

[20] Thelen E, Fisher DM. From spontaneous to instrumental behavior: kinematic analysis of movement changes during very early learning. Child Dev 1983;54(1):129–40.

[21] Hutt SJ, Hutt C. Stereotypy, arousal and autism. Hum Dev 1968;11(4):277–86.

[22] Bos KJ, Zeanah CH, Smyke AT, Fox NA, Nelson CA. Stereotypies in children with a history of early institutional care. Arch Pediatr Adolesc Med 2010;164(5):406–11.

[23] McCarty MJ, Brumback AC. Rethinking stereotypies in autism. Semin Pediatr Neurol 2021;38:100897.

[24] Mahone EM, Crocetti D, Tochen L, Kline T, Mostofsky SH, Singer HS. Anomalous putamen volume in children with complex motor stereotypies. Pediatr Neurol 2016;65:59–63.

[25] Kates WR, Lanham DC, Singer HS. Frontal white matter reductions in healthy males with complex stereotypies. Pediatr Neurol 2005;32 (2):109–12.

[26] Dean SL, Tochen L, Augustine F, Ali SF, Crocetti D, Rajendran S, et al. The role of the cerebellum in repetitive behavior across species: childhood stereotypies and deer mice. Cerebellum 2022;21(3):440–51.
[27] Harris AD, Singer HS, Horska A, Kline T, Ryan M, Edden RA, et al. GABA and glutamate in children with primary complex motor stereotypies: an 1H-MRS study at 7T. AJNR Am J Neuroradiol 2016;37(3):552–7.
[28] Maraganore DM, Lees AJ, Marsden CD. Complex stereotypies after right putaminal infarction: a case report. Mov Disord 1991;6(4):358–61.
[29] Mendez MF, Shapira JS, Miller BL. Stereotypical movements and frontotemporal dementia. Mov Disord 2005;20(6):742–5.
[30] Miller JM, Singer HS, Bridges DD, Waranch HR. Behavioral therapy for treatment of stereotypic movements in nonautistic children. J Child Neurol 2006;21(2):119–25.
[31] Specht MW, Mahone EM, Kline T, Waranch R, Brabson L, Thompson CB, et al. Efficacy of parent-delivered behavioral therapy for primary complex motor stereotypies. Dev Med Child Neurol 2017;59(2):168–73.
[32] Ghosh D, Rajan PV, Erenberg G. A comparative study of primary and secondary stereotypies. J Child Neurol 2013;28(12):1562–8.
[33] Goldman S, Wang C, Salgado MW, Greene PE, Kim M, Rapin I. Motor stereotypies in children with autism and other developmental disorders. Dev Med Child Neurol 2009;51(1):30–8.
[34] Michelotti J, Charman T, Slonims V, Baird G. Follow-up of children with language delay and features of autism from preschool years to middle childhood. Dev Med Child Neurol 2002;44(12):812–19.
[35] Fernandez TV, Williams ZP, Kline T, Rajendran S, Augustine F, Wright N, et al. Primary complex motor stereotypies are associated with de novo damaging DNA coding mutations that identify candidate risk genes and biological pathways. bioRxiv 2019;730952.
[36] Breen J, Hare DJ. The nature and prevalence of catatonic symptoms in young people with autism. J Intellect Disabil Res 2017;61(6):580–93.
[37] Shah A, Wing L. Psychological approaches to chronic catatonia-like deterioration in autism spectrum disorders. Int Rev Neurobiol 2006;72:245–64.
[38] Wing L, Shah A. A systematic examination of catatonia-like clinical pictures in autism spectrum disorders. Int Rev Neurobiol 2006;72:21–39.
[39] Hauptman AJ, Cohen D, Dhossche D, Raffin M, Wachtel L, Ferrafiat V. Catatonia in neurodevelopmental disorders: assessing catatonic deterioration from baseline. Lancet Psychiatry 2023;10(3):228–34.
[40] Mazzone L, Postorino V, Valeri G, Vicari S. Catatonia in patients with autism: prevalence and management. CNS Drugs 2014;28(3):205–15.
[41] Wachtel LE. Treatment of catatonia in autism spectrum disorders. Acta Psychiatr Scand 2019;139(1):46–55.
[42] Bodfish JW, Symons FJ, Parker DE, Lewis MH. Varieties of repetitive behavior in autism: comparisons to mental retardation. J Autism Dev Disord 2000;30(3):237–43.
[43] Melo C, Ruano L, Jorge J, Pinto Ribeiro T, Oliveira G, Azevedo L, et al. Prevalence and determinants of motor stereotypies in autism spectrum disorder: a systematic review and meta-analysis. Autism 2020;24(3):569–90.
[44] Lovaas OI, Litrownik A, Mann R. Response latencies to auditory stimuli in autistic children engaged in self-stimulatory behavior. Behav Res Ther 1971;9(1):39–49.
[45] Koegel RL, Covert A. The relationship of self-stimulation to learning in autistic children. J Appl Behav Anal 1972;5(4):381–7.
[46] Cunningham AB, Schreibman L. Stereotypy in autism: the importance of Function. Res Autism Spectr Disord 2008;2(3):469–79.
[47] Barry S, Baird G, Lascelles K, Bunton P, Hedderly T. Neurodevelopmental movement disorders – an update on childhood motor stereotypies. Dev Med Child Neurol 2011;53(11):979–85.
[48] Symons FJ, Sperry LA, Dropik PL, Bodfish JW. The early development of stereotypy and self-injury: a review of research methods. J Intellect Disabil Res 2005;49(Pt 2):144–58.
[49] Chavez B, Chavez-Brown M, Rey JA. Role of risperidone in children with autism spectrum disorder. Ann Pharmacother 2006;40(5):909–16.
[50] LeClerc S, Easley D. Pharmacological therapies for autism spectrum disorder: a review. P T 2015;40(6):389–97.
[51] Rooker GW, Hausman NL, Fisher AB, Gregory MK, Lawell JL, Hagopian LP. Classification of injuries observed in functional classes of self-injurious behaviour. J Intellect Disabil Res 2018;62(12):1086–96.
[52] Fazzi E, Lanners J, Danova S, Ferrarri-Ginevra O, Gheza C, Luparia A, et al. Stereotyped behaviours in blind children. Brain Dev 1999;21 (8):522–8.
[53] Zhang D, Poustka L, Marschik PB, Einspieler C. The onset of hand stereotypies in fragile X syndrome. Dev Med Child Neurol 2018;60 (10):1060–1.
[54] Carter P, Downs J, Bebbington A, Williams S, Jacoby P, Kaufmann WE, et al. Stereotypical hand movements in 144 subjects with Rett syndrome from the population-based Australian database. Mov Disord 2010;25(3):282–8.
[55] Dy ME, Waugh JL, Sharma N, O'Leary H, Kapur K, D'Gama AM, et al. Defining hand stereotypies in Rett syndrome: a movement disorders perspective. Pediatr Neurol 2017;75:91–5.
[56] Temudo T, Maciel P, Sequeiros J. Abnormal movements in Rett syndrome are present before the regression period: a case study. Mov Disord 2007;22(15):2284–7.
[57] Grados MA, Alvi MH, Srivastava S. Behavioral and psychiatric manifestations in Cornelia de Lange syndrome. Curr Opin Psychiatry 2017;30 (2):92–6.
[58] Moss J, Oliver C, Arron K, Burbidge C, Berg K. The prevalence and phenomenology of repetitive behavior in genetic syndromes. J Autism Dev Disord 2009;39(4):572–88.
[59] Smith AC, Boyd KE, Brennan C, Charles J, Elsea SH, Finucane BM, et al. Smith-Magenis syndrome GeneReviews [Internet]. 2001Seattle, WA. Available from. Available from: https://www.ncbi.nlm.nih.gov/books/NBK1310.

[60] Jinnah HA, Visser JE, Harris JC, Verdu A, Larovere L, Ceballos-Picot I, et al. Delineation of the motor disorder of Lesch-Nyhan disease. Brain 2006;129(Pt 5):1201–17.

[61] Tully HM, Dempsey JC, Ishak GE, Adam MP, Mink JW, Dobyns WB, et al. Persistent figure-eight and side-to-side head shaking is a marker for rhombencephalosynapsis. Mov Disord 2013;28(14):2019–23.

[62] Dean SL, Huisman TAGM, Poretti A, Singer HS. Figure of eight stereotypies in a young girl with a prenatal cerebellar injury. Mov Disord Clin Pract 2019;6(6):488–90.

[63] Reddy OJ, Gafoor JA, Suresh B, Prasad PO. Bobble head doll syndrome: a rare case report. J Pediatr Neurosci 2014;9(2):175–7.

[64] Crosland KA, Zarcone JR, Schroeder S, Zarcone T, Fowler S. Use of an antecedent analysis and a force sensitive platform to compare stereotyped movements and motor tics. Am J Ment Retard 2005;110(3):181–92.

[65] Reese HE, Scahill L, Peterson AL, Crowe K, Woods DW, Piacentini J, et al. The premonitory urge to tic: measurement, characteristics, and correlates in older adolescents and adults. Behav Ther 2014;45(2):177–86.

[66] Misirlisoy E, Brandt V, Ganos C, Tübing J, Münchau A, Haggard P. The relation between attention and tic generation in Tourette syndrome. Neuropsychology 2015;29(4):658–65.

[67] Scahill L, Specht M, Page C. The prevalence of tic disorders and clinical characteristics in children. J Obsessive Compuls Relat Disord 2014;3 (4):394–400.

[68] Cubo E, Gabriel Y, Galán JM, Villaverde VA, Velasco SS, Benito VD, et al. Prevalence of tics in schoolchildren in central Spain: a population-based study. Pediatr Neurol 2011;45(2):100–8.

[69] Kurlan R, McDermott MP, Deeley C, Como PG, Brower C, Eapen S, et al. Prevalence of tics in schoolchildren and association with placement in special education. Neurology 2001;57(8):1383–8.

[70] Black KJ, Black ER, Greene DJ, Schlaggar BL. Provisional tic disorder: what to tell parents when their child first starts ticcing. F1000Res 2016;5:696.

[71] Snider LA, Seligman LD, Ketchen BR, Levitt SJ, Bates LR, Garvey MA, et al. Tics and problem behaviors in schoolchildren: prevalence, characterization, and associations. Pediatrics 2002;110(2 Pt 1):331–6.

[72] Linazasoro G, Van Blercom N, de Zárate CO. Prevalence of tic disorder in two schools in the Basque country: results and methodological caveats. Mov Disord 2006;21(12):2106–9.

[73] Khalifa N, von Knorring AL. Tourette syndrome and other tic disorders in a total population of children: clinical assessment and background. Acta Paediatr 2005;94(11):1608–14.

[74] Diagnosing tic disorders: CDC. Available from: https://www.cdc.gov/ncbddd/tourette/diagnosis.html.

[75] Levine JLS, Szejko N, Bloch MH. Meta-analysis: adulthood prevalence of Tourette syndrome. Prog Neuropsychopharmacol Biol Psychiatry 2019;95:109675.

[76] Müller-Vahl KR, Sambrani T, Jakubovski E. Tic disorders revisited: introduction of the term "tic spectrum disorders". Eur Child Adolesc Psychiatry 2019;28(8):1129–35.

[77] Paulus T, Schappert R, Bluschke A, Alvarez-Fischer D, Naumann KER, Roessner V, et al. Questioning the definition of Tourette syndrome-evidence from machine learning. Brain Commun 2021;3(4):fcab282.

[78] Robertson MM, Eapen V, Singer HS, Martino D, Scharf JM, Paschou P, et al. Gilles de la Tourette syndrome. Nat Rev Dis Prim 2017;3:16097.

[79] Singer HS. Discussing outcome in Tourette syndrome. Arch Pediatr Adolesc Med 2006;160(1):103–5.

[80] Bloch MH, Peterson BS, Scahill L, Otka J, Katsovich L, Zhang H, et al. Adulthood outcome of tic and obsessive-compulsive symptom severity in children with Tourette syndrome. Arch Pediatr Adolesc Med 2006;160(1):65–9.

[81] Hirschtritt ME, Lee PC, Pauls DL, Dion Y, Grados MA, Illmann C, et al. Lifetime prevalence, age of risk, and genetic relationships of comorbid psychiatric disorders in Tourette syndrome. JAMA Psychiatry 2015;72(4):325–33.

[82] Kumar A, Trescher W, Byler D. Tourette syndrome and comorbid neuropsychiatric conditions. Curr Dev Disord Rep 2016;3(4):217–21.

[83] Marwitz L, Pringsheim T. Clinical utility of screening for anxiety and depression in children with Tourette syndrome. J Can Acad Child Adolesc Psychiatry 2018;27(1):15–21.

[84] Termine C, Grossi E, Anelli V, Derhemi L, Cavanna AE. Possible tics diagnosed as stereotypies in patients with severe autism spectrum disorder: a video-based evaluation. Neurol Sci 2021;42(4):1559–61.

[85] Wong DF, Singer HS, Brandt J, Shaya E, Chen C, Brown J, et al. D2-like dopamine receptor density in Tourette syndrome measured by PET. J Nucl Med 1997;38(8):1243–7.

[86] Mataix-Cols D, Isomura K, Pérez-Vigil A, Chang Z, Rück C, Larsson KJ, et al. Familial risks of Tourette syndrome and chronic tic disorders. A population-based cohort study. JAMA Psychiatry 2015;72(8):787–93.

[87] Browne HA, Hansen SN, Buxbaum JD, Gair SL, Nissen JB, Nikolajsen KH, et al. Familial clustering of tic disorders and obsessive-compulsive disorder. JAMA Psychiatry 2015;72(4):359–66.

[88] Yang Z, Wu H, Lee PH, Tsetsos F, Davis LK, Yu D, et al. Investigating shared genetic basis across Tourette syndrome and comorbid neurodevelopmental disorders along the impulsivity-compulsivity spectrum. Biol Psychiatry 2021;90(5):317–27.

[89] Carias KV, Wevrick R. Clinical and genetic analysis of children with a dual diagnosis of Tourette syndrome and autism spectrum disorder. J Psychiatr Res 2019;111:145–53.

[90] Lin WD, Tsai FJ, Chou IC. Current understanding of the genetics of tourette syndrome. Biomed J 2022;45(2):271–9.

[91] Chao T-K, Hu J, Pringsheim T. Prenatal risk factors for Tourette Syndrome: a systematic review. BMC Pregnancy Childbirth 2014;14(1):53.

[92] Lennington JB, Coppola G, Kataoka-Sasaki Y, Fernandez TV, Palejev D, Li Y, et al. Transcriptome analysis of the human striatum in Tourette syndrome. Biol Psychiatry 2016;79(5):372–82.

[93] Moore DP. Neuropsychiatric aspects of Sydenham's chorea: a comprehensive review. J Clin Psychiatry 1996;57(9):407–14.

[94] Swedo SE, Leonard HL, Garvey M, Mittleman B, Allen AJ, Perlmutter S, et al. Pediatric autoimmune neuropsychiatric disorders associated with streptococcal infections: clinical description of the first 50 cases. Am J Psychiatry 1998;155(2):264–71.

[95] Dean S, Wolf D, Singer H. PANDAS. The curated reference collection in neuroscience and biobehavioral psychology. Amsterdam, Netherlands: Elsevier Science Ltd; 2016. p. 369–73.

[96] Singer HS, Mascaro-Blanco A, Alvarez K, Morris-Berry C, Kawikova I, Ben-Pazi H, et al. Neuronal antibody biomarkers for Sydenham's chorea identify a new group of children with chronic recurrent episodic acute exacerbations of tic and obsessive compulsive symptoms following a streptococcal infection. PLoS One 2015;10(3):e0120499.

[97] Morris-Berry CM, Pollard M, Gao S, Thompson C, Singer HS, Group TSS. Anti-streptococcal, tubulin, and dopamine receptor 2 antibodies in children with PANDAS and Tourette syndrome: single-point and longitudinal assessments. J Neuroimmunol 2013;264(1–2):106–13.

[98] Singer HS, Gause C, Morris C, Lopez P, Group TSS. Serial immune markers do not correlate with clinical exacerbations in pediatric autoimmune neuropsychiatric disorders associated with streptococcal infections. Pediatrics 2008;121(6):1198–205.

[99] Garvey MA, Perlmutter SJ, Allen AJ, Hamburger S, Lougee L, Leonard HL, et al. A pilot study of penicillin prophylaxis for neuropsychiatric exacerbations triggered by streptococcal infections. Biol Psychiatry 1999;45(12):1564–71.

[100] Pavone P, Rapisarda V, Serra A, Nicita F, Spalice A, Parano E, et al. Pediatric autoimmune neuropsychiatric disorder associated with group a streptococcal infection: the role of surgical treatment. Int J Immunopathol Pharmacol 2014;27(3):371–8.

[101] Perlmutter SJ, Leitman SF, Garvey MA, Hamburger S, Feldman E, Leonard HL, et al. Therapeutic plasma exchange and intravenous immunoglobulin for obsessive-compulsive disorder and tic disorders in childhood. Lancet 1999;354(9185):1153–8.

[102] Bergink V, Gibney SM, Drexhage HA. Autoimmunity, inflammation, and psychosis: a search for peripheral markers. Biol Psychiatry 2014;75(4):324–31.

[103] Prata J, Machado AS, von Doellinger O, Almeida MI, Barbosa MA, Coelho R, et al. The Contribution of inflammation to autism spectrum disorders: recent clinical evidence. Methods Mol Biol 2019;2011:493–510.

[104] Dean SL, Singer HS. Treatment of Sydenham's chorea: a review of the current evidence. Tremor Other Hyperkinet Mov (N Y) 2017;7:456.

[105] Mejia NI, Jankovic J. Secondary tics and tourettism. Braz J Psychiatry 2005;27(1):11–17.

[106] Müller-Vahl KR, Edwards MJ. Mind the difference between primary tics and functional tic-like behaviors. Mov Disord 2021;36(12):2716–18.

[107] Pringsheim T, Ganos C, McGuire JF, Hedderly T, Woods D, Gilbert DL, et al. Rapid onset functional tic-like behaviors in young females during the COVID-19 pandemic. Mov Disord 2021;36(12):2707–13.

[108] Olvera C, K A. Functional tic disorder: a pandemic in a pandemic. Practical Neurol 2022;March–April:59–66.

[109] Paulus T, Bäumer T, Verrel J, Weissbach A, Roessner V, Beste C, et al. Pandemic tic-like behaviors following social media consumption. Mov Disord 2021;36(12):2932–5.

[110] Kim J, Kim JY, Lee JM, Kang DH, Lee CH, Park IS, et al. Progressive cervical spondylotic myelopathy caused by tic disorders in a young adult with Tourette syndrome. Kor J Neurotr 2019;15(2):199–203.

[111] Krauss JK, Jankovic J. Severe motor tics causing cervical myelopathy in Tourette's syndrome. Mov Disord 1996;11(5):563–6.

[112] Lehman LL, Gilbert DL, Leach JL, Wu SW, Standridge SM. Vertebral artery dissection leading to stroke caused by violent neck tics of Tourette syndrome. Neurology 2011;77(18):1706–8.

[113] Lim S, Rezai KA, Abrams GW, Eliott D. Self-induced, bilateral retinal detachment in Tourette syndrome. Arch Ophthalmol 2004;122(6):930–1.

[114] Pringsheim T, Okun MS, Müller-Vahl K, Martino D, Jankovic J, Cavanna AE, et al. Practice guideline recommendations summary: treatment of tics in people with Tourette syndrome and chronic tic disorders. Neurology 2019;92(19):896–906.

[115] Lavenstein B, Miyares L, Dodge L. Pain in childhood Tourette syndrome-retrospective analysis (P3.267). Neurology 2016;86(16 Supplement): P3.267.

[116] Charania SN, Danielson ML, Claussen AH, Lebrun-Harris LA, Kaminski JW, Bitsko RH. Bullying victimization and perpetration among US children with and without Tourette syndrome. J Dev Behav Pediatr 2022;43(1):23–31.

[117] Freeman RD, Zinner SH, Müller-Vahl KR, Fast DK, Burd LJ, Kano Y, et al. Coprophenomena in Tourette syndrome. Dev Med Child Neurol 2009;51(3):218–27.

[118] Eapen V, Črnčec R, McPherson S, Snedden C. Tic disorders and learning disability: clinical characteristics, cognitive performance and comorbidity. Austral J Spec Educ 2013;37(2):162–72.

[119] Group TsSS. Treatment of ADHD in children with tics: a randomized controlled trial. Neurology 2002;58(4):527–36.

[120] McGuire JF, Piacentini J, Brennan EA, Lewin AB, Murphy TK, Small BJ, et al. A meta-analysis of behavior therapy for Tourette syndrome. J Psychiatr Res 2014;50:106–12.

[121] McGuire JF, Piacentini J, Scahill L, Woods DW, Villarreal R, Wilhelm S, et al. Bothersome tics in patients with chronic tic disorders: characteristics and individualized treatment response to behavior therapy. Behav Res Ther 2015;70:56–63.

[122] Singer HS, McDermott S, Ferenc L, Specht M, Mahone EM. Efficacy of parent-delivered, home-based therapy for tics. Pediatr Neurol 2020;106:17–23.

[123] Seideman MF, Seideman TA. A review of the current treatment of Tourette syndrome. J Pediatr Pharmacol Ther 2020;25(5):401–12.

[124] Du YS, Li HF, Vance A, Zhong YQ, Jiao FY, Wang HM, et al. Randomized double-blind multicentre placebo-controlled clinical trial of the clonidine adhesive patch for the treatment of tic disorders. Aust N Z J Psychiatry 2008;42(9):807–13.

[125] Scahill L, Chappell PB, Kim YS, Schultz RT, Katsovich L, Shepherd E, et al. A placebo-controlled study of guanfacine in the treatment of children with tic disorders and attention deficit hyperactivity disorder. Am J Psychiatry 2001;158(7):1067–74.

[126] Murphy TK, Fernandez TV, Coffey BJ, Rahman O, Gavaletz A, Hanks CE, et al. Extended-release guanfacine does not show a large effect on tic severity in children with chronic tic disorders. J Child Adolesc Psychopharmacol 2017;27(9):762–70.

[127] Donegan S, Dixon P, Hemming K, Tudur-Smith C, Marson A. A systematic review of placebo-controlled trials of topiramate: how useful is a multiple-indications review for evaluating the adverse events of an antiepileptic drug? Epilepsia 2015;56(12):1910–20.

[128] Sallee F, Kohegyi E, Zhao J, McQuade R, Cox K, Sanchez R, et al. Randomized, double-blind, placebo-controlled trial demonstrates the efficacy and safety of oral aripiprazole for the treatment of Tourette's disorder in children and adolescents. J Child Adolesc Psychopharmacol 2017;27(9):771–81.

[129] Scahill L, Leckman JF, Schultz RT, Katsovich L, Peterson BS. A placebo-controlled trial of risperidone in Tourette syndrome. Neurology 2003;60(7):1130–5.

[130] Sandor P, Stephens RJ. Risperidone treatment of aggressive behavior in children with Tourette syndrome. J Clin Psychopharmacol 2000;20(6):710–12.

[131] Shapiro E, Shapiro AK, Fulop G, Hubbard M, Mandeli J, Nordlie J, et al. Controlled study of haloperidol, pimozide and placebo for the treatment of Gilles de la Tourette's syndrome. Arch Gen Psychiatry 1989;46(8):722–30.

[132] Mylan Institutional L. Haloperidol package insert. Rockford, IL; 2019.

[133] Rath JJ, Tavy DL, Wertenbroek AA, van Woerkom TC, de Bruijn SF. Botulinum toxin type A in simple motor tics: short-term and long-term treatment-effects. Parkinsonism Relat Disord 2010;16(7):478–81.

[134] Marras C, Andrews D, Sime E, Lang AE. Botulinum toxin for simple motor tics: a randomized, double-blind, controlled clinical trial. Neurology 2001;56(5):605–10.

[135] Martinez-Ramirez D, Jimenez-Shahed J, Leckman JF, Porta M, Servello D, Meng FG, et al. Efficacy and safety of deep brain stimulation in Tourette syndrome: the international Tourette syndrome deep brain stimulation public database and registry. JAMA Neurol 2018;75(3):353–9.

[136] Welter ML, Houeto JL, Thobois S, Bataille B, Guenot M, Worbe Y, et al. Anterior pallidal deep brain stimulation for Tourette's syndrome: a randomised, double-blind, controlled trial. Lancet Neurol 2017;16(8):610–19.

[137] Cubo E, González M, Singer H, Mahone EM, Scahill L, Müller-Vahl KR, et al. Impact of placebo assignment in clinical trials of tic disorders. Mov Disord 2013;28(9):1288–92.

Chapter 19

Spinal cord disorders

Philippines Cabahug[1,2], Janet Dean[1,2], Sarah Ann Korth[1,2] and Cristina Lavinia Sadowsky[1,2]

[1]*Kennedy Krieger Institute, International Center for Spinal Cord Injury, Baltimore, MD, United States,* [2]*Department of Physical Medicine and Rehabilitation, Kennedy Krieger Institute and Johns Hopkins School of Medicine, Baltimore, MD, United States*

Epidemiology of spinal cord injury

In the United States, traumatic spinal cord injury (TSCI) in childhood is rare, with only 3%–5% of SCIs each year occurring in children younger than 15 years of age [1,2]. Several studies from 1997 to 2010 estimate the incidence of SCI in individuals 21 years and younger to be between 15.5 and 24 cases per 1 million [3–5]. In the youngest children, males and females are affected at similar rates; however, as age at injury increases, the ratio of males to females increases. By adolescence, males are 2–4 times more likely to sustain an SCI than females [3,4]. Racial differences are found, with African Americans having the highest rate of injury at 15.3 per 1 million, followed by Native Americans (10), Hispanics (8.7), and Asians (3.6) per million [3].

The neurological level and degree of completeness varies with age [6,7]. Because of their disproportionately larger heads and underdeveloped neck musculature, infants and young children are more susceptible to upper cervical injuries. Children under 5 years of age are more likely to have an upper cervical lesion C1–4 (47%), have more severe injury, and concurrent brain injury (24%). Older children and adolescents more commonly sustain C4–8 lesions (53%) [4,7]. Among children injured prior to 12 years of age, approximately two-thirds are paraplegic and approximately two-thirds have complete lesions. Among adolescents, approximately 50% have paraplegia and 55% have complete lesions [6,8].

A study by New in 2019 mapped the global incidence rates of TSCI and nontraumatic SCI (NTSCI). The median TSCI rates differ per region, ranging from 3.3 to 13.2/million population per year, while median NTSCI rates ranged from 2.1 to 6.5/million population/year. Prevalence rates for TSCI ranged from 10.1 to 51.5 cases/million population; NTSCI prevalence rates were from 2.5 to 19.6 cases/million population [9].

Traumatic spinal cord injury

The most common cause of SCI in children and adolescents is motor vehicle crashes. In urban areas, violence such as gunshot or penetrating wounds is a common cause of SCI in all ages of children but is especially common in African American and Hispanic youth [7]. Falls, sport and recreation injuries including diving accidents are other common causes of SCI [4,6,7].

Physical and developmental characteristics predispose infants and young children to unique types of TSCI, including SCI without radiographic abnormality (SCIWORA), lap belt injuries, and birth-related trauma (Table 19.1) [2,4,10–13].

Pathophysiology of acute traumatic spinal cord injury

Acute SCI involves two phases: primary injury (occurring at time of impact) and secondary injury (due to ensuing pathophysiologic processes). Immediate mechanical injury to the cord may result in laceration, contusion, compression (permanent or temporary), shear, and traction of neural tissue. Secondary injury occurs hours to weeks following the primary injury due to pathophysiologic biochemical processes affecting the neuronal and glial cells. This includes vascular perfusion abnormalities, edema, free radical generation, lipid peroxidation, excitotoxicity, inflammation, and cell death [14,15]. Mechanism of injury associated with pediatric-onset cervical SCI include acceleration/deceleration, rotational injury, and flexion/extension [14,16] (Table 19.2). Thoracolumbar (TL) SCI include spine fractures that are

Capute and Accardo's Neurodevelopmental Disabilities in Infancy and Childhood. DOI: https://doi.org/10.1016/B978-0-12-824060-1.00006-7

TABLE 19.1 Anatomic characteristics of the pediatric spine.

Anatomic characteristic	Significance
Larger head in comparison to trunk (children <8 years of age)	Fulcrum of motion at the C2/C3 level, thus upper cervical spine is more prone to injury Occipital-atlantoaxial complex is at risk
Underdeveloped neck muscles Horizontally oriented facet joints Anterior wedging of the vertebra Underdeveloped uncinate processes Open ossification centers	Increased risk of ligamentous injury Increased risk of growth plate injuries Increased risk of dislocation injuries
Increased elasticity of spinal column compared to spinal cord	Spinal column can tolerate up to 5 cm distraction, whereas the spinal cord can tolerate only up to 5 mm, leading to SCIWORA.

SCIWORA, Spinal cord injury without radiographic abnormality.

TABLE 19.2 Mechanisms of injury: cervical spine.

Mechanism	Presentation	Imaging	Management
Acceleration/Deceleration (e.g., high-speed collisions, auto versus pedestrian, airbag injuries)	Occipito-atlantal, atlantoaxial injuries: partial or absent neurological deficits May be worsened by traction or cervical collar placement (distraction of tectorial membrane)	Sagittal CT and MRI Subluxation may be occult on X-rays	Surgical fusion if tectorial membrane is ruptured
	Odontoid injury in <7 years of age (forward facing car seat, falls): may have fatal outcome	Lateral X-rays Reconstructed CT images	Prompt immobilization with or without halo
	Subaxial ligamentous injury in >8 years of age: symptoms depend on severity of subluxation/angulation	MRI CT scan	Compression fractures: immobilization Burst/tear-drop fractures: surgical stabilization
Rotational injury (e.g., fall or collision)	Can lead to AARF Type I: atlas rotated on odontoid, no anterior displacement Type II: atlas rotated on one lateral articular process, minimal anterior displacement Type III: atlas rotated on both lateral articular processes, with >5 mm anterior displacement Type IV: rotation and posterior displacement of atlas	CT, MRI	Type I: soft collar with or without traction Type II-IV: surgical stabilization
Flexion/Extension (e.g., contact sports)	Cervical cord neuropraxia from neck lateral flexion Mild SCIWORA Temporary sensory and motor symptoms of some or all extremities Last 15 mins to 48 h		Immobilization for 2 weeks

AARF, Atlantoaxial rotatory fixation; *SCIWORA*, spinal cord injury without radiographic abnormality; *CT*, computed tomography; *MRI*, magnetic resonance imaging.

frequently associated with high-speed motor vehicle accidents or falls from significant heights. Compression fractures are most common, usually occurring at the TL junction. In comparison, burst fractures are associated with high energy trauma. Seatbelt injuries are flexion-distraction injuries and are associated with severe abdominal trauma. Sudden deceleration causes compression of the viscera between the seatbelt and the spine, along with a flexion fulcrum at the anterior vertebral body. This usually occurs at L2 and L3 levels. Slow vehicle crushing injuries (driveway injuries) occur when drivers back out slowly from a parking space. Hyperextension and thoracoabdominal injury occur when the upright thoracic spine is pinned between the bumper and another surface.

Traumatic spondylolisthesis/spondylolysis may be seen in pediatric athletes (gymnastics, weightlifting, football). Fatigue and fracture of the pars of the lower lumbar spine may result from repeated hyperextension and axial loading of the immature spine (Table 19.3) [17].

Spinal cord injury without radiographic abnormality

SCIWORA is described as a traumatic myelopathy without evidence of skeletal injury on plain or flexion-extension films of the spine, tomography, and occasionally myelography. Magnetic resonance imaging (MRI) abnormalities are subsequently found in 19%–23% of these children [18]. SCIWORA injury occurs more frequently in the cervical spine (46%) than the thoracic spine (14%). In young children, the most common site of injury is the upper cervical spine from C1 to C4, whereas the thoracic cord is more frequently injured in teenagers [19,20]. The cause of SCIWORA varies with age. Younger children are injured in pedestrian–motor vehicle accidents and falls, while older children are often injured as passenger in a motor vehicle crash or have sport-related injury [18]. The majority of patients are managed conservatively, typically in hard collar bracing for 12 weeks [18,21]. The main predictor of long-term outcomes in children with SCIWORA is the neurological status at admission [18].

Lap belt injury

Five- to 9-year-old motor vehicle passengers are at increased risk for lap belt injuries due to frequent nonuse of booster seats resulting in malpositioning of the seatbelt on the abdomen rather than the pelvic brim, providing a fulcrum for

TABLE 19.3 Mechanisms of injury: thoracolumbar spinal cord injury.

Thoracolumbar injury	Presentation	Management
Compression fracture (lower energy axial loading forces, e.g., minor to moderate falls or sports injuries)	May affect multiple levels. Consider intraabdominal injury. Most result in <30% loss of vertebral height	TLSO for 6–8 weeks
Burst fracture (axial loading injuries associated with high energy trauma)	Considered unstable and is associated with neurologic injury	Surgical decompression and stabilization if with neurologic deficit or spinal canal compromise. Nonsurgical management for stable fractures without neural compromise: TLSO 8–12 weeks
Vertebral apophysis fracture/slipped vertebral apophysis injury	Usually occurs at L4 or L5 back pain and radicular leg pain. Neurogenic claudication. Overweight children may be at risk.	Posterior laminar decompression
Seatbelt injury (compression of viscera between seatbelt and spine with flexion fulcrum at anterior vertebral body)	Flexion-distraction injury associated with severe abdominal trauma. Small bowel mesenteric tear or perforation and compression/chance-type fracture. Most common level at L2 and L3	Comprehensive trauma survey is warranted.
Traumatic spondylolisthesis/ spondylolysis	Pediatric athletes. Fatigue and fracture of the parts of the lower lumbar spine may result from repeated hyperextension and axial loading of the immature spine	Surgery: if pain persists >6 months, debilitating pain, high-grade listhesis, or presence of a neurologic deficit. Direct pars arthrodesis, or pedicle screw-laminar hook constructs preserve motion in affected segment and is preferred.

TLSO, Thoracolumbosacral orthosis.

flexion/distraction forces in the mid-lumbar spine [10,22]. While vertebral damage can be found from the mid-thoracic spine to the conus, L2–L4 is the most common level for damage. SCIWORA or evidence of vertebral damage is seen in 23%–30% of children presenting with lap belt injury [8].

Birth-related spinal cord injury

In neonates, the leading cause of SCI is obstetrical complications. Traumatic/ischemic SCIs occur in approximately 1/60,000 births. Injuries to the upper cervical cord are associated with forceps rotation during vertex delivery, whereas lower cervical and thoracic level injuries are associated with vaginal breech delivery [11,23]. SCI should be suspected in babies with decreased or absent spontaneous movement, absent deep tendon reflexes, atonic rectal sphincter, absent or periodic breathing, or lack of response to painful stimuli below the level of the lesion. If SCI is suspected in the delivery room, the head, neck, and spine should be immobilized and an MRI should be obtained to confirm diagnosis [23]. Neonatal SCI from thoracic or lumbar lesions are less common. They can result from vascular occlusion associated with umbilical artery catheters or paradoxical air embolism through transitional cardiovascular shunts.

One should screen for conditions that may occur with neonatal SCI such as brachial plexus injuries, phrenic nerve injuries, and hypoxic ischemic encephalopathy.

Management of acute traumatic spinal cord injury

Prehospital care

Suspect SCI in the following circumstances: high-risk injuries (head injury or multisystem trauma), unconsciousness, torticollis, cervical muscle rigidity, neck pain, or the presence of neurologic signs and symptoms. Immediate management should follow the Advanced Trauma Life Support ATLS guidelines. Severe hemodynamic instability, apnea, cardiac arrest, and concomitant traumatic brain injury may be present. Ensuring an adequate airway, resuscitation, provision of aggressive hemodynamic support and ventilation is critical. Immobilization in a neutral position is done to prevent further injury and progression of deficits in an unstable spine until the child is safely transported to a medical facility [12,13,16]. Children <8 years of age have a relatively large head, with the occiput extending beyond the shoulder when in supine. Using a standard spine board will place the neck in flexion. This may be avoided by using a spine board with an occipital recess or raising the trunk to align the auditory meatus with the shoulders. Very young children may not be able to tolerate the use of rigid cervical collars. Blocks or rolled-up towels placed at each side of the head and secured with tape may be used instead [13,14,16].

Acute management

Stabilization efforts continue in the emergency department. Immobilization using the spinal board is maintained until initial evaluation and imaging are completed. Patients are then taken off the spine board due to risk of developing pressure injuries [16].

Patients with acute SCI can present with neurogenic shock (NS), hemorrhagic shock, or mixed shock. NS is a form of distributive shock that presents as hypotension without compensatory tachycardia and hypothermia. This develops due to loss of all sympathetic and parasympathetic tone distal to the level of injury. In comparison, spinal shock refers to the transient depression and eventual return of spinal reflexes distal to the level of injury. This may last anywhere from 4 to 6 weeks [24].

NS shock is more common in cervical and high thoracic SCI. Impaired sympathetic outflow results in a drop in the peripheral resistance, blood pressure (BP), and pooling of blood in the peripheral vasculature. The unopposed vagal activity leads to bradycardia and reduced myocardial contractility. Thus there is loss of chronotropic and inotropic response of the heart to hypotension. In the dermis, loss of sympathetic tone leads to vasodilation, temperature dysregulation, and hypothermia. Aggressive fluid resuscitation with early initiation of vasopressors is recommended in the treatment of NS. The Consortium for Spinal Cord Medicine Clinical Practice Guidelines recommends monitoring and maintenance of mean arterial pressure above 85 mm Hg for a minimum of 7 days after injury to ensure adequate spinal cord perfusion and reduce secondary injury [16,25]. Vasopressors such as norepinephrine and dopamine are recommended for cervical and high thoracic injuries due to their alpha- and beta-adrenergic effects on the heart and vessels. Lower thoracic injuries, wherein sympathetic outflow is preserved, may benefit from the use of phenylephrine, a pure vasoconstrictor. Managing NS with fluids alone risks the development of pulmonary edema. Persons with NS are at increased risk for developing adrenal insufficiency and may benefit from hydrocortisone treatment [16].

Respiratory impairment following SCI correlates with the injury level. Injuries above C5 will present with decreased tidal volumes and vital capacity. Cervical and high thoracic injuries (T6 and above) can present with respiratory muscle weakness (decreased inspiration, expiration, and cough), autonomic dysfunction, changes in chest wall and abdominal wall dynamics, and respiratory fatigue. Autonomic dysfunction is due to unopposed vagal activity from loss of sympathetic nervous system. This leads to increased bronchial secretions, bronchoconstriction, and decreased tone, which in turn predisposes patient to atelectasis, pneumonia, and potential respiratory failure [26]. Aggressive secretion management and respiratory support are recommended. During spinal shock, patient may present with flaccid paralysis and areflexia. The increased compliance of the lung, chest, abdomen, and pelvic wall muscles can increase the work of breathing. A paradoxical breathing pattern may be observed. The upper rib cage retracts inward while the abdomen protrudes outward during inspiration due to paralysis of the intercostal muscles. Positioning can influence vital capacity. In a supine position, the vital capacity is greater as abdominal contents displace the diaphragm cephalad, lengthening it to a more favorable length tension position. Early intubation is preferred as children are more likely to experience progression of symptoms over the first few days due to the acute inflammatory response and development of edema. Initial ventilator support may be required in a child with a cervical level SCI. However, many of these children are weaned from this support. Children with high cervical injuries may require lifelong ventilatory support or phrenic nerve or diaphragm pacing [26–28].

Venous thromboembolism (VTE) is a potentially life-threatening complication following SCI. It is uncommon in pediatric-onset SCI between birth and 12 years of age. Adolescents with SCI have a higher risk for developing VTE. Risk factors include immobility, comorbid polytrauma with major vascular injury, pelvic or lower limb fractures, and use of central venous catheters. Mechanical prophylaxis in the acute setting is recommended for children of all ages. If indicated, antithrombotic prophylaxis with low molecular weight heparin should start if no active bleeding or high bleeding risk is present [29].

Neurogenic bladder, neurogenic bowel, autonomic dysfunction, effects of immobility, and risk for pressure injuries are some concerns that need to be addressed during the acute phase. Early consultation with physiatry is recommended to assist the primary team in managing these conditions (Table 19.4) [25,28,30–32].

Persons with cervical and high thoracic (T6 and above) SCI are at risk for autonomic dysreflexia (AD). It is more prevalent in those with complete tetraplegia and occurs more often in the chronic stage of SCI. Painful or irritating stimuli below the level of injury leads to massive reflex sympathetic outflow, splanchnic vasoconstriction, and hypertensive crisis, often with reflex bradycardia. AD can present as life-threatening episodes of hypertension, which may be accompanied by pounding headache, bradycardia, and upper body flushing. It may present with less severe symptoms (sweating above the level of SCI, nasal congestion, and piloerection) or may even be asymptomatic. If untreated, AD may result in severe consequences such as seizures, intracranial hemorrhage, retinal detachment, cardiac arrhythmias, and death [33].

Variability in cognitive and communication abilities from infancy to adolescents can affect identification of AD. Infants and young children may present with vague symptoms of AD, and young children may not be able to articulate their symptoms as clearly as adults. Suspect AD in infants or children with SCI who are unusually sleepy, irritable, or crying [33,34].

BP in children and adolescents with SCI is affected by age and neurologic level of injury. BP increases with advancing age and approximate adult norms in older teenagers. Those with injuries at T6 and above are expected to have lower baseline BPs compared to those without SCI. Accurate measurements for baseline BP determinations and during AD episodes are important. The measurement of BP in a reassuring environment (to alleviate anxiety) and use of appropriately sized BP cuffs are recommended. In a child or adolescent with SCI, the baseline BP should be established and followed annually as the child ages (or as needed). Elevation of BP that is 15 mm Hg (in children) and 20 mm Hg (in adolescents) or more above baseline should be considered as a sign of AD [34,35]. Initial treatment includes sitting the patient, loosening any tight clothing (e.g., abdominal binder, compression stockings), and identification and removal of noxious stimuli/triggers (Table 19.5) [33]. Bladder and bowel problems are the most common triggers of AD. Pharmacologic agents (Table 19.6) should be considered if systolic BPs are at or above 120 mm Hg in children under 5 years old, 130 mm Hg in children 6–12 years old, and 140 mm Hg in adolescents [28,33,34].

SCI can occur together with traumatic brain injury, and both may present with autonomic dysfunction. AD should be differentiated from paroxysmal sympathetic hyperactivity (PSH) or storming. Both present with elevated BP; however, etiology and management are distinct from each other (Table 19.6) [33,36]. PSH triggers include a wide range of stimuli (both noxious and nonnoxious) including pain (allodynia), endotracheal suctioning, passive movement, constipation or urinary retention, emotional stimuli, and loud noises. PSH resolves within weeks or months of injury and is usually self-limiting. AD does not typically appear during the acute phase (spinal shock). However, there is evidence of early episodes of AD in the first days/weeks after injury and may be underrecognized during the acute phase of SCI.

TABLE 19.4 Related complications in the acute setting.

Presentation/Condition	Possible diagnosis	Interventions
Urinary tract: retention or incontinence	Neurogenic bladder	Monitor fluid intake and output Bladder scans Foley catheter insertion Transition to intermittent catheterization once stable Identify if reflexic or areflexic bladder[a] Reflexic bladder: consider anticholinergics
Gastrointestinal: bowel constipation or incontinence	Neurogenic bowel	Identify if reflexic or areflexic bowel[a] Establish bowel program (regular and daily bowel movements) Consider stool softeners/laxatives Bowel emptying by digital stimulation or manual evacuation Abdominal X-ray for suspected fecal impaction/ bowel obstruction
Autonomic dysfunction (lesions T6 and above): elevated blood pressures Low blood pressures Poikilothermia Fever Bradycardia	Autonomic dysreflexia	AD management: see Table 19.5 Differentiate AD from storming
	Orthostatic hypotension	Abdominal binder, compression stockings, gradual assumption of upright position Premedication with midodrine
	Temperature dysregulation	Infants and young children at risk for hypothermia or hyperthermia Environmental awareness, dressing in layers
	Quadriplegia fever or SCI fever (fever that cannot be attributed to an infectious or other noninfectious cause in patients with high SCI)	Diagnosis of exclusion (rule out sepsis and other infections, hyperthyroidism, DVT, etc.) Treat with antipyretics
	Bradycardia	Caution with procedures that can increase vagal tone (e.g., tracheal suctioning, intubation) Atropine Pacers
Swelling of extremities	Venous thromboembolism (VTE)	Doppler ultrasound VTE prevention: Children of all ages with acute SCI receive mechanical prophylaxis with GCSs and or PCDs Chemoprophylaxis in adolescents for 8 weeks specially if with additional risk
	Fracture	X-rays If stable, recommend same surgical intervention as persons without SCI
	Heterotopic ossification	Triple phase bone scan. Consider indomethacin. ROM
Immobility	Pressure injuries	Turning schedule Monitor skin Appropriate seating and sleeping surfaces (wheelchair, mattress) Optimize nutrition Consult wound care

a, Age appropriate; *AD*, autonomic dysreflexia; *DVT*, deep venous thrombosis; *SCI*, spinal cord injury; *GCSs*, graduated compression stockings; *PCDs*, pneumatic compression devices; *ROM*, range of motion.
[a]*Upper motor neuron SCI can present with spastic or reflexic bladder and bowel; lower motor neuron SCI can present with flaccid or areflexic bladder and bowel.*

TABLE 19.5 Potential triggers of autonomic dysreflexia.

Urinary system
Distended bladder Renal or bladder stones Blocked catheter Catheterization Detrusor sphincter dyssynergia Urinary tract infection Urologic procedures (urodynamic studies, cystoscopy)
Gastrointestinal system
Bowel distention Bowel impaction Appendicitis Gallstones Hemorrhoids Gastric ulcers Hemorrhoids
Skin
Ingrown toenails Pressure injuries Constrictive clothing, shoes, or appliances (e.g., splints) Burns, blisters, insect bites Contact with hard, sharp or uncomfortable surfaces
Reproductive system
Sexual activity/intercourse Orgasm Females: menstruation, lactation, breastfeeding, labor and delivery, vaginitis, mastitis Males: ejaculation, scrotal compression (sitting on scrotum), epididymitis, prostatitis
Other
Boosting[a] Excessive intake of alcohol, caffeine, or diuretics Fractures or other trauma below level of injury Heterotopic ossification Deep vein thrombosis, pulmonary embolism Syringomyelia Surgical and invasive diagnostic procedures

[a]*Boosting: autonomic dysreflexia (AD) is intentionally caused by an athlete to enhance one's performance in sport.*

Neuroprotective therapies

The use of high-dose methylprednisolone (MPS) in pediatric TSCI is controversial. In the 1990s, the National Acute Spinal Cord Injury Surveys II and III studies showed beneficial effect of MPS in SCI if given within 8 hours after injury. However, these studies did not include patients <13 years of age. Re-analysis of these studies did not demonstrate clinically significant improvement, along with increased risk of adverse effects. Complications from MPS use include gastrointestinal hemorrhage, wound infection, severe sepsis, and pneumonia. In 2013 the Congress of Neurological Surgeons recommended against routine use of MPS in SCI [37]. Other therapies being studied include GM-1 gangliosides, riluzole, minocycline, and granulocyte colony stimulating factor [13,28].

Surgical management

Surgical decompression of neural elements and stabilization of the spine are done to prevent further injury. Considerations unique to pediatric SCI include the intrinsic elasticity and potential growth of the developing spine. The pediatric cervical spine may be more difficult to immobilize due to its intrinsic elasticity. External fixation may be done via a halo device (for upper cervical spine) or a Minerva body jacket (for mid to lower cervical spine). Surgical

TABLE 19.6 Autonomic dysreflexia versus paroxysmal sympathetic hyperactivity (storming).

Autonomic dysreflexia (AD)	Paroxysmal sympathetic hyperactivity (PSH)
Etiology	
Cervical SCI and high thoracic SCI (T6 and above)	TBI, hypoxic/anoxic injury, subarachnoid hemorrhage, stroke, encephalitis, vasculitis
Presentation	
Sudden increase in BP 15 mm Hg above baseline Sudden onset pounding headache Bradycardia (or tachycardia) Cardiac arrhythmias Profuse sweating or flushing of skin (usually face, neck, shoulders)[a] Piloerection Nasal congestion Blurred vision and/or spots in visual field Apprehension or anxiety Minimal or no symptoms	Systemic hypertension Tachycardia Tachypnea Hyperthermia Diaphoresis Posturing (decerebrate) Generalized rigidity
Management	
Sit up patient, loosen tight clothing Identify and remove noxious stimuli – Distended bladder: catheterize; unkink indwelling catheter – Fecal impaction: use lidocaine jelly, disimpact bowel[b] – Vent gastrostomy tube (if present) Pharmacologic intervention if AD does not respond to conservative measures or if severe elevation of BP above baseline – Topical nitropaste 2% (0.5–2 inches) – Nifedipine (0.25 mg/kg; adolescents ≥ 40 kg: 10 mg) – Recurrent AD: prazosin or terazosin	Minimize nociception Cooling blankets for hyperthermia Medications: Pain: NSAIDs, acetaminophen, opiate agonist (morphine), benzodiazepines; gabapentin (can also control autonomic symptoms/ paroxysms) Hypertension, tachycardia, diaphoresis: beta blockers (propranolol) PSH paroxysms: propranolol, clonidine Malignant hyperthermia: dantrolene Paroxysms and sympathetic overactivity: intrathecal baclofen Dopamine agonists: amantadine, bromocriptine

[a]*above or possibly below level of lesion.*
[b]*if systolic BP > 150 mm Hg, give pharmacologic agent.*

instrumentation and fusion may inhibit future growth and affect the curvature of the spine (development of excessive kyphosis). The extent of fusion should be limited when possible. Surgical intervention should be done within 8–24 hours postinjury. Early decompression may improve neurologic function, reduce cost of care and allow for earlier mobilization. This in turn may reduce morbidity associated with prolonged immobility [13,28].

Nontraumatic spinal cord injury disorders

Between 2010 and 2015, the global incidence of nontraumatic pediatric SCI ranged from 6 to 19.6 cases/million population with tumor (30%–63%) and inflammatory/autoimmune causes (28%–35%) being the most common [9]. A recent Australian retrospective case series of 103 consecutive admissions of children with SCI reported that more than one-half of the children (66%) sustained an NTSCI, with the majority (59%) the result of neoplasm or transverse myelitis with an additional 18% caused by other spinal cord disorders [38]. Causes of NTSCI unique to the pediatric population include high cervical injuries related to skeletal dysplasia, such as achondroplasia, juvenile rheumatoid arthritis, Down syndrome, and C1–C2 subluxation with pharyngitis and tonsillitis [19].

Spinal cord tumors

The central nervous system (CNS) is the most common site for cancer in children 0–14 years of age and the second most common cause in adolescents and young adults 15–39 years of age with an age adjusted incidence rate of 5.83

and 19.59/100,000 population, respectively. The spinal cord/cauda equina account for 10.1% of CNS tumors [20]. In the pediatric age group, extradural tumors account for two-thirds (66%) of all spinal tumors while intradural-extramedullary tumors account for 9% and intramedullary account for 25% [39]. Intramedullary tumors occur in children between 1 and 5 years of age, equally in males and females. These tumors are predominantly histologically benign, regardless of size with astrocytoma being the most common (82% of cases), followed by ganglioglioma. Ependymoma are uncommon in children outside the setting of neurofibromatosis Type 2 [40]. Laminectomy and radical resection of these tumors generally results in a good long-term survival and acceptable morbidity [41]. Adjuvant chemotherapy or radiation therapy is generally reserved for high-grade or malignant tumors with rapid postoperative tumor regrowth or for tumors that cannot be safely removed. The occurrence of paralysis following surgical resection of intramedullary tumors is related to the preoperative motor status. Patients having no or minimal preoperative motor deficits had less than 1% incidence of postoperative paralysis [42]. A multisite long-term follow-up study was reported on 164 children who underwent a radical resection of an intramedullary tumor. Postoperatively, 60% of children reported good quality of life and independent function, while 23% reported substantial weakness, most often in the legs, and with rehabilitation, were able to regain everyday independence. Urinary incontinence was reported in 26.6% of patients. Pain (6.9%) and dysesthesia (61.2%) were common sources of discomfort but only rarely necessitated medications [42]. Approximately two-third of children develop a spinal deformity following resection; however, only a third of these children required a stabilization procedure. Significant factors associated with progressive spinal deformity were cystic tumor, prior radiation, and age less than 7 years [41,42]. Malignant tumors have a very poor outcome, and surgery is usually limited to conservative debulking [43]. Children in the terminal or palliative phase of care will benefit from inpatient or outpatient rehabilitation interventions. Rehabilitation goals will focus on symptom control, family education, attainable functional goals, assistive devices, and medical equipment to facilitate independence and decrease caregiver burden.

Spina bifida

Spina bifida (SB) is a type of neural tube defect (NTD) resulting from failure of neural tube closure during early embryonic development resulting in incomplete formation of the vertebral column and/or spinal cord. The term "spina bifida" is often used interchangeably with its most common and severe form, myelomeningocele, also known as "open spina bifida," characterized by herniation of a dysplastic spinal cord, cerebral spinal fluid, and meninges through an open posterior vertebral column defect. SB differs from acquired SCIs in that it usually involves impairments of both the spinal cord and the brain.

The etiology of SB is multifactorial with genetic, environmental, and/or geographical contributors. Risk factors include history of NTDs in the mother, in a prior pregnancy, or in a family member; maternal diabetes; obesity; exposure to valproic acid, carbamazepine, or isotretinoin; insufficient folic acid during time of neural tube formation; maternal hyperthermia; and low socioeconomic status.

Estimates of prevalence of NTDs worldwide are highly variable, likely due to inconsistencies in surveillance systems. In the United States, birth prevalence of all NTDs is around 5.3–6.5 cases per 10,000 births [44,45], and that of SB is around 3.9 per 10,000 live births [46]. Periconceptional folic acid supplementation has been shown to lower the birth prevalence, with a notable 19% decline after a US mandate for folate fortification of flour and cereal grain products in 1998 [47,48], likely in combination with simultaneous advances in prenatal detection of NTDs and access to elective pregnancy terminations [49].

Early prenatal diagnosis is essential for parental decisions regarding termination, as well as for potential prenatal closure of the spinal defect. A high-resolution fetal anatomy ultrasound between 18 and 22 weeks of gestation and/or amniotic fluid alpha-fetoprotein are used for screening. Amniocentesis or chorionic villus sampling are reserved for high-risk pregnancies and for inconclusive screening results [50].

There is no standardized approach to categorizing levels of SB. Examples of common methods include (1) anatomic level, labeled based on the most cranial vertebral defect seen on spinal imaging [51] and (2) functional level, divided into thoracic (no movement in lower limbs), high-lumbar (hip flexion present only), mid-lumbar (knee extension present), low-lumbar (ankle dorsiflexion present), and sacral (foot plantar flexion present); with each muscle group having at least antigravity muscle strength. These two methods do not always correlate, and best approaches to categorize SB are still under debate [52].

Commonly associated diagnoses and sequelae of SB are discussed below.

Motor impairment: Weakness usually occurs in myotomes innervated by radicular levels below the spinal lesion. While many factors contribute to the ability to ambulate, generally speaking, people with thoracic and high-lumbar functional levels typically use a wheelchair for mobility; those with low-lumbar levels can ambulate community distances with ankle foot orthoses (AFOs) and an assistive device (such as forearm crutches), and people with sacral levels can ambulate at the community level with or without bracing, and no assistive device [52].

Sensory impairment: Occurs below the level of the spinal lesion in a dermatomal pattern, depending on which roots are impacted. Lack of sensation in the lower limbs leads to the high risk of pressure sores. Routine self or caregiver-assisted skin checks are essential.

Chiari II malformation (C2M): Occurs in 100% of cases of open SB. Involves elongation and caudal migration of the cerebellar vermis and medulla into the spinal canal, resulting in variable levels of brainstem dysfunction. Symptoms can include apnea, dysphagia, hoarseness, spasticity, and headaches. Frequency of surgical decompressions for symptomatic C2M have declined given the inconsistent clinical impact, as well as focus on management of decompensated hydrocephalus as a root cause of symptomatic C2M [53,54].

Hydrocephalus (HC): Occurs in 60%–95% of children with SB and is most common in higher (thoracic) than lower (sacral) lesions [53]. Common symptoms include headaches, vomiting, lethargy, irritability, and confusion. Severe HC can be fatal if not treated. Management often requires ventricular shunting; however, given shunt failures are a primary cause of morbidity and mortality in this population [53,55], a newer technique of endoscopic third ventriculostomy with choroid plexus coagulation is also being used in an attempt to avoid shunt placement.

Tethered cord syndrome (TCS): Incidence of TCS is 14%–32% [56]. Due to adhesions of MMC malformation and scar tissue formed at the initial myelomeningocele (MMC) closure site, the spinal cord is often tacked down to adjacent structures in people with SB, limiting its normal ability to rise with growth or stretch. Particularly during times of rapid growth, this can lead to a longitudinal traction on the cord, reducing blood flow to the spinal nerves and damaging the spinal cord. Signs and symptoms include new and/or progressive weakness, deterioration of gait, sensory changes, progressive pain, progressive scoliosis, and changes in bowel or bladder function. Most people with SB appear on imaging to have a tethered cord, thus the decision for an untethering surgery is based on the neurosurgeon's clinical judgment of symptoms and risks. After surgical repair, children are at high risk of recurrent or new symptoms consistent with tethering.

Syringomyelia: An accumulation of cerebrospinal fluid (CSF) within the spinal cord can result from shunt malfunction. Symptoms can include pain, weakness, stiffness, and loss of temperature sensation, usually impacting the hands. Management should always begin with evaluation of shunt function.

Neurogenic bowel dysfunction (NBD): NBD affects the majority of people with SB. The slow colonic transit, anal sphincter dysfunction, and decreased sensation all contribute to constipation and fecal incontinence. NBD has an enormous impact on health, affecting bladder function, skin integrity, and shunt function. It also has significant psychosocial impacts on the caregiver and the patient, affecting ability to socialize, develop relationships, as well as employability [57,58]. While there is not yet a standardized approach to bowel management, many people with SB can gain control of their bowel with a combination of dietary management, oral medications such as stool softeners and stimulants, and/or enemas for recto-sigmoid emptying [58].

Neurogenic bladder dysfunction: Over 90% of newborns with SB have normal upper urinary tracts. However, without intervention, 50% will have upper urinary tract damage in the first 5 years of life as a result of lower urinary tract dysfunction [59]. Lower urinary tract involvement can include partial or complete denervation of the bladder, poor compliance and poor contractility, internal sphincter incompetence, and detrusor sphincter dyssynergia, all putting health of the kidneys at risk. Patients should have close monitoring by a urologist, and some may require monitoring by a nephrologist as well. While there is not yet a standardized approach, maintenance of normal renal function is the core of urologic management, usually involving a combination of upper tract monitoring with ultrasounds and kidney function tracking with creatinine and/or cystatin C. Bladder function is managed with a combination of medical (catheters), pharmacologic, and, in certain cases, surgical interventions [59].

Orthopedic deformities: Club foot is the most common foot deformity. Hip deformities and lower limb contractures are common due to an imbalance of muscle strength across the joint or flaccid paralysis, depending on lesion level.

Cognition: Children with SB are very strong in skills that use rule-based processing. They tend to have a strong vocabulary and are good at math facts. They have difficulty with visual spatial reasoning, spoken and written language comprehension, and problem-solving in math [60–62]. Some children with SB may present with "cocktail party syndrome." The child is verbose with good articulation but with shallow intellect, poor social, academic, and daily living skills [63]. Children with SB should also be closely monitored for attention deficits. This often presents with difficulty with initiation, a sluggish cognitive tempo, and a lack of focus with difficulty orienting to external stimuli [64]. Unlike developmental attention deficit hyperactivity disorder (ADHD) involving the frontal lobes, these attention challenges in SB are a result of the Chiari malformation, tectal beaking, and hypogenesis of the corpus callosum [65].

Historically, the open spinal defect was surgically closed shortly after birth. However, prenatal closures of SB defects have been shown to decrease the need for ventriculoperitoneal shunting at 12 months, reduce Chiari II malformations, improve lower limb motor function at 30 months compared to predicted anatomical level [66,67], and improve composite scores of neurodevelopmental outcomes [53]. Maternal and fetal risks of the prenatal procedure include uterine wall thinning, peripartum bleeding, premature birth, as well as a potential for increased risk of future symptomatic tethering of the spinal cord compared to those with postnatal closures [56]. Risks have been reduced with improvements in surgical techniques and protocols, and a fetoscopic approach has also shown success and is being evaluated as a potential means to further minimize risks [66,68–71].

To maximize a child's potential for success, a family-centered approach to care is essential from the beginning. Better outcomes are seen in children whose parents have higher expectations, teach their kids skills to address their attention difficulties, allow them to do things for themselves to maximize their physical strengths and independence, and encourage language development [72]. Because of its complexity, medical management, education, and routine monitoring of physical, cognitive, communicative, and social development should be provided by a multidisciplinary team of specialists knowledgeable in SB beginning in utero and continuing across the lifespan.

Neuroimmune: transverse myelitis, multiple sclerosis

Acute transverse myelitis (ATM) has an incidence of one to four new cases per million people per year, affecting individuals of all ages with bimodal peaks between the ages of 10 and 19 years and 30 and 39 years. In children, the incidence of ATM has been estimated to be 1–2 per million children, and it is more than four times more common in adolescence as in childhood [73,74]. The disease presents as acute onset of motor, sensory, and sphincter deficits that can be localized to the spinal cord and has a monophasic course [75]. There is no specific sex preference [76], but males are more likely to present with ATM in general and females are more commonly affected in regions with known higher multiple sclerosis (MS) risk [77]. Clinically, ATM is characterized by acute or subacute onset of back or radicular pain and progressive neurologic dysfunction in the motor, sensory, and autonomic realm, resulting in para/tetraparesis/plegia, numbness, paresthesia, or band-like dysesthesias and bowel and bladder dysfunction [74,76,77]. Imaging is highly variable and up to half of all patients with clinical ATM do not show any imaging abnormalities [78]. If hyperintense signal on T2-weighted MR imaging extends over three or more adjacent vertebral segments, and involves more than two-thirds of its cross-sectional area, longitudinally extensive transverse myelitis is diagnosed [79]. The most common MRI feature of ATM is a centrally located hyperintensity causing cord enlargement in the cervical or thoracic spine [77,79,80]. A specific form of ATM is neuromyelitis optica, which presents with the triad of myelitis, optic neuritis, and positive anti-AQP4 antibody [81]. ATM can have a viral etiology, or it can be a presentation of acute disseminated encephalomyelitis (ADEM), MS, collagen vascular disease like systemic lupus erythematosus, Behçet disease, Sjogren syndrome, and paraneoplastic syndromes; in many cases, the cause is idiopathic [82]. CSF analysis may show mild pleocytosis, elevated protein levels, or oligoclonal bands [83,84]. In 2002 the Transverse Myelitis Group proposed a set of uniform diagnostic criteria and nosology for ATM in order to allow for development of multicenter clinical trials, given the rare occurrence of the clinical syndrome [76]:

Diagnostic criteria include:

- Sensory, motor, or autonomic dysfunction originating from the spinal cord.
- T2 hyperintense signal changes on MRI.
- No evidence of a compressive lesion.
- Bilateral signs/symptoms.
- Clearly defined sensory level.
- Evidence of inflammatory process demonstrated by gadolinium enhancement on MRI, CSF analysis showing pleocytosis, or elevated immunoglobulin G (IgG) index.
- Progression to nadir between 4 hours and 21 days.

The pharmacologic interventions used in ATM consist of immunomodulatory therapy, like high-dose intravenous steroids with/without plasmapheresis, intravenous immunoglobulin (IVIG), and azathioprine [79,85,86]. And for functional recovery, activity-based rehabilitation interventions are employed [87]. Most cases of ATM have full or at least partial recovery [88], with approximately one-third of patients recovering with little to no sequelae, one-third being left with moderate degree of permanent disability, and one-third having severe disabilities [88–91].

Multiple sclerosis

Pediatric-onset multiple sclerosis is an inflammatory autoimmune emyelinating disorder accounting for 3%–5% of all MS cases [92,93], with 3%–10% first presenting during the adolescent years, with less during childhood [94]. It manifests as white matter lesions spread over time and space in the CNS, with significant spinal cord burden [95]. There are several forms of MS, with relapsing remitting MS being the most common form in both adult and pediatric population (85% of MS patients); other forms include primary progressive MS, secondary progressive, and progressive-relapsing MS [96]. Clinically, it presents with a constellation of motor, sensory, and autonomic symptoms that usually have a patchier, more incomplete occurrence then ATM and are recurrent in nature. Gait abnormalities and spasticity are encountered frequently, especially after the initial episode of neurologic dysfunction. Compared to MS in adults, children have a higher rate of relapse [97], more extensive MRI lesions [98], and delayed onset of physical disability, often occurring after 10 years from diagnosis [99]. The disease pathology is characterized by both focal and diffuse white matter and gray matter damage. The diagnosis is confirmed by occurrence of specific MRI changes consisting of focal, well circumscribed, hyperintense plaques on T2-weighted images of the spine, best seen on short tau inversion recovery sequencing [98]. CSF findings consist of moderately elevated protein levels (with elevated IgG/albumin index), mild mononuclear pleocytosis, and presence of oligoclonal IgG bands. Treatment of the acute relapse is done with intravenous high-dose corticosteroids. Other considerations for refractory symptoms include adrenocorticotrophic hormone therapy and plasma exchange. Longer-term, disease-modifying therapies such as interferon-b, glatiramer [95,100], and fingolimod have been used to provide long-term control [100]. As the goal of intervention in MS is functional recovery, adjunct symptomatic treatment can include medications to control spasticity and pain and activity-based physical and occupational therapies to optimize mobility and activities of daily living.

Infectious: acute flaccid myelitis

Acute flaccid myelitis (AFM) is a polio-like illness primarily affecting children and characterized by acute onset of asymmetric flaccid paralysis due to anterior horn gray matter involvement. While it predominantly affects children (median age of onset is 6.3) [101], it was first reported to the California Department of Public Health in 2012 in a 29-year-old poliomyelitis unvaccinated man with no history of travel [102]. A viral etiology (primarily enterovirus (EV) D68) is supported by the temporal and geographic correlations between AFM cases and EV D68 circulation [103,104] and by the fact that EV D68 is the predominant pathogen identified in the biologic specimens (typically respiratory samples) collected from individuals with AFM [105,106]. Other viruses, like EV A71, rhinovirus, parechovirus, echovirus, and Coxsackievirus A16 were also isolated from CSF, respiratory and stool samples [101]. From 2012 to 2018, AFM had a biennial occurrence, with peak number of cases happening in even years, pattern disrupted by the SARS CoV2 pandemic [107].

Specific clinical diagnostic criteria have been proposed [108] and Centers for Disease Control and Prevention (CDC) established and updated AFM case definition (used for classification) since 2015 [109] (Table 19.7).

TABLE 19.7 2015 CDC Case defintion for Acute flaccid myelitis.

A1. Clinical criteria
• An illness with onset of acute flaccid weakness of one or more limbs, AND • Absence of a clear alternative diagnosis attributable to a nationally notifiable condition
A2. Laboratory/Imaging criteria
Confirmatory laboratory/imaging evidence: • MRI showing spinal cord lesion with predominant gray matter involvement and spanning one or more vertebral segments, AND • Excluding persons with gray matter lesions in the spinal cord resulting from physician diagnosed malignancy, vascular disease, or anatomic abnormalities.
Presumptive laboratory/imaging evidence: • MRI showing spinal cord lesion where gray matter involvement is present but predominance cannot be determined, AND • Excluding persons with gray matter lesions in the spinal cord resulting from physician diagnosed malignancy, vascular disease, or anatomic abnormalities.
Supportive laboratory/imaging evidence: • MRI showing a spinal cord lesion in at least some gray matter and spanning one or more vertebral segments, AND • Excluding persons with gray matter lesions in the spinal cord resulting from physician diagnosed malignancy, vascular disease, or anatomic abnormalities.

The clinical presentation of AFM starts with a prodromal illness characterized by fever (81%), respiratory tract (81%), and/or gastrointestinal (37%) symptoms [101,110], followed by the onset of asymmetric, flaccid weakness, typically starting as monoplegia and potentially progressing over hours to days to involvement of other limbs and bulbar structures, resulting in paralysis and sometimes respiratory insufficiency or failure. Cranial nerves involvement has also been described [101,111–113]. The onset of flaccid, areflexic paralysis is frequently followed by pain and autonomic nervous system involvement, consisting of BP and heart rate variability and bowel and bladder dysfunction [114]. The primary site of injury is the anterior horn cells of the spinal cord, resulting in a motor neuronopathy.

Diagnosis is confirmed by imaging, specifically MRI showing non/minimally enhancing T2 hyperintense lesions of the spinal cord gray matter involving, most commonly, the cervical spine but extending longitudinally to involve the thoracic spine and sometimes the brainstem (dorsal pons) [111]; and CSF analysis which shows pleocytosis and mildly or moderately elevated CSF protein [108].

Differential diagnosis includes infectious and noninfectious conditions that affect the central and peripheral nervous system like poliomyelitis, West Nile virus, tick-borne and Japanese encephalitis, varicella zoster virus myelitis; demyelinating myelitis (i.e., aquaporin-4-IgG seropositive or seronegative neuromyelitis optica spectrum disorder, myelin oligodendrocyte glycoprotein-antibody (MOG) associated myelitis, MS, and ADEM); spinal stroke, fibrocartilaginous emboli; Guillain-Barré syndrome (GBS) and other predominantly motor neuropathies; and disorders affecting the neuromuscular junction (like myasthenia gravis, botulism, and tick paralysis) [108,115].

Detection of the etiologic agent can be attempted by collecting respiratory (nasopharyngeal and oropharyngeal) and stool or rectal swab specimens, which can show the presence of EV D68, EV A71, or other enteroviruses, including polio by reverse transcription polymerase chain reaction [108].

Nerve conduction studies (NCS) and electromyography (EMG) can be used early to differentiate AFM from GBS or other acute neuromuscular disorders. The AFM-induced EMG/NCS changes emerge over several weeks and include diminished or absent compound motor action potentials, decreased or absent F waves, and normal sensory NCS associated with muscle denervation changes (reduced or absent recruitment and fibrillations and positive sharp waves) [108].

AFM is a newly described, polio-like disease affecting small children that has no specific medical treatment and induces significant long-term neurologic and functional consequences. The COVID-19 pandemic showed that typical measures that generally contain virus spread (hand hygiene, mask wearing, social distancing, and quarantine) are effective in preventing new AFM cases, but they are not sustainable for prolonged periods. Thus early diagnosis and development of effective antiviral interventions are essential in minimizing the significant consequences of paralysis.

The acute management of AFM is mainly supportive, with hospitalization and admission to an intensive care unit for impending respiratory insufficiency or failure. Optimization of cardiopulmonary parameters, management of pain and bowel and bladder dysfunction, and prevention of immobilization complications are targeted goals. IVIG could be administered early postdisease onset, as the presence of neutralizing antibodies against contemporary strains of EV D68 has been documented in the infusions [116]. If white matter involvement is seen on MRI, high-dose corticosteroids with or without plasma exchange are advocated by some [117]. There is no antiviral agent or vaccine for EV D68 and an open label fluoxetine trial showed no clinical benefit [118], despite its in vitro anti-enterovirus D68 activity [119].

Initiation of early rehabilitation strategies in the acute care setting and seamless transition to an inpatient setting (or directly outpatient setting if there is only minor or no persisting neurologic deficit) are strongly recommended to facilitate functional recovery [108,115,120]. Neurologic recovery occurs over an extended period of time and appears to progress in a distal to proximal pattern [113,121]. Bulbar innervated muscles' strength appears to recover sooner than limb muscle strength and the extent of recovery is highly variable, with full return to preparalysis state being rarely achieved [105,122–124]. Only a small percentage of children require ventilator support long term [111,113].

Nerve transfers have been performed in a limited number of cases in order to improve specific motor patterns, with more success achieved when targeting elbow, wrist, and hip function [125–128] and less improvements in shoulder function [129].

Long-term AFM sequelae are similar to sequelae of other spinal cord-related paralysis: chronic pain (neuropathic, musculoskeletal, and visceral), soft tissue contractures and limitations in range of motion, joint subluxations, limb length discrepancies, scoliosis, chest wall abnormalities, decreased bone mass and pathologic fractures [123,130], and psychological sequelae, such as anxiety and depression [113,123,124,131].

Spinal cord infarct and fibrocartilaginous embolism

Spinal cord ischemia and infarct/stroke have quite a heterogeneous etiology, including both systemic and local pathologies. Among the systemic conditions that could lead to spinal cord ischemia are those associated with prolonged hypotension (shock, iatrogenic, etc.) or those that increase thromboembolic states (leukemia, polycythemia, sickle cell disease, decompression sickness, etc.). Localized pathological conditions that could lead to spinal cord infarction include vascular abnormalities like arteriovenous malformations with subsequent steal phenomenon or venous congestion and fibrocartilaginous embolism (FCE). Spinal cord strokes occur more frequently in the adult than the pediatric population and they are estimated to make about 5%–8% of all acute myelopathies and 1%–2% of all vascular neurological pathologies [132]. FCE in itself is estimated to represent 5.5% of spinal cord infarctions but may be more common than initially presumed if underdiagnosed due to its vague clinical presentation [133,134]. The clinical presentation of spinal cord stroke depends on the infarcted territory, with anterior spinal cord syndrome (loss of motor function, pain and temperature sensation below infarcted level, with preservation of proprioception, autonomic dysfunction manifested as hypotension, sexual, bowel, and bladder dysfunction) being the most common [135]. Because the presumed etiology of FCE is increased intravertebral body pressure by axial loading forces applied to the spine, such as heavy lifting or straining causing dislodged components of the nucleus pulposus to migrate into the surrounding radicular arteries, acute onset of pain also accompanies the clinical picture if this entity [136]. Radiologic changes associated with spinal cord stroke are progressive spinal cord enlargement and increased T2 signal in the spinal cord on MRI, but early MRI can show no specific findings if diffusion-weighted imaging is not used to show spinal cord cytotoxic changes. In the case of FCE, disk disease can also be found on imaging [137]. Acute treatment is supportive, and neurological outcome varies [138].

Other causes of nontraumatic spinal cord injury

A total of 15%–20% of children with Down syndrome have atlantoaxial instability associated with increased ligamentous laxity. An atlantodens interval of greater than 4.5 mm is considered abnormal. These individuals may present with occipital atlanto instability. Many children with atlantoaxial instability are asymptomatic. Restriction of high-risk activities in asymptomatic individuals is controversial [139–141].

Idiopathic juvenile arthritis may present with increased risk of cervical fusion at C2–C3 and progression of fusion, predisposing to an increased risk of fracture. They are also at risk for atlantoaxial instability at C1–C2 due to synovitis of the facet and synovial joints surrounding the dens. Destruction of the dens may result from inflammatory processes [142,143].

Skeletal dysplasias may be associated with myelopathy. Infants with achondroplasia may have small foramen magnum which can result in compression of the upper cervical cord. They can also present with spinal stenosis which can lead to spastic paraplegia [144]. Osteogenesis imperfecta is associated with basilar impression and spinal cord impairment.

Assessment of spinal cord injury in the pediatric population

History

The evaluation of a child suspected with SCI offers unique challenges. Considering age and development, eliciting a history of neck pain may be difficult as children and may not be able to express their symptoms clearly or cooperate with the physical examination.

The examiner should ask for history of midline pain and tenderness, altered level of consciousness, and the presence of focal neurologic deficits (current or transient). One should evaluate for concomitant SCI in the following situations: high energy mechanisms of trauma (such as motor vehicle accidents, fall from a height >10 feet), injuries above the clavicle, loss of consciousness, loss of bowel or bladder function, and paresthesia. The presence of other injuries in multitrauma situations (e.g., fractures) can distract from assessing for SCI.

Physical examination

Initial evaluation should assess for airway, breathing, and circulation as patients with SCI are at risk for respiratory failure and NS. Determine the mental status using the Glasgow coma scale. Assess for posterior midline pain/tenderness of the spine. Check for limitation and pain of neck range of motion. Neurologic exam should include evaluation for motor weakness, sensory deficits (delineate a level of sensory loss if possible), and abnormal reflexes (areflexia/hyperreflexia). Note that if the patient is still in spinal shock, there is absence of spinal reflexes below the level of injury. Posterior column function is evaluated by testing proprioception and vibratory sense. The sacral reflexes (anal reflex and/or bulbocavernosus) and rectal tone should be tested. The presence of abdominal wall contusions and abdominal visceral injury should alert one to screen for TL spinal injuries.

In patients with suspected or diagnosed SCI, serial neurologic examination should be performed to evaluate for improvement or deterioration. A total of 25%–50% of children with SCI may experience delayed onset of neurologic dysfunction that ranges from 30 minutes to 4 days. They can present with transient and subtle neurologic symptoms such as paresthesia or subjective weakness. Posttraumatic occlusion of radicular arteries, natural expansion of the cord injury by inflammation, and repeated trauma to the spinal cord because of occult spinal instability are some mechanisms thought to be responsible for delayed onset of SCI [2].

Neurologic assessment and classification of SCI should be done as per the International Standards for Neurologic Classification of SCI (ISNCSCI). The ISNCSCI determines the neurologic level of injury, if the SCI is complete versus incomplete, aids in prognostication after injury, and assists in the evaluation of progressive neurologic symptoms during the acute phase and at follow-up. The worksheet can be downloaded for free from https://asia-spinalinjury.org/international-standards-neurological-classification-sci-isncsci-worksheet/. The WeeSTeP is the International Standards examination developed for the pediatric population. Training is available online for ISNCSCI and WeeSTeP (https://asia-spinalinjury.org/learning/). (Fig. 19.1).

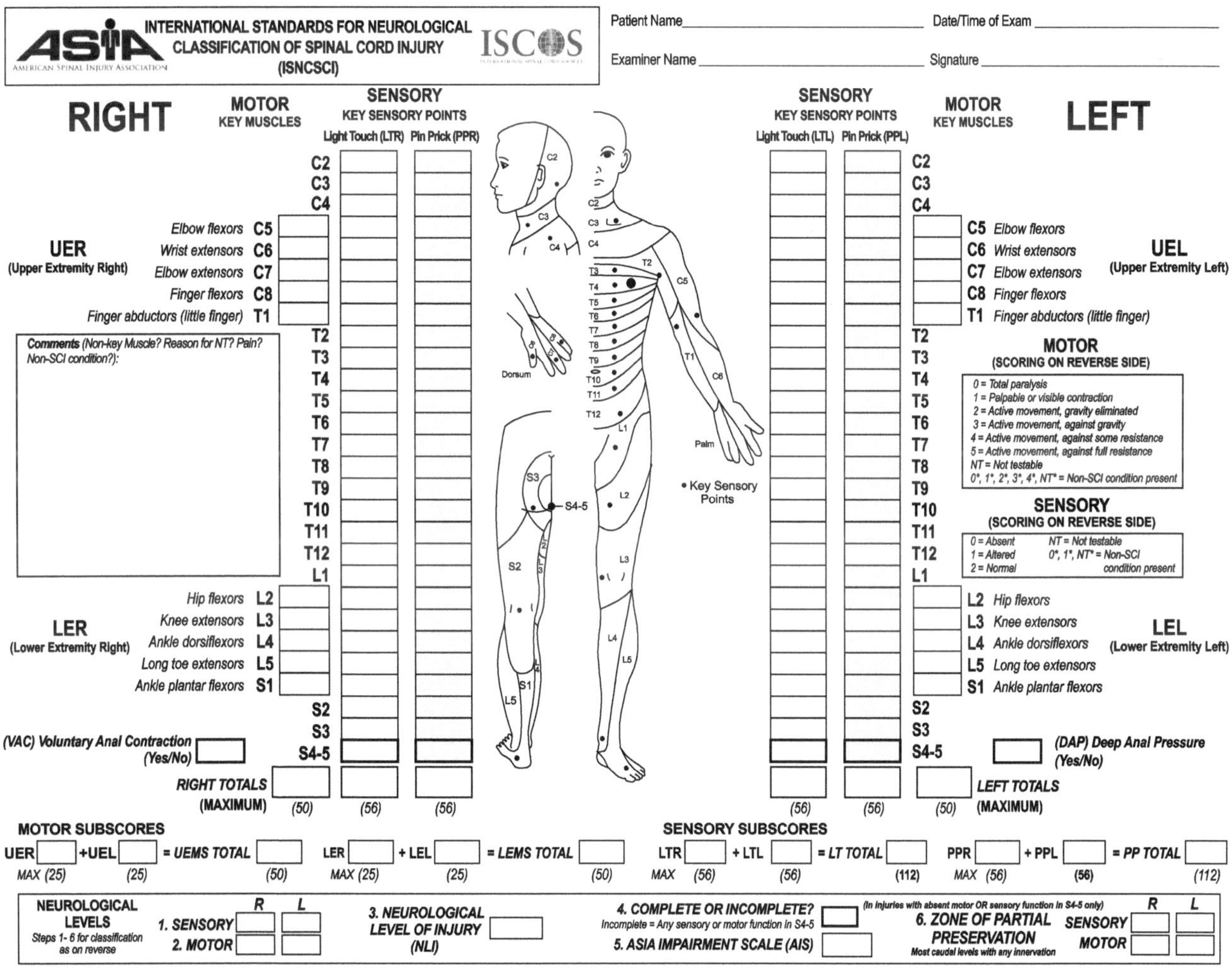
ASIA AMERICAN SPINAL INJURY ASSOCIATION
INTERNATIONAL STANDARDS FOR NEUROLOGICAL CLASSIFICATION OF SPINAL CORD INJURY (ISNCSCI)
ISCOS
Patient Name ______ Date/Time of Exam ______
Examiner Name ______ Signature ______

RIGHT — MOTOR KEY MUSCLES — SENSORY KEY SENSORY POINTS Light Touch (LTR) Pin Prick (PPR)
LEFT — MOTOR KEY MUSCLES — SENSORY KEY SENSORY POINTS Light Touch (LTL) Pin Prick (PPL)

C2, C3, C4
UER (Upper Extremity Right) / UEL (Upper Extremity Left):
C5 Elbow flexors
C6 Wrist extensors
C7 Elbow extensors
C8 Finger flexors
T1 Finger abductors (little finger)
T2, T3, T4, T5, T6, T7, T8, T9, T10, T11, T12, L1
LER (Lower Extremity Right) / LEL (Lower Extremity Left):
L2 Hip flexors
L3 Knee extensors
L4 Ankle dorsiflexors
L5 Long toe extensors
S1 Ankle plantar flexors
S2, S3, S4-5

Comments (Non-key Muscle? Reason for NT? Pain? Non-SCI condition?):

(VAC) Voluntary Anal Contraction (Yes/No)
(DAP) Deep Anal Pressure (Yes/No)

RIGHT TOTALS (MAXIMUM) (50) (56) (56)
LEFT TOTALS (MAXIMUM) (56) (56) (50)

MOTOR (SCORING ON REVERSE SIDE)
0 = Total paralysis
1 = Palpable or visible contraction
2 = Active movement, gravity eliminated
3 = Active movement, against gravity
4 = Active movement, against some resistance
5 = Active movement, against full resistance
NT = Not testable
0*, 1*, 2*, 3*, 4*, NT* = Non-SCI condition present

SENSORY (SCORING ON REVERSE SIDE)
0 = Absent; 1 = Altered; 2 = Normal; NT = Not testable; 0*, 1*, NT* = Non-SCI condition present

• Key Sensory Points; Dorsum; Palm

MOTOR SUBSCORES
UER ☐ + UEL ☐ = UEMS TOTAL ☐ — MAX (25) (25) (50)
LER ☐ + LEL ☐ = LEMS TOTAL ☐ — MAX (25) (25) (50)

SENSORY SUBSCORES
LTR ☐ + LTL ☐ = LT TOTAL ☐ — MAX (56) (56) (112)
PPR ☐ + PPL ☐ = PP TOTAL ☐ — MAX (56) (56) (112)

NEUROLOGICAL LEVELS (Steps 1-6 for classification as on reverse) R L
1. SENSORY
2. MOTOR
3. NEUROLOGICAL LEVEL OF INJURY (NLI)
4. COMPLETE OR INCOMPLETE? Incomplete = Any sensory or motor function in S4-5
5. ASIA IMPAIRMENT SCALE (AIS)
6. ZONE OF PARTIAL PRESERVATION (In injuries with absent motor OR sensory function in S4-5 only) Most caudal levels with any innervation — R L SENSORY, MOTOR

FIGURE 19.1 The International Standards for Neurologic Classification of SCI (ISNCSCI) Worksheet. *With permission from The International Standards for Neurological Classification of Spinal Cord Injury, (American Spinal Injury Association: International Standards for Neurological Classification of Spinal Cord Injury, revised 2019; Richmond, VA.).*

Muscle Function Grading

0 = Total paralysis
1 = Palpable or visible contraction
2 = Active movement, full range of motion (ROM) with gravity eliminated
3 = Active movement, full ROM against gravity
4 = Active movement, full ROM against gravity and moderate resistance in a muscle specific position
5 = (Normal) active movement, full ROM against gravity and full resistance in a functional muscle position expected from an otherwise unimpaired person
NT = Not testable (i.e. due to immobilization, severe pain such that the patient cannot be graded, amputation of limb, or contracture of > 50% of the normal ROM)
0*, 1*, 2*, 3*, 4*, NT* = Non-SCI condition present [a]

Sensory Grading

0 = Absent **1** = Altered, either decreased/impaired sensation or hypersensitivity
2 = Normal **NT** = Not testable
0*, 1*, NT* = Non-SCI condition present [a]

[a] Note: Abnormal motor and sensory scores should be tagged with a '*' to indicate an impairment due to a non-SCI condition. The non-SCI condition should be explained in the comments box together with information about how the score is rated for classification purposes (at least normal / not normal for classification).

When to Test Non-Key Muscles:

In a patient with an apparent AIS B classification, non-key muscle functions more than 3 levels below the motor level on each side should be tested to most accurately classify the injury (differentiate between AIS B and C).

Movement	Root level
Shoulder: Flexion, extension, adbuction, adduction, internal and external rotation **Elbow:** Supination	C5
Elbow: Pronation **Wrist:** Flexion	C6
Finger: Flexion at proximal joint, extension **Thumb:** Flexion, extension and abduction in plane of thumb	C7
Finger: Flexion at MCP joint **Thumb:** Opposition, adduction and abduction perpendicular to palm	C8
Finger: Abduction of the index finger	T1
Hip: Adduction	L2
Hip: External rotation	L3
Hip: Extension, abduction, internal rotation **Knee:** Flexion **Ankle:** Inversion and eversion **Toe:** MP and IP extension	L4
Hallux and Toe: DIP and PIP flexion and abduction	L5
Hallux: Adduction	S1

ASIA Impairment Scale (AIS)

A = Complete. No sensory or motor function is preserved in the sacral segments S4-5.

B = Sensory Incomplete. Sensory but not motor function is preserved below the neurological level and includes the sacral segments S4-5 (light touch or pin prick at S4-5 or deep anal pressure) AND no motor function is preserved more than three levels below the motor level on either side of the body.

C = Motor Incomplete. Motor function is preserved at the most caudal sacral segments for voluntary anal contraction (VAC) OR the patient meets the criteria for sensory incomplete status (sensory function preserved at the most caudal sacral segments S4-5 by LT, PP or DAP), and has some sparing of motor function more than three levels below the ipsilateral motor level on either side of the body. (This includes key or non-key muscle functions to determine motor incomplete status.) For AIS C – less than half of key muscle functions below the single NLI have a muscle grade ≥ 3.

D = Motor Incomplete. Motor incomplete status as defined above, with at least half (half or more) of key muscle functions below the single NLI having a muscle grade ≥ 3.

E = Normal. If sensation and motor function as tested with the ISNCSCI are graded as normal in all segments, and the patient had prior deficits, then the AIS grade is E. Someone without an initial SCI does not receive an AIS grade.

Using ND: To document the sensory, motor and NLI levels, the ASIA Impairment Scale grade, and/or the zone of partial preservation (ZPP) when they are unable to be determined based on the examination results.

AMERICAN SPINAL INJURY ASSOCIATION

INTERNATIONAL STANDARDS FOR NEUROLOGICAL CLASSIFICATION OF SPINAL CORD INJURY

ISCOS

INTERNATIONAL SPINAL CORD SOCIETY

Steps in Classification

The following order is recommended for determining the classification of individuals with SCI.

1. Determine sensory levels for right and left sides.
The sensory level is the most caudal, intact dermatome for both pin prick and light touch sensation.

2. Determine motor levels for right and left sides.
Defined by the lowest key muscle function that has a grade of at least 3 (on supine testing), providing the key muscle functions represented by segments above that level are judged to be intact (graded as a 5).
Note: in regions where there is no myotome to test, the motor level is presumed to be the same as the sensory level, if testable motor function above that level is also normal.

3. Determine the neurological level of injury (NLI).
This refers to the most caudal segment of the cord with intact sensation and antigravity (3 or more) muscle function strength, provided that there is normal (intact) sensory and motor function rostrally respectively.
The NLI is the most cephalad of the sensory and motor levels determined in steps 1 and 2.

4. Determine whether the injury is Complete or Incomplete.
(i.e. absence or presence of sacral sparing)
If voluntary anal contraction = ***No*** *AND all S4-5 sensory scores =* ***0*** *AND deep anal pressure =* ***No****, then injury is* ***Complete.***
Otherwise, injury is ***Incomplete****.*

5. Determine ASIA Impairment Scale (AIS) Grade.
Is injury Complete? If YES, AIS=A

NO ↓

Is injury Motor Complete? If YES, AIS=B

NO ↓ (No=voluntary anal contraction OR motor function more than three levels below the motor level on a given side, if the patient has sensory incomplete classification)

Are at least half (half or more) of the key muscles below the neurological level of injury graded 3 or better?

NO ↓ **AIS=C**

YES ↓ **AIS=D**

If sensation and motor function is normal in all segments, AIS=E
Note: AIS E is used in follow-up testing when an individual with a documented SCI has recovered normal function. If at initial testing no deficits are found, the individual is neurologically intact and the ASIA Impairment Scale does not apply.

6. Determine the zone of partial preservation (ZPP).
The ZPP is used only in injuries with absent motor (no VAC) OR sensory function (no DAP, no LT and no PP sensation) in the lowest sacral segments S4-5, and refers to those dermatomes and myotomes caudal to the sensory and motor levels that remain partially innervated. With sacral sparing of sensory function, the sensory ZPP is not applicable and therefore "NA" is recorded in the block of the worksheet. Accordingly, if VAC is present, the motor ZPP is not applicable and is noted as "NA".

FIGURE 19.1 (Continued).

Imaging

Plain radiographs should be obtained to screen patients for spine injury. If this is negative, and there is still concern for SCI, an MRI should be obtained, given the high frequency of SCIWORA in this age group. MRI is the most sensitive modality for SCI as it can evaluate bony, ligamentous, and cord injuries. MRI can also indicate patient prognosis [13]. Poor outcomes are seen in spinal cord hemorrhage, while patients with single-level edema tend to have better outcomes [13,31].

Outcomes

Neurologic measures

International Standards for Neurological Classification of Spinal Cord Injury

The most common scale employed in defining the neurologic outcome following TSCI is the ISNCSCI developed by American Spinal Injury Association (ASIA) in 1982, subsequently revised multiple times (last revision 2019) [90,145–149]. The ASIA Impairment Scale (AIS) is the result of systematic exam of dermatomes and myotomes in the body in order to determine the cord segments affected by the SCI. It is considered unreliable for children under 6 years of age. Several measures of neurological damage can be generated: neurological level, sensory level, and motor level (on right and left sides), sensory scores (pin prick and light touch), motor score, and zone of partial preservation. The

AIS grades the degree of impairment in five different subgroups: (1) Complete. No sensory or motor function is preserved in the sacral segments S4–S5. (2) Incomplete. Sensory but not motor function is preserved below the neurological level and includes the sacral segments S4–S5. (3) Incomplete. Motor function is preserved below the neurological level, and more than half of key muscles below the neurological level have a muscle grade less than 3 (grades 0–2). (4) Incomplete. Motor function is preserved below the neurological level and at least half of key muscles below the neurological level have a muscle grade greater than or equal to 3. (5) Normal. Sensory and motor functions are normal.

Functional measures

National Institutes of Health common data elements toolbox

In 2017 in the context of developing common data elements (CDE) to be used in standardizing data collection and data sharing in clinical studies (project undertaken under the auspices of National Institute of Neurologic Disorders and Stroke of the National Institutes of Health [NIH]), pediatric SCI CDE's were published [150]. Only a limited number of functional measures used for adults with SCI could be validated in children with SCIs and new, pediatric specific measures were recommended. A specific purpose of pediatric SCI CDEs was to place emphasis on longitudinal monitoring, thus instruments that could follow pediatric patients through childhood, adolescence, and adulthood and link pediatric SCI functional outcomes with adult ones were strongly considered. As a result, the NIH toolbox contains recommended outcomes for usage in pediatric SCI classified as: core, supplemental highly recommended, supplemental, and exploratory.

Patient-reported outcome measures

1. The Pediatric Spinal Cord Injury Activity Measure (PEDI-SCI AM) [151] was developed at Shriners Hospitals for Children and assesses different aspects of the ability of a child with SCI to perform everyday activities organized into four domains: general mobility, wheeled mobility (power and manual), ambulation, and daily routine. It can be administered as a computer adaptive test or as a short form and is available as a self-report for children aged 8–21 and a report from a primary caregiver for children aged 4–21.
2. The Spinal Cord Injury Functional Index (SCI-FI) [152] was developed in adults with SCI at Boston University and University of Michigan and contains five scales: basic mobility (54 items), fine motor function (36 items), self-care (90 items), ambulation (39 items), and wheelchair mobility (56 items); it has subsequently been linked to PEDI-SCI AM scores [153].

Functional Independence Measure for Children

The Functional Independence Measure for Children (WeeFIM) was adapted from the adult FIM in 1987 to provide uniform assessment of rehabilitation outcomes, the severity of disability, and burden of care in children over 6 months. The FIM focuses on six areas of functioning (self-care, sphincter control, mobility, locomotion, communication, and social cognition) with each of the items evaluated on level of assistance (in terms of dependence on others and assistive devices) using a 7-point scale, where 7 denotes independence and 1 complete dependence [154].

Physical Ability and Mobility Scale

Physical Ability and Mobility Scale contain 20 questions created to measure different motor skills, including tolerance to positioning, tolerance to sitting in a chair, tolerance to orthoses or splint, support for seating system, head control, trunk control, rolling supine to/from prone, transitioning from supine to sit, transitioning from sit to stand, standing, transitioning from floor to stand, environmental transfers, transfers into and out of a car, walking on level ground—assistive device, walking on level ground—distance, walking on level ground—level of assistance, community skills, wheelchair mobility, standing balance, and stairs; it has been shown to be a useful measure capturing incremental and granular functional motor skills changes occurring during inpatient rehabilitation for children with spinal cord-related paralysis [155].

Prognosis and functional recovery

ISNCSCI exam done at 72 hours after injury is determined to be more accurate compared to 24-hour examinations, as postinjury swelling and spinal cord edema may increase between 24 and 72 hours [156]. Other studies have shown that the neurologic exam performed 1 week following injury had prognostic value for determining recovery [157]. Motor level, completeness of injury, and AIS grade help predict functional outcomes (Table 19.8) [158].

TABLE 19.8 Selected functional outcomes after spinal cord injury.

	C1–3	C4	C5	C6	C7–8	T1–9	T10–L1	L2–S5
Impairment	Total paralysis of trunk, upper and lower limbs; vent dependent	Paralysis of trunk, upper and lower limbs	No elbow extension, pronation, wrist, and hand movement; paralysis of trunk, lower limbs	No wrist flexion, elbow extension, hand movement; paralysis of trunk, lower limbs	Limited grasp release and dexterity; paralysis of trunk, lower limbs	Paralysis of lower trunk and lower extremities	Paralysis of lower extremities	Partial paralysis of lower extremities
Respiratory	Vent dependent Cannot clear secretions	May be able to breathe without vent	↓Endurance and vital capacity; needs assist to clear secretions			Compromised vital capacity and endurance	Intact function	Intact function
Bowel	Total assist	Total assist	Total assist	Some to total assist	Some to total assist	Independent	Independent	Independent
Bladder			Total assist	Some to total assist	Independent to some assist			
Eating			Total assist for set-up; independent with equipment	Independent with or without equipment	Independent			
Bathing			Total assist	Upper body: independent Lower body: some to total assist	Upper body: independent Lower body: some assist to independent			
Grooming			Some to total assist	Some assist to independent with equipment	Some assist to independent			
Dressing			LE: Total assist UE: Some assist	Upper body: independent Lower body: some to total assist	UE: independent LE: independent to some assist			
Transfers			Total assist	Level: some assist to independent Uneven: some to total assist	Level: independent Nonlevel: independent to some assist			
Wheelchair	Manual: total assist Power: independent with equipment	Manual: total assist Power: independent	Manual: independent to some assist indoor, some to total assist outdoors Power: independent	Manual: independent indoors, some to total assist outdoors Power: independent	Manual: independent indoor and level outdoor terrain			
Standing (S) and Ambulation (A)	S: Total assist A: Not indicated	S: Total assist A: Not indicated	Total assist	S: Total assist A: Not indicated	S: Independent to some assist A: Not indicated	S: Independent A: Not functional	S: Independent A: Functional, some assist to independent	S: Independent A: Functional, independent to some assist

TABLE 19.9 Life expectancy after spinal cord injury.

Life expectancy in years for persons with SCI who survive at least 24 h postinjury						
Age at injury (in years)	**Neurologic level and extent of lesion**					
	Vent dependent any level		**AIS ABC**		**AIS D any level**	**No SCI[a]**
		C1–C4	**C5–C8**	**T1–S3**		
10	16.8	41.3	48.4	54.0	61.7	69.3
15	12.8	36.6	43.7	49.2	56.8	64.3
20	10.5	32.6	39.4	44.8	52.1	59.5
Life expectancy in years for persons with SCI who survive at least 1 year postinjury						
Current age (in years)	**Neurologic level and extent of lesion**					
	Vent dependent any level		**AIS ABC**		**AIS D any level**	**No SCI[a]**
		c1–c4	c5–c8	**T1–S3**		
10	24.5	42.4	49.3	54.5	62.1	69.3
15	20.3	37.7	44.5	49.7	57.2	64.3
20	17.6	33.7	40.1	45.3	52.5	59.5

AIS, American Spinal Injury Association Impairment Scale; *SCI*, Spinal cord injury.
[a]*From 2018 life tables for US general population.*
Source: National Spinal Cord Injury Statistical Center, University of Alabama at Birmingham, 2021 Annual Statistical Report—Complete Public Version.

Neurologic recovery is most rapid during the first 3 months postinjury, with most of the recovery occurring during the first 6 months. A plateau is reached 9–12 months postinjury, though later recovery can occur. A total of 20%–30% of persons initially classified as having complete (AIS A) injury will convert to an incomplete status by 1 year. And 50% will become sensory incomplete (AIS B), while the other 50% will be motor incomplete (AIS C or D). Conversion after 1 year is less likely. A total of 68% of persons who were initially classified with AIS B injury will improve to motor incomplete status by 1 year. And 65%–85% of persons with AIS C will improve to AIS D/E. Majority of those with AIS D will remain AIS D at 1 year. Persons who sustained complete SCI due to penetrating spinal injuries are less likely to convert to incomplete SCI [159].

Life expectancy following SCI is inversely proportional to the severity of injury, with it being almost normal with an AIS D injury. Life expectancy decreases as age increases within a neurologic category (Table 19.9) [160].

Medical and rehabilitation management of chronic spinal cord injury

Management of secondary health conditions following spinal cord injury

Latex allergy

Latex allergy, first identified in up to 60% of children with SB, has now been documented, although to a lesser degree in 6%–8% of children with traumatic SCI. Children diagnosed with allergies should be educated on avoiding latex, obtaining a medical alert bracelet and providing prescriptions for auto-injectable epinephrine [161,162].

Spasticity

Spasticity occurs in approximately 50% of children with SCI, most often in those with incomplete injury [6]. While increased spasticity has many negative consequences such as decreased function, orthopedic complications, pain and negative self-image, for some children, increased muscle tone has positive effects on functions such as the improved ability to stand and transfer. As a result, spasticity management must be individually tailored and consider the patient, family, and rehabilitation team goals. Management begins with prevention of common precipitating factors such as urinary tract infection, constipation, and skin breakdown. Physical modalities including range of motion exercises, stretching, orthotics, standing, locomotor training, and functional electrical stimulation (FES) cycling are the starting point and least invasive methods of spasticity management. Focal hypertonicity can be managed with intermuscular botulinum toxin injections and phenol/alcohol neurolysis [163]. Systemic antispastic medication such as baclofen, tizanidine, clonidine, and diazepam can be added for generalized spasticity. The major adverse effect of systemic antispastic medication is increased sedation with the potential for negative effects on cognition, which can be minimized with a slow uptitration of the dose [164]. Implantation of an intrathecal baclofen pump has shown good outcomes for the treatment of generalized spasticity unresponsive to other treatments; however, it requires refills every 3–6 months and has the potential for serious adverse complications [165].

Heterotopic ossification

Heterotopic ossification in children with SCI is uncommon, reported in only 10% of pediatric patients. The average time of onset postinjury is 14 months, and the hip is most commonly involved. Decreased range of motion in the effected extremity was the most common presenting sign. Treatment with etidronate or postoperative use of radiation therapy is contraindicated in children prior to puberty [166].

Osteopenia/Osteoporosis

For children with paralysis, lack of weight-bearing activity, essential for optimal bone mass accrual during childhood and adolescence, places them at risk for lower peak bone mass, decreased bone strength, and increased risk for osteoporosis in adulthood [167–169]. In children with SCI, bone mineral density (BMD) values are substantially lower than age-matched peers at the hip (60%) and a similar trend was found at the distal femur (30%–40%) [169]. Pathologic long bone fractures are reported in 14% of children and youth with SCI and in children with limited ambulation up to 50% of fractures occur in the femur [170,171]. The goal for management of bone density in children with SCI is to optimize bone mass accrual by preventing bone loss at acute phase and to preserve and increase bone mass during the growing and chronic phase [172]. Bone densitometry can identify and quantify decreased bone mass as well as guide and monitor treatment response [168]. Primary prevention strategies include a balanced diet with sufficient intake of calcium and vitamin D as well as regular physical exercise [172]. In children with SCI, physical exercise including a combination of cycling, standing, stepping, and FES may increase BMD by approximately 25% [169]. Bisphosphonates are the main pharmacologic treatment for pediatric osteoporosis, although criteria for instituting their use have not been standardized [172].

Hip instability and deformities

Hip subluxation/dislocation is common in children with SCI with incidence between 29% and 82% and is strongly related to age at injury [173,174]. Approximately 48%–93% of children sustaining an SCI less than 10 years of age had at least one subluxed hip with only 3%–9% of children greater than 10 years of age having a subluxed hip [173,175]. Treatment of hip dislocation or subluxation in children who have spastic SCI generally requires controlling the spasticity first. Surgical correction involves performing soft tissue releases of any contractures and then bony procedures [173]. Surgical correction is advocated for children with spasticity and response to FES, as they are candidates for FES-assisted ambulation and could benefit from future technological advances [176].

Spine deformities

Spine deformities can occur because of muscle weakness, muscle imbalances (muscles on one side of the spine are stretched and overused while those on the other side become tighter and weaker from underuse), residual deformity or can occur following an earlier laminectomy. Neuromuscular scoliosis is very common in children sustaining an SCI prior to skeletal maturity with 98% developing scoliosis and 67% requiring surgical fixation. In contrast, for those acquiring SCIs after skeletal maturity, the risk of scoliosis is reduced to 20% with only 5% requiring surgical correction [177]. Conservative treatment includes observation, bracing, and seating modifications. Therapies including locomotor training may be prescribed for posture, sitting balance, and trunk strengthening. Spinal fusion is recommended for children over 10 years of age with curves greater than 40 degrees [178]. Bracing with thoracic lumbar sacral orthosis may delay the need for surgery and is recommended for skeletally immature children with curves between 20 and 45 degrees [179]. Children under 10 years of age can tolerate up to 80 degrees of curve if the curve is flexible and decreased while braced; otherwise, surgical placement of a growing spinal system is recommended [177,178]

Neurogenic bladder

Acute management of neurogenic bladder in the setting of spinal shock and areflexic bladder involves placement of a Foley catheter to allow careful monitoring of fluid balance. As spinal shock resolves, children with upper motor neuron injury develop a hyperreflexic spastic bladder and sphincter characterized by detrusor sphincter dyssynergia (DSD) and high bladder pressure resulting in low volume urination. For children with SB or lower motor neuron injury, the bladder generally remains hyporeflexic, characterized by a hypotonic bladder and normal or hypotonic sphincter, resulting in urinary retention and overflow incontinence. With transfer to acute rehabilitation, a timed intermittent catheterization program is introduced in the beginning at every 4 hours. The goals of intermittent catheterization are to effectively empty the bladder, decrease urinary tract infection and renal and bladder calculi, prevent renal injury and AD, and, ultimately, to promote independent bladder continence. Urologic evaluation including urodynamic studies to evaluate lower urinary tract functions should begin shortly after birth or 3 months after acute injury and will be necessary throughout life [180,181]. A renal/bladder ultrasound is completed at least annually to evaluate the upper urinary tract for vesical-ureteral reflux and renal or bladder calculi. In children with SB or very young children with SCI, intermittent catheterization is initiated when urodynamic studies show DSD, with intravesical pressures greater than 40 cm H_2O, the inability to empty the bladder fully, or by 3–4 years of age [180,182,183]. Children with adequate upper extremity function typically accomplish self-catheterization by 5–7 years of age. For children with hyperreflexic bladder, anticholinergics medications such as oxybutynin are the mainstay of medical treatment, although there are growing concerns regarding the impact of long-term use of these medications on the growing brain. Newer targeted muscarinic receptor subtypes medication, such as mirabegron, have been tested and are now available in doses suitable for children [184]. Intervesicular botulinum toxin injections, which inhibit acetylcholine neurotransmitter release at the neuromuscular junction, are being administered in children as young as 2 years old resulting in increased bladder capacity, decreased intravesical pressure, and improvement is continence and is an effective alternative for decreasing the use of anticholinergic medication [185]. For children who fail to achieve bladder continence with medical management or for those with limited hand function, surgical procedures may be indicated [186,187]. A suprapubic catheter is placed by inserting a catheter through the lower abdomen into the bladder and is a simple and completely reversible procedure. The catheter can be continuously drained via a leg bag or can be capped and emptied every 4 hours similar to catheterization. Suprapubic catheters have fallen out of favor with the advent of the appendicovesicostomy, a more complicated surgical procedure involving the use of the appendix, if available, or other small bowel tissue to create a stoma with a channel to the bladder through the umbilicus. Bladder augmentation, done by attaching a piece of small bowel to the bladder, may be done with this surgery or independently to increase bladder capacity [186,187]. Self-catheterization through the umbilical stoma is easier to accomplish for those with decreased arm and hand function and eliminates the need for an indwelling catheter.

Neurogenic bowel

In the setting of spinal shock with decreased gastric motility and areflexic bowel, oral medication to maintain a soft stool consistency and laxative to promote gastrointestinal motility are initiated once the child has begun oral or enteral feedings. With transfer to acute rehabilitation, a daily bowel program should be started. The goal of a bowel program is to prevent constipation and impaction, optimize stool consistency, and to establish predictable and effective bowel emptying [188]. As spinal shock resolves, children with upper motor neuron SCI will develop a hyperreflexic bowel characterized by increased colonic and anal sphincter tone often resulting in constipation and stool retention. The addition of a daily rectal stimulant medication and manual digital stimulation of the rectal sphincter will facilitate reflex evacuation of stool. Children with SB or lower motor neuron SCI will remain with an areflexic bowel characterized by loss of centrally mediated motor activity leading to slow bowel transit and atonic anal sphincter, resulting in constipation and the risk of fecal incontinence. The use of a daily lubricating suppository and 1–2 times per day digital stimulation and manual disimpaction will facilitate evacuation. For older children, the introduction of daily enemas administered by cone or a transanal irrigation system may be initiated to decrease constipation and improve continence [189]. For children unable to achieve bowel continence with medical management or for those with limited upper extremity function, a surgical Antegrade Continence Enema procedure may be recommended using the appendix or other small bowel tissue to create a continent stoma and channel between the abdomen and bowel. Following the procedure, an enema can be administered with a catheter inserted through the stoma on the abdomen every 1–2 days [190]. For those with intractable bowel complication, a colostomy stoma has been shown to improve quality of life, reduces time spent on bowel care, and increases independence [191]. Once able, upright sitting on the toilet with adaptive equipment if needed, allows for gravity and activity to stimulate Valsalva, facilitating evacuation with any type of bowel program. The age of independence in completing a bowel program is variable.

Syringomyelia

Syringomyelia may be associated with Chiari malformation, tumors, or trauma and is reported in up to 51% of pediatric SCI cases. In a retrospective review of 88 children SCI, Backe et al. found that the cyst was always at the level of injury with no patients presenting with ascending myelopathy: likely representing posttraumatic myelopathy [192]. The onset of symptoms is usually insidious in nature and may present as increased spasticity, a decrease in neurologic function, rapid progression of scoliotic curve, ascending sensory level, and increase or new onset of pain [192,193]. Surgical shunting can be combined with curve repair, if the latter is indicated [192].

Skin

Children with SCI/disease are at risk for development of pressure injuries and burns due to lack of protective sensation. Younger children are at risk as they can sustain trauma from play activities, increased time spent in prone, and limited understanding. The risk can increase in adolescents due to unwillingness to do skin checks and body image issues. A developmental approach must be used when initiating preventive measures and management of skin injuries. As the child grows, new equipment (i.e., wheelchair and seat/cushion) must be matched to their size. As the child matures, the responsibility for monitoring pressure relief, good hygiene, and proper nutrition should be gradually shifted from the parents to the child [30,194].

Pain

Pain can be nociceptive, neuropathic, or a combination of both. Neuropathic pain may be central pain, dysesthesia, or radicular in nature. It may negatively affect daily activities, school, work, mood, and social interactions. Developmental level and communication abilities may limit appropriate evaluation of pain in infants and young children. Treatment of pain may include physical modalities (transcutaneous nerve stimulation, physical therapy) and medications (for neuropathic pain may consider gabapentin, pregabalin, amitriptyline, duloxetine) [194].

Rehabilitation

Pediatric SCI rehabilitation addresses the prevention of secondary complications associated with SCI, maintenance of good health, and promotion of age-appropriate functional independence. It should incorporate developmental and age-related functional goals, taking into consideration the dynamic nature of growth and development with regard to the physical, physiologic, cognitive, and psychological changes that occur from infancy through childhood to adolescence and eventual transition to adulthood. This will involve a diverse rehabilitation team (physicians, nursing, physical/occupational/speech therapists, social workers, teachers, psychologists, etc.) working with the family.

Therapy interventions should focus on age-appropriate mobility, basic and instrumental activities of daily living, establishment of bowel, bladder, and skin programs. Play, leisure, and recreation should be incorporated in therapy. Social services, psychological, and vocational counseling should be included.

Equipment will change and may need frequent replacement with increased size and evolving needs. For example, an infant or toddler with paraplegia may crawl at home or use a parapodium for standing and a wheeled stroller for mobility. The child eventually transitions to independent wheelchair mobility and/or orthotic use for ambulation. School age children may need different orthotics for ambulation and standing. Wheelchairs will need to be replaced as the child grows. Adolescents may need access to motor vehicles for community mobility as they negotiate school, social life, and eventual transition to adulthood.

Promotion of functional recovery: activity-based restorative therapy

As activity is known to promote functional recovery after spinal cord traumatic and nontraumatic insults in both adults and pediatric individuals with spinal cord-related paralysis, rehabilitative interventions have shifted toward implementation of activity-based (restorative) therapies (ABRT) [90,145,195–197]. The five key therapeutic components of ABRT are (1) weight-loading activities, (2) FES, (3) task-specific practice, (4) massed practice, and (5) locomotor gait training, applied in the context of optimal neurostimulation above and below neurologic injury level and aiming to make neuroplastic changes in the damaged CNS and prevent medical comorbidities [195,196].

A typical rehabilitative prescription should involve all components coordinated into a supervised program: stretching, strengthening through traditional, electrically stimulated, gravity-eliminated, or bio-feedback methods, balance training, mobility and ADL training, gait training utilizing the overground method, but also weight-supported or FES-assisted techniques and aquatherapy facilitated gait training and endurance building, employing methods like FES-assisted bicycling, aquatherapy, and wheelchair sports. Such a program applied into the acute/subacute SCI rehabilitative phase accomplishes functional training and limits the onset of secondary complications related to the neurologic deficit (muscle atrophy, contractures, spasticity, bone loss, glucose, and lipid-related metabolic changes). When applied in the chronic phase, it partially restores muscle mass, reverses or minimizes bone and metabolic changes, and even improves neurologic function [90,145–148].

Upper extremity function

Functional use of the upper extremities is important for performing independent activities of daily living and for independence in the community. In a study of adults with tetraplegia, most patients (76%) and caregivers (64%) considered hand and upper extremity function as the most important function to be restored [198]. During acute rehabilitation, therapy is initiated to obtain/maintain joint mobility and maximize the strength, endurance, and the balance of voluntary muscles. Orthotics may also be used to maintain range of motion and improve function but are often rejected by children [199]. Depending on the extent of limb paralysis, some children may be candidates for surgical reconstruction to improve function. These procedures include tendon transfer surgery and nerve transfer surgery. Tendon transfer surgeries, involving rerouting of a functional tendon or muscle to a different muscle that needs restoration, have been the primary strategy for restoration. An advantage of tendon transfers is a wide window of time postinjury to consider and complete the procedure [200]. More recently, nerve transfer surgeries have shown success in restoring function, especially in children with AFM [125]. In this procedure, a motor branch innervating a healthy, functioning, and voluntarily controlled muscle is cut and coapted onto the motor branch of a paralyzed variable muscle [200]. Nerve transfer surgery is most successful when completed prior to 15 months post onset before irreversible distal nerve and motor endplate degeneration occurs [125]. Goals for surgical reconstruction may include shoulder abduction, elbow flexion or extension, wrist extension, hand grasp, and pinch. For children who are not candidates for surgery, there are orthotic FES systems available which can provide a stimulated pinch and grasp. It is possible that implantable FES technology available in the past could become available in the future.

Ambulation

Interventions for standing and gait/walking address achieving control of upright trunk posture (sitting), standing, and walking. In following normal development, a child learns to sit, then stand, then walk. Orthotics and assistive devices maybe utilized, depending on the child's age, size, neurologic level, compliance, and preference. Children who are young have an L3 lesion or below and AIS D injury are more likely to become community ambulators.

Parapodiums are used in early childhood (starting at 9–12 months of age). They permit the child to stand, have free use of both hands, and see their peers at eye level. By 7–10 years of age, children may abandon use of parapodium in favor of other orthotics or wheeled mobility. Children can ambulate using a walker or forearm crutches together with the parapodium. Reciprocating gait orthosis (RGOs) is used by children with lower thoracic or upper lumbar paraplegia. Knee-ankle-foot orthosis or hip-knee-ankle-foot orthosis may also be utilized. RGOs provide gait that is energy efficient. Use of RGOs can be started as early as 15–18 months. Other devices include standing wheelchairs, standing frames, or mobile standers. As the child grows into adolescence, this may be replaced by wheeled mobility (whether manual, power-assisted, or power wheelchairs).

Other interventions focused on facilitating neuroplasticity to promote recovery include locomotor training (body-weight-supported overground training or treadmill training) and FES. These can benefit those with incomplete SCI.

Education

Attending school is a primary task of childhood and adolescence. Educational achievement is an important factor in determining employability after SCI; a smooth re-entry into a school is a major goal for the family, child, and the rehabilitation team [201–203].

Returning to school is important as it allows the child to reestablish peer interaction and friendships and is an important step in normalizing child and family life. Children with SCI, even those requiring ventilator assistance, should return to a regular education classroom at their home school with assistance from school nurses, educational aids, or private duty nurses to accommodate their impaired mobility and health. Children with paraplegia will need little modification to access educational materials, while children with tetraplegia will need evaluation and training for assistive technology and alternative computer access, such as dictation software or an alternative mouse [204]. Children with a concurrent brain injury will require additional academic and neuropsychological testing to make recommendations for school placement. To assist in the development of the written Individualized Education Program (IEP) or 504 plan, the hospital team can submit written reports regarding the child's anticipated educational, therapeutic, and health care needs during the school day. Related educational services such as physical and occupational therapy services, the need for an educational aid or private duty nurse provided by the school, as well as the flexibility to attend additional outpatient therapy during the school day are also written into the formal educational plan. If distance permits, members of the hospital rehabilitation team can offer to visit the home school to evaluate accessibility and provide SCI education to school professionals, as well as age-appropriate classroom peer education aimed at increasing injury-specific knowledge and decreasing anxiety surrounding the return of the child to the classroom [201]. Reported barriers to school re-entry include the lack of physical and functional accessibility of the school, including bathrooms, water fountains, and sometimes classrooms, lack of training for school personnel, and difficulty in obtaining appropriate health services such as a private duty nurse or a 1:1 aid during the school day [201–203]. Some adolescents expressed psychosocial barriers including concerns about self-image, peer reactions, and teacher expectations and some reported loosing friend after their injury. Many students reported that their peers and teachers were helpful with their return to school. Some students limited their activities to avoid asking for help and to not receive unwanted leniency from teacher [202,205]. Overall, students believed returning to school was positive part of the rehabilitation process [206].

Recreation, fitness, and leisure

Because of their relatively long lifespan and decreased mobility, children with SCI and SB are at a greater risk of the secondary health conditions experienced by adults, including obesity, deficits in lean mass, cardiovascular disease, and metabolic syndrome [207]. Assisting children to develop a healthy lifestyle that promotes exercise and nutrition may help to mitigate these conditions. As play is a primary activity for children, fitness and recreational activities need to be fun, developmentally appropriate, consistent with preinjury interests, and integrated into family activities. During inpatient rehabilitation, participation in child life and recreational therapy programs are important to educate families in new ways to engage with their child in leisure and recreational activities. Participation in adapted physical education should be part of the child's IEP. At school age, as with their typical siblings, children with SCI should be encouraged to try a variety of fitness, recreation, and leisure activities. With advancements in assistive technology and a better awareness of disability, the variety of sports and leisure activities available for children with SCI have increased tremendously for both individual and team sport, leading to the development of international sporting events including the Paralympics [207]. Some adolescents with SCI may choose to participate on their public-school athletic teams, and there is federal legislation that prohibits discrimination based on disabilities and requires that public schools ensure that they have equal access to all school programs and activities [208]. Some colleges now have competitive wheelchair sports teams, offer athletic scholarships, and serve as a conduit to participation in Paralympic sports.

Psychiatric concerns and mental health

The physical sequela following SCI may have psychological implications. High cervical injuries may result in respiratory dysfunction requiring mechanical ventilation and limitations in the ability to communicate. Issues regarding breathing and vent problems (e.g., hypoxia events, ventilator adjustments, weaning attempts) may trigger anxiety and hypervigilance in the child and caregivers. This can be compounded by the child's inability to express distress and/or have volitional movement. A child may take some time to be used to seeing the tracheostomy site. Chronic distress and low self-worth may develop in some youth with SCI [209].

In adolescents who were independent prior to SCI, the inability to perform activities of daily living independently (e.g., showering, grooming, dressing) can be challenging. The loss of control and autonomy may contribute to depression or low self-worth. Neurogenic bladder and bowel dysfunction are associated with loss of volitional control and issues with continence (urinary and bowel) and constipation. Incontinence can impact one's self-worth and affect social engagement. It can lead to distress, which in turn can cause irritability, anxiety, avoidance and can progress to an anxiety or mood disorder. Regular bladder and bowel regimens can help ensure continence, decrease social embarrassment, and allow for participation in school and community.

Individuals with pediatric-onset SCI are at risk for sleep and cognitive dysfunction. Persistent pain, spasms, sleep-disordered breathing, incontinence, anxiety, and depression can affect sleep and result in insomnia and nighttime awakenings. Cognitive deficits can result from the injury or complications after the injury. This can affect the understanding of impairments and management of complications of the SCI, ability to gain independence in activities of daily living (ADLs), and future scholastic performance and career planning. The presence of concomitant traumatic brain injury (TBI) can also affect attention, memory, executive function, mood, and judgment.

Skeletal problems, such as scoliosis and hip dislocation/subluxation, can develop following SCI. Physically apparent deformities and/or gait abnormalities may negatively impact self-esteem and social engagement. Children may experience painful muscle spasms. Anticipation of pain and discomfort may lead to anticipatory anxiety.

During the acute phase of initial hospitalization, psychiatric management should include monitoring for delirium, adjustment to the diagnosis, screen for and address acute anxiety, mood changes, and depression, including psychological trauma and acute stress reactions. Early involvement of mental health professionals is crucial for monitoring and managing the child's psychological distress and to support and educate the family during this period. Patient and family understanding regarding the meaning of the diagnosis, and its functional implications, and prognosis is important and may require ongoing discussions. To limit polypharmacy, consider psychiatric consult to optimize medications addressing anxiety and distress.

There is limited literature regarding the risk of suicide in pediatric-onset SCI. During the acute phase, a thorough safety assessment should be done. During hospitalization and rehabilitation, patients may encounter new functional limitations that may lead to periods of frustrations and hopelessness. It is important to reassess safety frequently. Those with concomitant TBI may have issues with impulse control and judgment and may be at higher risk for suicide; however, this risk is lesser in TBI sustained as a child/adolescent. This may be related to children having more resources and support and limited access to dangerous means of self-harm.

Youth with pediatric SCI are generally resilient and adapt well to their injury. Levels of depression and anxiety are similar to their noninjured peers. These may fluctuate over time and with changes in physical functioning [210]. Frequent monitoring is recommended.

Caregivers are at risk for developing acute stress symptoms. They may have difficulty processing and retaining information. Regular multidisciplinary care meetings with the caregivers/guardians/pertinent family members should be done to discuss diagnostic findings, its implications on function, and provide proactive education. These meetings serve as a venue to identify concerns, formulate and review plans of care, and answer questions. Multidisciplinary team participants can include the primary medical team, nursing, other subspecialty services, mental health, physiatry and rehabilitative services (physical therapy, occupational therapy, speech), social work, pain or palliative care, and spiritual care. Caregivers should be given the tools and correct knowledge so they may advocate and problem-solve for their child effectively. Caregiver problem-solving was related to patients' physical and psychosocial health-related quality of life. Effective caregiver problem-solving leads to better managed health issues and lower incidence of secondary complications following SCI. The team should regularly check in on how the caregivers are coping, provide resources and mental health referrals (if needed) [209].

The involvement of a peer mentor can be an invaluable resource to patients and families. This can be instituted before hospital discharge.

Psychosocial issues and sexual health

"Sexual health" is defined by the World Health Organization as "a state of physical, emotional, mental and social well-being in relation to sexuality... [requiring] a positive and respectful approach to sexuality and sexual relationships, as well as the possibility of having pleasurable and safe sexual experiences, free of coercion, discrimination and violence..." [211].

The development of sexual health occurs in the same stages for all children, beginning in the first years of life, and is formed by a combination of biological, psychological, social, and cultural influences. Youth with SCI face additional psychosocial and physical challenges, including lower self-esteem and body image, as well as potential motor and sensory deficits impacting sexual functioning [212]. Youths with SCI will have sexual experiences as adults; so, helping these youths achieve the same sexual developmental milestones as their peers should be a primary goal [213]. However, many studies report the availability of sexual education for people with disabilities to be inadequate [214,215]. Information regarding disability and sex tends to focus on function or fertility, rather than sexual satisfaction and intimacy.

Anticipatory guidance regarding sexual development as well as sexuality should be provided to children with SCI and their caregivers routinely throughout their childhood [213]. Acquisition of accurate and age-appropriate knowledge helps a child develop a healthy sexual balance while limiting negative outcomes such as sexually transmitted infections, unintended pregnancy, or sexual exploitation [216]. Health care providers should provide information to caregivers both on typical sexual development for child's age, as well as additional issues related to the SCI that may impact that stage. In early childhood, counseling should be offered on normalization of sexual exploration and appropriate showing of body parts, as the child may be used to less privacy than other children, given increased medical needs. In middle to late childhood, emphasis should be on the importance of acknowledgment of the child's sexuality, creating an open forum for discussion about feelings on sexuality and body image. Throughout middle to late childhood, adolescence, and emerging adulthood, information should be offered about how SCI may impact a teenager's sexual functioning as well as potential management options, including ability to experience erections, masturbation, sexual intercourse, orgasms, fertility, and SCI-related pregnancy issues including potential for AD. Focus should include identity development separate from disability. Discussions on risks of abuse, on safe sex practices, and sexually transmitted diseases are essential. Health care providers should also interview adolescents without their parents present so that the youth can freely discuss potentially sensitive issues [212]. Proactive effort and open communication between parent, child, and provider at each stage of sexual development maximize self-confidence and the ability to have healthy, satisfying, and safe intimate relationships throughout the lifespan.

References

[1] Huisman TAGM, Wagner MW, Bosemani T, Tekes A, Poretti A. Pediatric spinal trauma. J Neuroimaging 2015;25(3):337–53. Available from: https://doi.org/10.1111/jon.12201. Accessed Jun 8, 2022.

[2] Hamilton MG, Myles ST. Pediatric spinal injury: review of 174 hospital admissions. J Neurosurg 1992;77(5):700–4. Available from: https://doi.org/10.3171/jns.1992.77.5.0700.

[3] Vitale MG, Goss JM, Matsumoto H, Roye DPJ. Epidemiology of pediatric spinal cord injury in the united states: years 1997 and 2000. J Pediatr Orthop 2006;26(6):745–9. Available from: https://doi.org/10.1097/01.bpo.0000235400.49536.83.

[4] Selvarajah S, Schneider EB, Becker D, Sadowsky CL, Haider AH, Hammond ER. The epidemiology of childhood and adolescent traumatic spinal cord injury in the United States: 2007–2010. J Neurotrauma 2014;31(18):1548–60. Available from: https://doi.org/10.1089/neu.2014.3332.

[5] Piatt JHJ. Pediatric spinal injury in the US: epidemiology and disparities. J Neurosurg Pediatr 2015;16(4):463–71. Available from: https://doi.org/10.3171/2015.2.PEDS1515.

[6] Vogel L, Mulcahy MJ, Betz RI. The child with a spinal cord injury. Dev Med Child Neurol 1997;39(3):202–7. Available from: https://doi.org/10.1111/j.1469-8749.1997.tb07411.x, https://onlinelibrary.wiley.com/doi/abs/10.1111/j.1469-8749.1997.tb07411.x.

[7] DeVivo MJ, Vogel LC. Epidemiology of spinal cord injury in children and adolescents. J Spinal Cord Med 2004;27(Suppl 1):4. Available from: https://doi.org/10.1080/10790268.2004.11753778, https://www.ncbi.nlm.nih.gov/pubmed/15503696.

[8] Vogel LC, Betz RR, Mulcahey MJ. Spinal cord injuries in children and adolescents. Handb Clin Neurol 2012;109:131–48. Available from: https://doi.org/10.1016/B978-0-444-52137-8.00008-5.

[9] New PW, Lee BB, Cripps R, Vogel LC, Scheinberg A, Waugh M. Global mapping for the epidemiology of paediatric spinal cord damage: towards a living data repository. Spinal Cord 2019;57(3):183–97. Available from: https://doi.org/10.1038/s41393-018-0209-5, https://www-ncbi-nlm-nih-gov.proxy1.library.jhu.edu/pubmed/30552413.

[10] Achildi O, Betz RR, Grewal H. Lapbelt injuries and the seatbelt syndrome in pediatric spinal cord injury. J Spinal Cord Med 2007;30(Suppl 1):21. Available from: https://doi.org/10.1080/10790268.2007.11753964.

[11] McCall T, Fassett D, Brockmeyer D. Cervical spine trauma in children: a review. Neurosurg Focus 2006;20(2):5. Available from: https://doi.org/10.3171/foc.2006.20.2.6, https://www.ncbi.nlm.nih.gov/pubmed/16512656.

[12] Gopinathan NR, Viswanathan VK, Crawford AH. Cervical spine evaluation in pediatric trauma: a review and an update of current concepts. Indian J Orthop 2018;52(5):489–500. Available from: https://doi.org/10.4103/ortho.IJOrtho_607_17.

[13] Copley PC, Tilliridou V, Kirby A, Jones J, Kandasamy J. Management of cervical spine trauma in children. Eur J Trauma Emerg Surg 2019;45(5):777–89. Available from: https://doi.org/10.1007/s00068-018-0992-x.

[14] Benmelouka A, Shamseldin LS, Nourelden AZ, Negida A. A review on the etiology and management of pediatric traumatic spinal cord injuries. Adv J Emerg Med 2020;4(2):e28. Available from: https://doi.org/10.22114/ajem.v0i0.256.

[15] Oyinbo CA. Secondary injury mechanisms in traumatic spinal cord injury: a nugget of this multiply cascade. Acta Neurobiol Exp 2011;71(2):281–99.

[16] Lemley K, Bauer P. Pediatric spinal cord injury: recognition of injury and initial resuscitation, in hospital management, and coordination of care. J Pediatr Intensive Care 2015;4(1):27–34. Available from: https://doi.org/10.1055/s-0035-1554986.

[17] Srinivasan V, Jea A. Pediatric thoracolumbar spine trauma. Neurosurg Clin N Am 2017;28(1):103–14. Available from: https://doi.org/10.1016/j.nec.2016.07.003.

[18] Pang D. Spinal cord injury without radiographic abnormality in children, 2 decades later. Neurosurgery. 2004;55(6):1325. Available from: https://doi.org/10.1227/01.neu.0000143030.85589.e6.

[19] Zidek K, Srinivasan R. Rehabilitation of a child with a spinal cord injury. Semin Pediatr Neurol 2003;10(2):140–50. Available from: https://doi.org/10.1016/s1071-9091(03)00022-6.

[20] Ostrom QT, Patil N, Cioffi G, Waite K, Kruchko C, Barnholtz-Sloan J. CBTRUS statistical report: primary brain and other central nervous system tumors diagnosed in the united states in 2013-2017. Neuro Oncol 2020;22(12):iv1–iv96. Available from: https://doi.org/10.1093/neuonc/noaa200.

[21] Knox J. Epidemiology of spinal cord injury without radiographic abnormality in children: a nationwide perspective. J Child Orthop 2016;10 (3):255–60. Available from: https://doi.org/10.1007/s11832-016-0740-x.

[22] Eberhardt CS, Zand T, Ceroni D, Wildhaber BE, La Scala G. The seatbelt syndrome – do we have a chance? A report of 3 cases with review of literature. Pediatr Emerg Care 2016;32(5):318–22. Available from: https://doi.org/10.1097/PEC.0000000000000527.

[23] Tekes A, Pinto PS, Huisman TAGM. Birth-related injury to the head and cervical spine in neonates. Magn Reson Imaging Clin N Am 2011;19 (4):777–90. Available from: https://doi.org/10.1016/j.mric.2011.08.004.

[24] Ditunno JF, Little JW, Tessler A, Burns AS. Spinal shock revisited: a four-phase model. Spinal Cord 2004;42(7):383–95. Available from: https://doi.org/10.1038/sj.sc.3101603.

[25] Consortium for Spinal Cord Medicine. Early acute management in adults with spinal cord injury: a clinical practice guideline for health-care professionals. *J Spinal Cord Med* 2008;31(4):403–79. doi:10.1043/1079-0268-31.4.408.

[26] Berlowitz DJ, Wadsworth B, Ross J. Respiratory problems and management in people with spinal cord injury. Breathe. 2016;12(4):328–40. Available from: https://doi.org/10.1183/20734735.012616.

[27] Consortium for Spinal Cord Medicine. Respiratory management following spinal cord injury: a clinical practice guideline for health-care professionals. *J Spinal Cord Med* 2005;28(3):259–93. doi:10.1080/10790268.2005.11753821.

[28] Thomas AX, Riviello JJ, Davila-Williams D, et al. Pharmacologic and acute management of spinal cord injury in adults and children. Curr Treat Options Neurol 2022;24(7):285–304. Available from: https://doi.org/10.1007/s11940-022-00720-9.

[29] Consortium for Spinal Cord Medicine. Prevention of venous thromboembolism in individuals with spinal cord injury: clinical practice guidelines for health care providers, 3rd ed. *Top Spinal Cord Inj Rehabil* 2016;22(3):209–40. Available from: https://doi.org/10.1310/sci2203-209.

[30] Powell A, Davidson L. Pediatric spinal cord injury: a review by organ system. Phys Med Rehabil Clin N Am 2015;26(1):109–32. Available from: https://doi.org/10.1016/j.pmr.2014.09.002.

[31] Wang J, Yang M, Meng M, Li Z. Clinical characteristics and treatment of spinal cord injury in children and adolescents. Chin J Traumatol 2023;26(1):8–13. Available from: https://doi.org/10.1016/j.cjtee.2022.04.007.

[32] Consortium for Spinal Cord Medicine. Acute management of autonomic dysreflexia: individuals with spinal cord injury presenting to health-care facilities. *J Spinal Cord Med* 2002;25(Suppl 1):67.

[33] Krassioukov A, Linsenmeyer TA, Beck LA, et al. Evaluation and management of autonomic dysreflexia and other autonomic dysfunctions: preventing the highs and lows: management of blood pressure, sweating, and temperature dysfunction. Top Spinal Cord Inj Rehabil 2021;27 (2):225–90. Available from: https://doi.org/10.46292/sci2702-225.

[34] McGinnis KB, Vogel LC, McDonald CM, et al. Recognition and management of autonomic dysreflexia in pediatric spinal cord injury. J Spinal Cord Med 2004;27(Suppl 1):S61–74. Available from: https://doi.org/10.1080/10790268.2004.11753787.

[35] Hwang M, Zebracki K, Betz RR, Mulcahey MJ, Vogel LC. Normative blood pressure and heart rate in pediatric spinal cord injury. Top Spinal Cord Inj Rehabil 2013;19(2):87–95. Available from: https://doi.org/10.1310/sci1902-87.

[36] Valbuena Valecillos A,D, Gater DR, Alvarez G. Concomitant brain injury and spinal cord injury management strategies: a narrative review. J Pers Med 2022;12(7). Available from: https://doi.org/10.3390/jpm12071108.

[37] Hurlbert RJ, Hadley MN, Walters BC, et al. Pharmacological therapy for acute spinal cord injury. Neurosurgery 2013;72(Suppl 2):93–105. Available from: https://doi.org/10.1227/NEU.0b013e31827765c6.

[38] Galvin J, Scheinberg A, New PW. A retrospective case series of pediatric spinal cord injury and disease in Victoria, Australia. Spine. 2013;38 (14):E878–82. Available from: https://doi.org/10.1097/BRS.0b013e318294e839.

[39] Rossi A, Gandolfo C, Morana G, Tortori-Donati P. Tumors of the spine in children. Neuroimaging Clin N Am 2007;17(1):17–35. Available from: https://doi.org/10.1016/j.nic.2006.11.004.

[40] Tortori-Donati P., Rossi A., Biancheri R., Garrè M.L., Cama A. Tumors of the spine and spinal cord. Berlin, Heidelberg: Springer p. 1609–51. 10.1007/3-540-26398-5_40.

[41] Jallo GI, Freed D, Epstein F. Intramedullary spinal cord tumors in children. Child Nerv Syst 2003;19(9):641–9. Available from: https://doi.org/10.1007/s00381-003-0820-3.

[42] Constantini S, Houten J, Miller DC, et al. Intramedullary spinal cord tumors in children under the age of 3 years. J Neurosurg 1996;85 (6):1036–43. Available from: https://doi.org/10.3171/jns.1996.85.6.1036.

[43] Huisman TAGM. Pediatric tumors of the spine. Cancer Imaging 2009;9(Spec No A):S45–8. Available from: https://doi.org/10.1102/1470-7330.2009.9012.

[44] Zaganjor I, Sekkarie A, Tsang BL, et al. Describing the prevalence of neural tube defects worldwide: a systematic literature review. PLoS One 2016;11(4):e0151586. Available from: https://doi.org/10.1371/journal.pone.0151586.

[45] Kancherla V, Carmichael SL, Feldkamp ML, Berry RJ. Teratology society position statement on surveillance and prevalence estimation of neural tube defects. Birth Defects Res 2019;111(1):5–8. Available from: https://doi.org/10.1002/bdr2.1434.

[46] Mai CT, Isenburg JL, Canfield MA, et al. National population-based estimates for major birth defects, 2010–2014. Birth Defects Res 2019;111 (18):1420–35. Available from: https://doi.org/10.1002/bdr2.1589.

[47] Honein MA, Paulozzi LJ, Mathews TJ, Erickson JD, Wong LY. Impact of folic acid fortification of the US food supply on the occurrence of neural tube defects. JAMA 2001;285(23):2981–6. Available from: https://doi.org/10.1001/jama.285.23.2981.

[48] Cornel MC, de Smit DJ, de Jong-van den Berg LTW. Folic acid – the scientific debate as a base for public health policy. Reprod Toxicol 2005;20(3):411–15. Available from: https://doi.org/10.1016/j.reprotox.2005.03.015.

[49] Detrait ER, George TM, Etchevers HC, Gilbert JR, Vekemans M, Speer MC. Human neural tube defects: developmental biology, epidemiology, and genetics. Neurotoxicol Teratol 2005;27(3):515–24. Available from: https://doi.org/10.1016/j.ntt.2004.12.007.

[50] Flick A, Krakow D, Martirosian A, Silverman N, Platt LD. Routine measurement of amniotic fluid alpha-fetoprotein and acetylcholinesterase: the need for a reevaluation. Am J Obstet Gynecol 2014;211(2):139.e1–6. Available from: https://doi.org/10.1016/j.ajog.2014.02.005.

[51] Chao TT, Dashe JS, Adams RC, Keefover-Hicks A, McIntire DD, Twickler DM. Fetal spine findings on MRI and associated outcomes in children with open neural tube defects. Am J Roentgenol 2011;197(5):W956–61. Available from: https://doi.org/10.2214/AJR.11.6745.

[52] Dias LS, Swaroop VT, de Angeli L,RA, Larson JE, Rojas A, Karakostas T. Myelomeningocele: a new functional classification. J Child Orthop 2021;15(1):1–5. Available from: https://doi.org/10.1302/1863-2548.15.200248.

[53] Blount JP, Bowman R, Dias MS, Hopson B, Partington MD, Rocque BG. Neurosurgery guidelines for the care of people with spina bifida. J Pediatr Rehabil Med 2020;13(4):467–77. Available from: https://doi.org/10.3233/PRM-200782.

[54] Shoja MM, Johal J, Oakes WJ, Tubbs RS. Embryology and pathophysiology of the Chiari I and II malformations: a comprehensive review. Clin Anat 2018;31(2):202–15. Available from: https://doi.org/10.1002/ca.22939.

[55] Protzenko T, Bellas A, Pousa MS, et al. Reviewing the prognostic factors in myelomeningocele. Neurosurg Focus 2019;47(4):E2. Available from: https://doi.org/10.3171/2019.7.FOCUS19462.

[56] Bradko V, Castillo H, Janardhan S, Dahl B, Gandy K, Castillo J. Towards guideline-based management of tethered cord syndrome in spina bifida: a global health paradigm shift in the era of prenatal surgery. Neurospine. 2019;16(4):715–27. Available from: https://doi.org/10.14245/ns.1836342.171.

[57] Mosiello G, Safder S, Marshall D, Rolle U, Benninga MA. Neurogenic bowel dysfunction in children and adolescents. J Clin Med 2021;10(8). Available from: https://doi.org/10.3390/jcm10081669.

[58] Beierwaltes P, Church P, Gordon T, Ambartsumyan L. Bowel function and care: guidelines for the care of people with spina bifida. J Pediatr Rehabil Med 2020;13(4):491–8. Available from: https://doi.org/10.3233/PRM-200724.

[59] Joseph DB, Baum MA, Tanaka ST, et al. Urologic guidelines for the care and management of people with spina bifida. J Pediatr Rehabil Med 2020;13(4):479–89. Available from: https://doi.org/10.3233/PRM-200712.

[60] Dennis M, Barnes MA. The cognitive phenotype of spina bifida meningomyelocele. Dev Disabil Res Rev 2010;16(1):31–9. Available from: https://doi.org/10.1002/ddrr.89.

[61] Holck P, Nettelbladt U, Sandberg AD. Children with cerebral palsy, spina bifida and pragmatic language impairment: differences and similarities in pragmatic ability. Res Dev Disabil 2009;30(5):942–51. Available from: https://doi.org/10.1016/j.ridd.2009.01.008.

[62] Vachha B, Adams R. Language differences in young children with myelomeningocele and shunted hydrocephalus. Pediatr Neurosurg 2003;39(4):184–9. Available from: https://doi.org/10.1159/000072469.

[63] Hurley AD, Dorman C, Laatsch L, Bell S, D'Avignon J. Cognitive functioning in patients with spina bifida, hydrocephalus, and the "cocktail party" syndrome. Dev Neuropsychol 1990;6(2):151. Available from: https://doi.org/10.1080/87565649009540456.

[64] Burmeister R, Hannay HJ, Copeland K, Fletcher JM, Boudousquie A, Dennis M. Attention problems and executive functions in children with spina bifida and hydrocephalus. Child Neuropsychol 2005;11(3):265–83. Available from: https://doi.org/10.1080/092970490911324.

[65] Queally JT, Barnes MA, Castillo H, Castillo J, Fletcher JM. Neuropsychological care guidelines for people with spina bifida. J Pediatr Rehabil Med 2020;13(4):663–73. Available from: https://doi.org/10.3233/PRM-200761.

[66] Adzick NS, Thom EA, Spong CY, et al. A randomized trial of prenatal versus postnatal repair of myelomeningocele. N Engl J Med 2011;364(11):993–1004. Available from: https://doi.org/10.1056/NEJMoa1014379.

[67] Joyeux L, Danzer E, Flake AW, Deprest J. Fetal surgery for spina bifida aperta. Arch Dis Child 2018;103(6):F589–95. Available from: https://doi.org/10.1136/archdischild-2018-315143.

[68] Belfort MA, Whitehead WE, Shamshirsaz AA, et al. Fetoscopic open neural tube defect repair: development and refinement of a two-port, carbon dioxide insufflation technique. Obstet Gynecol 2017;129(4):734–43. Available from: https://doi.org/10.1097/AOG.0000000000001941.

[69] Kohn JR, Rao V, Sellner AA, et al. Management of labor and delivery after fetoscopic repair of an open neural tube defect. Obstet Gynecol 2018;131(6):1062–8. Available from: https://doi.org/10.1097/AOG.0000000000002577.

[70] Miller JL, Groves ML, Ahn ES, et al. Implementation process and evolution of a laparotomy-assisted 2-port fetoscopic spina bifida closure program. Fetal Diagn Ther 2021;48(8):603–10. Available from: https://doi.org/10.1159/000518507.

[71] Verweij EJ, de Vries M,C, Oldekamp EJ, et al. Fetoscopic myelomeningocoele closure: is the scientific evidence enough to challenge the gold standard for prenatal surgery? Prenat Diagn 2021;41(8):949–56. Available from: https://doi.org/10.1002/pd.5940.

[72] Landry SH, Smith KE, Swank PR, Guttentag C. A responsive parenting intervention: the optimal timing across early childhood for impacting maternal behaviors and child outcomes. Dev Psychol 2008;44(5):1335–53. Available from: https://doi.org/10.1037/a0013030.

[73] Berman M, Feldman S, Alter M, Zilber N, Kahana E. Acute transverse myelitis: incidence and etiologic considerations. Neurology. 1981;31(8):966–71. Available from: https://doi.org/10.1212/wnl.31.8.966.

[74] Absoud M, Greenberg BM, Lim M, Lotze T, Thomas T, Deiva K. Pediatric transverse myelitis. Neurology. 2016;87(9):S46–52. Available from: https://doi.org/10.1212/WNL.0000000000002820.

[75] Choi KH, Lee KS, Chung SO, et al. Idiopathic transverse myelitis: MR characteristics. Am J Neuroradiol 1996;17(6):1151–60.

[76] Transverse Myelitis Consortium Working Group. Proposed diagnostic criteria and nosology of acute transverse myelitis. *Neurology*. 2002;59(4):499–505. Available from: https://doi.org/10.1212/wnl.59.4.499.

[77] Lu VM, Niazi TN. Pediatric spinal cord diseases. Pediatr Rev 2021;42(9):486–99. Available from: https://doi.org/10.1542/pir.2020-000661.

[78] Griessenauer CJ, Raborn J, Foreman P, Shoja MM, Loukas M, Tubbs RS. Venous drainage of the spine and spinal cord: a comprehensive review of its history, embryology, anatomy, physiology, and pathology. Clin Anat 2015;28(1):75–87. Available from: https://doi.org/10.1002/ca.22354.

[79] Celik H, Aksoy E, Oztoprak U, et al. Longitudinally extensive transverse myelitis in childhood: clinical features, treatment approaches, and long-term neurological outcomes. Clin Neurol Neurosurg 2021;207:106764. Available from: https://doi.org/10.1016/j.clineuro.2021.106764.

[80] Barnett Y, Sutton IJ, Ghadiri M, Masters L, Zivadinov R, Barnett MH. Conventional and advanced imaging in neuromyelitis optica. Am J Neuroradiol 2014;35(8):1458–66. Available from: https://doi.org/10.3174/ajnr.A3592.

[81] Jarius S, Paul F, Weinshenker BG, Levy M, Kim HJ, Wildemann B. Neuromyelitis optica. Nat Rev Dis Prim 2020;6(1):85. Available from: https://doi.org/10.1038/s41572-020-0214-9.

[82] Huh Y, Park E, Jung J, Oh S, Choi S. Clinical insights for early detection of acute transverse myelitis in the emergency department. Clin Exp Emerg Med 2015;2(1):44–50. Available from: https://doi.org/10.15441/ceem.14.034.

[83] Cordonnier C, de Seze J, Breteau G, et al. Prospective study of patients presenting with acute partial transverse myelopathy. J Neurol 2003;250(12):1447–52. Available from: https://doi.org/10.1007/s00415-003-0242-x.

[84] De Seze J, Stojkovic T, Destee A, et al. Acute myelopathies: clinical, laboratory and outcome profiles in 79 cases. Brain. 2001;124:1509–21. Available from: https://doi.org/10.1093/brain/124.8.1509.

[85] Greenberg BM, Thomas KP, Krishnan C, Kaplin AI, Calabresi PA, Kerr DA. Idiopathic transverse myelitis: corticosteroids, plasma exchange, or cyclophosphamide. Neurology. 2007;68(19):1614–17. Available from: https://doi.org/10.1212/01.wnl.0000260970.63493.c8.

[86] Elsone L, Panicker J, Mutch K, Boggild M, Appleton R, Jacob A. Role of intravenous immunoglobulin in the treatment of acute relapses of neuromyelitis optica: experience in 10 patients. Mult Scler 2014;20(4):501–4. Available from: https://doi.org/10.1177/1352458513495938.

[87] Sadowsky CL, Becker D, Bosques G, et al. Rehabilitation in transverse myelitis. Continuum 2011;17(4):816–30. Available from: https://doi.org/10.1212/01.CON.0000403797.10612.d3.

[88] Defresne P, Hollenberg H, Husson B, et al. Acute transverse myelitis in children: clinical course and prognostic factors. J Child Neurol 2003;18(6):401–6. Available from: https://doi.org/10.1177/08830738030180060601.

[89] Simone C, Emmady P. Transverse myelitis. StatPearls 2022.

[90] Hammond ER, Recio AC, Sadowsky CL, Becker D. Functional electrical stimulation as a component of activity-based restorative therapy may preserve function in persons with multiple sclerosis. J Spinal Cord Med 2015;38(1):68–75. Available from: https://doi.org/10.1179/2045772314Y.0000000238.

[91] Christensen PB, Wermuth L, Hinge HH, Bømers K. Clinical course and long-term prognosis of acute transverse myelopathy. Acta Neurol Scand 1990;81(5):431–5. Available from: https://doi.org/10.1111/j.1600-0404.1990.tb00990.x.

[92] Chitnis T, Glanz B, Jaffin S, Healy B. Demographics of pediatric-onset multiple sclerosis in an MS center population from the northeastern United States. Mult Scler 2009;15(5):627–31. Available from: https://doi.org/10.1177/1352458508101933.

[93] Harding KE, Liang K, Cossburn MD, et al. Long-term outcome of paediatric-onset multiple sclerosis: a population-based study. J Neurol Neurosurg Psychiatry 2013;84(2):141–7. Available from: https://doi.org/10.1136/jnnp-2012-303996.

[94] Boiko A, Vorobeychik G, Paty D, Devonshire V, Sadovnick D. Early onset multiple sclerosis: a longitudinal study. Neurology. 2002;59(7):1006–10. Available from: https://doi.org/10.1212/wnl.59.7.1006.

[95] Sheremata W, Tornes L. Multiple sclerosis and the spinal cord. Neurol Clin 2013;31(1):55–77. Available from: https://doi.org/10.1016/j.ncl.2012.09.007.

[96] Ghasemi N, Razavi S, Nikzad E. Multiple sclerosis: pathogenesis, symptoms, diagnoses and cell-based therapy. Nashrīyah-i pizishkī-i yākhtah 2017;19(1):1–10. Available from: https://doi.org/10.22074/cellj.2016.4867.

[97] Benson LA, Healy BC, Gorman MP, et al. Elevated relapse rates in pediatric compared to adult MS persist for at least 6 years. Mult Scler Relat Disord 2014;3(2):186–93. Available from: https://doi.org/10.1016/j.msard.2013.06.004.

[98] Yeh EA, Weinstock-Guttman B, Ramanathan M, et al. Magnetic resonance imaging characteristics of children and adults with paediatric-onset multiple sclerosis. Brain. 2009;132:3392–400. Available from: https://doi.org/10.1093/brain/awp278.

[99] Renoux C, Vukusic S, Mikaeloff Y, et al. Natural history of multiple sclerosis with childhood onset. N Engl J Med 2007;356(25):2603–13. Available from: https://doi.org/10.1056/NEJMoa067597.

[100] Gontika M, Skarlis C, Markoglou N, et al. Fingolimod as a first- or second-line treatment in a mini-series of young hellenic patients with adolescent-onset multiple sclerosis: focus on immunological data. Neurol Sci 2022;43(4):2641–9. Available from: https://doi.org/10.1007/s10072-021-05623-2.

[101] Patel M. National overview of acute flaccid myelitis – United States, 2014–2018. Centers for Disease Control and Prevention Website. https://www.cdc.gov/ddid/bsc/afm-overview-2018.html. Updated 2018. [accessed September, 2021].

[102] Ayscue P, Van Haren K, Sheriff H, et al. Acute flaccid paralysis with anterior myelitis – California, June 2012–June 2014. Morb Mortal Wkly Rep 2014;63(40):903–6.

[103] Van Haren K, Ayscue P, Waubant E, et al. Acute flaccid myelitis of unknown etiology in california, 2012–2015. JAMA 2015;314(24):2663–71. Available from: https://doi.org/10.1001/jama.2015.17275.

[104] Messacar K, Robinson CC, Pretty K, Yuan J, Dominguez SR. Surveillance for enterovirus D68 in colorado children reveals continued circulation. J Clin Virol 2017;92:39–41. Available from: https://doi.org/10.1016/j.jcv.2017.05.009.

[105] Knoester M, Helfferich J, Poelman R, Van Leer-Buter C, Brouwer OF, Niesters HGM. Twenty-nine cases of enterovirus-D68-associated acute flaccid myelitis in europe 2016: a case series and epidemiologic overview. Pediatr Infect Dis J 2019;38(1):16–21. Available from: https://doi.org/10.1097/INF.0000000000002188.

[106] McLaren N, Lopez A, Kidd S, et al. Characteristics of patients with acute flaccid myelitis, United States, 2015–2018. Emerg Infect Dis 2020;26(2):212–19. Available from: https://doi.org/10.3201/eid2602.191453.

[107] AFM cases and outbreaks. Centers for Disease Control and Prevention Website. https://www.cdc.gov/acute-flaccid-myelitis/cases-in-us.html. Updated 2023. [accessed September, 2021].

[108] Murphy OC, Messacar K, Benson L, et al. Acute flaccid myelitis: cause, diagnosis, and management. Lancet 2021;397(10271):334–46. Available from: https://doi.org/10.1016/S0140-6736(20)32723-9.

[109] AFM case definitions. https://www.cdc.gov/acute-flaccid-myelitis/hcp/case-definitions.html. Updated 2023. [accessed September, 2023].

[110] Lopez A, Lee A, Guo A, et al. Vital signs: surveillance for acute flaccid myelitis – United States, 2018. Morb Mortal Wkly Rep 2019;68 (27):608–14. Available from: https://doi.org/10.15585/mmwr.mm6827e1.

[111] Gordon-Lipkin E, Muñoz LS, Klein JL, Dean J, Izbudak I, Pardo CA. Comparative quantitative clinical, neuroimaging, and functional profiles in children with acute flaccid myelitis at acute and convalescent stages of disease. Dev Med Child Neurol 2019;61(3):366–75. Available from: https://doi.org/10.1111/dmcn.14030.

[112] Hopkins SE. Acute flaccid myelitis: etiologic challenges, diagnostic and management considerations. Curr Treat Options Neurol 2017;19 (12):48. Available from: https://doi.org/10.1007/s11940-017-0480-3.

[113] Melicosta ME, Dean J, Hagen K, et al. Acute flaccid myelitis: rehabilitation challenges and outcomes in a pediatric cohort. J Pediatr Rehabil Med 2019;12(3):245–53. Available from: https://doi.org/10.3233/PRM-180549.

[114] Hardy D, Hopkins S. Update on acute flaccid myelitis: recognition, reporting, aetiology and outcomes. Arch Dis Child 2020;105(9):842–7. Available from: https://doi.org/10.1136/archdischild-2019-316817.

[115] Ide W, Melicosta M, Trovato MK. Acute flaccid myelitis. Phys Med Rehabil Clin North Am 2021;32(3):477–91. Available from: https://doi.org/10.1016/j.pmr.2021.02.004.

[116] Zhang Y, Moore DD, Nix WA, Oberste MS, Weldon WC. Neutralization of enterovirus D68 isolated from the 2014 US outbreak by commercial intravenous immune globulin products. J Clin Virol 2015;69:172–5. Available from: https://doi.org/10.1016/j.jcv.2015.06.086.

[117] Greenberg B, Plumb P, Cutter G, et al. Acute flaccid myelitis: long-term outcomes recorded in the CAPTURE study compared with paediatric transverse myelitis. BMJ Neurol Open 2021;3(1):e000127. Available from: https://doi.org/10.1136/bmjno-2021-000127.

[118] Messacar K, Sillau S, Hopkins SE, et al. Safety, tolerability, and efficacy of fluoxetine as an antiviral for acute flaccid myelitis. Neurology. 2019;92(18):e2118–26. Available from: https://doi.org/10.1212/WNL.0000000000006670.

[119] Tyler KL. Rationale for the evaluation of fluoxetine in the treatment of enterovirus D68-associated acute flaccid myelitis. JAMA Neurol 2015;72(5):493–4. Available from: https://doi.org/10.1001/jamaneurol.2014.4625.

[120] Hagen K, Porter C, Martin R, Dean J, Salorio C, Sadowsky C. Improvements in function following inpatient activity-based therapy for children with acute flaccid myelitis. Top Spinal Cord Inj Rehabil 2020;26(4):275–82. Available from: https://doi.org/10.46292/sci20-00010.

[121] Weber EL, Werner JM, Johnson MB, et al. Characteristics of upper extremity recovery in acute flaccid myelitis: a case series. Plast Reconstr Surg 2021;147(3):645–55. Available from: https://doi.org/10.1097/PRS.0000000000007583.

[122] Chong PF, Kira R, Mori H, et al. Clinical features of acute flaccid myelitis temporally associated with an enterovirus D68 outbreak: results of a nationwide survey of acute flaccid paralysis in japan, august-december 2015. Clin Infect Dis 2018;66(5):653–64. Available from: https://doi.org/10.1093/cid/cix860.

[123] Bove R, Rowles W, Carleton M, et al. Unmet needs in the evaluation, treatment, and recovery for 167 children affected by acute flaccid myelitis reported by parents through social media. Pediatr Neurol 2020;102:20–7. Available from: https://doi.org/10.1016/j.pediatrneurol.2019.08.009.

[124] Martin JA, Messacar K, Yang ML, et al. Outcomes of colorado children with acute flaccid myelitis at 1 year. Neurology. 2017;89(2):129–37. Available from: https://doi.org/10.1212/WNL.0000000000004081.

[125] Pino PA, Intravia J, Kozin SH, Zlotolow DA. Early results of nerve transfers for restoring function in severe cases of acute flaccid myelitis. Ann Neurol 2019;86(4):607–15. Available from: https://doi.org/10.1002/ana.25558.

[126] Doi K, Sem SH, Hattori Y, Sakamoto S, Hayashi K, De la Red-Gallego MA. Surgical reconstruction for upper-extremity paralysis following acute flaccid myelitis. JB & JS Open Access 2019;4(4):e0030. Available from: https://doi.org/10.2106/JBJS.OA.19.00030.

[127] Moore AM, Bettlach CR, Tung TT, West JM, Russo SA. Lower extremity nerve transfers in acute flaccid myelitis patients: a case series. Plast Reconstr Surg 2021;9(7):e3699. Available from: https://doi.org/10.1097/GOX.0000000000003699.

[128] Saltzman EB, Rancy SK, Sneag DB, Feinberg Md J,H, Lange DJ, Wolfe SW. Nerve transfers for enterovirus D68-associated acute flaccid myelitis: a case series. Pediatr Neurol 2018;88:25–30. Available from: https://doi.org/10.1016/j.pediatrneurol.2018.07.018.

[129] Paziuk TM, Tadley M, Aversano M, Kozin SH, Zlotolow DA. The utilization of nerve transfer for reestablishing shoulder function in the setting of acute flaccid myelitis: a single-institution review. Pediatr Neurol 2020;111:17–22. Available from: https://doi.org/10.1016/j.pediatrneurol.2020.06.016.

[130] Curley N, Yang Y, Dean J, Salorio C, Sadowsky C. Description of bone health changes in a cohort of children with acute flaccid myelitis (AFM). Top Spinal Cord Inj Rehabil 2022;28(1):42–52. Available from: https://doi.org/10.46292/sci21-00035.

[131] Yea C, Bitnun A, Robinson J, et al. Longitudinal outcomes in the 2014 acute flaccid paralysis cluster in canada. J Child Neurol 2017;32 (3):301–7. Available from: https://doi.org/10.1177/0883073816680770.

[132] Sandson TA, Friedman JH. Spinal cord infarction. Report of 8 cases and review of the literature. Medicine. 1989;68(5):282–92.

[133] AbdelRazek MA, Mowla A, Farooq S, Silvestri N, Sawyer R, Wolfe G. Fibrocartilaginous embolism: a comprehensive review of an understudied cause of spinal cord infarction and proposed diagnostic criteria. J Spinal Cord Med 2016;39(2):146–54. Available from: https://doi.org/10.1080/10790268.2015.1116726.

[134] AbdelRazek M, Elsadek R, Elsadek L. Case series of two patients with fibrocartilaginous embolism mimicking transverse myelitis of the spinal cord. J Clin Neurosci 2017;40:66–8. Available from: https://doi.org/10.1016/j.jocn.2017.02.003.

[135] Pearl N, Dubensky L. Anterior cord syndrome. StatPearls 2022.

[136] Raventhiranathan N, Petropoulou K, Sakonju A, Bakrukov D, Mirchia K. Pediatric fibrocartilaginous spine embolism induced by trauma. Radiol Case Rep 2021;16(8):1951–5. Available from: https://doi.org/10.1016/j.radcr.2021.04.068.

[137] Manara R, Calderone M, Severino MS, et al. Spinal cord infarction due to fibrocartilaginous embolization: the role of diffusion weighted imaging and short-tau inversion recovery sequences. J Child Neurol 2010;25(8):1024–8. Available from: https://doi.org/10.1177/0883073809355822.

[138] Reisner A, Gary MF, Chern JJ, Grattan-Smith J. Spinal cord infarction following minor trauma in children: fibrocartilaginous embolism as a putative cause. J Neurosurg 2013;11(4):445–50. Available from: https://doi.org/10.3171/2013.1.PEDS12382.

[139] Davidson RG. Atlantoaxial instability in individuals with Down syndrome: a fresh look at the evidence. Pediatrics. 1988;81(6):857–65.

[140] Cullen S, O'Connell E, Blake NS, Ward OC. Atlantoaxial instability in Down's syndrome: clinical and radiological screening. Ir Med J 1989;82(2):64–5.

[141] Ferguson RL, Putney ME, Allen BL. Comparison of neurologic deficits with atlanto-dens intervals in patients with Down syndrome. J Spinal Disord 1997;10(3):246–52.

[142] Espada G, Babini JC, Maldonado-Cocco J, García-Morteo O. Radiologic review: the cervical spine in juvenile rheumatoid arthritis. Semin Arthritis Rheum 1988;17(3):185–95. Available from: https://doi.org/10.1016/0049-0172(88)90019-4.

[143] Nathan FF, Bickel WH. Spontaneous axial subluxation in a child as the first sign of juvenile rheumatoid arthritis. J Bone Joint Surg 1968;50(8):1675–8.

[144] Yang SS, Corbett DP, Brough AJ, Heidelberger KP, Bernstein J. Upper cervical myelopathy in achondroplasia. Am J Clin Pathol 1977;68(1):68–72. Available from: https://doi.org/10.1093/ajcp/68.1.68.

[145] Hagen K, Porter C, Martin R, Dean J, Salorio C, Sadowsky C. Improvements in function following inpatient activity-based therapy for children with acute flaccid myelitis. Top Spinal Cord Inj Rehabil 2020;26(4):275–82. Available from: https://doi.org/10.46292/sci20-00010.

[146] Sadowsky CL, Hammond ER, Strohl AB, et al. Lower extremity functional electrical stimulation cycling promotes physical and functional recovery in chronic spinal cord injury. J Spinal Cord Med 2013;36(6):623–31. Available from: https://doi.org/10.1179/2045772313Y.0000000101.

[147] Hammond ER, Metcalf HM, McDonald JW, Sadowsky CL. Bone mass in individuals with chronic spinal cord injury: associations with activity-based therapy, neurologic and functional status, a retrospective study. Arch Phys Med Rehabil 2014;95(12):2342–9. Available from: https://doi.org/10.1016/j.apmr.2014.07.395.

[148] Mertins R, Selvarajah S, Sadoswky C. The efficacy of aquatic and land-based locomotor training on ambulatory function and quality of life in adults with incomplete spinal cord injuries: two case reviews. J Aquat Phys Ther 2017;25(2):52–61.

[149] ASIA and ISCoS International Standards Committee. The 2019 revision of the International Standards for Neurological Classification of Spinal Cord Injury (ISNCSCI) – what's new? *Spinal Cord* 2019;57(10):815–17. Available from: https://doi.org/10.1038/s41393-019-0350-9.

[150] Mulcahey MJ, Vogel LC, Sheikh M, et al. Recommendations for the national institute for neurologic disorders and stroke spinal cord injury common data elements for children and youth with SCI. Spinal Cord 2017;55(4):331–40. Available from: https://doi.org/10.1038/sc.2016.139.

[151] Calhoun Thielen C, Slavin MD, Ni P, Mulcahey MJ. Development and initial validation of ability levels to interpret pediatric spinal cord injury activity measure and pediatric measure of participation scores. J Pediatr Rehabil Med 2021;14(3):463–76. Available from: https://doi.org/10.3233/PRM-200774.

[152] Jette AM, Tulsky DS, Ni P, et al. Development and initial evaluation of the spinal cord injury-functional index. Arch Phys Med Rehabil 2012;93(10):1733–50. Available from: https://doi.org/10.1016/j.apmr.2012.05.008.

[153] Ni P, Mulcahey MJ, Slavin MD, et al. Tracking spinal cord injury functional outcomes across the lifespan: validation of linking coefficients. Arch Phys Med Rehabil 2019;100(10):1924–31. Available from: https://doi.org/10.1016/j.apmr.2019.05.022.

[154] Kidd D, Stewart G, Baldry J, et al. The functional independence measure: a comparative validity and reliability study. Disabil Rehabil 1995;17(1):10–14. Available from: https://doi.org/10.3109/09638289509166622.

[155] Salorio C, Rogers K, Neuland E, Cagney J, Sadowsky C. Use of the physical abilities and mobility scale (PAMS) in children receiving inpatient rehabilitation for spinal cord related paralysis. Dev Neurorehabil 2022;25(3):186–92. Available from: https://doi.org/10.1080/17518423.2021.1962424.

[156] Brown PJ, Marino RJ, Herbison GJ, Ditunno JF. The 72-hour examination as a predictor of recovery in motor complete quadriplegia. Arch Phys Med Rehabil 1991;72(8):546–8.

[157] Marino RJ, Ditunno JF, Donovan WH, Maynard F. Neurologic recovery after traumatic spinal cord injury: data from the model spinal cord injury systems. Arch Phys Med Rehabil 1999;80(11):1391–6. Available from: https://doi.org/10.1016/s0003-9993(99)90249-6.

[158] Consortium for Spinal Cord Medicine. Outcomes following traumatic spinal cord injury: clinical practice guidelines for health-care professionals. *J Spinal Cord Med* 2000;23(4):289–316. Available from: https://doi.org/10.1080/10790268.2000.11753539.

[159] Chay W, Kirshblum S. Predicting outcomes after spinal cord injury. Phys Med Rehabil Clin N Am 2020;31(3):331–43. Available from: https://doi.org/10.1016/j.pmr.2020.03.003.

[160] National spinal cord injury statistical center, University of Alabama at Birmingham, 2021 annual statistical report – complete public version; 2021.

[161] Konz KR, Chia JK, Kurup VP, Resnick A, Kelly KJ, Fink JN. Comparison of latex hypersensitivity among patients with neurologic defects. J Allergy Clin Immunol 1995;95(5):950–4. Available from: https://doi.org/10.1016/S0091-6749(95)70094-3.

[162] Vogel LC, Schrader T, Lubicky JP. Latex allergy in children and adolescents with spinal cord injuries. J Pediatr Orthop 1995;15(4):517–20. Available from: https://doi.org/10.1097/01241398-199507000-00021.

[163] Gooch JL, Patton CP. Combining botulinum toxin and phenol to manage spasticity in children. Arch Phys Med Rehabil 2004;85(7):1121–4. Available from: https://doi.org/10.1016/j.apmr.2003.09.032.

[164] Tilton A, Vargus-Adams J, Delgado MR. Pharmacologic treatment of spasticity in children. Semin Pediatr Neurol 2010;17(4):261–7. Available from: https://doi.org/10.1016/j.spen.2010.10.009.

[165] Armstrong RW, Steinbok P, Farrell K, Cochrane D, Norman MG, Kube S. Continuous intrathecal baclofen treatment of severe spasms in two children with spinal-cord injury. Dev Med Child Neurol 1992;34(8):731–8. Available from: https://doi.org/10.1111/j.1469-8749.1992.tb11509.x.

[166] Garland DE, Shimoyama ST, Lugo C, Barras D, Gilgoff I. Spinal cord insults and heterotopic ossification in the pediatric population. Clin Orthop Relat Res 1989;245:303–10.

[167] Sadowsky CL, Mingioni N, Zinski J. A primary care provider's guide to bone health in spinal cord-related paralysis. Top Spinal Cord Inj Rehabil 2020;26(2):128–33. Available from: https://doi.org/10.46292/sci2602-128.

[168] Khalatbari H, Binkovitz LA, Parisi MT. Dual-energy X-ray absorptiometry bone densitometry in pediatrics: a practical review and update. Pediatr Radiol 2021;51(1):25–39. Available from: https://doi.org/10.1007/s00247-020-04756-4.

[169] Lauer R, Johnston TE, Smith BT, Mulcahey MJ, Betz RR, Maurer AH. Bone mineral density of the hip and knee in children with spinal cord injury. J Spinal Cord Med 2007;30(Suppl 1):S10–14. Available from: https://doi.org/10.1080/10790268.2007.11753962.

[170] Betz RR. Unique management needs of pediatric spinal cord injury patients: orthopedic problems in the child with spinal cord injury. J Spinal Cord Med 1997;20(1):14–16.

[171] Henderson RC, Berglund LM, May R, et al. The relationship between fractures and DXA measures of BMD in the distal femur of children and adolescents with cerebral palsy or muscular dystrophy. J Bone Miner Res 2010;25(3):520–6. Available from: https://doi.org/10.1359/jbmr.091007.

[172] Ward LM, Konji VN, Ma J. The management of osteoporosis in children. Osteoporos Int 2016;27(7):2147–79. Available from: https://doi.org/10.1007/s00198-016-3515-9.

[173] McCarthy JJ, Betz RR. Hip disorders in children who have spinal cord injury. Orthop Clin N Am 2006;37(2):197–202. Available from: https://doi.org/10.1016/j.ocl.2005.09.004 vi.

[174] Rink P, Miller F. Hip instability in spinal cord injury patients. J Pediatr Orthop 1990;10(5):583–7. Available from: https://doi.org/10.1097/01241398-199009000-00002.

[175] McCarthy JJ, Chafetz RS, Betz RR, Gaughan J. Incidence and degree of hip subluxation/dislocation in children with spinal cord injury. J Spinal Cord Med 2004;27(Suppl 1):S80–3. Available from: https://doi.org/10.1080/10790268.2004.11753423.

[176] Betz RR, Mulcahey MJ, Smith BT, Triolo RJ, McCarthy JJ. Implications of hip subluxation for FES-assisted mobility in patients with spinal cord injury. Orthopedics. 2001;24(2):181–4. Available from: https://doi.org/10.3928/0147-7447-20010201-23.

[177] Dearolf WW, Betz RR, Vogel LC, Levin J, Clancy M, Steel HH. Scoliosis in pediatric spinal cord-injured patients. J Pediatr Orthop 1990;10(2):214–18.

[178] Lubicky JP, Betz R. Spinal deformities in children and adolescents after spinal cord injury. In: Bets R, Mulcahey MJ, editors. The child with a spinal cord injury. Rosemont IL: American Academy of Orthopedic Surgeons; 1996, p. 363–70.

[179] Betz R.R., Mulcahey M.J. Spinal cord rehabilitation. In: Weinstein SL, editor. The pediatric spine: principles and practice. New York: Raven Press; p. 7811-810.

[180] Eswara JR, Castellan M, González R, Mendieta N, Cendron M. The urological management of children with spinal cord injury. World J Urol 2018;36(10):1593–601. Available from: https://doi.org/10.1007/s00345-018-2433-1.

[181] Rawashdeh YF, Austin P, Siggaard C, et al. International children's continence society's recommendations for therapeutic intervention in congenital neuropathic bladder and bowel dysfunction in children. Neurourol Urodyn 2012;31(5):615–20. Available from: https://doi.org/10.1002/nau.22248.

[182] Mitsui T, Kitta T, Moriya K, Takeda M, Shinohara N. Traumatic pediatric spinal cord injury: long-term outcomes of lower urinary tract function. Spinal Cord Ser Cases 2015;1:15023. Available from: https://doi.org/10.1038/scsandc.2015.23.

[183] Hopps CV, Kropp KA. Preservation of renal function in children with myelomeningocele managed with basic newborn evaluation and close followup. J Urol 2003;169(1):305–8. Available from: https://doi.org/10.1016/S0022-5347(05)64112-2.

[184] Park JS, Lee YS, Lee CN, Kim SH, Kim SW, Han SW. Efficacy and safety of mirabegron, a β3-adrenoceptor agonist, for treating neurogenic bladder in pediatric patients with spina bifida: a retrospective pilot study. World J Urol 2019;37(8):1665–70. Available from: https://doi.org/10.1007/s00345-018-2576-0.

[185] Pascali MP, Mosiello G, Boldrini R, Salsano ML, Castelli E, De Gennaro M. Effects of botulinum toxin type a in the bladder wall of children with neurogenic bladder dysfunction: a comparison of histological features before and after injections. J Urol 2011;185(6):2552–7. Available from: https://doi.org/10.1016/j.juro.2011.01.019.

[186] Sorokin I, De E. Options for independent bladder management in patients with spinal cord injury and hand function prohibiting intermittent catheterization. Neurourol Urodyn 2015;34(2):167–76. Available from: https://doi.org/10.1002/nau.22516.

[187] Sylora JA, Gonzalez R, Vaughn M, Reinberg Y. Intermittent self-catheterization by quadriplegic patients via a catheterizable mitrofanoff channel. J Urol 1997;157(1):48–50.

[188] Qi Z, Middleton JW, Malcolm A. Bowel dysfunction in spinal cord injury. Curr Gastroenterol Rep 2018;20(10):47. Available from: https://doi.org/10.1007/s11894-018-0655-4.

[189] Mosiello G, Marshall D, Rolle U, et al. Consensus review of best practice of transanal irrigation in children. J Pediatr Gastroenterol Nutr 2017;64(3):343–52. Available from: https://doi.org/10.1097/MPG.0000000000001483.

[190] Sinha CK, Butler C, Haddad M. Left antegrade continent enema (LACE): review of the literature. Eur J Pediatr Surg 2008;18(4):215–18. Available from: https://doi.org/10.1055/s-2008-1038499.

[191] Waddell O, McCombie A, Frizelle F. Colostomy and quality of life after spinal cord injury: systematic review. BJS Open 2020;4(6):1054–61. Available from: https://doi.org/10.1002/bjs5.50339.

[192] Backe HA, Betz RR, Mesgarzadeh M, Beck T, Clancy M. Post-traumatic spinal cord cysts evaluated by magnetic resonance imaging. Paraplegia. 1991;29(9):607–12. Available from: https://doi.org/10.1038/sc.1991.89.

[193] Gower DJ, Pollay M, Leech R. Pediatric syringomyelia. J Child Neurol 1994;9(1):14–21. Available from: https://doi.org/10.1177/088307389400900103.

[194] Zebracki K, Melicosta M, Unser C, Vogel LC. A primary care provider's guide to pediatric spinal cord injuries. Top Spinal Cord Inj Rehabil 2020;26(2):91–9. Available from: https://doi.org/10.46292/sci2602-91.

[195] Sadowsky CL, McDonald JW. Activity-based restorative therapies: concepts and applications in spinal cord injury-related neurorehabilitation. Dev Disabil Res Rev 2009;15(2):112–16. Available from: https://doi.org/10.1002/ddrr.61.

[196] Dolbow DR, Gorgey AS, Recio AC, et al. Activity-based restorative therapies after spinal cord injury: inter-institutional conceptions and perceptions. Aging Dis 2015;6(4):254–61. Available from: https://doi.org/10.14336/AD.2014.1105.

[197] Musselman KE, Walden K, Noonan VK, et al. Development of priorities for a Canadian strategy to advance activity-based therapies after spinal cord injury. Spinal Cord 2021;59(8):874–84. Available from: https://doi.org/10.1038/s41393-021-00644-2.

[198] Hanson RW, Franklin MR. Sexual loss in relation to other functional losses for spinal cord injured males. Arch Phys Med Rehabil 1976;57(6):291–3.

[199] Peljovich A. Hand reconstruction in children with spinal cord injury. Phys Med Rehabil Clin North Am 2020;31(3):471–98. Available from: https://doi.org/10.1016/j.pmr.2020.04.008.

[200] Vova JA, Davidson LT. Nerve and tendon transfers after spinal cord injuries in the pediatric population: clinical decision making and rehabilitation strategies to optimize function. Phys Med Rehabil Clin N Am 2020;31(3):455–69. Available from: https://doi.org/10.1016/j.pmr.2020.04.006.

[201] Canter KS, Roberts MC. A systematic and quantitative review of interventions to facilitate school reentry for children with chronic health conditions. J Pediatr Psychol 2012;37(10):1065–75. Available from: https://doi.org/10.1093/jpepsy/jss071.

[202] Graham P, Weingarden S, Murphy P. School reintegration: a rehabilitation goal for spinal cord injured adolescents. Rehabil Nurs 1991;16(3):122–7. Available from: https://doi.org/10.1002/j.2048-7940.1991.tb01196.x.

[203] Massagli TL, Dudgeon BJ, Ross BW. Educational performance and vocational participation after spinal cord injury in childhood. Arch Phys Med Rehabil 1996;77(10):995–9. Available from: https://doi.org/10.1016/s0003-9993(96)90058-1.

[204] Reyes MR. Educational and vocational issues for children with spinal cord injuries. In: Betz RR, Mulcahey MJ, editors. The child with spinal cord injury. American Academy of Orthopedic Surgeons; 1996, p. 331–42.

[205] Mulcahey MJ. Returning to school after a spinal cord injury: perspectives from four adolescents. Am J Occup Ther 1992;46(4):305–12. Available from: https://doi.org/10.5014/ajot.46.4.305.

[206] Sandford PR, Falk-Palec D, Spears K. Return to school after spinal cord injury. Arch Phys Med Rehabil 1999;80(8):885–8. Available from: https://doi.org/10.1016/s0003-9993(99)90078-3.

[207] Johnston TE, McDonald CM. Health and fitness in pediatric spinal cord injury: medical issues and the role of exercise. J Pediatr Rehabil Med 2013;6(1):35–44. Available from: https://doi.org/10.3233/PRM-130235.

[208] Deepan KC, Zebracki K, Klass SJ, Cooper RA. Leisure and recreational activities. In: Betz RR, Mulcahey MJ, editors. The child with spinal cord injury. Rosemont IL: American Academy of Orthopedic Surgeons; 1996, p. 376–85.

[209] Harness J, Pierce J, Malas N. Psychiatric evaluation and management in pediatric spinal cord injuries: a review. Curr Psychiatry Rep 2021;23(7):40. Available from: https://doi.org/10.1007/s11920-021-01256-6.

[210] January AM, Zebracki K, Chlan KM, Vogel LC. Symptoms of depression over time in adults with pediatric-onset spinal cord injury. Arch Phys Med Rehabil 2014;95(3):447–54. Available from: https://doi.org/10.1016/j.apmr.2013.11.011.

[211] World Health Organization. Defining sexual health. sexual and reproductive health research. https://www.who.int/teams/sexual-and-reproductive-health-and-research/key-areas-of-work/sexual-health/defining-sexual-health; 2006 [accessed 05.15.23].

[212] Papadakis JL, Zebracki K, Chlan KM, Vogel LC. Sexuality in pediatric spinal cord injury. Top Spinal Cord Inj Rehabil 2017;23(1):42–8. Available from: https://doi.org/10.1310/sci2301-42.

[213] Alexander MS, Alexander CJ. Recommendations for discussing sexuality after spinal cord injury/dysfunction in children, adolescents, and adults. J Spinal Cord Med 2007;30(Suppl 1):S65–70. Available from: https://doi.org/10.1080/10790268.2007.11753971.

[214] Gatti C, Del Rossi C, Ferrari A, Casolari E, Casadio G, Scire G. Predictors of successful sexual partnering of adults with spina bifida. J Urol 2009;182(4):1911–16. Available from: https://doi.org/10.1016/j.juro.2009.02.065.

[215] Hess MJ, Hough S. Impact of spinal cord injury on sexuality: broad-based clinical practice intervention and practical application. J Spinal Cord Med 2012;35(4):211–18. Available from: https://doi.org/10.1179/2045772312Y.0000000025.

[216] Houtrow A, Roland M. Sexual health and education guidelines for the care of people with spina bifida. J Pediatr Rehabil Med 2020;13(4):611–19. Available from: https://doi.org/10.3233/PRM-200743.

Chapter 20

Neuromuscular disorders

Michael Alonzo Lopez[1], Jenna Klotz[2] and Meeta Cardon[3]

[1]*Division of Pediatric Neurology, Department of Pediatrics, Heersink School of Medicine, University of Alabama at Birmingham, Birmingham, AL, United States,* [2]*Department of Neurology, Division of Child Neurology and Neuromuscular Medicine, Stanford University School of Medicine, Palo Alto, CA, United States,* [3]*Division of Child Neurology, Department of Neurology, University of New Mexico, Albuquerque, NM, United States*

Approach to pediatric neuromuscular disorders

Neuromuscular disorders (NMDs) consist of diseases of the peripheral nervous system. This includes the anterior horn cells residing in the ventral spinal cord, spinal nerve roots and rootlets, cervical and lumbosacral plexus, peripheral nerves, neuromuscular junction, and muscle. The autonomic nervous system is also included but has more complex localization. Localizing disease to the right anatomical level can help with rapid diagnosis and appropriate treatment.

Common presenting symptoms for neuromuscular disorders include weakness, hypotonia, elevated creatine kinase (CK), and motor delays. In a child presenting with weakness, central nervous system (CNS) etiologies (pyramidal weakness) should be distinguished from peripheral nervous system (PNS) etiologies. Additionally, it is important to determine if the weakness is a result of an acquired or genetic etiology. Central weakness commonly results from acquired injuries such as infection, stroke, or trauma. Peripheral weakness often results from genetic etiologies that may have more systemic disease (CNS + PNS) but may disproportionately affect the PNS. Thus, NMD, especially genetic NMD, has a significant impact on the neurological development of the pediatric patient. Herein, we review the approach for the history, exam, and testing of common genetic NMD in pediatric patients.

The goal of history-taking is to assist with localization, followed by an accurate and timely diagnosis. A motor symptom history should ascertain the pattern and severity of weakness and whether these are accompanied by sensory problems such as numbness, tingling, or paresthesias. Patients with genetic NMD frequently have comorbid intellectual disabilities, developmental delays, attention deficit/hyperactivity disorders, autism, and neurodevelopmental disorders. Thus, history-taking should not focus exclusively on motor function [See chapter: The developmental history]. Family history is particularly helpful if there is a pattern of inheritance recognized. However, genetic NMD can arise de novo or go unrecognized in mildly affected patients. Exam features consistent with a peripheral etiology include depressed reflexes, atrophy of muscles, a symmetric pattern of bilateral weakness, and fasciculations (involuntary, rapid, and visually apparent muscle twitching that does not lead to movement across a joint). Exam findings that should prompt an evaluation for central etiologies include hyperreflexia, unilateral weakness, or focal weakness of one limb.

Occasionally, it is difficult to precisely localize, and there can be mixed central and peripheral involvement. In these cases, nerve conduction studies and electromyography (NCS/EMG) are helpful in defining the phenotype. NCS/EMG provides key information on localization within the peripheral nervous system (nerve, neuromuscular junction, or muscle), type of pathology (axonal, demyelinating), temporal features (acute, subacute, or chronic), distribution (symmetric vs asymmetric, bulbar vs limb, proximal vs distal, diffuse vs focal), and severity.

In addition, lab evaluations are important for the evaluation of acquired neuromuscular diseases such as autoimmune, demyelinating, infectious, or traumatic diseases. Acquired NMD is suspected by an abrupt change in the usual state of health or based on a history of relevant exposures (environmental exposures, drugs, toxins, fever, immune compromised). When a genetic etiology is suspected, targeted panel or single gene testing is preferred, especially for a clinically identifiable syndrome. Due to advancements in sequencing technology and growing knowledge of genotype-phenotype correlations, genetic testing has become first-line testing in many disorders. When there is not an obvious candidate disease, broader testing, such as panel or whole exome sequencing with parental variant testing, is helpful [1]. Consultation with a geneticist can guide more advanced testing such as methylation assays, repeat expansion testing, whole genome sequencing, RNA sequencing, or long-read genomic sequencing.

Capute and Accardo's Neurodevelopmental Disabilities in Infancy and Childhood. DOI: https://doi.org/10.1016/B978-0-12-824060-1.00039-0

Finally, a biopsy of a muscle or peripheral nerve can be utilized if the molecular diagnosis remains elusive. It can refine a genetic diagnosis based on histopathologic features, or it can clarify the interpretation of genetic variants of uncertain clinical significance. It can also help to exclude acquired etiologies like dermatomyositis or vasculitis.

There is an expanding number of motor function assessments that are important in monitoring patients with neuromuscular disorders. These tend to be disease-specific and are performed by persons with expertise and training, such as neuromuscular specialists or physical/occupational therapists.

While the development of molecular treatments for some neuromuscular conditions has begun revolutionizing the field, supportive care to address cardiac, pulmonary, and rehabilitative therapies remains a mainstay of management. Clinics with multidisciplinary specialty care access can help patients achieve maximum function and are the standard of care [1].

Multidisciplinary clinics include many specialists and require asynchronous visits to fit the needs of the patient population. Members of the team include the neuromuscular specialists who guide diagnosis and management. A cardiologist evaluates arrhythmias or cardiomyopathy in appropriate groups. A pulmonologist evaluates for evidence of respiratory dysfunction such as nocturnal hypoventilation and impaired airway clearance (weak cough) and assesses the need for noninvasive ventilation. Additional specialists are invaluable in supportive care, as patients often require expertise in orthopedics (scoliosis, contractures, or fractures), endocrinology (osteopenia, fractures, or pubertal delay), and gastroenterology (nutrition management, dysmotility, or severe constipation). Rehabilitative therapy and equipment are essential for maximizing function. Equipment recommendations are important and typically include orthotic bracing, standers, walkers, power wheelchairs, bathroom equipment, and home lift systems. These are made through collaboration between the physician, physical and occupational therapists, and medical equipment specialists.

The psychosocial impact of neuromuscular disease is high and should be routinely addressed [2–5]. Quality of life surveys for patients and their families have been used on a research basis. Age-appropriate depression and suicidality scales can be used, but they can be challenging to implement when communication is severely impaired and when the disease process affects intellectual capacity. Counseling is often helpful for families and patients coping with neuromuscular disease. Involvement of a child psychiatry specialist can be helpful for patients with more severe mood or behavioral issues. Neuropsychological assessments are helpful for school-age children to identify baseline function and direct strategies for maximizing learning in school.

Multi-disciplinary clinics are oftentimes sponsored by nonprofit advocacy and research organizations, which can help reduce the burden on patients and supplement institutional clinical resources. There are numerous advocacy organizations and supportive communities for patients and families, which are reliable resources for parents and patients. Overall, supportive care in a multidisciplinary setting reduces psychosocial burden, prolongs life span, and reduces morbidity by identifying and treating neuromuscular disease challenges proactively [6].

Muscle disorders

Muscular dystrophies

Muscular dystrophies are a group of muscle disorders with characteristic histopathological features arising from genetic mutations (pathogenic variants) in genes important to skeletal muscle function. The limb girdle muscular dystrophies (LGMD) describe a pattern of progressive proximal weakness affecting the lower limbs more than, the upper limbs. The number of genes involved in LGMD is expanding, and a comprehensive review has been detailed by others [2]. We focus on Duchenne muscular dystrophy (DMD), which is a prototypical example of muscular dystrophies. While DMD fits the pattern of LGMD weakness, it is considered separately in LGMD naming conventions, which are evolving. In practice, it is important to define the molecular etiology in order to understand phenotype for each disease.

DMD is the most common muscular dystrophy. It occurs in 1 in 3500–6000 live newborn males worldwide [3,6,7]. DMD is an X-linked recessive disorder usually manifesting as a severe progressive muscle wasting disease in boys. It is caused by loss-of-function mutations in the *DYSTROPHIN* gene, which lead to the absence of the dystrophin protein and compromise the integrity of the sarcolemmal membrane. Important notes about *DYSTROPHIN* include that it is the largest gene in the human genome. It is widely expressed in tissue-specific isoforms, including the brain, eye, gastrointestinal tract, heart, and skeletal muscle. Pathogenic changes lead to a spectrum of phenotypes that include asymptomatic female carriers, manifesting carriers, Becker muscular dystrophy, intermediate muscular dystrophy, and DMD. The mildest syndromes can include rhabdomyolysis, myalgias, exercise intolerance, or asymptomatic hyperCKemia.

On muscle biopsy, DMD and LGMD are defined by the presence of myofiber size variability, internal nuclei, hypertrophic fibers, and mild inflammatory infiltration that progresses to end-stage sequela of fibroadipocytic infiltration of skeletal muscle fibers. These changes manifest as unrelenting progressive weakness and severe physical disability with early death.

Clinical presentation

In a developing newborn, symptoms of muscle weakness may be absent or mild. Mild symptoms can include hypotonia, developmental delay, growth difficulty, or myalgias. Gross and fine motor delays become more apparent by 2–3 years old due to abnormal gait. The typical age of presentation of DMD is 3–5 years old for a child who is having trouble rising from the floor. However, the range of presentation can be quite broad, from newborn (due to high CK or positive family history) to late ambulatory age (8–13 years old without treatment). Patients may have a history of delayed independent ambulation, toe walking, and increased plantarflexion tone. DMD often manifests with nonmotor symptoms like failure to thrive and constipation. These associated symptoms commonly lead to diagnostic evaluation, which reveals an elevation in aspartate aminotransferase (AST) and alanine aminotransferase (ALT) emanating from muscle. The serum CK will be markedly elevated ($> 10\times$ normal). There is also an increased association of neurodevelopmental disorders with DMD, including autism, speech delay, and intellectual disability in boys with DMD.

Notable exam features include macroglossia, pseudohypertrophic muscles (usually calves ± scapular or limb muscles), increased plantarflexion tone, neck flexion weakness, proximal weakness worse in the lower limbs, reduced deep tendon reflexes (usually not absent), a lordotic waddling gait, and difficulty rising from the floor (i.e., the Gower sign). At presentation, facial strength, coordination, and sensation are normal.

Diagnosis

Muscular dystrophies are diagnosed by genetic testing using Next Generation Sequencing (NGS) panels of genes known to cause muscle disease. Single-candidate gene testing is preferred when a clinically identifiable syndrome is apparent, as is often the case with DMD. Mutations in *DMD* that preserve the reading frame can lead to a less severe phenotype, such as in Becker muscular dystrophy. Although genetic modifiers and some in-frame mutations can cause unexpected phenotypes.

While mutations in the *DYSTROPHIN* gene are a common cause of LGMD, other gene mutations that disrupt the sarcolemmal membrane can lead to similar monogenic LGMD phenotypes. There are over 27 causative genes leading to autosomal recessive and, more rarely, autosomal dominant or de novo LGMD [8]. Laboratory testing can be helpful to differentiate nongenetic or acquired myopathies, such as endocrine, immune-mediated/inflammatory, and drug/toxic myopathies.

Treatments

Treatments for muscular dystrophies are limited. The mainstay includes comprehensive multidisciplinary care under the auspices of a neuromuscular specialist aimed at maximizing supportive care for the patient and family. Tailored goals of supportive care evolve with the progression of disease. In the initial stages of a new diagnosis, supportive care is directed at counseling on disease course, etiology, and ascertainment of baseline function for monitoring of progression. Patients and families will need varying amounts of psychosocial support, often in consultation with genetic counselors [4].

Supportive care is further aimed at maximizing independence in collaboration with physical medicine and rehabilitation with physical and occupational therapies. In the later nonambulatory phase, patients with muscular dystrophies will start to develop neuromuscular respiratory complications and may require more frequent monitoring for nocturnal hypoventilation or sleep-disordered breathing problems. Cardiac screening is indicated as cardiomyopathy and arrhythmia are often comorbid with genetic myopathies. In DMD, for example, dilated cardiomyopathy occurs in the teenage years.

For DMD, antiinflammatory treatments with oral corticosteroids are the standard of care [9]. Corticosteroid therapies include daily oral prednisone, deflazacort, or vamorolone. There is little data to support their use in other muscular dystrophies. In DMD, corticosteroids have significant evidence showing modest benefits in prolongation of ambulation, slowing respiratory function decline, and delaying the development of cardiomyopathy. Treatment with corticosteroids begins at diagnosis, 3–5 years of age, and is life-long. With the long-term use of corticosteroids, there are numerous side effects that are often intolerable for patients and their families. Dosing adjustments should be made if mood, weight gain, or other behavioral problems occur.

Beyond corticosteroids and supportive care, there is a growing repertoire of antisense oligonucleotide (ASO) therapies for amenable mutations in DMD [10]. These include ASOs targeting specific exonic deletions that alter the pre-mRNA to expand the patient's native deletion, thereby facilitating read-through and translation of an internally truncated dystrophin protein. Current FDA-approved ASO or "exon-skipping" therapies include targeting of exons 45, 51, and 53. These therapies are administered weekly for life via intravenous infusion. The effectiveness of ASOs in DMD is seen by increases in dystrophin protein expression by muscle biopsy and long-term studies demonstrating additional improvements compared with standard of care in motor and pulmonary outcomes [11]. Side effects of DMD ASO therapies have included hypersensitivity reactions, proteinuria, and milder symptoms (headache, nausea, and abdominal pain). Renal function should be monitored as with other ASOs.

Gene replacement therapies with *DYSTROPIN* mimics, that is, microdystrophin and minidystrophin, are in active clinical trial investigations using a adeno-associated viral vectors and transgene strategies. Currently, delandistrogene moxeparvovec is approved for administration in boys ages 4 years and older with few contraindications. Treatment is associated with a trend toward improved timed motor metrics. However, serious side effects include acute liver injury, immune-mediated myositis, and myocarditis have been reported. The clinical trial pipeline is robust for DMD, and a variety of investigational products are being used to restore dystrophin expression or target disease-driving pathways.

Clinical trial studies routinely incorporate the North Star Ambulatory Assessment, which includes timed motor metrics such as the 10-meter walk run, supine to stand time, and 4-stair climb. Additional clinical endpoints include loss of ambulation, pulmonary function testing, cardiac function, and dystrophin expression in muscle biopsies.

Associated neurodevelopmental disorders

The function of dystrophin is still not fully defined. However, evidence is mounting for an increasing role in neurodevelopment independent of the impact of muscle dysfunction. DMD/BMD patients often have developmental delays, attention deficit hyperactivity disorders (ADHD), and intellectual disabilities. Screening for these disorders begins at the time of diagnosis. Neuropsychological testing can be helpful during the school year to improve support in school beyond mobility issues. Patients should have comprehensive individualized education and disability (i.e., 504) plans to address these issues. The diagnosis and treatment of ADHD are done in collaboration with pediatricians, who can provide standard diagnostic tools and medications. Stimulants can be used in consultation with a cardiologist, but caution is advised should there be evidence of cardiomyopathy.

Facioscapulohumeral muscular dystrophy

Facioscapulohumeral muscular dystrophy (FSHD) is an autosomal dominant muscular dystrophy with variable penetrance. It is usually less severe than DMD, with a pattern of weakness that includes facial, scapular, and humeral muscle groups. These patients often have a less severe pattern of proximal limb weakness but may still lose ambulation. Few studies report cognitive testing in FSHD, but there are reports of significant intellectual disability [12]. Patients are further at risk of developing retinal disease, that is, Coat's disease, and hearing loss.

The genetics are complex but generally caused by inappropriate activation of the *DUX4* gene in skeletal muscle. There are two types of FSHD: Type 1 and Type 2. Both Type 1 and Type 2 require the 4qA permissive allele in the D4Z4 macrosatellite repeat locus of the long arm of chromosome 4 (4q). Type 1 is associated with a deletion mutation leading to contraction of the D4Z4 allele. Type 2 is less common, representing 5% of mutations and associated with normal D4Z4 size, but mutations involving *SMCHD1*, *DNMT3B*, or *LRIF1* genes [13]. Both Type 1 and Type 2 are similar in clinical presentation.

Unique exam findings of FSHD include weakness of facial muscles and sparing extraocular movements in conjunction with proximal upper limb weakness of the scapular and humeral muscles. Importantly, FSHD is an example of a genetic myopathy that can have an asymmetric pattern of weakness. Characteristic findings include a transverse smile (flat smile), prominence of shoulder muscles (triple hump sign), and abdominal weakness (Beevor's sign). Treatments remain elusive, but current preclinical strategies target inappropriate expression of *DUX4* [14].

Congenital muscular dystrophies

The congenital muscular dystrophies (CMD) result from defects in genes encoding extracellular matrix structural or enzymatic functions (Table 20.1) [15–17]. They more often involve early-onset weakness (infancy or toddlerhood) with variable facial weakness, ophthalmoplegia, respiratory failure, cardiomyopathy, or congenital contractures. CNS involvement is common, including intellectual disability and seizures. Additional CNS abnormalities include microcephaly, macrocephaly, structural brain abnormalities such as brainstem defects, cerebellar abnormalities, cysts, neuronal migration defects, and white matter disease. Ocular abnormalities also occur, including retinal degeneration, cataracts, glaucoma, and coloboma. There are recognized CMD syndromes collectively termed dystroglycanopathies, which include Fukuyama congenital muscular dystrophy (FCMD), muscle-eye-brain disease (MEB), and Walker-Warburg syndrome (WWS). MEB is more common in Finland. There can be in utero weakness followed by neonatal

TABLE 20.1 Congenital muscular dystrophies.

Gene symbol	Type of CMD	Syndrome/characteristic features	Functional classification	Inheritance pattern
COL12A1	Collagen associated myopathies	Bethlem	ECM constituents	AR
COL6A1	Collagen associated myopathies	Ullrich CMD, Bethlem	ECM constituents	AD, AR
COL6A2	Collagen associated myopathies	Ullrich CMD, Bethlem	ECM constituents	AD, AR
COL6A3	Collagen associated myopathies	Ullrich CMD, Bethlem	ECM constituents	AD, AR
DAG1	Dystroglycanopathies	Muscle-eye-brain disease, Walker-Warburg syndrome	ECM receptor	AR
FKRP	Dystroglycanopathies	Muscle-eye-brain disease, Walker-Warburg syndrome	ECM receptor modifications	AR
FKTN	Dystroglycanopathies	Fukuyama congenital muscular dystrophy, Walker-Warburg syndrome	ECM receptor modifications	AR
LARGE1	Dystroglycanopathies	Muscle-eye-brain disease, Walker-Warburg syndrome	ECM modifications	AR
POMGNT1	Dystroglycanopathies	Muscle-eye-brain disease, Walker-Warburg syndrome	ECM receptor modifications	AR
POMT1	Dystroglycanopathies	Walker-Warburg syndrome	ECM receptor modifications	AR
POMT2	Dystroglycanopathies	Muscle-eye-brain disease, Walker-Warburg syndrome	ECM receptor modifications	AR
RXYLT1	Dystroglycanopathies	Muscle-eye-brain disease, Walker-Warburg syndrome	ECM receptor modifications	AR
LAMA2	Merosinopathy	Wide spread white matter changes involving U fibers. May develop seizures.	ECM constituents	AR
ITGA7	Other	Facial + LG, equivocal brain malformations	ECM receptor	AR
SEPN1	Other	LGMD, severe axial weakness	ER	AR

AD, Autosomal dominant; *AR*, autosomal recessive; *CMD*, congenital muscular dystrophy; *ECM*, extracellular matrix; *LG*, limb girdle; *ER*, endoplasmic reticulum.

hypotonia. Life span is shortened with early death in late teens or early twenties [16]. Diagnosis is determined by genetic testing. If needed, a muscle biopsy can be used to assess for abnormal expression patterns of disease-causing gene candidates via western blot or immunohistochemistry.

Congenital myopathies

Congenital myopathies are historically defined by muscle biopsy features with characteristic findings. The list of genes involved is expanding, but commonly includes *ACTA1*, *DNM2*, *MTM1*, *MYH7*, *NEB*, *RYR1*, *SEPN1*, *TPM2*, *TPM3*, and *TTN*. They do not typically include CNS involvement or an intellectual disability. There is a spectrum of severity, ranging from onset at birth with severe generalized weakness to mild limb weakness manifesting in adulthood. Congenital myopathies are usually more stable and not as progressive as muscular dystrophies. Ultrastructural skeletal muscle findings include nemaline bodies defined as the presence of rods, cores, caps, and zebra bodies. Based on ultrastructural and microscopic histopathologic findings, they can be further categorized as nemaline myopathy, core myopathy, centronuclear myopathy, congenital fiber-type disproportion, and myosin storage myopathy. Treatment is supportive, although *MTM1* clinical trials utilizing gene replacement therapies are ongoing.

Pompe disease (glycogen storage type II)

Pompe disease is a glycogen storage disease due to pathogenic variants resulting in a deficiency or absence of alpha-glucosidase (*GAA*). Pompe disease has many associated systemic comorbidities, including neurological problems with sensorineural hearing loss, autonomic dysfunction, and small-fiber neuropathy [18]. Other systemic issues include gastrointestinal, vascular, and musculoskeletal dysfunction. There is variable onset in infantile-onset (IOPD) and late-onset Pompe disease (LOPD). IOPD involves cardiomyopathy along with muscle weakness, whereas LOPD has onset after 12 months, with patients presenting with bulbar, respiratory, and proximal limb weakness and less commonly cardiomyopathy. There are three FDA-approved disease-modifying enzyme replacement therapies (ERT) consisting of alaglucosidase alfa (Lumizyme, Myozyme) and avalglucosidase alfa (Nexviazyme). Immunogenic reactions to the ERT are significant and can neutralize ERT replacement due to the generation of antibodies to recombinant human GAA protein. In IOPD, determination of cross-reactive immunologic material (CRIM) status is required prior to ERT to improve immunotolerance to ERT. Treatment has improved survival. Newborn screening for Pompe disease is currently being conducted in several U.S. states. The Pompe Disease Newborn Screening Working Group has published articles on the evaluation and management of Pompe and positive newborn screen results [19,20].

Neuromuscular junction disorders

Congenital myasthenic syndromes

Clinical presentation

Congenital myasthenic syndromes (CMS) are genetic disorders disrupting synaptic transmission between the motor neurons and muscle, that is, of the neuromuscular junction (Table 20.2) [21–31]. They arise from

TABLE 20.2 Genetic neuromuscular junction disorders (congenital myasthenic syndromes).

Gene symbol	Distinguishing features	Synaptic localization	Cognitive involvment	Inheritance pattern
COLQ	Can spare ocular muscles, some long-term benefit to ephedrine or albuterol treatment	Synaptic	−	AR
LAMA5	May have cerebral volume loss and periventricular T2 hyperintensities on brain MRI	Synaptic	−	AR
LAMB2	Possibly comorbid with renal disease	Synaptic	+	AR
CHAT	Episodic apnea, hypotonia	Pre	±	AR
MYO9A	Episodic apnea, may worsen with fluoxetine and 3,4-DAP	Pre	+	AR
PREPL	Might respond to pyridostigmine	Pre	−	AR
RPH3A	Hand tremors, paroxysmal hyperglycemia	Pre	+	AR
SLC18A3	Episodic apnea, paramyotonia features, some response to AChEI	Pre	±	AR
SLC5A7	Lethal antenatal arthrogryposis or severe neonatal hypotonia with episodic apnea, might respond to AChEI or salbutamol	Pre	±	AR
SNAP25	Some benefit with 3,4-DAP treatment	Pre	+	AD
SYB1	Infantile onset, severe with dysphagia and respiratory failure	Pre	Not reported	AR
SYT2	Earlier onset and more severe weakness in AR cases, reduced CMAP amplitudes and even distal polyneuropathy, may benefit from 3,4-DAP treatment	Pre	±	AD/AR

(Continued)

TABLE 20.2 (Continued)

Gene symbol	Distinguishing features	Synaptic localization	Cognitive involvment	Inheritance pattern
UNC13A	Microcephaly, facial dysmorphisms, abnormal electrographic activity	Pre	+	AR
VAMP1	Severe weakness in limbs and face	Pre	±	AR
AGRN	Fetal akinesia deformation sequence (FADS), responds to ephedrine	Post	–	AR
CHD8	Macrocephaly and autism, may respond to 3,4-DAP	Post	±	AD
CHRNA1	High lethality	Post	–	AD/AR
CHRNB1	High lethality	Post	–	AD/AR
CHRND	High lethality	Post	–	AD/AR
CHRNE	Fluctuating ophthalmoparesis and limb weakness, can positively respond to ice pack test, most benefit from cholinergic agonists, some benefit from salbutamol	Post	–	AD/AR
CHRNG	Multiple pterygia, arthrogryposis	Post	–	AD/AR
COL13A1	Possible benefit from salbutamol and 3,4-DAP	Post	±	AR
DES	Muscle biopsy may have myopathic features, may respond to salbutamol	Post	–	AR
DOK7	Limb-girdle weakness or early onset vocal cord paralysis requiring ventilation, can worsen with AChEI, can benefit from ephedrine and salbutamol	Post	–	AR
LRP4	Albuterol might be effective treatment	Post	–	AR
MACF1	May benefit from AChEI and salbutamol	Post	–	AR
MUSK	Adult-onset limb-girdle weakness or infantile limb weakness and head drop and vocal cord paralysis, may worsen with AChEI, may respond to salbutamol	Post	–	AR
RAPSN	Fluctuating ptosis and limb weakness ± bulbar symptoms, can benefit from AChEI and 3,4-DAP, can worsen with fluoxetine	Post	–	AR
SCN4A	Paroxysmal "attacks" of weakness, like periodic paralysis	Post	+	AR
TOR1AIP1	Normal early development followed by childhood-onset limb-girdle weakness and muscle atrophy	Post	–	AR
SLC25A1	Might respond to 3,4-DAP	Other, mitochondrial	±	AR
PLEC	Myasthenic ± muscular dystrophy symptoms ± epidermolysis bullosa	Other, crosslinks intermediate filaments to target tissues	–	AR
ALG14	Either adult-onset weakness or early-onset weakness with epilepsy and early mortality	Other (glycosylation)	+	AR
ALG2	Limb-girdle weakness since infancy, muscle biopsy may have features of mitochondrial myopathy, may respond to AChEI or salbutamol and ephedrine	Other (glycosylation)	±	AR
DPAGT1	Limb-girdle weakness, tubular aggregates in SR, typically benefit from AChEI and 3,4-DAP	Other (glycosylation)	±	AR
GFPT1	Limb-girdle weakness, benefit from pyridostigmine, tubular aggregates in SR	Other (glycosylation)	–	AR
GMPPB	Limb-girdle weakness, elevated serum CK, dystrophic features on muscle biopsy	Other (glycosylation)	±	AR

3,4 DAP, 3,4-Diaminopyridine; *AChEI*, acyetycholinesterase inhibitor; *AD*, autosomal dominant; *AR*, autosomal recessive; *SR*, sarcoplasmic reticulum.

pathogenic changes of structural and functional components of the neuromuscular junction. Identification of this diagnosis can be quite challenging because the symptoms vary widely in severity, as does the age at presentation. Limb and axial weakness, bulbar weakness, ptosis, and ophthalmoparesis are common complaints, along with a pattern of fatiguability or fluctuation throughout the day. Weakness, including that in the eyes, can become fixed over time, and some presentations are simply of a limb-girdle pattern of weakness without facial involvement or fluctuating severity.

Molecular pathophysiology demonstrates multiple foci of potential disruption in the process of neuromuscular signal transmission. Pre-synaptic dysfunction has been attributed to defects in acetylcholine production (choline acetyltransferase) and synaptic vesicle fusion at the terminal membrane (Synaptotagmin 2, SNAP25B, and Synaptobrevin SNARE proteins). Dysfunction within the synapse has been attributed to defects in basal lamina structure (Laminin β) as well as the anchoring of acetylcholinesterase to the basal lamina (Collagen-like tail, or ColQ). Post-synaptic dysfunction is attributed to deficiencies in specific acetylcholine receptor subunits (cholinergic receptor nicotinic ε, α1, β1, and δ subunits) and a cascade of signals essential for localization and clustering of acetylcholine receptors on the postsynaptic membrane (muscle-associated receptor tyrosine kinase, agrin, lipoprotein-related protein 4, docking protein 7, receptor-associated protein of the synapse). Disruption of the acetylcholine receptor subunits, which specifically shorten channel openings, can result in fast-channel congenital myasthenic syndrome (FCCMS). This typically presents early in life, as opposed to slow-channel congenital myasthenic syndrome (SCCMS), which presents after adolescence and has mild symptoms due to pathogenic variants that specifically prolong depolarization. FCCMS can be responsive to pyridostigmine and 3,4-diaminopyridine; however, SCCMS symptoms worsen with these treatments. Disorders of glycosylation affect the solubility and function of proteins throughout the cell, including some critical for neuromuscular junction transmission (Glutamine:fructose-6-phosphate amidotransferase 1, dolichyl-phosphate n-acetylglucosamine phosphotransferase, alpha-1,3/1,6-mannosyltranserase, and UDP-n-acetylglucosaminyltransferase subunit), and this can result in a CMS phenotype.

Of note, CMS are genetic disorders of the NMJ distinct from autoimmune myasthenia gravis, wherein pathology is due to autoantibodies that attack components of the NMJ and are treated with immunosuppressive and immunomodulatory therapies.

Diagnosis

The diagnosis of CMS can be made based on appropriate genetic testing. Nerve conduction study with repetitive nerve stimulation and electromyography may show hallmark findings such as a $\geq 10\%$ decrement of the compound muscle action potential (CMAP) amplitudes on low frequency 2–3 Hz repetitive nerve stimulation. At times, the decrement is only observed after tetanic activation for 1 minute. Single fiber electromyography (SFEMG) is more sensitive and may show increased jitter or blocking. A trial of oral pyridostigmine or intravenous neostigmine while in a high level of inpatient care, or an ice pack test may or may not show symptomatic benefit. Evidence is lacking on whether patients with CMS can be identified reliably with these approaches. Edrophonium is no longer recommended for diagnosis. Muscle biopsy and MRI of limb muscles may provide clues to diagnose a specific CMS subtype. However, they may also be normal or misleadingly dystrophic.

Prognosis

The prognosis of CMS varies widely based on the severity of the symptoms. The most common cause of death is respiratory failure and its associated complications. Because many CMS subtypes have just been identified and are still not well defined, further discussion is speculative.

Treatment and future directions

Currently, there is no cure for CMS. As reviewed in Table 20.2, identification of the specific subtype may suggest a likelihood of responding to oral treatments like pyridostigmine, 3,4-diaminopyridine, salbutamol, albuterol, fluoxetine, and ephedrine. Certain CMS types may worsen with pyridostigmine, and, if possible, identification of the CMS gene etiology would help direct specific treatments. Immunosuppressants and immunomodulating therapies are not treatment options for CMS.

Peripheral neuropathies

Neuropathies with pediatric presentations

Polyneuropathies are characterized by the onset of relatively symmetric, distal-predominant motor, sensory, and/or autonomic symptoms. There is usually associated hyporeflexia or areflexia of the affected limb(s). Polyneuropathies fall into two broad categories. The first is primary neuropathies, in which polyneuropathy is the major or sole feature. The prototypical example is Charcot-Marie-Tooth disease. The second is syndromic neuropathies, in which polyneuropathy is part of the presentation of a multisystemic neurodegenerative or metabolic disorder.

Charcot-Marie-tooth disease (hereditary motor and sensory neuropathy)

Charcot-Marie-Tooth disease (CMT) encompasses genetically heterogeneous neuropathies with variable age of onset, disease severity, and rate of disease progression. CMT is categorized into subtypes primarily based on the pattern of inheritance and/or electrophysiologic characteristics in nerve conduction studies: CMT1 – autosomal dominant, demyelinating or dysmyelinating; CMT2 – autosomal dominant, axonal; CMT4 – autosomal recessive; and CMTX – X-linked. CMT3 refers to severe, early-onset phenotypes known as Dejerine–Sottas syndrome (infantile onset) and congenital hypomyelinating neuropathy (prenatal/neonatal onset) [32]. Because CMT3 was ultimately found to be caused by several different genes associated with other CMT subtypes, it is no longer considered a separate category [33]. To date, more than 80 genes have been identified as causing CMT [32]. Many of these genes are important for appropriate myelination, axonal transport, energy production, or trophic support and, when dysfunctional, cause progressive nerve damage.

Clinical presentation

CMT is a length-dependent sensorimotor polyneuropathy. Patients present with distal numbness, weakness, and muscle atrophy affecting the lower more than the upper limbs. Although both motor and sensory fibers are involved, sensory symptoms can be minimally apparent to patients, and the primary manifestations are often related to motor deficits [34]. Toddlers and young children may present with toe walking due to early ankle dorsiflexion weakness. Older patients may present with the classic steppage gait with a drop foot. Clumsiness, poor balance, and increased tripping and falling, rather than frank motor delays, may be the initial presentation. Common complaints include difficulty with fine motor tasks such as handwriting, using a zipper, or buttoning clothing.

On exam, foot deformities, including high arched feet, hammer toes, and flat feet, are often seen, and hypo or areflexia, especially at the ankles, is common. Although some axonal forms may present with normal or increased muscle stretch reflexes [32].

The incidence of CMT is approximately 1 in 2500 [32,34]. In the most common form, CMT1A, due to *PMP22* duplication, early motor milestones are usually on track, including independent ambulation by 15 months of age. In its classic form, symptoms develop in the first two decades and though progressive, rarely lead to loss of ambulation. 90% of patients with CMT1A have a positive family history, but the severity of the phenotype can vary significantly even within the same family, such that it may be undiagnosed in the affected parent [34].

In contrast, the most common form of CMT2, CMT2A, due to pathogenic variants in *MFN2*, presents in infancy or early childhood with a more rapidly progressive course resulting in wheelchair dependence by age 20 [34,35]. CMT3 classically refers to a phenotype of severe hypotonia and sometimes respiratory failure in infancy, but milder phenotypes of CMT presenting at birth with foot deformity but with normal early development have also been described [36]. Despite having a severe presentation in infancy, patients with Dejerine–Sottas syndrome typically improve after this period and gradually achieve motor milestones with variable delays. Intellectual disability is generally uncommon in CMT but has occurred in some patients with Dejerine–Sottas syndrome, CMTX, and CMT2 [36,37]. While there is significant overlap in presentation between the different genetic causes of CMT, certain additional features, such as sensorineural hearing loss or vocal cord paresis, should increase suspicion for specific genes (Table 20.3) [33,35,37].

Diagnosis

With the improved availability and affordability of genetic testing, gene panels have superseded electrodiagnostic testing for many cases of CMT. However, when genetic testing is nondiagnostic, such as when testing results in variants of

TABLE 20.3 Phenotypic feastures associted with genetic peripheral neuropathies (Charcot-Marie-tooth syndrome).

Gene symbol	CMT subtype	Distinguishing feature
GJB1/Connexin-32	CMTX1	Sensorineural hearing loss
PRPS1	CMTX5	
PMP22	CMT1E	
NEFL	CMT1F, CMT2E	
MPZ	CMT2J	
PMP22, MPZ	CMT3	
KARS1, MTM2, SH3TC2, NDRG1	CMT4	
MFN2	CMT2A	Vocal cord paresis, white matter abnormalities on brain MRI
TRPV4	CMT2C	
GDAP1	CMT2K/4A	
MFN2	CMT2A	Optic atrophy
Xq24-26.1	CMTX4	
PRPS1	CMTX5	
GARS	CMT2D	Prominent upper limb involvement
HSPB8	CMT2L	
GARS, BSCL2	dHMN5	
RAB7A	CMT2B	Severe sensory loss with foot ulceration and mutilation
GJB1	CMTX1	
SLC12A6	Andermann Syndrome	
Xp22.2	CMTX2	Intellectual disability
Xp24.26.1	CMTX4	
DYNC1H1	CMT2O/SMA-LED	
MORC2	CMT2Z	
SLC12A6	Andermann Syndrome	
EGR2	CMT1D	Scoliosis
LMNA	CMT2B1	
GARS	CMT2D	
PMP22, MPZ, PRX, EGR2	CMT3	
GDAP1, MTMR2, SBF1, SBF2, FGD4, FIG4	CMT4	

Note: The table is not all inclusive.
CMT, Charcot-Marie-tooth; *dHMN*, distal hereditary motor neuropathy; *SMA-LED*, spinal muscular atrophy, lower extremity dominant.

uncertain significance or when the initial presentation is unclear, electrodiagnostic testing remains an important tool in diagnosing CMT [33].

Treatment

There are no specific treatments for CMT, but this is an active area of research within neuromuscular medicine [33]. Treatment is supportive, and patients with CMT should work with physical and occupational therapy regularly. Cardiovascular exercise is beneficial for CMT. Orthopedic surgeries may be needed for contractures and foot deformities that impact function or cause significant pain. Pain in CMT is more commonly musculoskeletal in origin and may respond best to orthopedic interventions and physical therapy rather than medications targeting neuropathic pain [32]. Disease severity and progression can be monitored with the CMT Neuropathy Score version 2 (CMTNSv2) and the CMT Pediatric Scale (CMTPedS).

Hereditary sensory and autonomic neuropathy

Like CMT, hereditary sensory and autonomic neuropathies (HSAN) represent a genetically heterogeneous group of disorders. However, HSAN involves selective degeneration of sensory and autonomic nerve fibers. Some forms may present with distal weakness. Autosomal dominant forms tend to have onset in the second and third decades, but autosomal recessive forms are present in the newborn period or infancy [37,38]. In Riley-Day syndrome (HSAN3 or familial dysautonomia, autosomal recessive mutations in *IKBKAP*), newborns may present with difficulty feeding due to gastrointestinal dysmotility, recurrent pneumonias from aspiration, alacrima, emotional lability, and marked blood pressure lability. Children typically have normal intelligence [37,38]. In contrast, children with congenital insensitivity to pain with anhidrosis (HSAN4, autosomal recessive mutations in *NTRK1*) have mild intellectual disability, hyperactivity, and self-mutilating behavior with painless ulcerations in the mouth and distal limbs [37,38]. HSAN5 (*NGF*) has a similar phenotype to HSAN4, but with normal intelligence [33,37]. HSAN6 (*DST*) presents with a severe phenotype with neonatal hypotonia, autonomic dysfunction, feeding and respiratory issues, global developmental delays, and increased mortality [33].

Giant axonal neuropathy

Giant axonal neuropathy (GAN) is a rare pediatric neurodegenerative disorder caused by autosomal recessive mutations in *GAN*, which encodes the protein gigaxonin. Gigaxonin regulates intermediate filament turnover and is highly expressed in the brain, heart, and skeletal muscle [39,40]. When gigaxonin is dysfunctional, there is a toxic accumulation of intermediate filaments within these tissues. On a nerve biopsy, there is reduced nerve fiber density with giant axons due to neurofilament accumulation, giving the disease its name.

Clinical presentation

In GAN, development is typically normal within the first year of life [39–41]. However, even prior to the onset of neurological symptoms, characteristic features of tightly curled, "kinky" hair and long, curly eyelashes are often seen [39,41]. The mean age of onset is 2–3 years old and characterized by regression in motor skills with progressive loss of ambulation by age 10 and death in the third decade, often from respiratory failure, though milder forms exist [39–41]. In contrast with most forms of CMT, there is early involvement of the proximal muscles and the CNS [41]. CNS signs include nystagmus, oculomotor apraxia, upper motor neuron findings (spasticity, hyperreflexia), seizures, white matter abnormalities on brain MRI, optic atrophy, and cognitive issues [39–41]. Cognitive issues range from attention deficit hyperactivity disorder to intellectual disability [39].

Diagnosis

Diagnosis is determined by genetic testing. Additional supportive testing includes nerve or skin biopsy showing "giant axons," brain and spine MRI showing progressive atrophy and white matter signal abnormalities, and NCS/EMG studies demonstrating an axonal sensorimotor polyneuropathy with secondary demyelination, all of which are supportive [39–41].

Treatment

There are no disease-specific treatments for GAN [39–41]. Treatment is supportive. Orthopedic screening and referral are important for monitoring scoliosis or severe contractures impairing function (finger contractures). School support services for learning issues are also essential.

Secondary and syndromic neuropathies

Storage disorders

Metabolic disorders with toxic accumulation of metabolites causing neurodegeneration may eventually involve the peripheral nervous system. There are several cases where peripheral neuropathy is a more prominent or early feature of the disease, as noted in Table 20.4 [42–49].

TABLE 20.4 Genetic peripheral neuropathies associated with cerebral white matter disease.

Gene symbol	Type of leukodystrophy	Characteristics of brain MRI findings	Specific serum or urine testing	Type of peripheral neuropathy
ARSA	Metachromatic leukodystrophy	Symmetric T2-hyperintense frontal and parietal periventricular while matter leading to diffuse and confluent subcortical white matter involvement, tigroid appearance	Decreased Arylsulfatase A activity, elevated sulfatides	Demyelinating
ABCD1	X-linked adrenoleukodystrophy	Symmetric, confluent posterior T2-hyperintense white matter with contrast enhancement of the leading edge	Elevated long chain fatty acids and plasmalogen	Axonal
MFN2	MFN2 cavitating leukodystrophy	Multifocal cavitating white matter lesions ± diffusion restriction	–	Axonal
GALC	Krabbe disease	Patchy T2-hyperintense white matter involving the U fibers and cerebellum, extends to include pyramidal tracts and splenium of the corpus callosum ± cranial nerve enhancement	Decreased Galactocerebrosidase activity, elevated psychosine	Demyelinating
PLP1	*PLP1* null syndrome	Diffuse T2-hyperintensity in white matter, T1-hypointense	–	Demyelinating
GAN	Giant axonal neuropathy	T2-hyperintensity of the cerebellar white matter, can extend supratentorially to anterior and posterior periventricular white matter	–	Axonal
CSA and CSB	Cockayne syndrome	White matter dysmyelination, cerebellar atrophy ± basal ganglia calcifications	–	Demyelinating ± conduction block

Metachromatic leukodystrophy can initially present with a CMT-like picture with demyelinating features on nerve conduction studies, and central nervous system pathology may not be apparent clinically or radiographically until later [33,38,43]. Hematopoietic stem cell transplants have symptomatic and survival benefits but primarily slow the progression of central and not peripheral nervous system disease [43].

In early infantile Krabbe disease, children are typically normal at birth but show developmental delays and regression starting at 6 months of age [50]. There appears to be simultaneous demyelination both centrally and peripherally, with the severity of the polyneuropathy correlating with the severity of the disease, and abnormal nerve conduction studies are seen even in presymptomatic neonates [51].

Painful acroparesthesias due to small fiber neuropathy are the initial presentation in ~80% of male patients with X-linked recessive Fabry disease, which can present in childhood [52]. This disorder is responsive to enzyme replacement therapy [52].

Congenital disorders of glycosylation

Congenital disorders of glycosylation (CDG) are multisystemic diseases with prominent neurological symptoms, global developmental delay, or developmental regression. More than 90% of patients have intellectual disabilities, cerebellar syndromes, and hypotonia with hyporeflexia [53]. Epilepsy and movement disorders are frequent. Autism spectrum disorder is uncommon in CDG, and social development is frequently a strength [53,54]. While nearly all CDG can have peripheral neuropathy, it is best described in *PMM2*- and *NGLY1*-associated CDG, which have motor predominant and sensorimotor polyneuropathy, respectively [53,54].

Plexopathies with childhood presentations

Neonatal brachial plexus injury

Neonatal brachial plexus palsy (NBPP) occurs in 0.4–4 per 1000 live births [55]. The primary risk factor is birthweight (specifically >4 kg), but shoulder dystocia, breech delivery, vacuum- or forceps-assisted delivery, prolonged labor, maternal obesity or diabetes, and advanced maternal age also contribute [55,56].

Clinical presentation and diagnosis

NBPP can be recognized on examination by reduced spontaneous and active movement of the affected limb and an asymmetric Moro reflex. In an Erb's palsy (upper trunk, C5–C6 nerve roots), which accounts for ~50% of NBPP, infants may have the classic "waiter's tip" arm positioning (arm adducted and internally rotated, elbow extended and wrist flexed), with reduced biceps reflex [55]. Klumpke's palsy (isolated lower trunk, C8-T1 nerve roots) accounts for ~2% of NBPP and is characterized by preserved movement in the proximal arm with claw hand (weakness of wrist and finger extensors) and reduced grasp reflex [55,56]. Extended Erb's palsy (C5–C7 involvement) and total plexopathy injuries also occur. The presence of Horner's syndrome can occur when the sympathetic chain is involved, which indicates a likely avulsion injury. Thus, there is no chance for spontaneous recovery [55,56]. Other signs of proximal nerve root injury or avulsion include scapular winging (involvement of the long thoracic nerve) and asymmetric chest rise due to hemidiaphragm paralysis (involvement of the phrenic nerve) [55].

Treatment

Early physical and occupational therapy with range-of-motion exercises is important for recovery [55,56]. The limb should not be restrained unless there are any associated fractures [56]. "Tummy time" is encouraged to prevent plagiocephaly secondary to torticollis, and children should be monitored for speech delay, which can cooccur [56].

Many NBPP, particularly those that are isolated neurapraxic injuries without axonal damage, spontaneously recover within days to weeks [55,56]. Any NBPP that does not fully recover by 1 month should be referred to a NBPP surgical specialist [56]. In the most severe injuries, surgery should be completed between 3 and 6 months of age to prevent permanent changes in the impacted neuromuscular junction and muscles [55,56]. The overall prognosis for NBPP is good, but it depends on the severity of the injury. 20%–30% of cases will have residual deficits [55,56]. Early referral and interventions are key to minimizing long–term disability.

Neuralgic amyotrophy

Neuralgic amyotrophy, also known as Parsonage–Turner syndrome or brachial neuritis, presents with severe-onset unilateral or bilateral shoulder pain. Patients subsequently develop weakness and muscle atrophy in the affected limb(s). Weakness and atrophy predominantly affect the periscapular muscles, though there can be more diffuse or distal limb involvement [57,58]. This syndrome is rare in children compared with adults but has a bimodal peak incidence in children at <8 weeks old, more commonly associated with osteomyelitis or septic arthritis of the humerus, and at 7–15 years old [57]. Steroids can be helpful for pain in the acute phase, but it is unclear if they impact the long-term functional outcome [33,57,58]. Many adult patients make a full recovery within 2–3 years of symptom onset, but in a single-center study of 22 pediatric patients, motor deficits persisted in 8 of 13 patients with follow-up (4–36 months) [57]. Patients with recurrent neuralgic amyotrophy or with a positive family history can be evaluated for pathogenic variants in *SEPTIN9*, which account for ~55% of familial cases in North America [58].

Motor neuron disease

Spinal muscular atrophy

Spinal muscular atrophy (SMA) is the most common pediatric motor neuron disorder, affecting 1 in 6,000 to 1 in 10,000 live births [59]. Autosomal recessive deletions and other pathogenic variants in the *SURVIVAL MOTOR NEURON 1 (SMN1)* gene cause loss of survival motor neuron (SMN) protein, which in turn leads to degeneration of motor neurons [60]. A small amount of functional SMN can be produced from the homologous *SMN2* gene, and patients with higher copy numbers of *SMN2* generally have less severe phenotypes [60]. *SMN* is ubiquitously expressed, and although the predominant symptoms are related to motor neuron loss, in the most severe phenotypes, extra-neuronal involvement, including brain atrophy and white and deep gray matter abnormalities, is seen [61].

Clinical presentation

Prior to disease-modifying treatments, SMA subtypes were classified according to age of onset and maximal motor milestone achieved. SMA type 0 is the most severe, with prenatal onset, associated with only 1 copy of *SMN2* and often fatal in utero [62]. SMA type 1 infants (2–3 copies of *SMN2*) are symptomatic by 6 months old and never achieve sitting. They present with hypotonia, marked proximal more than distal weakness, initially sparing the facial muscles, tongue fasciculations, and areflexia. Based on natural history studies, untreated SMA type 1 patients develop respiratory failure and die by 2 years old [63]. Even with supportive care, the median age for permanent mechanical ventilation or death is 10.5 months [64]. SMA type 2 patients sit but do not stand or walk. Their development is normal until 6–8 months, and weakness is apparent by 18 months. Based on motor function scales, they continue to make gains until ~5 years old, after which there is a more marked decline [65,66]. SMA type 3a patients develop weakness between 18 months and 3 years old and lose ambulation in the second decade, whereas type 3b patients are symptomatic after age 3 years and lose ambulation in the fourth or fifth decade [67,68]. Type 3 patients can present similarly to limb girdle myopathy, with a waddling gait and a Gower's sign. Loss of muscle stretch reflexes out of proportion to the degree of weakness and the presence of a tremor (polyminimyoclonus) may be clues. Type 4 patients are at more than 30 years old and often have four or more copies of *SMN2* [62]. In the posttreatment period of SMA, the long-term evolution of motor and other clinical outcomes remains to be defined. Hence, the classical types of SMA will need to be reassessed.

Cognitive development in spinal muscular atrophy

SMN protein is ubiquitously expressed, including in the brain and other extra-neuronal tissues. The more severe patients may also have associated brain malformations and dysautonomia. This suggests a direct impact of severe disease on the brain in addition to a secondary impact of impaired mobility, respiratory, and bulbar function on learning due to diminished interaction with their environment [61,69]. The impact of SMA on neurocognition in children is understudied. There are conflicting studies reporting average, above-average, and below-average performance on various cognitive tasks [61,69]. Studies that concluded below-average performance tended to include a greater number of more severely affected SMA patients.

Diagnosis

Genetic testing showing homozygous deletions in *SMN1* confirms the diagnosis of SMA. Approximately 5% of SMA is caused by single nucleotide variations in *SMN1* [70]. Genetic testing should also include the *SMN2* copy number, which informs prognosis and management.

Prenatal carrier testing is now routinely offered to expectant parents but typically fails to detect carriers with two copies of *SMN1* on one allele (silent carriers) and those with single nucleotide variants [71,72]. A prenatal diagnosis can be made by chorionic villi sampling and amniocentesis. SMA was added to the Recommended Uniform Screening Panel for newborn screens in the United States in July 2018 and is nearly adopted by all states in the United States. Newborn screening will miss 5% of SMA patients with point mutations. Thus, it is important to send genetic testing for SMA in patients presenting with weakness and hyporeflexia in those with negative newborn screens [70].

Treatment

Three breakthrough therapies are currently available for the treatment of SMA. These include nusinersen, onasemnogene abeparvovec-xioi, and risdiplam. All three treatments demonstrate increased survival, enhanced motor function outcomes, and increased functional SMN protein.

Onasemnogene abeparvovec-xioi is a one-time intravenous gene replacement therapy using adeno-associated virus 9 (AAV9) to transduce motor neurons with a *SMN1* transgene and is approved for term patients under 2 years old without end-stage weakness. Severe fatal immunogenic reactions can occur, leading to hepatic inflammation or failure, and require prophylactic immunosuppression with corticosteroids. Nusinersen is an antisense oligonucleotide administered intrathecally. It may lead to renal toxicity or proteinuria, coagulation abnormalities, or thrombocytopenia. Risdiplam is an oral small-molecule splice modifier. It may cause a rash, diarrhea, or fever. Pre-clinical data in animals suggests a toxic effect on spermatogenesis and male reproductive organs. Both nusinersen and risdiplam target *SMN2* pre-mRNA to promote inclusion of exon 7 in its transcript, which increases its stability and allows more functional, full-length

SMN protein to be produced from *SMN2*. Nusinersen and risdiplam are approved for all ages and subtypes of SMA. There are no head-to-head comparisons of efficacy, but all three treatments have proven effective [73–78]. Overall, the clinical benefits for survival and motor outcomes clearly favor early treatment with careful monitoring and management of adverse effects.

Indeed, there is ongoing research into combination therapy and treatment with dual therapy in young patients with suboptimal responses to gene replacement therapy (NCT04488133). Additional therapeutics targeting muscle rather than motor neurons are also under investigation (NCT05156320, NCT05337553).

Functional outcomes in both clinical trials and clinical practice for SMA patients are assessed using several different scales, including the Children's Hospital of Philadelphia Infant Test of Neuromuscular Disorders (CHOP INTEND), the Hammersmith Functional Motor Scale Expanded (HFMSE), the Hammersmith Infant Neurological Examination (HINE), and the Revised Upper Limb Module (RULM).

Systemic neuromuscular disorders

Myotonic dystrophy type 1

Clinical presentation

Congenital and childhood-onset myotonic dystrophy type 1 (DM1) are the predominant manifestations of this multisystem neuromuscular disorder in the pediatric population [79]. DM1 is an autosomal dominant trinucleotide repeat disorder. The most common presenting signs of congenital DM1 are hypotonia at birth with associated feeding difficulties, and respiratory insufficiency following pregnancy complications of polyhydramnios, and decreased fetal movements. A neuromuscular evaluation reveals severe facial weakness with distal more than proximal limb weakness, talipes equinovarus, and hyporeflexia. Clinical myotonia is absent in infants and toddlers with congenital DM1. Symptoms evolve into a global developmental delay requiring physical, occupational, speech, and feeding therapy. The natural history of congenital DM1 is biphasic, with neonates improving in strength within the first few weeks and then developing more typical signs and symptoms of adult DM1 later in life [80]. Childhood-onset DM1 or juvenile DM1 presents between 1 and 10 years with variable mild distal weakness, painful myotonia, difficulty with fine motor skills, and ankle contractures. Learning difficulties are a common reason for the patient to present for evaluation, and neuromuscular clues pointing to the diagnosis may be slim [80]. Retrospective studies of children with DM1 show high rates of intellectual disability and learning disabilities affecting schooling [79,81]. Moreover, impaired language skills, attention deficit/hyperactivity disorder (ADHD), and autism spectrum disorder are reported in congenital and childhood-onset DM1 [80,81].

Extra-muscular comorbidities are numerous and affect many organ systems. Smooth muscle dysfunction results in dysphagia, gastroesophageal reflux, and constipation. Cardiac arrhythmias are more prevalent than cardiomyopathy. Respiratory muscle weakness predisposes to obstructive sleep apnea and nocturnal hypoventilation. Systemic nonmuscle disease includes cataracts, insulin resistance, and excessive daytime sleepiness. Behavior and neurodevelopmental diseases include conduct disorder, anxiety, and depression, which present over a lifetime [79,82]. Adults with DM1 can experience cognitive decline, low fertility, and hypogonadism. Recent studies suggest a higher risk for cancer in adult DM1 patients, including thyroid, endometrial, ovarian, testicular, cutaneous melanoma, and colon cancer [83].

DM1 localizes to the 3′ noncoding region of the *DMPK* (myotonic dystrophy protein kinase) gene on chromosome 19q13.3 [84]. More than 50 CTG repeats in this region are diagnostic of DM1 and more likely to be adult-onset symptoms, whereas <30 is normal [85]. CTG repeat expansion in this region transmits from parent to child in an autosomal dominant pattern. The condition exhibits anticipation in affected females, in which the CTG repeats expand further in gametes. Thus, symptomatic infants who have inherited DM1 do so more often from their mothers. Repeat expansion greater than 1000 is typical for congenital and childhood-onset DM1 [85]. It is proposed that this repeat expansion leads to dysregulation of RNA metabolism, including aberrant RNA, alternative splicing, disruption of local chromatic structure, and aberrant sequestration of RNA-binding proteins [86,87]. A comparison of the clinical course of congenital and childhood-onset myotonic dystrophy type 1 is shown in Table 20.5 [88].

Diagnosis

Diagnostic testing might start with serum CK, which may be elevated, and then move to genetic testing on a whole blood sample [82]. A muscle biopsy shows nonspecific myopathic changes, such as increased central myonuclei. However, biopsy is no longer recommended due to the availability and precision of genetic testing for repeat expansions [86].

TABLE 20.5 Comparison of clinical course of congenital and childhood-onset myotonic dystrophy type 1.

Characteristic features	Congenital myotonic dystrophy	Childhood-onset myotonic dystrophy
CTG repeats	>1000	>800
Prenatal Symptoms	+	±
Age of onset	Birth	1–10 years
Motor delay	+	±
Learning disability	++	+
Progression	+	+
Early respiratory failure	+	–
Dysphagia	++	+

Prognosis

Mortality is reported to be 25% in the infant period due to respiratory failure and associated complications [89]. Life expectancy is reported to be in the 50s, and causes of death include respiratory complications, cardiac arrythmia, and cancer [82,89,90]. A younger age of onset and the presence of proximal weakness confer a greater risk of early mortality [90]. Sudden death in adult DM1 patients is reported in multiple studies of mortality and is suspected to be related to cardiac arrythmia.

Treatment and future directions

Currently, there is no cure for DM1, and care is supportive. Multidisciplinary care is required throughout life to monitor for and manage the associated morbidities. Mexiletine is used in nondystrophic myotonia and safely reduces myotonia in DM1 patients, but seems to affect neither gross motor outcomes nor quality of life measurements [91,92]. A phase 2 study of a small-molecule inhibitor of upregulated RNA regulatory kinase, GSKβ, showed improvements in motor and neuropsychiatric function in congenital and childhood-onset DM1 patients [93].

Myotonic dystrophy type 2

Although it shares much of the same phenotype, myotonic dystrophy type 2 (DM2) is a genetically distinct condition that is unlikely to present in the pediatric population [82]. It is characterized less often by facial and distal limb weakness and instead causes proximal limb weakness without cognitive impairment.

Friedreich ataxia

Clinical Presentation

Friedreich ataxia (FRDA) is an autosomal recessive neurodegenerative disease with characteristic neuromuscular impairments. It is characterized by a progressive gait and limb ataxia, which evolve to include distal greater than proximal limb weakness, peripheral sensory neuropathy, dorsal root ganglionopathy, severe dysarthria/dysphagia, and pyramidal and extrapyramidal dysfunction in bowel and bladder control [94]. Neurologic examination reveals square wave jerks, areflexia, pes cavus, loss of vibration sense, and extensor plantar responses [95]. Patients are divided into three categories: classical type, in which the age of symptom onset is 10–16 years of age; late-onset, in which symptoms appear at 25–40 years of age, and very late-onset, in which symptoms appear after 40 years of age. Although the acquisition of early developmental milestones may have been normal, patients with FRDA show slower decision time, poorer executive function, reduced verbal fluency, and impairments in visuoperceptive and visuoconstructive abilities, even when accounting for motor deficits such as limb control, dysarthria, and oculomotor control in standardized testing [96].

Beyond the neurologic symptoms, Friedreich ataxia patients have a higher rate of decreased visual acuity, hearing loss, scoliosis, and impaired glucose tolerance, often progressing to diabetes mellitus with a risk of acute diabetic ketoacidosis. Depression, anxiety, and auditory and visual hallucinations, in addition to possible Charles Bonnet syndrome, occur in some patients and seem to correlate with longer disease duration [97]. Most significant, however, is the development of concentric left ventricular cardiomyopathy and associated supraventricular tachyarrhythmias [94,97].

Friedreich ataxia arises from biallelic mutations of the *FRATAXIN* gene (*FXN*), formerly *X25*, on chromosome 9q13 [98]. In 95% of cases, it is caused by a GAA repeat expansion in the first intronic region of both *FXN* alleles. In other cases, point mutations play a role in gene disruption, such as a compound heterozygous mutation with a repeat expansion [98,99]. Greater than 70 GAA repeats are pathogenic, and most disease—causing alleles have 600—800 repeats. Repeat expansion is inversely correlated with the age of disease onset, and earlier disease onset is characterized by more severe symptoms [94,99,100]. The carrier rate of the pathologic trinucleotide repeat expansion has been estimated in an European population at 1 in 85 [101]. Frataxin protein is transported to the mitochondria, where it regulates an iron-sulfur cluster critical for electron transport chain enzyme function [102]. Reduced frataxin expression to less than 10% of normal is correlated with iron accumulation and cell death in the cerebellar dentate nucleus, spinocerebellar, and corticospinal tracts, correlating with the clinical phenotype [99,103]. It is proposed that the increased intracellular oxidative stress or possibly abnormal DNA heterochromatization affecting nearby gene expression accounts for the multisystem effects [94,102].

Diagnosis

The diagnostic journey for patients with FRDA culminates with genetic testing, specifically repeat and sequence analysis of the *FXN* gene [94]. Magnetic resonance imaging (MRI) of the brain and spinal cord may show mineralization and atrophy in the dentate nucleus of the cerebellum and spinal cord atrophy [103,104]. NCS/EMG may reveal an axonal sensory or sensorimotor polyneuropathy [95].

Prognosis

Life expectancy in Friedreich ataxia is variable, with the majority of deaths occurring from 16 to 45 years [105]. The most common causes of death are congestive heart failure and/or cardiac arrhythmia.

Treatment and future directions

Currently, there is no cure for Friedreich's ataxia. Treatment historically has focused on beta-adrenergic antagonism, angiotensin-converting enzyme inhibition, and angiotensin receptor blockade for hypertrophic cardiomyopathy and the associated heart failure. Defibrillator implantation for arrythmias and heart transplantation have been successful in some patients, and expert consensus guidelines recommend that they be considered. Supportive care with rehabilitative therapies and pharmacotherapy for possible spasticity and muscle spasms is recommended. Orthopedic, urologic, and ophthalmologic management of scoliosis, overactive bladder, and square wave jerks, respectively, should be considered. In severe cases of dysarthria, dietary modifications and even gastrostomy tubes should be managed by gastroenterology. Diabetes mellitus is treated with insulin, as metformin is thought to detrimentally inhibit the activity of complex I in the electron transport chain [106].

Omaveloxolone is the first FDA-approved drug for the treatment of FRDA. It is a potent activator of nuclear factor erythroid 2-related factor 2 (*NRF2*), a reducer of oxidative damage, and a restorer of mitochondrial function [107]. Ongoing clinical trials attempt to address variable aspects of the proposed pathogenesis of this condition, including reduced frataxin expression, iron accumulation, and associated ferroptosis in tissues, oxidative stress and mitochondrial dysfunction, and generalized inflammation [102].

Mitochondrial disorders

Clinical presentation

Mitochondrial disorders are multisystem diseases that have the potential to cause systemic diseases manifesting as neuromuscular disorders. This mainly includes mitochondrial myopathy, neuropathy, and neuromuscular junction symptoms, in addition to central nervous system deficits such as cranial neuropathy and other organ dysfunction. The specific pattern and severity of symptoms vary widely, as does the age of symptom onset, but the majority are progressive. Thus, mitochondrial neuromuscular disorders are likely underrecognized and can be subclinical or masked by other neurological symptoms [108]. There are several prototypical syndromes with neuromuscular and CNS involvement (Table 20.6) [33,108—113]. Historical features that could suggest an underlying mitochondrial disorder include muscle weakness, exercise intolerance, fatigue, ophthalmoplegia, sensorineural hearing loss, cardiomyopathy or

TABLE 20.6 Mitochondrial neuromuscular disorders.

Gene symbol	Condition	Central nervous system	Eye/ear findings	Muscle biopsy findings	Neuropathy	Cardiac involvement	Other organ involvement
MT-TL1	MELAS	Stroke-like episodes after normal early development, Epilepsy, migraine, early dementia	SNHL, ptosis	Ragged red fibers on modified Gomori trichrome, can be positive for cytochrome c oxidase	Mixed axonal/ demyelinating peripheral neuropathy	Concentric hypertrophic cardiomyopathy, conduction abnormalities (Wolff-Parkinson-White syndrome)	Nephropathy, Diabetes mellitus, short stature
Large deletion in mtDNA	Kearns-Sayre Syndrome	Intellectual disability, leukoencephalopathy on Brain MRI, cerebellar ataxia, dysphagia	Ptosis, ophthalmoplegia, pigmentary retinopathy, SNHL	Ragged red fibers on modified Gomori trichrome, negative for cytochrome c oxidase	± Peripheral neuropathy	Conduction block > cardiomyopathy	Renal tubular acidosis, endocrine dysfunction
MT-TK	MERRF	Myoclonus, generalized epilepsy, cerebellar ataxia, dementia, Parkinsonism	Deafness, ophthalmoparesis	Ragged red fibers, COX-negative fibers	Axonal or mixed axonal/demyelinating neuropathy, sensory ± motor involvement	Cardiomyopathy, WPW	Short stature, lipomatosis
POLG	*POLG*-related disorders	Epilepsy, migraine, Parkinsonism, intellectual disability	Ptosis, ophthalmoplegia, cataracts, retinitis pigmentosa, hearing loss	Ragged red fibers, dysmorphic mitochondria	Axonal or mixed axonal/demyelinating sensory ± motor polyneuropathy	Cardiomyopathy	GI dysmotility, hypogonadism, sudden respiratory failure
Large deletion in mtDNA	Leigh Syndrome	Severe developmental delay and regression, cerebellar ataxia, spasticity, chorea, dystonia, deep gray matter lesions	Ophthalmoparesis	± histologic abnormalities, etc deficiencies	Polyneuropathy	Cardiomyopathy leading to early death	Elevated lactate in serum and CSF
ATP6	NARP	Cerebellar atrophy and ataxia, learning disability, epilepsy	Optic atrophy or retinopathy, ophthalmoplegia, SNHL		Axonal Sensorimotor polyneuropathy	Arrythmia	Short stature, sleep apnea
TYMP	MNGIE	Progressive ophthalmoplegia	Deafness		Mixed axonal and demyelinating or just demyelinating polyneuropathy, autonomic dysfunction		Severe GI dysmotility

COX, Cyclooxygenase; *CSF*, cerebrospinal fluid, *etc*, electron transport chain; *GI*, gastrointestinal; *MELAS*, mitochondrial myopathy, encephalopathy, lactic acidosis, and stroke-like episodes; *MERRF*, myoclonic epilepsy with red ragged fibers; *MNGIE*, mitochondrial neurogastrointestinal encephalopathy; *MRI*, magnetic resonance imaging; *mtDNA*, mitochondrial DNA; *NARP*, neuropathy, ataxia, and retinitis pigmentosa; *POLG*, polymerase G; *SNHL*, sensorineural hearing loss; *WPW*, Wolff-Parkinson-White.

conduction defects, developmental regression, a combination of neuropathy and myopathy with or without rhabdomyolysis, and episodic worsening with fasting or a transient illness.

Mitochondrial disease can be maternally inherited through mitochondrial DNA (mtDNA), X-linked, autosomal recessive, or autosomal dominant. Heteroplasmy occurs when the pathogenic variant is in the mitochondrial genome but only affects a specific percentage of the mitochondria within each cell. Pathogenic variants affecting proteins of mitochondrial maintenance and repair may result in a mitochondrial depletion syndrome, also known as hepato-cerebral syndrome.

Diagnosis

The diagnosis of mitochondrial disorders can be challenging due to varied clinical presentations. A high index of suspicion and family history can assist in making the diagnosis. Subsequently, molecular and genetic testing of the mitochondrial genome in addition to nuclear-encoded DNA from blood or muscle tissue is helpful. NCS/EMG may reveal evidence of more than one neuromuscular localization, such as polyneuropathy and myopathy. Muscle biopsy can be characterized by ragged red fibers on modified Gomori Trichrome staining or succinate dehydrogenase histochemistry, which correlates with mitochondrial proliferation and aggregation in the subsarcolemmal region. Electron microscopy can identify dysmorphic mitochondria. Myofibers that do not stain with cytochrome C oxidase (COX) at a rate >5% can be diagnostic. Genetic testing of muscle samples may be an important diagnostic step if whole blood testing is nondiagnostic and mitochondrial myopathy is suspected. Immunofluorescent studies and functional analysis of electron transport chain enzymes may be helpful in diagnosis as well.

Prognosis

Because of the vast spectrum of phenotypic severity within this category of disease, prognosis also varies widely—from severe, degenerative, and fatal prenatally or in infancy to mildly disabling in adulthood.

Treatment

There is currently no cure for mitochondrial disorders. Symptom management and surveillance by a multidisciplinary team of specialists, oftentimes led by a metabolic specialist, is typical. Prevention of crises by avoiding infectious illnesses and not exercising while sick may be helpful. Supplementation with coenzyme Q10 or its synthetic analog idebenone, alpha-lipoic acid, vitamin E, creatine monohydrate, and even L-carnitine if levels are low has been suggested. Mitochondrial encephalopathy, lactic acidosis, and stroke-like episodes (MELAS) are treated with high-dose intravenous arginine during a stroke-like episode as well as low-dose daily arginine supplementation.

Conclusion

Pediatric neuromuscular disorders often result in severe motor impairments that disrupt neurodevelopment. The underlying etiology is frequently genetic, leading to lifelong impairments and comorbidities, which can be devastating for patients and families, often prolonging the diagnostic process. Advances in genetic sequencing and gene discovery have transformed clinical practice by improving access to genetic testing and accelerating the development of novel treatments. This progress has enhanced diagnostic success and significantly improved outcomes for patients. Given these advances and their broad impact on health and neurodevelopment, care for patients with neuromuscular disorders is best provided by an interdisciplinary team of therapists, social workers, psychologists, and families, led by a neuromuscular neurologist, to maximize support and treatment and mitigate the negative impact on neurodevelopment.

References

[1] Herman I, Lopez MA, Marafi D, Pehlivan D, Calame DG, Abid F, et al. Clinical exome sequencing in the diagnosis of pediatric neuromuscular disease. Muscle Nerve 2021;63(3):304–10. Available from: https://doi.org/10.1002/mus.27112.

[2] Nigro V, Aurino S, Piluso G. Limb girdle muscular dystrophies: update on genetic diagnosis and therapeutic approaches. Curr Opin Neurol 2011;24(5):429–36. Available from: https://doi.org/10.1097/WCO.0b013e32834aa38d.

[3] Emery AE. Population frequencies of inherited neuromuscular diseases—a world survey. Neuromuscul Disord 1991;1(1):19–29.

[4] Huang M, Chen T, Wang Y, Zhou C, Cao J, Lu X, et al. Chronic pain, psychological distress, and quality of life in males with Duchenne muscular dystrophy. Dev Med Child Neurol 2023;65(5):640–54. Available from: https://doi.org/10.1111/dmcn.15404.

[5] Inhestern L, Brandt M, Driemeyer J, Denecke J, Johannsen J, Bergelt C. Experiences of health care and psychosocial needs in parents of children with spinal muscular atrophy. Int J Env Res Public Health 2023;20(7). Available from: https://doi.org/10.3390/ijerph20075360.
[6] Mah JK, Korngut L, Dykeman J, Day L, Pringsheim T, Jette N. A systematic review and meta-analysis on the epidemiology of Duchenne and Becker muscular dystrophy. Neuromuscul Disord 2014;24(6):482–91. Available from: https://doi.org/10.1016/j.nmd.2014.03.008.
[7] Mendell JR, Lloyd-Puryear M. Report of MDA muscle disease symposium on newborn screening for Duchenne muscular dystrophy. Muscle Nerve 2013;48(1):21–6. Available from: https://doi.org/10.1002/mus.23810.
[8] Narasimhaiah D, Uppin MS, Ranganath P. Genetics and muscle pathology in the diagnosis of muscular dystrophies: an update. Indian J Pathol Microbiol 2022;65(Supplement):S259–70. Available from: https://doi.org/10.4103/ijpm.ijpm_1074_21.
[9] Birnkrant DJ, Bushby K, Bann CM, Apkon SD, Blackwell A, Brumbaugh D, et al. Diagnosis and management of Duchenne muscular dystrophy, part 1: diagnosis, and neuromuscular, rehabilitation, endocrine, and gastrointestinal and nutritional management. Lancet Neurol 2018;17(3):251–67. Available from: https://doi.org/10.1016/S1474-4422(18)30024-3.
[10] Rivera SR, Jhamb SK, Abdel-Hamid HZ, Acsadi G, Brandsema J, Ciafaloni E, et al. Medical management of muscle weakness in Duchenne muscular dystrophy. PLoS One 2020;15(10):e0240687. Available from: https://doi.org/10.1371/journal.pone.0240687.
[11] Mitelman O, Abdel-Hamid HZ, Byrne BJ, Connolly AM, Heydemann P, Proud C, et al. A combined prospective and retrospective comparison of long-term functional outcomes suggests delayed loss of ambulation and pulmonary decline with long-term eteplirsen treatment. J Neuromuscul Dis 2022;9(1):39–52. Available from: https://doi.org/10.3233/jnd-210665.
[12] Dos Santos VB, Saute JAM, Jacinto-Scudeiro LA, Ayres A, Rech RS, de Olivera AA, et al. Cognitive profile of patients with facioscapulohumeral muscular dystrophy. Dement Neuropsychol 2021;15(4):541–7. Available from: https://doi.org/10.1590/1980-57642021dn15-040015.
[13] Jia FF, Drew AP, Nicholson GA, Corbett A, Kumar KR. Facioscapulohumeral muscular dystrophy type 2: an update on the clinical, genetic, and molecular findings. Neuromuscul Disord 2021;31(11):1101–12. Available from: https://doi.org/10.1016/j.nmd.2021.09.010.
[14] Tihaya MS, Mul K, Balog J, de Greef JC, Tapscott SJ, Tawil R, et al. Facioscapulohumeral muscular dystrophy: the road to targeted therapies. Nat Rev Neurol 2023;19(2):91–108. Available from: https://doi.org/10.1038/s41582-022-00762-2.
[15] Saito K. Fukuyama congenital muscular dystrophy – Table 20.3. Distinguishing between the major phenotypes of the alpha-dystroglycanopathies: FCMD, MEBD, and WWS. Internet: GeneReviews®; 2006.
[16] Mendell JR, Boue DR, Martin PT. The congenital muscular dystrophies: recent advances and molecular insights. Pediatr Dev Pathol 2006;9(6):427–43. Available from: https://doi.org/10.2350/06-07-0127.1.
[17] Xia W, Ni Z, Zhang Z, Sang H, Liu H, Chen Z, et al. Case report: a boy from a consanguineous family diagnosed with congenital muscular dystrophy caused by integrin alpha 7 (ITGA7) mutation. Front Genet 2021;12:706823. Available from: https://doi.org/10.3389/fgene.2021.706823.
[18] Chan J, Desai AK, Kazi ZB, Corey K, Austin S, Hobson-Webb LD, et al. The emerging phenotype of late-onset Pompe disease: a systematic literature review. Mol Genet Metab 2017;120(3):163–72. Available from: https://doi.org/10.1016/j.ymgme.2016.12.004.
[19] Kronn DF, Day-Salvatore D, Hwu WL, Jones SA, Nakamura K, Okuyama T, et al. Management of confirmed newborn-screened patients with pompe disease across the disease spectrum. Pediatrics 2017;140(Suppl 1):S24–45. Available from: https://doi.org/10.1542/peds.2016-0280E.
[20] Burton BK, Kronn DF, Hwu WL, Kishnani PS, Pompe Disease Newborn Screening Working Group. The initial evaluation of patients after positive newborn screening: Recommended algorithms leading to a confirmed diagnosis of pompe disease. Pediatrics 2017;140(Suppl 1):S14–23. Available from: https://doi.org/10.1542/peds.2016-0280D.
[21] Finsterer J. Congenital myasthenic syndromes. Orphanet J Rare Dis 2019;14(1):57. Available from: https://doi.org/10.1186/s13023-019-1025-5.
[22] Abicht A, Muller JS, Lochmuller H. In: Adam MP, Mirzaa GM, Pagon RA, et al., editors. Congenital myasthenic syndromes overview. Internet: GeneReviews®; 1993.
[23] Cossins J, Webster R, Maxwell S, Rodriguez Cruz PM, Knight R, Llewelyn JG, et al. Congenital myasthenic syndrome due to a TOR1AIP1 mutation: a new disease pathway for impaired synaptic transmission. Brain Commun 2020;2(2):fcaa174. Available from: https://doi.org/10.1093/braincomms/fcaa174.
[24] Ehrstedt C, Liu WW, Frykholm C, Beeson D, Punga AR. Novel pathogenic ALG2 mutation causing congenital myasthenic syndrome: a case report. Neuromuscul Disord 2022;32(1):80–3. Available from: https://doi.org/10.1016/j.nmd.2021.11.012.
[25] Engel AG, Selcen D, Shen XM, Milone M, Harper CM. Loss of MUNC13-1 function causes microcephaly, cortical hyperexcitability, and fatal myasthenia. Neurol Genet 2016;2(5):e105. Available from: https://doi.org/10.1212/NXG.0000000000000105.
[26] Maselli RA, Vazquez J, Schrumpf L, Arredondo J, Lara M, Strober JB, et al. Presynaptic congenital myasthenic syndrome with altered synaptic vesicle homeostasis linked to compound heterozygous sequence variants in RPH3A. Mol Genet Genom Med 2018;6(3):434–40. Available from: https://doi.org/10.1002/mgg3.370.
[27] Maselli RA, Arredondo J, Vazquez J, Chong JX, Bamshad MJ, Nickerson DA, et al. A presynaptic congenital myasthenic syndrome attributed to a homozygous sequence variant in LAMA5. Ann N Y Acad Sci 2018;1413(1):119–25. Available from: https://doi.org/10.1111/nyas.13585.
[28] Maselli RA, Wei DT, Hodgson TS, Sampson JB, Vazquez J, Smith HL, et al. Dominant and recessive congenital myasthenic syndromes caused by SYT2 mutations. Muscle Nerve 2021;64(2):219–24. Available from: https://doi.org/10.1002/mus.27332.
[29] Oury J, Liu Y, Topf A, Todorovic S, Hoedt E, Preethish-Kumar V, et al. MACF1 links Rapsyn to microtubule- and actin-binding proteins to maintain neuromuscular synapses. J Cell Biol 2019;218(5):1686–705. Available from: https://doi.org/10.1083/jcb.201810023.
[30] Shen XM, Scola RH, Lorenzoni PJ, Kay CS, Werneck LC, Brengman J, et al. Novel synaptobrevin-1 mutation causes fatal congenital myasthenic syndrome. Ann Clin Transl Neurol 2017;4(2):130–8. Available from: https://doi.org/10.1002/acn3.387.
[31] Wang X, Xiao H, Su B, Ren Y, Ding J, Wang F. LAMB2 novel variant c.2885-9 C > A affects RNA splicing in a minigene assay. Mol Genet Genom Med 2021;9(7):e1704. Available from: https://doi.org/10.1002/mgg3.1704.

[32] Jani-Acsadi A, Ounpuu S, Pierz K, Acsadi G. Pediatric Charcot-Marie-tooth disease. Pediatr Clin North Am 2015;62(3):767–86. Available from: https://doi.org/10.1016/j.pcl.2015.03.012.

[33] Klein CJ. Charcot-Marie-tooth disease and other hereditary neuropathies. Contin (Minneap Minn) 2020;26(5):1224–56. Available from: https://doi.org/10.1212/CON.0000000000000927.

[34] Siskind CE, Shy ME. Genetics of neuropathies. Semin Neurol 2011;31(5):494–505. Available from: https://doi.org/10.1055/s-0031-1299788.

[35] Tazir M, Hamadouche T, Nouioua S, Mathis S, Vallat JM. Hereditary motor and sensory neuropathies or Charcot-Marie-tooth diseases: an update. J Neurol Sci 2014;347(1–2):14–22. Available from: https://doi.org/10.1016/j.jns.2014.10.013.

[36] Baets J, Deconinck T, De Vriendt E, Zimon M, Yperzeele L, Van Hoorenbeeck K, et al. Genetic spectrum of hereditary neuropathies with onset in the first year of life. Brain 2011;134(Pt 9):2664–76. Available from: https://doi.org/10.1093/brain/awr184.

[37] Wilmshurst JM, Ouvrier R. Hereditary peripheral neuropathies of childhood: an overview for clinicians. Neuromuscul Disord 2011;21(11):763–75. Available from: https://doi.org/10.1016/j.nmd.2011.05.013.

[38] Landrieu P, Baets J. Early onset (childhood) monogenic neuropathies. Handb Clin Neurol 2013;115:863–91. Available from: https://doi.org/10.1016/B978-0-444-52902-2.00049-7.

[39] Bharucha-Goebel DX, Norato G, Saade D, Paredes E, Biancavilla V, Donkervoort S, et al. Giant axonal neuropathy: cross-sectional analysis of a large natural history cohort. Brain 2021;144(10):3239–50. Available from: https://doi.org/10.1093/brain/awab179.

[40] Echaniz-Laguna A, Cuisset JM, Guyant-Marechal L, Aubourg P, Kremer L, Baaloul N, et al. Giant axonal neuropathy: a multicenter retrospective study with genotypic spectrum expansion. Neurogenetics 2020;21(1):29–37. Available from: https://doi.org/10.1007/s10048-019-00596-z.

[41] Johnson-Kerner BL, Roth L, Greene JP, Wichterle H, Sproule DM. Giant axonal neuropathy: an updated perspective on its pathology and pathogenesis. Muscle Nerve 2014;50(4):467–76. Available from: https://doi.org/10.1002/mus.24321.

[42] Adang L. Leukodystrophies. Contin (Minneap Minn) 2022;28(4):1194–216. Available from: https://doi.org/10.1212/CON.0000000000001130.

[43] Beerepoot S, Nierkens S, Boelens JJ, Lindemans C, Bugiani M, Wolf NI. Peripheral neuropathy in metachromatic leukodystrophy: current status and future perspective. Orphanet J Rare Dis 2019;14(1):240. Available from: https://doi.org/10.1186/s13023-019-1220-4.

[44] Opal P. In: Adam MP, Everman DB, Mirzaa GM, Pagon RA, Wallace SE, Bean LJH, et al., editors. GAN-related neurodegeneration. Internet: GeneReviews®; 2021. Available from: https://www.ncbi.nlm.nih.gov/books/NBK1136/.

[45] Gitiaux C, Blin-Rochemaure N, Hully M, Echaniz-Laguna A, Calmels N, Bahi-Buisson N, et al. Progressive demyelinating neuropathy correlates with clinical severity in Cockayne syndrome. Clin Neurophysiol 2015;126(7):1435–9. Available from: https://doi.org/10.1016/j.clinph.2014.10.014.

[46] Laugel V. In: Adam MP, Everman DB, Mirzaa GM, Pagon RA, Wallace SE, Bean LJH, et al., editors. Cockayne syndrome. Internet: GeneReviews®; 2019. Available from: https://www.ncbi.nlm.nih.gov/sites/books/NBK1342/.

[47] Zaman M, Shutt TE. The role of impaired mitochondrial dynamics in MFN2-mediated pathology. Front Cell Dev Biol 2022;10:858286. Available from: https://doi.org/10.3389/fcell.2022.858286.

[48] Manokaran RK, Mahalingam HV, Kewalramani D. Cavitating leukodystrophy as a manifestation of cerebral involvement in MFN2 neuropathy. Neurology 2020;95(1):42–3. Available from: https://doi.org/10.1212/WNL.0000000000009746.

[49] Huffnagel IC, Dijkgraaf MGW, Janssens GE, van Weeghel M, van Geel BM, Poll-The BT, et al. Disease progression in women with X-linked adrenoleukodystrophy is slow. Orphanet J Rare Dis 2019;14(1):30. Available from: https://doi.org/10.1186/s13023-019-1008-6.

[50] Bradbury AM, Bongarzone ER, Sands MS. Krabbe disease: New hope for an old disease. Neurosci Lett 2021;752:135841. Available from: https://doi.org/10.1016/j.neulet.2021.135841.

[51] Siddiqi ZA, Sanders DB, Massey JM. Peripheral neuropathy in Krabbe disease: Electrodiagnostic findings. Neurology 2006;67(2):263–7. Available from: https://doi.org/10.1212/01.wnl.0000230153.34613.84.

[52] El-Abassi R, Singhal D, England JD. Fabry's disease. J Neurol Sci 2014;344(1–2):5–19. Available from: https://doi.org/10.1016/j.jns.2014.06.029.

[53] Paprocka J, Jezela-Stanek A, Tylki-Szymanska A, Grunewald S. Congenital disorders of glycosylation from a neurological perspective. Brain Sci 2021;11(1). Available from: https://doi.org/10.3390/brainsci11010088.

[54] Lam C, Wolfe L, Need A, Shashi V, Enns G. In: Adam MP, Everman DB, Mirzaa GM, et al., editors. NGLY1-related congenital disorder of deglycosylation. Internet: GeneReviews®; 2018. Available from: https://www.ncbi.nlm.nih.gov/sites/books/NBK481554/.

[55] Govindan M, Burrows HL. Neonatal brachial plexus injury. Pediatr Rev 2019;40(9):494–6. Available from: https://doi.org/10.1542/pir.2018-0113.

[56] Smith BW, Daunter AK, Yang LJ, Wilson TJ. An update on the management of neonatal brachial plexus palsy-replacing old paradigms: A review. JAMA Pediatr 2018;172(6):585–91. Available from: https://doi.org/10.1001/jamapediatrics.2018.0124.

[57] Al-Ghamdi F, Ghosh PS. Neuralgic amyotrophy in children. Muscle Nerve 2018;57(6):932–6. Available from: https://doi.org/10.1002/mus.26060.

[58] Van Eijk JJ, Groothuis JT, Van Alfen N. Neuralgic amyotrophy: An update on diagnosis, pathophysiology, and treatment. Muscle Nerve 2016;53(3):337–50. Available from: https://doi.org/10.1002/mus.25008.

[59] D'Amico A, Mercuri E, Tiziano FD, Bertini E. Spinal muscular atrophy. Orphanet J Rare Dis 2011;6:71. Available from: https://doi.org/10.1186/1750-1172-6-71.

[60] Anderton RS, Meloni BP, Mastaglia FL, Boulos S. Spinal muscular atrophy and the antiapoptotic role of survival of motor neuron (SMN) protein. Mol Neurobiol 2013;47(2):821–32. Available from: https://doi.org/10.1007/s12035-013-8399-5.

[61] Masson R, Brusa C, Scoto M, Baranello G. Brain, cognition, and language development in spinal muscular atrophy type 1: A scoping review. Dev Med Child Neurol 2021;63(5):527–36. Available from: https://doi.org/10.1111/dmcn.14798.

[62] Zerres K, Rudnik-Schoneborn S. Natural history in proximal spinal muscular atrophy. Clinical analysis of 445 patients and suggestions for a modification of existing classifications. Arch Neurol 1995;52(5):518–23. Available from: https://doi.org/10.1001/archneur.1995.00540290108025.

[63] Kolb SJ, Kissel JT. Spinal muscular atrophy: A timely review. Arch Neurol 2011;68(8):979–84. Available from: https://doi.org/10.1001/archneurol.2011.74.

[64] Kolb SJ, Coffey CS, Yankey JW, Krosschell K, Arnold WD, Rutkove SB, et al. Natural history of infantile-onset spinal muscular atrophy. Ann Neurol 2017;82(6):883–91. Available from: https://doi.org/10.1002/ana.25101.

[65] Coratti G, Lucibello S, Pera MC, Duong T, Muni Lofra R, Civitello M, et al. Gain and loss of abilities in type II SMA: A 12-month natural history study. Neuromuscul Disord 2020;30(9):765–71. Available from: https://doi.org/10.1016/j.nmd.2020.07.004.

[66] Mercuri E, Lucibello S, Pera MC, Carnicella S, Coratti G, de Sanctis R, et al. Long-term progression in type II spinal muscular atrophy: A retrospective observational study. Neurology 2019;93(13):e1241–7. Available from: https://doi.org/10.1212/WNL.0000000000008166.

[67] Piepers S, van den Berg LH, Brugman F, Scheffer H, Ruiterkamp-Versteeg M, van Engelen BG, et al. A natural history study of late onset spinal muscular atrophy types 3b and 4. J Neurol 2008;255(9):1400–4. Available from: https://doi.org/10.1007/s00415-008-0929-0.

[68] Wadman RI, Wijngaarde CA, Stam M, Bartels B, Otto LAM, Lemmink HH, et al. Muscle strength and motor function throughout life in a cross-sectional cohort of 180 patients with spinal muscular atrophy types 1c-4. Eur J Neurol 2018;25(3):512–18. Available from: https://doi.org/10.1111/ene.13534.

[69] Polido GJ, de Miranda MMV, Carvas N, Mendonca RH, Caromano FA, Reed UC, et al. Cognitive performance of children with spinal muscular atrophy: A systematic review. Dement Neuropsychol Oct- 2019;13(4):436–43. Available from: https://doi.org/10.1590/1980-57642018dn13-040011.

[70] Phan HC, Taylor JL, Hannon H, Howell R. Newborn screening for spinal muscular atrophy: Anticipating an imminent need. Semin Perinatol 2015;39(3):217–29. Available from: https://doi.org/10.1053/j.semperi.2015.03.006.

[71] Serra-Juhe C, Tizzano EF. Perspectives in genetic counseling for spinal muscular atrophy in the new therapeutic era: early pre-symptomatic intervention and test in minors. Eur J Hum Genet 2019;27(12):1774–82. Available from: https://doi.org/10.1038/s41431-019-0415-4.

[72] Post AL, Mottola AT, Kuller JA. What's new in prenatal genetics? A review of current recommendations and guidelines. Obstet Gynecol Surv 2017;72(10):610–17. Available from: https://doi.org/10.1097/OGX.0000000000000491.

[73] Darras B.M., Mazurkiewicz-Beldzinska R., Rose M., Xiong K., Zanoteli H., Baranello E., et al. FIREFISH Part 2: 24-month Efficacy and Safety of Risdiplam in Infants with Type 1 Spinal Muscular Atrophy (SMA). presented at: *American Academy of Neurology 2021 Virtual Annual Meeting*; April 17–22, 2021; Session Session P6: Neuromuscular Disorders and Clinical Trials. https://n.neurology.org/content/96/15_Supplement/4126.

[74] Darras BT, Masson R, Mazurkiewicz-Beldzinska M, Rose K, Xiong H, Zanoteli E, et al. Risdiplam-treated infants with type 1 spinal muscular atrophy versus historical controls. N Engl J Med 2021;385(5):427–35. Available from: https://doi.org/10.1056/NEJMoa2102047.

[75] Finkel RS, Mercuri E, Darras BT, Connoly AM, Kuntz NL, Kirschner J, et al. Nusinersen versus sham control in infantile-onset spinal muscular atrophy. N Engl J Med 2017;377(18):1723–32. Available from: https://doi.org/10.1056/NEJMoa1702752.

[76] Mendell JR, Al-Zaidy S, Shell R, Arnold WD, Rodino-Klapac LR, Prior TW, et al. Single-dose gene-replacement therapy for spinal muscular atrophy. N Engl J Med 2017;377(18):1713–22. Available from: https://doi.org/10.1056/NEJMoa1706198.

[77] Mercuri E, Darras BT, Chiriboga CA, Day JW, Campbell C, Connolly AM, et al. Nusinersen versus sham control in later-onset spinal muscular atrophy. N Engl J Med 2018;378(7):625–35. Available from: https://doi.org/10.1056/NEJMoa1710504.

[78] Pane M, Palermo C, Messina S, Sansone VA, Bruno C, Catteruccia M, et al. Nusinersen in type 1 SMA infants, children and young adults: Preliminary results on motor function. Neuromuscul Disord 2018;28(7):582–5. Available from: https://doi.org/10.1016/j.nmd.2018.05.010.

[79] Ho G, Carey KA, Cardamone M, Farrar MA. Myotonic dystrophy type 1: Clinical manifestations in children and adolescents. Arch Dis Child 2019;104(1):48–52. Available from: https://doi.org/10.1136/archdischild-2018-314837.

[80] Echenne B, Rideau A, Roubertie A, Sebire G, Rivier F, Lemieux B. Myotonic dystrophy type I in childhood long-term evolution in patients surviving the neonatal period. Eur J Paediatr Neurol 2008;12(3):210–23. Available from: https://doi.org/10.1016/j.ejpn.2007.07.014.

[81] Douniol M, Jacquette A, Guile JM, Tanguy ML, Angeard N, Heron D, et al. Psychiatric and cognitive phenotype in children and adolescents with myotonic dystrophy. Eur Child Adolesc Psychiatry 2009;18(12):705–15. Available from: https://doi.org/10.1007/s00787-009-0037-4.

[82] Johnson NE, Abbott D, Cannon-Albright LA. Relative risks for comorbidities associated with myotonic dystrophy: A population-based analysis. Muscle Nerve 2015;52(4):659–61. Available from: https://doi.org/10.1002/mus.24766.

[83] Emparanza JI, Lopez de Munain A, Greene MH, Matheu A, Fernandez-Torron R, Gadalla SM. Cancer phenotype in myotonic dystrophy patients: Results from a meta-analysis. Muscle Nerve 2018;58(4):517–22. Available from: https://doi.org/10.1002/mus.26194.

[84] Fu YH, Pizzuti A, Fenwick Jr. RG, King J, Rajnarayan S, Dunne PW, et al. An unstable triplet repeat in a gene related to myotonic muscular dystrophy. Science 1992;255(5049):1256–8. Available from: https://doi.org/10.1126/science.1546326.

[85] Lanni S, Pearson CE. Molecular genetics of congenital myotonic dystrophy. Neurobiol Dis 2019;132:104533. Available from: https://doi.org/10.1016/j.nbd.2019.104533.

[86] Machuca-Tzili L, Brook D, Hilton-Jones D. Clinical and molecular aspects of the myotonic dystrophies: A review. Muscle Nerve 2005;32(1):1–18. Available from: https://doi.org/10.1002/mus.20301.

[87] López-Martínez A, Soblechero-Martín P, de-la-Puente-Ovejero L, Nogales-Gadea G, Arechavala-Gomeza V. An overview of alternative splicing defects implicated in myotonic dystrophy type I. Genes (Basel) 2020;11(9). Available from: https://doi.org/10.3390/genes11091109.

[88] Turner C, Hilton-Jones D. Myotonic dystrophy: Diagnosis, management and new therapies. Curr Opin Neurol 2014;27(5):599–606. Available from: https://doi.org/10.1097/WCO.0000000000000128.

[89] Reardon W, Newcombe R, Fenton I, Sibert J, Harper PS. The natural history of congenital myotonic dystrophy: Mortality and long term clinical aspects. Arch Dis Child 1993;68(2):177–81. Available from: https://doi.org/10.1136/adc.68.2.177.

[90] Mathieu J, Allard P, Potvin L, Prevost C, Begin P. A 10-year study of mortality in a cohort of patients with myotonic dystrophy. Neurology 1999;52(8):1658–62. Available from: https://doi.org/10.1212/wnl.52.8.1658.

[91] Heatwole C, Luebbe E, Rosero S, Eichinger K, Martens W, Hilbert J, et al. Mexiletine in myotonic dystrophy type 1: A randomized, double-blind, placebo-controlled trial. Neurology 2021;96(2):e228–40. Available from: https://doi.org/10.1212/WNL.0000000000011002.

[92] Mousele C, Matthews E, Pitceathly RDS, Hanna MG, MacDonald S, Savvatis K, et al. Long-term safety and efficacy of mexiletine in myotonic dystrophy types 1 and 2. Neurol Clin Pract 2021;11(5):e682–5. Available from: https://doi.org/10.1212/CPJ.0000000000001073.

[93] Horrigan J, Gomes TB, Snape M, Nikolenko N, McMorn A, Evans S, et al. A phase 2 study of AMO-02 (Tideglusib) in congenital and childhood-onset myotonic dystrophy type 1 (DM1. Pediatr Neurol 2020;112:84–93. Available from: https://doi.org/10.1016/j.pediatrneurol.2020.08.001.

[94] Cook A, Giunti P. Friedreich's ataxia: Clinical features, pathogenesis and management. Br Med Bull 2017;124(1):19–30. Available from: https://doi.org/10.1093/bmb/ldx034.

[95] Durr A, Cossee M, Agid Y, Campuzano V, Mignard C, Penet C, et al. Clinical and genetic abnormalities in patients with Friedreich's ataxia. N Engl J Med 1996;335(16):1169–75. Available from: https://doi.org/10.1056/NEJM199610173351601.

[96] Nieto A, Correia R, de Nobrega E, Monton F, Hess S, Barroso J. Cognition in Friedreich ataxia. Cerebellum 2012;11(4):834–44. Available from: https://doi.org/10.1007/s12311-012-0363-9.

[97] Fichera M, Castaldo A, Mongelli A, Marchini G, Gellera C, Nanetti L, et al. Comorbidities in Friedreich ataxia: Incidence and manifestations from early to advanced disease stages. Neurol Sci 2022;43(12):6831–8. Available from: https://doi.org/10.1007/s10072-022-06360-w.

[98] Campuzano V, Montermini L, Molto MD, Pianese L, Cossee M, Cavalcanti F, et al. Friedreich's ataxia: Autosomal recessive disease caused by an intronic GAA triplet repeat expansion. Science 1996;271(5254):1423–7. Available from: https://doi.org/10.1126/science.271.5254.1423.

[99] Rummey C, Corben LA, Delatycki M, Wilmot G, Subramony SH, Corti M, et al. Natural history of Friedreich's Ataxia: Heterogeneity of neurological progression and consequences for clinical trial design. Neurology 2022;99(14):e1499–510. Available from: https://doi.org/10.1212/WNL.0000000000200913.

[100] Reetz K, Dogan I, Costa AS, Dafotakis M, Fedosov K, Giunti P, et al. Biological and clinical characteristics of the European Friedreich's Ataxia consortium for translational studies (EFACTS) cohort: A cross-sectional analysis of baseline data. Lancet Neurol 2015;14(2):174–82. Available from: https://doi.org/10.1016/S1474-4422(14)70321-7.

[101] Cossee M, Schmitt M, Campuzano V, Reutenauer L, Moutou C, Mandel JL, et al. Evolution of the Friedreich's ataxia trinucleotide repeat expansion: founder effect and premutations. Proc Natl Acad Sci USA 1997;94(14):7452–7. Available from: https://doi.org/10.1073/pnas.94.14.7452.

[102] Delatycki MB, Bidichandani SI. Friedreich ataxia-pathogenesis and implications for therapies. Neurobiol Dis 2019;132:104606. Available from: https://doi.org/10.1016/j.nbd.2019.104606.

[103] Waldvogel D, van Gelderen P, Hallett M. Increased iron in the dentate nucleus of patients with Friedrich's ataxia. Ann Neurol 1999;46(1):123–5. Available from: https://doi.org/10.1002/1531-8249.

[104] Selvadurai LP, Harding IH, Corben LA, Georgiou-Karistianis N. Cerebral abnormalities in Friedreich ataxia: A review. Neurosci Biobehav Rev 2018;84:394–406. Available from: https://doi.org/10.1016/j.neubiorev.2017.08.006.

[105] Tsou AY, Paulsen EK, Lagedrost SJ, Perlman SL, Mathews KD, Wilmot GR, et al. Mortality in Friedreich ataxia. J Neurol Sci 2011;307(1–2):46–9. Available from: https://doi.org/10.1016/j.jns.2011.05.023.

[106] Corben LA, Lynch D, Pandolfo M, Schulz JB, Delatycki MB, Clinical Management Guidelines Writing G. Consensus clinical management guidelines for Friedreich ataxia. Orphanet J Rare Dis 2014;9:184. Available from: https://doi.org/10.1186/s13023-014-0184-7.

[107] Lynch DR, Chin MP, Delatycki MB, Subramony SH, Corti M, Hoyle JC, et al. Safety and efficacy of Omaveloxolone in Friedreich Ataxia (MOXIe study). Ann Neurol 2021;89(2):212–25. Available from: https://doi.org/10.1002/ana.25934.

[108] Lu JQ, Tarnopolsky MA. Mitochondrial neuropathy and neurogenic features in mitochondrial myopathy. Mitochondrion 2021;56:52–61. Available from: https://doi.org/10.1016/j.mito.2020.11.005.

[109] Velez-Bartolomei F, Lee C, Enns G. Merrf. In: Adam MP, Everman DB, Mirzaa GM, Pagon RA, Wallace SE, Bean LJH, et al., eds. Internet: GeneReviews((R)); 1993.

[110] Finsterer J. Inherited mitochondrial neuropathies. J Neurol Sci 2011;304(1–2):9–16. Available from: https://doi.org/10.1016/j.jns.2011.02.012.

[111] Ahmed ST, Craven L, Russell OM, Turnbull DM, Vincent AE. Diagnosis and treatment of mitochondrial myopathies. Neurother J Am Soc Exp NeuroTher 2018;15(4):943–53. Available from: https://doi.org/10.1007/s13311-018-00674-4.

[112] Olimpio C, Tiet MY, Horvath R. Primary mitochondrial myopathies in childhood. Neuromuscul Disord 2021;31(10):978–87. Available from: https://doi.org/10.1016/j.nmd.2021.08.005.

[113] Thorburn DRRJ, Rahman S, Adam MPED, Mirzaa GM, Pagon RA, Wallace SE, et al. *Mitochondrial DNA-associated leigh syndrome and NARP*. University of Washington; 2003 [Updated 28.09.2017]. https://www.ncbi.nlm.nih.gov/books/NBK1173/.

Chapter 21

Neurodegenerative diseases

Aram Kim[1] and Alexis L. Franks[2]
[1]*Division of Neurodevelopmental Medicine, Thompson Autism and Neurodevelopmental Center - Children's Hospital of Orange County, Orange, CA, United States,* [2]*Department of Pediatrics, Division of Child Neurology and Child Development, University of Pittsburgh School of Medicine, Pittsburgh, PA, United States*

Introduction

Neurodegenerative diseases are progressive disorders affecting the central nervous system (CNS) and/or peripheral nervous system (PNS), leading to loss of previously acquired skills. Pediatric neurodegenerative diseases are especially rare and consist of several broad categories of disorders such as leukodystrophies, genetic leukoencephalopathies, hereditary ataxias, and storage disorders. Incidence and prevalence of these uncommon diseases can be difficult to ascertain. The estimated incidence of leukodystrophies are approximately 1:7500 live births, and the estimated prevalence of hereditary ataxias—both autosomal dominant and recessive—are 2.7–3.3:100,000 live births [1,2]. However, with diagnostic improvements and increasing availability and affordability of next generation sequencing, as well as implementation of additional screening tools such as newborn screening, the prevalence, and the symptomatic spectrum, of these diseases will likely continue to increase. This chapter will review the general clinical presentation and recommended diagnostic workup for individuals suspected to have a neurodegenerative disease, with a focus on leukodystrophies, potentially treatable conditions, and select examples. Additional neurogenetic and neurometabolic conditions are reviewed in Chapters: Approach to diagnosis and management of genetic neurodevelopmental disorders and Neurometabolic disorders.

Clinical presentation overview and general approach

Initial presentations of neurodegenerative conditions often include neurobehavioral symptoms that not uncommonly trigger subspecialty neurodevelopmental referrals (Table 21.1). Thus, maintaining a working understanding of the various presentations and manifestations of neurodegenerative disease is essential when evaluating patients from a neurodevelopmental disabilities perspective. A clinician should suspect a neurodegenerative disease when a child presents with a stagnation or deceleration of development, a developmental regression, or inconsolable protracted irritability. However, new onset of executive dysfunction in school age children without prior evidence of such cognitive or behavioral tendencies should also trigger suspicion. For example, symptoms of inattention may be a harbinger of active disease in the cerebral form of X-linked adrenoleukodystrophy (ALD) [3]. Wilson disease (WD) may present with new onset learning problems in adolescents [18,19,26]. Children with megalencephalic leukoencephalopathy with subcortical cysts can come to medical attention due to symptoms of an autism spectrum disorder and macrocephaly [27]. Thus augmenting a careful clinical assessment with behavioral screenings and neuropsychological testing to identify disproportionately and newly affected neurocognitive areas, or sequential developmental screenings to trend developmental quotients, can be helpful for refining phenotypic symptomatology. In general, even without extensive screenings, red flags for neurodegenerative disease include loss of or failure to gain developmental milestones, refractory or unprecedented behavioral inattention and irritability, cognitive decline affecting executive function, memory, and/or learning, the new development of epilepsy or epileptic encephalopathies in previously neurotypical individuals, and the new onset of a movement disorder including dystonia or ataxia [28].

In childhood, primary neurodegenerative conditions often reflect clinical manifestations of underlying genetic conditions, including inborn errors of metabolism (IEMs). Except for severe IEMs presenting within the neonatal period before measurable neurodevelopment can occur, the majority of these conditions present in a subacute or chronic

Capute and Accardo's Neurodevelopmental Disabilities in Infancy and Childhood. DOI: https://doi.org/10.1016/B978-0-12-824060-1.00015-8

TABLE 21.1 Overview of most common ages of onset and presenting symptoms for selective pediatric neurodegenerative diseases.

Condition	Most common ages of onset[a]	Common presenting symptoms (primarily neurodevelopmental symptoms shown in bold)[b]
X-linked adrenoleukodystrophy (ALD)	Childhood—adrenal insufficiency 0–10 years; cerebral disease 3–11 years	• Adrenal crises in males • **Deficits in cognition involving visuospatial and visuomotor function, attention, reasoning** • **Decline in school performance**
Metachromatic leukodystrophy (MLD)	Late infantile <30 months (mean age of symptom onset: 18 months)	• **Gross motor delay: initially hypotonia and progression to hypertonia** • Feeding and swallowing difficulties
Pelizaeus–Merzbacher (PMD)	Infantile to early childhood	• **Nystagmus** • **Hypotonia and progression to hypertonia** • **Seizures**
Alexander disease	Infantile (<30 days of age)	• **Hypotonia** • **Myoclonus** • **Seizures** • **Megalencephaly** • **Loss of gross and fine motor skills**
Krabbe disease	Infantile	• **Irritability** • **Regression of development—mainly motor** • Difficulty with feeds • **Visual difficulties**
Aicardi–Gouteries syndrome (AGS)	Infantile	• **Appendicular hypertonia** • **Axial hypotonia** • Low grade fevers • **Ocular jerks** • Feeding issues • Thrombocytopenia at birth • Hepatosplenomegaly • Dermatologic manifestations
Canavan disease (CD)	Infantile	• **Irritability** • **Hypotonia** • **Visual difficulties** • **Macrocephaly**
Cerebrotendinous xanthomatosis (CTX)	Childhood to adolescence	• Childhood onset diarrhea • **Learning issues causing school difficulties**
Vanishing white matter disease (VWM)	Childhood	• **Abnormal gait, hypotonia, ataxia** • **Seizures** • **Stress-provoked episodes of neurocognitive decline, such as motor regression or encephalopathy**
Megalencephalic leukoencephalopathy with subcortical cysts (MLC)	Infantile	• **Macrocephaly** • **Autistic symptoms** • **Ataxia** • **Spasticity and progressive motor decline**
Peroxisomal biogenesis disorders/ Zellweger spectrum disorders (ZSD)	Variable	• Enamel abnormalities, **sensorineural hearing loss**, ocular abnormalities such as cataracts[b] • Neonatal-infantile: **hypotonia**, difficulty feeding, hepatomegaly, **seizures** • Childhood: adrenal insufficiency, liver dysfunction, **seizures**, **developmental regression** • Adolescent: **peripheral neuropathy**

(*Continued*)

TABLE 21.1 (Continued)

Condition	Most common ages of onset[a]	Common presenting symptoms (primarily neurodevelopmental symptoms shown in bold)[b]
Neurodegeneration with brain iron accumulation (NBIA)	Variable	• **New-onset movement disorder (e.g., dystonia, chorea, parkinsonism)** • **Progressive loss of motor skills** • **Cognitive decline, variably preceded by nonspecific developmental delay**
Hepatolenticular degeneration/ Wilson disease (WD)	Adolescence	• Hepatic dysfunction • Kaiser–Fleischer rings • **Mild learning difficulties, decline in school performance** • **Mild psychiatric/mood symptoms**
Ataxia-telangiectasia (A-T)	Toddler or preschool age	• **Failure to progress in gross motor development beyond immature gait** • **Ataxia and imbalance** • **Choreoathetosis, tremor, myoclonic jerks, and/ or dystonia and parkinsonism** • Scleral telangiectasias • **Abnormal eye movements (oculomotor apraxia, abnormal saccades, nystagmus)**
Friedrich's ataxia (FA)	Adolescence	• Scoliosis • **Ataxia and areflexia** • Cardiomyopathy, diabetes • **Vision and hearing impairment** • **Mild deficits in executive function and emotional lability**

[a]*Significant clinical heterogeneity exists for age of onset, presenting symptoms, disease severity, and prognosis. See text for additional review.*
[b]*Presentation is variable based on age; however, there are common symptoms. ALD [3–4], MLD [5], PMD [6], AD [7], Krabbe [8], AGS [9], Canavan [10,11], CTX [12], ZSD [13], NBIA [14–17], WD [18–20], A-T [21,22], FA [23–25].*

fashion. Underlying functional alterations in proteins may lead to gradual deterioration over time or permit normal development, prior to reaching a pathologic threshold where developmental plateau or regression is appreciated. Patients may experience an intermittent course. With IEMs or mitochondrial diseases, systemic stressors can provoke metabolic crises, and variable degrees of recovery may be possible with treatment and supportive care. The courses of some individual neurodegenerative conditions also can follow a well-characterized pattern of age of onset and speed of progression, such as Krabbe disease, where age of onset has significant ramifications for individual prognosis [8]. Other conditions, however, demonstrate significant variability based on the severity of the underlying disease, such as X-linked ALD where children with predisposing genetic variants still may or may not develop cerebral symptoms [4].

Definitive diagnosis of neurodegenerative conditions relies on genetic testing and magnetic resonance imaging (MRI) findings; however, given the frequency of non-specific MRI findings and variants of unclear significance detected on genetic testing, a thorough history and physical examination remains essential to narrow the differential diagnosis and guide confirmatory testing. Furthermore, it is crucial for the clinician to maintain a broad differential diagnosis and elicit features of the illness which may point to secondary neurodegeneration due to a disorder that presents with seemingly progressive neurologic symptoms (Table 21.2).

Diagnostic approach

Diagnosis for childhood neurodegenerative disorders is based on clinical suspicion along with ancillary testing such as neuroimaging, electroencephalograms (EEGs), biochemical testing, and confirmatory genetic testing (Fig. 21.1).

TABLE 21.2 Alternative etiologic considerations of progressive neurologic symptoms in childhood, with examples [29–35].

Systemic illnesses
• Sepsis-associated encephalopathy • Hypertensive encephalopathy • Thyrotoxicosis
Neuroinflammatory conditions
• Multiple sclerosis • Acute disseminated encephalomyelitis • Autoimmune encephalitis • Paraneoplastic encephalitis • Systemic rheumatologic illness
Metabolic and nutritional derangements
• Wernicke encephalopathy • Heavy metal poisoning
Chronic infections
• Human immunodeficiency virus • Fungal meningitis • Measles/subacute sclerosing panencephalitis
Epileptic encephalopathies
• Electrographic status epilepticus of sleep/Landau–Kleffner syndrome • Rasmussen encephalitis • Progressive myoclonic epilepsies
Traumatic brain injuries
Psychiatric illness
• Childhood-onset schizophrenia
Autism with regressive features (early in course)

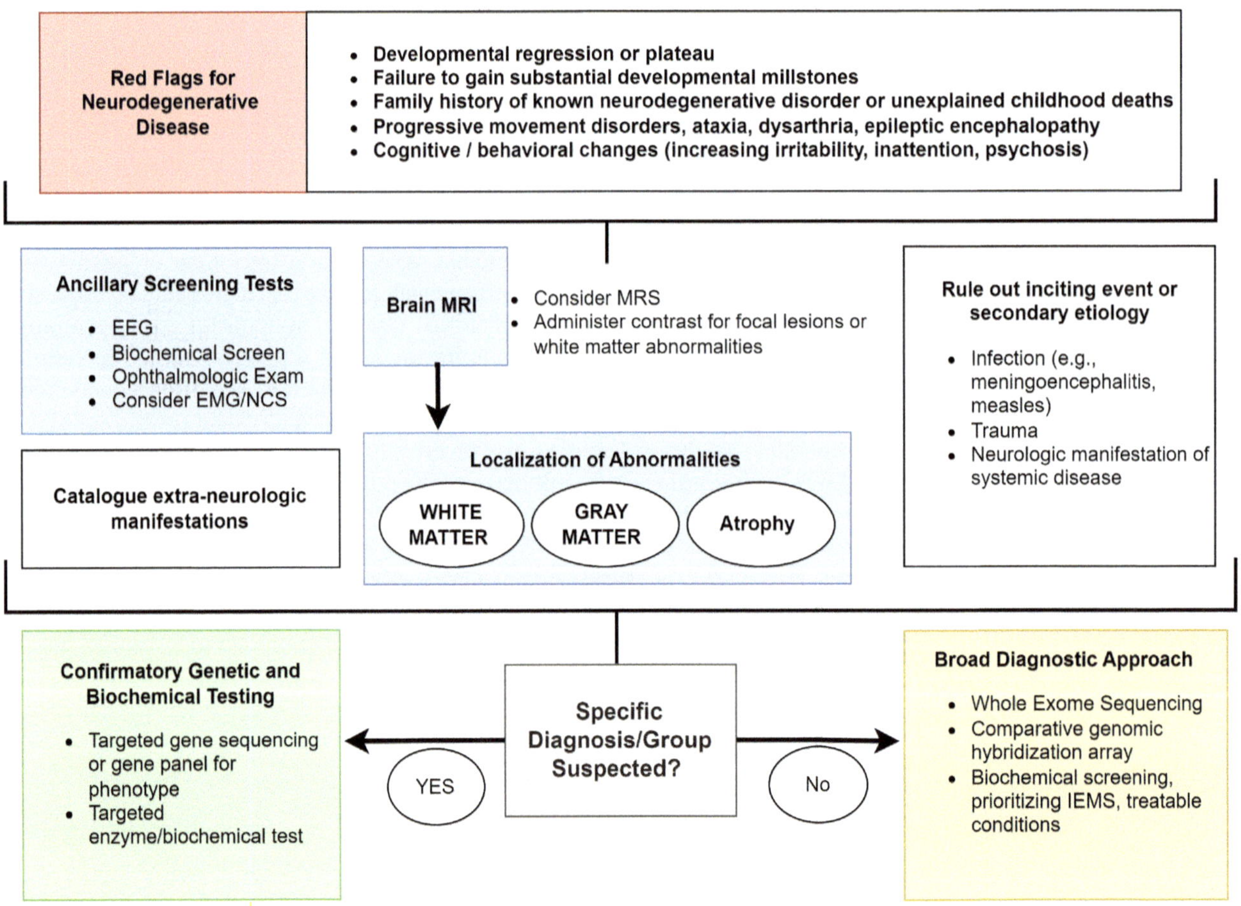

FIGURE 21.1 Diagnostic approach for neurodegenerative disease in childhood.

Clinical evaluation

Key historical features include the timing of symptom onset (neonatal period, <12 months of age, early or late childhood) along with the rate and pattern of progression (regression of skills, global developmental delays, episodic symptoms). Family history may point to heritable conditions. Comorbid medical conditions and nonneurologic findings on physical examination may narrow the differential diagnosis (Table 21.1).

Brain magnetic resonance imaging

Brain MRI is an exceedingly useful tool as it will not only assist with identifying differences in brain development, but also reveal abnormalities in the white matter (e.g., demyelination or hypomyelination), gray matter (e.g., gliosis after prior insult, abnormal deposition of compounds), and overall brain volume (e.g., global or focal atrophy). Additional specialized sequences, such as diffusion tensor imaging (including diffusion-weighted imaging) and administration of intravenous gadolinium contrast, can indicate areas of ongoing brain injury, inflammation, and cellular disruption. Finally, magnetic resonance spectroscopy (MRS) can also identify spectral peaks reflective of levels of certain metabolites in the brain (including lactate and N-acetylaspartate [NAA]), which can be helpful to better characterize nonspecific findings or support a working diagnosis.

Since many individual conditions demonstrate characteristic MRI patterns, along with the ubiquity of MRI in the workup of neurodevelopmental conditions, early brain MRI can be essential to guide workup of childhood neurodegenerative disease. Specifically, MRI findings can distinguish neurodegenerative conditions based on the pattern and location of any abnormalities. Confirmatory genetic and metabolic testing should be pursued; if the MRI is nondiagnostic, a broad genetic testing approach is important. See subchapter: Neuroimaging in neurodevelopmental disabilities.

Laboratory evaluation: metabolic and genetic testing

Clinical history, physical exam, and MRI findings can expedite laboratory evaluation by improving specificity of tests sent. If a specific condition is suspected, then directed genetic and metabolic testing should be prioritized. Nonspecific screening for easily reversible or treatable conditions is acceptable if MRI findings are nonspecific, and broad genetic approaches (e.g., whole exome sequencing or whole genome sequencing) are required. Screening for IEMs is frequently warranted, particularly if the course is episodic (see Chapter: Neurometabolic disorders), and enzymatic function assessment can also guide diagnosis. An initial screening approach should prioritize treatable conditions, including confirmation of newborn screening results, a comprehensive metabolic panel, complete blood count, lactate, ammonia, thyroid studies, lead screen, urinalysis, plasma amino acids, urine organic acids, acylcarnitine profile, plasma very long chain fatty acids (VLCFA), leukocyte lysosomal enzyme battery (must include galactocerebrosidase and arylsulfatase A), cholestanol, urine glycosaminoglycans, and urine sulfatides [36,37].

Genetic testing is useful in both targeted and broad approaches. Testing of a single gene or gene panel (e.g., leukodystrophy gene panel) based on initial findings may help delineate a diagnosis. However, if a single diagnosis is not clear, broad testing with comparative genomic hybridization array and/or whole exome (or genome) sequencing should be considered. Positive variants and variants of unclear significance should undergo confirmatory metabolic testing, if applicable. Variants of unclear significance should be considered within the context of an individual patient's presentation (see Chapter: Approach to diagnosis and management of genetic neurodevelopmental disorders).

Additional testing

An ophthalmologic examination may assist with characterization of eye movements, identification of any retinal abnormalities, or execution of specialized testing such as visual evoked potentials to hone a patient's diagnosis. EEG should also be pursued, even if a clinical history of seizures is lacking, in order to eliminate epileptic encephalopathies as an etiology. Electromyography and nerve conduction studies (EMG/NCS) can be useful to identify evidence of PNS involvement. Further testing such as cerebrospinal fluid profiles (e.g., protein and glucose, neurotransmitters, amino acids, lactate) can also aid in narrowing diagnoses [36,37].

General approach to management

Certain neurodegenerative conditions have been shown to benefit from targeted therapies or hematopoietic stem cell transplant (HSCT) if performed sufficiently early in the disease course. For other conditions, there may be clinical centers where ongoing research or clinical trials are under investigation (see http://www.clinicaltrials.gov). When applicable, individual disease-specific therapies will be described in the following sections. Unfortunately, the majority of childhood neurodegenerative diseases do not have a targeted disease-modifying or curative therapy. Therefore for many conditions, multidisciplinary supportive care is paramount and achieved through optimizing therapies, managing comorbidities, and minimizing or preventing complications. Even if a specific diagnosis is not attainable, such an approach should still be emphasized. Specific attention should be given to ensuring adequate nutrition and growth, monitoring respiratory function, assessing swallowing function, assessing seizure risk, managing behavior or neuroirritability, and screening for cardiac function if indicated. Involvement of palliative or supportive care is recommended early even if targeted therapy for the condition is available.

Abnormal white matter

Leukodystrophies

Both leukodystrophies and genetic leukoencephalopathies can present with white matter abnormalities; however, classification will be based on expert opinion with leukodystrophies being defined as heritable disorders where pathologic alterations of nonneuronal glial cells lead to detrimental effects on the white matter or the CNS, with or without involvement of the PNS (Table 21.3) [38].

TABLE 21.3 Disorders classified as leukodystrophies.

Condition
18q minus syndrome
X-linked adrenoleukodystrophy[a]
Adult onset leukodystrophy neuronal spheroids and pigmented glia
Aicardi–Goutieres syndrome
Alexander disease
Autosomal dominant leukodystrophy with autonomic disease
Canavan disease
Cerebrotendinous xanthomatosis[a]
Chloride ion channel 2-related leukoencephalopathy with intramyelinic edema
eIF2B-related disorder (vanishing white matter disease)
Fucosidosis
Krabbe disease[a]
Hypomyelination with atrophy of basal ganglia and cerebellum
Hypomyelination with brainstem and spinal cord involvement and leg spasticity
Hypomyelination with congenital cataract
Leukoencephalopathy with brainstem and spinal cord involvement and lactate elevation
Leukoencephalopathy thalamus and brainstem involvement with high lactate
Megalencephalic leukoencephalopathy with subcortical cysts
Metachromatic leukodystrophy[a]
Oculodentodigital dysplasia
Pelizaeus–Merzbacher disease
Pelizaeus–Merzbacher like disease
Peroxisomal biogenesis (or Zellweger spectrum) disorders
Pol-III related disorders
Polyglucosan body disease
RNAse T2-deficient leukoencephalopathy
Sialic acid storage disorders
Single enzyme deficiencies of peroxisomal fatty acid beta oxidation
D-Bifunctional protein deficiency peroxisomal acyl-CoA oxidase deficiency sterol carrier protein X deficiency
Sjogren–Larsson syndrome
SOX-10 associated PCWH: peripheral demyelinating neuropathy, central dysmyelinating leukodystrophy, Waardenburg syndrome, and Hirschsprung disease

[a]*Treatable entities; italicized = primarily adult onset [36,38].*

TABLE 21.4 Typical brain magnetic resonance imaging (MRI) findings in selected leukodystrophies.

Condition	Brain MRI findings
X-linked ALD (ALD)	• **T2 hyperintensities (posterior WM, CC splenium)**
Metachromatic leukodystrophy (MLD)	• T2 hyperintensities (parieto-occipital/central > fronto-central, callosal WM) • Cerebral atrophy
Pelizaeus–Merzbacher disease (PMD)	• Normal or hypomyelination in IC, MCP
Alexander disease	• **WM T2 hyperintensities (anterior > posterior, cerebellum, basal ganglia, thalamus, brainstem)** • **PV rim T1 hyperintensities, T2 hypointensities** • **Additional contrast enhancement ventricular lining, optic chiasm, fornix, dentate nucleus**
Krabbe disease	• WM T2 hyperintensities (PV, cerebellar, IC) • Enlargement of optic nerves and cauda equina, **cranial nerve enhancement**, spontaneous third ventriculostomy
Aicardi–Goutieres syndrome (AGS)	• WM T2 hyperintensities (lobar but spares PV, CC, optic radiations) • Global atrophy • Lateral ventricle: temporal horn dilatation, edema • Calcifications (thalami, putamen, dentate nucleus, subcortical WM)
Canavan disease (CD)	• WM T2 hyperintensities + edema and restricted diffusion (subcortical, extension to PV) • T2 hyperintensities in the thalamus and globus palladi • MRS: NAA peak
Cerebrotendinous xanthomatosis (CTX)	• T2 hyperintensities (dentate nucleus, WM–PV brainstem, IC) • Global atrophy • MRS: increased lactate and lipid, decreased NAA
Vanishing white matter disease (VWM)	• Diffuse abnormal WM T2 FLAIR signal—approximately isointense to CSF (relative sparing of temporal lobes) • *Lack* of contrast enhancement
Megalencephalic leukoencephalopathywith subcortical cysts (MLC)	• Diffuse WM T2 hyperintensities and edema; when severe evolves into subcortical temporal cysts • Relative sparing subcortical WM and optic radiations • Re-normalization possible in remitting subtype
Zellweger spectrum disorder (ZSD)	• Classic form: WM T2 hyperintensities (cerebellum, midbrain, hemispheres) • Infantile form: delayed myelination, decreased WM volume, polymicrogyria, pachygyria

Findings in **bold** are contrast enhancing.
CC, Corpus callosum; *WM*, white matter; *IC*, internal capsule; *MCP*, middle cerebellar peduncle; *PV*, periventricular; *FLAIR*, fluid-attenuated inversion recovery, *CSF*, cerebrospinal fluid. (ALD [39], MLD [40], PMD [41], Alexander [42], Krabbe [43–45], AGS [46], CD [10,47], CTX [48], VWM [49], MLC [27], ZSD [13]).

Most, if not all, are progressive in nature and of varying etiology. Not uncommonly, brain MRI demonstrates specific patterns of white matter abnormalities that offer the first clue to an underlying diagnosis, which are summarized in Table 21.4.

X-linked adrenoleukodystrophy

X-linked ALD is a rare disorder, affecting both men and women, with an estimated incidence of 1:14,700 live births [50]. It is characterized by mutations in the peroxisomal *ABCD1* gene, which encodes a transporter for VLCFA into the peroxisome for beta-oxidation and ultimately leads to toxic accumulation [4].

Clinical presentation

There are three forms of ALD—childhood cerebral ALD (cALD), adrenomyeloneuropathy (AMN), and isolated adrenal insufficiency. Childhood cALD is characterized by cerebral inflammatory demyelination with associated behavioral changes such as hyperactivity, cognitive changes such as school failure, and neurologic changes including seizures and complete disability and death within 4 years of diagnosis, occurring mostly in males between 3 and 11 years old [3]. There is a similar adolescent and adult cALD phenotype, though progression of disease may be slower. AMN is characterized by a progressive axonopathy, atrophy of the spinal cord, and peripheral neuropathy with associated limb weakness, spasticity, and bladder and bowel dysfunction. Most adults with ALD experience these symptoms, though in females, onset is later with milder symptoms [4].

Diagnosis

Individuals with X-linked ALD will show elevated VLCFA levels and variants in the *ABCD1* gene. Adrenal dysfunction should be monitored in all males with ALD soon after diagnosis [51]. With the addition of ALD to the newborn screen, diagnosis in the asymptomatic period can be made. Surveillance MRI should be undertaken to detect cerebral involvement early for possible therapy, as clear genotype–phenotype correlation is lacking. Noncontrasted MRI should be obtained between 12 and 18 months, a second MRI one year after first, followed by contrasted MRIs in the high-risk age range between 3 and 12 years old every 6 months, and yearly MRI thereafter barring development of any clinical deterioration which would prompt immediate additional neuroimaging [3,39].

Treatment

It is standard of care that individuals with adrenal insufficiency undergo replacement therapy. Though HSCT does not affect development of adrenal insufficiency or AMN, it can lead to improved survival and function for individuals with cALD who are neurologically asymptomatic to minimally symptomatic [52–54]. As not all individuals develop cerebral disease, close monitoring of individuals in an experienced center is recommended to determine the optimal timing and indication for HSCT. Other experimental therapies such as Lorenzo's oil, a combination of oleic and erucic acid triglycerides, has been tried with some evidence of modest reduction in the risk of developing cerebral ALD [55]. However, without clear improvements in endocrine and neurologic function and lack of placebo-controlled trials, it is not currently approved by the U.S. Food and Drug Administration [4]. Multiple clinical trials investigating novel therapies for patients with ALD are currently underway, including gene therapy, which has shown initial positive results with short-term follow-up [56].

Aicardi–Goutieres syndrome

Aicardi–Goutieres syndrome (AGS) is an interferonopathy that can be caused by variants in several known genes, with the most common being *TREX1*, *RNASEH2A*, *RNASEH2B*, and *RNASEH2C*. AGS is usually inherited in an autosomal recessive fashion; however, specific autosomal-dominant pathogenic variants have been described [57]. The estimated prevalence of AGS is unknown.

Clinical presentation

AGS typically presents with progressive encephalopathy before 4 months of age, though a small subset may present with global developmental regression and onset of spasticity between 4 and 12 months. Microcephaly, developmental delay, feeding issues and emesis, ocular jerks, intermittent low-grade fevers, and appendicular hypertonicity with axial hypotonia are common early symptoms and can be mistaken for a congenital infection; seizures are variably present. Extraneurologic symptoms may include hepatosplenomegaly and thrombocytopenia at birth, vascular necrotic skin lesions, edematous hands and feet, glaucoma, pulmonary hypertension, cardiomyopathy, and autoimmune thyroiditis [9,58]. Neurologic symptoms can be time-limited, though individuals can alternatively decompensate [59]. Patients can present with nonclassic forms of AGS where patients are asymptomatic during first year of life with minimal clinical and radiological evidence of neurologic disease [59].

Diagnosis and treatment

Chronic cerebrospinal fluid (CSF) lymphocytosis and elevated interferon-alpha are seen; however, these levels decrease with age [57]. Due to high genetic heterogeneity, specialist input is recommended for guidance on testing. Treatment is mainly supportive, requiring a multidisciplinary team to ensure close monitoring as multiple organ systems are involved.

Alexander disease

Alexander disease is an autosomal dominant, neurodegenerative disorder caused by de novo mutations in the *GFAP* gene, which encodes for glial fibrillary acidic protein (GFAP) and is histologically defined by formation of Rosenthal fibers in the brain [60]. Due to its rarity, its worldwide prevalence is unknown; however, in Japan it is reported to be approximately 1:2.7 million individuals [61].

Clinical presentation

Clinical presentation is heterogeneous and has been classically categorized based on age of presentation. The infantile form (0–2 years) is the most common and is associated with regression of developmental milestones, megalencephaly, and seizures with eventual progression to spastic quadriparesis. Survival is variable from a few weeks to several years after birth with rare cases of individuals living into their teen years. Patients with the juvenile form (onset 2–12 years) will usually present between 4 and 10 years of age with slowly progressive bulbar signs along with lower limb spasticity, failure to thrive, scoliosis, and cognitive decline. Patients with adult onset ($>$ 12 years) can present similarly to patients with juvenile onset disease with additional symptoms such as palatal myoclonus or abnormal movements [62,63].

Diagnosis and treatment

MRI plays a major role in diagnosis with confirmatory genetic testing [42] (Table 21.4). Treatment is supportive.

Canavan disease

Canavan disease (CD) is an autosomal recessive, neurodegenerative disease caused by mutations in the *ASPA* gene, which encodes for the enzyme aspartoacylase (ASPA). ASPA deficiency leads to accumulation of NAA and is characterized by both a lack of myelination and spongy degeneration of the white matter [10].

Clinical presentation

Based on the age of presentation, CD can be classified into three different groups—the congenital, infantile, and juvenile form. Patients with the congenital form present with symptoms within the first few weeks of life; whereas those with the infantile form, which is the most common, presents by 6 months of age [10]. Common initial symptoms are irritability, hypotonia with poor head control, poor visual tacking, developmental delay, and macrocephaly. Eventually patients will develop spasticity and seizures [10,11]. Patients with the juvenile form have a milder phenotype with a more variable age of presentation and symptomatology; they can present with intellectual disability and macrocephaly without profound regression [64,65].

Diagnosis and treatment

Patients with CD will have elevated urinary NAA, though due to low specificity, additional testing and genetic confirmation is recommended [10]. There is no definitive treatment for patients with CD, so supportive therapy is key.

Cerebrotendinous xanthomatosis

Cerebrotendinous xanthomatosis (CTX) is a rare, autosomal recessive disorder caused by mutations in the *CYP27A1* gene, which encodes for the mitochondrial enzyme sterol 27-hyroxylase. Mutations in this enzyme lead to decreased bile acid synthesis and excess production of cholestanol. CTX not only causes both CNS and PNS symptoms but also a wide range of systemic disease. Estimated worldwide prevalence is $<$5:100,000 [66].

Clinical presentation

CTX is clinically heterogeneous. Symptom onset ranges from childhood to adolescence (9–19 years); however, delay in diagnosis is usually about 30 years. Infantile or childhood onset diarrhea may be followed by development of epilepsy in the first decade and psychiatric symptoms along with cataracts and tendon xanthomas by the second decade of life. Most will meet developmental milestones appropriately, but later demonstrate difficulty with learning and school performance with poor adaptive skills in their childhood years. Axonal or demyelinating polyneuropathy and movement disorders may develop in the third to fourth decade of life [12]. CTX is also associated with premature atherosclerosis and cardiovascular disease, osteoporosis, and pulmonary disease [48].

Diagnosis

Patients with CTX have elevated plasma and CSF cholestanol and elevated bile alcohol concentration in plasma and urine. Apolipoprotein B may also be elevated in the CSF. Total cholesterol levels are usually low to normal with decreased chenodeoxycholic acid (CDCA) [67].

Treatment

Early replacement therapy with CDCA to decrease the synthesis of toxic bile acids may prevent disease progression and even improve neuropathy when initiated prior to development of neurologic symptoms [68]. Other bile acids, including ursodeoxycholic acid, cholic acid, or taurocholic acid have also been trialed [67,69]. Ongoing management of psychiatric comorbidities and continued monitoring of biochemical abnormalities is recommended.

Krabbe disease

Krabbe disease is an autosomal recessive disorder caused by mutations in the *GALC* gene, which encodes the enzyme galactocerebrosidase. Galactocerebrosidase dysfunction leads to toxic accumulation of the metabolite psychosine in oligodendrocytes and Schwann cells, causing symptoms in both the CNS and PNS [8]. The estimated incidence is approximately 1:100,000–400,000 live births [70].

Clinical presentation

Krabbe disease is categorized based on age of symptom presentation [8,71]. Patients with early infantile onset (<6 months) present with stiffness, extreme irritability, regression of development—especially motor skills—difficulty with feeds, seizures, visual difficulty due to optic atrophy with subacute decline, and an average lifespan of 2 years after symptom onset [8]. Disease progression is more variable in patients with late infantile (7–12 months) and juvenile (12 months to 10 years) disease. Patients with late-infantile disease on average present with symptoms by 8 months of age: loss of milestones, poor feeding, irritability, cortical fisting, and visual changes, with a 5-year survival rate of 50%. Patients with juvenile disease commonly present with progressively abnormal gait, developmental regression, and vision loss. Adolescent (11–20 years) and adult (> 20 years) onset is more rare and usually presents with pronounced peripheral neuropathy [71].

Diagnosis

Krabbe disease is part of several states' newborn screens, but diagnosis still relies on biochemical testing (psychosine levels and galactocerebrosidase activity) with confirmatory genetic testing in individuals with high clinical suspicion and supportive MRI findings [72,73]. Visual-evoked potentials and auditory brainstem-evoked potentials commonly show prolonged latencies, nerve conduction studies can show sensorimotor demyelinating polyneuropathy, and CSF testing shows elevated protein [71,74].

Treatment

Currently, HSCT is the only disease-modifying treatment available for Krabbe disease and should be pursued as early in the disease course as possible for better outcomes [75,76]. Thus all diagnosed cases should be referred to a center specializing in Krabbe disease.

Megalencephalic leukoencephalopathy with subcortical cysts

Megalencephalic leukoencephalopathy with subcortical cysts (MLC) is a rare, infantile onset disease caused by both recessive and dominant mutations either in the *MLC1* or *GLIALCAM* gene. MLC1, the product of the gene *MLC1*, is a membrane protein in astrocytes that regulates brain ion and water homeostasis; GlialCAM is a chaperone of MLC1 [27]. The incidence or prevalence of MLC is not known.

Clinical presentation

Patients with MLC can either present with the classic (autosomal recessive) or remitting (autosomal dominant) form. Mutations in *MLC1* more commonly cause classic MLC. In the classic form, patients present in the first year of life with macrocephaly, and subsequently they develop ataxia, spasticity, motor deterioration at a median age of 5 years followed by progressive cognitive decline. By teenage years, individuals often require a wheelchair, though retaining some degree of ambulation with or without support at age 15 years is a good prognostic marker. While patients with remitting MLC initially present similarly to the classic form with macrocephaly, they usually only experience mild developmental delays without significant neurologic deterioration or decline. Both classic and remitting MLC can be associated with symptoms of an autism spectrum disorders within the context of developmental delay and regression. Epilepsy can be seen in both classic and remitting MLC, though more commonly with the classic form [27].

Diagnosis and treatment

Diagnosis is based on neuroimaging and genetic testing. Therapy is limited to symptomatic management and supportive care [27].

Metachromatic leukodystrophy

Metachromatic leukodystrophy (MLD) is an autosomal recessive, lysosomal storage disorder caused by mutations in the *ARSA* gene, which encodes for the enzyme arylsulfatase A. Deficiency in arylsulfatase A leads to accumulation of sulfatides, which in turn causes demyelination of both the peripheral and central nerves [77]. The incidence of MLD is estimated to be around 1:100,000 live births [78].

Clinical presentation

There are three different forms of MLD based on age of onset for which initial symptoms and prognoses differ [77]. Patients with the late infantile form (<30 months) tend to have gross motor delay before 18 months and experience rapid regression initially developing hypotonia and depressed deep tendon reflexes with progression into hypertonia, spastic quadriplegia, and loss of gross motor function by 40 months. These symptoms are accompanied by regression in all domains of development with visual abnormalities and peripheral neuropathy [5,77]. Patients with the juvenile form (onset 30 months to 16 years) tend to have a more protracted course, and initial symptoms are commonly cognitive (individuals can present with symptoms of new-onset inattention and disruptive behaviors). Overall gross motor development remains stable until development of hypertonicity or spasticity, at which point rapid decline ensues. Patients with the adult form (>16 years) will develop slow intellectual decline and memory loss with emotional instability that may be initially misdiagnosed as a psychiatric illness [77].

Diagnosis

Laboratory measurements consist of blood enzyme activity and urine sulfatide levels along with confirmatory *ARSA* genetic testing. Variant MLD can present with abnormal urine sulfatides and diagnostic MRI findings (see Table 21.4), despite normal *ARSA* genetic and metabolic testing, due to defects of other proteins in the metabolic pathway (e.g., saposin B deficiency, multiple sulfatase deficiencies) [77].

Treatment

HSCT may be considered to modify the disease course for patients with juvenile or adult MLD who present with minimal or no symptoms. HSCT has not been shown to be beneficial at any stage in late infantile MLD [40,79–81].

Pelizaeus–Merzbacher disease

Pelizaeus–Merzbacher disease (PMD) is a X-linked disorder caused by mutations in the *PLP1* gene, which encodes for proteolipid protein (PLP), one of the most abundant proteins found in myelin. PMD has phenotypic variability, and mutations in the *PLP1* gene can also cause spastic paraplegia 2 [82]. The prevalence reported in the United States ranges 1:200,000–1:500,000 [6].

Clinical presentation

PMD is clinically heterogeneous but has historically been classified based on age at presentation. Connatal PMD (neonatal onset) is the most severe form and presents with feeding issues, optic atrophy, extrapyramidal symptoms, and severe hypotonia. Patients have global developmental delay, typically do not ambulate, and expressive language delay is disproportionately more severe than receptive skills. Classic PMD (onset <12 months) presents with horizontal or vertical nystagmus, delayed motor development, and hypotonia with eventual progression to appendicular spasticity, optic atrophy, and seizures. If ambulation is achieved, this skill is typically lost by adolescence. Patients with either connatal or classic PMD can develop seizures that are responsive to antiepileptic agents. Transitional PMD comprises of two forms—spastic paraplegia and the PLP1 null syndrome—which have clinical features of both connatal and classic PMD. Patients present with spastic diplegia before age 5 and can develop cognitive involvement later. PLP null syndrome instead is characterized by mild spastic quadriparesis and a variable amount of cognitive impairment with peripheral neuropathy [6,83].

Diagnosis and treatment

Genetic testing demonstrates a significant amount of heterogeneity; duplications are the most common and are usually associated with the classic phenotype, whereas point mutations can be associated with variable phenotypes [84,85]. Due to this heterogeneity, consultation with a geneticist may be helpful to guide specific testing. Treatment is supportive.

Peroxisomal biogenesis (or Zellweger spectrum) disorders

Peroxisomal biogenesis disorder, or the Zellweger spectrum disorders (ZSD), describes a multisystem disorder due to mutations in the *PEX* gene that cause defects in peroxisome assembly and metabolism. ZSD describes a group of clinical conditions previously thought to represent separate disorders (i.e., infantile Refsum disease, neonatal ALD, and Zellweger syndrome). Due to biochemical and molecular advances, these disorders are now known to represent part of a single clinical spectrum of ZSDs. The incidence of ZSDs in the United States is 1:50,000 newborns [13].

Clinical presentation

Presentation is dependent on age of onset, though some symptoms are common in all forms, including developmental delay, sensorineural hearing loss, and ocular abnormalities including cataracts. Patients who have neonatal-infantile onset (previously Zellweger syndrome) present with profound hypotonia, absent deep tendon reflexes, minimal spontaneous movement, hepatomegaly, and difficulty feeding. Seizures are common, and patients have atypical craniofacial features including high forehead, epicanthal folds, large anterior fontanel, and midface hypoplasia. Mortality is high and most do not survive past the first year [13,86].

Childhood presentations (previously neonatal ALD) are more variable in course and symptomatology. Symptoms include hepatomegaly and liver dysfunction, adrenal insufficiency, renal stones, seizures, and some patients can present with progressive loss of milestones with associated rapidly progressive leukodystrophy. Most patients die before reaching adolescence [13].

Patients with the adolescent–adult form have less severe symptoms and a highly variable course. The majority of these individuals still have symptoms of developmental delay/intellectual disability and a slow progression of disease. Many will have peripheral neuropathy, cerebellar ataxia, and adrenal insufficiency [13].

Diagnosis

Plasma levels of VLCFAs, peroxisomal bile acid intermediates—dihydroxycholestanoic acid (DHCA) and trihydroxycholestanoic acid (THCA), phytanic acid, pristanic acid, and pipecolic acid can be measured along with urinary bile and oxalic acids. Plasmalogen levels in erythrocytes and C26:0-lysoPC in dried blood spots can also be measured, and further testing using fibroblasts may provide clarity in ambiguous cases. Genetic testing can confirm diagnosis and provide information on genotype–phenotype correlations [13].

Treatment

There is no cure for the neurologic symptoms associated with ZSD, and therapy is mainly supportive. Cholic acid, however, has been approved by the Food and Drug Administration as a safe treatment for patients with ZSD and has been shown to improve growth with reduced urinary and plasma excretion of bile acids. However, caution should be used in patients with advanced liver disease [87–89].

Vanishing white matter disease

Vanishing white matter (VWM) disease is an autosomal recessive disease caused by mutations in the *EIF2B1–5* gene, which encodes for subunits of the eIFB2b or eukaryotic initiation factor 2b. eIF2B is important in initiation of translation and regulating protein synthesis under numerous conditions including stress [90]. The prevalence is unknown.

Clinical presentation

Clinical manifestations of disease are quite variable with age of onset representing an important prognostic marker. Disease onset prior to 1 year of age is associated with the most severe and rapidly progressive symptoms. These children develop encephalopathy, irritability, somnolence, and seizures with survival limited to mere months after symptom onset. Nonspecific antenatal symptoms of intrauterine growth restriction, oligohydramnios, and reduced fetal movements have been also reported in this group. Disease onset between 1 and 4 years demonstrates a more variable course, but tends to be associated with progressive spasticity, ataxia, and eventual wheelchair dependence. Survival into adulthood is possible. Later onset is typically accompanied by lower mortality with common presenting symptoms of cognitive and psychiatric issues. Vision and hearing are typically normal, but peripheral neuropathy can be seen [91]. Disease course may be episodic, where stressors such as febrile illnesses and head trauma may provoke symptom deterioration, including worsening motor issues or even language loss and coma in severe instances. Unfortunately, recovery is typically incomplete after these episodes, and younger patients are disproportionately affected [90]. Extraneurologic manifestations of VWM include ovarian failure, growth failure, cataracts, hepatosplenomegaly, pancreatitis, and renal hypoplasia. Severe neurologic disease is correlated with increased multisystem involvement [49].

Diagnosis and treatment

While biochemical testing including elevated CSF glycine supports the diagnosis of VWM, diagnosis relies on specific MRI criteria [49]. Genetic testing may provide further information based on described genotype–phenotype correlations [90]. Supportive measures are the mainstay including aggressive temperature control during illnesses and after vaccinations to help prevent episodes of deterioration.

Genetic leukoencephalopathies

Genetic leukoencephalopathies (gLE) are heritable disorders with significant white matter abnormalities; however, unlike leukodystrophies, the effects on white matter in gLE are not due to primary degeneration within neurons and systemic manifestations of disease predominate, in contrast to leukodystrophies [38]. Not all gLE are degenerative, and gLE encompass a wide range of disorders including IEMs such as disorders of branched chain amino acids and vascular disorders such as cerebral autosomal-dominant arteriopathy with subcortical infarcts. Common neurodegenerative disorders classified as gLE are listed in Table 21.5 along with typical MRI findings.

Abnormal gray matter

Examination of abnormal gray matter signal on brain MRI can also guide etiological elucidation in childhood neurodegenerative disease, particularly in the metabolically active basal ganglia. While ultimate diagnosis, similar to leukodystrophies, often requires genetic testing, familiarity with this group of conditions and their presentations can assist with expediting and improving diagnostic precision.

TABLE 21.5 Overview of select forms of genetic leukoencephalopathies.

Condition	Abbreviation	Affected gene	Inheritance	Brain MRI findings	Additional considerations[a]
Neuronal ceroid lipofuscinosis (Batten's disease)	CLN 1–14	*Variable*	Variable, mostly AR	Global cerebral volume loss and T2 hyperintensities in PV WM	Most are symptomatic by first decade of life with seizures, vision issues, progressive decline in cognitive and motor ability due to progressive neuronal death. ERT available for CLN2
GM1 gangliosidosis	GM1	*GLB1*	AR	Dilated perivascular spaces T2 hyperintensities in cerebral WM and hypointensities in thalami Global cerebral and cerebellar atrophy	Symptoms vary based on age of onset with infantile disease having more rapid progression Common symptoms are hypotonia, macular cherry red spots (infantile), coarse fascies, dysphagia, visceral findings—cardiomyopathy and HS, seizures, ataxia, scoliosis
GM2 gangliosidosis—Tay Sach's (TS) and Sandhoff disease (SD)	GM2	*HEXA* (TS) *HEXB* (SD)	AR	T2 hyperintensities in bilateral basal ganglia, thalami, and cerebral and cerebellar WM with speckled hypointensities in thalami	Symptoms vary based on age of onset Infantile disease: hypotonia, exaggerated startle reflex, seizures, developmental regression, macular cherry red spots, HSM (SD) Juvenile disease: gait issues, seizures, dementia, neuropathy, cerebellar dysfunction
Mucopolysaccharidoses[b]	MPS I-III	*IDUA* (MPSI) *IDS* (MPSII) *Variable* (MPSIII)	MPS I, III: AR MPS II: X-linked	T2 hyperintensities in PV WM Enlarged perivascular spaces and ventricular spaces with concern for hydrocephalus Global cerebral atrophy Cervical spinal canal stenosis	MPS I and II: dysostosis multiplex leading to limited range of motion, decreased pulmonary function, cardiac valve abnormalities, HSM, communicating hydrocephalus, carpal tunnel syndrome, vision issues, cognitive delays MPS III: minimal systemic findings with more pronounced cognitive and behavioral issues Therapy: ERT exists for MPS I and II; hematopoietic stem cell transplant is standard of care of individuals with MPS I (early)

(Continued)

TABLE 21.5 (Continued)

Condition	Abbreviation	Affected gene	Inheritance	Brain MRI findings	Additional considerations[a]
Molybdenum cofactor deficiency	MoCD	*MOSC1* (Type A) *MOSC2* (Type B) *GPHN* (Type C)	AR	Diffuse cerebral infarction of WM and cortex with edema and eventual multicystic WM lesions Edematous lesions in globus palladi and caudate Dandy walker malformations MRS: Choline peak	Normal at birth but soon after develops seizures, feeding issues, coarse fascies, ectopia lentis, nystagmus Therapy for MoCD Type A: cyclic pyranopterin monophosphate

AR, Autosomal recessive; *PV*, periventricular; *WM*, white matter; *HSM*, hepatosplenomegaly.
[a]*Approved, existing therapies are included. Numerous additional therapies not listed here are currently under investigation.*
[b]*Types I–III will only be discussed as they are the most common forms of MPS related with neurodegeneration. (NCL [92], GM1 [93], GM2 [94–96], MPS [97–100], MoCD [101,102]).*

Neurodegeneration with brain iron accumulation

Deposition of heavy metals, specifically iron, within the basal ganglia (i.e., globus pallidus and substantia nigra) is the pathognomonic finding in the neurodegeneration with brain iron accumulation (NBIA) family of otherwise clinically and genetically heterogeneous conditions [16]. Despite their pathologic link, the majority of implicated genes are not directly involved in iron metabolism and are widely expressed throughout the brain, even with the limitation of iron deposition to certain areas [17]. Subsequently, iron deposition in NBIA may be an epiphenomenon, since it is also observed in normal aging and other neurodegenerative conditions (e.g., Alzheimer's disease, Parkinson's disease) [17]. A relationship to mitochondrial dysfunction has been suggested. Prevalence of NBIA is rare and estimated at 1 case per 500,000 [16]. Currently, up to 15 distinct genetic forms of NBIA have been reported [17], though this number is increasing as additional genes are discovered and characterized. The most common forms comprising over 90% of NBIA include pantothenate kinase-associated neurodegeneration (PKAN), phospholipase A1 group VI-associated neurodegeneration [103], beta-propeller protein-associated neurodegeneration (BPAN), and mitochondrial membrane protein-associated neurodegeneration (MPAN) (Table 21.6) [15–17].

Clinical presentation

The age of onset, typical symptoms, and rate of progression of NBIA is highly variable, though typically initial findings include development of a movement disorder, such as dystonia, chorea, or parkinsonism, and can evolve into progressive loss of motor abilities and neuropsychiatric regression. Dysarthria and spasticity may follow. Cognition is variably intact at onset, and many individuals may have early signs of developmental delays [14–17].

Diagnosis

Diagnosis is typically prompted by detection of iron deposition in deep gray matter structures on brain MRI obtained early in the course of disease. Exact etiology is determined through genetic testing, either through a targeted gene panel or exome sequencing [15,16].

Treatment

Treatment options are limited, and supportive care is central for the majority of NBIA. Most targeted therapeutic studies have been for PKAN. Iron chelation therapy has been trialed, and deferiprone has demonstrated limited use in PKAN for slowing or stabilizing progression. Often, clinical symptoms do not improve, though burden of iron deposition on MRI may lessen. In other forms of NBIA, chelation therapy may actually worsen the clinical picture. Other pharmacologically targeted therapies have either been unsuccessful or remain in the preclinical phases, including gene therapy [17].

TABLE 21.6 Overview of common forms of neurodegeneration with brain iron accumulation (NBIA).

Condition	Abbreviation	Affected gene	Inheritance	Location of iron deposition/brain MRI findings	Additional considerations
Pantothenate kinase-associated neurodegeneration	PKAN	*PANK2*	AR	GP, "eye of the tiger sign" (bilateral T2 hyperintense signal surrounded by hypointensity)	Neuroferritinopathy. pigmentary retinopathy, neuropsychiatric symptoms, prominent speech symptoms. No peripheral nervous system involvement. Chelation therapy may be beneficial in some patients.
Phospholipase A2 group VI (PLA2G6)-associated neurodegeneration	PLAN	*PLA2G6*	AR	GP, SN, cerebellar atrophy	Synucleinopathy. Developmental delay, visual impairment, and epilepsy common.
Beta-propeller protein-associated neurodegeneration	BPAN	*WDR45*	X-linked dominant	SN > GP, T1 hyperintense "halo" adjacent to SN in cerebral peduncles	Early developmental delay and intellectual disability, epilepsy. Significant expressive language development limitations. Stereotypies common. Skewed X-inactivation may lead to affected symptoms in females.
Mitochondrial membrane protein-associated neurodegeneration	MPAN	*C19orf12*	AR	GP, SN	Synucleinopathy. Prominent neuropsychiatric features/ behavioral disturbances and later hallucinations. Early parkinsonism with dystonia and spasticity with dementia.

AR, Autosomal recessive; *GP*, globus pallidus; *SN*, substantial nigra [16,17].

Wilson disease/hepatolenticular degeneration

Hepatolenticular degeneration or WD is a rare (prevalence estimates range widely due to variability of frequency in populations and reduced penetrance, between 1:1,500 and 1:30,000 [18]) genetic condition due to biallelic variants in *ATP7B*. Classically, affected individuals develop hepatic failure and neuropsychiatric symptoms due to an inability to excrete hepatic copper into the bile causing toxic levels of copper deposition in tissues [20]. Evidence also supports *ATP7B* has a role in transporting copper out of the brain across the blood–brain barrier [18]. Traditionally, childhood onset WD is thought to primarily present with signs of hepatic dysfunction only, though researchers postulate this view likely leads to an underestimation of pediatric cases [19].

Clinical presentation

Age of onset ranges widely, with an average of approximately 13 years in children. Initial symptoms may include subtle neurodevelopmental or psychiatric symptoms—such as mild cognitive symptoms causing learning difficulties and declines in school performance, and/or depressive symptoms and neurotic behaviors—and markers of liver dysfunction are often discovered incidentally. Neurologic symptoms generally include movement disorders, including dystonia, parkinsonism, ataxia, and stereotypies. Classic examination findings include the postural, upper extremity "wing-beating" tremor and risus sardonicus, which is an abnormal, sustained spasm of the muscles of the face mimicking a grin. WD can eventually cause global effects on multiple aspects of executive function, including attention, reasoning, processing, and memory. Individuals may also be at risk for seizures, headaches, neuropathy, and autonomic dysfunction. Psychiatric symptoms may be prominent in adolescents and adults, including mood disorders, impulsive behavior, disinhibition, and may progress to symptoms of psychosis and schizophrenia when advanced. While neuropsychiatric and hepatic symptoms are considered most prominently, because copper deposition can occur systemically, there is associated risk for multisystem organ involvement and associated dysfunction. Without effective treatment, affected individuals progress into fulminant liver failure and neurological deterioration [18–20].

Diagnosis

Diagnosis of WD includes measurements of serum copper and ceruloplasmin levels, 24-hour urine copper excretion, and a hepatic function panel. Ophthalmologic evaluation can demonstrate Kaiser–Fleischer rings using a slit lamp examination, visible evidence of copper deposition. Definitive diagnosis may be elusive despite genetic testing and/or liver biopsy to measure dry copper concentration, and thus a diagnostic scoring system has been developed [20]. Brain MRI can demonstrate abnormalities in both the gray and white matter, with atrophy of basal ganglia structures and increased susceptibility [26]. Individuals can have T2 hyperintensities throughout the putamen, thalami, midbrain, and pons, with the putamen most prominently affected in younger children [104–106]. The classic MRI finding known as "face of the giant panda" sign is due to a characteristic T2 hyperintense appearance of the midbrain tegmentum surrounding normal but darker appearing red nuclei and substantial nigra [19,105,106].

Treatment

Therapeutic options are targeted at reducing copper overload through copper chelation therapy (e.g., D-penicillamine, trientine) and/or blocking copper absorption through zinc supplementation (zinc will be absorbed preferentially in lieu of the similarly divalent copper ions). Chelation therapy can reverse hepatic symptoms; however, neuropsychiatric symptoms cannot be reversed, and symptoms may continue to progress despite treatment. Liver transplantation, however, is curative and can even stabilize or potentially improve neurologic symptoms [18–20].

Inborn errors of metabolism and mitochondrial disease

Multiple IEMs and mitochondrial diseases can present in a neurodegenerative fashion if not effectively treated (when able) and are associated with abnormal findings in the metabolically active deep gray matter regions on brain MRI. Testing for underlying metabolic disease is warranted when a child presents with symptoms suggesting neurodegeneration, particularly when MRI demonstrates abnormal signal in the putamen, globus pallidus, or thalamus. Urea cycle disorders and methylmalonic aciduria can present with MRI involvement of the globus pallidus, while glutaric aciduria type 1 frequently affects the putamen. Stroke-like episodes or metabolic strokes can be seen in forms of mitochondrial disease, aminoacidopathies, organic acidemias, and urea cycle defects in the deep gray matter and/or brainstem. MRS of lesions may additionally guide diagnosis [106]. See chapter: Neurometabolic disorders.

Progressive volume loss/atrophy

MRI findings in certain neurodegenerative conditions are linked primarily to volumetric abnormalities rather than alterations in signal in the parenchymal tissue. General findings of atrophy may be nonspecific in later stages of neurodegenerative disease, though localized progressive atrophy early in a disease course may guide the evaluation.

Hereditary ataxias and cerebellar atrophy

Heritable ataxias make up many of the childhood neurodegenerative conditions and may present with progressive cerebellar atrophy on brain MRI. Clinical signs and symptoms of cerebellar dysfunction are frequently present, including ataxia, tremor, motor dysregulation, and dysarthria.

Ataxia telangiectasia

Ataxia telangiectasia (A-T) is due to autosomal recessive biallelic variants in *ATM*. With an estimated prevalence of 1:40,000–100,000, A-T is highly rare. Abnormal function of *ATM* leads to deficiencies in DNA repair, which results in a constellation of classic symptoms spanning multiple organ systems. Diagnostic testing in children is typically prompted by the detection of neurologic symptoms in toddler/preschool age children, or preclinically as a result of newborn screening for severe combined immune deficiency. Children may have normal neurodevelopment early in life, but then in toddlerhood, symptoms are detected when children gain initial walking skills but fail to refine them, remaining unsteady. Children often gradually develop worsening imbalance, incoordination, and movement disorders, and typically require a wheelchair for ambulation by age 10 years. Movement disorder symptoms can include choreoathetosis, tremors, myoclonic jerks, and/or dystonia and parkinsonism. Also around this age, telangiectasias are frequently noted in the sclera, along with abnormal eye movements (oculomotor apraxia, abnormal saccades, nystagmus). Extraneurologic symptoms can include immune deficiency, an increased risk of malignancy (especially hematologic), and chronic lung disease. Early in the course, brain MRI may be normal, however, subsequent studies demonstrate progressive cerebellar atrophy that inversely correlates with age [21,22]. Unfortunately, no treatments are currently available, and affected individuals rarely survive past their mid 20s. Given the increased risk of malignancy, exposure to ionizing radiation should be avoided when at all possible [22].

Friedrich ataxia

Friedrich's ataxia (FA) is the most common genetic cause of ataxia (global prevalence 1:40,000) and is due to biallelic trinucleotide repeat (GAA) expansion variants in frataxin (*FXN*), which has a role in mitochondrial energy production [24,25]. Onset of symptoms typically is between ages 10 and 15 years, though can be highly variable [22,23]. Typical symptoms include ataxia and clumsiness which may be accompanied by scoliosis. As the disease progresses, affected individuals may also develop wheelchair dependence, dysarthria, areflexia, cardiomyopathy, and/or diabetes as well as vision and hearing impairment. The ataxia in FA is multimodal with cerebellar, sensory, and vestibular contributions [24]. Psychologically, patients may experience mild deficits in executive function and emotional lability [25]. The number of GAA repeats correlates with severity of disease [24], and cause of death is typically due to hypertrophic cardiomyopathy [25]. MRI demonstrates decreased volume primarily within the brainstem and specific cerebellar structures, including the dentate nucleus and superior and inferior peduncles [23]. The mainstay of treatment is supportive, though the antioxidant idebenone is being studied as a treatment for cardiomyopathy [25].

Spinocerebellar ataxias and recessive ataxias

Spinocerebellar ataxias (SCAs) represent a clinically and genetically heterogeneous group of progressive heritable ataxias [107]. "SCA" is used to refer to the heritable ataxias with autosomal-dominant inheritance, while "recessive ataxias" or "spinocerebellar ataxias, recessive (SCARs)" refers to autosomal recessive conditions. There are over 40 SCAs, the majority of which are due to expansion of trinucleotide repeats in various genes. Likewise, over 50 SCARs have been defined, including both A-T and FA [108]. While a detailed review of all of these conditions is beyond the scope of this chapter, the vast number of identified genetic causes of progressive ataxia emphasizes the utility of MRI, which commonly will demonstrate variable degrees of olivopontocerebellar atrophy [109], to narrow the differential and allow for efficient diagnosis through genetic sequencing [110]. Many SCAs may have overlapping or nonspecific symptoms. Thus a targeted ataxia gene panel with broad coverage of autosomal recessive, autosomal dominant, and trinucleotide repeat disorders is typically the most appropriate first step (particularly because next generation sequencing and comparative genomic hybridization array will not detect trinucleotide repeat disorders) unless family history or clinical, biochemical, or radiological findings suggest a single specific etiology [110].

Cerebral atrophy

Rett syndrome

Rett syndrome is an X-linked dominant condition associated with variants in *MECP2* that presents early in females with a variable pattern of significant developmental regression. Within this context, affected individuals often also are diagnosed with an autism spectrum disorder and classically develop prominent hand stereotypies accompanied by the loss of fine motor skills. Brain MRI may demonstrate decreased global cerebral volumes or in specific regions (e.g., cerebellum early in the disease course), in line with the classic clinical sign of acquired microcephaly [111]. See Chapter Genetic neurodevelopmental disorders: Approach to diagnosis and management for additional review of Rett syndrome.

Childhood onset, neurodegeneration with brain atrophy

A recently described condition characterized by neurodegeneration and global developmental regression, childhood onset, neurodegeneration with brain atrophy, is associated with de novo autosomal-dominant variants in *UBTF*. Affected individuals may have a period of brief neurodevelopment until age 2–3 years, after which global developmental regression ensues and affected individuals can develop ataxia and hypotonia that progresses to spasticity. Epilepsy, acquired microcephaly, and severe intellectual disability are common. Difficulties with attention and autism-like behaviors may also be observed. MRI demonstrates progressive, diffuse global cerebral atrophy, which is typically severe. Treatment is supportive [112–115].

Conclusions

Neurodevelopmental care of children with neurodegenerative conditions is challenging, as therapeutic options are frequently limited. Given the inherent rarity of these conditions, understanding of the natural history and targeted disease-modifying treatments have been unavailable. However, advances in utilization of HSCT, liver transplantation, and trials of gene therapy, may have the potential to broadly change prognosis. Until these treatments are widely available, neurodevelopmentalists should pursue a diagnostic and neuroimaging evaluation, provide anticipatory guidance about long-term expectations tailored to the specific condition and age of onset, and help families connect to specialty clinics and/or clinical trials, as well as provide connections to disease foundations and support groups. Treatment should include maximizing supportive care while maintaining therapies and school supports to maximize functional quality of life. Early engagement of a multidisciplinary team to holistically address care is encouraged. Referrals to other physician specialists, in addition to neurodevelopmental disabilities, may include palliative care, physical medicine and rehabilitation, geneticists, gastroenterology, pulmonology, psychiatry, and cardiology or orthopedics if indicated. Care teams should include nonphysician members, such as genetic counselors (particularly if other family members are at risk or in anticipation of family planning needs), physical therapists, occupational therapists, speech therapists, nutritionists, and special education teachers. The need for therapies is anticipated to increase as neurodegenerative diseases progress, parents should be encouraged to develop an individualized education plan with their child's school in order to ensure appropriate special education and support services are in place. Parents may require the assistance of an educational advocate if inadequate supports are available. Home nursing may be needed to help families manage complex care needs at home, particularly as nutritional and ventilatory needs increase and as affected children get closer to the end of life. End-of-life planning conversations should take place with ample time for patients and families to have careful consideration of goals of care. Psychological counseling should also be encouraged for patients and families to assist those coping with these life-altering diseases.

References

[1] Bonkowsky JL, Nelson C, Kingston J, Filloux F, Mundorff M, Srivastava R. The burden of inherited leukodystrophies in children. Neurology 2010;75(8):718–25.

[2] Ruano L, Melo C, Silva MC, Coutinho P. The global epidemiology of hereditary ataxia and spastic paraplegia: a systematic review of prevalence studies. Neuroepidemiology 2014;42(3):174–83.

[3] Kemp S, Berger J, Aubourg P. X-linked adrenoleukodystrophy: clinical, metabolic, genetic and pathophysiological aspects. Biochim Biophys Acta 2012;1822(9):1465–74.

[4] Turk BR, Theda C, Fatemi A, Moser AB. X-linked adrenoleukodystrophy: pathology, pathophysiology, diagnostic testing, newborn screening and therapies. Int J Dev Neurosci 2020;80(1):52–72.

[5] Harrington M, Whalley D, Twiss J, Rushton R, Martin S, Huynh L, et al. Insights into the natural history of metachromatic leukodystrophy from interviews with caregivers. Orphanet J Rare Dis 2019;14(1):1–10.

[6] Wolf NI, van Spaendonk RML, Hobson GM, et al. PLP1 disorders. In: Adam MP, Feldman J, Mirzaa GM, et al., editors. GeneReviews® [Internet]. Seattle (WA): University of Washington, Seattle; 1999.

[7] Srivastava S, Waldman A, Naidu S. Alexander disease GeneReviews is a registered trademark of the University of Washington, Seattle. All rights reserved In: Adam MP, Ardinger HH, Pagon RA, Wallace SE, Bean LJH, Gripp KW, et al., editors. GeneReviews(®). Seattle (WA): University of Washington; 1993.

[8] Suzuki K. Globoid cell leukodystrophy (Krabbe's disease): update. J Child Neurol 2003;18(9):595–603.

[9] Goutières F. Aicardi–Goutières syndrome. Brain Dev 2005;27(3):201–6.

[10] Hoshino H, Kubota M. Canavan disease: clinical features and recent advances in research. Pediatr Int 2014;56(4):477–83.

[11] Bley A, Denecke J, Kohlschütter A, Schön G, Hischke S, Guder P, et al. The natural history of Canavan disease: 23 new cases and comparison with patients from literature. Orphanet J Rare Dis 2021;16(1):227.

[12] Mignarri A, Gallus GN, Dotti MT, Federico A. A suspicion index for early diagnosis and treatment of cerebrotendinous xanthomatosis. J Inherit Metab Dis 2014;37(3):421–9.

[13] Klouwer FC, Berendse K, Ferdinandusse S, Wanders RJ, Engelen M. Zellweger spectrum disorders: clinical overview and management approach. Orphanet J Rare Dis 2015;10(1):1–11.

[14] Di Meo I, Tiranti V. Classification and molecular pathogenesis of NBIA syndromes. Eur J Paediatr Neurol 2018;22(2):272–84.

[15] Gregory A, Hayflick S. Neurodegeneration with brain iron accumulation disorders overview GeneReviews is a registered trademark of the University of Washington, Seattle. All rights reserved In: Adam MP, Ardinger HH, Pagon RA, Wallace SE, Bean LJH, Mirzaa G, et al., editors. GeneReviews(®). Seattle (WA): University of Washington; 1993.

[16] Hayflick SJ, Kurian MA, Hogarth P. Neurodegeneration with brain iron accumulation. Handb Clin Neurol 2018;147:293–305.

[17] Iankova V, Karin I, Klopstock T, Schneider SA. Emerging disease-modifying therapies in neurodegeneration with brain iron accumulation (NBIA) disorders. Front Neurol 2021;12:629414.

[18] Bandmann O, Weiss KH, Kaler SG. Wilson's disease and other neurological copper disorders. Lancet Neurol 2015;14(1):103–13.

[19] Fernando M, van Mourik I, Wassmer E, Kelly D. Wilson disease in children and adolescents. Arch Dis Child 2020;105(5):499–505.

[20] Socha P, Janczyk W, Dhawan A, Baumann U, D'Antiga L, Tanner S, et al. Wilson's disease in children: a position paper by the hepatology committee of the European Society for paediatric gastroenterology, hepatology and nutrition. J Pediatr Gastroenterol Nutr 2018;66(2):334–44.

[21] Dineen RA, Raschke F, McGlashan HL, Pszczolkowski S, Hack L, Cooper AD, et al. Multiparametric cerebellar imaging and clinical phenotype in childhood ataxia telangiectasia. Neuroimage Clin 2020;25:102110.

[22] Rothblum-Oviatt C, Wright J, Lefton-Greif MA, McGrath-Morrow SA, Crawford TO, Lederman HM. Ataxia telangiectasia: a review. Orphanet J Rare Dis 2016;11(1):159.

[23] Harding IH, Chopra S, Arrigoni F, Boesch S, Brunetti A, Cocozza S, et al. Brain structure and degeneration staging in Friedreich ataxia: magnetic resonance imaging volumetrics from the ENIGMA-Ataxia Working Group. Ann Neurol 2021;90(4):570–83.

[24] Parkinson MH, Boesch S, Nachbauer W, Mariotti C, Giunti P. Clinical features of Friedreich's ataxia: classical and atypical phenotypes. J Neurochem 2013;126(Suppl 1):103–17.

[25] Williams CT, De Jesus O. Friedreich ataxia. Treasure Island (FL): StatPearls; 2021.

[26] Shribman S, Bocchetta M, Sudre CH, Acosta-Cabronero J, Burrows M, Cook P, et al. Neuroimaging correlates of brain injury in Wilson's disease: a multimodal, whole-brain MRI study. Brain 2021;.

[27] Hamilton EM, Tekturk P, Cialdella F, van Rappard DF, Wolf NI, Yalcinkaya C, et al. Megalencephalic leukoencephalopathy with subcortical cysts: characterization of disease variants. Neurology 2018;90(16):e1395–403.

[28] Rich J. Degenerative central nervous system (CNS) disease. Pediatr Rev 2001;22(5):175–6.

[29] Berkowitz AL. Approach to neurologic infections. Continuum 2021;27(4):818–35.

[30] Davies E, Connolly DJ, Mordekar SR. Encephalopathy in children: an approach to assessment and management. Arch Dis Child 2012;97 (5):452–8.

[31] Donald KA, Walker KG, Kilborn T, Carrara H, Langerak NG, Eley B, et al. HIV Encephalopathy: pediatric case series description and insights from the clinic coalface. AIDS Res Ther 2015;12(1):2.

[32] Fernandez A, Drozd MM, Thümmler S, Dor E, Capovilla M, Askenazy F, et al. Childhood-onset schizophrenia: a systematic overview of its genetic heterogeneity from classical studies to the genomic era. Front Genet 2019;10:1137.

[33] Kaur J, Singhi P, Singhi S, Malhi P, Saini AG. Neurodevelopmental and behavioral outcomes in children with sepsis-associated encephalopathy admitted to pediatric intensive care unit: a prospective case control study. J Child Neurol 2016;31(6):683–90.

[34] Khan S, Al Baradie R. Epileptic encephalopathies: an overview. Epilepsy Res Treat 2012;2012:403592.

[35] Vekatesan A. Encephalitis and brain abscess. Continuum 2021;27(4):855–86.

[36] Parikh S, Bernard G, Leventer RJ, van der Knaap MS, van Hove J, Pizzino A, et al. A clinical approach to the diagnosis of patients with leukodystrophies and genetic leukoencephelopathies. Mol Genet Metab 2015;114(4):501–15.

[37] Hoytema van Konijnenburg EMM, Wortmann SB, Koelewijn MJ, Tseng LA, Houben R, Stöckler-Ipsiroglu S, et al. Treatable inherited metabolic disorders causing intellectual disability: 2021 review and digital app. Orphanet J Rare Dis 2021;16(1):170.

[38] Vanderver A, Prust M, Tonduti D, Mochel F, Hussey HM, Helman G, et al. Case definition and classification of leukodystrophies and leukoencephalopathies. Mol Genet Metab 2015;114(4):494–500.

[39] Mallack EJ, Turk BR, Yan H, Price C, Demetres M, Moser AB, et al. MRI surveillance of boys with X-linked adrenoleukodystrophy identified by newborn screening: meta-analysis and consensus guidelines. J Inherit Metab Dis 2021;44(3):728–39.

[40] Groeschel S, Kühl J-S, Bley AE, Kehrer C, Weschke B, Döring M, et al. Long-term outcome of allogeneic hematopoietic stem cell transplantation in patients with juvenile metachromatic leukodystrophy compared with nontransplanted control patients. JAMA Neurol 2016;73 (9):1133–40.

[41] Van der Knaap MS, Valk J. Magnetic resonance of myelination and myelin disorders. Springer Science & Business Media; 2005.

[42] van der Knaap MS, Naidu S, Breiter SN, Blaser S, Stroink H, Springer S, et al. Alexander disease: diagnosis with MR imaging. Am J Neuroradiol 2001;22(3):541–52.

[43] Abdelhalim AN, Alberico RA, Barczykowski AL, Duffner PK. Patterns of magnetic resonance imaging abnormalities in symptomatic patients with Krabbe disease correspond to phenotype. Pediatr Neurol 2014;50(2):127–34.

[44] Morana G, Biancheri R, Dirocco M, Filocamo M, Marazzi M, Pessagno A, et al. Enhancing cranial nerves and cauda equina: an emerging magnetic resonance imaging pattern in metachromatic leukodystrophy and krabbe disease. Neuropediatrics 2009;40(06):291–4.

[45] Zuccoli G, Kim A, Poe M, Escolar ML. Spontaneous third ventriculostomy in Krabbe disease. Pediatr Neurol 2020;108:99–105.

[46] Uggetti C, La Piana R, Orcesi S, Egitto M, Crow Y, Fazzi E. Aicardi-Goutieres syndrome: neuroradiologic findings and follow-up. Am J Neuroradiol 2009;30(10):1971–6.

[47] Merrill ST, Nelson GR, Longo N, Bonkowsky JL. Cytotoxic edema and diffusion restriction as an early pathoradiologic marker in canavan disease: case report and review of the literature. Orphanet J Rare Dis 2016;11(1):169.

[48] Nie S, Chen G, Cao X, Zhang Y. Cerebrotendinous xanthomatosis: a comprehensive review of pathogenesis, clinical manifestations, diagnosis, and management. Orphanet J Rare Dis 2014;9(1):1–11.
[49] van der Knaap MS, Pronk JC, Scheper GC. Vanishing white matter disease. Lancet Neurol 2006;5(5):413–23.
[50] Moser AB, Jones RO, Hubbard WC, Tortorelli S, Orsini JJ, Caggana M, et al. Newborn screening for X-linked adrenoleukodystrophy. Int J Neonatal Screen 2016;2(4):15.
[51] Regelmann MO, Kamboj MK, Miller BS, Nakamoto JM, Sarafoglou K, Shah S, et al. Adrenoleukodystrophy: Guidance for adrenal surveillance in males identified by newborn screen. J Clin Endocrinol Metab 2018;103(11):4324–31.
[52] Shapiro E, Krivit W, Lockman L, Jambaque I, Peters C, Cowan M, et al. Long-term effect of bone-marrow transplantation for childhood-onset cerebral X-linked adrenoleukodystrophy. Lancet 2000;356(9231):713–18.
[53] Peters C, Charnas LR, Tan Y, Ziegler RS, Shapiro EG, DeFor T, et al. Cerebral X-linked adrenoleukodystrophy: the international hematopoietic cell transplantation experience from 1982 to 1999. Blood 2004;104(3):881–8.
[54] van Geel BM, Verrips A, Boelens J-J, Kemp S, Engelen M. Hematopoietic cell transplantation does not prevent myelopathy in X-linked adrenoleukodystrophy: a retrospective study. J Inherit Metab Dis 2015;38(2):359–61.
[55] Moser HW, Raymond GV, Lu S-E, Muenz LR, Moser AB, Xu J, et al. Follow-up of 89 asymptomatic patients with adrenoleukodystrophy treated with Lorenzo's oil. Arch Neurol 2005;62(7):1073–80.
[56] Eichler F, Duncan C, Musolino PL, Orchard PJ, De Oliveira S, Thrasher AJ, et al. Hematopoietic stem-cell gene therapy for cerebral adrenoleukodystrophy. N Engl J Med 2017;377(17):1630–8.
[57] Rice G, Patrick T, Parmar R, Taylor CF, Aeby A, Aicardi J, et al. Clinical and molecular phenotype of Aicardi-Goutieres syndrome. Am J Hum Genet 2007;81(4):713–25.
[58] Adang LA, Frank DB, Gilani A, Takanohashi A, Ulrick N, Collins A, et al. Aicardi goutières syndrome is associated with pulmonary hypertension. Mol Genet Metab 2018;125(4):351–8.
[59] Crow YJ, Vanderver A, Orcesi S, Kuijpers TW, Rice GI. Therapies in Aicardi-Goutières syndrome. Clin Exp Immunol 2014;175(1):1–8.
[60] Sawaishi Y. Review of Alexander disease: beyond the classical concept of leukodystrophy. Brain Dev 2009;31(7):493–8.
[61] Yoshida T, Sasaki M, Yoshida M, Namekawa M, Okamoto Y, Tsujino S, et al. Nationwide survey of Alexander disease in Japan and proposed new guidelines for diagnosis. J Neurol 2011;258(11):1998–2008.
[62] Messing A, Goldman JE, Johnson AB, Brenner M. Alexander disease: new insights from genetics. J Neuropathol Exp Neurol 2001;60 (6):563–73.
[63] Martidis A, Yee RD, Azzarelli B, Biller J. Neuro-ophthalmic, radiographic, and pathologic manifestations of adult-onset Alexander disease. Arch Ophthalmol 1999;117(2):265–7.
[64] Tacke U, Olbrich H, Sass JO, Fekete A, Horvath J, Ziyeh S, et al. Possible genotype-phenotype correlations in children with mild clinical course of Canavan disease. Neuropediatrics 2005;36(4):252–5.
[65] Çakar NE, Uzunhan TA. A case of juvenile Canavan disease with distinct pons involvement. Brain Dev 2020;42(2):222–5.
[66] Björkhem I, Hansson M. Cerebrotendinous xanthomatosis: an inborn error in bile acid synthesis with defined mutations but still a challenge. Biochem Biophys Res Commun 2010;396(1):46–9.
[67] Mondelli M, Sicurelli F, Scarpini C, Dotti MT, Federico A. Cerebrotendinous xanthomatosis: 11-year treatment with chenodeoxycholic acid in five patients. An electrophysiological study. J Neurol Sci 2001;190(1):29–33.
[68] Yahalom G, Tsabari R, Molshatzki N, Ephraty L, Cohen H, Hassin-Baer S. Neurological outcome in cerebrotendinous xanthomatosis treated with chenodeoxycholic acid: early versus late diagnosis. Clin Neuropharmacol 2013;36(3):78–83.
[69] Berginer VM, Berginer J, Korczyn AD, Tadmor R. Magnetic resonance imaging in cerebrotendinous xanthomatosis: a prospective clinical and neuroradiological study. J Neurol Sci 1994;122(1):102–8.
[70] Anderson S. Newborn screening for lysosomal storage disorders. J Pediatr Health Care 2018;32(3):285–94.
[71] Duffner PK, Barczykowski A, Kay DM, Jalal K, Yan L, Abdelhalim A, et al. Later onset phenotypes of Krabbe disease: results of the worldwide registry. Pediatr Neurol 2012;46(5):298–306.
[72] Wenger DA, Rafi MA, Luzi P, Datto J, Costantino-Ceccarini E. Krabbe disease: genetic aspects and progress toward therapy. Mol Genet Metab 2000;70(1):1–9.
[73] Guenzel AJ, Turgeon CT, Nickander KK, White AL, Peck DS, Pino GB, et al. The critical role of psychosine in screening, diagnosis, and monitoring of Krabbe disease. Genet Med 2020;22(6):1108–18.
[74] Bascou N, DeRenzo A, Poe MD, Escolar ML. A prospective natural history study of Krabbe disease in a patient cohort with onset between 6 months and 3 years of life. Orphanet J Rare Dis 2018;13(1):126.
[75] Wright MD, Poe MD, DeRenzo A, Haldal S, Escolar ML. Developmental outcomes of cord blood transplantation for Krabbe disease A 15-year study. Neurology 2017;. Available from: https://doi.org/10.1212/WNL.0000000000004418.
[76] Allewelt H, Taskindoust M, Troy J, Page K, Wood S, Parikh S, et al. Long-term functional outcomes after hematopoietic stem cell transplant for early infantile Krabbe disease. Biol Blood Marrow Transplant 2018;24(11):2233–8.
[77] Gieselmann V, Krägeloh-Mann I. Metachromatic leukodystrophy—an update. Neuropediatrics 2010;41(01):1–6.
[78] Poorthuis BJ, Wevers RA, Kleijer WJ, Groener JE, de Jong JG, Van Weely S, et al. The frequency of lysosomal storage diseases in The Netherlands. Hum Genet 1999;105(1–2):151–6.
[79] Boucher AA, Miller W, Shanley R, Ziegler R, Lund T, Raymond G, et al. Long-term outcomes after allogeneic hematopoietic stem cell transplantation for metachromatic leukodystrophy: the largest single-institution cohort report. Orphanet J Rare Dis 2015;10(1):1–19.

[80] Bredius R, Laan L, Lankester A, Poorthuis B, van Tol M, Egeler R, et al. Early marrow transplantation in a pre-symptomatic neonate with late infantile metachromatic leukodystrophy does not halt disease progression. Bone Marrow Transplant 2007;39(5):309–10.

[81] Peters C, Steward C. Hematopoietic cell transplantation for inherited metabolic diseases: an overview of outcomes and practice guidelines. Bone Marrow Transplant 2003;31(4):229–39.

[82] Pelizaeus-Merzbacher disease, Pelizaeus-Merzbacher-like disease 1, and related hypomyelinating disorders. In: Hobson GM, Garbern JY, editors. Seminars in neurology. Thieme Medical Publishers.; 2012.

[83] Osório MJ, Goldman SA. Neurogenetics of Pelizaeus–Merzbacher disease. Handb Clin Neurol 2018;148:701–22.

[84] Cailloux F, Gauthier-Barichard F, Mimault C, Isabelle V, Courtois V, Giraud G, et al. Genotype–phenotype correlation in inherited brain myelination defects due to proteolipid protein gene mutations. Eur J Hum Genet 2000;8(11):837–45.

[85] Garbern J, Cambi F, Shy M, Kamholz J. The molecular pathogenesis of Pelizaeus-Merzbacher disease. Arch Neurol 1999;56(10):1210–14.

[86] Steinberg SJ, Dodt G, Raymond GV, Braverman NE, Moser AB, Moser HW. Peroxisome biogenesis disorders. Biochim Biophys Acta 2006;1763(12):1733–48.

[87] Setchell KD, Bragetti P, Zimmer-Nechemias L, Daugherty C, Pelli MA, Vaccaro R, et al. Oral bile acid treatment and the patient with Zellweger syndrome. Hepatology 1992;15(2):198–207.

[88] Maeda K, Kimura A, Yamato Y, Nittono H, Takei H, Sato T, et al. Oral bile acid treatment in two Japanese patients with Zellweger syndrome. J Pediatr Gastroenterol Nutr 2002;35(2):227–30.

[89] Heubi JE, Bove KE, Setchell KDR. Oral cholic acid is efficacious and well tolerated in patients with bile acid synthesis and Zellweger spectrum disorders. J Pediatr Gastroenterol Nutr 2017;65(3):321–6.

[90] Hamilton EM, van der Lei HD, Vermeulen G, Gerver JA, Lourenço CM, Naidu S, et al. Natural history of vanishing white matter. Ann Neurol 2018;84(2):274–88.

[91] Federico A, Scali O, Stromillo M, Di Perri C, Bianchi S, Sicurelli F, et al. Peripheral neuropathy in vanishing white matter disease with a novel EIF2B5 mutation. Neurology 2006;67(2):353–5.

[92] Mole SE, Anderson G, Band HA, Berkovic SF, Cooper JD, Holthaus S-MK, et al. Clinical challenges and future therapeutic approaches for neuronal ceroid lipofuscinosis. Lancet Neurol 2019;18(1):107–16.

[93] Arash-Kaps L, Komlosi K, Seegräber M, Diederich S, Paschke E, Amraoui Y, et al. The clinical and molecular spectrum of GM1 gangliosidosis. J Pediatr 2019;215(152-7):e3.

[94] Smith NJ, Winstone AM, Stellitano L, Cox TM, Verity CM. GM2 gangliosidosis in a UK study of children with progressive neurodegeneration: 73 cases reviewed. Dev Med Child Neurol 2012;54(2):176–82.

[95] Mugikura S, Takahashi S, Higano S, Kurihara N, Kon K, Sakamoto K. MR findings in Tay-Sachs disease. J Comput Assist Tomography 1996;20(4):551–5.

[96] Beker-Acay M, Elmas M, Koken R, Unlu E, Bukulmez A. Infantile type Sandhoff disease with striking brain MRI findings and a novel mutation. Pol J Radiol 2016;81:86.

[97] Muenzer J. Overview of the mucopolysaccharidoses. Rheumatology 2011;50(suppl_5):v4–v12.

[98] Martins AM, Dualibi AP, Norato D, Takata ET, Santos ES, Valadares ER, et al. Guidelines for the management of mucopolysaccharidosis type I. J Pediatr 2009;155(4):S32–46.

[99] Transplant outcomes in mucopolysaccharidoses. In: Prasad VK, Kurtzberg J, editors. Seminars in hematology. Elsevier; 2010.

[100] Reichert R, Campos LG, Vairo F, de Souza CFM, Pérez JA, Duarte JÁ, et al. Neuroimaging findings in patients with mucopolysaccharidosis: what you really need to know. Radiographics 2016;36(5):1448–62.

[101] Atwal PS, Scaglia F. Molybdenum cofactor deficiency. Mol Genet Metab 2016;117(1):1–4.

[102] Vijayakumar K, Gunny R, Grunewald S, Carr L, Chong KW, DeVile C, et al. Clinical neuroimaging features and outcome in molybdenum cofactor deficiency. Pediatr Neurol 2011;45(4):246–52.

[103] Burger-Caplan R, Saulnier CA, Sparrow SS. Vineland adaptive behavior scales. Encyclopedia of clinical neuropsychology. Springer; 2018. p. 1–5.

[104] Yu XE, Gao S, Yang RM, Han YZ. MR imaging of the brain in neurologic Wilson disease. Am J Neuroradiol 2019;40(1):178–83.

[105] Kim TJ, Kim IO, Kim WS, Cheon JE, Moon SG, Kwon JW, et al. MR imaging of the brain in Wilson disease of childhood: findings before and after treatment with clinical correlation. Am J Neuroradiol 2006;27(6):1373–8.

[106] Paprocka J, Machnikowska-Sokołowska M, Gruszczyńska K, Emich-Widera E. Neuroimaging of basal ganglia in neurometabolic diseases in children. Brain Sci 2020;10:11.

[107] Sullivan R, Yau WY, O'Connor E, Houlden H. Spinocerebellar ataxia: an update. J Neurol 2019;266(2):533–44.

[108] Beaudin M, Matilla-Dueñas A, Soong BW, Pedroso JL, Barsottini OG, Mitoma H, et al. The classification of autosomal recessive cerebellar ataxias: a consensus statement from the society for research on the cerebellum and ataxias task force. Cerebellum 2019;18(6):1098–125.

[109] Meira AT, Arruda WO, Ono SE, Neto AC, Raskin S, Camargo CHF, et al. Neuroradiological findings in the spinocerebellar ataxias. Tremor Other Hyperkinet Mov 2019;9.

[110] Brandsma R, Verschuuren-Bemelmans CC, Amrom D, Barisic N, Baxter P, Bertini E, et al. A clinical diagnostic algorithm for early onset cerebellar ataxia. Eur J Paediatr Neurol 2019;23(5):692–706.

[111] Shiohama T, Levman J, Takahashi E. Surface- and voxel-based brain morphologic study in Rett and Rett-like syndrome with MECP2 mutation. Int J Dev Neurosci 2019;73:83–8.

[112] Edvardson S, Nicolae CM, Agrawal PB, Mignot C, Payne K, Prasad AN, et al. Heterozygous de novo UBTF gain-of-function variant is associated with neurodegeneration in childhood. Am J Hum Genet 2017;101(2):267–73.

[113] Toro C, Hori RT, Malicdan MCV, Tifft CJ, Goldstein A, Gahl WA, et al. A recurrent de novo missense mutation in UBTF causes developmental neuroregression. Hum Mol Genet 2018;27(4):691–705.

[114] Sedláčková L, Laššuthová P, Štěrbová K, Haberlová J, Vyhnálková E, Neupauerová J, et al. UBTF mutation causes complex phenotype of neurodegeneration and severe epilepsy in childhood. Neuropediatrics 2019;50(1):57–60.

[115] Ikeda C, Kawarai T, Setoyama C, Orlacchio A, Imamura H. Recurrent de novo missense variant E210K in UBTF causes juvenile dystonia-parkinsonism. Neurol Sci 2021;42(3):1217–19.

Chapter 22

Neurometabolic disorders

Andrea Lynne Gropman[1], Matthew Whitehead[2,3], Jennifer Harmon[4] and Kirkland A. Wilson[5,6]
[1]*Neurometabolic Translational Research, Center for Experimental Neurotherapeutics, St. Jude's Pediatric Research Hospital, Memphis, TN, United States,* [2]*Department of Radiology, Children's Hospital of Philadelphia, PA, United States,* [3]*Perelman School of Medicine, Philadelphia, PA, United States,* [4]*Department of Neurology, Wake Forest University School of Medicine, Winston Salem, NC, United States,* [5]*Rare Disease Institute, Children's National Hospital, Washington, DC, United States,* [6]*National Human Genome Research Institute, National Institutes of Health, Bethesda, MD, United States*

Introduction

Neurometabolic disorders, including inborn errors of metabolism (IEM), are a group of heritable single gene disorders that result in altered metabolism, and are defined by the *International Classification of Inherited Metabolic Disorders* as "any condition in which primary alteration of a biochemical pathway is intrinsic to specific biochemical, clinical, and/or pathophysiological features, regardless of whether there are abnormalities in currently available biochemical laboratory tests" [1]. In these disorders, altered metabolism causes deficiencies of metabolites essential for normal cellular and organ function. While individually rare, collective incidence is estimated to be 1:400 and comprises over 1400 individual disorders [2–4]. While each disorder has its own unique pattern of signs and symptoms, there are several common neurodevelopmental presentations that may indicate that a patient may have a neurometabolic disorder.

Neurometabolic disorders can be broadly separated into two categories: disorders presenting with acute metabolic crisis in the setting of an intoxication syndrome and disorders presenting with more subacute or chronic symptoms (Table 22.1) [5]. The former group is primarily comprised of disorders in which patients typically appear healthy at birth then present in the first few days of life with acute intoxication syndromes or in early childhood with acute metabolic decompensation in the setting of illness or fasting [5]. The latter group includes a heterogeneous array of disorders with subacute onset and slow progression of symptoms that can encompass all organ systems [5–7]. Although the

TABLE 22.1 Neurometabolic diseases by presentation.

Acute metabolic decompensation	Subacute progressive neurological dysfunction	Global developmental delay and intellectual disability
Carbamoyl phosphate synthetase 1 deficiency	Arginase deficiency	Phenylalanine hydroxylase deficiency
N-acetylglutamate synthase deficiency	Galactosemia	Smith–Lemli–Opitz syndrome
Ornithine transcarbamylase deficiency	Leigh disease	
Argininosuccinate synthase deficiency	Pyruvate dehydrogenase complex deficiency	
Argininosuccinate lyase deficiency	Mitochondrial encephalomyopathy, lactic acidosis, and stroke-like episodes (MELAS)	
Propionic acidemia	Mucopolysaccharidoses I, II, III, VII	
Methylmalonic acidemia	GM1 gangliosidosis	
Maple syrup urine disease	Tay–Sachs	
	Niemann–Pick type C disease	
	Neuronal ceroid lipofuscinoses (CLN 1, 2, 3 diseases)	
	Molybdenum cofactor deficiency A and B	

Capute and Accardo's Neurodevelopmental Disabilities in Infancy and Childhood. DOI: https://doi.org/10.1016/B978-0-12-824060-1.00046-8

majority of these disorders are typically progressive, metabolic stabilization of patients with some disorders may lead to a nonprogressive chronic phenotype, such as the stable encephalopathy observed in patients with well-treated phenylketonuria (PKU), described further in this chapter [6].

When to suspect a metabolic disorder

Acute metabolic decompensation

Because symptoms of neurometabolic disorders mimic other disease presentations, a high degree of clinical suspicion is required for prompt diagnosis of these disorders.

Rapid and accurate diagnosis is crucial in the setting of acute metabolic decompensation, especially in infants and young children in whom reduction and removal of neurotoxic metabolites is crucial to prevent severe neurological sequelae or death. Because toxic metabolites cross the placenta and are subsequently catabolized by maternal enzymes, infants with IEM appear phenotypically normal at birth. With establishment of feeding in the first few days of life, these patients become increasingly symptomatic from accumulation of toxic metabolic intermediates as feeding is established. Early symptoms of metabolic decompensation are often subtle and typically include poor feeding and lethargy; because these nonspecific symptoms also typical of more common neonatal emergencies, particularly sepsis, workup may initially focus on infectious etiologies of presentation without consideration of alternate etiologies [8]. Patients with untreated hyperammonemia or elevations in branched chain amino acids (BCAAs)—particularly leucine—are at high risk of cerebral edema and may rapidly progress from lethargy to coma and subsequent death with ongoing accumulation of these toxic metabolites; therefore, timely recognition of hyperammonemia, either as result of a urea cycle defect or organic acidemia, or maple syrup urine disease is tantamount [9,10].

Assessment for these most life-threatening IEM can be completed with a relatively standard biochemical workup, including basic metabolic panel assessing for acidosis, ammonia level, urinalysis to evaluate for ketosis, lactate level, and measurement of plasma amino and urine organic acids, although it is important to note that metabolite concentrations in plasma are not necessarily reflective of metabolite concentrations in cerebral spinal fluid (CSF) [3,9,11]. Brain imaging is often required to assess for life-threatening structural changes, such as massive cerebral edema with impending herniation. Magnetic resonance imaging (MRI) may show diffuse cytotoxic cerebral edema and/or areas of acute injury, including evidence of acute energy failure; in some disorders, magnetic resonance spectroscopy (MRS) can show accumulation of certain metabolites, such as lactate or glutamine, that may be diagnostic or suggestive of IEM. Treatment for acute metabolic crisis includes reduction of the substrate of the defective enzyme, strict avoidance of catabolism through administration of 10% dextrose-containing fluids, and therapy with scavengers for the toxic metabolites; patients with severe and life-threatening elevations in ammonia or leucine require dialysis to further accelerate clearance of these molecules from the central nervous system (CNS) [9]. While biochemical screening labs may yield diagnosis, genotyping through molecular techniques such as next-generation sequencing (NGS) through either specific gene, panel, whole-exome or whole-genome sequencing is typically completed to confirm diagnosis and specify the enzymatic defect, which may further guide treatment such as vitamin and cofactor supplementation [12]. Tables 22.2–22.4 summarize the clinical features of the selected neurometabolic conditions discussed in further detail in this chapter.

It should be noted that, although initial presentation of IEM causing acute metabolic decompensation occurs most commonly in the newborn period; some patients with more mild enzyme defects will present later in childhood or even into adulthood [9]. Although other acute disorders, such as infection or sepsis, remain more common in these older children, IEM should be considered in the differential for any patient presenting in extremis, with unexplained acute alterations in mental status or with unexpected laboratory abnormalities in the setting of acute catabolic stress that increases breakdown of fats, carbohydrates, and/or protein by the body, including the stress of birth and pregnancy, increased protein intake, fasting or starvation, fad diets, medications, surgery, and/or infection [9,10,38,39]. History and physical exam often reveal information that should prompt consideration and investigation of IEM; these red flags include decompensation with stress as above, unexplained intermittent symptoms, symptoms attributable to organs with high energy requirements, aversion and self-avoidance of food groups or atypical diets, particularly protein, and failure to thrive in the setting of at least one of these other red flags. A thorough three-generation family history should also be completed, including assessing for consanguinity in all generations; although family history will be negative in many patients due to autosomal recessive inheritance patterns of most IEM; consanguineous families may have multiple affected members over multiple generations, whereas pedigrees of patients with X-linked or mitochondrial disorders may show multiple generations of affected family members in the maternal lineage [40].

TABLE 22.2 Presenting symptoms of urea cycle disorders.

Name of disorder	Type of disorder	Incidence	Associated gene	Mode of inheritance	Cognitive symptoms	Behavioral symptoms	Other neurological symptoms	Common systemic symptoms
Carbamoyl phosphate synthetase I(CPSI) deficiency	Proximal urea cycle	1 in 1.3 million [13]	*CPS1*	AR	Motor > language delay Learning disability	Agitation Excitation Insomnia	Hypotonia Spasticity Epilepsy	
N-acetylglutamate synthase (NAGS) deficiency	Proximal urea cycle	1 in 2 million [13]	*NAGS*	AR	Motor and language delay Learning disability	Agitation Excitation Aggression Insomnia	Hypotonia Spasticity Epilepsy	
Ornitine transcarbamylase (OTC) deficiency	Proximal urea cycle	1 in 56,500 [13]	*OTC*	X-linked	Isolated language to GDD ADHD Dyslexia Variable ID Executive function deficits	Irritability Self-injury	Dystonia Spastic paresis	
Argininosuccinate synthase (ASS) deficiency	Distal urea cycle	1 in 250,000 [13]	*ASS1*	AR	Language delay Developmental regression ADHD Social impairment Variable ID	Irritability	Hypotonia Spasticity	Cataracts
Argininosuccinate lyase (ASL) deficiency	Distal urea cycle	1 in 218,750 [13]	*ASL*	AR	Isolated language to GDD Developmental regression ADHD Variable ID Variable LD	Paranoia Psychosis Aggression Self-injury	Dystonia Tone abnormalities Cerebellar ataxia	Hepatomegaly/ hepatic fibrosis Brittle hair Hypertension
Arginase deficiency	Distal urea cycle	1 in 950,000 [13]	*ARG1*	AR	Isolated motor to GDD Developmental regression ADHD Variable ID	Irritability	Spasticity (LE > UE) Hyperreflexia	

ADHR, attention deficit hyperactivity disorder; *AR*, Autosomal recessive; *GDD*, global developmental delay; *ID*, intellectual disability; *LD*, learning disability.

TABLE 22.3 Diagnosis and treatment of urea cycle disorders.

Name of disorder	MRI findings	Laboratory abnormalities	Tested on newborn screening	Treatment
Carbamoyl phosphate synthetase I (CPSI) deficiency	Cerebral edema Subarachnoid or subdural hemorrhage White matter and basal ganglia signal abnormality *Ex vacuo* ventriculomegaly MRS with elevated glutamine, reduced myoinositol and choline [14–16]	↑ NH_3 ↑ plasma glutamine ↓ plasma citrulline and arginine ↓ plasma orotic acid [17,18]	Yes	Protein restriction Avoidance of catabolism Nitrogen scavengers
N-acetylglutamate synthase (NAGS) deficiency	Similar to CPS1	Similar to CPS1	Yes	Carglumic acid Protein restriction Avoidance of catabolism Nitrogen scavengers
Ornitine transcarbamylase (OTC) deficiency	Cerebral edema Hypomyelination Cortical atrophy Deep gray matter injury and cystic degeneration *Ex vacuo* ventriculomegaly MRS with elevated glutamine, reduced myoinositol and choline [16,19–22]	↑ NH_3 ↑ plasma glutamine and orotic acid ↓ plasma citrulline and arginine [19]	Yes	Protein restriction Avoidance of catabolism Nitrogen scavengers Liver transplant
Argininosuccinate synthase (ASS) deficiency	Cerebral edema Signal abnormality of thalamus, basal ganglia, and corpus callosum *Ex vacuo* ventriculomegaly [23,24]	↑ NH_3 ↑ plasma glutamine, orotic acid, and citrulline ↓ plasma arginine and ornithine [25,26]	Yes	Protein restriction Avoidance of catabolism Nitrogen scavengers Liver transplant
Argininosuccinate lyase (ASL) deficiency	Gray matter lesions Cerebral atrophy	↑ NH_3 ↑ plasma argininosuccinate and citrulline ↓ plasma arginine [27]	Yes	Arginine supplementation Protein restriction Avoidance of catabolism Nitrogen scavengers Liver transplant
Arginase deficiency	Cerebral and cerebellar atrophy Demyelinating lesions MRS with elevated glutamine and glutamate, decreased myoinositol [16,28,29]	↑ NH_3 ↑ plasma arginine ↓ plasma ornithine [29,30]	Yes	Arginine supplementation Protein restriction Avoidance of catabolism Nitrogen scavengers Liver transplant

NH_3, Ammonia; *MRS*, magnetic resonance spectroscopy.

Subacute progressive neurological dysfunction

Progressive intellectual and neurological deterioration (PIND) is described as a family of severe childhood disorders defined as advancing deterioration over 3 months with progressive cognitive and/or developmental decline in patients who previously had normal intellectual ability and appropriate attainment of developmental milestones, accompanied by new onset of other neurological signs such as epilepsy or motor dysfunction, often associated with progressive dysfunction of other organ systems, and not attributable to other known extrinsic neurological injury [12]. Of the at least 85 known disorders that cause PIND, it is estimated that 62.5%–75% of all cases of PIND are caused by IEM. There are

TABLE 22.4 Presenting symptoms of organic acidemias and amino acidopathies.

Disorder name	Type of disorder	Incidence	Associated genes	Mode of inheritance	Cognitive symptoms	Behavioral symptoms	Other neurological symptoms	Common systemic symptoms
Propionic acidemia	Organic acidemia	1 in 50,000 [31]	*PCCA* *PCCB*	AR	Isolated motor to GDD; rarely other isolated delays ADHD Variable ID	Irritability Lethargy	Ataxia Hypotonia Progressive spastic quadriparesis Generalized epilepsy, including absence, myoclonic and GTC seizures Metabolic stroke-like episodes	Cardiomyopathy
Methylmalonic acidemia	Organic acidemia	1 in 50,000 [32]	*MMUT* *MMAA* *MMAB* *MMADHC* *MCEE*	AR	Isolated motor to GDD; rarely other isolated delays ADHD Variable ID	Irritability Lethargy	Hypotonia Progressive spastic quadriparesis Generalized epilepsy, including absence, myoclonic, and GTC seizures Metabolic stroke-like episodes	Chronic renal failure
Isovaleric acidemia	Organic acidemia	1 in 159,000 [33]	*IVD*	AR	Fine motor delay Delayed social development Inattention	Irritability	Hypotonia Progressive spastic quadriparesis Rarely infantile spasms Metabolic stroke-like episodes	"Sweaty feet" odor
Glutaric aciduria type 1	Organic acidemia	1 in 110,000 [34,35]	*GCDH*	AR	Sudden onset of central hypotonia with limb dystonia, dyskinesia, and/or choreoathetosis Motor delays Preserved cognition	Nonspecific, generally similar to other individuals with chronic disease	Sudden onset motor dysfunction and/or hypotonia with progression to dystonia and spasticity Epilepsy	Retinal hemorrhage
Maple syrup urine disease	Amino acidopathy	1 in 185,000 [36]	*BCKDHA* *BCKDHB* *DBT*	AR	Motor and/or speech delay Visuospatial impairment ADHD, isolated hyperactivity Variable ID Impaired executive function	Hallucinations	Dystonia that progresses to parkinsonian movements Epilepsy	
Phenylketonuria	Amino acidopathy	1 in 10,000 [37]	*PAH*	AR	Motor delay ADHD Variable ID Variable learning disability	Depression/anxiety	Seizures in untreated patients Blindness Quadraparesis Poor coordination	"Musty" odor Decreased pigmentation of the skin, eyes, and hair

AR, Autosomal recessive; *GDD*, global developmental delay; *ID*, intellectual disability; *LD*, learning disability.

two broad categories of IEM that lead to PIND: (1) storage disorders, in which molecules that are unable to be completely metabolized accumulate in organelles, tissues, and organs; and (2) disorders with ongoing energy failure, such as mitochondrial disorders [12].

History/family history and thorough physical exam should be completed; careful abdominal assessment for hepatosplenomegaly and thorough ophthalmic exam to assess for macular or retinal manifestations of IEM should be completed. MRI and MRS should be completed for all patients with subacute symptoms of neurodegeneration; many disorders in this family have MRI abnormalities that precede symptoms [41]. Laboratory evaluation may involve biochemical screening labs such as basic metabolic panel assessing for acidosis, ammonia level, urinalysis to evaluate for ketosis, lactate level, and measurement of plasma amino and urine organic acids; however, normal biochemical screening labs do not preclude many IEM in this category, for whom specific enzymatic testing and/or next-generation whole exome or whole genome sequencing will yield a molecular diagnosis [12].

Global developmental delay and intellectual disability

A third common presentation of IEM is described as a more indolent neurodevelopmental phenotype defined by global developmental delay (GDD) in early childhood and intellectual disability (ID) in later life. Although GDD and ID are common in the general population, affecting between 5%–10% and 1%–3% of individuals, respectively, IEM is a rare cause of nonsyndromic developmental or cognitive impairment, accounting for an estimated 1%–5% of all cases of developmental delay [42,43]. Although IEM represent this small percentage of patients with GDD and ID, accurate diagnosis of these disorders remains important, as early diagnosis may introduce treatment opportunities to prevent further delays or loss of cognition, to reduce risk of future acute metabolic decompensation at times of metabolic stress, and to provide support and counseling for the family [42,43].

IEM most likely to present as developmental delay or ID cover a broad variety of metabolic pathways and include PKU, galactosemia, lysosomal storage disorders (LSD), urea cycle defects (UCD), disorders of cholesterol synthesis, disorders of neurotransmitter synthesis, congenital disorders of glycosylation, disorders of creatine synthesis, and transport and disorders of purine and pyrimidine synthesis [42,44]. In a recent publication, Konijnenburg et al. evaluated an updated review of literature regarding treatable disorders causing ID and added 44 disorders from the literature to those obtained on their previous review in 2012, demonstrating that ongoing investigation regarding etiologies of ID and ongoing research to develop novel therapies for IEM continue to expand our ability to identify and treat patients who have metabolic etiologies of GDD/ID [4].

Similarly to patients with other presentations of IEM, thorough evaluation is required; those with IEM may have other subtle signs and symptoms of other neurological or systemic organ dysfunction that may increase clinical suspicion and further guide testing [42,43]. Although the diagnostic yield of specific biochemical tests in the evaluation of GDD/ID is not clearly defined, and although some tests in the typical biochemical workup for IEM as previously described above may have lower utility than other for the GDD/ID phenotype, the potentially high impact of clinical information compared to the relatively low cost, easy accessibility, and quick turnaround time of metabolic testing favors routine testing for IEM as part of workup for GDD/ID and was recommended in the American Academy of Pediatrics clinical report regarding evaluation of patients with GDD/ID [44]. Because basic metabolic screening labs will not detect all IEM causing GDD/ID, and because thorough biochemical/enzymatic testing for all known treatable causes of GDD/ID would be neither technically feasible nor cost-efficient, comprehensive genetic testing for all IEM that cause treatable GDD/ID, either as part of a comprehensive GDD/ID sequencing panel, whole-exome or whole-genome sequencing should be considered a critical part of evaluation for all patients with unexplained GDD/ID, especially if all other testing has been negative [4,44].

While these patients with either the PIND or GDD/ID phenotypes of IEM do not typically present in the setting of a life-threatening crisis requiring acute initiation of treatment, timely diagnosis remains important to prevent further neurological deterioration. For treatable IEM, disease-directed therapy such as enzyme replacement (ERT), hematopoietic stem cell transplant (HSCT), gene therapy, dietary modification—including initiation of the ketogenic diet for some patients—or the supplementation of deficient metabolic cofactors can alter the trajectory of the disease; therefore, early diagnosis and prompt initiation of therapies may arrest progression of disease, preserve cognitive function, and prevent further disability or mortality. For patients for whom no disease-directed therapies are available, timely diagnosis can aid in care planning, initiation of supportive therapies and genetic counseling for the family [4,12]. It is important for any physician who evaluates and treats patients with GDD and/or ID to be able to identify patients who symptoms suspicious for IEM and to be familiar with their common presentations. A comprehensive review of every IEM is beyond the scope of this chapter, but we aim to highlight those disorders most notable for their neurologic and neurodevelopmental symptoms.

Urea cycle disorders

Urea cycle disorders (UCD) are a family of disorders caused by disruption in urea cycle, the primary pathway by which nitrogen resulting from protein catabolism is detoxified and excreted (Fig. 22.1). These disorders are further divided into proximal UCD, affecting the urea cycle enzymes located in the mitochondria, and the distal UCD affecting those located in the cytosol [45].

Common across UCD are elevations in ammonia and glutamine, which are typically more pronounced in the proximal UCD. Hyperammonemia results from direct accumulation in the setting of continued breakdown of nitrogenous sources. Glutamate and glutamine build up secondary to ammonia accumulation, and these metabolites are independently toxic at high concentrations [46]. The majority of UCD present with variably severe acute metabolic decompensation with superimposed chronic neurodevelopmental disease, including developmental delay, ID, attention deficit, motor impairment, and epilepsy; presentation of the individual UCD are explored further below and summarized in Table 22.2.

Treatment of nitrogen imbalances is standard across UCD and long-term therapy involves use of nitrogen scavengers such as compounds containing phenylbutyrate, phenylacetate, and/or benzoate. Depending on disease type and severity, supplementation with arginine and/or citrulline can be used [19,45,47]. Dietary protein restriction is also commonly used to limit exogenous nitrogen [48]. Lactulose can also be used to encourage ammonia intestinal excretion, and antibiotic use can inhibit intestinal microflora growth and reduce intestinal ammonia production. Certain medications, including phenobarbital, valproate, carbamazepine, and phenytoin, should also be used with care as they increase the risk of hyperammonemic crises [49].

Acutely, treatment may also involve the cessation of protein intake and supplementation with glucose either orally or IV to halt protein catabolism. More aggressive ammonia management with hemodialysis has also been used [14,17,46]. Finally, orthotopic liver transplant is an option for patients with poor nitrogen management on prior therapies and/or frequent hyperammonemic crises, as the liver is the only organ with every enzyme of the urea cycle. While liver transplant can improve a patient's ammonia control and—if done in a timely fashion—does appear to improve neurocognitive function, neurological sequelae may develop despite appropriate therapy, likely because of the inability of the transplanted liver to correct CSF concentrations of toxic metabolites [17,48,50,51].

Carbamoyl phosphate synthetase I

Carbamoyl phosphate synthetase I (CPS1) deficiency is a rare autosomal recessive proximal UCD affecting the first urea cycle enzyme [17,52]. Patients with CPS1 have the most severe symptoms of all UCD and classically present in the neonatal period with emesis, lethargy, and encephalopathy that may progress to coma; patients with less severe deficiencies in this enzyme can present with symptoms as late as adulthood [14,17,18].

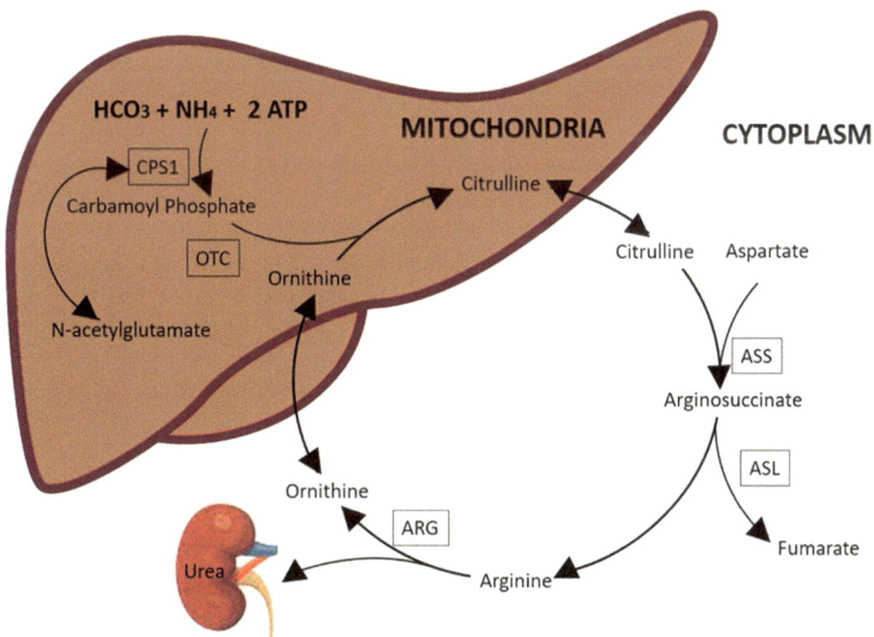

FIGURE 22.1 Schematic of the urea cycle. The urea cycle is the primary mechanism by which surplus nitrogen, usually from protein catabolism, is detoxified and excreted. The urea cycle can be divided into enzymes that are located in the mitochondrial and enzymes that are located in the cytosol. Disorders of the mitochondrial enzymes are called "proximal urea cycle disorders," while disorders of the cytosolic enzymes are called "distal urea cycle disorders."

Patients can develop intellectual and developmental delays, most prominent of which are gross motor deficits. Learning abilities range from severe academic difficulties to normal academic performance. Seizures are both a common complication of acute brain injury in the setting of metabolic decompensation as well as chronic sequelae of structural brain abnormalities caused by prior acute injury [13,15,53]. Both acute and chronic alterations in behavior, including psychosis, agitation, excitability, insomnia, aggression, and sleep disturbance; aggression is the most frequent chronic behavioral disturbance observed. As these neurodevelopmental symptoms may be the presenting symptoms in later-onset disease, patients may be misdiagnosed with personality disorders and/or learning deficits leading to delays in treatment [14,15,53].

Diagnostic abnormalities, including expected MRI and laboratory findings, are summarized in Table 22.3. Treatment of disease, both acute and chronic exacerbations, is per standard care of UCD as noted earlier in the chapter [14,17,18,46,50].

N-acetylglutamate synthase deficiency

N-acetylglutamate synthase (NAGS) deficiency is an autosomal recessive proximal UCD affecting the synthesis of a cofactor for CPS1; as a result, the presentation of patients with this deficiency is largely identical to the presentation of CPS1 deficiency [54–56]. It is important to definitively distinguish NAGS deficiency from CPS1 deficiency, usually via genetic testing, because it can be successfully treated with N-carbamylglutamate, a NAG analog [54,57].

Ornithine transcarbamylase deficiency

Ornithine transcarbamylase (OTC) deficiency is an X-linked proximal UCD. Presentation and onset of disease is variable due to the X-linked nature of this disorder: males, who are classically hemizygous, usually present in the neonatal or infancy period, but females, who are classically heterozygous for a pathologic variant, may remain asymptomatic or experience milder symptoms in the setting of a catabolic trigger [19].

Clinically, patients in a hyperammonemic crisis can display acute behavior changes including irritability, psychosis, and lethargy that may progress to coma. Milder and/or subacute neurocognitive dysfunction, such as ADHD and executive dysfunction and/or psychosis may be the presenting symptom in some patients, especially in hemizygous females. Learning deficits, such as ID, dyslexia, and deficits in executive function and speech are common in this patient population, as are gross motor delays and movement disorders, including dystonia and spasticity. Epilepsy has also been described in some patients [19,20,58,59].

Diagnostic abnormalities, including expected MRI and laboratory findings, are summarized in Table 22.3 [19–21,47]. Treatment is per standard urea cycle management; orthotopic liver transplant is a treatment option, but notably, developmental delay is still documented posttransplant [19].

Argininosuccinate synthase deficiency

Argininosuccinate synthase (ASS) deficiency, also known as citrullinemia type I, is an autosomal recessive disorder distal UCD [60]. There is a spectrum of disease from neonatal to adult onset of symptoms [25,61]. Patients present after catabolic stressors in hyperammonemic crisis marked by lethargy, failure to thrive, seizure, and/or coma that may progress permanent neurological impairment or death [61]. Intellectual and developmental delays and/or regression, especially in the language domain, are common. Poor eye contact and communication delays have also been seen in patients; as a result, these patients are often suspected to have autism spectrum disorder (ASD) but—upon testing—a true diagnosis of ASD is not especially common, differentiating this disorder from other IEM. Hyperactivity and/or inattention, are common behavioral changes in patients, as is aggression. ASS deficiency is associated with motor dysfunction, including hypotonia and spasticity, and patients may be initially incorrectly diagnosed with cerebral palsy (CP) leading to delays in care [25,47,62–64].

Diagnostic abnormalities, including expected MRI and laboratory findings, are summarized in Table 22.3. Treatment is per standard urea cycle management; orthotopic liver transplant is a treatment option, but transplant does not alter the localized CSF concentrations of toxic metabolites [47].

Argininosuccinate lyase deficiency

Argininosuccinate lyase (ASL) deficiency, also termed argininosuccinic aciduria, is an autosomal recessive distal UCD caused by due to defects in the enzyme ASL and is the second most common UCD after OTC deficiency [13,48,65,66].

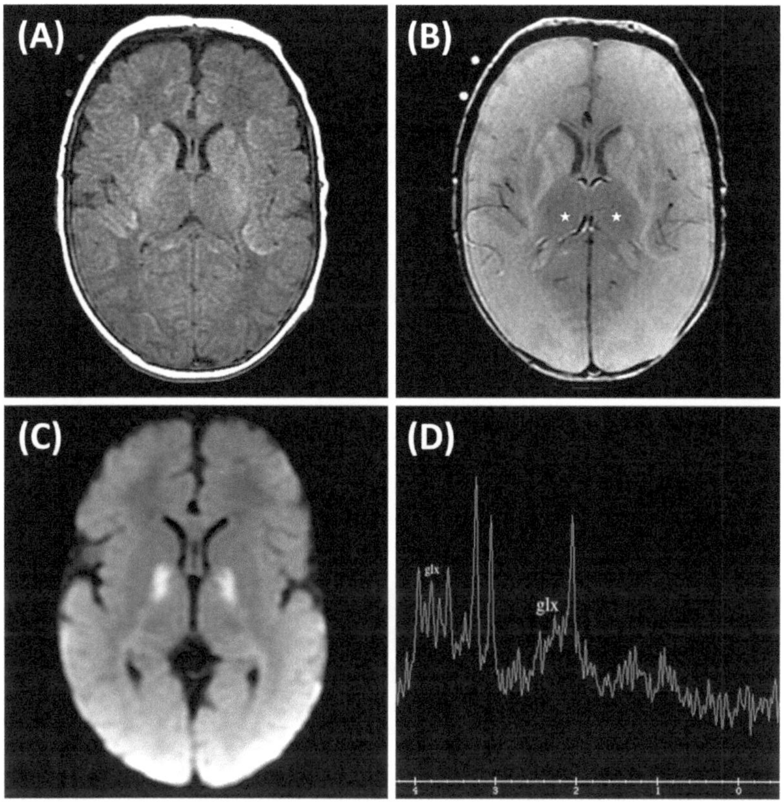

FIGURE 22.2 Neonatal male with a urea cycle disorder (arginosuccinate lyase deficiency). Axial T1WI (A) proton density images (B), and diffusion weighted imaging (DWI) (C) demonstrate typical MRI changes compatible with subacute hyperammonemic brain injury with hyperintense lesions involving the basal ganglia and perisylvian/periinsular tissues; the thalami are spared helping to distinguish the appearance from hypoxic-ischemic injury (B). PRESS short TE single voxel 1H magnetic resonance spectroscopy (MRS) (D) over the left basal ganglia reveals increased concentrations of glutamine and glutamate (glx).

Classically, ASL deficiency presents with neonatal onset of hyperammonemia causing emesis, lethargy, irritability, and coma; presentation at a later age has been described in the setting of catabolic stressors [27].

Patients with ASL deficiency have the highest incidence of neurocognitive and neurodevelopmental abnormalities of all the UCD. Abnormal development is usually noted by 2 years of age, and delays across all domains are common, even in the absence of documented hyperammonemia. Learning difficulties and ADHD are similarly common. Behavioral abnormalities, including psychosis, paranoia, aggression, and self-injurious behavior are seen as well. Motor abnormalities, including abnormal tone, abnormal reflexes, and cerebellar ataxia have also been seen. Patients can furthermore have seizures of variable semiology including absence seizures and generalized tonic-clinic seizures (GTC), which can abate with dietary modification [27,66,67]. Systemic symptoms include brittle hair, hypertension, hepatomegaly, and/or hepatic fibrosis [27,48,49,66–68].

Diagnostic abnormalities, including expected MRI and laboratory findings, are shown in Fig. 22.2 and summarized in Table 22.3. Treatment is per standard care of UCD with protein restriction and arginine supplementation [27,66]. Early implementation of dietary modification can decrease the development of intellectual and developmental issues. Liver transplant is also a therapeutic option to reduce the risk of hyperammonemia and reduces the need for protein-restricted diet, though arginine supplementation is continued in some cases [66,69].

Arginase deficiency

Arginase deficiency is an autosomal recessive distal UCD affecting the last enzyme in the urea cycle [28]. Neonatal onset is rare as is severe hyperammonemia, though both have been reported in the literature. Classically, early development is normal with initially presentation of psychomotor retardation occurring typically by age 3; progressive microcephaly has also been described [29].

Patients can develop intellectual and developmental delays and/or regressions, with language being most affected. Behavior changes can also occur, including increased irritability and attention difficulties. Patient develop seizures, gait disturbances, hyperreflexia, and spasticity more frequently and severely affecting the lower extremities [28–30]. Because spasticity is typically the most obvious presenting symptom, arginase deficiency is often misdiagnosed as CP; thus, in patients with a diagnosis of CP and progression of disease, arginase deficiency must be excluded [29,70].

Diagnostic abnormalities, including expected MRI and laboratory findings, are summarized in Table 22.3. Treatment is per standard therapy for UCD, and diet modification is reported to improve developmental delays in some patients [28,29]. Liver transplant can normalize ammonia and arginine concentrations in blood but not CSF [28].

Organic acidemias and amino acidopathies

Organic acidemias and amino acidopathies are disorders that involve altered fatty acid or amino acid metabolism which result in accumulation of toxic metabolites that cause disease: amino acid metabolites, in the case of amino acidopathies, or nonamino organic acid metabolites, in the case of organic acidemias. The metabolites most associated with the specific disease, which are often useful for diagnosis, are rarely useful for assessing clinical severity. Instead, assessment of patient acid-base balance, measurement of glucose, lactate, and ketone body accumulation as proxies for relative energy deficit, and evaluation of plasma ammonia predict disease severity and metabolic crisis [2,3,71–75].

Acute treatment of the metabolically decompensating patient is similar regardless of the underlying defect and includes reversal of catabolism through cessation of all dietary protein sources for a maximum of 36–48 hours with concomitant supplementation with nonprotein calories. Therapy also involves carnitine supplementation to promote urinary excretion of toxic metabolites and the use of ammonia scavengers such as sodium benzoate and/or sodium phenylacetate to allow for alternate excretion of nitrogen. Patient may require hemodialysis as necessary to further remove ammonia or other toxic metabolites. Long-term dietary management and avoidance is a common treatment strategy, though the individual diet is based on the specific defect [2,3,71–76]. The typical presentations of the most common defects are discussed in further detail below as well as in Table 22.4.

Propionic acidemia

Propionic acidemia is an autosomal recessive organic acidemia that results in the accumulation of toxic compounds due to defects in propionyl-CoA carboxylase [31]. Propionic acidemia patients are observed to be the most neurologically and neurodevelopmentally affected of the organic acidemias [74,77]. Many patients present shortly after birth, and the majority typically present by 90 days of life; however, it should be noted that there is a spectrum of disease and patients may not present with symptoms until several years after birth [38,74,78–80].

Developmental delays and regressions are common in this disorder; motor function is classically the most impaired domain, and patients may develop spasticity, hemiplegia, and/or hypotonia. Cognitive delays are usually present by age 2 years, and severity of deficits is associated with age of onset. Attention difficulties, distractibility, and true ADHD are also described and are not always amenable to stimulant medications. Characteristic neurologic findings of disease include ataxia, hypotonia, lethargy, and seizures, which are typically generalized or myoclonic in early childhood but progress to mild generalized or absence seizures in later childhood. Metabolic stroke-like episodes, including the presence of vegetative symptoms such as low body temperature, drowsiness, and bradycardia, can also occur [2,38,78,81–85].

Diagnostic abnormalities, including expected MRI and laboratory findings, are shown in Fig. 22.3 and summarized in Table 22.5. Treatment strategies are those described above, including a low-protein diet with supplemented caloric intake; patients usually also require additional nocturnal feedings to prevent fasting. Patients also benefit from carnitine

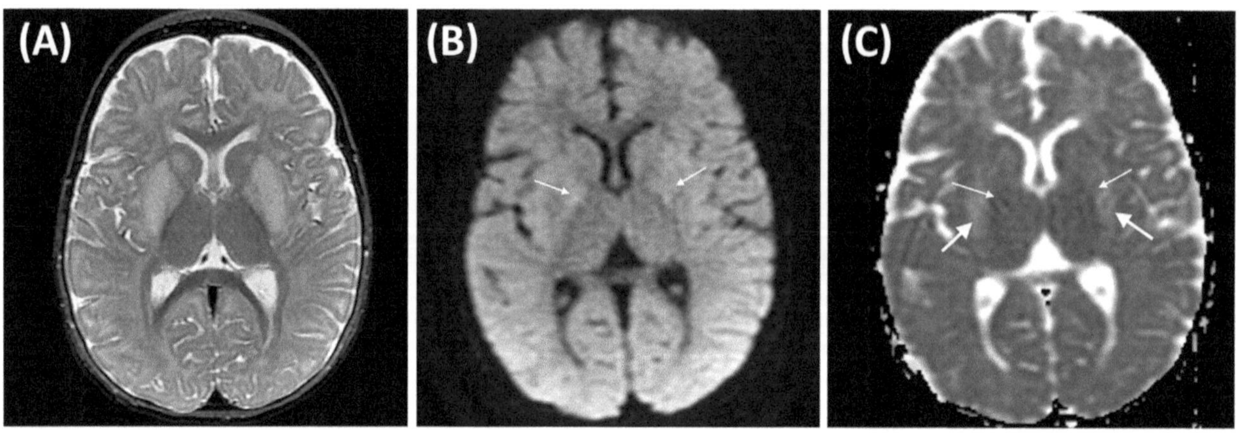

FIGURE 22.3 A 9-month-old female with propionic acidemia. Axial T2WI (A), diffusion weighted imaging (DWI) (B), and apparent diffusion coefficient (ADC) map (C) show swelling and edema in the striatum bilaterally compatible with active metabolic injury. Diffusion characteristics of the lesions are variable; facilitated diffusion reflecting vasogenic edema ± chronic lesions predominate (large arrows, C) but there are also small regions of reduced diffusion consistent with cytotoxic edema (small arrows, B and C).

TABLE 22.5 Diagnosis and treatment of organic acidemias and amino acidopathies.

Name of disorder	MRI findings	Laboratory abnormalities	Tested on newborn screening	Treatment
Propionic acidemia	Cerebral edema Gray matter lesions Gliosis	↑ NH_3 ↑ plasma propionate, 3-hydroxypropionate, and methylcitrate [78,79,91]	Yes	Protein restriction; specific restriction of dietary isoleucine, valine, threonine, and methionine Avoidance of catabolism Carnitine supplementation Nitrogen scavengers Carglumic acid Metronidazole to reduce endogenous propionate production by gut flora Liver transplant
Methylmalonic acidemia	Gray matter lesions, including subcortical lesions Altered and/or delayed myelination Optic nerve atrophy [92–95]	↑ NH_3 ↑ plasma propionate, 3-hydroxypropionate, methylcitrate, methylmalonate, and methylmalonylcarnitine [96]	Yes	Protein restriction; specific restriction of dietary isoleucine, valine, threonine, and methionine Avoidance of catabolism Carnitine supplementation Nitrogen scavengers Carglumic acid Cobalamin supplementation Liver transplant
Isovaleric acidemia	Cerebral edema Intracranial hemorrhage T2 hyperintensity of globus pallidus T2 hyperintensity of cerebral white matter [75,77,97]	↑ NH_3 ↑ plasma isovaleric acid ↑ urine isovalerylglycine, isovalerylcarnitine, and 3-hydroxyisovaleric acid [71,98,99]	Yes	Protein restriction; specific restriction of dietary leucine Avoidance of catabolism Carnitine and glycine supplementation
Glutaric aciduria type 1	Diffusion restriction and increased T2 signal intensity in bilateral basal ganglia, especially putamen [100,101]	↑ plasma and urine glutaric acid, 3-hydroxyglutaric acid, and glutarylcarnitine [100,102,103]	Yes	Lysine-restricted diet Carnitine supplementation Avoidance of catabolism Supportive management of motor disease, including medical tone management and orthotics
Maple syrup urine disease	Cerebral edema in acute metabolic crisis [104,105]	↑ plasma leucine, isoleucine, and valine [106,107]	Yes	Dietary restriction of branch chain amino acids Trial of thiamine Avoidance of catabolism Orthotopic liver transplant
Phenylketonuria	Decreased cerebral volume Periventricular and callosal white matter abnormalities [108,109]	↑ plasma phenylalanine [110]	Yes	Dietary restriction of phenylalanine Supplementation of synthetic cofactor (sapropterin or Kuvan) Enzyme substitution therapy (pegvaliase-pqpz or Palanziq)

supplementation to increase secretion of toxic metabolites as well as treatment with carglumic acid [76,82–84,86]. Periodic treatment with antibiotics, typically metronidazole, is also occasionally used to minimize propionate producing intestinal microflora, a large source of free propionate [83,87,88]. Orthotopic liver transplant may prolong survival and delay neurological deterioration; however, it is not a curative measure and, at best, shifts the trajectory of the disease to that of chronic insidious progression [76,81,84,89,90]. Even in the presence of optimal biochemical management, long-term chronic neurological complications occur [2,38,81].

Methylmalonic acidemia

Methylmalonic acidemia (MMA) is an autosomal recessive organic acidemia caused by a defect in either of two enzymes: methylmalonyl-CoA epimerase or methylmanoyl-CoA mutase [32,111–115]. There is significant overlap in clinical presentation with propionic acidemia, given that these enzymes are the direct next step in the shared pathway. The optic nerve is also susceptible to damage, resulting in optic nerve atrophy and blindness [114]. ID is a common neurocognitive outcome, with processing speed disproportionately affected compared to other etiologies of ID [116]. Reports suggest that the risk of metabolic stroke-like episodes in this patient population is between 17% and 30% [117,118].

Diagnostic abnormalities, including expected MRI and laboratory findings, are shown in Fig. 22.4 and summarized in Table 22.5. Treatment and outcomes do not differ greatly from propionic acidemia, though a trial of cobalamin supplementation, as cobalamin is the cofactor for methylmalonyl mutase, to test for patient responsiveness, is recommended [76,96].

Isovaleric acidemia

Isovaleric acidemia is an autosomal recessive organic acidemia resulting from defective leucine metabolism secondary to deficiency of the enzyme isovaleryl-CoA dehydrogenase [71,119]. The vast majority of patients present in the first year of life, and the common neonatal presentation involves a brief period of wellness after birth with subsequent metabolic crisis marked by development of emesis, failure to thrive, hypothermia, and lethargy that progresses to coma. Classically, acutely decompensating patients are said to have a smell like "dirty socks" or "sweaty feet" [119]. Despite this risk of acute metabolic decompensation leading to neurological injury, isovaleric acidemia is classically the least neurologically severe of the organic acidemias [75,77,97].

Isovolemic acidemia (IVA) patients have relatively mild neurodevelopmental findings but notably have an increased incidence of fine motor deficits that affect daily routines, decreased attention span, irritability, and a spectrum of social impairment including severe ASD [75,77,97]. Social deficits and abnormal peer interactions have also been seen and range in severity from severe ASD to those with minimal difficulties [120]. Rarely infantile spasms have been associated with isovaleric acidemia [121].

Diagnostic abnormalities, including expected MRI and laboratory findings, are summarized in Table 22.5. Management of these patients includes standard treatment for acute metabolic decompensations; long-term dietary management is summarized in Table 22.6 [122,123].

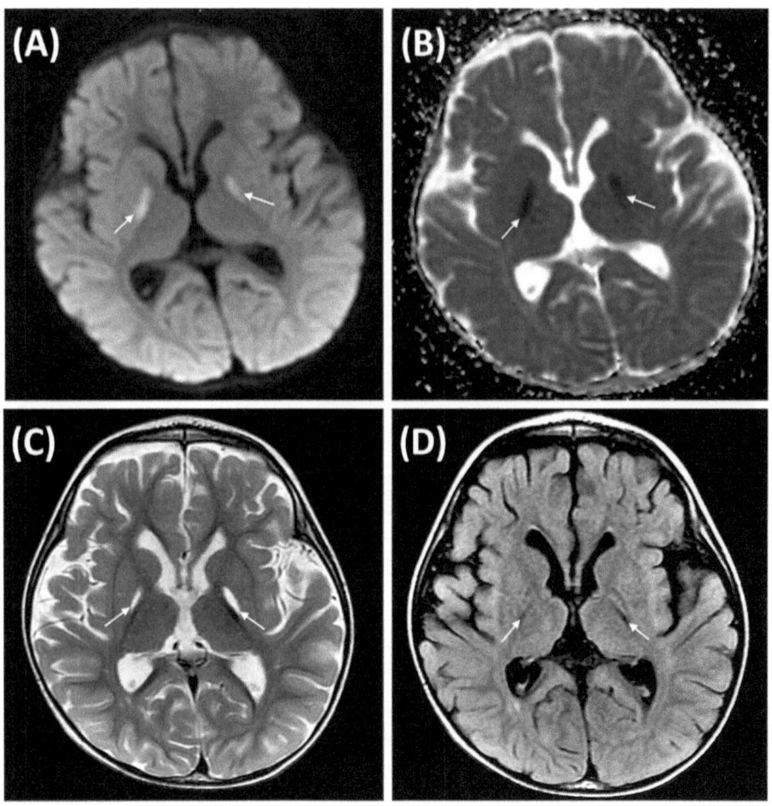

FIGURE 22.4 A 16-month-old male with methylmalonic acidemia. Axial diffusion weighted imaging (DWI) (A) and apparent diffusion coefficient (ADC) map (B) show reduced diffusion consistent with acute metabolic injury involving the globus pallidus bilaterally (arrows). T2WI (C) and T2 FLAIR images (D) from a follow-up MRI at 23 months show chronic, necrotic lesions in the globus pallidus (arrows) and progressive atrophy.

Glutaric aciduria type I

Glutaric aciduria type I (GA1) is an autosomal recessive disorder involving a defect in lysine metabolism [100,102,103]. Presentation of glutaric aciduria type I, unlike other organic acidemias, is classically thought of as primarily a neurological disorder [129]. The first presenting sign is often macrocephaly with increased CSF space causing stretching of bridging veins, increasing risk of subdural hemorrhage with relatively minor trauma. Retinal hemorrhage is often observed with intracranial hemorrhage and can lead to a misdiagnosis of nonaccidental head trauma [100].

Neonatal onset of crises is rare; instead, in the large majority of patients, encephalopathic crises occur between 3 and 24 months of life. Similarly, encephalopathy has not been described in asymptomatic patients greater than 6 years of age [130]. Patients in metabolic crisis initially present with sudden onset motor dysfunction, such as sudden onset hypotonia that progresses to tongue thrusting, dystonia, dyskinesia, and/or choreoathetosis, and even a single crisis can result in severe motor neurological deficits [34,35,103,130,131]. Clinically, some patients may present with a diagnosis of "progressive cerebral palsy," thus in patients with a diagnosis of dystonic CP without clear history of risk factors for hypoxic-ischemic injury, especially those with disease progression, GA1 must be excluded.

Imaging features of GA1 are shown in Fig. 22.5 and described in Table 22.5, along with typical laboratory abnormalities and standard treatment approaches. Notably, unlike other metabolic disorders, ketonuria, lactic acidosis, and hyperammonemia are usually absent even during crises [35,103]. Even with optimal management, treatment does not correct existing neurological disease, and patients may experience progressive neurological sequelae despite adequate nutritional therapy [34,35,130].

Maple syrup urine disease

Maple syrup urine disease (MSUD) is an autosomal recessive disorder affecting the metabolism of branched chain amino acids, notably leucine, which is the primary neurotoxicant, as well as isoleucine, and valine [36,132].

There is a range of severity of disease and while most patients present with symptoms in the neonatal period, first onset of symptoms may not occur until much later in life [104,133,134]. Classically, patients have a maple syrup odor to their cerumen and urine with neonates presenting with lethargy, opisthotonos, and seizures often associated with

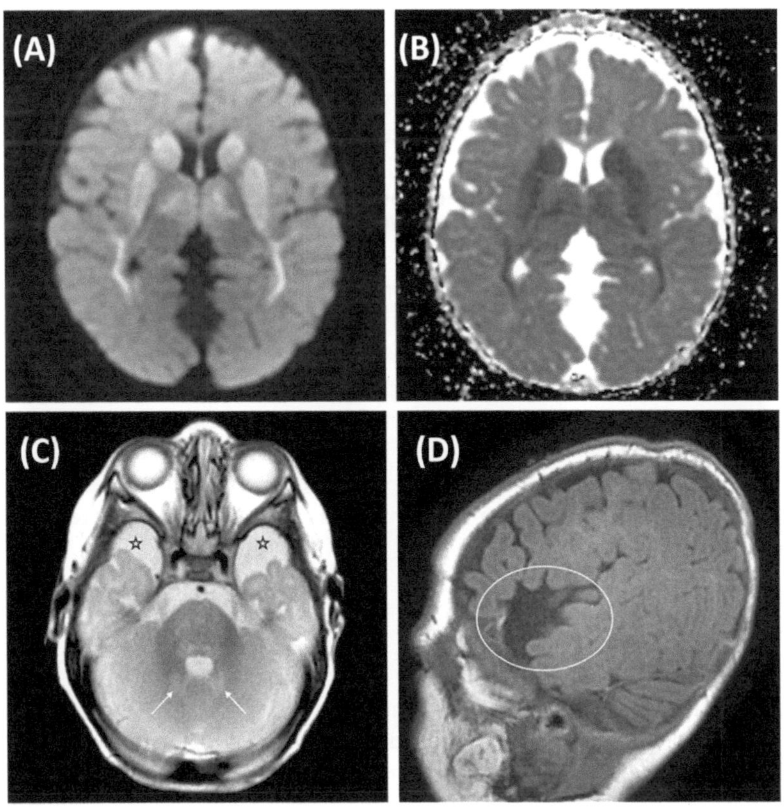

FIGURE 22.5 A 7-month-old female with glutaric aciduria type I. Axial diffusion weighted imaging (DWI) (A) and apparent diffusion coefficient (ADC) map (B) show signal abnormality with reduced diffusion affecting the basal ganglia > thalami and sagittal stratum reflecting metabolic injury. Abnormal signal is also present in the cerebellar deep gray nuclei on T2WI (C) (arrows). Concurrent structural abnormalities including temporopolar fluid space enlargement (stars, C) and underopercularization as seen on sagittal T1WI (D) (oval) clinch the diagnosis.

bicycling movements. GDD, with relative sparing of language, and attention deficits are common; dystonia, choreoathetosis, ataxia, cognitive impairment, and hallucinations can all occur [104,135,136]. Metabolic decompensations commonly occur with illness and are independently associated with neurodegeneration and Wernicke encephalopathy [104,133,134].

Imaging features of MSUD are described in Table 22.5, along with typical laboratory abnormalities and standard treatment approaches. Acute management is similar to other organic acidemias; patients with a rare variant are thiamine-responsive, so thiamine trial is recommended in all MSUD patients. With prompt initiation of treatment, patients with MSUD may achieve a normal IQ [134,137]. Orthotopic liver transplant is a successful treatment of MSUD, and it should be considered for patients with more severe and difficult to manage disease [104,106].

Phenylalanine hydroxylase deficiency

Phenylalanine hydroxylase deficiency, also known as PKU, is an autosomal recessive disorder of phenylalanine metabolism and is one of the most common IEM [37,110]. PKU has a range of severity which is defined primarily by the severity of accumulation of phenylalanine prior to onset of treatment and/or in times of dietary nonadherence. Notably, unlike other IEM true metabolic decompensation episodes are rare [138]. Classically, patients with PKU have musty odor, and some patients also have decreased pigmentation in their skin, eyes, and hair [139]. Untreated patients can develop seizures, visual impairment, motor deficits, poor coordination, learning disabilities, ID, and ADHD [21,140–144].

Maternal PKU can occur in patients who are inadequately treated during pregnancy and exposure to excessive phenylalanine as a teratogen can result in fetal loss or fetal anomalies, including cardiac, pulmonary, neurological, and facial abnormalities; these affected fetuses are at risk of developmental and intellectual disabilities [139].

Imaging features of PKU are described in Table 22.5, along with typical laboratory abnormalities and standard treatment approaches. The primary treatment of PKU is diet modification to restrict dietary phenylalanine intake [145]. A minority of patients are responsive to sapropterin, a synthetic form of the cofactor for phenylalanine hydroxylase, and an enzyme substitution therapy was recently approved [110,146]. It should be noted that treatment of symptomatic patients does result in improvement in behavioral outcomes, but does not reverse ID [147].

Galactosemia

Galactosemia is an inherited metabolic disorder of carbohydrate metabolism due to defects in galactose metabolism occurring in 1 in 50,000 individuals [148–152]. Patients with galactosemia clinically present with cataracts, neonatal jaundice, failure to thrive, *Escherichia coli* sepsis, and progressive hepatomegaly and renal failure. Subacute symptoms usually include primary ovarian insufficiency, intellectual delays, motor difficulties such as myoclonus, cerebellar ataxia, tremor, and dystonia affecting upper extremities more than lower [153]. Additionally, motor developmental delays and intellectual delays have been seen in this patient population, with noted impairments in activities of daily living and adaptive functioning. A range of social and communication deficits, from diagnosed ASD to isolated language defects have been seen. Finally, behavioral and psychiatric problems, most frequently, depression and anxiety, are also associated with disease, though acute episodes of psychosis have also been observed [149,151].

On imaging, patients often show cerebral atrophy, variable ventriculomegaly, white matter lesions with clustering around the lateral ventricle horns, and peripheral cerebral demyelination [154–156]. Laboratory abnormalities include reducing substances in the urine and increased levels of plasma galactose, galactose-1-phosphate, and galactitol. Reducing substances are commonly found in the urine of untreated patients [151,153]. It should be noted that there are variants that show only biochemical disease without clinical symptoms, the most common of which is "Duarte Galactosemia" [151,152]. Treatment primarily revolves around a lactose-free and galactose-restricted diet. Diet modification typically improves the neonatal complications, but it has also been shown that it is not sufficient to prevent neurologic sequelae, though it may lessen the severity [149,151,157].

Mitochondrial disorders

Mitochondrial disorders are a family of clinically heterogeneous genetic disorders characterized by mitochondrial dysfunction. The symptoms of mitochondrial disorders arise from disruption of the normal functions of the mitochondrial, which are shown in detail in Fig. 22.6 [158]. Mitochondrial genetics are unique due to dual contributors from both the mitochondrial DNA (mtDNA), which is solely maternally inherited, and the nuclear DNA (nDNA), which encodes for the vast majority of the approximately 1500 mitochondrial proteins [158]. As such, mitochondrial disorders can arise

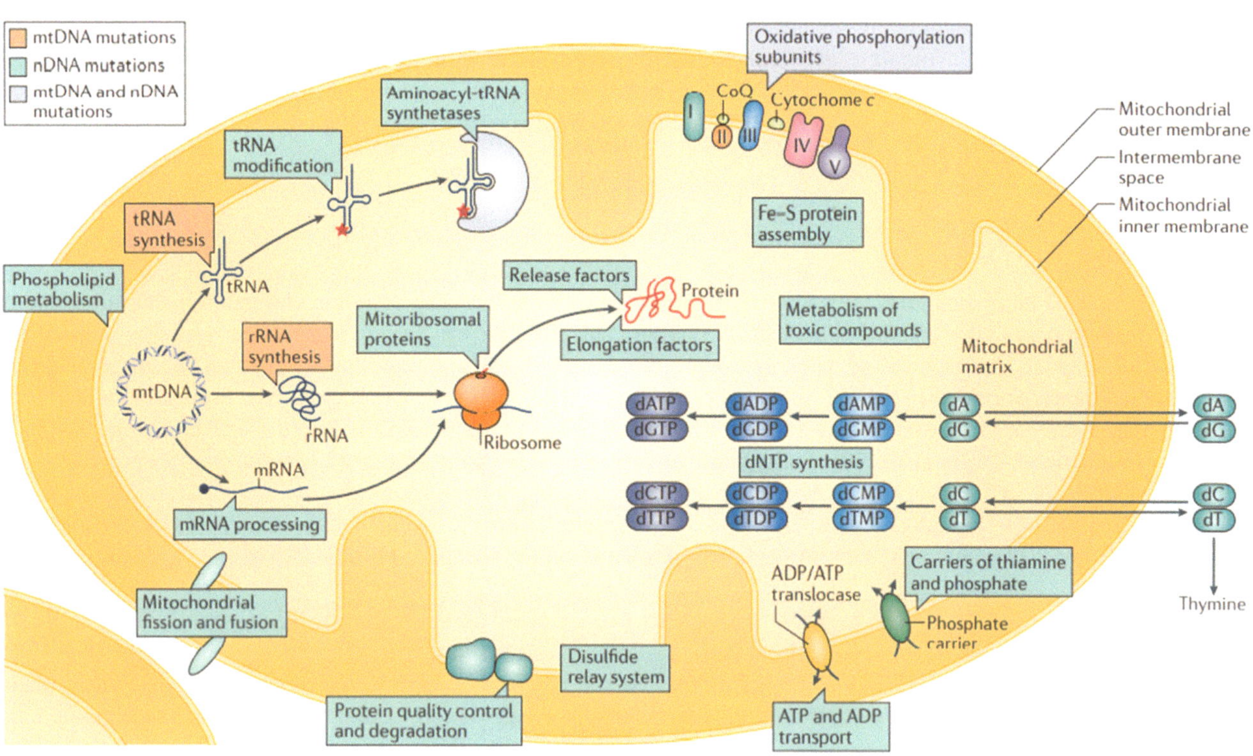

FIGURE 22.6 Genes associated with human mitochondrial diseases and their role in mitochondrial function. Defects in nuclear DNA (nDNA) can cause abnormalities in oxidative phosphorylation, mitochondrial DNA (mtDNA) maintenance, solute carriers of thiamine and phosphate, and respiratory chain deficiency, among others. Mutations in mtDNA can cause defects in structural proteins, defective mitochondrial RNA synthesis and respiratory chain deficiency. *CoQ*, Coenzyme Q10; *dNTP*, deoxynucleotide triphosphate; *Fe–S*, iron–sulfur [158].

from a variety of inheritance patterns, including autosomal recessive, X-linked, autosomal dominant, or maternal; furthermore, unique aspects of phenotypic penetrance occur in certain mitochondrial disorders, leading to unique inheritance patterns such as pseudodominant inheritance in pyruvate dehydrogenase complex (PDHc) deficiency due to skewed X-inactivation in female carriers and the predominance of affected males in the maternal lineage in patients with Leber's hereditary optic neuropathy that may give the specious appearance of X-linked inheritance [158–160]. Additionally, inheritance of mtDNA mutations can be heterogeneous between family members due to heteroplasmy, which is variable presence of mutated mtDNA within a mitochondrion and/or cell and which is related to the level of mitochondrial and cellular dysfunction [158]. Due to the deleterious effects of proteins affected by biallelic mutations in nDNA versus the effects of variable dysfunction caused by mtDNA, childhood mitochondrial disorders are more often attributable to nDNA mutations inherited in an autosomal recessive or X-linked pattern and occur with an estimated prevalence of 5–15 in 100,000 individuals under the age of 16; conversely, the majority of adult-onset mitochondrial disorders are caused by mtDNA mutations and are estimated to affect between 10 and 15 in 100,000 adults [158]. Although individual mitochondrial disorders may be rare, but as a group, they affect an estimated 1 in 5000 individuals and are one of the most common inherited metabolic disorders [161].

Because the mitochondrial are critical to cellular energy production, mitochondrial disorders disproportionately affect tissues with high energy demand [158]. Episodic encephalopathy, developmental delay and, especially, developmental regression occurring acutely with partial interim recovery between events leading to stepwise clinical deterioration are frequent hallmarks of mitochondrial disorders [158,162].

Diagnosis of these disorders can be challenging and often requires multiple testing modalities, including neuroimaging to assess for typical patterns of neurological injury, screening for biochemical markers of mitochondrial dysfunction, and NGS of both the nuclear and mitochondrial genomes; additionally, diagnosis may require histological, biochemical, and or molecular assessment of biopsied tissue, but these more invasive diagnostics have become less frequently utilized as NGS has made genetic testing more agile and accessible [158,163–167].

Therapies for mitochondrial disorders are typically twofold: treatment of the symptoms of the disorder and supportive management with supplemental vitamins and nutrients and/or specialized diet that aims to optimize mitochondrial function [162,168,169]. Additionally, management of these patients includes prevention of further decompensation by avoiding catabolism and avoidance of contraindicated medications such as valproic acid, statins, and propofol [162,165]. Given the vast array of identified mitochondrial disorders, it is not feasible to explore this topic fully in the scope of this chapter; however, we will focus in detail on several disorders with prominent neurodevelopmental phenotypes presenting in childhood.

Leigh disease

Leigh disease, also known as subacute necrotizing encephalopathy, is the most common childhood-onset mitochondrial disorder, occurring in 2.5 in 100,000 live births [158,160]. This disorder has broad genetic heterogeneity, with over 75 identified monogenic causes associated with multiple modes of inheritance, including autosomal recessive, X-linked, and maternal inheritance [160,170]. Clinically, Leigh disease presents as a neurodegenerative disorder with normal development in infancy followed by onset of developmental delay and regression before the age of 2, typically in the setting of infection or illness [160,170]. Following onset of symptoms, patients experience episodic progressive neurological dysfunction, including continued developmental regression, hypotonia, seizures, ataxia, dystonia, dysphagia, ophthalmologic manifestations such as nystagmus and optic atrophy, and/or respiratory dysfunction [160,171]. Following the onset of symptoms, neurological deterioration is typically rapid and typically results in death by 3 years of age [160,170].

Leigh disease is associated with a characteristic MRI pattern of bilateral symmetric T2 hyperintensities classically involving primarily basal ganglia and brainstem; MRS typically shows an elevated lactate peak (Fig. 22.7) [160,170–174]. Laboratory abnormalities include elevated serum lactate and plasma alanine levels, and identification of a causative gene mutation using NGS can provide molecular confirmation of diagnosis; formal diagnosis of Leigh disease requires the triad of neurodegeneration, characteristic symmetric MRI lesions, and evidence of mitochondrial dysfunction [160,170].

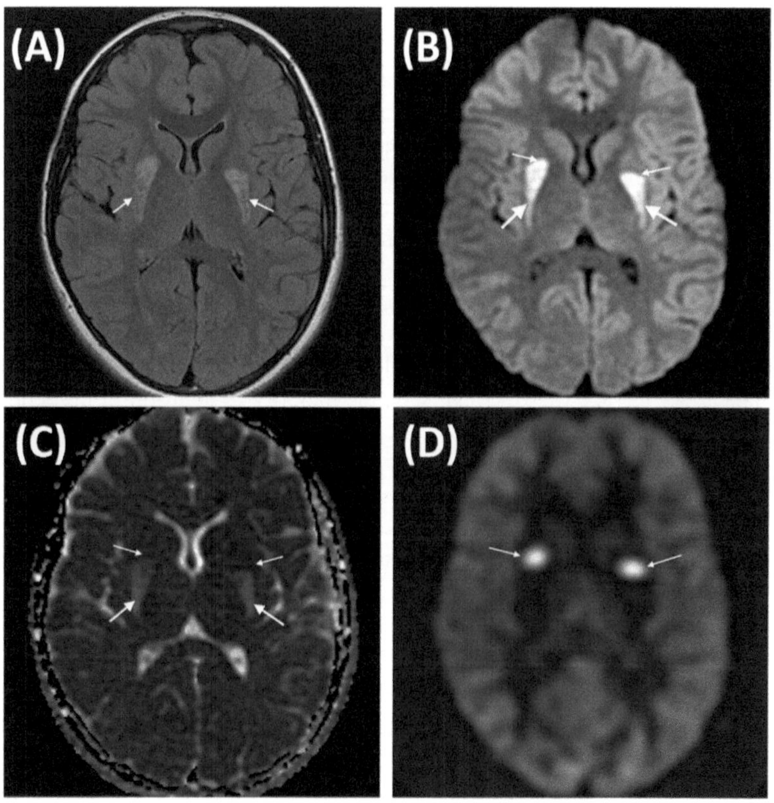

FIGURE 22.7 A 15-year-old male with Leigh syndrome (mitochondrial gene mutation). Axial T2 FLAIR (A), diffusion weighted imaging (DWI) (B), apparent diffusion coefficient (ADC) map (C), and argininosuccinate lyase (ASL) perfusion images (D) depict acute on chronic metabolic injury. Chronic, hyperintense lesions are present in the posterior putamen on T2 FLAIR and ADC images (large arrows). Acute, active putamen lesions with reduced diffusion (small arrows, B and C) and hyperperfusion (arrows, D) are also visible at the anterior margin of the chronic lesions.

Identification of a genetic etiology may direct supportive therapies, such as supplementation of Coenzyme Q10 (CoQ10) in patients with CoQ10 deficiency or initiation of ketogenic diet in patients with PDHc deficiency [170]. As with other mitochondrial diseases, a "mitochondrial cocktail" of vitamins and supplements, including CoQ10, vitamin E, L-carnitine, thiamine, may be prescribed to support energy production; otherwise, treatment is focused on alleviating symptom burden and optimizing comfort, often in the palliative care setting [170].

Pyruvate dehydrogenase complex deficiency

Dysfunction in PDHc is one etiology of Leigh disease, with 84% of cases occurring due to mutations in the *PDHA1* gene on the X chromosome [160,175]. Despite its location on the X chromosome, approximately equal males and females are diagnosed with *PDHA1* mutations, with skewed X-inactivation leading to a milder neurological phenotype in females [159,160]. Patients with PDHc deficiency often present with a Leigh disease phenotype [159]. In addition to the commonly described MRI findings characteristic of Leigh disease, MRI of patients with PDHc may also show structural abnormalities such as ventriculomegaly or dysgenesis of the corpus callosum [159]. Ketogenic diet has been shown to decrease lactic acidosis, slow neurodegeneration, and stabilize disease progression in some patients [159,176,177].

Mitochondrial encephalomyopathy, lactic acidosis, and stroke-like episodes

Mitochondrial encephalomyopathy, lactic acidosis, and stroke-like episodes (MELAS) is a progressive neurodegenerative disorder most commonly associated with a mutation in the mtDNA gene *MT-TL1* and is the most common mtDNA-associated childhood onset mitochondrial disorder, with the most common mutation occurring in 60 per 100,000 people [178]. Symptoms typically occur prior to the age of 40, with 65%–76% of patients presenting prior to the age of 20 and 5%–8% of patients presenting before the age of 2 [179]. The three invariable criteria of MELAS are stroke-like episodes before the age of 40, encephalopathy characterized by seizures and/or dementia, and myopathy characterized by presence of lactic acidosis and/or ragged red fibers [161,178,179].

The most common features of MELAS are progressive neurodegeneration associated with acute stroke-like episodes, headache, refractory focal seizures including *epilepsia partialis continua*, exercise intolerance, myopathy, dementia, and sensorineural hearing loss [161]. Stroke-like episodes occur in up to 99% of affected patients and typically present with variable symptoms, including partially reversible aphasia, cortical vision loss, focal motor weakness, headache, encephalopathy, and/or seizures, including acute onset of *epilepsia partialis continua* [178,179]. Slowly progressive dementia affecting all cognitive domains and refractory epilepsy are common symptoms [179]. Recurrent severe migrainous headaches occur in the majority of patients with MELAS and may precede, trigger, or cooccur with a stroke-like episode [161,179]. Although MELAS is defined by its neurological symptoms, multisystem organ dysfunction is a frequent phenotypic feature [178,179].

MRI will typically show a characteristic pattern of injury with acute diffusion restriction and T2 hyperintensity attributable to vasogenic edema corresponding with infarct in nonvascular territories, most commonly located either unilaterally or bilaterally in the parieto-occipital region (Fig. 22.8); MRS will typically demonstrate a lactate peak both in acutely injured areas as well as in noninfarcted areas, and N-acetyl aspartate (NAA) is often elevated as well (Fig. 22.8) [164,173,174]. Laboratory confirmation may include demonstration of lactic acidemia, which is present in 94% of affected patients, muscle biopsy and/or molecular confirmation of a pathogenic mtDNA mutation [179].

As with many other mitochondrial disorders, treatment is largely symptomatic; treatment with a "mitochondrial cocktail" of vitamins and supplements, including CoQ10, creatine, alpha-lipoic acid, thiamine, and L-carnitine may have some modest benefit for some patients [178]. IV arginine administered in the setting of an acute stroke-like episode has been shown to improve symptoms and should be initiated urgently in patients presenting with MELAS presenting acutely with sudden onset of new neurological symptoms, and oral arginine supplementation can reduce the severity and frequency of these stroke-like episodes [165,180].

Peroxisomal disorders

Peroxisomal disorders are a family of diseases caused by dysfunction of the peroxisomes, essential and ubiquitous cellular organelles containing over 50 different enzymes responsible for a multitude of cellular metabolic functions.

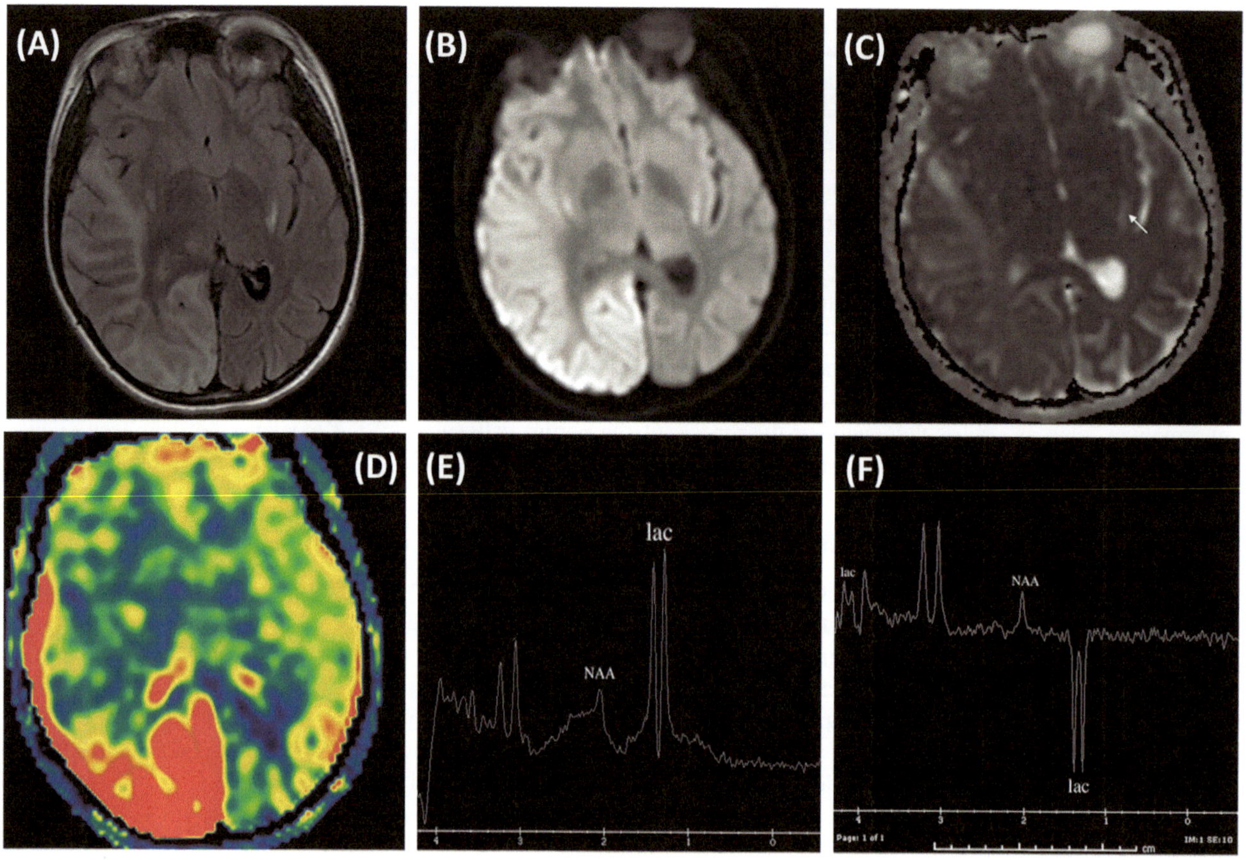

FIGURE 22.8 A 12-year-old female with mitochondrial encephalomyopathy, lactic acidosis, and stroke-like episodes (MELAS) syndrome. Axial T2 FLAIR (A), diffusion weighted imaging (DWI) (B), apparent diffusion coefficient (ADC) map (C), and argininosuccinate lyase (ASL) perfusion image (D) demonstrate right temporo-occipital swelling, mixed cortical/subcortical vasogenic and cytotoxic edema (A–C), and hyperperfusion (D) crossing large arterial territories consistent with acute metabolic stroke. A small chronic lesion in the posterior with facilitated diffusion left putamen is consistent with chronic sequela of prior metabolic injury (arrow, C). PRESS single voxel 1H magnetic resonance spectroscopy (MRS) over the right parietal white matter with 35 msec TE (E) and 144 msec TE (F) reveal marked lactate (lac) and diminished N-acetylaspartate (NAA).

Peroxisomal disorders as a whole have an incidence of 1 in 5000 individuals [181]. Due to the predominant neurodegenerative phenotype in the most common of these disorders, X-linked adrenoleukodystrophy and Zellweger syndrome, these disorders are covered in the chapters: Approach to diagnosis and management of genetic neurodevelopmental disorders; Neurodegenerative diseases.

Lysosomal storage disorders

LSDs are a broad family of diseases caused by the disruption of the function of the lysosomes, membrane-bound organelles which contain enzymes responsible for the degradation of myriad macromolecules [182]. Lysosome-mediated degradation of these molecules requires coordination of multiple cellular processes, and LSDs are caused by disruptions of this process, the end result of which is the accumulation of the nondegraded macromolecule inside the lysosome and subsequent lysosomal dysfunction [182].

As with most metabolic disorders, patients with LSDs are normal at birth and become symptomatic with age, displaying an array of phenotypes that are dependent on the macromolecule that is unable to be degraded [182]. Although not all LSDs cause neurologic injury, many disorders in this disease class do have a developmental phenotype. Table 22.6 summarizes presenting symptoms of lysosomal storage disorders.

TABLE 22.6 Presenting symptoms of lysosomal storage disorders.

Name of disorder	Incidence	Associated gene	Mode of inheritance	Cognitive symptoms	Behavioral symptoms	Other neurological symptoms	Common systemic symptoms
Mucopolysaccharidosis type I (Hurler syndrome)	1 in 100,000 [124]	*IDUA*	AR	Slowing of developmental milestones Progressive GDD Progressive ID	Psychiatric disorder		Thoracolumbar kyphosis Umbilical hernia Progressive corneal clouding Recurrent respiratory infections Mitral and aortic valvular insufficiency Coarsening of facial features, including frontal bossing, thickening of the lips and tongue, broadening of the nasal bridge, and dental abnormalities
Mucopolysaccharidosis type II (Hunter syndrome)	1 in 100,000 live male births [124]	*IDS*	X-linked	Motor Regression Progressive ID	Hyperactive behavior Aggression Poor attention	Generalized Epilepsy Sleep disturbances	Thoracolumbar kyphosis Umbilical hernia Recurrent respiratory infections Mitral and aortic valvular insufficiency Coarsening of facial features
Mucopolysaccharidosis type III (Sanfilippo syndrome)	1 in 70,000 [124]	*SGSH* *NAGLU* *NGSNAT* *GNS*	AR	Speech delay Developmental arrest Progressive language regression Progressive cognitive regression	Hyperactivity Restlessness Destructive behavior Anxiety Severe tantrums Aggression "Klüver-Bucy like" hyperoral activity and apparent reduced fear, risk-taking	Epilepsy	Mild coarsening of the facial features Hirsutism Joint stiffness Mild hepatosplenomegaly

(Continued)

TABLE 22.6 (Continued)

Name of disorder	Incidence	Associated gene	Mode of inheritance	Cognitive symptoms	Behavioral symptoms	Other neurological symptoms	Common systemic symptoms
Mucopolysaccharidosis type VII (Sly syndrome)	1 in 300,000–1,000,000 [124]	*GUSB*	AR	Developmental arrest Developmental regression Progressive language regression Progressive neurocognitive decline		Hearing impairment	Respiratory difficulties Sleep apnea Hepatosplenomegaly Coarse facial features Macrocephaly with scaphocephaly
GM1 gangliosidosis	1 in 100,000–200,000 [125]	*GLB1*	AR	Early GDD followed by rapid developmental regression		Increased startle Hypotonia Spasticity Hearing impairment Seizures Gait abnormalities	Cardiomyopathy Hepatosplenomegaly Ophthalmic manifestations including cherry-red macula and blindness Skeletal dysplasia leading to restrictive lung disease
Tay–Sachs disease	1 in 222,000 live births Increased carrier frequency in Eastern Quebecois and Ashkenazi Jew population [126]	*HEXA*	AR	Abrupt developmental arrest then regression		Hypotonia Exaggerated startle reflex Progressive psychomotor degeneration Seizures	Cherry-red macula
Niemann–Pick type C disease	1 in 120,000 live births [127]	*NPC1* *NPC2*	AR	GDD followed by developmental arrest then regression Learning disability	Juvenile: Attention deficit Adult-onset: Psychosis Paranoid delusions Auditory or visual hallucinations	Infantile: Hypotonia Spasticity Late infantile/ juvenile: Cerebellar ataxia Dysarthria Dysphagia, Seizures Dystonia Cataplexy, which is typically induced by laughter Vertical supranuclear gaze palsy	Hepatomegaly Cholestasis Jaundice Splenomegaly

CLN1 disease	0.1–5 in 100,000 [128]	*CLN1*	AR	Decreased social behavior followed by general developmental regression		Hypotonia Ataxia Myoclonus Epilepsy Vision loss leading to blindness Early background suppression on EEG	Optic atrophy Retinal disease without pigmentary aggregation
CLN2 disease	0.2–9 in 100,000 live births per year [128]	*CLN2*	AR	Early language delay Developmental arrest followed by motor regression Cognitive decline		Refractory epilepsy; myoclonic, tonic, atypical absence and/or generalized tonic-clonic seizures Vision loss Ataxia	Absence of retinal pigmentary accumulation
CLN3 disease	0.5–5 in 100,000 live births per year [128]	*CLN3*	AR	Loss of independent adaptive skills	Angry outbursts Anxiety Depression Physical violence	Vision loss progressing to blindness Seizures Parkinsonism Dysarthria	Retinal degneration

AR, Autosomal recessive; *EEG*, electroencephalogram; *GDD*, global developmental delay; *ID*, intellectual disability; *LD*, learning disability.

Mucopolysaccharidoses

The mucopolysaccharidoses (MPS) are LSDs caused by impaired degradation of glycoaminoglycans (GAGs) [124,183,184]. As with most LSDs, individuals with these disorders appear normal at birth but start to display a multitude of clinical manifestations with GAG accumulation, including coarsening of the facial features, corneal clouding, hepatosplenomegaly, skeletal deformity, joint rigidity, and cardiorespiratory dysfunction, with variable neurodevelopmental manifestations occurring primarily in MPS I, II, III, and VII (Fig. 22.9) [124,183]. Slowing of development, GDD, and behavioral abnormalities may be the earliest manifestations of these diseases for some children; prompt diagnosis of these patients is critical to delay or mitigate further clinical decompensation, although treatment does not typically reverse or arrest neurocognitive decline [183].

Mucopolysaccharidosis type I

Mucopolysaccharidosis type I (MPS I) is an autosomal recessive disorder caused by reduced alpha-L-iduronidase activity [124,183,185]. MPS I can generally be differentiated into two clinical phenotypes: a severe phenotype (MPS IH, or Hurler syndrome) that presents in the first months of life and a more variable attenuated phenotype that occurs on a spectrum between the intermediate Hurler–Scheie phenotype to the more mild Scheie phenotype [185].

Patients with Hurler syndrome generally present with slowing of developmental milestones by 1 year of age, progression to GDD by 2 years of age, slowing of cognitive development between 2 and 4 years of age followed by rapid progression to severe ID as the patient ages [183,185]. Somatic features of Hurler syndrome are described in Table 22.6 and shown in Fig. 22.9 [185].

Gold standard for laboratory diagnosis of MPS I is measurement of alpha-L-iduronidase activity, which serves as the basis for newborn screening in some states, and can be confirmed with NGS of the IDUA gene [185]. Treatments for MPS I include both HSCT and ERT. HSCT is considered the treatment of choice for Hurler syndrome and, if completed prior to 2.5 years of age and onset of cognitive decline, can arrest progression to ID and preserve cognition in the range of the normal full-scale IQ [183,185]. ERT with the recombinant enzyme laronidase is the preferred therapy for patients with Hurler–Scheie or Scheie syndromes but has only been shown to improve somatic symptoms [183,185,198].

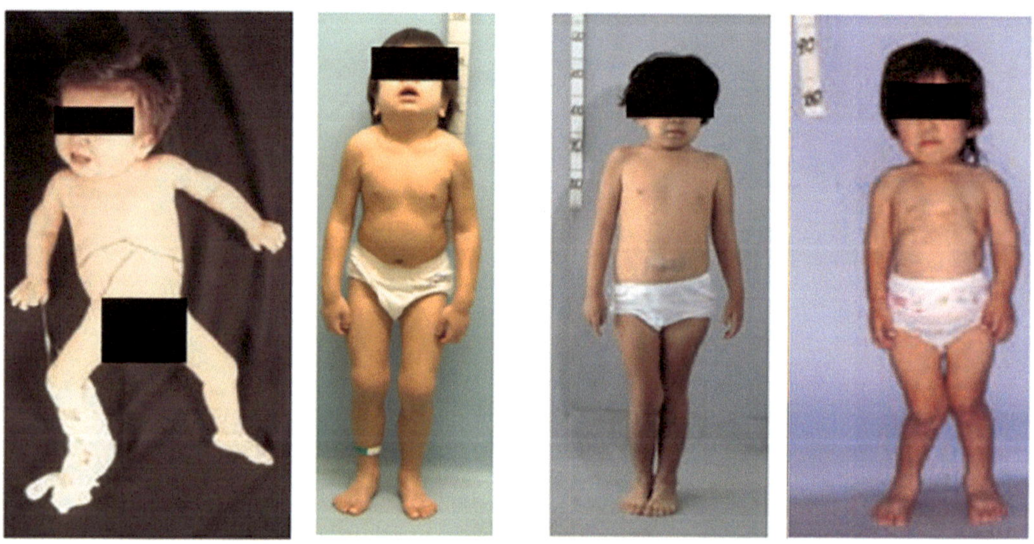

1.5 years (I: severe) 8 years (II: severe) 12 years (IIIB: severe) 7 years (VII: skeletal type)

FIGURE 22.9 Clinical features of patients with mucopolysaccharidosis (MPS) I, II, III, and VII. MPS I: coarse hair and face, short neck, hepatosplenomegaly, dysostosis multiplex, stiffed joints, and dysostosis metacarpal. MPS II: coarse hair and face, macrocephaly, short neck, hepatosplenomegaly, dysostosis multiplex, and stiffed joints. MPS IIIB: coarse facial feature. MPS VII: short stature, short neck, short trunk, dysostosis multiplex, kyphoscoliosis, knock-knees, joint instability and laxity, and dysostosis multiplex. *Images kindly provided and permitted by Carol Ann Foundation for MPS IVA and by Dr. Orii at Gifu University for other types of MPS; Stapleton M, Arunkumar N, Kubaski F, Mason RW, Tadao O, Tomatsu S. Clinical presentation and diagnosis of mucopolysaccharidoses. Mol Genet Metab 2018;125(1–2):4–17.*

Mucopolysaccharidosis type II

Mucopolysaccharidosis type II (MPS II, or Hunter syndrome) is an X-linked disorder caused by reduced iduronate sulfatase (IDS) [183,186,187,198]. The somatic manifestations of MPS II are similar to those described in MPS I, with exception of corneal clouding, which a differentiating feature not seen commonly in MPS II [124,183,186].

Patients with MPS II are classified as either severe (neuronopathic) or attenuated based on neurological involvement, which is outlined in Table 22.6. Infants with severe/neuronopathic MPS II appear normal at birth and may attain early developmental milestones; motor and speech delay tend to be apparent by 18–24 months of age and development plateaus between 3 and 5 years of age after which cognition declines to the point of severe ID in the second decade of life [124,183,186].

Imaging features of MPS II are described in Table 22.7, along with typical laboratory abnormalities. Carrier testing with NGS of the *IDS* gene is available for females with a family history of affected male relatives.

ERT with the recombinant enzyme idursulfase can halt progression of the somatic symptoms of MPS II; this enzyme does not cross the blood–brain barrier and therefore does not ameliorate the neurological manifestations of this disease, for which only symptomatic treatment is available [183,186,198]. Additionally, pharmacological treatment of behavioral symptoms should be considered in these patients. Hyperactive behaviors typically do not respond well to methylphenidate [186].

Mucopolysaccharidosis III

Mucopolysaccharidosis type III (MPS III or Sanfilippo syndrome) is a family of autosomal recessive disorders encompassing four subtypes that are delineated based on their genetic and biochemical etiology. They have fairly similar phenotypes with prominent neurological symptoms and more mild somatic manifestations [124,199].

Like other MPS discussed in this chapter, MPS III can present as either a severe or attenuated phenotype defined primarily by severity of neurological involvement. The majority of patients with the severe phenotype first exhibit speech delay, then developmental arrest around the age of 3 followed by subsequent regression in language and cognition. Due to the prominent language and behavioral symptoms, patients with MPS III are frequently misdiagnosed with ASD, although their social skills tend to be relatively preserved at the time of initial presentation [124,183]. Behavioral abnormalities and somatic features are outlined in Table 22.6 [183,200]. For patients with the severe phenotype, death typically occurs in the second or third decade of life as a result of severe neurological deterioration [183,188].

Imaging features of MPS III are described in Table 22.7, along with typical laboratory abnormalities. Molecular confirmation of diagnosis is completed with NGS of the four genes associated with the four subtypes of this disease [124]. There is no current disease-directed therapy approved for clinical use, and current treatment regimens focus on symptom management [183].

Mucopolysaccharidosis VII

Mucopolysaccharidosis type VII (MPS VII or Sly syndrome) is an autosomal recessive disorder caused by deficiency in the enzyme β-glucuronidase, with presentations ranging from severe to attenuated [201]. Due to the relative rarity of this type of MPS, it is the least clinically characterized. Somatic symptoms of MPS VII are similar to other MPS types [124]. In patients with the more severe phenotype, developmental arrest and subsequent developmental regression occurs between 1 and 3 years of age, with progressive neurocognitive decline and language difficulties that are frequently exacerbated by hearing impairment [124,183].

First line screening for MPS VII is indirect measurement of β-glucuronidase activity through biochemical quantification of urinary GAGs; diagnosis is then molecularly confirmed through NGS of the *GUSB* gene [124]. ERT with the recombinant enzyme vestronidase alfa is approved for both pediatric and adult patients with MPS VII; however, neurological symptoms of the disease are not targeted by this therapy [202].

Gangliosidoses

The gangliosidoses are a group of rare neurodegenerative autosomal recessive conditions caused by defective lysosomal hydrolase enzymes leading to impaired catabolism of gangliosides, the primary sphingolipid molecules comprising neuronal plasma membranes [189,203]. These disorders are designated as either GM1 or GM2 gangliosidoses based on the accumulated metabolites; while all diseases in this class entail progressive and irreversible neurological dysfunction, they represent a spectrum with presentations ranging from a severe infantile form to a milder attenuated form that presents in adulthood [203].

GM1 gangliosidosis

GM1 gangliosidosis is an autosomal recessive disorder caused by a deficiency in the hydrolytic enzyme β-galactosidase [125]. GM1 gangliosidosis is associated with progressive neurodegeneration and is further divided into three subtypes based on age of presentation and severity of disease which correlate with residual activity of the mutant β-galactosidase [125,203].

Type I (infantile) GM1 is the most severe subtype presenting between in the first 3 months of life with early developmental delay, followed by progressive neurological deterioration resulting in death between 2 and 4 years of age. While neurological dysfunction is the hallmark of this disease, there is concurrent progression of variable systemic symptoms, which are detailed in Table 22.6 [125,203].

Type II (late infantile and juvenile) GM1 is an intermediate phenotype with bimodal distribution of age of onset. The late infantile subtype presents between 7 months and 2 years of age with subsequent developmental regression that slowly progresses until death between 5 and 10 years of age. The juvenile subtype initially presents between 3 and 10 years of age, with very slowly progressive neurological dysfunction and loss of primarily motor milestones leading to death in the third decade of life. Type III (adult) GM1 is the mildest phenotype with primarily motor dysfunction between 3 and 30 years of age with prominent extrapyramidal signs progressing to more prominent Parkinsonian features [125,203].

Imaging features of GM1 are described in Table 22.7, along with typical laboratory abnormalities and treatment paradigms. Historically, diagnosis of GM1 gangliosidosis is made with enzyme activity assays; molecular diagnosis has become more common as first-line diagnostic testing [189].

GM2 gangliosidosis

GM2 gangliosidoses are autosomal recessive disorders caused by dysfunction of one of three lysosomal β-hexosaminidases. Tay–Sachs disease, the most common GM2 gangliosidosis, is caused by deficiency in the hydrolytic enzyme hexosaminidase A [126,189]. Tay–Sachs disease can be divided into three subtypes differentiated by age of onset. The infantile form of Tay–Sachs is the most severe form, presenting between 3 and 6 months of age with early symptoms of abrupt developmental arrest, hypotonia, and exaggerated startle reflex [203,204]. Following initial onset of symptoms, patients experience progressive psychomotor degeneration; death occurs in the majority of affected patients between 3 and 5 years of age [126,182].

Juvenile Tay–Sachs disease presents at an average age of 5 years with initial symptoms of impaired coordination followed by neurological deterioration with seizures, worsening spasticity, progressive speech and swallowing dysfunction, loss of continence, blindness, and sleep disruptions [126,204]. Patients eventually deteriorate to a vegetative state, and death typically occurs between the ages of 10 and 15 years, typically as a result of infection [204]. The late onset/adult subtype of Tay–Sachs has a variable presentation with onset typically beginning late in the second decade of life; motor dysfunction and psychosis are unique features of this subtype [6,203].

Imaging features of Tay–Sachs disease are described in Table 22.7, along with typical laboratory abnormalities and treatment paradigms. Historically, diagnosis of GM1 gangliosidosis is made with enzyme activity assays in leukocytes or skin fibroblasts; molecular diagnosis has become more common as first-line diagnostic testing [189]. There are currently no disease-directed therapies.

Niemann–Pick type C disease

Niemann–Pick type C disease (NPC) is an autosomal recessive disorder caused by impaired trafficking and subsequent accumulation of cholesterol and sphingolipids species within the lysosome [125,127,196,205,206]. As with other LSDs, age of presentation of NPC varies from perinatal onset to first presentation well in the adult years, with subtypes of this disease defined as follows: early infantile, late infantile, juvenile, and adolescent/adult. These categories typically are defined by the onset of neurological symptoms, whereas systemic manifestations are independent of neurological dysfunction, can precede neurological symptoms for years to decades, and are not reliable predictors of disease progression [127,196]. Regardless of age of neurological symptom onset, over 40% of patients with NPC present before 12 months of age with hepatomegaly and cholestasis causing jaundice, which typically resolve, and splenomegaly, which may persist [206].

Early infantile NPC is a visceral-neurodegenerative disease with typical onset between ages 2 months and 2 years and may present initially with an isolated hepatosplenomegaly [127,206]. Infants with this form have typical development from birth to the age of 8–9 months, at which time they start to have delays in attainment of milestones followed by developmental arrest then regression; following this loss of motor milestones, patients develop prominent spasticity. The majority of patients with this form die prior to 5 years of life, typically as a result of neurological disease progression leading to respiratory infections and subsequent respiratory failure [127,206].

Both the late infantile and juvenile forms of NPC are primarily neurodegenerative, with indolent onset of neurological dysfunction following normal development in infancy and early childhood; the most pathognomonic of these neurological findings is vertical supranuclear gaze palsy causing restriction of voluntary vertical eye movements that can be overcome by oculocephalic maneuver (Table 22.7) [127,206]. The most common presentation of NPC is the juvenile onset phenotype, which present between 6 and 15 years of age; onset may be preceded by splenomegaly, and learning and attentional difficulties that may be misdiagnosed as learning disabilities or attention deficit disorder [127,207]. As with infantile and late infantile NPC, patients with late stage juvenile NPC develop significant spasticity and severe dysphagia, requiring gastrostomy for nutritional support [127]. Life expectancy for this form is broad, and some patients survive past the age of 30 [127].

Adult-onset NPC is a primarily neuropsychiatric phenotype; patients typically develop symptoms after the age of 15 and may display a milder constellation of neurological symptoms similar to those observed in the juvenile form, including vertical supranuclear gaze palsy [127,192]. A total of 40% of patients with adult-onset NPC typically present with psychosis with paranoid delusions and auditory or visual hallucinations; these symptoms may predate motor symptoms by years; therefore, misdiagnosis and delayed diagnosis are common [127,191,192].

Imaging features of NPC are described in Table 22.7, along with typical laboratory abnormalities. Any suggestive results obtained by laboratory testing are confirmed with NGS [127,208]. Specific substrate reduction therapy with miglustat is recommended for treatment of patients with neuronopathic NPC [127,191,209].

Neuronal ceroid lipofuscinosis (Batten disease)

The neuronal ceroid lipofuscinoses (NCL) are a family of neurodegenerative disorders caused by intracellular accumulation of ceroid lipofuscin, with four subtype groups collectively under the designation "Batten disease" [193,196]. With the identification of 14 distinct genes responsible for the NCL phenotype, these diseases have been renamed CLN1−14 disease according to the affected genes [193,196]. With the exception of CLN4 disease, which has an autosomal dominant mode of inheritance, these disorders are inherited in an autosomal recessive pattern. This review will focus primarily on the most common childhood onset CLN diseases: CLN 1, 2, and 3.

CLN1 disease

CLN1 disease is an autosomal recessive disorder caused by deficiency in the lysosomal enzyme palmitoyl protein thioesterase (PPT1) [128,193,210]. CLN1 disease was historically designated as infantile onset Batten disease; this disease is defined by normal development in the early infant period, with onset of disease between 6 and 18 months marked by decreased social behavior and hypotonia followed by rapid neurodegeneration leading eventually to a vegetative state of consciousness [193,194]. Imaging features of NPC are described in Table 22.7. While diagnosis can be made with demonstration of deficient PPT1 enzyme activity, panel-based NGS panel provides definitive diagnosis [194,196].

CLN2 disease

CLN2 disease is an autosomal recessive disorder caused by deficiency in the soluble lysosomal enzyme tripeptidyl peptidase 1 (TPP1) [128,193,210]. CLN2 disease was historically designated as late infantile onset Batten disease; this disease is defined by normal development in the infant and early toddler period, with motor regression occurring between 2 and 3 years of age [193,194,211]. Medically refractory epilepsy is a prominent feature of CLN2 disease, with median age of onset typically at age 3 [193,194,211].

EEG is often the first neurodiagnostic study performed and typically demonstrates nonspecific abnormalities, including background slowing and posterior-predominant epileptiform discharges; more specific abnormalities include a "flash-per-flash" photic response at frequencies between 1 and 2 Hz [197]. Imaging and laboratory abnormalities are described in Table 22.7. Diagnosis is typically confirmed with identification of two pathogenic variants in the *CLN2* gene identified using NGS; for patients with phenotypic evidence of CLN disease, a sequencing panel or whole-exome/whole-genome sequencing may supersede enzymatic testing [197].

Cerliponase alfa, an intraventricular recombinant ERT therapy for the treatment of CLN2, was approved by the federal drug administration (FDA) in 2017; treatment with this disease-modifying therapy has been demonstrated to slow both motor and language regression compared to historical controls, which highlights the importance of timely diagnosis [212]. Notably, females have been described to have a later symptom onset but more severe disease course with more rapid loss of independence, which may impact their response to disease-directed therapy [213].

CLN3 disease

CLN3 disease is an autosomal recessive disorder caused by a deficiency in the CLN3 protein, the function of which is not fully understood [128,193,210]. CLN3 disease is the most common form of NCL worldwide and was historically designated as classic juvenile-onset Batten disease [210]. In this form, patients develop normally in infancy and early childhood; retinal degeneration marked by vision loss is the first symptom to present between the ages of 4 and 7 years and is followed by a neurocognitive decline, described further in Table 22.6; thus, diagnosis of CLN3 disease should be considered for any patients with precipitous vision loss in early childhood, especially if associated with other neurocognitive dysfunction. Life expectancy is typically in the third decade of life, and females will typically experience later symptom onset but more rapid progression of disease [193,194,214]. Imaging and laboratory abnormalities are described in Table 22.7. While diagnosis is made based on observation of vacuolated lymphocytes in a peripheral blood sample, demonstration of CLN3 mutation by NGS provides molecular confirmation of the diagnosis [194].

TABLE 22.7 Diagnosis and treatment of lysosomal storage disorders.

Name of disorder	MRI findings	Laboratory abnormalities	Tested on newborn screening	Treatment
Mucopolysaccharidosis type I		↓ alpha-L-iduronidase activity [185]	Some states	Hematopoietic stem cell transplant Enzyme replacement Therapy with laronidase (Aldurazyme)
Mucopolysaccharidosis type II	Enlarged perivascular and/or subarachnoid spaces White matter abnormalities Hydrocephalus [124,183,186]	↑ urinary heparan and dermatan sulfates ↓ iduronate sulfatase activity [186,187]	Some states	Enzyme replacement therapy with idursulfase (Elaprase)
Mucopolysaccharidosis type III (Sanfilippo syndrome)	General cerebral atrophy Severe atrophy of amygdala [183,188]	↓ activity of one of four enzymes associated with disease [124]	No	Supportive
Mucopolysaccharidosis type VII (Sly syndrome)		↑ urinary glycoaminoglycans [124]	No	Enzyme replacement therapy with vestronidase alfa (Mepsevii) for somatic symptoms
GM1 gangliosidosis	Mild T2 hyperintensities in the brainstem and white matter Nonspecific cerebral and/or cerebellar atrophy T2-weighted hyperintensities of the bilateral putamina (Type III) [182,189]	↓ β-galactosidase activity [189]	No	Supportive
Tay–Sachs disease	Multifocal T2-weighted hyperintensities in the cerebral white matter, deep gray structures, pyramidal tract, and cerebellum (infantile) Cerebellar atrophy Diffuse reduction in N-acetylaspartate (NAA) on MRS [126]	↓ leukocyte or skin fibroblast hexosaminidase activity [190]	No	Supportive

(Continued)

TABLE 22.7 (Continued)

Name of disorder	MRI findings	Laboratory abnormalities	Tested on newborn screening	Treatment
Niemann–Pick type C disease	Atrophy of the cerebellum, hippocampus, and deep gray structures Diffuse white matter abnormalities [191]	Abnormal filipin staining ↑ serum cholestane-3β, 5α, 6β-triol, lyso-sphingomyelin isoforms, and bile acid [127,191,192]	No	Miglustat Supportive treatment of neurological symptoms
CLN1 disease	T2 hypointensities in the thalami and basal ganglia Diffuse cerebral atrophy White matter with extreme hyperintensity on T2-weighted imaging [193–195]	↓ palmitoyl protein thioesterase activity [196]	No	Supportive
CLN2 disease	Cerebral and cerebellar atrophy with more significant volume loss in the gray matter Mild white matter hyperintensities [197]	↓ tripeptidyl peptidase 1 activity [197]	No	Intraventricular recombinant enzyme replacement therapy with cerliponase alfa (Brineura)
CLN3 disease		Presence of vacuolated lymphocytes in peripheral blood [194]	No	Supportive

Disorders of purine and pyrimidine metabolism

Disorders of purine and pyrimidine metabolism are a clinically heterogeneous family of over 30 rare diseases arising from mutations that lead to either deficient or accelerated activity of enzymes involved in purine or pyrimidine biosynthesis, with an unknown incidence in the population [215]. Due to their role in DNA and RNA biosynthesis as well as energy production, these disorders affect almost all tissues [42]. Few of these disorders have primarily neurological presentations, including Lesch–Nyhan syndrome, an X-linked disorder of the purine salvage pathway that is covered in further depth in another chapter of this text, and the molybdenum cofactor deficiencies.

Molybdenum cofactor deficiency

Molybdenum cofactor deficiency (MoCD) is a group of autosomal recessive disorders caused by biallelic mutations in one of four genes (*MOSC1*, *MOSC2*, MOSC3, and GPHN) involved in molybdenum cofactor biosynthesis, leading to a combined deficiency of molybdenum-dependent enzymes, occurring with an estimated prevalence of 1 in 100,000–200,000 live births [216–218]. MoCD is typically divided into two subtypes differentiated by causative gene: Type A, which is associated with biallelic mutations in MOSC1, and Type B, which is associated with biallelic mutations in MOSC2. subtypes differentiated by the causative gene. Mutations in the other two genes implicated in molybdenum cofactor biosynthesis, *MOSC3* and *GPHN*, have been described in 1 and 2 patients, respectively [218–221].

MoCD A and B are clinically indistinguishable: the majority of patients present in the first few days of life with intractable neonatal seizures that are often refractory to standard antiseizure therapies, exaggerated startle reactions, tone abnormalities, lethargy, feeding difficulties, autonomic dysfunction, and profound encephalopathy with rapid neurological deterioration typically leading to death in early childhood; additionally, dysmorphic facial features, including frontal bossing, widely spaced eyes, puffy cheeks, a small nose, thickened lips, and a long philtrum are variably observed [215,216,221–223].

Neurodiagnostic studies show early evidence of severe neurological dysfunction. EEG frequently shows a burst suppression pattern, indicative of profound encephalopathy [218]. MRI performed within the first few days of life shows temporal abnormalities, cerebral edema, and variable laminar necrosis at the gray–white junction progressing to cystic degeneration of subcortical structures in the subacute phase, leading to frequent misdiagnosis of hypoxic-ischemic encephalopathy (HIE); these two diseases may be differentiated by involvement of the caudate, which is frequently affected in MoCD and infrequently affected in HIE and/or by demonstration of significant MRI abnormalities, including diffuse encephalomalacia and early subcortical cystic changes, in the first few days of life, which would be inconsistent with a diagnosis of HIE (Fig. 22.10) [216,224].

Laboratory evaluation includes demonstration of decrease serum and urine uric acid and presence of sulfites in the urine; molecular confirmation with NGS should be rapidly performed to confirm MoCD subtype, as this determination will guide treatment [216]. Disease-directed therapy with fosdenopterin is FDA-approved for the treatment of MoCD Type A and has been shown to normalize disease biomarkers, improve neurodevelopmental outcome, and prolong lifespan in patients with this disorder [223,225]. Treatment is otherwise supportive, with most patients requiring multiple antiseizure medications for management of intractable epilepsy [216].

Disorders of cholesterol metabolism

Disorders of cholesterol metabolism are a family of disorders arising from the defects in the cholesterol biosynthesis pathway, which leads to impaired production of cell membranes, myelin, steroid hormones, and bile acids [226]. Disorders of cholesterol metabolism can occur at any step along the cholesterol biosynthesis pathway, with both autosomal recessive and X-linked modes of inheritance described in pathogenesis of various diseases in this family [226]. Most disorders in this family present with neurodevelopmental findings, behavioral difficulties, and/or structural brain anomalies [226]. Of the disorders of cholesterol metabolism that with a primarily neurodevelopmental phenotype, Smith–Lemli–Opitz syndrome (SLOS) is the most prevalent of these disorders as well as the second most prevalent treatable IEM causing ID [226,227].

Smith–Lemli–Opitz syndrome

SLOS is an autosomal recessive disorder caused by biallelic mutations in the gene *DHCR7* leading to a deficiency in the enzyme 7-dehydrocholestrol Δ7 reductase, with an estimated prevalence of 1 in 20,000–40,000 live births [226,228–230]. SLOS was initially defined as a multiple malformation syndrome associated with a wide variety of CNS malformations, limb defects including the classic finding of cutaneous syndactyly of the second and third toes that occurs in approximately 80% of patients with SLOS, urogenital anomalies, and dysmorphic facial features, including microcephaly, ptosis, midface hypoplasia, small upturned nose, micrognathia, and midline defects [226,227,230]. Neurodevelopmental manifestations include variable ID typically in the moderate to severe range, attention deficits and hyperactivity, language impairment, sleep disturbances, self-injurious behaviors, and ASD; patients with a mild phenotype of SLOS may display neurodevelopmental symptoms in the absence of any observable congenital malformations [226,229,231].

Biochemical screening for SLOS can be performed through evaluation for elevated levels of 7 dehydrocholesterol in plasma; molecular confirmation is typically performed through demonstration of biallelic mutations in the *DHCR7* gene through NGS [226]. Treatment is supportive, as there are no disease-directed therapies for SLOS; although dietary cholesterol supplementation is typically recommended and may improve some non-CNS manifestations of this disease, it is not demonstrated to improve neurodevelopmental outcomes, likely due to poor penetrance of exogenous cholesterol across the blood–brain barrier [226,227,229,230].

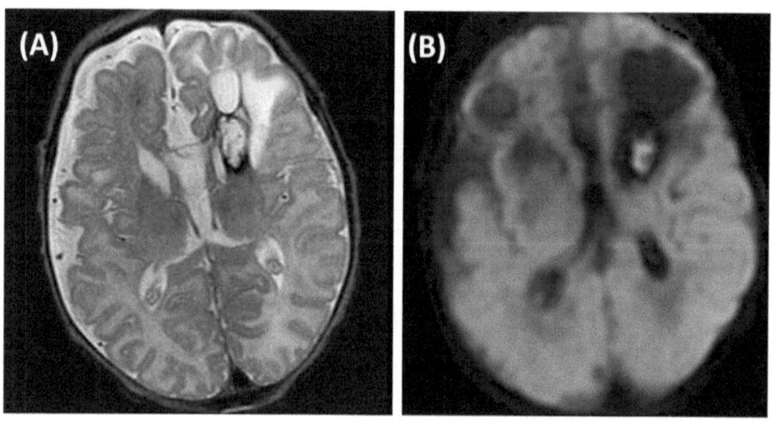

FIGURE 22.10 Neonatal female with molybdenum cofactor deficiency. Axial T2WI (A) and diffusion weighted imaging (DWI) (B) show chronic lesions in the basal ganglia and frontal lobes compatible with sequela of prior hemorrhage and metabolic injury, indicating a prenatal disease onset. Hyperintense signal on the DWI sequence represents blood products rather than active disease (B).

References

[1] Ferreira CR, Rahman S, Keller M, Zschocke J, ICIMD Advisory Group. An international classification of inherited metabolic disorders (ICIMD). J Inherit Metab Dis 2021;44(1):164–77.

[2] Scholl-Bürgi S, Haberlandt E, Gotwald T, Albrecht U, Baumgartner Sigl S, Rauchenzauner M, et al. Stroke-like episodes in propionic acidemia caused by central focal metabolic decompensation. Neuropediatrics 2009;40(2):76–81.

[3] Kölker S, Sauer SW, Surtees RAH, Leonard JV. The aetiology of neurological complications of organic acidaemias–a role for the blood-brain barrier. J Inherit Metab Dis 2006;29(6):701–4.

[4] Hoytema van Konijnenburg EMM, Wortmann SB, Koelewijn MJ, Tseng LA, Houben R, Stöckler-Ipsiroglu S, et al. Treatable inherited metabolic disorders causing intellectual disability: 2021 review and digital app. Orphanet J Rare Dis 2021;16(1):170.

[5] Levy PA. Inborn errors of metabolism: part 1: overview. Pediatr Rev 2009;30(4):131–7.

[6] Walterfang M, Bonnot O, Mocellin R, Velakoulis D. The neuropsychiatry of inborn errors of metabolism. J Inherit Metab Dis 2013; 36(4):687–702.

[7] Ibrahim M, Parmar HA, Hoefling N, Srinivasan A. Inborn errors of metabolism: combining clinical and radiologic clues to solve the mystery. Am J Roentgenol 2014;203(3):W315–27.

[8] Burton BK. Inborn errors of metabolism in infancy: a guide to diagnosis. Pediatrics 1998;102(6):E69.

[9] MacNeill EC, Walker CP. Inborn errors of metabolism in the emergency department (undiagnosed and management of the known). Emerg Med Clin North Am 2018;36(2):369–85.

[10] Saudubray J-M, Garcia-Cazorla À. Inborn errors of metabolism overview: pathophysiology, manifestations, evaluation, and management. Pediatr Clin North Am 2018;65(2):179–208.

[11] Scholl-Bürgi S, Sass JO, Zschocke J, Karall D. Amino acid metabolism in patients with propionic acidaemia. J Inherit Metab Dis 2012; 35(1):65–70.

[12] Warmerdam HAG, Termeulen-Ferreira EA, Tseng LA, Lee JY, van Eeghen AM, Ferreira CR, et al. A scoping review of inborn errors of metabolism causing progressive intellectual and neurologic deterioration (PIND). Front Neurol 2019;10:1369.

[13] Summar ML, Koelker S, Freedenberg D, Le Mons C, Haberle J, Lee H-S, et al. The incidence of urea cycle disorders. Mol Genet Metab 2013;110(1–2):179–80.

[14] Fan L, Zhao J, Jiang L, Xie L, Ma J, Li X, et al. Molecular, biochemical, and clinical analyses of five patients with carbamoyl phosphate synthetase 1 deficiency. J Clin Lab Anal 2020;34(4):e23124.

[15] Imataka G, Ishii J, Ando Y, Yoshihara S, Takagi Y, Nitta A, et al. Long-term survival of a patient with acute neonatal-onset metabolic encephalopathy with carbamoyl phosphate synthetase 1 deficiency. Eur Rev Med Pharmacol Sci 2020;24(19):10051–3.

[16] Sen K, Anderson AA, Whitehead MT, Gropman AL. Review of multi-modal imaging in urea cycle disorders: the old, the new, the borrowed, and the blue. Front Neurol 2021;12:632307.

[17] Diez-Fernandez C, Häberle J. Targeting CPS1 in the treatment of carbamoyl phosphate synthetase 1 (CPS1) deficiency, a urea cycle disorder. Expert Opin Ther Targets 2017;21(4):391–9.

[18] Nitzahn M, Lipshutz GS. CPS1: looking at an ancient enzyme in a modern light. Mol Genet Metab 2020;131(3):289–98.

[19] Choi J-H, Lee BH, Kim JH, Kim G-H, Kim Y-M, Cho J, et al. Clinical outcomes and the mutation spectrum of the OTC gene in patients with ornithine transcarbamylase deficiency. J Hum Genet 2015;60(9):501–7.

[20] Gropman A. Brain imaging in urea cycle disorders. Mol Genet Metab 2010;100(Suppl 1):S20–30.

[21] Harding BN, Leonard JV, Erdohazi M. Ornithine carbamoyl transferase deficiency: a neuropathological study. Eur J Pediatr 1984; 141(4):215–20.

[22] Sen K, Whitehead MT, Gropman AL. Multimodal imaging in urea cycle-related neurological disease - what can imaging after hyperammonemia teach us? Transl Sci Rare Dis 2020;5(1–2):87–95.

[23] Albayram S, Murphy KJ, Gailloud P, Moghekar A, Brunberg JA. CT findings in the infantile form of citrullinemia. Am J Neuroradiol 2002; 23(2):334–6.

[24] Karall D, Haberlandt E, Albrecht U, Rostasy K, Häberle J, Scholl-Bürgi S. Unrecognized citrullinemia mimicking encephalitis in a 14-year-old boy: unexpected result through the use of a standardized lumbar puncture protocol. Neuropediatrics 2012;43(2):59–63.

[25] Lin Y, Gao H, Lu B, Zhou S, Zheng T, Lin W, et al. Citrullinemia type I is associated with a novel splicing variant, c.773 + 4A > C, in ASS1: a case report and literature review. BMC Med Genet 2019;20(1):110.

[26] Saheki T, Nakano K, Kobayashi K, Imamura Y, Itakura Y, Sase M, et al. Analysis of the enzyme abnormality in eight cases of neonatal and infantile citrullinaemia in Japan. J Inherit Metab Dis 1985;8(3):155–6.

[27] Ficicioglu C, Mandell R, Shih VE. Argininosuccinate lyase deficiency: longterm outcome of 13 patients detected by newborn screening. Mol Genet Metab 2009;98(3):273–7.

[28] Sin YY, Baron G, Schulze A, Funk CD. Arginase-1 deficiency. J Mol Med Berl Ger 2015;93(12):1287–96.

[29] Prasad AN, Breen JC, Ampola MG, Rosman NP. Argininemia: a treatable genetic cause of progressive spastic diplegia simulating cerebral palsy: case reports and literature review. J Child Neurol 1997;12(5):301–9.

[30] Cederbaum SD, Shaw KN, Valente M. Hyperargininemia. J Pediatr 1977;90(4):569–73.

[31] Lehnert W. Long-term results of selective screening for inborn errors of metabolism. Eur J Pediatr 1994;153(7Suppl 1):S9–13.

[32] Coulombe JT, Shih VE, Levy HL. Massachusetts metabolic disorders screening program. II. Methylmalonic aciduria. Pediatrics 1981; 67(1):26–31.

[33] Therrell BL, Lloyd-Puryear MA, Camp KM, Mann MY. Inborn errors of metabolism identified via newborn screening: ten-year incidence data and costs of nutritional interventions for research agenda planning. Mol Genet Metab 2014;113(1–2):14–26.

[34] Kölker S, Burgard P, Okun JG, Schulze-Bergkamen A, Assmann B, Greenberg CR, et al. Looking forward–an evidence-based approach to glutaryl-CoA dehydrogenase deficiency. J Inherit Metab Dis 2004;27(6):921–6.

[35] Kölker S, Greenberg CR, Lindner M, Müller E, Naughten ER, Hoffmann GF. Emergency treatment in glutaryl-CoA dehydrogenase deficiency. J Inherit Metab Dis 2004;27(6):893–902.

[36] Chuang D, Shih V. Disorders of branched-chain amino acid and ketoacid metabolism. In: Scriver CR, Beaudet AL, Sly WS, Val D, editors. The metabolic and molecular bases of inherited disease. McGraw Hill, Inc.; 1995.

[37] Hillert A, Anikster Y, Belanger-Quintana A, Burlina A, Burton BK, Carducci C, et al. The genetic landscape and epidemiology of phenylketonuria. Am J Hum Genet 2020;107(2):234–50.

[38] Wolf B, Hsia YE, Sweetman L, Gravel R, Harris DJ, Nyhan WL. Propionic acidemia: a clinical update. J Pediatr 1981;99(6):835–46.

[39] Balakrishnan U. Inborn errors of metabolism-approach to diagnosis and management in neonates. Indian J Pediatr 2021;88(7):679–89.

[40] Ahrens-Nicklas RC, Slap G, Ficicioglu C. Adolescent presentations of inborn errors of metabolism. J Adolesc Health Off Publ Soc Adolesc Med 2015;56(5):477–82.

[41] Trakadis YJ, Fulginiti V, Walterfang M. Inborn errors of metabolism associated with psychosis: literature review and case-control study using exome data from 5090 adult individuals. J Inherit Metab Dis 2018;41(4):613–21.

[42] Sklirou E, Lichter-Konecki U. Inborn errors of metabolism with cognitive impairment: metabolism defects of phenylalanine, homocysteine and methionine, purine and pyrimidine, and creatine. Pediatr Clin North Am 2018;65(2):267–77.

[43] Cleary MA, Green A. Developmental delay: when to suspect and how to investigate for an inborn error of metabolism. Arch Dis Child 2005; 90(11):1128–32.

[44] Moeschler JB, Shevell M. Committee on genetics. Comprehensive evaluation of the child with intellectual disability or global developmental delays. Pediatrics 2014;134(3):e903–18.

[45] Brusilow SW, Horwich AL. Urea cycle enzymes. In: Valle DL, Antonarakis S, Ballabio A, Beaudet AL, Mitchell GA, editors. The online metabolic and molecular bases of inherited disease [Internet]. New York, NY: McGraw-Hill Education; 2019. Available from: ommbid.mhmedical.com/content.aspx?aid = 1181435469.

[46] Zimmermann M, Reichert AS. Rapid metabolic and bioenergetic adaptations of astrocytes under hyperammonemia - a novel perspective on hepatic encephalopathy. Biol Chem 2021;402(9):1103–13.

[47] Maestri NE, Clissold DB, Brusilow SW. Long-term survival of patients with argininosuccinate synthetase deficiency. J Pediatr 1995; 127(6):929–35.

[48] Nagamani SCS, Erez A, Lee B. Argininosuccinate lyase deficiency. In: Adam MP, Ardinger HH, Pagon RA, Wallace SE, Bean LJ, Mirzaa G, et al., editors. GeneReviews® [Internet]. Seattle (WA): University of Washington, Seattle; 1993. Available from: http://www.ncbi.nlm.nih.gov/books/NBK51784/.

[49] Matoori S, Leroux J-C. Recent advances in the treatment of hyperammonemia. Adv Drug Deliv Rev 2015;90:55–68.

[50] Tuchman M, Yudkoff M. Blood levels of ammonia and nitrogen scavenging amino acids in patients with inherited hyperammonemia. Mol Genet Metab 1999;66(1):10–15.

[51] Posset R, Gropman AL, Nagamani SCS, Burrage LC, Bedoyan JK, Wong D, et al. Impact of diagnosis and therapy on cognitive function in urea cycle disorders. Ann Neurol 2019;86(1):116–28.

[52] Metzenberg RL, Marshall M, Cohen PP. Carbamyl phosphate synthetase: studies on the mechanism of action. J Biol Chem 1958; 233(6):1560–4.

[53] Wakutani Y, Nakayasu H, Takeshima T, Mori N, Kobayashi K, Endo F, et al. A case of late-onset carbamoyl phosphate synthetase I deficiency, presenting periodic psychotic episodes coinciding with menstrual periods. Rinsho Shinkeigaku 2001;41(11):780–5.

[54] Kenneson A, Singh RH. Presentation and management of N-acetylglutamate synthase deficiency: a review of the literature. Orphanet J Rare Dis 2020;15(1):279.

[55] Hall LM, Metzenberg RL, Cohen PP. Isolation and characterization of a naturally occurring cofactor of carbamyl phosphate biosynthesis. J Biol Chem 1958;230(2):1013–21.

[56] Shigesada K, Tatibana M. Enzymatic synthesis of acetylglutamate by mammalian liver preparations and its stimulation by arginine. Biochem Biophys Res Commun 1971;44(5):1117–24.

[57] Grisolia S, Cohen PP. The catalytic rôle of carbamyl glutamate in citrulline biosynthesis. J Biol Chem 1952;198(2):561–71.

[58] Nicolaides P, Liebsch D, Dale N, Leonard J, Surtees R. Neurological outcome of patients with ornithine carbamoyltransferase deficiency. Arch Dis Child 2002;86(1):54–6.

[59] Tuchman M, Lee B, Lichter-Konecki U, Summar ML, Yudkoff M, Cederbaum SD, et al. Cross-sectional multicenter study of patients with urea cycle disorders in the United States. Mol Genet Metab 2008;94(4):397–402.

[60] Beaudet AL, O'Brien WE, Bock HG, Freytag SO, Su TS. The human argininosuccinate synthetase locus and citrullinemia. Adv Hum Genet 1986;15(161–96):291–2.

[61] Woo HI, Park H-D, Lee Y-W. Molecular genetics of citrullinemia types I and II. Clin Chim Acta 2014;431:1–8.

[62] Majoie CBLM, Mourmans JM, Akkerman EM, Duran M, Poll-The BT. Neonatal citrullinemia: comparison of conventional MR, diffusion-weighted, and diffusion tensor findings. Am J Neuroradiol 2004;25(1):32–5.

[63] Kadwa RA, Sankhyan N, Ahuja CK, Singhi P. Late-onset citrullinemia type I: a radiological mimic of herpes encephalitis. J Pediatr Neurosci 2019;14(1):36–7.

[64] Janwadkar A, Shirole N, Nagral A, Bakshi R, Vasanth S, Bagde A, et al. Citrullinemia type 1: behavioral improvement with late liver transplantation. Indian J Pediatr 2019;86(7):639–41.
[65] Ratner S. Enzymes of arginine and urea synthesis. Adv Enzymol Relat Areas Mol Biol 1973;39:1–90.
[66] Baruteau J, Diez-Fernandez C, Lerner S, Ranucci G, Gissen P, Dionisi-Vici C, et al. Argininosuccinic aciduria: recent pathophysiological insights and therapeutic prospects. J Inherit Metab Dis 2019;42(6):1147–61.
[67] Odent S, Roussey M, Journel H, Betremieux P, David V, Le Marec B. Argininosuccinic aciduria. A new case revealed by psychiatric disorders. J Genet Hum 1989;37(1):39–42.
[68] Fichtel JC, Richards JA, Davis LS. Trichorrhexis nodosa secondary to argininosuccinicaciduria. Pediatr Dermatol 2007;24(1):25–7.
[69] Ranucci G, Martinelli D, Maiorana A, Liguori A, Liccardo D, Candusso M, et al. P017 impact of liver transplant on plasma and cerebrospinal fluid amino acids in patients with argininosuccinic aciduria. Dig Liver Dis 2018;50(4):e365.
[70] Scheuerle AE, McVie R, Beaudet AL, Shapira SK. Arginase deficiency presenting as cerebral palsy. Pediatrics 1993;91(5):995–6.
[71] Tanaka K, Budd MA, Efron ML, Isselbacher KJ. Isovaleric acidemia: a new genetic defect of leucine metabolism. Proc Natl Acad Sci USA 1966;56(1):236–42.
[72] Zwickler T, Riderer A, Haege G, Hoffmann GF, Kölker S, Burgard P. Usefulness of biochemical parameters in decision-making on the start of emergency treatment in patients with propionic acidemia. J Inherit Metab Dis 2014;37(1):31–7.
[73] Scriver CR, Rosenberg LE. Amino acid metabolism and its disorders. Major Probl Clin Pediatr 1973;10:1–478.
[74] Kölker S, Garcia-Cazorla A, Cazorla AG, Valayannopoulos V, Lund AM, Burlina AB, et al. The phenotypic spectrum of organic acidurias and urea cycle disorders. Part 1: initial presentation. J Inherit Metab Dis 2015;38(6):1041–57.
[75] Sogut A, Acun C, Aydin K, Tomac N, Tomsac N, Demirel F, et al. Isovaleric acidaemia: cranial CT and MRI findings. Pediatr Radiol 2004; 34(2):160–2.
[76] Fraser JL, Venditti CP. Methylmalonic and propionic acidemias: clinical management update. Curr Opin Pediatr 2016;28(6):682–93.
[77] Nizon M, Ottolenghi C, Valayannopoulos V, Arnoux J-B, Barbier V, Habarou F, et al. Long-term neurological outcome of a cohort of 80 patients with classical organic acidurias. Orphanet J Rare Dis 2013;8:148.
[78] Sass JO, Hofmann M, Skladal D, Mayatepek E, Schwahn B, Sperl W. Propionic acidemia revisited: a workshop report. Clin Pediatr 2004; 43(9):837–43.
[79] Pena L, Franks J, Chapman KA, Gropman A, Ah Mew N, Chakrapani A, et al. Natural history of propionic acidemia. Mol Genet Metab 2012;105(1):5–9.
[80] Nguyen NHT, Morland C, Gonzalez SV, Rise F, Storm-Mathisen J, Gundersen V, et al. Propionate increases neuronal histone acetylation, but is metabolized oxidatively by glia. Relevance propionic acidemia. J Neurochem 2007;101(3):806–14.
[81] Schreiber J, Chapman KA, Summar ML, Ah Mew N, Sutton VR, MacLeod E, et al. Neurologic considerations in propionic acidemia. Mol Genet Metab 2012;105(1):10–15.
[82] Rafique M. Emerging trends in management of propionic acidemia. Arq Bras Endocrinol Metab 2014;58(3):237–42.
[83] Baumgartner MR, Hörster F, Dionisi-Vici C, Haliloglu G, Karall D, Chapman KA, et al. Proposed guidelines for the diagnosis and management of methylmalonic and propionic acidemia. Orphanet J Rare Dis 2014;9:130.
[84] Chapman KA, Summar ML, Enns GM. Propionic acidemia: to liver transplant or not to liver transplant? Pediatr Transpl 2012;16(3):209–10.
[85] Hayasaka K, Tada K. Effects of the metabolites of the branched-chain amino acids and cysteamine on the glycine cleavage system. Biochem Int 1983;6(2):225–30.
[86] Grünert SC, Müllerleile S, De Silva L, Barth M, Walter M, Walter K, et al. Propionic acidemia: clinical course and outcome in 55 pediatric and adolescent patients. Orphanet J Rare Dis 2013;8:6.
[87] Robinson WG, Coon MJ. Synthesis of malonic semialdehyde, β-hydroxypropionate, and β-hydroxyisobutyrate. Methods in enzymology [Internet]. Elsevier; 1963. p. 549–53. Available from: https://linkinghub.elsevier.com/retrieve/pii/0076687963062200.
[88] Thompson GN, Walter JH, Bresson JL, Ford GC, Lyonnet SL, Chalmers RA, et al. Sources of propionate in inborn errors of propionate metabolism. Metabolism 1990;39(11):1133–7.
[89] Romano S, Valayannopoulos V, Touati G, Jais J-P, Rabier D, de Keyzer Y, et al. Cardiomyopathies in propionic aciduria are reversible after liver transplantation. J Pediatr 2010;156(1):128–34.
[90] de Baulny HO, Benoist JF, Rigal O, Touati G, Rabier D, Saudubray JM. Methylmalonic and propionic acidaemias: management and outcome. J Inherit Metab Dis 2005;28(3):415–23.
[91] Surtees RA, Matthews EE, Leonard JV. Neurologic outcome of propionic acidemia. Pediatr Neurol 1992;8(5):333–7.
[92] Baker EH, Sloan JL, Hauser NS, Gropman AL, Adams DR, Toro C, et al. MRI characteristics of globus pallidus infarcts in isolated methylmalonic acidemia. Am J Neuroradiol 2015;36(1):194–201.
[93] Brismar J, Ozand PT. CT and MR of the brain in disorders of the propionate and methylmalonate metabolism. Am J Neuroradiol 1994; 15(8):1459–73.
[94] Harting I, Seitz A, Geb S, Zwickler T, Porto L, Lindner M, et al. Looking beyond the basal ganglia: the spectrum of MRI changes in methylmalonic acidaemia. J Inherit Metab Dis 2008;31(3):368–78.
[95] Radmanesh A, Zaman T, Ghanaati H, Molaei S, Robertson RL, Zamani AA. Methylmalonic acidemia: brain imaging findings in 52 children and a review of the literature. Pediatr Radiol 2008;38(10):1054–61.
[96] Fenton WA, Gravel RA, Rosenblatt DS. Disorders of propionate and methylmalonate metabolism. In: Valle DL, Antonarakis S, Ballabio A, Beaudet AL, Mitchell GA, editors. The online metabolic and molecular bases of inherited disease [Internet]. New York, NY: McGraw-Hill Education; 2019. Available from: ommbid.mhmedical.com/content.aspx?aid = 1181437326.

[97] Fischer AQ, Challa VR, Burton BK, McLean WT. Cerebellar hemorrhage complicating isovaleric acidemia: a case report. Neurology 1981; 31(6):746–8.

[98] Tanaka K. Isovaleric acidemia: personal history, clinical survey and study of the molecular basis. Prog Clin Biol Res 1990;321:273–90.

[99] Loots DT, Erasmus E, Mienie LJ. Identification of 19 new metabolites induced by abnormal amino acid conjugation in isovaleric acidemia. Clin Chem 2005;51(8):1510–12.

[100] Strauss KA, Puffenberger EG, Robinson DL, Morton DH. Type I glutaric aciduria, part 1: natural history of 77 patients. Am J Med Genet C Semin Med Genet 2003;121C(1):38–52.

[101] Bjugstad KB, Goodman SI, Freed CR. Age at symptom onset predicts severity of motor impairment and clinical outcome of glutaric acidemia type 1. J Pediatr 2000;137(5):681–6.

[102] Baric I, Wagner L, Feyh P, Liesert M, Buckel W, Hoffmann GF. Sensitivity and specificity of free and total glutaric acid and 3-hydroxyglutaric acid measurements by stable-isotope dilution assays for the diagnosis of glutaric aciduria type I. J Inherit Metab Dis 1999;22(8):867–81.

[103] Lindner M, Kölker S, Schulze A, Christensen E, Greenberg CR, Hoffmann GF. Neonatal screening for glutaryl-CoA dehydrogenase deficiency. J Inherit Metab Dis 2004;27(6):851–9.

[104] Blackburn PR, Gass JM, Vairo FPE, Farnham KM, Atwal HK, Macklin S, et al. Maple syrup urine disease: mechanisms and management. Appl Clin Genet 2017;10:57–66.

[105] Yudkoff M, Daikhin Y, Nissim I, Horyn O, Luhovyy B, Luhovyy B, et al. Brain amino acid requirements and toxicity: the example of leucine. J Nutr 2005;135(6 Suppl):1531S–1538SS.

[106] Strauss KA, Mazariegos GV, Sindhi R, Squires R, Finegold DN, Vockley G, et al. Elective liver transplantation for the treatment of classical maple syrup urine disease. Am J Transplant 2006;6(3):557–64.

[107] Strauss KA, Morton DH. Branched-chain ketoacyl dehydrogenase deficiency: maple syrup disease. Curr Treat Options Neurol 2003;5 (4):329–41.

[108] Clocksin HE, Hawks ZW, White DA, Christ SE. Inter- and intra-tract analysis of white matter abnormalities in individuals with early-treated phenylketonuria (PKU). Mol Genet Metab 2021;132(1):11–18.

[109] Palaiodimou L, Zompola C, Lachanis S, Mylona V, Theodorou A, Papagiannopoulou G, et al. True restriction in diffusion-weighted imaging in a mistreated patient with phenylketonuria. Neurologist 2020;26(1):20–1.

[110] Blau N, van Spronsen FJ, Levy HL. Phenylketonuria. Lancet 2010;376(9750):1417–27.

[111] Barness LA, Young D, Mellman WJ, Kahn SB, Williams WJ. Methylmalonate excretion in a patient with pernicious anemia. N Engl J Med 1963;268:144–6.

[112] Childs B, Nyhan WL, Borden M, Bard L, Cooke RE. Idiopathic hyperglycinemia and hyperglycinuria: a new disorder of amino acid metabolism. I. Pediatrics 1961;27:522–38.

[113] Oberholzer VG, Levin B, Burgess EA, Young WF. Methylmalonic aciduria. An inborn error of metabolism leading to chronic metabolic acidosis. Arch Dis Child 1967;42(225):492–504.

[114] Manoli I, Sloan JL, Venditti CP. Isolated methylmalonic acidemia. In: Adam MP, Ardinger HH, Pagon RA, Wallace SE, Bean LJ, Mirzaa G, et al., editors. GeneReviews® [Internet]. Seattle (WA): University of Washington; 1993. Available from: http://www.ncbi.nlm.nih.gov/books/NBK1231/.

[115] MacMahon M. Requesting vitamin B12 and folate assays. Lancet 1995;346(8980):973.

[116] O'Shea CJ, Sloan JL, Wiggs EA, Pao M, Gropman A, Baker EH, et al. Neurocognitive phenotype of isolated methylmalonic acidemia. Pediatrics 2012;129(6):e1541–51.

[117] Baumgarter ER, Viardot C. Long-term follow-up of 77 patients with isolated methylmalonic acidaemia. J Inherit Metab Dis 1995; 18(2):138–42.

[118] Hörster F, Baumgartner MR, Viardot C, Suormala T, Burgard P, Fowler B, et al. Long-term outcome in methylmalonic acidurias is influenced by the underlying defect (mut0, mut-, cblA, cblB). Pediatr Res 2007;62(2):225–30.

[119] Vockley J, Ensenauer R. Isovaleric acidemia: new aspects of genetic and phenotypic heterogeneity. Am J Med Genet C Semin Med Genet 2006;142C(2):95–103.

[120] Castorina M, Rigante D, Antuzzi D, Sciascia Cannizzaro G, Ricci R. Different outcome in isovaleric acidemia might be related to unsatisfactory diet compliance. Scand J Gastroenterol 2008;43(6):767–8.

[121] Sezer T, Balci O. Infantile spasms during acute metabolic decompensation in an infant with isovaleric acidemia. J Clin Neurol Seoul Korea 2016;12(3):376–7.

[122] Gibson KM, Breuer J, Nyhan WL. 3-Hydroxy-3-methylglutaryl-coenzyme A lyase deficiency: review of 18 reported patients. Eur J Pediatr 1988;148(3):180–6.

[123] Fries MH, Rinaldo P, Schmidt-Sommerfeld E, Jurecki E, Packman S. Isovaleric acidemia: response to a leucine load after three weeks of supplementation with glycine, L-carnitine, and combined glycine-carnitine therapy. J Pediatr 1996;129(3):449–52.

[124] Stapleton M, Arunkumar N, Kubaski F, Mason RW, Tadao O, Tomatsu S. Clinical presentation and diagnosis of mucopolysaccharidoses. Mol Genet Metab 2018;125(1–2):4–17.

[125] Lang FM, Korner P, Harnett M, Karunakara A, Tifft CJ. The natural history of Type 1 infantile GM1 gangliosidosis: a literature-based meta-analysis. Mol Genet Metab 2020;129(3):228–35.

[126] Cachon-Gonzalez MB, Zaccariotto E, Cox TM. Genetics and therapies for GM2 gangliosidosis. Curr Gene Ther 2018;18(2):68–89.

[127] Vanier MT. Niemann-Pick disease type C. Orphanet J Rare Dis 2010;5:16.

[128] Sleat DE, Gedvilaite E, Zhang Y, Lobel P, Xing J. Analysis of large-scale whole exome sequencing data to determine the prevalence of genetically-distinct forms of neuronal ceroid lipofuscinosis. Gene 2016;593(2):284–91.
[129] Boy N, Mühlhausen C, Maier EM, Heringer J, Assmann B, Burgard P, et al. Proposed recommendations for diagnosing and managing individuals with glutaric aciduria type I: second revision. J Inherit Metab Dis 2017;40(1):75–101.
[130] Kyllerman M, Skjeldal O, Christensen E, Hagberg G, Holme E, Lönnquist T, et al. Long-term follow-up, neurological outcome and survival rate in 28 Nordic patients with glutaric aciduria type 1. Eur J Paediatr Neurol 2004;8(3):121–9.
[131] Mühlhausen C, Ergün S, Strauss KA, Koeller DM, Crnic L, Woontner M, et al. Vascular dysfunction as an additional pathomechanism in glutaric aciduria type I. J Inherit Metab Dis 2004;27(6):829–34.
[132] Lang CH, Lynch CJ, Vary TC. BCATm deficiency ameliorates endotoxin-induced decrease in muscle protein synthesis and improves survival in septic mice. Am J Physiol Regul Integr Comp Physiol 2010;299(3):R935–44.
[133] Manara R, Del Rizzo M, Burlina AP, Bordugo A, Citton V, Rodriguez-Pombo P, et al. Wernicke-like encephalopathy during classic maple syrup urine disease decompensation. J Inherit Metab Dis 2012;35(3):413–17.
[134] Thompson GN, Francis DE, Halliday D. Acute illness in maple syrup urine disease: dynamics of protein metabolism and implications for management. J Pediatr 1991;119(1 Pt 1):35–41.
[135] Levin ML, Scheimann A, Lewis RA, Beaudet AL. Cerebral edema in maple syrup urine disease. J Pediatr 1993;122(1):167–8.
[136] Carecchio M, Schneider SA, Chan H, Lachmann R, Lee PJ, Murphy E, et al. Movement disorders in adult surviving patients with maple syrup urine disease. Mov Disord J Mov Disord Soc 2011;26(7):1324–8.
[137] Scriver CR, Clow CL, Mackenzie S, Delvin E. Thiamine-responsive maple-syrup-urine disease. Lancet 1971;297(7694):310–12.
[138] van Spronsen FJ, van Rijn M, Dorgelo B, Hoeksma M, Bosch AM, Mulder MF, et al. Phenylalanine tolerance can already reliably be assessed at the age of 2 years in patients with PKU. J Inherit Metab Dis 2009;32(1):27–31.
[139] Maillot F, Lilburn M, Baudin J, Morley DW, Lee PJ. Factors influencing outcomes in the offspring of mothers with phenylketonuria during pregnancy: the importance of variation in maternal blood phenylalanine. Am J Clin Nutr 2008;88(3):700–5.
[140] Bauman ML, Kemper TL. Morphologic and histoanatomic observations of the brain in untreated human phenylketonuria. Acta Neuropathol 1982;58(1):55–63.
[141] Huttenlocher PR. The neuropathology of phenylketonuria: human and animal studies. Eur J Pediatr 2000;159(Suppl 2):S102–6.
[142] Leuzzi V, Rinalduzzi S, Chiarotti F, Garzia P, Trasimeni G, Accornero N. Subclinical visual impairment in phenylketonuria. A neurophysiological study (VEP-P) with clinical, biochemical, and neuroradiological (MRI) correlations. J Inherit Metab Dis 1998;21(4):351–64.
[143] Bilder DA, Noel JK, Baker ER, Irish W, Chen Y, Merilainen MJ, et al. Systematic review and meta-analysis of neuropsychiatric symptoms and executive functioning in adults with phenylketonuria. Dev Neuropsychol 2016;41(4):245–60.
[144] Arnold GL, Vladutiu CJ, Orlowski CC, Blakely EM, DeLuca J. Prevalence of stimulant use for attentional dysfunction in children with phenylketonuria. J Inherit Metab Dis 2004;27(2):137–43.
[145] van Vliet D, van Wegberg AMJ, Ahring K, Bik-Multanowski M, Blau N, Bulut FD, et al. Can untreated PKU patients escape from intellectual disability? A systematic review. Orphanet J Rare Dis 2018;13(1):149.
[146] Burton BK, Longo N, Vockley J, Grange DK, Harding CO, Decker C, et al. Pegvaliase for the treatment of phenylketonuria: results of the phase 2 dose-finding studies with long-term follow-up. Mol Genet Metab 2020;130(4):239–46.
[147] Knox WE. An evaluation of the treatment of phenylketonuria with diets low in phenylalanine. Pediatrics 1960;26:1–11.
[148] Banford S, McCorvie TJ, Pey AL, Timson DJ. Galactosemia: towards pharmacological chaperones. J Pers Med 2021;11(2):106.
[149] Kuiper A, Grünewald S, Murphy E, Coenen MA, Eggink H, Zutt R, et al. Movement disorders and nonmotor neuropsychological symptoms in children and adults with classical galactosemia. J Inherit Metab Dis 2019;42(3):451–8.
[150] Cerone J, Rios A. Galactosemia. Pediatr Rev 2019;40(Suppl 1):24–7.
[151] Demirbas D, Coelho AI, Rubio-Gozalbo ME, Berry GT. Hereditary galactosemia. Metabolism 2018;83:188–96.
[152] Podskarbi T, Kohlmetz T, Gathof BS, Kleinlein B, Bieger WP, Gresser U, et al. Molecular characterization of Duarte-1 and Duarte-2 variants of galactose-1-phosphate uridyltransferase. J Inherit Metab Dis 1996;19(5):638–44.
[153] Lai K, Elsas LJ, Wierenga KJ. Galactose toxicity in animals. IUBMB Life 2009;61(11):1063–74.
[154] Nelson MD, Wolff JA, Cross CA, Donnell GN, Kaufman FR. Galactosemia: evaluation with MR imaging. Radiology 1992;184(1):255–61.
[155] Kaufman FR, McBride-Chang C, Manis FR, Wolff JA, Nelson MD. Cognitive functioning, neurologic status and brain imaging in classical galactosemia. Eur J Pediatr 1995;154(7Suppl 2):S2–5.
[156] Welsink-Karssies MM, Schrantee A, Caan MWA, Hollak CEM, Janssen MCH, Oussoren E, et al. Gray and white matter are both affected in classical galactosemia: An explorative study on the association between neuroimaging and clinical outcome. Mol Genet Metab 2020; 131(4):370–9.
[157] Welling L, Bernstein LE, Berry GT, Burlina AB, Eyskens F, Gautschi M, et al. International clinical guideline for the management of classical galactosemia: diagnosis, treatment, and follow-up. J Inherit Metab Dis 2017;40(2):171–6.
[158] Gorman GS, Chinnery PF, DiMauro S, Hirano M, Koga Y, McFarland R, et al. Mitochondrial diseases. Nat Rev Dis Prim 2016;2:16080.
[159] Patel KP, O'Brien TW, Subramony SH, Shuster J, Stacpoole PW. The spectrum of pyruvate dehydrogenase complex deficiency: clinical, biochemical and genetic features in 371 patients. Mol Genet Metab 2012;105(1):34–43.
[160] Lake NJ, Compton AG, Rahman S, Thorburn DR. Leigh syndrome: one disorder, more than 75 monogenic causes. Ann Neurol 2016; 79(2):190–203.
[161] Saneto RP, Sedensky MM. Mitochondrial disease in childhood: mtDNA encoded. Neurother J Am Soc Exp Neurother 2013;10(2):199–211.

[162] Parikh S, Goldstein A, Karaa A, Koenig MK, Anselm I, Brunel-Guitton C, et al. Patient care standards for primary mitochondrial disease: a consensus statement from the Mitochondrial Medicine Society. Genet Med J Am Coll Med Genet 2017;19(12).

[163] Friedman SD, Shaw DWW, Ishak G, Gropman AL, Saneto RP. The use of neuroimaging in the diagnosis of mitochondrial disease. Dev Disabil Res Rev 2010;16(2):129–35.

[164] Baertling F, Klee D, Haack TB, Prokisch H, Meitinger T, Mayatepek E, et al. The many faces of paediatric mitochondrial disease on neuroimaging. Childs Nerv Syst 2016;32(11):2077–83.

[165] Parikh S, Goldstein A, Koenig MK, Scaglia F, Enns GM, Saneto R, et al. Diagnosis and management of mitochondrial disease: a consensus statement from the Mitochondrial Medicine Society. Genet Med J Am Coll Med Genet 2015;17(9):689–701.

[166] Schon KR, Ratnaike T, van den Ameele J, Horvath R, Chinnery PF. Mitochondrial diseases: a diagnostic revolution. Trends Genet 2020; 36(9):702–17.

[167] McCormick EM, Zolkipli-Cunningham Z, Falk MJ. Mitochondrial disease genetics update: recent insights into the molecular diagnosis and expanding phenotype of primary mitochondrial disease. Curr Opin Pediatr 2018;30(6):714–24.

[168] Gropman AL. Diagnosis and treatment of childhood mitochondrial diseases. Curr Neurol Neurosci Rep 2001;1(2):185–94.

[169] Haas RH, Parikh S, Falk MJ, Saneto RP, Wolf NI, Darin N, et al. Mitochondrial disease: a practical approach for primary care physicians. Pediatrics 2007;120(6):1326–33.

[170] Baertling F, Rodenburg RJ, Schaper J, Smeitink JA, Koopman WJH, Mayatepek E, et al. A guide to diagnosis and treatment of Leigh syndrome. J Neurol Neurosurg Psychiatry 2014;85(3):257–65.

[171] Goldstein AC, Bhatia P, Vento JM. Mitochondrial disease in childhood: nuclear encoded. Neurotherapeutics 2013;10(2):212–26.

[172] Dimmock DP, Lawlor MW. Presentation and diagnostic evaluation of mitochondrial disease. Pediatr Clin North Am 2017;64(1):161–71.

[173] Saneto RP, Friedman SD, Shaw DWW. Neuroimaging of mitochondrial disease. Mitochondrion 2008;8(5–6):396–413.

[174] Gropman AL. Neuroimaging in mitochondrial disorders. Neurotherapeutics 2013;10(2):273–85.

[175] Gerards M, Sallevelt SCEH, Smeets HJM. Leigh syndrome: resolving the clinical and genetic heterogeneity paves the way for treatment options. Mol Genet Metab 2016;117(3):300–12.

[176] Vidali S, Aminzadeh S, Lambert B, Rutherford T, Sperl W, Kofler B, et al. Mitochondria: the ketogenic diet–A metabolism-based therapy. Int J Biochem Cell Biol 2015;63:55–9.

[177] Wexler ID, Hemalatha SG, McConnell J, Buist NR, Dahl HH, Berry SA, et al. Outcome of pyruvate dehydrogenase deficiency treated with ketogenic diets. Studies in patients with identical mutations. Neurology 1997;49(6):1655–61.

[178] Sproule DM, Kaufmann P. Mitochondrial encephalopathy, lactic acidosis, and strokelike episodes: basic concepts, clinical phenotype, and therapeutic management of MELAS syndrome. Ann NY Acad Sci 2008;1142:133–58.

[179] El-Hattab AW, Adesina AM, Jones J, Scaglia F. MELAS syndrome: clinical manifestations, pathogenesis, and treatment options. Mol Genet Metab 2015;116(1–2):4–12.

[180] Hirano M, Emmanuele V, Quinzii CM. Emerging therapies for mitochondrial diseases. Essays Biochem 2018;62(3):467–81.

[181] Waterham HR, Ferdinandusse S, Wanders RJA. Human disorders of peroxisome metabolism and biogenesis. Biochim Biophys Acta 2016;1863(5):922–33.

[182] Ferreira CR, Gahl WA. Lysosomal storage diseases. Transl Sci Rare Dis 2017;2(1–2):1–71.

[183] Barone R, Pellico A, Pittalà A, Gasperini S. Neurobehavioral phenotypes of neuronopathic mucopolysaccharidoses. Ital J Pediatr 2018;44 (Suppl 2):121.

[184] Fecarotta S, Tarallo A, Damiano C, Minopoli N, Parenti G. Pathogenesis of mucopolysaccharidoses, an update. Int J Mol Sci 2020;21(7):E2515.

[185] Kubaski F, de Oliveira Poswar F, Michelin-Tirelli K, Matte U, da S, Horovitz DD, et al. Mucopolysaccharidosis type I. Diagnostics 2020; 10(3):E161.

[186] Wraith JE, Scarpa M, Beck M, Bodamer OA, De Meirleir L, Guffon N, et al. Mucopolysaccharidosis type II (Hunter syndrome): a clinical review and recommendations for treatment in the era of enzyme replacement therapy. Eur J Pediatr 2008;167(3):267–77.

[187] Scarpa M. Mucopolysaccharidosis type II. In: Adam MP, Ardinger HH, Pagon RA, Wallace SE, Bean LJ, Mirzaa G, et al., editors. GeneReviews® [Internet]. Seattle (WA): University of Washington; 1993. Available from: http://www.ncbi.nlm.nih.gov/books/NBK1274/.

[188] Potegal M, Yund B, Rudser K, Ahmed A, Delaney K, Nestrasil I, et al. Mucopolysaccharidosis type IIIA presents as a variant of Klüver-Bucy syndrome. J Clin Exp Neuropsychol 2013;35(6):608–16.

[189] Sandhoff K, Harzer K. Gangliosides and gangliosidoses: principles of molecular and metabolic pathogenesis. J Neurosci 2013; 33(25):10195–208.

[190] Leal AF, Benincore-Flórez E, Solano-Galarza D, Garzón Jaramillo RG, Echeverri-Peña OY, Suarez DA, et al. GM2 gangliosidoses: clinical features, pathophysiological aspects, and current therapies. Int J Mol Sci 2020;21(17):E6213.

[191] Geberhiwot T, Moro A, Dardis A, Ramaswami U, Sirrs S, Marfa MP, et al. Consensus clinical management guidelines for Niemann-Pick disease type C. Orphanet J Rare Dis 2018;13(1):50.

[192] Di Lazzaro V, Marano M, Florio L, De Santis S. Niemann-Pick type C: focus on the adolescent/adult onset form. Int J Neurosci 2016; 126(11):963–71.

[193] Mink JW, Augustine EF, Adams HR, Marshall FJ, Kwon JM. Classification and natural history of the neuronal ceroid lipofuscinoses. J Child Neurol 2013;28(9):1101–5.

[194] Schulz A, Kohlschütter A, Mink J, Simonati A, Williams R. NCL diseases - clinical perspectives. Biochim Biophys Acta 2013;1832(11):1801–6.

[195] Vanhanen S-L, Puranen J, Autti T, Raininko R, Liewendahl K, Nikkinen P, et al. Neuroradiological findings (MRS, MRI, SPECT) in infantile neuronal ceroid-lipofuscinosis (Infantile CLN1) at different stages of the disease. Neuropediatrics 2004;35(01):27–35.

[196] Sun A. Lysosomal storage disease overview. Ann Transl Med 2018;6(24):476.

[197] Fietz M, AlSayed M, Burke D, Cohen-Pfeffer J, Cooper JD, Dvořáková L, et al. Diagnosis of neuronal ceroid lipofuscinosis type 2 (CLN2 disease): expert recommendations for early detection and laboratory diagnosis. Mol Genet Metab 2016;119(1–2):160–7.

[198] Parini R, Deodato F. Intravenous enzyme replacement therapy in mucopolysaccharidoses: clinical effectiveness and limitations. Int J Mol Sci 2020;21(8):E2975.

[199] Benetó N, Vilageliu L, Grinberg D, Canals I. Sanfilippo syndrome: molecular basis, disease models and therapeutic approaches. Int J Mol Sci 2020;21(21):E7819.

[200] Wagner VF, Northrup H. Mucopolysaccharidosis type III. In: Adam MP, Ardinger HH, Pagon RA, Wallace SE, Bean LJ, Mirzaa G, et al., editors. GeneReviews® [Internet]. Seattle (WA): University of Washington; 1993. Available from: http://www.ncbi.nlm.nih.gov/books/NBK546574/.

[201] Tomatsu S, Montaño AM, Dung VC, Grubb JH, Sly WS. Mutations and polymorphisms in GUSB gene in mucopolysaccharidosis VII (Sly Syndrome). Hum Mutat 2009;30(4):511–19.

[202] McCafferty EH, Scott LJ. Vestronidase alfa: a review in mucopolysaccharidosis VII. BioDrugs 2019;33(2):233–40.

[203] Regier DS, Proia RL, D'Azzo A, Tifft CJ. The GM1 and GM2 gangliosidoses: natural history and progress toward therapy. Pediatr Endocrinol Rev 2016;13(Suppl 1):663–73.

[204] Kolter T, Sandhoff K. Sphingolipid metabolism diseases. Biochim Biophys Acta 2006;1758(12):2057–79.

[205] Newton J, Milstien S, Spiegel S. Niemann-Pick type C disease: the atypical sphingolipidosis. Adv Biol Regul 2018;70:82–8.

[206] Seker Yilmaz B, Baruteau J, Rahim AA, Gissen P. Clinical and molecular features of early infantile niemann pick type C disease. Int J Mol Sci 2020;21(14):E5059.

[207] Patterson M, Niemann-Pick Disease Type C. In: Adam MP, Ardinger HH, Pagon RA, Wallace SE, Bean LJ, Mirzaa G, et al., editors. GeneReviews® [Internet]. Seattle (WA): University of Washington; 1993. Available from: http://www.ncbi.nlm.nih.gov/books/NBK1296/.

[208] Sitarska D, Ługowska A. Laboratory diagnosis of the Niemann-Pick type C disease: an inherited neurodegenerative disorder of cholesterol metabolism. Metab Brain Dis 2019;34(5):1253–60.

[209] Pineda M, Walterfang M, Patterson MC. Miglustat in Niemann-Pick disease type C patients: a review. Orphanet J Rare Dis 2018;13(1):140.

[210] Kyttälä A, Lahtinen U, Braulke T, Hofmann SL. Functional biology of the neuronal ceroid lipofuscinoses (NCL) proteins. Biochim Biophys Acta 2006;1762(10):920–33.

[211] Nickel M, Simonati A, Jacoby D, Lezius S, Kilian D, Van de Graaf B, et al. Disease characteristics and progression in patients with late-infantile neuronal ceroid lipofuscinosis type 2 (CLN2) disease: an observational cohort study. Lancet Child Adolesc Health 2018;2(8):582–90.

[212] Lewis G, Morrill AM, Conway-Allen SL, Kim B. Review of cerliponase alfa: recombinant human enzyme replacement therapy for late-infantile neuronal ceroid lipofuscinosis type 2. J Child Neurol 2020;35(5):348–53.

[213] Cialone J, Adams H, Augustine EF, Marshall FJ, Kwon JM, Newhouse N, et al. Females experience a more severe disease course in batten disease. J Inherit Metab Dis 2012;35(3):549–55.

[214] Adams HR, Mink JW, University of Rochester Batten Center Study Group. Neurobehavioral features and natural history of juvenile neuronal ceroid lipofuscinosis (Batten disease). J Child Neurol 2013;28(9):1128–36.

[215] Nyhan WL. Disorders of purine and pyrimidine metabolism. Mol Genet Metab 2005;86(1–2):25–33.

[216] Atwal PS, Scaglia F. Molybdenum cofactor deficiency. Mol Genet Metab 2016;117(1):1–4.

[217] Arican P, Gencpinar P, Kirbiyik O, Bozkaya Yilmaz S, Ersen A, Oztekin O, et al. The clinical and molecular characteristics of molybdenum cofactor deficiency due to MOCS2 mutations. Pediatr Neurol 2019;99:55–9.

[218] Mastrangelo M. Actual insights into treatable inborn errors of metabolism causing epilepsy. J Pediatr Neurosci 2018;13(1):13–23.

[219] Mendel RR. The molybdenum cofactor. J Biol Chem 2013;288(19):13165–72.

[220] Bayram E, Topcu Y, Karakaya P, Yis U, Cakmakci H, Ichida K, et al. Molybdenum cofactor deficiency: review of 12 cases (MoCD and review). Eur J Paediatr Neurol 2013;17(1):1–6.

[221] Scelsa B, Gasperini S, Righini A, Iascone M, Brazzoduro VG, Veggiotti P. Mild phenotype in molybdenum cofactor deficiency: a new patient and review of the literature. Mol Genet Genomic Med 2019;7(6). Available from: https://onlinelibrary.wiley.com/doi/10.1002/mgg3.657.

[222] Schwarz G. Molybdenum cofactor biosynthesis and deficiency. Cell Mol Life Sci 2005;62(23):2792–810.

[223] Schwahn BC, Van Spronsen FJ, Belaidi AA, Bowhay S, Christodoulou J, Derks TG, et al. Efficacy and safety of cyclic pyranopterin monophosphate substitution in severe molybdenum cofactor deficiency type A: a prospective cohort study. Lancet 2015;386(10007):1955–63.

[224] Topcu M, Coskun T, Haliloglu G, Saatci I. Molybdenum cofactor deficiency: report of three cases presenting as hypoxic-ischemic encephalopathy. J Child Neurol 2001;16(4):264–70.

[225] Kang C. Fosdenopterin: first approval. Drugs 2021;81(8):953–6.

[226] Kanungo S, Soares N, He M, Steiner RD. Sterol metabolism disorders and neurodevelopment-an update: sterol metabolism disorders and neurodevelopment. Dev Disabil Res Rev 2013;17(3):197–210.

[227] Nowaczyk MJ, Whelan DT, Heshka TW, Hill RE. Smith-Lemli-Opitz syndrome: a treatable inherited error of metabolism causing mental retardation. Can Med Assoc J 1999;161(2):165–70.

[228] Yu H, Patel SB. Recent insights into the Smith-Lemli-Opitz syndrome. Clin Genet 2005;68(5):383–91.

[229] Porter FD. Smith-Lemli-Opitz syndrome: pathogenesis, diagnosis and management. Eur J Hum Genet 2008;16(5):535–41.

[230] Jira PE, Waterham HR, Wanders RJA, Smeitink Ja M, Sengers RCA, Wevers RA. Smith-Lemli-Opitz syndrome and the DHCR7 gene. Ann Hum Genet 2003;67(Pt 3):269–80.

[231] DeBarber AE, Eroglu Y, Merkens LS, Pappu AS, Steiner RD. Smith-Lemli-Opitz syndrome. Expert Rev Mol Med 2011;13:e24.

Chapter 23

Approach to diagnosis and management of genetic neurodevelopmental disorders

Alisa Mo[1], Michael Oanea[1], Julie S. Cohen[2,3], Lance Rodan[1,4] and Siddharth Srivastava[1,5]

[1]*Department of Neurology, Boston Children's Hospital, Harvard Medical School, Boston, MA, United States,* [2]*Department of Neurology and Developmental Medicine, Kennedy Krieger Institute, Baltimore, MD, United States,* [3]*Department of Neurology, Johns Hopkins University School of Medicine, Baltimore, MD, United States,* [4]*Division of Genetics and Genomics, Boston Children's Hospital, Boston, MA, United States,* [5]*Rosamund Stone Zander Translational Neuroscience Center, Department of Neurology, Boston Children's Hospital, Boston, MA, United States*

Introduction

Genetic abnormalities are a major contributor to global developmental delay (GDD)/intellectual disability (ID), autism spectrum disorder (ASD), cerebral palsy (CP), and other neurodevelopmental disorders (NDDs). In the past decade, advances in genomics research, in particular next-generation sequencing (NGS), have rapidly increased the number of genes associated with NDDs and transformed our approach toward the diagnosis of suspected genetic conditions. This chapter presents information on the genetic basis of NDDs, focusing on the identification and evaluation of suspected genetic etiologies and highlighting the clinical phenotype of several selected disorders.

All individuals with GDD/ID, ASD, and CP should be offered genetic evaluation. There are a multitude of potential benefits of a genetic diagnosis, including reproductive counseling, changes in medical management, initiation of systemic surveillance, facilitation of family testing, access to clinical trials and/or advocacy groups, and termination of the diagnostic odyssey [1–4].

Approach to genetic testing

A genetic diagnosis may be suspected after a detailed medical history, family history, physical examination, and review of prior investigations including neuroimaging and laboratory studies. In some cases, the clinical presentation may suggest a specific neurogenetic disorder, but many individuals will have nonspecific findings.

Several types of genetic tests used in the evaluation of NDDs are described below, including *FMR1* repeat analysis, karyotyping, fluorescence in situ hybridization (FISH), Sanger sequencing, mitochondrial DNA (mtDNA) sequencing, chromosomal microarray (CMA), gene panels, exome sequencing (ES), and genome sequencing (GS). Some of these tests are used in specific contexts. For example, *FMR1* repeat analysis is used specifically to detect Fragile X syndrome (FXS). FISH is used primarily as a follow-up after the detection of certain types of cytogenetic variants. Sanger sequencing is used for single gene sequencing. mtDNA sequencing is used for detection of mitochondrial disorders outside of those due to variants in the nuclear genome. Other tests—specifically CMA, gene panels, ES, and GS—are often utilized in the agnostic (i.e., hypothesis-free) evaluation of NDDs when a specific genetic diagnosis does not come to mind. They each have varying abilities to detect different types of genetic variants and genome characteristics, as shown in Table 23.1.

Types of testing

FMR1 repeat analysis

FXS is primarily caused by unstable *CGG* expansions in the 5′ untranslated region of *FMR1* leading to DNA methylation of the region and subsequent gene silencing [5]. Detection of *CGG* repeat number is by polymerase chain reaction-based technology and Southern blotting.

Capute and Accardo's Neurodevelopmental Disabilities in Infancy and Childhood. DOI: https://doi.org/10.1016/B978-0-12-824060-1.00002-X

TABLE 23.1 Ability of different genetic tests to detect various types of genetic variants and genome characteristics.

	Genome sequencing	Exome sequencing	Gene panels	Chromosomal microarray
Sequence variants (coding)	Yes	Yes	Yes	No
Sequence variants (noncoding)	Yes	No	No	No
Copy number variants	≥1 kb	≥ 3 exons	≥1 exon	≥25 kb deletions; ≥50 kb duplications
Regions of homozygosity	Yes	Yes	No	Yes
Uniparental disomy	Yes with parent(s)	Yes with parent(s)	No	Isodisomy only
Mosaicism	Limited	Yes	Yes	Limited
Balanced chromosomal rearrangements	Yes	No	No	No
Repeat expansion disorders	Yes	No	No	No
Mitochondrial genome (mtDNA) variants	Yes	No	No	No

Karyotyping

Karyotyping involves use of a microscope to visualize size, shape, and number of chromosomes in order to detect alterations in structure that are several millions of base pairs or larger. Such alterations can include aneuploidies, chromosomal deletions, duplications, translocations, or inversions. Karyotyping can detect balanced and unbalanced chromosomal rearrangements as well as somatic mosaicism.

Fluorescence in situ hybridization

FISH involves the use of fluorescent probes composed of single-stranded DNA to bind to certain regions of the genome. Although FISH has various applications, in current practice, FISH is primarily useful to follow-up on certain cytogenetic variants, such as when a CMA result is suggestive of an unbalanced translocation.

Sanger sequencing

Developed in the 1970s, the Sanger method continues to be a reliable technology for sequencing genomic DNA [6], but it is costly and low-throughput, as it processes small targeted regions. The largest worldwide effort using Sanger sequencing was the Human Genome Project which sequenced the complete human genome over a period of 13 years and cost almost 3 billion dollars [7]. Sanger sequencing is used for single gene sequencing, detection of specific familial variants, and confirmation of specific types of variants detected by NGS.

Mitochondrial DNA sequencing

The human mitochondrial genome is separate from the nuclear genome. The human mitochondrial genome encodes 37 genes, encoding 2 ribosomal RNAs, 22 transfer RNAs, and 13 protein components of the respiratory chain. Variants in mtDNA include point mutations, large-scale deletions/duplications, and depletions. Such variants can lead to mitochondrial diseases, although it is noteworthy that mitochondrial diseases can also result from variants in the nuclear genome. Mitochondrial diseases can present quite heterogeneously with a range of neurological and systemic phenotypes [8].

Chromosomal microarray

Chromosomal duplications and deletions lead to gains or losses of genetic material, known as copy number variations (CNVs). Healthy individuals may carry benign CNVs that contribute to phenotypic variation [9]. In contrast, pathogenic CNVs can lead to NDDs. CNVs, which can be identified with CMA, range in size from one kilobase to many megabases. A CMA that is single-nucleotide polymorphism (SNP)-based can also detect regions of homozygosity due to common ancestry, consanguinity, or uniparental disomy (UPD) [10,11].

Current high-resolution CMAs can detect CNVs up to the level of a single exon, which increases the yield of diagnosing intragenic CNVs [12]. However, CMAs are limited in their ability to detect low-level somatic mosaicism (i.e., de novo CNVs that arose after fertilization and are only present in a small percentage of cells). In addition, CMAs cannot detect balanced chromosomal rearrangements, such as translocations and inversions, or triploidy.

Gene panels, exome sequencing, genome sequencing

The introduction of NGS technology in the mid-2000s allowed genomes to be sequenced at a substantially improved speed and drop in costs, leading to a revolution in the identification of new genes that cause human disease [13]. NGS is often combined with targeted enrichment methods that allow for selective sequencing of desired regions. ES uses targeted probes to detect variants in protein-coding regions. ES can detect single nucleotide variants (SNVs) and small insertions/deletions (indels) across exons and exon-intron boundaries [14]. However, ES may miss relevant SNVs at certain locations if those locations do not have sufficient coverage, defined as the number of independent reads mapped to a given location. Other limitations of ES include challenges in detecting trinucleotide repeat expansion disorders, imprinting disorders, variants in intronic/noncoding regions, and balanced rearrangements. In some cases, depending on the platform technology used, ES may be able to detect intragenic deletions/duplications and chromosomal CNVs, but sensitivity can be limited and dependent on the quality of the specific run of the exome.

Genome sequencing (GS) [15] is seeing increasing clinical use in the diagnosis of rare diseases. Advantages of GS include improved detection of structural variants and repeat expansions, as well as the ability to probe variants across both noncoding and coding genomic regions [16]. In essence, among NGS technologies, GS can capture the broadest range of variants, including SNVs, CNVs, UPD, repeat expansion disorders, complex structural rearrangements, and mitochondrial DNA variants.

Other testing

Other testing used in the neurogenetics evaluation include screening for metabolic disorders, DNA methylation assays [17], and RNA sequencing [18]. [See Chapter: Epigenetics in neurodevelopmental disabilities and chapter: Neurometabolic disorders].

Biochemical/metabolic testing

Biochemical/metabolic testing refers to the measurement of metabolite levels in body fluids (e.g., blood, urine, and cerebrospinal fluid [CSF]), intracellular metabolite levels (e.g., white blood cells coenzyme Q10 levels), and/or enzyme activity (e.g., to help diagnose lysosomal storage disorders). Biochemical testing does not provide information about DNA sequencing/molecular testing. For example, decreased CSF glucose and normal blood glucose are biochemical tests suggestive of glucose transporter type 1 deficiency syndrome, whereas a pathogenic variant in the *SLC2A1* gene is the corresponding molecular testing finding. In some cases, an individual may have biochemical evidence of a genetic disorder while molecular testing does not show causal variants in the associated disease gene (e.g., if the variant is an intronic variant but the testing used for the patient only evaluated coding variants).

Newborn screening

Newborn screening (NBS) is an example of biochemical testing to evaluate for multiple genetic conditions (as well as some nongenetic conditions) in a newborn. In the United States, each state has implemented NBS, but the list/number of screened conditions varies from state to state. There are certain caveats about NBS. NBS does not capture *every* inborn error of metabolism, as the number of screened conditions generally ranges from 30 to 40s (with some states lower than this range), which is a small subset of the >500 known inborn errors of metabolism (IEMs). Moreover, NBS can be associated with both false positives and false negatives. For example, some individuals with glutaric acidemia type 1 are low excretors of the metabolites glutaric acid and 3-hydroxyglutaric acid, leading to a normal (and hence false negative) NBS result. In essence, a normal NBS result does not completely rule out the presence of the screened disorders.

Diagnostic yields of genetic testing

The diagnostic yields of genetic testing vary in the literature, in part because of the heterogeneity of the studied population and because of technological advances with newer methods. The yield of *FMR1* repeat analysis is approximately 1% in males and females [19]. The diagnostic yield of metabolic testing is generally low (1%–5%) though estimates vary across studies; however, neurometabolic conditions are important to identify as they are a potentially treatable cause of NDD [20]. One study found that the diagnostic yield of CMA is up to 28% for ID/GDD and up to 20.5% in ASD [19], though lower in individuals with milder symptoms or ASD without systemic, syndromic features.

In a metaanalysis of 21 studies of individuals with isolated ID/GDD or ASD, ES had a 31% diagnostic yield, with a higher yield of 53% when NDD was associated with additional neurological or systemic features [21]. The yield is typically higher when the affected individual is sequenced together with parental nonaffected samples ("trio" sequencing). Two metaanalyses have examined the diagnostic yield of ES for CP, highlighting an overall diagnostic yield of 23% [22] to 31.1% [23].

The diagnostic yield of GS is even higher than ES for various indications. In a systematic review of the literature evaluating the diagnostic yield of GS versus ES for suspected genetic disorders in the pediatric population, the pooled diagnostic yield of GS was 38.6%, compared to a diagnostic yield of 37.8% for ES [24]. Another systematic review and metaanalysis focusing on a broad range of rare diseases demonstrated a higher clinical utility of GS compared to ES [25]. With respect to NDDs, one study identified a diagnostic yield of GS of 30% (n = 45/150) for individuals with a broad range of NDDs [26]. Another study compared the diagnostic yields of three different pipelines (GS first, GS as secondary test, CMA with or without FMR1 analysis) for evaluation of patients with ID. The GS-first pipeline was associated with the highest diagnostic yield (35%) [27]. For CP, one study performed GS in an unselected cohort of 150 individuals with CP (cases were not clinically selected for sequencing and were from a convenience cohort recruited from Botox clinics in Australia). The study found an overall diagnostic yield of 24.7% [28].

Choice of genetic testing

The selection and order of desired genetic tests varies considerably and depends on clinician judgment, practice factors, availability of genetic counseling, and cost or coverage by insurance. In the past, many organizations had recommended CMA as the first-tier test for children with ID/ASD [29–32], in conjunction with *FMR1* repeat analysis.

However, there is now ample evidence to suggest the use of GS (or ES if GS is not accessible) as a first-line test for children with unexplained NDDs including ID, ASD, and CP. In fact, a recent practice guideline from the American College of Medical Genetics and Genomics (ACMG) strongly recommended ES or GS as a first- or second-tier test for individuals with congenital anomalies or ID [33]. Although these guidelines do not yet apply for individuals with isolated ASD, we anticipate that GS (or ES if GS is not accessible) will be incorporated in future clinical testing guidelines as choices for first-line tests for isolated ASD as well.

It has been appreciated that sometimes genetic disorders can initially appear to present as CP, and GS (or ES if GS is not accessible) is suggested as a first-line test for CP, based on one metaanalysis [22]. Certain clinical scenarios may enrich the possibility of a genetic diagnosis in a person with CP: (1) individuals with a label of CP who have progressive or regressive symptoms (termed "CP masqueraders," as regression and/or progressive symptoms are incompatible with the definition of CP); (2) individuals who have unexplained CP (i.e., cryptogenic CP) without a suggestive perinatal history; (3) individuals with CP who have features heralding the presence of a genetic disorder such as systemic anomalies or hypotonic-ataxia type; (4) individuals with CP in whom there is a mismatch between the perinatal history, neuroimaging findings, and/or motor examination. However, we would suggest genetic testing for all individuals with CP, based on evidence citing a relatively high diagnostic yield of ES/GS among cohorts containing a mixture of explained and unexplained CP [22,23,28].

A suggested workflow for genetic testing for individuals with ASD, ID, and/or CP is presented in Fig. 23.1. If there are clues about a specific known genetic condition, such as FXS, Rett syndrome (RTT), Angelman syndrome (AS), or Prader–Willi syndrome (PWS), then it is reasonable to first perform targeted testing.

More commonly, the clinical presentation of an individual with NDD does not provide specific clues to a particular disease. Therefore a broad survey is recommended. It is recommended to use GS (or ES if GS is not accessible) as a first-line test given recent studies shower higher diagnostic yield than ES, CMA, and *FMR1* repeat analysis. When possible, "trio" GS that includes both parental samples should be performed to help interpret rare genetic variants that may be present in the affected child. If ES is performed as a first-line test and does not yield a diagnostic result, then it is recommended to perform CMA testing plus *FMR1* repeat analysis. If these results are negative, then we would recommend referral to a neurogenetics or genetics specialist to consider additional testing or enrollment in genetic testing on a research basis. For variants of uncertain significance (VUS), the clinician should review the findings with a clinical

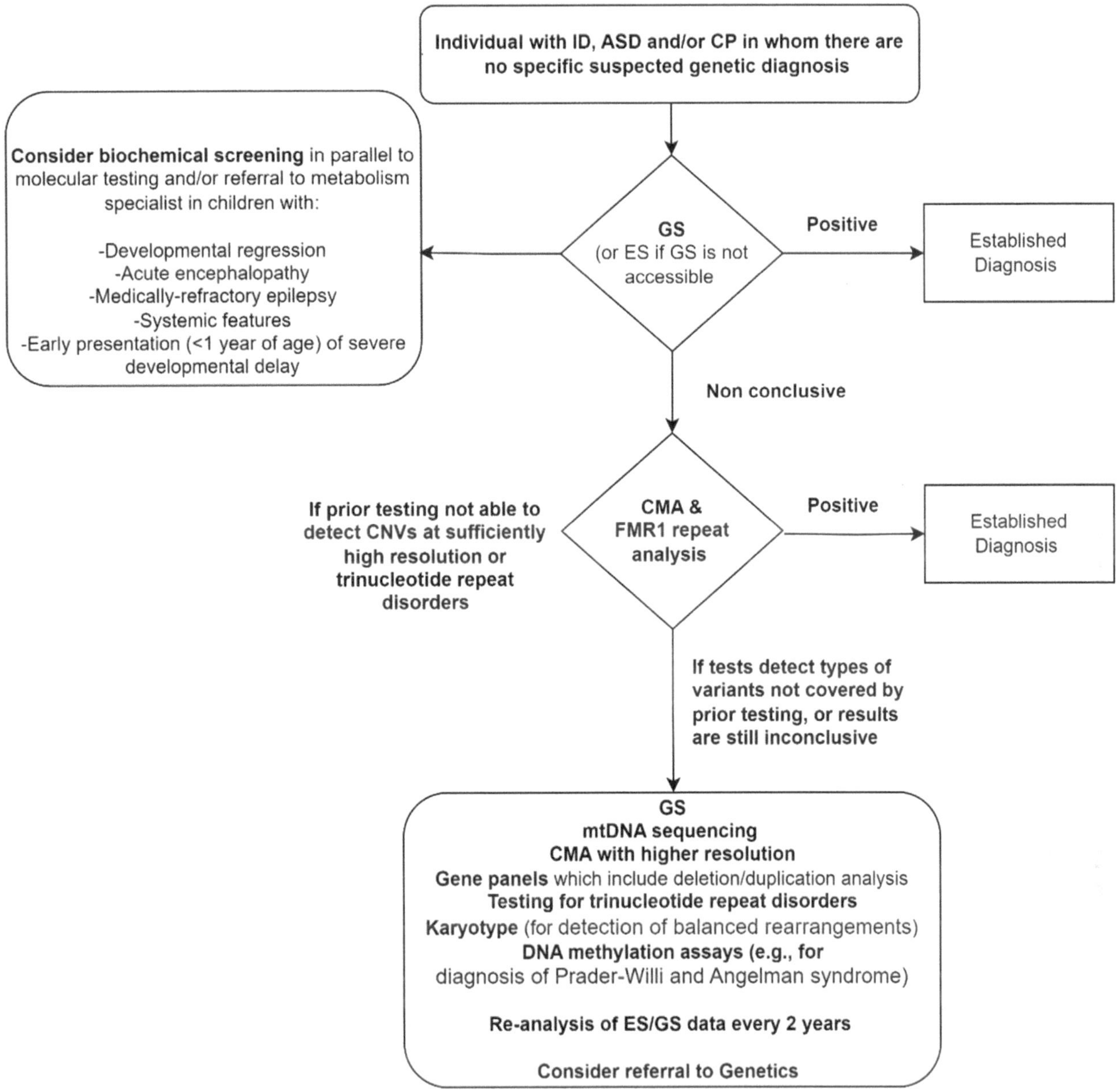

FIGURE 23.1 A proposed recommendation for genetic testing in individuals with intellectual disability, autism spectrum disorder, and/or cerebral palsy. This recommendation does not apply to individuals with other isolated neurodevelopmental disorders (NDDs) such as attention-deficit/hyperactivity disorder (ADHD) or specific learning disability. *ES*, Exome sequencing; *GS*, whole genome sequencing; *CNV*, copy number variant; *mtDNA*, mitochondrial DNA; *CMA*, chromosomal microarray.

geneticist, as some variants may have variable expressivity, incomplete penetrance, or additional support (or lack of support) for pathogenicity. These variants should then be reviewed every 2 years to reassess their classification, as some may later be determined to be pathogenic or benign.

Metabolic testing can sometimes pinpoint a disorder sooner than molecular testing. In Canada, there is an initiative termed the Treatable Intellectual Disability Endeavor (TIDE; https://www.treatable-id.org/) which protocolizes biochemical screening for a wide range of potentially treatable IEMs in individuals with global delay/ID. Therefore, although the diagnostic yield of biochemical testing in cases of ID is generally low, the rationale for the TIDE program is to identify an IEM as soon as possible to institute treatment and potentially preserve some degree of cognitive functioning (i.e., low yield but high impact of testing). A review identified 116 treatable genetic disorders (representing 139 unique genes) that could present as ID, as presented in Table 23.2 of work by van Konijnenburg et al. [34]. A similar approach has been used to perform biochemical screening to identify treatable IEMs in patients with mimics of CP [35].

TABLE 23.2 Examples of chromosomal disorders associated with neurodevelopmental disorders (NDDs).

Location	Incidence	Particular gene(s) of interest	Neurological features	Systemic features
1p36 deletion syndrome	1:5000	*DVL1, GABRD, GNB1*	Usually severe ID, speech more affected than motor skills, microcephaly, seizures, hypotonia	Dysmorphic features (straight eyebrows, wide depressed nasal bridge, pointed chin), cardiac defects, renal abnormalities, skeletal anomalies
2q37 deletion syndrome	>100 cases	*GPC1, GPR35, STK25, CENTG2, HTR2B*	Mild to severe ID, seizures, ASD, hypotonia	Dysmorphic features (frontal bossing, low frontal hairline, thin eyelids, thin lips, smooth philtrum), short stature, cardiac defects, limb defects
5q35 deletion syndrome	1:10,000	*NSD1*	ID, learning disabilities, hypotonia	Dysmorphic features (broad forehead, downslanting palpebral fissures, long narrow face with long chin), dolichocephaly, generalized overgrowth, cardiac anomalies, renal anomalies
9q34.3 deletion syndrome (Kleefstra syndrome)	>125 cases	*EHMT1*	Moderate to severe ID, ASD, sleep and behavioral disturbances, psychiatric disorders, hypotonia, seizures, dysgenesis of the corpus callosum, white matter anomalies	Dysmorphic features (brachycephaly, unusual eyebrows, widely spaced eyes, eversion of vermilion of the lower lip, prognathism), hearing loss, cardiac defects, renal and urological defects
15q24 deletion syndrome	1:42,000	*CYP11A1, SEMA7A, CPLX3, ARID3B, STRA6, SIN3A, CSK*	Mild to severe ID, ASD, behavioral dysregulation, hypotonia	Dysmorphic features (long face, high anterior hairline, downslanting palpebral fissures, broad nasal bridge, long smooth philtrum), growth restriction, genital abnormalities, digit abnormalities, skeletal defects
17q21.31 deletion syndrome (Koolen–de Vries syndrome)	1:16,000	*KANSL1*	Mild to moderate ID, friendly/cheerful behavior, hypotonia, seizures	Dysmorphic features (upslanting palpebral fissures, blepharophimosis, ptosis, tubular-shaped nose), ocular defects, congenital heart defects, genitourinary anomalies, skeletal anomalies
22q11.2 deletion syndrome (velocardiofacial syndrome/DiGeorge syndrome)	1:5000	*TBX1, DGC8, CRKL, SNAP29, PRODH*	ID, ASD, mild to moderate learning disabilities, seizures	Variable mild dysmorphic features, thymic/parathyroid hypoplasia, palate abnormality, conotruncal cardiac defects
22q13.3 deletion syndrome (Phelan–McDermid syndrome)	>600 cases	*SHANK3*	Severe to profound ID, nonverbal or severely delayed speech, ASD, hypotonia, subacute neuropsychiatric decompensation	Dysmorphic features (dolichocephaly, full brow, full cheeks, flat midface, wide nasal bridge, prominent ears, large fleshy hands, dysplastic toenails), congenital heart defects, renal defects, endocrine abnormalities
MECP2 duplication syndrome (Xq28 microduplication)	1:30,000 (males)	*MECP2*	Severe ID, nonverbal or severely affected speech, ASD, speech regression, hypotonia, progressive spasticity, ataxia, seizures, movement disorder, mild brain atrophy, dysgenesis of the corpus callosum	Mild dysmorphic features (brachycephaly, midface retrusion, depressed nasal bridge), short stature, recurrent infections, severe constipation

We suggest consideration of metabolic testing, or referral to a metabolic specialist, in children with features that include, but are not limited to, the following: developmental regression, acute encephalopathy, medically refractory epilepsy, systemic features (e.g., hepatosplenomegaly, coarse facial features, or skeletal anomalies), or early presentation (<1 year of age) of severe developmental delay (language and visual-motor developmental quotients <50%). In these scenarios, one could consider biochemical screening in parallel to molecular testing.

Interpretation of variants

According to the ACMG and Clinical Genome Resource (ClinGen) guidelines, CNVs and SNVs/indels are classified into five categories based on the level of evidence that they are disease causing: pathogenic, likely pathogenic, variant of uncertain significance, likely benign, and benign [36,37]. These classifications of pathogenicity can change over time as additional data arise about a variant. A genetic counselor or geneticist should be integrated into the NDD care team to help with selecting the appropriate genetic test, interpreting the result, addressing the potential for secondary findings (i.e., pathogenic variants in genes of medical value that are not related to the primary reason for genetic testing), and discussing the benefits, risks, limitations, and results of testing with the affected individual(s) and family.

Management of children with confirmed genetic disorders

After genetic testing results have returned, the next step is typically counseling the family about the interpretation of results. When a clinician has identified a pathogenic or likely pathogenic variant in a gene that fits the child's neurodevelopmental syndrome, it is critical to review literature reports of other individuals with the same genetic syndrome, including possibly the same genetic variant, to help counsel families about common neurodevelopmental features of the syndrome and any additional recommended evaluations. Many NDDs are associated with nonneurological abnormalities, and counseling the families about possible systemic conditions can change management decisions. For example, if the syndrome is associated with scoliosis, then a radiographic scoliosis survey may be indicated. If the condition has been linked to cardiac defects, then there can be a referral to cardiology for an echocardiogram. The family should also be counseled about recurrence risk depending on the inheritance pattern of the disorder. If the family desires, they can be connected with disease-specific family support groups. Families should also be presented information about disease-specific targeted treatments and, if interested, clinical trials. Longitudinal evaluations of these children are critical in order to periodically review the child's clinical history and management in light of new research findings. Furthermore, several NDDs can be associated with developmental regression and loss of functional skills, which in some cases may respond to therapy, and so establishing a longitudinal relationship with these patients is important.

Neurogenetic disorders

In the next section, we discuss chromosomal disorders and SNVs/indels that lead to NDDs and focus on a selection of classic NDD syndromes.

Chromosomal disorders

In 1959 Lejeune et al. showed by karyotyping that an extra copy of chromosome 21 resulted in Down syndrome, or Trisomy 21 [38], the first recognition that chromosomal abnormalities could lead to ID. Subsequently, advances in chromosome banding techniques in the 1980s led to the discovery of microdeletion syndromes including PWS and AS (15q11−q13 deletion) [39−41] and Smith−Magenis syndrome (SMS) (17p12 deletion) [42]. FISH and CMA technology allowed the discovery of even smaller CNVs. Below, we review several chromosomal disorders that lead to distinctive neurodevelopmental profiles. Further examples are listed in Table 23.2.

Trisomy 21 (Down syndrome)

Epidemiology: Trisomy 21 is the most common cause of ID caused by a chromosomal abnormality, occurring in approximately 1 in 700 live births. The incidence rises with increased maternal age, ranging from 1 in 1500 for ages 15−29 years to 1 in 50 for ages over 45 years [43].

Genetics: Approximately 95% of cases are caused by complete trisomy of chromosome 21 due to nondisjunction during gamete formation, which is generally of maternal origin [44]. It can also be associated with a Robertsonian translocation involving chromosome 21. Rarely, there can be an unbalanced translocation of chromosome 21 to another chromosome, or nondisjunction can occur after fertilization and result in somatic mosaicism. Prenatal screening allows for early detection of pregnancies affected by trisomy 21. Definitive genetic testing is by karyotyping on a blood sample.

Neurobiology: Chromosome 21 contains more than 500 genes including approximately 200 protein-coding genes [45], but the functions of the majority of genes are unknown. Furthermore, the precise mechanisms by which an extra dosage of these genes leads to the neurological consequences of trisomy 21 are poorly understood. Trisomy 21 is thought to affect multiple critical periods of brain development that control neuronal proliferation, differentiation, migration, and cortical organization [46]. The brain is smaller as a result of decreased cellular proliferation and increased apoptosis. Both excitatory and inhibitory neurons are decreased, and an imbalance of excitatory and inhibitory inputs is thought to contribute to epileptogenesis [46,47]. Synaptic density and dendritic spine numbers are reduced, and spines have abnormal morphology and function. Myelination is delayed [48]. In individuals aged 20–40 years, neuropathological hallmarks of early-onset Alzheimer disease are typical [48].

Neuroimaging: The brain may show globally decreased volume and simplified gyri [48]. There is a higher incidence of basal ganglia calcifications, particularly in the globus pallidus. A magnetic resonance imaging (MRI) or computerized tomography (CT) scan of the temporal bone can detect inner ear malformations in those with hearing loss, and spinal imaging can show evidence of craniocervical instability [49].

Clinical features: Trisomy 21 effects nearly every organ system, but clinical features exemplify variable expressivity and variation in severity across individuals [43,50]. Characteristic facial dysmorphic features include brachycephaly, upslanting almond-shaped palpebral fissures, epicanthic folds, and flat nasal bridge. Other features include Brushfield spots, single transverse palmar crease, and short fifth finger with hypoplasia of the mid-phalanx and clinodactyly. Feet may show a wide gap between the first and second toes. Congenital heart disease affects approximately 50% of individuals. Individuals should have early screens for congenital cataracts, refractive error, strabismus, amblyopia, and hearing loss. Additional complications include celiac disease, gastrointestinal reflux, duodenal atresia, and hematologic abnormalities. There is an increased risk of atlanto-axial instability, and a minority of individuals can develop signs of spinal cord compression [50].

Trisomy 21 is associated with varying degrees of ID, communication disorder, autistic symptomatology, memory difficulties, motor impairments, and behavioral and psychiatric problems [50]. Infants are hypotonic, and developmental delay commonly becomes apparent within the first year of life. The average age of milestones include head control at 6 months of age, sitting between 8.5 and 15.2 months of age, crawling between 13.1 and 23.1 months of age, and walking between 19.7 and 36.3 months of age [51]. Babbling is delayed, and the first word generally appears between 18 and 36 months of age. Adaptive skills are also delayed.

Approximately 80% of individuals have moderate ID, though IQ can also be in the average, mild ID, or severe-profound ID range [50,52]. Cognitive progress tends to slow with increased age [51]. Most individuals are verbal, and receptive language is stronger than expressive language [52]. Phonological and syntactic errors are common, and syntax is more impaired than both expressive and receptive vocabulary. Certain areas of language pragmatics such as initiating conversation or elaboration in maintaining topics are also impaired [52,53]. Social skills and daily living skills are generally strengths. However, up to 15% of individuals may meet criteria for ASD [52]. Rarely, children and young adults can show a subacute regression of language and cognitive skills and worsening ASD-like symptomatology of unclear underlying etiology, termed Down syndrome disintegrative disorder [54]. By age 40 years, neuropathological changes of Alzheimer disease are seen in nearly all individuals, and up to 75% of individuals older than 60 years of age show clinical signs of this disease [55].

In children with trisomy 21, psychiatric symptoms and disorders include aggressive behavior (7%), attention deficit hyperactivity disorder (6%), and conduct or oppositional disorder (5%) [56]. In adults, major depressive disorder (6%), aggressive behavior (6%), and stereotypic behavior (4%) are most frequent [56]. Sleep quality should be monitored, and an overnight sleep study should be considered for obstructive sleep apnea, which can occur in about one-third of individuals.

Up to 13% of individuals have epilepsy [57]. There is a trimodal age of onset, with 40% of individuals developing epilepsy before 1 year of age, 40% of individuals developing epilepsy in the third decade, and a subset of older individuals developing late-onset myoclonic epilepsy [47]. Infantile spasms are one of the most significant seizure types, affecting 2%–5% of individuals [58]. Compared to individuals with infantile spasms who do not have trisomy 21, individuals with spasms in the setting of trisomy 21 progress less often to Lennox–Gastaut syndrome and have relatively better seizure control. In the infant age group, tonic–clonic seizure with myoclonus is another significant seizure type [47].

Management: Detailed guidelines for medical monitoring and management of individuals with trisomy 21 have been developed by the Committee on Genetics of the American Academy of Pediatrics [50]. While this chapter has focused on neurodevelopmental features of trisomy 21, neurodevelopmental specialists should also be aware of general medical issues that are more prevalent in this population. When possible, individuals should be followed in multidisciplinary specialty programs focused on trisomy 21. As summarized in Ref. [50], medical issues include monitoring for growth (using trisomy 21-specific growth charts), evaluation for congenital heart disease, serial hearing and vision evaluations, screening for thyroid disease (due to increased risk of hypothyroidism), celiac disease, and obstructive sleep apnea, and serial monitoring for myeloproliferative disorders (due to increased risk of leukemia). Routine neurological evaluation of signs and symptoms of spinal cord injury is important to assess for symptomatic atlanto-axial instability.

Families should be referred to an early intervention program for infant feeding support, physical therapy, occupational therapy, and speech and language therapy. Information regarding parental support groups, community resources, and local and national advocacy groups should be provided. Later, depending on the individual's capabilities, families can be counseled on semiindependent living and employment. As Alzheimer disease is more common and occurs in younger individuals with trisomy 21, monitoring for cognitive decline starting in mid-adulthood is important.

Prader–Willi syndrome

Epidemiology: Estimates of the birth incidence of PWS range between 1 in 10,000 and 1 in 30,000 [59].

Genetics: In contrast to AS (described below), the majority of PWS cases arise from loss of paternally derived imprinted genes between chromosome 15q11 and 15q13 due to microdeletion (60%–70% of cases) or from maternal UPD (30%–40% of cases). About 3% of cases are due to defects in the PWS/AS imprinting center [59]. The first step in the genetic diagnosis is a DNA methylation analysis to detect the absence of paternally imprinted genes in the PWS/AS region. If there is abnormal DNA methylation consistent with PWS, further studies including FISH, microsatellite analysis, and imprinting center analysis can determine the molecular subtype of PWS [60].

Neurobiology: Multiple genes in the PWS locus contribute to the clinical phenotype. The *SNORD116* locus encodes multiple noncoding RNAs that are important for regulating sleep, appetite, and temperature. Loss of *Snord116* in the murine hypothalamus leads to hyperphagia [61]. Deletion of *MAGEL2* appears to lead to ASD symptomatology [62]. Other genes implicated in the PWS phenotype include genes encoding for gamma-aminobutyric acid (GABA) receptors and *NECDIN*.

Neuroimaging: Individuals with PWS can have reduced cortical gyrification compared to typical individuals [63].

Clinical features: Severe neonatal hypotonia is universal and contributes to infant feeding difficulties and failure to thrive [64]. Other clinical features include dysmorphic facies (narrow bifrontal diameter, almond-shaped palpebral fissures), strabismus, hypopigmentation, hypogonadism, small hands and feet, scoliosis, and short stature. Obesity arises from voracious food-seeking behaviors and ingestion of large quantities of food due to persistent hunger, impaired satiety, and decreased emesis. Insufficient physical activity is another contributor to obesity in this population [64,65].

Developmental milestones are delayed [65]. The average age of crawling is at 16 months, walking at 28 months, and talking with more than 10 words at 39 months. Individuals typically show mild ID with an average full-scale intellectual quotient (IQ) of 65 [65]. Arithmetic and writing are particularly affected in school-age children. Behavioral problems such as frequent temper tantrums, obsessive-compulsive behaviors, and skin-picking behavior are common and can be provoked by withholding food [65]. ASD symptomatology can be found in up to 27% of individuals and occurs more often in those with PWS caused by UPD [66]. Complications due to obstructive sleep apnea are common. Up to 26% of individuals with PWS have epilepsy [67]. Focal epilepsy is most common, but individuals can also show generalized tonic–clonic seizures, tonic seizures, atypical absence seizures, and myoclonic seizures.

Management: Individuals with PWS need strict limitation of food using behavioral modification, close supervision, and physical barriers [64]. There is some evidence for using topiramate to help with hyperphagia behaviors [68]. Polysomnography should be used to detect obstructive and central sleep apnea. Individuals with PWS also need close monitoring and management of multiple endocrine abnormalities such as short stature, hypogonadism, hypothyroidism, osteoporosis, and adrenal insufficiency [64]. Growth hormone treatment is often used to improve height, mobility, and body composition [69]. Epilepsy in individuals with PWS is generally well controlled with antiseizure medication [67].

Angelman syndrome

Epidemiology: AS occurs in approximately 1 in 15,000 births [70].

Genetics: The main molecular mechanism in the pathogenesis of AS is loss of the maternally inherited ubiquitin-protein ligase E3A (*UBE3A*) gene, which maps within the chromosome 15q11–q13 region [70]. *UBE3A* is an imprinted gene where the maternally inherited copy is functional, and the paternally inherited copy is silenced. About 70% of cases are caused by deletions of the maternally inherited copy of chromosome 15q11–q13. The rest of the cases are due to UPD where both copies of chromosome 15 are inherited from the father (2%); point mutations in *UBE3A* (10%); and defects in the AS imprinting center (3%) [70]. Similar to the diagnosis of PWS, the genetic diagnosis of AS is established by DNA methylation studies of chromosome 15. Further tests including CMA, UPD studies, imprinting center studies, and *UBE3A* sequencing may be needed to determine the subtype of AS.

Neurobiology: Mouse models of AS with a maternally inherited *Ube3a* deletion show memory deficits, abnormal motor skills, and seizures [71]. UBE3A functions by conjugating ubiquitin groups to other proteins that target them for degradation [72]. Targets of UBE3A include ECT2, a RhoA guanine nucleotide exchange factor, EPHEXIN5, a negative synaptic regulator, GAT1, a GABA transporter, and ARC, a protein that helps trafficking AMPA receptors. Furthermore, many proteins that are not known to be directly ubiquitinated by UBE3A also show altered levels of expression in AS mouse models.

Neuroimaging: Brain MRI can show subtle anatomical differences such as mild dysmyelination, gray matter abnormalities, and acquired microcephaly [73].

Clinical features: Classic facial dysmorphisms include mandibular prognathia, protruding tongue, brachycephaly, postnatal microcephaly, hypopigmentation, and deep-set eyes. Individuals may also have thoracic scoliosis and gastrointestinal problems [74].

Individuals with AS characteristically exhibit hypotonia with severe impairment in motor abilities, limited verbal abilities (nonverbal or up to a few words), and severe ID [74]. Receptive and nonverbal communication skills are stronger than expressive skills, and most affected individuals can understand simple commands. The average age of sitting is at 12 months, crawling is at 18–24 months, and walking is at 4 years [74]. Individuals with AS have a characteristic slow, stiff-legged ataxic gait with jerky arm movements. Arms are characteristically upheld with flexion at the wrists and elbows. Individuals may also have prolonged tremors of the arms and legs due to cortical myoclonus. The characteristic AS behavioral profile includes frequent smiling with an apparent happy demeanor, emotional lability with easy excitability, hand flapping behaviors, mouthing behaviors, and short attention span. AS individuals may also have an unusual fascination with water. ASD-like symptomatology is diagnosed in about half of individuals. Sleep disturbances are common.

Seizures are present in approximately 80% of individuals, generally start in infancy, and are typically medically refractory. Seizures range from febrile seizures to tonic, tonic–clonic, atypical absence, focal, and myoclonic seizures. Even in the absence of seizures, the EEG is usually abnormal and has three characteristic patterns: (1) persistent large-amplitude rhythmic 4–6 Hz activity; (2) prolonged frontal-predominant runs of triphasic 2–3 Hz activity; and (3) posterior-predominant spikes with 3–4 Hz components facilitated by eye closure [74].

Management: As with many other NDDs, early intervention services with physical therapy, occupational therapy, and speech/language therapy are essential. At this time, there are no treatments that are specific to AS. Many individuals require multiple antiseizure medications. Behavioral and sleep issues are treated with therapy and medications. Families of individuals with AS should be directed to national and international support groups.

Williams syndrome

Epidemiology: The incidence of Williams syndrome (WS) is approximately 1 in 7500 live births [75].

Genetics: A 1.5–1.8 Mb recurrent microdeletion of 7q11.23 causes WS and is usually diagnosed by CMA [75].

Neurobiology: The WS deletion region contains at least 25 genes. The gene *ELN* encodes elastin, a component of elastic fibers in connective tissues which is thought to drive the vascular, cardiac, gastrointestinal, orthopedic, and facial anomalies [75]. Haploinsufficiency of other genes in the microdeletion region have been linked to the neurocognitive phenotype. These genes include *GTF2I* and *GTF2IRD1* which contribute to visuospatial deficits in individuals with WS [76].

Neuroimaging: Brain imaging may show subtle anomalies including decreased brain size particularly in the parieto-occipital region [77]. There can be decreased posterior fossa size, which may predispose individuals to Chiari I malformation [75].

Clinical features: Individuals with WS have distinctive facies with narrow bitemporal diameter, broad forehead, periorbital fullness, epicanthal folds, flat nasal bridge, a short nose with bulbous tip, and a wide mouth with prominent lips [75]. A narrow face and long neck are typical in older children and adults. Linear height is decreased. Cardiovascular anomalies, most commonly supravalvar aortic stenosis, occur in 80% of individuals due to elastin arteriopathy and are the major cause of morbidity and mortality [75]. Hypertension occurs in half of individuals. Other systemic medical problems include strabismus, sensorineural hearing loss, dental problems, feeding difficulties, lax skin, hypercalcemia, hypothyroidism, progressive joint contractures, scoliosis, and congenital anomalies of the kidney and urinary tract (CAKUT). Joint laxity in children and contractures in older individuals contribute to gait difficulties.

Younger children are hypotonic but may subsequently develop spasticity and hyperreflexia. About 95% of individuals have developmental delay [75]. About 75% have ID, typically in the mild to moderate range, though up to 5% of individuals can have normal intelligence. Individuals with WS have relative strengths in language abilities and auditory rote memory [78]. In contrast, performance on visuospatial construction is a marked weakness. Adaptive behavior skills are also a relative weakness [75].

Individuals with WS have been described to have a "cocktail party" personality, characterized by an apparently overfriendly, loquacious demeanor. They may show lack of social inhibition and easily approach strangers. Many individuals have difficulty with emotional regulation, with approximately 50% of individuals requiring treatment for attention-deficit/hyperactivity disorder (ADHD) or anxiety disorder [75]. Many affected individuals also exhibit hypersensitivity to sound. Sleep disorders occur in approximately 50% of individuals. Epilepsy is uncommon [79].

Management: Guidelines for healthcare maintenance and supervision of individuals with WS have been developed by the American Academy of Pediatrics [75].

Smith–Magenis syndrome

Epidemiology: The incidence of SMS is approximately 1 in 15,000 to 1 in 25,000 [80].

Genetics: SMS is caused by a microdeletion of 17p11.2 (occurring in 90% of affected individuals) or an intragenic pathogenic variant affecting the retinoic acid-induced 1 (*RAI1*) gene in this region (occurring in 10% of affected individuals) [80]. Diagnosis commonly occurs with CMA. In individuals without a 17p11.2 deletion where clinical suspicion remains high for SMS, *RAI1* sequencing should be performed to detect heterozygous pathogenic SNVs/indels.

Neurobiology: A defect in RAI1 leads to many of the clinical characteristics of SMS due to its function as a transcriptional regulator that increases expression of CLOCK, a key regulator of the circadian oscillator [81]. Nearly all individuals with SMS show a shift of melatonin secretion leading to a sleep phase advance. In typically developing individuals, peak secretion of melatonin occurs in the middle of the night. In individuals with SMS, peak secretion of melatonin is around noon, which coincides with excessive sleepiness at that time [82]. Other genes in the microdeletion area contribute to the SMS phenotype. For example, individuals with large deletions that involve *PMP22*, a gene encoding a myelin protein, can lead to focal pressure neuropathies [80].

Neuroimaging: Brain anomalies can include ventriculomegaly, enlargement of the cisterna magna, partial cerebellar vermian agenesis, frontal lobe calcification, and hypoplasia of the cerebellar vermis [80,82].

Clinical features: Characteristic facial dysmorphisms include brachycephaly, frontal bossing, synophrys, midface hypoplasia with coarse facial features, hypertelorism, a tented appearance of the upper lip, teeth agenesis, and initial micrognathia progressing toward prognathism [80]. Truncal obesity and multiple skeletal anomalies are common. Systemic medical problems include feeding difficulties leading to failure to thrive in infancy, hearing loss, eye problems, hypercholesterolemia, and structural cardiac and renal problems.

Affected neonates are often hypotonic. Individuals typically have mild to moderate ID, and IQ tends to decrease with age [80]. Language delay is typical, with more prominent delays in expressive language compared to receptive language. Despite often being socially engaged and interactive, they can show deficits in social skills as well as restricted patterns of behaviors and interests, which can lead to a diagnosis of ASD. Long-term memory is typically stronger than working memory and short-term memory. There are also delays in motor skills and impaired adaptive function.

Self-injurious behaviors include head banging, biting, nose picking, onychotillomania (pulling out fingernails and toenails), and polyembolokoilamania (inserting objects in body orifices) and may in part reflect a decreased sensitivity to pain. Other behavioral problems include temper tantrums, aggression, and attention-seeking behaviors. When excited, individuals may show stereotypies such as self-hugging and page-flipping [80]. A hallmark feature of SMS is severe difficulty with sleep due to an inverted circadian secretion of melatonin. Infants are hypersomnolent, whereas older children have difficulty falling asleep, decreased rapid eye movement (REM) and total sleep, fragmented sleep cycles with frequent nocturnal awakenings, and excessive daytime sleepiness [80].

EEGs are often abnormal, and approximately 10%–30% of individuals have clinical seizures. Seizure types are variable, and catamenial seizures are a feature in female individuals [80]. A subset of individuals where the deletion affects *PMP22* show signs of peripheral neuropathy, including hyporeflexia, decreased lower extremity muscle mass, decreased sensitivity to pain, and pes planus or pes cavus. Other neurological problems include a festinating gait and balance problems [80].

Management: In addition to therapies targeting speech and language, feeding, and hypotonia, behavioral therapies should be implemented in early childhood. Psychotropic medications that target self-injurious and aggressive behaviors can be helpful. In small studies, beta-blockers may suppress daytime melatonin secretion and improve sleep [83]. Beta-blockers may also be combined with an evening dose of melatonin.

45X syndrome (Turner syndrome)

Epidemiology: Turner syndrome (TS) occurs in approximately 1 in 2500 live female births, making it the most common abnormality of the sex chromosome in females [84,85].

Genetics: Approximately 45% of cases of TS are caused by monosomy X (45,X karyotype). Most of the rest of the cases are due to mosaicism for various X-chromosome anomalies (such as 45,X/46,XX or 45,X/47,XXX) [84]. Additional implicated anomalies include a structurally abnormal X chromosome consisting of two copies of Xq (the X chromosome long arm), ring chromosome X, or partial deletions of the short or long arm. Definitive genetic diagnosis is made through karyotyping [84].

Neurobiology: Reduced dosage of X-linked genes and sex steroid defects from ovarian dysgenesis lead to the TS phenotype, but the precise contribution of each gene and interactions across genes are poorly understood. Dosage-sensitive X-linked genes include *SHOX*, *KDM6A*, *NLGN4X*, and *USP9X* [86].

Neuroimaging: Brain MRI may show decreased gray matter volume in the superior parietal and postcentral gyrus.

Clinical features: At birth, TS may be suspected due to congenital lymphedema of the hands and feet and webbed neck. Common systemic features include short stature, primary ovarian failure, structural heart defects, "shield chest" with widely spaced nipples, renal anomalies, hearing and ocular anomalies, and extremity deformities [85].

Early gross motor skills can be delayed, and individuals typically have below-average motor coordination. Fine motor delays can present with difficulty writing in school-age children [85]. The majority of individuals have a low-average full-scale IQ, with a mean of 94.6. Affected individuals are typically stronger in verbal skills and more severely impaired in nonverbal skills, such as mathematics, executive functioning, processing speed, and visuospatial reasoning, correlating with neuroimaging showing decreased brain volume in the parietal lobe [85]. Behavioral issues include difficulty in peer relationships, difficulty recognizing social cues, and ADHD. About half of affected individuals are diagnosed with anxiety or depression [85]. Individuals with TS associated with structural abnormalities of the X chromosome, such as ring chromosome X, may have more significant neurodevelopmental abnormalities.

Management: Guidelines for the management of TS have been published by the American Academy of Pediatrics [87]. Learning disabilities should be identified through regular developmental screenings, with neuropsychological assessments as needed. Stimulant medications for ADHD can be helpful, but physicians should consult with cardiology before their initiation.

Single gene disorders

The list of single gene disorders associated with NDDs is continuously expanding. Single gene variants can lead to ID via autosomal recessive, autosomal dominant, X-linked, or mitochondrial forms of inheritance. Table 23.3 lists examples of single gene disorders linked to NDDs.

Autosomal-dominant intellectual disability

Autosomal-dominant intellectual disability (ADID) is the predominant genetic cause of ID in nonconsanguineous populations, where de novo variants are observed about five times as frequently as autosomal recessive mutations [88]. As of July 2021, SysID, a database of published NDD-associated genes, includes 548 genes strongly linked to ADID and 493 candidate ADID genes [89]. In large population studies, several genes are overrepresented in ADID (e.g., *SYNGAP1*, *ARID1B*, *SCN1A*, *SATB2*, *DYRK1A*, *MED13L*, *ANKRD11*, *STXBP1*) [88]. In a recent study of individuals with severe ID, 42% of the cohort had pathogenic de novo variants; approximately half of these variants led to loss-of-function and half led to altered function [90].

Autosomal recessive intellectual disability

Autosomal recessive intellectual disability (ARID) is rare in nonconsanguineous individuals with a genetic diagnosis of ID, but the risk is several folds higher in consanguineous populations [88,91,92]. It is estimated that the number of

ARID genes is in the thousands, and the majority of genes are still undiscovered [92,93]. As of July 2021, SysID lists 946 genes strongly linked to ARID and 770 candidate genes [89]. Genes involved in ARID are highly heterogeneous and are involved in many different molecular pathways. With the exception of well-known syndromic disorders that include ARID as a feature occurring alongside epilepsy, structural brain defects, and other congenital abnormalities (e.g., metabolic disorders, ciliopathies, leukodystrophies), large family studies of ARID have not shown any particularly overrepresented genes. Many ARID genes are also pleiotropic with a broad spectrum of phenotypes [91].

X-linked intellectual disability

In the 1960s, it was suggested that pathogenic variants in genes located on the X-chromosome were overrepresented in NDDs, leading to the higher proportion of affected males compared to females [5]. In 1969 H. A. Lubs identified a family of four males with a cytogenetic abnormality of the X-chromosome leading to X-linked intellectual disability (XLID), later recognized to be FXS [94]. Currently, approximately 5%–10% of male individuals with ID are thought to have pathogenic variants in X-chromosome genes [95]. Based on SysID, there are 127 human genes strongly linked to XLID and 59 candidate genes [89].

Below, we highlight two NDDs linked to the X-chromosome.

Fragile X syndrome

Epidemiology: FXS is the most common form of inherited ID. The incidence varies but is approximately 1 in 2500 individuals. The premutation carrier state is substantially higher and occurs in about 1 in 251 males and 1 in 100 females [5].

Genetics: FXS is caused by reduced levels of fragile X mental retardation protein (FMRP) due to an unstable expansion of the *CGG* trinucleotide repeat at the 5′ promoter of the gene [5]. Expansion to >200 repeats in the full mutation leads to complete DNA methylation-dependent silencing of *FMR1* gene expression. In the premutation state (between 55 and 200 repeats), the gene is unstable and can be transmitted as a full mutation from a carrier mother to future generations.

Neurobiology: *FMR1* encodes FMRP, an RNA-binding protein that is abundantly expressed in neurons. FMRP binds other proteins and mRNAs to form messenger ribonucleoprotein complexes that are involved in the transport of RNAs from the nucleus to the cytoplasm, including into the dendrites. Loss of FMRP leads to abnormal translation of mRNAs involved in synaptic transmission and synaptic plasticity [96]. FMRP has been shown to inhibit metabotropic glutamate receptor (mGluR)-mediated local mRNA translation in dendrites [96,97]. Thus absence of FMRP induces excessive responses to mGluR signaling which then leads to altered protein expression, abnormally long and thin dendritic spines, abnormal synaptic transmission, and behavioral deficits.

Neuroimaging: FXS individuals can show increased volume of the caudate nucleus and decreased volume of the cerebellum [5].

Clinical features: In the full mutation, facial dysmorphisms include a long, narrow face, large ears, arched palate, prominent forehead, and prognathism [5]. Macroorchidism is found in 90% of males by adolescence. Other systemic features include mitral valve prolapse, joint hyperextension, recurrent otitis media, and strabismus.

Male infants with the full mutation are hypotonic, and GDD/ID is nearly universal. Sitting is generally achieved at 10 months, and walking and first word are generally achieved around 20 months [98]. In most affected individuals, ID is in the moderate to severe range, though 11% have IQs in the mild ID range [5]. Verbal skills are stronger than mathematical, visuospatial, and executive function. Expressive language is more impaired compared to receptive language, and affected individuals may show verbal pragmatic errors such as perseveration, repetitive speech, and echolalia [99]. Individuals also show deficits in verbal short-term memory, executive function, abstract reasoning, attention, and visuospatial processing [5]. Individuals can learn simple verbal and nonverbal tasks and information, but have difficulty with abstract, nonsequential information. Acquisition of verbal, mathematical, visual, and memory skills typically slows after early childhood, leading to increased discrepancy between age-equivalent scores and chronological age.

Behavioral problems include social anxiety, avoidance behavior, gaze aversion, mood instability, and mannerisms such as hand flapping [5,99]. A total of 86% of individuals meet diagnostic criteria for an anxiety disorder. Between 25% and 47% of males with FXS can have comorbid ASD [5]. Seizures develop in 10%–20% of individuals [100], with the risk highest in childhood, and are typically easily controlled with medications. The most common epilepsy type resembles benign childhood epilepsy with centrotemporal spikes.

Due to the presence of a normal allele and some degree of X-inactivation of the chromosome that contains the trinucleotide expansion, females with the full mutation have reduced expression of FMRP, leading to a less severe cognitive phenotype [101]. Most affected females have borderline to low-average IQ. Neuropsychiatric issues such as anxiety, social avoidance, and depression are seen at significantly higher rates than in the general population. 5% of females have seizures.

Individuals with a *CGG* premutation can also exhibit similar but less severe neurocognitive and behavioral features as those with the full mutation [5,101]. About 40%–45% of males and 8%–16% of females over 50 years of age who carry a premutation can develop fragile X-associated tremor/ataxia syndrome (FXTAS), a neurodegenerative disorder characterized by ataxia, progressive tremor, parkinsonism, cognitive decline, peripheral neuropathy, autonomic dysfunction, and behavioral problems [101]. About 20% of females with a premutation can develop primary ovarian insufficiency [101].

Management: Guidelines are provided by the American Academy of Pediatrics [101]. Management of FXS involves addressing systemic medical problems, treating seizures, and referring to early developmental programs, neuropsychological testing, and behavioral interventions. Pharmacological interventions for behavioral difficulties and seizures may be necessary. Trials of FXS-specific therapies such as mGluR antagonists have not yet shown success in improving behavior or cognition [102].

Rett syndrome

Epidemiology: RTT is estimated to have a prevalence of 1 in 10,000 to 1 in 15,000 and occurs almost exclusively in females [103].

Genetics: De novo dominant missense, nonsense, or frameshift pathogenic variants in the X-linked methyl CpG binding protein 2 (*MECP2*) gene cause the large majority of classic sporadic RTT [104]. Genotype–phenotype correlations show that R133C, R294X, and C-terminal nonstop variants are correlated with milder disease [105]. Severe disease phenotypes can be associated with certain specific variants such as R168X.

Neurobiology: MECP2 binds to methylated cytosines in DNA and interacts with other proteins to act as a transcriptional regulator [106]. *MECP2* is highly expressed in neurons in the brain starting from mid-gestation and remains high in adulthood [107]. Defects in *MECP2* disturb patterns of brain gene expression [106] and lead to abnormalities in synaptic development and maturation in the cerebral cortex [108].

Neuroimaging: MRI shows decreased gray and white matter of the brain, particularly in the frontal lobes [109].

Clinical features: The typical profile of RTT syndrome is a girl with typical development for the first 6–18 months of life who subsequently regresses in her motor skills, language, and social interaction, and develops stereotypic hand movements [110]. Classic RTT can be divided into four clinical stages [111]. The first stage ("early onset stagnation stage") occurs between 6 and 18 months, where there is developmental stagnation of verbal, cognitive, social, and fine and gross motor abilities. Decelerating head growth, which may begin a few months after birth, continues and leads to microcephaly. The second stage ("rapid destructive stage") is a period of rapid developmental regression, emergence of intense stereotypies and breathing irregularities, and loss of purposeful hand use in individuals between 1 and 4 years of age. Expressive language is generally completely lost, though some individuals may have preserved single words [112]. A degree of nonverbal communication with gestures and eye gaze is typically preserved. Hand stereotypies include flapping, wringing, finger kneading, and rubbing. Common motor disturbances include bruxism, oculogyric crises, dystonia, and other extrapyramidal dysfunctions. Gait is ataxic. The third stage ("pseudostationary stage") is a period of stabilization during preschool and early school years. The fourth stage ("late motor deterioration stage") is a period of further motor deterioration. Nonneurological clinical features of RTT include failure to thrive, poor growth, scoliosis, disorganized breathing (episodes of hyperventilation alternating with apnea during wakefulness), and autonomic dysfunction.

Seizures can be either focal or generalized, typically arise in late childhood, and eventually affect the majority of RTT individuals [113]. Seizures often follow a pattern of relapse and remission. EEG is often abnormal after age 3 years even in individuals without seizures. Nonepileptic spells such as breath-holding spells, movement disorders, and behavioral arrest are also common and may need to be characterized by EEG to rule out seizure.

Atypical RTT can either be more mild or more severe than typical RTT, and not all of the characteristic features are present. Atypical RTT syndromes are classified into the milder variant with preserved speech (Zappella variant), early-onset seizure variant (Hanefeld variant) due to pathogenic variants in *CDKL5*, and the congenital variant (Rolando variant) due to pathogenic variants in *FOXG1* with regression shortly after birth [114]. Given substantial clinical differences and molecular mechanisms, the latter two variants may be viewed as distinct syndromes from RTT. There are also several other genetic disorders that may mimic RTT [115].

TABLE 23.3 Examples of single gene disorders leading to neurodevelopmental disorder (NDD).

Inheritance pattern	Incidence	Implicated genes	Neurological features	Systemic features
Autosomal recessive				
Smith–Lemli–Opitz	1:20,000	*DHCR7*	Moderate to severe ID, microcephaly, neonatal hypotonia followed by hypertonia	Increased 7-dehydrocholesterol level ± decreased total cholesterol, dysmorphic features (narrow forehead, epicanthal folds, ptosis, short mandible), two-thirds toe syndactyly, single transverse palmar crease, genital abnormalities, short stature, cardiac defects
Meckel–Gruber syndrome	1:100,000	Ciliopathy genes (*MKS1*, *TMEM67*, *TMEM216*, *CEP290*, *CC2D2A*, *RRPGIP1L*, and *B9D1*)	Microcephaly, occipital encephalocele, cerebral and cerebellar hypoplasia, absence of corpus callosum, olfactory structures, and septum pellucidum	Orofacial clefts, often do not survive past a few weeks, polydactyly, kidney dysplasia, liver fibrosis, growth deficiency, cryptorchidism
Zellweger syndrome	1:50,000	Peroxisomal genes	Profound ID, hypotonia, vision/hearing impairment, seizures	Dysmorphic features (high forehead, large anterior fontanel, widely split sutures, flat face, broad nasal bridge), hepatomegaly, kidney disease, chondrodysplasia punctata at the patella
Autosomal dominant				
Cornelia de Lange syndrome	1:20,000	Cohesin-associated genes including pathogenic variants in *NIPBL* (50%), *SMC1L1* (5%; X-linked, milder phenotype), and rarely *SMC3*	Variable ID, initial hypertonicity, broad-based gait, behavioral disturbances, microbrachycephaly	Dysmorphic features (synophrys, thick eyebrows, short nasal bridge with anteverted nares), failure to thrive, hirsutism, micromelia, and other limb anomalies, genitourinary deficits
Rubinstein–Taybi syndrome	1:125,000	*CBP*, *P300*	Moderate to severe ID, language difficulties, seizures, unsteady gait, hypotonia, hyperreflexia, behavioral problems, microcephaly	Dysmorphic features (broad nasal bridge with overhanging columella, downslanting palpebral fissures, facial grimace), scoliosis, cryptorchidism, hirsutism, cardiac defects
Kabuki syndrome	1:50,000	*KMT2D* (80%) or *KDM6A*	Microcephaly, mild to moderate ID, hypotonia, seizures	Dysmorphic features (long palpebral fissures with eversion of lateral lower eyelid, highly arched eyebrows, broad nose with short columella, prominent ears), skeletal abnormalities, short stature
Noonan syndrome	1:1000	*PTPN1* (40%–50%), *SOS1* (10%–20%), *RAF1* (3%–17%), and other pathogenic variants in Ras-MAPK signaling pathway genes	Mild ID, learning disabilities, ASD, ADHD	Dysmorphic features (widely spaced, downslanting palpebral fissures, ptosis, low set ears), short stature, congenital heart disease

(Continued)

TABLE 23.3 (Continued)

Inheritance pattern	Incidence	Implicated genes	Neurological features	Systemic features
Tuberous sclerosis complex	1:6000	*TSC1, TSC2*	Variable ID, ASD, epilepsy, ADHD, mood disorders, subependymal nodules, cortical tubers, subependymal giant cell astrocytomas	Hypomelanotic macules, confetti skin lesions, facial angiofibromas, shagreen patches, fibrous cephalic plaques, ungual fibromas; angiomyolipomas, renal cysts, renal cell carcinomas; cardiac rhabdomyomas, lymphangioleiomyomatosis
Weaver syndrome	1:15,000	*EZH2*	Variable ID	Dysmorphic features (broad forehead, widely spaced eyes with almond-shaped palpebral fissures), generalized overgrowth
X-linked				
X-linked hydrocephalus spectrum	1:30,000 (males)	*L1CAM*	Hydrocephalus, ID, spastic paraparesis	Thumb abnormalities, intestinal pseudoobstruction
Menkes syndrome	1:35,000 (males)	*ATP7A*	Cerebral/cerebellar degeneration, seizures	Copper deficiency, growth retardation, hair abnormality

Management: Expert-panel consensus guidelines have recently been published for management of RTT [116]. Management in a multidisciplinary clinic focusing on RTT is ideal. Currently, there are no standard pharmacological options to treat the breathing abnormalities associated with RTT.

Primary mitochondrial disorders

Mutations in nuclear genes that code for mitochondrial proteins or in the maternally inherited mitochondrial genome can lead to a range of disorders with variable levels of neurological, developmental, and multisystemic organ involvement. Some classic examples of mitochondrial diseases due to variants in the mitochondrial genome include mitochondrial encephalopathy, lactic acidosis and stroke-like episodes (MELAS) syndrome, Leber hereditary optic neuropathy (LHON), and Kearns–Sayre syndrome (KSS). MELAS syndrome can present with encephalopathy, myopathy, and recurrent stroke-like episodes accompanied by focal neurological deficits. LHON can present with progressive vision loss. KSS can present with progressive limitation of extraocular muscle movement, myopathy, cardiac conduction block, and hearing loss [117]. See chapter: Neurometabolic disorders.

A greater number of mitochondrial diseases are due to defects in the nuclear genome [118], many of which can have neurodevelopmental manifestations [119]. Mitochondrial defects should be suspected in the individual with ID, multisystem organ involvement, certain neurological features (myoclonus, stroke in nonvascular territories), characteristic laboratory abnormalities (such as elevated lactate, anion-gap metabolic acidosis), or characteristic neuroimaging abnormalities (such as symmetric basal ganglia hyperintensities, cerebral or cerebellar atrophy) [119]. It is important to note that some genetic disorders (such as RTT and FXS) can have features of secondary mitochondrial dysfunction, although these disorders are not considered primary mitochondrial diseases themselves [120].

Conclusion

Individuals with genetic causes of NDD can be challenging for the clinician and need a multidisciplinary team for their diagnosis, evaluation, and management. All individuals with unexplained NDDs should be offered genetic evaluation. The establishment of a specific genetic etiology offers prognostic information, anticipatory guidance, community support, information of recurrence risk, and potential access to etiology-specific treatments. A genetic diagnosis can also offer closure to what is often a long diagnostic odyssey for a family. Although this chapter has primarily focused on diseases caused by a variant in a single gene or chromosomal anomaly, it is well recognized that NDDs often emerge via polygenic inheritance and gene−environment interactions [121]. Ongoing advances in clinical and basic science research will further our recognition of gene−phenotype associations, facilitate the rapid diagnosis of neurogenetic disorders in the clinic, and improve the treatment of affected individuals.

References

[1] Clark MM, Zornitza S, Lauge F, Tiong YT, Susan MW, David D, et al. Meta-analysis of the diagnostic and clinical utility of genome and exome sequencing and chromosomal microarray in children with suspected genetic diseases. NPJ Genomic Med 2018;3:16. Available from: https://doi.org/10.1038/s41525-018-0053-8.

[2] Malinowski J, Miller DT, Demmer L, Gannon J, Maria Pereira E, Schroeder MC, et al. Systematic evidence-based review: outcomes from exome and genome sequencing for pediatric patients with congenital anomalies or intellectual disability. Genet Med June 2020;22 (6):986−1004. Available from: https://doi.org/10.1038/s41436-020-0771-z.

[3] Niguidula N, Alamillo C, Shahmirzadi Mowlavi L, Powis Z, Cohen JS, Farwell Hagman KD. Clinical whole-exome sequencing results impact medical management. Mol Genet Genomic Med 2018;6(6):1068−78. Available from: https://doi.org/10.1002/mgg3.484.

[4] Ontario Health (Quality). Genome-wide sequencing for unexplained developmental disabilities or multiple congenital anomalies: a health technology assessment. Ont Health Technol Assess Ser 2020;20(11):1−178.

[5] Gallagher A, Hallahan B. Fragile X-associated disorders: a clinical overview. J Neurol 2012;259(3):401−13.

[6] Sanger F, Nicklen S, Coulson AR. DNA sequencing with chain-terminating inhibitors. Proc Natl Acad Sci USA 1977;74(12):5463−7.

[7] Lander ES, Linton LM, Birren B, Nusbaum C, Zody MC, Baldwin J, et al. Initial sequencing and analysis of the human genome. Nature 2001;409(6822):860−921.

[8] Parikh S, Goldstein A, Kay Koenig M, Scaglia F, Enns GM, Saneto R, Anselm I, et al. Diagnosis and management of mitochondrial disease: a consensus statement from the Mitochondrial Medicine Society. Genet Med 2015;17(9):689−701. Available from: https://doi.org/10.1038/gim.2014.177.

[9] Redon R, Ishikawa S, Fitch KR, Feuk L, Perry GH, Andrews TD, et al. Global variation in copy number in the human genome. Nature 2006;444(7118):444−54.

[10] Wiszniewska J, Bi W, Shaw C, Stankiewicz P, Kang S-HL, Pursley AN, et al. Combined array CGH plus SNP genome analyses in a single assay for optimized clinical testing. Eur J Hum Genet 2014;22(1):79−87.

[11] Mefford HC, Batshaw ML, Hoffman EP. Genomics, intellectual disability, and autism. N Engl J Med 2012;366(8):733−43.

[12] Jezkova J, Heath J, Williams A, Barrell D, Norton J, Collinson MN, et al. Exon-focused targeted oligonucleotide microarray design increases detection of clinically relevant variants across multiple NHS genomic centres. NPJ Genom Med 2020;5(1):28.

[13] Yang Y, Muzny DM, Reid JG, Bainbridge MN, Willis A, Ward PA, et al. Clinical whole-exome sequencing for the diagnosis of mendelian disorders. N Engl J Med 2013;369(16):1502−11.

[14] Retterer K, Scuffins J, Schmidt D, Lewis R, Pineda-Alvarez D, Stafford A, et al. Assessing copy number from exome sequencing and exome array CGH based on CNV spectrum in a large clinical cohort. Genet Med 2015;17(8):623−9.

[15] Bowling KM, Thompson ML, Amaral MD, Finnila CR, Hiatt SM, Engel KL, et al. Genomic diagnosis for children with intellectual disability and/or developmental delay. Genome Med 2017;9(1).

[16] Souche E, Beltran S, Brosens E, Belmont JW, Fossum M, Riess O, et al. Recommendations for whole genome sequencing in diagnostics for rare diseases. Eur J Hum Genet 2022;30(9):1017−21.

[17] Sadikovic B, Levy MA, Kerkhof J, Aref-Eshghi E, Schenkel L, Stuart A, et al. Clinical epigenomics: genome-wide DNA methylation analysis for the diagnosis of Mendelian disorders. Genet Med 2021;23(6):1065−74.

[18] Byron SA, Van Keuren-Jensen KR, Engelthaler DM, Carpten JD, Craig DW. Translating RNA sequencing into clinical diagnostics: opportunities and challenges. Nat Rev Genet 2016;17(5):257−71.

[19] Savatt JM, Myers SM. Genetic testing in neurodevelopmental disorders. Front Pediatr 2021;9:526779.

[20] Michelson DJ, Shevell MI, Sherr EH, Moeschler JB, Gropman AL, Ashwal S. Evidence report: genetic and metabolic testing on children with global developmental delay: report of the Quality Standards Subcommittee of the American Academy of Neurology and the Practice Committee of the Child Neurology Society. Neurology 2011;77(17):1629−35.

[21] Srivastava S, Love-Nichols JA, Dies KA, Ledbetter DH, Martin CL, Chung WK, et al. Meta-analysis and multidisciplinary consensus statement: exome sequencing is a first-tier clinical diagnostic test for individuals with neurodevelopmental disorders. Genet Med 2019;21(11):2413−21.

[22] Srivastava S, Lewis SA, Cohen JS, Zhang B, Aravamuthan BR, Chopra M, et al. Molecular diagnostic yield of exome sequencing and chromosomal microarray in cerebral palsy: a systematic review and meta-analysis. JAMA Neurol 2022;79(12):1287−95. Available from: https://doi.org/10.1001/jamaneurol.2022.3549.

[23] Gonzalez-Mantilla PJ, Hu Y, Myers SM, Finucane BM, Ledbetter DH, Martin CL, et al. Diagnostic yield of exome sequencing in cerebral palsy and implications for genetic testing guidelines: a systematic review and meta-analysis. JAMA Pediatr 2023;177(5):472−8. Available from: https://doi.org/10.1001/jamapediatrics.2023.0008.

[24] Nurchis MC, Altamura G, Teresa Riccardi M, Clementina Radio F, Chillemi G, Silvio Bertini E, et al. Whole genome sequencing diagnostic yield for paediatric patients with suspected genetic disorders: systematic review, meta-analysis, and GRADE assessment. Arch Public Health 2023;81(1):93. Available from: https://doi.org/10.1186/s13690-023-01112-4.

[25] Chung CCY, Shirley PYH, Nicole YTN, Phoenix HLD, Hong Kong Genome Project, Chu ATW, et al. Meta-analysis of the diagnostic and clinical utility of exome and genome sequencing in pediatric and adult patients with rare diseases across diverse populations. Genet Med 2023;25(9):100896. Available from: https://doi.org/10.1016/j.gim.2023.100896.

[26] van der Sanden BPGH, Schobers G, Corominas Galbany J, Koolen DA, Sinnema M, van Reeuwijk J, et al. The performance of genome sequencing as a first-tier test for neurodevelopmental disorders. Eur J Hum Genet 2023;31(1):81–8. Available from: https://doi.org/10.1038/s41431-022-01185-9.

[27] Lindstrand A, Marlene Ek MK, Britt-Marie Anderlid EB, Jonas Carlsten JE, et al. Genome sequencing is a sensitive first-line test to diagnose individuals with intellectual disability. Genet Med 2022;24(11):2296–307. Available from: https://doi.org/10.1016/j.gim.2022.07.022.

[28] Eyk CLvan, Webber DL, Minoche AE, Pérez-Jurado LA, Corbett MA, Gardner AE, et al. Yield of clinically reportable genetic variants in unselected cerebral palsy by whole genome sequencing. NPJ Genomic Med 2021;6(1):74. Available from: https://doi.org/10.1038/s41525-021-00238-0.

[29] Manning M, Hudgins L. Professional practice and guidelines committee. Array-based technology and recommendations for utilization in medical genetics practice for detection of chromosomal abnormalities. Genet Med 2010;12(11):742–5.

[30] Miller DT, Adam MP, Aradhya S, Biesecker LG, Brothman AR, Carter NP, et al. Consensus statement: chromosomal microarray is a first-tier clinical diagnostic test for individuals with developmental disabilities or congenital anomalies. Am J Hum Genet 2010;86(5):749–64.

[31] Moeschler JB, Shevell M. Committee on genetics. Comprehensive evaluation of the child with intellectual disability or global developmental delays. Pediatrics 2014;134(3):e903–18.

[32] Carter MT, Srour M, Au PYB, Buhas D, Dyack S, Eaton A, et al. Genetic and metabolic investigations for neurodevelopmental disorders: position statement of the Canadian College of Medical Geneticists (CCMG). J Med Genet 2023;60(6):523–32.

[33] Manickam K, McClain MR, Demmer LA, Biswas S, Kearney HM, Malinowski J, et al. Exome and genome sequencing for pediatric patients with congenital anomalies or intellectual disability: an evidence-based clinical guideline of the American College of Medical Genetics and Genomics (ACMG). Genet Med 2021;23(11):2029–37.

[34] Hoytema van Konijnenburg EMM, Saskia BW, Marina JK, Laura AT, Roderick H, et al. Treatable inherited metabolic disorders causing intellectual disability: 2021 review and digital app. Orphanet J Rare Dis 2021;16(1):170. Available from: https://doi.org/10.1186/s13023-021-01727-2.

[35] Leach EL, Shevell M, Bowden K, Stockler-Ipsiroglu S, van Karnebeek CDM. Treatable inborn errors of metabolism presenting as cerebral palsy mimics: systematic literature review. Orphanet J Rare Dis 2014;9:197. Available from: https://doi.org/10.1186/s13023-014-0197-2.

[36] Richards S, Aziz N, Bale S, Bick D, Das S, Gastier-Foster J, et al. Standards and guidelines for the interpretation of sequence variants: a joint consensus recommendation of the American College of Medical Genetics and Genomics and the Association for Molecular Pathology. Genet Med 2015;17(5):405–24.

[37] Riggs ER, Andersen EF, Cherry AM, Kantarci S, Kearney H, Patel A, et al. Technical standards for the interpretation and reporting of constitutional copy-number variants: a joint consensus recommendation of the American College of Medical Genetics and Genomics (ACMG) and the Clinical Genome Resource (ClinGen). Genet Med 2020;22(2):245–57.

[38] Lejeune J, Gauthier M, Turpin R. Les chromosomes humains en culture de tissus. CR Hebd Séances Acad Sci 1959;248:602–3.

[39] Butler MG, Meaney FJ, Palmer CG. Clinical and cytogenetic survey of 39 individuals with Prader-Labhart-Willi syndrome. Am J Med Genet 1986;23(3):793–809.

[40] Ledbetter DH, Riccardi VM, Airhart SD, Strobel RJ, Keenan BS, Crawford JD. Deletions of chromosome 15 as a cause of the Prader-Willi syndrome. N Engl J Med 1981;304(6):325–9.

[41] Magenis RE, Brown MG, Lacy DA, Budden S, LaFranchi S. Is Angelman syndrome an alternate result of del(15)(q11q13)? Am J Med Genet 1987;28(4):829–38.

[42] Smith AC, McGavran L, Robinson J, Waldstein G, Macfarlane J, Zonona J, et al. Interstitial deletion of (17)(p11.2p11.2) in nine patients. Am J Med Genet 1986;24(3):393–414.

[43] Jones K. Smith's recognizable patterns of human malformation: expert consult - online and print. 7th ed. London, England: W B Saunders; 2013.

[44] Sheets KB, Crissman BG, Feist CD, Sell SL, Johnson LR, Donahue KC, et al. Practice guidelines for communicating a prenatal or postnatal diagnosis of Down syndrome: recommendations of the national society of genetic counselors. J Genet Couns 2011;20(5):432–41.

[45] Moyer AJ, Gardiner K, Reeves RH. All creatures great and small: new approaches for understanding Down syndrome genetics. Trends Genet 2021;37(5):444–59.

[46] Haydar TF, Reeves RH. Trisomy 21 and early brain development. Trends Neurosci 2012;35(2):81–91.

[47] Arya R, Kabra M, Gulati S. Epilepsy in children with Down syndrome. Epileptic Disord 2011;13(1):1–7.

[48] Baburamani AA, Patkee PA, Arichi T, Rutherford MA. New approaches to studying early brain development in Down syndrome. Dev Med Child Neurol 2019;61(8):867–79.

[49] Rodrigues M, Nunes J, Figueiredo S, de Campos AM, Geraldo AF. Neuroimaging assessment in Down syndrome: a pictorial review. Insights Imag 2019;10(1):1–3.

[50] Bull MJ, Trotter T, Santoro SL, Christensen C, Grout RWCouncil on Genetics. Health supervision for children and adolescents with Down syndrome. Pediatrics 2022;149(5):e2022057010.

[51] Locatelli C, Onnivello S, Antonaros F, Feliciello A, Filoni S, Rossi S, et al. Is the age of developmental milestones a predictor for future development in Down Syndrome? Brain Sci 2021;11(5):655.

[52] Martin GE, Klusek J, Estigarribia B, Roberts JE. Language characteristics of individuals with Down syndrome. Top Lang Disord 2009;29 (2):112–32.

[53] Miller JF. Profiles of language development in children with Down syndrome. In: Miller JF, Leddy M, Leavitt LA, editors. Improving the communication of people with Down syndrome. Baltimore: Brookes; 1999. p. 11.

[54] Rosso M, Fremion E, Santoro SL, Oreskovic NM, Chitnis T, Skotko BG, et al. Down syndrome disintegrative disorder: a clinical regression syndrome of increasing importance. Pediatrics 2020;145(6):e20192939.

[55] Roizen NJ, Patterson D. Down's syndrome. Lancet 2003;361(9365):1281–9.

[56] Myers BA, Pueschel SM. Psychiatric disorders in persons with Down syndrome. J Nerv Ment Dis 1991;179(10):609–13.

[57] Stafstrom CE, Patxot OF, Gilmore HE, Wisniewski KE. Seizures in children with Down syndrome: etiology, characteristics and outcome. Dev Med Child Neurol 1991;33(3):191–200.

[58] Beatty CW, Wrede JE, Blume HK. Diagnosis, treatment, and outcomes of infantile spasms in the Trisomy 21 population. Seizure 2017;45:184–8.

[59] Bohonowych J, Miller J, McCandless SE, Strong TV. The global Prader-Willi syndrome registry: development, launch, and early demographics. Genes 2019;10(9):713.

[60] Ramsden SC, Clayton-Smith J, Birch R, Buiting K. Practice guidelines for the molecular analysis of Prader-Willi and Angelman syndromes. BMC Med Genet 2010;11(1):70.

[61] Polex-Wolf J, Lam BY, Larder R, Tadross J, Rimmington D, Bosch F, et al. Hypothalamic loss of Snord116 recapitulates the hyperphagia of Prader-Willi syndrome. J Clin Invest 2018;128(3):960–9.

[62] Mendiola AJP, LaSalle JM. Epigenetics in Prader-Willi syndrome. Front Genet 2021;12:624581.

[63] Lukoshe A, Hokken-Koelega AC, van der Lugt A, White T. Reduced cortical complexity in children with Prader-Willi syndrome and its association with cognitive impairment and developmental delay. PLoS ONE 2014;9(9):e107320.

[64] McCandless SECommittee on Genetics. Clinical report—health supervision for children with Prader-Willi syndrome. Pediatrics 2011;127 (1):195–204.

[65] Butler MG. Prader-Willi syndrome: current understanding of cause and diagnosis. Am J Med Genet 1990;35(3):319–32.

[66] Bennett JA, Germani T, Haqq AM, Zwaigenbaum L. Autism spectrum disorder in Prader-Willi syndrome: a systematic review. Am J Med Genet A 2015;167A(12):2936–44.

[67] Verrotti A, Soldani C, Laino D, d'Alonzo R, Grosso S. Epilepsy in Prader-Willi syndrome: clinical, diagnostic and treatment aspects. World J Pediatr 2014;10(2):108–13.

[68] Consoli A, Cabal Berthoumieu S, Raffin M, Thuilleaux D, Poitou C, Coupaye M, et al. Effect of topiramate on eating behaviors in Prader-Willi syndrome: TOPRADER double-blink randomised placebo-controlled study. Transl Psychiatry 2019;9(1):274.

[69] Deal CL, Tony M, Höybye C, Allen DB, Tauber M, Christiansen JS, et al. Growth Hormone Research Society workshop summary: consensus guidelines for recombinant human growth hormone therapy in Prader-Willi syndrome. J Clin Endocrinol Metab 2013;98(6):E1072–87.

[70] Wheeler AC, Sacco P, Cabo R. Unmet clinical needs and burden in Angelman syndrome: a review of the literature. Orphanet J Rare Dis 2017;12(1):164.

[71] Jiang YH, Armstrong D, Albrecht U, Atkins CM, Noebels JL, Eichele G, et al. Mutation of the Angelman ubiquitin ligase in mice causes increased cytoplasmic p53 and deficits of contextual learning and long-term potentiation. Neuron 1998;21(4):799–811.

[72] Sell GL, Margolis SS. From UBE3A to Angelman syndrome: a substrate perspective. Front Neurosci 2015;9:322.

[73] Aghakhanyan G, Bonanni P, Randazzo G, Nappi S, Tessarotto F, De Martin L, et al. From cortical and subcortical grey matter abnormalities to neurobehavioral phenotype of angelman syndrome: a voxel-based morphometry study. PLoS ONE 2016;11(9):e0162817.

[74] Clayton-Smith J, Laan L. Angelman syndrome: a review of the clinical and genetic aspects. J Med Genet 2003;40(2):87–95.

[75] Morris CA, Braddock SRCouncil on Genetics. Health care supervision for children with Williams syndrome. Pediatrics 2020;145(2): e20193761.

[76] Hirota H, Matsuoka R, Chen X-N, Salandanan LS, Lincoln A, Rose FE, et al. Williams syndrome deficits in visual spatial processing linked to GTF2IRD1 and GTF2I on chromosome 7q11.23. Genet Med 2003;5(4):311–21.

[77] Jackowski AP, Rando K, Maria de Araújo C, Del Cole CG, Silva I, Tavares de Lacerda AL. Brain abnormalities in Williams syndrome: a review of structural and functional magnetic resonance imaging findings. Eur J Paediatr Neurol 2009;13(4):305–16.

[78] Mervis CB, Robinson BF, Bertrand J, Morris CA, Klein-Tasman BP, Armstrong SC. The Williams syndrome cognitive profile. Brain Cogn 2000;44(3):604–28.

[79] Nicita F, Garone G, Spalice A, Savasta S, Striano P, Pantaleoni C, et al. Epilepsy is a possible feature in Williams-Beuren syndrome patients harboring typical deletions of the 7q11.23 critical region. Am J Med Genet A 2016;170A(1):148–55.

[80] Elsea SH, Girirajan S. Smith-Magenis syndrome. Eur J Hum Genet 2008;16(4):412–21.

[81] Williams SR, Zies D, Mullegama SV, Grotewiel MS, Elsea SH. Smith-Magenis syndrome results in disruption of CLOCK gene transcription and reveals an integral role for RAI1 in the maintenance of circadian rhythmicity. Am J Hum Genet 2012;90(6):941–9.

[82] Poisson A, Nicolas A, Cochat P, Sanlaville D, Rigard C, de Leersnyder H, et al. Behavioral disturbance and treatment strategies in Smith-Magenis syndrome. Orphanet J Rare Dis 2015;10(1):111.

[83] De Leersnyder H, de Blois MC, Vekemans M, Sidi D, Villain E, Kindermans C, et al. beta(1)-adrenergic antagonists improve sleep and behavioural disturbances in a circadian disorder, Smith-Magenis syndrome. J Med Genet 2001;38(9):586–90.
[84] Wolff DJ, Van Dyke DL, Powell CM. Working Group of the ACMG Laboratory Quality Assurance Committee. Laboratory guideline for Turner syndrome. Genet Med 2010;12(1):52–5.
[85] Hutaff-Lee C, Bennett E, Howell S, Tartaglia N. Clinical developmental, neuropsychological, and social-emotional features of Turner syndrome. Am J Med Genet C Semin Med Genet 2019;181(1):126–34.
[86] Farooqui A, Tazyeen S, Ahmed MM, Alam A, Ali S, Malik MZ, et al. Assessment of the key regulatory genes and their Interologs for Turner Syndrome employing network approach. Sci Rep 2018;8(1).
[87] Gravholt CH, Andersen NH, Conway GS, Dekkers OM, Geffner ME, Klein KO, et al. Clinical practice guidelines for the care of girls and women with Turner syndrome: proceedings from the 2016 Cincinnati International Turner Syndrome Meeting. Eur J Endocrinol 2017;177(3):G1–70.
[88] The Deciphering Developmental Disorders Study. Large-scale discovery of novel genetic causes of developmental disorders. Nature 2015;519 (7542):223–8.
[89] Kochinke K, Zweier C, Nijhof B, Fenckova M, Cizek P, Honti F, et al. Systematic phenomics analysis deconvolutes genes mutated in intellectual disability into biologically coherent modules. Am J Hum Genet 2016;98(1):149–64.
[90] Deciphering Developmental Disorders Study. Prevalence and architecture of de novo mutations in developmental disorders. Nature 2017;542 (7642):433–8.
[91] Jamra R. Genetics of autosomal recessive intellectual disability. Med Genet 2018;30(3):323–7.
[92] Reuter MS, Tawamie H, Buchert R, Hosny Gebril O, Froukh T, Thiel C, et al. Diagnostic yield and novel candidate genes by exome sequencing in 152 consanguineous families with neurodevelopmental disorders. JAMA Psychiatry 2017;74(3):293–9.
[93] Hu H, Kahrizi K, Musante L, Fattahi Z, Herwig R, Hosseini M, et al. Genetics of intellectual disability in consanguineous families. Mol Psychiatry 2019;24(7):1027–39.
[94] Lubs H. A marker X chromosome. Am J Hum Genet 1969;21:231–44.
[95] Tzschach A, Grasshoff U, Beck-Woedl S, Dufke C, Bauer C, Kehrer M, et al. Next-generation sequencing in X-linked intellectual disability. Eur J Hum Genet 2015;23(11):1513–18.
[96] Huber KM, Gallagher SM, Warren ST, Bear MF. Altered synaptic plasticity in a mouse model of fragile X mental retardation. Proc Natl Acad Sci USA 2002;99(11):7746–50.
[97] Bear MF, Huber KM, Warren ST. The mGluR theory of fragile X mental retardation. Trends Neurosci 2004;27(7):370–7.
[98] Maes B, Fryns JP, Ghesquière P, Borghgraef M. Phenotypic checklist to screen for fragile X syndrome in people with mental retardation. Ment Retard 2000;38(3):207–15.
[99] Kau ASM, Meyer WA, Kaufmann WE. Early development in males with Fragile X syndrome: a review of the literature. Microsc Res Tech 2002;57(3):174–8.
[100] Berry-Kravis E. Epilepsy in fragile X syndrome. Dev Med Child Neurol 2002;44(11):724–8.
[101] Hersh JH, Saul RA, Committee on Genetics. Health supervision for children with fragile X syndrome. Pediatrics 2011;127(5):994–1006.
[102] Erickson CA, Davenport MH, Schaefer TL, Wink LK, Pedapati EV, Sweeney JA, et al. Fragile X targeted pharmacotherapy: lessons learned and future directions. J Neurodev Disord 2017;9(1):7.
[103] Hagberg B. Rett's syndrome: prevalence and impact on progressive severe mental retardation in girls. Acta Paediatr 1985;74(3):405–8.
[104] Amir RE, Van den Veyver IB, Wan M, Tran CQ, Francke U, Zoghbi HY. Rett syndrome is caused by mutations in X-linked MECP2, encoding methyl-CpG-binding protein 2. Nat Genet 1999;23(2):185–8.
[105] Neul JL, Fang P, Barrish J, Lane J, Caeg EB, Smith EO, et al. Specific mutations in methyl-CpG-binding protein 2 confer different severity in Rett syndrome. Neurology 2008;70(16):1313–21.
[106] Chahrour M, Jung SY, Shaw C, Zhou X, Wong STC, Qin J, et al. MeCP2, a key contributor to neurological disease, activates and represses transcription. Science. 2008;320(5880):1224–9.
[107] Shahbazian MD, Antalffy B, Armstrong DL, Zoghbi HY. Insight into Rett syndrome: MeCP2 levels display tissue- and cell-specific differences and correlate with neuronal maturation. Hum Mol Genet 2002;11(2):115–24.
[108] Chao H-T, Zoghbi HY, Rosenmund C. MeCP2 controls excitatory synaptic strength by regulating glutamatergic synapse number. Neuron 2007;56(1):58–65.
[109] Naidu S, Kaufmann WE, Abrams MT, Pearlson GD, Lanham DC, Fredericksen KA, et al. Neuroimaging studies in Rett syndrome. Brain Dev 2001;23(Suppl 1):S62–71.
[110] Williamson SL, Christodoulou J. Rett syndrome: new clinical and molecular insights. Eur J Hum Genet 2006;14(8):896–903.
[111] Hagberg B, Witt-Engerström I. Rett syndrome: a suggested staging system for describing impairment profile with increasing age towards adolescence. Am J Med Genet Suppl 1986;1:47–59.
[112] Weaving LS, Ellaway CJ, Gécz J, Christodoulou J. Rett syndrome: clinical review and genetic update. J Med Genet 2005;42(1):1–7.
[113] Tarquinio DC, Hou W, Berg A, Kaufmann WE, Lane JB, Skinner SA, et al. Longitudinal course of epilepsy in Rett syndrome and related disorders. Brain 2017;140(2):306–18.
[114] Pini G, Bigoni S, Congiu L, Romanelli AM, Scusa MF, Di Marco P, et al. Rett syndrome: a wide clinical and autonomic picture. Orphanet J Rare Dis 2016;11(1):132.
[115] Srivastava S, Desai S, Cohen J, Smith-Hicks C, Baranano K, Fatemi A, et al. Monogenic disorders that mimic the phenotype of Rett syndrome. Neurogenetics 2018;19(1):41–7.

[116] Fu C, Armstrong D, Marsh E, Lieberman D, Motil K, Witt R, et al. Consensus guidelines on managing Rett syndrome across the lifespan. BMJ Paediatr Open 2020;4(1):e000717.

[117] Chinnery PF. Primary mitochondrial disorders overview. In: Margaret PA, Feldman J, Mirzaa GM, Pagon RA, Wallace SE, et al., editors. GeneReviews®. Seattle (WA): University of Washington; 1993. Available from: http://www.ncbi.nlm.nih.gov/books/NBK1224/.

[118] Angelini C, Bello L, Spinazzi M, Ferrati C. Mitochondrial disorders of the nuclear genome. Acta Myol 2009;28(1):16–23.

[119] Falk MJ. Neurodevelopmental manifestations of mitochondrial disease. J Dev Behav Pediatr 2010;31(7):610–21. Available from: https://doi.org/10.1097/DBP.0b013e3181ef42c1.

[120] Ortiz-González XR. Mitochondrial dysfunction: a common denominator in neurodevelopmental disorders? Dev Neurosci 2021;1–8.

[121] Abrahams BS, Geschwind DH. Advances in autism genetics: on the threshold of a new neurobiology. Nat Rev Genet 2008;9(5):341–55.

Chapter 24

Intellectual and developmental disabilities

Meghan O'Neill[1] and Patricia Ann Evans[2]

[1]*Department of Developmental Behavioral Pediatrics, Ann & Robert H. Lurie Children's Hospital of Chicago, Chicago, IL, United States,*
[2]*Departments of Neurology and Neurotherapeutics, Pediatrics, and Psychiatry, University of Texas Southwestern School of Medicine, Children's Medical Center, Dallas, TX, United States*

History of intellectual and developmental disabilities (IDD)

Historically, cultures have struggled with how to understand and respond to individuals with intellectual disabilities (ID). Individuals with IDD experienced scorn, neglect, and persecution worldwide for centuries, viewed as outcasts of society who were often mutilated, abandoned, segregated, or institutionalized. It was not until the 19th and 20th centuries when some societies began attempts at educating or "training" individuals with IDD, with a turning point coming in the 1950s on the heels of the civil rights movement in the United States as groups of parents of individuals with IDD began calling for recognition and rights for their children. The passage of the Rehabilitation Act of 1973 (Section 504), the Education for All Handicapped Children Act of 1975 (later reauthorized as the Individuals with Disabilities Education Act, or IDEA), and the Americans with Disabilities Act (ADA) of 1990 with subsequent amendments, court decisions, and administrative actions have cumulatively granted individuals with disabilities legal protections and access to educational supports and services that would have been unheard of in prior centuries. Concurrently, there has been a general shift in the way that individuals with IDD are viewed in our society, with a move toward greater community inclusion, amplified acceptance of individuals who are neurodiverse, and increased promotion of autonomy and self-advocacy among those with IDD. Recognizing that individuals with IDD have the right to a rich human experience that may necessitate specific educational and therapeutic supports and tailored medical care, this chapter focuses on how we diagnose, evaluate, and care for individuals with IDD.

Definition

Contemporary conceptualizations of ID emphasize functional and social abilities rather than test scores. Definitions of ID have been formulated by the World Health Organization (WHO) [1] *International Classification of Diseases, Tenth Edition* (ICD-10), the US Individuals with Disabilities Education Act (IDEA), the American Psychiatric Association (APA) *Diagnostic and Statistical Manual of Mental Disorders, Fifth Edition* (DSM-5) [2], and the American Association on Intellectual and Developmental Disabilities (AAIDD) [3], all requiring significant impairment in adaptive behavior and general intellectual function (reasoning, learning, problem solving), and social skills. This focus on conceptual, social, and practical skills enables the development of individual treatment plans designed to enhance functioning. Consistent across these definitions is onset of symptoms before age 22 years of age, separating deficits occuring during the developmental period from those acquired in later adulthood.

Significant impairment in general intellectual function refers to performance on an individually administered standardized test of intelligence that is approximately 2 standard deviations (SD) below the mean. Typically, tests measuring intelligence provide a standard score with a mean of 100 and a normal "bell curve" distribution with standard deviation of 15 points. Subsequently an intelligence quotient (IQ) score of less than 70 meets the criteria of having intellectual impairment. Given issues with standard error of measurement in addition to the cultural and other biases inherent to tests of intelligence, it is critical to correlate scores in the very low range (below 70) with measures of adaptive function so as to not over diagnose individuals with adequate adaptive functioning as having mild ID.

Capute and Accardo's Neurodevelopmental Disabilities in Infancy and Childhood. DOI: https://doi.org/10.1016/B978-0-12-824060-1.00019-5

TABLE 24.1 Types of measurable adaptive skills.

Adaptive skill	Examples
Conceptual	Language and communication, reading, writing, time concepts, math and number concepts, problem-solving, judgment, planning, executive functioning, self-direction
Social	Interpersonal skills, social problem-solving, capacity to follow rules, obey laws, and avoid victimization, participation in leisure activities, understanding social responsibilities
Practical	Personal care, toileting, hygiene, dressing, feeding/cooking, managing household duties, money management, shopping, taking medication and directing health care, participating in an occupation, coordinating and completing daily routines, managing a safe environment, handling travel and transportation, community navigation

Adaptive behavior refers to the skills required for people to function in their everyday lives. The AAIDD and DSM-5 classifications of adaptive behavior both address three broad sets of skills: conceptual, social, and practical (Table 24.1). For a deficit in adaptive behavior to be present, a significant delay in at least one of the three skill areas must be present. The rationale for requiring only one area is the empirically derived finding that people with IDD, especially in the mild range, can have varying patterns of ability and may not have deficits in all three areas. Additionally, requiring more than one area of deficit in adaptive behavior may delay identification and implementation of early intervention in young children. Though severity levels in IDD used to be classified based on IQ standard deviations, there has been a general move toward classifying severity level based on adaptive functioning deficits. A summary of basic adaptive abilities attained by adulthood for individuals with IDD within each severity group is detailed in Table 24.1 adapted from ref. [3].

In making the diagnosis of IDD, other co-occurring conditions that might distort assessment or provide alternative explanation for atypical functioning should be ruled out. Severe hearing and vision impairment, refractory seizure disorders, autism spectrum disorders, severe attentional or emotional dysregulation, significant mental health concerns, severe motor impairment, or progressive neurological conditions may mimic IDD or render conventional standardized assessments null. Additionally, it is important to acknowledge that standardized assessments are not without bias, such that some individuals who fall within a borderline range for cognitive functioning may have adaptive functioning that is still well within a normative range and thus should not be classified as having IDD. Historically, the inadequate assessment of adaptive functioning has likely contributed to overdiagnosis of IDD among black and Hispanic males in the United States, leading to significant racial disparity in IDD diagnoses from a epidemiologic standpoint. Conversely, individuals may present with significantly lower adaptive functioning than expected relative to cognitive functioning, often due to other prominent conditions (autism spectrum disorder, significant motor impairment, poor attention), and should also not be classified as having IDD. Given these complicating factors, making the diagnosis of IDD requires knowledge of the associated conditions that may mimic IDD, understanding of the potential limitations of both cognitive and adaptive assessments, and repeated evaluations over time to confirm validity and stability of the diagnosis.

Etiology

Though IDD is one of the most common causes of disability in children globally, reported prevalence of IDD in different parts of the world varies substantially depending on study design, definitional constructs, severity levels, and population characteristics such as location or age. The prevalence of IDD in the general population is approximately 1%, with 85% of cases involving mild IDD. Variable rates across populations have been reported, ranging from lows of around 300 cases per 100,000 people in high-income North America, Western Europe, and East Asia to highs of around 500 cases per 100,000 people in South Asia. Beyond definitional and study design differences, differences in these rates are largely attributable to variable rates of mild IDD, with differential rates of poverty, educational access, and environmental exposures potentially contributing to higher rates of mild IDD diagnoses in developing countries. Conversely, the prevalence of severe IDD is relatively stable across populations. IDD occurs more in boys than in girls, though the difference diminishes with increasing severity of disability [4].

IDD may result from prenatal, peri-natal, postnatal, or childhood-related events that lead to a cascade of atypical brain development. Infection, trauma, perinatal insults, prematurity, hypoxia-ischemia, toxic exposures, metabolic dysfunction, endocrine abnormalities, malnutrition, and genetic abnormalities (ranging from chromosomal level aberrations to single base pair variants to epigenetic changes) may all potentially cause or contribute to the development of IDD. Identifying a definitive underlying diagnosis based on pre- or perinatal history or dysmorphology is challenging, necessitating further medical and genetic evaluation as recommended by practice guidelines from the American College of Medical Genetics (ACMG) [5], the American Academy of Neurology (AAN), and the American Academy of Pediatrics (AAP). Details on etiologic workup in ID and GDD are addressed further, but in general, the yield of etiologic workup tends to be lower with more mild levels of IDD. Additionally, with the continued advancement of genetic technologies and data science (including sharing of data across large databases), the number of identified biologic and genetic causes of IDD is continuing to increase rapidly, though currently there are over 115 disorders (all of which are metabolic in nature) for which treatment may ameliorate the core symptoms of IDD [6] (see tidebc.org).

In children with severe IDD, a biologic cause (usually with prenatal onset) can be identified in about three-fourths of all cases. Causes include chromosomal abnormalities (e.g., Down syndrome, microdeletion syndromes), genetic and epigenetic disorders (e.g., fragile X, Rett syndrome, Angelman syndrome, and tuberous sclerosis), abnormalities of brain development (e.g., lissencephaly), and inborn errors of metabolism and mitochondrial disorders (e.g., mucopolysaccharidoses, mitochondrial respiratory chain complex disorders). IDD is sometimes classified as "syndromic" (when present with additional clinical symptoms or physical anomalies that are consistent with a known syndrome and causative biologic defect) or "nonsyndromic" (where constellations of cooccurring dysmorphic features are largely absent, requiring broader testing such as molecular sequencing to evaluate for genetic etiologies). Nonsyndromic severe IDD may be a result of inherited or de novo gene mutations, as well as microdeletions or microduplications not detected on standard chromosomal microarray. Currently, well over a thousand single genes have been associated with ID. Generally, the earlier the problem occurs in embryologic development, the more severe its consequences tend to be with symptoms and clinical presentations that tend to be detected earlier in childhood.

Beyond those factors mentioned above, there are several nongenetic risk factors that must be considered and are often associated more with mild IDD, including low socioeconomic status, low maternal education levels, residence in a developing country, malnutrition, and poor access to healthcare. Fetal environmental factors during pregnancy may lead to IDD, including maternal exposure to various toxins [Intrauterine alcohol and drug exposure: preventable causes of intellectual and behavioral disorders] or infectious agents, uncontrolled maternal health conditions, and birth complications. Finally, IDD can be acquired during early childhood via encephalitis or meningitis, head trauma, asphyxia, malnutrition, exposure to toxic substances (the most common being lead), or oncologic processes (either directly from a tumor itself or from the impact of chemotherapy, surgeries, etc.).

Evaluating for a cause of IDD is generally valued by parents [7], especially among cases of severe IDD where yield of testing tends to be higher. As such, genetic testing is generally recommended in the evaluation of individuals with IDD and identification of an underlying causative etiology can provide the family with more information about prognosis and anticipation of future needs, reproductive risk with future pregnancies, potential associated medical and mental health comorbidities, connection to family and advocacy groups, and the possibility for enrollment in condition-specific research studies and clinical trials. It is rare that identification of a genetic etiology will lead to specific treatment options to address the underlying cause of IDD or modify the course of symptoms aside from some treatable metabolic disorders (see tidebc.org), though targeted treatment options may expand throughout the modern genomic era of medicine. Finally, even though identification of a genetic cause may not substantially alter management in some cases, most parents consider identification of an etiologic diagnosis to be highly relevant for their own emotional relief and many adults with IDD have also reported both the desire and capability to learn more about genetic conditions and genetic testing.

Clinical manifestations

Presentation through the lifespan

The presentation of IDD is variable depending on severity level, age, and underlying etiology. An initial diagnostic evaluation is often initiated when a parent brings concerns that a child is not meeting age-appropriate developmental milestones. Earlier presentation of delays and more significant levels of delay relative to chronological age expectations both tend to correlate with more severe levels of IDD later on in life, an increased prevalence of neurologic comorbidities, and, thus, a need for higher levels of support.

The earliest identification of IDD now occurs during fetal life with the advent of more sophisticated prenatal screening and detailed radiologic techniques that may identify conditions known to be associated with the development of IDD. Central nervous system malformations and genetic conditions such as trisomy 13, 18, and 21 represent the more common prenatally recognized conditions and provide the opportunity for more advanced planning on the part of the family. Newborn screening initiatives may also identify many children with conditions that increase risk for intellectual disability. In the newborn period, clinical concerns for IDD arise when a child has dysmorphic facial features, significant hypotonia, and constellations of particular congenital anomalies that raise suspicions for a potential neurogenetic syndrome, such as Down syndrome, Prader–Willi syndrome, or Williams syndrome. Major organ system dysfunction including feeding- and breathing-related concerns may be a marker for more widespread neurologic dysfunction. In early infancy, around 2–4 months of age, parents may notice that their child fails to interact with the environment. There may be initial concerns about eye contact, tracking, response to sound or voice, and general awareness of the external world. There may also be concerns for either hypotonia or hypertonia.

By the time the child enters later infancy through about 18 months, delays in acquisition of motor milestones tend to be most apparent. Sometimes the child will fall within the later range of normal with sitting, standing, and walking, though more severely impacted children may have more obvious gross motor delays from very early on, starting with poor head control. The toddler-aged child may be clumsier than expected with difficulties progressing to running or jumping. By 18 months onward, language expectations increase and delays in this area are brought to attention, though failure to meet language milestones even earlier in the first year of life may be predictive of eventual intellectual disability. Receptive language delays raise the most concern for IDD, especially in the context of delays in other areas of development. A portion of children with more isolated language delays at this age will go on to have a developmental language disorder without larger cognitive concerns, but often these children will display typical play and adaptive behaviors and have fewer difficulties with fine and gross motor skills than those children who go onto have IDD. Parents may also report concerns during toddlerhood when the child's younger siblings begin to surpass the development of the affected child.

By the preschool years, concerns for IDD stem from continued difficulties with language, delayed emergence of play skills, behavioral issues that may arise when age-appropriate demands surpass the child's capabilities, and more widespread delays with learning and fine motor skills, such as mastering the alphabet, cutting, coloring, drawing, and basic number sense. By the school-age years, children will demonstrate broader academic underachievement and sometimes manifest behavioral difficulties ranging from inattention to anxiety to oppositional behaviors. For some children with mild IDD, the diagnosis remains uncertain during the early years. Because of this, the diagnosis of Global Developmental Delay was coined. This diagnosis is reserved for children 5 years of age and under who have significant delays ($\geq$ 2 or more standard deviations below the mean) in 2 or more areas of development. As the demands of the school setting increase, the diagnosis clarifies, often with the transition from "learning to read" to "reading to learn" which occurs around third grade.

By later school ages, those with more severe levels of IDD have usually been identified and supports have been established. For children with milder levels of IDD, limitations become more apparent as demands for independence at home and in the social arena increase with age. More nuanced concepts of time and money may present a challenge and executive functioning skills are reduced, leading to challenges with management of longer-term projects, planning ahead, organizing time, managing money, and budgeting skills. Fluid intelligence, which refers to the use of mental operations to think and reason abstractly and develop problem-solving strategies, is notably challenging for individuals with IDD at all levels of severity. This greatly influences learning in everyday life and in educational settings, rendering abstract concepts in language, reading, and math elusive into adulthood.

Socially, children and young adults with mild to moderate levels of IDD carry on meaningful conversation, may be up to date on common trends, and may not be recognized as having any notable difficulties at first glance. However, many will need support with navigating more complicated social landscapes and are at risk for emulating their social milieu to gain acceptance. Adolescents with ID are at high risk of being bullied and of being taken advantage of from a social perspective. Accordingly, these individuals should be monitored closely for mental health comorbidities such as anxiety and depression. Efforts to develop self-advocacy and self-agency skills should be initiated at an early age.

In adulthood, the ultimate level of functioning attained by an individual with ID will often correlate with the level of severity of symptoms. Table 24.2 details a sampling of skills and abilities often attained by individuals within each severity group, though it must be recognized that these generalizations will not apply precisely to every individual, each of whom will have their own specific pattern of strengths and challenges.

TABLE 24.2 Severity of intellectual disability and adult-age functioning.

Level	Mental age as adult	Adult adaptation
Mild	9–11 years	Reads at fourth to fifth grade level; simple multiplication and division; writes simple letter, lists; completes job application; basic independent job skills (arrive on time, stay at task, interact with coworkers); uses public transportation, might qualify for driver's license; keeps house, cooks using recipes; interested in romantic relationships; will often need more support in decision making surrounding health care, finances/budgeting, and raising a family.
Moderate	6–8 years	Sight-word reading; copies information (e.g., address from card to job application); matches written number to number of items; recognizes time on clock; communicates; some independence in self-care; housekeeping with supervision or cue cards; meal preparation, can follow picture recipe cards; job skills learned with much repetition; uses public transportation with some supervision; supported employment is possible; successful friendships and romantic relationships can be attained but social judgment and life decisions require continued supports.
Severe	3–5 years	Needs continuous support and supervision to complete basic activities of daily living and to participate in activities; little understanding of written language, number, time, or money concepts; often capable of communicating wants and needs, sometimes with augmentative communication techniques; capable of developing strong attachment to caregivers.
Profound	Less than 3 years	Limitations of self-care, continence, communication, and mobility; might need complete custodial or nursing care; limited understanding of symbolic communication but may understand and use some gestures and emotional cues; recognizes caregivers.

Regression and neurodegeneration

IDD is usually conceptualized as nonprogressive or static. Individuals with IDD will continue to make developmental progress over time, although at a slower rate than typically developing peers and sometimes with a somewhat decelerating or plateauing trajectory in late childhood to adulthood. It is important to distinguish this static entity from cases in which *regression* occurs. Regression is defined by true loss of previously acquired skills and should prompt additional and often more extensive etiologic workup, especially when loss of skills is widespread across language, motor, and cognitive domains [See Appendix 2.2: Regression or Plateau in Developmental Milestones]. Regression may occur because of evolution of medical complications (increasing spasticity/contracture over time with gait deterioration in cerebral palsy) or in the setting of a progressive encephalopathy due to an underlying inborn error of metabolism or genetically based neurodegenerative process. As such, it is important to follow the developmental trajectory closely in any child with identified delays, as true loss of developmental milestones should necessitate prompt efforts at identification of an underlying genetic or metabolic etiology which may alter management, prognosis, and counseling on reproductive risk. [See chapter: Neurodegenerative disorders].

Not all declines in function are true regression indicative of neurodegeneration. In younger children, over-programmed skills may no longer be manifested if they are not reinforced. A decline in IQ test scores may be the result of a slowing trajectory, difficulty progressing to abstraction and generalization that is typical of adolescents, or the child approaching their maximal adult functional level. Occasionally, individuals with IDD can also experience psychiatric conditions or catatonia which may manifest as profound changes in functional abilities, often during the teen or young adult years. In all cases of regression, medical causes should also be sought which might provide explanation for change in functioning.

Neurogenetic syndromes with more frequent medical and behavioral complications

Several neurogenetic syndromes are associated with particular behavioral and cognitive phenotypes and are characterized by well-studied established patterns of relative strengths and weaknesses. Some of the most recognizable and researched conditions include fragile X syndrome, Down syndrome, Prader–Willi syndrome, Smith–Magenis syndrome, Turner syndrome, Williams syndrome, and velocardiofacial syndrome [See chapter: Approach to diagnosis and management of genetic neurodevelopmental disorders]. Of note, many of these conditions are also associated with particular medical complications or conditions and may require individualized assessment of issues that are relevant to that child's disorder. This might include dental examinations in children who exhibit bruxism, nutritional and endocrine assessments in children with hyperphagia or obesity, thyroid function testing in children with Down syndrome, and cardiac evaluations in

Williams syndrome and velocardiofacial syndrome. For some of these conditions, the AAP has published series of guidelines and/or toolkits to guide medical screening (Down syndrome, fragile X syndrome, Williams syndrome, fetal alcohol spectrum disorder).

Importance of recognition and treatment

Early diagnosis of IDD facilitates earlier intervention, identification of strengths and challenges, realistic goal setting, monitoring for potential comorbid conditions, easing of parental anxiety through connection to family groups or larger networks of support, and greater inclusion of the child in the community. Although the core features of IDD itself are generally not responsive to any specific medical treatments, early and comprehensive therapy and educational supports can help a child to meet their maximal developmental potential and set the stage for a successful transition to adulthood. Additionally, astute recognition of associated impairments throughout the lifespan (hearing impairment, vision impairment, mobility-related issues, sleep apnea and other sleep dysfunction, attentional disorders, mood disorders, epilepsy and other medical comorbidities) is critical as many of these conditions are amenable to medical treatment which leads to improved overall outcomes.

Prevention of IDD and related problems

IDD cannot usually be prevented, though there are primary, secondary, and tertiary strategies that can be implemented to address the risk factors that may contribute to or cause IDD.

Genetic counseling

When IDD is identified in a family member and a specific etiology is determined, recurrence risks can be estimated which can facilitate decision-making regarding future reproductive options. This now includes the use of preimplantation genetic testing in concert with in vitro fertilization or prenatal diagnosis for some forms of genetically based IDD. Examples may include fragile X testing or conditions caused by balanced translocation.

Pregnancy-based initiatives

Given that critical periods of brain development occur prenatally, initiatives targeting fetal health during pregnancy are vital in efforts at preventing intellectual disability. Basic strategies include reduction of teen pregnancy, improvements in prenatal care, nutritional initiatives (folic acid), and prevention of poverty and malnourishment during pregnancy. Campaigns to promote safe sexual practices and screening for sexually transmitted diseases and other infections that may cause congenital infection represent additional prevention initiatives, including promotion of measures to prevent infection (postexposure prophylaxis in the setting of maternal human immunodeficiency virus, or HIV, exposures, maternal hygiene practices to prevent primary maternal cytomegalovirus, or CMV, etc.) or treat infection (antibiotics for syphilis or toxoplasmosis). Increasing public knowledge regarding the harms of alcohol or other drugs of abuse is another vital primary preventative strategy given that fetal alcohol exposure remains one of the most common preventable causes of IDD worldwide.

Preventing injury and poisoning

Prevention of accidental traumatic brain injury during childhood represents a postnatal strategy to prevent both death and serious disability including IDD. This includes use of age-appropriate car seat restraints, helmet use with biking, scootering, or skateboarding, and limiting exposure to firearms and overall reduction of gun violence. Home-safety measures include use of window locks and guard rails to prevent falls and proper securing of medications and poisons both through manufacturer-based safety mechanisms and through parental education regarding best practices.

Immunization

Immunization initiatives to prevent measles, mumps, rubella, hepatitis B, varicella, and *Neisseria meningitidis*, haemophilus influenzae, pneumococcus, and influenza can reduce encephalitis, meningitis, and congenital infection, thereby reducing rates of both prenatal and postnatally acquired IDD. Currently, vaccine development is underway for CMV, group B streptococcus, HSV (herpes simplex virus), HIV, syphilis, and zika virus, all of which may cause perinatal or congenital infections with long-term developmental impacts.

Presymptomatic detection

Presymptomatic detection includes initiatives such as state newborn screening (using tandem mass spectrometry to detect >60 rare genetic disorders and congenital hypothyroidism), newborn hearing screening, and widespread lead screening during toddlerhood, all of which can detect conditions prior to the development of more obvious clinical manifestations. Screening for comorbid conditions that may not have obvious symptoms and may negatively impact development in IDD represents a secondary preventative strategy. Conditions such as hearing and vision impairment, seizures, thyroid disease, sleep apnea, and so forth should be considered and treated appropriately, especially in more at-risk populations (such as individuals with Down syndrome and other neurogenetic syndromes) [8].

Treatment and supportive care from birth to age 22 years

A multidisciplinary approach: family, school, therapists, clinicians, and patient involvement

Management of intellectual disability involves maximizing a child's developmental potential and optimizing functioning in all settings including the home, school, and community. Individuals with IDD will require multiple and variable supports over time which require individualization and continuous re-evaluation [9]. Management of IDD is inherently multidisciplinary and hinges on establishment of a medical home, collaboration of involved treating professionals, and provision of support for family members. Care should be family-centered, culturally sensitive, and rooted in advocacy for the individual with IDD while also promoting the individual's capacity for self-advocacy, self-direction, and autonomy to the fullest extent possible. Consultant medical specialties often include neurodevelopmental disabilities, developmental-behavioral pediatrics, neurology, genetics, physical medicine and rehabilitation, orthopedics, psychiatry, ophthalmology, and audiology.

A strengths-based approach focusing on optimizing participation in all aspects of life may be more beneficial than an exclusive focus on ameliorating cognitive deficits or academic skill attainment [10,11]. Though each family and individual with IDD will have unique and personalized goals of treatment, often the most salient objectives include promoting effective communication abilities, fostering maximal independent functioning, and appropriately addressing any challenging behaviors, often with the larger goal of maximizing community inclusion and social connectedness. Preservation of physical well-being, support of positive relationship building, and promotion of recreational and leisure opportunities are also important. Detection and treatment of comorbid disorders may reduce the negative impacts of the disability and improve overall inclusion in both school and community-based programming.

Family support is equally important in the quest to assist those with IDD. Raising a child with IDD is most often perceived as rewarding but can be equally exhausting given the varied emotional aspects, the increased need for time, resources, and money to dedicate toward the child's various needs, the challenges of service navigation, and the knowledge that the need for supports will persist throughout both the life of the child and the parent. Family members may experience grief, resentment, disappointment, frustration, and guilt, all of which are normal and expected emotions that may ebb and flow throughout the caretaker's journey and are best managed with the support of providers and the community at large. Service coordinators or case managers can help a family to create, maintain, and execute a reasonable and appropriate support plans and can be critical in facilitating communication among providers, family, and the individual with IDD.

Finally, the social model of disability and the self-advocacy movement demands that the individual with IDD be central to the discussion of goals of care. Self-advocacy stresses that individuals with IDD should be involved as much as possible in decision making, should be treated equitably and fully included in society, and should be given as much control over their own self-direction as possible. This movement emphasizes the tenet of "nothing about us without us" and above all seeks to redefine the approach to disability, moving toward a social model of acceptance, equality, accommodation, and inclusion, putting the onus on society to change its attitudes and practices.

Therapies: the role of speech, occupational, physical, educational, behavioral, and other therapies

Many supports and therapeutic options exist to both enhance an individual's relative strengths or compensate for an individual's relative challenges. Table 24.3 references many therapeutic options that exist, all of which require goal

TABLE 24.3 List of disciplines involved in the care of children with intellectual disabilities, and the goals of their respective interventions.

Support	Goals of the intervention and other details
Speech therapy	• Build expressive language skills with incorporation of Augmentative and Alternative Communication (AAC) where needed • Further develop receptive language skills • Enhancing pragmatic and social communication skills • Expand vocabulary and length of utterance • Work on speech articulation and oral motor issues • Address abnormal feeding or swallowing patterns
Augmentative and alternative communication evaluation	• Provide methods of communication for those who have trouble with speech or language skills that either add to existing speech abilities or offer options to be used instead of speech (when minimal verbal abilities exist or significant articulation issues limit the communication partner's understanding) • May be used intermittently or throughout life • Range from no-tech/low-tech (sign language, picture boards) to high-tech options (speech-generating devices)
Occupational therapy	• Work toward functional gains in eye–hand coordination, fine motor manipulation, and grasping patterns • Address sensory processing/integration issues and self-regulation • Work on self-care skills (grooming, dressing, feeding, hygiene) • Help with participation in meaningful and purposeful activities, often with a focus on play among children and leisure activities in adults • Enhance involvement in specific employment skills and activities in adulthood • Address challenges with domestic activities (cooking, cleaning, laundry)
Physical therapy	• Maximize mobility and self-locomotion • Work on strengthening, range of motion, balance, and stability/postural support • Provide adaptive solutions to mobility problems • Improve overall sensory integration and awareness of body in space
Hearing services	• Provision of audiology services to identify hearing impairment • Hearing-based therapies to work on listening and spoken language development • Increase access to sound through use of devices (hearing aids, cochlear implants)
Vision services	• Ophthalmologic services to identify vision impairment • Use of surgery, glasses, magnifiers, binoculars, telescopes or tinted lenses as needed, as well as other adaptive equipment • Receipt of services from a teacher of the visually impaired • Services to address orientation and mobility-related issues • Orthoptic eye exercises (orthoptics) to improve binocular function • Of note, behavioral vision therapy is considered to be scientifically unproven and is not recommended by the AAP or AAPOS
Nutrition/dietician	• Ameliorate malnutrition and failure to thrive • Address overeating or excessive weight/obesity-related issues
Behavioral interventions	• Extinguish or reduce challenging behaviors through the use of positive behavioral strategies (token economies, positive praise), applied behavioral analysis techniques, functional behavioral analysis/assessment with the development of a formal behavioral intervention plan, and so forth • Address specific comorbid conditions (attention deficit hyperactivity disorder, autism spectrum disorder, anxiety, depression, sleep problems, irritability/aggression) and rule out contributing medical problems (constipation, dental pain, headache, fatigue, etc.) • Accommodations including sensory strategies, frequent breaks, shorter work periods, clearly laid out expectations, help with organization, etc. • Improve behavior-related functioning, performance, and participation in everyday activities • Inclusion of parent/caregiver training to maintain consistency in behavioral approaches
Psychotherapy	• Requires professionals with specialized skills and an understanding of psychological, biological, and social-cultural dimensions of IDD • Involves assessment of needs, administration of modified psychotherapy services, setting up positive behavior supports and effective environments, addressing cooccurring medical conditions, and management of psychopharmacology • May include dialectical behavior therapy, behavioral relaxation training, positive psychology approaches, and so forth

(*Continued*)

TABLE 24.3 (Continued)

Support	Goals of the intervention and other details

Special education services	• Provision of specialized instruction, or instruction that is specially designed to meet the unique needs of a child with IDD
	• Must be provided to students with disabilities in the least restrictive environment to ensure that children with disabilities are educated with nondisabled children, to the maximum extent appropriate. Removal should occur only if the nature or severity of the disability is such that education in regular classes with the use of supplementary aids and services cannot be achieved satisfactorily
	• Creation of an individualized curriculum that may be different from that of same-age, nondisabled peers with adaptations or modifications
	  ◦ Modifications refer to changes in what is being taught to or expected from the student
	  ◦ Accommodations refer to changes that helps a student overcome or work around the disability
	  ◦ Modifications and accommodations may be made in relation to scheduling, setting, materials, instruction, and student response
	• Use of educational strategies that tend to be more effective in IDD, including:
	  ◦ Breaking down learning tasks into small steps
	  ◦ Using a hands-on kinesthetic teaching approach where information is presented in a concrete and observable fashion
	  ◦ Using visual aids including charts, pictures, social stories, schedules, and graphs to present material and help students understand what is expected
	  ◦ Providing direct and immediate feedback to help form cause-effect connections

Vocational training	• Enhance life skills, including personal, financial, and household management • Foster personal social skills, including communication, self-awareness, and interpersonal skills, as well as understanding personal responsibilities, anxiety, and anger management • Increase employability skills, which includes specific job skills training • Develop digital skills, such as the basics of computer operations and the Internet, video and photo editing, etc. • Encourage involvement in community integration interventions to improve friendship development and inclusion in recreation with typical peers • Work toward a goal of competitive, integrated employment and community inclusion where possible • Provision of a job coach to support employment

Recreational therapies and other options	• Therapies "disguised as fun"
	  ◦ Therapeutic horseback riding (equine/hippotherapy)
	  ◦ Aquatic therapy
	  ◦ Music therapy
	  ◦ Art therapy
	  ◦ Dance therapy
	• Reduce isolation and increase community participation through formal and informal leisure and recreational activities
	  ◦ Church groups and communities of faith
	  ◦ Boy and Girl Scouts programs
	  ◦ Creative opportunities including choir and instrumental lessons, art lessons, and cooking classes
	  ◦ Sports programs and Special Olympics
	  ◦ Involvement in Best Buddies and other one-to-one friendship programs
	  ◦ Dances, opportunities for dating, and provision of education on reproductive health and sexual behavior, boundaries, and navigation of romantic relationships
	  ◦ Community-based programming and social groups
	  ◦ Volunteering opportunities (animal rescue, homeless shelter, park cleanup, etc.)

setting that must be tailored to the individual with progress that is measured over time. Frequent re-evaluations should occur to ensure that time, money, and resources dedicated toward these therapies prove to be of continued benefit to the individual.

Public school federal law

The early management of IDD hinges upon education services (See chapter: Legislative Mandates). In the United States, the IDEA, Part B, mandates a free and appropriate public education (FAPE) in the least restrictive environment

(LRE) for all school-aged children (3 through 21 years of age). Additionally, Part C mandates early intervention services from birth through 36 months of age to be delivered in a natural environment (usually home) for all qualifying infants and toddlers. Special education settings range from less restrictive (general education classroom with modification and support) to more restrictive (therapeutic schools), and while the setting should be tailored to the individual's needs, IDEA mandates placement for all students in the least restritive setting possible. [See chapter: Classroom].

Behavior

Although mental illness is generally of biologic origin and responds to medication, behavioral disorders may result from a mismatch between the child's abilities and the demands of the situation, psychosocial challenges, and family or parenting difficulties. These behaviors may represent attempts by the child to communicate, gain attention, gain access to desired tangibles, escape certain demands, exert control, or avoid frustration. Determining the antecedents, functions, and consequences of behavior can help in developing an effective behavior intervention plan. The most formal iteration of this is a functional behavioral assessment carried out by a well-trained behavior analyst. In assessing challenging behavior, one must also consider whether it is inappropriate for the child's *mental or developmental age*, rather than the child's *chronological age*. However, if the behavior is dangerous to the individual or the external world, intervention is required regardless of whether it is "appropriate" for developmental age. When intervention is needed, an environmental change, such as a more appropriate classroom setting, may improve certain behavior problems. Behavior management techniques and parent training are invaluable. Finally, psychopharmacologic agents may be appropriate in certain situations, such as aggression or self-injurious behaviors with high levels of intensity and frequency. [See chapters: Functional and Behavioral Assessment and Distruptive and self-injurious behaviors].

Medication management

There are currently no medications that are effective in improving the core symptoms of IDD. There are several precision-based agents being researched in specific disorders with known biologic mechanisms (e.g., mTOR inhibitors may augment defective neurodevelopmental signaling pathways in tuberous sclerosis and disorders associated with PTEN gene mutations), with the hope for future pharmacologic options that could alter the natural course of cognitive development in individuals with these disorders, perhaps limiting the severity of IDD. Currently, medications are most commonly used in the management of associated behavioral and psychiatric disorders that do not respond to initial behavioral approaches and environmental manipulation. Medications used among individuals with IDD generally target specific associated symptom complexes, including inattention, impulsivity, and hyperactivity (stimulant medications and alpha-agonists), self-injurious behavior and aggression (antipsychotics), anxiety, obsessive-compulsive disorder, and depression (selective serotonin reuptake inhibitors), and sleep-related dysfunction (melatonin, alpha-agonists, gabapentin, trazodone, etc.). Generally, medications should be started at low dose and increased slowly with frequent assessments of whether benefits outweigh any adverse effects and whether medication is still required for symptom management. [See chapter: Principles of pharmacotherapy].

Community supports

Beyond education-based services, families of children with IDD often are eligible for federal or state-provided social services. All states in the US offer developmental disabilities programs that provide home- and community-based services to eligible children and adults, potentially including in-home supports, care coordination services, residential living arrangements, behavior management services, vocational and employment support programs, and additional therapeutic options. Other supported services can include social skills training, legal and advocacy services, adult day-programs, and public transportation. A variety of Medicaid waiver programs are also offered for children with disabilities within each state. Children with IDD who live in low-income status households can qualify to receive supplemental security income (SSI), but many of these families are unaware of their eligibility and do not receive the benefits to which they are eligible. [See chapter: Community].

Transition to adulthood

Transition to adulthood in adolescents with IDD can be stressful for both the individual and the family, regardless of disabilities. A successful transition correlates with later improved quality of life but in the process, families tend to find that policies, systems, and services are more fragmented, less readily available, and more difficult to navigate in the adult world. Several domains of transition must be addressed, such as education and employment, health and living, finances and independence, and social and community life. Specific issues to manage include transitioning to an adult healthcare provider, enrolling in adult-based insurance plans, determining the need for decision-making assistance such as supported decision making, guardianship, or medical power of attorney, and securing government benefits after aging out of youth-based programs. State programs may also provide varying degrees of help with vocational planning and skill development as well as options for supportive housing, particularly in the context of a young adult who is unable to provide basic activities of daily living. [See chapter: Transition and Aging].

Summary

Intellectual disabilities are defined by significant limitations in both adaptive and intellectual functioning with varying levels of severity and may occur due to a multitude of different causes. While there are no medically based treatments that currently improve the core features of IDD broadly, therapies, educational services, and individualized supports can help to maximize competencies and limit challenges for the individual with IDD. Additionally, individuals with IDD benefit from appropriate recognition and medical management of cooccurring conditions through the lifespan, as these conditions often contribute more to limitations in function than the core features of IDD themselves. In conclusion, people with IDD can and should live dignified, productive, purposeful, satisfying, and connected lives within their own communities when they and their families are provided with appropriate and adequate supports and services. Management approaches should be person-centered and strengths-based, should discourage infantilization while promoting autonomy, and should emphasize inclusive practices throughout the lifespan.

References

[1] World Health Organization. Mental retardation. In: International Statistical Classification of Diseases and Related Health Problems, 10th revision Geneva: WHO; 2011.

[2] American Psychiatric Association. Diagnostic and statistical manual of mental disorders. 5th ed. Arlington, VA: American Psychiatric Publishing; 2013.

[3] Schalock RL, Luckasson R, Tassé MJ. Intellectual Disability: Definition, Diagnosis, Classification, and Systems of Supports. 12th Edition Washington, DC: American Association on Intellectual and Developmental Disabilities; 2021.

[4] Maulik PK, Mascarenhas MN, Mathers CD, et al. Prevalence of intellectual disability: a meta-analysis of population-based studies. Res Dev Disabil 2011;32(2):419–36.

[5] Manickam K, McClain MR, Demmer LA, Biswas S, Kearney HM, Malinowski J, et al. ACMG Board of Directors Exome and genome sequencing for pediatric patients with congenital anomalies or intellectual disability: an evidence-based clinical guideline of the American College of Medical Genetics and Genomics (ACMG). Genet Med 2021. Available from: https://doi.org/10.1038/s41436-021-01242-6. Epub ahead of print. PMID: 34211152.

[6] Hoytema van Konijnenburg EMM, Wortmann SB, Koelewijn MJ, Tseng LA, Houben R, Stöckler-Ipsiroglu S, et al. Treatable inherited metabolic disorders causing intellectual disability: 2021 review and digital app. Orphanet J Rare Dis 2021;16(1):170. Available from: https://doi.org/10.1186/s13023-021-01727-2; PMID: 33845862; PMCID: PMC8042729.

[7] Dikow N, Moog U, Karch S, Sander A, Kilian S, Blank R, et al. What do parents expect from a genetic diagnosis of their child with intellectual disability? J Appl Res Intellect Disabil 2019;32:10.

[8] O'Neill ME, Shapiro BK. Developmental delay and intellectual disability. In: Kliegman RK, St Geme III JW, Blum NJ, Tasker RC, Wilson KM, Schuh AM, et al., editors. Nelson Textbook of Pediatrics. 22 ed. Philadelphia: Elsevier; 2025, p. 345–62.

[9] Chow AJ, Saad A, Al-Baldawi Z, Iverson R, Skidmore B, Jordan I, et al. Family-centered care interventions for children with chronic conditions: A scoping review. Health Expect 2024;27(1):e13897. Available from: https://doi.org/10.1111/hex.13897.

[10] Andrews J, Falkmer M, Girdler S. Community participation interventions for children and adolescents with a neurodevelopmental intellectual disability: a systematic review. Disabil Rehabil 2015;37:825.

[11] Taheri A, Perry A, Minnes P. Examining the social participation of children and adolescents with Intellectual Disabilities and Autism Spectrum Disorder in relation to peers. J Intellect Disabil Res 2016;60:435.

Chapter 25

Specific learning disabilities

Sheryl Lynn Frierson
Luke Waites Center for Dyslexia and Learning Disorders, Scottish Rite for Children, Dallas, TX, United States

Overview

An estimated 5%–15% of school children struggle with a developmental disorder that disrupts their education by impairing their ability to learn skills specific to the academic domains of reading, writing, and/or math [1]. Specific learning disability (SLD) is a term frequently used to describe these developmental academic disorders. Under US educational law (Individuals with Disabilities Education Act [IDEA], originally enacted in 1975 as the Education for All Handicapped Children Act [Public Law 94–142]), IDEA provides free and appropriate public education [FAPE] to students with disabilities. In the 1990 reauthorization, the name was changed to IDEA. There are 13 eligibility categories of disability that a student can be identified with to receive special education services (see all 13 categories at https://www.understood.org/articles/en/conditions-covered-under-idea), but SLD is by far the most commonly used category. In the 2019/2020 school year, more than one-third of students (5–21 years old) receiving services under IDEA were identified with SLD [2] which translates into 2.4 million students identified with SLD as opposed to 2.2 million identified with speech and language impairment and other health impairment *combined* [3].

SLD ranges in severity from mild (e.g., a single area of deficit affecting only one academic domain that may require an accommodation) to severe (e.g., learning deficits affecting all three domains that may require extensive special education support). Reported estimates of male to female ratio range from nearly equal (1.15:1) to strong male predominance (5:1). The strong male predominance is particularly related to more severe symptoms [4]. Variability is also related to which type of measure is used to define dyslexia, for example, decoding versus fluency [4]. There is no agreement about the role of referral bias inflating the male prevalence estimates, but evidence suggests that a slight male predominance may be attributed to neurological differences [5]. A more consistent characteristic of SLD is comorbidity, especially with attention deficit hyperactivity disorder (ADHD) [6,7].

SLD is a developmental disorder typically considered a problem for school age children. However, longitudinal research has documented that SLD is a lifelong condition with recent findings of precursor symptoms detectable on event related potentials (ERP) in newborns [8] and well-established evidence of persistent impact continuing into adulthood [9]. Currently, identification of SLD is delayed until after a child has attended formal school instruction, yet there are early warning signs of increased risk. Family history of parental SLD confers four times increased risk of a child being identified with SLD. Other early warning signs before school entry are the presence of conditions commonly comorbid with SLD, particularly ADHD, speech sound disorder, and language disorder [10]. These early warning signs should not be ignored because waiting for the student to fail academically is costly in terms of lost self-confidence and missed opportunity for intervention. There is evidence that refractory achievement gaps, shown to persist into high school, develop as early as first grade [11].

The diagnosis of SLD is clinical. While there have been research advances in understanding the genetics and neuroanatomy of SLD, there are no current genetic tests or neuroimaging techniques to identify individuals with SLD [12,13]. Since diagnosis depends on clinical judgment, a clinician relying on arbitrary standard score thresholds risks misdiagnosis or missing the diagnosis. A comprehensive evaluation includes history from multiple sources, educational records, and relevant standardized testing with careful interpretation accounting for external factors (e.g., school setting, socioeconomic status, and other social determinants of health), familial factors (e.g., family history), and internal assets and challenges of the student (e.g., oral language, cognitive abilities, and comorbid disorders).

Capute and Accardo's Neurodevelopmental Disabilities in Infancy and Childhood. DOI: https://doi.org/10.1016/B978-0-12-824060-1.00040-7

Intervention is the goal of establishing the diagnosis of SLD. Four major goals of intervention are instruction, accommodations, rebuilding the student's positive self-image, and addressing comorbidities. Outcomes are as individual as the presentations. Many individuals with SLD go on to pursue and accomplish their chosen educational and professional goals. However, across the whole population of individuals with SLD, there is increased risk of lower academic achievement, lower rate of graduating and pursuing postsecondary education, lower rates of employment, and lower lifetime earning power [14]. Therefore, the goal of diagnosing and intervening in SLD is to minimize individual risk for these adverse outcomes.

Terminology

The terminology used to define and describe learning disorders can be confusing. There is a plethora of similar but not synonymous terms, which is a product of the involvement of many stakeholders representing different professions in both academic and clinical settings, as well as parent and community organizations.

In the US, educators and medical professionals have their own distinct definitions of SLD and distinct guidelines for identification and treatment recommendations. As mentioned above, SLD is one of 13 eligibility categories of disability that can qualify a student to receive special education services. Notice that the educational (IDEA) definition in Box 25.1 is broader in scope than the DSM-5 TR definition in Box 25.2. This chapter aligns with the more restricted DSM-5 TR definition of SLD, which is specific to learning impairments manifesting as deficits in reading, writing, and math. In practice, as well as in this chapter, the two terms are often used interchangeably using the abbreviation SLD.

BOX 25.1 IDEA specific learning disability.

1. In general

The term "specific learning disability" means a disorder in one or more of the basic psychological processes involved in understanding or in using language, spoken or written, which disorder may manifest itself in the imperfect ability to listen, think, speak, read, write, spell, or do mathematical calculations.

2. Disorders included

Such term includes conditions like perceptual disabilities, brain injury, minimal brain dysfunction, dyslexia, and developmental aphasia.

3. Disorders not included

Such term does not include a learning problem that is primarily the result of visual, hearing, or motor disabilities, of intellectual disabilities, of emotional disturbance, or of environmental, cultural, or economic disadvantage.

Section 1401 (30)—Individuals with Disabilities Education Act.

BOX 25.2 DSM-5 specific learning disorder.

Difficulties learning and using academic skills, as indicated by the presence of at least one of the following symptoms that have persisted for at least 6 months, despite the provision of interventions that target those difficulties:

- Inaccurate or slow and effortful word reading (e.g., reads single words aloud incorrectly or slowly and hesitantly, frequently guesses words, has difficulty sounding out words). Difficulty understanding the meaning of what is read (e.g., may read text accurately but not understand the sequence, relationships, inferences, or deeper meanings of what is read).
- Difficulties with spelling (e.g., may add, omit, or substitute vowels or consonants).
- Difficulties with written expression (e.g., makes multiple grammatical or punctuation errors within sentences; employs poor paragraph organization; written expression of ideas lacks clarity).
- Difficulties mastering number sense, number facts, or calculation (e.g., has poor understanding of numbers, their magnitude, and relationships; counts on fingers to add single-digit numbers instead of recalling the math fact as peers do; gets lost in the midst of arithmetic computation and may switch procedures). Difficulties with mathematical reasoning (e.g., has severe difficulty applying mathematical concepts, facts, or procedures to solve quantitative problems).

The learning difficulties are not better accounted for by intellectual disabilities, uncorrected visual or auditory acuity, other mental or neurological disorders, psychosocial adversity, lack of proficiency in the language of academic instruction, or inadequate educational instruction [15].

The SLD framework used outside of US public schools comes from the American Psychiatric Association (APA) via the Diagnostic and Statistical Manual, currently in its fifth edition (DSM-5 TR). In the DSM-5 TR, SLD is categorized as a neurodevelopmental disorder along with other developmental disorders including intellectual developmental disability (IDD, often shortened to ID), communication disorders, motor disorders, autism spectrum disorder (ASD), and attention deficit hyperactivity disorder. In previous versions of the Diagnostic Statistical Manual (e.g., DSM-IV), deficits in each domain were described as distinct, that is, reading disorders, math disorders, and disorders of written expression. In DSM-5 TR, the APA has reframed SLD as a single disorder, with different possible manifestations impairing reading, writing, and math. This current definition of SLD in the DSM-5 TR aligns with a neurodevelopmental disabilities spectrum and continuum framework [16] that eschews the idea that categorical diagnoses are independent disorders but are instead different patterns and combinations of impairment all arising from developmental brain dysfunction [17]. While evolving knowledge in the field appropriately leads to changes in definitions, changes can cause confusion when current and past definitions are both in use.

Other terms that can be sources of confusion include learning disability, learning disorder, and language-based learning disorder. The terms learning disability and learning disorder are often abbreviated as LD which is sometimes used as short-hand for SLD. However, it is important to understand the perspective of the person using the term LD. For example, in the United States, the DSM-5 TR defines SLD as "specific," in part, because the learning deficits are not better explained by intellectual disability (ID). Clinically, ID is distinguished from SLD by its more significant impact on the development of adaptive function and personal independence [18]. In contrast, in the United Kingdom, the term LD refers to the more pervasive cognitive disorder, known in the United States as ID. In UK terminology, *learning difficulties* would be the title of this chapter.

Even when international differences are not at work, some experts in learning and development use the term LD to refer to a much broader range of neurodevelopmental disorders including ASD, ADHD, and language disorders [10,19]. However, there are other chapters in this book dedicated to each of those diagnoses so this chapter is limited to the DSM-5 TR definition of SLD.

Language-based learning disorder is terminology used to highlight that scholastic tasks are language-based tasks, with reading, writing, and mathematics all revolving around written language whether it is letters or numbers [20].

Finally, the widely used terms dyslexia, dysgraphia, and dyscalculia describe specific presentations of SLD and are often used as stand-alone diagnoses. Later in this chapter, a framework for understanding these three terms in the context of DSM-5 TR taxonomy is offered.

Multidimensional DSM-5 TR definition

The DSM-5 TR definition of SLD appears in Box 25.2 and so is not repeated here except to call attention to the fact that the definition is multidimensional. In addition to identifying the affected academic domain(s), the clinician uses specifiers to document the affected skills and assigns severity (mild, moderate, severe) based on the degree of support the student will need to accomplish routine educational tasks.

Although not part of the DSM taxonomy, it can be conceptually helpful to cluster the deficits within each domain into four developmental levels of skill: readiness, basic, fluency, and application (see Table 25.1). Readiness skills are the precursor skills and abilities that are typically acquired prior to entering formal education. Basic skills are the foundational scholastic skills in a domain. Fluency is evidence that basic skills have been mastered to a level of automaticity that facilitates application. Application level skills integrate several foundational skills with other abilities (e.g., self-regulation, working memory) to accomplish more complex learning tasks.

The following four brief case descriptions illustrate the use of DSM-5 TR specifiers.

CASE A

Fourth grader with excellent math skills but who struggled with early reading. In second grade, she was identified with dyslexia and began receiving explicit, structured, sequential, multisensory, phonologically based reading intervention. Her decoding has improved but she is still not a good speller. Her oral reading is very slow and she has trouble understanding what she reads.

Diagnosis: SLD with impairment in reading (basic reading, reading fluency, and reading comprehension) is the DSM-5 TR diagnosis after a comprehensive diagnostic evaluation.

TABLE 25.1 Skills commonly assessed as part of comprehensive evaluation for specific learning disability are categorized according to complexity.

Level	Reading	Writing	Mathematics
Readiness	Phonological processing[a] Phonemic awareness Rapid naming Phonemic memory Alphabet (rote sequence) Letter recognition	Text conventions (e.g., left-to-right in English) Writing first letter in own name	Counting (rote sequence) Counting objects (1-to-1 correspondence) Number recognition[b,c]
Basic skills	Phonics (letter-sound correspondence) *Word reading accuracy* Word recognition[b,c] Decoding[b,c]	Accurately writing own name[b,c] Handwriting (accurate letter formation) Alphabet writing[b] *Spelling accuracy*[b,c]	Writing numbers accurately *Number sense*[b,c] Basic calculations[b,c]
Automaticity	*Reading rate or fluency* Word reading fluency[d] Decoding fluency[d] Oral reading fluency[b,c,e]	Quick, accurate copying from a model. Accurate writing from dictation *Grammar and punctuation accuracy*[f]	*Memorization of arithmetic facts*[b,c] *Accurate or fluent calculation*[b,c]
Application	*Reading comprehension*[b,c,e,f]	*Clarity or organization of written expression*[b,c,f,g]	*Accurate math reasoning*[b,c]

Readiness level skills can be assessed in preschool. Basic skills begin to be taught in early elementary school. Automaticity level is the transitional level between basic and application skills. Application level skills require integrating basic skills with other domain-general skills, for example, working memory. Bolded, italicized skills are specifying terms in DSM-5 taxonomy. Examples of commonly used standardized measures to assess specific skills are cited here and briefly described in Table 25.2. If there is no citation, the skill is typically assessed using informal measures or by examiner observations during standardized measures.
[a]*CTOPP-2: Comprehensive Test of Phonological Processing second edition [21].*
[b]*WIAT-4: Wechsler Individual Achievement Test fourth edition [22].*
[c]*WJ-IV: Woodcock-Johnson Tests of Achievement fourth edition [23].*
[d]*TOWRE-2: Test of Word Reading Efficiency second edition [24].*
[e]*Gray Oral Reading Test fifth edition [25].*
[f]*Gates MacGinitie Reading Tests fourth edition [26].*
[g]*TOWL-4: Test of Written Language fourth edition [27].*

CASE B

Fourth grader with excellent math skills and precocious oral vocabulary. She had initial difficulty with decoding but memorized enough sight words to meet expectations and quickly memorized the short books assigned in kindergarten giving impression that she was reading well. Now in fourth grade, her memorization strategy can no longer mask her reading difficulties. Her oral reading is often inaccurate and dysfluent. She appears to guess at words and then has to re-read sections that do not make sense. Despite stumbling through the passage, she is able to tell you what happened in the story.

Diagnosis: SLD with impairment in reading (basic reading and reading fluency) is the DSM-5 TR diagnosis after a comprehensive diagnostic evaluation.

CASE C

Fourth grader who had no difficulty learning to read and always maintained average grades, until this year. She has a lot of unfinished classwork and spends hours completing homework. She struggles to understand math word problems and is falling behind in social studies because she has to re-read assigned text several times to get the main idea.

Diagnosis: SLD with impairment in reading (reading comprehension) is the DSM-5 TR diagnosis after a comprehensive diagnostic evaluation.

CASE D

Student D is a fifth-grade student who is able to read okay but is a poor speller. Writing tasks are frustrating. His written work is brief and with short choppy sentences, inconsistent capitalization and punctuation, simple vocabulary, and poor sequencing making it difficult to follow his train of thought. In mathematics, despite conceptually understanding math, he still has not memorized multiplication facts forcing him to use serial addition and counting on his fingers to complete long division problems.

Diagnosis: SLD with impairment in written expression (spelling accuracy, capitalization and punctuation, and clarity or organization of written expression) and impairment in mathematics (memorization of arithmetic facts and accurate or fluent calculation) is the DSM-5 TR diagnosis after a comprehensive diagnostic evaluation.

These cases highlight the variability of SLD. The first three all have SLD with impairment in reading but with very different presentations. Case A is a classic profile of a student with dyslexia. Case B may also have dyslexia but is still able to extract meaning from text despite word reading errors, probably by relying on prior knowledge. Case C does not have dyslexia, but has hit a fourth grade slump due to an isolated, quantifiable deficit in reading comprehension that the literature describes as specific reading comprehension disorder (SRCD: [28,29]). The fourth case shows that SLD can manifest with impairment in more than one academic domain.

Dyslexia, dyscalculia, and dysgraphia

Dyslexia, dysgraphia, and dyscalculia are commonly used terms that are independent of DSM-5 TR taxonomy. Prevalence of each one is estimated somewhere between 4% and 9% [18] and each of them is highly comorbid with the other two [30–33]. These three diagnoses roughly approximate SLD with impairment in reading, writing, and mathematics, respectively. However, these are historical terms from the early 20th century that describe clinical syndromes, or groups of symptoms consistently observed to occur together. Symptoms within each syndrome may cross academic domains or may include skills that are not currently part of the SLD definition. This section describes these syndromes.

Dyslexia

Dyslexia is primarily a reading disorder and is recognized in DSM-5 TR as an alternative term to under the category of SLD with impairment in reading. It is characterized by deficits in recognition of common (high frequency) words, decoding (sounding out words), and/or reading fluently, which encompasses accuracy, pace, and prosody. However, it is also characterized by poor spelling which is typically considered a writing task, not a reading task. The key to understanding the reading–spelling connection occurs at the reading readiness skills of phonemic/phonological awareness (see Table 25.1).

Phonemic awareness is an auditory skill that is independent of written language. It is the ability to hear and manipulate individual sounds in spoken words. Current expert consensus is that phonemic awareness is a core deficit of dyslexia [34,35]. In an alphabetic language, spoken words are connected to written words through the correspondence between spoken sounds and written letters. Instruction that teaches the sound-letter correspondence is phonics. The application of phonics is decoding, also referred to as the "sounding out" process converting written words into spoken words. Spelling also requires application of phonics knowledge, but in the opposite direction; namely, it requires hearing the sounds in spoken words and converting that information into written letters. Dyslexia-related errors, whether in reading or spelling, often arise from sounds that have been omitted, inserted, misplaced, or associated with the incorrect written letter or letters. Converting the syndrome of dyslexia into DSM-5 TR taxonomy requires identifying SLD in both reading (basic reading) and written expression (spelling) and may require additional specifiers for reading fluency, reading comprehension, and other aspects of written expression. Thus dyslexia highlights the blurry distinction between the academic domains of reading and writing that some experts do not identify as distinct domains [19].

Dysgraphia

The syndrome of dysgraphia is characterized by illegible and/or slow, laborious handwriting despite conventional instruction. The clinical impact of dysgraphia is difficulty producing writing accurately and efficiently enough to complete tasks in the educational setting including copying from the board, note taking, and writing a report. Given the close relationship between written language output (writing) and written language input (reading), dyslexia and dysgraphia are closely related.

Compared to dyslexia, dysgraphia is not as well studied or understood and, to date, there is no clear consensus on dysgraphia's precise definition. Some investigators use the term dysgraphia to refer to the negative impact of developmental coordination disorder (DCD) on the motor aspects of handwriting [36]. Other research characterizes dysgraphia as primarily a spelling disorder [37,38]. A third approach to defining dysgraphia encompasses both handwriting and spelling and refers to this combination of skills as transcription [39]. This line of research has revealed the importance of cognitive abilities in the motor planning and execution of handwriting.

Orthography refers to conventions that govern how a spoken language is written. Encoding that information into memory and retrieving it on demand is referred to as orthographic memory. The "*orthographic loop*" refers to the connection between cognition (orthographic memory) and writing (accurate motor planning and execution). Importantly, orthography conventions occur at different levels of language. Berninger et al. [40] describe three levels of orthographic deficits corresponding to different presentations of SLD [40]. The first is the *text level* encompassing orthographic convention rules for organizing words into sentences, paragraphs, stories, essays, etc. These conventions include grammar, syntax, punctuation, and capitalization rules. In DSM-5 TR terminology, deficits in these skills would be described as SLD impairing written expression.

The second is the *word level* that would correspond to orthographic conventions that govern how letters are combined into words including recognizing and recalling from memory both regularly and irregularly spelled "sight" words. The authors propose that deficits at the word level correspond to dyslexia.

The third level is the "subword" level meaning individual letters and how they are placed on the page. Subword skills include recognizing letters, remembering how letters should look without referring to a model, and developing automatic motor memory of how to form each letter. Subtle features are important including letter orientation, relative size, accurate spacing, and proximity to the lines and margins of the page. Students with dysgraphia show delays in mastering these skills despite receiving the same instruction as typically developing writers. It is common to see persistent reversals beyond the developmentally typical age of achieving this milestone around 7 or 8 years of age. These students may not remember how to form letters that they have already been taught and often benefit from having a model from which to copy. This general lack of automaticity can result in letters that appear "drawn" instead of written. Legibility and efficiency are compromised.

Researchers describe the sequential steps that connect orthographic memory to the sensorimotor actions that accomplish the writing as the "orthographic loop." Thus breakdown anywhere on the loop, including orthographic recall, can lead to poor handwriting output even in the absence of motor deficits [39,40]. This framework for understanding dysgraphia identifies a distinct clinical population of individuals presenting with writing deficits that cannot be fully accounted for by motor impairments or poor spelling but stem from deficits in storing accurate subword orthographic information.

This understanding of dysgraphia helps distinguish it from dyslexia. Both syndromes have a negative impact on spelling but the nature of the spelling errors differs. Dyslexia's phonological awareness deficits lead to sound-based spelling errors, for example, omitting or inserting sounds or choosing the incorrect letter for a sound. Dysgraphia's orthographic deficits lead to phonetically plausible misspellings, for example, muther, or unpronounceable strings of letters, for example, dgkt. However, the orthographic and phonologic spelling errors do not segregate discretely so there can be significant overlap in symptoms [41].

Dyscalculia

The World Health Organization (2010) describes dyscalculia as a specific impairment in arithmetical skills which is not solely explicable on the basis of general mental retardation or of grossly inadequate schooling. It is estimated to affect about 4%–7% of the population depending on the country of study and criteria of diagnosis [42]. Dyscalculia is acknowledged in DSM-5 TR as an alternative term for SLD with impairment in mathematics. Clinically, math disorders present with poor memory for numbers, difficulty counting, telling time, recognizing sequences, and recalling math facts and procedures. However, DSM-5 TR does not explicitly include visual-spatial symptoms that some consider part of a broader syndrome of dyscalculia, for example, mental rotation and visualization and sequential memory, though these may be implied under math reasoning. Other symptoms identified in broad characterization of dyscalculia might not fit into the DSM-5 TR framework, for example, directional confusion and difficulty with wayfinding and navigating through unfamiliar spaces (http://www.dyscalculia.org).

Number sense or numeracy is the innate ability to understand how numbers represent quantities and how the quantities relate to one another. Numeracy has been proposed as a core deficit underlying dyscalculia akin to phonological processing being a core deficit in dyslexia [43] although this is not a universally held opinion [44]. However, counting skills are also early skills; counting skill in kindergarten has been correlated to math achievement in second grade [45]. Working memory, especially visual-spatial working memory, has also been associated with dyscalculia [46].

Dyscalculia is encompassed in the term language-based learning disorder. Math disorders require students to "read" and "write" number, assign meaning to math symbols and to solve math problems written as combinations of numbers and symbols [47]. Furthermore, the ability to solve math word problems requires reading comprehension, including understanding math vocabulary and the language structure of math word problems. Recent research suggests that explicitly teaching a strategic approach to deciphering the language structure of math word problems can improve math performance more than practice on basic math skills alone [48]. Indeed, there is evidence that the ability to solve math word problems is a stronger predictor of early algebraic reasoning than is the ability to do calculations [49].

Compared to other manifestations of SLD, both dyscalculia and SLD mathematics have a unique association with the internalizing symptoms of anxiety [50]. For people with dyscalculia, performing number-related tasks can be associated with anxiety and distress. While all reading and written expression manifestations of SLD are associated with increased prevalence of internalizing symptoms, including anxiety [51], the literature describes a math-specific form of anxiety. Math anxiety does not necessarily correlate with poor mathematics achievement and so cannot be explained solely as an effective response to poor achievement. It is hypothesized that math anxiety is rooted in cultural reinforcement that math is difficult and, in particular, that girls are not good at mathematics [50]. This is the opposite of a growth mindset that normalizes making errors as part of the process of learning, rather than as evidence of a lack of capability. The research suggests a preventive approach that includes adults, especially adult women, not reinforcing negative stereotypes when referring to themselves. Beyond changing how we talk about math, it will be important for educators to set goals that focus less on the outcome of correct answers in math and more on the process of exploration and learning that will encourage students rather than increase anxiety.

Comorbidity

In SLD, there is a 50-50 chance of impairment in more than one domain [31]. Using syndromic terms, having dyscalculia more than doubles the chances of having dyslexia [52] and students with writing problems are reported to also have dyslexia 30%–70% of the time [30,33].

The risk of SLD comorbidity with other neurodevelopmental disabilities is also high with the most common comorbid diagnoses being ADHD, speech and language disorders, and behavioral-emotional symptoms. SLD is comorbid with ADHD 30%–50% of the time [51] and the risk of ADHD is even higher when there is SLD in both math and reading [53]. Speech and language disorders, whether resolved in the past or persistent at the time of evaluation are commonly comorbid with SLD. A recent estimate identified comorbidity 20%–46% [54]. Both ADHD and speech/language disorders generally present prior to school entry and so can be useful early markers of a student who is at increased risk of SLD [34,55]. Anxiety and mood disorders overlap with SLD approximately 37% of the time [56].

Etiology

Neurocognitive

For dyslexia, the current consensus is that there is a core deficit in phonemic awareness which is in the ability to recognize, isolate, and manipulate the individual sounds that make up spoken words [34]. Phonemic awareness and letter recognition are prerequisites for phonics, learning to match spoken sounds in English to individual letters. Rapid naming of digits and letters is considered a phonological processing assessing fluency (speed with accuracy) of retrieving this phonological information memory. While phonemic awareness is one of the strongest early predictors of later reading in English [57], rapid naming has been identified as a strong predictor of reading deficits in many different languages [58]. One theory of dyslexia in English, the double deficit hypothesis, proposes that deficits in either phonemic awareness or rapid naming can result in the syndrome of dyslexia, but that the most impaired individuals may have deficits in both aspects of prereading language skill [59].

Phonological skills are also implicated in dysgraphia and dyscalculia. Distinctive neurocognitive findings in dyscalculia are referred to as numeracy, including difficulty with the innate awareness of quantity and size as well as basic understanding that numbers are sequential and represent quantities. Distinctive neurocognitive findings in dysgraphia include orthographic memory deficits and, for some students, fine motor deficits.

Genetics

SLD is highly heritable. An individual with a family history of SLD has quadruple the risk of having SLD [7]. Despite employing multiple genetic approaches (e.g., classic linkage, large pedigrees, chromosomal aberrations, genetic associations with comorbid diagnoses, and large-scale genome-wide association studies), no single "SLD" gene has been identified. Instead, there are a plethora of candidate genes that account for only a small amount of variance in phenotype. The current consensus is that SLD is polygenic and the etiology is multifactorial characterized by significant gene by environment interactions [12].

Neurobiology

SLD is not characterized by gross brain abnormalities on clinical neuroimaging studies and, despite rapid advances in neuroimaging, there are currently no clinical brain imaging methods that can reliably determine whether or not an individual has SLD. However, research methods of imaging (see Box 25.3 for some differences from clinical imaging) have identified structural and functional differences in individuals with various presentations of SLD. Heterogeneity of neurobiological findings is expected because SLD is heterogeneous. Yet even when researchers narrow the focus to one set of characteristics of SLD, results across studies remain variable due to, among other things, differences in imaging procedures, data analysis techniques and approaches, and differences in the sample size and criteria for inclusion. Nevertheless, research imaging has advanced our understanding of different manifestations of SLD and has recently detected evidence of neurobiological differences in infants who later present with SLD supporting the hypothesis that there are some congenital developmental differences [8].

Prior to the emergence of neuroimaging technology, indirect approaches gave hints about the neurobiological underpinnings of SLD. Structurally, autopsy data suggested an association between SLD and small areas of cortical heterotopia supporting a postulate of errant neuronal migration. Functionally, studies of individuals who lost skills in reading and writing after injuries to left hemisphere perisylvian language-related regions raised the question of whether the same distributed network of brain regions would be implicated in developmental learning disorders. This leads to intense interest in these cortical regions, primarily the inferior frontal gyrus (IFG), a temporo-parietal region, and an occipito-temporal region, and the white matter connections between them. Once neuroimaging became accessible, there was rapid development of a wide range of techniques to assess structural and functional correlates of SLD. The early hypotheses remain strong drivers of research and reflect some of the most consistent findings in the field.

SLD impairing reading has been the most widely researched. A consistent finding by multiple researchers is that SLD impairing reading is characterized by underactivation of left-hemisphere temporo-parietal and occipito-temporal

BOX 25.3 Challenges in interpreting functional neuroimaging of specific learning disability (SLD).

- Proxy measure: Functional magnetic resonance imaging (fMRI) is based on blood oxygenation-level dependent (BOLD) signal changes in response to a short stimulus. BOLD response in a region is a proxy for changes in local neuronal activity.
- Postacquisition processing: Raw imaging data undergoes extensive preprocessing to prepare it for statistical analysis. This includes aligning the map of activity (BOLD signal) with an anatomical map of the brain so that activity can be attributed to anatomical regions for interpretation.
- Cognitive subtraction: Since fMRI measures changes in BOLD signal, results reported to be associated with a task, for example, reading words during scanning, are actually contrast images subtracting the BOLD signal of interest from the BOLD signal of a control stimulus, for example, resting in the scanner. The assumption is that imaging signal that is not accounted for by the control stimulus can be attributed to the task of interest.
- Group level data: Results represent contrasts between pooled groups of controls and pooled groups of individuals displaying a selected characteristic of SLD. The assumption is that any other (non-SLD) neurobiological heterogeneity that might affect BOLD signal, for example, medications or comorbid disorders, will be similar enough in both groups that signal that is not accounted for by the control group can be attributed to the presence of SLD. Data are pooled to obtain sufficient data for statistical modeling.
- Regions of interest: Imaging studies can inspect data across the whole brain but this increases the risk of false positive results due to multiple comparisons. One way to address this is to limit data analysis to selected regions of interest; however, this approach increases risk of bias in data selection, which may lead to confirming expected findings while overlooking unexpected or unknown ones.

regions. This finding is not limited to English or even to alphabetic languages; there seems to be something universal about these regions that is associated with reading deficits [60]. These regions make sense with what is known about the psychological underpinnings of dyslexia. The temporo-parietal region has been linked to phonological processing and the occipito-temporal region has been linked to rapid association of "sight words" to meaning ([34] for a helpful figure). Another commonly reported finding is overactivation in the right inferior frontal gyrus that may represent a compensatory response and thus be secondary to SLD rather than part of the etiology. There has also been much research interest in the major white matter bundles that connect these two posterior regions with the inferior frontal gyrus which is linked to speech production. These investigations have identified some areas of group difference in structure of the white matter pathways, but there is limited consistency across studies. Recently, the literature has reflected a trend toward considering multiple systems of connectedness between noncontiguous brain regions. For example, a comparison of dyslexia and dysgraphia found differences in functional and structural connectivity within the left hemisphere language network, particularly in the inferior parietal cortex [61–63].

Math uniquely features number sense as a core cognitive skill and distinctive neurobiological correlates of math SLD. In particular, measures of numeracy are associated with function and structure of the right intraparietal sulcus (IPS) and dorsal aspects of the posterior parietal cortex (PPC). Overall, the distributed network associated with SLD math is more bilateral and more dorsal than the dyslexia network [13,64,65]. There are also significant areas of overlap between these two networks in the occipito-temporal cortex (fusiform gyrus), prefrontal regions (bilateral IFG), and the left hemisphere language network that is engaged when subject are retrieving basic math facts [13].

Identification

As per the DSM-5 TR, identification of SLD requires demonstrating three core features: (1) a persistent pattern of difficulty learning key academic skills despite receiving appropriate instruction, (2) symptoms that become readily apparent at some point during schooling, and (3) evidence of performance that is substantially and quantifiably below average (e.g., negative impact on school or work performance). The latter refers to a student's performance on psychometrically sound measures of academic achievement. Most often, these measures are standardized so that a standard score (SS) of 100 is average with a standard deviation (SD) of 15 points; thus, one SD below average would be SS = 85. DSM-5 TR requires scores within an academic domain at least 1.5 SD below the mean (SS = 78). This definition indicates a score in the lowest seventh percentile of scores on the normal curve. Defining SLD in this way raises the question of whether it is a distinct disorder or just represents the low end of normally distributed scores. However, within that left-hand tail of the normal distribution, there are individuals who demonstrate gaps in their academic learning for myriad reasons; for example, students who meet criteria for the more global disorder of ID or students who may have been distracted and missed important content during instruction (e.g., due to ADHD or life stressors like food insecurity). So, SLD does not encompass nor is it limited to students with a SS <85 in reading, writing, or math. Students with SLD show an unexpected pattern of persistent and specific difficulty learning of some aspect(s) of one or more of the key academic skills of reading, writing, or math. The DSM-5 TR describes four qualities that make SLD "specific" or "unexpected."

First, the student does not have a more global developmental disorder and specifically does not meet diagnostic criteria for ID. In some settings, examiners attempt to operationalize this form of unexpectedness by using discrepancy scores that define impairment by a certain number of standard score points discrepancy between achievement scores to IQ or oral language scores but, *used in isolation*, this has proven problematic for identification, for example, a student with IQ = 130 and average academic achievement would meet criteria but not a student with IQ = 90 despite below average academic achievement scores. Similar difficulties are encountered with any rigid discrepancy formula, for example, cross battery assessments compare scores across measures of underlying abilities and academic achievement to document specific patterns of discrepancy.

Second, the learning difficulty is unexpected because it is not secondary to a neurological disorder (e.g., trauma or stroke). Third, SLD is unique in that it may manifest as isolated deficits in just one domain (see cases above). Fourth, the student must have received adequate routine instruction without, for example, excessive absenteeism, and still make poor progress. The corollary to this is that continuing with routine instruction, even augmented with additional review and practice, will result in the student continuing to fall further behind. Thus SLD is specific because affected students require, not more of the same, but specific, different instruction. This characteristic is the basis for recent approaches to identification in schools called *response to intervention.* Namely, students are monitored for growth in reading. Those who make poor progress move to a higher "tier" of intervention. Continued poor response indicates that the student probably has SLD and requires more comprehensive evaluation.

Evaluation

An evaluation for SLD is called a psychological evaluation or a psycho-educational evaluation. To meet DSM-5 TR criteria, the evaluation documents quantifiable deficits in academic achievement using psychometrically sound, standardized measures. Evaluation frequently tests cognition, language and fine motor skills, academic readiness abilities (see Table 25.1). It also commonly includes behavioral rating scales to identify commonly comorbid conditions, including ADHD. However, interpreting test scores in isolation can be misleading. For example, below average scores can occur for reasons other than SLD and average scores can mask error patterns characteristic of SLD.

Therefore a comprehensive evaluation interprets test scores and rating scales in a larger context that includes extensive past history (medical, educational, social, and family history). The clinician will document observations of the student's behavior and approach to tasks during the assessment that factor into interpretation of scores.

A *medically* comprehensive evaluation further contributes information about coexisting neurological or genetic diagnoses, would consider the potential impact of chronic medical conditions (e.g., cardiac, rheumatological, or endocrine disorders), and identification of any previously undetected medical conditions that may be causing or exacerbating symptoms (e.g., thyroiditis).

Identification of SLD using DSM-5 TR includes documenting quantifiable deficits in academic achievement using standardized tests; however, tests scores alone are not sufficient to make the diagnosis. First, quantitative test scores do not always capture highly relevant qualitative aspects of the assessment. For example, average test scores may obscure qualitative evidence of SLD, for example, characteristic error patterns and observations of the examinee's effort. On the other hand, below average test scores may not reflect the student's best abilities due to a variety of factors, for example, fatigue, distractibility, or a less than ideal testing environment. A clinician skilled in administration and interpretation of standardized measures is necessary. Second, test scores at one point in time cannot establish whether any deficits are persistent over time, to what extent the deficits interfere with day-to-day educational functioning, whether the student has received appropriate instruction, or whether there are any comorbid diagnoses or predisposing conditions, for example, premature birth. Thus identification of SLD requires a comprehensive evaluation and synthesis of the test scores with other data obtained from the student, parents, teachers, school records, and any prior evaluations. This is consistent with the DSM-5 TR call for the clinician to synthesize multiple streams of data. See Box 25.4 for a summary of components of a comprehensive diagnostic evaluation. Examples of how the selected skills are commonly measured on various standardized tests are described in Table 25.2.

BOX 25.4 Components of a comprehensive diagnostic evaluation.

Comprehensive student history

Birth history

Developmental history: Past or present delays in language or motor milestones and developmental interventions (even if there is no longer a perceived problem)

Medical history: Including all medications and dietary supplements

Social history and context: Family unit (parents/guardians/caregivers, siblings, others in the home), success with friends and age-appropriate activities

Family history: With an emphasis on any neurodevelopmental or genetic diagnoses

Behavioral history: Past or present behavioral concerns or discipline problems and any past or present counseling or parenting training

Educational history: Prior schools, grade retention or delays in starting formal education ("red-shirting"), class size and setting (public, private, home school), any disruptions to instruction (e.g., family moves or infectious pandemic restricting in-person instruction), past or present tutoring, history of access to supplemental interventions in school

Questionnaires

Parent/Guardian observations: About each academic domain as well as executive function (e.g., working memory, organization, task completion, self-monitoring), and internalizing and externalizing behaviors, standardized questionnaires

Teacher and tutor observations: About each academic domain, executive function (e.g., working memory, organization, task completion, self-monitoring), and internalizing and externalizing behaviors, standardized questionnaires

School documentation

School and state benchmark testing results

Report cards

Prior school evaluations (sometimes referred to as full independent evaluation or FIE)

Prior school services including response to intervention supplemental instruction, other direct interventions

Prior formal documentation including Section 504 Plan or individualized education plan (IEP)

TABLE 25.2 Examples of how the selected skills are commonly measured on various standardized tests.

Brief descriptions of common formats of standardized tests for select skills	
Reading	
Phonological processing	• Phonemic awareness assesses isolating sounds (beginning, ending, and middle sounds) in spoken words, blending sounds to form spoken words, and removing sounds from spoken words to form new words. • Rapid naming of letters and digits or objects and colors as appropriate for age. • Phonological memory is repeating sequences of digits and blending sequences of letter sounds into words.
Word recognition	Accuracy of reading a list of common sight words.
Decoding	Accuracy of "sounding out" a list of nonsense words.
Word reading fluency	Timed reading of a list of common sight words.
Decoding fluency	Timed "sounding out" a list of nonsense words.
Oral reading fluency	Examinee reads passages aloud and the examiner scores based on accuracy, speed, and ability to answer verbally presented comprehension questions at the end.
Reading comprehension	• Tests vary in format. Formats include: • Matching pictures with words or phrases. • Supplying a missing word in a sentence (i.e., cloze format). • Silently reading short passages and answering questions read by the examiner. • Silently reading a series of short passages and responding to written multiple choice questions after each passage without examiner prompting.
Writing	
Visual-motor coordination	Shape-copying task requires integrating visual perceptual and fine motor (finger and hand movement) abilities. Optional subtests include measures of visual perception and fine motor coordination.
Alphabet writing	Accuracy of the sequence and speed of completing the task. Accuracy of letter forms is observed.
Spelling	Accuracy of spelling from dictation. For young children, accuracy of writing own name and letters from dictation.
Written expression	• Sentence writing, especially for young children. • Narrative story writing based on a picture prompt. • Essay writing for older children.
Mathematics	
Number sense	Accuracy of number and symbol recognition, one-to-one correspondence (counting objects), relationships between numbers, and understanding of sequential patterns.
Math calculation	Worksheet sampling addition, subtraction, multiplication, division, fractions, decimals, algebra, and geometry as appropriate for age.
Math fact fluency	Timed addition, subtraction, multiplication.
Math reasoning	Math word problems read aloud by examiner. Some items have visual supports. The use of paper and pencil for computation is allowed.

The list is not intended to be exhaustive.

Illustrating the importance of this comprehensive approach, a group of researchers used a Bayesian statistical approach to estimate risk of SLD for certain populations. Assuming baseline prevalence of SLD at 5% and a 2:1 prevalence in males, researchers estimated that the baseline risk jumps to 24% for males with ADHD and up to 76% for males with ADHD and a family history of SLD. Adding the psychoeducational evaluation shot the risk above 90% for males with ADHD, a family history of SLD, and very low reading achievement scores [66]. Adding past educational data would be important for confirming a consistent pattern of low reading achievement, not just at a one-time evaluation.

In addition to these intrinsic characteristics that modify SLD, the clinician must consider other factors that appear intrinsic (e.g., motivation, engagement with learning, and behavioral regulation) but may be traced back to external factors ranging in severity from mild and modifiable (e.g., limited prior exposure to a classroom environment) to significantly impairing, for example, adverse childhood experiences [67,68].

Treatment

Core treatment for SLD consists of remedial instruction, educational accommodations, and access to circumvention strategies.

Remedial instruction addresses deficits in the affected academic domain, for example, math deficits require specific math intervention. Although, intervention addresses domain-specific deficits, every level of instruction also relies on and practices use of more generalized skills, for example, working memory and self-monitoring for errors, that are referred to as domain-general skills. These domain-general skills become increasingly important as basic skills in a domain (e.g., word reading) come together into more complex application level skills (e.g., reading comprehension).

Effective basic reading instruction addresses sight word recognition as well as the underlying core deficit of phonological awareness as a step to foster effective decoding to sound out words when reading [69] and encoding to translate sounds in words into spelling. A structured approach to spelling also involves learning situational rules for decoding and encoding words; for example, different syllable types affect how the sounds of the letters, for example, short vowel syllables versus silent-e syllables. Comprehensive reading interventions incorporate text-level instruction to build oral reading fluency and develop strategies for effective reading comprehension. The basics of letter formation and handwriting are also relevant to reading instruction since writing reinforces the skills needed for reading [39].

Intervention for dysgraphia would require additional emphasis on letter formation and handwriting fluency. Research has shown that handwriting fluency and spelling predict written expression [70]. Therefore dysgraphia intervention would include spelling instruction. The phonological encoding (sound spelling) may seem less important in isolated dysgraphia; however, letter-sound correspondence contains orthographic information [71] to which a student with dysgraphia might benefit from repeated exposure. Other orthographic aspects of spelling include situational spelling rules and understanding word structure (morphology), for example, suffixes, prefixes, stems, and root words. Other aspects of writing instruction would address sentence structure, grammar, capitalization and punctuation conventions, and organization of ideas into clear written expression.

Math interventions may need to address deficits in number sense, for example, using number lines to reinforce the sequential nature of numbers. Interventions directed to math operations improve math fact understanding but math reasoning skills may require more domain-general language comprehension strategies [72].

Effective remedial instruction differs from conventional instruction in approach and intensity. This approach is explicit, not assuming prior mastery of foundational skills, and sequential meaning that it builds on skills that have already been explicitly taught. For example, effective dyslexia intervention explicitly teaches early literacy skills, for example, phonological awareness, letter naming, and letter-sound correspondences. Then, building on that foundation, letter combinations are introduced including digraphs (e.g., /sh/ or /ee/), diphthongs (e.g., /oi/ or /ow/), and syllable patterns (e.g., vowel-consonant-vowel or silent-e). Effective remedial approaches are often described as multisensory or multimodal. This is because each new concept is taught and practiced using different modalities, for example, reading, writing, and speaking, to develop robust neural networks and consolidation into long-term memory. Modified intensity may increase the time spent on instruction, that is, longer or more frequent lessons. In most cases, it also means instruction is delivered in a smaller group setting (one to six students) by an educator with specialized training. These qualities allow content and pacing to be adjusted to meet different students' needs.

An important aspect of remedial instruction is that it teaches skills that can be generalized. For example, learning to decode gives the student a strategic way to approach new vocabulary words encountered when reading. This is not quite the same as "catching up" to classmates. Instead, it arms the student with tools for effective reading practice; tools that the typically developing readers gained access to without the need for such explicit instruction.

There is evidence that reading achievement improves with remedial intervention and that the improvement is associated with normalizing of functional neuroimaging patterns [73]. Expert opinion about when to begin remedial instruction for SLD has evolved. In the past, identification and treatment were deferred to third grade. However, longitudinal research showed that students with dyslexia had already fallen behind their classmates in reading achievement by first grade and that the gap persisted into ninth grade [11]. This suggests that intervention needs to begin in kindergarten or preschool to prevent the learning gap from developing. Other authors are proponents of explicit intervention with struggling readers early in kindergarten so that it may prevent a significant proportion of struggling readers from ever needing special education services [69].

Other treatments for SLD are designed to facilitate learning in all subjects, not just in the affected domain. First, educational accommodations are strategies intended to level the playing field so the student with SLD has access to the same content as classmates. For example, a student with dyslexia may be able to solve math word problems if they are read out loud. Second, circumvention strategies give the student an alternative way to accomplish a task, for example, a student with dysgraphia may be allowed to do an oral report instead of a written report. Technologies can be helpful when students need to operate more independently, for example, text-to-speech software can read a math problem to a student and speech-to-text (dictation) can capture a student's ideas in written form bypassing slow or illegible handwriting.

It is also important to address the frequently cooccurring symptoms that may include anxiety, depressed mood, negative self-image, inattentiveness, and disruptive behavior. For many students, the combination of remediation, accommodations, and bypass strategies is sufficient to alleviate cooccurring symptoms. However, for others, these symptoms signal comorbid disorders that need to be treated in order for the student to benefit from the SLD treatment.

Future directions

- Consensus definition of dysgraphia will allow research on prevalence, and response to intervention to proceed.
- Effectiveness of early literacy (prereading instruction) intervention to prevent first grade achievement gap in reading and to prevent need of special education services.
- Can genetics and neuroimaging contribute to risk assessment of individuals who are struggling academically? To predicting which students will respond to current interventions and who may need something different?

References

[1] Wright JH. Specific learning disorder. Diagnostic and statistical manual of mental disorders (DSM-5). 5th ed. American Psychiatric Association Publishing; 2013. p. 66–74.

[2] Office of Special Education Programs. OSEP fast facts: race and ethnicity of children with disabilities served under IDEA part B. U.S. Department of Education. 2021. Available from: https://sites.ed.gov/idea/osep-fast-facts-race-and-ethnicity-of-children-with-disabilities-served-under-idea-part-b/.

[3] Office of Special Education Programs. OSEP fast facts: school aged children 5 (in kindergarten) through 21 served under part B, of the IDEA. U. S. Department of Education. 2021. Available from: https://sites.ed.gov/idea/osep-fast-facts-school-aged-children-5-21-served-under-idea-part-b-21/.

[4] Quinn JM, Wagner RK. Gender differences in reading impairment and in the identification of impaired readers: results from a large-scale study of at-risk readers. J Learn Disabil 2015;48(4):433–45. Available from: https://doi.org/10.1177/0022219413508323.

[5] Arnett AB, Pennington BF, Peterson RL, Willcutt EG, DeFries JC, Olson RK. Explaining the sex difference in dyslexia. J Child Psychol Psychiatry Allied Discip 2017;58(6):719–27. Available from: https://doi.org/10.1111/jcpp.12691.

[6] Moll K, Landerl K, Snowling MJ, Schulte-Körne G. Understanding comorbidity of learning disorders: task-dependent estimates of prevalence. J Child Psychol Psychiatry Allied Discip 2019;60(3):286–94. Available from: https://doi.org/10.1111/jcpp.12965.

[7] Wagner RK, Zirps FA, Edwards AA, Wood SG, Joyner RE, Becker BJ, et al. The prevalence of dyslexia: a new approach to its estimation. J Learn Disabil 2020;53(5):354–65. Available from: https://doi.org/10.1177/0022219420920377.

[8] Lohvansuu K, Torppa M, Ahonen T, Eklund K, Hämäläinen JA, Leppänen PHT, et al. Unveiling the Mysteries of Dyslexia—Lessons Learned from the Prospective Jyväskylä Longitudinal Study of Dyslexia. Brain Sciences 2021;11(4):427. Available from: https://doi.org/10.3390/brainsci11040427.

[9] Shaywitz SE, Fletcher JM, Holahan JM, Shneider AE, Marchione KE, Stuebing KK, et al. Persistence of dyslexia: the connecticut longitudinal study at adolescence. Pediatrics 1999;104(6):1351–9. Available from: http://publications.aap.org/pediatrics/article-pdf/104/6/1351/887790/pe129901351p.pdf.

[10] Pennington BF, McGrath LM, Peterson RL. *Diagnosing learning disorders: from science to practice*. 3rd ed. Guilford Press; 2019.

[11] Ferrer E, Shaywitz BA, Holahan JM, Marchione KE, Michaels R, Shaywitz SE. Achievement gap in reading is present as early as first grade and persists through adolescence. J Pediatr 2015;167(5):1121–5. Available from: https://doi.org/10.1016/j.jpeds.2015.07.045.

[12] Georgitsi M, Dermitzakis I, Soumelidou E, Bonti E. The polygenic nature and complex genetic architecture of specific learning disorder. Brain Sci 2021;11(5). Available from: https://doi.org/10.3390/brainsci11050631.

[13] Ashkenazi S, Black JM, Abrams DA, Hoeft F, Menon V. Neurobiological underpinnings of math and reading learning disabilities. J Learn Disabil 2013;46(6):549–69. Available from: https://doi.org/10.1177/0022219413483174.

[14] Newman L, Wagner M, Knokey A-M, Marder C, Nagle K, Shaver D, et al. The post-high school outcomes of young adults with disabilities up to 8 years after high school a report from the national longitudinal transition study-2 (NLTS2). 2011.

[15] Jesse HW, Gregory KB, Michael E T, Monica RB. Diagnostic and statistical manual of mental disorders (DSM-5). 5th ed. American Psychiatric Association; 2013. p. 810.

[16] Accardo PJ, Accardo JA, Capute AJ. A neurodevelopmental perspective on the continuum of developmental disabilities. In: 3rd ed. Accardo PJ, editor. Capute & Accardo's neurodevelopmental disabilities in infancy and childhood, 1. 2008. p. 3–25.

[17] Moreno-De-Luca A, Myers SM, Challman TD, Moreno-De-Luca D, Evans DW, Ledbetter DH. Personal View Developmental brain dysfunction: revival and expansion of old concepts based on new genetic evidence. Lancet Neurol 2013;. Available from: http://www.thelancet.com/neurology.

[18] American Psychiatric Association. Desk reference to the diagnostic criteria from DSM-5. American Psychiatric Association Publishing; 2013.

[19] Peterson RL, McGrath LM, Willcutt EG, Keenan JM, Olson RK, Pennington BF. How specific are learning disabilities? J Learn Disabil 2021;54(6):466–83. Available from: https://doi.org/10.1177/0022219420982981.

[20] Fuchs LS, Geary DC, Compton DL, Fuchs D, Hamlett CL, Seethaler PM, et al. Do different types of school mathematics development depend on different constellations of numerical versus general cognitive abilities. Dev Psychol 2010;46(6):1731–46. Available from: https://doi.org/10.1037/a0020662.

[21] Wagner RK, Torgesen JK, Rashotte CA, Pearson NA. *Comprehensive Test of Phonological Processing* (CTOPP-2). 2nd ed Austin, TX: ProEd; 2013.

[22] Breaux KC. WIAT-4: Wechsler Individual Achievement Test. Fourth Edition Pearson; 2020.

[23] Schrank FA, McGrew KS, Mather N. *Woodcock-Johnson IV Tests of Cognitive Abilities*. Rolling Meadows. IL: Riverside Publishing; 2014.

[24] Torgesen JK, Wagner RK, Rashotte CA. *TOWRE-2 Test of Word Reading Efficiency*. New York: Pearson; 2012.

[25] Wiederholt, JL, & Bryant, BR. Gray *Oral Reading Tests—Fifth Edition (GORT-5)*. Austin, TX: Pro-Ed, 2012.

[26] MacGinitie WH, MacGinitie RK, Maria K, Dreyer LG. *Gates-MacGinitie Reading Test*. 4th ed. Itasca, IL: Riverside Publishing; 2000.

[27] Hammill DD, Larsen SC. *TOWL 4: Test of Written Language*. Austin TX. Pro Ed 2009.

[28] Cutting LE, Clements-Stephens A, Pugh KR, Burns S, Cao A, Pekar JJ, et al. Not all reading disabilities are dyslexia: distinct neurobiology of specific comprehension deficits. Brain Connect 2013;3(2):199–211. Available from: https://doi.org/10.1089/brain.2012.0116.

[29] Wagner RK, Beal B, Zirps FA, Spencer M. A model-based meta-analytic examination of specific reading comprehension deficit: how prevalent is it and does the simple view of reading account for it? Ann Dyslexia 2021;71(2):260–81. Available from: https://doi.org/10.1007/s11881-021-00232-2.

[30] Ijeoma C J-A, Chinelo J, U. Dyslexia, neurodevelopmental conditions and comobidity: a rule rather than an exception. Arch Neurol Neurosci 2019;4(2). Available from: https://doi.org/10.33552/ann.2019.04.000585.

[31] Moll K, Kunze S, Neuhoff N, Bruder J, Schulte-Körne G. Specific learning disorder: prevalence and gender differences. PLoS ONE 2014;9(7). Available from: https://doi.org/10.1371/journal.pone.0103537.

[32] Haberstroh S, Schulte-Körne G. Clinical practice guideline: The diagnosis and treatment of dyscalculia. Dtsch Arztebl Int 2019;116:107–14. Available from: https://doi.org/10.3238/arztebl.2019.0107.

[33] Chung, PJ, Patel, DR, Nizami, I. Disorder of written expression and dysgraphia: definition, diagnosis, and management. *Transl Pediatr*, 2020; 9(Suppl 1):S46–S54. Available from: https://doi.org/10.21037/tp.2019.11.01

[34] Peterson RL, Pennington BF. Developmental dyslexia. Annu Rev Clin Psychol 2015;11:283–307. Available from: https://doi.org/10.1146/annurev-clinpsy-032814-112842.

[35] Snowling MJ, Hulme C, Nation K. Defining and understanding dyslexia: past, present and future. Oxf Rev Educ 2020;46(4):501–13. Available from: https://doi.org/10.1080/03054985.2020.1765756.

[36] Biotteau M, Danna J, Baudou É, Puyjarinet F, Velay JL, Albaret JM, et al. Developmental coordination disorder and dysgraphia: signs and symptoms, diagnosis, and rehabilitation. Neuropsychiatr Dis Treat 2019;15:1873–85. Available from: https://doi.org/10.2147/NDT.S120514.

[37] Kandel S, Lassus-Sangosse D, Grosjacques G, Perret C. The impact of developmental dyslexia and dysgraphia on movement production during word writing. Cognit Neuropsychol 2017;34(3–4):219–51. Available from: https://doi.org/10.1080/02643294.2017.1389706.

[38] Rapcsak SZ, Beeson PM, Henry ML, Leyden A, Kim E, Rising K, et al. Phonological dyslexia and dysgraphia: cognitive mechanisms and neural substrates. Cortex 2009;45(5):575–91. Available from: https://doi.org/10.1016/j.cortex.2008.04.006.

[39] Berninger VW, Abbott RD, Abbott SP, Graham S, Richards T. Writing and reading: connections between language by hand and language by eye. J Learn Disabil 2002;35(1):38–56.

[40] Berninger VW, Richards TL, Abbott RD. Differential diagnosis of dysgraphia, dyslexia, and OWL LD: behavioral and neuroimaging evidence. Read Writ 2015;28(8):1119–53. Available from: https://doi.org/10.1007/s11145-015-9565-0.

[41] Georgiou GK, Martinez D, Vieira APA, Guo K. Is orthographic knowledge a strength or a weakness in individuals with dyslexia? Evidence from a meta-analysis. Ann Dyslexia 2021;71(1):5–27. Available from: https://doi.org/10.1007/s11881-021-00220-6.

[42] Ogbogo S, Orluwene GW. Psychometric analysis of dyscalculia test. Br J Educ 2021;9:55–70. Available from: https://doi.org/10.37745/bje.2013.

[43] Mazzocco MMM, Feigenson L, Halberda J. Preschoolers' precision of the approximate number system predicts later school mathematics performance. PLoS ONE 2011;6(9). Available from: https://doi.org/10.1371/journal.pone.0023749.

[44] Evans TM, Ullman MT. An extension of the procedural deficit hypothesis from developmental language disorders to mathematical disability. Front Psychol 2016;7(SEP). Available from: https://doi.org/10.3389/fpsyg.2016.01318.

[45] Kaufmann L, Mazzocco MM, Dowker A, von Aster M, Göbel SM, Grabner RH, et al. Dyscalculia from a developmental and differential perspective. Front Psychol 2013;4:516. Available from: https://doi.org/10.3389/fpsyg.2013.00516.

[46] Menon V. Working memory in children's math learning and its disruption in dyscalculia. Current Opinion in Behavioral Sciences 2016;10:125–32. Available from: https://doi.org/10.1016/j.cobeha.2016.05.014.

[47] Geary DC, Hoard MK, Byrd-Craven J, Nugent L, Numtee C. Cognitive mechanisms underlying achievement deficits in children with mathematical learning disability. Child Dev 2007;78(4):1343–59.

[48] Fuchs, L.S., Fuchs, D., Seethaler, P.M., Cutting, L.E., & Mancilla-Martinez, J. (2019). Connections between reading comprehension and word-problem solving via oral language comprehension: implications for comorbid learning disabilities New directions for child and adolescent development (pp. 73–90). Wiley-Blackwell Publishing.

[49] Powell SR, Fuchs LS. Does early algebraic reasoning differ as a function of students' difficulty with calculations versus word problems? Learn Disabil Res Pract 2014;29(3):106–16.

[50] Suárez-Pellicioni M, Núñez-Peña MI, Colomé À. Math anxiety: a review of its cognitive consequences, psychophysiological correlates, and brain bases. Cogn Affect Behav Neurosci 2016;16(1):3–22. Available from: https://doi.org/10.3758/s13415-015-0370-7.

[51] Willcutt EG, Pennington BF. Psychiatric comorbidity in children and adolescents with reading disability. J Child Psychol Psychiat 2000;41 (8):1039–48.

[52] Joyner RE, Wagner RK. Co-occurrence of reading disabilities and math disabilities: a meta-analysis. Sci Stud Read 2020;24(1):14–22. Available from: https://doi.org/10.1080/10888438.2019.1593420.

[53] Willcutt EG, Petrill SA, Wu S, Boada R, DeFries JC, Olson RK, et al. Comorbidity between reading disability and math disability: concurrent psychopathology, functional impairment, and neuropsychological functioning. J Learn Disabil 2013;46(6):500–16. Available from: https://doi.org/10.1177/0022219413477476.

[54] Price KM, Wigg KG, Misener VL, Clarke A, Yeung N, Blokland K, et al. Language Difficulties in School-Age Children with Developmental Dyslexia. J Learn Disabil 2022;55(3):200–12. Available from: https://doi.org/10.1177/00222194211006207.

[55] Hulme C, Snowling MJ. Reading disorders and dyslexia. Curr Opin Pediatr 2016;28(6):731–5. Available from: https://doi.org/10.1097/MOP.0000000000000411.

[56] Aro T, Eklund K, Eloranta AK, Ahonen T, Rescorla L. Learning Disabilities Elevate Children's Risk for Behavioral-Emotional Problems: Differences Between LD Types, Genders, and Contexts. J Learn Disabil 2022;55(6):465–81. Available from: https://doi.org/10.1177/00222194211056297.

[57] Wagner RK, Torgesen JK, Rashotte CA, Hecht SA, Barker TA, Burgess SR, et al. Changing relations between phonological processing abilities and word-level reading as children develop from beginning to skilled readers: a 5-year longitudinal study. Dev Psychol 1997;33(3).

[58] Georgiou GK, Aro M, Liao CH, Parrila R. Modeling the relationship between rapid automatized naming and literacy skills across languages varying in orthographic consistency. J Exp Child Psychol 2016;143:48–64. Available from: https://doi.org/10.1016/j.jecp.2015.10.017.

[59] Wolf M. What time may tell: towards a new conceptualization of developmental dyslexia. Ann Dyslexia 1999;49.

[60] Pernet C, Andersson J, Paulesu E, Demonet JF. When all hypotheses are right: a multifocal account of dyslexia. Hum Brain Mapp 2009; 30(7):2278–92. Available from: https://doi.org/10.1002/hbm.20670.

[61] Richards TL, Grabowski TJ, Boord P, Yagle K, Askren M, Mestre Z, et al. Contrasting brain patterns of writing-related DTI parameters, fMRI connectivity, and DTI-fMRI connectivity correlations in children with and without dysgraphia or dyslexia. Neuroimage Clin 2015;8:408–21. Available from: https://doi.org/10.1016/j.nicl.2015.03.018.

[62] Aboud KS, Bailey SK, Petrill SA, Cutting LE. Comprehending text versus reading words in young readers with varying reading ability: distinct patterns of functional connectivity from common processing hubs. Dev Sci 2016;19(4):632–56. Available from: https://doi.org/10.1111/desc.12422.

[63] Fan Q, Anderson AW, Davis N, Cutting LE. Structural connectivity patterns associated with the putative visual word form area and children's reading ability. Brain Res 2014;1586:118–29. Available from: https://doi.org/10.1016/j.brainres.2014.08.050.

[64] Soares N, Evans T, Patel DR. Specific learning disability in mathematics: a comprehensive review. Transl Pediatr 2018;7(1):48–62. Available from: https://doi.org/10.21037/tp.2017.08.03.

[65] Bailey DH, Farkas G, Oh Y, Morgan P, Hillemeier M. Reciprocal effects of reading and mathematics? Beyond the cross-lagged panel model. Dev Psychol 2020;56(5):912–21. Available from: https://doi.org/10.1037/dev0000902.supp.

[66] Wagner, R.K., Edwards, A.A., Malkowski, A., Schatschneider, C., Joyner, R.E., Wood, S., et al. (2019). Combining old and new for better understanding and predicting dyslexia. New directions for child and adolescent development (pp. 11–23). Wiley-Blackwell Publishing.

[67] Shonkoff JP, Garner AS, Siegel BS, Dobbins MI, Earls MF, McGuinn L, et al. The lifelong effects of early childhood adversity and toxic stress. Pediatrics 2012;129(1). Available from: https://doi.org/10.1542/peds.2011-2663.

[68] Catts HW, Petscher Y. A cumulative risk and resilience model of dyslexia. J Learn Disabil 2021;. Available from: https://doi.org/10.1177/00222194211037062.

[69] Schatschneider C, Torgesen JK. Using our current understanding of dyslexia to support early identification and intervention. J Child Neurol 2004;19(10):759–65.

[70] Puranik CS, Alotaiba S. Examining the contribution of handwriting and spelling to written expression in kindergarten children. Read Writ 2012;25(7):1523–46. Available from: https://doi.org/10.1007/s11145-011-9331-x.

[71] Apel K, Henbest VS, Masterson J. Orthographic knowledge: clarifications, challenges, and future directions. Reading and writing, vol. 32. The Netherlands: Springer; 2019. p. 873–89. Available from: https://doi.org/10.1007/s11145-018-9895-9.

[72] Powell SR, Fuchs LS, Cirino PT, Fuchs D, Compton DL, Changas PC. Effects of a multitier support system on calculation, word problem, and prealgebraic performance among at-risk learners. Except Child 2015;81(4):443–70. Available from: https://doi.org/10.1177/0014402914563702.

[73] Simos PG, Fletcher JM, Sarkari S, Billingsley RL, Denton C, Papanicolaou AC. Altering the brain circuits for reading through intervention: a magnetic source imaging study. Neuropsychology 2007;21(4):485–96. Available from: https://doi.org/10.1037/0894-4105.21.4.485.

Chapter 26

Speech and language disorders

Mary L. Leppert
Department of Neurology and Developmental Medicine, Kennedy Krieger Institute and Department of Pediatrics, Johns Hopkins University School of Medicine, Baltimore, MD, United States

Perhaps one of the greatest marvels of the human condition is the capacity to learn to communicate through speech, gesture, sign, or written word so as to convey or understand information, inquiry, and emotion. In typical development, newborns lack the capacity to communicate, and all vocalizations are guttural or the consequence of physiological functions, but within a few short months, their vocalizations have communicative intent. By a year of age, children begin to use and understand single words and gestures, and by 2–3 years, children speak in sentences, follow commands, and begin to understand the pragmatic communications of body language, facial expression, tone, and inference.

The development of speech, or the production of vocal sounds, progresses through the guttural phase and the prelinguistic phase in the first year of life to the linguistic phase of communication with words or signs in the second year. Speech requires adequate hearing to emulate sound, and the motor capacity to produce clear, accurate, and fluent utterances. The components of speech include articulation, phonation, voice, and fluency. The majority of speech sounds are acquired by 4 years of age, although a few, such as "r" and "th," may not be fully accurate until 6–7 years of life [1]. Children may demonstrate errors of speech production including omissions, substitutions, phoneme rearrangements, or assimilations as well as developmental dysfluency during the first few years, but children's utterances should be completely understandable and fluent by 4 years and reach mature adult patterns of speech by 8–9 years of age [2].

The development of language is dependent on adequate hearing, cognition, attention, and interest of children from early infancy. As early as the prelinguistic phase, infants begin to use vocalizations with social intent and reciprocity. They attend to the verbal output of those around them and, with attention to the context, begin to lay visual, verbal, and often emotional representations in their lexical repositories. This lexicon becomes the basis of expression and understanding of language that is required for human communication. Robust communication consists of expressive, receptive, and pragmatic language, each of which has components that affect the accuracy and efficiency of communication. The most basic components of expressive language are the size and breadth of the vocabulary or lexicon and the length and completeness of sentence formation. The efficiency, effectiveness, and accuracy of expressive language require exacting word choices (semantics) and correct grammar, morphology, and syntax. In addition to the appropriate expressive language formulation, accurate delivery of organized, sequential thought is required for discourse. The pragmatic skill of reading the language recipient's understanding of the discourse allows the speaker to repair communication to provide clarity. Similarly, receptive language requires accurate decoding, efficient processing and comprehension of communications, and analysis of meaning in the context of the situation and the preceded communication [3]. Inaccuracy in decoding may result in misunderstanding, estimations, and loss of detail. A child with sufficient vocabulary who is asked to "go get the umbrella, it's raining" but returns with a raincoat or boots clearly understood the concept of foul weather but did not respond with accuracy. Alternatively, prolonged decoding or processing of verbally delivered information may be inefficient. A child with slower integration of verbally presented information may miss whole segments of language content as they listen, try to process, and listen again. When these intervening segments are lost, the listener may misunderstand, inaccurately estimate the meaning of the information, or become frustrated with the disconnected bits of information and cease listening. These children may appear inattentive, defiant, or less intelligent, despite good cognition, attention, and will to understand.

Pragmatic language skills provide an understanding of communication that is not at the word level. The interpretation of tone, facial expression, body posture, or gesture, and at higher levels, the ability to garner meaning from inference rather than pedantic understanding are some of the components of pragmatics that enrich communication.

Capute and Accardo's Neurodevelopmental Disabilities in Infancy and Childhood. DOI: https://doi.org/10.1016/B978-0-12-824060-1.00029-8

Children acquire pragmatic abilities across early and middle childhood, beginning with the interpretation of gesture, facial expression, and body language and progressing through understanding riddles, jokes, idioms, irony, and sarcasm. By early adolescence, the use and understanding of inference becomes central to learning and to social skills. Reading demands progress to the most sophisticated phase, having progressed from learning to read and reading to learn, students must learn to draw meaning between the lines of the story, make inferences, and understand theme and moral of the story. The social dynamics of early adolescence place additional demands on pragmatic understanding as sarcasm is honed at this age.

The synchronous acquisition of appropriate expressive, receptive, and pragmatic skills collectively enables the clear, concise communicative skills required for forming and understanding discourse and for learning. A mature communicator will listen, assimilate, understand spoken language and its accompanying body language, facial expression, and inference and comment or request for clarification with succinct, exact verbal response complete with appropriate facial and body expression. This same dynamic is present in the mature reader who "listens"—that is, reads, assimilates written words, and produces succinct, exact written summation or interpretation.

Given the complexity of speech and language development and the overlapping interdependency of its components, it is not surprising that disorders of language development are common. As in all streams of development, language acquisition may be delayed, deviant, or dissociated. Delayed language is simply a slower rate of acquisition of skills, including prelinguistic milestones, vocabulary, length of utterances, parts of speech used, and understanding. Deviant language is characterized by atypical sequence of milestone acquisition, and dissociation is manifested when the multiple domains of language develop at asynchronous rates.

Definitions

The classification of disorders of speech and language has been problematic, as terms have been too broad, too specific, or too similar. Recently, the CATALISE Consortium [4] presented consensus guided terms to describe disorders of speech and language. Language disorders are broadly defined as problems that "are likely to endure to middle childhood and beyond, with significant impact on everyday social interaction or educational progress." The consortium divided language disorders into two groups: language disorder due to "differentiating conditions" and "developmental language disorders." Language disorders due to differentiating conditions occur as a part of a constellation of findings that are common to an underlying condition such as cerebral palsy, genetic, or teratogenic disorders such as Fragile X, Angelman, or fetal alcohol syndromes, or known behavioral phenotypes such as autism. Developmental language disorders (DLD) are not explained by a unifying etiology or underlying condition. Further, the Consortium stated that language disorders may be diagnosed in the presence of delayed nonverbal skills, so a discrepancy between verbal and nonverbal abilities is not required to make the diagnosis [4]. Within the classification of language disorder, the broad category of speech sound disorders (SSD) is comprised of subclassification of disorders or specific mechanisms of speech most often defined by speech language pathologists. The spectrum of SSD includes disorders of articulation or phonation, childhood apraxia of speech, and disorders of fluency. Language disorders, broadly classified into expressive, receptive, and pragmatic or higher order language impairments, are also composed of subclassifications of more specific impairments (Table 26.1). The list of subclassifications of language disorders is as diverse as the components

TABLE 26.1 Developmental language disorders.

Speech sound disorders	Expressive language disorders	Receptive language disorders	Pragmatic language disorders
Dysarthria	Small vocabulary	Inaccurate understanding	Difficulty reading tone or body language
Phonological disorders	Poor connected language or length of utterance	Slow, inefficient processing or integration of verbally presented information	Difficulty with inferential thought
Childhood apraxia of speech	Morphological errors		Misinterpretation or misunderstanding of social cues
Disorders of voice	Syntactical errors		
Disorders of fluency	Semantical errors		

of language, which have confused the literature with classification systems suggested by authorities such as the APA, Rapin and Allen, and Wilson and Risucci [5]. The American Speech and Hearing Association (ASHA) dichotomizes language disorders as spoken or oral language disorders and written language disorders. A spoken language disorder (SLD) is defined as "a significant impairment in the acquisition and use of language across modalities due to deficits in comprehension and/or production across any of the five language domains, phonology, morphology, syntax, semantics and pragmatics" (ASHA.org). Similar to the CATALISE consortium, ASHA makes the distinction between SLDs which occur as a primary disorder from those that are secondary to other established diagnoses, such as intellectual disability, autism, hearing impairment, and developmental or mental health disorders. The Diagnostic and Statistical Manual, Fifth Edition (DSM-5) reclassified the broad class of "communication disorders" to include language disorder (formerly expressive and mixed receptive-expressive language disorder), SSD, child onset fluency disorders, and social pragmatic communication disorder [6]. For the purpose of this chapter, definitions will be grouped as SSDs, DLDs (including expressive and mixed receptive and expressive language disorders that are primary, or not associated with other conditions), and pragmatic language disorders or social pragmatic language disorders [6].

Prevalence

In early childhood, errors in speech production and developmental dysfluency are common and generally resolve without intervention. Disorders of speech production or SSD occur in 3.6% of children at 8 years of age, with a higher prevalence in boys (4.6% vs 2.5%) [7]. SSDs that persist to middle childhood are also more likely to persist into young adulthood, with a prevalence of 2.4% of college freshmen [8].

Language delays are also relatively common in early childhood, as 11%–18% of young children demonstrate early language delays, which often resolve by 3–4 years of age [9,10]. By school age, approximately 7% of children qualify for a diagnosis of DLD [11,12]. Social communication disorder (SCD), introduced as a diagnosis in DSM-5, had previously been referred to as pragmatic language disorder or pragmatic/semantic language impairment. Although the diagnosis is relatively new, prevalence estimates of SCD range from 7% to 11% [13].

Individually, each of these disorders of speech and language appear to be more common in early childhood. Many will resolve but are considered persistent if they continue to middle childhood. Each of these disorders may also cooccur with other speech and language disorders in children.

Etiology and risk for speech and language disorders

Speech and language disorders due to differentiating conditions are in many cases explained by or are a consequence of the underlying condition. SSDs may be directly attributable to the motor impairment of a child with cerebral palsy or structural changes of the oropharynx such as cleft palate. Language delays are explained by cognitive level in children with intellectual disability or the social communication deficits in a child with autism. The etiologies of these differentiating conditions are diverse and may include fetal exposures, genetic conditions, neurologic or neuromuscular disorders, and metabolic disorders. In genetic disorders, for example, Fragile X, Williams, or Angelman syndromes, speech and language impairments are a frequently known component of the constellation of findings that comprise the genetic disorder.

DLDs are, by definition, not explained by an underlying condition and therefore are without specific identifiable etiology. A number of risk factors are consistently associated with DLDs, many of which cooccur and perhaps suggest multifactorial contributions to the disorder. Language disorders most certainly run in families. A history of language disorder is very often found in first-degree relatives in children with DLD [14,15] and in children with family history of dyslexia [16]. The strong inheritance of language delay and the observations of higher concordance in monozygotic twins also suggest a genetic underpinning of DLD. The genetics of DLD remains unclear but specific genes associated with DLD include CNTNAP (7q35), CMIP (16q23.2), and ATP2C2 (16q24.1). Childhood apraxia of speech is associated with variants in FOXP2 (7q31.1) [12,17].

Males, children from lower socioeconomic backgrounds, and lower parental education levels are overrepresented in DLD cohorts [10,14]. The contribution of environment to DLD is also associated with early language development, as children's vocabulary and access to that vocabulary are dependent on the breadth and frequency of the words they hear [18]. Exposure to higher quantity and quality or complexity of words and the expectation and opportunity for reciprocal dialog or engagement are all associated with better language outcomes by school entry [9,19]. Preterm birth, low birth weight, and birth order (younger children of large sibling groups) are additional associated risks with less supporting evidence [10,14]. Neuroimaging studies do not show consistent differences in children with DLD as compared to typically developing controls, and data from functional imaging using functional magnetic resonance imaging (fMRI) and diffusion tensor imaging (DTI) imaging provide only suggestive findings [12].

Risks established by early or persistent speech sound and developmental language disorders

In addition to the conditions that establish a risk for disorders of speech production and language, SSD and DLD, particularly when persistent, establish risk for developmental, behavioral, and mental health disorders as well as lower educational and employment attainment. Language delays seen in very young children are likely to resolve without specific intervention. Children identified early with language impairment who resolve their language difficulties by school age have relatively good educational prognoses compared to children with persisting language deficits [16]. Children identified with SSD carried small but notable risk for lower literacy skills at school entry and in early elementary education, particularly if the SSD persisted until school entry [20]. In the preschool years, poor comprehension, poor gesture use, and lack of imitation of body movement at 3 years of age carry a poor prognosis, particularly if coupled with a family history of language delay or reading difficulties [4]. In 4-year-old children, the more domains of language function that are impaired, the greater the risk for poor outcome. By 5 years, children who continue to have language difficulties are likely to have persistent DLD and are at higher risk of associated deficits [4,16]. Persistent receptive language disorder and lower nonverbal cognitive skills carry poorer learning and literacy prognoses than does the persistence of speech sound or articulation disorder [21].

Cooccurring speech and language impairments, particularly persisting combinations of disorders, are consistently and strongly associated with poorer literacy outcomes [4,20–22]. The overall estimate of children who have comorbid SSDs and language impairments is extremely varied but estimated at 1.3%–4.6% [23,24]. The comorbidity of SSD and DLD varies with the index disorder, as 11%–15% of children with persisting speech disorders had language impairments, and 5%–8% of children with persistent language disorders had speech disorders [24]. Similarly, an estimated 30% of children with DLD also have SCD, and the prevalence of SCD in the absence of DLD is about 9% [13].

Associated disorders and comorbidities

DLDs do not occur in isolation but are associated with or comorbid with motor, learning, social, behavioral, and mental health disorders. These disorders vary in severity and in the age at which they present. Associated disorders may be underidentified or subclinical when mild. Motor weakness ascribed to clumsiness, weak reading fluency or comprehension insufficient to be considered a learning disability (LD), inattentiveness without hyperactivity that goes unnoticed, a preference to avoid socializing attributed to "introverted" personality, or subclinical anxiety all contribute to the impairment of individuals with DLD. Temporal variability in the age of presentation of associated disorders further confounds the assessment of comorbid conditions with DLD. Disorders of learning become apparent as academic demands exceed the ability of a child to perform in the classroom, but social impairment and many mental health disorders present later, often in adolescence.

Associated motor findings

Developmental coordination disorder (DCD) is characterized by suboptimal motor function in the absence of significant neurologic findings and is a common associated disorder in children with DLD [25]. Children with DCD are typically less coordinated than expected for age and may appear clumsy in gross and fine motor activity. Difficulty with manual dexterity is the most common finding in DCD [26], with common manifestations including trouble with buttoning, snapping, using scissors, or handwriting. (See chapter: Developmental coordination disorder and related neuromotor abnormalties).

Learning disabilities

Language is the basis of reading and learning. It is no surprise that children with SSD, DLD, or some combination have difficulties with language-based learning tasks: spelling, reading, and writing. The prerequisites of reading—phonological awareness, alphabetic awareness with sound-letter symbol correlation, decoding skills, and the size, accessibility, and understanding of the lexical repository—are all established in early childhood and are compromised in children with language impairment. Children with SSD diagnosed in early childhood (3½ years) were more likely to have disordered phonemic awareness and spelling at 5½, lower word reading at 8 years [20], and difficulty with multisyllabic words, poorer vocabulary, decoding, and spelling skills as adolescents [2]. Vocabulary size and understanding in children with DLD is constricted in comparison to the vocabulary of same-age peers and does not show any catch-up acquisition in vocabulary across education. The vocabulary growth, however, follows the same curve as typically developing peers [27]. Deficits in vocabulary size, decoding skills, grammar, and phonological skills persist in children with DLD, as do word recognition and comprehension. Longitudinal studies demonstrate that 42% of children with DLD at school entry meet criteria for LD in reading, 36% maintain LD in reading by fourth grade and continue to exhibit deficits in word recognition and reading comprehension throughout high school [21,22,28,29]. Like vocabulary, DLD readers follow the same reading trajectory as typically developing readers, albeit at a delayed rate [28,29].

Written langauge disorders are also more common in children with DLD. Dockrell et al. [30] describe the process of writing as a "writing burst" in which the author forms an idea, translates the idea to a string of words (inclusive of proper morphology syntax, semantics), and transcribes the words to written representation, with evaluation and revision at the translation and transcription steps [31]. Children with DLD have the same number of writing bursts as do children without DLD, indicating that they have the same number of ideas, but they produce text that is shorter, with fewer words, of lower quality, and with more pauses and spelling errors [31]. The transcribing of thought to word is also slower, more effortful, and of lower quality in children with DLD, perhaps due to difficulties with manual dexterity. The inefficiency of writing persists throughout educational years, as adolescents continue to demonstrate poor writing fluency, sentence structure, quality, and organization of text [30].

Math skill deficits are unexpectedly also more common in children with persistent DLD. Tasks that are verbally demanding (arithmetic, counting, and word problems) are more problematic for children with DLD than are tasks such as using a number line or math concepts in which children with DLD perform similar to peers [32]. LD in math is nearly four times higher in children with DLD than in language typical peers [33], suggesting that language deficits permeate all learning domains.

Behavioral disorders and attention deficit hyperactivity disorder

There is clear evidence that children with DLD have about a twofold increase in rates of internalizing and externalizing behaviors and have a greater severity of these disorders [34,35]. Further, these differences in behavior were evident in children as young as 18 months of age with language impairment and appeared to persist or become more evident with advancing age [22,34]. Teachers tend to rate more severe problem behaviors than do parents of children with DLD [34]. Conduct disorders appear to be associated with DLD. Teachers also tended to rate boys with language impairment higher than typically developing peers on ratings of aggression, while parents tended to rate their children with language impairment higher on problems with delinquency than their peers [36]. The association between DLD and attention deficit hyperactivity disorder (ADHD) is also clear. Children with DLD are at increased risk of ADHD, and children with ADHD are at increased risk of DLD [37]. The question as to whether these are independently associated diagnoses or each is the consequence of the other is an interesting one. Children require the ability to hear, to pay attention, or actively listen and to establish rapport in order to lay language, build their lexicon, study the context surrounding language, and the pragmatic communication with which it is delivered. It seems then that a child without the attention to listen or focus on verbally delivered information and the scaffolding with which it is relayed is likely to miss the opportunity to effectively build their lexicon and to learn appropriate semantics, syntax, and grammar. Thus it is possible that attention deficits contribute to DLD. Conversely, a child with inaccurate or slower understanding of language will become lost with longer or more complex discourse and cease paying attention when the information becomes confusing. In this scenario, the child with DLD will appear inattentive. The prevalence of ADHD in children with DLD varies in early studies from 16% to 33%, and the prevalence of DLD in children with ADHD ranges from 26% to 62% [37]. More recent analysis suggests that the relative risk of ADHD in a child with DLD is very similar to the relative risk of DLD in a child with ADHD. Both risks are significantly higher than that seen in control groups, but ADHD-inattentive type appears to have a stronger association with DLD [37].

Mental health disorders

Mental health disorders are also twice as common in children with DLDs and show evidence of increasing prevalence over time [38,39]. Any psychiatric disorder occurs in similar rates in children with SSD versus controls but increase substantially in children with DLD (42%) and even more so in those with DLD and SSD (57%) [22]. Children diagnosed with SSD at 5 years of age were twice as likely to have anxiety disorders, social phobia, and antisocial personality disorders as were their typically developing peers in early adulthood. Similarly, children diagnosed with DLD at 5 and followed to young adulthood were more than three times more likely to have anxiety disorder and 2½ times more likely to have social phobia and antisocial personality disorders than their peers [39]. Higher rates of anxiety and depression were reported independently by language-impaired adolescents and their parents in more recent studies and were not predicted by the language level or by gender as males and females were equally affected [40].

The frequency of the association of mental health disorders in youth with DLD mirrors the frequency of DLD in youth with mental health disorders. Excluding ADHD from mental health disorders, rates of affective, antisocial, anxiety, depressive, and substance use disorders range from 12% to 38% in youth with DLD, such that in sum, 40% of DLD youths experience psychiatric illness [39,41]. Similarly, 40% of youth with psychiatric disorders were found to have previously undetected language impairments [42]. It remains unclear whether the association between language impairment and mental health disorders is causal or a consequence of a larger phenotype of language disorder. Children and adolescents who lack the accuracy or efficiency of understanding, or who misread or misinterpret linguistic or social cues, may become anxious by the uncertainty. Social isolation, social phobia, or a tendency to be "introverted" or a loner may be an avoidant response to the discomfort of misinterpreting social cues rather than an inherent desire to be alone. Conversely, it may also be that children with anxiety, social phobia, or depression may be less able to attend to verbal discourse or social and pragmatic communications.

The cause-effect versus consequence question applies to learning, attention, behavioral, and emotional or mental health comorbidities in children with DLD. Additionally, each of the conditions carry risk for other, often overlapping associated disorders. In these associated conditions, children with SSD are comparable to speech typical peers, but children with DLD have much higher rates of cooccurring disorders. The consistency and strength of the associated disorders coupled with high rates of motor findings, which are unlikely a direct consequence of DLD, suggest that there may be a DLD phenotype. A DLD phenotype includes a tendency toward motor clumsiness, academic underachievement, internalizing behaviors, and emotional disorders including ADHD-inattentive type, anxiety, and depression.

Individually, DLD and each of its associated disorders influence the trajectory and outcomes of children and youth. The constellation of language, learning, motor, behavioral, and mental health difficulties collectively have significant impact on outcomes. Literature largely reports on outcomes using language as the index disorder and does not consider the weight of the many subclinical or clinical associated disorders.

Outcomes of developmental language disorders

Literacy and education

Children identified as having language impairment early in life and followed into adulthood showed poorer receptive language and literacy skills than did their age-matched peers. Deficits in speech production, receptive grammar, short-term verbal memory, phonological awareness, and lower overall literacy [43,44] were seen in young adults, with 75% performing below average range [45]. Many longitudinal follow-up studies report significant deficits in core language skills, with poor reading comprehension persisting into adulthood [43,46,47]. The average reading age of young adults was the equivalent of an 11–12 year old [43], yet 75% of young adults with a history of DLD reported reading books, magazines, or newspapers sometimes or often [45].

Self-reports of young adults in their 30s show that in those with DLD but normal intelligence, 36% reported continued difficulty reading, and 60% reported ongoing difficulty with spelling [47]. DLD young adults also reported more difficulty with reading timetables, forms, letters, and bank statements and with filling out checks, forms, and emails [45]. Collectively, this suggests that young children with DLD, despite the high rates of LD, become adults who read and write, albeit with greater difficulty.

The educational consequence of persisting language impairment is overall lower academic achievement. Individuals with DLD were less likely than same-age peers to finish high school or the equivalent compulsory education [46,48]. Those students with DLD who did complete compulsory education had lower grades and scored lower on standardized tests and qualifying exams [43,49]. In recent years, young adults with DLD were equally likely to pursue postsecondary education as same-age peers but more likely to pursue postsecondary vocational programs and less likely to attend college or university. Only 10% of the DLD group completed university degrees compared to 41% of age-matched peers [43]. Young adults who pursued vocational qualifications completed basic qualification courses at the same rate as their language typical peers but were far less likely to receive higher qualifications in vocational careers [43].

Employment and independence outcomes

The cumulative burden of lower literacy, language, and education carries into adult employment outcomes. Cohorts described in earlier literature found that individuals with DLD were much more likely than typical language peers to be unemployed in young adult years [38]. More recent studies show similar rates of employment among DLD and language typical peers, but a greater proportion of DLD individuals are in part-time employment [43,46,49]. DLD young adults are more likely to be in lower-skilled jobs, less likely to require submitting a curriculum vitae or interviewing for employment positions and less likely to be in managerial or professional occupations [43]. In a recent comparison of young adults with language impairment and pragmatic language impairment, individuals with only pragmatic language impairment had higher language and academic scores and a greater proportion of higher-skilled jobs [44]. Overall, employment income was lower in DLD adults than in their language typical peers [23]. Once in employment, DLD adults were paid similarly to their peers at equivalent job positions and with peers of similar educational achievement or qualifications [43,49]. Contrary to early cohort studies, employed DLD adults were also similar to their peers in job satisfaction, job security, and getting along with coworkers [43].

In addition to having lower-paying jobs on average, young adults with DLD had poor financial literacy skills and were more likely to require continuing financial support and oversight by their parents [49]. The rates of young adults with DLD living independently vary. Followed into their 30s, slightly more individuals with DLD were living with their parents [48], but the vast majority of individuals with DLD were independent [47,50].

Psychosocial outcomes

Twenty percent of young adults with DLD reported that they did not have any close friendships compared to 50% of young adults with pragmatic language impairment. Individuals with pragmatic language impairment had poorer quality friendships than did language typical peers and were less likely to be in romantic relationships [50]. Young adults with DLD, however, were as likely as typically developing peers to be married or to be living with a partner [46,49]. Individuals with DLD were slightly more likely to have children in their mid-20s and slightly less likely to have children in their 30s than were same-age peers [46,47]. The children of individuals with DLD had higher rates of SSD (8%–31%) and were far more likely to have reading difficulties [46,47].

In late adolescence, teens with DLD reported higher rates of victimization, bullying, and sadness than did their peers, though the rates of bullying and victimization decreased between middle childhood and later teen years [51]. Sadness and adolescent psychiatric illness can persist into adulthood for children with DLD. Depression occurs in 15%–26% of individuals with DLD [50,52]. Anxiety is also much higher in the DLD group with 16% of individuals reporting such diagnoses in their mid-20s [50].

Behavioral disorders also persist into adulthood. Parents of older teen boys with DLD report higher rates of delinquency and aggression, and DLD boys' self-reports indicate higher rates of arrest (41% DLD vs 20% in age-matched peers) and increased rates of criminal convictions (29% DLD vs 14.5% age-matched peers) [36]. To understand the association between DLD and criminal behavior, Snow [53] undertook a study of juvenile offenders. After controlling for IQ, they found that 50% of juveniles in the justice system had language impairment and poorer social skills than expected for age. Thirty-six percent of these juveniles reported that they were "not good" at reading or writing, compared to 14% of their age-matched peers, and only 50% reported that they were still in school at an average age of 15.8 years [53]. There was no difference in the type of crime, as DLD youth and age-matched peers had similar rates of arrest for violent and property offenses. Eighty percent of young offenders also reported substance use. In a study of 100 young offenders incarcerated at age 19, 46% had language impairment [54].

The association of DLD with later substance/alcohol use and trouble with the law appears to be mitigated by the early identification and intervention for language impairment. In contrast to the findings of Snow and Powell, Winstanley reported that children with DLD who received early intervention were 2.5 times less likely to be in trouble with the police and two times less likely to be arrested than age-matched peers. The same study showed that children with DLD and age-matched peers reported drinking alcohol at the same rates, but the DLD group were more likely to drink alone, while the age-matched peers were more likely to start drinking at earlier age and drink more per session than DLD peers [55].

The educational and psychosocial outcomes of children with language impairment must be understood in the context of the many confounders of the outcomes for these children and young adults. Children with language disorder are more likely to be born into families with lower educational attainment and lower occupational states and were less likely to have been read to in early childhood or have parents interested in their education [48]. Each of these disadvantages may individually, or in combination, influence the outcomes of children with language impairment. It may also be true that the heritability of language impairment perpetuates intergenerational risk for lower educational attainment and expectation.

The promise that early identification and intervention of DLD may mitigate outcomes is suggested by the improving outcomes of children with DLD. The rates of youth with a history of DLD leaving school before the end of compulsory education is falling, and the rates of those entering postsecondary education has improved substantially over the last three decades. As young adults, individuals with DLD do not differ from peers in their rates of employment or marriage. They also report salaries similar to non-DLD peers with the same qualifications, and most importantly, they report similar job satisfaction and quality of life [46]. The evidence of the improved outcomes of children identified early provides much hope regarding means to build resilience for future generations of children with DLD. Early intervention, early reading, and awareness and monitoring for learning differences, pragmatic differences, and mental health disorders may continue to optimize outcome and trajectory of children with DLD and reduce the number of young adults with DLD who leave the educational system early, get in trouble with the law, or engage in alcohol/substance use. These efforts may reduce, but not eliminate, the collective risks for individuals with DLD and the disparities that accompany those risks.

The role of the pediatrician

Given the frequency with which disorders of speech and language occur in childhood, pediatricians require a sound understanding of the disorders. The pediatrician is expected to identify speech disorders or language delays in the routine care of young children through observation, surveillance, and screening. When a disorder is suspected, it is incumbent on the primary care clinician to either establish a diagnosis or to refer to a speech language pathologist to confirm and further define the disorder. Once a diagnosis is established, clinicians should have heightened awareness of the trajectory of language competency and the substantial risk for language-based learning disorders. Anticipation of difficulties with reading, written language, and inferential thought provides an opportunity to intervene with services that may mitigate educational and social outcomes. The number of disorders that may cooccur with or may masquerade as language disorders is extensive and must also remain in the fore of the mind of the treating clinician. Anxiety, ADHD, central auditory processing disorder, autism, mood disorders, intellectual and learning disabilities may all share overlapping symptoms that may mimic DLD and may also complicate the care of children with DLD when these other disorders cooccur. The importance of identifying cooccurring disorders cannot be understated, as effective pharmacologic management of anxiety, ADHD, or mood disorders optimizes language learning and proficiency. Finally, the pediatrician caring for children with DLD can offer valuable counsel to families by providing informed anticipatory guidance, ensuring appropriate referrals, and advocating for children with DLD in educational, extracurricular, and employment settings.

Summary

Disorders of speech and language are common in early childhood, but many resolve before school entry. Language disorders are more likely to persist and carry greater risk for poor literacy and educational outcomes than are SSDs. Persisting language disorders substantially increase the risk for learning disorders in reading, writing, and even math. ADHD, behavioral disorders, and disorders of mental health are much more common in children with DLD than in the general pediatric population. Interestingly, DLD is also far more prevalent in individuals with ADHD, psychiatric disorders, and in those in the juvenile justice system. Perhaps the new classification system will inform or refine specific phenotypes or trajectories for children with SSDs or those with DLDs as opposed to those language disorders associated with other conditions. Mitigation strategies to improve the educational and psychosocial trajectories of children with DLD may continually improve the trajectories and reduce the intergenerational burden of DLD.

Language is the very basis of human communication. Attention to early language, vocabulary, and literacy are of inestimable value, as language is the medium through which we communicate, learn, inquire, negotiate, connect, and emote. To enrich a young child through verbal dialog, broad, repeated, and incrementally advancing vocabulary exposure, and early literacy skills is to provide a lifetime gift for the child and perhaps for the giver.

References

[1] Crowe K, McLoed S. Children's english consonant acquisition in the United States: a review. Am J Speech Lang Pathol 2020;1–15:1–1044/2020.

[2] Lewis BA, Freebairn L, Tag J, Ciesla AA, Iyengar SK, Stein CM, Taylor HG. Adolescent outcomes of children with early speech sound disorders with and without language impairment. Am J Speech Lang Pathol 2015;24:150–63.

[3] Rapin I. Practitioner review: developmental language disorders: a clinical update. J Child Psychol Psychiatry 1996;37(6):643–55.

[4] Bishop DVM, Snowling JM, Thompson PA, Greenhalgh T. The CATALISE-2 Consortium. Phase 2 of CATALISE: a multinational and multidisciplinary Delphi consensus study of problems with language development: Terminology. Journal of Child Psychology and Psychiatry 2017;58(10):1068–80.

[5] Korkman M, Hakkien-Rihu P. A new classification of developmental language disorders. Brain Lang 1994;47:96–116.

[6] American Psychiatric Association. Diagnostic and statistical manual of mental disorders, fifth edition (DSM-5). Arlington, VA: American Psychiatric Association; 2013.

[7] Wren YE, Miller LL, Peters T, Emond A, Roulstone SE. Prevalence and Predictors of persistent speech sound disorder at eight years old: findings from a population cohort study. J Speech Lang Hear Res 2016;59:647–73.

[8] Coulton GL. Speech disorders among college freshman: a 13 year survey. J Speech Hear Disord 1986;51:3–7.

[9] Feldman HM. How young children learn language and speech. Pediatr Rev 2019;40(8):398–411.

[10] Sansavini A, Favilla ME, Guasti MT, et al. Developmental language disorder: early predictors, age for the diagnosis and diagnostic tools. A scoping review. Brain Sci 2021;. Available from: https://doi.org/10.3390/brainsci11050654.

[11] Feldman HM. The importance of language-learning environments to child language outcomes. Pediatrics 2019;144. Available from: https://doi.org/10.1542/peds.2019-2157.

[12] Paul R. Language disorders. In: Gallagher A, Bulteau C, Cohen D, Michaud JL, editors. "Handbook of clinical neurology." Neurocognitive development: disorders and disabilities, 174. United Kingdom: Elsevier Publishing; 2020. p. 21–35. Available from: 10/1016/B978-0-444-64148-9.00002-8.

[13] Weismer SE, Tomblin JB, Durkin MS, Bolt D, Palta M. A preliminary epidemiologic study of social (pragmatic) communication disorder in the context of developmental language disorder. Int J Lang Commun Disord 2021;1–14. Available from: https://doi.org/10.1111/1460-6984,12664.

[14] Rudolph JM, Leonard LB. Early language milestones and specific language impairment. J Early Interv 2016;38:41–58.

[15] Snowling MJ, Duff FJ, Nash HM, Hulme C. Language profiles and literacy outcomes of children with resolving, emerging, or persisting language impairments. J Child Psychol Psychiatry 2016;57(12):1360–9.

[16] Snowling MJ, Melby-Lervåg M. Oral language deficits in familial dyslexia: a meta-analysis and review. Psychol Bull 2016;142(5):498–545. Available from: https://doi.org/10.1037/bul0000037.

[17] Deriziotis P, Fisher SE. Speech and language: translating the genome. Trends Genet 2017;33(9):642–56.

[18] Massaro DW. Two different communication genres and implications for vocabulary development and learning to read. J Lit Res 2015;47(4):505–27. Available from: https://doi.org/10.1177/1086296X15627528.

[19] Hart B, Risley TR. The early catastrophe: the 30 million word gap. Am Educator 2003;27:4–9.

[20] Hayiou-Thomas MD, Carroll JM, Leavett R, Hulme C, Snowling MJ. When does speech sound disorder matter for literacy? The role of disordered speech errors, co-occurring language impairment and family risk of dyslexia. J Child Psychol Psychiatry 2017;58(2):197–205.

[21] Beitchman JH, Wilson B, Brownlie EB, Walters H, Lancee W. Long-term consistency in speech/language profiles: I. Development and academic outcomes. J Am Acad Child Adolesc Psychiatry 1996;35(6):804–14.

[22] Beitchman JH, Wilson B, Brownlie EB, Walters H, Inglis A, Lancee W. Long-term consistency in speech/langauge profiles: II. Behavioral, emotional, and social Outcomes. J Am Acad Child Adolesc Psychiatry 1996;26(6):815–25.

[23] Beitchman JH, Nair R, Clegg M, Patel PG. Prevalence of speech and language disorders in 5-year-old kindergarten children in the Ottawa-Carleton region. J Speech Hear Disord 1986;51:8–110.

[24] Shrilberg LD, Tomblin JB, McSweeny JL. Prevalence of speech delay in 6-year-old children and comorbidity with language impairment. J Speech Lang Hear Res 1999;42:1461–81.

[25] Webster RI, Majnemer A, Platt RW, Shevell MI. Motor function in children with a preschool diagnosis of specific language impairment (SLI). J Pediatr 2005;146:80–5.

[26] Bolk J, Farooqi A, Hafström M, Åden U, Serenius F. Developmental coordination disorder and its association with developmental comorbidities at 6.5 years in apparently healthy children born extremely preterm. JAMA Pediatr 2018;172(8):765–74. Available from: https://doi.org/10.1001/jamapediatrics.2018.1394.

[27] Rice ML, Hoffman L. Predicting vocabulary growth in children with and without specific language impairment: a longitudinal study from 2;6 to 21 years of age. J Speech Lang Hear Res 2015;58:345–59.

[28] Catts HW, Bridges MS, Little TD, Tomblin JB. Reading achievement growth in children with language impairments. J Speech Lang Hear Res 2008;51:1569–79.

[29] Catts HW, Fey ME, Tomblin JB, Zhang X. A longitudinal investigation of reading outcomes in children with language impairments. J Speech Lang Hear Res 2002;45:1142–57.

[30] Dockrell JE, Lindsay G, Connelly V. The impact of specific language impairment on adolescents' written text. Except Child 2009;75(4):427–46.

[31] Connelly V, Dockrell JE, Walter K, Critten S. Predicting the quality of composition and written language bursts form oral language, spelling and handwriting skills in children with and without specific language impairment. Writ Commun 2012;29(3):278–302.

[32] Cross AM, Joanisse MF, Archibald LMD. Mathematical abilities in children with developmental language disorder. Lang Speech Hear Serv Sch 2019;50:150–63.

[33] Young AR, Beitchman JH, Johnson C, Douglas L, Atkinson L, Escobar M, Wilson B. Young adult academic outcomes in a longitudinal sample of early identified language impaired and control children. J Child Psychol Psychiatry 2002;43(5):635–43.

[34] Curtis PR, Frey JR, Watson CD, Hampton LH, Roberts MY. Language disorders and problem behaviors: a meta-analysis. Pediatrics 2018; 142(2):e20173551.

[35] Yew SGK, O'Kearney R. Emotional and behavioural outcomes later in childhood and adolescence for children with specific language impairmnets: meta-analyses of controlled prospective studies. J Child Psychol Psychiatry 2013;54(5):516–24.

[36] Brownlie EB, Beitchman JH, Escobar M, Young A, Atkinson L, Johnson C, Wilson B, Douglas L. Early language impairment and young adult delinquent and aggressive behavior. J Abnorm Child Psychol 2004;32(4):453–67.

[37] Mueller K, Tomblin JB. Examining the comorbidity of language and ADHD. Top Lang Disord 2012;32(3):228–46.

[38] Law J, Rush R, Schoon I, Parsons S. Modeling developmental language difficulties from school entry into adulthood: literacy, mental health, and employment outcomes. J Speech Lang Hear Res 2009;52:1401–16.

[39] Beitchman JH, Wilson B, Johnson CJ, Atkinson L, Young A, Adalf E, Escobar M, Douglas L. Fourteen-year follow-up of speech/language-impaired and control children: psychiatric outcome. J Am Acad Child Adolesc Psychiatry 2001;40(1):75–82.

[40] Conti-Ramsden G, Botting N. Emotional health in adolescents with and without a history of specific language impairment. J Child Psychol Psychiatry 2008;49(5):516–25.

[41] Conti-Ramsden G, Durkin K. Developmental language disorders. In: Skuse D, Bruce H, Dowdney L, editors. Child psychology and psychiatry: frameworks for clinical training and practice. John Wiley & Sons, Ltd; 2017. 3rd ed.

[42] Cohen NJ, Barwick MA, Horodezky NB, Vallance DD, Im N. Language, achievement and cognitive processing in psychiatrically disturbed children with previously identified and unsuspected language impairment. J Child Psychol Psychiatry Allied Discip 1998;39(6):865–77.

[43] Conti-Ramsden G, Durkin K, Toseeb U, Botting N, Pickles A. Education and employment outcomes of young adults with a history of developmental language disorder. Int J Lang Commun Disord 2018;53(2):237–55.

[44] Whitehouse AJO, Line EA, Watt HJ, Bishop DVM. Qualitative aspects of developmental language impairment relate to language and literacy outcome in adulthood. Int J Lang Commun Disord 2009;44(4):489–510.

[45] Botting N. Language, literacy and cognitive skills of young adults with developmental language disorder. Int J Commun Disord 2020;55(2): 255–65.

[46] Johnson CJ, Beitchman JH, Brownlie EB. Twenty-year follow-up of children with and without speech-language impairments: family, educational, occupational and quality of life outcomes. Am J Speech Lang Pathol 2010;19:51–65.

[47] Elbro C, Dalby M, Maarbjerg S. Language-learning impairments: a 30-year follow up of language-impaired children with and without psychiatric, neurological and cognitive difficulties. International Journal of Language & Communication Disorders 2011;46(4):437–48.

[48] Schoon I, Parsons S, Rush R, Law J. Children's language ability and psychosocial development: a 29 year follow-up study. Pediatrics. 2010;128(1):e73–e380.

[49] Dubois P, St. Pierre MC, Desmarais C, Guay F. Young adults with developmental language disorder: a systematic review of education, employment and independent living outcomes. J Speech Lang Hear Res 2020;63:3786–800.

[50] Whitehouse AJO, Watt HJ, Line EA, Bishop DVM. Adult psychosocial outcomes of children with specific language impairment, pragmatic language impairment and autism. Int J Lang Commun Disord 2009;44(11):511–28.

[51] Van den Bedem NP, Dockrell JE, van Alphen PM, Kalicharan SV, Rieffe C. Victimization, bullying, and emotional competence: longitudinal associations in (pre)adolescents with and without developmental language disorder. J Speech Lang Hear Res 2018;61:2028–44.

[52] Bottting N, Durkin K, Toseeb A, Conti-Ramsden G. Emotional health, support and self-efficacy in young adults with a history of language impairment. British Journal of Developmental Psychology 2016;34(4):538–54.

[53] Snow PC. Oral language competence, social skills and high-risk boys: what are juvenile offenders trying to tell us? Child Soc 2008;22:16–28.

[54] Snow PC, Powell MB. Oral language competence in incarcerated young offenders: links with offending severity. Int J Speech Lang Pathol 2011;13(6):480–9.

[55] Whitstanley M, Webb RT, Conti-Ramsden G. More or less likely to offend? Young adults with a history of identified language disorders. Int J Lang Commun Disord 2018;53(2):256–70.

Chapter 27

Attention deficit hyperactivity disorder

Steven M. Lazar[1] **and Robert G. Voigt**[2,3]

[1]*Section of Pediatric Neurology and Developmental Neuroscience, Meyer Center for Developmental Pediatrics & Autism, Baylor College of Medicine, Texas Children's Hospital, Houston, TX, United States,* [2]*Michael R. Boh Centers for Child Development, Developmental Pediatrics, Ochsner Children's Hospital, New Orleans, LA, United States,* [3]*Faculty of Medicine The University of Queensland Medical School, Brisbane, Australia*

Epidemiology

Attention-deficit/hyperactivity disorder (ADHD) is the most prevalent neurobehavioral disability. Early epidemiologic data showed a wide range of prevalence from 2% to 17% depending on the population studied [1]. While the diagnostic criteria for ADHD and further understanding of the condition evolved, large metaanalyses incorporating over 100 epidemiologic studies with child and adolescent populations from 35 countries across 6 continents estimated the global prevalence of ADHD to range from 5.3% to 7.2% [2,3]. Additional large metaanalytic studies confirmed that the large variance in reported ADHD prevalence was primarily due to methodological differences, including diagnostic criteria, sources of information, and the presence or absence of functional impairment due to symptoms [2,4]. Estimates of the prevalence of ADHD in preschool-aged children is similar to school-aged children, ranging from 2% to 5.7%, with older ages showing more stability in diagnosis [5].

ADHD prevalence in adulthood has an inverse relationship with age, ranging from approximately 5% to 9% in those from 18 to 25 years of age to 0.8% to 4.5% in those >60 years of age [5]. Longitudinal population studies have also shown decline in symptoms over age with estimates that 15% of patients with childhood and adolescent diagnosed ADHD will continue to reach full diagnostic criteria in adulthood, while up to 40%–60% will have persistent symptoms not reaching full diagnostic criteria [6]. Metaanalytic data have estimated that the prevalence of ADHD in adulthood ranges from approximately 2.5% with noted symptom onset in childhood to 6.7% irrespective of when symptoms were first noted, adding to the literature of changing symptomatology of ADHD over time and providing credence for an adolescent/adult-onset subtype [7].

Point prevalence of ADHD does not appear to vary geographically between Europe, Oceania, South America, Asia, Africa, the Middle East, or North America [2,4]. Within consistent methods, temporal prevalence of diagnosis also has not changed over time [4,8]. However, in contrast to large research population-based studies, there is significant variation in prevalence of clinical diagnosis. Within the United States, regional and state to state prevalence of ADHD varies significantly [8,9].

Within clinical samples of individuals with ADHD, there is a 4:1 male to female ratio, while larger population studies show a lower 2–2.5:1 male to female ratio [8]. This discrepancy is felt to be related to both males being more likely to manifest with externalizing (hyperactive-impulsive) symptoms as opposed to internalizing (inattentive) symptoms [10–12], and younger age of peak symptomatic incidence in boys as compared to girls. This combined with a dramatic decrease in this discrepancy with increasing age, with near equivalent rates in adulthood, suggest underidentification of girls with ADHD in childhood [13–15].

Ethnic disparities are also seen in the diagnosis of ADHD. Over a 20-year period in the United States, 12.8% of non-Hispanic black, 12% of non-Hispanic white, and 6.1% of Hispanic youth were diagnosed with ADHD [16]. Additionally, black and Latinx children are less likely to be diagnosed and medicated compared to their white peers when controlling for socioeconomic factors and symptom severity [17,18].

A large portion of the variation in prevalence estimates could be due to methodological differences [4]. Additionally, cultural differences may explain some variability, including over- and underreporting of symptoms by

Capute and Accardo's Neurodevelopmental Disabilities in Infancy and Childhood. DOI: https://doi.org/10.1016/B978-0-12-824060-1.00017-1

parents in some countries [19]. Given this, interpretation of symptom reporting should be taken in the context of the family's cultural beliefs and expectations about behavior.

Signs and symptoms

ADHD is a neurobehavioral syndrome that leads to specific impairments in interpersonal, academic, or occupational functioning (Fig. 27.1). ADHD is diagnosed clinically by the presence of developmentally inappropriate and functionally impairing symptoms within the spectrum of attentional regulation, activity level, and impulse control and self-regulation. There are no neuroimaging, biochemical, or neuropsychological tests or other biomarkers that confirm a diagnosis of ADHD.

Historical notes

The earliest mentions in the medical literature of a syndrome of disordered attention and impulsive behavior occurred in the writings of Weikard [20] and Crichton [21] in the late 18th century. Various references in popular literature expanded on these symptoms via children's books authored by Hoffman featuring the characters "Fidgeting Philip" and "Johnny Head-in-the-Air" [22]. At the turn of the 20th century, the constellation of behaviors consisting of restlessness, inattentiveness, and overarousal with an inability to internalize rules and limits were attributed to a deficit in a child's morals, despite normal intelligence and proper parenting [23,24]. In the 1920s, following the global influenza pandemic, an epidemic of postinfectious encephalitis (von Economo's encephalitis) in a subset of children and adults with similar patterns of aggressive and hyperactive behaviors was described [25]. In the 1940s, Alfred Strauss recognized that a subset of children with intellectual disability exhibited hyperactive behaviors that were discrepant from their underlying developmental ages. At that time, he also noted that a similar cluster of behaviors, including distractibility, perseveration, hyperactivity, impulsivity, low frustration tolerance, and emotional lability were characteristic of patients with posttraumatic brain injury. Noting these similar behavioral aberrations, Strauss hypothesized that both the subset of children with intellectual disability and those with direct brain injury were exhibiting behavior that characterized a "brain-injured child" [26]. This concept later evolved into the term "minimal brain dysfunction" in the 1960s and 1970s

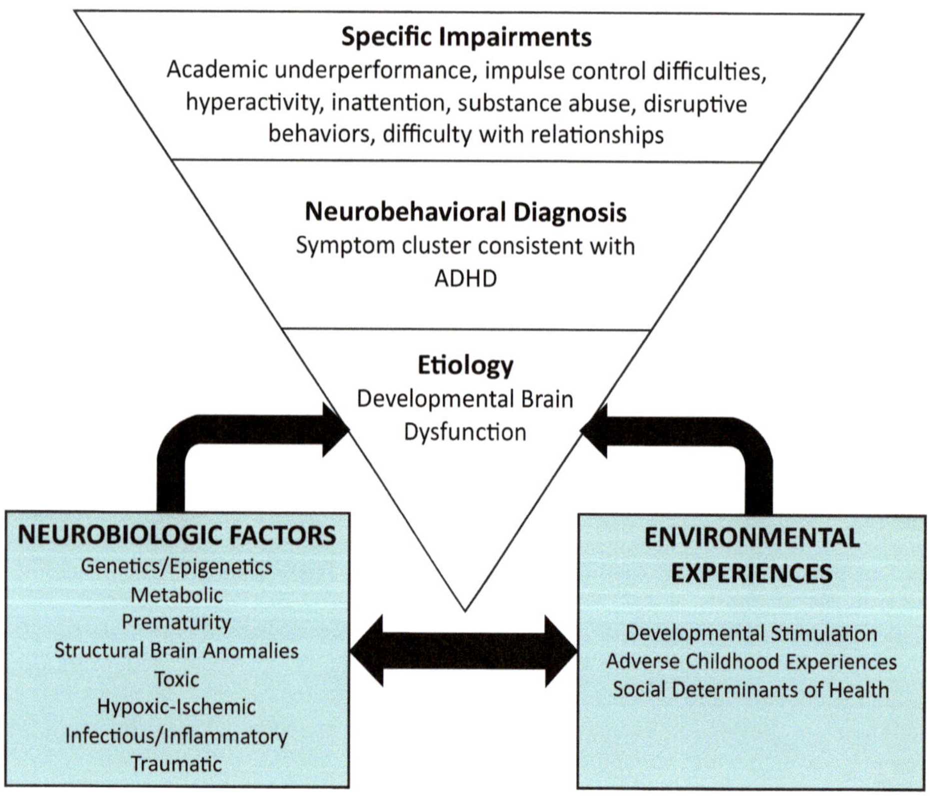

FIGURE 27.1 A neurodevelopmental framework of attention-deficit/hyperactivity disorder (ADHD) across diagnostic levels.

[27–29]. More recent genetic, neuroimaging, and clinical data have been used to help solidify this concept as a framework to better understand ADHD within the spectrum and continuum of neurodevelopmental disabilities (developmental brain dysfunction) [30,31].

Parallel to this broad etiologic framework, codified criteria for a symptomatic diagnosis of ADHD evolved. What is now known as ADHD was initially described as "Hyperkinetic Reaction of Childhood" in 1968 with the publication of the American Psychiatric Association's *Diagnostic and Statistical Manual (DSM) of Mental Disorders 2nd Edition* (*DSM-II*) [32]. This initial focus on hyperactive and disruptive behavior progressed by 1980 to the *DSM-III* diagnostic criteria for the newly defined "Attention-Deficit Disorder (with or without hyperactivity)" that involved varied impairments in attention, impulse control, and activity [33]. This categorization was made alongside other disruptive behavior disorders, such as conduct disorder and oppositional defiant disorder. By 1987 with the publication of the *DSM-III-Revised*, the combined term ADHD was first introduced [34]. To expound on the variable presentation of a categorical diagnosis of ADHD, the DSM-IV in 1994 [35] added predominantly inattentive, predominantly hyperactive, and combined types of ADHD [36]. As a broader approach to the functional impairments, the American Academy of Pediatrics (AAP) produced the *DSM-Primary Care* (*DSM-PC*) *Child and Adolescent Version* which emphasized utility of intervention for children who may not meet full diagnostic criteria for ADHD [37], though this has not been revised to meet the current *DSM* diagnostic criteria.

Current diagnostic criteria

The current edition of the *DSM-5* was published in 2013, and it presents the diagnosis of ADHD as a threshold diagnosis in which a child must have at least six of the described symptoms of inattention or impulsivity/hyperactivity with several symptoms occurring before the age of 12 years, persisting for at least 6 months, occurring across two or more environments, and reducing the quality of social, academic, or occupational functioning, while not being better accounted for by another disorder [38]. These diagnostic criteria have been reaffirmed in the 2022 DSM-5 Text Revision (DSM5-TR) [39]. Recent changes to the diagnosis include those over 17 years of age now only having to meet at least five symptoms in each category. In *DSM-5*, ADHD was also notably transitioned from categorization with behavioral and disruptive disorders in childhood to neurodevelopmental disorders. Additionally, subtypes of ADHD are now defined as the predominant presenting symptoms to better emphasize the variation of symptomatology both across individuals and within an individual over time. The age of onset of symptoms was also increased from before 7 years of age in *DSM-IV-TR* to before 12 years of age in *DSM-5*, allowing for inclusion of individuals with later onset symptoms. A population-based study showed this change led to an increased prevalence (7.4%–10.8%) of ADHD in childhood as well as improved catchment of lower resource and ethnic minority families [40], though there is growing evidence that a strict age cutoff at any level may limit the true prevalence and identification of ADHD through the lifespan [41].

Notably, an evaluation or determination of etiology is not required for a formal diagnosis of ADHD. Insight into the etiology of ADHD and risk factors has led to some groups distinguishing primary or secondary (acquired) ADHD. Primary ADHD is defined as an inherent neurodevelopmental condition, while secondary ADHD is used to describe those who develop clinically significant symptomatology after an acquired brain injury (particularly traumatic brain injury). Either group can receive a diagnosis of ADHD using the *DSM-5* criteria and both groups show relatively similar efficacy of standard treatments when controlling for other secondary disabilities due to injury [42–47].

Parallel to the DSM-5, the International Classification of Disease—11th edition presents a broad diagnostic description that gives significant weight to clinical judgment to guide a diagnosis without mention of etiology [48].

A checklist approach to a diagnosis of ADHD is problematic in that many of the single behavioral criteria can be seen in a variety of other neurodevelopmental disabilities. For example, some degree of inattention, impulsivity, and hyperactivity characterize a normal preschooler's behavior. Further, every child with autism spectrum disorder (ASD) exhibits a pattern of atypical attention [49]. In a school setting, several of the clinical criteria for ADHD could be met solely due demands and expectations for academic performance that exceed underlying abilities in children with below average intelligence, language disorders, or learning disabilities [50,51]. Additionally, there is variation in the reporting of ADHD symptoms among parents [52] and between parents and educators [53–56].

Another limitation of a threshold symptom checklist approach is that it neglects to recognize the continuous spectrum of inattentive, impulsive, and hyperactive behaviors. A checklist approach leads to a nonevidenced-based categorical diagnostic structure, whereas most current research favors a holistic, dimensional, spectrum-based approach to neurobehavior leveraging clinical judgment [57]. Approaching ADHD-related behaviors along a spectrum allows for a more fluid diagnostic structure that will allow for monitoring and individualized treatment strategies. Some additional diagnostic terms have been proposed such as "Sluggish Cognitive Tempo," but whether this is a separate subgroup or simply subthreshold inattentive symptomatology of ADHD remains to be determined [58–60].

Additionally, viewing neurobehavioral symptoms as continuous as opposed to categorical traits has significant implications for research and understanding of the neurobiological underpinnings of ADHD and other neurodevelopmental disorders [30,61,62]. Some theoretical frameworks have promoted that behaviors on the positive end of the neurobehavioral spectrum can have protective effects on other areas on the continuum of disability, highlighting the importance of comprehensive evaluation to evaluate neurodevelopmental and neurobehavioral strengths as well as weaknesses [63]. For example, children with a specific learning disability in reading with high levels of attentional regulation and impulse-control may be more adept to time-intensive instructional techniques and interventions than those with comorbid ADHD.

Hence, clinical judgment and evaluation of ADHD symptoms, including a complete neurodevelopmental history and evaluation, is essential both for making a diagnosis and isolating targeted treatment strategies across the spectrum and continuum of neurodevelopmental disabilities.

Clinical presentation of attention-deficit/hyperactivity disorder

The clinical presentation of ADHD tends to vary by age. Evaluation of infant behaviors emphasizes the early developmental origins of ADHD. Studies have evaluated the quality of early infant general movements as predictors of later neurodevelopmental disability, with significantly abnormal movements predicting later diagnosis of cerebral palsy while mildly abnormal general movements have been associated with minor neurologic dysfunction and ADHD [64]. Later diagnosis of ADHD is also associated with other infantile behaviors, including visual attention [57], activity level [58,59], and emotional regulation and temperament [65–67], although each of these is variably associated with the continuum of neurodevelopmental disabilities and not specific for ADHD [68].

During the preschool ages, hyperactivity and impulsivity tend to be the predominant symptoms. ADHD is significantly difficult to diagnose in this age group, as some level of inattention, impulsivity, and increased activity level are characteristic of typical preschoolers. Thus only the most severe of those on the spectrum of ADHD behaviors should meet criteria for a preschool-aged diagnosis of ADHD [69]. Additionally, children diagnosed with ADHD in this age group tend to also be diagnosed with other neurodevelopmental disorders, such as developmental delay, ASD, and language disorders [70–72] consistent with the basic neurodevelopmental tenet that the more severe the developmental disruption, the earlier and more comorbid diagnoses can be ascribed. Thus to make an accurate diagnosis, all preschool-aged children should undergo a comprehensive evaluation, including hearing and visual assessment, as well as a complete neurodevelopmental history and evaluation. If a diagnosis is made in the preschool age, most of the patients will continue to meet criteria into school age [6,73–76]. Finally, if a diagnosis is made in the preschool age, behavioral interventions should be pursued prior to a trial of psychotropic medication [77].

During school age, common presenting symptoms include problems with listening, task completion, work accuracy, and difficulty with social interactions. Additionally, overt signs of hyperactivity and impulsive behaviors are at their peak [6,78,79]. At this age, defining the presence of a learning disability or ADHD, or a combination of both, is imperative for targeted intervention. As with the preschool age, a comprehensive neurodevelopmental evaluation both assists in the initial diagnosis and the development of appropriate treatment recommendations. Approximately 65% of children diagnosed with ADHD in childhood will continue to carry a diagnosis into adolescence, while many of those who lose the ADHD diagnosis continue to have present but decreased symptom severity [50,80,81].

As children reach adolescence, there is a notable decline in reported hyperactivity [76,80,82]. Specific academic concerns tend to revolve around work completion and organization. Although a DSM-5 diagnosis of ADHD requires impairing symptoms by history, there is a likely subset of first-time adolescent diagnoses for individuals who were able to compensate for their ADHD symptoms through a variety of factors, including limited hyperactive/impulsive symptoms, strong intellectual or social skills, or familial and community support. However, this population's ability to adapt becomes overwhelmed by the increasing social and academic demands of adolescence, and their symptoms become more externally apparent. Factors contributing to persistence into adulthood are consistent with other age groups with severity of symptoms being the most significant [6,83].

Cross-sectional birth cohort studies have guided interest in a subset of individuals with ADHD who may have onset in late adolescence or adulthood [73,74,79,84–87]. However, more prospective, longitudinal studies have reported that a large proportion of late adolescent and adult patients meeting criteria for ADHD, in the absence of a childhood diagnosis, show significant fluctuation in their symptoms, have symptoms that tend to be transient in nature, or have symptoms deemed secondary to another condition, such as substance use disorder [86,88,89].

Across all age groups, the prominent factors for persistence of an ADHD diagnosis include the severity of ADHD symptoms, pharmacologic treatment for ADHD (which may be a marker for severe symptoms itself), cooccurring disruptive behaviors or mood disorders (i.e., oppositional defiant disorder, conduct disorder, depression), or a family history of persistent ADHD [73,75,85,90].

Assessment of attention-deficit/hyperactivity disorder

Inattention and hyperactivity as chief complaints have broad neurodevelopmental differential diagnoses. The assessment process of attentional concerns is not unifactorial, and a diagnosis of ADHD cannot be made based on a chief complaint, checklist of symptoms, or observation. The diagnosis must be made through clinical judgment based on a collection of historical information and direct, comprehensive neurodevelopmental evaluation to confirm both the developmental inappropriateness of and impairment caused by the presenting symptoms, and to ensure that the presenting symptoms are not secondary to demands and expectations that exceed an individual's underlying developmental abilities. Typical data that are collected, analyzed, and synthesized to arrive at a clinical diagnosis include but are not limited to: a targeted history from the patient and/or caregiver(s), a review of the patient's medical, social, family, educational, and most importantly, developmental histories, a review of all prior neuropsychological or psychoeducational testing reports, and physical and neurodevelopmental examinations. Through this evaluation, it is recommended that clinicians, regardless of background, consider five specific areas of inquiry as they relate to attentional dysregulation [91].

1. Symptoms: How many symptoms of attentional dysregulation are present? Does the patient meet the symptom count criteria for a DSM-5 or ICD-11 diagnoses of ADHD?
2. Impairment: How is the patient impaired? How severe are the symptoms and in which environments are the symptoms impairing? Are the impairments related to attentional dysregulation or secondary to other unrecognized neurodevelopmental concerns or due to inconsistent parenting?
3. Comorbidity: What comorbid conditions and/or associated problems are present by history or need to be evaluated further?
4. Mimics: Can the symptoms be explained fully by a typical developmental trajectory or another neurodevelopmental/neuropsychiatric diagnosis where secondary inattention results from inappropriate demands and expectations, treatment, or environmental situations that may mimic attentional dysregulation? Do these conditions exacerbate or even obscure an underlying diagnosis of ADHD?
5. Strengths: What are the patient's and their support system's strengths? How does the patient excel? Are they able to compensate for their symptoms?

Typically, a referral for evaluation of ADHD occurs based on a problematic behavior observed within the home or school environment as opposed to a dedicated symptom. It is thus recommended to first uncover the core impairments before directed questioning regarding ADHD symptomatology. This aims to avoid biasing the evaluator by assuming a diagnosis at the outset (anchoring bias) and providing leading questions that could miss mimics and alternative diagnoses. A complete neurodevelopmental assessment and relevant contributions to a diagnosis of ADHD is summarized in Table 27.1.

Accessory information to the neurodevelopmental evaluation typically includes parent and teacher completion of ADHD rating scales. However, caution needs to be taken when interpreting these rating scales, as they possess all the limitations inherent in instruments that rely on untrained, biased observers to make difficult judgments on complex behaviors [72,92]. Rating scales are collections of targeted questions in which a rater/observer gives a rating, typically on a Likert scale, describing the presence of and severity of behaviors. Several rating scales have been developed to help quantify ADHD symptoms in a more standardized way.

Rating scales have the benefit of being cost-effective and can provide a brief dimensional estimate of targeted symptoms, but they also should be used with caution as there is only modest agreement between different raters (self and parents, parents and teachers, and even between parents) [55,93,94]. Consequently, it is important for the clinician to determine the reason for this lack of agreement, which may include lack of context, failure to define answers (how many fidgets are a "very much"?), rater bias, unrealistic expectations for developmental level, and cultural perspectives of typical behaviors. ADHD rating scales may be more useful in evaluating treatment effects when administered prior to and after medication initiation or behavioral intervention [95]. Clinicians must understand the limitations of ADHD rating scales, as a high symptom score does not necessarily equate with a clinical diagnosis [96,97]. Most importantly, care must be made not to interpret symptom rating scales as diagnostic tools given that these measures are based on

TABLE 27.1 The neurodevelopmental assessment of attention-deficit/hyperactivity disorder (ADHD).

Component of the neurodevelopmental evaluation	Key evaluation strategies and examples that support a diagnosis of ADHD
Chief complaint	Failed behavioral screening test; specific impairment or primary symptoms related to inattention or impulsivity-hyperactivity (i.e., does not complete school-work, forgetfulness, staring spells)
History of present illness	Onset of symptoms typically prior to age 12 years; symptoms of inattention and/or hyperactivity-impulsivity result in impairment in more than one setting
Developmental and academic history	Pattern of developmental history and trajectory (fetal hyperactivity, difficult baby, "dare-devil" behaviors); developmental history indicates that behavioral symptoms are inappropriate for developmental level and are not secondary to inappropriate demands or expectations placed on child; further focus on academic performance and concerns (i.e., frequent mistakes, forgetful in homework, low grades secondary to incomplete assignments, seems to rush through assignments)
Birth history	Maternal history of miscarriages; maternal medications; prenatal toxin exposures; gestational hypertension/preeclampsia; gestational age; birthweight; other risk factors for neurologic impairment
Medical history	Personal history of frequent injuries due to impulsiveness/hyperactivity, cooccurring neurological disorders
Family history	Family history of ADHD or other neurodevelopmental disorders
Social history	Adverse childhood experiences; social determinants of health
Physical/ neurodevelopmental examination	Neurologic "soft signs" (i.e., behavioral toe walking, poor coordination, overflow movements); dysmorphology or neurocutaneous stigmata concerning for high risk neurogenetic disorders; easy distractibility, poor impulse control, excessive motor activity directly observed during developmental testing; developmental testing shows that levels of inattention, impulsivity, and excessive motor activity are inappropriate for developmental level and not secondary to alternative neurodevelopmental diagnosis (below average cognition, intellectual disability, language disorder, specific learning disability)
Review of prior neuropsychological and academic testing	Pattern of executive dysfunction; dissociated adaptive/functional skills from overall cognitive abilities
Review of prior laboratory testing	Assessing for risk factors such as lead poising, genetic syndromes associated with ADHD, metabolic disorders

chronological age and not on cognitive/developmental level, and thus, they do not evaluate whether the behaviors being rated are truly developmentally inappropriate.

Rating scales can be broadband measures, which evaluate behaviors beyond ADHD symptoms, such as the Child Behavior Checklist (CBCL) [98] (attention problems scale: sensitivity: 77%, specificity: 73% [99]), Strengths and Difficulties Questionnaire (SDQ) [100], Behavior Assessment Scale for Children—3rd Edition (BASC-3) [101], and Connors 3 [102]. Rating scales can alternatively be narrowband, typically having a direct relation to the DSM-IV-TR or DSM-5 symptom criteria. Examples include the ADHD Rating Scale 5 [103] and the NICHQ Vanderbilt Assessment Scales [104] (parent form: sensitivity: 69%, specificity: 84% [105]; teacher form: sensitivity: 80%, specificity: 75% [106]). Some rating measures are offered at no cost and are easily accessible, while others have fees associated with them.

Clinically, utilization of both a broadband measure of psychopathology and a narrowband scale of ADHD-related symptoms may be helpful with diagnostic formulation. Preferably, scales are completed by both a primary caregiver and a teacher who works directly with the child at school. Teacher behavioral rating scales can be particularly useful, as they show high levels of reproducibility and validity when compared to direct observations and clinical interviews [107]. Rating scales are ideally completed prior to the initial patient/caregiver interview to avoid biasing ratings based on discussion of symptoms and counseling on behaviors.

Neuropsychological tests, particularly those focusing on executive function, have shown associations with a diagnosis of ADHD. Even though there are associations, no neuropsychological test or task performance can reliably diagnose ADHD. Neurocognitive deficits associated with ADHD, such as reaction time variability, vigilance, working memory, and response inhibition, may lead to targeted treatment outcomes and progress, but they are nonspecific and cannot confer a clinical diagnosis of ADHD [108–111].

Throughout the neurodevelopmental assessment, care must be made to determine the source of the symptomatology. Medical problems that may mimic an ADHD diagnosis given their shared symptoms are varied and important to assess as the treatments vary considerably [91,112]. Table 27.2 summarizes some common ADHD mimic conditions, key historical components, and steps to evaluate. Note that "mimic" conditions are considered as separate from high risk predisposing factors, such as genetic disorders (see Table 27.4), prematurity, and traumatic brain injury.

Several evidence-based guidelines for the assessment of ADHD are available to clinicians. The AAP has published an evidence-based practice guideline for the diagnosis and treatment of ADHD in the primary care setting [113]. The Society for Developmental and Behavioral Pediatrics has published an evidence-based practice guideline for the assessment and treatment of children and adolescents with "complex ADHD," defined by age (presentation at <4 years or >12 years of age), or by presence of coexisting conditions (such as ASD), moderate to severe functional impairment, diagnostic uncertainty, or inadequate response to treatment [77].

TABLE 27.2 **Some Conditions that can mimic attention-deficit/hyperactivity disorder (ADHD) symptoms and their evaluation.**

Mimic condition	Red flags and associations	Evaluation
Visual/auditory impairment	Limited fixation on visual tracking; atypical localization to sounds; requires repeated visual or auditory cueing	Vision and hearing screen
Intellectual disability/borderline intelligence/below average cognitive abilities/"slow learner" (IQ <90)	Developmental history indicates static pattern of delays across developmental streams; developmental/school history indicates that demands and expectations at school and/or home are inconsistent with child's developmental level; symptoms most noticeable during more advanced cognitive tasks	Developmental testing; neuropsychological testing; genetic testing
Learning disabilities	Symptoms occur primarily in learning settings (at school; during homework time); symptoms notable with specific tasks (reading, arithmetic, writing tasks)	Developmental testing; psychoeducational testing
Psychiatric disorders (i.e., depression, anxiety, adjustment disorder)	Developmental history indicates age-appropriate acquisition of developmental milestones; history of trauma/social stressor prior to behavioral symptom onset; depressed mood; hypervigilance; excessive worry	Developmental testing; broad-band behavioral screening
Sleep disorders	Noisy breathing, snoring, and/or periods of breathing pauses; need for frequent daytime naps; daytime sleepiness	Sleep diary; polysomnogram
Thyroid disorders	Sudden new onset concerns Hyperthyroid: heat intolerance, hypertension, tachycardia, jitteriness, hyperreflexia Hypothyroid: cold intolerance, unexplained weight gain, fatigue	Thyroid function tests and antibody testing; endocrinology evaluation
Adrenal disorders	Persistent or paroxysmal hypertension/tachycardia; decompensation with mild illness	Cortisol and adrenal function tests; endocrinology evaluation
Seizures (i.e., childhood absence epilepsy)	Inability to regain attention by others; periods of lost time; paroxysmal stereotyped movement with or without maintained awareness	Neurological examination; electroencephalogram, and brain MRI if indicated
Medication effect	Timing of symptoms around peak effect of medicine; no concerns prior to initiation of medicine	Medication list; trial of stopping/adjusting current medications
Neurodegenerative disorders (i.e., leukodystrophy, demyelinating disorders, mitochondrial/metabolic disorders)	Developmental history indicates a progressive pattern of developmental delay (plateauing or regression of developmental milestones); progressive worsening of behavioral symptoms; decompensation with illness	Review of newborn screening tests; abnormal findings on neurological exam; brain MRI; developmental testing; genetic/metabolic testing

Neurobiology and etiology of attention-deficit/hyperactivity disorder

Current understanding of the etiology and neurobiology of ADHD continues to develop. As with all other neurodevelopmental disabilities, ADHD is a descriptive behavioral diagnosis with multiple etiologies and without a single underlying neuropathology. Numerous genetic, anatomic, and structural and functional neuroimaging methods have shown a wide variety of associations between ADHD symptoms and underlying neuropathology.

Neuroanatomy of attention

The foundation of attention and attentional dysregulation involves a complex interplay between numerous regional brain areas and interrelated circuits, each of which has been evaluated and implicated in the pathophysiology of ADHD. Though there are associations, caution must be taken in making direct neurocognitive and behavioral correlates to single brain regions or networks.

Principally, there has been significant interest in dopaminergic and noradrenergic systems in the pathophysiology of ADHD, particularly given the efficacy of neurostimulants, such as methylphenidate, that act on these systems [114–117]. Dopaminergic system dysfunction has been associated with ADHD regarding mesocortical, mesolimbic, and nigrostriatal pathways corresponding to cognitive deficits related to motivation, reward, and voluntary movement, respectively [117–119]. Noradrenaline system dysfunction has also been implicated due to the intrinsic relationship between dopamine and noradrenaline, as dopamine is hydroxylated into noradrenaline. Dysfunction in this system has led to atomoxetine as a potential therapy in ADHD [116,120].

Fig. 27.2 and Table 27.3 provide a brief review of regional anatomy and functional circuits along with their relationship to the neuroanatomy of attention [80].

Neuroimaging and attention-deficit/hyperactivity disorder

Over the last two decades, there has been an exponential growth in structural and functional neuroimaging studies of ADHD. Frequently, the goals of these studies have been to characterize structural and functional correlates of ADHD

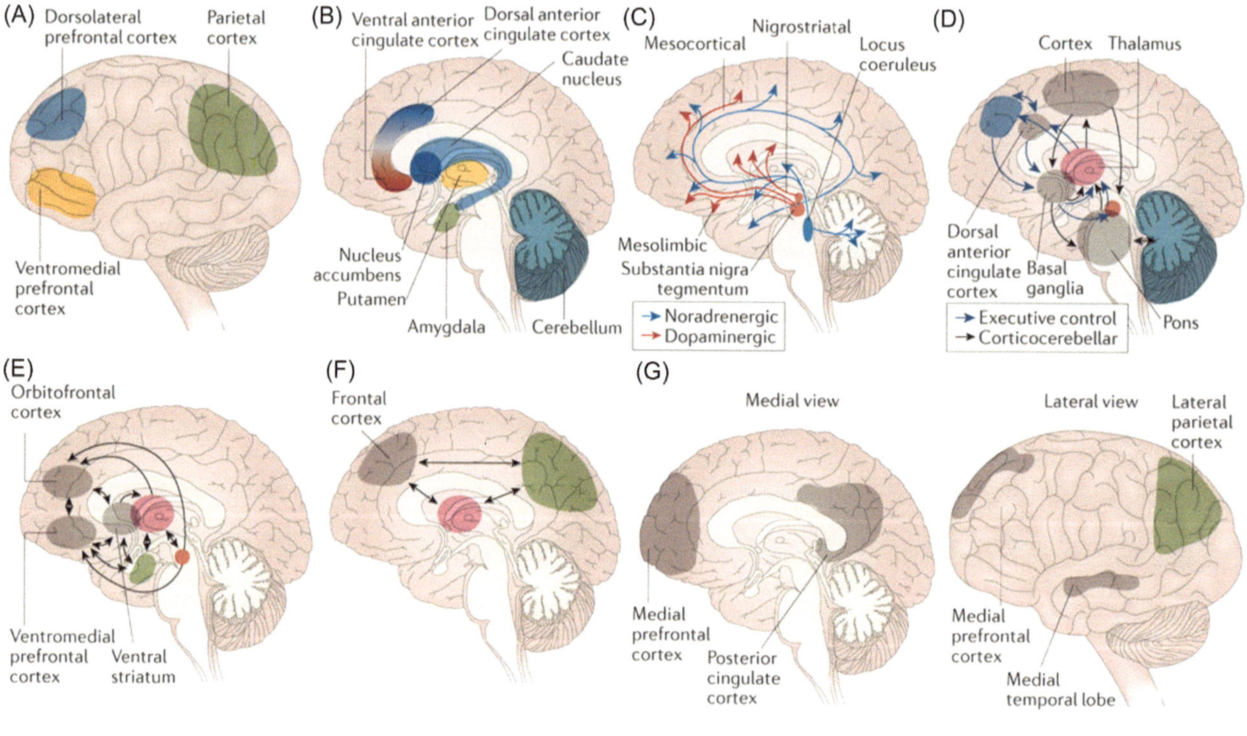

FIGURE 27.2 Neuroanatomic regions and functional networks associated with attention-deficit/hyperactivity disorder (ADHD). *Reprinted and adapted by permission from Faraone SV, et al. Attention-deficit/hyperactivity disorder. Nat Rev Dis Primers 2015. https://doi.org/10.1038/nrdp.2015.20.*

TABLE 27.3 Brain mechanisms of attention-deficit/hyperactivity disorder (ADHD).

Brian region		Neurocognitive association
Cerebral cortex (a)	Dorsolateral prefrontal cortex	Working memory
	Ventromedial prefrontal cortex	Complex decision making and strategic planning
	Parietal cortex	Spatial orientation of attention
Fronto-striatal circuit (b)	Central anterior cingulate cortex, dorsal anterior cingulate cortex, basal ganglia (nucleus accumbens, caudate, putamen)	Affective and cognitive components of executive function
Dopaminergic system (c)	Substantia nigra/tegmentum Mesolimbic, mesocortical, nigrostriatal pathways	Planning and initiation of motor responses Activation Switching Reaction to novelty Processing of reward
Noradrenergic system (c)	Locus coeruleus	Arousal modulation Signal to noise ratio in cortical areas State-dependent cognitive processes Cognitive preparation of urgent stimuli
Executive control network (d)	Corticocerebellar circuit	Planning Goal-directed behaviors Inhibition Working memory Flexible adaptation to context
Reward network (e)	Ventromedial prefrontal cortex, orbitofrontal cortex, ventral striatum, thalamus, amygdala, substantia nigra	Anticipation and receipt of reward
Alerting network (f)	Frontal and parietal cortical areas, thalamus	Attentional regulation and sustained alertness
Default mode network (g)	Medial prefrontal cortex, posterior cingulate gyrus, lateral parietal cortex, medial temporal lobe	Active at rest and diminished with externally oriented tasks

Source: Reprinted and adapted by permission from Faraone SV, et al. Attention-deficit/hyperactivity disorder. Nat Rev Dis Primers 2015. https://doi.org/10.1038/nrdp.2015.20.

and attentional dysregulation, as well as to assess the diagnostic utility of neuroimaging techniques. Through these numerous studies, many metaanalytic reviews with large data sets have shown significant heterogeneity and inconsistent findings. This is not unexpected, given the variability of symptomatic presentations, etiologic factors, and related cognitive impairments in individuals with ADHD [121]. While it is unlikely that there will be isolated diagnostic biomarkers for the spectrum of ADHD, these studies can help elucidate pathophysiologic changes that may predict treatment effects and consequences or suggest novel therapeutic interventions.

Structural neuroimaging

Magnetic resonance imaging (MRI) has allowed researchers to noninvasively evaluate anatomic images of the brain with increasingly higher spatial resolution as techniques advance. MRI studies have tended to focus on brain volumetrics, cortical surface measures, and white matter integrity. Many studies have used direct comparisons focusing on ADHD as a dichotomous trait, comparing those with a clinical diagnosis to those without. Only more recently has there been an increased interest in determining the relationship of structural imaging findings with attentional dysregulation as a more continuous trait.

Volumetric MRI studies have estimated the differences between total brain volume, made volume comparisons between gray and white matter, and evaluated targeted volumes of regional interest [122–124]. The most consistent finding among metaanalytic studies has been overall reduced (3%–5%) whole brain and gray matter volume in individuals with ADHD compared to typical controls [122,124–127]. Furthermore, whole brain volume has also been found to negatively correlate with attentional dysregulation symptoms in the general population [128]. The relative white matter volume among those with ADHD compared to controls has been more inconsistent, with some studies showing reductions in white matter volume [122], while others have not found a significant difference [125]. Region of interest-based studies has evaluated both cortical and subcortical gray matter structures. Cortical regions implicated in volumetric studies include the total cortical volume, multiple frontal areas, and occipital and parietal lobes [122,124]. Subcortical regions of interest studies have reported more consistent findings, including reduced volumes of portions of the basal ganglia, including the putamen, globus pallidus, and caudate nucleus, in children with ADHD compared to controls [125,126,129–131].

Further studies have attempted to illuminate the specific contribution of these volumetric changes, cortical thickness, cortical surface area, and gyrification (a measure of cortical folding). There is significant inconsistency in the findings of measurements of cortical thickness, with some studies showing varied regions of cortical thinning (including predominantly the frontal, prefrontal, parietal, temporoparietal, and occipital regions) in those with ADHD compared to controls [126,132–135], while other studies have failed to replicate this difference [136,137] or have even shown increased cortical thickness [138]. Relatedly, overall brain surface area has been shown to be reduced predominantly in the prefrontal, frontal, parietal, temporal, and left occipital lobes in those with ADHD compared to controls [134,136–138]. Studies of gyrification have been more inconsistent, with larger powered studies showing no difference between controls and those with ADHD [139].

Apart from cortical and subcortical gray matter structures, the implication of impaired regional interconnectivity in ADHD has led to investigations of white matter tract integrity utilizing diffusion tensor imaging (DTI). Using fractional anisotropy, a DTI-derived index that has been associated with myelination, axonal density, and fiber organization and integrity, both increased and decreased fractional anisotropy values have been reported in individuals with ADHD compared to controls. Atypical interhemispheric connectivity, particularly within the corpus callosum, is the consistent finding across multiple studies and metaanalyses [140–142]. Additionally, there does not appear to be a difference in white matter tractography between inattentive and combined type presentations of ADHD [143].

History of and current stimulant use in ADHD management also plays a role in brain volumetrics. Two metaanalyses have demonstrated that treatment with stimulants is associated with a less pronounced regional and total volumetric discrepancy between those with ADHD and controls [124,144–148]. This finding, however, has not been consistently replicated, and caution must be made in assuming "normalization" of structural changes with stimulant treatment as well as the causative implication of these findings [125–127].

Inconsistent data with structural imaging are not unexpected given the heterogeneity of ADHD and the high incidence of comorbidity that could account for these differences. This has led to an increasing interest in functional networks and interactions as opposed to just structural differences.

Functional neuroimaging

Functional neuroimaging is classically based on indirect evaluation of neuronal function through hemodynamic techniques, including functional MRI (fMRI), positron emission tomography (PET), single-photon emission computed tomography (SPECT), as well as combined techniques.

Initial studies focused on the use of SPECT and PET targeting dopaminergic systems, given their implication in pathophysiology and effective stimulant treatment in ADHD. Metaanalysis has shown up to a 14% higher density of dopamine transporters in ADHD groups compared to control groups in the striatum, but this has been felt to be a response to stimulant use as opposed to being pathogenic [149].

Utilizing task-based fMRI, studies have shown reduced neuronal activity in bilateral fronto-striatal networks, and to a lesser degree in occipital, parietal, and temporal areas, in ADHD groups compared to controls. Specific tasks have included motor response, interference inhibition, and set shifting tasks, all of which have relationships to known neurocognitive differences in ADHD compared to controls [131,150–152]. Studies targeting working memory tasks have shown decreased activation in bilateral frontal areas, frontoparietal areas, and the insula [151,153], while attentional tasks have shown decreased activation of the right prefrontal cortex (PFC), posterior basal ganglia, thalami, and parietal regions and increased activation of the cerebellum and precuneus [151,152] in those with ADHD compared to controls. Further investigations have shown age-dependent disruptions of frontal activation from inefficient nonspecific activation in childhood to discrete circuits in adulthood [154], which is consistent with studies showing delayed frontal maturation being related to a diagnosis of ADHD [135].

Aside from provoked and task-based activation procedures for fMRI, further interest has arisen in resting-state fMRI. Functional connectivity studies have focused on the Default Mode Network (DMN; regions typically reduced during tasks and increased during rest), cognitive control (frontoparietal regions), and attentional (dorsal, ventral, salience) networks, as well as their interrelationships [121,155,156]. Consistent findings have shown individuals with ADHD to have decreased synchrony between regions of the DMN and reduced anticorrelation between the DMN and cognitive control and attentional networks [157–159]. Spatial convergence has only trended to significance in a larger more recent metaanalysis [160]. Impairment in affective (limbic/amygdala related) networks and fronto-striatal motivational/reward circuits has also been implicated [158,161–163].

Functional neuroimaging studies have also provided insight into markers of treatment effect, as well as long-term treatment exposure, particularly with stimulant use. Stimulant use prior to task-based fMRI studies has been associated with normalization of both functional activation patterns on inhibition, timing, and attention tasks [147,152,164,165] and resting state functional connectivity [160,166,167].

As with structural MRI studies, higher powered functional MRI studies and metaanalyses frequently tend to have inconsistent findings and low reproducibility. Multiple factors related to these discrepancies have been postulated, including increased motion artifact in ADHD groups, treatment/medication effects, methodological differences, as well as heterogeneity of symptomatology, age, comorbidity, and etiologic factors [121,129,168]. Many of the reported findings have been noted in children, particularly in regions related to fronto-striatal circuits, but significant relationships have not been found for adolescents and adults [129], implicating the developmental origins of ADHD and varied prevalence by age. This can be further elucidated with longitudinal studies of individuals across symptom trajectories as well as with pharmacologic and nonpharmacologic interventions.

To address the limitations of previous research, focus has recently shifted to multifactorial studies, particularly within the Enhancing NeuroImaging Genetics Through Metaanalysis (ENIGMA) project [169,170]. The increasing availability of freely accessible open data sets can enhance power and effect size, as well as standardize methodology among international cohorts to determine within and across group variance among neurodevelopmental and psychiatric disorders [171]. Given the increased power of these large data sets, multimodal studies become a possibility to ascertain and integrate neuroimaging, genetic, environmental, and phenotypic data [129,172–175].

Genetics of attention-deficit/hyperactivity disorder

Through the multifactorial etiology of complex behaviors, genetic predisposition stands out to be a leading factor in the risk of developing ADHD. Studies using identical (monozygotic) twins have estimated heritability coefficients of approximately 80% [176,177] with similar rates found between dizygotic twins, full siblings, and maternal and paternal half siblings [176,178]. Accounting for reporting variance, heritability estimates for diagnosis across the lifespan [179–183] and persistence of a clinical diagnosis into adulthood [180,184,185] are also consistent regarding clinical diagnostics and caregiver reports, with lower estimates with isolated self-reports. When viewed as a quantitative trait, with a diagnosis ADHD being the extreme of a behavioral continuum, ADHD symptoms are equally heritable across varied levels of symptom severity [62,168,186,187]. Additionally, sibling studies have shown a ninefold increased risk of ADHD in siblings of probands with clinically diagnosed ADHD compared to siblings of controls [188].

Early molecular genetic studies looking to find common genetic variants among those with ADHD used a candidate-gene approach [168,186]. Most had focused on genes related to dopamine and norepinephrine pathways (given the therapeutic nature of stimulant medications), serotonergic pathways, neurotransmitter release pathways, as well as synaptogenesis and neuronal proliferation [114,168,189,190]. Results were frequently contradictory [191,192] but a metaanalysis found consistent implication of eight unique DNA variants associated with six genes, although only a small strength of association was reported for each of these genes [193]. These genes included a serotonin transporter (*5HTT*), dopamine transporter (*DAT1/SLC6A3*), D4 dopamine receptor (*DRD4*), D5 dopamine receptor (*DRD5*), serotonin 1B receptor (*HTR1B*), and a synaptic vesicle regulator (*SNAP25*). The dopamine transporter gene (*DAT1/SLC6A3*)'s common noncoding region regulatory alleles also show variation in tandem repeats, with 10 repeats conferring risk of ADHD in childhood [191] and 9 repeats conferring risk in adulthood [194]. Additionally, data have supported differential dopamine transporter function in PET studies in human adults with the 9-repeat allele [195] linking genetic associations to in vivo functional studies.

Further inquiries into the genetic underpinnings that define the heredity of ADHD have utilized genome-wide association studies (GWAS) to scan the entire genome to detect common population DNA variants (single nucleotide polymorphisms—SNPs) that have small etiologic effect. The cumulative effect of SNP-based heritability has been estimated to be approximately 22% [196], leaving a significant gap to reach the overall heritability of 80% in ADHD. Theories underlying this "hidden heritability" have included low predictive value of GWAS methods, low effect size of individual SNPs, and other genetic variance, including higher effect, rare variant SNPs, as well as common and rare copy number variants (CNVs) [189,197]. Though early GWAS studies had not found any individual SNPs reaching clinical significance, later analyses of higher-powered data sets focusing on predetermined risk loci related to groups of candidate genes and pathways involved in the regulation of neurotransmitter release, neurite outgrowth, and axon guidance have demonstrated statistically significant relationships [198–201].

Of more recent interest is the interplay of multiple common variants ("genetic load") in a process to develop polygenic risk scores (PRS) using results of GWAS studies to calculate clinically significant risk prediction [202–205]. This methodology has shown associations between subclinical attentional dysregulation in the general population as well in clinically diagnosed ADHD [206]. Polygenic risk assessment has also furthered the evidence behind shared genetic foundations among various neurodevelopmental and neuropsychiatric conditions, including ASD, depression, anxiety, and bipolar disorder [207–211]. PRS have also been implicated in prediction of persistence or remission of ADHD symptoms linking ADHD PRS to developmental trajectories [203]. However, clinical use of PRS related to clinical conditions must be used with caution, given that most GWAS studies have used a predominantly European ancestry and have only accounted for about 5.5% of the variance among groups [196,212].

While many population-based samples for high-powered genetic association studies focus on low effect, common variants (>1% prevalence in the general population), more rare variants and well-described genetic syndromes have been frequently associated with ADHD [189,213,214]. ADHD is frequently associated with rare genetic syndromes that have varied pathogenesis (Table 27.4) [214–229].

While each of these syndromes provide examples of rare variants with increased risk of ADHD, this is only one component of the diverse neurodevelopmental phenotypes related to developmental brain dysfunction in each of these genetic syndromes.

Given the mismatch between the high familial heritability of ADHD and low defined genetic contribution, gene–environment interactions have been evaluated, although consistent results and replications have been lacking [230–232]. Plausible mechanisms for such gene–environment interactions include epigenetic factors related to methylation, notably within catecholamine and GABAergic systems [233–236].

Targeted genetic testing does not assist in a clinical diagnosis of ADHD though a comprehensive neurodevelopmental evaluation can guide etiologic evaluation of broader systemic or syndromic phenotypes. Additionally, continued interest in multimodal studies evaluating the relative contributions of genetic factors and gene–environment interactions can help illuminate further research into the neurobiologic mechanisms that might yield targeted treatments and medication selection.

Environmental factors and attention-deficit/hyperactivity disorder

While genetic contributions account for the bulk of risk related to ADHD, environmental factors, including prenatal factors, toxic exposures, nutritional deficiencies, and social determinants of health, have been examined as relative risk (RR) factors.

TABLE 27.4 Various genetic syndromes confer increased risk for attention-deficit/hyperactivity disorder (ADHD).

Syndrome	Underlying genetics	Prevalence of ADHD	Relative rate compared to general population
Williams syndrome	Microdeletion of 7q11.23	~66%	11.6×
Klinefelter syndrome	47:XXY chromosomal aneuploidy	~63%	11.1×
Fragile X syndrome (FXS)	Loss of function of *FMR1* gene typically due to trinucleotide repeat expansion	~53%	9.3×
Down syndrome	Trisomy of chromosome 21	~44%	7.7×
Tuberous sclerosis complex (TSC)	Pathogenic variants in or deletion of *TSC1* or *TSC2* genes	~30%–60%	5.3–10.5×
DiGeorge/velo-cardio-facial syndrome	Microdeletion of 22q11	~40%	7.0×
Neurofibromatosis type 1 (NF1)	Pathogenic variants in or deletion of *NF1* gene	~33%	5.8×
Turner syndrome	45:XO chromosomal aneuploidy	~20%	3.5×

The most substantial (threefold increased) prenatal risk for a diagnosis of ADHD is among the very/extremely preterm or very/extremely low birth weight. This risk has been found to increase with decreasing birth weight and earlier gestational age, and there is an inverse relationship between attentional dysregulation symptomatology and birth weight irrespective of formal ADHD diagnosis [237–240]. Furthermore, maternal hypertension has shown to confer a 25% increased risk of ADHD [241], while preeclampsia has shown up to a 15% increased risk that rises to over 40% with associated intrauterine growth restriction [242]. Children have also been reported to have a 1.5–2 times increased risk of an ADHD diagnoses when born to obese mothers [243,244], with increasing risk for morbid obesity [245]. Maternal Type 1 or Type 2 diabetes has not been consistently associated with increased risk of ADHD in offspring, though gestational diabetes mellitus in Caucasian populations increased the risk of ADHD in offspring 1.6-fold [246]. Prior miscarriages confer a 9% increased risk of ADHD, with increasing risk of up to 22% with two or more prior miscarriages [247].

Prenatal toxic exposures also increase the risk of ADHD. Prenatal acetaminophen use has been associated with up to a 33% increased risk of ADHD [248,249] with a positive dose–response relationship [250]. In a longitudinal sibling control study, there was no increased risk in short term use of acetaminophen (≤28 days), but a two-fold increased risk of ADHD in long term use of acetaminophen (>28 days). However, though this association was believed to be influenced by multiple confounding factors when controlling for intrafamilial factors [251]. Measurable phthalate metabolite levels in pregnancy have been associated with a threefold increased risk of ADHD [252]. In Danish and Swedish national registry studies, the only prenatal antiepileptic medication exposure associated with an increased risk of ADHD was valproic acid, which conferred a 50% increased risk [253,254]. In utero tobacco exposure has been shown to attribute a 50% increased risk of ADHD [255,256], although larger population studies that control for family history of ADHD have not found a significant association, presuming shared genetic risk for smoking and ADHD [257–259]. Prenatal methamphetamine has not been shown to increase risk of later ADHD diagnosis when controlling for confounders, such as polysubstance use and environmental influences [260]. Low-moderate prenatal alcohol use has mixed evidence of increasing risk of ADHD [261,262]. With increasing prenatal alcohol exposure, including binge drinking, the risk of a later ADHD diagnosis increases, ranging from 5 to 10 times the general population rate [262,263]. Notably, 50%–93% of children with a fetal alcohol spectrum disorder carry a diagnosis of ADHD [264,265].

Postnatal exposure to secondhand smoking has also been linked to a 60% increased risk of ADHD, though it is unclear whether this association is confounding versus causal [266,267]. Other significant postnatal environmental toxicant exposures have included a 1.5-fold to 2-fold increased risk with organophosphate exposure (with measurable metabolites) [268]. There is mixed evidence associating lead [269] and mercury [270] exposure to later development of ADHD, although further longitudinal, prospective studies are needed to better understand these associations. No significantly increased risk of a diagnosis of ADHD has been linked with sugar consumption [271], perfluoroalkyl substances [272], air pollutants [273], or artificial food coloring (with minimal increase in symptom ratings noted on parent but not teacher reports) [274].

Nutritional deficiencies have also been implicated in ADHD risk. While serum iron levels have not been found to be significantly different, modestly lower ferritin levels have been reported in those with ADHD compared to those without [275,276]. Prenatal vitamin D deficiency has been associated with a 50% increased risk of ADHD in offspring in an isolated cohort [277]. Those with ADHD have also been reported to have moderately lower omega-3 fatty acid levels than typical controls with a small effect improvement in symptoms with supplementation [278].

Social determinants of health have noteworthy associations with risk of ADHD diagnosis. Lower family income has been correlated with increased risk of an ADHD diagnosis even when controlling for gender, other familial factors [279,280], and any bias toward diagnosing those in a lower socioeconomic status (SES) [281,282]. Additionally, a dose–response relationship has been found between cumulative effects of lower parental educational achievement, parental unemployment, and parental relative poverty and childhood diagnosis of ADHD [283]. Familial stress, including prenatal maternal bereavement due to the death of a child or spouse during pregnancy (72% increase) [284] or death of a family member (60% increase) [285], has an increased risk of diagnosis of ADHD. Other adverse childhood experiences, such as residential instability (including foster care) and parental substance abuse, criminality, and exposure to violence or familial mental health disorders, confer a greater than twofold risk of an ADHD diagnosis [285], though the shared heritability of ADHD within families may account for some of this risk.

Among all these environmental factors, it is difficult to conclude causation, and each risk in isolation is not necessarily sufficient to lead to a diagnosis of ADHD. The etiology of ADHD, like the etiology of all neurodevelopmental disabilities, is a complex interplay between genetic and environmental factors. Even though determination of these factors is not diagnostic, they can provide insight into the pathophysiology of ADHD and individual risk of current and future diagnosis, as well as help guide responsible treatment recommendations and counseling.

Outcomes and functional impairment in attention-deficit/hyperactivity disorder

Existence of functional impairment is a core feature of ADHD. Untreated ADHD has been associated with impairments throughout the lifespan. Key impairments are noted in academic, occupational, and social life [286]. Impairment goes beyond the individual, with strife extending to the family as well as societal levels, with ADHD also bearing significant economic burden on the individual, family, and society [287–290]. Treatment with medications tends to confer a robust protective effect on mood disorders, suicidality, criminality, substance use disorders, accidents and injuries, traumatic brain injuries, motor vehicle collisions, and educational outcomes [291].

Among the most studied impacts of ADHD on an individual is within academic achievement. One of the most common specific impairments leading to an evaluation for ADHD is academic underachievement. Impairments related to school include decreased academic success (typically measured as grade retention, classroom grades, years of school completed, and college attendance), as well as content related to skills and information learned (typically measured using standardized academic achievement tests) [292–295]. Compared to controls, those with ADHD tend to have lower grades, are more likely to be retained in grade, have a lower number of school years completed, and have lower admission rates to postsecondary education [293,294,296]. Academic impairment from ADHD symptoms also crosses subject matter affecting reading, writing, and mathematics [297], and the gap between academic performance among those with ADHD and their typical peers increases over time with up to three to five grade level lower scores on academic achievement tests in adulthood [295,298]. Some of this difference could be related to notable working memory and other associated neurocognitive differences as opposed to primary ADHD symptomatology [299]. Given the high comorbidity with learning disabilities and other psychopathology, there is felt to be an additive effect on academic underachievement [300,301].

Outside of academic achievement, the risk of substance abuse among those with ADHD remains one of the most highly studied outcomes, particularly due to the theoretical dysfunction of the dopaminergic and reward neural systems shared by both conditions [302,303]. Adults with ADHD show increased odds of comorbid substance use, including nicotine use, polysubstance use, and alcohol dependence [304]. Given the increased risk, substance use may itself be a form of self-medication and a presenting sign of ADHD in adolescents and adults.

Studies across multiple countries have found that at all levels of judicial systems, those with ADHD tend to be disproportionately involved compared to general population estimates [286,305]. Up to 20% of adolescents within juvenile correctional facilities have a diagnosis of ADHD [306]. This risk is even further compounded by comorbidity with conduct disorder [306,307]. Criminal convictions also show a direct correlation with symptom severity in some population studies [308], and individuals with ADHD are more than five times as likely to have a criminal conviction compared to their typical peers [309]. A diagnosis of childhood ADHD has been significantly associated with increased RR of adolescent and adulthood arrests (RR: 2.2), convictions (RR: 3.3), and incarcerations (RR: 2.9) [310]. This increased criminality has been felt to be in part related to impulsivity and risk-taking behaviors.

Related to risk-taking, individuals with ADHD are at a nearly twofold increased risk of physical injury [311,312] and have increased rates of traffic violations, motor vehicle collisions, and substance-impaired driving [313]. Given this increased risk, the relationship with ADHD and traumatic brain injury has garnered interest. In those with ADHD, there is a significant positive relationship between the severity of ADHD symptoms and risk of mild traumatic brain injury [314]. In a large population study, those with ADHD were found to have an increased risk of traumatic brain injury at all levels of severity [315]. Notably, severity of traumatic brain injury is also related to increased risk of a postinjury diagnosis of ADHD [42]. Symptoms of inattention and impulsivity following traumatic brain injury are felt to be within the spectrum of ADHD that is traditionally diagnosed without premorbid injury [43]. Furthermore, a diagnosis of ADHD portends an increased all-cause mortality rate compared to typical controls [286].

Attention-deficit/hyperactivity disorder associations

As with other neurodevelopmental disabilities, the cooccurrence of other neurodevelopmental and psychiatric disorders among individuals with ADHD is quite high. While other pathology may mimic the observed symptoms of ADHD (e.g., inattention due to depression), a comprehensive neurodevelopmental evaluation allows for determination of the primary cause of symptoms or coexistence of a secondary diagnosis. Within ADHD, comorbidity is the norm as opposed to the exception. Given the need for additional or alternative management plans, cooccurring neurodevelopmental disabilities (learning and intellectual disabilities, language disorders, ASD, motor incoordination, and cerebral palsy), neurological disorders (tic disorders, epilepsy, sleep disorders), and psychiatric disorders (internalizing disorders [mood and anxiety disorders], externalizing disorders [oppositional defiant disorder and conduct disorder], substance use disorders) must be recognized to avoid incomplete diagnosis and management.

Other neurodevelopmental disabilities and attention-deficit/hyperactivity disorder

Within the spectrum and continuum of neurodevelopmental disabilities, intellectual disability, language disorders, and learning disabilities have been most frequently associated with an ADHD diagnosis. Borderline intellect and intellectual disability have been associated with an increased risk of ADHD with a notable indirect correlation between IQ and rate of ADHD diagnosis [316]. Among those diagnosed with ADHD, population-based studies have estimated a cooccurrence with speech and language disorders of 25%–50% [317,318].

ADHD and the relationship to specific learning disabilities (SLD) in reading (retrieval of linguistic information, verbal working memory), mathematics (calculation, problem-solving, visuospatial working memory), and writing (verbal and visual-spatial working memory integration) has been appraised with an overall median comorbid prevalence of 33% (with studies ranging from 7% to 92% depending on method), with a threefold increased risk of an SLD in those with ADHD compared to typical controls [300,319]. The most frequently studied SLD in relation to ADHD are disorders of reading, such as dyslexia, with a comorbid prevalence of 25%–48% within those with ADHD [11,300,320]. A total of 11%–30% of those with ADHD will also have a mathematics disorder, while approximately 55%–64% of children with ADHD have a written language disorder [321]. There are shared neuropsychologic factors among ADHD and learning disabilities but also independent factors that help determine their relative contributions. This common comorbidity emphasizes the critical importance of a comprehensive neurodevelopmental evaluation for all children presenting with concerns about ADHD [83]. This is especially imperative given the functional impairment of academic underachievement as a common presenting complaint for evaluation for both learning disabilities and ADHD.

As all children with ASD exhibit atypical attention (e.g., overfocus/perseveration on visually stimulating activities [such as spinning objects] and underfocus/inattention to language stimuli [such as a lack of response to name being called]), prior to the DSM-5, ADHD and ASD were considered mutually exclusionary diagnoses. However, within the most recent addition of the DSM, should both full sets of symptom criteria be met, diagnoses of both ADHD and ASD can be given simultaneously. ADHD has been estimated to be present in 30%–80% of children with ASD, while ASD is estimated to be present in 20%–65% of children with ADHD [322–325]. Consistent with other evaluations of comorbid neurodevelopmental disabilities, the more severe the symptoms, the more likely there is to be comorbidity [323,324]. This cooccurrence seems to increase significantly with increasing age, likely due to the increasing demands related to social adaptability and executive skills as well as academic pressures [326]. There are also shared neuropsychological findings (e.g., deficits in executive functioning), structural and functional neuroimaging findings, and genetic origins among individuals with ADHD and ASD, supporting the concept of multifactorial developmental brain dysfunction leading to shared endophenotypes with varied expression among neurodevelopmental and neuropsychiatric disorders [186,327–329].

In terms of motor development, among those diagnosed with cerebral palsy, 19%–35% have also been diagnosed with a comorbid ADHD [330]. Developmental coordination disorder or other subtle motor incoordination can be observed in 40%–50% of children with ADHD [331,332]. Determining whether the coordination and motor deficits seen in ADHD are secondary to attentional dysregulation or a cooccurring motor impairment is an area of active debate [333,334].

Neurological disorders and attention-deficit/hyperactivity disorder

Tic disorders, ranging from transient or chronic motor or vocal tic disorders to Tourette syndrome, are highly associated with ADHD. Children with ADHD have a fourfold to sixfold increased risk of a chronic tic disorder than typical controls [335]. Inversely, approximately 50% of children with tic disorders have a diagnosis of ADHD [336,337]. In children with both ADHD and Tourette syndrome, the ADHD typically is first diagnosed, and ADHD symptoms tend to cause greater functional impairment than tics [337]. Again, increasing severity of tics and persistence are associated with higher rates of comorbidity [336]. Notably, treating ADHD symptoms with stimulant medication has not been shown to cause tics nor has it been shown to worsen them [338]. This high incidence of comorbidity may be related to the shared neurobiology of impaired dopaminergic systems.

An increased incidence of ADHD had also been reported among other neurologic disorders. Children with epilepsy have a 2.5–5 times higher risk of being diagnosed with ADHD [339,340]. Further, symptoms of ADHD may be exacerbated by or present as an adverse effect of certain antiseizure medications, such as phenobarbital, topiramate, and valproic acid [341]. Additionally, one's risk of ADHD has been found to be increased with a family history of epilepsy in their mother (85% increase), father, sibling (50%–60% increase), and first-degree cousins (15%) [342].

Sleep-disordered breathing has been modestly associated with an ADHD diagnosis in children and adolescents [343], though the shared neurobiology of these conditions and causative directionality remains unclear. Those with ADHD have been reported to have greater sleep onset latency and subjectively worse sleep efficiency, though polysomnography has not been found to be significantly different between adults with ADHD and controls [344,345]. Restless leg syndrome has been inconsistently associated with ADHD as well, with up to 44% of children with ADHD having symptoms of restless leg syndrome. This comorbidity is believed to be an effect of shared dopaminergic dysregulation and a component of iron deficiency [346].

Psychiatric disorders and attention-deficit/hyperactivity disorder

Nearly half to two-thirds of children and adolescents presenting for specialty evaluation of attentional dysregulation will also have a coexisting psychiatric disorder [347–349], while approximately a quarter will have more than one coexisting psychiatric disorder [322]. The most common coexisting conditions in clinical samples of children and adolescents include externalizing behavior disorders, such as oppositional defiant disorder (26%–60%) [350–353] and conduct disorder (up to 20%) [322,351]. Hyperactive and impulsive symptoms are most strongly correlated with oppositionality [354].

Depending on the study and population, ADHD has also been variably associated with internalizing disorders, such as anxiety and depression. Across anxiety disorders, including separation anxiety, generalized anxiety, specific phobias, and obsessive-compulsive disorder, cooccurrence occurs in 15%–35% of children and adults with ADHD [348,350,355–358]. Associations between ADHD and specific mood disorders are less consistent among varied populations and studies, though the rate of comorbid ADHD diagnosis ranges from 73% in childhood to 17% in adulthood for bipolar disorder, and from 28% in childhood and 7% in adulthood for depression, corresponding to a threefold increased risk of ADHD in those diagnosed with any mood disorder [359]. The increasing comorbid incidence with younger ages is believed to be due to the overlap in symptom presentation at younger ages. Given this high incidence of comorbidity, targeted psychopharmacology interventions must account for the overlap in symptomatology and avoidance of polypharmacy. When ineffective pharmacologic management occurs, an alternative diagnosis or comorbid psychopathology should be considered.

Management of attention-deficit/hyperactivity disorder

The management of ADHD typically includes a combination of family and patient education, pharmacotherapy, behavioral or family counseling, and educational interventions. Numerous specialty society, national, and international evidence-based guidelines have been developed to help guide best practices in management of ADHD [77,113,360]. Treatment of the core symptoms leading to functional impairment is imperative, given the individual and societal costs associated with ADHD. Fig. 27.3 provides a sample framework for the longitudinal management of ADHD in children with an emphasis on identifying and measuring targeted behaviors for intervention and assessment of efficacy.

Family and patient education

To help develop an effective treatment plan, patients and families must be informed and educated about ADHD. Direct education at the time of diagnosis, particularly regarding the endogenous neurodevelopmental origins of ADHD (including reassurance that ADHD is "no one's fault"), symptom management options with risks and benefits of each, and further resources (including information about advocacy organizations, books, and online resources) can all be beneficial. In addition, families should be provided with emotional support, given the known burden on caregivers and individuals from an academic, financial, and emotional perspective [287].

Pharmacotherapy

While many parents and healthcare providers can be hesitant to start medications in children with ADHD, the medications used are safe and as efficacious, or more efficacious, than many medications used for nonpsychiatric disorders [361]. Pharmacologic treatment for symptoms of ADHD has a greater than 70-year history of evidence from numerous trials and metaanalyses [362,363]. Stimulant medications have been consistently shown to be the most effective class of medication for ADHD [364,365] with nearly 70% of children diagnosed with ADHD taking a stimulant medication [366]. Nonstimulant medications, including atomoxetine, sustained-release guanfacine, and sustained released-clonidine, are also approved by the US Food and Drug Administration (FDA) for treatment of ADHD.

Stimulant medications

The first described use of stimulant medications in the treatment of ADHD symptoms occurred in the 1930s, when Bradley found that Benzedrine, an amphetamine derivative, decreased hyperactivity in children with behavioral issues

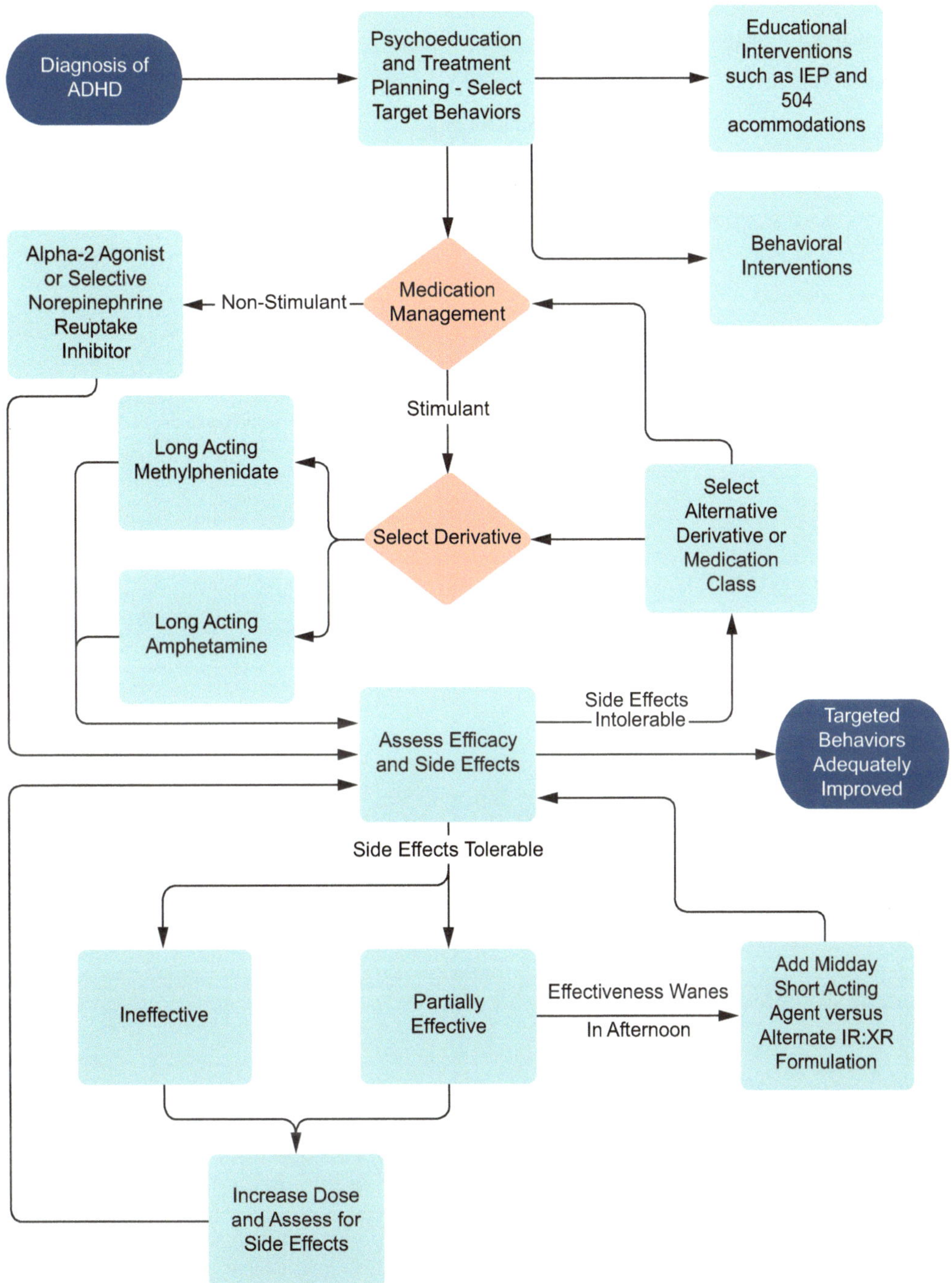

FIGURE 27.3 Management and intervention algorithm for attention-deficit/hyperactivity disorder (ADHD) including medication titration.

in a school setting [363]. In the 1950s, the FDA first approved methylphenidate for use in depression and narcolepsy. In the 1960s, methylphenidate approval was expanded to children with behavioral disorders. Stimulant medications consist of formulations of methylphenidate or amphetamine that have similar mechanisms of action, with the end effect of increasing catecholamine transmission (both block presynaptic dopamine and norepinephrine transporters, while amphetamine also increases the presynaptic efflux of dopamine) [115,167]. A dichotomous relationship also exists in which low-moderate levels of stimulation to dopaminergic systems (notably D1, D2, and D5 receptors) and noradrenergic systems (α-2 adrenoreceptors) increase neuronal firing, while higher levels of stimulation have a paradoxical diminishing effect [354]. Even with similar mechanisms of action, should a person with ADHD not respond to one stimulant derivative, they may have an acceptable response to an alternate stimulant derivative. A list of FDA-approved short- and long-acting stimulant medications is provided in Tables 27.5 and 27.6.

TABLE 27.5 Long-acting stimulants.

Mechanism of delivery	Brand name	Formulation and dose (mg)	Duration of action	FDA approved (years)	IR:XR ratio
Methylphenidate (MPH) derivatives					
ER capsule	Metadate CD/ER	Capsule: 10, 20, 30 (max: 60/day)	8–10 h	6 and up	30:70
	Ritalin LA	Capsule: 20, 30, 40 (max: 60/day)	8–10 h	6 and up	50:50
SR capsule	Ritalin SR	Capsule: 20 (max: 60/day)	4–8 h with low predictability	6 and up	N/A
OROS	Concerta	Capsule: 18, 27, 36, 54, 72 (max: age 6–12: 54/day age 13 + : 72/day)	10–12 h	6 and up	22:78
ER transdermal patch	Daytrana	Patch: 10, 15, 20, 30 (max: 60/day)	9 h with patch on; 3 + hours after removal	6 and up	N/A
ER oral suspension	Quillivant XR	Soln: 25 mg/5 mL (max: 60/day)	10–12 h	6 and up	20:80
ER chewable tablets	Quillichew ER	Tab: 20, 30, 40 (max: 60/day)	8–10 h	6 and up	20:80
ER and multilayer release	Aptensio XR	Capsule: 10, 15, 20, 30, 40, 50, 60 (max: 60/day)	8–12 h	6 and up	40:60
	Adhansia XR	Capsule: 25, 35, 45, 55, 70, 85 (max: 85–100/day)	16 h	6 and up	20:80
ER ODT	CoTempla XR-ODT	Tab: 8.6, 17.3, 25.9 (max: 51.8/day)	10–12 h	6 and up	25:75
Delayed ER	Jornay PM	Capsule: 20, 30, 40, 50, 60, 80, 100 (max: 100/day)	13 h	6 and up	20:80
ER capsule	Focalin XR	Capsule: 5, 10, 15, 20, 25, 30, 40 (max: 30–40/day)	8–12 h	6 and up	50:50
Amphetamine (AMP) derivatives					
ER capsule 3:1 D:L formulation	Adderall XR	Capsule: 5, 10, 15, 20, 25, 30 (max: 30/day)	8–12 h	6 and up	50:50
ER ODT 3:1 D:L formulation	Adzenys-XR ORT	ODT Tab: 3.1, 6.3, 9.4, 12.5, 15.7, 18.8 (max: 18.8/day)	8–10 h	6 and up	50:50
ER oral suspension 3:1 D:L formulation	Adzenys-ER	Soln: 1.25 mg/mL (max: age 6–12: 18.8/day age 13 + : 12.5/day)	8–10 h	6 and up	50:50
	Dyanavel XR	Soln: 2.5 mg/mL (max: 20/day)	10–12 h	6 and up	20:80
ER triple bead (1 IR, 2 ER) 3:1 D:L formulation	Mydayis	Capsule: 12.5, 25, 37.5, 50 (max: ages 13–17: 25/day; ages 18 + : 50/day)	12–16 h	13 and up	33:66
Prodrug (Lis-dexamfetemine) 3:1 D:L formulation	Vyvanse	Capsule: 10, 20, 30, 40, 50, 60, 70 Chewable: 20, 30, 40, 50, 60, 70 (max: 70/day)	10–12 h	6 and up	N/A
ER capsule 1:0 D:L formulation	Dexedrine spansules	Capsule: 5, 10, 15 (max: 40/day)	8–10 h	6–16	N/A

Shared adverse effects

Common side effects: Loss of appetite, upset stomach, insomnia, and headache.
Less common side effects: Rebound irritability, dysphoria, agitation, and growth impairment.
Rare adverse events: Cardiovascular risk in those with preexisting conditions (sudden death, stroke, and myocardial infarction), peripheral vasculopathy, priapism (methylphenidate), permanent skin color loss (Daytrana patch), rhabdomyolysis, worsening psychiatric symptoms including psychosis and mania.
FDA black box warnings: Potential for abuse (amphetamines) and dependence (methylphenidates).

SR, Sustained release; *ER/XR*, extended release; *LA*, long acting; *IR*, immediate release; *D*, dextro; *L*, levo; *ODT*, oral dissolvable tablet; *OROS*, osmotic-controlled release oral delivery system; *Soln*, solution.

TABLE 27.6 Short-acting stimulants.

Generic name	Brand name	Formulation and dose (mg)	Duration of action	FDA approved (years)	Dosing/notes
Methylphenidate (MPH) derivatives					
D,L-methylphenidate	Ritalin	Tab: 5, 10, 20 (max: 60/day)	4–5 h	6 and up	Given two to three times daily or as single booster dose with an extended release
	Methylin chew	Chewable: 2.5, 5, 10 (max: 60/day)	4–5 h	6 and up	
	Methylin liquid	Soln: 5 mg/mL, 10 mg/mL (max: 60/day)	4–5 h	6 and up	
D-methylphenidate	Focalin	Tab: 2.5, 5, 10 (max: 60/day)	4–5 h	6 and up	1.5–1.67 × potency than D,L methylphenidate
Amphetamine (AMP) derivatives					
Amphetamine	Evekeo	Tab: 5, 10 (max: 40/day)	4–6 h	3 and up	1:1; D:L racemic mixture
		ODT: 5, 10, 15, 20 (max: 40/day)	4–6 h	6 and up	
Mixed amphetamine salts	Adderall	Tab: 5, 10, 15, 20, 25, 30 (max: 40/day)	3–6 h	3 and up	3:1; D:L racemic mixture
D-amphetamine	Dexedrine	Tab: 5, 10 (max: 40/day)	4–6 h	3 and up	1:0; D:L racemic mixture
	Zenzedi	Tab: 2.5, 5, 7.5, 10, 15, 20, 30 (max: 40/day)	4–6 h	3 and up	
	Procentra	Soln: 5 mg/mL (max: 40/day)	4–6 h	3 and up	

Shared adverse effects

Common side effects: Loss of appetite, upset stomach, insomnia, and headache
Less common side effects: Rebound irritability, dysphoria, agitation, and growth impairment
Rare adverse events: Cardiovascular risk in those with preexisting conditions (sudden death, stroke, and myocardial infarction), peripheral vasculopathy, priapism (methylphenidates), rhabdomyolysis, worsening psychiatric symptoms including psychosis and mania
FDA black box warnings: Potential for abuse (amphetamines) and dependence (methylphenidates) [a]

D, Dextro; *L*, levo; *ODT*, oral dissolvable tablet; *Soln*, solution.

[a] *Immediate-release formulations have an increased risk of abuse/dependence and diversion compared to extended-release formulations.*

Stimulant medications are the first-line pharmacotherapy for ADHD internationally [77,113,360]. Stimulants may reduce symptoms in up to 70%–90% of children with clinically diagnosed ADHD in a dose-response fashion [367,368]. Theoretical underpinnings of this effect include the "inverted U-shape" of D1/D5 dopaminergic receptor and noradrenergic receptor function in the PFC. This framework is based in the findings of neurophysiologic studies that a moderate amount of stimulation to dopamine and noradrenergic receptors is required for ideal neural regulation in the PFC [369]. Individuals with ADHD are assumed to have a low basal level of stimulation and the use of exogenous psychostimulants allows for homeostatic neuronal regulation. This "inverted-U shape" response is thought to translate to dose effects in which increasing dosage past an ideal therapeutic response does not necessarily provide additional benefit [370].

Use of stimulants has been shown to lead to a rapid improvement in attention and distractibility, along with a decrease in impulsivity and hyperactivity. Teacher and parent ratings of behavior show marked improvement with stimulant use with a moderate-to-large effect size [364], with some reports of slightly higher effects of amphetamine relative to methylphenidate [368]. While most data are related to childhood and school-aged samples, efficacy and safety have also been demonstrated in preschool [10,371] and adult [368] ages. Academic productivity is shown to improve, but inconsistent improvement has been seen in accuracy and overall achievement [372], which is likely a product of high comorbidity of learning disabilities and ADHD. Self-reported quality of life measures also show improvement with stimulant use in children, adolescents, and adults with ADHD [365,373]. During periods of medication adherence, stimulants decrease the risk of motor vehicle collisions [374]. Notably, use of stimulant medication in ADHD has also shown a significant protective effect on risk of comorbid mood disorders, accidents and injuries, and academic achievement, with trending protective effects across studies for suicidality, substance use disorders, criminality, and traumatic brain injury [291,375].

While there are numerous benefits to treatment with stimulant medication in ADHD, there are also side effects. Within amphetamine and methylphenidate derivatives, the most common side-effects include appetite suppression and insomnia, with some reports of amphetamine being more prone to these side effects. Younger children (<5 years of age) and those with intellectual disability or ASD are more prone to intolerable side effects [368]. Other concerns related to stimulant use include influence on height and weight trajectories, cardiovascular effects, and substance abuse, dependence, and diversion.

Evaluations of the effect of stimulant use on growth trajectories have demonstrated mixed results, although most suggest that consistent use over years can lead to an approximate 1–3 cm reduction in adult height. While initial treatment tends to confer modest weight loss [371], long-term stimulant use has been associated with a substantial increase in weight and body-mass index [376,377]. These effects on growth may be attenuated through drug holidays (periods of time off mediation) [378]. While these differences in growth in association with stimulant medication treatment are statistically significant, they have not been found to be clinically significant [379,380].

Given the sympathomimetic function of stimulant medications, there is concern for an increase in adverse cardiovascular events. While there has been a noted modest increase in basal heart rate and blood pressure in those taking stimulants, the overall risk of adverse cardiac events is not significantly different from control populations [381,382]. Additionally, those using stimulants for ADHD management do not show a higher risk for all-cause mortality [381]. Given the known mechanisms of stimulant medication and some reports of increased risk of cardiovascular effects in certain populations, those who are at high risk (i.e., known personal or family history of arrhythmias, underlying structural heart defects, etc.) should use stimulants with caution and close clinical monitoring [383–385].

While stimulant medications have great efficacy for clinical use, there remains risk related to abuse, dependence, and diversion of medications. Given their potential for abuse, stimulants are currently classified as controlled substances by the US Drug Enforcement Agency. When stimulants are injected intravenously or inhaled intranasally, they can produce a feeling of euphoria or "high" through a rapid, phasic dopaminergic stimulation. However, when used in oral formulations as prescribed, this effect does not occur due to the delayed passage from the gastrointestinal system to the central nervous system, leading to slow steady-state increase in dopamine leading to tonic firing of the receptors [386]. Use of slow or delayed release formulations and prodrugs, such as lisdexamfetamine, makes abuse also less likely to occur [387].

Even though abuse potential is low, diversion of stimulant medication for secondary uses, such as increased wakefulness, is quite common among high school and college students, with nearly 20% of college students reporting use of stimulant medications that are not prescribed to them [388,389]. The most cited reasons for nonmedical use include academic and occupational performance enhancement, despite little evidence that academic performance is improved by use of stimulants in individuals without ADHD [390]. Furthermore, nonmedical use of stimulant medication has been associated with lower educational attainment [391]. Anticipatory guidance for adolescents prior to entering college, as well as medication contracts to encourage appropriate use, are ways the prescriber can help deter diversion [392].

While a clinical diagnosis of ADHD is related to higher rates of substance abuse [303,393], treatment with stimulant medication does not increase this risk and likely even lowers the long-term risk [394]. When there is a high risk of comorbidity between substance use disorders and ADHD, psychosocial treatments, monitoring, and nonstimulant medication use is considered first line. Should a nonstimulant agent be unsuccessful, a long-acting stimulant agent with low abuse potential should be considered [302].

Initiating and titrating stimulants

The decision to initiate treatment of ADHD with stimulants occurs when the diagnosis is confirmed and targeted symptoms or impairments are defined. Determining target symptoms allows for continual evaluation of response to treatment. Some example behaviors to target include level of (hyper)activity, independence in self-care or schoolwork, completion of chores at home, decreased periods of "daydreaming," and safer behavior (decreased impulsiveness). Target outcomes should be realistic, something the child is able to do, and something that can be measured. Standardized behavior rating scales are commonly used to monitor treatment effect. The selection of which class of stimulant is typically aided by the mode of delivery (is the child able to swallow whole capsules or is a liquid form needed?), age of approved use, insurance coverage and costs, and shared decision making with the patient and family regarding risks and benefits.

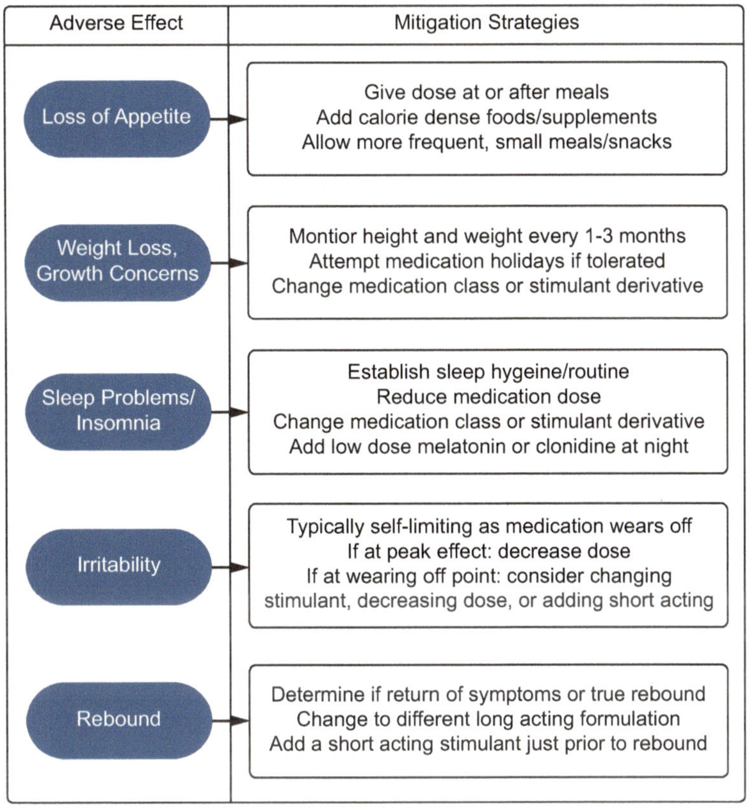

FIGURE 27.4 Common stimulant adverse effects and mitigation strategies.

In children between 3 and 5 years of age with severe ADHD symptoms refractory to behavioral interventions, immediate release dextroamphetamine and mixed-amphetamine salts are the only FDA-approved stimulants currently available. In this population, start at the lowest dose once a day, and at a weekly interval the dose may be increased in increments equivalent to the starting dose for effect. Alternatively, additional midday doses may be added for a total of two to three doses a day should the effect wear off. Clinical trials are currently in progress for extended-release stimulants in the preschool age group [395].

A trial of extended-release formulation is recommended to start in children 6 years of age and older. Start with the lowest available dose followed by regular check-ins on a weekly basis to determine dose titration (typically in increments equivalent to the starting dose every 1–2 weeks). The dose can be increased in this fashion until target outcomes are met, side effects become unbearable, or maximum dosing is met without significant effect. If outcome measures are met in the early part of the day but the effect weans by midday, addition of a short-acting stimulant at half the daily stimulant dose, increasing the dose of extended-release formulation, or changing to a higher extended:immediate release formulation can be considered. Stimulants should be given daily regardless of school status, as ADHD symptoms cross environments though individual treatment plans may vary based on side effect profiles and shared decision making with patients and families. Recommendations for strategies to approach common side effects are reviewed in Fig. 27.4.

Nonstimulant medications

Up to nearly a third of children with ADHD will not have an obvious benefit from treatment with stimulant medications. Additionally, there are many families who are hesitant to start a stimulant medication given perceived safety risks, even in the setting of appropriate education and counseling. Nonstimulant medications with the highest level of evidence in the treatment of ADHD include alpha-2-adrenergic agonists (clonidine and guanfacine) and norepinephrine reuptake inhibitors (atomoxetine and viloxazine) (Table 27.7).

Alpha-2-adrenergic agonists, including clonidine and guanfacine, have been in use for the treatment of ADHD since the 1980s. Currently, only the long-acting formulations of these medications have been approved by the FDA, though the short-acting formulations are often used off-label. Clinical trials have supported the safety and efficacy of long-

TABLE 27.7 Nonstimulant medications for attention-deficit/hyperactivity disorder (ADHD).

Generic name	Brand name	Formulation and dose (mg)	Duration of action	FDA approved ages (years)	Dosing/notes
Alpha-2 agonists					
Clonidine-XR	Kapvay	Tab: 0.1	Effects seen in 2–4 weeks	6 and up	Given once or twice daily Max: 0.4 mg/day
Common side effects: Hypotension or orthostatic hypotension, dry mouth, abdominal pain, drowsiness-rebound hypertension with abrupt discontinuation **Less common side effects:** Headache, sleep difficulties **Rare adverse events:** Depressed mood, anxiety, nightmares					
Guanfacine-XR	Intuniv	Tab: 1, 2, 3, 4	Effects seen in 2–4 weeks	6 and up	Max: 4 mg/day
Common side effects: Dry mouth, somnolence, headache, fatigue, dizziness, abdominal pain, constipation, appetite decreased, sedation **Less common side effects:** Weight gain, agitation, rebound hypertension with abrupt discontinuation **Rare adverse events:** Hallucinations, QTc prolongation					
Selective norepinephrine reuptake inhibitors					
Atomoxetine	Strattera	Capsule: 10, 18, 25, 40, 60, 80, 100	Effects seen in 1–4 weeks	6 and up	Starting dose: 0.5 mg/kg, goal dose 1.2–1.4 mg/kg Max: 100 mg/day
Common side effects: Nausea, upset stomach, dry mouth, loss of appetite, drowsiness **Less common side effects:** Urinary retention, mood swings, dizziness/fainting **Rare adverse events:** Cardiovascular risk in those with preexisting conditions (sudden death, stroke, and myocardial infarction) and severe liver injury that resolves with discontinuation of medicine **FDA black box warnings:** Suicidal ideation in the pediatric and adolescent population					
Viloxazine ER capsule	QELBREE	Capsule: 100, 150, 200	Effects seen >7 days	6–17	Capsules can be opened and sprinkled Max: 400 mg/day
Common side effects: Drowsiness or insomnia, nausea, loss of appetite, headache **Less common side effects:** Irritability, mood swings **Rare adverse events:** Paradoxical mania **FDA black box warnings:** Suicidal ideation in the pediatric and adolescent population					

ER/XR, Extended release.

acting clonidine formulations in reduction of ADHD symptoms both as monotherapy and adjunct therapy along with stimulants [396–398], although some metaanalyses have not reached significant difference from placebo [368]. Extended release guanfacine has also been shown to reduce ADHD symptoms with once-a-day dosing, regardless of morning or evening dosing [368,399,400]. Common side effects of alpha-2-adrenergic agonists include somnolence, headache, and fatigue, but these are noted to diminish with continued use [368,401]. Reductions in heart rate and blood pressure have also been noted, but these are typically not clinically meaningful [402], though rebound hypertension has been reported with abrupt discontinuation of clonidine [403].

Atomoxetine (Strattera) was the first FDA-approved selective norepinephrine reuptake inhibitor (SNRI) for the treatment of ADHD in those older than 6 years of age. It is effective in reducing symptoms of ADHD [368,404–406], and it may be particularly useful in those with coexisting conditions such as tics and anxiety [338,407]. However, atomoxetine is still not considered a first-line agent for ADHD, given its lower effect size when compared to stimulants [404,406,408]. The most frequent side effects reported are gastrointestinal side effects and fatigue. The gastrointestinal upset can typically be prevented by coadministration with food, especially foods with high protein or fat content. Other reported side effects include dizziness, irritability, somnolence, and rare allergic reactions [409]. The most concerning adverse effect of SNRIs is an increase in suicidal ideation in some children [410], and cases have been reported of severe hepatotoxicity [411], and increased heart rate and blood pressure, thus increasing cardiovascular risk in those with known cardiac disease [382]. Atomoxetine is metabolized through the cytochrome P450 system, and those with genetic variations related to the CYP2D6 or those using other medications that affect this system may have decreased efficacy or increased side effects that require dose adjustments [412].

Nonpharmacologic treatments

Alongside pharmacologic treatments for ADHD symptoms, numerous behavioral interventions have been described and promoted, though there is a limited evidence base for many of these, and outside of the preschool age, stimulant medication use is considered first-line treatment [77,113,360]. Large prospective studies, including the Multimodal Treatment for ADHD (MTA) Study, showed that stimulant treatment alone was as effective as combined stimulant and behavioral treatment for core ADHD symptoms, but the combination of stimulant and behavioral treatment improved anxiety symptoms, academic performance, parent–child relations, and social skills better than stimulant medication alone [413,414].

There is no strong evidence in support of restrictive or elimination diets for ADHD treatment, given high levels of confounding and lack of blinding in most studies [415]. There is no consistent evidence that supports the use of St. John's wort, tyrosine, multivitamins, tryptophan, Ginkgo biloba, magnesium, or phenylalanine supplementation in treating of ADHD symptoms. Mixed evidence exists for carnitine, pycnogenol, essential fatty acids, and zinc supplementation on ADHD symptoms, and more research is required before drawing conclusions [416,417]. There is some moderate evidence that omega-3 fatty acid supplementation may lead to mild improvement in ADHD symptoms [278].

One of the leading factors for mixed evidence of nonpharmacologic treatments for ADHD is the diverse nature and implementation of these interventions. Some behavioral methods have focused on a wide range of outcomes and skills, including academic, social, and practical adaptive skills, while others focus primarily on core ADHD symptoms. Additionally, behavioral interventions vary based on the age and population studied. Like pharmacologic therapy, while symptom measures of ADHD may be shown to improve, a practical and generalizable effect on interpersonal, academic, and adaptive function is not guaranteed. For a recent detailed review, see [418].

Cognitive and behavioral interventions

For preschool-aged children, a parent training program is recommended as first-line intervention for ADHD [77,113]. A recent metaanalysis reported that parent training, including parent management training (PMT), is associated with a moderate reduction in parent-reported ADHD symptoms, although this effect did not persist when using independent raters [419,420]. Direct cognitive training of preschoolers with a focus on working memory, inhibitory control, and cognitive flexibility has been shown to result in a modest improvement in executive function measures, but not in core ADHD symptom or functional status improvement [421]. Additionally, metaanalyses have not shown functional improvement or reduction in ADHD symptoms with neurofeedback or direct cognitive training with appropriate blinding and controls [422,423]. There is insufficient evidence to make conclusions on the benefit of meditation-based therapies [424] or social skills training in ADHD [425]. Organizational skills training has shown moderate reduction in parent-reported inattentive symptoms, though this was not confirmed with independent raters [426].

In adults with ADHD, cognitive-behavioral therapy (CBT) has shown moderate to significant improvements in symptoms and functional status in self report, but this effect is diminished with independent observer ratings [427]. Further evaluations with active controls showed moderate improvement in ADHD symptoms for those receiving CBT [428,429].

Educational interventions

In the public education system of the United States, intervention for ADHD is typically provided through one of two venues. Special Education services as detailed in the Individuals with Disabilities Education Act (IDEA) can be provided to some children with ADHD under the special education categorical label of "Other Health Impairment" (OHI). Students with OHI must both have a qualifying medical diagnosis, and that diagnosis must lead to impairments in educational achievement. The OHI classification entitles students to receive an Individualized Education Program (IEP) that includes special educational services that allow the structure and content of the educational curriculum to be adapted to fit the needs of the individual student.

Another venue in which accommodations can be provided to students with ADHD, if they do not qualify under IDEA, is through Section 504 of the Rehabilitation Act of 1973. To qualify for these accommodations, a student's disabilities must limit major life activities, including learning, concentrating, thinking, and communication with others. These accommodations typically adjust the regular curriculum without modifying the content, such as reducing the number of homework problems, providing preferred seating in a classroom, providing individual or small group instruction, changing formatting of material presented, and providing more frequent breaks and more time to complete assignments. A more detailed discussion of educational interventions in neurodevelopmental disabilities can be reviewed in Chapters (Classroom Setting) and (Legislative Mandates) of this book.

In the school environment, common interventions include token reinforcement systems, focused teacher or paraprofessional aide for attention and instruction, and environmental modifications, such as preferred seating and smaller class size. Even with symptom improvement with standard medication intervention, academic difficulties can persist. Most targeted training and teaching methods focus on study skills, organization, planning, and memory strategies to bridge this gap to assist in functional skill areas [430,431]. Overall, given the heterogeneity of instructional strategies and implementation, the evidence-base for these interventions is limited [432].

References

[1] Scahill L, Schwab-Stone M. Epidemiology of ADHD in school-age children. Child Adolesc Psychiatr Clin North Am 2000;9:541–55. Available from: https://doi.org/10.1016/s1056-4993(18)30106-8.

[2] Polanczyk G, de Lima MS, Horta BL, Biederman J, Rohde LA. The worldwide prevalence of ADHD: a systematic review and metaregression analysis. Am J Psychiatry 2007;164:942–8. Available from: https://doi.org/10.1176/ajp.2007.164.6.942.

[3] Thomas R, Sanders S, Doust J, Beller E, Glasziou P. Prevalence of attention-deficit/hyperactivity disorder: a systematic review and meta-analysis. Pediatrics 2015;135:e994–e1001. Available from: https://doi.org/10.1542/peds.2014-3482.

[4] Polanczyk Gv, Willcutt EG, Salum GA, Kieling C, Rohde LA. ADHD prevalence estimates across three decades: an updated systematic review and meta-regression analysis. Int J Epidemiol 2014;43:434–42. Available from: https://doi.org/10.1093/ije/dyt261.

[5] Dobrosavljevic M, Solares C, Cortese S, Andershed H, Larsson H. Prevalence of attention-deficit/hyperactivity disorder in older adults: a systematic review and meta-analysis. Neurosci Biobehav Rev 2020;118:282–9. Available from: https://doi.org/10.1016/j.neubiorev.2020.07.042.

[6] Faraone Sv, Biederman J, Spencer T, Mick E, Murray K, Petty C, et al. Diagnosing adult attention deficit hyperactivity disorder: are late onset and subthreshold diagnoses valid. Am J Psychiatry 2006;163:1720–9. Available from: https://doi.org/10.1176/ajp.2006.163.10.1720.

[7] Simon V, Czobor P, Bálint S, Mészáros Á, Bitter I. Prevalence and correlates of adult attention-deficit hyperactivity disorder: meta-analysis. Br J Psychiatry 2009;194:204–11. Available from: https://doi.org/10.1192/bjp.bp.107.048827.

[8] Polanczyk Gv, Salum GA, Sugaya LS, Caye A, Rohde LA. Annual research review: a meta-analysis of the worldwide prevalence of mental disorders in children and adolescents. J Child Psychol Psychiatry 2015;56:345–65. Available from: https://doi.org/10.1111/jcpp.12381.

[9] National Center on Birth Defects and Developmental Disabilities. State-based prevalence of ADHD diagnosis and treatment | CDC. Centers for Disease Control. <https://www.cdc.gov/ncbddd/adhd/data/diagnosis-treatment-data.html>; 2020 (accessed 10.09.21).

[10] Franz AP, Bolat GU, Bolat H, Matijasevich A, Santos IS, Silveira RC, et al. Efficacy and safety of immediate-release methylphenidate treatment for preschoolers with ADHD. Arch Clin Neuropsychol 2018;33:143–57. Available from: https://doi.org/10.1093/arclin/acx055.

[11] Levy F, Hay DA, Bennett KS, McStephen M. Gender differences in ADHD subtype comorbidity. J Am Acad Child Adolesc Psychiatry 2005;44:368–76. Available from: https://doi.org/10.1097/01.chi.0000153232.64968.c1.

[12] Loyer Carbonneau M, Demers M, Bigras M, Guay MC. Meta-analysis of sex differences in ADHD symptoms and associated cognitive deficits. J Atten Disord 2020. Available from: https://doi.org/10.1177/1087054720923736.

[13] Murray AL, Booth T, Eisner M, Auyeung B, Murray G, Ribeaud D. Sex differences in ADHD trajectories across childhood and adolescence. Dev Sci 2019;22:1–11. Available from: https://doi.org/10.1111/desc.12721.

[14] Mowlem FD, Rosenqvist MA, Martin J, Lichtenstein P, Asherson P, Larsson H. Sex differences in predicting ADHD clinical diagnosis and pharmacological treatment. Eur Child Adolesc Psychiatry 2019;28:481–9. Available from: https://doi.org/10.1007/s00787-018-1211-3.

[15] Arnett AB, Pennington BF, Willcutt EG, Defries JC, Olson RK. Sex differences in ADHD symptom severity. J Child Psychol Psychiatry 2015;56:632–9. Available from: https://doi.org/10.1111/jcpp.12337.

[16] Xu G, Strathearn L, Liu B, Yang B, Bao W. Twenty-year trends in diagnosed attention-deficit/hyperactivity disorder among US children and adolescents, 1997-2016. JAMA Netw Open 2018;1:1–9. Available from: https://doi.org/10.1001/jamanetworkopen.2018.1471.

[17] Cuffe SP, Moore CG, McKeown RE. Prevalence and correlates of ADHD symptoms in the National Health Interview Survey. J Atten Disord 2005;9:392–401. Available from: https://doi.org/10.1177/1087054705280413.

[18] Coker TR, Elliott MN, Toomey SL, Schwebel DC, Cuccaro P, Emery ST, et al. Racial and ethnic disparities in ADHD diagnosis and treatment. Pediatrics 2016;138. Available from: https://doi.org/10.1542/peds.2016-0407.

[19] MacDonald B, Pennington BF, Willcutt EG, Dmitrieva J, Samuelsson S, Byrne B, et al. Cross-country differences in parental reporting of symptoms of ADHD. J Cross Cult Psychol 2019;50:806–24. Available from: https://doi.org/10.1177/0022022119852422.

[20] Barkley RA, Peters H. The earliest reference to ADHD in the medical literature? Melchior Adam Weikard's description in 1775 of "attention deficit" (Mangel der Aufmerksamkeit, Attentio Volubilis). J Atten Disord 2012;16:623–30. Available from: https://doi.org/10.1177/1087054711432309.

[21] Palmer ED, Finger S. An early description of ADHD (inattentive subtype): Dr Alexander Crichton and "mental restlessness" (1798). Child Psychol Psychiatry Rev 2001;6:66–73. Available from: https://doi.org/10.1017/s1360641701002507.

[22] Thome J, Jacobs KA. Attention deficit hyperactivity disorder (ADHD) in a 19th century children's book. Eur Psychiatry 2004;19:303–6. Available from: https://doi.org/10.1016/j.eurpsy.2004.05.004.

[23] Still GF. Some abnormal psychical conditions in children. Lancet 1902;1077–82.

[24] Barkley RA. The relevance of the Still lectures to attention-deficit/hyperactivity disorder: a commentary. J Atten Disord 2006;10:137–40. Available from: https://doi.org/10.1177/1087054706288111.

[25] Dickman MS. Von Economo encephalitis. Arch Neurol 2001;58:1696–8. Available from: https://doi.org/10.1001/archneur.58.10.1696.

[26] Strauss AA, Lehtinen LE. Psychopathology and education of the brain-injured child. Oxford, England: Grune & Stratton; 1947.

[27] Wender PH. Minimal brain dysfunction in children. Diagnosis and management. Pediatr Clin North Am 1973;20:187–202. Available from: https://doi.org/10.1016/s0031-3955(16)32819-x.

[28] Wender PH. The minimal brain dysfunction syndrome. Annu Rev Med 1975;26:45–62. Available from: https://doi.org/10.1146/annurev.me.26.020175.000401.

[29] Menkes MM, Rowe JS, Menkes JH. A twenty-five year follow-up study on the hyperkinetic child with minimal brain dysfunction. Pediatrics 1967;39:393–9.

[30] Moreno-De-Luca A, Myers SM, Challman TD, Moreno-De-Luca D, Evans DW, Ledbetter DH. Developmental brain dysfunction: revival and expansion of old concepts based on new genetic evidence. Lancet Neurol 2013;12:406–14. Available from: https://doi.org/10.1016/S1474-4422(13)70011-5.

[31] Accardo PJ, Shapiro BK, Capute AJ, editors. Behavior belongs in the brian: neurobehavioral syndromes. Baltimore, MD: York Press; 1997.

[32] American Psychiatric Association. Diagnostic and statistical manual of mental disorders. 2nd Edition, Washington DC: American Psychiatric Association; 1968.

[33] American Psychiatric Association. Diagnostic and statistical manual of mental disorders. third edition (DSM-III), Washington, DC: American Psychiatric Association; 1980.

[34] American Psychiatric Association. Diagnostic and statistical manual of mental disorders, 3rd edition - Revised (DSM-III - Revised). Washington, DC. 1987.

[35] American Psychiatric Association. Diagnostic and statistical manual. 4th edition (DSM-IV), Washington, DC: American Psychiatric Association; 1994.

[36] American Psychiatric Association. Diagnostic and statistical manual of mental disorders, fourth edition - text revision (DSMIV-TR). American Psychiatric Association; 2000.

[37] Wolraich ML, Felice ME, Drotar D. The classification of child and adolescent mental diagnoses in primary care: diagnostic and statistical manual for primary care (DSM-PC) child and adolescent version. Am Acad Pediatr 1996.

[38] American Psychiatric Association. Diagnostic and statistical manual of mental disorders: DSM-5. 5th edition, Arlington, VA, US: American Psychiatric Publishing, Inc; 2013. Available from: https://doi.org/10.1176/appi.books.9780890425596.

[39] American Psychiatric Association. Diagnostic and statistical manual of mental disorders (5th ed., text rev.) (DSM5-TR). American Psychiatric Association; 2022.

[40] Voort vande JL, He J-P, Jameson ND, Merikangas KR. Impact of the DSM-5 attention-deficit/hyperactivity disorder age-of-onset criterion in the US adolescent population. J Am Acad Child Adolesc Psychiatry 2014;53:736–44. Available from: https://doi.org/10.1016/J.JAAC.2014.03.005.

[41] Asherson P, Agnew-Blais J. Annual research review: does late-onset attention-deficit/hyperactivity disorder exist? J Child Psychol Psychiatry 2019;60:333–52. Available from: https://doi.org/10.1111/JCPP.13020.

[42] Asarnow RF, Newman N, Weiss RE, Su E. Association of attention-deficit/hyperactivity disorder diagnoses with pediatric traumatic brain injury: a meta-analysis. JAMA Pediatr 2021. Available from: https://doi.org/10.1001/JAMAPEDIATRICS.2021.2033.

[43] Eme R. ADHD: an integration with pediatric traumatic brain injury. Expert Rev Neurother 2012;12:475–83. Available from: https://doi.org/10.1586/ern.12.15.

[44] Levin H, Hanten G, Max J, Li X, Swank P, Ewing-Cobbs L, et al. Symptoms of attention-deficit/hyperactivity disorder following traumatic brain injury in children. J Dev Behav Pediatr 2007;28:108–18. Available from: https://doi.org/10.1097/01.DBP.0000267559.26576.cd.

[45] Narad ME, Riemersma J, Wade SL, Smith-Paine J, Morrison P, Taylor HG, et al. Impact of secondary ADHD on long-term outcomes after early childhood traumatic brain injury. J Head Trauma Rehabil 2020;35:E271–9. Available from: https://doi.org/10.1097/HTR.0000000000000550.

[46] Adeyemo BO, Biederman J, Zafonte R, Kagan E, Spencer TJ, Uchida M, et al. Mild traumatic brain injury and ADHD: a systematic review of the literature and meta-analysis. J Atten Disord 2014;18:576–84. Available from: https://doi.org/10.1177/1087054714543371.

[47] Max JE, Schachar RJ, Levin HS, Ewing-Cobbs L, Chapman SB, Dennis M, et al. Predictors of secondary attention-deficit/hyperactivity disorder in children and adolescents 6 to 24 months after traumatic brain injury. J Am Acad Child Adolesc Psychiatry 2005;44:1041–9. Available from: https://doi.org/10.1097/01.chi.0000173292.05817.f8.

[48] International Classification of Diseases, Eleventh Revision (ICD-11), World Health Organization (WHO) 2019/2021, https://icd.who.int/browse11. Licensed under Creative Commons Attribution-NoDerivatives 3.0 IGO licence (CC BY-ND 3.0 IGO).

[49] Voigt RG. Making developmental-behavioral diagnoses. In: Voigt RG, Macias MM, Myers SM, Tapia CD, editors. AAP developmental and behavioral pediatrics. 2nd ed. American Academy of Pediatrics; 2018. p. 187–222.

[50] Bélanger SA, Andrews D, Gray C, Korczak D. ADHD in children and youth: part 1-etiology, diagnosis, and comorbidity. Paediatr Child Health 2018;23:447–53. Available from: https://doi.org/10.1093/pch/pxy109.

[51] Mattfeld AT, Gabrieli JDE, Biederman J, Spencer T, Brown A, Kotte A, et al. Brain differences between persistent and remitted attention deficit hyperactivity disorder. Brain 2014;137:2423–8. Available from: https://doi.org/10.1093/brain/awu137.

[52] de Reyes AL, Kazdin AE. Informant discrepancies in the assessment of childhood psychopathology: a critical review, theoretical framework, and recommendations for further study. Psychol Bull 2005;131:483–509. Available from: https://doi.org/10.1037/0033-2909.131.4.483.

[53] Tandon M, Pergjika A. Attention deficit hyperactivity disorder in preschool-age children. Child Adolesc Psychiatr Clin North Am 2017;26:523–38. Available from: https://doi.org/10.1016/j.chc.2017.02.007.

[54] Slobodin O, Davidovitch M. Gender differences in objective and subjective measures of ADHD among clinic-referred children. Front Hum Neurosci 2019;13:1–14. Available from: https://doi.org/10.3389/fnhum.2019.00441.

[55] Narad ME, Garner AA, Peugh JL, Tamm L, Antonini TN, Kingery KM, et al. Parent-teacher agreement on ADHD symptoms across development. Psychol Assess 2015;27:239–48. Available from: https://doi.org/10.1037/a0037864.

[56] Murray AL, Booth T, Ribeaud D, Eisner M. Disagreeing about development: an analysis of parent-teacher agreement in ADHD symptom trajectories across the elementary school years. Int J Methods Psychiatr Res 2018;27:1–8. Available from: https://doi.org/10.1002/mpr.1723.

[57] Arildskov TW, Sonuga-Barke EJS, Thomsen PH, Virring A, Østergaard SD. How much impairment is required for ADHD? No evidence of a discrete threshold. J Child Psychol Psychiatry 2021. Available from: https://doi.org/10.1111/JCPP.13440.

[58] Becker SP. Systematic review: assessment of sluggish cognitive tempo over the past decade. J Am Acad Child Adolesc Psychiatry 2021;60:690–709. Available from: https://doi.org/10.1016/j.jaac.2020.10.016.

[59] Servera M, Sáez B, Burns GL, Becker SP. Clinical differentiation of sluggish cognitive tempo and attention-deficit/hyperactivity disorder in children. J Abnorm Psychol 2018;127:818–29. Available from: https://doi.org/10.1037/abn0000375.

[60] Becker SP, Mossing KW, Zoromski AK, Vaughn AJ, Epstein JN, Tamm L, et al. Assessing sluggish cognitive tempo and ADHD inattention in elementary students: empirical differentiation, invariance across sex and grade, and measurement precision. Psychol Assess 2020;32:1047–56. Available from: https://doi.org/10.1037/pas0000946.

[61] Mill J, Xu X, Ronald A, Curran S, Price T, Knight J, et al. Quantitative trait locus analysis of candidate gene alleles associated with attention deficit hyperactivity disorder (ADHD) in five genes: DRD4, DAT1, DRD5, SNAP-25, and 5HT1B. Am J Med Genet B Neuropsychiatr Genet 2005;133 B:68–73. Available from: https://doi.org/10.1002/ajmg.b.30107.

[62] Larsson H, Anckarsater H, Råstam M, Chang Z, Lichtenstein P. Childhood attention-deficit hyperactivity disorder as an extreme of a continuous trait: a quantitative genetic study of 8,500 twin pairs. J Child Psychol Psychiatry 2012;53:73–80. Available from: https://doi.org/10.1111/j.1469-7610.2011.02467.x.

[63] Greven CU, Buitelaar JK, Salum GA. From positive psychology to psychopathology: the continuum of attention-deficit hyperactivity disorder. J Child Psychol Psychiatry 2018;59:203–12. Available from: https://doi.org/10.1111/jcpp.12786.

[64] Hadders-Algra M, Groothuis AMC. Quality of general movements in infancy is related to neurological dysfunction, ADHD, and aggressive behaviour. Dev Med Child Neurol 1999;41:381–91. Available from: https://doi.org/10.1017/S0012162299000845.

[65] Sullivan EL, Holton KF, Nousen EK, Barling AN, Sullivan CA, Propper CB, et al. Early identification of ADHD risk via infant temperament and emotion regulation: a pilot study. J Child Psychol Psychiatry 2015;56:949–57. Available from: https://doi.org/10.1111/JCPP.12426.

[66] Miller Nv, Hane AA, Degnan KA, Fox NA, Chronis-Tuscano A. Investigation of a developmental pathway from infant anger reactivity to childhood inhibitory control and ADHD symptoms: interactive effects of early maternal caregiving. J Child Psychol Psychiatry 2019;60:762–72. Available from: https://doi.org/10.1111/JCPP.13047.

[67] Gurevitz M, Geva R, Varon M, Leitner Y. Early markers in infants and toddlers for development of ADHD. J Atten Disord 2014;18:14–22. Available from: https://doi.org/10.1177/1087054712447858.

[68] Johnson MH, Gliga T, Jones E, Charman T. Annual research review: infant development, autism, and ADHD – early pathways to emerging disorders. J Child Psychol Psychiatry 2015;56:228–47. Available from: https://doi.org/10.1111/JCPP.12328.

[69] Posner K, Melvin GA, Murray DW, Gugga SS, Fisher P, Skrobala A, et al. Clinical presentation of attention-deficit/hyperactivity disorder in preschool children: the preschoolers with attention-deficit/hyperactivity disorder treatment study (PATS). J Child Adolesc Psychopharmacol 2007;17:547–62. Available from: https://doi.org/10.1089/cap.2007.0075.

[70] Curchack-Lichtin JT, Chacko A, Halperin JM. Changes in ADHD symptom endorsement: preschool to school age. J Abnorm Child Psychol 2014;42:993–1004. Available from: https://doi.org/10.1007/s10802-013-9834-9.

[71] Chacko A, Wakschlag L, Hill C, Danis B, Espy KA. Viewing preschool disruptive behavior disorders and attention-deficit/hyperactivity disorder through a developmental lens: what we know and what we need to know. Child Adolesc Psychiatr Clin North Am 2009;18:627–43. Available from: https://doi.org/10.1016/j.chc.2009.02.003.

[72] Wakschlag LS, Tolan PH, Leventhal BL. Research review: "ain't misbehavin": towards a developmentally-specified nosology for preschool disruptive behavior. J Child Psychol Psychiatry 2010;51:3–22. Available from: https://doi.org/10.1111/j.1469-7610.2009.02184.x.

[73] Agnew-Blais JC, Polanczyk Gv, Danese A, Wertz J, Moffitt TE, Arseneault L. Evaluation of the persistence, remission, and emergence of Attention-deficit/hyperactivity disorder in young adulthood. JAMA Psychiatry 2016;73:713–20. Available from: https://doi.org/10.1001/jamapsychiatry.2016.0465.

[74] Franke B, Michelini G, Asherson P, Banaschewski T, Bilbow A, Buitelaar JK, et al. Live fast, die young? A review on the developmental trajectories of ADHD across the lifespan. Eur Neuropsychopharmacol 2018;28:1059–88. Available from: https://doi.org/10.1016/j.euroneuro.2018.08.001.

[75] Holbrook JR, Cuffe SP, Cai B, Visser SN, Forthofer MS, Bottai M, et al. Persistence of parent-reported ADHD symptoms from childhood through adolescence in a community sample. J Atten Disord 2016;20:11–20. Available from: https://doi.org/10.1177/1087054714539997.

[76] Riddle MA, Yershova K, Lazzaretto D, Paykina N, Yenokyan G, Greenhill L, et al. The preschool attention-deficit/hyperactivity disorder treatment study (PATS) 6-year follow-up. J Am Acad Child Adolesc Psychiatry 2013;52:264–78. Available from: https://doi.org/10.1016/j.jaac.2012.12.007.

[77] Barbaresi WJ, Campbell L, Diekroger EA, Froehlich TE, Liu YH, O'Malley E, et al. Society for developmental and behavioral pediatrics clinical practice guideline for the assessment and treatment of children and adolescents with complex attention-deficit/hyperactivity disorder. J Dev Behav Pediatr 2020;41(Suppl 2):S35–57. Available from: https://doi.org/10.1097/DBP.0000000000000770.

[78] Caye A, Petresco S, de Barros AJD, Bressan RA, Gadelha A, Gonçalves H, et al. Relative age and attention-deficit/hyperactivity disorder: data from three epidemiological cohorts and a meta-analysis. J Am Acad Child Adolesc Psychiatry 2020;59:990–7. Available from: https://doi.org/10.1016/j.jaac.2019.07.939.

[79] Caye A, Rocha TBM, Anselmi L, Murray J, Menezes AMB, Barros FC, et al. Attention-deficit/hyperactivity disorder trajectories from childhood to young adulthood evidence from a birth cohort supporting a late-onset syndrome. JAMA Psychiatry 2016;73:705–12. Available from: https://doi.org/10.1001/jamapsychiatry.2016.0383.

[80] Faraone SV, Asherson P, Banaschewski T, Biederman J, Buitelaar JK, Ramos-Quiroga JA, et al. Attention-deficit/hyperactivity disorder. Nat Rev Dis Prim 2015;1. Available from: https://doi.org/10.1038/nrdp.2015.20.

[81] Sibley MH, Kuriyan AB. DSM-5 changes enhance parent identification of symptoms in adolescents with ADHD. Psychiatry Res 2016;242:180–5. Available from: https://doi.org/10.1016/j.psychres.2016.05.036.

[82] Chan E, Fogler JM, Hammerness PG. Treatment of attention-deficit/hyperactivity disorder in adolescents: a systematic review. J Am Med Assoc 2016;315:1997–2008. Available from: https://doi.org/10.1001/jama.2016.5453.

[83] Biederman J, Petty CR, Woodworth KY, Lomedico A, Hyder LL, Faraone SV. Adult outcome of attention-deficit/hyperactivity disorder: a controlled 16-year follow-up study. J Clin Psychiatry 2012;73:941–50. Available from: https://doi.org/10.4088/JCP.11m07529.

[84] Bonvicini C, Faraone Sv, Scassellati C. Attention-deficit hyperactivity disorder in adults: a systematic review and meta-analysis of genetic, pharmacogenetic and biochemical studies. Mol Psychiatry 2016;21:872–84. Available from: https://doi.org/10.1038/mp.2016.74.

[85] Breda V, Rohde LA, Menezes AMB, Anselmi L, Caye A, Rovaris DL, et al. The neurodevelopmental nature of attention-deficit hyperactivity disorder in adults. Br J Psychiatry 2021;218:43–50. Available from: https://doi.org/10.1192/bjp.2020.200.

[86] Kosaka H, Fujioka T, Jung M. Symptoms in individuals with adult-onset ADHD are masked during childhood. Eur Arch Psychiatry Clin Neurosci 2019;269:753–5. Available from: https://doi.org/10.1007/s00406-018-0893-3.

[87] Moffitt TE, Houts R, Asherson P, Belsky DW, Corcoran DL, Hammerle M, et al. Is adult ADHD a childhood-onset neurodevelopmental disorder? Evidence from a four-decade longitudinal cohort study. Am J Psychiatry 2015;172:967–77. Available from: https://doi.org/10.1176/appi.ajp.2015.14101266.

[88] Sibley MH, Yeguez CE. The impact of DSM-5 a-criteria changes on parent ratings of ADHD in adolescents. J Atten Disord 2018;22:83–91. Available from: https://doi.org/10.1177/1087054714526040.

[89] Sibley MH. Empirically-informed guidelines for first-time adult ADHD diagnosis. J Clin Exp Neuropsychol 2021. Available from: https://doi.org/10.1080/13803395.2021.1923665.

[90] Cabral MDI, Liu S, Soares N. Attention-deficit/hyperactivity disorder: diagnostic criteria, epidemiology, risk factors and evaluation in youth. Transl Pediatr 2020;9:S104–13. Available from: https://doi.org/10.21037/tp.2019.09.08.

[91] Stein MA, Shin D. Disorders of attention: diagnosis. In: 3rd ed. Accardo PJ, editor. Capute & Accardo's neurodevelopmental disabilities in infancy and childhood: the spectrum of neurodevelopmental disabilities, vol. 2. Brookes Publishing; 2007. p. 639–56.

[92] Accardo PJ. The misdiagnosis of the hyperactive child. In: Accardo PJ, Blondis TA, Whitman BY, editors. Attention deficit disorders and hyperactivity in children. 1st ed. New York, NY: Marcel Dekker; 1991.

[93] Mandal RL, Olmi DJ, Wilczynski SM. Behavior rating scales: concordance between multiple informants in the diagnosis of attention-deficit/hyperactivity disorder. J Atten Disord 1999;3:97–103. Available from: https://doi.org/10.1177/108705479900300204.

[94] Danforth JS, DuPaul GJ. Interrater reliability of teacher rating scales for children with attention-deficit hyperactivity disorder. J Psychopathol Behav Assess 1996;18:227–37. Available from: https://doi.org/10.1007/BF02229046.

[95] Shapiro BK. Use and abuse of behavioral rating scales. In: Rogers BT, Montgomery TR, Lock TM, Accardo PJ, editors. Attention deficit hyperactivity disorder: the clinical spectrum. Baltimore, MD: York Press; 2001. p. 73–86.

[96] Conners CK. Rating scales in attention-deficit/hyperactivity disorder: use in assessment and treatment monitoring. J Clin Psychiatry 1998;59:24–30.

[97] Conners CK. Clinical use of rating scales in diagnosis and treatment of attention- deficit/hyperactivity disorder. Pediatr Clin North Am 1999;46:857–70. Available from: https://doi.org/10.1016/S0031-3955(05)70159-0.

[98] Achenbach TM, Ruffle TM. The child behavior checklist and related forms for assessing behavioral/emotional problems and competencies. Pediatr Rev 2000;21:265–71. Available from: https://doi.org/10.1542/pir.21-8-265.

[99] Chang LY, Wang MY, Tsai PS. Diagnostic accuracy of rating scales for attention-deficit/hyperactivity disorder: a meta-analysis. Pediatrics 2016;137. Available from: https://doi.org/10.1542/peds.2015-2749.

[100] Goodman R. The strengths and difficulties questionnaire: a research note. J Child Psychol Psychiatry 1997;38:581–6. Available from: https://doi.org/10.1111/J.1469-7610.1997.TB01545.X.

[101] Reynolds CR, Kamphaus RW. Behavior assessment system for children—third edition (BASC-3). Bloomington, MN: Pearson; 2015.

[102] Conners CK. Conners third edition (Conners 3). Bloomington, MN: Pearson; 2008.

[103] DuPaul GJ, Power TJ, Anastopoulos AD, Reid R. ADHD rating scale - 5 for children and adolescents: checklists, norms, and clinical interpretation. Guilford Publications; 2016.

[104] Wolraich ML, Lambert W, Doffing MA, Bickman L, Simmons T, Worley K. Psychometric properties of the Vanderbilt ADHD diagnostic parent rating scale in a referred population. J Pediatr Psychol 2003;28:559–67. Available from: https://doi.org/10.1093/JPEPSY/JSG046.

[105] Bard DE, Wolraich ML, Neas B, Doffing M, Beck L. The psychometric properties of the vanderbilt attention-deficit hyperactivity disorder diagnostic parent rating scale in a community population. J Dev Behav Pediatr 2013;34:72–82. Available from: https://doi.org/10.1097/DBP.0b013e31827a3a22.

[106] Wolraich ML, Bard DE, Neas B, Doffing M, Beck L. The psychometric properties of the vanderbilt attention-deficit hyperactivity disorder diagnostic teacher rating scale in a community population. J Dev Behav Pediatr 2013;34:83–93. Available from: https://doi.org/10.1097/DBP.0b013e31827d55c3.

[107] Staff AI, Oosterlaan J, van der Oord S, Hoekstra PJ, Vertessen K, de Vries R, et al. The validity of teacher rating scales for the assessment of ADHD symptoms in the classroom: a systematic review and meta-analysis. J Atten Disord 2020;25:1578–93. Available from: https://doi.org/10.1177/1087054720916839.

[108] Crisci G, Caviola S, Cardillo R, Mammarella IC. Executive functions in neurodevelopmental disorders: comorbidity overlaps between attention deficit and hyperactivity disorder and specific learning disorders. Front Hum Neurosci 2021;15. Available from: https://doi.org/10.3389/fnhum.2021.594234.

[109] Pievsky MA, McGrath RE. The neurocognitive profile of attention-deficit/hyperactivity disorder: a review of meta-analyses. Arch Clin Neuropsychol 2018;33:143–57. Available from: https://doi.org/10.1093/arclin/acx055.

[110] Sjöwall D, Bohlin G, Rydell AM, Thorell LB. Neuropsychological deficits in preschool as predictors of ADHD symptoms and academic achievement in late adolescence. Child Neuropsychol 2017;23:111–28. Available from: https://doi.org/10.1080/09297049.2015.1063595.

[111] Perugini EM, Harvey EA, Lovejoy DW, Sandstrom K, Webb AH. The predictive power of combined neuropsychological measures for attention-deficit/hyperactivity disorder in children. Child Neuropsychol 2000;6:101–14. Available from: https://doi.org/10.1076/chin.6.2.101.7059.

[112] Pearl PL, Weiss RE, Stein MA. Mimics of ADHD: medical and neurological conditions. Attention-deficit hyperactivity disorder in adults and children. Cambridge University Press; 2015. p. 150–60. Available from: https://doi.org/10.1017/CBO9781139035491.014.

[113] Wolraich ML, Hagan JF, Allan C, Chan E, Davison D, Earls M, et al. Clinical practice guideline for the diagnosis, evaluation, and treatment of attention-deficit/hyperactivity disorder in children and adolescents. Pediatrics 2019;144:20192528. Available from: https://doi.org/10.1542/peds.2019-2528.

[114] Mehta TR, Monegro A, Nene Y, Fayyaz M, Bollu PC. Neurobiology of ADHD: a review. Curr Dev Disord Rep 2019;6:235–40. Available from: https://doi.org/10.1007/s40474-019-00182-w.

[115] Faraone SV. The pharmacology of amphetamine and methylphenidate: relevance to the neurobiology of attention-deficit/hyperactivity disorder and other psychiatric comorbidities. Neurosci Biobehav Rev 2018;87:255–70. Available from: https://doi.org/10.1016/j.neubiorev.2018.02.001.

[116] del Campo N, Chamberlain SR, Sahakian BJ, Robbins TW. The roles of dopamine and noradrenaline in the pathophysiology and treatment of attention-deficit/hyperactivity disorder. Biol Psychiatry 201169. Available from: https://doi.org/10.1016/j.biopsych.2011.02.036.

[117] Jenson D, Yang K, Acevedo-Rodriguez A, Levine A, Broussard JI, Tang J, et al. Dopamine and norepinephrine receptors participate in methylphenidate enhancement of in vivo hippocampal synaptic plasticity. Neuropharmacology 2015;90:23–32. Available from: https://doi.org/10.1016/j.neuropharm.2014.10.029.

[118] Volkow ND, Wang GJ, Kollins SH, Wigal TL, Newcorn JH, Telang F, et al. Evaluating dopamine reward pathway in ADHD: clinical implications. J Am Med Assoc 2009;302:1084–91. Available from: https://doi.org/10.1001/jama.2009.1308.

[119] Aarts E, van Holstein M, Hoogman M, Onnink M, Kan C, Franke B, et al. Reward modulation of cognitive function in adult attention-deficit/hyperactivity disorder: a pilot study on the role of striatal dopamine. Behav Pharmacol 2015;26:227–40. Available from: https://doi.org/10.1097/FBP.0000000000000116.

[120] Arnsten AFT, Pliszka SR. Catecholamine influences on prefrontal cortical function: relevance to treatment of attention deficit/hyperactivity disorder and related disorders. Pharmacol Biochem Behav 2011;99:211–16. Available from: https://doi.org/10.1016/j.pbb.2011.01.020.

[121] Albajara Sáenz A, Villemonteix T, Massat I, Sáenz AA, Villemonteix T, Massat I. Structural and functional neuroimaging in attention-deficit/hyperactivity disorder. Dev Med Child Neurol 2019;61:399–405. Available from: https://doi.org/10.1111/dmcn.14050.

[122] Valera EM, Faraone Sv, Murray KE, Seidman LJ. Meta-analysis of structural imaging findings in attention-deficit/hyperactivity disorder. Biol Psychiatry 2007;61:1361–9. Available from: https://doi.org/10.1016/j.biopsych.2006.06.011.

[123] Ellison-Wright I, Ellison-Wright Z, Bullmore E. Structural brain change in attention deficit hyperactivity disorder identified by meta-analysis. BMC Psychiatry 20088. Available from: https://doi.org/10.1186/1471-244X-8-51.

[124] Nakao T, Radua J, Rubia K, Mataix-Cols D. Gray matter volume abnormalities in ADHD: voxel-based meta-analysis exploring the effects of age and stimulant medication. Am J Psychiatry 2011;168:1154–63. Available from: https://doi.org/10.1176/appi.ajp.2011.11020281.

[125] Greven CU, Bralten J, Mennes M, O'Dwyer L, van Hulzen KJE, Rommelse N, et al. Developmentally stable whole-brain volume reductions and developmentally sensitive caudate and putamen volume alterations in those with attention-deficit/hyperactivity disorder and their unaffected siblings. JAMA Psychiatry 2015;72:490–9. Available from: https://doi.org/10.1001/jamapsychiatry.2014.3162.

[126] Hoogman M, Bralten J, Hibar DP, Mennes M, Zwiers MP, Schweren LSJ, et al. Subcortical brain volume differences in participants with attention deficit hyperactivity disorder in children and adults: a cross-sectional mega-analysis. Lancet Psychiatry 2017;4:310–19. Available from: https://doi.org/10.1016/S2215-0366(17)30049-4.

[127] Boedhoe PSW, van Rooij D, Hoogman M, Twisk JWR, Schmaal L, Abe Y, et al. Subcortical brain volume, regional cortical thickness, and cortical surface area across disorders: findings from the ENIGMA ADHD, ASD, and OCD working groups. Am J Psychiatry 2020;177:834–43. Available from: https://doi.org/10.1176/appi.ajp.2020.19030331.

[128] Hoogman M, Rijpkema M, Janss L, Brunner H, Fernandez G, Buitelaar J, et al. Current self-reported symptoms of attention deficit/hyperactivity disorder are associated with total brain volume in healthy adults. PLoS ONE 2012;7:e31273. Available from: https://doi.org/10.1371/journal.pone.0031273.

[129] Pereira-Sanchez V, Castellanos FX. Neuroimaging in attention-deficit/hyperactivity disorder. Curr Opin Psychiatry 2021;34:105–11. Available from: https://doi.org/10.1097/YCO.0000000000000669.

[130] Chen S, Guan L, Tang J, He F, Zheng Y. Asymmetry in cortical and subcortical structures of the brain in children and adolescents with attention-deficit/hyperactivity disorder. Neuropsychiatr Dis Treat 2021;17:493–502. Available from: https://doi.org/10.2147/NDT.S292444.

[131] Samea F, Soluki S, Nejati V, Zarei M, Cortese S, Eickhoff SB, et al. Brain alterations in children/adolescents with ADHD revisited: a neuroimaging meta-analysis of 96 structural and functional studies. Neurosci Biobehav Rev 2019;100:1–8. Available from: https://doi.org/10.1016/j.neubiorev.2019.02.011.

[132] Narr KL, Woods RP, Lin J, Kim J, Phillips OR, Del'Homme M, et al. Widespread cortical thinning is a robust anatomical marker for attention-deficit/hyperactivity disorder. J Am Acad Child Adolesc Psychiatry 2009;48:1014–22. Available from: https://doi.org/10.1097/CHI.0b013e3181b395c0.

[133] Almeida LG, Ricardo-Garcell J, Prado H, Barajas L, Fernández-Bouzas A, Ávila D, et al. Reduced right frontal cortical thickness in children, adolescents and adults with ADHD and its correlation to clinical variables: a cross-sectional study. J Psychiatr Res 2010;44:1214–23. Available from: https://doi.org/10.1016/j.jpsychires.2010.04.026.

[134] Silk TJ, Beare R, Malpas C, Adamson C, Vilgis V, Vance A, et al. Cortical morphometry in attention deficit/hyperactivity disorder: contribution of thickness and surface area to volume. Cortex 2016;82:1–10. Available from: https://doi.org/10.1016/j.cortex.2016.05.012.

[135] Shaw P, Eckstrand K, Sharp W, Blumenthal J, Lerch JP, Greenstein D, et al. Attention-deficit/hyperactivity disorder is characterized by a delay in cortical maturation. Proc Natl Acad Sci USA 2007;104:19649–54. Available from: https://doi.org/10.1073/pnas.0707741104.

[136] Ambrosino S, de Zeeuw P, Wierenga LM, van Dijk S, Durston S. What can cortical development in attention-deficit/ hyperactivity disorder teach us about the early developmental mechanisms involved? Cereb Cortex 2017;27:4624–34. Available from: https://doi.org/10.1093/cercor/bhx182.

[137] Wolosin SM, Richardson ME, Hennessey JG, Denckla MB, Mostofsky SH. Abnormal cerebral cortex structure in children with ADHD. Hum Brain Mapp 2009;30:175–84. Available from: https://doi.org/10.1002/hbm.20496.

[138] Almeida Montes LG, Prado Alcántara H, Martínez García RB, de La Torre LB, Ávila Acosta D, Duarte MG. Brain cortical thickness in ADHD: age, sex, and clinical correlations. J Atten Disord 2013;17:641–54. Available from: https://doi.org/10.1177/1087054711434351.

[139] Forde NJ, Ronan L, Zwiers MP, Alexander-Bloch AF, Faraone SV, Oosterlaan J, et al. No association between cortical gyrification or intrinsic curvature and attention-deficit/hyperactivity disorder in adolescents and young adults. Front Neurosci 2017;11:218. Available from: https://doi.org/10.3389/fnins.2017.00218.

[140] Aoki Y, Cortese S, Castellanos FX. Research review: diffusion tensor imaging studies of attention-deficit/hyperactivity disorder: meta-analyses and reflections on head motion. J Child Psychol Psychiatry 2018;59:193–202. Available from: https://doi.org/10.1111/jcpp.12778.

[141] Chen L, Hu X, Ouyang L, He N, Liao Y, Liu Q, et al. A systematic review and meta-analysis of tract-based spatial statistics studies regarding attention-deficit/hyperactivity disorder. Neurosci Biobehav Rev 2016;68:838–47. Available from: https://doi.org/10.1016/j.neubiorev.2016.07.022.

[142] van Ewijk H, Heslenfeld DJ, Zwiers MP, Buitelaar JK, Oosterlaan J. Diffusion tensor imaging in attention deficit/hyperactivity disorder: a systematic review and meta-analysis. Neurosci Biobehav Rev 2012;36:1093–106. Available from: https://doi.org/10.1016/J.NEUBIOREV.2012.01.003.

[143] Saad JF, Griffiths KR, Kohn MR, Braund TA, Clarke S, Williams LM, et al. No support for white matter connectivity differences in the combined and inattentive ADHD presentations. PLoS ONE 2021;16:e0245028. Available from: https://doi.org/10.1371/journal.pone.0245028.

[144] Frodl T, Skokauskas N. Meta-analysis of structural MRI studies in children and adults with attention deficit hyperactivity disorder indicates treatment effects. Acta Psychiatr Scand 2012;125:114–26. Available from: https://doi.org/10.1111/j.1600-0447.2011.01786.x.

[145] Villemonteix T, de Brito SA, Kavec M, Balériaux D, Metens T, Slama H, et al. Grey matter volumes in treatment naïve vs. chronically treated children with attention deficit/hyperactivity disorder: a combined approach. Eur Neuropsychopharmacol 2015;25:1118–27. Available from: https://doi.org/10.1016/j.euroneuro.2015.04.015.

[146] Chang JC, Lin HY, Lv J, Tseng WYI, Gau SSF. Regional brain volume predicts response to methylphenidate treatment in individuals with ADHD. BMC Psychiatry 2021;21:1–14. Available from: https://doi.org/10.1186/s12888-021-03040-5.

[147] Fotopoulos NH, Devenyi GA, Guay S, Sengupta SM, Chakravarty MM, Grizenko N, et al. Cumulative exposure to ADHD medication is inversely related to hippocampus subregional volume in children. NeuroImage 2021;31:102695. Available from: https://doi.org/10.1016/J.NICL.2021.102695.

[148] Pretus C, Ramos-Quiroga JA, Richarte V, Corrales M, Picado M, Carmona S, et al. Time and psychostimulants: opposing long-term structural effects in the adult ADHD brain. A longitudinal MR study. Eur Neuropsychopharmacol 2017;27:1238–47. Available from: https://doi.org/10.1016/j.euroneuro.2017.10.035.

[149] Fusar-Poli P, Rubia K, Rossi G, Sartori G, Balottin U. Striatal dopamine transporter alterations in ADHD: pathophysiology or adaptation to psychostimulants? A meta-analysis. Am J Psychiatry 2012;169:264–72. Available from: https://doi.org/10.1176/appi.ajp.2011.11060940.

[150] Lei D, Du M, Wu M, Chen T, Huang X, Du X, et al. Functional MRI reveals different response inhibition between adults and children with ADHD. Neuropsychology 2015;29:874–81. Available from: https://doi.org/10.1037/neu0000200.

[151] Cortese S, Kelly C, Chabernaud C, Proal E, di Martino A, Milham MP, et al. Toward systems neuroscience of ADHD: a meta-analysis of 55 fMRI sudies. Am J Psychiatry 2012;169:1038–55. Available from: https://doi.org/10.1176/appi.ajp.2012.11101521.

[152] Hart H, Radua J, Nakao T, Mataix-Cols D, Rubia K. Meta-analysis of functional magnetic resonance imaging studies of inhibition and attention in attention-deficit/hyperactivity disorder: exploring task-specific, stimulant medication, and age effects. JAMA Psychiatry 2013;70:185–98. Available from: https://doi.org/10.1001/jamapsychiatry.2013.277.

[153] McCarthy H, Skokauskas N, Frodl T. Identifying a consistent pattern of neural function in attention deficit hyperactivity disorder: a meta-analysis. Psychol Med 2014;44:869–80. Available from: https://doi.org/10.1017/S0033291713001037.

[154] Yap KH, Abdul Manan H, Sharip S. Heterogeneity in brain functional changes of cognitive processing in ADHD across age: a systematic review of task-based fMRI studies. Behav Brain Res 2021;397:112888. Available from: https://doi.org/10.1016/j.bbr.2020.112888.

[155] Cortese S. Understanding the network bases of ADHD: an overview of the fMRI evidence. Brain network dysfunction in neuropsychiatric illness. Cham: Springer; 2021. p. 333–43. Available from: https://doi.org/10.1007/978-3-030-59797-9_16.

[156] Damiani S, Tarchi L, Scalabrini A, Marini S, Provenzani U, Rocchetti M, et al. Beneath the surface: hyper-connectivity between caudate and salience regions in ADHD fMRI at rest. Eur Child Adolesc Psychiatry 2020;30(4):619–31. Available from: https://doi.org/10.1007/S00787-020-01545-0.

[157] Posner J, Park C, Wang Z. Connecting the dots: a review of resting connectivity MRI studies in attention-deficit/hyperactivity disorder. Neuropsychol Rev 2014;24:3–15. Available from: https://doi.org/10.1007/s11065-014-9251-z.

[158] Castellanos FX, Aoki Y. Intrinsic functional connectivity in attention-deficit/hyperactivity disorder: a science in development. Biol Psychiatry 2016;1:253–61. Available from: https://doi.org/10.1016/j.bpsc.2016.03.004.

[159] Sutcubasi B, Metin B, Kurban MK, Metin ZE, Beser B, Sonuga-Barke E. Resting-state network dysconnectivity in ADHD: a system-neuroscience-based meta-analysis. World J Biol Psychiatry 2020;21:662–72. Available from: https://doi.org/10.1080/15622975.2020.1775889.

[160] Cortese S, Aoki YY, Itahashi T, Castellanos FX, Eickhoff SB. Systematic review and meta-analysis: resting-state functional magnetic resonance imaging studies of attention-deficit/hyperactivity disorder. J Am Acad Child Adolesc Psychiatry 2021;60:61–75. Available from: https://doi.org/10.1016/j.jaac.2020.08.014.

[161] Hulvershorn LA, Mennes M, Castellanos FX, di Martino A, Milham MP, Hummer TA, et al. Abnormal amygdala functional connectivity associated with emotional lability in children with attention-deficit/hyperactivity disorder. J Am Acad Child Adolesc Psychiatry 2014;53:351–61. Available from: https://doi.org/10.1016/j.jaac.2013.11.012.

[162] Posner J, Rauh V, Gruber A, Gat I, Wang Z, Peterson BS. Dissociable attentional and affective circuits in medication-naïve children with attention-deficit/hyperactivity disorder. Psychiatry Res 2013;213:24–30. Available from: https://doi.org/10.1016/j.pscychresns.2013.01.004.

[163] Tomasi D, Volkow ND. Abnormal functional connectivity in children with attention-deficit/hyperactivity disorder. Biol Psychiatry 2012;71:443–50. Available from: https://doi.org/10.1016/j.biopsych.2011.11.003.

[164] Norman LJ, Carlisi C, Lukito S, Hart H, Mataix-Cols D, Radua J, et al. Structural and functional brain abnormalities in attention-deficit/hyperactivity disorder and obsessive-compulsive disorder: a comparative meta-analysis. JAMA Psychiatry 2016;73:815–25. Available from: https://doi.org/10.1001/jamapsychiatry.2016.0700.

[165] Rubia K, Halari R, Cubillo A, Mohammad AM, Brammer M, Taylor E. Methylphenidate normalises activation and functional connectivity deficits in attention and motivation networks in medication-naïve children with ADHD during a rewarded continuous performance task. Neuropharmacology 2009;57:640–52. Available from: https://doi.org/10.1016/j.neuropharm.2009.08.013.

[166] Pereira-Sanchez V, Franco AR, Vieira D, de Castro-Manglano P, Soutullo C, Milham MP, et al. Systematic review: medication effects on brain intrinsic functional connectivity in patients with attention-deficit/hyperactivity disorder. J Am Acad Child Adolesc Psychiatry 2021;60:222–35. Available from: https://doi.org/10.1016/j.jaac.2020.10.013.

[167] Rubia K, Alegria AA, Cubillo AI, Smith AB, Brammer MJ, Radua J. Effects of stimulants on brain function in attention-deficit/hyperactivity disorder: a systematic review and meta-analysis. Biol Psychiatry 2014;76:616–28. Available from: https://doi.org/10.1016/j.biopsych.2013.10.016.

[168] Posner J, Polanczyk Gv, Sonuga-Barke E. Attention-deficit hyperactivity disorder. Lancet 2020;395:450–62. Available from: https://doi.org/10.1016/S0140-6736(19)33004-1.

[169] Hoogman M, Muetzel R, Guimaraes JP, Shumskaya E, Mennes M, Zwiers MP, et al. Brain imaging of the cortex in ADHD: a coordinated analysis of large-scale clinical and population-based samples. Am J Psychiatry 2019;176:531–42. Available from: https://doi.org/10.1176/appi.ajp.2019.18091033.

[170] Hoogman M, van Rooij D, Klein M, Boedhoe P, Ilioska I, Li T, et al. Consortium neuroscience of attention deficit/hyperactivity disorder and autism spectrum disorder: the ENIGMA adventure. Hum Brain Mapp 2020;23:27. Available from: https://doi.org/10.1002/hbm.25029.

[171] Thompson PM, Stein JL, Medland SE, Hibar DP, Vasquez AA, Renteria ME, et al. The ENIGMA consortium: large-scale collaborative analyses of neuroimaging and genetic data. Brain Imaging Behav 2014;8:153–82. Available from: https://doi.org/10.1007/s11682-013-9269-5.

[172] Bacanlı A, Unsel-Bolat G, Suren S, Yazıcı KU, Callı C, Aygunes Jafari D, et al. Effects of the dopamine transporter gene on neuroimaging findings in different attention deficit hyperactivity disorder presentations. Brain Imaging Behav 2021;15:1103–14. Available from: https://doi.org/10.1007/s11682-020-00437-w.

[173] Yadav SK, Bhat AA, Hashem S, Nisar S, Kamal M, Syed N, et al. Genetic variations influence brain changes in patients with attention-deficit hyperactivity disorder. Transl Psychiatry 2021;11. Available from: https://doi.org/10.1038/s41398-021-01473-w.

[174] Durston S. Imaging genetics in ADHD. NeuroImage 2010;53:832–8. Available from: https://doi.org/10.1016/j.neuroimage.2010.02.071.

[175] Klein M, Walters RK, Demontis D, Stein JL, Hibar DP, Adams HH, et al. Genetic markers of ADHD-related variations in intracranial volume. Am J Psychiatry 2019;176:228–38. Available from: https://doi.org/10.1176/appi.ajp.2018.18020149.

[176] Faraone Sv, Larsson H. Genetics of attention deficit hyperactivity disorder. Mol Psychiatry 2019;24:562–75. Available from: https://doi.org/10.1038/s41380-018-0070-0.

[177] Cortese S. The neurobiology and genetics of attention-deficit/hyperactivity disorder (ADHD): what every clinician should know. Eur J Paediatr Neurol 2012;16:422–33. Available from: https://doi.org/10.1016/j.ejpn.2012.01.009.

[178] Chen Q, Brikell I, Lichtenstein P, Serlachius E, Kuja-Halkola R, Sandin S, et al. Familial aggregation of attention-deficit/hyperactivity disorder. J Child Psychol Psychiatry 2017;58:231–9. Available from: https://doi.org/10.1111/jcpp.12616.

[179] Larsson H, Chang Z, D'Onofrio BM, Lichtenstein P. The heritability of clinically diagnosed attention deficit hyperactivity disorder across the lifespan. Psychol Med 2014;44:2223–9. Available from: https://doi.org/10.1017/S0033291713002493.

[180] Kan KJ, Dolan Cv, Nivard MG, Middeldorp CM, van Beijsterveldt CEM, Willemsen G, et al. Genetic and environmental stability in attention problems across the lifespan: evidence from the netherlands twin register. J Am Acad Child Adolesc Psychiatry 2013;52:12–25. Available from: https://doi.org/10.1016/j.jaac.2012.10.009.
[181] Brikell I, Kuja-Halkola R, Larsson H. Heritability of attention-deficit hyperactivity disorder in adults. Am J Med Genet B Neuropsychiatr Genet 2015;168:406–13. Available from: https://doi.org/10.1002/ajmg.b.32335.
[182] Freitag CM, Rohde LA, Lempp T, Romanos M. Phenotypic and measurement influences on heritability estimates in childhood ADHD. Eur Child Adolesc Psychiatry 2010;19:311–23. Available from: https://doi.org/10.1007/s00787-010-0097-5.
[183] Schultz MR, Rabi K, Faraone SV, Kremen W, Lyons MJ. Efficacy of retrospective recall of attention-deficit hyperactivity disorder symptoms: a twin study. Twin Res Hum Genet 2006;9:220–32. Available from: https://doi.org/10.1375/183242706776382374.
[184] Rietveld MJH, Hudziak JJ, Bartels M, van Beijsterveldt CEM, Boomsma DI. Heritability of attention problems in children: longitudinal results from a study of twins, age 3 to 12. J Child Psychol Psychiatry 2004;45:577–88. Available from: https://doi.org/10.1111/j.1469-7610.2004.00247.x.
[185] Kuntsi J, Rijsdijk F, Ronald A, Asherson P, Plomin R. Genetic influences on the stability of attention-deficit/hyperactivity disorder symptoms from early to middle childhood. Biol Psychiatry 2005;57:647–54. Available from: https://doi.org/10.1016/j.biopsych.2004.12.032.
[186] Thapar A. Discoveries on the genetics of ADHD in the 21st century: new findings and their implications. Am J Psychiatry 2018;175:943–50. Available from: https://doi.org/10.1176/appi.ajp.2018.18040383.
[187] Hinshaw SP. Attention deficit hyperactivity disorder (ADHD): controversy, developmental mechanisms, and multiple levels of analysis. Annu Rev Clin Psychol 2018;14:291–316. Available from: https://doi.org/10.1146/annurev-clinpsy-050817-084917.
[188] Chen W, Zhou K, Sham P, Franke B, Kuntsi J, Campbell D, et al. DSM-IV combined type ADHD shows familial association with sibling trait scores: a sampling strategy for QTL linkage. Am J Med Genet B Neuropsychiatr Genet 2008;147:1450–60. Available from: https://doi.org/10.1002/ajmg.b.30672.
[189] Grimm O, Kranz TM, Reif A. Genetics of ADHD: what should the clinician know? Curr Psychiatry Rep 2020;22. Available from: https://doi.org/10.1007/s11920-020-1141-x.
[190] Kambeitz J, Romanos M, Ettinger U. Meta-analysis of the association between dopamine transporter genotype and response to methylphenidate treatment in ADHD. Pharmacogenom J 2014;14:77–84. Available from: https://doi.org/10.1038/tpj.2013.9.
[191] Faraone Sv, Mick E. Molecular genetics of attention deficit hyperactivity disorder. Psychiatr Clin North Am 2010;33:159–80. Available from: https://doi.org/10.1016/j.psc.2009.12.004.
[192] Franke B, Faraone Sv, Asherson P, Buitelaar J, Bau CHD, Ramos-Quiroga JA, et al. The genetics of attention deficit/hyperactivity disorder in adults, a review. Mol Psychiatry 2012;17:960–87. Available from: https://doi.org/10.1038/mp.2011.138.
[193] Gizer IR, Ficks C, Waldman ID. Candidate gene studies of ADHD: a meta-analytic review. Hum Genet 2009;126:51–90. Available from: https://doi.org/10.1007/s00439-009-0694-x.
[194] Franke B, Vasquez AA, Johansson S, Hoogman M, Romanos J, Boreatti-Hümmer A, et al. Multicenter analysis of the SLC6A3/DAT1 VNTR haplotype in persistent ADHD suggests differential involvement of the gene in childhood and persistent ADHD. Neuropsychopharmacology 2010;35:656–64. Available from: https://doi.org/10.1038/npp.2009.170.
[195] Faraone Sv, Spencer TJ, Madras BK, Zhang-James Y, Biederman J. Functional effects of dopamine transporter gene genotypes on in vivo dopamine transporter functioning: a meta-analysis. Mol Psychiatry 2014;19:880–9. Available from: https://doi.org/10.1038/mp.2013.126.
[196] Demontis D, Walters RK, Martin J, Mattheisen M, Als TD, Agerbo E, et al. Discovery of the first genome-wide significant risk loci for attention deficit/hyperactivity disorder. Nat Genet 2019;51:63–75. Available from: https://doi.org/10.1038/s41588-018-0269-7.
[197] Williams NM, Zaharieva I, Martin A, Langley K, Mantripragada K, Fossdal R, et al. Rare chromosomal deletions and duplications in attention-deficit hyperactivity disorder: a genome-wide analysis. Lancet 2010;376:1401–8. Available from: https://doi.org/10.1016/S0140-6736(10)61109-9.
[198] Mooney MA, McWeeney SK, Faraone SV, Hinney A, Hebebrand J, Faraone SV, et al. Pathway analysis in attention deficit hyperactivity disorder: an ensemble approach. Am J Med Genet B Neuropsychiatr Genet 2016;171:815–26. Available from: https://doi.org/10.1002/ajmg.b.32446.
[199] Aebi M, van Donkelaar MMJ, Poelmans G, Buitelaar JK, Sonuga-Barke EJS, Stringaris A, et al. Gene-set and multivariate genome-wide association analysis of oppositional defiant behavior subtypes in attention-deficit/hyperactivity disorder. Am J Med Genet B Neuropsychiatr Genet 2016;171:573–88. Available from: https://doi.org/10.1002/ajmg.b.32346.
[200] Poelmans G, Pauls DL, Buitelaar JK, Franke B. Integrated genome-wide association study findings: identification of a neurodevelopmental network for attention deficit hyperactivity disorder. Am J Psychiatry 2011;168:365–77. Available from: https://doi.org/10.1176/appi.ajp.2010.10070948.
[201] Groenman AP, Greven CU, van Donkelaar MMJ, Schellekens A, van Hulzen KJE, Rommelse N, et al. Dopamine and serotonin genetic risk scores predicting substance and nicotine use in attention deficit/hyperactivity disorder. Addict Biol 2016;21:915–23. Available from: https://doi.org/10.1111/adb.12230.
[202] Li JJ, He Q. Polygenic scores for ADHD: a meta-analysis. Res Child Adolesc Psychopathol 2021;49:297–310. Available from: https://doi.org/10.1007/s10802-021-00774-4.
[203] Riglin L, Collishaw S, Thapar AK, Dalsgaard S, Langley K, Smith GD, et al. Association of genetic risk variants with attention-deficit/hyperactivity disorder trajectories in the general population. JAMA Psychiatry 2016;73:1285–92. Available from: https://doi.org/10.1001/jamapsychiatry.2016.2817.

[204] Chatterjee N, Wheeler B, Sampson J, Hartge P, Chanock SJ, Park JH. Projecting the performance of risk prediction based on polygenic analyses of genome-wide association studies. Nat Genet 2013;45:400–5. Available from: https://doi.org/10.1038/ng.2579.

[205] Martin AR, Daly MJ, Robinson EB, Hyman SE, Neale BM. Predicting polygenic risk of psychiatric disorders. Biol Psychiatry 2019;86:97–109. Available from: https://doi.org/10.1016/j.biopsych.2018.12.015.

[206] Groen-Blokhuis MM, Middeldorp CM, Kan KJ, Abdellaoui A, van Beijsterveldt CEM, Ehli EA, et al. Attention-deficit/hyperactivity disorder polygenic risk scores predict attention problems in a population-based sample of children. J Am Acad Child Adolesc Psychiatry 2014;53:1123–9. Available from: https://doi.org/10.1016/j.jaac.2014.06.014.

[207] Yang L, Neale BM, Liu L, Lee SH, Wray NR, Ji N, et al. Polygenic transmission and complex neuro developmental network for attention deficit hyperactivity disorder: genome-wide association study of both common and rare variants. Am J Med Genet B Neuropsychiatr Genet 2013;162:419–30. Available from: https://doi.org/10.1002/ajmg.b.32169.

[208] Stergiakouli E, Martin J, Hamshere ML, Langley K, Evans DM, St Pourcain B, et al. Shared genetic influences between attention-deficit/hyperactivity disorder (ADHD) traits in children and clinical ADHD. J Am Acad Child Adolesc Psychiatry 2015;54:322–7. Available from: https://doi.org/10.1016/j.jaac.2015.01.010.

[209] du Rietz E, Coleman J, Glanville K, Choi SW, O'Reilly PF, Kuntsi J. Association of polygenic risk for attention-deficit/hyperactivity disorder with co-occurring traits and disorders. Biol Psychiatry 2018;3:635–43. Available from: https://doi.org/10.1016/j.bpsc.2017.11.013.

[210] Andersson A, Tuvblad C, Chen Q, du Rietz E, Cortese S, Kuja-Halkola R, et al. Research review: the strength of the genetic overlap between ADHD and other psychiatric symptoms – a systematic review and meta-analysis. J Child Psychol Psychiatry 2020;61:1173–83. Available from: https://doi.org/10.1111/jcpp.13233.

[211] Lichtenstein P, Carlström E, Råstam M, Gillberg C, Anckarsäter H. The genetics of autism spectrum disorders and related neuropsychiatric disorders in childhood. Am J Psychiatry 2010;167:1357–63. Available from: https://doi.org/10.1176/appi.ajp.2010.10020223.

[212] Palk AC, Dalvie S, de Vries J, Martin AR, Stein DJ. Potential use of clinical polygenic risk scores in psychiatry - ethical implications and communicating high polygenic risk. Philos Ethics Humanit Med 2019;14:1–12. Available from: https://doi.org/10.1186/s13010-019-0073-8.

[213] Martin J, O'Donovan MC, Thapar A, Langley K, Williams N. The relative contribution of common and rare genetic variants to ADHD. Transl Psychiatry 20155. Available from: https://doi.org/10.1038/tp.2015.5.

[214] Lo-Castro A, D'Agati E, Curatolo P. ADHD and genetic syndromes. Brain Dev 2011;33:456–61. Available from: https://doi.org/10.1016/j.braindev.2010.05.011.

[215] Leyfer OT, Woodruff-Borden J, Klein-Tasman BP, Fricke JS, Mervis CB. Prevalence of psychiatric disorders in 4 to 16-year-olds with Williams syndrome. Am J Med Genet B Neuropsychiatr Genet 2006;141:615–22. Available from: https://doi.org/10.1002/ajmg.b.30344.

[216] Green T, Avda S, Dotan I, Zarchi O, Basel-Vanagaite L, Zalsman G, et al. Phenotypic psychiatric characterization of children with Williams syndrome and response of those with ADHD to methylphenidate treatment. Am J Med Genet B Neuropsychiatr Genet 2012;159 B:13–20. Available from: https://doi.org/10.1002/ajmg.b.31247.

[217] Rhodes SM, Riby DM, Matthews K, Coghill DR. Attention-deficit/hyperactivity disorder and Williams syndrome: shared behavioral and neuropsychological profiles. J Clin Exp Neuropsychol 2011;33:147–56. Available from: https://doi.org/10.1080/13803395.2010.495057.

[218] Cederlöf M, Ohlsson Gotby A, Larsson H, Serlachius E, Boman M, Långström N, et al. Klinefelter syndrome and risk of psychosis, autism and ADHD. J Psychiatr Res 2014;48:128–30. Available from: https://doi.org/10.1016/j.jpsychires.2013.10.001.

[219] Tartaglia N, Cordeiro L, Howell S, Wilson R, Janusz J. The spectrum of the behavioral phenotype in boys and adolescents 47,XXY (Klinefelter syndrome). Pediatr Endocrinol Rev 2010;8(Suppl 1):151–9.

[220] Crawford H, Abbeduto L, Hall SS, Hardiman R, Hessl D, Roberts JE, et al. Fragile X syndrome: an overview of cause, characteristics, assessment and management. Paediatr Child Health 2020;30:400–3. Available from: https://doi.org/10.1016/j.paed.2020.08.007.

[221] Savatt JM, Myers SM. Genetic testing in neurodevelopmental disorders. Front Pediatr 2021;9:1–24. Available from: https://doi.org/10.3389/fped.2021.526779.

[222] Ekstein S, Glick B, Weill M, Kay B, Berger I. Down syndrome and attention-deficit/hyperactivity disorder (ADHD). Dev Med Child Neurol 2011;26:1290–5. Available from: https://doi.org/10.1177/0883073811405201.

[223] de Vries PJ, Belousova E, Benedik MP, Carter T, Cottin V, Curatolo P, et al. Tuberous sclerosis complex-associated neuropsychiatric disorders (TAND): new findings on age, sex, and genotype in relation to intellectual phenotype. Front Neurol 2020;11:603. Available from: https://doi.org/10.3389/fneur.2020.00603.

[224] D'Agati E, Moavero R, Cerminara C, Curatolo P. Attention-deficit hyperactivity disorder (ADHD) and tuberous sclerosis complex. J Child Neurol 2009;24:1282–7. Available from: https://doi.org/10.1177/0883073809341272.

[225] Niklasson L, Gillberg C. The neuropsychology of 22q11 deletion syndrome. A neuropsychiatric study of 100 individuals. Res Dev Disabil 2010;31:185–94. Available from: https://doi.org/10.1016/j.ridd.2009.09.001.

[226] Niklasson L, Rasmussen P, Óskarsdóttir S, Gillberg C. Attention deficits in children with 22q.11 deletion syndrome. Dev Med Child Neurol 2005;47:803–7. Available from: https://doi.org/10.1017/S0012162205001702.

[227] Niarchou M, Martin J, Thapar A, Owen MJ, van den Bree MBM. The clinical presentation of attention deficit-hyperactivity disorder (ADHD) in children with 22q11.2 deletion syndrome. Am J Med Genet B Neuropsychiatr Genet 2015;168:730–8. Available from: https://doi.org/10.1002/ajmg.b.32378.

[228] Payne JM, Haebich KM, MacKenzie R, Walsh KS, Hearps SJC, Coghill D, et al. Cognition, ADHD symptoms, and functional impairment in children and adolescents with neurofibromatosis type 1. J Atten Disord 2021;25:1177–86. Available from: https://doi.org/10.1177/1087054719894384.

[229] Green T, Bade Shrestha S, Chromik LC, Rutledge K, Pennington BF, Hong DS, et al. Elucidating X chromosome influences on attention deficit hyperactivity disorder and executive function. J Psychiatr Res 2015;68:217–25. Available from: https://doi.org/10.1016/j.jpsychires.2015.06.021.

[230] Gould KL, Coventry WL, Olson RK, Byrne B. Gene-environment interactions in ADHD: the roles of SES and chaos. J Abnorm Child Psychol 2018;46:251–63. Available from: https://doi.org/10.1007/s10802-017-0268-7.

[231] Nigg J, Nikolas M, Burt SA. Measured gene-by-environment interaction in relation to attention-deficit/hyperactivity disorder. J Am Acad Child Adolesc Psychiatry 2010;49:863–73. Available from: https://doi.org/10.1016/j.jaac.2010.01.025.

[232] Palladino VS, McNeill R, Reif A, Kittel-Schneider S. Genetic risk factors and gene-environment interactions in adult and childhood attention-deficit/hyperactivity disorder. Psychiatr Genet 2019;29:63–78. Available from: https://doi.org/10.1097/YPG.0000000000000220.

[233] Neumann A, Walton E, Alemany S, Cecil C, González JR, Jima DD, et al. Association between DNA methylation and ADHD symptoms from birth to school age: a prospective meta-analysis. Transl Psychiatry 2020;10:1–11. Available from: https://doi.org/10.1038/s41398-020-01058-z.

[234] Rijlaarsdam J, Barker ED, Caserini C, Koopman-Verhoeff ME, Mulder RH, Felix JF, et al. Genome-wide DNA methylation patterns associated with general psychopathology in children. J Psychiatr Res 2021;140:214–20. Available from: https://doi.org/10.1016/j.jpsychires.2021.05.029.

[235] Fageera W, Chaumette B, Fortier MÈ, Grizenko N, Labbe A, Sengupta SM, et al. Association between COMT methylation and response to treatment in children with ADHD. J Psychiatr Res 2021;135:86–93. Available from: https://doi.org/10.1016/j.jpsychires.2021.01.008.

[236] Barker ED, Walton E, Cecil CAM. Annual research review: DNA methylation as a mediator in the association between risk exposure and child and adolescent psychopathology. J Child Psychol Psychiatry 2018;59:303–22. Available from: https://doi.org/10.1111/jcpp.12782.

[237] Franz AP, Bolat GU, Bolat H, Matijasevich A, Santos IS, Silveira RC, et al. Attention-deficit/hyperactivity disorder and very preterm/very low birth weight: a meta-analysis. Pediatrics 2018;141:e20171645. Available from: https://doi.org/10.1542/peds.2017-1645.

[238] Sucksdorff M, Lehtonen L, Chudal R, Suominen A, Joelsson P, Gissler M, et al. Preterm birth and poor fetal growth as risk factors of attention-deficit/hyperactivity disorder. Pediatrics 2015;136:e599–608. Available from: https://doi.org/10.1542/peds.2015-1043.

[239] Lindström K, Lindblad F, Hjern A. Preterm birth and attention-deficit/hyperactivity disorder in schoolchildren. Pediatrics 2011;127:858–65. Available from: https://doi.org/10.1542/peds.2010-1279.

[240] Momany AM, Kamradt JM, Nikolas MA. A meta-analysis of the association between birth weight and attention deficit hyperactivity disorder. J Abnorm Child Psychol 2018;46:1409–26. Available from: https://doi.org/10.1007/s10802-017-0371-9.

[241] Maher GM, O'Keeffe GW, Kearney PM, Kenny LC, Dinan TG, Mattsson M, et al. Association of hypertensive disorders of pregnancy with risk of neurodevelopmental disorders in offspring a systematic review and meta-analysis. JAMA Psychiatry 2018;75:809–19. Available from: https://doi.org/10.1001/jamapsychiatry.2018.0854.

[242] Maher GM, Dalman C, O'Keeffe GW, Kearney PM, McCarthy FP, Kenny LC, et al. Association between preeclampsia and attention-deficit hyperactivity disorder: a population-based and sibling-matched cohort study. Acta Psychiatr Scand 2020;142:275–83. Available from: https://doi.org/10.1111/acps.13162.

[243] Jenabi E, Bashirian S, Khazaei S, Basiri Z. The maternal prepregnancy body mass index and the risk of attention deficit hyperactivity disorder among children and adolescents: a systematic review and meta-analysis. Korean J Pediatr 2019;62:374–9. Available from: https://doi.org/10.3345/kjp.2019.00185.

[244] Sanchez CE, Barry C, Sabhlok A, Russell K, Majors A, Kollins SH, et al. Maternal pre-pregnancy obesity and child neurodevelopmental outcomes: a meta-analysis. Obes Rev 2018;19:464–84. Available from: https://doi.org/10.1111/obr.12643.

[245] Andersen CH, Thomsen PH, Nohr EA, Lemcke S. Maternal body mass index before pregnancy as a risk factor for ADHD and autism in children. Eur Child Adolesc Psychiatry 2018;27:139–48. Available from: https://doi.org/10.1007/s00787-017-1027-6.

[246] Zhao L, Li X, Liu G, Han B, Wang J, Jiang X. The association of maternal diabetes with attention deficit and hyperactivity disorder in offspring: a meta-analysis. Neuropsychiatr Dis Treat 2019;15:675. Available from: https://doi.org/10.2147/NDT.S189200.

[247] Wang H, Li F, Miao M, Yu Y, Ji H, Liu H, et al. Maternal spontaneous abortion and the risk of attention-deficit/ hyperactivity disorder in offspring: a population-based cohort study. Hum Reprod 2020;35:1211–21. Available from: https://doi.org/10.1093/humrep/deaa035.

[248] Chen MH, Pan TL, Wang PW, Hsu JW, Huang KL, Su TP, et al. Prenatal exposure to acetaminophen and the risk of attention-deficit/hyperactivity disorder: a nationwide study in Taiwan. J Clin Psychiatry 2019;80:0. Available from: https://doi.org/10.4088/JCP.18m12612.

[249] Alemany S, Avella-García C, Liew Z, García-Esteban R, Inoue K, Cadman T, et al. Prenatal and postnatal exposure to acetaminophen in relation to autism spectrum and attention-deficit and hyperactivity symptoms in childhood: meta-analysis in six European population-based cohorts. Eur J Epidemiol 2021;8:1–12. Available from: https://doi.org/10.1007/s10654-021-00754-4.

[250] Ystrom E, Gustavson K, Brandlistuen RE, Knudsen GP, Magnus P, Susser E, et al. Prenatal exposure to acetaminophen and risk of ADHD. Pediatrics 2017;140:20163840. Available from: https://doi.org/10.1542/peds.2016-3840.

[251] Gustavson K, Ystrom E, Ask H, Ask Torvik F, Hornig M, Susser E, et al. Acetaminophen use during pregnancy and offspring attention deficit hyperactivity disorder – a longitudinal sibling control study. JCPP Adv 2021;1:e12020. Available from: https://doi.org/10.1002/jcv2.12020.

[252] Engel SM, Villanger GD, Nethery RC, Thomsen C, Sakhi AK, Drover SSM, et al. Prenatal phthalates, maternal thyroid function, and risk of attention-deficit hyperactivity disorder in the Norwegian mother and child cohort. Environ Health Perspect 2018;126:057004. Available from: https://doi.org/10.1289/EHP2358.

[253] Christensen J, Pedersen L, Sun Y, Dreier JW, Brikell I, Dalsgaard S. Association of prenatal exposure to valproate and other antiepileptic drugs with risk for attention-deficit/hyperactivity disorder in offspring. JAMA Netw Open 2019;2:e186606. Available from: https://doi.org/10.1001/jamanetworkopen.2018.6606.

[254] Wiggs KK, Rickert ME, Sujan AC, Quinn PD, Larsson H, Lichtenstein P, et al. Antiseizure medication use during pregnancy and risk of ASD and ADHD in children. Neurology 2020;95:e3232–40. Available from: https://doi.org/10.1212/WNL.0000000000010993.

[255] Huang L, Wang Y, Zhang L, Zheng Z, Zhu T, Qu Y, et al. Maternal smoking and attention-deficit/hyperactivity disorder in offspring: a meta-analysis. Pediatrics 2018;141:2021. Available from: https://doi.org/10.1542/peds.2017-2465.

[256] Dong T, Hu W, Zhou X, Lin H, Lan L, Hang B, et al. Prenatal exposure to maternal smoking during pregnancy and attention-deficit/hyperactivity disorder in offspring: a meta-analysis. Reprod Toxicol 2018;76:63–70. Available from: https://doi.org/10.1016/j.reprotox.2017.12.010.

[257] Joelsson P, Chudal R, Talati A, Suominen A, Brown AS, Sourander A. Prenatal smoking exposure and neuropsychiatric comorbidity of ADHD: a Finnish nationwide population-based cohort study. BMC Psychiatry 2016;16:1–10. Available from: https://doi.org/10.1186/s12888-016-1007-2.

[258] Obel C, Zhu JL, Olsen J, Breining S, Li J, Grønborg TK, et al. The risk of attention deficit hyperactivity disorder in children exposed to maternal smoking during pregnancy - a re-examination using a sibling design. J Child Psychol Psychiatry 2016;57:532–7. Available from: https://doi.org/10.1111/jcpp.12478.

[259] Skoglund C, Chen Q, D'Onofrio BM, Lichtenstein P, Larsson H. Familial confounding of the association between maternal smoking during pregnancy and ADHD in offspring. J Child Psychol Psychiatry 2014;55:61–8. Available from: https://doi.org/10.1111/jcpp.12124.

[260] Perez FA, Blythe S, Wouldes T, McNamara K, Black KI, Oei JL. Prenatal methamphetamine—impact on the mother and child—a review. Addiction 2021. Available from: https://doi.org/10.1111/add.15509.

[261] San Martin Porter M, Maravilla JC, Betts KS, Alati R. Low–moderate prenatal alcohol exposure and offspring attention-deficit hyperactivity disorder (ADHD): systematic review and meta-analysis. Arch Gynecol Obstet 2019;300(2):269–77. Available from: https://doi.org/10.1007/S00404-019-05204-X.

[262] Mattson SN, Bernes GA, Doyle LR. Fetal alcohol spectrum disorders: a review of the neurobehavioral deficits associated with prenatal alcohol exposure. Alcohol Clin Exp Res 2019;43:1046–62. Available from: https://doi.org/10.1111/acer.14040.

[263] Pagnin D, Zamboni Grecco ML, Furtado EF. Prenatal alcohol use as a risk for attention-deficit/hyperactivity disorder. Eur Arch Psychiatry Clin Neurosci 2019;269:681–7. Available from: https://doi.org/10.1007/s00406-018-0946-7.

[264] Weyrauch D, Schwartz M, Hart B, Klug MG, Burd L. Comorbid mental disorders in fetal alcohol spectrum disorders: a systematic review. J Dev Behav Pediatr 2017;38:283–91. Available from: https://doi.org/10.1097/DBP.0000000000000440.

[265] Fryer SL, McGee CL, Matt GE, Riley EP, Mattson SN. Evaluation of psychopathological conditions in children with heavy prenatal alcohol exposure. Pediatrics 2007;119:e733–41. Available from: https://doi.org/10.1542/peds.2006-1606.

[266] Huang A, Wu K, Cai Z, Lin Y, Zhang X, Huang Y. Association between postnatal second-hand smoke exposure and ADHD in children: a systematic review and meta-analysis. Environ Sci Pollut Res 2021;28:1370–80. Available from: https://doi.org/10.1007/s11356-020-11269-y.

[267] Lin LZ, Xu SL, Wu QZ, Zhou Y, Ma HM, Chen DH, et al. Association of prenatal, early postnatal, or current exposure to secondhand smoke with attention-deficit/hyperactivity disorder symptoms in children. JAMA Netw Open 2021;4:e2110931. Available from: https://doi.org/10.1001/jamanetworkopen.2021.10931.

[268] Bouchard MF, Bellinger DC, Wright RO, Weisskopf MG. Attention-deficit/hyperactivity disorder and urinary metabolites of organophosphate pesticides. Pediatrics 2010125. Available from: https://doi.org/10.1542/peds.2009-3058.

[269] Donzelli G, Carducci A, Llopis-Gonzalez A, Verani M, Llopis-Morales A, Cioni L, et al. The association between lead and attention-deficit/hyperactivity disorder: a systematic review. Int J Environ Res Public Health 2019;16:382. Available from: https://doi.org/10.3390/IJERPH16030382.

[270] Yoshimasu K, Kiyohara C, Takemura S, Nakai K. A meta-analysis of the evidence on the impact of prenatal and early infancy exposures to mercury on autism and attention deficit/hyperactivity disorder in the childhood. Neurotoxicology 2014;44:121–31. Available from: https://doi.org/10.1016/J.NEURO.2014.06.007.

[271] Farsad-Naeimi A, Asjodi F, Omidian M, Askari M, Nouri M, Pizarro AB, et al. Sugar consumption, sugar sweetened beverages and attention deficit hyperactivity disorder: a systematic review and meta-analysis. Complement Therap Med 2020;53:102512. Available from: https://doi.org/10.1016/j.ctim.2020.102512.

[272] Forns J, Verner MA, Iszatt N, Nowack N, Bach CC, Vrijheid M, et al. Early life exposure to perfluoroalkyl substances (PFAS) and ADHD: a meta-analysis of nine European population-based studies. Environ Health Perspect 2020;128:20. Available from: https://doi.org/10.1289/EHP5444.

[273] Zhang M, Wang C, Zhang X, Song H, Li Y. Association between exposure to air pollutants and attention-deficit hyperactivity disorder (ADHD) in children: a systematic review and meta-analysis. Int J Environ Health Res 2020. Available from: https://doi.org/10.1080/09603123.2020.1745764.

[274] Nigg JT, Lewis K, Edinger T, Falk M. Meta-analysis of attention-deficit/hyperactivity disorder or attention-deficit/hyperactivity disorder symptoms, restriction diet, and synthetic food color additives. J Am Acad Child Adolesc Psychiatry 2012;51:86–97. Available from: https://doi.org/10.1016/j.jaac.2011.10.015.

[275] Tseng PT, Cheng YS, Yen CF, Chen YW, Stubbs B, Whiteley P, et al. Peripheral iron levels in children with attention-deficit hyperactivity disorder: a systematic review and meta-analysis. Sci Rep 2018;8:1–11. Available from: https://doi.org/10.1038/s41598-017-19096-x.

[276] Wang Y, Huang L, Zhang L, Qu Y, Mu D. Iron status in attention-deficit/hyperactivity disorder: a systematic review and meta-analysis. PLoS ONE 2017;12:e0169145. Available from: https://doi.org/10.1371/journal.pone.0169145.

[277] Sucksdorff M, Brown AS, Chudal R, Surcel HM, Hinkka-Yli-Salomäki S, Cheslack-Postava K, et al. Maternal vitamin D levels and the risk of offspring attention-deficit/hyperactivity disorder. J Am Acad Child Adolesc Psychiatry 2021;60:142–51. Available from: https://doi.org/10.1016/j.jaac.2019.11.021.

[278] Hawkey E, Nigg JT. Omega-3 fatty acid and ADHD: blood level analysis and meta-analytic extension of supplementation trials. Clin Psychol Rev 2014;34:496–505. Available from: https://doi.org/10.1016/j.cpr.2014.05.005.

[279] Larsson H, Sariaslan A, Langström N, D'Onofrio B, Lichtenstein P. Family income in early childhood and subsequent attention deficit/hyperactivity disorder: a quasi-experimental study. J Child Psychol Psychiatry 2014;55:428–35. Available from: https://doi.org/10.1111/jcpp.12140.

[280] Choi Y, Shin J, Cho KH, Park EC. Change in household income and risk for attention deficit hyperactivity disorder during childhood: a nationwide population-based cohort study. J Epidemiol 2017;27:56–62. Available from: https://doi.org/10.1016/j.je.2016.09.004.

[281] Russell AE, Ford T, Russell G. Socioeconomic associations with ADHD: findings from a mediation analysis. PLoS ONE 2015;10:e0128248. Available from: https://doi.org/10.1371/journal.pone.0128248.

[282] Russell G, Ford T, Rosenberg R, Kelly S. The association of attention deficit hyperactivity disorder with socioeconomic disadvantage: alternative explanations and evidence. J Child Psychol Psychiatry 2014;55:436–45. Available from: https://doi.org/10.1111/jcpp.12170.

[283] Keilow M, Wu C, Obel C. Cumulative social disadvantage and risk of attention deficit hyperactivity disorder: results from a nationwide cohort study. SSM Popul Health 2020;10:100548. Available from: https://doi.org/10.1016/j.ssmph.2020.100548.

[284] Li J, Olsen J, Vestergaard M, Obel C. Attention-deficit/hyperactivity disorder in the offspring following prenatal maternal bereavement: a nationwide follow-up study in Denmark. Eur Child Adolesc Psychiatry 2010;19:747–53. Available from: https://doi.org/10.1007/s00787-010-0113-9.

[285] Björkenstam E, Björkenstam C, Jablonska B, Kosidou K. Cumulative exposure to childhood adversity, and treated attention deficit/hyperactivity disorder: a cohort study of 543 650 adolescents and young adults in Sweden. Psychol Med 2018;48:498–507. Available from: https://doi.org/10.1017/S0033291717001933.

[286] Barbaresi WJ, Colligan RC, Weaver AL, Voigt RG, Killian JM, Katusic SK. Mortality, ADHD, and psychosocial adversity in adults with childhood ADHD: a prospective study. Pediatrics 2013;131:637–44. Available from: https://doi.org/10.1542/peds.2012-2354.

[287] Zhao X, Page TF, Altszuler AR, Pelham WE, Kipp H, Gnagy EM, et al. Family burden of raising a child with ADHD. J Abnorm Child Psychol 2019;47:1327–38. Available from: https://doi.org/10.1007/s10802-019-00518-5.

[288] Daley D, Jacobsen RH, Lange AM, Sørensen A, Walldorf J. The economic burden of adult attention deficit hyperactivity disorder: a sibling comparison cost analysis. Eur Psychiatry 2019;61:41–8. Available from: https://doi.org/10.1016/j.eurpsy.2019.06.011.

[289] Le HH, Hodgkins P, Postma MJ, Kahle J, Sikirica V, Setyawan J, et al. Economic impact of childhood/adolescent ADHD in a European setting: the Netherlands as a reference case. Eur Child Adolesc Psychiatry 2014;23:587–98. Available from: https://doi.org/10.1007/s00787-013-0477-8.

[290] Doshi JA, Hodgkins P, Kahle J, Sikirica V, Cangelosi MJ, Setyawan J, et al. Economic impact of childhood and adult attention-deficit/hyperactivity disorder in the United States. J Am Acad Child Adolesc Psychiatry 2012;51:990–1002. Available from: https://doi.org/10.1016/j.jaac.2012.07.008.

[291] Boland H, DiSalvo M, Fried R, Woodworth KY, Wilens T, Faraone SV, et al. A literature review and meta-analysis on the effects of ADHD medications on functional outcomes. J Psychiatr Res 2020;123:21–30. Available from: https://doi.org/10.1016/j.jpsychires.2020.01.006.

[292] Raggi VL, Chronis AM. Interventions to address the academic impairment of children and adolescents with ADHD. Clin Child Family Psychol Rev 2006;9:85–111. Available from: https://doi.org/10.1007/s10567-006-0006-0.

[293] Langberg JM, Molina BSG, Arnold LE, Epstein JN, Altaye M, Hinshaw SP, et al. Patterns and predictors of adolescent academic achievement and performance in a sample of children with attention-deficit/hyperactivity disorder. J Clin Child Adolesc Psychol 2011;40:519–31. Available from: https://doi.org/10.1080/15374416.2011.581620.

[294] Arnold LE, Hodgkins P, Kahle J, Madhoo M, Kewley G. Long-term outcomes of ADHD: academic achievement and performance. J Atten Disord 2020;24:73–85. Available from: https://doi.org/10.1177/1087054714566076.

[295] Voigt RG, Katusic SK, Colligan RC, Killian JM, Weaver AL, Barbaresi WJ. Academic achievement in adults with a history of childhood attention-deficit/hyperactivity disorder: a population-based prospective study. J Dev Behav Pediatr 2017;38:1–11. Available from: https://doi.org/10.1097/DBP.0000000000000358.

[296] Loe IM, Feldman HM. Academic and educational outcomes of children with ADHD. J Pediatr Psychol 2007;32:643–54. Available from: https://doi.org/10.1093/jpepsy/jsl054.

[297] Massetti GM, Lahey BB, Pelham WE, Loney J, Ehrhardt A, Lee SS, et al. Academic achievement over 8 years among children who met modified criteria for attention-deficit/hyperactivity disorder at 4-6 years of age. J Abnorm Child Psychol 2008;36:399–410. Available from: https://doi.org/10.1007/s10802-007-9186-4.

[298] Lawrence D, Houghton S, Dawson V, Sawyer M, Carroll A. Trajectories of academic achievement for students with attention-deficit/hyperactivity disorder. Br J Educ Psychol 2021;91:755–74. Available from: https://doi.org/10.1111/bjep.12392.

[299] Simone AN, Marks DJ, Bédard AC, Halperin JM. Low working memory rather than ADHD symptoms predicts poor academic achievement in school-aged children. J Abnorm Child Psychol 2018;46:277–90. Available from: https://doi.org/10.1007/s10802-017-0288-3.

[300] DuPaul GJ, Gormley MJ, Laracy SD. Comorbidity of LD and ADHD: implications of DSM-5 for assessment and treatment. J Learn Disabil 2013;46:43–51. Available from: https://doi.org/10.1177/0022219412464351.

[301] Visser L, Linkersdörfer J, Hasselhorn M. The role of ADHD symptoms in the relationship between academic achievement and psychopathological symptoms. Res Dev Disabil 2020;97:103552. Available from: https://doi.org/10.1016/j.ridd.2019.103552.

[302] Harstad E, Levy S, Ammerman SD, Gonzalez PK, Ryan SA, Siqueira LM, et al. Attention-deficit/hyperactivity disorder and substance abuse. Pediatrics 2014;134:e293–301. Available from: https://doi.org/10.1542/peds.2014-0992.

[303] Groenman AP, Janssen TWP, Oosterlaan J. Childhood psychiatric disorders as risk factor for subsequent substance abuse: a meta-analysis. J Am Acad Child Adolesc Psychiatry 2017;56:556–69. Available from: https://doi.org/10.1016/j.jaac.2017.05.004.

[304] Capusan AJ, Bendtsen P, Marteinsdottir I, Larsson H. Comorbidity of adult ADHD and its subtypes with substance use disorder in a large population-based epidemiological study. J Atten Disord 2019;23:1416–26. Available from: https://doi.org/10.1177/1087054715626511.

[305] Knecht C, de Alvaro R, Martinez-Raga J, Balanza-Martinez V. Attention-deficit hyperactivity disorder (ADHD), substance use disorders, and criminality: a difficult problem with complex solutions. Int J Adolesc Med Health 2015;27:163–75. Available from: https://doi.org/10.1515/ijamh-2015-5007.

[306] Beaudry G, Yu R, Långström N, Fazel S. An updated systematic review and meta-regression analysis: mental disorders among adolescents in juvenile detention and correctional facilities. J Am Acad Child Adolesc Psychiatry 2021;60:46–60. Available from: https://doi.org/10.1016/j.jaac.2020.01.015.

[307] Erskine HE, Norman RE, Ferrari AJ, Chan GCK, Copeland WE, Whiteford HA, et al. Long-term outcomes of attention-deficit/hyperactivity disorder and conduct disorder: a systematic review and meta-analysis. J Am Acad Child Adolesc Psychiatry 2016;55:841–50. Available from: https://doi.org/10.1016/j.jaac.2016.06.016.

[308] Anker E, Ginsberg Y, Heir T. Prevalence of criminal convictions in Norwegian adult ADHD outpatients and associations with ADHD symptom severity and emotional dysregulation. BMC Psychiatry 2021;21. Available from: https://doi.org/10.1186/s12888-021-03223-0.

[309] Dalsgaard S, Mortensen PB, Frydenberg M, Thomsen PH. Long-term criminal outcome of children with attention deficit hyperactivity disorder. Crim Behav Ment Health 2013;23:86–98. Available from: https://doi.org/10.1002/cbm.1860.

[310] Mohr-Jensen C, Steinhausen HC. A meta-analysis and systematic review of the risks associated with childhood attention-deficit hyperactivity disorder on long-term outcome of arrests, convictions, and incarcerations. Clin Psychol Rev 2016;48:32–42. Available from: https://doi.org/10.1016/j.cpr.2016.05.002.

[311] Amiri S, Sadeghi-Bazargani H, Nazari S, Ranjbar F, Abdi S. Attention deficit/hyperactivity disorder and risk of injuries: a systematic review and meta-analysis. J Inj Violence Res 2017;9:95.

[312] Ruiz-Goikoetxea M, Cortese S, Aznarez-Sanado M, Magallón S, Alvarez Zallo N, Luis EO, et al. Risk of unintentional injuries in children and adolescents with ADHD and the impact of ADHD medications: a systematic review and meta-analysis. Neurosci Biobehav Rev 2018;84:63–71. Available from: https://doi.org/10.1016/j.neubiorev.2017.11.007.

[313] Curry AE, Yerys BE, Metzger KB, Carey ME, Power TJ. Traffic crashes, violations, and suspensions among young drivers with ADHD. Pediatrics 2019;143:20182305. Available from: https://doi.org/10.1542/peds.2018-2305.

[314] Karic S, DesRosiers M, Mizrahi B, Zevallos J, Rodriguez P, Barengo NC. The association between attention deficit hyperactivity disorder severity and risk of mild traumatic brain injury in children with attention deficit hyperactivity disorder in the United States of America: a cross-sectional study of data from the National. Child Care Health Dev 2019;45:688–93. Available from: https://doi.org/10.1111/cch.12684.

[315] Liou YJ, Wei HT, Chen MH, Hsu JW, Huang KL, Bai YM, et al. Risk of traumatic brain injury among children, adolescents, and young adults with attention-deficit hyperactivity disorder in Taiwan. J Adolesc Health 2018;63:233–8. Available from: https://doi.org/10.1016/j.jadohealth.2018.02.012.

[316] Voigt RG, Barbaresi WJ, Colligan RC, Weaver AL, Katusic SK. Developmental dissociation, deviance, and delay: occurrence of attention-deficit-hyperactivity disorder in individuals with and without borderline-to-mild intellectual disability. Dev Med Child Neurol 2006;48:831–5. Available from: https://doi.org/10.1017/S0012162206001782.

[317] Mueller KL, Tomblin JB. Examining the comorbidity of language impairment and attention-deficit/ hyperactivity disorder. Top Lang Disord 2012;32:228–46. Available from: https://doi.org/10.1097/TLD.0b013e318262010d.

[318] Tirosh E, Cohen A. Language deficit with attention-deficit disorder: a prevalent comorbidity. J Child Neurol 1998;13:493–7. Available from: https://doi.org/10.1177/088307389801301005.

[319] Harris MN, Voigt RG, Barbaresi WJ, Voge GA, Killian JM, Weaver AL, et al. ADHD and learning disabilities in former late preterm infants: a population-based birth cohort. Pediatrics 2013;132:e630–6. Available from: https://doi.org/10.1542/peds.2012-3588.

[320] Hendren RL, Haft SL, Black JM, White NC, Hoeft F. Recognizing psychiatric comorbidity with reading disorders. Front Psychiatry 2018;0:101. Available from: https://doi.org/10.3389/FPSYT.2018.00101.

[321] Pham Av, Riviere A. Specific learning disorders and ADHD: current issues in diagnosis across clinical and educational settings. Curr Psychiatry Rep 2015;17. Available from: https://doi.org/10.1007/s11920-015-0584-y.

[322] Jensen CM, Steinhausen HC. Comorbid mental disorders in children and adolescents with attention-deficit/hyperactivity disorder in a large nationwide study. Atten Defic Hyperact Disord 2015;7:27–38. Available from: https://doi.org/10.1007/s12402-014-0142-1.

[323] Sokolova E, Oerlemans AM, Rommelse NN, Groot P, Hartman CA, Glennon JC, et al. A causal and mediation analysis of the comorbidity between attention deficit hyperactivity disorder (ADHD) and autism spectrum disorder (ASD). J Autism Dev Disord 2017;47(6):1595–604. Available from: https://doi.org/10.1007/S10803-017-3083-7.

[324] Visser JC, Rommelse NNJ, Greven CU, Buitelaar JK. Autism spectrum disorder and attention-deficit/hyperactivity disorder in early childhood: a review of unique and shared characteristics and developmental antecedents. Neurosci Biobehav Rev 2016;65:229–63. Available from: https://doi.org/10.1016/j.neubiorev.2016.03.019.

[325] Rau S, Skapek MF, Tiplady K, Seese S, Burns A, Armour AC, et al. Identifying comorbid ADHD in autism: attending to the inattentive presentation. Res Autism Spectr Disord 2020;69:101468. Available from: https://doi.org/10.1016/j.rasd.2019.101468.

[326] Hartman CA, Geurts HM, Franke B, Buitelaar JK, Rommelse NNJ. Changing ASD-ADHD symptom co-occurrence across the lifespan with adolescence as crucial time window: Illustrating the need to go beyond childhood. Neurosci Biobehav Rev 2016;71:529–41. Available from: https://doi.org/10.1016/j.neubiorev.2016.09.003.

[327] Rommelse N, Buitelaar JK, Hartman CA. Structural brain imaging correlates of ASD and ADHD across the lifespan: a hypothesis-generating review on developmental ASD–ADHD subtypes. J Neural Transm 2017;124:259–71. Available from: https://doi.org/10.1007/s00702-016-1651-1.

[328] Martin J, Hamshere ML, Stergiakouli E, O'Donovan MC, Thapar A. Genetic risk for attention-deficit/hyperactivity disorder contributes to neurodevelopmental traits in the general population. Biol Psychiatry 2014;76:664–71. Available from: https://doi.org/10.1016/j.biopsych.2014.02.013.

[329] Solberg BS, Zayats T, Posserud MB, Halmøy A, Engeland A, Haavik J, et al. Patterns of psychiatric comorbidity and genetic correlations provide new insights into differences between attention-deficit/hyperactivity disorder and autism spectrum disorder. Biol Psychiatry 2019;86:587–98. Available from: https://doi.org/10.1016/j.biopsych.2019.04.021.

[330] Craig F, Savino R, Trabacca A. A systematic review of comorbidity between cerebral palsy, autism spectrum disorders and attention deficit hyperactivity disorder. Eur J Paediatr Neurol 2019;23:31–42. Available from: https://doi.org/10.1016/j.ejpn.2018.10.005.

[331] Athanasiadou A, Buitelaar JK, Brovedani P, Chorna O, Fulceri F, Guzzetta A, et al. Early motor signs of attention-deficit hyperactivity disorder: a systematic review. Eur Child Adolesc Psychiatry 2020;29:903–16. Available from: https://doi.org/10.1007/s00787-019-01298-5.

[332] Kaiser ML, Schoemaker MM, Albaret JM, Geuze RH. What is the evidence of impaired motor skills and motor control among children with attention deficit hyperactivity disorder (ADHD)? Systematic review of the literature. Res Dev Disabil 2015;36:338–57. Available from: https://doi.org/10.1016/j.ridd.2014.09.023.

[333] Lee J, Mayall LA, Bates KE, Hill EL, Leonard HC, Farran EK. The relationship between motor milestone achievement and childhood motor deficits in children with attention deficit hyperactivity disorder (ADHD) and children with developmental coordination disorder. Res Dev Disabil 2021;113:103920. Available from: https://doi.org/10.1016/j.ridd.2021.103920.

[334] Goulardins JB, Rigoli D, Licari M, Piek JP, Hasue RH, Oosterlaan J, et al. Attention deficit hyperactivity disorder and developmental coordination disorder: two separate disorders or do they share a common etiology. Behav Brain Res 2015;292:484–92. Available from: https://doi.org/10.1016/j.bbr.2015.07.009.

[335] Poh W, Payne JM, Gulenc A, Efron D. Chronic tic disorders in children with ADHD. Arch Dis Child 2018;103:847–52. Available from: https://doi.org/10.1136/archdischild-2017-314139.

[336] Scahill L, Specht M, Page C. The prevalence of tic disorders and clinical characteristics in children. J Obsessive Compuls Relat Disord 2014;3:394–400. Available from: https://doi.org/10.1016/j.jocrd.2014.06.002.

[337] Cooke T, So T-Y. Attention deficit hyperactive disorder and occurrence of tic disorders in children and adolescents—what is the verdict. Curr Pediatr Rev 2016;12:230–8. Available from: https://doi.org/10.2174/1573396312666160728113443.

[338] Osland ST, Steeves TD, Pringsheim T. Pharmacological treatment for attention deficit hyperactivity disorder (ADHD) in children with comorbid tic disorders. Cochrane Database Syst Rev 2018;2018. Available from: https://doi.org/10.1002/14651858.CD007990.pub3.

[339] Bertelsen EN, Larsen JT, Petersen L, Christensen J, Dalsgaard S. Childhood epilepsy, febrile seizures, and subsequent risk of ADHD. Pediatrics 2016138. Available from: https://doi.org/10.1542/peds.2015-4654.

[340] Chou IC, Chang YT, Chin ZN, Muo CH, Sung FC, Kuo HT, et al. Correlation between epilepsy and attention deficit hyperactivity disorder: a population-based cohort study. PLoS ONE 2013;8. Available from: https://doi.org/10.1371/journal.pone.0057926.

[341] Verrotti A, Moavero R, Panzarino G, di Paolantonio C, Rizzo R, Curatolo P. The challenge of pharmacotherapy in children and adolescents with epilepsy-ADHD comorbidity. Clin Drug Investig 2017;38(1):1–8. Available from: https://doi.org/10.1007/S40261-017-0585-1.

[342] Brikell I, Ghirardi L, D'Onofrio BM, Dunn DW, Almqvist C, Dalsgaard S, et al. Familial liability to epilepsy and attention-deficit/hyperactivity disorder: a nationwide cohort study. Biol Psychiatry 2018;83:173–80. Available from: https://doi.org/10.1016/j.biopsych.2017.08.006.

[343] Sedky K, Bennett DS, Carvalho KS. Attention deficit hyperactivity disorder and sleep disordered breathing in pediatric populations: a meta-analysis. Sleep Med Rev 2014;18:349–56. Available from: https://doi.org/10.1016/j.smrv.2013.12.003.

[344] Díaz-Román A, Mitchell R, Cortese S. Sleep in adults with ADHD: systematic review and meta-analysis of subjective and objective studies. Neurosci Biobehav Rev 2018;89:61–71. Available from: https://doi.org/10.1016/j.neubiorev.2018.02.014.

[345] Lugo J, Fadeuilhe C, Gisbert L, Setien I, Delgado M, Corrales M, et al. Sleep in adults with autism spectrum disorder and attention deficit/hyperactivity disorder: a systematic review and meta-analysis. Eur Neuropsychopharmacol 2020;38:1–24. Available from: https://doi.org/10.1016/j.euroneuro.2020.07.004.

[346] Wajszilber D, Santiseban JA, Gruber R. Sleep disorders in patients with ADHD: impact and management challenges. Nat Sci Sleep 2018;10:453. Available from: https://doi.org/10.2147/NSS.S163074.

[347] Pliszka SR. Patterns of psychiatric comorbidity with attention-deficit/hyperactivity disorder. Child Adolesc Psychiatr Clin North Am 2000;9:525–40. Available from: https://doi.org/10.1016/s1056-4993(18)30105-6.

[348] Reale L, Bartoli B, Cartabia M, Zanetti M, Costantino MA, Canevini MP, et al. Comorbidity prevalence and treatment outcome in children and adolescents with ADHD. Eur Child Adolesc Psychiatry 2017;26:1443–57. Available from: https://doi.org/10.1007/s00787-017-1005-z.

[349] Anker E, Bendiksen B, Heir T. Comorbid psychiatric disorders in a clinical sample of adults with ADHD, and associations with education, work and social characteristics: a cross-sectional study. BMJ Open 2018;8:e019700. Available from: https://doi.org/10.1136/bmjopen-2017-019700.

[350] Mohammadi MR, Zarafshan H, Khaleghi A, Ahmadi N, Hooshyari Z, Mostafavi SA, et al. Prevalence of ADHD and its comorbidities in a population-based sample. J Atten Disord 2021;25:1058–67. Available from: https://doi.org/10.1177/1087054719886372.

[351] Burke JD, Loeber R, Birmaher B. Oppositional defiant disorder and conduct disorder: a review of the past 10 years, part II. J Am Acad Child Adolesc Psychiatry 2002;41:1275–93. Available from: https://doi.org/10.1097/00004583-200211000-00009.

[352] Connor DF, Doerfler LA. ADHD with comorbid oppositional defiant disorder or conduct disorder: discrete or nondistinct disruptive behavior disorders? J Atten Disord 2008;12:126–34. Available from: https://doi.org/10.1177/1087054707308486.

[353] Connor DF, Steeber J, McBurnett K. A review of attention-deficit/hyperactivity disorder complicated by symptoms of oppositional defiant disorder or conduct disorder. J Dev Behav Pediatr 2010;31:427–40. Available from: https://doi.org/10.1097/DBP.0B013E3181E121BD.

[354] Wood AC, Rijsdijk F, Asherson P, Kuntsi J. Hyperactive-impulsive symptom scores and oppositional behaviours reflect alternate manifestations of a single liability. Behav Genet 2009;39:447–60. Available from: https://doi.org/10.1007/s10519-009-9290-z.

[355] Chen Q, Hartman CA, Haavik J, Harro J, Klungsøyr K, Hegvik TA, et al. Common psychiatric and metabolic comorbidity of adult attention-deficit/hyperactivity disorder: a population-based cross-sectional study. PLoS ONE 201813. Available from: https://doi.org/10.1371/journal.pone.0204516.

[356] Larson K, Russ SA, Kahn RS, Halfon N. Patterns of comorbidity, functioning, and service use for US children with ADHD, 2007. Pediatrics 2011;127:462–70. Available from: https://doi.org/10.1542/peds.2010-0165.

[357] Abramovitch A, Dar R, Mittelman A, Wilhelm S. Comorbidity between attention deficit/hyperactivity disorder and obsessive-compulsive disorder across the lifespan: a systematic and critical review. Harv Rev Psychiatry 2015;23:245–62. Available from: https://doi.org/10.1097/HRP.0000000000000050.

[358] Yoshimasu K, Barbaresi WJ, Colligan RC, Voigt RG, Killian JM, Weaver AL, et al. Childhood ADHD is strongly associated with a broad range of psychiatric disorders during adolescence: a population-based birth cohort study. J Child Psychol Psychiatry 2012;53:1036–43. Available from: https://doi.org/10.1111/j.1469-7610.2012.02567.x.

[359] Sandstrom A, Perroud N, Alda M, Uher R, Pavlova B. Prevalence of attention-deficit/hyperactivity disorder in people with mood disorders: a systematic review and meta-analysis. Acta Psychiatr Scand 2021;1–12. Available from: https://doi.org/10.1111/acps.13283.

[360] NICE. Attention deficit hyperactivity disorder: diagnosis and management. Nice Guideline no 87. National Institute for Health and Clinical Excellence; 2018. p. 62.

[361] Leucht S, Hierl S, Kissling W, Dold M, Davis JM. Putting the efficacy of psychiatric and general medicine medication into perspective: review of meta-analyses. Br J Psychiatry 2012;200:97–106. Available from: https://doi.org/10.1192/bjp.bp.111.096594.

[362] Baumeister AA, Henderson K, Pow JL, Advokat C. The early history of the neuroscience of attention-deficit/hyperactivity disorder. J Hist Neurosci 2012;21:263–79. Available from: https://doi.org/10.1080/0964704X.2011.595649.

[363] Bradley C. The behavior of children receiving benzedrine. Am J Psychiatry 1937;94:577–85. Available from: https://doi.org/10.1176/ajp.94.3.577.

[364] Storebø OJ, Ramstad E, Krogh HB, Nilausen TD, Skoog M, Holmskov M, et al. Methylphenidate for children and adolescents with attention deficit hyperactivity disorder (ADHD). Cochrane Database Syst Rev 2015;2015. Available from: https://doi.org/10.1002/14651858.CD009885.pub2.

[365] Coghill DR, Banaschewski T, Soutullo C, Cottingham MG, Zuddas A. Systematic review of quality of life and functional outcomes in randomized placebo-controlled studies of medications for attention-deficit/hyperactivity disorder. Eur Child Adolesc Psychiatry 2017;26:1283–307. Available from: https://doi.org/10.1007/s00787-017-0986-y.

[366] Visser SN, Danielson ML, Bitsko RH, Holbrook JR, Kogan MD, Ghandour RM, et al. Trends in the parent-report of health care provider-diagnosed and medicated attention-deficit/hyperactivity disorder: United States, 2003-2011. J Am Acad Child Adolesc Psychiatry 2014;53:34–46. Available from: https://doi.org/10.1016/j.jaac.2013.09.001.

[367] Wigal SB. Efficacy and safety limitations of attention-deficit hyperactivity disorder pharmacotherapy in children and adults. CNS Drugs 2009;23:21–31. Available from: https://doi.org/10.2165/00023210-200923000-00004.

[368] Cortese S, Adamo N, del Giovane C, Mohr-Jensen C, Hayes AJ, Carucci S, et al. Comparative efficacy and tolerability of medications for attention-deficit hyperactivity disorder in children, adolescents, and adults: a systematic review and network meta-analysis. Lancet Psychiatry 2018;5:727–38. Available from: https://doi.org/10.1016/S2215-0366(18)30269-4.

[369] Arnsten AFT. Stimulants: Therapeutic Actions in ADHD. Neuropsychopharmacology 2006;31(11):2376–83. Available from: https://doi.org/10.1038/sj.npp.1301164.

[370] Levy F. Dopamine vs noradrenaline: inverted-U effects and ADHD theories. Aust N Z J Psychiatry 2009;43:101–8. Available from: https://doi.org/10.1080/00048670802607238.

[371] Greenhill L, Kollins S, Abikoff H, McCracken J, Riddle M, Swanson J, et al. Efficacy and safety of immediate-release methylphenidate treatment for preschoolers with ADHD. J Am Acad Child Adolesc Psychiatry 2006;45:1284–93. Available from: https://doi.org/10.1097/01.chi.0000235077.32661.61.

[372] Kortekaas-Rijlaarsdam AF, Luman M, Sonuga-Barke E, Oosterlaan J. Does methylphenidate improve academic performance? A systematic review and meta-analysis. Eur Child Adolesc Psychiatry 2019;28:155–64. Available from: https://doi.org/10.1007/s00787-018-1106-3.

[373] Bernardi S, Faraone Sv, Cortese S, Kerridge BT, Pallanti S, Wang S, et al. The lifetime impact of attention deficit hyperactivity disorder: results from the National Epidemiologic Survey on Alcohol and Related Conditions (NESARC). Psychol Med 2012;42:875–87. Available from: https://doi.org/10.1017/S003329171100153X.

[374] Chang Z, Quinn PD, Hur K, Gibbons RD, Sjölander A, Larsson H, et al. Association between medication use for attention-deficit/hyperactivity disorder and risk of motor vehicle crashes. JAMA Psychiatry 2017;74:597–603. Available from: https://doi.org/10.1001/jamapsychiatry.2017.0659.

[375] Pringsheim T, Hirsch L, Gardner D, Gorman DA. The pharmacological management of oppositional behaviour, conduct problems, and aggression in children and adolescents with attention-deficit hyperactivity disorder, oppositional defiant disorder, and conduct disorder: a systematic review and meta-analysi. Can J Psychiatry 2015;60:42–51. Available from: https://doi.org/10.1177/070674371506000202.

[376] Greenhill LL, Swanson JM, Hechtman L, Waxmonsky J, Arnold LE, Molina BSG, et al. Trajectories of growth associated with long-term stimulant medication in the multimodal treatment study of attention-deficit/hyperactivity disorder. J Am Acad Child Adolesc Psychiatry 2020;59:978–89. Available from: https://doi.org/10.1016/j.jaac.2019.06.019.

[377] Faraone Sv, Biederman J, Morley CP, Spencer TJ. Effect of stimulants on height and weight: a review of the literature. J Am Acad Child Adolesc Psychiatry 2008;47:994–1009. Available from: https://doi.org/10.1097/chi.obo13e31817eoea7.

[378] Ibrahim K, Donyai P. Drug holidays from ADHD medication: international experience over the past four decades. J Atten Disord 2015;19:551–68. Available from: https://doi.org/10.1177/1087054714548035.

[379] Carucci S, Balia C, Gagliano A, Lampis A, Buitelaar JK, Danckaerts M, et al. Long term methylphenidate exposure and growth in children and adolescents with ADHD. A systematic review and meta-analysis. Neurosci Biobehav Rev 2021;120:509–25. Available from: https://doi.org/10.1016/j.neubiorev.2020.09.031.

[380] McCarthy S, Neubert A, Man KKC, Banaschewski T, Buitelaar J, Carucci S, et al. Effects of long-term methylphenidate use on growth and blood pressure: results of the German Health Interview and Examination Survey for Children and Adolescents (KiGGS). BMC Psychiatry 2018;18:1–10. Available from: https://doi.org/10.1186/s12888-018-1884-7.

[381] Liu H, Feng W, Zhang D. Association of ADHD medications with the risk of cardiovascular diseases: a meta-analysis. Eur Child Adolesc Psychiatry 2019;28:1283–93. Available from: https://doi.org/10.1007/s00787-018-1217-x.

[382] Liang EF, Lim SZ, Tam WW, Ho CS, Zhang MW, McIntyre RS, et al. The effect of methylphenidate and atomoxetine on heart rate and systolic blood pressure in young people and adults with attention-deficit hyperactivity disorder (ADHD): systematic review, meta-analysis, and meta-regression. Int J Environ Res Public Health 2018;15:1789. Available from: https://doi.org/10.3390/ijerph15081789.

[383] Fay TB, Alpert MA. Cardiovascular effects of drugs used to treat attention-deficit/hyperactivity disorder part 2: impact on cardiovascular events and recommendations for evaluation and monitoring. Cardiol Rev 2019;27:173–8. Available from: https://doi.org/10.1097/CRD.0000000000000234.

[384] Fay TB, Alpert MA. Cardiovascular effects of drugs used to treat attention-deficit/hyperactivity disorder: part 1: epidemiology, pharmacology, and impact on hemodynamics and ventricular repolarization. Cardiol Rev 2019;27:113–21. Available from: https://doi.org/10.1097/CRD.0000000000000233.

[385] Dalsgaard S, Kvist AP, Leckman JF, Nielsen HS, Simonsen M. Cardiovascular safety of stimulants in children with attention-deficit/ hyperactivity disorder: a nationwide prospective cohort study. J Child Adolesc Psychopharmacol 2014;24:302–10. Available from: https://doi.org/10.1089/cap.2014.0020.

[386] Volkow ND, Swanson JM. Variables that affect the clinical use and abuse of methylphenidate in the treatment of ADHD. Am J Psychiatry 2003;160:1909–18. Available from: https://doi.org/10.1176/appi.ajp.160.11.1909.

[387] Faraone SV, Upadhyaya HP. The effect of stimulant treatment for ADHD on later substance abuse and the potential for medication misuse, abuse, and diversion. J Clin Psychiatry 200768. Available from: https://doi.org/10.4088/jcp.1107e28.

[388] Clemow DB, Walker DJ. The potential for misuse and abuse of medications in ADHD: a review. Postgrad Med 2014;126:64–81. Available from: https://doi.org/10.3810/pgm.2014.09.2801.

[389] Benson K, Flory K, Humphreys KL, Lee SS. Misuse of stimulant medication among college students: a comprehensive review and meta-analysis. Clin Child Family Psychol Rev 2015;18:50–76. Available from: https://doi.org/10.1007/s10567-014-0177-z.

[390] Faraone Sv, Rostain AL, Montano CB, Mason O, Antshel KM, Newcorn JH. Systematic review: nonmedical use of prescription stimulants: risk factors, outcomes, and risk reduction strategies. J Am Acad Child Adolesc Psychiatry 2020;59:100–12. Available from: https://doi.org/10.1016/j.jaac.2019.06.012.

[391] McCabe SE, Veliz P, Wilens TE, Schulenberg JE. Adolescents' prescription stimulant use and adult functional outcomes: a national prospective study. J Am Acad Child Adolesc Psychiatry 2017;56:226–33. Available from: https://doi.org/10.1016/j.jaac.2016.12.008.

[392] Colaneri N, Keim S, Adesman A. Physician practices to prevent ADHD stimulant diversion and misuse. J Subst Abuse Treat 2017;74:26–34. Available from: https://doi.org/10.1016/j.jsat.2016.12.003.

[393] Rucklidge JJ, Downs-Woolley M, Taylor M, Brown JA, Harrow SE. Psychiatric comorbidities in a New Zealand sample of adults with ADHD. J Atten Disord 2016;20:1030–8. Available from: https://doi.org/10.1177/1087054714529457.

[394] Quinn PD, Chang Z, Hur K, Gibbons RD, Lahey BB, Rickert ME, et al. ADHD medication and substance-related problems. Am J Psychiatry 2017;174:877–85. Available from: https://doi.org/10.1176/appi.ajp.2017.16060686.

[395] Young JR, Yanagihara A, Dew R, Kollins SH. Pharmacotherapy for preschool children with attention deficit hyperactivity disorder (ADHD): current status and future directions. CNS Drugs 2021;35(4):403–24. Available from: https://doi.org/10.1007/S40263-021-00806-Z.

[396] Scahill L. Alpha-2 adrenergic agonists in children with inattention, hyperactivity and impulsiveness. CNS Drugs 2009;23:43–9. Available from: https://doi.org/10.2165/00023210-200923000-00006.

[397] Kollins SH, Jain R, Brams M, Segal S, Findling RL, Wigal SB, et al. Clonidine extended-release tablets as add-on therapy to psychostimulants in children and adolescents with ADHD. Pediatrics 2011;127:e1406–13. Available from: https://doi.org/10.1542/peds.2010-1260.

[398] Jain R, Segal S, Kollins SH, Khayrallah M. Clonidine extended-release tablets for pediatric patients with attention-deficit/hyperactivity disorder. J Am Acad Child Adolesc Psychiatry 2011;50:171–9. Available from: https://doi.org/10.1016/j.jaac.2010.11.005.

[399] Huss M, Dirks B, Gu J, Robertson B, Newcorn JH, Ramos-Quiroga JA. Long-term safety and efficacy of guanfacine extended release in children and adolescents with ADHD. Eur Child Adolesc Psychiatry 2018;27:1283–94. Available from: https://doi.org/10.1007/s00787-018-1113-4.

[400] Newcorn JH, Stein MA, Childress AC, Youcha S, White C, Enright G, et al. Randomized, double-blind trial of guanfacine extended release in children with attention-deficit/hyperactivity disorder: morning or evening administration. J Am Acad Child Adolesc Psychiatry 2013;52:921–30. Available from: https://doi.org/10.1016/j.jaac.2013.06.006.

[401] Faraone Sv, Glatt SJ. Effects of extended-release guanfacine on ADHD symptoms and sedation-related adverse events in children with ADHD. J Atten Disord 2010;13:532–8. Available from: https://doi.org/10.1177/1087054709332472.

[402] Biederman J, Melmed RD, Patel A, McBurnett K, Donahue J, Lyne A. Long-term, open-label extension study of guanfacine extended release in children and adolescents with ADHD. CNS Spectr 2008;13:1047–55. Available from: https://doi.org/10.1017/S1092852900017107.

[403] Elbe D. Switching from clonidine immediate-release to guanfacine extended-release. J Can Acad Child Adolesc Psychiatry 2020;29:121.

[404] Cheng JYW, Chen RYL, Ko JSN, Ng EML. Efficacy and safety of atomoxetine for attention-deficit/hyperactivity disorder in children and adolescents - meta-analysis and meta-regression analysis. Psychopharmacology 2007;194:197–209. Available from: https://doi.org/10.1007/s00213-007-0840-x.

[405] Gayleard JL, Mychailyszyn MP. Atomoxetine treatment for children and adolescents with attention-deficit/hyperactivity disorder (ADHD): a comprehensive meta-analysis of outcomes on parent-rated core symptomatology. Atten Defic Hyperact Disord 2017;9:149–60. Available from: https://doi.org/10.1007/s12402-017-0216-y.

[406] Shang CY, Pan YL, Lin HY, Huang LW, Gau SSF. An open-label, randomized trial of methylphenidate and atomoxetine treatment in children with attention-deficit/hyperactivity disorder. J Child Adolesc Psychopharmacol 2015;25:566–73. Available from: https://doi.org/10.1089/cap.2015.0035.

[407] Coughlin CG, Cohen SC, Mulqueen JM, Ferracioli-Oda E, Stuckelman ZD, Bloch MH. Meta-analysis: reduced risk of anxiety with psychostimulant treatment in children with attention-deficit/hyperactivity disorder. J Child Adolesc Psychopharmacol 2015;25:611–17. Available from: https://doi.org/10.1089/cap.2015.0075.

[408] Patra S, Nebhinani N, Viswanathan A, Kirubakaran R. Atomoxetine for attention deficit hyperactivity disorder in children and adolescents with autism: a systematic review and meta-analysis. Autism Res 2019;12:542–52. Available from: https://doi.org/10.1002/aur.2059.

[409] Solmi M, Fornaro M, Ostinelli EG, Zangani C, Croatto G, Monaco F, et al. Safety of 80 antidepressants, antipsychotics, anti-attention-deficit/hyperactivity medications and mood stabilizers in children and adolescents with psychiatric disorders: a large scale systematic meta-review of 78 adverse effects. World Psychiatry 2020;19:214–32. Available from: https://doi.org/10.1002/WPS.20765.

[410] Bangs ME, Tauscher-Wisniewski S, Polzer J, Zhang S, Acharya N, Desaiah D, et al. Meta-analysis of suicide-related behavior events in patients treated with atomoxetine. J Am Acad Child Adolesc Psychiatry 2008;47:209–18. Available from: https://doi.org/10.1097/CHI.0B013E31815D88B2.

[411] Bangs ME, Jin L, Zhang S, Desaiah D, Allen AJ, Read HA, et al. Hepatic events associated with atomoxetine treatment for attention-deficit hyperactivity disorder. Drug Saf 2008;31(4):345–54. Available from: https://doi.org/10.2165/00002018-200831040-00008.

[412] Trzepacz PT, Williams DW, Feldman PD, Wrishko RE, Witcher JW, Buitelaar JK. CYP2D6 metabolizer status and atomoxetine dosing in children and adolescents with ADHD. Eur Neuropsychopharmacol 2008;18:79–86. Available from: https://doi.org/10.1016/J.EURONEURO.2007.06.002.

[413] Fernández de la Cruz L, Simonoff E, McGough JJ, Halperin JM, Arnold LE, Stringaris A. Treatment of children with attention-deficit/hyperactivity disorder (ADHD) and irritability: results from the multimodal treatment study of children with ADHD (MTA). J Am Acad Child Adolesc Psychiatry 2015;54:62–70. Available from: https://doi.org/10.1016/j.jaac.2014.10.006.

[414] Pelham WE, Altszuler AR. Combined treatment for children with attention-deficit/hyperactivity disorder: brief history, the multimodal treatment for attention-deficit/hyperactivity disorder study, and the past 20 years of research. J Dev Behav Pediatr 2020;41(Suppl 2):S88–98. Available from: https://doi.org/10.1097/DBP.0000000000000777.

[415] Nigg JT, Holton K. Restriction and elimination diets in ADHD treatment. Child Adolesc Psychiatr Clin North Am 2014;23:937. Available from: https://doi.org/10.1016/J.CHC.2014.05.010.

[416] Rucklidge JJ, Johnstone J, Kaplan BJ. Nutrient supplementation approaches in the treatment of ADHD. Expert Rev Neurother 2014;9:461–76. Available from: https://doi.org/10.1586/ERN.09.7.

[417] Rucklidge JJ, Frampton CM, Gorman B, Boggis A. Vitamin–mineral treatment of attention-deficit hyperactivity disorder in adults: double-blind randomised placebo-controlled trial. Br J Psychiatry 2014;204:306–15. Available from: https://doi.org/10.1192/BJP.BP.113.132126.

[418] Shrestha M, Lautenschleger J, Soares N. Non-pharmacologic management of attention-deficit/hyperactivity disorder in children and adolescents: a review. Transl Pediatr 2020;9:S114. Available from: https://doi.org/10.21037/TP.2019.10.01.

[419] Rimestad ML, Lambek R, Zacher Christiansen H, Hougaard E. Short- and long-term effects of parent training for preschool children with or at risk of ADHD: a systematic review and meta-analysis. J Atten Disord 2019;23:423–34. Available from: https://doi.org/10.1177/1087054716648775.

[420] Abikoff HB, Thompson M, Laver-Bradbury C, Long N, Forehand RL, Miller Brotman L, et al. Parent training for preschool ADHD: a randomized controlled trial of specialized and generic programs. J Child Psychol Psychiatry 2015;56:618–31. Available from: https://doi.org/10.1111/jcpp.12346.

[421] Scionti N, Cavallero M, Zogmaister C, Marzocchi GM. Is cognitive training effective for improving executive functions in preschoolers? A systematic review and meta-analysis. Front Psychol 2020;10:2812. Available from: https://doi.org/10.3389/fpsyg.2019.02812.

[422] Cortese S, Ferrin M, Brandeis D, Buitelaar J, Daley D, Dittmann RW, et al. Cognitive training for attention-deficit/hyperactivity disorder: meta-analysis of clinical and neuropsychological outcomes from randomized controlled trials. J Am Acad Child Adolesc Psychiatry 2015;54:164–74. Available from: https://doi.org/10.1016/j.jaac.2014.12.010.

[423] Cortese S, Ferrin M, Brandeis D, Holtmann M, Aggensteiner P, Daley D, et al. Neurofeedback for attention-deficit/hyperactivity disorder: meta-analysis of clinical and neuropsychological outcomes from randomized controlled trials. J Am Acad Child Adolesc Psychiatry 2016;55:444–55. Available from: https://doi.org/10.1016/j.jaac.2016.03.007.

[424] Zhang J, Díaz-Román A, Cortese S. Meditation-based therapies for attention-deficit/hyperactivity disorder in children, adolescents and adults: a systematic review and meta-analysis. Evid Based Ment Health 2018;21:87–94. Available from: https://doi.org/10.1136/ebmental-2018-300015.

[425] Storebø OJ, Andersen ME, Skoog M, Hansen SJ, Simonsen E, Pedersen N, et al. Social skills training for attention deficit hyperactivity disorder (ADHD) in children aged 5 to 18 years. Cochrane Database Syst Rev 20192019. Available from: https://doi.org/10.1002/14651858.CD008223.pub3.

[426] Bikic A, Reichow B, McCauley SA, Ibrahim K, Sukhodolsky DG. Meta-analysis of organizational skills interventions for children and adolescents with attention-deficit/hyperactivity disorder. Clin Psychol Rev 2017;52:108–23. Available from: https://doi.org/10.1016/j.cpr.2016.12.004.

[427] Knouse LE, Teller J, Brooks MA. Meta-analysis of cognitive-behavioral treatments for adult ADHD. J Consult Clin Psychol 2017;85:737–50. Available from: https://doi.org/10.1037/ccp0000216.

[428] Young Z, Moghaddam N, Tickle A. The efficacy of cognitive behavioral therapy for adults with ADHD: a systematic review and meta-analysis of randomized controlled trials. J Atten Disord 2020;24:875–88. Available from: https://doi.org/10.1177/1087054716664413.

[429] Nimmo-Smith V, Merwood A, Hank D, Brandling J, Greenwood R, Skinner L, et al. Non-pharmacological interventions for adult ADHD: a systematic review. Psychol Med 2020;50:529–41. Available from: https://doi.org/10.1017/S0033291720000069.

[430] Dahl V, Ramakrishnan A, Spears AP, Jorge A, Lu J, Bigio NA, et al. Psychoeducation interventions for parents and teachers of children and adolescents with ADHD: a systematic review of the literature. J Dev Phys Disabil 2020;32:257–92. Available from: https://doi.org/10.1007/s10882-019-09691-3.

[431] DuPaul GJ, Gormley MJ, Laracy SD. School-based interventions for elementary school students with ADHD. Child Adolesc Psychiatr Clin North Am 2014;23:687–97. Available from: https://doi.org/10.1016/j.chc.2014.05.003.

[432] DuPaul GJ, Evans SW, Mautone JA, Owens JS, Power TJ. Future directions for psychosocial interventions for children and adolescents with ADHD. J Clin Child Adolesc Psychol 2020;49:134–45. Available from: https://doi.org/10.1080/15374416.2019.1689825.

Chapter 28

Autism spectrum disorder

Scott M. Myers[1,2] and Thomas D. Challman[1,2]
[1]Geisinger Autism & Developmental Medicine Institute, Danville, PA, United States, [2]Geisinger Health System and Geisinger Commonwealth School of Medicine, Lewisburg, PA, United States

Introduction

Autism spectrum disorder (ASD) is a common but complex and heterogeneous neurodevelopmental disorder that requires personalized assessments and intervention strategies; coordination among medical, educational, and community intervention and support services; and longitudinal care. Many substantial changes have occurred since the publication of the third edition of this textbook 15 years ago, including a rise in measured prevalence of ASD, changes to the most commonly accepted diagnostic criteria and classification systems, elucidation of the importance of associated features and cooccurring disorders, expansion of the evidence base for therapeutic interventions, rapid progress in identifying causal genomic factors, and substantial increases in research funding and publications related to ASD [1–5]. Clinical practice guidelines for diagnosis and management are available from professional organizations such as the American Academy of Child and Adolescent Psychiatry (AACAP), American Academy of Pediatrics (AAP), Canadian Paediatric Society, and European Society of Child and Adolescent Psychiatry [6–11].

The term *ASD* is used throughout this chapter for uniformity and to emphasize reference to the clinical disorder, as opposed to *autism* or the adjective form, *autistic*, which might refer to a broader phenotype, including individuals with traits that are subthreshold for clinical diagnosis. The authors acknowledge the value of both person-first terminology (e.g., "person with ASD") and identity-first terminology (e.g., "autistic person") and the preference for the latter among many in the ASD community [12]. Because of the decision to use the term *ASD* and lack of an equivalent adjective, person-first terminology is used more frequently in this chapter.

Definition and classification

Historical aspects

In 1943 Leo Kanner eloquently described 11 children with innate "autistic disturbances of affective contact" characterized by profoundly deficient social engagement and interaction; communication disturbance ranging from mutism to echolalia, pronoun reversal, and literalness; and unusual interactions with the environment, including an "anxiously obsessive desire for the maintenance of sameness" that was not explained by general cognitive impairment [13]. The next year, Hans Asperger contributed a report of four boys with a similar constellation of characteristics constituting "autistic psychopathy," including lack of reciprocity in social interactions and a restricted range of intense and unusual interests, but normal or precocious language acquisition, with above-average linguistic skills but subtle abnormalities of verbal and nonverbal communication and motor clumsiness [14,15]. Asperger's work, published in German, remained virtually unknown to the English-speaking scientific community until it was described by Wing in 1981 [14] and a translation published by Frith in 1991 [15]. Both Kanner and Asperger were likely influenced by George Frankl's work on emotional (or affective) language, and the importance of gestures and other nonverbal communication for the development of normal affective contact [16,17].

It should be recognized that although Kanner and Asperger are the names universally associated with the earliest academic recognition of the condition, the first clinical description of six children with a constellation of traits easily recognizable as what is currently known as ASD was published by Russian psychiatrist Grunya Sukhareva almost two decades earlier, in 1925 [18,19]. The descriptions she provided are strikingly similar to the diagnostic criteria currently

Capute and Accardo's Neurodevelopmental Disabilities in Infancy and Childhood. DOI: https://doi.org/10.1016/B978-0-12-824060-1.00038-9

in use [20], and she foreshadowed contemporary issues of sex differences in the presentation of what is now known as ASD in her 1927 paper describing five girls with this array of symptoms [21]. Sukhareva continued to contribute to the ASD literature for several more decades [19]. It is striking that although her initial description was published in Russian in 1925, republished in German in 1926, and translated into English in 1996, Sukhareva's pioneering, forward thinking, and sustained contributions are infrequently acknowledged.

Evolution of diagnostic criteria and classification

Autism was first included as an official diagnostic category in the American Psychiatric Association's *Diagnostic and Statistical Manual of Mental Disorders, Third Edition* (DSM-III) in 1980 as a class of conditions, pervasive developmental disorders (PDDs), which included five categorical diagnoses: infantile autism (IA), IA residual state, childhood onset pervasive developmental disorder (COPDD), COPDD residual state, and atypical PDD [22,23]. The diagnosis of IA required pervasive lack of responsiveness to other people, gross deficits in language development, peculiar speech patterns (if speech is present), and "bizarre" responses to various aspects of the environment, and onset before age 30 months [22]. This definition came to be viewed as too restrictive because it was monothetic (i.e., required that every criterion must be met) and focused on infants and toddlers [22,23].

The 1987 revision, DSM-III-R, instituted a more flexible and developmentally oriented approach that would be useful across ages and developmental levels, and the PDD category was reduced to two diagnoses: autistic disorder and PDD not otherwise specified (PDD-NOS) [23,24]. Sixteen detailed criteria for autistic disorder were provided, organized into three major domains of dysfunction: (1) qualitative impairments in reciprocal social interaction, (2) impairments in verbal and nonverbal communication, and (3) markedly restricted repertoire of activities and interests. A total of at least eight criteria, including at least two from the social domain and at least one from each of the other two categories, was required to make the diagnosis of autistic disorder. However, when implemented in practice and subjected to clinical research, the DSM-III-R criteria were found to be too broad, leading to many false-positive diagnoses [25].

Following an elaborate drafting process, DSM-IV (1994) introduced the categorical diagnoses of Asperger's disorder, childhood disintegrative disorder (CDD), Rett disorder, and PDD-NOS in addition to maintaining the autistic disorder term [23,26]. The three-domain symptom categorization model of DSM-III-R (reciprocal social interaction, communication, restricted/repetitive behavior) was maintained, and DSM-IV was closely aligned with the *International Classification of Diseases, 10th edition* (ICD-10) [27]. In the DSM-IV text revision (DSM-IV-TR), published in 2000, the text description of PDD-NOS was edited slightly to correct a copyediting error and increase specificity, but the diagnostic categories and specific criteria were otherwise unchanged [28].

The evolution of DSM-IV-TR to DSM-5 involved three major changes: (1) elimination of the PDD umbrella term and the five subcategories in favor of a single diagnosis, "autism spectrum disorder"; (2) restructuring of the three-domain symptom model into a two-domain symptom model, essentially by combining the communication and social behavior domains into a single social-communication domain; and (3) introduction of core symptom domain severity levels based on the level of support needed for individual functioning [23,29]. It is also notable that, in contrast to DSM-IV-TR, it is specified in DSM-5 that behavioral criteria can be met on the basis of historical report, and sensory-related behaviors are included as symptoms within the restricted/repetitive behavior domain. Diagnostic restrictions that had previously prevented the codiagnosis of other disorders, such as attention-deficit/hyperactivity disorder (ADHD), in the setting of autism were eliminated in DSM-5, and a new diagnosis, social (pragmatic) communication disorder (SCD), was introduced [29]. It was thought that the diagnosis of SCD would apply to some individuals who would have been classified as having PDD-NOS by DSM-IV-TR criteria but do not meet the DSM-5 criteria for ASD due to lack of threshold restricted/repetitive behaviors, but questions remain about whether SCD is qualitatively distinct from ASD [30–34]. ICD-11 also adopted the conceptualization of ASD as a single umbrella diagnosis and the two-domain symptom model [35].

Current definition and classification: DSM-5-TR, ICD-11

Almost 80 years after the seminal papers by Kanner and Asperger, and nearly a century after the initial prescient description by Sukhareva, the ASD diagnostic criteria remain remarkably consistent with the original observations of these astute clinicians.

DSM-5 and ICD-11 are well-aligned in their conceptualization of ASD as a single diagnosis defined by clinically significant, persistent deficits in social communication and interaction and atypically restricted and repetitive behaviors and interests that cause significant impairment in adaptive functioning (Table 28.1) [29,35]. The specific manifestations

TABLE 28.1 Diagnostic criteria and codes for autism spectrum disorder: DSM-5/DSM-5-TR and ICD-11.

DSM-5, DSM-5-TR[a]	ICD-11
A. Persistent deficits in social communication and social interaction across multiple contexts, as manifested by all of the following, currently or by history (examples are illustrative, not exhaustive): 1. Deficits in social-emotional reciprocity, ranging, for example, from abnormal social approach and failure of normal back-and-forth conversation; to reduced sharing of interests, emotions, or affect; to failure to initiate or respond to social interactions. 2. Deficits in nonverbal communicative behaviors used for social interaction, ranging, for example, from poorly integrated verbal and nonverbal communication; to abnormalities in eye contact and body language or deficits in understanding and use of gestures; to a total lack of facial expressions and nonverbal communication. 3. Deficits in developing, maintaining, and understanding relationships, ranging, for example, from difficulties adjusting behavior to suit various social contexts; to difficulties in sharing imaginative play or in making friends; to absence of interest in peers.	A. Persistent deficits in initiating and sustaining social communication and reciprocal social interactions that are outside the expected range of typical functioning given the individual's age and level of intellectual development. Specific manifestations of these deficits vary according to chronological age, verbal and intellectual ability, and disorder severity. Manifestations may include limitations in the following: • Understanding of, interest in, or inappropriate responses to the verbal or nonverbal social communications of others. • Integration of spoken language with typical complementary nonverbal cues, such as eye contact, gestures, facial expressions, and body language. These nonverbal behaviors may also be reduced in frequency or intensity. • Understanding and use of language in social contexts and ability to initiate and sustain reciprocal social conversations. • Social awareness, leading to behavior that is not appropriately modulated according to the social context. • Ability to imagine and respond to the feelings, emotional states, and attitudes of others. • Mutual sharing of interests. • Ability to make and sustain typical peer relationships.
B. Restricted, repetitive patterns of behavior, interests, or activities, as manifested by at least two of the following, currently or by history (examples are illustrative, not exhaustive): 1. Stereotyped or repetitive motor movements, use of objects, or speech (e.g., simple motor stereotypies, lining up toys or flipping objects, echolalia, idiosyncratic phrases). 2. Insistence on sameness, inflexible adherence to routines, or ritualized patterns of verbal or nonverbal behavior (e.g., extreme distress at small changes, difficulties with transitions, rigid thinking patterns, greeting rituals, need to take same route or eat food every day). 3. Highly restricted, fixated interests that are abnormal in intensity or focus (e.g., strong attachment to or preoccupation with unusual objects, excessively circumscribed or perseverative interest). 4. Hyper- or hyporeactivity to sensory input or unusual interests in sensory aspects of the environment (e.g., apparent indifference to pain/temperature, adverse response to specific sounds or textures, excessive smelling or touching of objects, visual fascination with lights or movement).	B. Persistent restricted, repetitive, and inflexible patterns of behavior, interests, or activities that are clearly atypical or excessive for the individual's age and sociocultural context. These may include: • Lack of adaptability to new experiences and circumstances, with associated distress, that can be evoked by trivial changes to a familiar environment or in response to unanticipated events. • Inflexible adherence to particular routines; for example, these may be geographic such as following familiar routes, or may require precise timing such as mealtimes or transport. • Excessive adherence to rules (e.g., when playing games). • Excessive and persistent ritualized patterns of behavior (e.g., preoccupation with lining up or sorting objects in a particular way) that serve no apparent external purpose. • Repetitive and stereotyped motor movements, such as whole-body movements (e.g., rocking), atypical gait (e.g., walking on tiptoes), unusual hand or finger movements and posturing. These behaviors are particularly common during early childhood. • Persistent preoccupation with one or more special interests, parts of objects, or specific types of stimuli (including media) or an unusually strong attachment to particular objects (excluding typical comforters). • Lifelong excessive and persistent hypersensitivity or hyposensitivity to sensory stimuli or unusual interest in a sensory stimulus, which may include actual or anticipated sounds, light, textures (especially clothing and food), odors and tastes, heat, cold, or pain.
C. Symptoms must be present in the early developmental period (but may not become fully manifest until social demands exceed limited capacities, or may be masked by learned strategies in later life).	C. The onset of the disorder occurs during the developmental period, typically in early childhood, but characteristic symptoms may not become fully manifest until later, when social demands exceed limited capacities.

(Continued)

TABLE 28.1 (Continued)

DSM-5, DSM-5-TR[a]	ICD-11
D. Symptoms cause clinically significant impairment in social, occupational, or other important areas of current functioning. E. These disturbances are not better explained by intellectual disability (intellectual developmental disorder) or global developmental delay. Intellectual disability and autism spectrum disorder frequently cooccur; to make comorbid diagnoses of autism spectrum disorder and intellectual disability, social communication should be below that expected for general developmental level.	D. The symptoms result in significant impairment in personal, family, social, educational, occupational or other important areas of functioning. Some individuals with autism spectrum disorder are able to function adequately in many contexts through exceptional effort, such that their deficits may not be apparent to others. A diagnosis of autism spectrum disorder is still appropriate in such cases.
Specify the current severity of social communication impairment (criterion A) • Level 1—Requiring support: Social communication deficits cause noticeable functional impairment; difficulty initiating social interactions and atypical or unsuccessful responses to overtures of others; may appear to have decreased interest in social interactions. • Level 2—Requiring substantial support: Marked deficits in verbal and nonverbal social communication skills cause social impairment that is apparent even with supports in place; limited initiation of social interactions and reduced or abnormal responses to social overtures from others. • Level 3—Requiring very substantial support: Severe deficits in verbal and nonverbal social communication skills cause severe functional impairment; very limited initiation of social interactions and minimal response to social overtures from others. Specify the current severity of restricted, repetitive behavior (criterion B) • Level 1—Requiring support: Significant interference with functioning in one or more contexts due to inflexibility of behavior; difficulty switching between activities; problems with organization and planning hamper independence. • Level 2—Requiring substantial support: Inflexibility of behavior, difficulty coping with change, or other restricted/repetitive behaviors appear frequently enough to be obvious to the casual observer and interfere with functioning in a variety of contexts; distress and/or difficulty changing focus or action. • Level 3—Requiring very substantial support: Inflexibility of behavior, extreme difficulty coping with change, or other restricted/repetitive behaviors markedly interfere with functioning in all spheres; great distress/difficulty changing focus or action. *Specify* if: With or without accompanying intellectual impairment With or without accompanying language impairment Associated with a known medical or genetic condition or environmental factor Associated with another neurodevelopmental, mental, or behavioral problem With catatonia Codes: 299.00 Autism spectrum disorder	Specifiers: Without disorder of intellectual development With disorder of intellectual development With mild or no impairment of functional language With impaired functional language (i.e., not able to use more than single words or simple phrases) With complete, or almost complete, absence of functional language Qualifier: Without loss of previously acquired skills With loss of previously acquired skills Codes: 6A02 Autism spectrum disorder (ASD) 6A02.0—without disorder of intellectual development and with mild or no impairment in functional language 6A02.1—with disorder of intellectual development and with mild or no impairment in functional language 6A02.2—without disorder of intellectual development and with impaired functional language 6A02.3—with disorder of intellectual development and with impaired functional language 6A02.5—with disorder of intellectual development and with absence of functional language 6A02.Y—used if above parameters do not apply 6A02.Z—used if above parameters are unknown An additional digit can be applied to the above codes to indicate the presence or absence of regression: 6A02.x0—without loss of previously acquired skills 6A02.x1—with loss of previously acquired skills

DSM-5, Diagnostic and statistical manual of mental disorders, fifth edition; *DSM-5-TR*, diagnostic and statistical manual of mental disorders, fifth edition, text revision; *ICD-11*, international classification of diseases, eleventh revision.
[a]*The criteria as stated in the table reflect the minor text revisions to DSM-5 (DSM-5-TR).*
Source: American Psychiatric Association (2013) [29], World Health Organization (2019/2021) [35], and American Psychiatric Association (2022) [36].

of the core deficits vary with age, language and intellectual ability, and disorder severity. Symptoms begin in the developmental period but may not become fully manifest until social demands exceed limited capacities. The deficits in social communication and interaction must be present to a degree that is outside of the expected range of functioning for the individual's age and level of intellectual development.

The DSM-5 includes guidelines for severity level classification for both the social-communication and restricted/repetitive behavior core symptom domains based on the level of support required for functioning at home and in the community and specifiers regarding the presence or absence of common cooccurring diagnoses (e.g., intellectual impairment, structural language impairment, catatonia, associated medical conditions, or other neurodevelopmental, mental, or behavioral disorders) (Table 28.1) [29]. These specifiers provide clinicians with an opportunity to individualize the diagnosis and communicate a richer description of the affected individual's clinical course and current symptomatology [29]. The DSM-5 specifiers for cooccurring diagnoses are recorded but, in contrast to ICD-11 specifiers and qualifiers, are not coded differently. The 2022 text revision to DSM-5 (DSM-5-TR) did not include any major changes to the ASD criteria, although two small adjustments were made [36]. The text description of criterion A was modified slightly from "as manifested by the following" to "as manifested by **all of** the following," clarifying that all three components of the social communication and interaction criteria must be met. Additionally, the specifier "associated with another neurodevelopmental, mental or behavioral disorder," was altered to read "associated with a neurodevelopmental, mental, or behavioral **problem**," allowing clinicians to indicate cooccurring problems that do not necessarily rise to the level of disorder [36].

Although the ICD-11 recognizes ASD as the single umbrella term, it retained a multicategorical coding system describing various combinations of intellectual and language impairments and the presence or absence of loss of previously acquired skills (Table 28.1) [35]. There are other ways in which the two diagnostic classification systems differ. For example, whereas DSM-5 specifies that all three social communication criteria and at least two out of the four restricted/repetitive behavior criteria must be met, ICD-11 does not require a set number or combination of features. Seven symptom groups are provided for each of the two core domains, and it is up to the clinician to decide whether a patient's features satisfy the definition.

Profound autism

"Profound autism" has been proposed as an administrative term to apply to children (age ~8 years or older) and adults with ASD who have, or are likely to have as adults, the need for extensive functional supports including 24-hour access to an adult who can care for them if concerns arise, due to being unable to take care of their basic daily adaptive needs or to be left completely alone in a residence [3]. In most cases, these individuals will have severe intellectual disability (ID) (e.g., IQ <50), very limited language (e.g., limited ability to communicate with a stranger using comprehensible sentences), or both (i.e., meeting ICD-11 code 6A02.5, ASD with disorder of intellectual development and with absence of functional language; see Table 28.1). Profound autism is commonly associated with complex cooccurring medical issues (e.g., epilepsy) and problem behaviors such as self-injury and aggression, but it is defined by the intellectual and/or language impairment and not by the core ASD features or the associated impairments [3].

Neurodiversity

Some, but not all, individuals with ASD and other stakeholders who engage in advocacy identify with the neurodiversity movement and/or neurodiversity theoretical perspective and aligned approaches. Although there is not a single, agreed-upon definition of the term *neurodiversity*, it can reasonably be said that proponents of neurodiversity approaches consider ASD and other neurodevelopmental disabilities, such as ADHD, dyslexia, and developmental coordination disorder (DCD), to be neurological differences rather than disorders representing pathology [37,38]. The neurodiversity framework tends to favor social models of disability that explain disablement and distress in terms of societal barriers, as opposed to individual medical problems [37]. Therefore proponents reject the medical model in which disorders, such as ASD, are defined by clusters of symptoms representing specific deficits or impairments, and especially the idea that these symptoms represent targets for intervention [37,39]. From the neurodiversity perspective, interventions to correct or ameliorate core deficits and reduce autistic traits are not desirable or appropriate; instead, the focus should be on promoting societal education, acceptance, and accommodations. Proponents who do accept a role for some interventions directed at the individual rather than just society (e.g., teaching adaptive skills) assert that intervention goals for children with ASD should be based on autistic rather than neurotypical behavior, so the goals must not only be developmentally and functionally relevant but also congruent with the development of an "autistic way of being"

[37,39] Chapman and Botha [38] proposed the development of neurodivergence-informed therapy, described as "therapy that resists default normalization, is sensitive to neurodivergent perspectives, understands disablement as relational and political, and considers disability as a potential source of community and pride."

Neurodiversity proponents and other advocates have been a driving force behind efforts to promote acceptance and societal changes that advance the rights and welfare of neurologically atypical people, especially those with disability. The neurodiversity perspective has been important in the evolution of understanding, for example, that social challenges do not originate solely with an individual with ASD but are relational, meaning that the interactional partner shares responsibility for preventing and remedying misunderstandings and social rejection [39]. The preference for identity-first language (e.g., "autistic people") rather than person-first language (e.g., "people with autism") has been well-received by many, but calls to eliminate the use of terminology such as disorder, symptom, impairment, deficit, risk, intervention, and treatment are more controversial [40–42]. There is also ongoing controversy about how well individuals with the highest support needs, including those with substantial ID and limited functional communication, are represented by a nonmedical model [3].

Epidemiology

There has been a marked increase in the number and complexity of epidemiological studies reporting ASD prevalence estimates in recent decades. However, methodological differences, including variable case definitions, ascertainment strategies, and source population considerations (e.g., age, socioeconomic, and race/ethnicity differences), make it difficult to interpret and compare published ASD prevalence data [43–47]. Most studies use datasets derived from existing registries or administrative databases (e.g., health insurance or educational databases), cross-sectional surveys performed at one point in time in a particular population, or active public health surveillance programs. Data sources vary, even within cross-sectional surveys and surveillance programs. Some contemporary surveys include large populations, multiple sites, and stratified samples, and employ intricate sets of screening procedures followed by some form of diagnostic confirmation, whereas others rely on parent report of ASD diagnosis [43–45]. Epidemiological surveys that rely on passive ascertainment of clinically diagnosed individuals, such as large national registries or administrative databases, tend to underestimate ASD prevalence, whereas studies that utilize active case ascertainment, especially those that rely solely on parent report of ASD diagnosis, tend to provide overestimates [43,44].

Prevalence

The population prevalence of ASD is currently estimated to be approximately 1%–3%, based on studies from North America and Europe, primarily [45,48,49]. A rigorous review of 31 recent epidemiological surveys from high-income countries reported a median ASD prevalence of 1.14% in school-age children [45]. This review focused on cross-sectional, population-based studies of more than 5000 individuals and excluded surveys relying exclusively on parent report of ASD diagnosis. The male preponderance in ASD is widely recognized; large reviews and metaanalyses describe an average male-to-female (M/F) sex ratio of 4:1 to 4:2 [44–46,50].

However, there is substantial variation in ASD prevalence even among high socioeconomic index countries. For example, within the multisite, population registry-based Autism Spectrum Disorders in the European Union (ASDEU) study, the prevalence of ASD among 7- to 9-year-old children in 2015 ranged from 0.48% in South-East France to 3.13% in Iceland, with intermediate rates in South-West France (0.73%), Finland (0.77%), and Denmark (1.26%) [51]. ASD prevalence estimation in the ASDEU study seemed to be very sensitive to local clinical practice effects and investment in diagnostic training. In Europe, higher prevalence of ASD is often found among racial/ethnic minorities and those with immigrant backgrounds. For example, geographic and racial/ethnic variation were prominent in a large case-control prevalence study that used 2017 English National Pupil Database and reported an ASD prevalence estimate of 1.76% among school children aged 5–19 years [52]. In this study, racial/ethnic differences in ASD prevalence were found to be mediated by socioeconomic disadvantage.

ASD prevalence is monitored longitudinally by several public health surveillance programs in the United States (US). The longest running and most well-established public health surveillance program is the Centers for Disease Control and Prevention (CDC) Autism and Developmental Disabilities Monitoring (ADDM) Network, a multisite, multiple-source, record-based surveillance system that includes systematic screening of records related to health care, service provision for developmental disabilities, and special education services within the chosen geographic areas. Information extracted from the child's records is then used to evaluate whether criteria for ASD are met based on DSM-based criteria. The most recent ADDM Network surveys yielded prevalence estimates of 2.8% among 8-year-olds and 2.2% among 4-year-olds in 2020, with M/F ratios of 3.8:1 and 3.1:1, respectively [49,53]. Rates of parent-reported lifetime diagnosis of ASD in children aged 3–17 years in recent national health surveys were 2.7% in the 2020 National Survey of Children's Health and 3.5% in the much smaller National Health Interview Survey [54,55]. Although these parent surveys have the advantage of utilizing large, nationally representative samples, they lack diagnostic confirmation by a professional.

The prevalence and characteristics of profound autism were described for the first time in early 2023, based on ADDM Network surveillance data from surveillance years 2000–2010 and 2016. The 2012 and 2014 surveillance years could not be included because information on nonverbal status was not captured. Among 20,135 children with ASD at age 8 years, 26.7% were classified as having profound autism because they had an overall IQ score or IQ subdomain score of <50 or were nonverbal or minimally verbal [56]. Lower adaptive scores, seizures, self-injurious behavior, prematurity or low birth weight, female sex, racial and ethnic minority background, and low socioeconomic status (SES) were overrepresented in the profound autism subgroup of children with ASD. In the 2016 surveillance year, in which the overall ASD prevalence was 1.85% [57], the prevalence of profound autism was 0.46% [56].

Until recently, racial and ethnic disparities in ASD prevalence in the US have tended to be in the opposite direction of the situation in Europe—racial/ethnic minorities and those with lower SES tended to have lower rates of ASD in the US [58,59]. However, in ADDM Network surveillance years 2016 and 2018, there was no overall difference in ASD prevalence between non-Hispanic White, non-Hispanic Black, and Hispanic 8-year-olds, and in 2020 the prevalence in non-Hispanic White children was lower [49,57,60]. This reversal in the pattern of association of race/ethnicity and the prevalence of ASD in the US has also been demonstrated using California administrative data [61].

Time trends

There has been a substantial increase in the prevalence of ASD over time in high-income countries, even when age-standardized estimates are considered [43–47,62]. For example, ADDM Network estimates of the prevalence of ASD among 8-year-old children in the US have increased gradually from 0.67% in surveillance years 2000 and 2002 to 2.3% in 2018 and 2.8% in 2020 [49,60]. Interpretation of temporal trends within ADDM Network data is complicated by changes in participating sites/catchment areas, case definitions, and diagnostic practices, however [60]. It is also clear that as the prevalence of ASD has increased, the proportion of individuals with known concurrent ID has decreased [44,45,63]. Among children with ASD and available IQ data in the ADDM Network, the proportion with IQ ≤ 70 declined from approximately one half in 2000 and 2002 to just over one-third in 2016, 2018, and 2020 (33%, 35.2%, and 37.9%, respectively). However, it should be noted that even in 2018 and 2020, cognitive data were available for only 60% and 67% of the children with ASD, respectively [49,60].

The sex ratio of approximately four males to one female has been a robust characteristic of ASD across epidemiological and clinical study populations. There is conflicting information about whether there has been a decline in the M/F ratio over time. For example, evidence of a decrease in M/F ratio between 1990 and 2019 was reported in the Global Burden of Diseases, Injuries, and Risk Factors Study 2019 [47], whereas the recent comprehensive reviews by Fombonne and colleagues did not find epidemiological evidence that the sex ratio has changed or that the magnitude of male preponderance has been overestimated [44,45]. Assertions that females are underdiagnosed, the true M/F ratio is significantly less than 4:1, and that this is due, at least in part, to female "camouflaging" (i.e., masking or compensating for their autistic characteristics as an attempt to fit in and cope) are controversial and require further investigation [44,50,64–67].

Potential reasons for increasing prevalence

Nonetiologic factors, such as variation in study methods and changes in public and professional awareness, screening and diagnostic practices, and availability of services, have contributed to the temporal increase in ASD prevalence estimates [43,45,46,68]. As case definitions and ascertainment strategies have evolved, ASD has been increasingly identified in previously underdiagnosed populations defined by sex, geography, race/ethnicity, or SES.

Although there is some evidence that changes in ICD and DSM diagnostic criteria have contributed, they were not the primary driver of the widespread increase in ASD prevalence estimates, especially in recent decades [68–70]. In fact, the transition from DSM-IV-TR to DSM-5 was expected to result in a decrease in ASD prevalence. Studies comparing the two sets of criteria suggest approximately a 20% reduction in ASD diagnoses when the DMS-5 criteria were used, with fewer than one-third of those who met DSM-IV-TR criteria but not DSM-5 criteria being captured instead by SCD [34]. However, the prevalence of ASD has not declined since the DSM-5 was published in 2013, it has continued to increase [45,49,60]. There is also evidence of diagnostic substitution, or diagnostic migration, which is a temporal shift in diagnostic category assignment rather than an increase in disability overall, particularly from studies using special education system administrative data, which were not intended for use as a public health surveillance tool [71,72].

Perhaps most importantly, there has been a temporal change over the last two decades in the threshold at which clinicians make a diagnosis of ASD, with a shift toward diagnosing individuals with fewer and less-severe symptoms. Significant decreases in the number and severity of ASD symptoms and in Autism Diagnostic Observation Schedule (ADOS) total scores have been observed in clinical populations [73–75]. For example, in the Child and Adolescent Twin Study in Sweden, the mean number of ASD symptoms on the Autism-Tics, ADHD, and other comorbidities inventory for individuals diagnosed at age 7–12 years decreased by 50% from 2004 to 2014 [74]. Furthermore, effect sizes obtained from ASD versus control group comparison studies on emotion recognition, theory of mind, and executive function tasks as well as P300 (P3b) event-related potential amplitude and brain size have decreased significantly (relative decrease in mean effect size of 45%–80%, $P < .05$) [76]. These decreases over time in the magnitude of differences between individuals with ASD and controls suggest that changes in diagnostic practices may have contributed substantially to the observed increase in prevalence. Additional evidence of temporal change in diagnostic threshold comes from retrospective epidemiological studies in Sweden and the US showing stability of ASD symptoms despite an increase in clinical diagnosis of ASD during the same period [77,78].

The temporal change in diagnostic threshold accompanying the increase in ASD prevalence could potentially be interpreted as evidence of improved recognition of the milder end of the spectrum or as evidence of overdiagnosis. It seems likely that both are involved to some degree, although overdiagnosis is difficult to define and the possibility of widespread overdiagnosis has not received much attention in the literature. However, Fombonne and colleagues have reported that among 232 children and adolescents with a community diagnosis of ASD who were referred for a neuroimaging research study, only 47% actually met criteria for ASD after an extensive diagnostic re-evaluation process [79,80]. Another study reported a lower, but still substantial, rate of community overdiagnosis of ASD. Hausman-Kedem and colleagues [81] evaluated 87 individuals with a DSM-IV-based community diagnosis of ASD (mean age 7.6 years, range 2.2–25.9 years), and 23% did not meet ASD criteria based on consensus diagnoses made using standardized tools, including the ADOS-2 and measures of cognition and adaptive functioning, plus clinical observation. Many clinicians working in referral centers have made similar observations (Challman and Myers, unpublished observations).

It is not clear whether there has been a true secular increase in ASD due to etiologic factors in addition to the contributions of nonetiologic factors. Some biological factors associated with ASD risk, such as advanced maternal age, hypertension during pregnancy, and maternal obesity, have also increased in prevalence in several parts of the world, making it plausible that such factors could have influenced change in ASD prevalence [47]. Similarly, an increase in environmental risk factors such as chemical exposures cannot be ruled out. However, the lack of change over time in the proportion of variance in heritability explained by nonshared environment suggests that it is unlikely that environmental factors can explain the rising prevalence of ASD, at least in Sweden [82]. Research aimed at understanding the role of environmental exposures in the etiology of ASD is complex and still in its infancy and, to date, there is very limited evidence that differences in certain environmental exposures may be related to temporal or geographic variation in prevalence [83–85].

Etiology and pathobiology

The last two decades have brought tremendous progress in our understanding of the complex genetic architecture of ASD. Despite remarkable progress with gene discovery, there are major challenges to understanding the biological mechanisms underlying ASD, including substantial phenotypic and genetic locus heterogeneity and overlap with other neurodevelopmental and psychiatric disorders. Because ASD is clinically and etiologically heterogeneous, it is very unlikely that a single unifying pathobiology will be identified for the disorder as a whole or even for its core behavioral components. However, there is emerging evidence from convergent neuroscience approaches that multiple implicated rare genetic variants converge on a smaller number of highly interconnected core molecular pathways, and several neuropathological abnormalities have been reported in postmortem brain tissue with some consistency, including altered size of neuronal cell bodies, changes in neuronal density, and abnormal distribution of neurons during development. Investigators strive to clarify the role of prenatal and postnatal environmental factors in ASD risk through increasingly sophisticated epidemiological studies and statistical analyses, identify valid and reliable biomarkers, and understand the role of epigenetic mechanisms. This work is driven by the promise that elucidating the genetic and other etiologies will lead toward an understanding of the pathobiological mechanisms and ultimately enable development of targeted treatments and prevention strategies.

Genetics and genomics

Although clinically and etiologically heterogeneous, ASD is strongly genetic. Twin and family studies have demonstrated that the heritability of ASD is high (~70%–90%), risk increases with increasing genetic relatedness, and shared environmental factors contribute minimally [82,86–88]. Rapid progress has been made in discovery and delineation of the genetic architecture of ASD in recent years, largely due to (1) widespread use of new high-throughput, cost-effective, and genome-wide testing methods; (2) successful patient recruitment and prioritization of collaborative projects and rapid data sharing, leading to assembly of very large data sets; and (3) advances in genomic data processing and statistical analytic methods [89–91]. Broadly, the two major components of the genetic architecture of ASD identified to date are rare deleterious variants of large effect size [92–94] and polygenic risk conferred by additive effects of a large number of common variants of individually very small effect size but major collective importance [94–96]. Emerging evidence suggests that rare and common variation act additively to create ASD risk [97–99]. Genetic influences on ASD risk—whether common or rare, inherited or de novo—also influence typical variation in social and communication ability in the general population [100]. At the population level, common variation is currently thought to play the dominant role in ASD liability, whereas a rare pathogenic variant may make the largest contribution to liability for an individual [98,101]. Despite the remarkable progress, important aspects of the genetics of ASD are still poorly understood, including the role of epistasis (gene–gene interactions) and environmental modification of genotype effects.

Rare variation

In some cases, individuals with ASD have clinically recognizable, monogenic syndromes such as fragile X syndrome, tuberous sclerosis complex, Smith-Lemli-Opitz syndrome, or *PTEN* hamartoma tumor syndrome. Similarly, some individuals with aneuploidy (e.g., Down syndrome, Turner syndrome, Klinefelter syndrome) or large, microscopically visible deletions or duplications meet criteria for ASD [102–105]. These conditions are typically suspected clinically and confirmed by laboratory testing. More frequently, pathogenic rare variants of large effect size are identified by genome-wide clinical or research testing, especially chromosomal microarray analysis (CMA), exome sequencing (ES; also referred to as whole exome sequencing, WES), or genome sequencing (GS; also referred to as whole genome sequencing, WGS). Large-effect rare single-gene variants implicated in ASD to date occur primarily within the coding portions of the genome and are most often de novo, although they may also be inherited in a dominant or recessive fashion [92,94,99,106,107]. Recent rare variant discovery efforts that include analyses of both de novo and inherited coding variants have also begun to identify genes associated with ASD risk of moderate effect size [107]. Sequence-level variants, including single-nucleotide variants (SNVs) and short insertions and deletions (indels), are the most commonly identified large-effect rare variants identified, followed by copy number variants (CNVs), which are deletions or duplications of longer DNA segments [93,94]. Approximately a dozen high-confidence CNVs and more than 100 high-confidence ASD genes enriched with de novo protein truncating variants or damaging missense variants have been identified to date [92–94,108]. Examples of clinical genetic syndromes and high-confidence ASD-associated genes and CNVs are provided in Table 28.2 [91,93,94,103,108]. In addition to CNVs, other types of structural variants (e.g., inversions, larger insertions, uniparental disomy, and tandem repeat expansions) contribute to the genetic architecture of ASD [94,109]. Evidence of enrichment of rare noncoding variants in association with ASD is quite limited currently, even among studies focusing on putatively functional elements of the noncoding genome, but this is likely to change as WGS studies proliferate [91,94,99,110]. The contribution of mitochondrial DNA variants appears to be small based on available evidence, which is currently limited or conflicting [94]. In addition to the various types of germline variants, pathogenic somatic (mosaic) variants, including SNVs and CNVs detectable in blood or saliva, contribute to ASD etiology in approximately 3%–5% of simplex families [111–113]. Cases of tissue-specific causal variants have also been described in postmortem brain tissue [113,114].

Common variation

Common variant-associated risk is conferred by inherited and de novo single nucleotide polymorphisms (SNPs), mostly within noncoding regions of the genome, which are discovered primarily in genome-wide association

TABLE 28.2 Genetic etiologies: examples of ASD-associated clinical syndromes, genes, and CNVs.

Clinical genetic syndromes (implicated genes)	High-confidence ASD-associated genes and CNVs		
	Genes		Recurrent CNVs
Often suspected clinically and confirmed by targeted genetic testing, but may also be found on genome-wide testing (ES or GS)[a]	SNVs and small insertions/deletions identified by ES or GS[b]		Recurrent deletions and duplications identified by CMA, ES with CNV calling, or GS
Angelman syndrome (*UBE3A*)	*ADNP*	*GRIN2B*	1q21.1 duplication
CHARGE syndrome (*CHD7*)	*ANK2*	*KDM6B*	3q29 deletion
Down syndrome/Trisomy 21	*ANKRD11*	*KMT5B*	7q11.23 duplication
Fragile X syndrome (*FMR1*)	*ARID1B*	*NRXN1*	15q11.2q13.1 (BP1-BP3) duplication
Klinefelter syndrome (47, XXY)	*ASH1L*	*MED13L*	15q13.2q13.3 (BP4-BP5) deletion
MECP2 duplication syndrome (*MECP2*)	*CHD2*	*POGZ*	16p11.2 proximal (BP4-BP5) deletion
Neurofibromatosis Type 1 (*NF1*)	*CHD8*	*SCN2A*	16p11.2 proximal (BP4-BP5) duplication
PTEN hamartoma tumor syndrome (*PTEN*)	*CTNNB1*	*SETD5*	16p11.2 distal (BP2-BP3) deletion
Rett syndrome (*MECP2*)	*DEAF1*	*SHANK2*	17q12 deletion
Smith-Lemli-Opitz (*DHCR7*)	*DSCAM*	*SHANK3*	17q12 duplication
Timothy syndrome (*CACNA1C*)	*DYRK1A*	*STXBP1*	22q11.2deletion
Tuberous sclerosis (*TSC1, TSC2*)	*FOXP1*	*SYNGAP1*	22q11.2 duplication
Turner syndrome (45,X)	*GIGYF1*	*WDFY3*	22q13.33 deletion

ASD, Autism spectrum disorder; *CMA*, chromosomal microarray analysis; *CNV*, copy number variant; *ES*, exome sequencing; *GS*, genome sequencing; *SNV*, single nucleotide variant.
[a]*Fragile X syndrome could be detected on GS, but not CMA or ES; Down syndrome and sex chromosome aneuploidies could be detected on CMA, ES, or GS.*
[b]*In the case of* NRXN1, *a very large gene (1.1 Mb, 24 exons), large intragenic deletions may be detected by CMA.*
Source: Searles Quick et al. (2021) [91]; Satterstrom et al. (2020) [92]; Fu et al. (2022) [93]; Trost et al. (2022) [94]; Richards et al. (2015) [102]; Sztainberg and Zoghbi (2016) [103]; Sanders et al. (2015) [108].

studies (GWAS) of large research populations and often summarized as a polygenic risk score (PRS) [96,97,99,115]. The largest ASD GWAS metaanalysis published to date, which included over 18,000 individuals with ASD and nearly 28,000 controls, identified for the first time five common variant loci that met the statistical standard for genome-wide significance [96]. In addition to identifying genome-wide significance of individual loci, GWAS data can also be leveraged to compute an ASD PRS. A PRS is a quantifiable genetic risk score that summarizes the cumulative impact of thousands of SNPs, each weighted by the effect size of its association with a target trait derived from GWAS [116]. Although they cannot predict the clinical outcome of interest with precision at the individual patient level, potential uses of PRS include stratifying risk to facilitate earlier diagnosis (especially when combined with other risk information for an individual), allowing more nuanced assessment of recurrence risk, forecasting disease trajectory, and informing pharmacological and other interventions. Studies have shown clinical validity in some conditions for which PRS-based stratification has the potential to improve patient outcomes, but unequivocal clinical predictive utility has not yet been demonstrated [117]. At this time, the difference in mean PRS between groups of individuals with and without ASD is small, and even a top decile ASD PRS is associated with only a modest increase in ASD odds relative to a bottom decile score (OR $\sim$1.5–2.8) [94,96]. Despite the current limitations, ASD PRS is likely to become an important component of multifactorial models of ASD risk in the future [91,94,115].

Etiologic heterogeneity and variable expressivity

It has become clear that genetic risk associated with neurodevelopmental and psychiatric diagnoses, including ASD, is not strictly disorder specific. In fact, each categorical clinical diagnosis is associated with multiple genetic variants (etiologic heterogeneity), and each implicated variant is associated with multiple categorical diagnoses (variable expressivity). The latter is often referred to as pleiotropy, but we prefer the term variable expressivity to describe the variation in related cognitive and behavioral phenotypes (e.g., neurodevelopmental and psychiatric diagnoses) associated with a single genetic variant or gene [118,119]. Here, we reserve the term pleiotropy for the association of a variant or gene with seemingly unrelated phenotypes, often involving more than one organ system (e.g., ASD and congenital heart defects). Rare pathogenic variants in the same genes have been identified across diagnostic categories including ASD, ID, epilepsy, ADHD, schizophrenia, OCD, and others [92,93,120–128]. Polygenic risk is also shared across neurodevelopmental and psychiatric disorders and traits [96,124,129–134] and has been demonstrated to affect phenotypic outcomes even in individuals who carry large-effect rare variants [135–138].

No rare sequence level variants or CNVs that confer only ASD risk have been identified [139–141]. However, several studies have reported relative ASD specificity, or predominance of the ASD phenotype, for certain genes based on comparison of the distribution of likely gene-disruptive de novo variants in cohorts ascertained for ASD to those ascertained for ID/severe neurodevelopmental disorder phenotypes [92,93]. Other studies using a similar approach in terms of comparison groups, but different analytic strategies, concluded that there was no statistically significant evidence of relative ASD specificity [141,142]. Though convenient, this approach has been criticized for several reasons, including potential bias introduced by unequal opportunity for each participant to receive each diagnosis (i.e., ASD and ID) due to the lack of uniform phenotyping across studies [139]. It is certainly possible that for a given effect size of deleterious impact on IQ, rare variants in particular genes may have a substantially greater or lesser effect size on core ASD features such as social communication and interaction [118,139]. However, it is also possible that among individuals whose developmental brain dysfunction is primarily attributable to a rare variant of large effect size, factors such as background polygenic risk conferred by common variants, additional rare variants of more modest effect size ("second hits"), and stochastic developmental variation may be more important determinants of the specific neurodevelopmental phenotypes expressed than the gene(s) involved in the primary genetic etiology [139,143,144]. Large-scale studies that cross diagnostic boundaries and include adequate phenotyping of all affected individuals will be required to clarify these issues.

Parental age effects

Advanced maternal and paternal age are both associated with an increased risk of ASD in the offspring. A dose-response metaanalysis indicated that an increase of 10 years in maternal age was associated with an 18% higher risk of ASD and an increase of 10 years in paternal age was associated with a 21% higher risk of ASD [145]. Genomic mechanisms involved in parental age effects on ASD risk in offspring include (1) accumulation of de novo pathogenic rare variants with age in the paternal germline, (2) inherited rare variants that directly contribute to parental age behavior in fathers, and (3) a polygenic (common variant) mechanism that influences parental age in both mothers and fathers, mediated largely by educational attainment polygenic scores [115,146]. In addition to genomic mechanisms, ASD risk associated with older maternal age may also be affected by age-related increased risk of environmental factors such as obesity, gestational diabetes mellitus (DM), preterm delivery, low birth weight, autoimmune disease, and possibly increased opportunity for exposure to pollutants or medications [145,147,148].

From genes to pathobiology

Genetic etiologic research is guided by the promise that identifying the genetic and genomic causes will lead to a mechanistic understanding of the pathobiology of ASD which, in turn, will drive tangible advances in clinical care including not only development of targeted therapies but also improvements in diagnostic approaches and ability to predict natural history and clinical course [91,149]. Advances in gene discovery provide a foundation of knowledge that forms the basis for efforts to elucidate the pathobiological mechanisms underlying ASD and identify molecular, cellular, or circuit level targets for intervention [90,149–151]. However, translating the remarkable progress in gene discovery to an understanding of the pathobiology of ASD remains largely unrealized, and mapping biological pathways onto specific dimensions of human cognition and behavior is a daunting task [90,149–151].

The number of implicated genes far exceeds the number of clinically defined phenotypes, suggesting that multiple rare genetic variants are likely to converge on a smaller number of shared causal pathways at some point on the path from genes to behavior [149,152,153]. Convergent neuroscience approaches, rooted in functional genomics and systems biology, seek to identify intersections among molecular, cellular, and circuit level functions, anatomical brain regions, and developmental time periods across multiple ASD-associated genes [149,152]. Complementary methodologies are used to analyze high-confidence ASD genes, including gene ontology and gene set enrichment analyses and comparisons to various pathway databases; comparisons to spatiotemporally defined gene coexpression networks derived from normal human brain tissue (compiled from early gestation to late adulthood) and maps of interactions between proteins and/or genes; comparisons of postmortem brain tissue transcriptome data from individuals with ASD to matched controls; functional and transcriptional profiling in patient-derived model systems such as induced pluripotent stem cell (iPSC)-derived monolayer models and brain organoids; and in vitro and in vivo investigations of the effects of perturbation of multiple ASD risk genes in parallel using clustered regularly interspaced short palindromic repeats (CRISPR)-

based technology [152]. Gene expression analyses have been completed at the level of bulk tissue from major brain regions and more finely dissected cortical areas and also at the level of single cells, allowing identification of convergence on specific anatomical locations and cell types [152]. Together, these studies provide evidence for convergence of a larger number of ASD risk genes on a smaller number of biological pathways, cell types, brain regions, and developmental stages, although it is not yet clear whether the specific profiles of convergence are distinct from those involved in other neurodevelopmental disorders.

The available data provide evidence that ASD risk genes converge on biological pathways related to regulation of gene expression (including chromatin modification and transcription regulation, impacting key signaling pathways such as PI3K/AKT/mTOR, ERK/MAPK, and WNT/β-catenin) and synaptic structure and function (including ion channels, receptors, scaffolding proteins, and cytoskeletal and vesicle machinery), causing dysregulation of processes including cell proliferation, neurogenesis, cell fate determination, migration, neurite outgrowth, synaptogenesis, and synaptic wiring [92,152,154–157]. Nongenetic risk factors such as viral infection or immune activation and exposure to valproic acid may converge on some of the same signaling pathways [154,156,158]. In their synthesis of the data, Courchesne and colleagues propose a multistage process of ASD pathobiology involving two epochs of gene expression that together span much of prenatal and early postnatal neural development and many brain-specific mechanisms and functions [154,155]. According to this model, high-confidence ASD genes underlie the regulation of core prenatal processes. In trimesters one to three, variants in one set of ASD risk genes (mostly broadly expressed regulatory genes, but also some brain-specific genes) disrupt cell proliferation, neurogenesis, migration, and cell fate, while in trimester three and early postnatal life variants in another set of mostly brain-specific genes disrupts neurite outgrowth, synaptogenesis, and the "wiring" of the cortex [155].

Hypothesis-free system biological approaches initially identified developing excitatory neurons in human midgestational prefrontal cortex as an important pathobiological nexus in ASD; subsequently, other cortical areas and cell types (including inhibitory neurons and nonneuronal cells) have been implicated and convergent transcriptional changes have been demonstrated among cell types with shared developmental lineages in postmortem brain tissue from adults with ASD [90,92,152,159,160]. Recent studies have identified more widespread gene and transcript expression changes across frontal, temporal, parietal, and occipital cerebral cortex, affecting many neural cell types and specific biological processes and extending beyond higher-order association areas to include primary sensory areas [161]. In the neurotypical brain, cortical regions can be distinguished on the basis of differences in gene expression (reflecting variability in the cytoarchitecture, connectivity, and function of each region); however, there is marked attenuation of these typical gene-expression differences among individuals with ASD [161]. Rare and common variant enrichment point to cortex-wide downregulation of genes involved in synaptic signaling plasticity, blood–brain barrier membrane transport, and neurite morphogenesis and regionally pronounced downregulation of neuronal energy processes and oligodendrocyte organelle regulation in posterior regions, especially primary visual cortex [161]. ASD-associated common genetic variation has also been observed to be enriched in an upregulated coexpression module comprising genes involved in proteostasis, cytokine signaling, and reactive astrocytes throughout the cortex, with pronounced upregulation impacting reactive microglia and neuronal signal transduction in primary visual cortex [161].

Efforts are underway to take advantage of gene discovery directly by developing novel therapies, including gene replacement, gene editing, and antisense oligonucleotide strategies for large-effect rare loss-of-function variants in certain ASD-associated genes [162,163]. Identifying points of convergence has the potential to support development of more traditional therapeutics by identifying biological mechanisms and targets in common pathways that are downstream of individual genes and, therefore, applicable to more than one gene at a time [152,163,164].

Neuropathology

Neuropathological research may help to bridge the gap between molecular abnormalities and clinical phenotypes but has been limited by the small number of postmortem brains available for study. The existing literature must be interpreted with caution not only because of the small sample sizes but also because of potential bias related to factors such as the paucity of studies focused on the whole brain (most focus on one or more specific regions), availability of predominantly adult brain tissue, manner of death (including frequent hypoxia/ischemia and reperfusion injury), postmortem interval, brain bank methodologies, cooccurring conditions (e.g., high rates of ID and epilepsy), and medication effects [165–167]. It is difficult to establish causality and to exclude, for example, the possibility that observed changes are secondary consequences of ASD or cooccurring conditions such as epilepsy or drug therapy, or some combination of these factors [165,166].

Abnormalities in neuronal morphology and cytoarchitectural organization have been described in a variety of brain regions, including the prefrontal cortex, fronto-insular cortex, cingulate cortex, fusiform gyrus, hippocampus, amygdala, and cerebellum [165,168,169]. The vast majority of abnormalities described originate prenatally. Some seem to be present in postmortem brain tissue from young children with ASD, but not in samples from older children or adults, suggesting abnormal developmental trajectory. Several studies from one laboratory group have found that ASD is consistently associated with decreased cortical minicolumn width and increased number of minicolumns per cortical area, most prominently in the dorsolateral prefrontal cortex [165,168,170]. These findings await independent replication. Although patches of abnormal laminar cytoarchitecture and cortical disorganization have been identified in prefrontal and superior temporal cortical neurons based on reduced or abnormal marker expression [171], cortical layering was largely undisturbed in most studies and there is little overall evidence of dysregulated layer formation in ASD [165]. Other studies have reported axonal abnormalities, including an excess of thin axons, which form short-range pathways, and absence of the normal increase from childhood to adulthood in thick axons, which form long-range pathways [165,172,173]. Increased density of dendritic spines has also been reported in several cortical regions and the amygdala in childhood [165,174]. Findings such as increased number of neurons, decreased neuronal size, and increased neuropil in certain regions are not consistent and it is difficult to draw firm conclusions [165]. The most consistent cerebellar finding is decreased size and number of Purkinje neurons, which was initially thought to be of prenatal onset but may actually be an acquired process that occurs postnatally, possibly related to seizures, medications, and/or ischemia near the time of death or even an artifact of postmortem factors rather than being part of the core pathology of ASD [167]. Most, but not all, studies examining markers of neuroinflammation in the postmortem brains of people with ASD have found abnormalities such as activation of astroglia and microglia and, in some cases, alterations in levels of cytokines and chemokines [166]. Microglial densities normally decrease throughout childhood and early adulthood in neurotypical subjects, but this seems not to be the case among individuals with ASD.

Environmental risk

In addition to genetic and genomic variation, there is evidence that some environmental factors may be associated with elevated risk of ASD [84,147,148,175,176]. These include advanced parental age, prenatal exposures (e.g., medications, toxins), maternal factors (e.g., cardiometabolic pregnancy complications, short interpregnancy interval), antenatal infection/maternal immune activation, and perinatal insults (e.g., preterm birth, low birth weight, neonatal encephalopathy, perinatal ischemic stroke). However, many of these are confounded by genetics, as previously discussed for parental age effects [115,145,177,178]. Importantly, epidemiological research has also shown that certain factors are *not* associated with increased ASD risk; these include vaccines given to mothers during pregnancy, exposure to epidural analgesia during labor, delivery by cesarean section, and childhood vaccines, among others [84,147,148,179]. Interestingly, neither maternal smoking nor mild-moderate alcohol consumption during pregnancy appear to be associated with ASD risk in offspring, despite their association with other neurodevelopmental disorders [84,180,181].

Prenatal exposures

ASD risk is significantly elevated in children born to women treated with the antiseizure and mood stabilizing medication, valproic acid [182–185]. The pooled results of case-control studies and cohort studies also suggest a significant association between selective serotonin reuptake inhibitor (SSRI) exposure during pregnancy and risk of ASD in children [186–189]. However, a metaanalysis of cohort studies showed that the presence of a maternal psychiatric disorder without SSRI exposure was also associated with ASD risk in the offspring, with a similar effect size and largely overlapping confidence interval [186], corroborating an earlier metaanalysis in which the association between SSRI use and increased ASD risk did not persist when analysis was confined to children born to mothers with psychiatric disorders [188]. These finding suggest that maternal psychiatric disorder is likely to be a strong confounder. Because of likely confounding by indication, the importance of the association between SSRI exposure during pregnancy and ASD is not yet clear, and any potential role of SSRIs (and other antidepressant medications) in the pathogenesis of ASD remains speculative [186–189].

A potential link between acetaminophen and ASD is another active topic of investigation. Several systematic reviews and metaanalyses of epidemiological studies have reported an association between prenatal exposure to acetaminophen and both ASD and ADHD in childhood [190–192]. A metaanalysis of six European population-based birth cohorts including almost 74,000 mother-child pairs found that children prenatally exposed to acetaminophen were 19% more likely to have ASD symptoms within the clinical or borderline range compared to nonexposed children [191].

Although potential indications for acetaminophen use (e.g., maternal fever or infection during pregnancy, maternal chronic illnesses) were included as covariates, confounding by indication could not be ruled out. One study found cord plasma acetaminophen metabolite measurements to be associated with significantly increased risk of ASD and ADHD in a dose-response fashion, providing some support for the findings of studies based on historical exposure, but the validity has been questioned because metabolites were detected in 100% of samples and an unusually large proportion of children were diagnosed with ASD or ADHD (37%), whereas only 33% were classified as having no developmental disability [193,194]. Despite the considerable scientific and media attention the topic has received, the relationship between in utero acetaminophen exposure and ASD remains uncertain due to methodological issues that plague many of the published studies, such as measurement errors and potential confounding by indication and genetic factors [194,195]. In addition, when the broader relationship between prenatal acetaminophen exposure and neurodevelopmental outcomes is considered, the findings are conflicting and inconclusive. For example, a recent prospective study employing a battery of cognitive outcome measures found no significant association between maternal acetaminophen intake any time before or during pregnancy and adverse cognitive effects at age 4 years [196].

Concerns about environmental chemicals such as air pollutants, pesticides, plastics derivatives, and metals have also been investigated [84,176,197]. This is a challenging research to complete, since it is difficult to estimate the burden of real-life exposure to a wide variety of chemical substances, which occur in various combinations over different periods of time in people with different genetic backgrounds [84,176]. There is evidence supporting an association between prenatal and early postnatal exposure to the fine particulate matter component of ambient air pollution ($PM_{2.5}$) and ASD [198–201]. The available evidence is weak and inconsistent, however, for larger particulate matter and other pollutants, including nitrogen dioxide, ozone, polycyclic aromatic hydrocarbons, metals, solvents, styrene, and pesticides [84,148,198–200,202]. Diet is the main source of exposure to most pesticides in use today, and a study of mother-child pairs with increased familial probability of ASD did not find a relationship between pesticide residues and child ASD-related outcomes at the age of 3 years [202].

It is also possible that some prenatal exposures may reduce ASD risk. There is evidence, for example, of a protective effect of maternal folic acid supplementation in the prepregnancy period and during pregnancy [203,204], possibly due to amelioration of the impact of toxic exposures [197]. Several studies suggest that associations between ASD or ASD-related traits and prenatal exposure to pesticides, air pollutants, and phthalates are stronger when mothers did not receive folic acid supplementation or had higher folate requirements during pregnancy [197,205–207].

Maternal factors

In addition to older age at the time of pregnancy (especially age >35 years) intrinsic and extrinsic maternal factors including cardiometabolic status and infectious and immunological processes are associated with significantly increased probability of ASD in offspring [145,148,175,208,209]. Maternal cardiometabolic conditions associated with increased ASD risk in children include obesity/overweight before or during pregnancy, excessive gestational weight gain, type 1 and type 2 DM, gestational DM, chronic and gestational hypertension, and preeclampsia [148,175,210–219]. Findings regarding the use of assistive reproductive technology (ART) and ASD risk are inconclusive; the most recent metaanalysis found a significant overall association, with reduced effect size when controlling for confounders and no association when the analysis was limited to singletons [220]. Several studies have demonstrated a U-shaped association between interpregnancy interval (the time from the birth of one child to conception of the next child), with both short (<12 months) and long (≥60 months) intervals being associated with increased risk of ASD in offspring [221–224].

Maternal immune activation during pregnancy, indicated by infection, fever, elevated inflammatory markers, or autoimmune disease, is often reported to be associated at the population level with elevated risk of ASD in offspring [148,208,209,225–235]. Maternal infection has received much attention as a potentially preventable and treatable ASD risk factor. A causal relationship has not been established, however, and some studies have not found a significant association. Some specific maternal infections, such as influenza, are not associated with significantly increased ASD risk based on the best available information at this time [236]. However, the weight of the current evidence supports an association—a recent metaanalysis found a modest but statistically significant association of prenatal maternal infection/fever across all three trimesters with ASD risk in offspring [208]. An association between maternal infection during pregnancy and ASD was also observed at the population level in a very large Swedish cohort study [237]. However, maternal infections in the year before pregnancy (negative control exposures) were equally strongly associated with ASD, suggesting that the gestational infection itself was not responsible for the increased likelihood of ASD in the offspring [237]. In an analysis restricted to full sibling pairs (10,030 with ASD and 11,834 without ASD), the association between maternal infection during pregnancy and ASD was completely attenuated. Together, these findings suggest confounding of the associations by unmeasured familial factors, implying that, in general, maternal infections during pregnancy are not promising targets for ASD prevention strategies [237].

Perinatal and neonatal factors

The prevalence of ASD is significantly increased in association with exposure to perinatal complications (e.g., hypoxia/birth asphyxia, low Apgar scores, and neonatal encephalopathy), preterm birth, low birth weight, and being small for gestational age (birth weight < 10th percentile for gestational age), with the highest risk in those born extremely preterm (22–27 weeks) [147,238–245]. A large, Danish population-based cohort study found that children with ASD who were born preterm had elevated rates of common complications of prematurity such as bronchopulmonary dysplasia, intraventricular hemorrhage, necrotizing enterocolitis, retinopathy of prematurity, and neonatal seizures, and those born at term had elevated rates of sepsis and hypoglycemia [246]. However, there was no evidence of a distinctive neonatal morbidity profile in ASD compared to other neurodevelopmental disorders. There is also evidence of an association between postterm birth and ASD outcome, at least among males [247].

ASD risk is also significantly increased in association with perinatal (≤28 days of age) and early childhood ischemic stroke, and this is not explained by being born prematurely or small for gestational age, or by having a first-degree relative with ASD [248–250]. This risk is not entirely attributable to structural damage caused by the stroke; however, since there is also an increased risk of ASD in siblings of children with ischemic stroke who do not have ASD, suggesting that there are shared genetic risk factors between ischemic stroke and ASD [249].

Vaccines

Assertions that toxic effects of the measles-mumps-rubella (MMR) vaccine, mercury exposure from thimerosal-containing vaccines, or excessive burden on the immune system caused by a larger number of vaccines are responsible for the increase in prevalence of ASD have been soundly refuted by epidemiological studies, and the weight of all of the scientific evidence overwhelmingly favors rejection of the hypothesis that there is a causal association between childhood vaccines and ASD [251–261]. Similarly, there is no significant association between maternal Tdap (tetanus-diphtheria-acellular pertussis) or influenza vaccines given during pregnancy and ASD outcome in offspring [262–264]. Evidence supporting the hypothesis that vaccinations act as a "second hit," conferring ASD risk in genetically susceptible subgroups, is also lacking. Even among children at elevated risk due to having an affected older sibling, the MMR vaccine is not associated with increased risk of ASD [265].

Gene x environment interaction

As recognition that both genetic and environmental factors contribute to the risk of ASD and other neurodevelopmental and psychiatric disorders has increased, the need for studies of environmental exposures in the context of genetic risk has become apparent, and a limited number of epidemiological investigations of ASD-related gene-by-environment interactions (GxE) have been published [85,176,197,266–268]. These studies provide preliminary evidence that certain genetic variants, including SNPs and CNVs, increase susceptibility to certain environmental exposures, such as air pollution [269–271]. Identifying important GxE effects on risk and/or severity of ASD and also on cooccurring conditions will be a complex and difficult undertaking, but there is optimism about new opportunities made possible by increasing availability of prospective population-based birth cohorts and exposure-based cohorts, higher resolution environmental data, in vitro model systems such as human iPSC-derived neural systems for mechanistic and screening studies, and transgenic animal models to confirm GxE interactions identified in simpler systems [176,267,268,272]. It will be important for this line of research to avoid common pitfalls of candidate GxE research in psychiatry such as inadequate statistical power (and need for large sample sizes), high false discovery rates, and publication bias, and to strive to ensure accurate exposure measurement and account for gene–environment correlation [266,273].

Epigenetics

Molecular epigenetic mechanisms include (1) DNA methylation and posttranslational modifications of histones that act as epigenetic "marks" that control the activity of a gene locus and (2) chromatin organization and remodeling by proteins recruited by transcription factors [274–278]. Epigenetic changes do not alter the DNA sequence but have important influences on gene expression, tissue differentiation and development, and disease susceptibility [277–279].

The epigenome is largely shaped by DNA sequence variation and transcription, but also influenced by environmental exposure, providing buffering and helping to stabilize a particular profile of gene expression [277–280].

There has been considerable interest in epigenetics as a potential way to understand the role of gene–environment interactions in the pathogenesis of ASD, and numerous DNA methylation studies comparing ASD and control groups have been published, including investigations of specific targeted genes and genome-wide analyses performed on various postmortem brain tissue regions and peripheral tissues (e.g., blood, buccal cells, placental tissue, umbilical cord blood) [281–286]. Two main lines of evidence suggest that epigenetic mechanisms may be important in ASD: (1) Abnormal DNA methylation has been documented in patients with ASD who have pathogenic variants in genes that encode proteins involved in the epigenetic machinery, and (2) ASD is associated with altered DNA methylation at specific loci and broadly throughout the genome in the absence of any identified pathogenic change in the DNA sequence [281,283].

ASD is among the phenotypic manifestations of Mendelian disorders of the epigenetic machinery (MDEMs), which are a subset of a broader group of chromatinopathies caused by pathogenic germline variants in genes encoding components of the epigenetic machinery [287–289]. Epigenetic machinery genes encode enzymatic writers, erasers, and remodelers of chromatin marks and/or nonenzymatic readers which bind to specific chromatin modifications and architectural proteins [287,288]. Pathogenic variants in MDEM genes result in disruption of chromatin and transcription states at downstream target genes and global aberrant DNA methylation [288]. Although they disrupt the epigenetic machinery, these are germline DNA variants that are fully captured within the previous discussion of rare genetic variants of large effect size. The epigenomic signature associated with a MDEM (e.g., abnormal DNA methylation pattern) is not the cause of the disorder; it is a secondary event in the disease pathogenesis, downstream of the causative DNA variant [280,283].

Although many differences have been described, the evidence of epigenetic dysregulation in ASD is less robust in the absence of a pathogenic germline DNA variant affecting the epigenetic machinery, and causality has not been established [281,284,290]. Targeted gene-specific analyses, primarily using postmortem brain tissue, have identified abnormal methylation of promoter regions or transcription start sites for several genes in people with ASD compared to controls, although the sample sizes are small [165,281]. Epigenome-wide association studies evaluating the relationship between differentially methylated regions (DMRs) or differentially methylated positions (DMPs) in ASD cohorts and controls have also revealed evidence of epigenetic alterations in ASD-associated genes in brain, blood, and other tissues, but most studies have not identified DMRs meeting rigorous statistical significance [282,291–293]. For example, in a large analysis of DNA methylation in ASD, the authors found little evidence to support an association between DNA methylation at birth and ASD diagnosis in a cross-cohort metaanalysis, but they discovered that polygenic risk (ASD PRS) was associated with differential methylation at two sites on chromosome 8 and, when using a more relaxed "discovery" *P*-value threshold, 49 DMPs were identified [292]. Some other significant outcomes are reported, however. For example, DNA methylation signatures in cord blood differentiated groups with ASD from typically developing controls, and some specific regions were consistently differentially methylated [285].

Issues including technological and biological factors make the results of both individual candidate gene and genome-wide DNA methylation profiling studies difficult to interpret and preclude drawing definitive conclusions based on currently available data [281]. DNA methylation findings from peripheral (nonbrain) tissues must be interpreted with caution and require not only replication but also support from orthogonal methods [294].

Biomarkers

A biomarker is a defined molecular, histologic, radiographic, or physiologic characteristic that is measured as an indicator of a normal biological process, pathogenic process, or biological response to an exposure or intervention [295]. The US Food and Drug Administration (FDA) and the National Institutes of Health (NIH) jointly sponsor the online *Biomarkers, EndpointS, and other Tools* (BEST) resource, created by the FDA-NIH Biomarker Working Group to harmonize biomarker-related nomenclature. The BEST resource categorizes biomarkers based on their proposed utility: susceptibility/risk, diagnosis, monitoring, prognostication, prediction, response, or safety [295]. Identification of biomarkers that are valid, reliable, and usable at scale is an important step toward understanding heterogeneity in outcomes (including response to treatment); developing mechanism-based therapies; and implementing practices at the level of individual patients to allow more rapid diagnosis, earlier and more accurate matching to treatments with the highest likelihood of response, and improved prediction and prevention of risks such as cooccurring conditions and relapse [296,297].

A wide variety of proposed molecular/biochemical, neuroimaging, electrophysiological, and eye tracking, and other biomarkers have been studied in relation to ASD [298–307]. However, comprehensive systematic reviews of the literature have not identified any biomarkers with sufficient evidence for clinical use for diagnosis or monitoring or to inform clinical trials, and no biomarker has yet been qualified by the FDA Center for Drug Evaluation and Research Biomarker Qualification Program [148,308,309]. Fortunately, large-scale international efforts to identify and rigorously evaluate ASD-related biomarkers are underway, including the EU-AIMS Longitudinal European Autism Project [310] and the Autism Biomarkers Consortium for Clinical Trials (ABC-CT) [311,312].

The authors of a systematic review of response biomarkers to guide interventions in ASD noted that in most cases, biomarkers that were assessed in multiple studies were found to have both inconsistent evidence of association with ASD and variable direction of effect [309]. The biochemical/molecular biomarker literature in ASD, which includes markers of oxidative stress (e.g., decreased glutathione) and mitochondrial dysfunction, autoantibodies (e.g., maternal fetal brain directed, various other brain directed, and folate receptor alpha autoantibodies), cytokines, and various neurotransmitters, hormones, vitamins, and minerals, is particularly broad but shallow and difficult to interpret [148,308,309,313]. Even the extensive body of neuroimaging research has not yielded a valid and reproducible stratification biomarker for ASD, but there are some findings that merit additional discussion, as do some promising neurophysiological markers and recent advances involving eye tracking paradigms and multiple-task digital phenotyping.

Structural neuroimaging

Magnetic resonance imaging (MRI) studies employing voxel-based morphometry and diffusion tensor imaging (DTI) have most consistently reported that ASD is associated with age-related alterations in cortical surface area, thickness, and gyrification; brain lateralization; white matter connectivity; corpus callosum morphology; gray matter volume in the prefrontal and temporal cortex, cerebellum, amygdala, and caudate nucleus; and extra-axial cerebrospinal fluid (CSF) volume [314,315]. A metaanalysis of DTI studies found evidence of pervasive hypoconnectivity across language-associated tracts in people with ASD compared to controls, and subanalyses by age group indicated that children, but not adults, with ASD showed remarkable fractional anisotropy reduction among all tracts [316]. Typical brain lateralization, such as asymmetries in language and motor regions, tend to be reduced among those with ASD but the effect sizes are small [317]. In general, although there is evidence of group differences between individuals with ASD and controls for various structural imaging findings, the effect sizes are generally small and there is large variation within ASD groups.

Cross-sectional and longitudinal studies suggest that as a group, children later diagnosed with an ASD have average head circumferences at birth, followed by abnormal acceleration in brain growth during the second year of life leading to significantly above-average head circumferences and MRI brain volumes in toddlers, and then a plateau in brain growth leading to brain volumes in adolescence and adulthood that are not significantly different from controls [318–322]. This is commonly referred to as early brain overgrowth, although a systematic review and metaanalysis concluded that the prevalence of macrocephaly, defined as HC >97th percentile, in individuals with ASD is only 15.7%, so most do not experience true overgrowth [322].

Longitudinal studies of high-risk infants that include serial structural MRI exams have clarified the nature of early brain growth differences and demonstrated quite accurate individual-level prediction of ASD status from imaging performed at 6–12 months of age [314,323]. A prospective MRI study of infants at high familial risk of ASD identified hyperexpansion of cortical surface area between 6 and 12 months of age, followed by a significant increase in total brain volume growth rate in the second year of life, in those who went on to be diagnosed with ASD at age 24 months [324]. A deep-learning algorithm using primarily surface area information from MRI at 6–12 months of age predicted the diagnosis of autism in individual high-risk children at 24 months (positive predictive value 81%, sensitivity of 88%) [324]. A similar study confirmed an earlier finding [321] of excessive extra-axial CSF in the subarachnoid space in infants who were later diagnosed with ASD as compared to infants who did not have ASD and showed that MRI at 6 months of age predicted ASD diagnosis at age 24 months based on a machine learning algorithm using extra-axial fluid volume (sensitivity 66%, specificity 68%) [325].

Neurophysiological measures

The available neurophysiological information comes from studies utilizing functional MRI (fMRI), electroencephalography (EEG), event-related potentials (ERPs) and, to a lesser extent, magnetoencephalography (MEG), magnetic resonance spectroscopy (MRS), positron emission tomography (PET), single photon emission computed tomography

(SPECT), functional near-infrared spectroscopy (fNIRS), repetitive transcranial magnetic stimulation (rTMS), pupillometry, and auditory brainstem response (ABR) [303,307,314,326,327].

Information about structural connectivity from DTI is complemented by functional connectivity data, largely from resting state fMRI, EEG, and MEG. A comprehensive connectome-wide mega-analysis of resting state fMRI data revealed robust alterations in functional connectivity in ASD, primarily involving hypoconnectivity of primary sensory and attention networks and hyperconnectivity between the default mode network, subcortex, and the rest of the brain [327]. Task-dependent fMRI studies have elucidated differences in neural activation in people with ASD compared to controls on a wide variety of tasks such as face processing, executive functioning, and reward processing [328–332].

Resting state EEG and event-related potentials are among the most promising candidate biomarkers, and the ABC-CT recently competed a large-scale rigorous evaluation of a battery of well-studied assays of resting-state activity, face perception (N170 ERP latency), biological motion perception (P1, N2, and P3 ERP components), and visual evoked potentials (VEPs) [302,333]. The N170 ERP response to faces emerged as a promising biomarker for continued evaluation based on psychometrics (acquisition rate, construct performance, 6-week stability), group discriminations, and phenotype correlations and will move on to the next stage of biomarker development efforts focused on context of use [333]. Resting-state activity and VEP yielded mixed results, whereas the biological motion perception assay failed to display adequate construct performance.

Analogous to the infant structural MRI studies discussed previously, abnormal functional connectivity on resting state fMRI in high-risk infants at 6 months of age predicted ASD diagnosis at age 24 months [334], as did frontal EEG spectral power at age 3–12 months [335] and nonlinear EEG features (e.g., entropy) at age 3 months [336].

Eye tracking and digital phenotyping

Eye tracking tools allow precise measurement of visual attention to social information, which is known to be altered in ASD across a variety of experimental modalities and tasks [337–340]. Existing evidence suggests that eye tracking measures may be useful as diagnostic, predictive, prognostic, and/or response biomarkers [340,341].

The ABC-CT evaluated five eye tracking assays (activity monitoring, social interactive, static social scenes, biological motion preference, and pupillary light reflex) and an oculomotor index of gaze to human faces (OMI) derived from gaze to faces during the activity monitoring, social interactive, and static social scenes tasks [340]. All tasks met expectations for construct validity and acquisition rates had moderate or high 6-week stability. Pupillary light response and biological motion preference tasks exhibited suboptimal group discrimination and correlation with child characteristics. Among participants with ASD, higher OMI was associated with increased memory for faces, decreased ASD symptom severity, and higher verbal IQ and social communication skills [340]. Like the N170 ERP response to faces, the OMI has been chosen for further evaluation as an ASD biomarker by the ABC-CT.

Klin and colleagues developed an eye tracking-based measure of social visual engagement to supplement informed and experienced clinician judgment in the assessment and diagnosis of ASD [342,343]. Development included two prospective double-blind studies (discovery and replication studies including 719 and 370 children, respectively) of children aged 16–30 months who were referred to specialty centers for evaluation [343]. High sensitivity and specificity compared to the reference standard which was specialist diagnosis using best-practice standardized protocols. The measure was then assessed in an independent sample of 475 referred children in the same age range in six specialty clinics and was found to be predictive of ASD diagnosis by clinical experts (sensitivity 71.0%, specificity 80.7%) [342]. In a subset of 335 children whose ASD diagnosis was rated by clinicians as certain, the sensitivity was 78.0% and specificity was 85.4%.

Another recently developed multitask digital phenotyping tool that uses computer vision technology and was developed using machine learning also involves eye tracking and social attention but also captures differences in facial expression, head movements, response to name, blink rates, and motor behaviors [344]. This tool was designed for screening and assessed in a general pediatric population of children aged 17–36 months at well-child visits, including 49 children diagnosed with ASD and 98 diagnosed with developmental delay without ASD. The algorithm based on multiple digital phenotypes achieved high diagnostic accuracy (sensitivity 87.8%, specificity 80.8%, positive predictive value 40.6%, area under the receiver operating characteristic curve 0.90) [344].

Clinical features

ASD is defined by the presence of early onset, persistent core deficits in social communication and interaction, and restricted, inflexible, and repetitive patterns of behavior and interests that interfere with adaptive functioning.

Associated clinical features such as intellectual and motor deficits, maladaptive behaviors, and cooccurring medical and mental health conditions do not appear in the ASD diagnostic criteria but frequently have a substantial impact on adaptive functioning and quality of life and are also important aspects of the clinical presentation of ASD.

Core impairments

Clinical diagnosis using DSM-5 or ICD-11 criteria hinges upon impairments in the two core domains: (1) social communication and interaction and (2) restricted, repetitive, and inflexible patterns of behavior, interests, or activities (Table 28.1) [29,35,36]. These impairments must interfere substantially with adaptive functioning and cannot simply reflect coexisting ID. The specific manifestations of impairment within the two core domains vary according to chronological age, language and cognitive ability, and disorder severity, as outlined in Tables 28.3 and 28.4.

TABLE 28.3 Social communication, social interaction, and language impairment in autism spectrum disorder.

Younger or more severely impaired children	Older or less severely impaired children
Marked lack of interest in other people, deficits in social-emotional reciprocity • Often aloof and indifferent, may prefer to be alone • More interested in objects than people • May bump into people as if they did not see them or climb on them as if they were furniture • Approach to others may seem abrupt, aggressive, or disruptive • Indifference or aversion to physical contact and affection • Fewer cuddling, smiling, and laughing interactions • Rather than being distant, some children display indiscriminate affection • Decreased sharing of joyful affect, objects, interests • Decreased imitation of others (motor, vocal, social behaviors) • Failure to notice distress and offer comfort to others • Lack of adjustment of behavior according to environment or social context **Delays/deficits in joint attention** • Lack of response when name is called • Failure to follow the gaze and pointing gestures of others • Failure to spontaneously alternate gaze between an object and another person • Absent or limited attempts to draw the attention of others to objects or events for the purpose of sharing experiences (including lack of pointing or showing for the purpose of sharing interest or commenting) **Deficits in nonverbal social communicative behavior** • Reduced or atypical use of eye contact (sometimes gaze aversion) • Limited or absent responsive smiling • Delayed, decreased, or absent use of gestures – Deictic gestures (reaching, giving, showing, pointing) – Conventional representational gestures (e.g., waving hi/bye, shaking head for "no," nodding head for "yes") • Persistent use of primitive motor gestures to communicate (e.g., leading or pulling another's hand, using others' hands as tools) • Failure to coordinate verbal and nonverbal communication (e.g., protodeclarative pointing with vocalization and referential gaze shift) **Spoken language impairment** • Greater proportion of syllables with atypical phonation (e.g., squeals, growls, yells), fewer speech-like vocalizations • Atypical intonation (e.g., cartoonish, singsong) • Delayed receptive and expressive language milestones • May lack meaningful speech	**Difficulty forming developmentally appropriate friendships that involve a mutual sharing of interests, activities, and emotions** • May prefer solitary play or interaction with much younger or older people • May desire friendship but lack complete or realistic understanding of what friendship entails **Socially immature and less independent than peers** • Unusual quality of social overtures and responses, may be viewed as odd or eccentric by peers • May initiate interactions in inappropriate, awkward, or stilted ways • Lack of awareness of personal space • Gullible, naïve, lacking "common sense" • Difficulty understanding social conventions • May prefer to interact with adults rather than peers • Lack of modulation of behavior according to social context **Poor perspective-taking ability and insight into other people that may contribute to inappropriate, offensive behavior and vulnerability** • Difficulty anticipating how another person will feel or what they might think • Deficits in cognitive aspects of empathy • Difficulty knowing how to react to another person's behavior or emotions • Difficulty accepting that there might be multiple perspectives, not just a single correct perspective • Difficulty recognizing deceit or socially dangerous situations **Deficits in nonverbal social communicative behavior** • Decreased or atypical eye contact • Absent or limited/reduced use of appropriate facial expression, gestures, and body postures to facilitate communication (and diminished ability to recognize and understand these behaviors in others) **Semantic language impairment** • Unusual vocabulary for age/social group • Idiosyncratic word use, neologisms • Difficulty answering open-ended questions • Difficulty with comprehending and accepting multiple meanings • Overly literal; difficulty understanding figurative language (e.g., idioms, metaphors, humor, sarcasm, and irony)[a]; misinterpretation of contextual meaning[a] • Good memory for details, but difficulty recognizing and understanding themes

(Continued)

TABLE 28.3 (Continued)

Younger or more severely impaired children	Older or less severely impaired children
• May label objects or actions but not use words to make requests, answer questions, or comment • Nonsequential acquisition of milestones (e.g., early labeling of letters and numbers despite having a small vocabulary and not using words to make requests) • Increased atypical, idiosyncratic, or inappropriate means of communication – Idiosyncratic or metaphorical language (language that has meaning only to those familiar with the individual's communication style) – Neologisms (made up words) – Pronoun reversal (especially substituting "you" for "I" or "me") – Referring to self by name rather than using pronouns – Self-injurious behavior, aggression, tantrums **Play deficits/atypicality** • Early lack of interest in toys or preoccupation with elementary sensory features (feeling, mouthing, lining, spinning, arranging, hoarding, carrying) • Later appreciation of appropriate functional use – Sequences of appropriate actions – Impoverished, lacks typical creativity or variability – Repetitious, mechanical – Often precisely imitated from videos • Delayed onset, decreased frequency, diminished diversity, and atypical quality of symbolic play (use of objects in imaginative, pretend ways) • Solitary and parallel play, often some interactive physical play • Delayed and impaired development of pretend play • Difficulty with turn-taking in play	**Dysprosody (atypical intonation, stress/inflection, rhythm of voice)** • Monotone, robotic • Exaggerated intonation (singsong, cartoonish) **Pragmatic language impairment, difficulty engaging in conversation** • Difficulty with initiating or entering a conversation and with closing a conversation • Abrupt topic changes, tangential remarks • Failure to provide the right amount of background information – Not enough (expecting the listener to already have the same background knowledge and perspective) – Too much (excessive detail, unimportant aspects) • Difficulty maintaining conversation by elaborating or requesting more information • Difficulty recognizing and resolving communication breakdowns (unresponsive to partner's cues) • Tendency to engage in one-sided discourse that revolves almost exclusively around the individual's intense interest or includes irrelevant detail – Lack of reciprocal exchange (monologue-like) – Topic preoccupation, perseveration • Asks socially inappropriate questions or makes socially inappropriate comments • Difficulty adapting style of communication to social situation – Inability to infer expected degree of formality – Difficulty changing to match the situational context or needs of the listener – Excessively formal or pedantic style (e.g., content may be inappropriately philosophical, pompous, or quaint) **Play deficits/atypicality** • Difficulty with flexible, cooperative imaginative play (e.g., interactive role-playing with peers) • May insist on playing by rigidly fixed rules • May have difficulty tolerating losing at games or sports • May lack novelty, creativity, and imagination in play

[a]*There is overlap between some aspects of pragmatics and higher-order semantics; some would consider these deficits to be examples of pragmatic, rather than semantic impairment.*

Social communication and social interaction

Pervasive and persistent impairment in reciprocal social behavior that is not accounted for by developmental level/mental age is a defining feature of ASD; deficits in the development and understanding of social and emotional reciprocity, nonverbal communicative behaviors, and interpersonal relationships are required for DSM-5 diagnosis (Table 28.1) [29,36]. The earliest signs of ASD often appear as skill deficits or decreased frequency of age-appropriate behaviors rather than excesses of atypical behavior (Table 28.3) [345]. Differences in social interaction and communication that may emerge within the first year of life include delayed or diminished orienting and attending to social stimuli, sharing of positive affect, and directing of eye gaze, facial expressions, and vocalizations to others during interactions [323,346]. Fewer canonical (speech-like) vocalizations and reduced frequency of babbling, deictic gesture use (e.g., pointing, reaching, showing, giving), and coordination of gestures and vocalizations may also be noted within the first 12–14 months of life [323,347,348].

Infants and young children with ASD typically lack developmentally appropriate joint attention (JA), which is coordinating attention with another individual to an external focus (e.g., an object or event), showing social engagement and awareness of the partner's mutual interest [346,349]. JA behaviors can be receptive, (e.g., responding by following the direction of eye gaze and gestures of others) or expressive (e.g., initiating JA by using eye contact and gestures to direct

TABLE 28.4 Repetitive behavior, restricted interests/activities, and atypical interactions with sensory stimuli in autism spectrum disorder.

Younger or more severely impaired children

Stereotyped, repetitive motor behaviors
- Hand flapping, finger flicking, rocking, spinning, jumping, prancing on toes, whole body twisting or posturing, bruxism, etc.
- Incorporation of objects (e.g., lining, spinning, flipping, or tapping various objects; dangling and jiggling or twirling strings or cords; opening and closing doors or drawers, turning lights on and off)
- Self-injurious behavior (independent of situation, automatically reinforced)

Stereotyped, repetitive speech
- Immediate echolalia ("parroting"), delayed echolalia (reciting memorized dialog, scripted verses)
- Ritualized patterns of verbal behavior (e.g., greeting rituals)

Insistence on sameness, behavioral rigidity/inflexibility
- Difficulty coping with change; excessive distress with deviation from routines or small changes in the environment (e.g., a change in packaging of a preferred food, taking a different route in the car, or someone sitting in a different place than usual)
- Difficulty with transitions among activities or locations
- Extreme dietary selectivity (including restrictive food intake disorder)
- Insisting that others do or say things a certain way (often the way they first experienced the action or verbiage)

Restricted interests and activities, unusual preoccupations
- Reduced exploratory behaviors
- Lack of playing with a variety of toys
- Preoccupying interest in or attachment to unusual objects or parts of objects (e.g., a battery or piece of string)

Unusual sensory interests and exploration
- Visual fascination with lights or objects that move or spin
- Atypical visual inspection, lateral eye gaze, squinting
- Excessive smelling, licking, or touching of objects

Aversion or hyperreactivity to certain sensory stimuli
- Distress, avoidance, ear-covering in response to certain sounds
- Gagging in response to smells, tastes, or sight of foods of certain textures
- Tactile defensiveness, aversion to social touch or certain textures

Hyporeactivity to sensory stimuli
- Lack of response to certain sounds, especially human voice
- Apparent indifference to pain and/or temperature
- Increased tolerance of spinning or being upside-down

Older or less severely impaired children

Repetitive movements and compulsions
- Complex stereotyped motor mannerisms
- Touching, tapping, rubbing
- Ordering, arranging
- Pacing

Repetitive and/or ritualized patterns of verbal behavior
- Perseverative telling or asking
- Reciting of memorized dialog, scripted discourse
- Idiosyncratic phrases

Strong desire to maintain sameness
- Inflexible adherence to schedules, routines, rituals
- Rigid thinking patterns; excessively rule-bound, inflexible
- Difficulty accepting unexpected change
- Difficulty with transitions
- Overly selective, restricted diet
- Insisting that others adhere to specific verbal scripts

Excessively narrow or circumscribed scope of interests and activities
- Special interests of abnormal intensity and/or content (topic/focus)
  - Restricted, intense interests; inflexibility about deviating from them
  - Unusual, all-encompassing interests/preoccupations (often with topics that are not salient to the general population)
- Obsession with facts, details, or collections
- May be excessively perfectionistic and intolerant of mistakes
- Excessive focus on nonrelevant or nonfunctional parts of objects

Unusual sensory interests
- Craving of and sensory experiences that are prolonged or intense
- Excessive smelling or touching of objects or people

Extreme reaction/aversion to and avoidance of certain sensations
- Sounds
- Odors
- Touch (e.g., tags in clothing, social touch)
- Food textures, tastes
- Temperature (hot or cold)

Hyporeactivity to sensory stimuli
- Interoceptive signals, such as pain, temperature extremes, hunger
- Scents, tastes, ambient sounds

the attention of others to an object or event of interest for the purpose of sharing their experience). Both are commonly delayed in children with ASD; however, initiation of JA for the purpose of sharing interest and enjoyment/positive affect rather than requesting is particularly atypical and persistent [349,350]. Initiation of JA evolves during the second year of life, with increasing frequency, quality, and variety of social communicative bids [346]. JA ability is an important precursor and predictor of language and core aspects of social cognition such as theory of mind (mindreading, perspective-taking) [346,349,351].

Social behavior in children and adolescents with ASD can range from aloofness and indifference toward others to active but indiscriminate social interactions that are awkward, inappropriate, and/or one-sided [352]. The latter may be quantitatively appropriate but qualitatively atypical. Social conventions may go unnoticed and ability to appropriately modulate behavior according to social context may be lacking. Reciprocal social behaviors such as seeking to share enjoyment with others, recognizing and reacting to the mental states of others, and forming friendships that involve a mutual sharing of interests, activities, and emotions are often limited, even in older and intellectually capable children (Table 28.3) [353,354]. Even highly functioning adolescents and adults with ASD typically have difficulty navigating social interactions when it is necessary to make accurate inferences about other people's beliefs, motivations, goals, and emotional states [355,356]. Social naiveté and gullibility can lead to difficulty recognizing socially dangerous situations and vulnerability to exploitation by others, especially during adolescence.

Deficits in pragmatics, which is the ability to use language in context for social purposes, are universal in individuals with ASD, although not all parts of pragmatics are impaired and not all individuals with ASD have difficulty with the same pragmatic phenomena [357]. Even the most able communicators with ASD typically exhibit atypicality in conversational discourse, including poor topic maintenance, difficulty with turn-taking, failure to respond adequately to questions or comments, inclusion of noncontextual or socially inappropriate utterances, and difficulty recognizing and repairing a misunderstanding or unclear referent [358]. Other common deficits involve understanding figurative and metaphorical language, making inferences, and resolving ambiguous language [357–359]. Speech may be characterized by pedantic precision and lack normal prosody and emotional tone [357].

Children with ASD typically exhibit deficits in the development of symbolic play, including delayed onset, decreased frequency, reduced diversity, and qualitative atypicality [360–362]. Qualitatively, their play tends to be solitary, repetitive, stereotyped, and excessively focused on sensorimotor manipulation of objects rather than interactive, varied, flexible, imaginative, and creative (Table 28.3). When playing with others, children with ASD may have difficulty with turn-taking, letting others choose the activity, accepting novelty introduced by playmates, and tolerating losing.

Repetitive behavior and restricted interests

The second core domain of impairment required for ASD diagnosis involves persistent restricted, repetitive, or sensory stimuli-related behaviors and interests that are clearly atypical or excessive for the individual's age and sociocultural context [29,35,36]. Groups of repetitive behaviors and restricted interests (RBRI) contributing to the DSM-5 criteria include (1) repetitive sensorimotor actions (e.g., stereotyped or repetitive movements, manipulation of objects, or speech), (2) inflexible insistence on sameness/behavioral rigidity (e.g., resistance to minor changes in routine or personal environment, compulsive adherence to nonfunctional rituals or routines), (3) circumscribed interests (e.g., fixation/preoccupation with topics, objects, or activities that is abnormal in intensity or focus), and (4) unusual response to or interest in sensory stimuli in the environment (e.g., hypo- or hyperreactivity to sensory input, sensory-seeking behaviors) (Table 28.3) [29,36]. Examples of these behavior and their expression in children of different ages and degrees of impairment are provided in Table 28.4.

Repetitive behaviors, including stereotyped movements and behaviors related to desire for sameness, also occur in normal development and in the course of other neurodevelopmental and psychiatric conditions, but most are more prevalent and/or severe in children with ASD [363–368]. Although RBRI such as unusual visual inspection of objects, repetitive object use, and stereotyped motor mannerisms and head movements have been noted to be more frequent by age 12 months in prospective studies of high-risk infants who go on to have ASD, parental concerns about these behaviors do not differentiate between ASD and non-ASD groups until after age 15 months [369,370]. RBRI become increasingly evident in the second and third years of life, and most types peak in later childhood/preadolescence and then decline, although more severe repetitive motor behaviors tend to peak earlier and the evidence for decline by adolescence is mixed for circumscribed interests [371–373].

Stereotypies and other motor behaviors, including repetitive self-injurious behavior, are more common and intense in younger children and those with lower IQ and language abilities [372,374]. These are sometimes conceptualized as "lower-order" repetitive behaviors and distinguished from "higher order" repetitive behaviors related to insistence on sameness or circumscribed interests [375]. Behaviors related to inflexible insistence on sameness, which conflict with the need to adapt to changing environments, are associated with anxiety and executive function capabilities and have been hypothesized to play a role in reducing fears by constraining the environment and unpredictability, thereby serving as an early form of self-regulation [372,376]. They tend to increase along with fears in typical development and decline as executive function abilities and more mature forms of self-regulation develop [367,372]. Circumscribed interests can be divided into two subtypes: those that, although characterized by unusually high intensity and/or inflexibility, are otherwise common in terms of their content, labeled as restricted interests (RI), and those that are generally not salient outside of ASD, labeled as unusual interests (UI) [377]. Emerging evidence suggests that RI and UI have distinct patterns of association with age and IQ; older age and higher IQ are associated with increased RI and decreased UI [377,378]. Interests tend to shift to more typical content in older and more cognitively able children, but they remain unusual in their intensity, and even among intellectually very capable adolescents and adults, excessive and/or unusual preoccupations and behavioral rigidity often interfere with social and adaptive functioning [355,378,379]. Conversely, circumscribed interests are often inherently rewarding and can be adaptive, for example, when leveraged for facilitating learning and knowledge acquisition, as themes for socialization with others with similar interests, and potentially in the workplace [377,380–383].

Sensory modulation impairments have been defined as difficulties in regulating and organizing the type and intensity of behavioral responses to sensory input to match environmental demands and are categorized as follows: (1) overresponsivity (hyperreactivity), which is exaggerated, rapid onset, and/or prolonged reactions to sensory stimulation; (2) underresponsivity (hyporeactivity), which is unawareness or slow response to sensory input; and (3) sensation seeking, which describes atypical interest in and craving of sensory experiences that are prolonged or intense [384,385]. Individuals with ASD often exhibit behavior fitting one or more of these patterns of atypical sensory modulation related to sight, sound, smell, taste, touch, and/or movement, and these symptoms are important throughout the lifespan and across levels of functioning (Table 28.4) [385–388]. In 2013, hyperreactivity and hyporeactivity to sensory input were added to the DSM-5 diagnostic criteria for ASD, grouped with unusual interests in sensory aspects of the environment (sensation seeking) as the fourth criterion within the RBRI category [29]. Sensation seeking was captured in DSM-IV within the stereotyped and restricted patterns of interest criterion, but this was the first time that overresponsivity and underresponsivity were captured within the diagnostic criteria. All three categories of sensory modulation impairment are significantly more common in children with ASD than in typically developing children, and a metaanalysis showed that the mean effect size is greatest for underresponsivity ($d = 1.38$), followed by overresponsivity ($d = 1.28$) and sensation seeking ($d = 0.66$) [385]. However, only overresponsivity symptoms differentiated individuals with ASD from those with other developmental disorders and other clinical comparison groups regardless of age, IQ, and study methodology.

Neuropsychological features

The search for a primary underlying deficit that is necessary and sufficient to explain the wide variety of cognitive and behavioral features of ASD has largely failed, and the weight of the available evidence suggests that ASD is characterized by combinations of impairments in a range of higher-order cognitive abilities rather than one core mechanistic impairment [389–391]. Although no single theory or unitary deficit adequately explains all aspects of ASD, impairments in social motivation and reward processing, perspective-taking/theory of mind, cognitive flexibility, and ability to process complex information and form a coherent global picture are common, as is a profile of relatively high systemizing and low cognitive empathy traits [392–398].

Despite the lack of evidence that a single deficit underlies the core and associated clinical features of ASD, neuropsychological characteristics are important at the group and individual levels. Abilities vary widely among individuals with ASD, and features such as IQ and language are correlated with important outcomes. Assessment of language, intellectual and adaptive functioning, and other domains such as attention, executive function, and achievement is important for differentiating ASD from other disorders, identifying cooccurring conditions, informing intervention planning, and monitoring progress toward adaptive, academic, and occupational goals.

Language

The severity of language impairment ranges from severe (e.g., nonverbal or minimally verbal) to relatively mild (e.g., isolated pragmatic impairment), and the clinical features vary with age and developmental level (Table 28.3) [357,358]. Communicative speech is often delayed or absent in young children with ASD, and comprehension is impaired. Nonverbal communication is also impaired, so there is often little or no ability to compensate by using facial expressions, gestures, or pantomime, especially in young children [359,399]. Approximately 30% of school-age children, adolescents, and even adults with ASD remain minimally verbal, meaning that they have only a few words or phrases, or lack functional spoken language [357]. This literature is difficult to interpret, however, due to large inconsistencies in the definitions of "nonverbal" and "minimally verbal" and variability in the measures used and ages of study participants [400].

Pragmatic language impairment, involving deficits in using language for social purposes, is a hallmark of ASD and is included in the discussion of the core impairment in social communication and social interaction (Table 28.3). Although structural language skills (phonology, morphology, syntax, compositional semantics) are extremely variable and are normal in some individuals [357], atypicalities such as echolalia, jargon, neologisms, and other unconventional speech and unusual semantic and syntactic error patterns are more frequent among children with ASD than typically developing children or children with ID [358,401]. Even among individuals with higher verbal skills, the language profile that emerges in childhood and persists into adulthood is often characterized by unevenness and idiosyncrasy [358,402].

Intelligence and adaptive behavior

Although it was once thought that ID was present in the vast majority (~70%) of children with ASD [403], more recent large epidemiological studies suggest that the prevalence of ID is substantially lower (26%–45%) [49,404–406]. Among prevalence studies published between 2012 and 2020, the median prevalence of ID was 33% [46]. Estimates of the prevalence of ID among individuals with ASD are generally higher (~45%–55%) in earlier studies, those that were specifically designed to assess the prevalence of ID in ASD as opposed to the epidemiology of ASD in general, and those focused on preschool-age children [45,50,53,407,408]. A review of ASD prevalence surveys conducted between 1966 and 2020 found that the median proportion of individuals with ID was 47% and had decreased substantially over time [45]. Similarly, a systematic review and metaanalysis of M/F ratio in ASD included 24 epidemiological studies published between 1996 and 2015 that provided sufficient information to determine that the mean proportion of participants with IQ < 70 was 48% [50]. Despite the lower prevalence of ID in recent studies, the IQ distribution among individuals with ASD is clearly shifted substantially to the left. Among the two-thirds of 8-year-old autistic children with available data on cognitive ability in ADDM Network surveillance year 2020, 61.4% had an IQ one standard deviation (SD) or more below the mean; 37.9% were classified as having ID and 23.5% were classified in the borderline range (IQ 71–85) [49].

The distribution of full-scale IQ (FSIQ) scores in ASD is broader than the normal distribution in the general population. For example, in a representative sample of 671 children with ASD, the average SD of FSIQ was almost 26 points (1.7 times as great as the SD in the normative population) [409]. Individuals with ASD often display an unusual degree of unevenness in cognitive abilities, frequently including a significant discrepancy between verbal (V) and nonverbal (N) abilities, which can be in either direction (N > V or V > N) and/or isolated strengths that significantly exceed not only their own general level of ability but sometimes general population norms as well [410–412]. A strong correlation between uneven intellectual development and autistic-like symptoms, including those outside of the clinical range for ASD, has also been reported [413]. Wechsler Intelligence Scale for Children subtest performance profiles tend to be characterized by relatively high block design and low coding subtest scores across intellectual levels, whereas the classic "islets of ability" profile (low verbal comprehension scores, high visual-spatial and working memory scores) is more frequent in those with FSIQ scores ≤ 85 and the "right-descending" profile (high verbal comprehension and visual-spatial scores, low processing speed scores) is more likely to be found in those with FSIQ ≥ 86 [412]. A metaanalysis of visuospatial task performance did not confirm the often-reported superior performance of individuals with ASD [414]. ASD was associated with enhanced performance on block design and figure disembedding tasks, but the effect size was small, heterogeneity was very large, and there was little or no evidence of superiority on other visuospatial tasks examined. Overall, although a high degree of variability between domains and subtests is common and some subtest performance patterns are observed at the group level, a single prototypical profile of cognitive strengths and weaknesses characteristic of ASD does not exist [410,412,415].

At the group level, mean IQ typically increases through early childhood in children with ASD [416,417]. However, there is evidence that there are subgroups of children with different developmental trajectories. In a study of children with ASD followed from early childhood through preadolescence using an empirical data-driven approach to isolate trajectories of intellectual functioning, Solomon and colleagues identified three distinct IQ trajectory groups: (1) a group with ID from early childhood through preadolescence (45%), (2) a group whose IQs increased substantially during early childhood (39%), and (3) a group whose IQs were in the average or above range through the study period (16%) [418]. Other recent longitudinal studies have found that group mean IQ (particularly nonverbal and perceptual scores) tends to increase throughout adolescence despite considerable individual variability [419,420]. This seems to be true even in the absence of significant improvement of core ASD symptoms [419].

Although IQ is one of the most robust predictors of outcomes for individuals with ASD, it must be recognized that low IQ is more reliably predictive of a poorer outcome than high IQ is of a better outcome and that IQ, especially among higher-ability individuals, often significantly exceeds real-world adaptive functioning [421–427]. Deficits in adaptive functioning tend to be prominent across IQ levels and degrees of core ASD symptoms, and there is often a significant discrepancy between level of cognitive ability and adaptive functioning, particularly among higher-ability individuals [423,425,427–430]. Older age, higher IQ, and higher levels of social-communication symptoms are associated with greater IQ-adaptive functioning discrepancy [423], and emotion perception has also been found to play an important role in functional adaptive outcomes [431,432].

Executive function and other cognitive domains

Deficits in executive functions that are important for goal-oriented behavior such as identifying goals for a task, planning and organizing an approach to problem solving, inhibiting distractions, initiating a task, using abstract reasoning, self-monitoring, and being flexible (e.g., trying a new approach, innovation) are also common among individuals with ASD [433]. A metaanalysis identified reduced executive function abilities of moderate effect size across all major domains, including concept formation, mental flexibility, fluency, planning, response inhibition, and working memory, among individuals with ASD [395]. These deficits contribute to difficulties with independent, goal-oriented behaviors that are important for social and adaptive functioning. Attentional abnormalities are also common in individuals with ASD, particularly the orienting and executive components of attention [390,395,433,434].

Neuropsychological deficits associated with ASD are not limited to children or those with general intellectual impairment. A systematic review and metaanalysis showed that even among adults with ASD and intact general intellectual ability (mean IQ 108.2, SD 9.1), there are deficits of medium effect size in processing speed, verbal learning and memory, and reasoning and problem solving relative to neurotypical controls (mean IQ 109.8, SD 7.7) [391]. The ASD group means were significantly lower than those of neurotypical controls in all domains of nonsocial cognition assessed; deficits of small effect size were found for the domains of visual learning and memory, verbal fluency, verbal comprehension, attention and vigilance, and working memory. Unsurprisingly, effect sizes were large for social cognition domains (theory of mind, emotion perception and processing) [391].

Academic achievement

Academic outcomes are widely variable, and uneven patterns of academic achievement are common among individuals with ASD. Achievement in basic academic domains may be commensurate with or above what would be expected for the child's level of cognitive ability [435,436]. Skills requiring primarily rote, mechanical, or procedural abilities are often intact, but skills requiring more abstract reasoning, conceptualization, or interpretive analysis are often deficient [353,437]. In a longitudinal follow-up study, most children and adolescents with ASD (74% at age 9; 92% at age 18) were found to have a discrepancy between IQ and achievement of 15 or more points in at least one academic domain [435]. Reading comprehension is often a relative weakness, and individuals with ASD are significantly worse at comprehending highly social text [435,436,438]. Deficits in both oral language and word reading contribute to reading comprehension difficulty [439]. Even among high-functioning individuals with ASD, difficulties with reading comprehension and mathematical problem solving are common, and weaknesses in graphomotor skills, attention, and processing speed can interfere substantially with academic performance [353,440,441]. Because the types of skills emphasized in the early elementary grades play to their strengths in memorization and concrete tasks, such as decoding and numerical operations, young children with ASD may perform well academically. In later elementary and middle school grades, when comprehension, analytic interpretation, and abstract reasoning skills are emphasized, these children often struggle academically and require more supports [353,436,442].

Motor impairments

In addition to motor stereotypies, children with ASD often exhibit less distinctive motor abnormalities, including deficits in motor planning, control, and coordination that lead to functional impairment. An increasingly extensive body of literature, including numerous systematic reviews and metaanalyses, has documented the wide range of motor aberrations associated with ASD, including differences in gait and other locomotor skills, balance, postural control, manual dexterity, bilateral coordination, object control/manipulation, handwriting, visual-motor integration, praxis/imitation, motor learning, and interpersonal coordination [441,443–456]. At the group level, gross and fine motor deficits can be detected as early as 6 months of age and can distinguish between high-risk and low-risk infants, but no motor markers have been identified that differentiate children who later receive an ASD diagnosis from those with developmental differences that do not meet criteria for ASD, suggesting that motor deficits are early-emerging but nonspecific [457,458]. Clinically, parents may be concerned about delayed attainment of gross and/or fine motor milestones, clumsiness, toe-walking, or difficulty with drawing or handwriting. A range of minor neurological dysfunctions, or neurological subtle signs, such as hypotonia, associated movements (e.g., during sequential finger opposition and/or stressed gait maneuvers), coordination and balance difficulties, toe-walking, and elicited involuntary movements (e.g., choreiform movements of the fingers or face when arms are extended) are common among children with ASD and may be observed on neurological examination [459–462].

It is well-established that there are group-level differences between individuals with ASD and typically developing controls across a variety of motor skill domains; however, the true prevalence of motor impairments among people with ASD is unclear due to variability in study design, specific aspects of motor function assessed, case definitions and outcome measures used, and ages of participants included [443–445]. Population-based epidemiological data using comparable case definitions across a range of ages are lacking. However, a systematic review and metaanalysis estimated that 50%–88% of 5- to 12-year-old children with ASD had significant motor impairments on standardized motor assessments and/or functional measures consistent with DCD [444]. Individuals identified as having significant functional motor deficits in research studies often do not carry clinical diagnoses reflecting this impairment [463–465]. For example, in a population-based cohort study of over 2000 young children with ASD (age ≤ 6 years), 79% of children had some degree of motor impairment based on the Vineland Adaptive Behavior Scales motor domain, and 35% had standard scores below 70, yet motor difficulties were reported by diagnosing clinicians in only 1.3% [465]. Among a large sample ($n = 11,814$) of 5- to 15-year-old autistic children without significant cognitive impairment, 87% were found to be at risk for DCD based on scoring below the 10th percentile on the parent-completed Developmental Coordination Disorder Questionnaire, but only 15% carried a diagnosis of DCD/motor delay and 32% were receiving physical therapy [463].

Although no ubiquitous motor signature of ASD has been identified, praxis (i.e., performance of skilled motor gestures during imitation, on verbal command, and during tool use), object control/manipulation, and postural control are motor skill domains that may be more selectively impaired in ASD relative to other neurodevelopmental disorders [445,466]. There is strong evidence of motor imitation impairment among individuals with ASD, and the severity of impairment is correlated with measures of core social communication and interaction deficits [453,466]. Excessive reliance on proprioceptive feedback and weak dynamic integration of visual feedback and perception-action coupling seems to be a core characteristic of motor dysfunction in ASD [445,453,467,468]. Many children with ASD may have a specific impairment in efficient use of visual information to form and update internal representations of the self and of the external world for predictive motor control [453]. These children demonstrate not only impaired performance of skilled motor gestures but also impaired ability to recognize these gestures as performed by others [469].

Savant skills

Savant syndrome is usually defined as a rare and remarkable condition in which an individual with a developmental disability, brain injury, or acquired brain disease has one or more exceptional "islands" of skill or ability that stand in marked contrast to overall handicap [470,471]. However, there is no universally agreed upon definition or set of diagnostic criteria, and it is sometimes defined simply as a condition in which an individual exhibits exceptional ability in one or more areas that is inconsistent with their overall level of functioning, without the disability requirement [472]. Overlapping terms used to describe such skills or the individuals who possess them include "extraordinary talents," "special isolated skills," and "twice-exceptional," among others. Savant skills can exist in a variety of areas, such as music (e.g., absolute pitch, reproducing a piece of music after hearing it once or a few times), art (e.g., producing extremely detailed drawings), mathematical computation (e.g., fast mental arithmetic, generating prime numbers), calendar calculation, mechanical/visual-spatial skills, feats of memory (e.g., recall of facts, hyperthymesia), precise appreciation of passing time without access to a time-keeping instrument, and facility with foreign language acquisition (e.g., learning multiple languages rapidly and speaking with no accent) [470,471,473,474].

Robust, population-based epidemiological data are lacking, but the available information suggests that 90% of savant syndrome is congenital, 50%–75% of people with congenital savant syndrome have ASD, and perhaps 10%–26% of individuals with ASD have parent-reported savant level skills [471,474–477]. The Autism Diagnostic Interview-Revised (ADI-R) includes questions about the presence of "special isolated skills," and several studies that used these questions as the outcome measure and included past and/or current skills have reported higher lifetime prevalence rates of 46%–63% [478,479]. Among individuals with ASD, those with savant syndrome as a group tend to have higher IQ, better working memory, and less social impairment, and to exhibit higher sensory sensitivity, obsessive behavior, and systemizing and technical/spatial traits than those without savant skills [471,474,477,480].

Hyperlexia is defined as the cooccurrence of advanced reading skills relative to comprehension skills or general intelligence, early acquisition of reading skills without explicit teaching, and a strong orientation toward written material, generally in the context of a neurodevelopmental disorder [481]. Most, but not all, children with hyperlexia have ASD. Estimates of the prevalence of hyperlexia among children with ASD range from 6% to 20%, with the lower estimates coming from the studies that used the most stringent criteria for defining case status [481]. Children with ASD and hyperlexia may spontaneously begin to read as early as 18–24 months of age, and there is often a period of intense, almost exclusive interest in all types of printed material. Hyperlexia appears to be mediated by a different process for the acquisition of reading skills; typical reading acquisition relies on previously acquired language abilities, whereas hyperlexic reading frequently emerges before any communicative oral language appears [481].

Autistic regression

Autistic regression refers to loss of previously established language and/or social interest and engagement skills. However, there is no consensus, standardized definition of exactly how regression should be operationally defined or measured. Many studies define regression as a period of apparently typical development for the first one or two years of life, followed by an abrupt or gradual loss of previously acquired skills with or without recovery [482,483]. However, regression can occur in children with preexisting atypical development, and developmental plateau/stagnation (without overt loss of existing skills) is included in some operational definitions of regression [482–484]. In fact, the evidence for a distinct regressive ASD subgroup characterized by completely typical development prior to the loss of skills is much more limited than the evidence that there is a subgroup of very young children who regress in the setting of preexisting developmental delays and atypicalities [483]. Within the ICD-11 diagnostic framework, regression is included as a dichotomous qualifier ("without loss of previously acquired skills" or "with loss of previously acquired skills") that can be added to any of the five specifiers and coded using an additional digit (Table 28.1) [35]. The DSM-5 ASD criteria do not include a specifier or other marker for regression, but the phenomenon is discussed as a potential part of the developmental course of the disorder [29,36].

A large systematic review and metaanalysis of studies published from 2010 through 2019 estimated the prevalence of regression in children with ASD to be 20%–40%, depending on the developmental domains included in the definition of regression, with a pooled mean proportion of 30% [482]. The mean prevalence of regression was lower in subgroups of studies that only considered loss of language skills in the definition of regression (20%), was prospective (24%), used population-based samples (25%), or had low risk of bias (26%) [482]. The weighted average age of onset was 19.8 months. These findings were similar to a previous metaanalytic review of studies published between 1980 and 2010, which reported an overall prevalence of 32% and mean age of onset of 21.4 months [485]. However, prospective studies of infants at high risk due to family history have shown that parental observations and eye tracking measures can identify more gradual, subtle changes and declining trajectories of development such as declining fixation on eyes, gaze to faces, social engagement, and increasing difficulty disengaging visual fixation [483,486,487]. Convergent evidence across studies suggests that ASD groups are not reliably distinguishable from controls at 9 months of age, and that statistically significant group differences start at approximately 12 months and widen over time [486,488]. These studies suggest that the onset of ASD involves declines in the rates of key social and communication behaviors, including loss of previously attained skills, and some report evidence of regression in over 85% of the children who go on to have ASD [486,487,489,490]. This body of work suggests that regression could be the norm in the development of children with ASD, and some investigators have proposed that ASD onset should be conceptualized not as a dichotomy (regression vs early onset/no regression) but as a continuum of age at which social and communication behaviors begin to diverge or decline [483,486]. Interestingly, parents do not typically identify very early skill losses explicitly as regression, even though they are able to implicitly identify the changes in their child's development over time when rating the frequencies of current behaviors [486,490]. However, at the other end of the continuum, the declines in some children are experienced late enough that more numerous and sophisticated skills have been acquired and losses are more easily perceived, corresponding to the traditional concept of regression reported in retrospective studies.

The literature addressing the natural history of autistic regression is conflicting and difficult to interpret due to variability in case definition and study methods, including reliance on parent-reported skill loss without verification by rigorous clinical information gathering [483,491]. Even when regression is defined by parent-reported loss of previously acquired communication and other social skills and treated as a dichotomous variable (present/absent), the literature describing the relationship with important outcomes is mixed. Most, but not all, studies have found that a history of regression is associated at the group level with increased severity of core ASD symptoms and more significant cognitive and language impairment, especially when outcomes are measured after the preschool period [483,492–498]. Association of regression with other adverse outcomes including lower adaptive functioning and increased prevalence of sleep problems, psychiatric symptoms in later childhood or adolescence, gastrointestinal symptoms, immunological function, epilepsy, and atypical epileptiform EEG is more equivocal [483,492,493,499–505]. Overall, there is little evidence that parent-reported autistic regression is associated with significantly increased severity of psychiatric symptoms in school-age children and adolescents with ASD, especially when accounting for IQ [483,487,495]. Gadow and colleagues found elevated schizophrenia spectrum symptoms (especially disorganized behavior and avolition) among youth with a history of regression, but the difference was no longer statistically significant when only individuals with IQ ≥70 were considered [495]. A metaanalytic review found that children with a history of autistic regression were at a slightly elevated risk for epilepsy (OR 1.59; 95% CI 1.31, 1.95) and having an epileptiform EEG (OR 1.29; 95% CI 1.09, 1.52) when compared to children with ASD but no regression [504].

Loss of established skills within the first two years of life is often appropriately noted to be a "red flag" for ASD [8,29,354]. However, it should be recognized that parent-reported loss of communication and/or social skills is not unique to ASD [506,507]. For example, in a prospective population cohort study, skill loss in the second year of life was reported in a substantial proportion of children with ASD (41%), language impairment (30%), and typical development (26%), as measured using the Communication and Symbolic Behavior Scales Developmental Profile Infant-Toddler Checklist (CSBS-ITC) [507]. Skill loss was most common in the domain of "emotion and eye gaze" for all three groups, although it was significantly more frequent in the ASD group [507]. Children with ASD were also significantly more likely than those with language impairment or typical development to lose skills in more than one of the seven cluster domains on the CSBS-ITC. It is also important to understand that the presence of regression does not imply the existence of an environmental cause or trigger. In fact, very high rates of regression are reported in children with ASD caused by certain genetic disorders, such as Phelan-McDermid syndrome (due to either 22q13 deletion or a pathogenic nonsense, frameshift, or missense variant in *SHANK3*), *SYNGAP1*-related disorder, *MECP2* duplication syndrome, and mitochondrial disorders [508–513]. A preliminary study did not find a significant difference in the rate of regression between autistic individuals with and without de novo, likely gene disrupting variants [514].

In summary, it is currently unclear whether autistic regression is a pervasive process with some variation in age of presentation or a qualitatively different phenomenon that occurs in a minority of individuals, potentially representing a distinct ASD subtype in terms of etiology and natural history. It is clear, however, that dramatic, catastrophic loss of skills in ASD is rare [506,507,515]. Previously, in DSM-IV-TR and ICD-10, the diagnosis CDD captured children who seemed to develop typically for the first 2 years of life, then regressed in multiple developmental domains before age 10 years (mean age 3 years 2 months) [516]. CDD was estimated to be at least 100 times less common than DSM-IV-TR autism spectrum disorders (autistic disorder, Asperger's disorder, PDD-NOS). Regression in CDD was typically sudden in onset, rapid, included prominent loss of adaptive skills (e.g., loss of previously acquired bowel/bladder control) and often functional motor skills and was frequently associated with symptoms of emotional distress (e.g., fear, anxiety), challenging behavior (e.g., hyperactivity, aggression, fecal smearing), more frequent epilepsy, and greater global developmental deficit than autistic disorder [516]. These patients are now captured under the ASD diagnosis in DSM-5 and ICD-11. Regression that involves loss of motor and other skills beyond the social and communication domains occurs after age 24–30 months, progresses rapidly, is associated with seizures, or involves other organ systems (e.g., loss of vision or hearing, development of cardiomyopathy) raises the possibility of one of the many neurodegenerative disorders captured by the umbrella term "progressive intellectual and neurological deterioration" (PIND) and warrants medical investigation (see Appendix 2.2: Approach to regression or plateau in developmental milestones) [7,29,516,517]. Etiologies of PIND include metabolic and other genetic disorders, developmental and epileptic encephalopathies (including Landau-Kleffner syndrome), autoimmune encephalopathies, viral infections (e.g., subacute sclerosing panencephalitis), and prion diseases [516,517].

Associated conditions

In addition to the core impairments that define ASD and the common neuropsychological features already described, there are other conditions that frequently impact clinical presentation, prognosis, and management. These associated neurodevelopmental, psychiatric, and somatic medical conditions occur among individuals with ASD at a higher frequency than expected based on background population rates, often cause significant impairment or distress, and are associated with poorer outcomes [518–523]. In some cases, especially in later childhood or adolescence, it is a cooccurring condition that first brings an individual with ASD to clinical attention. A large longitudinal study of an insured population found that the burden of common psychiatric and somatic medical conditions (anxiety, ADHD, depression, sleep disorders, obesity, cardiovascular disorders, and neurological disorders) increased throughout adolescence, and by age 22, all conditions assessed were more prevalent among females with ASD than among males with ASD [519]. Apart from cardiovascular disease, the prevalence of all of these conditions was higher in the ASD group than in controls. In some cases, both the ASD and the associated conditions can be attributed to an identified underlying genetic etiology [524], but more often the reason for the cooccurrence is unclear.

Here, we divide these associated conditions into brain-based (i.e., neurodevelopmental, neurological, and psychiatric) and nonneurological somatic medical conditions for organizational purposes, recognizing that there is significant interaction and overlap between the central nervous system and other organ systems and that both groups of conditions are profoundly influenced by environment.

Neurodevelopmental, neurological, and psychiatric disorders

Cooccurrence among neurodevelopmental disorders is the rule rather than the exception; most children with one diagnosis meet criteria for at least one additional disorder [29,525,526]. ID was previously discussed in this chapter as a neuropsychological feature sometimes associated with ASD, as were deficits in language, academic achievement, and motor functioning that in some cases warrant additional diagnoses (e.g., language disorder, specific learning disorder, DCD). Among the DSM-5 communication disorders, only SCD is precluded because the diagnosis of ASD supersedes that of SCD whenever the criteria for ASD are met [29]. Similarly, motor stereotypies are among the diagnostic characteristics of ASD, so an additional diagnosis of stereotypic movement disorder is not given when the repetitive behaviors are better explained by the presence of ASD [29].

In addition to the large proportion of children with ASD who exhibit motor deficits suggestive of DCD described previously [444], a small fraction have cerebral palsy (2.9%–4.3% in available population-based studies) [527]. There is also some evidence that macrocephaly, hydrocephalus, congenital abnormalities of the nervous system, and migraine/headache occur more frequently among people with ASD as compared to population controls, although there is significant heterogeneity, and the range of prevalence reported by individual studies is wide [528]. Tic disorders, including Tourette's disorder, also seem to cooccur with ASD more frequently than expected based on the general population frequency. However, population-based prevalence data are not currently available for tic disorders among people with ASD. Individual studies of clinical samples have reported tics in 22%–34% of patients with ASD and Tourette's disorder in 6.5%–11% [529,530]. In the largest and most recent study published to date, tics were identified in 18.4% of 679 children with ASD who were enrolled in a genetics study [531].

Epilepsy

ASD is associated with increased risk of epilepsy, including all seizure onset types and semiologies, etiologic categories, and electroclinical epilepsy syndromes, although most large studies suggest that focal onset is more common than generalized onset [532–534]. The most recent, large systematic review and metaanalysis of the prevalence of epilepsy in people with ASD reported an overall pooled prevalence estimate of 10% (9% in population-based samples and 19% in clinic-based samples) [535]. The pooled prevalence was 7% in children and 19% in adults. The authors of another recent, large systematic review did not complete a metaanalysis to obtain pooled estimates because of between-study heterogeneity, inconsistent reporting of important demographic- and disease-related data, and variability in diagnostic criteria used [536]. Instead, they reported descriptive statistics, including the median overall period prevalence, which was 12.1% (19% in females, 11% in males). When the analyses were confined to population-based studies, the median overall period prevalence was 10.8% (15.5% in females, 8.8% in males) [536]. Some studies suggest bimodal peaks in the age of onset in early childhood and adolescence, but the most recent epidemiological evidence indicates that main trend is gradually increasing prevalence with age, and children and youth with ASD are more likely to develop seizures in the second and third decades of life than in the preschool years [535,537].

Increasing age, female sex, and the presence of ID are associated with increased epilepsy risk in people with ASD [505,535–537]. Although the prevalence of epilepsy is highest in individuals with severe ID and those with identified genetic etiologies, even the subgroup with idiopathic ASD and normal IQ has a rate of epilepsy that is substantially higher than the rate in the general population [532]. Ewen and colleagues found that although ID was the predictor of greatest magnitude, language function, ASD-specific symptoms, and motor function all contributed independently to the risk of epilepsy in individuals with ASD [505]. People with ASD experience the full range of epilepsy severity. Some studies suggest that the rate of medically refractory epilepsy may be increased in this population, but existing epidemiological data are insufficient to answer this question [532,538]. Many children with ASD have interictal epileptiform activity on EEG, especially during sleep, even though they have not had definite clinical seizures [532,539]. The clinical significance of the increased rate of interictal EEG abnormalities is unclear, however, and there is little evidence to support the idea that epilepsy or epileptiform EEG abnormalities cause ASD or have treatment implications [532,537,540,541].

Neurobehavioral symptoms and psychiatric disorders

Neurobehavioral features that are commonly associated with ASD and cause significant impairment and distress include inattention/distractibility, impulsivity, hyperactivity, anxiety, mood disturbances/instability, irritability, agitation, anger outbursts, aggression, self-injurious behavior, property destruction, oppositional behavior, elopement or wandering, sleep dysfunction, feeding and eating disorders, and other maladaptive behaviors and manifestations of emotional dysregulation. Although these problem behaviors are not core features of ASD, they commonly interfere with functioning in school, at home, and in the community; are associated with worse long-term outcomes, including reduced quality of life; and contribute substantially to the burden on families [3,523,542–544]. Estimates of the prevalence of frequently reported mental health conditions based on two metaanalyses and two subsequent large studies are provided in Table 28.5 [523,545–547]. These conditions occur significantly more frequently in children and adolescents with ASD than in the general population. Prevalence estimates from clinical sample-based studies are higher than those from population-based and registry-based studies.

Metaanalyses indicate that ADHD is the psychiatric condition that cooccurs most commonly with ASD [523,545], and children and adolescents with both ASD and ADHD have an increased risk of anxiety and mood disorders [547,548]. Cooccurring anxiety disorders are also common and associated with poorer social functioning, increased RBRI, and more parent-reported self-injurious behavior, depressive symptoms, and parental stress in youth with ASD

TABLE 28.5 Estimated prevalence of cooccurring psychiatric disorders among individuals with autism spectrum disorder.

DSM-5 category	Lai et al. (2019) Metaanalysis[a] pooled prevalence (population samples, clinical samples)	Mutluer et al. (2022) Population studies metaanalysis[b] pooled prevalence	Khachadurian et al. (2023) SPARK[c] prevalence	Casseus et al. (2023) NSCH[d] prevalence
Attention-deficit hyperactivity disorder	28% (22%, 34%)	26%	35%	44%
Anxiety disorders	20% (15%, 26%)	11%	18%	60%
Sleep-wake disorders	13% (11%, 16%)	19.7%	21%	
Disruptive, impulse-control, and conduct disorders	12% (7%, 22%)	7%	2% CD 7.5% ODD	
Depressive disorders	11% (8%, 14%)	2.7%	6% MDD or dysthymia 3% DMDD	24%
Obsessive-compulsive and related disorders	9% (4%, 12%)	1.8%	8% OCD 2% hoarding	
Bipolar and related disorders	5% (3%, 7%)	2%	2%	
Schizophrenia spectrum and psychotic disorders	4% (2%, 7%)	0.6%		

CD, Conduct disorder; *DMDD*, disruptive mood dysregulation disorder; *MDD*, major depressive disorder (depression); *NCHS*, National Survey of Children's Health; *OCD*, obsessive-compulsive disorder; *ODD*, oppositional defiant disorder; *SPARK*, Simons Foundation Powering Autism Research for Knowledge.
[a]*Metaanalysis including 96 studies (1993–2019) reporting confirmed clinical diagnoses of ASD and 8 cooccurring psychiatric/mental health conditions by DSM or ICD criteria; substantial heterogeneity and wide 95% prediction intervals noted for prevalence estimates of all cooccurring conditions [523].*
[b]*Metaanalysis including 39 population-based studies (2015–2020); age 6–18; primary diagnosis of ASD, reporting prevalence of a wide range of DSM-5 disorders [545].*
[c]*SPARK cohort; birth year 1999–2019, age <18 years at the time of study registration; n = 40,582 with ASD, 79% male; data based on parental report of professional diagnosis [546].*
[d]*NSCH pooled 2016–2018 data, age 3–17; data based on parental report of professional diagnosis. Prevalence of conditions that could not be directly mapped with confidence to one of the eight DSM-5 categories included "behavior or conduct problems" (70%) and "other mental health condition" (60%) [547].*

[549,550]. Diagnosing anxiety disorders can be challenging because of cognitive and language impairments, compromised reporting of emotions, overlapping symptoms between anxiety and ASD, and unique behavioral expressions of anxiety in the ASD population [550–552]. In general, autistic children with higher IQs have higher rates of anxiety and depression, but there is much heterogeneity [553,554]. Cooccurring depressive disorders are associated with poorer adaptive functioning, diminished quality of life, and increased behavior problems and caregiver burden [543,554–556]. Catatonia also cooccurs with ASD; the estimated prevalence is 10.4%, with onset peaking between 15 and 19 years of age, and it is one of the specifiers in the DSM-5 [557]. Depressive, bipolar, and schizophrenia spectrum disorders become more prevalent with increasing age, similar to the temporal pattern in the general population [523,555]. Among adults with ASD, recent pooled prevalence estimates are 7.5% for bipolar disorder and 9.4% for psychosis [558]. The metaanalytic pooled prevalence of self-injurious behavior among individuals with ASD is 42% (95% CI, 38%–47%), with hand hitting, skin picking, and hitting self against objects occurring most commonly [559].

Suicidality is a major public health issue among autistic children, adolescents, and young adults, who are more likely than their nonautistic peers to attempt suicide and to die [560–563]. A 2022 systematic review and metaanalysis of studies reporting on suicidality in people with ASD under the age of 25 years found the pooled prevalence of suicidal ideation to be 25.2%, suicide attempts 8.3%, and suicide deaths 0.2% [564]. Suicidality risk is amplified by cooccurring depressive, bipolar, and psychotic disorders. There is evidence that ASD is associated with increased likelihood of gender dysphoria [565,566]. Gender dysphoria among young people with ASD is associated with increased prevalence of suicidal ideation when compared to both autistic and nonautistic peers without gender dysphoria, and preliminary evidence suggests that transgender autistic youth have a higher risk of suicidal ideation and attempts in comparison to cisgender autistic youth and transgender/cisgender nonautistic youth [564,567–569].

Sleep dysfunction

Significant sleep differences between individuals with ASD and typically developing controls have been demonstrated on both objective measures (polysomnography, actigraphy) and subjective measures (validated sleep questionnaires) [570–572], and sleep problems are common at all ages and levels of intellectual ability [573–577]. ASD is associated with significantly lower total sleep duration and higher levels of bedtime resistance, sleep onset delay, sleep anxiety, night awakenings, parasomnias, sleep-disordered breathing, daytime sleepiness, and general sleep problems on subjective measures [571]. Children with ASD also differ significantly from controls on several objective polysomnographic parameters. Specifically, lower total sleep time, longer sleep onset latency, more time spent in stage 1 sleep, less time spent in REM sleep, lower sleep efficiency, and more time awake after sleep onset are associated with ASD [571,572].

Although rates of parent-reported sleep problems range from 50% to 80% in children with ASD (vs 9%–50% in comparison groups), studies using chart review methods and ICD billing codes generally arrive at substantially lower prevalence estimates for sleep disorders [570,578,579]. For example, a large, retrospective study of 48,762 children with ASD and 243,819 controls who were enrolled in the military health system found that 31% of children with ASD and 14% of controls had a diagnosis of a sleep disorder based on ICD codes [578]. In a registry study of over 1500 4- to 10-year-old children with ASD, 71% had significant sleep problems as defined by a threshold total score on the Children's Sleep Habits Questionnaire, but only 30% had a sleep disorder diagnosis [579]. Large metaanalyses have reported lower, but still substantial, rates of sleep disorders among individuals with ASD, ranging from 13% (11% in population-based or registry studies, 16% in studies of clinical samples) [523] to 19.7% [545] (Table 28.5). These rates are significantly higher than the general pediatric population 1-year prevalence estimate of 3.7% [523,580]. Sleep disturbances associated with ASD are important because of their potential adverse effects on adaptive functioning, learning/academic performance, regulation of emotions and behavior, physical health, and quality of life [573,574,579,581–583].

Feeding problems/Atypical eating behaviors

Feeding problems, or atypical eating behaviors, are very common in children with ASD. These include behaviors related to selective intake ("picky eating"; often referred to as food selectivity, neophobia, refusal, avoidance, or aversion), hypersensitivity to food textures, holding food in the mouth without swallowing, rumination, and pica [584–587]. A metaanalysis estimated the prevalence to be 59% among individuals with ASD [588]. Atypical eating behaviors were significantly more common in children with ASD (70.4%) than in children with other neurodevelopmental disorders (13.1%) and typically developing children (4.8%) based on data collected using a standardized parent interview, the Checklist for Autism Spectrum Disorder [585]. Among children with ASD who had atypical eating behaviors, the most common behavior was food selectivity (88%), followed by hypersensitivity to food textures (46%), other peculiar patterns such as eating only one brand of food (27%), pocketing food in the mouth without swallowing (19%), and pica (12%) [585]. Some children with ASD meet the DSM-5 criteria for avoidant/restrictive food intake disorder, but the prevalence is unknown [588,589].

Food selectivity is characterized by food refusal, limited food repertoire, and high-frequency single food intake [590]. Food refusal is often based on hypersensitivity to texture, but aversion based on taste, temperature, or color/appearance is also common and restriction of intake is sometimes related to fear-based avoidance (e.g., fear of choking or contamination) [585,587]. Problematic mealtime behaviors such as gagging, spitting, vomiting, self-injury, and aggression sometimes arise, which can be quite stressful for families. Feeding difficulties related to selective intake are correlated with sensory processing abnormalities, cognitive and behavioral rigidity, challenging behaviors, and constipation [586,591,592]. Most studies suggest that feeding problems persist throughout childhood and adolescence and are unlikely to resolve spontaneously, although some aspects improve with advancing age [586]. Although there are cases of protein-energy malnutrition as well as scurvy and other micronutrient deficiencies, the feeding problems and associated nutritional (especially calcium and protein) intake suboptimality in children with ASD do not typically result in greater risk for compromised growth [584,586,587,591]. In contrast to eating disorders, feeding problems associated with ASD are usually unrelated to weight, shape, and/or body image concerns [591,593]. There is some evidence of increased ASD prevalence in people with anorexia nervosa, but population-based studies evaluating the prevalence of anorexia nervosa in people with ASD are lacking [594].

Somatic medical conditions

In addition to the brain-based conditions that were reviewed in the preceding sections, there is evolving evidence of an excess burden of nonneurological somatic medical conditions associated with ASD. The association between childhood gastrointestinal conditions and ASD is perhaps the most well-known example. However, the relationship between nonneurological somatic medical issues and ASD is not confined to early childhood. For example, the burden of somatic conditions is substantial in transition-age autistic youth; independent adults with ASD self-reported high rates of current symptoms including fatigue (73%), sleep problems (69%), menstrual problems (61% of females), and pain (back pain 48%, joint pain 44%, abdominal pain 43%), all of which were more prevalent in women [520]. The total symptom burden and individual symptoms were associated with higher levels of depression, anxiety, and autistic traits, as well as lower quality of life [520].

Gastrointestinal disorders

Systematic literature reviews and metaanalyses have confirmed that gastrointestinal (GI) problems are common among individuals with ASD and that overall/aggregate GI symptoms and specific symptoms including constipation, diarrhea, and abdominal pain are significantly more prevalent among those with ASD versus typically developing controls [588,595–597]. A recent metaanalysis estimated the pooled prevalence of overall/aggregate GI symptoms to be 48.7% [588]. Among the most common specific GI symptoms are constipation (26%), diarrhea (20%), and abdominal pain/discomfort (21%) [588]. Relatively few studies have included children with other developmental disabilities as comparison groups, but those that have typically find that the prevalence of aggregate and specific GI symptoms among those with other developmental disabilities falls between the ASD and typically developing groups [502,598,599]. For example, in a report from the Study to Explore Early Development, the prevalence of any current GI symptom in 2- to 5-year-old children was significantly higher in those with ASD (34.6%) as compared to those with other developmental delays (DD; 22.1%) and children from the general population (POP; 12.0%) [502]. Similar statistically significant differences were found for constipation (ASD 23.5%, DD 16.3%, POP 8.5%) and diarrhea (ASD 10.1%, DD 2.6%, POP 1.5%) [502].

There is substantial evidence linking GI symptoms and behavior. Among children with ASD, cooccurring GI symptoms are associated with increased sensory issues, sleep problems, anxiety and other internalizing symptoms, and challenging behaviors [591,597,600–602]. In a well-characterized sample of children who participated in two Research Units on Pediatric Psychopharmacology Autism Network medication trials, the subgroup of children with moderate or greater GI problems (23%) had greater symptom severity on measures of irritability, anxiety, and social withdrawal than those without GI problems, and their irritability was less likely to respond to treatment with risperidone [603]. GI symptoms may not be obvious in some children with ASD due to communication deficits and/or differences in responses to sensory input but should be considered in the case of unexplained food refusal, irritability, agitation, self-injury, aggression, sleep disturbance, or other problem behavior that is new or represents a significant exacerbation of baseline status [595,604]. The contribution of behavioral factors such as delayed bowel training, rigidity about toileting routine, sensorimotor issues, and food selectivity to the observed higher prevalence of symptoms in children with ASD is unclear but plausibly substantial [595].

There is no evidence suggesting a unique GI pathophysiology in ASD [595,604]. In some cases, GI problems may be directly related to underlying genetic etiology [605,606]. Most GI-related symptoms that cooccur with ASD are functional problems, but there is now substantial evidence that inflammatory bowel disease is also overrepresented among children with ASD [607]. Among 1307 patients with ASD evaluated for celiac disease, 21 (1.6%) had evidence of celiac disease or celiac autoimmunity [608], which is not significantly different than the pooled global prevalence (1.4%) [609]. The summed ASD prevalence of 0.55% (210/38,426) among individuals with celiac disease is also unremarkable [608]. Numerous studies have described stool microbial differences among individuals with ASD in comparison to controls, but these studies fail to show consistent or reproducible microbial signatures, and a causal relationship has not been established [610,611]. A recent analysis using a Bayesian differential ranking algorithm provided some evidence for consistent alterations in the gut microbiome in ASD which, if confirmed, could potentially be a cause, effect, or confounder of ASD [611]. However, a large stool metagenomics study found negligible evidence for direct associations between the stool microbiome composition and ASD diagnostic status [612]. Instead, ASD was robustly associated with dietary variables which were linked to reduced microbiome diversity and looser stool consistency. These findings challenge the suggestion that the microbiome may be causally related to ASD traits and a potential target for intervention [612].

Cardiometabolic disease and obesity

ASD is associated with cardiometabolic risk, including increased risk of type 1 and type 2 DM, dyslipidemia (elevated triglycerides and high-density lipoprotein), and atherosclerotic heart disease [613]. There is also increased risk of DM and hypertension in children with ASD (vs children without ASD) [613], which may be partially mediated by the increased risk of overweight and obesity [519,614]. A 2019 metaanalysis reported high prevalence of overweight (37%) and obesity (22%) among children with ASD [614] and a subsequent large study found obesity in 33.5% of individuals with ASD (vs 21% of controls) at age 22 years [519]. The implications of these findings are substantial, since early-onset cardiometabolic disease is associated with increased morbidity, health care needs, and mortality in the general population [615,616]. The contributions of ASD-associated risk factors such as psychotropic medication use, selective diet, sedentary behavior, sleep problems, and genetic factors to obesity, type 2 DM, dyslipidemia, and atherosclerotic heart disease are not yet clear but warrant investigation.

Other somatic conditions

Ophthalmological problems are more common among children with ASD compared to control populations [617–619]. A large metaanalysis found significantly higher prevalence of general/unspecified vision problems, strabismus, accommodative deficits, and refractive error (but not disorders of refraction or visual acuity) among individuals with ASD versus those without ASD [619]. ASD was also associated with worse peripheral vision, poorer stereoacuity, and impairments in color vision and contrast sensitivity [619]. The evidence of overrepresentation in ASD is strongest for strabismus [618,619], which occurs in approximately 8%–10% of children with ASD [617,618]. Currently, there is no conclusive evidence that children with ASD are at increased risk for peripheral hearing loss [620,621].

The prevalence of atopic dermatitis is significantly higher among children with ASD in comparison to controls, whereas the prevalence of asthma is not, despite some evidence of familial coaggregation [622–625]. Associations between food allergies, allergic rhinitis, autoimmune diseases (as a group), and psoriasis and ASD are supported by a large case-control study but overall, the epidemiological evidence for an excess of immune-mediated conditions in ASD is limited [624].

Identification

Timely identification of ASD depends on community and professional awareness of the signs and symptoms, implementation of developmental surveillance and screening in primary care, and access to expert clinical diagnostic evaluation. Children typically present for diagnostic evaluation because of concerns about attainment of developmental milestones or the presence of aberrant behaviors, either raised spontaneously by parents or other caregivers or elicited by health care providers through developmental surveillance and screening (also see Chapter Developmental surveillance and screening). A comprehensive diagnostic evaluation should be completed regardless of the primary concern prompting referral; children referred for non-ASD concerns, such as language delay or disruptive behaviors, may be determined to have ASD and, conversely, those presenting with a specific concern about ASD are often found to have other neurodevelopmental disorders but not ASD [3].

Surveillance and screening

The AAP has established guidelines for developmental surveillance and screening, including ASD-specific recommendations [7,626]. Developmental surveillance, which should take place at every health supervision visit, is a flexible, longitudinal, and cumulative process through which knowledgeable clinicians identify children who may have developmental problems [626]. The key components include eliciting and attending to parental concerns, obtaining a developmental history, making accurate and informed observations of the child, identifying risk factors and protective factors, documenting the process and findings, and communicating with other professionals involved in the child's care to share and obtain information [626]. Developmental screening is the use of standardized tools to identify, confirm, and/or refine risk. The AAP recommends general developmental screening using a standardized broadband measure for all children at the 9-, 18-, and 30-month preventive care visits and when surveillance identifies concerns about a child's development [626,627]. In addition, an ASD-specific screening test is recommended at the 18- and 24-month visits [7,626]. An ASD-specific screening test may also be administered at any time if a parent or caregiver expresses concern, or the clinician observes or elicits a history of behavioral signs that may suggest ASD, and children may be screened earlier and more frequently if there is increased historical risk (e.g., prematurity or having an older sibling with ASD) [626].

Although the AAP recommends universal ASD screening at the 18- and 24-month visits, there is disagreement about whether this is warranted. The US Preventive Services Task Force (USPSTF) completed a literature review through August 2014 and concluded that there was insufficient evidence to assess the balance of benefits and harms of screening for ASD in children aged 18–30 months for whom no concerns of ASD have been raised [628,629]. The USPSTF did not recommend for or against universal screening; they advocated for more research and stated that in the meantime, clinicians should (1) use their clinical judgment about screening in the absence of concern about ASD, especially when caring for populations in which there are barriers to identification through surveillance, and (2) use validated tools to assess whether further diagnostic evaluation or services are needed to follow-up on concerns expressed by parents [629]. They did not address screening of children at elevated familial risk (e.g., due to having a sibling with ASD).

Available tools for early childhood screening for ASD have been thoroughly reviewed elsewhere [630–634]. A scoping review identified 26 different instruments (in 75 adaptations) that were potentially suitable for primary care screening for ASD in the first 3 years of life [635]. Examples of parent-completed and clinician-administered ASD-specific screening tools are presented in Table 28.6. There is substantial evidence that screening of unselected children between the ages of 16 and 40 months using available instruments can accurately identify many children with ASD [628,629,634,636–638]. Attempts to screen for ASD at younger ages have met with mixed success; prediction is less accurate and stable in younger children [630,639–641]. Separate bodies of evidence suggest potential benefits of screening on initiation of services and benefits of early intervention on positive outcomes, but studies directly evaluating the impact of screening (vs no screening) on clinical outcomes are lacking [629–634].

Information on universal ASD screening for older children in primary care or primary care-like settings is scarce [634–642]. One study provided some evidence that the Social Communication Questionnaire (SCQ), completed by early childhood care providers in Head Start and daycare/preschool settings, accurately identified ASD among 848 children aged 30–76 months [643]. Most of the evidence supporting a role for the SCQ (Lifetime Version) in identifying ASD comes from referred clinical populations of children aged 4 years or older [644]. There is a substantial reduction in diagnostic accuracy when children younger than 4 years are included and when the Current Version of the SCQ is used [644,645].

The Modified Checklist for Autism in Toddlers (M-CHAT) and the M-CHAT Revised With Follow-up (M-CHAT-R/F) are the most thoroughly studied and widely used ASD-specific screening tools. The M-CHAT-R/F includes a brief, structured interview administered by a health care professional when the score on the questionnaire is elevated. A meta-analysis of 50 studies showed that the version (M-CHAT vs M-CHAT-R/F) did not significantly affect sensitivity and specificity, so the data were summarized collectively and the abbreviation M-CHAT(-R/F) was used to include both versions [638]. The pooled sensitivity of M-CHAT(-R/F) was 0.83, and the pooled specificity was 0.94. However, the sensitivity and specificity estimates varied widely, and it is concerning that low sensitivity estimates were found in two recent, large primary care studies in Pennsylvania (0.39) [646] and Utah (0.33) [647]. Some of the variance is explained by study characteristics that affected diagnostic accuracy of the M-CHAT(-R/F) [638]. For example, sensitivity was higher when a concurrent, as opposed to prospective, case confirmation strategy was used, and specificity was higher in populations with low versus high likelihood of ASD. The M-CHAT(-R/F) was initially validated for the age range of 16 to 30 months, but it has been studied and used for older children, and this systematic review and metaanalysis provides evidence of utility up to age 48 months [638]. Although including the follow-up portion does not affect sensitivity, it does significantly reduce false-positive rates, which is important because this reduces the referral burden on diagnostic clinics and intervention services [638].

TABLE 28.6 Examples of autism spectrum disorder screening instruments designed for use in early childhood.

Screening instrument	Age range (months)
Parent-report questionnaires	
Early screening for autism and communication disorders (ESAC)	12–36
First-year inventory (FYI)	12
Infant-toddler checklist (ITC)[a]	9–24
Modified checklist for autism in toddlers (M-CHAT) Modified checklist for autism in toddlers, revised, with follow-up (M-CHAT-R/F)[b]	16–30
Parent observation of early milestones scale (POEMS)	3–24
Parent's observation of social interaction (POSI)[c]	16–36
Psychological development questionnaire-1 (PDQ-1)	18–36
Quantified checklist for autism in toddlers (Q-CHAT) Quantified checklist for autism in toddlers, 10-item version (Q-CHAT-10)	18–24
Direct observation measures	
Autism detection in early childhood (ADEC)	12–36
Autism observation scale for infants (AOSI)	6–18
Joint attention observation schedule (JA-OBS)	20–48
Screening tool for autism in toddlers and young children (STAT)[d]	12–36
Social attention and communication surveillance-revised (SACS-R)	11–30
Social attention and communication surveillance-preschool (SACS-PR)	31–60
Systematic observation of red flags of autism (SORF)	18–24

[a] *The ITC is a component of the communication and symbolic behavior scales-developmental profile (CSBS-DP); although designed as a broad screen risk for communication and/or social impairment (age 6–24 months), not an ASD screening tool, it includes social and symbolic behavior scales and has shown promise as an ASD screen from age 9 to 24 months.*
[b] *M-CHAT-R/F includes a brief, structured interview administered by a health care professional when the score on the questionnaire is elevated.*
[c] *The POSI is a component of the survey for well-being of young children (SWYC), a comprehensive primary care developmental screening instrument.*
[d] *Initially called the screening tool for autism in 2-year-olds (STAT).*
Source: Towle and Patrick (2016) [630]; Thabtah and Peebles (2019) [631]; Sánchez-García et al. (2019) [632]; Petrocchi et al. (2020) [633]; Levy et al. (2020) [634]; Sobieski et al. (2022) [635].

Diagnostic evaluation

Due to the challenges inherent in differentiating ASD from other developmental, behavioral, and emotional problems, the clinical standard of care for ASD diagnosis is expert clinician judgment, informed by data from a comprehensive evaluation and application of the current DSM or ICD diagnostic criteria (Table 28.1) [3,6,7,9,648]. Children and adolescents are most often diagnosed by licensed psychologists with expertise in neurodevelopmental disorders or by medical subspecialists in neurodevelopmental disabilities, developmental-behavioral pediatrics, child and adolescent psychiatry, or child neurology, either working independently or as part of a multidisciplinary team. Published guidelines often suggest that a multidisciplinary diagnostic team is preferred, but there is little supporting evidence available in the literature, and recommendations about practical issues such as distribution of roles and resolution of discrepant conclusions are lacking [3,649,650]. Some authors have asserted that in this context, "multidisciplinary" should be interpreted to mean that multiple areas of functioning are considered [651].

The goal of the diagnostic evaluation should not be to simply decide whether the child has ASD or not, but to determine the child's diagnoses, whatever they may be, and identify domain-specific deficits and strengths to inform intervention and service planning [3,11,651,652]. A comprehensive evaluation includes assessment of the child's level of functioning across domains and identification of cooccurring conditions, thereby avoiding the need for sequential disorder-specific evaluations (e.g., one assessment pathway for ASD, followed by another for ADHD) [3]. It is also important to ascertain patient and family needs, resources, and preferences as part of the diagnostic process and consider this information when formulating treatment recommendations.

A medical diagnostic evaluation should incorporate the following elements (also see Chapters, The developmental history, The physical evaluation, Developmental testing for the physician, The etiological evaluation): [6,7,9,651–653]:

1. *Parent/caregiver interview*, including a thorough medical history and review of systems, developmental and behavioral history (including past milestone attainment and current abilities), social history, and three-generation family history. Standardized rating scales completed by parents/caregivers and/or professionals (e.g., teachers, therapists, behavioral intervention staff) are commonly used to quantify symptoms and compare them to normative populations. Examples of commonly used ASD-related measures of this type are provided in Table 28.7. Structured interview tools, such as those described in Table 28.8, are informative but are often impractical in clinical settings because of time constraints.
2. *Review of existing data* such as pertinent medical and educational records, including any available standardized testing and other pertinent data collected by other professionals (e.g., early intervention evaluators, psychologists, speech-language pathologists, occupational therapists, and others). The results of any previous etiologic investigations and other pertinent studies should be noted (e.g., genetic testing, audiometry, brain imaging).
3. *Direct clinical assessment*, including physical examination, standardized developmental and/or psychological testing (appropriate for age and level of ability), and neurobehavioral observation/elicitation. It is important to measure current cognitive, language, and adaptive functioning. In the case of team evaluations, this testing is often completed by someone other than the physician. In some cases, physician diagnosticians who work independently may review and interpret the results of current standardized testing performed by professionals in other disciplines in lieu of directly administering tests themselves. Administration of a standardized, ASD-specific direct observation measure, such those described in Table 28.8, may be helpful to support and enhance expert clinical judgment. Physical examination may identify clues to etiology or important comorbidities and thereby help to guide future investigations and treatments.
4. *Integration of information and determination of categorical diagnoses*, most often using current criteria from the DSM-5-TR [36] or the ICD-11 [35]. Clinical judgment is informed by the information gathered in Steps 1 through 3. Other conditions in the differential diagnosis must be considered (Table 28.9). If ASD is diagnosed using DSM-5 criteria, the current severity levels of social communication impairment and restricted, repetitive behavior are recorded, and diagnostic specifiers are addressed (Table 28.1). Similarly, for ICD-11-based ASD diagnosis, specifiers and qualifiers are noted. Beyond enumeration of categorical diagnoses, the diagnostic formulation should include information about the individual child's profile of pertinent cognitive and functional strengths and weaknesses, which are relevant to short-term and long-term outcomes and care decisions.

TABLE 28.7 Examples of standardized, quantitative parent-report measures of autism spectrum disorder-related behaviors.

Instrument	Forms available (age range in years)	Number of items
Autism spectrum rating scales (ASRS)	Parent form (2–5) Parent form (6–18) Teacher form (2–5) Teacher form (6–18) Parent/Teacher short form (2–18)	70 71 70 71 15
Children's communication checklist (CCC-2)[a,b]	Single form (4–16)	70
Gilliam autism rating scale (GARS-3)	Single form (3–22)	58
Social responsiveness scale (SRS-2)	Preschool (2.5–4.5) School-age (4–18) Adult (≥19) Adult self-report (≥19)	65 65 65 65

[a]*The CCC-2 is not an ASD-specific measure; it assesses general language and pragmatics. However, the social-interaction difference index (SIDI) identifies children with disproportionate pragmatic and social difficulties in relation to their structural language impairments.*
[b]*The CCC-2 should be administered only if the child is able to speak in simple sentences.*
Source: Charman and Gotham (2013) [654]; Evers et al. (2020) [655]; Constantino and Gruber (2012) [656]; Frazier et al. (2014) [657]; Goldstein and Naglieri (2013) [658]; Goldstein and Naglieri (2014) [659]; Gilliam (2014) [660]; Bishop (2003) [661]; Nowell et al. (2022) [662].

TABLE 28.8 Autism spectrum disorder diagnostic tools: examples of structured interviews and observational measures.

Diagnostic instrument	Comments
Structured caregiver interviews	
Autism diagnostic interview-revised (ADI-R)	Semistructured caregiver interview Age: Mental age >24 months for published version; existing algorithms extend the age range down to 12 months (and nonverbal mental age ≥10 months) Administration time: 90–150 min
Developmental, dimensional and diagnostic interview (3Di) Developmental diagnostic dimensional interview—short version (3Di-sv)	Standardized, computer-based diagnostic interview 3Di (full version) includes mandatory modules covering ASD symptoms and optional modules on comorbid symptoms Age ≥2 years Administration time: • 3Di: 90–180 min • 3Di-sv: 45 min
Diagnostic interview for social and communication disorders (DISCO-11)	Semistructured interview, but ratings can be based on any available information, including direct observation and reports from teachers or other caregivers Age: All Administration time: 120–240 min (including scoring); short version 45–60 minu
Observational measures	
Autism diagnostic observation schedule (ADOS-2)	Semistructured standardized observation/elicitation Administration time: 40–60 min Form selection: Toddler module: Age 12–30 months, no consistent phrase speech Module 1: Age >30 months, no consistent phrase speech Module 2: Any age, use phrase speech but not verbally fluent Module 3: Verbally fluent children and young adolescents Module 4: Verbally fluent older adolescents and adults
Childhood autism rating scale, standard (CARS2-ST)[a] Childhood autism rating scale, high-functioning (CARS2-HF)[a]	Combination interview and unstructured observation/elicitation Yields cutoff scores as well as T-scores and percentile ranks (derived from ASD populations) Administration time: ~30–45 min (5–10 min after the information needed to make the ratings has been collected) Form selection: • CARS2-ST: Age <6 years **or** estimated IQ <80 **or** notable communication impairment • CARS2-HF: Age ≥6 years **and** IQ≥80 **and** fluent communication
Childhood autism rating scale adaptation based exclusively on patient observation (CARS2^{obs})	Adaptation of the CARS2-ST or CARS2-HF with scoring based exclusively on a 15-min protocol-based observation for expedited assessmentIncludes the first eight items of the CARS2-ST or CARS2-HF
Telehealth tools	
Naturalistic observation diagnostic assessment (NODA)	Store-and-forward telehealth approach that relies on developmental history and parent-collected videos Four 10-min scenarios: Family mealtime, playtime with others, playtime alone, and parent concerns Instructions to introduce specific social presses included in the application
Brief observation of symptoms of autism (BOSA)	Adapted from the brief observation of social communication change (BOSCC) and the ADOS-2 naturalistic social context with standardized materials and activities Four versions: Minimally verbal (MV); phrase speech-young fluent (PSYF); fluent, age 6–10 years (F1); fluent, age 11-adult (F2)

[a]*Information from parents can be obtained with the CARS2-QPC (questionnaire of parent concerns), an unscored form in which parents/caregivers answer questions about their observations.*

Source: Charman and Gotham (2013) [654]; Randall et al. (2018) [663]; Evers et al. (2020) [655]; Lebersfeld et al. (2021) [664]; Rutter, Le Couteur, et al. (2003) [665]; Kim and Lord (2012) [666]; Kim et al. (2013) [667]; de Bildt et al. (2015) [668]; Wing et al. (2002) [669]; Wing (2006) [670]; Skuse et al. (2004) [671]; Santhosh et al. (2009) [672]; Slappendel et al. (2016) [673]; Schopler et al. (2010) [674]; Moon et al. (2019) [675]; Sanchez and Constantino (2020) [676]; Lord, Rutter, et al. (2012) [677]; Lord, Luyster, et al. (2012) [678]; Smith, Rozga, et al. (2017) [679], Lord, Holbrook, et al. (2020) [680]; Dow et al. (2022) [681].

TABLE 28.9 Differential diagnosis: conditions that may mimic ASD.[a,b]

DSM-5 neurodevelopmental disorders Global developmental delay Intellectual disability Social (pragmatic) communication disorder Developmental language disorder Attention-deficit/hyperactivity disorder Stereotypic movement disorder Tourette's disorder
Learning differences (not captured in DSM-5) Language-based learning disorder Nonverbal learning disorder Intellectual giftedness
Other DSM-5 disorders Attachment disorders Anxiety disorders (especially selective mutism, social anxiety disorder) Obsessive-compulsive disorder Depressive disorders Schizophrenia spectrum and other psychotic disorders Personality disorders
Other neurological/sensory disorders Hearing impairment Acquired epileptic aphasia

ASD, Autism spectrum disorder; *DSM-5*, diagnostic and statistical manual of mental disorders, fifth edition.
[a]*Because of shared behavioral and cognitive features, many disorders may mimic ASD, especially when they appear in combination.*
[b]*With the exception of social (pragmatic) communication disorder, each of these conditions may also cooccur with ASD.*

5. *Consideration of etiologic possibilities* and determination of appropriate investigations to evaluate for etiologic diagnoses and/or associated medical conditions (see Management section).
6. *Recommendations for intervention and support*, based on the individual child's diagnoses and profile of abilities and clinician knowledge of the literature regarding evidence-based treatment (see Section IV: Management of neurodevelopmental disorders).

The task of gathering and interpreting historical information is much more complicated than simply taking an inventory of ASD symptoms and mapping them to the DSM-5 or ICD-11 criteria. The diagnostician must determine whether relevant behaviors are situation-specific or pervasive and attempt to establish whether these behaviors have an autistic quality while accounting/adjusting for the confounding influences of comorbid symptoms [79,682]. This requires clinical experience and expertise not only with ASD but also with other neurodevelopmental and psychiatric disorders as well as age-dependent normal variation in development and behavior (Table 28.9). When evaluating young children, expertise in global developmental delay, ID, language disorders, ADHD, and anxiety disorders is particularly important. Additional clinical experience and proficiency with mood disorders, obsessive-compulsive disorder, and psychotic disorders becomes increasingly important for assessment of older children and adolescents. Personality disorders are added to the list of necessary psychopathology expertise for diagnosis of older adolescents and adults [683].

A variety of standardized instruments have been developed to aid information gathering and support clinical judgment [654,655,663,664]. These include standardized dimensional measures that are completed by parent, teachers, or other caregivers and help to structure and quantify information about ASD-related traits (Table 28.7). Measures such as the Social Responsiveness Scale, Second Edition, Autism Spectrum Rating Scales, and Gilliam Autism Rating Scale are commonly used for this purpose [656–660]. However, caution is warranted, especially regarding cutoff scores, because multiple studies suggest favorable sensitivity but fair to poor specificity in clinical populations [648,684–692]. Although not an ASD-specific measure, the Children's Communication Checklist is useful for quantifying pragmatic and higher-order semantic aspects of language and detecting deficits that are discrepant from structural language functioning [661,662]. These scales cannot be used as the sole or primary basis for making a diagnosis, but they do facilitate quantitative documentation of ASD features in both core domains and provide benchmarking for assessment of change over time [3].

Clinician-administered diagnostic tools based on structured interviews and/or direct observations are also available (Table 28.8). Semistructured interview tools include the ADI-R [664–668], Diagnostic Interview for Social and Communication Disorders [655,669,670], and Developmental, Dimensional and Diagnostic Interview [655,671–673]. All are valid and reliable tools that are lengthy to administer and require substantial examiner training. SCQ is a parent-completed questionnaire consisting of 40 dichotomous items based on the ADI-R [644,645,693]. Like the ADI-R, the SCQ Lifetime Version focuses on past behavior (specifically, the period between their fourth and fifth birthdays) in addition to current symptoms. The SCQ has been widely used as a screener for entry into research studies and for clinical screening of patients referred for non-ASD concerns.

The Childhood Autism Rating Scale (CARS2) "standard" and "high-functioning" forms are five-item questionnaires that are completed by trained clinicians based on interview, unstructured observation of the patient, and collateral information to quantify ASD symptom burden (Table 28.8) [674,675]. Each item is scored from 1 to 4 at half-point intervals using narrative scoring anchors, and training requirements are minimal as long as the clinician has expertise in ASD. It is not appropriate to ask parents or teachers to complete the CARS2 forms, but an unscored form, the CARS2-QPC (questionnaire of parent concerns), is available to collect information from parents and other caregivers. Both forms of the CARS2 provide severity grouping based on cutoff scores (minimal-to-no, mild-to-moderate, or severe symptoms of ASD) as well as T-scores and percentile ranks [674]. It should be noted that T-scores and percentile ranks on the CARS2 are based on clinical reference samples of individuals with ASD, not the general population, so a T-score of 50 indicates that the child scored at the mean for similar-age children with ASD. Initial reports suggest that the CARS2^{obs}, an adaptation based on a brief observation and scoring of the first eight CARS2 items, may be an effective component of expedited clinical assessment [676].

The ADOS-2, widely considered to be the "gold standard" of ASD observational measures, is a clinician-administered interactive assessment designed to elicit behaviors relevant to the diagnosis of ASD using specific planned social occasions, prompts, and materials [654,664,677,678]. Five modules are available, based on age and level of communication (Table 28.8). Module 1–4 scoring algorithms apply cut-offs to provide a categorical ADOS-2 classification of "autism," "autism spectrum," or "non-spectrum," whereas the Toddler Module outcome is expressed as an ADOS-2 range of concern ("moderate-to-severe concern," "mild-to-moderate concern," or "little-to-no-concern") [677,678]. For Modules 1–3, a dimensional comparison score (or calibrated severity score) ranging from 1 to 10 provides an indicator of severity and tool for comparison across modules or time [694]. Comparison score algorithms for the Toddler Module and Module 4 have been published but are still undergoing investigation and are not currently part of the version of the ADOS-2 available commercially [695,696].

Telehealth is the use of telecommunications and information technology to provide access to health assessment, diagnosis, intervention, consultation, supervision, and information at a distance [697]. Reimbursable telehealth involves two-way, real-time interactive communication between the patient and the physician or practitioner at the distant site. Telehealth evaluation expanded greatly during the COVID-19 pandemic, and several ASD measures have been adapted or created for this purpose (Table 28.8) [679–681,698,699]. Although there is emerging evidence in support of these tools and of telehealth as an effective alternative to in-person diagnostic evaluation, additional rigorous empirical research is needed [698,699].

The *Lancet* Commission report recommends use of at least one standardized instrument for documentation of the severity of ASD signs and symptoms and to provide a benchmark for later reassessments [3]. Use of these tools is also supported, but not mandated, by professional society guidelines [6,7,9,11]. However, it is important to keep in mind that these instruments were developed to structure, support, and enhance expert clinical judgment, not to replace it or stand alone as arbiters of diagnostic validity [79,700]. As emphasized by Bishop and Lord [700], ***instruments do not make diagnoses; clinicians make diagnoses***. Among the various diagnostic tools, there is substantial variability in the number and nature of items applicable to DSM-5 subcriteria included, and the algorithms generally do not adequately address the DSM-5 C ("early onset"), D ("impact on daily functioning"), or E ("not better explained by ID") criteria [655]. Determination of whether these criteria are met requires additional information gathering and clinician judgment. Moreover, the discriminative abilities of both parent-completed scales and clinician-administered measures are impacted by individual characteristics beyond ASD signs and symptoms, such as IQ and cooccurring emotional or behavioral problems (e.g., anxiety, ADHD) [440,648,685]. No tool or combination of tools can make up for lack of clinical expertise in ASD and related disorders, and the practice of requiring the use of specific tools in order to access services directly contradicts best practice recommendations for individualizing assessment procedures [3,79,700]. A prospective diagnostic study found that clinical diagnoses of ASD in young children by developmental-behavioral pediatricians with and without the ADOS were consistent in 90% of cases, and clinicians were able to identify children for who the ADOS may contribute to accurate diagnosis [701]. This suggests that the ADOS is generally not required for accurate ASD diagnosis in preschool-age children by expert clinicians.

Potentially problematic diagnostic practices encountered in research and sometimes in clinical practice include relying on self-report ASD symptom checklists and/or unverified self-reported diagnosis, standardized rating scale scores (e.g., SCQ or SRS), or isolated administration of a single instrument as the sole determinant of ASD status; collecting information only about current symptoms, often from a single reporter; and failing to adjust for the effect of cooccurring conditions [79,648]. Some authorities have expressed concern that many clinicians who trained within the current era of lower ASD diagnostic threshold and increased proportion of referrals with borderline or higher IQ may have less experience and expertise with ID (especially in the severe to profound range) than previous generations, particularly if their training had an ASD focus [702]. Another concern is that large numbers of professionals from various disciplines are being trained to administer the ADOS (e.g., in a 2-day introductory training session) despite a lack of broader training in assessment and differential diagnosis of ASD and other neurodevelopmental and psychiatric disorders [700]. This was not the intended use of the ADOS, and the validity data do not reflect administration by examiners with this level of training and expertise.

Children who have been diagnosed with ASD also require periodic formal reevaluation because their strengths, weaknesses, and educational needs often change even if the ASD diagnosis does not [3,651]. Due to the rapid developmental changes in preschool years, the *Lancet* Commission recommends focused reevaluation, ideally, within a year of the first diagnosis in this age group [3]. Focused or more comprehensive reassessment may be helpful when specific concerns are raised, at times of transition, and for periodic monitoring of progress throughout childhood and adolescence. Tracking developmental trajectories over time helps to refine intervention needs, guide etiologic considerations and investigations, and inform counseling regarding prognosis.

Management

Medical management of patients after the diagnosis of ASD includes evaluating for underlying etiologic diagnosis and/or associated medical conditions; providing recurrence risk counseling; guiding families to effective psychosocial and medical interventions; providing family and individual support, education, and access to resources; and promoting general health and well-being through effective longitudinal health care within a medical home [3,7,10,11]. Strengths and needs of individuals with ASD are heterogeneous and evolve over time, so periodic reassessment is important to guide intervention. Effective management requires coordination across systems of care (e.g., health care, education, social care, employment, financial, and safety net services) and navigation of transitions within and between these systems of care due to age or other determinants of service eligibility [3].

Etiologic evaluation

When the diagnosis of ASD is made by a physician or a diagnostic team that includes a physician, they typically also take responsibility for the etiologic evaluation. If the diagnosis is made by a psychologist or by a physician who does not provide this service, referral to a medical geneticist is appropriate. In some cases, the pediatrician or other primary health care provider may take on this role. There is currently no single, consensus approach to etiologic evaluation for children diagnosed with ASD. There is general agreement that genetic testing should be offered, but not about which test(s) should be first tier or other details of the approach to testing [3,703–706]. In recent years, the availability of new methods of genetic testing has progressed rapidly and the publication of state-of-the-art guidelines by professional organizations has lagged [707]. Therefore the genetic testing recommendations in this chapter reflect the authors' assessment of the current literature rather than strict adherence to guidelines published previously by various professional organizations.

Etiologic evaluation begins with the history and physical examination, which guide decision making about clinical laboratory tests and neurodiagnostic studies, such as neuroimaging and electrophysiological studies, to evaluate for underlying etiologic diagnosis and/or associated medical conditions. Examples of potential indications for clinical testing are outlined in Table 28.10. A more detailed discussion of neuroimaging, neurophysiologic, metabolic, and genetic testing in the etiologic investigation of neurodevelopmental disabilities can be found in Chapter (The etiological evaluation).

Clinical genetic testing

A causal genetic variant can be identified in at least 15%–20% of affected individuals, and all patients with an ASD diagnosis should be offered genetic etiologic evaluation [705,707–709]. For patients with ASD, the molecular diagnostic yield is approximately 16%–17% for exome sequencing (also referred to as WES) [705,708], 7%–9% for CMA [710–713], and 0.5% for *FMR1* CGG repeat analysis [707,710,711,714,715]. Factors associated with increased

TABLE 28.10 Potential indications for clinical testing in the evaluation of patients with ASD.

Genetic testing	Metabolic testing	Neuroimaging[a]	Electroencephalography
All patients with ASD	Acute or intermittent decompensations Developmental regression (especially progressive, degenerative symptoms) Lethargy Seizures (especially focal or intractable) Extrapyramidal movement disorder Severe hypotonia Ataxia Hearing loss Oculomotor abnormalities, ptosis Retinopathy, cataracts, corneal clouding or depositions, lens dislocation Cyclic vomiting Failure to thrive Hepatomegaly, splenomegaly Multisystem organ dysfunction (e.g., cardiac, hepatic, renal) Parental consanguinity	Focal neurologic findings (e.g., cranial nerve abnormalities, spasticity, ataxia, focal weakness, or sensory loss) Skin findings suggestive of neurocutaneous syndromes Epilepsy Signs of elevated intracranial pressure Unexplained micro/macrocephaly Developmental regression, especially progressive, degenerative symptoms Genetic diagnosis with recognized CNS abnormalities Historical evidence of an acquired brain injury or stupor/coma	Clinical concern for seizures Developmental regression (particularly marked regression in receptive and expressive language) Cortical CNS malformations

ASD, Autism spectrum disorder; *CNS*, central nervous system; *MR*, magnetic resonance; *MRI*, magnetic resonance imaging.
[a]*MRI of the brain, often with MR spectroscopy when a metabolic disorder is suspected.*

molecular diagnostic yield include low IQ, epilepsy, congenital anomalies, dysmorphic features, female sex, simplex status, and larger number of unaffected siblings [93,108,153,712,716]. However, the yield is substantial even in the absence of ID, dysmorphic features, or congenital anomalies and in multiplex families [92,99,712]. Identification of an underlying genetic etiology is important because it may allow more accurate prognostic and recurrence risk counseling, prevention of morbidity through timely identification and treatment of medical conditions known to be associated with the genetic variant, refinement of treatment options (including the potential for targeted treatment in some cases), and access to condition-specific family support groups and clinical trials [7,707,717].

Standard genetic etiologic evaluation should be within the purview of medical subspecialists who diagnose and treat children with ASD [707]. However, clinicians should only order tests that they are capable of explaining and interpreting; full interpretation of abnormal or equivocal results may involve referral to an appropriate subspecialist. The process of obtaining informed consent for testing includes describing the nature of the test, including potential outcomes (including secondary findings), and addressing points such as possible impact on other family members. Some neurodevelopmental disabilities and pediatric neurology clinics have integrated genetic counselors to provide support and expertise around the issue of genetic testing, including pretest and posttest counseling.

The following stepwise general approach to genetic diagnostic testing in ASD, based on the authors' review of the literature and clinical experience, is offered as a practical guideline, recognizing that the approach is certain to evolve over time (adapted from Ref. [707]):

Step 1. The etiologic investigation begins with a careful personal and family medical history and physical examination, including assessment for abnormal growth (including head circumference), dysmorphic features, evidence of visceral storage, skin manifestations of neurocutaneous disorders or mosaicism, and neurological abnormalities [703]. **If a specific etiology is suspected based on history and physical examination, pursue the appropriate specific genetic testing** (e.g., *FMR1* CGG repeat analysis, *PTEN* sequencing and deletion/duplication analysis, *MECP2* sequencing and deletion/duplication analysis, etc.). **If a specific etiology is not suspected, proceed to Step 2.**

Step 2. Pursue broad evaluation for causal variants: Either exome sequencing (ES) with CNV calling or genome sequencing (GS). ES with CNV calling detects exonic sequence variants (SNVs, indels) and CNVs. GS detects exonic and noncoding sequence variants (SNVs, indels), mitochondrial DNA variants, and structural variants including CNVs, tandem repeat expansions, inversions, and larger insertions. Trio analysis is preferred; this informs variant classification and is associated with increased diagnostic yield [718–720]. If GS is not feasible and insurance restrictions or laboratory offerings dictate a stepwise approach to CNV analysis and ES, CMA and ES can be pursued individually. If an etiology is not determined, proceed to Step 3.

Step 3. Consider additional testing and/or referral to medical genetics and facilitate periodic reanalysis of the exome or genome sequence. Additional testing, such as fragile X analysis (*FMR1* CGG repeat analysis) if the test performed in Step 2 was not GS and/or mitochondrial genome sequencing, should be considered. If GS was not performed in Step 2, it can be considered at this point. Periodic reanalysis of the exome or genome sequence obtained in Step 2 is recommended, perhaps every 2 years (although the optimal interval is not known), because it increases the diagnostic yield due to novel gene-disease discoveries, updated clinical information, and improved bioinformatics tools [721]. At this point, referral to a medical genetics service for further evaluation should be considered. In some cases, other test modalities, such as genome-wide DNA methylation "episignature" or RNA-sequencing studies may be considered to support variant reclassification or to suggest a specific condition in people with previously nondiagnostic testing [722–724].

These recommendations differ from those published by professional organizations, some of which have not been updated recently, but they reflect the current literature and evidence-based recommendations of an interdisciplinary group of experts in clinical genetics, laboratory genetics, genetic counseling, child neurology, neurodevelopmental disabilities, and developmental-behavioral pediatrics [705,708,725,726]. The AAP guidelines recommend offering CMA and fragile X analysis to all patients with ASD, consideration of *MECP2* sequence analysis for girls, and then potentially ES if those are unrevealing [7]. In 2017 the Autism and Intellectual Disability Committee of the AACAP also recommended ES as a second-tier test to be considered if CMA and fragile X analysis are negative [704]. Based on a systematic evidence-based review of the literature, the ACMG updated their recommendations for diagnostic testing in children with ID in 2021 to include ES/GS as a first- or second-tier test [727,728], but the 2013 ASD guideline has not yet been updated [703].

Targeted ASD gene panels are available for clinical use but were not included in our recommendations. Commercial targeted ASD panels vary tremendously in the number of genes included (e.g., from 27 to 2641 genes in the panels assessed by Ní Ghrálaigh and colleagues [729]) and the percentage overlap with the Simons Foundation Autism Research Initiative Gene database [729]. Compared to ES, targeted gene panels have a lower diagnostic yield and higher rate of inconclusive results due to variants of uncertain significance [729,730]. Unlike gene panels, ES includes genes not yet known to be associated with a specific phenotype, such as ASD. As a result, when new gene-disease relationships are discovered, they can be readily incorporated into variant interpretation algorithms and captured on exome reanalysis, whereas gene panel reanalysis typically can only incorporate new information about the genes originally included in the panel [139,707].

ES with CNV calling is more widely utilized currently, but it is anticipated that as the cost decreases, GS will be used more often and eventually become the preferred first-tier genetic test for etiologic evaluation of ASD and other neurodevelopmental disorders. As described in Step 2, GS can detect more types of genetic variation, so it is not surprising that the molecular diagnostic yield is higher than that of ES for the evaluation of ASD and related neurodevelopmental disorders as well as for other pediatric indications [731–734].

Clinical metabolic testing, neuroimaging, and electroencephalography

Although metabolic testing, neuroimaging, and EEG are not recommended for everyone with ASD, certain clinical findings should prompt consideration of one or more studies, as outlined in Table 28.10.

Clinical metabolic testing

Most metabolic disorders (inborn errors of metabolism), including 85% of the approximately 300 disorders that were newly described between 2011 and 2016, display predominantly neurological manifestations, and all categories of inborn errors have been implicated in neurodevelopmental disorders, including ASD [735–738]. These are most commonly autosomal recessive disorders that present early in life with symptoms such as those listed in Table 28.10 [703,706,737,739]. Mitochondrial disorders are among the metabolic disorders implicated in a small percentage of individuals with ASD, and multiple lines of evidence suggest mitochondrial dysfunction, even in the absence of an identifiable disorder [740]. One study suggested that individuals with ASD caused by a mitochondrial disorder were not clinically distinguishable from those with idiopathic ASD [741], but another demonstrated that careful assessment identified clinical findings that differentiated them from children with idiopathic autism [742].

Although many different metabolic disorders can present with an ASD phenotype, collectively they are responsible for only a small fraction of ASD cases [703,737–739]. Population-based estimates of the prevalence of metabolic disorders among individuals with ASD are lacking, but the existing literature suggests that the yield of routine metabolic testing in ASD is very low [703,743,744]. For example, one study of 274 patients with "nonsyndromic" ASD (although it did include individuals with ASD and ID) failed to detect any creatine deficiency syndromes, mucopolysaccharidoses, amino acid metabolism defects, or abnormalities of purine and pyrimidine metabolism but did identify two individuals with nondiagnostic persistent biochemical abnormalities [743]. Another study did not identify any metabolic disorders among 406 children and young adults with "nonsyndromic" ASD whose urine samples were analyzed for evidence of the same groups of disorders [744]. The prevalence of inborn errors of metabolism is much higher among patients with ASD in populations in which consanguineous marriages are common, so it is important to maintain a high index of suspicion when there is a family history of consanguinity [745,746]. Many of these metabolic disorders can be detected by newborn screening using tandem mass spectrometry and also by ES/GS in Step 2 of the genetic etiologic evaluation suggested above [747,748].

Based on the available evidence, no metabolic tests are routinely recommended for all children with ASD at this time [7,703,706,749]. However, the potential impact of diagnosis is high because of the availability of treatment, in some cases, and high recurrence risk within families, so testing should be considered in certain circumstances, as outlined in Table 28.10 [703,706,750]. Bolstered by the development of analyte databases, untargeted urine and plasma metabolomic analyses have emerged as a promising tool for screening for biochemical abnormalities and evaluating them as potential biomarkers, but the prospect of clinical use remains speculative at this time [751,752].

Clinical neuroimaging

The clinical neuroimaging literature has consistently demonstrated that there are no specific abnormalities or patterns of findings that are unique to ASD or meaningfully correlated with clinical, etiologic, or pathophysiological aspects of the disorder [753,754]. Some clinical MRI studies have found that neuroradiological findings are more common in children with ASD than in typically developing controls [754–756], whereas others have not found significant differences [757–759]. Studies comparing ASD to other neurodevelopmental disorders such as ADHD [758] and ID [754,755] have not demonstrated significant differences in the prevalence of neuroradiological findings. The vast majority of clinical neuroimaging findings described in ASD cohorts are incidental findings, defined as "unsought asymptomatic brain findings generated while seeking other information of interest," [759] which are common in children undergoing MRI of the brain for any reason and are rarely clinically meaningful [753,760,761]. Mega cisterna magna, qualitative abnormalities of the opercular formation, macrocephaly, ventriculomegaly, and hypoplasia of the corpus callosum are among the most common incidental findings described in ASD cohorts more commonly than in controls [754,756]. However, in a large study with systematic, comprehensive assessment of neuroradiological findings, only mega cisterna magna and opercular abnormalities remained significant after application of a false discovery rate correction to control for multiple comparisons [754].

Although structural neuroimaging research has identified morphometric differences in groups of individuals with ASD relative to controls, reliable markers for diagnosis or clinical subtyping have not been identified, and clinical neuroimaging is not recommended for all individuals with ASD [6,7,703]. However, MRI of the brain should be considered on an individual basis in the presence of any of the historical and examination findings described in Table 28.10 [6,7,703,753,762].

Clinical electroencephalography

As previously described, ASD is associated with increased risk of epilepsy; the prevalence is estimated to be about 10% in population-based samples and closer to 20% in clinical samples [535], with higher rates among those with cooccurring ID. Additionally, many more children with ASD who have never had a clinical seizure have been found to have interictal EEG abnormalities of unclear significance [532,537,539–541,763]. In clinical practice, EEG is not routinely indicated for all children with ASD [7,532,764]. Instead, EEG should be used as a tool in the overall clinical evaluation for epilepsy, in the context of a careful clinical history. When seizures are suspected clinically, the EEG may provide supporting evidence for or against a diagnosis of epilepsy and help with epilepsy syndrome classification [765]. Although the evidence is mixed, some authorities consider unexplained developmental regression, especially involving language, to warrant a sleep-deprived EEG with appropriate sampling of slow-wave sleep (often an overnight study) [532,764,766].

Recurrence risk counseling

Parents of a child with ASD should be counseled regarding recurrence risk in subsequent pregnancies. When a specific genetic etiology has been identified through testing, the family can be provided with accurate information about the risk of recurrence in each subsequent pregnancy. When genetic testing has not been completed or has not revealed the etiology of the child's ASD, recurrence risk counseling is based on estimates derived from empirical data in the ASD epidemiological literature.

Relative recurrence risk (RRR) is the risk of an ASD diagnosis in a sibling of a child with ASD compared to the risk of ASD diagnosis in a sibling of an unaffected child, so it accounts for the general ASD risk in the base population at the same time and can be compared between groups. In the largest and most recent population-based epidemiological study of recurrence risk, using population-based data from five countries, the RRR was 9.3 for full siblings [767], which is similar to the findings of other large studies (RRR 7.5–11.8) [768,769]. Absolute recurrence risk estimates published in the last decade range from 10% to 20% [768,770–773]. The highest rate of recurrence (19.5%) was found in a large, prospective study of 1241 younger siblings of children with ASD who were recruited in infancy and monitored closely [773]. There may be bias inherent to volunteering to participate in this type of study, so it is unclear whether this can be generalized to the broader population. However, an ASD recurrence range estimate of 10%–20% has been recommended for counseling purposes when a couple has already had one child with ASD [774]. When a couple already has two or more children with ASD, the recurrence risk increases to 30%–50% [772–775].

Because of the increased risk of ASD, all younger siblings of an affected child should be monitored through developmental surveillance and routine administration of ASD screening tools to facilitate early identification and intervention. However, the risk is not limited to ASD. Studies have documented not only increased ASD risk but also increased risk for other neurodevelopmental disorders, which further complicates recurrence risk counseling [771,776–778].

When the genetic etiology is known, the added risk (beyond the background population rate) of each subsequent child having ASD or a related phenotype of similar severity may range from as high as 50% in the case of an inherited maternally derived chromosome 15q11-q13 interstitial duplication to 25% in the case of an autosomal recessive disorder (e.g., Smith-Lemli-Opitz syndrome), to 1%–2% if the affected child has a de novo pathogenic CNV or SNV (the estimated chance of parental germline/gonadal mosaicism for the variant). Since the most commonly identified pathogenic variants are de novo SNVs and CNVs, the recurrence risk estimate provided to the family (1%–2% plus the background population risk of 2%–3%) is lower than the empiric recurrence risk of 10%–20% that is provided when no pathogenic variant is identified or when genetic testing is declined.

Approach to intervention

Meeting the varied and complex needs of children and adolescents with ASD can be challenging for families and for pediatric service and health care providers. Among the most important roles of clinicians who diagnose and manage children with ASD are educating families about empirically supported interventions, guiding them to evidence-based options and accurate information resources, and helping them to navigate the multiple systems of care and transitions that are typically involved [3,7].

Goals and therapeutic targets

Ultimately, the goal of intervention is to enhance quality of life by preparing individuals to make and communicate decisions about their own lives, live as independently as possible, achieve academically and vocationally to the best of their potential, develop satisfying social relationships, and participate as fully in the community as they desire. Broadly, the more proximal goals of intervention include building developmentally appropriate adaptive skills and eliminating or minimizing maladaptive behaviors and emotional states that have deleterious effects on learning, functional independence, community participation, and quality of life (Table 28.11) [3,779–782]. These are not completely independent concepts since successful reduction of maladaptive behaviors usually involves teaching and reinforcing other skills and behaviors. Both skill-building and reduction of problem behaviors often involve remediation of deficits and enhancement of compensatory processes. Psychosocial interventions, such as behavioral and developmental interventions, social skills groups, cognitive-behavioral therapy (CBT), and other educational and mental health practices, are commonly used for both skill-building and reduction of interfering symptoms or cooccurring conditions. In contrast, the use of

TABLE 28.11 Autism spectrum disorder intervention: broad goals and specific targets.

1. Building developmentally appropriate skills
   - Learning readiness skills (not directly related to task content; e.g., on task behavior, engagement)
   - Verbal and nonverbal communication
   - Interpersonal social skills needed to interact with others
   - Higher cognitive functions (e.g., intellectual performance, executive function, problem-solving, information processing, reasoning, theory of mind, memory, creativity, attention)
   - Motor skills (fine motor, gross motor, visual-motor coordination)
   - Play and leisure skills
   - Preacademic, academic, and vocational/employment preparation skills
   - Adaptive/self-care skills needed for functional independence
   - Self-determination, including self-directed actions and regulation of one's own behavior to set and achieve goals, make decisions, and solve problems
2. Reducing maladaptive behaviors and emotional states that have negative effects
   - Problem behaviors that can harm the individual or others, result in property destruction, or interfere with learning and functioning as independently as possible in the community
   - Difficulties with sensory and emotional regulation (i.e., ability to flexibly modify one's level of arousal or response in order to function effectively in the environment)
   - Functionally impairing or ego-dystonic/distressing repetitive behaviors and restricted interests
   - Cooccurring mental health and medical conditions that cause significant impairment or distress

Source: Lord et al. (2022) [3]; Hume et al. (2021) [779]; Steinbrenner et al. (2020) [780]; National Autism Center (2015) [781]; Lai et al. (2020) [782].

medical interventions (predominantly pharmacotherapy) is currently substantiated only for addressing certain interfering symptoms/behaviors or co-occurring somatic and mental health conditions.

The specific targets of intervention vary with age, ASD severity, current abilities (e.g., language, cognitive, and functional adaptive skill levels), needs related to cooccurring mental health conditions, and family circumstances and preferences [3,782]. Key skill-building objectives of intervention in early childhood may include promoting language, cognitive, and motor development, teaching basic daily living skills (e.g., toilet training), and enhancing social communication and interaction by facilitating caregiver-child synchrony and establishing skills such as JA, imitation, and symbolic play. At the same time, it may be important to reduce distress caused by inflexible adherence to rituals or routines, marked aversions to sensory stimuli, dangerously restrictive food selectivity, or emotional outbursts associated with aggressive, self-injurious, or destructive behaviors. Early intervention practices increasingly reflect the importance of involving and training parents and other caregivers [3,782–784]. At school age, the emphasis may shift toward facilitating learning and independence across domains including pragmatic language, academics, daily living skills, organization and self-regulation, general social behavior (e.g., respecting boundaries and personal space), and building peer relationships (e.g., collaboration, developing friendships) as well as ameliorating anxiety, ADHD symptoms, and/or other cooccurring mental health conditions. Interventions in later childhood and adolescence may focus increasingly on higher cognitive functions (e.g., problem solving, making inferences, executive functions, and aspects of social cognition), self-determination, vocational preparation, and mental health and emotional well-being in addition to general functional independence across environments.

Lai and colleagues [782] refer to skill-building objectives as "maximizing potential" and reducing challenging behaviors related to sensory experiences, ego-dystonic RBRI, or cooccurring health conditions as "minimizing barriers." The authors include augmentative and alternative communication (AAC) interventions, including speech generating devices and the manualized picture exchange communication system (PECS) within the minimizing barriers category, although PECS is clearly a skill-building program as well, and the same could be argued for other AAC interventions [785]. They also describe a third goal of ASD intervention, "optimizing person-environment fit," which includes strategies aimed at changing the knowledge, expectations, and behavior of others (e.g., caregivers, peers, and society in general), making environmental adaptations, and providing vocational supports [782]. The thematic aim of optimizing person-environment fit is to address socio-cultural issues and create autism-friendly environments. This aspect of intervention has not been as well studied, and evidence of effectiveness is preliminary.

Personalized stepped care approach

As emphasized by the *Lancet* Commission on the future of care and clinical research in autism, the needs of individuals with ASD are heterogeneous and evolve over the life course [3]. The Commission proposed a modified stepped care approach in delivering and monitoring the effectiveness of services across time, requiring personalized treatment plans that focus on individual strengths and difficulties, changes in contexts and expectations across the lifespan, and the priorities and preferences of individuals and families at each step [3]. Stepped-care treatment delivery models provide highly accessible, less resource-intensive interventions first, with the goal of helping a meaningful proportion of patients, then gradually stepping up to more resource-intensive services for those individuals who need them [3,786–788]. Inherent to any stepped-care approach is the need for systematic, measurement-based monitoring to assess symptom status and update intervention needs, which is necessary to avoid individuals and families getting stuck in an early step of care without recognition of needs that should be addressed in more resource-intensive steps [3,786].

Although the least resource-intensive services are typically those with the lowest financial cost to care delivery systems and society, the personalized approach advocated by the *Lancet* Commission also considers the burden on individuals and families beyond financial cost, such as time, effort, and stress and emphasizes participatory decision making [3]. Accessibility and cost/burden depend, in part, on where and when interventions are delivered, intensity (i.e., number of hours per day, week, or month), amount of parent/caregiver time and effort required, whether they are provided individually or in groups, and the types of professionals involved in delivering the services. One difficulty with this approach is the lack of data about the relationships between baseline characteristics of the individual with ASD (e.g., age, cognitive and language abilities, severity of core ASD symptoms, behavior challenges, and cooccurring conditions), initial rate of progress, type and intensity of intervention, and eventual outcome to guide clinical decision making. Tasks such as choosing among available intervention options, determining the most appropriate intensity of services, and deciding when to step up to more intensive services are left to clinicians, families, and individuals with ASD, with limited specific information available about how these decisions are likely to affect outcomes.

Evidence-based psychosocial intervention

Psychosocial interventions, including educational and behavioral services, are the mainstay of management of ASD and associated impairments and conditions. The intervention literature reflects substantial variation in underlying theoretical principles (e.g., behavior analytic, developmental, or combined), study design (e.g., randomized controlled trials, quasiexperimental group designs, and single case experimental designs), and practice elements such as scope (focused/targeted vs comprehensive), delivery modality (individual vs group/classroom, delivered by a professional vs a trained parent), setting (e.g., home, inclusive or specialized classroom, clinic, community), and specific intervention targets (e.g., social interaction, communication, cognition, daily living skills, reduction of problem behavior) [779–781,789–791]. Despite the clinical heterogeneity of ASD and these sources of additional variation, many interventions have become increasingly well-studied and numerous systematic reviews and metaanalyses have been published, as illustrated by a recent umbrella review that examined the results of 128 metaanalyses synthesizing the evidence from more than 190 controlled clinical trials [792]. It stands to reason that all ASD interventions should be based on sound theoretical constructs and objective scientific evidence of efficacy and effectiveness. However, there is not complete agreement on what constitutes evidence-based intervention for ASD, leading to discrepancies among systematic reviews and metaanalyses [793]. For example, six recent large-scale systematic reviews of ASD psychosocial interventions, some including quantitative metaanalyses, all used different parameters to determine the strength of evidence for specific interventions [779–781,789–791,794]. The disagreement is largely about whether to include or exclude evidence from studies utilizing single case experimental designs, which made up 83% of the ASD focused intervention literature between 1990 and 2017 [779,780]. Although single case experimental design studies are a valid scientific approach, they do have significant weaknesses [779,789,791,795,796]. Exclusion of studies using single subject experimental designs has the largest impact on behavioral interventions because most of the evidence supporting their efficacy comes from small studies that use these designs. Reviews that include these studies conclude that there is strong evidence supporting behavioral interventions [779–781,789], whereas those that exclude this type of study design conclude that the strength of evidence is weak [791,794,797].

The National Clearinghouse for Autism Evidence and Practice (NCAEP) [779,780] has completed the third iteration of a systematic review initially conducted under the auspices of the National Professional Development Center on Autism Spectrum Disorder (NPDC) [798–800]. The NCAEP review, which reflects the literature from 1990 through 2017, included group and single case experimental design (also called single-subject design or within-subject experimental design) studies evaluating behavioral, developmental, academic, and/or vocational focused intervention practices that measured outcomes related to communication, social skills, challenging behaviors, academic/preacademic skills, play, school readiness, adaptive/self-help skills, mental health, vocational skills, or self-determination in individuals with ASD from birth through age 21 years [779,780]. As outlined in Table 28.12, this review identified 28 evidence-

TABLE 28.12 Evidence-based focused intervention practices: the national clearinghouse for autism evidence and practice (NCAEP) systematic review.[a]

Evidence-based practice	Description
Antecedent-based interventions MIMC[b]: • Picture exchange communication system (PECS)	Arrangement of events or circumstances that precede an activity or demand in order to increase the occurrence of a behavior or lead to the reduction of the challenging/interfering behaviors.
Augmentative and alternative communication	Interventions using and/or teaching the use of a system of communication that is not verbal/vocal which can be aided (e.g., device, communication book) or unaided (e.g., sign language).
Ayres sensory integration	Interventions that target a person's ability to integrate sensory information (visual, auditory, tactile, proprioceptive, and vestibular) from their body and environment in order to respond using organized and adaptive behavior.
Behavioral momentum intervention	The organization of behavior expectations in a sequence in which low probability, or more difficult, responses are embedded in a series of high probability, or less effortful, responses to increase persistence and the occurrence of the low probability responses.
Cognitive behavioral/instructional strategies	Instruction on management or control of cognitive processes that lead to changes in behavioral, social, or academic behavior.
Differential reinforcement of alternative, incompatible, or other behavior	A systematic process that increases desirable behavior or the absence of an undesirable behavior by providing positive consequences for demonstration/nondemonstration of such behavior. These consequences may be provided when the learner is: (1) engaging in a specific desired behavior other than the undesirable behavior (DRA), (2) engaging in a behavior that is physically impossible to do while exhibiting the undesirable behavior (DRI), or (3) not engaging in the undesirable behavior (DRO).
Direct instruction	A systematic approach to teaching using a sequenced instructional package with scripted protocols or lessons. It emphasizes teacher and student dialog through choral and independent student responses and employs systematic and explicit error corrections to promote mastery and generalization.
Discrete trial training	Instructional approach with massed or repeated trials with each trial consisting of the teacher's instruction/presentation, the child's response, a carefully planned consequence, and a pause prior to presenting the next instruction.
Exercise and movement	Interventions that use physical exertion, specific motor skills/techniques, or mindful movement to target a variety of skills and behavior.
Extinction	The removal of reinforcing consequences of a challenging behavior in order to reduce the future occurrence of that behavior.
Functional behavioral assessment	A systematic way of determining the underlying function or purpose of a behavior so that an effective intervention plan can be developed.
Functional communication training	A set of practices that replace a challenging behavior that has a communication function with more appropriate and effective communication behaviors or skills.
Modeling	Demonstration of a desired target behavior that results in use of the behavior by the learner and that leads to the acquisition of the target behavior.
Music-mediated intervention	Intervention that incorporates songs, melodic intonation, and/or rhythm to support learning or performance of skills/behaviors. It includes music therapy, as well as other interventions that incorporate music to address target skills.
Naturalistic intervention MIMC[b]: • Joint attention, symbolic play, engagement, and regulation (JASPER) • Milieu teaching • Pivotal response treatment (PRT)	A collection of techniques and strategies that are embedded in typical activities and/or routines in which the learner participates to naturally promote, support, and encourage target skills/behaviors.

(Continued)

TABLE 28.12 (Continued)

Evidence-based practice	Description
Parent-implemented intervention MIMC[b]: • Project ImPACT (improving parents as communication teachers) • Stepping stones triple P	Parent delivery of an intervention to their child that promotes their social communication or other skills or decreases their challenging behavior.
Peer-based instruction and intervention	Intervention in which peers directly promote autistic children's social interactions and/or other individual learning goals, or the teacher/ other adult organizes the social context (e.g., play groups, social network groups, recess) and when necessary, provides support (e.g., prompts, reinforcement) to the autistic children and their peer to engage in social interactions.
Prompting	Verbal, gestural, or physical assistance given to learners to support them in acquiring or engaging in a targeted behavior or skill.
Reinforcement	The application of a consequence following a learner's use of a response or skills that increases the likelihood that the learner will use the response/skills in the future.
Response interruption/redirection	The introduction of a prompt, comment, or other distractors when an interfering behavior is occurring that is designed to divert the learner's attention away from the interfering behavior and result in its reduction.
Self-management	Instruction focusing on learners discriminating between appropriate and inappropriate behaviors, accurately monitoring and recording their own behaviors, and rewarding themselves for behaving appropriately.
Social narratives MIMCb: • Social stories	Interventions that describe social situations in order to highlight relevant features of a target behavior or skill and offer examples of appropriate responding.
Social skills training MIMC[b]: • Program for the education and enrichment of relational skills (PEERS)	Group or individual instruction designed to teach learners ways to appropriately and successfully participate in their interactions with others.
Task analysis	A process in which an activity or behavior is divided into small, manageable steps in order to assess and teach the skill. Other practices, such as reinforcement, video modeling, or time delay, are often used to facilitate acquisition of the smaller steps.
Technology-aided instruction and intervention MIMC[b]: • FaceSay • Mindreading	Instruction or intervention in which technology is the central feature and the technology is specifically designed or employed to support the learning or performance of a behavior or skill for the learner.
Time delay	A practice used to systematically fade the use of prompts during instructional activities by using a brief delay between the initial instruction and any additional instructions or prompts.
Video modeling	A video-recorded demonstration of the targeted behavior or skill shown to the learner to assist learning in or engaging in a desired behavior or skill.
Visual supports	A visual display that supports the learner engaging in a desired behavior or skills independent of additional prompts.

DRA, Differential reinforcement of alternative behavior; *DRI*, differential reinforcement of incompatible behavior; *DRO*, differential reinforcement of other behavior; *MIMC*, manualized interventions meeting criteria.
[a]*Sources: Hume et al. (2021) [779], Steinbrenner et al. (2020) [780]. The NCAEP systematic review included group [randomized controlled trial, sequential multiple assignment randomized trial (SMART), quasiexperimental design, regression discontinuity design] and single case experimental design [withdrawal of treatment (ABAB), concurrent multiple baseline, multiple probe, alternating treatments, and changing criterion design] studies evaluating behavioral, developmental, academic, and/or vocational focused intervention practices that measured outcomes related to communication, social skills, challenging behaviors, academic/preacademic skills, play, school readiness, adaptive/self-help skills, mental health, vocational skills, or self-determination.*
[b]*Manualized interventions meeting criteria (MIMC) for evidence-based practice: a set of manualized interventions within established EBP categories that independently have sufficient evidence to be classified as evidence-based practices.*

based practices (EBPs) and 10 manualized interventions within established EBP categories that independently had sufficient evidence to be classified as EBPs [779,780]. This review updates the previous NPDC work [799,800], which came to conclusions that were very similar to those of the independent National Standards Project systematic review [781]. These evidence-based focused intervention practices can be assembled in various combinations to address specific learning needs in educational programs for children and adolescents with ASD [779,780,801,802]. The NCAEP review is a valuable information resource and an important contribution to the field of ASD intervention, although there are some problems with the methodology (e.g., not including studies with null findings, which may inflate confidence about intervention effectiveness) that should be addressed in subsequent analyses or future iterations of the study [803].

Project AIM (autism intervention metaanalysis) evaluated six categories of interventions for young children with ASD (age 0–8 years) using robust variance estimation with small sample adjustments when synthesizing effect sizes and conducting metaregressions: behavioral, developmental, naturalistic developmental behavioral interventions (NDBI), sensory-based interventions, technology-based interventions, and TEACCH (treatment and education of autistic and related communication-handicapped children) [791]. The analysis was limited to randomized clinical trial (RCT) and quasiexperimental study designs; studies using single case experimental designs were excluded. The metaanalytic approach required that at least five studies contribute to the generation of effect sizes, so animal-assisted therapy, CBT, and other approaches were excluded from summary effect size estimates. This study found evidence of effectiveness for NDBI, developmental, and behavioral intervention types based on summary effect sizes for various outcomes, as described in Table 28.13 [791,804]. The authors concluded that NDBI and developmental were "promising" intervention types based on significant summary effects for at least one outcome when important quality indicators were considered. Studies supported the effectiveness of traditional behavioral interventions for a wide range of skill-building and behavior reduction outcomes (Table 28.13), but there were few RCTs and outcome measures that were often based on caregiver report, so confidence in the accuracy of the summary effect size estimates was limited and they were described as showing "some evidence of effectiveness" [791]. NDBIs, which combine behavior analytic strategies and developmental approaches, were clearly the intervention type most supported by evidence from RCTs, particularly for social communication, language, and play skills outcomes (Table 28.13) [791,804]. The Project AIM NDBI intervention category largely corresponds to the naturalistic intervention category in the NCAEP systematic review and also includes some interventions from other categories such as parent-implemented intervention (Table 28.12) [779,780]. There were no effect size estimates with 95% CIs that did not include zero for any outcomes in studies of sensory-based interventions, technology-based interventions, or TEACCH, so these were considered to be intervention types with "little evidence of effectiveness" [791]. Although there were not enough studies to permit summary effect size estimates, animal-assisted interventions were also included in this category based on the review of the four available studies. It should be noted that the Project AIM sensory intervention category was quite heterogeneous and sensory integration intervention approaches were not considered separately, as they were in the NCAEP review [779,780]. A subsequent umbrella review of 58 systematic reviews of nonpharmacological interventions for children with ASD under age 13 years found evidence for positive effects on a range of child outcomes for behavioral interventions, developmental interventions, and NDBIs, as well as CBT and technology-based interventions [805].

Most of the evidence-based practices listed in Tables 28.12 and 28.13 (including all of the behavioral interventions and most of the NDBIs) are applied behavior analysis (ABA)-based practices [780,791,806]. NDBIs represent the integration of ABA and developmental science, and the authors of the 2015 consensus statement designating these approaches as a new intervention category stated that NDBIs are "based firmly in ABA," "fully meet criteria as ABA techniques," and should be covered by funding that provides coverage of ABA-based interventions [806]. The efficacy of NDBIs and traditional behavioral interventions for multiple outcomes is supported by numerous analyses in addition to the primary Project AIM analysis [804,807–815]. Although significant short-term benefits of developmental interventions have been established only for social communication outcomes, a well-designed study of an intervention in this category has provided some of the best evidence of lasting benefit several years after the end of the treatment period. Findings from the methodologically rigorous Preschool Autism Communication Trial (PACT, although this abbreviation has subsequently been used by the authors to denote the intervention, Pediatric Autism Communication Therapy, rather than the trial name) provide evidence that intervention-related proximal changes in parent-child dyadic social communication can facilitate long-term change in developmentally distal outcomes (in this case, core ASD symptom severity both in the short term and almost 6 years after the end of the intervention), even in the absence of continued intervention [791,816–818]. This study extended the findings of a smaller NDBI study which showed that benefits of intervention using the Early Start Denver Model (ESDM) beginning at 18–30 months of age were maintained at follow-up 2 years after the intervention ended, and that core ASD symptoms improved in the ESDM group relative to the community-treatment-as-usual group during that time interval [819,820]. However, it should be noted that the more

TABLE 28.13 Project AIM: significant autism spectrum disorder intervention outcomes in children based on studies using quasiexperimental and RCT designs.

Intervention category	Description[a]	Examples[a,b]	Outcomes with summary effect size estimate 95% CIs that do not include zero[a,c]
Behavioral interventions	• Approaches derived from operant learning theory • Characterized by discrete presentation of information (a stimulus), the prompted exhibition of target responses (e.g., desired academic, adaptive, and communicative behaviors), and the provision of extrinsic positive reinforcement in the presence of those responses • Target skills are chosen based on functional areas of need • Skills tend to be initially targeted in highly structured interactions (e.g., one-on-one with a therapist) within isolated clinical contexts, but more natural settings and interaction partners are gradually integrated as a child demonstrates progress	• Autism partnership • Discrete trial training (DTT) • Early intensive behavioral intervention (EIBI) • Lovaas model • Picture exchange communication system (PECS) • Verbal behavior	• Adaptive • Cognitive • Language • Motor • Social communication • Socio-emotional and challenging behaviors
Developmental interventions	• Derived from developmental theories of learning, rooted in constructivist theory • Early deficits in social processes (e.g., joint attention) are posited to lead to difficulties in early caregiver-child social interactions, disrupting the primary context for subsequent language and social communication development • Focus on improving the synchrony, reciprocity, and duration of parent–child or child–child interactions as a pathway for ameliorating deficits in social communication and generating cascading improvements in developmentally related skills • Primarily delivered in the context of everyday routines such as play; goals are chosen based on the typical sequences of social communication and language development	• Developmental, individual-differences, and relationship-based model (DIR/floortime) • Hanen models • Pediatric autism communication therapy (PACT) • Responsive teaching	• Social communication
Naturalistic developmental behavioral interventions (NDBI)	• Theoretical underpinnings in both behavioral and developmental theories of learning and development • Involve the use of behavioral principles of learning to teach skills chosen from a developmental sequence in naturalistic environments and using natural rewards • Skills selected as relevant for intervention are those that allow the child to participate more fully within reciprocal interactions with the adult • Interventions are delivered primarily in the context of play, but control of interactions within this context is shared by both the child and the adult, through balanced turn-taking	• Early Start Denver Model (ESDM) • Enhanced milieu teaching (EMT) • Incidental teaching (IT) • Joint attention, symbolic play, engagement, and regulation (JASPER) • Pivotal response treatment (PRT) • Prelinguistic milieu teaching (PMT) • Project ImPACT (improving parents as communication teachers) • Reciprocal imitation training (RIT) • Social communication/ emotional regulation/ transactional supports (SCERTS)	• Cognitive • Language • Social communication • Play

CI, Confidence interval; *RCT*, randomized controlled trial.
Project AIM evaluated six categories of interventions in young children with ASD (age 0–8 years) by metaanalysis: behavioral, developmental, NDBI, sensory-based interventions, technology-based interventions, and TEACCH (treatment and education of autistic and related communication-handicapped children). Focused intervention practices and comprehensive models were eligible for inclusion, but the analysis was limited to RCT and quasiexperimental study designs; studies using single case experimental designs were excluded. The metaanalytic approach required that at least five studies contribute to the generation of effect sizes, so animal-assisted therapy, cognitive-behavioral therapy, and other approaches were excluded from summary effect size estimates.
[a] *Sandbank et al. (2020) [791].*
[b] *Crank et al. (2021) [804].*
[c] *There were no effect size estimates with 95% CIs that did not include zero for any outcomes in studies of sensory-based interventions, technology-based interventions, or TEACCH.*

distal outcome in the original PACT study, improvement in overall ASD symptom severity relative to the control group, was not found in the adaptation of the intervention delivered across home and educational settings (Pediatric Autism Communication Therapy-Generalized; PACT-G), despite improvement in proximal social communication outcomes [818]. The effect size of the improvement in child-parent social communication associated with PACT-G intervention was substantially smaller than that observed in the initial study, but the intervention was associated with significant improvement in parental well-being and reduction in child disruptive behavior at home and in the educational setting.

ASD interventions have been studied most thoroughly in preschool-age and elementary school-age children [780], and there is substantial evidence supporting the efficacy of comprehensive intervention models such as early intensive behavioral intervention (EIBI) and the ESDM, as well as focused practices such as pivotal response treatment (PRT) and joint attention, symbolic play, engagement, and regulation (JASPER) (Tables 28.12 and 28.13) [779,781,789,791,805,821]. Some parent-mediated interventions (also called parent-implemented interventions or parent training) have been shown in systematic reviews and metaanalyses to improve positive behaviors/social skills, language/communication, maladaptive behaviors, and adaptive behavior/daily living skills, although the pooled effect size for the latter outcome is small [783,790,809–811,814,822–828]. These interventions involve intensive training of parents by professionals so that they can assist their children skill acquisition or behavior change. This approach is not limited to one philosophical category of intervention; there are parent-mediated interventions rooted in ABA principles (e.g., JASPER, PRT, Social ABCs), developmental theory (e.g., PACT, Play Project Home), social skills training (e.g., Program for the Education and Enrichment of Relational Skills [PEERS] for Preschoolers), and within the TEACCH structured learning approach [811,816,827,829–836]. Parent education, sometimes referred to as parent support, is often distinguished from parent training/parent-mediated intervention. Whereas the emphasis of parent training is on teaching and coaching parents to implement specific intervention strategies (technique-focused), parent education/support involves care coordination and psychoeducation at varying levels of intensity (e.g., handouts and other resources for self-guided reading, didactic presentations, and structured group educational programs) and often addresses child and parent needs and outcomes [837–839]. The literature reflects emerging, but still limited, empirical support for parent education approaches [837,840]. For example, although a 24-week parent training program was superior to parent education in a clinical trial for disruptive behavior in children with ASD, both were associated with substantial improvement [839]. A community-based education program for parents/caregivers of children newly diagnosed with ASD, consisting of five 2-hour group sessions distributed over 4–12 months, was effective in improving parents' knowledge, increasing parent empowerment, and reducing stress related to parent–child interactions [837].

The few trials comparing ASD interventions to each other, rather than to treatment-as-usual controls, have generally found improvement across interventions, without demonstrating significant differences between intervention types [841–843]. For example, a multisite, controlled trial randomized toddlers with ASD to either ESDM or EIBI (mainly discrete trial training; DTT), each delivered for either 15 or 25 hours per week for 12 months by trained research staff [842]. Neither intervention type nor intensity had main effects on autism symptom severity, expressive communication, receptive language, or nonverbal ability measured by examiners who were naïve to treatment assignment [842]. Similarly, Kasari and colleagues [843] showed that some young children with ASD who have significant intellectual impairment and little or no speech can improve substantially with high-quality intervention, but they did not find a difference between JASPER and DTT. No single intervention is effective for all children, and even when an intervention is effective overall, the response of individual children varies greatly. Investigators are increasingly exploring predictors of response to the interventions they study, which is an important step toward developing methods to personalize interventions [844–846]. For example, among preschool children with ASD and limited language, improvement in expressive language was used to retrospectively identify "super responders" (47% of children treated) and "slow responders" to JASPER intervention [846]. The model indicated that children who played functionally with a wider variety of toys (i.e., more play diversity) had increased odds of being a super responder to JASPER intervention [846]. Conversely, the combination of lower play diversity and delayed fine motor skills increased the odds of being slow responders. Although the evidence of intervention efficacy in preschoolers with ASD is very substantial overall, there is limited evidence at this time that preemptive interventions for infants or toddlers who are showing early signs of or have been diagnosed with ASD before 24 months of age improve child outcomes [847,848]. However, a trial comparing iBASIS-Video Interaction to Promote Positive Parenting (iBASIS-VIPP) preemptive intervention plus usual care to usual care only in very young children (age 9–15 months at entry screening; mean 12 months) indicated that iBASIS-VIPP was associated with reduced ASD symptom severity and odds of ASD diagnosis at age 3 years, and this is an active area of investigation [849].

Older children and adolescents tend to rely increasingly on school-based interventions, often including special education services and supports for goals related to academic achievement and social skills [850–856]. The percentage of autistic students qualifying for special education who were included in general education settings more than 80% of the time increased from 9% in 1992–1993 to 40% in 2018 [850]. ASD interventions have increasingly been studied within specialized and inclusive educational settings and found to be beneficial [780,857,858]. Interventions to improve social outcomes include adult-led individualized skill-building, peer-mediated approaches, and group-based social skills instruction [779,780,850,858,859]. Group-based social skills interventions are widely used for building social competence among children and adolescents with ASD in clinical and educational settings, and several systemic reviews and metaanalyses have found significant positive effects [858–863]. UCLA PEERS includes curricula for a wide range of ages and abilities, from preschool through adolescence, and is among the most well-evaluated and evidence-based group social skills interventions [862,864]. Potential alternative and/or complementary delivery methods for social skills interventions are being actively investigated. A preliminary metaanalysis found no significant differences between face-to-face social skills training and intervention using behavioral intervention technologies (e.g., smartphone application games and other computer software, human-like avatars, and therapeutic robots) to target social skills deficits [865]. Despite the progress with development and implementation of social skills instruction strategies, which are most effective for increasing social knowledge, emerging evidence suggests that there is a knowledge-performance distinction in ASD, social knowledge deficits are not reliably related to outcomes, and performance-related factors may be more important for addressing social challenges [866].

Although there is substantial evidence of intervention effects on language outcomes in young children with ASD [779,867–869], the efficacy of interventions in later childhood and adolescence, when pragmatic language is the primary target for verbal individuals, is much less clear [870–872]. In addition to their importance in teaching and maintaining new skills and desirable behaviors, behavior analytic strategies are the primary treatment for problematic maladaptive behaviors such as self-injury, aggression toward others, elopement, and other disruptive and noncompliant behaviors [839,873–875]. The goal of these problem-focused interventions is to reduce the problem behavior while also establishing and strengthening adaptive behaviors using function-based treatment [873,874,876]. There is also support in the literature for Parent–Child Interaction Therapy [877,878] and Stepping Stones Triple P [879,880] for reducing disruptive behaviors, and CBT is efficacious for treating anxiety in verbal children and adolescents with ASD, especially when the CBT is adapted for ASD [881–884]. Finally, there is a growing body of literature establishing evidence-based practices to enhance transition to adulthood, including successful postsecondary education and employment outcomes [885–891].

For many psychosocial interventions, there is insufficient evidence in the scientific literature to determine efficacy or to rule out the possibility that they are harmful [781,791]. In the case of facilitated communication, however, the intervention has been proven ineffective and potentially harmful [892–897]. Facilitated communication and its derivatives, rapid prompting method and spelling to communicate, should be strictly avoided [896–899].

Common elements

Although there are differences, there are also common elements among empirically supported intervention practices. In fact, there is some evidence that the style/category and particular "brand" of intervention may be less important than more general characteristics of high-quality interventions, such as using an evidence-based approach that involves frequent assessment of the child, addresses developmental and behavioral needs across all domains, incorporates developmental and behavioral science in treatment content and delivery, is implemented with fidelity, and integrates intervention into everyday activities by coaching caregivers regularly [842]. These and other features shared by many empirically supported interventions and practice guidelines are presented in Table 28.14 [7,652,782,789,842,900,901]. ASD intervention should be assessment-based and individualized, and many children with ASD can benefit from environment-related interventions and supports, including not only structuring of the physical environment but also education and support of others who are part of the child's environment, such as parents and other family members, professionals who provide services, peers, and communities (Table 28.14).

Research gaps

Although the number of published intervention trials, including RCTs, has increased dramatically in the last decade [780,791], there are still many gaps in the literature and deficiencies in our understanding of ASD interventions. Effect sizes are generally modest for proximal outcomes (i.e., those that are directly addressed by the intervention), and little is known about whether they lead to positive distal outcomes (i.e., outcomes that are developmentally downstream from

TABLE 28.14 Psychosocial interventions for autism spectrum disorder in childhood: common elements of empirically supported interventions and suggested practices.

1. Assessment-based and individualized interventions:
   - Selection of individualized, measurable goals, and instructional procedures based on systematic, objective assessment of the child's existing skills
   - Use of empirically supported instructional methods (e.g., Tables 28.12 and 28.13) to build, generalize, and maintain skills and reduce problem behaviors
   - Consideration of patient and family preferences, shared decision making regarding intervention priorities and methods
   - Incorporation of strategies to maximize adherence to the intervention methods, such as manualized practices, ongoing training, and monitoring of fidelity of implementation
   - Inclusion of specific intervention content to address the core impairments in social communication and interfering or ego-dystonic repetitive behaviors/restricted interests
   - Evaluation of the child's interests and preferences to guide development of reinforcement systems and incorporation of preferred activities to increase engagement
   - Functional assessment of problem behaviors to identify the functions/outcomes of the behaviors and selection of treatment strategies based on the assessment (e.g., interventions designed to change the outcomes of the problem behaviors and procedures to teach and differentially reinforce acceptable behaviors to serve the original functions)
   - Integration of intervention into everyday activities by coaching caregivers regularly to promote reliable implementation throughout the day and across settings and contexts, enhancing generalization and increasing learning opportunities
   - Systematic measurement and documentation of the individual child's progress and adjustment of instructional strategies as necessary to enable acquisition of target skills and/or reduction of interfering behaviors
2. Environment-related interventions and supports[a]:
   - Structured learning environments, including elements such as predictable routine and use of visual activity schedules to help children to anticipate transitions between activities
   - Organized workspaces with features such as physical boundaries to minimize distractions and facilitate task completion
   - Arrangement of physical environments to control access to materials of interest and set up communicative temptations to facilitate child initiation and interaction
   - Education within the least restrictive environment (school setting) that is appropriate for meeting the child's needs
   - Parent/caregiver education about ASD and associated difficulties and how they are managed
   - Explicit transition planning and use of evidence-based transition practices

[a]*Environment-related interventions and supports include not only practices involving adaptation of physical environments but also strategies aimed at changing the knowledge and behavior of people other than the child with ASD (e.g., parents/caregivers, other family members, professionals who provide services, peers, and communities).*
Source: Lord et al. (2022) [3]; Hyman et al. (2020) [7]; Ip et al. (2019) [10]; Fuentes et al. (2021) [11]; Myers and Challman (2018) [652]; Lai et al. (2020) [782]; Smith and Iadarola (2015) [789]; Schreibman et al. (2015) [806]; Rogers et al. (2021) [842]; Iovannone et al. (2003) [900]; National Institute for Health and Clinical Excellence (2013) [901].

what is directly addressed by the intervention) [782,791,805,902,903]. Similarly, relatively little is known about the relationship between intensity/dosage of intervention and outcomes, which subcomponents of multicomponent interventions are responsible for efficacy (the "active ingredients"), or what factors are mediators or moderators of intervention effects [782,791,805]. Few studies have evaluated important outcomes related to child and family well-being and quality of life. It is also clear that there is a need for valid, reliable, and feasible measure that quantify meaningful changes in response to treatment. There has been a lack of consensus about which outcome measures should be used in ASD intervention research and, therefore, very little consistency in outcome measures is used across studies, making synthesis and generalization of results very difficult [904–906]. Most measures of core ASD symptoms were designed as diagnostic tools and have not been proven to be sensitive to change with intervention (Tables 28.7 and 28.8), but efforts to develop new measures are underway [680,904,907,908]. Examples include the Autism Impact Measure, Brief Observation of Social Communication Change, Brief Observation of Symptoms of Autism, Early Social Communication Scales, and Social Skills Improvement System. Implementation strategies to close the gap in intervention effect sizes between effectiveness in community settings and efficacy trials in university and other research settings are also needed [909].

Medical care

Children with ASD have the same needs as other children for health supervision and acute care visits but typically have additional needs for care coordination, direction to community resources, and assistance with navigating developmental and behavioral challenges such as sleep hygiene, feeding issues, toilet training, and challenging behaviors [7,910,911].

Some have additional health care needs related to associated medical problems and/or underlying etiologic conditions and may be medically complex [7,912]. More time is often required for office visits, which may be challenging due to the patients' deficits in social interaction and communication, difficulty accepting novelty, and sensory aversions [652,913]. Strategies such as familiarizing the patient with the office environment, staff, and routine during a "practice" visit; using visual schedules to outline the steps or events that will be occurring; slowing down the pace of the visit; keeping instructions simple; using visual cues and supports; exaggerating social cues; allowing ample time for talking before touching the patient; allowing the child to manipulate instruments and materials; and having family and/or familiar staff available may be helpful in reducing resistance to physical examination and increasing family and patient comfort with the clinician [652,913–915]. Although children with ASD would be very likely to benefit from receiving care through the medical home model (see Chapter 41, Medical home), they are less likely to receive such care compared to other children with special health care needs [911].

As previously reviewed, individuals with ASD often have cooccurring medical and neurodevelopmental or psychiatric conditions, and clinicians should be aware of and monitor for these conditions, some of which may have atypical presentations, especially in children with limited communication skills. They may have difficulty communicating and/or localizing discomfort, and those with GI distress, for example, may present with nonspecific signs or symptoms such as irritability, aggressive or self-injurious behavior, sleep disturbance, anxiety, or other change in behavior [604,916,917]. In general, cooccurring conditions are treated the same way in autistic individuals as they are in those with typical development. For example, when a child with ASD also has epilepsy, the treatment is based on seizure type, epilepsy syndrome (if applicable), causal genetic variant (if one has been identified), and risk level, with consideration of cooccurring medical and psychiatric disorders [918–920]. Similarly, common GI problems such as constipation, diarrhea, or abdominal pain should be evaluated and treated just as they would be in any other child, with some adaptations to existing guidelines as necessary (e.g., related to ability to tolerate certain tests) [604,916,921,922]. It is not clear how frequently effective treatment of GI problems is associated with improvement in problem behavior; there are anecdotal reports in the literature, but quantitative data are lacking. In a few cases, ASD-specific recommendations that include adaptation of existing guidelines have been formalized and published. Examples include the guideline for management of constipation from the Gastroenterology Committee of the Autism Treatment Network (now the Autism Care Network) [922] and the Healthy Weight Research Network recommendations for managing overweight and obesity in children with ASD, which include adaptations to the 2007 guidelines endorsed by the AAP [923,924].

Psychopharmacology

Pharmacological treatments have not been demonstrated to correct the core social and communication impairments that characterize ASD, and their impact on restricted and repetitive behaviors is very limited, but they are commonly used to target cooccurring mental health conditions or associated neurobehavioral symptoms that cause significant impairment and distress (e.g., ADHD, anxiety, irritability, sleep dysfunction) [164,925–927]. According to recent estimates, approximately 40%–50% of children and adolescents with ASD are treated with psychotropic medication, and polypharmacy is common (approximately 20%–30%) [928–931]. Higher rates of psychotropic medication use (64%–69%) and polypharmacy (31%–33%) were described in a large study utilizing claims data from Medicaid and a commercial insurance plan [932]. Psychotropic medication use increases with increased age, lower IQ/presence of ID, and higher prevalence levels of challenging behavior or coexisting psychiatric diagnoses [928,932].

A detailed discussion of pharmacological management of ASD is beyond the scope of this chapter, but numerous literature reviews have summarized the evidence for conventional and novel agents, some of them quite comprehensively [164,299,925,926,933–935], and the principles of psychotropic medication management are outlined in Chapter 33: Principles of pharmacologic treatment and also in ASD-specific review articles [7,910,936,937]. However, five clusters of behaviors/symptoms associated with ASD (ADHD symptoms, irritability/severe disruptive behavior, RBRI, anxiety, and sleep problems) that are commonly targeted with medication are presented in Table 28.15, with a summary of the evidence of efficacy for various psychotropic medications. Although these problem behaviors are not core features of ASD, they commonly interfere with functioning in school, at home, and in the community and contribute substantially to the burden on families, as previously reviewed in the Associated Conditions section of this chapter. Several evidence-informed clinical practice guidelines that are pertinent to ASD-related psychopharmacologic management are available (Table 28.15). Currently, only two medications have a US FDA indication for use in ASD; risperidone and aripiprazole are FDA-approved for use in ASD for the symptomatic treatment of irritability, including aggressive behavior, deliberate self-injury, and temper tantrums in children and adolescents with autism. Other medications may

TABLE 28.15 Potential conventional psychotropic medication options for common target symptoms in autism spectrum disorder.[a]

Target symptoms	Published treatment guidelines/ recommendations	Medication class	Examples[a]	Selected supporting references
ADHD symptoms – Hyperactivity, impulsivity – Inattention, distractibility	*ASD-specific guidelines:* Rodrigues et al. (2021) [938] Young et al. (2020) [939] Mahajan et al. (2012) [940] *Other pertinent guidelines:* Barbaresi et al. (2020) [941]	Psychostimulants	Methylphenidate	**SR/MA**: Rodrigues et al. (2021) [938]; Sturman et al. (2017) [942] **RCT**: RUPP Autism Network (2005) [943], Posey et al. (2007) [944]; Pearson et al. (2013) [945]
		Selective norepinephrine reuptake inhibitors	Atomoxetine	**SR/MA:** Rodrigues et al. (2021) [938]; Patra et al. (2019) [946] **RCT**: Handen et al. (2015), [947] Smith et al. (2016) [948]; Harfterkamp et al. (2012) [949]; Arnold et al. (2006) [950]
		Alpha-2 adrenergic agonists	Guanfacine	**SR/MA**: N/A **RCT**: Scahill et al. (2015) [951]
		Atypical antipsychotics[b] (second-generation antipsychotics)	Aripiprazole Risperidone	**SR/MA:** Rodrigues et al. (2021) [938]; Wang et al. (2021) [952] **RCT:** McCracken et al. (2002), [953] Shea et al. (2004), [954] Marcus et al. (2009), [955] Owen et al. (2009) [956]
Irritability, emotion dysregulation, and related severely disruptive behavior – Outbursts of anger, frustration, and distress – Impulsive aggression, self-injurious behavior, property destruction – Sometimes referred to by caregivers as agitation, tantrums, meltdowns, or rages	*ASD-specific guidelines:* McGuire et al. (2016) [937] *Other pertinent guidelines:* N/A	Atypical antipsychotics (second-generation antipsychotics)	Aripiprazolec Risperidone[c]	**SR/MA:** Fung et al. (2016) [957]; Salazar de Pablo et al. (2023) [958]; Iffland et al. (2023) [959] **RCT:** McCracken et al. (2002) [953]; Shea et al. (2004) [954]; Pandina et al. (2007) [960]; Aman et al. (2009) [961]; Arnold et al. (2012) [962]; Kent et al. (2013) [963]; Marcus et al. (2009) [955], Owen et al. (2009) [956]; Ichikawa et al. (2017) [964]; DeVane et al. (2019) [965]
		Selective serotonin reuptake inhibitors[d]	Fluvoxamine (adults)	**SR/MA**: N/A **RCT**: McDougle et al. (1996) [966] (adults)
		Anticonvulsant mood stabilizers	Valproic acid/ Divalproex sodium[e]	**SR/MA**: N/A **RCT**: Hollander et al. (2010) [967]
Repetitive behavior/ restricted interests that interfere with functioning or cause distress – Stereotyped, repetitive movements, manipulation of objects, or speech – Inflexible insistence on sameness/behavioral rigidity, compulsions	*ASD-specific guidelines:* N/A *Other pertinent guidelines:* N/A	Atypical antipsychotics (second-generation antipsychotics)	Aripiprazole Risperidone	**SR/MA:** Zhou et al. (2021) [927] **RCT:** McDougle et al. (1998) [968] (adults); McCracken et al. (2002) [953]; Shea et al. (2004) [954]; Marcus et al. (2009) [955]; Owen et al. (2009) [956]; Ichikawa et al. (2017) [964]

(*Continued*)

TABLE 28.15 (Continued)

Target symptoms	Published treatment guidelines/ recommendations	Medication class	Examples[a]	Selected supporting references
		Selective serotonin reuptake inhibitors	Fluoxetine (children, adults)[f] Fluvoxamine (adults)	**SR/MA:** N/A[g] **RCT:** Hollander et al. (2005) [969]; Hollander et al. (2012) (adults) [970]; McDougle et al. (1996) [966] (adults)
		Anxiolytic serotonin 5-HT_{1A} receptor partial agonist	Buspirone[h]	**SR/MA:** N/A **RCT:** Chugani et al. (2016) [971]
		Alpha-2 adrenergic agonists	Guanfacine	**SR/MA:** N/A **RCT:** Scahill et al. (2015) [951], Politte et al. (2018) [972]
Anxiety	*ASD-specific guidelines*[i]: Vasa et al. (2016) [973] *Other pertinent guidelines:* Walter et al. (2020) [974]	Selective serotonin reuptake inhibitors	None	**SR/MA:** N/A **RCT:** N/A
		Atypical antipsychotics (second-generation antipsychotics)	Risperidone (adults)	**SR/MA:** N/A **RCT:** McDougle et al. (1998) [968]
Sleep dysfunction/insomnia	*ASD-specific guidelines:* Williams Buckley et al. (2020) [975] Malow et al. (2012) [976] *Other pertinent guidelines:* N/A	Endogenous indoleamine chronobiotic and hypnotic	Melatonin	**SR/MA:** Parker et al. (2019) [977]; Cuomo et al. (2017) [978] (metasynthesis) **RCT:** Wright et al. (2011) [979]; Cortesi et al. (2012) [980]; Gringras et al. (2017) [981]; Malow et al. (2021) [982]

ADHD, Attention-deficit/hyperactivity disorder; *AIR-P*, autism intervention research network on physical health; *ACN*, autism care network (formerly autism treatment network); *MA*, metaanalysis; *N/A*, not available; *RCT*, randomized clinical trial; *RUPP*, research units on pediatric psychopharmacology; *SR*, systematic review.
[a]*Supported by at least one double-blind, placebo controlled RCT in ASD with* N $\geq$ *20* and *FDA approved for some mental health condition.*
[b]*Hyperactivity and related behaviors were secondary outcomes in RCTs assessing atypical antipsychotics for the treatment of irritability; atypical antipsychotics have not been evaluated for the treatment of ADHD symptoms in patients who were not selected for significant irritability.*
[c]*Risperidone and aripiprazole are currently the only medications with US Food and Drug Administration-approved labeling specific to autism (for the symptomatic treatment of irritability, including aggressive behavior, deliberate self-injury, and temper tantrums in children and adolescents with autism).*
[d]*Irritability also improved on citalopram relative to placebo in one controlled trial in children, but this was a secondary finding of small effect size that was judged by the authors not to be clinically significant [983].*
[e]*Another valproate trial did not show significant improvement in irritability/aggression relative to placebo in adults with ASD using the same outcome measure (n = 30) [984].*
[f]*A single fluoxetine trial in children indicated significant benefit for RBRI (n = 39) [969], whereas two larger fluoxetine trials in children did not (n = 109; n = 158) [985,986].*
[g]*SR/MA by Zhou et al. [927] shows lack of benefit of SSRIs as a group for RBRI (across all ages).*
[h]*Chugani et al. [971] found improvement in RBRI in very young children (age 2–6 years) treated with 2.5 mg/day but not in those treated with 5 mg/day or placebo.*
[i]*Due to the lack of data regarding the pharmacological treatment of anxiety in children and adolescents with ASD, the medication component of the treatment recommendations is based data in typically developing pediatric populations and expert consensus [973].*

be used for FDA-approved indications that are cooccurring diagnoses in children with ASD (e.g., ADHD or generalized anxiety disorder) or off-label for symptomatic treatment.

Several studies, including double-blind and placebo-controlled RCTs, have demonstrated significant reduction in ADHD symptoms and superiority to placebo for ADHD medications including methylphenidate, atomoxetine, and extended-release guanfacine (Table 28.15) [938,942–955,987–989]. Smaller RCTs, not included in Table 28.5 due to sample size, provide additional support for guanfacine [987] and some evidence that clonidine is also superior to placebo for ADHD symptoms [988,989]. However, extended-release clonidine, which is approved for the treatment of ADHD, has not yet been studied in patients with ASD. Similarly, despite ample evidence of efficacy for ADHD, amphetamine products such as mixed amphetamine salts and lisdexamfetamine have not been examined RCTs in children with ASD. Atypical antipsychotics have also been repeatedly demonstrated to improve hyperactivity and impulsivity as secondary outcomes when they are studied for the treatment of irritability in children with ASD [952–956]. These medications have not been studied for ADHD symptoms in children with ASD who do not have significant irritability, since the potential benefits would not outweigh the potential adverse effects. Several useful guidelines and practice pathways for the management of ADHD symptoms among children and adolescents with ASD are available, as noted in Table 28.15 [938–941].

There is strong evidence that two second-generation antipsychotic medications, risperidone and aripiprazole, are effective for reducing irritability and associated problem behaviors in children with ASD, as documented by numerous RCTs and several systematic reviews and metaanalyses indicating not only statistical significance but large effect size (Table 28.15) [953–965]. Metaanalyses also suggest that ADHD medications may reduce irritability slightly [958,959]. Overall, there is no clear evidence that antidepressants, including SSRIs, have an effect on irritability in children with ASD [958,959,990], although a small trial in adults did find significant improvement with fluvoxamine treatment [966]. Irritability also improved on citalopram relative to placebo in one controlled trial in children, but this was a secondary finding of small effect size that was judged by the authors not to be clinically meaningful [983]. The evidence for valproic acid/divalproex sodium is limited and mixed—significant improvement in irritability was noted in one small trial ($n = 27$) [967] but not in another ($n = 30$) [984]—and overall, there is no convincing evidence that anticonvulsant mood stabilizers are useful for treating irritability associated with ASD [957–959,991]. The practice pathway developed by the Autism Intervention Research Network on Physical Health (AIR-P) and Autism Treatment Network (ATN; now the Autism Care Network, ACN) for the evaluation and management of irritability and problem behavior in ASD includes a stepwise approach and detailed checklist that can be applied to other behavioral symptoms as well [937].

A metaanalysis showed that there is a small, statistically significant improvement in RBRI outcomes associated with atypical antipsychotics (risperidone and aripiprazole, but not lurasidone) [927]. As a group, SSRIs were not superior to placebo for this outcome, although there is some evidence for fluvoxamine in adults [966], and for the 5-HT_{1A} receptor partial agonist, buspirone, in very young children [971]. Two small fluoxetine studies, one in children [969] and one in adults [970], suggested benefit, but much larger RCTs in children showed no superiority to placebo for RBRI outcomes for citalopram [983] or fluoxetine [985,986]. In the trial targeting primarily ADHD symptoms, extended release guanfacine was secondarily associated with modest but statistically significant improvement on measures of stereotypy and inappropriate (including repetitive) speech [951,972]. Although anxiety is common in children with ASD with and without cooccurring depression, a systematic review and metaanalysis did not identify any RCTs of antidepressant or antianxiety medication that used outcome measures for anxiety or depression; instead, they focused on outcomes such as RBRI and associated problem behaviors such hyperactivity and aggression [990]. In one small double-blind, placebo-controlled RCT in adults, risperidone treatment resulted in significant changes in clinician-rated visual analog scales including "anxious or nervous" and "depressed" [968]. As previously reviewed, there is substantial evidence that modified CBT is effective for reducing anxiety symptoms in children with ASD who have sufficient cognitive abilities, and nonpharmacologic interventions are the first-line approach [883,884,973,992]. Due to the lack of data regarding the pharmacological treatment of anxiety in children and adolescents with ASD, the medication component of the treatment recommendations is based on data in typically developing pediatric populations and expert consensus [973,974,992,993]. For core anxiety symptoms, the AIR-P/ACN clinical guidelines suggest considering sertraline, fluoxetine, citalopram, or escitalopram and note that alpha-2 agonists, such as guanfacine and clonidine, and beta-blockers, such as propranolol, may be useful for anxiety-related physiologic symptoms and behavioral dysregulation, and a short-acting benzodiazepine such as lorazepam or a beta-blocker can be considered for predictable episodic situational anxiety [973].

A metasynthesis of eight previously published systematic reviews examining the efficacy of sleep interventions in children with ASD found that parent education, behavioral interventions, and melatonin appear to be the most effective for multiple domains of sleep problems [978], and several comprehensive yet practical reviews have been published [994–996]. Although more research is needed, there is a solid evidence base supporting the use of melatonin (dose range 1–10 mg) for certain aspects of sleep dysfunction, including sleep latency and duration (Table 28.15) [977–982]. Several ASD-specific guidelines for managing sleep disturbances are available, including an AIR-P/ACN pathway published in 2012 [976] and a 2020 practice guideline from the American Academy of Neurology (AAN) [975]. The AAN practice guideline concludes that clinicians should assess for and address coexisting medical conditions and concomitant medications that could contribute to the sleep disturbance and counsel families about strategies to improve sleep habits, with behavioral strategies as a first-line treatment approach for sleep disturbance either alone or in combination with pharmacological intervention (melatonin, starting at a low dose and using a pharmaceutical-grade product if available) [975].

Evaluation of problem behaviors begins with defining and characterizing the specific behaviors, prioritizing based on severity and threat to the safety of the patient with ASD and others, and identifying potential contributing factors such as medical problems, adverse effects of current medications, coexisting psychiatric disorders, discrepancy between educational or other demands and the patient's abilities, functional communication deficits, psychosocial stressors (including parental stress), and maladaptive reinforcement patterns [910,936,937]. In some cases, medical conditions may cause or exacerbate maladaptive behaviors, and recognition and treatment of these medical factors may eliminate or reduce the need for psychotropic medications or other interventions [7,652,937]. For example, an individual may present with disrupted sleep due to obstructive sleep apnea or gastroesophageal reflux or an acute onset or exacerbation of self-injurious behavior due to discomfort caused by headache, otitis media, constipation, dental pain, a medication side effect, or menstruation-related pain or dysphoria [652]. Maintaining a relatively high index of suspicion and completing a careful history and physical examination may help the clinician to detect these and other occult sources of distress in a patient who has limited ability to communicate and who may exhibit aberrant behaviors at baseline. In most cases, problem-focused behavioral interventions informed by functional behavioral assessment are an important and effective component of treatment [873,874,937]. After treatable medical causes and modifiable behavioral and environmental factors have been addressed, a therapeutic trial of medication directed at specific target symptoms or behaviors may be considered. Selected options for common target symptom clusters are outlined in Table 28.15. Occasionally, nonpharmacological medical interventions may be appropriate. For example, children and adolescents with ASD are at increased risk for catatonia, and electroconvulsive therapy (ECT) can be effective, especially if the patient is unresponsive to treatment with a benzodiazepine [557,997,998]. rTMS is another nonpharmacological biomedical intervention that may eventually have a role in ASD intervention, but the available safety and efficacy data are limited at this time [999].

In addition to conventional psychotropic medications, many novel agents have been investigated, whether to try to improve core ASD deficits or associated conditions, symptoms, or impairments. Some lines of pharmacotherapy research focus on novel mechanisms based on preclinical data. For example, interest in the evidence suggesting that an imbalance between excitatory and inhibitory signaling contributes to ASD pathobiology has led to active investigation of pharmacological manipulation of glutamatergic and GABAergic systems [164,925,926]. Glutamate modulators that have been tested in RCTs in ASD include amantadine, memantine, ketamine, D-cycloserine, N-acetylcysteine (NAC), and riluzole, all with ultimately negative or mixed results [164,925,934]. Enhancing inhibitory GABAergic neurotransmission by $GABA_B$ receptor agonism (arbaclofen), $GABA_A$ receptor modulation (R07017773), or indirectly using bumetanide has also met with mixed results [164,925]. For example, a large ($n = 125$), double-blind, placebo-controlled RCT did not demonstrate superiority of arbaclofen versus placebo on the primary outcome measure, the Aberrant Behavior Checklist Social Withdrawal/Lethargy subscale, in children and adolescents with ASD [1000]. Recognition of convergence of ASD risk genes on biological pathways related to regulation of gene expression prompted another focus of pharmacological interest: manipulation of transcriptional regulators, such as mTOR and MAPK, using insulin-like growth factor 1 (IGF-1) or its analogs [164]. There has also been great interest in the neuropeptides oxytocin and vasopressin. The results of oxytocin trials have been mixed, especially in children, with several metaanalysis finding no significant benefit [1001,1002]. Balovaptan, a selective vasopressin 1a receptor antagonist, showed promise initially but recent large RCTs in adults ($n = 322$) and children ($n = 339$) did not demonstrate improvement in social communication and interaction [1003,1004].

There is great interest in exploring the endocannabinoid system, which modulates the activity of other neurotransmitters and has a role in social behavior. Aran and colleagues completed a RCT showing that disruptive behavior improved significantly more on one of two primary outcome measures in patients taking a whole-plant cannabis extract containing cannabidiol and tetrahydrocannabidiol (THC) than in those receiving placebo [1005]. The authors concluded that the evidence of efficacy was mixed and insufficient and that further testing of cannabinoids is needed. Others have expressed great concern about exposing children to THC rather than pure, pharmaceutical grade cannabidiol [1006]. It is clear that much more research is needed to determine whether there is a role for cannabidiol in the treatment of ASD and associated conditions.

Complementary health approaches

The National Center for Complementary and Integrative Health now uses the term "complementary health approaches" (CHAs) to refer to treatments that fall outside the mainstream of standard medical care, and "integrative health" to convey the idea of using standard and nonstandard therapies in a coordinated manner [1007,1008]. CHAs were previously referred to as "complementary and alternative medicine." CHAs commonly used for children with ASD include "natural products" such as special diets, vitamins, minerals, and omega-3 fatty acids; mind-body practices such as auditory integration training, chiropractic acupuncture, and therapeutic touch; and nonstandard biomedical interventions such as hyperbaric oxygen, antifungal agents, and heavy metal chelation therapy [652,1007]. Potential risks of CHA treatments include direct toxic effects of biological agents, adverse physical effects of manipulative techniques, presence of contaminants, interactions with prescribed medications, interference with appropriate nutrition, and interruption or postponement of empirically validated therapies, as well as unwarranted expenditure of time, effort, and financial resources [1007,1009,1010].

Use of CHAs is common among children with ASD, and this can be a source of conflict between clinicians and families [1009,1011,1012]. The authors of the 2022 *Lancet* Commission report noted that families are often vulnerable to false claims of effectiveness and unstudied treatments that might have substantial adverse effects, and they recommended that health care providers learn about the evidence regarding potential benefits and harms of complementary health approaches, ask families about their use of or interest in these approaches, and partner with families by providing information about the evidence base and potential risks and harms to enable responsible decision making [3]. This is similar to the 2007 AAP statement that pediatricians should (1) critically evaluate the scientific evidence of efficacy and risk of harm of various treatments and convey this information to families, (2) help families understand how to evaluate scientific evidence and recognize unsubstantiated treatments and pseudoscience, and (3) insist that studies that examine CHA treatments be held to the same scientific standards as all clinical research [910].

Outcomes

As is true of all of the neurodevelopmental disabilities, the management and outcome of ASD are determined by the primary symptoms and the associated dysfunctions and their interaction with the individual's environment. Within the continuum of neurologic dysfunction, each person has conditions/traits that are of variable severity, some rising to the level of a diagnosis and some falling within the broad range of typical. Consequently, the outcomes for people with ASD are quite variable [1013]. Although specific features change over time and outcomes vary, most children diagnosed with ASD using DSM criteria remain within the spectrum as adolescents and adults and, regardless of their level of intellectual functioning, continue to experience lifelong difficulties with social relationships, independent living, employment and other community participation, and mental/behavioral health [424,1014,1015].

Approximately 10%–20% of children with well-documented ASD ultimately lose the diagnosis because they no longer fulfill diagnostic criteria and are reported to function within the range of normal cognitive and adaptive functioning [1016,1017]. As a group, these individuals with "loss of autism diagnosis" have a different pattern of developmental trajectories on standardized measures than most individuals with ASD and similar average IQ. It is not yet clear whether this is attributable to interventions, the nature of the original clinical presentation, intrinsic/genetic predisposition to a highly atypical developmental trajectory, resolution of an interfering process, or some combination of these.

While outcomes are highly variable, most individuals with ASD show improvements in communication, social skills, and adaptive functioning and reduction in severity of autism symptoms as they transition from childhood to adolescence and young adulthood [426,1014,1018]. Bal and colleagues [1018] found that verbal children showed improvement in nonverbal communication and the development and maintenance of social relationships as they entered adulthood while nonverbal children showed improvement in nonverbal communication but little change in development and maintenance of social relationships. Some longitudinal studies document modest increases in IQ [419]. More individuals with ASD are gaining academic skills, participating in higher education, and living independently as adults than in the past, but this is likely due to more people with average or higher cognitive ability being diagnosed [3,1019,1020].

Although IQ is one of the most robust predictors of outcome for individuals with ASD, high IQ does not guarantee a good outcome, since deficits in adaptive functioning are common across all levels of cognitive functioning [423–428]. Recent metaanalyses suggest that only 18%–20% of individuals with ASD studied could be considered to be well integrated within society (good outcome), and 48%–50% were described as having a poor or very poor outcome [424,1014,1021]. Many young adults with ASD, including those with average intellectual ability, continue to rely on their families for financial support, housing, activities, and social relationships [1022]. Relatively few develop or sustain long-term intimate relationships, including marriages or stable partnerships [1014,1020]. Quality of life of adults with ASD is lower than that of typically developing adults, when measured with tools designed for the general population, which may not be the best way to assess it for autistic individuals [1023].

As previously discussed, mental and physical health challenges are common and are associated with poorer outcomes [518–523]. Longitudinal studies show that ADHD symptoms tend to decrease over time, depressive symptoms fluctuate but peak at clinically significant levels in young adulthood, and anxiety symptoms tend to occur at a stable high rate [1024]. The trajectory of cooccurring medical disorders (including epilepsy, obesity, and dyslipidemia) is also one of increasing prevalence through adolescence and young adulthood [519,521]. ASD is associated with increased risk of premature mortality, including death from natural causes and from unnatural causes, and lower IQ, epilepsy, and psychiatric comorbidity are associated with higher risk [1025–1028].

The information presented represents one view of success and focuses on the population level. There are multiple stories of individuals who have led fulfilling lives in atypical manners. True success depends upon the characteristics of the individual, their family, the environment in which the person lives, and the accommodations and supports that can be provided. As societal approaches to people with neurodevelopmental disabilities change, the definition of success may require revision.

References

[1] Whyatt CP, Torres EB. Autism research: an objective quantitative review of progress and focus between 1994 and 2015. Front Psychol. 2018;9:1526.

[2] Lord C, Brugha TS, Charman T, Cusack J, Dumas G, Frazier T, et al. Autism spectrum disorder. Nat Rev Dis Primers. 2020;6(1):5.

[3] Lord C, Charman T, Havdahl A, Carbone P, Anagnostou E, Boyd B, et al. The *Lancet* Commission on the future of care and clinical research in autism. Lancet 2022;399(10321):271–334.

[4] Hirota T, King BH. Autism spectrum disorder: a review. JAMA 2023;329(2):157–68.

[5] Interagency Autism Coordinating Committee. Interagency Autism Coordinating Committee (IACC) 2021–2023 strategic plan for autism research, services, and policy draft. January 2023. Retrieved on 2/28/2023 from the U.S. Department of Health and Human Services Interagency Autism Coordinating Committee website: https://iacc.hhs.gov/meetings/iacc-meetings/2023/full-committee-meeting/january18/strategic_plan_draft.pdf.

[6] Volkmar F, Siegel M, Woodbury-Smith M, King B, McCracken J, State M. Practice parameter for the assessment and treatment of children and adolescents with autism spectrum disorder American Academy of Child and Adolescent Psychiatry (AACAP) Committee on Quality Issues (CQI) J Am Acad Child Adolesc Psychiatry 2014;53(2):237–57.

[7] Hyman SL, Levy SE, Myers SM. Identification, evaluation, and management of children with autism spectrum disorder Council on Children With Disabilities, Section on Developmental and Behavioral Pediatrics Pediatrics 2020;145(1):e20193447.

[8] Zwaigenbaum L, Brian JA, Ip A. Early detection for autism spectrum disorder in young children. Paediatr Child Health. 2019;24(7):424–43.

[9] Brian JA, Zwaigenbaum L, Ip A. Standards of diagnostic assessment for autism spectrum disorder. Paediatr Child Health. 2019;24(7):444–60.

[10] Ip A, Zwaigenbaum L, Brian JA. Post-diagnostic management and follow-up care for autism spectrum disorder. Paediatr Child Health. 2019;24 (7):461–77.

[11] Fuentes J, Hervás A, Howlin P. ESCAP practice guidance for autism: a summary of evidence-based recommendations for diagnosis and treatment ESCAP ASD Working Party Eur Child Adolesc Psychiatry 2021;30(6):961–84.

[12] Taboas A, Doepke K, Zimmerman C. Preferences for identity-first versus person-first language in a US sample of autism stakeholders. Autism 2023;27(2):565–70.

[13] Kanner L. Autistic disturbances of affective contact. Nervous Child. 1943;2:217–50.

[14] Wing L. Asperger's syndrome: a clinical account. Psychol Med 1981;11(1):115–29.

[15] Frith U. Translation and annotation of "Autistic psychopathy in childhood" by Hans Asperger. In: Frith U, editor. Autism and Asperger syndrome. Cambridge: Cambridge University Press; 1991, p. 36–92.

[16] Frankl G. Language and affective contact. Nervous Child. 1943;2(3):251–62.

[17] Muratori F, Calderoni S, Bizzari V. George Frankl: an undervalued voice in the history of autism. Eur Child Adolesc Psychiatry 2021;30 (8):1273–80.

[18] Ssucharewa GE, Wolff S. The first account of the syndrome Asperger described? Translation of a paper entitled "Die schizoiden Psychopathien im Kindesalter" by Dr. G.E. Ssucharewa; scientific assistant, which appeared in 1926 in the Monatsschrift für Psychiatrie und Neurologie 60:235–261. Eur Child Adolesc Psychiatry 1996;5(3):119–32.

[19] Sher DA, Gibson JL. Pioneering, prodigious and perspicacious: Grunya Efimovna Sukhareva's life and contribution to conceptualising autism and schizophrenia. Eur Child Adolesc Psychiatry 2023;32(3):475–90.

[20] Manouilenko I, Bejerot S. Sukhareva-Prior to Asperger and Kanner. Nord J Psychiatry 2015;69(6):1761–4.

[21] Simmonds C, Sukhareva GE. The first account of the syndrome Asperger described? Part 2: the girls. Eur Child Adolesc Psychiatry 2020;29 (4):549–64.

[22] Volkmar FR, Cohen DJ, Paul R. An evaluation of DSM-III criteria for infantile autism. J Am Acad Child Psychiatry 1986;25(2):190–7.
[23] Rosen NE, Lord C, Volkmar FR. The diagnosis of autism: from Kanner to DSM-III to DSM-5 and beyond. J Autism Dev Disord 2021;51 (12):4253–70.
[24] Waterhouse L, Wing L, Spitzer R, Siegel B. Pervasive developmental disorders: from DSM-III to DSM-III-R. J Autism Dev Disord 1992;22 (4):525–49.
[25] Volkmar FR, Cicchetti DV, Bregman J, Cohen DJ. Three diagnostic systems for autism: DSM-III, DSM-III-R, and ICD-10. J Autism Dev Disord 1992;22(4):483–92.
[26] Volkmar FR, Klin A, Siegel B, Szatmari P, Lord C, Campbell M, et al. Field trial for autistic disorder in DSM-IV. Am J Psychiatry 1994;151 (9):1361–7.
[27] World Health Organization. *(ICD-10)* The international classification of diseases. 10th ed. Geneva, Switzerland: American College of Physicians. World Health Organization; 1992.
[28] American Psychiatric Association. *Text Revision (DSM-IV-TR)* Diagnostic and statistical manual of mental disorders. 4th ed. Washington, DC: American Psychiatric Association; 2000.
[29] American Psychiatric Association. Diagnostic and statistical manual of mental disorders. 5th ed. Arlington, VA: American Psychiatric Association; 2013.
[30] Baird G, Norbury CF. Social (pragmatic) communication disorders and autism spectrum disorder. Arch Dis Child 2016;101:745–51.
[31] Mandy W, Wang A, Lee I, Skuse D. Evaluating social (pragmatic) communication disorder. J Child Psychol Psychiatry 2017;58:1166–75.
[32] Flax J, Gwin C, Wilson S, Fradkin Y, Buyske S, Brzustowicz L. Social (pragmatic) communication disorder: another name for the broad autism phenotype? Autism. 2019;23(8):1982–92.
[33] Kulage KM, Goldberg J, Usseglio J, Romero D, Bain JM, Smaldone AM. How has DSM-5 affected autism diagnosis? A 5-year follow-up systematic literature review and meta-analysis. J Autism Dev Disord 2020;50(6):2102–27.
[34] Ellis Weismer S, Rubenstein E, Wiggins L, Durkin MS. A preliminary epidemiologic study of social (pragmatic) communication disorder relative to autism spectrum disorder and developmental disability without social communication deficits. J Autism Dev Disord. 2021;51 (8):2686–96.
[35] World Health Organization (WHO). International classification of diseases, eleventh revision (ICD-11), 2019/2021. https://icd.who.int/browse11. Licensed under Creative Commons Attribution-NoDerivatives 3.0 IGO licence (CC BY-ND 3.0 IGO) [accessed 27.8.22].
[36] American Psychiatric Association. Text Revision Diagnostic and statistical manual of mental disorders. 5th ed. Arlington, VA: American Psychiatric Association; 2022.
[37] Dwyer P. The neurodiversity approach(es): what are they and what do they mean for researchers? Hum Dev 2022;66(2):73–92.
[38] Chapman R, Botha M. Neurodivergence-informed therapy. Dev Med Child Neurol 2023;65(3):310–17.
[39] Schuck RK, Tagavi DM, Baiden KMP, Dwyer P, Williams ZJ, Osuna A, et al. Neurodiversity and autism intervention: reconciling perspectives through a naturalistic developmental behavioral intervention framework. J Autism Dev Disord 2022;52(10):4625–45.
[40] Monk R, Whitehouse AJO, Waddington H. The use of language in autism research. Trends Neurosci 2022;45(11):791–3.
[41] Dwyer P, Ryan JG, Williams ZJ, Gassner DL. First do no harm: suggestions regarding respectful autism language. Pediatrics 2022;149(Suppl 4):e2020049437N.
[42] Singer A, Lutz A, Escher J, Halladay A. A full semantic toolbox is essential for autism research and practice to thrive. Autism Res. 2023;16 (3):497–501.
[43] Chiarotti F, Venerosi A. Epidemiology of autism spectrum disorders: a review of worldwide prevalence estimates since 2014. Brain Sci. 2020;10(5):274.
[44] Fombonne E, MacFarlane H, Salem AC. Epidemiological surveys of ASD: advances and remaining challenges. J Autism Dev Disord 2021;51 (12):4271–90.
[45] Fombonne E, MacFarlane H, Salem AC, Zuckerman KE. Epidemiological surveys of ASD: current findings and new directions. In: Matson JL, Sturmey P, editors. Handbook of autism and pervasive developmental disorder: assessment, diagnosis, and treatment. Cham, Switzerland: Springer; 2022, p. 135–84.
[46] Zeidan J, Fombonne E, Scorah J, Ibrahim A, Durkin MS, Saxena S, et al. Global prevalence of autism: a systematic review update. Autism Res. 2022;15(5):778–90.
[47] Solmi M, Song M, Yon DK, Lee SW, Fombonne E, Kim MS, et al. Incidence, prevalence, and global burden of autism spectrum disorder from 1990 to 2019 across 204 countries. Mol Psychiatry 2022;27(10):4172–80.
[48] Bougeard C, Picarel-Blanchot F, Schmid R, Campbell R, Buitelaar J. Prevalence of autism spectrum disorder and co-morbidities in children and adolescents: a systematic literature review. Front Psychiatry. 2021;12:744709.
[49] Maenner MJ, Warren Z, Williams AR, Amoakohene E, Bakian AV, Bilder DA, et al. Prevalence and characteristics of autism spectrum disorder among children aged 8 years – Autism and Developmental Disabilities Monitoring Network, 11 sites, United States, 2020. MMWR Surveill Summ 2023;72(2):1–14.
[50] Loomes R, Hull L, Mandy WPL. What is the male-to-female ratio in autism spectrum disorder? A systematic review and meta-analysis. J Am Acad Child Adolesc Psychiatry 2017;56(6):466–74.
[51] Delobel-Ayoub M, Saemundsen E, Gissler M, Ego A, Moilanen I, Ebeling H, et al. Prevalence of autism spectrum disorder in 7-9-year-old children in Denmark, Finland, France and Iceland: a population-based registries approach within the ASDEU Project. J Autism Dev Disord 2020;50(3):949–59.

[52] Roman-Urrestarazu A, van Kessel R, Allison C, Matthews FE, Brayne C, Baron-Cohen S. Association of race/ethnicity and social disadvantage with autism prevalence in 7 million school children in England. JAMA Pediatr. 2021;175(6):e210054.

[53] Shaw KA, Bilder DA, McArthur D, Williams AR, Amoakohene E, Bakian AV, et al. Early identification of autism spectrum disorder among children aged 4 years – Autism and Developmental Disabilities Monitoring Network, 11 sites, United States, 2020. MMWR Surveill Summ 2023;72(1):1–15.

[54] Li Q, Li Y, Liu B, Chen Q, Xing X, Xu G, et al. Prevalence of autism spectrum disorder among children and adolescents in the United States from 2019 to 2020. JAMA Pediatr. 2022;176(9):943–5.

[55] Lebrun-Harris LA, Ghandour RM, Kogan MD, Warren MD. Five-year trends in US children's health and well-being, 2016–2020. JAMA Pediatr. 2022;176(7):e220056.

[56] Hughes MM, Shaw KA, DiRienzo M, Durkin MS, Esler A, Hall-Lande J, et al. The prevalence and characteristics of children with profound autism, 15 sites, United States, 2000–2016. Public Health Rep 2023;138(6):971–80.

[57] Maenner MJ, Shaw KA, Baio J, Washington A, Patrick M, DiRieno M, et al. Prevalence of autism spectrum disorder among children aged 8 years – Autism and Developmental Disabilities Monitoring Network, 11 Sites, United States, 2016. MMWR Surveill Summ 2020;69(4):1–12.

[58] Durkin MS, Maenner MJ, Meaney FJ, Levy SE, DiGuiseppi C, Nicholas JS, et al. Socioeconomic inequality in the prevalence of autism spectrum disorder: evidence from a US cross-sectional study. PLoS One 2010;5(7):e11551.

[59] Durkin MS, Maenner MJ, Baio J, Christensen D, Daniels J, Fitzgerald R, et al. Autism spectrum disorder among US children (2002–2010): socioeconomic, racial, and ethnic disparities. Am J Public Health 2017;107(11):1818–26.

[60] Maenner MJ, Shaw KA, Bakian AV, Bilder DA, Durkin MS, Esler A, et al. Prevalence and characteristics of autism spectrum disorder among children aged 8 years – Autism and Developmental Disabilities Monitoring Network, 11 Sites, United States, 2018. MMWR Surveill Summ 2021;70(11):1–16.

[61] Winter AS, Fountain C, Cheslack-Postava K, Bearman PS. The social patterning of autism diagnoses reversed in California between 1992 and 2018. Proc Natl Acad Sci USA 2020;117(48):30295–302.

[62] Russell G, Stapley S, Newlove-Delgado T, Salmon A, White R, Warren F, et al. Time trends in autism diagnosis over 20 years: a UK population-based cohort study. J Child Psychol Psychiatry 2022;63(6):674–82.

[63] Shenouda J, Barrett E, Davidow AL, Sidwell K, Lescott C, Halperin W, et al. Prevalence and disparities in the detection of autism without intellectual disability. Pediatrics 2023;151(2):e2022056594.

[64] Fombonne E. Camouflage and autism. J Child Psychol Psychiatry 2020;61(7):735–8.

[65] Lai MC, Hull L, Mandy W, Chakrabarti B, Nordahl CW, Lombardo MV, et al. Commentary: 'Camouflaging' in autistic people – reflection on Fombonne (2020). J Child Psychol Psychiatry 2021;62(8):1037–41.

[66] Williams ZJ. Commentary: The construct validity of 'camouflaging' in autism: psychometric considerations and recommendations for future research – reflection on Lai et al. (2020). J Child Psychol Psychiatry 2022;63(1):118–21.

[67] Cook J, Hull L, Crane L, Mandy W. Camouflaging in autism: a systematic review. Clin Psychol Rev 2021;89:102080.

[68] Hansen SN, Schendel DE, Parner ET. Explaining the increase in the prevalence of autism spectrum disorders: the proportion attributable to changes in reporting practices. JAMA Pediatr. 2015;169(1):56–62.

[69] Miller JS, Bilder D, Farley M, Coon H, Pinborough-Zimmerman J, Jenson W, et al. Autism spectrum disorder reclassified: a second look at the 1980s Utah/UCLA Autism Epidemiologic Study. J Autism Dev Disord 2013;43(1):200–10.

[70] Rice CE, Rosanoff M, Dawson G, Durkin MS, Croen LA, Singer A, et al. Evaluating changes in the prevalence of the autism spectrum disorders (ASDs). Public Health Rev 2012;34(2):1–22.

[71] Polyak A, Kubina RM, Girirajan S. Comorbidity of intellectual disability confounds ascertainment of autism: implications for genetic diagnosis. Am J Med Genet B Neuropsychiatr Genet 2015;168(7):600–8.

[72] Cardinal DN, Griffiths AJ, Maupin ZD, Fraumeni-McBride J. An investigation of increased rates of autism in US public schools. Psychol Sch 2021;58:124–40.

[73] Whitehouse AJ, Cooper MN, Bebbington K, Alvares G, Lin A, Wray J, et al. Evidence of a reduction over time in the behavioral severity of autistic disorder diagnoses. Autism Res. 2017;10(1):179–87.

[74] Arvidsson O, Gillberg C, Lichtenstein P, Lundström S. Secular changes in the symptom level of clinically diagnosed autism. J Child Psychol Psychiatry 2018;59(7):744–51.

[75] Avlund SH, Thomsen PH, Schendel D, Jørgensen M, Clausen L. Time trends in diagnostics and clinical features of young children referred on suspicion of autism: a population-based clinical cohort study, 2000–2010. J Autism Dev Disord 2021;51(2):444–58.

[76] Rødgaard EM, Jensen K, Vergnes JN, Soulières I, Mottron L. Temporal changes in effect sizes of studies comparing individuals with and without autism: a meta-analysis. JAMA Psychiatry 2019;76(11):1124–32.

[77] Lundström S, Reichenberg A, Anckarsäter H, Lichtenstein P, Gillberg C. Autism phenotype versus registered diagnosis in Swedish children: prevalence trends over 10 years in general population samples. BMJ 2015;350:h1961.

[78] Myers SM, Voigt RG, Colligan RC, Weaver AL, Storlie CB, Stoeckel RE, et al. Autism spectrum disorder: incidence and time trends over two decades in a population-based birth cohort. J Autism Dev Disord 2019;49(4):1455–74.

[79] Fombonne E. Editorial: is autism overdiagnosed? J Child Psychol Psychiatry 2023;64(5):711–14.

[80] Duvall S, Greene R, Phelps R, Markwardt S, Rutter T, Grieser Painter J. et al. Factors associated with unconfirmed/inaccurate community-based ASD diagnosis in a research referred sample. Austin, TX, USA: International Society for Autism Research Annual Meeting; 2022.

[81] Hausman-Kedem M, Kosofsky BE, Ross G, Yohay K, Forrest E, Dennin MH, et al. Accuracy of reported community diagnosis of autism spectrum disorder. J Psychopathol Behav Assess 2018;40(3):367–75.
[82] Taylor MJ, Rosenqvist MA, Larsson H, Gillberg C, D'Onofrio BM, Lichtenstein P, et al. Etiology of autism spectrum disorders and autistic traits over time. JAMA Psychiatry 2020;77(9):936–43.
[83] Nevison CD. A comparison of temporal trends in United States autism prevalence to trends in suspected environmental factors. Environ Health 2014;13:73.
[84] Bölte S, Girdler S, Marschik PB. The contribution of environmental exposure to the etiology of autism spectrum disorder. Cell Mol Life Sci 2019;76(7):1275–97.
[85] Pugsley K, Scherer SW, Bellgrove MA, Hawi Z. Environmental exposures associated with elevated risk for autism spectrum disorder may augment the burden of deleterious de novo mutations among probands. Mol Psychiatry 2022;27(1):710–30.
[86] Tick B, Bolton P, Happé F, Rutter M, Rijsdijk F. Heritability of autism spectrum disorders: a meta-analysis of twin studies. J Child Psychol Psychiatry 2016;57(5):585–95.
[87] Sandin S, Lichtenstein P, Kuja-Halkola R, Hultman C, Larsson H, Reichenberg A. The heritability of autism spectrum disorder. JAMA 2017;318(12):1182–4.
[88] Bai D, Yip BHK, Windham GC, Sourander A, Francis R, Yoffe R, et al. Association of genetic and environmental factors with autism in a 5-country cohort. JAMA Psychiatry. 2019;76(10):1035–43.
[89] Lehner T, Senthil G, Addington AM. Convergence of advances in genomics, team science, and repositories as drivers of progress in psychiatric genomics. Biol Psychiatry 2015;77(1):6–14.
[90] Willsey AJ, Morris MT, Wang S, Willsey HR, Sun N, Teerikorpi N, et al. The psychiatric cell map initiative: a convergent systems biological approach to illuminating key molecular pathways in neuropsychiatric disorders. Cell. 2018;174(3):505–20.
[91] Searles Quick VB, Wang B, State MW. Leveraging large genomic datasets to illuminate the pathobiology of autism spectrum disorders. Neuropsychopharmacology 2021;46(1):55–69.
[92] Satterstrom FK, Kosmicki JA, Wang J, Breen MS, De Rubeis S, An JY, et al. Large-scale exome sequencing study implicates both developmental and functional changes in the neurobiology of autism. Cell. 2020;180(3):568–84.
[93] Fu JM, Satterstrom FK, Peng M, Brand H, Collins RL, Dong S, et al. Rare coding variation provides insight into the genetic architecture and phenotypic context of autism. Nat Genet 2022;54(9):1320–31.
[94] Trost B, Thiruvahindrapuram B, Chan AJS, Engchuan W, Higginbotham EJ, Howe JL, et al. Genomic architecture of autism from comprehensive whole-genome sequence annotation. Cell. 2022;185(23):4409–27 e18.
[95] Autism Spectrum Disorders Working Group of the Psychiatric Genomics Consortium. Meta-analysis of GWAS of over 16,000 individuals with autism spectrum disorder highlights a novel locus at 10q24.32 and a significant overlap with schizophrenia. Mol Autism 2017;8:21.
[96] Grove J, Ripke S, Als TD, Mattheisen M, Walters RK, Won H, et al. Identification of common genetic risk variants for autism spectrum disorder. Nat Genet 2019;51(3):431–44.
[97] Weiner DJ, Wigdor EM, Ripke S, Walters RK, Kosmicki JA, Grove J, et al. Polygenic transmission disequilibrium confirms that common and rare variation act additively to create risk for autism spectrum disorders. Nat Genet 2017;49(7):978–85.
[98] Klei L, McClain LL, Mahjani B, Panayidou K, De Rubeis S, Grahnat AS, et al. How rare and common risk variation jointly affect liability for autism spectrum disorder. Mol Autism. 2021;12(1):66.
[99] Cirnigliaro M, Chang TS, Arteaga SA, Pérez-Cano L, Ruzzo EK, Gordon A, et al. The contributions of rare inherited and polygenic risk to ASD in multiplex families. Proc Natl Acad Sci USA 2023;120(31):e2215632120.
[100] Robinson EB, St Pourcain B, Anttila V, Kosmicki JA, Bulik-Sullivan B, Grove J, et al. Genetic risk for autism spectrum disorders and neuropsychiatric variation in the general population. Nat Genet. 2016;48(5):552–5.
[101] Gaugler T, Klei L, Sanders SJ, Bodea CA, Goldberg AP, Lee AB, et al. Most genetic risk for autism resides with common variation. Nat Genet 2014;46(8):881–5.
[102] Richards C, Jones C, Groves L, Moss J, Oliver C. Prevalence of autism spectrum disorder phenomenology in genetic disorders: a systematic review and meta-analysis. Lancet Psychiatry 2015;2(10):909–16.
[103] Sztainberg Y, Zoghbi HY. Lessons learned from studying syndromic autism spectrum disorders. Nat Neurosci 2016;19(11):1408–17.
[104] Tartaglia NR, Wilson R, Miller JS, Rafalko J, Cordeiro L, Davis S, et al. Autism spectrum disorder in males with sex chromosome aneuploidy: XXY/Klinefelter syndrome, XYY, and XXYY. J Dev Behav Pediatr 2017;38(3):197–207.
[105] Björlin Avdic H, Butwicka A, Nordenström A, Almqvist C, Nordenskjöld A, Engberg H, et al. Neurodevelopmental and psychiatric disorders in females with Turner syndrome: a population-based study. J Neurodev Disord. 2021;13(1):51.
[106] Wilfert AB, Turner TN, Murali SC, Hsieh P, Sulovari A, Wang T, et al. Recent ultra-rare inherited variants implicate new autism candidate risk genes. Nat Genet 2021;53(8):1125–34.
[107] Zhou X, Feliciano P, Shu C, Wang T, Astrovskaya I, Hall JB, et al. Integrating de novo and inherited variants in 42,607 autism cases identifies mutations in new moderate-risk genes. Nat Genet 2022;54(9):1305–19.
[108] Sanders SJ, He X, Willsey AJ, Ercan-Sencicek AG, Samocha KE, Cicek AE, et al. Insights into autism spectrum disorder genomic architecture and biology from 71 risk loci. Neuron 2015;87(6):1215–33.
[109] Trost B, Engchuan W, Nguyen CM, Thiruvahindrapuram B, Dolzhenko E, Backstrom I, et al. Genome-wide detection of tandem DNA repeats that are expanded in autism. Nature 2020;586(7827):80–6.

[110] Werling DM, Brand H, An JY, Stone MR, Zhu L, Glessner JT, et al. An analytical framework for whole-genome sequence association studies and its implications for autism spectrum disorder. Nat Genet 2018;50(5):727–36.

[111] Lim ET, Uddin M, De Rubeis S, Chan Y, Kamumbu AS, Zhang X, et al. Rates, distribution and implications of postzygotic mosaic mutations in autism spectrum disorder. Nat Neurosci 2017;20(9):1217–24.

[112] Krupp DR, Barnard RA, Duffourd Y, Evans SA, Mulqueen RM, Bernier R, et al. Exonic mosaic mutations contribute risk for autism spectrum disorder. Am J Hum Genet 2017;101(3):369–90.

[113] D'Gama AM. Somatic mosaicism and autism spectrum disorder. Genes 2021;12(11):1699.

[114] Rodin RE, Dou Y, Kwon M, Sherman MA, D'Gama AM, Doan RN, et al. The landscape of somatic mutation in cerebral cortex of autistic and neurotypical individuals revealed by ultra-deep whole-genome sequencing. Nat Neurosci 2021;24(2):176–85.

[115] Antaki D, Guevara J, Maihofer AX, Klein M, Gujral M, Grove J, et al. A phenotypic spectrum of autism is attributable to the combined effects of rare variants, polygenic risk and sex. Nat Genet 2022;54(9):1284–92.

[116] Wray NR, Lin T, Austin J, McGrath JJ, Hickie IB, Murray GK, et al. From basic science to clinical application of polygenic risk scores: a primer. JAMA Psychiatry. 2021;78(1):101–9.

[117] Kumuthini J, Zick B, Balasopoulou A, Chalikiopoulou C, Dandara C, El-Kamah G, et al. The clinical utility of polygenic risk scores in genomic medicine practices: a systematic review. Hum Genet 2022;141(11):1697–704.

[118] Moreno-De-Luca A, Myers SM, Challman TD, Moreno-De-Luca D, Evans DW, Ledbetter DH. Developmental brain dysfunction: revival and expansion of old concepts based on new genetic evidence. Lancet Neurol 2013;12(4):406–14.

[119] Kingdom R, Wright CF. Incomplete penetrance and variable expressivity: from clinical studies to population cohorts. Front Genet. 2022;13:920390.

[120] Wang W, Corominas R, Lin GN. *De novo* mutations from whole exome sequencing in neurodevelopmental and psychiatric disorders: from discovery to application. Front Genet. 2019;10:258.

[121] Sanders SJ, Sahin M, Hostyk J, Thurm A, Jacquemont S, Avillach P, et al. A framework for the investigation of rare genetic disorders in neuropsychiatry. Nat Med 2019;25(10):1477–87.

[122] Satterstrom FK, Walters RK, Singh T, Wigdor EM, Lescai F, Demontis D, et al. Autism spectrum disorder and attention deficit hyperactivity disorder have a similar burden of rare protein-truncating variants. Nat Neurosci 2019;22(12):1961–5.

[123] Rees E, Kirov G. Copy number variation and neuropsychiatric illness. Curr Opin Genet Dev 2021;68:57–63.

[124] Lee PH, Feng YA, Smoller JW. Pleiotropy and cross-disorder genetics among psychiatric disorders. Biol Psychiatry 2021;89(1):20–31.

[125] Epi25 Collaborative. Ultra-rare genetic variation in the epilepsies: a whole-exome sequencing study of 17,606 individuals. Am J Hum Genet 2019;105(2):267–82.

[126] Rees E, Creeth HDJ, Hwu HG, Chen WJ, Tsuang M, Glatt SJ, et al. Schizophrenia, autism spectrum disorders and developmental disorders share specific disruptive coding mutations. Nat Commun. 2021;12(1):5353.

[127] Lin GN, Song W, Wang W, Wang P, Yu H, Cai W, et al. De novo mutations identified by whole-genome sequencing implicate chromatin modifications in obsessive-compulsive disorder. Sci Adv. 2022;8(2):eabi6180.

[128] Andreassen OA, Hindley GFL, Frei O, Smeland OB. New insights from the last decade of research in psychiatric genetics: discoveries, challenges and clinical implications. World Psychiatry 2023;22(1):4–24.

[129] Brainstorm Consortium Anttila V, Bulik-Sullivan B, Finucane HK, Walters RK, Bras J, et al. Analysis of shared heritability in common disorders of the brain. Science 2018;360(6395):eaap8757.

[130] Sullivan PF, Geschwind DH. Defining the genetic, genomic, cellular, and diagnostic architectures of psychiatric disorders. Cell. 2019;177(1):162–83.

[131] Cross-Disorder Group of the Psychiatric Genomics Consortium. Genomic relationships, novel loci, and pleiotropic mechanisms across eight psychiatric disorders. Cell 2019;179(7):1469–82.

[132] Serdarevic F, Tiemeier H, Jansen PR, Alemany S, Xerxa Y, Neumann A, et al. Polygenic risk scores for developmental disorders, neuromotor functioning during infancy, and autistic traits in childhood. Biol Psychiatry 2020;87(2):132–8.

[133] Lu H, Qiao J, Shao Z, Wang T, Huang S, Zeng P. A comprehensive gene-centric pleiotropic association analysis for 14 psychiatric disorders with GWAS summary statistics. BMC Med 2021;19(1):314.

[134] Mallard TT, Grotzinger AD, Smoller JW. Examining the shared etiology of psychopathology with genome-wide association studies. Physiol Rev 2023;103(2):1645–65.

[135] Oetjens MT, Kelly MA, Sturm AC, Martin CL, Ledbetter DH. Quantifying the polygenic contribution to variable expressivity in eleven rare genetic disorders. Nat Commun. 2019;10(1):4897.

[136] Niemi MEK, Martin HC, Rice DL, Gallone G, Gordon S, Kelemen M, et al. Common genetic variants contribute to risk of rare severe neurodevelopmental disorders. Nature 2018;562(7726):268–71.

[137] Davies RW, Fiksinski AM, Breetvelt EJ, Williams NM, Hooper SR, Monfeuga T, et al. Using common genetic variation to examine phenotypic expression and risk prediction in 22q11.2 deletion syndrome. Nat Med 2020;26(12):1912–18.

[138] Alver M, Mancini V, Läll K, Schneider M, Romano L. Contribution of schizophrenia polygenic burden to longitudinal phenotypic variance in 22q11.2 deletion syndrome Estonian Biobank Research Team, et al. Mol Psychiatry 2022;27(10):4191–200.

[139] Myers SM, Challman TD, Bernier RA, Bourgeron T, Chung WK, Constantino JN, et al. Insufficient evidence for "autism-specific" genes. Am J Hum Genet 2020;106:587–95.

[140] Schaaf CP, Betancur C, Yuen RKC, Parr JR, Skuse DH, Gallagher L, et al. A framework for an evidence-based gene list relevant to autism spectrum disorder. Nat Rev Genet 2020;21(6):367–76.

[141] Wang T, Kim CN, Bakken TE, Gillentine MA, Henning B, Mao Y, et al. Integrated gene analyses of de novo variants from 46,612 trios with autism and developmental disorders. Proc Natl Acad Sci USA 2022;119(46):e2203491119.

[142] Coe BP, Stessman HAF, Sulovari A, Geisheker MR, Bakken TE, Lake AM, et al. Neurodevelopmental disease genes implicated by de novo mutation and copy number variation morbidity. Nat Genet 2019;51(1):106–16.

[143] Bassett AS, Lowther C, Merico D, Costain G, Chow EWC, van Amelsvoort T, et al. Rare genome-wide copy number variation and expression of schizophrenia in 22q11.2 deletion syndrome. Am J Psychiatry 2017;174(11):1054–63.

[144] Pizzo L, Jensen M, Polyak A, Rosenfeld JA, Mannik K, Krishnan A, et al. Rare variants in the genetic background modulate cognitive and developmental phenotypes in individuals carrying disease-associated variants. Genet Med 2019;21(4):816–25.

[145] Wu S, Wu F, Ding Y, Hou J, Bi J, Zhang Z. Advanced parental age and autism risk in children: a systematic review and meta-analysis. Acta Psychiatr Scand 2017;135(1):29–41.

[146] Gratten J, Wray NR, Peyrot WJ, McGrath JJ, Visscher PM, Goddard ME. Risk of psychiatric illness from advanced paternal age is not predominantly from de novo mutations. Nat Genet 2016;48(7):718–24.

[147] Modabbernia A, Velthorst E, Reichenberg A. Environmental risk factors for autism: an evidence-based review of systematic reviews and meta-analyses. Mol Autism 2017;8:13.

[148] Kim JY, Son MJ, Son CY, Radua J, Eisenhut M, Gressier F, et al. Environmental risk factors and biomarkers for autism spectrum disorder: an umbrella review of the evidence. Lancet Psychiatry. 2019;6(7):590–600.

[149] Manoli DS, State MW. Autism spectrum disorder genetics and the search for pathological mechanisms. Am J Psychiatry 2021;178(1):30–8.

[150] Sestan N, State MW. Lost in translation: traversing the complex path from genomics to therapeutics in autism spectrum disorder. Neuron 2018;100(2):406–23.

[151] Iakoucheva LM, Muotri AR, Sebat J. Getting to the cores of autism. Cell. 2019;178(6):1287–98.

[152] Willsey HR, Willsey AJ, Wang B, State MW. Genomics, convergent neuroscience and progress in understanding autism spectrum disorder. Nat Rev Neurosci 2022;23(6):323–41.

[153] Kostic A, Buxbaum JD. The promise of precision medicine in autism. Neuron 2021;109(14):2212–15.

[154] Courchesne E, Pramparo T, Gazestani VH, Lombardo MV, Pierce K, Lewis NE. The ASD living biology: from cell proliferation to clinical phenotype. Mol Psychiatry 2019;24(1):88–107.

[155] Courchesne E, Gazestani VH, Lewis NE. Prenatal origins of ASD: the when, what, and how of ASD development. Trends Neurosci 2020;43 (5):326–42.

[156] Eyring KW, Geschwind DH. Three decades of ASD genetics: building a foundation for neurobiological understanding and treatment. Hum Mol Genet 2021;30(20):R236–44.

[157] Liao C, Moyses-Oliveira M, De Esch CEF, Bhavsar R, Nuttle X, Li A, et al. Convergent coexpression of autism-associated genes suggests some novel risk genes may not be detectable in large-scale genetic studies. Cell Genom. 2023;3(4):100277.

[158] Kumar S, Reynolds K, Ji Y, Gu R, Rai S, Zhou CJ. Impaired neurodevelopmental pathways in autism spectrum disorder: a review of signaling mechanisms and crosstalk. J Neurodev Disord. 2019;11(1):10.

[159] Velmeshev D, Schirmer L, Jung D, Haeussler M, Perez Y, Mayer S, et al. Single-cell genomics identifies cell type-specific molecular changes in autism. Science 2019;364(6441):685–9.

[160] Velmeshev D, Magistri M, Mazza EMC, Lally P, Khoury N, D'Elia ER, et al. Cell-type-specific analysis of molecular pathology in autism identifies common genes and pathways affected across neocortical regions. Mol Neurobiol 2020;57(5):2279–89.

[161] Gandal MJ, Haney JR, Wamsley B, Yap CX, Parhami S, Emani PS, et al. Broad transcriptomic dysregulation occurs across the cerebral cortex in ASD. Nature 2022;611(7936):532–9.

[162] Chen GT, Geschwind DH. Challenges and opportunities for precision medicine in neurodevelopmental disorders. Adv Drug Deliv Rev 2022;191:114564.

[163] Hong D, Iakoucheva LM. Therapeutic strategies for autism: targeting three levels of the central dogma of molecular biology. Transl Psychiatry. 2023;13(1):58.

[164] Baribeau D, Anagnostou E. Novel treatments for autism spectrum disorder based on genomics and systems biology. Pharmacol Ther 2022;230:107939.

[165] Fetit R, Hillary RF, Price DJ, Lawrie SM. The neuropathology of autism: a systematic review of post-mortem studies of autism and related disorders. Neurosci Biobehav Rev 2021;129:35–62.

[166] Liao X, Liu Y, Fu X, Li Y. Postmortem studies of neuroinflammation in autism spectrum disorder: a systematic review. Mol Neurobiol 2020;57(8):3424–38.

[167] Casanova MF. The neuropathology of autism. In: 4th ed. Volkmar FR, Rogers SJ, Paul R, Pelphrey KA, editors. Handbook of autism and pervasive developmental disorders, I. Hoboken, NJ: John Wiley & Sons; 2014.

[168] Varghese M, Keshav N, Jacot-Descombes S, Warda T, Wicinski B, Dickstein DL, et al. Autism spectrum disorder: neuropathology and animal models. Acta Neuropathol 2017;134(4):537–66.

[169] Chen JA, Peñagarikano O, Belgard TG, Swarup V, Geschwind DH. The emerging picture of autism spectrum disorder: genetics and pathology. Annu Rev Pathol 2015;10:111–44.

[170] Casanova MF. The neuropathology of autism. Brain Pathol 2007;17(4):422–33.

[171] Stoner R, Chow ML, Boyle MP, Sunkin SM, Mouton PR, Roy S, et al. Patches of disorganization in the neocortex of children with autism. N Engl J Med 2014;370(13):1209–19.

[172] Zikopoulos B, Liu X, Tepe J, Trutzer I, John YJ, Barbas H. Opposite development of short- and long-range anterior cingulate pathways in autism. Acta Neuropathol 2018;136(5):759–78.

[173] Trutzer IM, García-Cabezas MÁ, Zikopoulos B. Postnatal development and maturation of layer 1 in the lateral prefrontal cortex and its disruption in autism. Acta Neuropathol Commun. 2019;7(1):40.

[174] Tang G, Gudsnuk K, Kuo SH, Cotrina ML, Rosoklija G, Sosunov A, et al. Loss of mTOR-dependent macroautophagy causes autistic-like synaptic pruning deficits. Neuron 2014;83(5):1131–43.

[175] Katz J, Reichenberg A, Kolevzon A. Prenatal and perinatal metabolic risk factors for autism: a review and integration of findings from population-based studies. Curr Opin Psychiatry 2021;34(2):94–104.

[176] Cheroni C, Caporale N, Testa G. Autism spectrum disorder at the crossroad between genes and environment: contributions, convergences, and interactions in ASD developmental pathophysiology. Mol Autism. 2020;11(1):69.

[177] Carlsson T, Molander F, Taylor MJ, Jonsson U, Bölte S. Early environmental risk factors for neurodevelopmental disorders – a systematic review of twin and sibling studies. Dev Psychopathol 2021;33(4):1448–95.

[178] Nudel R, Thompson WK, Børglum AD, Hougaard DM, Mortensen PB, Werge T, et al. Maternal pregnancy-related infections and autism spectrum disorder – the genetic perspective. Transl Psychiatry. 2022;12(1):334.

[179] Mikkelsen AP, Greiber IK, Scheller NM, Lidegaard Ø. Association of labor epidural analgesia with autism spectrum disorder in children. JAMA 2021;326(12):1170–7.

[180] Rosen BN, Lee BK, Lee NL, Yang Y, Burstyn I. Maternal smoking and autism spectrum disorder: a meta-analysis. J Autism Dev Disord 2015;45(6):1689–98.

[181] Tang S, Wang Y, Gong X, Wang G. A meta-analysis of maternal smoking during pregnancy and autism spectrum disorder risk in offspring. Int J Environ Res Public Health 2015;12(9):10418–31.

[182] Christensen J, Grønborg TK, Sørensen MJ, Schendel D, Parner ET, Pedersen LH, et al. Prenatal valproate exposure and risk of autism spectrum disorders and childhood autism. JAMA 2013;309(16):1696–703.

[183] Wiggs KK, Rickert ME, Sujan AC, Quinn PD, Larsson H, Lichtenstein P, et al. Antiseizure medication use during pregnancy and risk of ASD and ADHD in children. Neurology 2020;95(24):e3232–40.

[184] Bjørk MH, Zoega H, Leinonen MK, Cohen JM, Dreier JW, Furu K, et al. Association of prenatal exposure to antiseizure medication with risk of autism and intellectual disability. JAMA Neurol. 2022;79(7):672–81.

[185] Dreier JW, Bjørk MH, Alvestad S, Gissler M, Igland J, Leinonen MK, et al. Prenatal exposure to antiseizure medication and incidence of childhood- and adolescence-onset psychiatric disorders. JAMA Neurol. 2023;80(6):568–77.

[186] Kaplan YC, Keskin-Arslan E, Acar S, Sozmen K. Maternal SSRI discontinuation, use, psychiatric disorder and the risk of autism in children: a meta-analysis of cohort studies. Br J Clin Pharmacol 2017;83(12):2798–806.

[187] Kaplan YC, Keskin-Arslan E, Acar S, Sozmen K. Prenatal selective serotonin reuptake inhibitor use and the risk of autism spectrum disorder in children: a systematic review and meta-analysis. Reprod Toxicol 2016;66:31–43.

[188] Kobayashi T, Matsuyama T, Takeuchi M, Ito S. Autism spectrum disorder and prenatal exposure to selective serotonin reuptake inhibitors: a systematic review and meta-analysis. Reprod Toxicol 2016;65:170–8.

[189] Leshem R, Bar-Oz B, Diav-Citrin O, Gbaly S, Soliman J, Renoux C, et al. Selective serotonin reuptake inhibitors (SSRIs) and serotonin norepinephrine reuptake inhibitors (SNRIs) during pregnancy and the risk for autism spectrum disorder (ASD) and attention deficit hyperactivity disorder (ADHD) in the offspring: a true effect or a bias? A systematic review and meta-analysis. Curr Neuropharmacol 2021;19(6):896–906.

[190] Masarwa R, Levine H, Gorelik E, Reif S, Perlman A, Matok I. Prenatal exposure to acetaminophen and risk for attention deficit hyperactivity disorder and autistic spectrum disorder: a systematic review, meta-analysis, and meta-regression analysis of cohort studies. Am J Epidemiol 2018;187(8):1817–27.

[191] Alemany S, Avella-García C, Liew Z, García-Esteban R, Inoue K, Cadman T, et al. Prenatal and postnatal exposure to acetaminophen in relation to autism spectrum and attention-deficit and hyperactivity symptoms in childhood: meta-analysis in six European population-based cohorts. Eur J Epidemiol 2021;36(10):993–1004.

[192] Khan FY, Kabiraj G, Ahmed MA, Adam M, Mannuru SP, Ramesh V, et al. A systematic review of the link between autism spectrum disorder and acetaminophen: a mystery to resolve. Cureus. 2022;14(7):e26995.

[193] Ji Y, Azuine RE, Zhang Y, Hou W, Hong X, Wang G, et al. Association of cord plasma biomarkers of in utero acetaminophen exposure with risk of attention-deficit/hyperactivity disorder and autism spectrum disorder in childhood. JAMA Psychiatry. 2020;77(2):180–9.

[194] Damkier P, Cleary B, Weber-Schoendorfer C, Shechtman S, Cassina M, Panchaud A, et al. Handle with care – interpretation, synthesis and dissemination of data on paracetamol in pregnancy. Nat Rev Endocrinol. 2022;18(3):191.

[195] Liew Z, Ernst A. Intrauterine exposure to acetaminophen and adverse developmental outcomes: epidemiological findings and methodological issues. Curr Environ Health Rep. 2021;8(1):23–33.

[196] Lye JM, Knight JA, Arneja J, Seeto RA, Wong J, Adel Khani N, et al. Maternal acetaminophen use and cognitive development at 4 years: the Ontario birth study. Pediatr Res 2023;93(4):959–63.

[197] Volk H.E., Ames J.L., Chen A., Fallin M.D., Hertz-Picciotto I., Halladay A., et al. Considering toxic chemicals in the etiology of autism. Pediatrics. 2022. 149(1):e2021053012.

[198] Chun H, Leung C, Wen SW, McDonald J, Shin HH. Maternal exposure to air pollution and risk of autism in children: a systematic review and meta-analysis. Environ Pollut 2020;256:113307.

[199] Dutheil F, Comptour A, Morlon R, Mermillod M, Pereira B, Baker JS, et al. Autism spectrum disorder and air pollution: a systematic review and meta-analysis. Environ Pollut 2021;278:116856.

[200] Rahman MM, Shu YH, Chow T, Lurmann FW, Yu X, Martinez MP, et al. Prenatal exposure to air pollution and autism spectrum disorder: sensitive windows of exposure and sex differences. Environ Health Perspect 2022;130(1):17008.

[201] Lin LZ, Zhan XL, Jin CY, Liang JH, Jing J, Dong GH. The epidemiological evidence linking exposure to ambient particulate matter with neurodevelopmental disorders: a systematic review and meta-analysis. Environ Res 2022;209:112876.

[202] Joyce EE, Chavarro JE, Rando J, Song AY, Croen LA, Fallin MD, et al. Prenatal exposure to pesticide residues in the diet in association with child autism-related traits: results from the EARLI study. Autism Res. 2022;15(5):957–70.

[203] Liu X, Zou M, Sun C, Wu L, Chen WX. Prenatal folic acid supplements and offspring's autism spectrum disorder: a meta-analysis and meta-regression. J Autism Dev Disord 2022;52(2):522–39.

[204] Levine SZ, Kodesh A, Viktorin A, Smith L, Uher R, Reichenberg A, et al. Association of maternal use of folic acid and multivitamin supplements in the periods before and during pregnancy with the risk of autism spectrum disorder in offspring. JAMA Psychiatry. 2018;75 (2):176–84.

[205] Schmidt RJ, Kogan V, Shelton JF, Delwiche L, Hansen RL, Ozonoff S, et al. Combined prenatal pesticide exposure and folic acid intake in relation to autism spectrum disorder. Environ Health Perspect 2017;125(9):097007.

[206] Goodrich AJ, Volk HE, Tancredi DJ, McConnell R, Lurmann FW, Hansen RL, et al. Joint effects of prenatal air pollutant exposure and maternal folic acid supplementation on risk of autism spectrum disorder. Autism Res. 2018;11(1):69–80.

[207] Oulhote Y, Lanphear B, Braun JM, Webster GM, Arbuckle TE, Etzel T, et al. Gestational exposures to phthalates and folic acid, and autistic traits in Canadian children. Environ Health Perspect 2020;128(2):27004.

[208] Tioleco N, Silberman AE, Stratigos K, Banerjee-Basu S, Spann MN, Whitaker AH, et al. Prenatal maternal infection and risk for autism in offspring: a meta-analysis. Autism Res 2021;14(6):1296–316.

[209] Estes ML, McAllister AK. Maternal immune activation: implications for neuropsychiatric disorders. Science 2016;353(6301):772–7.

[210] Lei XY, Li YJ, Ou JJ, Li YM. Association between parental body mass index and autism spectrum disorder: a systematic review and meta-analysis. Eur Child Adolesc Psychiatry 2019;28(7):933–47.

[211] Gardner RM, Lee BK, Magnusson C, et al. Maternal body mass index during early pregnancy, gestational weight gain, and risk of autism spectrum disorders: results from a Swedish total population and discordant sibling study. Int J Epidemiol 2015;44:870–83.

[212] Tian ZX, Wan M, Gao YL, Wu BF, Xie Y, Liu J, et al. Gestational weight gain and risk of autism spectrum disorders in offspring: a systematic review and meta-analysis. J Obstet Gynaecol 2020;40(7):953–60.

[213] Lyall K, Ning X, Aschner JL, Avalos LA, Bennett DH, Bilder DA, et al. Cardiometabolic pregnancy complications in association with autism-related traits as measured by the social responsiveness scale in ECHO. Am J Epidemiol 2022;191(8):1407–19.

[214] Wan H, Zhang C, Li H, Luan S, Liu C. Association of maternal diabetes with autism spectrum disorders in offspring: a systemic review and meta-analysis. Medicine 2018;97(2):e9438.

[215] Kong L, Nilsson IAK, Brismar K, Gissler M, Lavebratt C. Associations of different types of maternal diabetes and body mass index with offspring psychiatric disorders. JAMA Netw Open. 2020;3(2):e1920787.

[216] Persson M, Reichenberg A, Andersson Franko M, Sandin S. Maternal type 1 diabetes, pre-term birth and risk of autism spectrum disorder – a prospective cohort study. Int J Epidemiol 2023;52(2):377–85.

[217] Rowland J, Wilson CA. The association between gestational diabetes and ASD and ADHD: a systematic review and meta-analysis. Sci Rep 2021;11(1):5136.

[218] Maher GM, O'Keeffe GW, Dalman C, et al. Association between preeclampsia and autism spectrum disorder: a population-based study. J Child Psychol Psychiatry 2020;61:131–9.

[219] Sun BZ, Moster D, Harmon QE, Wilcox AJ. Association of preeclampsia in term births with neurodevelopmental disorders in offspring. JAMA Psychiatry 2020;77:823–9.

[220] Andreadou MT, Katsaras GN, Talimtzi P, Doxani C, Zintzaras E, Stefanidis I. Association of assisted reproductive technology with autism spectrum disorder in the offspring: an updated systematic review and meta-analysis. Eur J Pediatr 2021;180(9):2741–55.

[221] Cheslack-Postava K, Suominen A, Jokiranta E, Lehti V, McKeague IW, Sourander A, et al. Increased risk of autism spectrum disorders at short and long interpregnancy intervals in Finland. J Am Acad Child Adolesc Psychiatry 2014;53(10):1074–81.

[222] Durkin MS, DuBois LA, Maenner MJ. Inter-pregnancy intervals and the risk of autism spectrum disorder: results of a population-based study. J Autism Dev Disord 2015;45(7):2056–66.

[223] Zerbo O, Yoshida C, Gunderson EP, Dorward K, Croen LA. Interpregnancy interval and risk of autism spectrum disorders. Pediatrics 2015;136(4):651–7.

[224] Schieve LA, Tian LH, Drews-Botsch C, Windham GC, Newschaffer C, Daniels JL, et al. Autism spectrum disorder and birth spacing: findings from the study to explore early development (SEED). Autism Res. 2018;11(1):81–94.

[225] Jiang HY, Xu LL, Shao L, Xia RM, Yu ZH, Ling ZX, et al. Maternal infection during pregnancy and risk of autism spectrum disorders: a systematic review and meta-analysis. Brain Behav Immun 2016;58:165–72.

[226] Croen LA, Qian Y, Ashwood P, Zerbo O, Schendel D, Pinto-Martin J, et al. Infection and fever in pregnancy and autism spectrum disorders: findings from the study to explore early development. Autism Res. 2019;12(10):1551–61.

[227] Antoun S, Ellul P, Peyre H, Rosenzwajg M, Gressens P, Klatzmann D, et al. Fever during pregnancy as a risk factor for neurodevelopmental disorders: results from a systematic review and meta-analysis. Mol Autism 2021;12(1):60.

[228] Han VX, Patel S, Jones HF, Dale RC. Maternal immune activation and neuroinflammation in human neurodevelopmental disorders. Nat Rev Neurol. 2021;17(9):564–79.

[229] Hall MB, Willis DE, Rodriguez EL, Schwarz JM. Maternal immune activation as an epidemiological risk factor for neurodevelopmental disorders: considerations of timing, severity, individual differences, and sex in human and rodent studies. Front Neurosci. 2023;17:1135559.

[230] Zawadzka A, Cieślik M, Adamczyk A. The role of maternal immune activation in the pathogenesis of autism: a review of the evidence, proposed mechanisms and implications for treatment. Int J Mol Sci. 2021;22(21):11516.

[231] Lyall K, Ames JL, Pearl M, Traglia M, Weiss LA, Windham GC, et al. A profile and review of findings from the early markers for autism study: unique contributions from a population-based case-control study in California. Mol Autism. 2021;12(1):24.

[232] Che X, Hornig M, Bresnahan M, Stoltenberg C, Magnus P, Surén P, et al. Maternal mid-gestational and child cord blood immune signatures are strongly associated with offspring risk of ASD. Mol Psychiatry 2022;27(3):1527–41.

[233] Chen SW, Zhong XS, Jiang LN, Zheng XY, Xiong YQ, Ma SJ, et al. Maternal autoimmune diseases and the risk of autism spectrum disorders in offspring: a systematic review and meta-analysis. Behav Brain Res 2016;296:61–9.

[234] Nielsen TC, Nassar N, Shand AW, Jones HF, Han VX, Patel S, et al. Association of maternal autoimmune disease and early childhood infections with offspring autism spectrum disorder: a population-based cohort study. Autism Res. 2022;15(12):2371–80.

[235] Sun CK, Cheng YS, Chen IW, Chiu HJ, Chung W, Tzang RF, et al. Impact of parental rheumatoid arthritis on risk of autism spectrum disorders in offspring: a systematic review and meta-analysis. Front Med 2022;9:1052806.

[236] Zerbo O, Qian Y, Yoshida C, Fireman BH, Klein NP, Croen LA. Association between influenza infection and vaccination during pregnancy and risk of autism spectrum disorder. JAMA Pediatr. 2017;171(1):e163609.

[237] Brynge M, Sjöqvist H, Gardner RM, Lee BK, Dalman C, Karlsson H. Maternal infection during pregnancy and likelihood of autism and intellectual disability in children in Sweden: a negative control and sibling comparison cohort study. Lancet Psychiatry. 2022;9(10):782–91.

[238] Getahun D, Fassett MJ, Peltier MR, Wing DA, Xiang AH, Chiu V, et al. Association of perinatal risk factors with autism spectrum disorder. Am J Perinatol 2017;34(3):295–304.

[239] Modabbernia A, Mollon J, Boffetta P, Reichenberg A. Impaired gas exchange at birth and risk of intellectual disability and autism: a meta-analysis. J Autism Dev Disord 2016;46(5):1847–59.

[240] Modabbernia A, Sandin S, Gross R, Leonard H, Gissler M, Parner ET, et al. Apgar score and risk of autism. Eur J Epidemiol 2019;34 (2):105–14.

[241] Badawi N, Dixon G, Felix JF, Keogh JM, Petterson B, Stanley FJ, et al. Autism following a history of newborn encephalopathy: more than a coincidence? Dev Med Child Neurol 2006;48(2):85–9.

[242] Crump C, Sundquist J, Sundquist K. Preterm or early term birth and risk of autism. Pediatrics 2021;148(3):e2020032300.

[243] Agrawal S, Rao SC, Bulsara MK, Patole SK. Prevalence of autism spectrum disorder in preterm infants: a meta-analysis. Pediatrics 2018;142 (3):e20180134.

[244] Laverty C, Surtees A, O'Sullivan R, Sutherland D, Jones C, Richards C. The prevalence and profile of autism in individuals born preterm: a systematic review and meta-analysis. J Neurodev Disord 2021;13(1):41.

[245] Guo BQ, Li HB, Zhai DS, Yang LQ. Prevalence of autism spectrum disorder diagnosis by birth weight, gestational age, and size for gestational age: a systematic review, meta-analysis, and meta-regression. Eur Child Adolesc Psychiatry 2022. Available from: https://doi.org/10.1007/s00787-022-02078-4. Epub ahead of print.

[246] Atladóttir HÓ, Schendel DE, Parner ET, Henriksen TB. A descriptive study on the neonatal morbidity profile of autism spectrum disorders, including a comparison with other neurodevelopmental disorders. J Autism Dev Disord 2015;45(8):2429–42.

[247] Jenabi E, Farashi S, Salehi AM, Parsapoor H. The association between post-term births and autism spectrum disorders: an updated systematic review and meta-analysis. Eur J Med Res 2023;28(1):316.

[248] Hamner T, Shih E, Ichord R, Krivitzky L. Children with perinatal stroke are at increased risk for autism spectrum disorder: prevalence and co-occurring conditions within a clinically followed sample. Clin Neuropsychol 2022;36(5):981–92.

[249] Sundelin H, Söderling J, Bang P, Bolk J. Risk of autism after pediatric ischemic stroke: a nationwide cohort study. Neurology 2022;98(19): e1953–63.

[250] Wu CM, Zhang B, Trenor CC, Rivkin MJ, Danehy A, Lehman LL. Autism spectrum disorder in children with perinatal ischemic stroke varies by stroke type. J Child Neurol 2023; 8830738231188395.

[251] Taylor LE, Swerdfeger AL, Eslick GD. Vaccines are not associated with autism: an evidence-based meta-analysis of case-control and cohort studies. Vaccine 2014;32(29):3623–9.

[252] Uno Y, Uchiyama T, Kurosawa M, Aleksic B, Ozaki N. Early exposure to the combined measles-mumps-rubella vaccine and thimerosal-containing vaccines and risk of autism spectrum disorder. Vaccine 2015;33(21):2511–16.

[253] Gidengil C, Goetz MB, Newberry S, Maglione M, Hall O, Larkin J, et al. Safety of vaccines used for routine immunization in the United States: an updated systematic review and meta-analysis. Vaccine 2021;39(28):3696–716.

[254] Hviid A, Hansen JV, Frisch M, Melbye M. Measles, mumps, rubella vaccination and autism: a nationwide cohort study. Ann Intern Med 2019;170(8):513–20.

[255] DeStefano F, Shimabukuro TT. The MMR vaccine and autism. Annu Rev Virol. 2019;6(1):585–600.

[256] Di Pietrantonj C, Rivetti A, Marchione P, Debalini MG, Demicheli V. Vaccines for measles, mumps, rubella, and varicella in children. Cochrane Database Syst Rev 2021;11(11):CD004407.
[257] Parker SK, Schwartz B, Todd J, Pickering LK. Thimerosal-containing vaccines and autistic spectrum disorder: a critical review of published original data. Pediatrics 2004;114(3):793–804.
[258] Schechter R, Grether JK. Continuing increases in autism reported to California's developmental services system: mercury in retrograde. Arch Gen Psychiatry 2008;65(1):19–24.
[259] Gerber JS, Offit PA. Vaccines and autism: a tale of shifting hypotheses. Clin Infect Dis 2009;48(4):456–61.
[260] DeStefano F, Price CS, Weintraub ES. Increasing exposure to antibody-stimulating proteins and polysaccharides in vaccines is not associated with risk of autism. J Pediatr 2013;163(2):561–7.
[261] DeStefano F, Bodenstab HM, Offit PA. Principal controversies in vaccine safety in the United States. Clin Infect Dis 2019;69(4):726–31.
[262] Becerra-Culqui TA, Getahun D, Chiu V, Sy LS, Tseng HF. Prenatal tetanus, diphtheria, acellular pertussis vaccination and autism spectrum disorder. Pediatrics 2018;142(3):e20180120.
[263] Ludvigsson JF, Winell H, Sandin S, Cnattingius S, Stephansson O, Pasternak B. Maternal Influenza A (H1N1) immunization during pregnancy and risk for autism spectrum disorder in offspring: a cohort study. Ann Intern Med 2020;173(8):597–604.
[264] Becerra-Culqui TA, Getahun D, Chiu V, Sy LS, Tseng HF. Prenatal influenza vaccination or influenza infection and autism spectrum disorder in offspring. Clin Infect Dis 2022;75(7):1140–8.
[265] Jain A, Marshall J, Buikema A, Bancroft T, Kelly JP, Newschaffer CJ. Autism occurrence by MMR vaccine status among US children with older siblings with and without autism. JAMA 2015;313(15):1534–40.
[266] Musci RJ, Augustinavicius JL, Volk H. Gene-environment interactions in psychiatry: recent evidence and clinical implications. Curr Psychiatry Rep 2019;21(9):81.
[267] Hollander JA, Cory-Slechta DA, Jacka FN, Szabo ST, Guilarte TR, Bilbo SD, et al. Beyond the looking glass: recent advances in understanding the impact of environmental exposures on neuropsychiatric disease. Neuropsychopharmacology 2020;45(7):1086–96.
[268] Keil-Stietz K, Lein PJ. Gene × environment interactions in autism spectrum disorders. Curr Top Dev Biol 2023;152:221–84.
[269] Volk HE, Kerin T, Lurmann F, Hertz-Picciotto I, McConnell R, Campbell DB. Autism spectrum disorder: interaction of air pollution with the MET receptor tyrosine kinase gene. Epidemiology 2014;25(1):44–7.
[270] Kim D, Volk H, Girirajan S, Pendergrass S, Hall MA, Verma SS, et al. The joint effect of air pollution exposure and copy number variation on risk for autism. Autism Res. 2017;10(9):1470–80.
[271] Alemany S, Vilor-Tejedor N, García-Esteban R, Bustamante M, Dadvand P, Esnaola M, et al. Traffic-related air pollution, *APOEε4* status, and neurodevelopmental outcomes among school children enrolled in the BREATHE Project (Catalonia, Spain). Environ Health Perspect 2018;126(8):087001.
[272] Lipkin WI, Bresnahan M, Susser E. Cohort-guided insights into gene-environment interactions in autism spectrum disorders. Nat Rev Neurol. 2023;19(2):118–25.
[273] Duncan LE, Keller MC. A critical review of the first 10 years of candidate gene-by-environment interaction research in psychiatry. Am J Psychiatry 2011;168(10):1041–9.
[274] Huang S. Towards a unification of the 2 meanings of "epigenetics". PLoS Biol 2022;20(12):e3001944.
[275] Talbert PB, Henikoff S. The yin and yang of histone marks in transcription. Annu Rev Genomics Hum Genet 2021;22:147–70.
[276] Li S, Peng Y, Panchenko AR. DNA methylation: precise modulation of chromatin structure and dynamics. Curr Opin Struct Biol 2022;75:102430.
[277] Cavalli G, Heard E. Advances in epigenetics link genetics to the environment and disease. Nature 2019;571(7766):489–99.
[278] Feinberg AP. The key role of epigenetics in human disease prevention and mitigation. N Engl J Med 2018;378(14):1323–34.
[279] Farsetti A, Illi B, Gaetano C. How epigenetics impacts on human diseases. Eur J Intern Med 2023;114:15–22.
[280] Horsthemke B. A critical appraisal of clinical epigenetics. Clin Epigenetics. 2022;14(1):95.
[281] Tremblay MW, Jiang YH. DNA methylation and susceptibility to autism spectrum disorder. Annu Rev Med 2019;70:151–66.
[282] Bakulski KM, Dou JF, Feinberg JI, Aung MT, Ladd-Acosta C, Volk HE, et al. Autism-associated DNA methylation at birth from multiple tissues is enriched for autism genes in the early autism risk longitudinal investigation. Front Mol Neurosci. 2021;14:775390.
[283] LaSalle JM. Epigenomic signatures reveal mechanistic clues and predictive markers for autism spectrum disorder. Mol Psychiatry 2023;28(5):1890–901.
[284] Stoccoro A, Conti E, Scaffei E, Calderoni S, Coppedè F, Migliore L, et al. DNA methylation biomarkers for young children with idiopathic autism spectrum disorder: a systematic review. Int J Mol Sci. 2023;24(11):9138.
[285] Mordaunt CE, Jianu JM, Laufer BI, Zhu Y, Hwang H, Dunaway KW, et al. Cord blood DNA methylome in newborns later diagnosed with autism spectrum disorder reflects early dysregulation of neurodevelopmental and X-linked genes. Genome Med. 2020;12(1):88.
[286] Ravaei A, Emanuele M, Nazzaro G, Fadiga L, Rubini M. Placental DNA methylation profile as predicting marker for autism spectrum disorder (ASD). Mol Med 2023;29(1):8.
[287] Harris JR, Gao CW, Britton JF, Applegate CD, Bjornsson HT, Fahrner JA. Five years of experience in the epigenetics and chromatin clinic: what have we learned and where do we go from here? Hum Genet 2023;1–18.
[288] Fahrner JA, Bjornsson HT. Mendelian disorders of the epigenetic machinery: postnatal malleability and therapeutic prospects. Hum Mol Genet 2019;28(R2):R254–64.
[289] Ciptasari U, van Bokhoven H. The phenomenal epigenome in neurodevelopmental disorders. Hum Mol Genet 2020;29(R1):R42–50.

[290] Dall'Aglio L, Muka T, Cecil CAM, Bramer WM, Verbiest MMPJ, Nano J, et al. The role of epigenetic modifications in neurodevelopmental disorders: a systematic review. Neurosci Biobehav Rev 2018;94:17–30.

[291] Andrews SV, Sheppard B, Windham GC, Schieve LA, Schendel DE, Croen LA, et al. Case-control meta-analysis of blood DNA methylation and autism spectrum disorder. Mol Autism 2018;9:40.

[292] Hannon E, Schendel D, Ladd-Acosta C, Grove J, iPSYCH-Broad ASD GroupHansen CS, et al. Elevated polygenic burden for autism is associated with differential DNA methylation at birth. Genome Med. 2018;10(1):19.

[293] Ellis SE, Gupta S, Moes A, West AB, Arking DE. Exaggerated CpH methylation in the autism-affected brain. Mol Autism. 2017;8:6.

[294] Havdahl A, Niarchou M, Starnawska A, Uddin M, van der Merwe C, Warrier V. Genetic contributions to autism spectrum disorder. Psychol Med 2021;51(13):2260–73.

[295] FDA-NIH Biomarker Working Group. BEST (Biomarkers, EndpointS, and other Tools) resource [Internet]. Silver Spring (MD): Food and Drug Administration (US); 2016. Glossary. 2016 Jan 28 [Updated 2021 Nov 29]. Available from: https://www.ncbi.nlm.nih.gov/books/NBK338448/. Co-published by National Institutes of Health (US), Bethesda (MD).

[296] Sahin M, Sweeney JA, Jones SR. Editorial: biomarkers to enable therapeutics development in neurodevelopmental disorders. Front Integr Neurosci. 2020;14:616641.

[297] García-Gutiérrez MS, Navarrete F, Sala F, Gasparyan A, Austrich-Olivares A, Manzanares J. Biomarkers in psychiatry: concept, definition, types and relevance to the clinical reality. Front Psychiatry. 2020;11:432.

[298] Vargason T, Grivas G, Hollowood-Jones KL, Hahn J. Towards a multivariate biomarker-based diagnosis of autism spectrum disorder: review and discussion of recent advancements. Semin Pediatr Neurol 2020;34:100803.

[299] McCracken JT, Anagnostou E, Arango C, Dawson G, Farchione T, Mantua V, et al. Drug development for autism spectrum disorder (ASD): progress, challenges, and future directions. Eur Neuropsychopharmacol 2021;48:3–31.

[300] Frye RE, Vassall S, Kaur G, Lewis C, Karim M, Rossignol D. Emerging biomarkers in autism spectrum disorder: a systematic review. Ann Transl Med. 2019;7(23):792.

[301] Jensen AR, Lane AL, Werner BA, McLees SE, Fletcher TS, Frye RE. Modern biomarkers for autism spectrum disorder: future directions. Mol Diagn Ther 2022;26(5):483–95.

[302] Kang E, Keifer CM, Levy EJ, Foss-Feig JH, McPartland JC, Lerner MD. Atypicality of the N170 event-related potential in autism spectrum disorder: a meta-analysis. Biol Psychiatry Cogn Neurosci Neuroimaging 2018;3(8):657–66.

[303] Clairmont C, Wang J, Tariq S, Sherman HT, Zhao M, Kong XJ. The value of brain imaging and electrophysiological testing for early screening of autism spectrum disorder: a systematic review. Front Neurosci. 2022;15:812946.

[304] Pagnozzi AM, Conti E, Calderoni S, Fripp J, Rose SE. A systematic review of structural MRI biomarkers in autism spectrum disorder: a machine learning perspective. Int J Dev Neurosci 2018;71:68–82.

[305] Hiremath CS, Sagar KJV, Yamini BK, Girimaji AS, Kumar R, Sravanti SL, et al. Emerging behavioral and neuroimaging biomarkers for early and accurate characterization of autism spectrum disorders: a systematic review. Transl Psychiatry. 2021;11(1):42.

[306] Zhang-James Y, Buitelaar JK, The ENIGMA-ASD Working Groupvan Rooij D, Faraone SV. Ensemble classification of autism spectrum disorder using structural magnetic resonance imaging features. JCPP Advances 2021;1:e12042.

[307] Zhang F, Roeyers H. Exploring brain functions in autism spectrum disorder: a systematic review on functional near-infrared spectroscopy (fNIRS) studies. Int J Psychophysiol 2019;137:41–53.

[308] Cortese S, Solmi M, Michelini G, Bellato A, Blanner C, Canozzi A, et al. Candidate diagnostic biomarkers for neurodevelopmental disorders in children and adolescents: a systematic review. World Psychiatry 2023;22(1):129–49.

[309] Parellada M, Andreu-Bernabeu Á, Burdeus M, San José Cáceres A, Urbiola E, Carpenter LL, et al. In search of biomarkers to guide interventions in autism spectrum disorder: a systematic review. Am J Psychiatry 2023;180(1):23–40.

[310] Loth E, Charman T, Mason L, Tillmann J, Jones EJH, Wooldridge C, et al. The EU-AIMS Longitudinal European Autism Project (LEAP): design and methodologies to identify and validate stratification biomarkers for autism spectrum disorders. Mol Autism. 2017;8:24.

[311] McPartland JC, Bernier RA, Jeste SS, Dawson G, Nelson CA, Chawarska K, et al. The Autism Biomarkers Consortium for Clinical Trials (ABC-CT): scientific context, study design, and progress toward biomarker qualification. Front Integr Neurosci. 2020;14:16.

[312] Webb SJ, Shic F, Murias M, Sugar CA, Naples AJ, Barney E, et al. Biomarker acquisition and quality control for multi-site studies: the autism biomarkers consortium for clinical trials. Front Integr Neurosci. 2020;13:71.

[313] Carvalho AF, Solmi M, Sanches M, Machado MO, Stubbs B, Ajnakina O, et al. Evidence-based umbrella review of 162 peripheral biomarkers for major mental disorders. Transl Psychiatry. 2020;10(1):152.

[314] Li X, Zhang K, He X, Zhou J, Jin C, Shen L, et al. Structural, functional, and molecular imaging of autism spectrum disorder. Neurosci Bull 2021;37(7):1051–71.

[315] Hoogman M, van Rooij D, Klein M, Boedhoe P, Ilioska I, Li T, et al. Consortium neuroscience of attention deficit/hyperactivity disorder and autism spectrum disorder: the ENIGMA adventure. Hum Brain Mapp 2022;43(1):37–55.

[316] Li M, Wang Y, Tachibana M, Rahman S, Kagitani-Shimono K. Atypical structural connectivity of language networks in autism spectrum disorder: a meta-analysis of diffusion tensor imaging studies. Autism Res 2022;15(9):1585–602.

[317] Postema MC, van Rooij D, Anagnostou E, Arango C, Auzias G, Behrmann M, et al. Altered structural brain asymmetry in autism spectrum disorder in a study of 54 datasets. Nat Commun. 2019;10(1):4958.

[318] Courchesne E, Pierce K, Schumann CM, Redcay E, Buckwalter JA, Kennedy DP, et al. Mapping early brain development in autism. Neuron 2007;56(2):399–413.

[319] Courchesne E, Campbell K, Solso S. Brain growth across the life span in autism: age-specific changes in anatomical pathology. Brain Res 2011;1380:138–45.

[320] Hazlett HC, Poe MD, Gerig G, Styner M, Chappell C, Smith RG, et al. Early brain overgrowth in autism associated with an increase in cortical surface area before age 2 years. Arch Gen Psychiatry 2011;68(5):467–76.

[321] Shen MD, Nordahl CW, Young GS, Wootton-Gorges SL, Lee A, Liston SE, et al. Early brain enlargement and elevated extra-axial fluid in infants who develop autism spectrum disorder. Brain 2013;136(Pt 9):2825–35.

[322] Sacco R, Gabriele S, Persico AM. Head circumference and brain size in autism spectrum disorder: a systematic review and meta-analysis. Psychiatry Res 2015;234(2):239–51.

[323] Dawson G, Rieder AD, Johnson MH. Prediction of autism in infants: progress and challenges. Lancet Neurol 2023;22(3):244–54.

[324] Hazlett HC, Gu H, Munsell BC, Kim SH, Styner M, Wolff JJ, et al. Early brain development in infants at high risk for autism spectrum disorder. Nature 2017;542(7641):348–51.

[325] Shen MD, Kim SH, McKinstry RC, Gu H, Hazlett HC, Nordahl CW, et al. Increased extra-axial cerebrospinal fluid in high-risk infants who later develop autism. Biol Psychiatry 2017;82(3):186–93.

[326] McPartland JC, Lerner MD, Bhat A, Clarkson T, Jack A, Koohsari S, et al. Looking back at the next 40 years of ASD neuroscience research. J Autism Dev Disord 2021;51(12):4333–53.

[327] Ilioska I, Oldehinkel M, Llera A, Chopra S, Looden T, Chauvin R, et al. Connectome-wide mega-analysis reveals robust patterns of atypical functional connectivity in autism. Biol Psychiatry 2023;94(1):29–39.

[328] Goodwill AM, Low LT, Fox PT, Fox PM, Poon KK, Bhowmick SS, et al. Meta-analytic connectivity modelling of functional magnetic resonance imaging studies in autism spectrum disorders. Brain Imaging Behav 2023;17(2):257–69.

[329] Costa C, Cristea IA, Dal Bò E, Melloni C, Gentili C. Brain activity during facial processing in autism spectrum disorder: an activation likelihood estimation (ALE) meta-analysis of neuroimaging studies. J Child Psychol Psychiatry 2021;62(12):1412–24.

[330] Zhang Z, Peng P, Zhang D. Executive function in high-functioning autism spectrum disorder: a meta-analysis of fMRI studies. J Autism Dev Disord 2020;50(11):4022–38.

[331] Voorhies W, Dajani DR, Vij SG, Shankar S, Turan TO, Uddin LQ. Aberrant functional connectivity of inhibitory control networks in children with autism spectrum disorder. Autism Res. 2018;11(11):1468–78.

[332] Janouschek H, Chase HW, Sharkey RJ, Peterson ZJ, Camilleri JA, Abel T, et al. The functional neural architecture of dysfunctional reward processing in autism. Neuroimage Clin. 2021;31:102700.

[333] Webb SJ, Naples AJ, Levin AR, Hellemann G, Borland H, Benton J, et al. The autism biomarkers consortium for clinical trials: initial evaluation of a battery of candidate EEG biomarkers. Am J Psychiatry 2023;180(1):41–9.

[334] Emerson RW, Adams C, Nishino T, Hazlett HC, Wolff JJ, Zwaigenbaum L, et al. Functional neuroimaging of high-risk 6-month-old infants predicts a diagnosis of autism at 24 months of age. Sci Transl Med. 2017;9(393):eaag288.

[335] Gabard-Durnam LJ, Wilkinson C, Kapur K, Tager-Flusberg H, Levin AR, Nelson CA. Longitudinal EEG power in the first postnatal year differentiates autism outcomes. Nat Commun. 2019;10(1):4188.

[336] Bosl WJ, Tager-Flusberg H, Nelson CA. EEG analytics for early detection of autism spectrum disorder: a data-driven approach. Sci Rep. 2018;8(1):6828.

[337] Klin A, Jones W, Schultz R, Volkmar F, Cohen D. Visual fixation patterns during viewing of naturalistic social situations as predictors of social competence in individuals with autism. Arch Gen Psychiatry 2002;59(9):809–16.

[338] Klin A, Lin DJ, Gorrindo P, Ramsay G, Jones W. Two-year-olds with autism orient to non-social contingencies rather than biological motion. Nature 2009;459(7244):257–61.

[339] Frazier TW, Strauss M, Klingemier EW, Zetzer EE, Hardan AY, Eng C, et al. A meta-analysis of gaze differences to social and nonsocial information between individuals with and without autism. J Am Acad Child Adolesc Psychiatry 2017;56(7):546–55.

[340] Shic F, Naples AJ, Barney EC, Chang SA, Li B, McAllister T, et al. The autism biomarkers consortium for clinical trials: evaluation of a battery of candidate eye-tracking biomarkers for use in autism clinical trials. Mol Autism. 2022;13(1):15.

[341] Frazier TW, Klingemier EW, Parikh S, Speer L, Strauss MS, Eng C, et al. Development and validation of objective and quantitative eye tracking-based measures of autism risk and symptom levels. J Am Acad Child Adolesc Psychiatry 2018;57(11):858–66.

[342] Jones W, Klaiman C, Richardson S, Aoki C, Smith C, Minjarez M, et al. Eye-tracking-based measurement of social visual engagement compared with expert clinical diagnosis of autism. JAMA 2023;330(9):854–65.

[343] Jones W, Klaiman C, Richardson S, Lambha M, Reid M, Hamner T, et al. Development and replication of objective measurements of social visual engagement to aid in early diagnosis and assessment of autism. JAMA Netw Open. 2023;6(9):e2330145.

[344] Perochon S, Di Martino JM, Carpenter KLH, Compton S, Davis N, Eichner B, et al. Early detection of autism using digital behavioral phenotyping. Nat Med 2023;29(10):2489–97.

[345] Tanner A, Dounavi K. The emergence of autism symptoms prior to 18 months of age: a systematic literature review. J Autism Dev Disord 2021;51(3):973–93.

[346] Franchini M, Armstrong VL, Schaer M, Smith IM. Initiation of joint attention and related visual attention processes in infants with autism spectrum disorder: literature review. Child Neuropsychol 2019;25(3):287–317.

[347] Yankowitz LD, Petrulla V, Plate S, Tunc B, Guthrie W, Meera SS, et al. Infants later diagnosed with autism have lower canonical babbling ratios in the first year of life. Mol Autism. 2022;13(1):28.

[348] Manwaring SS, Stevens AL, Mowdood A, Lackey M. A scoping review of deictic gesture use in toddlers with or at-risk for autism spectrum disorder. Autism Dev Lang Impair. 2018;3:1–27.

[349] Mundy P. A review of joint attention and social-cognitive brain systems in typical development and autism spectrum disorder. Eur J Neurosci 2018;47(6):497–514.

[350] Nyström P, Thorup E, Bölte S, Falck-Ytter T. Joint attention in infancy and the emergence of autism. Biol Psychiatry 2019; 86(8):631–8.

[351] Brooks R, Meltzoff AN. Connecting the dots from infancy to childhood: a longitudinal study connecting gaze following, language, and explicit theory of mind. J Exp Child Psychol 2015;130:67–78.

[352] Wing L, Gould J. Severe impairments of social interaction and associated abnormalities in children: epidemiology and classification. J Autism Dev Disord 1979;9:11–29.

[353] Bauminger-Zviely N. School-age children with ASD. In: 4th ed. Volkmar FR, Rogers SJ, Paul R, Pelphrey KA, editors. Handbook of autism and pervasive developmental disorders, I. Hoboken, NJ: John Wiley & Sons; 2014, p. 148–75.

[354] National Institute for Health and Clinical Excellence. Autism spectrum disorder in under 19s: recognition, referral and diagnosis. Published Sept 28, 2011. Updated Dec. 20, 2017. https://www.nice.org.uk/guidance/cg128 [accessed 16.06.23].

[355] Howlin P, Magiati I. Autism spectrum disorder: outcomes in adulthood. Curr Opin Psychiatry 2017;30:69–76.

[356] Poon KK, Sidhu DJ. Adults with autism spectrum disorders: a review of outcomes, social attainment, and interventions. Curr Opin Psychiatry 2017;30:77–84.

[357] Schaeffer J, Abd El-Raziq M, Castroviejo E, Durrleman S, Ferré S, Grama I, et al. Language in autism: domains, profiles and co-occurring conditions. J Neural Transm 2023;130(3):433–57.

[358] Eigsti IM, de Marchena AB, Schuh JM, Kelley E. Language acquisition in autism spectrum disorders: a developmental review. Res Autism Spectr Disord. 2011;2:681–91.

[359] Tager-Flusberg H, Caronna E. Language disorders: autism and other pervasive developmental disorders. Pediatr Clin North Am 2007;54 (3):469–81.

[360] Kasari C, Chang Y. Play development in children with autism spectrum disorders: skills, object play, and interventions. In: 4th ed. Volkmar FR, Rogers SJ, Paul R, Pelphrey KA, editors. Handbook of autism and pervasive developmental disorders, I. Hoboken, NJ: John Wiley & Sons; 2014.

[361] Thiemann-Bourque K, Johnson LK, Brady NC. Similarities in functional play and differences in symbolic play of children with autism spectrum disorder. Am J Intellect Dev Disabil. 2019;124(1):77–91.

[362] González-Sala F, Gómez-Marí I, Tárraga-Mínguez R, Vicente-Carvajal A, Pastor-Cerezuela G. Symbolic play among children with autism spectrum disorder: a scoping review. Children 2021;8(9):801.

[363] Leekam SR, Prior MR, Uljarevic M. Restricted and repetitive behaviors in autism spectrum disorders: a review of research in the last decade. Psychol Bull 2011;137(4):562–93.

[364] Joseph L, Thurm A, Farmer C, Shumway S. Repetitive behavior and restricted interests in young children with autism: comparisons with controls and stability over 2 years. Autism Res. 2013;6(6):584–95.

[365] Harrop C, McConachie H, Emsley R, Leadbitter K, Green JPACT Consortium. Restricted and repetitive behaviors in autism spectrum disorders and typical development: cross-sectional and longitudinal comparisons. J Autism Dev Disord 2014;44(5):1207–19.

[366] Keating J, Van Goozen S, Uljarevic M, Hay D, Leekam SR. Restricted and repetitive behaviors and their developmental and demographic correlates in 4-8-year-old children: a transdiagnostic approach. Front Behav Neurosci. 2023;17:1085404.

[367] Evans DW, Leckman JF, Carter A, Reznick JS, Henshaw D, King RA, et al. Ritual, habit, and perfectionism: the prevalence and development of compulsive-like behavior in normal young children. Child Dev 1997;68(1):58–68.

[368] Jiujias M, Kelley E, Hall L. Restricted, repetitive behaviors in autism spectrum disorder and obsessive-compulsive disorder: a comparative review. Child Psychiatry Hum Dev 2017;48(6):944–59.

[369] Miller M, Sun S, Iosif AM, Young GS, Belding A, Tubbs A, et al. Repetitive behavior with objects in infants developing autism predicts diagnosis and later social behavior as early as 9 months. J Abnorm Psychol 2021;130(6):665–75.

[370] Sacrey LA, Zwaigenbaum L, Bryson S, Brian J, Smith IM, Roberts W, et al. Can parents' concerns predict autism spectrum disorder? A prospective study of high-risk siblings from 6 to 36 months of age. J Am Acad Child Adolesc Psychiatry 2015;54(6):470–8.

[371] Chaxiong P, Dimian AF, Wolff JJ. Restricted and repetitive behavior in children with autism during the first three years of life: a systematic review. Front Psychol. 2022;13:986876.

[372] Uljarević M, Frazier TW, Jo B, Billingham WD, Cooper MN, Youngstrom EA, et al. Big data approach to characterize restricted and repetitive behaviors in autism. J Am Acad Child Adolesc Psychiatry 2022;61(3):446–57.

[373] Courchesne V, Bedford R, Pickles A, Duku E, Kerns C, Mirenda P, et al. Non-verbal IQ and change in restricted and repetitive behavior throughout childhood in autism: a longitudinal study using the Autism Diagnostic Interview-Revised. Mol Autism. 2021;12(1):57.

[374] Richler J, Huerta M, Bishop SL, Lord C. Developmental trajectories of restricted and repetitive behaviors and interests in children with autism spectrum disorders. Dev Psychopathol 2010;22(1):55–69.

[375] Turner M. Annotation: repetitive behaviour in autism: a review of psychological research. J Child Psychol Psychiatry 1999;40(6):839–49.

[376] Uljarević M, Richdale AL, Evans DW, Cai RY, Leekam SR. Interrelationship between insistence on sameness, effortful control and anxiety in adolescents and young adults with autism spectrum disorder (ASD). Mol Autism. 2017;8:36.

[377] Spackman E, Smillie LD, Frazier TW, Hardan AY, Alvares GA, Whitehouse A, et al. Characterizing restricted and unusual interests in autistic youth. Autism Res. 2023;16(2):394–405.
[378] Spackman E, Smillie LD, Frazier TW, Hardan AY, Alvares GA, Whitehouse A, et al. Profiles of circumscribed interests in autistic youth. Front Behav Neurosci. 2023;17:1037967.
[379] Grove R, Hoekstra RA, Wierda M, Begeer S. Special interests and subjective wellbeing in autistic adults. Autism Res. 2018;11(5):766–75.
[380] Harrop C, Amsbary J, Towner-Wright S, Reichow B, Boyd BA. That's what I like: the use of circumscribed interests within interventions for individuals with autism spectrum disorder. A systematic review. Res Aut Spectr Disord 2019;57:63–86.
[381] Cope R, Remington A. The strengths and abilities of autistic people in the workplace. Autism Adulthood. 2022;4(1):22–31.
[382] Kirchner JC, Dziobek I. Toward the successful employment of adults with autism: a first analysis of special interests and factors deemed important for vocational performance. Scand J Child Adolesc Psychiatr Psychol. 2014;2(2):77–85.
[383] Krieger B, Kinebanian A, Prodinger B, Heigl F. Becoming a member of the work force: perceptions of adults with Asperger syndrome. Work 2012;43(2):141–57.
[384] Miller LJ, Anzalone ME, Lane SJ, Cermak SA, Osten ET. Concept evolution in sensory integration: a proposed nosology for diagnosis. Am J Occup Ther 2007;61(2):135–40.
[385] Ben-Sasson A, Gal E, Fluss R, Katz-Zetler N, Cermak SA. Update of a meta-analysis of sensory symptoms in ASD: a new decade of research. J Autism Dev Disord 2019;49(12):4974–96.
[386] Kirby AV, Bilder DA, Wiggins LD, Hughes MM, Davis J, Hall-Lande JA, et al. Sensory features in autism: findings from a large population-based surveillance system. Autism Res. 2022;15(4):751–60.
[387] McCormick C, Hepburn S, Young GS, Rogers SJ. Sensory symptoms in children with autism spectrum disorder, other developmental disorders and typical development: a longitudinal study. Autism 2016;20(5):572–9.
[388] MacLennan K, O'Brien S, Tavassoli T. In our own words: the complex sensory experiences of autistic adults. J Autism Dev Disord 2022;52(7):3061–75.
[389] Brunsdon VE, Colvert E, Ames C, Garnett T, Gillan N, Hallett V, et al. Exploring the cognitive features in children with autism spectrum disorder, their co-twins, and typically developing children within a population-based sample. J Child Psychol Psychiatry 2015;56(8):893–902.
[390] Vivanti G, Yerys B, Salomone E. Psychological factors in autism spectrum disorders. In: Volkmar F, editor. Autism and pervasive developmental disorders. Cambridge: Cambridge University Press; 2019. p. 61–88.
[391] Velikonja T, Fett AK, Velthorst E. Patterns of nonsocial and social cognitive functioning in adults with autism spectrum disorder: a systematic review and meta-analysis. JAMA Psychiatry 2019;76(2):135–51.
[392] Clements CC, Zoltowski AR, Yankowitz LD, Yerys BE, Schultz RT, Herrington JD. Evaluation of the social motivation hypothesis of autism: a systematic review and meta-analysis. JAMA Psychiatry 2018;75(8):797–808.
[393] Peñuelas-Calvo I, Sareen A, Sevilla-Llewellyn-Jones J, Fernández-Berrocal P. The "Reading the Mind in the Eyes" test in autism-spectrum disorders comparison with healthy controls: a systematic review and meta-analysis. J Autism Dev Disord 2019;49(3):1048–61.
[394] Gao S, Wang X, Su Y. Examining whether adults with autism spectrum disorder encounter multiple problems in theory of mind: a study based on meta-analysis. Psychon Bull Rev 2023. Available from: https://doi.org/10.3758/s13423-023-02280-8 Epub ahead of print.
[395] Demetriou EA, Lampit A, Quintana DS, Naismith SL, Song YJC, Pye JE, et al. Autism spectrum disorders: a meta-analysis of executive function. Mol Psychiatry 2018;23(5):1198–204.
[396] Booth RDL, Happé FGE. Evidence of reduced global processing in autism spectrum disorder. J Autism Dev Disord 2018;48(4):1397–408.
[397] Neufeld J, Hagström A, Van't Westeinde A, Lundin K, Cauvet É, Willfors C, et al. Global and local visual processing in autism – a co-twin-control study. J Child Psychol Psychiatry 2020;61(4):470–9.
[398] Greenberg DM, Warrier V, Allison C, Baron-Cohen S. Testing the empathizing-systemizing theory of sex differences and the extreme male brain theory of autism in half a million people. Proc Natl Acad Sci USA 2018;115(48):12152–7.
[399] Rapin I, Dunn M. Update on the language disorders of individuals on the autistic spectrum. Brain Dev 2003;25(3):166–72.
[400] Koegel LK, Bryan KM, Su PL, Vaidya M, Camarata S. Definitions of nonverbal and minimally verbal in research for autism: a systematic review of the literature. J Autism Dev Disord 2020;50(8):2957–72.
[401] Luyster RJ, Zane E, Wisman Weil L. Conventions for unconventional language: revisiting a framework for spoken language features in autism. Autism Dev Lang Impair. 2022;7:1–19.
[402] Boucher J. Research review: structural language in autistic spectrum disorder – characteristics and causes. J Child Psychol Psychiatry 2012;53(3):219–33.
[403] Fombonne E. Epidemiological surveys of autism and other pervasive developmental disorders: an update. J Autism Dev Disord 2003;33(4):365–82.
[404] Idring S, Lundberg M, Sturm H, Dalman C, Gumpert C, Rai D, et al. Changes in prevalence of autism spectrum disorders in 2001–2011: findings from the Stockholm youth cohort. J Autism Dev Disord 2015;45(6):1766–73.
[405] Saemundsen E, Magnússon P, Georgsdóttir I, Egilsson E, Rafnsson V. Prevalence of autism spectrum disorders in an Icelandic birth cohort. BMJ Open. 2013;3(6):e002748.
[406] Zhou H, Xu X, Yan W, Zou X, Wu L, Luo X, et al. Prevalence of autism spectrum disorder in China: a nationwide multi-center population-based study among children aged 6 to 12 years. Neurosci Bull 2020;36(9):961–71.
[407] Charman T, Pickles A, Simonoff E, Chandler S, Loucas T, Baird G. IQ in children with autism spectrum disorders: data from the special needs and autism project (SNAP). Psychol Med 2011;41:619–27.

[408] Postorino V, Fatta LM, Sanges V, Giovagnoli G, De Peppo L, Vicari S, et al. Intellectual disability in autism spectrum disorder: investigation of prevalence in an Italian sample of children and adolescents. Res Dev Disabil 2016;48:193–201.

[409] Billeiter KB, Froiland JM. Diversity of intelligence is the norm within the autism spectrum: full scale intelligence scores among children with ASD. Child Psychiatry Hum Dev 2023;54(4):1094–101.

[410] Joseph RM. The significance of IQ and differential cognitive abilities for understanding ASD. In: Fein D, editor. The neuropsychology of autism. New York, NY: Oxford University Press; 2011, p. 281–94.

[411] Nowell KP, Schanding Jr GT, Kanne SM, Goin-Kochel RP. Cognitive profiles in youth with autism spectrum disorder: an investigation of base rate discrepancies using the Differential Ability Scales-Second Edition. J Autism Dev Disord. 2015;45(7):1978–88.

[412] Takayanagi M, Kawasaki Y, Shinomiya M, Hiroshi H, Okada S, Ino T, et al. Review of cognitive characteristics of autism spectrum disorder using performance on six subtests on four versions of the Wechsler Intelligence Scale for Children. J Autism Dev Disord 2022;52(1):240–53.

[413] Melling R, Swinson JM, Birak KS. The relationship between Autism Spectrum Quotient (AQ) and uneven intellectual development in school-age children. Cogent Psychol. 2016;3:1149136.

[414] Muth A, Hönekopp J, Falter CM. Visuo-spatial performance in autism: a meta-analysis. J Autism Dev Disord 2014;44(12):3245–63.

[415] Charman T, Jones CR, Pickles A, Simonoff E, Baird G, Happé F. Defining the cognitive phenotype of autism. Brain Res 2011;1380:10–21.

[416] Anderson DK, Liang JW, Lord C. Predicting young adult outcome among more and less cognitively able individuals with autism spectrum disorders. J Child Psychol Psychiatry 2014;55(5):485–94.

[417] Flanagan HE, Smith IM, Vaillancourt T, Duku E, Szatmari P, Bryson S, et al. Stability and change in the cognitive and adaptive behaviour scores of preschoolers with autism spectrum disorder. J Autism Dev Disord 2015;45(9):2691–703.

[418] Solomon M, Cho B, Iosif AM, Heath B, Srivastav A, Nordahl C, et al. IQ trajectories in autistic children through preadolescence. JCPP Adv. 2023;3(1):e12127. Available from: https://doi.org/10.1002/jcv2.12127 Epub 2023 Jan 31.

[419] Simonoff E, Kent R, Stringer D, Lord C, Briskman J, Lukito S, et al. Trajectories in symptoms of autism and cognitive ability in autism from childhood to adult life: findings from a longitudinal epidemiological cohort. J Am Acad Child Adolesc Psychiatry 2020;59(12):1342–52.

[420] Prigge MBD, Bigler ED, Lange N, Morgan J, Froehlich A, Freeman A, et al. Longitudinal stability of intellectual functioning in autism spectrum disorder: from age 3 through mid-adulthood. J Autism Dev Disord 2022;52(10):4490–504.

[421] Bishop SL, Farmer C, Thurm A. Measurement of nonverbal IQ in autism spectrum disorder: scores in young adulthood compared to early childhood. J Autism Dev Disord 2015;45(4):966–74.

[422] Kirby AV, Baranek GT, Fox L. Longitudinal predictors of outcomes for adults with autism spectrum disorder: systematic review. OTJR 2016;36(2):55–64.

[423] Tillmann J, San José Cáceres A, Chatham CH, Crawley D, Holt R, Oakley B, et al. Investigating the factors underlying adaptive functioning in autism in the EU-AIMS Longitudinal European Autism Project. Autism Res. 2019;12(4):645–57.

[424] Mason D, Capp SJ, Stewart GR, Kempton MJ, Glaser K, Howlin P, et al. A meta-analysis of outcome studies of autistic adults: quantifying effect size, quality, and meta-regression. J Autism Dev Disord 2021;51(9):3165–79.

[425] Alvares GA, Bebbington K, Cleary D, Evans K, Glasson EJ, Maybery MT, et al. The misnomer of 'high functioning autism': intelligence is an imprecise predictor of functional abilities at diagnosis. Autism 2020;24(1):221–32.

[426] Pickles A, McCauley JB, Pepa LA, Huerta M, Lord C. The adult outcome of children referred for autism: typology and prediction from childhood. J Child Psychol Psychiatry 2020;61(7):760–7.

[427] McQuaid GA, Pelphrey KA, Bookheimer SY, Dapretto M, Webb SJ, Bernier RA, et al. The gap between IQ and adaptive functioning in autism spectrum disorder: disentangling diagnostic and sex differences. Autism. 2021;25(6):1565–79.

[428] Chatham CH, Taylor KI, Charman T, Liogier D'ardhuy X, Eule E, Fedele A, et al. Adaptive behavior in autism: minimal clinically important differences on the Vineland-II. Autism Res. 2018;11(2):270–83.

[429] Bal VH, Kim SH, Cheong D, Lord C. Daily living skills in individuals with autism spectrum disorder from 2 to 21 years of age. Autism 2015;19(7):774–84.

[430] Pathak M, Bennett A, Shui AM. Correlates of adaptive behavior profiles in a large cohort of children with autism: the autism speaks Autism Treatment Network registry data. Autism 2019;23(1):87–99.

[431] Hudepohl MB, Robins DL, King TZ, Henrich CC. The role of emotion perception in adaptive functioning of people with autism spectrum disorders. Autism 2015;19(1):107–12.

[432] Otsuka S, Uono S, Yoshimura S, Zhao S, Toichi M. Emotion perception mediates the predictive relationship between verbal ability and functional outcome in high-functioning adults with autism spectrum disorder. J Autism Dev Disord 2017;47(4):1166–82.

[433] Braconnier ML, Siper PM. Neuropsychological assessment in autism spectrum disorder. Curr Psychiatry Rep 2021;23(10):63.

[434] Ng R, Heinrich K, Hodges EK. Brief Report: neuropsychological testing and informant-ratings of children with autism spectrum disorder, attention-deficit/hyperactivity disorder, or comorbid diagnosis. J Autism Dev Disord 2019;49(6):2589–96.

[435] Kim SH, Bal VH, Lord C. Longitudinal follow-up of academic achievement in children with autism from age 2 to 18. J Child Psychol Psychiatry 2018;59(3):258–67.

[436] Keen D, Webster A, Ridley G. How well are children with autism spectrum disorder doing academically at school? An overview of the literature. Autism 2016;20(3):276–94.

[437] Prior M, Ozonoff S. Psychological factors in autism. In: Volkmar FR, editor. Autism and pervasive developmental disorders. 2nd ed. Cambridge, UK: Cambridge University Press; 2007, p. 69–128.

[438] Brown HM, Oram-Cardy J, Johnson A. A meta-analysis of the reading comprehension skills of individuals on the autism spectrum. J Autism Dev Disord 2013;43(4):932–55.

[439] Sorenson Duncan T, Karkada M, Deacon SH, Smith IM. Building meaning: meta-analysis of component skills supporting reading comprehension in children with autism spectrum disorder. Autism Res. 2021;14(5):840–58.

[440] Troyb E, Orinstein A, Tyson K, Helt M, Eigsti IM, Stevens M, et al. Academic abilities in children and adolescents with a history of autism spectrum disorders who have achieved optimal outcomes. Autism 2014;18(3):233–43.

[441] Kushki A, Chau T, Anagnostou E. Handwriting difficulties in children with autism spectrum disorders: a scoping review. J Autism Dev Disord 2011;41:1706–16.

[442] Randi J, Newman T, Grigorenko EL. Teaching children with autism to read for meaning: challenges and possibilities. J Autism Dev Disord 2010;40:890–902.

[443] Fournier KA, Hass CJ, Naik SK, Lodha N, Cauraugh JH. Motor coordination in autism spectrum disorders: a synthesis and meta-analysis. J Autism Dev Disord 2010;40(10):1227–40.

[444] Kangarani-Farahani M, Malik MA, Zwicker JG. Motor impairments in children with autism spectrum disorder: a systematic review and meta-analysis. J Autism Dev Disord 2023;. Available from: https://doi.org/10.1007/s10803-023-05948-1 Epub ahead of print.

[445] Zampella CJ, Wang LAL, Haley M, Hutchinson AG, de Marchena A. Motor skill differences in autism spectrum disorder: a clinically focused review. Curr Psychiatry Rep 2021;23(10):64.

[446] Wang LAL, Petrulla V, Zampella CJ, Waller R, Schultz RT. Gross motor impairment and its relation to social skills in autism spectrum disorder: a systematic review and two meta-analyses. Psychol Bull 2022;148(3-4):273–300.

[447] Lum JAG, Shandley K, Albein-Urios N, Kirkovski M, Papadopoulos N, Wilson RB, et al. Meta-analysis reveals gait anomalies in autism. Autism Res 2020;14:733–47.

[448] Stins JF, Emck C. Balance performance in autism: a brief overview. Front Psychol. 2018;9:901.

[449] Lim YH, Partridge K, Girdler S, Morris SL. Standing postural control in individuals with autism spectrum disorder: systematic review and meta-analysis. J Autism Dev Disord 2017;47(7):2238–53.

[450] Khoury E, Carment L, Lindberg P, Gaillard R, Krebs MO, Amado I. [Sensorimotor aspects and manual dexterity in autism spectrum disorders: a literature review]. English abstract. Encephale. 2020;46(2):135–45.

[451] Handle HC, Feldin M, Pilacinski A. Handwriting in autism spectrum disorder: a literature review. NeuroSci. 2022;3(4):558–65.

[452] Edwards LA. A meta-analysis of imitation abilities in individuals with autism spectrum disorders. Autism Res 2014;7(3):363–80.

[453] Lidstone DE, Mostofsky SH. Moving toward understanding autism: visual-motor integration, imitation, and social skill development. Pediatr Neurol 2021;122:98–105.

[454] de Moraes ÍAP, Massetti T, Crocetta TB, da Silva TD, de Menezes LDC, Monteiro CBM, et al. Motor learning characterization in people with autism spectrum disorder: a systematic review. Dement Neuropsychol. 2017;11(3):276–86.

[455] Cerullo S, Fulceri F, Muratori F, Contaldo A. Acting with shared intentions: a systematic review on joint action coordination in autism spectrum disorder. Brain Cogn 2021;149:105693.

[456] McNaughton KA, Redcay E. Interpersonal synchrony in autism. Curr Psychiatry Rep 2020;22(3):12.

[457] Iverson JM, Shic F, Wall CA, Chawarska K, Curtin S, Estes A, et al. Early motor abilities in infants at heightened versus low risk for ASD: a baby siblings research consortium (BSRC) study. J Abnorm Psychol. 2019;128(1):69–80.

[458] West KL. Infant motor development in autism spectrum disorder: a synthesis and meta-analysis. Child Dev 2019;90:2053–70.

[459] Ming X, Brimacombe M, Wagner GC. Prevalence of motor impairment in autism spectrum disorders. Brain Dev 2007;29(9):565–70.

[460] Lopez-Espejo MA, Nuñez AC, Moscoso OC, Escobar RG. Clinical characteristics of children affected by autism spectrum disorder with and without generalized hypotonia. Eur J Pediatr 2021;180(10):3243–6.

[461] Tripi G, Roux S, Carotenuto M, Bonnet-Brilhault F, Roccella M. Minor neurological dysfunctions (MNDs) in autistic children without intellectual disability. J Clin Med. 2018;7(4):79.

[462] Leyden J, Fung L, Frick S. Autism and toe-walking: are they related? Trends and treatment patterns between 2005 and 2016. J Child Orthop 2019;13(4):340–5.

[463] Bhat AN. Is motor impairment in autism spectrum disorder distinct from developmental coordination disorder? A report from the SPARK study. Phys Ther 2020;100(4):633–44.

[464] Ketcheson LR, Pitchford EA, Wentz CF. The relationship between developmental coordination disorder and concurrent deficits in social communication and repetitive behaviors among children with autism spectrum disorder. Autism Res. 2021;14(4):804–16.

[465] Licari MK, Alvares GA, Varcin K, Evans KL, Cleary D, Reid SL, et al. Prevalence of motor difficulties in autism spectrum disorder: analysis of a population-based cohort. Autism Res. 2020;13(2):298–306.

[466] Kaur M, Srinivasan SM, Bhat AN. Comparing motor performance, praxis, coordination, and interpersonal synchrony between children with and without autism spectrum disorder (ASD). Res Dev Disabil 2018;72:79–95.

[467] Lidstone DE, Miah FZ, Poston B, Beasley JF, Mostofsky SH, Dufek JS. Children with autism spectrum disorder show impairments during dynamic versus static grip-force tracking. Autism Res. 2020;13(12):2177–89.

[468] Cook J. From movement kinematics to social cognition: the case of autism. Philos Trans R Soc Lond B Biol Sci 2016;371(1693):20150372.

[469] Dowell LR, Mahone EM, Mostofsky SH. Associations of postural knowledge and basic motor skill with dyspraxia in autism: implication for abnormalities in distributed connectivity and motor learning. Neuropsychology 2009;23(5):563–70.

[470] Treffert DA, Rebedew DL. The Savant Syndrome Registry: a preliminary report. WMJ 2015;114(4):158–62.

[471] Hughes JEA, Ward J, Gruffydd E, Baron-Cohen S, Smith P, Allison C, et al. Savant syndrome has a distinct psychological profile in autism. Mol Autism. 2018;9:53.

[472] Park HO. Autism spectrum disorder and savant syndrome: a systematic literature review. Soa Chongsonyon Chongsin Uihak. 2023;34 (2):76–92.

[473] Treffert DA. The savant syndrome: an extraordinary condition. A synopsis: past, present, future. Philos Trans R Soc Lond B Biol Sci 2009;364(1522):1351–7.

[474] Howlin P, Goode S, Hutton J, Rutter M. Savant skills in autism: psychometric approaches and parental reports. Philos Trans R Soc Lond B Biol Sci 2009;364(1522):1359–67.

[475] Rimland B. Savant capabilities of autistic children and their cognitive implications. In: Serban G, editor. Cognitive defects in the development of mental illness. New York, NY, USA: Brunner/Mazel; 1978, p. 43–65.

[476] Bölte S, Poustka F. Comparing the intelligence profiles of savant and nonsavant individuals with autistic disorder. Intelligence 2004;32 (2):121–31.

[477] Daniel E, Menashe I. Exploring the familial role of social responsiveness differences between savant and non-savant children with autism. Sci Rep. 2020;10(1):2255.

[478] Meilleur AA, Jelenic P, Mottron L. Prevalence of clinically and empirically defined talents and strengths in autism. J Autism Dev Disord 2015;45(5):1354–67.

[479] Bal VH, Wilkinson E, Fok M. Cognitive profiles of children with autism spectrum disorder with parent-reported extraordinary talents and personal strengths. Autism 2022;26(1):62–74.

[480] Bennett E, Heaton PF. Defining the clinical and cognitive phenotype of child savants with autism spectrum disorder. Curr Pediatr Res. 2017;21(1):140–7.

[481] Ostrolenk A, Forgeot d'Arc B, Jelenic P, Samson F, Mottron L. Hyperlexia: systematic review, neurocognitive modelling, and outcome. Neurosci Biobehav Rev 2017;79:134–49.

[482] Tan C, Frewer V, Cox G, Williams K, Ure A. Prevalence and age of onset of regression in children with autism spectrum disorder: a systematic review and meta-analytical update. Autism Res 2021;14:582–98.

[483] Boterberg S, Charman T, Marschik PB, Bölte S, Roeyers H. Regression in autism spectrum disorder: a critical overview of retrospective findings and recommendations for future research. Neurosci Biobehav Rev 2019;102:24–55.

[484] Thurm A, Powell EM, Neul JL, Wagner A, Zwaigenbaum L. Loss of skills and onset patterns in neurodevelopmental disorders: understanding the neurobiological mechanisms. Autism Res. 2018;11(2):212–22.

[485] Barger BD, Campbell JM, McDonough JD. Prevalence and onset of regression within autism spectrum disorders: a meta-analytic review. J Autism Dev Disord 2013;43:817–28.

[486] Ozonoff S, Iosif AM. Changing conceptualizations of regression: what prospective studies reveal about the onset of autism spectrum disorder. Neurosci Biobehav Rev 2019;100:296–304.

[487] Pearson N, Charman T, Happé F, Bolton PF, McEwen FS. Regression in autism spectrum disorder: reconciling findings from retrospective and prospective research. Autism Res. 2018;11(12):1602–20.

[488] Zwaigenbaum L, Bauman ML, Stone WL, Yirmiya N, Estes A, Hansen RL, et al. Early identification of autism spectrum disorder: recommendations for practice and research. Pediatrics 2015;136(Suppl 1):S10–40.

[489] Ozonoff S, Iosif AM, Baguio F, Cook IC, Hill MM, Hutman T, et al. A prospective study of the emergence of early behavioral signs of autism. J Am Acad Child Adolesc Psychiatry 2010;49(3):256–66 e1-2.

[490] Ozonoff S, Gangi D, Hanzel EP, Hill A, Hill MM, Miller M, et al. Onset patterns in autism: variation across informants, methods, and timing. Autism Res. 2018;11(5):788–97.

[491] Barbaresi WJ. The meaning of "regression" in children with autism spectrum disorder: why does it matter? J Dev Behav Pediatr 2016;37 (6):506–7.

[492] Goin-Kochel RP, Esler AN, Kanne SM, Hus V. Developmental regression among children with autism spectrum disorder: onset, duration, and effects on functional outcomes. Res Autism Spectr Disord. 2014;8(7):890–8.

[493] Zachor DA, Ben-Itzchak E. Specific medical conditions are associated with unique behavioral profiles in autism spectrum disorders. Front Neurosci. 2016;10:410.

[494] Bradley CC, Boan AD, Cohen AP, Charles JM, Carpenter LA. Reported history of developmental regression and restricted, repetitive behaviors in children with autism spectrum disorders. J Dev Behav Pediatr 2016;37(6):451–6.

[495] Gadow KD, Perlman G, Weber RJ. Parent-reported developmental regression in autism: epilepsy, IQ, schizophrenia spectrum symptoms, and special education. J Autism Dev Disord 2017;47(4):918–26.

[496] Thompson L, Gillberg C, Landberg S, Kantzer AK, Miniscalco C, Barnevik Olsson M, et al. Autism with and without regression: a two-year prospective longitudinal study in two population-derived Swedish cohorts. J Autism Dev Disord 2019;49(6):2281–90.

[497] Pickles A, Wright N, Bedford R, Steiman M, Duku E, Bennett T, et al. Predictors of language regression and its association with subsequent communication development in children with autism. J Child Psychol Psychiatry 2022;63(11):1243–51.

[498] Hu C, Yang F, Yang T, Chen J, Dai Y, Jia F, et al. A multi-center study on the relationship between developmental regression and disease severity in children with autism spectrum disorders. Front Psychiatry. 2022;13:796554.

[499] Giannotti F, Cortesi F, Cerquiglini A, Vagnoni C, Valente D. Sleep in children with autism with and without autistic regression. J Sleep Res 2011;20(2):338–47.

[500] Baird G, Charman T, Pickles A, Chandler S, Loucas T, Meldrum D, et al. Regression, developmental trajectory and associated problems in disorders in the autism spectrum: the SNAP study. J Autism Dev Disord 2008;38(10):1827–36.
[501] Ferguson BJ, Marler S, Altstein LL, Lee EB, Akers J, Sohl K, et al. Psychophysiological associations with gastrointestinal symptomatology in autism spectrum disorder. Autism Res. 2017;10(2):276–88.
[502] Reynolds AM, Soke GN, Sabourin KR, Croen LA, Daniels JL, Fallin MD, et al. Gastrointestinal symptoms in 2- to 5-year-old children in the study to explore early development. J Autism Dev Disord 2021;51(11):3806–17.
[503] Scott O, Shi D, Andriashek D, Clark B, Goez HR. Clinical clues for autoimmunity and neuroinflammation in patients with autistic regression. Dev Med Child Neurol 2017;59(9):947–51.
[504] Barger BD, Campbell JM, Simmons C. The relationship between regression in autism spectrum disorder, epilepsy, and atypical epileptiform EEGs: a meta-analytic review. J Intellect Dev Disabil 2017;42(1):45–60.
[505] Ewen JB, Marvin AR, Law K, Lipkin PH. Epilepsy and autism severity: a study of 6,975 children. Autism Res. 2019;12(8):1251–9.
[506] Thurm A, Manwaring SS, Luckenbaugh DA, Lord C, Swedo SE. Patterns of skill attainment and loss in young children with autism. Dev Psychopathol 2014;26(1):203–14.
[507] Brignell A, Williams K, Prior M, Donath S, Reilly S, Bavin EL, et al. Parent-reported patterns of loss and gain in communication in 1- to 2-year-old children are not unique to autism spectrum disorder. Autism 2017;21(3):344–56.
[508] Reierson G, Bernstein J, Froehlich-Santino W, Urban A, Purmann C, Berquist S, et al. Characterizing regression in Phelan McDermid Syndrome (22q13 deletion syndrome). J Psychiatr Res 2017;91:139–44.
[509] De Rubeis S, Siper PM, Durkin A, Weissman J, Muratet F, Halpern D, et al. Delineation of the genetic and clinical spectrum of Phelan-McDermid syndrome caused by *SHANK3* point mutations. Mol Autism. 2018;9:31.
[510] Tammimies K. Genetic mechanisms of regression in autism spectrum disorder. Neurosci Biobehav Rev 2019;102:208–20.
[511] Peters SU, Hundley RJ, Wilson AK, Carvalho CM, Lupski JR, Ramocki MB. Brief report: regression timing and associated features in MECP2 duplication syndrome. J Autism Dev Disord 2013;43(10):2484–90.
[512] Peters SU, Hundley RJ, Wilson AK, Warren Z, Vehorn A, Carvalho CM, et al. The behavioral phenotype in MECP2 duplication syndrome: a comparison with idiopathic autism. Autism Res. 2013;6(1):42–50.
[513] Shoffner J, Hyams L, Langley GN, Cossette S, Mylacraine L, Dale J, et al. Fever plus mitochondrial disease could be risk factors for autistic regression. J Child Neurol 2010;25(4):429–34.
[514] Goin-Kochel RP, Trinh S, Barber S, Bernier R. Gene disrupting mutations associated with regression in autism spectrum disorder. J Autism Dev Disord 2017;47(11):3600–7.
[515] Ozonoff S, Young GS, Brian J, Charman T, Shephard E, Solish A, et al. Diagnosis of autism spectrum disorder after age 5 in children evaluated longitudinally since infancy. J Am Acad Child Adolesc Psychiatry 2018;57(11):849–57 e2.
[516] Mehra C, Sil A, Hedderly T, Kyriakopoulos M, Lim M, Turnbull J, et al. Childhood disintegrative disorder and autism spectrum disorder: a systematic review. Dev Med Child Neurol 2019;61(5):523–34.
[517] Verity C, Baker E, Maunder P, Pal S, Winstone AM. Differential diagnosis of progressive intellectual and neurological deterioration in children. Dev Med Child Neurol 2021;63(3):287–94.
[518] Rydzewska E, Dunn K, Cooper SA. Umbrella systematic review of systematic reviews and meta-analyses on comorbid physical conditions in people with autism spectrum disorder. Br J Psychiatry 2021;218(1):10–19.
[519] Malow BA, Qian Y, Ames JL, Alexeeff S, Croen LA. Health conditions in autism: defining the trajectory from adolescence to early adulthood. Autism Res. 2023;16(7):1437–49.
[520] Williams ZJ, Gotham KO. Current and lifetime somatic symptom burden among transition-aged autistic young adults. Autism Res 2022;15(4):761–70.
[521] Davignon MN, Qian Y, Massolo M, Croen LA. Psychiatric and medical conditions in transition-aged individuals with ASD. Pediatrics 2018;141(Suppl 4):S335–45.
[522] Hossain MM, Khan N, Sultana A, Ma P, McKyer ELJ, Ahmed HU, et al. Prevalence of comorbid psychiatric disorders among people with autism spectrum disorder: an umbrella review of systematic reviews and meta-analyses. Psychiatry Res 2020;287:112922.
[523] Lai MC, Kassee C, Besney R, Bonato S, Hull L, Mandy W, et al. Prevalence of co-occurring mental health diagnoses in the autism population: a systematic review and meta-analysis. Lancet Psychiatry 2019;6(10):819–29.
[524] Kurtz-Nelson EC, Beighley JS, Hudac CM, Gerdts J, Wallace AS, Hoekzema K, et al. Co-occurring medical conditions among individuals with ASD-associated disruptive mutations. Child Health Care 2020;49(4):361–84.
[525] Willcutt EG. Behavior and molecular genetic approaches to comorbidity. Curr Dev Disord Rep. 2019;6(2):31–6.
[526] Ismail FY, Shapiro BK. What are neurodevelopmental disorders? Curr Opin Neurol 2019;32(4):611–16.
[527] Craig F, Savino R, Trabacca A. A systematic review of comorbidity between cerebral palsy, autism spectrum disorders and attention deficit hyperactivity disorder. Eur J Paediatr Neurol 2019;23(1):31–42.
[528] Pan PY, Bölte S, Kaur P, Jamil S, Jonsson U. Neurological disorders in autism: a systematic review and meta-analysis. Autism 2021;25(3):812–30.
[529] Canitano R, Vivanti G. Tics and Tourette syndrome in autism spectrum disorders. Autism 2007;11(1):19–28.
[530] Baron-Cohen S, Scahill VL, Izaguirre J, Hornsey H, Robertson MM. The prevalence of Gilles de la Tourette syndrome in children and adolescents with autism: a large scale study. Psychol Med 1999;29(5):1151–9.
[531] Kim YR, Song DY, Bong G, Han JH, Kim JH, Yoo HJ. Clinical characteristics of comorbid tic disorders in autism spectrum disorder: exploratory analysis. Child Adolesc Psychiatry Ment Health 2023;17(1):71.

[532] El Achkar CM, Spence SJ. Clinical characteristics of children and young adults with co-occurring autism spectrum disorder and epilepsy. Epilepsy Behav 2015;47:183–90.

[533] Alaimo H, Geller E, Mahalingam R, Rodriguez A, Goldberg R, Bojko A, et al. Ictal EEG in patients with autistic spectrum disorder and epilepsy. Epilepsy Res 2020;168:106482.

[534] Jokiranta E, Sourander A, Suominen A, Timonen-Soivio L, Brown AS, Sillanpää M. Epilepsy among children and adolescents with autism spectrum disorders: a population-based study. J Autism Dev Disord 2014;44(10):2547–57.

[535] Liu X, Sun X, Sun C, Zou M, Chen Y, Huang J, et al. Prevalence of epilepsy in autism spectrum disorders: a systematic review and meta-analysis. Autism 2022;26(1):33–50.

[536] Lukmanji S, Manji SA, Kadhim S, Sauro KM, Wirrell EC, Kwon CS, et al. The co-occurrence of epilepsy and autism: a systematic review. Epilepsy Behav. 2019;98(Pt A):238–48.

[537] Tuchman R. What is the relationship between autism spectrum disorders and epilepsy? Semin Pediatr Neurol 2017;24(4):292–300.

[538] Sansa G, Carlson C, Doyle W, Weiner HL, Bluvstein J, Barr W, et al. Medically refractory epilepsy in autism. Epilepsia 2011;52 (6):1071–5.

[539] Ghacibeh GA, Fields C. Interictal epileptiform activity and autism. Epilepsy Behav 2015;47:158–62.

[540] Precenzano F, Parisi L, Lanzara V, Vetri L, Operto FF, Pastorino GMG, et al. Electroencephalographic abnormalities in autism spectrum disorder: characteristics and therapeutic implications. Medicina (Kaunas) 2020;56(9):419.

[541] Sánchez Fernández I, Loddenkemper T, Galanopoulou AS, Moshé SL. Should epileptiform discharges be treated? Epilepsia 2015;56 (10):1492–504.

[542] McConachie H, Livingstone N, Morris C, Beresford B, Le Couteur A, Gringras P, et al. Parents suggest which indicators of progress and outcomes should be measured in young children with autism spectrum disorder. J Autism Dev Disord 2018;48(4):1041–51.

[543] Menezes M, Mazurek MO. Associations between domains of health-related quality of life and comorbid emotional and behavioral problems in youth with autism spectrum disorder. Res Autism Spectr Disord. 2021;82:101740.

[544] Kuhlthau KA, McDonnell E, Coury DL, Payakachat N, Macklin E. Associations of quality of life with health-related characteristics among children with autism. Autism 2018;22(7):804–13.

[545] Mutluer T, Aslan Genç H, Özcan Morey A, Yapici Eser H, Ertinmaz B, Can M, et al. Population-based psychiatric comorbidity in children and adolescents with autism spectrum disorder: a meta-analysis. Front Psychiatry 2022;13:856208.

[546] Khachadourian V, Mahjani B, Sandin S, Kolevzon A, Buxbaum JD, Reichenberg A, et al. Comorbidities in autism spectrum disorder and their etiologies. Transl Psychiatry. 2023;13(1):71.

[547] Casseus M, Kim WJ, Horton DB. Prevalence and treatment of mental, behavioral, and developmental disorders in children with co-occurring autism spectrum disorder and attention-deficit/hyperactivity disorder: a population-based study. Autism Res. 2023;16(4):855–67.

[548] Gordon-Lipkin E, Marvin AR, Law JK, Lipkin PH. Anxiety and mood disorder in children with autism spectrum disorder and ADHD. Pediatrics 2018;141(4):e20171377.

[549] Hunsche MC, Cervin M, Storch EA, Kendall PC, Wood JJ, Kerns CM. Social functioning and the presentation of anxiety in children on the autism spectrum: a multimethod, multiinformant analysis. J Psychopathol Clin Sci. 2022;131(2):198–208.

[550] Kerns CM, Kendall PC, Berry L, Souders MC, Franklin ME, Schultz RT, et al. Traditional and atypical presentations of anxiety in youth with autism spectrum disorder. J Autism Dev Disord 2014;44(11):2851–61.

[551] Vasa RA, Mazurek MO. An update on anxiety in youth with autism spectrum disorders. Curr Opin Psychiatry 2015;28(2):83–90.

[552] Kerns CM, Kendall PC, Zickgraf H, Franklin ME, Miller J, Herrington J. Not to be overshadowed or overlooked: functional impairments associated with comorbid anxiety disorders in youth with ASD. Behav Ther 2015;46(1):29–39.

[553] Mingins JE, Tarver J, Waite J, Jones C, Surtees AD. Anxiety and intellectual functioning in autistic children: a systematic review and meta-analysis. Autism 2021;25(1):18–32.

[554] Hudson CC, Hall L, Harkness KL. Prevalence of depressive disorders in individuals with autism spectrum disorder: a meta-analysis. J Abnorm Child Psychol 2019;47(1):165–75.

[555] Pezzimenti F, Han GT, Vasa RA, Gotham K. Depression in youth with autism spectrum disorder. Child Adolesc Psychiatr Clin N Am 2019;28 (3):397–409.

[556] Greenlee JL, Mosley AS, Shui AM, Veenstra-VanderWeele J, Gotham KO. Medical and behavioral correlates of depression history in children and adolescents with autism spectrum disorder. Pediatrics 2016;137(Suppl 2):S105–14.

[557] Vaquerizo-Serrano J, Salazar De Pablo G, Singh J, Santosh P. Catatonia in autism spectrum disorders: a systematic review and meta-analysis. Eur Psychiatry 2021;65(1):e4.

[558] Varcin KJ, Herniman SE, Lin A, Chen Y, Perry Y, Pugh C, et al. Occurrence of psychosis and bipolar disorder in adults with autism: a systematic review and meta-analysis. Neurosci Biobehav Rev 2022;134:104543.

[559] Steenfeldt-Kristensen C, Jones CA, Richards C. The prevalence of self-injurious behaviour in autism: a meta-analytic study. J Autism Dev Disord 2020;50(11):3857–73.

[560] Kõlves K, Fitzgerald C, Nordentoft M, Wood SJ, Erlangsen A. Assessment of suicidal behaviors among individuals with autism spectrum disorder in Denmark. JAMA Netw Open. 2021;4(1):e2033565.

[561] Blanchard A, Chihuri S, DiGuiseppi CG, Li G. Risk of self-harm in children and adults with autism spectrum disorder: a systematic review and meta-analysis. JAMA Netw Open 2021;4(10):e2130272.

[562] Lai MC, Saunders NR, Huang A, Artani A, Wilton AS, Zaheer J, et al. Self-harm events and suicide deaths among autistic individuals in Ontario, Canada. JAMA Netw Open. 2023;6(8):e2327415.
[563] Kirby AV, Bakian AV, Zhang Y, Bilder DA, Keeshin BR, Coon H. A 20-year study of suicide death in a statewide autism population. Autism Res. 2019;12(4):658–66.
[564] O'Halloran L, Coey P, Wilson C. Suicidality in autistic youth: a systematic review and meta-analysis. Clin Psychol Rev 2022;93:102144.
[565] Kallitsounaki A, Williams DM. Autism spectrum disorder and gender dysphoria/incongruence. A systematic literature review and meta-analysis. J Autism Dev Disord 2023;53(8):3103–17.
[566] Maggio MG, Calatozzo P, Cerasa A, Pioggia G, Quartarone A, Calabrò RS. Sex and sexuality in autism spectrum disorders: a scoping review on a neglected but fundamental issue. Brain Sci. 2022;12(11):1427.
[567] Chang JC, Lai MC, Tai YM, Gau SS. Mental health correlates and potential childhood predictors for the wish to be of the opposite sex in young autistic adults. Autism 2022;26(1):146–59.
[568] Strauss P, Cook A, Watson V, Winter S, Whitehouse A, Albrecht N, et al. Mental health difficulties among trans and gender diverse young people with an autism spectrum disorder (ASD): findings from trans pathways. J Psychiatr Res 2021;137:360–7.
[569] Strang JF, Anthony LG, Song A, Lai MC, Knauss M, Sadikova E, et al. In addition to stigma: cognitive and autism-related predictors of mental health in transgender adolescents. J Clin Child Adolesc Psychol 2023;52(2):212–29.
[570] Elrod MG, Hood BS. Sleep differences among children with autism spectrum disorders and typically developing peers: a meta-analysis. J Dev Behav Pediatr 2015;36(3):166–77.
[571] Díaz-Román A, Zhang J, Delorme R, Beggiato A, Cortese S. Sleep in youth with autism spectrum disorders: systematic review and meta-analysis of subjective and objective studies. Evid Based Ment Health 2018;21(4):146–54.
[572] Petruzzelli MG, Matera E, Giambersio D, Marzulli L, Gabellone A, Legrottaglie AR, et al. Subjective and electroencephalographic sleep parameters in children and adolescents with autism spectrum disorder: a systematic review. J Clin Med. 2021;10(17):3893.
[573] Cohen S, Conduit R, Lockley SW, Rajaratnam SM, Cornish KM. The relationship between sleep and behavior in autism spectrum disorder (ASD): a review. J Neurodev Disord. 2014;6(1):44.
[574] Schreck KA, Richdale AL. Sleep problems, behavior, and psychopathology in autism: inter-relationships across the lifespan. Curr Opin Psychol. 2020;34:105–11.
[575] Baker EK, Richdale AL. Examining the behavioural sleep-wake rhythm in adults with autism spectrum disorder and no comorbid intellectual disability. J Autism Dev Disord 2017;47(4):1207–22.
[576] Ballester P, Martínez MJ, Javaloyes A, Inda MD, Fernández N, Gázquez P, et al. Sleep problems in adults with autism spectrum disorder and intellectual disability. Autism Res. 2019;12(1):66–79.
[577] Bianca B, Silvia G, Elisa F, Deny M, Giovanni V, Lino N, et al. Insomnia in children with autism spectrum disorder: a cross-sectional study on clinical correlates and parental stress. J Autism Dev Disord 2022. Available from: https://doi.org/10.1007/s10803-022-05793-8 Epub ahead of print.
[578] Elrod MG, Nylund CM, Susi AL, Gorman GH, Hisle-Gorman E, Rogers DJ, et al. Prevalence of diagnosed sleep disorders and related diagnostic and surgical procedures in children with autism spectrum disorders. J Dev Behav Pediatr 2016;37(5):377–84.
[579] Malow BA, Katz T, Reynolds AM, Shui A, Carno M, Connolly HV, et al. Sleep difficulties and medications in children with autism spectrum disorders: a registry study. Pediatrics 2016;137(Suppl 2):S98–104.
[580] Meltzer LJ, Johnson C, Crosette J, Ramos M, Mindell JA. Prevalence of diagnosed sleep disorders in pediatric primary care practices. Pediatrics 2010;125(6):e1410–18.
[581] Han GT, Trevisan DA, Abel EA, Cummings EM, Carlos C, Bagdasarov A, et al. Associations between sleep problems and domains relevant to daytime functioning and clinical symptomatology in autism: a meta-analysis. Autism Res 2022;15(7):1249–60.
[582] Favole I, Davico C, Marcotulli D, Sodero R, Svevi B, Amianto F, et al. Sleep disturbances and emotional dysregulation in young children with autism spectrum, intellectual disability, or global developmental delay. Sleep Med 2023;105:45–52.
[583] Medic G, Wille M, Hemels ME. Short- and long-term health consequences of sleep disruption. Nat Sci Sleep. 2017;9:151–61.
[584] Sharp WG, Berry RC, McCracken C, Nuhu NN, Marvel E, Saulnier CA, et al. Feeding problems and nutrient intake in children with autism spectrum disorders: a meta-analysis and comprehensive review of the literature. J Autism Dev Disord 2013;43(9):2159–73.
[585] Mayes SD, Zickgraf H. Atypical eating behaviors in children and adolescents with autism, ADHD, other disorders, and typical development. Res Autism Spectr Disord. 2019;64:76–83.
[586] Page SD, Souders MC, Kral TVE, Chao AM, Pinto-Martin J. Correlates of feeding difficulties among children with autism spectrum disorder: a systematic review. J Autism Dev Disord 2022;52(1):255–74.
[587] Bourne L, Mandy W, Bryant-Waugh R. Avoidant/restrictive food intake disorder and severe food selectivity in children and young people with autism: a scoping review. Dev Med Child Neurol 2022;64(6):691–700.
[588] Wang J, Ma B, Wang J, Zhang Z, Chen O. Global prevalence of autism spectrum disorder and its gastrointestinal symptoms: a systematic review and meta-analysis. Front Psychiatry 2022;13:963102.
[589] Kozak A, Czepczor-Bernat K, Modrzejewska J, Modrzejewska A, Matusik E, Matusik P. Avoidant/restrictive food disorder (ARFID), food neophobia, other eating-related behaviours and feeding practices among children with autism spectrum disorder and in non-clinical sample: a preliminary study. Int J Environ Res Public Health 2023;20(10):5822.
[590] Bandini LG, Anderson SE, Curtin C, Cermak S, Evans EW, Scampini R, et al. Food selectivity in children with autism spectrum disorders and typically developing children. J Pediatr 2010;157(2):259–64.

[591] Zickgraf H, Mayes SD. Psychological, health, and demographic correlates of atypical eating behaviors in children with autism. J Dev Phys Disabil 2019;31:399–418.

[592] Leader G, O'Reilly M, Gilroy SP, Chen JL, Ferrari C, Mannion A. Comorbid feeding and gastrointestinal symptoms, challenging behavior, sensory issues, adaptive functioning and quality of life in children and adolescents with autism spectrum disorder. Dev Neurorehabil 2021;24 (1):35–44.

[593] Baraskewich J, von Ranson KM, McCrimmon A, McMorris CA. Feeding and eating problems in children and adolescents with autism: a scoping review. Autism 2021;25(6):1505–19.

[594] Westwood H, Tchanturia K. Autism spectrum disorder in anorexia nervosa: an updated literature review. Curr Psychiatry Rep 2017;19(7):41.

[595] McElhanon BO, McCracken C, Karpen S, Sharp WG. Gastrointestinal symptoms in autism spectrum disorder: a meta-analysis. Pediatrics 2014;133(5):872–83.

[596] Holingue C, Newill C, Lee LC, Pasricha PJ, Daniele Fallin M. Gastrointestinal symptoms in autism spectrum disorder: a review of the literature on ascertainment and prevalence. Autism Res. 2018;11(1):24–36.

[597] Leader G, Abberton C, Cunningham S, Gilmartin K, Grudzien M, Higgins E, et al. Gastrointestinal symptoms in autism spectrum disorder: a systematic review. Nutrients. 2022;14(7):1471.

[598] Chaidez V, Hansen RL, Hertz-Picciotto I. Gastrointestinal problems in children with autism, developmental delays or typical development. J Autism Dev Disord 2014;44(5):1117–27.

[599] Bresnahan M, Hornig M, Schultz AF, Gunnes N, Hirtz D, Lie KK, et al. Association of maternal report of infant and toddler gastrointestinal symptoms with autism: evidence from a prospective birth cohort. JAMA Psychiatry. 2015;72(5):466–74.

[600] McCue LM, Flick LH, Twyman KA, Xian H. Gastrointestinal dysfunctions as a risk factor for sleep disorders in children with idiopathic autism spectrum disorder: a retrospective cohort study. Autism 2017;21(8):1010–20.

[601] Neuhaus E, Bernier RA, Tham SW, Webb SJ. Gastrointestinal and psychiatric symptoms among children and adolescents with autism spectrum disorder. Front Psychiatry. 2018;9:515.

[602] Restrepo B, Angkustsiri K, Taylor SL, Rogers SJ, Cabral J, Heath B, et al. Developmental-behavioral profiles in children with autism spectrum disorder and co-occurring gastrointestinal symptoms. Autism Res. 2020;13(10):1778–89.

[603] Nikolov RN, Bearss KE, Lettinga J, Erickson C, Rodowski M, Aman MG, et al. Gastrointestinal symptoms in a sample of children with pervasive developmental disorders. J Autism Dev Disord 2009;39(3):405–13.

[604] Buie T, Campbell DB, Fuchs 3rd GJ, Furuta GT, Levy J, Vandewater J, et al. Evaluation, diagnosis, and treatment of gastrointestinal disorders in individuals with ASDs: a consensus report. Pediatrics. 2010;125(Suppl 1):S1–18.

[605] Mitchel MW, Myers SM, Heidlebaugh AR, Taylor CM, Rea H, Neuhaus E, et al. *CHD8*-related neurodevelopmental disorder with overgrowth. In: Adam MP, Mirzaa GM, Pagon RA, Wallace SE, Bean LJH, Gripp KW, et al., editors. GeneReviews® [Internet]. Seattle (WA): University of Washington; 2022, p. 1993–2023.

[606] Beck VC, Isom LL, Berg AT. Gastrointestinal symptoms and channelopathy-associated epilepsy. J Pediatr 2021;237:41–9 e1.

[607] Kim JY, Choi MJ, Ha S, Hwang J, Koyanagi A, Dragioti E, et al. Association between autism spectrum disorder and inflammatory bowel disease: a systematic review and meta-analysis. Autism Res 2022;15(2):340–52.

[608] Quan J, Panaccione N, Jeong J, Underwood FE, Coward S, Windsor JW, et al. Association between celiac disease and autism spectrum disorder: a systematic review. J Pediatr Gastroenterol Nutr 2021;72(5):704–11.

[609] Singh P, Arora A, Strand TA, Leffler DA, Catassi C, Green PH, et al. Global prevalence of celiac disease: systematic review and meta-analysis. Clin Gastroenterol Hepatol 2018;16(6):823–36.

[610] Alharthi A, Alhazmi S, Alburae N, Bahieldin A. The human gut microbiome as a potential factor in autism spectrum disorder. Int J Mol Sci. 2022;23(3):1363.

[611] Morton JT, Jin DM, Mills RH, Shao Y, Rahman G, McDonald D, et al. Multi-level analysis of the gut-brain axis shows autism spectrum disorder-associated molecular and microbial profiles. Nat Neurosci 2023;26(7):1208–17.

[612] Yap CX, Henders AK, Alvares GA, Wood DLA, Krause L, Tyson GW, et al. Autism-related dietary preferences mediate autism-gut microbiome associations. Cell. 2021;184(24):5916–31 e17.

[613] Dhanasekara CS, Ancona D, Cortes L, Hu A, Rimu AH, Robohm-Leavitt C, et al. Association between autism spectrum disorders and cardiometabolic diseases: a systematic review and meta-analysis. JAMA Pediatr. 2023;177(3):248–57.

[614] Kahathuduwa CN, West BD, Blume J, Dharavath N, Moustaid-Moussa N, Mastergeorge A. The risk of overweight and obesity in children with autism spectrum disorders: a systematic review and meta-analysis. Obes Rev 2019;20(12):1667–79.

[615] Pool LR, Aguayo L, Brzezinski M, Perak AM, Davis MM, Greenland P, et al. Childhood risk factors and adulthood cardiovascular disease: a systematic review. J Pediatr 2021;232:118–26.

[616] Jacobs Jr DR, Woo JG, Sinaiko AR, Daniels SR, Ikonen J, Juonala M, et al. Childhood cardiovascular risk factors and adult cardiovascular events. N Engl J Med. 2022;386(20):1877–88.

[617] Butchart M, Long JJ, Brown M, McMillan A, Bain J, Karatzias T. Autism and visual impairment: a review of the literature. Rev J Autism Dev Disord. 2017;4:118–31.

[618] Chang MY, Doppee D, Yu F, Perez C, Coleman AL, Pineles SL. Prevalence of ophthalmologic diagnoses in children with autism spectrum disorder using the Optum Dataset: a population-based study. Am J Ophthalmol 2021;221:147–53.

[619] Perna J, Bellato A, Ganapathy PS, Solmi M, Zampieri A, Faraone SV, et al. Association between autism spectrum disorder (ASD) and vision problems. A systematic review and meta-analysis. Mol Psychiatry 2023;. Available from: https://doi.org/10.1038/s41380-023-02143-7 Epub ahead of print.

[620] Mishaal RA, Weikum WM, Brooks B, Derry K, Lanphear NE. Appraising the need for audiological assessment before autism spectrum disorder referral. Paediatr Child Health. 2022;27(3):176–82.
[621] Beers AN, McBoyle M, Kakande E, Dar Santos RC, Kozak FK. Autism and peripheral hearing loss: a systematic review. Int J Pediatr Otorhinolaryngol 2014;78(1):96–101.
[622] Tsai TY, Chao YC, Hsieh CY, Huang YC. Association between atopic dermatitis and autism spectrum disorder: a systematic review and meta-analysis. Acta Derm Venereol 2020;100(10):adv00146.
[623] Zheng Z, Zhang L, Zhu T, Huang J, Qu Y, Mu D. Association between asthma and autism spectrum disorder: a meta-analysis. PLoS One 2016;11(6):e0156662.
[624] Zerbo O, Leong A, Barcellos L, Bernal P, Fireman B, Croen LA. Immune mediated conditions in autism spectrum disorders. Brain Behav Immun 2015;46:232–6.
[625] Gong T, Lundholm C, Lundström S, Kuja-Halkola R, Taylor MJ, Almqvist C. Understanding the relationship between asthma and autism spectrum disorder: a population-based family and twin study. Psychol Med 2023;53(7):3096–104.
[626] Lipkin PH, Macias MM. Council On Children With Disabilities, Section On Developmental And Behavioral Pediatrics Promoting optimal development: identifying infants and young children with developmental disorders through developmental surveillance and screening. Pediatrics 2020;145(1):e20193449.
[627] Hagan JF, Shaw JS, Duncan PM. Bright futures: guidelines for health supervision of infants, children, and adolescents. 4th ed. Elk Grove Village, IL: American Academy of Pediatrics; 2017.
[628] McPheeters ML, Weitlauf A Vehorn A, Taylor C, Sathe NA, Krishnaswami S, et al. Screening for autism spectrum disorder in young children: a systematic evidence review for the US preventive services task force [Internet]. Rockville (MD): Agency for Healthcare Research and Quality (US); 2016. Report No.: 13-05185-EF-1. (Evidence Syntheses, No. 129.) Available from: https://www.ncbi.nlm.nih.gov/books/NBK349703/.
[629] Siu AL, US Preventive Services Task Force (USPSTF)Bibbins-Domingo K, Grossman DC, Baumann LC, Davidson KW, et al. Screening for autism spectrum disorder in young children: US Preventive Services Task Force recommendation statement. JAMA 2016;315(7):691–6.
[630] Towle PO, Patrick PA. Autism spectrum disorder screening instruments for very young children: a systematic review. Autism Res Treat. 2016;2016:4624829.
[631] Thabtah F, Peebles D. Early autism screening: a comprehensive review. Int J Environ Res Public Health 2019;16(18):3502.
[632] Sánchez-García AB, Galindo-Villardón P, Nieto-Librero AB, Martín-Rodero H, Robins DL. Toddler screening for autism spectrum disorder: a meta-analysis of diagnostic accuracy. J Autism Dev Disord 2019;49(5):1837–52.
[633] Petrocchi S, Levante A, Lecciso F. Systematic review of level 1 and level 2 screening tools for autism spectrum disorders in toddlers. Brain Sci. 2020;10(3):180.
[634] Levy SE, Wolfe A, Coury D, Duby J, Farmer J, Schor E, et al. Screening tools for autism spectrum disorder in primary care: a systematic evidence review. Pediatrics 2020;145(Suppl 1):S47–59.
[635] Sobieski M, Sobieska A, Sekułowicz M, Bujnowska-Fedak MM. Tools for early screening of autism spectrum disorders in primary health care – a scoping review. BMC Prim Care. 2022;23(1):46.
[636] Wetherby AM, Guthrie W, Hooker JL, Delehanty A, Day TN, Woods J, et al. The early screening for autism and communication disorders: field-testing an autism-specific screening tool for children 12 to 36 months of age. Autism 2021;25(7):2112–23.
[637] Sturner R, Howard B, Bergmann P, Attar S, Stewart-Artz L, Bet K, et al. Autism screening at 18 months of age: a comparison of the Q-CHAT-10 and M-CHAT screeners. Mol Autism. 2022;13(1):2.
[638] Wieckowski AT, Williams LN, Rando J, Lyall K, Robins DL. Sensitivity and specificity of the modified checklist for autism in toddlers (original and revised): a systematic review and meta-analysis. JAMA Pediatr. 2023;177(4):373–83.
[639] Lee HY, Vigen C, Zwaigenbaum L, Bryson S, Smith I, Brian J, et al. The Performance of the First Year Inventory (FYI) screening on a sample of high-risk 12-month-olds diagnosed with autism spectrum disorder (ASD) at 36 months. J Autism Dev Disord 2019;49(12):4957–73.
[640] Sacrey LR, Zwaigenbaum L, Bryson S, Brian J, Smith IM, Roberts W, et al. Screening for behavioral signs of autism spectrum disorder in 9-month-old infant siblings. J Autism Dev Disord 2021;51(3):839–48.
[641] Zwaigenbaum L, Bryson SE, Brian J, Smith IM, Sacrey L, Armstrong V, et al. Assessment of autism symptoms from 6 to 18 months of age using the Autism Observation Scale for Infants in a prospective high-risk cohort. Child Dev 2021;92(3):1187–98.
[642] Hirota T, So R, Kim YS, Leventhal B, Epstein RA. A systematic review of screening tools in non-young children and adults for autism spectrum disorder. Res Dev Disabil 2018;80:1–12.
[643] Janvier YM, Harris JF, Coffield CN, Louis B, Xie M, Cidav Z, et al. Screening for autism spectrum disorder in underserved communities: early childcare providers as reporters. Autism 2016;20(3):364–73.
[644] Chesnut SR, Wei T, Barnard-Brak L, Richman DM. A meta-analysis of the Social Communication Questionnaire: screening for autism spectrum disorder. Autism 2017;21(8):920–8.
[645] Marvin AR, Marvin DJ, Lipkin PH, Law JK. Analysis of Social Communication Questionnaire (SCQ) screening for children less than age 4. Curr Dev Disord Rep. 2017;4(4):137–44.
[646] Guthrie W, Wallis K, Bennett A, Brooks E, Dudley J, Gerdes M, et al. Accuracy of autism screening in a large pediatric network. Pediatrics 2019;144(4):e20183963.

[647] Carbone PS, Campbell K, Wilkes J, Stoddard GJ, Huynh K, Young PC, et al. Primary care autism screening and later autism diagnosis. Pediatrics 2020;146(2):e20192314.

[648] Havdahl KA, Hus Bal V, Huerta M, Pickles A, Øyen AS, Stoltenberg C, et al. Multidimensional influences on autism symptom measures: implications for use in etiological research. J Am Acad Child Adolesc Psychiatry 2016;55(12):1054–63.

[649] Penner M, Anagnostou E, Andoni LY, Ungar WJ. Systematic review of clinical guidance documents for autism spectrum disorder diagnostic assessment in select regions. Autism 2018;22(5):517–27.

[650] Hayes J, Ford T, Rafeeque H, Russell G. Clinical practice guidelines for diagnosis of autism spectrum disorder in adults and children in the UK: a narrative review. BMC Psychiatry 2018;18(1):222.

[651] Huerta M, Lord C. Diagnostic evaluation of autism spectrum disorders. Pediatr Clin North Am 2012;59:103–11.

[652] Myers SM, Challman TD. Autism spectrum disorders. In: Voigt RG, Macias MM, Myers SM, Tapia CD, editors. American academy of pediatrics developmental and behavioral pediatrics. 2nd ed. Itasca, IL: American Academy of Pediatrics; 2018, p. 199–223.

[653] Myers SM. Diagnosing developmental disabilities. In: Batshaw ML, Roizen NJ, Pellegrino L, editors. Children with disabilities. 8th ed. Baltimore: Brookes; 2019, p. 199–223.

[654] Charman T, Gotham K. Measurement issues: screening and diagnostic instruments for autism spectrum disorders – lessons from research and practise. Child Adolesc Ment Health. 2013;18(1):52–63.

[655] Evers K, Maljaars J, Carrington SJ, Carter AS, Happé F, Steyaert J, et al. How well are DSM-5 diagnostic criteria for ASD represented in standardized diagnostic instruments? Eur Child Adolesc Psychiatry 2021;30(1):75–87.

[656] Constantino JN, Gruber CP. Social responsiveness scale. 2nd ed. Los Angeles, CA: Western Psychological Services; 2012.

[657] Frazier TW, Youngstrom EA, Embacher R, Hardan AY, Constantino JN, Law P, et al. Demographic and clinical correlates of autism symptom domains and autism spectrum diagnosis. Autism 2014;18(5):571–82.

[658] Goldstein S, Naglieri JA. Autism spectrum rating scales (ASRS) technical manual. New York: Multi-Health Systems; 2013.

[659] Goldstein S, Naglieri JA. Autism spectrum rating scales (ASRS) technical report #2. New York: Multi-Health Systems; 2014.

[660] Gilliam JE. Gilliam autism rating scale. 3rd ed. Austin, TX, USA: PRO-ED Inc.; 2014.

[661] Bishop DV. *The children's communication checklist (CCC-2): CCC-2 manual*. 2nd ed. London: Harcourt Assessment; 2003.

[662] Nowell SW, Tomaszewski B, Steinbrenner JR, Sam AM, Odom SL. Use of the Children's Communication Checklist-2 in school-aged students with autism: a psychometric analysis. J Autism Dev Disord 2022;52(9):4089–99.

[663] Randall M, Egberts KJ, Samtani A, Scholten RJ, Hooft L, Livingstone N, et al. Diagnostic tests for autism spectrum disorder (ASD) in preschool children. Cochrane Database Syst Rev 2018;7(7):CD009044.

[664] Lebersfeld JB, Swanson M, Clesi CD, O'Kelley SE. Systematic review and meta-analysis of the clinical utility of the ADOS-2 and the ADI-R in diagnosing autism spectrum disorders in children. J Autism Dev Disord 2021;51(11):4101–14.

[665] Rutter M, Le Couteur A, Lord C. Autism diagnostic interview-revised (ADI-R). Los Angeles, CA: Western Psychological Services; 2003.

[666] Kim SH, Lord C. New autism diagnostic interview-revised algorithms for toddlers and young preschoolers from 12 to 47 months of age. J Autism Dev Disord 2012;42(1):82–93.

[667] Kim SH, Thurm A, Shumway S, Lord C. Multisite study of new autism diagnostic interview-revised (ADI-R) algorithms for toddlers and young preschoolers. J Autism Dev Disord 2013;43(7):1527–38.

[668] de Bildt A, Sytema S, Zander E, Bölte S, Sturm H, Yirmiya N, et al. Autism Diagnostic Interview-Revised (ADI-R) algorithms for toddlers and young preschoolers: application in a non-US sample of 1,104 children. J Autism Dev Disord 2015;45(7):2076–91.

[669] Wing L, Leekam SR, Libby SJ, Gould J, Larcombe M. The diagnostic interview for social and communication disorders: background, inter-rater reliability and clinical use. J Child Psychol Psychiatry 2002;43(3):307–25.

[670] Wing L. Diagnostic interview for social and communication disorders. 11th ed Bromley, UK: Centre for Social and Communication Disorders; 2006.

[671] Skuse D, Warrington R, Bishop D, Chowdhury U, Lau J, Mandy W, et al. The developmental, dimensional and diagnostic interview (3Di): a novel computerized assessment for autism spectrum disorders. J Am Acad Child Adolesc Psychiatry 2004;43(5):548–58.

[672] Santosh PJ, Mandy WP, Puura K, Kaartinen M, Warrington R, Skuse DH. The construction and validation of a short form of the developmental, diagnostic and dimensional interview. Eur Child Adolesc Psychiatry 2009;18(8):521–4.

[673] Slappendel G, Mandy W, van der Ende J, Verhulst FC, van der Sijde A, Duvekot J, et al. Utility of the 3Di short version for the diagnostic assessment of autism spectrum disorder and compatibility with DSM-5. J Autism Dev Disord 2016;46(5):1834–46.

[674] Schopler E, Van Bourgondien ME, Wellman GJ, Love SR. *The childhood autism rating scale (CARS2)*. 2nd ed. Los Angeles, CA: Western Psychological Services; 2010.

[675] Moon SJ, Hwang JS, Shin AL, Kim JY, Bae SM, Sheehy-Knight J, et al. Accuracy of the childhood autism rating scale: a systematic review and meta-analysis. Dev Med Child Neurol 2019;61(9):1030–8.

[676] Sanchez MJ, Constantino JN. Expediting clinician assessment in the diagnosis of autism spectrum disorder. Dev Med Child Neurol 2020;62 (7):806–12.

[677] Lord C, Rutter M, DiLavore PC, Risi S, Gotham K, Bishop SL. Autism diagnostic observation schedule. Part I: modules 1-4. *2nd ed.* Los Angeles, CA: Western Psychological Services; 2012.

[678] Lord C, Luyster RJ, Gotham K, Guthrie W. Autism diagnostic observation schedule. Part II: toddler module. *2nd ed.* Los Angeles, CA: Western Psychological Services; 2012, p. 2012.

[679] Smith CJ, Rozga A, Matthews N, Oberleitner R, Nazneen N, Abowd G. Investigating the accuracy of a novel telehealth diagnostic approach for autism spectrum disorder. Psychol Assess 2017;29(3):245–52.
[680] Lord C, Holbrook A, Dow D, Byrne K, Grzadzinksi R, Sterrett K, et al. Brief observation of symptoms of autism (BOSA). Los Angeles, CA: Western Psychological Services; 2020.
[681] Dow D, Holbrook A, Toolan C, McDonald N, Sterrett K, Rosen N, et al. The brief observation of symptoms of autism (BOSA): development of a new adapted assessment measure for remote telehealth administration through COVID-19 and beyond. J Autism Dev Disord 2022;52(12):5383–94.
[682] Lord C, Bishop SL. Let's be clear that "autism spectrum disorder symptoms" are not always related to autism spectrum disorder. Am J Psychiatry 2021;178(8):680–2.
[683] Rinaldi C, Attanasio M, Valenti M, Mazza M, Keller R. Autism spectrum disorder and personality disorders: comorbidity and differential diagnosis. World J Psychiatry 2021;11(12):1366–86.
[684] Aldridge FJ, Gibbs VM, Schmidhofer K, Williams M. Investigating the clinical usefulness of the Social Responsiveness Scale (SRS) in a tertiary level, autism spectrum disorder specific assessment clinic. J Autism Dev Disord 2012;42(2):294–300.
[685] Hus V, Bishop S, Gotham K, Huerta M, Lord C. Factors influencing scores on the social responsiveness scale. J Child Psychol Psychiatry 2013;54(2):216–24.
[686] Cholemkery H, Mojica L, Rohrmann S, Gensthaler A, Freitag CM. Can autism spectrum disorders and social anxiety disorders be differentiated by the social responsiveness scale in children and adolescents? J Autism Dev Disord 2014;44:1168–82.
[687] Cholemkery H, Kitzerow J, Rohrmann S, Freitag CM. Validity of the social responsiveness scale to differentiate between autism spectrum disorders and disruptive behaviour disorders. Eur Child Adolesc Psychiatry 2014;23(2):81–93.
[688] Capriola-Hall NN, McFayden T, Ollendick TH, White SW. Caution when screening for autism among socially anxious youth. J Autism Dev Disord 2021;51(5):1540–9.
[689] Camodeca A. Description of criterion validity of the autism spectrum rating scales 6-18 parent report: initial exploration in a large community sample. Child Psychiatry Hum Dev 2019;50(6):987–1001.
[690] Hong JS, Perrin J, Singh V, Kalb L, Cross EA, Wodka E, et al. Psychometric evaluation of the autism spectrum rating scales (6-18 years parent report) in a clinical sample. J Autism Dev Disord 2022;1–12.
[691] Li C, Zhou H, Wang T, Long S, Du X, Xu X, et al. Performance of the autism spectrum rating scale and social responsiveness scale in identifying autism spectrum disorder among cases of intellectual disability. Neurosci Bull 2018;34(6):972–80.
[692] Camodeca A. Diagnostic utility of the Gilliam Autism Rating Scales-3rd edition parent report in clinically referred children. J Autism Dev Disord 2023;53(5):2112–26.
[693] Rutter M, Bailey A, Lord C. Social communication questionnaire (SCQ). Los Angeles, LA: Western Psychological Services; 2003.
[694] Janvier D, Choi YB, Klein C, Lord C, Kim SH. Brief report: examining test-retest reliability of the Autism Diagnostic Observation Schedule (ADOS-2) calibrated severity scores (CSS). J Autism Dev Disord 2022;52(3):1388–94.
[695] Esler AN, Bal VH, Guthrie W, Wetherby A, Ellis Weismer S, Lord C. The Autism diagnostic observation schedule, toddler module: standardized severity scores. J Autism Dev Disord 2015;45(9):2704–20.
[696] Morrier MJ, Ousley OY, Caceres-Gamundi GA, Segall MJ, Cubells JF, Young LJ, et al. Brief report: relationship between ADOS-2, module 4 calibrated severity scores (CSS) and social and non-social standardized assessment measures in adult males with autism spectrum disorder (ASD). J Autism Dev Disord 2017;47(12):4018–24.
[697] Centers for Medicare & Medicaid Services. Telehealth. Available at: https://www.medicaid.gov/medicaid/benefits/telehealth/index.html. [accessed 18.06.23.]
[698] Jang J, White SP, Esler AN, Kim SH, Klaiman C, Megerian JT, et al. Diagnostic evaluations of autism spectrum disorder during the COVID-19 pandemic. J Autism Dev Disord 2022;52(2):962–73.
[699] Liu M, Ma Z. A systematic review of telehealth screening, assessment, and diagnosis of autism spectrum disorder. Child Adolesc Psychiatry Ment Health 2022;16(1):79.
[700] Bishop SL, Lord C. Commentary: Best practices and processes for assessment of autism spectrum disorder – the intended role of standardized diagnostic instruments. J Child Psychol Psychiatry 2023;64(5):834–8.
[701] Barbaresi W, Cacia J, Friedman S, Fussell J, Hansen R, Hofer J, et al. Clinician diagnostic certainty and the role of the Autism Diagnostic Observation Schedule in autism spectrum disorder diagnosis in young children. JAMA Pediatr 2022;176(12):1233–41.
[702] Thurm A, Farmer C, Salzman E, Lord C, Bishop S. State of the field: differentiating intellectual disability from autism spectrum disorder. Front Psychiatry 2019;10:526.
[703] Schaefer GB, Mendelsohn NJProfessional Practice and Guidelines Committee. Clinical genetics evaluation in identifying the etiology of autism spectrum disorders: 2013 guideline revisions. Genet Med 2013;15(5):399–407.
[704] Muhle RA, Reed HE, Vo LC, Mehta S, McGuire K, Veenstra-VanderWeele J, et al. Clinical diagnostic genetic testing for individuals with developmental disorders. J Am Acad Child Adolesc Psychiatry 2017;56(11):910–13.
[705] Srivastava S, Love-Nichols JA, Dies KA, Ledbetter DH, Martin CL, Chung WK, et al. Meta-analysis and multidisciplinary consensus statement: exome sequencing is a first-tier clinical diagnostic test for individuals with neurodevelopmental disorders. Genet Med 2019;21(11):2413–21.
[706] Carter MT, Srour M, Au PB, Buhas D, Dyack S, Eaton A, et al. Genetic and metabolic investigations for neurodevelopmental disorders: position statement of the Canadian College of Medical Geneticists (CCMG). J Med Genet 2023;60(6):523–32.

[707] Savatt JM, Myers SM. Genetic testing in neurodevelopmental disorders. Front Pediatr. 2021;9:526779.

[708] Stefanski A, Calle-López Y, Leu C, Pérez-Palma E, Pestana-Knight E, Lal D. Clinical sequencing yield in epilepsy, autism spectrum disorder, and intellectual disability: a systematic review and meta-analysis. Epilepsia 2021;62(1):143–51.

[709] Miyake N, Tsurusaki Y, Fukai R, Kushima I, Okamoto N, Ohashi K, et al. Molecular diagnosis of 405 individuals with autism spectrum disorder. Eur J Hum Genet 2023. Available from: https://doi.org/10.1038/s41431-023-01335-7 Epub ahead of print.

[710] Shen Y, Dies KA, Holm IA, Bridgemohan C, Sobeih MM, Caronna EB, et al. Clinical genetic testing for patients with autism spectrum disorders. Pediatrics 2010;125(4):e727–35.

[711] Eriksson MA, Liedén A, Westerlund J, Bremer A, Wincent J, Sahlin E, et al. Rare copy number variants are common in young children with autism spectrum disorder. Acta Paediatr. 2015;104(6):610–18.

[712] Tammimies K, Marshall CR, Walker S, Kaur G, Thiruvahindrapuram B, Lionel AC, et al. Molecular diagnostic yield of chromosomal microarray analysis and whole-exome sequencing in children with autism spectrum disorder. JAMA 2015;314(9):895–903.

[713] Harris HK, Sideridis GD, Barbaresi WJ, Harstad E. Pathogenic yield of genetic testing in autism spectrum disorder. Pediatrics 2020;146(4): e20193211.

[714] Roesser J. Diagnostic yield of genetic testing in children diagnosed with autism spectrum disorders at a regional referral center. Clin Pediatr 2011;50:834–43.

[715] Tassone F, Choudhary NS, Tassone F, Durbin-Johnson B, Hansen R, Hertz-Picciotto I, et al. Identification of expanded alleles of the FMR1 gene in the childhood autism risks from genes and environment (CHARGE) study. J Autism Dev Disord 2013;43:530–9.

[716] Robinson EB, Samocha KE, Kosmicki JA, McGrath L, Neale BM, Perlis RH, et al. Autism spectrum disorder severity reflects the average contribution of de novo and familial influences. Proc Natl Acad Sci U S A. 2014;111(42):15161–5.

[717] Blesson A, Cohen JS. Genetic counseling in neurodevelopmental disorders. Cold Spring Harb Perspect Med. 2020;10(4):a036533.

[718] Richards S, Aziz N, Bale S, Bick D, Das S, Gastier-Foster J, et al. Standards and guidelines for the interpretation of sequence variants: a joint consensus recommendation of the American College of Medical Genetics and Genomics and the Association for Molecular Pathology. Genet Med 2015;17(5):405–24.

[719] Retterer K, Juusola J, Cho MT, Vitazka P, Millan F, Gibellini F, et al. Clinical application of whole-exome sequencing across clinical indications. Genet Med 2016;18(7):696–704.

[720] Wright CF, Fitzgerald TW, Jones WD, Clayton S, McRae JF, van Kogelenberg M, et al. Genetic diagnosis of developmental disorders in the DDD study: a scalable analysis of genome-wide research data. Lancet 2015;385(9975):1305–14.

[721] Dai P, Honda A, Ewans L, McGaughran J, Burnett L, Law M, et al. Recommendations for next generation sequencing data reanalysis of unsolved cases with suspected Mendelian disorders: a systematic review and meta-analysis. Genet Med 2022;24(8):1618–29.

[722] Wojcik MH, Reuter CM, Marwaha S, Mahmoud M, Duyzend MH, Barseghyan H, et al. Beyond the exome: what's next in diagnostic testing for Mendelian conditions. Am J Hum Genet 2023;110(8):1229–48.

[723] Murdock DR, Dai H, Burrage LC, Rosenfeld JA, Ketkar S, Müller MF, et al. Transcriptome-directed analysis for Mendelian disease diagnosis overcomes limitations of conventional genomic testing. J Clin Invest 2021;131(1):e141500.

[724] Aref-Eshghi E, Kerkhof J, Pedro VP, Groupe DI FranceBarat-Houari M, Ruiz-Pallares N, et al. Evaluation of DNA methylation episignatures for diagnosis and phenotype correlations in 42 Mendelian neurodevelopmental disorders. Am J Hum Genet 2020;106 (3):356–70.

[725] Arteche-López A, Gómez Rodríguez MJ, Sánchez Calvin MT, Quesada-Espinosa JF, Lezana Rosales JM, Palma Milla C, et al. Towards a change in the diagnostic algorithm of autism spectrum disorders: evidence supporting whole exome sequencing as a first-tier test. Genes 2021;12(4):560.

[726] Sheth F, Shah J, Jain D, Shah S, Patel H, Patel K, et al. Comparative yield of molecular diagnostic algorithms for autism spectrum disorder diagnosis in India: evidence supporting whole exome sequencing as first tier test. BMC Neurol 2023;23(1):292.

[727] Manickam K, McClain MR, Demmer LA, Biswas S, Kearney HM, Malinowski J, et al. Exome and genome sequencing for pediatric patients with congenital anomalies or intellectual disability: an evidence-based clinical guideline of the American College of Medical Genetics and Genomics (ACMG). Genet Med 2021;23(11):2029–37.

[728] Malinowski J, Miller DT, Demmer L, Gannon J, Pereira EM, Schroeder MC, et al. Systematic evidence-based review: outcomes from exome and genome sequencing for pediatric patients with congenital anomalies or intellectual disability. Genet Med 2020;22(6):986–1004.

[729] Ní Ghrálaigh F, McCarthy E, Murphy DN, Gallagher L, Lopez LM. Brief report: evaluating the diagnostic yield of commercial gene panels in autism. J Autism Dev Disord 2023;53(1):484–8.

[730] Rehm HL, Alaimo JT, Aradhya S, Bayrak-Toydemir P, Best H, Brandon R, et al. The landscape of reported VUS in multi-gene panel and genomic testing: time for a change. Genet Med 2023;25(12):100947.

[731] Clark MM, Stark Z, Farnaes L, Tan TY, White SM, Dimmock D, et al. Meta-analysis of the diagnostic and clinical utility of genome and exome sequencing and chromosomal microarray in children with suspected genetic diseases. NPJ Genom Med 2018;3:16.

[732] Lowther C, Valkanas E, Giordano JL, Wang HZ, Currall BB, O'Keefe K, et al. Systematic evaluation of genome sequencing for the diagnostic assessment of autism spectrum disorder and fetal structural anomalies. Am J Hum Genet 2023;110(9):1454–69.

[733] van der Sanden BPGH, Schobers G, Corominas Galbany J, Koolen DA, Sinnema M, van Reeuwijk J, et al. The performance of genome sequencing as a first-tier test for neurodevelopmental disorders. Eur J Hum Genet 2023;31(1):81–8.

[734] Nurchis MC, Altamura G, Riccardi MT, Radio FC, Chillemi G, Bertini ES, et al. Whole genome sequencing diagnostic yield for paediatric patients with suspected genetic disorders: systematic review, meta-analysis, and GRADE assessment. Arch Public Health 2023;81(1):93.

[735] Saudubray JM, Garcia-Cazorla A. An overview of inborn errors of metabolism affecting the brain: from neurodevelopment to neurodegenerative disorders. Dialogues Clin Neurosci 2018;20(4):301–25.

[736] van de Burgt N, van Doesum W, Grevink M, van Niele S, de Koning T, Leibold N, et al. Psychiatric manifestations of inborn errors of metabolism: a systematic review. Neurosci Biobehav Rev 2023;144:104970.

[737] Simons A, Eyskens F, Glazemakers I, van West D. Can psychiatric childhood disorders be due to inborn errors of metabolism? Eur Child Adolesc Psychiatry 2017;26(2):143–54.

[738] Žigman T, Petković Ramadža D, Šimić G, Barić I. Inborn errors of metabolism associated with autism spectrum disorders: approaches to intervention. Front Neurosci. 2021;15:673600.

[739] Zecavati N, Spence SJ. Neurometabolic disorders and dysfunction in autism spectrum disorders. Curr Neurol Neurosci Rep 2009;9(2):129–36.

[740] Hollis F, Kanellopoulos AK, Bagni C. Mitochondrial dysfunction in autism spectrum disorder: clinical features and perspectives. Curr Opin Neurobiol 2017;45:178–87.

[741] Oliveira G, Diogo L, Grazina M, Garcia P, Ataíde A, Marques C, et al. Mitochondrial dysfunction in autism spectrum disorders: a population-based study. Dev Med Child Neurol 2005;47(3):185–9.

[742] Weissman JR, Kelley RI, Bauman ML, Cohen BH, Murray KF, Mitchell RL, et al. Mitochondrial disease in autism spectrum disorder patients: a cohort analysis. PLoS One 2008;3(11):e3815.

[743] Schiff M, Benoist JF, Aïssaoui S, Boespflug-Tanguy O, Mouren MC, de Baulny HO, et al. Should metabolic diseases be systematically screened in nonsyndromic autism spectrum disorders? PLoS One 2011;6(7):e21932.

[744] Campistol J, Díez-Juan M, Callejón L, Fernandez-De Miguel A, Casado M, Garcia Cazorla A, et al. Inborn error metabolic screening in individuals with nonsyndromic autism spectrum disorders. Dev Med Child Neurol 2016;58(8):842–7.

[745] İnci A, Özaslan A, Okur İ, Biberoğlu G, Güney E, Ezgü FS, et al. Autism: screening of inborn errors of metabolism and unexpected results. Autism Res. 2021;14(5):887–96.

[746] Moravej H, Inaloo S, Nahid S, Mazloumi S, Nemati H, Moosavian T, et al. Inborn errors of metabolism associated with autism among children: a multicenter study from Iran. Indian Pediatr 2023;60(3):193–6.

[747] Adhikari AN, Gallagher RC, Wang Y, Currier RJ, Amatuni G, Bassaganyas L, et al. The role of exome sequencing in newborn screening for inborn errors of metabolism. Nat Med 2020;26(9):1392–7.

[748] Delanne J, Bruel AL, Huet F, Moutton S, Nambot S, Grisval M, et al. The diagnostic rate of inherited metabolic disorders by exome sequencing in a cohort of 547 individuals with developmental disorders. Mol Genet Metab Rep. 2021;29:100812.

[749] Asato MR, Goldstein AC, Schiff M. Autism and inborn errors of metabolism: how much is enough? Dev Med Child Neurol 2015;57(9):788–9.

[750] Márquez-Caraveo ME, Ibarra-González I, Rodríguez-Valentín R, Ramírez-García MÁ, Pérez-Barrón V, Lazcano-Ponce E, et al. Brief report: delayed diagnosis of treatable inborn errors of metabolism in children with autism and other neurodevelopmental disorders. J Autism Dev Disord 2021;51(6):2124–31.

[751] Glinton KE, Elsea SH. Untargeted metabolomics for autism spectrum disorders: current status and future directions. Front Psychiatry. 2019;10:647.

[752] Siracusano M, Arturi L, Riccioni A, Noto A, Mussap M, Mazzone L. Metabolomics: perspectives on clinical employment in autism spectrum disorder. Int J Mol Sci. 2023;24(17):13404.

[753] Cooper AS, Friedlaender E, Levy SE, Shekdar KV, Bradford AB, Wells KE, et al. The implications of brain MRI in autism spectrum disorder. J Child Neurol 2016;31(14):1611–16.

[754] Ambrosino S, Elbendary H, Lequin M, Rijkelijkhuizen D, Banaschewski T, Baron-Cohen S, et al. In-depth characterization of neuroradiological findings in a large sample of individuals with autism spectrum disorder and controls. Neuroimage Clin. 2022;35:103118.

[755] Erbetta A, Bulgheroni S, Contarino VE, et al. Low-functioning autism and nonsyndromic intellectual disability: magnetic resonance imaging (MRI) findings. J Child Neurol 2015;30:658–63.

[756] Rochat MJ, Distefano G, Maffei M, Toni F, Posar A, Scaduto MC, et al. Brain magnetic resonance findings in 117 children with autism spectrum disorder under 5 years old. Brain Sci. 2020;10(10):741.

[757] Gupta S, Kanamalla U, Gupta V. Are incidental findings on brain magnetic resonance imaging images in children merely incidental? J Child Neurol 2010;25:1511–16.

[758] Vasa RA, Ranta M, Huisman TA, Pinto PS, Tillman RM, Mostofsky SH. Normal rates of neuroradiological findings in children with high functioning autism. J Autism Dev Disord 2012;42:1662–70.

[759] Monterrey JC, Philips J, Cleveland S, Tanaka S, Barnes P, Hallmayer JF, et al. Incidental brain MRI findings in an autism twin study. Autism Res. 2017;10(1):113–20.

[760] Dangouloff-Ros V, Roux CJ, Boulouis G, Levy R, Nicolas N, Lozach C, et al. Incidental brain MRI findings in children: a systematic review and meta-analysis. AJNR Am J Neuroradiol 2019;40(11):1818–23.

[761] Jansen PR, Dremmen M, van den Berg A, Dekkers IA, Blanken LME, Muetzel RL, et al. Incidental findings on brain imaging in the general pediatric population. N Engl J Med 2017;377(16):1593–5.

[762] Moeschler JB, Shevell M. Comprehensive evaluation of the child with intellectual disability or global developmental delays Committee on Genetics Pediatrics 2014;134:e903–18.

[763] Santarone ME, Zambrano S, Zanotta N, Mani E, Minghetti S, Pozzi M, et al. EEG features in autism spectrum disorder: a retrospective analysis in a cohort of preschool children. Brain Sci. 2023;13(2):345.

[764] Jeste SS, Tuchman R. Autism spectrum disorder and epilepsy: two sides of the same coin? J Child Neurol 2015;30:1963–71.

[765] Benbadis SR, Beniczky S, Bertram E, MacIver S, Moshé SL. The role of EEG in patients with suspected epilepsy. Epileptic Disord 2020;22(2):143–55.

[766] Besag FM. Epilepsy in patients with autism: links, risks and treatment challenges. Neuropsychiatr Dis Treat 2017;14:1–10.

[767] Hansen SN, Schendel DE, Francis RW, Windham GC, Bresnahan M, Levine SZ, et al. Recurrence risk of autism in siblings and cousins: a multinational, population-based study. J Am Acad Child Adolesc Psychiatry 2019;58(9):866–75.

[768] Sandin S, Lichtenstein P, Kuja-Halkola R, Larsson H, Hultman CM, Reichenberg A. The familial risk of autism. JAMA 2014;311(17):1770–7.

[769] Grønborg TK, Schendel DE, Parner ET. Recurrence of autism spectrum disorders in full- and half-siblings and trends over time: a population-based cohort study. JAMA Pediatr. 2013;167(10):947–53.

[770] Risch N, Hoffmann TJ, Anderson M, Croen LA, Grether JK, Windham GC. Familial recurrence of autism spectrum disorder: evaluating genetic and environmental contributions. Am J Psychiatry 2014;171(11):1206–13.

[771] Jokiranta-Olkoniemi E, Cheslack-Postava K, Sucksdorff D, Suominen A, Gyllenberg D, Chudal R, et al. Risk of psychiatric and neurodevelopmental disorders among siblings of probands with autism spectrum disorders. JAMA Psychiatry. 2016;73(6):622–9.

[772] Ozonoff S, Young GS, Carter A, Messinger D, Yirmiya N, Zwaigenbaum L, et al. Recurrence risk for autism spectrum disorders: a baby siblings research consortium study. Pediatrics 2011;128(3):e488–95.

[773] Messinger DS, Young GS, Webb SJ, Ozonoff S, Bryson SE, Carter A, et al. Early sex differences are not autism-specific: a baby siblings research consortium (BSRC) study. Mol Autism. 2015;6:32.

[774] Schaefer GB. Clinical genetic aspects of ASD spectrum disorders. Int J Mol Sci. 2016;17(2):180.

[775] Werling DM, Geschwind DH. Recurrence rates provide evidence for sex-differential, familial genetic liability for autism spectrum disorders in multiplex families and twins. Mol Autism 2015;6:27.

[776] Constantino JN, Zhang Y, Frazier T, Abbacchi AM, Law P. Sibling recurrence and the genetic epidemiology of autism. Am J Psychiatry 2010;167:1349–56.

[777] Miller M, Musser ED, Young GS, Olson B, Steiner RD, Nigg JT. Sibling recurrence risk and cross-aggregation of attention-deficit/hyperactivity disorder and autism spectrum disorder. JAMA Pediatr. 2019;173(2):147–52.

[778] Finucane B, Myers SM. Genetic counseling for autism spectrum disorder in an evolving theoretical landscape. Curr Genet Med Rep 2016;. Available from: https://doi.org/10.1007/s40142-016-0099-9.

[779] Hume K, Steinbrenner JR, Odom SL, Morin KL, Nowell SW, Tomaszewski B, et al. Evidence-based practices for children, youth, and young adults with autism: third generation review. J Autism Dev Disord 2021;51(11):4013–32.

[780] Steinbrenner J.R., Hume K., Odom S.L., Morin K.L., Nowell S.W., Tomaszewski B., et al. Evidence-based practices for children, youth, and young adults with autism. The University of North Carolina at Chapel Hill, Frank Porter Graham Child Development Institute, National Clearinghouse on Autism Evidence and Practice Review Team; 2020. https://ncaep.fpg.unc.edu/sites/ncaep.fpg.unc.edu/files/imce/documents/EBP%20Report%202020.pdf.

[781] National Autism Center. Findings and conclusions: national standards project, phase 2. Randolph, MA: Author; 2015.

[782] Lai MC, Anagnostou E, Wiznitzer M, Allison C, Baron-Cohen S. Evidence-based support for autistic people across the lifespan: maximising potential, minimising barriers, and optimising the person-environment fit. Lancet Neurol 2020;19(5):434–51.

[783] Cheng WM, Smith TB, Butler M, Taylor TM, Clayton D. Effects of parent-implemented interventions on outcomes of children with autism: a meta-analysis. J Autism Dev Disord 2023;53(11):4147–63.

[784] Reichow B, Kogan C, Barbui C, Maggin D, Salomone E, Smith IC, et al. Caregiver skills training for caregivers of individuals with neurodevelopmental disorders: a systematic review and meta-analysis. Dev Med Child Neurol 2023. Available from: https://doi.org/10.1111/dmcn.15764 Epub ahead of print.

[785] Bondy A, Frost L. The picture exchange communication system. Behav Modif 2001;25(5):725–44.

[786] Patel V, Saxena S, Lund C, Thornicroft G, Baingana F, Bolton P, et al. The *Lancet* Commission on global mental health and sustainable development. Lancet 2018;392(10157):1553–98.

[787] Green J. Editorial perspective: delivering autism intervention through development. J Child Psychol Psychiatry 2019;60(12):1353–6.

[788] Storch EA, Schneider SC, Olsen SM, Ramirez AC, Berry LN, Goin-Kochel RP, et al. Stepped-care cognitive behavioral therapy in children on the autism spectrum with co-occurring anxiety. J Autism Dev Disord 2022;1–16. Available from: https://doi.org/10.1007/s10803-022-05775-w Epub ahead of print.

[789] Smith T, Iadarola S. Evidence base update for autism spectrum disorder. J Clin Child Adolesc Psychol 2015;44(6):897–922.

[790] French L, Kennedy EMM. Annual research review: early intervention for infants and young children with, or at-risk of, autism spectrum disorder: a systematic review. J Child Psychol Psychiatry 2018;59(4):444–56.

[791] Sandbank M, Bottema-Beutel K, Crowley S, Cassidy M, Dunham K, Feldman JI, et al. Project AIM: autism intervention meta-analysis for studies of young children. Psychol Bull 2020;146(1):1–29.

[792] Gosling CJ, Cartigny A, Mellier BC, Solanes A, Radua J, Delorme R. Efficacy of psychosocial interventions for autism spectrum disorder: an umbrella review. Mol Psychiatry 2022;27(9):3647–56.

[793] Vivanti G. What does it mean for an autism intervention to be evidence-based? Autism Res. 2022;15(10):1787–93.

[794] Reichow B, Hume K, Barton EE, Boyd BA. Early intensive behavioral intervention (EIBI) for young children with autism spectrum disorders (ASD). Cochrane Database Syst Rev 2018;5(5):CD009260.

[795] What Works Clearinghouse. What works clearinghouse standards handbook, Version 4.1. US Department of Education, Institute of Education Sciences, National Center for Education Evaluation and Regional Assistance; 2020. https://ies.ed.gov/ncee/wwc/Docs/referenceresources/WWC-Standards-Handbook-v4-1-508.pdf.

[796] Odom SL. Education of students with disabilities, science, and randomized controlled trials. Res Pract Persons Severe Disabil. 2021;46(3):132–45.

[797] Weitlauf AS, McPheeters ML, Peters B, Sathe N, Travis R, Aiello R, et al. Therapies for children with autism spectrum disorder: behavioral interventions update. Comparative Effectiveness Review No. 137. [Internet]. Rockville (MD): Agency for Healthcare Research and Quality (US); 2014. https://www.ncbi.nlm.nih.gov/books/NBK241444/.

[798] Odom SL, Collet-Klingenberg L, Rogers SJ, Hatton DD. Evidence-based practices in interventions for children and youth with autism spectrum disorders. Prevent School Failure 2010;54(4):275–82.

[799] Wong C, Odom SL, Hume K, et al. Evidence-based practices for children, youth, and young adults with autism spectrum disorder. Chapel Hill, NC: The University of North Carolina, Frank Porter Graham Child Development Institute, Autism Evidence-Based Practice Review Group; 2014. Available from: https://autismpdc.fpg.unc.edu/sites/autismpdc.fpg.unc.edu/files/2014-EBP-Report.pdf.

[800] Wong C, Odom SL, Hume KA, Cox AW, Fettig A, Kucharczyk S, et al. Evidence-based practices for children, youth, and young adults with autism spectrum disorder: a comprehensive review. J Autism Dev Disord 2015;45(7):1951–66.

[801] Sam AM, Cox AW, Savage MN, Waters V, Odom SL. Disseminating information on evidence-based practices for children and youth with autism spectrum disorder: AFIRM. J Autism Dev Disord 2020;50(6):1931–40.

[802] Sam AM, Odom SL, Tomaszewski B, Perkins Y, Cox AW. Employing evidence-based practices for children with autism in elementary schools. J Autism Dev Disord 2021;51(7):2308–23.

[803] Leaf JB, Sato SK, Javed A, Arthur SM, Creem AN, Cihon JH, et al. The evidence-based practices for children, youth, and young adults with autism report: concerns and critiques. Behav Interv 2021;36(2):457–72.

[804] Crank JE, Sandbank M, Dunham K, Crowley S, Bottema-Beutel K, Feldman J, et al. Understanding the effects of naturalistic developmental behavioral interventions: a Project AIM meta-analysis. Autism Res 2021;14(4):817–34.

[805] Trembath D, Varcin K, Waddington H, Sulek R, Bent C, Ashburner J, et al. Non-pharmacological interventions for autistic children: an umbrella review. Autism 2023;27(2):275–95.

[806] Schreibman L, Dawson G, Stahmer AC, Landa R, Rogers SJ, McGee GG, et al. Naturalistic developmental behavioral interventions: empirically validated treatments for autism spectrum disorder. J Autism Dev Disord 2015;45(8):2411–28.

[807] Tiede G, Walton KM. Meta-analysis of naturalistic developmental behavioral interventions for young children with autism spectrum disorder. Autism 2019;23(8):2080–95.

[808] Uljarević M, Billingham W, Cooper MN, Condron P, Hardan AY. Examining effectiveness and predictors of treatment response of pivotal response treatment in autism: an umbrella review and a meta-analysis. Front Psychiatry 2022;12:766150.

[809] Fuller EA, Oliver K, Vejnoska SF, Rogers SJ. The effects of the Early Start Denver Model for children with autism spectrum disorder: a meta-analysis. Brain Sci. 2020;10(6):368.

[810] Wang Z, Loh SC, Tian J, Chen QJ. A meta-analysis of the effect of the Early Start Denver Model in children with autism spectrum disorder. Int J Dev Disabil. 2021;68(5):587–97.

[811] Waddington H, Reynolds JE, Macaskill E, Curtis S, Taylor LJ, Whitehouse AJ. The effects of JASPER intervention for children with autism spectrum disorder: a systematic review. Autism 2021;25(8):2370–85.

[812] Rodgers M, Marshall D, Simmonds M, Le Couteur A, Biswas M, Wright K, et al. Interventions based on early intensive applied behaviour analysis for autistic children: a systematic review and cost-effectiveness analysis. Health Technol Assess 2020;24(35):1–306.

[813] Eckes T, Buhlmann U, Holling HD, Möllmann A. Comprehensive ABA-based interventions in the treatment of children with autism spectrum disorder – a meta-analysis. BMC Psychiatry 2023;23(1):133.

[814] Rodgers M, Simmonds M, Marshall D, Hodgson R, Stewart LA, Rai D, et al. Intensive behavioural interventions based on applied behaviour analysis for young children with autism: an international collaborative individual participant data meta-analysis. Autism 2021;25(4):1137–53.

[815] Gitimoghaddam M, Chichkine N, McArthur L, Sangha SS, Symington V. Applied behavior analysis in children and youth with autism spectrum disorders: a scoping review. Perspect Behav Sci. 2022;45(3):521–57.

[816] Green J, Charman T, McConachie H, Aldred C, Slonims V, Howlin P, et al. Parent-mediated communication-focused treatment in children with autism (PACT): a randomised controlled trial. Lancet 2010;375(9732):2152–60.

[817] Pickles A, Le Couteur A, Leadbitter K, Salomone E, Cole-Fletcher R, Tobin H, et al. Parent-mediated social communication therapy for young children with autism (PACT): long-term follow-up of a randomised controlled trial. Lancet 2016;388(10059):2501–9.

[818] Green J, Leadbitter K, Ellis C, Taylor L, Moore HL, Carruthers S, et al. Combined social communication therapy at home and in education for young autistic children in England (PACT-G): a parallel, single-blind, randomised controlled trial. Lancet Psychiatry. 2022;9(4):307–20.

[819] Dawson G, Rogers S, Munson J, Smith M, Winter J, Greenson J, et al. Randomized, controlled trial of an intervention for toddlers with autism: the Early Start Denver Model. Pediatrics 2010;125(1):e17–23.

[820] Estes A, Munson J, Rogers SJ, Greenson J, Winter J, Dawson G. Long-term outcomes of early intervention in 6-year-old children with autism spectrum disorder. J Am Acad Child Adolesc Psychiatry 2015;54(7):580–7.

[821] Shi B, Wu W, Dai M, Zeng J, Luo J, Cai L, et al. Cognitive, language, and behavioral outcomes in children with autism spectrum disorders exposed to early comprehensive treatment models: a meta-analysis and meta-regression. Front Psychiatry 2021;12:691148.

[822] Deb SS, Retzer A, Roy M, Acharya R, Limbu B, Roy A. The effectiveness of parent training for children with autism spectrum disorder: a systematic review and meta-analyses. BMC Psychiatry 2020;20(1):583.

[823] Ratliff-Black M, Therrien W. Parent-mediated interventions for school-age children with ASD: a meta-analysis. Focus Autism Other Dev Disabl 2021;36(1):3–13.

[824] Conrad CE, Rimestad ML, Rohde JF, Petersen BH, Korfitsen CB, Tarp S, et al. Parent-mediated interventions for children and adolescents with autism spectrum disorders: a systematic review and meta-analysis. Front Psychiatry 2021;12:773604.

[825] Nevill RE, Lecavalier L, Stratis EA. Meta-analysis of parent-mediated interventions for young children with autism spectrum disorder. Autism 2018;22(2):84–98.

[826] Tarver J, Palmer M, Webb S, Scott S, Slonims V, Simonoff E, et al. Child and parent outcomes following parent interventions for child emotional and behavioral problems in autism spectrum disorders: a systematic review and meta-analysis. Autism 2019;23(7):1630–44.

[827] Ona HN, Larsen K, Nordheim, Brurberg KG. Effects of pivotal response treatment (PRT) for children with autism spectrum disorders (ASD): a systematic review. Rev J Autism Dev Disord. 2019;7(1):78–90.

[828] Postorino V, Sharp WG, McCracken CE, Bearss K, Burrell TL, Evans AN, et al. A systematic review and meta-analysis of parent training for disruptive behavior in children with autism spectrum disorder. Clin Child Fam Psychol Rev 2017;20(4):391–402.

[829] Rojas-Torres LP, Alonso-Esteban Y, Alcantud-Marín F. Early intervention with parents of children with autism spectrum disorders: a review of programs. Children 2020;7(12):294.

[830] Iadarola S, Levato L, Harrison B, Smith T, Lecavalier L, Johnson C, et al. Teaching parents behavioral strategies for autism spectrum disorder (ASD): effects on stress, strain, and competence. J Autism Dev Disord 2018;48(4):1031–40.

[831] Brian JA, Smith IM, Zwaigenbaum L, Bryson SE. Cross-site randomized control trial of the Social ABCs caregiver-mediated intervention for toddlers with autism spectrum disorder. Autism Res. 2017;10(10):1700–11.

[832] Brian J, Drmic I, Roncadin C, Dowds E, Shaver C, Smith IM, et al. Effectiveness of a parent-mediated intervention for toddlers with autism spectrum disorder: evidence from a large community implementation. Autism 2022;26(7):1882–97.

[833] Solomon R, Van Egeren LA, Mahoney G, Quon Huber MS, Zimmerman P. PLAY Project Home Consultation intervention program for young children with autism spectrum disorders: a randomized controlled trial. J Dev Behav Pediatr 2014;35(8):475–85.

[834] Ozonoff S, Cathcart K. Effectiveness of a home program intervention for young children with autism. J Autism Dev Disord 1998;28 (1):25–32.

[835] Virues-Ortega J, Julio FM, Pastor-Barriuso R. The TEACCH program for children and adults with autism: a meta-analysis of intervention studies. Clin Psychol Rev 2013;33(8):940–53.

[836] Tripathi I, Estabillo JA, Moody CT, Laugeson EA. Long-term treatment outcomes of PEERS® for preschoolers: a parent-mediated social skills training program for children with autism spectrum disorder. J Autism Dev Disord 2022;52(6):2610–26.

[837] Milgramm A, Corona LL, Janicki-Menzie C, Christodulu KV. Community-based parent education for caregivers of children newly diagnosed with autism spectrum disorder. J Autism Dev Disord 2022;52(3):1200–10.

[838] Bearss K, Burrell TL, Stewart L, Scahill L. Parent training in autism spectrum disorder: what's in a name? Clin Child Fam Psychol Rev 2015;18(2):170–82.

[839] Bearss K, Johnson C, Smith T, Lecavalier L, Swiezy N, Aman M, et al. Effect of parent training vs parent education on behavioral problems in children with autism spectrum disorder: a randomized clinical trial. JAMA 2015;313(15):1524–33.

[840] Dawson-Squibb JJ, Davids EL, Harrison AJ, Molony MA, de Vries PJ. Parent education and training for autism spectrum disorders: scoping the evidence. Autism 2020;24(1):7–25.

[841] Boyd BA, Hume K, McBee MT, et al. Comparative efficacy of LEAP, TEACCH and non-model-specific special education programs for preschoolers with autism spectrum disorders. J Autism Dev Disord 2014;44(2):366–80.

[842] Rogers SJ, Yoder P, Estes A, Warren Z, McEachin J, Munson J, et al. A multisite randomized controlled trial comparing the effects of intervention intensity and intervention style on outcomes for young children with autism. J Am Acad Child Adolesc Psychiatry 2021;60 (6):710–22.

[843] Kasari C, Shire S, Shih W, Landa R, Levato L, Smith T. Spoken language outcomes in limited language preschoolers with autism and global developmental delay: RCT of early intervention approaches. Autism Res. 2023;16(6):1236–46.

[844] Asta L, Persico AM. Differential predictors of response to Early Start Denver Model vs. Early Intensive Behavioral Intervention in young children with autism spectrum disorder: a systematic review and meta-analysis. Brain Sci 2022;12(11):1499.

[845] Bent C, Glencross S, McKinnon K, Hudry K, Dissanayake C, Victorian ASELCC TeamVivanti G. Predictors of developmental and adaptive behaviour outcomes in response to Early Intensive Behavioural Intervention and the Early Start Denver Model. J Autism Dev Disord 2023. Available from: https://doi.org/10.1007/s10803-023-05993-w Epub ahead of print.

[846] Panganiban J, Kasari C. Super responders: predicting language gains from JASPER among limited language children with autism spectrum disorder. Autism Res. 2022;15(8):1565–75.

[847] Hampton LH, Rodriguez EM. Preemptive interventions for infants and toddlers with a high likelihood for autism: a systematic review and meta-analysis. Autism 2022;26(6):1364–78.

[848] McGlade A, Whittingham K, Barfoot J, Taylor L, Boyd RN. Efficacy of very early interventions on neurodevelopmental outcomes for infants and toddlers at increased likelihood of or diagnosed with autism: a systematic review and meta-analysis. Autism Res 2023;16(6):1145–60.

[849] Whitehouse AJO, Varcin KJ, Pillar S, Billingham W, Alvares GA, Barbaro J, et al. Effect of preemptive intervention on developmental outcomes among infants showing early signs of autism: a randomized clinical trial of outcomes to diagnosis. JAMA Pediatr. 2021;175(11):e213298.

[850] Odom SL, Hall LJ, Morin KL, Kraemer BR, Hume KA, McIntyre NS, et al. Educational interventions for children and youth with autism: a 40-year perspective. J Autism Dev Disord 2021;51(12):4354–69.

[851] Plavnick JB, Marchand-Martella NE, Martella RC, Thompson JL, Wood AL. A review of explicit and systematic scripted instructional programs for students with autism spectrum disorder. Rev J Autism Dev Disord. 2015;2(1):55–66.

[852] Fleury VP, Hedges S, Hume K, Browder DM, Thompson JL, Fallin K, et al. Addressing the academic needs of adolescents with autism spectrum disorder in secondary education. Remedial Spec Educ. 2014;35(2):68–79.

[853] Knight VF, Sartini E. A comprehensive literature review of comprehension strategies in core content areas for students with autism spectrum disorder. J Autism Dev Disord 2015;45(5):1213–29.

[854] King SA, Lemons CJ, Davidson KA. Math interventions for students with autism spectrum disorder: a best-evidence synthesis. Except Child 2016;82(4):443–62.

[855] Asaro-Saddler K. Writing instruction and self-regulation for students with autism spectrum disorders. Top Lang Disord. 2016;36(3):266–93.

[856] Taylor JC, Rizzo KL, Hwang J, Hill D. A review of research on science instruction for students with autism spectrum disorder. Sch Sci Math 2020;120(2):116–25.

[857] Watkins L, Ledbetter-Cho K, O'Reilly M, Barnard-Brak L, Garcia-Grau P. Interventions for students with autism in inclusive settings: a best-evidence synthesis and meta-analysis. Psychol Bull 2019;145(5):490–507.

[858] Dean M, Chang YC. A systematic review of school-based social skills interventions and observed social outcomes for students with autism spectrum disorder in inclusive settings. Autism 2021;25(7):1828–43.

[859] McKeithan GK, Sabornie EJ. Social-behavioral interventions for secondary-level students with high-functioning autism in public school settings: a meta-analysis. Focus Autism Other Dev Disabl 2020;35(3):165–75.

[860] Gates JA, Kang E, Lerner MD. Efficacy of group social skills interventions for youth with autism spectrum disorder: a systematic review and meta-analysis. Clin Psychol Rev 2017;52:164–81.

[861] Wolstencroft J, Robinson L, Srinivasan R, Kerry E, Mandy W, Skuse D. A systematic review of group social skills interventions, and meta-analysis of outcomes, for children with high functioning ASD. J Autism Dev Disord 2018;48(7):2293–307.

[862] Zheng S, Kim H, Salzman E, Ankenman K, Bent S. Improving social knowledge and skills among adolescents with autism: systematic review and meta-analysis of UCLA PEERS® for adolescents. J Autism Dev Disord 2021;51(12):4488–503.

[863] Gilmore R, Ziviani J, Chatfield MD, Goodman S, Sakzewski L. Social skills group training in adolescents with disabilities: a systematic review. Res Dev Disabil 2022;125:104218.

[864] Laugeson EA, Park MN. Using a CBT approach to teach social skills to adolescents with autism spectrum disorder and other social challenges: the PEERS (R) method. J Ration Emot Cogn Behav Ther. 2014;32(1):84–97.

[865] Soares EE, Bausback K, Beard CL, Higinbotham M, Bunge EL, Gengoux GW. Social skills training for autism spectrum disorder: a meta-analysis of in-person and technological interventions. J Technol Behav Sci 2021;6(1):166–80.

[866] Gates JA, McNair ML, Richards JK, Lerner MD. Social knowledge and performance in autism: a critical review and recommendations. Clin Child Fam Psychol Rev 2023;26(3):665–89.

[867] Fuller EA, Kaiser AP. The effects of early intervention on social communication outcomes for children with autism spectrum disorder: a meta-analysis. J Autism Dev Disord 2020;50(5):1683–700.

[868] Sandbank M, Bottema-Beutel K, Crowley S, Cassidy M, Feldman JI, Canihuante M, et al. Intervention effects on language in children with autism: a Project AIM meta-analysis. J Speech Lang Hear Res 2020;63(5):1537–60.

[869] Hampton LH, Kaiser AP. Intervention effects on spoken-language outcomes for children with autism: a systematic review and meta-analysis. J Intellect Disabil Res 2016;60(5):444–63.

[870] Parsons L, Cordier R, Munro N, Joosten A, Speyer R. A systematic review of pragmatic language interventions for children with autism spectrum disorder. PLoS One 2017;12(4):e0172242.

[871] Adams C, Lockton E, Freed J, Gaile J, Earl G, McBean K, et al. The social communication intervention project: a randomized controlled trial of the effectiveness of speech and language therapy for school-age children who have pragmatic and social communication problems with or without autism spectrum disorder. Int J Lang Commun Disord 2012;47(3):233–44.

[872] Parsons L, Cordier R, Munro N, Joosten A. A randomized controlled trial of a play-based, peer-mediated pragmatic language intervention for children with autism. Front Psychol. 2019;10:1960.

[873] Newcomb ET, Hagopian LP. Treatment of severe problem behaviour in children with autism spectrum disorder and intellectual disabilities. Int Rev Psychiatry 2018;30(1):96–109.

[874] Kurtz PF, Leoni M, Hagopian LP. Behavioral approaches to assessment and early intervention for severe problem behavior in intellectual and developmental disabilities. Pediatr Clin North Am 2020;67(3):499–511.

[875] Scahill L, Bearss K, Lecavalier L, Smith T, Swiezy N, Aman MG, et al. Effect of parent training on adaptive behavior in children with autism spectrum disorder and disruptive behavior: results of a randomized trial. J Am Acad Child Adolesc Psychiatry 2016;55(7):602–9.

[876] Beavers GA, Iwata BA, Lerman DC. Thirty years of research on the functional analysis of problem behavior. J Appl Behav Anal 2013;46(1):1–21.

[877] Vetter JA. Parent-child interaction therapy for autism spectrum and attention-deficit/hyperactivity disorders: a review of the literature. Child Fam Behav Ther. 2018;40:204–32.

[878] Allen K, Harrington J, Quetsch LB, Masse J, Cooke C, Paulson JF. Parent-child interaction therapy for children with disruptive behaviors and autism: a randomized clinical trial. J Autism Dev Disord 2023;53(1):390–404.

[879] Kasperzack D, Schrott B, Mingebach T, Becker K, Burghardt R, Kamp-Becker I. Effectiveness of the Stepping Stones Triple P group parenting program in reducing comorbid behavioral problems in children with autism. Autism 2020;24(2):423–36.

[880] Ruane A, Carr A. Systematic review and meta-analysis of Stepping Stones Triple P for parents of children with disabilities. Fam Process 2019;58(1):232–46.

[881] Wood JJ, Kendall PC, Wood KS, Kerns CM, Seltzer M, Small BJ, et al. Cognitive behavioral treatments for anxiety in children with autism spectrum disorder: a randomized clinical trial. JAMA Psychiatry. 2020;77(5):474–83.

[882] White SW, Simmons GL, Gotham KO, Conner CM, Smith IC, Beck KB, et al. Psychosocial treatments targeting anxiety and depression in adolescents and adults on the autism spectrum: review of the latest research and recommended future directions. Curr Psychiatry Rep 2018;20(10):82.

[883] Perihan C, Bicer A, Bocanegra J. Assessment and treatment of anxiety in children with autism spectrum disorder in school settings: a systematic review and meta-analysis. School Ment Health 2022;14(1):153–64.

[884] Sharma S, Hucker A, Matthews T, Grohmann D, Laws KR. Cognitive behavioural therapy for anxiety in children and young people on the autism spectrum: a systematic review and meta-analysis. BMC Psychol 2021;9(1):151.

[885] Shattuck PT, Lau L, Anderson KA, Kuo AA. A national research agenda for the transition of youth with autism. Pediatrics 2018;141(Suppl 4):S355–61.

[886] Anderson AH, Stephenson J, Carter M, Carlon S. A systematic literature review of empirical research on postsecondary students with autism spectrum disorder. J Autism Dev Disord 2019;49(4):1531–58.

[887] Capriola-Hall NN, Brewe AM, Golt J, White SW. Anxiety and depression reduction as distal outcomes of a college transition readiness program for adults with autism. J Autism Dev Disord 2021;51(1):298–306.

[888] White SW, Smith IC, Miyazaki Y, Conner CM, Elias R, Capriola-Hall NN. Improving transition to adulthood for students with autism: a randomized controlled trial of STEPS. J Clin Child Adolesc Psychol 2021;50(2):187–201.

[889] Fong CJ, Taylor J, Berdyyeva A, McClelland AM, Murphy KM, Westbrook JD. Interventions for improving employment outcomes for persons with autism spectrum disorders: a systematic review update. Campbell Syst Rev. 2021;17(3):e1185.

[890] Baker-Ericzén MJ, ElShamy R, Kammes RR. Current status of evidence-based practices to enhance employment outcomes for transition age youth and adults on the autism spectrum. Curr Psychiatry Rep 2022;24(3):161–70.

[891] National Technical Assistance Center on Transition: The Collaborative. Effective practices in secondary transition: operational definitions. Charlotte, NC: Author; 2022. https://transitionta.org/wp-content/uploads/docs/Updated_EBP_Definitions_June_2022.pdf [accessed 26.11.23].

[892] Lilienfeld SO, Marshall J, Todd JT, Shane HC. The persistence of fad interventions in the face of negative scientific evidence: facilitated communication for autism as a case example. Evid Based Commun Assess Interv. 2014;8:62–101.

[893] Schlosser RW, Balandin S, Hemsley B, Iacono T, Probst P, von Tetzchner S. Facilitated communication and authorship: a systematic review. Augment Altern Commun 2014;30:359–68.

[894] Hemsley B, Bryant L, Schlosser RW, Shane HC, Lang R, Paul D, et al. Systematic review of facilitated communication 2014–2018 finds no new evidence that messages delivered using facilitated communication are authored by the person with disability. Autism Dev Lang Impair. 2018;3:1–8.

[895] Saloviita T. Does linguistic analysis confirm the validity of facilitated communication? Focus Autism Other Dev Disabl 2018;33(2):91–9.

[896] International Society for Augmentative and Alternative Communication. ISAAC position statement on facilitated communication. Augment Altern Commun 2014;30:357–8.

[897] American Speech-Language-Hearing Association. *Facilitated communication* [position statement]. Rockville, MD: American Speech-Language-Hearing Association; 2018. Available from: http://www.asha.org/policy/.

[898] American Speech-Language-Hearing Association. *Rapid prompting method* [position statement]. Rockville, MD: American Speech-Language-Hearing Association; 2018. Available from: http://www.asha.org/policy/.

[899] Schlosser RW, Hemsley B, Shane H, Todd J, Lang R, Lilienfeld SO, et al. Rapid prompting method and autism spectrum disorder: systematic review exposes lack of evidence. Rev J Autism Dev Disord. 2019;6:403–12.

[900] Iovannone R, Dunlap G, Huber H, Kincaid D. Effective educational practices for students with autism spectrum disorders. Focus Autism Other Dev Disabl 2003;18:150–65.

[901] National Institute for Health and Clinical Excellence. Autism spectrum disorder in under 19s: support and management. Published Aug 28, 2013. Updated June 14, 2021. https://www.nice.org.uk/guidance/cg170 [accessed 29.11.23].

[902] Green J, Garg S. Annual research review: the state of autism intervention science: progress, target psychological and biological mechanisms and future prospects. J Child Psychol Psychiatry 2018;59(4):424–43.

[903] Franz L, Goodwin CD, Rieder A, Matheis M, Damiano DL. Early intervention for very young children with or at high likelihood for autism spectrum disorder: an overview of reviews. Dev Med Child Neurol 2022;64(9):1063–76.

[904] Provenzani U, Fusar-Poli L, Brondino N, Damiani S, Vercesi M, Meyer N, et al. What are we targeting when we treat autism spectrum disorder? A systematic review of 406 clinical trials. Autism 2020;24(2):274–84.

[905] McConachie H, Parr JR, Glod M, Hanratty J, Livingstone N, Oono IP, et al. Systematic review of tools to measure outcomes for young children with autism spectrum disorder. Health Technol Assess 2015;19(41):1–506.

[906] Anagnostou E, Jones N, Huerta M, Halladay AK, Wang P, Scahill L, et al. Measuring social communication behaviors as a treatment endpoint in individuals with autism spectrum disorder. Autism 2015;19(5):622–36.

[907] Grzadzinski R, Janvier D, Kim SH. Recent developments in treatment outcome measures for young children with autism spectrum disorder (ASD). Semin Pediatr Neurol 2020;34:100806.

[908] Kim SH, Grzadzinski R, Martinez K, Lord C. Measuring treatment response in children with autism spectrum disorder: applications of the brief observation of social communication change to the autism diagnostic observation schedule. Autism 2019;23(5):1176–85.

[909] Nahmias AS, Pellecchia M, Stahmer AC, Mandell DS. Effectiveness of community-based early intervention for children with autism spectrum disorder: a meta-analysis. J Child Psychol Psychiatry 2019;60(11):1200–9.

[910] Myers SM, Johnson CP. Management of children with autism spectrum disorders American Academy of Pediatrics Council on Children With Disabilities Pediatrics 2007;120(5):1162–82.

[911] Todorow C, Connell J, Turchi RM. The medical home for children with autism spectrum disorder: an essential element whose time has come. Curr Opin Pediatr 2018;30(2):311–17.

[912] Kuo DZ, Houtrow AJCouncil On Children With Disabilities. Recognition and management of medical complexity. Pediatrics 2016;138(6): e20163021.

[913] Volkmar FR, Wiesner LA, Westphal A. Healthcare issues for children on the autism spectrum. Curr Opin Psychiatry 2006;19:361–6.

[914] Bultas MW, Johnson NL, Burkett K, Reinhold J. Translating research to practice for children with autism spectrum disorder: Part 2: behavior management in home and health care settings. J Pediatr Health Care 2016;30(1):27–37.

[915] Straus J, Coburn S, Maskell S, Pappagianopoulos J, Cantrell K. Medical encounters for youth with autism spectrum disorder: a comprehensive review of environmental considerations and interventions. Clin Med Insights Pediatr. 2019;13:1179556519842816.

[916] Madra M, Ringel R, Margolis KG. Gastrointestinal issues and autism spectrum disorder. Child Adolesc Psychiatr Clin N Am 2020;29 (3):501–13.

[917] Ferguson BJ, Dovgan K, Takahashi N, Beversdorf DQ. The relationship among gastrointestinal symptoms, problem behaviors, and internalizing symptoms in children and adolescents with autism spectrum disorder. Front Psychiatry. 2019;10:194.

[918] Watkins LV, O'Dwyer M, Shankar R. A review of the pharmacotherapeutic considerations for managing epilepsy in people with autism. Expert Opin Pharmacother 2022;23(7):841–51.

[919] Fisher RS, Cross JH, French JA, Higurashi N, Hirsch E, Jansen FE, et al. Operational classification of seizure types by the international league against epilepsy: position paper of the ILAE commission for classification and terminology. Epilepsia 2017;58(4):522–30.

[920] Nevitt SJ, Sudell M, Cividini S, Marson AG, Tudur Smith C. Antiepileptic drug monotherapy for epilepsy: a network meta-analysis of individual participant data. Cochrane Database Syst Rev 2022;4(4):CD011412.

[921] Buie T, Fuchs 3rd GJ, Furuta GT, Kooros K, Levy J, Lewis JD, et al. Recommendations for evaluation and treatment of common gastrointestinal problems in children with ASDs. Pediatrics. 2010;125(Suppl 1):S19–29.

[922] Furuta GT, Williams K, Kooros K, Kaul A, Panzer R, Coury DL, et al. Management of constipation in children and adolescents with autism spectrum disorders. Pediatrics 2012;130(Suppl 2):S98–105.

[923] Curtin C, Hyman SL, Boas DD, Hassink S, Broder-Fingert S, Ptomey LT, et al. Weight management in primary care for children with autism: expert recommendations. Pediatrics 2020;145(Suppl 1):S126–39.

[924] Spear BA, Barlow SE, Ervin C, et al. Recommendations for treatment of child and adolescent overweight and obesity. Pediatrics 2007;120 (suppl 4):S254–88.

[925] Henneberry E, Lamy M, Dominick KC, Erickson CA. Decades of progress in the psychopharmacology of autism spectrum disorder. J Autism Dev Disord 2021;51(12):4370–94.

[926] Persico AM, Ricciardello A, Lamberti M, Turriziani L, Cucinotta F, Brogna C, et al. The pediatric psychopharmacology of autism spectrum disorder: a systematic review – part I: the past and the present. Prog Neuropsychopharmacol Biol Psychiatry 2021;110:110326.

[927] Zhou MS, Nasir M, Farhat LC, Kook M, Artukoglu BB, Bloch MH. Meta-analysis: pharmacologic treatment of restricted and repetitive behaviors in autism spectrum disorders. J Am Acad Child Adolesc Psychiatry 2021;60(1):35–45.

[928] Jobski K, Höfer J, Hoffmann F, Bachmann C. Use of psychotropic drugs in patients with autism spectrum disorders: a systematic review. Acta Psychiatr Scand 2017;135(1):8–28.

[929] Madden JM, Lakoma MD, Lynch FL, Rusinak D, Owen-Smith AA, Coleman KJ, et al. Psychotropic medication use among insured children with autism spectrum disorder. J Autism Dev Disord 2017;47(1):144–54.

[930] Shurtz L, Schwartz C, DiStefano C, McPartland JC, Levin AR, Dawson G, et al. Concomitant medication use in children with autism spectrum disorder: data from the autism biomarkers consortium for clinical trials. Autism 2023;27(4):952–66.

[931] Feroe AG, Uppal N, Gutiérrez-Sacristán A, Mousavi S, Greenspun P, Surati R, et al. Medication use in the management of comorbidities among individuals with autism spectrum disorder from a large nationwide insurance database. JAMA Pediatr. 2021;175(9):957–65.

[932] Houghton R, Ong RC, Bolognani F. Psychiatric comorbidities and use of psychotropic medications in people with autism spectrum disorder in the United States. Autism Res. 2017;10(12):2037–47.

[933] Aishworiya R, Valica T, Hagerman R, Restrepo B. An update on psychopharmacological treatment of autism spectrum disorder. Neurotherapeutics 2022;19(1):248–62.

[934] Baribeau D, Vorstman J, Anagnostou E. Novel treatments in autism spectrum disorder. Curr Opin Psychiatry 2022;35(2):101–10.

[935] Hellings J. Pharmacotherapy in autism spectrum disorders, including promising older drugs warranting trials. World J Psychiatry. 2023;13(6):262–77.

[936] Myers SM. The status of pharmacotherapy for autism spectrum disorders. Expert Opin Pharmacother 2007;8(11):1579–603.

[937] McGuire K, Fung LK, Hagopian L, Vasa RA, Mahajan R, Bernal P, et al. Irritability and problem behavior in autism spectrum disorder: a practice pathway for pediatric primary care. Pediatrics 2016;137(Suppl 2):S136–48.

[938] Rodrigues R, Lai MC, Beswick A, Gorman DA, Anagnostou E, Szatmari P, et al. Practitioner review: pharmacological treatment of attention-deficit/hyperactivity disorder symptoms in children and youth with autism spectrum disorder: a systematic review and meta-analysis. J Child Psychol Psychiatry 2021;62(6):680–700.

[939] Young S, Hollingdale J, Absoud M, Bolton P, Branney P, Colley W, et al. Guidance for identification and treatment of individuals with attention deficit/hyperactivity disorder and autism spectrum disorder based upon expert consensus. BMC Med 2020;18(1):146.

[940] Mahajan R, Bernal MP, Panzer R, Whitaker A, Roberts W, Handen B, et al. Clinical practice pathways for evaluation and medication choice for attention-deficit/hyperactivity disorder symptoms in autism spectrum disorders. Pediatrics 2012;130(Suppl 2):S125–38.

[941] Barbaresi WJ, Campbell L, Diekroger EA, Froehlich TE, Liu YH, O'Malley E, et al. The society for developmental and behavioral pediatrics clinical practice guideline for the assessment and treatment of children and adolescents with complex attention-deficit/hyperactivity disorder: process of care algorithms. J Dev Behav Pediatr 2020;41(Suppl 2S):S58–74.

[942] Sturman N, Deckx L, van Driel ML. Methylphenidate for children and adolescents with autism spectrum disorder. Cochrane Database Syst Rev 2017;11(11):CD011144.

[943] Research Units on Pediatric Psychopharmacology Autism Network. Randomized, controlled, crossover trial of methylphenidate in pervasive developmental disorders with hyperactivity. Arch Gen Psychiatry 2005;62:1266–74.

[944] Posey DJ, Aman MG, McCracken JT, et al. Positive effects of methylphenidate on inattention and hyperactivity in pervasive developmental disorders: an analysis of secondary measures. Biol Psychiatry 2007;61:538544.

[945] Pearson DA, Santos CW, Aman MG, et al. Effects of extended release methylphenidate treatment on ratings of attention-deficit/hyperactivity disorder (ADHD) and associated behavior in children with autism spectrum disorders and ADHD symptoms. J Child Adolesc Psychopharmacol 2013;23:337–51.

[946] Patra S, Nebhinani N, Viswanathan A, Kirubakaran R. Atomoxetine for attention deficit hyperactivity disorder in children and adolescents with autism: a systematic review and meta-analysis. Autism Res 2019;12(4):542–52.

[947] Handen BL, Aman MG, Arnold LE, et al. Atomoxetine, parent training, and their combination in children with autism spectrum disorder and attention-deficit/hyperactivity disorder. J Am Acad Child Adolesc Psychiatry 2015;54:905–15.

[948] Smith T, Aman MG, Arnold LE, Silverman LB, Lecavalier L, Hollway J, et al. Atomoxetine and parent training for children with autism and attention-deficit/hyperactivity disorder: a 24-week extension study. J Am Acad Child Adolesc Psychiatry 2016;55(10):868–76.

[949] Harfterkamp M, van de Loo-Neus G, Minderaa RB, et al. A randomized double-blind study of atomoxetine versus placebo for attention-deficit/hyperactivity disorder symptoms in children with autism spectrum disorder. J Am Acad Child Adolesc Psychiatry 2012;51:733–41.

[950] Arnold LE, Aman MG, Cook AM, et al. Atomoxetine for hyperactivity in autism spectrum disorders: placebo-controlled crossover pilot trial. J Am Acad Child Adolesc Psychiatry 2006;45:1196–205.

[951] Scahill L, McCracken JT, King BH, Rockhill C, Shah B, Politte L, et al. Extended-release guanfacine for hyperactivity in children with autism spectrum disorder. Am J Psychiatry 2015;172(12):1197–206.

[952] Wang F, Wen F, Yu L, Yan J, Liu J, Li Y, et al. The efficacy and safety in attention deficit hyperactivity disorder of second-generation antipsychotics and other medications for hyperactivity in children and adolescents with autism: a meta-analysis. Int Clin Psychopharmacol 2021;36(3):109–16.

[953] McCracken JT, McGough J, Shah B, Cronin P, Hong D, Aman MG, et al. Risperidone in children with autism and serious behavioral problems. N Engl J Med 2002;347(5):314–21.

[954] Shea S, Turgay A, Carroll A, Schulz M, Orlik H, Smith I, et al. Risperidone in the treatment of disruptive behavioral symptoms in children with autistic and other pervasive developmental disorders. Pediatrics 2004;114(5):e634–41.

[955] Marcus RN, Owen R, Kamen L, Manos G, McQuade RD, Carson WH, et al. A placebo-controlled, fixed-dose study of aripiprazole in children and adolescents with irritability associated with autistic disorder. J Am Acad Child Adolesc Psychiatry 2009;48(11):1110–19.

[956] Owen R, Sikich L, Marcus RN, Corey-Lisle P, Manos G, McQuade RD, et al. Aripiprazole in the treatment of irritability in children and adolescents with autistic disorder. Pediatrics 2009;124(6):1533–40.

[957] Fung LK, Mahajan R, Nozzolillo A, Bernal P, Krasner A, Jo B, et al. Pharmacologic treatment of severe irritability and problem behaviors in autism: a systematic review and meta-analysis. Pediatrics 2016;137(Suppl 2):S124–35.

[958] Salazar de Pablo G, Pastor Jordá C, Vaquerizo-Serrano J, Moreno C, Cabras A, Arango C, et al. Systematic review and meta-analysis: efficacy of pharmacological interventions for irritability and emotional dysregulation in autism spectrum disorder and predictors of response. J Am Acad Child Adolesc Psychiatry 2023;62(2):151–68.

[959] Iffland M, Livingstone N, Jorgensen M, Hazell P, Gillies D. Pharmacological intervention for irritability, aggression, and self-injury in autism spectrum disorder (ASD). Cochrane Database Syst Rev 2023;10(10):CD011769.

[960] Pandina GJ, Bossie CA, Youssef E, Zhu Y, Dunbar F. Risperidone improves behavioral symptoms in children with autism in a randomized, double-blind, placebo-controlled trial. J Autism Dev Disord 2007;37(2):367–73.

[961] Aman MG, McDougle CJ, Scahill L, Handen B, Arnold LE, Johnson C, et al. Medication and parent training in children with pervasive developmental disorders and serious behavior problems: results from a randomized clinical trial. J Am Acad Child Adolesc Psychiatry 2009;48(12):1143–54.

[962] Arnold LE, Aman MG, Li X, Butter E, Humphries K, Scahill L, et al. Research Units of Pediatric Psychopharmacology (RUPP) autism network randomized clinical trial of parent training and medication: one-year follow-up. J Am Acad Child Adolesc Psychiatry 2012;51(11):1173–84.

[963] Kent JM, Kushner S, Ning X, Karcher K, Ness S, Aman M, et al. Risperidone dosing in children and adolescents with autistic disorder: a double-blind, placebo-controlled study. J Autism Dev Disord 2013;43(8):1773–83.

[964] Ichikawa H, Mikami K, Okada T, Yamashita Y, Ishizaki Y, Tomoda A, et al. Aripiprazole in the treatment of irritability in children and adolescents with autism spectrum disorder in Japan: a randomized, double-blind, placebo-controlled study. Child Psychiatry Hum Dev 2017;48(5):796–806.

[965] DeVane CL, Charles JM, Abramson RK, Williams JE, Carpenter LA, Raven S, et al. Pharmacotherapy of autism spectrum disorder: results from the randomized BAART clinical trial. Pharmacotherapy 2019;39(6):626–35.

[966] McDougle CJ, Naylor ST, Cohen DJ, Volkmar FR, Heninger GR, Price LH. A double-blind, placebo-controlled study of fluvoxamine in adults with autistic disorder. Arch Gen Psychiatry 1996;53(11):1001–8.

[967] Hollander E, Chaplin W, Soorya L, Wasserman S, Novotny S, Rusoff J, et al. Divalproex sodium vs placebo for the treatment of irritability in children and adolescents with autism spectrum disorders. Neuropsychopharmacology 2010;35(4):990–8.

[968] McDougle CJ, Holmes JP, Carlson DC, Pelton GH, Cohen DJ, Price LH. A double-blind, placebo-controlled study of risperidone in adults with autistic disorder and other pervasive developmental disorders. Arch Gen Psychiatry 1998;55(7):633–41.

[969] Hollander E, Phillips A, Chaplin W, Zagursky K, Novotny S, Wasserman S, et al. A placebo controlled crossover trial of liquid fluoxetine on repetitive behaviors in childhood and adolescent autism. Neuropsychopharmacology 2005;30(3):582–9.

[970] Hollander E, Soorya L, Chaplin W, Anagnostou E, Taylor BP, Ferretti CJ, et al. A double-blind placebo-controlled trial of fluoxetine for repetitive behaviors and global severity in adult autism spectrum disorders. Am J Psychiatry 2012;169(3):292–9.

[971] Chugani DC, Chugani HT, Wiznitzer M, Parikh S, Evans PA, Hansen RL, et al. Efficacy of low-dose buspirone for restricted and repetitive behavior in young children with autism spectrum disorder: a randomized trial. J Pediatr 2016;170:45–53 e1-4.

[972] Politte LC, Scahill L, Figueroa J, McCracken JT, King B, McDougle CJ. A randomized, placebo-controlled trial of extended-release guanfacine in children with autism spectrum disorder and ADHD symptoms: an analysis of secondary outcome measures. Neuropsychopharmacology 2018;43(8):1772–8.

[973] Vasa RA, Mazurek MO, Mahajan R, Bennett AE, Bernal MP, Nozzolillo AA, et al. Assessment and treatment of anxiety in youth with autism spectrum disorders. Pediatrics 2016;137(Suppl 2):S115–23.

[974] Walter HJ, Bukstein OG, Abright AR, Keable H, Ramtekkar U, Ripperger-Suhler J, et al. Clinical practice guideline for the assessment and treatment of children and adolescents with anxiety disorders. J Am Acad Child Adolesc Psychiatry 2020;59(10):1107–24.

[975] Williams Buckley A, Hirtz D, Oskoui M, Armstrong MJ, Batra A, Bridgemohan C, et al. Practice guideline: treatment for insomnia and disrupted sleep behavior in children and adolescents with autism spectrum disorder: report of the Guideline Development, Dissemination, and Implementation Subcommittee of the American Academy of Neurology. Neurology 2020;94(9):392–404.

[976] Malow BA, Byars K, Johnson K, Weiss S, Bernal P, Goldman SE, et al. A practice pathway for the identification, evaluation, and management of insomnia in children and adolescents with autism spectrum disorders. Pediatrics 2012;130(Suppl 2):S106–24.

[977] Parker A, Beresford B, Dawson V, Elphick H, Fairhurst C, Hewitt C, et al. Oral melatonin for non-respiratory sleep disturbance in children with neurodisabilities: systematic review and meta-analyses. Dev Med Child Neurol 2019;61(8):880–90.

[978] Cuomo BM, Vaz S, Lee EAL, Thompson C, Rogerson JM, Falkmer T. Effectiveness of sleep-based interventions for children with autism spectrum disorder: a meta-synthesis. Pharmacotherapy 2017;37(5):555–78.

[979] Wright B, Sims D, Smart S, Alwazeer A, Alderson-Day B, Allgar V, et al. Melatonin versus placebo in children with autism spectrum conditions and severe sleep problems not amenable to behaviour management strategies: a randomised controlled crossover trial. J Autism Dev Disord 2011;41(2):175–84.

[980] Cortesi F, Giannotti F, Sebastiani T, Panunzi S, Valente D. Controlled-release melatonin, singly and combined with cognitive behavioural therapy, for persistent insomnia in children with autism spectrum disorders: a randomized placebo-controlled trial. J Sleep Res 2012;21(6):700–9.

[981] Gringras P, Nir T, Breddy J, Frydman-Marom A, Findling RL. Efficacy and safety of pediatric prolonged-release melatonin for insomnia in children with autism spectrum disorder. J Am Acad Child Adolesc Psychiatry 2017;56(11):948–57.

[982] Malow BA, Findling RL, Schroder CM, Maras A, Breddy J, Nir T, et al. Sleep, growth, and puberty after 2 years of prolonged-release melatonin in children with autism spectrum disorder. J Am Acad Child Adolesc Psychiatry 2021;60(2):252–61.

[983] King BH, Hollander E, Sikich L, McCracken JT, Scahill L, Bregman JD, et al. Lack of efficacy of citalopram in children with autism spectrum disorders and high levels of repetitive behavior: citalopram ineffective in children with autism. Arch Gen Psychiatry 2009;66(6):583–90.

[984] Hellings JA, Weckbaugh M, Nickel EJ, Cain SE, Zarcone JR, Reese RM, et al. A double-blind, placebo-controlled study of valproate for aggression in youth with pervasive developmental disorders. J Child Adolesc Psychopharmacol 2005;15(4):682–92.

[985] Reddihough DS, Marraffa C, Mouti A, O'Sullivan M, Lee KJ, Orsini F, et al. Effect of fluoxetine on obsessive-compulsive behaviors in children and adolescents with autism spectrum disorders: a randomized clinical trial. JAMA 2019;322(16):1561–9.

[986] Herscu P, Handen BL, Arnold LE, Snape MF, Bregman JD, Ginsberg L, et al. The SOFIA study: negative multi-center study of low dose fluoxetine on repetitive behaviors in children and adolescents with autistic disorder. J Autism Dev Disord 2020;50(9):3233–44.

[987] Handen BL, Sahl R, Hardan AY. Guanfacine in children with autism and/or intellectual disabilities. J Dev Behav Pediatr 2008;29(4):303–8.

[988] Fankhauser MP, Karumanchi VC, German ML, Yates A, Karumanchi SD. A double-blind, placebo-controlled study of the efficacy of transdermal clonidine in autism. J Clin Psychiatry 1992;53(3):77–82.

[989] Jaselskis CA, Cook Jr EH, Fletcher KE, Leventhal BL. Clonidine treatment of hyperactive and impulsive children with autistic disorder. J Clin Psychopharmacol. 1992;12(5):322–7.

[990] Deb S, Roy M, Lee R, Majid M, Limbu B, Santambrogio J, et al. Randomised controlled trials of antidepressant and anti-anxiety medications for people with autism spectrum disorder: systematic review and meta-analysis. BJPsych Open 2021;7(6):e179.

[991] Limbu B, Deb S, Roy M, Lee R, Roy A, Taiwo O. Randomised controlled trials of mood stabilisers for people with autism spectrum disorder: systematic review and meta-analysis. BJPsych Open 2022;8(2):e52.

[992] Vasa RA, Carroll LM, Nozzolillo AA, Mahajan R, Mazurek MO, Bennett AE, et al. A systematic review of treatments for anxiety in youth with autism spectrum disorders. J Autism Dev Disord 2014;44:3215–29.

[993] Stefánsdóttir ÍH, Ivarsson T, Skarphedinsson G. Efficacy and safety of serotonin reuptake inhibitors (SSRI) and serotonin noradrenaline reuptake inhibitors (SNRI) for children and adolescents with anxiety disorders: a systematic review and meta-analysis. Nord J Psychiatry 2023;77(2):137–46.

[994] Bruni O, Angriman M, Calisti F, Comandini A, Esposito G, Cortese S, et al. Practitioner review: treatment of chronic insomnia in children and adolescents with neurodevelopmental disabilities. J Child Psychol Psychiatry 2018;59(5):489–508.

[995] Cortese S, Wang F, Angriman M, Masi G, Bruni O. Sleep disorders in children and adolescents with autism spectrum disorder: diagnosis, epidemiology, and management. CNS Drugs 2020;34(4):415–23.

[996] Rana M, Kothare S, DeBassio W. The assessment and treatment of sleep abnormalities in children and adolescents with autism spectrum disorder: a review. J Can Acad Child Adolesc Psychiatry 2021;30(1):25–35.

[997] Wachtel LE. Treatment of catatonia in autism spectrum disorders. Acta Psychiatr Scand 2019;139(1):46–55.

[998] Moore S, Amatya DN, Chu MM, Besterman AD. Catatonia in autism and other neurodevelopmental disabilities: a state-of-the-art review. NPJ Mental Health Res. 2022;1:12. Available from: https://doi.org/10.1038/s44184-022-00012-9.

[999] Cole EJ, Enticott PG, Oberman LM, Gwynette MF, Casanova MF, Jackson SLJ, et al. The potential of repetitive transcranial magnetic stimulation for autism spectrum disorder: a consensus statement. Biol Psychiatry 2019;85(4):e21–2.

[1000] Veenstra-VanderWeele J, Cook EH, King BH, Zarevics P, Cherubini M, Walton-Bowen K, et al. Arbaclofen in children and adolescents with autism spectrum disorder: a randomized, controlled, phase 2 trial. Neuropsychopharmacology 2017;42(7):1390–8.

[1001] Ooi YP, Weng SJ, Kossowsky J, Gerger H, Sung M. Oxytocin and autism spectrum disorders: a systematic review and meta-analysis of randomized controlled trials. Pharmacopsychiatry 2017;50(1):5–13.

[1002] Wang Y, Wang MJ, Rong Y, He HZ, Yang CJ. Oxytocin therapy for core symptoms in autism spectrum disorder: an updated meta-analysis of randomized controlled trials. Res Autism Spectr Disord 2019;64:63–75.

[1003] Jacob S, Veenstra-VanderWeele J, Murphy D, McCracken J, Smith J, Sanders K, et al. Efficacy and safety of balovaptan for socialisation and communication difficulties in autistic adults in North America and Europe: a phase 3, randomised, placebo-controlled trial. Lancet Psychiatry. 2022;9(3):199–210.

[1004] Hollander E, Jacob S, Jou R, McNamara N, Sikich L, Tobe R, et al. Balovaptan vs placebo for social communication in childhood autism spectrum disorder: a randomized clinical trial. JAMA Psychiatry. 2022;79(8):760–9.

[1005] Aran A, Harel M, Cassuto H, Polyansky L, Schnapp A, Wattad N, et al. Cannabinoid treatment for autism: a proof-of-concept randomized trial. Mol Autism. 2021;12(1):6.

[1006] Pedrazzi JFC, Ferreira FR, Silva-Amaral D, Lima DA, Hallak JEC, Zuardi AW, et al. Cannabidiol for the treatment of autism spectrum disorder: hope or hype? Psychopharmacology (Berl) 2022;239(9):2713–34.

[1007] Challman TD, Myers SM. Complementary health approaches in developmental and behavioral pediatrics. In: Voigt RG, Macias MM, Myers SM, Tapia CD, editors. American academy of pediatrics developmental and behavioral pediatrics. 2nd ed. Itasca, IL: American Academy of Pediatrics; 2018. p. 575–90.

[1008] National Center for Complementary and Integrative Health (NCCIH). Complementary, alternative, or integrative health: what's in a name? Last updated April 2021. Available at: https://www.nccih.nih.gov/health/complementary-alternative-or-integrative-health-whats-in-a-name# [accessed 11.12.23].

[1009] Levy SE, Hyman SL. Complementary and alternative medicine treatments for children with autism spectrum disorders. Child Adolesc Psychiatr Clin N Am 2015;24(1):117–43.

[1010] Challman TD. Complementary and alternative medicine in autism: promises kept? In: Shapiro BK, Accardo PJ, editors. Autism: clinical and research frontiers. Baltimore, MD: Brookes; 2008, p. 177–90.

[1011] Levy SE, Frasso R, Colantonio S, Reed H, Stein G, Barg FK, et al. Shared decision making and treatment decisions for young children with autism spectrum disorder. Acad Pediatr. 2016;16(6):571–8.

[1012] Salomone E, Charman T, McConachie H, Warreyn P. Working Group 4, COST Action 'Enhancing the Scientific Study of Early Autism'. Prevalence and correlates of use of complementary and alternative medicine in children with autism spectrum disorder in Europe. Eur J Pediatr 2015;174(10):1277–85.

[1013] Waizbard-Bartov E, Fein D, Lord C, Amaral DG. Autism severity and its relationship to disability. Autism Res. 2023;16(4):685–96.

[1014] Howlin P. Adults with autism: changes in understanding since DSM-III. J Autism Dev Disord 2021;51(12):4291–308.

[1015] McCauley JB, Pickles A, Huerta M, Lord C. Defining positive outcomes in more and less cognitively able autistic adults. Autism Res. 2020;13:1548–60.

[1016] Whiteley P, Carr K, Shattock P. Is autism inborn and lifelong for everyone? Neuropsychiatr Dis Treat 2019;15:2885–91.

[1017] Elias R, Lord C. Diagnostic stability in individuals with autism spectrum disorder: insights from a longitudinal follow-up study. J Child Psychol Psychiatry 2022;63(9):973–83.

[1018] Bal VH, Kim SH, Fok M, Lord C. Autism spectrum disorder symptoms from ages 2 to 19 years: implications for diagnosing adolescents and young adults. Autism Res. 2019;12(1):89–99.

[1019] Kim SH, Bal VH, Lord C. Longitudinal follow-up of academic achievement in children with autism from age 2 to 18. J Child Psychol Psychiatry 2018;59(3):258–67.

[1020] Farley M, Cottle KJ, Bilder D, Viskochil J, Coon H, McMahon W. Mid-life social outcomes for a population-based sample of adults with ASD. Autism Res. 2018;11(1):142–52.

[1021] Steinhausen HC, Mohr Jensen C, Lauritsen MB. A systematic review and meta-analysis of the long-term overall outcome of autism spectrum disorders in adolescence and adulthood. Acta Psychiatr Scand 2016;133(6):445–52.

[1022] Lord C, McCauley JB, Pepa LA, Huerta M, Pickles A. Work, living, and the pursuit of happiness: vocational and psychosocial outcomes for young adults with autism. Autism 2020;24(7):1691–703.

[1023] Ayres M, Parr JR, Rodgers J, Mason D, Avery L, Flynn D. A systematic review of quality of life of adults on the autism spectrum. Autism 2018;22(7):774–83.

[1024] McCauley JB, Elias R, Lord C. Trajectories of co-occurring psychopathology symptoms in autism from late childhood to adulthood. Dev Psychopathol 2020;32(4):1287–302.

[1025] Catalá-López F, Hutton B, Page MJ, Driver JA, Ridao M, Alonso-Arroyo A, et al. Mortality in persons with autism spectrum disorder or attention-deficit/hyperactivity disorder: a systematic review and meta-analysis. JAMA Pediatr 2022;176(4):e216401.

[1026] Schendel DE, Overgaard M, Christensen J, et al. Association of psychiatric and neurologic comorbidity with mortality among persons with autism spectrum disorder in a Danish population. JAMA Pediatr 2016;170:243–50.

[1027] Hirvikoski T, Mittendorfer-Rutz E, Boman M, Larsson H, Lichtenstein P, Bölte S. Premature mortality in autism spectrum disorder. Br J Psychiatry 2016;208:232–8.

[1028] Jokiranta-Olkoniemi E, Gyllenberg D, Sucksdorff D, Suominen A, Kronström K, Chudal R, et al. Risk for premature mortality and intentional self-harm in autism spectrum disorders. J Autism Dev Disord 2021;51(9):3098–108.

Chapter 29

Vision and hearing impairment

Clay Smith[1,2] and Bruce K. Shapiro[1]
[1]*Department of Neurology and Developmental Medicine, Kennedy Krieger Institute, Baltimore, MD, United States,* [2]*Department of Pediatrics, Johns Hopkins University School of Medicine, Baltimore, MD, United States*

Introduction

All living things rely on their senses to learn about their environment. Hearing and vision are particularly important to human development and learning. Proper assessment and treatment of visual and hearing impairment is of utmost importance. Cortical processing of auditory and visual signals is subject to experience-dependent plasticity. As a result, unrecognized sensory deprivation can have profound long-term consequences.

Assessments of visual and auditory anatomy and function begin in the newborn period. However, because of functional development as well as acquired dysfunction, serial assessments are necessary. This may entail simple screenings all the way to complex diagnostics. Although the fundamentals of visual and auditory assessment and treatment are the same for all children, those with developmental disabilities may present particular challenges. The assessments may rely more heavily on objective tests when communication and feedback are limited. It may not be clear to caregivers whether problems with hearing or vision function are due to disorder of the sensory organs, neurological system, or other causes.

Individuals with neurodevelopmental disabilities are much more likely than their typically developing peers to have hearing loss or visual impairment. This is particularly the case for individuals with cerebral palsy where rates of severe hearing and visual impairment are estimated to be 4% and 10%, respectively [1]. A normal assessment does not preclude dysfunction in later life and because of communicative and cognitive issues in children with neurodevelopmental disorders, we must have a high index of suspicion for co-occurring visual and auditory disorders.

Vision

The assessment of the visual apparatus begins essentially at birth and is a serial process throughout development. Knowing key principles of vision and its development are essential for recognizing normal from abnormal. With early identification and intervention, we can sometimes avert permanent deviations of visual development.

It is estimated that nearly 23% (1.62 million), of people with visual acuity loss or blindness in the United States are younger than 40 years [2]. Etiology and prevalence data for childhood visual impairment are highly influenced by regional and socioeconomic factors. The prevalence of childhood blindness (20/400 or worse) in low-income countries is more than three times that of high-income countries [3]. Within the United States, social determinants of health play a significant role in visual health. Age, sex, income, adverse childhood events, and insurance status are all correlated with rates of pediatric vision testing [4]. Co-occurring neurodevelopmental disabilities (NDDs) are also quite common among children with visual impairment. Among those with acuity ≤ 20/70 in better eye with correction, it is estimated that 55% have co-occurring intellectual disability, 42% with cerebral palsy, 8% with autism spectrum disorder, and 5% with hearing loss [5].

Visual anatomy and physiology

The anterior structures of the eye serve to focus images on the retina. As light traverses different media of the eye, it is subject to refraction. Most refraction is a result of the corneal curvature which can undergo some conformational change for accommodation. Emmetropization refers to changes in the eye to harmonize corneal curvature, lens power, and axial length for refraction. This process occurs predominantly in the first 3 years of life but continues up to school-age. Real-time dynamic adjustments to the lens, accommodation, modify the refractive power in order to focus the image on the retina. Other dynamic processes in our vision include adjustments to the pupillary aperture as a result of ambient light.

Capute and Accardo's Neurodevelopmental Disabilities in Infancy and Childhood. DOI: https://doi.org/10.1016/B978-0-12-824060-1.00059-6

Beyond the cornea, anterior chamber, and lens, light traverses the vitreous and meets the retina. Photoreceptor cells, rods and cones, serve to transduce the light signal into nerve impulses that travial via ganglion cell axons in the optic nerve. The fovea has a higher concentration of color-sensitive cones and the thinner retinal periphery contains more rods for low-light, or scotopic, vision. Signals from the nasal retina decussate at the optic chiasm and thus optic tracts carry information from our contralateral visual fields.

Subcortical sites of visual processing include the pulvinar and lateral geniculate nuclei in the thalami and the superior colliculus. From the optic tract, the signals continue via the tectopulvinar or genticulostriate pathways. Two major types of cells predominate each pathway, the magnocellular and parvocelluar, respectively. The tectopulvinar pathway is phylogenetically older and predominantly magnocellular. It conveys information with high temporal resolution and directs visuospatial attention while the genticulostriate serves visual processing with high spatial resolution.

The primary visual cortex in the occipital lobe maintains retinotopic organization. Visual information is further processed in visual association cortical areas and often divided into dorsal and ventral streams. The dorsal and ventral streams are colloquially referred to as the where and what streams, respectively. Dorsal visual association cortex includes parietal as well as superior and middle temporal areas. It serves to process information on orientation and movement. The ventrally located inferior temporal cortex aids in processing information on form, color, and general object recognition.

Visual development

The temporal sequence of visual development and milestones is essential to identifying abnormal exam features. The pupillary light reflex may be absent in neonates born <30 weeks gestational age. It emerges around the 30-week period and is well established by 35 weeks [6,7]. A blink response to bright light emerges concomitantly. Sometimes this is referred to as the dazzle reflex though this is a term more often used in veterinary medicine. Newborns tend to fixate on objects in the range of 6–12 inches from their face, or generally the distance to a caregiver's face when held. Neonates also show a preference high contrast, black and white images with color vision evolving over the first 6 months of life. Tracking of objects develops in stages. Around 2 months of age, babies can track objects in the vertical and horizontal planes. At 3 months of age, they can track things in circular arcs. Intermittent misalignment of the eyes is common until around 4 months of age when they are better yoked and followed by substantive development in depth perception. Around this time, at 4–5 months, infants may start reaching and grasping at objects. There are many aspects to functional vision and they mature at different rates. Visual acuity matures rapidly before 6 months with a range of 3–6 years or later for mature visual acuity abilities [8,9] (Table 29.1).

TABLE 29.1 Visual developmental milestones.

Pupillary response	30 weeks gestational age
Blink to bright light	30 weeks gestational age
Turn to diffuse light	32–36 gestational age
Fixate	Birth
Track vertical/horizontal	2 months
Recognize mother	2 months
Looks at hands	3 months
Track circular	3 months
Visual threat	3 months
Accommodation	4 months
Color vision	5 months
Pats self in mirror	6 months
Depth perception	7 months
Inspects details	7 months

Physical exam

The exam should begin with a brief observation of the eyes and associated anatomy. Are the eyebrows thick or sparse, highly arched, broad, or tapered. Do they meet in the midline (synophrys), or extend laterally? Are the lashes remarkable? Do the eyelid margins evert or invert (ectropion and entropion)?

The presence and size of the eye, or globe, should be assessed as microphthalmia-anophthalmia may be associated with an underlying genetic condition. Buphthalmos, or enlarged globe, may indicate increased intraocular pressure such as in congenital glaucoma. Palpebral fissures should be assessed for size and inclination. Blepharophimosis is a fixed reduction in the horizontal palpebral fissures. Blepharoptosis, a reduction of the vertical palpebral fissures may be congenital or acquired. The inclination of the palpebral fissures is assessed as an imaginary line between the medial and lateral canthi in reference to another between the medial canthi. Thus upslanted palpebral fissures give the appearance of lateral canthi above the medial canthi. Epicanthus is a fold of skin that stretches from the medial aspect of the upper eyelid and stretches downward to cover the medial canthus. Epicanthi, just as many of these morphological findings, are often nonspecific findings but in context may increase suspicion for a syndrome. The distance between the pupils should be assessed for hypo- and hypertelorism. Telecanthus should not be confused with hypertelorism and can occur along with widely spaced eyes. It refers to an increased distance between the inner canthi.

The alignment of the eyes should be assessed with the cover–uncover test when able. A pseudostrabismus may appear in the case of epicanthal folds, giving the appearance of eye misalignment. This can also be assessed with the corneal light reflex. In pseudostrabismus, the light reflection will be symmetric in contrast with a tropia. Congenital or infantile nystagmus presents in the first 6 months of life and children should be assessed for accompanying sensory deficit. Acute nystagmus is less common but may indicate an underlying urgent condition and should be evaluated as such.

The appearance of the sclerae, irides, and pupils should be observed. Discrete lesions or discoloration of the sclerae should be noted. The size, shape, and colors of the irides should be examined. Aniridia may be a complete or partial absence of the iris. Heterochromia may suggest Waardenburg syndrome and a stellate iris Williams syndrome. Lisch nodules, or iris hamartomas, may be seen by the naked eye but should be examined by slit lamp. Failure of the embryonic fissure to close properly may lead to colobomas that affect various parts of the eye. Defects of the lid or iris are outwardly apparent and the latter classically gives a keyhole appearance to the pupil. Colobomas may be isolated, associated with other abnormalities of the eye, or other organ systems as in Wolf–Hirschhorn or CHARGE syndromes.

The red reflex should be assessed in infancy and regularly thereafter. The clinician should assess for the presence of the red reflex, asymmetry, or leukocoria. Retinoblastoma or cataract are commonly feared, although the differential is broader and abnormalities in the red reflex should prompt comprehensive ophthalmic examination.

Fundoscopy can be challenging, particularly with the developmentally young, anxious, or hyperactive individual. Incorporating it into assessments allows for continued refinement and comfort with technique and can be integral in identifying pathology. The optic nerve should be assessed with attention to the color, contour, and size of the disk and cup. The periphery and fovea should be assessed and may reveal pigmentary or vascular abnormalities. The classic cherry red spot is associated with Tay–Sachs disease and other storage diseases like Sandhoff, Niemann–Pick, Krabbe, Farber, and metachromatic leukodystrophy. This results from vascular choroid that can be seen through the foveola in contrast with surrounding ganglion cell layers that have opacified as a result of lipid deposition, or ischemia in central retinal artery occlusion (Table 29.2).

TABLE 29.2 Eye exam findings and related conditions.

Findings	Related conditionsa
Anophthalmia-microphthalmia	SOX2 disorder Waardenburg anophthalmia syndrome (*SMOC1*) Lenz microphthalmia syndrome (*NAA10*) Galloway–Mowat syndrome (*WDR73*)
Buphthalmos	Congenital glaucoma
Blepharophimosis	3MC syndrome
Blepharoptosis	Retraction syndrome of Duane 3MC syndromes Turner syndrome

(Continued)

TABLE 29.2 (Continued)

Findings	Related conditionsa
	Cardiofaciocutaneous (CFC) syndrome Intellectual disability with dysmorphic facies and ptosis; IDDDFP (*BRPF1*) Galloway–Mowat syndrome
Epicanthal folds	Cri du chat Down syndrome Galloway–Mowat syndrome Chorioretinal dysplasia, lymphedema, and microcephaly (*KIF11*)
Upslanting palpebral fissures	Down syndrome Prader–Willi syndrome
Downslanting palpebral fissures	Cri du chat CHARGE syndrome Coffin Lowry Coffin Siris Cardiofaciocutaneous syndrome Rubinstein–Taybi syndrome ZTTK Syndrome (*SON*)
Hypertelorism	Cardiofaciocutaneous syndrome Turner syndrome Cri du chat Apert syndrome Di George syndrome Galloway–Mowat syndrome
Hypotelorism	Holoprosencephaly Coffin–Siris syndrome Williams syndrome
Telecanthus	Cri du chat Wardenburg syndrome
Cataracts	Congenital rubella Down syndrome Galactosemia Other metabolic disorders
Aniridia	PAX6 disorders, e.g., WAGR syndrome
Heterochromia	Wardenburg syndrome
Coloboma (may involve iris, crystalline, choroid, retina, or optic nerve)	CHARGE syndrome Wolf–Hirschhorn syndrome
Cherry red spot	Tay–Sachs disease (GM2 gangliosidosis type 1) Sandhoff disease (GM2 gangliosidosis type 2) Niemann–Pick disease Farber disease Metachromatic leukodystrophy Mucolipidosis type 1 Galactosialidosis
Pigmented retinopathy	Retinitis pigmentosa with or without skeletal anomalies (*CWC27*) Chorioretinal dysplasia, lymphedema, and microcephaly
Optic atrophy	ZTTK syndrome Chorioretinal dysplasia, lymphedema, and microcephaly Peroxisomal biogenesis disorders

[a]*Many findings are nonspecific and conditions may not be exhaustive.*

Visual screening

Regular vision screening and intervention may reduce the risk of persistent amblyopia at 7 years old by more than half [10]. The American Academy of Pediatrics (AAP) Section on Ophthalmology produces multilateral recommendations for visual assessment by pediatricians. Universal to screening across the ages are ocular history, external exam, red reflex testing, and pupillary responses. Family history should be reviewed as congenital cataracts, retinoblastoma, and heritable disease involving the eye should prompt referral to an ophthalmologist. Children whose first-degree relatives had strabismus or amblyopia should be referred for assessment when able to cooperate with exam or sooner if symptoms are present. Strabismus in the neonatal period is not unusual and intermittent strabismus in early infancy is often seen as well. However, persistent horizontal strabismus after 4 months should prompt referral [11]. Assessment of ocular motility begins at 6–12 months then regularly thereafter. In assessing visual acuity, the fixate and follow response in infants should be present by 6 months. Developmentally appropriate optotypes can be used as early as 3 years old, though typically later.

Instrument-based screening should begin at 6–12 months or 1–3 years. The discrepancy in recommendation exists because of concerns about false-positive results and a low-likelihood of intervention. Reliability progressively improves with age after the first year of life with increases in readability and positive predictive values [12,13]. Yearly repeats should be done through age 5 years or until assessment with optotypes can be performed reliably. Instrument-based screening may detect amblyopia, high refractive error, and strabismus, the most common causes of visual impairment in children [14–16]. Though it is not a direct measure of visual acuity, it can be particularly helpful method of vision screening for those who are unable to participate in optotype testing.

Prematurity

Retinopathy of prematurity (ROP) is the leading cause of blindness in premature and low birth weight individuals. Decreases in neonatal mortality have inversely correlated with increased incidence of ROP. In 2012 the incidence of ROP was nearly 20%. Predictors for the development of ROP include female gender, birth weight, and gestational age [17].

Rates of severe ROP are inversely correlated with gestational age, increasing from 3% to 34% as gestational age decreases from 27 to $\leq$ 24 weeks [18]. Severe ROP can also be a predictor of nonvisual disability. In a cohort study at 5 years, the rates of motor, cognitive, and severe hearing impairments were three to four times more common than in those without severe ROP [19].

In ROP, normal angiogenesis is disrupted and the neovascularization that takes place can lead to edema, hemorrhage, and even retinal detachment. It is unclear what predicts resumption of normal angiogenesis or development of ROP after initial injury but the dynamics between insulin-like growth factor-1 and vascular endothelial growth factor (VEGF) have been implicated. ROP generally begins and ends between 34 and 45 weeks postmenstrual age. Most cases resolve spontaneously. However, severe untreated ROP can lead to long-term structural damage and poor visual acuity.

Current recommendations include screening for children with birthweight $\leq$ 1500 g or gestational age $\leq$ 30 weeks as well as those deemed high-risk by professional opinion to be screened [20]. While many cases may resolve spontaneously, treatment is initiated when an infant develops a threshold severity based on classification that includes the zone in relation to the disk, the stage of vascular change, and the extent of involvement.

Treatment modalities generally include laser photocoagulation or intravitreal injection of anti-VEGF agents. Infants treated with anti-VEGF therapies may require longer follow-up due to risk of late recurrence. The risk of strabismus, myopia, astigmatism, and anisometropia are increased in infants with ROP. Late retinal complications can occur in patients with ROP that did not meet criteria for treatment and thus follow-up may be warranted.

Ophthalmological problems associated with premature birth are not restricted to ROP. Errors of refraction, strabismus, cerebral visual impairment (CVI), deficits in color vision, contrast sensitivity, visual fields, and acuity have all been associated with prematurity. It is difficult to segregate risk factors and establish causality but these conditions are not universally secondary to ROP or known neurological insult [21].

Higher order visual processing and cerebral visual impairment

The eyes are thought of foremost when we think of vision but after the sensory information is transduced, it must be interpreted and processed in the brain. While the primary visual cortex lies in the occipital lobes, many other parts of the brain serve associated visual processes. These networks help in top-down processing to help us contextualize visual stimuli. As there is much more to vision than the eyes, there is much more to visual function than acuity and fields.

Visual processing is often separated into the dorsal and ventral streams. The dorsal stream is associated with directionality, stereopsis, and motion processing. The ventral stream is associated with perception and recognition. Individuals with William syndrome often demonstrate a particular pattern of impaired visuospatial abilities consistent with dorsal stream dysfunction [22,23]. Studies of other genetic and acquired developmental disorders have suggested a particular vulnerability of the dorsal stream while specific disorders of the ventral stream are rare [24–26]. However, many patients with higher order visual impairment are not limited to an either/or in this dichotomy and broader connectivity issues are implicated [27].

Like much of clinical neurology, function has been best studied by correlating acquired dysfunction with focal lesions. Acquired deficits described include inability to recognize faces (prosopagnosia), objects (semantic agnosia), movement (akinetopsia), words (alexia), or multiple objects simultaneously (simultanagnosia) just to name a few. These visuo-cognitive deficits have largely been described in previously healthy and mature individuals.

Identifying similar deficits in an individual with neurodevelopmental disabilities or a child who has not demonstrated prior mastery can be difficult. Assessing higher-order visual skills cannot be done with a single test. Therefore it is often a battery of measurements and those tests must be accessible and valid for individuals of different cognitive, communicative, and physical abilities. Developmental CVI is not always associated with classic localization derived from mature patients. In neonatal brain injury, the posterior putamen is more consistently implicated in abnormal visual performance when compared to occipital cortex or optic radiations [28] (Table 29.3).

CVI is a leading cause of significant visual impairment in the industrialized world and it is almost certainly underrecognized, particularly in the pediatric and NDD population [29–33]. CVI has been described as disturbances of vision or visual perception that is secondary to dysfunction or damage of visual association territories and their pathways posterior to the optic chiasm [34]. The retrochiasmatic visual apparatus is complex and responsible for a number of visual processes. When we discuss and diagnose CVI, then we are talking about a heterogenous group consisting of varying deficit(s) of varying etiologies.

The prevalence of CVI is increasing. This may be due in part to increased awareness but a significant driver is likely medical advances decreasing mortality and morbidityin neonatology. In a retrospective study, Huo et al. found that the

TABLE 29.3 Cerebral visual impairment.

Causative or related conditions that convey risk
Prematurity
Cerebral palsy
Severe epilepsy
Developmental delay
Stroke
Metabolic disorders
Periventricular leukomalacia
Hypoxic ischemic encephalopathy
Meningoencephalitis
Traumatic brain injury
Areas of atypical behavior elicited by history
Abnormal light response—gazing or aversion
Fixation and tracking issues
Preference for stationary versus moving objects
Preference for color versus black and white
Issues with contrast sensitivity and crowded fields
Visual field preference

four most common causes of CVI were perinatal hypoxia, cerebrovascular accident, meningitis, and acquired hypoxia. Other neurological abnormalities were common and most prevalent were epilepsy, cerebral palsy, hemiparesis, and hypotonia [35].

At current, there is a lack of consensus in diagnosing CVI. Common elements of diagnosis in the literature include some combination of medical history, assessments of ocular health and functional vision, visual perceptual tests, structured inventories, intelligence testing, neurodevelopmental assessment, clinical electrophysiology, and neuroimaging though often retrospective. McConnell et al. noted that the current diagnosis of exclusion approach is likely to lead to diagnostic overshadowing with attribution to preexisting impairment rather than CVI because isolated lesions are less common [36]. Multidisciplinary evaluations will continue to be relied upon in diagnosis and awareness creating an index of suspicion will be needed to ensure diagnosis and intervention.

Reading

Reading and vision are naturally linked and there are proponents that reading disability is associated with disorders of higher order visual processing. One such theory is the magnocellular theory of developmental dyslexia. Broadly speaking, the magnocellular system is responsible for processing high temporal frequency while the parvocellular system is responsible for high spatial resolution. The magnocellular deficit has been implicated in producing visual instability or impaired simultaneous recognition that interferes with reading [37,38]. That deficits of the magnocellular system contribute significantly to dyslexia has been refuted by many in the field [39–45]. Patterns of saccades and fixation in those with dyslexia have been noted to be similar to beginning readers but normalize when content is adjusted for ability [46,47]. When primary difficulties with decoding and comprehension interfere with fluent reading, there are increased fixations and backward saccades [48]. Issues with convergence insufficiency and poor accommodation may interfere with reading comfort and stamina but do not directly affect decoding [49]. Joint statements from leading professional societies like the AAP have been unequivocal that the evidence is insufficient for vision therapy as an intervention for dyslexia [50].

Hearing

Neurobiology of audition

To understand hearing and its impairment requires familiarity with the principal anatomy and physiology involved in audition and speech. Sound waves enter the external acoustic meatus and the signal is transduced via the ossicles to the cochlea where neural signals are generated. Just as the cochlea is organized in a tonotopic configuration, the primary auditory cortex is as well. Before the signal reaches the auditory cortex, there are a number of relays in the cochlear nuclei, superior olivary complex, inferior colliculus, and medial geniculate body. These neuronal relays help decode important information like duration, intensity, frequency, localization and even begin integration with other processes like motor response.

The brain generally has two systems for dealing with audition: one which focuses on attaching meaning to the sound and the other which is involved with speech. The cortical response in the temporal lobe is to identify and categorize the stimuli, cross modal matching, or pairing of visual and auditory stimuli, connect emotional responses, and with the hippocampus, aid in spatial navigation. Within the superior temporal gyrus (STG), activation in response to phonemes occurs in the left mid-STG and integration into words in the left anterior STG. There is a preference for phonetic sounds relative to artificial control sounds or environmental noise [51]. These signals are further decoded in categorical invariant fashion in the inferior frontal cortex. In the premotor cortex, the signals are transformed into articulatory representations. Further processing through the inferior parietal lobule and posterior superior temporal region where they are compared with auditory and other sensory information [52]. Generally, the auditory ventral pathway plays a role in auditory object recognition and the dorsal stream plays a role in sensorimotor integration.

Types of hearing loss

A basic knowledge of auditory structures and their function allows us to localize hearing deficits and inform intervention. Conceptually, it is important to consider hearing issues by their anatomical correlates, moving from the periphery to central structures. Conductive hearing loss occurs in the outer to middle ear, sensorineural hearing loss (SNHL) occurs in the inner ear (cochlea) or auditory nerve, and cortical deafness, or central auditory processing disorder (CAPD) are the most central deficits.

Conductive hearing loss

A conductive hearing loss occurs when there is interference with sound waves traveling through the outer or middle ear and impeding its signal to the inner ear. This type of loss may be transient (e.g., middle ear effusion) or permanent in the case of congenital malformation.

Conductive hearing loss due to congenital malformations may occur anywhere along the course of the auditory structures. Moving peripherally to centrally, conditions may include microtia, atretic canals, ossicular chain abnormalities, or malformations of the oval or round windows. These congenital issues may occur in isolation but should always prompt scrutiny.

Conductive hearing loss may be acquired in the cases of cerumen impaction, foreign bodies, tympanic membrane (TM) perforation, middle ear effusions, trauma, or cholesteatomas.

Sensorineural hearing loss

SNHL occurs with dysfunction of the cochlea and/or auditory nerve. Cases may be congenital, acquired, static, progressive, hereditary, environmental, or idiopathic.

Genetic etiologies may be responsible for 50%–60% of the cases of pediatric SNHL [53]. Hearing loss may be the lone feature in about 70% of genetic hearing loss while others are classified as syndromes [54]. More than 120 unique genes have been identified in nonsyndromic hearing loss including *GJB2* (connexin 26) and *STRC* (stereocilin) [55]. Examples that are not confined to the auditory system include Waardenburg (pigmentation), Pendred (thyroid), Usher (retinal/vestibular), Alport (renal), Jervell and Lange-Nielsen syndromes (arrhythmogenic), and more. As many as 19%–30% of cases thought to be isolated may actually be nonsyndromic mimics without readily apparent involvement of other organ systems [56–58]. It is therefore important to consider genetics as other organ symptoms may be involved, for example, thyroid dysfunction or cardiac dysrhythmias. Screening electrocardiograms should be considered in cases of severe-to-profound bilateral SNHL. Other etiologies included in congenital SNHL but nonhereditary may be congenital infection or dysplasias of the inner ear structures (Table 29.4).

Cases of congenital SNHL, both unilateral and bilateral, should prompt evaluation for congenital cytomegalovirus infection (cCMV). In pediatric hearing impairment, as many as 10%–20% of cases may be attributable to cCMV [59]. cCMV may be otherwise asymptomatic and the hearing loss vexingly progressive, sometimes allowing evasion of newborn screening. To establish the diagnosis, cytomegalovirus testing is performed on urine or saliva samples obtained in the first 3 weeks of life. polymerase chain reaction using dried blood spot may be done but sensitivity does suffer as not all infected newborns are viremic.

TABLE 29.4 Selected nonsyndromic mimics.

Inheritance	Condition	Features	Hearing loss			Pearls
			Type	Onset	Severity	
AR	Usher	Retinitis pigmentosa ± vestibular dysfunction	SN	Prelingual	Varies by subtype	Progressive vision loss beginning with nyctalopia
	Pendred	Goiter	SN	Prelingual	Severe-to-profound	Goiter often euthyroid Mondini malformation
	Jervell and Lange-Nielsen	Long QT	SN	Prelingual	Profound	Risk of tachyarrhythmia, sudden death. May have history of syncope.
AD	Waardenburg	Pigmentary abnormalities	SN	Prelingual	Variable	White forelock Heterochromia irides
	Branchiooto-renal Spectrum Disorders	Branchial fistulae/cysts Renal hypo-/aplasia Ear malformations	Mix	Variable	Variable	Variable expression

SNHL may also be acquired. There are a number of risk factors for acquired SNHL including noise exposure, ototoxic drugs, infection, trauma, or tumors. Prematurity is a risk factor for a number of reasons such as hyperbilirubinemia, infection, and/or need for ototoxic drugs.

Auditory neuropathy spectrum disorder (ANSD) refers to a hearing deficit with intact cochlear outer hair cells but a defect in inner hair cell or cochlear nerve function. This can be identified by normal otoacoustic emission (OAE) but abnormal auditory brainstem response (ABR). There is a lot of clinical and etiological heterogeneity within ANSD. The majority of which are suspected to be genetic with syndromic, nonsyndromic, and mitochondrial contributors [60].

Central auditory disorders

Central auditory disorders include cortical deafness and CAPD. Cortical deafness may be the consequence of brain injury be it diffuse or lesions from the cochlear nucleus to the primary auditory cortex. As with many developmental conditions, CAPD can be conceptually and diagnostically variable. The American Speech-Language-Hearing Association describes CAPD as a group of problems in processing auditory information within the central nervous system leading to demonstrable issues in sound localization and lateralization; auditory discrimination; auditory pattern recognition; temporal aspects of audition, including temporal integration, temporal discrimination (e.g., temporal gap detection), temporal ordering, and temporal masking; auditory performance in competing acoustic signals (including dichotic listening); and auditory performance with degraded acoustic signals. These deficits are not due to a deficit in higher order language, cognitive, or related factors [61]. Diagnosis must incorporate audiological assessment, speech-language abilities, and potentially other psychometrics. Treatment or intervention must be tailored to the individual given the heterogeneity of deficits and may include direct skill remediation, compensatory strategies, or environmental modifications.

Hearing assessment

There are a number of different methods to assess auditory status but it is important to know which components of audition are assessed by each of these methods.

Tympanometry uses a tone generator, microphone, air pump, and manometer to measure acoustic impedance at different pressures thus gleaning information on TM compliance and middle ear pressures. A seal is created in the external ear canal (EAC) and the pump insufflates or deflates to create pressure changes in the EAC. It is an important tool in the audiologist's assessment but is not a direct measure of audition. Tympanometry patterns can indicate middle ear effusions, eustachian tube dysfunction, or issues with the ossicular chain and acoustic reflex.

OAEs are sounds generated by the outer hair cells of the cochlea in response to sound. OAEs may also be utilized in newborn nurseries. They indicate normal cochlear function but not more.

OAEs are akin to a reflex arc to the cochlea while ABR reflect signal conduction that is retrocochlear or more central including the auditory nerve, cochlear nucleus, superior olivary complex, lateral lemniscus, and inferior colliculus. These tests measure the action potentials from the cochlear nerve in response to a click stimulus and thus may also be referred to as brainstem auditory evoked response (BAER). There are both screening and diagnostic forms. Automated auditory brainstem response (AABR) will generate a pass/fail response. Diagnostic ABRs provide quantitative data that require specialized interpretation. AABRs are often utilized in newborn nursery screening.

The standard for assessing functional hearing abilities is essentially a call and response. Pure tone audiometry tests standard frequencies from 250 to 8000 Hz at varying intensities or volume in decibels (dB). Thresholds are determined by the intensity at which the child can detect the tone 50% of the time. Ideally, this is tested using headphones in order to test each ear individually. The above methods are of course air conductive (AC) measures but bone conduction (BC) may be measured as well by placing an oscillator over the mastoid and bypassing the middle ear. Contrasting results of bone and air conduction measures can help localize the deficit and a disparity between them (BC > AC) indicates conductive hearing loss while equal hearing loss (BC = AC) indicates SNHL.

In young children and those with neurodevelopmental differences, comprehension and compliance may be difficult. For those with sensory issues, this may need to be done without headphones in what is referred to as a sound field measure. Audiologists then may report on thresholds "in at least the better ear" as they are unable to determine any interaural asymmetry.

Behavioral observation audiometry is a subjective measure by observing response to sound field stimuli. This is typically performed in children who are at a developmental equivalent of 6–8 months or have other disabilities that impair their ability to respond. The responses may be startle, cessation of behavior, or even more subtle. Visual reinforcement audiometry uses an intermittent reinforcement schedule to condition a child to turn toward audible stimuli. This can often be performed in children with a developmental age of 6–24 months of age.

Degree of hearing loss

The degree of hearing loss is based on the intensity required to perceive sound. This measure can be reported as decibels hearing loss (dB HL) (Table 29.5).

Early hearing detection and intervention

The Early Hearing Detection and Intervention (EHDI) Program is a federal initiative with programs in each state and territory. The 1-3-6 Benchmarks include "1" receiving a hearing screening by 1 month of age; "3" receiving a follow-up diagnostic test by 3 months of age if screening was not passed; and "6" enrollment in early intervention services by 6 months of age if diagnosed with permanent hearing loss. EHDI Program Coordinators annually report data via The Hearing Screening Follow-Up Survey to the Centers for Disease Control and Prevention. In 2019 nearly 6000 babies were identified as deaf or hard of hearing, more than twice the number identified in 2005.

Over that same time period, the reported incidence of hearing loss in those screened increased from 0.9 per 1000 to 1.7 per 1000. Notably, the number of failed screenings with no documented diagnosis decreased from 70.5% to 38% in that time. Documented enrollment in early intervention services for those with hearing loss increased from 57.7% to 61.7%.

These numbers are encouraging but there are still improvements to be made. Only 49.1% of those who failed their screen were diagnosed by 3 months of age in 2019 [62]. Delayed amplification or cochlear implantation is associated with poorer language outcomes [63]. Comprehensive data during the COVID-19 pandemic has not been released. It is anticipated that the pandemic have impacted each stage of screening, diagnosis, and intervention [64].

Intervention

The intervention for hearing loss is influenced by its permanency, severity, etiology, and age-of-onset. General principles include the involvement of a multidisciplinary clinical team and early intervention or local educational agencies. Clinical teams should include audiologists, speech-language pathologists, otolaryngologists, and often clinical geneticists and genetic counselors. Individuals with clinically significant hearing loss should also be referred to ophthalmologists because of an increased reliance on sight for communication. Co-occurring developmental disabilities in hearing loss are common.

TABLE 29.5 Degree of hearing loss and impact of hearing loss on communication.

Minimal hearing loss: Average hearing threshold levels in dB: 16–25 dB	• Difficulty with fast-paced verbal instruction, softly spoken or distant speech • Extra effort to follow spoken instruction may lead to fatigue • May present as attention and behavior problems
Mild hearing loss: Average hearing threshold levels in dB: 26–40 dB	• Background noise may create difficulty at conversational levels • Difficulty with portions of speech spectrum at conversational levels, particularly unstressed consonant sounds • Phonological and early reading skills may be impacted
Moderate hearing loss: Average hearing threshold levels in dB: 41–55 dB	• May miss 50%–80% of speech at conversational levels • Without amplification, may rely on visual cues • Absence of early amplification may impact speech, vocabulary, and syntax development
Moderate to severe hearing loss: Average hearing threshold levels in dB: 56–70 dB	• Heavy reliance on amplification and lip reading • The quality of speech production is likely to be affected
Severe hearing loss: Average hearing threshold levels in dB: 71–90 dB	• Benefit from amplification but may rely on lip reading and sign language for communication
Profound hearing loss: Average hearing threshold levels in dB: 91 + dB	• Benefits of amplification may be limited • Oral speech skills will be severely delayed

Source: Modified from Educational Audiology Association. Hearing loss counseling handouts. http://www.edaud.org.

Among individuals with hearing loss ≥ 40 dB unaided in the better ear, common conditions are intellectual disability (23%), cerebral palsy (10%), autism spectrum disorder (7%), and visual impairment (5%) [5].

Clinical intervention

Hearing loss amenable to surgical interventions tends to be conductive in nature though not uniformly. Tympanostomy tubes may aid in recurrent acute otitis media, middle ear effusions, or eustachian tube dysfunction. Surgical excision is required for tumors and cholesteatomas. Abnormalities of the ossicular chain may be amenable to amplification and/or surgical intervention. Perilymph fistulas, a cause of SNHL, may require a surgical patch if conservative approaches fail.

Amplification is the next consideration in cases that either are not candidates for surgical intervention, or whose issues were not resolved by surgery. Hearing aids may not normalize hearing deficits but should be expected to significantly augment hearing ability. Hearing aids come in different styles including behind-the-ear, in-the-ear, in-the-canal, or bone conduction devices. Behind-the-ear devices are most common for young children as the ear molds are easily revised with growth. Bone conduction devices are helpful for children with malformations of the ear that preclude the use of other devices or in cases of unilateral deafness. These devices can be held against the skull with a headband or surgically implanted. Children generally are not candidates for bone-anchored devices until around school age or they have developed sufficient skull bone thickness.

Some children may readily welcome hearing aids while others may actively resist wearing them. Considerations when a child is resistant to wearing their device are comfort and fit, volume and other detailed settings, or feedback issues. Parent training and, when available, behavioral psychology services should be part of the treatment plan to help with tolerance. Older children may resist hearing aids due to social stigma, perceived or real. Preparing children in how to respond to questions about their hearing device can help them confront these issues and foster self-confidence.

For those with limited benefit from hearing aid amplification, cochlear implants (CI) may be a consideration. CI transduce sound into electrical signals that stimulate the cochlear nerve. While they cannot fully replicate natural hearing, they are associated with improved speech and language skills [65]. Historically, delays in cognitive and adaptive skills have been used to deny CI, though individuals with CI show faster gains in cognitive, adaptive, language, and auditory skills as well as less family stress relative to their hearing aid only peers [66].

In cases of severe cochlear or auditory nerve dysfunction, congenital or acquired, an auditory brainstem implant (ABI) may help to improve environmental sound awareness and speech discrimination. Originally pioneered for individuals with *NF2*-related schwannomatosis (NF2), use has since expanded for pediatric and nontumor consideration. Individuals with NF2 have a predisposition to develop schwannomas. Vestibular schwannomas (acoustic neuromas) can lead to retrocochlear hearing loss due to direct invasion of the tumor or iatrogenic injury from surgery or radiation. Such individuals usually are not good candidates for CI given the lesion is more proximal to the intervention and there may be little to no residual function. ABI utilizes an external processor and internal implant. ABI systems are similar to CI in that they transduce sound into electrical signal, but ABI bypass the cochlea, transmitting electrical signal directly to the second order neurons in the cochlear nucleus complex [67]. Candidates for ABI, in addition to individuals with NF2, include those with cochlear or cochlear nerve aplasia and postmeningitic cochlear ossification. Nearly half of nontumor pediatric patients may develop the ability to understand common phrases without lip reading or better, though individuals with concurrent nonauditory disabilities may not reach equivalent outcomes [68].

Amplification and cochlear implants are an effective means of circumventing hearing deficits but there are many efforts to directly address specific biological deficits. Monogenic etiologies of SNHL make relatively good candidates for specific gene therapies. There are numerous preclinical efforts utilizing the principles of gene replacement, suppression, and editing that have shown varying degrees of promise [69]. There are also nonspecific efforts to generate hair cells using cell-based therapies and transcription factor-mediated reprogramming [70,71]. Translational studies are highly anticipated but candidacy will be very selective with highly specific targets. Otolaryngologists' recommendations for intervention should not be deferred for the prospect of gene therapy.

Educational intervention

Early intervention and local education agencies also play an integral role in supporting children who are deaf or hard of hearing. Early intervention services help families understand their child's hearing loss and communication needs as well as foster development in these realms. Based on an individual's needs, schools may emphasize auditory/oral, cued speech, manual or total communication development. Cued speech utilizes eight hand gestures to represent consonant phonemes and four vowel group positions that help distinguish sounds that look alike on the lips. Total communication is a philosophy that incorporates a variety of communication techniques such as spoken language, signs and gestures, and lipreading. Updated rates of American Sign Language use among children are needed but recent assessments of adult use are substantial, approaching 3%. The vast majority of these individuals have good-to-excellent hearing acuity. The rate of sign language usage was almost sevenfold greater (more than 23%) among deaf respondents than any other hearing acuity group [72].

Classroom supports and services include environmental modifications, staff support, assistive technology and adaptations of information delivery or assessment. Environmental modifications may include seating preferences that help in the classroom as well as the cafeteria or bus. Staff may serve as aides, interpreters, or transcriptionists. Hearing assistive technology such as wireless or FM systems can help individuals with residual hearing abilities who otherwise may miss spoken instruction due to competing background noise or insufficient volume.

Local educational agencies and parents must emphasize the individualization in developing an Individualized Education Program for a child who is deaf or hard of hearing. Considerations must be given to a child's mode of expressive and receptive communication, their language attainment compared their hearing peers, their ability to communicate with their peers and educators using their preferred mode of communication, and their academic level. The provision of free and appropriate public education should be based on the child's unique needs and the least restrictive environment may range from the regular classroom to a center or school tailored to the needs of deaf and hard of hearing individuals.

Debate persists around the optimal approach to early language and communication instruction with deaf children [73]. There is evidence that children with cochlear implants and no sign language exposure fare better on spoken language and reading metrics when compared to their peers with sign language exposure [74]. The Deaf community may not even consider those as being valid outcome measures when deafness is viewed as an ethnolinguistic identity rather than disability. Among the many variables to be considered when trying to determine the most successful interventions consideration are is the outcomes among for deaf children of deaf parents versus deaf children of hearing parents. The current state of evidence should inform but not dictate approaches and they should be personalized and incorporate shared decision making.

The hearing world can and should do better to include deaf culture rather than coerce the deaf and hard of hearing (D/HH) community to assimilate. Deaf culture is "not simply a camaraderie" but a cultural identity among individuals with shared language and history [75]. Ethical issues in consent and research exist and contribute to disparities in service delivery and the advancement of research for the d/Deaf community [76]. Some institutions have begun explicit training in Deaf culture to ameliorate this problem [77]. Clinicians, scientists, and educators should work to increase sensitivity and awareness regarding the Deaf community to combat this risk.

Conclusion

Recognizing differences in hearing and vision quality is crucial in providing individuals with the respect they deserve. Individuals with NDDs are at increased risk for delay or under recognition of atypical hearing and vision status. Screening, diagnosing, and addressing hearing and vision differences can alter academic performance, social and emotional development, adaptive performance, and most importantly, quality of life. As such, stakeholders should be strong advocates for the appropriate assessment of hearing and vision in all youth and particularly those with NDDs.

References

[1] Novak I, Hines M, Goldsmith S, Barclay R. Clinical prognostic messages from a systematic review on cerebral palsy. Pediatrics 2012;130(5): e1285–312. Available from: https://doi.org/10.1542/peds.2012-0924.

[2] Flaxman AD, Wittenborn JS, Robalik T, et al. Prevalence of visual acuity loss or blindness in the US: a Bayesian meta-analysis. JAMA Ophthalmol 2021;139(7):717–23. Available from: https://doi.org/10.1001/jamaophthalmol.2021.0527.

[3] Burton MJ, Ramke J, Marques AP, et al. The Lancet Global Health Commission on global eye health: vision beyond 2020. Lancet Glob Health 2021;9(4):e489–551. Available from: https://doi.org/10.1016/S2214-109X(20)30488-5.

[4] Shelton E, Robich M, VanNasdale D. Assessing the role of social determinants of health in children receiving a vision test using national data from the National Survey of Children's Health. Invest Ophthalmol Vis Sci 2022;63(7):2154–0182.

[5] Van Naarden Braun K, Christensen D, Doernberg N, et al. Trends in the prevalence of autism spectrum disorder, cerebral palsy, hearing loss, intellectual disability, and vision impairment, metropolitan atlanta, 1991-2010. PLoS ONE 2015;10(4):e0124120. Available from: https://doi.org/10.1371/journal.pone.0124120.

[6] Osorio MJ, Hertle RW, Painter M, Hinch K. Pupillary light reflex in premature infants prior to 30 weeks postmenstrual age. J AAPOS 2009;13:608–9.

[7] Robinson J, Fielder AR. Pupillary diameter and reaction to light in preterm neonates. Arch Dis Child 1990;65:35–8.

[8] Chandna A. Natural history of the development of visual acuity in infants. Eye 1991;5(Pt 1):20–6. Available from: https://doi.org/10.1038/eye.1991.4.

[9] Leat SJ, Yadav NK, Irving EL. Development of visual acuity and contrast sensitivity in children. J Optom 2009;2(1):19–26. Available from: https://doi.org/10.3921/joptom.2009.19.

[10] de Koning HJ, Groenewoud JH, Lantau VK, et al. Effectiveness of screening for amblyopia and other eye disorders in a prospective birth cohort study. J Med Screen 2013;20(2):66–72. Available from: https://doi.org/10.1177/0969141313497355.

[11] Birch E, Stager D, Wright K, Beck R. The natural history of infantile esotropia during the first six months of life. Pediatric Eye Disease Investigator Group. J AAPOS 1998;2(6):325–9. Available from: https://doi.org/10.1016/s1091-8531(98)90026-x.

[12] Donahue SP, Johnson TM. Age-based refinement of referral criteria for photoscreening. Ophthalmology 2001;108(12):2309–15. Available from: https://doi.org/10.1016/s0161-6420(01)00810-7.

[13] Longmuir SQ, Pfeifer W, Leon A, Olson RJ, Short L, Scott WE. Nine-year results of a volunteer lay network photoscreening program of 147 809 children using a photoscreener in Iowa. Ophthalmology 2010;117(10):1869–75. Available from: https://doi.org/10.1016/j.ophtha.2010.03.036.

[14] Chia A, Lin X, Dirani M, et al. Risk factors for strabismus and amblyopia in young Singapore Chinese children. Ophthalmic Epidemiol 2013;20(3):138–47. Available from: https://doi.org/10.3109/09286586.2013.767354.

[15] Donahue SP, Nixon CN. Section on Opthamology, American Academy of Pediatrics; visual system assessment in infants, children, and young adults by pediatricians. Pediatrics 2016;137(1):28–30. Available from: https://doi.org/10.1542/peds.2015-3596.

[16] Donnelly UM. Horizontal strabismus worldwide–what are the risk factors? Ophthalmic Epidemiol 2012;19(3):117–19. Available from: https://doi.org/10.3109/09286586.2012.681002.

[17] Ludwig CA, Chen TA, Hernandez-Boussard T, Moshfeghi AA, Moshfeghi DM. The epidemiology of retinopathy of prematurity in the United States. Ophthalmic Surg Lasers Imaging Retina 2017;48(7):553–62. Available from: https://doi.org/10.3928/23258160-20170630-06.

[18] Darlow BA, Hutchinson JL, Henderson-Smart DJ, et al. Prenatal risk factors for severe retinopathy of prematurity among very preterm infants of the Australian and New Zealand Neonatal Network. Pediatrics 2005;115(4):990–6. Available from: https://doi.org/10.1542/peds.2004-1309.

[19] Schmidt B, Davis PG, Asztalos EV, Solimano A, Roberts RS. Association between severe retinopathy of prematurity and nonvisual disabilities at age 5 years. JAMA 2014;311(5):523–5. Available from: https://doi.org/10.1001/jama.2013.282153.

[20] Fierson WMAmerican Academy of Pediatrics Section on OphthalmologyAmerican Academy of OphthalmologyAmerican Association for Pediatric Ophthalmology and StrabismusAmerican Association of Certified Orthoptists. Screening examination of premature infants for retinopathy of prematurity. Pediatrics 2018;142(6):e20183061. Available from: https://doi.org/10.1542/peds.2018-3810.

[21] O'Connor AR, Wilson CM, Fielder AR. Ophthalmological problems associated with preterm birth. Eye 2007;21(10):1254–60. Available from: https://doi.org/10.1038/sj.eye.6702838.

[22] Atkinson J, King J, Braddick O, Nokes L, Anker S, Braddick F. A specific deficit of dorsal stream function in Williams' syndrome. Neuroreport 1997;8(8):1919–22. Available from: https://doi.org/10.1097/00001756-199705260-00025.

[23] Bellugi U, Lichtenberger L, Mills D, Galaburda A, Korenberg JR. Bridging cognition, the brain and molecular genetics: evidence from Williams syndrome. Trends Neurosci 1999;22(5):197–207. Available from: https://doi.org/10.1016/s0166-2236(99)01397-1.

[24] Dalrymple KA, Corrow S, Yonas A, Duchaine B. Developmental prosopagnosia in childhood. Cogn Neuropsychol 2012;29(5-6):393–418. Available from: https://doi.org/10.1080/02643294.2012.722547.

[25] Atkinson J. The Davida Teller Award Lecture, 2016: visual brain development: a review of "dorsal stream vulnerability"-motion, mathematics, amblyopia, actions, and attention. J Vis 2017;17(3):26. Available from: https://doi.org/10.1167/17.3.26.

[26] Braddick O, Atkinson J, Wattam-Bell J. Normal and anomalous development of visual motion processing: motion coherence and 'dorsal-stream vulnerability'. Neuropsychologia 2003;41(13):1769–84. Available from: https://doi.org/10.1016/s0028-3932(03)00178-7.

[27] Martín MB, Santos-Lozano A, Martín-Hernández J, et al. Cerebral versus ocular visual impairment: the impact on developmental neuroplasticity. Front Psychol 2016;7:1958. Available from: https://doi.org/10.3389/fpsyg.2016.01958.

[28] Mercuri E, Atkinson J, Braddick O, et al. Basal ganglia damage and impaired visual function in the newborn infant. Arch Dis Child Fetal Neonatal Ed 1997;77(2):F111–14. Available from: https://doi.org/10.1136/fn.77.2.f111.

[29] Rahi JS, Cable NBritish Childhood Visual Impairment Study Group. Severe visual impairment and blindness in children in the UK. Lancet 2003;362(9393):1359–65. Available from: https://doi.org/10.1016/S0140-6736(03)14631-4.

[30] Boonstra N, Limburg H, Tijmes N, van Genderen M, Schuil J, van Nispen R. Changes in causes of low vision between 1988 and 2009 in a Dutch population of children. Acta Ophthalmol 2012;90(3):277–86. Available from: https://doi.org/10.1111/j.1755-3768.2011.02205.x.

[31] Hatton DD, Schwietz E, Boyer B, Rychwalski P. Babies count: the national registry for children with visual impairments, birth to 3 years. J AAPOS 2007;11(4):351–5. Available from: https://doi.org/10.1016/j.jaapos.2007.01.107.

[32] Flanagan NM, Jackson AJ, Hill AE. Visual impairment inchildhood: insights from a community-based survey. Child Care Health Dev. 2003;29(6):493–9. Available from: https://doi.org/10.1046/j.1365-2214.2003.00369.x.

[33] Kong L, Fry M, Al-Samarraie M, Gilbert C, Steinkuller PG. An update on progress and the changing epidemiology of causes of childhood blindness worldwide. J AAPOS 2012;16(6):501–7. Available from: https://doi.org/10.1016/j.jaapos.2012.09.004.

[34] Philip SS, Dutton GN. Identifying and characterising cerebral visual impairment in children: a review. Clin Exp Optom 2014;97(3):196–208. Available from: https://doi.org/10.1111/cxo.12155.

[35] Huo R, Burden SK, Hoyt CS, Good WV. Chronic cortical visual impairment in children: aetiology, prognosis, and associated neurological deficits. Br J Ophthalmol 1999;83(6):670–5. Available from: https://doi.org/10.1136/bjo.83.6.670.

[36] McConnell EL, Saunders KJ, Little JA. What assessments are currently used to investigate and diagnose cerebral visual impairment (CVI) in children? A systematic review. Ophthalmic Physiol Opt 2021;41(2):224–44. Available from: https://doi.org/10.1111/opo.12776.

[37] Stein J. The magnocellular theory of developmental dyslexia. Dyslexia 2001;7(1):12–36. Available from: https://doi.org/10.1002/dys.186.

[38] Werth R. Is developmental dyslexia due to a visual and not a phonological impairment? Brain Sci 2021;11(10):1313. Available from: https://doi.org/10.3390/brainsci11101313.

[39] Skottun BC, Parke LA. The possible relationship between visual deficits and dyslexia: examination of a critical assumption. J Learn Disabil 1999;32(1):2–5. Available from: https://doi.org/10.1177/002221949903200101.

[40] Vellutino FR, Fletcher JM, Snowling MJ, Scanlon DM. Specific reading disability (dyslexia): what have we learned in the past four decades? J Child Psychol Psychiatry. 2004;45(1):2–40. Available from: https://doi.org/10.1046/j.0021-9630.2003.00305.x.

[41] Victor JD, Conte MM, Burton L, Nass RD. Visual evoked potentials in dyslexics and normals: failure to find a difference in transient or steady-state responses. Vis Neurosci 1993;10(5):939–46. Available from: https://doi.org/10.1017/s0952523800006155.

[42] Victor JD. Defective visual pathway in reading-disabled children. N Engl J Med 1993;329(8):579. Available from: https://doi.org/10.1056/NEJM199308193290815.

[43] Conlon E, Sanders M, Zapart S. Temporal processing in poor adult readers. Neuropsychologia 2004;42(2):142–57.

[44] Skottun BC, Skoyles JR. Attention, dyslexia, and the line-motion illusion. Optom Vis Sci 2006;83(11):843–9. Available from: https://doi.org/10.1097/01.opx.0000233645.60607.0b.

[45] Skottun BC, Skoyles J. Yellow filters, magnocellular responses, and reading. Int J Neurosci 2007;117(2):287–93. Available from: https://doi.org/10.1080/0020745050053407610.1016/j.neuropsychologia.2003.07.004.

[46] Beauchamp GR, Kosmorsky G. Learning disabilities: update comment on the visual system. Pediatr Clin North Am 1987;34(6):1439–46. Available from: https://doi.org/10.1016/s0031-3955(16)36366-0.

[47] Olitsky SE, Nelson LB. Reading disorders in children. Pediatr Clin North Am 2003;50(1):213–24. Available from: https://doi.org/10.1016/s0031-3955(02)00104-9.

[48] Hoyt CS. Visual training and reading. Am Orthop J 1999;49:23–5. Available from: https://doi.org/10.1080/0065955X.1999.11982187.

[49] Granet DB, Castro EF, Gomi CF. Reading: do the eyes have it? Am Orthop J 2006;56(1):44–9. Available from: https://doi.org/10.3368/aoj.56.1.44.

[50] Handler SM, Fierson WM, Section on Ophthalmology, et al. Learning disabilities, dyslexia, and vision. Pediatrics 2011;127(3):e818–56. Available from: https://doi.org/10.1542/peds.2010-3670.

[51] DeWitt I, Rauschecker JP. Phoneme and word recognition in the auditory ventral stream. Proc Natl Acad Sci USA 2012;109(8):E505–14. Available from: https://doi.org/10.1073/pnas.1113427109.

[52] Rauschecker JP. Ventral and dorsal streams in the evolution of speech and language. Front Evol Neurosci 2012;4:7. Available from: https://doi.org/10.3389/fnevo.2012.00007.

[53] Sheffield AM, Smith RJH. The epidemiology of deafness. Cold Spring Harb Perspect Med 2019;9(9):a033258. Available from: https://doi.org/10.1101/cshperspect.a033258.

[54] Kalatzis V, Petit C. The fundamental and medical impacts of recent progress in research on hereditary hearing loss. Hum Mol Genet 1998;7(10):1589–97. Available from: https://doi.org/10.1093/hmg/7.10.1589.

[55] Van Camp G, Smith R. Hereditary hearing loss. https://hereditaryhearingloss.org/. Updated 2021; (accessed January 1, 2023).

[56] Shearer AE, Shen J, Amr S, Morton CC, Smith RJNewborn Hearing Screening Working Group of the National Coordinating Center for the Regional Genetics Networks. A proposal for comprehensive newborn hearing screening to improve identification of deaf and hard-of-hearing children [published correction appears in Genet Med. 2019 Jun 18;:]. Genet Med 2019;21(11):2614–30. Available from: https://doi.org/10.1038/s41436-019-0563-5.

[57] Vona B, Nanda I, Hofrichter MA, Shehata-Dieler W, Haaf T. Non-syndromic hearing loss gene identification: A brief history and glimpse into the future. Mol Cell Probes 2015;29(5):260–70. Available from: https://doi.org/10.1016/j.mcp.2015.03.008.

[58] Xiang J, Jin Y, Song N, et al. Comprehensive genetic testing improves the clinical diagnosis and medical management of pediatric patients with isolated hearing loss. BMC Med Genom 2022;15(1):142. Available from: https://doi.org/10.1186/s12920-022-01293-x.

[59] Goderis J, De Leenheer E, Smets K, Van Hoecke H, Keymeulen A, Dhooge I. Hearing loss and congenital CMV infection: a systematic review. Pediatrics 2014;134(5):972–82. Available from: https://doi.org/10.1542/peds.2014-1173.

[60] Manchaiah VK, Zhao F, Danesh AA, Duprey R. The genetic basis of auditory neuropathy spectrum disorder (ANSD). Int J Pediatr Otorhinolaryngol 2011;75(2):151–8. Available from: https://doi.org/10.1016/j.ijporl.2010.11.023.

[61] Working Group on Auditory Processing Disorders. (Central) Auditory processing disorders—the role of the audiologist. American Speech-Language-Hearing Association. April 2005; (accessed December 20, 2022). https://www.asha.org/policy/ps2005-00114/.

[62] Centers for Disease Control and Prevention. 2020 summary of national CDC EHDI data. National center on birth defects and developmental disabilities: hearing loss in children. September 29, 2022; (accessed November 30, 2022). https://www.cdc.gov/ncbddd/hearingloss/2020-data/01-data-summary.html.

[63] Ching TYC, Dillon H, Button L, et al. Age at intervention for permanent hearing loss and 5-year language outcomes. Pediatrics 2017;140(3):e20164274. Available from: https://doi.org/10.1542/peds.2016-4274.

[64] Jenks CM, DeSell M, Walsh J. Delays in infant hearing detection and intervention during the COVID-19 pandemic: commentary. Otolaryngol Head Neck Surg 2022;166(4):603–4. Available from: https://doi.org/10.1177/01945998211067728.

[65] Tomblin JB, Spencer L, Flock S, Tyler R, Gantz B. A comparison of language achievement in children with cochlear implants and children using hearing aids. J Speech Lang Hear Res 1999;42(2):497–509. Available from: https://doi.org/10.1044/jslhr.4202.497.

[66] Oghalai JS, Bortfeld H, Feldman HM, Chimalakonda N, Emery C, Choi JS, et al. Cochlear implants for deaf children with early developmental impairment. Pediatrics 2022;149(6). Available from: https://doi.org/10.1542/peds.2021-055459 e2021055459.

[67] Wong K, Kozin ED, Kanumuri VV, Vachicouras N, Miller J, Lacour S, et al. Auditory brainstem implants: recent progress and future perspectives. Front Neurosci 2019;13:10. Available from: https://doi.org/10.3389/fnins.2019.00010.

[68] Noij KS, Kozin ED, Sethi R, Shah PV, Kaplan AB, Herrmann B, et al. Systematic review of nontumor pediatric auditory brainstem implant outcomes. Otolaryngol Head Neck Surg 2015;153(5):739−50. Available from: https://doi.org/10.1177/0194599815596929.

[69] Omichi R, Shibata SB, Morton CC, Smith RJH. Gene therapy for hearing loss. Hum Mol Genet 2019;28(R1):R65−79. Available from: https://doi.org/10.1093/hmg/ddz129.

[70] Iyer AA, Hosamani I, Nguyen JD, et al. Cellular reprogramming with ATOH1, GFI1, and POU4F3 implicate epigenetic changes and cell-cell signaling as obstacles to hair cell regeneration in mature mammals. Elife 2022;11:e79712. Available from: https://doi.org/10.7554/eLife.79712.

[71] Bergman JE, Davies C, Denton AJ, Ashman PE, Mittal R, Eshraghi AA. Advancements in stem cell technology and organoids for the restoration of sensorineural hearing loss. J Am Acad Audiol 2021;32(10):636−45. Available from: https://doi.org/10.1055/s-0041-1728677.

[72] Mitchell RE, Young TA. How many people use sign language? A national health survey-based estimate. J Deaf Stud Deaf Educ 2022;28(1):1−6. Available from: https://doi.org/10.1093/deafed/enac031.

[73] White KR, Cooper LZ. Opportunities and shared decision-making to help children who are deaf to communicate. Pediatrics 2017;140(1): e20171287. Available from: https://doi.org/10.1542/peds.2017-1287.

[74] Geers AE, Mitchell CM, Warner-Czyz A, Wang NY, Eisenberg LSCDaCI Investigative Team. Early sign language exposure and cochlear implantation benefits. Pediatrics 2017;140(1):e20163489. Available from: https://doi.org/10.1542/peds.2016-3489.

[75] Padden C, Humphries T. Deaf in America. Harvard University Press; 1988.

[76] McKee M, Schlehofer D, Thew D. Ethical issues in conducting research with deaf populations. Am J Public Health 2013;103(12):2174−8. Available from: https://doi.org/10.2105/AJPH.2013.301343.

[77] Epstein S, Johnson LM, Sie KCY, Norton SJ, Ou HC, Horn DL. Sensitivity to deaf culture among otolaryngology and audiology trainees [published online ahead of print, 2022 Jul 13]. Ann Otol Rhinol Laryngol 2022;. Available from: https://doi.org/10.1177/00034894221111248.

Chapter 30

Systemic diseases with increased risk of neurodevelopmental disabilities

Darrah Haffner[1], Jessica Durkee-Shock[2], Tanjala T. Gipson[3], Meghan O'Neill[4] and Eliza Gordon-Lipkin[5]

[1]*Division of Pediatric Neurology, Nationwide Children's Hospital, The Ohio State University, Columbus, OH, United States,* [2]*Medical Virology Section, National Institute of Allergy and Infectious Diseases, National Institutes of Health, Bethesda, MD, United States,* [3]*The University of Tennessee Health Science Center, Boling Center for Developmental Disabilities, Le Bonheur Children's Hospital, Memphis, TN, United States,* [4]*Department of Developmental Behavioral Pediatrics, Ann & Robert H. Lurie Children's Hospital of Chicago, Chicago, IL, United States,* [5]*Metabolism, Infection and Immunity Section, National Human Genome Research Institute, National Institutes of Health, Bethesda, MD, United States*

Introduction

Human physiology relies on a network of interwoven systems that impact human function. The cardiovascular, respiratory, gastrointestinal, endocrine, immune, renal, nervous systems, and others all rely on one another to orchestrate life. Through this complex interplay, dysfunction of a single system may impact another. As the nervous system plays a key role in human cognition, communication, and mobility, changes or disruptions in its function may lead to disability. Well-known examples of secondary neurologic dysfunction include stroke, where a cardiovascular insult causes focal brain injury and, hepatic encephalopathy, where liver dysfunction causes ammonia accumulation and altered mental status.

Neurodevelopment also depends on a complicated orchestration of cellular and biochemical processes as neuronal cell layers migrate, synapses form and prune, axons are myelinated, and glia form neuronal support networks. Disruption of any or many of these processes may result in neurodevelopmental disabilities (NDDs). NDDs are a group of conditions that are caused by neurologic dysfunction during early periods of brain development and lead to varying degrees of long-term disability. NDD may manifest as impairments in cognitive, motor, sensory, social, and behavioral function and key examples include intellectual disability, cerebral palsy, autism spectrum disorder, attention-deficit/hyperactivity disorder (ADHD), learning disabilities, communication disorders, and hearing or vision impairment.

When considering NDD in the context of systemic disease, we recognize that disruptions in systemic physiology may have downstream effects on neurodevelopment during critical developmental windows. This secondary neurologic injury may occur by several mechanisms (Table 30.1). These include deficiency of vital elements (e.g., hypoxia, malabsorption), neurotoxicity by biochemical intermediates (e.g., uremia, hyperammonemia), neurotoxicity of therapy (e.g., chemotherapy or radiation), shared embryologic origins (e.g., neurocutaneous disorders), and vulnerability to environmental insults (e.g., primary immunodeficiencies).

Perturbations in neurodevelopment may also occur temporally either as a single event or as recurrent/chronic events over time (Table 30.1). A single insult may have long lasting consequences by changing the trajectory of CNS development, but does not continually affect neurophysiology over time. In contrast, recurrent or chronic systemic disease may continuously affect the nervous system, leading to NDD that may evolve as the systemic disease evolves.

NDDs associated with systemic diseases may also have a broad range of manifestations. This spectrum of manifestations may impact different functions of the nervous system and may range from mild to severe (Fig. 30.1). Additionally, while neurologic injury may occur early in brain development, NDD may not be detected until late childhood or sometimes adulthood. While NDDs such as cerebral palsy or severe/profound intellectual disability are frequently symptomatic in the first year of life, other NDDs such as learning disabilities, executive function issues, or minor neuromotor abnormalities may not manifest until school age or beyond. Some may be asymptomatic or

Capute and Accardo's Neurodevelopmental Disabilities in Infancy and Childhood. DOI: https://doi.org/10.1016/B978-0-12-824060-1.00023-7

TABLE 30.1 Key examples by which different mechanisms of systemic dysfunction may cause neurodevelopmental disabilities (NDDs).

	Single event	Recurrent or chronic
Deficiency of vital elements	Hypoxic ischemic encephalopathy (oxygen deprivation)	Congenital hypothyroidism (thyroid hormone deficiency) Complex congenital heart disease (decreased blood oxygenation)
Toxicity by inappropriate accumulation of biochemical intermediates	Kernicterus (accumulation of bilirubin)	Inborn errors of metabolism (e.g., Phenylketonuria (PKU); accumulation of phenylalanine)
Toxicity of therapy	Aminoglycosides (gentamicin-associated hearing loss)	Chemotherapy (methotrexate-associated leukoencephalopathy)
Shared embryologic origins	Neural tube defects (disruption of closure of the neural tube)	Neurocutaneous disorders (skin and neural tissue both arising from ectoderm)
Vulnerability to environmental insults	Immunodeficiency (vulnerability to neuroinfectious disease)	

Nonexhaustive.

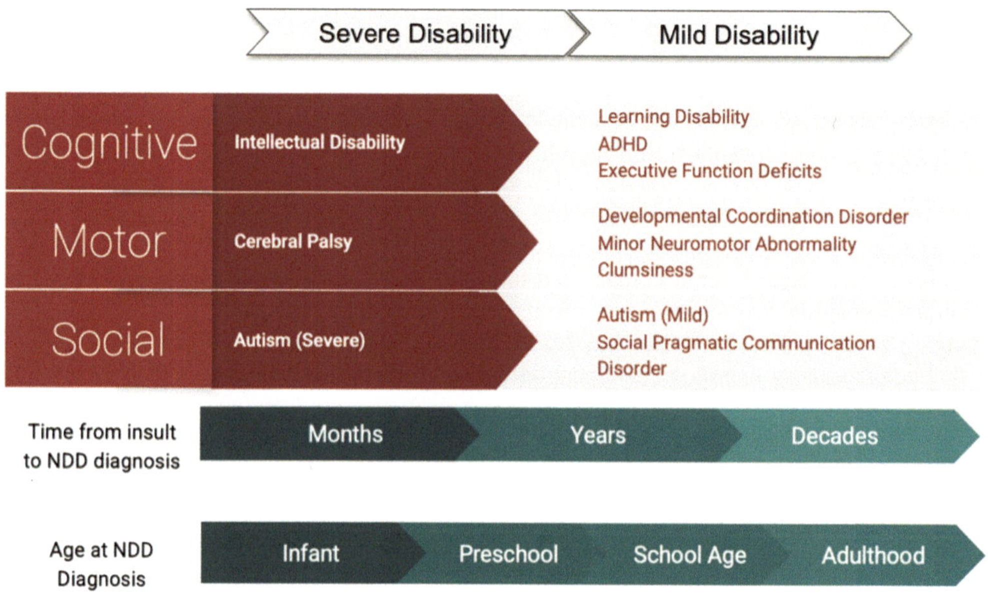

FIGURE 30.1 The spectrum of neurodevelopmental disabilities (NDDs) associated with systemic disease.

misattributed to the sociodynamic stressors of chronic disease and require investigation, such as neuropsychological testing, to identify.

The following chapter will take a systems-based approach to understand the relationship between chronic systemic disease and NDD with key examples from each system. This chapter will address conditions that primarily affect physiology *outside* of the nervous system. While many genetic or metabolic conditions have both systemic and neurodevelopmental manifestations, these are primarily addressed in other chapters of this text.

Congenital heart disease

Each year, 1 in 100 infants are born with congenital heart disease (CHD) [1,2]. Moderate to severe CHD occurs in 6–8 infants/1000 live births [2]. Over the last few decades, survival has significantly increased [3]. There has been increasing emphasis on the long-term neurodevelopmental outcomes of survivors. Up to half of all infants with CHD have some degree of neurodevelopmental impairment [4,5].

CHD's impact on the brain begins in fetal life. Aberrant systemic circulation in congenital heart disease may result in reduced blood flow and subsequently oxygen to the brain [6]. Imaging studies report delayed brain maturation compared to age-matched peers [7]. Grossly, fetuses and neonates with CHD are more likely microcephalic [8]. Smaller head sizes were associated with CHD types with compromised blood delivery to the brain such as transposition of the great arteries (TGA) and left outflow tract obstruction [9]. Total brain volume is smaller in third trimester fetuses with more severe forms of CHD than fetuses without CHD; these differences increase as the third trimester progresses [10]. Magnetic resonance (MR) spectroscopy and diffusion tensor imaging at term demonstrate delayed brain metabolism and microstructure compared to normal controls, with an equivalent maturation to preterm infants born at 34–36 weeks gestation [11,12]. On a macroscopic level, delays in sulcation, myelination, germinal matrix involution, and migration of glial cells correlate with a 4- to 5-week delay in maturation [12]. These structural delays correlate with delayed oligodendrocyte maturation [12]. Early work points to altered functional brain connectivity even prior to surgery [13]. All of these differences may be compounded by underlying syndromes or genetic disorders.

In addition to delayed maturations, infants with CHD are at high risk of acquired injury both before and after surgery [11,14]. Preoperative injury is reported in 17%–41% of term infants with CHD and 25%–67% postoperatively [11,14,15]. In comparison, Miller et al. reported all control neonates had normal scans [11]. Full-term infants with CHD have a high incidence of white matter injury similar to the white matter injury and periventricular leukomalacia seen in preterm infants [11,12]. Risk factors for preoperative brain injury include balloon atrial septostomy [14]. Risk factors for postoperative brain injury include cardiopulmonary bypass and low mean blood pressures in the immediate postoperative period [14].

Brain volumes remain smaller into adolescence and adulthood, particularly in those with cyanotic CHD [16]. Decreased volume correlated with worse cognitive, motor, and executive functions [16]. Myelination remains abnormal into adolescence [17]. There is emerging evidence for persistent abnormalities across multiple resting state networks, including those integral to executive function [18]. Persistent white matter injury and volume loss in adolescence correlates with working memory [19].

Neurologic and motor impairments are common in early childhood with up to 29% of children affected, most commonly hypotonia and gross motor delay [20]. In general, children with surgically repaired CHD have similar intellectual functioning to the general population [21]. However, there are specific types of CHD with significant differences. Multiple studies have reported that children with single ventricle physiology and/or cyanotic CHD tend to have lower intelligence quotients (IQs), with some studies reporting a mean difference of 10 points [22]. Mild cognitive impairment, impaired social interaction and impairments in core communication skills, inattention and impulsivity, and impaired executive functioning are common [23–25]. Specific deficits in motor planning, visuo-spatial skills, and higher order language are also common [12]. These differences persist in a significant proportion, who ultimately may need specialized services into adulthood [12].

Increased risk for neurodevelopmental impairment is associated with a host of different factors including the direct influence of the CHD on the brain, pre- and postoperative conditions, and the environmental milieu throughout childhood (Table 30.2). Multiple studies report the association with palliative surgery versus corrective surgery, microcephaly, and deep hypothermic arrest time [20,26]. Cooccurring genetic and extra-cardiac anomalies further increase risk of neurodevelopmental disabilities [27].

In 2012 the American Heart Association (AHA) with the endorsement of the American Academy of Pediatrics (AAP) published guidelines for optimizing neurodevelopmental outcomes in pediatric CHD [28]. The guidelines state "surveillance, screening, evaluation, and re-evaluation of developmental disability and developmental delays in the pediatric CHD population are essential steps to obtain appropriate interventions to maximize these children's potential

TABLE 30.2 Risk factors for neurodevelopmental impairment in congenital heart disease (CHD).

- Complications due to heart disease (abnormal fetal cerebral blood flow, type of CHD, postnatal diagnosis, lower birth rate)
- Perioperative factors (balloon atrial septostomy, cardiopulmonary bypass, deep hypothermic circulatory arrest, type of surgery)
- Postoperative factors (hemodynamics, need for extracorporeal membrane oxygenation (ECMO), longer intensive care stay, seizures)
- Genetic factors
- Socioeconomic/environmental factors

overall development, quality of life, and opportunity to become productive, responsible adults" [28]. The guidelines recommended developmental screening and surveillance within the medical home with appropriate referral for more detailed assessment and services. Yet many primary care providers are unaware of these recommendations. A 2016 survey of pediatric primary caregivers reported that only 21% of providers were aware of the guidelines and less than 10% received guidance from cardiology regarding neurodevelopment in children with CHD [29]. Many large medical centers have set up cardiac neurodevelopmental follow-up programs in order to centralize more comprehensive care. These programs often mirror or are integrated with neonatal follow-up programs in early childhood and have more extensive neuropsychological assessments in school age and adolescents. Despite these guidelines, further work is needed to ensure optimal follow-up as earlier identification allows for earlier intervention and improved outcomes.

Congenital diaphragmatic hernia

Congenital diaphragmatic hernia (CDH) occurs in 1/2500 live births. There is incomplete closure of the pleuroperitoneal membrane with subsequent herniation of the abdominal viscera into the thoracic cavity [30]. Lung compression by the herniated abdominal contents impedes lung development at a critical developmental window. Clinical course and outcomes vary widely, highly influenced by degree of pulmonary hypoplasia and pulmonary hypertension [30]. In addition, children with familial CDH, bilateral CDH, those with a syndromic/genetic etiology, and liver herniation are more likely to have poor outcomes. A total of 40% infants with CDH have other congenital anomalies, with congenital heart disease accounting for at least two-thirds [31–33]. Survival has increased significantly with better prenatal diagnosis and advancements in neonatal care including ECMO and more gentle ventilation. However, medical comorbidities including neurodevelopmental impairment in survivors remains common.

Early studies on CDH outcomes report developmental delay in more than one-third of CDH survivors [34]. More recent reports suggest rates of neurodevelopmental delay and/or impairment may be present in 40%–80% of survivors in early childhood [35–37]. Similar to CHD, infants with CDH are often hemodynamically unstable with significant hypoxemia. Delayed brain development, maturation, and overall decreased brain growth are also reported [38–40]. Specific to CDH anatomy, large diaphragmatic defects and intrathoracic liver position are associated with persistent developmental delays [41]. Preterm delivery, periventricular leukomalacia, and hypotonia on presentation also contribute [41]. ECMO and need for respiratory support at 30 days of life are often cited as a risk factor for neurodevelopmental impairment [42,43]. However, evidence is mixed, with some studies reporting no association and suggesting increased risk for NDDs in CDH is not solely related to ECMO [41].

As children with CDH age, delays do appear to improve. By school age, overall neurocognitive scores are not significantly different than the general population [44]. However, specific deficits do become more apparent, including difficulties with executive function, attention and social communication, motor delay, and planning [45–47]. Sensorineural hearing loss is also common [48].

The AAP recommends brain imaging prior to discharge for neonates with CDH and abnormal head ultrasound, seizures or abnormal neurologic exam, or need for ECMO or patch repair [48]. Similar to CHD, high level neurodevelopmental screening and surveillance is essential for optimal developmental outcomes.

Biliary atresia

Biliary atresia is a congenital liver disease that occurs due to obstruction of the bile ducts that exit the liver, leading to jaundice which, if untreated, may progress to liver failure and death. In the 1950s, a Japanese surgeon developed a surgical procedure to allow for bile drainage known as the hepatoportoenterostomy or Kasai procedure. Most infants with biliary atresia will undergo this procedure in the first months of life and up to 70% may also require liver transplantation [49]. These children may require frequent and prolonged hospitalizations in the first year of life and complications of liver failure include cirrhosis, bleeding, and hyperammonemia.

As early as the 1980s, developmental differences in this population of patients have been recognized [50]. More recently, large multicenter studies such as the Childhood Liver Disease Research Network (ChiLDReN) have been organized to formally define outcomes and identify opportunities for intervention. In general, risk for NDD is correlated with the severity of liver disease. A 2018 study followed 148 children with biliary atresia and native liver at 12 and 24 months of age and on average, children performed below average, particularly on motor scales at 1 year of age [51]. Unsuccessful hepatoportoenterostomy increased the risk of neurodevelopmental impairment by four times. For those patients who require liver transplant, follow-up studies have shown average to below average scores prior to transplant with a dramatic drop by one standard deviation in the months posttransplant, particularly in the motor domain followed by recovery to baseline at 1 year. Despite this recovery to baseline, at 1-year posttransplant, 35%–50% were diagnosed as developmentally delayed [52,53].

However, whether these delays persist later in childhood remains unclear. One study of 93 children ages 3–12 years with biliary atresia and native liver showed on average, normal IQ scores, but, markers of advanced liver disease were associated with lower full scale IQ [54]. Two separate European studies of 46 Dutch children and 39 Finnish children identified low motor scores in 40%–50% and a need for special education in 26% of the Dutch group [55,56]. In both studies, children with biliary atresia had on average lower IQ scores (mean full scale IQ of 91) than the normative population, regardless of transplant status [55,56].

Chronic kidney disease

Chronic kidney disease (CKD) is defined as decline in glomerular filtration rate below 60 mL/min/1.73 m^2 of body surface area for more than 3 months and is caused by abnormalities in the structure or function of the kidneys. While CKD is a continuum of renal function, the most severe form, end stage renal disease (ESRD), where an individual is dependent on renal replacement therapy, affects 100 per million children in the United States [57]. Adolescents are more likely to be affected than young children and black and Pacific Islanders are more likely to be affected than white [57]. Individuals with CKD frequently have chronic progressive renal function and become increasingly at risk for complications, which include hypertension, uremia, and anemia. It has also long been recognized that CKD may have long-term consequences on brain function.

As early as the 1980s, there have been reports of developmental delays in children with ESRD [58–60]. A 2000 clinical trial of 62 CKD patients in comparison to their siblings identified that children with CKD tended to have lower IQs than their siblings and that younger age, longer time to dialysis, and low maternal education predicted lower scores [61]. In general, while IQ scores vary widely in this patient population, approximately 5%–25% have intellectual disability [62–64]. Greater disease severity is associated with lower IQ. For those with mild to moderate CKD, cognitive function tends to fall in the below average range [65–67]. Specifically, measures of attention, memory, and executive function fall on average, below their peers [65,68].

Approximately 18% of patients with CKD have hearing loss which may affect language development, if undetected [69]. However, in those with intact hearing, language deficits are not present. While motor impairment may also occur in CKD, cerebral palsy has been reported in less than 10% of patients and for those with intact gross motor function, fine motor function appears to be spared [67,69].

The underlying neuropathology of neurocognitive changes in CKD is not well understood. Neuroimaging studies have revealed cerebral volume changes and atrophy [70]. Cerebrovascular changes which can occur secondary to blood pressure fluctuations may also contribute to cognitive differences in these patients. In some cases, underlying genetic disorders, such as Galloway–Mowat syndrome or Wilms Tumor Aniridia Genitourinary Abnormalities and Range of Developmental Delays Syndrome (WAGR syndrome), which affect both kidney and brain development, may also play a role [71].

It is also unclear whether renal transplant improves cognitive outcomes and evidence is mixed. While it may be beneficial for some aspects of neurocognition, other deficits may persist [72,73].

Overall, subtle cognitive deficits may occur in children affected with CKD, affecting their emotional-behavioral outcomes and quality of life [67]. Investigation by neuropsychological evaluation may be indicated and lead to supportive interventions.

Hypothyroidism and diabetes

The endocrine system plays a key role in human brain development as it influences the growth and differentiation of neurons and glial cells. Endocrine disorders therefore influence the developing brain and may be associated with varying degrees of NDDs.

Hypothyroidism leads to deregulation of many cellular processes including altered dendritic cell morphology, abnormal cell migration and proliferation, and impaired synaptic transmission in the hippocampus, among others. Children with congenital or early hypothyroidism that goes untreated are at risk for impaired brain development and subsequent cognitive and motor deficits [74–76]. Other clinical manifestations of hypothyroidism in infancy include poor feeding, poor growth, jaundice, edema, constipation, goiter, and hypothermia.

Historically, hypothyroidism due to insufficient dietary intake of iodine was a major cause of intellectual disability, commonly referred to as cretinism. With the introduction of iodized salts to major food supplies in the 20th century, rates of cretinism have declined in developed countries. However, it remains a continued public health concern in developing countries around the world, particularly land-locked countries [77,78].

Hypothyroidism may also be congenital, caused by idiopathic thyroid dysgenesis in the majority of cases, or, by inborn errors of thyroid hormone metabolism (in <5% of cases) [78]. Efficient identification and treatment of congenital hypothyroidism during the newborn period is critical for optimizing neurologic outcomes [75,79]. For this reason, this condition was added to the universal newborn screen in most developed countries as early as the 1970s [78]. Long-term outcomes of children with congenital hypothyroidism, who are treated within the first 3 months of life, reveal that while these children ultimately have intellectual quotients that fall within the normal range, they perform below their peers in multiple cognitive arenas including attention, memory, and visuospatial processing [80,81]. Of note, common comorbidities in these patients also include CHD in 10% and hearing loss in 10%, both of which may impact developmental outcomes [78].

In contrast to congenital hypothyroidism, children with juvenile-onset diabetes, an autoimmune disorder causing insulin deficiency and hyperglycemia, may have more subtle neurodevelopmental differences since metabolic changes occur later in brain development. While these children may have typical early cognitive development, they are at higher risk for psychiatric disorders in adolescence [82,83]. Those with cooccurring psychiatric disorders may also be particularly vulnerable to academic underachievement [84]. Poorer glycemic control has also been associated with poor performance on cognitive tasks [85–88]. Recent clinical trials are examining if improved insulin delivery systems that lead to better glycemic control may also lead to improved cognitive performance [89]. Early data suggest that these systems may be associated with higher perceptual reasoning score and more neurotypical functional MRI measures [89].

Sickle cell disease

Sickle cell disease (SCD) is a genetic disorder affecting the structure of hemoglobin within red blood cells, resulting in "sickled" cells that are rigid, fragile, and have short survival. People with SCD suffer from chronic anemia, frequent pain, lung disease, avascular necrosis, and increased risk for infection. These individuals also have high risk for stroke, when sickled cells block vessels within the CNS. Stroke in SCD may be overt, with clear clinical symptoms such as weakness, paralysis or aphasia, or, they may be silent, where lesions are identified on MRI in the absence of any clear symptoms. By age 15 years, approximately 38% of children with SCD have silent strokes on MRI [90].

CNS vascular disease in children with SCD is associated with neurocognitive deficits. Several studies have demonstrated a stepwise decline in IQ, where SCD with normal MRIs have IQ in the 90s, those with silent infarcts have IQ in the 80s and those with overt infarcts have IQ in the 70s [90,91]. Not only do children with SCD who stroke have lower IQ, but one study of 19 children showed that 80% have deficiencies in various domains on neuropsychological testing [92]. Anemia may also independently contribute to lower IQ in these patients [93].

Stroke, anemia, and several other factors including school absenteeism due to hospitalization/pain, socioeconomic status, and parental education contribute to multifactorial poorer neurodevelopmental outcomes in this patient population [94]. Neurodevelopmental delays in both motor and cognitive domains may manifest as early as preschool age [95,96]. By school age, one metaanalysis showed that 38% of students with SCD received special education services and were 10 times more likely than their peers to experience grade retention [96]. Another study of 276 children with SCD identified NDD in 24%, with the most common diagnoses being language disorder, ADHD, and specific learning disability [97]. Given this high-risk population, the American Society of Hematology has recommended surveillance for NDD by hematologists. However, efforts are underway to establish formal neurodevelopmental screening guidelines for children with SCD, similar to those for children with CHD [98].

Achondroplasia

Skeletal dysplasias are a group of congenital conditions affecting bone development. The most common form of skeletal dysplasia is achondroplasia, an autosomal-dominant genetic condition marked by pathogenic variants in the *FGFR3* gene. Achondroplasia is most notably marked by short stature (dwarfism) and occurs in approximately 4.6 in every 1,000,000 live births [99]. People with achondroplasia typically have normal cognitive function [100]. Early motor delays, however, are common [101–103]. Developmental concerns and neurodevelopmental disabilities in this population primarily arise secondary to comorbidities that commonly occur [102,103]. Most notably, hydrocephalus may occur in one-third of children with achondroplasia although shunting is required in only 5% [99]. Developmental outcomes of those with hydrocephalus are poorer [104]. Other comorbidities that may affect development if untreated include recurrent otitis media in up to 40%, hearing loss in up to 50%, obstructive sleep apnea in over half, and myelopathy in up to

20% [99]. In particular, obstructive sleep apnea has been reported to affect attention, memory, and executive function [100].

Similar profiles have been described in other forms of skeletal dysplasias such as osteogenesis imperfecta [105,106]. Therefore when considering the potential neurodevelopmental outcomes of patients with skeletal dysplasias, identification and treatment of comorbidities that impact neurologic function and development is of utmost importance.

Neurocutaneous disorders

Neurocutaneous disorders (also known as phakomatoses) are a group of genetic conditions characterized by both skin and neurologic abnormalities. The cooccurrence of both dermatologic and neurologic findings may be explained in some of these disorders by shared embryologic origins as both skin and neural tissue are derived from the ectodermal germ layer during early stages of embryonic development. Characteristic hyper- or hypopigmented skin lesions may be the earliest or most outwardly visible manifestations of disease. In this respect, a careful skin exam by primary pediatricians during early childhood is important because these unusual skin findings may be indicative of a genetic disorder with neurodevelopmental implications. Examples include tuberous sclerosis complex (TSC), neurofibromatosis (NF), and Sturge–Weber syndrome, among others. The incidence of NDD among patients with neurocutaneous disorder is high with variable manifestations depending on the part of the nervous system affected.

TSC is a neurocutaneous disorder with characteristic skin findings which include hypomelanotic macules (ash-leaf spots), forehead plaques, shagreen patches, facial angiofibromas, and periungual fibromas. The earliest manifestation is cardiac rhabdomyomas which can be detected prenatally on third trimester ultrasound. A fetal brain MRI can then be ordered to look for cortical dysplasias. If both of these major features of TSC are present, the infant would meet criteria for clinically definite TSC [107]. After delivery, often the earliest signs are skin manifestations, most commonly hypopigmented macules, and seizures [107]. Importantly, prevention of infantile spasms has been associated with improved epilepsy outcomes [108]. However, there are many individuals with TSC who are not diagnosed until later in life and this diagnosis may be incidental after an unexpected seizure or a skin manifestation that was not recognized as being related to TSC. The neurodevelopmental and neuropsychiatric manifestations of TSC are collectively termed TAND (TSC-associated neuropsychiatric disorders), and there are consensus guidelines for the evaluation and management of these conditions in TSC [109–111]. Therefore failure to recognize and treat TSC early can lead to poor medical and neurodevelopmental outcomes [112–114].

NF is another phakomatosis which causes tumors to grow on nerves throughout the body. There are three types of NF: NF1, NF2, and schwannomatosis. NF1 is the most common type of NF, affecting approximately 1 in 3000 people worldwide. While all types of NF may experience NDD, individuals with NF1 are more likely to be affected [115,116. NF1 is caused by mutations in the NF1 gene, which makes a protein called neurofibromin that regulates cell growth and division. In NF1, there is uncontrolled growth of cells, resulting in neurofibromas which grow on the skin, brain, and spinal cord. In addition to tumors, individuals with NF1 may also have café-au-lait spots (light brown patches on the skin), freckling in the armpits and groin, and bone abnormalities. Additionally, aberrant interactions between glial cells and neurons, altered neurotransmitter systems, and imbalances in neurotrophic factors have been implicated in the neurobiological mechanisms underlying neurodevelopmental disorders in NF1 [117]. Children with NF1 have a high prevalence of ADHD, with over half having attention difficulties [116,118,119]. They have also been found to have learning disabilities and deficits in social communication with symptoms that frequently overlap with autism [116]. However, whether these deficits truly meet criteria for ASD is not consistently agreed upon [120].

Primary immunodeficiencies

NDDs are common in many inborn errors of immunity (IEI) or genetic defects leading to deficits or dysregulation of the immune system. Neurologic disease can develop via multiple mechanisms in this population and can be broadly divided into three groups: those due to the underlying genetic syndrome, those due to autoimmune/autoinflammatory manifestations, and those due to invasive infections. Key examples include purine nucleoside phosphorylase (PNP) deficiency and adenosine deaminase (ADA) deficiency. These disorders poison the immune system through toxic metabolic products, leading to profound or severe combined immunodeficiency (SCID) with markedly decreased-to-absent T-cell mediated immunity. Due to the accumulation of toxic metabolites, patients with PNP and ADA deficiencies also suffer neurologic complications, including hypotonia, spasticity, ataxia, cognitive impairment, and behavioral problems [121,122]. Counterintuitively, many IEI result in substantial immune dysregulation, leading to autoimmune and autoinflammatory manifestations often involving the central nervous system. Aicardi–Gutierrez syndrome, an interferonopathy, presents with encephalopathy in the first year of life with concurrent neuroinflammation, and later microcephaly, leukodystrophy, central nervous system calcifications, intellectual disability, spasticity, dystonia, and hypotonia while other autoinflammatory disorders in this family may result in more subtle neurodevelopmental deficits [123].

Invasive infections of the CNS also cause substantial short-term and long-term neurologic complications in patients with IEI. For these patients, their neurodevelopmental outcome is largely dependent on the severity and clinical course of their infection. Patients with defects in terminal complement (C5-C9) have a predisposition to meningitis caused by the encapsulated organism *Neisseria meningitidis*. Particular susceptibility to invasive bacterial infections, including meningitis, is observed in IRAK4 deficiency and MYD88 deficiency. Defects in the Toll-like receptor 3, an innate immune sensor for foreign, endosomal, double-stranded DNA, leads to isolated herpes simplex virus encephalitis without other notable infectious susceptibility. Invasive fungal central nervous system infections, caused by candida and aspergillus and others, are seen with increased frequency in individuals with certain immunodeficiencies including CARD9 deficiency and chronic granulomatous disease [124].

One of the greatest challenges to clinicians is management of syndromic IEI with prominent NDD. In these individuals, while immunodeficiency and immune dysregulation may be greatly improved by definitive therapy, such as enzyme replacement, gene therapy or hematopoietic stem cell transplantation (HSCT), similar correction of the neurologic complications is rare. For example, in patients with ataxia telangiectasia who underwent HSCT, immune reconstitution was successful but did not change the course of neurologic disease [125–127]. Thus appropriate joint decision making with family members to convey the long-term prognosis in an individual with a specific syndromic IEI is essential.

Hematopoietic stem cell transplantation

HSCT for the management of numerous malignant and nonmalignant processes continues to expand. Of children who undergo HSCT, early neurologic complications are present in 11%–59% of individuals, while late complications are present in about 5% of individuals [128,129]. These early and late complications of HSCT may lead to NDD long after recovery from the acute transplant period. Early neurologic complications of HSCT include infectious, metabolic, medication, immune dysregulation, and vasculopathy [128]. Medications are also a common cause of neurologic symptoms during HSCT. The most common culprits include ganciclovir, dimethylsufoxide, amphotericin B, narcotics, benzodiazepines, and cyclosporin and tacrolimus causing calcineurin inhibitor neurotoxicity. Immune dysregulation impacting the CNS peri-HSCT is also common, manifesting as microangiopathy, graft-versus-host disease, and acute disseminated encephalomyelitis [128].

Given the increased use of HSCT for treatment of a variety of conditions, long-term neurologic complications of HSCT are of increasing importance. Risk factors associated with increased rates of long-term complications include early memory loss, acute graft-versus-host disease, immunosuppressive therapy, history of delirium, increase length of hospital stay, unrelated donor HSCT, single dose total body irradiation, and cognitive decline in the first year after HSCT [130]. Radiation therapy is particularly important as a driver of late neurologic complications, as total body irradiation is commonly used as a component of conditioning for HSCT and radiation-related deficits may progress over time. Radiation results in higher rates of long-term neurologic complications as compared to chemotherapy alone, especially when administered at ages less than 3 years; those who receive total body irradiation also have more neuropsychiatric deficits and poorer educational attainment compared to those who receive chemotherapy alone [131,132]. In addition to significant early toxicity, calcineurin inhibitors can also lead to long-term complications including seizures, occipital blindness, and hemiparesis [133].

Other long-term problems that may contribute to disability include psychiatric disorders, hearing and vision deficits, chronic pain, and motor difficulties. These deficits may not be observable at 1- to 2-year follow-ups, but become more prominent at 5-year or longer follow-ups. Anxiety, depression, posttraumatic stress disorder, disruptive behavior, and problems with social function post-HSCT are frequent [131]. At over 10 years of follow-up, one study reported cataracts in 36% and chronic pain in 21% of patients post-HSCT in childhood. The same study also reported coordination problems (7%), problems with chewing and swallowing (8%), and muscle weakness (6%)—most prominent in those with chronic graft-versus-host disease [133].

Increasingly HSCT is being used for the treatment of certain metabolic and genetic conditions where HSCT may improve long-term neurologic outcome and mitigate NDD, although response is mixed. In patients with SCD, improvements in hemiparesis and stable central nervous system imaging are observed peri-HSCT. Furthermore, HSCT may stabilize SCD-associated IQ loss [134]. Among the IEI, ADA deficiency enzyme replacement therapy appears to be more effective in attenuating CNS symptoms than HSCT [121,135]. HSCT is used to treat Gaucher, Farber's, attenuated Hunters, and Hurler syndrome, although studies have reported somewhat mixed clinical outcomes for neuromuscular, neurocognitive, and neurodevelopmental symptoms [136]. Other metabolic disease for which HSCT is used for the treatment of neurologic symptoms include X-linked adrenoleukodystrophy, metachromatic leukodystrophy, gangliosidosis, and Niemann–Pick disease [136,137].

Implications of anesthesia for children with systemic disease

Medical or surgical procedures requiring anesthesia are frequently necessary for children with systemic disease. In the last decade, there has been increased research and concern for the neurodevelopmental consequences of anesthesia during childhood, particularly during early periods of critical brain development. In 2016 the Food and Drug Administration issued a warning for 11 drugs commonly used in general anesthesia and sedation that they "may affect the development of children's brains," specifically pointing to their use for more than 3 hours or repeated use in children less than age 3 years or pregnant women in the third trimester [138]. The list of drugs included all anesthetic gases, propofol, ketamine, barbiturates, and benzodiazepines. Unfortunately, there remain no good alternatives to these medications in children under 3 years and most procedures for children with systemic diseases cannot safely be delayed. Evidence to support the recommendation primarily stems from in vitro and animal models. Clinical data support that a single exposure to anesthesia for minor procedures such as hernia repair, do not have lasting neurodevelopmental consequences [139–142]. However, long-term outcomes studies for procedures requiring repeated or prolonged anesthesia suggest increased risk for ADHD, learning disability, and other neurodevelopmental differences, these studies are confounded by the complex clinical scenarios for which those procedures are indicated [141–143].

Sociobehavioral concerns and school performance in children with chronic disease

Many chronic health conditions of childhood (e.g., celiac disease, cystic fibrosis, and hemophilia A) have not shown an association with atypical neurocognitive development [144,145]. However, these children may nevertheless be impacted emotionally and academically by their disease. Children with frequent hospitalizations, doctor's appointments, or therapies may have increased school absenteeism. Healthcare-related stressors may also contribute to mental health issues such as anxiety or depression which can impact academic performance. Children with chronic health conditions are also at increased risk of peer victimization (bullying) [146].

Identification and clinical care

As illustrated in the above examples, for children with systemic medical conditions, primary dysfunction of one physiologic system may impact the central nervous system. In some patients, this impact may be subtle and require investigation for diagnosis via neuropsychological testing. While developmental screening is recommended for all children at specific intervals, these recommendations are not always adhered to and special attention should be paid to children with vulnerable conditions [147,148]. In some communities, chronic medical conditions will automatically qualify for early intervention monitoring or services. In these cases, referrals should be made either by the family themselves, healthcare professionals, or educational professionals.

Additionally, some subspecialities have adapted consensus statements that include specific neurodevelopmental assessments as part of routine and ongoing disease monitoring [28]. To dovetail these recommendations, some clinical care centers have established multidisciplinary follow-up programs that provide both subspeciality and neurodevelopmental care for children with systemic diseases that are particularly at risk (e.g., CHD, CDH, and SCD). However, these programs are typically at academic or tertiary care centers and may not be accessible to all families. In these cases, neurodevelopmental surveillance typically falls on the primary care providers as the medical home.

Ultimately, identification and treatment of NDD in children with systemic disease can improve quality of life of the individual, of the family and optimize function in adulthood [149–151].

References

[1] Ribera I, Ruiz A, Sánchez O, Eixarch E, Antolín E, Gómez-Montes E, et al. Multicenter prospective clinical study to evaluate children short-term neurodevelopmental outcome in congenital heart disease (children NEURO-HEART): study protocol. BMC Pediatr 2019;19(1):326.

[2] Hoffman JIE, Kaplan S. The incidence of congenital heart disease. J Am Coll Cardiol 2002;39(12):1890–900.

[3] Moons P, Bovijn L, Budts W, Belmans A, Gewillig M. Temporal trends in survival to adulthood among patients born with congenital heart disease from 1970 to 1992 in Belgium. Circulation 2010;122(22):2264–72.
[4] Bellinger DC, Jonas RA, Rappaport LA, Wypij D, Wernovsky G, Kuban KC, et al. Developmental and neurologic status of children after heart surgery with hypothermic circulatory arrest or low-flow cardiopulmonary bypass. N Engl J Med 1995;332(9):549–55.
[5] Massaro AN, Glass P, Brown J, Chang T, Krishnan A, Jonas RA, et al. Neurobehavioral abnormalities in newborns with congenital heart disease requiring open-heart surgery. J Pediatr 2011;158(4):678–81.
[6] McQuillen PS, Miller SP. Congenital heart disease and brain development. Ann N Y Acad Sci 2010;1184:68–86.
[7] Feldmann M, Guo T, Miller SP, Knirsch W, Kottke R, Hagmann C, Latal B, Jakab A. Delayed maturation of the structural brain connectome in neonates with congenital heart disease. Brain Commun. 2020 Nov 27;2(2):fcaa209. Available from: https://doi.org/10.1093/braincomms/fcaa209. PMID: 33381759; PMCID: PMC7756099.
[8] Limperopoulos C, Majnemer A, Shevell MI, Rosenblatt B, Rohlicek C, Tchervenkov C. Neurologic status of newborns with congenital heart defects before open heart surgery. Pediatrics 1999;103(2):402–8.
[9] Masoller N, Martínez JM, Gómez O, Bennasar M, Crispi F, Sanz-Cortés M, et al. Evidence of second-trimester changes in head biometry and brain perfusion in fetuses with congenital heart disease. Ultrasound Obstet Gynecol 2014;44(2):182–7.
[10] Limperopoulos C, Tworetzky W, McElhinney DB, Newburger JW, Brown DW, Robertson Jr RL, et al. Brain volume and metabolism in fetuses with congenital heart disease: evaluation with quantitative magnetic resonance imaging and spectroscopy. Circulation 2010;121(1):26–33.
[11] Miller SP, McQuillen PS, Hamrick S, Xu D, Glidden DV, Charlton N, et al. Abnormal brain development in newborns with congenital heart disease. N Engl J Med 2007;357(19):1928–38.
[12] Marelli A, Miller SP, Marino BS, Jefferson AL, Newburger JW. Brain in congenital heart disease across the lifespan: the cumulative burden of injury. Circulation 2016;133(20):1951–62.
[13] De Asis-Cruz J, Donofrio MT, Vezina G, Limperopoulos C. Aberrant brain functional connectivity in newborns with congenital heart disease before cardiac surgery. Neuroimage Clin 2018;17:31–42.
[14] McQuillen PS, Barkovich AJ, Hamrick SEG, Perez M, Ward P, Glidden DV, et al. Temporal and anatomic risk profile of brain injury with neonatal repair of congenital heart defects. Stroke 2007;38(2 Suppl):736–41.
[15] Mahle WT, Tavani F, Zimmerman RA, Nicolson SC, Galli KK, Gaynor JW, et al. An MRI study of neurological injury before and after congenital heart surgery. Circulation 2002;106(12 Suppl 1):I109–14.
[16] von Rhein M, Buchmann A, Hagmann C, Huber R, Klaver P, Knirsch W, et al. Brain volumes predict neurodevelopment in adolescents after surgery for congenital heart disease. Brain 2014;137(Pt 1):268–76.
[17] Easson K, Gilbert G, Rohlicek CV, Saint-Martin C, Descoteaux M, Deoni SCL, et al. Altered myelination in youth born with congenital heart disease. Hum Brain Mapp 2022;43(11):3545–58.
[18] Enguix V, Easson K, Gilbert G, Saint-Martin C, Rohlicek C, Luck D, et al. Altered resting state functional connectivity in youth with congenital heart disease operated during infancy. PLoS ONE 2022;17(4):e0264781.
[19] Ehrler M, Latal B, Kretschmar O, von Rhein M, O'Gorman Tuura R. Altered frontal white matter microstructure is associated with working memory impairments in adolescents with congenital heart disease: a diffusion tensor imaging study. Neuroimage Clin 2020;25:102123.
[20] Majnemer A, Limperopoulos C, Shevell M, Rosenblatt B, Rohlicek C, Tchervenkov C. Long-term neuromotor outcome at school entry of infants with congenital heart defects requiring open-heart surgery. J Pediatr 2006;148(1):72–7.
[21] Goldberg CS, Schwartz EM, Brunberg JA, Mosca RS, Bove EL, Schork MA, et al. Neurodevelopmental outcome of patients after the fontan operation: a comparison between children with hypoplastic left heart syndrome and other functional single ventricle lesions. J Pediatr 2000;137(5):646–52.
[22] Forbess JM, Visconti KJ, Bellinger DC, Jonas RA. Neurodevelopmental outcomes in children after the fontan operation. Circulation 2001;104(12 Suppl 1):I127–32.
[23] Shillingford AJ, Glanzman MM, Ittenbach RF, Clancy RR, Gaynor JW, Wernovsky G. Inattention, hyperactivity, and school performance in a population of school-age children with complex congenital heart disease. Pediatrics 2008;121(4):e759–67.
[24] Cassidy AR, White MT, DeMaso DR, Newburger JW, Bellinger DC. Executive function in children and adolescents with critical cyanotic congenital heart disease. J Int Neuropsychol Soc 2015;21(1):34–49.
[25] Bellinger DC. Are children with congenital cardiac malformations at increased risk of deficits in social cognition? Cardiol Young 2008;18(1):3–9.
[26] Sadhwani A, Wypij D, Rofeberg V, Gholipour A, Mittleman M, Rohde J, et al. Fetal brain volume predicts neurodevelopment in congenital heart disease. Circulation 2022;145(15):1108–19.
[27] Gaynor JW, Stopp C, Wypij D, Andropoulos DB, Atallah J, Atz AM, et al. Neurodevelopmental outcomes after cardiac surgery in infancy. Pediatrics 2015;135(5):816–25.
[28] Marino BS, Lipkin PH, Newburger JW, Peacock G, Gerdes M, Gaynor JW, et al. Neurodevelopmental outcomes in children with congenital heart disease: evaluation and management: a scientific statement from the American Heart Association. Circulation 2012;126(9):1143–72.
[29] Knutson S, Kelleman MS, Kochilas L. Implementation of developmental screening guidelines for children with congenital heart disease. J Pediatr 2016;176:135–41.
[30] Hedrick HL. Management of prenatally diagnosed congenital diaphragmatic hernia. Semin Fetal Neonatal Med 2010;15(1):21–7.
[31] Graziano JNCongenital Diaphragmatic Hernia Study Group. Cardiac anomalies in patients with congenital diaphragmatic hernia and their prognosis: a report from the Congenital Diaphragmatic Hernia Study Group. J Pediatr Surg 2005;40(6):1045–9.

[32] Qiao L, Wynn J, Yu L, Hernan R, Zhou X, Duron V, et al. Likely damaging de novo variants in congenital diaphragmatic hernia patients are associated with worse clinical outcomes. Genet Med 2020;22(12):2020–8.

[33] Cohen MS, Rychik J, Bush DM, Tian ZY, Howell LJ, Adzick NS, et al. Influence of congenital heart disease on survival in children with congenital diaphragmatic hernia. J Pediatr 2002;141(1):25–30.

[34] Nobuhara KK, Lund DP, Mitchell J, Kharasch V, Wilson JM. Long-term outlook for survivors of congenital diaphragmatic hernia. Clin Perinatol 1996;23(4):873–87.

[35] Cortes RA, Keller RL, Townsend T, Harrison MR, Farmer DL, Lee H, et al. Survival of severe congenital diaphragmatic hernia has morbid consequences. J Pediatr Surg 2005;40(1):36–45.

[36] Chen C, Jeruss S, Terrin N, Tighiouart H, Wilson JM, Parsons SK. Impact on family of survivors of congenital diaphragmatic hernia repair: a pilot study. J Pediatr Surg 2007;42(11):1845–52.

[37] Nijhuis-van der Sanden MWG, van der Cammen-van Zijp MHM, Janssen AJWM, Reuser JJCM, Mazer P, van Heijst AFJ, et al. Motor performance in five-year-old extracorporeal membrane oxygenation survivors: a population-based study. Crit Care 2009;13(2):R47.

[38] Danzer E, Zarnow D, Gerdes M, D'Agostino JA, Siegle J, Bebbington MW, et al. Abnormal brain development and maturation on magnetic resonance imaging in survivors of severe congenital diaphragmatic hernia. J Pediatr Surg 2012;47(3):453–61.

[39] Moran MM, Gunn-Charlton JK, Walsh JM, Cheong JLY, Anderson PJ, Doyle LW, et al. Associations of neonatal noncardiac surgery with brain structure and neurodevelopment: a prospective case-control study. J Pediatr 2019;212:93–101.e2.

[40] Van der Veeken L, Russo FM, Litwinska E, Gomez O, Emam D, Lewi L, et al. Prenatal cerebellar growth is altered in congenital diaphragmatic hernia on ultrasound. Prenat Diagn 2022;42(3):330–7.

[41] Danzer E, Gerdes M, D'Agostino JA, Hoffman C, Bernbaum J, Bebbington MW, et al. Longitudinal neurodevelopmental and neuromotor outcome in congenital diaphragmatic hernia patients in the first 3 years of life. J Perinatol 2013;33(11):893–8.

[42] Wynn J, Aspelund G, Zygmunt A, Stolar CJH, Mychaliska G, Butcher J, et al. Developmental outcomes of children with congenital diaphragmatic hernia: a multicenter prospective study. J Pediatr Surg 2013;48(10):1995–2004.

[43] Church JT, Mon R, Wright T, Coughlin MA, Ladino-Torres M, Tapley C, et al. Neurodevelopmental outcomes in CDH survivors: a single institution's experience. J Pediatr Surg 2018;53(6):1087–91.

[44] Tureczek I, Caflisch J, Moehrlen U, Natalucci G, Bernet V, Latal B. Long-term motor and cognitive outcome in children with congenital diaphragmatic hernia. Acta Paediatr 2012;101(5):507–12.

[45] Madderom MJ, Toussaint L, van der Cammen-van Zijp MHM, Gischler SJ, Wijnen RMH, Tibboel D, et al. Congenital diaphragmatic hernia with(out) ECMO: impaired development at 8 years. Arch Dis Child Fetal Neonatal Ed 2013;98(4):F316–22.

[46] Snoek KG, Reiss IKM, Tibboel J, van Rosmalen J, Capolupo I, van Heijst A, et al. Sphingolipids in congenital diaphragmatic hernia; results from an international multicenter study. PLoS ONE 2016;11(5):e0155136.

[47] Frisk V, Jakobson LS, Unger S, Trachsel D, O'Brien K. Long-term neurodevelopmental outcomes of congenital diaphragmatic hernia survivors not treated with extracorporeal membrane oxygenation. J Pediatr Surg 2011;46(7):1309–18.

[48] American Academy of Pediatrics Section on Surgery, American Academy of Pediatrics Committee on Fetus and NewbornLally KP, Engle W. Postdischarge follow-up of infants with congenital diaphragmatic hernia. Pediatrics 2008;121(3):627–32.

[49] McDiarmid SV, Anand R, Lindblad ASSPLIT Research Group. Studies of pediatric liver transplantation: 2002 update. An overview of demographics, indications, timing, and immunosuppressive practices in pediatric liver transplantation in the United States and Canada. Pediatr Transpl 2004;8(3):284–94.

[50] Burgess DB, Martin HP, Lilly JR. The developmental status of children undergoing the Kasai procedure for biliary atresia. Pediatrics 1982;70(4):624–9.

[51] Ng VL, Sorensen LG, Alonso EM, Fredericks EM, Ye W, Moore J, et al. Neurodevelopmental outcome of young children with biliary atresia and native liver: results from the ChiLDReN study. J Pediatr 2018;196:139–47.

[52] Patterson C, So S, Rogers A, Ng VL. Motor outcomes in young children pre-and one-year post-liver transplant. Pediatr Transpl 2022;26(3):e14200.

[53] Wayman KI, Cox KL, Esquivel CO. Neurodevelopmental outcome of young children with extrahepatic biliary atresia 1 year after liver transplantation [Internet]. J Pediatr 1997;131:894–8. Available from: https://doi.org/10.1016/s0022-3476(97)70039-8.

[54] Squires JE, Ng VL, Hawthorne K, Henn LL, Sorensen LG, Fredericks EM, et al. Neurodevelopmental outcomes in preschool and school aged children with biliary atresia and their native liver. J Pediatr Gastroenterol Nutr 2020;70(1):79–86.

[55] Rodijk LH, den Heijer AE, Hulscher JBF, Alizadeh BZ, de Kleine RHJ, Verkade HJ, et al. Long-term neurodevelopmental outcomes in children with biliary atresia [Internet]. J Pediatr 2020;217:118–24. Available from: https://doi.org/10.1016/j.jpeds.2019.10.054.

[56] Ruuska S, Lähteenmäki M, Häyrinen T, Kanerva K, Jahnukainen T, Haataja L, et al. Neurocognitive and motor functions in biliary atresia patients: a cross-sectional, prospective national cohort study. J Pediatr Gastroenterol Nutr 2021;73(4):491–8.

[57] Available from Centers for Disease Control and Prevention. Chronic kidney disease surveillance system—United States. http://www.cdc.gov/ckd.

[58] Polinsky MS, Kaiser BA, Stover JB, Frankenfield M, Baluarte HJ. Neurologic development of children with severe chronic renal failure from infancy. Pediatr Nephrol 1987;1(2):157–65.

[59] Schärer K. Growth and development of children with chronic renal failure [Internet]. Acta Paediatrica 1990;90–2. Available from: https://doi.org/10.1111/j.1651-2227.1990.tb11605.x.

[60] Nevins TE, Rotundo A, Lockman LA, Michael Mauer S, Michael AF. A progressive encephalopathy in children with renal failure in infancy [Internet]. Chronic Ren Dis 1985;587–90. Available from: https://doi.org/10.1007/978-1-4684-4826-9_60.

[61] Brouhard BH, Donaldson LA, Lawry KW, McGowan KR, Drotar D, Davis I, et al. Cognitive functioning in children on dialysis and post-transplantation. Pediatr Transpl 2000;4(4):261–7.

[62] Madden SJ, Ledermann SE, Guerrero-Blanco M, Bruce M, Trompeter RS. Cognitive and psychosocial outcome of infants dialysed in infancy. Child Care Health Dev 2003;29(1):55–61.

[63] Warady BA, Belden B, Kohaut E. Neurodevelopmental outcome of children initiating peritoneal dialysis in early infancy. Pediatr Nephrol 1999;13(9):759–65.

[64] Hulstijn-Dirkmaat GM, Damhuis IH, Jetten ML, Koster AM, Schröder CH. The cognitive development of pre-school children treated for chronic renal failure. Pediatr Nephrol 1995;9(4):464–9.

[65] Hooper SR, Gerson AC, Butler RW, Gipson DS, Mendley SR, Lande MB, et al. Neurocognitive functioning of children and adolescents with mild-to-moderate chronic kidney disease. Clin J Am Soc Nephrol 2011;6(8):1824–30.

[66] Alhamed AA, Toly VB, Hooper SR, Dell KM. The link between executive function, socio-emotional functioning and health-related quality of life in children and adolescents with mild to moderate chronic kidney disease [Internet]. Child Care Health Dev 2022;455–64. Available from: https://doi.org/10.1111/cch.12946.

[67] Hooper SR, Johnson RJ, Gerson AC, Lande MB, Shinnar S, Harshman LA, et al. Overview of the findings and advances in the neurocognitive and psychosocial functioning of mild to moderate pediatric CKD: perspectives from the Chronic Kidney Disease in Children (CKiD) cohort study. Pediatr Nephrol 2022;37(4):765–75.

[68] Gipson DS, Hooper SR, Duquette PJ, Wetherington CE, Stellwagen KK, Jenkins TL, et al. Memory and executive functions in pediatric chronic kidney disease. Child Neuropsychol 2006;12(6):391–405.

[69] Qvist E, Pihko H, Fagerudd P, Valanne L, Lamminranta S, Karikoski J, et al. Neurodevelopmental outcome in high-risk patients after renal transplantation in early childhood. Pediatr Transpl 2002;6(1):53–62.

[70] Hartung EA, Erus G, Jawad AF, Laney N, Doshi JJ, Hooper SR, et al. Brain magnetic resonance imaging findings in children and young adults with CKD [Internet]. Am J Kidney Dis 2018;349–59. Available from: https://doi.org/10.1053/j.ajkd.2017.11.024.

[71] Verbitsky M, Kogon AJ, Matheson M, Hooper SR, Wong CS, Warady BA, et al. Genomic disorders and neurocognitive impairment in pediatric CKD. J Am Soc Nephrol 2017;28(8):2303–9.

[72] Mendley SR, Zelko FA. Improvement in specific aspects of neurocognitive performance in children after renal transplantation. Kidney Int 1999;56(1):318–23.

[73] Icard P, Hooper SR, Gipson DS, Ferris ME. Cognitive improvement in children with CKD after transplant. Pediatr Transpl 2010;14(7):887–90.

[74] Bernal J, Guadaño-Ferraz A, Morte B. Perspectives in the study of thyroid hormone action on brain development and function. Thyroid 2003;13(11):1005–12.

[75] Klein AH, Meltzer S, Kenny FM. Improved prognosis in congenital hypothyroidism treated before age three months. J Pediatr 1972;81 (5):912–15.

[76] Salazar P, Cisternas P, Martinez M, Inestrosa NC. Hypothyroidism and cognitive disorders during development and adulthood: implications in the central nervous system. Mol Neurobiol 2019;56(4):2952–63.

[77] Delange F, Dunn JT, Glinoer D. Iodine deficiency in Europe: a continuing concern. Springer Science & Business Media; 2013. p. 492.

[78] Wassner AJ. Congenital hypothyroidism. Clin Perinatol 2018;45(1):1–18.

[79] Virtanen M, Mäenpää J, Santavuori P, Hirvonen E, Perheentupa J. Congenital hypothyroidism: age at start of treatment versus outcome. Acta Paediatr Scand 1983;72(2):197–201.

[80] Rovet JF, Ehrlich R. Psychoeducational outcome in children with early-treated congenital hypothyroidism. Pediatrics 2000;105(3 Pt 1):515–22.

[81] Rovet JF. Long-term neuropsychological sequelae of early-treated congenital hypothyroidism: effects in adolescence. Acta Paediatr Suppl 1999;88(432):88–95.

[82] Dybdal D, Tolstrup JS, Sildorf SM, Boisen KA, Svensson J, Skovgaard AM, et al. Increasing risk of psychiatric morbidity after childhood onset type 1 diabetes: a population-based cohort study. Diabetologia 2018;61(4):831–8.

[83] Butwicka A, Frisén L, Almqvist C, Zethelius B, Lichtenstein P. Risks of psychiatric disorders and suicide attempts in children and adolescents with type 1 diabetes: a population-based cohort study. Diabetes Care 2015;38(3):453–9.

[84] Liu S, Ludvigsson JF, Lichtenstein P, Gudbjörnsdottir S, Taylor MJ, Larsson H, et al. Educational outcomes in children and adolescents with type 1 diabetes and psychiatric disorders. JAMA Netw Open 2023;6(4):e238135.

[85] Kar Ş, Er E, Ata A, İnal-Kaleli İ, Özcan T, Köse S, et al. Effect of metabolic control on cognitive functions in children and adolescents with type 1 diabetes mellitus J Pediatr Endocrinol Metab 2023;Available from. Available from: http://doi.org/10.1515/jpem-2023-0027.

[86] Perantie DC, Lim A, Wu J, Weaver P, Warren SL, Sadler M, et al. Effects of prior hypoglycemia and hyperglycemia on cognition in children with type 1 diabetes mellitus. Pediatr Diabetes 2008;9(2):87–95.

[87] He J, Li S, Liu F, Zheng H, Yan X, Xie Y, et al. Glycemic control is related to cognitive dysfunction in Chinese children with type 1 diabetes mellitus. J Diabetes 2018;10(12):948–57.

[88] Semenkovich K, Patel PP, Pollock AB, Beach KA, Nelson S, Masterson JJ, et al. Academic abilities and glycaemic control in children and young people with Type 1 diabetes mellitus. Diabet Med 2016;33(5):668–73.

[89] Reiss AL, Jo B, Arbelaez AM, Tsalikian E, Buckingham B, Weinzimer SA, et al. A pilot randomized trial to examine effects of a hybrid closed-loop insulin delivery system on neurodevelopmental and cognitive outcomes in adolescents with type 1 diabetes. Nat Commun 2022;13 (1):4940.

[90] DeBaun MR, Armstrong FD, McKinstry RC, Ware RE, Vichinsky E, Kirkham FJ. Silent cerebral infarcts: a review on a prevalent and progressive cause of neurologic injury in sickle cell anemia. Blood 2012;119(20):4587–96.

[91] Armstrong FD, Thompson Jr RJ, Wang W, Zimmerman R, Pegelow CH, Miller S, et al. Cognitive functioning and brain magnetic resonance imaging in children with sickle cell disease. Neuropsychology Committee of the cooperative study of sickle cell disease. Pediatrics 1996;97(6 Pt 1):864–70.

[92] Schatz J, Brown RT, Pascual JM, Hsu L, DeBaun MR. Poor school and cognitive functioning with silent cerebral infarcts and sickle cell disease. Neurology 2001;56(8):1109–11.

[93] Steen RG, Miles MA, Helton KJ, Strawn S, Wang W, Xiong X, et al. Cognitive impairment in children with hemoglobin SS sickle cell disease: relationship to MR imaging findings and hematocrit. Am J Neuroradiol 2003;24(3):382–9.

[94] King AA, Strouse JJ, Rodeghier MJ, Compas BE, Casella JF, McKinstry RC, et al. Parent education and biologic factors influence on cognition in sickle cell anemia. Am J Hematol 2014;89(2):162–7.

[95] Knight LMJ, King AA, Strouse JJ, Tanabe P. Pediatric neurodevelopmental delays in children 0 to 5 years of age with sickle cell disease: a systematic literature review. J Pediatr Hematol Oncol 2021;43(3):104–11.

[96] Heitzer AM, Cohen DL, Okhomina VI, Trpchevska A, Potter B, Longoria J, et al. Neurocognitive functioning in preschool children with sickle cell disease. Pediatr Blood Cancer 2022;69(3)e29531.

[97] Lance EI, Cannon AD, Shapiro BK, Lee LC, Johnston MV, Casella JF. Co-occurrence of neurodevelopmental disorders in pediatric sickle cell disease. J Dev Behav Pediatr 2021;42(6):463–71.

[98] Lance EI, Cannon AD, Casella JF, Shapiro BK. Neurocognitive screening in sickle cell disease. Pediatr Blood Cancer 2022;69(9):e29861.

[99] Stender M, Pimenta JM, Cheung M, Irving M, Mukherjee S. Comprehensive literature review on the prevalence of comorbid conditions in patients with achondroplasia. Bone 2022;162:116472.

[100] Wigg K, Tofts L, Benson S, Porter M. The neuropsychological function of children with achondroplasia. Am J Med Genet A 2016;170 (11):2882–8.

[101] Hecht JT, Thompson NM, Weir T, Patchell L, Horton WA. Cognitive and motor skills in achondroplastic infants: neurologic and respiratory correlates. Am J Med Genet 1991;41(2):208–11.

[102] Hecht JT, Bodensteiner JB, Butler IJ. Neurologic manifestations of achondroplasia. Handb Clin Neurol 2014;119:551–63.

[103] Pauli RM. Achondroplasia: a comprehensive clinical review. Orphanet J Rare Dis 2019;14(1):1.

[104] Thompson NM, Hecht JT, Bohan TP, Kramer LA, Davidson K, Brandt ME, et al. Neuroanatomic and neuropsychological outcome in school-age children with achondroplasia. Am J Med Genet 1999;88(2):145–53.

[105] Ablon J. Personality and stereotype in osteogenesis imperfecta: behavioral phenotype or response to life's hard challenges? Am J Med Genet A 2003;122A(3):201–14.

[106] Engelbert RH, Uiterwaal CS, Gerver WJ, van der Net JJ, Pruijs HE, Helders PJ. Osteogenesis imperfecta in childhood: impairment and disability. A prospective study with 4-year follow-up. Arch Phys Med Rehabil 2004;85(5):772–8.

[107] Roach ES, DiMario FJ, Kandt RS, Northrup H. Tuberous Sclerosis Consensus Conference: recommendations for diagnostic evaluation. National Tuberous Sclerosis Association. J Child Neurol 1999;14(6):401–7.

[108] Kotulska K, Kwiatkowski DJ, Curatolo P, Weschke B, Riney K, Jansen F, et al. Prevention of epilepsy in infants with tuberous sclerosis complex in the EPISTOP trial. Ann Neurol 2021;89(2):304–14.

[109] Gipson TT, Johnston MV. New insights into the pathogenesis and prevention of tuberous sclerosis-associated neuropsychiatric disorders (TAND). F1000Res 2017;6. Available from: https://doi.org/10.12688/f1000research.11110.1.

[110] Vanclooster S, Bissell S, van Eeghen AM, Chambers N, De Waele L, Byars AW, et al. The research landscape of tuberous sclerosis complex-associated neuropsychiatric disorders (TAND)-a comprehensive scoping review. J Neurodev Disord 2022;14(1):13.

[111] de Vries P, Humphrey A, McCartney D, Prather P, Bolton P, Hunt A, et al. Consensus clinical guidelines for the assessment of cognitive and behavioural problems in tuberous sclerosis. Eur Child Adolesc Psychiatry 2005;14(4):183–90.

[112] Gipson TT. Consequences of delay in screening, monitoring, and treatment of angiomyolipoma and tuberous sclerosis: a case report. Clin Nephrol 2018;90(1):71–5.

[113] Gipson TT, Gerner G, Wilson MA, Blue ME, Johnston MV. Potential for treatment of severe autism in tuberous sclerosis complex. World J Clin Pediatr 2013;2(3):16–25.

[114] Gipson TT, Gerner G, Srivastava S, Poretti A, Vaurio R, Hartman A, et al. Early neurodevelopmental screening in tuberous sclerosis complex: a potential window of opportunity. Pediatr Neurol 2014;51(3):398–402.

[115] Mautner VF, Lindenau M, Baser ME, Hazim W, Tatagiba M, Haase W, et al. The neuroimaging and clinical spectrum of neurofibromatosis 2. Neurosurgery 1996;38(5):880–5.

[116] Hyman SL, Shores A, North KN. The nature and frequency of cognitive deficits in children with neurofibromatosis type 1. Neurology 2005;65 (7):1037–44.

[117] Tadini G, Legius E, Brems H. Multidisciplinary approach to neurofibromatosis type 1. Springer Nature; 2020. p. 313.

[118] Garg S, Lehtonen A, Huson SM, Emsley R, Trump D, Evans DG, et al. Autism and other psychiatric comorbidity in neurofibromatosis type 1: evidence from a population-based study. Dev Med Child Neurol 2013;55(2):139–45.

[119] Morotti H, Mastel S, Keller K, Barnard RA, Hall T, O'Roak BJ, et al. Autism and attention-deficit/hyperactivity disorders and symptoms in children with neurofibromatosis type 1. Dev Med Child Neurol 2021;63(2):226–32.

[120] Klein-Tasman BP. Are the autism symptoms in neurofibromatosis type 1 actually autism? Dev Med Child Neurol 2021;63(2):132.

[121] Hönig M, Albert MH, Schulz A, Sparber-Sauer M, Schütz C, Belohradsky B, et al. Patients with adenosine deaminase deficiency surviving after hematopoietic stem cell transplantation are at high risk of CNS complications. Blood 2007;109(8):3595–602.

[122] Schejter YD, Even-Or E, Shadur B, NaserEddin A, Stepensky P, Zaidman I. The broad clinical spectrum and transplant results of PNP deficiency. J Clin Immunol 2020;40(1):123–30.

[123] Crow YJ, Manel N. Aicardi-Goutières syndrome and the type I interferonopathies. Nat Rev Immunol 2015;15(7):429–40.

[124] Bousfiha A, Jeddane L, Picard C, Ailal F, Bobby Gaspar H, Al-Herz W, et al. The 2017 IUIS phenotypic classification for primary immunodeficiencies. J Clin Immunol 2018;38(1):129–43.

[125] Castagnoli R, Delmonte OM, Calzoni E, Notarangelo LD. Hematopoietic stem cell transplantation in primary immunodeficiency diseases: current status and future perspectives. Front Pediatr 2019;7:295.

[126] Barton LL, Friedman NR. The neurological manifestations of pediatric infectious diseases and immunodeficiency syndromes. Springer Science & Business Media; 2008. p. 421.

[127] Rothblum-Oviatt C, Wright J, Lefton-Greif MA, McGrath-Morrow SA, Crawford TO, Lederman HM. Ataxia telangiectasia: a review. Orphanet J Rare Dis 2016;11(1):159.

[128] Shin H, Yum MS, Kim MJ, Suh JK, Im HJ, Kim H, et al. Hematopoietic stem cell transplantation-associated neurological complications and their brain MR imaging findings in a pediatric population. Cancers 2021;13(12). Available from: https://doi.org/10.3390/cancers13123090.

[129] Abrahão R, Huynh JC, Benjamin DJ, Li QW, Winestone LE, Muffly L, et al. Chronic medical conditions and late effects after acute myeloid leukaemia in adolescents and young adults: a population-based study. Int J Epidemiol 2021;50(2):663–74.

[130] Bevans M, El-Jawahri A, Tierney DK, Wiener L, Wood WA, Hoodin F, et al. National institutes of health hematopoietic cell transplantation late effects initiative: the patient-centered outcomes working group report. Biol Blood Marrow Transpl 2017;23(4):538–51.

[131] Simms S, Kazak AE, Golomb V, Goldwein J, Bunin N. Cognitive, behavioral, and social outcome in survivors of childhood stem cell transplantation. J Pediatr Hematol Oncol 2002;24(2):115–19.

[132] Smedler AC, Winiarski J. Neuropsychological outcome in very young hematopoietic SCT recipients in relation to pretransplant conditioning. Bone Marrow Transpl 2008;42(8):515–22.

[133] Gurney JG, Ness KK, Rosenthal J, Forman SJ, Bhatia S, Baker KS. Visual, auditory, sensory, and motor impairments in long-term survivors of hematopoietic stem cell transplantation performed in childhood: results from the bone marrow transplant survivor study. Cancer 2006;106(6):1402–8.

[134] King AA, McKinstry RC, Wu J, Eapen M, Abel R, Varughese T, et al. Functional and radiologic assessment of the brain after reduced-intensity unrelated donor transplantation for severe sickle cell disease: blood and marrow transplant clinical trials network study 0601. Biol Blood Marrow Transpl 2019;25(5):e174–8.

[135] Scott O, Kim VHD, Reid B, Pham-Huy A, Atkinson AR, Aiuti A, et al. Long-term outcome of adenosine deaminase-deficient patients-a single-center experience. J Clin Immunol 2017;37(6):582–91.

[136] Gassas A, Raiman J, White L, Schechter T, Clarke J, Doyle J. Long-term adaptive functioning outcomes of children with inherited metabolic and genetic diseases treated with hematopoietic stem cell transplantation in a single large pediatric center: parents' perspective. J Pediatr Hematol Oncol 2011;33(3):216–20.

[137] Chen BT, Ortiz AO, Dagis A, Torricelli C, Parker P, Openshaw H. Brain imaging findings in symptomatic patients after allogeneic haematopoietic stem cell transplantation: correlation with clinical outcome. Eur Radiol 2012;22(10):2273–81.

[138] Andropoulos DB, Greene MF. Anesthesia and developing brains - implications of the FDA warning. N Engl J Med 2017;376(10):905–7.

[139] Sun LS, Li G, Miller TLK, Salorio C, Byrne MW, Bellinger DC, et al. Association between a single general anesthesia exposure before age 36 months and neurocognitive outcomes in later childhood. JAMA 2016;315(21):2312–20.

[140] Davidson AJ, Morton NS, Arnup SJ, de Graaff JC, Disma N, Withington DE, et al. Apnea after awake regional and general anesthesia in infants: the general anesthesia compared to spinal anesthesia study—comparing apnea and neurodevelopmental outcomes, a randomized controlled trial. Anesthesiology 2015;123(1):38–54.

[141] Wilder RT, Flick RP, Sprung J, Katusic SK, Barbaresi WJ, Mickelson C, et al. Early exposure to anesthesia and learning disabilities in a population-based birth cohort. Anesthesiology 2009;110(4):796–804.

[142] Warner DO, Zaccariello MJ, Katusic SK, Schroeder DR, Hanson AC, Schulte PJ, et al. Neuropsychological and behavioral outcomes after exposure of young children to procedures requiring general anesthesia: the mayo anesthesia safety in kids (MASK) study. Anesthesiology 2018;129(1):89–105.

[143] Sprung J, Flick RP, Katusic SK, Colligan RC, Barbaresi WJ, Bojanić K, et al. Attention-deficit/hyperactivity disorder after early exposure to procedures requiring general anesthesia. Mayo Clin Proc 2012;87(2):120–9.

[144] Namatovu F, Strandh M, Ivarsson A, Nilsson K. Effect of childhood coeliac disease on ninth grade school performance: evidence from a population-based study. Arch Dis Child 2018;103(2):143–8.

[145] Moser JJ, Veale PM, McAllister DL, Archer DP. A systematic review and quantitative analysis of neurocognitive outcomes in children with four chronic illnesses [Internet]. Pediatr Anesth 2013;1084–96. Available from: https://doi.org/10.1111/pan.12255.

[146] Sentenac M, Arnaud C, Gavin A, Molcho M, Gabhainn SN, Godeau E. Peer victimization among school-aged children with chronic conditions. Epidemiol Rev 2012;34:120–8.

[147] Lipkin PH, Macias MM, Baer Chen B, Coury D, Gottschlich EA, Hyman SL, et al. Trends in pediatricians' developmental screening: 2002-2016. Pediatrics 2020;145(4). Available from: https://doi.org/10.1542/peds.2019-0851.

[148] Lipkin PH, Macias MMCouncil on Children with Disabilities, Section on Developmental and Behavioral Pediatrics. Promoting optimal development: identifying infants and young children with developmental disorders through developmental surveillance and screening. Pediatrics 2020;145(1). Available from: http://doi.org/10.1542/peds.2019-3449.

[149] Campbell F, Conti G, Heckman JJ, Moon SH, Pinto R, Pungello E, et al. Early childhood investments substantially boost adult health. Science 2014;343(6178):1478–85.

[150] Whitehouse AJO, Varcin KJ, Pillar S, Billingham W, Alvares GA, Barbaro J, et al. Effect of preemptive intervention on developmental outcomes among infants showing early signs of autism: a randomized clinical trial of outcomes to diagnosis. JAMA Pediatr 2021;175(11): e213298.

[151] Estes A, Swain DM, MacDuffie KE. The effects of early autism intervention on parents and family adaptive functioning. Pediatr Med 2019;2. Available from: https://doi.org/10.21037/pm.2019.05.05.

Section IV

Management of neurodevelopmental disorders

Chapter 31

Developing a management program

Sherry Vinson[1] and Bruce K. Shapiro[2]

[1]*Department of Pediatrics, Baylor College of Medicine, Houston, TX, United States,* [2]*Department of Pediatrics, Johns Hopkins University School of Medicine, Baltimore, MD, United States*

Introduction

The management of neurodevelopmental disabilities (NDD) is challenging for various reasons, including the need to address multiple dysfunctions. The definition of NDD itself—a subset of special healthcare needs that have their onset in childhood, affect more than three areas of life function, and are expected to last for an extended period of time—suggests the complexity of NDD management.

While an NDD diagnosis such as cerebral palsy, intellectual disability, learning disability, and communication disorders e.g., receptive-expressive language disorder or autism spectrum disorder describes the most evident manifestation of brain dysfunction, the etiology of the NDD affects the entire brain, resulting in diminshed ability and inreased behavior difficulties. Whether the NDD is caused by a genetic or an acquired etiology, comorbidities are common.

Obtaining treatments for NDD often requires time-consuming advocacy. NDD treatments extend beyond medicine and require partnerships with education, nutrition, nursing, occupational therapy, psychology, physical therapy, speech therapy, social services, and vocational rehabilitation professionals. While such partnerships allow for shared responsibilities, their necessity demands more effort than simply prescribing a medication.

Unlike an acute health condition that resolves with one treatment, NDD requires lifelong monitoring for intervention needs. Maturation affects the expression of brain dysfunction over time, with different challenges arising at different ages throughout the lifetime of the individual with NDD. Furthermore, the chronicity of NDD makes individuals with NDD and their families susceptible to questionable interventions that promise cures. Appropriate management of NDD considers the patient, the family, and the larger community. Additionally, it requires taking the time to help the patient's family understand the patient's condition and needs while connecting them with supports to navigate helpful systems.

Background/foundations

A useful way of approaching the management of NDD is based on the World Health Organization's International Classification of Functioning, Disability, and Health (ICF) [1]. The ICF views disability, at the individual level, as impairment, functional limitations, and restrictions to participation. Impairment refers to the anatomical or physiological (including psychological functions) derangement. Activity limitations are difficulties an individual may have in executing a task or action (activities). Participation restrictions are problems an individual may experience in interacting in the world around them. (Previously, activity limitation was called disability and restriction to participation was handicap.) An example may be a young adult who was born prematurely and experienced hypoxic ischemic encephalopathy that resulted in periventricular leukomalacia and spastic diplegic cerebral palsy. His current impairment is his spasticity. His impairment limits his activities by preventing him from walking and requires the use of a wheelchair. The lack of a functional elevator at his local subway station restricts his participation because he is unable to travel to work. Another example may be a teenager who has impaired attention which gives rise to academic underachievement (activity limitation) and, without intervention, may cause him to not graduate from high school, thereby limiting his future employment opportunities. Distinguishing between the terms, impairment, activity, and restriction can be difficult. At times the distinction between increasing the ability to perform tasks (diminishing activity limitations) and increasing the ability to participate in everyday life situations overlap. However, understanding the distinction between these terms provides a

Capute and Accardo's Neurodevelopmental Disabilities in Infancy and Childhood. DOI: https://doi.org/10.1016/B978-0-12-824060-1.00011-0

structure for optimal NDD management combining treatment, interventions, and environmental modifications to help individuals with NDD. Using the WHO ICF approach facilitates asking three questions to address the needs of the individual with NDD: What can be done to cure or ameliorate the impairment? What can be done to enable the individual to be more successful in undertaking activities? What are the obstacles that society creates that limit the individual's ability to participate in the greater society and how do we minimize their effect?

Planning an NDD management program necessitates that the planner understands the difference between ability, capacity, capability, and performance. Ability is the actual skill. Capacity is when someone can perform that skill in a standardized, controlled environment. Capability describes what a person can do in their daily environment. Performance refers to what a person actually does in their daily environment. An adult without an impairment has the capability and the capacity to crawl but typically does not perform this skill. However, an 8-month-old without an impairment would have the capacity, capability, and performance with regard to crawling. An 8-month- old with an impairment that keeps him from getting into a crawling position but who can crawl if placed in the crawling position would have the capacity but not the capability or performance function of crawling. A person with slow processing speed may have the capacity to exhibit a skill when given double the typical time for exhibiting the skill but might not have the capability or performance ability for exhibiting the skill.

An understanding of the different types of therapy, what the various therapists do, and where there are overlaps between the therapy types is essential before developing an NDD management program. Knowing that some speech therapists specialize in the motor processes of speech/feeding, while others specialize in helping patients use language to socially connect, and others specialize in training patients to understand word meanings, is essential to avoid wasted referrals. Recognizing that some occupational therapists are trained in feeding issues, while others work on skills of daily living and fine motor issues, provide planners an alternative pathway if a speech therapist is available for feeding therapy. Realizing that physical therapists work on mobility and movement aids in decisions of when to choose a physical therapist.

Management and therapy are often used interchangeably but are quite different. Most therapy focuses on remediation: reducing impairments, increasing the ability to undertake activities, or enhancing the individuals's ability to participate in daily life. The interaction primarily involves the individual, the therapist, and the family. Management, on the other hand, is not a singular entity. It may have different goals addressed by a team (see Table 31.1) or multiple goals addressed by a single therapist.

Circumvention therapy bypasses the challenges imposed by the impairment and capitalizes on the individual's strengths, often through the integration of technology. Motor disorders may be circumvented by power wheel chairs, speech synthesizers, and nonoral feeding techniques. Advances in technology have seamlessly integrated many circumvention devices with smart phones. Executive dysfunction may be mitigated by multiple alarms to ensure medication compliance, calendars to aid in organization, and multiple approaches to enhance classroom performance. [See chapters: Principles of neurorehabilitation and Circumvention therapy].

Rehabilitation therapists aim to restore the individual to their premorbid state. Rehabilitative goals are often appropriate for individuals with acquired conditions or who have undergone surgical intervention for their cerebral palsy. The key outcome question of rehabilitation therapy is clear: Is the person functioning as well (or better) as they were before the intervention?

TABLE 31.1 Goals of therapy [2].

Goal	Intensity	Length	Outcome
Circumvention	High	Short	Clear
Rehabilitation	High	Moderate	Clear
Acquisition of new skills	High	Moderate	Clear
Improve skill quality	Moderate to high	Long	Unclear
Maintenance	Moderate to low	Long	Unprovable
Prevention of adverse outcome	Moderate to low	Variable	Unprovable
Development of prerequisite skills	Moderate	Long	Unclear

Source: Adapted from Accardo J, Shapiro B. Neurodevelopmental disabilities: beyond the diagnosis. Seminars in pediatric neurology. Elsevier; 2006. p. 242–49.

Acquisition of new skills is an appropriate goal for those who have the necessary prerequisite skills but who require an extra "push" to get to the next level. An example of this is the 21-month-old girl with prerequisites for walking but whose progress toward ambulation has stalled. In her case, a short course of physical therapy often will achieve the desired goal of ambulation.

Therapy to improve quality seeks to improve the way an activity is performed. For instance, quality improvement therapy for children with developmental coordination disorder who show dysgraphia seeks to bridge the gap between capacity and performance. While a child may demonstrate excellent penmanship within a controlled environment (treatment session), they might revert to previous writing style when confronted with the distractions and pressures of a typical classroom (e.g., 2-minute math tests), and the need for active learning. Additionally, focusing on whether the child is adhering to a "normal" way of performing an activity may be incorrect and unfair. For an individual child, doing an activity in the "normal" way may not be the most mechanically efficient approach. Finally, there may be instances of therapy to improve quality, for example, gait training for genu recurvatum, that may prevent long-term damage.

Maintenance therapy aims to preserve function and prevent the loss of skills. It is widely employed and has significant appeal. Examples of maintenance therapy include range of motion exercises in rehabilitation settings and summer reading programs in schools. However, since it is challenging to prove a negative outcome, attributing the result solely to the intervention can be difficult.

Preventive therapy: The effectiveness of therapy in preventing an adverse outcome is clearest when the adverse outcome is almost universal. This is most applicable to metabolic disorders like phenylketonuria (PKU) or congenital hypothyroidism. While the outcome might lend support to the therapy in some conditions, it may not be as applicable to others. For example, bisphosphonates effectively prevent fractures in children with osteogenesis imperfecta but their impact on cerebral palsy is less clear. Outside the biomedical sphere, positioning may be an intervention to prevent plagiocephaly. Consequently, a general statement cannot be made about the outcome of therapy to prevent adversity.

Therapy to develop prerequisite skills is the most controversial among therapeutic goals. The controversy arises from two issues. First, the therapy is often applied to developing children, making it unclear whether the changes seen are due to the intervention or due to maturation and development is unclear. Second, the end point of this type of therapy is often nebulous. When does therapy reach a point where it should end? If it is when a child achieves typical functionality, then the quest to develop prerequisite skills is destined to fail.

Establishing therapeutic goals can be short term, intermediate term, and long term. An individual child may have goals of varying time frames. Long term goals serve as a road map to what the individual, family, and others may achieve. Conversely, short term goals focus on the immediate application of therapy. Communicating the goals of therapy when developing an NDD management plan is more likely to lead to successful advocacy for patients.

Management programs are inherently complex as they entail a comprehensive approach that focuses on the individual within the context of their family, home, and community. Such programs require multiple iterations and therapeutic strategies, thus necessitating thoughtful prioritization and coordination. Successful management programs require communication among all the participants: individuals, families, teachers, physicians, therapists, coaches, and others.

Insight into who sets the goals for someone with NDD is essential for effective planning. The process of goal setting starts with the individual who has NDD and involves their parents/guardians. Professionals from the relevant disciplines may recommend specific goals, assess their feasibility, and propose techniques that facilitate the achievement of those goals.

Developing a neurodevelopmental disabilities management program

Development of a comprehensive and effective NDD management program for an individual must consider approximately 10 domains to thoroughly address that individual's needs. These domains are:

1. *Maintenance of physiologic stability* such as ensuring hydration and adequate caloric intake; addressing seizures, sleep apnea, gastroesophageal reflux; and monitoring for toxic effects of medication management.
2. *Nursing care* to ensure appropriate medication administration; bowel/bladder hygiene; appropriate food/liquid intake; optimal positioning, maintenance of skin health/integrity, proper application of bracing/splinting.
3. *Activities of daily living (ADL)* focusing on the individual's ability to master control of feeding, dressing, bathing, toileting, and other self-care skills.
4. *Communication* to ensure the individual has the optimal chance to communicate through verbal/nonverbal/augmentative technology techniques since communication allows individuals to control their environment.
5. *Mobility* to ensure that even individuals who cannot ambulate have appropriate mobility.

6. *Transportation* to ensure safe transport with suitable adaptations needed for individuals with motor impairments such as spasticity/rigidity/hypotonia or intellectual disability.
7. *Education/Work* to ensure the individual has appropriate activities designed to help them progress academically and functionally to their full potential using the Individuals with Disabilities Education Act (IDEA), Americans with Disability Act (ADA), and Section 504 of the Rehabilitation Act.
8. *Leisure* activities to ensure the individuals have play and socialization opprtunities that are appropriate to their capabilities, thereby enhancing their quality of life.
9. *Living* arrangement to ensure the individual has an appropriate residence with suitable care, guidance, and supervision.
10. *Behavior* to ensure appropriate cooperation with instructors and avoidance of problems with peers or self-injury.

Ignoring a domain will most likely interfere with successful interventions proposed to address needs in another domain. For example, neglecting the behavior domain will most likely interfere with appropriate plans in all other nine domains.

There are five questions that need to be asked when developing an individual's NDD management program. These questions are: (1) What is the person's capacity, capability, performance in language, problem solving, motor, and social behavioral areas? (2) What are the needs that the patient may have in each of the domains? (3) What interventions are needed to address the short term issues? (4) What will the patient look like when maturity is reached? (5) What interventions are needed to get them there?

The choice of optimal interventions by NDD management planners should be based on research demonstrating effectiveness. it is imperative to consider what is required to administer the intervention including family's capacity, cost, intensity, burden of care (including multiple interventions), and length of time needed for the intervention to achieve success. Defining the criteria for success is essential when selecting interventions. These decisions guide which interventions to choose based on the needs and when to stop or change the intervention.

Implementing a neurodevelopmental disabilities management program

The most difficult part of helping individuals with NDD reach their optimal potential and quality of life is implementing the developed program and assisting them or their families in navigating the involved systems. Assembling information from medical, education, psychology, therapy, and social service professionals regarding best practices in implementation *before* moving forward with an individual's service plan is the key to successful execution. Naturally, gathering information takes time, and individuals may require services before finding "the perfect" way to implement the plan. However, professionals involved in the NDD management program for an individual patient should build upon their knowledge of best implementation practice for certain needs/types of NDD and collaborate in an interdisciplinary manner to help their patients.

Implementing an NDD management program requires consideration of the different delivery models. Should therapy be done through the public school or through private medical therapists or a combination of both? Should special education be done through public or private or home school? Should therapy use a coaching model or direct therapy delivery? One size does not fit all and different models might be necessary to address different goals. The correct choices depend on the answers to the five aforementioned questions, alongside research on effective interventions, the practical aspects of administering the intervention, and the definition of success for the chosen interventions.

Therapist factors should not be underestimated. The therapist's ability to deliver therapy, interact with the child and family, work with other team members, show enthusiasm, and demonstrate flexibility are important considerations when selecting therapists. Therapists should actively participate in setting the overall management program. They should think about long-term outcomes and how short-term activities enable the patient to progress toward the long-term goals. The therapists should also consider what is required to administer the intervention, and advocate for the patient and the family to the rest of the team. The choice of therapists is influenced by NDD management planners' understanding of each therapist's role and potential overlaps, the objectives of the selected therapies, and the individual's impairment, disability, handicap, capacity, capability, and performance.

Coordination and prioritization of the chosen interventions is important for effective implementation of the NDD management program. As alluded to previously, interdisciplinary and multidisciplinary approaches are needed. The benefits of partnerships across disciplines include improved communication, setting priorities, effective coordination, information regarding novel approaches, and reciprocal assistance "horse trading" (i.e., "if you do this for me and my patient, I'll do this for you and your patient").

Every person with NDD needs a medical home. For the more common NDDs, the primary care provider is assigned. However, the primary care provider for someone with NDD *must* have sufficient staff to liaise with the individual's other medical providers as well as the patient's education, psychology, therapy, and social service professionals. In the case of the individual with rare or complex disorders, the primary care provider might lack the resources or the expertise to serve as the coordinator of the medical home. In that case, it may fall to the subspecialist team to provide the medical home and coordinate primary care activities as part of the management program. [See chapter: Medical home]

Parents are the keystone of the therapeutic arch for children with NDD. Often they serve as case manager as well as parent. The success of these parents depends on the training they receive from each member of the interdisciplinary team. Appropriate parental training starts with establishing the definition of success, providing the parent (s) with a comprehensive view of their child, and initiating the process of long-term planning. This training also helps the parent connect with support groups. The learning program for a parent is extensive and complicated. It may include knowing the best practices for interventions, school advocacy, lifespan planning and transitions, understanding Medicaid waivers, social security income (SSI) disability, exploring guardianship and alternatives to guardianship, and postsecondary training and work, establishing adult medical care, and considering residential facility use. All of this training should aim at helping and empowering the parent(s) to successfully navigate complex systems.

Monitoring

Once an individual's NDD management program is being implemented, monitoring of the program is required. The goals of this monitoring should be: (1) ensuring the therapies are working in concert to achieve the goals; (2) addressing previously unanswered or unaddressed questions; (3) sharing new information, such as emerging interventions with proven efficacy; and (4) providing anticipatory guidance, including the next steps to take.

Recommendations for the frequency of monitoring include seeing the individual with NDD at times of transition: (1) initial diagnosis; (2) initial implementation of management program; (3) school entry; (4) mid elementary school; (5) middle school; (6) early high school when transition to adulthood planning is formalized, although long term planning should proceed from the time of the initial diagnosis; (7) late high school to address vocational and higher education, decision making, living and other issues related to the individuals potential [See chapter: Transition and Aging]. A recommended monitoring schedule is as follows: (1) every 3–6 months in the first 3 years of life; (2) every 6–12 months during the elementary and middle school years; (3) yearly in high school and postsecondary stages. Additionally, the individual with NDD should be seen whenever there are concerns about loss of skills, new challenges, or that the planned program is not working as desired.

The content of each monitoring session should include assessment of the individual's current functional performance. Each monitoring session should also ask questions about the individual's communication and related concerns. Inquiries about the individual's health, behavior, skill gains and losses since the last monitoring session, as well as any new challenges. The professional conducting the monitoring needs to ask the following question with each monitoring session, "Are additional evaluations required to address specific questions?"

Among the questions that arise during monitoring sessions are questions about when to start therapy, no matter the type of therapy. Is starting earlier better? How early is considered early enough? The best answer to these questions is to intervene with therapy early enough that harm does not accrue.

Other questions about therapy should aim to reveal red flags hinting at the need to change therapists or stop a certain therapy. Such red flags are that therapy is a chore, the child show aversion toward the therapist, the therapist does not respect or regard the child's cues, the therapy goals are not clearly stated, the therapist cannot forecast what the child's abilities will be when therapy is completed, gives no advice regarding how to determine progress, or suggests that treatment failures are the parents' fault.

Summary

Management of NDD, no matter the type, involves many disciplines, a central coordinator, and partnerships between the person with NDD, the parents and caretakers, and all professionals involved. NDD management needs to be based on an initial history, physical, ability, and functional assessment in the language and problem-solving, motor and social domains with consideration of physiological needs, nursing care, activities of daily living training, communication, mobility, transportation, education/work, leisure, living, and behavior domains. It requires ongoing monitoring and much advocacy aided by use of laws and helpful support groups. Effective NDD management aims to help individuals with NDD reach their full potential by supporting families, addressing impairments, diminishing functional limitations, and ehancing the ability of the individual to participate in the world around them.

References

[1] World Health Organization. The international classification of impairments, disabilities, and handicaps. World Health Organization; 2002.

[2] Accardo J, Shapiro B. Neurodevelopmental disabilities: beyond the diagnosis. Seminars in pediatric neurology. Elsevier; 2006. p. 242–9.

Chapter 32

Early Intervention

Thomas D. Challman[1,2] and Scott M. Myers[1,2]
[1]Geisinger Autism & Developmental Medicine Institute, Danville, PA, United States, [2]Geisinger Health System and Geisinger Commonwealth School of Medicine, Lewisburg, PA, United States

Introduction

The first few years of life are a time of dynamic change in a child's communication, motor, social, cognitive, and adaptive skills. Pediatricians and other primary care providers are tasked with monitoring a child's development across these domains, with a goal of identifying children who are not making the expected progress. Interest in interventions that might favorably alter the rate of acquisition of developmental skills emerged in the early decades of the 20th century, leading to the creation of programs designed to provide enrichment for children with a history of environmental deprivation or biological risk factors. As the research base supporting the value of these types of programs matured, there was an increase in legislative activities and mandates supporting the formation of systems of Early Intervention (EI) services. This chapter provides an overview of EI history and effectiveness, as well as discussion of some practical considerations for pediatricians and specialists who care for children who have, or are at risk for, developmental delays.

History of Early Intervention

Infant stimulation programs in the United States originated in several centers by the mid-20th century, directed toward children with evidence of, or who were at increased risk for, motor or intellectual disabilities. These programs were designed to be multidisciplinary in nature, combining early intensive therapy with support from medical and surgical specialists. Over subsequent years, interest grew in providing intervention to children with a wider array of developmental conditions and risks [1,2]. As more programs were established, there were attempts to measure the effectiveness of these activities. In 1974 the United Cerebral Palsy Association and the Bureau of the Handicapped of the United States Department of Health, Education, and Welfare organized a study of over 1000 infants participating in infant stimulation programs at five centers [3]. While this study was not able to readily evaluate outcomes due to the uncontrolled design, several principles emerged which formed a template for later EI programs: parent training, services provided within the home, and availability of a multidisciplinary team supporting the family. Concurrent efforts studied the effectiveness of other intervention programs, such as the Ypsilanti Perry Preschool Project, designed to study the effects of systematic early educational assistance to economically disadvantaged children [4,5]. Unfortunately, the logistical, financial, and geographic realities were such that access to any EI program at that time was significantly limited for most children who could potentially benefit.

Prior to the mid-20th century, the US federal government played a limited role in supporting public school education or in providing educational benefits for individuals with special developmental needs. In 1912 the Children's Bureau (whose first chief was Julia Lathrop—the first woman to lead a US federal bureau) was established for the purpose of investigating and reporting on "all matters pertaining to the welfare of children and child life among all classes of our people" [6]. This agency played an important role in early efforts to reduce infant mortality and to improve children's living conditions. The Social Security Act of 1935, which included several provisions geared toward children and individuals with disabilities, expanded the federal footprint of the Children's Bureau [7]. Title V of the Social Security Act continues to provide block grants to states to support maternal and child health programs, including programs for children with disabilities. The National Defense Education Act (NDEA) of 1958, spurred by the launch of the Sputnik satellite by the Soviet Union and concern that the United States was lagging behind geopolitical rivals in scientific

Capute and Accardo's Neurodevelopmental Disabilities in Infancy and Childhood. DOI: https://doi.org/10.1016/B978-0-12-824060-1.00057-2

expertise, provided funding for science and technology education in US schools and was the first major legislative effort at the federal level to provide financial support for public schools [8]. Shortly after the passage of NDEA, Public Law 85–926 was signed by President Dwight Eisenhower [9]. This law, through the provision of grants to institutions of higher learning and state educational agencies, encouraged the expansion of training programs for teachers of children with intellectual disabilities. The scope of this legislation was later expanded in 1963 via Public Law 88–164 to include programs for teachers being trained in the education of children with other disabilities (vision, speech, hearing, motor, emotional impairment) [10]. The Elementary and Secondary Education Act of 1965 provided additional funding for primary and secondary education, with a goal of reducing gaps in achievement among students from disadvantaged backgrounds [11]. Title 1 of this act directed funding to school districts with an increased percentage of pupils from low-income families. Unfortunately, children with special educational needs continued to face substantial barriers in most areas of the United States. Until the 1970s, most states were not providing adequate educational opportunities for these children, either placing children in programs inappropriate for their needs, or denying them entry into a school program altogether.

Special education legislation

Persistent efforts by parents of children with disabilities and other advocates led to a series of legislative successes in the last decades of the 20th century, creating the modern special education and EI landscape (Fig. 32.1). In 1971, an advocacy group for children with intellectual disabilities, PARC, sued the Commonwealth of Pennsylvania over a statute that allowed Pennsylvania schools to deny children entry into an educational program if they had not achieved a mental age of 5 years by the time they were 8 years old [12]. This case is widely regarded as the first "right to education" lawsuit in the United States. The suit was settled in favor of PARC and led to an agreement that the state would provide free educational services to all children without regard to their disability status. Another landmark lawsuit filed in 1971, *Mills v. Board of Education of District of Columbia* [13], resulted in a similar outcome—that the District of Columbia was compelled to provide a free public education and various procedural protections to children with disabilities.

Numerous lawsuits in other states followed, leading to recognition by the US Congress that federal intervention was needed to provide a suitable framework for educational services and protections for children with special needs. A series of amendments to the Elementary and Secondary Education Act, and the passage of the Education of the Handicapped Act of 1970, laid incremental groundwork in expanding the federal role in ensuring access to educational programs for individuals with disabilities. Despite these efforts, congressional hearings revealed troubling evidence that the educational needs of many children with disabilities were not being met [14]:

- More than half of the children with disabilities in the United States were not receiving appropriate educational services.
- One million children with disabilities in the United States were excluded entirely from the public school system.
- Many children participating in regular school programs were prevented from having a successful educational experience because their disabilities were undetected.
- Because of the lack of adequate services within the public school system, families were often forced to find services outside the public school system, often at great distance and expense.
- Financial resources were inadequate to meet the special educational needs of children with disabilities.

It became clear that a more robust federal response was needed to address these inequities. Two statutes were subsequently passed that have had far-reaching implications—the Rehabilitation Act of 1973 [15] and the Education for All Handicapped Children Act of 1975 (EAHCA) [14]. Section 504 of the Rehabilitation Act created additional civil rights

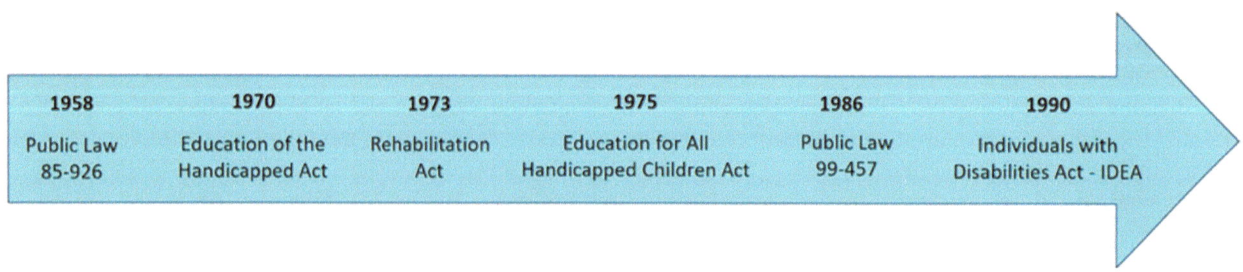

FIGURE 32.1 Timeline of selected special education and Early Intervention legislation.

protections for individuals with disabilities, specifically prohibiting discrimination based on disability by any program receiving federal financial assistance. This particular statute created a foundation for future disability rights legislation, including the Americans with Disabilities Act. EAHCA mandated that all public schools receiving federal funding provide equal access to services and one free meal per day for children with disabilities age 5 and older. EAHCA also required that schools have a mechanism for identifying and evaluating children with disabilities; perform annual meetings for the purpose of creating and monitoring an individualized education program; and ensure that students were provided educational services in the least restrictive environment. Services for children younger than age 3 were not initially addressed by this legislation.

While EAHCA initially did not specifically mandate the provision of educational services for preschool children with disabilities, incentives were offered to states to create intervention programs for children in the 3- to 5-year age range. As evidence grew that EI programs provided measurable value to young children with developmental delays or risk factors, there was interest in incorporating these programs into the existing special education legislation to improve access to these beneficial programs for young children. In 1986 PL 99−457 expanded EAHCA to require states to develop and implement comprehensive, coordinated, multidisciplinary, interagency programs of EI services for infants and toddlers with developmental delays from birth to age 3 [16].

In 1990 the EAHCA was reauthorized and renamed the Individuals with Disabilities Act (IDEA) [17]. IDEA, which was reauthorized and amended in 2004, is comprised of four parts. Part A outlines foundational principles for the Act; Part B contains provisions for special education for children with disabilities ages 3 through 21, including the state grants program for preschool children with disabilities ages 3 through 5; Part C details provisions for the grant program for infants and toddlers with disabilities; and Part D governs grant activities geared toward improving the education of children with disabilities at a national level. Core elements of the original EAHCA have been retained through various amendments and reauthorizations. Overarching principles of IDEA include:

- Free appropriate public education (FAPE)
- Appropriate evaluation
- Individualized education plan (IEP) / Individualized Family Service Plan (IFSP)
- Least restrictive environment (LRE)
- Parent and teacher participation
- Procedural safeguards

IDEA Part C—the program for infants and toddlers with disabilities

Beginning with the 1986 amendment to EAHCA (Part H), and subsequently in Part C of IDEA, the US Congress recognized several significant needs not addressed in previous legislation [18]:

- To enhance the development of infants and toddlers with disabilities, to minimize their potential for developmental delay, and to recognize the significant brain development that occurs during a child's first 3 years of life.
- To reduce the educational costs to society, including schools, by minimizing the need for special education and related services after infants and toddlers with disabilities reach school age.
- To maximize the potential for individuals with disabilities to live independently in society.
- To enhance the capacity of families to meet the special needs of their infants and toddlers with disabilities.
- To enhance the capacity of State and local agencies and service providers to identify, evaluate, and meet the needs of all children, particularly minority, low-income, inner city, and rural children, and infants and toddlers in foster care.

The main goals of Part C of IDEA are to develop statewide systems for the provision of intervention services for infants and toddlers with disabilities and their families; to assist in the coordination of payment for EI services from federal, state, local, and private sources; to improve existing programs and increase capacity to provide quality EI services to all children, including historically underrepresented populations; and to encourage states to expand opportunities for children under 3 years of age deemed "at-risk" of experiencing a significant developmental delay if EI services were not provided. Minimum requirements for each state's EI program for infants and toddlers outlined in Part C of IDEA include the following [19]:

- Rigorous definition of the term "developmental delay" to be used by the *state* in carrying out EI programs.
- State policy ensuring that EI services are based on scientifically based research and are available to all infants and toddlers with disabilities and their families.

- Timely, comprehensive, multidisciplinary evaluation.
- Individualized family service plan.
- Comprehensive Child Find system to identify infants and toddlers with disabilities.
- Public awareness program focusing on early identification of infants and toddlers with disabilities.
- Central directory that includes information on EI services, resources, available experts, and research projects.
- Comprehensive system of personnel development.
- Policies and procedures relating to the establishment and maintenance of qualifications for EI personnel.
- Establishment of a lead agency.
- Policy pertaining to the contracting or making of other arrangements with service providers to provide EI services in the state.
- Procedures for securing timely reimbursements of funds.
- Procedural safeguards.
- System for compiling data.
- State interagency coordinating council.
- Policies that ensure EI services are provided in natural environments, or in the most appropriate setting if services cannot be provided in a natural environment.

Part C of IDEA mandates the provision of EI services for infants and toddlers exhibiting delays in one or more of the following areas of development: physical, cognitive, communication, social or emotional, or adaptive. Principal EI services include physical therapy, occupational therapy, speech-language therapy, and special instruction—although a wider array of services and supports can be provided through IDEA Part C (Table 32.1). Importantly, each state or territory is permitted to develop their own criteria for what constitutes "developmental delay," resulting in some geographic variation in eligibility for EI programs. For example, several states require at least a 50% delay on a standardized measure in one or more developmental domains, whereas in others, a 25% delay is sufficient (Table 32.2). Children with specific diagnosed conditions likely to result in developmental delays (such as trisomy 21) are also eligible for EI services. Additionally, while Part C of IDEA encourages states to offer services to "at-risk" infants and toddlers (i.e., those children exposed to certain biological or environmental factors such as low birth weight, nutritional deprivation, or a history of abuse or neglect that can be associated with increased risk for developmental delays), most states currently do not [20].

Effectiveness of Early Intervention

Variability between EI programs and differences in the populations of children receiving these services create methodological difficulties in the rigorous analysis of the effectiveness of these programs. EI programs typically provide

TABLE 32.1 Early Intervention services included in IDEA Part C [21].

- Assistive technology
- Audiology services
- Family training, counseling, and home visits
- Health services (necessary to enable an otherwise eligible child to benefit from the other Early Intervention services)
- Medical services (for diagnostic or evaluation purposes to determine a child's developmental status and need for Early Intervention services)
- Nursing services
- Nutrition services
- Occupational therapy
- Physical therapy
- Psychological services (including administering and interpreting developmental tests, obtaining and interpreting information about a child's behavior, managing counseling and parent training)
- Service coordination
- Sign language and cued language services
- Social work services
- Special instruction
- Speech-language pathology services
- Transportation and related costs
- Vision services

TABLE 32.2 IDEA Part C eligibility by state/territory [22].

State/ territory	Delay criteria (in one or more developmental area, unless otherwise specified)	"At-risk" infant and toddler eligibility for IDEA Part C services
Alabama	25%	No
Alaska	50%	No
American Samoa	Any delay in one or more developmental areas	Yes
Arizona	50%	No
Arkansas	25%	No
California	33%	Yes
Colorado	33%	No
Connecticut	2 SD below the mean in one developmental area or 1.5 SD below the mean in two or more developmental areas	No
Delaware	25% or 1.75 SD below the mean	No
District of Columbia	25%	
Florida	1.5 SD below the mean in two or more developmental areas or 2 SD below the mean in one or more developmental areas	Yes
Georgia	2 SD below the mean in one or more developmental areas; or at least 1.5 SD (moderate delay) below the mean in two or more developmental areas; and/or the child's developmental issues interfere with their functional ability when compared with peers	No
Guam	2 SD or more below the mean or 30% or greater delay in one or more developmental areas or 1.5 SD or 22%–29% below the mean in two or more developmental areas	Yes
Hawaii	1.4 SD or more below the mean in at least one developmental area or subarea; or 1 SD or more below the mean in at least two or more developmental areas or subareas	No
Idaho	30% below age norm or exhibits a 6-month delay, whichever is less; or at least 2 SD below the mean in one developmental area; or at least 1.5 SD below the mean in two or more developmental areas	No
Illinois	30%	No
Indiana	At least 25% in function below child's chronological age or 2 SD below the mean in one or more developmental areas; or at least 20% in function below child's CA or 1.5 SD below the mean in two or more developmental areas adjusted for prematurity as applicable and on an assessment instrument that yields scores in months	No
Iowa	25%	No
Kansas	Discrepancy of 25% or more between CA and developmental age, after correction for prematurity, in any one developmental area or at least 20% discrepancy between CA and developmental age, after correction for prematurity, in two or more developmental areas	No
Kentucky	2 SD below the mean in one developmental area or at least 1.5 SD below the mean in two developmental areas	No
Louisiana	1.5 SD below the mean on the Battelle Developmental Inventory, second edition (BDI-2) in two developmental areas	No
Maine	2 or more SD below the mean in at least one developmental area or 1.5 or more SD below the mean in at least two developmental areas	No

(Continued)

TABLE 32.2 (Continued)

State/ territory	Delay criteria (in one or more developmental area, unless otherwise specified)	"At-risk" infant and toddler eligibility for IDEA Part C services
Maryland	25% delay in at least one or more of five developmental areas or manifests atypical development or behavior in one or more of the five developmental areas, interferes with current development, and is likely to result in a subsequent delay (even when diagnostic instruments and procedures do not document a 25% delay)	No
Massachusetts	1.5 SD below the mean, as measured by the Battelle Developmental Inventory, second edition (BDI-2), in one or more developmental areas	Yes
Michigan	20% or 1 SD below the mean in one or more developmental areas	No
Minnesota	1.5 SD below the mean in one or more developmental areas	No
Mississippi	33% delay in one area of development or a 25% delay in two or more areas of development; or 2 SD below the mean in one developmental area or 1.5 SD below the mean in each of the two developmental areas	No
Missouri	50%	No
Montana	50% delay in one developmental area or 25% delay in two or more developmental areas	
Nebraska	2 SD below the mean in one developmental area or 1.3 SD below the mean in two developmental areas	No
Nevada	50% delay of a child's chronological age in any one developmental area or 25% delay of child's chronological age in any two developmental areas	No
New Hampshire	33% delay in one or more developmental areas; or atypical behavior	Yes
New Jersey	2 SD below the mean in one developmental area or 1.5 SD below the mean in two or more of the developmental areas	No
New Mexico	25% or greater delay in one or more developmental areas using the IDA (infant toddler developmental assessment) or another approved tool; a domain-specific tool may be used to establish eligibility if the score is 1.5 SD below the mean or greater	Yes
New York	12-month delay in one domain; or 33% delay in one domain; or 25% delay in each of two domains; or if standardized instruments used, a score of at least 2 SD below the mean in one domain; or at least 1.5 SD below the mean in each of two domains; or for children with delay only in the communication domain, a score of 2 SD below the mean in the area of communication	No
North Carolina	2 SD below the mean of the composite score (total test score) on standardized tests in at least one of the areas of development; or 30% delay on instruments which determine scores in months in at least one of the areas of development; or 1.5 SD below the mean of the composite score (total test score) on standardized tests in at least two of the areas of development; or 25% delay on instruments which determine scores in months in at least two of the areas of development	No
North Dakota	25% below age norms in two or more areas of development; or 50% below age norms in one or more areas	No
Northern Mariana Islands	25%	No
Ohio	1.5 SD below the mean in at least one area of development as indicated by the use of one of the two approved evaluation tools	No

(Continued)

TABLE 32.2 (Continued)

State/ territory	Delay criteria (in one or more developmental area, unless otherwise specified)	"At-risk" infant and toddler eligibility for IDEA Part C services
Oklahoma	A delay in developmental age compared to CA of 50% or a score 2 SD below the mean in one of the developmental domains or subdomains; OR a delay in developmental age compared to CA of 25% or score 1.5 SD below the mean in two or more of the developmental domains or subdomains	No
Oregon	2 SD or more below the mean in one or more developmental areas, or 1.5 SD below the mean in two or more developmental areas	No
Pennsylvania	25% delay or 1.5 SD below the mean in one area of development	No
Puerto Rico	Quantitative and qualitative criteria listed for each area, including: Motor, cognitive, and language skills: 2 SD below the mean or 33% delay; 1.5 SD below the mean or 25% delay with other delays Social-emotional and adaptive skills: Informed clinical opinion	No
Rhode Island	2 SD below the mean in at least one area of development or 1.5 SD below the mean in two or more areas of development	No
South Carolina	Delay of 40% (2 SD below the mean) in one area of development or a delay of 25% (1.5 SD below the mean) in two areas of development	No
South Dakota	At least 1.5 SD below the mean in one or more areas of development	No
Tennessee	25% delay in two developmental areas or a 40% delay in one area	No
Texas	Delay of at least 25% in one or more of areas of development; or a delay of at least 33% if the child's only delay is in expressive language; or a qualitative determination of delay, as indicated by responses or patterns that are disordered or qualitatively different from what is expected for the child's age, and significantly interfere with the child's ability to function in the environment	No
Utah	1.5 SD at or below the mean, or at or below the 7th percentile in one or more of the areas of development on approved instrument	No
Vermont	The child is experiencing a clearly observable and measurable delay in one or more of developmental areas (as stated above) and the delayed development shall be at the level that the child's future success in home, school, or community cannot be assured without the provision of Early Intervention services	No
Virgin Islands	25% delay in one or more areas of development when comparing functional age to CA or standardized test scores of 1.5 SD below the mean	No
Virginia	At least 25% below chronological or adjusted age in one or more areas of development or atypical development (even in the absence of 25% delay)	No
Washington	1.5 SD or 25% delay from chronological age in one or more developmental areas	No
West Virginia	40% delay in functional abilities/developmental skills in one or more areas of development; or 25% delay in functional abilities/developmental skills in two or more areas of development; or substantially atypical development in two or more developmental areas, even when evaluation does not document a 25% delay	Yes
Wisconsin	25% or 1.3 SD below the mean in one or more areas of development	No
Wyoming	1.5 SD below the mean or 25% delay in one or more areas of developmental	No

services in varying combinations and quantities, making it difficult to interpret the impact of individual elements. Despite these challenges, a body of research developed over past decades provides evidence of the effectiveness of EI programs. In economically disadvantaged children, early education programs have a positive impact on cognitive and behavioral outcomes [23]. What does the evidence say about the effectiveness of EI in children in other developmental delay risk groups? Evidence developed in the later part of the 20th century supported the perspective that EI is effective in enhancing developmental progress in children with (or at risk for) developmental disabilities [24–26]. More recent data have largely agreed that EI programs are beneficial in specific subgroups of children with, or at risk for, developmental delays. The methodological quality of studies of EI in children at high risk of cerebral palsy is variable [27], although there is good evidence supporting the involvement of parents in intervention programs and the provision of task- and context-specific therapies to improve motor and cognitive outcomes [28]. Young children at high risk for cerebral palsy should be referred promptly for intervention. An overview of systematic reviews and recent controlled trials suggests that earlier intervention is superior to later intervention in this population, although more rigorous research is needed [29]. In young children with developmental language disorder, early intensive speech-language intervention is effective in improving phonological expressive skills [30]. In children born prematurely, EI programs have beneficial effects on cognitive and motor outcomes during infancy, with some benefits persisting to age 3–5 years [31,32]. There is also evidence that psychosocial interventions are helpful in young children in developing countries who have poor nutritional status. For instance, early childhood stimulation has been shown to have a range of long-term benefits, including higher adult IQ, higher educational attainment, fewer symptoms of depression, and reductions in violent behavior in a population of growth-restricted Jamaican children [33]. Beyond the goal of improvements in children's developmental skills, additional caregiver benefits are observed in families participating in EI programs. These can include increased advocacy ability and capacity to access supports, as well as enhanced overall belief that the family is better off as a result of participation in EI [34]. Additional discussion regarding the use of specific interventions, such as speech-language (see subshapter: speech and language therapy) and physical therapies (see subchapter: Physical therapy), in children with developmental disabilities is presented in greater detail elsewhere in this volume.

The role of healthcare providers in the Early Intervention process

Pediatricians and other healthcare professionals are important partners for families in the successful provision of EI and special education services to eligible children [35]. Valuable assistance can be provided to families in the following areas [36,37]:

- Providing an appropriate medical home with regular developmental screening and surveillance.
- Referring children for services.
- Identifying eligible diagnoses in children.
- Participating in assessments and sharing information with EI or school programs.
- Providing advice to families.
- Assisting in the creation of the individualized family service plan (IFSP).
- Utilizing the information in the IFSP to help develop and deliver coordinated medical services.
- Serving in other advocacy capacities.

Developmental screening and surveillance in the context of a medical home and referrals to EI programs for children who may be eligible for these services are essential responsibilities of primary care providers (see chapter: Developmental surveillance and screening). The pediatric medical home is a process for the provision of care to children that is "accessible, family-centered, continuous, comprehensive, coordinated, compassionate, and culturally effective" [38] (see chapter: Medical home). Close, mutually respectful partnerships between families and the healthcare community create environments in which developmental concerns are more likely to be articulated by caregivers, and appropriate developmental screening and referral for EI services by primary care providers is more likely to occur. Evidence-based guidelines for developmental screening are available [39]. While completion of developmental screening in practice has not always been optimal, recent improvements are evident, with the use of developmental screening tools by pediatricians increasing from 21% in 2002 to 63% in 2016 [40]. Primary care providers can also play a role in identifying specific diagnoses associated with a high risk of developmental delays that may automatically qualify children for EI services in certain states. In some circumstances, pediatricians may be able to participate directly as part of a multidisciplinary team in the evaluation of children for EI services. In many cases that degree of involvement is not possible for busy clinicians, but families can still benefit from the advice and guidance of a trusted healthcare professional in navigating what can be an unfamiliar and sometimes confusing EI process.

For a child who is determined to be eligible for EI services, an IFSP is created. This document lays out the specific intervention plan for the child, outlining the supports the family and child will receive to help address the identified areas of developmental need. Pediatricians and other primary care providers should be familiar with the elements of the IFSP [37], as they can help to review the document, determine if appropriate resources are being offered, and clarify any questions the family has regarding the EI plan. Information derived in the EI evaluation process and contained in the IFSP can sometimes be helpful in developing medical treatment plans. Specific medical services may be included as part of the IFSP for certain children, usually those with complex medical needs, and the primary care provider may serve as the main point of contact for delivering or coordinating these services. Lastly, primary care providers can serve in other advocacy roles within the EI system—for instance, as consultants to local EI programs and school districts or as members of state agencies and councils.

Primary care providers need to be aware of the contact information for their local EI programs so that families of potentially eligible children can be appropriately directed to these resources. The Centers for Disease Control and Prevention maintains a list of *Early Intervention Contact Information by State* which provides current information [41]. The Center for Parent Information and Resources [42] is another useful resource for families that contains an extensive amount of information regarding EI and special education services.

Access to Early Intervention services: challenges and opportunities

Successful EI programs should be family-centered, be based in local communities, incorporate multiple disciplines, and be able to effectively coordinate services among multiple public and private agencies [26]. In 2021, 3.2% of children 0–3 years of age were receiving Part C EI services [43], although there is considerable state-to-state variability. Due to differences in the definition of what constitutes "delay," the percentage of US children potentially eligible for EI services can vary considerably depending on location—in 2013, ranging from 2% to 78% in different states [44]. These variations can create the situation in which a particular child receiving Part C EI services in one state might be ineligible if the family moves to a different state. To date, there has been no agreed upon uniform national definition of what constitutes a sufficient degree of delay to warrant the provision of EI services.

The two primary avenues that children enter the EI system through Part C of IDEA are by either meeting the state-specific developmental delay criteria or by having a diagnosed condition likely to result in developmental delays (also referred to as "Category 1" conditions). As with the developmental delay criteria, each state and territory determine the specific list of disorders used to determine eligibility under the "diagnosed conditions" provision. In a study of variability between states in these diagnosed conditions, a total of 620 listed disorders were identified [45]. No single diagnosed condition was listed as a qualifying disorder across the entire study group of 49 states, the District of Columbia, and 4 territories. Hearing impairment and fetal alcohol syndrome were the most commonly specified "Category 1" conditions.

Regardless of the criteria used, however, many children who are eligible for EI do not actually receive these services. Several factors have been identified that contribute to disparities in the receipt of EI services. In a cohort of 900 infants and toddlers derived from the 2005 to 2006 National Survey of Children with Special Health Care Needs (NS-CSHCN), less than half were receiving EI services [46]. Children with special health care needs (SHCN) receiving EI services were also less likely to be Hispanic, live in a multiracial family, or be living in poverty. Among poor children with SCHN (<100% of the federal poverty level), the adjusted probability of receiving EI services was 0.18, compared to 0.36 for nonpoor children (>185% federal poverty level). Latent class analysis using data from same 2005 to 2006 NS-CSHCN survey identified four subgroups of children with developmental vulnerability: those with developmental disability (Group 1), mild developmental delay (Group 2), socially at risk with behavior problems (Group 3), and socially at risk with vision difficulties (Group 4) [47]. Children who were black, non-Hispanic were more likely to fall into Group 3, and children in Groups 1, 3, and 4 were significantly less likely than the children with mild developmental delay to receive EI services. Geographic variation is evident in access to EI services. In a cohort of children with low birth weight in South Carolina, only 54% were enrolled in EI at some point between birth and age 3 [48], whereas 80% of a sample of 2-year-old children with very low birth weight and developmental delay in Wisconsin were receiving therapy by parent report [49]. In an analysis using data from the 2009/2010 NS-CSHCN [50], children who lived in a state with narrow eligibility criteria were less likely to receive EI services, and children living in states whose eligibility criteria narrowed after 2005–06 (a trend largely related to state budget pressures) were also less likely to receive services compared to those children living in states with consistent EI eligibility criteria over this period. Even more surprising and distressing was that this finding was most pronounced in the group of children with more severe impairments.

Similar access challenges are found in groups of children in specific risk categories. Developmental delays are a risk factor for child maltreatment [51], and vice versa, underscoring the importance of heightened attention to developmental needs in the population of children referred because of concerns of maltreatment. While the Keeping Children and Families Safe Act of 2003 required states to develop "provisions and procedures for referral of a child under the age of 3 who is involved in a substantiated case of child abuse or neglect to EI services funded under part C of the Individuals with Disabilities Education Act" [52], 5 years after the passage of this legislation only 18.2% of children in a nationally representative sample who had experienced substantiated child maltreatment had been referred to Part C EI programs [53].

There are multiple steps involved in the process of delivering EI therapies to children: from initial concern raised by caregiver or primary care provider, referral to the appropriate EI program, evaluation to determine eligibility, and ultimately to the actual receipt of services by the child. Each of these steps poses a potential barrier to access and effective delivery of EI services. A systematic review in 2018 highlighted limitations in the literature evaluating the steps involved in connecting children with Part C services [54]. The authors identified 43 studies that included at least two steps of the EI service delivery process, but few that evaluated the entire process from start (community surveillance) to finish (delivery of services). Another systematic review of factors related to Part C enrollment identified five themes across studies: policies related to "at-risk" children, funding of programs, eligibility criteria, referral source, and administrative structure of the EI programs [55]. These areas are all potential sources of variability and disparity among children who are potentially eligible to receive services under Part C of IDEA. An additional retrospective cohort study utilizing primary care and EI records of 14,710 children with developmental delays or disability in a large, safety net health system showed gaps at each stage of the EI process [56]. In this study, only 18.7% of eligible children received an EI referral, and only about a quarter of those (5% of eligible children) actually received services. Children who were Black, non-Hispanic were less likely to receive an EI referral, as were children with a diagnosed condition. Disparities in service provision related to income were also found, with children from higher income families more likely to receive physical therapy and occupational therapy. For those families who are successful in accessing EI services, experiences are largely favorable [57].

Practice-specific characteristics have the potential to impact the likelihood that an eligible child will appropriately access EI services. In one study, the use of electronic health record templates and phone follow-up following abnormal developmental screening improved EI referral rates and completed evaluations, although a minority of children who failed developmental screening in this study were actually referred to EI programs [58]. Similarly, in an urban primary care population, 58% of children identified with a developmental concern were referred to EI,but only 30% completed the EI evaluation process [59]. Electronic transmittal of EI referrals (as opposed to relying solely on caregivers to initiate the contact) was associated with an increased probability of completion of the EI evaluation process. Surprisingly, data from children with special healthcare needs and a developmental condition impacting daily functioning in the 2009/2010 NS-CSHCN indicated that those children with a patient-centered medical home (PCMH) were no more likely to be receiving Part C EI services than those without a PCMH [60]. Additionally, and perhaps counterintuitively, children who had access to effective care coordination were actually less likely to be receiving Part C services. Careful attention does need to be given to health beliefs and expectations [61], as well as social and financial stressors, in parents of children with developmental delays since these factors can influence the successful completion of the EI referral process. Clearly, further research analyzing all steps of the EI process and associated potential barriers is needed to help inform the continued development of best practices for the effective delivery of these services.

Conclusions

Through IDEA and other legislation, US states and territories have a robust mechanism to deliver a range of services for children with, or at risk for, developmental delays. Part C of IDEA supports the provision of intervention services for infants and toddlers through age 3. There is evidence that these intervention services strengthen skill development in children with developmental delays and enhance family capacity to care for children and access available supports. There continues to be a need for rigorous scientific study of the effectiveness of specific elements of EI programs with particular attention to the development of a better understanding of the optimal "dose" (intensity, frequency) of therapy in children with specific needs. Disparities in the ability of certain populations of children to access appropriate EI services remains an area of concern, highlighting the essential role healthcare providers need to play in identifying children with developmental delays, making necessary referrals to EI programs, and partnering with families to ensure that these helpful resources are successfully implemented.

References

[1] Tjossem TD, editor. Intervention strategies for high risk infants and children. Baltimore: University Park Press; 1976.

[2] Denhoff E. Current status of infant stimulation or enrichment programs for children with developmental disabilities. Pediatrics 1981;67(1):32–7.

[3] Schilling M, Siepp J, Patterson EG, et al. The first three years: programming for atypical infants and their families. A united cerebral palsy nationally-organized collaborative project to provide comprehensive services. New York: United Cerebral Palsy Associations; 1974.
[4] Weikart DP, Bond JT, McNeil JT. The Ypsilanti Perry Preschool Project: preschool years and longitudinal results through fourth grade. Issue 3 of monographs of the high/scope educational research foundation. Ypsilanti: High/Scope Educational Research Foundation; 1978.
[5] Schweinhart LJ. Long-term follow-up of a preschool experiment. J Exp Criminol 2013;9:389–409.
[6] United States Code. Ch. 73, §2, 37 Stat. 79; 1912.
[7] PL 74-271. Available from: https://www.govinfo.gov/content/pkg/CPRT-113WPRT85308/pdf/CPRT-113WPRT85308.pdf; 1935 (accessed 7/24/2022).
[8] PL 85-864. Available from: https://www.govinfo.gov/content/pkg/STATUTE-72/pdf/STATUTE-72-Pg1580.pdf; 1958 (accessed 7/24/2022).
[9] PL 85-926. Available from: https://www.congress.gov/85/statute/STATUTE-72/STATUTE-72-Pg1777.pdf; 1958 (accessed 7/24/2022).
[10] PL 88-164. Available from: https://www.govinfo.gov/content/pkg/STATUTE-77/pdf/STATUTE-77-Pg282.pdf; 1963 (accessed 7/24/2022).
[11] PL 89-10. Available from: https://www.govinfo.gov/content/pkg/STATUTE-79/pdf/STATUTE-79-Pg27.pdf; 1965 (accessed 7/24/2022).
[12] PARC v. Commonwealth of Pa., 334 F. Supp. 1257; 1971.
[13] Mills v. Board of Education of District of Columbia, 348 F. Supp. 866; 1972.
[14] PL 94-142. Available from: https://www.govinfo.gov/content/pkg/STATUTE-89/pdf/STATUTE-89-Pg773.pdf; 1975 (accessed 7/24/2022).
[15] PL 93-112. Available from: https://www.govinfo.gov/content/pkg/COMPS-799/pdf/COMPS-799.pdf; 1973 (accessed 7/24/2022).
[16] PL 99-457. Available from: https://www.govinfo.gov/content/pkg/STATUTE-100/pdf/STATUTE-100-Pg1145.pdf; 1986 (accessed 7/24/2022).
[17] PL 101-476. Available from: https://www.govinfo.gov/content/pkg/STATUTE-104/pdf/STATUTE-104-Pg1103.pdf; 1990 (accessed 7/24/2022).
[18] PL 108-446 Sec 631. Available from: https://www.congress.gov/108/plaws/publ446/PLAW-108publ446.pdf; 2004 (accessed 7/24/2022).
[19] Individuals with Disabilities Education Act. Available from: https://sites.ed.gov/idea/statute-chapter-33/subchapter-iii/1435 (accessed 7/24/2022).
[20] Early Childhood Technical Assistance Center. Available from: https://ectacenter.org/topics/earlyid/state-info.asp (accessed 7/24/2022).
[21] Individuals with Disabilities Education Act. Available from: https://sites.ed.gov/idea/regs/c/a/303.13 (accessed 7/24/2022).
[22] Data derived from: Available from: https://ectacenter.org/topics/earlyid/state-info.asp (accessed 7/24/2022).
[23] Barnett WS. Effectiveness of early educational intervention. Science 2011;333(6045):975–8.
[24] Casto G, Mastropieri MA. The efficacy of Early Intervention programs: a meta-analysis. Except Child 1986;52(5):417–24.
[25] Shonkoff JP, Hauser-Cram P. Early Intervention for disabled infants and their families: a quantitative analysis. Pediatrics 1987;80(5):650–8.
[26] Guralnick M, editor. The effectiveness of Early Intervention. Baltimore: Paul H. Brookes Publishing Co; 1997.
[27] Hadders-Algra M, Boxum AG, Hielkema T, Hamer EG. Effect of Early Intervention in infants at very high risk of cerebral palsy: a systematic review. Dev Med Child Neurol 2017;59(3):246–58.
[28] Morgan C, Fetters L, Adde L, et al. Early Intervention for children aged 0 to 2 years with or at high risk of cerebral palsy: international clinical practice guideline based on systematic reviews. JAMA Pediatr 2021;175(8):846–58.
[29] Damiano DL, Longo E. Early Intervention evidence for infants with or at risk for cerebral palsy: an overview of systematic reviews. Dev Med Child Neurol 2021;63(7):771–84.
[30] Rinaldi S, Caselli MC, Cofelice V, et al. Efficacy of the treatment of developmental language disorder: a systematic review. Brain Sci 2021;11 (3):407.
[31] Spittle A, Orton J, Anderson PJ, Boyd R, Doyle LW. Early developmental intervention programmes provided post hospital discharge to prevent motor and cognitive impairment in preterm infants. Cochrane Database Syst Rev 2015;2015(11):CD005495.
[32] Spittle A, Treyvaud K. The role of early developmental intervention to influence neurobehavioral outcomes of children born preterm. Sem Perinatol 2016;40(8):542–8.
[33] Walker SP, Chang SM, Vera-Hernández M, Grantham-McGregor S. Early childhood stimulation benefits adult competence and reduces violent behavior. Pediatrics 2011;127(5):849–57.
[34] Bailey DB, Hebbeler K, Spiker D, Scarborough A, Mallik S, Nelson L. Thirty-six-month outcomes for families of children who have disabilities and participated in Early Intervention. Pediatrics 2005;116(6):1346–52.
[35] Rose L, Herzig LD, Hussey-Gardner B. Early Intervention and the role of pediatricians. Pediatr Rev 2014;35(1):e1–e10.
[36] American Academy of Pediatrics. Committee on Children with Disabilities. The pediatrician's role in development and implementation of an individual education plan (IEP) and/or an individual family service plan (IFSP). Pediatrics 1999;104(1 Pt 1):124–7.
[37] Lipkin PH, Okamoto JCouncil on Children with Disabilities, & Council on School Health. The Individuals with Disabilities Education Act (IDEA) for children with special educational needs. Pediatrics 2015;136(6):e1650–62.
[38] Adams RC, Tapia CCouncil on Children with Disabilities. Early Intervention, IDEA Part C services, and the medical home: collaboration for best practice and best outcomes. Pediatrics 2013;132(4):e1073–88.
[39] Lipkin PH, Macias MMCouncil on Children with Disabilities, Section on Developmental and BehavioralPediatrics. Promoting optimal development: identifying infants and young children with developmental disorders through developmental surveillance and screening. Pediatrics 2020;145(1):e20193449.
[40] Lipkin PH, Macias MM, Baer Chen B, et al. Trends in pediatricians' developmental screening: 2002-2016. Pediatrics 2020;145(4):e20190851.
[41] Centers for Disease Control and Prevention. Available from: https://www.cdc.gov/ncbddd/actearly/parents/state-text.html.
[42] Center for Parent Information and Resources. Available from: https://www.parentcenterhub.org (accessed 7/24/2022).
[43] Early Childhood Technical Assistance Center. Available from: https://ectacenter.org/partc/partcdata.asp (accessed 7/24/2022).
[44] Rosenberg SA, Robinson CC, Shaw EF, Ellison MC. Part C Early Intervention for infants and toddlers: percentage eligible versus served. Pediatrics 2013;131(1):38–46.

[45] Barger B, Squires J, Greer M, et al. State variability in diagnosed conditions for IDEA Part C eligibility. Infants Young Child 2019;32(4):231–44.

[46] McManus B, McCormick MC, Acevedo-Garcia D, Ganz M, Hauser-Cram P. The effect of state Early Intervention eligibility policy on participation among a cohort of young CSHCN. Pediatrics 2009;124(Suppl 4):S368–74.

[47] McManus BM, Carle AC, Rapport MJ. Classifying infants and toddlers with developmental vulnerability: who is most likely to receive Early Intervention? Child Care Health Dev 2014;40(2):205–14.

[48] Wang CJ, Elliott MN, Rogowski J, Lim N, Ratner JA, Schuster MA. Factors influencing the enrollment of eligible extremely-low-birth-weight children in the Part C Early Intervention program. Acad Pediatr 2009;9(4):283–7.

[49] McManus BM, Robert S, Albanese A, Sadek-Badawi M, Palta M. Predictors of receiving therapy among very low birth weight 2-year olds eligible for Part C Early Intervention in Wisconsin. BMC Pediatr 2013;13:106.

[50] McManus BM, Magnusson D, Rosenberg S. Restricting state Part C eligibility policy is associated with lower Early Intervention utilization. Matern Child Health J 2014;18(4):1031–7.

[51] van IJzendoorn MH, Bakermans-Kranenburg MJ, Coughlan B, Reijman S. Annual research review: umbrella synthesis of meta-analyses on child maltreatment antecedents and interventions: Differential susceptibility perspective on risk and resilience. J Child Psychol Psychiatry 2020;61(3):272–90.

[52] PL 108-36. Available from: https://www.govinfo.gov/content/pkg/PLAW-108publ36/pdf/PLAW-108publ36.pdf; 2003 (accessed 7/24/2022).

[53] Johnson-Motoyama M, Moses M, Conrad-Hiebner A, Mariscal ES. Development, CAPTA Part C referral and services among young children in the U.S. child welfare system: implications for Latino children. Child Maltreat 2016;21(3):186–97.

[54] Barger B, Rice C, Simmons CA, Wolf R. A systematic review of Part C Early Identification studies. Top Early Child Spec Educ 2018;38(1):4–16.

[55] Twardzik E, Cotto-Negrón C, MacDonald M. Factors related to Early Intervention Part C enrollment: a systematic review. Disabil Health J 2017;10(4):467–74.

[56] McManus BM, Richardson Z, Schenkman M, et al. Child characteristics and Early Intervention referral and receipt of services: a retrospective cohort study. BMC Pediatr 2020;20(1):84.

[57] Bailey DB, Hebbeler K, Scarborough A, Spiker D, Mallik S. First experiences with Early Intervention: a national perspective. Pediatrics 2004;113(4):887–96.

[58] Talmi A, Bunik M, Asherin R, et al. Improving developmental screening documentation and referral completion. Pediatrics 2014;134(4):e1181–8.

[59] Jimenez ME, Fiks AG, Shah LR, et al. Factors associated with Early Intervention referral and evaluation: a mixed methods analysis. Acad Pediatr 2014;14(3):315–23.

[60] Ross SM, Smit E, Twardzik E, Logan SW, McManus BM. Patient-centered medical home and receipt of Part C Early Intervention among young CSHCN and developmental disabilities versus delays: NS-CSHCN 2009-2010. Matern Child Health J 2018;22(10):1451–61.

[61] Magnusson DM, Minkovitz CS, Kuhlthau KA, Caballero TM, Mistry KB. Beliefs regarding development and Early Intervention among low-income African American and Hispanic mothers. Pediatrics 2017;140(5):e20172059.

Chapter 33

Principles of pharmacotherapy

Benjamin Joffe Schindel[1] and Bruce K. Shapiro[2]

[1]Department of Neurology and Developmental Medicine, Kennedy Krieger Instittute, Johns Hopkins University School of Medicine, Baltimore, MD, United States, [2]Department of Pediatrics, Johns Hopkins University School of Medicine, Baltimore, MD, United States

Introduction

The care of patients with neurodevelopmental disabilities (NDDs) requires a unique perspective. Unlike conditions treated by many other medical specialties, neurodevelopmental disabilities can be treated but, at the time of this writing, not cured. The underlying pathophysiology is often unknown, and those rare conditions with well-elucidated mechanisms often evade curative therapy by the nature of the nervous system and by the current understanding of neurobiology and molecular neuroscience. In fact, to approach management of NDDs from the perspective of *curing* them is to fail to acknowledge the complex and full spectrum of human experience that is exemplified by neurodiversity.

Rather than cure, the current goal in managing NDDs is intervention on specific symptoms. Successful intervention, therefore, requires identifying target symptoms and prescribing therapies meant to address those specific symptoms. It also requires specific and accurate diagnosis, which is a substantial challenge for several reasons. The Diagnostic and Statistical Manual of Mental Disorders, the gold standard for mental health diagnosis which is currently in its fifth edition, defines its diagnoses by symptom description rather than by the underlying etiology. This is perhaps best illustrated by examining the criteria for autism spectrum disorder, which describe behaviors of various origins. In clinical practice, autism unsurprisingly presents with heterogeneous symptoms and behaviors.

As the criteria for NDDs lack specificity, there is a fair amount of diagnostic confusion. How much social impairment distinguishes autism spectrum disorder from attention-deficit/hyperactivity disorder (ADHD)? How much perseveration is required to qualify interests as "repetitive" or "restrictive"? Whether a child has a social (pragmatic) communication disorder or autism spectrum disorder may be in the eye of the beholder.

Furthermore, in the world of NDDs, multiple descriptive diagnoses are the rule, not the exception. Those tasked with managing patients with NDDs have to carefully consider how best to approach individual patients who often have multiple comorbidities. How does one identify a primary disorder to treat? Should the patient's ADHD or anxiety be addressed first?

Successful treatment requires using the right intervention for the right diagnosis at the right time. Most NDDs are best managed by a combination of pharmacology and nonpharmacologic treatments [1]. Nonpharmacologic treatment options include therapies—physical, occupational, speech, behavioral, cognitive behavioral, and others—as well as environmental modifications (see Section IV of this book). The goal is not to resolve the condition but to alleviate the symptoms , enabling patients to participate in their daily activities and engage in their communities.

The role of the prescriber in managing NDDs is to facilitate fmedical and nonmedical therapies through judicious prescribing. This is not to minimize the role of the prescriber in managing NDDs: pharmacotherapy is an important part of the comprehensive management of patients with disability, and the principles of prudent management apply to all parties involved in a patient's care plan.

Sometimes a single treatment may address multiple symptoms—alpha adrenergic agents, for example, are effective for both ADHD and tics, which frequently co-occur. However, the vast majority of therapies are effective for one symptom at a time. In the spirit of identifying specific target behaviors, medications should generally be prescribed for individual symptoms. This approach allows for titration to achieve maximal effect while minimizing confounding and side effects.

It is also important to note that dosing of medications may be atypical in patients with NDD. While many medications prescribed in pediatric populations are dosed based on the patients' weight, patients with NDDs often respond

Capute and Accardo's Neurodevelopmental Disabilities in Infancy and Childhood. DOI: https://doi.org/10.1016/B978-0-12-824060-1.00031-6

atypically— either too strongly or not strongly enough—to standard dosing of neuropsychiatric medications. Consequently, dosing needs to be adjusted based on patient's response.

Comorbidity often creates clinical situations that require atypical, individualized pharmacotherapy. Diagnostic overshadowing may lead to a failure to recognize the comorbid condition, resulting in the omission of appropriate pharmacotherapy. Alternatively, the patient's constellation of symptoms may require more than one pharmacologic agent in addition to nonpharmacologic approaches. Second-line drugs may be employed in the hope of 'killing two birds with one stone,' although using two first-line agents might be more successful.

The medications that are generally prescribed for symptom management are of the following categories:

- Stimulants
- Norepinephrine reuptake inhibitors
- Alpha-adrenergic agonists
- Neuroleptics
- Antidepressants
- Anxiolytics
- Hormones

Other interventions for specific treatment management, such as a phenylalanine-avoidant diet for phenylketonuria or the ketogenic diet for some intractable epilepsies, fall outside of the purview of this chapter. Furthermore, while targeted genetic therapies hold promise for preventing manifestations of some conditions, they present a number of problems:

- The body of literature is not robust;
- these treatments are often prohibitively expensive;
- as they are relatively new, most of these therapies are still in research phases;
- they are limited to a few NDDs for which specific mechanisms have been identified; and
- there are ethical dilemmas about the need to "prevent" or "cure" many NDDs.

Perhaps one day a future edition of this textbook will include a list of specific, targeted therapies for individual genetic mutations. In the meantime, this chapter aims to provide the prescribing practitioner caring for neurodevelopmentally disabled patients with a foundation in principles of pharmacology, familiarity with some of the most commonly prescribed medications used to manage NDDs, and evidence-based suggestions for the use of these medications.

Review of principles of pharmacology

In brief summary, drugs are absorbed, distributed, metabolized, and then excreted. Elimination is the combined result of metabolism and excretion. In most cases, the desired and undesired effects of a drug occur during this phase.

Absorption describes how a drug is delivered to the bloodstream. Most drugs prescribed for NDD symptom management are introduced enterally, either through the mouth or through alternate administration methods like gastric tubes, and travel to the small intestine before being absorbed into the blood. On their journey through the gastrointestinal tract, the drugs encounter multiple types of semipermeable membranes, which act as physiological barriers to control the rate of absorption. Ultimately, these drugs passively diffuse or are actively transported across the cell membranes and into the bloodstream. The proportion of a drug that reaches the bloodstream relative to the amount consumed is known as bioavailability. The small intestine, with its relatively great surface area, is the site of much drug absorption. This means that people with anatomical or functional abnormalities of the gastrointestinal system, particularly of the small intestine, may have altered responses to medication dosages that are typical of the larger population.

Drug manufacturers have developed different medication formulations to control the rate of absorption. Practically, this helps to control the dosage frequency and, in principle, improve patient adherence. The Food and Drug Administration (FDA) has defined four forms of drug release [2]:

1. Immediate release: Allows the drug to dissolve in the gastrointestinal contents, with no intention of delaying or prolonging the dissolution or absorption of the drug.
2. Modified release: Dosage forms whose drug release characteristics of time course and/or location are chosen to accomplish therapeutic or convenience objectives not offered by conventional dosage forms such as a solution or an immediate-release dosage form. Modified-release solid oral dosage forms include both delayed- and extended-release drug products.

3. Delayed release: Release of a drug (or drugs) at a time other than immediately following oral administration.
4. Extended release: A dosage form that allows a reduction in dosing frequency as compared to that presented by a conventional dosage form, for example, a solution or an immediate release dosage form.

The FDA further breaks down this fourth category, extended release, into sustained release and controlled release. Sustained release formulations control the *rate* of drug release over a period of time, while controlled release formulations control the *plasma concentration* of the drug. To accomplish this, some long-acting or extended release formulations are coated in polymers, which provide an additional physical barrier between the compound and the bodily enzymes that metabolize it, or are enmeshed in a matrix in which the drug is dispersed, providing an additional chemical barrier. These barriers slow down the rate of absorption such that much of it occurs in the large intestine. A discussion of some mechanisms of release is included in the section on stimulants.

Once absorbed into the bloodstream, drugs are distributed throughout the body, where they meet their target receptors and have their effects. The drugs are transported through the blood either unbound, and therefore available for pharmacologic effect, or reversibly bound to plasma proteins. The volume of distribution, the ratio of the total amount of the drug in the body to the plasma drug concentration, reflects these two modes of transportation. Drugs like amphetamines, which are readily taken up by body tissues, and therefore exist minimally in the plasma, have alarge volume of distribution.

Pharmacologic management of central nervous system conditions requires that the drugs distribute sufficiently across the largely impermeable blood–brain barrier, which is composed of tightly joined cerebral capillary endothelial cells and astrocytes. Primarily, psychoactive medications exert their effects by entering cerebrospinal fluid through the choroid plexus and then diffusing into the brain.

Elimination refers to the way a drug is removed from the body. It occurs both by removal of the drug from the body through excretion and by changes in the properties of the drug by metabolism. Excretion is a similar concept to that of clearance, which refers to the plasma volume cleared of a drug per unit time. Note that excretion refers to the *amount* of the drug removed per unit time, usually measured in units of mass over time, while clearance is measured in units of volume over time. Metabolism occurs primarily by liver enzymes, and excretion occurs primarily through the renal system, although certainly there are other ways in which these processes occur: many tissues contain the enzymes responsible for medication metabolism but in lower concentrations than the liver, and many drugs are excreted, in part, in the stool.

Metabolism refers to the various types of specific reactions that change the drug from its initial form into other forms. These reactions include oxidation, reduction, hydrolysis, conjugation, and isomerization, among others. Metabolism usually occurs in two phases, which are defined by the reactions. In Phase I reactions, which are usually oxidation reactions, most drugs are converted to forms with some residual pharmacologic activity. Some drugs are introduced into the body in active forms, while others, termed prodrugs, are introduced into the body, absorbed, and distributed in inert form before being activated by Phase I reactions. Phase II reactions conjugate the product from the Phase I reaction with another molecule (often through methylation, glucuronidation, acetylation, or sulfation) and inactivate it. Often, this phase makes the conjugated metabolite water-soluble to allow for reabsorption into the serum and eventual elimination. The liver is the most important site of these metabolic reactions, and the hepatic cytochrome P450 (CYP450) enzymes are particularly important for metabolism of many drugs, including some used for management of NDD symptoms, like amphetamines, selective serotonin reuptake inhibitors (SSRIs), and neuroleptics [3]. Due to the common use of this superfamily of enzymes—as much as 90% of drugs are metabolized by the CYP450 system—interactions are somewhat common and need to be checked before initiating pharmacotherapy in patients who are already on medications [4] However, routine measurement of enzyme activity before initiating pharmacotherapy is not universally endorsed.

In pharmacology, elimination typically proceeds by first-order kinetics, which refers to the fact that the clearance rate, the plasma volume cleared of a drug per unit time, is dependent upon the concentration of the drug. A useful metric of clearance rate is half-life, which is the amount of time it takes for half of a drug's initial concentration to be cleared. In first-order kinetics, half-life is calculated by the following formula:

$$t_{1/2} = (0.693 \times V_d)/CL$$

In this formula, $t_{1/2}$ is the half-life, V_d is the volume of distribution, and CL is the clearance, which in turn is equal to the quotient of the rate of drug elimination over the plasma drug concentration. Rearranging this formula, we see that $t_{1/2}$ is directly proportional to the plasma drug concentration and will decrease as the concentration progressively decreases. With some specific exceptions, including ethanol and phenytoin, drugs are cleared by first-order kinetics unless their concentration is so great that the enzymes responsible for the metabolic processes are overwhelmed, in which case clearance shifts to zero-order kinetics and the clearance rate is constant.

These principles are important for the prescribing practitioner to understand and consider. Frequently, these principles must be inferred from the clinical effectiveness of a clinician's medication regimen on an individual patient. An understanding of the above pharmacology distinguishes the clinician who is able to adapt and modify treatment regimens based on the known scientific principles of the drugs.

Evidence-based pharmacologic management of frequent target behaviors treated in neurodevelopmental disabilities

The most common target symptom clusters encountered in the care of people with NDDs include:

- Inattention, impulsivity, and hyperactivity;
- aggression and disruptive behaviors, including self-injury;
- anxiety;
- depressed mood;
- obsessive-compulsive behaviors; and
- sleep disturbances.

While most pharmacologic interventions address symptoms, understanding the mechanism of the symptoms and the role of comorbid conditions is critical for successful management with medications.

Consider the following example: While evaluating a child referred for academic underachievement, a clinician discovers that the child does not appear to pay attention during class. They prescribe stimulants, the first-line treatment for ADHD. Several weeks later, the parents call the clinician back with concerns that not only has the child's attention not improved, but they now have a loss of appetite and daily abdominal pain from hunger. Re-evaluation of the child is reveals a specific learning disorder in reading, and the child's academic performance and attention improve with cessation of stimulant medication and provision of appropriate accommodations.

A close, detailed analysis of patients' presenting symptoms, with particular attention to identifying the causes of the patients' presentations, is crucial for managing these symptoms with medications. Like other conditions, many of the presenting symptoms of NDDs are "great mimickers" with low specificity and multiple possible etiologies. If a treatment regimen is not working, the clinician is advised to reconsider the etiology of the target symptom, set realistic expectations for changes in this symptom, and collaborate with parents, therapists, and educators to find the most appropriate and effective way to address the symptom.

Inattention, impulsivity, and hyperactivity

According to the DSM definitions , inattention is failing to pay attention to detail or sustain attention, making careless mistakes, and failing to finish tasks. Hyperactivity and impulsivity are characterized by fidgeting, squirming, acting "as if driven by a motor," interrupting, etc. These somewhat opposite manifestations are united by their response, in the appropriate clinical setting, to similar medications (see Chapter: ADHD).

Difficulties with attention and impulse control are very common presentations in many neurodevelopmental disabilities. The DSM-5 criteria for ADHD diagnosis hinge on "a persistent pattern of inattention and/or hyperactivity-impulsivity that interferes with functioning or development." In situations in which inattention, hyperactivity, and impulsivity—the 'core symptoms' of ADHD—exist as primary deficits, pharmacologic management may be helpful. When associated deficits in social and academic functioning are present, pharmacologic management is more strongly indicated. For example, a child with difficulty paying attention may exhibit situational avoidance, and may not identify this behavior as rooted in their difficulties with inattention.

Stimulants—amphetamine and methylphenidate formulations—are prescribed as the first-line medications for treatment initiation when targeting the core symptoms of ADHD. This applies across all age groups, except for children 4-6 years, where behaviroal therapy is the first line treatment.

The American Academy of Pediatrics (AAP's) 2019 guidelines divide treatments by age group, as follows:

- Age 4 years to sixth birthday: Methylphenidate may be prescribed in this age group, but only after failure to respond to a course of behavioral therapy. There is strong evidence that this age group responds well to behavioral therapy alone, but financial, social, and geographic limitations may preclude adequate intervention. Due to its gentler side effects when compared with amphetamines, methylphenidate is the only medication studied in this age group. However, it is not currently FDA-approved in this age group.

- Age 6 years to 12th birthday: Behavioral interventions, both parent-directed and academic (through an Individualized Education Program or 504 Plan), should be prescribed alongside pharmacologic interventions. In this age group, the AAP endorses a sequence to specific medications and medication categories based on amount, if not always strength, of evidence: long-acting stimulants should be prescribed first, then atomoxetine, extended-release guanfacine, and extended-release clonidine, in that order. In other countries, medications like bupropion, modafinil, and imipramine are sometimes prescribed [5].
- Age 12 years to 18th birthday: The recommendations for this age group are the same as for the 6–12 age group, although the emphasis on behavioral interventions is less. This is due to weaker evidence that behavioral interventions are as effective in this age group.

Stimulants are the first-line medication category for management of inattention, hyperactivity, and impulsivity. While these symptoms are the 'core features' of ADHD, it is important to remember that they are also present in many other NDDs. Failure of these symptoms to respond to stimulant medication may indicate the need for reassessment of symptom etiology.

According to the FDA, a stimulant is a drug which helps restore mental alertness or wakefulness during fatigue or drowsiness (FDA, 21CFR340).

Stimulants come in two categories: amphetamines and methylphenidate. The primary effects of these medications, in summary, are to increase dopamine and norepinephrine activity, although there are some differences in mechanism and associated effects [6]. Their most common side effects include appetite suppression, irritability, headaches, insomnia, and nervousness. People with preexisting tics may experience an exacerbation of these movements. Stimulants have a Black Box Warning due to their high potential for abuse.

Amphetamines are taken up by the presynaptic nerve terminals in competition with endogenous monoamines like norepinephrine, dopamine, and serotonin. This both increases the synaptic presence of these neurotransmitters and induces their further release from the presynaptic neuron. It exerts the latter effect by binding to and reversing the effect of vesicular monoamine transporter 2 (VMAT2) and by activating trace amine-associated receptor 1 (TAAR1). VMAT2 is typically responsible for uptake of intracellular monoamines into vesicles for later release, and TAAR1 is responsible for release of monoamines into the synaptic cleft. The dextroamphetamine (D-amphetamine) isomer is a more potent agonist of TAAR1, so amphetamine formulations containing greater proportions of D-amphetamine tend to have a stronger psychopharmacologic effect than those with greater proportions of L-amphetamine. Lisdexamfetamine, marketed as Vyvanse, is an inactive prodrug that converts to L-lysine and D-amphetamine in the body. Oral amphetamine has a bioavailability of greater than 75% and usually reaches its peak therapeutic effect 1–3 hours after ingestion. It is metabolized by the CYP2D6 system in the liver and eliminated in urine, with about 10% of the initial dose remaining after 3 days. The half-life for D-amphetamine is 9–11 hours, and for L-amphetamine 11–14 hours, although the phenomenon of "rapid metabolizers" suggests that there may be variability to this reported duration of effect [7]. Similar variability has been noted with methylphenidate.

Like amphetamines, methylphenidate acts by blocking presynaptic reuptake of both norepinephrine and dopamine. It may also increase synaptic norepinephrine and dopamine by increasing neuronal firing, although this mechanism is not yet elucidated. Methylphenidate is an agonist at the 5-HT_{1A} receptor [8]. Oral methylphenidate has a bioavailability of up to 50%, half-life of 2–3 hours, and 97% renal excretion. While the exact mechanism underlying this clinical observation has not been fully elucidated, documented evidence shows notable discrepancies between serum and brain levels of methylphenidate and its metabolites in mice. This may suggest a broad range of metabolic rates in patients taking this medication. [9]. The side effect profile is similar to that of amphetamines but methylphenidate tends to be better tolerated, with fewer and milder side effects.

When prescribing stimulant medications, it is often desirable to control the duration of effect by prescribing long- and short-acting formulations. Long-acting methylphenidate sold under the trade name Ritalin LA, for example, uses the spheroidal oral drug absorption system (SODAS). A capsule of Ritalin LA contains small beads of methylphenidate, half of which are protected by a polymer coat. When the capsule dissolves in the gastrointestinal tract, the beads of methylphenidate are released, allowing access to the medication contained inside. The noncoated methylphenidate is able to be used immediately, but the half of the beads with the polymer coat cannot be accessed until that polymer is broken down, a process that takes about 4 hours. This means that when the body's fluids create sufficient pores in the polymer layer to allow the methylphenidate to be released, a second effect may occur, almost as if the patient dosed the immediate-acting methylphenidate twice.

One mechanism of drug delivery is the osmotic-release oral system (OROS) methylphenidate, marketed under the trade name Concerta. Concerta Capsules are made of an external layer of immediate release methylphenidate surrounding a semipermeable membrane with an exit pore on one end that allows for slow release of the methylphenidate contained inside. As the body's water enters the capsule through the semipermeable membrane, it expands a compartment inside—the so-called "push layer"—that forces more of the methylphenidate out through the exit pore. This allows for a slow, prolonged release of methylphenidate; Concerta is apparently active for up to 12 hours with nearly continuous effect [10].

Although not classified as a stimulant, another medication used to treat the inattention, hyperactivity, and impulsivity seen in ADHD and other conditions is atomoxetine. This norepinephrine reuptake inhibitor acts primarily by blocking the presynaptic norepinephrine transporter, with an almost 300-fold greater affinity for this transporter than for the presynaptic dopamine transporter.

Several studies have attempted to clarify clinical response to stimulant therapy for ADHD. These studies have been dogged by issues with regimen adherence and poorly standardized titration. One study that carefully monitored clinical response to stimulants in a standardized fashion found the rate of response to stimulants, as measured by frequent patient and parent standardized interviews, increased from 44% to 67% [11]. A similar study of aggressive behaviors in 6- to 12-year olds with ADHD found that almost 64% met criteria for remission of these behaviors after optimization of monotherapy with a stimulant, with the implication that this was likely due to optimized control of the core ADHD symptoms [12].

In general, about 90% of patients will respond to stimulants. A study from 2013 demonstrated that about half of children respond to either stimulant class, while 20% of the study population responded to methylphenidate and another 20% responded to amphetamine. In this study, consistent with previously published data, only about 10% did not respond at all [13].

Of note, while there is recent interest in pharmacogenetic studies to elucidate and anticipate differential response, this emerging field is not yet established enough to qualify for standard recommendation.

Nonstimulant medications like alpha agonists and atomoxetine are also used for ADHD symptom management, both as monotherapeutic agents and in combination with stimulants. In children, effect sizes for these nonstimulant medications are lower than effect sizes for stimulants. In adults, however, these medications may have comparable efficacy to stimulants, though the data are not yet conclusive [14]. There are more and larger studies using guanfacine than clonidine, hence the AAP's recommendation for guanfacine use before clonidine in children. The data, in general, do not suggest significant differences in tolerance and perhaps only a small difference in effect, as below. Head-to-head studies of the alpha agonists are lacking in the literature.

Nonstimulant medications are approved as additions to stimulant medications or may be used as monotherapies. However, literature suggests significant improvement in ADHD symptomatology following stimulant optimization. To avoid polypharmacy, it is advisable to maximize the first-line regimen before adding a second agent or switching to a second-line agent.

Evidence for nonstimulant medications suggests a potential difference in effectiveness. For example, guanfacine has been shown to improve ADHD core symptoms both on its own and in combination with stimulants, although its efficacy as a monotherapy was less than that of methylphenidate in children [15]. Similarly, clonidine appears to have a similar effect on ADHD symptoms when combined with stimulants. However, when used as monotherapy, it seems to particularly impact symptoms of conduct disorder without significantly affecting ADHD core symptoms. [16].

Stimulant therapy should be started when a patient's inability to self-regulate has substantial effects on their school performance, social engagement, self-worth, safety, or ability to function in the home or in society. While behavioral therapy may be effective in some cases and for specific behaviors, medications are meant to allow patients to 'access' the parts of their brains that enable them to engage in such therapies—and, indeed, in the activities of their lives. Medication may be discontinued as environmental factors, like school accommodations and employment workarounds, allow the patient to meet their needs. ADHD pharmacology must be discontinued if side effects become impairing or if there is no response to a reasonable trial of several medications, as determined by the clinician and patient/caregivers., Such treatment failure suggests a misdiagnosis.

As with all NDDs, when starting, changing, or adding medications to treat ADHD symptoms, one must identify a specific therapeutic goal. Once a stimulant has been initiated, close collaboration between the prescriber, caregiver, and patient will help to 'fine-tune' the treatment regimen. In general, 'less is more' when it comes to stimulants: the goal should be to maximize effect with minimal, if any, side effects, which requires prescribing the lowest effective dose and weighing the costs of these side effects with the benefits of management of ADHD core symptoms. However, prescribing the lowest effective dose increases the risk of underdosing the patient and therefore failing to adequately address their symptoms. Such fine-tuning will involve timing, redosing, and identification of both effects attributable to the medication as well as side effects, both subtle and well-known.

However, underdosing defeats the point of medicating in the first place. Optimally dosed medication will maximize the intended effect while minimizing the unintended side effects. In order to optimally dose medications, dosages should be gradually increased until there is either no additional benefit, significantly impairing or interfering side effects, or deterioration in functioning, an unusual phenomenon known to experienced prescribers of stimulants. In these situations, some desired therapeutic effects increase with increasing stimulant dose to a certain point, after which it appears not to plateau but to decrease, while side effects may continue to accumulate. This is reported in the literature as far back as 1977, when Sprague and Sleator identified improvement in memory tasks in children with ADHD treated with methylphenidate administered at low doses, but significant decrease in performance on these tasks as the dosage was increased [17,18]. When faced with this inverted U of stimulant responsiveness clinically, it may be prudent to decrease to a previous dose; switch to another medication, either of the same class or another; or to re-evaluate the diagnosis.

Another unusual use of stimulant medications is sleep assistance. As discussed in the section on sleep, sleep resistance is a very common complaint in the NDD practitioner's office. This sleep resistance is often behavioral in origin and is a common finding in patients with ADHD. In many cases, sleep refusal in the patient with ADHD is due to the same difficulties with organization that may have been present throughout the day. However, there are some patients who experience 'reactivation' or 'rebounding' after their stimulant dosage has worn off. These patients may become more irritable and disorganized than they were prior to their initial daily stimulant dosing or when compared with weekends or holidays, when they might not take their ADHD medications at all. At night, this manifests as particular difficulty organizing the mind for sleeping. Such patients may benefit from a short-acting evening dose of a stimulant.

Long-acting formulations can help to reduce frequency of dosing and stigma associated with having to disrupt the day for medication administration. Some children metabolize stimulants rapidly, greatly reducing the effective duration. In these cases, it may be to the child's benefit to prescribe a long-acting formulation, and here it becomes important to recall the mechanisms by which these longer-acting formulations are made. As examples, Concerta's OROS delivery system allows a slow 'trickle' of stimulant medication, while Ritalin LA is designed to have two separate timed serum peaks. This means that rapid stimulant metabolizers prescribed the latter drug may experience discontinuous clinical effect as one peak will wear off to a subtherapeutic level before the other activates.

While consistent pharmacology should be encouraged, some families feel that it is not necessary to take stimulants every day of the week, and some may reduce doses or skip afternoon doses. Many younger children will have 'medication holidays' on the weekends or on breaks from school, when the demands on their executive function abilities are less. Older children, teenagers, and adults with prescriptions for stimulants often "self-titrate" in order to optimize the effect when they most need it, such as during school or when studying.

While this is not the intended use of the prescription, sometimes it is prudent to allow modifications in order to achieve adherence in these patients. Parents often report concern about the addictive potential of stimulant medications, and adolescents may report stigma about having to take medication. Some patients may follow the prescription regimen more faithfully if given the freedom to partner with the prescriber and maintain a degree of ownership over the management of their ADHD.

However, it is critical that parents and adolescents understand the utility of the stimulant prescriptions. While improvement in academic achievement may be the most concrete outcome goal of these medications, stimulants, by managing the pervasive difficulties with self-regulation that necessarily occur in multiple contexts, are also helpful for improving the quality of social interactions, extracurricular activity involvement, and home and family dynamics. Patients who struggle with other adaptive functions besides academic performance should consider taking their medication daily. Furthermore, people who exhibit potentially dangerous behaviors due to their inattention or high activity levels should be encouraged to take their medications daily without breaks.

Prescribers must also be aware of the risk of diversion, particularly in adolescents and adults with stimulant prescriptions. Monitoring refill rates, which can be accomplished in many electronic medical record systems, may provide insights into the rate at which a prescription is being used.

Anxiety

Anxiety presents several significant challenges for treatment. Its manifestations are myriad—it may be generalized, specific (phobic), associated with panic, socially oriented—and treatment evidence is lacking for some of these manifestations. Adjustment disorders, which may present with anxious mood, are usually not appropriately managed by pharmacotherapy, but distinguishing these conditions from primary anxiety disorders can present a significant diagnostic challenge. Compounding these challenges, anxiety is a very common symptom. A high percentage of children with

ADHD manifest anxiety, either as a primary comorbidity or secondary to their primary attention difficulties. Furthermore, there are not sufficient studies that compare specific drugs to each other.

Adjustment disorders are conditions that occur as responses to identifiable stressors. The DSM specifies that they must resolve within 6 months of the "termination of the stressor or its consequences," but prolonged, continuous stressors will not have this same timeline requirement. For many people with NDDs, the demands of school, home life, work, or society may be sources of stress, as the nature of the NDD may make it difficult for the person to participate as expected, demanded, required, or desired.

Adjustment disorders are often underappreciated. Failure to recognize the neurodevelopmental foundation for this mismatch will commonly result in misdiagnosis and inappropriate therapy. Young children who refuse to undertake a task may be diagnosed with oppositional defiant disorder when, in fact, the reason for their oppositionality is inability to understand or perform the task or the demand. Mismatches in demand or expectation and capacity, as above, often result in behavioral disturbances. This may be enhanced by deficits in language abilities that typically allow verbal outlet of internal stresses. Adjusting expectations to meet the person's capacities frequently resolves the behavioral disturbance. (See chapter: Mood and anxiety disorders)

Treatment of adjustment disorders involves the removal of the stressor.

While treatment of primary anxiety with cognitive behavioral therapy (CBT) is an acceptable first-line approach, CBT is not always a clinically reasonable option for many patients with NDDs (See subchapter: cognitive and behavioral therapy). If pharmacotherapy is required, first-line therapy is selective serotonin reuptake inhibitors (SSRI).

Like the stimulant medications discussed above, the SSRIs and serotonin norepinephrine reuptake inhibitors (SNRIs) act primarily on presynaptic receptors responsible for neurotransmitter reuptake. SSRIs target serotonin receptors, while SNRIs target both serotonin and norepinephrine receptors, with a 30:1 proportion in venlafaxine and 10:1 in duloxetine. [19]. In contrast, bupropion primarily acts as a dual dopamine and norepinephrine reuptake inhibitor [20].

The serotonin-increasing effect of SSRIs is accomplished in two phases. The first is the direct blockade of serotonin reuptake by presynaptic neurons. This has the effect of increasing synaptic serotonin, thereby increasing the postsynaptic uptake of this neurotransmitter and enabling its antidepression and antianxiety effects. Over about 2–4 weeks, postsynaptic 5-HT_{1A} receptors become desensitized to this barrage of serotonin and are downregulated, which in turn causes increased serotonin release from the presynaptic neuron. Antidepressant effects are typically not demonstrated until this final step. The process takes several weeks to develop, which likely explains the long time to therapeutic effect observed clinically.

There are six SSRIs: fluoxetine, paroxetine, sertraline, citalopram, escitalopram, and fluvoxamine. Their pharmacokinetic properties differ, although they share a final common pathway, as outlined above. Half-lives tend to be around 24 hours, although fluoxetine's $t_{1/2}$ is closer to 48–96 hours when it has been taken for a prolonged time. The SSRIs are metabolized hepatically by CYP450 enzymes and they tend to be eliminated in urine and feces in approximately a 2:1 ratio [21].

Evidence for the treatment of anxiety with CBT, medications, or a combination of both is limited by a few factors. For one, the diversity of anxious pathologies, which range from specific phobias to generalized anxiety, limit the ability to study homogeneous populations. Secondly, researcher bias toward a few medications has restricted the availability of studies on other medication classes for anxiety treatment. CBT is a scarce resource depending on geography, which skews the results of these studies to specific populations [22]. Lastly, language abilities and IQ impact the ability to benefit from CBT, which means that young children as well as children and adults with impaired language or cognition may not be advanced enough to gain benefit.

In comparing pharmacotherapy to psychotherapy for anxiety disorders, currently available research demonstrates mixed results. In one study [23], children with social anxiety disorder responded more strongly to a 12-week course of social effectiveness therapy for children (SET-C), a type of behavioral therapy, than to a 12-week course of an SSRI, in this case fluoxetine.

Another study [24] investigated the effect of SSRI therapy with sertraline compared with CBT for a sample of 488 children suffering from generalized anxiety disorder, social anxiety disorder, and separation anxiety disorder. They found no significant difference in response between the two groups. However, this study also included subjects assigned to receive both CBT and sertraline and found that the group receiving both therapies responded much more strongly, on clinician rating scales of anxiety symptoms, compared with those receiving either intervention in isolation. To date, this is the largest study available in the literature of comparative effectiveness of the therapies for anxiety.

Initiation of an SSRI, the first-line pharmacotherapy for primary anxiety disorders, is clinician dependent and should be made on an individualized basis, with some of the considerations being:

- Severity of anxiety symptoms;

- access to CBT near the patient's home;
- patient and parental preference;
- patient and parental ability to adhere to medication regimen, which is particularly important given the prolonged titration and daily dosing required for SSRI therapy; and
- patient and parental ability to implement and support nonpharmacologic interventions.

SNRIs, such as venlafaxine, may offer useful alternatives to SSRIs. Because they have not been studied as extensively. SNRIs are reserved for medically refractory cases, as demonstrated by failure to respond to at least two SSRIs. While venlafaxine is more commonly prescribed, data suggest that duloxetine may also be helpful for children with generalized anxiety disorder [25].

In making the determination of treatment success or failure, it is crucial to allow a sufficient time for the SSRI to take effect. Treatment guidelines suggest starting at a low dose, with expected initial improvement, as measured by standardized questionnaires, noted after about 2 weeks and more significant clinical improvement, as measured by patient or parent self-report, after about 8 weeks. Maximal improvement is expected at 12 weeks.[26] (citation: printed psych guidelines paper). Failure suggests underdosing or that an alternative medication may prove more useful, although, as noted above, head-to-head trials of specific SSRIs are lacking in the current literature.

Guanfacine, an alpha-agonist that is sometimes used for management of ADHD symptoms (see above section), has shown limited efficacy in the management of generalized anxiety disorder (GAD), separation anxiety disorder, and social anxiety disorder [27]. The improvement was noted exclusively in scores on the Clinical Global Impression of Improvement Scale. This should be carefully considered in patients with dual diagnoses of ADHD and an anxiety disorder.

Anxiolytic medications, primarily benzodiazepines, are indicated for short-term use due to their significant side effect profiles and the capacity for dependence. Long-term pharmacologic management of anxiety is accomplished with SSRIs, as discussed above.

Benzodiazepines modulate $GABA_A$ receptors, which are synaptic protein complexes that, under the influence of gamma-aminobutyric acid (GABA), permit influx of chloride ions and subsequent neuronal hyperpolarization, leading to a reduction in neuron firing frequency. Benzodiazepines bind to the GABAA_AA receptor and enhance chloride influx by increasing the frequency of channel opening. This is in contrast to the barbiturate class of anxiolytics, which increase the duration of GABAA_AA channel opening to achieve the same increase in chloride efflux. This mechanism has recently come under question. Benzodiazepines bind to the $GABA_A$ receptor and enhance chloride influx by increasing the frequency of channel opening. This is in contrast to barbiturate class of anxiolytics, which increase the duration of $GABA_A$ channel opening to achieve the same increase in chloride efflux. This mechanism has recently come under question [28].

Benzodiazepines are not recommended for long-term therapy due to their side effect profile, including drowsiness, abuse potential, and respiratory depression. However, their rapid onset makes them useful for panic disorder or other cases of debilitating anxiety. These should be started at doses low enough to achieve therapeutic effect with minimal risk of side effects. Long-acting benzodiazepines, like diazepam and clonazepam, are preferred to reduce the frequency of dosing while maintaining their antianxiety effects.

As with SSRIs, tricyclic antidepressants (TCAs) have mixed efficacy in the treatment of anxiety disorders.: The largest study to date showed no difference when comparing TCAs with placebo in treatment of generalized anxiety disorder and separation anxiety disorder [29]. Furthermore, TCAs have very notable and dangerous anticholinergic, behavioral, and cardiovascular side effects, precluding their widespread use in the treatment of pediatric anxiety disorders.

Depression

Depression is a common diagnosis, both as a primary disability and as a comorbidity, in people with NDDs. Management of pediatric depression may reasonably fall under the purview of the NDD clinician but severe or refractory depression, particularly when associated with suicidal ideation, may be better managed by colleagues in psychiatry. Furthermore, like anxiety, depressed mood may be due to a stressor and qualify as an adjustment disorder. As above, identification and removal of the stressor frequently resolves the mood difficulties. (See chapter: Mood and anxiety disorders).

The pharmacologic treatment of choice for depression in children and adolescents starts with a SSRI. Of the SSRIs, fluoxetine has been the most widely studied and is often used as the initial drug of choice. While there are few studies in the literature comparing fluoxetine to other SSRIs, studies have demonstrated SSRI superiority in reducing symptoms of major depression when compared with placebo [30,31], and with TCAs [32].

The other SSRIs are of less certain effect and appear to have more significant withdrawal and side effects. Sertraline, for example, appears to have similar efficacy in treating the symptoms of depression but has more significant withdrawal effects, such as agitation and gastrointestinal upset [33]. Citalopram and escitalopram are stereoisomers of each other: escitalopram contains only the psychoactive S-citalopram enantiomer, while citalopram is a racemic mixture of both enantiomers, requiring approximately twice the dose to achieve equivalent effects. Their findings when compared with placebo have been inconsistent.[34–37]. Paroxetine and fluvoxamine are generally not prescribed for pediatric depression due to their apparent inefficacy, although this is likely at least partly due to lack of robust literature [34].

SSRIs require up to 8 weeks to reach maximal effect. Recall that over the first 4 weeks of treatment, postsynaptic 5-HT_{1A} receptors are downregulated in order to increase serotonin release from presynaptic neurons, and more time is often required in order to see clinical improvement in symptoms, with maximal improvement expected at 12 weeks. Patients must be counseled on the importance of patience and regimen adherence when initiating treatment with an SSRI.

Typical side effects of SSRIs are mild and usually short-lived; they include agitation, gastrointestinal upset, and headaches [38]. Activation, which may manifest as impulsivity, akathisia, and irritability, typically disappears when dosage is decreased; formulation is switched to a longer acting one, as higher serum levels of the pharmacoactive ingredients are more likely to trigger side effects; or medication is stopped [39]. People who have had activating responses to one SSRI may be more likely to have activation if switched to another SSRI. Citalopram and escitalopram appear to have a slightly increased risk of rare cardiac effects, such as ventricular arrhythmia and sudden cardiac death, when compared with the other SSRIs [40].

SSRIs inhibit the cytochrome P450 system and may have interactions with other medications [41]. Although fluvoxamine appears to have the most potent CYP450 inhibitory properties it is also rarely prescribed, as discused above. However, the frequently prescribed fluoxetine may have more clinically relevant CYP450 inhibition, as this can last for several weeks due to the drug's prolonged half-life.

One of the most well-known associations of SSRIs are their increased risk of suicidal and homicidal ideation and behavior. In fact, the FDA has issued a Black Box Warning for SSRIs due to this apparent association [42]. The evidence suggests a possible slightly increased risk of suicidal ideation but without an associated increased rate of completed suicide when compared with placebo [43]. Furthermore, when compared with the risk of suicidality in untreated depression, the benefits of prescribing SSRIs greatly outweigh the possible risk of increased suicidal thoughts. Of course, this decision must be made clinically on a case-by-case basis, and collaboration with colleagues in psychiatry may be appropriate when patient suicide is a consideration.

Venlafaxine, a SNRI, is typically reserved for children who have failed two SSRIs. Venlafaxine has proven to have similar efficacy as other SSRIs when used as a second-line therapy in children and adolescents who had not responded to a first-line treatment with an SSRI. However, the response was less sustained when compared with the SSRIs, and the side effects were greater [44,45]. Duloxetine, another SNRI, has not demonstrated effect in clinical trials [46].

If two SSRIs or two SSRIs and venlafaxine fail, or based on clinical determination, the next treatment of choice may be another medication, such as bupropion (a dual norepinephrine and dopamine reuptake inhibitor) or duloxetine, or more intensive therapies which might require inpatient admission. However, data on the aforementioned pharmacotherapy in children are lacking. Depression that fails to remit after multiple appropriate trials of the above treatments may require the guidance of a psychiatrist.

Aggression

Aggression and aggressive behaviors are frequently managed with neuroleptic medications. The word "neuroleptic" has roots in the Greek *neuron*, relating to the nervous system, and *lepsis*, meaning "to take hold off"; therefore, a neuroleptic medication is one that "takes hold of the nervous system" [47]. These medications are also called antipsychotics.

This class of medications has a long and fraught history. The first, chlorpromazine, was used as an alternative to surgical frontal lobotomy for behavioral disorders in the 1950s. Research into its mechanism of action, dopamine D_2 receptor blocking, led to the development of the class of medications that became known as neuroleptics, all of which have similar effects on dopamine. Subsequently, the neuroleptic medications were divided into "first-generation," or "typical," and "second-generation," or "atypical," classes based on their perceived ability to induce the extrapyramidal symptoms of akathisia, dystonia, pseudoparkinsonism, and tardive dyskinesia. There remains some debate about the risk of extrapyramidal symptom development with typical versus atypical neuroleptic medications, but common practice is to prescribe the atypical medications as first-line for psychotic disorders and bipolar disorder.

To that end, second-generation neuroleptic medications like risperidone and olanzapine are approved for irritability, aggression, mood variability, and disruptive behaviors in children with intellectual disability, mood disorders, and other causes of disruptive behavior disorders [48].

The neuroleptic medications act by binding to and blocking dopamine D_2 receptors. While the antipsychotic effects appear to be due to blockage of the mesolimbic dopamine receptors, the presence of D_2 receptors throughout the neurocircuitry—they are present in mesocortical, tuberoinfundibular, and nigrostriatal pathways in addition to the mesolimbic pathway—and the nonselective nature of neuroleptic D_2 blockade underlie the common and significant side effects. In general, this nonselectivity applies to all of the neuroleptics with variability dependent upon individual molecular characteristics; however, as a group, the atypical (second-generation) neuroleptics have additional pharmacokinetic characteristics that are thought to temper the extrapyramidal symptoms and hyperprolactinemia. Clozapine and quetiapine, for example, bind to and block D_2 receptors for a brief enough period of time to prevent psychosis but not long enough to induce extrapyramidal symptoms (EPS). Furthermore, the atypical neuroleptics have 5-HT_{2A} and 5-HT_{2C} blocking effects, which are known to modulate dopamine activity in the various dopaminergic pathways of the nervous system.

Risperidone and aripiprazole, second-generation or "atypical" antipsychotics, are FDA-approved to treat aggression in children greater than 5 years old, but only those with diagnoses of autism spectrum disorder. Specifically, risperidone is approved for aggression, tantrum behaviors, and self-injury in children with autism spectrum disorder (ASD), while aripiprazole is approved only for treatment of irritability in children with ASD. Indications for pharmacotherapy for aggressive behavior include very severe behaviors that endanger the life of the patient and those in close contact with the patient, as well as failure to identify an underlying cause that might be the target of environmental modification or behavioral therapy.

Antipsychotic medications approved for use in patients with NDDs are the following:

- Aripiprazole, approved for "irritability associated with autistic disorder" and "treatment of Tourette's disorder" [49].
- Risperidone, approved for "treatment of irritability associated with autistic disorder in children and adolescents aged 5–16 years" [50].

A 2017 Cochrane Review showed some evidence that risperidone reduced aggression compared with placebo; there were not sufficient data for assessing other antipsychotics [51].

Children treated with risperidone have shown decreased irritability as measured by the corresponding section of the Clinical Global Impression-Improvement scale [52].

The other antipsychotic medications, including olanzapine, haloperidol, quetiapine, clozapine, and ziprasidone are not approved for treatment of symptoms of NDDs; however, they are sometimes used off-label for symptom management.

A population-wide study of Canadian adults with neurodevelopmental disabilities investigated the rates of antipsychotic use [53]. They found that almost 40% of the adult population with intellectual and developmental disabilities had received at least one antipsychotic prescription during the study period, with just over two-thirds of that group carrying a diagnosis of a psychiatric illness. These findings were similar to another study from the United Kingdom in 2015, in which about 30% of patients with NDDs who were prescribed antipsychotics carried mental health diagnoses in a population-wide study [54].

In evaluating the ideal management of aggressive behaviors in people with NDDs, it is of particular importance to investigate underlying mechanisms. For instance, a child with poor verbal language who has a toothache may not be able to communicate the need for dental care, and situational frustration may lead to behavioral outbursts. Menstrual cramps, constipation, earaches, headaches, hunger, boredom, sexual frustration, and many other causes may actually be underlying the behavior, and children and young adults with intellectual disability may demonstrate behaviors that are actually appropriate for their developmental age.

The dopamine-blocking antipsychotic medications are associated with a number of significant side effects, including weight gain, insulin resistance, hyperlipidemia, metabolic syndrome, transaminitis, cardiac conduction anomalies (specifically QTc prolongation), extrapyramidal symptoms (dystonia, akathisia, parkinsonism, and tardive dyskinesia—less common with the second-generation antipsychotics), agranulocytosis, and neuroleptic malignant syndrome. Prescribing clinicians and those encountering children with aggressive behaviors treated with antipsychotic medications must recall these side effects and monitor appropriately with regular laboratory evaluations of blood glucose, serum lipid levels, hepatic function, and blood counts, as well as electrocardiogram (EKGs).

Other medications frequently used off-label include other second-generation antipsychotics, like olanzapine, quetiapine, and ziprasidone; some antiseizure medications; alpha-2 agonists like clonidine and guanfacine; mood stabilizers like lithium; and beta-2 antagonists [55].

Sleep disturbances

Difficulty with sleep is reported frequently by patients with NDDs [56] and is associated with difficult behaviors, obesity, and school problems in both general pediatric populations and populations of patients with NDDs [57–59] (See chapter: Sleep disorders). Some NDDs are particularly and specifically associated with sleep difficulties: Angelman syndrome, for example, is thought to have a neurobiologic basis for notoriously abnormal sleep-wake cycles [60].

There are a few categories of sleep disturbances [61].

- *Sleep latency* refers to the time that elapses between turning off the lights and the onset of any stage of sleep. Formal measurement typically requires EEG and other metrics by polysomnography.
- *Total sleep time* is the complete amount of time a patient spends in sleep phases during a given monitoring period.
- Related to total sleep time, *sleep efficiency* refers to the total amount of sleep expressed as the percentage of total time in bed that is spent in one of the sleep stages.
- *Wake after sleep onset*, abbreviated WASO, is the total amount of time patients are awake after having initially fallen asleep; this may be compared with *total wake time* which, by including sleep latency period, is the total amount of time in bed that is spent awake during a given monitoring period. WASO, then, is a component of total wake time and gives an indication of *night awakenings*.
- *Sleep continuity* consolidates sleep efficiency, total sleep time, and WASO to report an overall metric of quality of sleep.

Bedtime resistance, an important concept in assessing sleep difficulties, refers to the inability to wind down for sleep. This may be due to strong sleep associations, like parental presence or a specific soothing toy, that make initiating sleep, particularly after a nighttime awakening, difficult. It may also be due to a lack of bedtime rules, like having lights off and no access to screens. In both cases, it manifests with opposition to the behaviors associated with sleep onset. This may appear as crying, fussing, or arguing instead of initiating sleep onset behaviors. An associated, biologically based condition called delayed sleep phase syndrome, in which endogenous melatonin secretion is delayed relative to desired sleep-wake cycle rhythms, may be a consideration in some cases.

Insomnia is defined by the International Classification of Sleep Disorders, Third Edition, as "trouble initiating or maintaining sleep which is associated with day-time consequences and is not attributable to environmental circumstances or inadequate opportunity to sleep" [62]. All of the above categories of sleep disturbances, not just difficulty with sleep latency, must be considered for a diagnosis of insomnia. This definition requires symptom frequency of at least three times a week and duration of at least 3 months.

Treatment for sleep disturbance in the NDD practitioner's office may involve referral for behavioral therapy and recommendations of melatonin. Oral melatonin is a synthetic form of the endogenous hormone produced by the pineal gland that is thought to help regulate circadian rhythm by initiating sleep onset. Over the counter melatonin is a supplement that is not monitored by the FDA, and dosages may be very hard to make exact, as concentrations vary by brand. For this reason, melatonin should be prescribed when exact dosing is required.

There is evidence that melatonin may be effective in initiating sleep by advancing sleep-wake rhythm and might lengthen sleep, although the heterogeneous etiologies of sleep onset disorders may preclude accurate and precise studies [63]. Dosing is widely variable in the literature, with no dose–response relationship in sleep onset noted for a range of 0.05–0.15 mg/kg melatonin in one study [64]. Typically, dosages of 1–5 mg are appropriate, and the medication should be administered as much as 3–4 hours before actual desired sleep onset time [63].

There have not been significant side effects of melatonin reported in the literature, although long-term studies are lacking. Some studies report minor, transient side effects like headache and diarrhea, although these tend to remit after a few days [65].

An American Academy of Neurology literature review to develop guidelines for treating sleep disturbances in children with autism spectrum disorders found the following evidence [66]:

- *Bedtime resistance*: The authors found evidence that melatonin, with or without behavioral therapy (specifically CBT), is "probably effective." CBT alone is considered "possibly effective."
- *Sleep onset latency*: Similarly to bedtime resistance, melatonin, both with or without CBT, is thought to be "probably effective," and CBT on its own is "possibly effective." Parental education, including behavioral training, does not have sufficient support in the literature to date.
- *Sleep continuity*: The authors found that for the various metrics of sleep continuity—sleep efficiency, total sleep time, and WASO—melatonin, again with or without CBT, was "probably effective," and CBT alone was "possibly effective."

The authors note, however, that there remains insufficient evidence in the literature for other potential medical approaches to sleep management which are not typically included in the roster of medications but are known to often have effects on sleep. These include alpha-adrenergic agonists and antiseizure, antidepressant, and antipsychotic medications. Diphenhydramine, a histamine antagonist with strong sedating effects, is not indicated for sleep.

It is also important to note that ASD diagnosis was somewhat heterogeneous in the literature, so it is possible that some of the diagnoses would not carry over. As a result, the findings of this review may be more generalizable than just to the pediatric ASD population.

Conclusion

NDDs are varied in their presentations and require equally varied management approaches. This may take the form of nonpharmacologic therapies like environmental modification, CBT and behavioral therapy, pharmacologic intervention, or most commonly a combination of pharmacologic and nonpharmacologic approaches. The goal is never 'resolution' of the NDD but, rather, symptom alleviation that enables adequate participation in the activities of the patient's life. Therefore the role of the prescriber in managing NDDs is facilitation of these activities—school, work, social interaction, therapy, etc.—through judicious prescribing based on identification of specific target symptoms. It is important for the prescriber to understand not only the best available evidence for use of the medications, but also to understand the pharmacologic foundations and mechanisms of these medications in order to prescribe safely, adequately, and with confidence. Establishing close communication with caregivers and, as developmentally appropriate, patients themselves will allow the greatest, most sustainable, and longest-term therapeutic alliance.

References

[1] Wachtel LE, Hagopian LP. Psychopharmacology and applied behavioral analysis: tandem treatment of severe problem behaviors in intellectual disability and a case series. Isr J Psychiatry Relat Sci 2006;43(4):265–74.

[2] Perrie Y, Rades T. FASTtrack Pharmaceutics: Drug Delivery and Targeting. 2nd ed. London: Pharmaceutical Press; 2012. p. 1–24.

[3] Ogu CC, Maxa JL. Drug interactions due to cytochrome P450. Proc (Bayl Univ Med Cent) 2000;13(4):421–3. Available from: https://doi.org/10.1080/08998280.2000.11927719.

[4] Lynch T, Price A. The effect of cytochrome P450 metabolism on drug response, interactions, and adverse effects. Am Fam Phys 2007;76(3): 391–6.

[5] Cortese S, Newcorn JH, Coghill D. A practical, evidence-informed approach to managing stimulant-refractory attention deficit hyperactivity disorder (ADHD). CNS Drugs 2021;35(10):1035–51. Available from: https://doi.org/10.1007/s40263-021-00848-3.

[6] Faraone SV. The pharmacology of amphetamine and methylphenidate: relevance to the neurobiology of attention-deficit/hyperactivity disorder and other psychiatric comorbidities. Neurosci Biobehav Rev 2018;87:255–70. Available from: https://doi.org/10.1016/j.neubiorev.2018.02.001.

[7] Weiss MD, Surman CBH, Elbe D. Stimulant 'rapid metabolizers': wrong label, real phenomena. Atten Defic Hyperact Disord 2018;10(2): 113–18. Available from: https://doi.org/10.1007/s12402-017-0242-9.

[8] Markowitz JS, DeVane CL, Pestreich LK, Patrick KS, Muniz R. A comprehensive in vitro screening of d-, l-, and dl-threo-methylphenidate: an exploratory study. J Child Adolesc Psychopharmacol 2006;16(6):687–98. Available from: https://doi.org/10.1089/cap.2006.16.687.

[9] Balcioglu A, Ren JQ, McCarthy D, Spencer TJ, Biederman J, Bhide PG. Plasma and brain concentrations of oral therapeutic doses of methylphenidate and their impact on brain monoamine content in mice. Neuropharmacology 2009;57(7-8):687–93. Available from: https://doi.org/10.1016/j.neuropharm.2009.07.025.

[10] Pelham WE, Gnagy EM, Burrows-Maclean L, Williams A, Fabiano GA, Morrisey SM, et al. Once-a-day Concerta methylphenidate versus three-times-daily methylphenidate in laboratory and natural settings. Pediatrics 2001;107(6):E105. Available from: https://doi.org/10.1542/peds.107.6.e105.

[11] Coghill D, Seth S. Effective management of attention-deficit/hyperactivity disorder (ADHD) through structured re-assessment: the Dundee ADHD Clinical Care Pathway. Child Adolesc Psychiatry Ment Health 2015;9:52. Available from: https://doi.org/10.1186/s13034-015-0083-2.

[12] Blader JC, Pliszka SR, Kafantaris V, Foley CA, Carlson GA, Crowell JA, et al. Stepped treatment for attention-deficit/hyperactivity disorder and aggressive behavior: a randomized, controlled trial of adjunctive risperidone, divalproex sodium, or placebo after stimulant medication optimization. J Am Acad Child Adolesc Psychiatry 2021;60(2):236–51. Available from: https://doi.org/10.1016/j.jaac.2019.12.009.

[13] Ramtvedt BE, Røinås E, Aabech HS, Sundet KS. Clinical gains from including both dextroamphetamine and methylphenidate in stimulant trials. J Child Adolesc Psychopharmacol 2013;23(9):597–604. Available from: https://doi.org/10.1089/cap.2012.0085.

[14] Leucht S, Hierl S, Kissling W, Dold M, Davis JM. Putting the efficacy of psychiatric and general medicine medication into perspective: review of meta-analyses. Br J Psychiatry 2012;200(2):97–106. Available from: https://doi.org/10.1192/bjp.bp.111.096594.

[15] McCracken JT, McGough JJ, Loo SK, Levitt J, Del'Homme M, Cowen J, et al. Combined stimulant and guanfacine administration in attention-deficit/hyperactivity disorder: a controlled, comparative study. J Am Acad Child Adolesc Psychiatry 2016;55(8):657–66. Available from: https://doi.org/10.1016/j.jaac.2016.05.015.

[16] Hazell PL, Stuart JE. A randomized controlled trial of clonidine added to psychostimulant medication for hyperactive and aggressive children. J Am Acad Child Adolesc Psychiatry 2003;42(8):886–94. Available from: https://doi.org/10.1097/01.CHI.0000046908.27264.00.

[17] Levy F. Stimulant side effects and inverted-U: implications for ADHD guidelines. Aust N Z J Psychiatry 2012;47(3):217–21.

[18] Sprague RL, Sleator EK. Methylphenidate in hyperkinetic children: differences in dose effects on learning and social behavior. Science 1977;198(4323):1274–6. Available from: https://doi.org/10.1126/science.337493.

[19] Sansone RA, Sansone LA. Serotonin norepinephrine reuptake inhibitors: a pharmacological comparison. Innov Clin Neurosci 2014;11(3-4):37–42.

[20] Stahl SM, Pradko JF, Haight BR, Modell JG, Rockett CB, Learned-Coughlin S. A review of the neuropharmacology of bupropion, a dual norepinephrine and dopamine reuptake inhibitor. Prim Care Companion J Clin Psychiatry 2004;6(4):159–66. Available from: https://doi.org/10.4088/pcc.v06n0403.

[21] Preskorn SH. Clinically relevant pharmacology of selective serotonin reuptake inhibitors. An overview with emphasis on pharmacokinetics and effects on oxidative drug metabolism. Clin Pharmacokinet 1997;32(Suppl 1):1–21. Available from: https://doi.org/10.2165/00003088-199700321-00003.

[22] Payne KA, Myhr G. Increasing access to cognitive-behavioural therapy (CBT) for the treatment of mental illness in Canada: a research framework and call for action. Healthc Policy 2010;5(3):e173–85.

[23] Beidel DC, Turner SM, Sallee FR, Ammerman RT, Crosby LA, Pathak S. SET-C versus fluoxetine in the treatment of childhood social phobia. J Am Acad Child Adolesc Psychiatry 2007;46(12):1622–32. Available from: https://doi.org/10.1097/chi.0b013e318154bb57.

[24] Ginsburg GS, Kendall PC, Sakolsky D, Compton SN, Piacentini J, Albano AM, et al. Remission after acute treatment in children and adolescents with anxiety disorders: findings from the CAMS. J Consult Clin Psychol 2011;79(6):806–13. Available from: https://doi.org/10.1037/a0025933.

[25] Strawn JR, Prakash A, Zhang Q, Pangallo BA, Stroud CE, Cai N, et al. A randomized, placebo-controlled study of duloxetine for the treatment of children and adolescents with generalized anxiety disorder. J Am Acad Child Adolesc Psychiatry 2015;54(4):283–93. Available from: https://doi.org/10.1016/j.jaac.2015.01.008.

[26] American Psychiatric Association. Practice guideline for the treatment of patients with major depressive disorder. 3rd ed. American Psychiatric Publishing; 2010.

[27] Strawn JR, Compton SN, Robertson B, Albano AM, Hamdani M, Rynn MA. Extended release guanfacine in pediatric anxiety disorders: a pilot, randomized, placebo-controlled trial. J Child Adolesc Psychopharmacol 2017;27(1):29–37. Available from: https://doi.org/10.1089/cap.2016.0132.

[28] Bianchi MT, Botzolakis EJ, Lagrange AH, Macdonald RL. Benzodiazepine modulation of GABA(A) receptor opening frequency depends on activation context: a patch clamp and simulation study. Epilepsy Res 2009;85(2–3):212–20. Available from: https://doi.org/10.1016/j.eplepsyres.2009.03.007.

[29] Berney T, Kolvin I, Bhate SR, Garside RF, Jeans J, Kay B, et al. School phobia: a therapeutic trial with clomipramine and short-term outcome. Br J Psychiatry 1981;138:110–18. Available from: https://doi.org/10.1192/bjp.138.2.110.

[30] Gibbons RD, Hur K, Brown CH, Davis JM, Mann JJ. Benefits from antidepressants: synthesis of 6-week patient-level outcomes from double-blind placebo-controlled randomized trials of fluoxetine and venlafaxine. Arch Gen Psychiatry 2012;69(6):572–9. Available from: https://doi.org/10.1001/archgenpsychiatry.2011.2044.

[31] Varigonda AL, Jakubovski E, Taylor MJ, Freemantle N, Coughlin C, Bloch MH. Systematic review and meta-analysis: early treatment responses of selective serotonin reuptake inhibitors in pediatric major depressive disorder. J Am Acad Child Adolesc Psychiatry 2015;54(7):557–64. Available from: https://doi.org/10.1016/j.jaac.2015.05.004.

[32] Qin B, Zhang Y, Zhou X, Cheng P, Liu Y, Chen J, et al. Selective serotonin reuptake inhibitors versus tricyclic antidepressants in young patients: a meta-analysis of efficacy and acceptability. Clin Ther 2014;36(7):1087–95. Available from: https://doi.org/10.1016/j.clinthera.2014.06.001.

[33] Wagner KD, Ambrosini P, Rynn M, Wohlberg C, Yang R, Greenbaum MS, et al. Efficacy of sertraline in the treatment of children and adolescents with major depressive disorder: two randomized controlled trials. JAMA 2003;290(8):1033–41. Available from: https://doi.org/10.1001/jama.290.8.1033.

[34] Hetrick SE, McKenzie JE, Cox GR, Simmons MB, Merry SN. Newer generation antidepressants for depressive disorders in children and adolescents. Cochrane Database Syst Rev 2012;11(11):CD004851. Available from: https://doi.org/10.1002/14651858.CD004851.pub3.

[35] Wagner KD, Jonas J, Findling RL, Ventura D, Saikali K. A double-blind, randomized, placebo-controlled trial of escitalopram in the treatment of pediatric depression. J Am Acad Child Adolesc Psychiatry 2006;45(3):280–8. Available from: https://doi.org/10.1097/01.chi.0000192250.38400.9e.

[36] Bridge JA, Iyengar S, Salary CB, Barbe RP, Birmaher B, Pincus HA, et al. Clinical response and risk for reported suicidal ideation and suicide attempts in pediatric antidepressant treatment: a meta-analysis of randomized controlled trials. JAMA 2007;297(15):1683–96. Available from: https://doi.org/10.1001/jama.297.15.1683.

[37] Wagner KD, Robb AS, Findling RL, Jin J, Gutierrez MM, Heydorn WE. A randomized, placebo-controlled trial of citalopram for the treatment of major depression in children and adolescents. Am J Psychiatry 2004;161(6):1079–83. Available from: https://doi.org/10.1176/appi.ajp.161.6.1079.

[38] Birmaher B, Brent D, AACAP Work Group on Quality IssuesBernet W, Bukstein O, Walter H, et al. Practice parameter for the assessment and treatment of children and adolescents with depressive disorders. J Am Acad Child Adolesc Psychiatry 2007;46(11):1503–26. Available from: https://doi.org/10.1097/chi.0b013e318145ae1c.

[39] Luft MJ, Lamy M, DelBello MP, McNamara RK, Strawn JR. Antidepressant-induced activation in children and adolescents: risk, recognition and management. Curr Probl Pediatr Adolesc Health Care 2018;48(2):50–62. Available from: https://doi.org/10.1016/j.cppeds.2017.12.001.

[40] Czaja AS, Valuck RJ, Anderson HD. Comparative safety of selective serotonin reuptake inhibitors among pediatric users with respect to adverse cardiac events. Pharmacoepidemiol Drug Saf 2013;22(6):607–14. Available from: https://doi.org/10.1002/pds.3420.

[41] Hemeryck A, Belpaire FM. Selective serotonin reuptake inhibitors and cytochrome P-450 mediated drug-drug interactions: an update. Curr Drug Metab 2002;3(1):13–37. Available from: https://doi.org/10.2174/1389200023338017.

[42] Friedman RA, Leon AC. Expanding the black box - depression, antidepressants, and the risk of suicide. N Engl J Med 2007;356(23):2343–6. Available from: https://doi.org/10.1056/NEJMp078015.

[43] Hammad TA, Laughren T, Racoosin J. Suicidality in pediatric patients treated with antidepressant drugs. Arch Gen Psychiatry 2006;63 (3):332–9. Available from: https://doi.org/10.1001/archpsyc.63.3.332.

[44] Vitiello B, Emslie G, Clarke G, Wagner KD, Asarnow JR, Keller MB, et al. Long-term outcome of adolescent depression initially resistant to selective serotonin reuptake inhibitor treatment: a follow-up study of the TORDIA sample. J Clin Psychiatry 2011;72(3):388–96. Available from: https://doi.org/10.4088/JCP.09m05885blu.

[45] Brent D, Emslie G, Clarke G, Wagner KD, Asarnow JR, Keller M, et al. Switching to another SSRI or to venlafaxine with or without cognitive behavioral therapy for adolescents with SSRI-resistant depression: the TORDIA randomized controlled trial. JAMA 2008;299(8):901–13. Available from: https://doi.org/10.1001/jama.299.8.901.

[46] Emslie GJ, Prakash A, Zhang Q, Pangallo BA, Bangs ME, March JS. A double-blind efficacy and safety study of duloxetine fixed doses in children and adolescents with major depressive disorder. J Child Adolesc Psychopharmacol 2014;24(4):170–9. Available from: https://doi.org/10.1089/cap.2013.0096.

[47] King C, Voruganti LN. What's in a name? The evolution of the nomenclature of antipsychotic drugs. J Psychiatry Neurosci 2002;27(3):168–75.

[48] Zuddas A, Zanni R, Usala T. Second generation antipsychotics (SGAs) for non-psychotic disorders in children and adolescents: a review of the randomized controlled studies. Eur Neuropsychopharmacol 2011;21(8):600–20. Available from: https://doi.org/10.1016/j.euroneuro.2011.04.001.

[49] FDA: Abilify. https://www.accessdata.fda.gov/drugsatfda_docs/label/2014/021436s038,021713s030,021729s022,021866s023lbl.pdf.

[50] FDA: Risperidone. https://www.accessdata.fda.gov/drugsatfda_docs/label/2009/020272s056%2C020588s044%2C021346s033%2C021444s03lbl.pdf.

[51] Loy JH, Merry SN, Hetrick SE, Stasiak K. Atypical antipsychotics for disruptive behaviour disorders in children and youths. Cochrane Database Syst Rev 2017;8(8):CD008559. Available from: https://doi.org/10.1002/14651858.CD008559.pub3.

[52] McCracken JT, McGough J, Shah B, Cronin P, Hong D, Aman MG, et al. Risperidone in children with autism and serious behavioral problems. N Engl J Med 2002;347(5):314–21. Available from: https://doi.org/10.1056/NEJMoa013171.

[53] Lunsky Y, Khuu W, Tadrous M, Vigod S, Cobigo V, Gomes T. Antipsychotic use with and without comorbid psychiatric diagnosis among adults with intellectual and developmental disabilities. Can J Psychiatry 2018;63(6):361–9. Available from: https://doi.org/10.1177/0706743717727240.

[54] Deb S, Unwin G, Deb T. Characteristics and the trajectory of psychotropic medication use in general and antipsychotics in particular among adults with an intellectual disability who exhibit aggressive behaviour. J Intellect Disabil Res 2015;59(1):11–25. Available from: https://doi.org/10.1111/jir.12119.

[55] Huffman LC, Sutcliffe TL, Tanner IS, Feldman HM. Management of symptoms in children with autism spectrum disorders: a comprehensive review of pharmacologic and complementary-alternative medicine treatments. J Dev Behav Pediatr 2011;32(1):56–68. Available from: https://doi.org/10.1097/DBP.0b013e3182040acf.

[56] Hodge D, Carollo TM, Lewin M, Hoffman CD, Sweeney DP. Sleep patterns in children with and without autism spectrum disorders: developmental comparisons. Res Dev Disabil 2014;35(7):1631–8. Available from: https://doi.org/10.1016/j.ridd.2014.03.037.

[57] Gottlieb DJ, Vezina RM, Chase C, Lesko SM, Heeren TC, Weese-Mayer DE, et al. Symptoms of sleep-disordered breathing in 5-year-old children are associated with sleepiness and problem behaviors. Pediatrics 2003;112(4):870–7. Available from: https://doi.org/10.1542/peds.112.4.870.

[58] Rangan A, Zheng M, Olsen NJ, Rohde JF, Heitmann BL. Shorter sleep duration is associated with higher energy intake and an increase in BMI z-score in young children predisposed to overweight. Int J Obes 2018;42(1):59–64. Available from: https://doi.org/10.1038/ijo.2017.216.

[59] Fallone G, Acebo C, Seifer R, Carskadon MA. Experimental restriction of sleep opportunity in children: effects on teacher ratings. Sleep 2005;28(12):1561–7. Available from: https://doi.org/10.1093/sleep/28.12.1561.

[60] Goldman SE, Bichell TJ, Surdyka K, Malow BA. Sleep in children and adolescents with Angelman syndrome: association with parent sleep and stress. J Intellect Disabil Res 2012;56(6):600–8. Available from: https://doi.org/10.1111/j.1365-2788.2011.01499.x.

[61] Shrivastava D, Jung S, Saadat M, Sirohi R, Crewson K. How to interpret the results of a sleep study. J Commun Hosp Intern Med Perspect 2014;4(5):24983. Available from: https://doi.org/10.3402/jchimp.v4.24983.

[62] Sateia MJ. International classification of sleep disorders-third edition: highlights and modifications. Chest 2014;146(5):1387–94. Available from: https://doi.org/10.1378/chest.14-0970.

[63] Bruni O, Alonso-Alconada D, Besag F, Biran V, Braam W, Cortese S, et al. Current role of melatonin in pediatric neurology: clinical recommendations. Eur J Paediatr Neurol 2015;19(2):122–33. Available from: https://doi.org/10.1016/j.ejpn.2014.12.007.

[64] van Geijlswijk IM, van der Heijden KB, Egberts AC, Korzilius HP, Smits MG. Dose finding of melatonin for chronic idiopathic childhood sleep onset insomnia: an RCT. Psychopharmacology 2010;212(3):379–91. Available from: https://doi.org/10.1007/s00213-010-1962-0.

[65] Besag FMC, Vasey MJ, Lao KSJ, Wong ICK. Adverse events associated with melatonin for the treatment of primary or secondary sleep disorders: a systematic review. CNS Drugs 2019;33(12):1167–86. Available from: https://doi.org/10.1007/s40263-019-00680-w.

[66] Williams Buckley A, Hirtz D, Oskoui M, Armstrong MJ, Batra A, Bridgemohan C, et al. Practice guideline: treatment for insomnia and disrupted sleep behavior in children and adolescents with autism spectrum disorder: report of the Guideline Development, Dissemination, and Implementation Subcommittee of the American Academy of Neurology. Neurology 2020;94(9):392–404. Available from: https://doi.org/10.1212/WNL.0000000000009033.

Chapter 34

Disruptive behaviors/self-injury in neurodevelopmental disabilities

Michelle A. Frank-Crawford, Samantha L. Hardesty and Louis P. Hagopian
Kennedy Krieger Institute, and Johns Hopkins University School of Medicine, Baltimore, MD, United States

Description, prevalence, and course

Individuals with neurodevelopmental disabilities (NDD) are at greater risk for engaging in problem behaviors than their neurotypical peers (e.g., [1]). Problem behavior in this population is quite heterogeneous and can vary in terms of its form as well as its degree of intensity/severity (e.g., likelihood of causing injury), frequency (e.g., daily, weekly, monthly, episodic [2]), and its controlling variables. Cyclical patterns of problem behavior characterized by alternating periods of worsening behavior and behavioral stability are not unusual among individuals with NDD [3–5]. Commonly reported forms of problem behavior include aggression (e.g., physical aggression including hitting and kicking others; verbal aggression including swearing and threatening others), self-injurious behavior (SIB; e.g., head banging and hitting, pinching, scratching, eye poking, self-biting), disruptive or destructive behaviors (e.g., property destruction, tipping furniture, throwing items), pica (i.e., ingestion of inedible objects), and elopement (e.g., moving away from a caregiver without consent and not complying with prompts to return; Table 34.1). When the problem behaviors are severe and persistent, they can pose significant risks of injury to the individual themselves or to others, can be destructive to the environment, and impair functioning. For example, SIB can cause contusions and lacerations, loss of tissue, blindness, and irreversible disfigurement [10–13]. Aggression not only places the individual and others around them in danger for experiencing severe injury, but is also associated with increased risk of physical abuse from caregivers [14] and residential staff [15]. Elopement is also a particularly dangerous form of problem behavior; research has suggested it is relatively common [16], and places individuals at increased risk for injury, such as death by drowning [17].

Prevalence

The emergence and prevalence of severe problem behavior in individuals with disabilities has been the focus of research across many decades [12,18,19]. Although prevalence estimates of problem behavior vary widely (likely owing to differences in study methodology, measurement, and reporting periods), many studies suggest that somewhere between 30% and 50% of individuals with NDD engage in at least one form of problem behavior [1,6,20]. Severity of problem behavior for individuals with NDD has been evaluated using several methods, including rating scales [21] and assessments that quantify the severity of injuries related to SIB based on the number of injuries, injury location, and injury type (i.e., the self-injury trauma scale [22]. Results have generally suggested that although the prevalence of problem behavior among individuals with NDD is high, a smaller subset of this population, 5%–10%, engaged in particularly severe problem behavior that poses the greatest risks [20,21,23]. Furthermore, severity of injury produced by problem behavior has been linked to increasing age and size of the individual engaging in the behaviors [24], and certain subtypes of self-injury [9,25].

Emergence and maintenance

Problem behaviors likely emerge because of dysfunction resulting from intellectual and/or developmental disabilities, and experiences that occasion and reinforce the behaviors [26–28]. For example, individuals with NDD typically

Capute and Accardo's Neurodevelopmental Disabilities in Infancy and Childhood. DOI: https://doi.org/10.1016/B978-0-12-824060-1.00030-4

TABLE 34.1 Examples of commonly reported problem behaviors exhibited by those with neurodevelopmental disabilities (NDD) and autism spectrum disorder (ASD).

Problem behavior	Examples	Prevalence (based on literature)	Population (based on literature)
Self-injurious behavior	Head-hitting, self-biting	82% [2]	Adults with profound intellectual and multiple disabilities
		27.7% [6]	Children with ASD
Verbal aggression	Yelling at others, threatening others	66.1% [7]	Adults with NDD
Aggression	Hitting others, kicking others	47.8% [7]	Adults with NDD
		32% [8]	Children with ASD
Destructive behavior	Breaking objects, throwing objects	38.1% [7]	Adults with NDD
Pica	Ingestion of nonnutritive or inedible objects	0.3%–25% [9]	Individuals with NDD in community and living in institutions

experience deficits in adaptive skills, social skills, emotion regulation, and appropriate functional communication. Children with limited adaptive and leisure skills may also experience difficulty effectively engaging with stimuli in their environment, leading to diminished access to external forms of reinforcement. This may result in an increase in repetitive behavior as a means to produce sensory stimulation. As these behaviors increase in frequency, they may be more likely to persist, and even contact social contingencies provided by caregivers. Children with NDD who have little to no way to communicate theirs wants and needs may experience frustration which can be associated with distress and problem behavior. This may result in reactions on the part of caregivers that may inadvertently reinforce the behavior—such as consoling the child, giving the child access to things that have previously calmed them (e.g., food or toys), or removing possible aversive stimuli (e.g., instructions to complete nonpreferred tasks). Although the caregiver may intend to simply reduce the occurrence of the behavior in the moment, such responses may actually reinforce and strengthen the problematic behaviors through operant reinforcement processes. As a result, the problem behavior (and caregiver responses that temporarily reduce problem behavior) is more likely to persist and continue occurring in the future under similar conditions [29,30].

Operant reinforcing functions of problem behavior

Problem behavior is most often maintained by operant reinforcement processes [31,32]. Although not readily apparent based on brief observation or caregiver report, the events in the environment that occasion and reinforce it can be identified via behavioral assessment procedures. Problem behavior can be classified into two broad functional classes based on whether its controlling variables are mediated by others (socially maintained), or it occurs independent of social interaction (automatically maintained). Within these broad functional classes of socially and automatically maintained self-injury, there are also subclasses. With the class of socially maintained problem behavior, the most common subclasses include behavior maintained by attention, escape from nonpreferred situations, and access to preferred items [31]. With automatically maintained problem behavior, the behavior itself is thought to produce its own reinforcing consequences, such as sensory reinforcement. The prevalence of different operant reinforcing functions of problem behavior may also vary according to the form of problem behavior. Aggressive behaviors are primarily maintained by social consequences, pica is almost exclusively automatically maintained, and self-injury is automatically maintained in approximately 25% of cases. Recent research has identified subtypes of automatically maintained self-injury have been identified that include treatment responsive and treatment-resistant subtypes—and it has also been posited that some types of automatically maintained self-injury may be a function of motor or sensory dysfunction rather than merely sensory reinforcement processes [33].

Course over the lifespan

The presence of problem behavior is not unique to individuals with NDD. In fact, repetitive and sometimes harmful behaviors (e.g., self-injury) are not uncommon in neurotypically developing infants and toddlers (e.g., [34]) and preschoolers [35,36]. However, unlike individuals with NDD, problem behavior typically resolves during early childhood for neurotypical individuals as more refined and complicated motor development takes the place of less adaptive repetitive behaviors [34,37].

Problem behavior exhibited by those with NDD often emerges early in life [18], and may persist across the individual's lifespan [38,39]. Research on the prevalence of problem behavior has sometimes taken the form of a snapshot, with researchers recording the prevalence of problem behavior in a specific population at one point in time [1]. Other studies have attempted to evaluate the course of problem behavior over the lifespan by retrospectively examining records for children across multiple years [6] or evaluating the presence of problem behavior in cohorts of individuals [40,41]. Results from each of these lines of research have generally painted a dire picture: very young children and adolescents with NDD are at a high risk for developing problem behavior, some forms of which progress into severe problem behavior [28,42–44]. It may begin as self-stimulatory stereotypy or proto-injurious behavior, before developing into a topography that produces actual tissue damage [28,42]. Research on the course of problem behavior has suggested that there might be a modest decline into adulthood [35,36]; however, a large percentage of adults with NDD are still reported to engage in some form of problem behavior. For example, a study conducted by Hove and Havik [45] revealed that roughly 20.2% of the adults with NDD living in Norway who were receiving community care were reported to have problem behavior, with physical and verbal aggression being the most prevalent topographies. Thus it appears as if problem behavior often develops early, and may also persist throughout the lifespan for some individuals.

Risk factors and sequelae

Risk factors

A growing body of research has pointed to a number of variables associated with an increased risk of behavioral problems in children and adolescents with autism and other NDDs. It is important to note that while specific variables have been linked to problem behavior such as aggression and SIB, the causal direction of some of these relations are not well understood. In general, some risk factors have been more consistently identified across studies and are typically tied to the psychological and psychiatric health of the population. For example, it is well documented that certain diagnoses increase the risk of an individual developing problem behavior. Notably, level of intellectual disability [46–50], presence of an autism spectrum disorder [38], and certain genetic disorders such as Prader–Willi syndrome [51] and Lesch–Nyhan syndrome [52] are all associated with a greater risk of developing severe problem behavior. Neurological dysfunction (e.g., traumatic brain injury) has also been related to increases in aggression [53], as have psychiatric disorders, such as depression, paranoia, psychosis, or personality disorders (e.g., [47,54]). Deficits in skills such as social interactions, adaptive/functional life abilities, and communication are also associated with an increased likelihood of engaging in problem behavior [20,55]. It is not surprising that deficits in early social and communication skills are associated with increases in the presence of problem behavior given the importance of such skills on social development. Competent social behavior provides individuals with access to important learning experiences throughout the lifespan, leading to the development of age-appropriate skills.

Familial, physical, social, and economic sequelae

The societal and personal costs associated with caring for individuals who engage in severe problem behavior can be immense. For both service providers and families, severe problem behavior is associated with monetary losses as well as other burdens of unknown monetary value. Service providers experience monetary costs to protect against lawsuits and to provide workers' compensation benefits, to support additional staffing needs, and to provide personal protective equipment to staff. Abstract costs include low staff morale, high levels of staff turnover, absenteeism, reduced productivity, and problems associated with staff recruitment [56]. The management and treatment of severe problem behaviors often requires highly specialized services and environments, and individuals employed in such settings require frequent and on-going training that can be costly to the organization [47]. Organizations serving individuals with the most severe behavior also often come under criticism from the media and other groups for their frequent use of crisis management procedures and protective procedures including the use of restraint.

Families of children with NDD who engage in problem behavior also face numerous burdens, such as disruptions to daily routines, harm to well-being of immediate family members, limited access to supports and services, and general diminished quality of life. Families often report feeling isolated as they are unable to participate in community activities, socialize, or take vacations. Respite or other child care providers can be reluctant to take on or maintain services with a client who engages in severe problem behavior. Aggressive and disruptive behaviors are also one of the strongest predictors of stress among parents of children with NDD [26,57,58]. These families also report experiencing enormous financial burdens related to frequent repairs to or replacement of personal property, safety adaptations made to the home, and loss of employment income associated with difficulty in obtaining and maintaining a job that can accommodate the unpredictable needs of their child [47].

Problem behaviors also have deleterious effects on the individuals who exhibit these behaviors themselves. Severe problem behavior interferes with learning, development, and, depending on the social and cultural context, social participation and interactions [47]. If left untreated, problem behaviors are likely to persist over time, becoming more severe in adulthood [59]. Furthermore, individuals who exhibit severe problem behaviors may be at an increased risk for experiencing trauma [60], more restrictive environments, overmedication, or inflicting physical injury on themselves [61]. With respect to overmedication, a recent systematic review of medication use among youth and adolescents diagnosed with NDD found that the majority of individuals were prescribed two or more medications (polypharmacy [62]. Several factors seem to be closely associated with polypharmacy, including the presence of externalizing disorders (i.e., oppositional defiant disorder) and more specifically the presence of aggressive behaviors [62,63]. Given the significant risks to the well-being of individuals who engage in problem behavior as well as to that of their caregivers, it is important that families be directed to the necessary resources and supports in their community.

Behavior analytic approaches to assessment and treatment

Applied behavior analysis and behavioral assessment

Empirically supported behavioral assessment and treatment procedures are rooted in the discipline of applied behavior analysis (ABA). ABA applies principles of learning and behavioral sciences based on laboratory research for the purpose of improving socially important issues such as behavior problems and learning [64]. ABA-based interventions relevant to individuals with NDD have two broad domains of application, mainly distinguished by targets of treatment: "comprehensive" and "focused." The goal of comprehensive ABA interventions is to produce changes in adaptive skills that impact global measures of functioning including IQ, adaptive skills, and social functioning. To achieve this goal, comprehensive ABA interventions use a variety of procedures to target skills such as attention, communication, discrimination, and socialization, as well as more advanced educational skills. These types of interventions are applied over extended periods of time (i.e., years). When used with younger children, these interventions are often referred to as "early intensive behavioral intervention" (EIBI [65]. It is well documented that comprehensive ABA interventions are effective in improving IQ, adaptive skills, and social functioning when programming is delivered early in life and on an intensive schedule [66–68]. It appears as if gains in functioning (i.e., IQ and adaptive and social skills) are more robust when behavioral intervention is initiated earlier during the child's development.

Focused ABA interventions are generally more time-limited (i.e., months) because they are designed to address specific behavior concerns, such as aggression and SIB. The goals of these programs are to reduce the undesirable problem behavior while strengthening socially appropriate, functional alternatives to it. The first step in addressing problem behavior is to conduct a functional behavioral assessment (FBA) to identify its controlling variables (i.e., the antecedents that occasion it and the consequences that reinforce it). The current discussion aims to provide a brief overview of the FBA process for the purpose of understanding the behavioral analytic approach to assessing and treating challenging behavior. Chapter The functional and behavioral assessment of this text also provides a more in-depth discussion of the functional and behavioral assessment process.

Although it is generally understood that for any given individual, the learning history leading to the development and maintenance of severe problem behavior cannot be known, the variables presently maintaining it can be identified through FBA. FBA represents the gold standard approach to identifying the variables that occasion and reinforce problem behavior. FBA is a client-driven process for identifying the antecedents to and consequences for problem behavior. Importantly, this assessment information is then used to guide the development of individualized, function-based treatments that are directly linked to the variables that maintain the problem behavior. FBA includes three general types of methods, indirect assessments, descriptive assessments, and controlled functional analysis [69]. Indirect assessments include questionnaires and open-needed interviews through which clinicians gather information from informants who know the target individual and have observed them engage in problem behavior [70,71]. Descriptive assessments include direct observations of the individual in their natural context without manipulation of the environment; descriptive data are collected on the events that occurred before and after the problem behavior [72]. For some individuals, particularly those with recently emerging or less severe problem behavior, the use of indirect assessments represents a reasonable starting point, and may provide sufficient information to develop an efficacious intervention.

For those with more longstanding and severe problem behavior, more intensive assessment is often indicated. For example, research examining the validity of indirect and descriptive assessments have generally suggested that they can lead to ambiguous outcomes and are less accurate that controlled functional analyses that incorporate both direct observation and systematic manipulation of environmental conditions [71,73]. Thus indirect and descriptive assessments may be most useful in generating hypotheses regarding the variables maintaining problem behavior which can then be tested in a controlled functional analysis [66,74].

Controlled functional analyses involve the systematic manipulation of the antecedents and consequences hypothesized to give rise to and maintain problem behavior, respectively. This is generally regarded as the most rigorous approach to the assessment of problem behavior because it represents a controlled analysis where antecedents and consequences are manipulated and their effects on problem behavior are directly observed and recorded in a quantifiable and objective manner. Unlike the use of rating scales or informal observations, functional analysis requires considerably more resources including an area to conduct the assessment where the environment can be controlled, more precise data collection (ideally using computerized data collection systems), additional time to obtain multiple samples of behavior across several conditions being tested, additional staff, and specialized expertise to design the conditions and interpret the data (and in for cases with particularly dangerous behavior, additional training and experience).

Behavioral treatment

The second step in achieving the goals of focused ABA interventions is to use the information gathered during the FBA process to guide the development of an individualized treatment. Typically, function-based treatments involve altering the environment to minimize problem behavior, establishing and reinforcing adaptive behaviors, and withholding reinforcement for problem behavior (i.e., extinction). The hallmarks of effective behavioral treatments include objective measurement of behavior, use of procedures based on scientifically established principles of behavior, and precise control of the environment to allow for the objective evaluation of outcomes. Nearly half a century of research has shown that focused ABA-based treatment approaches are effective in reducing problem behavior and establishing appropriate skills [13,44,75–79].

Although behavioral interventions are highly individualized based on identification of the controlling variables relevant to each case, they universally include components that establish and reinforce appropriate alternative responses, extinction for problem behavior (withholding reinforcement when problem behavior occurs), and a number of other procedures to signal events and expectations. One of the most commonly used reinforcement-based interventions to address socially maintained problem behavior includes differential reinforcement of alternative behavior (DRA). Generally, DRA involves withholding reinforcers for problem behavior (i.e., extinction) while reinforcing an appropriate alternative behavior. For example, in the case of problem behavior motivated by escape from instruction, DRA might include providing a brief break following compliance with the academic task [80]. One frequently used variation of DRA is functional communication training [81]. Functional communication training involves providing access to the functional reinforcer following a socially appropriate communicative response, such as "excuse me" to access attention [82]. DRA, and functional communication training specifically, has been identified as well-established treatments for severe problem behavior according to APA's criteria for empirically supported treatments [44,83], metaanalysis [84], and consecutive controlled case series studies [13,85]. Noncontingent reinforcement is another empirically supported intervention for problem behavior, according to APA's criteria for empirically supported treatments [76], metaanalysis [86], and consecutive controlled case series [87]. Although some specific interventions were described above, it should be noted that many other behaviorally based interventions exist. Furthermore, in many cases, particularly when problem behavior is severe or occurs for multiple reasons, behavioral treatments often increase in complexity and include many different components including those that are aimed at preventing injury to the individual and others.

Interdisciplinary approach

Diagnostic and functional classification systems

Identifying the reinforcing function of problem behavior is critical for developing individualized behavioral interventions that directly address its controlling variables relevant for each case. In addition to the functional classification, problem behavior can also be classified diagnostically. The Diagnostic and Statistical Manual 5 (DSM-5) allows for the classification of problem behavior in terms of its clinical presentation including its form and cooccurring problems, as well as part of the presentation of comorbid psychiatric conditions. Stereotypic movement disorder with SIB is diagnosed when self-injury is present; whereas intermittent explosive disorder may be diagnosed in cases where problem behavior has a sudden and intense onset. When problem behavior is thought to be part of the presentation or directly due to some other psychiatric condition, then an additional diagnosis would be indicated. For example, if problem behavior is associated with other signs such as dysregulated and cyclical mood, then the diagnosis of bipolar disorder or some other type of mood disorder may be applicable. It has been noted that the use of both functional and diagnostic classification systems is complementary. When used together, they can characterize the environmental and biological variables that are evoking and/or maintaining the problem behavior. Understanding of these different type of determinants can guide interdisciplinary treatment where behavioral and pharmacological interventions are more fully integrated and applied in a more targeted fashion [88,89]. Detailed discussion on the use of medication for problem behavior in individuals with NDD is beyond the scope of the current chapter (see Chapter Principles of pharmacotherapy in NDD).

Education, surveillance, early identification, and coordination of care

There are several roles that pediatricians can serve in the assessment and treatment of problem behavior, including surveillance of emerging problems (particularly in high-risk populations), delivery of first-line behavioral and pharmacological interventions, and coordination of services by identifying needs and initiating referrals [90]. Given the importance of early identification, pediatricians identify potential risk factors for problematic behaviors (i.e., the presence of specific genetic syndromes or diagnoses, skill deficits, caregiver stress) and can monitor for the emergence of problem behavior. Surveillance strategies should include inquiring about problematic behaviors during routine visits, including monitoring behaviors that although low in severity currently, could evolve into more serious problems (i.e., repetitive or proto-injurious behaviors, emotional reactivity, and frequent tantrums that do not include severe problem behavior). Pediatric health care providers can provide caregivers of high-risk populations with training and recommendations on optimal caregiver–child interactions as a preventative strategy; and inform them about programs offered by public health and education systems children who are at risk.

Identification of emerging problem behavior as well as surveillance of these risk factors can facilitate more rapid access to community-based early intervention services. Comprehensive and focused ABA services are empirically supported, and following landmark insurance reform, they are now widely covered and accessible children diagnosed with autism spectrum disorders through both private and public insurances. Early identification and access to services is key to successful intervention.

Pediatric health care providers are also in an excellent position to directly gain insight into caregiver–child interactions around problem behavior based on interviews and direct observations (in the examination room or in the waiting area of the office). This enables these providers to begin to assess the behavior, identify possible biological and behavioral determinants. Although providers vary in terms of their time and expertise, many can provide first-line behavioral and pharmacological interventions to address these problems. As pediatric care providers are among the first to identify problems and have a wide degree of knowledge with respect to the wide range of needs of children with NDD, they play a key role in service coordination [90].

Summary

Individuals with NDD are at an increased risk of developing severe problem behavior, which can manifest in many forms and vary in severity and persistence over time. Prevalence estimates suggest that somewhere between 30% and 50% of those diagnosed with NDD will develop at least one form of problem behavior over their lifetime, with roughly 5%–10% being at risk of developing more severe forms of problem behavior. The personal and financial costs incurred by those impacted by problem behavior can be immense. Research suggests that much of the problem behavior exhibited by those with NDD is the product of deficits associated with the primary disability, events in the environment that occasion and reinforce those behaviors, and comorbid psychiatric conditions. Empirically supported behavioral treatments for problem behavior are rooted in the field of ABA and decades of research demonstrate ABA's effectiveness in reducing maladaptive behaviors and promoting socially appropriate behaviors—and now widely available through landmark insurance reform. Pediatricians serve a vital role in the identification of emerging problem behavior, ongoing surveillance, delivery of first-line behavioral and pharmacological interventions, and coordination of services through identifying needs and initiating referrals.

References

[1] Dekker MC, Koot HM, Ende JVD, Verhulst FC. Emotional and behavioral problems in children and adolescents with and without intellectual disability. J Child Psychol Psychiatry 2002;43(8):1087–98. Available from: https://doi.org/10.1111/1469-7610.00235.

[2] Poppes P, Van der Putten AJJ, Vlaskamp C. Frequency and severity of challenging behaviour in people with profound intellectual and multiple disabilities. Res Dev Disabil 2010;31(6):1269–75. Available from: https://doi.org/10.1016/j.ridd.2010.07.017.

[3] Fisher WW, Piazza CC, Roane HS. Sleep and cyclical variables related to self-injurious and other destructive behaviors. In: Schroeder SR, Oster-Granite ML, Thompson T, editors. Self-injurious behavior: gene-brain-behavior relationships. American Psychological Association; 2002. p. 205–21. Available from: https://doi.org/10.1037/10457-014.

[4] Iwata BA, Dorsey MF, Slifer KJ, Richman GS. Toward a functional analysis of self-injury. J Appl Behav Anal 1994;27(2):197–209. Available from: https://doi.org/10.1901/jaba.1994.27-197.

[5] Mazaleski JL, Iwata BA, Rodgers TA, Vollmer TR, Zarcone JR. Protective equipment as treatment for stereotypic hand mouthing: sensory extinction or punishment effects? J Appl Behav Anal 1994;27(2):345–55. Available from: https://doi.org/10.1901/jaba.1994.27-345.

[6] Soke GN, Rosenberg SA, Hamman RF, Fingerlin T, Robinson C, Carpenter L, et al. Brief report: prevalence of self-injurious behaviors among children with autism spectrum disorder—a population-based study. J Autism Dev Disord 2016;46(11):3607–14. Available from: https://doi.org/10.1007/Fs10803-016-2879-1.

[7] O'Dwyer C, McCallion P, Burke É, Carroll R, O'Dwyer M, McCarron M. Prevalence and associated factors of problem behaviours among older adults with intellectual disabilities in Ireland. Res Dev Disabil 2018;80:192–204. Available from: https://doi.org/10.1016/j.ridd.2018.05.007.

[8] Ming X, Brimacombe M, Chaaban J, Zimmerman-Bier B, Wagner GC. Autism spectrum disorders: concurrent clinical disorders. J Child Neurol 2008;23(1):6–13.

[9] Ali Z. Pica in people with intellectual disability: a literature review of etiology, epidemiology and complications. J Intellect Dev Disabil 2001;26:205–15. Available from: https://doi.org/10.1080/13668250020054486.

[10] Hyman SL, Fisher W, Mergcugliano M, Cataldo MF. Children with self-injurious behavior. Pediatrics 1990;85(3):437–41.

[11] Kuhn DE, Hagopian L, Terlonge C. Treatment of life-threatening self-injurious behavior secondary to hereditary sensory and autonomic neuropathy type II: a controlled case study. J Child Neurol 2008;23(4):381–8. Available from: https://doi.org/10.1177/0883073807309236.

[12] McClintock K, Hall S, Oliver C. Risk markers associated with challenging behaviours in people with intellectual disabilities: a meta-analytic study. J Intellect Disabil Res 2003;47(6):405–16. Available from: https://doi.org/10.1046/j.1365-2788.2003.00517.x.

[13] Rooker GW, Bonner AC, Dillon CM, Zarcone JR. Behavioral treatment of automatically reinforced SIB: 1982–2015. J Appl Behav Anal 2018;51(4):974–97. Available from: https://doi.org/10.1002/jaba.492.

[14] Stith SM, Liu T, Davies C, Boykins EL, Alder MC, Harris JM, et al. Risk factors in child maltreatment: a meta-analytic review of the literature. Aggress Violent Behav 2009;14(1):13–29. Available from: https://doi.org/10.1016/j.avb.2006.03.006.

[15] Stormshak EA, Bierman KL, Bruschi C, Dodge KA, Coie JDThe Conduct Problems Prevention Research Group. The relation between behavior problems and peer preference in different classroom contexts. Child Dev 1999;70(1):169–82.

[16] Anderson C, Law JK, Daniel A, Rice C, Mandell DS, Hagopian L, et al. Occurrence and family impact of elopement in children with autism spectrum disorders. Pediatrics 2012;130(5):870–7. Available from: https://doi.org/10.1542/peds.2012-0762.

[17] Guan J, Li G. Characteristics of unintentional drowning deaths in children with autism spectrum disorder. Injury Epidemiol 2017;4(1):32. Available from: https://doi.org/10.1186/s40621-017-0129-4.

[18] Fodstad JC, Rojahn J, Matson JL. The emergence of challenging behaviors in at-risk toddlers with and without autism spectrum disorder: a cross-sectional study. J Dev Phys Disabil 2012;24(3):217–34. Available from: https://doi.org/10.1007/s10882-011-9266-9.

[19] Rojahn J. Self-injurious and stereotypic behaviour of noninstitutionalised mentally retarded people: prevalence and classification. Am J Ment Deficiency 1986;91(3):268–76.

[20] Emerson E, Kiernan C, Alborz A, Reeves D, Mason H, Swarbrick R, et al. The prevalence of challenging behaviors: a total population study. Res Dev Disabil 2001;22(1):77–93. Available from: https://doi.org/10.1016/s0891-4222(00)00061-5.

[21] Crocker J, Brook AT, Niiya Y, Villacorta M. The pursuit of self-esteem: contingencies of self-worth and self-regulation. J Personal 2007;74 (6):1749–71. Available from: https://doi.org/10.1111/j.1467-6494.2006.00427.x.

[22] Iwata BA, Pace GM, Kissel RC, Nau PA, Farber JM. The self-injury trauma (SIT) scale: a method for quantifying surface tissue damage caused by self-injurious behavior. J Appl Behav Anal 1990;23(1):99–110. Available from: https://doi.org/10.1901/jaba.1990.23-99.

[23] Sturmey P, Seiverling L, Ward-Horner J. Assessment of challenging behaviors in people with autism spectrum disorders. Baton Rouge: Academic Press; 2008. p. 131–63. Available from: https://doi.org/10.1016/B978-012373606-2.50007-3.

[24] Mazurek MO, Vasa RA, Kalb LG, Kanne SM, Rosenberg D, Keefer A, et al. Anxiety, sensory over-responsivity, and gastrointestinal problems in children with autism spectrum disorders. J Abnorm Child Psychol 2013;41(1):165–76. Available from: https://doi.org/10.1007/s10802-012-9668-x.

[25] Rooker W Griffin, Hagopian P Louis, Becraft L Jessica, Javed Noor, Fisher B Alyssa, Finney S Katharine. Injury characteristics across functional classes of self-injurious behavior J Appl Behav Anal 2020;53(2):1042–57In press. Available from: https://doi.org/10.1002/jaba.664.

[26] Furniss F, Biswas AB. Recent research on a etiology, development and phenomenology of self-injurious behaviour in people with intellectual disabilities: a systematic review and implications for treatment. J Intellect Disabil Res 2012;56(5):453–75. Available from: https://doi.org/10.1111/j.1365-2788.2012.01534.x.

[27] Romanczyk RG, Kistner JA, Plienis A. Self-stimulatory and self-injurious behavior: etiology and treatment. Adv Child Behav Anal Ther 1982;2:189–254.

[28] Richman DM, Lindauer SE. Longitudinal assessment of stereotypic, proto-injurious, and self-injurious behavior exhibited by young children with developmental delays. Am J Ment Retard 2005;110(6):439–50. Available from: https://doi.org/10.1352/0895-8017(2005)110[439:laospa]2.0.co;2.

[29] Kurtz PF, Leoni M, Hagopian LP. Behavioral approaches to assessment and early Intervention for severe problem behavior in intellectual and developmental disabilities. Pediatr Prev 2020;67(3):499–511. Available from: https://doi.org/10.1016/j.pcl.2020.02.005.

[30] Stocco CS, Thompson RH. Contingency analysis of caregiver behavior: implications for parent training and future directions. J Appl Behav Anal 2015;48(2):417–35. Available from: https://doi.org/10.1002/jaba.206.

[31] Beavers GA, Iwata BA, Lerman DC. Thirty years of research on the functional analysis of problem behavior. J Appl Behav Anal 2013;46(1):1–21. Available from: https://doi.org/10.1002/jaba.30.

[32] Iwata BA, Kahng SW, Wallace MD, Lindberg JS. The functional analysis model of behavioral assessment. In: Austin J, Carr JE, editors. Handbook of applied behavior analysis. Context Press/New Harbinger Publications; 2000. p. 61–89.

[33] Hagopian LP, Frank-Crawford MA. Classification of self-injurious behaviour across the continuum of relative environmental–biological influence. J Intellect Disabil Res 2018;62(12):1108–13. Available from: https://psycnet.apa.org/doi/10.1111/jir.12430.

[34] Baghdadli A, Pascal C, Grisi S, Aussilloux C. Risk factors for self-injurious behaviours among 222 young children with autistic disorders. J Intellect Disabil Res 2003;47(8):622–7. Available from: https://doi.org/10.1046/j.1365-2788.2003.00507.x.

[35] Egger HL, Angold A. Common emotional and behavioral disorders in preschool children: presentation, nosology, and epidemiology. J Child Psychol Psychiatry 2006;47(3-4):313–17. Available from: https://doi.org/10.1111/j.1469-7610.2006.01618.x.

[36] Fahmie A Tara, Garcia R Amanda, Poetry H Jasmine, Tierman M Emily, Hamawe Rima, Marks T Sarah, et al. Topographies and functions of emerging problem behavior and appropriate requests in neurotypical preschoolers. J Appl Behav Anal 2020;53(4):2199–214. Available from: https://doi.org/10.1002/jaba.741.

[37] MacDonald R, Green G, Mansfield R, Geckeler A, Gardenier N, Anderson J, et al. Stereotypy in young children with autism and typically developing children. Res Dev Disabil 2007;28(3):266–77. Available from: https://doi.org/10.1016/j.ridd.2006.01.004.

[38] Murphy GH, Beadle-Brown J, Wing L, Gould J, Shah A, Holmes N. Chronicity of challenging behaviors in people with severe intellectual disabilities and/or autism: a total population sample. J Autism Dev Disord 2005;35(4):405–18. Available from: https://doi.org/10.1007/s10803-005-5030-2.

[39] Taylor Lorne, Oliver Chris, Murphy Glynis. The Chronicity of Self-Injurious Behaviour: A Long-Term Follow-Up of a Total Population Study. J Appl Res Intellect Disabil 2011;24(2):105–17. Available from: https://doi.org/10.1111/j.1468-3148.2010.00579.x.

[40] Matson JL, Mahan S, Hess JA, Fodstad JC, Neal D. Progression of challenging behaviors in children and adolescents with autism spectrum disorders as measured by the Autism Spectrum Disorders-Problem Behaviors for Children (ASD-PBC). Res Autism Spectr Disord 2010;4(3):400–4. Available from: https://doi.org/10.1016/j.rasd.2009.10.010.

[41] Richards C, Davies L, Oliver C. Predictors of self-injurious behavior and self-restraint in autism spectrum disorder: towards a hypothesis of impaired behavioral control. J Autism Dev Disord 2017;47(3):701–13. Available from: https://doi.org/10.1007/s10803-016-3000-5.

[42] Berkson G, Tupa M, Sherman L. Early development of stereotyped and self-Injurious behaviors: I. incidence. Am J Ment Retard 2001;106(6):539–47. Available from: https://doi.org/10.1352/0895-8017(2001)106%3C0539:EDOSAS%3E2.0.CO;2.

[43] Hall S, Oliver C, Murphy G. Early development of self-injurious behavior: an empirical study. Am J Ment Retard 2001;106(2):189–99. Available from: https://doi.org/10.1352/0895-8017(2001)106%3C0189:EDOSIB%3E2.0.CO;2.

[44] Kurtz PF, Boelter EW, Jarmolowicz DP, Chin MD, Hagopian LP. Identifying empirically supported treatments for pica in individuals with intellectual disabilities. Res Dev Disabil 2011;32(6):2114–20. Available from: https://doi.org/10.1016/j.ridd.2011.07.042.

[45] Hove OR, Havik OE. Mental disorders and problem behavior in a community sample of adults with intellectual disability: three-month prevalence and comorbidity. J Ment Health Res Intellect Disabil 2008;1(4):223–37. Available from: https://doi.org/10.1080/19315860802269198.

[46] Borthwick-Duffy SA. Epidemiology and prevalence of psychopathology in people with mental retardation. J Consult Clin Psychol 1994;62(1):17–27. Available from: https://doi.org/10.1037/0022-006X.62.1.17.

[47] Holden B, Gitlesen JP. A total population study of challenging behaviour in the county of Hedmark, Norway: prevalence, and risk markers. Res Dev Disabil 2006;27(4):456–65. Available from: https://doi.org/10.1016/j.ridd.2005.06.001.

[48] Lahey BB, Waldman ID, McBurnett K. The development of antisocial behavior: an integrative causal model. J Child Psychol Psychiatry 2008;40(5):669–82. Available from: https://psycnet.apa.org/doi/10.1111/1469-7610.00484.

[49] Moffitt TE. Adolescence-limited and life-course-persistent antisocial behavior: a developmental taxonomy. Psychol Rev 1993;100(4):674–701. Available from: https://doi.org/10.1037/0033-295X.100.4.674.

[50] Tremblay RE, Nagin DS, Seguin JR, Zoccolillo M, Zelaro PD, Bolvin M, et al. Physical aggression during early childhood: trajectories and predictors. Pediatrics 2004;114(1):43–50. Available from: https://doi.org/10.1542/peds.114.1.e43.

[51] Dykens E, Shah B. Psychiatric disorders in prader-willi syndrome. CNS Drugs 2003;17(3):167–78. Available from: https://doi.org/10.2165/00023210-200317030-00003.

[52] Olson L, Houlihan D. A review of behavioral treatments used for lesch-nyhan syndrome. Behav Modif 2000;24(2):202–22. Available from: https://doi.org/10.1177/F0145445500242003.

[53] Davison GC, Neale JM. Abnormal psychology: an experimental clinical approach. John Wiley & Sons; 1974.

[54] Szymanski L, King BH. Practice parameters for the assessment and treatment of children, adolescents, and adults with mental retardation and comorbid mental disorders. J Am Acad Child Adolesc Psychiatry 1999;38(12):5–31. Available from: https://doi.org/10.1016/S0890-8567(99)80002-1.

[55] Walz NC, Benson BA. Labeling and discrimination of facial expressions by aggressive and nonaggressive men with mental retardation. Am J Ment Retard 1996;101(3):282–91.

[56] Hunter M, Carmel H. The cost of staff injuries from inpatient violence. Psychiatry Serv 1992;43(6):586. Available from: https://doi.org/10.1176/ps.43.6.586.

[57] Baker L Bruce, Blacher Jan, Crnic A Keith, Edelbrock Craig. Behavior Problems and Parenting Stress in Families of Three-Year-Old Children With and Without Developmental Delays. Am J Ment Retard 2002;107(6):433–44. Available from: https://doi.org/10.1352/0895-8017(2002)107%3C0433:BPAPSI%3E2.0.CO;2.

[58] Lecavalier L, Leone S, Wiltz J. The impact of behaviour problems on caregiver stress in young people with autism spectrum disorders. J Intellect Disabil Res 2006;50(3):172–83. Available from: https://doi.org/10.1111/j.1365-2788.2005.00732.x.

[59] Murphy O, Healy O, Leader G. Risk factors for challenging behaviors among 157 children with autism spectrum disorder in Ireland. Res Autism Spectr Disord 2009;3(2):474–82. Available from: https://doi.org/10.1016/j.rasd.2008.09.008.

[60] Wigham S, Hatton C, Taylor JL. The effects of traumatizing life events on people with intellectual disabilities: a systematic review. J Ment Health Res Intellect Disabil 2011;4(1):19–39. Available from: https://doi.org/10.1080/19315864.2010.534576.

[61] Sturmey P. Correlates of restraint use in an institutional population. Res Dev Disabil 1999;20(5):339–46. Available from: https://doi.org/10.1016/S0891-4222(99)00015-3.

[62] Ritter C, Hewitt K, McMorris CA. Psychotropic polypharmacy among children and youth with autism: a systematic review. J Child Adolesc Psychopharmacol 2021;31(4):244–58. Available from: https://doi.org/10.1089/cap.2020.0110.

[63] Wink L, Pedapati E, Horn P, McDougle C, Erickson C. Multiple antipsychotic medication use in autism spectrum disorder. J Child Adolesc Psychiatry 2018;27(1):91–4.

[64] Baer DM, Wolf MW, Risley TR. Some current dimensions of applied behavior analysis. J Appl Behav Anal 1968;1(1):91–7. Available from: https://doi.org/10.1901/jaba.1968.1-91.

[65] Eikeseth S, Klintwall L, Jahr E, Karlsson P. Outcome for children with autism receiving early and intensive behavioral intervention in mainstream preschool and kindergarten settings. Res Autism Spectr Disord 2012;6(2):829–35. Available from: https://doi.org/10.1016/j.rasd.2011.09.002.

[66] Dawson G, Burner K. Behavioral interventions in children and adolescents with autism spectrum disorder: a review of recent findings. Curr Opin Pediatr 2011;23(6):616–20. Available from: https://doi.org/10.1097/mop.0b013e32834cf082.

[67] Reichow B, Barton EE, Boyd BA, Hume K. Early intensive behavioral intervention (EIBI) for young children with autism spectrum disorders (ASD). Cochrane Database Syst Rev 2012;10(1):1465–858. Available from: https://doi.org/10.1002/14651858.CD009260.pub2.

[68] Warren Z, McPheeters ML, Sathe N, Foss-Feig JH, Glasser A, Veenstra-VanderWeele J. A systematic review of early intensive intervention for autism spectrum disorders. Pediatrics 2011;127(5):1303–11. Available from: https://doi.org/10.1542/peds.2011-0426.

[69] Rooker W Griffin, DeLeon G Iser, Borrero S W Carrie, Frank-Crawford A Michelle, Roscoe M Eileen. Reducing Ambiguity in the Functional Assessment of Problem Behavior. Behav Interv 2015;30(1):1–35. Available from: https://doi.org/10.1002/bin.1400.

[70] Hanley GP, Jin CS, Vanselow NR, Hanratty LA. Producing meaningful improvements in problem behavior of children with autism via synthesized analyses and treatments. J Appl Behav Anal 2014;47(1):16–36. Available from: https://doi.org/10.1002/jaba.106.

[71] Iwata BA, DeLeon IG, Roscoe EM. Reliability and validity of the functional analysis screening tool. J Appl Behav Anal 2013;46(1):271–84. Available from: https://doi.org/10.1002/jaba.31.

[72] Lalli JS, Browder DM, Mace FC, Brown DK. Teacher use of descriptive analysis data to implement interventions to decrease students' problem behaviors. J Appl Behav Anal 1993;26(2):227–38. Available from: https://doi.org/10.1901/jaba.1993.26-227.

[73] Mace FC, Lalli JS. Linking descriptive and experimental analyses in the treatment of bizarre speech. J Appl Behav Anal 1991;24(3):553–62. Available from: https://doi.org/10.1901/jaba.1991.24-553.

[74] Roscoe EM, Schlichenmeyer KJ, Dube WV. Functional analysis of problem behavior: a systematic approach for identifying idiosyncratic variables. J Appl Behav Anal 2015;48(2):289–314. Available from: https://doi.org/10.1002/jaba.201.

[75] Asmus JM, Ringdahl JE, Sellers JA, Call NA, Andelman MS, Wacker DP. Use of a short-term inpatient model to evaluate aberrant behavior: outcome data summaries from 1996 to 2001. J Appl Behav Anal 2004;37(3):283–304. Available from: https://doi.org/10.1901/jaba.2004.37-283.

[76] Carr JE, Severtson JM, Lepper TL. Noncontingent reinforcement is an empirically supported treatment for problem behavior exhibited by individuals with developmental disabilities. Res Dev Disabil 2009;30(1):44–57. Available from: https://doi.org/10.1016/j.ridd.2008.03.002.

[77] Hagopian LP, Fisher WW, Legacy SM. Schedule effects of noncontingent reinforcement on attention-maintained destructive behavior. J Appl Behav Anal 1994;27(2):317–25. Available from: https://doi.org/10.1901/jaba.1994.27-317.

[78] Plauche Johnson C, Myers SMThe Council on Children with Disabilities. Management of children with autism spectrum disorders. Pediatrics 2007;120(5):1162–82. Available from: https://doi.org/10.1542/peds.2007-2362.

[79] Wolf MM, Risley TR, Mees HL. Application of operant conditioning procedures to the behavior problems of an autistic child. Behav Res Ther 1963;1(2-4):305–12. Available from: https://doi.org/10.1016/0005-7967(63)90045-7.

[80] Roberts ML, Mace FC, Daggett JA. Preliminary comparison of two negative reinforcement schedules to reduce self-injury. J Appl Behav Anal 1995;28(4):579–80. Available from: https://doi.org/10.1901/jaba.1995.28-579.

[81] Carr EG, Durand VM. Reducing behavior problems through functional communication training. J Appl Behav Anal 1985;18(2):111–26. Available from: https://doi.org/10.1901/jaba.1985.18-111.

[82] Fisher WW, Kuhn DE, Thompson RH. Establishing discriminative control of responding using functional and alternative reinforcers during functional communication training. J Appl Behav Anal 1998;31(4):543–60. Available from: https://doi.org/10.1901/jaba.1998.31-543.

[83] Petscher ES, Rey C, Bailey JS. A review of empirical support for differential reinforcement of alternative behavior. Res Dev Disabil 2008;30 (3):409–25. Available from: https://doi.org/10.1016/j.ridd.2008.08.008.

[84] Chezan LA, Wolfe K, Drasgow E. A meta-analysis of functional communication training effects on problem behavior and alternative communicative responses. Focus Autism Other Dev Disabil 2018;33(4):195–205. Available from: https://doi.org/10.1177/F1088357617741294.

[85] Hagopian LP, Fisher WW, Sullivan MT, Acquisto J, LeBlanc LA. Effectiveness of functional communication training with and without extinction and punishment: a summary of 21 inpatient cases. J Appl Behav Anal 1998;31(2):211–35. Available from: https://doi.org/10.1901/jaba.1998.31-211.

[86] Richman DM, Barnard-Brak L, Grubb L, Bosch A, Abby L. Meta-analysis of noncontingent reinforcement effects on problem behavior. J Appl Behav Anal 2015;48(1):131–52. Available from: https://doi.org/10.1002/jaba.189.

[87] Phillips CL, Iannaccone JA, Rooker GW, Hagopian LP. Noncontingent reinforcement for the treatment of severe problem behavior: an analysis of 27 consecutive applications. J Appl Behav Anal 2017;50(2):357–76. Available from: https://doi.org/10.1002/jaba.376.

[88] Wachtel LE, Hagopian LP. Psychopharmacology and applied behavioral analysis: tandem treatment of severe problem behaviors in intellectual disability and a case series. Isr J Psychiatry Relat Sci 2006;43(4):265–74.

[89] Hagopian P Louis, Caruso-Anderson E M. Integrating behavioral and pharmacological interventions for severe problem behavior displayed by children with neurogenetic and developmental disorders.. In: Shapiro, B. K., & P. J. Accardo (Eds.), Neurogenetic syndromes: Behavioral issues and their treatment (p. 217–239). Baltimore, MD: Paul H. Brookes Publishing. 2010.

[90] McGuire K, Fung LK, Hagopian L, Vasa RA, Mahajan R, Bernal P. Irritability and problem behavior in autism spectrum disorder: a practice pathway for pediatric primary care. Pediatrics 2016;137(2):136–48. Available from: https://doi.org/10.1542/peds.2015-2851L.

Chapter 35

Sleep disorders

Jennifer Accardo
Departments of Pediatrics and Neurology, Virginia Commonwealth University, Richmond, VA, United States

Introduction

Children with neurodevelopmental disabilities (NDD) often have difficulty sleeping. Although in the United States, sleep medicine has become the purview of pulmonologists, internationally, neurologists have led the way in its study. Without taking away from the importance of sleep-related breathing disorders such as obstructive sleep apnea, neurologically and behaviorally based sleep disorders like insomnia are even more common and impactful.

Just as with eating, many consider sleeping an activity that comes naturally, rather than a skill to acquire. Yet no sane person wants to literally "sleep like a baby." Similar to how children's patterns of eating develop over time, their sleep evolves in the context of neurologic maturation and behavioral acquisition. Sleep even has a few milestones of its own: sleeping through the night, falling asleep independently, and giving up naps.

Children with NDD face a variety of obstacles to sleep: physical, behavioral, and psychological. Conditions comorbid with disabilities can lead to pain and discomfort, which they may not be able to communicate. Across disabilities, children more frequently resist bedtime, prolong routines, evidence distress at small changes, ignore social cues to sleep, and perseverate on other activities (e.g., electronics use) to the detriment of sleep initiation. Anxiety often further delays sleep [1].

Sufficient quantity and quality of sleep are increasingly recognized as necessary substrates for learning, attention, and emotional self-regulation. For the clinician, asking about sleep opens a window into a child and family's daily routine and presents an opportunity to intervene in a meaningful way.

Background

The epidemiology of sleep disorders in children with NDD is a challenging topic. First, there is the question of definition: where does one place the boundaries for a sleep variant versus a problem, versus a disorder? Next, problems and disorders are varied in their causes and manifestations, to say nothing of the variety of expressions of NDD among children. Finally, the baseline prevalence of sleep problems and disorders tends to be high among typically developing children. One study of children with autism found both parent-reported and actigraphic sleep problems in 66% of children with autism spectrum disorder—but in 45% of typically developing controls as well [2]. Sleep problems are perhaps more commonly reported in 10%–25% of typically developing children [3].

Sleep is a state through which animals cycle at regular, and ideally predictable, intervals. The hypothalamic suprachiasmatic nucleus, or master circadian pacemaker, governs transcription of clock proteins that feedback on each other to maintain circadian timing. Hormones modulate it internally, and zeitgebers (literally, time givers), such as light and food, externally. Light travels directly via the retinohypothalamic tract and in particular, entrains the organism to the cycle of days and nights [4]. Sleeping organisms show decreased responses to external stimuli such as sound and touch. In contrast to coma, sleep is easily reversible. Research on the purpose of sleep has used sleep deprivation to demonstrate its uses. Physiologic functions, such as metabolism, blood pressure, and seizure thresholds, can be affected. Also, depriving a creature of sleep can result in loss of ability to monitor the environment, react efficiently, learn new information, and modulate responses.

How did this frequently vulnerable state evolve, and to what end? What is sleep for? This seemingly simple question eluded answers for many years [5]. Recent research implicates sleep's role in the glymphatic system (the lymphatic

Capute and Accardo's Neurodevelopmental Disabilities in Infancy and Childhood. DOI: https://doi.org/10.1016/B978-0-12-824060-1.00051-1

drainage system of the brain), which filters wastes from the brain differentially during sleep [6]. But where this theory points toward accumulation of toxins and amyloid over time, it seems less likely to account for shorter-term effects of sleep loss or effects of sleep deprivation in children and young adults. Ahmad et al. reflect that "sleep should be conceptualized as a complex state, behavior, and process without a monolithic function but rather an omniphenomenon differentially focused on the needs of the organism as determined by developmental period" [3].

Suffice it to say, sleep or an analogous state can be identified in almost all animal life. Circadian rhythms are evident across not only animal life but also plants [7]. The elegant circuitry that controls human sleep propels organisms between sleep and wake with opposing signals—the cholinergic neurons in the ascending reticular activating system and the sleep activating neurons of the ventrolateral preoptic nucleus—that create a flip flop switch, also between non-rapid eye movement (non-REM) and rapid eye movement (REM) sleep stages [8]. While in theory, one is either asleep or awake, in non-REM sleep or in REM sleep, increasingly, mixed states are being recognized and studied.

The study of NDD depends on several key concepts. Developmental milestones, the orderly progression of skills building upon skills, from simpler to more complex, within specific time windows, form the benchmarks by which to identify patterns of delay, dissociation, and deviance (see Chapter Typical and atypical brain development). One could conceptualize developmental milestones for sleep as for language, fine motor, and gross motor skills. Having a child sleep through the night uninterrupted is memorable and satisfying for families. Consolidation of daytime naps from two naps to one at 12–18 months, with 94% giving up naps entirely by age 5 years, to the regret of parents, also constitute predictable benchmarks [9]. Likewise, the deviance of the "never nappers" causes familial distress.

Other progressions evolve more subtly over time. Sleep shifts from a longer to a shorter duration over the course of childhood, and indeed, adult life. Borbely hypothesized a two-process model to govern the duration and timing of sleep [10,11]. Process S (homeostatic or sleep debt) and Process C (circadian) interact to regulate sleep and wakefulness. Circadian signals help maintain wakefulness during the day, even as sleep need gradually increases, leading to predictable dips in alertness. During adolescence, there is a circadian shift toward later sleep onset and wake times, which becomes most pronounced in late adolescence, particularly if school start times allow for this shift [12].

Duration and timing of sleep are intuitively understandable concepts, despite challenges to measurement, given that self-report and family observation can be highly subjective, and actigraphic recordings can require a surprising amount of interpretation. While quantity is (somewhat) easier to measure, quality is harder to define [13]. At a more complex level, electroencephalographic (EEG) activity during sleep yields not only sleep architecture, the orderly progression through different sleep stages, but also microarchitecture, with microarousal phenomena such as cyclic alternating patterns [14]. Both levels of architecture have tempted researchers with the promise of potential biomarkers that could yield pathognomonic signatures for different disabilities [15]. While abnormalities are present in children with NDD, the promise of pathognomonic patterns has proved elusive.

Expectations for sleep vary considerably across cultures. Historically, people observed a night watch, with first and second sleep, bringing into question the idea that normal sleep involves sleeping straight through the night [16]. Many cultures cosleep, in contrast to reactive cosleeping often seen with families of children with disabilities (bedsharing as survival strategy rather than heritage). Middle Eastern families often keep later hours regardless of age. Some European and South American countries continue the tradition of midday siestas. There is more research emerging on how Black children's sleep problems and patterns are different than their white peers [17,18].

Impact

Sleep intimately affects physical functions such as cardiovascular health, seizure threshold, immune function, and appetite and metabolism, the latter via the neurotransmitter orexin/hypocretin [19–22].

Studies evaluating sleep's impact on neuropsychological function have been surprisingly inconclusive. Children with obstructive sleep apnea often present as irritable and hyperactive rather than sleepy [23]. Past studies have been confounded by socioeconomic status and IQ. Intermittent hypoxia may injure vulnerable neurons such as in the hippocampi and cerebellum.

Sleep deprivation for families as well as children drives many requests for sleep evaluation. While parents often apologize for asking for help, the reality is that parenting a child with or without a disability is difficult on a full night's sleep, let alone without one. A child who stays awake until 6 a.m. and requires supervision during the night may have a parent who left a job in order to care for him after several near-miss collisions while driving sleep-deprived. Families worry about how neighbors and landlords will react when children wake up and make noise during the night. Quality of life for families can be significantly affected [24].

Approach to clinical concern

History

A good sleep history may be directed by a question, or questions, about a child's sleep. What is the child's family worried about? The clinician often must look further, as the chief complaint may be driven by issues that the family does not initially recognize. Commonly, deficits in self-soothing to sleep independently manifest as difficulty staying asleep (a frequent complaint) rather than difficulty falling asleep.

Context for sleep history includes length of commute to school, school start times, and daycare/afterschool programs. Bus rides to specialized programs or private school placements can necessitate early wake times due to the length of ride, cutting into sleep time. In the recent past in the context of the COVID-19 pandemic, asking about in person versus virtual school also helps clarify how much structure is placed on a child's life. Children with special health care needs or significant behavior challenges may qualify for homebound instruction, resulting in a nightmarish lack of structure in which sleep-wake schedules spiral. Homeschooling has become increasingly popular and requires questioning to elicit duration and timing of academics, as well as what additional activities comprise the day. Daycare before school may prompt very early wake times for parents to drop children off before they start their own work days.

The COVID-19 pandemic exposed many cracks in the foundation of society, not only in school systems but also in children's health and leisure. The national epidemic of sedentary electronics addiction ran rampant, further encouraged by social isolation. Even before the pandemic, children who stay indoors, often in dim light, with constant screen exposure and few zeitgebers, such as regular meals, sunlight, or physical activity, tend to lose their circadian rhythmicity. Such situations, fueled by social anxiety, school avoidance, or amotivational depression, manifest sleep as a symptom of a larger disorder, rather than the primary cause. Regular exercise is important for patients to dig out, though deconditioned young people are hard to mobilize.

How much sleep do children need? Several recent consensus statements have addressed this question. The American Academy of Sleep Medicine made recommendations for sleep need per 24 hours on a regular basis among multiple age groups, including 10–13 hours (including naps) for 3–5 year olds, 9–12 hours for 6–12 year olds, and 8–10 hours for 13–18 year olds [25]. Putting children to bed earlier and earlier if they are not falling asleep quickly, to try to maximize sleep time, can quickly backfire if they are not sleepy and learn to associate being in bed with being awake and bored.

Defining terms is important. Is bedtime the time that children start getting ready for bed (commonly referred to as "going upstairs"?), the time that they are in bed with lights out and intended to fall asleep, or the time they actually fall asleep? Considerable lags may exist between these three entities, and a clinician should try to account for the sequence of events. Sleep onset latency is the more technical term for the amount of time between getting into bed with the intention of falling asleep and actually falling asleep. If sleep onset latency takes more than 20–30 minutes, ask what the child is doing during that time and whether screens are involved. Many young people, especially with attention deficit hyperactivity disorder (ADHD), will "go until they drop," energetically going on about their business until they suddenly lose steam. They have no wind-down period.

Night wakings should be characterized by whether they are true wakings (fully awake) or parasomnias, during which the child is actually in a mixed state: brain asleep, but performing some activities as though awake. This can be hard for parents to discern late at night, in dark environments, especially with children with limited verbal abilities. Though there are two types of parasomnias, REM parasomnias are not well characterized in children, and this discussion will address non-REM parasomnias. Like autism, parasomnias exist on a spectrum, and the phenomena of sleep talking, sleepwalking, confusional arousals, and night terrors can be considered as related—and highly heritable [26]. Classically, non-REM parasomnias do not interrupt sleep, and the literature suggests that their occurrence should not be associated with sleepiness the next day. Anecdotally, however, families have reported that their children are more tired the day after.

Classic questions for snoring are frequency, intensity, and triggers (e.g., supine position, nasal congestion, excessive tiredness). Associated questions to address obstructive sleep apnea syndrome include noisy breathing (which parents sometimes distinguish from snoring), mouth breathing, snorting, gasping, night sweats, and bruxism. Having witnessed apneas is not required for obstructive sleep apnea to be present. Apnea often worsens during REM sleep, a stage that typically increases as the sleep period progresses. Unless parents observe throughout the night, they may miss significant apneas. Sleep with neck in hyperextension or in other unusual positions, such as sitting with trunk folded forward, may signal obstructive sleep apnea (OSA), especially in children with Down syndrome [27].

Snoring at least three times a week meets the threshold for ordering PSG. Adolescents, who have their own spaces and are less closely supervised than younger children, may fly under the radar for snoring. Clinical suspicion of obstructive sleep apnea may be sufficient in the absence of witnesses. Family members with hearing impairment may not be aware of symptoms of snoring [28]. Clinicians often ask whether a child's snoring can be heard outside a closed door as a way to quantify loudness, though there is no specific data to support the use of this question.

Restless sleep has recently been proposed as a disorder, rather than a symptom, as a way to better study it [29]. It is difficult to quantify how restless a child is, both for parents and for professionals.

Asking when a child wakes for the day, should be accompanied by asking how they wake for the day. Do they wake spontaneously or to an alarm, or do they require a family member to wake them, sometimes repeatedly? How long does is take them to wake, and what kind of mood are they in? Some children wake early in order to access electronics before others in the household are awake.

Review of systems focuses on issues that sleep impacts but also that can impact sleep (see Table 35.1). Weight gain in particular can trigger or exacerbate obstructive sleep apnea. Need for orthodonture suggests upper airway effects, while history of orthodonture can change oropharyngeal anatomy in a way that affects breathing during sleep. OSA may worsen asthma via interleukin release. Gastroesophageal reflux disorder can spark central sleep apnea as well as worsen OSA and cause discomfort overnight, disrupting sleep. The relationship between epilepsy and sleep is complex and bidirectional [19]. Pain and discomfort from poor dentition, fecal impaction, eczema, or muscle spasm may be difficult for patients to communicate and can disrupt sleep.

Sleep-focused physical examination

In terms of physical exam, it is no longer typical to find young children with failure to thrive, and even congestive heart failure, because of severe OSA, as when the condition was initially described in pediatrics. But it is dismayingly common to see histories of snoring in children with obesity and potentially large neck circumferences. Children with developmental disabilities can have a higher risk of obesity than typically developing peers [30]. While obesity increases risk, it is not the only factor contributing to OSA, as not all obese children have OSA. Obesity and sleep have a complicated relationship beyond the boundaries of breathing [31]. Acanthosis nigricans points to impaired glucose tolerance and can be shown to families as an objective marker to substantiate concern.

Oropharyngeal anatomy is not limited to tonsils but their size should be graded or surgical absence noted. Children with NDD, especially autism, can have difficulty cooperating with oropharyngeal exam; if they are disinclined to either follow directions or imitate opening the mouth, the examiner may want to defer this portion of the exam or postpone until the end of the visit. At minimum, parents should be discouraged from prying open unwilling mouths. While visualization of adenoids is not typically possible for the non-otolaryngologist, adenoidal facies (the phenomenon of drooping eyelids, slack facial features, and open-mouthed posture) is easily observed. In contrast to tonsils, surgically removed adenoids can regrow over time. Retrognathia or micrognathia, overbite or crossbite, and any evidence of bruxism (including audible demonstration in office) are other risk factors worth documenting. Evidence of environmental allergies such as allergic shiners and allergic salute suggest congestion and discomfort, while excoriation and dry patches

TABLE 35.1 Pediatric sleep medicine review of systems.

Symptom	Risk for sleep-disordered breathing	Source of discomfort
Weight gain	Present	–
Orthodonture	Present	–
Nasal congestion (seasonal allergies or chronic rhinitis)	Present	Yes
Asthma	Present	Yes
Gastroesophageal reflux disorder	Present	Yes
Dental caries	–	Yes
Constipation	–	Yes
Eczema	–	Yes
Epilepsy	Present	Yes
Muscle spasms/clonus	–	Yes

round out the allergic triad with eczema. Severe scoliosis and/or kyphosis can contribute to restrictive lung disease as they progress. Bitten fingernails and for that matter, toenails, suggest anxiety, as can evidence of trichotillomania in the form of sparse and missing patches of hair, especially affecting eyebrows and hairline.

Other evaluations

Use of questionnaires can supplement clinical judgment. Sleep questionnaires often show different results from "objective" devices noted below, but they may capture different concerns than those devices. The Children's Sleep Habits Questionnaire, originally developed for 4–10 year olds, is widely used in a variety of populations and freely available, providing a broad survey of sleep symptoms [32]. The Pediatric Sleep Questionnaire focuses more on sleep-disordered breathing and restless legs syndrome (RLS) [33]. One of its subscales addresses symptoms of ADHD, but this makes its use in populations with ADHD diagnoses circular [1]. The Brief Infant Sleep Questionnaire addresses primarily behavioral sleep in younger children, from 1 to 36 months [34]. The Epworth Sleepiness Scale is commonly used by sleep medicine practitioners to quantify daytime sleepiness, including narcolepsy, with a series of questions about likelihood of falling asleep in certain situations. A pediatric modification of this questionnaire changes out the question about falling asleep in a car while driving [35]. More recently, the Patient Reported Outcome Measurement Information System developed by the National Institutes of Health for clinical and research use have item banks related both to pediatric sleep disturbance and also sleep-related impairment [36]. Meanwhile, the Pediatric Sleep Practices Questionnaire asks patients of 8–17 year olds to self-report their sleep timing, routines, environment, technology use before bedtime, and need for parental presence [37].

Actigraphy is a way to document sleep-wake schedules. Wearable motion sensors record movement as a proxy for wake and generate an estimate of sleep-wake schedules over time (usually continuously over 1–2 weeks), often based on proprietary algorithms. Actigraphy is most accurate in documenting time of sleep onset, less accurate with night wakings. Its use does require experienced interpretation, and insurance companies still consider it experimental after years of clinical experience. It is less accurate with children with disorders of mobility or movement and with very young children. Ideally, parents should keep sleep diaries in tandem with the actigraphy, which requires consistent documentation over at least a week.

Overnight polysomnography (PSG) is also known as the overnight sleep study. To refer rationally for PSG, understanding what overnight sleep studies can and cannot show is crucial. Having a sleep *problem* does not necessarily equal need for sleep *study*. Overnight polysomnograms are indicated to diagnose sleep-disordered breathing, including obstructive sleep apnea and central sleep apnea syndrome. They are also indicated to test for resolution of obstructive sleep apnea following interventions such as adenotonsillectomy in the case of moderate-severe obstructive sleep apnea or effective treatment with continuous positive airway pressure (CPAP) titration. They can identify periodic limb movements during sleep (PLMS), though contrary to popular opinion cannot form the sole basis for diagnosing RLS, a clinical diagnosis. The presence of PLMS can support a diagnosis of RLS. PSG is classically not indicated for evaluation of insomnia, although if behavioral strategies are tried and ineffective, testing for sleep disruptors such as sleep-disordered breathing may be a reasonable option. Also, in certain high-risk conditions (e.g., Down syndrome), PSG should be performed even in the absence of snoring. While not painful, the monitoring process is a sensorily intense process for children with NDD, especially the nasal cannula. If available, a process of desensitization ahead of time can be invaluable [38].

Multiple sleep latency test (MSLT) is a series of four to five nap opportunities recorded with a limited montage of monitoring focusing on sleep staging. Patients are asked to try to fall asleep in a quiet, dark room over a 15 minute period. Information obtained is whether they fall asleep, average amount of time to fall asleep (mean sleep latency), and number of naps in which REM stage sleep is attained (sudden onset REM periods or SOREMPs). The goal of this test is to objectively document degree of daytime sleepiness, with less than 8 minutes mean sleep latency and at least two naps with REM sleep, potentially indicative of narcolepsy. Factors other than underlying sleep disorders can affect MSLT results. Potential for sleep deprivation preceding the test is the rationale for actigraphy leading up to the date of MSLT, when available. Use of substances that affect sleep provides a rationale for drug screening prior to MSLT and request to avoid caffeine the day of testing, as well as weaning psychoactive medications off 2 weeks prior to testing. This can be challenging for young people with significant comorbid psychiatric conditions, such that unmedicated testing is not always feasible. Anecdotally, narcolepsy has spiked in younger children. Normal values for mean sleep latency are not available for children who are still developmentally expected to nap, though recurrence of requirement for nap after it has been given up may be telling.

Several specific NDD have sleep phenotypes worthy of individual discussion (Table 35.2).

TABLE 35.2 Common sleep disturbances in selected neurodevelopmental disorders.

Neurodevelopmental disorder	Common sleep disturbances	References
22q11.2 deletion syndrome	Increased need for sleep; anatomic risks for sleep-disordered breathing (e.g., velopharyngeal insufficiency); nocturnal leg pain.	[50]
Angelman syndrome	Severe night wakings and early morning wakings, potentially associated with epilepsy and gastroesophageal reflux disease; apparent decreased need for sleep; altered melatonin secretion.	[51,52]
Attention-deficit hyperactivity disorder (ADHD)	Insomnia; irregular sleep schedules varying night to night; daytime sleepiness; delayed sleep phase disorder; significant overlap with restless leg syndrome.	[53,54]
Autism spectrum disorder (ASD)	Multiple types of sleep disorders, especially sleep onset and maintenance insomnia, but also restless sleep, obstructive sleep apnea, parasomnias, question of circadian rhythm sleep disorders (most frequently delayed sleep phase disorder); question of short sleep need.	[55]
Cerebral palsy	Sleep onset and maintenance insomnia with wakings potentially related to discomfort, caregiving interventions; likely increased risk for sleep-disordered breathing.	[56]
Down syndrome	High risk of obstructive sleep apnea increasing with age and weight, potentially associated with regression, does not consistently resolve with adenotonsillectomy.	[57]
Prader-Willi syndrome	Central and obstructive sleep apnea, potentially severe, with risk of exacerbation due to treatment with growth hormone; excessive daytime sleepiness and narcolepsy-like symptoms; insomnia; ventilatory control abnormalities.	[58]
Rett syndrome	Irregular breathing with intermittent hyperventilation during wakefulness as well as sleep; breath holding spells; obstructive and central sleep apnea; ventilatory control abnormalities.	[59]
Smith-Magenis syndrome	Fragmented sleep cycles with short total sleep time potentially due to circadian dysregulation and abnormal melatonin secretion; daytime sleepiness; increased risk of obstructive sleep apnea due to facial structure, obesity, and hypotonia.	[60]

Down syndrome places children and adults at substantial risk for sleep-disordered breathing, primarily obstructive sleep apnea. Lifetime risk approaches 40%–60%, and OSA may contribute to developmental regression in adulthood. Anatomy (relative macroglossia), hypotonia, and obesity all contribute to risk. PSG to assess for sleep-disordered breathing, even in the absence of snoring, is indicated as a baseline as well as in the event of symptoms.

Cerebral palsy has a variety of effects on sleep as of yet poorly characterized. Sleep-disordered breathing may be related to tone abnormalities. Comorbid conditions such as muscle spasms, epilepsy, and gastroesophageal reflux contribute to sleep disruption and discomfort. Challenges with mobility may necessitate caregiver intervention (and potential sleep disruption) to achieve position change during the night. Any accompanying visual impairment may weaken circadian signals such that regular sleep schedules are difficult to maintain.

Autism spectrum disorder is a perfect storm for sleep. Insomnia affects many children and adults, based on medical and psychiatric comorbidities as well as characteristics of autism itself that make sleep more difficult (e.g., lack of perception of social signals, sensory sensitivities, perseveration on preferred activities). Prevalence of obstructive sleep apnea is unknown but may be higher than suspected.

Prader-Willi syndrome, with its characteristic severe hypotonia and morbid obesity, creates an ideal setup for obstructive sleep apnea. In addition, treatment with growth hormone injections has been associated with sudden death for reasons that are unclear but potentially associated with sleep-disordered breathing. Growth hormone-induced lymphoid hyperplasia may be a mechanism. PSG prior to initiation of growth hormone treatments has been recommended.

Angelman syndrome, in which children are thin, hypertonic, and have significant epilepsy, has its own sleep manifestations. Day-night reversal can stymie tired parents and uncontrolled seizures as well as hyperactivity can exacerbate sleep disruption [39]. Mouse models suggest that deficiency of UBE3A expression due to haploinsufficiency lengthens circadian periods, affecting sleep-wake cycles [40]. Treatment with melatonin and control of seizures are mainstays.

Breathing during Rett syndrome can be dysfunctional during wake as well as sleep, with disorganized breathing followed by compensatory hyperpnea. Additionally, epilepsy and EEG abnormalities likely disrupt sleep further [41].

Sleep disorders

The International Classification of Sleep Disorders-third edition catalogs a multitude of separate sleep disorders, a triumph of splitting over lumping. At the end of the day, however, there are four basic sleep disorders: not sleeping (well) enough, that is, insomnia; sleeping too much (hypersomnia); sleeping at the wrong times (circadian rhythm sleep disorders); or doing something unusual during sleep, usually involving breathing or movements (parasomnias, movement disorders, and sleep-related breathing disorders).

Insomnia refers to chronic difficulty falling asleep, staying asleep, waking too early, or getting restful sleep that impacts daytime function. The first line for insomnia is ensuring adequate sleep opportunity, in conditions that support a good night's sleep. Control of light is crucial, and cool temperatures tend to promote sleep.

Behavior therapy, when available, can be quite helpful. If ongoing follow-up is needed, an experienced child psychologist has the skills to guide behavioral sleep strategies. Amid the current mental health crisis, this service can be hard to find. Some families benefit from telehealth opportunities, though others find it more difficult to connect on a personal level, and internet connections are inconsistent across regions. Some occupational therapists have the skill set to provide direction for sleep hygiene, as well as explore sensory and calming techniques that can support sleep. Sleep coaches are increasingly present in the community, and their training and experience vary widely. Clinicians should be aware that parents may seek out their services.

Some families benefit from book recommendations. *Sleeping Through the Night: How Infants, Toddlers, and Their Parents Can Get a Good Night's Sleep,* and *Taking Charge of Your Child's Sleep: The All-In-One Resource for Solving Sleep Problems in Kids and Teens*, get parents of various age ranges through the basics and beyond in sleep [42,43]. *Sleep Better! A Guide to Improving Sleep in Children with Special Needs* by V. Mark and *Solving Sleep Problems in Children with Autism Spectrum Disorder* by Terry Katz and Beth Malow are both practical, readable guidebooks for families of children with NDD from which clinicians would benefit from reading as well [44,45]. Often parents want to search the internet. The Pediatric Sleep Council website is geared toward infant and toddler sleep, while Start School Later focuses on teenagers and advocacy for later school start times [46,47]. Better Days, Better Nights is an evidence-based online program for Canadian families to work on sleep skills [48].

Melatonin is often used to shorten sleep onset; it rarely helps with night wakings, despite hopes to the contrary. Other medications can be considered when prolonged sleep onset or night wakings interfere with a child's health, behavior, or family life. Medication cannot overcome inadequate opportunity for sleep, inappropriate sleep environment, or counterproductive sleep habits, particularly electronics use. Table 35.3 is not a comprehensive list. Recall too that there are no FDA-approved medications for insomnia for children. Other medications can be considered, such as mirtazapine, but there is limited guidance about dosing available.

Narcolepsy is a neurologically based sleep disorder that may have an autoimmune component and stems from a deficiency of the neurotransmitter hypocretin (orexin). It not only results in excessive daytime sleepiness but also in fragmented nighttime sleep, as well as more specific symptoms such as sleep paralysis, hypnogogic and hypnopompic

TABLE 35.3 Pharmacotherapy for sleep.

Medication	Starting dosage	Indication	Side effects
Clonidine	0.05–0.1 mg	Sleep onset; sometimes helps sleep maintenance	Dry mouth
Hydroxyzine	12.5–25 mg	Sleep onset; anxiety	Anticholinergic
Trazodone	25–50 mg	Sleep onset; anxiety/depression	
Gabapentin	10 mg/kg or 300 mg, whichever is smaller × 1 week, then increase to 20 mg/kg or 600 mg	Sleep onset; sleep maintenance; restless legs syndrome; mild anxiolytic	Potential irritability

hallucinations, and potentially cataplexy. Although narcolepsy is underdiagnosed, more children have hypersomnia for reasons such as inadequate nighttime sleep than have narcolepsy. This neurologic disorder does benefit from nonmedication strategies: regular routine, exercise, control of light, and sleep hygiene (Table 35.4).

Pharmacotherapy options include stimulant medications and nonstimulant wakefulness promoting agents (modafinil and its variants), as well as venlafaxine and other antidepressants to treat cataplexy, and sodium oxybate as a powerful sleep consolidator and anticataplexy agent together. Anecdotally sleep medicine specialists may shy away from stimulant medication such as is used in ADHD in favor of more sleep-specific agents. Histamine anatagonists as well as dopamine and norepinephrine reuptake inhibitors have been added recently to the armamentarium.

Obstructive sleep apnea in children is often treated by surgical removal of tonsils and adenoids. Sometimes it is helpful to have an overnight sleep study to establish whether disease is severe, suggesting inpatient procedure to monitor for higher risk of postoperative bleeding rather than same-day surgery with discharge to home. Children with autism and other sensory issues can become highly averse to oral intake following surgery, due to discomfort, and also deserve close monitoring.

Obesity is a risk for OSA, and weight loss, a potential treatment. Guidance about weight loss in children with NDD is beyond the scope of this chapter. In adults, however, even modest weight loss can have a positive effect on breathing during sleep.

Medication treatment of mild-moderate obstructive sleep apnea has some evidence base. Agents studied are nasal steroid sprays and montelukast. Nasal steroid sprays must be used chronically over a period of months and not as needed as families sometimes do during allergy season. Cooperation with nasal sprays may be a rate limiting factor and strong smell/taste aversive to children.

CPAP can be frightening to children and families. However, it is an effective treatment that should not be used as a threat. Children can learn to tolerate it, and some children with disabilities such as Down syndrome, especially when used to frequent medical procedures, may cooperate better with the intervention than typically developing peers. Children with different facial anatomy may need different mask interfaces to optimize fit. Children with sensory sensitivities may need a more gradual introduction, with practice first with the mask without pressure to acclimate to the sensation, before proceeding with pressurized air. Clinicians should respect the family's right to decline this specific therapy once risks and benefits have been explained and discussed. While CPAP is not a medication and lacks drug interactions, it entails significant and ongoing effort and is not without side effects.

Restless leg syndrome (RLS) (also known as Ekbom-Willis syndrome) affects 2% of children, with 0.5%–1% having severe symptoms. It is a diagnosis made by history. Children with NDD often cannot communicate a history of feelings of discomfort relieved by movement. Families may identify history of restlessness that are most prominent at bedtime, or need to have legs rubbed, and other affected family members. Growing pains may have significant overlap with RLS. There is substantial overlap between RLS and ADHD. Having children draw pictures of their symptoms can be revealing. Symptoms have been associated with serum ferritin levels that may not be below the laboratory cutoff for normal range but are below 50 ng/mL. Oral iron supplementation can be helpful for symptoms, though side effects such

TABLE 35.4 Sleep hygiene recommendations.

1. Physiologic
   a. Avoid caffeine
   b. Avoid alcohol
   c. Avoid nicotine
   d. Be mindful of medication effects (alerting, sedating)
   e. Limit daytime sleep opportunity to allow for adequate sleep pressure for nighttime
2. Structural (bedtime routine)
   a. Consistent timing from night to night, also between school nights and weekend nights
   b. Calm, familiar activities
   c. Steps leading inexorably toward bedroom
   d. Predictable enforcement of limits to avoid bargaining/stalling
3. Environmental
   a. Low light; crucial due to light being the most powerful zeitgeber
   b. Cool room temperature consonant with drop in body temperature during sleep
   c. Quiet
   d. Room exclusively for sleep and limited morning/bedtime routines in order to strengthen associations between bed and sleep

as constipation and GI distress can be difficult to tolerate. Iron transfusions are used in select centers but are not widely available for RLS. Medications can mitigate symptoms: clonidine and gabapentin shorten sleep onset. Dopamine agonists such as ropinirole and pramipexole are widely used in adults, but the side effect of disinhibition carries extra concerns for children.

Areas for further research/investigation

Home sleep apnea testing (HSAT) is widely used in adult sleep medicine patients. Its use in children has been met with skepticism based on children's lower likelihood of tolerance of or cooperation with monitoring. However, access to pediatric PSG is limited compared to demand, and parents who have themselves had HSAT may ask if it is an option for their child.

Glossopharyngeal nerve stimulation is a surgical procedure implanting a device that stimulates that cranial nerve, enervating the tongue and altering tone. It was initially studied in select candidates who could not tolerate CPAP. Research to evaluate its use in young people with Down syndrome is emerging with some success for refractory OSA [49].

Dental sleep medicine interventions can be offered to adults with OSA as an alternative to CPAP but are not usually appropriate for children as their jaws and faces continue to grow and develop. Mandibular advancement, however, is a surgical option for some children, particularly those with overbites.

Melatonin for insomnia is coming under increasing scrutiny for impurities of production and also side effects. There is a push to modify cognitive behavioral therapy for insomnia (CBT-i), a method which has been successfully used in adults with insomnias of all sorts, for use in children. CBT-i takes a two-pronged approach, combining stimulus control (associating the bed with sleep only) and sleep restriction. An adaptation for children with NDD is being developed and may look somewhat different. In the context of limited availability in many areas, telehealth options are also in process. Introduction of apps would be ironic considering the destructive impact of electronic on sleep.

Conclusion

Good sleep benefits from external structure and/or self-discipline; calm, ordered surroundings; and sensitivity to social and environmental cues as well as functioning neurological substrate and appropriate anatomy.

References

[1] Accardo JA, Marcus CL, Leonard MB, Shults J, Meltzer LJ, Elia J. Associations between psychiatric comorbidities and sleep disturbances in children with attention-deficit/hyperactivity disorder. J Dev Behav Pediatr 2012;33:97–105.

[2] Souders MC, Mason TB, Valladares O, Bucan M, Levy SE, et al. Sleep behaviors and sleep quality in children with autism spectrum disorders. Sleep 2009;32(12):1566–78.

[3] Ahmad SF, Buckley AW, Glaze DG. Neurology of sleep. Neurol Clin 2021;39:867–82.

[4] Sollars PJ, Pickard GE. The neurobiology of circadian rhythms. Psychiatr Clin North Am 2015;38(4):645–65.

[5] Frank M, Heller HC. The function(s) of sleep. Handb Exp Pharmacol 2019;253:3–34.

[6] Xie L, Kang H, Xu Q, Chen MJ, Liao Y, et al. Sleep drives metabolite clearance from the adult brain. Science 2013;342(6156):373–7.

[7] Anafi RC, Kayser MS, Raizen DM. Exploring phylogeny to find the function of sleep. Nat Rev Neurosci 2019;20(2):109–16.

[8] Fuller PM, Gooley JJ, Saper CB. Neurobiology of the sleep-wake cycle: sleep architecture, circadian regulation, and regulatory feedback. J Biol Rhythms 2006;21(6):482–93.

[9] Staton S, Rankin PS, Harding M, Smith SS, Westwood E, et al. Many naps, one nap, none: a systematic review and meta-analysis of napping patterns in children 0–12 years. Sleep Med Rev 2020;50:101247.

[10] Borbely AA, Daan S, Wirz-Justice A, DeBoer T. The two-process model of sleep regulation: a reappraisal. J Sleep Res 2016;25(2):131–43.

[11] Borbely AA. A two process model of sleep regulation. Hum Neurobiol 1982;1(3):195–204.

[12] Crowley SJ, Acebo C, Carskadon MA. Sleep, circadian rhythms, and delayed phase in adolescence. Sleep Med 2007;8(6):602–12.

[13] Krystal AD, Edinger JD. Measuring sleep quality. Sleep Med 2008;9(Suppl 1):S10–17.

[14] Bruni O, Novelli L, Miano S, Parrino L, Terzano MG, et al. Cyclic alternating pattern: a window into pediatric sleep. Sleep Med 2010;11(7):628–36.

[15] Gorgoni M, Scarpelli S, Reda F, DeGennaro L. Sleep EEG oscillations in neurodevelopmental disorders without intellectual disabilities. Sleep Med Rev 2020;49:101224.

[16] Ekirch AR. At day's close: night in times past. New York, NY: Norton; 2005.

[17] Crosby B, LeBourgeois MK, Harsh J. Racial differences in reported napping and nocturnal sleep in 2- to 8-year-old children. Pediatrics 2005;.

[18] Williamson AA, Rubens SL, Patrick KE, Moore M, Mindell JA. Differences in sleep patterns and problems by race in a clinical sample of black and white preschoolers. J Clin Sleep Med 2017;13(11):1281–8.
[19] Accardo JA, Malow BA. Sleep, epilepsy, and autism. Epilepsy Behav 2015;47:202–6.
[20] Haspel JA, Anafi R, Brown MK, Cermekian N, Depner C, et al. Perfect timing: circadian rhythms, sleep, and immunity—an NIH workshop summary. JCI Insight. 2020;5(1):e131487.
[21] Leproult R, Van Cauter E. Role of sleep and sleep loss in hormonal release and metabolism. Endocr Dev 2010;17:11–21.
[22] Baker-Smith CM, Isaiah A, Melendres MC, Mahgerefteh J, Lasso-Pirot A, et al. Sleep disordered breathing and cardiovascular disease in children and adolescents. J Am Heart Assoc. 2021;10:e022427.
[23] Halbower A, Mahone EM. Neuropsychological morbidity linked to childhood sleep-disordered breathing. Sleep Med Rev 2006;10:97–107.
[24] Martin CA, Sciberras E, Papadapoulos N, Engel L, Hiscock H, et al. Associations between child sleep problem severity and maternal well-being in children with autism spectrum disorder. J Autism Dev Disord 2021;51:2500–10.
[25] Paruthi S, Brooks LJ, D'Ambrosio C, Hall WA, Kotagal S, et al. Recommended amount of sleep for pediatric populations: a consensus statement of the American Academy of Sleep Medicine. J Clin Sleep Med 2016;12(6):785–6.
[26] Mason TB, Pack AI. Pediatric parasomnias. Sleep 2007;30(2):141–51.
[27] Senthilvel E, Krishna J. Body position and obstructive sleep apnea in children with Down syndrome. J Clin Sleep Med 2011;7(2):158–62.
[28] Accardo J, Reesman J. Can you hear me snore? J Clin Sleep Med 2013;9(11):1221–3.
[29] DelRosso L, Picchietti DL, Spruyt K, Bruni O, Garcia-Borreguero D, et al. Restless sleep in children: a systematic review. Sleep Med Rev 2021;56:101406.
[30] Segal M, Eliasziw M, Phillips S, Bandini L, Curtin C, et al. Intellectual disability is associated with increased risk for obesity in a nationally representative sample of U.S. children. Disabil Health J. 2016;9(3):392–8.
[31] Vgontzas AN, Bixler EO, Basta M. Obesity and sleep: a bidirectional association? Sleep 2010;33(5):573–4.
[32] Owens JA, Spirito A, McGuinn M. The Children's Sleep Habits Questionnaire (CSHQ): psychometric properties of a survey instrument for school-aged children. Sleep 2000;23(8):1041–51.
[33] Chervin RD, Hedger K, Dillon JE, Pituch KJ. Pediatric Sleep Questionnaire (PSQ): validity and reliability of scales for sleep-disordered breathing, snoring, sleepiness, and behavioral problems. Sleep Med 2000;1(1):21–32.
[34] Sadeh A. A brief screening questionnaire for infant sleep problems: validation and findings for an Internet sample. Pediatrics 2004;113: e570–7.
[35] Melendres MC, Lutz JM, Rubin ED, Marcus CL. Daytime sleepiness and hyperactivity in children with suspected sleep-disordered breathing. Pediatrics 2004;114(3):768–75.
[36] Meltzer L, Forrest CB, de la Motte A, Bevans KB. Clinical validity of the PROMIS pediatric sleep measures across populations of children with chronic illnesses and neurodevelopment disorders. J Pediatr Psychol 2020;45(3):319–27.
[37] Meltzer L, Forrest CB, de la Motte A, Mindell JA, Bevans KB. Development and validation of the Pediatric Sleep Practices Questionnaire: a self-report measure for youth ages 8–17 years. Behav Sleep Med 2021;19(1):126–43.
[38] Paasch V, Hoosier TM, Accardo J, Ewen JB, Slifer KJ. Technical tips: performing EEGs and polysonograms on children with neurodevelopmental disabilities. Neurodiagn J. 2012;52:333–48.
[39] Bakke KA, Howlin P, Helverschou SB. Hyperactive behaviour in Angelman syndrome: the association with sleep problems and age of epilepsy onset. J Int Disabil Res. 2021;65(7):666–74.
[40] Shi S, Mahoney CE, Houdek P, Zhao W, Anderson MP, et al. Circadian rhythms and sleep are dependent upon expression levels of key ubiquitin ligase Ube3a. Front Behav Neurosci 2022;16:837523.
[41] Glaze DG, Frost JD, Zoghbi HY, Percy AK. Rett's syndrome: characterization of respiratory patterns and sleep. Ann Neurol 1987;21:377–82.
[42] Mindell JA. Sleeping Through the night: how infants, toddlers, and their parents can get a good night's sleep. New York, NY: HarperResource,; 2005.
[43] Owens JA, Mindell JA. Taking charge of your child's sleep: the all-in-one resource for solving sleep problems in kids and teens. New York, NY: Marlowe & Co.; 2005.
[44] Durand VM. Sleep better! a guide to improving sleep in children with special needs. 2nd ed. Baltimore, MD: Paul H. Brookes Publishing Co.; 2013.
[45] Katz T, Malow BA. Solving sleep problems in children with autism spectrum disorders: a guide for frazzled families. Bethesda, MD: Woodbine House; 2014.
[46] Pediatric Sleep Council website. https://www.babysleep.com/; [accessed 07.31.22].
[47] Start School Later. https://www.startschoollater.net/; [accessed 07.31.22].
[48] Better Nights, Better Days website. https://ndd.betternightsbetterdays.ca/; [accessed 07.31.22].
[49] Caloway CL, Diercks GR, Kamy D, de Guzman V, Soose R, et al. Update on hypoglossal nerve stimulation in children with Down syndrome and obstructive sleep apnea. Laryngoscope 2020;130(4):E263–7.
[50] Goldenberg P. 22q11.2 Deletion syndrome. In: Accardo JA, editor. Sleep in children with neurodevelopmental disabilities. Switzerland: Springer Nature; 2019.
[51] Trickett J, Heald M, Oliver C, Richards C. A cross-syndrome cohort comparison of sleep disturbance in children with Smith-Magenis syndrome, autism spectrum disorder and tuberous sclerosis complex. J Neurodev Disord 2018;10(1):9.

[52] Buonfiglio D, Hummer DL, Armstrong A, Ehlen C, DeBruyne JP. Angelman syndrome and melatonin: what can they teach us about sleep regulation. J Pineal Res 2020;69(4):e12697.
[53] Becker SP. ADHD and sleep: recent advances. Curr Opin Psychol. 2020;34:50–6.
[54] Gruber R, Sadeh A, Raviv A. Instability of sleep patterns in children with attention-deficit/hyperactivity disorder. J Am Acad Child Adolesc Psychiatry 2000;39(4):495–501.
[55] Schwichtenberg AJ, Janis A, Lindsay A, Desai H, Sahi A, et al. Sleep in children with autism spectrum disorder: a narrative review and systematic update. Curr Sleep Med Rep. 2022;8(4):51–61.
[56] Hulst RY, Gorter JW, Voorman JM, Kolk E, Van Der Vossen S, et al. Sleep problems in children with cerebral palsy and their parents. Dev Med Child Neurol 2021;63(11):1344–50.
[57] Ingram DG, Ruiz AG, Gao D, Friedman NR. Success of tonsillectomy for obstructive sleep apnea syndrome. J Clin Sleep Med 2017;13(8):975–80.
[58] Duis J, Pullen LC, Picone M, Friedman N, Hawkins S, et al. Diagnosis and management of sleep disorders in Prader-Willi syndrome. J Clin Sleep Med 2022;18(6):1687–96.
[59] Ramirez JM, Karlen-Amarante M, Wang JJ, Huff A, Burgraff N. Breathing disturbances in Rett syndrome. Handb Clin Neurol 2022;189:139–51.
[60] Shayota BJ, Elsea SH. Behavior and sleep disturbance in Smith-Magenis syndrome. Curr Opin Psychiatry 2019;32(2):73–8.

Chapter 36

Dysphagia and feeding disorders

Kevin Senn

Department of Pediatrics, Institute on Development and Disability, Oregon Health and Sciences University, Portland, OR, United States

Overview

The rationale behind screening, diagnosing, and treating feeding disorders is to optimize the nutritional status of the child with the parallel goals of adequate nutrient intake, safe intake of food, and preservation of quality of life. The sequelae of undernutrition may include inadequate energy for daily activities, secondary organ system dysfunction, poor bone growth, and a compromised immune system as well as inadequate reserve when additional physiologic demands arise. The correlation between undernutrition and suboptimal cognitive and behavior skills is strong in studies from the developing world but less clear in industrialized and wealthier countries [1]. These studies have shown a need for a strong focus on the first year of life, as a correlation was found between smaller head size and subsequent lower cognitive scores in the undernourished [2]. Older autopsy and animal studies and current in-depth neuroimaging studies have shown the micro and macro changes in the brain exposed to early undernutrition. There is decreased neuronal proliferation and myelinization, altered dendritic pruning patterns, and general brain undergrowth, in addition to molecular and structural changes in hippocampal and cortical regions after prenatal protein malnutrition [3]. There remain substantial gaps in knowledge that were summarized by Galler et al. [4], exploring current neuroimaging of the sequelae of undernutrition. Newer understanding of the effects of the environment and nutrition on epigenetic modulation and long-term health has substantiated the relationship between early nutrition and the long-term outcomes of obesity and metabolic syndrome [5].

The effect of early life malnutrition persists into adulthood. Adolescents in a low-income cohort in Chile had significantly lower IQ and smaller brain volume despite recovery from malnutrition in their first year of life, compared to matched controls [6]. While determining how strongly this research relates to a child with a neurodevelopmental concern in an otherwise well-resourced situation, enabling any young child to reach their best potential in developmental skills and growth for their condition should be a goal of all treatment programs. Special syndromic growth charts enable the monitoring of the growth trajectory of some children with neurodevelopmental disabilities.

Epidemiology

Pediatric feeding disorders have many definitions depending on the professionals evaluating them with some focusing on the mechanical aspects of swallowing, while others focus on restricted nutritional intake. The diagnostic and statistical manual of mental disorders (DSM) and international classification of diseases (ICD) systems have only recently attempted to refine the developmental and physical issues involved in pediatric feeding disorders and maintain separation from psychogenic eating disorders.

In 2019, a consensus was built around a new and comprehensive definition using the *International Classification of Functioning, Disability, and Health* framework. Pediatric feeding disorders are defined as "impaired oral intake that is not age-appropriate, and is associated with medical, nutritional, feeding skill, and/or psychosocial dysfunction." This framework focuses on four main integral and interdependent domains that all practitioners should address [7].

Definitions and accurate surveys are necessary to find the true prevalence of feeding difficulties. This is challenging when there has been a wide variety of terms and definitions with evolving names over time. For example, many neurotypical children may be classified as having feeding problems based on parental reports [8]. Some may be of a developmental or transitory nature. Parents will report 15%−50% of the time that their typically developing child is a

Capute and Accardo's Neurodevelopmental Disabilities in Infancy and Childhood. DOI: https://doi.org/10.1016/B978-0-12-824060-1.00047-X

challenging eater—often termed a "picky eater," and a smaller group of them might actually be "problem eaters." The majority of these children will not eventually have any disability.

The challenge is compounded by the spectrum of severity seen in many neurodevelopmental conditions associated with dysphagia or feeding difficulty. Therefore it would make little sense to investigate and provide percentages of a level of need if we used an undefined denominator. For example, the percent of children with cerebral palsy (CP) needing tube feeding would only have value if the CP was of the same type and severity level. Children with neurodevelopmental disorders (NDD) may develop a feeding disorder before a clear neurocognitive or motor dysfunction is diagnosed. Many medical causes of feeding difficulty in both neurotypical children and children with NDD can be remediated.

Feeding difficulty can be physiological as in the case of extreme prematurity or a manifestation of brain injury related to prematurity. Feeding disorder may also be the first sign of developmental delay in a child. For example, the risk of aspiration is not always readily evident from signs and symptoms and may dynamically change over time and when new challenges arise for the child. A feeding evaluation program may be tasked with evaluating infants with feeding challenges who have no obvious medical conditions and require resources for a comprehensive evaluation.

Evaluation and treatment by specialists in feeding disorders can be an opportunity to pursue the goals of a healthier outcome despite a neurodevelopmental difference. A thorough understanding of the developmental anatomy and physiological maturation sequence of the functional skills of eating is key to a comprehensive assessment.

Anatomy and innervation

In young children, the esophagus and trachea are closely adjacent and quite compact. The pharyngeal structures and muscles must be well formed and sturdy enough to keep their form under inspiratory pressures and perform the complex pattern of peristalsis together.

Swallowing is a complex function that requires sensory and motor integration. The act of swallowing of amniotic fluid occurs prenatally. Ultrasound measures of amniotic fluid amounts can be the first sign of a functional or anatomic concern for dysphagia.

The first step is the preparation of the bolus which requires intact innervation and coordinated function of oropharyngeal muscles, which is discussed in more details below. Once a bolus is in place, it has to be sensed to prepare for the rapid transfer past the airway inlets and the protective systems should prevent aspiration.

Oropharyngeal musculature and innervation

The motor section of the neural apparatus of swallow has to activate in order to propel the bolus to the esophagus, and begin the peristalsis to its intestinal tract destination, The following tables are provided as references to the more specific motor and sensory innervation and related functions (Tables 36.1 and 36.2).

Developmental aspects and maturation

Magendie, in 1825, is credited with first describing the action and neurology of swallowing in three phases: oral, pharyngeal, and esophageal [9], although others may split the oral phase into oral preparatory and oral to distinguish the sucking and chewing actions during the oral time before the swallow. Another way is to separate into two phases—oral and pharyngoesophageal—with voluntary control on oral and a reflexive pharyngeal phase, with subsequent esophageal portions under mixed somatic and autonomic controls.

The first patterns of swallowing are the result of reflexive brainstem activities, often preserved even when higher cortical injury has occurred. The fetus shows suckling actions by 18 weeks of gestation typically. After birth, muscular activation of the lips, tongue, palate, and buccal pads strip milk with compression and suction until a large enough bolus accumulates in the center of the tongue to trigger a reflexive swallow activation of multiple muscles. After breathing is paused, and if the upper esophageal sphincter is relaxed sufficiently, peristalsis moves the bolus forward. The newborn oral cavity is made smaller by relative tongue size, buccal pads, and absence of teeth—so swallows are small volume at the start. The bolus spends very little time in the pharynx and relatively longer time in the esophagus.

Prior to term, babies often have a one suck to swallow ratio—likely the in utero pattern of swallowing amniotic fluid, and as they mature, more sucks are needed before a swallow. The rule of thumb in neonatal units is expectation of full oral feeding by 34 weeks postmenstrual age—in a healthy preterm, perhaps later in situation of neurological dysfunction.

TABLE 36.1 The motor cranial nerves and related functions.

Motor cranial nerves	Innervation	Function
Trigeminal (V)	Temporalis, masseter, and lateral pterygoid muscles	Primarily for jaw motion and eventually for chewing
Facial (VII)	Orbicularis oris, zygomaticus, and buccinators	Close up the lips, lift the corners of the mouth, and tense the cheeks to prevent food loss, allow for sucking pressure and prevent cheek pocketing
Glossopharyngeal and vagus (IX and X)	Activate the upper pharyngeal constrictors, with IX activating the stylopharyngeus muscle to and X adducting the vocal folds to protect the airway and activating the palatoglossal musculature	Narrows the oropharynx by elevating the tongue, drawing the faucial pillars and soft palate inferiorly during the swallow, and regulates the tone in the cricopharyngeus and levator veli palatine muscles for closing the nasopharynx. The right and left recurrent laryngeal nerve branches from the vagus to the posterior cricoarytenoid muscles and innervate the vocal cords. These are particularly prone to surgical trauma due to the long route through the chest near and around heart vessels.
Spinal accessory (XI)	Palatopharyngeus muscle, the muscularis uvula, and the levator veli palatini (with vagus)	Assists with closure and propulsion
Hypoglossal (XII)	All of the intrinsic and extrinsic tongue muscles	Tongue movements

TABLE 36.2 The sensory cranial nerves involved in taste and proprioception.

Cranial nerve	Function
Trigeminal (V)	For sensation (not taste) in the anterior 2/3 of tongue
Facial (VII)	For taste from the anterior 2/3 of tongue
Glossopharyngeal (IX)	Taste and sensation from the posterior 1/3 of tongue. The soft palate and superior portion of the pharynx
Vagus (X)	Sensation of the palate, larynx, and inferior portions of pharynx

The growth of facial and airway structures is significant even in the first 4 years and much different after full growth has completed. The buccal pads will disappear, areas for oral pockets of food will enlarge, and the pharynx and larynx move deeper into the neck. The muscle activations in swallowing are stronger over time to contract the longer distance. Some babies with safe reflexive sucking and swallowing may progress on to a higher aspiration risk after this natural anatomical change occurs.

The musculature of front to back tongue motion at first will later progress to effective lateral tongue movement, and jaw movements will progress from vertical munch to rotary motions over time. Understanding the expected time for these transitions also helps to judge what textures a child may be ready for in an oral preparatory sense. Eventually, even mixed textures of liquids and solids together can be sorted by these actions voluntarily prior to presentation of the ready bolus.

Early feeding problems

The evaluation of the infant with feeding difficulty is complex. It encompasses cardiopulmonary, structural malformations of the esophagus, craniofacial conditions, and metabolic or neurological dysfunction. Isolated feeding problems are rare, and therefore a comprehensive history and physical examination should be part of the evaluation of the infant with feeding difficulty.

Anatomic orofacial malformations can range from obvious clefts in the lips, nares, and palate, and the soft palate can also be disrupted by the tongue, as with Pierre Robin Sequence or weakly supported, such as in 22q11 deletions. Low tone in general, as in Trisomy 21, may lead to oral preparatory challenges. Protective reflexes—notably the diving reflex to inhibit inhalation, may not be as protective if the child is tachypneic from bronchopulmonary dysplasia (BPD) or challenged by the desensitization of chronic gastroesophageal reflux disease (GERD).

Volpe provides a systematic categorization of feeding disorders due to various neurologic causes: encephalopathies, issues in the deep brain and cranial nerve nuclei, more distal disorders of the cranial nerves, the neuromuscular junction, and primary muscle disorders. Examination, imaging, and labs could elucidate the problem and consultation with experts in newborn nursing, lactation specialists, Neonatal therapists, and medical providers may provide important information. Observation of feeding and response to manipulation of feeding techniques are valuable parts of the exam and history. Several examples of etiology are givein in Table 36.3 for some of the underlying pathologies of feeding difficulties.

Etiologies of newborn feeding difficulty

Signs of dysphagia may include jaw thrust, bite reflex (tonic or phasic), tongue thrust, lip retraction, jaw clenching, and persistence of suckling pattern. Among the symptoms that are often seen are difficulty with latch and lip closure, tonic bite reflex, lingual incoordination, inability to be satisfied with intake, poor weight gain, noisy congested breathing, coughing, clicking, gasping, medical events of pneumonia, cyanosis, stridor, and wheezing. The symptoms may mimic sudden acute apnea or unresponsiveness events also known as a Brief Resolved Unexplained Event. These symptoms may be texture-specific, and failure to advance to developmentally appropriate textures is a common chief complaint. There will be some infants that gather the skills of eating more slowly than expected, and sustenance by nasogastric or gastrostomy tube as they improve is often necessary.

Feeding team evaluation

A multidisciplinary feeding team can address the varied etiology of the feeding problem and assess the nutritional risks. When staffed with feeding therapists, dieticians, and medical providers, the global evaluation can review the oral motor function, the nutritional status, and the medical/developmental and behavior concerns. The oral motor specialist addresses safe and effective methods of feeding. The nutritional objective is to supply the necessary substrates for growth and consider special dietary needs.

For outpatients, many tertiary care centers have organized teams for feeding support, sometimes in combination with neonatology follow-up, gastroenterology, or neurodevelopmental specialties. If preliminary measures for improvement of the feeding are not fully sufficient, a referral for a team approach could help achieve the goals.

Interdisciplinary collaboration is beneficial in that each provider can share their special expertise and communicate efficiently [11]. Observation of feeding gives the opportunity for trial manipulations, perhaps of the bottles, nipples, flow rates, positions, and behavioral techniques. Family education and collaboration are inherent and critical for success. Follow-up after evaluation is frequently needed to support successful home feeding and growth.

Monitoring and charting the growth accurately are important in infants and in later years as well, when body types may not be easy to measure. Myelomeningocele, cerebral palsy, and genetic conditions complicate growth assessment.

TABLE 36.3 Possible etiologies of feeding difficulties [10].

Pathophysiology	Possible diagnoses
Encephalopathy	Metabolic, infarction, ischemia, seizures, withdrawal
Deep brain/nuclei	Moebius sequence, hypoxia, Chiari malformation type 2
Nerves	Injury, dysautonomia, tumor
Neuromuscular junction	Botulism, myasthenia gravis
Muscular	Congenital muscular dystrophy, spinal muscular atrophy

Source: Adapted from Joseph J. Neurology of the newborn. Volpe, 5th ed. Saunders Elsevier; 2008.

With electronic record and growth charts published by patient registries, we can often find a valid chart to track growth. Getting accurate lengths, weights, and body composition measures not only requires consistency but may include special equipment, such as long measuring tables, skin fold calipers, wheelchair scales, and the commitment to consistent accurate measurement.

Breast milk for babies is the first priority. Modifications in the manner in which breast milk is supplied to the infant may be necessary for babies with NDD. Many times, the milk or formula will need modification for issues of reflux, motility, constipation, and allergies in children with NDD conditions, just as in the typical population. Calorie needs can vary in genetic and metabolic situations. Micronutrient needs can also be altered by certain medications, and recommended dietary intake may need to be revised for those individuals.

Oral motor therapists may be credentialed or simply experienced, trained professionals who go beyond typical occupational therapy and speech language pathology practices but remain true to their scope of practice. Focused evaluation of the oral anatomy, the sensory profile pertaining to feeding, and the muscular control for oral feeding are systematically observed to create the feeding plan. When the imaging or the observations have suggested that a thicker liquid meal is safer, the mechanics of delivery have to be shown to work for the family. In the end, the bottle, nipple, formula or food, and the oral skills have to work together for efficient delivery.

Imaging may include the standard modified barium swallow (or videofluoroscopy study) and/or endoscopic approaches. The former does not lend itself to breast feeding—but all the phases of the swallow are seen and the risk of aspiration and penetration of liquids into the airway are documented. Direct endoscopy could also give the dynamic view of vocal fold and cord movements and could be used with breast feeding—but cannot always show the full swallow and the residue down in the airway—if there is any. An experienced evaluator and proper equipment can add value to these studies [12].

Medical providers may not always have the largest role on this team but are necessary to exclude the cardiopulmonary, genetic, metabolic, and structural malformations etiologies. They can coordinate treatment plans and provide treatments and studies as needed. Ankyloglossia can be a frequently encountered hypothesis, and detection can be somewhat subjective. Growth, safety of swallowing, sleep disturbances, primary GI issues of reflux and constipation, and quality of diet are medical areas to be monitored. With additional neurodevelopmental assessments, the provider can help the behavioral plans by reviewing the overall motor, cognitive, communicative, and social skills of the child. As an example, different approaches are used for a highly sensory-defensive child with autism than for a motor-delayed child with Down syndrome.

Special considerations in NDD associated dysphagia

Growth

For each child, there may be a optimal growth pattern to meet their unique needs. A set rule of so many grams/day for acceptable growth may not always be correct. In the small for gestational age and the term baby, some evidence has suggested that excessively fast growth may lead to further metabolic stress in the adult life [13,14]. Formulae that mimic the lower protein content of breast milk appear to be better for prevention of obesity and adding certain fatty acids in the milk showed better visual-spatial skills [15]. For children challenged in movement and ambulation from moderate to severe CP, a better growth trend led to fewer days of health care use and more social participation [16]. Overweight status, however, may hinder motor progress. When a biomedical demand is present, such as surgery, illness, or the need to foster early pulmonary growth in a premature infant, the nutrition plan may require individualized modification.

In early life, the current WHO growth standards have been preferred over the CDC reference charts for a variety of reasons. The WHO charts are based on a broader and more diverse set of populations, which helps avoid unwanted trends like obesity that might be present in charts solely derived from one population. Moreover, the inclusion of multicultural populations and a higher rate of neonates and infants who are breastfeeding, coupled with efforts to minimize outliers, make the current WHO charts a preferable choice.

Tube feeding

Temporary feeding by the nasogastric route and long-term feeding by gastrostomy or jejunostomy device insertion are safer options for children with aspiration and inability or unwillingness to consume sufficient nutrients. A comprehensive review of the current state of alternate enteral nutrition delivery is given by ESPGHAN in 2010 [17]. Tube feeding bypasses the aberrant swallowing and supports the nutritional objectives of growth. However, it is a substantial deviation from what parents expect. Along with other medical interventions, tube feeding may be a major factor in promoting a better life expectancy and a better quality of life, despite the medicalization of feeding. Parents report that they are freed from the work of oral feeding to focus on other developmental needs of the child [18]. Practices may vary by socioeconomic status, cultural, and racial or ethnic backgrounds, and these should be further studied [19,20].

Special diets that may not be as palatable can be delivered by tube, for example hypoallergenic formulas, formulas with easier to digest proteins or altered sugars, or ketogenic and related diets for epilepsy [21].

Complete nutrition—including all necessary nutrients in the right amounts and proportions—can be more reliably given by tube [22]. The placement procedures and the daily care are reasonably safe and well tolerated. With small gastric buttons and pumps, meals can occur in the setting of the family table, promoting time for oral feeding practice, when that is a safe goal. In some situations, weaning from the enteral tube feeding can occur when medical and neurological development is more optimal for feeding. With care, parents can often move to blended home food by tube or opt for one of the growing numbers of specialty products available. Reflux, retching, and constipation may be good reasons to try new food products, so medications can be reduced.

When an undernourished child is given a rapid increase in caloric intake, it is done carefully and often established in a monitored environment due to the potential of an insulin release reaction termed refeeding syndrome, with hypomagnesemia, hypokalemia, or hypophosphatemia with lethal outcome at worst [23]. Recognized refeeding reactions of any severity occur as low as 5%, but it is often unrecognized [24].

Many sites have reported on the experience of gastrostomy surgery, and complication rates are fairly low for percutaneous endoscopic gastrostomy tubes and low-profile skin surface button devices; the most common methods. Local cellulitis and other stomal problems rose to nearly 10% in one large hospital cohort [25]. In that series, the rate of conversion to jejunal feeding tube (14%) or a second surgery for a reflux procedure (4%) are notable, highlighting the concern of feeding intolerance after gastric tube placement. Neurologically impaired children have long been seen as at risk for pathological GERD when converted to gastrostomy feeding, but preemptive fundoplication is not supported by current studies [26,27].

Medication effects

It is well known that certain medications affect appetite, but they also may affect aspiration risks and micronutrient needs.

Some antiepileptic drugs affect vitamin D bioavailability. Valproate may interfere with bone mineralization but with additional intake of calcium and vitamin D, it may be avoided [28]. This effect is especially concerning for children who are nonweight bearing.

Sedating medications, benzodiazepines, and muscle relaxants may interfere with the alertness and awareness for safer swallowing patterns and functional peristalsis [29]. Additionally, implanted vagal nerve stimulation devices for epilepsy may occasionally have an effect on the larynx and give symptoms of cough or swallow dyscoordination in some modes [30].

Proton pump inhibitor medications for GERD have been associated with bone density and fracture concerns in adults, and current protocols attempt to limit prolonged use. However, when they can reduce painful reflux, the appetite can return and sensory awareness of the bolus may be restored.

Appetite promotion is not easily a medication target, and children who are not on the cycle of hunger and satiation with the resultant hormonal swings may have a poor appetite. For children, we cannot use steroids, but prokinetics and cyproheptadine may promote better interest in eating. The literature is sparse, and the evidence base is more anecdotal than desired, but often a safe trial can be undertaken in selected children [31].

Neurocognitive conditions

The classic patterns of feeding in the first few months have been well described as a reflexive, homeostatic drive to self-nourish, provided the ability and environment are right for this to occur. This is appropriately followed by patterns of collaboration, then individuation and separation from the feeder, provided the dyad of feeding can still be maintained for effective delivery. Complete independence in food intake is usually at a relatively young age, when provided the right food items, the tools and textures, and the physical ability.

At any stage, feeding aversion can begin, triggered by an internal or external/relational conflict with food or feeding. Medical aversion can come about from illness and internal discomforts, such as pain, procedures, maldigestion, and allergic reactions. Psychological aversion can be internal, reactive, or mixed. It is well known that children with autism often have selectivity, rigidity, and prefer routines and sameness [32]. When applied to food, the selectivity in autism can range to the extremes in ways that are hard to change with feeding therapy. Parents can have maladaptive feeding styles and misperceptions of healthy nutrition that a mental health provider may be well suited to address and modify. Many situations are mixed, as the family resorts to maladaptive measures in response to the early difficulties.

Motor conditions

Conditions with altered tone and movements can clearly increase the difficulty and risk of oral feeding, particularly when the whole body is involved. When thoroughly evaluated, over 90% of preschool children who eventually received a CP diagnosis had been having oral motor dysfunction. Children with cerebral palsy may evidence abnormal oral motor patterns, including jaw thrusting, tongue thrusting, tongue retraction, lip retraction, and hyperactive gag reflexes. Bite reflexes may interfere with mature sucking or chewing patterns and may affect the child's ability to grade jaw opening. This results in prolonged time for feeding meals and actual undernutrition in more than a third [33]. In children with athetosis, positioning is more difficult because of excessive movement, variable tone, and neck extension, which opens the airway for potential aspiration.

There are no good measures that predict the caloric needs of the child. The number of calories that a child requires is defined empirically and can be a wide range. Caloric needs are not always predicted by the muscle tone of the patient and should be individualized.

Positioning

While the evidence is limited proper positioning is among the most commonly employed interventions to improve safety and effectiveness of swallowing and feeding in children with oral motor dysphagia [34]. In children with cerebral palsy, correct positioning may provide a stable foundation that aids in controlling muscle movements to improve coordination during swallowing and feeding and limit the influence of abnormal extensor responses or athetosis.

Treatment considerations

Once an understanding is gained of the skills of feeding and the environment of eating, the interventions can be tailored to that child's situation and needs.

Gradual changes have been used for children with highly sensitive and rigid temperaments. A skilled therapist may intentionally guide the child and family on a path that eventually gets the new foods and textures acceptable in the mouth for consumption. The many steps along the way may include precontemplation and involvement in food preparation in the older child, repeated exposures, inclusion in the choices and preparation of foods and socially rewarding experiences to promote the intake [35]. A younger child may be approached with chaining and linking methods to expand variety, which makes new foods look, feel, and smell familiar to accepted foods. When parents see success for others, they may be empowered to adopt the techniques. Family stress of feeding may be reduced in a mutually helpful manner [36].

Feeding programs that maintain success have determined the individual underpinnings of that child's problem as best as possible and work to transition the helpful changes into the family approach at home.

Conclusion

Providing quality nutrition for a child with a neurodevelopmental condition can be challenging, but there is a broad base of research that supports the methods of remediation and circumvention. A comprehensive, coordinated, and interdisciplinary approach that addresses the strengths and challenges of the child and family and utilizes the full range of available community resources is likely to achieve the best outcomes.

References

[1] Gorman KS. Malnutrition and cognitive development: evidence from experimental/quasi-experimental studies among the mild-to-moderately malnourished. J Nutr 1995;125(8 Suppl):2239S–44S. Available from: https://doi.org/10.1093/jn/125.suppl_8.2239S.

[2] Galler JR. The behavioral consequences of malnutrition in early life. In: Galler JR, editor. Nutrition and behavior. New York: Springer Science & Business Media; 1984. p. 63–117. Available from: 10.1007/978-1-4615-7219-0_4.

[3] Barra R, Morgan C, Sáez-Briones P, Reyes-Parada M, Burgos H, Morales B, et al. Facts and hypotheses about the programming of neuroplastic deficits by prenatal malnutrition. Nutr Rev 2019;77(2):65–80. Available from: https://doi.org/10.1093/nutrit/nuy047.

[4] Galler JR, Bringas-Vega ML, Tang Q, Rabinowitz AG, Musa KI, Chai WJ, et al. Neurodevelopmental effects of childhood malnutrition: a neuroimaging perspective. Neuroimage 2021;231:117828. Available from: https://doi.org/10.1016/j.neuroimage.2021.117828.

[5] Worthman CM, Kuzara J. Life history and the early origins of health differentials. Am J Hum Biol 2005;17(1):95–112. Available from: https://doi.org/10.1002/ajhb.20096.

[6] Ivanovic DM, Leiva BP, Perez HT, Inzunza NB, Almagia AF, Toro TD, et al. Long-term effects of severe undernutrition during the first year of life on brain development and learning in Chilean high school graduates. Nutrition. 2000;11-12:1056–63.

[7] Goday PS, Huh SY, Silverman A, Lukens CT, Dodrill P, Cohen SS, et al. Pediatric feeding disorder: consensus definition and conceptual framework. J Pediatr Gastroenterol Nutr 2019;68(1):124–9. Available from: https://doi.org/10.1097/MPG.0000000000002188.

[8] Kerzner B, Milano K, MacLean Jr WC, Berall G, Stuart S, et al. A practical approach to classifying and managing feeding difficulties. Pediatrics. 2015;135(2):344–53. Available from: https://doi.org/10.1542/peds.2014-1630.

[9] Ertekin C, Aydogdu I. Neurophysiology of swallowing. Clin Neurophysiol 2003;114(12):2226–44. Available from: https://doi.org/10.1016/s1388-2457(03)00237-2.

[10] Neurology of the Newborn. Volpe, Joseph J. 5th ed. Saunders Elsevier; 2008

[11] Silverman AH. Interdisciplinary care for feeding problems in children. Nutr Clin Pract 2010;25(2):160–5. Available from: https://doi.org/10.1177/0884533610361609.

[12] Arvedson JC, Lefton-Greif MA. Instrumental assessment of pediatric dysphagia. Semin Speech Lang 2017;38(2):135–46. Available from: https://doi.org/10.1055/s-0037-1599111.

[13] Koletzko B, Demmelmair H, Grote V, Totzauer M. Optimized protein intakes in term infants support physiological growth and promote long-term health. Semin Perinatol 2019;43(7):151153. Available from: https://doi.org/10.1053/j.semperi.2019.06.001.

[14] Clayton PE, Cianfarani S, Czernichow P, Johannsson G, Rapaport R, Rogol A. Management of the child born small for gestational age through to adulthood: a consensus statement of the International Societies of Pediatric Endocrinology and the Growth Hormone Research Society. J Clin Endocrinol Metab 2007;92(3):804–10. Available from: https://doi.org/10.1210/jc.2006-2017.

[15] Alexander DD, Yan J, Bylsma LC, Northington RS, Grathwohl D, Steenhout P, et al. Growth of infants consuming whey-predominant term infant formulas with a protein content of 1.8 g/100 kcal: a multicenter pooled analysis of individual participant data. Am J Clin Nutr 2016;104(4):1083–92. Available from: https://doi.org/10.3945/ajcn.116.130633.

[16] Stevenson RD, Conaway M, Cameron Chumlea W, Rosenbaum P, Fung EB, Henderson RC, et al. Growth and health in children with moderate-to-severe cerebral palsy. Pediatrics 2006;118(3):1010–18. Available from: https://doi.org/10.1542/peds.2006-0298.

[17] Braegger C, Decsi T, Dias JA, Hartman C, Kolacek S, Koletzko B, et al. Practical approach to paediatric enteral nutrition: a comment by the ESPGHAN committee on nutrition. J Pediatr Gastroenterol Nutr 2010;51(1):110–22. Available from: https://doi.org/10.1097/MPG.0b013e3181d336d2.

[18] Åvitsland TL, Birketvedt K, Bjørnland K, Emblem R. Parent-reported effects of gastrostomy tube placement. Nutr Clin Pract 2013;28(4):493–8. Available from: https://doi.org/10.1177/0884533613486484.

[19] Faigle R, Bahouth MN, Urrutia VC, Gottesman RF. Racial and socioeconomic disparities in gastrostomy tube placement after intracerebral hemorrhage in the United States. Stroke. 2016;47(4):964–70. Available from: https://doi.org/10.1161/STROKEAHA.115.011712.

[20] Garcia RM, Prabhakaran S, Richards CT, Naidech AM, Maas MBR. Socioeconomic status, and gastrostomy after spontaneous intracerebral hemorrhage. J Stroke Cerebrovasc Dis 2020;29(2):104567. Available from: https://doi.org/10.1016/j.jstrokecerebrovasdis.2019.104567.

[21] Hosain SA, La Vega-Talbott M, Solomon GE. Ketogenic diet in pediatric epilepsy patients with gastrostomy feeding. Pediatr Neurol 2005;32(2):81–3. Available from: https://doi.org/10.1016/j.pediatrneurol.2004.09.006.

[22] Edwards S, Davis AM, Bruce A, Mousa H, Lyman B, Cocjin J, et al. Caring for tube-fed children: a review of management, tube weaning, and emotional considerations. JPEN J Parenter Enter Nutr 2016;40(5):616–22. Available from: https://doi.org/10.1177/0148607115577449.

[23] da Silva JSV, Seres DS, Sabino K, Adams SC, Berdahl GJ, Citty SW, et al. ASPEN consensus recommendations for refeeding syndrome. Nutr Clin Pract 2020;35(2):178–95. Available from: https://doi.org/10.1002/ncp.10474.

[24] Walmsley RS. Refeeding syndrome: screening, incidence, and treatment during parenteral nutrition. J Gastroenterol Hepatol 2013;28(Suppl 4):113–17. Available from: https://doi.org/10.1111/jgh.12345.

[25] McSweeney ME, Jiang H, Deutsch AJ, Atmadja M, Lightdale JR. Long-term outcomes of infants and children undergoing percutaneous endoscopy gastrostomy tube placement. J Pediatr Gastroenterol Nutr 2013;57(5):663–7. Available from: https://doi.org/10.1097/MPG.0b013e3182a02624.

[26] Franken J, Stellato RK, Tytgat SHAJ, Van der Zee DC, Mauritz FA, Lindeboom MYA. Gastro-esophageal reflux after laparoscopic gastrostomy placement in children. J Pediatric Gastroenterol Nutr 2020;70(2):e41–7. Available from: https://doi.org/10.1097/MPG.0000000000002530.

[27] Thomson M, Rao P, Rawat D, Wenzl TG. Percutaneous endoscopic gastrostomy and gastro-oesophageal reflux in neurologically impaired children. World J Gastroenterol 2011;17(2):191–6. Available from: https://doi.org/10.3748/wjg.v17.i2.191.

[28] Durá-Travé T, Gallinas-Victoriano F, Malumbres-Chacón M, Moreno-Gónzalez P, Aguilera-Albesa S, Yoldi-Petri ME. Vitamin D deficiency in children with epilepsy taking valproate and levetiracetam as monotherapy. Epilepsy Res 2018;139:80–4. Available from: https://doi.org/10.1016/j.eplepsyres.2017.11.013.

[29] Wyllie E, Wyllie R, Cruse RP, Rothner AD, Erenberg G. The mechanism of nitrazepam-induced drooling and aspiration. N Engl J Med 1986;314(1):35–8. Available from: https://doi.org/10.1056/NEJM198601023140107.

[30] Lundgren J, Ekberg O, Olsson R. Aspiration: a potential complication to vagus nerve stimulation. Epilepsia. 1998;39(9):998–1000. Available from: https://doi.org/10.1111/j.1528-1157.1998.tb01450.x.

[31] Sant'Anna AM, Hammes PS, Porporino M, Martel C, Zygmuntowicz C, Ramsay M. Use of cyproheptadine in young children with feeding difficulties and poor growth in a pediatric feeding program. J Pediatr Gastroenterol Nutr 2014;59(5):674–8. Available from: https://doi.org/10.1097/MPG.0000000000000467.

[32] Hubbard KL, Anderson SE, Curtin C, Must A, Bandini LG. A comparison of food refusal related to characteristics of food in children with autism spectrum disorder and typically developing children. J Acad Nutr Diet 2014;114(12):1981–7. Available from: https://doi.org/10.1016/j.jand.2014.04.017 Epub 2014 Jun 11.

[33] Reilly S, Skuse D, Poblete X. Prevalence of feeding problems and oral motor dysfunction in children with cerebral palsy: a community survey. J Pediatr 1996;129(6):877–82. Available from: https://doi.org/10.1016/s0022-3476(96)70032-x.

[34] Parr J, Pennington L, Taylor H, Craig D, Morris C, McConachie H, et al. Parent-delivered interventions used at home to improve eating, drinking and swallowing in children with neurodisability: the FEEDS mixed-methods study. Health Technol Assess 2021;25(22):1–208. Available from: https://doi.org/10.3310/hta25220.

[35] Morgan AT, Dodrill P, Ward EC. Interventions for oropharyngeal dysphagia in children with neurological impairment. Cochrane Database Syst Rev 2012;17(10):CD009456. Available from: https://doi.org/10.1002/14651858.CD009456.pub2.

[36] Nelson KE, Lacombe-Duncan A, Cohen E, Nicholas DB, Rosella LC, Guttmann A, et al. Family experiences with feeding tubes in neurologic impairment: a systematic review. Pediatrics. 2015;136(1):e140–51. Available from: https://doi.org/10.1542/peds.2014-4162.

Chapter 37

Mood and anxiety disorders in neurodevelopmental disorders

Carmen Lopez-Arvizu[1] and Elaine Tierney[2]
[1]*Center for Developmental Behavioral Health, Kennedy Krieger Institute, Department of Psychiatry and Behavioral Science, Johns Hopkins School of Medicine, Baltimore, MD, United States,* [2]*Department of Psychiatry and Behavioral Sciences, Johns Hopkins School of Medicine, Kennedy Krieger Institute, Baltimore, MD, United States*

Introduction

Individuals with neurodevelopmental disorders (NDD) often experience multiple comorbid mental health diagnoses, which can significantly impact their educational and vocational engagement and lead to increased social isolation [1]. The presence of a developmental disability does not exclude an individual from having a psychiatric disorder, but NDD actually increases the risk, with prevalence rates ranging from 30% to 50% [2]. The prevalence of mental health disorders in children and adolescents with NDD varies between 10% and 60%, with a relative risk ranging from 2.8 to 4.5, depending on the studied population [2,3].

Dual diagnosis, which refers to the cooccurrence of psychiatric illnesses, mental disorders, and intellectual disability (ID) or NDD, is associated with earlier and more frequent hospitalizations compared to individuals diagnosed solely with a psychological disorder [4]. Psychiatric disorders occur three times more frequently in children and adolescents with ID compared to those with typical development [5,6]. Individuals with a dual diagnosis experience greater levels of disability and higher mortality rates compared to those with psychiatric illness alone [4]. Only one in ten children and adolescents with dual diagnoses receives specialized mental health services, and a lack of social support is associated with a higher likelihood of mental illness [7].

In individuals with NDD, the presentation of psychopathology may vary greatly depending on factors such as age, NDD diagnosis, severity of the disorder, whether ID is also present, and available supports. For example, children with Down syndrome are less likely to have cooccurring mental health disorders, while anxiety in individuals with severe degrees of ID may manifest as hypervigilance or imaginary friends [8]. Limited language abilities in individuals with NDD, or language disorders, can hinder the communication of symptoms related to anxiety, mood, psychosis, and other phenomena, making diagnosis challenging. Emotional responses and maladaptive behaviors are common in individuals with NDD, particularly during times of stress, and determining the underlying cause can be difficult for caregivers and clinicians [2].

The diagnosis of mental health disorders in individuals with NDD can be further complicated by baseline maladaptive behavior and the failure to recognize changes in frequency, intensity, or form of behavior. This delay in diagnosis, known as diagnostic overshadowing, can impede effective treatment [8]. Adjustment disorders, which result from a mismatch between the individual's capacity and the demands of the situation, are common and can be associated with anxiety or mood symptoms. Stress plays a significant role in mental illness and behavior problems in individuals with NDD [7]. Attention deficit hyperactivity disorder (ADHD) often coexists with other NDDs but may be challenging to diagnose due to overlapping symptoms with autism spectrum disorder (ASD) and difficulties in quantifying inappropriate attention or activity levels in children with ID or cerebral palsy [2].

Certain behavioral disorders, such as self-injury in individuals with severe to profound ID, are more frequently observed in individuals with NDD than in the general population. Very often, self-injury is for attention or for communication functions. Functional communication programs and augmentative communication devices can help reduce self-injury by providing a more effective means of requesting attention [9]. Psychosis is also seen but is difficult to diagnose in individuals with severe ID, while anxiety and mood disorders are common [2].

Capute and Accardo's Neurodevelopmental Disabilities in Infancy and Childhood. DOI: https://doi.org/10.1016/B978-0-12-824060-1.00005-5

To assist clinicians in diagnosing psychiatric disorders in individuals with ID, the National Association for the Dually Diagnosed, in collaboration with the American Psychiatric Association, published the Diagnostic Manual-Intellectual Disability (DM-ID) in 2008 and a revised version, DM-ID-2, in 2017. These manuals closely correspond to the DSM-5 classification system, providing adapted diagnostic criteria for individuals with severe to profound ID and additional information to aid in diagnosis [10,11]. Diagnosis is based on the presence of symptoms that cause emotional distress or interfere with daily activities (school, work, and social life) [2].

Psychiatric evaluation of individuals with NDD

A psychiatric evaluation consists of a clinical interview of the individual that involves an assessment of mental status, medical history, psychosocial environment, family history, and educational history. Often, collateral information is obtained from family members and caregivers. The interview should be tailored to take into consideration the communication and intellectual abilities of the individual while respecting their autonomy.

When communicating with patients, it is important to follow certain guidelines to ensure effective and respectful interaction. Always address the patient directly, acknowledging their presence and involvement in the conversation. Assume that the patient can hear and understand what you are saying, even if they have communication difficulties.

To establish a rapport, greet the patient by asking how they prefer to be called and use that name throughout the conversation. Additionally, familiarize yourself with any clinical information and be aware of important communication barriers, such as the need for an interpreter or any hearing or visual impairments. Also, take note if the patient utilizes alternative forms of nonverbal communication, such as the Picture Exchange Communication System (PECS) or an electronic device.

In cases where self-reporting is not possible, it is crucial to gather collateral information from sources like school, caregivers, and parents to determine the presence of symptoms. When communicating, use language that is simple, clear, and appropriate for the patient's developmental age. Avoid using complicated medical terms that may confuse or overwhelm the patient or his/her caregivers.

Allow the patient sufficient time to respond to questions and avoid rushing them. If the individual appears confused or is not fully engaged in the interview, consider repeating the question in its entirety or simplifying the vocabulary to their developmental level to aid comprehension. Emphasize the difference between responses in a clear and simplified manner, keeping in mind that there is a tendency for patients to answer "yes" to leading questions.

Visual techniques can be helpful in communication. For example, you can use markings to indicate the number of days or symptoms on a visual timeline. Additionally, incorporating significant markers such as the patient's birthday or major holidays they celebrate can aid in understanding and recalling information. To ensure understanding, periodically confirm the patient's comprehension and, if necessary, ask them to repeat the question back to you in his/her own language.

It is also important to assess for any physical problems that may be causing pain or distress for the patient. Finally, modify your language and approach to be developmental age-appropriate.

Screening measures developed with a typically developing sample may not be valid for individuals with NDD (the questions may not be able to elicit behavioral symptoms or internalizing feelings).

Treatment decision-making should be a collaborative process wherever possible, involving the patient, other health professionals, and a parent or legal guardian if they are involved. Factors specific to the individual patient should be taken into account in any subsequent therapeutic approach, including the acceptability of blood test monitoring and comorbid illness, among other factors. In the case of adults with intellectual impairment, there needs to be a clear process for informed consent and the presence of a surrogate decision-maker if appropriate. Mental health laws might vary depending on the country and state of residence.

The following section summarizes mood and anxiety disorders in individuals with NDD, with a special focus on unique aspects of diagnosis and management.

Mood disorders in NDD

The prevalence of bipolar disorder in the general population has been estimated to be between 3% and 5% of the population when both bipolar disorder types 1 and 2 are included [12]. In the typically developing population, younger individuals with bipolar disorder present most commonly with a mixed-state presentation (mania and depression symptoms are present at the same time or within a short amount of time) and will often have a more chronic course and more severe presentation.

Individuals diagnosed with moderate to profound ID are more likely to be diagnosed with bipolar disorder than any other psychiatric disorder [13], and bipolar disorder is diagnosed twice as frequently in individuals who have ID [14]. Agitation and irritability are symptoms that are found in mania and may also present in individuals with NDD, and individuals who have been diagnosed with NDD plus bipolar disorder frequently have aggression [15].

History/diagnosis/epidemiology

In individuals with NDD, the clinical presentation of mood disorders varies depending on the level of intellectual disability (ID). Diagnostic criteria used for the general population may have clinical utility for those with milder ID, though ID-specific classification systems, such as the DS-ID-2, may be more useful for those with moderate-severe ID [1].

People with NDD may be prone to difficulties with mood regulation due to poorly developed socialization and coping skills. Environmental, medical conditions, and medications can cause mood outbursts [2]. Factors that predicted depression in individuals with NDD predicted life events and problem behaviors [16]. Children with NDD are at increased risk of depression later in life [12].

Diferential diagnosis

For individuals with NDD, the process of diagnosing depression can be challenging. It is crucial to carefully consider and exclude clinical confounders that may lead to pain or distress, as these factors can contribute to a phenotype that mimics depressive symptoms. Physical pain or distress can have a detrimental effect on focused attention, sleep patterns, eating habits, and even contribute to feelings of agitation and irritability.

To assist clinicians in navigating these challenges, the DM-ID-2 guidelines for diagnosing a major depressive episode provide valuable information. This guidance aims to help clinicians account for potential clinical confounders related to physical pain or distress that may influence the presentation of depressive symptoms in individuals with ID. Understanding the specific patterns and presentations of the range of depressive disorders in the ID population is essential for accurate diagnosis and effective intervention.

Table 37.1 lists the different depressive disorders included in the DM-ID-2 and their distinct diagnostic criteria in individuals with ID.

TABLE 37.1 Different depressive disorders included in diagnostic manual-intellectual disability-2.

Depressive disorder	Description: distinct diagnostic criteria in individuals with ID
Major depressive disorder	• 4 or more symptoms were present during the same 2-week period, representing a change from previous functioning. • At least one of the symptoms is either (1) depressed mood, (2) loss of interest or pleasure, or (3) irritable mood.
Disruptive mood dysregulation disorder	• Severe recurrent temper outbursts that are out of proportion, intensity, or duration to the situation. • Inconsistent with developmental level and occur on average three or more times per week. • Persistently irritable or angry mood most of the day, between temper outbursts. • Symptoms present for 12 or more months. • Onset of the disorder before 10 years of age.
Persistent depressive disorder (dysthymia)	• A depressed mood present most of the day, more days than not, for two years. • In youths, mood can be irritable, with a duration of at least one year. • Two or more physical symptoms present, such as abnormal appetite, sleep, energy, self-esteem, concentration, or hopelessness.
Substance/medication-induced depressive disorder	• Persistent depressed mood that developed during or after substance intoxication or withdrawal, or after exposure to a medication.
Premenstrual dysphoric disorder	• Presence of at least five symptoms in most menstrual cycles during the week before the onset of menses. • Symptoms start to improve within a few days after the onset of menses and become minimal or absent in the week postmenses.

(Continued)

TABLE 37.1 (Continued)

Depressive disorder	Description: distinct diagnostic criteria in individuals with ID
	• Symptoms include affective lability, irritability, or anger; increased interpersonal conflicts; a depressed mood; feelings of hopelessness or self-deprecating thoughts; anger; tension; or feelings of being keyed up or on the edge. • Symptoms cause significant distress or interference with work, school, or social activities.
Bipolar I disorder	• Episodes of elevated mood for a minimum of seven days, occurring most of the day. • Depressive episodes last two weeks or longer. • Mania symptoms may include abnormalities in self-esteem, sleep, talkativeness, racing thoughts, distractibility, increased goal-directed activity or psychomotor agitation, engaging in risky activities, an abnormal appetite, and increased energy. • The depressive component includes symptoms of depressed mood, loss of interest or pleasure, changes in weight or activities, concentration difficulties, decreased energy, and feelings of hopelessness, worthlessness, guilt, and suicidal thoughts.
Bipolar II disorder	• Depressive and hypomanic episodes that alternate. • Hypomania, a mood lasting at least four consecutive days, is present most of the day, almost every day. • Episodes are typically less severe and do not inhibit function.
Cyclothymic disorder	• Cyclic disorder is characterized by brief episodes of hypomania and depression. • Symptoms lasting at least two years in adults and one year in children and adolescents.
Depressive disorder due to another medical condition	• Depression occurs as a result of another medical condition, such as hypothyroidism.

Screening tools

Table 37.2 lists screening tools that are used commonly in depression and other mood disorders.

TABLE 37.2 Commonly used screening tools to assess depression mood and anxiety disorders.

Tool	Domain	Age	Availability
Beck depression inventory 2 (BDI 2)	Behavioral, cognitive, and somatic components of depression focus on negative attitudes toward oneself.	13 years and older	Available for purchase: http://www.pearsonassessments.com
Mood Disorders Questionnaire (MDQ)	Identifies current or prior mania/hypomania (bipolar disorder).	12 years. and older	In public domain; https://www.ohsu.edu/sites/default/files/2019-06/cms-quality-bipolar_disorder_mdq_screener.pdf
Child depression inventory (CDI)	The CDI has two forms: the original 27-item version and the 10-item short-form version, which take between 5 and 15 minutes for the child to complete. The CDI is designed to detect symptoms of depression and to distinguish between depression and other psychiatric disorders. It can also be used as an instrument to monitor changes in depression symptoms over time.	7–17 years.	Available for purchase: http://www.pearsonassessments.com
Nine item depression scale of the patient health questionnaire. (PHQ-9)	Focus on symptoms and functional impairment.	12 years. and older	In public domain; https://aims.uw.edu/resource-library/phq-9-depression-scale

(Continued)

TABLE 37.2 (Continued)

Tool	Domain	Age	Availability
Patient health questionnaire modified for adolescents. (PHQ-A)	Focus on symptoms and functional impairment.	11–17 years.	In public domain; http://www.phqscreeners.com/instructions/instructions.pdf.
Pediatric symptom checklist	Identified anxiety, depression, and ADHD symptoms.	Proxy 7 and up Self-report 11 and older	In public domain
Screen for child anxiety related disorders (SCARED)	Parent and child version.	8–18 years.	In public domain
Reiss scales for children's dual diagnosis (RSCDD)	Empirically driven and developed from the DSM-III-R taxonomy, 60 items are distributed on 10 subscales: anxiety disorder; anger/self-control; attention deficit; autism/PDD; conduct disorder; depression; poor self-esteem; psychosis; somatoform behaviors; withdrawn/isolated; total score. Provides information on how scores correspond to DSM-IV diagnosis when appropriate.	4–21 years.	Available for purchase: http://www.idspublishing.com
Developmental behavior checklist (DBC)	96 items are distributed on 5 subscales total score; disruptive/antisocial; self-absorbed; communication disturbance; anxiety; social relating; and depressive (adults only) symptom cluster information available to calculate depression and attention deficit/hyperactivity scores for the parent version. Separate versions for parents and teachers Short versions are available; a daily monitoring version available good psychometric properties and empirical support.	Child and adult version	Available for purchase: https://www.monash.edu
Nisonger child behavior rating form (NCBRF)	10 social competence items are distributed on two subscales. 66 problem behavior items were distributed over six subscales. parent and teacher versions good psychometric properties but not as much empirical support as the DB.	3–16 year-old intellectual or developmental disabilities and normally developing children	In public domain: https://nisonger.osu.edu/nisonger-child-behavior-rating-form/
Aberrant behavior checklist (ABC)	56 caregiver completed the survey. Subtests: (I) Irritability, Agitation, Crying; (II) Lethargy, Social Withdrawal; (III) Stereotypic Behavior; (IV) Hyperactivity, Noncompliance; and (V)	Full range of ID (particularly severe to profound) Separately normed community and residential versions	Available for purchase: http://www.slosson.com

Interventions

Behavioral

Psychological and behavioral therapy can help people with NDD identify the challenges they face and help them deal with the day-to-day conflicts that could be contributing to their difficulties coping with anxiety and mood disorder symptoms. For example, the therapist can help individuals with NDD identify activities that decrease stress and improve mood, and then this can be incorporated into their daily routine. Therapists can also educate the people caring for individuals with NDD about mood disorders and how to effectively help them.

Cognitive behavioral therapy (CBT) is a form of psychotherapy that has been found to decrease depressive thoughts in individuals with mild to moderate depressive disorders. CBT focuses on education about mood and self-monitoring to identify thoughts that result in behavioral responses. CBT is supported by empirical evidence and is established as the gold standard psychological treatment for addressing anxiety and mood disorders in typically developing youths and adults. There have been several attempts to modify and adapt CBT to cater to special populations. These adaptations involve using developmentally appropriate and simplified language, employing flexible methods, adopting a directive style, including developmentally appropriate activities, and including caregivers [17,18]. CBT has been studied in individuals with mild ID who have the skills to undergo adapted CBT. However, studies in individuals with ASD are limited to individuals with higher functioning [19], as the evidence of effectiveness on moderate and severe ID remains unclear [20].

Applied behavioral analysis (ABA) is a therapy based on the science of learning and behavior. It uses a scientific approach to understand behavior by assessing how environmental changes affect behavior. ABA therapy can introduce relevant skills to help children make cause-and-affect associations and provide immediate rewards upon mastering a task. Social skills in daily living can be taught this way and mean greater independence for the individual and decreased problematic behavior [21], possibly due to decreased stress and improved mood.

Psychopharmacology

The American Academy of Child and Adolescent Psychiatry has published A Practice Parameter for the Assessment and Treatment of Psychiatric Disorders in Children and Adolescents with Intellectual Disability (Intellectual Developmental Disorder), which is available for download at https://www.jaacap.org/article/S0890-8567(19)32223-3/fulltext [22]. It summarizes the randomized, controlled trials of individuals with ID that have occurred since 1999. Due to the difficulty in defining psychiatric disorders in individuals with ID, the criteria used for diagnostic purposes for entry into one study might not be the same as for another study that is addressing a similar symptom profile.

Antidepressants: There are multiple classes of medications that are used for the treatment of depressive disorders (see Table 37.3). The selective serotonin reuptake inhibitor class of antidepressants (SSRIs), such as fluoxetine, are usually the first choice of antidepressant due to their low incidence of side effects in comparison with older antidepressants, and SSRIs are the most commonly used antidepressant class by individuals with ID [23]. Antidepressants, particularly SSRIs, showed improvement of aggression and self-injury behavior on average in less than 50% of cases, with the remainder of the individual showing no improvement or deterioration [24].

TABLE 37.3 Commonly used medications to treat mood and anxiety disorders.

	Pediatric		Adult	
Medications	Typical dose range (mg/d)	FDA indications	Typical dose range (mg/d)	FDA indications
Selective serotonin reuptake inhibitors				
Citalopram	20–40	None	40	MDD
Escitalopram	10–40	MDD 12 +	20–40	MDD, GAD
Fluoxetine	20–80	MDD 8 + , OCD 6 +	20–80	MDD, OCD, PD
Fluvoxamine	50–300	OCD 8 +	100–300	OCD
Paroxetine	20–60	None	40–60	MDD,OCD, GAD, SAD,PD
Sertraline	100–200	OCD 6 +	150–250	MDD,OCD,SAD,PD
Serotonin-norepinephrine reuptake inhibitors				
Venlafaxine	150–225	None	75–375	MDD,GAD,SAD,PD
Duloxetine	40–60	GAD 7 +	20–80	MDD,GAD
Desvenlafaxine	25–100	None	50–400	MDD
Atypical antidepressants				
Bupropion	150–300	GAD 6 +	150–300	MDD
Duloxetine	30–60	GAD 7 +	30–120	MDD,GAD

(Continued)

TABLE 37.3 (Continued)

	Pediatric		Adult	
Medications	**Typical dose range (mg/d)**	**FDA indications**	**Typical dose range (mg/d)**	**FDA indications**
Mirtazapine	15–45	None	15–45	MDD
Vilazodone	–	None	10–40	MDD
Vortioxetine	5–20	None	10–80	MDD
Buspirone	5–60	None	25–60	GAD
Tricyclic antidepressants				
Clomipramine	50–200	OCD 10 +	100–250	OCD
Desipramine	50–200	None	150–300	MDD
Nortriptyline	30–150	None	75–150	MDD
Imipramine	50–150	Enuresis 6 +	150–300	MDD

MDD, Major depressive disorder; *GAD*, gen. anxiety disorder; *OCD*, obsessive-compulsive disorder; *PD*, panic disorder; *SAD*, seasonal affective disorder.

Treatment of individuals with NDD who have bipolar disorder is approached in the same way as for children without NDD, namely, the use of mood stabilizers (i.e., valproic acid, lithium, etc.). See Table 37.4. At times, antipsychotics (neuroleptics) are used for the treatment of bipolar disorder, either as a sole agent or as an adjuvant to a mood stabilizer (see Table 37.5). Newer atypical antipsychotics are generally preferred over older, first-generation antipsychotics because of the possible increased sensitivity to extrapyramidal symptoms in the ID population. Few antipsychotics have FDA indication for treatment of mood disorders in typically developing children, and only two antipsychotics have FDA indication for the treatment of a NDD disorder, risperidone and aripiprazole, both of which were shown to improve symptoms of irritability in individuals with autism spectrum disorder [25–27].

TABLE 37.4 Commonly used medications to treat anxiety and mood disorders (Benzodiazepines and mood stabilizers).

	Pediatric		adults	
Benzodiazepines	**Typical dose range (mg/d)**	**FDA**	**Typical dose range (mg/d)**	**FDA indications**
Alprazolam	0.25–8	None	0.25–8	GAD, PD
Clonazepam	0.25–6	None	0.25–6	PD
Diazepam	2–5	None	2–10	GAD, PD
Lorazepam	0.25–3	12 +	0.25–8	GAD
Mood Stabilizers				
Carbamazepine	Blood level 4–12	None	Blood level 4–12	BP
Divalproex/Valproate	Blood level 75–125	None	Blood level 7.5–125	BP
Oxcarbazepine	600–2400	None	600–2400	None
Lithium	Blood level 0.8–1.2	BP 12 +	Blood level 0.8–1.2	BP
Lamotrigine	By weight	None	By weight	None

GAD, Gen. anxiety disorder; *PD*, panic disorder; *BP*, bipolar disorder.

TABLE 37.5 Commonly used antipsychotic (neuroleptic) medications.

	Pediatric		adults	
Medications	**Typical dose range (mg/d)**	**FDA indications**	**Typical dose range (mg/d)**	**FDA indications**
Atypical Antipsychotics				
Aripiprazole	1–20	BP 10 + Tourette 6 +	2–30	BP
Asenapine	5–20	BP 10 +	5–20	BP
Cariprazine	3–6	None	3–6	BP
Lurasidone	40–160	BP 10 +	40–160	
Olanzapine	1.5–20	BP 13 +	1.5–20	BP
Quetiapine	25–800	BP 10 +	25–800	BP
Risperidone	0.25–8	BP 10 +	2–16	BP
Ziprasidone	20–160		20–160	
Typical antipsychotics				
Clozapine	300–900	None	300–900	BP
Haloperidol	1–3	Tourette 3 +	1–3	Tourette
Loxapine	1–3	None	1–3	BP
Brexpiprazole	2–4	None	2–4	BP
Iloperidone	2–24	None	12–24	None

BP, Bipolar disorder; *Tourette*, Tourette disorder.

Skill generalization or maintenance of improvement in mood disorders and other psychiatric symptoms across time or settings. Once medication is initiated for someone with NDD, the clinician should monitor the response. The most commonly used outcome measure in psychopharmacology studies in NDD is the Aberrant Behavior Checklist-Community, with subscales on hyperactivity, social withdrawal, lethargy, stereotypy, and irritability [28].

Complementary and alternative medicine approaches

Various approaches have been explored, including alternative and complementary therapies, to manage mood disorders in individuals with NDD. Table 37.6 summarizes current evidence for commonly used approaches.

TABLE 37.6 The evidence behind commonly used complementary and alternative approaches used to manage mood disorders in individuals with NDD.

Approach	Description and supporting evidence
Communication interventions	A metaanalysis of 44 studies of alternative and assistive communication interventions for challenging behavior in children, adolescents, and adults demonstrated decreases in the levels of problem behavior and increases in the levels of alternative communicative responses [29].
Meditation	Mindfulness-informed interventions have decreased depressive symptoms as well as aggressive behaviors in individuals with ID. A study of children who had a 20-minute yoga practice found a decrease in disruptive behavior and improved social adaptive abilities.
Acupuncture	There is weak evidence for treating major depression, with no definitive studies in the NDD population.

(*Continued*)

TABLE 37.6 (Continued)

Approach	Description and supporting evidence
Exercise	Demonstrated good effect for the treatment of mild depressive disorders, but there is no evidence that there are effective treatments for moderate to severe depressive disorders. However, exercise is routinely recommended to lessen stress, elevate mood as much as possible, and decrease the potential metabolic side effects that frequently occur with psychiatric medications.
Cannabis	Safety studies have been completed, showing variability in the efficacy, safety, and tolerability of medical cannabis administration. It is known that acute administration of THC activates stress-response systems, and tolerance can develop. Chronic use might be associated with a blunted stress response, but there are no randomized control studies on the use of cannabis with anxiety disorder in individuals with NDD. As such, the efficacy or safety of such treatment is not advised for the treatment of anxiety in this population [30]. Furthermore, the use of cannabis is an independent risk factor for developing schizophrenia [31].
Cannabidiol (CBD)	CBD is an FDA-approved agent for the treatment of epilepsy. No clinical trials have been completed on its use as a treatment for depressive disorders in humans, and it is not effective in animal models of depression. CBD might have an antistress effect, which can reduce depressive symptoms related to stress. CBD can affect the levels of other medications by changing P450 metabolism.
St. John's Wort, Ginkgo Biloba, and Kava-Kava	Not found to be an effective treatment for depressive disorders in NDD.

Anxiety disorders

History/diagnosis/epidemiology

Anxiety disorders include generalized anxiety disorder, specific phobia, agoraphobia, social anxiety disorder, separation anxiety disorder, panic disorder, obsessive-compulsive disorder (OCD), trauma-related disorders, and selective mutism. Symptoms of diagnoses are in multiple disorders in both the DSM-5 and the DS-ID-2. When considering an anxiety diagnosis, a complete review of symptoms should be obtained.

Diferential diagnosis

Anxiety disorders are seen commonly in individuals with NDD. The lifetime prevalence of anxiety disorders in the general population is 29%, and the prevalence in adolescents is 1.8% [32]. Anxiety disorders occur in individuals with NDD as often as in the general population. Studies have shown that an anxiety disorder diagnosis was present in approximately 15%–20% of individuals with NDD [33], which is similar to the general population [34]. The median age of onset for anxiety disorders in the typical population is 11 years [32]. Anxiety disorder symptoms can overlap and make the diagnostic process complicated.

Specific anxiety disorders in individuals with NDD are more difficult to identify than in the general population due to a lack of self-reporting. However, anxiety symptoms might be inferred from fear and avoidance behaviors. Fear responses are a part of evolution and are protective for the individual. They involve physical responses triggered by neurotransmitters and are shaped by life experiences [10]. Fear responses are not considered pathological unless the response is out of proportion to the stimulus and leads to avoidance, distress, or impairment in daily functioning that interferes with quality. Individuals might not be aware of the triggers for their anxiety but still physically and emotionally feel distress.

The course of anxiety tends to be chronic and potentially fluctuating in intensity. A careful longitudinal evaluation of variation in expression, severity, and response to treatment is important for successful management.

Alongside anxiety disorders, consideration should be given to adjustment disorders that might result from environmental stressors like bullying, change in routines, or the environment.

The transition from DSM-IV to DSM-5 removed OCD, hoarding disorder and posttraumatic stress disorder (PTSD) from the anxiety disorders group. However, it is important to note that those three diagnoses include anxiety-related features and should be considered when assessing anxiety symptoms in individuals with NDD.

OCD often involves the presence of obsessions, compulsions, or both. Obsessions are defined as thoughts, urges, or images that are intrusive, unwanted, and cause distress, while compulsions are repetitive behaviors or mental acts that the individual is driven to perform with the goal of reducing anxiety or distress. It is important to differentiate compulsions from the restricted, repetitive patterns of behavior, interests, or activities that are part of ASD. Restricted behaviors are often driven by sensory experiences, focused on specific topics, and tend to prefer rigid adherence to established routines. Disruptions of routines, however, may lead to anxiety.

PTSD involves recurrent, involuntary, and intrusive memories, dreams, and dissociative reactions with marked physical reactions to internal or external cues, avoidance of stimuli associated with the traumatic event, and a negative alternation in cognition and mood. PTSD is associated with exposure to actual or threatened death, injury, or sexual violence. Identifying traumatic events is key to distinguishing PTSD from anxiety disorder.

The diagnosis of anxiety should consider the possibility of any comorbid medical conditions (hyperthyroidism, substance abuse disorder, atrial fibrillation, etc.), the evaluation of any other medications being administered (steroids, stimulant medications, etc.), and any other substance that might increase heart rate (caffeine, amphetamines) that can be contributing to the clinical presentation.

Self-reported symptoms of anxiety can manifest as worries and/or fear of an identifiable trigger, though individuals with NDD might not be able to identify a specific trigger. Anticipatory anxiety, characterized by thoughts like ("what if"), can also be experienced.

Observed symptoms of anxiety or fear can manifest as crying episodes, avoidance, tantrums, failing to speak (selective mutism), restlessness, irritability, muscle tension, and sleep disturbances. These symptoms may lead to disruptive behaviors and aggression, causing disruptions to the individual's routine, school refusal, or other maladaptive behaviors. These externalizing behaviors are usually what trigger medical evaluation and can exhibit symptoms that overlap with mood disorders, emphasizing the imperativeness of a careful and thorough review of systems.

Screening measures for anxiety

See Table 37.2.

Interventions

Behavioral

Cognitive behavioral therapy (CBT) in the context of anxiety focuses on educating the individuals about anxiety symptoms, self-monitoring to identify triggering thoughts (worries, beliefs) leading to behaviors (avoidance, fear), and incorporating relaxation techniques and guided imagery, as well as cognitive restructuring strategies to change the initial responses or reactions. It is assumed that the individual who participates in CBT has the skills necessary to understand and implement these concepts to address anxiety symptoms (including understanding language and identifying emotions). For younger children, the focus of CBT shifts towards family interventions [22].

ABA therapy can introduce relevant skills to help children make cause-and-affect associations and provide immediate rewards upon mastering a task. Social skills in daily living can be taught this way and mean greater independence for the individual and decreased problematic behavior [21], possibly due to decreased stress and anxiety.

Psychopharmacology

Individuals with NDD might be more sensitive to adverse effects of medications or might not be able to report them accurately. It is important to have close monitoring and thoughtful discussion with the individual and caregivers to discuss the risk versus benefit of the use of any psychopharmacological interventions.

There are minimal studies to address the treatment approach to anxiety in individuals with NDD. Selective serotonin reuptake inhibitors (SSRIs) continue to be the treatment of choice because of their evidence of efficacy in the general population for individuals with NDD. However, buspirone is the treatment of choice for anxiety in individuals with ASD [35]. Serotonin-norepinephrine reuptake inhibitors (SNRIs) have also been recommended for social anxiety, generalized anxiety disorder (GAD), separation anxiety, and panic disorder. SNRIs have shown some empirical results as an additional option when failing an SSRI trial. SNRI is usually a second-line if the first-line use is contraindicated, ineffective, or cannot be tolerated due to adverse effects (see Table 37.3). At times, benzodiazepines are used in the treatment of anxiety (see Table 37.4).

Other agents have shown some empiric results in the treatment of anxiety, including alpha-agonists and B-blockers. But there are no clinical trials investigating the use of these medications for this indication in children with NDD.

Complementary and alternative medicine

Table 37.7 summarizes the current evidence in the treatment of anxiety disorders in NDD using complementary and alternative medicine approaches.

TABLE 37.7 The evidence behind commonly used complementary and alternative medicine approaches to manage anxiety disorders in NDD.

Approach	Description and supporting evidence
Meditation and yoga	Meditation can decrease anxiety symptoms and improve sleep.
Acupuncture	There are no definitive studies in the NDD population. In some studies, individuals with ASD show some improvement in overall (not specific to anxiety) behavior, but acupuncture was better than behavioral therapy. A combination of acupuncture plus behavioral therapy was better than either treatment individually [36].
Musical therapy	Music as an intervention to reduce procedural anxiety in children [37,38].
Exercise	Demonstrated medium effect in anxiety disorders.
Cannabis	Safety studies have been completed, showing variability in the efficacy, safety, and tolerability of medical cannabis administration. It is known that acute administration of THC activates stress-response systems, and tolerance can develop. Chronic use might be associated with a blunted stress response, but there is no RCT on the use of cannabis with anxiety disorder in individuals with NDD. As such, the efficacy or safety of such treatment is not advised for the treatment of anxiety in this population [30]. Furthermore, the use of cannabis is an independent risk factor for developing schizophrenia [31].
Cannabidiol (CBD)	No studies have been completed on its use as a treatment for anxiety in NDD.

There is a clear need for further research in anxiety disorders in the context of NDD, including well-designed randomized controlled studies that take into consideration the unique features of anxiety in NDD. Such research endeavors should aim to clarify the specific skill sets required for individuals to actively participate in and comprehend the concepts of cognitive behavioral therapy (CBT). Moreover, they should emphasize the importance of the involvement of competent therapists who possess the necessary skills and techniques to effectively assess and manage anxiety disorders in individuals with NDD.

Behavioral phenotypes of selected genetic syndromes and other NDDs

There are distinct behavioral phenotypes that have been identified in certain genetic and neurodevelopmental disorders. These phenotypes, although not exclusive to the disorder, show a range of distinctive behaviors commonly observed in individuals with specific genetic syndromes, or NDD.

The underlying biological mechanisms of the disease contribute to the phenotypic expression of the cognitive and psychiatric features. As discussed in chapter Biological basis of NDD), there is a complex interplay between genetic mutations impaired plasticity mechanisms that hinder normal brain maturation and development and compromise brain function.

Table 37.8 summarizes common phenotypic behaviors observed in specific genetic syndromes and NDD. The information is compiled from the American Academy of Child and Adolescent Psychiatry Practice Parameter for the Assessment and Treatment of Psychiatric Disorders in Children and Adolescents with Intellectual Disability [22]. These phenotypic behaviors can be considered potential symptoms of psychiatric disorders, and a corresponding psychiatric diagnosis may be made if these symptoms impair function or cause significant distress.

TABLE 37.8 Common phenotypic behaviors observed in specific genetic syndromes and NDD.

	Cognitive features	Psychiatric features	Specific strategies (diagnosis/ management)
Down syndrome (trisomy 21)	Mild-to-severe intellectual disability. A strength is grammar, weakness is expressive language. Visual processing is better than auditory processing.	Greater than 50% of individuals have psychiatric features of hyperactivity, impulsiveness, inattention, and stubbornness; 30% have anxiety, 30% have depression, and 37% meet criteria for diagnosis of ASD [39].	See information on the National Down Syndrome Society webpage https://ndss.org/resources/mental-health-down-syndrome and Ref. [40].

(Continued)

TABLE 37.8 (Continued)

	Cognitive features	Psychiatric features	Specific strategies (diagnosis/ management)
Fragile X syndrome	Mild-to-severe intellectual disability. They have difficulty with abstract thinking, sequential cognitive processing, short-term memory, mathematics, and visual motor processing.	Psychiatric features include attention dysfunction, hyperarousal, social anxiety, social cognition, and communication challenges; 25% use sign language, and 25% have ASD.	See GeneReviews' *FMR1 Disorders* https://www.ncbi.nlm.nih.gov/books/NBK1384/ and Refs. [41], [42], and [43].
Rett syndrome	Severe to profound intellectual disability, limited language acquisition and use.	Psychiatric features: Stage I: Decreased interactions. Stage II: Social withdrawal, irritability, autistic-like behaviors, sleep disturbance. Stage III: Improvement in alertness, interactions, ongoing sleep disturbance. Stage IV: Persistence of poor communication, irritability.	See the *Consensus Guidelines on Managing Rett Syndrome across the Lifespan* [44] and the International Rett Syndrome Foundation for information on cognitive assessment https://www.rettsyndrome.org/nec-soleat-concludaturque/.
Prader-Willi syndrome (15q11Lq13)	Mild to borderline intellectual disability	Extreme hyperphagia, self-injurious behavior (skin picking), obsessive-compulsive disorder, social cognition deficits, cognitive inflexibility with poor affect regulation and explosiveness, depression/mood disorders and psychosis.	See the *Foundation for Prader-Willi Research Mental Health Guidebook* at https://www.fpwr.org/pws-mental-health-guidebook and Ref. [45].
Angelman syndrome	Severe to profound intellectual disability with minimal expressive language and better receptive language.	They are social and happy, frequent inappropriate and unexpected laughter, little stranger anxiety; fear of crowds and noise, hyperactivity and inattention, and sleep disturbance.	See GeneReviews' *Angelman Syndrome* https://www.ncbi.nlm.nih.gov/books/NBK1144/ and Ref. [46].
Williams syndrome	Mild to moderate intellectual disability; 75% strengths in auditory rote memory and language but severe deficits in visual-spatial construction. Adaptive behavior is less than expected for their IQ.	Superficial sociability and can have temper tantrums. At times they have obsessions, preoccupations, fears, anxiety sadness and depression: 50% are diagnosed with ADHD and they frequently have sleep disturbance.	See GeneReviews' *Williams Syndrome* https://www.ncbi.nlm.nih.gov/books/NBK1249/.
Deletion 22q11.2	Borderline to mild intellectual disability, strength is language abilities and weakness is receptive and high order language skills with abstract reasoning and visual-spatial deficits.	Emotional dysregulation, ADHD, anxiety and phobia, poor social adaptation with withdrawal autism, 30% have psychotic symptoms.	See GeneReviews' *22q11.2 Deletion Syndrome* https://www.ncbi.nlm.nih.gov/books/NBK1523/.
Smith–Magenis syndrome	Mild-to-severe intellectual disability, weakness is short-term memory, visual motor coordination, sequencing and response speed. Receptive language is better than expressive language.	Sleep disturbance, self-injurious behavior, self-hug, onychotillo-mania, lick and flip melting objects, teeth grinding, body rocking and spinning. They demand much adult attention and have delayed empathic skills.	See GeneReviews' *Smith-Magenis Syndrome* https://www.ncbi.nlm.nih.gov/books/NBK1310/.
Lesch–Nyhan syndrome	Mild to moderate intellectual disability	Chronic, compulsive self-injurious behavior resulting in self-mutilation via biting, eye poking, fingernail pulling, psychogenic vomiting, arching, head snapping and head banging. The language pattern consists of repeated ambivalent statements with anxiety and vulgarity, and they have frequent compulsive aggression toward others with grabbing and pinching.	See GeneReviews' *HPRT1 Disorders* https://www.ncbi.nlm.nih.gov/books/NBK1149/.

(*Continued*)

TABLE 37.8 (Continued)

	Cognitive features	Psychiatric features	Specific strategies (diagnosis/ management)
Smith–Lemli–Opitz syndrome	Mild to profound intellectual disability, receptive language is better than expressive.	Self-injury occurs in 50% at some point in their life and 50% meet diagnostic criteria for autism [47].	See Refs. [47] and [48].
Sturge–Weber syndrome	Variable	Mood disorder diagnoses were in 31% of the individuals, disruptive behavior disorder and adjustment disorder and 25%. Substance related disorder was the most frequent disorder in adults (67%) [49].	See GeneReviews' *Sturge-Weber Syndrome* and Ref. [49].
Nail–patella syndrome	Average IQ	They have an increase in symptoms of attention deficit hyperactivity disorder and depressive disorders [50].	See the National Health Service https://www.nhs.uk/conditions/nail-patella-syndrome/ and Ref. [50].

Summary and conclusion

Recognizing the unique features of anxiety and mood disorders in individuals with NDD plays a crucial role in early recognition, diagnosis, and management. It is essential to adopt specific diagnostic criteria and customize behavioral interventions based on the cognitive and developmental level for successful management.

References

[1] Munir KM. The co-occurrence of mental disorders in children and adolescents with intellectual disability/intellectual developmental disorder. Curr Opin Psychiatry 2016;29(2):95–102.

[2] Noel J. Recognition and treatment of mood dysregulation in adults with intellectual disability. Ment Health Clin 2018;8(6):264–74.

[3] Koskentausta T, Iivanainen M, Almqvist F. Psychiatric disorders in children with intellectual disability. Nord J Psychiatry 2002.

[4] Morgan VA, Leonard H, Bourke J, Jablensky A. Psychiatric disorders in children with intellectual disability. Nord J Psychiatry 2002.

[5] Emerson E, Einfeld S. Emotional and behavioural difficulties in young children with and without developmental delay: a bi-national perspective. J Child Psychol Psychiatry 2010.

[6] Dekker MC, Koot HM. DSM-IV disorders in children with borderline to moderate intellectual disability. I: prevalence and impact. J Am Acad Child Adolesc Psychiatry 2003;42(8):915–22. Available from: https://doi.org/10.1097/01.CHI.0000046892.27264.1A.

[7] Scott HM, Havercamp SM. Mental health for people with intellectual disability: the impact of stress and social support. Am J Intellect Dev Disabil 2014.

[8] Reiss S, Levitan GW, Szyszko J. Emotional disturbance and mental retardation: diagnostic overshadowing. Am J Ment Defic 1982.

[9] Carr EG, Durand VM. Reducing behavior problems through functional communication training. J Appl Behav Anal 1985.

[10] Bailey NM. Diagnostic manual@ intellectual disability: a textbook of diagnosis of mental disorders in persons with intellectual disability (Dm ID). Advances in Mental Health and Learning Disabilities 2008;2:60–1.

[11] Fletcher RJ, Cooper SA, Lunsky Y. Diagnostic Manual – Intellectual Disability: A Clinical Guide for Diagnosis of Mental Disorders in Persons with Intellectual Disability. 2nd ed.; DM-ID-2 Kingston NY: NADD Press; 2017.

[12] Akiskal HS, Bourgeois ML, Angst J, Post R, Möller H, Hirschfeld R. Re-evaluating the prevalence of and diagnostic composition within the broad clinical spectrum of bipolar disorders. J Affect Disord 2000;59(Suppl 1):S5–s30.

[13] Hurley A, D F, Lam N. Patients with and without intellectual disability seeking outpatient psychiatric services: diagnoses and prescribing pattern. Journal of Intellectual Disability Research: JIDR 2003;47:39–50.

[14] Cain N, Davidson P, Burhan A, Andolsek M, Baxter J, Sullivan L, et al. Identifying bipolar disorder in individuals with intellectual disability. Journal of Intellectual Disability Research: JIDR 2003;47:31–8.

[15] Tsiouris JA, Kim SY, Brown WT, Cohen IL. Association of aggressive behaviours with psychiatric disorders, age, sex and degree of intellectual disability: a large-scale survey. J Intellect Disabil Res 2011;55(7):636–49.

[16] American Psychiatric Association. Diagnostic and statistical manual of mental disorders. Washington, D.C: American Psychiatric Publishing; 2013.

[17] Dagnan D, Jackson I, Eastlake L. A systematic review of cognitive behavioural therapy for anxiety in adults with intellectual disabilities. J Intellect Disabil Res 2018;62(11):974–91.

[18] Surley L, Dagnan D. A review of the frequency and nature of adaptations to cognitive behavioural therapy for adults with intellectual disabilities. J Appl Res Intellect Disabil 2019;32(2):219–37.

[19] Sukhodolsky DG, Bloch MH, Panza KE, Reichow B. Cognitive-behavioral therapy for anxiety in children with high-functioning autism: a meta-analysis. Pediatrics 2013;132(5):e1341–50.

[20] Hassiotis A, Serfaty M, Azam K, Strydom A, Blizard R, Romeo R, et al. Manualised individual cognitive behavioural therapy for mood disorders in people with mild to moderate intellectual disability: a feasibility randomised controlled trial. J Affect Disord 2013;151(1):186–95.

[21] Wachtel LE, Hagopian LP. Psychopharmacology and applied behavioral analysis: tandem treatment of severe problem behaviors in intellectual disability and a case series. Isr J Psychiatry Relat Sci 2006;43(4):265–74.
[22] Siegel M, McGuire K, Veenstra-VanderWeele J, Stratigos K, King B, Bellonci C, et al. Practice parameter for the assessment and treatment of psychiatric disorders in children and adolescents with intellectual disability (intellectual developmental disorder). J Am Acad Child Adolesc Psychiatry 2020;59(4):468–96.
[23] Cooper SA, Smiley E, Allan L, Morrison J. Incidence of unipolar and bipolar depression, and mania in adults with intellectual disabilities: prospective cohort study. Br J Psychiatry 2018;212(5):295–300.
[24] Sohanpal SK, Deb S, Thomas C, Soni R, Lenôtre L, Unwin G. The effectiveness of antidepressant medication in the management of behaviour problems in adults with intellectual disabilities: a systematic review. J Intellect Disabil Res 2007;51(Pt 10):750–65.
[25] Marcus RN, Owen R, Kamen L, Manos G, McQuade RD, Carson WH, et al. A placebo-controlled, fixed-dose study of aripiprazole in children and adolescents with irritability associated with autistic disorder. J Am Acad Child Adolesc Psychiatry 2009;48(11):1110–19.
[26] McCracken JT, McGough J, Shah B, Cronin P, Hong D, Aman MG, et al. Risperidone in children with autism and serious behavioral problems. N Engl J Med 2002;347(5):314–21.
[27] Owen R, Sikich L, Marcus RN, Corey-Lisle P, Manos G, McQuade RD, et al. Aripiprazole in the treatment of irritability in children and adolescents with autistic disorder. Pediatrics 2009;124(6):1533–40.
[28] Aman MG, Singh NN, Stewart AW, Field CJ. The aberrant behavior checklist: a behavior rating scale for the assessment of treatment effects. Am J Ment Defic 1985;89(5):485–91.
[29] Ganz JB, Earles-Vollrath TL, Heath AK, Parker RI, Rispoli MJ, Duran JB. A meta-analysis of single case research studies on aided augmentative and alternative communication systems with individuals with autism spectrum disorders. J Autism Dev Disord 2012;42(1):60–74.
[30] al'Absi M, Allen AM. Impact of acute and chronic cannabis use on stress response regulation: challenging the belief that cannabis is an effective method for coping. Front Psychol 2021;12:687106.
[31] Andreasson S, Allebeck P, Engstrom A, Rydberg U. Cannabis and schizophrenia. A longitudinal study of Swedish conscripts. Lancet 1987;2(8574):1483–6.
[32] Kessler RC, Berglund P, Demler O, Jin R, Merikangas KR, Walters EE. Lifetime prevalence and age-of-onset distributions of DSM-IV disorders in the national comorbidity survey replication. Arch Gen Psychiatry 2005;62(6):593–602.
[33] Ezell J, Hogan A, Fairchild A, Hills K, Klusek J, Abbeduto L, et al. Prevalence and predictors of anxiety disorders in adolescent and adult males with autism spectrum disorder and fragile X syndrome. J Autism Dev Disord 2019;49(3):1131–41.
[34] Munir S, Takov V. Generalized anxiety disorder. StatPearls. Treasure Island (FL) ineligible companies. Disclosure: Veronica Takov declares no relevant financial relationships with ineligible companies. StatPearls Publishing LLC; 2023.
[35] Ceranoglu TA, Wozniak J, Fried R, Galdo M, Hoskova B, DeLeon Fong M, et al. A retrospective chart review of buspirone for the treatment of anxiety in psychiatrically referred youth with high-functioning autism spectrum disorder. J Child Adolesc Psychopharmacol 2019;29(1):28–33.
[36] Ming X, Chen X, Wang XT, Zhang Z, Kang V, Zimmerman-Bier B. Acupuncture for treatment of autism spectrum disorders. Evid Based Complement Alternat Med 2012;2012:679845.
[37] Freitas C, Fernandez-Company JF, Pita MF, Garcia-Rodriguez M. Music therapy for adolescents with psychiatric disorders: an overview. Clin Child Psychol Psychiatry 2022;27(3):895–910.
[38] Geretsegger M, Fusar-Poli L, Elefant C, Mossler KA, Vitale G, Gold C. Music therapy for autistic people. Cochrane Database Syst Rev 2022;5(5):CD004381.
[39] Dimachkie Nunnally A, Nguyen V, Anglo C, Sterling A, Edgin J, Sherman S, et al. Symptoms of autism spectrum disorder in individuals with Down syndrome. Brain Sci 2021;11(10).
[40] Vicari S, Pontillo M, Armando M. Neurodevelopmental and psychiatric issues in Down's syndrome: assessment and intervention. Psychiatr Genet 2013;23(3):95–107.
[41] Wadell PM, Hagerman RJ, Hessl DR. Fragile X syndrome: psychiatric manifestations, assessment and emerging therapies. Curr Psychiatry Rev 2013;9(1):53–8.
[42] Hagerman RJ, Polussa J. Treatment of the psychiatric problems associated with fragile X syndrome. Curr Opin Psychiatry 2015;28(2):107–12.
[43] Hagerman RJ, Berry-Kravis E, Hazlett HC, Bailey Jr. DB, Moine H, Kooy RF, et al. Fragile X syndrome. Nat Rev Dis Primers 2017;3:17065.
[44] Fu C, Armstrong D, Marsh E, Lieberman D, Motil K, Witt R, et al. Consensus guidelines on managing Rett syndrome across the lifespan. BMJ Paediatr Open 2020;4(1):e000717.
[45] Skokauskas N, Sweeny E, Meehan J, Gallagher L. Mental health problems in children with prader-willi syndrome. J Can Acad Child Adolesc Psychiatry 2012;21(3):194–203.
[46] Pelc K, Cheron G, Dan B. Behavior and neuropsychiatric manifestations in Angelman syndrome. Neuropsychiatr Dis Treat 2008;4(3):577–84.
[47] Tierney E, Nwokoro NA, Porter FD, Freund LS, Ghuman JK, Kelley RI. Behavior phenotype in the RSH/Smith-Lemli-Opitz syndrome. Am J Med Genet 2001;98(2):191–200.
[48] Thurm A, Tierney E, Farmer C, Albert P, Joseph L, Swedo S, et al. Development, behavior, and biomarker characterization of Smith-Lemli-Opitz syndrome: an update. J Neurodev Disord 2016;8:12.
[49] Turin E, Grados MA, Tierney E, Ferenc LM, Zabel A, Comi AM. Behavioral and psychiatric features of Sturge-Weber syndrome. J Nerv Ment Dis 2010;198(12):905–13.
[50] López-Arvizu C, Sparrow EP, Strube MJ, Slavin C, DeOleo C, James J, et al. Increased symptoms of attention deficit hyperactivity disorder and major depressive disorder symptoms in Nail-patella syndrome: potential association with LMX1B loss-of-function. Am J Med Genet B Neuropsychiatr Genet 2011;156b(1):59–66.

Chapter 38

Neurorehabilitation and specific therapies

Chapter 38.1

Neurorehabilitation

William Ide[1,2], Anisa Tatini[2], Kavita Nadendla[2] and Melissa Trovato[1,2]

[1]Kennedy Krieger Institute, Baltimore, MD, United States, [2]Department of Neurology and Developmental Medicine, Kennedy Krieger Institute, Department of Physical Medicine & Rehabilitation, Johns Hopkins University School of Medicine, Baltimore, MD, United States

Introduction

The goal of rehabilitation for a child with a neurodevelopmental disability is to optimize function and enhance the quality of life for the patient and their family. Rehabilitation is best achieved through a multidisciplinary process consisting of a spectrum of interventions tailored to the unique needs of the child, helping them achieve their fullest functional, psychological, emotional, social, and educational potential [1]. In this chapter, the role of rehabilitation in various settings will be discussed, followed by a review of spasticity management, specific therapies, equipment, and orthotic devices.

Rehabilitation in acute care

Over the past three decades, admissions to the pediatric intensive care unit (PICU) in developed countries have achieved a marked reduction in mortality with an associated increase in moderate to severe disability among survivors [2]. Accordingly, the proportion of children admitted to intensive care with chronic conditions and/or disabilities is also on the rise. Cremer et al. reported a 67% prevalence of chronic conditions in neonatal and pediatric ICUs in 23 centers located in Europe and North America [3]. With children who have baseline and acquired disabilities accounting for a greater proportion of the modern hospital case mix, the initiation of rehabilitation services in the acute care setting is gaining emphasis. Comorbidities such as neuromuscular weakness, soft tissue pressure injuries, delirium, and contractures can develop at this stage and have an impact on the success of rehabilitation efforts later in the recovery phase [4]. The early initiation of bedside therapies can be a key component of the overall success of treatment and should be achieved in parallel with the primary goal of maintaining medical stability and optimizing neurologic recovery.

Early rehabilitation is gaining momentum in the neonatal ICU (NICU). With advances in reproductive and perinatal medicine, survival rates of preterm and very low-birth-weight infants have increased, along with a higher incidence of neurological disability [5]. Increasing evidence supports an adverse impact of the NICU environment on the trajectory of infant brain development due to excess exposure to noise, light, painful stimuli, and a lack of meaningful social interaction, language, touch, and sleep [6]. Studies have proposed the implementation of early rehabilitation in the neonatal ICU to include visual, auditory, oral motor, respiratory, and neuromotor interventions [7,8]. Liu [9] found preterm infants receiving this early combined rehabilitation intervention had shorter durations of oxygen supplementation and indwelling gastric tube use, fewer hospitalization days, and lower incidences of retinopathy of prematurity, bronchopulmonary dysplasia, and necrotizing enterocolitis.

Literature has described improved outcomes in adult patient with early mobilization and rehabilitation in the critical care setting, and pediatric programs have demonstrated early mobilization in the critically ill pediatric population to be

Capute and Accardo's Neurodevelopmental Disabilities in Infancy and Childhood. DOI: https://doi.org/10.1016/B978-0-12-824060-1.00016-X

safe and feasible [10,11]. Daily assessments of sedation, delirium, nutrition, pain, sleep, and barriers to mobilization are essential for optimizing rehabilitation for children in the critical care setting. Mobility goals are determined daily and should progress from bed-level to out-of-bed activities as appropriate. Nurses and family members should be empowered to facilitate and engage in patient mobilization. An interdisciplinary rehabilitation team consisting of the critical care medical team, physical therapists, occupational therapists, speech language pathologists, nurses, respiratory therapists, child life specialists, and nutritionists is vital to sustaining a rehabilitative culture in the PICU. A pediatric rehabilitation medicine (PRM) physician may be consulted in the critical care stage. These are physicians who have completed a primary residency in physical medicine and rehabilitation (PM&R) as well as subspecialty training in pediatric rehabilitation. This nonsurgical neuromuscular field focuses on restoring function and improving quality of life for children with a wide range of disabilities. The PRM consultant may assist patients and families in understanding and utilizing the specialties and resources they may encounter throughout their recovery process.

A priority of rehabilitation consultation during the acute care phase of a neurological injury is to provide recommendations that aid in preventing medical complications that have the potential to adversely impact functional recovery. This includes advice about skin protection, bowel and bladder programs, sleep hygiene, gastric protection, bracing, and spasticity management. The PRM consultant may assist primary teams in the anticipation, identification, and mitigation of triggers for serious complications such as autonomic dysreflexia in spinal cord injury or paroxysmal sympathetic hyperactivity in the brain-injured patient.

A PRM consultation often involves requests to provide recommendations for therapeutic disposition once the patient is medically ready to be discharged from acute care. Options include in-home or outpatient therapies, intensive outpatient programs, or acute inpatient rehabilitation facilities. There are facilities that also offer a continuum of care model, allowing the patient to transition from acute inpatient rehabilitation to an intensive day program, then to home-based or outpatient services based on their needs. Intensive day programs allow the patient to continue to receive 5 days per week of therapy as an outpatient. These programs allow for ease of transition home, support families ability to care for the child, and promote ongoing education as the child is healing and continuing to progress with functional skills. The determination of postacute rehabilitation needs should involve consideration of multiple factors and input from members of the child's care team. The patient's baseline and current functional status, social circumstances, medical stability, cognitive and behavioral status, and the goals of the child and family must all be carefully considered. The recommendation of acute care therapists from each involved discipline will be of utmost value; therefore, it may be necessary for the PRM consultant to suggest evaluations from certain therapists if they have not previously been obtained.

Inpatient rehabilitation

Children with traumatic brain injuries admitted to inpatient rehabilitation have demonstrated consistent improvements in function, correlating with the severity of their impairments on admission [12]. A systematic review by Königs et al. concluded early neurorehabilitation and intensive inpatient rehabilitation programs promote functional recovery compared to usual care in adults with moderate to severe traumatic brain injury (TBI) [13]. A retrospective chart review by Lee et al. of 44 children aged 6 months to 10 years with moderate to severe bilateral spastic cerebral palsy (CP) found that intensive inpatient rehabilitation was more effective for improving gross motor function than intermittent outpatient rehabilitation [14]. In children with brain tumors, inpatient rehabilitation following resection has demonstrated functional benefits at discharge and three-month follow-up [15]. Despite these advantages, children commonly have unmet inpatient and outpatient rehabilitation needs following brain injury; data suggests that more children would qualify for inpatient rehabilitation than are referred [16,17]. Receiving acute care at a facility with an onsite rehabilitation unit, increased length of critical care stay, and need for mechanical ventilation are all factors associated with an increased likelihood of admission to rehabilitation at the end of hospitalization in children [17,18].

In the early stages of treatment for a serious neurological injury, a PRM physician should meet with the family to discuss potential issues that may arise during the rehabilitation phase. This may include an explanation of the expected sequence of returning skills, with an approximation of timing as best as can be determined. The rehabilitation physician's perspective can aid families in gaining a more comprehensive understanding of the challenges in the postintensive care setting. In acutely disabling conditions such as TBI, families will likely have had no previous experience with the multidisciplinary rehabilitation model of care. Thus, it may be necessary to explain the roles of therapists and other team members.

When a patient is admitted to a neurorehabilitation unit, well-organized scheduling of initial evaluations and feedback to the family by the entire team is an important step in facilitating the transition. A crucial issue to address soon after admission is the discharge plan, which often requires time and effort to secure the correct resources, equipment, family training, and finances.

Spasticity management

Spasticity is a motor disorder characterized by a velocity-dependent increase in tonic stretch reflex (muscle tone) with exaggerated deep tendon jerks resulting from hyperexcitability of the stretch reflexes, as one component of the upper motor neuron syndrome [19]. Chronic spasticity leads to secondary changes in the muscle actin and myosin filaments and connective tissue elements, resulting in muscle fibrosis and atrophy [20]. Abnormal muscle tone impairs function, interferes with growth and development, and causes pain.

The decision to treat spasticity should be made after a thorough evaluation of its severity and its effects on the patient's function. Treatment should be driven by clearly stated goals created with the patient's family. In severely affected patients, treatment goals may aim to improve hygiene, daily care, and the child's comfort.

When selecting interventions, it is important to consider any benefits the patient may experience from spasticity. Increased tone may assist with sitting, transferring, and ambulation that the patient may lose if tone is reduced. Further, reducing spasticity may reveal underlying weakness, poor endurance, and impaired selective muscle control. In a child with significant weakness and a lack of selective muscle control, a reduction in muscle tone without other intervention does not improve function [20].

The reduction of spasticity often involves a multimodal approach that may include the elimination of aggravating factors (i.e., pain or infection), rehabilitative therapies, use of orthotic and adaptive equipment, oral pharmacotherapy, intrathecal baclofen pump, orthopedic surgery, selective dorsal rhizotomy, and chemical denervation. The selection of therapeutic modalities is dependent on the degree of spasticity, the number of limbs affected, and the side effect profile of the intervention.

Spasticity rehabilitation interventions

Rehabilitation interventions are an important component in the management of spasticity and are used in conjunction with pharmacologic management. Performing functional activities positively affects muscle tone [21]. Prolonged stretching and range of motion exercises have long been a mainstay of treatment; however, there is little evidence that this intervention is helpful. It is still popular because it is easy for therapists and families to perform [22]. Early introduction of orthotic devices (e.g., ankle foot orthosis (AFO)) should be considered to secure the functional position of affected muscles and joints and to decrease energy expenditure in ambulatory patients. Orthoses with dynamic properties (e.g., articulated AKA, "hinged" AFO) can assist in balancing muscle forces across joints during therapeutic and daily activities [23]. Braces are further detailed in the "Orthoses" section below. Fig. 38.1.1 features different types of lower extremity orthoses commonly utilized in children.

If a spastic contracture occurs, serial casting may be used. Serial casting can be used with or without concurrent botulinum toxin injections prior to casting. If used with botulinum toxin injections, it is recommended to wait 4 weeks for the maximum effect of the toxin injection [22].

Pharmacological treatment

Oral agents are recommended if a general reduction in muscle tone is desired. They may be most beneficial when all four limbs are affected, but they can be considered even if one body limb is affected. The most common agents are benzodiazepines, baclofen, alpha-2 agonists, and dantrolene. Benzodiazepines (usually diazepam) are considered the oral medications with the strongest evidence of efficacy in the cerebral palsy population. However, medications should be chosen based on patient clinical presentation as well as medication side effect profile [22]. Apart from dantrolene, all oral agents are centrally acting to decrease excitability in the CNS. As such, these centrally-acting medications can cause sedation, generalized weakness, and cognitive side effects. It is important for the clinician to titrate the medication to maximize therapeutic use while also limiting adverse effects. Dantrolene is unique among oral agents as it works on skeletal muscle to inhibit calcium release at the sarcoplasmic reticulum, which prevents muscle contraction. Weakness and hepatotoxicity are known adverse effects, and routine monitoring of liver function tests is required for patients on this medication [24].

Surgical interventions

Intrathecal Baclofen (ITB) was first introduced in the 1980s as a treatment for adults with spasticity from a spinal cord injury. ITB is often recommended for patients who have global spasticity that cannot be adequately managed with oral medications. Baclofen acts at the spinal cord level, primarily by inhibiting the release of excitatory neurotransmitters. ITB decreases spasticity via direct effect on the CNS and, as such, does not have the systemic side effects observed in oral therapy. A continuous supply of medication is delivered by an infusion system that uses a catheter in the intrathecal space, a pump in the abdomen, and telemetry programming. The required doses of intrathecal baclofen are a fraction of the effective doses used in oral therapy [25]. When assessing if a patient is an appropriate candidate for ITB pump placement, it is important to consider patient adherence with refills and dose adjustments, potential complications during and after placement, and financial factors. Patient adherence may be impeded by geographic location and financial status. Complications of ITB include pump function, integrity of the catheter, overdosing or withdrawal of medication, and displacement of the pump, all of which can be life-threatening and are considered medical emergencies [26].

Selective Dorsal Rhizotomy (SDR) is a neurosurgical procedure in which dorsal rootlets at the lumbosacral level are cut to disrupt afferent input to the monosynaptic reflex arc. The Dorsal Rhizotomy Procedure was first used in the early 1900s but fell out of favor due to the nonspecific nature of the rootlets being severed. The procedure was refined in the 1980s to focus only on severing dorsal rootlets [27]. These innovations in surgical procedures eliminated sensory and bladder adverse events of the surgery [28]. The number of rootlets cut can range from 20% to 80% [29]. There is no international standardized selection criteria for patients who most benefit from SDR [28]. Since SDR is irreversible, appropriate patient selection is critical, and a multidisciplinary team should work with families to determine patient eligibility. Traditionally, it was felt that children ages 4–10 with diplegic CP who were ambulatory, had pure spasticity, had good muscle control and strength, and were cognitively intact were the best candidates for their procedure. These children generally fall under the Gross Motor Function Classification System (GMFCS) I–III [27]. However, there are some studies that explore SDR as an alternative to ITB therapy for patients who fall under GMFCS IV–V. The goals of treatment for these patients would be to improve ease of care [30]. Another consideration for patients with a GMFCS of V is ventral and dorsal rhizotomy, which is helpful if the patient has both spasticity and dystonia. SDR is contraindicated if dystonia is present, as it may worsen dystonia [31]. Post-operatively, the patient will participate in an intensive therapy program that will gradually taper over the course of the year. An intensive therapy program may occur in inpatient or outpatient rehabilitation, depending on the needs of the patient. The goals of therapy are to strengthen, improve endurance, and train gait [27]. There is mixed evidence on long-term functional outcomes for patients after SDR, and high-quality studies are needed to fully determine the clinical impact [32].

Chemical denervation

For focal spasticity, chemical denervation with phenol alcohol or botulinum neurotoxin may be considered.

Phenol in a concentration of 3%–6% denatures proteins and causes tissue necrosis. Transcutaneous phenol injection may target the supplying motor nerve (obturator, musculocutaneous nerves) or motor points of spastic muscles using electrical stimulation. After the introduction of botulinum toxin, phenol use for the management of spasticity has been limited. Phenol injections are technically challenging due to the precision needed to perform them and are often not well tolerated by pediatric patients. This procedure is often done for children with sedation or anesthesia. Common adverse reactions include pain during injection and local tenderness at the injection site, believed to be secondary to local tissue necrosis and inflammation. If a sensory or mixed nerve is injected, there is a significant risk of pain and paresthesia. In a study comparing botulinum toxin and phenol, botulinum toxin was found to be more effective in the management of lower extremity spasticity. However, phenol is substantially less expensive compared to botulinum toxin, and its effects can last up to 18 months [33]. Often, phenol is used as an adjunct to botulinum toxin for easier identification of nerves.

Botulinum neurotoxin (BoNT) is the treatment of choice for local chemodenervation in children [22]. Botulinum A toxin (BTX-A), the most popular serotype used in the USA, interacts at presynaptic cholinergic terminals in the neuromuscular junction (NMJ). It prevents presynaptic acetylcholine-containing vesicles from fusion with the synaptic membrane, functionally producing chemical denervation. Neuromuscular blockade is achieved by a direct injection into the target muscle. Localization may be aided by using anatomic landmarks and palpation, electromyography, electric stimulation, ultrasonography, or a combination of these methods [24]. The selectivity of botulinum neurotoxin to the NMJ and the fact that it does not affect axons or peripheral nerves enhance its safety and minimize side effects. The injection can be performed using reassurance and restraint, topical cold spray or lidocaine cream, conscious sedation, or general anesthesia. The number of injections, their site, and the comfort level of the patient and the practitioner should be considered.

In clinical trials, botulinum toxin has been effective in reducing muscle tone, painful spasms, and improving limb function. Current indications for use of Botulinum neurotoxin are equinus deformity, crouched gait, enhancing range of motion, improvement of upper extremity deformity, elimination of pain associated with spasticity, and as an alternative or adjunct to serial casting. Side effects are rare but may include pain during injection, infection, allergic reactions, flu-like symptoms, weakness, and fatigue [34]. BoNT holds a "black box" warning from the FDA due to the higher risk of mortality in GMFCS V patients. The therapeutic effect after injection is usually observed within a few days of injection, with maximum results seen in 3–4 weeks. Reduction in spasticity will usually last 3 months but may vary between patients [35]. An important consideration with continuous repetition of BoNT injection is the potential for the formation of autoantibodies against the toxins, which will decrease the efficacy of the medication [36]. More recent literature has also investigated whether continuous use of BoNT may result in increased fibrosis of injected muscles over time; however, the consensus of such research is still to be determined [34]. Intensive rehabilitative therapy consisting of stretching, strengthening, functional training, orthotics use, and electric stimulation can enhance the results of BoNT.

Effective muscle hypertonia management coupled with rehabilitative therapies can improve a child's tolerance and response to therapeutic modalities, leading to more satisfactory goals' achievement.

Rehabilitation therapy interventions

The goals of rehabilitation therapy interventions for children with disorders of the brain are to promote functional skill acquisition. The choice of therapy interventions should be determined by the patient's family and therapy team (physical therapist, occupational therapist, speech therapist).

There is no one therapy that is considered superior, and often children will use different therapy types throughout their lives depending on their functional goals. If the goal is to improve gross motor skills, treadmill training and task-specific training are the most evidence-supported therapy interventions. If the goal is to improve hand function, constraint-induced movement therapy, manual therapy, and environmental enrichment have the most supporting evidence. For communication and cognition goals, literacy-based therapy, as well as augmentative and alternative communication, had the strongest evidence of efficacy [22].

Orthoses

An orthosis is any externally applied device, static or dynamic, used to modify the structural and functional characteristics of the neuromuscular skeletal system [37]. Static orthoses, such as a resting hand splint, elbow extension splint, or solid ankle-foot orthosis (AFO), can be utilized for protection and improved positioning, immobilization of affected joints to prevent or correct deformities and relieve pain, and modification of abnormal tone by providing gentle stretch. A dynamic orthosis, such as an elbow Dynasplint, an articulated AFO, or a reciprocating gait orthosis (RGO), can assist with function by compensating for deformity and muscle weakness. A supramalleolar orthosis (SMO) can be used to control pronation and increase medial-lateral stability of the ankle while allowing for dorsiflexion and plantarflexion. The University of California Berkeley Laboratories Shoe Insert (UCBL) is worn inside the shoe and provides support for the longitudinal arch and calcaneus in patients with deformity or instability of the hindfoot and/or midfoot (Fig. 38.1.1).

When prescribing an orthosis for a child, careful consideration should be given to the differences in orthotic devices and their potential effect on outcome. A child's age, functional level, motor control, type of deformity, and commitment to use, as well as the patient's and family's goals, can impact which device is prescribed. Any device should be simple, lightweight, durable, easy to use, and optimized for functional independence.

Mobility and adaptive equipment

There are a wide variety of assistive devices available to improve positioning, maximize functional mobility and independence, and facilitate daily care by families and caregivers.

Transfer aids assist with the transition between positions to protect the caregiver from overexertion injuries that can occur from improper or repeated heavy lifting while keeping the child safe during the transfer. Overhead trapeze bars help with bed mobility (i.e., rolling from supine to sitting position). A transfer board provides a bridge between two surfaces, such as bed, chair, car, or commode transfers. Different types of commercially available mechanical lifts facilitate safe transfers.

Children who are nonambulatory or minimally so may benefit from the use of passive or mobile *standing devices* to provide therapeutic benefits of weight bearing and a positive effect on bone mineral density, sustained muscle stretch, improvement in head and trunk control, and maintenance of trunk and lower limb passive range of motion. The supine stander, prone stander, or multiposition stander help support the child in an upright position by providing different levels of support at the trunk, hips, and knees [38]. A systematic review by Paleg et al. proposed specific clinical dosing

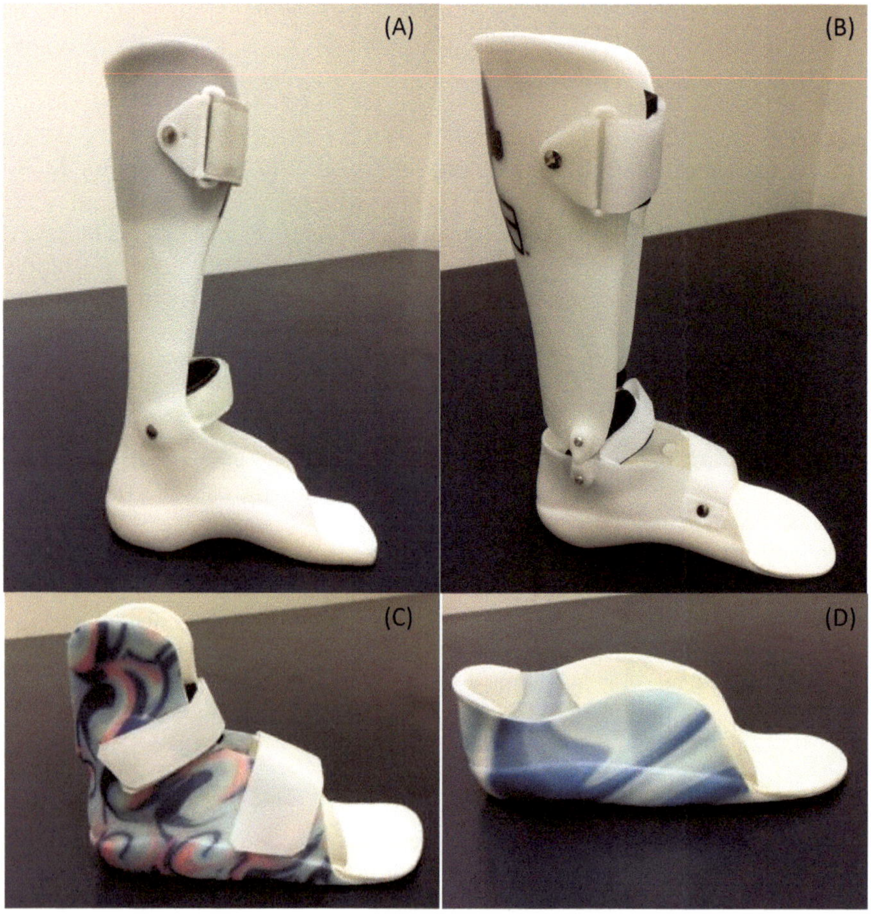

FIGURE 38.1.1 (A) Custom molded ankle foot orthosis, (B) Custom molded articulated (hinged) ankle foot orthosis, (C) custom molded supramalleolar orthosis, (D) University of California Berkeley Laboratories Shoe Insert. *Photo credit: David R. Harris, P.T., C.O.*

recommendations for pediatrically supported standing programs. Standing 5 days per week for 30–45 min/day can minimize the effects of spasticity, 45–60 min/day can increase range of motion, 60 min/day in hip abduction can improve hip positioning, and 60–90 min/day can improve bone mineral density [39].

Gait aids, including canes, crutches, walkers, and gait trainers, facilitate functional independence, improve balance and posture, and decrease energy expenditure and pain. When recommending ambulatory aids, factors of tone, volitional muscle control, balance, and prognosis must be considered to determine a child's ability to utilize the device functionally or therapeutically. Canes and axillary or forearm (Lofstrand) crutches provide the lowest level of support. Posterior wheeled walkers are widely used in children for their ability to promote an erect posture and the ability to add modifications individualized to the child's needs. For children who need more trunk and postural support for ambulation, gait trainers are both functionally and physiologically beneficial [38].

Mobility devices, including scooters, strollers, and manual or power wheelchairs, allow children to be more independent, explore their environment, and enhance their social skills. Children with adequate cognitive, visual, and spatial discrimination abilities can learn how to operate a motorized device at 3 years of age [40]. Many factors should be taken into consideration when choosing a wheelchair, including type of disability, potential for growth and development, specific children's needs, and functional and cognitive status.

The achievement of independence in *self-care skills*, an essential part of normal development, is often delayed in children with disabilities. Environmental modifications, the use of adaptive devices, and compensatory techniques for feeding, writing, dressing, hygiene, and other daily activities help children to be more successful in achieving functional independence.

Adjunctive interventions

Interventions such as electrical stimulation, adaptive recreation, locomotor therapy, and constraint-induced movement therapy may be used in conjunction with traditional therapies in pediatric rehabilitation practice.

Electrical stimulation may improve motor function in children with poor selective motor control, where strengthening can be a challenge. Neuromuscular electrical stimulation (NMES) consists of the application of electric current to elicit a visible muscle contracture with the goal of strengthening and spasticity reduction. When applied to support muscle contraction during a specific functional activity, the treatment is termed functional electrical stimulation (FES). A review of 12 studies revealed two randomized controlled trials demonstrating statistically significant improvement in strength and range of motion [41]. Threshold electrical stimulation (TES) is a low-level electrical stimulus of subcontracture intensity that is often applied during sleep with the proposed mechanism to treat muscle atrophy. However, research reported no statistically significant effects [41].

Adaptive recreation (AR) for individuals with disabilities is an important aspect of any rehabilitation program. AR not only increases physical activity but confers positive psychosocial impacts as well. This includes, but is not limited to, adapted sports. For example, hippotherapy, or horseback riding with integrated skilled therapy, has long been a popular intervention that has demonstrated clinical improvement in walking, running, and jumping in a heterogeneous group of children with mild to severe cerebral palsy [42]. Aquatic therapy is commonly utilized in children with muscle weakness, as the elimination of gravity in the water reduces effort in movement. Pool therapy can improve muscle weakness, increase range of motion, alter abnormal tone, and enhance endurance and cardiopulmonary fitness by providing multiple sensory stimuli. However, evidence-based research is limited to the pediatric population [43]. Music therapy is another therapeutic tool to meet the physical, emotional, social, psychological, and spiritual needs of people of all ages [44]. In the pediatric setting, music creates an enjoyable atmosphere to promote gross and fine motor skills, speech and language, and attention in conjunction with traditional therapies.

Locomotor therapy provides a task-specific approach that enables severely affected patients to repetitively practice complex gait cycles [45]. Several studies have shown its potential in patients after stroke, SCI, and CP to improve gait speed, endurance, and mechanics. This can be accomplished by treadmill training with or without body weight support and robotic gait training. Treadmill therapy with bodyweight support was compared with physical therapy using early therapy-assisted ambulation with the support of orthotics and assistive devices in stroke. Intensity, frequency, and overall length of treatment in both groups were matched (45 min/day, 5 days per week, until patient could walk overground unassisted). The overall outcome did not differ, except for patients with major hemispheric stroke who were difficult to mobilize using physiotherapy alone [46].

Constraint-Induced Movement Therapy (CIMT), developed by Taub et al., was initially applied to adults with hemiparesis after stroke, combining restraint of the unaffected limb with intensive motor training of the affected limb. Since the first published trials on adult chronic stroke patients in the 1990s, CIMT has been applied to patients with acute stroke, traumatic brain injury, and children with cerebral palsy (Fig. 38.1.2). A randomized controlled trial of 18 children with CP showed significant

FIGURE 38.1.2 A child with hemiplegic cerebral palsy in constraint-induced movement therapy. *Photo credit: Christina Ide, PA-C.*

improvement in the skills of the treated upper extremity. The CIMT group underwent immobilization of the unaffected (or less affected) upper limb for a period of 3 weeks using a bi-valved cast while receiving intensive therapy of 6 h per day for motor learning. The control group was assigned to conventional OT for a mean of 2.2 h per week. At the completion of the 3-week period and at 6-month follow-up, the CIMT group showed significant improvement in the use of the trained extremity [47]. Several elements of CIMT may have contributed to its success, one of which is the intensity of the therapy. However, CIMT in this traditional paradigm of at least 20 h per week may be difficult to realize in the pediatric population.

Interdisciplinary approach to rehabilitation

The importance of a team-based approach in the care of children cannot be overstated and is paramount to the specialty of pediatric rehabilitation medicine. An Accreditation Council for Graduate Medical Education (ACGME) requirement for all physicians trained in PM&R is to build skills in team leadership to ensure optimal communication and coordination between members. Having worked closely with an interdisciplinary team throughout training, the pediatric rehabilitation medicine physician has a comprehensive understanding of the roles of the other members of the team and is able to more effectively coordinate the efforts of the rehabilitation team. Ramsdell et al. [48]. For example, the rehabilitation team may identify hypertonia as a barrier to patient functional progression with therapies. The pediatric physiatrist may coordinate with the therapists to address range of motion, bracing, and modalities, which may be carried over with the nursing staff and patient family members. The pediatric physiatrist will be in communication with all members of the team and plan for either targeted or global medication management, depending on the feedback provided by the team. The communication is ongoing for the entirety of the patient's rehabilitation course and leads to a unified effort to address patient care.

Summary

The rehabilitation plan for a child with CNS dysfunction requires a multidisciplinary approach customized with succinct, definable, and achievable therapeutic goals. These goals should address short- and long-term neuromuscular concerns such as tone management and motor function, functional mobility, independence with self-care, and cognitive and social development, all of which contribute to a better quality of life for patients and their families.

References

[1] Jaffe KM. Pediatric trauma rehabilitation: a value-added safety net. J Trauma Acute Care Surg 2008;64(3):819–23.
[2] Namachivayam P, Shann F, Shekerdemian L, Taylor A, van Sloten I, Delzoppo C, Daffey C, Butt W. Three decades of pediatric intensive care: who was admitted, what happened in intensive care, and what happened afterward. Pediatr Crit Care Med 2010;11(5):549–55.
[3] Cremer R, Leclerc F, Lacroix J, Ploin D. Children with chronic conditions in pediatric intensive care units located in predominantly french-speaking regions: prevalence and implications on rehabilitation care need and utilization, and the GFRUP/RMEF chronic diseases in PICU study group. Crit Care Med 2009;37(4):1456.
[4] Hopkins RO, Choong K, Zebuhr CA, Kudchadkar SR. Transforming PICU culture to facilitate early rehabilitation. J Pediatr Intensive Care 2015;4(04):204–11.
[5] Cheong JLY, Spittle AJ, Burnett AC, Anderson PJ, Doyle LW. Have outcomes following extremely preterm birth improved over time? Sem Fetal Neonat Med 2020;25(3):101114.
[6] Letzkus L, Conaway M, Miller-Davis C, Darring J, Keim-Malpass J, Zanelli S. A feasibility randomized controlled trial of a NICU rehabilitation program for very low birth weight infants. Sci Rep 2022;12(1):1729. Available from: https://doi.org/10.1038/s41598-022-05849-w PMID: 35110644; PMCID: PMC8810863.
[7] Filippa M, Lordier L, De Almeida JS, et al. Early vocal contact and music in the NICU: new insights into preventive interventions. Pediatr Res 2020;87(2):249–64.
[8] Ghomi H, Yadegari F, Soleimani F, Knoll BL, Noroozi M, Mazouri A. The effects of premature infant oral motor intervention (PIOMI) on oral feeding of preterm infants: a randomized clinical trial. Int J Pediatr Otorhinolaryngol 2019;120:202–9.
[9] Liu Y, Zheng-feng L, Zhong Y, Zhao Z, Deng W, Chen L, Liu B, Du T, Zhang Y. BMC Pediatr 2021;21:269.
[10] Tipping CJ, Harrold M, Holland A, Romero L, Travis N, Hodgson CL. The effects of active mobilisation and rehabilitation in ICU on mortality and function: a systematic review. Intensive Care Med 2017;43(2):171–83.
[11] Wieczorek B, Ascenzi J, Kim Y, Lenker H, Potter C, Shata N, Mitchell L, Haut C, Berkowitz I, Pidcock F, Hoch J, Malamed C, Kravitz T, Kudchadkar S. PICU Up!: impact of a quality improvement intervention to promote early mobilization in critically ill children. Pediatr Crit Care Med 2016;17(12):e559–66.
[12] Rice SA, Blackman JA, Braun S, Linn RT, Granger CV, Wagner DP. Rehabilitation of children with traumatic brain injury: descriptive analysis of a nationwide sample using the WeeFIM. Arch Phys Med Rehabil 2005;86(4):834–6.
[13] Königs M, Beurskens EA, Snoep L, Scherder EJ, Oosterlaan J. Effects of timing and intensity of neurorehabilitation on functional outcome after traumatic brain injury: a systematic review and meta-analysis. Arch Phys Med Rehabil 2018;99(6):1149–59 e1.

[14] Lee SH, Shim JS, Kim K, Moon J, Kim M. Gross motor function outcome after intensive rehabilitation in children with bilateral spastic cerebral palsy. Ann Rehabil Med 2015;39(4):624.

[15] Day AM, Slomine BS, Salama C, Quinton TL, Suskauer SJ, Salorio CF. Functional Gains in Children Receiving Inpatient Rehabilitation After Brain Tumor Resection. Arch Phys Med Rehabil 2021;102(11):2134–40. Available from: https://doi.org/10.1016/j.apmr.2021.05.001 Epub 2021 May 25. PMID: 34044002; PMCID: PMC8572161.

[16] Fuentes MM, Wang J, Haarbauer-Krupa J, Yeates KO, Durbin D, Zonfrillo MR, Jaffe KM, Temkin N, Tulsky D, Bertisch H. Unmet rehabilitation needs after hospitalization for traumatic brain injury. Pediatrics 2018;141(5).

[17] Greenspan AI, MaCkenzie EJ. Use and need for post-acute services following paediatric head injury. Brain Injury 2000;14(5):417–29.

[18] Horvat C, Au A, Kantawala S, Houtrow A, Clark R, Fink E. 862: Factors associated with disposition to rehabilitation among children admitted to intensive care. Crit Care Med 2018;46(1):415.

[19] Lance JW. Disordered muscle tone and movement. Clin Exp Neurol 1981;18:27–35.

[20] Tilton AH. Approach to the rehabilitation of spasticity and neuromuscular disorders in children. Neurol Clin 2003;21(4):853–81.

[21] Damiano DL. Activity, activity, activity: rethinking our physical therapy approach to cerebral palsy. Phys Ther 2006;86(11):1534–40.

[22] Novak I, Morgan C, Fahey M, Finch-Edmondson M, Galea C, Hines A, Langdon K, Mc Namara M, Paton MC, Popat H. State of the evidence traffic lights 2019: systematic review of interventions for preventing and treating children with cerebral palsy. Curr Neurol Neurosci Rep 2020;20(2):1–21.

[23] Aboutorabi A, Arazpour M, Bani MA, Saeedi H, Head JS. Efficacy of ankle foot orthoses types on walking in children with cerebral palsy: a systematic review. Ann Phys Rehabil Med 2017;60(6):393–402.

[24] Tilton A, Vargus-Adams J, Delgado MR. Pharmacologic treatment of spasticity in children. Semin Pediatr Neurol 2010;17(4):261–7. Available from: https://doi.org/10.1016/j.spen.2010.10.009 PMID: 21183133.

[25] Hasnat MJ, Rice JE. Intrathecal baclofen for treating spasticity in children with cerebral palsy. Cochrane Database Syst Rev 2015;2015(11): CD004552. Available from: https://doi.org/10.1002/14651858.CD004552.pub2 PMID: 26563961; PMCID: PMC9361875.

[26] Awaad Y, Rizk T, Siddiqui I, Roosen N, McIntosh K, Waines GM. Complications of intrathecal baclofen pump: prevention and cure. ISRN Neurol 2012;2012:575168. Available from: https://doi.org/10.5402/2012/575168 Epub 2012 Mar 24. PMID: 22548189; PMCID: PMC3323842.

[27] Wang KK, Munger ME, Chen B, Novacheck TF. Selective dorsal rhizotomy in ambulant children with cerebral palsy. J Child's Orthop 2018;12(5):413–27.

[28] Grunt S, Fieggen AG, Vermeulen RJ, Becher JG, Langerak NG. Selection criteria for selective dorsal rhizotomy in children with spastic cerebral palsy: a systematic review of the literature. Develop Med Child Neurol 2014;56(4):302–12.

[29] Nicolini-Panisson RD, Tedesco AP, Folle MR, Donadio MVF. Selective dorsal rhizotomy in cerebral palsy: selection criteria and postoperative physical therapy protocols. Rev Paul Pediatr 2018;36:100–8.

[30] D'Aquino D, Moussa AA, Ammar A, Ingale H, Vloeberghs M. Selective dorsal rhizotomy for the treatment of severe spastic cerebral palsy: efficacy and therapeutic durability in GMFCS grade IV and V children. Acta Neurochir 2018;160(4):811–21.

[31] Ahluwalia R, Bass P, Flynn L, Martin E, Riordan H, Lawrence A, Naftel RP. Conus-level combined dorsal and ventral lumbar rhizotomy for treatment of mixed hypertonia: technical note and complications. J Neurosurg Pediatr 2020;27(1):102–7.

[32] Tedroff K, Hägglund G, Miller F. Long-term effects of selective dorsal rhizotomy in children with cerebral palsy: a systematic review. Develop Med Child Neurol 2020;62(5):554–62.

[33] Gonnade N, Lokhande V, Ajij M, Gaur A, Shukla K. Phenol versus botulinum toxin a injection in ambulatory cerebral palsy spastic diplegia: a comparative study. J Pediatr Neurosci 2017;12(4):338.

[34] Multani I, Manji J, Hastings-Ison T, Khot A, Graham K. Botulinum toxin in the management of children with cerebral palsy. Pediatr Drugs 2019;21(4):261–81.

[35] Chen Y, Zhang C, Liu Y, Magat E, Verduzco-Gutierrez M, Francisco GE, Zhou P, Zhang Y, Li S. The effects of botulinum toxin injections on spasticity and motor performance in chronic stroke with spastic hemiplegia. Toxins 2020;12(8):492.

[36] Delgado MR, Bonikowski M, Carranza J, Dabrowski E, Matthews D, Russman B, Tilton A, Velez JC, Grandoulier A, Picaut P. Safety and efficacy of repeat open-label abobotulinum toxin A treatment in pediatric cerebral palsy. J Child Neurol 2017;32(13):1058–64.

[37] Cifu DX. Braddom's physical medicine & rehabilitation. 6th ed. Philadelphia: Elsevier Inc; 2021.

[38] Murphy KP, McMahon MA, Houtrow AJ. Pediatric rehabilitation: principles and practice. Springer Publishing Company; 2020.

[39] Paleg GS, Smith BA, Glickman LB. Systematic review and evidence-based clinical recommendations for dosing of pediatric supported standing programs. Pediatr Phys Ther 2013;25(3):232–47.

[40] Molnar GE. Rehabilitation in cerebral palsy. West J Med 1991;154(5):569.

[41] Kerr C, McDowell B, McDonough S. Electrical stimulation in cerebral palsy: a review of effects on strength and motor function. Dev Med Child Neurol 2004;46(3):205–13.

[42] Sterba JA, Rogers BT, France AP, Vokes DA. Horseback riding in children with cerebral palsy: effect on gross motor function. Dev Med Child Neurol 2002;44(5):301–8.

[43] Dumas H, Francesconi S. Aquatic therapy in pediatrics: annotated bibliography. Phys Occup Ther Pediatr 2001;20(4):63–78.

[44] Kennelly J. The specialist role of the music therapist in developmental programs for hospitalized children. J Pediatr Health Care 2000;14(2):56–9.

[45] Schindl MR, Forstner C, Kern H, Hesse S. Treadmill training with partial body weight support in nonambulatory patients with cerebral palsy. Arch Phys Med Rehabil 2000;81(3):301–6.

[46] Kosak MC, Reding MJ. Comparison of partial body weight-supported treadmill gait training versus aggressive bracing assisted walking post stroke. Neurorehabil Neural Repair 2000;14(1):13–19.

[47] Taub E, Ramey SL, DeLuca S, Echols K. Efficacy of constraint-induced movement therapy for children with cerebral palsy with asymmetric motor impairment. Pediatrics 2004;113(2):305–12.

[48] Ramsdell K., Nitu-Marquise A.D., Turk M.A. Rehabilitation team functioning. PM&R KnowledgeNow. Published April 21, https://now.aapmr.org/rehabilitation-team-functioning/; 2017 [accessed 09.05.23].

Chapter 38.2

Physical therapy

Janice Laux and Deanna Johnson
Kennedy Krieger Institute, Department of Physical Therapy, Baltimore, MD, United States

Introduction

Pediatric physical therapy's role is to optimize gross motor capacity so as to support their environmental interactions, whether that is at home, school or their community. Their need for our services does not end in childhood but rather continues through adolescence and into adulthood, each with its own set of nuances. The focus of this chapter is to: (1) define pediatric physical therapy and our role within an interprofessional team; (2) when to refer to physical therapy; (3) explain the physical therapy examination process using the International Classification of Functioning (ICF) model; (4) review various service delivery models; and (5) discuss transitional periods/lifespan issues and future needs.

What is pediatric physical therapy?

Pediatric physical therapists (PTs) work with individuals with childhood-onset conditions who are experiencing illnesses, injuries, or other conditions that limit their ability to move or function in their daily lives. Provided services aim to improve postural and mobility skills, develop or restore function, alleviate pain, prevent or decrease permanent physical disabilities or secondary complications, and promote overall health and wellness (Academy of Pediatric Physical Therapy fact sheet). Simply put, physical therapy works with the child and their families to assist each child in reaching their maximum potential to function independently for active participation in home, school, and community environments. Physical therapy interventions include prescribed individualized therapeutic exercise for strengthening, balance and coordination training, manual (hands-on) therapy, gait training, adaptive equipment prescription, and patient/caregiver education [1]. As primary health care providers, PTs work in collaboration with families, communities, and other medical, educational, developmental, and rehabilitation specialists [1].

Pediatric physical therapy education and practice is grounded in neuroscience, neuro-motor development, developmental anatomy and physiology, and infant/child development. Shaping how physical therapy interventions themselves are conducted involves the construct of engagement, known to be a key factor in the effectiveness of therapy outcomes. In the complex dynamics that exist in neurodevelopmental pediatric therapy, the fundamental importance of engaging children and caregivers in the therapy process is crucial to service delivery and clinical practice. Pediatric physical therapists strive to acquire expertise in the core interpersonal and communication skills needed to engage and motivate clients and caregivers. One national model for family engagement is Erikson's FAN, Facilitating Attuned Interactions. This psychologically informed communication tool, developed by Lisa Gilkerson, is applicable for use with families across a variety of settings and from infancy onward. King et al. [2] describe engagement as intentionally entering the world of the child and family to be aware of, anticipate, and be responsive to engagement and disengagement. The intent is to create a space for enjoyable and meaningful interactions for both clients and caregivers, develop good interpersonal and supportive relationships, collaborate on aligned goals, plans, and expectations, and discuss perceptions of progress [2]. These intentional practices optimize the likelihood of change and contribute to the perceived value and success of therapy, especially when therapy is likely a lifespan need, as with childhood neurodevelopmental conditions.

Pediatric physical therapists treat a diverse patient population with varying diagnoses, ages, and complexities. Several diagnoses, such as cerebral palsy, Down syndrome, and developmental delay, have extensive evidence describing developmental courses and outcomes. For example, the work of Palisano, Rosenbaum, Bartlett, and Livingston established the five-level Gross Motor Function Classification System (GMFCS). This system provides a description of gross motor function and guidance for future levels of assistance that a child will likely need. The related motor growth curves developed from a large population-based sample [3], help understand how the gross motor abilities of children with CP at each GMFCS level change with age, estimate future motor capacities, and, over how time, compare an individual child's function to expectations. The work by Pat Winders, PT, has captured information about how children with Down syndrome learn gross motor skills, best practices for teaching motor skills, and average ages for gross motor skill acquisition that serves as a guide to understanding gross motor development in children with Down syndrome. Additionally, typical infant and childhood motor development has been well studied and serves as the basis for numerous developmental checklists. There is limited evidence-based guidance and prognostic information available for more complex and varied neurological disorders, such as neurogenetic conditions. Without such guidelines for gross motor development, the therapist must rely on their clinical knowledge and reasoning to address the questions of concerned parents who are anxious to know when their child may sit, crawl, or walk.

When to refer to physical therapy?

PTs contribute information on gross motor ability that enables the interprofessional team to fully understand the child's challenges in order to create a comprehensive, coordinated care plan. Indications for referral for physical therapy will primarily fall into the neuromuscular and musculoskeletal domains. Age of the child is a consideration in interpreting findings and assigning a level of concern. Neuromuscular impairments may be postural and/or motor control, static, or progressive in nature. During the first 6 months of life (corrected for prematurity), these concerns may include difficulty lifting their head or poor head control, floppiness, stiffness or little to no movement in the legs, asymmetric movement and posture of the extremities, pushing back into extension when lying on their back, and lack of arm movement or hands fisted predominantly. Between 6 and 15 months of corrected age, neuromotor concerns may include ongoing poor head control, the inability to straighten their back for sitting, asymmetries in posture, difficulty rolling, crawling, or floor mobility, being unable to sit independently, and limited weight bearing on the legs. Between 15 and 24 months of corrected age, neuromotor concerns may include walking on toes or the inability to take steps independently. At five years, a child with gross and/or fine motor dyscoordination and a motor skill deficit not otherwise explained by an intellectual disability, muscle or neurologic condition, or disease is a concern.

Musculoskeletal impairments may be asymmetry in neck posture and motion and/or cranial asymmetry in a neonate to 3 months and older infant. In the older infant, LE alignment impairments warranting a physical therapy referral may include altered foot alignment, torsional profile, angular conditions of the knee, joint pain, or limping. Developmental orthopedics defines age ranges for typical and atypical LE joint alignment, of which pediatric physical therapists are aware to assist with framing levels of concern in such conditions of altered alignment. The toe walking pattern from the time of independent walking, without ever being full sole, is atypical. Excessive joint mobility, such as increased lumbar lordosis, knee hyperextension, and foot pronation in standing, can be indicative of a connective tissue disorder. Inflammatory joint swelling and pain can be indicative of an injury or disease process, such as juvenile arthritis. Scoliosis can be present in an infant or young child (Table 38.2.1).

Any identified or diagnosed conditions of the neurologic system, including neurogenetic, musculoskeletal system, primary muscle disease, or conditions of progressively declining neuromuscular functions.

Pediatric physical therapists are not alone in evaluating and treating children to reach their full potential; they are part of an interprofessional team. In addition to family and caregivers, teams can include medical providers (i.e., physicians, physician associates, nurse practitioners), occupational therapists, speech and language pathologists, social workers, psychologists, psychiatrists, nutritionists, educational professionals, orthotists, and equipment vendors that work collaboratively in providing care. This interprofessional team forms a dynamic partnership, promoting coordination of care toward goal achievement. Children and their caregivers must be intimately involved and engaged in the development and implementation of the plan of care, with a focus on the child's participation in their natural setting (i.e., home, school, community). Table 38.2.2 (Inter-professional team).

Physical therapy examination and its contribution in the management of infants and children with a neuromotor disability

Physical therapy examination process

Physical therapy examination and clinical reasoning required to understand a child's activity and participation capacities are complex processes guided by the identification of impairments that exist in the setting of intrinsic and extrinsic factors relative to the child, family, larger environment, and community. The ever-changing needs of a child as an individual, a member of a family, and part of a community guide physical therapy care during infancy, childhood, school years, and into adulthood.

Motor disabilities affect and are affected by the development and functioning of other developmental domains such as cognition, language, sensation, perception, and behavior, as well as external factors such as family, community, and

TABLE 38.2.1 Indications for physical therapy referral.

Head/neck	Decreased control
Trunk	Excessive extension/Inadequate extension
Extremities	Asymmetric use Distal posturing Unable to lift against gravity Poor repertoire, quality and quantity of movements (i.e., jerky, dystonic, sluggish) Poor weight bearing
Primitive reflexes	Obligatory Retained > 6 mos
Gross motor skills	Delayed rolling, sitting
Foot Deformities	Kidney-shaped Banana-shaped Flat foot with deviation
Torsional profile	Excessive in- or out-toeing
Angular, knee	Excessive bow or knock
Joints, ankle/knee/hip	Swelling Pain Limping
Trunk	Scoliosis, congenital Scoliosis, torticollis linked Scoliosis, idiopathic Asymmetry
Lower extremity	Toe walking Leg length discrepancy

TABLE 38.2.2 Inter-professional team.

Inter-professional team members	
• Assistive Technology • Audiology • Augmentative communication • Behavioral psychology • Case management • Child life • Education • Law • Nutrition • Neuropsychology • Occupational therapy • Orthotics and prosthetics • Physical therapy • Seating • Social work • Speech and language pathology	• Therapeutic recreation • Dentistry • Endocrine • Gastroenterology • Genetic counseling • Genetics • Medicine • Neurology • Neurosurgery • Nursing • Ophthalmology • Pediatrics • Orthopedics • Physical medicine and rehabilitation • Prenatal medicine • Urology • Pharmacy • Psychiatry

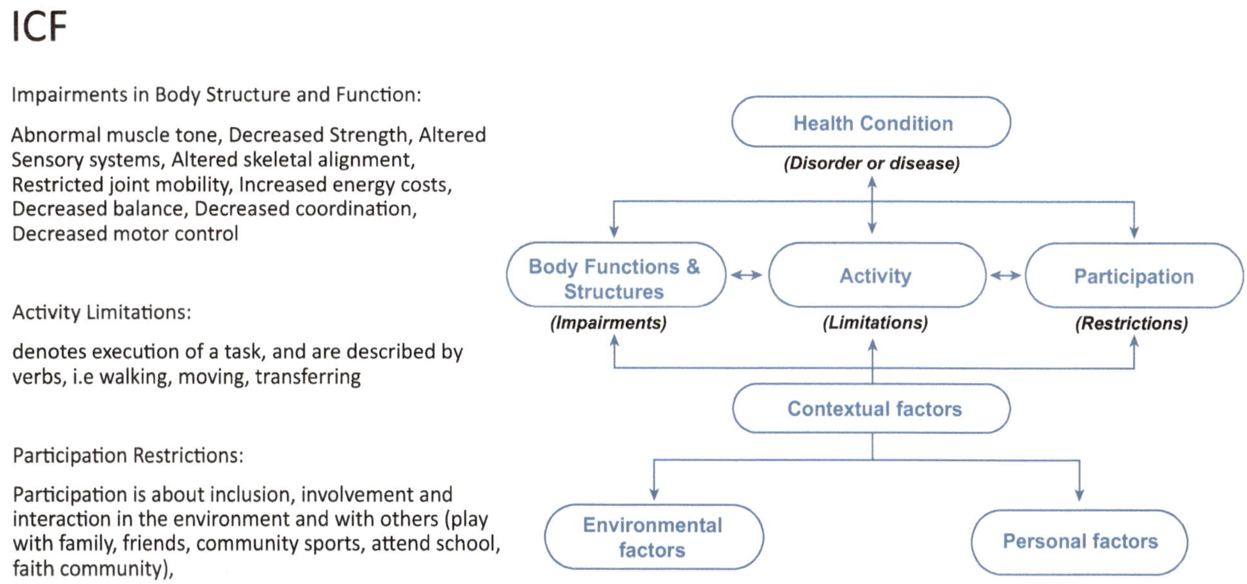

FIGURE 38.2.1 International Classification of Functioning, Disability, and Health—Children and Youth (ICF-CY).

medical resources. To assist with conceptualizing the complexity of childhood disabilities for optimal care and management within this population, pediatric physical therapy examination and interventions are typically framed by the International Classification of Functioning, Disability, and Health—Children and Youth (ICF-CY) developed by the World Health Organization. The ICF model considers that the child's functioning is determined by the multidirectional relationship among the domains of body function and structure, activity and participation, and contextual factors, as well as environmental and personal factors [4]. This model emphasizes the positivity of participation and activity versus functional limitations and disabilities (Fig. 38.2.1).

Contextual factors, defined as personal and environmental, interface with the three domains to influence functional performance. Environmental factors include physical, social, and attitudinal characteristics and are barriers or facilitators in time, place, and support. Personal factors are intrinsic to the person, such as age, gender and cultural factors, temperament, attitudes, beliefs, motivation, and engagement. The environment plays a key role in shaping all aspects of human development, with the family providing the most direct influence on a child's growth and development [5]. The ICF conceptual framework acknowledges the influence that the physical, social, and attitudinal environments have on a person's ability to participate in everyday activities. For infants and young children, the family plays a key role in shaping the influential environment that impacts on their motor development.

The medical language of the ICF-CY framework was modified by Rosenbaum and Gorter [6] into people's first language of the six "F" words in childhood disability: function, fun, family, friends, fitness, and future. These F-words focus on six key areas of child development and shift attitudes toward what individuals with childhood disabilities can do, the benefits of fun and physical activity on health and quality of life, where supports are found, and how futures are not defined by impairments (Fig. 38.2.2).

Test and measures for physical therapy diagnosis and prognosis

Typical motor development is represented by periods of adjustment and variability in motor skills performed by the infant [7]. Dynamic and functional movements are an essential part of development across multiple developmental domains. For example, infants with mobility restrictions, such as those that might be present following prematurity, may face adversity in proper perceptual motor development [8]. Motor delays are frequently the first or most obvious sign of an underlying medical condition in infancy. Parents and pediatricians keep a close eye on developing milestones since there is an expected pattern of achievement. The observation of altered achievement of gross motor milestones is likely to lead to early intervention services for evaluation.

Physicians and physical therapists alike rely on examinations that include the taking of birth and medical histories, system reviews, clinical observation, and tests and measures to identify children at risk. Physical therapists, with a focus on a child's motor capacity, take the subjective inquiry to understand social history, home setting, typical day, equipment used,

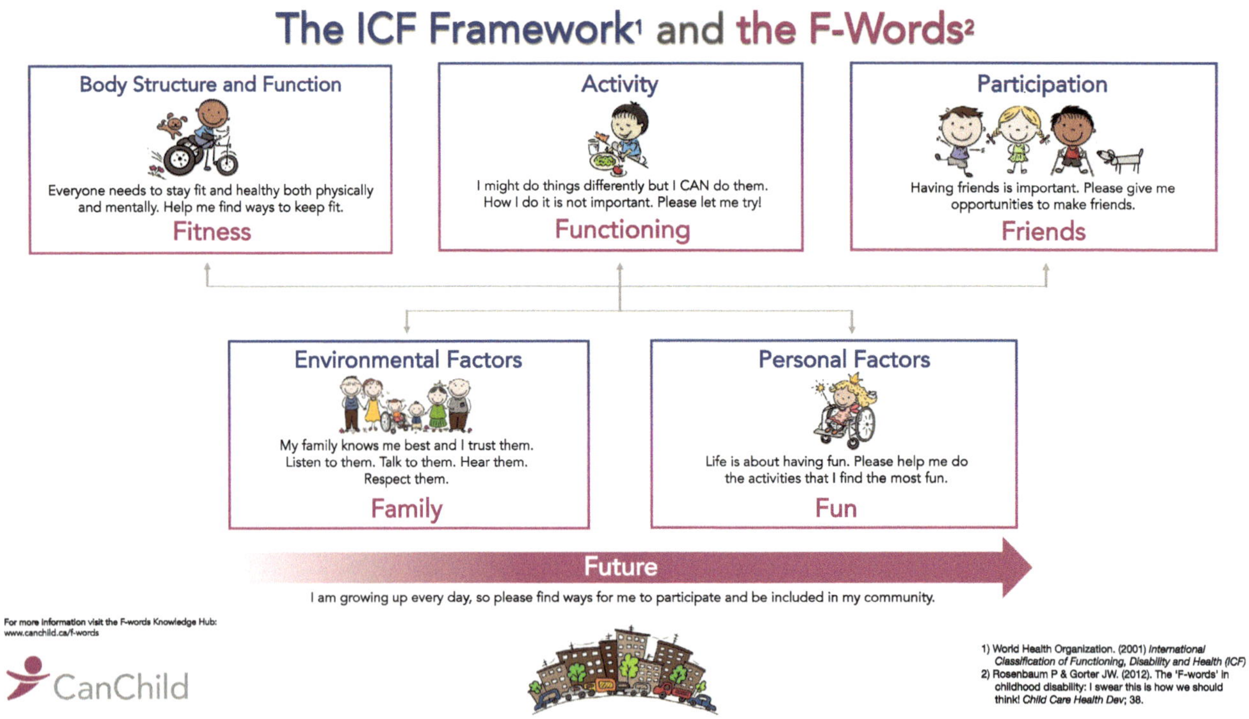

FIGURE 38.2.2 The ICF framework and the six key areas of child development.

the presence of pain, and patient/caregiver goals. The objective section of the PT examination may include range of motion, strength, sensation, pain, muscle tone, posture, balance, coordination, motor control, skin integrity, functional movement, and developmental skills. Listed in Table 38.2.3 are some of the many standardized outcome measures that are reliable and validated for use within the pediatric population for both diagnostic and prognostic assessment.

Dosing

The term dosing as applied to physical therapy means finding the right amount of therapy at the right time and includes determining the frequency (how often), the duration (how long), and the intensity of services (how hard the patient works). Dosing requires thoughtful consideration of the child's age, level of severity, readiness for change, and comorbidities. The frequency and duration of services may change as the child moves through various phases of development. Physical therapy services are goal-based, with a focus on functional activity and participation. Dosing is determined to meet the defined functional goals.

How dosing is determined is a complex process. According to Gannotti et al. [9], there are several guiding principles in determining the dosing of services. These include consideration of critical ages, lifespan activities, intensity and type of intervention, engagement, and supportive environments. The optimal dose should create sustainable changes at the structural and functional levels (ICF Model). Achieving movement in the patient's natural environment is the most sustainable. Considered by PT when recommending and dosing services are the known established critical periods for muscle development, brain development, and skeletal development [9]. Gannotti et al. [9] described the complexities of dosing using a pathway model as a framework to advance consideration of mediating and moderating factors that shape dosing parameters of interventions for children with CP (Fig. 38.2.3).

In this model, intervention type (focus on body structures, activity, or the environment) acts on a child first through the family, then through the dose (frequency, intensity, and time) to yield structural and behavioral changes. As a result, these changes are linked to improvements in functional independence. Family cultural values relate to outcomes for function and participation and need to be considered when deciding on physical therapy interventions and dosing.

Habilitative, rehabilitative, and supportive services as categories are different in respect to dosing. Habilitative implies teaching a new skill or skills. Rehabilitative means working to regain a lost skill. Both habilitative and rehabilitation services rely on learning-induced brain plasticity via activity-dependent cortical reorganization. Two of the key elements of motor learning are repetition and positive feedback. Dosing, therefore, strives to meet the repetition and intensity requirements.

TABLE 38.2.3 Outocme measures and tests.

Measure	Ages	Target	Main purpose	ICF category
Gross motor function measure (GMFM)	6 months–18 + years	CP, TBI, DS; other children whose motor function is less than 5 year level	Evaluative	Activity, body functions
Functional independence measure for children (WeeFIM)	6 months-21 years with disabilities; 6 months–7 years without disabilities	Children with disabilities, including head injury and CP	Discriminative; document rehabilitation outcomes	Activity
Alberta infant motor scale (AIMS)	0–18 months	Infants at risk for motor delay	Evaluative, predictive	Activity
Movement assessment of infants (MAI)	0–12 months	High risk infants	Assessment, detection	Activity, body structures and function
Batelle developmental inventory, 2nd edition	0–7 years	Infants and children with suspected or with developmental delay	Screening, evaluative, measurement, multidomain	Activity
Peabody developmental motor scale, revised (PDMS-2)	0–60 months	Children with suspected fine and/or gross motor delay	Developmental "diagnosis"; discriminative and evaluative	Activity, body structures and function
Bruininks-Oseretsky test of motor performance (BOTMP)	4.5–15.5 years	Children with motor problems	Discriminative; may "diagnose" coordination disorders	Body structures and function, activity
Movement assessment battery for children, 2nd edition (M-ABC2)	3–16 years	Children with motor function impairment	Evaluative, identification, descriptive	Body structures and function, activity
Hammersmith infant neurological examination	2–24 months adjusted age	Infants at high risk	Diagnostic	Body structures and function, activity
Bayley scales of infant and toddler development III (BSID-III)	1–42 months	Children with developmental delays	Diagnostic	Activity
Pediatric clinical test of sensory interaction for balance (CTSIB)	6–12 years	Children with balance deficits	Evaluative	Body structures and function
Pediatric Berg balance scale (PBBS)	5–15 years	Children with balance deficits	Evaluative	Activity
Beighton scale of hypermoblity	6 yrs - adult	Children with suspected joint hypermobility, joint pain	Evaluative, Diagnostic, quantify	Body structures and function
Barry Albright Dystonia Scale (BADS)	4 years, however, designed for adults	Children with cerebral palsy, secondary dystonia	Evaluative of severity	Body structures and function
Hypertonia assessment tool (HAT)	4–19 years	Children with increased muscle tone to discriminate between spasticity, dystonia and rigidity.	Evaluative	Body structures and function

(Continued)

TABLE 38.2.3 (Continued)

Measure	Ages	Target	Main purpose	ICF category
Selective control assessment of the lower extremity (SCALE)	4–25 years	Children with increased muscle tone	Evaluative providing a grading scale	Body structures and function
Pediatric quality of life inventory (Peds QL)	5–18 years	Children with disabilities	Evaluative measure of quality of life	Activity and participation
School function assessment (FSA)	Kindergarten through grade 6	Children with disabilities	Evaluative	Activity and participation
Pediatric evaluation of disability inventory (PEDI) and	0–8 years	Children with physical and/or behavioral conditions who may require caregiver assistance and modifications	Measurement/ Identification of function, Multidomain, document rehabilitation outcomes	Activity and participation (PEDI-CAT)
Pediatric evaluation of disability inventory – computer adapted test (PEDI-CAT)	0–20 years			
Pediatric outcomes data collection instrument (PODCI)	2–18 years	Children with musculoskeletal disorder/ injury	Outcome measure, patient reported	Activity and participation
Canadian occupational performance measure (COPM)	7+ years	Children with disability, adults	Outcome measure, self or parent reported	Activity and participation
Children's assessment of participation and enjoyment (CAPE) and preference for participation (PAC)	6–21 years	Children with disability	Outcome measure, self-report	Participation
Activities scale for kids – capacity (ASKc) and performance (ASKp)	5–15 years	Children with musculoskeletal limitations	Measure, self or parent report	Activity and participation

Practice is crucially important in motor learning. The means to achieve the requisite practice can be especially challenging for infants and children with neuromotor disabilities. The use of adaptive equipment and technology serves to create and facilitate the means for this practice to occur. Increasingly, as the fields of biomedical engineering and rehabiltation medicine collaborate, robotics, smart, and interactive equipment is being developed to bolster means of practice such as early kicking, reaching, and supported mobility for infants and children with neuromotor conditions.

In habilitative care, for the child who has never experienced typical development, dosing may be higher at younger ages, less than seven years of age, to maximize skill acquisition. Developmental literature describes that gross motor maturation is typically reached by 7 years, with further development manifesting in coordination for play and sport. Early intervention is, therefore, important because learning-induced brain plasticity at an early age seems to have a unique impact on brain development. Dosing for children with CP can be guided by the gross motor function curve. This curve describes the ages of maximal gross motor level by GMFCS level, helping to direct service dosing to maximize skill development before it is expected to level off. Within these early years of childhood, dosing frequency increases may follow a demonstration of emerging skills and, therefore, neuromotor readiness to benefit from intense, skilled practice for motor learning and control.

Rehabilitative services, on the other hand, are provided to assist a child in regaining skills lost secondary to acquired injury or illness, and for those children with neurodevelopmental conditions following surgery, growth with the associated increased body anthropometrics, range of motion loss, altered muscle function for force production, muscle weakness, and neurological changes such as sequelae from seizures. Dosing in rehabilitative cases relies on the principles of

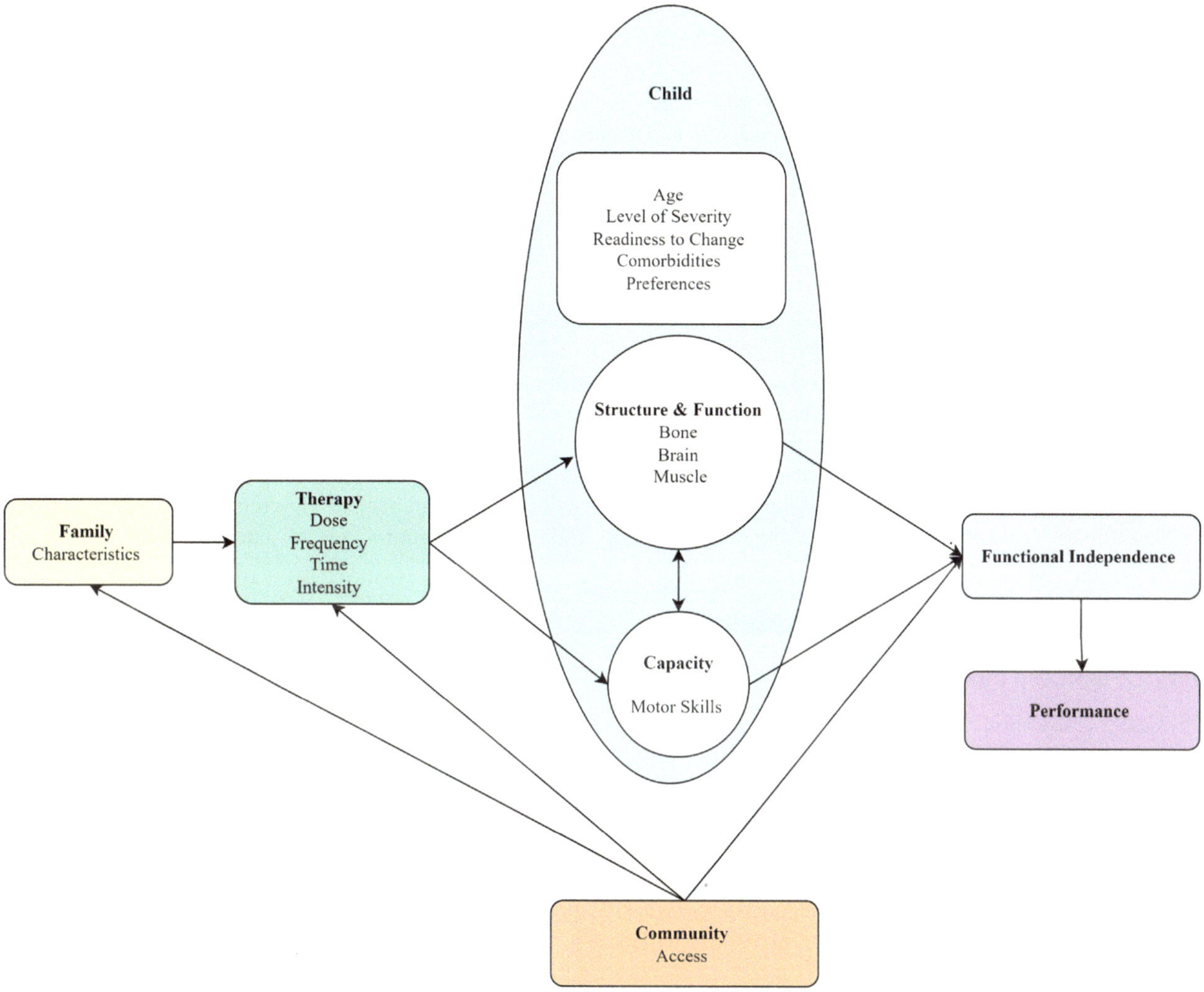

FIGURE 38.2.3 Pathway model for dosing.

neural plasticity and the body's innate healing processes while considering the same factors as with habilitative service. Rehabilitation dosing likewise depends on the severity of the presenting body structure and function impairments, activity and participation abilities, and the motor learning response to treatment.

In either a habilitative or rehabilitative service, physical therapy care may include or move to focus on the use of adaptive equipment and technology that serve to augment skills or compensate for functional challenges. This supportive service employs adaptive equipment prescription and training and a home exercise program to meet needs without expectation for new learning or regaining lost skills. This level of service also addresses potential or present secondary complications. The duration of the physical therapy care is likely shorter in these cases.

Physical therapy in progressive neuromotor disorders (neurodegenerative diseases) carries separate goals, expectations regarding response to intervention, and secondary or mitigating complications. The general approach to physical therapy remains defined by the ICF model and 6-Fs. Depending on the nature of the progressive neurologic condition or muscle disease, physical therapy will strive to maximize a child's strength, endurance, and function. As the degenerative process unfolds, physical therapy will work with the family and caregivers within the interprofessional team to anticipate physical management care needs and quality of life. This focus will include recommending and assisting with the procurement of indicated adaptive equipment for child engagement, seating and mobility, and ADLs.

Dosing may be reduced for several reasons, such as the child mastering the necessary motor skills for their participation goals, no longer needing skilled therapy to practice those skills, the establishment of a home exercise program or participation in a community-based fitness program, or due to family constraints. Additionally, if motor learning has plateaued despite continued efforts, this may indicate the need for discharge from skilled physical therapy, shifting the focus to providing education and other resources for the family.

Caregiver and patient education on home exercise programs and lifestyle routines is important for overall development and wellness. This may include education on proper sleep hygiene, nutrition, mindfulness, and physical activity. Similar to a typically developing child, daily wellness practices and regular exercise are recommended for children with neuromotor impairment. The patient may benefit from routine wellness exams a physical therapist for exercise prescriptions for musculoskeletal health, physical activity, and fitness. For the older child, coordinating efforts with athletic trainers and community programs to increase the frequency, timing, and duration of exercise for children with special needs is indicated across the lifespan. Recreational programs and exercise groups have been created to fill this need, including aquatic therapy, hippotherapy, yoga, dance, team sports, sailing, and skiing for children with special needs. All of these programs are valuable resources to promote movement activity, especially after discharge from physical therapy services.

Discharge planning begins during the initial evaluation, with the patient and their caregiver playing an active role in developing the plan of care, goals, and discharge plan. Measuring success or progress versus plateau in skill acquisition or lack of progress is ongoing during treatment and through regular re-evaluations. Physical therapists are trained to use critical reasoning in collaboration with the family and patient. The complexity inherent in the development of the plan of care stems from the numerous patient-specific variables that influence care, such as the setting, family dynamics, the availability of resources, and the innate variability in the growth and development of children (APTA FACT SHEET). Critical thinking guides the PT in generating, implementing, and evaluating approaches for quality patient care.

Service delivery systems and models

Service delivery models

Physical therapy services are delivered in a variety of public and private settings (hospitals, clinics, homes, schools, and communities). Questions often arise as to the role physical therapists (PTs) play in each setting, especially regarding public schools as defined by the Individuals with Disabilities Education Act (IDEA) and private clinics. Differences between school-based and clinic-based physical therapy services follow. This list should serve as a general guideline; the requirements and regulations themselves may vary by state or local education agency (LEA).

The Individuals with Disabilities Education Act (IDEA) was enacted "to ensure that all children with disabilities have available to them a free, appropriate public education that emphasizes special education and related services designed to meet their unique needs and prepare them for further education, employment, and independent living" (IDEA §300.1). Physical therapy, as a related service, is provided "to assist a child with a disability to benefit from special education" (IDEA §300.34). Physical therapy services are provided in several different models according to the age of the child and include the Infants and Toddlers Program (0–3 years, typically home-based), Child Find Preschool Services (center-based services), and school-based (to age 21 years). As indicated, providing additional medically necessary private home-based outpatient and inpatient services serves to fill gaps in care outside of the definition of IDEA services. Telehealth PT is a recently added service delivery model in response to the COVID-19 pandemic, and its future use remains to be determined.

The earliest setting where physical therapy might contribute to an infant's care is in the NICU. Collaboration between families and pediatric physical therapists is essential to achieving the most favorable developmental outcomes for high-risk infants [10]. Physical therapists working with infants in the NICU and their families have advanced postgraduate training to assume the role of supporting parent-infant interaction to promote bonding, caregiving, feeding, and positive neurodevelopmental outcomes [11]. Evidence supports that an early, intentional focus on caregiver education and family-centered services improves child outcomes in both the NICU setting and in home and community environments.

Early intervention (EI) services provided as Part C of IDEA may follow a NICU stay, or any infant or toddler at risk for developmental delay can be referred for eligibility. The focus of EI physical therapy is to enhance the development of infants and toddlers with disabilities through age 3 years. EI services are family-centered, occur in the context of a family's natural environments and routines, employ a coaching model, and are implemented in accordance with the Individualized Family Service Plan (IFSP).

Chronologically, school-based physical therapy (IDEA Part B) might follow. Part B serves the primary purpose of "ensuring that children with disabilities have available to them a free appropriate public education (FAPE) that emphasizes special education and related services designed to meet their unique needs and prepare them for further education, employment, and independent living" [12]. Under IDEA, physical therapy is considered a related service. Intervention in the educational setting under IDEA is not provided to remediate medical diagnoses or physical impairments, but instead to help the student achieve his/her learning and participation goals despite the existing diagnosis-driven impairments. Schools are responsible only for those services necessary for the student to make progress in his/her special education program. Discontinuation of related services may occur when skills needed have been incorporated into school routines, when needed supports are in place without the addition of services, or when the student no longer needs related service support to progress on goals/objectives (FACT Sheet).

School-based physical therapists (SBPTs) use a variety of adaptive equipment within their practice to help support students in the educational setting. The provision of equipment used by students during the school day is determined and documented in the student's Individual Education Plan (IEP). The team PTs are often responsible for providing needed materials and adaptive equipment for full access to the student's educational setting during the school day and to make progress toward participation goals [13].

Many school-based PTs have a limited or no budget allocation to obtain equipment to meet students' needs. Therefore, much coordination occurs between school-based, hospital-based, or outpatient-based PTs and their families to determine and assist with the attainment of optimal equipment. Insurance coverage of equipment for purchase requires documented trials to demonstrate efficacy. PT and the medical team work with insurance companies to purchase durable medical equipment for the individual student. Sometimes a child uses their personal piece of equipment within both the school and home; at other times, the school may provide the item from their existing inventory or purchase an item for their inventory for use by the student. At other times, a student may use one piece of equipment at school and another at home or in the community, which can be less than optimal considering motor control challenges that may impede learning with the use of two separate devices, such as walkers and gait trainers.

Along this age spectrum, from infancy into and including school age, a child might receive PT both through IDEA and privately. Similarities and differences between school-based and clinic-based physical therapy services are often misunderstood. Table 38.2.4 outlines some of the main differences. Please note that these are general guidelines; readers

TABLE 38.2.4 Differences between school-based and clinic-based physical therapy services.

What is the focus of the PT?	**School based PT Services**	**Clinic based PT services**
	Assist a student to achieve educational goals developed by the Individualized Education Program (IEP) team Promote access to academic curriculum and participation in other school activities Improve access to the school environment Address post-secondary transition goals.	Assist a child to achieve functional goals that enhance performance at home and/or in the community. Address medical continuum of needs, including impairments and functional limitations. Improve access to the home or community environment.
Who decides need & scope of physical therapy?	The IEP team decides by consensus, with consideration of a school-based PT's recommendation. The IEP team prioritizes to create discipline-free IEP goals and decides which services are necessary to achieve goals. Physicians may be a member of the IEP team. However, the physician referral alone does not drive decisions.	The PT, in partnership with the patient and family (and physician, if appropriate), agree upon a plan of care that is carried out by a PT or physical therapist assistant (PTA). Reimbursement for services may be limited by insurance coverage.
Where do physical therapy services occur?	In the least restrictive environment in the school where a student is expected to perform the task Worksites, buses, community, or other instructional settings. In a separate location, only when an intervention requires privacy or the student needs intense remediation that cannot be carried out in the natural environment.	Clinic, hospital, home, community, and other settings.
How are physical therapy services delivered?	Integration of physical therapy into the student's actual classroom or school activities. Consultation and collaboration with school staff and parent/guardian. Direct intervention in groups or individually, where necessary.	Direct intervention, telehealth/remote or consultation. Typically individual but may also be provided in groups.

Source: Adapted from the subcommittee on Physical Therapy for Educational Benefit (chaired by Carlo Vialu, PT) and the School-Based Physical Therapy Special Interest Group, (chaired by Marcia K. Kaminker, PT, DPT, Ms, Board-Certified Pediatric Clinical Specialist).

should familiarize themselves with the policies of their state or local education agency (LEA), as requirements and regulations may vary.

Transitional periods, lifespan issues and future needs

Transitional periods

Healthcare transition is a purposeful, uninterrupted process that addresses the medical, psychological, and educational/vocational needs of the individual and their family [15]. In the realm of neurodevelopmental practice, the transition from early school age into adolescence and then into young adulthood can be particularly challenging. Transitional pathways are well defined from EI to early childhood and into school-aged services. However, in adolescence and the teenage years, participation in community-based endeavors can be challenging due to limited programs that can accommodate children with special needs and access to available programs. Furthermore, during the transition into adulthood, finding age-appropriate, high-quality, adult-oriented health care systems can be rife with barriers such as a lack of information on resources available, a lack of adult specialists with experience, poor coordination between pediatric and adult providers, and a lack of reimbursement by insurance companies [15].

Transitional programs are being developed with solutions to improve the current system. Pediatric physical therapists can play a crucial role in interprofessional management, promoting patient's transitions as they age. PTs can empower children and adolescents, include patient perspectives early in the transition to adult care, assist families in finding providers, provide lists of community resources in their area, foster communication between the pediatric and adult healthcare teams, and lastly, advocate for improved healthcare coverage. The interprofessional team must have an early focus on the transition to adulthood.

Wellness across the lifespan

The life expectancy of children with intellectual and/or developmental disabilities (I/DD) such as cerebral palsy, spina bifida, and other neuro-developmental disabilities has improved dramatically over the last 30 years. Young adults with neuromuscular disabilities are living longer lives, with the majority of children with IDD surviving well into adulthood [16]. Although many diagnoses are "static" or "nonprogressive," adults with I/DD experience increased risk for health issues as compared to adults without IDD [17,18]. PT coordinates with the team and addresses, directly or indirectly, areas of potential secondary complications impacting wellness. Included in this focus on future health, wellness, and quality of life are (1) musculoskeletal/orthopedic, (2) pain, (3) fatigue and poor sleep, (4) metabolic, endocrine, and obesity, (5) nutritional intake and elimination, (6) genitourinary, and (7) psychosocial adaptation [19].

Musculoskeletal problems may include progressively altered muscle function and musculoskeletal deformities of the extremities and spine. These can be further compounded by decreased bone mineral density, osteopenia, and osteoporosis (Murphy KP). Evidence describes the altered biomechanical adaptations in muscle structure and function present in the spastic/hypertonic muscles of children with CP [20]. The force production aspect of muscle strength is decreased in spastic muscles and can become functionally more pronounced during and following a child's anthropometric growth. Longer lever arms and greater load (body weight) tax already weak muscles. Additionally, abnormal and restricted movement patterns and postures likely change the developing skeletal framework, potentially leading to abnormal orthopedic development, joint contracture, and altered lines of force production. PT evaluates for these impairments, anticipates future deformities, and, in coordination with the medical team, primarily orthopedics and physical medicine, intervenes accordingly with patient/parental/caregiver education such as activity-based strengthening, range of motion, bracing and splinting, and positioning protocols. Lastly, altered and reduced activity in infancy and childhood can result in the underdevelopment of appropriate bone mineral density in early childhood, potentially leading to osteoporosis or osteopenia (Henderson et al.). Medication interactions and side effects can impact bone growth and health and the PT will consider these effects in the treatment plan of care.

Perceived chronic pain and fatigue are very prevalent and impactful on quality of life. Pain can be secondary to early onset osteoarthritis in the back, neck, hips, and upper and lower extremities related to overuse, repetitive use patterns, and/or prolonged abnormal joint alignment and movement [21,22], chronic hypertonicity, or muscle spasm [23]. Fatigue may be due to the level of exertion necessary to perform daily activities owing to the structural and functional impairments of decreased muscle force production, altered joint range of motion, decreased aerobic capacity and cardiorespiratory endurance, limited access and/or opportunity to partake in physical activity, poor sleep, and pain. PT coordinates with the medical team and specifically addresses impairments contributing to pain and fatigue, working to remediate or provide compensatory or adaptive strategies to help manage pain.

Metabolic or endocrine issues can be defined as early-onset (second or third decade) "metabolic syndrome," such as early-onset type II diabetes, obesity, coronary artery disease, and high cholesterol [24]. Risk factors may include a lack of muscle mass and a high ratio of adipose tissue to muscle [24]. PT strives from infancy and early childhood to maximize muscle development and strength for physical activity that can be sustained at some level through the lifespan. Nutritional intake, poor hydration, and elimination problems, although not primarily addressed by PT, fall within the scope of PT practice and are an area of care coordination with nutrition and medicine.

Genitourinary problems may newly arise or those present from childhood, such as incontinence, may be exacerbated by early-onset aging and compounded by changes in muscle tone, deformity, and loss of muscle strength [17,18,25]. PT can help address these conditions by promoting physical activity, improving posture and positioning, and strengthening muscles for bladder and bowel control [25].

A decline in functional status [26] secondary to pain, fatigue, deconditioning, and a lack of muscular development from early infancy through adolescence is a key area addressed and monitored by physical therapy and collaboratively with the interprofessional team. Beyond these physical domain contributions to functional decline are challenges with psychosocial adaptation. These can be further compounded by a lack of social interaction, challenges with successful employment, a lack of independent living opportunities, and a risk for depression and substance abuse [16,27]. Physical therapists, understanding that infancy and childhood are the times to lay the foundation for physical function, work to maximize early physical capacity. Affecting even a small change in motor function early in life may tilt the slope of developmental curves, both physical and psychological, to see much larger impacts into adulthood [28]. PT throughout the lifespan is typically indicated to address the physical domain of the multidimensional activity and participation constructs for future health, wellness, and quality of life. Clearly, interprofessional collaboration and care are required from infancy into adulthood.

Future needs

The evolution of habilitative and rehabilitative care will continue to follow the clinical translation of research in neuroplasticity, robotics and other technologies, dosing and service delivery models, including telehealth technology, and physical inactivity sequelae.

Neuroplasticity research tells us that stimulation, especially experience-dependent, provided early in development or after injury or illness matters, and the intensity of practice is vitally important [29,30]. Most PT inventions for infants and children with neurodevelopmental conditions strive to tap into neuroplasticity, working through sensori-motor systems and motor learning theory. Increasingly, evidence-informed models of motor system plasticity are propelling advances in therapeutic interventions. Constraint-induced movement therapy (CIMT) is more recently being adapted to infants and babies under the age of 18 months who are at risk of developing hemiplegic cerebral palsy (CP). Virtual reality or active video gaming (VR/AVG) systems augmented, when appropriate, with neuromuscular stimulation are other examples. VR/AVG are systems, hardware, software, and/or setups that create computer-generated interactive simulations, presenting users with opportunities to use body movements to engage in environments that appear to be and feel similar to real-world objects and events [31,32]. Increasingly, neuromuscular electric stimulation (NMES) and functional electric stimulation (FES) modalities, through which muscles can be selectively activated and strengthened during functional tasks, are being delivered wirelessly and algorithmically specific to the motor task being performed. Evidence-informed models of motor system plasticity are propelling advances in manual therapy [33,34]. The precise use of the therapists' hands to improve body structures and physiology. Increasingly, engineering and physical therapy are collaborating in areas of research and development of rotobics, virtual reality (VR), computer-assisted rehabilitation, and neuromuscular electrical stimulation to further treatment and human performance measures. Devices to assist with early motor skill development, such as infant kicking [35,36], smart garments to promote reaching [37], and crawling [38], are in active research and development. Devices for movement's sake, such as Go Baby Go cars [39], versus destination-focused movement using powered mobility, demonstrate inclusivity for children with disabilities. Further research and development are ongoing with these collaborative efforts.

Research is demonstrating the potential for noninvasive brain stimulation (NIBS) to optimize developmental trajectories toward better outcomes, such as with the use of transcranial direct current stimulation (tDCS) combined with virtual reality (VR) [40,41] or treadmill training to improve balance and mobility in children with cerebral palsy [42].

Ongoing study on the mediation and modulation of dosing is needed. The complex and influential relationships between types of intervention and intervention responses owing to the constructs of family characteristics, child characteristics along the ICF domains, and community access warrant consideration from researchers who are designing intervention studies to guide clinical practice. Specific to neuroplasticity, further development of methods to promote and assist physical activity in daily routines, including the daily use of technology, and evidence of its benefit to improved function is warranted.

In response to the 2020 COVID-19 pandemic, telehealth PT quickly expanded its role in healthcare delivery. Used across service delivery models in IDEA and private outpatient clinics, telehealth demonstrates promise in improving healthcare access and equity, furthering parental and caregiver adoption of home exercise programs, and grounding interventions into the home and community routines important to the child or family. Telehealth outpatient services, conceptualized as a bridge between home-based and clinic-based care, have the setting in the patient's natural environment, but the therapist is remote. To support pediatric physical therapy telehealth practice, it is essential to develop training resources specific to telehealth and conduct further research to demonstrate its efficacy and effectiveness in this unique patient population.

The CDC lists childhood obesity in the United States as a serious problem that puts children and adolescents at risk for poor health. Children and adolescents with cerebral palsy and other neuromuscular conditions often do not meet the 60-min daily physical activity level recommended by the WHO [43,44]. Promoting wellness, physical fitness, and, most importantly, reducing the sedentary behavior of children and adults with neuromuscular disabilities is a critical step toward adopting a healthy lifestyle for long-term health benefits. Evaluation of sedentary behavior and interventions to decrease sedentary behavior in children with physical disabilities is needed [45]. Research has found that the relationship between physical activity and sleep can be bidirectional, where regular physical activity can improve sleep and sleep duration and quality will likely influence physical activity. Research is defining physical activity in individuals with neuromuscular impairment. For example, Verschuren et al. [46] have shown that transitioning from seated to standing positions, which is feasible for children with CP classified in GMFCS Levels IV and V, contributes to the accumulation of light activity, reducing sedentary behavior.

These future trends, tapping into neuroplasticity, embracing technology, understanding physical activity and sedentary behaviors, and designing optimal programs including dosing, are interrelated in providing children with neurodevelopmental disabilities with services that will improve their quality of life.

References

[1] APTA Academy of Pediatric PT FACT SHEET. Clinical reasoning in pediatric physical therapist practice; February, 2021.

[2] King G, Chiarello LA, Ideishi Roger, Ziviani Jenny, Phoenix Michelle, McLarnon Matthew JW, et al. The complexities and synergies of engagement: an ethnographic study of engagement in outpatient pediatric rehabilitation sessions. Disabil Rehabil 2021;43(16):2353–65. Available from: https://doi.org/10.1080/09638288.2019.1700562.

[3] Rosenbaum P, Walker S, Hanna S, Palisano R, Russell D, Raina P, et al. Prognosis for gross motor function in cerebral palsy: creation of motor development curves. J Am Med Assoc 2002;288(11):1357–63.

[4] World Health Organization. Toward a common language for functioning, health, and disability: the international classification of functioning, disability and health. Geneva, Switzerland: World Health Organization; 2002:9.

[5] Bronfenbrenner U. Children and families: 1984? Society 1981;18:38–41. Available from: https://doi-org.proxy1.library.jhu.edu/10.1007/BF02694667.

[6] Rosenbaum P, Gorter JW. The 'F-words' in childhood disability: I swear this is how we should think!. Child Care Health Dev 2012;38(4):457–63. Available from: https://doi.org/10.1111/j.1365-2214.2011.01338.x. Epub 2011 Nov 1. PMID: 22040377.

[7] Case-Smith J. Analysis of current motor development theory and recently published infant motor assessments. Infants Young Child 1996;9:29–41.

[8] Valentini NC, Gereira KRG, dos Santos Chiquetti EM, et al. Motor trajectories of preterm and full-term infants in the first year of life. Jpn Pediatric Soc 2019;61:967–77.

[9] Gannotti ME, Christy JB, Heathcock JC, Kolobe TH. A path model for evaluating dosing parameters for children with cerebral palsy. Phys Ther 2014;94(3):411–21. Available from: https://doi.org/10.2522/ptj.20130022. Epub 2013 Nov 14. PMID: 24231231; PMCID: PMC3967121.

[10] Sweeney JK, Heriza HB, Blanchard Y. Neonatal physical therapy: clinical competencies and NICU clinical training models. Part I. Pediatr Phys Ther 2009;21(4):296–307.

[11] Sweeney JK, Heriza HB, Blanchard Y, Dusing S. Neonatal physical therapy. Part II: practice frameworks and evidence-based practice guidelines. Pediatr Phys Ther 2010;2(1):2–16 3.

[12] Individuals with Disabilities Education Act. 20 USC §1400(d). School based paragraph section III; 2004.

[13] APTA Academy of Pediatric PT FACT SHEET. Management of equipment used by physical therapists in school-based practice; March 2021.

[14] ©2015 by the APTA Academy of Pediatric Physical Therapy, 1020 N Fairfax St, Suite 400, Alexandria, VA 22314-1488, http://www.pediatricapta.org (Section III similarities and differences between school based and clinic based PT).

[15] Lilly H, Bitzel M, Pejnovic T, et al. Barriers and characteristics for successful transition to adult healthcare for individuals with cerebral palsy: a systematic review. Phys Ther Rev 2019;24(5):195–207.

[16] Roebroeck ME, Jahnsen R, Carona C, Kent RM, Chamberlain MA. Adult outcomes and lifespan issues for people with childhood-onset physical disability. Dev Med Child Neurol 2009;51(8):670–8.

[17] Turk MA. Health, mortality, and wellness issues in adults with cerebral palsy. Dev Med Child Neurol 2009;51(suppl 4):24–9.

[18] Turk MA, Scandale J, Rosenbaum PF, Weber RJ. The health of women with cerebral palsy. Phys Med Rehabil Clin N Am 2001;12(1):153–68.

[19] Haak P, Lenski M, Hidecker MJ, Li M, Paneth N. Cerebral palsy and aging. Dev Med Child Neurol 2009;51(suppl 4):16–23.

[20] Moreau N, et al. Rapid force generation is impaired in cerebral palsy and is related to decreased muscle size and funcritonal mobility. Gait Posture 2012;35(1):154–8.

[21] Jahnsen R, Villien L, Aamodt G, Stanghelle JK, Holm I. Musculoskeletal pain in adults with cerebral palsy compared with the general population. J Rehabil Med 2004;36(2):78–84 10.

[22] Jahnsen R, Villien L, Egeland T, Stanghelle JK, Holm I. Locomotion skills in adults with cerebral palsy. Clin Rehabil 2004;18(3):309–16.

[23] Tuzson AE, Granata KP, Abel MF. Spastic velocity threshold constrains functional performance in cerebral palsy. Arch Phys Med Rehabil 2003;84(9):1363–8.

[24] Peterson MD, Gordon PM, Hurvitz EA. Chronic disease risk among adults with cerebral palsy: the role of premature sarcopoenia, obesity and sedentary behaviour. Obes Rev 2013;14(2):171–82.

[25] Henderson CM, Rosasco M, Robinson LM, et al. Functional impairment severity is associated with health status among older persons with intellectual disability and cerebral palsy. J Intellect Disabil Res 2009;53(11):887–97.

[26] Jahnsen R, Villien L, Aamodt G, Stanghelle J, Holm I. Physiotherapy and physical activity - Experiences of adults with cerebral palsy with implications for children. Advances in Physiotherapy 2003;5(1):21–32.

[27] Vogtle LK. Employment outcomes for adults with cerebral palsy: an issue that needs to be addressed. Dev Med Child Neurol 2013;55(11):973.

[28] Grohs MN, Hilderley A, Kirton A. The therapeutic potential of non-invasive neurostimulation for motor skill learning in children with neurodevelopmental disorders. Curr Dev Disord Rep 2019;6:19–28.

[29] Kolb B, et al. Brain plasticity in the developing brain. Prog Brain Res 2013;207:35–64.

[30] Mateos-Aparcio P, Rodriguez-Moreno A. The impact of the study of brain plasticity. Front Cell Neurosci 2019;13:66.

[31] Levac D, Glegg S, Colquhoun H, Miller P, Noubary F. Virtual reality and active videogame-based practice, learning needs, and preferences: a cross-Canada survey of physical therapists and occupational therapists. Games Health J 2017;6(4):217–28.

[32] Banerjee-Guénette P, Bigford S, Glegg SMN. Facilitating the implementation of virtual reality-based therapies in pediatric. Rehabil Phys Occup Ther Pediatr 2020;40(2):201–16.

[33] Ponzo V, Cinnera AM, Mommo F, Caltagirone C, Koch G, Tramontano M. Osteopathic manipulative therapy potentiates motor cortical plasticity. J Osteopath Med 2018;118(6):396–402.

[34] Tramontano M, Cerritelli F, Piras F, Spanò B, Tamburella F, Piras F, et al. Brain connectivity changes after osteopathic manipulative treatment: a randomized manual placebo-controlled trial. Brain Sci 2020;10(12):969.

[35] V. Emeli, K.E. Fry and A. Howard, Towards infant kick quality detection to support physical therapy and early detection of cerebral palsy: a pilot study, 29th IEEE international conference on robot and human interactive communication (RO-MAN), 2020a, p. 1069–1074, Available from: https://doi.org/10.1109/RO-MAN47096.2020.9223571.

[36] V. Emeli, K.E. Fry and A. Howard, Robotic system to motivate spontaneous infant kicking for studies in early detection of cerebral palsy: a pilot Study, 8th IEEE RAS/EMBS international conference for biomedical robotics and biomechatronics (BioRob), 2020b, pp. 175–180, Available from: https://doi.org/10.1109/BioRob49111.2020.9224409.

[37] Greenspan B, Cunha AB, Lobo MA. Design and validation of a smart garment to measure positioning practices of parents with young infants. Infant Behav Dev 2021;62:101530. Available from: https://doi.org/10.1016/j.infbeh.2021.101530. Epub 2021 Feb 4. PMID: 33548894; PMCID: PMC7927978.

[38] Tripathi T, Dusing S, Pidcoe PE, Xu Y, Shall MS, Riddle DL. A motor learning paradigm combining technology and associative learning to assess prone motor learning in infants. Phys Ther 2019;99(6):80716. Available from: https://doi.org/10.1093/ptj/pzz066. PMID: 31155666.

[39] Huang H, Ragonesi CB, Stoner T, Peffley T, Galloway JC. Modified toy cars for mobility and socialization. Pediatr Phys Ther Spr 2014;26 (1):76–84.

[40] Llorens R, Fuentes MA, Borego A, et al. Effectiveness of a combines transcranial direct current stimulation and virtual reality-based interventiom on upper limb function in chronic individuals post-stroek with persistent severe hemiparesis: a randomized control trial. J Neuroeng Rehabil 2021;18:108.

[41] Silva TDd, Fontes AMGG, Oliveira-Furlan BSd, Roque TT, Lima AII, Souza BMMd, Alberissi CAdO, Silveira AC, Moraes ÍAPd, Collett J, Silva RP, Airoldi MJ, Ribeiro-Papa DC, Dawes H, Monteiro CBdM. Effect of combined therapy of virtual reality and transcranial direct current stimulation in children and adolescents with cerebral palsy: a study protocol for a triple blinded tandomized controlled crossover trial. Front Neurol 2020;11:953.

[42] Grecco LAC, Duarte NdAC, Mendonça ME, Cimolin VO, Galli M, Fregni F, Oliveira CS. Transcranial direct current stimulation during treadmill training with children with cerebral palsy: a randomized controlled double-blind trial. Res Dev Disabil 2014;35(11):2840–8.

[43] Johansen M, Rasmussen HM, Lauruschkus K, Laugesen B. Measurement of physical activity in children and adolescents with cerebral palsy: a scoping review protocol. JBI Evid Synth 2021;19(9):2339–49.

[44] Carlon SL, et al. Differences in habitual physical activity levels of young people with cerebral palsyand their typically developing peers: a systematic review. Disab Rehab 2013;35(8):647–55.

[45] Ganz F, et al. Sedentary behavior and children with physical disabilities: a scoping review. Disabil Rehabil 2020.

[46] Verschuren O, Peterson MD, Leferink S, Darrah J. Muscle activation and energy-requirements for varying postures in children and adolescents with cerebral palsy. Pediatrics 2014;165:1011–16.

Chapter 38.3

Occupational therapy

Scott Frampton
Department of Occupational Therapy, Kennedy Krieger Institute, Baltimore, MD, United States

Introduction

Pediatric occupational therapists receive training that is based on occupation and activity analysis. Their education is based on infant and child development, anatomy and physiology, neuro-motor development, neurosciences, and psychosocial education. Neurodevelopmental occupational therapy is rooted in engaging children and their families in meaningful occupations in their service delivery model.

In occupational therapy, *occupations* refer to the everyday activities that people do as individuals, in families, and within communities to occupy time and bring meaning and purpose to life. Occupations include things people need to, want to, and are expected to do [1]. Occupations have significant meaning to the child and their caregivers. This adds value to their identity and sense of self. Meaningful engagement in daily activities is important for development, learning, practice, and mastery of skills.

Occupational therapists focus on the child's engagement in meaningful occupations. They use their skills to assess the relationship between the child, social and environmental context, performance skills, and patterns. A holistic approach is utilized when evaluating a child and planning treatment interventions.

The occupational therapy domain includes occupations, context, performance patterns, performance skills, and client factors. Occupations are activities such as activities of daily living (ADL's), instrumental activities of daily living (IADL's), health management, rest and sleep, education, play, leisure, and social participation. The contexts are both environmental and personal factors. Performance patterns include habits, roles, routines, and rituals that a child engages in. Performance skills such as motor skills, process skills, and social interaction skills are crucial for occupational engagement. Client factors include values, beliefs, and spirituality, as well as body functions and body structures [2].

Occupational therapists take all of the aspects of our domains to form an occupational profile of the child and their family or caregivers. "The occupational profile is a summary of a client's (person's, group's, or population's) occupational history and experiences, patterns of daily living, interests, values, needs, and relevant contexts" [3]. Information is obtained through observation, discussion with the child and/or family, and formal evaluations. The occupational profile and treatment interventions are individualized and tailored to that child or family.

Occupational therapy, in the context of neurodevelopmental disabilities, is charged with communicating the importance of occupations to families and caregivers and explaining their role, as well as the role of other team members, in acheiving successful engagement. Treatment focuses on the use of occupations and the grading of tasks up or down (making them harder or easier based on performance) to work toward the attainment of skills and increased occupational performance.

Multidisciplinary care

There are various terms used to describe groups of professionals working with children and their caregivers. This can include interprofessional, interdisciplinary, and multidisciplinary. A range of terms—interprofessional, multiprofessional, interdisciplinary, and multidisciplinary—are used to describe health care teams. Multidisciplinary is most frequently used to describe health care teams [4]. Within these various terms, occupational therapists may work in coordination with other members of the team to identify an appropriate plan of care, or they may determine a plan of care for occupational therapy services and communicate this with the child's pediatrician. However, communication amongst professionals and the physician is important to ensure that all of the needs of the child and family are being met. This allows for open dialog and the ability to make referrals to other disciplines when they see issues that are outside of their scope of practice.

Occupational therapists do not work alone but rather are part of a multidisciplinary team to evaluate a child and then provide treatment geared toward the child reaching their potential and successful engagement in occupations in their natural environment. This can be in the home, school, community, clinic, etc.—any place that has meaning and in which the child is participating. Communication and coordination of care amongst team members are important to provide a holistic approach to treating the child. The multidisciplinary team may include a few or many members, depending on the needs of the child. However, it should be noted that communication amongst disciplines is key to looking at the child and the family in a holistic manner. Multidisciplinary teams for children may include, but are not limited to, the following:

Assistive technology	Audiology	Behavior psychology
Child life specialists	Education	Endocrine
Neuropsychology	Neurology	Neurosurgery
Nursing	Nutrition	Occupational therapists
Ophthalmology	Orthopedics	Orthotists
Pediatrics	Physical medicine/rehab	Social work
Speech language pathology		

Occupational therapy evaluations

Neuromotor dysfunction can manifest in children in many areas. There may be impairments in fine motor, visual motor, visual perceptual skills, postural control, functional mobility, muscle tone, oral motor/swallowing skills, and appropriate responses to sensory input. These may limit a child's performance in preferred occupations. Basic activities of daily living (ADL), such as dressing, grooming, and being able to feed one's self, may all be impacted. A child may have difficulty participating in their roles and preferred leisure activities due to any of these impairments. A referral to occupational therapy may be appropriate when there are concerns in any of the above areas that may be affecting a child's performance or participation in activities.

Occupational therapists may use both norm-referenced and criterion-referenced assessments. A norm-referenced motor assessment provides information regarding the function of an individual in comparison to a normative sample. This allows the clinician to rate the child's performance and compare it to that of children of a similar age. In comparison, a criterion-referenced assessment analyzes the child's performance in regards to a specific task. This allows the clinician to assess skills in relation strictly to the task versus how they do compared to their peers.

There are many different assessments available to occupational therapists, in addition to observation and clinical reasoning. Using their training, an occupational therapist will determine which assessment will provide the information needed to best serve the child and their caregivers in determining strengths and limitations in occupational performance. The focus may fall within many areas, including fine motor/visual motor skills, visual perceptual skills, sensory integration, general development, handwriting, and feeding. Both normative and criterion-referenced assessments may be used. The following tables list some possible assessments that may be utilized, though this is not an exhaustive list (Tables 38.3.1 and 38.3.2).

Information obtained through these assessments, in conjunction with direct observation and caregiver interviews, assists the occupational therapist in identifying the child's occupational profile. Strengths and limitations are noted in order to create an individualized plan to assist the child with the successful performance of their occupations (Table 38.3.2).

General principles in pediatric occupational therapy treatment

Pediatric occupational therapy practice draws on developmental anatomy and physiology, infant and child development, neuroscience, and neuro-motor development. Their training in psychosocial assessment and interactions allows them to take a holistic approach to working with the child and caregivers. Occupational therapists utilize activity analysis to plan a treatment program. There are many steps involved in all tasks. The occupational therapist analyzes all of the components of a task and the environment. They grade the task up or down (making it harder or easier) to improve the child's success, based on their current performance. Through frequent task analysis and observation of performance, the occupational therapist strategizes how to advance the child's performance with the ultimate goal of performing the task or activities with the least amount of assistance, in order to promote greater independence.

This allows activities to be adjusted to ensure success and the ability to improve the skills that are needed for improved participation in a child's occupations. Occupational therapists use frames of reference to guide their treatment. Frames of reference provide a theoretical basis for the therapist to use when working with the child. There are many frames of reference that may be utilized by the occupational therapist. Several examples are the biomechanical

TABLE 38.3.1 Norm referenced assessments.

Name of assessment	Age group	Domains of evaluation
Bayley Scales of Infant Development, (BSID, 4th ed.)	Ages 1–42 months	Evaluates 5 domains: cognitive, language (receptive & expressive), motor (gross & fine), social-emotional and adaptive
Beery- Buktenica developmental test of visual-motor integration (Beery VMI, 6th ed.)	Ages 2–18 years	Evaluates the extent to which individuals can integrate their visual and motor abilities.
Bruininks-Oseretsky test of- motor proficiency (BOT-2)	Evaluates ages 4 years through 21 years	Evaluates fine manual control, manual coordination, body coordination, and strength and agility
Developmental test of visual perception (3rd ed.)	4–12 years, 11 months	Evaluation of visual perception and visual-motor integration
Motor-free visual perception test-4 (4th ed.)	4 years to 80 + years	Evaluates visual perceptual skills without requiring a motor response
Peabody developmental motor scales (PDMS-2)	Birth through 5 years of age	Evaluates fine and gross motor skills
Pediatric evaluation of disability inventory (PEDI)	6 months–7 years	Evaluates self-care, social function, and mobility
Sensory integration and praxis test (SIPT)	Ages 4 years–8 years, 11 months	Evaluates sensory integration processes that affect learning and behavior
Sensory profile 2	Birth to 14 years, 11 months	Evaluates a child's sensory processing patterns in the context of home, school, and community-based activities
Vineland adaptive behavior scales (Vineland 3)	Birth to 90 years old	Evaluates communication, daily living skills, socialization, motor skills, and maladaptive behavior

TABLE 38.3.2 Criterion referenced assessments.

Name of assessment	Age group	Domains of evaluation
Hawaii early learning profile (HELP)	Ages 0–6 years	Evaluates 7 domains: regulatory/sensory organization, cognition, language, gross motor, fine motor, social emotional, and self help
Evaluation tool of children's handwriting	Ages 2 years, 9 months–5 years, 8 months	Evaluates foundation index (assesses basic sensory-motor abilities), coordination index (consists of more complex gross, fine and oral motor tasks), verbal index (language items testing verbal memory, sequencing, comprehension, association and expression), nonverbal index (nonspoken items examining memory, sequencing, visualization and mental manipulations), and complex tasks index (the interpretation of visual-spatial information)
School functional assessment (SFA)	Kindergarten through 6th grade	Evaluates functional tasks that affect the academic and social aspects of an elementary school program

frame of reference, Developmental frame of reference, Rehabilitation frame of reference, and Model of Human Occupation (MOHO), though there are many more. The occupational therapist will use this knowledge in determining the most appropriate frame(s) of reference for analyzing and treating the child in order to modify their environment, improve skills, and increase their overall participation in daily occupations.

Service delivery models

Occupational therapy services may be delivered in a variety of settings (homes, outpatient clinics, schools, and hospitals). Therefore, this will be discussed to provide a better understanding of the purpose and operation within these settings.

Infant care in the Neonatal Intensive Care Unit (NICU) may be a child's first experience with occupational therapy. "Occupational therapists must be able to understand, assimilate, and use advanced clinical reasoning and appropriate intervention to promote the organization of the infant while supporting the parent- or caregiver-infant dyad" [5].

The clinician uses their knowledge and clinical reasoning skills to assess for both typical and atypical performance. Adapting or structuring the environment to enhance function is a well-accepted approach to caring for infants in the NICU [5]. Occupational therapists have a good understanding of the development of infants who are born prematurely and are able to support the team and family in working toward enhancing the development of the child. A knowledge of neurodevelopment and neurobehavioral organization assists them with working with a child who may have difficulty with all of the sensory stimulation and demands for which they are not yet prepared.

The Individuals with Disabilities Education Act is a law that entitles every child, from infancy to young adulthood, to a free, appropriate, public education through early education (EI) and special education services. Within this delivery model, occupational therapy is a "related" service and is designed to be provided "to assist a child with a disability to benefit from special education [6]. The focus is on education versus a rehabilitative model.

Part C of IDEA deals with early intervention services (birth through 36 months of age), while Part B, also known as Child Find, applies to services for school-aged children (3 through 21 years of age) [7]. Under this law, occupational therapy services through EI would be provided to children who qualify and are aged birth through 3 years. An infant or toddler can begin the process of receiving Part C early intervention services by having a condition with a high probability of developmental delay [8].

Two concepts are increasingly prioritized when translating evidence-based neuroscience into functional application for "best practice" provision of EI services: (1) creating frequent opportunities that allow for "learning in the natural environment" rather than in simulated "treatment" situations; and (2) utilizing methods of "coaching" as a model for families, medical homes, and EI programs providing services to infants [9]. An Individualized Family Service Plan (IFSP) is developed in coordination with the child's caregivers and members of the child's team. This is a family-centered approach that utilizes the child being in their natural environment and engaging in occupation. Occupational therapists use their knowledge of development, skill acquisition, and context to work with a child and their family. In this delivery model, the focus is on coaching caregivers in techniques that advance their child's development. The active participation of the caregivers and child is crucial to achieving their goals. Using task analysis and by modifying the environment to enhance the success of the child, the occupational therapist works with the family to incorporate learned strategies into their daily routines. This allows for opportunities for practice within their natural environment.

Following EI services, a child may move to school-based occupational therapy services through IDEA Part B. In this service model, also known as Child Find, occupational therapy is a related service with the goal of enabling the child with special needs to benefit from special education. An individualized education plan is created by an interdisciplinary team to ensure that the child is able to fully participate in the special education program. The focus of occupational therapy is not to repair or remediate the impairments of the child but to assist with the skills needed for participation in a special education setting. These services may be either direct or consultative.

School occupational therapists (and occupational therapy assistants, under the supervision of the occupational therapist) support academic and nonacademic outcomes, including social skills, math, reading and writing (i.e., literacy), behavior management, recess, participation in sports, self-help skills, prevocational/vocational participation, transportation, and more [10]. Occupational therapists provide treatment and strategies that are focused on the child's participation within the school environment. This may include working on fine motor, visual motor, and visual perceptual skills for better manipulation of classroom materials and handwriting. Occupational therapists may also address issues such as clothing management and self-feeding as they relate to the child in the school.

An occupational therapist may also address issues with sensory integration as they pertain to a child's ability to participate in activities in an educational setting. Issues with sensory integration disorder manifest in many ways. Children may be sensory-seeking or sensory-avoiding. Each makes it difficult to focus on tasks, both at school and in the home environment. They may be overly sensitive to light, sound, the feeling of their clothing, etc. This can make it difficult to focus on the task at hand, as they may be overwhelmed by sensory stimuli. At the other end, they may be under-sensitive and require frequent movement, deep pressure, etc. in order to regulate their sensory systems. Difficulty maintaining one's sensory regulation, no matter where they fall on the continuum, can significantly impact a child's performance of daily activities, limiting their occupational performance. An occupational therapist can identify the issues and make recommendations to ensure the child's success.

In addition to addressing the barriers to a student's participation, such as those listed above, and providing recommendations to enhance the student's performance, the occupational therapist may identify any physical barriers that may need to be addressed, such as adaptive seating or adaptations for better manipulation of items utilized in the educational setting. An occupational therapist will also look for other instruments to assist students, such as augmentative communication.

Outpatient occupational therapy services in a clinic are another delivery model for children with neurodevelopmental disabilities. These services may occur in conjunction with school-based services or separately. In this setting, children are seen for a variety of reasons, including, but not limited to: fine motor, visual motor, visual perceptual skills, sensory integrative dysfunction, self-help skills, and oral motor/swallowing difficulties. The occupational therapist will work with the child and family to determine the challenges the child is having with their participation in preferred occupations. This will help decide what tools will be utilized in the assessment of the child.

A "therapeutic triad" exists in pediatric care in which there must be communication between parent-child, child-practitioner, and parent-practitioner. This will lead to better engagement from all parties and promote follow-through on recommendations from the occupational therapist. Working together, taking what is learned during therapy sessions, and incorporating this into a child's daily roles, and routines can enhance their performance in occupations.

Inpatient hospital occupational therapy services may be required at times for children with neurodevelopmental disabilities. This may be in concert with a surgical procedure, a prolonged medical admission, or as part of an acute rehabilitation program associated with a newly acquired brain injury. In this model, children receive intensive services from an interdisciplinary team in a medically based model. In addition to hospital-based therapies, occupational therapists will assist with transitioning the child to the appropriate community-based services.

There are many similarities between school-based and clinic-based occupational therapy services:

- Adhere to state licensure laws, regulations, and professional standards. Apply the best evidence and best practices in pediatric occupational therapy.
- Provide services based on the needs of the child.
- Utilize principles in both occupation and neuroscience.
- Work with family and other team members to provide treatment and interventions to increase a child's participation in their occupation.
- Monitor progress and adjust treatments or interventions as needed.
- Discontinue services when appropriate, though they may be restarted based on changes in the child's performance, which may warrant another bout of therapy (Table 38.3.3).

TABLE 38.3.3 Occupational therapy services: clinic-based compared to school-based.

	Clinic-based occupational therapy services	School-based occupational therapy services
Focus of occupational therapy	Emphasis on helping the child to improve function and performance for success in the home and community environments. Address medical continuum of needs. This includes: Occupations, Content & Environment, Performance Patterns, Performance Skills, and Client Factors.	Provide services that help the student to achieve goals developed by the Individualized Education Program (IEP) team. The goals are based on educationally relevant needs. Address the child's skills to foster greater participation, management of school materials, and success in school related activities. Address transition from school-based services.
Determination of occupational therapy services	The Occupational Therapist works with the patient and family to determine the goals and plan of care. Services are carried out by an Occupational Therapist or Certified Occupational Therapy Assistant. Insurance coverage may affect reimbursement for services.	Services and goals are determined by the IEP team, with input from the Occupational Therapist. Goals are based on the needs of the student. The team determines who will provide assistance, where, and how often to assist with success of the student.
Setting for occupational therapy services	Hospital, therapy clinic, home, as well as community settings	In following federal regulations, services are provided in the least restrictive environment. Services may be provided within the classroom setting, or the student may be pulled out and be seen individually, as necessary.
Delivery of occupational therapy services	Occupational Therapist or COTA provide direct intervention with the patient and/or family. Services are generally "in-person," though a TeleHealth model may also be used. Generally, services are individual, but may also be cotreats with other disciplines, or group sessions, depending on the setting, and needs of the patient.	Occupational therapy services may direct, or consultative. They may take place in the classroom, and/or the student may be pulled out for an individual session.

Seating and mobility

Children with neurodevelopmental disabilities may require specialized seating systems to meet their positioning and mobility needs. Proper positioning is of the utmost importance for these children to prevent possible deformities that may occur over time and to promote function and mobility. A proper seating system is crucial for proper alignment of the trunk, head, and lower extremities and allows for optimal positioning and use of the upper extremities. Adequate positioning will allow for greater interaction with the environment and participation in occupations, whether visually or through upper-extremity engagement. It also improves pulmonary function and decreases the likelihood of skin breakdown (Table 38.3.3).

The team to assess the seating and mobility system may consist of an occupational therapist and/or physical therapist, a child, a caregiver, and an assistive technology professional (ATP). The team assesses the child's position, motor function, cognition, mobility, social, and environmental factors. The occupational therapy assessment seeks to promote a seating system that will maximize a child's participation in daily occupations.

Occupational therapists work with the child and family to determine what occupations are important to them and how they engage in them. Assessing both strengths and limitations to the child's participation is important in order to identify a seating and mobility system that will optimize the child's participation in preferred occupations.

A mat evaluation is completed to provide the necessary information for determining which type of seating or mobility system is appropriate. The occupational therapist assesses both active and passive ranges of motion as they relate to a child's sitting posture. This includes the hips, knees, ankles, and upper extremities. The position of the pelvis is assessed both in a supine and a seated position. The available range of motion, pelvis obliquities, tilt, and rotation are noted. The spine is assessed for any signs of scoliosis, kyphosis, or lordosis. If any of these are noted, they will be assessed to see if these spinal limitations are fixed or flexible. The child's sitting balance and posture, along with their head control, are assessed. The muscle tone and its impact on positioning are assessed. This may be in any part of the body and may be hypertonic, hypotonic, or variable. Skin integrity is also assessed. Any current or prior issues with skin integrity or decreased sensation are noted. This will assist in determining a proper, pressure-relieving seat cushion.

An environmental evaluation of mobility also occurs. Many factors are taken into consideration. In what environments will the child be using the wheelchair? This may include the home environment, paved roads (cut-out or no cutouts), grass or gravel, or rural environments. Also, a big factor is the bulk of the seating system. Things to take into consideration are if the person will be using a wheelchair-accessible vehicle or if and how it can be transported in the family's vehicle.

There are many options available for seating and mobility systems. Seating and mobility systems can take the form of a special needs stroller, a manual upright wheelchair, a tilt-in-space wheelchair, or a power wheelchair. Each seating system takes into account what is required to maintain proper trunk, head alignment, skin integrity, means of mobility, and the ability to promote engagement in occupation. The occupational therapist, along with the team, assesses the supports that are needed for each child. There are many specialized options available to meet a child's needs. This can range from a standard seating system with things such as trunk lateral supports and a head rest to a custom-molded seat and seat back.

The means of mobility are also assessed. The child may be dependent or have some upper extremity function to allow them to propel a wheelchair. For those who have the appropriate cognitive, visual, and motor skills, power mobility may be another option that could lead to greater independence.

Transportation for children with special needs

In a policy statement, the American Academy of Pediatrics states, "All children, including those with special health care needs, should have access to proper resources for safe transportation" [11]. The use of child safety seats is important for all children when riding in a vehicle. Studies found that the use of child safety seats reduces the risk of injury by 71%–82% [12] and the risk of death by 28% when compared to children of similar ages with only seat belts [13].

Child safety seats (CSS) are essential for maintaining the safety of children when riding in a vehicle. Vehicle safety is designed for adults, not children. CSS is utilized to provide safety protections for children. In order for it to function and work as intended, there are three things that must be followed: appropriate CSS, the child positioned properly, and the CSS installed according to the manufacturer's instructions.

An appropriate child safety restraint is one that meets or exceeds FMVSS 213. This is a federal standard set by the National Highway Traffic Safety Administration (NHTSA) that regulates the design and performance of child restraint

TABLE 38.3.4 Stages of car seats.

Stage 1: Rear facing – until at least 2 years of age; best practice says until the child has outgrown the height or weight limits based on the manufacturer's guidelines.
Stage 2: Forward facing – recommended to keep children harnessed in a forward facing positon until at least 5 years of age.
Stage 3: Booster seat – used until 4 ft 9 in, or the seat belt fits appropriately; at least 8 years of age.
Stage 4: Seat belt – the shoulder strap should rest on the shoulder (not against the neck), and the lap portion should rest on the upper thighs; the child's knees should be able to bend when sitting all the way back in the seat; feet can be positioned flat on the floor; should ride in the back seat until at least 13 years of age.

systems for children weighing up to 80 lb. For the seat to be appropriate, the child should meet the height and weight guidelines as set by the manufacturer for either rear-facing or forward-facing. In addition, the seat should not have been recalled (see the NTHSA website at https://www.nhtsa.gov/recalls#car-seats) or expired. Because car seats are made of petroleum-based products and can be exposed to extreme temperatures, both hot and cold, it is possible that the plastic may become compromised over time. The owner's manual will provide information on the time from manufacture to the time of expiration. This information may also be found on the CSS. Most car seats typically have expiration dates of six to eight years from the date of manufacture.

Positioning the child in the CSS appropriately means following the manufacturer's instructions. This is important, especially for rear-facing car seats, as there are often head supports or body positioning supports that have specific weights for when they may or may not be used. When a child is riding in a rear-facing position, the straps should be at or below the height of the child's shoulders, the chest clip should be at arm pit level, and the harness should be tightened so that the straps cannot be pinched off of the child's shoulders. For those appropriate to ride in a forward-facing position, the straps should be at or above the height of the child's shoulders, the chest clip should at armpit level, and the harness should be tightened so that the straps cannot be pinched off of their shoulders.

The CSS should also be installed properly, according to the manufacturer's recommendations. Not all car seats work in all vehicles. Families that require more than one car seat will need to make sure that they can all be installed properly and fit within the vehicle. Rear-facing car seats have specifications for the correct angle that is allowed based on the weight and/or age of the child. There are also requirements about how much of the base of the car seat is on the seat of the vehicle. All car seats, once installed, should not move more than one inch along the belt path.

In a position statement by the American Academy of Pediatrics (AAP), "All infants and toddlers should ride in a rear-facing car safety seat (CSS) as long as possible, until they reach the highest weight or height allowed by their CSS's manufacturer. Most convertible seats have limits that will permit children to ride rear-facing for 2 years or more" [14].

The rear-facing limit on some convertible car seats is now 50 lb. While the child may be older than 2 years of age, this position may be appropriate for older children with neurodevelopmental disabilities. The best practice is to remain rear-facing as long as possible, based on the manufacturer's instructions.

There are many CSS on the market, and they are ever-evolving. In choosing an appropriate restraint, there are general guidelines to consider when deciding which type of car seat to use (see Table 38.3.4). It is important to keep in mind that the child fits appropriately in the seat and that it can be installed in the family's vehicle.

For some children with special health care needs, a standard car safety seat (CSS) provides the best protection for most travel needs. Best practice states that if a commercially available car seat is appropriate, this is the best option. However, there are children with neurodevelopmental disabilities for whom these seats do not work.

Children with special needs may require a specialized CSS in order to be transported safely. When assessing for a special needs CSS, it is important to have knowledge of the presentation of the child, including head control, trunk control, posture, muscle tone, range of motion limitations, as well as any behavioral issues. Collaboration with a child passenger safety technician (CPST) can be helpful in order to determine how various CSS work with the child and the family's vehicle. This is a person who has received and completed a training program to ensure proper use of a CSS. A CPST with special needs training will also incorporate knowledge of medical conditions and positioning, as well as how various special needs CSS work. Certified personnel can be found at: http://www.cert.safekids.org.

When identifying an appropriate CSS for children with special health needs, many factors must be taken into consideration. This includes: will the recommended seat appropriately position the child; can this seat be appropriately installed into the family's vehicle; what type and how much traveling does the family do; does the child have any specialized equipment; and are there other children riding in the vehicle as well?

It is beneficial to assess the child with special health care needs and trial them in the potential CSS. Assessing the seat in the family's vehicle to ensure it fits appropriately is also recommended. There are no "one size fits all" solutions because of the variables that must be considered. Choosing a CSS based strictly on the height and weight guidelines from the car seat manufacturer can be risky, as these measurements do not take body proportion into account. For example, two children may be the same height, but one may have more height in the trunk versus the legs. They may both be within the height guidelines, but one could have a higher seated have a higher seated shoulder height, which makes the car seat an inappropriate match. That is why it is important to have an understanding of physical issues as well as the intricacies of the seats available and how they may or may not work with the family's vehicle.

Some children with neurodevelopmental disabilities may require riding in a wheelchair-accessible vehicle. "The wheelchair must be effectively secured to the vehicle in a manner that limits its movement in a crash or during vehicle maneuvers. This is accomplished by using a crash-tested tie-down system or securement device that "anchors" the wheelchair to the vehicle floor in a manner that will prevent the mass of the wheelchair from loading the wheelchair-seated occupant in a crash" [15]. This is to ensure that the wheelchair remains in place and does not become a projectile with the child in the seat.

In order to safely transport a child in a wheelchair-accessible vehicle, the wheelchair should be secured via tie-downs or special locking systems. Once the wheelchair is secured, a lap and shoulder belt should also be utilized. The anterior chest harness and seat belt in a wheelchair are designed for positioning the child in the wheelchair and are not designed for safety transportation. The lap or shoulder belt is what will protect the child in the event of an accident.

Dosing, frequency, and scope of therapy

The frequency and determination of the treatment plans will vary depending on the setting in which occupational therapy services are provided. There is a large variation. For example, in the school setting, services may be direct or consultative, meaning possibly seeing a child one time per week to one time per quarter. Inpatient services have a more intense focus, and patients may be seen six days per week. The setting and focus of the therapy will provide a guideline for the frequency of services. However, the occupational therapist will use their clinical reasoning skills to determine this, within the guidelines of their setting.

Therapy success is measured on an individual basis and varies depending on the setting. Making progress toward goals is a major factor in determining success. Boosts in therapy may be warranted when a new, functional skill is achieved. On the other hand, when progress is plateauing, it may be appropriate to discontinue services. The focus should be on improving "functional" performance.

Red flags/when to make occupational therapy referrals

Occupational therapy referrals should be made when children are not meeting their developmental milestones in regards to their fine motor/visual motor/visual perceptual skills, as well as their performance in daily activities. Neurodevelopmental physicians may identify some possible delays in the child's development. Early identification of possible deficits and a referral to occupational therapy can help assess the strengths and challenges that a child has and determine a plan of care to address these issues.

Summary

Pediatric occupational therapists are trained in anatomy and physiology, general infant and child development, neuroscience, and neuro-motor development. They focus on a child's occupations and occupational performance by analyzing skills and the environment in order to provide treatment that uses occupation both as a means and an end. They play a major part in the advancement of skills for those with neuro-developmental disabilities.

References

[1] Wfot.org. Available from: https://www.wfot.org/about-occupational-therapy, [cited 14.11.21].

[2] Occupational therapy practice framework: domain and process—fourth edition. Am J Occup Ther. 2020;74(Supplement_2):1–87. Available from: https://doi.org/10.5014/ajot.2020.74S2001.

[3] Guidelines for Supervision, Roles, and Responsibilities During the Delivery of Occupational Therapy Services. Am J Occup Ther. 2020;74 (Supplement_3):1–6. Available from: https://doi.org/10.5014/ajot.2020.74S3004.

[4] Chamberlain-Salaun J, Mills J, Usher K. Terminology used to describe health care teams: an integrative review of the literature. J Multidiscip Healthc 2013;6:65–74.

[5] Sandra Carroll, OTR, CLC C, Sue Ludwig, OTR, Chrysty Sturdivant, OTR, CLC C. Occupational therapy's role in the neonatal intensive care unit. Am J Occup Ther. 2018;72:1–9.

[6] The Individuals With Disabilities Education Act (IDEA) for children with special educational needs, In: Lipkin PH, Okamoto J, (editors). FAAP, the Council on Children with Disabilities and Council on School Health.

[7] Sec. 300.34. Related services – individuals with disabilities education act. Sites.ed.gov, available from: https://sites.ed.gov/idea/regs/b/a/300.34; 2017 [cited 14.11.21].

[8] CDC. Individuals with Disabilities Education Act (IDEA) services. Cdc.gov, Available from: https://www.cdc.gov/ncbddd/cp/treatment.html; 2021 [cited 14.11.21].

[9] Adams RC, Tapia C. Council on children with disabilities. Early intervention, IDEA Part C services, and the medical home: collaboration for best practice and best outcomes. Pediatrics 2013;132(4):e1073–88.

[10] Occupational Therapy in the School Setting. Aota.org, available from: https://www.aota.org/About-Occupational-Therapy/Professionals/CY/school-settings.aspx; [cited 14.11.21].

[11] O'Neil J, Hoffman B. Council on injury, violence, and poison prevention. Transporting children with special health care needs. Pediatrics 2019;143(5)e20190724.

[12] Durbin DR, Hoffman BD. Council on injury, violence, and poison prevention. Child passenger safety. Pediatrics 2018;142(5):e20182460.

[13] Yonkman J, O'Neil J, Talty J, Bull MJ. Transporting children in wheelchairs in passenger vehicles: a comparison of best practice to observed and reported practice in a pilot sample. Am J Occup Ther 2010;64(5):804–8.

[14] Muller VM, Burke RV, Arbogast H, Ruiz PC, Nunez NM, San Mateo KR, et al. Evaluation of a child passenger safety class in increasing parental knowledge. Accid Anal Prev 2014;63:37–40.

[15] Elliott MR, Kallan MJ, Durbin DR, Winston FK. Effectiveness of child safety seats vs seat belts in reducing risk for death in children in passenger vehicle crashes. Arch Pediatr Adolesc Med 2006;160(6):617–21.

Chapter 38.4

Speech and language therapy

Natalie Schmitt and Kirby Matthews
Speech Language Department, Kennedy Krieger Institute, Baltimore, MD, United States

Speech-language therapy

According to the National Health Interview Survey (NHIS) in 2012, 7.7% of children aged 3–17 years had a communication disorder [1]. Of those children, 55% of them received an intervention service [1]. The goal of this chapter is to give an overview of what pediatric speech-language therapy services are, what areas they address, and best practices in those areas. Pediatric speech-language pathologists (SLPs) provide speech-language therapy services to treat speech, language, social communication, cognitive communication, and swallowing disorders in children [2]. The services provided may be habilitative or rehabilitative. Habilitative speech-language therapy is provided to "help a person learn, keep, or improve skills and functional abilities that they may not be developing normally" (e.g., language delay), whereas rehabilitative therapy is provided to "help a person improve skills that have been lost after a stroke, head injury, illness, or other cause" [3]. The focus of this chapter will be on habilitative speech-language treatment services and will highlight best practices in the following areas: speech sound production, fluency, voice, oral language, and written language. In addition, it will also review best practices for children who are multilingual. Habilitative speech-language therapy services may be provided in a variety of settings, including, but not limited to, the child's home, community (e.g., library and playground), daycare, school (preschool, elementary, middle, and high), or outpatient clinic.

The four major purposes of intervention include: changing or eliminating the underlying impairment; changing the disorder; teaching compensatory strategies; and modifying the child's environment. Paul et al. [4] discussed how intervention often involves the fourth option of modifying the child's environment combined with one of the other three. The most common purpose of language intervention is to modify the language disorder [4]. When this is the case, the goal of therapy is to attempt to improve the child's language function by teaching specific behaviors to help the child become a better communicator [4]. An example of this would be helping the child to expand their mean length of utterance.

No matter the purpose of intervention, applying evidence-based practice (EBP) is of utmost importance. According to the American Speech-Language-Hearing Association (ASHA), EBP "is the integration of clinical expertise/expert opinion, evidence (external and internal), and client/patient/caregiver perspectives" [5]. Furthermore, ASHA states, "When all three components of EBP are considered together, clinicians can make informed, evidence-based decisions and provide high-quality services reflecting the interests, values, needs, and choices of individuals with communication disorders" [5].

Service delivery methods

If a child is found to have a speech or language disorder, there are both private and public options for speech-language treatment. At any age, families may seek speech-language treatment through insurance or self-pay at a private practice or outpatient medical center. Coverage is dependent on the family's insurance policy, and the treating clinician determines treatment methods.

Through the Individuals with Disabilities Education Improvement Act, families are additionally entitled to public speech-language treatment that does not require insurance [6]. Children born through age three years receive early intervention (EI) services under IDEA (Individuals with Disabilities Education Act) Part C. Children three through twenty-one receive special education and related services under IDEA Part B. Receiving services through IDEA can help the child more easily generalize speech-language gains to their natural environments (e.g., home and school) and also allows for fluid collaboration between the speech-language pathologist and adults that are integral to the child's life (i.e., teachers, caregivers, and daycare providers).

While different states have specific guidelines for EI services, research has supported a shift from direct intervention toward consultative and collaborative approaches. In this model, the provider indirectly works with the child by training the caregiver in communication and play strategies [7]. Additionally, many Part C programs have adopted a "primary service provider" model, which involves a transdisciplinary team with one service provider acting as the primary point of contact and treatment for the family. The intent of this model is to address the child holistically and help the family see how all aspects of the child's development overlap in the child's natural environment [8]. Ideally, the service provider would align with the child's most significant area of weakness. For example, a speech-language pathologist may be a primary service provider if the child demonstrates severe delays in language skills but mild delays in cognitive and motor development. The transdisciplinary team would meet frequently to discuss goal progress.

At 3 years, the caregiver has the option to extend the IFSP for one year or transition services to an Individualized Education Program (IEP), covered by Part B of IDEA. When receiving services through an IEP, the child typically receives speech-language treatment at a public school. The child's IEP includes whether services are individual or in a small group and if they occur inside or outside of the general education classroom. See chapter: Legislative mandates. Traditionally, speech-language services are provided in a separate room outside of the general education environment. This service delivery allows for focused individual or small group intervention; however, it removes the child from their peers and risks missing classroom instruction. See chapter: Classroom.

Traditionally, speech-language services are provided in a separate room outside of the general education environment. This service delivery allows for focused individual or small group intervention; however, it removes the child from their peers and risks missing classroom instruction.

For language therapy, current research supports collaborative speech-language treatment, where speech-language pathologists closely collaborate with classroom and/or special education teachers. Classroom-based services may be more effective in generalization than more traditional pullout therapy approaches for some areas of language intervention, such as curricular vocabulary [9–11]. The speech-language pathologist may want to combine service delivery models (e.g., one session outside of general education, one within general education) to best meet the needs of the individual student.

For school-based speech production therapy, shorter, more frequent sessions are gaining more support in research. For example, Byers et al. (2021) found that students with mild-moderate speech-sound disorders benefited equally from short, more frequent individual sessions (5 min individual sessions, 3 × per week) compared to the more traditional group therapy (30 min, 2 × per week). This service delivery model allows for more individualized treatment and less time removed from the general education environment [12].

If a child does not have significant enough speech-language needs to qualify for speech-language services at school, they may qualify for a 504 plan [13]. A 504 plan provides the students with accommodations to support their disability. Common accommodations for children with speech-language disorders include: visual supports, repetition of instruction, word banks, and speech-to-text or text-to-speech (for written language needs). The speech-language pathologist may consult with teachers of children with 504 plans to ensure their communication needs are supported through accommodations. For children with more significant medical or behavioral needs, home-based intervention may be warranted.

Criterion for success in treatment

In any type of speech-language treatment, measurable goals are created and monitored frequently to determine progress. Goals should be selected based on the child's specific areas of need, functional impact, and academic impact (if implemented at school). No matter the setting, it is important for the therapist to gather a baseline of the child's skill in each goal to measure progress appropriately. Success in therapy can be measured by both progress in specific goal areas as well as functional communication (e.g., increases in intelligibility and use of language targets in spontaneous speech). Factors, independent of treatment methods and initial speech or language skills, that can increase success in therapy include higher child engagement, higher nonverbal cognition, lower externalizing behaviors, and age [14,15].

Bilingual and english language learner considerations

Before detailing best practices for specific speech-language interventions, it is important to note considerations for bilingual children and children who are learning English as a second language. Children that are exposed to and/or use multiple languages should be carefully evaluated to determine whether speech and language concerns are due to language differences or a true language disorder. Only those speech sounds and language structures that are present across both languages should be targeted during speech-language intervention [16]. For example, in Spanish, there is no voiced or voiceless /th/ therefore articulation of /th/ would not be an appropriate target for speech-language intervention [17]. If speech-language differences are a concern for caregivers, teachers, or medical professionals, the child may benefit from targeting these areas with bilingual or English as a Second Language services. The child's school, dependent on performance on state-approved oral language proficiency tests, may provide these services.

Speech production treatment

The overall goal of speech production treatment is to increase the intelligibility of the child. Treatment methods and frequency are dependent on the type of speech production disorder. Of note, the most evidence-based way to target improvements in intelligibility is by practicing speech itself. While implemented by some speech-language pathologists, nonspeech oral motor exercises (e.g., tongue wagging, blowing, horns, lip pucker, or smile) have little to no evidence for improving speech production [18].

The most common types of speech production disorders are organic in etiology and include articulation and phonological disorders. When selecting which sounds to target, clinicians can choose a developmental or complex approach. A developmental target selection selects speech-sound errors in the order they develop. For instance, target earlier-developing sounds such as /p/, /b/, and /m/ before targeting later-developing sounds such as /s/ and /r/. Contrastingly, a complexity target selection approach focuses first on more complex targets that are not in the child's phonological repertoire to encourage generalized learning of sounds. Proponents of this approach argue that by targeting later-developing, more complex sounds, such as fricatives, earlier-developing sounds, such as stops, will improve spontaneously. Similarly, targeting the more complex consonant clusters (i.e., spoon, brush) will help improve the singletons (i.e., /s/, /p/, /b/, /r/) [19].

Children with single articulation errors (e.g., lateral or frontal lips) may benefit best from a traditional articulation approach, as the deficit is in the motor movements. In the traditional articulation approach, the child first learns how to produce the sound in isolation and then practices the target sound in all parts of the word (i.e., beginning, middle, and end) in increasing language complexity (e.g., words, phrases, sentences, conversation). Children with phonological disorders (i.e., patterns of phonological errors) often have both motor (physical production of the sound) and linguistic components. As such, treatment for this type of disorder should focus on teaching the child to produce the target sounds with accuracy, as well as when to use that sound in context.

For young, highly unintelligible children with multiple error patterns, an SLP may use the cycle approach, created by Hodson. The cycles approach's goal is to facilitate efficient change in the phonological system by engaging in amplified stimulation of phonological patterns and practicing 4–5 high-frequency words following one phonological pattern for 2–6 h of therapy. Once all patterns are targeted, a "cycle" is complete, the child's phonological skills are re-evaluated, and new targets are selected [20].

For children with fewer phonological processes, contrastive approaches are the most efficient. Contrastive articulation approaches such as minimal pairs and maximal oppositions aim to improve intelligibility by highlighting sound contrasts necessary to differentiate words [21,22]. These approaches use word pairs as targets instead of individual sounds. For example, if a child is always substituting /t/ for /k/, a contrastive word pair could be "tea" and "key," as these words only differ in the target sound. The multiple oppositions approach is similar to the above approaches; however, it contrasts the child's error with three to four sounds that are maximally different (e.g., "door," "four," "chore," and "store"). This type of approach is best suited for children that have phoneme collapse, substituting one sound for many sounds (e.g., saying "doh" for "go," "sew," "show," and "foe") [23]. Children with motor speech disorders (i.e., dysarthria and childhood apraxia of speech) will often not respond as well to phonological or articulation approaches and will likely need higher-intensity speech-language treatment with specialized intervention.

Dysarthria, a speech disorder due to motor weaknesses in the speech mechanism, is common in children with global motor involvement, such as cerebral palsy (CP) and Down syndrome. Intervention for these children depends on the type and severity of the dysarthria and the speech subsystems (i.e., respiratory, laryngeal, velopharyngeal, and articulatory) involved. With that, intervention should focus on the subsystems that are impacting intelligibility. There are documented improvements in intelligibility following direct intervention on respiration (e.g., positioning, breath control, phrasing, hypernasality (e.g., combination of surgical-pharyngoplasty, palatal lift, and speech-language treatment), articulation precision (e.g., with tactile cues of PROMPT intervention, oral placement therapy, instrumental biofeedback, and contrastive phonological approaches), and loudness (e.g., Lee Silverman Voice Treatment) [24–30]. Contrasting to children with phonological disorders, children with dysarthria will likely need explicit intervention on generalizing target sounds to words with higher motor complexity, sentences, and conversation.

For children with dysarthria with subtler oral motor weakness, learning and practicing intelligibility strategies in therapy may be the most beneficial [31]. For example, Sakash et al. found that children with CP with moderate-severe intelligibility concerns who speak in longer utterances benefit from a slow speech strategy to improve intelligibility [32]. Common intelligibility strategies include slowing the rate of speech and using overarticulation. Often, a child will need to learn how to recognize when the listener does not understand them and independently use the strategies to resolve communication breakdowns. If a child's speech remains unintelligible, the use of assistive technology should be strongly considered.

For childhood apraxia of speech (CAS), the deficit is in motor programming and planning. As such, treatment should focus on the movement between sounds rather than specific sounds and incorporate explicit targeting of prosody, as this is a major area of weakness for children with CAS. The following treatment methods have been shown to be effective for treating the core deficits in children with CAS.

Dynamic Temporal and Tactile Cueing (DTTC), created by Strand, is a treatment technique developed for children with severe CAS who are young or beginning their CAS treatment journey [33]. With appropriate dosage (i.e., three to four times per week), children should progress to a point where DTTC is no longer needed, which is often less than one year. In DTTC, treatment targets are selected based on functional use and the child's phonetic inventory. Targets may be as few as three to four words for children beginning therapy and increase in number as the child has success. The acronym DTTC describes the cueing continuum that is implemented during treatment. Cues in DTTC are "temporal" in that the clinician slowly increases the time between the model and the child's imitation. The temporal hierarchy includes simultaneous production, direct imitation, delayed imitation, and spontaneous production. Cues are "dynamic" in that the clinician is constantly adjusting based on the needs of the child, providing visual, auditory, and tactile cueing as needed. Additionally, prosody is explicitly targeted at each level of the temporal continuum. For example, the child may practice saying target words as if they were making a statement, asking a question, or being angry [33].

The Nuffield Dyspraxia Program is a motor skills learning approach that aims to build accurate speech from core units. It first targets single speech sounds and simple syllables and combines them into words. Cues, feedback, frequent practice, and repetitive sequencing exercises help develop new motor programs. Minimal word pairs are additionally utilized to incorporate phonological skills [34].

ReST Rapid Syllable Transition Training, best for school-age children with milder CAS, was designed to target the three core features of CAS, including consistency and accuracy of sounds, sequencing, and prosody. ReST uses 2–3 syllable pseudo-words that are carefully selected based on the child's phonetic inventory (only sounds that are consistently within or emerging) to help improve the syllable shape complexity and prosody of the child [35].

Similar to ReST, the Treatment for Establishing Motor Program Organization (TEMPO) uses pseudo-words to help improve intelligibility and prosody. Treatment targets consist of 3–4 syllable pseudo-words that have either a strong-weak (SW) or weak-strong (WS) stress pattern over the first two syllables. During treatment, correct sound production, fluent transition between syllables, and accurate lexical stress are targeted. A recent study found reduced segmentation and improved stress contrasts in children that received TEMPO intervention four times a week for four weeks and retention of these effects one month after treatment [36].

Fluency disorders

Treatment for children with fluency disorders (i.e., stuttering and cluttering) should be highly individualized and multidimensional, considering all aspects of the child's disfluency (i.e., observable disfluencies, negative thoughts related to disfluencies) and concomitant communication disorders [37].

For preschool children, the first treatment decision is whether ongoing treatment is warranted at all! Most children recover from stuttering, with or without speech-language therapy, within 3–4 years, usually by age seven; many recover within 12 months of onset [38]. Singer et al. completed a metaanalysis to rank six risk factors for persistent stuttering. These include a family history of stuttering, being male, poor articulation skills, a higher percentage of stuttering-like disfluencies, poor receptive language, and poor expressive language [38]. No matter the risk to the child of persistent stuttering, all families with children with disfluencies benefit from indirect treatment to create a more fluency-enriching environment and learn how to react during stuttering events. This may help give caregivers confidence as well as prevent young children from developing negative feelings about their speech, as children who stutter have self-reported negative feelings about their communication as early as preschool, independent of the severity of the stutter and the time since the onset of stuttering [39]. If treatment is initiated, treatment for preschoolers can include indirect and direct methods. Indirect treatment, such as phase one of the Palin Parent-Child Interaction Therapy, involves counseling families to create a fluency-enhancing environment by using strategies such as reducing the communication rate, reducing demands, and recasting or rephrasing to model fluency techniques [40]. Caregivers may additionally require explicit instruction and practice on how to react during the child's stuttering events. During moments of disfluency, caregivers should maintain eye contact and allow the child to finish his or her thought. It is best that the caregiver model slow speech rather than telling the child to "slow down" [41].

If stuttering persists following indirect intervention, the clinician may guide the parent in operant treatment (e.g., Lidcombe Program). This therapy model trains the parent to give frequent positive reinforcement for fluent speech (e.g., "Smooth talking!" or "No bumps!") and sometimes (about a 5:1 ratio) redirect disfluent speech by making a comment or asking for correction (e.g., "That word got stuck!" or "Try that one again") [42].

Syllable-timed speech is another technique that, when practiced daily, can help preschoolers decrease stuttering events by 96% over one year. It uses rhythm to induce fluent speech by having the child practice speaking one syllable at a time (e.g., "Ice-cream is the most de-li-cious dess-ert."). Syllable timed speech has the best outcomes when practiced 5–10 min, 4–5 times per day. The frequency of practice decreases along with decreases in the child's stuttering events [43].

As a child enters school age, there is a shift to direct intervention to help minimize the negative impacts of the child's disfluencies. Yarus et al. describe how a comprehensive treatment program should include: increasing acceptance and openness in stuttering; reducing secondary behaviors and minimizing avoidances; improving social communication; improving self-confidence and self-efficacy; managing bullying effectively; and minimizing the impact of stuttering [37]. By taking a comprehensive treatment approach to stuttering, success is measured by reductions in the child's negative reactions, reductions in communication avoidance, and improvements in the child's communication abilities, regardless of whether he/she stutters. School-age children have likely been stuttering and receiving therapy for stuttering for many years. It is important to consider the child's specific communication goals (e.g., ordering from a restaurant, initiating a conversation with a specific peer) when developing a treatment plan.

Direct treatment for reducing disfluencies can help a child learn strategies to help them communicate more effectively. Treatment strategies can include speech modification and/or stuttering modification. Speech modification helps teach the child to change the timing and tension of his or her speech to increase fluency. Techniques can include reduced rates, easy onsets, prolonged syllables, and continuous phonation to increase fluency. Stuttering modification teaches strategies to help the child stutter more easily, moving through and past stuttering moments with decreased tension. The clinician may also address any secondary behaviors (e.g., eye blinking, limb movement) that coincide with stuttering events [44].

Common methods that can help target negative feelings associated with the presence of stuttering include desensitization of stuttering fears (by practicing stuttering in a safe environment), pseudostuttering (practicing stuttering on purpose to feel in control of one's speech), self-disclosure (talking to others about your stuttering), and attending therapy in a group with other children who stutter (to feel less isolated and provide a safe space to talk about concerns) [37].

There is a paucity of research regarding treatment for the less frequent fluency disorder, cluttering. Theoretically, people who clutter are thought to speak at a rate that is too fast for their motor system to handle. Therefore, techniques that help regulate the rate of speech, such as pausing and overarticulation, may be beneficial [45]. Additionally, similar to children with poor intelligibility due to speech production errors, children who clutter may benefit from explicit practice in identifying when listeners do not understand them and implementing strategies to resolve the communication breakdowns. Auditory-visual feedback training is a helpful tool to increase self-awareness of cluttering moments and produce natural speech [46]. Strategies to increase fluency for children who clutter should be individualized for the child, but commonly include slowing rate, pausing, increasing volume, and using overarticulation.

Voice treatment

The reported prevalence of pediatric voice disorders has ranged from 1.4% to 6.0%, with vocal fold nodules being the most frequently diagnosed, possibly as high as 16.9% in the school-age population [47]. Prior to a speech-language pathologist treating an identified voice disorder, a referral to other medical professionals, such as an otolaryngologist (ENT) or allergist, may be appropriate to determine if there is an underlying cause (e.g., vocal polyps, allergies) that may benefit from medical intervention [47]. According to Hartnick et al. (2017), 95% of ENTs recommend voice therapy as a treatment for pediatric vocal fold nodules [48]. Therapy approaches to pediatric voice disorders may be direct, indirect, or a combination of the two [47]. Indirect therapy approaches focus on education (e.g., vocal hygiene, physiology of voice production, etc.) and counseling (e.g., stress management strategies), whereas direct therapy focuses on altering vocal production, including phonation and respiration [47,48].

Vocal hygiene tips may be different depending on the patient and what their daily activities involve, but in general, vocal hygiene involves staying hydrated, minimizing caffeine, resting your voice periodically, and avoiding unhealthy vocal practices (e.g., shouting, coughing) [47,49]. Hartnick et al. (2017) determined that both indirect and direct treatment approaches were effective for school-age (6 years old to 10 years old) children with vocal fold nodules, but that there was no significant difference between the two [48]. Many direct voice treatments that were originally developed for adults can be modified to meet a child's developmental level for voice therapy [50]. Common direct voice treatments that can be adapted for children include, but are not limited to, resonant voice therapy, flow phonation, and semioccluded vocal tract exercises [50,51]. These voice treatments aim to teach the child how to use voicing most efficiently with the least amount of vocal strain.

Language treatment

According to ASHA, a spoken language disorder (SLD) refers to "a significant impairment in the acquisition and use of language across modalities due to deficits in comprehension and/or production across any of the five language domains (i.e., phonological, morphological, syntax, semantics, and pragmatics)" [51]. A child with the primary disability of SLD that is not accompanied by other conditions or disabilities (e.g., intellectual disability) is classified as having a specific language impairment (SLI) [51]. The estimated total population prevalence of language disorders is 9.92%, with 7.58% for language disorders of unknown etiology and 2.34% for language disorders associated with intellectual disability and/or a medical diagnosis [52].

Language is comprised of three major aspects: form (phonology, morphology, and syntax), content (semantics), and use (pragmatics) [51]. However, it should be noted that not all children with language disorders will have impairments in all three of these areas [4]. The main goal of treating a language disorder is to improve overall communicative effectiveness (understanding and expression) for that child. This may look very different for a toddler or preschooler as compared to a school-aged child or adolescent. When generating language goals for a child, Paul et al. suggest prioritizing goals and objectives that target behaviors within the child's zone of proximal development (ZPD), which is described as "the distance between a child's current level of independent functioning and potential level of performance" [4].

There are three types of approaches to treating language disorders: clinician-directed, child-centered, and a hybrid of the two [4,51]. There is not one right approach for each child, and often a clinician may end up using a combination of all three approaches [4]. During clinician-directed therapy, the clinician determines all aspects of therapy, including the setting, materials to be used, and how the child will use them [4,51]. Although clinician-directed therapy tends to be less naturalistic, it maximizes opportunities for the child to produce target responses [4]. In contrast, child-centered or child-oriented approaches follow the child's lead and often occur in more natural, everyday settings and activities [4,51]. When using a child-centered approach, the clinician chooses activities that provide opportunities for the child to produce target responses as a part of play or conversation [4]. This type of approach is most often used with younger children but can also be altered for use with older children as well [51]. Paul et al. (2018) emphasize how important it is to wait and respond to the child when using this approach [4]. Examples of child-centered approaches include self-talk (the caregiver describing their own actions), parallel talk (describing the child's actions), expansions (taking what the child says and adding grammatical markers or semantic details), and extensions (commenting on what the child said by adding semantic information) [4,51]. Finally, the combination of clinician-directed and child-centered approaches allows the clinician to select activities that are more natural while at the same time increasing opportunities for the child's spontaneous production of targeted language forms [4,51]. While the options for treatment approaches may be the same across ages, treatment goals and methods will differ.

Late language emergence

Toddlers with late language emergence (LLE) are often described as "late talkers." LLE is a "delay in language onset with no other diagnosed disabilities or developmental delays in other cognitive or motor domains." [53] While some children with LLE may catch up to their peers (approximately 50%–70%), others do not and may be at risk for developing language and/or literacy difficulties [53]. At this age, language therapy may be indirect, direct, or a hybrid of both. Determining the type of service (direct vs indirect) as well as the frequency and duration of therapy should be individualized for each child and family depending on the nature and severity of the language delay [53]. Indirect therapy consists of parent/caregiver education on language stimulation techniques (e.g., communication temptations, parallel-talk, self-talk), language enrichment ideas (e.g., book sharing), and use of multimodal communication (e.g., gestures, speech, signs, pictures) [53]. Direct intervention involves the SLP implementing activities with the child as a part of working toward goals and objectives generated together with the child's caregivers [53].

Preschool language

For preschoolers (ages 3–5) with language difficulties, language is not progressing at the same rapid pace as their typically developing peers [51]. Intervention at this age is often a combination of play-based activities (e.g., more child-centered) and structured activities (e.g., shared book reading, doing a craft, clinician-directed). In the area of semantics, intervention may include increasing receptive and expressive vocabulary (including basic concepts such as spatial concepts) and increasing understanding and use of semantic relationships [51]. Therapy may also include improving morphology and syntax, such as using age-appropriate morphemes (e.g., pronouns, verbs), increasing sentence length and complexity, and increasing the variety of sentence types used (e.g., statement, question, command, exclamation) [51]. Pragmatic language, or the use of language in different settings, is also important for this age group, and skills addressed may include increasing flexibility in language use in various situations, improving conversational skills, and developing narrative and storytelling skills [51]. Many studies have demonstrated that oral narrative ability is related to later language and academic outcomes for children of all ages, including written language outcomes [54].

Parent/caregiver training is also important for preschool language intervention, as parents/caregivers are often the ones who are around the child the most and who are providing language models for them. Roberts et al. conducted a systematic review and metaanalysis that found that training parents to implement language and communication intervention techniques with children with a mean age of less than 6 years was associated with improved language outcomes for the child [55]. Examples of language stimulation techniques for this age group that can be taught to parents or caregivers may include expansion (e.g., the child says, "Car" and the caregiver repeats with added words, "A red car") and modeling correct grammar (e.g., the child says, "Her goed fast" and the caregiver repeats, "Yes, she went fast"). Research is scarce in recommending the frequency of language intervention for this age group. In a pilot study with preschool children, Bellon-Harn (2012) compared a concentrated treatment schedule (4 times per week for 6 weeks) to a distributed treatment schedule (2 times per week for 12 weeks), where treatment addressed semantic and morphologic skills. There were no significant differences between the two groups, and both groups had positive outcomes [56].

School-age language

ASHA's practice portal on spoken language disorders states that "the focus of language intervention for elementary school children with language difficulties is to help the child acquire the language skills needed to learn and succeed in a classroom environment." Treatment with school-aged children (5–10 years old) may incorporate improving vocabulary knowledge and use, including knowledge of curriculum vocabulary, abstract vocabulary, subtle differences in meaning, changes in vocabulary meaning given different contexts, and figures of speech [51]. It may also include improving understanding of abstract language concepts such as figurative language and multiple meaning words, as well as targeting improved comprehension. An important part of improving comprehension may involve improving self-advocacy skills [51]. If a child does not understand but does not ask for clarification, they may become even more confused, and this could lead to reduced participation and reduced learning. Language treatment with elementary school children with language disorders may also include increasing use and understanding of more advanced morphology and syntax, including understanding and using complex sentence structures, as well as using those advanced skills to communicate a variety of pragmatic functions (e.g., story retelling, persuasion, and opinion) in a variety of settings [51]. These settings include school interactions (e.g., classroom discussions, presentations), social interactions (e.g., peers at school, peers outside of school), community interactions, and family interactions. Within all these different settings, children at this age participating in language therapy may also be working to improve conversational skills (e.g., repairing breakdowns, initiating conversation, maintaining conversation) as well as learning what to say, what not to say, and when and where to do so [51].

The most recent systematic review of language intervention with school-age children with spoken language disorders by Cirrin and Gillam found limited research evidence concerning the effectiveness of language intervention practices that were currently being used with school-age children [57]. This review revealed gaps in the current literature and suggests a need for future research in this area, especially with elementary-age children [57]. While there is also not much research on what is an appropriate amount of treatment for children with language impairment (LI), Schmitt et al.'s [58] findings suggested that more intensive language treatment in the public-school setting is not necessarily associated with better treatment outcomes. Their research on speech-language treatment in U.S. public schools indicated that the best outcomes for children with LI resulted from high-frequency, low-dose, or low-frequency, high-dose treatment [58]. Meaning that children had better treatment outcomes when they were seen for more than one session per week for shorter sessions or for fewer sessions per week for longer sessions [58]. In this study, children who were seen for high-frequency and high-dose had worse treatment outcomes than these other two groups [58]. In addition, Schmitt et al. also discussed how research suggests that treatments distributed over time lead to better retention and access as stronger neural connections are made [58].

Adolescent language

According to Nippold and Schwarz, multiple studies have demonstrated that for most children with SLI, symptoms change over time; however, the underlying difficulties remain and negatively impact both spoken and written language [59]. In contrast to the rapid language changes that occur in younger children, change occurs much more gradually in older children and adolescents [59]. Prior to the 1970s, it was widely accepted that language development was basically complete by the ages of 5 or 6 [59]. After elementary school, the language problems of secondary students were not considered developmental, and therefore, intervention strategies were only aimed at how to compensate for their difficulties [60,61]. However, once researchers realized that reading and writing had strong linguistic bases and that there was significant growth in those areas during elementary school, they started to question this assumption [61]. As a result of research in this area, it is now known that language continues to develop throughout childhood, adolescence, and into adulthood [61].

Children in their adolescent years (ages 11 through high school) face a variety of challenges, including increasing academic and social demands. Adolescents with language disorders often have difficulty meeting these increased demands, and children who were able to keep up in elementary school now struggle in secondary school [51,62]. Language disorders not only impact adolescents' ability to understand and speak, but they also have a great impact on their academic, social, emotional, and vocational success [62]. And unfortunately, it is documented throughout the literature that adolescents with language disorders have historically been and still are an underserved population [62].

Service delivery methods that can be considered for this population include individual treatment, group treatment, collaboration with teachers, and supplemental treatment. No matter the service delivery method, an important aspect of intervention with this age group is teaching ways to compensate for language difficulties [51]. Many adolescents with identified language disorders who have received years of intervention aimed at changing their disorder [4]. While the goal of changing the disorder is still a possibility, it is likely time to consider how to compensate for it [4]. Strategy-based intervention involves teaching how to use external aids such as graphic organizers or spell check as well as how to use strategies such as using context clues to understand unfamiliar vocabulary [51]. Additionally, self-advocacy is an important skill for adolescents with language disorders to develop. When a child recognizes they are having trouble and has the skills to ask for help, repetition, slow rate, accommodations, etc., they are empowered to have more successful interactions and learning experiences. Explicit teaching and role-playing situations in speech-language therapy can have a positive impact on a child's self-advocacy [63]. It is also important to involve the child at this age as much as possible in developing goals to foster feelings of collaboration and responsibility as well as maximize chances for success [4,51].

Although there remains a need for further research, many recent studies have demonstrated that adolescents have made significant improvements using targeted intervention programs (as compared to broader intervention topics), including the following areas: narrative discourse, word-finding ability, semantics, verb argument structure, and vocabulary [62]. There are many variables to consider when determining which treatment delivery method is best for an older child with a language disorder. Individual language therapy may be best for a child that has very specific goals that need to be targeted or a child that has secondary concerns such as attention or behavioral difficulties. According to Ebbels et al., direct 1:1 intervention with an SLP can be effective in all areas of language for older children with language disorders, regardless of their gender, receptive language, autism spectrum disorder status, or age [64]. Although this study had limitations and acknowledged that more research in this area is needed, it demonstrated that direct speech-language therapy (SLT) services for older children with language disorders should be available and can make a difference [64]. If generalization is a goal, a group may be beneficial for the child to be able to practice skills learned with peers, especially with social-pragmatic language goals. In general, it is important to not only address academic language skills with this group but functional language skills as well [4]. Functional language skills may include following directions, asking questions, initiating and maintaining a conversation, identifying and solving interpersonal conflicts, gaining information from basic writing, and using basic writing to communicate basic information (e.g., on a form or questionnaire) [4].

Starling et al. conducted one of the first research studies that evaluated a collaborative intervention that linked changes in mainstream secondary teachers' instructional language practices with improvements in the language abilities of adolescents with language impairments [65]. They concluded that offering professional development in language modification techniques resulted in positive benefits for students with language impairments [65]. The SLP can provide education to the general education or special education teacher about language disorders as well as teach techniques to implement in the classroom. Examples of techniques include chunking information presented (e.g., breaking down assignments into more manageable amounts), providing notes ahead of the lecture for the student to follow along and take notes on, and using visual aids in the classroom (e.g., graphic organizers) [65]. Finally, a child may receive supplemental therapy, additional speech-language therapy outside of services received through school, to address areas not addressed in school therapy. With this type of therapy, it is important for all SLPs involved in the child's care to collaborate.

Written language

According to ASHA, "a disorder of written language involves a significant impairment in fluent word recognition (i.e., reading decoding and sight word recognition), reading comprehension, written spelling, or written expression (i.e., written composition)" [66]. Just as with spoken language disorders, written language disorders can involve any of the five language domains (i.e., phonology, morphology, syntax, semantics, and pragmatics) [66]. While reading specialists and special educators typically perform literacy intervention in the public school system, SLPs are more frequently involved in literacy instruction and responsiveness to intervention (RTI) models [4]. For the purpose of this chapter, we will be examining the relationship between spoken and written language as well as interventions for children with written language difficulties.

The simple view of reading that Kilpatrick describes shows that reading comprehension abilities are dependent on decoding skills and language comprehension abilities [67]. This demonstrates that we must have data on decoding skills and language comprehension abilities to determine the most appropriate intervention for the child. There is a close relationship between spoken and written languages [66]. Children with spoken language difficulties often have difficulties learning to read and write, and children with difficulties in reading and writing often have spoken language deficits [66].

Intervention options for children with written language disorders vary significantly depending on the child's needs. In general, though, effective intervention includes the following principles: ongoing assessment and adjustments to goals, individualized instruction, and targeting improvement of the child's skills one step beyond their current level [66]. For children with decoding difficulties due to a phonological core deficit, Kilpatrick (2015) lists three elements that the most successful studies on reading intervention contained: addressed phonological awareness deficits and taught phonological awareness to the advanced level (e.g., phonemic manipulation), provided instruction in phonic decoding, and provided opportunities to apply these skills in reading connected text [67]. It is also important to understand the relationship between reading and writing, as it is important to address both aspects of written language during intervention. According to ASHA, "reading and writing are highly interrelated, and it is difficult to isolate any aspect of reading development that does not have a writing counterpart" [66].

References

[1] Black LI, Vahratian A, Hoffman HJ. Communication disorders and use of intervention services among children aged 3–17 years: United States, 2012. NCHS Data Brief 2015;(205):1–8.

[2] Speech-Language Pathologists. Asha.org. American Speech-Language-Hearing Association, available from: https://www.asha.org/students/speech-language-pathologists/; [cited 28.09.21].

[3] Hasselkus A. Habilitation – what it is and why it matters to you. Leader live. Available from: https://leader.pubs.asha.org/do/10.1044/habilitation-what-it-is-and-why-it-matters-to-you/full/; 2012 [cited 28.09.21].

[4] Paul R, Norbury C, Gosse C. Language disorders from infancy through adolescence: listening, speaking, reading, writing, and communicating. St. Louis, MO: Mosby; 2018.

[5] Evidence-based practice (EBP). Asha.org. American Speech-Language-Hearing Association, Available from: https://www.asha.org/research/ebp/; [cited 28.09.21].

[6] Individuals with Disabilities Education Improvement Act of 2004. 2004.

[7] Kemp P, Turnbull AP. Coaching with parents in early intervention: an interdisciplinary research synthesis. Infants Young Child 2014;27 (4):305–24.

[8] Marturana E, McComish C, Woods J, Crais E. Early intervention teaming and the primary service provider approach: who does what, when, why, and how? Perspect Lang Learn Educ 2011;18(2):47–52.

[9] Archibald LMD. SLP-educator classroom collaboration: a review to inform reason-based practice. Autism Dev Lang Impair 2017;2: 239694151668036.

[10] Cirrin FM, Schooling TL, Nelson NW, Diehl SF, Flynn PF, Staskowski M, et al. Evidence-based systematic review: effects of different service delivery models on communication outcomes for elementary school-age children. Lang Speech Hear Serv Sch 2010;41(3):233–64.

[11] Throneburg RN, Calvert LK, Sturm JJ, Paramboukas AA, Paul PJ. A comparison of service delivery models: effects on curricular vocabulary skills in the school setting. Am J Speech Lang Pathol 2000;9(1):10–20.

[12] Byers BA, Bellon-Harn ML, Allen M, Saar KW, Manchaiah V, Rodrigo H. A comparison of intervention intensity and service delivery models with school-age children with speech sound disorders in a school setting. Lang Speech Hear Serv Sch 2021;52(2):529–41.

[13] Education USD. Office for Civil Rights, Free appropriate public education for students with disabilities: requirements under Section 504 of the Rehabilitation Act of 1973. Washington, D. C; 2010.
[14] Justice LM, Jiang H, Logan JA, Schmitt MB. Predictors of language gains among school-age children with language impairment in the Public Schools. J Speech Lang Hear Res 2017;60(6):1590–605.
[15] Schmitt MB. Children's active engagement in public school language therapy relates to greater gains. Am J Speech Lang Pathol 2020;29 (3):1505–13.
[16] McLeod S, Verdon S, Baker E, Ball MJ, Ballard E, David AB, et al. Tutorial: speech assessment for multilingual children who do not speak the same language(s) as the speech-language pathologist. Am J Speech Lang Pathol 2017;26(3):691–708.
[17] Gildersleeve-Neumann CE, Kester ES, Davis BL, Peña ED. English speech sound development in preschool-aged children from bilingual English–Spanish environments. Lang Speech Hear Serv Sch 2008;39(3):314–28.
[18] Watson MM, Lof GL. Epilogue: what we know about nonspeech oral motor exercises. Semin Speech Lang 2008;29(4):339–44.
[19] Storkel HL. The complexity approach to phonological treatment: how to select treatment targets. Lang Speech Hear Serv Sch 2018;49 (3):463–81.
[20] Hodson BW. Enhancing phonological patterns of young children with highly unintelligible speech. ASHA Lead 2011;16(4):16–19.
[21] Blache SE, Parsons CL, Humphreys JM. A minimal-word-pair model for teaching the linguistic significance of distinctive feature properties. J Speech Hear Disord 1981;46(3):291–6.
[22] Gierut JA. The conditions and course of clinically induced phonological change. J Speech Lang Hear Res 1992;35(5):1049–63.
[23] Williams AL. Multiple oppositions: theoretical foundations for an alternative contrastive intervention approach. Am J Speech Lang Pathol 2000;9(4):282–8.
[24] Pennington L, Roelant E, Thompson V, Robson S, Steen N, Miller N. Intensive dysarthria therapy for younger children with cerebral palsy. Dev Med Child Neurol 2013;55(5):464–71.
[25] McLeod S, Baker E. Children's speech: an evidence-based approach to assessment and intervention. Upper Saddle River, NJ: Pearson; 2016.
[26] Ward R, Leitão S, Strauss G. An evaluation of the effectiveness of PROMPT therapy in improving speech production accuracy in six children with cerebral palsy. Int J Speech Lang Pathol 2014;16(4):355–71.
[27] AAPPSPA position statement – oral-motor therapy. Talktools.com. Available from: https://talktools.com/blogs/from-the-experts/aappspa-position-statement-oral-motor-therapy. [cited 24.09.21].
[28] Bahr D, Rosenfeld-Johnson S. Treatment of children with speech oral placement disorders (OPDs): a paradigm emerges. Commun Disord Q 2010;31(3):131–8.
[29] Nordberg A, Carlsson G, Lohmander A. Electropalatography in the description and treatment of speech disorders in five children with cerebral palsy. Clin Linguist Phon 2011;25(10):831–52.
[30] Levy ES. Implementing two treatment approaches to childhood dysarthria. Int J Speech Lang Pathol 2014;16(4):344–54.
[31] Levy E.S. Say it strong—and say it again. Leader Live, available from: https://leader.pubs.asha.org/do/10.1044/leader.FTR2.26012021.54/full/; 2021 [cited 24.09.21].
[32] Sakash A, Mahr TJ, Natzke PEM, Hustad KC. Effects of rate manipulation on intelligibility in children with cerebral palsy. Am J Speech Lang Pathol 2020;29(1):127–41.
[33] Strand EA. Dynamic temporal and tactile cueing: a treatment strategy for childhood apraxia of speech. Am J Speech Lang Pathol 2020;29 (1):30–48.
[34] Murray E, McCabe P, Ballard KJ. A randomized controlled trial for children with childhood apraxia of speech comparing Rapid Syllable Transition treatment and the Nuffield Dyspraxia Programme–Third Edition. J Speech Lang Hear Res 2015;58(3):669–86.
[35] McCabe P, Thomas DC, Murray E. Rapid syllable transition treatment—a treatment for childhood apraxia of speech and other pediatric motor speech disorders. Perspect ASHA Spec Interest Groups 2020;5(4):821–30.
[36] Miller HE, Ballard KJ, Campbell J, Smith M, Plante AS, Aytur SA, et al. Improvements in speech of children with apraxia: the efficacy of treatment for Establishing Motor Program Organization (TEMPOSM). Dev Neurorehabil 2021;24(7):494–509.
[37] Yaruss JS, Coleman CE, Quesal RW. Stuttering in school-age children: a comprehensive approach to treatment. Lang Speech Hear Serv Sch 2012;43(4):536–48.
[38] Yairi E, Ambrose N. Epidemiology of stuttering: 21st century advances. J Fluen Disord 2013;38(2):66–87.
[39] Singer CM, Hessling A, Kelly EM, Singer L, Jones RM. Clinical characteristics associated with stuttering persistence: a meta-analysis. J Speech Lang Hear Res 2020;63(9):2995–3018.
[40] Millard SK, Zebrowski P, Kelman E. Palin parent-child interaction therapy: the bigger picture. Am J Speech Lang Pathol 2018;27 (3S):1211–23.
[41] Stutteringhelp.org. Available from: https://www.stutteringhelp.org/7-tips-talking-your-child-. [cited 29.09.21].
[42] Kingston M, Huber A, Onslow M, Jones M, Packman A. Predicting treatment time with the Lidcombe Program: replication and meta-analysis. Int J Lang Commun Disord 2003;38(2):165–77.
[43] Trajkovski N, Andrews C, Onslow M, O'Brian S, Packman A, Menzies R. A phase II trial of the Westmead Program: syllable-timed speech treatment for pre-school children who stutter. Int J Speech Lang Pathol 2011;13(6):500–9.
[44] Max L, Caruso AJ. Contemporary techniques for establishing fluency in the treatment of adults who stutter. Contemp Issues Commun Sci Disord 1997;24(Spring):39–46.
[45] St. Louis KO, Raphael LJ, Myers FL, Bakker K. Cluttering updated. ASHA Lead 2003;8(21):4–22.

[46] van Zaalen Y, Reichel I. Clinical success using the audio-visual feedback training for cluttering. Perspect ASHA Spec Interest Groups 2019;4(6):1589–94.
[47] Voice Disorders. Asha.Org; American Speech-Language-Hearing Association. Retrieved 25.09.21, from https://www.asha.org/practice-portal/clinical-topics/voice-disorders/. 2021.
[48] Hartnick C, Ballif C, De Guzman V, Sataloff R, Campisi P, Kerschner J, et al. Indirect vs direct voice therapy for children with vocal nodules: a randomized clinical trial. JAMA Otolaryngol Head Neck Surg 2018;144(2):156–63.
[49] U.S. Department of Health and Human Services. Taking care of your voice. National Institute of Deafness and Other Communication Disorders. Retrieved 25.09.21, from https://www.nidcd.nih.gov/health/taking-care-your-voice. 2021.
[50] Braden M, Verdolini Abbott K. Advances in pediatric voice therapy. Perspect ASHA Spec Interest Groups 2018;3(3):68–76.
[51] Spoken Language Disorders. Asha.Org; American Speech-Language-Hearing Association. Retrieved 25.09.21, from https://www.asha.org/practice-portal/clinical-topics/spoken-language-disorders/. 2021.
[52] Norbury CF, Gooch D, Wray C, Baird G, Charman T, Simonoff E, et al. The impact of nonverbal ability on prevalence and clinical presentation of language disorder: evidence from a population study. J Child Psychol Psychiatry Allied Discip 2016;57(11):1247–57.
[53] Late Language Emergence. Asha.Org; American Speech-Language-Hearing Association. Retrieved 25.09.21, from https://www.asha.org/practice-portal/clinical-topics/late-language-emergence/. 2021.
[54] Pico DL, Prahl AH, Biel CH, Peterson AK, Biel EJ, Woods C, et al. Interventions designed to improve narrative language in school-age children: a systematic review with meta-analyses. Lang Speech Hear Serv Sch 2021;1–18.
[55] Roberts MY, Curtis PR, Sone BJ, Hampton LH. Association of parent training with child language development: a systematic review and meta-analysis: a systematic review and meta-analysis. JAMA Pediatr 2019;173(7):671–80.
[56] Bellon-Harn ML. Dose frequency: comparison of language outcomes in preschool children with language impairment. Child Lang Teach Ther 2012;28(2):225–40.
[57] Cirrin FM, Gillam RB. Language intervention practices for school-age children with spoken language disorders: a systematic review. Lang Speech Hear Serv Sch 2008;39(1):S110–37.
[58] Schmitt MB, Justice LM, Logan JAR. Intensity of language treatment: contribution to children's language outcomes. Int J Lang Commun Disord 2017;52(2):155–67.
[59] Nippold MA, Schwarz IE. Do children recover from specific language impairment? Adv Speech-Lang Pathol 2002;4(1):41–9.
[60] Kumpula-Lacey K. A comparison of two language screening instruments in two populations (Paper 3178) [Master's thesis, Portland State University], Retrieved from Dissertations and Theses. 1982.
[61] Nippold MA. Developmental markers in adolescent language: syntax, semantics, and pragmatics. Lang Speech Hear Serv Sch 1993;24(1):21–8.
[62] Joffe VL, Nippold MA. Progress in understanding adolescent language disorders. Lang Speech Hear Serv Sch 2012;43(4):438–44.
[63] Martin GE, Barstein J, Hornickel J, Matherly S, Durante G, Losh M. Signaling of noncomprehension in communication breakdowns in fragile X syndrome, Down syndrome, and autism spectrum disorder. J Commun Disord 2017;65:22–34.
[64] Ebbels SH, Wright L, Brockbank S, Godfrey C, Harris C, Leniston H, et al. Effectiveness of 1:1 speech and language therapy for older children with (developmental) language disorder. Int J Lang Commun Disord 2017;52(4):528–39.
[65] Starling J, Munro N, Togher L, Arciuli J. Training secondary school teachers in instructional language modification techniques to support adolescents with language impairment: a randomized controlled trial. Lang Speech Hear Serv Sch 2012;43(4):474–95.
[66] Written language disorders. Asha.Org; American Speech-Language-Hearing Association. Retrieved 25.09.21, from https://www.asha.org/Practice-Portal/Clinical-Topics/Written-Language-Disorders/. 2021.
[67] Kilpatrick DA. Essentials of assessing, preventing, and overcoming reading difficulties. 1st ed. John Wiley & Sons; 2015.

Chapter 38.5

Cognitive and behavioral therapy

Valerie Paasch[1,2], Lindsay Cirincione[1,2], Margaret A. Tunney[1,2] and Keith J. Slifer[1,2]

[1]*Department of Behavioral Psychology, Kennedy Krieger Institute, Baltimore, MD, United States,* [2]*Department of Physical Medicine & Rehabilitation, Johns Hopkins University School of Medicine, Baltimore, MD, United States*

Introduction

The 20th century brought the emergence and, by mid-century, the dominance of a behavioral conceptualization of psychology. The meaning of "behavior" to modern psychologists has its origins in the work of pioneering behaviorists such as B.F. Skinner [1]. For Skinner, behavior was defined as overt movements of individuals or organisms through the environment as directly observed visually or, by extension, through mechanical measurement. Behaviorism specifically did not attempt to define or describe internal physiological events and especially did not characterize mental or cognitive constructs, eschewing them as explanatory fiction. In contrast, the 1960s gave rise to what is now termed cognitive behavior therapy (CBT), beginning with the work of Aaron Beck [2], who introduced cognitive therapy as a conceptualization that acknowledged the impact of private, subjective phenomena such as thoughts and emotions on one's overt, observable behavior [3,4]. "Cognitive" refers to mental processes that are known primarily through language used by individuals to describe subjective perceptions. Cognitive psychology uses verbal constructs such as thoughts, emotions, sensation, and schema to describe private, subjective phenomena that are also associated with measurable physiological processes. Cognitive-behavioral therapy involves a synthesis of the two approaches to describe and understand how individuals function in the world.

In his work as a psychotherapist with patients diagnosed with depression, Beck noticed that patients' thoughts often lacked validity from the therapist's perspective. He began to recognize patterns of "cognitive distortions" in his patients' thinking. Through his observations, he began to regard depression as a cognitive disorder, and he developed a cognitive theory of depression [5,6]. He went on to develop a treatment protocol that provided a manual of clinical interventions that included cognitive, emotional, and behavioral components. CBT is structured, didactic, and goal-directed. It involves a collaborative interaction between therapist and client with the goal of helping the client recognize and subsequently modify, unhelpful patterns of thinking, emotions, and behavior to promote improved mood, functioning, and quality of life [6].

CBT employs a conceptual model characterizing the relationships among thoughts, emotions, and behaviors. The primary cognitive components of the model include "automatic thoughts," "cognitive distortions," and "underlying beliefs." In the CBT conceptualization, a person's underlying beliefs shape perceptions and interpretations of events in life. Belief systems are formed based on life experiences but are not necessarily objectively true and may be rigid and overgeneralized. These belief systems (also referred to as schema) are "rules" by which information is processed at the most reflexive and superficial level, resulting in automatic thoughts. Beliefs may be core beliefs involving primary ideas about oneself, others, and the world (e.g., I am unlovable; the world is dangerous). Beliefs may also be intermediate, involving assumptions, attitudes, and rules that are influenced by core beliefs in their development. Examples of intermediate beliefs that can be dysfunctional include "I must please others to be accepted" or "I must be excellent at everything at all times."

Automatic thoughts are immediate, unprocessed thoughts arising from an event or situation that influence an individual's emotional response and behavior. In effect, one interprets and reacts to events based on these immediate thoughts, which may be inaccurate or distorted. Such dysfunctional thoughts play an important role in manifestations of anxiety and mood disorders.

Cognitive distortions are errors in logic that lead to erroneous conclusions, which have a negative impact on psychological and behavioral functioning. Examples of cognitive distortions include: (1) dichotomous thinking, emphasizing extreme and mutually exclusive categories with no room for nuance in between; (2) overgeneralization, focusing on extreme examples to make broad generalizations about a situation; (3) selective abstraction, focusing on negative, upsetting aspects of a situation that confirm one's negative interpretation while ignoring or minimizing other neutral or positive aspects; (4) mind reading, assuming to know the thoughts and intentions of others; (5) fortune telling, predicting negative events before they occur; (6) minimization by ignoring or discounting some experiences as insignificant; (7) catastrophizing, envisioning the worst possible outcomes though they are unlikely, or perceiving nonpreferred or uncomfortable experiences as impossible or unbearable; (8) emotional reasoning, in which arguments and decisions are based on how one feels rather than an objective assessment of reality; (9) rigidity, following rigid rules about how things ought to be; and (10) personalization, involving self-blame or blaming others as directly responsible for a negative event [6,7].

Research on CBT has resulted in numerous outcome studies demonstrating its effectiveness for treating psychological and psychiatric disorders such as anxiety, depression, eating, substance use, and personality disorders. It has been applied to understanding and treating behavioral and psychological difficulties in children and youth and various social relationship difficulties across the life span. CBT has also been effective for improving the coping and adjustment of adults and youth with health conditions such as irritable bowel syndrome, chronic fatigue, insomnia, migraine headaches, and chronic pain. There is now an extensive body of research on CBT conducted by a host of investigators, which has established it as the predominant form of evidence-based psychotherapy in contemporary practice. Along with research on the experimental analysis of behavior and applied behavior analysis, CBT has helped give rise to the current emphasis among mental health professionals on empirically validated psychological treatments.

The literature includes many examples of specific CBT protocols that are effective in improving the functioning and outcomes of individuals with certain types of disorders. However, most therapists who employ CBT individualize the therapy for the unique and specific needs of individual patients. The individualized conceptualization is formed as the therapist and patient engage in a process of assessment and rapport-building. As the model is applied, the approach is explained, and the patient's presenting problems are characterized and prioritized. The problems of greatest concern to the patient and their family are addressed first, unless there are prerequisite concepts or skills the patient needs to acquire before the prioritized problem can be addressed.

Each CBT session is structured, including a brief check on symptoms and mood since the prior session and a review of the results of any specific homework or assignment that was previously agreed upon. This is followed by active presentations of new concepts and/or participation in therapist-led activities selected to reinforce those concepts. The didactic part of the session either introduces or rehearses content and skills. The emphasis is placed on the consistent presentation of generalizable concepts and the practice of skills that can be applied across multiple situations and experiences. The session concludes with feedback on observed performance, a summarization of the objective of the session, and agreement on homework assignments for the patient to complete between sessions.

CBT has been used alone or in combination with medications, such as antidepressants or anxiolytics, to address a host of behavioral and psychological difficulties. The primary core symptoms that are addressed are anxiety, depression, panic, anger, inattention, and overactivity. For anxiety, the focus of CBT is on identifying and ultimately helping the patient to recognize and replace negative or otherwise unhelpful automatic thoughts and unsupported beliefs. Another component involves teaching and helping the patient to apply relaxation responses that can compete with or help countercondition anxious autonomic nervous system arousal. In cases involving panic responses, there is an additional component of identifying situations and stimuli that elicit the panic symptoms, then conducting planned, strategic, stepwise exposures during which the patient does not leave the situation until anxiety has peaked and begun to diminish.

In CBT treatment of depression, in addition to the process of cognitive restructuring described above, an approach described as "behavioral activation" is conducted with patients who have stopped participating in previously preferred activities. These individuals are encouraged and assisted with activity scheduling to help them re-experience the positive aspects of participating in these activities. Experiencing the positive aspects of the activity often reinforces participation and makes it more likely to continue. As a result, the patient may once again experience that their behavior can have positive results, elevating their mood and helping them overcome their belief that they are helpless to change the circumstances of their lives [8].

In CBT aimed at anger and other self-regulation difficulties, emphasis is placed on recognizing and modifying environmental triggers. This is combined with strategies for modifying rigid, overgeneralized, automatic thoughts and related beliefs and attitudes that have been consistently associated with increased sympathetic nervous system arousal and become "conditioned" stimuli that trigger emotional arousal, angry postures, verbalizations, gestures, and disruptive behavior.

Finally, with symptoms of inattention and overactivity, emphasis during CBT is placed on identifying environmental variables that exacerbate or mitigate the symptoms. To the extent possible, antecedent conditions are arranged that have been associated with increased attention, reduced distraction, and lower activity levels. With children, CBT intervention also involves training caregivers to help structure the child's environment and routine to increase attention and decrease activity levels. Caregivers are assisted in establishing a set of rules and contingencies that reward self-regulation and remove reinforcement for overactive and inattentive behavior. In school-age children, the behavioral approach is often used in combination with psychostimulant medications that can increase attention to environmental stimuli and contingencies.

The average "dose" and course of CBT are 40–60 min per week for 8–12 weeks. However, these vary based on a multitude of psychosocial factors impacting the patient and family. As presented above, CBT is a valid and effective treatment for a wide range of mental and behavioral health conditions and symptoms. Whether these conditions are seen as symptoms of a disorder is largely dependent on their severity, interference with successful functioning, level of distress to the person, and statistical probability relative to the general population. The applicability and utility of the CBT conceptualization are robust, generalizable, and, in many respects, universal, even when a mental health diagnosis is not assigned to the patient. It is the conceptual and practical approach that currently dominates psychological and behavioral interventions in the 21st century.

When encountering the kinds of challenges described above in their patients, physicians and allied health professionals are advised to refer the patient to a psychologist or other trained behavioral health professional for consultative assessment and treatment. Ideally, the behavioral health therapist will maintain a collaborative relationship with the referring healthcare professional so they can together monitor progress and share recommendations. As emphasized elsewhere in this volume, there is additional value in an interdisciplinary approach to patients, particularly for developing children and youth and individuals with cognitive and developmental differences.

The cognitive behavior therapy + model

In this variation or elaboration of CBT, there is an expanded emphasis on the interactive and dynamic relationships among thoughts, physiological responses, and associated emotions and behaviors. CBT + hypothesizes that the relationships among these components are complex, interactive, and multidirectional in causation [9]. In working with pediatric populations, CBT has in many respects merged over time with earlier psychological conceptualizations and theories such as operant and respondent conditioning, behavior modification, the experimental analysis of behavior, and applied behavior analysis.

Assessment and outcome measures

In CBT, cognitive refers to subjective, private internal events such as thoughts, feelings, emotions, cognitions, and cognitive social constructs such as beliefs, attitudes, and opinions. These cognitive phenomena have been demonstrated to impact the observable and measurable behavior of individuals. This can be accomplished by observing and recording operationally defined behaviors or by administering standardized psychological or behavioral measures such as questionnaires about behavior, mood, anxiety, etc. Furthermore, CBT also describes and measures physiological processes that occur within the body that may or may not be detected by human observation. Therefore, assessment and outcome measurement may also involve mechanical or electronic instruments to detect and record autonomic (sympathetic nervous system) responses to environmental situations, thoughts, and emotions. These physiological responses may include respiration, heart rate, blood pressure, skin conductance, body temperature, electromyography, and associated biological measures of chemical and hormonal changes in the body and central and peripheral nervous systems (e.g., endocrine functions, neurotransmitter production, gastrointestinal processes).

Indications and contraindications for cognitive behavior therapy

In differentiating between behavioral therapy, applied behavior analysis (ABA), and CBT when used in the context of children and youth across the developmental spectrum, CBT is indicated for individuals who (1) are school age or older, (2) present with no more than moderate neurodevelopmental disorders (NDD), (3) exhibit less severe behavior problems, and (4) who can communicate verbally (or functionally using sign language, pictures, symbol systems, or electronic adaptive communication devices). Children and youth who are likely to benefit from CBT include those with problems such as mild to moderate acquired brain injury (without current posttraumatic amnesia [PTA]), functional neurological symptoms, functional deficits in daily living, mood, anxiety, coping or adjustment difficulties, sleep and arousal difficulties, chronic pain, and those who are experiencing challenges with cooperation, participation, or adherence with medical care. Each of these indications and associated protocols is elaborated in the recent literature on child and adolescent psychology, child psychiatry, and pediatric psychology.

Although applications are individualized, contraindications for CBT include: (1) lack of established instructional control; (2) severe inattention or overactivity; (3) severe disruptive behavior; (4) active suicidal ideation and intent; and

(5) severe psychiatric disorders such as active psychosis. Children whose behavior or cognitive capacity precludes attending to and participating in didactic presentations, reciprocal interactions, and structured activities are not likely to benefit from CBT. For patients with these behavioral contraindications, alternative approaches are used that are less dependent on verbal interaction and didactic explanations and better suited for teaching by arranging environmental experiences and contingencies. Behavioral approaches such as ABA or caregiver intervention are more feasible and effective with patients with the above contraindications for CBT. ABA techniques focus on identifying motivating variables that can be changed via environmental modification, controlling antecedent stimuli, and assessing and modifying environmental and social contingencies. ABA assessment and treatment typically begins with implementation by a behavioral therapist or team and subsequently provides parent, teacher, and other caregiver training.

Strategies and intervention components included in cognitive behavior therapy with children and youth

The concept of motivation is of fundamental importance, along with strategies to assess and optimize it. In behavioral therapy and CBT with children and youth, motivation is reflected by the amount of behavior invested in seeking access to items and events that are desired, preferred, and, if available, are interacted with frequently (collectively referred to as "reinforcers"). A reinforcer can be positive if it is provided contingent on a specified behavior, and that behavior occurs more frequently. A reinforcer can also be negative if a behavior prevents or eliminates something that is unwanted or not preferred, and that behavior subsequently occurs more frequently.

Common reinforcer assessment methods involve observations or questionnaires to determine how the youth spends free time, what they report to be enjoyable, and what they actively request, refuse, or avoid. In most applications, CBT employs positive reinforcement to strengthen (increase the frequency of) a target behavior that the individual, caregiver, and therapist agree to be desirable because it may improve mood, activity level, learning, or daily functioning. The primary approach to decreasing less desirable or dysfunctional behavior is to attempt to reduce or eliminate variables that maintain or strengthen the undesirable behavior. In the behavioral literature, discontinuing reinforcement of a behavior leads to the reduction and eventual elimination of a target behavior because it is no longer reinforced.

A primary motivational strategy in CBT is to arrange for a target behavior to reliably produce positive consequences. For example, conditions could be arranged in which a child who wants to play with a favorite toy receives greater access to that toy if they follow a parent's verbal directive. Another strategy would be for a child to be allowed to stay up later (delay bedtime) by cooperating during bath time. Conversely, a problem behavior might decrease when it reliably leads to the loss of something that the child really wants, such as social attention, favorite activities, or tangible items.

Another way to motivate behavior through the manipulation of environmental circumstances (antecedent stimuli) that are observed to be associated with the occurrence of the behavior. For example, being seated in front of a tray of food might increase the probability that eating will occur. Other environmental conditions might be arranged that impact a child's appetite or level of deprivation for a certain type of food and make it more or less likely to be eaten.

Other motivational variables commonly employed in CBT involve assessments of the individual's reported goals or personal values that they believe are likely to impact their future behavior. A value might be "feeling appreciated," "being successful," or "helping others." If a targeted behavior can be associated with one of these values, it may be more likely to be performed by the individual. In daily practice, homework assignments that involve behavioral diaries or charts for self-monitoring can be used to help patients identify antecedents, consequences, and value-congruent goals that could impact their future behavior.

The cognitive component of CBT interventions involves examining and modifying the patient's automatic thoughts and core beliefs. This can include examining situations and identifying accompanying automatic thoughts that increase anxiety, depress mood, or interfere with coping. The clinician begins by describing "cognitive distortions" (as discussed above) and engaging patients in an evaluation of their own thoughts to determine their objective accuracy. Clinicians often use Socratic questioning to assist patients in identifying the objective accuracy of automatic thoughts. By doing so, patients can also identify why some automatic thoughts may be unhelpful, as well as replacement thoughts that result in less overall distress. Patients are guided in understanding the impact of automatic thoughts, any problematic core beliefs they may hold related to the automatic thoughts, and identifying helpful alternatives to compete with unhelpful thoughts and core beliefs. For example, a school-aged child who has automatic thoughts of "I have to get an A on my test, or I will be in trouble" may hold a core belief of "I am only lovable if I do things perfectly." In this way, identification and replacement of such automatic thoughts ultimately leads to the deconstruction of problematic core beliefs, allowing the child to recognize that they have many outstanding, loveable qualities, even when perfection is not achieved.

In the discussion above, we have seen that behavior can be impacted in CBT by identifying and modifying unhelpful thoughts or by discovering and altering environmental conditions that can impact the probability of more functional and effective thoughts and behaviors to achieve desirable outcomes.

A related strategy in CBT involves helping patients develop skills to induce states of comfort or relaxation that can be purposely employed to counter or compete with physiological arousal that is distressing. For example, if a patient is experiencing anxiety, teaching them to use relaxation exercises (diaphragmatic breathing, progressive muscle relaxation, visual imagery, and autogenic phrases) can help them learn to induce relaxation to compete with and counteract their anxiety symptoms. Patients learn and practice these strategies during planned exercises, then apply them in vivo according to a planned set of situational exposures. For example, a patient who experiences debilitating fear or anxiety when undergoing a certain type of medical care may be taught to employ techniques that induce a competing relaxation response during simulations of that care, so the strategies can later be employed when confronted with that real medical care. They can learn to exert volitional control over their physiological reactions, resulting in more comfortable and successful medical care. This often requires effort, trial and error, modification, and repetition to achieve improvement and proficiency. Skilled therapists assist with the arrangement of circumstances and opportunities for this rehearsal to occur and find strategies to help motivate the patient to participate in the process [10].

Individualization of cognitive behavior therapy

While there are proponents of using manualized treatment protocols resulting from successful clinical trials, in much of daily clinical practice, the strategies and tactics reviewed in this chapter are applied in myriad combinations to meet the unique presentation of each patient, particularly children and youth with neurodevelopmental disabilities. Each patient presents with a particular and unique combination of behavioral and psychological symptoms, cognitive strengths, vulnerabilities, and patterns of coping (usually a mix of successful and maladaptive coping behaviors), necessitating some "flexibility within fidelity" of intervention techniques, design, and application. In addition, each person lives in a unique cultural, environmental, and psychosocial context, and it is imperative that any therapeutic approach take the family's values, needs, and life circumstances into consideration.

Working with parents and caregivers

The parent-child relationship involves a developing child whose role is to explore, learn, grow, and mature cognitively and emotionally, and a caregiver whose role is to nurture, protect, encourage, and teach. Over time, children and youth develop new competencies and greater autonomy. For youth with neurodevelopmental differences, this process may take a different course, and caregivers often need additional guidance and support from medical and mental health professionals. As behavioral health professionals, we can help caregivers recognize that their expressed emotions and behaviors have observable and predictable influences on their child's thoughts, emotions, and behavior. Caregivers who understand this will be more open to changing their own behavior and maintaining hope that their own efforts and those of the healthcare team can support their child's optimal functioning and quality of life.

In the context of CBT, therapists need the input and collaboration of caregivers to help individualize and implement CBT strategies. Caregivers can serve as models for demonstrating and learning the use of cognitive-behavioral techniques. They can help motivate their child to apply skills learned in therapy in their everyday environments. By structuring the social environment with organizational cues such as schedules, routines, prompts, and rewards for practice, they can communicate to their child the importance of trying and rehearsing new ways of thinking and behaving that will help them function and cope more effectively with life's challenges. To the extent that caregivers control their child's access to preferred items and activities, these things can be provided contingent on the child's positive and healthy behavior. With guidance from a skilled therapist, caregiver reinforcement can provide powerful motivation. Through changes in their own expressed thoughts and emotions, the caregiver can model that learning and change are life-long processes and thereby encourage their child's openness, insight, and flexibility concerning the malleability of thoughts, emotions, and patterns of behavior. Ultimately, for older children, teens, and young adults, self-motivation becomes critical to success in CBT. Often, this involves learning ways to organize one's own environment and routines to prompt and support positive functioning and behavior. An important role of the CBT professional is to help family caregivers allow their children to explore their own thoughts, beliefs, and personal values. Giving up some aspects of control of the home, school, and community environments to allow children to explore, learn, change, and make choices (and sometimes fail) can be difficult for conscientious and well-meaning caregivers. However, these opportunities to learn from new experiences are key to healthy development and more independent functioning.

Discussing at length the myriad specific techniques a CBT therapist can use with caregivers and families is beyond the scope of this chapter. The literature on behavioral caregiver training and family therapy is extensive, reflecting the diversity of skills required of cognitive-behavioral therapists in partnering with children and their caregivers to achieve socially and functionally significant changes in behavioral health.

Specialized applications of cognitive behavior therapy

In this section, we will discuss specialized protocols that combine aspects of operant and respondent conditioning from the behavior therapy literature.

Desensitization and counterconditioning

These terms are used to describe protocols that involve systematic exposure to a nonpreferred stimulus or situation. This approach has its origins in the behavior therapy literature, beginning with the work of Joseph Wolpe [11,12] who coined the term "systematic desensitization." This therapy involves the treatment of fears or phobias that limit an individual's functioning because the behavior results in the escape or avoidance of that fear. As a result, the person lacks both ongoing exposure and coping skills that would allow them to adapt and gain mastery over the situation. Systematic desensitization is a type of counterconditioning that involves working with the patient to develop a hierarchy of feared stimuli or situations, teaching them alternative behaviors that will create a competing physiological response, and then exposing them to the stimuli from their fear hierarchy. Exposures progress one at a time, beginning with the least feared and gradually progressing through imagined and then in vivo exposure. As the patient can cope with one level of their hierarchy, they are challenged and encouraged to move on to the next. In this protocol, coping refers to the patient's ability to expose themselves to the feared stimulus long enough for their physiological arousal to reach its zenith and begin to diminish. When the person can approach the stimulus with little to no arousal or desire to avoid or escape, they are determined to have gained mastery over that step in the hierarchy. When the patient and therapist can agree that the patient has gained mastery over the entire hierarchy, the therapy sessions can be reduced in frequency and ultimately discontinued.

Desensitization and counterconditioning (also referred to as exposure therapy) have been applied successfully to a host of problems in adults and pediatric populations, ranging from simple phobias (spiders, needles, and dogs) to specialized protocols for school phobias, social phobias, and obsessive-compulsive disorder. They can also be used for uncomfortable and feared medical procedures [10,13–16].

Comprehensive behavioral intervention for tics

In the medical and behavioral literature, a "tic" is a frequently occurring vocalization or body movement for which the person experiences a premonitory urge or compulsion such that the behavior is difficult to control or suppress. Tics may manifest alone or as part of a more complex pattern of symptoms or characteristics, such as Tourette syndrome, which is characterized by multiple motor tics (e.g., blinking, coughing, sniffing, and facial twitching) and at least one vocal or verbal tic that is often preceded by a premonitory urge or sensation in the affected muscles. The tics can sometimes be delayed or suppressed temporarily; thus, they are not completely involuntary. The symptoms may change in intensity, location, and frequency over time and across situations. The individual has varying degrees of conscious awareness of their tics depending on their activities, level of stress, and degree of fatigue (see Chapter 18 Tics and Stereotypies).

Both comprehensive behavioral intervention for tics (CBIT) and habit reversal training (HRT) are specialized treatment protocols for tics. These therapies include several primary components: (1) teaching the person to be more aware of their tics, urges to tic, and potential contributors to the occurrence of tics such as stress, anxiety, and fatigue; (2) strategies to decrease physiological arousal (relaxation training); (3) teaching them to engage in an alternative motor behavior that is incompatible with or competes with the tic; (4) teaching them to monitor and be more aware of the environments, physical states, or activities associated with greater or lesser frequency of tics and strength of urges to tic; (5) systematically planning for the application of these strategies in new or unfamiliar settings (generalization training). CBIT and HRT can be used to treat primary motor tics, Tourette disorder, and other repetitive habitual behaviors (except stereotypies in neurodevelopmental disorders such as self-injurious behavior, flapping, rocking, etc.). CBIT and HRT have been used successfully with children, youth, and adults [17].

Dialectical behavioral therapy

Dialectical Behavioral Therapy (DBT) is an evidence-based treatment developed for working with patients presenting with difficult-to-treat problems with more chronic and severe symptoms and interpersonal conflicts (personality disorders, suicidal ideation, self-harm, and severe mood disorders). DBT is a modified and elaborated type of CBT that seeks to teach patients to live in the present moment, develop healthy ways to cope with stress, regulate their own (often intense) emotions, and have more satisfying interpersonal relationships. Key intervention components include mindfulness, radical acceptance, and self-regulation [18,19].

Acceptance and commitment therapy

Acceptance and commitment therapy (ACT) uses concepts of "acceptance," "mindfulness," and "commitment" to offer strategies for coping with intense emotions and facilitating behavior change. Techniques stemming from these concepts are used to teach the person to have more cognitive flexibility. Hayes [20] first developed this approach in the early 1980s with the goal of integrating cognitive behavioral therapy and applied behavior analysis. The emphasis is placed on using data recording to help patients identify the negative effects of overly rigid adherence to verbal rules. In ACT, the goal is not necessarily to eliminate difficult emotions but to be present with them and the situations that induce them, to be more accepting of them by not trying to avoid them, and to be more open to experiencing those feelings and situations in order to pursue one's personal values in life.

Trauma focused cognitive behavioral therapy

Trauma-Focused Cognitive Behavioral Therapy (TF-CBT) is another evidence-based treatment for people who have experienced and been negatively impacted by traumatic events. It was developed to help treat the behavioral and psychological difficulties of people with posttraumatic stress disorder due to prior trauma. Developed by Cohen, Mannarino, and Deblinger [21], this approach aims to help patients and caregivers recognize, cope with, and better regulate emotions, thoughts, and behavior patterns that stem from or relate to their past traumatic experiences. TF-CBT has a growing research literature supporting its efficacy with children who have experienced or been witnesses to traumatic events such as physical or sexual abuse, community violence, accidents, war, and natural disasters [21,22]. The emphasis is placed on techniques to educate the child about trauma and its effects on a person, as well as teaching individualized coping and self-regulation skills. All the specialized applications of CBT introduced above have substantive and replicated supporting studies. However, a comprehensive review of each of these is beyond the scope of this chapter.

References

[1] Skinner BF. The behavior of organisms: an experimental analysis. Appleton-Century; 1938. p. 1–457.

[2] Beck AT. Thinking and depression: I. Idiosyncratic content and cognitive distortions. Arch Gen Psychiatry 1963;9:324–35.

[3] Beck AT. Cognitive therapy and the emotional disorders. International Universities Press, Inc.; 1976. ISBN 10: 0823609901/ISBN 13: 9780823609901.

[4] Beck AT. Theoretical perspectives on clinical anxiety. In: AH Tuma, JD Maser, editors. Anxiety and the anxiety disorders. Lawrence Erlbaum Associates, Inc; 1985. p. 183–96. Available from: https://doi.org/10.4324/9780203728215.

[5] Beck AT, Rush A, Shaw B, Emery G, DeRueis RJ, Hollon SD. Cognitive therapy of depression. New York: The Guilford Press; 2024. p. 1–406. eText ISBN 9781462555963, 1462555969.

[6] Chand SP, Kuckel DP, Huecker MR. (2023). Cognitive behavior therapy. StatPearls Publishing, Treasure Island, FL; https://www.ncbi.nlm.nih.gov/books/NBK470241/.

[7] Sullivan MO, Watson S, Butler B. A practical guide to working with depression; a cognitive behavioral approach for Mental Health Workers. Pavilion Publishing; 2016. p. 1–91. ISBN: 9781910366905.

[8] Martell CR, McCauley E, Gudmundsen GR, Schloredt CA, Dimidjian S. Behavioral activation with adolescents: a clinician's guide. New York: Guilford Publications; 2016. p. 1–226. ISBN 9781462523986.

[9] Kennedy F, Pearson D. Integrating CBT and third wave therapies: distinctive features. 1st ed Routledge; 2020. p. 1–164. Available from: https://doi.org/10.4324/9780429442926.

[10] Slifer KJ. A clinician's guide to helping children cope and cooperate with medical care: An applied behavioral approach. Baltimore, Maryland: Johns Hopkins University Press; 2014. p. 1–288. ISBN 142141113X (electronic).

[11] Wolpe J. Psychotherapy by reciprocal inhibition. Conditional Reflex 1968;3(4):234–40. Available from: https://doi.org/10.1007/BF03000093. PMID: 5712667.

[12] Wolpe J. Behavior therapy in complex neurotic states. Br J Psychiatry 1964;110(464):28–34. Available from: https://doi.org/10.1192/bjp.110.464.28. PMID: 14103366.

[13] Slifer KJ, Babbitt RL, Cataldo MD. Simulation and counterconditioning as adjuncts to pharmacotherapy for invasive pediatric procedures. J Dev Behav Pediatr 1995;16:133–41. PMID: 7560116.

[14] Slifer KJ, Eischen SE, Busby S. Using counterconditioning to treat behavioral distress during subcutaneous injections in a pediatric rehabilitation patient. Brain Injury 2002;16(10):901–16. Available from: https://doi.org/10.1080/02699050210131948. PMID: 12419003.

[15] Slifer KJ, Hankinson JC, Zettler MA, Frutchey RA, Hendricks MC, Ward CM, Reesman J. Distraction, exposure therapy, counterconditioning, and topical anesthetic for acute pain management during needle sticks in children with intellectual and developmental disabilities. Clin Pediatr 2011;50(8):688–97. Available from: https://doi.org/10.1177/0009922811398959.

[16] Paasch V, Leibowitz L, Accardo J, Slifer K. Preparing children with autism spectrum disorders for overnight sleep studies: a case series. Clin Pract Pediatr Psychol 2016;4(2):153–63. Available from: https://psycnet.apa.org/doi/10.1037/cpp0000139.

[17] Scotti-Degnan CM, Ford HA. Comprehensive behavioral intervention for tics: a practical review for clinicians. Pract Neurol 2020;20:42–6.

[18] Linehan MM. DBT skills training manual. second edition New York: Guilford Press; 2014. ISBN-10: 1462516998.

[19] Rathus JH, Miller AL. DBT skills manual for adolescents. New York: Guilford Press; 2014. ISBN-13978-1572307810.

[20] Hayes SC, Strosahl KD, Bunting K, Twohig M, Wilson KG. What is acceptance and commitment therapy? In: Hayes SC, Strosahl KD, editors. A practical guide to acceptance and commitment therapy. Boston, MA: Springer; 2004. Available from: https://doi.org/10.1007/978-0-387-23369-7_1.

[21] Cohen JA, Mannarino AP, Berliner L, Deblinger E. Trauma-focused cognitive behavioral therapy for children and adolescents: an empirical update. J Interpers Violence 2000;15(11):1202–23. Available from: https://doi.org/10.1177/088626000015011007.

[22] de Arellano MA, Lyman DR, Jobe-Shields L, George P, Dougherty RH, Daniels AS, et al. Trauma-focused cognitive-behavioral therapy for children and adolescents: assessing the evidence. Psychiatr Serv 2014;65(5):591–602. Available from: https://doi.org/10.1176/appi.ps.201300255. PMID: 24638076; PMCID: PMC4396183.

Chapter 38.6

Circumvention and assistive technology

Brooke Kimbrell[1,2] and Tara Johnson[3]

[1]*Department of Neurology and Developmental Medicine, The Kennedy Krieger Institute, Baltimore, MD, United States,* [2]*Department of Pediatrics, Johns Hopkins University School of Medicine, Baltimore, MD, United States,* [3]*Arkansas Children's Biomedical Research Institute, Arkansas Children's Hospital, Department of Pediatrics, University of Arkansas for Medical Sciences, Little Rock, AR, United States*

Introduction

The human mind and body are independently elegant in their complexity and efficiency. The interaction between mind and body creates the physical representation of each person's individuality. The infinite combinations of individuals in a culture create society, which provides a moderately adaptable framework for individual interactions, employment, education, and leisure. Given the richness of human diversity, it is somewhat surprising that limitations so often disrupt individuals' engagement with society. In fact, physical and mental differences that limit societal engagement are labeled disabilities. When medical and rehabilitative efforts cannot fully restore an individual's capacity to function efficiently, technology can often expand people's abilities and support integration into society. These sorts of interventions circumvent an individual's unique limitation or disability by adapting the process or the environment of a task. Thus, circumvention therapy and assistive technology have the explicit goal of improving the function of tasks, which benefits societal integration and personal independence.

One example of well-established assistive technology (AT) is corrective lenses for near-or far-sighted vision impairments. This technology restores functional vision without directly healing the misaligned anatomy of the eye. With vision restored, the wearer can participate in normal activities without restriction. This assistive technology is so well accepted by individuals and by society that most people with myopia and hyperopia would not even describe their vision problems as a disability. These devices are so ubiquitously available that they are sometimes even used for the purely social purpose of fashion by people with normal vision. It is encouraging to note this example of how society can adjust to include and even celebrate people with disabilities and their assistive technology. This is the ultimate goal of rehabilitative engineering, not only to accommodate for an individual's limitations but also to expand the boundaries of societal norms so that the secondary harm of stigma is abolished.

The focus of this chapter is a brief introduction to types of assistive technology with a few practical examples of devices that may support a person's achievements and integration into society by adjusting how an individual performs the task. It is worth mentioning briefly that the modification of society to reduce barriers to disabled individual's participation and success is an ongoing and worthwhile effort. Since the 1960s, there has been a movement in architectural design to explore universal design [1]. Systems built according to these principles pursue the ideal of being equally accessible to all physical abilities. The premise of universal design is that society can be consciously as well as unconsciously shaped to be inclusive, and in doing so, society benefits from increased diversity. This is different from assistive technology, which strives to meet the needs of an individual. The two fields meet in the middle, and rehabilitative engineering addresses both assistive technology and universal design.

Milestones in the United States of America's societal evolution toward inclusivity can be marked by several major policy changes over the past century, such as the Individuals with Disabilities Education Act (IDEA) of 1975, reauthorized in 2004, and the Americans with Disabilities Act (ADA) of 1990, reauthorized in 2008 [1]. These laws affirmed society's obligation to make reasonable accommodations for people with disabilities and to reduce barriers to employment, education, and leisure. Assistive technology was specifically addressed and prioritized in the Assistive Technology Act as amended in 2004 (ATA2004). Section 2(b) [2] states that AT increases involvement in and reduces expenditures associated with programs and activities that facilitate communication, ensure independent functioning, enable early childhood development, support educational achievement, provide and enhance employment options, and enable full participation in community living for individuals with disability. Access to such devices can also reduce expenditures associated with early childhood intervention, education, rehabilitation and training, health care, employment, residential living, independent living, recreation opportunities, and other aspects of daily living [3]. This language reflects an emerging emphasis on the use of technology to empower individuals to achieve their best potential in society. This chapter will introduce types of assistive technology used to support disabilities in vision, hearing, language, and motor function.

Assistive technology for vision impairment

Assistive technology for people with visual impairment or blindness includes enhancement of visual stimuli and circumvention of visual systems to audible and tactile inputs. As with all AT, the goal is to promote autonomy, independence, and self-expression in society. In addition, there is a particular focus on access to information and education for these technologies. This emphasis was a foundational belief of Louise Braille, who invented braille in the 1820s [4]. This tactile system of reading and writing has remained at the core of education for blind people for over a century. As technology has developed, braille reading and writing devices have evolved to facilitate screen-based written language and even translate spoken language in real time with refreshable braille displays. These adaptations improve accessibility and communication options for people who read and write braille. Braille has been a powerful tool for the blind community, but learning and using it could be compared to learning and using another language with the need for translation to and from mainstream written language. This laborious process means that some people with cognitive impairments cannot master the use of braille, and some people with low vision would rather use their residual vision than rely on braille. For these individuals and many braille-literate people, the advent of text-to-speech and voice-to-text technology has enabled easier access to digital written information. Kurzweil invented optical character recognition (OCR) technology in the 1970s, and Kurzweil systems were the first to be able to scan and convert written documents to speech. In partnership with the National Federation for the Blind, Kurzweil released a hand-held OCR device in 2006 called the Kurzweil National Federation for the Blind Reader commonly known as the KNFB Reader. This technology was released as a smart phone application, also known as KNFB Reader, in 2014 [5].

Over recent decades, many other subtypes of assistive technology have evolved. This process has accelerated thanks to the broad commercial availability of smart and internet connected technology. Since most of these products and platforms are commercially developed, they usually do not have any published data or research regarding efficacy. We will briefly summarize some of the AT commonly used at home, in the community, and at school or work. Of course, these technologies cross domains, and well-suited AT can be used universally. We will forgo discussion of vision restoration technology and treatments and assume that these avenues have been optimized. We will also forgo discussion of living assistants, both human and animal.

Around the house, AT should help enable comfortable, safe, and efficient living. Occupational therapists have used low-tech strategies for years to help with daily tasks. There are many tricks, like stitching shaped buttons to the inside of clothes, to help with the identification and matching of outfits. Color reader wands and apps are a high-tech option for this and other uses. Bump dots are raised stickers that can be placed anywhere to help orient a user. For example, on buttons on an appliance. Stickers with braille labels or with RFID chips that can be read aloud by a wand can help with identifying tactilely nondescript objects like cans or makeup products. Talking clocks can be used to announce the time with the push of a button. Smart home technology also has a role, with many people enjoying the convenience of asking their internet-connected device to announce the time, play a song, adjust the thermostat, lock the door, adjust the lighting, or do many other things [2].

Recreation at home includes many typical activities such as reading, crafting, watching movies and shows, and even video gaming, which are enjoyed by people with vision impairments or blindness. Many enjoyable activities, such as reading, crafting, and handheld gaming, can be facilitated with appropriate magnification and lighting for some people with residual vision. These magnifiers are usually called close circuit television (CCTV) devices and can be handheld, desktop, or wearable devices. Smartphone apps like WeZoom are convenient for many people, especially for short texts. Written material can also be made available via braille or audio formats. The national library service makes audio format books available free to any person with documented vision impairment via a program called Brail and Audio Reading Download (BARD). These audiobooks can be mailed via cassettes for use with a provided reader device or downloaded from the internet [6]. Many other databases for audiobooks include Audible, Librivox, and Overdrive. Enjoyment of shows and movies is becoming more accessible thanks to features such as audio descriptions of the actions and expressions, which are being embedded by many platforms like Netflix and Amazon. For more social recreation at home, there are many braille-adapted board games, playing cards, dice, and dominos [1].

In the community, assistive technology can help with navigation, shopping, education, and employment. The socially identifiable white cane is a well-known example of AT for navigation. Even this venerable symbol is getting "smart" with technology such as ultrasound and laser distance sensors [7]. A notable example is the WeWalk white cane, which vibrates and provides audible cues for obstacles ahead of the cane tip and objects above waist level. This device also interfaces with its app for turn-by-turn navigation and clock-style orientation. Electronic white canes fall into the class of electronic travel aids (ETAs), along with wearable devices and smart phone apps for navigation. Wearable ETA's such as Buzzclip can sense obstacles above the waist that a cane would not detect. More complex wearable technology is being developed, with several in the prototype phase. These devices are striving for features such as object recognition and guidance for navigation near moving objects [8]. Smart phone navigation apps, including Apple Maps and Google Maps offer turn-by-turn directions. In addition to these features, several apps, such as Blindsquare, Lazarillo, Soundscape, and WeWalk, offer more information about orientation and the surrounding landscape, such as nearby businesses and intersections. Many of these apps also offer directions via public transportation [9]. Of course, ridesharing apps like Uber and Lyft offer excellent convenience and are easily accessible via smart phone.

Day-to-day tasks in the community may also need assistive technology for success. Any person with a vision impairment may request a free talking banknote reader device from the US Treasury Department. Object identification in the environment is a challenge that several smart phone apps attempt to address. BeMyEyes is a free app that connects a person with vision impairment to a seeing volunteer who can identify, read, or explain material presented through the smart phone's camera. Fostering even more independence, although likely less accurate, apps such as Seeing AI (free) and Envision AI (subscription service) use the smart phone's camera to narrate the visible environment, including object identification, face recognition, barcode scanning, money identification, scene and photo description, and read-aloud modes. Short text mode will read aloud any text in view. A document scanner will capture pages of text that can be read aloud, stored, or optimized for best readability with font size, contrast, and color. These optical character recognition (OCR) apps and the relatively venerable KNFB reader app are tailored for people with vision impairments in that they will announce when the borders of the page are visible. KNFB readers can interface with braille displays [10]. Another OCR app is the Voice Dream scanner, which can manage more complex layouts like multiple pages or columns and has customizable voices and reading speeds. The companion app, VoiceDream Reader, takes advantage of this customization for text already in digital form.

Head wearable technologies with similar features of object identification, facial recognition, and reading are becoming commercially available at significantly higher costs than the previously mentioned mobile applications. A notable example is the Orcam brand MyEye products, which are low-profile devices that can be clipped to eyeglass frames and activated with gestures or button taps. Recently re-released, Google Glass has collaborated with Envision AI to offer the app's features "hands-free." Other wearable devices, such as NuEyes, eSight, IrisVision, and Acesight, combine some of these capabilities with vision augmentation for people with visual impairments [11]. These goggle-like devices remain somewhat bulky-looking and cost and battery life must be considered.

Many of the technologies we have already discussed can assist participation in work or school. In addition, laptop and desktop computers are accessible to people with vision impairments and blindness thanks to built-in accessibility features and third-party software. Even in the early days of personal computers, manufacturers included screen-reading technology. Today, there are many different software packages with different features. A few to note include VoiceOver for iOS, Job Access with Speech (JAWS) for Windows, COBRA, which interfaces with over 70 refreshable braille displays, and Orca, which is open source software. Refreshable braille displays can be used for both reading and inputting data. They range from Bluetooth-connected pocket-sized devices with 14 cells, like the Focus 14 Blue Braille Display, to the VarioPro 80 Cell unit, which is integrated with full keyboard and cursor navigation. For people with some sight, screen displays can be adjusted to optimize vision by changing font size, screen brightness, contrast, color choices, icon size, and many other options. Screen magnification systems are software programs that can further adjust the visible presentation to optimize accessibility. In some cases, these technologies may be acquired through the school or workplace as accommodations under the Americans with Disabilities Act [12].

Combinations of these assistive technologies should be tailored to the needs, interests, and social environment of each individual. Use of technology itself must not become the end goal, but practice and skill are required to use most of these systems efficiently. For this reason, low vision specialists as well as therapists are important resources to guide the selection of technology.

Assistive technology for hearing impairment

We began this chapter with the example of corrective lenses as an assistive technology that society has fully embraced to support people with vision impairments. Another such device is the smart phone, which has been particularly relevant to people with hearing impairments. This section will briefly overview some categories of AT for people with hearing impairments. In many cases, we will describe the individual technology items and at least one of the apps or features available on smart phones that can also fill that need. By the time this comes to print, it will be out of date, but hopefully this snapshot will offer a starting point to explore options. Not surprisingly, the video platform of YouTube is a good place to learn about up-to-date technology for people with hearing impairments. This section will not focus on technology that directly supports or replaces hearing, such as hearing aids or cochlear implants. This discussion presumes that patients have worked with their medical team to optimize their unique hearing capacity. Our goal is to identify technology to assist with everyday activities or unique challenges that an individual with residual hearing impairment may have.

Most hearing individuals unconsciously rely on auditory input around the home. Doorbells, smoke alarms, appliance buzzers, alarm clocks, baby monitors, security systems, televisions, and telephones are just a few of the household items that default to audible notifications. These systems may need to be adapted or replaced with accessible technology for a hearing-impaired person's home to be comfortable and safe for them. Historically, doorbells could be updated with systems that included flashing lights, or a hearing dog could be trained to run back and forth from the door (or any other audible notification in the home) to the home occupant to alert them. Recently, smart doorbells such as Ring have provided a visible, tactile, or even video notification to a smart phone when someone is at the door. Smoke alarms should be installed with flashing lights to notify occupants of danger. Advanced systems can link with smart phones to augment the alert if the person is in a place where the alarm light may not be noticed, such as asleep or in a different room. Smart home security systems with video and/or sound detection can be used to improve awareness of the environment for both safety and convenience. This can help a hearing-impaired person know when a household appliance beeps, when a baby wakes up from their nap, or when another person moves around the house. BraciPro is a smart phone app that can be "trained" to recognize household sounds and notify the user. Many televisions are smart technology in their own right, with helpful features including closed captioning and Bluetooth connections to deliver sound directly to paired hearing aids [2].

At home and in the community, telecommunication is an integral part of many people's day. This field has benefited from many generations of innovative technology. Most of the telecommunication systems and devices for the hearing impaired are being naturally supplanted by smart phone technology. It is worth reviewing other systems since each was developed to address a specific need. In the 1960s, some deaf individuals were able to obtain teletype machines for their homes. The earliest of these were decommissioned devices from World War II. A teletype machine could relay messages only to another teletype machine. To communicate with a hearing person via telephone, a relay service was needed. These services began sporadically but were mandated under the Americans with Disabilities Act (ADA) of 1990. This free 24/7 service is still accessible to anyone in the USA by dialing 711. Teletype devices are now much smaller and more convenient, and they include brail reader and writer options, but they still largely rely on this human relay service. Similarly, there are services available for video translation of sign language into spoken language called Video Relay Service (VRS). These services commonly incur a cost for the user. People with hearing impairments who use their voice to speak may be able to use commercially available telephones if they are equipped with amplification, Bluetooth connections to hearing aids, visual or tactile notifications, or voice-to-text technology. Voice-to-text technology is most commonly used for captioned telephone calls. Some states offer Captioned Telephone Service (CTS) through CapTell, where the addition of a second phone allows the captions to appear on a screen on the phone. CaptionCall is a company that offers this service privately via a combination of a phone and internet connection and also offers apps for iOS and Android. Smart phones have built-in features for amplification, blue tooth connection to hearing aids, tactile and visual notifications for alerts, and video communication. Apps can be selected and added based on individual needs. Useful apps include Zoom for video calls, Purple VRS or Convo for translation relay between sign and speech, and CaptionCall for text-to-speech captioning.

In the community, there are many needs for amplification or filtering of auditory information, and there are many devices intended to meet these needs. Three amplification systems with historical and ongoing applications are loop systems, FM systems, and pocket talkers [13]. Loop systems are well suited for capturing sounds in large spaces, such as classrooms. They work by using a loop of wire as an antenna, then delivering amplified sound to a hearing aid, earpiece, or headphones. The loop may be small enough to wear around a user's neck like a lanyard or embed in a clipboard, or it may be permanently installed under the flooring in a large conference room or classroom. FM systems detect the speech of a single speaker, like a teacher with a worn microphone, and direct it to a hearing device via a radio receiver. Pocket talkers use a similar approach, but often the microphone is embedded into a handheld device that can be placed near the speaker. Smart phones come preprogrammed with assistive technology settings for hearing impairments and can be used much like a pocket talker. The use of Bluetooth microphones can extend the physical range of these settings. Many hearing-impaired people have a particularly difficult time when multiple people are talking at once. Amplification and filtering technology can help, but it is limited. Another helpful option is the Ava app, which provides real-time captioning of ambient conversations.

Hearing impairment can present a barrier to social interactions, classroom education, professional development, and even comfort and safety in the home. These limitations can be circumvented by creative and thoughtfully employed technology.

Although the array of potential technologies can be dizzying, ideally accessible assistive technology can be integrated into individual areas of need in a person's daily life. The goal is to allow people with hearing impairments to live without being limited by their hearing capacity. The advent of smart phones changed the way that technology interacts with everyone's daily life and will continue to change the way that people with hearing impairments interact with the world.

Assistive technology for language impairment

A large part of society revolves around communication through spoken language that requires complex cognitive processing and motor output coordinated with swallow and respiratory functions. These processes can be disrupted due to developmental dysfunction, motor impairment, or brain injury to motor speech areas. Lack of verbal communication is socially isolating and often frustrating for the affected individual and their contacts. Facilitating functional communication in a nonverbal individual can be accomplished with augmentative and alternative communication (AAC). Augmentative and alternative communication ranges from gestural and facial expression to picture boards to sophisticated digital technology. The underlying principle is to offer the ability for self-expression, including requesting, answering, and social interaction. This can be as simple as using a button or combination of switches for basic expressions or requests. A picture exchange system is a low-tech way for a person to make requests and answer queries. High-tech tools such as Proloquo2go on an iPad can offer a readily adaptable lexicon of icons to request or express ideas and have the advantage of producing an audible spoken output. If cognitively appropriate, spelling boards are a low-tech option to allow people to write their thoughts. A high-tech text-based system can use text-to-speech technology, which allows a user to write their thoughts and have the computer speak their words. Word prediction software can help make this process more efficient. Some softwares have features that allow the user to customize their computer-generated voice. A commercially available example is VocaliD, which blends vocal signatures from the user with a matched speaker from a voicebank to create a unique synthetic voice. Often, nonverbal individuals who can spell and write are nonverbal due to motor disorders and are unable to write or type with traditional methods. Fortunately, modified inputs such as a large ball computer mouse, a one-handed keyboard, and eye-gaze technology allow access to these technologies.

Many verbal individuals have receptive or expressive language disorders, learning disabilities in reading or writing, or writing difficulty due to fine motor abnormalities. Assistive technology can be used to improve access to education and ease of expression. Stories and books with diverse vocabulary nurture language skills, but reading these stories can be laborious for children with reading difficulties. The task of reading can be circumvented by text-to-speech technology for digital material and optical computerized reader (OCR) devices or applications for printed text. Writing difficulty can also be circumvented by voice-to-text technology, which is integrated into some word processing software or can be added to products such as Dragon. Word processors also offer the benefits of spelling correction, grammar checking, and word prediction.

Assistive technology for motor impairment

Loss of strength, lack of coordination, or deformity can limit independence and participation in society. Assistive technology can help ease the daily effort burden and reduce barriers to participation for these common and diverse motor disabilities, whether congenital or acquired across the life span.

Lower-extremity impairment primarily impacts walking and standing. Common societally accepted assistive technology such as braces, canes, and walkers can support stability and safety for minor impairments. Orthoses can support regional limb weakness, such as a drop foot. Functional Electrical Stimulation has also been used for drop foot with a decrease in energy consumption and improvement in walking sleek for people with multiple sclerosis and slow self-selected walking speed [14]. Crutches and prostheses can bear weight for more significant lower body weakness or deformity. Standers facilitate hands-free upright positioning and weight bearing in people with severe lower extremity or truncal impairment.

When walking is not possible or is prohibitively inefficient, wheelchairs and exoskeletons can facilitate mobility and participation. Wheelchairs offer a broad spectrum of features, including power mobility, seat elevation, chair reclining, or head and trunk support, depending on the needs and budget of the consumer. Power mobility navigation options are diverse and include power propulsion assist wheels, joystick control, head control, and sip and puff control depending on upper extremity function. There are many attachments, ranging from cup holders to robotic arms, that can customize wheelchairs to optimize their function for an individual's needs. Kinova's Jaco is a notable robotic arm that is controlled with a joystick. Exoskeletons have been developed and are being refined to facilitate walking with powered robotic assistance. These devices support short-distance mobility at home and in the community. Other assistive technology needs may include lifts to assist with turns and transfers, shower chairs, and tools to grasp high or low objects not accessible from a seated or supported standing position.

Upper extremity impairment causes decreased functional arm and hand use for activities of daily living, such as writing, accessing technology, and infinite other functions. For individuals with partial arm function, many household items, such as utensils and writing implements, may be modified by adding grips suited to individual biomechanics. Specially designed cups and dishes can help with some grip or coordination difficulties. High-tech utensils such as Gyenno or Liftware products can dampen tremor or adjust the utensil angle to minimize coordination difficulties for some people with movement disorders. Robotic feeders such as Obi can restore mealtime independence for people who are able to use button switches to select serving and command delivery of a bite. Devices for dressing and doing fasteners can be helpful, or better yet, adaptive clothing has been developed to support independence for people with limited hand use, such as the Tommy Hilfiger line. Dynamic arm supports use stored energy from elastics or counterbalances and can increase function for people with limited shoulder and elbow strength while seated. Upper extremity exoskeletons and wearable powered arm supports, such as the Abilitech Assist, have been developed for adults with disabilities or working in industry. Writing can be facilitated by adaptive keyboards or circumvented by speech-to-text technology, which is further discussed in the language section of this chapter. Voice commands and internet-connected technology can be helpful for diverse functions around the house, as reviewed in the vision and hearing impairment sections of this chapter. Technology access can be facilitated by alternative inputs such as voice commands, switches, joy sticks, head control, sip and puff, and eye gaze technology. These modalities may also be used for transportation in power wheelchairs and automobiles.

Practicalities

Selection, acquisition, and application of assistive technology is a challenging but rewarding process that should be centered on the needs and goals of the patient and family. A child with a disability will not consistently use technology to do something unless the technology makes it easier than undertaking the same task naturally. Furthermore, most children will avoid assistive technology if it is socially "uncool," even if the tasks are more challenging to complete unaided [15]. This effect is well documented, and device abandonment is one of the best-researched aspects of assistive technology science. For example, Copley and Ziviani reviewed the literature on barriers to the integration of assistive technology in the classroom and summarized that they include "lack of appropriate staff training and support, negative staff attitudes, inadequate assessment and planning processes, insufficient funding, difficulties procuring and managing equipment, and time constraints" [16].

Although it is relatively easy to identify factors in device abandonment in retrospect, it is much more challenging to develop a formula to avoid them when prescribing the technology. One valiant attempt by Brian Hoppestad used the Delphi method to curate 39 elements experts in the field thought were necessary to assess when considering new adaptive technology for computer access [17]. Although all 39 elements are likely important, remembering to assess that number of factors is overwhelming. I humbly suggest the following principles for identifying successful assistive devices: (1) Listen to the patient; (2) Partner with the family and teachers; (3) Match goals and maturity with technology; (4) Try before you buy; and (5) Do not get discouraged (Fig. 38.6.1).

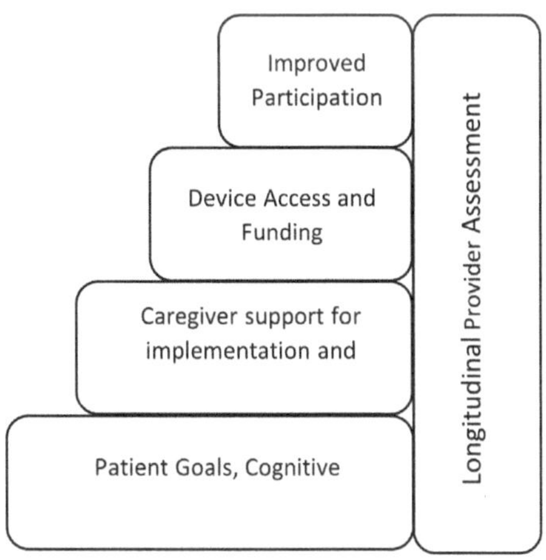

FIGURE 38.6.1 Framework for assistive technology assessment.

Listen to your patients: Start by asking, "What are you good at?" This is the perfect starting point for identifying strengths that an adaptive device will leverage to circumvent limitations. Eventually, asking "What is hard for you?" or "If you have three wishes, what would they be?" may identify a target area for adaptive technology. Avoid limiting the focus of the conversation to disabilities, and challenges or opportunities may be missed to better align patient and provider goals. Children often have imaginative and exaggerated expectations, so discussing device goals with them directly is important to promote buy-in.

Partner with the family and the teachers: As whimsical as some of our patients' wishes may be, their families and teachers are usually reliable sources regarding what factors are most limiting to the child's success. Screening for tasks that need help, prompt tantrums, and add stress to daily life is a good place to start. Asking about missed opportunities in the community for the child or family be a high yield approach. Discussing long-term goals may allow for the identification of barriers that assistive technology should target. Cultural perspectives, socioeconomics, physical space, and group dynamics all influence the application of a device in the home or classroom. The adults in these environments will be responsible for supporting the application of assistive technology and should be integral in its selection. Training and supporting parents and education staff on the use of assistive technology is of paramount importance in facilitating access to assistive technology for children with disabilities.

Match the Goals and Technology to Maturity: At best, technology can enable skills that are within a child's cognitive capacity, and goals should be set accordingly. For example, providing a nonverbal child who's intellectual skills are approximately 2 years old with a communication device should have the goal of facilitating two-word sentences, which would be a typical language expectation for that mental age. A family member may be hoping for performance closer to their chronological age and may become discouraged and abandon the device unless appropriate counseling is given by the prescriber. This child will benefit more from a simple device interface with good portability, which reflects the skills and needs of a toddler. As a child matures, he or she may find that a previously adopted system no longer suits their needs and abandon it. This does not reflect the failure of the device, but rather the success of the child! Periodical reassessment of the child's readiness for new skill emergence should be performed, and their assistive technology should adjusted to help facilitate progress.

Try before you buy: many states have adaptive equipment libraries, which allow families to try commercial products before they are purchased. The opportunity to interact with a device directly is an invaluable test of its fit for integration into the child's daily activities. This step can screen for many of the barriers to adoption, including limitations of physical space, social acceptance, and developmental readiness. Assistive technology is notoriously difficult to acquire through insurance, which may be related to the limited availability of scientific evidence of efficacy and the high rate of device abandonment. However, a letter with information about how a device has shown its potential to affect a child's life can support the effort. If insurance approval is not obtained, it is even more important for a family to try a device before paying for it out of pocket.

Don't get discouraged: identifying and supplying assistive technology that enables a child to express their potential is richly rewarding; however, the process can be tedious with some degree of educated guesswork involved. Devices may not be adopted because they are not well suited to a patient's skills, environment, or available support, and they may be abandoned as a child develops new skills. These outcomes are expressions of a unique child's needs rather than the success or failure of the prescriber's efforts. Assessment of prior assistive technology use and the reason for discontinuation provides invaluable information for guiding the selection of future devices. Identifying assistive devices for children with multiple severe disabilities is particularly challenging, and the adoption of the devices may take more time than anticipated. Borgestig et al. studied nonverbal children with severe physical impairment who were provided with eye gaze technology for use in daily activities and 9–10 months of multidisciplinary support to the child, their parents, and school personnel. They found that all 10 children adopted the technology in daily life and that there was improvement in time-on-task after 5 months and improved accuracy in selecting targets after 15–20 months [18]. These positive effects may not have been captured with less comprehensive support or a briefer attempt.

Future directions

Improving access to quality assistive technology needs substantial and sustained effort in several domains. First, the devices must be developed and tested. Device design is plagued by challenges including poorly defined needs, unclear safety standards, a lack of categorization systems, and access for the target population. As a result, there is a somewhat comical effect of re-inventing the same technology every few years as devices fall out of vogue. One study undertook a review and classification of dynamic arm support devices and found a total of 104 such devices invented since the polio era of the 1940s, most of which had fallen out of use and been replaced by similar devices [19]. The Rehabilitation Engineering and Assistive Technology Society of North America (RESNA) is drafting and publishing safety standards and some classification schemes, which will aid future device development. The World Health Organization has published a classification system [3]. The International Classification of Functioning (ICF) framework may be well suited to evaluate assistive technology. However, a systematic review of eye-gaze assistive technology for children with complex disabilities by Perfect et al. found that none of the 11 studies included utilized the ICF framework to assess outcomes, and the majority used nonvalidated outcome measures [20,21]. Some work has been done to develop tools and metrics to guide selection and evaluate responses to assistive technology. This groundwork may facilitate RTCs in the future. The strengthening of condition-specific patient advocacy groups may assist in device design and testing by identifying specific needs and facilitating consumer marketing.

Once designed, assistive technology must reach its target population through diverse pathways, including direct consumer marketing, assistive technology professionals, rehabilitation specialists, the school system, and the physician's office. This heterogeneous system makes it somewhat uncertain that disabled children will have access to the full range of devices available for their needs. In the past, the National Institute on Disability, Independent Living, and Rehabilitation Research sponsored a website called Abledata.com, which provided information on 40,000 assistive technology products and rehabilitation-related equipment. At the time this chapter is being written, this rich source of information has been taken offline after posting an announcement stating, "Due to a realignment of the National Institute on Disability, Independent Living, and Rehabilitation Research (NIDILRR) priorities, AbleData will be discontinued on September 27th, 2020." Hopefully, this database will be restored or replaced with another centralized repository of assistive technology information for consumers in the United States. In the meantime, the organization Global Public Inclusive Infrastructure (GPII) offers Unified Listing (ul.gpii.net), a searchable website consolidating the listings for communication, computer, and digital devices from assistive technology databases for 12 countries around the globe. Another searchable AT repository is Eastin.eu, which is managed by the Global Assistive Technology Information Network and accesses general AT listings from European and Australian databases.

Once a suitable device has been identified for a child's needs, cost often presents a barrier, as these devices are rarely covered by insurance and may be outside the budget of many families. Some devices are provided through the school system as part of the Individuals with Disabilities Education Act, but these devices are not always allowed to go home with the child. I expect that insurance approval will become more feasible for devices that have been validated in randomized controlled trials. These trials are quite rare in the literature currently. For example, a Cochran review of assistive technology for children and young people with low vision in 2015 found zero RTCs or quasi-RTCs to review [20]. Adapting commercial devices to meet AT needs is a relatively low-cost option. The use of AT applications on personal computers and smart devices is a good example of this approach, and it is worthwhile for providers to be aware of a few applications that may be useful to their patient populations.

Finally, although specialty training and certification are available for providers of assistive technology, access to this training should not be viewed as a prerequisite for connecting patients with assistive technology. There are many low-tech and low-cost AT options that can be suggested and tried safely and easily. These suggestions can significantly improve patient quality of life and provide a starting point for more formal assistive technology evaluation and selection.

Summary

The use of assistive technology to improve participation in society is an exciting and nuanced field. Assistive technology can circumvent many disabilities that limit people's success and facilitate expression of their potential. Society benefits from increased diversity when AT is accepted. The technologies available are ever-evolving and are as diverse as the population that uses them. Thus, this chapter is already out of date by the time this text is printed. We offer an overview with some useful examples of technology for common disabilities so that providers and families have a foundation to start exploring AT that is suited to their needs. A practical approach for assessing AT incorporates patient goals, caregiver insights, cognitive maturity, device access and suitability, and longitudinal assessment. Many AT needs can be met with applications of commercially available and societally integrated technologies such as smart phones and accessibility settings on computers. Familiarity with a few applications that are relevant to the disabilities encountered in a clinical setting can make a meaningful impact. Systematic device development, categorization, prescription, and access are ongoing areas of need and a current focus of the World Health Organization.

References

[1] Story M, Mace R, Meuller J. The Universal Design File: Designing for people of all ages and abilities. NC State University, Center for Universal Design; 1998. https://files.eric.ed.gov/fulltext/ED460554.pdf.

[2] Exentrquita C. Youtube: live accessible channel. Retrieved 04.12.21, from Youtube: https://www.youtube.com/watch?v = 9huTVJYBFdU. 2019.

[3] Bauer SM, Elsaesser L-J, Arthanat S. Assistive technology device classification based upon the World Health Organization's, International Classification of Functioning, Disability and Health (ICF). Disabil Rehabil Assist Technol 2011;6(3):243–59. Available from: https://doi.org/10.3109/17483107.2010.529631.

[4] Newman R. Remarks by Louis Braille. In: Frye D, Editor. Braille monitor. Retrieved from https://nfb.org/sites/default/files/images/nfb/publications/bm/bm10/bm1001/bm100113.htm, 2010; 53:1.

[5] Gashel J. National Federation for the Blind. Retrieved from https://nfb.org//sites/default/files/images/nfb/publications/bm/bm14/bm1411/bm141107.htm, 2014.

[6] Library of Congress. National library sercive for the blind and print disabled. Retrieved from https://nfb.org/sites/default/files/images/nfb/publications/bm/bm10/bm1001/bm100113.htm.

[7] Motta G, Ma T, Liu K, Pissaloux E, Yusro M, Ramli K, et al. Overview of smart white canes. In: Pissaloux E, Velázquez R, editors. Mobility of visually impaired people:fundamentalss and ICT assistive technologies. Springerlink; 2018. p. 469–536. Available from: https://do.org/10.10007/978-3-319-54446-5.

[8] Tapu R, Mocanu B, Zaharia T. Wearable assistive devices for visually impaired: a state of the art survey. Patt Recognit Lett 2020;37–52.

[9] Consult a Blind Guy. Blind – visually impaired GPS apps – orientation & mobility – navigation apps, Android - iOS Apps. Retrieved from Youtube: https://www.youtube.com/watch?v = grGcQ4OWTjY, 2020.

[10] National Federation of the Blind. (2023). KNFB Reader. Retrieved from National Federation of the Blind: https://nfb.org/programs-services/knfb-reader.

[11] The Blind Life. Retrieved from Youtube: https://www.youtube.com/@theblindlife, 2023.

[12] National Federation of the Blind. Higher Education Accessibility Online Resource Center. Retrieved from National Federation of the Blind: https://nfb.org/programs-services/center-excellence-nonvisual-access/higher-education-accessibility-online-resource.

[13] National Institute on Deafness and Other Communication Disorders. Assistive Devices for People with Hearing, Voice, Speech, or Language Disorders. Retrieved from https://www.nidcd.nih.gov/health/assistive-devices-people-hearing-voice-speech-or-language-disorders, 2023.

[14] Springer S, Khamis S. Effects of functional electrical stimulation on gait in people with multiple sclerosis – A systematic review. Mult Scler Relat Disord 2017;13:4–12. Available from: https://doi.org/10.1016/j.msard.2017.01.010.

[15] Soderstrom S, Ytterhus B. The use and non-use of assistive technologies from the world of information and communication technology by visually impaired young people: a walk on the tightrope of peer inclusion. Disabil Soc 2010;303–15.

[16] Copley J, Ziviani J. Barriers to the use of assistive technology for children with disabilities. Occup Ther Int 2004;229–43.

[17] Hoppestad B. Essential elements for assessment of persons with severe neurological impairments for computer access utilizing assistive technology devices: a Delphi study. Disability Rehabilitation: Assistive Technol 2006;3–16.

[18] Borgestig M, Sandqvist J, Ahlsten G, Falkmer T, Henningsson H. Gaze-based assistive technology in daily activities in children with severe physical impairments – an intervention study. Dev Neurorehabil 2017;129–41.

[19] Van der Heide L, van Ninhuijs B, Bergsma A, Gelderblom G, van der Pijl D, de Witte L. An overview and categorization of dynamic arm supports for people with decreased arm function. Prosthet Orthot Int 2014;287–302.

[20] Thomas R, Barker L, Rubin G, Dahlmann-Noor A. Assistive technology for children and young people with low vision. Cochrane Database of Systematic Reviews 2015;2015(6):CD011350. Available from: https://doi.org/10.1002/14651858.CD011350.pub2.

[21] Perfect J, Hoskin E, Noyek S, Davies C. A systematic review investigating outcome measures and uptake barriers when children and youth with complex disabilities use eye gaze assistive technology. Dev Neurorehabil 2020;145–59.

Further reading

Yochimowitz L. Pathstoliteracy.org. Retrieved from Games for Students with Visual Impairments: https://www.pathstoliteracy.org/games-for-students-with-visual-impairments/, 2023.

Chapter 39

Surgical interventions in cerebral palsy: orthopedics and neurosurgery

Jeremy P. Bauer[1] **and Peter A. Blasco**[2]

[1]*Shriners Children's Hospital, Oregon Health and Science University, Portland, OR, United States,* [2]*Department of Pediatrics, Division of Neurodevelopmental Pediatrics, Institute on Development and Disability, Oregon Health and Science University, Portland, OR, United States*

Introduction

Among the surgical subspecialists involved in the care of children with cerebral palsy (CP), the pediatric orthopedist is the one most likely to be sharing care with the neurodevelopmentalist. This chapter is an attempt to describe the child with cerebral palsy as seen through the eyes of the orthopedist, but it is written from the neurodevelopmental point of view. The intent is to provide a primer in orthopedic evaluation and management for the developmental specialist and for the general pediatrician or family practitioner interested in the care of children with neurodevelopmental disabilities. As an assist, terminology that is largely unique to orthopedics has been summarized in a glossary at the end of the chapter.

Orthopedic consultation is best initiated prior to any need for intervention in order to establish care and begin to screen for hip subluxation, scoliosis, and contracture. The mainstay of orthopedics intervention early on is nonoperative, with the goal of maintaining and gaining function and preserving joint motion and mobility. Orthopedics intervention occurs along the life span of the child and works to complement the need for bracing, casting, assistive devices, tone management, and developmental therapies. Surgical timing may be dictated by progressive deformity but, in the walking child, is generally timed at an age where there is less chance of recurrence. Nonetheless, as children continue to grow, they can have recurrences of deformity and contracture if other treatments are not ongoing after surgery.

Clinical assessment

Many view clinical orthopedics as a matter of simple logic—if a tendon is contracted, it is cut or lengthened. However, the mechanics of movement are too complicated to allow for this simplistic view. Most muscles of ambulation are two-joint muscles. If you alter a tendon (or a bone), usually *two joints* are directly influenced, and often a third is impacted by the domino effect. Gage [1] viewed all the muscles and joints of ambulation as a system in balance, and each must be accounted for clinically when contemplating intervention, even for an apparently isolated abnormality.

The orthopedic clinical evaluation is based on (1) a detailed visual and hands-on inspection of joint mobility and alignment, (2) the subjective and sometimes objective analysis of the mechanical characteristics of posture and gait, and (3) the examination of radiographic data—X-rays for bony alignment, neuroimaging for bone and soft tissue characteristics, and dexa scans for bone density. Motor developmental milestones are accorded appropriate attention, and the neurologic examination, focusing on strength, tone, reflexes, and coordination, is thorough but less rigorous than that performed by the child neurologist or neurodevelopmentalist.

Subtypes of cerebral palsy are classified based on the extremities involved (e.g., hemiplegic, diplegic) and on the predominant movement disorder or tone pattern. The orthopedist will be exact in describing which limbs are involved and tends to be less focused on the nature of the underlying neurologic substrate. The orthopedic clinical approach to diagnosis complements information generated by professionals from other disciplines, especially the neurodevelopmentalist and the physical therapist. Expertise in the mechanics of movement is unique to the orthopedist and can rarely be provided by another discipline. Hence, close working relationships often develop between the pediatric orthopedist, the pediatric physical therapist, and the neurodevelopmental subspecialist. The orthopedist is particularly attuned to understanding the goals of the family and child for treatment.

Capute and Accardo's Neurodevelopmental Disabilities in Infancy and Childhood. DOI: https://doi.org/10.1016/B978-0-12-824060-1.00001-8

Examination of gait

The evaluation of station and gait is primarily done by visual inspection of body alignment, joint movement, and functional movement patterns. Objective measures of gait have become available through the use of the gait laboratory and gait analysis. While a formal gait analysis study is not always available or feasible, a careful and thorough visual inspection of gait is critical.

Station is defined as the position or posture assumed when standing or sitting. Human stance is assessed from anterior, lateral, and posterior viewpoints with the subject standing erect. It is also valuable to view the individual from behind while he or she is bending forward (Fig. 39.1). *Gait* refers to walking. Typically, the infant begins to walk without support around 1 year of age. He or she steps with a wide base and relative flexion of the hips and knees, holds the arms in abduction and the elbows in flexion, and moves in a staccato manner. The width of the base diminishes gradually, movements become smoother, reciprocal arm swing appears, step length and walking velocity both increase, and an adult pattern of walking emerges by about 3 years of age. The 3-year-old's typical gait pattern shows a vigorous walk, reciprocal arm swing, well-developed heel strike, and smooth movements, all of which differ only slightly from those of an adult. By age 7, the child's gait is indistinguishable from that of an adult. *Step length*—the distance covered in one step—is obviously related to leg length but also depends on muscle tone and range of joint motion. *Stride length* is the distance covered in two steps (i.e., one complete *gait cycle)*. The *base* of the gait, measured as the perpendicular distance between the heels, is normally 5–10 cm, although it can be narrower. A wide-based gait is often seen when a patient is trying to compensate for difficulties with either balance or equilibrium.

Gait is assessed in progress rather than beginning from a stationary position [1,2] (Fig. 39.2). With the trunk slightly forward because of forward progression, the normal gait cycle begins with *heel strike* (better referred to as *initial contact)*, whereby the leading foot first makes contact with the floor. It proceeds through *stance phase*, then the *swing phase*, and then ends with heel strike of the same foot. Stance phase represents approximately 60% of the gait cycle, and swing phase represents 40%. Single limb stance phase begins just after heel strike, when the ankle is flexed at a 90-degree angle (referred to as neutral), the knee is nearly fully extended, and the hip is flexed about 40 degrees. Ankle motion in stance is divided into three stages, called rockers. In the first rocker, the ankle plantar flexes to get the foot flat on the floor. As the body mass progresses forward over the stance-phase foot, the tibialis anterior muscle is working eccentrically to slow progression. In midstance, or second rocker, the foot is still flat on the floor, but the ankle is slightly dorsiflexed and the knee is slightly flexed as well. As the body continues forward, the foot remains flat, and the ankle goes into increasing dorsiflexion while the knee progressively extends. The gastrocnemius-soleus muscle complex works in an eccentric manner to control the advancement of the tibia over the planted foot. As the contralateral foot

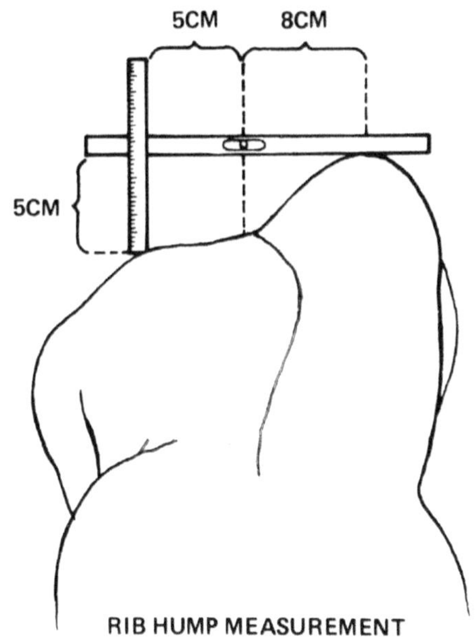

FIGURE 39.1 Posterior view of trunk in a person with scoliosis. With the patient bent over, the hump associated with scoliosis becomes quite striking. *From Horstmann H, Bleck E. Orthopedic management in cerebral palsy, Clinics in developmental medicine. London: Mac Keith Press; reprinted by permission; 2007.*

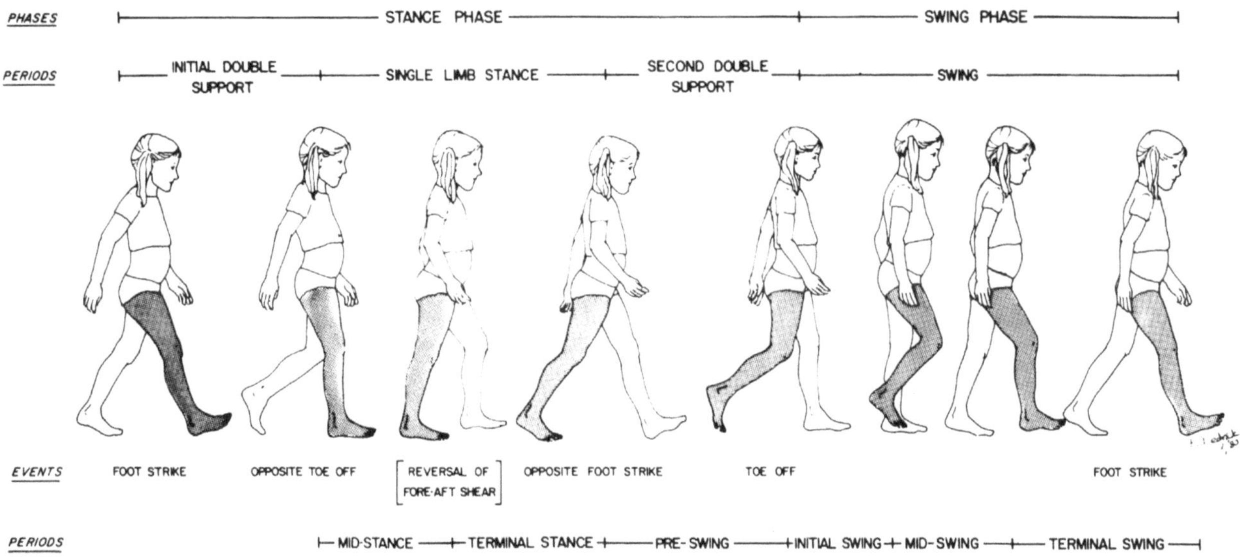

FIGURE 39.2 Normal gait cycle of a 7-year-old child. *From Sutherland DH. Gait disorders in childhood and adolescence. Baltimore: Wolters Kluwer [Williams & Wilkins]; reprinted by permission; 1984.*

makes initial floor contact (heel strike), the double-limb phase of stance begins. In third rocker, the stance foot rocks forward (i.e., the heel comes up from the floor) and prepares for *toe-off* as weight is transferred to the contralateral leg. Just before leaving the ground (toe-off, also called *terminal contact)* to enter swing phase, the ankle is actively plantar flexed 10–15 degrees by concentric contraction of the gastroc-soleus complex, and the knee is flexed 35 degrees.

Swing phase is separated into three periods: initial swing, midswing, and terminal swing. During initial swing, the hip, knee, and ankle must flex in order for the foot to clear the ground. This is followed by full extension of the knee as the leg swings through to terminal swing. Swing phase ends with heel strike, completing one gait cycle. Velocity during gait is not constant but varies slightly, speeding up through swing phase and slowing during stance phase.

As gait proceeds, the head moves up and down in a low sinusoidal curve. If one marks the center of gravity, located at about midpelvis, it describes essentially the same curve. The pelvis makes very little lateral rotational movement during gait. The more bobbing (up-and-down) movement that is created, the less efficient the gait becomes, leading to increased oxygen consumption.

In lower extremity spasticity, the typical pattern is a tendency toward flexion and adduction at the hip, flexion at the knee, plantar flexion *(equinus)* at the foot, and internal rotation of the entire lower extremity. This pattern produces a gait with increased bobbing of the pelvis, anterior pelvic tilt with lumbar hyperlordosis, increased side-to-side sway, decreased step length (because the knee and hip are restricted in extending), a wide base, abrupt changes in velocity, and a tendency to strike on the toe or forefoot rather than on the heel. Instead of forces being distributed evenly over the foot, this dynamic of the spastic limb produces extreme forces on the foot. Over time, instability of the subtalar joint frequently leads to the collapse of the foot into a *valgus* position (Fig. 39.3). In spastic diplegia, this occurs bilaterally. In hemiplegia, there tends to be less in the way of spastic hip adduction, more involvement in terms of knee and hip flexion, and an equinovarus foot. Over time, a heel cord contracture often develops.

Ataxic and choreoathetoid/dyskinetic gaits have their own unique characteristics, which are both more complex and more variable than those of the spastic syndromes. The major features are a wide base and excessive body sway in ataxia, and variable postural stability with erratic, jerky extremity movements in dyskinetic CP. The hallmark of these patterns is their nonreproducibility in gait analysis.

Instrumental gait analysis

The objective description of normal gait and gait disorders involves a thorough knowledge of force plate values, movement measurements, and electromyography. These measurements provide the foundation for kinematic description of movement (i.e., its geometry) and kinetic analysis of gait (i.e., the underlying forces). In children with cerebral palsy, there is an exceedingly complex interplay of tone abnormality, contracture, bone deformity, weakness, and multiple-

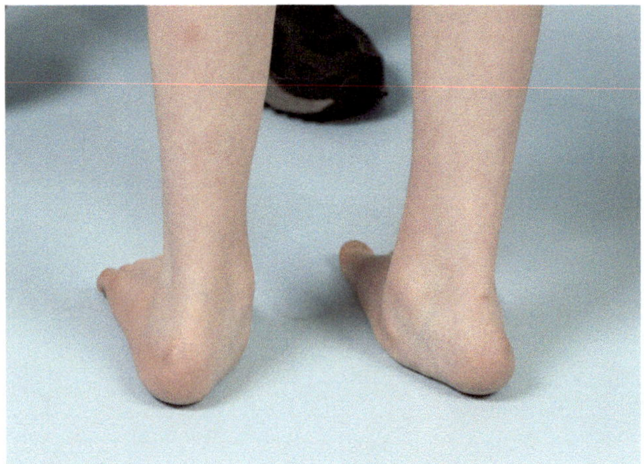

FIGURE 39.3 Feet depicted in severe planovalgus. The heels do not touch the ground due to Achilles tendon contracture. *Courtesy Jeremy P. Bauer, M.D.*

joint dynamics that leads to gait aberrations and increased energy expenditure. With the advent of gait analysis technology, objective criteria began to play a more prominent role in analyzing problems of locomotion, allowing the clinician to make rational decisions about surgery, orthotics, drugs, and physical therapy and to understand the distinctions between the underlying pathology and the compensations that are produced as a result of the pathology. If the true pathology is addressed, the compensations will be resolved. Because surgical procedures such as tendon lengthening can cause muscle weakness, identifying and operating only where there is pathology allows the orthopedist to leave untouched muscles that contribute power to gait. Experienced clinicians appreciate that the decision not to treat is of equal, if not more, value than the decision to treat. In addition, gait analysis information allows the clinician to make objective comparisons before and after treatment to better understand the efficacy of therapeutic maneuvers.

The sophisticated gait analysis laboratory requires a biotelemetry system consisting of three major subsystems: optical, electronic, and computer. The *optical system* requires motion picture recording with two or more cameras for true three-dimensional measurement. The subject to be studied is prepared with attached surface markers (sticks or light reflectors) placed at various critical points, especially joints. In addition, a digital time display is linked with the video.

The *electronic system* consists of a force plate and electromyography (EMG) components. The force plate is a foot pressure sensor that is mounted on the floor. It measures forces in three planes: vertical, fore-and-aft, and medial-lateral. Muscle group electrical activity is monitored with surface electrodes for large muscle groups (e.g., hip adductors). Deeper muscles and isolated muscles cannot be monitored without the use of internal (i.e., needle) electrodes. These fine-wire intramuscular electrodes have limited use for clinical evaluations, especially in children. In addition, a sensor, called a *foot switch*, may be placed on each heel to indicate the instant of the heel strike. Battery-operated radio transmitter units are used to relay these electrical pressure and EMG signals to a computer. A *computer system* performs all calculations from the data gathered and presents them in both graphic and tabular form.

There are five major determinants of mature gait: duration of single-limb stance, walking velocity (i.e., the average distance traveled per second), cadence (i.e., the number of steps per minute), step length, and the ratio of pelvic span to ankle spread. Mean values for gait events in a group of normal 7-year-olds are shown in Fig. 39.4. Energy is required to walk, and the measurement of oxygen consumption during gait is the ultimate test of the cost of walking in terms of energy expenditure [1].

In the absence of a gait analysis laboratory, it is imperative that a thorough examination of the child's gait be performed with the minimum of a video recording. The Edinburgh Visual Gait Score has been used to help quantify gait deviations without the use of an instrumented gait analysis [3].

Examination of the extremities

Leg length should be measured with the patient supine using a tape measure. The measurement is from the anterior superior iliac spine to the medial malleolus. If the patient has a flexion contracture of the knee joint, the measurement should follow the contour of the leg itself. It can be divided into two measurements that are added in order to get the actual limb length measurement. *Orthoroentgenograms*, which are basically plain radiographs, show the entire limb

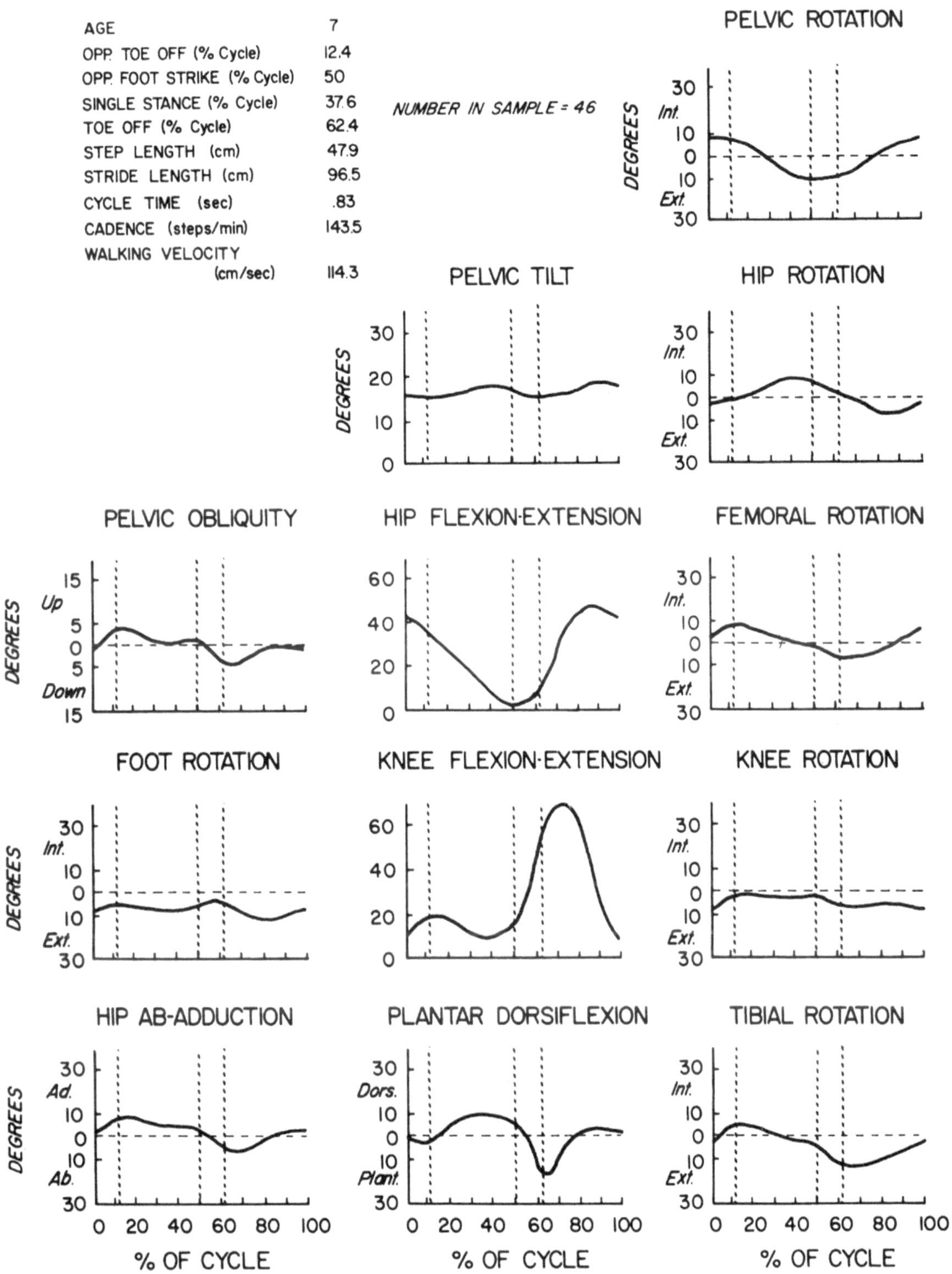

FIGURE 39.4 Mean values for gait events in a group of normal 7-year-old children. The column in the upper left-hand corner presents numerical data on events related to time and distance during walking. The graphs describe specific joint angles and their changes during the gait cycle. The first hashed vertical line represents initial toe-off, the second heel strike, and the third contralateral toe-off. *From Sutherland DH. Gait disorders in childhood and adolescence. Baltimore: Wolters Kluwer [Williams & Wilkins]; reprinted by permission; 1984.*

from the hip to the ankle and can demonstrate deformity in the limbs responsible for a limb length discrepancy. This view is not accurate for measuring limb length as the X-ray beam is centered at the knee and may cause distortion at the hip and the ankle. *Scanograms* are most frequently used to measure bone length radiographically, particularly in hemiplegic children, and involve shooting a series of hip, knee, and ankle films that focus on a long ruler adjacent to the limb being examined. This allows easy measurement of the length of the femur, tibia, and entire limb. More recently, the EOS imaging system [4] has been used to allow for simultaneous biplanar imaging with the option of 3-D reconstructions to determine torsional values. Lower extremity radiographs require the child to be able to stand independently. For children who are unable to do so, a specialized chair is used to obtain spine radiographs in the seated position.

Passive range of motion is intended to be a maximum (but not painful) measure and therefore requires that the patient be maximally relaxed, something not always possible in an awake infant or younger child. The hip range of motion is defined in terms of abduction and adduction. Hip abduction may be described in terms of the angle of the femurs when both are abducted or in terms of each femur individually as the angle of the abducted hip with a vertical plane. The latter method is preferable in order to identify and clearly document any asymmetry. Observations are made both with the hips flexed at 90 degrees and with them extended. The full range in the flexed position is normally 80–90 degrees on each side. With the legs extended, the normal range is 45–50 degrees on each side. Hip flexion contracture is common and can be measured in two ways: using either the Thomas test with the patient in supine (which is easier to accomplish) or the more accurate but more difficult method of [5] with the patient in prone (Fig. 39.5). With the Thomas test, it is important that the limb being assessed is able to hang over the edge of the exam table, as knee contracture is commonly present as well, which would limit hip extension. For the Staheli test, with the child in prone and the hips flush with the end of the table, the examiner places one hand on the posterior superior iliac spines while the other hand brings the lower limb slowly into extension. The point at which the examiner feels the pelvis begin to rotate forward is the point of deformity. Limitation of extension can be caused by a tight iliopsoas, a tight rectus femoris, or a tight tensor fascia lata (or its tendon, the iliotibial band). The normal range of hip extension is 0–20 degrees past neutral except in newborns, who have physiologic flexion contractures usually in the neighborhood of 25 degrees. Muscle imbalance in the spastic lower extremity causes the hip to be pulled into adduction and flexion, which leads to hip subluxation in CP. While it does not obviate the need for radiographic hip screening, determining the presence of limited or asymmetric hip range of motion is an important part of the orthopedic examination in order to screen for impending hip dislocation.

Internal (medial) and external (lateral) hip rotation are measured with the patient in prone position and the knees flexed to 90 degrees (Fig. 39.6). The pelvis must be firmly stabilized to prevent hip rotation caused by lateral rolling of the pelvis. These values change substantially during childhood. Internal rotation is least in infants and increases with age, reaching values of 20–60 degrees in adolescence. External rotation in infants averages 70 degrees and declines in adolescence

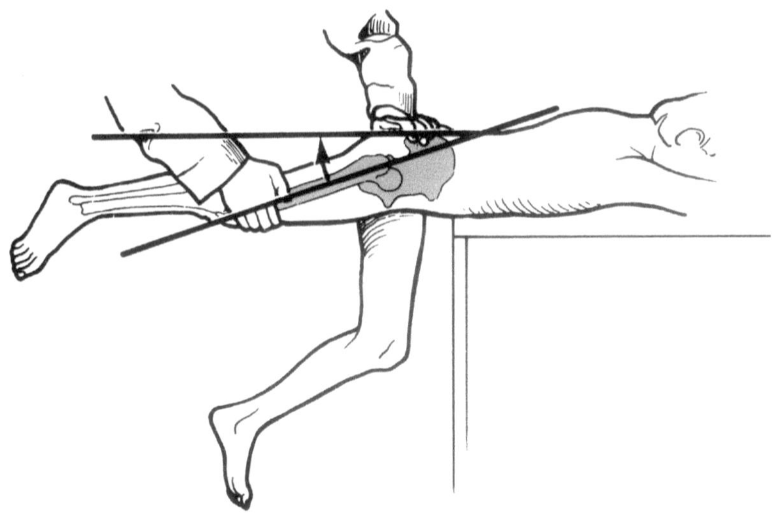

FIGURE 39.5 Staheli's test for hip flexion contracture. The angle subtended by the femur and the horizontal plane represents the angle of hip *extension*. Therefore, angles above the horizontal are positive and those below are negative. *From Horstmann H, Bleck E. Orthopedic management in cerebral palsy, Clinics in developmental medicine. London: Mac Keith Press; reprinted by permission; 2007.*

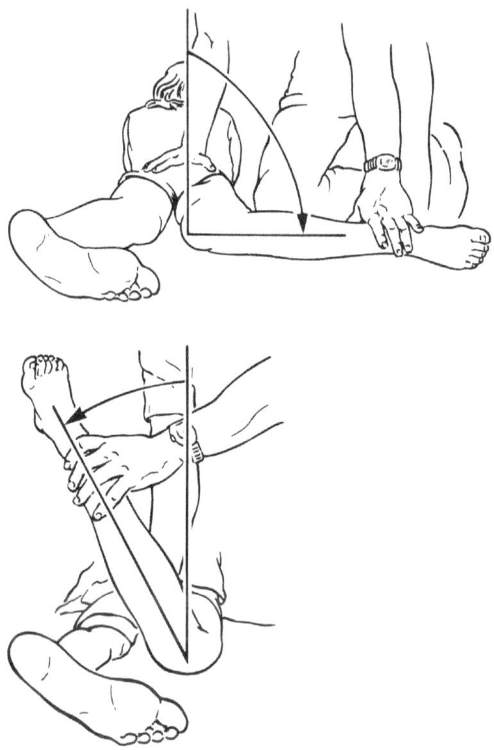

FIGURE 39.6 Measurement of hip internal rotation (top) and external rotation (bottom). *From Horstmann H, Bleck E. Orthopedic management in cerebral palsy, Clinics in developmental medicine. London: Mac Keith Press; reprinted by permission; 2007.*

to a range of 25–65 degrees. Internal rotation and external rotation should be fairly equal in range. A child with more internal than external rotation has what is known as *internal femoral torsion* or *femoral anteversion*. If external rotation is in excess of internal, the hip is in retroversion (or *external femoral torsion*). Internal femoral torsion occurs normally in the fetus and is expected to decrease with development but often fails to remodel with growth in children with CP.

During any of these examinations, further flexion of the knee (or abduction of the hip) done in a rapid manner can detect tightness (either spasticity or a contracture) in the quadriceps and adductor muscle groups. Knee flexor tightness in the hamstrings can be reliably measured with the patient supine. With the leg lying flat, the knee should extend fully to 0 degrees. The upper leg is then raised to flex the hip at 90 degrees, and the lower leg is gradually extended to its limit. Normal extension in the adult decreases to about 10–20 degrees shy of a straight line, and this quantity is referred to as the *popliteal angle*. (This is not actually the angle of the popliteal space but rather the complement to it.) Newborns demonstrate physiologic knee flexion contractures, with a popliteal angle range of 20–40 degrees. This decreases to 0 degrees by 9–10 months of age [6]. At the same time, the position of the patella should be noted in relation to its normal location in the patellofemoral groove. Flexion contracture within the knee joint itself can be distinguished to some extent from hamstring contracture by simply observing the knee in relaxed extension flat on the table (i.e., with the hip extended). If firm pressure on the anterior aspect of the knee joint fails to extend the knee completely, a contracture of the joint itself is likely present. In the presence of a severe hamstring contracture, it can be practically impossible to identify additional contracture within the joint capsule.

The rectus femoris muscle is a bi-articular muscle, crossing both the hip and the knee. Spasticity of the rectus femoris may lead to a stiff knee gait. To examine the rectus, the child is placed prone, and the knee is flexed. If the pelvis raises during the examination, this indicates either contracture or spasticity of the muscle as the rectus pulls on the pelvis when the knee is bent [7].

Lower leg deformities are generally described in terms of tibial-fibular torsion and foot rotation. If the tibia is twisted inward *(internal tibial torsion)*, the feet turn in. If the femurs are turned inward *(internal femoral torsion)*, the feet and the knees turn in as well. Precise measurements of tibial–fibular torsion are difficult and time-consuming, but good approximations are easy and quick [8]. Bleck recommended observing the patient sitting comfortably with knees flexed over the edge of the table (Fig. 39.7). The angle from an imaginary line through the malleoli that intersects the plane of the table edge estimates the amount of tibial rotation, which is normally around 20 degrees external (range

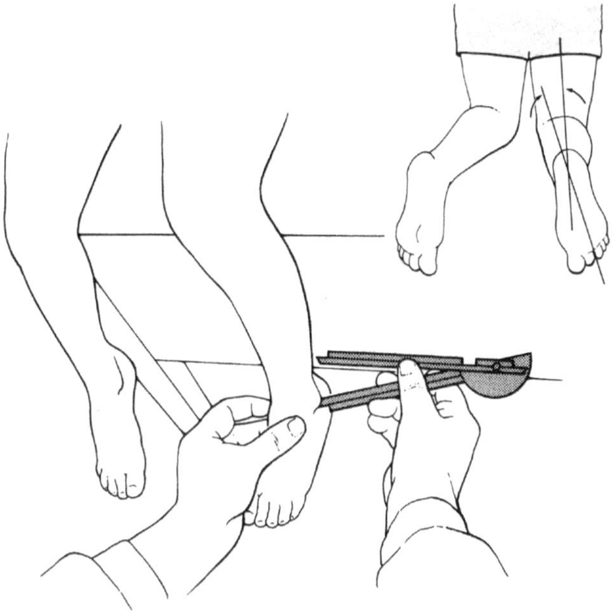

FIGURE 39.7 Measurement of tibial-fibular torsion. *Left*: A goniometer is used to measure the angle of the transmalleolar line and the vertical plane of the tibia (represented by the table edge). *Right*: The thigh-foot angle is determined from imaginary lines bisecting each. *From Horstmann H, Bleck E. Orthopedic management in cerebral palsy, Clinics in developmental medicine. London: Mac Keith Press; reprinted by permission; 2007.*

0–45 degrees). The thigh-foot angle is a close approximation of the same thing and is also easily measured. With the patient in prone and the knee flexed at 90 degrees, an imaginary line is drawn bisecting the thigh (femur) through the heel, and another is drawn bisecting the plantar aspect of the foot. The intersection of these lines represents a slightly narrower angle than the one obtained by the previously described method, and the normal range is from −3 to +20 degrees. The foot must be at rest when carrying out either of these measurements (for details, see [8] and [9]).

A foot that is positioned with the plane of the toes pointing down (i.e., with more than a 90-degree angle) is in *equinus*, and a foot that is positioned with the plane of the toes pointing up is in *calcaneus*. When the *hindfoot* (commonly called the heel) rolls in, it is called *varus*. When it rolls out, the foot appears to collapse medially, referred to as *valgus*. Varus is due to an overpull of the tibialis posterior or tibialis anterior muscles. It is to be distinguished from *in-toeing*, which is due to the rotation of the foot inward but with the heel still aligned flat on the floor. If these positions can be passively corrected, they are referred to as *dynamic*. If not, they are *fixed*. Equinus can also occur starting from the point of the midfoot, termed a *cavus deformity*. This excessive forefoot equinus is usually a fixed deformity and is also referred to as a *high arch*. Dorsiflexion of the ankle is measured with the foot held in slight inversion in order to lock the subtalar joint. It is important to check passive dorsiflexion with the knee both flexed and extended in order to distinguish soleus from gastrocnemius contracture (the Silfverskiold test). With the knee flexed at 90 degrees, the gastrocnemius is relaxed, so any equinus deformity is due to contracture of the soleus. If deformity appears or worsens (i.e., *less* dorsiflexion) when the knee is extended, then the gastrocnemius is the main site of contracture, or both muscles are implicated.

Toe deformities come in the form of flexion contractures and *hallux valgus*, an abduction deformity of the great toe. The former are common in children without cerebral palsy. In persons with cerebral palsy, they rarely cause a clinical problem in childhood. Hallux valgus, however, is associated with valgus foot deformity and can lead to significant pain at the first metatarsophalangeal joint in children with cerebral palsy.

Much of the passive orthopedic examination must be done on a floor mat or on a large examining table. Appreciating the ranges of normal and the subtle variations that depend on age and sex in typical children—let alone in children with various types of neuromuscular, connective tissue, or skeletal abnormalities—clearly depends on extensive clinical experience. In addition, reliance on physical exam parameters alone is not always enough to determine what surgery or intervention will be needed. Passive range of motion (ROM) measures do not always correlate with true range due to both body weight and spasticity. Spasticity and resulting range limitations may become more pronounced as the child walks. For some children who are unable to cooperate with passive examination or who have significant spasticity, the surgeon has the unique opportunity to supplement the clinical examination with an examination under anesthesia. This may help to better evaluate the true range of motion and degree of contracture at the hip, knee, and ankle.

Common clinical conditions

The following are some commonly encountered conditions and minor deformities that may appear in isolation or may be associated with encephalopathy. In the latter circumstance, they may be more likely to require surgical intervention.

1. *Metatarsus adductus is* a term denoting adduction of the forefoot (i.e., curvature in a medial direction). In congenital metatarsus adductus, in contrast to congenital clubfoot, the heel is not in varus, and there is no fixed equinus. In children with spastic cerebral palsy, spastic metatarsus adductus consists of forefoot adduction. It is a late-developing phenomenon and appears to be due to the spasticity of the intrinsic muscles of the foot. Although initially dynamic, it tends to evolve into a fixed deformity [8].
2. Congenital *clubfoot deformities (talipes equinovarus* and others) consist of varying degrees of fixed foot deformity, with equinovarus being the most common. Neither congenital nor acquired clubfeet are seen more commonly among children with cerebral palsy than in children who are otherwise typically developing.
3. *Excessive tibial torsion* occurs more frequently in the presence of neurologic impairment. Surgical correction is rarely needed, except in children with cerebral palsy or other neuromuscular disorders. Although internal tibial torsion may appear quite dramatic in the typical child, the natural history is for the tibia to rotate externally throughout development into adolescence [9]. Self-correction may occur in children with cerebral palsy as well.
4. *Excessive femoral internal torsion* fails to correct in most children with spastic cerebral palsy. It can be thought of as the maintenance of the fetal (newborn) angle rather than the creation of a new deformity. Internal femoral torsion can be measured radiographically by CT or EOS but is just as well monitored clinically. The femur has a normal angle of torsion, so the abnormal circumstance should be called *increased* femoral internal torsion. As with tibial torsion, even dramatic femoral torsion almost always resolves spontaneously in a typically developing child.
5. *Genu varum* and *genu valgum* are commonly referred to as "bow legs" and "knock knees," respectively. They rarely require surgery and are not seen more frequently in children with cerebral palsy than in typically developing children. Genu varum is physiologic in the newborn, with the angle of the femur and tibia averaging +17 degrees. This progresses to a maximum valgus angle of −10 degrees at 3 years of age, then returns to the stable adult range of 0–10 degrees by 6 or 7 years of age [10]. This is to be distinguished from Blount's disease, which is a pathologic entity consisting of progressive varus deformity of the proximal tibia (therefore, *tibia vara)* and which does require surgical intervention.
6. What used to be known as *congenital hip dysplasia and dislocation* is now called *developmental dysplasia of the hip* (DDH) to better reflect the evolving nature of the disorder. The hip is not really dysplastic at birth but becomes so in susceptible individuals. Unilateral DDH is more likely to be left-sided than right-sided. Bilateral involvement is present about one-third of the time, and DDH is six times more common in girls than in boys. Almost all cases are considered idiopathic and occur in isolation. If coincidentally present in an infant with cerebral palsy, DDH is greatly aggravated by the presence of spasticity after birth. Acquired hip dislocations secondary to spasticity are addressed in the following section.

Treatment

General considerations

The orthopedic surgeon's conceptual approach to treatment is no different from that of the neurodevelopmentalist. The ideal treatment of the child hinges on the early involvement of all team players with periodic follow-up evaluations. Ideally, all members of the interdisciplinary team approach the patient within the same general framework, which defines the elements of treatment as

1. Prerequisite data: an accurate diagnosis of
   a. Primary disability
   b. Associated problems
2. Treatment modalities
   a. Hands-on therapy
   b. Orthoses
   c. Adaptive and assistive equipment
   d. Medical treatments
   e. Surgery

Accurate diagnosis, including some appreciation for severity and associated problems, is the essential first step toward understanding the patient's current problems and potential future difficulties. Treatment follows from this and begins with hands-on therapy interventions that may have an immediate impact or are at least worth a trial for some period of time. Adaptive equipment (e.g., a seat insert), orthoses, and numerous other assistive devices (e.g., canes or an augmentative communication system) are provided as adjuncts.

At one time, it was mistakenly believed that cumbersome metal and leather braces could "fully control" the limbs of children with cerebral palsy, allowing them to develop walking skills naturally without acquiring deformities. (These were the so-called Phelps braces, named for Winthrop Phelps, an orthopedist who pioneered many treatment approaches for children with cerebral palsy.) Orthotic devices are currently used to prevent deformity (i.e., through control of spasticity), to improve function (i.e., maintaining foot position in stance or swing), and occasionally to impose partial postoperative immobilization. They are now made primarily of lightweight plastic and are attached by means of Velcro straps. The AFO (ankle-foot orthosis) is particularly useful because, depending on its design, it can control all motions of the foot and ankle: equinus, varus, and valgus. It is also convenient because it can be worn over a sock in an ordinary shoe. The primary indications for AFO use are spastic equinus or hypotonia, and AFO construction has a number of variations to suit specific purposes [1]. Knee hyperextension (*genu recurvatum*) can sometimes be managed with a rigid AFO molded to keep the foot in a few degrees of dorsiflexion, thereby mechanically driving the knee toward more flexion [11]. As another example, after hamstring lengthening and posterior capsulectomy of the knee joint, the knee-ankle-foot orthosis (KAFO) can be employed to control excessive flexion for a variable length of time until the patient regains quadriceps strength and competence during gait [8]. Orthoses seem to have little benefit for choreoathetoid patients. They may be somewhat useful to help maintain a stable, plantar-grade foot position. The smaller shoe insert-type orthoses—heel cups or UCBLs (named for the original brand, University of California Biomechanics Laboratory)—have limited utility, although some advocate their use to stabilize the symptomatic valgus hindfoot.

Medication management consists primarily of drug treatment intended specifically to diminish spasticity or lessen extraneous movements. The list of medications to choose from is very long, but the likelihood of gaining substantial clinical benefit (beyond modest tone reduction) is rather small. Diazepam and baclofen are the drugs most commonly employed to reduce spasticity and/or extrapyramidal rigidity. With all drug options, however, the balance between benefits and side effects is often quite narrow, and predicting whether a given drug will help or make matters worse is almost impossible. The use of baclofen intrathecally to relieve spasticity and dystonia has proven to be greatly effective [12,13]. The main risks relate to pump system malfunctions. Delivery using a programmable pump system allows for varying delivery during the course of the day. (see Chapter Cerebral Palsy) General pediatric care for associated medical problems (such as seizures) and intercurrent acute problems, to which children with disabilities are more susceptible, is an important element of the overall management plan.

Surgical intervention is not a last resort but an adjunct to other modalities of treatment. Surgery should be judiciously applied when the need becomes apparent. The general goals of surgery are to maximize function, minimize deformity, and consolidate interventions when possible, thereby keeping surgery and anesthesia episodes to a minimum. The first goal of surgery is to diminish muscle imbalance. This is accomplished by weakening muscles and decreasing spasticity (tendon lengthening and tendon release procedures) and by altering the direction of muscle force (specific muscle transfer procedures). Most muscle procedures are muscle shortening procedures via either fascial lengthening or tendon lengthening. This weakens the muscle by reducing the resting length (Fig. 39.8) and diminishes tension on the muscle spindle and tendon organs, thereby reducing spasticity [14]. The second goal is to prevent bone and soft tissue deformities. Where deformity has already developed, correcting bony deformity and releasing soft tissue contracture become necessary. Most bony procedures involve realignment or angulation changes in order to achieve better position and a better mechanical advantage. The resulting bone realignment and joint stabilization lead once again to maximal function.

There are several reasons for pursuing these operative procedures. Simple mechanical disadvantages may be the main factor impeding developmental progress. Pain may be a significant, immediate (or predictable future problem) that can be eliminated. Ease of management and reduction of general care needs are potential benefits even in those situations in which independent function is unlikely. Improved positioning with assistive devices can benefit the patient both in functional ways and in minimizing the long-term development or progression of deformity. Cosmesis, although almost never a primary indication for limb surgery, is often an associated benefit (e.g., improved functional gait). The ultimate long-term goal is to avoid the pain and superimposed disability associated with degenerative arthritis as a result of chronic and often unappreciated joint trauma.

The timing of surgical intervention varies greatly depending on the pathology being treated. For hip and spine pathology (to be addressed later), the decision to intervene is based more on the progression of deformity than on actual age. For children at the GMFCS I, II, and III levels, an attempt is made to consolidate the surgical interventions into one setting, followed by early rehabilitation. This is to avoid the so-called "Birthday Syndrome," as

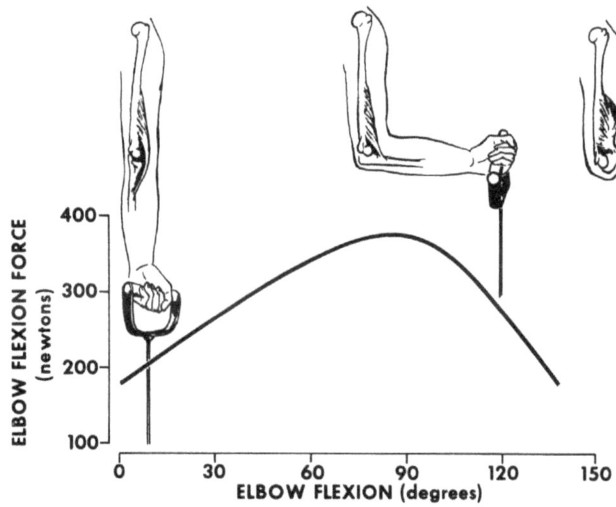

FIGURE 39.8 Muscle force graph. Maximal force is plotted against elbow flexion, which represents muscle length. Muscle strength is greatest at midlength and diminishes as the muscle is lengthened or shortened (as occurs with tendon release surgery). This is analogous to the Starling curve for myocardial contractility. *From Inman VT, Ralston JH, Todd F. Human walking. Baltimore: Wolters Kluwer [Williams & Wilkins]; reprinted by permission; 1981.*

described by Rang [15], where a child is treated with yearly surgical interventions as problems arise or are uncovered by the previous surgical intervention. Consolidating multiple surgical procedures is commonly referred to as single event multilevel surgery (SEMLS). Even with an experienced team and excellent preoperative planning, children may need more than one intervention [16]. In addition to the psychologic and financial burden put on the child and family by multiple interventions, there is a risk of recurrence with intervention at an early age, especially with rotational osteotomies. However, if a child has a severe deformity, surgery may be necessary earlier. A thorough review of the child's clinical examination and gait is imperative for preoperative planning. Determining what joints are causing gait dysfunction is difficult, and the surgeon must distinguish when contracture, malalignment, spasticity, weakness, or some combination is the cause.

Bleck [8] wrote cogently about preoperative management of the child and family. He maintained that the orthopedic surgeon who is conscientious and cautious in terms of preoperative analysis, selection of surgical procedure, surgical technique, and good follow-up care and who clearly defines the aims of surgery can expect an 80%–90% "good result." Promoting a trusting relationship is the fundamental basis for minimizing the physiologic and psychological discomfort surrounding the surgical procedure. To foster this attitude, sufficient time should be allowed for the physical therapist and patient to become acquainted, and a close working relationship should exist between the therapist and surgeon. Ensuring that parents have realistic expectations for the results of surgery is a critical part of preoperative counseling.

Specific surgical procedures

Procedures about the hip

Fixed and dynamic problems with the hip are the second most common orthopedic deformity in cerebral palsy. In children who have had progressive hip subluxation containment surgery is indicated to prevent further hip deformity and contracture. Indications for hip surgery include:

1. Hip subluxation and dislocation. Acquired hip dysplasia is rare before 6 months of age. Progressive hip subluxation initially is difficult to determine either by history or exam, as it is relatively painless early on and range of motion is preserved. With worsening deformity, a limited range develops. As a late consequence of a long-standing hip dislocation, the head becomes aspherical and becomes painful.
2. Asymmetrical sitting base and pelvic obliquity, caused by decreased mobility and hip stiffness from dislocation. Scoliosis is associated with pelvic obliquity, and it is debated whether the curve is aggravated by this posture.
3. Difficult perineal care.
4. Decubitus ulcers and skin problems.

All of these problems are interrelated. With persistent muscle imbalance, both the acetabulum and the femoral head will become dysplastic—that is, deformed to such an extent that they do not structurally fit together. The femoral head generally migrates posteriorly and laterally over time, and when 40% of the femoral head has migrated out of the acetabulum, the eventual consequence will be dislocation. Dislocation frequency increases with the severity of spasticity. The overall incidence of hip displacement (defined as 30% of the hip uncovered) is 35%; in children at GMFCS level V, it is 90% [17]. With close surveillance, frank hip dislocation should rarely happen because the indicators of progressive dysplasia are easy to recognize both clinically and radiographically, and preventive surgery is both safe and highly successful. In general, earlier surgery is associated with better long-term results. Non-operative measures have not been found to prevent subluxation, but attention to keeping the hips in abduction in the wheelchair and the use of a standing frame may help. Both the American Academy for Cerebral Palsy and Developmental Medicine (AACPDM) and the Pediatric Orthopedics Society of North America (POSNA) have recommended CP hip screening guidelines [18,19].

Adductor tenotomy surgery involves cutting the tendons of the adductor longus, gracilis, and (in severe cases) adductor brevis muscles (Fig. 39.9). There is some controversy as to whether to address the anterior branch of the obturator nerve by either sectioning or chemodenervation, but only in more severely involved (i.e., nonambulatory) children. The adductor magnus is always left intact to prevent an abduction contracture from developing. The iliopsoas, although not an adductor, is often lengthened as well, at least in children with hip dysplasia, preventing a hip flexion contracture that can lead to posterior dislocation [20]. The recurrence rate of contractures with adductor surgery is small, but children who have unilateral operations may ultimately have problems with windswept hips. The windswept hip deformity, a combination of adduction contracture on one side (often with subluxation of the femoral head) and abduction posture (usually with contracture of the iliotibial band) on the other, often gives the mistaken impression of a leg length discrepancy. Windswept hips are not uncommon among more severely involved children with cerebral palsy. The functional benefits of hip surgery are improved gait as a result of diminished scissoring and crouching and improved sitting as a result of abolished pelvic tilt. Children do not need to have hip flexion or adduction contractures to benefit from this procedure.

Complications of adductor surgery include the failure to halt the progression of hip dysplasia, excessive abduction resulting from overweakening of the adductor muscle group, insufficient release so that the child still scissors, windswept hip deformity, and heterotopic ossification. Treatment of windswept hips involves surgical release of the

FIGURE 39.9 Releasing spastic muscles about the hip. Two adductor muscles (the adductor longus and the gracilis) have been sectioned and the iliopsoas tendon is being cut. *Courtesy of Michael D. Sussman, M.D.*

adductors on one side and the iliotibial band contracture on the other, or bone realignment. Heterotopic ossification refers to the unexplained deposition of bone in muscle tissue. A final issue gets back to patient selection. Children with mixed quadriplegia with a predominant choreoathetoid component are at risk for a poor functional outcome if their lower extremity spasticity is reduced.

A neglected hip dislocation becomes painful and interferes with sitting and perineal care. Late treatment often requires salvage-type surgical options with unfavorable results. Surgery involves a varus de-rotational osteotomy (VDRO) of the involved hip and is generally performed on both hips to maintain symmetry in order to avoid a windswept hip and additional surgical intervention. The acetabulum often needs an osteotomy as the chronic subluxation leads to dysplasia, and this requires autologous bone grafting. The VDRO procedure will adduct the limb and require shortening of the femur. Adductor releases are performed if the hip has inadequate abduction after the associated shortening. The VDRO is fixed with hardware, and the child is maintained in an abduction pillow for 3–6 weeks. Some centers with access to inpatient rehab may start weight-bearing early; otherwise, it is allowed at the 3–6-week mark. Younger children have a high rate of recurrence, but intervention is still warranted to prevent worsening femoral and acetabular deformities.

Torsional malalignment procedures include those for femoral internal torsion and for tibial torsion. Increased internal femoral torsion leads to significant dysfunction in some children with CP, as they lack sufficient motor control to rotate the femur outward and therefore walk with their femurs rotated inward. Corrective surgery can be performed at any level: proximal, mid-shaft, or distal. The level of surgery depends on the concomitant procedures planned as well as the surgeon's preference. With proximal femoral VDRO, if the child has concomitant hip dysplasia, it can be addressed at the same time. If only rotational malalignment without hip dysplasia is present, a distal procedure can be performed. In distal femoral derotational osteotomy, the distal shaft of the femur is cut, rotated, and held in place with plate and screw fixation. The distal osteotomy also allows correction of a knee contracture at the same time. With more advanced fixation options available now, less immobilization in casts is required, and earlier weight-bearing mobilization is allowed. Some centers will begin weight-bearing as soon as it is tolerated by the child. Non-weight-bearing time can vary anywhere from 1–6 weeks, depending on available resources (e.g., in-patient rehab) and surgeon judgment as to when it is safe to do so.

Procedures about the knee

Before a SEMLS procedure, it is important to determine what issue is primary and what is secondary. A flexed knee gait could be secondary to either an incompetent gastroc-soleus complex or to a primary knee problem such as hamstring tightness, spasticity, or contracture. Surgical intervention at the level of the hamstrings has become more controversial because of the lack of correlation between popliteal angle and the flexed knee gait as well as the rate of recurrence after surgery. Indications for hamstring contracture surgery are a popliteal angle measurement of greater than 45 degrees and excessive knee flexion in mid- and terminal stance. Kyphosis in sitting and inability to use a standing frame, which can be a consequence of hamstring spasticity and/or contracture, are other indications for hamstring lengthening. Thus, children who are not ambulators can also benefit from surgery. The operation consists of fractional lengthening of the hamstring muscles (the biceps femoris, gracilis, semimembranosus, and semitendinosus). As noted previously, lengthening the hamstring muscles works by weakening each muscle and thereby reducing spasticity. Some surgeons prefer to lengthen only the medial hamstrings, leaving the biceps femoris intact. Additionally, many surgeons will transfer the hamstring or lengthen it in a way that avoids both overlengthening and excessive anterior pelvic tilt. When the hamstrings are lengthened, the sciatic nerve may remain tight and be excessively stretched. Children who have higher GMFCS levels, more contractures, and less communication ability are at increased risk for neuropraxia [21]. Care must be taken intraoperatively to not extend the knee with the hip flexed and in the postoperative period to avoid the child sitting with the knees extended. Postoperative immobilization is usually done with removable braces.

The recurrence rate of hamstring contractures is greater than that for hip adductor or heel cord contractures and is greatly influenced by the type and severity of involvement. Recurrence may be largely secondary to positioning (chair, sleep), hence the utility of knee immobilizers and standing devices following cast removal. If the hamstrings are overlengthened, genu recurvatum and anterior pelvic tilt may result. Because the hamstring muscles are powerful hip extensors in the first half of the stance phase, there may be a loss of propulsive power during gait if the hamstrings are not transferred or are overlengthened [22].

A long-standing crouch knee gait may lead to or be caused by extensor mechanism failure. As the degree of contracture worsens, hamstring surgery alone may not be sufficient to improve a static contracture. For moderate contractures (~15 degrees), some have advocated stretch casting to improve the range of motion [23], while the author will use a custom-hinged knee ankle foot orthosis (KAFO). When the contracture is larger, more intervention is needed. In the skeletally immature child, growth modulation may be performed by various methods to improve ROM and may be accompanied by a patellar tendon advancement or shortening. This slows the growth of the anterior aspect of the knee and allows the posterior portion to grow, reducing the amount of contracture. For larger contractures or in children with inadequate growth remaining, a more extensive surgery to improve ROM is a distal femoral extension osteotomy with a concomitant shortening of the patellar tendon or advancement of the patellar tendon or tibial tubercle [24].

Children who have limited motion of their knee in swing phase, also known as a stiff knee gait, may benefit from a rectus tendon release or tendon transfer. The child commonly has a positive Duncan-Ely test, and on kinematics, they will show delayed and diminished peak knee flexion in the swing phase. Either procedure helps improve the fluidity of gait.

Internal tibial torsion (ITT) in the absence of cerebral palsy rarely requires surgical correction. In children with cerebral palsy, it is more likely to persist, although spontaneous correction can still occur. Severe torsional malalignment at the tibia and foot may lead to decreased push-off power by decreasing the lever arm. In some cases, derotational osteotomy of the tibia and fibula is required. Most surgeons favor a distal procedure because of the risk of neurovascular complications with proximal tibia surgery. Postoperative care includes a below-knee cast for 4–6 weeks, followed by either an AFO or walking boot.

External tibial torsion (ETT) is more common than ITT in cerebral palsy, possibly as a compensation for femoral anteversion. Surgical intervention is often required for severe ETT in ambulatory persons with cerebral palsy, as it decreases push-off power and can cause a varus moment at the knee. It is especially important if correction of femoral anteversion is planned. Severe ETT may cause the child to walk with the femurs rotated inward as compensation for the torsion. A distal osteotomy is performed as for ITT.

Procedures about the ankle and foot

Ankle and foot problems represent the most common deformities in patients with cerebral palsy. In order to function as an adjustable base of support for standing, the foot must be flat and stable. The most common ankle deformity is equinus, which may be aligned straight but is most commonly associated with medial (varus) or lateral (valgus) deviation of the foot. In general, children with pure spastic diplegic cerebral palsy will develop planovalgus foot deformities, while those with hemiplegia or triplegia will develop an equinovarus deformity.

Any equinus deformity destabilizes the base of support and probably reduces or alters sensory input as well. Dynamic equinovarus is generally treated with an AFO. If the child cannot tolerate the ankle in a neutral position, then the AFO may be designed with slight plantarflexion. When a fixed contracture occurs, Achilles tendon lengthening or fascial lengthening of the gastrocnemius are the two principal surgical choices [25]. Great care must be taken to not overlengthen the gastroc-soleus complex, as this may lead to weakness and a crouched gait. In diplegic children, most surgeons advocate not performing an achilles lengthening and favor surgery more proximal, which has less risk of overlengthening. Weakening the gastroc-soleus complex leads to excessive dorsiflexion in stance and a compensatory crouch gait. There is no proven surgery to augment the incompetent plantar flexors. The child will often need a rigid orthosis, such as a ground reaction, or a solid AFO.

Postoperative management consists of short-leg casting for 4 weeks, rapid resumption of ambulation, and long-term protection with night splints (AFOs). The AFOs may reduce recurrence rates, which are reported as between 5% and 20% in various series. The rate of recurrence is associated with the age of the child, so the use of appropriate modalities such as serial casing, botulinum toxin, or bracing may help maintain foot position, thus avoiding surgery until the child is older and less likely to have a recurrence.

Varus deformity in children with spastic hemiplegic cerebral palsy is due to an imbalance of the two muscles governing medial and lateral ankle movement and is usually associated with equinus. The peroneal on the lateral side is overpowered as a result of overpull of the tibialis posterior or tibialis anterior. It is usually difficult to determine on clinical examination which of these muscles is the culprit. The use of foot pressure plates (pedobarography) and fine-wire EMG during gait may help distinguish between the two. In a child without contracture, a combination of tendon transfer or lengthening is performed. Any tendon transfer in a child with spasticity risks iatrogenic overcorrection. There are four surgical considerations:

1. If the child has dynamic hindfoot varus and equinus, posterior tibial tendon lengthening at the muscle-tendon junction or, in more severe cases, at the tendon level.
2. Supination of the forefoot is diagnosed by clinical exam, gait analysis, and pedobarography. A split anterior tibial tendon transfer (SPLATT) moves half of the tendon to the lateral cuneiform to control the supination. Some advocate full transfer of the tendon. With either procedure, care must be taken to not transfer too far laterally or with too much tension that causes the foot to be everted.

3. The decision to transfer the posterior tibialis muscle is controversial. Proponents of the procedure believe that by removing the deforming force and transferring it to the lateral side of the foot, you may improve eversion and possibly dorsiflexion. Some evidence has shown that in children with CP, the transfer works more like an internal brace than a dynamic everter.
4. If the hindfoot varus becomes rigid, then tendon lengthening or transfers alone will not correct the deformity. In that case, an osteotomy of the calcaneus may be performed to address the hindfoot varus, or a midfoot osteotomy may be used to correct rigid supination. With a more severe deformity, a triple arthrodesis may be needed to fully correct the foot.

In children who develop a severe planovalgus foot deformity, the initial treatment is often bracing. As the foot goes into progressive midfoot abduction and hindfoot valgus, the lever arm of the foot is reduced. With more severe deformities, it may not be possible to control the foot in a brace without the development of pain or sores. In the foot that is no longer amenable to bracing, two main options are available: fusion or osteotomy. With either option, the goal is to improve the lever arm of the foot. Children with significant spasticity or weak plantar flexors may still need an AFO after surgery, so the goal for these children is to make the foot more tolerant of the brace. Proponents of flatfoot reconstruction hope to improve the lever arm through a combination of osteotomies and soft tissue balancing procedures. This hopes to preserve joint motion and avoid adjacent joint degeneration and arthritis. With time, the foot may lose some correction, though it is unclear whether this is clinically significant [26]. After the reconstruction, the foot is reexamined for equinus and corrected with a gastrocnemius lengthening at the same surgery. Postoperative immobilization is required with a cast for 6–8 weeks, followed by an AFO.

Some advocate fusion, as it may be a more durable and reliable procedure. Options include subtalar fusion, talonavicular fusion, or triple arthrodesis (fusion of the talocalcaneal joint, talonavicular joint, and calcaneocuboid joint). The decision as to which joints to fuse depends on the level of severity of the deformity and may be performed in addition to the osteotomies above. Most surgeons avoid doing a fusion on the very young, as this may interrupt the growth of the foot. Any fusion surgery requires immobilization in a cast for 6–8 weeks to obtain fusion, followed by AFO use. While fusion in higher demand individuals may lead to increased joint forces at the adjacent joints and eventual arthritis, this has not been found to be the case in children with CP [27].

In summary, the benefit of obtaining a plantar-grade foot is the enhancement of sensory feedback and increased stability of the base of support, resulting in better equilibrium, improved lever arm function, better brace compliance, and decreased overall muscle tone.

Upper extremity procedures

Treatment of upper extremity deformities in CP is complex, and there is no clear consensus on how to best manage them at the various sites: shoulder, elbow, forearm, wrist, and fingers. Upper extremity surgery is infrequent compared to the number of procedures done on the lower extremities. There are two main categories of treatment for children with CP. The first group is children with severe elbow flexion, forearm pronation, wrist flexion, and a thumb-in-palm deformity. The mainstay of treatment for these deformities is to improve the resting position of the hand for hygiene, dressing, and pain. The second category consists of children with spastic hemiplegia or mild spastic quadriplegia, where the surgical goal is to improve the function of the involved hand to help with activities of daily living.

Initial treatment for both of these groups is occupational therapy (OT) and splinting or bracing. The goals are to improve extremity range of motion, prevent contracture, and help with hygiene. When the deformity can no longer be impacted by nonoperative measures, consideration of surgery is an option. Surgery involves both releasing tight, deforming tendons or their muscle bellies as well as transferring tendons to help improve positioning. For severe wrist flexion deformities, a fusion of the wrist is performed along with releases [28,29].

In spastic hemiplegia, the upper extremity typically exhibits more functional impairment than the lower extremity. the initial treatment is OT and splinting as above. Constraint-induced movement therapy (CIMT) is often employed to help improve the function of the affected arm by reducing the use of the contralateral arm. This may be particularly useful for children who have limited access to ongoing OT [30]. If sensation is relatively preserved, surgical treatment may well produce substantial functional gains. However, in the face of severe sensory loss, the lack of sensory feedback largely nullifies potential benefits from surgical procedures aimed at enhancing motor function [31]. Subjects must be old enough and cognitively intact enough (with a mental age of 5 years or better) to cooperate with sophisticated sensory testing [32]. Evaluations of pain, touch, and temperature sensation are not sufficient to discriminate the best surgical candidates; rather, it is stereognosis, two-point discrimination, and other "parietal lobe" functions that are more telling [33]. Dynamic EMG, videotaped evaluation, nerve blocks (lidocaine and related agents), and/or neuromuscular blocks (botulinum toxin) are relied on to help determine potential for surgical benefit [34]. Although it is difficult to predict which children are the best candidates for surgery, the best indicators are sensory testing, spontaneous hand use, and the absence of growth delay on the involved side. The Shriners Hospital Upper Extremity Evaluation (SHUEE) has been found to help predict those children who will benefit most [35]. The hand surgeon and occupational therapist outline surgical options and discuss the goals of the intervention with the family and child. The surgery may improve hand use and cosmesis, but the preexisting motor planning impairment persists and may result in less function than the parent or child would hope for [35].

Postoperative management

General principles of postoperative care—intensive monitoring, attention to fluids and electrolytes, pulmonary toilet, and pain management—are the same for orthopedic surgery as they are for other types of surgery. However, a few special procedures that relate to the orthopedic patient are worth noting.

Caudal anesthesia using bupivacaine (Marcaine) to decrease postoperative pain and reduce opioid usage has been advocated. Risks include masking a compartment syndrome or skin pressure necrosis. Intravenous and oral narcotics have a definite place in pain control. Studies of self-administered narcotic analgesia have shown the advantages of this approach [36]. The parents of the surgical patient actually control the administration based on their perception of the child's discomfort. Diazepam to decrease spasticity plays a major role in promoting comfort, and its muscle relaxant effect may prove to benefit the child even after the postoperative period. Physical therapy to achieve general mobilization is initiated as soon as possible after surgery. At first, it must be limited and gentle, with a plan for gradually increasing intensity in order to work up to the desired functional gains that engendered the surgery in the first place. Michael Sussman conceives of the process in terms of "prephysical therapy surgery" as opposed to "postoperative physical therapy."

Immobilization produces obvious muscle degeneration (atrophy) histologically after about 3 weeks. Because of this, the earliest mobilization possible is best to avoid the associated weakness. In centers with access to inpatient rehab, this is initiated early, whereas it may be delayed until the early healing of osteotomies has occurred in other centers. Early mobilization has become possible mostly as a result of advances in solid internal fixation methods that make osteotomy sites stable and resistant to intrinsic and external forces. Appropriate positioning and handling techniques can be initiated on an inpatient basis 3–4 days after surgery, followed by weight bearing about 1 week postoperatively. Temporary equipment—for example, an abduction pillow to maintain leg separation at night, knee immobilizers to maintain knee extension, casts or AFOs to maintain the foot and ankle position, or an adaptive seating system for use until ambulation can be resumed—may need to be employed.

Problems associated with surgery

Difficulties encountered with surgery fall into two major categories: complications common to surgical practice in general and errors in judgment related to specific orthopedic procedures. Pitfalls related to surgical technique revolve around the judgment as to how much to alter a tendon, muscle, or bone. For example, too much hip adductor release results in excessive abduction. Over-lengthening the hamstrings results in genu recurvatum. Over-lengthening the heel cord results in a crouch-type gait secondary to insufficient power in the plantar flexors. In the other direction, insufficient release leaves the child with residual deformity and results in recurrence or persistence of the original problems: scissoring through the hips, knee flexion, or equinus deformity at the ankle. The functional consequence of these errors can be a worsened (or unimproved) gait or a loss of sitting support. Another consideration is timing. By waiting too long, the child loses developmental opportunities. Conversely, operating too soon may increase the likelihood of unnecessary procedures being done or the need for repeat surgeries as the child grows and changes.

Perhaps the single biggest reason for surgical failure relates to the mismatch between the parental expectations and the surgeon's expectations of what can be accomplished. The most perfect, complication-free procedure is not satisfactory to the family or the child if it does not produce the functional outcome expected [8].

Children with CP can have complex medical issues that make them more prone to complications, and these comorbidities are correlated with the risk of postoperative complications. In a retrospective study of children with CP GMFCS levels IV and V, those with a combination of four comorbidities (gastrointestinal impairment, respiratory impairment, seizures, and communication impairment) had a 95% probability of a major complication. The rate of minor complications was also quite high, at 91% [37]. Careful preoperative assessment focusing on the management of the entire child, both before and after surgery, is the best strategy to mitigate these complications. Among the undesirable events to consider are the following:

1. Anesthesia always carries risks. There is an increased risk for malignant hyperthermia in neuromuscular diseases but not in cerebral palsy.
2. Wound infections are rare for most orthopedic procedures, but the rate is relatively high for spinal fusion procedures (see section on Scoliosis).
3. Other infections (pulmonary, urinary tract, decubiti, etc.) are more likely to develop in children with physical disabilities than in those without disabilities. The greater the motor impairment and the worse the oral-motor dysfunction, the greater the risk for these problems. Feeding, nutritional support, and respiratory care are the biggest challenges in the postoperative period. While gastrostomy tubes carry risk and are often a surrogate for the level of neurological involvement, they help maximize nutritional status. Feeding is started slowly postoperatively as gastrointestinal motility may be initially slow, and a complete bowel regimen is often started before surgery to avoid constipation. Attention to pulmonary toilet, both preoperatively and postoperatively, is particularly important.
4. Immobilization, although essential, works against the child in several ways. Contracture formation and muscle atrophy will start to develop. In addition, hypercalciuria is uniformly present. Secondary stone formation (renal, gallbladder) occurs, but the frequency is unknown. Symptomatic renal or gallbladder stones are quite rare. One study [38] reported a 20% incidence of cholelithiasis, mostly asymptomatic, following immobilization after spine stabilization surgery. These subjects, however, were mostly older children and young adults.
5. Heterotopic ossification (HO) represents a rare and unexplained phenomenon in which osteogenesis takes place in nonskeletal tissue, most commonly muscle. It occurs in inherited and acquired forms, the latter being associated with a number of inciting events, including orthopedic surgery [39]. The incidence following surgery is very low, and the ossification is seen exclusively around the hip. HO appears to be related to the extent of surgery and perhaps to overly aggressive or too early physical therapy following surgery. It can occur in children with or without CP. A history of head injuries is an aggravating factor for unexplained reasons.
6. Loss of gross motor skills is anticipated following surgery, but rapid recovery is to be expected. Developmental regression in nongross motor areas also occurs. Oral-motor function, for example, may suffer substantially in a child kept unfed for days, and recovery may be a very slow process.
7. Disturbances in associated problem areas (e.g., seizure control) may occur. Careful attention must be paid to antiseizure medications and potential medication interactions. Preoperative drug levels become extremely valuable if seizure activity ensues in the postoperative period. Thrombocytopenia and platelet dysfunction associated with valproic acid and unpredictable bleeding during surgery have been reported [40,41].

Although not really a complication of surgery, the orthopedist will be called upon to help with the management of children with CP who acquire fractures. Osteopenia is common among children with CP who are nonambulatory and/or have complex medical problems. This results in bone fragility, and pathologic fractures, although infrequent, can occur [42]. They are generally seen in the context of risk factors such as poor nutrition, seizure medications, decreased sun exposure (with associated low vitamin D), and decreased weight-bearing. Prevention strategies include: (1) optimization of nutrition with attention to calcium intake; (2) vitamin D supplementation; and (3) weight-bearing [43]. Among children who have experienced a pathologic fracture, some advocate treatment with bisphosphonates [44]. With the increased use of vitamin D, standing frames, and gait trainers, the rate of pathologic fractures has been reduced.

Scoliosis

Scoliosis is defined as a lateral curvature of the spine associated with rotation of the posterior vertebral elements into the concavity of the curve. This type of structural curvature is never "normal," although only a curve of greater than a 10-degree Cobb angle (see below) is designated as scoliosis. Nonstructural scoliosis refers to a visible curvature that does not have the true rotational component described above and is generally flexible back to normal when the offending cause resolves. Examples would be *splinting* (i.e., from thoracic or pleural pain); irritation of back soft tissues, especially the paraspinous muscles on one side; intervertebral disk disease; or simply poor posture. The term *fixed* scoliosis is sometimes used to indicate structural curvature, and *supple* or *postural* may be used to designate nonstructural curves. Any deformity may have a postural component added to a fixed degree of curvature.

Assessment

Screening for scoliosis on clinical examination is easy and reliable if the child can stand. Screening is unreliable for nonambulatory children. Visual inspection of the back with the child standing erect ("at attention") allows several important observations:

1. The height of the shoulders (or extended arms)
2. The obvious position of the spine
3. Symmetry of the scapulae and the flanks
4. The level of the pelvis relative to the plane of the floor

By asking the child to bend forward, one can then note the height of the scapulae, or "rib hump" (Fig. 39.1). The *scoliometer is* a device constructed like a water bubble level that is useful for quantifying the rib hump in degrees, especially in the screening setting. One can also measure it in centimeters, as depicted in Fig. 39.1. Many children with cerebral palsy will need to be viewed in a sitting position (often somewhat supported) to obtain a clinical impression of how straight the spine is. Objective measurement of the curve is only reliable when based on an upright standing (or minimally supported sitting) posteroanterior (PA) radiograph of the thoracolumbar spine. The angle of curvature is determined by the Cobb method (Fig. 39.10). The uppermost and lower vertebrae delineating the curve (the ones that are most tilted) are identified, and perpendiculars are constructed to them. The intersection of these lines reflects the curve's angle of concavity. (The concavity of the straight spine could be thought of as 180 degrees.) The Cobb angle is then the complement to the angle of concavity (e.g., 0 degrees in the straight spine). Scoliosis is a three-dimensional deformity, and while the coronal plane is measured on the PA radiograph, it does not fully elucidate the degree of rotational deformity, concomitant kyphosis, or lordosis that is commonly associated with more severe curves.

Kyphosis refers to the posterior angulation of the spine normally seen in the thoracic region, whereas *lordosis* is the normal anterior curvature of the lumbar spine. As seen in Fig. 39.11, a straightedge normally runs vertically between the sacrum and the normal thoracic kyphosis. An exaggerated hump (hyperkyphosis) can be visualized in terms of degrees of deviation from the vertical. The measurement is analogous to that of scoliosis, utilizing the lateral spine radiograph. Normal thoracic kyphosis ranges from 15 to 50 degrees. Thoracic kyphosis may be exaggerated as a result of tight hamstrings in children with cerebral palsy.

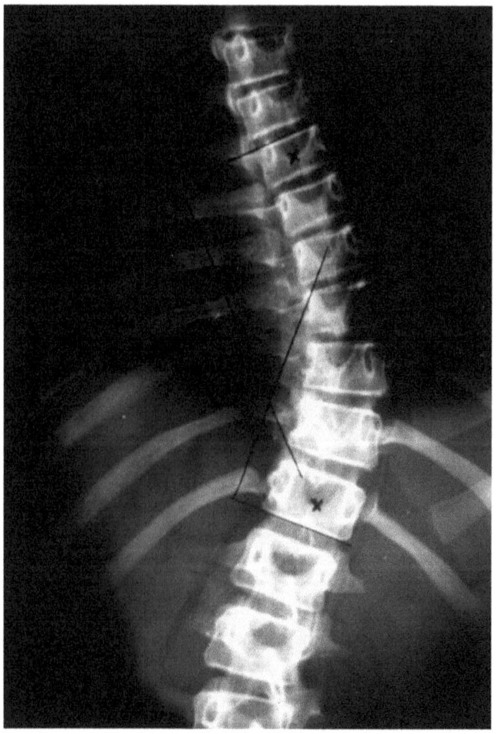

FIGURE 39.10 Posteroanterior spine film depicting construction of the Cobb angle. *Courtesy of John E. Lonstein, M.D.*

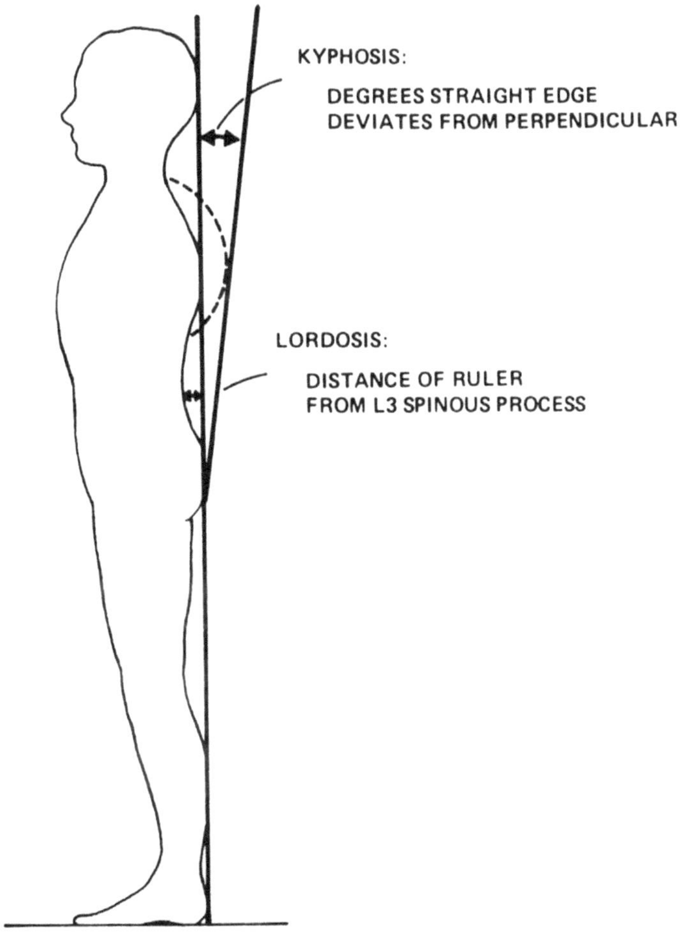

FIGURE 39.11 Assessment of standing posture. *From Horstmann H, Bleck E. Orthopedic management in cerebral palsy, Clinics in developmental medicine. London: Mac Keith Press; reprinted by permission; 2007.*

Lordosis is measured radiographically in the same manner as kyphosis. Clinically, it can be viewed in terms of distance from the L3 spinous process to the vertical, normally 3–4 cm at a maximum (Fig. 39.11). Lordosis may be exaggerated in certain conditions, especially cerebral palsy and peripheral neuromuscular diseases that cause weakness about the pelvic girdle. In children with cerebral palsy, it usually develops in compensation for excessive hip flexion with contractures and forward pelvic tilt. In pelvic girdle weakness, one sees progressive lordotic (or hyperlordotic) posturing as a natural mechanism to maintain the center of gravity because, once again, the pelvis tilts forward, in this instance as a result of weakness (as opposed to spasticity). Initially flexible, it becomes increasingly fixed as the child grows. Quite disabling problems with low back pain may develop in young adults. Treatment for people with cerebral palsy would be hip flexor contracture release. Unfortunately, there is no method to control hyperlordosis in people with neuromuscular disease.

Classification

There are four main categories of scoliosis: idiopathic, early-onset, congenital, and neuromuscular. Most children with CP who develop scoliosis will have neuromuscular scoliosis. Neuromuscular curves can be either double ("S"-type) or single ("C"-type) curves. Those that are primarily thoracic are seen in ambulatory children and during the adolescent growth spurt. Those that are primarily lumbar, as a rule, start earlier, tend to be more common in nonambulatory children, and are typically associated with pelvic obliquity. In ambulatory children with a double curve, there is generally a primary thoracic curve compensated by an opposite lumbar curve with little resultant pelvic obliquity. Interestingly, although scoliosis is common in hemiplegia, progression to the point of requiring intervention with bracing or surgery is uncommon.

Prevalence

The prevalence of scoliosis is best thought of in terms of the base population. In the general population, scoliotic curves (10 degrees or greater) have a prevalence of 3%–4%. Most of these are minor and untreated. Idiopathic scoliotic curves greater than 20 degrees occur in about 0.25% of adolescents. Girls have a great predisposition for curves to progress. Whereas the female-to-male ratio for the mildest curves is about 1:1, for curves greater than 30 degrees, it is 7:1. Among children with cerebral palsy, scoliosis is quite common. The prevalence ranges from 6% to 64%, depending on the type and severity of the disorder and the particular clinic population being surveyed. The incidence of scoliosis is directly related to the GMFCS level. Children at GMFCS level V have a 75% chance of developing a curve >40 degrees by age 20 [45], while children at GMFCS level I only develop such a curve at a rate of 5%. Because scoliosis has a relationship with the development of hip disorders and contractures, care must be taken to diagnose it early.

Etiology and progression

From a pathophysiologic standpoint, the exact mechanism that triggers spinal curvature is not clear. Asymmetric spasticity, equilibrium reaction deficits, primitive reflexes (especially the Galant), sensory deficits, and positional factors have all been implicated in various theories. The time and causea-and-effect relationships between scoliosis, pelvic obliquity, and hip dislocation—which often coexist—are controversial. Scoliosis may develop regardless of the hip and pelvic positions. In simple terms, it is probably due to the inability of the back muscles to support the flexible spine against gravity. Muscle imbalance is probably a second important factor.

Regardless of etiology, scoliosis curves tend to progress most rapidly during the adolescent growth spurt. In most circumstances, curves less than 40 degrees at skeletal maturity will be stable in adulthood, whereas those greater than 50 degrees are likely to progress, albeit at a slower rate, after skeletal maturity [46]. Thus, degree of curvature, proximity to skeletal maturity, and linear growth rate are the key elements the surgeon considers in determining scoliosis management. The Risser sign (degree of capping of the iliac crest apophyses) is a convenient marker of skeletal maturity and can be readily evaluated on spine radiographs that include the pelvis. Newer methods to assess skeletal maturity and, more importantly, peak growth velocity include the use of a lateral elbow radiograph or a PA hand radiograph [47]. Signs of puberty (Tanner staging and other characteristics) are also very helpful indicators of expected growth rate and growth potential.

Treatment

Treatment of scoliosis in children with cerebral palsy is based on the following indications:

1. Maintenance of balance for sitting or walking
2. Pain, most often at the apex of the curve or at the pelvis, as a result of discomfort in sitting
3. Pressure sores, especially in the presence of pelvic obliquity
4. Cardiopulmonary function, which is compromised in children with curves greater than 50 degrees. Pulmonary compromise does not become significant until the curve exceeds 100 degrees.

Evaluation of the spine is continued throughout childhood and adolescence, as small curves may rapidly progress during the adolescent growth spurt. A clinical evaluation is performed at 6–12 month intervals until maturity, and, when a curve is identified, radiographic monitoring is begun. Initial management of a child with a small curve less than 20–25 degrees is done by observation and regular radiographic follow-up. Postural curves are initially managed by wheelchair modifications. If the curve progresses, then the decision to use a brace is founded on the goal of helping improve function. There is no evidence that it actually decreases the rate of progression. Bracing for functional improvements with sitting balance, head control, postural control, and arm function is a reasonable goal [48]. Delaying the onset of a curve allows for greater lung development and reduces the risks associated with early fusion. These include limited growth of the chest cavity as well as continued growth of the anterior column of the spine, which may lead to a rotational deformity. Quality evidence for reducing curve progression is limited, but for a child who can tolerate the brace without discomfort, sores, or restriction in breathing or feeding, it is a reasonable option. It should be made clear to the family that the main goal of bracing is functional improvement.

If a curve is less than 40 degrees at skeletal maturity, it may still progress but will do so at a much slower rate [49]. Surgery is rarely required. Curves greater than 40 degrees have a high rate of progression [50,51] of up to 4 degrees per year and will require close observation and often surgical intervention.

Curve progression causes sitting intolerance, pain, and pressure sores and, when severe, may lead to feeding issues and respiratory compromise [52]. Surgical stabilization improves sitting balance, feeding, and care and has been shown to improve HRQOL [53]. Caregivers rated spine surgery as one of the most beneficial of all treatments impacting the lives of their children. Most orthopedists feel spine stabilization should not be delayed in a child with a large progressive curve. In contrast, a systematic review highlighted the lack of quality evidence for surgical fusion [54], although functional change was the outcome showing the least improvement.

Prior to surgery, multidisciplinary team involvement is imperative in order to optimize nutrition and respiratory status. Seizure control and shunt function need to be attended to. The surgeon must be prepared to deal with baclofen pump hardware in the vicinity of the fusion. The goal of surgery is to improve sitting balance, pelvic obliquity, and overall posture of the child, in addition to preventing the progression of scoliosis. This is achieved primarily by posterior fusion of the spine. While correction of the curve is important, it is arguably more important to improve the overall pelvic obliquity and the coronal and sagittal balance of the child. Fusion techniques have evolved, and in nonambulatory children, fusion to the pelvis is most often undertaken to correct pelvic obliquity. Some children who have large neglected curves may need adjuncts to posterior fusion, including preoperative or intraoperative traction, as well as anterior releases combined with posterior fusion and/or osteotomies. Fusion is obtained by rigid fixation and either pedicle screw fixation or sublaminar wiring. A cohort study showed similar outcomes of balance, complications, and fusions but a shorter hospital stay and lower blood loss in the pedicle screw fixation group [55]. The trend has shifted toward a more pedicle-screw-based fixation, although either option is reasonable [56].

Technical complications of spine surgery have improved dramatically, mainly due to the advent of segmental instrumentation, and medical complications have improved due to continuing advances in pediatric anesthesiology and preoperative care. Infection prevention measures have helped reduce infection risk, but the rate is still much higher than in children undergoing fusion for idiopathic scoliosis. In a recent multicenter study, the complication rate of children undergoing fusion was 46%, with the most common complications being pneumonia and surgical site infection [57]. Mortality rates reported to be 3% in the past, seen mostly in the more severely involved cases, are now at or below 0.5%. Pseudarthroses represent failure of fusion and vary widely in different series, but are now reported to occur in less than 5% of all cases. Neurologic complications caused by spinal cord stretching or vascular interruptions occur infrequently since the advent of intraoperative spinal cord monitoring, with less than 0.5% having permanent sequelae. Postoperative hyperesthesia is a focal, transient neurologic phenomenon that always resolves in about 2 weeks.

With spine surgery, the back always loses flexibility in terms of anterior–posterior and lateral flexion but gains stability. Because of the vertebral fusion, linear growth through the spine is severely limited and is a function of the number of segments fused. The loss in spinal growth can be roughly calculated before surgery [8]. Modern techniques can minimize blood loss, including the use of antifibrinolytics perioperatively, but commonly, children still reach the threshold for transfusion of blood products. The first choice is autologous blood that is collected from bleeding sites intraoperatively with the use of a cell saver, and, as a second option, blood bank products are transfused. The use of autolgous or related donor supplies has fallen out of favor. Given the overall high complication rate associated with fusion and the difficulty in demonstrating its clinical efficacy, it is important that the child be optimized medically beforehand and an in-depth discussion of the risks and benefits be had with the family.

Neurosurgery

Neurosurgery is appealing because of the hope that the neurosurgeon can get to the source of the disorder and "fix" the brain, thereby resolving all the joint and muscle problems. But the brain malformations and injuries that underlie CP are not really fixable, and most of what is done surgically involves destructive lesions or electrical stimulation strategies to counter the dysfunctional neurophysiology of CP. For the purposes of this chapter, neurosurgical involvement is covered in two areas—motor function and hydrocephalus.

Motor: spasticity

Neurosurgical techniques to improve motor function had little application to children with cerebral palsy until the surge of interest in selective dorsal rhizotomy (SDR) aimed at eliminating spasticity, especially in the lower limbs, which ensued 30–40 years ago. Based on Sherrington's experimental work in cats done in the 1890s and early 20th century, SDR—cutting the dorsal nerve roots of the spinal cord—is one of a number of spinal cord and spinal root ablative procedures that have been used in limited settings for many years, mostly in adults with spinal cord injury or disease. Rhizotomy in humans was first reported in 1908 but was popularized by Fasano in the 1970s [58]. By completely interrupting the spinal reflex arc, spasticity can be abolished. The technique involves gaining access to the lumbar spinal cord or cauda equina by performing a laminectomy. The dorsal roots are exposed, and the point at which the dorsal root enters the cord is dissected. Each root consists of 5–10 smaller rootlets rather than a single bundle. "Selectivity" refers to the rootlets; once they have been teased apart, they are then individually electrically stimulated under constant EMG surveillance. Although all the rootlets will produce a muscle contraction when stimulated, a few tend to produce an exaggerated response [59]. Presumably, these rootlets are the ones mainly carrying reflex arc fibers that are implicated in spasticity, whereas the others are carrying different afferent fibers. Only those rootlets producing the exaggerated response are sectioned; the others are left intact. Apparently, the stimulation procedure is extremely fickle, and there continues to be controversy about the validity of this selection process [60–62]. The total number of rootlets that should be sectioned (regardless of electrical characteristics) is also a matter of unresolved debate.

The application of selective dorsal rhizotomy to children can be attributed to Warwick Peacock's work in South Africa. In his hands, the procedure was tremendously successful [63,64]. He was careful to point out that the ideal candidate for rhizotomy appeared to be the child with pure spastic diplegia. The more truncal weakness, extrapyramidal involvement, and asymmetry the child exhibited, the less acceptable the results. Rhizotomy will worsen the weakness and unmask the movement disorder. Patient selection is of paramount importance (probably more than rootlet selection); however, the criteria are fairly subjective. The evaluation for patient selection should be carried out by at least a three-member team of experienced clinicians [65]. Age at surgery is a factor. In early childhood, spasticity is the main feature compromising motor function in cerebral palsy, but in later childhood, musculoskeletal features become more important. Rhizotomy is more effective in the younger group (probably up to age 7) [66]. In later years, orthopedic procedures seem to provide greater benefit [67]. Recent orthopedic surgery is a relative contraindication, and the need for later orthopedic intervention to correct fixed alignment problems is not uncommon [68]. Rhizotomy has not provided particular benefit for children with hemiplegia and has been abandoned for that population. Children with spastic quadriplegia without ambulatory potential are offered the procedure for the purpose of improved sitting and easier care [69].

Postoperatively, the children are arreflexic, hypotonic, and quite weak. They initially lose functional ground. There is occasionally some hyperesthesia of the legs, but this rapidly wears off. Strength returns gradually, but reflexes, as a rule, do not reappear, and there is typically a need for ankle support (AFOs) when ambulation is resumed. An extended, intense rehabilitation program is needed and is standard protocol in most institutions performing this procedure [68]. Limiting factors can include cognitive deficits, compliance problems, and the restrictions of insurance companies and other funding agencies.

Rhizotomy works in the sense of effectively eliminating or reducing lower extremity spasticity in the short term [70], and there has been very little recurrence in the long term [71,72]. Numerous follow-up studies support a wide variety of benefits from SDR [73,74]. Verbeek et al. [75] followed their cohort of spastic diplegia patients who underwent SDR in childhood for over 25 years. The subjects were stable over time in terms of daily activities and social participation, they did not show declines in terms of functional mobility, and they had no progression of spinal deformities and no increase in pain [75,76]. Munger et al. [71]. reported a 16–25 year follow-up comparing children with spastic diplegia who underwent SDR versus a matched control group who did not have SDR. Spasticity significantly decreased in the SDR group but not in the control (non-SDR) group, and gait quality improved in both groups but more so in the controls. The non-SDR group underwent significantly more interventions (orthopedic surgeries, intrathecal baclofen, botulinum toxin, and phenol) over the years. Survey measures of ICF domains for activity and participation (including pain) did not differ between the groups. In contrast, MacWilliams et al. [77]. showed improved gait in an SDR versus non-SDR study but no other significant differences. Energy consumption, while remaining high, was actually lower than baseline in both groups.

Long-term concerns about the operation have been addressed over the past several decades of experience.

1. Spinal column instability as a result of the laminectomy and the potentiation of scoliosis development. The modification of the original laminectomy procedure that allowed for replacement of the laminae and reattachment of the paraspinal musculature appears to have obviated this concern. Scoliosis development has not been an issue with long-term follow-up [68,78] Pronounced low lumbar hyperlordosis has been observed in some children, and its long-term effect appears to be of no consequence.
2. Recurrence of spasticity. This is rarely a problem. Long-term studies show very little recurrence [68,72].
3. Rapid hip dislocation soon after rhizotomy. This was reported mainly in children with spastic quadriplegia by Greene et al. in 1991 [79] and was very concerning. Later follow-up studies have found neither a positive nor a negative effect of SDR on hip dysplasia [80].
4. Weakness and contractures in the lower extremities. Children can exhibit impressive ankle valgus and a marked propensity for knee flexion when they begin to ambulate postoperatively. Long-term problems with ankle instability and crouch gait have not developed [71,78]. Joint movement limitations due to established contractures do not

improve with SDR, and SDR does not prevent the development of contractures [72], but it may reduce the need for soft-tissue surgeries that are aimed at decreasing spasticity [71].

5. Potential for long-term degenerative joint arthropathy as a result of lost proprioceptive and other sensory feedback. Some sensory loss can occur transiently, but it has not been found to be an issue in young adult patients [71]. SDR may have benefits in terms of the pain experienced by older patients [72]. Chronic joint changes are probably yet to be determined until later adulthood.
6. Bladder dysfunction. Post-operative bowel and bladder problems can appear but are transient. The longest-term studies to date have not found neurogenic bowel or bladder problems as long as judicious rootlet selection procedures are employed [68,71,78].

In conclusion, selective dorsal rhizotomy is widely offered for the management of spasticity in children with cerebral palsy. Its value compared to more traditional orthopedic approaches remains contentious [71,81,82]. Information regarding longer term complications to date has been reassuring, but potential problems in later adulthood remain to be determined.

Motor: movement disorders

Ventral posterolateral thalamotomy and dentate nucleus ablation represent old procedures pioneered in adults with movement disorders, particularly Parkinson's disease, that came to be applied to children with cerebral palsy. These stereotactic surgeries have an impressive history of success in adults with Parkinson's disease. They achieved little long-term benefit for spasticity, and, in terms of CP, they were felt to be more appropriate for cases with extrapyramidal features in which marked adventitious movement was the problem [83].

Cerebellar stimulation, based on the century-old observation in animals that electrical stimulation of the anterior lobe of the cerebellum decreased spastic and rigid hypertonus, was advocated as a major therapeutic breakthrough in the early 1970s by Irving Cooper, the neurosurgeon who introduced thalamotomy for Parkinson's disease. Reduction in spasticity, increased ease of management, and improvements in speech (apparently as a result of better breath control) were claimed but were clouded by the very subjective nature of the assessments. Spinal cord stimulation, a variation on the theme of cerebellar stimulation, later became favored over cerebellar stimulation, but the limited, if any, benefits of this procedure were brought to light by several studies [84,85].

Neurosurgical treatment of movement disorders in children, specifically dystonia, utilizing deep brain stimulation (DBS) found new enthusiasm around the turn of the century. In adults with Parkinson's disease, targeting the globus pallidus internus (GPi) produced a reduction of dyskinesias and, when applied to children with some primary dystonias, caused a dramatic resolution of symptoms.

Dystonia can be classified into two categories: primary, which are all genetic, and secondary, which are all acquired via some insult. CP largely falls into the secondary category, although about 10%–15% of CP is due to CNS malformations that may have genetic underpinnings. These malformations reflect a static situation, very different from the primary genetic dystonias that are usually progressive in nature. The universal experience has been that primary dystonias respond to DBS much better than secondary dystonias [86–89], although many questions remain regarding the details. Among patients with differing etiologies or even the same apparent etiology, the response to stimulation can be variable and difficult to predict [86] Inherited dystonia without apparent structural CNS pathology (e.g., the primary/idiopathic dystonias, as opposed to degenerative conditions like pantothenate kinase-associated neurodegeneration, Batten's disease, etc.) do better.

Later age of onset of dystonia, not the age at surgery, is associated with a better response [86]. The duration of life with dystonia and the shorter time between diagnosis and DBS treatment may be significant [88]; primary and secondary dystonias both do worse with a longer duration of symptoms. Children who already have contractures do not do as well, and Tsering [90] attributes this to a longer duration of dystonia, but you have to wonder if perhaps it points to some underlying spastic component. The benefits of DBS in secondary dystonia can be difficult to quantify using formal measures like the Burke-Fahn-Marsden (BFM) dystonia rating scale or Barry-Albright Dystonia (BAD) scale. Sanger [89] points out that patients will assert subtle, yet important (to them) changes with DBS (easier movement, episodes of relaxation, easier dressing, toileting, and equipment donning, better sleeping) that are not picked up on these scales. Other outcome elements, like pain and quality of life, are not addressed by the BFM and BAD scales.

Electrode location makes a difference—the thalamus is better for hyperkinetic movements, whereas the globus pallidus is better for hypertonus. The globus pallidus internus has become the standard target for treatment of secondary dystonia, and bilateral stimulation seems to be necessary [89]. Precise targeting using stereotactic techniques remains a substantial challenge. Complication rates for infection and equipment breakdown/mechanical failure are highly variable. Infection rates of 10% and an even higher rate of technical problems are reported [86,91]. Although speech and swallowing improve in many patients, stimulus-induced dysarthria is fairly common [92]. Infection almost always requires removal of the hardware. At present, DBS is reserved for children with major functional deficits who have not responded to other treatments, including failure to respond to oral medications and intrathecal baclofen, and have no other options [90].

Hydrocephalus

Cerebral palsy is strongly associated with preterm birth; roughly a third to half of children with CP are former preterm infants. Intraventricular hemorrhage (IVH) and the subsequent development of post-hemorrhagic hydrocephalus (PHH) are common among preterm infants. Among infants with grade I IVH, PHH occurs in 1%, whereas with grades III and IV, it is in the neighborhood of 25% [93].

Traditionally, the placement of a ventricular shunt has been used to divert obstructed cerebrospinal fluid (CSF) flow. The peritoneal cavity is by far the preferred location for the distal drainage tube, but atrial or pleural locations can also be used. In order to accomplish shunt placement, an infant weight of 1500–2500 g is required, depending on neurosurgeon preference [94,95]. For infants weighing less than the cut-off weight, a temporizing procedure in the form of a ventriculosubgaleal shunt (VSGS) or a ventricular reservoir (VR) is required. Most will then go on to require a permanent shunt [96].

Shunts are prone to failure due to obstruction, infection, and mechanical problems (tubing fractures, component disconnections, tubing migration, and overdrainage). The failure rate is about 50% after 2 years in pediatric patients [97]. Younger patients experience higher rates of shunt failure, particularly infants <6 months old. Additional risk factors for failure include prematurity, IVH, brain tumors, and postinfection. Over-drainage can occur acutely, causing extra-axial fluid collections and/or subdural hematomas due to rapid decompression of the ventricular system, or it can be chronic. With chronic overdrainage, slit ventricle syndrome can develop. Adjustable valves and antisiphon devices at the distal catheter are employed to manage overdrainage.

As an alternative to shunting, endoscopic third ventriculostomy (ETV) has gained interest for this population. It involves inserting an endoscope into the frontal horn of the lateral ventricle, threading it through the foramen of Monro into the third ventricle, and creating an opening in the floor in order to permit CSF drainage into the basal cisterns. Compared to VP shunting, ETV offers a lower risk of hardware-related complications, and infection risk appears to be lower [98]. Success of ETV alone varies with age, whereby younger patients (especially <2 years) do not do as well as older patients. The failure rate is high and, over time, appears to be similar to that for VP shunting [99]. For obstructive hydrocephalus due to aqueductal stenosis, the success rate at 12 months postprocedure for ETV was 66%, and for VP shunting, it was 83%. Infants younger than 6 months fared worse than older ones [100].

ETV and choroid plexus cauterization (CPC) were developed separately, and doing them in combination started in 2000. ETV plus CPC has shown greater success rates than ETV alone, overall 50%–60%. ETV/CPC success is highly dependent upon risk factors such as age and underlying etiology. Particularly positive results have been obtained among myelomeningococele and noninfectious causes of hydrocephalus, like CNS tumor-associated hydrocephalus, congenital aqueductal stenosis, and Dandy-Walker malformation. ETV failures tend to manifest early (< 6 months) as opposed to shunt failures, which appear later [101]. Some failures are due to stoma closure secondary to scarring, and re-opening these ("redo ETV") is considered a worthwhile endeavor. Infection with ETV or CPC is less common than with shunting.

In myelo-related hydrocephalus, ETV + CPC is well established and is 72%–78% effective [102]. Among <2-year-olds, myelomeningocele and aqueductal stenosis are most likely to achieve success (83%), while PHH of prematurity is least likely (40%). Long-term outcomes comparing ETV/CPC versus shunting are not well defined. In myelomeningocoele, neurocognitive outcome, and ventricular size are similar [103].

GLOSSARY

Antalgic to avoid pain (e.g., when there is a rock in one's shoe). The result is a shortened stance phase (quick and light step) on the affected side.

Cadence in gait analysis, the number of steps per minute.

Calcaneus largest of the seven tarsal bones; also used in reference to a deformity of the ankle whereby the foot is directed upward relative to its usual position (i.e., in dorsiflexion [*talipes calcaneus*]). Calcaneus gait is a short-step gait in which weight is borne primarily on the heel.

Cavus deformity equinus posture of the foot only from the point of the midfoot, also referred to as a high-arched foot or *pes cavus*.

Cobb angle the angle used to measure deformity in scoliosis. The measurement involves choosing the vertebrae at the top and at the bottom of the curve, drawing a line to mark their horizontal plane, and constructing the perpendiculars to those lines. The complement to the intersection of the perpendiculars represents the Cobb angle (see Fig. 39.10).

Contracture fixed limitation of passive muscle stretch as a result of soft tissue fibrosis (i.e., fibrosis of the muscle, tendon, or joint capsule).
Coxa Latin for "hip," as in *coxa valga*, the straightening of the angle formed by the femoral neck and shaft.
Dysplasia abnormal development with resultant distorted architecture.
Equinus positioning of the foot whereby the plane of the toes is pointing downward.
Femoral Anteversion (better termed Internal Femoral Torsion) internal rotation of the leg from the hip as a result of torsion of the femoral shaft such that the distal end (the intercondylar plane) is twisted medially relative to the proximal end (the plane of the femoral neck–shaft).
Gait walking.
Gait cycle two steps (or one stride).
Genu Latin for "knee," as in *genu varum* (bow legs) or *genu valgum* (knock knees).
Goniometer a device resembling two rulers hinged together that is used to measure joint angles.
Hallux Latin for the great toe, as in *hallux valgus* (a bunion deformity).
Hamstrings the knee flexor muscles that originate from the ischium: the biceps femoris, semimembranosis, and semitendinosis; functionally, the gracilis could be included as well.
Heel cord the Achilles tendon from the triceps surae inserting into the calcaneus.
Kinematics the description of the geometry of movement without consideration for the underlying forces involved. Typically, linear and angular displacements, velocity, and acceleration are the primary measures.
Kinetics the description of movement in terms of the internal and external forces involved. These moment and power measurements of energy are not possible to appreciate by visual inspection alone.
Obliquity being slanted, as in pelvic (also termed pelvic tilt).
Orthosis a support (or brace) for correcting and stabilizing the position of a body structure. Nomenclature is derived from the portion of the body that is covered by the device: for example, AFO, ankle–foot–orthosis; KAFO, knee–ankle–foot orthosis; HKAFO, hip–knee–ankle-foot orthosis; TLSO, thoracolumbosacral orthosis.
Planus flat, as in *pes planus*, a low-arched (or flat) foot.
Recurvatum curved backward; mostly used exclusively with *genu*. Therefore, *genu recurvatum* is hyperextension of the knee joint, also known as back-kneeing.
Station the position or posture assumed when standing (or sitting) erect.
Step length the distance covered in one step; the distance between right foot initial contact and left foot initial contact in double-limb stance.
Stride the distance covered in two steps; the distance from heel strike (or initial contact) to the following heel strike with the same foot.
Subluxation displacement of a joint from normal position, with some maintenance of continuity, that is to say, some direct contact. In contrast, dislocation indicates there is no contact.
Talipes Latin for a deformed foot ("ankle-foot") (i.e., a clubfoot). Although sometimes used synonymously with pes, pes more correctly refers simply to the foot.
Talus the tarsal bone that articulates with the tibia and fibula. Vertical talus refers to a congenital rigid flat (or "rocker-bottom") foot deformity.
Torsion being twisted, as in tibial or femoral.
Trendelenburg gait with each step the upper trunk sways toward the side on which the patient puts weight. This is due to insufficient hip abductor muscle function (i.e., weakness) or to hip pain.
Trendelenburg test the subject is asked to stand on one leg. Normally the powerful hip abductors of the stance leg will maintain the pelvis horizontal and parallel to the floor. If these abductors are weak, the pelvis will tilt, with the opposite side dropping below the stance leg side. This would be a positive Trendelenburg test indicating hip abductor weakness.
Valgus (feminine form *valga*, neuter *valgum*) Latin for "turned or bent out"; that is, the angulation of the distal segment of the body part being described is away from the midline of the body.
Varus (feminine form *vara*, neuter *varum*) Latin for "turned in"; that is, the angulation is pointed toward the midline of the body.
Windswept hips a deformity consisting of adduction contracture on one side and abduction posture, usually with an iliotibial band contracture, on the other side.

References

[1] Gage JR. Gait analysis in cerebral palsy (clinics in developmental medicine), 121. New York: Cambridge University Press; 1991.
[2] Sutherland DH. Gait disorders in childhood and adolescence. Baltimore: Williams & Wilkins; 1984.
[3] Read HS, Hazlewood ME, Hillman SJ, Prescott RJ, Robb JE. Edinburgh visual gait score for use in cerebral palsy. J Pediatr Orthop 2003;23:296–301.
[4] Melhem E, Assi A, El Rachkidi R, Ghanem I. EOS biplanar X-ray imaging: concept, developments, benefits, and limitations. J Child Orthop 2016;10:1–14.
[5] Staheli LT. The prone hip extension test. Clin Orthoped Relat Res 1977;123:12–15.
[6] Reade R, Hom L, Hallum A, Lopopolo R. Change in popliteal angle measurement in infants up to one year of age. Devel Med Child Neurol 1984;26:774–80.

[7] Lee SY, Sung KH, Chung CY, Lee KM, Kwon S-S, Kim TG, et al. Reliability and validity of the Duncan-Ely test for assessing rectus femoris spasticity in patients with cerebral palsy. Devel Med Child Neurol 2015;57:963–8.
[8] Bleck EE. *Orthopedic management in cerebral palsy* (clinics in developmental medicine, No. 99/100). London: MacKeith Press; 1987.
[9] Staheli LT. Lower-extremity-specific rotational problems in children: normal values to guide management. J Bone Jt Surg 1985;67A:39–47.
[10] Salenius P, Vankka E. The development of the tibiofemoral angle in children. J Bone Jt Surg 1975;57A:256–61.
[11] Rosenthal RK, Deutsch SD, Miller W, Schumann W, Hall JE. A fixed-ankle, below-the-knee orthosis for the management of genu recurvatum in spastic cerebral palsy. J Bone Jt Surg 1975;57A:545–7.
[12] Albright AL, Cervi A, Singletary J. Intrathecal baclofen for spasticity in cerebral palsy. JAMA 1991;265:1418–22.
[13] Rice J, O'Donnell M, Armstrong R. Intrathecal baclofen use in childen with spasticity: a physicians' survey. Devel Med Child Neurol 2006;48:1006–7.
[14] Inman VT, Ralston HJ, Todd F. Human walking. Baltimore: Williams & Wilkins; 1981.
[15] Rang M. Cerebral palsy. In: Morrissey RT, editor. Lovell and Winter's pediatric orthopedics. 3rd ed. Philadelphia: J.B. Lippincott; 1990. p. 465–506.
[16] McGinley JL, Dobson F, Ganeshalingam R, Shore BJ, Rutz E, Graham HK. Single-event multilevel surgery for children with cerebral palsy: a systematic review. Devel Med Child Neurol 2012;54:117–28.
[17] Soo B, Howard JJ, Boyd RN, Reid SM, Lanigan A, Wolfe R, et al. Hip displacement in cerebral palsy. J Bone Jt Surg 2006;88A:121–9.
[18] O'Donnell M, Mayson T, Miller S, Cairns R, Graham K, Love S, et al. AACPDM care pathways: hip surveillance. http://www.aacpdm.org/publications/care-pathways/hip-surveillance; 2017.
[19] Shore BJ, Mulpuri K, Shrader MW. Radiographic hip screening in cerebral palsy: developing POSNA-wide consensus. J POSNA 2021;3:1–13.
[20] Kalen V, Bleck EE. Prevention of spastic paralytic dislocation of the hip. Devel Med Child Neurol 1985;27:17–24.
[21] Karol LA, Chambers C, Popejoy D, Birch JG. Nerve palsy after hamstring lengthening in patients with cerebral palsy. J Pediatr Orthop 2008;28:773–6.
[22] De Mattos C, Do KP, Pierce R, Feng J, Aiona M, Sussman M. Comparison of hamstring transfer with hamstring lengthening in ambulatory children with cerebral palsy: further follow-up. J Child Orthop 2014;8:513–20.
[23] Westberry DE, Davids JR, Jacobs JM, Pugh LI, Tanner SL. Effectiveness of serial stretch casting for resistant or recurrent knee flexion contractures following hamstring lengthening in children with cerebral palsy. J Pediatr Orthop 2006;26:109–14.
[24] Novacheck TF, Stout JL, Gage JR, Schwartz MH. Distal femoral extension osteotomy and patellar tendon advancement to treat persistent crouch gain in cerebral palsy. J Bone Jt Surg 2009;91(suppl 2, part 2):271–86.
[25] Rose SA, DeLuca PA, Davis RB, Ounpuu S, Gage JR. Kinematic and kinetic evaluation of the ankle after lengthening of the gastrocnemius fascia in children with cerebral palsy. J Pediatr Orthop 1993;13:727–32.
[26] Rethlefsen SA, Hanson AM, Wren TAL, Kay RM. Calcaneal sliding osteotomy versus calcaneal lengthening osteotomy for valgus foot deformity correction in children with cerebral palsy. J Pediatr Orthop 2021;41:e433–8.
[27] Trehan SK, Ihekweazu UN, Root L. Long-term outcomes of triple arthrodesis in cerebral palsy patients. J Pediatr Orthop 2015;35:751–5.
[28] Koman LA, Gelberman RH, Toby EB, Poehling GG. Cerebral palsy: management of the upper extremity. Clin Orthoped Relat Res 1990;253:62–74.
[29] Louwery A, Warnink-Kavelaars J, Daams J, Deelen A. Effects of upper extremity surgery on activities and participation of children with cerebral palsy: a systematic review. Devel Med Child Neurol 2020;62:21–7.
[30] Hoare BJ, Wallen MA, Thorley MN, Jackman ML, Carey LM, Imms C. Constraint-induced movement therapy in children with unilateral cerebral palsy. The cochrane collaboration. Jon Wiley & Sons, Ltd; 2019.
[31] Goldner J. Upper extremity tendon transfers in cerebral palsy. Orthop Clin N Am 1974;5:389–414.
[32] Bolanos AA, Bleck EE, Firestone P, Young L. Comparison of stereognosis and two-point discrimination testing of the hands of children with cerebral palsy. Devel Med Child Neurol 1989;31:371–6.
[33] Kinnucan E, Van Heest A, Tomhave W. Correlation of motor function and stereognosis impairment in upper limb cerebral palsy. J Hand Surg 2010;35A:1317–22.
[34] Koman LA, Sarlikiotis T, Smith BP. Surgery of the upper extremity in cerebral palsy. Orthop Clin N Am 2010;41:519–29.
[35] Van Heest AE, Bagley A, Molitor, James MA. Tendon transfer surgery in upper-extremity cerebral palsy is more effective than botulinum toxin injections or regular, ongoing therapy. J Bone Jt Surg Am 2015;97:529–36.
[36] Rodgers BM, Webb CJ, Stergios D, Newman BM. Patient-controlled analgesia in pediatric surgery. J Pediatr Surg 1988;23:259–62.
[37] Shea J, Nunally KD, Miller PE, Difazio R, Matheney TH, Snyder B, et al. Hip reconstruction in nonambulatory children with cerebral palsy: identifying risk factors associated with postoperative complications and prolonged length of stay. J Pediatr Orthop 2020;40:e972–7.
[38] Teele RL, Nussbaum AR, Wyly JB, Allred EN, Emans J. Cholelithiasis after spinal fusion for scoliosis in children. J Pediatr 1987;111:857–60.
[39] Smith R, Triffitt JT. Bones in muscles: the problems of soft tissue ossification. Quart J Med 1986;61:985–90.
[40] Chambers HG, Mubarak SJ, Wenger DR. The effect of valproic acid on blood loss in patients with cerebral palsy. Devel Med Child Neurol 1993;35(Suppl. 69):18.
[41] Tetzlaff JE. Intraoperative defect in hemostasis in a child receiving valproic acid. Can J Anesth 1991;38:222–4.
[42] Henderson RC, Lark RK, Gurka MJ, Worley G, Fung EB, Conaway M, et al. Bone density and metabolism in children and adolescents with moderate to severe cerebral palsy. Pediatrics 2002;110:1–10 e5.
[43] Fehlings D, Switzer L, Stevenson R, Gaebler-Spira D, Dalziel B, Ozel S. AACPDM care pathways: osteoporosis. http://www.aacpdm.org/publications/care-pathways/osteoporosis; 2016.

[44] Bachrach SJ, Kecskemethy HH, Harcke HT, Lark RK, Miller F, Henderson RC. Pamidronate treatment and posttreatment bone density in children with spastic quadriplegic cerebral palsy. J Clin Densitom 2006;9:167–74.

[45] Hagglund G, Pettersson K, Czuba T, Persson-Bunke M, Rodby-Bousquet E. Incidence of scoliosis in cerebral palsy. Acta Orthop 2018;89:443–7.

[46] Thometz JG, Simon SR. Progression of scoliosis after skeletal maturity in institutionalized adults who have cerebral palsy. J Bone Jt Surg 1988;70A:1290–6.

[47] Sanders JO, Khoury JG, Kishan S, Browne RH, Mooney III JF, Arnold KD, et al. Predicting scoliosis progression from skeletal maturity: a simplified classification during adolescence. J Bone Jt Surg Am 2008;90:540–53.

[48] Pettersson K, Rodby-Bousquet E. Prevalence and goal attainment with spinal orthoses for children with cerebral palsy. J Pediatr Rehabil Med 2019;12:197–203.

[49] Miller A, Temple T, Miller F. Impact of orthoses on the rate of scoliosis progression in children with cerebral palsy. J Pediatr Orthop 1996;16:332–5.

[50] Majd ME, Muldowny DS, Holt RT. Natural history of scoliosis in the institutionalized adult cerebral palsy population. Spine 1997;22:1461–6.

[51] Saito N, Ebara S, Ohotsuka K, Kumeta H, Takaoka K. Natural history of scoliosis in spastic cerebral palsy. Lancet 1998;351:1687–92.

[52] Koop SE. Scoliosis in cerebral palsy. Devel Med Child Neurol 2009;51(suppl 4):92–8.

[53] Jain A, Sullivan BT, Shah SA, Samdani AF, Yaszay B, Marks MC, et al. Caregiver perceptions and health-related quality-of-life changes in cerebral palsy patients after spinal arthrodesis. Spine 2018;43:1052–6.

[54] Toovey R, Harvey A, Johnson M, Baker L, Williams K. Outcomes after scoliosis surgery for children with cerebral palsy: a systematic review. Devel Med Child Neurol 2017;59:690–8.

[55] Furhop SK, Keeler KA, Oto M, Miller F, Dabney KW, Bridwell KH, et al. Surgical treatment of scoliosis in non-ambulatory spastic quadriplegic cerebral palsy patients: a matched cohort comparison of unit rod technique and all-pedicle screw constructs. Spine Deform 2013;1:389–94.

[56] Rumalla K, Yarbrough CK, Pugely AJ, Koester L, Dorward IG. Spinal fusion for pediatric neuromuscular scoliosis: national trends and complications in-hospital outcomes. J Neurosurg Spine 2016;25:500–8.

[57] Miyanji F, Nasto LA, Sponseller PD, Hhah SA, Samdani AF, Lonner B, et al. Assessing the risk-benefit ratio of scoliosis surgery in cerebral palsy: surgery is worth it. J Bone Jt Surg Am 2018;100:556–63.

[58] Fasano VA, Broggi G, Barolat-Romana G, Sguazzi A. Surgical treatment of spasticity in cerebral palsy. Child's Brain 1978;4:289–305.

[59] Phillips LH, Park TS. Electrophysiologic studies of selective posterior rhizotomy patients. In: Park TS, Phillips LH, Peacock WJ, editors. Neurosurgery state of the art review: management of spasticity in cerebral palsy and spinal cord injury, 4. Philadelphia: Hanley and Belfus; 1989. p. 459–69.

[60] Georgoulis G, Brinzeu A, Sindou M. Dorsal rhizotomy for children with spastic diplegia of cerebral palsy origin: usefulness of intraoperative monitoring. J Neurosurg: Pediatr 2018;22:89–101.

[61] Hays RM, McLaughlin JF, Bjornson KF, Stephens K, Roberts TS, Price R. Electrophysiological monitoring during selective dorsal rhizotomy, and spasticity and GMFM performance. Devel Med Child Neurol 1998;40:233–8.

[62] Steinbok P, Tidemann AJ, Miller S, Mortenson P, Bowen-Roberts T. Electrophysiologically guided versus non-electrophysiologically guided selective dorsal rhizotomy for spastic cerebral palsy: a comparison of outcomes. Childs Nerv Syst 2009;25:1091–6.

[63] Peacock W, Arens LJ. Selective posterior rhizotomy for the relief of spasticity in cerebral palsy. S Afr Med J 1982;62:119–24.

[64] Peacock W, Arens LJ, Berman B. Cerebral palsy spasticity: selective posterior rhizotomy. Pediatr Neurosci 1987;13:61–6.

[65] Grunt S, Fieggen AG, Vermuelen RJ, Becher JG, Langerak NK. Selection criteria for selective dorsal rhizotomy in children with spastic cerebral palsy: a systematic review of the literature. Devel Med Child Neurol 2014;56:302–12.

[66] Baker R, Graham K. Functional decline in children undergoing selective dorsal rhizotomy after age 10. Devel Med Child Neurol 2011;53:677.

[67] MacWilliams BA, Johnson BA, Shuckra AL, D'Astous JL. Functional decline in children undergoing selective dorsal rhizotomy after age 10. Devel Med Child Neurol 2011;53:717–23.

[68] Wang KK, Munger ME, Chen BP-J, Novacheck TF. Selective dorsal rhizotomy in ambulant children with cerebral palsy. J Child Orthop 2018;12:413–27.

[69] D'Aquino D, Moussa AA, Ammar A, Ingale H, Vloeberghs M. Selective dorsal rhizotomy for the treatment of severe spastic cerebral palsy: efficacy and therapeutic durability in GMFCS grade IV and V children. Acta Neurochir (Wein) 2018;160:811–21.

[70] McLaughlin J, Bjornson K, Temkin N, Steinbok P, Wright V, Roberts T, et al. Selective dorsal rhizotomy: meta-analysis of three randomized controlled trials. Devel Med Child Neurol 2002;44:17–25.

[71] Munger ME, Aldahondo N, Krach LE, Novacheck TF, Schwartz MH. Long-term outcomes after selective dorsal rhizotomy: a retrospective matched cohort study. Devel Med Child Neurol 2017;59:1196–203.

[72] Tedroff K, Lowing K, Astrom E. A prospective cohort study investigating gross motor function, pain, and health-related quality of life 17 years after selective dorsal rhizotomy in cerebral palsy. Devel Med Child Neurol 2015;57:484–90.

[73] Novak I, McIntyre S, Morgan C, Campbell L, Dark L, Morton N, et al. A systematic review of interventions for children with cerebral palsy: state of the evidence. Devel Med Child Neurol 2013;55:885–910.

[74] Park TS, Dobbs MB, Cho J. Evidence supporting selective dorsal rhizotomy for treatment of spastic cerebral palsy. Cureus 2018;10:1–12.

[75] Verbeek BE, Lamberts RP, Fieggen AG, Mankahla N, de Villiers RVP, Botha E, et al. A long-term follow-up study of spinal abnormalities and pain in adults with cerebral palsy and spastic diplegia more than 25 years after selective dorsal rhizotomy. J Neurosurg Spine 2021;34:228–35.

[76] Verbeek BE, Lamberts RP, Fieggen AG, Verkoeijen PPJL, Langerak NG. Daily activities, participation, satisfaction, and functional mobility of adults with cerebral palsy more than 25 years after selective dorsal rhizotomy: a long-term follow-up during adulthood. Disabil Rehabil 2021;43:2191–9.

[77] MacWilliams BA, McMulkin ML, Duffy EA, Mungert ME, Chen BP-J, Novacheck TF, Schwartz MH for the Selective Dorsal Rhizotomy Outcomes Research Team. Long-term effects of spasticity treatment, including selective dorsal rhizotomy, for individuals with cerebral palsy. Devel Med Child Neurol 2021;Nov 10. Available from: https://doi.org/10.1111/dmcn.15075.

[78] Langerak NG, Vaughn CL, Hoffman EB, Figaji AA, Fieggen AG, Peter JC. Incidence of spinal abnormalities in patients with spastic diplegia 17 to 26 years after selective dorsal rhizotomy. Childs Nerv Syst 2009;25:1593–603.

[79] Greene WB, Dietz FR, Goldberg MJ, Gross RN, Miller F, Sussman MD. Rapid progression of hip subluxation in cerebral palsy after selective posterior rhizotomy. J Pediatr Orthop 1991;11:494–7.

[80] Miller SD, Juricic M, Hesketh K, McLean S, Magnuson S, Gasior S, et al. Prevention of hip displacement in children with cerebral palsy: a systematic review. Devel Med Child Neurol 2017;59:1130–8.

[81] Park TS. In support of selective dorsal rhizotomy in cerebral palsy: the strength of clinical experience. Devel Med Child Neurol 2019;62:654–5.

[82] Tedroff K, Hagglund G, Miller F. Long-term effects of selective dorsal rhizotomy in children with cerebral palsy: a systematic review. Devel Med Child Neurol 2019;62:554–62.

[83] Spelman JD, van Manen J. Cerebral palsy and stereotactic neurosurgery: long term results. J Neurol Neurosurg Psychiat 1989;52:23–30.

[84] Gottlieb GL, Myklebust BM, Stefoski D, Groth K, Kroin J, Penn RD. Evaluation of cervical stimulation for chronic treatment of spasticity. Neurology 1985;35:699–704.

[85] Ratusnick DL, Wolfe VL, Penn RD, Schewitz S. Effects on speech of chronic cerebellar stimulation in cerebral palsy. J Neurosurg 1978;48:876–82.

[86] Elkaim LM, Alotaibi NM, Sigal A, Alotaibi HM, Lipsman N, Kalia SK, et al.and the North American Pediatric DBS Collaboration Deep brain stimulation for pediatric dystonia: a meta-analysis with individual participant data. Devel Med Child Neurol 2018;61:49–56.

[87] Koy A, Hellmich M, Pauls KA, Marks W, Lin J-P, Fricke O, et al. Effects of deep brain stimulation in dyskinetic cerebral palsy: a meta-analysis. Mov Disord 2013;28:647–54.

[88] Lumsden DE, Kaminska M, Gimeno H, Tustin K, Baker L, Perides S, et al. Proportion of life lived with dystonia inversely correlates with response to pallidal deep brain stimulation in both primary and secondary childhood dystonia. Devel Med Child Neurol 2013;55:567–74.

[89] Sanger TD. Deep brain stimulation for cerebral palsy: where are we now? Devel Med Child Neurol 2020;62:28–33.

[90] Tsering D, Tochen L, Lavenstein B, Reddy SK, Granader Y, Keating RF, et al. Considerations in deep brain stimulation (DBS) for pediatric secondary dystonia. Childs Nerv Syst 2013;33:631–7.

[91] Kaminska M, Perides S, Lumsden DE, Nakou V, Selway R, Ashkan K, et al. Complications of deep brain stimulation – the challenges and 10 year experience in a large paediatric cohort. Eur J Paediatr Neurol 2017;21:168–75.

[92] Romito LM, Zorzi G, Marras CE, Frnanzini A, Nardocci N, Albanese A. Pallidal stimulation for acquired dystonia due to cerebral palsy: beyond 5 years. Eur J Neurol 2015;22:426–33.

[93] Christian EA, Jin DL, Attenello F, Wen T, Cen S, Mack WJ, et al. Trends in hospitalization of preterm infants with intraventricular hemorrhage and hydrocephalus in the United States, 2000–2010. J Neurosurg Pediatr 2016;17:260–9.

[94] Mazzola CA, Choudhri AF, Autuste KI, Librick Jr. DD, Rogido M, Mitchell L, et al. Pediatric hydrocephalus: systematic literature review and evidence-based guidelines. Part 2: Management of posthemorrhagic hydrocephalus in premature infants. J Neurosurg Pediatr 2014;14 (Suppl):8–23.

[95] Sandoval PV, Rosales PH, Hernandez DGQ, Naranjo EAC, Navarro VG. Intraventricular hemorrhage and post hemorrhagic hydrocephalus in preterm infants: diagnosis, classification, and treatment options. Childs Nerv Syst 2019;35:917–27.

[96] Wellons JC, Shannon CN, Holubkov R, Riva-Cambrin J, Kulkarni AV, Limbrick Jr. DD, et al.for the Hydrocephalus Clinical Research Network Shunting outcomes in posthemorrhagic hydrocephalus: results of a Hydrocephalus Clinical Research Network prospective cohort study. J Neurosurg Pediatr 2017;20:19–29.

[97] Hanak BW, Bonow RH, Harris CA, Browd SR. Cerebrospinal fluid shunting complications in children. Pediatr Neurosurg 2017;52:381–400.

[98] Sherrod BA, Iyer RR, Kestle JRW. Endoscopic third ventriculostomy for pediatric tumor-associated hydrocephalus. J Neurosurg/Neurosurg Focus 2020;48:1–9.

[99] Pande A, Lamba N, Mammi M, Gebrehiwet P, Trenary A, Doucette J, et al. Endoscopic third ventriculostomy versus ventriculoperitoneal shunt in pediatric and adult population: a systematic review and meta-analysis. Neurosurg Rev 2021;44:1227–41.

[100] Kulkarni AV, Sgouros S, Constantini S, Investigators IIHS. International infant hydrocephalus study: initial results of a prospective, multicenter comparison of endoscopic third ventriculostomy (ETV) and shunt for infant hydrocephalus. Childs Nerv Syst 2016;32:1039–48.

[101] Kulkarni AV, Drake JM, Kestle JRW, Mallucci CL, Sgouros S, Constantini, S.for the Canadian Pediatric Neurosurgery Study Group. Endoscopic third ventriculostomy vs cerebrospinal fluid shunt in the treatment of hydrocephalus in children: a propensity score-adjusted analysis. Neurosurgery 2010;67:588–93.

[102] Warf BC, Campbell JW. Combined endoscopic third ventriculostomy and choroid plexus cauterization as primary treatment of hydrocephalus for infants with myelomeningocele: long-term results of a prospective intent-to-treat study in 115 East African infants. J Neurosurg Pediatr 2008;2:310–16.

[103] Warf B, Ondoma S, Kulkarni A, Donnelly R, Ampiere M, Akona J, et al. Neurocognitive outcome and ventricular volume in children with myelomeningocele treated for hydrocephalus in Uganda. J Neurosurg Pediatr 2009;4:564–70.

Chapter 40

Innovative therapies

Mariam Hull[1], Kristen Fisher[1], Thomas D. Challman[2,3] and Mered Parnes[1]

[1]*Division of Pediatric Neurology and Developmental Neuroscience, Texas Children's Hospital, Baylor College of Medicine, Houston, TX, United States,* [2]*Geisinger Autism & Developmental Medicine Institute, Danville, PA, United States,* [3]*Geisinger Health System and Geisinger Commonwealth School of Medicine, Lewisburg, PA, United States*

Introduction

Children's neurologic and developmental outcomes can be adversely affected by genetic and acquired conditions. In recent years, there have been a number of advances in the development and implementation of innovative technologies to reduce morbidity and maximize function. Deep brain stimulation surgery, immunomodulatory therapies, stem cell therapies, and genetic therapies provide examples of both the promise and potential pitfalls that can be seen in novel and innovative treatment approaches for children with neurodevelopmental disorders.

Deep brain stimulation

In the 1980s, Benabid et al. demonstrated the utility of deep brain stimulation (DBS) in the treatment of Parkinson disease [1]. In 1997, its use was approved by the FDA for use in Parkinson disease and essential tremor, and since that time, it has expanded to include Tourette syndrome and dystonia (the latter through a humanitarian exemption) [2,3]. Use of DBS is also under investigation in adults for the treatment of depression, obsessive-compulsive disorder, obesity, chronic pain, and addiction [2,4,5].

The device delivers electrical current from an implantable pulse generator (IPG) by means of electrodes inserted into one or more targets in the brain. Target selection depends on the clinical indication and disease pathophysiology. An extension wire connects the electrode to the IPG, which is placed beneath the skin in the chest or abdomen. The IPG itself is a programmable battery-powered device similar to a pacemaker, which allows the clinician to adjust a number of stimulation parameters to modulate the device's effects with the goal of maximizing clinical efficacy and minimizing adverse effects. The clinician can select the specific area within the target to stimulate by choosing which contacts along the electrode to activate and by adjusting stimulation parameters, including amplitude, frequency, and pulse width (Fig. 40.1).

Proposed mechanisms

The underlying mechanism of action of DBS is not yet fully understood, and research has generated multiple hypotheses, some of which conflict [6]. Initially, it was supposed that DBS mimicked lesioning the target, as in an ablative procedure, but in a reversible and modulable way [4]. Electrical current has been shown to depolarize axons and generate action potentials by means of opening voltage-gated sodium channels. Downstream effects would then be mediated depending on whether the stimulated neuron is inhibitory or excitatory. However, studies have since shown that high-frequency versus low-frequency stimulation of the same target can produce either overall inhibitory or excitatory effects [4,6]. Currently, the leading hypotheses include (1) disruption of abnormal signals, (2) modulation of brain networks, (3) improvement of the regularity of signals, and (4) dissociation of input and output signals [7]. However, these proposed mechanisms are likely not exclusive to one another, and there is significant overlap [8].

Indications

Given the longstanding success observed in the treatment of movement disorders in adults, the technology was then used in the pediatric population based largely on adult data. Pediatric DBS studies are sparse and have often been limited by sample size, and therefore outcomes are not always as well understood. Despite this, compelling evidence for

Capute and Accardo's Neurodevelopmental Disabilities in Infancy and Childhood. DOI: https://doi.org/10.1016/B978-0-12-824060-1.00028-6

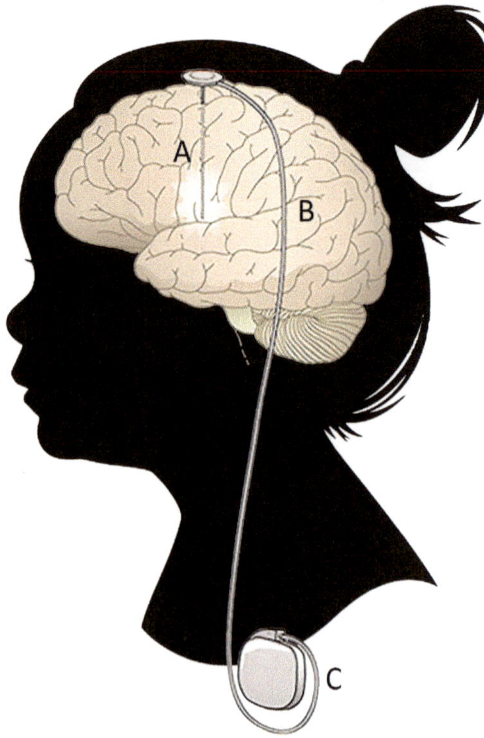

FIGURE 40.1 Deep brain stimulation. (A) Lead, (B) extension, (C) implantable pulse generator (IPG).

the use of DBS for the treatment of dystonia, Tourette syndrome, and juvenile parkinsonism is available [9]. While essential tremor does have an FDA-approved indication in adults, to date there is less robust evidence for its use in pediatric tremor, although this is an evolving subject.

Dystonia

The most common therapeutic application of DBS in the pediatric population is dystonia [9]. Dystonia is defined as sustained or intermittent muscle contractions causing abnormal, often repetitive movements, postures, or both, which are often patterned, twisting, or tremulous. It worsens with voluntary action and is often associated with overflow muscle activation [10].

Stimulation of the globus pallidus interna (GPi) is well-established as an effective treatment for medically-refractory dystonia in adults [11]. The subthalamic nucleus (STN) has also been shown to be an effective target for management of dystonia, and other targets are currently being explored, such as the ventral intermediate nucleus of the thalamus (VIM) [12].

No clear consensus data regarding appropriate patient selection in pediatrics has been established; however, younger patients with shorter disease courses tend to have more clinical benefit, as effective treatment of dystonia early on can reduce the development of musculoskeletal deformities. In general, patients who are considered for this intervention are medically refractory and have significant functional impairment from their dystonia [12–14].

Favorable outcomes in pediatric patients with several genetic dystonias have been well documented [9,15]. Furthermore, DBS in genetic dystonia produces better functional outcomes and symptomatic improvement than those with acquired forms of dystonia, such as those seen with anoxic brain injury and some cerebral palsy [9,15–17].

In a systematic review of pediatric cases of dystonia treated with DBS, dystonia rating scales improved the most among those with genetic etiologies, with *DYT-TOR1A* being the most responsive [18]. Other forms of genetic dystonia that have been shown to respond well to DBS (particularly GPi-DBS) include *DYT-THAP1*, *DYT-SGCE*, and *DYT-KMT2B* [8].

Pathogenic variants in *GNAO1* cause a syndrome of hypotonia, developmental delay, and a progressive movement disorder with recurrent episodes of severe chorea, which can be life-threatening. GPi-DBS has been shown to be particularly effective in reducing the frequency and severity of these events [19–21].

Tourette syndrome

Tourette syndrome (TS) is a neurobehavioral disorder characterized by motor and phonic tics. Tics are brief, patterned, repetitive, purposeless movements and sounds that wax and wane and change over time [22]. Onset is typically between 4 and 7 years of age. On average, tics tend to reach peak frequency and severity at roughly 10 to 12 years of age, and for many, they begin to decrease in late adolescence or early adulthood, though it is common for some tics to remain [23]. Comorbid obsessive-compulsive disorder and attention deficit hyperactivity disorder are common [23,24].

Targets vary and include GPi, centromedian-parafasicular nuclear complex of the thalamus (CM-Pf), STN, ventral capsule/ventral striatum, and nucleus accumbens; more data is required in order to further delineate the most ideal target in any given patient [7].

Patient selection in pediatric patients is more complicated than in adults, in particular due to the tendency for tics to decrease with age in many patients. Important factors to consider for patients with TS are that the tics are severe and impairing, that they are refractory to multiple classes of medications, that comorbid psychiatric disorders are stable, and that no substance abuse or suicidal ideation has been present in the six months leading up to the surgical evaluation [3,25].

DBS in TS has been associated with improvement in tics as well as obsessive-compulsive symptoms, depending on the target [3,25–27]. However, reports of improvement have largely been found in adults, with few published reports of adolescents undergoing implantation [27,28]. Among these adolescents with DBS, long-term outcomes are not yet known, and it is unclear whether these patients had the same improvement in symptoms as might be expected in the natural course of TS [7,11].

Juvenile Parkinsonism

Parkinson disease (PD) is a neurodegenerative process whose classic clinical features include rest tremor, bradykinesia, rigidity, and postural instability [29]. Typically, the age of onset is around 60, while young-onset PD symptoms arise between 21 and 40 years, and juvenile PD occurs under 21 years of age [30]. Juvenile PD is often genetic in origin, with implicated genes including *PINK1, PRKN*, and *DJ1*, among others [31]. Dopamine replacement is the standard treatment; however, adverse effects include dyskinesias and motor fluctuations. These often occur within years of treatment and can be quite limiting, especially in the genetic forms of PD. Although use of DBS has not been not reported for the management of juvenile PD, adults with the same genetic causes of juvenile PD have responded well to STN-DBS, which could potentially expand its indication in the pediatric population of these same genetic causes [11].

Immunomodulatory therapies

Immunomodulators are treatments that act to stimulate or suppress the immune system and include a wide range of medications, from oral corticosteroids to more novel targeted agents. Immunotherapies have been used for many years in neurologic conditions, and with newer treatments have come improved clinical results. The most common applications of these treatments are those used in the fields of rheumatology, immunology, neuroimmunology, and neuroinflammatory conditions, with the aim of controlling an inflammatory process to treat a disease and to prevent disease recurrence. Additionally, immunotherapies can be used to aid in the treatment of genetic conditions that are related to immune dysregulation and have neurologic sequelae, such as auto-inflammatory conditions and interferonopathies.

Commonly used immunomodulating treatments

Steroids

Corticosteroids have a role in immunomodulation by acting to treat ongoing inflammation. In neurology and neurodevelopment, they are used across a spectrum of conditions and various administration methods. High-dose intravenous (IV) steroids have been shown to be beneficial in the treatment of acute presentations of neuroinflammatory and demyelinating conditions, including acute disseminated encephalomyelitis (ADEM), multiple sclerosis (MS), neuromyelitis optica (NMO), and myelin oligodendrocyte glycoprotein antibody-associated disease (MOGAD).

High-dose steroids have also been shown to be beneficial in patients with epilepsy syndromes, particularly electrical status epilepticus in sleep (ESES) and infantile spasms. While there has been no specific mechanism or pathway identified in these patient populations, there is evidence of immune dysregulation with the presence of cytokine abnormalities in serum and cerebrospinal fluid (CSF) [32,33]. Immune dysregulation may further be supported by responses to steroid treatments. In ESES, patients demonstrate epileptiform activity that is significantly activated by sleep. Clinical syndromes associated with ESES include continuous spike wave in slow-wave sleep (CSWS), which commonly presents with global regression, and Landau-Kleffner syndrome (LKS), which commonly presents with an acquired auditory agnosia. In both CSWS and LKS, high-dose steroids, among other treatments, have been shown to improve cognitive performance [34–36]. Infantile spasms, or West syndrome, are a triad of epileptic spasms, developmental delay or regression, and findings of hypsarrhythmia on electroencephalogram. The development of epileptic spasms can be associated with underlying genetic disorders, such as tuberous sclerosis, but can also be of unclear etiology (cryptogenic). In patients with cryptogenic infantile spasms, the use of ACTH and oral corticosteroids has demonstrated improved development [37]. Further studies are needed to help better quantify immune dysfunction in these patient populations, which may further drive the opportunities for more effective treatment options.

Intravenous Immunoglobulin

Intravenous immunoglobulin (IVIG), pooled IgG from the plasma of individuals, has been shown to be beneficial in the treatment of various neurologic and autoimmune conditions. IVIG can act as an activator of innate immune cells or complement activation. Potential mechanisms include antibody neutralization, increased antibody clearance, suppression of immunoglobulin production, and modulation of antigen-presenting cell activation [38]. In pediatric neurology, IVIG has been shown to be effective for a wide array of autoimmune and immune-mediated conditions, including pediatric onset multiple sclerosis, and has demonstrated positive effects on the stabilization of disease or reducing the duration of recovery.

Rituximab

Rituximab is an anti-CD20 monoclonal antibody that results in B-cell depletion and can be used to treat various autoimmune and neuroinflammatory conditions. B-cells play a role in the production of antibodies, antigen presentation, and regulation of the immune response through the production of proinflammatory and antiinflammatory cytokines. Given this role, B-cells have been implicated in autoimmune neurologic conditions including myasthenia gravis, NMO, antibody-positive autoimmune encephalitis, and MS. While there have not been randomized clinical trials of rituximab to treat these conditions, case reports and retrospective studies have demonstrated efficacy with regard to reducing symptoms and decreasing time to recovery.

Common clinical diagnoses requiring immunomodulating treatments

Multiple sclerosis

MS is an autoimmune inflammatory disease that affects the central nervous system, leading to demyelination of the brain and spinal cord. Pediatric onset multiple sclerosis (POMS) makes up approximately 5%–10% of all patients diagnosed with multiple sclerosis [39]. Patients with POMS tend to have a comparatively early age of disability accumulation and increased cognitive impairment, thought to be due largely to the POMS patients having a more inflammatory disease course at the time of diagnosis, with increased relapses compared to adult-onset MS.

The treatment approach to patients with MS has varied over the years. In the past, there was an escalation approach, starting with less vigorous treatment and escalating if relapse were to occur. More recent data has supported the use of a de-escalation approach, as patients treated early with more highly effective therapies have been shown to have a decreased degree of cognitive impairment compared to those who remained on less effective treatments [40].

Currently, there is only one FDA-approved medication, fingolimod, for POMS. Fingolimod binds to the sphingosine-1-phosphate receptor, causing lymphocyte sequestration and preventing activated lymphocytes from crossing into the CNS. In clinical trials for POMS, it demonstrated an 82% reduction in annualized relapse rate in comparison to those patients treated with interferon beta-1a, and additionally, these patients had lower disability scores over time [41]. While fingolimod remains the only approved treatment for affected children, there have been case studies on the use of rituximab, interferons, glatiramer acetate, natalizumab, and cyclophosphamide. Clinical trials are currently ongoing for the use of additional therapies in POMS.

Neuromyelitis optica

Neuromyelitis optica (NMO) is an autoimmune neurologic disorder that is associated with aquaporin-4 antibodies. Aquaporin 4 has been demonstrated to be preferentially expressed in the retina, optic nerve, hypothalamus, cerebellum, periventricular and periaqueductal regions, and spinal cord. Aquaporin-4 antibodies mediate complement-dependent necrosis of astrocytes and lead to demyelination and neuronal loss, preferentially in the areas where aquaporin-4 is more prevalent. This can lead to characteristic demyelinating syndromes, most commonly optic neuritis and transverse myelitis. Patients with NMO have been shown to have significant residual disability following relapses. Optic neuritis related to NMO can be severe and have poor recovery, leading to significant visual impairment and the need for accommodations, especially in school-age children. Additionally, myelitis in NMO can lead to significant motor impairment due to the extent of spinal cord involvement, which leads to early disability accumulation.

Aggressive treatment at acute presentation is warranted for patients with suspected NMO, including high-dose IV steroids and the use of plasmapheresis. The use of plasmapheresis has been shown to help improve visual outcomes [42]. IVIG has been shown to be beneficial for acute presentations [43]. After acute treatment, patients require long-term immunomodulatory treatment to help prevent further relapses. In pediatric NMO, rituximab is commonly used due to its effectiveness. Additional case studies have described the use of mycophenolate mofetil, azathioprine, and tocilizumab in the treatment of pediatric NMO. Clinical trials are currently ongoing using FDA-approved treatments for NMO and expanding into pediatric NMO.

Autoimmune encephalitis

Autoimmune encephalitis is a collection of antibody-mediated disorders that result in neuroinflammation. Neurologic symptoms related to autoimmune encephalitis can vary, including altered mental status, seizures, movement disorders, developmental regression, and neuropsychiatric symptoms. These disorders typically have a subacute onset, and many patients require inpatient hospital care to varying degrees. There are commonly found antibodies linked with autoimmune encephalitis; in pediatrics, most commonly anti-NMDA receptor antibodies.

Treatment of autoimmune encephalitis includes IV steroids, IVIG, and, in more severe cases, the use of plasmapheresis. More recently, the use of rituximab has been introduced, which has been shown to be associated with improved outcomes [44,45]. The data on rituximab in pediatric autoimmune encephalitis is limited, but it is recommended as a second-line treatment for anti-NMDA receptor encephalitis in patients with refractory symptoms [46].

Opsoclonus myoclonus ataxia syndrome

Opsoclonus myoclonus ataxia syndrome (OMA) is an immune-mediated encephalopathy associated with a triad of neurologic symptoms including opsoclonus, myoclonus, and ataxia, although often not all of these symptoms will be present. It can present as a paraneoplastic condition as it is associated with neuroblastoma, but also can be cryptogenic. Patients typically have varying degrees of neurologic symptoms at presentation that progress, and there is additionally a strong association with developmental delay or regression and with behavioral symptoms. Patients with OMA typically undergo immunomodulating treatments with the aim of decreasing ongoing neuroinflammation, including IV steroids, IVIG, and rituximab. In more severe cases, the use of ACTH and cyclophosphamide is warranted. Patients with OMA have been followed long-term, and neuropsychological evaluation often shows speech and language dysfunction, anxiety, and ADHD. The use of immunomodulating treatments has been shown to improve long-term outcomes, but most patients continue to have ongoing neurocognitive and behavioral deficits [47].

Febrile infection-related epilepsy syndrome

Febrile infection-related epilepsy syndrome (FIRES) is an immune-mediated disorder with an acute onset of refractory epilepsy and status epilepticus associated with febrile illness. Patients typically require ICU care due to seizure severity and the need for anesthetic treatments for seizure control. Following the acute phase, children have ongoing refractory epilepsy and varying degrees of neuropsychological impairment. Children with FIRES have been found to have elevated pro-inflammatory and epileptogenic cytokines present in the CSF. Due to this, various immunotherapies have been trialed for patients who meet the diagnostic criteria for FIRES, including IV steroids, IVIG, plasmapheresis, and rituximab, with little benefit. More recently, the use of anakinra, an interleukin-1 receptor antagonist, has been utilized in patients with FIRES. In case reports, the use of anakinra has demonstrated improved seizure control, which in turn could potentially improve the neuropsychological outcomes of these patients, though more long-term follow-up is needed [48,49].

Immunomodulating treatments for genetic disorders with immune mediated pathologies

As genetic testing, including exome and genome sequencing, becomes more readily available, more novel conditions with neurologic and developmental sequelae are identified in patients. Included in this are immune-mediated conditions that can have a neurologic and developmental phenotype, such as autoinflammatory disorders and interferonopathies. CNS findings in such disorders may include stroke, inflammatory brain and spinal cord lesions, and sensorineural hearing loss.

Deficiency of adenosine deaminase 2 (DADA2) is an autosomal recessive autoinflammatory monogenic vasculitis syndrome that can lead to pediatric stroke. Due to cerebral vascular disease, patients can experience focal neurologic deficits and cognitive deficits related to areas of infarct. These patients have been shown to have benefits from the use of tumor necrosis factor (TNF) inhibitors, including prevention of further strokes and preservation of their level of function [50].

Aicardi-Goutieres syndrome (AGS) is a type 1 interferonopathy presenting in infancy characterized by CSF lymphocytosis and elevated interferon-alpha. Neurologic features include spasticity, dystonia, seizures, cortical blindness, and microcephaly, and children experience progressive neurodegeneration. Typical imaging findings include cerebral calcifications, leukoencephalopathy, and cerebral atrophy [51]. Treatment for AGS is generally supportive care, but more recently, there has been the development of novel therapies, including JAK kinase inhibitors, which prevent interferon activation. A current open-label study with baricitinib in patients with AGS suggests improvement in developmental outcomes in a small cohort of patients, and further studies are ongoing [52].

Stem cell therapy

Stem cells are the building blocks of the body. They are progenitor cells that can develop into the body's various tissues, including muscle, fat, bone, blood and immune cells, skin, and neural tissues [53]. There is considerable interest in investigating whether stem cells could have a therapeutic role in a range of diseases [54]. There are multiple types of stem cells, including hematopoietic, mesenchymal, neural, and embryonic. Hematopoietic stem cells (HSCs) are found in bone marrow and umbilical cord blood. Mesenchymal stem cells (MSC) can be derived from a variety of tissues, including bone marrow, adipose tissue, umbilical cord blood, and other birth tissues. Neural stem cells are found in the fetal brain, and embryonic stem cells originate in the preimplantation blastocyst.

HSC and MSC are the stem cell types most widely used in clinical research and therapeutic applications. The most common uses of stem cells are in conditions affecting the hematopoietic system that are inherited, acquired, or that result from myeloablative treatment. A bone marrow transplant, in the context of the treatment of a hematologic malignancy, for instance, is a familiar example of stem cell use. HSC transplantation is also an established approach to mitigating the effects of certain inborn errors of metabolism, including Hurler syndrome, metachromatic leukodystrophy, and X-linked leukodystrophy [55]. HSC and MSC have traditionally been obtained from bone marrow, although in recent years, umbilical cord blood has become a more popular source for these types of cells. Stem cells can be delivered to a recipient in either an autologous (using cells derived from the same individual) or allogeneic (using cells derived from a different individual) fashion, and there is an international network of umbilical cord blood banks that can provide these cells for use. Of note, a distinction should be made between stem cell *transplantation* and stem cell *therapy*. The intent of stem cell transplantation is for cells to engraft in the host and create new tissue, whereas in stem cell therapy, the goal is not cell engraftment but rather other direct or indirect effects of the stem cells on one or more organ systems.

Because of the potential that stem cells have in contributing to the regeneration of tissues, there has been interest in investigating whether stem cells could have a beneficial effect in neurologic diseases, including spinal cord injury, Parkinson's disease, traumatic brain injury, stroke, retinal disorders, cerebral palsy (CP), and autism spectrum disorder (ASD) [56–58]. While some studies have used stem cells for the primary therapeutic goal of engraftment of stem cells and the generation of new neural tissue, other studies have used an approach that hypothesizes that the stem cells act to secrete substances that have a positive effect on the target tissue via basic processes such as cell proliferation, angiogenesis, apoptosis, or oxidative stress. The use of stem cells is not without risk. Hypersensitivity and infusion reactions, undesirable immunological side effects, tumor development, and infection are all potential adverse effects [59,60].

What is the evidence supporting stem cell use in neurodevelopmental disorders? There is a large body of research investigating the effects of stem cells in animal (most often rodent) models of brain injury [61]. Varying routes of administration have been used in these studies, including intraventricular, intraperitoneal, intranasal, and intravenous. A number of these studies have shown positive effects, including improved functional outcome, increased blood vessel formation, elevated growth factor levels, improved cell growth, and reductions in areas of brain injury. As in all animal studies, a primary question is how generalizable these findings are to humans—particularly to children with neurodevelopmental disorders.

Stem cell therapy in cerebral palsy

The ClinicalTrials.gov website recently listed 21 completed stem cell therapy trials in children with CP, of which eight had randomized designs. Seven additional studies were in progress [62]. A 2021 review summarized the current state of the research [63]. At that time, a total of 2247 children had participated in trials of stem cell therapy for CP, with roughly half of these (1088) participating in early phase trials (case series or Phase 1). MSC and umbilical cord blood were the types of cells most commonly used in these studies. One randomized, controlled trial studied 63 children aged 1–6 years with CP [64]. Participants were allocated to receive a single dose of autologous cord blood or placebo at baseline, with crossover to the other treatment at 12 months. Several measures were obtained at baseline, 12, and 24 months, including the Gross Motor Function Measure-66 (GMFM-66), Peabody Developmental Motor Scales-2, and brain MRI with whole brain connectome analysis. No difference in mean change in GMFM-66 was seen between the treatment and placebo groups at 12 months. On post hoc analysis, however, there was some evidence of a potential dose effect, with individuals receiving a higher stem cell dose demonstrating greater improvements on the GMFM-66 as well as the MRI measures of brain connectivity. Another study from the same group enrolled 90 children aged 2–4 years with hypertonic CP [65]. These participants were randomized to one of three arms: a single dose of allogeneic cord blood, three doses of human cord-tissue-derived mesenchymal stromal cells, or a natural history group in which a single dose of umbilical cord blood was given at 12 months. At 6 months, there was no difference in the GMFM-66 score between the three groups. At 12 months, the mean GMFM-66 score of the mesenchymal stromal cell group was 1.4 points higher than the Natural History group ($P = .27$), and the mean GMFM-66 score of the cord blood group was 3.3 points higher than the Natural History group ($P = .02$). Several systematic reviews have concluded that there may be at least short-term positive effects of stem cells on motor function in children with CP [66–69], although the magnitude of this benefit appears to be modest. In aggregate, published studies suggest that there may be some degree of beneficial effect of stem cells in children with CP, although at present, stem cell therapy is not yet ready for clinical use. Key information regarding this therapy remains incomplete, including an understanding of who are the optimal candidates for its use, the most effective cells, dose, and mode of delivery, and the best methods for measuring functional improvement [70].

Stem cell therapy in autism spectrum disorder

Stem cell therapy has also garnered attention for children with ASD, although (as can often be the case with novel therapies), the desire of families to use this approach can sometimes outpace the current level of scientific support. The research base for stem cell therapy in ASD is even weaker than that for cerebral palsy. As of February 2022, the ClinicalTrials.gov website listed nine completed trials, with only three having randomized designs. Seven additional studies were underway [62]. In one small controlled trial, 29 children with ASD aged 2 to 6 years were randomized to receive autologous cord blood or placebo [71]. Participants were evaluated at baseline, 12, and 24 weeks, then crossed over to the other treatment. Assessments included safety, the Expressive One Word Picture Vocabulary Test, the Receptive One Word Picture Vocabulary Test, Clinical Global Impression, Stanford-Binet Fluid Reasoning and Knowledge, and the Vineland Adaptive Behavior and Socialization Subscales. No statistically significant differences for any endpoints were found, and there were no serious adverse events. In a larger randomized trial, 180 children with ASD aged 2 to 7 years were randomized to receive either a single dose of autologous cord blood, allogeneic cord blood, or placebo [72]. The primary outcome was the Vineland Adaptive Behavior Scales-Socialization Domain, and secondary outcomes included the PDD Behavior Inventory, Expressive One-Word Picture Vocabulary Test, and Clinical Global Impression Scale. There was no evidence that cord blood was associated with improvements in the primary or secondary outcome measures (and a large expectancy effect was noted for these measures). A sub-analysis of participants with a nonverbal IQ ≥ 70 suggested the possibility of some benefit among these higher functioning individuals in the allogeneic group, prompting the researchers to conclude that additional study was warranted. When all the available data is considered, however, at the present time, there is no convincing evidence that stem cell therapy is an effective treatment for children with ASD [73].

Stem cell therapy has become a potentially lucrative business, with hundreds of stem cell clinics operating in the United States and many more elsewhere in the world. There has been criticism of some of these clinics for offering stem cell therapy for unproven indications [74]. The U.S. Food and Drug Administration (FDA) has increased their oversight of stem cell clinics and, since 2017, has stepped up regulatory activities around stem cell therapy [60,75]. It is important for consumers to recognize that the FDA plays a role in stem cell treatment oversight and to become knowledgeable about whether there is sufficient evidence to support the use of stem cell therapy for specific conditions. While it is always important to be open to the possibilities that novel therapies such as stem cell therapy might offer to children with neurodevelopmental disorders [76], rigorous scientific evidence needs to be developed to justify any claims of therapeutic benefit.

Genetic therapies

Genetic therapies are another group of novel treatments that offer considerable potential for the treatment of neurodevelopmental disabilities. The basic processes of gene expression in cells include the transcription of genes into RNA, the splicing of RNA into messenger RNA (mRNA), and the translation of mRNA into protein. Genetic therapies operate at different points along this pathway, with the aim being to modify the expression of a gene to achieve a particular therapeutic goal. Genetic therapies utilize numerous strategies, including silencing a disease-causing gene, replacing a dysfunctional gene with a functional copy, directly editing the sequence of a particular gene, or altering the translation of mRNA. This group of treatments has its own set of issues to consider, including the duration and magnitude of the benefit as well as the cost.

Genetic therapies are being used or studied in a growing list of conditions, including various types of cancer, cardiovascular disease, infectious diseases, immunological disorders, and other genetic and metabolic disorders [77]. These approaches can achieve either a transient or permanent change in the gene expression of cells. Complementary DNA (cDNA) or mRNA delivery (in which these molecules are delivered directly into a cell to increase the formation of a particular protein) and antisense oligonucleotides (ASOs) are two methods that can be used to temporarily alter gene expression. ASOs are small segments of synthetic DNA designed to bind to a specific sequence on a mRNA molecule. ASOs can be used in several ways to alter mRNA translation, including binding to an unwanted (pathogenic) mRNA, causing it to be degraded, or binding to a specific area on an abnormal mRNA, allowing it to be spliced and translated correctly [78]. Two strategies that can permanently alter the genome of a cell include direct gene replacement and CRISPR ("clustered regularly interspaced short palindromic repeats"). CRISPR is a genetic editing technology that can be used to directly manipulate the genome to either correct or "knock out" a disease-causing gene [79]. There are a number of ways in which genes or oligonucleotides can be delivered into cells, including via liposomes (fatty acid vesicles that can enclose the desired genetic material) and viral vectors such as adeno-associated viruses and lentiviruses [80]. Duchenne muscular dystrophy and spinal muscular atrophy are two disorders relevant to the field of neurodevelopmental disabilities in which genetic therapies are currently being utilized.

Genetic therapies in Duchenne muscular dystrophy

ASOs are currently being used in the treatment of Duchenne muscular dystrophy, a disorder caused by mutations in the *DMD* gene and subsequent decreased levels of the protein dystrophin. Specific ASOs have been developed to treat children with particular types of *DMD* mutations. These ASOs are delivered into muscle cells and can bind to the abnormal *DMD* mRNA, permitting it to be translated into a shortened but still functional dystrophin—an approach referred to as "exon-skipping" [81]. There are several FDA-approved ASOs for the treatment of Duchenne muscular dystrophy, including eteplirsen, casimersen, viltolarsen, and golodirsen. Each of these ASOs is specific to individual types of *DMD* mutations, and so not all individuals with Duchenne muscular dystrophy are eligible to receive these treatments. Currently available ASOs have a modest impact on dystrophin levels but do have clinically meaningful benefits [82]. Gene therapy for Duchenne muscular dystrophy using the adeno-associated viral vector (AAV) is also under investigation [83].

Genetic therapies in spinal muscular atrophy

Genetic therapies are also being used in spinal muscular atrophy (SMA). SMA is characterized by progressive muscle weakness and atrophy caused by mutations in the gene *SMN1*. Nusinersen is an ASO that is FDA-approved for the treatment of SMA [84]. This ASO increases the function of another gene, *SMN2*, ameliorating the effects of the dysfunctional *SMN1* gene. Nusinersen is effective in improving motor function in early and later-onset SMA [85–88]. It is administered intrathecally, and multiple on-going doses are required. Another genetic therapy approach in SMA uses the AAV vector to deliver a functional version of the *SMN1* gene to motor neurons. Onasemnogene abeparvovec is FDA-approved for the treatment of SMA in children aged 0–2 years and consists of an AAV capsid containing an *SMN1* transgene [88]. This viral vector transgene is given intravenously as a one-time dose. It is taken up into motor neurons, where it can be transcribed into RNA—ultimately increasing the amount of SMN1 protein in the cells. Clinically meaningful benefits for individuals with SMA have been demonstrated [89–92].

Studies of genetic therapies are also underway in other disorders relevant to the field of neurodevelopmental disabilities, including Fragile X syndrome, tuberous sclerosis, Rett syndrome, Tay-Sachs disease, and Angelman syndrome [93]. Several additional issues warrant consideration. The magnitude and duration of the clinical benefits of genetic therapies remain incompletely delineated. Cost is also an area of concern. Genetic therapies are among the most expensive medication treatments ever developed. Antisense oligonucleotide therapy for Duchenne muscular dystrophy can cost $1,000,000 or more per patient per year [94]. Viral vector therapies are also quite costly—for example, onasemnogene abeparvovec for SMA currently costs approximately $2,100,000 per dose, prompting early efforts to analyze cost-effectiveness [95]. Nevertheless, genetic therapies do show considerable potential to benefit individuals with disorders that have been challenging to treat, and we can look forward to additional progress toward bringing these approaches into more widespread clinical use.

Summary

Caregivers of children with developmental disabilities are highly motivated to seek out therapies they hope will be beneficial. Clinicians should be aware of the types of therapies that are being promoted in their communities and need to understand the evidence base for and against various treatments to give families the most accurate and helpful advice. As these therapies grow in application, longitudinal studies are needed to establish their long-term effects on neurodevelopmental outcomes.

References

[1] Miocinovic S, Somayajula S, Chitnis S, Vitek JL. History, applications, and mechanisms of deep brain stimulation. JAMA Neurol 2013;70(2):163–71. Available from: https://doi.org/10.1001/2013.jamaneurol.45.

[2] Dougherty D.D. Deep brain stimulation clinical applications. https://doi.org/10.1016/j.psc.2018.04.004.

[3] Deeb W, Malaty I. Deep brain stimulation for tourette syndrome: potential role in the pediatric population. https://doi.org/10.1177/0883073819872620.

[4] Aum DJ, Tierney TS. Deep Brain Stimul Found Future Trends 2018;23.

[5] Engracia Valenti V, Tisch S, Vincent S, et al. Deep brain stimulation initiative: toward innovative technology, new disease indications, and approaches to current and future clinical challenges in neuromodulation therapy. Front Neurol http://www.frontiersin.org 2021;1:597451. Available from: https://doi.org/10.3389/fneur.2020.597451.

[6] Montgomery EB, Gale JT. Mechanisms of action of deep brain stimulation (DBS). Neurosci Biobehav Rev 2008;32(3):388–407. Available from: https://doi.org/10.1016/j.neubiorev.2007.06.003.

[7] Deeb W, Malaty I. Deep brain stimulation for Tourette syndrome: potential role in the pediatric population. J Child Neurol 2020;35(2):155–65. Available from: https://doi.org/10.1177/0883073819872620.

[8] Larsh T, Wu SW, Vadivelu S, Grant GA, O'malley JA. Deep brain stimulation for pediatric dystonia. SemPediatric Neurol 2021;38:100896. Available from: https://doi.org/10.1016/j.spen.2021.100896.

[9] Difrancesco MF, Halpern CH, Hurtig HH, Baltuch GH, Heuer GG. Pediatric indications for deep brain stimulation. https://doi.org/10.1007/s00381-012-1861-2.

[10] Albanese A, Bhatia K, Bressman SB, et al. Phenomenology and classification of dystonia: a consensus update. Mov Disord 2013;28(7):863–73. Available from: https://doi.org/10.1002/mds.25475.

[11] Difrancesco MF, Halpern CH, Hurtig HH, Baltuch GH, Heuer GG. Pediatric indications for deep brain stimulation. Childs Nerv Syst 2012;28:1701–14. Available from: https://doi.org/10.1007/s00381-012-1861-2.

[12] Larsh T, Wu SW, Vadivelu S, Grant GA, O'malley JA. Deep brain stimulation for pediatric dystonia. Published online 2021. https://doi.org/10.1016/j.spen.2021.100896.

[13] Apetauerova D, Schirmer CM, Shils JL, Zani J, Arle JE. Successful bilateral deep brain stimulation of the globus pallidus internus for persistent status dystonicus and generalized chorea: report of 2 cases. J Neurosurg 2010;113(3):634–8. Available from: https://doi.org/10.3171/2010.1.JNS091127.

[14] Allen NM, Lin JP, Lynch T, King MD. Status dystonicus: a practice guide. Dev Med Child Neurol 2014;56(2):105–12. Available from: https://doi.org/10.1111/dmcn.12339.

[15] Hale AT, Monsour MA, Rolston JD, Naftel RP, Englot DJ. Deep brain stimulation in pediatric dystonia: a systematic review. https://doi.org/10.1007/s10143-018-1047-9.

[16] Haridas A, Tagliati M, Osborn I, et al. Pallidal deep brain stimulation for primary dystonia in children. Neurosurgery 2011;68(3):738–43. Available from: https://doi.org/10.1227/NEU.0b013e3182077396.

[17] Borggraefe I, Mehrkens JH, Telegravciska M, Berweck S, Bö Tzel CK, Heinen F. Bilateral pallidal stimulation in children and adolescents with primary generalized dystonia – report of six patients and literature-based analysis of predictive outcomes variables. https://doi.org/10.1016/j.braindev.2009.03.010.

[18] Hale AT, Monsour MA, Rolston JD, Naftel RP, Englot DJ. Deep brain stimulation in pediatric dystonia: a systematic review. Neurosurg Rev 2020;43:873–80. Available from: https://doi.org/10.1007/s10143-018-1047-9.

[19] Koy A, Cirak S, Gonzalez V, et al. Deep brain stimulation is effective in pediatric patients with GNAO1 associated severe hyperkinesia. J Neurol Sci 2018;391(May 2018):31–9. Available from: https://doi.org/10.1016/j.jns.2018.05.018.

[20] Yilmaz S, Turhan T, Ceylaner S, Gökben S, Tekgul H, Serdaroglu G. Excellent response to deep brain stimulation in a young girl with GNAO1-related progressive choreoathetosis. https://doi.org/10.1007/s00381-016-3139-6.

[21] Waak M, Mohammad SS, Coman D, et al. GNAO1-related movement disorder with life-threatening exacerbations: movement phenomenology and response to DBS. J Neurol Neurosurg Psychiatry 2018;89(2):221–2. Available from: https://doi.org/10.1136/JNNP-2017-315653.

[22] Singer HS. Tics and tourette syndrome. CONTIN (MINNEAP MINN) 2019;25(4):936–58. Available from: https://doi.org/10.1007/978-3-7091-1628-9_28.

[23] Bloch MH, Leckman JF. Clinical course of Tourette syndrome. J Psychosom Res 2009;67:497–501. Available from: https://doi.org/10.1016/j.jpsychores.2009.09.002.

[24] Blackburn JS, Parnes M. Tics, tremors and other movement disorders in childhood. Curr Probl Pediatric Adolesc Health Care 2021;51 (3):100983. Available from: https://doi.org/10.1016/j.cppeds.2021.100983.

[25] Martino D, Deeb W, Jimenez-Shahed J, et al. The 5 pillars in Tourette Syndrome deep brain stimulation patient selection: present and future. Neurology 2021;96(14). Available from: https://doi.org/10.1212/WNL.0000000000011704.

[26] Massano J, Sousa C, Foltynie T, Ludvic MZ, Rui Vaz H. Successful pallidal deep brain stimulation in 15-year-old with Tourette syndrome: 2-year follow-up. https://doi.org/10.1007/s00415-013-7049-1.

[27] Coulombe MA, Elkaim LM, Alotaibi NM, et al. Deep brain stimulation for Gilles de la Tourette syndrome in children and youth: a meta-analysis with individual participant data. J Neurosurg Pediatr 2018;23(2):236–46. Available from: https://doi.org/10.3171/2018.7.PEDS18300.

[28] Massano J, Sousa C, Foltynie T, Ludvic, Marwan Z, Rui Vaz H. Successful pallidal deep brain stimulation in 15-year-old with Tourette syndrome: 2-year follow-up. J Neurol 2013;260:2417–19. Available from: https://doi.org/10.1007/s00415-013-7049-1.

[29] Kalia LV, Lang AE. Parkinson's Dis 2015;386. Available from: https://doi.org/10.1016/S0140-6736(14)61393-3, http://www.thelancet.com.

[30] Muthane UB, Swamy HS, Satishchandra P, Subhash MN, Rao S, Subbakrishna D. Early onset Parkinson's disease: are juvenile- and young-onset different? Mov Disord 1994;9(5):539–44. Available from: https://doi.org/10.1002/mds.870090506. PMID: 7527483.

[31] Lunati A, Lesage S, Brice A. The genetic landscape of Parkinson's disease. Neurol SOFMA 2018;174(9). Available from: https://doi.org/10.1016/j.neurol.2018.08.004ï.

[32] Türe E, Kamaşak T, Cora M, Karakaya D, Uçar T, Bebek N, Tüzün E, Demir K. Comparison of the serum cytokine levels before and after adrenocorticotropic hormone (ACTH) therapy in patients with infantile spasm. Seizure 2016;41:112–15. Available from: https://doi.org/10.1016/j.seizure.2016.06.021.

[33] Van Den Munckhof B, De Vries EE, Braun KPJ, et al. Serum inflammatory mediators correlate with disease activity in electrical status epilepticus in sleep (ESES) syndrome. Epilepsia 2016;57(2):e45–50. Available from: https://doi.org/10.1111/epi.13274.

[34] Hempel A, Frost M, Agarwal N. Language and behavioral outcomes of treatment with pulse-dose prednisone for electrical status epilepticus in sleep (ESES. Epilepsy Behav 2019;94:93–9. Available from: https://doi.org/10.1016/j.yebeh.2019.02.016.

[35] van den Munckhof B, Alderweireld C, Davelaar S, et al. Treatment of electrical status epilepticus in sleep: clinical and EEG characteristics and response to 147 treatments in 47 patients. Eur J Paediatr Neurol 2018;22(1):64–71. Available from: https://doi.org/10.1016/j.ejpn.2017.08.006.

[36] Van Den Munckhof B, Van Dee V, Sagi L, et al. Treatment of electrical status epilepticus in sleep: a pooled analysis of 575 cases. Epilepsia 2015;56(11):1738–46. Available from: https://doi.org/10.1111/epi.13128.

[37] Kivity S, Lerman P, Ariel R, Danziger Y, Mimouni M, Shinnar S. Long-term cognitive outcomes of a cohort of children with cryptogenic infantile spasms treated with high-dose adrenocorticotropic hormone. Epilepsia 2004;45(3):255–62. Available from: https://doi.org/10.1111/j.0013-9580.2004.30503.x.

[38] Lünemann JD, Quast I, Dalakas MC. Efficacy of intravenous immunoglobulin in neurological diseases. Neurotherapeutics 2016;13(1):34–46. Available from: https://doi.org/10.1007/s13311-015-0391-5.

[39] Renoux C, Vukusic S, Confavreux C. The natural history of multiple sclerosis with childhood onset. Clin Neurol Neurosurg 2008;110 (9):897–904. Available from: https://doi.org/10.1016/j.clineuro.2008.04.009.

[40] Gross RH, Corboy JR. Monitoring, switching, and stopping multiple sclerosis disease-modifying therapies. CONTLifelong Learn Neurol 2019;25(3):715–35. Available from: https://doi.org/10.1212/CON.0000000000000738.

[41] Chitnis T, Arnold DL, Banwell B, et al. Trial of Fingolimod versus Interferon Beta-1a in pediatric multiple sclerosis. N Engl J Med 2018;379 (11):1017–27. Available from: https://doi.org/10.1056/nejmoa1800149.

[42] Abboud H, Petrak A, Mealy M, Sasidharan S, Siddique L, Levy M. Treatment of acute relapses in neuromyelitis optica: Steroids alone versus steroids plus plasma exchange. Mult Scler 2016;22(2):185–92. Available from: https://doi.org/10.1177/1352458515581438.

[43] Elsone L, Panicker J, Mutch K, Boggild M, Appleton R, Jacob A. Role of intravenous immunoglobulin in the treatment of acute relapses of neuromyelitis optica: experience in 10 patients. Mult Scler J 2014;20(4):501–4. Available from: https://doi.org/10.1177/1352458513495938.

[44] Lee WJ, Lee ST, Byun JI, et al. Rituximab treatment for autoimmune limbic encephalitis in an institutional cohort. Neurology 2016;86 (18):1683–91. Available from: https://doi.org/10.1212/WNL.0000000000002635.

[45] Nepal G, Shing YK, Yadav JK, et al. Efficacy and safety of rituximab in autoimmune encephalitis: a meta-analysis. Acta Neurol Scand 2020;142(5):449–59. Available from: https://doi.org/10.1111/ane.13291.

[46] Nosadini M, Thomas T, Eyre M, et al. International Consensus Recommendations for the Treatment of Pediatric NMDAR Antibody Encephalitis. Published online 2021. Available from: https://doi.org/10.1212/NXI.0000000000001052.

[47] Mitchell WG, Wooten AA, O'Neil SH, Rodriguez JG, Cruz RE, Wittern R. Effect of Increased Immunosuppression on Developmental Outcome of Opsoclonus Myoclonus Syndrome (OMS). https://doi.org/10.1177/0883073814549581.

[48] Kenney-Jung DL, Vezzani A. Kahoud RJ, et al. Febrile infection-related epilepsy syndrome treated with anakinra. https://doi.org/10.1002/ana.

[49] Lai YC, Muscal E, Wells E, et al. Anakinra usage in febrile infection related epilepsy syndrome: an international cohort. Ann Clin Transl Neurol 2020;7(12):2467–74. Available from: https://doi.org/10.1002/acn3.51229.

[50] Meyts I, Aksentijevich I. Deficiency of adenosine deaminase 2 (DADA2): updates on the phenotype, genetics, pathogenesis, and treatment. https://doi.org/10.1007/s10875-018-0525-8.

[51] Eleftheriou D., Brogan P.A. Genetic interferonopathies: an overview. Published online 2017. https://doi.org/10.1016/j.berh.2017.12.002.

[52] Mao D, Reuter CM, Ruzhnikov MRZ, et al. De novo EIF2AK1 and EIF2AK2 variants are associated with developmental delay, leukoencephalopathy, and neurologic decompensation. Am J Hum Genet 2020;106(4):570–83. Available from: https://doi.org/10.1016/j.ajhg.2020.02.016.

[53] Zakrzewski W, Dobrzyński M, Szymonowicz M, Rybak Z. Stem cells: past, present, and future. Stem Cell Res Ther 2019;10(1):68. Available from: https://doi.org/10.1186/s13287-019-1165-5.

[54] Aly RM. Current state of stem cell-based therapies: an overview. Stem Cell Investig 2020;7:8. Available from: https://doi.org/10.21037/sci-2020-001.

[55] Tan EY, Boelens JJ, Jones SA, Wynn RF. Hematopoietic stem cell transplantation in inborn errors of metabolism. Front Pediatr 2019;7:433. Available from: https://doi.org/10.3389/fped.2019.00433.

[56] Andrzejewska, A, Dabrowska S, Lukomska B, Janowski M. Mesenchymal stem cells for neurological disorders. Adv Sci (Weinheim, Baden-Wurttemberg, Ger) 2021;8(7):2002944. Available from: https://doi.org/10.1002/advs.202002944.

[57] Sharma A, Jaganathan BG. Stem cell therapy for retinal degeneration: the evidence to date. Biol Targets Ther 2021;15:299–306. Available from: https://doi.org/10.2147/BTT.S290331.

[58] Sun JM, Kurtzberg J. Stem cell therapies in cerebral palsy and autism spectrum disorder. Dev Med Child Neurol 2021;63(5):503–10. Available from: https://doi.org/10.1111/dmcn.14789.

[59] Mousavinejad M, Andrews PW, Shoraki EK. Current biosafety considerations in stem cell therapy. Cell J 2016;18(2):281–7. Available from: https://doi.org/10.22074/cellj.2016.4324.

[60] Warns F.D.A. About stem cell therapies, https://www.fda.gov/consumers/consumer-updates/fda-warns-about-stem-cell-therapies, [accessed 20.02.22].

[61] Serrenho I, Rosado M, Dinis A, M Cardoso C, Grãos M, Manadas B, et al. Stem cell therapy for neonatal hypoxic-ischemic encephalopathy: a systematic review of preclinical studies. Int J Mol Sci 2021;22(6):3142. Available from: https://doi.org/10.3390/ijms22063142.

[62] U.S. National Library of Medicine, https://www.clinicaltrials.gov, [accessed 20.02.22].

[63] Paton M, Finch-Edmondson M, Fahey MC, London J, Badawi N, Novak I. Fifteen years of human research using stem cells for cerebral palsy: a review of the research landscape. J Paediatr Child Health 2021;57(2):295–6. Available from: https://doi.org/10.1111/jpc.15329.

[64] Sun JM, Song AW, Case LE, Mikati MA, Gustafson KE, Simmons R, et al. Effect of autologous cord blood infusion on motor function and brain connectivity in young children with cerebral palsy: a randomized, placebo-controlled trial. Stem Cell Transl Med 2017;6(12):2071–8. Available from: https://doi.org/10.1002/sctm.17-0102.

[65] Sun J., Case L., McLaughlin C., Burgess A., Skergan N., Cash J., et al. Phase 2 randomized study of allogeneic unrelated donor umbilical cord blood or umbilical cord tissue-derived mesenchymal stromal cells in 90 children with cerebral palsy. Presented at the transplantation & cellular therapy meetings, Feb 8–12, 2021.

[66] Paprocka J, Kaminiów K, Kozak S, Sztuba K, Emich-Widera E. Stem cell therapies for cerebral palsy and autism spectrum disorder – a systematic review. Brain Sci 2021;11(12):1606. Available from: https://doi.org/10.3390/brainsci11121606.

[67] Eggenberger S, Boucard C, Schoeberlein A, Guzman R, Limacher A, Surbek D, et al. Stem cell treatment and cerebral palsy: systemic review and meta-analysis. World J Stem Cell 2019;11(10):891–903. Available from: https://doi.org/10.4252/wjsc.v11.i10.891.

[68] Kułak-Bejda A, Kułak P, Bejda G, Krajewska-Kułak E, Kułak W. Stem cells therapy in cerebral palsy: a systematic review. Brain Dev 2016;38 (8):699–705. Available from: https://doi.org/10.1016/j.braindev.2016.03.002.

[69] Novak I, Walker K, Hunt RW, Wallace EM, Fahey M, Badawi N. Concise review: stem cell interventions for people with cerebral palsy: systematic review with meta-analysis. Stem Cell Transl Med 2016;5(8):1014–25. Available from: https://doi.org/10.5966/sctm.2015-0372.

[70] Jantzie LL, Scafidi J, Robinson S. Stem cells and cell-based therapies for cerebral palsy: a call for rigor. Pediatr Res 2018;83(1-2):345–55. Available from: https://doi.org/10.1038/pr.2017.233.

[71] Chez M, Lepage C, Parise C, Dang-Chu A, Hankins A, Carroll M. Safety and observations from a placebo-controlled, crossover study to assess use of autologous umbilical cord blood stem cells to improve symptoms in children with autism. Stem Cell Transl Med 2018;7(4):333–41. Available from: https://doi.org/10.1002/sctm.17-0042.

[72] Dawson G, Sun JM, Baker J, Carpenter K, Compton S, Deaver M, et al. A phase II randomized clinical trial of the safety and efficacy of intravenous umbilical cord blood infusion for treatment of children with autism spectrum disorder. J Pediatr 2020;222:164–173.e5. Available from: https://doi.org/10.1016/j.jpeds.2020.03.011.

[73] Price J. Cell therapy approaches to autism: a review of clinical trial data. Mol Autism 2020;11(1):37. Available from: https://doi.org/10.1186/s13229-020-00348-z.

[74] Master Z, Matthews K, Abou-El-Enein M. Unproven stem cell interventions: a global public health problem requiring global deliberation. Stem Cell Rep 2021;16(6):1435–45. Available from: https://doi.org/10.1016/j.stemcr.2021.05.004.

[75] Marks PW, Witten CM, Califf RM. Clarifying stem-cell therapy's benefits and risks. N Engl J Med 2017;376(11):1007–9. Available from: https://doi.org/10.1056/NEJMp1613723.

[76] Ballen K, Kurtzberg J. Exploring new therapies for children with autism: "Do no harm" does not mean do not try. Stem Cell Transl Med 2021;10(6):823–5. Available from: https://doi.org/10.1002/sctm.20-0548.

[77] Sayed N, Allawadhi P, Khurana A, Singh V, Navik U, Pasumarthi SK, et al. Gene therapy: comprehensive overview and therapeutic applications. Life Sci 2022;294:120375. Available from: https://doi.org/10.1016/j.lfs.2022.120375.

[78] Scoles DR, Minikel EV, Pulst SM. Antisense oligonucleotides: a primer. Neurol Genet 2019;5(2)e323. Available from: https://doi.org/10.1212/NXG.0000000000000323.

[79] Adli M. The CRISPR tool kit for genome editing and beyond. Nat Commun 2018;9(1):1911. Available from: https://doi.org/10.1038/s41467-018-04252-2.

[80] Dzierlega K, Yokota T. Optimization of antisense-mediated exon skipping for Duchenne muscular dystrophy. Gene Ther 2020;27(9):407–16. Available from: https://doi.org/10.1038/s41434-020-0156-6.

[81] Verhaart I, Aartsma-Rus A. Therapeutic developments for Duchenne muscular dystrophy. Nat Rev Neurol 2019;15(7):373–86. Available from: https://doi.org/10.1038/s41582-019-0203-3.

[82] Elangkovan N, Dickson G. Gene therapy for Duchenne muscular dystrophy. J Neuromuscul Dis 2021;8(s2):S303–16. Available from: https://doi.org/10.3233/JND-210678.

[83] Claborn MK, Stevens DL, Walker CK, Gildon BL. Nusinersen: a treatment for spinal muscular atrophy. Ann Pharmacother 2019;53(1):61–9. Available from: https://doi.org/10.1177/1060028018789956.

[84] Finkel RS, Mercuri E, Darras BT, Connolly AM, Kuntz NL, Kirschner J, et al. Nusinersen versus Sham control in infantile-onset spinal muscular atrophy. N Engl J Med 2017;377(18):1723–32. Available from: https://doi.org/10.1056/NEJMoa1702752.

[85] Mercuri E, Darras BT, Chiriboga CA, Day JW, Campbell C, Connolly AM, et al. Nusinersen versus sham control in later-onset spinal muscular atrophy. N Engl J Med 2018;378(7):625–35. Available from: https://doi.org/10.1056/NEJMoa1710504.

[86] Darras BT, Chiriboga CA, Iannaccone ST, Swoboda KJ, Montes J, Mignon L, et al.ISIS-396443-CS2/ISIS-396443-CS12 Study Groups Nusinersen in later-onset spinal muscular atrophy: long-term results from the phase 1/2 studies. Neurology 2019;92(21):e2492–506. Available from: https://doi.org/10.1212/WNL.0000000000007527.

[87] Acsadi G, Crawford TO, Müller-Felber W, Shieh PB, Richardson R, Natarajan N, et al. Safety and efficacy of nusinersen in spinal muscular atrophy: the EMBRACE study. Muscle Nerve 2021;63(5):668–77. Available from: https://doi.org/10.1002/mus.27187.

[88] McMillan HJ, Proud CM, Farrar MA, Alexander IE, Muntoni F, Servais L. Onasemnogene abeparvovec for the treatment of spinal muscular atrophy. Expert Opin Biol Ther 2022;1–16. Available from: https://doi.org/10.1080/14712598.2022.2066471.

[89] Lowes LP, Alfano LN, Arnold WD, Shell R, Prior TW, McColly M, et al. Impact of age and motor function in a phase 1/2a study of infants with sma type 1 receiving single-dose gene replacement therapy. Pediatr Neurol 2019;98:39–45. Available from: https://doi.org/10.1016/j.pediatrneurol.2019.05.005.

[90] Mercuri E, Muntoni F, Baranello G, Masson R, Boespflug-Tanguy O, Bruno C, et al. Onasemnogene abeparvovec gene therapy for symptomatic infantile-onset spinal muscular atrophy type 1 (str1ve-eu): an open-label, single-arm, multicentre, phase 3 trial. Lancet Neurol 2021;20:832–41. Available from: https://doi.org/10.1016/S1474-4422(21)00251-9.

[91] Day JW, Finkel RS, Chiriboga CA, Connolly AM, Crawford TO, Darras BT, et al. Onasemnogene abeparvovec gene therapy for symptomatic infantile-onset spinal muscular atrophy in patients with two copies of smn2 (str1ve): an open-label, single-arm, multicentre, phase 3 trial. Lancet Neurol 2021;20:284–93. Available from: https://doi.org/10.1016/S1474-4422(21)00001-6.

[92] Mendell JR, Al-Zaidy SA, Lehman KJ, McColly M, Lowes LP, Alfano LN, et al. Five-year extension results of the phase 1 start trial of onasemnogene abeparvovec in spinal muscular atrophy. JAMA Neurol 2021;78:834–41. Available from: https://doi.org/10.1001/jamaneurol.2021.1272.

[93] Weuring W, Geerligs J, Koeleman B. Gene therapies for monogenic autism spectrum disorders. Genes 2021;12(11):1667. Available from: https://doi.org/10.3390/genes12111667.

[94] Agboola F, Lin GA, Fluetsch N, Walton SM, Rind DM, Pearson SD. The effectiveness and value of deflazacort and exon-skipping therapies for the management of Duchenne muscular dystrophy. J Manag Care Special Pharm 2020;26(4):361–6. Available from: https://doi.org/10.18553/jmcp.2020.26.4.361.

[95] Broekhoff TF, Sweegers C, Krijkamp EM, Mantel-Teeuwisse AK, Leufkens H, Goettsch WG, et al. Early Cost-Effectiveness of Onasemnogene Abeparvovec-xioi (Zolgensma) and Nusinersen (Spinraza) treatment for spinal muscular atrophy I in The Netherlands with relapse scenarios. Value Health: J Int Soc Pharmacoecon Outc Res 2021;24(6):759–69. Available from: https://doi.org/10.1016/j.jval.2020.09.021.

Chapter 41

The medical home

Amy Jo Merchant Newmeyer[1] and Wilfreda Lindsey[2]

[1]Department of Pediatrics, Baylor College of Medicine-Temple Campus, McLane Chidren's Hospital, Temple, TX, United States, [2]Department of Neurology and Developmental Medicine, Kennedy Krieger Institute, Johns Hopkins University School of Medicine, Baltimore, MD, United States

Introduction

For the past century, the federal government has set forth special protections for children with complex and/or chronic medical conditions. There has been an evolution from the first "crippled children's programs" established by the Federal Children's Bureau in the early 20th century to the development of the Title V Maternal and Child Health Block Grant Program, which provides states funds to create service systems that deliver comprehensive care to women and children, especially those who are low income or have special health care needs [1–3]. In 1998, the Health Resources and Services Administration's Maternal and Child Health Bureau (HRSA MCHB) coined and defined the phrase children and youth with special health care needs (CYSHCN) as "those who have or are at increased risk for a chronic physical, developmental, behavioral, or emotional condition and who also require health and related services of a type or amount beyond that required by children generally" [4].

The children with special health care needs (CSHCN) screener is a five-item, parent survey-based tool developed through the Child and Adolescent Health Measurement Initiative to be an efficient and standardized way to identify children with chronic or special health care needs [5]. Since 2000, it has been widely used nationally, at the state level, and locally, but notably, it is embedded in the National Survey of Children's Health [6] and is the anchor assessment detailing both the prevalence and impact of special health care needs in the United States [1,3,7]. The National Survey of Children's Health [6] is a nationally distributed questionnaire that provides data on children's physical, mental, and emotional health, access to quality health care, family structure, neighborhood structure, school and after-school experiences, and social context. It is funded and directed by the HRSA MCHB and conducted by the US Census Bureau. The NSCH is designed to collect data at the national and state level on children between the ages of 0 and 17 years old [8,9].

The 2019–2020 NSCH found that 14.1 million children in the United States meet the definition for CYSHCN—accounting for a national prevalence of 19.4%—and that 28% of United States households with children had at least one child with SHCN. This is an increase from the 18.5% (13.6 million children) reported in the 2017–2018 NSCH [10,11]. Compared to children without special healthcare needs, CYSHCN are more likely to live in a single-parent household, be a part of families that receive public assistance, and experience adverse childhood events [12].

Though useful as a designation, the comprehensive definition of CYSHCN encompasses a varied population with a wide range of diagnoses, functional limitations, healthcare needs, and health statuses. Children with medical complexity (CMC) are a subpopulation of CYSHCN who are medically fragile, have significant functional limitations, and have increased health care service needs, leading to increased health care utilization [13]. Literature suggests that CMC are among those at the highest risk for adverse outcomes medically, developmentally, and psychosocially [14]. While the estimated prevalence of CMC in the United States is 0.4%–1.6%, CMC represents 6% of children on Medicaid or the Children's Health Insurance Program and comprises nearly 40% of pediatric health spending [15].

Children with intellectual and developmental disabilities are another subgroup of CYSHCN. Based on data from the 2009–2017 National Health Interview Survey, the overall national prevalence of any developmental disability amongst children between the ages of 3 and 17 years old was 16.93% [16]. Many studies have shown that health care costs are significantly higher for these individuals than for the general population. Health-related expenses for individuals with developmental disabilities, including cerebral palsy, intellectual disability, autism spectrum disorder, and genetic conditions, often include community resources, residential placement, nonmedical devices, and other nontraditional health needs in addition to the costs of traditional medical care [17].

Capute and Accardo's Neurodevelopmental Disabilities in Infancy and Childhood. DOI: https://doi.org/10.1016/B978-0-12-824060-1.00018-3

The heterogeneous nature of CYSHCN contributes to the difficulty of establishing policies, developing programs, and allocating resources to effectively serve this population [18]. Previously, there was a particular focus on further specifying and defining this population in an effort to deliver the most effective care in both medical and community settings [19]. However, there has been acknowledgment that a well-functioning system of care will benefit all children and families, including CYSHCN [18]. A well-functioning system, defined by HRSA, MCHB, and its partners, has six foundational components:

1. Early and continued screening for special health care needs.
2. Families as shared partners in the decision-making process at all levels.
3. Organized community-based services that are easily accessible by families.
4. CYSHCN receives care in the medical home.
5. CYSHCN has adequate medical insurance.
6. CYSHCN obtains the necessary services to transition to adulthood.

In spite of extensive descriptions of such systems, federal policy change—such as the passage of the Affordable Care Act—and advances in healthcare, based on the 2019–2020 NSCH, only 14% of CYSHCN receive all six components of a well-functioning system of care [10].

Health care spending and utilization

Due to medical and technological advances, more infants and children are surviving what were once considered life-threatening conditions. Children with special health care needs and those with medical complexity often receive fragmented and/or costly care or have poor access to care altogether. There has been increased recognition of the broad, comprehensive spectrum of care necessary for this population [20,21]. For any child with special needs, families experience additional costs and stress surrounding their health. Along with these increased needs, families often report increased rates of unemployment as well as unmet healthcare needs. This is especially difficult as CYSHCN often require multiple subspecialists, outpatient visits, hospitalizations, therapies, and frequently medical equipment. Due to the nature and impact of the health needs of CYSHCN, both clinicians and caretakers must dedicate an abundance of time to communicating and coordinating with all involved entities [22].

Using the 2007, 2008, and 2009 Medical Expenditure Panel Surveys of children between the ages of 0 and 17 years old, Kuo and colleagues found that compared to children without SHCN, CYSHCN had significantly higher mean inpatient visits, outpatient visits, and ED visits per year (0.03 vs 0.26, 2.6 vs 19.0, and 0.21 vs 0.51, respectively). Whereas children without SHCN had an annual mean of $804 in health care expenses, CYSHCN had a mean of $11,388 total. Children with the highest level of medical complexity had approximately 9 times more inpatient stays, approximately 7 times more office visits, and nearly 2.5 times more emergency department visits compared to children without SHCN [23].

In 2013, Medicaid spent more than 7 times more per child for children in the disability pathway compared to those in the nondisability pathway [24]. A 2020 study utilizing the 2000–2017 Medical Expenditure Panel Surveys found that, compared to children without intellectual and developmental disabilities (IDD), children with IDD had significantly higher health care utilization and spending in the number of inpatient admissions, medication expenses, office visit expenses, and home health care spending. Children with IDD were also found to have a 1.63 times higher incidence of hospital admissions than those without IDD [17].

Though healthy children only receiving preventive care make up 74.5% of the population, they only account for 5% of pediatric medical expenses. Children with complex needs and chronic conditions only make up 0.5% and 25% of the population, respectively, and account for a total of 95% of pediatric medical spending [25]. It is apparent from this data that CYSHCN account for the majority of healthcare dollars spent on children. This, paired with their increased health care service requirements, leaves them especially vulnerable to weaknesses everyone in the United States can experience, such as poor access to health care, high costs of health care, poor quality care, and inadequate insurance coverage [5]. Because of this, the American Academy of Pediatrics has consistently recommended that all children, but especially CYSHCN, should receive comprehensive, family-centered, coordinated, accessible care within a medical home model [26].

The medical home

The terminology for and concept of the medical home was introduced by the American Academy of Pediatrics (AAP) in 1967 and was initially designed to be the primary location to house a child's medical record with the goal of helping families with care coordination and improving care [27]. However, in 2002, the AAP released a policy statement expanding the medical home model in which it called for all children to have "accessible, continuous, comprehensive, family-centered, coordinated, compassionate, and culturally effective care" [26].

Five years later, the major primary care physician associations (the American Academy of Family Physicians, the American College of Physicians, the American Osteopathic Association, and the AAP) developed and endorsed the *Joint Principles of the Patient-Centered Medical Home*, detailing five characteristics of a practice model that would support the nuclear functions of primary care and chronic disease management more effectively:

1. Each patient should have a continuous relationship with a personal physician who would serve as their first contact and provide comprehensive care.
2. The personal physician should serve as the leader of the team that collectively provides care for the patient.
3. The personal physician provides for all health care needs and refers to appropriate professionals when necessary.
4. Care must be coordinated across the health care system and the patient's community.
5. Quality and safety were hallmark features of the patient-centered medical home [28,29].

Per the American College of Physicians, the medical home should function as the "central hub of patient information, primary care provision, and care coordination," and within this patient-centered model, the delivery of care should be evidence-based and the care team should extend to all health care professionals [30].

While there are many similarities between adult and pediatric medical homes, the American Pediatric Association, through a policy statement, addressed the intrinsic unique problems children and adolescents face in health care compared to adults and how these issues influence how pediatric medical homes operate differently from medical homes primarily (or only) providing care for adults. The differences were termed the "Five D's" and endorsed by the American Academy of Pediatrics' Council on Children with Disabilities and Medical Home Implementation Project Advisory Committee. The Five D's include developmental trajectory, dependency on adults, differential epidemiology of chronic disease, demographic patterns of poverty, and dollars spent on children compared with adults [31,32].

Intrinsic to both pediatric and adult medicine are challenges with creating the medical home, but the overarching concept remains the same. The medical home is not a place or a specific clinical practice; it is the way medical care is provided. It can be best defined as a patient-centered, team-based primary care philosophy that focuses on developing partnerships with patients, their families and caretakers, clinicians, and resources within the community.

Care coordination

From the outset, the concept of the medical home placed coordination of care as a key element. Care coordination is critical to a well-performing health system. Given the need for multiple care providers and health care services from different disciplines and in varying settings, care coordination is also essential for the best possible outcomes for CYSHCN and their families, whose needs are often complicated and complex, requiring extensive skill, time, and effort.

Within the medical home, care coordination encourages the integration of care for CYSHCN by creating and promoting collaboration between health care providers, educators, community entities, and any other professional involved in the child's care [33]. Care coordination should be a "patient- and family-centered, assessment-driven, team-based activity designed to meet the needs of children and youth while enhancing the caregiving capabilities of families" [34]. The four defining characteristics are:

1. Family-centeredness.
2. Planned, proactive, and comprehensive focus.
3. Promotion of self-care skills and independence.
4. Emphasis on cross-organizational relationships, including education and social services [35].

The Agency for Healthcare Research and Quality (AHRQ) developed an atlas for care coordination measures in which it provided a visual definition of care coordination with the central goal illustrated as meeting "patient needs and preferences in delivery of high-quality, high-value care" [36] (Fig. 41.1).

The concept of care coordination is important to distinguish from related concepts such as care management and care integration [37]. Care coordination has been characterized as what takes place in the gaps between providers, visits, and hospital stays, as well as the activities across outpatient visits, inpatient admissions, and accessing community resources [37,38]. Care management is a system-level approach proposed to aid patients, their families, and caregivers in managing their medical and mental health conditions. Care management includes both case management and disease management [36]. Care integration is the "seamless provision of health care services, from the perspective of the patient

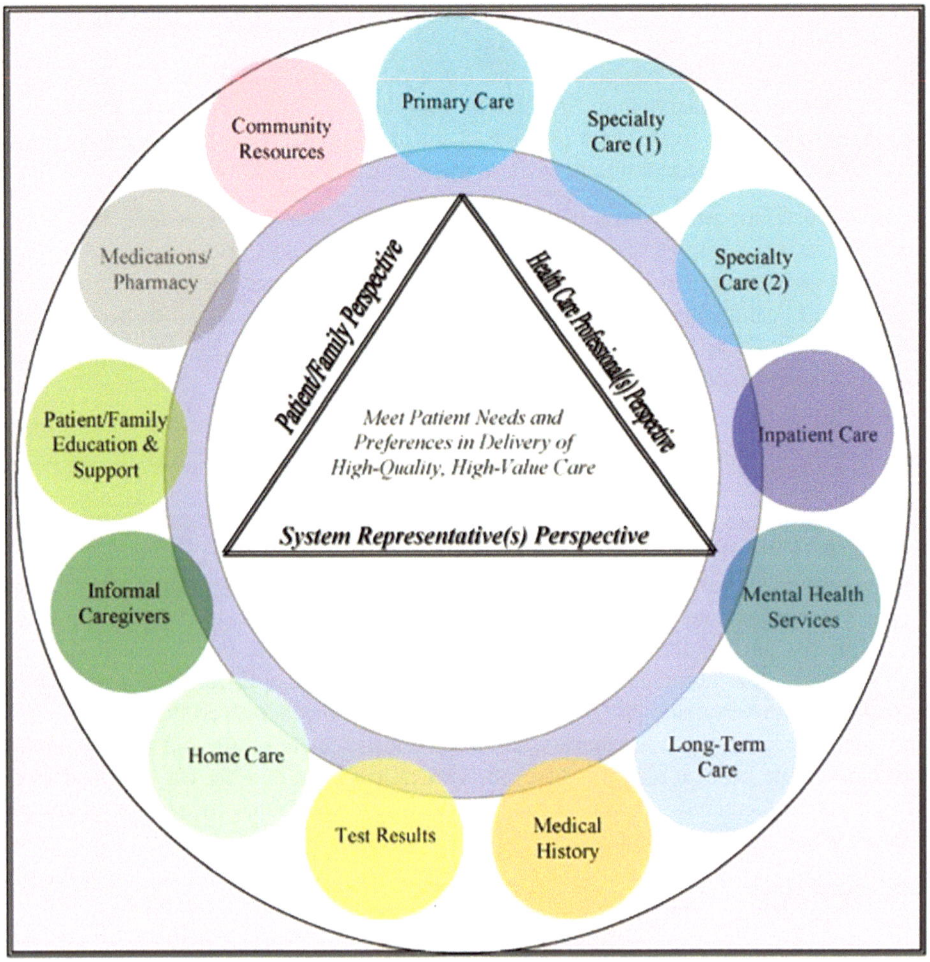

FIGURE 41.1 Care coordination ring (AHRQ)[36]).

and family, across the entire care continuum. It results from coordinating the efforts of all providers, irrespective of institutional, departmental, or community-based organizational boundaries" [25]. Care integration is the outcome when there has been effective care coordination.

Successful care coordination considers all aspects of a child's care as well as the associated costs, while recognizing the importance of the child's language and culture (AAP, 2014). By addressing interrelated needs across different domains, including but not limited to medicine, development and behavior, education, and social circumstance, care coordination's aim is to achieve "optimal health and wellness outcomes" [35].

While most, if not all, patients can benefit from care coordination, CYSHCN often require more services and are more likely to receive fragmented and duplicative care [39]. Based on the 2019–2020 National Survey of Children's Health, CYSHCN are also more likely than children without special health care needs to have unmet health care needs (8.8% vs 2.3%) and thus would likely benefit most from care coordination [10]. A systematic review of studies utilizing the National Survey of Children with Special Health Care Needs to report pediatric care coordination outcomes found that care coordination is correlated with decreased financial burdens and loss of work, which historically significantly impact families with CYSHCN. Those who received care coordination had decreased unmet needs (e.g., higher odds of unmet need for prescribed medications). Sufficient care coordination was also associated with decreased utilization of the emergency department and fewer missed school days [9]. However, according to the 2017–2018 NSCH, less than half (48.2%) of children from birth to 17 years old had a medical home, with a smaller percentage of CYSHCN having a medical home than children without special healthcare needs (42.7% vs 49.4%) [40].

Care coordination must be supported by adequate staffing, technology, evaluation tools, and proper training, with the systemic support of the family as a shared decision-maker and partner. Coordinated care for achieving all of the goals above necessitates leadership, adequate time, and careful, orderly planning that is inclusive of all members of the team [41].

The successful care team is a multidisciplinary team that includes the family as an equal member and functions as a part of the expanded medical home [42] Care coordinators promote and simplify communication amongst providers, facilitate relationships between families and necessary support services, and make certain the care needed is received. However, care coordination, especially for those with the highest needs, is inherently difficult as they often require care from multiple providers and various systems [33]. It is important that all team members have defined roles and responsibilities.

In 2020, after a comprehensive literature review, the National Academy for State Health Policy established evidence-based National Care Coordination Standards for CYSHCN, citing the following domains as critical to ensuring comprehensive, high-quality care coordination:

1. Screening, identification, and assessment.
2. Shared plan of care.
3. Team-based communication.
4. Child and family empowerment.
5. Skills development.
6. Care coordination workforce.
7. Care transitions [42].

Implementing care coordination through care planning

A well-functioning care team inclusive of patients and their families, caregivers, community partners, primary and specialty providers, therapists, pharmacists, educators, and mental health providers is crucial for care planning [43]. Members of the care team should have specific, well-defined roles with dedicated time to care plans and should work to the "top of their license," ensuring cost-effectiveness [37]. When care coordinators act independently, care planning is negatively impacted. The roles and contributions of all members should also be understood by the team as a whole [44]. Shared responsibility encourages efficiency in care planning at every stage. For the team members who care for CYSHCN, the shared responsibility must be agreed upon by all members. Each member should also clearly understand whom the family will contact for various needs or concerns, which provider will write and refill prescriptions, who will provide "supportive justification" for different needs (e.g., medications, medical supplies, medical equipment, therapies, home health care), and who will be responsible for keeping portions of the medical record current [45].

Care coordination is often supported through a written shared plan of care, which is a dynamic document guided by the patient and family's input that is comprehensive, family-centered, and allows for communication across the child's entire care team—clinical and nonclinical. The shared plan of care is developed with the family, "captures a child and family's medical and social needs, goals, and planned activities to achieve their desired outcomes, and is meant to be continuously reviewed and updated to reflect the child's progress and evolving care needs" [46]. It provides a well-rounded view of the family's needs and aims to coordinate and align all services with the family's goals. When appropriately used, the plan of care should not only increase the quality and efficiency of care but should also lead to better outcomes medically, functionally, socially, and financially [47].

The 2014 report, *Achieving a Shared Plan of Care with Children and Youth with Special Health Care Needs*, set forth ten principles for the successful use of a shared plan of care, four key elements of the care planning process, and ten steps toward the implementation of a care plan [47]. Using these as guides, care plans can achieve their purpose of producing optimal child and family outcomes through communication, collaboration, and comanagement. The NICHQ care plan template is customizable and available at https://static.nichq.org/care-plan/.

Barriers to effective care coordination and approaches to overcoming them

There are many challenges to providing coordinated care for primary care physicians, including, but not limited to, poor knowledge of certain medical conditions, a lack of information regarding appropriate community resources, the need for services that extend beyond health, poor communication amongst the other health care providers providing care, poorly defined roles and responsibilities for team members, a lack of time to provide coordinated care, insufficient reimbursement for effort and time required for care coordination, poorly organized systems, and cultural barriers [48] Katkin and colleagues suggest that the ideal pediatric care team acknowledges the family and children as key members of the team and recognizes that the team leaders should be pediatricians. Providers should assess not only family and child needs but also the availability of resources. Other roles include establishing the appropriate partnerships to address all requirements of the family and child, communicating effectively and efficiently to all members of the team, and incorporating all of the key elements of a high-functioning team (e.g., "shared goals, clear roles, mutual trust, effective communication, and measurable processes and outcomes") [45].

Even with the presence of well-functioning teams, inadequate reimbursement for care coordination exists. Historically, payment for care coordination has not been well supported or reimbursed by most payer models. The fee-for-service model, the most utilized payment model in the United States, compensates the medical office for visit time and the volume of care delivered rather than the quality of services rendered. Though diagnostic and management complexity as well as time spent are taken into consideration, many of the care coordination activities performed by physicians take place outside of direct encounters with patients and families. Though there is a trend toward value-based payments, monthly bundled payments, and other alternative models, there has not yet been a model that rewards the nonface-to-face, outside-of-clinic activities that are involved in care coordination. There has been consideration of an at-risk payment model where payment is linked to population numbers rather than service volume [37].

For adults, health care that is considered "high value" is generally characterized as achieving the best possible health outcome for the lowest cost. However, pediatric care is more often centered on preventive care, screening for developmental differences, and providing anticipatory guidance. This decreases the ability to lower short-term costs in the same way that is possible for adults. Most children with Medicaid coverage infrequently use inpatient services. Because of this, the value-based payment models that are based on adult care likely will not produce considerable savings [49].

In January 2015, the Centers for Medicare and Medicaid Services (CMS) made an effort to improve the way providers are compensated by making the shift to rewarding "value and care coordination—rather than volume and care duplication." In conjunction with the Department of Health and Human Services, CMS developed four categories to be reflective of the way in which clinicians receive payment for the care they provide (Fig. 41.2) [50].

Payment Taxonomy Framework

	Category 1: *Fee for Service–No Link to Quality*	Category 2: *Fee for Service–Link to Quality*	Category 3: *Alternative Payment Models Built on Fee-for-Service Architectutre*	Category 4: *Population-Based Payment*
Description	*Payments are based on volume of services and not linked to quality or efficiency*	*At least a portion of payments vary based on the quality or efficiency of health care delivery*	*Some payment is linked to the effective managemnt of a population or an episode of care. Payments still triggered by delivery of services, but opportunities for shared savings or 2-sided risk*	*Payment is not directly triggered by service delivery so volume is not linked to payment. Clinicians and organizations are paid and responsible for the care of a beneficiary for a long period (e.g.≥1 yr)*
Medicare FFS	• Limited in Medicare fee-for-service • Majority of Medicare payments now are linked to quality	• Hospital value-based purchasing • Physician Value-Based Modifier • Readmissions/Hospital Acquired Condition Reduction Program	• Accountable care organizations • Medical homes • Bundled payments • Comprehensive primary care initiative • Comprehensive ESRD • Medicare-Medicaid Financial Alignment Initiative Fee-For-Service Model	• Eligible Pioneer accountable care organizations in year 3-5

FIGURE 41.2 Categories delineating how clinicians receive pay for the care they provide [50].

The CMS's Innovative Center creates and evaluates different types of alternative payment models that reward effectively delivered, value-based care. "Alternative Payment Models are payment approaches, developed in partnership with the clinician community, that provide added incentives to deliver high-quality and cost-efficient care. APMs can apply to a specific clinical condition, a care episode, or a population" [51].

One such track that rewards value and outcomes and supports primary clinicians is merit-based incentive payment systems (MIPS). MIPS makes payment adjustments based on quality and cost performance and is available to Medicare-participating, eligible physicians. This payment system places emphasis on use of electronic health records, quality objectives, and evidence-based standards. [51]. However, a review of physician performance revealed only modest pay improvements. In 2019, performance bonuses remained low (maximum 1.88%), as did positive pay adjustments (maximum 1.68%) [52].

The concept for a new medical model, the Accountable Care Organization (ACO), was developed as a system-level approach to decrease the differences in the quality of care and to decelerate the ever-increasing health care costs. ACOs are physicians, hospitals, and health care professionals who volunteer to collectively assume responsibility and provide for a specific population's care [53,54]. ACOs seek to achieve the quadruple aim of improving the patient experience, improving population health, reducing health care costs, and ensuring providers also have a positive experience [54]. ACOs are responsible for the cost and quality of care provided and are held accountable through a pay-for-performance model in which providers share the savings accrued by achieving population-based performance standards on quality measures. The structure of ACOs incentivizes low-cost, high-quality care [53,55]. Though there is much heterogeneity in execution, ACOs have been found to lower costs without decreasing the quality of care and will likely be a framework to continue building upon in order to make more effective and efficient high-quality care models [54]. Members of an ACO often make up what is called a medical neighborhood.

The medical neighborhood

It is the responsibility of the primary care provider to direct, coordinate, and supervise the patient's care while also educating both the patient and his or her support team. To achieve medical needs, "whether they are primary, secondary, acute, chronic, preventive, or even end-of-life, ongoing relationships must be established" [56].

One can consider the integration of the medical home through two lenses. In one view, there is the integration of multiple health care services within the primary care practice. In the other view, primary care is integrated into the "medical neighborhood." A key tenet of the medical home is providing well-rounded, patient-centered, team-based care [57]. Medical care has become and continues to become increasingly complex. The medical home aims to recognize how health is impacted by physical, social, and cultural factors and to assemble a comprehensive team to address patients' needs. Gilfillan et al. [58]. Medical homes encourage primary teams to "share the care" by including all members of a patient's team in care delivery and encouraging shared accountability for patients' care quality and overall experience [59].

While the goal of the medical home is to strengthen and optimize the quality of primary care for patients, the medical neighborhood includes all practicing physicians. Medical expenditure continues to rise, and the medical home aims to control costs while still providing quality care. However, national spending related to primary care accounts for just approximately 6% of total health care costs [60]. In the United States, particularly where specialists outnumber primary care providers, care is often fragmented and patients are seen by multiple clinicians [61]. The number of pediatric providers with subspecialty certifications increased by 76.8% from 2003 to 2019 [62].

In 2010, the American College of Physicians published a position paper introducing the "patient-centered medical home neighbor"—a specialty or subspecialty practice that interacts with and aligns their processes with the patient's medical home—and provided guidelines for the facilitation of collaboration and integration between practices to provide care that is high-quality but also efficient [30,63].

The medical neighborhood is a set of principles and expectations that ensures coordinated care by bridging those specialists and subspecialists with primary care physicians and other health care providers to provide comprehensive, high-value care [61]. The Agency for Healthcare Research and Quality (AHRQ) illustrated a visual representation of the flow within a patient's medical neighborhood [64] (Fig. 41.3). The goal of the medical neighborhood is to improve patient safety and experience, provide quality evidence-based care, enhance collaboration and coordination of all involved entities, reduce duplicated services and fragmented care, and increase adherence to treatment recommendations [30,60,61]. In the medical neighborhood, the medical home serves as the home base, but each patient also receives care from a team of specialty providers at various locations [65].

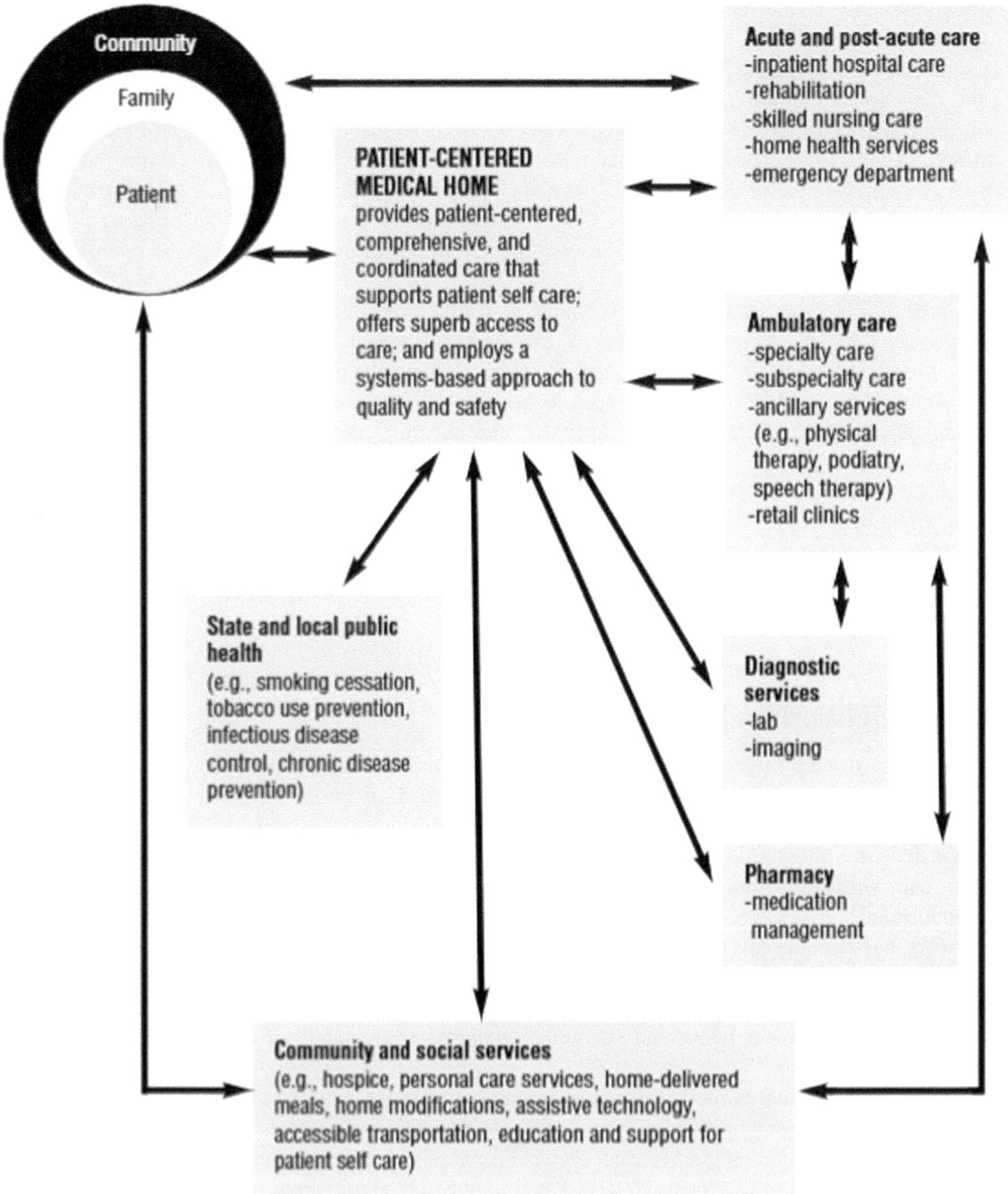

FIGURE 41.3 AHRQ 2011 [64].

Key features of a highly functioning medical neighborhood include:

1. Clarity and specificity of the roles of the specialty neighbors in the system.
2. Shared clinical information between neighbors to make informed decisions and reduce unnecessary duplication of services.
3. Individualized care plans are created by interdisciplinary care teams that include anticipatory guidance and proactive care interventions.
4. Continuity of care through setting transitions facilitated by active communication of all involved parties, including providers, patients, and families.
5. Focus on the patient's preferences, which are to be included in the decision-making process.
6. Strong community relationships, both clinical and nonclinical.

The goal outcomes of medical neighborhoods mirror those previously presented for ACOs—better health outcomes, improved patient safety, reduction of health costs, reduction of duplicated services, and delivery of high-quality, evidence-based care. Ultimately, many of the goals of the medical home rely on a high-functioning medical neighborhood that not only allows for but also encourages and expects the flow of communication between clinicians and patients [64].

Advantages, challenges, and limitations of the medical home

Advantages

Medical homes have a positive effect on healthcare spending as well as the quality of care received, health care outcomes, and medical needs met. Evidence shows that CYSHCN and their families benefit from having a medical home. Having a medical home leads to many positive outcomes, including the perception of higher quality of care [66], reduction of the family burden regarding coordination of care, reduced burden of care coordination by families [67], a reduction of unmet needs, improved access to care, decreased avoidable or preventable emergency department visits [68], fewer missed school days [69], decreased missed work days for parents, decreased hospitalizations [70], increased adherence with health-promoting behaviors such as daily reading, helmet use, and decreased screen time [71], and increased patient and family satisfaction [72]. Data suggests that medical homes lead to reduced overall health care spending through more efficient service delivery and reduced emergency department utilization, hospital admissions (acute and intensive care), readmissions, and the total number of days spent inpatient [21,57,73–75]. In Pennsylvania, it was found that medical homes generate the most cost-effective savings for patients who are in the highest risk categories [74,76].

Challenges and limitations

While some earlier studies reflected only marginal gains in quality and/or cost reduction with the implementation of patient-centered medical homes, it was discovered that many times those entities used outdated standards, attempted to draw conclusions out of sync with previously established goals, and/or lacked standardized metrics and goals [74,77–80]. There remains insufficient data on medical home implementation.

There are many barriers that keep primary care practices from fulfilling the requirements to be a well-functioning medical home. Knowledge of rare conditions can be a barrier to primary care clinicians' implementation of medical home care. In those cases, a specialty medical home may be most appropriate. In addition, extra time and effort are needed to establish a medical home that is appropriately integrated with the community and other clinicians. Providers can take more time but will see less or the same number of patients, with fewer but longer office visits and an increased workload outside of the actual encounter [56]. While there are increasingly more alternative payment models, the fee for service is still the most commonly used model in the United States.

Even when there are medical homes, there are disparities in access and utilization. Approximately 15% of CYSHCN receive care in a well-functioning system, but those who require the most care due to more complex needs and/or more activity limitations receive this level of care at even lower rates [3,18].

Future directions

Recently, with the recognition that children with chronic and complex conditions are living longer, the MCHB, with input from a diverse group of key stakeholders, developed the *Blueprint for Change: Guiding Principles for a System of Services for CYSHCN and Their Families*—a set of fundamental principles and actionable strategies to improve the current delivery of health care and to prioritize well-being and optimal health for CYSHCN. This publication addressed the comprehensive needs of CYSHCN within this changing health care system and the disparities many patients experience. Organized into 4 critical areas—health equity, family and child well-being and quality of life, access to services, and financing of services—*The Blueprint for Change* provides key principles and actionable strategies to attain its overall vision that CYSHCN "enjoy a full life… and thrive in a system that supports their social, health, and emotional needs, ensuring dignity, autonomy, independence, and active participation in their communities" [18]. Though the publication recognizes the implementation of the principles might be a lengthy process, it provides a framework to guide entities through the process of providing high-quality care.

Summary

Children and youth with special health care needs account for nearly 20% of the population and are a heterogeneous group with variable access to high-quality care delivered in well-functioning systems [10]. Children with neurodevelopmental disabilities are a portion of this group who have lifelong conditions that require complex but coordinated and integrated care between their families, primary and specialty providers, and communities. Because many specialists, subspecialists, services, and community resources are typically needed, the coordination of this care often requires much time and effort for everyone involved, and many times care is fragmented and services are unnecessarily duplicated. Because of this, it is critical to have a medical home within a medical neighborhood of multidisciplinary and interdisciplinary team members that includes the family and is committed to providing family-centered, coordinated, comprehensive, and culturally effective care.

References

[1] Schiff J, Manning L, VanLandeghem K, Langer CS, Schutze M, Comeau M. Financing care for CYSHCN in the next decade: reducing burden, advancing equity, and transforming systems. Pediatrics 2022;149(Supplement 7).

[2] US Department of Health and Human Services. Title V Maternal and Child Health Services Block Grant to States: Program Guidance and Forms for the Title V Application/Annual Report. 2020.

[3] Ghandour R.M., Hirai A.H., Kenney M.K. Children and youth with special health care needs: a profile. Pediatrics 2022;149(Supplement 7).

[4] McPherson M, Arango P. A new definition of children with special health care needs. Pediatrics 1998;102(1):137–40.

[5] Bethell CD, Read D, Stein RE, Blumberg SJ, Wells N, Newacheck PW. Identifying children with special health care needs: development and evaluation of a short screening instrument. Ambul Pediatrics 2002;2(1):38–48.

[6] Lichstein JC, Ghandour RM, Mann MY. Access to the medical home among children with and without special health care needs. Pediatrics 2018;142(6).

[7] Bethell CD, Blumberg SJ, Stein RE, Strickland B, Robertson J, Newacheck PW. Taking stock of the CSHCN screener: a review of common questions and current reflections. Academic pediatrics 2015;15(2):165–76.

[8] Ghandour RM, Jones JR, Lebrun-Harris LA, Minnaert J, Blumberg SJ, Fields J, et al. The design and implementation of the 2016 National Survey of Children's Health. Matern Child Health J 2018;22(8):1093–102.

[9] Swann-Thomsen HE, Lindsay R, Rourk S, Hofacer R, Nguyen E. National survey data to evaluate case management services: a systematic review on care coordination using the national survey of children with special health care needs. Prof. Case Manag 2022;27(3):124–40.

[10] US Department of Health and Human Services. (2020). Children with special health care needs: NSCH data brief July 2022.

[11] Warren M.D., McLellan S.E., Mann M.Y., Scott J.A., Brown T.W. Progress, persistence, and hope: building a system of services for CYSHCN and their families. Pediatrics 2022;149(Supplement 7).

[12] Abdi FM, Seok D, Murphey D. Children with special health care needs face challenges accessing information, support, and services. Child Trends 2020;.

[13] Cohen E, Kuo DZ, Agrawal R, Berry JG, Bhagat SK, Simon TD, et al. Children with medical complexity: an emerging population for clinical and research initiatives. Pediatrics 2011;127(3):529–38.

[14] Kuo DZ, Houtrow AJ, Norwood KW, Adams RC, Brei TJ, Davidson LF, et al. Recognition and management of medical complexity. Pediatrics 2016;138(6).

[15] Bergman DA, Keller D, Kuo DZ, Lerner C, Mansour M, Stille C, et al. Costs and use for children with medical complexity in a care management program. Pediatrics 2020;145(4).

[16] Zablotsky B, Black LI, Maenner MJ, Schieve LA, Danielson ML, Bitsko RH, et al. Prevalence and trends of developmental disabilities among children in the United States: 2009–2017. Pediatrics 2019;144(4).

[17] Kim J, Stevens P, Carbone PS, Jones KB. Health care use and spending of pediatric patients with an intellectual or developmental disability. Med Care 2020;58(5):468–73.

[18] McLellan SE, Mann MY, Scott JA, Brown TW. A blueprint for change: guiding principles for a system of services for children and youth with special health care needs and their families. Pediatrics 2022;149(Supplement 7).

[19] Cohen E, Berry JG, Sanders L, Schor EL, Wise PH. Status complexicus? The emergence of pediatric complex care. Pediatrics 2018;141 (Supplement_3):S202–11.

[20] Kuo DZ. The medical home for children with medical complexity: back to basics. J Pediatr 2019;206:8–9.

[21] Mosquera RA, Avritscher EB, Samuels CL, Harris TS, Pedroza C, Evans P, et al. Effect of an enhanced medical home on serious illness and cost of care among high-risk children with chronic illness: a randomized clinical trial. Jama 2014;312(24):2640–8.

[22] Kuo DZ, Cohen E, Agrawal R, Berry JG, Casey PH. A national profile of caregiver challenges among more medically complex children with special health care needs. Arch Pediatr Adolesc Med 2011;165(11):1020–6.

[23] Kuo DZ, Melguizo-Castro M, Goudie A, Nick TG, Robbins JM, Casey PH. Variation in child health care utilization by medical complexity. Matern Child Health J 2015;19(1):40–8.

[24] Musumeci M, Chidambaram P. Medicaid's role for children with special health care needs: a look at eligibility, services, and spending. Kaiser Family Foundation. 2019.

[25] Antonelli R. Walking before running: developing care coordination capacity to achieve high value outcomes for patients with behavioral health needs 2015. Available from: https://www.masspartnership.com/pdf/3-2-Walking-Before-Running-Antonelli.pdf.

[26] Medical Home Initiatives for Children With Special Needs Project Advisory Committee. The medical home. Pediatrics 2002;110(1):184–6.

[27] American Academy of Pediatrics Council on Pediatric Practice. (1967). Pediatric records and a "medical home." In Standards of child care (pp. 77–79). Evanston, IL: American Academy of Pediatrics.

[28] Fisher ES. Building a medical neighborhood for the medical home. N Engl J Med 2008;359(12):1202.

[29] American Academy of Family Physicians. American Academy of Pediatrics, American College of Physicians, American Osteopathic Association. Joint principles of the patient-centered medicalhome. February. Available from: http://www.aafp.org/online/etc/medialib/aafp_org/documents/policy/fed/jointprinciplespcmh0207.Par.0001.File.dat/022107medicalhome.pdf.

[30] Kirschner N, Greenlee MC. The patient-centered medical home neighbor: the interface of the patient-centered medical home with specialty/subspecialty practices. Philadelphia: American College of Physicians; 2010.

[31] Forrest CB, Simpson L, Clancy C. Child health services research: challenges and opportunities. Jama 1997;277(22):1787–93.

[32] Stille C, Turchi RM, Antonelli R, Cabana MD, Cheng TL, Laraque D, et al. The family-centered medical home: specific considerations for child health research and policy. Acad Pediatr 2010;10(4):211–17.

[33] Lipkin PH, Alexander J, Cartwright JD, Desch LW, Duby JC, Edwards DR, et al. Care coordination in the medical home: integrating health and related systems of care for children with special health care needs. Pediatrics 2005;116(5):1238–44.

[34] Turchi RM, Berhane Z, Bethell C, Pomponio A, Antonelli R, Minkovitz CS. Care coordination for CSHCN: associations with family-provider relations and family/child outcomes. Pediatrics 2009;124(Supplement_4):S428–34.

[35] Antonelli R.C., McAllister J.W., Popp, J. Making care coordination a critical component of the pediatric health system: a multidisciplinary framework, 2009.

[36] McDonald K.M., Schultz E., Albin L., Pineda N., Lonhart J., Sundaram V., et al. Care Coordination Atlas Version 4 (Prepared by Stanford University under subcontract to American Institutes for Research on Contract No. HHSA290-2010-00005I). AHRQ Publication No. 14-0037-EF. Rockville, MD: Agency for Healthcare Research and Quality, 2014.

[37] Kuo DZ, McAllister JW, Rossignol L, Turchi RM, Stille CJ. Care coordination for children with medical complexity: whose care is it, anyway? Pediatrics 2018;141(Supplement_3):S224–32.

[38] Turchi RM, Antonelli RC, Norwood, Adams RC, Brei TJ. Patient-and family-centered care coordination: a framework for integrating care for children and youth across multiple systems. Pediatrics 2014;133(5):e1451–60.

[39] Huang ZJ, Kogan MD, Stella MY, Strickland B. Delayed or forgone care among children with special health care needs: an analysis of the 2001 national survey of children with special health care needs. Ambul Pediatr 2005;5(1):60–7.

[40] US Department of Health and Human Services. (2020). Children with special health care needs: NSCH data brief July 2020

[41] Peikes D, Genevro J, Scholle SH, Torda P. The patient-centered medical home: Strategies to put patients at the center of primary care. Rockville, MD: AHRQ Publication No. 11-0029; 2011.

[42] National Care Coordination Standards for Children and Youth with Special Health Care Needs. The National Academy for State Health Policy, 2020. https://www.nashp.org/wp-content/uploads/2020/10/care-coordination-report-v5.pdf.

[43] Murphy NA, Carbone PS, Council on Children With Disabilities. Parent-provider-community partnerships: optimizing outcomes for children with disabilities. Pediatrics 2011;128(4):795–802.

[44] Phillips C. Care coordination for primary care practice. The Journal of the American Board of Family Medicine 2016;29(6):649–51.

[45] Katkin JP, Kressly SJ, Edwards AR, Perrin JM, Kraft CA, Richerson JE, et al. Guiding principles for team-based pediatric care. Pediatrics 2017;140(2).

[46] Normile B, VanLandeghem K. State strategies to advance medical homes for children and youth with special health care needs. June. National Academy for State Health Policy; 2018.

[47] McAllister JW. Achieving a shared plan of care with children and youth with special health care needs. *Lucille Packard Foundation for.* Children's Health 2014;13.

[48] American Academy of Pediatrics Council on Children with Disabilities. Care coordination in the medical home: integrating health and related systems of care for children with special health care needs. Pediatrics. 2005 Nov;116(5):1238-44. Available from: https://doi.org/10.1542/peds.2005-2070.

[49] Breslin, K. (2016). Opportunities to improve children's health by focusing on value.Behavioral-Health-News-article-by-Kate-Breslin-Oct-2016.pdf (scaany.org)

[50] Centers for Medicare & Medicaid Services. (2015). *Better Care. Smarter Spending. Healthier People: Paying Providers for Value, Not Volume.*[Fact sheet]. Available from: https://www.cms.gov/newsroom/fact-sheets/better-care-smarter-spending-healthier-people-paying-providers-value-not-volume.

[51] Centers for Medicare & Medicaid Services (CMS), HHS. Medicare Program; Merit-Based Incentive Payment System (MIPS) and Alternative Payment Model (APM) Incentive Under the Physician Fee Schedule, and Criteria for Physician-Focused Payment Models. Final rule with comment period. Fed Regist. 2016;81(214):77008-77831. PMID: 27905815.

[52] Rathi VK. Physician performance in the medicare merit-based incentive payment system. JAMA 2021;325(3):308.

[53] Tu T, David I, Lawrence I, Ross I. Origins and future of accountable care organizations. Washington, DC: Brookings Institution; 2015.

[54] Wilson M, Guta A, Waddell K, Lavis J, Reid R, Evans C. The impacts of accountable care organizations on patient experience, health outcomes and costs: a rapid review. J Health Serv Res Policy 2020;25(2):130–8.

[55] Abodunde B, Slater C, Coustasse A. MACRA and accountable care organizations: is it working? J Ambul Care Manag 2021;44(2):148–54.

[56] Budgen J, Cantiello J. Advantages and disadvantages of the patient-centered medical home: a critical analysis and lessons learned. Health Care Manag 2017;36(4):357–63.

[57] Nielsen M, Langner B, Zema C, Hacker T, Grundy P. Benefits of implementing the primary care patient-centered medical home. *Washington: Patient-Centered Primary Care Collaborative*; 2012.

[58] Gilfillan RJ, Tomcavage J, Rosenthal MB, Davis DE, Graham J, Roy JA, et al.). Value and the medical home: effects of transformed primary care. Am J Manag Care 2010;16(8):607–14.

[59] Ghorob AMPH, Bodenheimer T. Sharing the care to improve access to primary care. N Engl J Med 2012;366(21):1955–7.

[60] Feuerstein JD, Sheppard V, Cheifetz AS, Ariyabuddhiphongs K. How to develop the medical neighborhood. J Med Syst 2016;40(9):1–6.

[61] Greenberg JO, Barnett ML, Spinks MA, Dudley JC, Frolkis JP. The "medical neighborhood": integrating primary and specialty care for ambulatory patients. JAMA Intern Med 2014;174(3):454–7.

[62] Turner A, Ricketts T, Leslie LK. Comparison of number and geographic distribution of pediatric subspecialists and patient proximity to specialized care in the US between 2003 and 2019. JAMA Pediatr 2020;174(9):852–60.

[63] Laine C. Welcome to the patient-centered medical neighborhood. Ann Intern Med 2011;154(1):60 -60.

[64] Taylor EF, Lake T, Nysenbaum J, Peterson G, Meyers D. Coordinating care in the medical neighborhood: critical components and available mechanisms. White Paper (Prepared by Mathematica Policy Research under Contract No. HHSA290200900019I TO2). In: AHRQ Publication No, . 11-0064. Rockville, MD: Agency for Healthcare Research and Quality; June 2011.

[65] Wexler R, Hefner J, Welker MJ, McAlearney AS. Health care reform: possibilities & opportunities for primary care. J Fam Pract 2014;63 (6):298–304.

[66] Han B, Yu H, Friedberg MW. Evaluating the impact of parent-reported medical home status on children's health care utilization, expenditures, and quality: a difference-in-differences analysis with causal inference methods. Health Serv Res 2017;52(2):786–806.

[67] Miller JE, Nugent CN, Russell LB. Which components of medical homes reduce the time burden on families of children with special health care needs? Health Serv Res 2015;50(2):440–61.

[68] Butcher JS, Wolraich ML, Gillaspy SR, Martin VG, Wild RC. The impact of a medical home for children with developmental disability within a pediatric resident continuity clinic. J Okla State Med Assoc 2014;107(12):632–8.

[69] Knapp CA, Hinojosa M, Baron-Lee J, Fernandez-Baca D, Hinojosa R, Thompson L. Factors associated with a medical home among children with attention-deficit hyperactivity disorder. Matern Child Health J 2012;16(9):1771–8.

[70] Palfrey JS, Sofis LA, Davidson EJ, Liu J, Freeman L, Ganz ML. The pediatric alliance for coordinated care: evaluation of a medical home model. Pediatrics 2004;113(Supplement_4):1507–16.

[71] Long WE, Bauchner H, Sege RD, Cabral HJ, Garg A. The value of the medical home for children without special health care needs. Pediatrics 2012;129(1):87–98.

[72] Overview of Data Related to the Pediatric Medical Home. American Academy of Pediatrics, 2016. Available from: https://www.aap.org/en/practice-management/medical-home.

[73] Homer CJ, Klatka K, Romm D, Kuhlthau K, Bloom S, Newacheck P, et al. A review of the evidence for the medical home for children with special health care needs. Pediatrics 2008;122(4):e922–37.

[74] NCQA. Benefits of NCQA patient-centered medical home recognition. NCQA, 2019. Available from: https://www.ncqa.org/wp-content/uploads/2019/09/20190926_PCMH_Evidence_Report.pdf.

[75] Rosenberg CN, Peele P, Keyser D, McAnallen S, Holder D. Results from a patient-centered medical home pilot at UPMC Health Plan hold lessons for broader adoption of the model. Health Aff 2012;31(11):2423–31.

[76] Pines JM, Keyes V, van Hasselt M, McCall N. Emergency department and inpatient hospital use by medicare beneficiaries in patient-centered medical homes. Ann Emerg Med 2015;65(6):652–60.

[77] Cole ES, Campbell C, Diana ML, Webber L, Culbertson R. Patient-centered medical homes in Louisiana had minimal impact on Medicaid population's use of acute care and costs. Health Aff 2015;34(1):87–94.

[78] Jackson GL, Powers BJ, Chatterjee R, Prvu Bettger J, Kemper AR, Hasselblad V, et al. The patient-centered medical home: a systematic review. Ann Intern Med 2013;158(3):169–78.

[79] Friedberg MW, Schneider EC, Rosenthal MB, Volpp KG, Werner RM. Association between participation in a multipayer medical home intervention and changes in quality, utilization, and costs of care. Jama 2014;311(8):815–25.

[80] Vest JR, Bolin JN, Miller TR, Gamm LD, Siegrist TE, Martinez LE. Medical homes:"where you stand on definitions depends on where you sit. Med Care Res Rev 2010;67(4):393–411.

Chapter 42

Classroom setting

Sherry Vinson
Department of Pediatrics, Baylor College of Medicine, Houston, TX, United States

Introduction

Educating individuals with neurodevelopmental disabilities (NDDs) is associated with frequent questions and issues. How to achieve an optimal outcome is a complex process that varies according to the individual child, their pattern of strengths and challenges, classroom and teacher factors, and the many different interventions that can be undertaken. Knowledge of the educational choices available and the research based evidence regarding those choices aid professionals who guide parents in their quest to achieve a successful outcome for their child. This chapter covers topics aimed at providing such knowledge.

Understanding school options: private versus public versus charter versus home school

Parents in the United States have a variety of school options for their children. It is important to understand the differences between these options when searching for a school for their child with NDD.

Private schools are supported by a private organization or private individuals. Students attending private schools must pay tuition to attend, although some private schools offer scholarships. Private schools come in all assortments as to their cost, overall size, number of students per classroom, teacher educational and certification requirements, and purpose. Some private schools exclude children with NDD, some have programs that address the needs of children with milder NDD, and some focus on children with a specific NDD such as autism or language delay or intellectual disability or cerebral palsy or a learning disability or a variety of these conditions. When considering a private school placement, parents need to become knowledgeable about the programs offered by the school and gather as much information as possible to determine whether the school meets their child's needs. Professionals who care for children with NDD may be a useful source for referrals. Parents should visit any school they are considering for their child to ensure that the school follows research shown best practice instruction and to determine whether the school is a good fit for their child.

Private schools do not have to accept a child as a student. They may dismiss a child if they decide they cannot meet the student's needs. Finally, parents should know that private schools are not required to follow Individuals with Disabilities Education Act (IDEA). By contrast, the local education agency (LEA) of the public school system is required to evaluate a student who is attending a private school and whose parent requests an evaluation. After an evaluation, the LEA is also required to make recommendations if it is determined that special education services are needed. The LEA is also required to inform the parent whether the LEA special education service can be provided as an "add-on" to the private school (Section 612(a)(10)A of the IDEA, 34 Code of Federal Regulations/CFR 300.111, and 34 CFR 300.130–300).

While the cost of private schools may be prohibitive for many families, the IDEA mandates that a public school district pay for a private school that could provide an appropriate education when that public school district cannot do this [1,2].

Public schools are supported by their state's government and must follow federal school laws including IDEA as well as the state school laws. The public school must follow the state mandates regarding class size, length of the school day and year, teacher certification, curriculum, textbooks, and the like. Public education is free to each student living

Capute and Accardo's Neurodevelopmental Disabilities in Infancy and Childhood. DOI: https://doi.org/10.1016/B978-0-12-824060-1.00050-X

within the local public school zone. The public school must accept and educate each child of school age who resides within the specific public school district.

While education must take place in the least restrictive environment, there is a continuum of services to meet the student's needs. Some public school districts have magnet schools for specific interests like music or specific learning differences like giftedness or multihandicapped or blindness. Magnet schools may enroll students from several zones. They may have admission criteria or use systems such as a lottery to choose their students. Most states allow their public schools to connect with their state's school for the blind or school for the deaf to evaluate local students to make sure the local students have an appropriate education if deaf or blind. A very few students may have needs that cannot be met in the public school system and in those cases, the public school system will provide the education in a private setting. It may require a legal proceeding to adjudicate whether the child's public school program is appropriate. Parents and professionals should be aware that many states rank their public schools based on yearly test scores and other characteristics. Such rankings may not accurately depict a school, and parents who are seeking a public education for their child with NDD should visit the proposed school and talk to the principal and special education coordinator. Parents in the neighborhood are another source of information. Such conversations are a better way to determine if a school would be a good fit for their child than basing a decision solely on the state's ranking of the school.

Charter schools are supported by the state government where they are chartered and, like public schools, are free to all students and must follow federal school laws including IDEA. However, unlike public schools, charter schools are allowed flexibility regarding class size, length of the school day and year, teacher certification requirements, curriculum, and other areas to which public schools must strictly adhere to state mandates. As long as the charter school meets the achievement accountability standards of the state, it is allowed to remain open. However, parents need to understand that most states only hold accountability reviews of their charter schools every 3–5 years. As with private and traditional public schools, parental best practice should involve parents visiting the schools they are considering and talking to parents already using those schools.

Homeschooling has become more popular in recent years and is allowed by all states with varying accountability requirements. Parents choosing such an option for their child with an NDD are usually individuals with access to private specialized instruction, therapies, and special equipment since such school based supports are not required to be available for homeschooled children. Parents should know that 34 CFR 300.111 requires the LEA of the public school system to evaluate a student living in the LEA but being homeschooled upon parental request for the evaluation and after the evaluation give the recommendations for special education services if any were determined. However, it is up to the state if the LEA is allowed to use the small percent of IDEA funds allocated for nonpublic school students to give special education services to a child in the LEA who is homeschooled. Some states treat homeschooling like a private school but others do not allocate IDEA money for a homeschooler's public school special education service.

Ensuring appropriate classroom placement, instruction, and accommodations

The chapter of this book covering legislative mandates gives information about the various federal laws, such as IDEA and Section 504 of the Rehabilitation Act of 1973(504) and the Americans with Disabilities Act (ADA), which aim to ensure appropriate classroom placement, instruction, and accommodations. However, before using any of these laws or other means, the student's abilities in language, motor, adaptive/problem solving, and social development domains, as well as any academic abilities in reading, math, and written expression must be determined.

The prelude to appropriate placement, instruction, and accommodations is defining a student's ability levels and developing a plan that synchronizes short term goals with the student's current abilities. The plan needs to consider how the student's abilities in each of the above domains will impact the student's learning. The plan will identify areas for which aids and accommodations will minimize interference with the student's learning. For the student with NDD, consideration of any aspect of the student's condition for which an aid or accommodation would allow the student to learn and then apply age appropriate information is very important. For example, a blind student whose blindness is the single reason for delay in developmental domains will certainly need aids and accommodations, but the student should be able to progress in an age appropriate fashion. However, due to the issues mentioned in this book's chapters on the evolution of NDD, patterns of typical and atypical neurodevelopment, the biological basis of NDDs, genetic and epigenetic abnormalities in NDDs, and distorted neuroplasticity in NDDs, it is often difficult to determine if a student has a singular issue which needs to be addressed or if the student has multiple issues needing to be addressed. Regardless of the difficulty, the starting point is finding the student's ability level in each developmental domain and in each basic academic area. Failure to look at all domains often leads to failed management programs for the student, dissatisfied teachers, and resentful classmates. The chapter of this book that covers developmental assessment (Chapter 9) gives information regarding best practice to determine an individual's abilities in all developmental domains and basic academic areas.

The public school full and individual evaluation (FIE) mentioned in IDEA [3] should be requested in writing by the parent of a child with a known or suspected NDD. This should be initiated upon the child entering school, whether the school is private, home school, charter, or public. In addition, a written request for an FIE should be given to the zone public school anytime that grade retention is suggested or if a student makes little progress after six weeks of intervention in tier two or three of the public school's response to intervention (RTI) program. The FIE is an excellent complement to the evaluations done by NDD specialists and is required under IDEA to qualify for special education services. Even if after testing, the student does not qualify for a handicapping condition, the FIE will document the student's cognitive, communication, motor, social and behavioral, and academic abilities compared to their sameage peers. Although the student may not qualify for IDEA services, a weakness may be identified that qualifies the student for 504 accommodations. In addition, having a handicapping condition recorded on school records is useful for obtaining accommodations in other settings with the ADA [4] and the Rehabilitation Act [5] as well as state codes protecting against discrimination due to the handicapping condition.

Once the public school FIE results are available, the public school will invite the parents to a meeting to review the FIE results and join a team of school personnel to develop their child's individualized education plan (IEP). This IEP team must meet yearly per IDEA but can have emergency meetings called by any team member, including the parent, as needed during the year. Once the IEP is agreed upon and signed, it must be followed until another IEP meeting is called to make revisions. Of course, a private school does not have to follow the IEP but an initial one would be a good guide for that school or for the homeschooling parents.

It is of utmost importance for the parents to work as equal partners with the school personnel in developing the IEP of their child. Equal partnership is required by IDEA but sometimes parents are reluctant to speak up at IEP meetings. NDD specialists can be helpful in teaching parents about their rights as IEP team members and training them how to use knowledge of the requirements of IDEA, 504, ADA, and other relevant laws to advocate for their child. IEP development for a student with an NDD is optimal when school personnel and parents equally partner in giving input regarding what each has seen works best for the student's learning and participation in the classroom. Such input should then be considered along with the student's strengths and weaknesses in each domain of ability on the triannual FIE plus interim measurements of language understanding, communication, adaptive/problem solving, fine motor, gross motor, social and behavioral, academic abilities.

The IEP team should always keep in mind the student's language understanding and problem solving abilities, regardless of the student's NDD. Such a practice will allow for: (1) planning for someone with a motor affected condition who cannot speak or move but who could handle general education curriculum with an eye tracking device to have an appropriate education; (2) avoidance of placement of a child with a receptive language delay and autism with hyperlexia in a general curriculum reading classroom with no supports; (3) a student with dyslexia and no other delays to keep up with grade appropriate literature and historical information with use of books on tape; (4) an appropriate decision on whether to give a blind child significant Braille training with an otherwise appropriate for age/grade general curriculum or a deaf child significant American Sign Language training with an otherwise appropriate for age/grade general curriculum; and (5) a reasonable prediction for when a student with cognitive dysfunction will need increased individualized and small group (6 students:1 teacher) concrete instruction.

The best IEP teams are composed of school personnel and parents who keep updated with best practice in instruction and aid usage for the student's area of difficulty and advocate for these best practices. Chapters and appendices of this textbook covering specific clinical conditions give useful information regarding best practice instruction and aid usage for individuals with the specific condition covered in those chapters and appendices. Effective IEP teams also know where to find special advisors or instructors or equipment needed when not available in their specific school district such as knowing that most states have a type of advisement/equipment cooperative agreement like a regional education service center or advisement services from the state's School for the Blind or Deaf.

Overall, ensuring appropriate classroom placement, instruction, and accommodations for individuals with NDDs can be time consuming and difficult, especially when such students have multiple ability deficits. However, the parents and educators who follow the guidance given in this section will have an easier time with their given task.

Inclusion

Inclusion is the term used when students with special needs spend most or all of their time in the general education setting. The most recent data from the National Center for Educational Statistics show that 64% of all students with

TABLE 42.1 Percentage distribution of students 6–21 years old served under IDEA, Part B, by educational environment and type of disability: fall 2018.

Type of disability	<40% Time in general education	40%–70% Time in general education	>80% Time in general education
Autism	33.2	18.5	40.0
Deaf-blindness	35.7	12.6	25.7
Developmental delay	14.4	18.5	65.6
Emotional disturbance	17.2	17.4	49.6
Hearing impairment	10.5	14.7	63.3
Intellectual disability	48.5	27.5	17.3
Multiple disabilities	44.9	17.6	14.2
Orthopedic impairment	22.1	15.6	54.3
Other health impairment	8.5	20.2	67.3
Specific learning disability	4.6	21.4	72.2
Speech or language impairment	3.9	4.6	87.8
Traumatic brain injury	19.7	21.5	51.0
Visual impairment	8.9	12.4	68.2

Source: Digest of Education Statistics, 2019, Table 204.60.

disabilities in the US public schools spend 80% or more of their time in general education class. (see Table 42.1 for the percentage of students in each IDEA disability category spending 80% or more time in general education.) The principle underlying inclusion is that individuals with disabilities have the same value to society as those without disabilities and have the right to every opportunity given those without disabilities, including educational opportunities. The principle that each person has different circumstances, such as a disability, causing them to need specific interventions to reach equal outcomes brings about tension within the idea of inclusion, as does the principle that an individual's optimal ability may be limited due to that person's disability despite receiving appropriate specific interventions.

Most parents and educators agree that individuals with NDDs need experiences among peers without disabilities in order to adjust to real world environments and to have typical models for social skills and academics [6]. In addition, most parents and educators agree that school age peers without disabilities need experiences among peers with NDDs to gain insight and compassion [7]. However, research data to address inclusion questions is incomplete.

A review of the literature shows overwhelming statistical support for the academic and social benefits of inclusion for students with disabilities and, at least, no harm and more often benefits for the nondisabled peers with whom students with disabilities are included [8–12]. Theories about why students with disabilities have done better academically when in a general education class as compared to students with disabilities in a class of only students with a disability usually point to studies showing more instructional time versus noninstructional time in the general education classroom compared to the self contained special education classroom [13–15]. Another theory to explain the greater academic success of students with disabilities included in the general education classroom as opposed to their segregated peers with disabilities is that those in the general education classroom are exposed more to grade level information than those in the segregated special education classroom [16]. Theories to explain students without disabilities performing better academically when their general education classroom includes students with disabilities center on the idea that the teachers in those classrooms must differentiate instruction to meet all learners' needs instead of simply presenting information one way [17].

Despite the overwhelming statistical support for inclusion, Allison Gilmore in her 2018 publication "Has Inclusion Gone Too Far?" notes that there may be selection bias in inclusion studies since students with disabilities who have higher academic abilities and fewer behavioral challenges are more likely to be placed in inclusive settings as opposed to their peers with greater academic or behavior difficulties. Gilmore points out that the difficulty of accessing pure data sets separating out these differing students within the same handicapping group interferes with estimating the true causal effects of inclusion on student outcomes. However, she does mention three studies showing negative outcomes for students without disabilities whose general education classroom included students with an emotional/behavioral disorder [18–21].

Even the supporters of inclusion warn that simply placing a student with a disability in a general education classroom is apt to fail without optimal administrative support, behavioral support, teacher training in instructional techniques and classroom organization to meet each student's learning and socialization needs, and much curriculum planning involving all involved school personnel from both special education and general education departments [22–24]. Such supporters warn that without planning, students with disabilities who are in general education classrooms are sometimes segregated within that classroom, defeating the purpose of socialization and exposure to the general curriculum [25]. A recent review finds that some students with disabilities do not feel a sense of belonging to the group despite their general education classroom placement, exposing the need for curriculum design that allows for working together as equal partners at times without interfering with each students' instructional needs [26].

Disability rights advocates and the National Council on Disability continue to encourage an appropriate development of inclusion education, instead of using the excuse that the task is too difficult or saying it is the student's job to "fit in" [27]. The real question is not if inclusion is the better choice in the education of students with disabilities but rather "how can we improve inclusive education?" [28]. Rogers and Johnson released "Strategies to Include Students with Severe/Multiple Disabilities Within the General Education Classroom" in 2018 with their review of the literature to determine effective ways to include and support students with severe/multiple disabilities, mentioning the need for all general education teachers to be taught best practices, such as giving these students more time to respond ("wait time") and embedding several learning trials within an activity to ensure learning if these teachers are to be expected to include these students successfully [29]. The National Council on Disability released a 2018 report that included research based practices that lead to positive academic and/or behavioral outcomes for all students involved in inclusive education. The research based practices include (1) having a universal design for learning that understands the barriers presented by curricula not designed for the variety of student learning needs in a diverse school community; (2) giving opportunities for cooperative learning requiring students to work with their peers to accomplish a shared common goal within a classroom lesson; (3) using differentiated instruction which is a process to teach students of differing abilities in the same class to maximize each student's growth and individual success by meeting each student at his/her current skill level rather than expecting students to catch up to the curriculum; (4) informing instruction to promote learning through appropriate use of student data so that when a student is not seeming to learn, instruction can be quickly changed; (5) implementing a proactive, positive schoolwide approach to teaching appropriate social behavior and understanding the messages behind the problem behavior with the use of a variety of strategies for supporting and responding to student behavior, using data to determine appropriate behavior interventions, and developing an equitable school discipline restorative practices approach as found on the Positive Behavior Intervention and Support website at http://www.pbis.org; (6) putting into place peer assisted learning, which involves reciprocal tutoring roles, opportunities to respond and experience success, structured activities, and supplemental practice of skills taught in the core curriculum; (7) training of teachers in culturally responsive teaching that requires that educators value students' cultural and linguistic resources and build upon this understanding rather than view it as a barrier to learning; and (8) using a multitiered system of supports to address learning and behavior needs within a data informed decision model such that all learners including those with disabilities have access to a variety of academic interventions alongside their peers with removal for specific specialized instruction as needed to address specific disability related educational needs [30].

Parents trying to decide whether to put their children with special needs into some or all general education classrooms should ask questions of school officials including those on their child's IEP team and other parents whose children with disabilities have experienced general education at the school to determine if the school follows the best practices given in the previous paragraph. Parents should know their child's language understanding level to make certain that the general classroom teacher(s) provide instruction at that level and, as needed, have a designated well trained paraprofessional available to reexplain material to their child. They should know their child's accommodation, equipment, and intensive intervention needs to make sure these are provided in the general education classroom. Of special note, if a child has a research based intensive intervention need, such as phonological awareness training given to individuals with dyslexia or children with cognitive problems, the parent should ask for that to be given in a pull out class with accommodations such as having oral testing and use of books on tape for the general education classroom. This special pull out training should be done during the time the subject addressed in the pull out session is being covered in general education and not during recess or a special class such as music/art/PE which children tend to enjoy. If a child has never eloped from class or had an attention problem in a segregated classroom but has such in general education, the parent should submit a written request for an emergency IEP meeting to explore for a cause or whether an adjustment in the IEP is needed. Above all, parents should consider their child's academic, emotional, and social needs, advocating for the amount of inclusion and type of inclusion that is best for their individual child's needs.

Inclusion is not a "one size fits all" concept. Successful inclusion allows each student, no matter his or her disability, to feel valued and is different per individual. Would a segregated special education music class where the students with autism who were nonverbal learn to play autoharps to match general education peers' songs and then participate sometimes in the general education music class as contributors be better than their permanent general education music class placement where their frequent disruptions make them unaccepted by general education peers? Would a boy with moderate intellectual disability making academic progress at his 40% of age appropriate rate of learning in a pull out life skills class with optimal appropriate instruction time, and presentation of science/social studies information in a way he could understand, but who is included on his school's cross country team where he, as the fastest runner, feels more included than if he were put in an all day general education class where a paraprofessional sat with him to give 1:1 instruction? Should an eighth grade girl with moderate intellectual disability be made to stay in her general education history class when she is asking to go down the hall to teach the Preschool Program for Children with Disabilities students their letters and letter sounds? These are questions that continue to be asked.

Appropriate use of response to intervention

RTI is a process for early identification of K-2 students at risk for academic failure that employs a multitiered service delivery system to give high quality instruction to all students and provides struggling learners increasingly intensive academic and behavioral interventions starting in general education [31]. While the RTI idea preceded the 2004 amendment to IDEA, that 2004 amendment authorized the use of RTI while still not eliminating the requirement that a FIE be done for struggling students upon parental written request. Since that time, most public schools have used RTI as described in the IDEA amendment. RTI was thought to be a quicker way to get the correct instruction to struggling students who traditionally had to wait for an FIE. RTI in public schools involves testing all of the students at the beginning of school, placing nonstruggling students in the first tier, students at risk for problems in the second tier for targeted group interventions, and students with definite academic problems in the third tier for intensive individual interventions. If the third tier students do not improve with intervention, they are considered for the traditional FIE to evaluate for an increased level of service with appropriate special education that is supported by research to be effective in meeting the student's learning needs.

The RTI plan makes sense and sounds like a formulation of what every good teacher does: identify a struggling learner and change the method of instruction to the way that learner can learn. However, in 2015 the National Center for Evaluation and Regional Assistance released the results from a study commissioned by the US Department of Education focusing on 146 elementary schools using RTI. Unfortunately, the report concluded that the reading achievement of first graders was worse than before RTI was implemented and that the use of RTI had made no difference for second and third graders [32]. In carefully reviewing the study results, Fuchs and Fuchs saw some problems with evaluation methods causing flawed results and they also noted that many of the 146 schools were not consistently implementing RTI correctly [33]. Nancy Bailey reviewed the study and agreed that RTI had not been implemented appropriately and pointed out that the appropriate screening tools for detecting dyslexia were not being used by the schools in the study [34]. Others agreed with Bailey and said that inappropriate interventions for the Tier Two students were being used by many of the schools [35,36].

In his response to the 2015 report of RTI's ineffectiveness, Fletcher points out that the way to improve RTI is to make sure reliable and valid assessments are used, effective and research based instruction are used in Tier One, core instruction is supplemented with more intensive interventions based on students' needs, and highly qualified personnel are used to provide instructions and interventions. He urges the shifting of resources by local public school districts and states to give support to what will improve RTI [37].

One might read Fletcher's recommendations and argue that it is easy to say to use valid and reliable assessments along with research based instructions and interventions taught by highly qualified teachers but ask "Do we know what is valid and reliable and effective?" In the meantime, what are parents of individuals with NDDs to do? For parents whose children already have had an FIE and have an IEP in place, they should continue to make certain the FIE is updated every 3 years and they should continue to work with their child's IEP team as equal partners to adjust instruction if their child is not learning as predicted. For parents whose children have entered school without an identified handicapping condition but are struggling in school enough for the school to say, "Your child is in danger of grade retention," they should immediately submit a written request for an FIE. For those whose child has been identified at the beginning of school for Tier Two or Tier Three of RTI but are not told their child is in danger of grade retention, the parents should meet with the teacher(s) to learn the exact problem the child is having and make certain that the intervention that will be used for their child is known to be an effective intervention for the problem displayed by their child. (Parents and the professionals helping them should use a search engine to ask the question "What does ERIC say about using __ for __?," filling in the blanks with the intervention the school proposes and the problem identified in their child's learning. ERIC stands for Education Resources Information Center.) If the intervention is supposed to be best practice for the problem displayed by the child, the parent should agree to the intervention for six weeks but if the child has had minimal improvement within those six weeks, the parent should submit a written request for an FIE. On a special note, if a beginning reader is struggling with decoding words but can understand a grade level paragraph read to him, that child might benefit from the type of phonemic awareness training known to help students with dyslexia [38]. The parents should advocate that this method be used with their child. Of course, as with all learning, parents of all students should know what is being taught to their child and practice with their child at home to make certain the child is truly applying what is presented at school.

Use of consultation to the classroom

Upon reviewing their clients' IEP, professionals caring for students with NDDs are often questioning the use of once per month school based therapy services such as speech therapy or occupational therapy or of the only once per week auditory or vision instruction session. These services are supposed to be "consultative." The infrequent times the student sees the school based service provider is so the provider can reevaluate the student to adjust recommendations to the classroom teacher on the best way to instruct the student or on the student's needed accommodations or equipment for the classroom. These brief sessions with the student are not meant to give therapy or instruction. What is important for the parents and the professionals is that they monitor these recommendations to ensure they the recommendations are appropriate for the student's learning needs and based on the current research for best practices. It is also important to monitor the student's classroom(s) to make sure the recommendations are being followed and implemented. This monitoring can be difficult, especially the monitoring of classroom implementation due to privacy laws protecting all students in the classroom. The best way to do this monitoring is for the parents to ask at their child's annual IEP meeting about the consultant(s) recommendations and classroom implementation, also asking how these recommendations can also be implemented at home and asking that updated recommendations be provided to the parents. Simply asking shows the teacher(s) and consultant(s) parental interest, partnership, and support.

Grade retention

Grade retention refers to having a student repeat the same grade for a second year when that student is not meeting the standards for passing a grade. The suggestion of grade retention to "catch up" for someone who entered kindergarten with a known severe NDD is rare. However, individuals with mild NDDs not yet recognized, such as a mild cognitive impairment or a specific learning disability, are most at risk for grade retention suggestions by school officials. A typical suggestion from school officials regarding students later found to have borderline intellectual ability levels related to genetic conditions is that they repeat kindergarten or first grade to "gain maturity" but, often they continue to not meet promotion standards in future grades despite their kindergarten or first grade retention. Parents, at times, also request grade retention for such students, hoping their child will "catch up" during the extra year. An understanding of grade retention outcomes is useful for parents of individuals with NDDs and the professionals advising them.

The majority of studies show that unless a child has been physically absent from instruction during most of the academic year, grade retention to simply sit through the same instruction a second year does not bring long term academic benefits, with any short term gains dissipating within 3 years [39–43]. Only a few studies show academic benefit for retained students [44,45]. An interesting finding, however, in one study where first grade retention had no long term effect on students' academic ability and definitely did not have a negative effect on their academics, was that their probability of not graduating from high school or getting a general education diploma (GED) was greater than students of their same ability level who had no grade retention [46].

Due to the lack of demonstrated benefit of grade retention, the National Association of School Psychologists (NASP) issued a white paper in 2011 advising that early identification of learners' struggles and use of intense evidence based interventions to instruct them should replace the conundrum of whether to retain or promote a failing student. The white paper further advised that if a struggling student continues to perform below grade standards, investigation to rule out causes like a NDD should be undertaken and, if found, individualized, accelerated evidence based instruction should be provided. The white paper emphasizes that if the parents and education team do decide grade retention is in the student's best interest, the retention intervention should be more than simply a repeat of the previous year [47]. Long time reviewer of retention data, Shane R. Jimerson, along with Tyler L. Renshaw, made the following statement summing up their findings, "Like so much in education, what is most effective is a targeted approach that addresses students' academic, social, and mental health issues and links specific evidence based interventions to a student's individual needs" [48].

Following the recommendations given in the NASP white paper would seem to preclude retention without appropriate interventions for individuals who later are found to have NDDs not recognized upon school entrance [49]. Following the NASP recommendations would also have implications for students who already have IEPs. Whenever a student with an IEP is being considered for grade retention, the student's parents should work with the student's IEP team to develop an improved IEP. There is no need for grade retention for a student with an IEP since the student showing a gap large enough to suggest a need for grade retention would require a small group, intense, research based intervention, which could be provided in the next grade as well as in the same grade. However, in the rarer cases in which grade retention would be best for a student, that student's IEP should be revised to give the student different instruction since the previous instruction dwas not effective.

In the case of a student not being physically present for the majority of the school year, grade retention could be helpful since the student never had the instruction in the first place. However, parents should work with their child's school to arrange for the child to have appropriate research based homebound instruction given by the school when a child's condition prevents in person school attendance. A unique situation causing students to miss being present for instruction occurred with the COVID-19 pandemic when schools used online instruction with special needs students whose learning needs required in person instruction. Use of targeted intense research based small group instruction once in person school was possible along with extended year services, namely summer school, was a method used by IEP teams to avoid grade retention for these students [50].

Building self esteem in students with a neurodevelopmental disability

Every person deals with perceived ability weaknesses in themselves for which they feel insecure. Students with and without disabilities fear asking questions which might make them look "dumb." Unfortunately, peers, often the very ones who themselves are insecure, are glad to turn the spotlight on others, respond unkindly, and reinforce the questioner's insecurity. Teachers and other school personnel unwittingly at times respond in a way that reinforces the questioner's insecurity. An example would be the teacher who attempts to praise the child for listening to the previous day's instruction responds to a question by saying, "Now I know you know that." The fear of asking questions limits learners young and old, causing instructors to think the foundational information has been understood and not realize that the presentation of new information is not connected to the foundational information.

Students with disabilities often feel less capable due to handicapping conditions that affect appearance, motor skills, learning ability, articulation, vision, or hearing. Desiring not to be viewed as different, they stay in the background, avoid participation, or do not use their needed equipment. Even college students sometimes do not go to the college's office of disability services to gain available accommodations for fear that others will know they need help.

Assisting individuals to overcome their insecurities is an on going process that requires help from teachers and the varying professionals, peers, and family members with whom individuals interact. Monitoring for counseling needs is important and parents should use the school counselors and social workers as needed.

For school age individuals with disabilities, a good practice at the beginning of every school year is to have their classmates, teachers, and aides briefly talk about their own challenges for which they need help. Frequent teacher reminders to the entire class not to be embarrassed to ask questions, noting that other students may have the same questions but are afraid to ask them, encourage openness and motivate student questions. Having older students with disabilities return from high school and postsecondary experiences to talk to general education classes which include individuals with special needs as well as to segregated special education classes is a helpful way to show examples of people with disabilities who are successful. Having the teachers attempt to play an instrument or sing a song or play a game on the playground to show one does not have to be perfect to enjoy an activity is also helpful for students. Teachers giving all students jobs appropriate to their ability and praising them for doing the job builds self esteem. Involving students in extracurricular activities at the school, including music or sports or theater activities, leadership and planning committees, rearing and showing animals, and robotics clubs helps students to find new talents and connects them as valued members of the school community. (An older student mentor for a student with a disability in an extracurricular activity may be helpful.) Finally, successful completion of an academic task and allowing the student to teach the teacher how the task is to be done builds self-esteem.

Behavior in the classroom

Maladaptive behaviors in the classroom interfere with learning. The misbehaving child is not focused on the instruction. In addition, classmates often focus on the misbehaving child rather than on the instruction and the instructors cut short their instruction to deal with the misbehaving child. Other chapters of this book and the appendix give information about specific behavior problems and how to approach them. However, this section aims to give some practical points to remember for the classroom.

When a new behavior problem occurs, first determine if something has changed in the child's life. Does the child have a new acute or chronic health problem? Has the child's home situation changed, such as the organizing parent having to go out of town away from the family for work or the care of an elderly parent? Is there a new student in the classroom subtly hitting or bullying the student? Has the curriculum become more difficult to understand? While the answer to reduce the problem behavior may still be difficult, knowing the answer to "What has changed?" gives targeted guidance.

Sometimes the trigger for a child's misbehavior is not obvious. At other times, the cause of the misbehavior is so general, such as "He gets upset when he must do his work," that development of a plan to reduce the misbehavior seems impossible. At such a time, the parent of the child with the behavior problem should put in writing a request for a functional behavior assessment (FBA). The FBA is an organized way of collecting data to determine the triggers and consequences of a child's maladaptive behaviors to develop an effective positive behavior plan. The 1997 amendment to IDEA requires an FBA when a change in school placement is being considered due to the child's behavior [51] but an FBA may be useful in less severe situations. A review of collected data may reveal that the student does cooperate with doing certain work which would suggest a strength that can be used in the positive behavior plan. For example, if the child will do math calculation problems but will not even try math word problems, having in the behavior plan that the reward for trying a word problem will be getting to do a math calculation may motivate the student to try the word problem. The observed behaviors should alert the teacher to work with the student 1:1 to show how the word problem could be made into a calculation, thereby providing the student a strategy to use for success. Success in getting an answer triggers the child's brain reward system so that the student will want to continue. It is important, however, that parents participate in the development of the positive behavior plan since they will need to reinforce this plan at home to ensure consistency between home and school.

Students with NDDs that affect language understanding often show decreased attention to instruction. While some of these students may have attention deficit hyperactivity disorder that could be helped with medication, some may simply lose attention due to not comprehending. Teachers and parents should ensure that instruction is aimed at the student's level of understanding. Instruction should reinforce verbal presentation with visual cues. New information should be connected to established knowledge and supervised practice should be provided to ensure appropriate application of the instruction before leaving the child. (Well trained paraprofessionals are helpful in providing this supervised practice with reinstruction as needed. Sometimes new information is assumed to be linked to established knowledge but language delayed students may need a "bridge step" between the old information and the new information so that these learners will see the connection.) Teachers and parents should also make certain that instruction is given within the expected attention span for a student's language understanding level. While literature suggests that the attention spans of typical children are 2 to 3 minutes per year of their chronological age, in practice, it is best to assume that true attention span for learning novel information is 1 minute for every year of one's receptive language level before having to redirect.

Conclusion

The educational environment for children with NDD is difficult to navigate. Ongoing changes in educational goals and needs, variations in the delivery of educational services, and uninformed expectations from both teachers and parents and caregivers, among other factors mentioned above, can complicate the process. Therefore, equipping parents and caregivers with essential data and practical advice is vital for advocating effectively and ensuring a supportive and inclusive educational experience.

References

[1] 20 U.S. Code §1412 (a) (10).

[2] Endrew F.V. Douglas County School Dist. RE−1, 580 US; 2017.

[3] Individuals with Disabilities Act, Subchapter II (Part B) 1414.
[4] Americans with Disabilities Act, Title I.
[5] Rehabilitation Act, Section 501.
[6] US Dept of Education, National Center for Educational Statistics. (2021). Digest of Education Statistics, 2019 (NCES 2021-009), Chapter 2; "Percentage distribution of students 6 to 21 years old served under IDEA, Part B, by educational environment and type of disability: Fall 2018."
[7] UNESCO. The Salamanca statement and framework for action on special needs education. Adopted by the World Conference on Special Needs Education: Access and Quality. Salamanca, Spain: UNESCO; 1994.
[8] Salend, Spencer J, Duhaney, Garrik LM. what do families have to say about inclusion? Teach Exceptional Child 2001;35(1):62–6.
[9] Dudley-Marling C, Burns MB. Two perspectives on inclusion in the United States. Glob Educ Rev 2014;1(1):14–31.
[10] Oh-Young C, Filler, J. A meta-analysis of the effects of placement on academic and social skill outcome measures of students with disabilities. Res Dev Disabil 2015;47:80–92. Available from: https://doi.org/10.1016/j.ridd.2015.08.014.
[11] Heihir T, Grindal T, Freeman B, Lamoreau R, Borquaye Y, Burke S. A summary of the evidence on inclusive education. Abt Associates; 2016.
[12] Wagner M, Newman L, Cameto R, Levine P. The academic achievement and functional performance of youth with disabilities: a report from the National Longitudinal Transition Study-2 (NLTS2) (NCSER 2006-3000). Menlo Park, CA: SRI International; 2006.
[13] McGregor G, Vogelsber RT. Inclusive schooling practices: pedagogical and research foundations. A synthesis of the literature than informs best practices about inclusive schooling. University of Montana, Rural Institute on Disabilities; 1998.
[14] Theobald RJ, et al. Career and technical education, inclusion, and postsecondary outcomes for students with learning disabilities. J Learn Disabil 2019.
[15] Hunt P, Farron-Davis F, Beckstead S, Curtis D, Goetz L. Evaluating the effects of placement of students with severe disabilities in general education versus special education. J Assoc Pers Severe Handicap 1994;19(3):200–14.
[16] Helmstetter E, Curry CA, Brennan M, Sampson-Saul M. Comparison of general and special education classrooms of students with severe disabilities. Educ Train Ment Retard Dev Disabil 1998;33:216–27.
[17] McDonnell J, Thorson N, McQuivey C. Comparison of the instructional contexts of students with severe disabilities and their peers in general education classes. J Assoc Pers Severe Handicap 2000;25:54–8.
[18] Ryndak D, Jackson LB, White JM. Involvement and progress in the general curriculum for students with extensive support needs: K-12 inclusive-education research and implications for the future. Inclusion 2013;1(1):28–49.
[19] Waldron N, Cole C, Majd M. The academic progress of students across inclusive and traditional settings: a two year study Indiana inclusion study. Bloomington, IN: Indiana Institute on Disability and Community; 2001.
[20] Gilmour Allison F. Has inclusion gone too far? Weighing its effects on students with disabilities, their peers, and teachers. Educ Next 2018;18 (4):8–16.
[21] Carrell S, Hoekstra M. Externalities in the classroom: how children exposed to domestic violence affect everyone's kids. Am Econ J 2010;2 (1):211–28.
[22] Jason F. The effects of inclusion on classmates of students with special needs: the case of serious emotional problems. Educ Financ Policy 2008;4(3).
[23] Gottfried MA, Egalite A, Kirksey J. Does the presence of a classmate with emotional/behavioral disabilities link to other students' absences in kindergarten? Early Child Res Q 2016;.
[24] Obiakor F, Harris M, Mutua K, Rotatori A, Algozzine B. Making inclusion work in general education classrooms. Educ Treat Child 2012;35 (3):477–90.
[25] Jerome Sarah K. An analysis of inclusion in the field of special education. Selected Honors Theses; 2017.
[26] Wayne S. Equity as a basis for inclusive educational systems change. Swift Center Project using Grant from US Education Dept., Office of Special Education Programs Grant No. H326Y 120005. Australasian Journal of Special Education; 2016. p. 1–17. Available from: https://doi.org/10.1017//jse.2016.12.
[27] Hicks-Monroe S. A review of research on the educational benefits of the inclusive model of education for special education students, JAASEP; 2011. p. 61–69.
[28] Hergott J. Effects of special education and feeling of inclusion, special education and teaching commons. Iowa: Northwestern College; 2020.
[29] National Council on Disability. The segregation of students with disabilities, February 2018 report; 2018.
[30] Schifter L, Hehir T. The better question: how can we improve inclusive education? Educ Next 2018;.
[31] Rogers W, Johnson N. Strategies to include students with severe/multiple disabilities within the general education classroom. Phys Disabil Educ Relat Serv 2018;37(2):1–12.
[32] National Council on Disability. The segregation of students with disabilities, 2018 report; 2018.
[33] Jack F, Sharon V. Response to intervention: preventing and remediating academic difficulties. Child Dev Perspect 2009;3(1):30–7.
[34] Balu R, Zhu P, Doolittle F, Schiller E, Jenkins J, Gersten R. Evaluation of response to intervention practices for elementary school reading. Washington, DC: National Center for Education Evaluation and Regional Assistance; 2015.
[35] Fuchs D, Fuchs L. Critique of the national evaluation of response to intervention: a case for simpler frameworks. Exceptional Children; 2017.
[36] Bailey N. Why are schools still using response to intervention? Nancy Bailey's EduCation Website; 2017.
[37] Sparks S. Study: RTI practice falls short of promise. Education Week; 2015.
[38] Eide F. Response to intervention RTI hurts students with dyslexia. Dyslexic Advantage; 2016.

[39] Fletcher J. Response to RTI evaluation report. Building Capacity; 2015.
[40] Fletcher JM, Lyon GR, Fuchs LS, Barnes MA. Learning disabilities: from identification to intervention. 2nd ed. Guilford Press; 2018.
[41] Jimerson S. A synthesis of grade retention research: looking backward and moving forward. Calif Sch Psychol 2001;6:47–59.
[42] Hong G, Raudenbush S. Effects of kindergarten retention policy on children's cognitive growth in reading and mathematics. Educ Eval Policy Anal 2005;27(3):205–24.
[43] Wei Wu, et al. Effect of retention in first grade on children's achievement trajectories over 4 years: a piecewise growth analysis using propensity score matching. J Educ Pschol 2008.
[44] Hwang S, Cappella E. Rethinking early elementary grade retention: examining long-term academic and psychosocial outcomes. J Res Educ Effect 2018;11(4):559–87.
[45] Marcico Institute for Early Learning and Literacy. Does retention (repeating a grade) help struggling learners? University of Denver; 2012.
[46] Hughes JN, Chen Q, Thoemmes F, Kwok O. An investigation of the relationship between retention in first grade and performance on high stakes tests in third grade. Educ Eval Policy Anal 2010;32(2):166–82.
[47] Alexander KA, Entwisle DR, Dauber SL. On the success of failure. 2nd ed Cambridge University Press; 2003.
[48] Hughes JN. Affect of early grade retention on school completion. APA PsycNET; 2018.
[49] National Association of School Psychologists. Grade retention and social promotion (Position Statement); 2011.
[50] Jimerson S, Renshaw T. Retention and social promotion. ResearchGate; 2012.
[51] McLeskey J, Grizzle KL. Grade retention rates among students with learning disabilities. Child 1992;58(6):548–54.

Chapter 43

Transition and aging

Miya R. Asato[1,2] and J. Seth Van Zant[3,4]

[1]*Department of Neurology and Developmental Medicine, Kennedy Krieger Institute, Baltimore, MD, United States,* [2]*Departments of Neurology and Pediatrics, Johns Hopkins University School of Medicine, Baltimore, MD, United States,* [3]*Department of Pediatrics/Division of Pediatric Neurology, Baylor College of Medicine, Houston, TX, United States,* [4]*Department of Pediatric Neurology/Neurodevelopment, Christus Children's Hospital, San Antonio, TX, United States*

Illustrative case: Tyler is a 12-year-old boy with Down Syndrome. As an infant, his health concerns focused on nutrition, growth, and recurrent otitis media. As an older child, Tyler has been generally healthy and making developmental progress. Tyler's parents bring him to the neurodevelopmental clinic and raise questions and concerns about what to expect as Tyler grows up and as he transitions to middle school. They are particularly concerned about ensuring he has social opportunities with peers and ask what to expect as he gets older. You note that Tyler is shy during the visit and will answer questions with one-word answers when you encourage him to participate in the visit.

With advances in medical care, children with many different congenital and early childhood conditions thrive and progress within pediatric care. When they reach adult ages, there is a necessary move to adult systems of care. Medical transition is being increasingly recognized as an important component of routine visits, and medical team members are recognizing the need for preparing families and youth for transition at earlier stages of care as opposed to just prior to the actual transfer to an adult medicine physician. Patient age is often an obvious prompt for the physician to bring up transition as part of the visit. In neurodevelopmental practice, however, transition should be considered as a broad, longitudinal concept impacting all aspects of medical care, as well as social and educational planning, which should be regularly incorporated during each medical visit. Beginning with the delivery of the primary neurodevelopmental diagnosis and providing anticipatory guidance about the natural history and prognosis, the physician and family should be increasingly engaged to develop plans to optimize health, build independence, and promote positive physical and social outcomes in adult life, with increasing focus on the youth assuming a more active role to the greatest extent possible.

Transition represents a natural and developmentally guided process focused on the planning and preparation to leave pediatric care. This is in contrast to a *transfer* to adult care which includes the end of the care responsibility of the pediatric specialist, the preplanned move to an adult medicine setting, with acceptance by all parties involved. On a broader systems-level scale, health care transition for individuals with complex health care needs has been defined as "the movement of adolescents and young adults with chronic physical and medical conditions from child-centered to adult-oriented health care systems" [1,2]. The consensus statement on health care transition for young adults with special health care needs states the goal of transition "is to maximize lifelong functioning and potential through the provision of high quality, developmentally appropriate, health care services that continue uninterrupted as the individual moves from adolescence to adulthood" [3,4].

Transition for children and adolescents with intellectual and neurodevelopmental disabilities is complex and needs to take into consideration the youth's capabilities and expected capacity as an adult (Table 43.1). In addition to ensuring successful health care transition, individuals with neurodevelopmental disabilities and their families often navigate concurrent transitions in their education, social, vocational, and living environments. During adolescence, family members should shift from a direct caregiver role to that of a coach by promoting independence in daily activities and self-care. As the child ages, families need to assess the capacity for their child's future medical decision-making abilities as well as consideration of financial planning for the future. Physicians need to provide anticipatory guidance, identify needs, provide resources, and engage other health care team members—such as social workers and community support organizations.

The consequences of poor transition planning and execution can result in neglect or missed identification of medical problems resulting in complications or disease exacerbation, fragmented care, higher acute care utilization, and potentially higher health care costs. Compromised health status can impact health-related quality of life and can affect other

Capute and Accardo's Neurodevelopmental Disabilities in Infancy and Childhood. DOI: https://doi.org/10.1016/B978-0-12-824060-1.00053-5

TABLE 43.1 Health care transition steps and roles for individuals with intellectual and neurodevelopmental disabilities.

Health care transition steps	Age range	Physician role	Youth-centered participation to the greatest extent possible	Family centered
1. Partnership between patient, family, and health care team Ensure presence of a health care professional who can attend to the unique challenges of health care transition (HCT) and assumes responsibility for the process	Preadolescent (<12 years)	– Promotes concept and establishment of medical home, defines roles of different team members with child and family – Promotes education concerning diagnosis, natural history, and establishes a plan for therapeutic approaches – Allocates supportive resources (e.g., engages social worker, community agencies, support groups)	– Familiarity with care team, diagnosis, treatments, following medication or therapy program – Practices preparing for visits with family member (how to answer questions, explore questions about their care, and/or diagnosis)	– Partners with the health care team to increase communication among multidisciplinary team members – Supports child to prepare and participate in visits – Gains increasing awareness of expected natural history and outcomes based on child's diagnoses
2. Awareness of transition expected to occur at adult age Identify core knowledge and skills required to provide appropriate transition services and ensure they are part of training requirements for physicians who assume care of patients with intellectual and neurodevelopmental disabilities	Young adolescent (12–14 years)	– Conveys need to transition at agreed upon age to adult medical system – Reviews institutional policy, if applicable – Establishes goals for future visits to promote condition-specific self-care mastery and self-advocacy – Provision of anticipatory guidance to care issues or changes related to puberty; address or refer to specialist to address reproductive health – Initiate private meetings with youth during visits to promote communication and rapport building – Provision of informational resources	– Works on increasing self-care skills relative to medical condition (e.g., participates in medication management), knows emergency plans, and side effects of medication – Can articulate goals for future education and activities	– Structures and reinforces learning in home environment that foster adolescent's self-care skill mastery – Utilizes school and community resources to explore postgraduate options related to educational and vocational programs

(Continued)

TABLE 43.1 (Continued)

Health care transition steps	Age range	Physician role	Youth-centered participation to the greatest extent possible	Family centered
3. Assessment of transition-related skills and identification of needs	Early adolescence (14–16 years)	– Assessment of youth's functional abilities and transition skill level using structured or standardized testing (see Resource List at end of chapter) – Address and discuss family/caregiver's ability to manage evolving care needs over time – Provide anticipatory guidance about medical decision making – In cases of significant disability, discuss ongoing and future care needs with family	– Takes more active role in leading health care visits and providing information (verbally or creating written/digital summaries) – Is able to take increasing responsibility for medications or remembering appointments	– Family/caregivers shift to role of coach; foster behavior and habits that promote health – Increase youth's participation in health care-related tasks (e.g., calling in prescriptions)
4. Intentional transition planning, education, and assessment of readiness at annual visits	Early and mid-adolescence (14–18 years)	– Review progress in transition skill building – Address reproductive health issues – Address driving if applicable – Address alcohol and substance use, screen for mental health concerns, and make appropriate referrals – Create a written HCT plan, in conjunction with patient and their family/caregiver by age 14 and update plan annually	– Increases participation in activities in #3 – Contributes to HCT plan	– See #3 – Applies for state based developmental services programs, estate accounts, applies for adult Medicaid, SSI – Contributes to HCT plan
5. Establishment of medical decision-making capacity prior to reaching age of majority	Mid adolescence (16–18 years)	– Establishing medical decision making, need for health care proxy, provision of resources to patient and family	– Increasingly participates in discussions about his/her future (e.g., during health care visits, educational planning meetings, service coordinators, etc.).	– Is aware of need to explore and address medical decision making and community resources
6. Each specialist develops their component of the transition plan and reviews it regularly during visits	Mid- to late adolescence (16–21 years)	– Medical team seeks updates, clarifications during discussions around medical decision making, education and social planning	– Can identify progress in building transition skills – Is able to articulate health care needs, diagnoses, and/or participates in creating an accessible summary	– Works with health care, education, and community services teams in transition planning, seeks support through clinic, school, and community resources to greatest extent possible

(Continued)

TABLE 43.1 (Continued)

Health care transition steps	Age range	Physician role	Youth-centered participation to the greatest extent possible	Family centered
7. Transfer of care: identification of adult medicine provider(s)	Late adolescence/ early adulthood (18 years or older)	– Addresses any diagnostic testing needed – Discusses details around timing of transition for different specialists – Identifies adult medicine provider(s) – Prepares written summary to accepting provider	– Brings information to new provider and prepares to take a primary role in visit with new adult medicine providers if appropriate	– Addresses insurance coverage continuation or makes alternate plans – Supports or assists in making appointment with new adult medicine providers
8. Pediatric medicine provider reviews transition with young adult, family, and adult medicine provider	Time of transition (18 years or older)	– Addresses any systems-based barriers to transfer of care (e.g., provision of medical summaries) – Documents care transition in the medical record	– Learns how to communicate with new adult medicine provider (e.g., virtual messaging on health app, office phone)	– Provides support to young adult to participate to the highest level in first visits with new adult medicine providers – Assists with systems-based barriers (e.g., insurance, helping with forms as needed)

functional domains, such as the ability to participate in work and community activities [5,6]. Despite years of increased awareness of the growing importance of transition within pediatric care, data suggest continued unmet needs within neurodevelopmental populations and families feeling unprepared for leaving pediatric care systems [7].

While many general and diagnosis-specific approaches and models exist to promote successful medical transition, there are some common principles which have been provided in a consensus statement endorsed by the American Academy of Neurology, Child Neurology Society, and the American Academy of Pediatrics [8]. Based on age, these principles can guide anticipatory guidance and topics to discuss during visits. The framework for health care transition for children with special health care needs proposes six conceptual elements and steps to ensure successful transition, which are summarized below and modified for youth with intellectual and developmental disabilities, and which are discussed in greater detail throughout the chapter [3,4,9].

1. *Conveying the need to transition at adult age to the youth and family*. Identifying a health care professional on the team who can act as the main point person for guiding the youth and the family through the transition period is often helpful to streamline the transition process. This professional can clearly communicate any office policies or preferred timetables for transition.
2. *Ensuring that the health care team members have the necessary knowledge and skills to provide developmentally appropriate care throughout the transition process*. The main point person and key team members should have a working knowledge of the transition process, community resources, and referral options. Practitioners should document topics discussed pertaining to transition and assess the youth's knowledge of their health condition, medications, and emergency procedures.
3. *Creating and maintaining a portable medical summary that is accessible to the health care team, family, and youth and is updated at regular visits*. This document promotes tracking and monitoring of progress during the adolescent years. Digital formats are easily accessible by smartphone technology or portal health care apps and are not as prone

to loss as paper copies. Youth can enter information in the health section of their smartphone or tablet to use as a reminder of key medical team members and their contact information to facilitate medical information sharing. In addition, the portable medical summary will also serve as an important piece of the individual's medical history and provide their new physician with a succinct synopsis of their care prior to transferring providers.

4. *Developing and updating a written health care transition plan during medical visits for each specialist who is actively providing care*. Families often benefit from guidance around applying for insurance and other social supports, referrals for adult medicine practitioners, and timing of different specialty transitions. Focusing on transitioning specialists who address low acuity and complex medical problems, or primary care, can be a good transition strategy to avoid families and youth feeling none of their practitioners have a good grasp of the overall medical history.
5. *Addressing age-appropriate care issues, such as reproductive health care and medical decision-making issues and providing referrals and resources to support the youth and their family as needed*. Sexual education may be limited in school and home. Opportunities to discuss and refer for sexual and reproductive health needs during adolescence should not be missed or postponed. For youth with intellectual and developmental disabilities, assessment of functional abilities in young adolescence is an important step to prompt discussions around medical decision making and financial planning.
6. *Working to identify resources for affordable and maintainable insurance coverage into adulthood and applying for programs which provide adult-based services to support health*. Based on the state of residence, families may need reminders to apply for continued insurance which may expire and to consider application for adult care supports as appropriate through state-based programs, such as medical assistance. Adequate health insurance will ensure the patient's ability to access medications, continue to see medical team members for appointments, and seek emergency care. Supplemental state-based and disability funding may cover expenses such as therapies, transportation, technology for communication, and health expenses not covered by insurance. Families may experience long wait times to receive this funding due to limited state resources.

Illustrative case continued: Tyler is now a 10th grader and participates in an integrated high school program. He receives special education for reading and math but is in inclusive classes for the rest of his subjects. He enjoys lifting weights with the football team and helping with equipment at games. He does weightlifting and badminton for adapted competitive sports. Tyler is interested in working in a sports-oriented job after high school. He participates in his education planning meetings. He uses a timer to remember to take his medication. As an adolescent, he is able to hand his medical history to his health care team, which is recorded on his cell phone and answer questions about his health.

As an adolescent, Tyler is progressing with his self-care and community skills. He appears to be well-engaged during his visits. He and his parents hope he can participate in a job training program with a local sports store, although he would like to work at a sports stadium eventually. You note that Tyler expresses self-determination and has goals for his life as an adult and takes responsibility for his health and chores at home. Given his current status, he may be best served by a supported medical decision-making option where he designates a trusted adult to help him make decisions. Although he has just turned 15 years of age, you discuss the need to ensure continuation of health insurance into adulthood and let Tyler and his family know that sometime after his 18th birthday, he could transition to adult care. You review the hospital transition policy and mention that many of your patients choose to transition to the adult Down syndrome program at the university adult internal medicine clinic.

While the actual age of transfer to an adult medicine physician may vary according to a patient's specialized needs, practitioner, or institution specific policies, as well as according to family and youth readiness, the need to convey expectations for timing is essential. A clear discussion of timing should occur with the aim of addressing the goals of transition within a reasonable frame of time. Transitioning prior to skill development or preparation, or during a medical crisis or transition (e.g., changing seizure medications), is not recommended. The National Health Care Transition Center [9] has developed a general timeline for transition discussions to begin between the ages of 12 and 14 years with the goal of completing the transition process and transfer to adult care by the age of 26 years. This age coincides with provisions from the Affordable Care Act that allows for dependent coverage under parental insurance. Periodic assessment of transition-specific skills (e.g., knowing names and side effects of medication, who to call in the event of an emergency, etc.) can be done through clinic-specific questionnaires or a more general one such as the Transition Readiness Assessment Questionnaire [10], which has been utilized for several clinical populations of youth. This type of tool provides specific skill assessment that can be useful for goal setting and reviewing progress and can be incorporated into the electronic health record.

Preparing for adulthood: medical decision making

Due to the wide spectrum of intellectual and developmental disabilities, decision making and legal needs vary for each individual. Specifically, the degree of intellectual disability plays a large role in what types of decision making and supported decision making may be appropriate. Previous research has shown that individuals who have the ability to practice greater self-determination have greater levels of independence and may be able to play an integral role in medical care decision making. Use of school-based assessments, neuropsychology testing, adaptive skill assessment, and information obtained during visits from the youth and family can help the physician make recommendations about what level of support may be needed. Consideration of dual diagnosis such as mental health diagnoses may also necessitate input from psychiatry to evaluate whether a mental health condition could impact medical decision making.

Ideally, the educational law mandating transition as part of the educational plan can keep the youth and family engaged in the planning process and can offer the youth greater participation in planning of his or her future. Multiple opportunities to discuss the future and review of the youth's strengths and needs can provide an emerging understanding of values and needs to guide and support decision making for the young adult with intellectual and developmental disabilities. The health care team should regularly encourage the family and the young adult to discuss and revisit advance care planning and consider engaging the clinic social worker to help facilitate these discussions. Hospitalizations and acute illness can also offer discussion opportunities to advance medical decision making further.

Prior to a greater awareness of the importance of self-determination and promoting autonomy to individuals with intellectual and developmental disabilities, guardianship was routinely sought by families, particularly plenary or full guardianship, compared to limited guardianship [11] with the intention of protecting their family member. Full guardianship is the most restrictive form of legal and medical decision making and removes self-determination rights. Guardianship allows for the legal system to appoint a surrogate to make decisions on behalf of another individual who is deemed by the court system to need protection or to be cared for and is incapable of making some, or all, decisions on their own. Varying degrees of guardianship are determined depending on an individual's capacity to make decisions for themselves. Families may not always be aware of other options, such as power of attorney and supported decision-making options, which may vary according to the state of residence. Discussions of the many options of supported decision making can be reviewed with the family in consultation with clinic social workers, local advocacy groups, and attorneys with experience in disability law.

In addition to addressing medical decision making, families need to be aware that once their child reaches their 18th birthday, the young adult has the ability to make decisions and has access to their health records. Families may retain direct ability to communicate with the health care team and access medical records if consent is obtained.

Addressing social determinants of health and other domains of transition to adulthood

Transition to adulthood can be considered conceptually by categorical domains (Table 43.2). While medical transition is broadly recognized in the health care system, other domains of transition are also critically important and can often have a significant impact on the health and well-being of adolescents and young adults—particularly those with intellectual and neurodevelopmental disabilities. Therefore in addition to health care transition, educational, vocational, legal, financial, and social transitions also must be considered as key social determinants of health.

Education and employment

Transitioning from secondary school to postsecondary options can be an overwhelming process for adolescents and young adults, as well as their family members and caregivers. Many of the supports, therapies, and programs that individuals and families have come to rely on are no longer available once the individual completes secondary school. Ideally, stress associated with losing high school supports would be alleviated through enrollment and participation in postsecondary education, vocational rehabilitation, employment, or through other programs for individuals with intellectual and neurodevelopmental disabilities [12].

The underlying diagnosis and the presence and degree of intellectual disability will inform options when working toward graduation from secondary school. Some may choose to pursue a standard diploma and plan to graduate around the age of 18. Others may choose to pursue a nonacademic track and remain in school until the end of the school year of their 21st year to build additional skills and experiences.

TABLE 43.2 Transition to adulthood across multiple domains.

Health care
- Transition from pediatric to adult provider (or to adult model of care)
- Continue age-appropriate preventive care, health maintenance, and screenings
- Transition of therapy and rehabilitation services outside of the educational system

Behavioral/Mental health
- May require transition to new provider or model of care
- Co-occurrence of anxiety and/or depression with many neurodevelopmental disabilities; increased prevalence during times of transition and major adjustment periods
- Continued need for behavioral therapy and supports (particularly for individuals with autism spectrum disorder and related conditions)

Legal/Decision making
- Options for guardianship, supported decision making, living will, power of attorney, etc.
- Certain methods of decision making (particularly guardianship) require planning/obtaining prior to reaching the age of majority (age 18 in most states)

Education and employment
- IEP in place with accompanying transition plan while in secondary school
- Decisions on completing secondary school (regular diploma or alternate diploma with options for free appropriate public education (FAPE))—determines graduation at 18 compared to 22
- Vocational rehabilitation, life skills courses, or other types of volunteer and work experiences during secondary education
- Postsecondary options: trade school, college, integrated/competitive employment, supported employment, or sheltered employment/workshops

Social and community
- Options for independent and supported living
- Wellness and social activities (weight management, organized activities, etc.)
- Relationships and sexuality

Financial
- Establishing own finances (obtaining independent bank account, using debit card, etc.)
- Applying for/maintaining SSI benefits
- Qualifying for SSDI
- Creating an ABLE account and/or Special Needs Trust

SSI, Supplemental security income; *SSDI*, social security disability insurance; *ABLE*, achieving a better life experience.

Students in special education will receive more supports to plan postsecondary school plans and enrollment in preemployment and adult-based services programs [13]. Students with individualized educational plans (IEPs) should be supported to apply for preemployment programs, support from state-based vocational rehabilitation, and other postsecondary schooling programs. While the IEP process should involve discussion of the most appropriate programs as part of transition planning (by age 14 or 16 in most states)—as mandated under the Individuals with Disabilities Education Act—families and youth need to be continuously actively engaged to receive maximal benefit before they exceed the age limits to apply for these programs. Unfortunately, this does not always occur.

Underemployment adversely impacts adult outcomes of individuals with developmental disabilities. Youth should begin to build their work skills while in high school by volunteering and seeking short-term employment to build career awareness and interests. Gaining competitive work experience prior to graduation can help negate poor employment outcomes by helping the student develop professional relationships, learn valuable work and communication skills, and create a strong work ethic [14]. Employment training and job exposure opportunities should be integrated into the IEP. By the time the youth completes secondary education, the young adult can consider further education or employment. Based on prior employment experiences and level of skill, a young adult can consider integrated versus supported employment settings. Sheltered workshops are now considered restrictive and segregated due to concerns of poor compensation and lack of training leading to job advancement.

More college programs are offering specialized learning supports and programs for individuals with developmental disabilities, and students can pursue further education and build independence in these settings. College level education can also be degree or nondegree driven, depending on the goals and abilities of the student. Despite the growing number of programs, individuals with neurodevelopmental conditions are overall less likely to pursue postsecondary education. Unfortunately, with fewer personal development and social opportunities, they also experience higher degrees of un- or underemployment and participate in fewer leisure and social activities. Without an age-appropriate peer group once out of high school, young adults with neurodevelopmental disabilities rely more on their families for living arrangements and health maintenance. In individuals with cerebral palsy, patients experience poorer health than their peers without disabilities even at young ages. In addition to potentially having lower income due to employment status and less health care options, poorer health may also reflect the challenges many families face of having to support and manage all the medical and social well-being needs of their adult child [15].

Financial planning

Individuals with disabilities are at increased risk for poverty. With concerns about loss of family support over time, families and caregivers need to consider financial planning for the future as part of the transition process.

Legislation to allow the creation of tax-free savings accounts for individuals with disabilities was passed into law in 2013 as the Achieving a Better Life Experience (ABLE) Act. ABLE accounts can be used to pay for "qualified disability expenses," which can include basic costs of living, education and educational supports, food, employment, transportation, technology, supportive services, and other needs. Importantly, ABLE savings and investment accrual do not negatively impact eligibility for public benefits such as Medicaid. Special needs trust accounts can also be used to pay for additional items and comfort measures including assistive technology and therapy that is not covered by Medicaid, home furnishings, and other entertainment or leisure-related activities.

Supplemental security income (SSI) provides minimum basic financial assistance to older adults and persons with disabilities (regardless of age) with very limited income and resources. Children younger than 18 with certain disabilities and/or medical conditions can qualify to receive payments based on their personal income, as well as their family's income and resources. At age 18, as the individual becomes an adult, the need for continued payments and resources are reviewed. Family resources and income are not considered after the adolescent reaches age 18. Of note, eligibility for SSI can be jeopardized by marital status and the spouse's resources. Social security disability insurance (SSDI) is another federally supported program that supports individuals who are disabled and have a qualifying work history.

In addition to federal and state-funded programs like ABLE account programs, SSI/SSDI, other options, such as establishing a Special Needs Trust, can also help support and provide for the individual as they continue to move into adulthood. A Special Needs Trust can be used to help protect a person's assets—like money received through a legal settlement, an inheritance, or monetary gifts that allow an individual to retain eligibility for SSI and other benefits. Trusts often require legal assistance to establish, and attorneys who are experienced in Disability Law can provide necessary guidance.

Social and living arrangement transition

Young adults with intellectual and neurodevelopmental disabilities may have greater difficulty forming friendships, understanding and developing sexual relationships, and accessing and engaging meaningfully in their local community compared to their nondisabled peers [16,17]. This can be especially difficult during the transition after secondary education and is accentuated by the significant social loss experienced when leaving the school setting and being away from many of their previous social supports.

Transition planning as part of health care visits and as part of the IEP process should include consideration and discussion of current and future living arrangements. Many individuals with intellectual and neurodevelopmental disabilities express interest in living in independent or semi-independent settings. Transition Readiness Assessments can focus on skills necessary for independent or supported living arrangements, if desired. These skills can include self-care and personal hygiene practices, safety principles (e.g., during cooking), transportation and public safety skills, independent decision making, and time/resource management skills. Social and leisure activity interests can help promote independence in daily life. Alternatively, some individuals will choose to continue to reside in their family home for a period and then transition to another setting over time as independence in skills advance and when housing options become available.

One area that is often overlooked by both health care providers and often by family members and caregivers is providing appropriate sexual education and counseling. Individuals with intellectual and neurodevelopmental disabilities commonly have similar interests in relationships and sexual activities, although sometimes at slightly older ages than their peers without disabilities [18]. Individuals with disabilities may experience disparities, and in some cases, oppression and/or discrimination when it comes to their sexuality and sexual health. Some US states have laws in place that dictate that individuals with certain diagnoses are unable to consent to sexual relationships. Many schools and special education programs either do not include students with disabilities or do not offer developmentally appropriate health and sexual education courses. Families and caregivers can have difficulty recognizing, or accepting, the individual's sexuality, and health care providers often overlook, or are uncomfortable with, discussing these topics and providing appropriate health counseling and screenings related to sexual health and reproduction [19].

The need to have meaningful relationships in adulthood should be addressed during health visits, and the neurodevelopmental physician should have referral suggestions available to facilitate addressing reproductive health and sexual education. The physician should also be prepared to act as a mediator between patients and their families when discussing these topics.

In addition to romantic relationships and sexuality, preparation for navigating other social interactions and relationships is equally important. Focus is often placed on the need for opportunities to socialize and build community; conversely, it is also imperative that individuals with neurodevelopmental and related disorders receive coaching and instruction in recognizing and responding to unhealthy social interactions. For example, the ability to recognize when they are being bullied or exploited, how to end conversations and interactions they no longer wish to be a part of, recognizing dangerous situations, and understanding how others may view their responses and coping mechanisms. Coaching individuals early on can lead to improved social interactions and emotional safety, as well as physical safety.

Sensory seeking, avoidant, impulsive, or aggressive behaviors can all potentially be misinterpreted by others and may endanger individuals with disabilities when interacting with police, first responders, or the general public. Coaching individuals with disabilities to interact appropriately with those in authority and helping them understand how behavior may be misinterpreted is essential. Some individuals may benefit from a health alert card or phone ID providing pertinent information or registration with the local police station and may help promote safety in the community, particularly, if there are communication deficits.

Transition-related adjustments

Families may experience feelings of being overwhelmed with the many issues that need to be addressed during the adolescent years. Multiple transitions occurring during this phase may overwhelm the youth and their families (Fig. 43.1). It is critical for the health care team to appreciate how many potential concurrent transitions that may be happening for the young adult and family, such as insurance coverage changes, graduation from high school, as well as parent and family changes. In addition, many services that were covered within pediatric care, such as rehabilitation therapies,

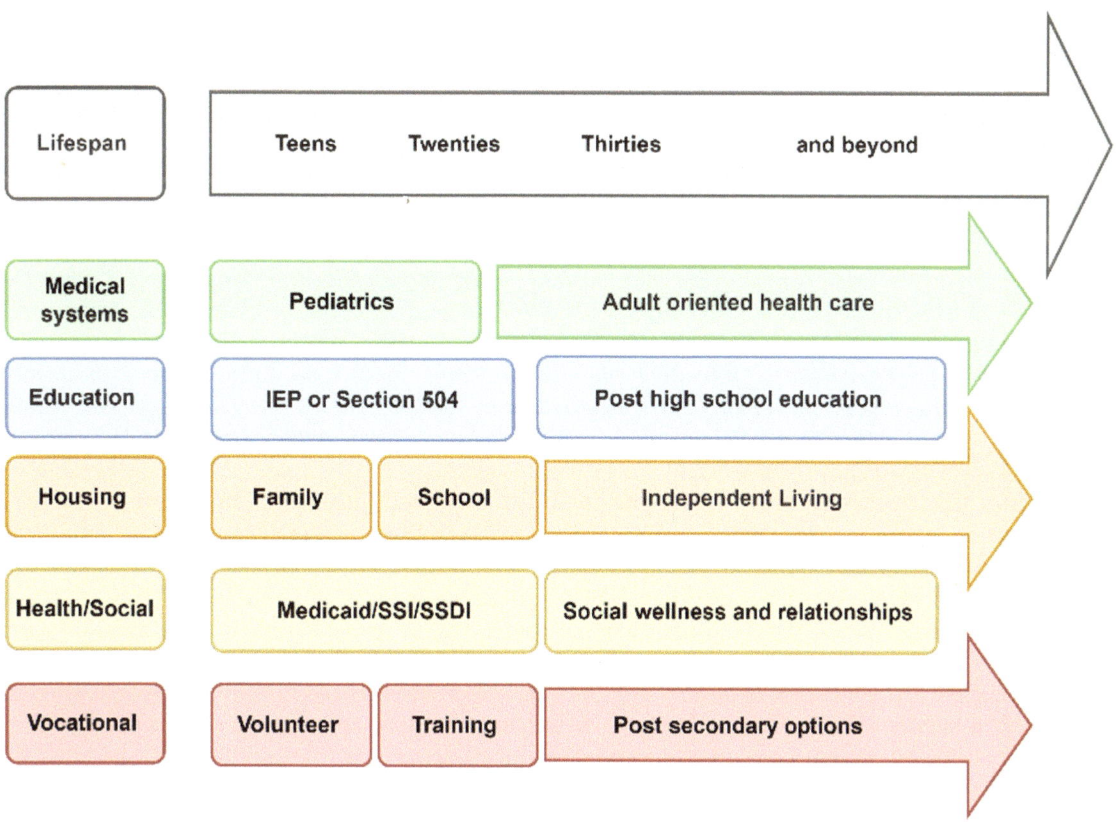

FIGURE 43.1 Multiple transitions occurring during time of late adolescence and early adulthood across different systems.

may no longer be covered by adult level insurance or after school graduation. Lower levels of physical activity due to not attending school, limited opportunities for social interaction, and therapies may be associated with weight gain and decline of mobility status. Increasing awareness of this during adolescent years and encouraging regular exercise is an important health maintenance topic.

Parents may need time and support to adjust to the role of a supportive coach rather than a primary caretaker to promote independence of their adolescent or young adult. They may also experience shifts in family structure and need to juggle care for multiple family members or experience changes in their own health which may accentuate concerns for their adult child's future. Families may anticipate a sense of impending loss when thinking of leaving a care team who has supported them for years to unknown health care settings.

Young adults may also experience significant adjustments when school-based programming ends. The social opportunities that school provides may not be fully realized until school is completed. Community programs to support young adults and accessing transportation to keep in touch with friends may be limited. Physicians can help youth and families prepare for the dramatic shifts occurring after school ends by encouraging community engagement and participation in activities that include social interaction during adolescence. Case coordinators through state and community agencies may be an important resource for information and access to programs, transportation supports, and housing options.

Physicians may feel unprepared to support the many psychosocial needs of transition age youth and their families. To appreciate the incremental steps that should be taken to prepare their patients and families, it may be conceptually helpful to consider a more holistic perspective of ongoing brain maturation and protracted cognitive development in typical development. The importance of focusing on transition issues within the context of social determinants of health (e.g., working knowledge of community, social support programs, and accessing work opportunities) may help re-frame the various complex needs transition of care requires. Using a team approach and checklist (Fig. 43.2) to address community-based resources, consulting the clinic social worker, and collecting community organization brochures can shift transition tasks to those with a working knowledge of resources. Some specialized transition programs offered through medical specialties such as adolescent medicine or medicine-pediatrics clinics may have increased capacity to focus on complex transition needs and also address sexual and reproductive health and socially appropriate interaction issues.

Once a young adult is ready to transition to an adult health care practitioner who has been previously contacted to accept the transfer, a medical summary letter should be sent to the accepting physician detailing pertinent history, the ability of the youth to participate in medical decision making, prior diagnostic tests, and active issues. Feedback should be sought from the patient and practice after the transition to ensure that the patient came to the appointment and to see if there are questions. The patient should have contact information for both offices, and refills of medication should be available in case of delays between the pediatric to new adult practitioner appointment. Decision making and legal issues should be documented in the medical record and transfer summary letters.

Aging in neurodevelopmental disabilities

Illustrative case continued: Tyler is now in his mid-thirties. He has successfully transitioned to the adult Down syndrome program and his family physician comanages his health care needs with the specialty clinic. Tyler works 10–15 hours per week at the local sports store in inventory management and assists customers. The store manager mentioned to Tyler's father that Tyler seems to be forgetting where some things are located and does not seem to be himself. Upon further questioning, it is revealed that Tyler's maternal uncle suddenly passed away and Tyler has been unable to attend the adaptive sports activities in which his uncle faciltiated Tyler's participation. On physical examination, Tyler had a slightly enlarged thyroid, so thyroid function tests were ordered.

There are many health conditions that become more prevalent as an individual ages. For individuals with neurodevelopmental disabilities, these same conditions can affect them with age but may present differently than in the general population. Primary care practitioners caring for aging individuals with intellectual and developmental disabilities should be aware of these differences and screen patients appropriately. Specifically, differences in cognitive abilities, speech and language, other cooccurring conditions, and overall medical complexity can make screening for, diagnosing, and managing conditions such as depression, anxiety, or cognitive regression challenging. Reports and observations from those who know the individual the best are critical to identify clinically significant changes and identify associated symptoms.

Individuals with intellectual disability are living to older ages, with some specific populations having life expectancies at near parity with the general population. Individuals with moderate and severe intellectual disability with well-managed health care conditions are commonly living into their fifth and sixth decades, including adults with Down syndrome. However, adults with Down syndrome may experience conditions commonly associated with senescence at accelerated rates such as cataracts, presbycusis, osteoporosis, and dementia. Thyroid disease is thought to occur at a

Age to address	*Transition step*	*Comment*
14 yrs	☐ Youth should attend and participate in IEP meetings	Youth plan and prepare for meeting with teachers and family to promote increasing contribution based on ability. IEP should include goals for education, employment, and living arrangements after high school.
	☐ Transition planning for after high school ends – vocational training, other community-based programs or post-secondary education	Exploration of diploma and non-diploma pathways available; foster youth's interest in different work and educational options
	☐ Application to state developmental disability	Eligibility and timing of application for state funded programs vary from state to state
	☐ Assess and practice independent living skills	Youth practice using transportation, budgeting, shopping, cooking, and housekeeping
	☐ Identify and address learning needs around practical money management	
16-17 yrs	☐ Apply to state vocational training programs or other pre-employment programs	Vocational rehabilitation services or program personnel can participate in IEP; explore community options for supported or independent employment
	☐ Obtaining an ID card or driver's license application	Youth should collect key health documents, education documents in a centralized secure place.
	☐ Volunteer or work experiences in community; create career development documents	Short term experiences can expand youth interest in different occupational fields or course of study Career development document development includes resume, cover letter, reference list writing; practice interviewing skills and identify accommodations to support success in work environments
	☐ If attending post-secondary education, explore school options and prepare to take standardized testing as appropriate	
	☐ Identify needs for financial support (e.g. Supplemental Security Income, Medicaid, other state systems)	
	☐ Addressing legal issues around advance directives, power of attorney, or guardianship	Consultation w/ attorney specialized in disability law
18-21 yrs	☐ Youth apply for Supplemental Security Income from the Social Security Administration	
	☐ Youth apply for Medicaid/Medical Assistance	
	☐ Voter registration	
	☐ Establishment of ABLE funding, trusts, and other funding resources to cover expenses into adulthood	Consultation w/ advocacy organizations and/or attorney to establish fund
	☐ If attending post-secondary education, initiate accommodation process with school's office of disability services.	

FIGURE 43.2 Checklist of transition tasks focused on medical and nonmedical domains by approximate recommended age.

higher rate in adult populations compared to children with Down syndrome. Increased musculoskeletal complications negatively impacting functional mobility and causing pain can be noted in adults with cerebral palsy. Chronic cardiac conditions commonly seen in younger patients with Prader Willi or Williams syndrome need ongoing monitoring for symptom progression. Given potentially limited experience with health surveillance for individuals with rare disorders, health supervision recommendations should be detailed in transition documentation for adult practitioners. For those with conditions where there may be a specialized adult clinic available within reasonable geographic proximity, this may be an option for specialized health surveillance [20].

For patients with dual diagnoses including mental health conditions or substance use, finding practitioners who have experience with adults with disabilities may be challenging. Families may need to find multiple adult medicine practitioners to treat different symptoms that may have been addressed by a single physician in pediatric care. Training needs to provide physicians and other health care practitioners with experience in disabilities is an ongoing need.

Patients and their families should be alerted that routine surveillance for cancer screening, risk factors for heart disease, and other routine primary care should be pursued. Maintaining exercise and avoidance of excessive sedentary habits should be highlighted as an important health priority.

Across different intellectual and neurodevelopmental conditions, the process of aging—and accelerated aging—varies and can be linked to the underlying condition. For instance, in Down syndrome, the excess genetic material associated with the additional chromosome 21 underlies much of the neurobiological changes. Specifically, the excess genetic material encoding amyloid precursor protein is thought to be directly associated with increased levels of amyloid beta and the increased rate and earlier onset of Alzheimer's disease in individuals with Down syndrome [21,22].

As individuals with intellectual and neurodevelopmental disabilities age, it is imperative that they continue to have a close relationship with their health care providers. In recent literature of individuals with disabilities and other forms of chronic disease/childhood diseases (such as congenital heart disease, sickle cell disease, and childhood cancers), the need for health screening, continued medical services, and specialized care into adulthood has been highlighted [23]. For example, in a recent review of individuals with cerebral palsy, patients, their family members, and caregivers reported difficulties accessing needed services and providers, despite having increased health needs, and in some instances, higher utilization of emergency and inpatient hospital services. In their review, difficulties in transition were cited as barriers to continuing to receive needed health services [24]. Other barriers may include lack of adult specialists with experience in childhood onset conditions, including developmental disabilities and limits in care coordination across multiple specialists.

Areas for future research

Given ongoing unmet needs for patients with intellectual and developmental disabilities and families, there is a need to provide consistent assessment of transition readiness and ongoing planning, resource allocation, and psychosocial support. Each transition may have different components based on the individual's needs, and plans may reflect regional and state laws and available programs. Successful transition is when patients and families do not get lost in the process and arrive in adult health care systems having already addressed medical decision making and with the young adult having some predefined ability to participate in meeting their health care needs and direct their care. Medical systems can increase their capacity to provide more systematic education and guidance starting in young adolescence, and adult and pediatric systems can continue to develop ways to freely communicate to ensure a safe transfer and ability to check back with the pediatric practitioners for guidance on issues. Transition takes time and adequate health reimbursement for building transition into medical visits can pose significant barriers for practitioners.

Given low priority of health insurance coverage for transition specific care and providing evidence of the importance and effectiveness of transition is an ongoing need. While a consensus definition of what the health care transition process involves has been published, there remains tremendous variation in approach. For example, there are several models of transition, different end points specific to individual patients and for specialized populations, and many systemic factors to consider in studies of transition success. In order to conduct high-quality research that is impactful to the field of health care transition, it is essential to determine key elements and outcomes to be measured and reported [14,15] in response to targeted interventions. Impact of transition-focused interventions on long-term health care outcomes and cost are other future areas of investigation and may benefit from focusing on specific populations of patients which would take natural history and prognosis into consideration.

Future research identifying successful strategies and programs are necessary to advance care and promote healthy outcomes. In addition, the creation of outcome measures that can be measured and tracked after transition to the adult health care system are also needed [25,26]. Given concerns for health inequities in health maintenance and screenings in adult disability populations, studies linking transition interventions and adult health outcomes are a future area of investigation, as well as tracking health satisfaction from the young adult, family, and practitioner perspectives.

In summary, transition is an important developmental step for patients and families. Health care during this time needs to be intentional and should make use of the time during adolescence to ensure that assessment, skill building, and discussions take place. Setting expectations early in adolescence and regularly providing care during this phase can help smooth the transition process. Transition is but one step in the ongoing care of people with neurodevelopmental and related disabilities. Continued monitoring and adjustment of the program based on the individual, family, and community are important pieces of the process. More systematic efforts to study and measure outcomes can advance health care in patients with neurodevelopmental disabilities.

Resources

https://www.gottransition.org/

Gottransition is a federally funded web-based resource to provide information about health care transition directed at youth consumers, medical professionals, and families. There are checklists and links to nonhealth transition resources.

https://www.childneurologyfoundation.org/free-downloads-transitions-tools-use-practice/

The Child Neurology Foundation has tools for download including youth and family checklists, templates for medical summaries and transfer letters, as well as a sample transition policy.

References

[1] Blum RW, Garell D, Hodgman CH, Jorissen TW, Okinow NA, Orr DP, et al. Transition from child-centered to adult health-care systems for adolescents with chronic conditions. A position paper of the Society for Adolescent Medicine. J Adolesc Health 1993;14:570–6. Available from: https://doi.org/10.1016/1054-139X(93)90143-D.

[2] Fair C, Cuttance J, Sharma N, Maslow G, Wiener L, Betz C, et al. International and interdisciplinary identification of health care transition outcomes. JAMA Pediatr 2016;170:205–11. Available from: https://doi.org/10.1001/jamapediatrics.2015.3168.

[3] American Academy of Pediatrics, American Academy of Family Physicians and American College of Physicians-American Society of Internal Medicine. A consensus statement on health care transitions for young adults with special health care needs. Pediatrics 2002;110:1304–6.

[4] American Academy of Pediatrics, American Academy of Family Physicians, American College of Physicians-American Society of Internal Medicine. Supporting the health care transition from adolescence to adulthood in the medical home. Pediatrics 2011;128:182–200. Available from: https://doi.org/10.1542/peds.2011-0969.

[5] Nugent J, Gorman G, Erdie-Lalena CR. Disparities in access to healthcare transition services for adolescents with Down syndrome. J Pediatr 2018;197:214–20. Available from: https://doi.org/10.1016/j.peds.2018.01.072.

[6] McManus MA, Pollack LR, Cooley WC, McAllister JW, Lotstein D, Strickland B, et al. Current status of transition preparation among youth with special needs in the United States. Pediatrics 2013;131:1090–7. Available from: https://doi.org/10.1542/peds.2012-3050.

[7] Lebrun-Harris LA, McManus MA, Ilango SM, Cyr M, McLellan SB, Mann MY, et al. Transition planning among US youth with and without special health care needs. Pediatrics 2018;142:e20180194. Available from: https://doi.org/10.1542/peds.2018-0194.

[8] Brown LW, Camfield P, Capers M, Cascino G, Ciccarelli M, de Gusmao CM, et al. The neurologist's role in supporting transition to adult health care. Neurology 2016;87(8):835–40. Available from: https://doi.org/10.1212/WNL.0000000000002965.

[9] The National Alliance to Advance Adolescent Health. Six core elements of health care transition. Got transition; 2014–2021. https://www.gottransition.org/six-core-elements/.

[10] Wood DL, Sawicki GS, Miller MD, Smotherman C, Lukens-Bull K, Livingood WC, et al. The transition readiness assessment questionnaire (TRAQ): its factor structure, reliability, and validity. Acad Pediatr 2014;14(4):415–22.

[11] Jameson JM, Riesen T, Polychronis S, Trader B, Mizner S, Martinis J, et al. Guardianship and the potential of supported decision making with individuals with disabilities. Res and Pract Pers Sev Dis 2015;1:36–51. Available from: https://doi.org/10.1177/1540796915586189.

[12] Lounds J, Seltzer MM, Greenberg JS, Shattuck PT. Transition and change in adolescents and young adults with autism: longitudinal effects on maternal well-being. Am J Mental Retard 2007;112:401–17.

[13] Gillian D, Coughlan B. Transition from special education into post-school services for young adults with intellectual disability: Irish parents' experience. Pol Pract Intellect Disabil 2010;7:196–203.

[14] Targett PS. Finding jobs for young people with disabilities. In: Wehman P, editor. Life beyond the classroom: transition strategies for young people with disabilities. 4th ed. Baltimore: Brookes; 2006. p. 255–88.

[15] Palisano RJ, Di Rezze B, Steward D, Rosenbaum PL, Hlyva O, Freeman M, et al. Life course health development of individuals with neurodevelopmental conditions. DMCN 2017;59:470–6. Available from: https://doi.org/10.1111/dmcn.13402.

[16] Cory L, Dattilo J, Williams R. Effects of a leisure education program on social knowledge and skills of youth with cognitive disabilities. Ther Recreat J 2006;40:144–64.

[17] Foley KR, Dyke P, Girdler S, Bourke J, Leonard H. Young adults with intellectual disability transitioning from school to post–school: a literature review from within the ICF. Disab Rehabil 2012;34(20):1747–64. Available from: https://doi.org/10.3109/09638288.2012.660603.

[18] Khan NF, Halpern CT. The relationship between cognitive ability and experience of vaginal, oral, and anal sex in the United States. J Sex Res 2018;55(1):99–105. Available from: https://doi.org/10.1080/00224499.2016.1247149.

[19] Schmidt EK, Hand BN, Havercamp S, Sommerich C, Weaver L, Darragh A. Sex education practices for people with intellectual and developmental disabilities: a qualitative study. Am J Occup Ther 2021;75(3). Available from: https://doi.org/10.5014/ajot.2020.044859.

[20] American Bar Association. PRACTICAL tool for lawyers: steps in supporting decision-making. American Bar Association; 2016. Accessed from: http://www.ambar.org/practicaltool.

[21] Perkins EA, Moran JA. Aging adults with intellectual disability. JAMA 2010;304(1):91–2. Available from: https://doi.org/10.1001/jama.2010.906.

[22] Cipriani G, Danti S, Carlesi C, Di Fiorino M. Aging with Down syndrome: the dual diagnosis: Alzheimer's disease and Down syndrome. Am J Alz Dis 2018;33(4):253–62. Available from: https://doi.org/10.1177/1533317518761093.

[23] Yeh JM, Ward ZJ, Chaudhry A, Liu Q, Yasui Y, Armstrong GT, et al. Life expectancy of adult survivors of childhood cancer over 3 decades. JAMA Oncol. 2020;6(3):350–7. Available from: https://doi.org/10.1001/jamaoncol.2019.5582.

[24] Manikandan M, Kerr C, Lavelle G, Walsh M, Walsh A, Ryan JM. Health service use among adults with cerebral palsy: a mixed methods systematic review. Dev Med Child Neurol 2021. Available from: https://doi.org/10.1111/dmcn.15097.

[25] Gabriel P, McManus M, Rogers K, White P. Outcome evidence for structured pediatric to adult health care transition interventions: a systematic review. J Pediatr 2017;188:263–9. Available from: https://doi.org/10.1016/j.jpeds.2017.05.066.

[26] Campbell F, Biggs K, Aldiss SK, O'Neill PM, Clowes M, McDonagh J, et al. Transition of care for adolescents from paediatric services to adult health services. Cochrane Database Syst Rev 2016;4. Available from: https://doi.org/10.1002/14651858.CD009794.pub2.

Chapter 44

Management in low-resource settings

Ryan Lee[1,2] and Andres Jimenez Gomez[3,4]

[1]*Department of Pediatrics, University of Hawaii - Manoa, Hawaii, HI, United States,* [2]*Shriners Hospitals for Children - Honolulu, Hawaii, HI, United States,* [3]*Neurodevelopmental Disabilities Program, Joe Dimaggio Children's Hospital, Hollywood, FL, United States,* [4]*Stiles-Nicholson Brain Institute, Florida Atlantic University, Jupiter, FL, United States*

Introduction

The latter half of the 20th century brought substantial changes to the global economy. Following the recovery and reconstruction after the Second World War, there was explosive population growth. Over the past 50 years, this growth has been particularly pronounced in nations and regions outside of "industrialized" or high-income countries (HICs) in Europe and North America. For instance, China's population grew from about 800 million in 1970 to over 1.4 billion by 2020; India similarly increased from 500 million to 1.3 billion citizens during the same period; and a similar trend of exponential growth has been observed in Latin America and former colonial states in Africa [1].

This demographic explosion occurred alongside a marked socioeconomic transformation that has shaped healthcare resources, individual financial capacity, and access to nutrition, preventive care, and childcare. However, this transformation process has also led to significant inequities. The structure of healthcare systems and resources—from infrastructure to personnel training and education—has resulted in substantial differences in key morbidity and mortality indicators across societies (Fig. 44.1). Data from the Global Burden of Disease Report show that the life expectancy of a person born in one of the world's poorer countries has consistently been more than 30 years shorter than that of a person born in one of the world's wealthiest nations [2]. These data highlight stark health disparities, with a disproportionate burden of disease on the poorest individuals in the world's poorest countries.

Notwithstanding these challenges, a number of pre-, perinatal, and early life interventions have significantly impacted the global disease burden across different societies, particularly by reducing mortality [3]. Low-cost, widely adopted interventions, such as hand washing, access to clean water, prenatal care, nutritional supplementation (e.g., vitamin and folic acid during pregnancy), fortification (e.g., iodine-fortified salt), vaccination, and premorbid screening (e.g., neonatal dried blood spot tests for thyroid function, pulse oximetry for congenital heart disease), have contributed to the decrease in global maternal and infant mortality in recent decades. Special attention has been paid to the resulting increase in chronic, noncommunicable diseases (NCDs) and disabilities, as well as their impact on developmental

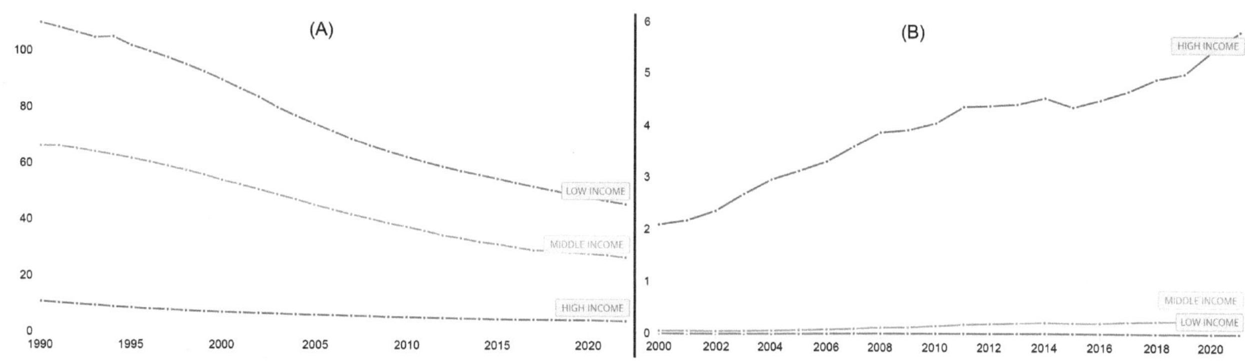

FIGURE 44.1 (A) Infant mortality trends per 1000 live births between 2000 and 2022 in each World Bank region, compared to (B) Health expenditure per capita in Thousand US$, 2000 to 2022. *From World Bank Data http://data.worldbank.org/. License CC BY-4.0.*

Capute and Accardo's Neurodevelopmental Disabilities in Infancy and Childhood. DOI: https://doi.org/10.1016/B978-0-12-824060-1.00024-9

potential. Longitudinal studies have demonstrated variability in this potential across societies, highlighting the need for interventions not only in healthcare on a global scale but also in educational and societal domains [4]. Care is no longer limited to the prevention of death; instead, longitudinal interventions must be applied, rather than just cross-sectional points of care, when assessing capacity and infrastructure needs. For example, when measuring the health needs of a community, long-term therapy, equipment, and support services must be considered in addition to immunization rates. Healthcare infrastructure in all communities relies on resource measurement and data to address disparities.

How can we bridge the divide in low-resource settings (LRSs)? In the following discussion, we outline key concepts related to the care of people with neurodevelopmental disabilities (NDDs) in LRSs, drawing on a global perspective. From defining the problem to providing examples of successful interventions, we propose a framework for establishing, developing, and maintaining care. This framework emphasizes the importance of considering socioeconomic, demographic, and cultural-ethnic factors in the delivery of care.

Understanding the low-resource setting

The establishment and maturation of local programs for managing NCDs are often significantly supported, mediated, and influenced by external actors. Foreign institutions, whether academic, medical, or financial, frequently lead these efforts. While many of these initiatives have reported short- or long-term success, it is crucial to pay special attention to those with mixed or negative outcomes, particularly from the perspective of the intended beneficiaries [5]. Often, projects aimed at supporting healthcare development in LRSs are introduced by foreign parties, with special interests that may conflict with those of the local stakeholders. To mitigate detrimental effects of such interventions, it is essential to clearly define key concepts, principles and axioms that underpin our approaches.

What constitutes a low-resource setting?

The concept of "low resource" is relative to the infrastructure and resources available in a given location. LRSs are defined as health systems that do not meet the accepted "norms" for healthcare as described by the World Health Organization, particularly in terms of quality of care, availability, and access [6]. Importantly, what defines a setting as "low resource" depends on many factors beyond economy and expenditure of the local government. Therefore, LRSs exist regardless of geographical location, including within HICs defined by the World Bank, which in 2022 classified HIC as those with gross national income of US $12,696 or more per capita . The intrinsic variability in services, particularly staffing, within the United States, has long been rcognized by the Health Resources and Services Administration through the designation of Healthcare Professional Shortage Areas, where primary care physicians are critically needed ro address inequities in care delivery. On a global scale, the categorization of LRSs requires clarification, as there are significant resource differences between low-income countries (LICs) and middle-income countries (MICs). In the United States, regions such as Appalachia and the Ozarks, as well as inner city and rural populations, also exemplify LRSs.

What resources are essential for a service setting? Assessing community needs beyond financial and infrastructural resources is paramount, as these concepts are highly complex. Van Zyl and colleagues [7] describe several major themes of a healthcare LRS, which include individual and societal financial shortages, underdeveloped infrastructure in both public and private facilities, limitations in telecommunications and connectivity, human resource limitations, restricted social resources leading to exclusion, geographic and environmental challenges , and a lack of knowledge and training influenced by local beliefs and practices. It is crucial to understand how theese factors and the plurality of feasible interventions are intertwined. In every situation, there are opportunities for impact and sustainability.

Resource poverty

It is important to note that the characteristics defined above stem from a HIC-centric perspective. One may consider these themes descriptive of a healthcare setting, rather than specifically defining a LRS, where one aspect may be lacking (e.g., financial, infrastructural), another may emerge to fill an apparent gap. The scarcity of knowledge, for instance,may lead to the development of novel approaches to care. Resourcefulness becomes a critical phenomenon, driven by both ingenuity and scarcity, in overcoming challenges. Regardless of the burden, resourcefulness is essential and can manifest in varying degrees of sophistication and effectiveness.

There are many examples of resourcefulness inspired by adversity in different practice settings. A well kown example is Kangaroo Mother Care. Developed in Colombia in the 1970s due to a shortage of incubator space for the sickest newborns, clinicians emulated the child-rearing practices of marsupials in an effort to reduce neonatal mortality.

This method, which involves skin-to-skin contact between mother and infant, is still widely used today, even in the most resource-rich settings. Research on Kangaroo Mother Care has expanded beyond its impact on mortality to explore its effects on neurodevelopment, attachment, and other outcomes [8]. Another example is the Jaipur leg, an orthopedic prosthetic developed in North India as an affordable alternative to traditional prosthetics. Traditional prosthetics are extremely costly and often do not provide the necessary mobility for users in areas with challenging topography and infrastructure. Due to its low cost and ease of manufacture, the Jaipur leg has been adopted and produced in other parts of the world [9,10].

A framework for disability and disability care

The above examples highlight interventions that have long-term impact on outcomes. However, the disability framework is much more extensive and requires sustained investment and commitment. This need has long been recognized by the World Health Organization and emphasized in multiple iterations of the International Classification of Functioning, Disability, and Health (ICF) [11]. This ICF framework for disability assessment and classification is grounded in a bio-psycho-social understanding of health and well-being. It recognizes multiple variables that impact disability and functionality, representing domains for potential intervention by healthcare providers across various disciplines (Fig. 44.2). A notable aspect of the ICF system is the development of independent "core sets" for highly prevalent disabilities. In the context NDDs, specific core sets have been developed for conditions such as autism spectrum disorder, attention deficit/hyperactivity disorder, and cerebral palsy provding a framework to facilitate local development and collaborative endeavors [12–14].

Defining resource needs in neurodevelopmental disabilities and the role of the team

Based on the above considerations, needs assessments are the sine qua non of initiatives in LRSs. Any resource inventory should necessarily begin by defining roles, including leadership roles from administrative, medical, and even political perspective. The role of the neurodevelopmental specialist is as valuable as any other, but it is crucial to recognize that this role must be assumed with humility and flexibility. In LRSs, knowledge and training often transcend disciplinary boundaries, leading to transdisciplinary collaboration. Roles in a LRS are fulfilled in ways that differ from how they are typically accomplished in a resource-rich environment. For example, an occupational therapist may also function as a physical therapist, or a general physician may take on key aspects of longitudinal neurodevelopmental surveillance.

While needs assessments are integral in planning, clinicians should be mindful of the scope and limitatons when developing work plans. Realistic initiatives often scale up from smaller projects. Starting with a narrower scope that serves as a proof-of-concept, with clearly defined short-, medium-, and long-term goals, can be effective. Tracking results and establishing accountability are crucial for achieving these goals as the project adapts to changes. Additionally, cultural nuances and socio-political dynamics are components of the process and require ongoing reassessment.

When conducting a needs assessment, it is important to consider how initiatives will be incorporated into the existing healthcare system. Models of practice, from referral mechanisms to the physical location of services, may vary significantly. Mapping these models can help in understanding the complex relationships between different entities. The development of initiatives aimed that improving system efficiency should be seen as an opportunity rather than an obstacle. For example, among the many regional variations of the Integrated Management of Childhood Illness (IMCI), (IMCI, a core set of acute care provisions for primary care settings), including the management of NCDs and disabilities emerged from initiatives that were tailored to regional needs [15].

Challenges in low-resource settings

As outlined above, the complexity of preventive, interventional, and surveillance services required to care for NCDs and disabilities in LRSs is immense. This complexity is further influenced by environmental factors, sometimes making meaningful change, despite the necessity and genuine desire, nearly impossible. Teams are often forced to adapt to survive and must build sustainability. Consequently, sustainability can take on many forms.

Systems of care, healthcare infrastructure, and politics

There are many essential components of stable infrastructure development, including understanding, incorporation, administration, and politics. The heterogeneity in the application from legislation to administration and resource utilization,

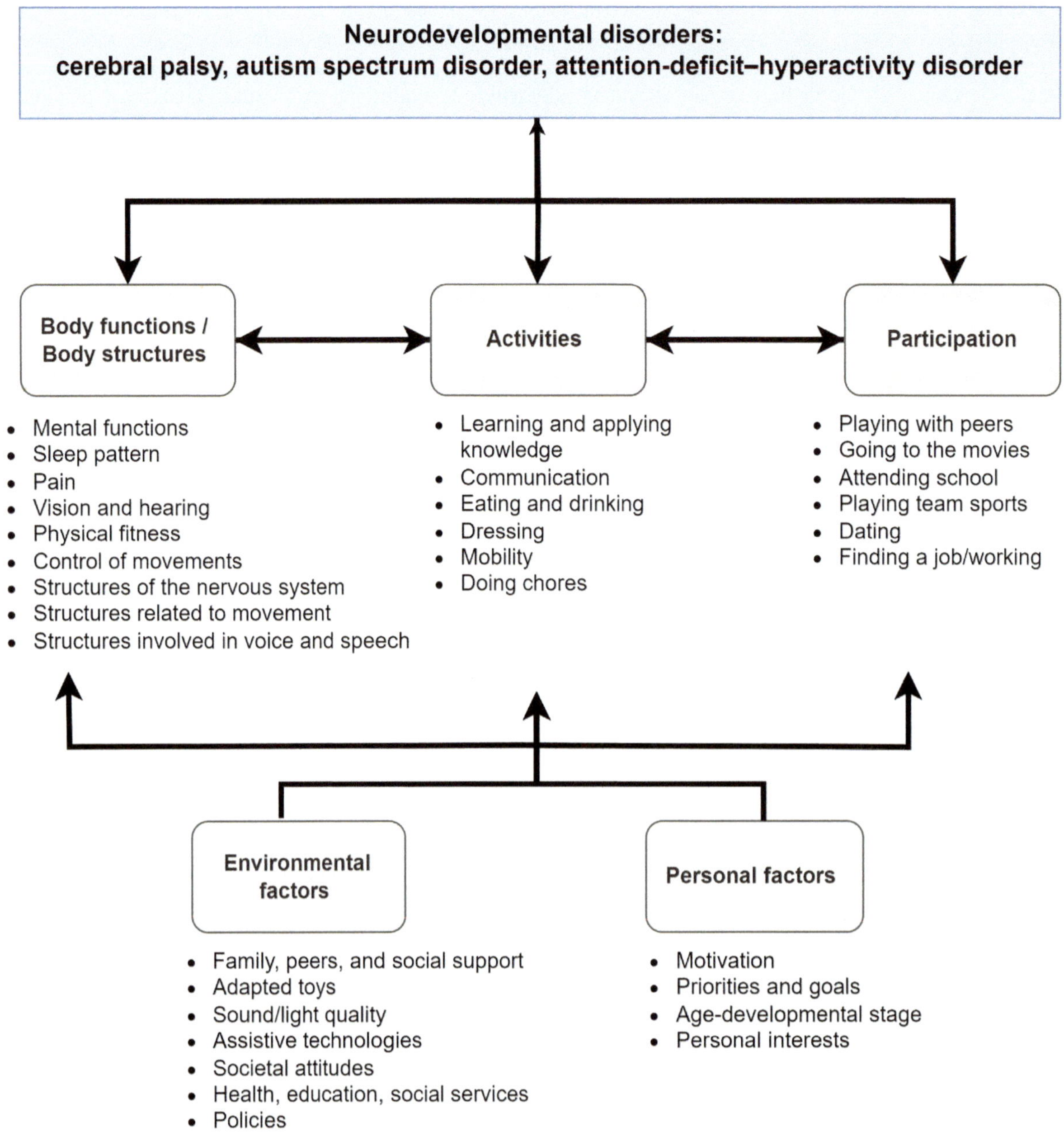

FIGURE 44.2 The International Classification of Functioning, Disability, and Health (ICF) biopsychosocial model. *From Schiariti V, Mahdi S, Bolte S. International classification of functioning, disability and health core sets for cerebral palsy, autism spectrum disorder, and attention deficit-hyperactivity disorder. Dev Med Child Neurol 60:933–41.*

suggests that the health systems—whether in low-, middle-, or high-income countries—face unique opportunities and challenges. In regions where the emerging burden of disability and its care is more pronounced (e.g., upper MICs), heavily structured systems often encouter conditions that require careful financing and resource allocation, as well as development of academic and research capacity. Legal aspects of practice may also hinder the flexibiltiy needed to structure effective disability care. An example of this can be seen in the health systems of Latin America, as reported in the Lancet's study on Universal Health Coverage in Latin America. Despite the precarious nature of resources and system structures, corruption, haphazard allocation based on kickbacks, and other illegal activities may exist. Minimizing these activities is challenging and relies on developing partnerships that enforce participation and accountability. One example of such an approach is the work done by the United States Agency for International Development (USAID).

Ethno-cultural perspectives on health, disease, and disability

Cultural and societal perceptions of disease, disability, and care significantly influence implementation of strategies for program development. Understanding the values by which a society and its families operate is imperative. While there may be universal values for child rearing, the prioritization of these varies widely. For example, the definitions happiness and health can differ significantly between a traditional family in Mexico City and one in Nairobi, Kenya, even when their socioeconomic profiles are similar. Understanding what happiness means to a family and the role of culture in their lives is essential to establishing an individualized plan of care. The definition of successful treatment also varies across cultures. When developing a program's vision and mission, it is crucial to consider social domains such as academic achievement, economic stability, marriage, and family dynamics. [16].

Understanding the effects of centuries of social conditioning on the parent-child relationship is essential for a clinician to be accepted by local communities. Child-rearing practices from nourishment to descipline, vary widely across cultures.. For example, while one person may be perplexed by the sight of a father's affectionate kiss, they may simultaneously view certain forms of punishment—criticized elsewhere—as culturally acceptable. It is the clinician's role to understand and respect these cultural practices, while also gently redirecting those that may threaten individual well-being. Communicating such recommendations in a nonjudgmental manner is crucial to maintaining a working relationship based on mutual trust.

The above considerations are particularly important in understanding the social and cultural perceptions of disease and disability. Ethno-cultural perceptions are evolve over time. For example. Epilepsy was once considered a royal disease in Europe, hypothyroid cretinism was folklorically vewed in Switzerland as a manifestation of God's anger meant to protect the community. However, common threads have emerged across different settings. Beliefs about functionality and beauty are prevalent across cultures and significantly influence how parents or caregivers adapt to any deviation from these ideals. Misunderstandings about the origins of neurodevelopmental disabilities, often based on cultural or religious belief systems, can further complicate this adaptation. For instance, genetic conditions may be perceived as a punishment from God, a result of bewitchment or possession by spirits, or the "evil eye" cast by a neighbor. The perceived burden and social stigma associated with disabilities often linger from past descriptions and terminologies, such as "Mongol" or "cripple", which have largely been retired [17]. While developed societies historically leaned toward institutionalizing individuals with disabilities, societies with fewer resources have often resorted to isolation, abandonment, or selective neglect and abuse [18]. In some cases, there has even been profitable exploitation of people with disabilities through exposure and public humiliation.

Caregivers are heavily influenced by their own health literacy and understanding, shaped by their personal beliefs. In a traditionally binary relationship between the care team and family, certain societies welcome a paternalistic approach by the care team, with little questioning of intent and goals. Other communities prefer to establish a more collaborative partnership with the care team and expect active participation. Still others view the relationship as customer-oriented, with interactons reflecting this perspective. These differing views are further complicated by traditional healing strategies, which have qualitative, not just quantitative, differences in their understanding of health. In many communities, traditional healing care is not merely seen as complementary or alternative medicine practitioners; they are essental community liaisons, serving as guides and counselors. The care team often needs to partner with these individuals and incorporate them into descision-making processes due to their expertise and influence within the community [19]. One such example is found in the Colombo-Venezuelan indigenous Wayuus' leaders, the Piachis. They partner with local and foreign staff deployed by the Fundacion Baylor Colombia to promote community dialog and primary care in a culturally sensitive manner. The approach shared the burden of care and understanding, ultimately enhancing trust within the community. This approach shares the burden of care and understanding, ultimately enhancing trust within the community.

At a more systemic level, involving cultural and legal infrastructure, disability profoundly impacts the relationship between affected individuals and their living environment. Pervasive disparities in access to care and services persist, often based on race, ethnicity, and socioeconomic status, disproportionately affecting people with disabilities. While some countries have introduced legislation to safeguard vulnerable populations, such as the Americans with Disabilities Act, many countries lack such laws andfind the practical application of these principles nearly impossible. Progress in this area requires a more purposeful and expansive approach, often involving adaptation. Traditionally, initiatives have focused on helping individuals adapt to existing environments. However, recent measures inspired by the growing visibility of NCDs and disability in LRSs, have inspired regional and global care initiatives aimed at achieving large-scale social impact. Examples include the NCD alliance and the Colombian Fundacion para la Investigacion y el Desarrollo de la Educacion Especial.

Examples of the systematic effects of discriminatory and neglectful policies are abundant and deeply rooted in the history of many nations. These policies have influenced the provision of care across societies, particularly when foreign interests are involved. From top-down interventions disguised as well-meaning projects and religious missions to clearly unethical interventions such as the Guatemala or Tuskegee syphilis studies, history has marked its pages [5,20,21]. Such actions have fostered distrust, often leading to resistance. Healing these wounds requires addressing fundamental misperceptions and building trusting relationships with local stakeholders.

Design and implementation of interventions

Attitudes and approaches

The attitude of a clinician embarking on the journey to provide neurodevelopmental services in a LRS often, though not always, involves a pioneering spirit and a deep understanding of available resources and potential liabilities. The challenges of serving in a LRS are best met with a focus on resourcefulness. A clinician does not require an abundance of technology and supplies to achieve good outcomes. Priorities for the clinician in an LRS include understanding NDDs, strengthening the physician-patient relationship, and maintaining flexibility. Much like the beak of the finch adapting to environmental challenges, a clinician in a LRS "bends" to get the most out of resources and may develop greater resilience and creativity as a result.

The design and implementation of support services for people with developmental disabilities require careful consideration of service delivery, resource allocation, training and education, and sustainability. For the clinical program to succeed, it must meet its goals. When setting goals or benchmarks, it is advised to consider feasibility, measurement, and monitoring. All too often, lofty goals are set that are vague and unmeasurable. Stakeholder participation during the process of goal development may be beneficial, though it may require training and education. However, feedback from stakeholders and self-advocates should be a mandatory part of the process. Inviting stakeholders and patients to participate in setting program goals not only strengthens relationships but also often leads to the discovery of new resources.

Goal setting should involve the clinician. Consideration should be given to clinic operations, particularly issues such as patient access. The time a clinician spends in the clinic and responding to patient questions outside of regular hours will require oversight, especially if there is no an additional clinician available for support during off hours or when the primary physician is on leave. Systems must be established early in the planning process to address patient concerns when the clinician is unavailable. Thoughtful planning for coverage and other operational elements will enhance the experience for all parties and increase the likelihood of sustainability.

Implementing strategies for continuity of care is important for sustaining efforts in LRSs. These communities often perceive themselves as revolving doors for providers, with high turnover and few clinicians willing to make long-term commitments to improving the community. Many underserved, low-resource communities have a long historys of failed attempts by providers who sought to bring meaningful change. This history can create a significant barrier for clinician trying to establish trust among patients and the community at large.

An effective way to implement continuity of care is to inquire about patient's value systems and the challenges they face. Such inquiries are most effective when made with cultural humility. Cultural humility is a process of self-reflection and discovery that builds honest and trustworthy relationships. It holds promise for helping people understand and eliminate health disparities [22]. Cultural humility involves challenging one's own cultural biases. Clinicians can take several steps toward cultural humility. These may include conducting home visits to observe the natural environment that influences the patient, attending community cultural events, sharing meals with people in the community, studying pivotal historic events that shaped the culture, and engaging in other activities that deepenthe clinician's appreciation of cultural influences on the patient. In many cases, knowledge about society and culture may be just as valuable as knowledge of pathophysiology.

Another strategy for establishing continuity of care is assessing system-wide services and supports while providing training and education to develop the workforce. Nearly every clinician serving a LRS faces workforce challenges. Partnering with state government and private agencies to develop the workforce is a priority. No single clinician can sustain a community indefinitely. Sustainability depends on the community's ability to define and strengthen effective systems of care. State government agencies, national programs, private agency partnerships, self-advocates, and comitted care providers are just some of the stakeholders who should be identified when seeking to improve infrastructure. The initial steps towards progress include identifing these stakeholders and establishing workgroups for strategic planning and implementation. The most effective processes are are driven by data, involve community-wide representation, and lead to policy development.

Disproportionate economics

Economics is the study of how humans make choices under conditions of scarcity. Scarcity arises when wants exceeds the availabile resources to satisfy them. LRSs are characterized by disproportionate economics, where unjust resource allocation creates dynamic interplay between poverty, child development, and opportunity. Understanding the cost of healthcare delivery is a prerequisite for allocating health resources—such as staff and medicines, in a way that is effective, efficient, just, and fair [23]. However, the true cost of health is often poorly understood and effective, efficient service delivery remains elusive in many countries. Judicious resource allocation can mean the difference between life and death, not just for individual patients but entire patient populations.

The impact of low resources on child development in communities around the world has received much attention. It is well established that living in low-resource environments early in life predisposes children to adverse health outcomes. Poor nutrition, inflammation, and traumatic stress during childhood increase risk for chronic disease health conditions [24–28]. This is supported by landmark studies on adverse childhood events (ACEs) and recent advances in understanding infant mental health [29,30]. There is a strong graded relationship between the extent of exposure to ACEs —such as abuse or household dysfunction— and multiple risk factors for several of the leading causes of death in adults [31].

Early brain development is impacted by social relationships and the environment [32]. High-risk children and youth are not receiving treatment that can help reduce their risk. In school-aged children, inadequate early childhood learning disproportionately affects low-income children [33]. A large study involving juveniles with either high levels of ACEs and/or high levels of mental health symptoms demonstrated that 41%–63% are not receiving services from behavioral health professionals [34]. In the same study, disparities were found in Black and Hispanic children receiving the least treatment when compared to non-Hispanic White children. This highlights the disproportionate burden faced by low-income and certain ethnic populations.

In a shift from the more traditional disease-focused model of global health interventions, increasing attention is now being placed on the importance of strengthening healthcare systems as a key component for achieving improved health outcomes [35]. By focusing on service delivery, a system can be evaluated based on its ability to provide patients with access to the high-quality care that they deserve. Service delivery is commonly used to evaluate the effectiveness of care in high-resource settings. Many available tools for measuring childhood outcomes were developed and validated in HICs and thus reflect the expectations for children of those cultures [36]. Clinicians and researchers cannot assume that translating such measurement tools into a local language also transfers its validity, since cross-cultural differences in concepts, norms, beliefs, and values for children's behavior are considerable [37–39]. In addition to the problem of how to measure outcomes in LRSs, there is the problem of which outcomes to measure. Reaching a common understanding and framework of what should be measured — and how— is challenging. A one-size-fits-all approach might facilitate some cross-country comparisons but fails to reflect important culture-specific and disease-specific outcomes [40]. while a consistent framework for measurement is necessary to support analysis across different settings, this consistency must be balanced with the needs to ensure that all tools are culturally appropriate and contextually relevant for validity [41]. There is an urgent need to establish guidelines for healthcare of individuals with NDDs in LRSs.

Clinical evaluation and management

The approach to clinical evaluation and management in LRSs does not differ markedly from that in high-resource settings. Accurate information about the child's disability, clear explanation of the child's limitations and behavior, and a realistic prognosis can serve as excellent therapeutic tools, even in the absence of other resources. The foundations of neurodevelopmental assessment —such as a comprehensive history and physical examination, combined with knowledge of typical human development and disease processes within the context of a strong physician-patient relationship—remain the cornerstone of quality care. In LRSs, however, the neurodevelopmental assessment should include a more extensive exploration of trauma, economic circumstance, culture, and available support resources. The skills of observation, listening, and effective communication become even more critical when social determinants of health impact a patient's condition. Observing the patient and their family or caregivers provides valuable information about the stability of relationships, attachment, caregiver capacity, and needs. Understanding these social determinants listening, attentive and careful observation is essential. It is often necessary to invest more time than anticipated in listening to a patient's circumstance. Through this listening, nonstandard care resources may be identified, reasons for missed appointments may be understood, and the physician-patient relationship can be strengthened. It's important to recognize that patients and their families may need multiple sessions before trusting the clinician with sensitive information pertaining to their circumstance. Be patient, listen, and observe. There is perhaps no greater asset for the clinician than an individualized understanding of the determinants of health when discussing a management plan.

Establishing practices that address NCDs in a LRS requires ingenuity and coordination. The clinical evaluation must involve processes that effectively address questions pertaining to the health and well-being of the patient. This task may appear straightforward, but in most cases, achieving these goals requires longterm commitments and strong relationship building skills. The latter may be particularly challenging for clinicians who lack, or are unwilling to embrace cultural humility.

The clinical interaction requires resourcefulness and an appreciation for the impact of culture on the understanding of disease. The history and physical examination are cornerstones of the clinical evaluation. When a language barrier exists, interpretation services should be arranged, with careful attention to local dialects to ensure accurate communication. For example, when arranging for a Filipino interpreter, it is important to clarify whether the patient speaks Ilocano or Tagalog; similarly, for the Chinese language, it is crucial to confirm whether Mandarin or Cantonese is spoken. Understanding should not be assumed unless the dialect is confirmed. The approach to history-taking and physical examination should not differ markedly between LRS and other environments.

The WHO has increasingly recognized the importance of NCD such as NDDs, trauma and their impact on the brain, as well as the need for global health equity. Screening practices are an effective way for the community to identify where to allocate resources for diagnosis and management. This is particularly valuable in LRS, where recources must be allocated judiciously. The WHO has published principles on screening, as well as guidelines on resources, and cost considerations [42]. Challenges to effective screening include community acceptance of public health efforts and utilization of data. In nearly every community, high or low resourced, there are groups of people who have a general distrust of mainstream medical practices. Total participation is not required for screening to be effective. Screening is not the same as sampling. Screening efforts, for example, may start with applying a tool to assess learning or behavior in children at the local school. Transparency to the public about intentions of screening and benefits to the community to identify and support needs will enhance public participation. Furthermore, screening tools must be culturally and linguistically appropriate. Several LRSs have established clinics for premature neonates at high risk for neurodevelopmental disability by applying culturally adapted scales and tools within a controlled setting [43]. More commonly, however, screening takes place in settings where longitudinal disease monitoring does not occur, such as a health fair or short-term missions. Regardless of whether screening occurs in a primary or acute care setting, a public health facility or a pop-up tent, cultural competency and literacy are essential. Adaptations should be reviewed by local cultural and linguistic experts for ease of understanding. When measures like these are implemented, screening provides a more accurate representation of disease burden.

Etiologic testing is challenging in LRSs. In middle-income countries (MICs) and low-income countries (LICs), the availability of first- and second-tier testing is gradually increasing. Historically, the utilization of these tests has been limited by factors such as distance to testing facilities, cost, administrative hurdles, and, at times, cultural reluctance to undergo testing. Identifying local modalities for testing, such as MRI or EEG, allows clinicians to adjust their expectations regarding diagnostic and etiologic evaluations. Recognizing these limitations should encourage clinicians to seek alternative, more attainable methods or models of care that address the needs of their patients.

In settings with limited access to advanced testing or treatment, clinicians should focus on the burdens expressed by patients and structure their recommendation around these concerns. For example, genetic testing to determine the cause of a neurodevelopmental diagnosis may not be easily accessible. The clinician must inform the patient about the therapeutic benefits versus the diagnostic curiosity, while primarily allowing space to listen to the patient's considerations and priorities. Patients often appreciate discussions about recent advances in medicine. Families, on the other hand, are frequently deprived of accurate information or may have been misinformed about their loved one's condition and the best care options. They may also have unspoken concerns about the cost of healthcare, which they might not voice unless asked by the clinician. A clinician who invites discussion about standards of care, while recognizing the patient's unique circumstances, can significantly strengthen the physician–patient relationship.

Screening and diagnostic tools adapted for use in local languages have been developed to meet the needs for diverse populations. One cornerstone of neurodevelopmental evaluation is the Capute Scales. This validated tool is highly regarded for its simplicity, low-cost materials, and ease of administration in several languages, making it a prime example of a well-recognized resource both in the United States and internationally [44–46]. In LRS, low-cost and uncopyrighted tools are commonly used in clinical practice. Notable examples include the Amiel-Tison scales for neuromotor assessment and the Hammersmith neonatal and infant neuromotor assessment scales. [47,48].

The delivery of a clinical diagnosis is just as important as its accurate determination. Cultural empathy must be paired with caregiver literacy to achieve effective communication of a diagnosis. When clinicians attempts to explain complex pathophysiology and genetics, they may encounter quiet submission from caregivers, which does not necessarily indicate misundestanding. To ensure clarity, crucial conversations may require the presence of a spiritual or cultural guide, the use of images, and simplified explanations. Using language that focuses on essential components or key take-home points is often more effective. Additionally, allowing caregivers and patients time to respond and ask questions in a back-and-forth, serve-and-return manner can further facilitate understanding. Recognizing cultural nuances such as feelings of guilt, mysticism, and other beliefs is essential for establishing a comprehensive clinical relationship.

Appropriate treatment and follow-up are the shared responsibility of both the clinical team and the patient, caregiver, or family unit. In LRSs, clinical interactions may be more limited than in a non-LRS. Clinic appointments may only occur once every 6 months or annually due to clinician availability. To address this challenge, an effective strategy is to recruit the parent or caregiver as an ad hoc therapist. The clinic setting thus becomes an opprtunity for education and training. Parents often practice the skills that therapists and other clinicians teach them while at home or in the community. Virtual or printed materials can provide step-by-step guidance for performing these maneuvers. Clinician feedback and verification may be less formal than standard skills assessments but, when combined with routine monitoring and advice, this model is effective in LRSs. Telehealth platforms have also been established to support LRSs. One notable example is Project Extension for Community Healthcare Outcomes (ECHO) (https://hsc.unm.edu/echo/), which establishes a virtual community for care, education,mentoring, and ongoing support for families, caregivers, and professionals. This model is currently being used to support patients with autism and other NDD in the United States and internationally.

Pitfalls in resource development

Pitfalls will be present for those that embark on the journey. Clinicians are often eager to succeed in a LRS and may become overly focused on the potential impact and acceptance of their work. This enthusiasm can sometimes blind them to the many challenges that lie ahead. In the following sections, we discuss several common pitfalls and offer recommendations for navigating them successfully.

The first and most significant pitfall is burnout. The burnout rate among physicians is about 50% [49]. Clinicians in LRSs are often expected to bridge a care gap that is too wide to fill on their own. The combination of system-wide needs, limited resources, and unrealistic expectations puts these clinicians at a high risk for burnout. Symptoms of burnout include emotional exhaustion, depersonalization, and a diminished sense of personal accomplishment. Burnout is more frequently reported among female and younger physicians. Several factors contribute to burnout, including excessive workloads, inefficient work processes, clerical burdens, work-home conflicts, lack of input or control over work-related issues, inadequate organizational support structures, and leadership culture. Addressing these contributors is essential for mitigating burnout and ensuring the sustainability of care in LRSs. [49,50]. The Agency for Healthcare Research and Quality (AHRQ) identifies several contributors to clinician burnout, including family responsibilities, time pressure, electronic health record (EHR), unfavorable organizational culture, chaotic work environment, and low control of work pace. The systemic impacts of burnout represent a public health crisis that begins with physicians and extends to healthcare organizations and systems. Burned-out clinicians are more likely to leave practice, leading to reduced patient access to continuity of care. Additionally, there is an increased risk of medical errors. Compassion fatigue, often described as a work-related stress response in healthcare providers, is considered a "cost of caring" and is a key contributor to the loss of compassion in healthcare [51].

An often-overlooked aspect of burnout is the transition from HIC to LRS. Clinicians moving from HICs and to a LRS are often confronted with sociopolitical and environmental crises that are entirely new to them. Conflicts such as war, famine, and corruption are prevalent in LRSs. The overwhelming exposure to death and disability can lead to feelings of despair and helplessness, causing clinicians to question their role and impact [52]. Addressing trauma and secondary trauma, along with integrated self-care practices is key for supporting the clinical team as theynavigate these challenges.

Burnout is best addressed when viewed as a shared responsibility between healthcare systems and individual clinicians [49]. Both the clinician and the healthcare setting must be proactive in supporting a practice model that is effective and sustainable. Successful strategies include locally developed practice modifications, increased support for clinical work, the promotion of community, connectedness, and a sense of meaning. Clinicians experience less burnout when they are not compensated based on individual productivity, are free from time pressures, have more control over clinical decisions, and can balance family life with their work (AHRQ). Additionally, implementing a patient-centered medical home can improve clinician satisfaction and burnout rates.

The second pitfall is dwelling on a lack of resources. By definition, LRSs have fewer dresources than high-resource settings. This disparity can lead to frustration for clinicians accustomed to working in more high-resource environments, as they may struggle to provide care with less. While a certain level of frustration can be healthy and help maintain motivation to improve conditions, clinicians must avoid becoming immobilized by it.

In LRSs, the most effective therapy sessions often involve the simplest, most available tools. For example, a caregiver using a stick and a torn piece of paper can create an engaging therapy session by practicing blowing out a birthday candle with three children of varied ages, etiologies, and developmental abilities. Songs and cultural traditions can also be used to teach identity and build relationships. Given that the greatest asset in neurodevelopmental care provision is human resources, many care needs can be met through creative use of available resources in any setting. Consider an orphanage in a rural LRS area that provides care for children with developmental disabilities. Medical staff perform evaluations using developmental testing and assign therapy based on needs. The staff conduct therapy sessions such as feeding the children in a prompt and safe manner using body cues, positioning, and fun engagement to complete each feeding session. Staff members are cross-trained to provide physical, occupational, speech-language, and play therapies, allowing them to develop understanding of each child's needs. The determinants of quality in this context are training, time, dedication, and compassion. In LRSs, there may be a temptation to abandon innovation when working under time constraints. However, innovation—across all levels of care and disciplines—is a critical tool for fostering resourcefulness and meeting community needs.

The clinician must be aware of the patient-specific, local, state, federal, and international resources available to support each patient's unique needs. Often these services are not coordinated and provided from different sources. For example, in the United States, Medicaid is a jointly funded Federal-State health insurance program for people with limited income and resources that provides an array of services for individuals with IDD. State schools provide school based supports and accommodations for students with developmental disabilities. The Social Security Act, Home and Community Waiver for individuals with IDD, provides person-centered alternatives to long-term institutionalization. Although disparities continue to exist in healthcare, advances such as Medicaid funding for state developmental disability programs serve as valuable resources for many individuals with IDD in the United States.

The promise

What is the promise we make to the people we serve? What is our promise to the community? The oaths that define professionalism for clinicians are numerous and varied, but they share common themes. These include pledges to provide treatments that benefit the patient according to one's abilities and judgment, refrain from voluntary acts of mischief or corruption, and live and practice the art of medicine with integrity, purity and holiness. While these oaths have been rewritten overtime to reflect changing cultural landscape and values, but the core tenets remain. The promise of every clinician is to dedicate one's life to helping people through the art of medicine, keeping the physician-patient relationship at the heart of medicine. This promise holds true regardless of the settings and resources available. Opportunities to fulfill this promise abound, including international collaboration with universities, the enhanced feasibility of parternships through telehealth, and the creation of global health fellowship training. The experience of caring for people with NDDs in LRSs are uniquely rewarding. With time, outcomes that differ from those typically seen in highly resourced settings emerge—such as a memorable walk beside a family supporting a child with a disability or the collective effort to build something meaningful within a community. It is essential to remember and practice the timeless aphorism by medieval French physician Ambroise Paré: "La médecine, c'est guérir parfois, soulager souvent, consoler toujours." ("Medicine is to cure occasionally, relieve often, and comfort always.")

References

[1] World Bank. World development indicators, http://datatopics.worldbank.org/world-development-indicators [accessed 10.12. 21].

[2] GBD, 2017 Mortality Collaborators. Global, regional, and national age-sex-specific mortality and life expectancy, 1950–2017: a systematic analysis for the Global Burden of Disease Study 2017. Lancet 2018;392(10159):1684–735. Available from: https://doi.org/10.1016/S0140-6736(18)31891-9.

[3] Institute for Health Metrics and Evaluation (IHME). http://healthdata.org/ [accessed 20.12. 21].

[4] Grantham-McGregor S, Cheung YB, Cueto S, Glewwe P, Richter L, Strupp B. Developmental potential in the first 5 years for children in developing countries. Lancet 2007;369:60–70.

[5] Edwards S. Toxic aid: economic collapse and recovery in Tanzania. Oxford University Press; 2014.

[6] WHO. Country-specific DALY estimates; 2016. http://www.who.int/healthinfo/global_burden_disease/estimates/en/.

[7] Van Zyl C, Bandenhorst M, Hanekorn S, Heine M. Unraveling 'low-resource settings': a systematic scoping review with qualitative content analysis. BMJ Glob Health 2021;6:e005190.

[8] Charpak N, Tessier R, Ruiz JG, Hernandez JT, Uriza F, Villegas J. Twenty-year follow-up of Kangaroo Mother Care versus traditional care. Pediatrics 2017;139(1):e20162063.

[9] Arya AP, Klenerman L. The Jaipur foot. J Bone Jt Surg 2008;90(11):1414–16.

[10] Mysore H. The "Jaipur Foot": India's most popular prosthetic for amputees is not the latest in technology, but it's still the most suitable option for many patients almost 50 years after its development. IEEE Pulse 2016;7(3):30–3.

[11] World Health Organisation. WHO I social determinants of health. WHO; 2020. Available: http://www.who.int/social_determinants/en/ [accessed 31.08.20].

[12] Schiariti V, Selb M, Cieza A, O'Donnell M. International classification of functioning, disability and health core sets for children and youth with cerebral palsy: a consensus meeting. Dev Med Child Neurol 2015;57:149–58.

[13] Bölte S, de Schipper E, Holtmann M, et al. Development of ICF core sets to standardize assessment of functioning and impairment in ADHD: the path ahead. Eur Child Adolesc Psychiatry 2017;23:1139–48.

[14] Bölte S, Mahdi S, de Vries PJ, et al. The gestalt of functioning in autism spectrum disorder: results of the international conference to develop final consensus ICF core sets. Autism 2019;23(2):449–67.

[15] Figueiras AC, Neves de Sousa IC, Rios VG, Benguigi Y. Manual para la vigilancia del desarrollo infantil (0-6 anos) en el contexto del AIEPI. 2nd ed. PAHO/WHO; 2011.

[16] Maldonado-Duran JM, Lecannelier F. What parents want for their children. In: Maldonado-Duran JM, Jimenez-Gomez A, Maldonado-Morales MX, Lecannelier F, editors. Clinical handbook of transcultural infant mental health. 1st ed. Springer Nature; 2019. p. 17–28.

[17] Jimenez-Gomez A. Cultural issues in response to young children's illnesses, chronic conditions, malformations and their management. In: Maldonado-Duran JM, Jimenez-Gomez A, Maldonado-Morales MX, Lecannelier F, editors. Clinical handbook of transcultural infant mental health. 1st ed. Springer Nature; 2019. p. 217–29.

[18] Nelson CA, Fox NA, Zeanah CH. Romania's abandoned children: deprivation, brain development, and the struggle for recovery. Harvard University Press; 2014.

[19] Jimenez-Gomez A, Munjuluri S. Use of traditional and folk remedies and traditional therapeutic strategies. In: Maldonado-Duran JM, Jimenez-Gomez A, Maldonado-Morales MX, Lecannelier F, editors. Clinical handbook of transcultural infant mental health. 1st ed. Springer Nature; 2019. p. 273–84.

[20] Reverby S. Compensation and reparations for victims and bystanders of the U.S. Public Health Service research studies in Tuskegee and Guatemala: who do we owe what? Bioethics 2020;34(9):893–8.

[21] Sullivan N. The trouble with medical "voluntourism." Sci Am 2017. https://blogs.scientificamerican.com/observations/the-trouble-with-medical-voluntourism/.

[22] Yeager KA, Bauer-Wu S. Cultural humility: essential foundation for clinical researchers. Appl Nurs Res 2013;26(4). Available from: https://doi.org/10.1016/j.apnr.2013.06.008.

[23] McBain RK, Jerome G, Warsh J, Browning M, Mistry B, Faure PAI, et al. Rethinking the cost of healthcare in low-resource settings: the value of time-driven activity-based costing. BMJ Glob Health 2016;1(3):e000134. Available from: https://doi.org/10.1136/bmjgh-2016-000134.

[24] Galler JR, Koethe JR, Yolken RH. Neurodevelopment: the impact of nutrition and inflammation during adolescence in low-resource settings. Pediatrics 2017;139(Suppl 1):S72–84. Available from: https://doi.org/10.1542/peds.2016-2828I.

[25] Jensen SKG, Tofail F, Haque R, Petri Jr WA, Nelson 3rd CA. Child development in the context of biological and psychosocial hazards among poor families in Bangladesh. PLoS One 2019;14(5):e0215304. Available from: https://doi.org/10.1371/journal.pone.0215304.

[26] John CC, Black MM, Nelson 3rd CA. Neurodevelopment: the impact of nutrition and inflammation during early to middle childhood in low-resource settings. Pediatrics 2017;139(Suppl 1):S59–71. Available from: https://doi.org/10.1542/peds.2016-2828H.

[27] Krebs NF, Lozoff B, Georgieff MK. Neurodevelopment: the impact of nutrition and inflammation during infancy in low-resource settings. Pediatrics 2017;139(Suppl 1):S50–8. Available from: https://doi.org/10.1542/peds.2016-2828G.

[28] Vohr BR, Poggi Davis E, Wanke CA, Krebs NF. Neurodevelopment: the impact of nutrition and inflammation during preconception and pregnancy in low-resource settings. Pediatrics 2017;139(Suppl 1):S38–49. Available from: https://doi.org/10.1542/peds.2016-2828F.

[29] Hughes K, Bellis MA, Hardcastle KA, Sethi D, Butchart A, Mikton C, et al. The effect of multiple adverse childhood experiences on health: a systematic review and meta-analysis. Lancet Public Health 2017;2(8):e356–66. Available from: https://doi.org/10.1016/S2468-2667(17)30118-4.

[30] Reichman NE, Corman H, Noonan K, Jiménez ME. Infant health and future childhood adversity. Matern Child Health J 2018;22(3):318–26. Available from: https://doi.org/10.1007/s10995-017-2418-5.

[31] Felitti VJ, Anda RF, Nordenberg D, Williamson DF, Spitz AM, Edwards V, et al. Relationship of childhood abuse and household dysfunction to many of the leading causes of death in adults. The Adverse Childhood Experiences (ACE) Study. Am J Prev Med 1998;14(4):245–58. Available from: https://doi.org/10.1016/s0749-3797(98)00017-8.

[32] Bick J, Nelson CA. Early adverse experiences and the developing brain. Neuropsychopharmacology 2016;41(1):177–96. Available from: https://doi.org/10.1038/npp.2015.252.

[33] Steinberg JR, Bruce JS, Marin-Nevarez P, Phan K, Merrell SB, Chamberlain LJ. Early childhood learning and the pediatrician: a qualitative study among diverse, low-income caregivers. J Dev Behav Pediatr 2018;39(5):376–86. Available from: https://doi.org/10.1097/DBP.0000000000000564.

[34] Finkelhor D, Turner H, LaSelva D. Receipt of behavioral health services among US children and youth with adverse childhood experiences or mental health symptoms. JAMA Netw Open 2021;4(3):e211435. Available from: https://doi.org/10.1001/jamanetworkopen.2021.1435.

[35] Bergquist HB, Burkholder TW, Muhammad Ali OA, Omer Y, Wallis LA. Considerations for service delivery for emergency care in low resource settings. Afr J Emerg Med 2020;10(Suppl 1):S7–11. Available from: https://doi.org/10.1016/j.afjem.2020.07.002.

[36] Henrich J, Heine SJ, Norenzayan A. The weirdest people in the world? Behav Brain Sci 2010;33:61–83.

[37] Lancy DF. The anthropology of childhood: cherubs, chattel and changelings. Cambridge, UK: Cambridge University Press; 2015.

[38] Sabanathan S, Wills B, Gladstone M. Child development assessment tools in low-income and middle-income countries: how can we use them more appropriately? Arch Dis Child 2015;100:482–8.

[39] Sternberg R. Culture and intelligence. Am Psychol 2004;59:325–38.

[40] McCoy DC, Peet ED, Ezzati M, Danaei G, Black MM, Sudfeld CR, et al. Early childhood developmental status in low- and middle-income countries: national, regional, and global prevalence estimates using predictive modeling. PLoS Med 2016;13:e1002034.

[41] Gladstone M, Abubakar A, Idro R, Langfitt J, Newton CR. Measuring neurodevelopment in low-resource settings. Lancet Child Adolesc Health 2017;1(4):258–9. Available from: https://doi.org/10.1016/S2352-4642(17)30117-7 Epub 2017 Oct 9. PMID: 30169180.

[42] Wilson J, Junger G. Principles and practice of screening for disease. Geneva: World Health Organization; 1968.

[43] Brito Moreno J.A., Perdomo Fonseca M.P., Mancilla N.I., Sanchez G. Seguimiento al desarrollo motor en niños con factores de risgo neurologico y peos menor a 1.500 gramos en el Programa Madre Canguro del Hospital Universitario Clínica San Rafael en el periodo comprendido entre Enero de 2009 a Abril de 2012. Repository Universidad Militar Nueva Granada; 2014. Accessed at: https://repository.unimilitar.edu.co/handle/10654/10786.

[44] Romo-Pardo B, Liendo-Vallejos S, Vargas-Lopez G, Rizzoli-Cordoba A, Buenrostro-Marquez G. Global neurodevelopmental screening tests for children under 5 years of age in the United States and Latin America: systematic review and comparative analysis. Biol Med Hosp Infant Mex 2012;69(6):450–62.

[45] Semrud-Clikerman M. The Capute scales. Pediatr Neurol 2006;34(1):79–80.

[46] Vericat A, Orden AB. Herramientas de screening del desarrollo psicomotor en Latinoamerica. Rev Chil Pediatr 2010;81(5):391–401.

[47] Gosselin J., Amiel-Tison C. Évaluation neurologique de la naissance à 6 ans. 2nd ed. Motréal: Éditions du CHU Sainte-Justine; 2007.

[48] Haataja L, Mercuri E, Regev R, Cowan F, Rutherford M, Dubowitz V, et al. Optimality score for the neurologic examination of the infant at 12 and 18 months of age. J Pediatr 1999;135:153–61.

[49] West CP, Dyrbye LN, Shanafelt TD. Physician burnout: contributors, consequences and solutions. J Intern Med 2018;283(6):516–29. Available from: https://doi.org/10.1111/joim.12752.

[50] Dillon EC, Tai-Seale M, Meehan A, Martin V, Nordgren R, Lee T, et al. Frontline perspectives on physician burnout and strategies to improve well-being: interviews with physicians and health system leaders. J Gen Intern Med 2020;35(1):261–7. Available from: https://doi.org/10.1007/s11606-019-05381-0.

[51] Sinclair S, Raffin-Bouchal S, Venturato L, Mijovic-Kondejewski J, Smith-MacDonald L. Compassion fatigue: a meta-narrative review of the healthcare literature. Int J Nurs Stud 2017;69:9–24. Available from: https://doi.org/10.1016/j.ijnurstu.2017.01.003.

[52] Balmer DF, Marton S, Gillespie SL, Schutze GE, Gill A. Reentry to pediatric residency after global health experiences. Pediatrics 2015;136(4):680–6.

Chapter 45

Community

Meghna Rajaprakash
Department of Neurology and Developmental Medicine, Kennedy Krieger Institute, Department of Neurology Johns Hopkins University School of Medicine, Baltimore, MD, United States

Introduction

Most discussions of the management of neurodevelopmental disabilities focus on the school or health setting; however, the person with neurodevelopmental disability (NDD) also functions within their community. Communities are where the person with NDD lives. Communities include the service providers and school but also provide the supports for the family and opportunities for socialization and independence. Communities are strongly associated with the social determinants of health and may influence life expectancy. This chapter summarizes important aspects related to families and communities involved in the care of a person with NDD.

Neurodevelopmental disability and the family unit

Within the community framework, children with NDD are managed first and foremost by their family unit. As such, there is a key association between the well-being of the family and that of the child. Family can take on many forms including units with two parents (married or unmarried), single-parent households, families of divorce, blended families, multigenerational families, adoptive families, or foster families [1]. Families can live within the same household or be separated geographically in different socioeconomic settings. Within a family, there can be diversity in ethnicity, culture or traditions, language, and religions [2].

The complexity and diversity of family backgrounds need to be considered in order to deliver the most optimal care for children with NDD [3]. This includes paying attention to the influences of socioeconomic status, family dynamics, and mental health on neurodevelopment. When there is dysfunction in family functioning, the scale tips toward risk and must be balanced with greater community support to optimize resiliency, as illustrated in Fig. 45.1.

Socioeconomic status

Family socioeconomic status (SES) or alternatively socioeconomic position is one of the strongest predictors of neurodevelopmental outcomes. SES is the most studied construct in relation to neurodevelopment and has been broadly defined as involving factors representing material resources and social status [4]. SES is used to collectively describe resources encompassing a family's material goods, money, power, social networks, healthcare, leisure time, or educational opportunities [5].

SES predicts cognitive performance as measured by intelligence quotient, with those from lower SES families performing more poorly than those from higher SES [6]. Lower family SES also correlates with poorer outcomes in specific domains including executive functioning and language [7]. Neuroimaging research has provided supporting evidence, showing an association between SES and the functional and structural differences in the prefrontal cortex, hippocampus, and amygdala [8,9].

Family dynamics

Another factor that contributes to neurodevelopment is family dynamics, or the interaction between the child with NDD and their caregivers. A large body of evidence suggests that synchrony of parental interactions with the child and the adaptation of behaviors have a profound impact on outcomes [10]. Parent responsiveness, in particular, has been shown

Capute and Accardo's Neurodevelopmental Disabilities in Infancy and Childhood. DOI: https://doi.org/10.1016/B978-0-12-824060-1.00012-2

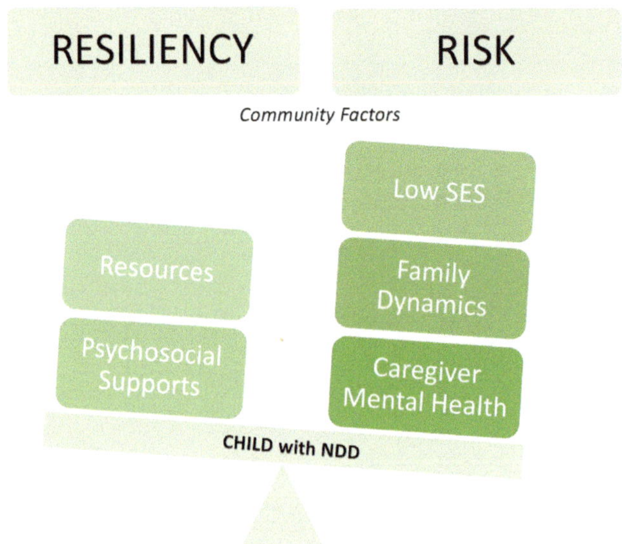

FIGURE 45.1 Diagram depicting factors affecting risk and resiliency in children with neurodevelopmental disabilites.

to be associated with better attachment and predicts higher academic performance in childhood [11]. This may explain why children who live in smaller families fare better in terms of cognitive, social, and emotional well-being [12].

Caregiver mental health

Exposure to early-life stressors such as acute events and caregiver-perceived stress adversely affects cognitive outcomes [13]. Stress can also lead to the development of caregiver psychopathology, including depression, which further places children at risk for neurodevelopmental disorders [14]. It is therefore of importance to consider the impact of the child's NDD on the mental health of caregivers and other family members.

Families can be impacted positively and negatively by the child's disability [15]. A systematic review of the lived experiences of US families living with children with autism spectrum disorder (ASD) revealed challenges, including the emotional strain of learning about the diagnosis, managing financial costs for medical care, difficulty accessing services, and loss of time with typically developing siblings. Positive themes of increased cohesion, adaptation, and appreciating small triumphs also emerged from the study [16].

Impact on parents

Compared to parents of typically developing children, parents of children with neurodevelopmental disabilities report higher levels of parental stress [17] that is related primarily to the child's maladaptive behaviors [18] but not necessarily correlated with severity of disability [19]. Mothers of children with disabilities experience the greatest stress and depressive symptoms as compared to other family members [20]. In some cases, there can be disagreements about approach to parenting, which can cause marital discord, especially if there are other issues in the relationship. However, other couples experience strengthened ties in caring for a child with disability and rely on the partnership for financial and emotional support [16]. The quality of marital relationship ultimately predicts lower parenting stress and depressive symptoms even when controlling for SES, child characteristic, and social supports [21]. As such, it is important to ensure that parents with children with NDD are provided with respite care, time for each other outside of the home, and the opportunity to strengthen their relationship.

Siblings of children with neurodevelopmental disability

The impact of neurodevelopmental disabilities on siblings is also highly variable. Though initial studies suggested negative impacts on siblings [22], more recent evidence suggests that sibling adjustment is associated more with overall family functioning rather than the disability itself or the sibling's resiliency [23,24]. Poor adjustment in siblings is related to parent and family factors such as family conflict [25], poor family adaptability [26], family size, low socioeconomic and immigrant status [27], and poor problem-solving and communication [28].

As a result of having a sibling with a disability, some children exhibit internalizing behaviors and withdraw [29]. Others are more resilient and the experience of having a sibling with a disability makes them sensitive, responsible, and independent individuals [30]. Various sibling support groups exist to provide siblings of individuals with neurodevelopmental disabilities with peer support and education.

Impact on grandparents

Like parents, grandparents with children with disabilities can experience stress and go through a mourning process as the expectations for their grandchild are challenged [31]. The impact on grandparents may be exacerbated by a lack of information and support that is usually provided to parents of the child with disability. As grandparents adapt and gain a better understanding of their grandchild's disability, they typically demonstrate a sense of responsibility and appreciation for the family's ability to cope [32]. At this stage, grandparents become a source of support and respite for the family, particularly when they have a good relationship with their own children [33]. However, grandparents' contributions can also significantly impact their own health and well-being [34].

Regardless of the family's reaction to the disability, special care must be taken to ensure that all members of the family are fully educated on the disability and provided with resources and psychological support to best optimize functioning within the family context.

Out-of-home placements

Despite the advantages that the family unit affords individuals with disability, it may be difficult to support them in the home environment due to medical complexity, lack of community resources, and funding issues [35]. Healthcare providers are uniquely positioned to assess these stressors and make referrals for community supports when required. Based on the local resources, families can be connected to a social worker or case manager to address particular areas where they need assistance. Providers can also educate parents about support groups, respite care, and medical waiver subsidies specific to each state.

In cases where community supports are inadequate, individuals are often transitioned to community settings such as skilled nursing facilities, intermediate care facilities, acute care or specialty hospitals providing long-term care, transitional facilities, rehabilitation facilities, residential schools, foster care, and medical group homes. As outlined in Table 45.1, these facilities differ in terms of the duration of stay, availability of skilled nursing, assistance with activities of daily living, and facilities to support mechanical ventilation and tube feeding [36]. Factors that contribute to placement of individuals with disabilities in out-of-home placements include lack of resources to support their complex health issues, need for technological care, and issues within the family unit including SES, family dynamics, and caregiver mental health [36].

Research has shown that 90% of families report that out-of-home placements are advantageous for family members [37]. However, compared to families with adults or adolescents in care, families with children under 15 years reported the highest levels of caregiver burden and lowest marital adjustment due to the placement of their child. Thus careful evaluation of the benefits and adverse impacts on the family need to be considered prior to considering an out-of-home placement.

In cases where there is child abuse, neglect, or parental mental illness or incarceration, children may be placed in foster care in a group facility or home. Foster children may be placed with a relative or nonrelative and transitioned through multiple foster homes until they can be reunited with their families or are adopted. Approximately 60% of preschool children in foster care have a NDD [38]. Older children show high rates of mental health issues, most commonly conduct disorder, oppositional defiant disorder, attention deficit hyperactivity disorder (ADHD), and anxiety disorders [39]. Children with NDD in the foster care system are more vulnerable to maltreatment [40], which further predisposes them to developmental and psychiatric issues. Some communities offer specialized training programs to potential foster families to address behavioral and developmental challenges.

Maltreatment and neurodevelopmental disability

Child abuse and neglect

Maltreatment includes but is not limited to physical abuse, sexual abuse, neglect, and emotional/psychological abuse and neglect [41]. In the United States, approximately 35.5 per 1000 children with disabilities are exploited in the form

TABLE 45.1 Characteristics of out-of-home placements.

Type of placement	Duration	Facilities/Services offered
Skilled nursing facilities	Long-term or short-term respite care	- 24-hour skilled nursing including skilled assessments, medication administration, tube feeds - Physicians provide support - Technological care including mechanical ventilation and parental nutrition
Hospitals	Short term	- Skilled nursing support - Physicians provide daily evaluation and adjust medications - Technological support - Parental nutrition and medication options
Transitional facilities	Short-term during transition from hospital to home	- Skilled nursing support - Parental training and discharge planning
Rehabilitation hospitals	Short term	- Skilled nursing support - Medication management - Rehabilitation therapies including physical therapy, occupational therapy, behavioral therapy, and rehabilitation
Intermediate care facilities	Long term	- Assistance with activities of daily living - Medication administration - Skilled nursing not available
Residential schools	Long term	- Educational and behavioral support for children with developmental and psychiatric issues - Skilled nursing services vary
Medical group homes	Long term	- Support for activities of daily living - Skilled nursing support limited

of child abuse and neglect as compared to 21.3 per 1000 children without disabilities [42]. A recent study of 4500 maltreated children showed that children with disabilities were 3.76 times more likely to be neglected, 3.79 times more likely to be physically abused, and 3.14 times more likely to be sexually abused with behavioral disorders showing the greatest prevalence [43]. Children with intellectual disabilities are particularly vulnerable to sexual abuse as they have difficulty with judgment and pragmatics [44]. However, challenges remain in quantifying the extent of abuse and determining whether the disability was preexisting or resulting from maltreatment [45].

The factors that predispose children with disabilities to abuse include those affecting the family unit, namely SES, family dynamics, and caregiver mental health. Families who do not have the community resources or respite to cope with the challenges of taking care of a child with disabilities are particularly vulnerable to maltreating the child with disability. The medical and behavioral complexity of the child's condition can also predispose them to neglecting their medical care [46]. Family dynamics and caregiver stress can also increase the likelihood of abuse for children with disability, particularly when there are maladaptive behavioral issues [46]. It is therefore of utmost importance to provide families who have a child with a disability with community supports.

Socialization and inclusion

One of the areas that has been met with much debate is how best to integrate individuals with NDD into the school and work environments in a process called socialization. Proponents of socialization state that it allows individuals with NDD to learn social norms, form relationships, and decreases social stigmas around the disability [47].

The Individuals with Disabilities Education Act mandates that children with NDD be placed in the least restrictive environment appropriate to their educational needs [48]. Inherent in this statement is the recognition that children should spend as much time as possible in the general education class with typically developing children in order to successfully integrate. Successful inclusion in educational settings is defined in many ways. One study suggested that the child is successfully included when (1) the child makes progress on their individualized outcomes, (2) makes gains in personal development and acquisition of skills anticipated for all children, (3) interacts with their typical peers and are accepted as full members of the group, and (4) parents are satisfied with the progress being made [49]. Many educators have positive attitudes toward inclusion, but there are challenges including lack of resources, difficulties with integrating special education into a stringent academic curriculum, lack of professional training or planning time, and large class sizes [50].

Another challenge is bullying of school-age children with NDD. Research shows that students in special education programs report greater rates of bullying and victimization than students in a general education class without disabilities [51]. Overt features of the disability including stuttering, gait abnormalities, use of sign language, hearing aids, or brailed texts may also lead to bullying [52]. Personal attributes including poor social skills, inability to recognize bullies, and behavioral differences further increase the risk of bullying for these individuals [53]. The school setting also plays a role with students in self-contained classrooms being more likely to be bullied than students with or without disabilities in inclusive classrooms [51]. As such, when integrating children with disabilities with their typically developing peers, special attention must be given to prevent victimization of individuals with disabilities.

Upon graduating from school, many adolescents with NDD find it difficult to find sustainable employment [54]. Transitional programs that offer vocational training during high school have been effective in helping individuals with disabilities integrate while providing their employers with the confidence to hire them more permanently [55]. Employment further provides children with NDD with life skills and helps boost their self-esteem as they contribute meaningfully to society.

Participation in leisure activities is another mechanism for involving children with NDD in the greater community. Participation in both spontaneous and structured forms of leisure ultimately improves the quality of life of children with NDD [56]. Involvement in activities such as noncompetitive sports, such as the Special Olympics, arts, entertainment, social groups, and religious activities, can help children with NDD learn social norms, make contributions to the community, and gain skills [57]. However, as adolescents with NDD age, opportunities for leisure decrease and most of their socialization is limited to the workplace or routine engagements such as religious activities. Same-aged peers who are meeting their own developmental challenges may lack time or interest to interact with individuals with NDD, and parents of typically developing children may object to their children playing with an older individual. As a result, adolescents and older individuals with NDD are at risk for isolation, excessive screen time, lack of physical fitness, and obesity.

Safety

Children and adolescents with cognitive disorders are prone to exploitation and require close supervision. Due to their trusting nature, poor pragmatic skills, and inability to predict consequences, they may become "innocent offenders," performing an illegal act while doing someone a favor; for example, complying with the request "I forgot my keys, could you please go through the window and open the door for me?" Individuals with NDD also may not report sexual abuse or bullying for fear of losing their "friend" or the perpetrator threating to harm family members. Close supervision is the intervention for these situations.

Environmental hazards also pose a safety risk for children with motor or behavioral disorders. Stairs, rugs, bathtubs, pools, and stoves are not uncommon threats to safety. Children with behavioral issues such as ADHD may have accidents and suffer from trauma in the face of impulsivity or temper tantrums. Exploration of dangerous objects including light sockets and consumption of toxic products due to pica by children with severe cognitive impairments may have untoward results. Finally, some behaviors associated with severe cognitive dysfunction may result in self-injury or aggression and, if not recognized, may result in catastrophic consequences.

Another safety issue is wandering or leaving a supervised space or care of a responsible person. Children with ASD are more prone to wandering, with a recent study showing that 60.4% of children with autism are affected compared to 22.3% of children with other developmental delay and 12.4% in typically developing peers [58]. Children with young and less educated mothers and those with comorbidities including ADHD, anxiety, depression, or oppositional behaviors are at a higher risk of wandering and safety events. A 2017 National Autism Association of children with ASD showed that of the 808 children that went missing between 2011 and 2016, 17% suffered death with 71% involved in drowning and 18% involved in a traffic incident [59]. There are also negative consequences of wandering for the children's families who are less likely to attend out-of-home activities due to fear of elopement. Healthcare providers must screen for risk of wandering and provide parents with education on safety risks and preventative guidance regarding managing behaviors, environmental modifications, and precautions such as GPS trackers, identification bracelets and locks.

Concluding remarks

Children with NDD are impacted by their community environment just as much as the community is impacted by them. Positive outcomes of resiliency and inclusion result when there is access to resources, community supports, and sound family dynamics and caregiver well-being. Unfavorable issues such as maltreatment, bullying, and mental health issues can be mitigated by focusing on the functioning of the community that surrounds the child, which first and foremost includes the family unit.

References

[1] Christian LG. Understanding families: applying family systems theory to early childhood practice. Young Child 2006;61(1):12–20.

[2] Caretelege G, Kea C, Hensley LD. State efforts to identify maltreated children with disabilities: a follow-up study. Child Maltreatment 1997;2:52–60.

[3] Xu Y. Empowering culturally diverse families of young children with disabilities: the double ABCX model. Early Child Educ J 2007;34 (6):431–7.

[4] Bradley RH, Corwyn RF. Socioeconomic status and child development. Annu Rev Psychol 2002;53:371–99.

[5] Oakes JM, Rossi PH. The measurement of SES in health research: current practice and steps toward a new approach. Soc Sci Med 2003;56 (4):769–84.

[6] von Stumm S, Plomin R. Socioeconomic status and the growth of intelligence from infancy through adolescence. Intelligence 2015;48:30–6.

[7] Noble KG, Norman MF, Farah MJ. Neurocognitive correlates of socioeconomic status in kindergarten children. Dev Sci 2005;8(1):74–87. Available from: https://doi.org/10.1111/j.1467-7687.2005.00394.x2.

[8] Brito NH, Noble KG. Socioeconomic status and structural brain development. Front Neurosci 2014;8. Available from: https://doi.org/10.3389/fnins.2014.002763.

[9] Johnson SB, Riis JL, Noble KG. State of the Art Review: Poverty and the Developing Brain. Pediatrics 2016;137(4):e20153075. Available from: https://doi.org/10.1542/peds.2015-30754.

[10] Leclere C, Viauz S, Avril M, Archard C, Chetouani M, Missionnier S, et al. Why synchrony matters during mother-child interactions: a systematic review. PLoS One 2014;9:113571.

[11] Pinquart M, Kauser R. Do the associations of parenting styles with behavior problems and academic achievement vary by culture? Results from a meta-analysis. Cultural Diversity Ethnic Minority Psychol 2017;24(1):75–100.

[12] Sandberg J, Rafail P. Family size, cognitive outcomes, and familial interaction in stable, two-parent families: United States, 1997–2002. Demography 2014;51:1895–931.

[13] Demir-Lira OE, Voss JL, O'Neil JT, Briggs-Gowan MJ, Wakschlag LS, Booth JR. Early life stress exposure associated with altered prefrontal resting-state fMRI connectivity in young children. Dev Cognit Neurosci 2016;19:107–14.

[14] Zhang T, Luo ZC, Ji Y, et al. The impact of maternal depression, anxiety, and stress on early neurodevelopment in boys and girls. J Affect Disord 2023;321:74–82. Available from: https://doi.org/10.1016/j.jad.2022.10.0305.

[15] Reichman NE, Corman H, Noonan K. Impact of child disability on the family. Matern Child Health J 2008;12:679–83.

[16] Corcoran J, Berry A, Hill S. The lived experience of US parents of children with autism spectrum disorders: a systematic review and meta-synthesis. J Intellect Disabil 2015;19(4):356–66.

[17] Dyson L. Fathers and mothers of school-age children with developmental disabilities: parental stress, family functioning, and social support. Am J Ment Retard 1997;102:267–79.

[18] Valicenti-McDermott M, Lawson K, Hottinger K, Seijo R, Schechtman M, Shulman L, et al. Parental stress in families of children with autism and other developmental disabilities. J Child Neurol 2015;30(13):1728–35.

[19] Stoneman Z, Gavida-Payne S. Marital adjustment in families of young children with disabilities: associations with daily hassles and problem-focused coping. Am J Ment Retard 2006;111:1–14.

[20] Emerson E. Mothers of children and adolescents with intellectual disability: social and economic situation, mental health status, and the self-assessed social and psychological impact of the child's difficulties. J Intellect Disabil Res 2003;98:23–4.

[21] Kersh J, Hesvat TT, Hauser-Cram P, Warfield ME. The contribution of marital quality to the well-being of parents of children with developmental disabilities. J Intellect Disabil Res 2006;50(12):883–93.

[22] Fowle CM. The effect of the severely mentally retarded child on his family. Am J Ment Defic 1969;73:468–73.

[23] Emerson E, Hatton C, Blacher J, Llewellyn G, Graham H. Socio-economic position, household composition, health status, and indicators of the well-being of mother of children with and without intellectual disabilities. J Intellect Disabil Res 2006;50:862–73.

[24] Giallo R, Gavidia-Payne S. Child, parent, and family factors as predictors of adjustment for siblings of children with disability. J Intellect Disabil Res 2006;50:937–48.

[25] Madeleco B, Olsen S, Dyches T, Marshall E. The relationship between family and sibling functioning in families raising a child with a disability. J Family Nurs 2003;9:365–96.

[26] Williams PD, Williams A, Hanson S, Graff C, Ridder L, Curry H, et al. Maternal mood, family functioning, and perceptions of social support, self-esteem, and mood among siblings of chronically ill children. Child Health Care 1999;28:297–310.

[27] Macks RJ, Reeve RE. The adjustment of non-disabled siblings of children of children with autism. J Autism Dev Disord 2007;37:1060–7.

[28] Van Riper M. Family variables associated with well-being in siblings of children with Down Syndrome. J Family Nurs 2000;6:267–86.
[29] Rossiter L, Sharpe D. The siblings of individuals with mental retardation: a quantitative integration of the literature. J Child Family Stud 2001;10:65–84.
[30] Walton KM, Ingersoll BR. Psychological adjustment and sibling relationships in siblings of children with autism spectrum disorder: risk and protective factors. J Autism Dev Disord 2015;45:2764–78.
[31] Hillman J. Grandparents of children with autism: a review with recommendations for education, practice, and policy. Educ Gerontol 2007;33:513–27.
[32] Mitchell W. The role of grandparents in inter-generational support for families with disabled children: a review of the literature. Child Family Soc Work 2007;12.
[33] Lee M, Gardner J. Grandparents' involvement and support in families with children with disabilities. Educ Gerontol 2010;36(6):467–99.
[34] Miller E, Buys L, Woodbridge S. Impact of disability on families: grandparents' perspectives. J Intellect Disabil Res 2012;56(1):487–500.
[35] Wang KW, Bernard A. Technology-dependent children and their families: a review. J Adv Nurs 2004;45(1):36–46.
[36] Friedman SL, Kalichman MD. Out-of-home placement for children and adolescents with disabilities. Pediatrics 2014;134(4):836–46.
[37] Baker BL, Blacher J. For better or worse? Impact of residential placement on families. Ment Retard 2002;40(1):1–13.
[38] Szilagyi M. The pediatrician and the child in foster care. Pediatrics 1998;19(2):39–50.
[39] Stein E, Rae-Grant N, Ackland S, Avison W. Psychiatric disorders of children "in care": methodology and demographic correlates. Can J Psychiatry 1994;39:341–7.
[40] Slayter E. Youth with disabilities in the United States Child Welfare System. Children and Youth Services Review 2016;64:155–65. Available from: https://doi.org/10.1016/j.childyouth.2016.03.012.
[41] Palusci VJ, Nazer D, Greydanus DE, Athens HC, Merrick J. Children, disability, and abuse. Int J Child Health Hum Dev 2017;10(3):205–14.
[42] Sedlak A.J., Mettenburg J., Besena M., Petta I., McPherson K., Greene A., et al. Fourth national incidence study of child abuse and neglect (NIS-4): report to congress, executive summary. Washington, DC: US Department of US Health and Human Services, Administration for Children and Families; 2010.
[43] Sullivan PM, Knutson JF. Maltreatment and disabilities: a population-based epidemiologic study. Child Abuse Negl 2000;24:1257–73.
[44] Kucuk S, Platin N, Erdem E. Increasing awareness of protection from sexual abuse in children with mild intellectual disabilities: An education study. Appl Nurs Res 2017;38:153–8. Available from: https://doi.org/10.1016/j.apnr.2017.10.0166.
[45] Bonner BL, Crow SM, Corwyn RF. Socioeconomic status and child development. Annu Rev Psychol 2002;53:371–99.
[46] Hibbard RA, Desch LW. Maltreatment of children with disabilities. Pediatrics 2007;119:1018–25.
[47] Kampert AL, Goreczny AJ. Community involvement and socialization among individuals with mental retardation. Res Dev Disabil 2007;28:278–86.
[48] Individuals with Disabilities Education Improvement Act of 2004. PL108-446, 20 USC; 2004.
[49] Cross AF, Traub EK, Hutter-Pishgahi L, Shelton G. Elements of successful inclusion for children with significant disabilities. Top Early Child Spec Educ 2004;24(3):169–83.
[50] Wilcox D, Hoon A. Inclusion – in whose best interest? How pediatricians can effectively advocate for children with neurodevelopmental disabilities. Dev Med Child Neurol 2013;56(3):206–9.
[51] Rose CA, Espelage DL, Monda-Amaya LE. Bullying and victimization rates among students in general and special education: a comparative analysis. Educ Psychol 2009;29:761–76.
[52] Rose CA, Allison S, Simpson CG. Addressing bullying among students with disabilities within a multi-tiered educational environment. In: Hollar DW, editor. Handbook of children with special health care needs. New York: Springer; 2012. p. 383–97.
[53] Flynt SW, Morton RC. Bullying and children with disabilities. J Instruct Psychol 2013;31(4):330–3.
[54] Winn S, Hay I. Transition from school for youths with a disability: Issues and challenges. Disabil Soc 2009;24(1):103–15. Available from: https://doi.org/10.1080/09687590802535725.
[55] Shandra CL, Hogan D. School-to-work program participation and the post-high school employment of young adults with disabilities. J Vocation Rehabil 2008;20:117–30.
[56] Dahan-Oliel N, Shaikako-Thomas K, Majnemer A. Quality of life and leisure participation in children with neurodevelopmental disabilities: a thematic analysis of the literature. Qual Life Res 2012;21:427–39.
[57] King G, Law M, King S, Rosenbaum P, Ketoy MK, Young N. A conceptual model of the factors affecting the recreation and leisure participation of children with disabilities. Dev Med Child Neurol 2003;48(5):337–42.
[58] Wiggins LD, DiGuiseppi C, Schieve L, Moody E, Soke G, Giarelli E, et al. Wandering among preschool children with and without autism spectrum disorder. J Dev Behav Pediatr 2020;41(4):251–7.
[59] McIlwain L., Fournier W. Mortality and risk in ASD wandering/elopement, 2011–2016 (p. 18); 2017. Retrieved from National Autism Association website: http://nationalautismassociation.org/wpcontent/uploads/2017/04/NAAMortalityRiskASDElopement.pdf.

Chapter 46

Telemedicine in neurodevelopmental disorders

Deepa Unnikrishnan Menon

Department of Neurology & Developmental Medicine, Johns Hopkins University School of Medicine, Kennedy Krieger Institute, Baltimore, MD, United States

In its simplest terms, telemedicine is the exchange of medical information from one site to another via electronic communication to support medical diagnosis, remote patient monitoring, and ongoing patient care [1].

Definition

The Health Resources and Services Administration, defines *telemedicine* as the use of electronic information and telecommunication technologies to support long-distance clinical healthcare, patient and professional health-related education, health administration, and public health. Modalities that make up telehealth include internet, videoconferencing, store-and-forward imaging, streaming media, and land and wireless communication [2].

The term *telehealth* refers to a broader scope of remote healthcare services and in addition to remote clinical care also includes remote nondirect patient care services, such as provider training, administrative meetings, and continuing medical education in addition to clinical services [2]. According to the Center for Connected Health Policy (CCHP), telehealth is a more universal term for the current broad array of applications and includes most healthcare disciplines including dentistry, counseling, physical and occupational therapy, home health, chronic disease monitoring, and management and disaster management [3].

Currently, telehealth is provided via four distinct applications. These include:

1. Live videoconferencing (synchronous)

The services are provided via a live two-way interaction between a person and a clinical provider using audiovisual telecommunication technology.

2. Store-and-forward (asynchronous)

This includes the transmission of recorded health history through an electronic communication system to a practitioner, usually a specialist, who uses the information to evaluate a case or render a service outside of a real-time or live interaction.

3. Remote patient monitoring

Personal health information and medical data are collected from an individual in one location via electronic communication devices and technologies, which are then transmitted to a provider in a different location for use in clinical care and support.

Capute and Accardo's Neurodevelopmental Disabilities in Infancy and Childhood. DOI: https://doi.org/10.1016/B978-0-12-824060-1.00035-3

4. Mobile health

Healthcare and public health activities and education are supported via mobile communication devices such as cell phones and tablet computers. The applications range from targeted text messages to promote healthy behavior to wide-scale alerts about disease outbreaks and other public service announcements [3].

Originating Site: The location where the patient receives medical services provided by the healthcare practitioner through a telecommunication system.

Distant Site: The location where the healthcare provider is present to deliver services using a telecommunication device or platform.

Hub and Spoke Model: Refers to a distribution method in which a centralized "hub" exists. Everything either originates in the hub or is sent to the hub for distribution to consumers. In the ECHO (Extension for Community Health Outcomes) case-based education model, the experts are in the "Hub" while community providers are the "Spokes."

History of telemedicine

Our recent brush with SARS-Cov-2 has highlighted the use of telehealth to better serve patients in a safe and equitable way and is a testament to new innovation. However, the dawn of telehealth may be traced to the mid-19th century when, during the American Civil War, casualty lists and medical supply orders were transmitted via telegraph. With the advent of the telephone, voice consultations were made possible, and then in 1910, amplified sounds of the stethoscope were transmitted via telephone networks. This then helped to create a method for transmissions of electrocardiograms and electroencephalograms [4].

With the advent of Morse code and two-way radio transmission, in the 1920s, the Seaman's Church Institute of New York became one of the first organizations to provide medical care for sick seaman and sailors onboard ships. The International Radio Medical Center, headquartered in Rome, Italy, was established in 1935 to provide services for sailors from five maritime nations via telemedicine. With the advent of air travel, telemedicine, and radio, medical advice for passengers on long-distance flights soon became a reality for in-flight medical assistance using on-call healthcare providers on the ground [4].

In the 1950s, the introduction of closed-circuit television made video communication for consultation between providers possible. In 1964 a two-way closed-circuit videoconferencing between specialists at the Nebraska Psychiatric Institute in Omaha and a state medical hospital in Norfolk helped to facilitate interactive and facilitated training and education at a distant site consultation [4]. In 1967, Massachusetts General Hospital/Logan International Airport Medical Station was established for two-way audiovisual consultation for passengers and airport employees, 24 hours a day. In the 1970s, mobile phones and satellite connections helped in the creation of mobile telemedicine. An example of this is the Alaska ATS-6 Satellite Biomedical Demonstration that lasted from 1971 to 1975, which improved village healthcare in Alaska [5].

Technological innovations have helped telehealth to further improve over the years. In 1969 the Department of Defense created the Advanced Research Projects Agency which was the precursor of the internet. In the 1970s, internet protocols were created, which then morphed into Web 2.0 in the 21st century. This paved the way for social networking and Voice over Internet protocols that made synchronous and asynchronous audiovisual communication possible. Innovations in the mobile technology and cell phone industry have also improved access to audio and video methods to facilitate real-time consultation between clinicians [6].

Benefits and barriers to telehealth

Numerous studies and papers have looked at the advantages and disadvantages of telepractice and telehealth. Benefits to the patient and families include improved access to specialty services that are not available in their medical homes and close to their residences. Telemedicine also benefits families by decreasing travel time and expenditure. In addition, telehealth has been especially beneficial in the management of "time-sensitive" disorders like acute management of stroke. Research has shown that live audiovisual teleconsultation by community hospital emergency room providers, working in tandem with stroke specialists in academic centers at a distance, has better outcomes than a phone consultation [7]. For providers, telehealth provides an opportunity for increased outreach and improved efficiency of the practice. Telehealth is also a venue for keeping abreast of new research and innovation by fostering educational seminars and training between academic centers and community physicians [7]. Innovative case-based learning models, like the Tele-ECHO model, have also helped to create local experts in areas that previously were underserved. For health systems, the increase in the use of telehealth models, like e-visits and teleconsultations, has served to keep patients within their medical homes and helped to decrease costs by reducing transfer to higher-level services remote from the patient's residence. For academic centers and tertiary hospitals, the use of telehealth models has helped to improve efficiency by reducing unnecessary care, improving service utilization and efficiency, trimming long waitlists, and improving patient health by providing intermediate-level care locally with improved consultations with community providers [8].

Interview-based specialties like behavioral health, nutrition, mental health/psychiatry, surgery (pre- and postop monitoring), and care coordination have fared best in terms of implementation and return on investment for both initial and follow-up patient visits, especially for the more stable medical patients. More complex specialties that may need the presence of a trained examiner/physician extender at the same site as the patient, include dermatology, orthopedics, cardiology, ophthalmology, and inpatient consultations in the ICU. Specialties that need specific training, like neurology and rheumatology, have been slower to adopt telehealth models for initial evaluations and patient care. However, recent papers have shown the utility of teleneurology consultation models for the management of epilepsy and neurodevelopmental disorders (NDD) [9–12]. There is also an increased interest in the use of telemedicine for the management of chronically ill children with medical complexity [13,14]. The Patient Protection and Affordable Care Act had four provisions that were focused on electronic medical records (EMRs) and the use of telemedicine which again helped to foster interest in and adoption of telemedicine practices more broadly [5].

Barriers to greater adoption of telehealth have included limitations of billing and reimbursement for the time spent with the patient, the complexity of malpractice coverage and licensing requirements that vary from state to state, the cost of technology and equipment, and the unequal distribution of broadband services that are necessary for audiovisual consultations (256–384 kB). Personal barriers include physician reluctance to adopt novel technology in practice. Provider acceptance is better with perceived ease of use and benefit to the practice and integration into existing workflow and EMRs [6]. Perceived limitations to the adaptation of telemedicine into routine medical practices include a more limited physical examination, the need for specialized equipment, and perceived difficulty in understanding body language, and nonverbal communication cues.

Patient barriers include concerns about confidentiality and privacy, concern about the quality of the patient-provider relationship ease of use of the technology, lack of equipment, and lack of adequate bandwidth to support the session.

Over the 2019–21 period, with the presence of the SARS-Cov-2 virus pandemic, emergency provisions that included the passage of Public Law 116–123 with Section 1135 waiver, allowed the Secretary of Health and Human Services to temporarily waive certain Medicare restrictions that included licensing requirements and widening the scope of practice. This allowed for the wholesale adoption of telehealth by practices both in the United States and globally to enable continuity of care and provision of healthcare services to patients within the safety of their homes [15]. This period also led to innovation in terms of equipment and technology to enhance remote patient monitoring (RPM) and grant-funded distribution of needed tablets, computers, and equipment to families and patients in need to further address healthcare disparities. This has been of particular benefit to families with children with special needs and complex medical issues.

Ethics and safety concerns

Telemedicine involves the utilization of electronic formats for gathering and maintaining medical records, documenting sensitive patient health information (PHI), and transfer of this information between providers and care management teams and third-party insurers. Thus there is a critical need to maintain high standards of confidentiality and maintenance of patient privacy. Federal Privacy laws like Health Insurance Portability and Accountability Act (HIPAA), which passed in 1996, ensure legal protections for health information [16]. The rapid adoption of telehealth brings with it certain ethical and legal concerns that need to be addressed to ensure patient satisfaction, privacy protection for patients, and liability protection for clinicians. During the telehealth visit, there is a virtual interaction between the clinician and the patient and documentation and transfer of PHI via EMRs. Regarding privacy and confidentiality, patients and families should be given information on how to contact the Compliance Officer if they have concerns about a breach of confidentiality [6,17].

Legal and ethical factors to consider include establishing the clinician-patient relationship, establishing privacy and confidentiality for the interaction, and ensuring adequate communication strategies either via telephone or EMR to ensure that the patient does not feel concerned about abandonment. In addition to the regular informed consent language, it is important to highlight the pros and cons of telemedicine. This should include a description of the elements of a telemedicine consultation and any associated risks, consequences, and benefits. Risks include interruption of transmission, equipment failure, and the inherent risk of possible breach of medical records and privacy. Explicit signed informed consent and permission should be obtained from the patient or family before any information is shared with a third party. All information and reports discussed should be part of the medical record. The technology used for telemedicine should be HIPAA-compatible to ensure privacy and confidentiality and security for the visit.

Ethical considerations include maintaining the same standard of care for face-to-face and in-person visits. This would ensure the accuracy of evaluation and documentation. The risk of misdiagnosis or delayed diagnosis due to the quality of the virtual visit should also be explained. When in doubt, a reasonable attempt should be made for an in-person evaluation to ensure the accuracy of findings.

Another concern that has been raised, during the social isolation due to the pandemic, includes concerns about possible suspected abuse or neglect or any reportable issue observed during a telehealth visit. Historically, stressors like economic instability, social isolation, and social distancing have precipitated a spike in child abuse with increased reports of spanking and head trauma. With the SARS-Cov-2 pandemic, reports of increased domestic violence to 30% in the United States were noted along with a reported 30%–60% increase in reports of child abuse. Data published from emergency room visit analysis demonstrated increased hospitalizations secondary to child abuse and neglect, with hospitalized children reportedly sustaining more severe trauma [18]. Telehealth may provide an opportunity for the clinician to identify and report signs of physical abuse (bruises) or neglect or may be a witness to worrisome living conditions or partner violence.

It is also important to screen for the possibility of cyberbullying and inappropriate internet usage or cyber victimization. Children with NDD are at a higher risk for cyberbullying and victimization [19].

If abuse or neglect is suspected, in addition to making a report to child protective services (CPS), an in-person visit should be scheduled, or more frequent telehealth visits can also be scheduled while also reporting to concerned authorities as a mandatory reporter. The provider may also have the advantage of triaging for potential mental health crises and preemptively provide appropriate resources to help the child and family [18].

Legal and regulatory concerns

Practices and clinicians interested in pursuing telehealth services should ensure that they take appropriate steps to ensure cybersecurity. Physicians need to evaluate their networks, EMRs, and all equipment used in the practice of telemedicine, and all portals whereby information is transmitted both into and out of the medical system are secure [20,21]. All hardware and software should be HIPAA-compliant and password-protected. Any medical device that is used should also be evaluated to ensure that appropriate safeguards exist [22].

Healthcare practitioners and clinicians wanting to practice telemedicine should be credentialed and be licensed in the state in which the patient resides. Medicare and Medicaid require that the clinician/healthcare provider be licensed in the patient's state and conduct an in-person medical evaluation prior to prescribing controlled substances [22,23].

Prior to the SARS-Cov-2 pandemic, practitioners could only provide telemedicine services to patients residing in the state in which the practitioner was licensed. With the Public Health Emergency (PHE) declaration (CMS 1135), providers were given more flexibility to use technologies such as Skype and equipment like cell phones to conduct the visit rather than have the patient travel to a hospital or clinic for the visit. Medicare and Medicaid regulations were also temporarily changed to allow providers to contact and take care of patients in the safety of their homes. Billing and reimbursement guidelines were modified to cover services that had previously not been covered under telehealth and also to allow reimbursement for audio-only visits. The telehealth state licensure requirements were also changed so that providers could deliver services across state lines and for the duration of the PHE. Temporary licenses were also granted to providers wishing to practice across state lines [15,20,22–24] (Table 46.1).

Interstate Compacts are agreements between two or more states to mutually recognize licenses and are a way of expediting the licensing process and allowing members to practice under a single multistate license. The larger compacts are recognized by well-established regulatory boards and meet the federal licensing requirements of the Centers for Medicare and Medicaid Services (CMS). A physician can access the Interstate Medical Licensing Compact, which streamlines the licensing process to enable practice in multiple states. The Nurse License Compact authorizes eligible nurses to practice across multiple states while maintaining a single license. It is recommended that physicians and nurses check their appropriate state regulatory site to confirm the status of Interstate Compacts (IMLCC.org).

In comparison, the Psychology Interjurisdictional Compact authorizes eligible psychologists to practice telepsychology across member states. Similar Interstate Compacts also exist for physical therapists (PT Compact), Speech and Language therapists (Audiology and Speech-Language Pathology Interstate Compact), and Emergency Medical Services (EMS Compact) [22–24].

TABLE 46.1 Comparison of CMMS telehealth regulations before and after Centers for Medicare and Medicaid Services (CMS) waiver 1135 (Public Health Emergency).

	Before March 2020	After CMS 1135 waiver
Who can perform telehealth	Only certain licensed providers	Any licensed medical and allied health clinician
Who can receive telehealth services	Patients and providers with a preexisting clinical relationship	No preexisting clinical relationship required
Site of services	Prespecified sites: designated rural areas, certain medical facilities	Telehealth providers and patients can be present from any site, including the patient's home
Site of provider	The provider should be in their designated clinic or office	Providers can conduct telehealth from home
State licensing requirements	The provider can provide services to patients in the state where they are licensed	Telehealth can now be provided to patients in another state (state-specific restrictions may still apply)
Telehealth format	Must be audiovisual (i.e., video technology)	Audiovisual OR audio-only allowed
Equipment requirements	Only approved telehealth platforms	Expanded approved platforms, including FaceTime, Skype, and Zoom
Telehealth reimbursement	Medicare coinsurance and deductibles apply to telehealth visits. Reimbursements for telehealth services is lower than for in-person services	Providers may waive cost-sharing for telehealth paid for by federal programs. All telehealth visits, including audio-only, will be reimbursed as if the service was furnished in person

Adapted from CARES Act AMA Covid-19 Fact Sheet (https://www.ama-assn.org/delivering-care/publichealth/cares-act-ama-covid-19-pandemic-telehealth-fact-sheet) and Medicare Telemedicine Healthcare Provider Fact Sheet (https://www.cms.gov/newsroom/fact-sheets/medicare-telemedicinehealth-care-provider-fact-sheet).

Reimbursement

CMS has a list of covered charges and reimbursement rates that are reviewed annually. Asynchronous telehealth services are marked by a GT modifier with the professional service current procedural terminology (CPT) or healthcare common procedure coding system (HCPS) code. If the originating site is a clinic, an originating site facility fee is reimbursed.

Telehealth reimbursement payment parity laws generally require commercial health carriers and benefit plans to reimburse healthcare providers at similar or equal rates for the same services provided in person. The reimbursement parity laws differ between states. Physicians can keep abreast of their state's parity laws by consulting the CCHP website (http://www.cchpca.org/topic/parity).

Currently, 50 states and the District of Columbia provide reimbursement for live video in Medicaid Fee-for-service. Twenty-two state Medicaid programs currently reimburse for an asynchronous store-and-forward visit. Twenty-nine state Medicaid programs reimburse for RPM monitoring. Twenty-two states continue to reimburse for audio-only visits. Eleven state Medicaid programs (AZ, CA, MA, MI, MN, NC, OH, OR, SC, TX, WA) reimburse all four modalities [25,26].

Service models

Teleconsultation

Teleconsultations can occur both in an inpatient as well as an outpatient setting. Teleconsultations involve establishing a communication/consult between physicians at the originating site requesting a consult on a patient in their practice and connecting to the expert located at an academic medical center or tertiary hospital which is the distant site.

These consults could be synchronous or asynchronous and could work for both acute and chronic disease management. This model improves access to specialists for practices and patients living in more remote, rural, or underserved areas; enhances care by decreasing wait time through early assessment; and could potentially be cost saving for families and may avoid more expensive hospitalizations by earlier diagnosis and treatment. For urgent inpatient consultations, specialists at a tertiary center can provide useful and timely consultation to their colleagues in emergency rooms and inpatient ICU setting to provide diagnostic and treatment advice (teleneurology—epilepsy; stroke). With the advent of FDA-approved medical equipment (stethoscope, ophthalmoscope, ultrasound) adapted for telemedicine, the consultant at the distant site would be able to provide concise diagnosis, guidelines, and management techniques that would be of benefit to their patient at the originating site. Thus the patient is able to stay in the local community hospital close to family, and value-added and less disruptive care is provided. For outpatient consultations on nonurgent cases, like a second opinion for a dermatological rash or a consult to enhance medication management, teleconsultations provide an ideal framework for busy primary care physicians in the originating site [6].

Telepractice

Telepractice is a way of extending the patients' medical home services with the provider evaluating and managing the patient in an alternate site like a child care center or school-based clinic. Benefits include appropriate healthcare by a clinician/physician already providing care to the patient in the medical home, fewer lost school days to illness, and cost saving to parents as they do not have to miss work or travel to a tertiary hospital or emergency room for more expensive care [6].

Remote patient monitoring for children with medical complexity

Children with medical complexity have multiple medical complexities, are dependent on medical technology, have severe functional limitations and often have disjointed/fragmented medical care, and are connected to multiple providers. They often tend to be high utilizers of medical health services and are at a risk for frequent hospitalizations [14]. Currently, due to innovation in medical devices and technology, these children could use peripheral devices like stethoscopes, high-resolution cameras, and otoscopes in their medical home or their own home, and using telehealth modalities can have fewer hospitalizations, lower cost of care, and better patient and family satisfaction [14]. Better care coordination by an Advanced Registered Nurse Practitioner who was involved in discharge planning for the child while an inpatient, and then ongoing connection and monitoring using telemedicine compliant tools, helped to decrease readmission rates and improve overall health for children and youth with special healthcare needs (CYSHCN) and medical complexity [11].

Teleresearch

Telemedicine can be used to recruit community participants for studies, outcome-based research, and dissemination of translational research findings into the community. It has also been studied extensively in health services research as a way to address healthcare disparities and to improve patient outcomes in a low-resource rural area [6]. An exciting area for use of telehealth has been the use in recruitment of patients with rare neurodevelopmental or metabolic or genetic disorders and conducting both recruitment and evaluation and to monitor disease progression or outcomes using the telehealth modalities [27,28]. There is also interest in the utilization and clinical trials of innovative devices and technology both for diagnosis and also disease-modifying potential, particularly in the rehabilitation arena [6]. There is an active body of ongoing research involving Health IT and analytical tools to mine big data and to search for clues to improve the health economy.

Neurodevelopmental disorders

CYSHCN constitute 13%–18% of the pediatric population and often have multiple medical and behavioral challenges and also often have difficulty accessing much-needed specialty healthcare. For children living in rural, remote, and underresourced areas, this is further compounded by a lack of money and transportation issues. In addition, 10%–15% of children with NDD also have significant communication and cognitive delays, severe behavioral challenges, self-injurious behavior, and need for medical equipment like specialized wheelchairs, which further complicate their issues in accessing needed healthcare. Telemedicine is beneficial in enabling families of children with NDD to access healthcare services in their medical homes. Using an asynchronous teleconsultation model, Langkamp et al. [29] demonstrated the utility of improving access, reducing transportation burden, and worsening behaviors associated with travel, by providing care in the child's school health clinic with the help of the school nurse. Thus both by providing synchronous audiovisual live consultation in the patient's home, as seen during the recent SARS-Cov-2 pandemic, and by providing asynchronous teleconsultation to children with special healthcare needs and neurodevelopmental disabilities, access to subspecialty care in rural and other underresourced areas is improved [29]. Teletherapies that are currently provided by speech therapists, care coordinators and social workers, mental health professionals, and psychiatrists have also proved very effective in improving the quality of life for children with special healthcare needs.

New frontiers and innovation

Tele-education

Telehealth has long been used for educational activity at academic centers that host webinars, continuing medical education virtual sessions, and meetings. Innovation in medical education that includes a mixed method of education/flipped classroom approach has also helped to revolutionize medical education by incorporating self-study, use of free online learning platforms like Coursera, and disease-specific novel case-based learning methods like the Project ECHO has helped to democratize translational research and academic guidelines and helped to create local experts, who then are able to better serve children with special healthcare needs in their medical homes. Tele-ECHO model is based on four tenets: (1) use technology to leverage scarce resources, (2) sharing "best practices" to reduce disparities, (3) case-based learning to master complexity, and (4) a web-based database to monitor outcomes. The goal of the ECHO model is to develop knowledge and capacity among community clinicians through telementoring and education.

Telerehabilitation

Innovation in the use of robotics, exoskeleton, virtual reality headsets, Google glass, and AI-based gait trainers have also been shown to be very effective in improving mobility and function in children and adults with traumatic brain injury, spinal cord injury, and other cognitive and motor delays. Using telemedicine virtual clinics to monitor the progress of patients within their own homes and communities also helps to enhance function [30].

Artificial intelligence and machine learning

With the universal adoption of EMRs and innovation in telediagnostics and increased sophistication of health informatics and analytical systems, currently, large datasets are now available that are being used to create machine learning algorithms. AI can help in the telediagnostic process by providing prompts for the clinician during history taking, building physician decision support, and providing best practices and evidence-based guidelines to help the clinician/healthcare provider become more efficient in providing care. Machine learning methods when applied to a large clinical dataset can also help with the predictive modeling of disease processes. This would be beneficial in the management of rare metabolic, genetic, and NDD. In patients with complex medical needs and children with NDD, judicious use of Information and Communication Technology and artificial intelligence could potentially enhance RPM and improve outcomes. In the realm of mental health and behavioral counseling, the development of computer-based generation and understanding of human-to-human interaction has led to the creation of virtual assistants and conversational agents. These can serve as a reminder/motivational message system; a way to create and monitor and provide routine checks using personal monitoring devices, create ways to address social isolation, and with the help of chatbots, even provide mental health counseling and motivational support for stable patients. In children with NDD who have fine motor or handwriting issues, these assistants and AI tools could bypass the neurodevelopmental deficits (speech—text). Chatbots and Avatars can also be used as a learning tools for behavioral support for children and youth with neurodevelopmental disabilities [31]. Thus AI techniques and tools could help to improve functionality and quality of life in individuals with neurodevelopmental disabilities in the future. AI is still in its infancy and though studies show promising results, ongoing monitoring of outcomes is necessary to ensure ongoing quality care for individuals with neurodevelopmental disabilities and complex medical needs [32].

Summary and future directions

In summary, telehealth is a promising methodology to enhance physicians' and clinicians' abilities to help children with special healthcare needs. Our recent experience due to the pandemic has accelerated the development of promising tools, techniques, and modalities of service delivery and also spurred new innovations in the telehealth realm of both clinical care and also the education of the future workforce. Some of the benefits of universal adoption of telemedicine include decreasing access burden to timely and convenient care, overcoming geographic barriers, and improving cost saving for patients and families. However, in the face of workforce shortage issues and as a way to curb rising healthcare costs, telehealth is also a useful alternative for RPM of stable patients who are in nontraditional housing and educational settings. Telehealth is a useful tool in a clinician toolbox to better address the issues of medically fragile and chronically ill children and youth in the safety of their own homes where more optimum interactions may be possible with the help of the extended families and local supports; and as a tool to extend the reach of the medical home by enabling subspecialists to work closely with the patient's primary care teams. Telehealth does not minimize the benefits of the in-person visits to evaluate a patient for their initial visit and for clarifying issues brought up during virtual visits. Most telehealth practitioners and researchers have shown that telemedicine will continue to play an adjunct role and not be a total replacement of traditional in-person care. Areas for future study should include developing best practices for the telehealth evaluation and management of children with complex healthcare needs and NDD, monitoring healthcare outcomes, and the economic impact of using a hybrid version of care to benefit both patients and families and the healthcare system.

References

[1] American Telemedicine Association. What is telemedicine? [accessed 1.10.21].
[2] Telemedicine and Telehealth. Available at healthit.gov/topic/health-it-health-care-settings/telemedicine-and telehealth [accessed 1.10.21].
[3] Center for Connected Health Policy. What is telehealth? [accessed 15.11.21].
[4] Craig J, Patterson V. Introduction to the practice of telemedicine. J Telemed Telecare 2005;11:3–9.
[5] Foote DR. Satellite communication for rural health care in Alaska. J Commun 1977;27:173–82.
[6] Burke BL, Hall RWSECTION ON TELEHEALTH CARE. Telemedicine: pediatric Applications. Pediatrics 2015;136:e293. Available from: https://doi.org/10.1542/peds.2015-1517.
[7] Weschler LR, Tsao JW, Levine SR, et al. Tele neurology applications: report of the telemedicine workgroup of the American Academy of Neurology. Neurology. 2013;80:670–6.
[8] McDowell M. Here and there. Children's Hospitals Today. 2013. p. 18–21.
[9] Velasquez SE, Chaves-Carballo E, Nelson E. Pediatric teleneurology: a model of epilepsy care for rural populations. Pediatr Neurol 2016;64:32–7.
[10] Pearl PL, Sable C, Evans S, et al. International telemedicine consultations for neurodevelopmental disabilities. Telemed J E Health 2014;20 (6):559–62.
[11] Hooshmand M, Foronda C. Comparison of Telemedicine to traditional face-face care for children with special needs: a quasi-experimental study. Telemed J E Health 2018;24(6). Available from: https://doi.org/10.1089/tmj.2017.0116.
[12] Soares NS, Langkamp DI. Telehealth in developmental-behavioral pediatrics. J Dev Behav Pediatr 2012;33:656–65.
[13] Ferro F, Tozzi AG, Ebra I, Gawronski O, et al. Impact of telemedicine on health outcomes in children with medical complexity: an integrative review. Eur J Pediatr 2021;. Available from: https://doi.org/10.1007/s00431-021-04164-2.
[14] Notario PM, Gentile E, Amidon M, Angst D, Lefaiver C, Webster K. Home-based telemedicine for children with medical complexity. Telemed J E Health 2019;25(11):1123–32. Available from: https://doi.org/10.1089/tmj.2018.0186.
[15] Menon DU, Belcher HME. Covid 19 pandemic health disparities and pediatric health care – the promise of telehealth. JAMA Pediatr 2021;175 (4):345–6. Available from: https://doi.org/10.1001/jamapediatrics2020.5097.
[16] US Department of Health and Human Services. Summary of privacy rule of 2003. https://www.hhs.gov/hipaa/for-professionals/privacy/laws-regulations/index.html [accessed 15.10.21].
[17] Nittari G, Khuman R, Baldoni S, et al. Telemedicine practice: review of the current ethical and legal challenges. Telemed J E Health 2020;26 (12):1427–37. Available from: https://doi.org/10.1089/tmj.2019.0158.
[18] Repine A, Macaulay J, Deutsch SA. Role of the pediatric health care provider in identifying and preventing child abuse during Covid-19 and beyond. Pediatr Consultant 2021;61(9):e1–6. Available from: https://doi.org/10.25270/con.2021.03.00010.
[19] Beckman L, Hellstrom L, vonKobyLetzki L. Cyberbullying among children with neurodevelopmental disorders: a systematic review. Scan J Psychol 2020;61(1):54–7. Available from: https://doi.org/10.1111/sjop.12525.
[20] American Medical Association. Medical cybersecurity: a patient safety issue. https://www.ama-assn.org/delivering-care/patient-suppott-advocacy/medical-cybersecurity-patient-safety-issue.
[21] U.S Food & Drug Administration. FDA fact sheet: the FDA's role in medical device cybersecurity, https://www.fda.gov/downloads/MedicalDevices? [accessed 15.10.21].

[22] Health and Human Services. Policy changes during Covid-19, https://telehealth.hhs.gov/policy-changes-during-during-the-covid-19-public-health-emergency [accessed 20.11.21].

[23] Center for Connected Health Policy. Telehealth in the time of Covid-19: how the pandemic has impacted policy, https://cchpca.org/covid-19-actions/; 2020 [accessed 20.10.21].

[24] Center for Connected Health Policy. State telehealth laws and reimbursement policies (Spring 2021), https:cchpca.org/2021/11/Spring2021_COVIDPolicies-1.pdf [accessed 20.10.21].

[25] Reger A. Telehealth reimbursement parity. Office of Legislative Research, 2020. https://www.cga.ct.gov/2020/rpt/pdf/2020-R-0161.pdf.

[26] Center for Connected Health Policy. State telehealth laws and reimbursement policies report, Fall 2021, https://www.cchpca.org/resources/state-teehealth-laws-and-reimbursement-policies-report-fall-2021; [accessed 1.11.21].

[27] Farmer JE, Muhlenbruck L. Telehealth for children with special healthcare needs: promoting comprehensive systems of care. Clin Pediatr 2001;40:93–8.

[28] Ragbeer SN, Augustine EF, Mink JW, Tatcher AR, Virhele A, Adams H. Remote assessment of cognitive function in juvenile neuronal ceroid lipofuscinosis (Batten disease): a pilot study of feasibility and reliability. J Child Neurol 2016;31(4):481–7.

[29] Langkamp DL, McManus M, Blakemore S. Telemedicine for children with developmental disabilities: a more effective clinical process than office-based care. Telemed J E Health 2015;21(2).

[30] Hsu N, Monasterio E, Rolin O. Telehealth in pediatric rehabilitation. Phys Med Rehab Clin N Am 2021;32:307–17. Available from: https://doi.org/10.1016/.pmr.2020.12.010.

[31] Hopkins IM, Gower MW, Perez TW, Smith DS, Amthor FR, Wimsatt CF, et al. Avatar assistant: improving social skills in students with ASD through computer-based intervention. J Autism Dev Disord 2011;41(11):1543–55. Available from: https://doi.org/10.1007/s10803-011-1179-z.

[32] Kuziemsky C, Maeder A, John O, Gogla S, Basu A, Meher S, et al. Role of artificial intelligence within the telehealth domain. Yearb Med Inform 2019;28(01):035–40. Available from: https://doi.org/10.1055/s-0039-1677897.

Section V

Special considerations

Chapter 47

Legislative mandates

Celanie K. Christensen, Kerry M. White, Kristin Fauntleroy-Love and Thomas Lock
Department of Pediatrics, Indiana University School of Medicine, Indianapolis, IN, United States

A physician working with people with disabilities needs practical knowledge of legislation that provides rules for how our society supports, cares for, and allows children with disabilities to reach their full potential. The purpose and interpretation of relevant legislation must be understood so that clinicians can help patients and their families in practical ways. There is often a difference between what is ideal and what is possible. At times we are constrained by what the system allows. At other times, not knowing what the system allows may lead the physician to be a suboptimal advocate. The provider who is not aware of current legislation may limit their patients' access to available services and opportunities. In this chapter, we will review relevant current legislation, describe how the laws are interpreted and implemented, apply this to clinical practice, and finally introduce some practical issues and challenges.

There are three main areas of support for people with disabilities: health care, education, and community inclusion. In this chapter, we focus on the main legislative mandates in these areas. Additional legislation applying to individuals with disabilities is outlined in Table 47.1.

Health care

Medicaid

Medicaid was established in 1965 as part of the Social Security Act (Title XIX) [1]. It provides health care coverage to millions of Americans including low-income families, and in 1990, coverage was expanded to include qualified pregnant women, children, and individuals receiving social security income (SSI). Medicaid is jointly funded by federal and state governments and is the single largest source of health coverage in the United States [2]. As of early 2022, more than 81 million people were covered by Medicaid [3]. It is implemented by individual states according to federal requirements, leading to variability in eligibility and provided services [3].

There are 56 Medicaid programs, one for each state, territory, and the District of Columbia [3]. Currently, over 10 million people qualify for coverage based on disability [4]. These individuals may have physical disabilities, developmental and intellectual disabilities, or serious mental health issues [5]. Clinicians must be knowledgeable about the Medicaid program covering their patients: who is covered, what services Medicaid provides, and how to qualify and obtain services. Understanding the benefits and limitations of financial support can help lead families to the most relevant and impactful resources.

Waivers to Medicaid

As recently as the 1960s, care for children with disabilities was implemented in institutional settings and outside of the community. In the 1970s and 1980s, efforts were made to deinstitutionalize children and allow for more care at home with additional services. To achieve this, changes to existing programs such as Medicaid were necessary. Standard Medicaid was created to provide health care based on need and is calculated solely via family income [1].

Medicaid waivers are programs that allow certain Medicaid rules to be bypassed [4]. This allows for broader eligibility and increased services to meet the needs of children with disabilities. Waivers are state-run and will vary both in practice and name. One of the first waivers developed in the early 1980s was the Beckett waiver. It allows for children with mainly physical disabilities and extensive medical needs to move home and still receive Medicaid benefits. This exception was codified in the Tax Equity and Fiscal Responsibility Act (TEFRA) of 1982 and allowed

Capute and Accardo's Neurodevelopmental Disabilities in Infancy and Childhood. DOI: https://doi.org/10.1016/B978-0-12-824060-1.00037-7

TABLE 47.1 Additional relevant federal disability rights legislation in the United States.

Law	Year	Purpose	Agency
Health care			
Family Medical Leave Act (FMLA)	1993	Allows eligible employees to take up to 12 weeks of unpaid leave within any 12-month period for certain medical and family reasons without losing their job. Requires continuation of group health care coverage as if the employee had not taken leave.	Wage and Hour Division of the US Department of Labor—https://www.dol.gov/agencies/whd/fmla
Genetic Information Nondiscrimination Act (GINA)	2008	Prevents discrimination based on and misuse of genetic information in health insurance and employment.	National Human Genome Research Institute of the National Institutes of Health—http://www.ginahelp.org/GINAhelp.pdf
Health Insurance Portability and Accountability Act (HIPAA)	1996	Defines personal health information (PHI) and states how it must be maintained and transferred.	Office for Civil Rights Headquarters of the US Department of Health and Human Services—https://www.hhs.gov/hipaa/
Community inclusion			
21st Century Communications and Video Accessibility Act (CVVA)	2010	Ensures accessibility of telecommunication devices as well as telephone, television, and broadband services including emerging digital and mobile innovations.	Federal Communications Commission, Disability Rights Office—https://www.fcc.gov/general/disability-rights-office
Achieving a Better Life Experience Act (ABLE)	2014	Enables people with disabilities to possess savings accounts that can pay for qualified expenses with tax advantages while still receiving benefits.	ABLE now in cooperation with the IRS—http://www.able-now.com/
Architectural Barriers Act (ABA)	1968	Requires buildings that are designed, constructed, or altered with federal funds, or at least by a federal agency comply with federal physical accessibility standards.	US Architectural and Transportation Barriers Compliance Board—https://www.access-board.gov/
Air Carrier Access Act (ACAA)	1986	Prohibits discrimination in publicly available domestic and international air carriers.	US Department of Transportation—https://www.transportation.gov/airconsumer/passengers-disabilities
Civil Rights Act of 1964	1964	Prohibits discrimination based on race, color, religion, sex, national origin, and later sexual orientation and gender identity.	US Department of Justice—Civil Rights Division, Department of Justice
Civil Rights of Institutionalized Persons Act (CRIPA)	1997	Authorizes the US Attorney General to investigate conditions of confinement at state and local government institutions including publicly operated nursing homes, and institutions for people with psychiatric and/or developmental disabilities to uncover and correct widespread violations jeopardizing health and safety of residents.	US Department of Justice, Civil Rights Division, Special Litigation Section—https://www.justice.gov/crt/civil-rights-institutionalized-persons
Fair Housing Act (FHA)	1968	Prohibits discrimination in private, federally funded, or state and local government housing, as well as in any aspect of selling or renting.	US Department of Housing and Urban Development—https://www.hud.gov/program_offices/fair_housing_equal_opp
Rehabilitation Act Section 508	1998	Ensures people with disabilities have access to electronic and information technologies of all federal agencies.	US Access Board—https://www.access-board.gov/ada/
Telecommunications Act	1996	Ensures accessibility of telecommunication devices and services.	Federal Communications Commission, Disability Rights Office—https://www.fcc.gov/general/disability-rights-office
Workforce Innovation and Opportunity Act (WIOA)	2014	Helps job seekers access employment, education, training, and support services to succeed in the labor market.	Employment and Training Administration of the Department of Labor—https://www.dol.gov/agencies/eta/wioa

children with certain types of disabilities requiring institutional care to receive financial support allowing for nursing level of care at home if care would be less costly than hospital or institutional care [6].

TEFRA resources primarily support children with physical disabilities and the equipment needed to care for them. There cannot be a waiting list, and eligibility is based only on child's income and resources. The disadvantage to TEFRA is that states cannot offer more than Medicaid would cover for another participant without disabilities and it is voluntary for states to offer it [6,7].

To provide wrap-around support beyond what Medicaid may offer most participants, 1915(c) Home and Community-Based Services (HCBS) waivers were introduced [8]. They may provide resources such as respite care, behavior services, modifications to homes and vehicles, and other therapy services such as music therapy [5]. 1915(c) waivers make up most waivers used in the country currently, and the levels of support differ by state. Forty-seven states and the District of Columbia are operating at least one 1915(c) waiver. Unlike TEFRA, waiting lists are often a barrier to access [4]. It is important to anticipate future needs for HCBS waivers so that application can be started early and waiting time for needed services is minimized.

Section 1115 of the Social Security Act gives authority to approve experimental pilot or demonstration projects to assist in promoting the objectives of Medicaid [9]. A small number of states use 1115 Demonstration Waivers which can streamline access to services but also allow other rules of Medicaid to be waived, potentially decreasing protections [4].

It is important for physicians working with patients with disabilities to know the eligibility requirements for the various waivers in their state. It is important to clarify the appropriateness of ICD10 codes, utilizing those that will help patients meet eligibility requirements. For example, learning disability (F81.9) often does not qualify but global developmental delay (F88) usually does. It is helpful to have ancillary staff available or know your family support organizations well to help families navigate the system that is often complicated and full of barriers that can prevent families from receiving appropriate services. If a family moves to a different state, they will have to apply for a new waiver because waivers are run on a state-by-state basis.

Patient Protection and Affordable Care Act of 2010

The main goal of Affordable Care Act (ACA) is to make affordable health care accessible to more Americans. It offers subsidies ("premium tax credits") that lower costs for households with income between 100% and 400% of the federal poverty level. Also, it allows states to expand their Medicaid program, though details of this expansion vary state by state [9,10].

The ACA prohibits discrimination based on preexisting conditions including disability. Individuals and families cannot be denied nor lose health insurance due to disability. Insurance plans cannot set premiums based on health conditions or disability. The ACA ends lifetime and yearly limits on health care coverage. Children can join or remain on their parents' plan until the age of 26 years [9,10].

In ACA-eligible plans, 10 essential health benefits must be covered. The required benefits include ambulatory services, emergency services, hospitalization, pregnancy, maternity and newborn care, prescription drugs, laboratory services, mental health and substance use disorders services, preventive and wellness services, pediatric oral and vision care, and rehabilitative and habilitative services and devices [10]. This bill has not been fully implemented nationwide. Clinicians need to be aware that other health care plans still exist in the United States that do not have to follow the ACA requirements.

Case study part 1

Anthony is a 6-month-old boy initially presenting to developmental clinic after a 2-month stay in the neonatal intensive care unit (NICU). His medical history is significant for prematurity (32 weeks gestational age), intraventricular hemorrhage (grade 3 on the right and grade 2 on the left), and feeding dysfunction, leading to gastrostomy tube placement prior to NICU discharge. Clinical evaluation reveals global developmental delay and high risk for spastic hemiplegic cerebral palsy. His family has concerns about health care financing to ensure access to therapies, medical services, and procedures he may need going forward. Parental income prevents Anthony from qualifying for Medicaid and the family's private insurance has a high deductible and limits on therapy services. Given Anthony's medical complexity and global delays, he qualifies for an HCBS Medicaid waiver. His family is provided application assistance in clinic so that he can be placed on the wait list for this resource.

Education

Individuals with Disabilities Education Act 1990, updated 2004

Individuals with Disabilities Education Act (IDEA) is the nation's special education law that provides financial aid to the states [11]. It requires states and public schools to serve the developmental and educational needs of children with disabilities. The law contains four parts. Part A created the federal Office of Special Education Programs that administers the program. Part B focuses on special education and related services for individuals 3–22 years of age. Part C is early intervention (EI) for children with disabilities ages 0–36 months. Part D provides grants and resources to improve education services and support programs geared toward maximizing outcomes for students with disabilities [12].

EI services are provided through an individualized family service plan (IFSP) and special education services are provided through an individualized education plan (IEP). These plans are created after an evaluation using suitable materials has been completed by an interdisciplinary team of professionals. Families, and when appropriate, students must be given the opportunity to participate in the plan's creation. For IFSPs and IEPs, families must give consent prior to starting services. Input from the family and child must be considered, and families can challenge the IEP through due process [12,13].

IDEA Part C, otherwise known as EI, serves several purposes. There is ample data showing EI improves developmental outcomes which leads to decreased need for services later in life, including institutionalization. Services are provided at the location of the infant/toddler such as home, day care, and other caregiver's location. Therapists and other providers work directly with the families/other caregivers as well as the child. Another important purpose of EI is to support the ability of families to interact with and meet the needs of the infant/toddler [12–14].

Once children enter the public school system at age 3 years, services are provided as part of an IEP. If the child has an IFSP, ideally parents or guardians will be guided through the transition process. Children are reevaluated by the public school system team to determine if ongoing services are necessary. There are 14 categories of IEP eligibility for children ages 3–9 years and 13 categories of IEP eligibility for people ages 9–22 years. They include specific learning disabilities, other health impairment, autism spectrum disorders, emotional disturbance, speech or language impairment, visual impairment, deafness, hearing impairment, deaf-blindness, orthopedic impairment, intellectual disability, traumatic brain injury, and multiple disabilities. The additional eligibility category for children ages 3–9 years is developmental delay [11,14].

Medical teams and schools use overlapping terms to describe children with disabilities. The two systems often develop different goals and provide complementary services. Medicine seeks to support the developmentally delayed child with extensive evaluation to determine diagnosis and best treatment plan. Schools must offer educational value for children of all developmental levels. IDEA states a free appropriate public education (FAPE) is a right of all children, and children must receive education in the least restrictive environment (LRE) required for them to further their education and seek employment as an adult with the goal of living independently. FAPE and LRE both illustrate differences in approach. Medical needs might require physical therapy; however, schools may not provide this intervention if physical abilities do not affect educational attainment. Schools may have certain resources available only in a special education classroom, but the child's doctor may feel a mainstream classroom with an aide would lead to the best outcome [11,14].

Clinicians are often faced with questions related to the quality of educational plans and may be asked to advocate in school matters. Parents often benefit from a discussion related to both their child's rights within the public school setting and realistic expectations within the scope of FAPE. It is important to stress that the standard for IEPs is appropriate, not ideal. Resources within the clinic should be readily available for support of parents navigating the school system.

Every Student Succeeds Act 2015

Every Student Succeeds Act (ESSA) reauthorizes the Elementary and Secondary Education Act of 1965 and supersedes the No Child Left Behind Act of 2002 [15]. The stated goal "is to provide all children significant opportunity to receive a fair, equitable, and high-quality education, and to close educational achievement gaps" [16]. It includes additional protection for high need students.

ESSA requires public schools be held accountable for educational achievement of students with disabilities. Standards must be aligned to the state's challenging existing content, promote access to the general education curriculum, and reflect professional judgment of the highest possible standards that can be achieved by individuals with disabilities [16].

For students with the most significant cognitive disabilities, ESSA establishes key provisions around alternate achievement standards, alternate assessments, and alternate diplomas. Standards must align to ensure students are on track to pursue postsecondary education or competitive integrated employment. An alternate diploma can be created for students who cannot meet the requirements of a regular diploma. ESSA clarifies that parents are to be fully apprised by the IEP team of the impact of choosing alternate standards and assessments for their child [16].

States are required to describe how they will support school districts "to improve school conditions for student learning, including through reducing—(i) incidences of bullying and harassment; (ii) the overuse of discipline practices that remove students from the classroom; and (iii) the use of aversive behavioral interventions that compromise student health and safety" [16]. Aversive behavioral interventions include seclusion, restraint, and exclusionary discipline.

The clinician should have a working knowledge of the alternate assessment and achievement options in their state. They must be able to offer counseling themselves or connect families to established and knowledgeable groups. If parents report suspicion of bullying or inappropriate discipline, they should be advised about state-specific reporting requirements.

Case study part 2

Anthony is now 8 years old and returns to developmental clinic for follow-up after several years. He has learning problems but no formal diagnosis of intellectual disability. He also has cerebral palsy, requiring a walker for ambulation most of the time and a manual wheelchair for longer distances. His mother has some concerns about appropriate therapies and classroom placement now that he is going into the third grade in a new school building. He benefited from EI services through an IFSP providing physical therapy (PT), occupational therapy (OT), and speech therapy (ST). The transition to the public school system at age 3 years went smoothly. He was placed in special education preschool and received school-based PT, OT, and ST. For first and second grades, he was placed in a general education classroom with pull outs to a resource room for math and reading as well as receiving PT, OT, and ST. He also had an aide to help him move between the classroom and lunchroom. The school has suggested moving him to a special education classroom for third grade and not providing an aide for transition between the lunchroom and classroom. In the clinic, his family is counseled to request formal school psychology assessment to help determine appropriate classroom setting and accommodations. His family is also connected with a local parent support resource focusing on public school system navigation.

Community inclusion

Section 504 of the Rehabilitation Act (Section 504) 1973

The Rehabilitation Act of 1973 is the first civil rights law relating to people with disabilities and is the predecessor to the Americans with Disabilities Act (ADA) [17,18]. This law was the first to address the needs of individuals with disabilities outside of basic medical care and moved toward community inclusion. The legislation attempts to address some of the societal barriers faced by individuals with disabilities, including isolation by placement in institutions, limited access to buildings, and discrimination in education and employment.

Section 504 states that "no qualified individual with a disability in the United States shall be excluded from, denied the benefits of, or be subjected to discrimination under any program or activity that either receives Federal financial assistance or is conducted by any Executive agency" [17]. While signed into law in 1973, the mandates and regulations of Section 504 did not begin implementation until 1977 after four years of intense activism by the disability community [19].

Requirements of Section 504 related to children and young adults are particularly relevant to public schools. Section 504 mandates that a student with a disability receives appropriate education even if they do not require special education services. Qualified students in general education classrooms must still receive reasonable accommodation through a 504 plan. This is especially important for students whose disability, such as attention deficit hyperactivity disorder (ADHD), does not fit under one of the IDEA categories. IDEA funds may not be used to implement 504 plans [17,20].

While Section 504 applies to many other services and programs, it explicitly states that individuals with disabilities do not have a right to accommodations that would fundamentally alter the activity or create an "undue burden." Each federal agency has its own set of Section 504 regulations that it is responsible for enforcing, and violations may be addressed through private lawsuits. It is not necessary to file a complaint with a federal agency or to receive a "right-to-sue" letter before going to court [17].

Pediatricians should know the differences between an IEP and a 504 plan and the diagnoses relevant to each. They should be able to discuss the types of accommodations appropriate for the individual child. See Table 47.2 for comparisons.

TABLE 47.2 Comparison of IFSPs, IEPs, and Section 504 plans.

	IFSP	IEP	504
Mandate	Individuals with Disabilities Education Act (IDEA), part C	Individuals with Disabilities Education Act (IDEA), part B	Section 504 of the Rehabilitation Act
Funding	Parents may pay for services based on state program, states receive funding from the federal government	Parents do not pay for services, states receive funding from the federal government	Parents do not pay for services, states do not receive additional funds, IEP funds cannot be used for 504 accommodations
Description	Family-centered and multidisciplinary service plan to maximize developmental outcomes	Special education plan for services and accommodations to meet a student's needs	Plan for accommodations to maximize learning in general education for a student with a variety of disabilities who does not require an IEP
Timing	0–36 months	3–22 years	K-12th grade and college
Eligibility	Documented developmental delay, diagnosis with high risk of developmental delay, and in some states, biological or environmental risk factors for delay	One of 13 IDEA disability categories—see text	Any disability that impairs education in a general classroom
Plan creation and review	Parent or guardian must consent, plan is reviewed every 6 months and updated yearly	Parent or guardian must consent and be given opportunity to be an active team member	Parental input not required; does not have to be written down, no progress monitoring
Example accommodations	Therapies in natural environment, nutrition services, counseling and family training	Therapies in school setting, transportation, special education services, classroom aides	Preferential seating, larger text, extended time for testing

IFSP, Individualized family service plan; *IEP*, individualized education plan.

Americans with Disabilities Act 1990

The ADA of 1990 is an unfunded mandate modeled after the Civil Rights Act of 1964. Its goal is to provide equal opportunities to people with disabilities and addresses employment, transportation, state and local services, and accommodations by private businesses [21].

ADA protects people with a disability and people with a relationship or association with an individual with a disability [18]. An individual with a disability is defined by the ADA as a person who has a physical or mental impairment that substantially limits one or more major life activities, a person who has a history or record of such an impairment, or a person who is perceived by others as having such an impairment. The ADA does not specifically name all the impairments that are covered [21].

Title I addresses employment and applies to organizations with 15 or more employees, all state and local governments, employment agencies, and labor unions. Employers must provide equal opportunity and access to employment and employment-related activities such as hiring, promotion, and pay for individuals with disabilities.

Title II addresses state and local government activities such as public education, employment, transportation, recreation, health care, social services, and architectural standards.

Title III applies to private and nonprofit organization activities including restaurants, retail stores, hotels, private schools, doctors' offices, day care centers, and recreation facilities including sports stadiums and fitness clubs. Religious organizations are exempt.

Title IV focuses on telephone and television access for people with hearing and speech disabilities. Telecommunications companies are required to establish interstate and intrastate telecommunications relay services (TRS). TRS allows callers with hearing and speech disabilities who use a telecommunications device for the deaf, and callers who use voice telephones to communicate with each other through a third-party communications assistant 24 hours a day, 7 days a week. Title IV also requires closed captioning of federally funded public service announcements.

Title V prohibits retaliation, intimidation, coercion, threats, or interference with people who seek to exercise their rights or who encourage or help others to do so. These provisions also protect people without disabilities if they advocate or testify on behalf of an individual with disability. Claims of "reverse discrimination," where an individual seeks remedies if they feel they were discriminated against because they do not have a disability, are also prohibited. Title V states that aids and accommodations do not have to be accepted by the individual and legal fees can be recovered if the lawsuit is successful [18,22].

Transition resources are limited, and community inclusion efforts are largely unfunded [21]. It can be difficult to access employment opportunities for individuals with neurodevelopmental disabilities. Discussions about transitioning to adult care and for independent living, if possible, need to start in the preteen years to prepare families and children. (See Chapter Aging and Transition)

Case study part 3

While transitioning to adult care, Anthony returns to developmental clinic at age 17 years. He was diagnosed with a mild intellectual disability by school psychology testing. His motor abilities remain stable. Anthony has been part of his IEP discussions since the age of 14 years. In clinic, discussion includes his ability to remain in school through the age of 22 years for vocational training. His family is encouraged to start looking at housing options as he becomes a young adult in need of adaptations for accessibility.Discussions were undertaken about supported decision making.

Using the support offered through federal and local mandates and expert guidance by his developmental specialist and clinical staff, Anthony's family maximized his potential and prepared him for the best outcome as an adult. He was covered through Medicaid allowing necessary medical care; school accommodations were in place to make the experience beneficial and useful; and community inclusion set the stage for a fulfilling job and the highest level of independence for Anthony.

Advocacy

There are 61 million adults and over 3 million children in America with disabilities. This means 25% of adults and 4.3% of children are at risk of facing barriers to equal rights and privileges [23]. In 2008 the United Nations adopted the Convention on the Rights of Persons with Disabilities (CRPD) [23,24]. Its purpose was to promote, protect, and ensure equal enjoyment of all human rights for individuals with disabilities and to promote respect for their inherent dignity. CRPD addresses barriers hindering full and effective participation in society. It comprises eight general principles that summarize its 50 articles. The articles address several topics including accessibility, privacy, education, employment, and work. Table 47.3 outlines the eight ethical principles of CRPD and a summary of additional protected rights discussed in its articles that are most relevant to physician advocates. CRPD was not the first national or international declaration that guides ethical practices. Despite it being the fastest negotiated human treaty, as disability advocacy is studied today, several infringements on the rights outlined in CRPD exist [25].

Individuals with disabilities remain at risk of poorer health outcomes and face discrimination due to the hindrance to their human rights. Disability advocacy is necessary to defend the rights of people with disabilities. Physicians have the responsibility to improve the health and well-being of their patients. CRPD specifically states health care professionals should raise awareness of the human rights, dignity, and autonomy for individuals with disabilities, calling clinicians to also act as advocates [25]. Becoming a physician advocate requires understanding human rights and the barriers that exist both in the health care system and society.

Medical and social models of disability

People with differing abilities have been viewed in many ways in diverse cultures. In the United States, disability is often viewed through one of two models, the medical model and the social model. The medical model focuses on an individual's physical or mental medical condition and regards them as "issues" needing to be fixed. It views disability as a personal impairment and permits the individual's body to be the sole target of intervention. This model prioritizes values and opinions of professionals over the views of individuals with disabilities when defining disability. It can allow society to avoid its social responsibility to provide equitable access and eliminate marginalization. The disability community has adopted "nothing for us without us" to stress the need for inclusion when defining disability, directing research, and adopting societal norms for the treatment of people with disabilities [26].

TABLE 47.3 Convention on the rights of persons with disabilities: eight general principles.

Respect for inherent dignity	Persons with disabilities are entitled to respect for inherent dignity and individual autonomy. All should have the freedom to make choices and the right to independence.
Nondiscrimination	All persons are equal under the law and entitled without discrimination. To promote equality and eliminated discrimination, appropriate steps should be made to ensure reasonable accommodations.
Full and effective participation and inclusion	Individuals with disabilities have the right to full and effective participation and inclusion in society. Community facilities shall be available on an equal basis. Measures should be taken to ensure the freedom to seek, receive, and impart information. This promotes the use of accessible communication formats.
Disability defined as a part of human diversity	Disability should be respected and accepted as a part of human diversity and humanity.
Equality of opportunity	Individuals with disabilities should have equal opportunity in all aspects of life including but not limited to work, education, political involvement, recreational activities, and choice in place of residence.
Accessibility	To enable individuals to live independently and participate fully in all aspects of life, appropriate measures should be made to ensure physical environments, transportation, information, and communication are accessible on an equal basis.
Equality between genders	Women with disabilities are subject to multiple disparities. Action should be taken to ensure full and equal participation. Measures should be taken to ensure full development, enhancement, and empowerment.
Respect for children with disabilities	Children with disabilities have the right to express their views freely on matters that affect them, with consideration given for age. The best interest of the child is the primary consideration. All measures should be taken to ensure enjoyment and freedoms that are equal with other children.

The social model regards disability as the result of bias, oppression, and lack of accommodation. In the social model, having a disability itself is a state of being, one with cultural diversity and pluralism. The impact of disability is a consequence of social and systemic barriers. It is the result of interactions between an individual and their environment [26].

The social aspects of disability have varied over time and place. In early Native American societies, the concept of disability did not exist. Everyone was seen to have different "gifts," and these gifts determined how the individual functioned within the community. For example, the widespread use of sign language among Native Americans who spoke different oral languages meant that deafness was not a barrier to community participation. Initially, European settlers were selected to fit certain roles in the new settlements and families were responsible for care of all family members. If a family could not care for a member, the community would usually take responsibility. However, there are documented cases of community conflict over the care of individuals with disabilities. As urbanization and industrialization progressed, the pressure for individuals to fit certain social roles increased. These produced definitions of disability that were centered around ability to perform wage earning jobs. The modern Disability Rights movement strives to counteract this pressure [19].

Barriers to disability inclusion

Accessibility is the degree to which an environment, item, or service is available for use by all members of its community. A discussion of all barriers to access would be extensive; however, many fall under the following broad categories [27].

Attitudinal barriers include prejudice, bias, and stigma toward individuals with disabilities. In health care, it infringes on the right to service without discrimination. When bias and stigma influence employment, earning potential is limited which decreases one's ability to access quality care, accommodations, and daily activities. Preexisting beliefs can impact social involvement and employment opportunities, thus negatively affecting one's quality of life, financial stability, and mental health.

Environmental barriers include inaccessible physical infrastructure and communication impediments that limit participation and access to information. The ADA addresses environmental barriers, though they have not been fully eliminated. For example, in 2021 65% of curb ramps and 48% of sidewalks in the United States are still not accessible [28].

Institutional barriers are laws and policies that negatively impact or exclude individuals with disabilities. These barriers lead to systemic disadvantages. For example, clinic policies that punish late arrivals unduly impact patients reliant on less dependable means of transit such as public and Medicaid transport.

Informational barriers are rooted in the definition of disability as the view that medical intervention is the sole solution. This allows government agencies and society to avoid the responsibility of improving inclusion. A prevalent example is significant government funding for extensive medical research to find cures for disability while providing limited funds for removal of environmental or institutional barriers.

The true influence of these barriers on fundamental rights of the individual with disability may not be readily apparent. Every person, regardless of ability, should have the opportunity to work, learn new things, and develop healthy relationships in a safe environment [27].

Employment barriers: In 2020 only 17.9% of individuals with disabilities were employed. They experience a higher rate of unemployment across all educational obtainment groups and all age groups. Barriers to employment include employer or coworker attitudes, need for specific features at work, lack of transportation, loss of governmental assistance, lack of training, or lack of education [29].

Education barriers: IDEA, as discussed previously in this chapter, aims to make appropriate public education free to eligible children with disabilities. It ensures special education and related services for eligible children [11]. It commits to providing 40% of the average per pupil expenditure (APPE) multiplied by the number of students in special education in each state. Unfortunately, IDEA has been underfunded. In 2019 the amount funded was about 14.3% of the national APPE, not even half of the 40% commitment [30].

Safety barriers: Individuals with disabilities are more likely to experience physical and sexual abuse. Current statistics are thought to be underestimated due to underreporting. Underreporting can be secondary to communication barriers, the relationship dynamics that exist in dependent care settings and the lack of recognition of warning signs [31].

Marriage barriers: While there are no laws prohibiting individuals with disabilities from getting married, SSI determinations impose an unofficial "marriage penalty." After marriage, spousal income is considered for SSI which reduces the amount received. If the combined income is over the SSI eligibility limit, benefits can be discontinued completely. In 2021 the individual limit for social security benefit is $794/mo. The couple benefit limit is $1191/month. If both partners depend on SSI, after marriage they will experience a combined income loss of $397 monthly or $4764 annually. This policy is an institutional barrier [32].

Health barriers: In 2021 Health Affairs published an article titled "Physicians Perceptions of People with Disability and Their Health Care." This article surveyed 714 practicing physicians and found that physicians' attitudes and lack of knowledge in caring for individuals with disabilities contribute toward health disparities. Of the physicians surveyed, 79.8% strongly agree that it is valuable to understand patients with disabilities, 40.7% felt "very confident" with their ability to provide the same quality care for individuals with disabilities, 56.6% "strongly" agreed that they welcome patients with disability in their practice, and 82.4% of physicians reported people with significant disability have a worse quality of life than people without disability [33]. Overall, these results suggest physicians practicing in the United States might hold strong biases toward individuals with disabilities. The lack of confidence in treating individuals with disabilities and assumption about their quality of life can negatively impact physicians' goals of care. This serves as an example of how the CRPD principles have not been achieved in our health care systems.

Dual role of neurodevelopmental and developmental behavioral physicians as clinician and advocate

While the medical model contributes to barriers for individuals with disabilities, it also provides valuable support. Governmental funding and physician participation in research increases the understanding of clinical conditions, provides new diagnostic technology, and uncovers prevention and treatment options. Indirectly, the medical model allows patients to qualify for additional support through waivers and other services. However, neglect of the social model limits the lives of persons with disabilities. Innovation in and implementation of strategies that reduce barriers and facilitate inclusion can create systems that support neurodevelopmental diversity.

Physician advocates are required to understand the rights of their patients and the barriers that they encounter in the health care system, including bias held by other professionals. Physicians can increase their political and community involvement promoting disability as a valuable form of human diversity rather than an individualized and devastating problem to be solved. Although medical education tends to stress the medical model over the social model of disability

TABLE 47.4 Convention on the rights of persons with disabilities: articles relevant to physician advocates.

Awareness-raising	Raise awareness through society. Combat stereotypes, prejudices, and harmful practices. Promote awareness of the capacities and contributions provided by individuals with disabilities.
Right to life	Every human being has the inherent right to life and right to effective enjoyment on an equal basis.
Protection from inhumane treatment	No one shall be subject to torture or cruel inhuman or degrading treatment. No one shall be subject to medical or scientific experimentation without their consent.
Freedom from abuse	Appropriate measures shall be taken to protect individuals with disabilities from exploitation, violence, and abuse.
Personal mobility	Measures should be taken to ensure mobility with the greatest possible independence. It should be the individual's choice and at an affordable cost. This may include mobility aids, devices, or assistive technology.
Respect for privacy	Protect the privacy of personal, health, and rehabilitation information.

in its curriculum, medical professionals should try to view disability through the social model and have an active role in addressing barriers. Care should emphasize improving quality of life and accessibility. Treatment plans should be grounded in the assumption that life with a disability has inherent value. Table 47.4 describes the CRPD articles most relevant for physician advocacy.

Physicians are in a unique position to act as advocates due to their respected role in society.

There are many levels of advocacy including the patient level, community level, and national level. Advocacy is defined as supporting or recommending something on behalf of a person or cause, thus the below examples do not fully encompass the many opportunities for advocacy.

Patient-level advocacy

The basis of patient-level advocacy includes providing high-quality care, protecting the rights of patients, and protecting patients from harm [34]. It should be guided by the patient's strengths, unique qualities, challenges, and goals. As neurodevelopmental pediatricians, there is an additional responsibility of addressing systemic barriers and connecting patients to all resources they qualify for. This may include connecting patients with clinical social workers, informing families of additional financial waivers, providing letters to schools to provide support for individualized educational plans, and writing letters to support the need for medical equipment [35].

Community-level advocacy

Community-level advocacy is the act of involving the community around issues related to disability. Physicians can accomplish this through knowledge distribution and participation in community organizations. This is demonstrated through educating others with continuing education courses or by providing education to family groups. Participation in the community through board memberships, especially in organizations that serve children with disabilities or the educational system allows physicians to advocate for their patient's interests within the community [36]. Membership and participation in local American Academy of Pediatrics (AAP) Chapters will provide additional opportunities for physicians to remain up to date on legislation and to advocate for specific issues which impact children with disabilities.

National-level advocacy

National-level advocacy is the act of advocating to influence change throughout the country. This can be accomplished through involvement in professional organizations. The AAP Council on Children with Disabilities comprises physicians who care for or have an interest in children and youth with disabilities. Their mission is to develop clinical guidelines, educate other pediatricians, collaborate with partner organizations, and provide expertise and advocacy both locally and nationally. The Association of University Centers on Disability works to advance policy for individuals with disabilities and is an additional resource for national involvement. Social media can also be considered as it now provides a unique opportunity for physicians to influence change beyond the scope of their immediate proximity [36].

In summary, while legislative mandates that dictate societal support for health care options, educational opportunities, and the ability to successfully function in one's community have improved, there remain significant barriers for

TABLE 47.5 Practical disability advocacy resources for the clinician.

Resource	Description	Website
Ten regional ADA centers	Central and comprehensive resource on ADA issues that provides information, training, and technical assistance	https://adata.org/
US Access Board	Federal agency promoting access for people with disabilities through developmental accessibility guidelines and standards	https://www.access-board.gov/
Medicaid.gov	General info about Medicaid, state-specific resources, waiver descriptions	https://www.medicaid.gov/
Medicaid and CHIP Payment and Access Commission (MACPAC)	Nonpartisan legislative branch agency and independent source of information on Medicaid and CHIP	https://www.macpac.gov/medicaid-101/
Department of Education—IDEA website	General and state-specific information about IDEA	https://sites.ed.gov/idea/
Center for Parent Information and Resources	Central hub of information and products created for the network of Parent Centers serving families of children with disabilities	https://www.parentcenterhub.org/find-your-center/
Understood.org	Advocate group for individuals with learning and thinking differences	https://www.understood.org/

ADA, Americans with Disabilities Act; *IDEA*, Individuals with Disabilities Education Act.

individuals with disabilities. Further advocacy is critical. Table 47.5 offers additional online resources for further education and updates regarding federal legislation in this area. Physicians should utilize current legislative mandates so their patients and families can be connected to available resources.

References

[1] Title XIX of the Social Security Act (the Act, P.L. 89–97). (1965). Public Law 89-97.

[2] American Academy of Pediatrics, Council on Children with Disabilities Policy statement. Supplemental Security Income (SSI) for children and youth with disabilities. Pediatrics 2009;124:1702–8.

[3] Centers for Medicare & Medicaid. (n.d.). By population. Retrieved from Medicaid.gov https://www.medicaid.gov/medicaid/program-information/medicaid-and-chip-enrollment-data/report-highlights/index.html

[4] Medicaid and CHIP Payment and Access Commission. (n.d.). People with disabilities. Retrieved from MACPAC website, https://www.macpac.gov/subtopic/people-with-disabilities/#: ~ :text = Over%2010%20million%20people%20qualify,do%20not%20have%20Medicare%20coverage.

[5] McLean KJ, Hoekstra AM, Bishop L. United States Medicaid home and community-based services for people with intellectual and developmental disabilities: a scoping review. J Appl Res Intellect Disabil. 2021;34(3):684–94. Available from: https://doi.org/10.1111/jar.12837.

[6] Tax Equity and Fiscal Responsibility Act of 1982 (1982). Pub.L. 97-248. Retrieved from GovTrack.us

[7] Musumeci M. Modernizing Medicaid eligibility criteria for children with significant disabilities: moving from a disabling to an enabling paradigm. Am J Law Med. 2011;37(1):81–127. Available from: https://doi.org/10.1177/009885881103700103.

[8] Omnibus Budget Reconciliation Act of 1981 (1981). Pub.L. 97-35. Retrieved from The United States Social Security Administration.

[9] U.S. Department of Health and Human Services. Available from: http://www.hhs.gov.

[10] United States. (2010). Patient Protection and Affordable Care Act, Pub. L. No. 111-148, 124 Stat. 119 (2010).

[11] United States. (2004). Individuals With Disabilities Education Act, 20 U.S.C. § 1400.

[12] U.S. Department of Education. Individuals with Disabilities Education Act (IDEA). U.S. Department of Education. (2020). Available from: https://sites.ed.gov/idea/.

[13] Adams RC, Tapia CCouncil on Children With Disabilities. . Early intervention, IDEA Part C services, and the medical home: collaboration for best practice and best outcomes. Pediatrics 2013;132(4):e1073–88. Available from: https://doi.org/10.1542/peds.2013-2305.

[14] Lipkin PH, Okamoto J, the Council on Children with Disabilities and Council on School HealthNorwood KW, Adams RC, Brei TJ, Burke RT, Davis BE, et al. The Individuals With Disabilities Education Act (IDEA) for children with special educational needs. Pediatrics 2015;136(6):e1650–62. Available from: https://doi.org/10.1542/peds.2015-3409.

[15] United States. (2015). Every Student Succeeds Act, 20 U.S.C. § 6301, https://www.congress.gov/bill/114th-congress/senate-bill/1177.

[16] U.S. Department of Education. Individuals with Disabilities Education Act (IDEA). U.S. Department of Education. (2020). Available from: https://www.everystudentsucceedsact.org.

[17] U.S. Congress. (1973). Section 504 of the 1973 Rehabilitation Act, Pub. L. No. 93-112, 87 Stat. 394 (September 26, 1973), codified at 29 U.S. C.§ 701 et seq.

[18] Americans With Disabilities Act of 1990, Pub. L. No. 101-336, § 1, 104 Stat. 328 (1990).

[19] Nielsen K.E. 2012. A disability history of the United States, Beacon Press, Boston.

[20] Understood. The Difference Between IEPs and 504 Plans Understood. 2020, November 5. Available from: https://www.understood.org/en/articles/the-difference-between-ieps-and-504-plans?_sp = f86e06f4-c3c8-408b-bff6-5c7720fca8b8.1665430937438.

[21] National Network | Information ADA. Guidance and Training on the Americans with Disabilities Act. (n.d.). Adata.org. Available from: https://adata.org.

[22] A desktop guide to disability rights laws in the United States. ADA Rocky Mountain ADA Center; 2021.

[23] Disability impacts all of us infographic [Internet]. Centers for Disease Control and Prevention. National Center on Birth Defects and Developmental Disabilities; 2020.

[24] Young, N.A. Childhood disability in the United States: 2019. ACSBR-006. Washington, DC: American Community Survey Briefs, US Census Bureau; 2021.

[25] United Nations. Convention on the rights of persons with disabilities; 2006.

[26] Hogan AJ. Social and medical models of disability and mental health: evolution and renewal. Can Med Assoc J 2019;191(1).

[27] Cable C. Partnerships, advocacy, and communication for social change [Internet]. The United Nations International Children's Emergency Fund; 2014. Available from: https://www.unicef.org/eca/sites/unicef.org.eca/files/IE_Webinar_Booklet_7.pdf

[28] Eisenberg Y, Heider A, Gould R, Jones R. Are communities in the United States planning for pedestrians with disabilities? Findings from a systematic evaluation of local government barrier removal plans. Cities 2020;102:102720.

[29] Persons with a disability: labor Force characteristics—2020 [Internet]. US Bureau of Labor Statistics. United States Department of Labor; 2021. Available from: https://www.bls.gov/news.release/pdf/disabl.pdf.

[30] Dragoo K.E.. The Individuals with Disabilities Education Act (IDEA) funding: a primer; 2019.

[31] Basile K, Breiding M, Smith S. Disability and risk of recent sexual violence in the United States. Am J Public Health 2016;106(5):928–33.

[32] A guide to supplemental security income (SSI) for groups [Internet]. The United States Social Security Administration; 2021. Available from: https://www.ssa.gov/pubs/EN-05-11015.pdf

[33] Iezzoni LI, Rao SR, Ressalam J, Bolcic-Jankovic D, Agaronnik ND, Donelan K, et al. Physicians' perceptions of people with disability and their health care. Health Aff 2021;40(2):297–306.

[34] Choi PP. Patient advocacy: the role of the nurse. Nurs Stand 2015;29(41):52–8. Available from: https://doi.org/10.7748/ns.29.41.52.e9772.

[35] Sullivan WF, Heng J. Supporting adults with intellectual and developmental disabilities to participate in health care decision making. Can Fam Physician 2018;64(Suppl 2):S32–6.

[36] Pettett G. Obligation or option? The physician's role in advocacy. Mo Med 2011;108(6):386–8.

Chapter 48

Neuroethics: a brief overview

Patricia Ann Evans
Departments of Neurology and Neurotherapeutics, Pediatrics, and Psychiatry, University of Texas Southwestern School of Medicine, Children's Medical Center, Dallas, TX, United States

Introduction

Neuroethics is a critically important component to the field of neurodevelopmental disorders, and yet, it is perhaps among the most challenging aspects of this clinical discipline. Neuroethics is best described as an interdisciplinary field which focuses on ethical issues for specialists who work with individuals with a wide range of cognitive and emotional impairment [1]. This is particularly important in the field of neurodevelopmental disorders, and the wide range of moral and ethical challenges which our patients and their families often face.

The following is a brief introduction to a biomedical ethics framework, with particular attention to autonomy, beneficence, nonmaleficence, and justice, applied specifically to the neurodevelopmentally disabled population. Samples of clinical applications are provided only as guidelines, as the application of ethical issues in caring for individuals with special needs is often as unique and individual as the patient.

Basic concepts of bioethics

This chapter borrows heavily from the excellent text, Principles of Biomedical Ethics, eighth edition, by Beauchamp and Childress, and their elucidation of four broad concerns: autonomy, beneficence, nonmaleficence, and justice (Table 48.1).

TABLE 48.1 Concepts and definitions.

Term	Definition	Special considerations
Autonomy	*Autonomy requires three elements*: Intentionality: actions require plans Understanding: actions require understanding to be truly autonomous Noncontrol: a person is free of control in order to execute the intended and understood plan	Balancing autonomy requires consideration of what is best for the individual's dignity and sense of worth and what is safest for both the individual and the community
Nonmaleficence	Do no harm	Considered the most passive of requirements, nonmaleficence is the bare bones minimum of medical ethics
Beneficence	One ought to *prevent* evil or harm One ought to *remove* evil or harm One ought to *do or promote* good	How do we as health care providers and as a community promote health and inclusion within the community?
Justice	Allocating, setting priorities, and rationing of health care: how do we as a community define what is just and what is unjust?	Health care distribution is inherently inequitable among minorities in the United States and requires extra care and concern to promote equal access to care

Capute and Accardo's Neurodevelopmental Disabilities in Infancy and Childhood. DOI: https://doi.org/10.1016/B978-0-12-824060-1.00036-5

Autonomy

The word autonomy derived from the Greek *autos* "self" and *nomos* "rule, governance or law," originally referred to the self-rule or self-governance of independent city-states. A person who is autonomous is said to be able to "act freely in accordance with a self-chosen plan" [2]. By contrast, those individuals who have diminished or no autonomy may be controlled to varying degrees by others. Within the context of neurodevelopmental disabilities (NDD) practice, our patients, regardless of age, often have at least some diminished autonomy. How do we as health care providers help those individuals and their families identify and support both the individuality and personhood of the client, as well as provide structure for support and care?

In seeking an ideal model of autonomy in the context of health care, [2] Beauchamp presents a three-condition theory:

1. *Intentionality*. For an act to be intentional, it must correspond to the individual's understanding of the action involved. Clearly, even good planning may result in undesirable outcomes; however, the individual needs intention to act autonomously.
2. *Understanding*. An individual must have a reasonable, if not full, understanding of the action to be taken, or it is not considered autonomous. Conditions that limit understanding include cognitive disabilities as well as maturity and communication disabilities, common findings in the NDD population.
3. *Noncontrol*. An individual must be free of controls, typically exerted by either external sources, such as guardians, or internal sources, such as comprehension, communication, or other cognitive impairments. Noncontrol may not be an option for many patients in the NDD population.

Clearly, a continuum may exist for any individual facing choices, from fully possessing clear intentions, understanding, and noncontrol to having no capacity for intention, understanding, or control. More typically, many individuals in the NDD population typically fall between these two extremes. Taking time to identify each individual's wants, needs, and wishes in this context can be invaluable in establishing long-term care. Identifying specifically those wants and needs typically takes careful conversations over time, as an individual matures and acquires better insight into his or her desires for the lifetime. The lines between adequate and inadequate degrees of understanding and degrees of control must be determined in light of specific objectives such as deciding whether to stay in the public school system until age 22 or attempt online college work; whether it is reasonable to have partial or complete guardianship; housing and jobs after high school training.

Decision-making capacity: issues of best versus substitute interest in informed consent

Who is competent to decide important issues in an individual's life? If an individual is not capable of making decisions, which impact daily living such as living arrangements, banking, and self-care, who should make those decisions on that person's behalf? Substituted judgment is often invoked as a guide for decision-making when a patient lacks decision-making capacity. The surrogate exercising substituted judgment is essentially being asked to reproduce the exact decision the patient would have made by applying the patient's values and preferences to the question at hand. Subsequently, true substituted judgment is usually impossible in many of the patients seen in pediatric neurology settings, particularly when their developmental disorder has deprived them of the capacity to possess or express health care preferences and values [3].

Subsequently, the best interests standard is the default surrogate decision-making standard when substituted judgment is impossible, which is why it would need to be used by parents or other surrogates in these cases. In short, the "Best Interests standard asks what the surrogate decision-maker *believes* is in the patient's best interest by balancing the benefits against the burdens of the proposed therapy or intervention" [4].

Finally, where appropriate, the use of narrative techniques can be very helpful in elucidating specific concerns. Such an approach allows the individual if possible, or close family or friends if needed, to create a narrative that describes the short- and long-term desires of the individual. The goal is to focus on each patient's dignity and individuality in the context of others who must make decisions on behalf of the patient [5].

Clinical examples of issues of autonomy

1. Matthew is a 16-year-old young man with intellectual disability with a full-scale IQ of 65, with moderate autism spectrum disorder. His parents plan to obtain full legal guardianship by the time he achieves legal age in 2 years;

however, they have also created a separate apartment in their home where he is able to live while he pursues life skills training in the public education system until his 22nd birthday. During this time, they plan to create a bank account for him, so that he has access to funds that they supply on a monthly basis, as well as whatever money he may earn in a jobs he hopes to have after graduating at age 22.
Matthew has friends at his local church, and he is able to participate in social activities as well as school and work independently but with support for his living and financial arrangements. Matthew's parents are finding ways to support his autonomy without risking his health, such as placing him in an apartment living alone, taking public transportation to a job where he may or may not be valued, and coming home alone.
Autonomy has been protected in that Matthew has a paid job that he enjoys and which he can control; as well as support from family and community in that he is allowed to live on his own, but with immediate help with finances and health from his parents, as well as socialization with church and his friends.

2. Melissa is an 18-year-old young woman whose grandparents have raised her from infancy. Melissa has a full-scale IQ of approximately 40–50, is nonverbal, has well-controlled epilepsy, and mild type 2 diabetes mellitus. She is obese and eats to self-soothe. She requires assistance in self-care such as hygiene, feeding, or laundry. Her periods are regular, and she is potentially fertile. It is anticipated that her grandparents, who are now in their 80s and fragile, will need to place her in a state facility, having no money to find a private group home or adult foster care. Her grandparents are understandably very worried about her future if they are not present to watch her and are especially worried about her capacity to bear children if she were taken advantage of in their absence.
The family has stayed on the 20-year long waiting list for a state group home to become available, but none has opened thus far. Time is spent reviewing methods for birth control, as well as placement, with a primary care physician who could potentially manage her medications, as well as an adult neurology service to manage her epilepsy and adult endocrinology to manage her diabetes. Time is spent with social work to provide respite care to the elderly grandparents until such time that Melissa can be placed in a group home safely.
Autonomy is quite limited in this situation but is focused primarily on the patient's safety and best living arrangements as her own guardians are increasingly unable to care for her. By providing support systems within the medical community, and ideally within the social community, Melissa will be known by many caring individuals prior to entering a state home when she leaves her grandparents' care.

3. Richard is a 28-year-old young adult who is mobile, but nonverbal, and currently resides in an assisted living facility with other similarly impaired adults. The residents are able to participate in a wide range of activities; however, Richard has been known to leave group activities and rattle the door knobs of other residents. This has led to frequent upset, particularly with one older, frail resident. The care team spends a great deal of time strategizing on how to best protect Richard's autonomy, without infringing upon the privacy of other residents. Ultimately, a decision is made to put group activities in an alternate large room, with a door that prevents Richard from wandering to the residential side of the home.

Beneficence versus nonmaleficence

Many ethical theories recognize a principle of beneficence as well as nonmaleficence. For the purposes of this brief chapter, the concepts of nonmaleficence are that one ought not to inflict harm; and that beneficence, that one ought to prevent harm; remove harm; and where possible, promote good. Each of the three principles of beneficence requires taking action by helping, that is, preventing harm, removing harm, and promoting good, whereas nonmaleficence requires only *intentional avoidance* of actions that cause harm. Historically, the conflicts between autonomy and beneficence have often been paternalistic/maternalistic in nature, particularly prior to modern medicine when patients were routinely not involved with information and decision-making in their health care. Paternalistic/maternalistic beneficence occurs in the medical setting when the health care provider must decide whether patient autonomy should be secondary to beneficence. In the context of NDD, suggestions by Beauchamp and Childress [2] provide when paternalism/maternalism may be justified:

1. A patient is at risk of a significant, preventable harm or failure to receive a benefit.
2. The paternalistic/maternalistic action will probably prevent the harm or secure the benefit.
3. The intervention to prevent harm or to secure a benefit for the patient probably outweighs the risks to the patient of the action taken.
4. There is no morally better alternative to the limitation of autonomy that will occur.
5. The least autonomy-restrictive alternative that will prevent the harm or secure the benefit is adopted.

Clinical examples of beneficence in the context of paternalism/maternalism

1. Anna is a 7-year-old nonverbal, autistic child, who demonstrates significant hand flapping throughout the day. She is otherwise a quiet child and uses headphones and an electronic tablet to help augment communication. She sleeps reasonably well and is working with a feeding therapist to help broaden her choice of foods. The parents are very embarrassed when they go out in public because her hand flapping can draw attention and unwanted comments. The family is seeking medication to reduce the frequency of the hand flapping. Recognizing that the hand flapping is not in itself causing harm to the patient, and that the use of medication to sedate the hand flapping has potential side effects, some time is spent discussing the pros and cons of the use of medication for this purpose. Parents are initially frustrated that medication is not the best answer in the specific context. However, parents agree to seek behavioral therapies to find ways to distract or to otherwise provide sensory input that might reduce the hand flapping without medication. Three months later, Anna's family is pleased that the use of a weighted vest with other sensory techniques have reduced her hand flapping and have found that her language has also increased with the same techniques.
   The use of nonmaleficence, in avoiding the use of medication, as well as beneficence, in promoting behavioral therapy to reduce undesirable behaviors was successful in achieving an optimal outcome in this clinical setting. By avoiding medication, autonomy has been preserved in that Anna can now elect to choose better ways to self-soothe without the risk of being overly sedated.
2. An older couple come for an initial visit with their very engaging, hyperactive, bright 7-year-old son. The child has had extensive psychometric testing, with no evidence of any autism. However, he does have evidence for a superior intellectual quotient with high verbal indices, accompanied by what the neuropsychologist is describing as oppositional behaviors. The family is seeking a diagnosis of autism spectrum disorder for their son since they believe he will receive more services within the public schools. In reviewing specific concerns of the family, the clinician shows them how avoiding the label of autism in a child who does not qualify as having the condition will be far healthier in all aspects of his social and academic life. Referrals are made to family counseling for help with parenting gifted children with oppositional traits, as well as social skills training for the child himself. Additionally, materials are provided for a range of academically challenging topics such as jazz camps, solar car engineering projects, and other possible projects to alleviate academic boredom. Family is willing to approach their challenging but non-autistic child in a far more beneficent, less patriarchal manner.

Justice

This world is unjust. This often feels especially true in the context of individuals born with serious NDD. Consequently, the medical ethical principle of justice refers not to attempting to correct those medical injuries which may not be correctable but rather ensuring that the individual has his or her dignity respected in their own reality at this given moment. The term, "distributive justice" refers to fair, equitable, and appropriate distribution of benefits and burdens determined by norms that structure the terms of social cooperation [2]. Its scope includes policies that allot benefits and burdens such as property, resources, taxation, privileges, opportunities, food distribution, jury service, and services. Subsequently, the application of distributive justice in the context of NDD must apply not only to the protection of the individual's worth and dignity but the provision of care within a society that profoundly underserves this community, especially in minority and low-income settings. In a day where the term "autism" can be still be weaponized, there exists a valid concern about when an individual on the autistic spectrum should actually be identified and/or medically managed. As always, each individual will need to be considered in the balance of what is most beneficent for that individual against autonomy and self-determination.

Clinical examples of justice distribution issues in the neurodevelopmental disabilities population

According to the Center for Disease Control, the prevalence of any developmental disorder in 1997–2008 was 13.87% [6]. The prevalence of developmental disabilities from 2006 to 2008 increased nearly by 2 million children from the decade before, with the prevalence of individuals with autism increasing by nearly 300% [6]. Children insured by Medicaid had twice the prevalence of any developmental disorder than those children with private insurance. Children from families with income below the federal poverty level, which in 2021 is $21,000 a year or less for a family of four, have increased prevalnce of any developmental disorder [7].

While some states now provide funding for Medicaid to pay for applied behavioral analysis (ABA) therapies, many still do not. Additionally, if a family is above the federal poverty level but does not make enough money to have private insurance, they are caught between have too much for Medicaid but not enough to pay for therapies. The minimum of 10 hours weekly of ABA therapy typically runs $1200 a week or $62,000 a year [8]. Clearly a family of four making $30,000 a year will not qualify for either Medicaid assistance for ABA or be able to pay for it privately. By not uniformly providing this critical service to all children, regardless of income, we continue to create severe disparities in the potential growth and development of the individual. This becomes more magnified as these individuals then mature into adulthood, with the exponential costs of housing and caring becoming dependent upon their families and communities.

Over the last few decades, the trend has been to create housing that allows individuals to stay within a community, rather than being institutionalized. To that end, the US Department of Housing and Urban Development (HUD) has created Section 8 vouchers, which allow people with mild or moderate special needs and low incomes to live on their own in the community. However, it usually takes years to obtain a Section 8 voucher, and once acquired, there may not be any available Section 8 units for rent in the individual's community. Section 8 housing is also not appropriate for people with more complex special needs who cannot live on their own [9]. As a rule, a Section 8 recipient is expected to contribute approximately one-third of his or her monthly income toward the rent, with the remainder being paid by the Section 8 voucher. Clearly, if we have a booming population of children on the autistic spectrum, living in poverty without access to adequate therapies early in life, we can certainly expect a housing crisis for these individuals as they reach their adult years [10].

Summary

Health care providers are privileged to care for individuals who find themselves blessed with many gifts but perhaps also challenged with significant emotional, cognitive, or physical disabilities. It is often a labor- and time-intensive task, which requires seeing each individual with all the strengths, weaknesses, and wants and needs. We are called to find creative ways to not only support the individual's gifts and personal dignity but also to support their health and incorporation into the society, as able with proactive ways to care in ways that are needed. Embracing the complexities and humanity of ethical principles will help create a path in which the individual is both celebrated and incorporated into the community.

References

[1] Taylor RM. Chapter 1 – ethical principles and concepts in medicine. In: Bernat James L, Richard Beresford H, editors. Handbook of clinical neurology, 118. Elsevier; 2013. p. 1–9. Available from: https://doi.org/10.1016/B978-0-444-53501-6.00001-9.

[2] Beauchamp TL, Childress JF. Principles of biomedical ethics. 8th ed. New York, NY: Oxford Press; 2019.

[3] Buchanan A. Mental capacity, legal competence and consent to treatment. J R Soc Med 2004;97(9):415–20. Available from: https://doi.org/10.1258/jrsm.97.9.415.

[4] Bernat JL. Emeritus Professor, Department of Neurology, Dartmouth College. Personal communication; 2021.

[5] Torke AM, Alexander GC, Lantos J. Substituted judgment: the limitations of autonomy in surrogate decision making. J Gen Intern Med 2008;23(9):1514–17. Available from: https://doi.org/10.1007/s11606-008-0688-8.

[6] Boyle CA, Boulet S, Schieve L, Cohen RA, Blumberg SJ, Yeargin-Allsopp M, et al. Trends in the prevalence of developmental disabilities in US children, 1997–2008. Pediatrics. 2011;127(6):1034–42.

[7] Federal Poverty Level (FPL). HHS poverty guidelines for 2021. Retrieved October 11, 2021 from https://aspe.hhs.gov/topics/poverty-economic-mobility/poverty-guidelines.

[8] Autism Parenting Magazine. How much does ABA therapy for autism cost? Retrieved October 11, 2021 from https://www.autismparentingmagazine.com/aba-therapy-autism-cost/.

[9] US Department of Housing and Urban Development (HUD). Housing choice voucher program (Section 8). https://www.benefits.gov/benefit/710#:~:text=What/20is/20Housing/20Choice/20Voucher/20Program/20/28Section/208/29/3F,subsidized/20housing/20projects./20Housing/20choice/20vouchers/20are/20administere.

[10] Pope TM. The best interest standard: both guide and limit to medical decision-making on behalf of incapacitated patients. J Clin Ethics 2011;22(2):134–8.

Chapter 49

Neurodevelopmental disabilities: a look at the future

Miya R. Asato[1,2], Fatima Y. Ismail[3,4] and Bruce K. Shapiro[1,5]

[1]*Department of Neurology and Developmental Medicine, Kennedy Krieger Institute, Baltimore, MD, United States,* [2]*Departments of Neurology and Pediatrics, Johns Hopkins University School of Medicine, Baltimore, MD, United States,* [3]*Pediatric Neurology and Developmental Medicine, Department of Pediatrics, College of Medicine and Health Sciences, United Arab Emirates University, United Arab Emirates,* [4]*Department of Neurology, Johns Hopkins University School of Medicine, Baltimore, MD, United States,* [5]*Department of Pediatrics, Johns Hopkins University School of Medicine, Baltimore, MD, United States*

Introduction

The field of neurodevelopmental disabilities (NDD) and its related disciplines has seen strides in medical knowledge, human rights, and education over the past century. At the present time, we are at an inflection point where scientific advances may bring about marked changes and potentially impact quality of life for individuals with NDD. This chapter addresses some of the trends that will shape the field.

Reconceptualizing the field

The care of people with NDD has mirrored the advances in other fields of medicine. These achievements are not solely within the discipline of NDD nor exclusively within the province of medicine. All disciplines that interact with people with NDD, individually and collectively, work to improve the outcomes for people with NDD. Increased knowledge of the disorders, recognition of new entities, new technology, and social changes have all contributed to our current state of care and will continue to shape future progress.

New knowledge through new technologies

Neurodevelopmental disorders reflect an evolving understanding of how disruptions in the developing brain alter maturational processes and critical neurological functions, as well as how mediators and moderators of brain plasticity enhance or diminish brain functions and ultimately influence behavior.

It is, therefore, imperative that we systematically gain deeper understanding of how the nervous system matures, attains specialized functions, and compensates and adapts following a congenital or an acquired injury. The tools for inquiry are many but advanced neuroimaging and access to affordable next generation genetic testing provide a foundation upon which future work will progress and where new approaches to both diagnosis and treatment will develop and potentially new disciplines will emerge.

Major advances in neuroimaging and genetics often provide the answer to the clinical question "why did my child develop this condition?" MRI is a now an indispensable tool for diagnosis and charting disease progression for many NDDs, especially those with neurodegenerative phenotypes. Timing of lesions can often be established and prognostic information can be informed by longitudinal MRI studies, both structural and functional. The field is moving from anatomical/structural lesion localization to local and global network analyses with higher spatial and temporal precision.

The advent of whole exome sequencing enables the identification of an etiology for approximately one-third of those with NDD [1], and this is expected to increase as analytic tools and functional studies expand. Genetic evaluation has not only provided etiologic diagnoses but has also identified new diseases with similar manifestations through

Capute and Accardo's Neurodevelopmental Disabilities in Infancy and Childhood. DOI: https://doi.org/10.1016/B978-0-12-824060-1.00008-0

incorporation of complex approaches like epigenetics, variable penetrance, and pleiotropy. This advancement holds potential to understand the biologic mechanisms underlying NDD and holds hope for the development of agents that may ameliorate or cure these disorders.

Just as the echocardiogram has supplanted the stethoscope, new technology that quantifies behavior will come into common usage. For example, gait analysis can be used to quantify kinematics of movements, guide surgical interventions, and gage response to therapy. Similarly, home-based technology for sleep and electroencephalography studies is increasingly available to enable patients to undergo diagnostic testing in a more natural environment.

An important limit to advancing progress in the care of people with NDD is that each person has their own disease. While some disorders are commonly recognized, the various co-occurring and comorbid conditions that have variable expressions result in an almost endless number of variables to measure. Artificial intelligence using machine learning algorithms holds promise for charting natural history, identifying prognostic factors, evaluating therapeutic efficacy, and guiding therapy.

Finally, the most direct effect on care has been the explosion in technology to circumvent the limitations imposed by the disorder and facilitate the performance of necessary tasks. Currently, there are technologies available to address motor, language, feeding, and academic issues. The proliferation of "smart phones" use has led to the development of applications that are inexpensive, easily accessible, and customizable. Future refinements to existing technologies or development of new ones may cater to the needs of people with complex care requirements, enhance current capacities, or broaden the range of individuals who can benefit from circumvention therapy.

New populations and wider recognition through screening

As medicine has progressed, NDDs have become more prevalent. Populations that once faced fatal outcomes now survive and do well, but some develop dysfunction that contributes to the spectrum and continuum of NDD. Perhaps the most obvious examples are in the care of the infant born prematurely or those with inborn errors of metabolism.

With the availability of gene-based therapies for progressive pediatric neuromuscular and gene or stem cell-based treatments for adrenoleukodystrophies and other storage disorders, the imperative of earlier identification has led many states to expand their newborn screening programs, and enable identified children to be evaluated for treatments at earlier stages. As the critical windows of interventions are defined, earlier and more expansive neurodevelopmental screening programs may be important for primary prevention. An example is the new time-sensitive therapeutic interventions in spinal muscular atrophy and muscular dystrophy that may alter the phenotype and severity of disability.

More recently, there has been a wider recognition of the developmental outcomes of systemic diseases of childhood and conditions that were traditionally not linked to adverse neurodevelopmental consequences. Screening for developmental concerns in children with conditions like congenital heart disease, chronic renal disease, and transplantation has become a new standard of care and a vital metric for assessing clinical outcomes.

New classification through new knowledge

A new classification of NDD that does not depend solely on function is needed. That is not to diminish the utility of the current classification, but an expanded classification could better distinguish disorders, encompass etiology, guide interventions, and serve as a foundation where the mechanism of the functional limitation can be studied. As our knowledge base is shifting from data based on grouped experiences to individualized approaches of precision medicine, existing classification systems will need to be revisited. Newer knowledge has highlighted the shortcomings of our current taxonomy. The Diagnostic and Statistics Manual (DSM) third edition was a multiaxial classification system that sought to ensure reliable psychiatric diagnoses. It has undergone several revisions and the multiaxial approach is no longer the focus. NDD described in the current DSM are categorized by symptoms and behavioral checklists and do not address etiology, mechanism, prognosis, or severity.

Defining disorders by behavior is limited by our language. Semantics often fail to distinguish conditions. The same behavior may be labeled with different terms and the same term can describe different behaviors. For example, hand flapping may be seen as overflow movement by some professionals or stereotypies by others. One person's persistence is another's perseveration. Similarly, restrictive behaviors may be seen in both autism spectrum disorder (ASD) and intellectual disability (ID), although they may differ in triggers, nature, course, and interference with daily activity. Repetitive motor activity can represent a feature of obsessive-compulsive disorder, tics, or stereotypies. Another example is hyperactivity. Is hyperactivity in the presence of intellectual disability the same as it is in attention deficit/hyperactivity disorder (ADHD)? Is ADHD an expression along a spectrum of a cognitive function or is it truly a distinct disorder?

Additionally, there is a greater recognition of the "less severe" expressions of impairment in cognitive, language, and motor functions that may or may not adversely affect execution of daily activities or quality of life. An example is handwriting difficulties in developmental coordination disorder. A keyboard or assisted technology of speech to text can easily and conveniently overcome the challenge imposed by impairment in fine motor skills.

The field should contemplate the following questions: Is the neurobiology distinct for behavioral manifestations of different NDDs, or are shared similar brain pathways responsible for their common coexistences? Should the treatments be based on these potential commonalities? Is the prognosis different based on neurobiology, or its association with an overarching diagnosis? To what degree does each of the diagnoses contribute to the outcome? How do we define treatment success? Future classifications of NDD will build on the DSM and incorporate the International Statistical Classification of Diseases and Related Health Problems and the International Classification of Function and Disability to facilitate tracking, diagnostic, clinical, and research purposes. These systems may eventually shift away from descriptive symptoms and move toward a brain-based model [2].

Shifting from a function-based to an etiological classification offers the advantage of biologically informed diagnosis and treatment. While the new genetics has been proposed as an alternate classification, the variation in expression of genetic disorders make it unlikely that a classification based solely on genetic etiologies will be sufficient. A successful example of this shift is seen in the field of epilepsy. Epilepsy diagnosis and classification transitioned from description of semiology to etiological classification based on underlying mechanisms of seizures. This was possible by aggregating and analyzing data pooled from deep phenotyping, genetics, brain imaging, and animal models. This approach allowed for targeted therapy that is mechanism-based and personalized.

Whether the current DSM classification for NDD survives will depend on its utility for clinicians, scientists, patients and families, and its applicability for transitioning toward precision medicine.

Care for individuals with neurodevelopmental disabilities

Interdisciplinary care is the rule

Advances in care in NDD are not solely discipline-based nor exclusively within the province of medicine. All disciplines that interact with people with NDD, individually and collectively, work to improve the outcomes for people with NDD. Increased knowledge of the disorders, recognition of new entities, new technology, and social changes have all contributed to our current state of care and will continue to shape future progress. The specialty of NDD is truly interdisciplinary as it intertwines multiple medical specialties and a number of new medical fields, allied health, psychology, law, and public health to address the complex care of individuals with NDD.

Medical home and care coordination

The need for care coordination remains an unmet need. For the medical home concept to be successful, the populations that can be served through primary care settings need to be defined, the primary care provider needs to develop the competencies needed to manage a NDD team, and the team needs to be compensated adequately for its time. Center-based care must be conceptualized to address more than the impairment. While the concept of medical home is laudable, it has not met its promise. The reasons for this may be listed as follows:

1. Primary care physicians and advanced practice providers will need broader and more in-depth training in the care of children with neurodevelopmental disorders. Comorbidities, added complexity of mental health disorders, psychopharmacology, education law and advocacy, and nonmedical disciplines are some of the areas of knowledge necessary to provide a medical home.
2. Medical care team members will need to define which children with NDD will be served in a primary care-based medical home. Many of the NDDs occur too rarely for a primary care physician to feel comfortable in delivering the needed care. While the prevalence of ADHD, language disorders, specific learning disabilities, and mild degrees of intellectual disability is relatively common, disorders such as spina bifida and cerebral palsy are not frequently seen in a general practice. Systems should be developed that augment the primary care physician care for children with NDD with accessibility to expert care input and collaboration. Training, consultation, and programs that share care via telehealth or ECHO (Extension for Community Health Outcomes) projects may enable the family to maintain care that is closer to home.
3. Primary care practices and tertiary centers will need to establish linkages to successfully provide care for children with medical complexity. Children with medical complexity often have needs that exceed the ability of a primary

care practice. To meet those needs may require a readily accessible team at a tertiary care facility. Whether the primary care physician will have the coordinating role or act as a member of the team will be determined by the team and based on the abilities of the child, family provider, and community.

4. Establishment of systems to adequately compensate the time that is required to coordinate and link with medical and nonmedical services are needed to make coordinated care a reality. An enhanced payment system could provide incentive to all parties to participate in and support medical homes. Compatible electronic records to promote electronic communication are also vital.

Harmonizing the electronic record

The medical record should narrate the patient's story. It should document the patient's progress toward meeting their goals. Within it, etiologies, comorbidities, intercurrent events, challenges, and successes should be documented. Neurodevelopmental disorders are ideal models for electronic records as each patient has a unique experience. The interactions of the large number of variables in each child are almost infinite. The advent of the electronic record has the potential to bring together data from multiple sources and create the large datasets that are required to refine interventions, enhance communication across care settings, and provide guidelines that incorporate the natural history and important "milestones" of the disorders that we manage. Unfortunately, the current versions of the electronic record are not governed by the clinicians who use them and are not customized for patients with NDD. The current system routinely queries about the tobacco use but does not systematically record and track information about the patient's level of function or capacity for medical decision making. Professional organizations outline guidelines about the etiologic evaluation and management for intellectual disability, global developmental delay, and autism spectrum disorder. However, similar guidelines for profiling and charting the patient's progress are lacking. Similarly, these organizations could advocate for the development and subsequent implementation of templates that are relevant to the management of NDD. These templates can then be customized. A good example of standardizing data collection is the National Institute of Health common data elements (https://cde.nlm.nih.gov/home). While it is a research-based initiative to harmonize data collection and combines datasets for optimal analysis, the model can be used for medical records to collate clinically meaningful data. In this textbook, we provided several templates that can be adapted, refined, and used. It is not until there is agreement about the data to be collected that the electronic medical record will meet its potential and quality improvement can be realized.

Transition and adult care

Most people with NDD have similar needs as everyone else and are able to be included into the mainstream of society with modest accommodations. These accommodations include additional time for visits, communication that is modified to ensure understanding, and identification of resources to assist decision making and facilitate successful integration. Future models of care may consider approaches to services modeling those created for senior citizens. Social supports and therapeutic services are best delivered in the person's natural environment.

As youth with NDD age out of pediatric care, the number of health care professionals and centers with expertise in disabilities can be even more limited. Support systems will need to be modified to achieve the goal of full participation in society. People with autism spectrum disorder, intellectual disability, or cerebral palsy require different social resources and approaches to participate fully in society. Coordinated, comprehensive teams to meet the needs of adults with complex needs will need to be established. Employment and accessibility laws will need ongoing advocacy and development. Expansion of funding for Medicaid medical and therapeutic services, development of alternative payment models that reward providers for delivering care based on quality and cost effectiveness, and better coordination of care between medical and other community-based sectors are needed.

New social forces shaping the field

Patient-centered versus family-centered

Families became active partners as the concept of family-centered care evolved and took priority in health care settings. During this time, we witnessed the growth of diagnosis-based parent groups that advocated for their children, sometimes unintentionally excluding others. Concurrent with the increased participation of families in the schools, pediatric medicine adopted a family-centered approach in hospital settings which is slowly altering adult care environments as well. By placing the patient and family at the center of health care decisions, communication, transparency, and information sharing is embedded into patient care. Through recognition and empowerment of families, organizations and resources have proliferated to provide additional supports and resources. Virtual communities have offered families whose children have rare conditions the opportunity to connect and decrease social isolation. While the growth of rare disorder communities has many advantages, power to leverage legislative reform and limited community funding can be compromised by multiple smaller-sized communities all trying to effect similar but unique changes and access finite and limited resources.

As more children with NDDs age into adulthood, a growing self-advocacy community is gaining momentum. The diversity of voices and conditions representing NDD offers a broad range of perspectives—ranging from a disability first versus a person first self-identification. Greater representation of self-advocates is a growing priority for research agencies and advocacy organizations, reflecting an increasing sentiment of "nothing about us without us" [3]. While pediatric care in the last couple of decades has focused on recognizing the vital role of parents and families, the transition from pediatric to adult care serves as an important reminder that a dynamic shift should occur toward patient-centered care during this time. At earlier stages of life, the focus during developing management plans and providing guidance is heavily oriented toward the caregivers. As the child grows older, there needs to be a steady shift to assessment of patient capacity and increasing patient engagement in care. This shift can be challenging for parents to navigate after years of working hard to optimize their child's health and taking primary responsibility in care, especially for individuals with more severe levels of disability. Increasing youth participation in visits and promoting opportunities and education to facilitate adult life mirrors the educational transition process. Families often still need to provide support, yet the intentional shift to prepare for adult care models is a critical component of NDD care.

Inclusion, equity, and access

Despite the common occurrence of disability in society, health disparities continue to persist for individuals with disability. For those with complex needs, accessing a specialized care team and basic primary, emergency, and dental care close to home can be challenging. Sparse transportation options in communities can hinder families from getting to appointments. Care that is located closer to home may limit the inclusion of patients with public insurance. Barriers within care environments can range from physical access to a lack of linguistic aids to facilitate communication. Discrimination, poverty, and disability are commonly intertwined and can occur across multiple generations within a single family. Children with NDDs and their families often face struggles impacting social determinants of health, such as obtaining quality education, stable and safe housing and neighborhoods, and meeting other basic needs. Advances in medical care are just a component of improving outcomes or preventing future NDDs. Advancing health in NDDs requires leadership across health care, government, and community systems.

Access is a recurring barrier to obtaining expert care. During the COVID pandemic and mandatory isolation, telehealth became an important tool. While telehealth could overcome barriers imposed by geography and time, it has not diminished the number of practitioners required or comes close to meeting the need for care. Telehealth may be useful for following established patients but initial evaluations of young children with NDDs often are limited because of the child's decreased ability to interact with a virtual image instead of a person. Physical findings, particularly affecting the motor system, may not be obtained with use of telehealth.

Ableism and neurodiversity

Ableism and neurodiversity are parallel constructs, as both deal with how a person with a disability is viewed. Both are dependent on a comprehensive view of the person and require an appreciation of an individual's characteristics.

Ableism is the intentional or unintentional discrimination against people with disabilities. It centers around the belief that individuals with disabilities are imperfect and need fixing. It is expressed in many forms ranging from personal to institutional. Much of the social movements behind rights for those with disabilities were based on minimizing ableism and empowering individuals with disabilities.

Neurodiversity describes the idea that people experience and interact with the world around them in many different ways; there is no one "right" way of thinking, learning, and behaving. Differences or deviations from "average" are not viewed as deficits needing repair. As a result of this social movement, some communities have called for eliminating designation of "autism" in research [4,5]. Advocates argue that the traditional medical model pathologizes certain features of disabilities and thereby perpetuates stigma and ableism. The neurodiversity movement rejects the idea that disability is a condition that needs "fixing." Within this framework, achieving "normality" should not be the therapeutic goal for individuals with NDD, and that society should provide more inclusive measures to ensure integration and inclusion and consideration of a broader view of quality of life.

If neurodiversity is misapplied, it may minimize the significance of dysfunction and prevent individuals with severe impairments from accessing care and support. However, who decides what to treat, what treatment to apply, and to whom to apply it requires a collaborative and thoughtful approach; identification of treatment priorities promotes a type of care management that is person-centered.

Global and low-resource settings

The limited breadth and depth of services for those with NDD in many international and low-resource locations presents a need to provide training in NDD beyond our geographic borders. There is increasing interest in establishing clinical skill competencies in global and low-resource settings. Creating communities of learning, such as ECHO projects, is a potential way to increase technical assistance and collaboration across international borders. While there are significant barriers for international physicians and other specialists to receive training in the United States, advances in telecommunications provide ongoing opportunities to provide remote consultation and expertise.

Working closely to develop training programs and systems of care in global settings require individualization that is based on understanding the local requirements and partnership with local stakeholders. While the US model of NDD training and care is well-developed, it may not be suitable in other global settings. We need to be mindful of the cultural variations that influence both the definition of a good training program and the criteria for measuring the success and effectiveness of services for individuals with NDD. A successful training program will have to customize its curriculum to meet the needs of the local NDD population. For example, certain geographical settings with a higher prevalence of genetic disorders may require more emphasis on the assessment and treatment of neurogenetic and neurometabolic disorders. Similarly, while some cultures prioritize the attainment of independent living by adulthood as a suitable goal, in other cultures, this may not be an objective for the families.

Training promises, pitfalls, and opportunities

In March 1999, the American Board of Medical Specialties approved the proposal for subspecialization in NDD as a joint certification of the American Board of Pediatrics and the American Board of Psychiatry and Neurology (ABPN). Training consists of: (1) 24 months of general pediatrics, and (2) 48 months of an integrated program of education, training, and experience in the management of neurological and neurodevelopmental disorders. The new subspecialty was to use its knowledge of NDD, the interdisciplinary approach to the challenges faced by people with NDD, and their ability to manage multiple coexisting conditions to improve the lives of people with NDD and their families through clinical care, teaching, consultation, and research.

Access to knowledgeable practitioners is a major contributor to unmet needs of patients and their families. With 20% of children having a neurodevelopmental disability, the current number of neurodevelopmentalists, child neurologists, developmental-behavioral pediatricians, child psychiatrists, and pediatric rehabilitation specialists are insufficient to meet the need. As a result, there are long waits for families to obtain confirmatory testing for children who are identified as "at risk" by screening tests. Additionally, ineffective, unnecessary, and expensive interventions may be sought by parents who are not guided by physicians with the clinical experience and the knowledge to create a comprehensive management program and to effectively advocate with community programs.

The short-term approach is to try to train additional providers who have expertise in the evaluation, diagnosis, and management of infants, children, youth, and adults with NDD. There are practical impediments to achieving this for the program and the trainee. Exposure to disability in medical school curricula is still insufficient. Many prospective trainees are not aware of NDD or disability-related fields because of lack of faculty and lack of clinical experiences or exposure.

For those who are aware of the field, the length of training, currently 6 years, serves as a disincentive. The cost of the training is high, both to the program and to the trainee. Neurodevelopmentalists are on the lower end of the physician pay scale. The prolonged time of training delays the trainees' ability to decrease their debt load and the relatively low compensation serve as disincentives for trainees to enter the field. Similarly, graduate medical educational programs may not be able to support the costs associated with 6 years of training.

Recently, in an effort to increase the number of neurodevelopmental physicians, an alternative pathway into NDD has been approved by the Accreditation Council for Graduate Medical Education (ACGME). This pathway is designed for trainees who have completed 2 years of core Pediatrics and a Child Neurology Residency in an ACGME program. It consists of an additional year of training in NDD and related disorders.

Most neurodevelopmentalists practice in university settings in order to meet their tripartite mission of patient care, training, and research. University hospital settings are uniquely able to provide the scope of interdisciplinary resources that are needed for the comprehensive management of children with NDD and their families, the research structure to conduct meaningful investigations, and the trainees who are seeking to develop clinical expertise. However, this may require families to travel long distances to obtain the resources that are not available in their local community.

Once patients graduate from pediatric to adult care, internal medicine physicians, neurologists, and psychiatrists do not have extensive training to address NDD. Advanced practice nurses or physician assistants have been suggested to meet the manpower needs. While they can be trained to develop expertise in specific areas, such as gastrostomy tube, bowel or bladder management, skin care, or monitoring an existing management program, their current training may be limited in making new diagnoses of NDDs or re-evaluating a program that is not working. Furthermore, psychologists or social workers are trained to provide mental health counseling but their experience with the special issues relating to those with limited communication or cognition may be insufficient.

At the current time, there are not enough NDD specialists being trained to replace those who are retiring. In addition to increasing visibility of the specialty, enhancing recruitment strategies, and developing new pathways, there are other interventions that may be effective. For example, loan forgiveness programs to reduce the debt incurred during training, the broadening of federal support for NDD training programs, and the development of state programs to support the advanced training of professionals in NDD are strategies that have helped address practitioner shortages in other fields of medicine.

NDD training programs continue to evolve. Each program seeks to provide activities that are relevant and maximize the yield to the trainee. Program directors will need to work with the certifying agencies to modify their programs to meet the current and future needs of people with NDD. Some of the questions that might be addressed are: Is 6 years of training too long? Should new experiences be added to the program and current ones be shortened? Is 2 core years of pediatrics still necessary? What does the experience in adult neurology add to the training? Should programs develop a research track where rigorous research training programs for fellows who are pursuing careers in academia?

As noted above, it is unlikely that there will ever be enough NDD specialists to meet the needs of the population. Expanding the curricula for primary care physicians and providing accessible continuing medical education opportunities for practicing physicians will increase the number of physicians who are able to care for people with the more common NDD. This will decrease the number of children who are being managed primarily by neurodevelopmentalists and expand their ability to consult on issues relating to diagnosis and management. Developing curricula for related fields of child neurology, psychiatry, neurology, and physiatry will increase the number of practitioners who are knowledgeable about NDD and increase access to care. Providing training in developmental disabilities for adult medicine specialists will also be crucial to meet longitudinal care needs across the lifespan.

Finally, medical schools should be encouraged to provide an NDD perspective as a core part of the curriculum. The NDD contribution to the medical field is unique and stresses the value of the individual, regardless of their capacities, and utilizes an interdisciplinary approach to maximize function. While it currently focuses on conditions that impact childhood and the developing central nervous system (CNS), this approach to care can be applied to all chronic disorders and disabling conditions.

In order to meet the needs of people with NDD, future approaches will require collaboration across medical disciplines of NDD, child neurology, child psychiatry, developmental-behavioral pediatrics, palliative and complex care, and pediatric rehabilitation. Some disciplines will need to expand their scope of practice while others will need to be more focused. In addition, primary care physicians in pediatrics and family practice will need to develop competence in the evaluation, diagnosis, and ongoing management of those at the milder end of the spectrum of NDD.

The simple answer to the access question is to increase the number of practitioners. To do this requires greater visibility of the field and more effective recruitment. Given the number of undergraduate students who are neuroscience majors and then go on to medical school, there is a potential pool of candidates with interest in the developing nervous system. Providing clinical and research experiences to these students may serve to increase the number of people training in NDD.

Making other health professionals aware of the needs of people with NDD and providing them with training on how to interact with people with NDD are ways to increase awareness. This would start in medical school and be added to the residency curricula. Similar training could be incorporated into nursing and pharmacy curricula. Mental health professionals would have NDD as a required part of their training.

A group that is often not considered as a target for training includes educational professionals. Children with NDDs spend a large portion of their time in school and many disorders are identified by teachers, yet communication between the health care team and the school is often difficult and limited. Educational approaches that train cross-disciplinary teams to collaborate can maximize resource allocation and decrease the unnecessary repetition and communication barriers that confront many families. This is a training opportunity that deserves further exploration and development.

Research challenges and priorities

At a fundamental level, research in NDD seeks to identify and link the contributions of neurobiological and environmental influences on the developing nervous system. To attain that goal, an ecosystem should develop to integrate the many contributors to the field, including, but not limited to, patients and their families, researchers in biological sciences, behavioral and social sciences, public health and education, advocacy groups, policymakers, and funding agencies.

In order for the field to advance, the focus of the field will have to progress beyond the current focus on function. These advances will not only come from within the NDD community. The developing brain does not belong to any single discipline. Each discipline is guided by its philosophy, tools, and constructs. Collaboration across disciplines is necessary to utilize the specialized knowledge needed to answer a question. Complex projects require collaboration from multiple disciplines. The achievements in molecular neurobiology, biochemistry/metabolism, genetics, and neuroimaging have identified a large number of etiologies of NDD. However, the mechanisms by which etiologies impede function is a gap in existing knowledge.

Understanding the mechanisms of normal developmental processes like vision, hearing, thought, language, learning, movement, attention, memory, and emotion are avenues for future research and the basis for understanding mechanisms of dysfunction. Studies of fetal development and its perturbations, prevention of prematurity, neuroprotection strategies, identifying critical/sensitive periods, exploration of plasticity and its limitations following injury, maturation and aging, socioeconomic status, nutrition, parenting, education, and their contributions to the pathological processes may have significant impact on the developmental disabilities.

Technological advances will change the nature of NDD-related research. Artificial intelligence with machine learning algorithms will use clinical registries that contain large datasets. These approaches will be able to identify which of the many variables that affect the trajectory of each individual with an NDD and enable prognostication that is relevant for the individual. Similar approaches will allow fine grain analysis that will lead to better classification of the NDD. If longitudinal data collection is employed, discerning risk factors might lead to effective prevention of long-term complications of the NDD and its associated disorders.

A second research focus would address therapeutic efficacy of currently employed interventions for NDD at population and individual levels. There is a pressing need to answer the question of which therapies to use, when to use them, to whom they should be applied, and are they more effective than alternative approaches. A parallel set of investigation would evaluate the mechanisms by which effective therapies exert their effects. For example, in the context of dyslexia treatment, how do therapies impact the function and organization of word recognition center in the brain? And can these mechanisms be adapted to other brain regions or functions?

A third research focus is precision medicine approach to treatment which holds the promise of amelioration or cure of certain NDD. NDDs are disorders with wide expression. Varying severities, comorbidities, and external factors limit prognostication to the extreme cases in the most general of terms. Intervening on clearly identified aberrant mechanisms of dysfunction has promise in some metabolic and genetic disorders like phenylketonuria, tuberous sclerosis, and fragile X syndrome. These research efforts are still in their infancy but should be supported.

Being able to make a prognosis specific to an individual would guide interventions, set long-term outcomes, and facilitate inclusion. Some have followed the path to define and identify biomarkers that would enable forecasting. Alternatively, the use of standard data collection strategies across multiple sites would provide a population of sufficient size to allow for accounting the many different variables that make prognostication difficult.

Future research should promote the inclusion of those with NDD in clinical trials in pediatric and adult populations. Promoting the use of innovative technology will bypass the limits imposed by geography and time to enable larger numbers of patients to participate in clinical trials. The development of multicenter collaborative groups will be required to enroll enough subjects to answer the questions of efficacy.

Finally, the inclusion of NDD specialists in the conceptualization and conduct of research in other fields will increase the internal and external validity of research and would advance the inclusion of people with NDD in research. NDD specialists, patients, and families will ensure that the measures used are relevant to the population included and suggest accommodations that would maximize the generalizability of the results [6].

Summary

Future directions for NDD include ongoing research, advancements in clinical care using an interdisciplinary model, and a greater commitment to training on multiple levels. On a broader systems level, funding for health care, community programs, medical and professional training, and research are necessary in order to optimize the health and outcomes of our patients with NDD.

References

[1] Srivastava S, Love-Nichols JA, Dies KA, Ledbetter DH, Martin CL, Chung WK, et al.NDD Exome Scoping Review Work Group Meta-analysis and multidisciplinary consensus statement: exome sequencing is a first-tier clinical diagnostic test for individuals with neurodevelopmental disorders. Genet Med 2019;21(11):2413–21.

[2] Doernberg E, Hollander E. Neurodevelopmental disorders (ASD and ADHD): DSM-5, ICD-10, and ICD-11. CNS Spectr 2016;21(4):295–9. Available from: https://doi.org/10.1017/S1092852916000262.

[3] Kover ST, Abbeduto L. Toward equity in research on intellectual and developmental disabilities. Am J Intellect Dev Disabil 2023;128 (5):350–70. Available from: https://doi.org/10.1352/1944-7558-128.5.350.

[4] Müller RA, Amaral DG. Editorial: time to give up on autism spectrum disorder? Autism Res 2017;10(1):10–14. Available from: https://doi.org/10.1002/aur.1746.

[5] Waterhouse L, London E, Gillberg C. The ASD diagnosis has blocked the discovery of valid biological variation in neurodevelopmental social impairment. Autism Res 2017;10(7):1182. Available from: https://doi.org/10.1002/AUR.1832.

[6] Perfetto EM, Burke L, Oehrlein EM, Epstein RS. Patient-Focused drug development: a new direction for collaboration. Med Care 2015;53 (1):9–17. Available from: https://doi.org/10.1097/MLR.0000000000000273.

Appendices

1. **Neurodevelopmental Concepts**
   1.1. Receptive language development milestones for the first two years
   1.2. Expressive language development milestones for the first two years
   1.3. Social skills milestones
   1.4. Developmental dissociation in neurodevelopmental disorders
2. **Approach to common clinical presentations in NDD**
   2.1. Delay in multiple areas of development
   2.2. Regression or plateau in developmental milestones
   2.3. Delay in motor development (0–2 years of age)
   2.4. Delay in language development in preschool children
   2.5. Delay in social skills
   2.6. Academic underachievement
   2.7. Hyperactivity
   2.8. Aggressive or self-injurious behaviors
   2.9. Pain and discomfort in a child with severe neurological impairment
3. **Clinic Templates**
   3.1. Developmental History
   3.2. New Evaluation for Sleep
   3.3. Physical Examination
   3.4. Follow up—Attention Deficit Hyperactivity Disorder
   3.5. Follow up—Autism Spectrum Disorder
   3.6. Follow up—Language and Communication Disorders
   3.7. Follow up—Cerebral Palsy
   3.8. Follow up—Early neonatal brain injury
   3.9. Follow up—Spina Bifida
   3.10. Elements of a successful management program

Capute and Accardo's Neurodevelopmental Disabilities in Infancy and Childhood. DOI: https://doi.org/10.1016/B978-0-12-824060-1.00043-2

1. Neurodevelopmental Concepts

1.1 Receptive language development milestones for the first two years

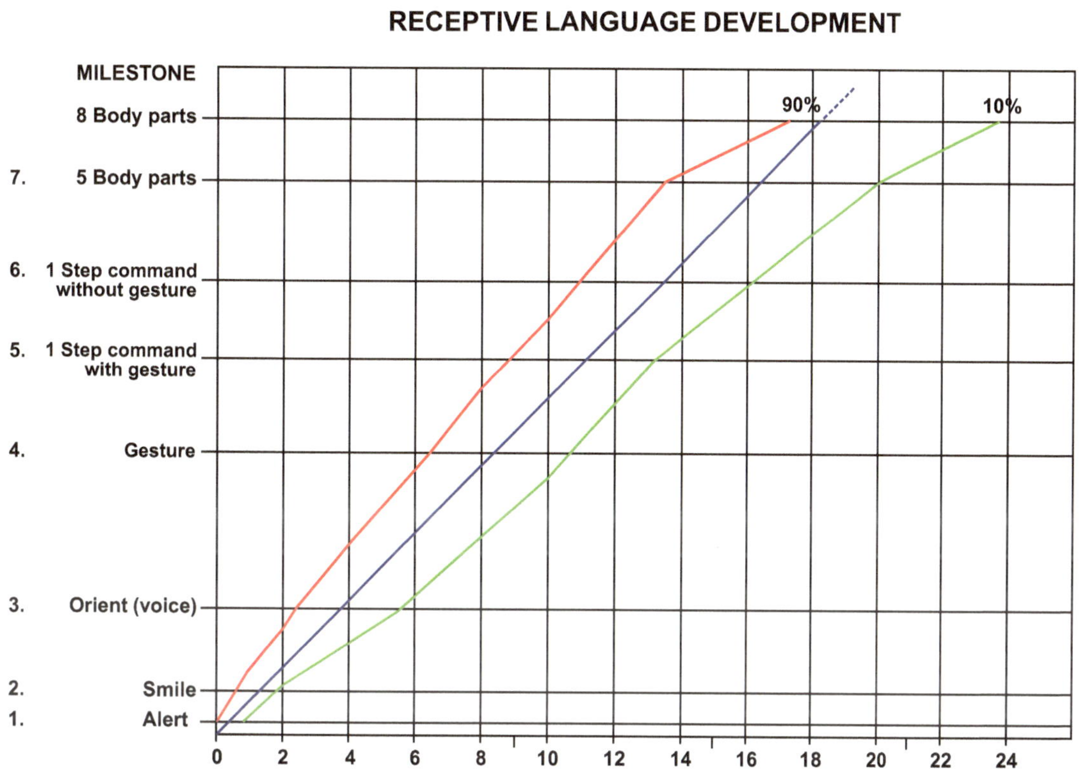

Receptive Milestones Survey

Receptive Milestones Survey

1. Does your baby respond to sound by blinking, startling or moving? (Response must be visible) Y N
   When was the first time? ______
2. Does your baby smile at you in response to your voice or touch? (a social interaction.) Y N
   When was the first time? ______
3. Does your baby turn cirrectly and immediately to find your voice? (Reliability of response needed) Y N
   When was the first time? ______
4. Does your baby wave "bye-bye," play "pat-a-cake" or "peek-a-boo?" (Spontaneous or imitative response) Y N
   When was the first time? ______
5. Does your baby follow simple commands such as:"Give me--," said with a gesture? (Do not include "NO!") Y N
   When was the first time? ______
6. Does your baby follow simple spoken commands without gestures? Y N
   When was the first time? ______
7. Does your baby point to body parts? Y N
   When did he point to 1 ? ______
   When did he point to 5 ? ______
   When did he point to 8 ? ______

Rules of Recording

1. Use survey questions as guides for standard assessment.
2. Request milestones until 2 failures (N) are recorded.
3. Record date of visit below graph.
4. Record age of acquisition in spaces provided; Record oldest age of an estimate to nearest half (½) month.
5. Plot milestone acquisition age on chart using code:
   X -- Milestone acquired by history
   O -- Milestone not acquired
   S -- Milestone not remembered but seen on this visit.

1.2 Expressive language development milestones for the first two years

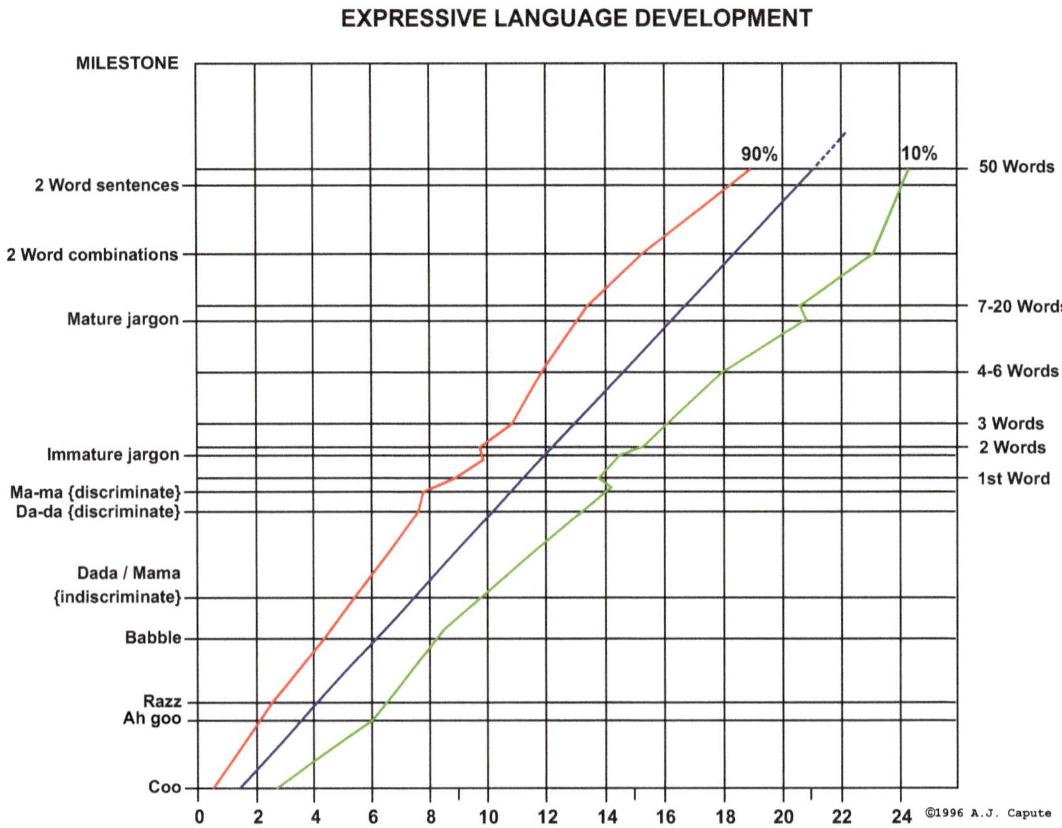

Expressive Milestones Survey

. Does your baby make long musical vowel sounds? (Distinguish from grunts, cries, short vowels and nonrepetitive noises.) Y N
When was the first time? ______

. Does your baby say "Ah-Goo"? (A specific vocalization.) Y N
When was the first time? ______

. Does your baby do the "raspberry"? (Noise made by tongue between lips blowing spray.) Y N
When was the first time? ______

. Does your baby babble? (Parental example, playful repetition of syllables with several different consonants and short vowels.) Y N
When was the first time? ______

. Does your baby say "dada" or "mama" evan if incorrect? (Indiscriminate.) Y N
When was the first time? ______

. Does your baby say "dada" or "mama" correctiy? Y N
When was the first time? ______

. Does your baby say any words? Y N
When did he say his first? ______
When did he say 2 words? ______
When did he say 3 words? ______
When did he say 4-6 words? ______
When did he say 7-20 words? ______
When did he say 50 words? ______
(Do not include proper names or nonspecific sounds. ______

8. Does your child talk in sentences like a foreign language in which no words can be understood? (Parental imitation helpful) Y N
When was the first time? ______

9. Does your child talk in sentences like a foreign language in which some words can be understood? (Parental imitation helpful) Y N
When was the first time? ______

10. Does your child put two words together (Noun + Noun)? Y N
When was the first time? ______

11. Does your child combine a noun or pronoun with a verb? Y N
When was the first time? ______

Vocabulary

date	word	date	word

1.3 Social Skills Milestones

2 months	Social smile
5 months	Recognizes caregiver
5–6 months	Reaches out to be picked up
6 months	Stranger anxiety
5–6 months	Pats image in a mirror
8 months	Follow's caregivers gaze
9 months	Enjoys interactive games (e.g., Peekaboo)
	Object constancy
10 months	Turns to speaker when name called
12–14 months	Points to request something (protoimperative pointing)
15 months	Temper tantrums
16 months	Points to indicate interest (protodeclarative pointing)
18 months	Brings objects or toys to caregivers to show or share the experience
	Early pretend play ("feed the baby")
	Parallel play
24 months	Imitates others, especially adults and older children
	Shows defiance, such as doing what he/she was told not to do.
30 months	Complex pretend play (uses a generic item to represent another object e.g., remote control as a phone)
	Plays dress up
3 years	Imaginative play
	Enjoys routines and may be upset when they are changed.
	Cooperative play
	Begins to take turns
4 years	Likes to be a Helper
	Comforts others who are sad or hurt
	Has a special friend
	Changes behavior relative to setting (Playground vs church)
	Distinguishes real from imaginary
	Pretends to be something or somebody else (dog, superhero, teacher)
	Trickster

References

- Gerber RJ, Wilks T, Erdie-Lalena C. Developmental milestones 3: social-emotional development. Pediatr Rev. 2011 Dec;32(12):533-6. doi: http://dx.doi.org/10.1542/pir.32-12-533. PMID: 22135423.
- Morin, A. Social and emotional skills at different ages. https://www.understood.org/articles/social-and-emotional-skills-what-to-expect-at-different-ages.
- Zubler JM, Wiggins LD, Macias MM, Whitaker TM, Shaw JS, Squires JK, Pajek JA, Wolf RB, Slaughter KS, Broughton AS, Gerndt KL, Mlodoch BJ, Lipkin PH. Evidence-Informed Milestones for Developmental Surveillance Tools. Pediatrics. 2022 Mar 1;149(3):e2021052138. doi: http://dx.doi.org/10.1542/peds.2021-052138. PMID: 35132439; PMCID: PMC9680195.

1.4 Developmental dissociation in neurodevelopmental disorders

Developmental Dissociation is defined as differences of 15% in developmental functioning between streams of development. The pattern of developmental dissociation differs among developmental disorders. While developmental disorders are defined by their most obvious level of dysfunction (e.g., motor dysfunction in cerebral palsy), dysfunctions in other streams of development are often observed and should be evaluated and managed. The table below depicts the most "typical" patterns of developmental dissociation in selected neurodevelopmental disorders. Variations are expected at the individual patient level.

Disorder	**Domains**						
	Gross Motor	**Fine Motor**	**Problem Solving**	**Language**	**Self-Help**	**Social**	**Executive functioning**
Cerebral Palsy	↓	↓	+/−	+/−	+/−	+/−	+/−
Intellectual Disabilities	Typical	+/−	↓	+/−	↓	+/−	↓
Autism Spectrum Disorder	Typical	+/−	+/−	↓	+/−	↓	+/−
Attention Deficit Hyperactivity Disorder	Typical	+/−	Typical	Typical	Typical	+/−	↓
Developmental Language Disorders	Typical	+/−	Typical	↓	Typical	+/−	+/−
Hearing Impairment / Deafness	Typical	Typical	Typical *with modifications	↓	+/−	+/−	+/−
Vision Impairment/ Blindness	+/−	+/−	Typical *with modifications	Typical *with modifications	+/−	+/−	Typical
Orthopedic impairments	↓	↓	+/−	Typical	+/−	+/−	+/−

Typical refers to range and level of skills expected for age; **+/−** reflects the range of skills within 15% of the lowest area; ↓ refers to most decreased or delayed (by more than 2 Standard Deviation) compared to typical development.

2. Approach to common clinical presentations in neurodevelopmental disabilities

2.1 Delay in multiple areas of development

This algorithm serves as a guide for assessing a child with delay in more than one domain of development. The primary goal is to diagnose and treat reversible causes promptly. The assessment should be individualized to the patient. *GDD: global developmental delay; ID: intellectual disability*

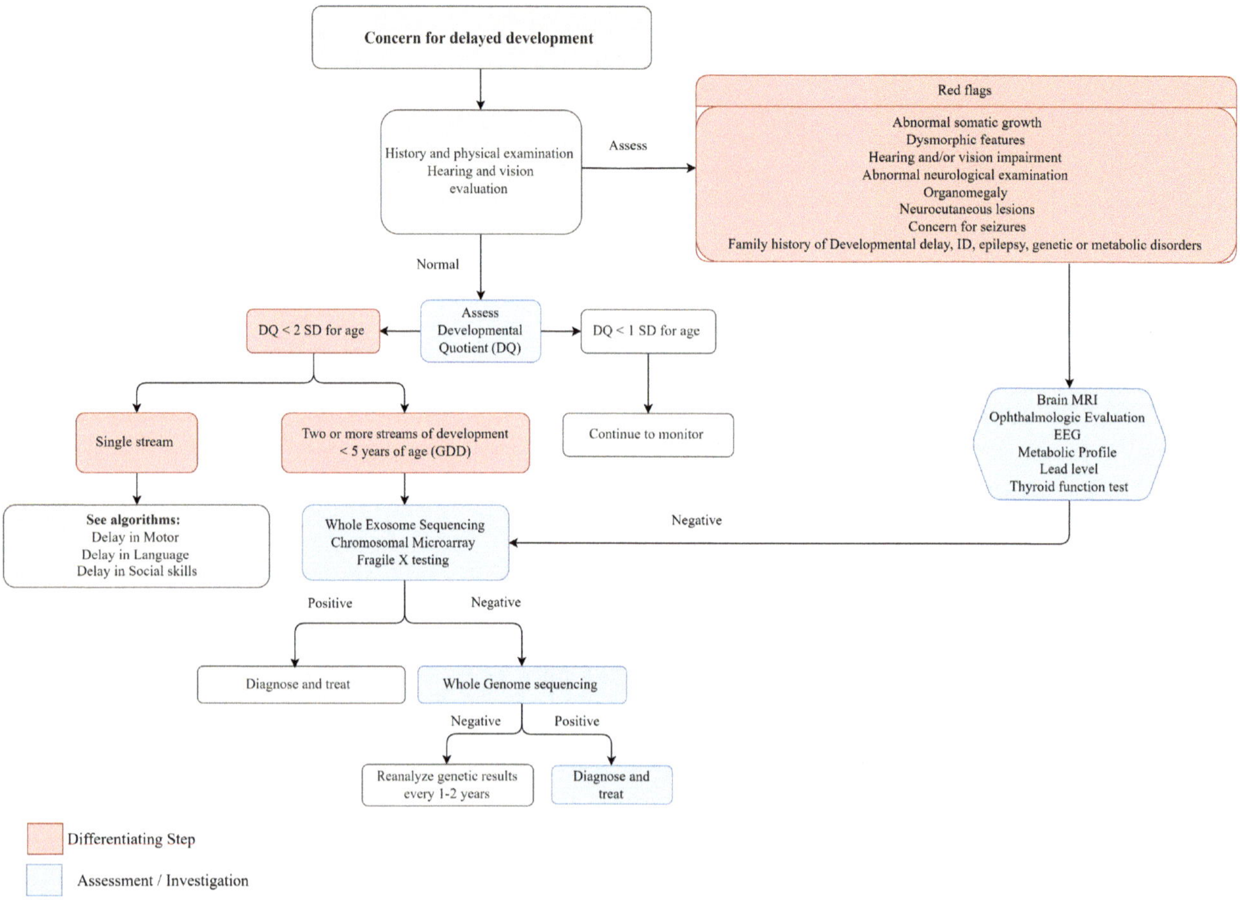

2.2 Regression or plateau in developmental milestones

This algorithm serves as a guide for assessing a child with a concern for loss of previously acquired skills, which may present as *persistent and unyielding decline in developmental skills or persistent decrease rate in attaining milestones or a prolonged plateau*. Pseudo-regression is defined as loss of skills that have not been firmly established or failure to progress. There are no standard measurements to capture regression at its onset. The key goal is to confirm developmental regression and differentiate between regression in one or more streams of development. The assessment should be individualized to the individual patient. This algorithm does not address acute loss of developmental skills (acute encephalopathy). *DD: developmental delay, ID: intellectual disability; LMN: lower motor neuron; UMN: upper motor neuron; NCS: nerve conduction studies; ESES: electrical status epilepticus in sleep; LKS: Landau-Kleffner syndrome*.

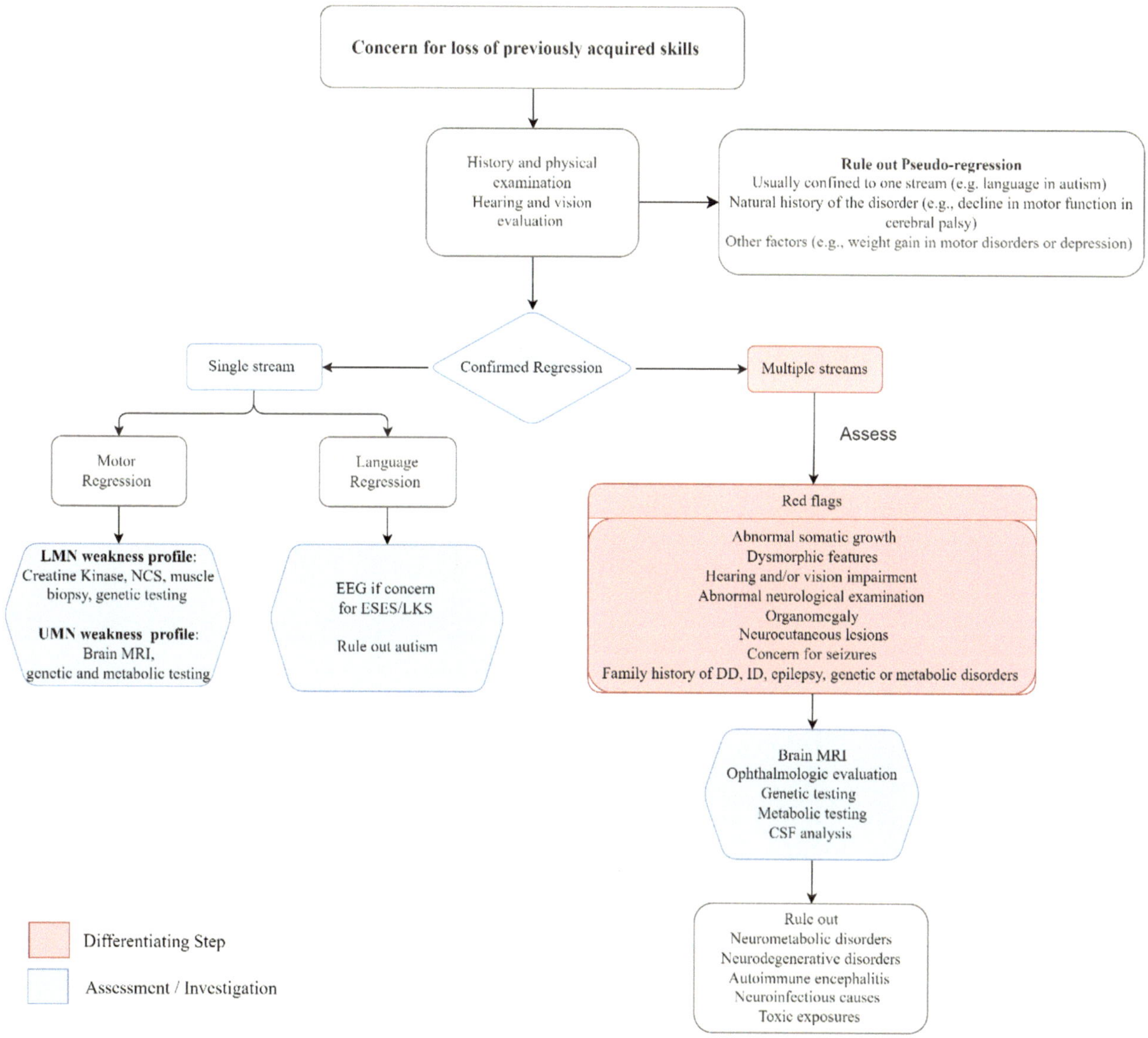

2.3 Delay in motor development (0–2 years of age)

This algorithm serves as a guide for assessing a child with delays in motor development. The primary goal is to rule out developmental regression, quantify motor quotient and assess neurological integrity. The assessment should be individualized to the patient.

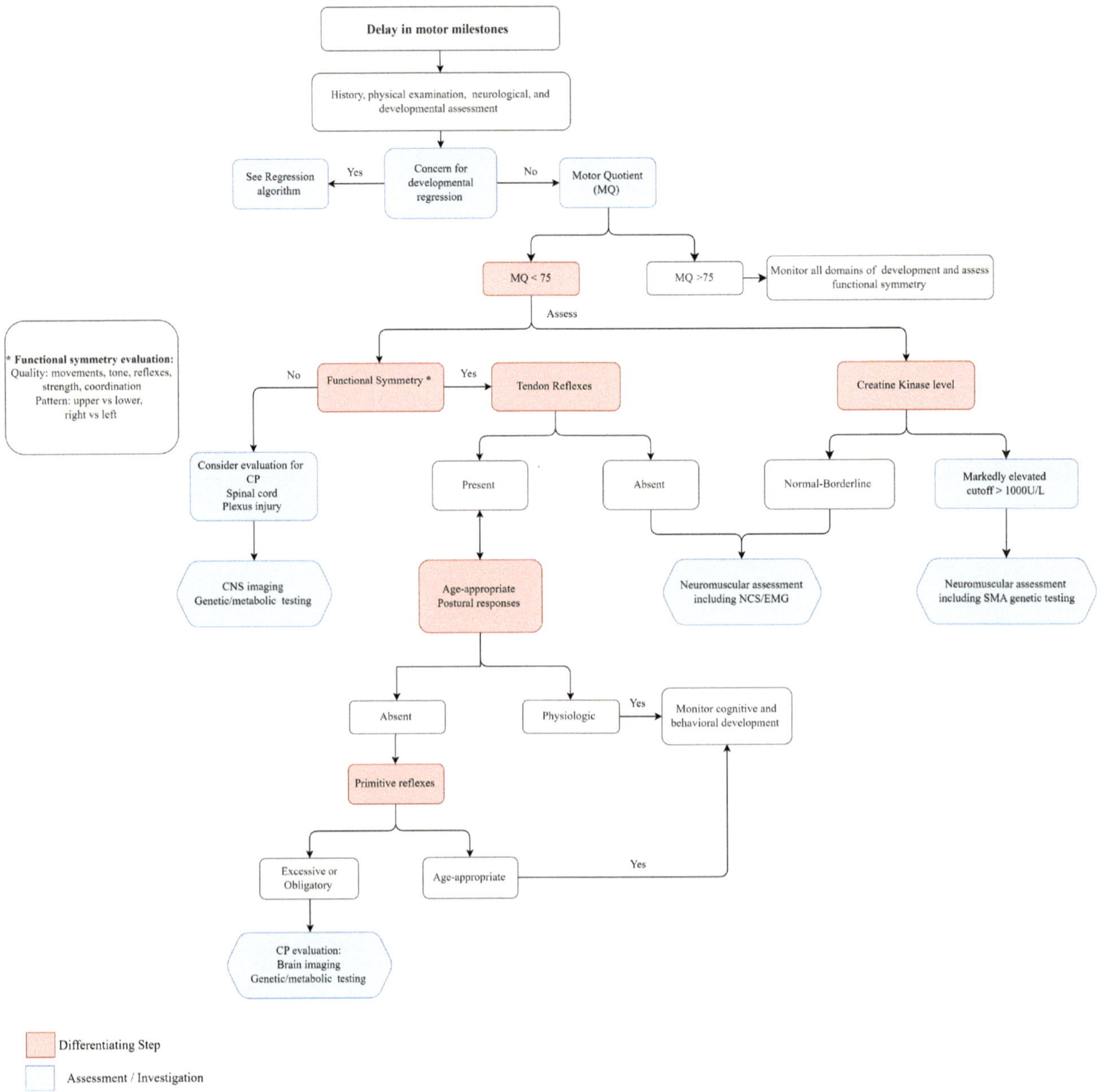

2.4 Delay in language in preschool children

This algorithm serves as a guide for assessing a child with delay in language development. The primary goal is rule out speech impairment, quantify language delay and identify specific domain(s) of impairment. The assessment should be individualized to the patient. *GDD: global developmental delay, ID: intellectual disability*

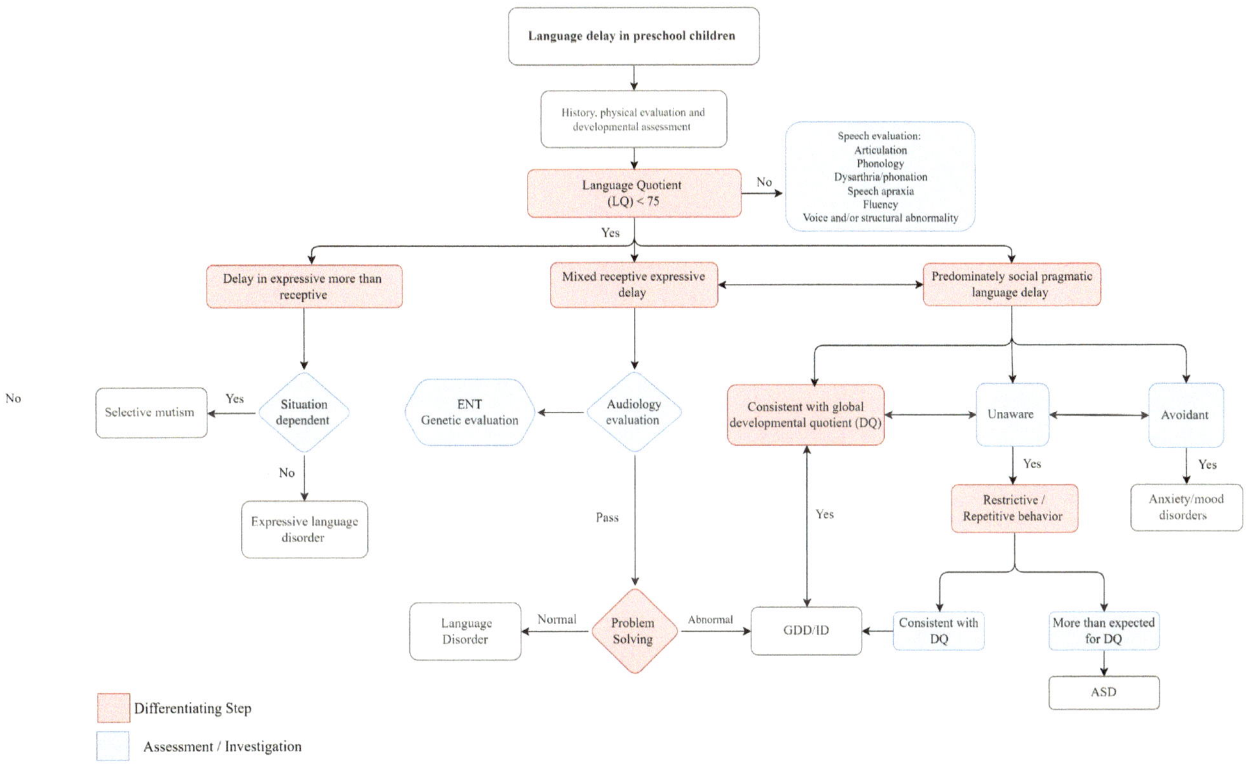

2.5 Delay in social skills

This algorithm serves as a guide for assessing a child with delay in social skills. The primary goal is to rule out global developmental delay and to characterize the predominant behavior associated with age-inappropriate social skills. The assessment should be individualized to the patient. *DMDD: Disruptive Mood Dysregulation Disorder. NDD: Neurodevelopmental disability*

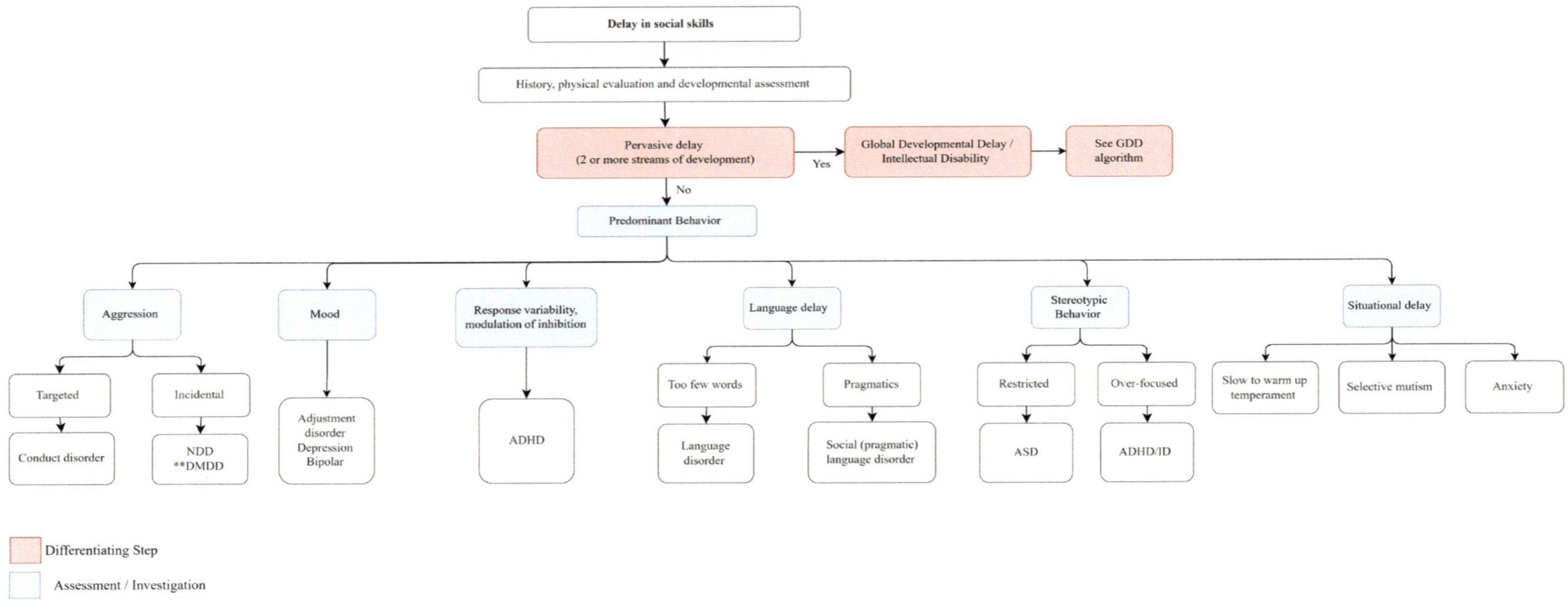

2.6 Academic underachievement

This algorithm serves as a guide for assessing a child with academic challenges. The primary goal is to rule out environmental or health conditions that may negatively impact academic skills, and identify whether academic challenges are mainly related to the global process vs. subject-specific mastery. The assessment should be individualized to the patient. *ADHD: attention deficit hyperactivity disorder; ASD: autism spectrum disorder; DCD: developmental coordination disorder.* * Processes include: Incomplete work, homework assignment not recorded, not completed, or not submitted, poor test performance but seems to know the material.

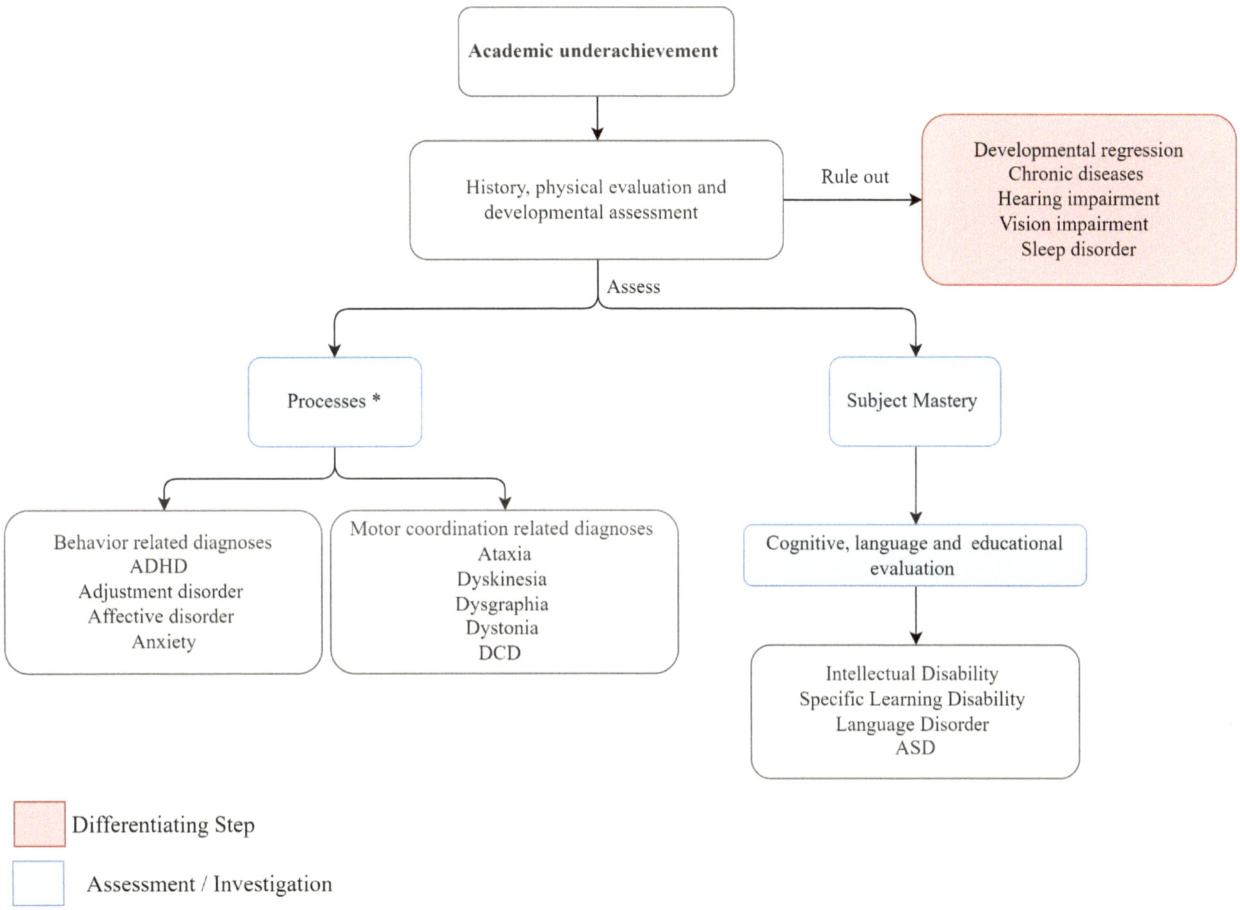

2.7 Hyperactivity

This algorithm serves as a guide for assessing a child with academic challenges. The primary goal is to rule out ADHD mimics and identify context of hyperactivity, and assess language and cognitive functions. The assessment should be individualized to the patient.

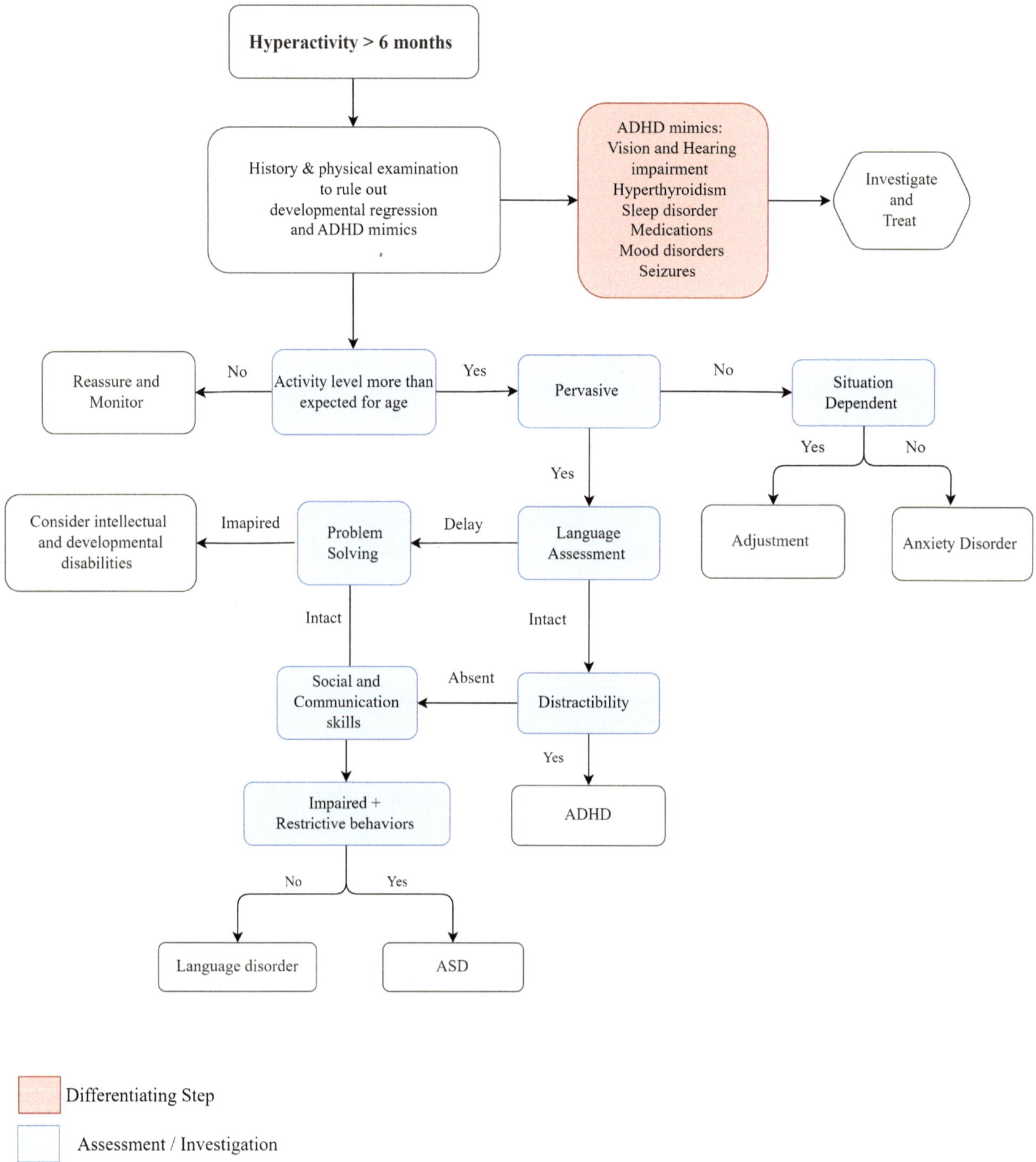

2.8 Aggressive or self-injurious behaviors

This algorithm serves as a guide for assessing a child for self-injurious behaviors (e.g., head banging, biting, excessive skin picking). The primary goal is to provide an approach to the identification of diagnoses associated with self-injurious behaviors (e.g., Lesch-Nyhan syndrome). The assessment should be individualized to the patient.

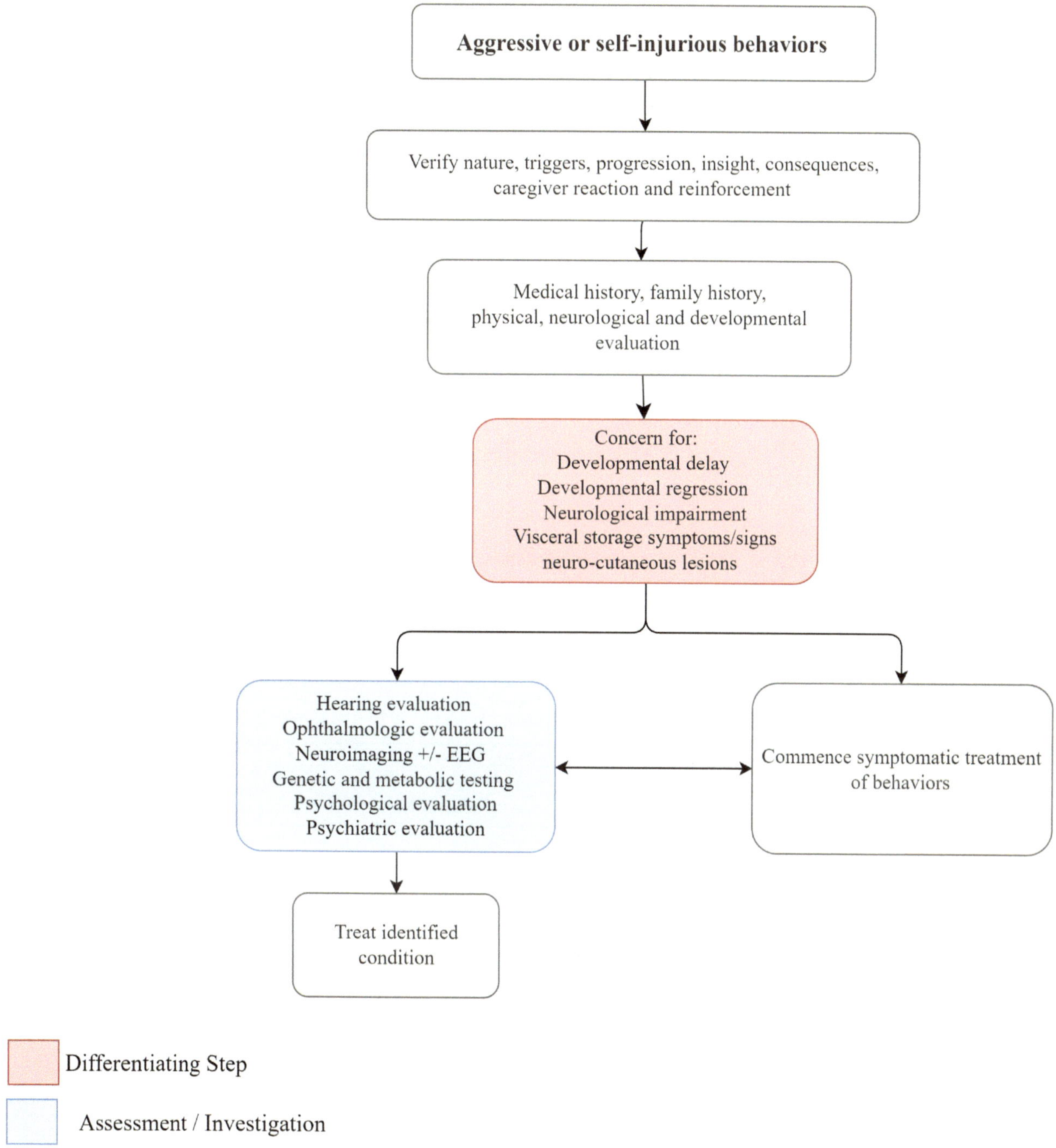

2.9 Pain and discomfort in children with severe neurological impairment who can not communicate

Nervous System	Seizures Increased Intracranial Pressure Neuro-irritability Autonomic Storming Dysautonomia Sympathetic Hyperactivity Dystonia
Eyes	Ocular Pain Corneal Abrasions
Head and Neck	Dental Pain Otitis Media Sinusitis
Respiratory System	Aspiration
Gastrointestinal System	Hunger Aerophagia Gastroesophageal Reflux Disease (GERD) Esophagitis Peptic or Ulcer G or J tube leakage Dysmotility Constipation Intestinal obstruction or malrotation
Renal and Urinary System	Kidney Stones Urinary Tract infection Urinary retention Urinary Tract Infection
Reproductive System	Mittelschmerz (ovulation pain) Menstrual pain Ovarian torsion Testicular torsion
Musculoskeletal System	Fracture Hip dislocation
Skin	Diaper rash Decubiti (Pressure Sores) Hair tourniquet syndrome
Endocrine System	Adrenal insufficiency

3. Clinic Templates

The templates serve as a guide to assist clinicians address important and relevant issues related to patients with neurodevelopmental disabilities. They can be individualized to the patient and the scope of the clinician's practice.

3.1 Developmental History

Name:
Date of Birth:
Date of Examination:
Chronological Age:
Referring Physician:
Record Number:

School

- Grade:
- Support Services: Individualized Educational Plan / 504 Plan / Early Intervention / Special Education / Speech Therapy / Occupational Therapy / Physical Therapy

Parental Concerns
Previous Diagnoses
Records Reviewed

- Individualized Educational Plan (IEP)
- Child Study Team
- Teacher comments
- Discipline evaluations
- Test data
- Classroom Teacher Observations

Birth and Medical History

- Maternal Parity
- Pregnancy/Labor/Delivery: Single or multiple gestations / Preterm or full term / Delivery: Spontaneous vaginal delivery or c-section / Indications for C-section: Repeat / Emergency / Complications
- Fetal Activity: Normal / Hyperactive / Hypoactive
- Birth Weight:
- Perinatal Course: Respiratory support / Seizures / Intracranial bleeding
- Hospitalizations:
- Surgery: PE tubes / T&A / Other
- Illnesses:
- Medications: Type/ Indications / Unusual responses to medications
- Lead Screening

Review of Systems

- HEENT: Vision / Glasses / Hearing Audiology / Recurrent otitis
- Dermatology: Chronic rashes
- GI: Diarrhea / Constipation / Rectal bleeding / Abdominal pain
- Endocrine: Growth (average - small – large) / FTT / Change in weight
- Cardiac: Heart murmur / Heart disease / Cyanosis / Shortness of breath
- Respiratory: Pneumonia / Bronchitis / Asthma / Chronic cough / Chest pain / Wheezing
- Bones and Joints: Limitation in movements / Pain / Swelling / Fractures / Scoliosis
- GU: Bladder infection / Dysuria / Enuresis (nocturnal or diurnal)
- Allergies: Hives / Eczema / Seasonal: Pollen, mold, dust, animal dander / Reactions to food, drugs
- Neurological: Tics / Paralysis / Weakness / (Febrile) seizures
- Feeding Problems: None / Oral motor difficulties (gagging, choking, drooling)
- Picky Eater: Texture / Position on plate / Stuffs mouth / No new foods
- Infancy: Pleasant / Colicky/ Irritable / Demanding / Cuddly/ Aloof / Vocal/ Quiet
- Day Diary: Describe child's activities of daily living (e.g., dressing, toileting, self-care)

Motor Milestones:
Walking: / Running: / Ride a bike:
Toe Walking: Yes/No / Past / Occasionally / Never / Falls a lot
Coordination: Good / Poor / Clumsy / Attentional when rushes
Tie shoe laces:
Handedness: Right / Left / Ambidextrous / Switches
Handwriting: Neat / Sloppy / Neater when slows down / Reverses letters / Numbers
Draws: Line / Circle / Square / Triangle / Face / Person
Toileting: Completed at: / Partial: wants to be changed / Daytime/ No accidents / Nocturnal Enuresis

Language Milestones:
Acquisition: On time / Early / Late
Sentence Length: Long/short/ Sentence structure: Simple complex (with subordinate clauses)
Vocabulary: Number of words or word approximations
Articulation: Clear / Mostly intelligible / Difficult to understand / Unintelligible / Stutter/Stammer
Language Deviance: Word finding / Word substitutions / Word opposites / Talks tangentially / Conversational Pragmatics (talks at rather that to/with)
Responds to Name: Yes/No /Sometimes
Follows Commands: Single/Multiple step / With or without repetition / Says 'huh' mishears directions
Points To: Wants / Body parts / Pictures / Interesting things

Attention: Short attention span / Variable / Extra-long / Daydreamer / Does better one on one / Small groups / Easily distractible / Doesn't listen / Needs repetition / Forgetful / doesn't always finish

Impulsivity: Impulsive / Cautious / Loud / Daredevil / Impatient / Cannot wait turn / Blurts out in class / Disruptive / Out of seat / Class clown / Interrupts / Easily frustrated / Gives up easily / Accident-prone / Spills / Drops

Hyperactivity: Always on the go / High energy level / Fidgety / Cannot sit still / Bounces in place / out of seat / Logorrhea ("chatterbox", "motor mouth") / Cannot keep hands to self / Busy hands / Extraneouis noises

School Performance:
Better at: Reading / Math / Calculation / Science / Social studies / Spelling / Writing
Worse at: Reading / Math / Calculations (careless errors) / Science / Social studies / Spelling / Writing
Erratic performance across subject areas/ Fails tests for incompleteness / Loses recess
Homework: Disorganized approach / Incomplete assignments / Takes longer than should / Loses / Rushes

Sleep Problems: Late onset / Less sleep / Tired during day / Snores / Bruxism / Enuresis / Talks in sleep / Sleepwalking / Nightmares / Night terrors / Moves around a lot / Falls out of bed / Deep sleeper

Behavior:
Response to change: Flexible / Goes with the flow / Does better with preparation / Needs preparation
Tantrums/meltdowns: Scream / Fall out / Kick / Hit / Bite / Throw
Duration: Frequency: Triggers: (obvious)
Mood lability: Explosive temperament
Aggressive: Bully/bullied arson (plays with matches) / Lies / Steals / Cruel to/rough with animals / Destructive / Oppositional / (Im)mature
Socially inappropriate / Nail biting
Fears: Loud noises / Vacuum / other
Common sense / Empathy / Remorse
Incontinence: Bowel / Bladder / Both / None
Somatization: Headaches / Stomach aches / Leg pains
Affect: Happy / Sad/ Depressed / Angry / Anxious/ Worrier / Talks down about self
Social: Loner / Gullible / Leader / Follower with friends
Group Activities: Scouts/ Church group / Cheerleading / Dance / Musical instrument / Clubs
Sports: Baseball / Basketball / Football / Karate / Tae kwan doe / Soccer / Swimming / Gymnastics
Hobbies / Collections
Why do they get into trouble? What do you do about it?

Deviance
Echoing: Immediate / Delayed / Movie scripts / Books
Pronominal Reversal: Yes / No / Refers to self by name / Third person
Talks to self: Yes / No
Regression: Yes / No
Motor: Rocking / Headbanging / Self-injurious behavior
Preservation of sameness: Routines / Rituals / Flexible/ Rigid
Concrete: Literal / Doesn't get jokes
Eye Contact: Good/Poor
Reads Faces: Yes / No
Emotions: Reads faces / Inappropriate facial expressions / Flat affect / Laughs inappropriately
Pain Sensitivity: Normal / High threshold
Joint Attention: Yes / No / Protodeclarative pointing
Selective Menu: Food color, smell, taste, position on plate
Peer Exposure: Yes / Limited / None
Play: Solitary / Pretend/ Imaginative / Odd / Lines up / Twirls / Spins toys / Spins self / Likes fans / Likes water / Treats people like furniture
Baby: Cuddly or not / Arching
Perseveration: Light switches / Doors/drawers / Flush toilets / Water faucets
Stereotypies: Flapping / Hand regard
Tactile aversive: Tags / Hats / Cloth texture / Seams / Dirty/ Sticky
Pica: Yes / No
In Own Little World: Acts deaf / Stares off / Busy
Memory: Directions / Maps
Splinter Skills: Blocks / Puzzles / Electronics / Drawing / Numbers / Letters / Shapes /colors hyperlexia
Fixations:

Family History
Home make up:
Housing: Education / Employment / Finances / Marriage/ Divorce / Separation / Moves / Deaths / Other stresses
Medical conditions
Intellectual disability / Cerebral palsy / Seizures / Learning problems / School problems / ADHD / Speech language delay / Autism / Depression / Bipolar disorder (manic depression) / Schizophrenia / Obsessive compulsive disorder / Anxiety / Panic / Celiac disease / Connective tissue disease / Thyroid / Heart disease / Hypertension / Cancer / Diabetes / Brain tumor / Degenerative disease / Alzheimer / Dementia

Therapy/Counseling
Type
Duration
Goals
Efficacy
Complementary therapies

3.2 New Evaluation for Sleep

By Jennifer Accardo, MD, MSCE

Name
Age
School
Grade:

- Time home from school:
- Activities once home:

Physical exercise: Type / Frequency / duration:
Dinner time:
Activities after dinner:
Activities leading up to bedtime, including start time:
Bedtime school nights (lights out, intend to fall asleep):
Length of time to fall asleep:

- What happens between time in bed and time actually asleep:
- If prolonged, worried/anxious? Annoyed about being awake?

BEDROOM

- Where does the child sleep?
- Does he/she fall asleep with someone else in room? Who?
- If so, conditions (talking, sharing bed, physical contact)
- Electronics in room? List:
- Conditions: Dark / Cool / Quiet / Nightlight / Ambient noise
- Physical complaints while trying to fall asleep
- Leg discomfort/urge to move

SLEEP

- Night wakings?
    - Frequency/duration
    - Time of night
    - What kind of mood is child in?
    - Is child fully awake during episodes?
    - How do parents become aware of wakings, and how do they respond?

Narcolepsy:

- Sleep paralysis
- Hypnogogic & hypnopompic hallucinations
- Cataplexy: emotional about something or laughing really hard, does anything unusual happen physically

Snoring:

- How loud?
- Frequency (nights/week)
- Triggers
- Associated symptoms
    - Noisy breathing
    - Mouth breathing
    - Labored breathing
    - Snort, gasp, any witnessed pauses in breathing?
    - Sweating during sleep
    - Bruxism
    - Bedwetting

- Sleep with neck in hyperextension/other unusual sleep positions
- First morning headache

Restlessness before falling asleep
Restlessness during sleep
Sleepwalking
Sleep talking
Night terrors
Frequent nightmares

Morning Routine

- At what time is the child supposed to wake up for school?
- At what time does the child actually wake up for school?
- Does the child wake up spontaneously, to an alarm, or does a parent wake them?
- How long does the waking-up process take?
- How much time does it take for the child to be fully awake?
- What is the child's mood upon waking?

Napping

- Napping or sleep during the day?
  - Timing/duration
  - Frequency
- Sleep during school?
- Sleep in car/how quickly?
- Does the child seem tired or sleepy? (explain difference between tired/fatigued and sleepy; potential for overlap)
- Signs for child of seeming tired or sleepy?

Weekend Routine:

- Estimated bedtime
- Estimated wake time

Ideal Sleep Schedule

- What is child's goal for sleep, if able to express this?
- What is parent's goal for child's sleep?

PAST MEDICAL HISTORY

SURGERIES

- Any prior adenotonsillectomy

REVIEW OF SYSTEMS

- Ophthalmologic: Blind / Light perception status
- Integumentary: Pruritis
- HEENT: Chronic rhinitis / Orthodontics
- Respiratory: Asthma
- Cardiovascular: Anemia / Iron deficiency
- Gastrointestinal: Picky eater / Constipation / Gastroesophageal reflux
- Genitourinary: Enuresis / Encopresis / Pubertal
- Musculoskeletal: Scoliosis
- Neurologic: Epilepsy / Migraine
- Psychiatric: Anxiety / Depression / PTSD/ Trauma

SOCIAL HISTORY

- Multiple households?
- Schedule differences between parents?

- Atypical parent work schedules-especially rotating/night shifts, parental travel
- Recent changes at home/stressors
- Caffeinated beverages

FAMILY HISTORY

- Snoring / OSA / CPAP use (successful?)
- Restless leg syndrome / Periodic limb movement disorder
- Anemia or iron deficiency
- Atypical sleep schedules (morning lark/night owl?) / insomnia

FOCUSED PHYSICAL EXAMINATION:

- Obesity
- Acanthosis nigricans
- Allergic shiners
- Allergic salute
- Deviated nasal septum
- Micrognathia or retrognathia
- Overbite / other bite abnormalities
- Evidence of bruxism
- Macroglossia / Tongue ridging
- Tonsils / size
- Palate—high arched?
- Adenoidal facies-mouth breathing / Slack jaw

3.3 Physical Examination

Vital signs
Growth parameters
General physical examination

- Dysmorphology
- Head and neck
- Cardiac
- Respiratory
- Gastrointestinal
- Musculoskeletal
- Skin

Neurological examination

- Mental status
- Cranial nerves
- Motor assessment
- Coordination and cerebellar functions
- Sensory assessment
- Tendon reflexes
- Postural reflexes
- Primitive reflexes
- Gait

Areas of focus in infant, preschool, and school age neurodevelopmental physical evaluation

Infant	General physical examination	Birthmarks Growth parameters Cardiovascular status (pulse, heart rate, respirations)
	Mental Status	Level of alertness Interaction with caregiver and provider
	Motor examination	Spontaneous movement of extremities Asymmetry in movement or tone
	Reflexes	Presence or absence of primitive and postural reflexes
Preschool	General physical examination	Nutrition and Growth
	Mental status	Communication Ability to follow commands Interaction with caregiver and provider
	Motor examination	Coordinated movements and balance Demonstration of motor milestones Excessive activity level
School Age	Mental status	Mood Understanding of complex language
	Motor examination	Strength testing Muscle tone and bulk Subtle neurological signs Coordinated movements Excessive activity level

Physical examination areas of focus in follow-up visits

ADHD	General physical examination	Nutrition and growth Cardiovascular examination Blood pressure Activity level
	Mental Status and cognitive functions	Affect / Attention Following commands Academic performance
	Motor examination	Subtle neurological signs "Motor on" behaviors (fidgeting, constant movement) Coordinated movement difficulties Repetitive motor movements
ASD	General physical examination	Dysmorphology examination Nutrition and Growth Abdominal examination Activity level Cutaneous examination
	Mental Status	Interaction with examiner Attention Use of Language (including higher level skills if applicable) Following commands
	Motor examination	Subtle neurological signs Coordinated movement difficulties Repetitive motor movements
GDD/IDD	General physical examination	Dysmorphology examination Nutrition and Growth Activity level Cutaneous examination
	Mental Status	Cognitive status Attention Use of Language Following commands
	Motor examination	Muscle tone and bulk

3.4 Follow up—Attention Deficit Hyperactivity Disorder

Patient Demographics
ADHD Diagnosis

- Initial parental concerns
- Age at diagnosis

Co-existing conditions
Reason for visit

- New/unresolved issues
- Current medications
  - List
  - Adherence
  - Efficacy
  - Symptoms aided
  - Duration of effect
  - Side effects

Previous medications and supplements: Type / Target symptoms / Schedule / Adherence / Side Effects
Current Function
Behaviors

- Anxiety, depression, conduct, oppositional, antisocial
- Stereotypic behaviors
- Discipline
- Sleep

School

- Grade / Academic Progress
- Subject Mastery
  - Reading
  - Math
  - Written Language
- Accommodations
- Special Education
- Tutoring
- Homework

Socialization

- Extracurricular Activities / Friends / Interests / Chores

Family History/dynamics
- Current Family Constellation / Changes in environment / Siblings / Stressors

Previous evaluations

- Cognitive and Educational assessment
- Etiologic investigation (if indicated): Neuroimaging / EEG / Genetic and metabolic testing

Interventions

- Educational (504 plan, IEP, RTI, tutoring)
- Behavioral (FBA, counseling)
- Assistive Technology
- Other interventions

Home strategies

- Communication strategies/ Behavior management techniques

Nutrition and growth
Parental Education and Support
Community Resources: Support Groups / Community Programs / Respite Care Services

PHYSICAL EXAMINATION

- Growth Parameters (height, weight, head circumference)
- Cardiovascular including blood pressure
- Neurological
  - Symmetry / Tone / Coordination / Cognitive functions / Academics and language (if indicated)

IMPRESSIONS

RECOMMENDATIONS

- **Management of Target Symptoms (Primary and co-existing)**
  - Medication changes
- **Outcomes and Long-Term Planning**
  - Family Goals and Concerns
  - Community Resource Identification and Access
  - Support Services Coordination
  - Transition Planning (Adolescence to Adulthood)
- **Communication and Documentation**
  - Regular Interdisciplinary Meetings
  - Patient and Caregiver Education Materials
  - Communication with Primary Care Physician

3.5 Follow up—Autism Spectrum Disorder

Patient Demographics
Autism Diagnosis

- Initial parental concerns
- Age at diagnosis

Co-existing conditions
Reason for visit

- New/unresolved issues

Current Function
Previous evaluations

- Cognitive and Educational assessment
- Etiologic investigation (if indicated): Neuroimaging / EEG / Genetic and metabolic testing

Functional and behavioral assessment

- Communication
  - Expressive Language: difficulties in verbal communication, including limited vocabulary, speech delays, or echolalia.
  - Receptive Language: challenges in understanding and responding to spoken language.
  - Non-Verbal Communication: the use of gestures, facial expressions, and body language.
- Social Interaction
  - Peer Relationships: how the child interacts with peers, initiates or responds to social cues, and engages in play.
  - Explore the quality of interactions with family members, teachers, and peers.
- Joint Attention
  - Evaluate the ability to share attention with others and follow someone else's gaze.
- Behavior and Emotional Regulation
  - Stereotypic Behaviors: Document repetitive movements or actions.
  - Emotional Regulation: Assess how the child manages and expresses emotions, noting any episodes of aggression, self-injury, or emotional meltdowns.
  - Attention and Hyperactivity: Observe attention span, focus, and levels of hyperactivity or impulsivity.

Cognitive and educational assessment
Adaptive Functioning

- Daily Living Skills: Evaluate independence in activities of daily living, such as dressing, grooming, and feeding.
- Self-Care: Assess the child's ability to perform tasks independently.

Interventions

- Applied Behavioral Analysis (ABA)
- Behavioral Therapy
- Occupational Therapy
- Speech and Language Therapy
- Pharmacologic
- Others
  - Medications / Target symptoms / Schedule / Adherence / Side Effects

Home strategies

- Communication strategies / Behavior management techniques

Nutrition

- Oral aversions / Food preferences / Growth

Behaviors

- Anxiety, depression, conduct, oppositional, antisocial
- Behavioral outbursts, irritability, defiance
- Stereotypic Behavior
- Discipline
- Sleep

School

- Grade /Academic Progress
- Subject Mastery
  - Reading
  - Math
  - Written Language
- Accommodations
- Special Education
- Tutoring
- Homework

Socialization

- Extracurricular Activities / Friends / Interests / Chores

Family History/dynamics

- Current Family Constellation / Changes in environment / Siblings / Stressors

Parental Education and Support

Community Resources

- Support Groups / Community Programs / Respite Care Services

PHYSICAL EXAMINATION

- Growth parameters (height, weight, head circumference)
- Dysmorphic features
- Neurocutaneous findings
- Neurological
  - Cranial Nerve Assessment
  - Motor Assessment including neuromotor abnormalities
  - Tone and reflexes
  - Gait and coordination
  - Sensory evaluation (hypersensitivity, hyposensitivity)
  - Vision and Hearing Assessment

IMPRESSIONS

RECOMMENDATIONS

- **Management of Target Symptoms** (Primary and co-existing)
- **Outcomes and Long-Term Planning**
  - Family Goals and Concerns
  - Community Resource Identification and Access
  - Support Services Coordination
  - Transition Planning (Adolescence to Adulthood)
- **Communication and Documentation**
  - Regular Interdisciplinary Meetings
  - Patient and Caregiver Education Materials
  - Communication with Primary Care Physician

3.6 Follow up—Language and Communication Disorders

Patient Demographics
Diagnoses

- Initial parental concerns
- Age at diagnosis

Co-existing conditions
Reason for visit

- New/unresolved issues

Current Function
Previous Evaluations

- Functional assessment
- Cognitive assessment
- Speech and language assessment
  - Communication
  - Expressive Language
  - Receptive Language
  - Non-Verbal Communication
  - Higher language function (humor, irony, metaphors, idioms)
  - Academics (decoding, reading comprehension, and written language)
- Audiology evaluation
- Etiologic investigation (if indicated): Neuroimaging / EEG / Genetic and metabolic testing

Interventions

- Speech and Language Therapy
- Educational Interventions
- Behavioral Therapy
- Assistive Technology
  - Type
  - Outcomes
- Pharmacological management of coexisting conditions
- others
  - Medications / Target symptoms / Schedule / Adherence / Side Effects

Home strategies

- Communication strategies
- Behavior management techniques

Nutrition and growth
Behaviors

- Anxiety, depression, conduct, oppositional, antisocial
- Stereotypic Behavior
- Discipline
- Sleep

School

- Grade / Academic Progress
- Subject Mastery
  - Reading
  - Math
  - Written Language

- Accommodations
- Special Education
- Review IEP

Socialization

- ○ Extracurricular activities / Friends / Interests / Chores

Family History/dynamics

- ○ Current Family Constellation / Changes in environment / Siblings / Stressors

Parental Education and Support
Community Resources: Support Groups / Community Programs / Respite Care Services

PHYSICAL EXAMINATION

- Growth Parameters
- Dysmorphic Features
- Neurocutaneous Findings

Neurological Examination

- Cranial Nerve Assessment
- Motor Assessment
- Tone and Reflexes
- Gait and Coordination
- Sensory Evaluation

Vision and Hearing Screening

IMPRESSIONS

RECOMMENDATIONS

- **Management of Target Symptoms (Primary and co-existing)**
- **Outcomes and Long-Term Planning**
  - ○ Family Goals and Concerns
  - ○ Community Resource Identification and Access
  - ○ Support Services Coordination
  - ○ Transition Planning (Adolescence to Adulthood)
- **Communication and Documentation**
  - ○ Regular Interdisciplinary Meetings
  - ○ Patient/Caregiver Education Materials
  - ○ Communication with Primary Care Physician

3.7 Follow up Clinic Template—Cerebral Palsy

Patient Demographic
Cerebral Palsy Diagnosis

- Age at diagnosis
- Type of CP: Spastic / Dyskinetic / Ataxic / Mixed / Hypotonic
- Distribution: Unilateral / Bilateral / Quadriplegic / Hemiplegic / Diplegic
- Etiology

Associated dysfunction(s) and coexisting conditions (e.g., seizures, intellectual disability, failure to thrive, scoliosis, etc.)
Reason for visit

- New/unresolved issues

Summary of previous assessments and Interventions

- Motor and gait
- Nutrition and Feeding
- Medications
    - Type/ Target symptoms / Schedule / Adherence / Side Effects
- Surgeries and orthoses
- Assistive Technology
- Cognitive
- Speech and Language

Diagnostic evaluations and their indications/results

- MRI, CT scans, X-rays
- Genetic/metabolic tests
- EEG

Current Functional Abilities

- Gross Motor Function Classification System (GMFCS) Level
- Manual Ability Classification System (MACS) for fine motor skills
- Communication Function Classification System (CFCS)
- Eating and Drinking Abilities Classification System (EDACS)
- Sleep
- Bowel and Bladder function
- Vision and Hearing
- Pain

PHYSICAL EXAMINATION
General Examination

- Vital Signs
- Height, Weight, Head Circumference
- Head and Neck
- Respiratory and cardiac function
- Abdominal examination
- Musculoskeletal assessment: Scoliosis / Hip dislocation / Pathological fractures / Contractures
- Skin
- Oral hygiene

Neurological and Developmental Evaluation

- Cognitive Function Evaluation
- Speech and Language Evaluation

- Cranial nerves
- Bulbar (oromotor) function/Feeding/Communication
- Sensory Function Assessment
- Motor Function Evaluation
    - Muscle Tone
    - Strength
    - Reflexes
    - Primitive reflexes
    - Postural responses
    - Gross and Fine Motor Skills

CEREBRAL PALSY CARE

Pain Assessment

- Pain Triggers
- Pain Management: Pharmacological Interventions / Non-pharmacological Techniques / Multidisciplinary Approach for Pain Relief

Nutrition

- Safe feeding
- Diet modifications

Musculoskeletal Management

- Interventions
    - Handling and positioning
    - Physical and Occupational Therapy
    - Pharmacotherapy (e.g., oral medications, botulinum toxin injections, baclofen pump)
    - Orthopaedic and neurological surgeries
    - Orthotics

Seating and Positioning

- Postural Assessment
- Wheelchair/Seating Evaluation
- Seating Interventions
    - Custom Seating/Positioning Devices / Postural Management Strategies / Occupational Therapy for Activities of Daily Living (ADLs)

Mobility Management

- Mobility and Assessment
- Assistive Devices Evaluation
- Mobility Interventions
    - Physical Therapy / Gait Training / Assistive Devices (e.g., wheelchairs, walkers) / Power mobility/

Communication

- Communication Interventions
    - Speech Therapy / Augmentative and Alternative Communication (AAC) Devices/Strategies / Educational Support for Communication Development

Educational Supports

- Academic Evaluation / Educational goals / Educational services and support

Psychological/Emotional Supports

- Emotional and Behavioral Assessment / Risk for depression, anxiety and mood disorders / Counselling / Pharmacological management

Family Supports

- Caregiver Education Programs / Caregiver Support Programs / Patient Support Groups / Parent mental health support

IMPRESSIONS

RECOMMENDATIONS

- **Management of Target Symptoms** (primary and co-existing)
- **Outcomes and Long-Term Planning**
  - Family Goals and Concerns
  - Functional Improvement Tracking
  - Community Resource Identification
  - Support Services Coordination
  - Transition Planning (Adolescence to Adulthood)
- **Communication and Documentation**
  - Regular Interdisciplinary Meetings
  - Patient/Caregiver Education Materials
  - Communication with Primary Care Physician

3.8 Follow up—Early brain injury

Patient Demographics
Primary brain injury diagnosis
Coexisting conditions
Reason for visit

- New/unresolved issues
- Interval history

Current Function

Developmental and functional skills (Interval change – gains or losses)

Feeding and Nutrition
Sleep
Interventions

- Physical therapy
- Occupational therapy
- Language therapy
- Pharmacologic
  - Medications / Target symptoms / Schedule / Adherence / Side Effects

Evaluations

- Etiologic investigation
  - Neuroimaging (Head US and/or Brain MRI)
  - EEG
  - Genetic and metabolic testing
- Vision and hearing assessment

Perinatal history (As relevant):

- Gravida Para / Group B Streptococcus (GBS) status / Serology / Gestational diabetes / Other maternal conditions
- Prenatal testing: QUAD screen (AFP, hCG, Inhibin A, unconjugated estriol)
- Fetal US
- Labor and Delivery
- Gestational Age (GA): , Confirmed by: US/LMP/physical examination
- Delivery mode:
- Apgar Scores: @ 1 min, @5 min, @ 10 min
- Birth Weight

NICU /PICU course

- Metabolic abnormalities: Hypoglycemia / Hypocalcemia / Other
- Gastrointestinal: Necrotizing enterocolitis
- Respiratory Support: Intubation / Other
- Infection(s): Sepsis / Pneumonia / Other
- Brain injury
  - Traumatic brain injury
  - Hypoxic Ischemic Injury
  - Intracranial bleeding (IVH, intraparenchymal hemorrhage, vascular malformation)
  - CNS infection: Meningitis / Encephalitis
  - Autoimmune Encephalitis
  - Other
- Complications
- NICU/PICU Discharge Physical, Neurological and Developmental Examination
- Length of hospitalization

Family History/dynamics

- Current Family Constellation / Changes in environment / Siblings / Stressors

Parental Education and Support
Community Resources

  - Support Groups / Community Programs / Respite Care Services

PHYSICAL EXAMINATION
Growth parameters (Correct for prematurity if applicable)
Dysmorphism
General physical examination

Neurological and Developmental Examination
Mental status and cognitive assessment including level of consciousness
Developmental assessment (Developmental Quotient – Instrument used)
Cranial Nerves: Pupils / Eye movements / Facial asymmetry / Sucking and swallowing
Muscle Tone / Muscle power / Tendon Reflexes / Coordination and Gait

Primitive Reflexes

- Moro Reflex (Startle Reflex) (Birth to 4 Months):
- Rooting Reflex (Birth to 4 Months):
- Sucking Reflex (Birth to 4 Months):
- Grasp Reflex (Birth to 6 Months):
- Tonic Neck Reflex (Fencing Reflex) (Birth to 6 Months):
- Babinski Reflex (Birth to 4 Months):
- Stepping Reflex (Birth to 2 Months):
- Galant Reflex (Birth to 2 Months):

Postural Reactions

- Pull to Sit Reaction (6 Months and Older):
- Parachute Reflex (6 Months and Older):
- Landau Reflex (3 Months and Older):

Sensory Examination:
Specialized Assessments:

- Hammersmith Neonatal / Infant Neurological Examination
- Prechtl's General Movement Assessment
- Other assessments as indicated

IMPRESSIONS

- Risk Factors from history and examination for long Term Neurodevelopmental Disabilities

RECOMMENDATIONS

- **Current management of symptoms** (Primary and co-existing)
- **Outcomes and Long-Term Planning**
  - Family Goals and Concerns
  - Functional Improvement Tracking
  - Community Resource Identification and Access
  - Support Services Coordination
- **Communication and Documentation**
  - Regular Interdisciplinary Meetings
  - Patient/Caregiver Education Materials
  - Communication with Primary Care Physician

3.9 Follow up—Spina Bifida

Patient Demographics
Spina Bifida Diagnosis

- Type: Myelomeningocele / Meningocele / Occulta
- Age at diagnosis
- Spina Bifida Lesion Level
- Location of Spina Bifida Lumbar / Sacral / Thoracic

Associated Conditions: (e.g., hydrocephalus, Arnold-Chiari malformation, epilepsy)
Previous Surgeries or Interventions
Reason for visit

- New/unresolved issues

Current function

- Mobility
- Continence (Bowel/Bladder)
- Decubitus prevention
- Cognitive function and communication

Evaluations

- Functional assessment
- Cognitive / neuropsychological assessment
- Hydrocephalus Screening
- Orthopedic Assessments
- Urological Evaluation
- Renal functions

Interventions

- Educational
- Behavioral
- Physical and occupational therapy
- Assistive Technology for mobility and communication
- Pharmacological treatment
  - Indication/ Target symptom
  - Response
  - Side effects
  - Adherence
- Continence Care

Nutrition and growth
Behaviors

- Anxiety, depression, conduct, oppositional, antisocial
- Stereotypic Behavior
- Discipline
- Sleep

School

- Grade / Academic Progress
- Subject Mastery
  - Reading
  - Math
  - Written Language

- Accommodations
- Special Education
- Tutoring

Socialization

- Extracurricular Activities / Friends / Interests / Chores

Family History/dynamics

- Current Family Constellation / Changes in environment / Siblings / Stressors

Parental Education and Support
Community Resources

- Support Groups / Community Programs / Respite Care Services

PHYSICAL EXAMINATION
Growth parameters including head circumference
Skin Inspection: ulcers, pressure sores, surgical site of repair

Orthopedic Assessment: Scoliosis, Hip dysplasia, Clubfoot, or Contractures
Neurological Examination

- Head circumference
- Mental status and cognitive assessment
- Cranial Nerve Assessment
- Gross and Fine Motor Assessment
- Tone and Reflexes
- Gait and Coordination
- Sensory Evaluation: above and below the level of the lesion
- Vision and Hearing Assessment

Developmental Assessment

- Gross Motor Skills
- Fine Motor Skills
- Speech and Language
- Cognitive Development
- Social and emotional

Spinal Cord Function

- Signs of tethered cord syndrome (changes in reflexes or muscle weakness).
- Sensory changes
- Changes in bowel and bladder functions
- Change in hand function (syrinx)

IMPRESSIONS
RECOMMENDATIONS

- **Management of Target Symptoms**
- **Outcomes and Long-Term Planning**
  - Family Goals and Concerns
  - Functional Improvement Tracking
  - Community Resource Identification and Access
  - Support Services Coordination
  - Transition Planning (Adolescence to Adulthood)
- **Communication and Documentation**
  - Regular Interdisciplinary Meetings
  - Patient/Caregiver Education Materials
  - Communication with Primary Care Physician

3.10 Elements of a management program

The goals of a management program are to minimize activity limitations and to maximize participation in the larger society. Most comprehensive and rigorous management programs are designed for children with severe physical (cerebral palsy) or cognitive dysfunctions (intellectual disability). In this context, we have formulated a detailed template. This template can be adapted to the child's specific disability based on patient and caregivers' goals.

Domains	Areas	Foci/Interventions	Related Disciplines
Physiologic Stability	1. Nutrition 2. Allergies 3. Medications 4. Primary care 5. Specialized monitoring	Adequate calories for growth Hydration Drug interactions Chronologic age appropriate activities (vaccinations, screenings, etc. Epilepsy/Genetic/Metabolic/Cardiovascular/Pulmonary/ Renal/ GI /Bone density (as indicated) Reproductive Health	Medicine
Nursing Care	1. Feeding 2. Positioning 3. Skin health 4. Elimination	Oral, Tube (exclusive or supplemental, bolus or continuous) Adaptive seating (if not sitting independently), Enhancing symmetry, hand splinting Decubitus prevention, pressure relief (especially heels, sacrum, occiput, scapula) Bowel and bladder programs	Medicine Nursing Occupational Therapy Physical Therapy
Activities of Daily Living	1. Feeding 2. Dressing 3. Bathroom	Adaptive silverware or dishes if needed Adaptations that bypass buttons and snaps and tie shoes Environmental adaptations to ensure safety on toilet or in bathtub. Electronic toothbrushes or water piks Electric razors with adaptations as needed	Medicine Occupational Therapy
Communication	1. Communicate needs and wants 2. Social interaction 3. Information transfer (learning, abstraction)	Speech: oral or augmented Augmented communication (cognitive, motor and social functions determine the best device) Language and cognition Pragmatics and higher language function	Medicine Speech Language Pathology
Mobility	Ambulation	Community Household Exercise Non-walker-Power mobility Wheelchair	Medicine Physical Therapy
Transportation	1. Safe transportation needs 2. Change with growth and motor function	Driving (self or others) Adapted vehicles Public Transportation	Physical Therapy Occupational therapy
Education/Work	1. Academic Achievement 2. Work	Reading, Mathematics, Written Language (direct instruction, circumvention techniques) IDEA, Section 504, ADA ADHD Multimodal approach Supported employment	Medicine Education Psychology Vocational Rehabilitation
Leisure and community engagement	Participation	Hobbies Friends Social groups (clubs, courses, scouts, sports, etc)	Social Work
Behavior	Management of maladaptive behaviors	Adjustment/depression/anxiety/ NDD specific-self-injury, stereotypies	Psychology Social Work Psychiatry

Domains	Areas	Foci/Interventions	Related Disciplines
Living	Establishing safe and stable living environment	Family Out of Home placements (Foster care, Kinship care, congregate living) Supported living Independent Alternative living situations	Community Agencies
Transition	Continuity of care into adulthood	Healthcare management Legal and financial management Accessing disability-related benefits Consent Guardianship Vocation	Medicine Adult Specialty clinics Social work Community Agencies Legal team

Index

Note: Page numbers followed by "*f*" indicate figures, "*t*" indicate tables, and "*b*" indicate boxes.

A

B

F

G

N

O

P

Q

R

S

T

U

V

W

X

Y

Z